# DIRECTORY OF GEOSCIENCE DEPARTMENTS

52nd Edition

Carolyn Wilson, Editor

American Geosciences Institute
Alexandria, Virginia

Directory of Geoscience Departments 2017, 52nd Edition

Edited by Carolyn Wilson

ISBN-13: 978-0-913312-85-8
ISSN: 0364-7811



Typeset in Times New Roman and Arial using Adobe InDesign CC.
Layout and programming: Christopher M. Keane
Advertising: John P. Rasanen

For more information on the American Geosciences Institute and its publications check us out at https://www.americangeosciences.org



Cover
Lava flowing into ocean from the Makaopuhi crater of Hawaii Volcanoes National Park, © Courtesy USGS Hawaiian Volcano Observatory/R.L. Christiansen

# Introduction

Thank you for using the 52nd edition of the Directory of Geoscience Departments. We, at the American Geosciences Institute (AGI), continue to be grateful to all the departments who took the time this year to update their information and provide additional historical data of enrollments and awarded degrees for their department.

We would also like to thank Kelsey Watson, the AGI Workforce Program fall intern, for her work updating the two-year college listings and further enhancing the completeness and quality of the department listings. Thanks to the efforts of all the data interns hired by the AGI Workforce Program since 2013, the international department listings have been massively updated with appropriate contact information allowing for department representatives at many of these schools to provide more details to their listings.

For this addition, individual faculty and staff members were able to edit their personal information and provide more specificity in their specialties than maybe previously recorded. Typically, departments have one or two people that edit the department listing annually. However, it can be difficult for that individual to receive all the up-to-date information in detail from all the faculty and staff within the department. Therefore, this year each person listed in the Directory received an individualized email listing out their information as it appeared in our database and allowed for them to update as needed. Approximately 3,000 faculty and staff made updates on their listing providing more accurate information in this version of the Directory.

This edition includes a listing of U.S. and Canadian student theses and dissertations from 2014 that have been reported to GeoRef Information Services. We would like to thank GeoRef for their diligent work compiling this list for use in this edition.

The data used to compile the directory is provided by the individuals and departments listed. AGI only edits the information for consistency and format. Any errors or omissions most likely reflect the entries of the individuals or the primary department contact. If you identify any issues, please email dgd@americangoesciences.org with information about a current responsible contact for the department in question, so that we may further encourage updates to the next edition.

The American Geosciences Institute does not warrant the accuracy of any of the self-reported information, but we do encourage individuals and departments to update the material they believe is out-of-date.

Some basic statistics for the 52nd edition of the Directory of Geoscience Departments: 1,964 academic departments and programs globally, with 971 of those departments in the United States. Three hundred thirty-three departments in the United States are two-year institutions. The overall number of academic departments and programs decreased from previous years due to the extensive cleaning and removing of duplicates in the international department listings.

Carolyn Wilson
Editor

# DIRECTORY OF GEOSCIENCE DEPARTMENTS
# 52nd Edition 2017

### Table of Contents

Directory of Geoscience Departments
U.S. Section ........ 1
Non-U.S. Section ........ 170
Theses and Dissertations ........ 267
U.S. State Geological Surveys ........ 286
U.S. Federal Agencies and National Laboratories ........ 303
List of Faculty Specialty Codes ........ 313
Faculty by Specialty Index ........ 314
Faculty Index ........ 405

## Usage Key

### Degrees Offered

A - Associate's or 2-year degree
B - Bachelor's or equivalent undergraduate degree
M - Master's Degree
D - Doctorate

Basic enrollment and degrees-granted data for the most recent year reported are shown below department information: degree level, enrollment, and degrees granted (in parentheses).

❐ indicates thesis and dissertations were provided.
● indicates department offers a field camp only for their majors.
○ indicates department offers a field camp with open enrollment.

The letters following individual faculty listings indicates their research specialties. Capital letters indicate general area, lowercase letters indicate subspecialty. Two or more consecutive lowercase letters refer to multiple subspecialties in the same major focus area referred to by the preceding capital letter.

# DEPARTMENTS AND FACULTIES

This section contains a global listing of academic geoscience departments. Universities and colleges in the United States are arranged alphabetically by state. Non-U.S. institutions are listed alphabetically by country following the U.S. section.

All data is as reported by the departments and faculty themselves and current as of January 28, 2016. If you are aware of updates, corrections, or other means to ensure continued improvement of this listing, please email dgd@agiweb.org with the appropriate information so that it can be addressed during the next revision.

## Alabama

### Alabama A&M University

**Dept of Biological and Environmental Sciences** (A,B,M,D) (2015)
P.O. Box 1208
Normal, AL 35762
p. (256) 372-4214
anthony.overton@aamu.edu
http://www.aamu.edu

Chair:
Anthony Overton

Professor:
Tommy L. Coleman, (D), Iowa State, 1980, So
Florence A. Okafor, (D), Nigeria, 1995, OnEn
Govind Sharma, (D), Kansas State, 1970, On
James W. Shuford, (D), Penn State, 1975, So

Associate Professor:
Monday O. Mbila, (D), Iowa State, 2000, Sd

Secretary:
Martha Palmer

### Auburn University

**Dept of Agronomy & Soils** (M,D) (2014)
201 Funchess Hall
Auburn University, AL 36849
p. (334) 844-4100
touchjt@auburn.edu
*Enrollment (2006): M: 8 (2) D: 6 (3)*

Head:
Joseph T. Touchton, (D), Illinois, 1977, Sc

Extension Agronomist-Soils:
Charles C. Mitchell, (D), Florida, 1980, Sc

Professor:
Elizabeth A. Guertal, (D), Oklahoma State, 1993, Sc
Joey N. Shaw, (D), Georgia, 1997, Sd
Charles W. Wood, (D), Colorado State, 1990, Sc

Associate Professor:
Yucheng Feng, (D), Penn State, 1995, Sb
John W. Odom, (D), Purdue, 1977, Sc

Assistant Professor:
Julie Howe, Wisconsin, 2005, Sc

**Dept of Geosciences** (B,M) (2017)
210 Petrie Hall
Auburn University, AL 36849-5305
p. (334) 844-4282
bdm0018@auburn.edu
http://www.auburn.edu/academic/science_math/geology/docs
*Enrollment (2016): B: 62 (17) M: 30 (8)*

Chair:
Mark G. Steltenpohl, (D), North Carolina, 1985, Gtc

Professor:
Willis E. Hames, (D), Virginia Tech, 1990, CcGpt
David T. King, Jr., (D), Missouri, 1980, GrOn
Ming-Kuo Lee, (D), Illinois, 1993, HwGe
Charles E. Savrda, (D), S California, 1986, GdPeGr
Lorraine W. Wolf, (D), Alaska (Fairbanks), 1989, Ysg

Associate Professor:
Phil L. Chaney, (D), Louisiana State, 1999, OynOn
Ronald D. Lewis, (D), Texas, 1982, Pi
Luke J. Marzen, (D), Kansas State, 2001, Oir
Karen S. McNeal, (D), Texas A&M, 2007, OeGeCm
Martin A. Medina Elizalde, (D), California (Santa Barbara), 2007, PeHyCl
Ashraf Uddin, (D), Florida State, 1996, GdOn
Haibo Zou, (D), Florida State, 1999, CcGiv

Assistant Professor:
Christopher G. Burton, (D), South Carolina, 2012, Oi
Chandana Mitra, (D), Georgia, 2011, OgaOi
Stephanie L. Shepherd, (D), Arkansas, 2010, GmOi

Postdoctoral Fellow:
Chong Ma, (D), Florida, 2015, GctGg

Instructor:
James Norwood, Oy

Lecturer:
Stefanie M. Brueckner, (D), Memorial, 2016, EmCag
Carmen P. Brysch, (D), Texas State Univ, 2014
John F. Hawkins, (M), Auburn, 2013, GcOe

Emeritus:
Robert B. Cook, (D), Georgia, 1971, Eg
Cyrus B. Dawsey, (D), Florida, 1975, Oy
James A. Saunders, (D), Colorado Mines, 1986, Cl

### Dauphin Island Sea Lab

**Marine Science Program** (M,D) (2014)
P.O. Box 369
101 Bienville Boulevard
Dauphin Island, AL 36528
p. (334) 861-7528
langelo@disl.org
http://www.disl.org/aboutus.html
Department Secretary: Carolyn F. Wood

Director:
William W. Schroeder, (D), Texas A&M, 1971, Og

Professor:
Thomas S. Hopkins, (D), California (San Diego), 1967, Ob

Associate Professor:
Jonathan R. Pennock, (D), Delaware, 1983, Oc

Librarian:
Dennis Patronas, On

Cooperating Faculty:
George F. Crozier, (D), California (San Diego), 1966, On

### University of Alabama

**Dept of Geological Sciences** (B,M,D) O (2016)
Box 870338
201 7th Avenue
Room 2003 Bevill Bldg.
Tuscaloosa, AL 35487-0338
p. (205) 348-5095
geology@geo.ua.edu
http://www.geo.ua.edu
*Enrollment (2016): B: 82 (19) M: 47 (5) D: 20 (2)*

Chair:
C. Fred T. Andrus, (D), Georgia, 2000, PeGaCs

Professor:
Ibrahim Çemen, (D), Penn State, 1983, Gct
Rona J. Donahoe, (D), Stanford, 1984, Cl
Delores Robinson, (D), Arizona, 2001, Gc
Harold H. Stowell, (D), Princeton, 1987, GpCcGt
Nick Tew, (D), Alabama, 1999, EoGro
Chunmiao Zheng, (D), Wisconsin, 1988, Hw

Associate Professor:
Andrew Goodliffe, (D), Hawaii, 1998, YgeGt
Samantha E. Hansen, (D), California (Santa Cruz), 2007, YsGt
Alberto Perez-Huerta, (D), Oregon, 2004, PgClGz
Ellen Spears, (D), Emory, 2006, On
Geoffrey Tick, (D), Arizona, 2003, Hw

Yong Zhang, (D), Nanjing, 1998, Hqw
Assistant Professor:
Julia Cartwright, (D), Manchester, 2010, XmgCc
Natasha T Dimova, (D), Florida State, 2010, HwCcm
Kimberly Genareau, (D), Arizona State, 2009, Gvi
Deborah A. Keene, (D), Georgia, 2002, Gga
Yuehan Lu, (D), Michigan, 2008, GgOuGe
Rezene Mahatsente, (D), Clausthal Tech, 1998, YgmYv
Marcello Minzoni, Kansas, 2007, GosEo
Rebecca T. Minzoni, (D), Rice Univ, 2015, GsPme
Tom S. Tobin, (D), Washington, 2014, PsCsPe
Matthew Wielicki, (D), California (Los Angeles), CcgGt
Bo Zhang, (D), Oklahoma, 2014, YeGoEo
Stable Isotope Research Scientist:
W. Joseph Lambert, (D), Alabama, 2010, CsPeCg
Emeritus:
Paul Aharon, (D), Australian National, 1980, ClmPy
Donald J. Benson, (D), Cincinnati, 1976, Gd
Richard H. Groshong, Jr., (D), Brown, 1971, Gc
W. Gary Hooks, (D), North Carolina, 1961, Gm
Ernest A. Mancini, (D), Texas A&M, 1974, Gro
Carl W. Stock, (D), North Carolina, 1977, Pi
Geochemical Research Lab Manager:
Sidhartha Bhattacharyya, (D), Alabama, 2010, CatGe

## University of South Alabama

**Dept of Earth Sciences** (B) ○ (2017)
5871 USA Dr. N.
Room 136
Mobile, AL 36688-0002
p. (251) 460-6381
dbeebe@southalabama.edu
http://www.usouthal.edu/earthsci
*Enrollment (2016): B: 135 (13)*

Chair:
Sytske K. Kimball, (D), Penn State, 2000, Ow
Professor:
Roy Ryder, (D), Florida, 1989, SdOry
Associate Professor:
David T. Allison, (D), Florida State, 1992, Gc
Keith G. Blackwell, (D), Texas A&M, 1990, Ow
Murlene W. Clark, (D), Florida State, 1983, PiGrg
Douglas W. Haywick, (D), James Cook, 1990, Gs
Carol F. Sawyer, (D), Texas State, 2007, GmOye
Assistant Professor:
Alex Beebe, (D), Clemson, 2013, Ge
Wes Terwey, (D), Colorado State, 2007, Ow
Instructor:
Karen J. Jordan, (M), Alabama, 2005, Oyr
Andrew Murray, Florida State, 2010, Ow
Diana Sturm, (D), Alabama, 2000, Geg
Sam Stutsman, (M), Alabama, 1996, Oiy
Emeritus:
Aaron Williams, (D), Oklahoma, 1971, Ow

**Dept of Marine Sciences** (M,D) (2015)
307 University Drive
Mobile, AL 36688-0002
p. (334) 460-7136
keyser@southalabama.edu
Administrative Assistant: Jan Keyser

Chair:
Sean Powers, (D), Ob
Professor:
Ronald P. Kiene, (D), SUNY (Stony Brook), 1986, Oc
Adjunct Professor:
Douglas W. Haywick, (D), James Cook, 1990, Gs
Erich M. Mueller, (D), Miami, 1983, Ob

## University of West Alabama

**Dept of Biological & Environmental Sciences** (B,M) (2017)
UWA Station 7
Livingston, AL 35470
p. (205) 652-3416
arindsberg@uwa.edu
http://www.uwa.edu/Biological_and_Environmental_Sciences.aspx

Director, Black Belt Museum:
James P. Lamb, (B), Alabama, PvePg
Associate Professor:
Andrew K. Rindsberg, (D), Colorado Mines, 1986, PeGePi

# Alaska

## Prince William Sound Community College

**Dept of Natural Sciences** (2011)
P.O. Box 97
Valdez, AK 99686
p. (907) 822-3673
rsbenda@pwscc.edu
http://www.pwscc.edu

## University of Alaska, Anchorage

**Dept of Geological Sciences** (B) ● (2016)
3211 Providence Drive
CPISB 101
Anchorage, AK 99508-4670
p. (907) 786-1298
uaa_mns@uaa.alaska.edu
http://www.uaa.alaska.edu/geology/
f: www.facebook.com/UAAGeologicalSciences
t: @UAAGeology
*Enrollment (2015): B: 60 (8)*

Chair:
Kristine J. Crossen, (D), Washington, 1997, Gl
Professor:
LeeAnn Munk, (D), Ohio State, 2001, Cle
Associate Professor:
Jennifer Aschoff, (D), Colorado Mines, 2008, GrdGo
Donald Matt Reeves, (D), Nevada, 2006, GeHgOn
Term Assistant Professor:
Terry R. Naumann, (D), Idaho, 1998, Gi
Assistant Professor:
Erin Shea, (D), MIT, 2014, CcGcCg
Term Instructor:
Peter J. Oswald, (M), Idaho, Gg
Mark Rivera, (M), New Mexico State, 2000, Gg

## University of Alaska, Fairbanks

**Alaska Quaternary Center** (2014)
907 Yukon Drive
Fairbanks, AK 99775-1200
p. (907) 474-7758
rmtopp@alaska.edu
Administrative Assistant: Leicha Welton

Professor:
W. Scott Ambruster, (D), California (Davis), 1981, On
Bruce P. Finney, (D), Oregon State, 1987, Ou
David M. Hopkins, (D), Harvard, 1955, Gg
Roger W. Powers, (D), Wisconsin, 1973, On
Dan L. Wetzel, (B), Alaska (Fairbanks), 1975, On
Research Associate:
Wendy H. Arundale, (D), Michigan State, 1976, On
Peter Bowers, (M), Washington State, 1980, Ga
Owen Mason, (D), Alaska, 1990, Ga
Adjunct Professor:
Daniel H. Mann, (D), Washington, 1983, Pe
Geology Librarian:
Judy Triplehorn, On

**Dept of Geosciences** (B,M,D) ○ (2016)
900 Yukon Drive
P.O. Box 755780
Fairbanks, AK 99775-5780
p. (907) 474-7565
pjmccarthy@alaska.edu
http://www.uaf.edu/geology/
*Enrollment (2010): B: 48 (11) M: 30 (9) D: 37 (6)*

Dean, CNSM:
Paul Layer, (D), Stanford, 1986, Ym
Chair:
Paul McCarthy, (D), Guelph, 1995, Gs

Associate Dean, CNSM :
Anupma Prakash, (D), Roorkee (India), 1966, Or
Professor:
James E. Beget, (D), Washington, 1981, Gm
Douglas Christensen, (D), Michigan, 1987, Ys
Hajo Eicken, (D), Bremen, 1990, Og
Jeffrey T. Freymueller, (D), South Carolina, 1991, Yd
Catherine L. Hanks, (D), Alaska (Fairbanks), 1991, Go
Regine M. Hock, (D), Swiss Fed Inst Tech, 1997, Ol
Jessica F. Larsen, (D), California (Santa Cruz), 1996, Gv
Rainer J. Newberry, (D), Stanford, 1980, Em
Vladimir Romanovsky, (D), Alaska (Fairbanks), 1996, Yg
Michael T. Whalen, (D), Syracuse, 1993, Gs
Associate Professor:
Bernard Coakley, (D), Columbia, 1991, Yg
Cary de Wit, (D), Kansas, 1997, Og
Patrick S. Druckenmiller, (D), Calgary, 2006, Pv
Sarah Fowell, (D), Columbia, 1994, Pl
Mary J. Keskinen, (D), Stanford, 1979, Gxz
Dan Mann, (D), Washington, 1983, Oy
Franz J. Meyer, (D), Tech (Munich), 2004, Or
Erin C. Pettit, (D), Washington, 2003, Ol
Assistant Professor:
Chris Maio, (D), Massachusetts (Boston), 2014, Oy
Elisabeth S. Nadin, (D), Caltech, 2007, Gtc
Carl Tape, (D), Caltech, 2009, Ys
Instructor:
Jochen E. Mezger, (D), Alberta, 1997, GcpCc
Emeritus:
Kenneth P. Severin, (D), California (Davis), 1987
David B. Stone, (D), Newcastle, 1963, Ym
Donald M. Triplehorn, (D), Illinois, 1961, Gs

**Dept of Mining & Geological Engineering** (B,M,D) (2014)
P.O. Box 755800
Fairbanks, AK 99775-5800
p. (907) 474-7388
fnedw@uaf.edu
http://www.alaska.edu/uaf/cem/ge/
Administrative Assistant: Judy L. Johnson
*Enrollment (2011): B: 10 (10) M: 2 (2) D: 2 (2)*
Professor:
Sukumar Bandopadhyay, (D), Penn State, 1982, Nm
Gang Chen, (D), Virginia Tech, 1989, Nm
Scott L. Huang, (D), Missouri (Rolla), 1981
Hsing K. Lin, (D), Utah, 1985, On
Paul A. Metz, (D), Imperial Coll (UK), 1991
Debsmita Misra, (D), HwOrNg
Daniel E. Walsh, (D), Alaska (Fairbanks), 1984, On
Assistant Professor:
Margaret M. Darrow, (D), Onn
Sabry Sabour Hafez, (D), Nm
Emeritus:
Rajive Ganguli, (D), Kentucky, 1999, Nm

**Museum** (1999)
Box 756960
Fairbanks, AK 99775-6960
p. (907) 474-7505
psdruckenmiller@alaska.edu
Administrative Assistant: Marie D. Ward
Curator:
Roland A. Gangloff, (D), California (Berkeley), 1975, Pi
Associate Scientist:
Paul E. Mattheus, (D), Alaska, 1997, Po
Adjunct Professor:
William A. Clemens, (D), California (Berkeley), 1960, Pv
Dale Guthrie, (D), Chicago, 1963, Pv
Donald M. Triplehorn, (D), Illinois, 1961, Gs

## University of Alaska, Southeast
**Dept of Natural Sciences** (2015)
11120 Glacier Hwy
Juneau, AK 99801
p. (907) 796-6163
cathy.connor@uas.alaska.edu
http://www.uas.alaska.edu

# Arizona

## Arizona State University
**School of Earth and Space Exploration** (B,M,D) (2014)
Box 871404
Tempe, AZ 85287-1404
p. (480) 965-5081
sese@asu.edu
http://sese.asu.edu
*Enrollment (2007): B: 84 (0) M: 37 (0) D: 43 (0)*
Director:
Kip V. Hodges, (D), MIT, 1982, Gc
Director, Astrobiology Program:
Jack D. Farmer, (D), California (Davis), 1978, Py
Professor:
Ariel Anbar, (D), Caltech, 1996, Cgs
Ramon Arrowsmith, (D), Stanford, 1995, GcmGt
Erik Asphaug, (D), Arizona, 1993, Xgy
Donald M. Burt, (D), Harvard, 1972, EgCgGz
Peter R. Buseck, (D), Columbia, 1962, Cg
Philip R. Christensen, (D), California (Los Angeles), 1981, Xg
Edward Garnero, (D), Caltech, 1994, Ys
Richard Hervig, (D), Chicago, 1979, Cg
Lawrence Krauss, (D), MIT, 1982, On
Stephen J. Reynolds, (D), Arizona, 1982, GctOe
Mark Robinson, (D), Hawaii, 1993, On
Steven Semken, (D), MIT, 1989, OenGg
Everett Shock, (D), California (Berkeley), 1987, On
Sumner Starrfield, (D), California (Los Angeles), YhsYv
James A. Tyburczy, (D), Oregon, 1983, GyzYx
Meenakshi Wadhwa, (D), Washington, 1994, On
Kelin Whipple, (D), Washington, 1999, Gt
Stanley N. Williams, (D), Dartmouth, 1983, Gv
Associate Professor:
Steven Desch, (D), Illinois (Urbana Champaign), 1998, On
Hilairy E. Hartnett, (D), Washington, 1998, ComOc
Arjun Heimsath, (D), California (Berkeley), 1999, Gc
James Rhoads, (D), Princeton, 1997, On
Thomas G. Sharp, (D), Arizona State, 1990, Gz
Assistant Professor:
Amanda Clarke, (D), Penn State, 2002, Gv
Christopher Groppi, (D), Arizona, 2003, On
Allen McNamara, (D), Michigan, 2002, Yx
Srikanth Saripalli, (D), S California, 2007, On
Evan Scannapieco, (D), California (Berkeley), 2001, On
Patrick Young, (D), Arizona, 2004, On
Research Associate:
Mikhail Y. Zolotov, (D), Vernadsky Inst (Russia), 1990, Cg
Emeritus:
L. Paul Knauth, (D), Caltech, 1973, Cs
Robert F. Lundin, (D), Illinois, 1962, Pm

## Arizona Western College
**Dept of Geosciences** (A) (2016)
2020 S. Avenue 8E
Yuma, AZ 85365
p. (928) 317-6000
fred.croxen@azwestern.edu
*Enrollment (2016): A: 3 (0)*
Professor:
Fred W. Croxen III, (M), N Arizona, 1977, HwPvOi
Catherine B. Hill, (M), GgOag
Adjunct Professor:
Earl E. Burnett, (B), Arizona, 1965, HwYgGe
Kelly L. Esslinger, (M), OapGg
Maureen Garrett, (M), GgeOg
Tia McCraley, (M), Gge
Todd Pinnt, (M), Oyi

## Central Arizona College
**Dvision of Science** (2011)
Coolidge, AZ 85228
p. (602) 426-4444
diane.beecroft@centralaz.edu

Instructor:
Allan E. Morton, (M), Brigham Young, 1975, Oe

## Cochise College
**Geology Dept** (A) (2016)
901 North Colombo Avenue
Sierra Vista, AZ 85635
p. (520) 515-5425
priddise@cochise.edu
http://skywalker.cochise.edu/wellerr/aawellerweb.htm
Instructor:
Joann Deakin, (M), Mast, Ggg

## Dine' College
**Science Dept** (A,B) (2017)
#1 Circle Drive
Tsaile, Navajo Nation, AZ 86556
p. (928) 724-6721
cacate@dinecollege.edu
http://www.dinecollege.edu
*Enrollment (2002): A: 18 (0)*
Chair:
Don Robinson, (D), Maharishi Management
Instructor:
Margaret Mayer, (M), Rhode Island, GeOn

## Mesa Community College
**Dept of Physical Science** (A) (2011)
1833 W. Southern Avenue
Mesa, AZ 85202
p. (480) 461-7015
kelli.santistevan@mesacc.edu
http://www.mesacc.edu/dept/d43/glg/
Professor:
Donna M. Benson, (M), Arizona State, 1996, Gg
Kaatje Kraft, (M), Arizona State, 1999, Gg
Robert S. Leighty, (D), Arizona State, 1997, Og
Adjunct Professor:
Zack Bowles, (M), Arizona State, Gg
Chloe Branciforte, (M), SD Mines, Gg
Mike Grivois, (M), Texas Christian
Steve Guggino
Jack Kepper
Jill Lockard
Tony Occhiuzzi, (M), Arizona State, On
Donna Pollard, (M)
Joanna Scheffler, (M), Washington State, Gg
Melinda Shimizu, (M), Arizona State, 2008, Gg
Carolyn Taylor, (M), Georgia State, Gg
Kelli Wakefield, (M), Arizona State, Gg
Emeritus:
Armand J. Lombard, (M), Arizona, 1977, Gg

## Northern Arizona University
**Dept of Geography, Planning, and Recreation** (B,M) (2013)
P.O. Box 15016
Flagstaff, AZ 86011
p. (928) 523-2650
geog@nau.edu
http://nau.edu/sbs/gpr/
*Enrollment (2013): B: 39 (15) M: 44 (17)*
Park Ranger Training Program Director:
Mark J. Maciha, (M), N Arizona, 2009, OnnOn
Professor and Chair:
Pamela Foti, (D), Wisconsin, 1988, On
Coordinator, Parks and Recreation Management Program:
Charles Hammersley, (D), New Mexico, 1988, On
Professor:
Alan A. Lew, (D), Oregon, 1986, On
Associate Professor:
Rebecca D. Hawley, (D), Arizona State, 1994, On
Ruihong Huang, (D), Wisconsin (Milwaukee), 2003
Assistant Professor:
Erik Schiefer, (D), British Columbia, 2004, GmOyi
Senior Lecturer:
Judith Montoya, (M), New Mexico, 1985, Oy
Lecturer - Outdoor Leadership and Education:
Aaron Divine, (M), N Arizona, 2004, OnnOn
Distance Learning Lecturer:
R. Marieke Taney, (M), N Arizona, 2002, OnnOn
Emeritus:
Graydon L. Berlin, (D), Tennessee, 1970, Or
Robert O. Clark, (D), Denver, 1970, On
Carolyn M. Daugherty, (D), Arizona State, 1987, On
Leland R. Dexter, (D), Colorado, 1986, Oy
Christina B. Kennedy, (D), Arizona, 1989, On
Stanley W. Swarts, (D), California (Los Angeles), 1975, Oy
George A. Van Otten, (D), Oregon State, 1977, On

**School of Earth Sciences and Environmental Sustainability** (B,M) (2014)
P O Box 4099
Building 12, Room 100
Knoles Drive
Flagstaff, AZ 86011-4099
p. (928) 523-4561
mary.reid@nau.edu
http://www.cefns.nau.edu/Academic/Geology
*Enrollment (2009): B: 108 (15) M: 29 (8)*
Chair:
Mary R. Reid, (D), MIT, 1987, Gi
Professor:
Ronald C. Blakey, (D), Iowa, 1973, Gr
David S. Brumbaugh, (D), Indiana, 1972, Ys
Ernest M. Duebendorfer, (D), Wyoming, 1986, Gct
David K. Elliott, (D), Bristol (UK), 1979, Pv
Thomas D. Hoisch, (D), S California, 1985, Gp
Darrell S. Kaufman, (D), Colorado, 1991, Gm
Michael H. Ort, (D), California (Santa Barbara), 1991, Gv
Roderic A. Parnell, (D), Dartmouth, 1982, Cl
Nancy Riggs, (D), California (Santa Barbara), 1991, Gv
James C. Sample, (D), California (Santa Cruz), 1986, Cl
Abraham E. Springer, (D), Ohio State, 1994, Hw
Paul J. Umhoefer, (D), Washington, 1989, Gt
Associate Professor:
Larry T. Middleton, (D), Wyoming, 1979, Gs
Research Associate:
Joseph E. Hazel, Jr., (M), N Arizona, 1991, Gs
Matthew A. Kaplinski, (M), N Arizona, 1990, Yg
Adjunct Professor:
Wendell A. Duffield, (D), Stanford, 1967, Gv
David D. Gillette, (D), S Methodist, 1974, Pv
John H. Sass, (D), Australian National, 1965, Yg
Wesley Ward, (D), Washington, 1978, Gs
Emeritus:
Charles W. Barnes, (D), Wisconsin, 1965, GcXg
Stanley S. Beus, (D), California (Los Angeles), 1963, Pi
Augustus S. Cotera, Jr., (D), Texas, 1962, Gg
Richard F. Holm, (D), Washington, 1969, Gv
James I. Mead, (D), Arizona, 1983, Po
Paul Morgan, (D), London, 1973, Yh
Jack D. Nations, (D), California (Berkeley), 1969, Pm
Laboratory Director:
James Wittke, (D), Texas, 1984, Gi
Office Manager:
Susan M. Sabala-Foreman, (A), N Arizona, 1981, On

## Phoenix College
**Dept of Physical Sciences** (A) (2016)
1202 West Thomas Road
Phoenix, AZ 85013
p. (602) 285-7244
abeer.hamdan@phoenixcollege.edu
Chair:
James J. White, (D), Arizona, 1986
Professor:
Richard Cups, (M), Arizona State, 1983, Gg
Don Speed, (M), Arizona, 1980, Gg

## Pima Community College, Community Campus

**Geology Dept** (1999)
401 North Bonita Ave.
Tucson, AZ 85709-5000
p. (520) 206-4500
wrcavanaugh@pima.edu
http://www.pima.edu

## Pima Community College, West Campus

**Geology Dept** (1999)
2202 West Anklam Rd.
Tucson, AZ 85709-0001
p. (520) 206-4500
jrasmussen@pima.edu
http://www.pima.edu

## Prescott College

**Dept of Environmental Studies** (B) (2014)
220 Grove Avenue
Prescott, AZ 86301
p. (928) 350-2256
kurt.refsnider@prescott.edu
http://www.prescott.edu

Professor:
Kurt Refsnider, (D), Colorado, 2012, GmlPe

## Scottsdale Community College

**Dept of Mathematics and Sciences** (1999)
9000 East Chaparral Road
Scottsdale, AZ 85256-2626
p. (480) 423-6111
roberto.ribas@scottsdalecc.edu
http://www.scottsdalecc.edu/academics/departments/math-sciences

## South Mountain Community College

**Geology Program** (1999)
7050 South 24th Street
Phoenix, AZ 85042-5806
p. (602) 243-8290
sian.proctor@southmountaincc.edu
http://www.southmountaincc.edu/math-science-engineering/program/geology/

## University of Arizona

**Dept of Geosciences** (B,M,D) ○ (2015)
Gould-Simpson Building
1040 E. Fourth Street
Tucson, AZ 85721-0077
p. (520) 621-6000
achase@email.arizona.edu
http://www.geo.arizona.edu
Administrative Assistant: Anne Chase
Administrative Assistant: Sylvia Quintero
*Enrollment (2006): B: 70 (0) M: 23 (0) D: 52 (0)*

Head:
Peter Reiners, (D), Washington, Cg

Professor:
Victor R. Baker, (D), Colorado, 1971, GmXgGh
Mark D. Barton, (D), Chicago, 1981, Eg
Susan L. Beck, (D), Michigan, 1987, Ys
Richard Bennett, (D), MIT, 1995, YdGt
Jon Chorover, California (Berkeley), 1993, Sc
Andrew S. Cohen, (D), California (Davis), 1982, Ps
Julia E. Cole, (D), Columbia, 1992, PeCsOa
Owen K. Davis, (D), Minnesota, 1981, Pl
Peter G. DeCelles, (D), Indiana, 1984, Gd
Robert T. Downs, (D), Virginia Tech, 1992, Gi
Mihai N. Ducea, (D), Caltech, 1998, Gt
Karl W. Flessa, (D), Brown, 1973, Po
Jiba Ganguly, (D), Chicago, 1967, Gx
George E. Gehrels, (D), Caltech, 1986, Gt
Vance Holliday, (D), Colorado, 1982, GamSa
Roy A. Johnson, (D), Wyoming, 1984, Ys
A. J. Timothy Jull, (D), Bristol, 1976, CclCg
Paul Kapp, (D), California (Los Angeles), 2001, GtcGm
Alfred S. McEwen, (D), Arizona State, 1988, Xg
Jonathan T. Overpeck, (D), Brown, 1985, Oa
Jon D. Pelletier, (D), Cornell, 1997, Gm
Mary Poulton, (D), Arizona, 1990, Ng
Jay Quade, (D), Utah, 1990, Sc
Randall M. Richardson, (D), MIT, 1978, Yg
Joaquin Ruiz, (D), Michigan, 1983, Cgc
Timothy Swindle, (D), Washington, 1986, Xc
Connie Woodhouse, (D), Arizona, 1996, Pe

Associate Professor:
Barbara Carrapa, (D), Vrije (Amsterdam), 2002, Gs
Joellen Russell, (D), Scripps, 1999, Oc
Eric Seedorff, (D), Stanford, 1987, Eg
Jessica Tierney, (D), Brown, 2010, Co
Marek Zreda, (D), New Mexico Tech, 1994, Cc

Assistant Professor:
Jennifer McIntosh, (D), Michigan, 2004, Hy
Matthew Steele-MacInnes, (D), Virginia Tech, 2013, Eg
Jianjun Yin, (D), Illinois (Urbana-Champaign), Og

Lecturer:
Paul Goodman, Washington, 2000, Oa
Jessica Kapp, (D), California (Los Angeles), Gg

Adjunct Professor:
Robert J. Kamilli, (D), Harvard, 1976, EgmCg
Charles Prewitt, (D), MIT, 1969, On
Marc Sbar, (D), Columbia, 1972, Yes

Emeritus:
William B. Bull, (D), Stanford, 1960, Gm
Clement G. Chase, (D), California (San Diego), 1970, Yg
George H. Davis, (D), Michigan, 1971, Gc
William R. Dickinson, (D), Stanford, 1958, Gt
John M. Guilbert, (D), Wisconsin, 1962, Em
DeVerle P. Harris, (D), Penn State, 1965, Eg
C. Vance Haynes, Jr., (D), Arizona, 1965, Ga
Everett H. Lindsay, (D), California (Berkeley), 1967, Pv
Edgar J. McCullough, Jr., (D), Arizona, 1963, Ge
Jonathan Patchett, (D), Edinburgh (UK), 1976
Joseph Schreiber, (D)
Spencer R. Titley, (D), Arizona, 1958, Eg
George Zandt, (D), MIT, 1978, Ys

Researcher:
David Dettman, (D), Michigan, 1994, Cs
Chris J. Eastoe, (D), Tasmania, 1979, CsGg

**Dept of Hydrology & Atmospheric Sciences** (B,M,D) ● (2014)
226A Harshbarger Building #11
PO Box 210011
Tucson, AZ 85721-0011
p. (520) 621-5082
programs@hwr.arizona.edu
http://www.hwr.arizona.edu
Administrative Assistant: Erma Santander
*Enrollment (2014): B: 0 (2) M: 23 (2) D: 27 (1)*

Head:
Thomas Maddock, III, (D), Harvard, 1972, Hq

Regents Professor:
Victor R. Baker, (D), Colorado, 1971, GmXgGh
W. James Shuttleworth, (D), Manchester (UK), 1971, Hs

Director, SAHRA Center:
Juan B. Valdes, (D), MIT, 1976, Hs

Professor:
Paul D. Brooks, (D), Colorado, 1995, Cg
Paul A. Ferre, (D), Waterloo, 1997, Hw
Hoshin V. Gupta, (D), Case Western, 1984, HqOga
Thomas Meixner, (D), Arizona, 1999, Cg
Peter A. Troch, (D), Ghent, 1993, Hs
Tian-Chyi Jim Yeh, (D), New Mexico Tech, 1983, Hq
Marek G. Zreda, (D), New Mexico Tech, 1994, Hw

Associate Professor:
Michael D. Bradley, (D), Michigan, 1971, On
Jennifer C. McIntosh, (D), Michigan, 2004, Cg

Associate Director, SAHRA Center:
James C. Washburne, (D), Arizona, 1994, HsOge

Associate Director, SAHRA Center:
Gary C. Woodard, (D), Michigan, 1981, On

Adjunct Associate Professor:
David C. Goodrich, (D), Arizona, 1990, Hs
Robert H. Webb, (D), Arizona, 1985, Gm
Adjunct Assistant Professor:
James E. Smith, (D), Waterloo, 1995, Hw
Adjunct Professor:
Roger C. Bales, (D), Caltech, 1984, Cg
David Hargis, (D), Arizona, Hwq
Leo S. Leonhart, (D), Arizona, 1978, Hw
Robert MacNish, (D), Michigan, 1966, Hw
Soroosh Sorooshian, (D), California (Los Angeles), 1978, Hg
Donald W. Young, (D), Arizona, 1994, On
Regents Professor:
Shlomo P. Neuman, (D), California (Berkeley), 1968, HwqGq
Emeritus:
Robert A. Clark, (D), Texas A&M, 1964, Hs
Lucien Duckstein, (D), Colorado State, 1962, Hq
Simon Ince, (D), Iowa, 1953, Hs
Austin Long, (D), Arizona State, 1966, Cl
Ernest T. Smerdon, (D), Missouri State, 1959, Hs
Arthur W. Warrick, (D), Iowa State, 1967, Sp
Peter J. Wierenga, (D), California (Davis), 1968, Sp
Lorne G. Wilson, (D), California (Davis), 1962, Hw
Cooperating Faculty:
Mark L. Brusseau, (D), Florida, 1989, Hw
Jonathan D. Chorover, (D), California (Berkeley), 1993, Cg
Bonnie C. Colby, (D), Wisconsin, 1983, On
R. B. Hawkins, (D), Colorado State, 1968, Hs
Katherine K. Hirschboeck, (D), Arizona, 1985, Pe
Kevin E. Lansey, (D), Texas, 1987, Hs
Sharon B. Megdal, (D), Princeton, 1981, Hg
Steven L. Mullen, (D), Washington, 1985, Oa
Robert G. Varady, (D), Arizona, 1981, On
Marvin Waterstone, (D), Rutgers, 1983, On

**Dept of Mining & Geological Engineering** (B,M,D) (2017)
1235 E. James E. Rogers Way
Mines Building, Room 229
PO Box 210012
Tucson, AZ 85721-0012
p. (520) 621-6063
ENGR-Mining@email.arizona.edu
http://www.mge.arizona.edu
f: https://www.facebook.com/UAMining/
*Enrollment (2012): B: 75 (18) M: 25 (12) D: 20 (4)*
Professor of Practice:
Victor O. Tenorio, (D), Arizona, 2012, Nmx
Head:
John M. Kemeny, (D), California (Berkeley), 1986, Nr
Director, Lab for Advanced Subsurface Imaging:
Ben K. Sternberg, (D), Wisconsin, 1977, Ye
Professor:
J. Brent Hiskey, (D), Utah, 1973, Nx
Pinnaduwa H. S. W. Kulatilake, (D), Ohio State, 1981, Nr
Mary M. Poulton, (D), Arizona, 1990, Ng
Associate Professor:
Sean Dessureault, (D), British Columbia, 2002, Nx
Jaeheon Lee, (D), Arizona, 2004, Nx
Moe Momayez, (D), McGill, NmYxNr
Jinhong Zhang, (D), Virginia Tech, 2006, Nm
Assistant Professor:
Kwangmin Kim, (D), Arizona, 2012, NmrNx

**Dept of Soil, Water & Environmental Science** (B,M,D) (2014)
PO Box 210038
Shantz 429
Tucson, AZ 85721-0038
p. (520) 621-1646
sleavitt@ltrr.arizona.edu
https://swes.cals.arizona.edu
*Enrollment (2001): B: 49 (18) M: 17 (3) D: 42 (15)*
Head:
Jonathan Chorover, (D), California (Berkeley), 1993, ScClo
Associate Director, SAHRA Center:
Katharine L. Jacobs, (M), California (Berkeley), 1981, Hg
Adjunct Professor:
Paul (Ty) Ferre, (D), So
Jim Yeh, (D), Sp
Professor:
Mark L. Brusseau, (D), Florida, 1989, Hw
Kevin Fitzsimmons, (D), Arizona, 1999, On
Charles P. Gerba, (D), Miami, 1973, Sb
Martha C. Hawes, (D)
Raina M. Maier, (D), Rutgers, 1988, On
Ian L. Pepper, (D), Ohio State, 1975, On
Craig Rasmussen, (D), GmCl
Charles Sanchez, (D), Iowa State, 1986, On
Jeffrey C. Silvertooth, (D), Oklahoma State, 1986, So
Markus Tuller, (D), On
James Walworth, (D), Georgia, 1985, Soc
Arthur W. Warrick, (D), Iowa State, 1967, Sp
Associate Professor:
Joan E. Curry, (D), California (Davis), 1992, Sc
Marcel Schaap, (D), So
Adjunct Assistant Professor:
Susan Moran, So
Assistant Professor:
Karletta Chief, (D), Arizona, 2007, On
Associate Specialist:
Michael Crimmins, (D), Arizona, On
Associate Research Scientist:
Janick F. Artiola, (D), Arizona, 1980, ScHsCa
Associate Professor of Practice:
Thomas B. Wilson, (D), On
Adjunct Professor:
Floyd Adamsen, (D), Colorado State, 1983, So
Emeritus:
Ed Glenn, (D), Hawaii, 1978, On
Donald F. Post, (D), Purdue, 1967, Sd
James Riley, (D), Arizona, 1968, On
Peter J. Wierenga, (D), California, 1968, Sp
Extension Specialist:
Paul W. Brown, (D), Wisconsin, 1981, Ow

**Laboratory of Tree Ring Research** O (2017)
Bryant Bannister Tree-Ring Bldg
1215 E. Lowell St.
Tucson, AZ 85721
p. (520) 621-1608
webmaster@ltrr.arizona.edu
http://ltrr.arizona.edu/
f: Laboratory of Tree-Ring Research
t: @TreeRingLabUA
Professor:
David C. Frank, (D), Bern, 2005
Professor:
Malcolm K. Hughes, (D), Durham (UK), 1970, Pe
Steven W. Leavitt, (D), Arizona, 1982, Csc
Russ Monson, (D), Washington State, 1982
Associate Professor:
Katherine K. Hirschboeck, (D), Arizona, 1985, OaHsg
Paul Sheppard, (D), 1995
Ronald Towner, (D), Arizona, 1997, Gae
Valerie Trouet, (D), KU Leuven (Belgium), 2004, PeOya
Assistant Professor:
Charlotte Pearson, (D), Reading, 2003
Research Associate:
Matthew W. Salzer, (D), Arizona, 2000, PeGav
Richard L. Warren, (B), Arizona, 1962, On
Emeritus:
Jeffrey S. Dean, (D), Arizona, 1967, On
Harold C. Fritts, (D), Ohio State, 1956, Pe
Charles W. Stockton, (D), Arizona, 1971, Hq
Thomas W. Swetnam, (D), Arizona, 1987, Pe
Research Professor:
David M. Meko, (D), Arizona, 1981, Hq
Ramzi Touchan, (D), Arizona, 1991
Research Associate:
Peter Brewer, (D), Reading, 2004
Curator:
Pearce Paul Creasman, (D), Texas A&M, 2008
Assoc. Research Professor:
Irina Panyushkina, (D), Sukachev Inst of Forest, 1997
Tomasz Wazny, (D), Hamburg, 1990

Asistant Research Professor:
Margaret Evans, (D), Arizona, 2003

**Lunar and Planetary Laboratory** (M,D) (2010)
1629 E. University Boulevard
Tucson, AZ 85721-0092
p. (520) 621-6963
acad_info@lpl.arizona.edu
http://www.lpl.arizona.edu
*Enrollment (2001): D: 33 (6)*
Head:
Michael J. Drake, (D), Oregon, 1972, Xc
Research Professor:
Martin G. Tomasko, (D), Princeton, 1969, Oa
Professor:
Victor R. Baker, (D), Colorado, 1971, XgGmh
William V. Boynton, (D), Carnegie Mellon, 1971, XcOr
Robert H. Brown, (D), Hawaii, 1982, Xc
Uwe Fink, (D), Penn State, 1965, Oa
Tom Gehrels, (D), Chicago, 1956, On
Richard J. Greenberg, (D), MIT, 1972, Xy
William B. Hubbard, (D), California (Berkeley), 1967, XyYvOa
Donald M. Hunten, (D), McGill, 1950, Oa
Jack R. Jokipii, (D), Caltech, 1965, Xy
Harold P. Larson, (D), Purdue, 1967, On
Dante Lauretta, (D), Washington Univ (St. Louis), 1997, XcmXg
John S. Lewis, (D), California (San Diego), 1968, Xcm
Jonathan I. Lunine, (D), Caltech, 1985, Xc
Renu Malhotra, (D), Cornell, 1988
Alfred McEwen, (D), Arizona State, 1988, Xg
Henry J. Melosh, (D), Caltech, 1972, Gt
George H. Rieke, (D), Harvard, 1969, On
Timothy D. Swindle, (D), Washington (St. Louis), 1986, Xm
Roger Yelle, (D), Wisconsin, 1984, On
Senior Research Scientist:
Lonnie ("Lon") L. Hood, (D), California (Los Angeles), 1979, XyOa
Larry A. Lebofsky, (D), MIT, 1974, OngXm
Senior Scientist:
Jozsef Kota, (D), Roland Eotvos, 1980, Oa
David Kring, (D), Harvard, 1989, XcGiXg
Associate Professor:
Caitlin Griffith, (D), SUNY (Stony Brook), 1991, On
Robert Kursinski, (D), Caltech, 1997, Oa
Associate Research Scientist:
Robert S. McMillan, (D), Texas (Austin), 1977, OnnOn
Assistant Professor:
Joe Giacalone, (D), Kansas, 1991, Xy
Adam Showman, (D), Caltech, 1998, On
Senior Research Scientist:
Lyle A. Broadfoot, (D), Saskatchewan, 1963, Or
Jay B. Holberg, (D), California (Berkeley), 1974, Oa
Bill R. Sandel, (D), Rice, 1972, Oa
Research Associate:
Alexander Dessler, (D), Duke, 1956, On
Ralph Lorenz, (D), Kent (UK), 1994, Xg
Elizabeth Turtle, (D), Arizona, 1998, Xg
Associate Research Scientist:
Peter Smith, (M), Arizona, 1977, Or
Emeritus:
Elizabeth Roemer, (D), California (Berkeley), 1955, On
Charles P. Sonett, (D), California (Los Angeles), 1954, Yg
Robert G. Strom, (M), Stanford, 1957, Xg
Graduate Secretary:
Mary Guerrieri, (N), Pima Comm Coll, 1990, On

**School of Geography and Development** (B,M,D) (2014)
Harvill Building
PO Box 210076
Tucson, AZ 85721
p. (520) 621-1652
robbins@email.arizona.edu
http://geog.arizona.edu/
*Enrollment (2010): B: 350 (0) M: 20 (0) D: 54 (0)*
Director:
Paul Robbins, (D), Clark, 1996
Dean:
John Paul Jones, III, (D), Ohio State, 1984, On
Co-Director of the University of Arizona Institute of the Environment:
Diana Liverman, (D), California (Los Angeles), 1984
Professor:
Michael E. Bonine, (D), Texas, 1975, Ou
Andrew Comrie, (D), Pennsylvania, 1992, Oy
Sallie A. Marston, (D), Colorado, 1986, Ou
Beth A. Mitchneck, (D), Columbia, 1990, On
David A. Plane, (D), Pennsylvania, 1981, On
Stephen R. Yool, (D), California (Santa Barbara), 1985, Or
Lecturer:
Dereka Rushbrook, (D), Arizona, 2005
Associate Professor:
Keiron D. Bailey, (D), Kentucky, 2002, Oi
Carl J. Bauer, (D), California (Berkeley), 1995, Hs
Sandy Dall'Erba, (D), Pau, 2004
Elizabeth Oglesby, (D), California (Berkeley), 2002, On
Christopher A. Scott, (D), Cornell, 1998, Hw
Willem van Leeuwen, (D), Arizona, 1995, Oy
Marvin Waterstone, (D), Rutgers, 1983, On
Margaret Wilder, (D), Arizona, 2002, On
Director of GIST:
Christopher Lukinbeal, (D), San Diego State, 2000, Oi
Assistant Professor:
Jeffrey Banister, (D), Arizona, 2010
Gary Christopherson, (D), Arizona, 2000, Oi
Sarah Moore, (D), Kentucky, 2006
Tracey Osborne, (D), California (Berkeley), 2010
Daoqin Tong, (D), Ohio State, 2007, Oi
Adjunct Professor:
Julio Betancourt, (D), Arizona, 1989, On
Vance Holliday, (D), Colorado, 1982, Ga
Laura Huntoon, (D), Pennsylvania, 1991, On
Charles F. Hutchinson, (D), California (Riverside), 1978, Or
Miranda Joseph, (D), Stanford, 1995, On
Barbara Morehouse, (D), Arizona, 1993, On
Thomas W. Swetnam, (D), Arizona, 1987, On
Director, International Studies-External Affairs:
Wayne Decker, (D), Johns Hopkins, 1979

# Arkansas

## Arkansas Tech University
**Dept of Physical Sciences-Geology** (B) (2015)
1701 N. Boulder Ave
McEver Science Building
Russellville, AR 72801
p. (479) 968-0293
jmusser@atu.edu
http://www.atu.edu/physci
*Enrollment (2014): B: 44 (10)*
Professor:
Cathy Baker, (D), Iowa, 1986, Gg
Assistant Professor:
Genet Duke, (D), South Dakota Mines
Jason A. Patton, (D), Arkansas, 2008, GgeGc

## Northwest Arkansas Community College
**Northwest Arkansas Community College** (1999)
One College Drive
Bentonville, AR 72712
p. (479) 631-8661
dandroes@nwacc.edu
http://www.nwacc.edu

## Ozarka College
**Math, Science, and Education** (1999)
218 College Drive
Melbourne, AR 72556
hayers@ozarka.edu
http://www.ozarka.edu/

## University of Arkansas at Little Rock
**Dept of Earth Sciences** (B) O (2016)
2801 South University Avenue

California

Little Rock, AR 72204-1099
p. (501) 569-3546
memcmillan@ualr.edu
http://www.ualr.edu/earthsciences/
Administrative Assistant: Ginny L. Oswalt
*Enrollment (2016): B: 48 (8)*
Chair:
Margaret E. McMillan, (D), Wyoming, 2003, Gm
Professor:
Jeffrey B. Connelly, (D), Tennessee, 1993, GcNg
Assistant Professor:
Michael T. DeAngelis, (D), Tennessee, 2011, GpCp
Laura S. Ruhl, (D), Duke, 2012, GeClGb
Rene A. Shroat-Lewis, (D), Tennessee, 2011, PeOe
Instructor:
Joshua C. Spinler, (D), Arizona, 2014, YdGt
Emeritus:
Philip L. Kehler, (D), S Methodist, 1970, Gr
Michael T. Ledbetter, (D), Rhode Island, 1978, Og

## University of Arkansas, Fayetteville

**Arkansas Water Resources Center** (2004)
113 Ozark Hall
Fayetteville, AR 72701
p. (501) 575-4403
haggard@uark.edu
http://www.uark.edu/depts/awrc
Administrative Assistant: Theresa J. Russell
Director:
Kenneth F. Steele, (D), North Carolina, 1971, Cg
Research Associate:
Terry E. Nichols, (D), Chicago, 1973, On
Laboratory Director:
Marc A. Nelson, (D), Arkansas, 1992, Cl

**Dept of Geosciences** (B,M,D) O (2017)
340 N. Campus Drive
216 Gearhart Hall
Fayetteville, AR 72701
p. (479) 575-3355
lmilliga@uark.edu
http://fulbright.uark.edu/departments/geosciences/
f: https://www.facebook.com/uageosciences/0396/
*Enrollment (2016): B: 202 (43) M: 66 (24) D: 14 (0)*
Maurice Storm Endowed Chair:
Christopher Liner, (D), Colorado Mines, 1989, YseGg
Distinguished Professor:
David W. Stahle, (D), Arizona State, 1990, GegGg
Assistant Professor:
Edward C. Holland, (D), Colorado (Boulder), unkn, Ou
Professor:
Stephen K. Boss, (D), North Carolina, 1994, GeYgGg
John C. Dixon, (D), Colorado, 1983, GmSdGg
Margaret J. Guccione, (D), Colorado, 1982, GmaGg
John G. Hehr, (D), Michigan State, 1971, Owy
Thomas R. Paradise, (D), Arizona State, 1993, OgyGm
University Professor:
W. Fred Limp, (D), Indiana, 1983, Oi
Director of the Center for Advanced Spatial Technologies:
Jackson D. Cothren, (D), Ohio State, 2004, Oi
Chair:
Ralph K. Davis, (D), Nebraska, 1992, HwGgg
Associate Professor:
Fiona M. Davidson, (D), Nebraska, 1991, Ou
Jason A. Tullis, (D), South Carolina, 2003, Ori
Assistant Professor:
Mohamed H. Aly, (D), Texas A&M, 2006, OriGg
Matthew Covington, (D), California, 2008, HqGmg
Gregory Dumond, (D), Massachusetts (Amherst), 2008, GtcGg
Song Feng, (D), Chinese Acad of Sci, 1999, OayOw
Adriana Potra, (D), Florida International, 2011, EgCcGg
John B. Shaw, (D), Texas (Austin), 2013, GsrGg
Xuan Shi, (D), West Virginia, 2007, Ori
Celina Suarez, (D), Kansas, 2010, ClPeGg
Associate Research Professor:
Phillip D. Hays, (D), Texas A&M, 1996, HwCsGg

Instructor:
Paula Anderson, (M), Arkansas, 2012, Gg
Rashauna Hintz, (M), Arkansas, Ou
Henry Turner, III, (D), Arkansas, 2010, GgtGc
Adjunct Professor:
Fred Paillet, (D), Rochester, unkn
Emeritus:
J. Van Brahana, (D), Missouri, 1973, HwGe
Malcolm Cleaveland, (D), Arizona, 1983, Pe
Thomas O. Graff, (D), Kansas, 1973, Ou
Walter Manger, (D), Iowa, Gr
Ken Steele, (D), North Carolina, HgCgGg
Doy L. Zachry, (D), Texas, 1969, Gr

# California

## American River College

**Dept of Earth Science** (A) (2014)
4700 College Oak Drive
Sacramento, CA 95841
p. (916) 484-8107
aubertj@arc.losrios.edu
http://arc.losrios.edu/~earthsci/
Chair:
Melissa H. Levy, (M), E Tennessee State, 1994, GgOgy
Professor:
John E. Aubert, (M), California (Davis), 1994, Oy
Charles E. Thomsen, (M), Cal State (Chico), 1994, Oy
GIS Coordinator:
Hugh H. Howard, (D), Kansas, 2003, Oi
Assistant Professor:
Glenn Jaecks, (D), California (Davis), 2002, GgOgPg
Adjunct:
Paul M. Veisze, (M), California (Berkeley), 1985, Oi
Adjunct Professor:
Terry J. Boroughs, (M), Ohio State, 1992, GgXgCg
Beth Dushman, (M), California (Davis), 2007, Gg
Robert Earle, (M), Oi
Nathan Jennings, (M), Oi
Tom Lupo, (M), Oi
Richard L. Oldham, (M), Nevada, 1972, Gg
Kimberly Olson, (M), Cal State (Chico), 2008, Oye
Steven C. Smith, (M), Cal State (Chico), 1994, OynOn

## Antelope Valley College

**Math, Science, and Engineering Div-Geosciences Prog** (A) (2016)
3041 West Avenue K
Lancaster, CA 93536
p. 661-722-6300
mpesses@avc.edu
https://www.avc.edu/sites/default/files/catalog/geosciences.pdf
Associate Professor:
Michael Pesses, Oyi
Instructor:
Aurora Burd, (D), GgOg

## Bakersfield College

**Physical Science Dept** (2010)
1801 Panorama Drive
Bakersfield, CA 93305
p. (661) 395-4391
yadira.guerrero@bakersfieldcollege.edu
Acting Chair:
Robert A. Schiffman, (M), California (Santa Barbara), 1971, Ga
Professor:
John C. Lyman, (B), Cal State (San Diego), 1962, Gg
Instructor:
Robert Lewy, (M), Cal State (Bakersfield), 1992, Gx
Michael Oldershaw, (M), California (Davis), 1987, Eo

## Cabrillo College

**Dept of Earth Science** (A) (2015)
6500 Soquel Drive
Aptos, CA 95003
p. (408) 479-6495

daschwar@cabrillo.edu
http://www.cabrillo.edu/academics/earthscience/
*Enrollment (2015): A: 35 (0)*

Chair:
David Schwartz, (M), San Jose State, 1983, Gu

## California Institute of Technology

**Div of Geological & Planetary Sciences** (B,M,D) (2015)
1200 East California Boulevard
MC 170-25
Pasadena, CA 91125
p. (626) 395-6111
shechet@gps.caltech.edu
http://www.gps.caltech.edu
*Enrollment (2015): B: 19 (4) M: 0 (16) D: 113 (11)*

Chair:
John P. Grotzinger, (D), Virginia Tech, 1985, GsPeXg

Professor:
Jess F. Adkins, (D), MIT, 1997, Cm
Paul D. Asimow, (D), Caltech, 1997, CpGzy
Jean-Philippe Avouac, (D), Inst Physique du Globe de Paris, 1991, GtYs
Geoffrey A. Blake, (D), Caltech, 1985, Xc
Michael E. Brown, (D), California (Berkeley), 1994, Xy
Robert W. Clayton, (D), Stanford, 1981, Ye
John M. Eiler, (D), Wisconsin, 1994, Cg
Charles Elachi, (D), Caltech, 1971, Xg
Kenneth A. Farley, (D), California (San Diego), 1991, Cg
Woodward W. Fischer, (D), Harvard, 2007, Py
Michael C. Gurnis, (D), Australian National, 1987, Ys
Thomas H. Heaton, (D), Caltech, 1979, Ne
Donald V. Helmberger, (D), California (San Diego), 1967, Ys
Andrew P. Ingersoll, (D), Harvard, 1966, Oa
Jennifer M. Jackson, (D), Illinois, 2005, Gy
Joseph L. Kirschvink, (D), Princeton, 1979, Pg
Shrinivas R. Kulkarni, (D), California (Berkeley), 1983, Xy
Michael P. Lamb, (D), California (Berkeley), 2008, Gm
Nadia Lapusta, (D), Harvard, 2001, GcYs
Jared R. Leadbetter, (D), Michigan State, 1997, Pm
Dianne K. Newman, (D), MIT, 1997, Py
Victoria Orphan, (D), California (Santa Barbara), 2001, Py
George R. Rossman, (D), Caltech, 1971, GzCa
Tapio Schneider, (D), Princeton, 2001, OawOg
Alex L. Sessions, (D), Indiana, 2001, Co
Mark Simons, (D), MIT, 1995, Ys
David J. Stevenson, (D), Cornell, 1976, Xy
Joann M. Stock, (D), MIT, 1988, Gt
Edward M. Stolper, (D), Harvard, 1979, Cp
Paul O. Wennberg, (D), Harvard, 1994, Oa
Brian P. Wernicke, (D), MIT, 1982, Gtc
Yuk L. Yung, (D), Harvard, 1974, Oa

Associate Professor:
Christian Frankenberg, (D), Ruprecht-Karls Univ, 2005, Or

Assistant Professor:
Jean-Paul Ampuero, (D), Paris VII, 2002, Ys
Konstantin Batygin, (D), Caltech, 2012, Xg
Simona Bordoni, (D), California (Los Angeles), 2007, Op
Bethany Ehlmann, (D), Brown, 2010, Xg
Heather Knutson, (D), Harvard, 2009, Xg
Andrew F. Thompson, Scripps, 2006, Op
Victor Tsai, (D), Harvard, 2009, Yg
Zhongwen Zhan, Caltech, 2013, Ys

Research Professor:
Egill Hauksson, (D), Columbia, 1981, Ys

Research Associate:
D.A. (Dimitri) Papanastassiou, (D), Caltech, 1970, Cc

Emeritus:
Arden L. Albee, (D), Harvard, 1957, GizXg
Clarence R. Allen, (D), Caltech, 1954, Ys
Donald S. Burnett, (D), California (Berkeley), 1963, Xc
Peter M. Goldreich, (D), Cornell, 1963, Xy
David G. Harkrider, (D), Caltech, 1963, Ys
Hiroo Kanamori, (D), Tokyo Univ, 1964, Ysg
Duane O. Muhleman, (D), Harvard, 1963, Xy
Jason B. Saleeby, (D), California (Santa Barbara), 1975, Gc
Leon T. Silver, (D), Caltech, 1955, Cc
Hugh P. Taylor, (D), Caltech, 1959, Cs
Gerald J. Wasserburg, (D), Chicago, 1954, Cc
Peter J. Wyllie, (D), St. Andrews, 1958, Cp

## California Lutheran University

**Dept of Geology** (B) (2016)
60 West Olsen Road #3700
Thousand Oaks, CA 91360-2787
p. (805) 493-3264
bilodeau@callutheran.edu
http://www.callutheran.edu/admission/undergraduate/majors/geology/
*Enrollment (2016): B: 16 (3)*

Chair:
William L. Bilodeau, (D), Stanford, 1979, GcsGg

Professor:
Linda A. Ritterbush, (D), California (Santa Barbara), 1990, Pi

## California Polytechnic State University

**Dept of Natural Resources Management and Environmental Sciences** (B,M) (2014)
One Grand Avenue
San Luis Obispo, CA 93407
p. (805) 756-2261
jstevens@calpoly.edu
http://nres.calpoly.edu
*Enrollment (2002): B: 126 (19) M: 3 (0)*

Professor:
Delmar D. Dingus, (D), Oregon State, 1975, So
Brent G. Hallock, (D), California (Davis), 1979, Sf
William L. Preston, (D), Oregon, 1979, Oe
Thomas A. Ruehr, (D), Colorado State, 1975, Sb
Terry L. Smith, (D), Nebraska, 1980, So

Assistant Professor:
Christopher (Chip) S. Appel, (D), Florida, 2001, Sc

Emeritus:
Thomas J. Rice, (D), North Carolina State, 1981, Sd
Ronald D. Taskey, (D), Oregon State, 1977, Sf

Administrative Coordinator:
Joan M. Stevens, On

Instructional Tech:
Craig Stubler, (B), California Polytech, 1996, So

**Department of Physics** (B) (2017)
1 Grand Ave
180 204
San Luis Obispo, CA 93407
p. (805) 756-2448
physics@calpoly.edu
http://physics.calpoly.edu
Administrative Assistant: Kathy Simon

Professor:
Antonio F. Garcia, (D), California (Santa Barbara), 2001, Gmg

Associate Professor:
John J. Jasbinsek, (D), Wyoming, 2008, Ysg
Scott Johnston, (D), California (Santa Barbara), 2006, GcpCc

Emeritus:
David H. Chipping, (D), Stanford, 1970, Grt

## California State Polytechnic University, Pomona

**Dept of Geological Sciences** (B,M) (2015)
3801 West Temple Avenue
Pomona, CA 91768
p. (909) 869-3454
janourse@csupomona.edu
http://geology.csupomona.edu
Administrative Assistant: Monica Baez
*Enrollment (2014): B: 130 (18) M: 28 (5)*

Chair:
Jonathan A. Nourse, (D), Caltech, 1989, GctEm

Professor:
Jeffrey S. Marshall, (D), Penn State, 2000, GmtHs
Jascha Polet, (D), Caltech, 1999, Ysg

Associate Professor:
Stephen G. Osborn, (D), Arizona, 2010, HwClSp

Assistant Professor:
Nicholas J. Van Buer, (D), Stanford, 2012, GxtGz

Emeritus:
David R. Berry, (D), California (Los Angeles), 1983, PgGo
David R. Jessey, (D), Missouri (Rolla), 1981, Eg
John A. Klasik, (D), Louisiana State, 1976, Gu
Equipment Technician:
Mike McAtee, (B), Cal Poly Pomona, 1978, On



## California State University, Bakersfield

**Dept of Geological Sciences** (B,M) (2015)
9001 Stockdale Hwy, 62 Sci
Bakersfield, CA 93311
p. (661) 654-3027
dbaron@csub.edu
http://www.csub.edu/Geology/
f: https://www.facebook.com/groups/103146842589/
Administrative Assistant: Sue Holt
*Enrollment (2014): B: 102 (24) M: 23 (5)*
Professor:
Dirk Baron, (D), Oregon Inst of Tech, 1995, Cl
Janice Gillespie, (D), Wyoming, 1991, Go
Robert A. Horton, Jr., (D), Colorado Mines, 1985, GdsGo
Assistant Professor:
Junhua "Adam" Guo, (D), Missouri, Gs
W Chris Krugh, Swiss Fed Inst Tech, 2008, GcmCc
Lecturer:
David Miller, (D), Stanford, GgtGc
Adjunct Professor:
Sarah Brown, (D), Simon Fraser, 2010, GtCac
Emeritus:
Robert M. Negrini, (D), California (Davis), 1986, YmgGn
Department IT Support:
Elizabeth Powers, (B), N Arizona, 1998, Pe

## California State University, Chico

**Dept of Geological and Environmental Sciences** (B,M) ● (2017)
400 W First Street
Chico, CA 95929-0205
p. (530) 898-5262
geos@csuchico.edu
http://www.csuchico.edu/geos/
f: https://www.facebook.com/CSUChicoGEOS/
*Enrollment (2014): B: 200 (50) M: 30 (6)*
Chair:
Russell Shapiro, (D), California (Santa Barbara), 1998, PyGdPg
Professor:
David L. Brown, (D), California (Berkeley), 1995, Hg
Ann Bykerk-Kauffman, (D), Arizona, 1990, GcOe
Julie Monet, (D), Rutgers, 2006, GgNg
Associate Professor:
Todd Greene, (D), Stanford, 2000, Gsr
Karin A. Hoover, (D), Johns Hopkins, 1998, Hw
Rachel Teasdale, (D), Idaho, 2001, Gv
Assistant Professor:
Hannah M. Aird, (D), Duke Univ, 2014, GxEg
Shane D. Mayor, (D), Wisconsin, 2001, Oa
Lecturer:
Carrie Monohan, (D), Washington, 2004, Hg
Jochen Nuester, (D), Max Planck Inst, 2005, Cm
Emeritus:
Jerold J. Behnke, (D), Nevada (Reno), 1968, Hw
Victor A. Fisher, (D), Florida State, 1968, Gr
Richard A. Flory, (D), Oregon State, 1975, Ps
Rolland K. Hauser, (D), Chicago, 1967, Oa
K. R. Gina Johnston, (D), Florida, 1970, Hs
Terence T. Kato, (D), California (Los Angeles), 1976, Gt
William M. Murphy, (D), California (Berkeley), 1985, Cl
James L. Regas, (D), Harvard, 1969, Xg
Howard L. Stensrud, (D), Washington, 1970, Em
Gregory R. Taylor, (D), Washington, 1987, Oa

## California State University, Dominguez Hills

**Earth Sciences** (B) (2014)
1000 E. Victoria Street
NSM B202
Carson, CA 90747
p. (310) 243-3377
cmtrujillo@csudh.edu
Administrative Assistant: Virginia L. Knauss
*Enrollment (2014): B: 55 (9)*
Dean, Natural & Behavioral Sciences:
Rodrick Hay, (D), Arizona State, 1996, Or
Chair:
Ashish Sinha, (D), S California, 1997, PeCs
Professor:
Brendan A. McNulty, (D), California (Santa Cruz), 1996, Gc
Associate Professor:
John Keyantash, (D), California (Los Angeles), 2001, HgOae
Ralph Saunders, (D), Arizona, 1996, Og
Lecturer:
Michael H. Ferris, (M), Cal State (Long Beach), 2012, Or
Judith A. King-Rundel, (M), California (Los Angeles), 2003, Oe
Emeritus:
David Sigurdson, (D), California (Riverside), 1974, GizEg

## California State University, East Bay

**Dept of Anthropology, Geography, and Environmental Studies** (2010)
25800 Carlos Bee Blvd
Hayward, CA 94542
p. (510) 885-3193
david.larson@csueastbay.edu
http://class.csueastbay.edu/geography
*Enrollment (2005): B: 0 (22) M: 14 (2)*
Chair:
David J. Larson, (D), California (Berkeley), 1994, Ou
Professor:
Scott W. Stine, (D), California (Berkeley), 1987, Gm
Associate Professor:
Karina Garbesi, (D), California (Berkeley), 1994, Cg
Michael D. Lee, (D), London Schl of Econ, 1990, Hg
Gary Li, (D), SUNY (Buffalo), 1997, Oi
David Woo, (D), California (Santa Barbara), 1991, Or

**Dept of Earth & Environmental Sciences** (B,M) (2016)
25800 Carlos Bee Blvd.
Hayward, CA 94542-3088
p. (510) 885-3486
geology@csueastbay.edu
http://www.sci.csueastbay.edu/earth
*Enrollment (2016): B: 77 (16) M: 20 (7)*
Professor:
Jean E. Moran, (D), Rochester, 1994, HwCl
Professor:
Mitchell S. Craig, (D), Georgia Tech, 1990, Yg
Jeffery C. Seitz, (D), Virginia Tech, 1994, CgOeGx
Associate Professor:
Luther M. Strayer, (D), Minnesota, 1998, Gc
Assistant Professor:
Michael Massey, (D), Stanford, 2013, OnGeCg
Patricia Oikawa, (D), Virginia, 2011, SbCoSf

## California State University, Fresno

**Dept of Earth & Environmental Sciences** (B,M) (2014)
2345 E. San Ramon Avenue
MH24
Fresno, CA 93740-8031
p. (559) 278-3086
slewis@csufresno.edu
http://www.csufresno.edu/ees/
*Enrollment (2009): B: 51 (13) M: 30 (5)*
Chair:
Stephen D. Lewis, (D), Columbia, 1982, Yg
Professor:
Keith D. Putirka, (D), Columbia, 1997, GviGp
Zhi (Luke) Wang, (D), Leuven (Belgium), 1997, SpHwOi
Associate Professor:
Chris Pluhar, (D), California (Santa Cruz), 2003, GtYmNg
John Wakabayashi, (D), California (Berkeley), GmcGt
Assistant Professor:
Robert G. Dundas, (D), California (Berkeley), 1994, PvGr
Mathieu Richaud, (D), N Illinois Univ, 2006, GusCs
Peter Van de Water, (D), Arizona, GrPbGf

Lecturer:
Jeff Anglen, (M), Texas Tech, 2001, GsdPg
Susan Bratcher, (M), Cal State (Fresno), 2008, Hw
Kerry Workman-Ford, (M), Cal State (Fresno), 2003, Gc
Adjunct Professor:
Jerry DeGraff, (M), Utah State, 1976, On
Dong Wang, (D), HwSpOr
Emeritus:
Jon C. Avent
Bruce A. Blackerby, (D), California (Los Angeles), Giz
Roland H. Brady, (D), California (Davis), 1986, NgOuGa
Seymour Mack, YrHgOn
Robert D. Merrill, (D), Texas, 1974, GsmPe
C. John Suen, (D), MIT, 1978, HwCsGe

## California State University, Fullerton

**Dept of Geological Sciences** (B,M) ● (2016)
800 N. State College Boulevard
Geological Sciences, MH 254
Fullerton, CA 92831
p. (657) 278-3882
geology@fullerton.edu
http://geology.fullerton.edu
f: https://www.facebook.com/CalStateFullerton.Geology
*Enrollment (2014): B: 145 (9) M: 16 (0)*

Chair:
Phillip A. Armstrong, (D), Utah, 1996, GcYg
Professor:
Diane Clemens-Knott, (D), Caltech, 1992, Gig
Matthew E. Kirby, (D), Syracuse, 2001, GnPeGg
Jeffrey R. Knott, (D), California (Riverside), 1998, Gmg
Adam D. Woods, (D), S California, 1998, GsPe
Associate Professor:
Nicole Bonuso, (D), S California, 2005, Pie
W. Richard Laton, (D), W Michigan, 1997, HwGeHy
Assistant Professor:
Joe Carlin, (D), Texas A&M, 2013, GuOuGs
Sean Loyd, (D), S California, 2010, CgGe
Valbone Memeti, (D), S California, 2009, GivCa
James Parham, (D), California (Berkeley), 2003, PvoGg
Lecturer:
Freddi Jo Bruschke, (D), Gg
Patricia M. Butcher, (M), Utah, 1993, Oeg
Robert deGroot, (D), Gg
Joanna Fantozzi, (M), Gg
Emily A. Hamecher, (D), Caltech, 2013, GgzGi
Wayne G. Henderson, (M), George Washington, 1997, PgGg
Scott Mata, (D), Gg
Carolyn Rath, (M), Gg
Kelly R. Ruppert, (M), California (Riverside), 2001, Gg
Kristin Weaver Bowman, (M), S California, 1999, Oe
Emeritus:
Galen R. Carlson, (D), Og
John H. Foster, (D)
Prem K. Saint, (D), Minnesota, 1973, Hw
Margaret S. Woyski, (D), Minnesota, 1946, Gi

## California State University, Long Beach

**Dept of Geological Sciences** (B,M) (2013)
1250 Bellflower Boulevard
Long Beach, CA 90840-3902
p. (562) 985-4809
rfrancis@csulb.edu
http://www.csulb.edu/colleges/cnsm/departments/geology/
Administrative Assistant: Margaret Costello
*Enrollment (2009): B: 67 (11) M: 16 (3)*

Chair:
Richard J. Behl, (D), California (Santa Cruz), 1992, GsPeGd
Stanley C. Finney, (D), Ohio State, 1977, Ps
Conrey Endowed Chair:
Matthew Becker, (D), Texas, 1996, Hwg
Professor:
Robert D. Francis, (D), California (San Diego), 1980, Co
Roswitha B. Grannell, (D), California (Riverside), 1969, Yv
Jack Green, (D), Columbia, 1953, Gv
Nate Onderdonk, (D), California (Santa Barbara), 2003, Gtm
Associate Professor:
Gregory J. Holk, (D), Caltech, 1997, CsGxEm
Thomas Kelty, (D), California (Los Angeles), 1998, Gc
Assistant Professor:
Lora R. Stevens (Landon), (D), Minnesota, 1997, Gn
Lecturer:
Bruce Perry, (M), Cal State (Long Beach), 1993, Gc
Emeritus:
Kwan M. Chan, (D), Liverpool, 1966, Oc
Paul J. Fritts, (D), Colorado, 1969, Pm
Charles T. Walker, (D), Leeds, 1952, Cl
Robert E. Winchell, (D), Ohio State, 1963, Gz

## California State University, Los Angeles

**Dept of Geosciences and Environment** (B,M) ○ (2016)
5151 State University Drive
Los Angeles, CA 90032
p. (323) 343-2400
mmurill@calstatela.edu
http://www.calstatela.edu/academic/geos
*Enrollment (2016): B: 65 (0) M: 13 (0)*

Department Chair:
Hengchun Ye, (D), Delaware, 1995, OyHs
Associate Dept Chair:
Steven Mulherin, (D), Ohio State Univ, 1999
Professor:
Qiu Hong-lie, (D), Louisiana State, 1994
Steve LaDochy, (D), Manitoba, 1985, On
Pedro C. Ramirez, (D), California (Santa Cruz), 1990, Gs
Associate Dept Chair:
Jennifer M. Garrison, (D), California (Los Angeles), 2004, GiCcGv
Associate Professor:
Andre Ellis, (D), U of Illinois, 2003, HwCg
Barry Hibbs, (D), Texas, 1993, Hw
Assistant Professor:
Kris Bezdecny, (D), South Florida, 2011, OniOn
Jingjing Li, (D), UC Irvine, 2013
Instructor:
Yoshie Hagiwara, (M)
Angel Hamane, (D), Pepperdine Univ, 2015, Oe
Adjunct Professor:
Mohammad Hassan Rezaie-Boroon, (D), Erlangen (Germany), 1997, GeCg
Emeritus:
Kim Bishop, (D), S California, 1994, Gc
Ivan P. Colburn, (D), Stanford, 1961, Gs
Robert J. Stull, (D), Washington, 1969, Gi

## California State University, Northridge

**Dept of Geography** (B,M) (2016)
18111 Nordhoff Street
Northridge, CA 91330-8249
p. (818) 677-3532
geography@csun.edu
http://www.csun.edu/social-behavioral-sciences/geography

Institute for Social and Behavioral Sciences:
Shawna J. Dark, (D), California (Los Angeles), 2003, Oi
Professor:
Helen M. Cox, (D), California (Los Angeles), 1998, OwgOr
Julie E. Laity, (D), California (Los Angeles), 1982, Oy
Amalie Orme, (D), California (Los Angeles), 1983, Gm
Yifei Sun, (D), SUNY (Buffalo), 2000, Oi
Associate Professor:
Mario A. Giraldo, (D), Georgia, 2007, OriOg
Assistant Professor:
Sanchayeeta Adhikari, (D), Florida, 2011, Oi
Soheil Boroushaki, (D), UWO, 2010, Oi
Regan Maas, (D), California (Los Angeles), 2010, Oi
Emeritus:
Gong-yuh Lin, (D), Hawaii, 1974, Oa
Eugene Turner, (D), Washington, 1977, Or

**Dept of Geological Sciences** (B,M) (2013)
18111 Nordhoff Street
MD: 8266
Northridge, CA 91330-8266

p. (818) 677-3541
doug.yule@csun.edu
http://www.csun.edu/geology/
Administrative Assistant: Mari C. Flores-Garcia
*Enrollment (2012): B: 72 (0) M: 23 (0)*

California

Professor:
Kathleen M. Marsaglia, (D), California (Los Angeles), 1989, Gs
Gerald W. Simila, (D), California (Berkeley), 1979, Yg
M. Ali Tabidian, (D), Nebraska, 1987, Hw
J. Douglas Yule, (D), Caltech, 1996, Gt
Associate Professor:
Matthew d'Alessio, (D), California (Berkeley), 2004, OeGt
Richard V. Heermance, (D), California (Santa Barbara), 2007, GrCc
Elena A. Miranda, (D), Wyoming, 2006, Gc
Jon R. Sloan, (D), California (Davis), 1980, Pm
Dayanthie Weeraratne, (D), Carnegie Inst, 2005, Yg
Assistant Professor:
M. Robinson Cecil, (D), Arizona, 2009, GiCcGz
Joshua J. Schwartz, (D), Wyoming, 2007, GiCc
Emeritus:
Herbert G. Adams, (D), California (Los Angeles), 1971, Ng
Lorence G. Collins, (D), Illinois, 1959, Gzx
George C. Dunne, (D), Rice, 1972, Gc
A. Eugene Fritsche, (D), California (Los Angeles), 1969, Gd
Vicki A. Pedone, (D), SUNY (Stony Brook), 1990, GdCs
Richard L. Squires, (D), Caltech, 1973, Pi

## California State University, Sacramento

**Dept of Geology** (B,M) O (2015)
6000 J Street
Placer Hall
Sacramento, CA 95819-6043
p. (916) 278-6337
geology@csus.edu
http://www.csus.edu/geology/
*Enrollment (2010): B: 77 (16) M: 18 (3)*
Chair:
Tim C. Horner, (D), Ohio State, 1992, HwGs
Professor:
Kevin J. Cornwell, (D), Nebraska, 1994, Gm
David G. Evans, (D), Louisiana State, 1989, HqYg
Lisa Hammersley, (D), California (Berkeley), 2003, Gig
Brian Hausback, (D), California (Berkeley), 1984, Gv
Judith E. Kusnick, (D), California (Davis), 1996, Oe
Lecturer:
Barbara J. Munn, (D), Virginia Tech, 1997, Ggp
Adjunct Professor:
Brian Bergamaschi, (D), Washington, 1995, HwOc
Emeritus:
Diane H. Carlson, (D), Washington State, 1984, Gc
Charles C. Plummer, (D), Washington, 1969, Gp
Greg Wheeler, (D), Washington, Em
Secretary:
Stacy Lindley, (B), California (Sacramento), 2010, On
Instructional Support Tech:
Steven W. Rounds, (B), Cal State (Sacramento), 1990, Gg

## California State University, San Bernardino

**Dept of Geological Sciences** (B,M) (2016)
5500 University Parkway
San Bernardino, CA 92407-2397
p. (909) 537-5336
smcgill@csusb.edu
http://www.geology.csusb.edu
Administrative Assistant: Christina Palmer
*Enrollment (2016): B: 51 (7) M: 4 (0)*
Professor:
Sally F. McGill, (D), Caltech, 1992, GtYdGm
Professor:
Joan E. Fryxell, (D), North Carolina, 1984, Gct
Erik Melchiorre, (D), Washington, 1997, EmHwCl
Alan L. Smith, (D), California (Berkeley), 1969, GviGz
Associate Professor:
W. Britt Leatham, (D), Ohio State, 1987, PsmOg
Assistant Professor:
Kerry Cato, (D), Texas A&M, 1991, Ng
Codi Lazar, (D), California (Los Angeles), 2010, Cp
Emeritus:
Louis A. Fernandez, (D), Syracuse, 1969, Gi
Cooperating Faculty:
Stuart S. Sumida, (D), California (Los Angeles), 1987, Pv

## California State University, Stanislaus

**Dept of Physics and Geology** (B) ● (2016)
One University Circle
Turlock, CA 95382
p. (209) 667-3466
HFerriz@csustan.edu
https://www.csustan.edu/geology
Administrative Assistant: Dawn McCulley
*Enrollment (2016): B: 40 (5)*
Chair:
Horacio Ferriz, (D), Stanford, 1985, NgHwYe
Professor:
Mario J. Giaramita, (D), California (Davis), 1989, Gpz
Julia Sankey, (D), Louisiana State, 1998, Pve
Associate Professor:
Robert D. Rogers, (D), Texas, 2003, GcmGt
Lecturer:
Garry F. Hayes, (M), Nevada (Reno), 1988, Gg
Roger Putnam, (M), North Carolina (Chapel Hill), 2013, Ggi
Michael Whittier, (M), CSU Stanislaus, 2005, Ggz

## Cerritos College

**Earth Science Dept** (A) (2013)
11110 E. Alondra Boulevard
Norwalk, CA 90650
p. (562) 860-2451
ddekraker@cerritos.edu
http://www.cerritos.edu/earth-science/
*Enrollment (2012): A: 9 (0)*
Earth Science Dept Chair:
Crystal LoVetere, (D), S California, Oy
Assistant Professor:
Dan DeKraker, (M), Cal State (Fullerton), 2004, OepGg
Aline Gregorio, (M), Cal State (Fullerton), 2011
Tor Lacy, (M), Cal State (Long Beach), 2005, Gg
Instructor:
Gary D. Johnpeer, (M), Arizona State, 1977, Ng

## Chabot College

**Div of Mathematics & Science** (A) (2004)
25555 Hesperian Boulevard
Hayward, CA 94545
p. (415) 786-6865
Instructor:
Adolph A. Oliver, (M), Stanford, 1974, Ys
David J. Perry, (M), San Jose State, 1967, Op

## Chaffey College

**Earth Science/Geology Dept** (2010)
5885 Haven Avenue
Rancho Cucamonga, CA 91737-3002
p. (909) 652-6402
henry.shannon@chaffey.edu
http://www.chaffey.edu/mathandscience/
Professor:
Jane Warger, (M), Columbia, 1990, OgGg

## City College of San Francisco

**Earth Sciences Dept** (A) (2014)
Box S50
50 Phelan Avenue
San Francisco, CA 94112
p. (415) 452-7014
cjlewis@ccsf.edu
http://www.ccsf.edu/Earth
*Enrollment (2011): A: 10 (0)*
Professor:
Darrel E. Hess, (M), California (Los Angeles), 1990, Oy
Chris Lewis, (M), California (Berkeley), 1993, GgEmOs

Katryn Wiese, (M), Oregon State, 1992, GiOg
Adjunct Professor:
Ian Duncan, (D), San Francisco State, Oy
James Kuwabara, (D), Caltech, 1980, Og
Joyce Lucas-Clark, (D), Stanford, PgGg
Russell McArthur, (M), California (Berkeley), Gg
Elizabeth Proctor, (M), San Francisco State, Oi
Kirstie L. Stramler, (D), Columbia, OagGg
Gordon Ye, (M), California (Berkeley), 1993, Oi

## College of Marin
**Geology Dept** (A) (2014)
835 College Ave
SC 192
Kentfield, CA 94904
p. (415) 457-8811
dfoss@marin.edu
http://marin.edu/~geology
Professor:
Donald J. Foss, (M), Boise State, 1980, Og
Emeritus:
James L. Locke, (M), San Jose State, 1971, Og

## College of San Mateo
**Geological Sciences** (A) (2016)
1700 W. Hillsdale Blvd.
San Mateo, CA 94402
p. (650) 574-6633
hand@smccd.edu
http://www.collegeofsanmateo.edu/geologicalsciences/index.asp
*Enrollment (2016): A: 10 (2)*
Professor:
Linda M. Hand, (M), Texas A&M, 1988, GgPgOg

## College of the Canyons
**Dept of Earth, Space, and Environmental Sciences** (A) (2014)
26455 Rockwell Canyon Road
Santa Clarita, CA 91355
p. (661) 362-3658
Vincent.Devlahovich@canyons.edu
http://www.canyons.edu/departments/ESES/
*Enrollment (2012): A: 4 (0)*
Chair:
Vincent A. Devlahovich, (D), Cal State (Northridge), 2012, GgOye
Professor:
Mary Bates, (M), Cal State (Northridge), Oyg

## College of the Desert
**Dept of Science** (A) (2016)
43-500 Monterey Avenue
Palm Desert, CA 92260
p. (760) 776-7272
nmoll@collegeofthedesert.edu
http://www.collegeofthedesert.edu/students/ss/ap/physsci/Pages/Geology.aspx
*Enrollment (2013): A: 3 (1)*
Professor:
Nancy E. Moll, (D), Washington, 1981, Gge
Adjunct Professor:
Brian Koenig, (M), Arizona, 1978, Gge
Robert Pellenbarg, (D), Delaware, 1976, OcGu

## College of the Sequoias
**Science Div** (A) (2004)
Visalia, CA 93277
p. (209) 730-3812
erich@giant.sequoias.cc.ca.us
*Enrollment (2000): A: 2 (0)*
Chair:
Eric D. Hetherington, (D), Minnesota, 1991, Gc
Emeritus:
John R. Crain, (M), Nevada, 1961, Gg

## College of the Siskiyous
**Dept of Biological & Physical Sciences** (A) (2016)
800 College Avenue
Weed, CA 96094
p. (530) 938-5255
hirt@siskiyous.edu
http://www.siskiyous.edu/class/ess/
*Enrollment (2015): A: 1 (0)*
Instructor:
William H. Hirt, (D), California (Santa Barbara), 1989, Gi

## Contra Costa College
**Astronomy/Engineering/Geology/Physics Dept** (A) (2015)
2600 Misson Bell Drive
San Pablo, CA 94806
p. (510) 235-7800
jsmithson@contracosta.edu
http://www.contracosta.edu/home/programs-departments/engineering/
Professor:
Mary Lewis, Gg
Jayne Smithson, Gg

## Cosumnes River College
**Dept of Science, Mathematics & Engineering** (A) (2015)
8401 Center Parkway
Sacramento, CA 95823
p. (916) 691-7210
MuranaB@crc.losrios.edu
Department Secretary: Sue McCoy
Chair:
Debra Sharkey, (M), California (Davis), 1994, Oy
Assistant Professor:
Hiram Jackson, (M), California (Davis), 1992, Gg
Adjunct Professor:
Gerry Drobny, (M), Washington State, 1981, Gg

## Crafton Hills College
**Geology Dept** (A) (2016)
11711 Sand Canyon Road
Yucaipa, CA 92399
p. (909) 794-2161
rihughes@craftonhills.edu
http://www.craftonhills.edu/courses_&_programs/Physical_Science/Geology/
*Enrollment (2013): A: 16 (4)*
Professor:
Richard O. Hughes III, (M), Ohio, 1994, GlOe

## Cuesta College
**Physical Sciences Div** (A) (2014)
P. O. Box 8106
San Luis Obispo, CA 93403
p. (805) 546-3230
jgrover@cuesta.edu
http://academic.cuesta.edu/physci/Geology/index.html
*Enrollment (2014): A: 6 (0)*
Professor:
Jeffrey A. Grover, (M), Arizona, 1982, Gg
Debra Stakes, (D), Gg

## Cuyamaca College
**Dept of Science and Engineering** (A) (2014)
900 Rancho San Diego Parkway
El Cajon, CA 92019
p. (619) 660-4345
Glenn.Thurman@gcccd.edu
http://www.cuyamaca.edu/
Instructor:
Lisa Chaddock, Oy
Michael Farrell, Gg
Bryan Miller-Hicks, Gg
Agatha Wein, Gg
Ray Wolcott, Og
Emeritus:
Waverly Ray, Gg

## Cypress College

**Geology Dept** (A) (2010)
9200 Valley View Street
Cypress, CA 90630
p. (714) 484-7153
RArmale@CypressCollege.edu
Professor:
Russell L. Flynn, (M), San Diego State, 1971, Op
Adjunct Professor:
Hank Wadleigh, (B), Cal State (Long Beach), 1957, Gg
Visiting Professor:
Curtis J. Williams, (M), Cal State (Los Angeles), 1996, Gg
Emeritus:
Keith E. Green, (M), S California, 1958, Pg
Altus Simpson, (M), S California, 1958, Go

## Diablo Valley College

**Div. of Physical Sciences** (A) (2013)
321 Golf Club Road
Pleasant Hill, CA 94523
p. (925) 685-1230 x46
JHetherington@dvc.edu
http://www.dvc.edu/org/departments/physics/
Chair:
Jean Hetherington, (M), Washington, 1983, Gg

## El Camino College

**Dept of Earth Sciences** (A) (2014)
16007 Crenshaw Blvd.
Torrance, CA 90506
p. (310) 660-3593
jholliday@elcamino.edu
http://www.elcamino.edu/academics/naturalsciences/earth/
Professor:
Jerry Brothen, (M), California (Los Angeles), Oy
Sara Di Fiori, (M), California (Los Angeles), GgOg
Matt Ebiner, (M), California (Los Angeles), Oy
Joseph W. Holliday, (M), Oregon State, 1982, Og
Associate Professor:
Chuck Herzig, (D), California (Riverside), GgOg
Instructor:
Gary Booher, Og
Robin Bouse, Gg
Charles Dong, Og
Lynn Fielding, Og
Patricia Neumann, Og
Douglas Neves, Og
Jim Noyes, (D), Scripps, Og
Ebenezer Peprah, Og

## Feather River College

**Feather River College** (A) (2010)
570 Golden Eagle Ave.
Quincy, CA 95971
dlerch@frc.edu
http://www.frc.edu/
Instructor:
Derek Lerch, Og

## Folsom Lake College

**Geosciences** (A) ● (2015)
10 College Parkway
Folsom, CA 95630
p. (916) 608-6668
pittmaj@flc.losrios.edu
http://www.flc.losrios.edu/Academics/Geology.htm
*Enrollment (2015): A: 7 (1)*
Professor:
Jason Pittman, (M), Oregon State, 1999, OgiOi

## Fresno City College

**Earth & Physical Science Dept** (A) (2017)
Math, Science & Engineering Div
1101 E. University Avenue
Fresno, CA 93741
p. (559) 442-4600
craig.poole@fresnocitycollege.edu
*Enrollment (2015): A: 15 (0)*
Instructor:
T. Craig Poole, (M), Cal State (Fresno), 1987, Gg

## Fullerton College

**Div of Natural Sciences (Geology)** (A) (2016)
321 E. Chapman Ave
Fullerton, CA 92832
p. (714) 992-7445
rlozinsky@fullcoll.edu
http://natsci.fullcoll.edu/
*Enrollment (2016): A: 5 (2)*
Professor:
William S. Chamberlin, (D), S California, 1989, OgbOe
Carolyn Heath, (D), California (Santa Cruz), Ob
Richard P. Lozinsky, (D), New Mexico Tech, 1988, GreGm
Associate Professor:
Roman DeJesus, (D), Og
Assistant Professor:
Marc Willis, (M), Gg

## Gavilan College

**Dept of Physical Sciences** (2011)
5055 Santa Theresa Boulevard
Gilroy, CA 95020
p. (408) 848-4701
rlee@gavilan.edu
http://www.gavilan.edu/natural_sciences/
Instructor:
Duane Willahan, (M), Santa Fe State, 1992, Gg

## Golden West College

**Physical Science Program** (A) (2011)
15744 Golden West Street
Huntington Beach, CA 92647
p. (714) 892-7711 x51116
msouto@gwc.cccd.edu
http://www.goldenwestcollege.edu/campus/physicalscience.html
*Enrollment (2005): A: 2 (0)*
Professor:
Ronald C. Gibson, (M), California (Riverside), 1964, Gc
Bernard J. Gilpin, (M), California (Riverside), 1976, Ys

## Grossmont College

**Dept of Earth Sciences** (A) (2013)
8800 Grossmont College Drive
El Cajon, CA 92020
p. (714) 644-7887
gary.jacobson@gcccd.edu
http://www.grossmont.edu/earthsciences/
*Enrollment (2004): A: 15 (0)*
Professor:
Chris Hill, (D), Gg
Instructor:
Gary L. Jacobson, (M), San Diego State, 1982, GucEm

## Hartnell College

**Div of Math and Science** (A) (2014)
411 Central Avenue
Salinas, CA 93901
aramirez@hartnell.edu
http://www.hartnell.edu/academics/math.html
Instructor:
Robert Barminski, (M), Moss Landing Marine Lab, GgOg

## Humboldt State University

**Dept of Geology** (B,M) ○ (2017)
1 Harpst Street
Arcata, CA 95521-8299

p. (707) 826-3931
geology@humboldt.edu
http://www.humboldt.edu/geology/
*Enrollment (2016): B: 85 (23) M: 5 (2)*
Chair:
Mark Hemphill-Haley, (D), Oregon, 2000, Gt
Professor:
Susan M. Cashman, (D), Washington, 1977, Gc
William C. Miller, (D), Tulane, 1984, Pi
Associate Professor:
Brandon L. Browne, (D), Alaska (Fairbanks), 2005, GivGz
Assistant Professor:
Melanie Michalak, (D), California (Santa Cruz), 2013
Jasper Oshun, (D), California (Berkeley), 2015, HyGmg
Research Associate:
Eileen Hemphill-Haley, (D), California (Santa Cruz), 1991, Pm
Harvey M. Kelsey, (D), California (Santa Cruz), 1977, Gmt
Robert C. McPherson, (M), Humboldt State, 1989, Gt
Dallas D. Rhodes, (D), Syracuse, 1973
Lecturer:
Amanda R. Admire, (M), Humboldt State, 2013, Og
Jason R. Patton, (D), Oregon State, 2014, GuOg
Adjunct Professor:
David Bazard, (D), Arizona, 1991
Thomas Lisle, (D), California (Berkeley), 1976, Hs
Mary Ann Madej, (D), Washington, Gm
Jon R. Pedicino, (D), Arizona, 1996
Brandon E. Schwab, (D), Oregon, 2000, GizGv
Robert R. Ziemer, (D), Colorado State, 1978, HgGmOn
Emeritus:
Kenneth R. Aalto, (D), Wisconsin, 1970, Gr
Raymond M. Burke, (D), Colorado, 1976, Gm
Gary A. Carver, (D), Washington, 1972, Gmt
Lorinda Dengler, (D), California (Berkeley), 1978, Yg
Andre K. Lehre, (D), California (Berkeley), 1982, GmHs
Alistair W. McCrone, (D), Kansas, 1961, Gs

## Imperial Valley College

**Science, Math & Engineering Div** (A) (2016)
380 E. Aten Rd.
Imperial, CA 92251
p. (760) 355-6304
ofelia.duarte@imperial.edu
http://www.imperial.edu/index.php?pid=343
Professor:
Kevin G. Marty, (M), New Orleans, 1994, GcOg
Instructor:
Kevin Marty, Gg

## Irvine Valley College

**Dept of Environnmental Sciences** (A) (2010)
5500 Irvine Center Drive
Irvine, CA 92618
p. (949) 451-5561
astinson@ivc.edu
http://www.ivc.edu
Administrative Assistant: Liz Nichols
*Enrollment (2007): A: 20 (5)*
Professor:
George Brogan, (M), San Diego State, Gc
Amy L. Stinson, (M), San Diego State, 1990, Gc
Adjunct Professor:
Mark Bordelon, (M)

## Laney College

**Dept of Geography/Geology** (A) (2010)
900 Fallon Street
Oakland, CA 94607
p. (510) 464-3233
dwoodrow@peralta.edu
http://www.laney.peralta.edu/apps/comm.asp?Q=30123

## Las Positas College

**Dept of Geology** (A) (2017)
3000 Campus Hill Drive
Livermore, CA 94551
p. (925) 424-1319
rhanna@laspositascollege.edu
http://www.laspositascollege.edu/geology/
*Enrollment (2016): A: 4 (4)*
Instructor:
Ruth L. Hanna, (M), California (Davis), 1988, Gg
Geology Lab Technician:
Carol Edson, Gg

## Loma Linda University

**Dept of Earth and Biological Sciences** (B,M,D) ● (2017)
Loma Linda University
Loma Linda, CA 92350
p. (909) 824-4530
pbuchheim@llu.edu
https://medicine.llu.edu/research/department-earth-and-biological-sciences
f: https://www.facebook.com/lomalindageology/?fref=ts
*Enrollment (2016): B: 1 (0) M: 1 (4) D: 11 (2)*
Professor:
Leonard R. Brand, (D), Cornell, 1970, Pv
Paul Buchheim, (D), Wyoming, 1978, GnsPe
Leroy Leggitt, (D), Loma Linda, 2005, Pi
Senior Scientist:
Benjamin L. Clausen, (D), Colorado, 1987, GiCtYg
Associate Professor:
Kevin Nick, (D), Oklahoma, 1990, GsbGr
Assistant Professor:
Raul Esperante, (D), Loma Linda, 2003, Pv
Ronald Nalin, (D), Padova, 2006, Gsr

## Los Angeles County Museum of Natural History

**Research and Collections Branch** (2016)
900 Exposition Boulevard
Los Angeles, CA 90007
p. (213) 763-3360
acelesti@nhm.org
https://www.nhm.org/site/research-collections/mineral-sciences
Curator:
Kenneth E. Campbell, (D), Florida, 1973, Pv
Associate Curator:
Luis M. Chiappe, (D), Buenos Aires, 1992, Pv
Curator:
Anthony R. Kampf, (D), Chicago, 1976, Gz
Associate Curator:
Aaron Celestian, (D)
Nathan Smith, (D)
Collections Manager:
Samuel A. McLeod, (D), California (Berkeley), 1981, Pv
Christopher A. Shaw, (M), Cal State (Long Beach), 1981, Pv

## Los Angeles Harbor College

**Dept of Earth Science** (A) (2013)
1111 Figueroa Place
Wilmington, CA 90744
p. (310) 233-4000
munasit@lahc.edu
http://www.lahc.cc.ca.us/
Instructor:
Patricia Kellner, (D), Og
John Mack, Og
Tissa Munasinghe, (D), Og
Melanie Renfrew, (D), Og
Susan White, Og

## Los Angeles Pierce College

**Physics and Planetary Sciences** (A) (2013)
6201 Winnetka Ave
Woodland Hills, CA 91371
p. (818) 710-2218
zayacjm@piercecollege.edu
*Enrollment (2012): A: 11 (0)*
Professor:
John M. Zayac, (M), California (Santa Barbara), 2006, Ggv

Professor:
Jason P. Finley, (M), California (Los Angeles), Ow
Stephen C. Lee, (B), Illinois, 1971, Og
W. Craig Meyer, (M), S California, 1973, GeuPm
Adjunct Professor:
Harry Filkorn, (D), Kent State, GgPi
James P. Krohn, (M), S California, 1974, NgGg
Donald Prothero, (D), Columbia, Gg
Emeritus:
Ruth Y. Lebow, (M), Chicago, 1941, Op
Mark L. Powell, (M), Cal State (Northridge), 1967, Oa
William H. Russell, (M), Cal State (Northridge), 1970, Ow
James Y. Vernon, (M), California (Los Angeles), 1951, Ow

California

## Los Angeles Southwest College

**Geology Dept** (A) (2011)
1600 West Imperial Highway
Los Angeles, CA 90047
p. (323) 241-5297
doosepr@lasc.edu
http://www.lasc.edu
Chair:
Glenn Yoshida, Gg
Professor:
Paul R. Doose, (D), California (Los Angeles), 1980

## Los Angeles Valley College

**Earth Science** (A) (2017)
5800 Fulton Ave.
Valley Glen, CA 91401
p. (818) 778-5566
hamsje@lavc.edu
http://lavc.edu/earthscience/index.aspx
*Enrollment (2014): A: 3 (0)*
Chair:
Jacquelyn E. Hams, (M), Cal State (Los Angeles), 1987, GgOue
Academic Senate President:
Donald J. Gauthier, (M), California (Los Angeles), 1993, OiyOw
Assistant Professor:
Meredith L. Leonard, (M), Cal State (Northridge), OweOi

## Mendocino College

**Earth Science** (A) (2014)
1000 Hensley Creek Road
Ukiah, CA 95482
p. (707) 468-3002
scardimo@mendocino.edu
*Enrollment (2010): A: 29 (3)*
Professor:
Steve Cardimona, (D), Texas, 1992, Ys

## Merced College

**Science, Mathematics & Engineering** (A) (2016)
3600 M Street
Merced, CA 95348
p. (209) 384-6293
kain.d@mccd.edu
http://www.mccd.edu/academics/sme/index.html

## MiraCosta College

**Dept of Geography** (A) (2017)
1 Barnard Drive
Oceanside, CA 92056
lmiller@miracosta.edu
https://www.miracosta.edu/Instruction/Geography/
Chair:
Herschel I. Stern, (D), Oregon, 1988, Oy

**Dept of Physical Sciences** (A) (2013)
1 Barnard Drive
Oceanside, CA 92056
p. (760) 944-4449 x7738
lmiller@miracosta.edu
https://www.miracosta.edu/Instruction/Geology/
Professor:
Keith H. Meldahl, (D), Arizona, 1990, Gg
Christopher V. Metzler, (D), California (San Diego), 1987, Gg
John Turbeville, GgOg
Adjunct Professor:
Phil Farquharson, Gg
Laboratory Director:
Larry Hernandez, Gg

## Modesto Junior College

**Dept of Science, Mathematics & Engineering** (A) (2010)
435 College Avenue
Modesto, CA 95350
p. (209) 575-6172
hayesg@mjc.edu
https://www.mjc.edu/instruction/sme/earthscience.php
Instructor:
Donald C. Ahrens, (D), N Colorado, Og
Garry F. Hayes, (M), Nevada (Reno), 1985, Gg

## Monterey Peninsula College

**Earth Sciences Dept** (A) (2014)
980 Fremont St.
Monterey, CA 93940
p. (831) 646-4000
ahochstaedter@mpc.edu
http://www.mpc.edu/academics/academic-divisions/physical-science/earth-science-eart
Chair:
Alfred Hochstaedter, (D), California (Santa Cruz), 1991, GgOgg

## Moorpark College

**Geology Program** (A) (2016)
7075 Campus Road
Moorpark, CA 93021
p. (805) 553-4161
rharma@vcccd.edu
http://www.moorparkcollege.edu/geology
*Enrollment (2016): A: 24 (2)*
Instructor:
Roberta L. Harma, (M), Hawaii, 1982, Gg

## Moss Landing Marine Laboratories

**CSU Consortium for Marine Sciences** (M) O (2014)
8272 Moss Landing Road
Moss Landing, CA 95039
p. (831) 771-4400
frontdesk@mlml.calstate.edu
http://www.mlml.calstate.edu
*Enrollment (2014): M: 7 (25)*
Director:
James T. Harvey, (D), Oregon, 1987, Ob
Librarian:
Joan Parker, (M), California (Los Angeles), 1986, On
Professor:
Kenneth H. Coale, (D), California (Santa Cruz), 1988, CtcOc
Jonathan Geller, (D), California (Berkeley), 1988, On
Michael Graham, (D), Scripps, 2000, Ob
Nicholas A. Welschmeyer, (D), Washington, 1982, Ob
Research Faculty:
John S. Oliver, (D), Scripps, 1980, Ob
Associate Professor:
Ivano Aiello, (D), Bologna, 1997, Gu
Assistant Professor:
Scott Hamilton, (D), California (Santa Barbara), 2008
Birgitte McDonald, (D)
Research Faculty:
Simona Bartl, (D), California (San Diego), 1989, On
Laurence Breaker, (D), Naval Postgrad Sch, 1983, Op
David Ebert, (D), Rhodes, 1990, On
Stacy Kim, (D), WHOI, 1996, On
Valerie Loeb, (D), Scripps, 1979, On
Richard Starr, (D), Autó Baja California Sur, 2002, On
Emeritus:
Gregor M. Cailliet, (D), California (Santa Barbara), 1972, On

Michael Foster, (D), California (Santa Barbara), 1971, On
H. Gary Greene, (D), Stanford, 1977, Gu
Diving Safety Officer:
Diana Steller, (D), California (Santa Cruz), 2003, On

## Mt. San Antonio College

**Dept of Earth Sciences & Astronomy** (A) (2017)
1100 North Grand Avenue
Walnut, CA 91789
p. (909) 594-5611
cwebb@mtsac.edu
*Enrollment (2010): A: 16 (0)*
Acting Chair:
Julie Ali-Bray, (M), S California, 1999, On
Professor:
Micol Christopher, (D), Caltech, 2007, On
Craig A. Webb, (M), Duke, 1996, OgwOg
Instructor:
Mark Boryta, (D), New Mexico Tech, 1997
Barbara Grubb, (M), Cal State (Long Beach), 1991, Ps
Larry Mendenhall, (D), Oregon State, 1961, Ow
Charles Roberts, (M), Ohio, 1972, Pm
Emeritus:
Hallock J. Bender, (D), San Gabriel, 1959, Gg
John W. Burns, (M), Pittsburgh, 1960, On
Damon P. Day, (M), Michigan Tech, 1965, Gg
Ron N. Hartman, (M), Cal State (Los Angeles), 1973, Xm
Kazimierz M. Pohopien, (M), McGill, 1951, Ge
Harold V. Thurman, (M), Cal State (Los Angeles), 1966, Og
Geotechnician:
Mark Koestel, (B), Arizona, 1978, Em

## Naval Postgraduate School

**Dept of Meteorology** (M,D) (2013)
589 Dyer Road
Root Hall, Room 254
Monterey, CA 93943-5114
p. (831) 656-2516
nuss@nps.edu
http://www.nps.edu/Academics/Schools/GSEAS/Departments/Meteorology/
*Enrollment (2012): M: 45 (36) D: 9 (2)*
Chair:
Wendell A. Nuss, (D), Washington, 1986, Ow
Distinguished Professor:
Michael T. Montgomery, (D), Harvard, 1990, OwaOg
Dean:
Philip A. Durkee, (D), Colorado State, 1984, Owr
Professor:
Patrick A. Harr, (D), Naval Postgrad Sch, 1993, Ow
Qing Wang, (D), Penn State, 1993, Ow
Associate Professor:
Joshua P. Hacker, (D), British Columbia, 2001, Ow
Assistant Professor:
Richard W. Moore, (D), Colorado State, 2004, Ow
Barbara V. Scarnato, (D), ETH, 2008, Ow
Research Associate:
Hway-Jen Chen, (M), California (Los Angeles), 1993, Ow
Paul A. Frederickson, (M), Maryland, 1989, Ow
Mary S. Jordan, (M), Naval Postgrad Sch, 1985, Ow
Kurt E. Nielsen, (M), Oklahoma, 1988, Ow
Andrew Penny, (M), Arizona, 2009, Ow
Emeritus:
Robert Haney, (D), California (Los Angeles), 1971, Ow
Robert J. Renard, (D), Ow
Carlyle H. Wash, (D), Wisconsin, 1978, Ow
Forrest Williams, (M), MIT, 1972, Ow
Roger T. Williams, (D), California (Los Angeles), 1963, Ow
Research Professor:
Kenneth L. Davidson, (D), Michigan, 1970, Ow
Peter S. Guest, (D), Naval Postgrad Sch, 1992, Ow
Research Associate Professor:
James Thomas Murphree, (D), California (Davis), 1989, Ow
NRC Postdoctoral Fellow:
Myung-Sook Park, (D), Ow
Distinguished Research Professor:
Russell L. Elsberry, (D), Colorado State, 1968, Ow
Distinguished Professor:
Chih-Pei Chang, (D), Washington, 1972, Ow
Meteorologist:
Robert L. Creasey, (M), Ow

**Dept of Oceanography** (M,D) (2014)
833 Dyer Road, Room 328
Monterey, CA 93943-5122
p. (831) 656-2673
pcchu@nps.edu
http://www.nps.edu/Academics/GSEAS/Oceanography/
*Enrollment (2014): M: 14 (14) D: 2 (2)*
Chair:
Peter C. Chu, (D), Chicago, 1985, Opr
Professor:
Ching-Sang Chiu, (D), MIT/WHOI, 1985, Op
Jeffrey D. Paduan, (D), Oregon State, 1987, Op
Associate Professor:
Jamie MacMahan, (D), Onp
Timour Radko, Florida State, Op
Emeritus:
Robert H. Bourke, (D), Oregon State, 1972, Op
Curtis A. Collins, (D), Oregon State, 1967, Op
Roland W. Garwood, (D), Washington, 1976, Op
Eugene C. Haderlie, (D), California (Berkeley), 1950, Ob
Thomas H. Herbers, (D), California (San Diego), 1990, Op
Albert J. Semtner, (D), Princeton, 1973, Op
Edward B. Thornton, (D), Florida, 1970, On
Eugene D. Traganza, (D), Miami, 1966, Oc
Stevens P. Tucker, (D), Oregon State, 1972, Og
Joseph J. Von Schwind, (D), Texas A&M, 1968, Op
Jacob B. Wickham, (M), Scripps, 1949, Op
Research Professor:
Wieslaw Maslowski, (D), Alaska (Fairbanks), 1994, Op
Timothy P. Stanton, (M), Auckland, 1978, Op
Research Associate Professor:
Robin T. Tokmakian, (D), Naval Postgrad Sch, 1997, Op

## Occidental College

**Dept of Geology** (B) (2016)
1600 Campus Road
Los Angeles, CA 90041
p. (323) 259-2823
jangarcia@oxy.edu
http://www.oxy.edu/Geology.xml
Administrative Assistant: Tracy Mikuriya
*Enrollment (2015): B: 15 (8)*
Professor:
Scott W. Bogue, (D), California (Santa Cruz), 1982, Ym
Margaret E. Rusmore, (D), Washington, 1985, Gc
James L. Sadd, (D), South Carolina, 1987, Oi
Related Staff:
Jan Garcia, On

## Ohlone College

**Dept of Geology** (A) (2014)
43600 Mission Boulevard
Fremont, CA 94539
p. (510) 979-7938
pbelasky@ohlone.edu
http://www.ohlone.edu/instr/geology/
Professor:
Paul Belasky, (D), California (Los Angeles), 1994, PqiGs

## Orange Coast College

**Div of Mathematics & Science** (A) (2016)
2701 Fairview Road
Box 5005
Costa Mesa, CA 92626-5005
p. (714) 432-5647
ebender@occ.cccd.edu
http://www.orangecoastcollege.edu/academics/divisions/math_science/geology/Pages/default.aspx
*Enrollment (2016): A: 8 (1)*

Chair:
E. Erik Bender, (D), Univ. of Southern California, 1994, Giz
Instructor:
Jim Schneider, (M), Cal State (Sacramento), Geu
Lecturer:
Joanna Fantozzi, (M), Cal State (Fullerton), 2011, GsPeHg
Diana Pomeroy, (M), Cal State (Long Beach), 2013, Pv
Michael Van Ry, (M), Cal State (Fullerton), 2010, Gv

## Palomar College

**Dept of Earth, Space, and Aviation Sciences** (A) (2016)
1140 West Mission Road
San Marcos, CA 92069
p. (760) 744-1150
sfigg@palomar.edu
http://www.palomar.edu/earthscience/
Administrative Assistant: Brenda Morris
*Enrollment (2014): A: 35 (0)*
Professor:
Doug Key, (M), San Diego State, Oy
Alan P. Trujillo, (M), N Arizona Univ, 1984, OguOe
Lisa Yon, (D), Brown, 1994, Ogg
Associate Professor:
Wing Cheung, (M), Indiana, 2007, OirOy
Patricia A. Deen, (M), San Diego State, 1984, Ogg
Cathy Jain, (M), San Diego State, 2000, Oy
Mark Lane, (M), San Diego State, 1996, On
Assistant Professor:
Sean Figg, (M), N Colorado, 2012, Gg
Emeritus:
Jim Pesavento, (D), GgOg

## Pasadena City College

**Dept of Geology** ● (2017)
Natural Sciences Division
1570 E. Colorado Boulevard
Pasadena, CA 91106
p. (626) 585-7138
naturalsciences@pasadena.edu
http://pasadena.edu/academics/divisions/natural-sciences/areas-of-study/geology.php
Dean, Natural Sciences:
David N. Douglass, (D), Dartmouth, 1987, Ym
Acting Chair:
Martha House, (D), MIT, Gg
Professor:
Elizabeth Nagy-Shadman, (D), CalTech, Gg
Yuet-Ling O'Connor, (D), S California, Ge
Bryan Wilbur, (D), California (Los Angeles), Gg
Instructor:
MIchael Vendrasco, (D), California (Los Angeles)
Emeritus:
Gerald L. Lewis, (M), Cal State (Long Beach), 1965, Ps
Geology Lab Tech:
Debra A. Cantarero, (A), Pasadena City Coll, 2000, On

## Pomona College

**Geology Dept** (B) (2017)
185 East Sixth Street
Claremont, CA 91711-6339
p. (909) 621-8675
LKeala@pomona.edu
http://www.geology.pomona.edu
Administrative Assistant: Lori Keala
*Enrollment (2016): B: 15 (7)*
Professor:
Robert R. Gaines, (D), California (Riverside), 2003, GsClPi
Eric B. Grosfils, (D), Brown, 1995, XgGvq
Chair:
Jade Star Lackey, (D), Wisconsin, 2005, GipCs
Associate Professor:
Linda A. Reinen, (D), Brown, 1993, Gct
Emeritus:
Richard W. Hazlett, (D), S California, 1986, Gv

## Riverside City College

**Dept of Physical Science: Geology** (A) (2017)
4800 Magnolia Avenue
Riverside, CA 92506-1299
p. (951) 222-8350
william.phelps@rcc.edu
*Enrollment (2016): A: 11 (0)*
Assistant Professor:
William Phelps, (D), GgOpPg

## Sacramento City College

**Dept of Physics, Astronomy, & Geology** (A) (2017)
3835 Freeport Blvd.
Sacramento, CA 95822
p. (916) 558-2343
stantok@scc.losrios.edu
http://www.scc.losrios.edu/pag/
*Enrollment (2016): A: 10 (0)*
Professor:
Kathryn Stanton, (D), California (Davis), 2006, GgPg

## Saddleback Community College

**Dept of Earth and Ocean Sciences** (A) (2016)
28000 Marguerite Parkway
Mission Viejo, CA 92692
p. (949) 582-4820
jrepka@saddleback.edu
http://www.saddleback.edu/mse/geo/
f: https://www.facebook.com/Saddleback-College-Geology-731491460197935/
*Enrollment (2015): A: 7 (1)*
Chair:
James Repka, (D), California (Santa Cruz), 1998, GgmOe
Professor:
Kalon Morris, (M), Scripps, 2001, OgpOw

## San Bernardino County Museum

**Geological Sciences Div** (2014)
2024 Orange Tree Lane
Redlands, CA 92374
p. (909) 307-2669
kspringer@sbcm.sbcounty.gov
http://www.co.san-bernardino.ca.us/museum/
Senior Curator:
Kathleen B. Springer, (M), California (Riverside), Og
Curator:
J. Chris Sagebiel, (M), Texas, 1998, Pv

## San Bernardino Valley College

**Earth and Spatial Sciences Program** (A) (2010)
701 South Mount Vernon Ave.
San Bernardino, CA 92410
p. (909) 384-8638
theibel@valleycollege.edu
Head:
Todd Heibel, Oyi
Professor:
Stephen H. Sandlin, Oy
Instructor:
Gary M. Croft, Oy
Vanessa Engstrom, Oy
Walter Grossman, Og
Jeffrey Krizek, Oi
William Muir, Og
Solomon Nana Kwaku Nimako, Oi
Edmund Jekwu Ogbuchiekwe, Oy
Lisa Schmidt, Oy
Adjunct Professor:
Donald G. Buchanan, (M), Naval Postgrad Sch, 1975, Og

## San Diego State University ❐

**Dept of Geological Sciences** (B,M,D) ● (2015)
5500 Campanile Drive
San Diego, CA 92182-1020
p. (619) 594-5586

dkimbrough@mail.sdsu.edu
http://www.geology.sdsu.edu
f: https://www.facebook.com/SDSU.Geology
t: @sdsugeology
Administrative Assistant: Irene Occhiello
Administrative Assistant: Pia Parrish
*Enrollment (2015): B: 72 (9) M: 25 (12) D: 9 (0)*

Chairman:
David L. Kimbrough, (D), California (Santa Barbara), 1982, Cc
Associate Dean, Division of Undergraduate Studies:
Stephen A. Schellenberg, (D), S California, 2000, Pe
Professor:
Eric G. Frost, (D), S California, 1983, Or
Gary H. Girty, (D), Columbia, 1983, Gc
Kim B. Olsen, (D), Utah, 1994, Ye
Thomas K. Rockwell, (D), California (Santa Barbara), 1983, Gm
Senior Scientist:
Barry B. Hanan, (D), Virginia Tech, 1980, Cc
Associate Professor:
Shuo Ma, (D), California (Santa Barbara), 2006, Ys
Kathryn W. Thorbjarnarson, (D), California (Los Angeles), 1990, Hq
Lecturer:
Victor E. Camp, (D), Washington State, 1976, Gv
Kevin Robinson, (M), San Diego State, 1996, Gc
Isabelle Sacramentogrilo, (M), San Diego State, 1999, Gg
Adjunct Professor:
Mario V. Caputo, (D), Cincinnati, 1988, GsdOn
Emeritus:
Patrick L. Abbott, (D), Texas, 1973, Gs
Kathe K. Bertine, (D), Yale, 1970, Cl
Steven M. Day, (D), California (San Diego), 1977, Ys
Clive E. Dorman, (D), Oregon State, 1974, Op
George R. Jiracek, (D), California (Berkeley), 1972, Ye
J. Philip Kern, (D), California (Los Angeles), 1968, Pg
Daniel Krummenacher, (D), Geneva, 1959, Cc
Claude Monte Marshall, (D), Stanford, 1971, Ym
Richard H. Miller, (D), California (Los Angeles), 1975, Ps
Gary L. Peterson, (D), Washington, 1963, Gr
Anton D. Ptacek, (D), Washington (St. Louis), 1965, Gx

## San Francisco State University

**Dept of Earth & Climate Sciences** (B,M) (2013)
1600 Holloway Avenue TH 509
San Francisco, CA 94132
p. (415) 338-2061
geosci@sfsu.edu
http://tornado.sfsu.edu
*Enrollment (2013): B: 56 (0) M: 22 (0)*

Chair:
Karen Grove, (D), Stanford, 1989, Gs
Professor:
David P. Dempsey, (D), Washington, 1985, Oa
Oswaldo Garcia, (D), SUNY (Albany), 1976, Oa
John P. Monteverdi, (D), California (Berkeley), 1977, Oa
David A. Mustart, (D), Stanford, 1972, GizGv
Raymond Pestrong, (D), Stanford, 1965, Ng
Associate Professor:
John Caskey, (D), Nevada (Reno), 1996, Gc
Petra Dekens, (D), California (Santa Cruz), 2007, Pe
Newell (Toby) Garfield, (D), Rhode Island, 1990, Op
Mary L. Leech, (D), Stanford, 1999, GptGz
Leonard Sklar, (D), California (Berkeley), 2003, Gm
Assistant Professor:
Jason Gurdak, (D), Colorado, 2006, Hw
Alexander Stine, (D), California (Berkeley), Gg
Adjunct Professor:
E. Jan Null, (D), California (Davis), 1974, Ge
Emeritus:
Charles E. Bickel, (D), Harvard, 1971, Gx
York T. Mandra, (D), Stanford, 1958, Pg
Erwin Seibel, (D), Michigan, 1972, On
Raymond Sullivan, (D), Glasgow, 1960, Go
Lisa D. White, (D), California (Santa Cruz), 1989, Pg
Geosciences Tech:
Russell McArthur, On

## San Joaquin Delta College

**Dept of Geology** (A) (2013)
Stockton, CA 95204
p. (209) 954-5354
gfrost@deltacollege.edu
http://www.deltacollege.edu/div/scimath/geology.html

Professor:
Gina Marie Frost, (D), California (Santa Cruz), Gg

## San Jose City College

**Dept of Physical Science** (2016)
2100 Moorpark Ave
San Jose, CA 95128
p. (408) 288-3716
jessica.smay@sjcc.edu
http://www.sjcc.edu/academics/departments-divisions/physical-sciences

Associate Professor:
Jessica J. Smay, (M), UC Santa Barbara, 2002, OeGm
Emeritus:
John W. Martin, (M), San Jose State, 1959, Og

## San Jose State University

**Dept of Geology** (B,M) O (2017)
One Washington Square
San Jose, CA 95192-0102
p. (408) 924-5050
leslie.blum@sjsu.edu
http://www.sjsu.edu/geology/
*Enrollment (2009): B: 26 (7) M: 24 (6)*

Emeritus Professor:
Calvin H. Stevens, (D), S California, 1963, PsGrs
John W. Williams, (D), Stanford, 1970, Ng
Chair:
Jonathan S. Miller, (D), North Carolina, 1994, GiCc
Professor:
David W. Andersen, (D), Utah, 1973, GsCl
Emmanuel Gabet, (D), California (Santa Barbara), 2002, Gm
Paula Messina, (D), CUNY, 1998, OeGm
Ellen P. Metzger, (D), Syracuse, 1984, GpOe
Robert B. Miller, (D), Washington, 1980, Gct
June A. Oberdorfer, (D), Hawaii, 1983, HwGeHy
Donald L. Reed, (D), California (San Diego), 1985, GuYg
Assistant Professor:
Kimberly Blisniuk, (D), California (Davis), 2011, Gmt
Ryan Portner, (D), Macquarie Univ, 2010, Gdv

## Santa Barbara City College

**Dept of Earth & Planetary Sciences** (A) (2015)
721 Cliff Drive
Santa Barbara, CA 93109
p. (805) 965-0581
schultz@sbcc.edu

Chair:
Michael A. Robinson, (D), California (Santa Barbara), 2009, OyiOw
Professor:
Jeffrey W. Meyer, (D), California (Santa Barbara), 1992, GgzGx
Jan Schultz, (M), California (Santa Barbara), 1991, GgePg
Assistant Professor:
William Dinklage, (D), California (Santa Barbara), Gge
Emeritus:
Robert S. Gray, (D), Arizona, 1965, PvGzx

## Santa Monica College

**Earth Science Dept** (A) O (2015)
1900 Pico Blvd.
Santa Monica, CA 90405
p. (310) 434-8652
drake_vicki@smc.edu
http://www.smc.edu/AcademicPrograms/EarthScience/Pages/default.aspx

Adjunct Professor:
Alessandro Grippo, (D), Gs
Michelle Hopkins, (D), Colorado (Boulder), 2014
Bryan Murray, (D), California (Santa Barbara), 2013

Richard Robinson, (M), Gg

## Santa Rosa Junior College

**Earth and Space Sciences** (A) (2011)
1501 Mendocino Avenue
Santa Rosa, CA 95401-4395
p. (707) 527-4365
llupa@santarosa.edu
http://online.santarosa.edu/presentation/?2992



## Santiago Canyon College

**Dept of Earth, Space, & Physical Sciences** (2010)
8045 East Chapman Avenue
Orange, CA 92869-4512
p. (714) 564-4788
hannes_susan@sccollege.edu

Chair:
Debra A. Brooks, (M), Texas A&M, 1989, Yg
Adjunct Professor:
Gail F. Montwill, (M), Cal State (Los Angeles), 1988, Gg
Lizanne V. Simmons, (B), Brigham Young, 1979, Gr
Amy L. Stinson, (M), San Diego State, 1990, Gc
Adam D. Woods, (D), S California, 1998, Gs

## Solano Community College

**School of Mathematics and Science** (1999)
4000 Suisun Valley Road
Fairfield, CA 94534
p. (707) 864-7211
John.Yu@solano.edu
http://www.solano.edu/

## Sonoma State University

**Dept of Geology** (B) (2015)
1801 East Cotati Avenue
Rohnert Park, CA 94928
p. (707) 664-2334
james@sonoma.edu
http://www.sonoma.edu/geology/
Administrative Assistant: Elizabeth Kettmann
*Enrollment (2015): B: 70 (12)*

Department Chair:
Matthew J. James, (D), California (Berkeley), 1987, PiGhPo
Professor:
Matty Mookerjee, (D), Rochester, 2005, GcYgOi
Assistant Professor:
Owen Anfinson, (D), Calgary, 2012, GsdGg
Emeritus:
Thomas B. Anderson, (D), Colorado, 1969, GsrGd
Rolfe C. Erickson, (D), Arizona, 1968, Gx
Retired:
Daniel B. Karner, (D), California (Berkeley), 1997, Cc
Instructional Support Technician:
Phillip Mooney, (M), California (Davis), 2010, OnGc

## Stanford University

**Dept of Geological and Environmental Sciences** (B,M,D) (2014)
Stanford, CA 94305-2115
p. (650) 723-0847
maslin@stanford.edu
http://pangea.stanford.edu/GES/
Administrative Assistant: Elaine Andersen
*Enrollment (2010): B: 21 (9) M: 11 (5) D: 45 (8)*

Chair:
Jonathan F. Stebbins, (D), California (Berkeley), 1983, Cg
Research Professor:
J. Michael Moldowan, (D), Michigan, 1972, Co
Consulting Professor:
Richard Bernknopf, (D), George Washington, 1980, On
Alan K. Cooper, Yr
Francois Farges, Yr
Professor:
Atilla Aydin, (D), Stanford, 1978, Gc
Dennis K. Bird, (D), California (Berkeley), 1978, Cg
Gordon E. Brown, Jr., (D), Virginia Tech, 1970, Gz
C. Page Chamberlain, (D), Harvard, 1985, Yr
Robert B. Dunbar, (D), California (San Diego), 1981, On
Marco T. Einaudi, (D), Harvard, 1969, Em
W. Gary Ernst, (D), Johns Hopkins, 1959, Cp
Rodney C. Ewing, (D), Stanford, 1974, GzeCl
Steven M. Gorelick, (D), Stanford, 1981, Hw
Stephan A. Graham, (D), Stanford, 1976, Go
James C. Ingle, Jr., (D), S California, 1966, Pm
Andre G. Journel, (D), Nancy (France), 1977, Gq
Juhn G. Liou, (D), California (Los Angeles), 1970, Gp
Keith Loague, (D), British Columbia, 1986, Hq
Donald R. Lowe, (D), Illinois, 1967, Gd
Gail A. Mahood, (D), California (Berkeley), 1979, Gi
Pamela A. Matson, (D), Oregon State, 1983, Sb
Michael O. McWilliams, (D), Australian National, 1978, CcYme
Elizabeth L. Miller, (D), Rice, 1977, Gc
David D. Pollard, (D), Stanford, 1969, Gc
Consulting Associate Professor:
Trevor Dumitru, (D), Melbourne, 1990, Gc
Thomas Holzer, (D), Stanford, 1970, Ng
Joseph W. Ruetz, (M), Stanford, 1977, Gg
Joseph Wooden, (D), North Carolina, 1975, Cg
Associate Professor:
Christopher F. Chyba, (D), Cornell, 1991, Gh
Assistant Professor:
Scott E. Fendorf, (D), Delaware, 1992, Sc
Adina Paytan, (D), California (San Diego), 1996, Oc
Courtesy Professor:
Peter G. Brewer, (D), Liverpool, 1967, Oc
David A. Clague, (D), California (San Diego), 1974, Ob
Peter K. Kitanidis, (D), Yr
Stephen G. Monismith, Yr
Courtesy Associate Professor:
James P. Barry, Yr
Ronaldo I. Borja, Yr
David L. Freyberg, Yr
Anders Nilsson, Yr
Alfred Spormann, Yr
Debra S. Stakes, (D), Oregon State, 1978, Ob
Courtesy Assistant Professor:
Kevin Arrigo, (D), S California, 1992, Og
Courtesy Professor:
James O. Leckie, (D), Harvard, 1970, Cl
Emeritus:
Robert G. Coleman, (D), Stanford, 1957, Gi
Robert R. Compton, (D), Stanford, 1949, Gx
John W. Harbaugh, (D), Wisconsin, 1955, Gq
Ronald J. P. Lyon, (D), California (Berkeley), 1954, Or
Irwin Remson, (D), Columbia, 1954, Hw
Tjeerd H. Van Andel, (D), Groningen (Neth), 1950, Gu

**Dept of Geophysics** (B,M,D) (2016)
397 Panama Mall
Stanford, CA 94305-2215
p. (650) 497-3498
mbrunner@stanford.edu
https://earth.stanford.edu/geophysics/
*Enrollment (2013): B: 2 (1) M: 2 (1) D: 67 (7)*

Chair:
Howard A. Zebker, (D), Stanford, 1984, Yx
Research Professor:
Bill Ellsworth, (D), MIT, 1978, Ys
Courtesy Associate Professor:
Simon L. Klemperer, (D), Cornell, 1985, GtYes
Professor:
Gregory C. Beroza, (D), MIT, 1989, Ys
Biondo L. Biondi, (D), Stanford, 1990, Ye
Jerry M. Harris, (D), Caltech, 1980, Ys
Simon L. Klemperer, (D), Cornell, 1985, Gt
Rosemary J. Knight, (D), Stanford, 1985, Hy
Gerald M. Mavko, (D), Stanford, 1977, Yg
Paul Segall, (D), Stanford, 1981, Yg
Norman H. Sleep, (D), MIT, 1972, Yg
Mark D. Zoback, (D), Stanford, 1975, Yg
Associate Professor:
Eric M. Dunham, (D), California (Santa Barbara), 2005, Ys
Tapan Mukerji, (D), Yg

Assistant Professor:
Dustin M. Schroeder, (D), Texas (Austin), 2014, Gl
Jenny Suckale, (D), Yg
Tiziana Vanorio, (D), Yx
Adjunct Professor:
Steven Gorelick, Yr
Emeritus:
Jon F. Claerbout, (D), MIT, 1967, Ye
Antony C. Fraser-Smith, (D), Auckland, 1966, Ym
Robert L. Kovach, (D), Caltech, 1962, Ys
Amos M. Nur, (D), MIT, 1969, Yg
Joan Roughgarden, (D), Harvard, 1971, Ob
George A. Thompson, (D), Stanford, 1949, Yg
Department Manager:
Csilla Csaplar, On

## Taft College

**Math/Science Div** (1999)
29 Emmons Park Drive
Taft, CA 93268
p. (661) 763-7932
ggolling@taftcollege.edu

## University of California, Berkeley

**Dept of Earth & Planetary Science** (B,M,D) (2015)
307 McCone
Berkeley, CA 94720-4767
p. (510) 642-3993
bbuffett@berkeley.edu
http://eps.berkeley.edu
*Enrollment (2002): B: 43 (18) M: 1 (1) D: 6 (6)*
Chair:
Hans-Rudolf Wenk, (D), Zurich, 1965, Gz
Professor:
Walter Alvarez, (D), Princeton, 1967, Grh
Jillian Banfield, (D), Johns Hopkins, 1990, CoGz
James K. Bishop, (D), MIT/WHOI, 1977, Oc
George H. Brimhall, (D), California (Berkeley), 1972, Eg
Roland Burgmann, (D), Stanford, 1993, Gt
Donald J. DePaolo, (D), Caltech, 1978, Cc
William E. Dietrich, (D), Washington, 1982, Gm
Inez Fung, (D), MIT, 1977, Oa
B. Lynn Ingram, (D), Stanford, 1992, Cs
Raymond Jeanloz, (D), Caltech, 1979, Gy
James W. Kirchner, (D), California (Berkeley), 1990, Ge
Michael Manga, (D), Harvard, 1994, GvXy
Mark A. Richards, (D), Caltech, 1986, Yg
Barbara A. Romanowicz, (D), Paris, 1979, Ys
Associate Professor:
Kristie Boering, (D), Stanford, 1991, Oa
Douglas S. Dreger, (D), Caltech, 1992, Ys
Assistant Professor:
Richard M. Allen, (D), Princeton, 2001, Ys
Burkhard Militzer, (D), Illinois (Urbana), 2000, Gy
Adjunct Associate Professor:
David L. Alumbaugh, (D), California (Berkeley), 1993, Ye
Paul Renne, (D), California (Berkeley), 1987, Cc
Adjunct Professor:
Steven Pride, (D), Texas A&M, 1991, Ys
Visiting Professor:
William D. Collins, (D), Chicago, 1988, Oa
Emeritus:
Mark S. Bukowinski, (D), California (Los Angeles), 1975, Gy
Ian S. Carmichael, (D), London, 1958, Gi
Garniss H. Curtis, (D), California (Berkeley), 1951, Cc
Lane R. Johnson, (D), Caltech, 1966, Ys
Doris Sloan, (D), California (Berkeley), 1981, Pme
Chi-Yuen Wang, (D), Harvard, 1964, Yg

**Dept of Environmental Science, Policy and Management** (B,M,D) (2012)
140 Mulford Hall
Berkeley, CA 94720-3110
p. (510) 643-3788
earthy@berkeley.edu
http://espm.berkeley.edu/

**Dept of Integrative Biology** (B,M,D) (2016)
3040 Valley Life Sciences Building #3140
Berkeley, CA 94720-3140
p. (510) 502-5887
johnh@berkeley.edu
http://ib.berkeley.edu/
Curator:
Carole S. Hickman, (D), Stanford, 1975, PoePy
Professor:
F. Stuart Chapin, (D), Stanford, 1973, Pe
William A. Clemens, (D), California (Berkeley), 1960, Pv
Robert J. Full, (D), SUNY (Buffalo), 1984, Pi
Harry W. Green, (D), Tennessee, 1977, Pv
Ned K. Johnson, (D), California (Berkeley), 1961, Pv
Mimi A. R. Koehl, (D), Duke, 1976, Po
Brent Mishler, (D), Harvard, 1984, Pe
Kevin Padian, (D), Yale, 1980, Pv
Wayne P. Sousa, (D), California (Santa Barbara), 1977, Ob
Associate Professor:
Mary E. Power, (D), Washington, 1981, Pe
Emeritus:
Roy L. Caldwell, (D), Iowa, 1969, Pi
Carole S. Hickman, (D), Stanford, 1975, Po
William Z. Lidicker, Jr., (D), Illinois, 1957, Pv
David R. Lindberg, (D), California (Santa Cruz), Pi
James L. Patton, (D), Arizona, 1968, Pv
Thomas M. Powell, (D), California (Berkeley), 1970, Op
Montgomery Slatkin, (D), Harvard, 1970, Po
Glennis Thompson, (D), Melbourne, 1974, On
James W. Valentine, (D), California (Los Angeles), 1958, Po
David B. Wake, (D), S California, 1964, Po
Marvalee H. Wake, (D), S California, 1968, Po

**Dept of Materials Science and Engineering** (B,M,D) (2014)
577 Evans Hall
Berkeley, CA 94720-1760
p. (510) 642-3801
hfmorrison@berkeley.edu
http://www.mse.berkeley.edu/
Chair:
Robert O. Ritchie, (D), Cambridge, 1973, Om
Professor:
Alex Becker, (D), McGill, 1963, Yg
George A. Cooper, (D), Cambridge, 1967, Np
Didier deFontaine, (D), Northwestern, 1967, Om
Lutgard DeJonghe, (D), California (Berkeley), 1970, Om
Thomas M. Devine, (D), MIT, 1974, Om
Fiona M. Doyle, (D), Imperial Coll (UK), 1983, Nx
James W. Evans, (D), SUNY (Buffalo), 1970, Nx
Douglas W. Fuerstenau, (D), MIT, 1953, Nx
Andreas Glaeser, (D), MIT, 1981, Om
Ronald Gronsky, (D), California (Berkeley), 1977, Om
Eugene Haller, (D), Basel, 1970, Om
J. W. Morris, Jr., (D), MIT, 1969, Om
H. Frank Morrison, (D), California (Berkeley), 1967, Yg
T. N. Narasimhan, (D), California (Berkeley), 1975, Hw
Timothy Sands, (D), California (Berkeley), 1984, Om
Kalanadh V. S. Sastry, (D), California (Berkeley), 1970, Nx
Eicke Weber, (D), Cologne, 1976, Om
Associate Professor:
Tad W. Patzek, (D), Silesian Tech, 1979, Np
James W. Rector, III, (D), Stanford, 1990, Ys
Assistant Professor:
Daryl Chrzan, (D), California (Berkeley), 1989, Om
Mauro Ferrari, (D), California (Berkeley), 1989, Om

**GeoEngineering Program** (B,M,D) (2011)
Berkeley, CA 94720
p. (510) 642-3157
pestana@ce.berkeley.edu
http://www.ce.berkeley.edu/geo/
Chair:
Lisa Alvarez-Cohen
Professor:
Alex Becker, (D), McGill, 1964, Ye
Huntly Frank Morrison, (D), California (Berkeley), 1967, Ye

Senior Scientist:
Ki Ha Lee, (D), California (Berkeley), 1978, Ye
Assistant Professor:
James W. Rector, (D), Stanford, 1990, Ye



**Museum of Paleontology** (D) (2016)
1101 Valley Life Sciences Bldg #4780
Berkeley, CA 94720-4780
p. (510) 642-1821
crmarshall@berkeley.edu
http://www.ucmp.berkeley.edu
*Enrollment (2016): D: 25 (3)*
Professor:
Charles R. Marshall, (D), Chicago, 1989, PqoPi
Curator:
Kevin Padian, (D), Yale, 1980, Pv
Principal Museum Scientist:
Mark B. Goodwin, (D), California (Davis), 2008, Pv
Museum Scientist:
Diane Erwin, (D), Alberta, 1990, Pb
Pat Holroyd, (D), Duke, 1994, Pv
Curator:
Leslea Hlusko, (D), Penn State, 2000, Onn
Associate Professor:
Cynthia Looy, (D), Utrecht Univ, 1995, Pb
Assistant Professor:
Seth Finnegan, (D), UC Riverside, 1994, Pio
Curatorial Associate:
Walter Alvarez, (D), Princeton, 1967, Gr
Curator:
Anthony D. Barnosky, (D), Washington, 1983, Pv
William A. Clemens, (D), California (Berkeley), 1960, Pv
David R. Lindberg, (D), California (Santa Cruz), 1983, Po
Emeritus:
Roy Caldwell, (D), Iowa, 1969, Po
James W. Valentine, (D), California (Los Angeles), 1958, Po
Assistant Director:
Lisa D. White, UC Santa Cruz, 1989, OePim

## University of California, Davis

**Dept of Earth and Planetary Sciences** (B,M,D) ● (2016)
2119 Earth & Physical Sciences Building
One Shields Ave.
Davis, CA 95616-5270
p. (530) 752-0350
geology@ucdavis.edu
http://geology.ucdavis.edu
f: https://www.facebook.com/pages/Geology-Department-at-UC-Davis/87869598212
*Enrollment (2016): B: 98 (25) M: 15 (8) D: 34 (2)*
Chair:
Dawn Y. Sumner, (D), MIT, 1995, Gs
Professor:
Magali I. Billen, (D), Caltech, 2001, Yg
Sandra J. Carlson, (D), Michigan, 1986, Po
Kari M. Cooper, (D), California (Los Angeles), 2001, Gi
Eric S. Cowgill, (D), California (Los Angeles), 2001, Gtc
Louise H. Kellogg, (D), Cornell, 1988, Yg
Charles E. Lesher, (D), Harvard, 1985, Gi
James S. McClain, (D), Washington, 1979, Yr
Isabel P. Montañez, (D), Virginia Tech, 1990, Gd
Ryosuke Motani, (D), Toronto, 1997, Pv
Sujoy Mukhopadhyay, (D), Caltech, 2002, Cg
Michael Oskin, (D), Caltech, 2002, GmtGc
Nicholas Pinter, (D), California (Santa Barbara), 1992, Gm
John Rundle, (D), California (Los Angeles), 1976, Yd
Howard J. Spero, (D), California (Santa Barbara), 1986, PeCs
Sarah T. Stewart, (D), Caltech, 2002
Geerat J. Vermeij, (D), Yale, 1971, Pe
Kenneth L. Verosub, (D), Stanford, 1973, Ym
Qing-zhu Yin, (D), Max Planck, 1995, Cc
Robert A. Zierenberg, (D), Wisconsin, 1983, Cs
Associate Professor:
Tessa M. Hill, (D), California (Santa Barbara), 2004, Pe
Research Associate:
Irina Delusina, (D), Tallinn Inst of Geology (Estonia), 1989, Pl
Oliver Kreylos, (D), California (Davis), 2003, Ng
Ann D. Russell, (D), Washington, 1994, Cm
Peter Thy, (D), Aarhus (Denmark), 1982, Gx
Burak Yikilmaz, California (Davis), 2010, Gc
Vice-Chair:
David A. Osleger, (D), Virginia Tech, 1990, Gr
Emeritus:
Cathy J. Busby, (D), Princeton, 1983, Gt
Richard Cowen, (D), Cambridge, 1966, Po
Howard W. Day, (D), Brown, 1971, Gp
John Dewey, (D), London (UK), 1960, Gt
James A. Doyle, (D), Harvard, 1970, PblPs
Charles G. Higgins, (D), California (Berkeley), 1950, Gm
Eldridge M. Moores, (D), Princeton, Gt
Jeffrey F. Mount, (D), California (Santa Cruz), 1980, Gm
Sarah M. Roeske, (D), California (Santa Cruz), 1988, GtcGp
James R. Rustad, (D), Minnesota, 1992, Cl
Peter Schiffman, (D), Stanford, 1978, Gp
Donald L. Turcotte, (D), Caltech, 1958, Yg
Robert J. Twiss, (D), Princeton, 1970, Gc
Cooperating Faculty:
William H. Casey, (D), Penn State, 1985, Cl
Graham E. Fogg, (D), Texas (Austin), 1986, Hw
Alexandra Navrotsky, (D), Chicago, 1967, Om

**Dept of Land, Air & Water Resources** (B,M,D) O (2017)
1110 Plant and Environmental Sciences Building
One Shields Avenue
Davis, CA 95616
p. (530) 752-1130
rjsouthard@ucdavis.edu
http://lawr.ucdavis.edu/
*Enrollment (2014): B: 5 (5) M: 54 (19) D: 55 (8)*
Specialist in Cooperative Extension:
Toby O'Geen, (D), Idaho, 2002, Os
Professor:
Cort Anastasio, (D), Duke, 1994, Oa
Shu-Hua Chen, (D), Purdue, 1999, OraOw
Randy A. Dahlgren, (D), Washington, 1987, Os
Graham E. Fogg, (D), Texas (Austin), 1986, Hw
Mark E. Grismer, (D), Colorado State, 1971, Hw
Richard Grotjahn, (D), Florida State, 1979, Oa
Peter J. Hernes, (D), Washington, 1999, Hg
Jan W. Hopmans, (D), Auburn, Os
William R. Horwath, (D), Michigan State, 1993, OsSb
Louise E. Jackson, (D), Washington, 1982, Os
Terrence R. Nathan, (D), SUNY (Albany), 1985, Oaw
Gregory B. Pasternack, (D), Johns Hopkins, 1998, GmHg
Kate Scow, (D), Cornell, 1988, Os
Randal J. Southard, (D), North Carolina State, 1983, Os
Susan L. Ustin, (D), California (Davis), 1983, Or
Associate Professor:
Ben Houlton, (D), Princeton, So
Sanjai Parikh, (D), Arizona, Os
Assoc Proj Scientist:
Martin Burger, (D), California (Davis), 2002, Spb
Specialist in Cooperative Extension:
Daniel Geisseler, (D), California (Davis), 2009, Os
20% Cooperative Extension:
Samuel Sandoval Solis, (D), Texas (Austin), 2011, Hg
Assistant Professor:
Helen E. Dahlke, (D), Cornell, 2011, Hgs
Yufang Jin, (D), Boston, 2002, Or
Adjunct Professor:
Minghua Zhang, (D), California (Davis), 1993, Hg
Emeritus:
James H. Richards, (D), Alberta, 1981, Os
Asst Proj Scientist:
Michael L. Grieneisen, (D), North Carolina (Chapel Hill), 1992, Hg
Asst Professional Researcher:
Lucas CR Silva, (D), Guelph, 2011, Sb
Cooperating Faculty:
Stephen R. Grattan, (D), California (Riverside), 1984, Sp
Thomas Harter, (D), Arizona, 1994, Hg
Richard L. Snyder, (D), Iowa State, 1980, Oa
Daniele Zaccaria, (D), Utah State, 2011, Hg

**Dept of Land, Air, & Water Resources - Hydrology Program**

(B,M,D) (2013)
1110A Plant Environmental Sciences
One Shields Avenue
Davis, CA 95616-8628
p. (530) 752-3060
radahlgren@ucdavis.edu
http://lawr.ucdavis.edu/
*Enrollment (2012): B: 25 (0) M: 16 (0) D: 21 (0)*

Chair:
Randy Dahlgran, (D), Washington, 1987, Sc
Jan W. Hopmans, (D), Auburn, 1985, Sp
Professor:
William H. Casey, (D), Penn State, 1985, Sc
Graham E. Fogg, (D), Texas (Austin), 1986, Hwg
Mark E. Grismer, (D), Colorado State, 1984, Spf
Miguel A. Marino, (D), California (Los Angeles), 1972, Hw
Gregory B. Pasternack, (D), Johns Hopkins, 1998, GmHg
Carlos E. Puente, (D), MIT, 1984, Hq
Susan L. Ustin, (D), California (Davis), 1983, Or
Wesley W. Wallender, (D), Utah State, 1982, Hg
Water Management Specialist:
Terry L. Prichard, (D), California (Davis), 1975, On
Plant-Water Relations Specialist:
Stephen R. Grattan, (D), California (Riverside), 1984, On
Irrigation Specialist:
Larry J. Schwankl, (D), California (Davis), 1991, On
Irrigation & Soil Specialist:
David A. Goldhamer, (D), California (Davis), 1980, On
Groundwater Hydrology Specialist:
Thomas L. Harter, (D), Arizona, 1993, Hw
Associate Professor:
Peter Hernes, (D), Washington, Og
Assistant Professor:
Samuel Sandoval, (D), Texas, 2011, Hg
Irrigation & Drainage Specialist:
Blaine R. Hanson, (D), Colorado State, 1977, Hg
Emeritus:
Robert M. Hagan, California (Davis), 1948, Hg
Theodore C. Hsiao, (D), Illinois (Urbana Champaign), 1963, SpOwn
Allen W. Knight, (D), Utah, 1965, Hs
Donald R. Nielsen, (D), Iowa State, 1958, Sp
Verne H. Scott, Colorado State, 1959, Hs

**Graduate Group in Hydrologic Sciences** (M,D) (2013)
1152 PES
One Shields Ave
Davis, CA 95616
p. (530) 752-1669
lawrgradadvising@ucdavis.edu
http://hydscigrad.ucdavis.edu/
Administrative Assistant: Diane Swindall
*Enrollment (2010): M: 14 (1) D: 14 (0)*

Chair:
Graham E. Fogg, (D), Texas, 1986, Hy
Professor:
William H. Casey, (D), Pennsylvania, 1985, Cl
Jeannie Darby, (D), Texas, 1988, Hg
Mark E. Grismer, (D), Colorado State, 1984, Hw
Jan W. Hopmans, (D), Auburn, 1985, So
Theodore C. Hsiao, (D), Illinois, 1964, On
B. E. Larock, (D), Stanford, 1966, On
Jay R. Lund, (D), Washington, 1986, Hs
M. A. Marino, (D), California (Los Angeles), 1972, Hw
James S. McClain, (D), Washington, 1979, Yr
Eldridge M. Moores, (D), Princeton, 1963, Gt
Jeffrey F. Mount, (D), California (Santa Cruz), 1980, Gs
Dennis E. Rolston, (D), California (Davis), 1971, So
K. K. Tanji, (M), California (Davis), 1961, On
Wes W. Wallender, (D), Utah State, 1982, Hg
B. C. Weare, (D), SUNY (Buffalo), 1974, Oa
Associate Professor:
Bruce Kutter, (D), Cambridge, 1983, Ne
Carlos E. Puente, (D), MIT, 1984, Hg
Emeritus:
Robert A. Matthews, (B), California (Berkeley), 1953, Ge

## University of California, Irvine

**Dept of Earth System Science** (B,D) (2015)
School of Physical Sciences
Irvine, CA 92697-3100
p. (949) 824-8794
essinfo@ess.uci.edu
http://www.ess.uci.edu
*Enrollment (2013): B: 166 (64) D: 49 (5)*

Chair:
Michael L. Goulden, (D), Stanford, 1992, Og
Advance Professor:
Ellen R. M. Druffel, (D), California (San Diego), 1980, Oc
Professor:
James Famiglietti, (D), Princeton, Hw
Gudrun Magnusdottir, (D), Colorado State, 1989, Oa
Michael J. Prather, (D), Yale, 1975, Oa
Eric Rignot, (D), S California, 1991, OlpOr
Eric S. Saltzman, (D), Rosenstiel, 1986, Oa
Soroosh Sorooshian, (D), California (Los Angeles), 1978, Hg
Susan E. Trumbore, (D), Columbia, 1989, Sc
Jin-Yi Yu, (D), Washington, OapOw
Charles Zender, (D), Colorado, 1996, Oa
Associate Professor:
Steven J. Davis, (D), Stanford, 2008, GeCsOu
Jefferson Keith Moore, (D), Oregon State, 1999, Op
Francois Primeau, (D), MIT/WHOI, 1998, Op
Assistant Professor:
Claudia Czimczik, (D)
Todd Dupont, (D), Penn State, 2004, Ol
Julie Ferguson, (D), Oxford, 2008, Oe
Kathleen Johnson, (D), California (Berkeley), 2004, Og
Saewung Kim, (D), Georgia Tech, 2007, Oa
Adam Martiny, (D), Tech (Denmark), 2003, Ob
Michael Pritchard, (D)
Isabella Velicogna, (D), Trieste, 1999, Or
Laboratory Director:
John Southon, (D), Auckland Univ, 1976, CcOcPe

## University of California, Los Angeles

**Dept of Atmospheric and Oceanic Sciences** (B,M,D) (2016)
7127 Math Sciences
Box 951565
Los Angeles, CA 90095
p. (310) 825-1217
deptinfo@atmos.ucla.edu
http://www.atmos.ucla.edu
f: https://www.facebook.com/AOS.UCLA

Chair:
Jochen P. Stutz, (D), Heidelberg, 1996, Oa
Professor:
Rong Fu
Kuo Nan Liou, (D), New York, 1970, Oa
Lawrence Lyons, (D), California (Los Angeles), 1972, Oa
James McWilliams, (D), Harvard, 1971, Oga
Carlos R. Mechoso, (D), Princeton, 1978, Oa
J. David Neelin, (D), Princeton, 1987, Oa
Suzanne Paulson, (D), Caltech, 1991, Oa
Richard M. Thorne, (D), MIT, 1968, Oa
Yongkang Xue, (D)
Associate Professor:
Jacob Bortnik, (D)
Marcelo Chamecki
Gang Chen
Alex Hall, (D)
Qinbin Li
Tina Treude
Assistant Professor:
Daniele Bianchi , (D)
Rob Eagle
Jasper Kok , (D)
Ulli Seibt , (D)
Andrew Stewart, (D)
Arahdna Tripati, (D)
Lecturer:
Jeffrey Lew, (D), California (Los Angeles), 1985, Oa

Adjunct Professor:
Yi Chao
Emeritus:
Akio Arakawa, (D), Tokyo, 1961, Oa
Robert Fovell, (D), Illinois, 1988, Oa
Michael Ghil, (D), Courant Inst, 1975, Oa
Richard Turco, (D), Illinois, 1971, Oa

**Dept of Earth, Planetary, and Space Sciences** (B,M,D) ○ (2016)
595 Charles E Young Drive East
3806 Geology Building, Box 951567
Los Angeles, CA 90095-1567
p. 310.825.3880
info@epss.ucla.edu
http://epss.ucla.edu
f: https://www.facebook.com/uclaepss
t: uclaepss
*Enrollment (2011): B: 69 (17) M: 11 (14) D: 61 (15)*
Chair:
Kevin D. McKeegan, (D), Washington Univ (St. Louis), 1987, XcCsc
Professor:
Vassilis Angelopoulos, (D), California (Los Angeles), 1993, XyOaYm
Jonathan M. Aurnou, (D), Johns Hopkins, 1999, XyYmx
Paul M. Davis, (D), Queensland, 1974, YsmYg
T. Mark Harrison, (D), Australian National, 1981, CcGt
Raymond V. Ingersoll, (D), Stanford, 1976, Gst
David Jewitt, (D), Caltech, 1983, XyOnn
Abby Kavner, (D), California (Berkeley), 1997, GyYg
Craig E. Manning, (D), Stanford, 1989, GxCp
Jean-Luc Margot, (D), Cornell, 1999, XyYdv
William I. Newman, (D), Cornell, 1979, Xy
David A. Paige, (D), Caltech, 1985, Xy
Gilles Peltzer, (D), Paris VII, 1987, OrGt
Christopher T. Russell, (D), California (Los Angeles), 1968, Xy
J. William Schopf, (D), Harvard, 1968, Po
Laurence C. Smith, (D), Cornell, 1996, Hs
Marco Velli, (D), Pisa, 1985, Yg
An Yin, (D), S California, 1988, Gt
Edward D. Young, (D), S California, 1990, CsXc
Associate Professor:
Caroline Beghein, (D), Utrecht, 2003, Ysg
Jonathan Mitchell, (D), Chicago, 2007, OaXg
Edwin A. Schauble, (D), Caltech, 2002, Cgs
Hilke Schlichting, (D), Caltech, 2009, Xy
Tina Treude, (D), Max Planck Inst, 2004, ObPyCm
Aradhna Tripati, (D), California (Santa Cruz), 2002, CmPeCs
Assistant Professor:
Lingsen Meng, (D), Caltech, 2012, Ys
Ulrike Seibt, (D), Hamburg, 2003, PyCg
Instructor:
Stephen L. Salyards, (D), Caltech, 1989, Ys
Adjunct Professor:
Robert C. Newton, (D), California (Los Angeles), 1963, CpGp
Edward J. Rhodes, (D), Oxford, 1990, CcGma
Emeritus:
G. Peter Bird, (D), MIT, 1976, YgGtg
Donald Carlisle, (D), Wisconsin, 1950, Eg
Paul J. Coleman, (D), California (Los Angeles), 1966, Xy
Wayne A. Dollase, (D), MIT, 1966, Gz
Clarence A. Hall, (D), Stanford, 1956, Gt
David D. Jackson, (D), MIT, 1969, YsdYg
Isaac R. Kaplan, (D), S California, 1962, Csg
Margaret G. Kivelson, (D), Radcliffe, 1957, Xy
Robert L. McPherron, (D), California (Berkeley), 1968, Xy
Paul M. Merifield, (D), Colorado, 1963, Ng
John L. Rosenfeld, (D), Harvard, 1954, Gp
Bruce Runnegar, (D), Queensland, 1967, Po
Gerald Schubert, (D), California (Berkeley), 1964, YgXy
Raymond J. Walker, (D), California (Los Angeles), 1973, Xy
John T. Wasson, (D), MIT, 1958, Xmc

**Inst of Geophysics & Planetary Physics** (2014)
3845 Slichter Hall
Los Angeles, CA 90095-1567
p. (310) 825-1664
jnakatsu@igpp.ucla.edu
http://www.igpp.ucla.edu
Professor:
Vassilis Angelopoulos, (D), California (Los Angeles), 1993, Yg
Maha Ashour-Abdalla, (D), Imperial Coll (UK), 1971, On
Friedrich H. Busse, (D), Munich, 1967, On
Richard E. Dickerson, (D), Minnesota, 1957, On
Michael Ghil, (D), New York Univ, 1975, OaYdOp
T. Mark Harrison, (D), Australian National, 1980, Cc
Charles F. Kennel, (D), Princeton, 1964, On
Margaret G. Kivelson, (D), Radcliffe, 1957, Xy
Robert L. McPherron, (D), California (Berkeley), 1968, Xy
James C. McWilliams, (D), Harvard, 1971, Oa
Paul H. Roberts, (D), Cambridge, 1967, Ym
Bruce Runnegar, (D), Queensland, 1967, Po
Christopher T. Russell, (D), California (Los Angeles), 1968, Yg
J. William Schopf, (D), Harvard, 1968, Po
Gerald Schubert, (D), California (Berkeley), 1964, Yg
Karl O. Stetter, (D), Tech (Munich), 1973, Po
Richard Turco, (D), Illinois, 1971, Oa
Raymond J. Walker, (D), California (Los Angeles), 1973, On
John T. Wasson, (D), MIT, 1958, Xm
Edward Young
Senior Scientist:
Stanislav I. Braginsky, (D), Moscow Inst, 1948, Ym
Robert J. Strangeway, (D), London, 1978, On
M I. Venkatesan, (D), Madras, 1973, Co
Paul Warren, (D), California (Los Angeles), 1979, Xm
Associate Professor:
Abby Kavner, (D), California (Berkeley), 1997, Yg
J. David Neelin, (D), Princeton, 1987, Oa
Associate Research Scientist:
Jean Berchem, (D), California (Los Angeles), 1986, On
Gregory Kallemeyn, (D), California (Los Angeles), 1982, Cg
Frank T. Kyte, (D), California (Los Angeles), 1983, Ct
Alan E. Rubin, (D), New Mexico, 1982, Xm
David Schriver, (D), California (Los Angeles), 1988, Yg
Fred Schwab, (B), California (Los Angeles), 1960, Ys
Assistant Research Scientist:
Kayo Ide, (D), Caltech, 1990, Og
Krishan Khurana, (D), Durham, 1984, On
Robert Richard, (D), California (Los Angeles), 1988, Yg
William Smythe, (D), California (Los Angeles), 1979, Ys
Ferenc D. Varadi, (D), California (Los Angeles), 1989, On
Emeritus:
Orson L. Anderson, (D), Utah, 1951, Yx
Paul J. Coleman, Jr., (D), California (Los Angeles), 1966, Gc
Isaac R. Kaplan, (D), S California, 1962, Cg
Leon Knopoff, (D), Caltech, 1949, Ys
Ronald L. Shreve, (D), Caltech, 1958, Gm
Department Manager:
James Nakatsuka, On

## University of California, Riverside

**Dept of Earth Sciences** (B,M,D) ● (2015)
University of California Riverside
Earth Sciences Department
Riverside, CA 92521
p. (951) 827-3182
david.oglesby@ucr.edu
http://earthscience.ucr.edu
*Enrollment (2013): B: 51 (13) M: 15 (8) D: 38 (4)*
Chair:
David D. Oglesby, (D), California (Santa Barbara), 1999, Ys
Distinguished Professor:
Timothy W. Lyons, (D), Yale, 1992, CgOcCl
Professor:
Mary L. Droser, (D), S California, 1987, Po
Nigel C. Hughes, (D), Bristol (UK), 1990, Po
Gordon Love, (D), Strathclyde, 1995, Cog
Richard A. Minnich, (D), California (Los Angeles), 1978, Oy
Andy Ridgwell, (D), East Anglia, 2001, On
Peter M. Sadler, (D), Bristol (UK), 1973, GrPs
Associate Professor:
Gareth J. Funning, (D), Oxford, 2005, YdgYs
Michael A. McKibben, (D), Penn State, 1984, Cg
Assistant Professor:
Robert J. Allen, (D), Yale, 2009, Oa
Nicolas Barth, (D), Otago, 2013, Gt

Andrey Bekker, (D), Virginia Tech, 2001, GsrCg
Heather Ford, (D), Ys
Abhijit Ghosh, (D), Washington, YsGt
Sandra Kirtland Turner, (D), Scripps, Cs
Adjunct Assistant Professor:
Katherine J. Kendrick, (D), California (Riverside), 1999, Gm
Thomas A. Scott, (D), California (Berkeley), 1987, Oy
Adjunct Professor:
Elizabeth Cochran, California (Los Angeles), 2005, Ys
Larissa F. Dobrzhinetskaya, (D), Inst of Physics of Earth (Moscow), 1978, Gx
Douglas M. Morton, (D), California (Los Angeles), 1966, Gp
Professor of the Graduate Division:
James H. Dieterich, (D), Yale, 1968, Yg
Harry W. Green, II, (D), California (Los Angeles), 1968, Yx
Emeritus:
Wilfred A. Elders, (D), Durham, 1961, Gi
Michael A. Murphy, (D), California (Los Angeles), 1954, Ps
Stephen K. Park, (D), MIT, 1984, Ym
Michael O. Woodburne, (D), California (Berkeley), 1966, Pv

## University of California, San Diego

**Scripps Institution of Oceanography** (B,M,D) (2017)
Graduate Office
9500 Gilman Dr.
Mail Code 0208
La Jolla, CA 92093-0208
p. (858) 534-3206
siodept@sio.ucsd.edu
https://scripps.ucsd.edu/education/
*Enrollment (2012): M: 7 (0) D: 62 (13)*
Vice Chancellor:
Margaret Leinen, (D), Rhode Island, 1980, OgPe
Director:
T. Guy Masters, (D), Cambridge (UK), 1979, YsGy
Professor:
Brian Palenik, (D), MIT/WHOI, 1989, Ob
Vice-Chair:
Catherine G. Constable, (D), California (San Diego), 1987, Ym
Director of MPL:
William A. Kuperman, (D), Maryland, 1972, Op
Professor:
Duncan C. Agnew, (D), California (San Diego), 1979, Ys
Lihini I. Aluwihare, (D), MIT/WHOI, 1999, Gu
Laurence Armi, (D), California (Berkeley), 1975, Op
Farooq Azam, (D), Czech Acad of Sci, 1968, Ob
Katherine A. Barbeau, (D), MIT/WHOI, 1998, Oc
Douglas H. Bartlett, (D), Illinois, 1985, Ob
Kevin M. Brown, (D), Durham (UK), 1987, Gu
Michael J. Buckingham, (D), Reading, 1971, Op
Ronald S. Burton, (D), Stanford, 1981, Ob
Paterno R. Castillo, (D), Washington Univ (St. Louis), 1987, Giu
Paola Cessi, (D), MIT, 1987, Op
Christopher D. Charles, (D), Columbia, 1991, Pe
Steven C. Constable, (D), Australian National, 1983, Yr
Andrew G. Dickson, (D), Liverpool (UK), 1978, Cm
Leroy M. Dorman, (D), Wisconsin, 1970, Yg
Neal W. Driscoll, (D), Columbia, 1992, Gu
William H. Fenical, (D), California (Riverside), 1968, Oc
Yuri A. Fialko, (D), Princeton, 1998, GqtGv
Peter J. S. Franks, (D), MIT/WHOI, 1990, Ob
Helen A. Fricker, (D), Tasmania, 1999, Yr
Terry Gaasterland, (D), Maryland, 1992, Ob
Jeffrey S. Gee, (D), California (San Diego), 1991, Gu
William H. Gerwick, (D), California (San Diego), 1981, Cm
Carl H. Gibson, (D), Stanford, 1962, Oo
Sarah T. Gille, (D), MIT/WHOI, 1995, Op
Vicki Grassian, (D), Berkeley, 1987, Cm
Philip A. Hastings, (D), Arizona, 1987, Ob
Myrl C. Hendershott, (D), Harvard, 1966, Op
John A. Hildebrand, (D), Stanford, 1983, Yr
David R. Hilton, (D), Cambridge, 1985, Cs
William S. Hodgkiss, Jr., (D), Duke, 1975, Oo
Nicholas D. Holland, (D), Stanford, 1965, Ob
Miriam Kastner, (D), Harvard, 1970, CmlCm
Ralph F. Keeling, (D), Harvard, 1988, Oa
Michael R. Landry, (D), Washington, 1976, Ob
Gabi Laske, (D), Karlsruhe, Germany, 1993, Ys
James J. Leichter, (D), Stanford, 1997, Ob
Lisa A. Levin, (D), California (San Diego), 1982, Ob
Peter F. Lonsdale, (D), California (San Diego), 1974, Gu
W. Kendall Melville, (D), Southampton (UK), 1974, Op
J. Bernard H. Minster, (D), Caltech, 1974, Ys
Mario J. Molina, (D), California (Berkeley), 1972, Oa
Bradley S. Moore, (D), Zurich, 1995, Cm
Joel R. Norris, (D), Washington, 1997, Oa
Richard D. Norris, (D), Harvard, 1990, Gu
Mark D. Ohman, (D), Washington, 1983, Ob
John A. Orcutt, (D), California (San Diego), 1976, YrsYd
Kimberly A. Prather, (D), California (Davis), 1990, OaCmHg
V. Ramanathan, (D), SUNY (Stony Brook), 1973, Oa
Dean H. Roemmich, (D), MIT/WHOI, 1980, Op
Gregory W. Rouse, (D), Sydney, 1991, Ob
Daniel L. Rudnick, (D), California (San Diego), 1987, Op
Lynn M. Russell, (D), Caltech, 1995, Oa
David T. Sandwell, (D), California (Los Angeles), 1981, Yr
John G. Sclater, (D), Cambridge, 1966, Yh
Uwe Send, (D), California (San Diego), 1988, Op
Jeffrey P. Severinghaus, (D), Columbia, 1995, Pe
Peter M. Shearer, (D), California (San Diego), 1986, Ysg
Len Srnka, (D), New Castle upon Tyne, 1974, Ym
Dariusz Stramski, (D), Gdansk (Poland), 1985, Op
George Sugihara, (D), Princeton, 1983, Ob
Lynne D. Talley, (D), MIT/WHOI, 1982, Op
Lisa Tauxe, (D), Columbia, 1983, Ym
Bradley T. Werner, (D), Caltech, 1987, Gm
William R. Young, (D), MIT/WHOI, 1981, Op
Associate Professor:
Eric E. Allen, (D), California (San Diego), 2002, Ob
Andreas Andersson, (D), Hawaii (Manoa), 2006, Oc
James Day, (D), Durham, England, Cg
Kerry Key, (D), Scripps, 2003, Yx
Todd Martz, (D), Montana, 2005, Gg
Jennifer E. Smith, (D), Hawaii, 2003, Obn
Assistant Professor:
Adrian Borsa, (D), Scripps, 2005, Hg
Chambers Hughes, (D), California (Berkeley), 2004, Oc
Anne Pommier, (D), Orleans, France, 2009, Cg
Dave Stegman, (D), California (Berkeley), Yd
Jennifer K. Vanos, (D), Guelph (Canada), 2012, Oa
Adjunct Professor:
Jay P. Barlow, (D), California (San Diego), 1982, Ob
Emeritus:
Gustaf Arrhenius, (D), Stockholm, 1952, ClGzCm
George E. Backus, (D), Chicago, 1956, YmsOp
Jeffrey L. Bada, (D), California (San Diego), 1968, Co
Wolfgang H. Berger, (D), California (San Diego), 1968, Gs
Steven C. Cande, (D), Columbia, 1976, Yr
David M. Checkley, (D), California (San Diego), 1978, Ob
Paul Crutzen, (D), Oc
Joseph R. Curray, (D), Scripps, 1959, Gu
Paul K. Dayton, (D), Washington, 1970, Ob
Horst Felbeck, (D), Muenster, 1979, Ob
Joris M. Gieskes, (D), Manitoba, 1965, Oc
Robert T. Guza, (D), California (San Diego), 1974, On
James W. Hawkins, (D), Washington, 1963, OuGip
Anthony DJ Haymet, (D), Chicago, 1981, OcgOn
Robert R. Hessler, (D), Chicago, 1960, Ob
Glenn R. Ierley, (D), MIT, 1982, Op
Jeremy B. C. Jackson, (D), Yale, 1971, PoePi
Charles F. Kennel, (D), Princeton, 1964, Oa
Gerald L. Kooyman, (D), Arizona, 1966, Ob
J. Douglas Macdougall, (D), California (San Diego), 1972
John A. McGowan, (D), California (San Diego), 1960, Ob
Walter H. Munk, (D), Scripps, 1947, Op
William A. Newman, (D), California (Berkeley), 1962, Ob
Robert L. Parker, (D), Cambridge, 1966, Ym
Robert Pinkel, (D), California (San Diego), 1974, Op
Richard L. Salmon, (D), California (San Diego), 1976, Op
Richard C. J. Somerville, (D), New York Univ, 1966, Oa
Victor D. Vacquier, (D), California (Berkeley), 1968, Ob
Ray F. Weiss, (D), California (San Diego), 1970, Cm
Clinton D. Winant, (D), S California, 1972, On

California

## University of California, Santa Barbara

**Dept of Earth Science** (B,M,D) (2014)
Room 1006
Webb Hall
Santa Barbara, CA 93106-9630
p. (805) 893-4688
gs-mso@geol.ucsb.edu
http://www.geol.ucsb.edu
Administrative Assistant: Hannah Smit
*Enrollment (2012): B: 81 (22) M: 12 (2) D: 36 (3)*

Chair:
Doug Burbank, (D), Dartmouth, 1982, Gmt
Professor:
Stanley M. Awramik, (D), Harvard, 1973, Po
Cathy J. Busby, (D), Princeton, 1983, Gs
Jordan F. Clark, (D), Columbia, 1995, ClHg
Bradley R. Hacker, (D), California (Los Angeles), 1998, Gp
Edward A. Keller, (D), Purdue, 1973, Gm
David W. Lea, (D), MIT, 1990, PeCm
Frank J. Spera, (D), California (Berkeley), 1977, Gi
Toshiro Tanimoto, (D), California (Berkeley), 1982, Ys
Bruce H. Tiffney, (D), Harvard, 1977, Pb
David Valentine, (D), Cm
Andre R. Wyss, (D), Columbia, 1989, Pv
Associate Professor:
Phillip B. Gans, (D), Stanford, 1987, Gt
Chen Ji, (D), Caltech, 2002, Ys
Lorraine Lisiecki, (D), Brown, 2005, Gn
Susannah Porter, (D), Harvard, 2002, Gn
Alexander Simms, (D), Rice, 2006, GsuGm
Assistant Professor:
John Cottle, (D), Oxford, 2007, Gi
Syee Weldeab, (D), Tubingen, 2002, PeCm
Undergraduate Staff Advisor:
Alice MaCall, (B)
Emeritus:
Ralph J. Archuleta, (D), California (San Diego), 1976, Ys

**Earth Research Institute** (2015)
Mail Code 1100
Santa Barbara, CA 93106-1100
p. (805) 893-8231
davey@eri.ucsb.edu
http://www.crustal.ucsb.edu
Administrative Assistant: Kathy J. Scheidemen

Director:
Douglas W. Burbank, (D), Dartmouth, 1982, Gm
Associate Director:
Bradley R. Hacker, (D), California (Los Angeles), 1988, Gx
Professor:
Ralph J. Archuleta, (D), California (San Diego), 1976, Ys
Cathy Busby, (D), Princeton, 1983, Gs
Edward A. Keller, (D), Purdue, 1973, Gt
Bruce P. Luyendyk, (D), California (San Diego), 1969, Yr
Frank J. Spera, (D), California (Berkeley), 1977, Cp
Toshiro Tanimoto, (D), California (Berkeley), 1982, Ys
Associate Professor:
Jordan Clark, (D), Columbia, 1995, Cg
Phillip B. Gans, (D), Stanford, 1987, Gt
Assistant Research Scientist:
Christopher C. Sorlien, (D), California (Santa Barbara), 1994, Yr
Assistant Professor:
Bodo Bookhagen, (D), Potsdam, 2005, Gm
Research Seismologist:
Jamison H. Steidl, (D), California (Santa Barbara), 1995, YsNeYg

## University of California, Santa Cruz

**Center for the Dynamics & Evolution of the Land-Sea Interface** (2014)
1156 High Street
Room A234, Earth & Marine Sciences Building
Santa Cruz, CA 95064
p. (831) 459-4089
acr@es.ucsc.edu

Professor:
Robert S. Anderson, (D), Washington, 1986, Gm
Kenneth W. Bruland, (D), California (San Diego), 1974, Oc
Margaret L. Delaney, (D), MIT/WHOI, 1983, Oc
Russell Flegal, (D), Oregon State, Cg
Laurel R. Fox, (D), California (Santa Barbara), On
Dianne Gifford-Gonzalez, (D), California (Berkeley), On
Mark S. Mangel, (D), British Columbia, 1978, On
Donald C. Potts, (D), California (Santa Barbara), Ob
Mary W. Silver, (D), California (San Diego), 1971, Ob
Jonathan P. Zehr, (D), California (Davis), 1985, Ob
Associate Professor:
Mark Carr, (D), California (Santa Barbara), Ob
Brent Haddad, (D), California (Berkeley), 1996, On
Karen D. Holl, (D), Virginia Tech, 1994, Ou
Christina Ravelo, (D), Columbia, 1991, Cs
Donald R. Smith, (D), California (Santa Cruz), On
Assistant Professor:
Don Croll, (D), California (Santa Cruz), Ob
Raphael M. Kudela, (D), S California, 1995, Or
Margaret A. McManus, (D), Old Dominion, 1996, On

**Center for the Origin, Dynamics, & Evolution of Planets** (2010)
1156 High Street
Room A234, Earth & Marine Sciences Building
Santa Cruz, CA 95064
p. (831) 459-4089
dkorycan@ucsc.edu

Professor:
Robert S. Anderson, (D), Washington, 1986, Gm
Peter Bodenheimer, (D), California (Berkeley), 1965, Xc
Frank Bridges, (D), California (San Diego), 1968, Yx
Douglas Lin, (D), Cambridge, 1976, Xc
Steven Vogt, (D), Texas, 1978, Xc
Other:
Don Korycansky, (D), California (Santa Cruz), Xc

**Earth & Planetary Sciences Dept** (B,M,D) ● (2016)
1156 High Street
Earth & Marine Sciences Bldg, Rm A232
Santa Cruz, CA 95064
p. (831) 459-4089
qwilliams@pmc.ucsc.edu
http://www.eps.ucsc.edu
f: https://www.facebook.com/UcscEPS
t: @EpsUcsc
*Enrollment (2016): B: 264 (63) M: 6 (4) D: 47 (7)*

Distinguished Professor, Department Chair:
Quentin Williams, (D), California (Berkeley), 1988, Gy
Distinguished Professor:
Gary B. Griggs, (D), Oregon State, 1968, On
Thorne Lay, (D), Caltech, 1983, Ys
Professor:
Emily Brodsky, (D), Caltech, 2001, YsgHw
Patrick Y. Chuang, (D), Caltech, 1999, Oa
Andrew T. Fisher, (D), Miami, 1989, Hw
Elise Knittle, (D), California (Berkeley), 1988, Gy
Paul L. Koch, (D), Michigan, 1989, Pv
Susan Y. Schwartz, (D), Michigan, 1988, Ys
Eli A. Silver, (D), California (San Diego), 1969, Yr
Othmar T. Tobisch, (D), Imperial Coll (UK), 1963, Gct
Slawek Tulaczyk, (D), Caltech, 1998, GlmNg
James C. Zachos, (D), Rhode Island, 1988, Cm
Associate Professor:
Matthew E. Clapham, (D), S California, 2006, Pe
Noah J. Finnegan, (D), Washington, 2007, Gm
Ian Garrick-Bethell, (D), MIT, 2009, Xy
Jeremy Hourigan, (D), Stanford, 2002, Cc
Assistant Professor:
Terrence Blackburn, (D), MIT, 2012, Cc
Xi Zhang, (D), CalTech, 2013, OaXy
Senior Lecturer:
Hilde Schwartz, (D), California (Santa Cruz), 1983, Pv
Emeritus:
Erik Asphaug, (D), Arizona, 1993, Xg
Kenneth L. Cameron, (D), Virginia Tech, 1971, Gx
Robert S. Coe, (D), California (Berkeley), 1966, Ym
Robert E. Garrison, (D), Princeton, 1964, Gs
James B. Gill, (D), Australian National, 1972, Gi

Gary A. Glatzmaier, (D), Colorado, 1980, YmXy
Leo F. Laporte, (D), Columbia, 1960, Pg
Karen C. McNally, (D), California (Berkeley), 1976, Ys
J. Casey Moore, (D), Princeton, 1971, Gc
Lisa C. Sloan, (D), Penn State, 1990, Pe
Gerald E. Weber, (D), California (Santa Cruz), 1980, NgGmo
Graduate Program Advisor:
Jennifer M. Fish, (B), California (Santa Cruz), On
Front Office Manager:
Amy Kornberg, On

**Inst of Geophysics & Planetary Physics** (2010)
1156 High Street
Earth & Marine Sciences Building, Room A234
Santa Cruz, CA 95064
p. (831) 459-4089
shalevst@gmail.com
http://igpp.ucsc.edu
Director:
Ana Christina Ravelo, (D), OcGe

## University of San Diego

**Dept of Marine Science & Environmental Studies** (B,M) (2014)
Alcala Park
San Diego, CA 92110
p. (619) 260-4795
boum@sandiego.edu
http://www.sandiego.edu/mars_envi/
Chair:
Michel A. Boudrias, (D), California (San Diego), 1992, Ob
Professor:
Hugh I. Ellis, (D), Florida, 1976, Ob
Associate Professor:
Ronald S. Kaufmann, (D), California (San Diego), 1992, Ob
Anne A. Sturz, (D), California (San Diego), 1991, Cm
Assistant Professor:
Richard Gonzalez, (D), Penn State, On
Sarah Gray, (D), California (Santa Cruz), On
Adjunct Professor:
Ann B. Bowles, (D), California (San Diego), On
Joseph R. Jehl, (D), Michigan, 1967, Ob
Donald B. Kent, (M), San Diego State, 1980, Ob
Brent S. Stewart, (D), California (Los Angeles), 1988, Ob
Pam Yochem, (D), California (Davis), 1992, On
Cooperating Faculty:
Gerald N. Estberg, (D), Cornell, 1966, Ow
Mary Sue Lowery, (D), California (San Diego), 1987, Ob
LeeAnn Otto, (D), British Columbia, 1981, On
Marie Simovich, (D), California (Riverside), 1985, On

## University of Southern California

**Dept of Earth Sciences** (B,M,D) ● (2016)
3651 Trousdale Parkway ,ZHS117
Los Angeles, CA 90089-0740
p. (213) 740-6106
waite@usc.edu
http://www.usc.edu/dept/earth
Administrative Assistant: Vardui Ter-Simonian
Administrative Assistant: Cynthia Waite
*Enrollment (2016): B: 30 (12) M: 0 (1) D: 61 (9)*
Chair:
William M. Berelson, (D), S California, 1986, Cm
Research Professor:
David A. Okaya, (D), Stanford, 1985, Ys
Professor:
Jan Amend, (D), California (Berkeley), 1995, ClmCo
Yehuda Ben-Zion, (D), S California, 1990, Ys
David J. Bottjer, (D), Indiana, 1978, Pe
Frank A. Corsetti, (D), California (Santa Barbara), 1998, Gs
James F. Dolan, (D), California (Santa Cruz), 1988, Gt
Douglas E. Hammond, (D), Columbia, 1975, Cm
Thomas H. Jordan, (D), Caltech, 1972, Ys
Steven P. Lund, (D), Minnesota, 1981, Ym
James Moffett, (D), Miami, 1986, Oc
Kenneth Nealson, (D), Chicago, 1969, Py
Scott R. Paterson, (D), California (Santa Cruz), 1986, Gc
John P. Platt, (D), California (Santa Barbara), 1973, GctGp
Charles G. Sammis, (D), Caltech, 1971, Yg
Sergio Sanudo, (D), California (Santa Cruz), 1993, Py
Lowell D. Stott, (D), Rhode Island, 1989, Pm
Ta-liang Teng, (D), Caltech, 1966, YsGtYe
Associate Professor:
Julien Emile-Geay, (D), Columbia, 2006, Oc
Sarah Feakins, (D), Columbia, 2006, CoGuCs
A. Joshua (Josh) West, (D), Cambridge (UK), 2007, ClGmHs
Research Associate Professor:
Yong-Gang Li, (D), S California, 1988, Ys
Ellen S. Platzman, (D), ETH, 1990
Emeritus:
Gregory A. Davis, (D), California (Berkeley), 1961, Gc
Alfred G. Fischer, (D), Columbia, 1950, Gs
Thomas L. Henyey, (D), Caltech, 1968, Yg
Teh-Lung Ku, (D), Columbia, 1966, Cg
Bernard W. Pipkin, (D), Arizona, 1964, Ng

## University of the Pacific

**Dept of Geological & Environmental Sciences** (B) (2017)
3601 Pacific Avenue
Stockton, CA 95211
p. (209) 946-2482
lrademacher@pacific.edu
http://pacific.edu/EES
f: https://www.facebook.com/pacificees/
t: @PacificEES
*Enrollment (2016): B: 48 (5)*
Chair:
Laura K. Rademacher, (D), California (Santa Barbara), 2002, ClHyOs
Professor:
Eugene F. Pearson, (D), Wyoming, 1972, GsPgOg
Associate Professor:
Kurtis C. Burmeister, (D), Illinois, 2005, GctEg
Lydia K. Fox, (D), California (Santa Barbara), 1989, GizOe
Visiting Professor:
Karrigan Bork, (D), California, Davis, 2011, Hg
Emeritus:
Roger T. Barnett, (D), California (Berkeley), 1973, Oy
J. Curtis Kramer, (D), California (Davis), 1976, Gg

## Ventura College

**Dept of Geology** (2017)
4667 Telegraph Road
Ventura, CA 93003
spalladino@vcccd.edu
Chair:
Luke D. Hall, (M), W Kentucky, 1975, Oy
Assistant Professor:
Steve D. Palladino, (M), California (Santa Barbara), 1994, Oi
Instructor:
William Budke, (M), California Polytech, 2001, Os

## Victor Valley College

**Victor Valley College** (2014)
18422 Bear Valley Road
Victorville, CA 92395
p. (415) 506-0234
carol.delong@vvc.edu
http://www.vvc.edu/

## West Valley College

**Dept of Geology** (A) (2014)
14000 Fruitvale Avenue
Saratoga, CA 95070-5698
p. (408) 741-2437
robert_lopez@westvalley.edu
Instructor:
Harry Shade, (M), Miami (Ohio), 1959, Gg
Emeritus:
Theodore C. Herman, (M), Michigan, 1961, Oe

## Whittier College

**Environmental Science Program** (B) (2011)

Whittier, CA 90608
p. (562) 907-4220
cswift@whittier.edu
http://www.whittier.edu/Academics/EnvironmentalSciences/
Chair:
Cheryl Swift
Visiting Professor:
Andrew H. Wulff, (D), Massachusetts, 1998, Gi
Emeritus:
William B. Wadsworth, (D), Northwestern, 1966, Gi

# Colorado

## Adams State University

**Biology and Earth Sciences** (B) O (2016)
208 Edgemont Blvd
Suite 3060
Alamosa, CO 81101
p. (719) 587-7256
babrink@adams.edu
http://www.adams.edu/academics/earthscience
f: https://www.facebook.com/groups/231799753584196/
*Enrollment (2014): B: 34 (7)*
Chair:
Benita A. Brink, (D), Marquette, 1989
Professor:
Robert G. Benson, (D), Colorado Mines, 1997, Gg
Associate Professor:
Jared M. Beeton, (D), Kansas, 2007, OySd

## Aims Community College

**Dept of Sciences** (A) (2013)
5401 West 20th Street
Greeley, CO 80634
p. (970) 339-6637
jim.stone@aims.edu
http://www.aims.edu/academics/sciences/index.php
Instructor:
Jim Stone, Og

## Arapahoe Community College

**Geology Program** (A) (2015)
5900 South Santa Fe Drive
Littleton, CO 80120
p. (303) 797-5831
henry.weigel@arapahoe.edu
http://www.arapahoe.edu/departments-and-programs/a-z-offerings/geology
Program Chair:
Henry Weigel

## Colorado College

**Geology Dept** (B) (2016)
14 E Cache La Poudre
Colorado Springs, CO 80903
p. (719) 389-6621
geology@coloradocollege.edu
http://www.coloradocollege.edu/academics/dept/geology
f: https://www.facebook.com/geodept.coloradocollege
*Enrollment (2016): B: 25 (19)*
Professor:
Eric M. Leonard, (D), Colorado, 1981, Gl
Paul M. Myrow, (D), Memorial, 1987, Gs
Jeffrey B. Noblett, (D), Stanford, 1980, Gxv
Christine S. Siddoway, (D), California (Santa Barbara), 1993, Gc
Department Chair:
Henry C. Fricke, (D), Michigan, 1997, Cs
Associate Professor:
Megan L. Anderson, (D), Arizona, 2005, YgGt
Technical Director:
Stephen G. Weaver, (D), Colorado Mines, 1988, Gx

## Colorado Mesa University

**Dept of Physical & Environmental Sciences** (A,B) O (2017)
1100 North Avenue
Grand Junction, CO 81501-3122
p. (970) 248-1993
rwalker@coloradomesa.edu
http://www.coloradomesa.edu/geosciences/
*Enrollment (2015): A: 21 (3) B: 83 (13)*
Professor:
Andres Aslan, (D), Colorado, 1994, Gm
Rex D. Cole, (D), Utah, 1975, GsrGo
Verner C. Johnson, (D), Tennessee, 1975, Yg
Richard F. Livaccari, (D), New Mexico, 1994, Gc
Gigi Richard, (D), Colorado State, 2001, Hsg
Instructor:
Cassandra Fenton, (D), Utah, 2002, CcGmCl
Lecturer:
Lawrence S. Jones, (D), Wyoming, 1996, GsmGr
Adjunct Professor:
William C. Hood, (D), Montana, 1964, Gg
Julia McHugh, (D), Iowa, 2012, Pgv
Dave Wolny, (B), Mesa State, 1992, Ys
Emeritus:
James B. Johnson, (D), Colorado, 1979, Gl
Jack E. Roadifer, (D), Arizona, 1966, Gx

## Colorado Mountain College

**Geology Div** (A) (2004)
3000 County Road 114
Glenwood Springs, CO 81601
p. (303) 945-7481 x263
Professor:
Garrett E. Zabel, (M), Houston, 1977, Gg

## Colorado School of Mines

**Dept of Chemistry** (B,M,D) (2016)
1012 14th Street, CO 204
Golden, CO 80401
p. (303) 273-3610
chemistry@mines.edu
http://chemistry.mines.edu/
f: https://www.facebook.com/csmchemistry/
t: @MinesChemistry
*Enrollment (2016): B: 78 (27) M: 14 (5) D: 52 (10)*
Head:
David T. Wu, (D), California (Berkeley), 1991, On
Professor:
Mark E. Eberhart, (D), MIT, 1983, Om
Mark Jensen, (D), Florida State, 1994
Daniel M. Knauss, (D), Virginia Tech, 1994
James Ranville, (D), Colorado Mines, Ca
Ryan M. Richards, (D), Michigan State, 2000
Bettina Voelker, (D), Swiss Fed Inst Tech, 1994, Ca
Kim R. Williams, (D), Michigan State, 1986, Ca
Associate Professor:
Stephen G. Boyes, (D), New South Wales (Australia), 2000
Renee L. Falconer, (D), South Carolina, 1994
Matthew C. Posewitz, (D), Dartmouth, 1995
Mark R. Seger, (D), Colorado State
Alan Sellinger, (D), Michigan, 1997
Angela C. Sower, (D), New Mexico
Assistant Professor:
Jenifer C. Braley, (D), Washington State, 2010
Allison Caster, (D), California (Berkeley), 2010
Svitlana Pylypenko, (D), New Mexico, Omn
Brian G. Trewyn, (D), Iowa State, 2006
Shubham Vyas, (D), Ohio State
Yongan Yang, (D), Chinese Acad of Sci, 1999
Emeritus:
Scott W. Cowley, (D), S Illinois, 1975, Cog
Dean W. Dickerhoof, (D), Illinois, 1961
Donald L. Macalady, (D), Wisconsin, 1969, Cl
Patrick MacCarthy, (D), Cincinnati, 1975, Co
Craig Simmons, (D), SUNY (Stony Brook), 1976, Cp
Kent J. Voorhees, (D), Utah State, 1970, Co
Thomas R. Wildeman, (D), Wisconsin, 1967, Ct

**Dept of Geology & Geological Engineering** (B,M,D) ● (2016)
1516 Illinois Street

Golden, CO 80401-1887
p. (303) 273-3800
cmedford@mines.edu
http://geology.mines.edu/
*Enrollment (2015): B: 131 (36) M: 121 (44) D: 54 (10)*
Professor:
Merritt S. Enders, (D)
Professor:
Wendy J. Harrison, (D), Manchester, 1979, Cl
Reed M. Maxwell, (D), California (Berkeley), 1998, Hs
Paul M. Santi, (D), Colorado Mines, 1995, Ng
Stephen A. Sonnenberg, (D), Colorado Mines, 1981, Go
Richard F. Wendlandt, (D), Penn State, 1978, GizCp
Lesli Wood, (D), Colorado State, 1992, GsoGm
Associate Professor:
David A. Benson, (D), Nevada (Reno), 1998, Hq
Thomas Monecke, (D), Germany, 2003, Em
Piret Plink-Bjorklund, (D), Goteborg, 1998, Gs
Kamini Singha, (D), Stanford, 2005, Hw
Bruce D. Trudgill, (D), Imperial Coll (UK), 1989, Gc
Wendy W. Zhou, (D), Missouri Univ of Sci & Tech, 2001, NgOir
Assistant Professor:
Alex Gysi, (D), Iceland, 2011, Cg
Yvette D. Kuiper, (D), New Brunswick, 2003, Gc
Alexis Navarre-Sitchler, (D), Penn State, 2008, Ca
Gabriel Walton, (D), Queens, 2014, NgrYg
Lecturer:
Christian V. Shorey, (D), Iowa, 2002, Gg
Emeritus:
L. Graham Closs, (D), Queen's, 1973, Ce
John B. Curtis, (D), Ohio State, 1989, GoCo
Jerry D. Higgins, (D), Missouri (Rolla), 1980, Ng
Keenan Lee, (D), Stanford, 1969, Or
Eileen P. Poeter, (D), Washington State, 1980, Ng
Samuel B. Romberger, (D), Penn State, 1968, Em
A. Keith Turner, (D), Purdue, 1969, Ng
John E. Warme, (D), California (Los Angeles), 1966, Pi
Robert J. Weimer, (D), Stanford, 1953, Gr

**Dept of Geophysics** (B,M,D) (2010)
1500 Illinois Street
Golden, CO 80401
p. (303) 273-3450
rsnieder@mines.edu
http://geophysics.mines.edu/
*Enrollment (2009): B: 69 (0) M: 26 (0) D: 25 (0)*
Professor:
Michael L. Batzle, (D), MIT, 1978, Yg
Dave Hale, (D), Stanford, 1997, Ye
Yaoguo Li, (D), British Columbia, 1992, Yev
Gary R. Olhoeft, (D), Toronto, 1975, Yg
Roel Snieder, (D), Utrecht (Neth), 1987, Ye
Ilya D. Tsvankin, (D), Moscow State, 1982, Ye
Senior Scientist:
Warren Hamilton, (D), California (Los Angeles), 1951, Ys
Misac Nabighian, (D), Columbia, 1967, Ye
Associate Research Professor:
Robert D. Benson, (M), Colorado Mines, 1984, Ye
Associate Professor:
Brandon Dugan, (D), Penn State, 2003, GuYrHw
Assistant Professor:
Paul Sava, (D), Stanford, 2005, Ye
Emeritus:
Thomas L. Davis, (D), Colorado Mines, 1974, Ye
Kenneth L. Larner, (D), MIT, 1970, Ye
Research Professor:
Norman Bleistein, (D), Courant Inst, 1965, Yg

**Dept of Mining Engineering** (B,M,D) ● (2016)
1600 Illinois
Golden, CO 80401
p. (303) 273-3700
csmmining@mines.edu
http://mining.mines.edu/
*Enrollment (2015): B: 100 (41) M: 18 (18) D: 14 (6)*
Department Head:
Priscilla P. Nelson, (D), Cornell, 1983, NrgNm
Professor:
Kadri Dagdelen, (D), Colorado Mines, 1985, Nm
M. Ugur Ozbay, (D), Witwatersrand, 1988, Nr
Associate Professor:
Mark Kuchta, (D), Lulea Univ of Tech, 1990, Nm
Hugh Miller, (D), Colorado Mines, 1996, Nmg
Masami Nakagawa, (D), Cornell, 1988, Oe
Assistant Professor:
Elizabeth Holley, (D), Colorado Mines, 2012, Eg
Rennie Kaunda, (D), W Michigan, 2007, Nm
Eunhye Kim, (D), Penn State, 2010, Nm
Research Professor:
Karl Zipf, (D), Penn State, 1988, Nm
Research Assistant Professor:
Vilem Petr, (D), Colorado Mines, 2000, Nm
Manager of Earth Mechanics Institute:
Brian Asbury, Nr
Research Associate:
Jürgen F. Brune, (D), Technische Univ Clausthal, 1994, Nm

## Colorado State University

**Dept of Atmospheric Science** (M,D) (2016)
1371 Campus Delivery
Fort Collins, CO 80523-1371
Fort Collins, CO 80523-1371
p. (970) 491-8360
info@atmos.colostate.edu
http://www.atmos.colostate.edu
*Enrollment (2013): M: 39 (19) D: 49 (6)*
Director:
Kristen L. Rasmussen, (D), Washington, 2014, Oa
Head:
Jeffrey L. Collett, Jr., (D), Caltech, 1989, Oa
Professor:
A. Scott Denning, (D), Colorado State, 1994, Oag
Sonia M. Kreidenweis, (D), Caltech, 1989, Oa
Eric Maloney, (D), Washington, 2000, Oa
David A. Randall, (D), California (Los Angeles), 1976, Oa
A.R. Ravishankara, (D), Florida, 1975, Oag
Steven A. Rutledge, (D), Washington, 1983, Oa
Associate Professor:
Michael M. Bell, (D), Naval Postgraduate School, 2010, Oa
Thomas Birner, (D), LMU Munich (Germany), 2003, Oa
Christian D. Kummerow, (D), Minnesota, 1987, Yr
Jeff Pierce, (D), Carnegie Mellon Univ, 2008, Oa
Russ Schumacher, (D), Colorado State Univ, 2008, Oa
Assistant Professor:
Elizabeth A. Barnes, (D), Washington, 2012, Oa
Emily Fischer, (D), Washington, 2010, Oa
David W. J. Thompson, (D), Washington, 2000, Yr
Sue van den Heever, (D), Colorado State, 2001, Oa

**Dept of Geosciences** (B,M,D) O (2015)
1482 Campus Delivery
Fort Collins, CO 80523-1482
p. (970) 491-5661
WCNR_GEO_Info@mail.colostate.edu
http://warnercnr.colostate.edu/departments/geosciences
Administrative Assistant: Sharon Gale
*Enrollment (2015): B: 164 (36) M: 43 (11) D: 32 (1)*
Head:
Rhicard C. Aster, (D), California (San Diego), 1991, YsgGv
Head:
Richard Aster, (D), Scripps, Ys
Professor:
Sven O. Egenhoff, (D), Technical Berlin, 2000, Gs
Judith L. Hannah, (D), California (Davis), 1980, GiCc
Dennis L. Harry, (D), Texas (Dallas), 1989, Yg
Ellen E. Wohl, (D), Arizona, 1988, Gm
Associate Professor:
Jerry F. Magloughlin, (D), Minnesota, 1993, Gpz
Sara L. Rathburn, (D), Colorado State, 2001, Ggm
John R. Ridley, (D), Edinburgh, 1982, Eg
Michael J. Ronayne, (D), Stanford, 2008, Hwq
William E. Sanford, (D), Cornell, 1992, Hw
Derek L. Schutt, (D), Oregon, 2000, Ysg
Sally J. Sutton, (D), Cincinnati, 1987, GdCl

Assistant Professor:
John Singleton, (D), Texas, 2011, Gct
Lisa Stright, (D), Stanford, 2011, GoNpYe
Research Associate:
James R. Chappell, (B), Colorado State, 2000, Ge
Svetoslav Georgiev, (D), ETH (Switzerland), 2008, CcGo
Ronald J. Karpilo, Jr, (M), Denver, 2004, Oy
Richard Markey, (M), N Illinois, 1990, Cca
Stephanie O'Meara, (M), Colorado State, 1997, Gc
Trista L. Thornberry-Ehrlich, (M), Colorado State, 2001, GgcGx
Gang Yang, (D), Sci & Tech (China), 2005, Cc
Aaron Zimmerman, (M), Colorado State, 2006, Cc
Emeritus:
Eric A. Erslev, (D), Harvard, 1981, GctGg
Frank G. Ethridge, (D), Texas A&M, 1970, Gs
Academic Success Coordinator:
Jill Putman, (M), Georgia, 2009



## Community College of Aurora

**Dept of Science** (1999)
16000 East Centretech Parkway
Aurora, CO 80011
p. (303) 361-7398
jim.weedin@ccaurora.edu
http://www.ccaurora.edu

## Denver Museum of Nature & Science

**Dept of Earth Sciences** (2016)
2001 Colorado Boulevard
Denver, CO 80205-5798
p. (303) 370-6000
Taylor.Foreman@dmns.org
http://www.dmns.org
Director of Earth and Space Sciences Branch, Department Chair of Earth Sciences, Curator of Paleobotany:
Ian Miller, (D), Yale, 2007, PbGg
Tim and Kathryn Ryan Curator of Geology:
James W. Hagadorn, (D), S California, 1998, GgsPg
Curator of Vertebrate Paleontology:
Tyler Lyson, (D), Yale, 2012, Pv
Joseph Sertich, (D), Stony Brook, 2011, Pv
Collections Manager:
Logan D. Ivy, (D), Colorado, 1993, Pv
Chief Preparator:
Mike Getty, (M), Calgary, Pv

## Fort Lewis College

**Dept of Geosciences** (B) (2013)
1000 Rim Drive
Durango, CO 81301
p. (970) 247-7278
gonzales_d@fortlewis.edu
http://geo.fortlewis.edu
*Enrollment (2012): B: 120 (22)*
Chair:
David A. Gonzales, (D), Kansas, 1997, Gx
Professor:
James D. Collier, (D), Colorado Mines, 1982, Cg
Gary Gianniny, (D), Wisconsin, 1995, Gs
Kimberly Hannula, (D), Stanford, 1993, GctGp
Ray Kenny, (D), Arizona State, 1991, Gm
Associate Professor:
Scott White, (D), Utah, 2001, Oy
Instructor:
Lauren Heerschap, (M), Colorado, GgEgGt
Adjunct Professor:
Charles Burnham, (D), MIT, 1961, Gz
Mary L. Gillam, (D), Colorado, 1998, Gm
Laboratory Director:
Andrea J. Kirkpatrick

## Front Range Community College, Larimer

**Natural Sciences** (A) (2016)
4616 S. Shields Street
Fort Collins, CO 80526
p. (970) 204-8607
stephanie.irwin@frontrange.edu
http://www.frontrange.edu/Academics/Academic-Departments/Larimer-Campus/Natural-Applied-Environment-Science/
Instructor:
Andy Caldwell, (M), N Colorado, 1999, Gg
Mike Smith, (M), N Colorado, 1998, GgOe

## Front Range Community College, Westminster

**Science and Technology** (A) (2017)
3645 West 112th Avenue
Westminster, CO 80031
p. (303) 404-5279
Fran.Goszdak@frontrange.edu
http://www.frontrange.edu/programs-and-courses/academic-departments/westminster-campus-departments/westminster-science

## Metropolitan State College of Denver

**Earth & Atmospheric Sciences Dept** (B) (2014)
P.O. Box 173362, Campus Box 22
Denver, CO 80217-3362
p. (303) 556-3143
cronoblj@mscd.edu
Administrative Assistant: Diane Hollenbeck
Chair:
James M. Cronoble, (D), Colorado Mines, 1977, Gg
Ken Engelbrecht
Professor:
John R. Kilcoyne, (D), Washington, 1973, Oy
Anthony A. Rockwood, (M), Colorado State, 1976, Ow
Roberta A. Smilnak, (D), Clark, 1973, Og
Associate Professor:
Thomas J. Corona, (M), Colorado State, 1978, Ow
Assistant Professor:
Robert E. Leitz, (M), California (Berkeley), 1974, Gz
Rafael Moreno, (D), Colorado State, 1992, Og
Emeritus:
James MacLachlan, (D), Princeton, Gg
H. Dixon Smith, (D), Minnesota, 1960, Og

## Northeastern Junior College

**Math, Science, and Health** (1999)
100 College Avenue
Sterling, CO 80751
p. (970) 521-6753
david.coles@njc.edu
http://njc.edu

## Red Rocks Community College

**Geology Program, Science Dept** (A) ● (2016)
13300 West Sixth Avenue
Campus Box 20
Lakewood, CO 80228
p. (303) 914-6290
eleanor.camann@rrcc.edu
http://rrcc.edu/geology/
Associate Professor:
Eleanor J. Camann, (D), North Carolina, 2005, OneGs

## United States Air Force Academy

**Dept of Economics & Geosciences** (B) (2014)
HQ USAFA/DFEG
2354 Fairchild Drive, Suite 6K110
USAF Academy, CO 80840-5701
p. (719) 333-3080
matthew.tracy@usafa.edu
http://www.usafa.edu/df/dfeg/?catname=dfeg
*Enrollment (2013): B: 90 (27)*
Professor:
Terry W. Haverluk, (D), Minnesota, 1993, Og
Associate Professor:
Steven J. Gordon, (D), Arizona State, 1999, Gm
Thomas Koehler, (D), Wisconsin, Oa

Lt Col :
Matthew Tracy, (D), Arizona State, 2008, On
Assistant Professor:
Glen Gibson, (D), Virginia Tech, 2012, Ori
Brett Machovina, (D), Denver, 2010, Oir
Evan Palmer, (D), Arizona State, 2014, Oig
Sarah Robinson, (D), Arizona State, GgOi
Instructor:
Scott Dubsky, (M), North Dakota, 2006, OyGm
Related Staff:
Danny Portillo, (B), Oi

## University of Colorado

**Dept of Geography** (B,M,D) (2017)
Campus Box 260
Boulder, CO 80309-0260
p. (303) 492-8312
emily.yeh@colorado.edu
*Enrollment (2016): B: 231 (0) M: 35 (0) D: 28 (0)*
Professor:
Waleed Abdalati, (D), Colorado, 1996, Or
Suzanne P. Anderson, (D), California (Berkeley), 1995, GmCl
Peter D. Blanken, (D), British Columbia (Canada), 1997, HsOaw
John Pitlick, (D), Colorado State, 1988, Gm
Mark Serreze, (D), Colorado, 1989, Oa
Thomas T. Veblen, (D), California (Berkeley), 1975, Og
Mark W. Williams, (D), California (Santa Barbara), 1990, Og
Associate Professor:
Holly R. Barnard, (D), Oregon State, 2009, Hg
Noah Molotch, (D), 2004, Univ, Hg
Assistant Professor:
Jennifer K. Balch, (D), Yale, 2008, Og

**Dept of Geological Sciences** (B,M,D) (2015)
Campus Box 399
Boulder, CO 80309-0399
p. (303) 492-8141
shemin.ge@Colorado.edu
http://www.colorado.edu/geolsci/
Administrative Assistant: Carmen Juszczyk
*Enrollment (2014): B: 248 (29) M: 13 (7) D: 45 (10)*
Associate Dean:
Mary J. Kraus, (D), Colorado, 1983, GsSa
Professor:
Robert Anderson, (D), Washington, 1986, Gm
David A. Budd, (D), Texas, 1984, Gd
G. Lang Farmer, (D), California (Los Angeles), 1983, Cg
Shemin Ge, (D), Johns Hopkins, 1990, HwEoYg
Bruce M. Jakosky, (D), Caltech, 1982, Xg
Craig H. Jones, (D), MIT, 1987, YsGtYg
Gifford H. Miller, (D), Colorado, 1975, Cc
Stephen J. Mojzsis, (D), California (San Diego), 1997, XcCcGp
Peter Molnar, (D), Columbia, 1970, Gt
Karl J. Mueller, (D), Wyoming, 1992, GtcGm
Anne F. Sheehan, (D), MIT, 1991, Yse
Joseph R. Smyth, (D), Chicago, 1970, Gz
Charles R. Stern, (D), Chicago, 1973, Gi
Jai P. Syvitski, (D), British Columbia, 1978, GsuGq
Eric E. Tilton, (D), California (Santa Cruz), 1998, Hw
Gregory E. Tucker, (D), Penn State, 1996, Gmt
Paul Weimer, (D), Texas (Austin), 1989, Gor
James W. C. White, (D), Columbia, 1983, Cs
Associate Professor:
Karen Chin, (D), California (Santa Barbara), 1996, Po
Jaelyn J. Eberle, (D), Wyoming, 1996, Pv
Rebecca M. Flowers, (D), MIT, 2005, CgcGt
Brian M. Hynek, (D), Washington, 2003, Xg
Thomas M. Marchitto, (D), MIT, 1999, Oc
Dena M. Smith, (D), Arizona, 2000, Pg
Alexis Templeton, (D), Stanford, 2002, ClGe
Assistant Professor:
Kevin H. Mahan, (D), Massachusetts, 2005, Gpc
Julio C. Sepúlveda, (D), Bremen, 2008, Co
Instructor:
Lon Abbott, (D), California (Santa Cruz), 1993, GgtGm
Emeritus:
John T. Andrews, (D), Nottingham (UK), 1965, GluGs
William W. Atkinson, Jr., (D), Harvard, 1973, EmCg
Peter W. Birkeland, (D), Stanford, 1961, Sd
William C. Bradley, (D), Stanford, 1956, Gm
Don L. Eicher, (D), Yale, 1958, Pm
Alexander Goetz, (D), Caltech, 1967, Or
Edwin E. Larson, (D), Colorado, 1965, Ym
James L. Munoz, (D), Johns Hopkins, 1966, Cp
Peter Robinson, (D), Yale, 1960, Pv
Donald D. Runnells, (D), Harvard, 1964, Cl
Hartmut A. Spetzler, (D), Caltech, 1969, Ys
Theodore R. Walker, (D), Wisconsin, 1952, Gd

**Univ of Colorado Museum** (2014)
Campus Box 265
Boulder, CO 80309-265
p. (303) 492-6165
linda.cordell@colorado.edu
http://cumuseum.colorado.edu/
Associate Professor:
Jaelyn J. Eberle, (D), Wyoming, 1996, Pve
Assistant Professor:
Karen Chin, (D), California (Santa Barbara), 1996, Po
Dena M. Smith, (D), Arizona, 2000, Po
Adjunct Curator:
Kenneth Carpenter, Pv
Mary Dawson, Pv
Trihn Dzanh, Pv
Emmett Evanoff, Pi
Jeff Indeck, Pv
Jonathan Marcot, Pv
Greg McDonald, Pv
Karen Sears, Pv
Emeritus:
Judith A. Harris, (D), Cambridge, 1972, Pv
Peter Robinson, (D), Yale, 1960, Pv
Museum Associate:
Emily Bray, Pg
Frank Fisher, Pg
Pat Monaco, Pg
Steve Wallace, Pg
Collection Manager:
Amy P. Moe-Hoffman, (M), Colorado, 2002, Pg
Tonia Superchi-Culver, (M), SD Mines, 2001, Pg
Associate Curator, Micropaleontology:
Donald Eicher, (D), Yale, 1958, Pm
Associate Curator, Fossil Primates:
Herbert Covert, Pv

## University of Colorado, Denver

**Dept of Geography, & Environmental Science** (2015)
CB 172
P.O. Box 173364
Denver, CO 80217-3364
p. (303) 556-2276
sue.eddleman@ucdenver.edu
Associate Professor:
John W. Wyckoff, (D), Utah, 1980, Oy
Emeritus:
Wesley E. LeMasurier, (D), Stanford, 1965, Gv
Martin G. Lockley, (D), Birmingham (UK), 1977, Pi

## University of Denver

**Dept of Geography and the Environment** (B,M,D) (2015)
2050 E. Iliff Avenue
Boettcher Center West, Room 120
Denver, CO 80208
p. (303) 871-2654
agoetz@du.edu
http://www.du.edu/geography
f: https://www.facebook.com/DUGeography/
*Enrollment (2015): B: 175 (53) M: 69 (26) D: 12 (2)*
Director, Graduate Program:
Matthew Taylor, (D), Arizona State, 2003, Og
Professor:
Andrew R. Goetz, (D), Ohio State, 1987, Ogu
Paul C. Sutton, (D), California (Santa Barbara), 1999, Oi

Teaching Associate Professor:
Hillary Hamann, (D), Colorado (Boulder), 2002, HgOy
Director, Undergraduate Program:
Donald G. Sullivan, (D), California (Berkeley), 1988, Og
Director, Environmental Science:
Michael W. Kerwin, (D), Colorado, 2000, Gg
Associate Professor:
Eric Boschmann, (D), Ohio State, 2008
Michael Daniels, (D), Wisconsin, 2002, Gm
Michael J. Keables, (D), Wisconsin, 1986, Oa
Rebecca Powell, (D), California (Santa Barbara), 2006
Assistant Professor:
Jing Li, (D), George Mason, 2012
Adjunct Professor:
Joseph Berry, (D), Colorado State, 1976
Andrea Gelfuso, (D), Denver, 1992
Michelle Moran-Taylor, (D), Arizona State, 2003
Martha A. Narey, (D), Denver, 1999, Og
Sean Tierney, (D), Denver, 2009
Emeritus:
David Longbrake, (D), Iowa, 1972
Terrence J. Toy, (D), Denver, 1973, Gm

## University of Northern Colorado

**Earth and Atmospheric Sciences** (B,M) (2017)
Campus Box 100
Greeley, CO 80639
p. (970) 351-2647
william.hoyt@unco.edu
http://esci.unco.edu
Administrative Assistant: Vicki L. Ouellette
*Enrollment (2016): B: 110 (14) M: 16 (4)*

Chair:
William H. Hoyt, (D), Delaware, 1982, OgGsOe
Director, Mathematics & Science Teaching Inst.:
Steve W. Anderson, (D), Arizona, 1990, GvOeXg
Professor:
Lucinda Shellito, (D), California (Santa Cruz), 2004, OaPeOw
Byron Straw, (M), N Colorado, 2010, GlmYg
Associate Professor:
Graham Baird, (D), Minnesota, 2006, GcgGz
Joe T. Elkins, (D), Georgia, 2002, OeGgCl
Emmett Evanoff, (D), Colorado, 1990, GrPgGs
Assistant Professor:
Wendilyn Flynn, (D), Illinois, 2012, Owa
David G. Lerach, (D), Colorado State, 2012, Owa
Lecturer:
Sarah M. Hirner, (M), Colorado, 2013, Gg
Carolyn D. Lambert, (M), Arizona, 2003, Hw
Amy Nicholl, (M), N Colorado, 2004, GgOge
Adjunct Professor:
Todd A. Dallegge, (D), Alaska (Fairbanks), 2002, GosGr
Walter A. Lyons, (D), Chicago, 1970, Oaw
Emeritus:
Richard D. Dietz, (D), Colorado, 1965, Xg
Kenneth D. Hopkins, (D), Washington, 1976, Gml
William D. Nesse, (D), Colorado, 1977, Gxz
K. Lee Shropshire, (D), Colorado, 1974, Pg

## Western State Colorado University

**Dept of Geology** (B) O (2016)
Gunnison, CO 81231
p. (303) 943-2015
astork@western.edu
http://www.western.edu/geology
*Enrollment (2016): B: 95 (16)*

Professor:
Robert P. Fillmore, (D), Kansas, 1994, Gs
David W. Marchetti, (D), Utah, 2006, GmClc
Allen L. Stork, (D), California (Santa Cruz), 1984, Giv
Rady Chair:
Bradford R. Burton, (D), Wyoming, 1997, GocGt
Moncrief Chair:
Elizabeth S. Petrie, (D), Utah State, 2014, GcoYe
Lecturer:
Holly Brunkal, (D), Colorado Mines, 2015, NgGm

Adjunct Professor:
James Coogan, (D), Wyoming, 1992, Gc

# Connecticut

## Central Connecticut State University

**Dept of Physics & Earth Sciences** (B,M) (2014)
1615 Stanley Street
New Britain, CT 06050-4010
p. (860) 832-2930
antar@CCSU.EDU
http://www.physics.ccsu.edu/
Department Secretary: Sandra O'Day

Chair:
Ali A. Antar, (D), Connecticut, On
Professor:
Sandra Burns, (D), Connecticut, 1972, Oe
Steven B. Newman, (D), SUNY (Albany), Oa
Associate Professor:
Marsha Bednarski, (D), Connecticut, 1997, Oe
Mark Evans, (D), Pittsburgh, 1989, GciGz
Kristine Larsen, (D), Connecticut, 1988, Xy
Jennifer L. Piatex, (D), Pittsburgh, XgOr
Michael Wizevich, (D), Virginia Tech
Emeritus:
Charles W. Dimmick, (D), Tulane, 1969, Ge
Related Staff:
R. Craig Robinson, (B), Millersville, 1971, Xy

## Eastern Connecticut State University

**Environmental Earth Science Dept** (B) (2017)
83 Windham Street
Willimantic, CT 06226
p. (860) 465-4317
drzewieckip@easternct.edu
http://www1.easternct.edu/environmentalearthscience/
f: https://www.facebook.com/groups/181927638678890/
*Enrollment (2010): B: 102 (22)*

Chair:
Peter A. Drzewiecki, (D), Wisconsin, 1996, GrsGd
Professor:
Catherine A. Carlson, (D), Michigan State, 1994, Hwg
William D. Cunningham, (D), Texas (Austin), 1993, GtcGp
James A. Hyatt, (D), Queens, 1993, Gml
Associate Professor:
Meredith Metcalf, (D), Connecticut, 2013, OirHw
Paul Torcellini, (D), Purdue, 1993, On
Assistant Professor:
Stephen Nathan , (D), Massachusetts (Amherst), 2005, PmEo
Bryan Oakley, (D), Rhode Island, 2012, OnGml
Adjunct Professor:
Susan Bruening, (M), E Connecticut State, 1988, Og
Heath Carlson, (M), Eastern Connecticut State, 2013, Og
Lynn-Ann DeLima, (M), S Connecticut State, 2000, Og
Vishnu R. Khade, (D), Cincinnati, 1987, Og
Emile Levasseur, (M), Sacred Heart, 1992, Og
Bruce Morton, (M), Connecticut, 1983, Og
James Motyka, (M), E Connecticut State, 1987, Og
Julie Sandeen, (M), Connecticut, 1991
Wesley Winterbottom, (M), Connecticut, 1988, Og
Emeritus:
Sherman M. Clebnik, (D), Massachusetts, 1975, Gl
Fred Loxsom, (D), Dartmouth, 1969, On
Henry I. Snider, (D), New Mexico, 1966, Ge
Roy R. Wilson, (D), Oregon State, 1984, Oi

## Middlesex Community College

**Div of Science, Allied Health and Engineering** (A) (2013)
100 Training Hill Road
Middletown, CT 06457
p. (860) 343-5779
MBusa@mxcc.commnet.edu
http://www.mxcc.commnet.edu/Content/Environmental_Science_1.asp

Professor:
Mark Busa, (D), Connecticut, OgEgYg

Associate Scientist:
Christine Witkowski, (M), Connecticut, 2002, GeOg

## Naugatuck Valley Community College
**Science, Technology, Engineering, and Mathematics (STEM)** (1999)
750 Chase Parkway
Waterbury, CT 06708
p. (203) 596-8690
cdonaldson@nv.edu
http://www.nvcc.commnet.edu

## Norwalk Community College
**Sciences Deptartment** (2014)
188 Richards Avenue
Norwalk, CT 06854
p. (203) 857-7275
mbarber@ncc.commnet.edu
http://www.ncc.commnet.edu/dept/science/default.asp

## Quinebaug Valley Community College
**Dept of Environmental Science** (1999)
742 Upper Maple Street
Danielson, CT 06239
p. (860) 412-7230
mvesligaj@gvcc.commnet.edu
http://www.qvcc.commnet.edu

## Southern Connecticut State University
**Dept of Earth Sciences** (B) O (2016)
501 Crescent Street
New Haven, CT 06515
p. (203) 392-5835
flemingt1@southernct.edu
http://www.southernct.edu/earthscience/
*Enrollment (2014): B: 46 (7)*

Professor:
Cynthia R. Coron, (D), Toronto, 1982, Eg
Thomas H. Fleming, (D), Ohio State, 1995, GiCgc
Associate Professor:
James W. Fullmer, (D), MIT, 1979, Ow
Assistant Professor:
Michael J. Knell, (D), Montana State, 2012, PvGsOg
Instructor:
Christopher Balsley, (M), Wesleyan, 1972, Gg
Daniel Coburn, (M), C Connecticut State, 2003, Oe
Jennifer Cooper, (M), Missouri (Columbia), 2006, Ggi
Yolanda Lee-Gorishti, (M), Connecticut, 2006, GaOe
Julie Rumrill, (M), Vermont, 2009, GgeGl
Emeritus:
John W. Drobnyk, (D), Rutgers, 1962, Ghs
Robert Radulski, (D), Rhode Island, Og
William Tolley, (M), Syracuse, Gg

## University of Connecticut
**Center for Integrative Geosciences** (B,M,D) (2014)
354 Mansfield Road
U-1045
Storrs, CT 06269-1045
p. (860) 486-4432
geology@uconn.edu
http://www.geosciences.uconn.edu
*Enrollment (2012): B: 31 (0) M: 10 (0) D: 7 (0)*

Program Director:
Pieter Visscher, (D), Groningen, 1991, Co
Professor:
Vernon F. Cormier, (D), Columbia, 1976, YsGt
William F. Fitzgerald, (D), MIT/WHOI, 1970, Oc
Gary A. Robbins, (D), Texas A&M, 1983, Hw
Robert M. Thorson, (D), Washington, 1970, Gm
Associate Professor:
Andrew M. Bush, (D), Harvard, 2005, PoqPi
Timothy Byrne, (D), California (Santa Cruz), 1981, Gc
Jean M. Crespi, (D), Colorado, 1985, Gc
Lanbo Liu, (D), Stanford, 1993, Yg
Assistant Professor:
Christophe Dupraz, (D), Fribourg, 1999, Py
Michael Hren, (D), Stanford, 2007, CslCo
William Ouimet, (D), MIT, 2007, GmOy
Emeritus:
Larry Frankel, (D), Nebraska, 1956, Pm
Alfred J. Frueh, (D), MIT, 1949, Gz
Norman H. Gray, (D), McGill, 1971, Gx
Raymond Joesten, (D), Caltech, 1970, Gp
Homer C. Liese, (D), Utah, 1962, Ct
Anthony R. Philpotts, (D), Cambridge, 1963, Gi

**Dept of Marine Sciences** (B,M,D) (2016)
1080 Shennecossett Road
Groton, CT 06340
p. (860) 405-9152
marinesciences@uconn.edu
http://www.marinesciences.uconn.edu/
*Enrollment (2015): B: 78 (19) M: 12 (2) D: 28 (3)*

Head:
James B. Edson, (D), Penn State, 1989, OwaOp
Professor:
Peter Auster, (D), National (Ireland), 2000, Ob
Ann Bucklin, (D), California (Berkeley), 1980, Ob
Timothy Byrne, (D), California (Santa Cruz), 1981, Ou
Hans G. Dam, (D), SUNY (Stony Brook), 1989, Ob
Heidi Dierssen, (D), California, 2000, Op
Senjie Lin, (D), SUNY (Stony Brook), 1995, Ob
Robert Mason, (D), Connecticut, 1991, OcCtm
George B. McManus, (D), SUNY (Stony Brook), 1986, Ob
James O'Donnell, (D), Delaware, 1986, Op
Sandra Shumway, (D), Coll of North Wales, 1976, Ob
Pieter T. Visscher, (D), Groningen, 1991, Co
J. Evan Ward, (D), Delaware, 1989, Ob
Associate Professor:
David Lund, (D), MIT/WHOI, 2006, Oc
Annelie Skoog, (D), Göteborg (Sweden), Oc
Craig Tobias, (D), William & Mary, 1999, Oc
Penny Vlahos, (D), Massachusetts, 2000, OcGeu
Michael Whitney, (D), Delaware, 2003, Op
Huan Zhang, (D), Tokyo Fisheries, 1995, Ob
Assistant Professor:
Hannes Baumann, (D), Hamburg, 2006
Hannes Baumann, (D), Hamburg, 2006, Ob
Melanie Fewings, (D), MIT/WHOI, 2007, Op
Julie Granger, (D), British Columbia, 2006, Oc
Kelly Lombardo, (D), Stony Brook, 2011, Op
Jamie Vaudrey, (D), Connecticut, 2007, Ob
Research Associate:
Zofia Baumann, (D), SUNY (Stony Brook), 2011, ObcCm
Emeritus:
Walter F. Bohlen, (D), MIT/WHOI, 1969, Op
William F. Fitzgerald, (D), MIT/WHOI, 1970, Oc
Edward C. Monahan, (D), MIT, 1966, Oa

## University of New Haven
**Dept of Environmental Sciences** (B,M) O (2016)
300 Boston Post Rd
West Haven, CT 06516
p. (203) 932-7101
rldavis@newhaven.edu
http://www.newhaven.edu
*Enrollment (2016): B: 28 (5) M: 29 (8)*

Undergraduate Director:
R. Laurence Davis, (D), Rochester, 1980, GemHg
Graduate Director:
Roman N. Zajac, (D), Connecticut, 1985, Oni
Provost and Senior Vice President of Academic Affairs:
Daniel J. May, (D), California (Santa Barbara), 1986, GtxGe
Professor:
Carmela Cuomo, (D), Yale, CmmGu
Associate Professor:
John Kelly, (D), California (Davis), Obn
Assistant Professor:
Amy L. Carlile, (D), Washington, Obn
Lecturer:
Jean-Paul Simjouw, (D), Old Dominion, 2004, OcCma

Practitioner-in-Residence:
Paul Bartholemew, (D), British Columbia, GeOiGz

## Wesleyan University

**Dept of Earth & Environmental Sciences** (B,M) (2016)
265 Church Street
Room 455
Middletown, CT 06459-0139
p. (860) 685-2244
vharris@wesleyan.edu
http://www.wesleyan.edu/ees
Administrative Assistant: Virginia M.. Harris
*Enrollment (2015): B: 42 (24) M: 5 (4)*

Chair:
Martha S. Gilmore, (D), Brown, 1997, XgGmOr
Professor:
Barry Chernoff, (D), Michigan, 1983, Ge
Suzanne B. OConnell, (D), Columbia, 1986, GsuGe
Peter C. Patton, (D), Texas, 1976, Gm
Dana Royer, (D), Yale, 2002, PeoPb
Johan C. Varekamp, (D), Utrecht (Neth), 1979, CgGzv
Associate Professor:
Timothy C.W. Ku, (D), Michigan, 2001, Cl
Phillip G. Resor, (D), Stanford, 2003, Gct
Assistant Professor:
James P. Greenwood, (D), Brown, 1997, Xc
Emeritus:
Jelle Z. De Boer, (D), Utrecht (Neth), 1963, YmGta
James T. Gutmann, (D), Stanford, 1972, Gv
Facilities Manager:
Joel LaBella, (B), S Connecticut, 1987, On

## Yale University

**Dept of Geology & Geophysics** (B,D) (2017)
210 Whitney Avenue
P.O. Box 208109
New Haven, CT 06520-8109
p. (203) 432-3114
rebecca.pocock@yale.edu
http://earth.yale.edu
*Enrollment (2009): B: 13 (7) D: 55 (5)*

Curator:
Derek E. G Briggs, (D), Cambridge (UK), 1976, PioPy
Professor:
Jay J. Ague, (D), California (Berkeley), 1987, Gx
David Bercovici, (D), California (Los Angeles), 1989, Yg
Ruth E. Blake, (D), Michigan, 1997, Cg
Mark T. Brandon, (D), Washington, 1984, Gc
Derek E G Briggs, (D), Cambridge, 1976, Yg
David A D Evans, (D), Caltech, 1998, YmGt
Alexey V. Fedorov, (D), Scripps, 1997, Op
Jacques Gauthier, (D), California (Berkeley), 1984, Pv
Shun-ichiro Karato, (D), Tokyo, 1977, Gy
Jun Korenaga, (D), MIT, 2000, YgsCg
Jeffrey J. Park, (D), California (San Diego), 1985, YsOg
Danny M. Rye, (D), Minnesota, 1972, Cs
Brian J. Skinner, (D), Harvard, 1955, CgEmGz
Ronald B. Smith, (D), Johns Hopkins, 1975, Oa
Mary-Louise Timmermans, (D), Cambridge, 2000, Oc
John S. Wettlaufer, (D), Washington, 1991, Yg
Senior Research Scientist:
Edward W. Bolton, (D), California (Los Angeles), 1985, Gq
Ellen Thomas, (D), Utrecht (Neth), 1979, Pm
Associate Professor:
William Boos, (D), MIT, 2008
Kanani K.M. Lee, (D), California (Berkeley), 2003, Gy
Maureen D. Long, (D), MIT, 2006, Ys
Trude Storelvmo, (D), Oslo, 2006
Assistant Professor:
Bhart-Anjan Bhullar, (D), Harvard, 2014
Pincelli Hull, (D), California (San Diego), 2010
Noah Planavsky, (D), California (Riverside), 2012
Alan Rooney, (D), Durham Univ, 2011
Emeritus:
Robert B. Gordon, (D), Yale, 1955, Nr
George Veronis, (D), Brown, 1954, Op
Elisabeth S. Vrba, (D), Cape Town, 1974, Pv
Research Affiliate:
William C. Graustein, (D), Yale, 1981, Cl
Related Staff:
Laurent Bonneau, (M), Connecticut, 1997, Oa
James O. Eckert, (D), Texas A&M, 1988, Gp
Francis J. Robinson, (D), Hong Kong Sci & Tech, 1999, Gg
H. Catherine W. Skinner, (D), Adelaide, 1959, Gz

**Peabody Museum of Natural History** (2016)
PO Box 208118
170 Whitney Avenue
New Haven, CT 06520-8118
p. (203) 432-3752
peabody.director@yale.edu
http://www.peabody.yale.edu/

Emeritus Curator:
Leo W. Buss, (D), Johns Hopkins, 1979, Pi
Elisabeth S. Vrba, (D), Cape Town, 1974, Pvo
Curator:
Jay J. Ague, (D), California (Berkeley), 1987, Gzx
Michael J. Donoghue, (D), Harvard, 1982, Pbo
Jacques A. Gauthier, (D), California (Berkeley), 1984, Pv
Senior Collections Manager:
Susan H. Butts, (D), Idaho, 2003, PiGsPe
Christopher A. Norris, (D), Oxford, 1992, Pv
Collections Manager:
Shusheng Hu, (D), Florida, 2006, PblGs
Stefan Nicolescu, (D), Gothenburg, 1998, CtGpz

# Delaware

## University of Delaware

**Dept of Geological Sciences** (B,M,D) ● (2016)
103 Penny Hall
Newark, DE 19716
p. (302) 831-2569
smcgeary@udel.edu
http://www.ceoe.udel.edu/schools-departments/department-of-geological-sciences
f: https://www.facebook.com/UDCEOE
t: @udceoe
Administrative Assistant: Cheryl Doherty
*Enrollment (2016): B: 47 (12) M: 16 (3) D: 12 (4)*

Chair:
Neil C. Sturchio, (D), Washington Univ (St. Louis), 1983, ClcCa
Professor:
Ronald E. Martin, (D), California (Berkeley), 1981, PmePg
James E. Pizzuto, (D), Minnesota, 1982, Gm
Associate Professor:
A. Scott Andres, (M), Lehigh, 1984, HwOi
Clara Chan, (D), California (Berkeley), 2006, PyCmGz
John A. Madsen, (D), Rhode Island, 1987, YrOue
Susan McGeary, (D), Stanford, 1984, Oe
Holly A. Michael, (D), MIT, 2005, Hwq
Michael ONeal, (D), Washington, 2005, Gm
William S. Schenck, (M), Delaware, 1997, GgpGc
Arthur C. Trembanis, (D), William & Mary, 2004, On
Assistant Professor:
Adam F. Wallace, (D), Virginia Tech, 2008, ClGzOm
Jessica Warren, (D), MIT/WHOI, 2007, GxCpGt
Visiting Professor:
Ed Kohut, (D), Oregon State, GvxGg
Emeritus:
Billy P. Glass, (D), Columbia, 1968, XmGsz
Robert R. Jordan, (D), Bryn Mawr, 1964, Gr
John C. Kraft, (D), Minnesota, 1955, Gs
Peter B. Leavens, (D), Harvard, 1967, Gz
Allan M. Thompson, (D), Brown, 1968, Gx
John F. Wehmiller, (D), Columbia, 1971, Cl
Laboratory Director:
Bill Parnella, (B), Buffalo, 1984, Gg
Cooperating Faculty:
Katharina Billups, (D), California (Santa Cruz), 1998, Ou
Shreeram Inamdar, (D), Virginia Tech, 1996, Hg
Thomas E. McKenna, (D), Texas, 1997, Hy
Peter P. McLaughlin, Jr., (D), Louisiana State, 1989, Gr

Kelvin W. Ramsey, (D), Delaware, 1988, Gs
Christopher K. Sommerfield, (D), SUNY (Stony Brook), 1997, On
John H. Talley, (M), Franklin & Marshall, 1974, Eg
William J. Ullman, (D), Chicago, 1982, Oc

**Oceanography Program** (M,D) (2014)
700 Pilottown Road
Lewes, DE 19958
p. (302) 645-4279
kbillups@udel.edu
http://www.ocean.udel.edu

Professor:
Thomas M. Church, (D), California (San Diego), 1970, Cm
Victor Klemas, (D), Braunschweig (Germany), 1965, Or
George W. Luther, III, (D), Pittsburgh, 1972, Cm
Jonathan H. Sharp, (D), Dalhousie, 1972, Oc
Christopher K. Sommerfield, (D), SUNY (Stony Brook), 1997, OuGsCc
William J. Ullman, (D), Chicago, 1982, Cm
Ferris Webster, (D), MIT, 1961, Op
Xiao-Hai Yan, (D), SUNY (Stony Brook), 1989, Or
Associate Professor:
Katharina Billups, (D), California (Santa Cruz), 1998, Pe
Douglas C. Miller, (D), Washington, 1985, Ob
Associate Scientist:
Charles H. Culberson, (D), Oregon State, 1972, Oc
Richard T. Field, (D), Delaware, 1994, Og
Assistant Professor:
Matthew J. Oliver, (D), Rutgers, 2006, ObrOg
Adjunct Professor:
Richard B. Coffin, (D), Delaware, 1986, On
James Crease, (D), Cambridge, 1951, Op
Marilyn L. Fogel, (D), Texas, 1977, Oc
Norden E. Huang, (D), Johns Hopkins, 1967, On
David E. Krantz, (D), South Carolina, Cs
Kamlesh Lulla, (D), Indiana State, 1963, Or
Donald B. Nuzzio, (D), Rutgers, 1982, Oc
Manmohan Sarin, (D), Gujarat (India), 1984, On
Alain J. Veron, (D), Paris, 1988, Oc
John F. Wehmiller, (D), Columbia, Gu
Emeritus:
Jin Wu, (D), Iowa, 1964, Op

# District of Columbia

## Carnegie Institution of Washington

**Dept of Terrestrial Magnetism** (2016)
5241 Broad Branch Road, N.W.
Washington, DC 20015-1305
p. (202) 478-8820
jdunlap@carnegiescience.edu
https://dtm.carnegiescience.edu

Director:
Richard W. Carlson, (D), California (San Diego), 1980, Cc
Senior Scientist:
Conel M. O'D Alexander, (D), Essex (UK), 1987, Xc
Alan P. Boss, (D), California (Santa Barbara), 1979, On
R. Paul Butler, (D), Maryland, 1993, On
John E. Chambers, (D), Manchester, 1994, On
Peter E. Driscoll, (D), Johns Hopkins, 2010, GtYm
Erik H. Hauri, (D), MIT/WHOI, 1992, Cc
Larry R. Nittler, (D), Washington (St. Louis), 1996, Xc
Diana C. Roman, (D), Oregon, 2004, Gv
Scott S. Sheppard, (D), Hawaii, 2004, Xg
Steven B. Shirey, (D), SUNY (Stony Brook), 1984, CcGiz
Peter E. vanKeken, (D), Utrecht, 1993, GtYsCg
Lara S. Wagner, (D), Arizona, 2005, YsgGt
Alycia J. Weinberger, (D), Caltech, 1998, Xg
Emeritus:
David E. James, (D), Stanford, 1967, YsGtYg
Alan T. Linde, (D), Queensland, 1972, Ys
Fouad Tera, (D), Vienna, 1962, CcgXm
Senior Scientist and former Director on Leave:
Sean C. Solomon, (D), MIT, 1971, YgsGt
Senior Research Scientist and SIMS Lab Manager:
Jianhua Wang, (D), Chicago, 1995, Cs
Senior Fellow:
I. Selwyn Sacks, (D), Witwatersrand, 1961, Ys
Mass Spectrometry Laboratory Manager:
Timothy D. Mock, (M), Vermont, 1989, Cc
Geochemistry Laboratory Manager:
Mary F. Horan, (M), SUNY (Stony Brook), 1984, Cc
Fiscal Officer:
Terry L. Stahl, On
Librarian:
Shaun J. Hardy, (M), SUNY (Buffalo), 1987, On

**Geophysical Laboratory** (2016)
5251 Broad Branch Road, N.W.
Washington, DC 20015-1305
p. (202) 478-8900
gcody@carnegiescience.edu
http://www.gl.ciw.edu/

Senior Scientist:
Ronald E. Cohen, (D), Harvard, 1985, Gy
Yingwei Fei, (D), City Univ of New York, 1989, Cp
Alexander F. Goncharov, (D), Russian Academy of Sciences, 1983, Gy
Robert M. Hazen, (D), Harvard, 1975, Gz
Ho-kwang Mao, (D), Rochester, 1968, Yx
Bjorn O. Mysen, (D), Penn State, 1974, Cp
Douglas Rumble, III, (D), Harvard, 1969, Gp
Anat Shahar, (D), California (Los Angeles), 2008, CgYxCs
Andrew Steele, (D), Portsmouth, 1996, Xy
Timothy A. Strobel, (D), Colorado Mines, 2008, Om
Viktor V. Struzhkin, (D), Moscow Inst of Physics & Tech, 1991, Gy
Senior Scientist:
T. Neil Irvine, (D), Caltech, 1959, Gi
Research Scientist:
Jinfu Shu, (M), Wuhan, 1981, Yx
Director:
Wesley T. Huntress, Jr., (D), Stanford, 1968, Xc
Research Scientist:
Muhetaer Aihaiti, (D), Electrocommunications Tokyo, 1996, Yg
Reinhard Boehler, (D), Tubingen, 1974, Yg
Xiaojia Chen, (D), Zhejiang, 1997, Ym
Dionysis Foustoukos, (D), Minnesota, 2005, Cg
Stephen Gramsch, (D), Chicago, 1994, Yg
Maddury Somayazulu, (D), Bombay, 1992, Yx
Changsheng Zha, (D), Beijing Inst Tech, 1969, YgOm
Director, HPCAT:
Guoyin Shen, (D), Uppsala, 1994, GzCg
Beamline Scientist, NSLS:
Zhenxian Liu, (D), Jilian (China), 1990, Gy
Beamline Scientist:
Paul Chow, (D), Illinois, 1988, Yg
Yue Meng, (D), Yg
Changyong Park, (D), Tohoku, 1998, Yg
Yuming Xiao, (D), California (Davis), 2007, Yg
Associate Director, HPCAT:
Stanislav Sinogeikin, (D), Yg
Acting Director:
George D. Cody, (D), Penn State, 1992, Co
Librarian:
Shaun J. Hardy, (M), SUNY (Buffalo), 1987, On

## George Washington University

**Dept of Geography** (B,M) (2011)
619 21st Street, NW
Washington, DC 20052
p. (202) 994-6185
mprice@gwu.edu
Department Secretary: Karin Johnston

Chair:
Elizabeth Chacko, (D), California (Los Angeles), 1997, Oy
Professor:
John C. Lowe, (D), Clark, 1969, Og
Assistant Professor:
Doult O. Fuller, (D), Maryland, 1994, Oy
Marie D. Price, (D), Syracuse, 1990, Oy
Visiting Professor:
Micheal Brewer, (D), Delaware, 1997, Or
Thomas Foltltin, (D), Montana, 1980, Oy
Emeritus:
Dorn C. McGrath, Jr., (M), Harvard, 1959, Oy

District of Columbia

## National Academy of Sciences\ National Research Council

**Board on Earth Sciences and Resources** (2016)
The National Academies Keck Center-6th Floor
500 5th Street, NW
Washington, DC 20001
p. (202) 334-2744
besr@nas.edu
http://dels.nas.edu/besr

Director:
Elizabeth A. Eide, (D), Stanford, 1993, GogGt
Senior Program Officer:
Sammantha L. Magsino, (M), Florida Intl, 1993, GvNg
Scholar:
Anne M. Linn, (D), California (Los Angeles), 1991, Gs
Senior Program Officer:
Deborah Glickson, (D), Washington, 2007, GutGg

## Smithsonian Institution/ National Air & Space Museum

**Center for Earth & Planetary Studies** (2016)
MRC 315, P.O. Box 37012
6th and Independence Ave., SW
Washington, DC 20013-7012
p. (202) 633-2470
campbellb@si.edu
http://airandspace.si.edu/research/earth-and-planetary/

Geophysicist:
Bruce A. Campbell, (D), Hawaii, 1991, Or
Planetary Geologist:
James R. Zimbelman, (D), Arizona State, 1984, Xg
Geologist:
Robert A. Craddock, (D), Virginia, 1999, Xg
John A. Grant, (D), Brown, 1990, Xg
Rossman P. Irwin, (D), Virginia, 2005
Thomas R. Watters, (D), George Washington, 1985, Xg
Geologist:
Ted A. Maxwell, (D), Utah, 1977, Xg
Program Manager:
Priscilla L. Strain, (B), Smith, 1974, Or
Photo Librarian:
Rosemary Aiello

## Smithsonian Institution/ National Museum of Natural History

**Dept of Mineral Sciences** (2015)
NHB MRC 119
10th & Constitution Avenue, NW
PO Box 37012
Washington, DC 20013-7012
p. (202) 633-1860
gvp@si.edu
http://mineralsciences.si.edu/

Chair:
Timothy J. McCoy, (D), Hawaii, 1994, Xmg
Research Geologist:
Benjamin Andrews, (D), Texas, 2009, Gv
Catherine Corrigan, (D), Case Western, 2004, XmcXg
Elizabeth Cottrell, (D), Columbia, 2004, Cp
Glenn J. MacPherson, (D), Princeton, 1981, XcGi
Jeffrey E. Post, (D), Arizona State, 1981, Gz
Cara Santelli, (D), MIT/WHOI, 2007, Py
Sorena S. Sorensen, (D), California (Los Angeles), 1984, Gp
Michael A. Wise, (D), Manitoba, 1987, Gz
Museum Specialist (Labs):
Timothy Gooding, (B), Hampshire Coll, 1990
Museum Specialist (IT):
Adam Mansur, (M), Maryland, 2008
Museum Specialist (GVP):
Sally K. Sennert, (M), Pittsburgh, 2003, GvOr
Edward Venzke, (M), Minnesota (Duluth), 1993, Gv
Richard Wunderman, (D), Michigan Tech, 1988, Gv
Museum Specialist (Education):
Adam Blankenbicker, (M), Michigan Tech, 2009, Oe
Geochemist:
Emma Bullock, (D), Open, 2006, Xm
Yulia Goreva, (D), Xm
Collection Manager:
Cathe Brown, (M), Maryland, 1996, Gzi
Russell Feather, (B), George Mason, 1982, Gz
Linda Welzenbach, (M), Bowling Green, 1992, Xm
Analytical Laboratories - Manager:
Timothy Rose, (M), Maryland, 1991, CaGv
Research Collaborator:
Steve Lynton, (D), Maryland, 2003
Postdoctoral Fellow:
Andrew Beck, (D), Tennessee, 2011, Xm
Karen R. Cahill, (D), Tennessee, 2005, Xg
Dominique Chaput, (D), Oxford, 2011, Py
Fred Davis, (D), Minnesota, 2012, Gv
Kathryn Gardner-Vandy, (D), Arizona, 2012, Xm
Brent Grocholski, (D), California (Berkeley), 2008, Xm
Marion Le Voyer, (D), Blaise Pascal, 2009, Gz
Christoph Popp, (D), Bern, 2008, Gv
Contract Geologist:
Rob Dennen, (M), SUNY (Buffalo), 2011, Gv
B. Carter Hearn Jr, (D), Johns Hopkins, 1959, Gic
Julie Herrick, (M), Michigan Tech, 2011, Gv
Sheryl Singerling, (M), Tennessee, 2012, Xm
Research Associate:
Maryjo Brounce, (B), Penn State, 2009
Emeritus:
Roy S. Clarke, Jr., (D), George Washington, 1976, Xm
Richard S. Fiske, (D), Johns Hopkins, 1960, Gv
William G. Melson, (D), Princeton, 1964, GvtGa
Management Support:
Phyllis McKenzie, On
Contractor:
Christine Webb, (B), Penn State, GzgOn
Collection Manager:
Leslie J. Hale, (B), Maryland, 1989, GgOnGh

**Dept of Paleobiology** (2010)
Dept. of Paleobiology, MRC121
NMNH, Smithsonian Institution
P.O. Box 37012
Washington, DC 20013-7012
p. (202) 633-1320
cloydd@si.edu
http://www.nmnh.si.edu/paleo/

Curator:
Scott L. Wing, (D), Yale, 1981, Pb
Curator:
Anna K. Behrensmeyer, (D), Harvard, 1973, Pv
William A. DiMichele, (D), Illinois, 1979, Pb
Brian T. Huber, (D), Ohio State, 1988, Pm
Conrad C. Labandeira, (D), Chicago, 1991, Pi
Ian G. Macintyre, (D), McGill, 1967, Gs
Curator:
Martin A. Buzas, (D), Yale, 1963, Pm
Douglas H. Erwin, (D), California (Santa Barbara), 1985, Pi
Daniel J. Stanley, (D), Grenoble, 1964, Ou
Curator:
Matthew T. Carrano, (D), Chicago, 1998, Pv
Curator:
Gene Hunt, (D), Chicago, 2003, Pm
Geologist:
Thomas Dutro, (D), Pi
Bevan French, (D), Xm
John Pojeta, (D), Cincinnati, 1963, Pi
Curator:
Alan H. Cheetham, (D), Columbia, 1959, Po
Robert Emry, (D), Columbia, 1970, Pv
Clayton E. Ray, (D), Harvard, 1962, Pv
Thomas R. Waller, (D), Columbia, 1966, Pi
Associate Chair & Collections Manager:
Jann W. M. Thompson, (B), George Washington, 1970, Pg

# Florida

## Broward College

**Natural Science Dept** (A) (2010)
111 East Las Olas Boulevard
Fort Lauderdale, FL 33301
p. (954) 201-7650
mdugan@broward.edu

Instructor:
Xenia Conquy, (M), Florida Atlantic, Gg
Henri L. Liauw, (M), South Florida, Gg
William Opperman, (M), Florida, Gg

## Broward College, Central Campus

**Dept of Physical Sciences (Oceanography & Geology)** (A,B) (2016)
3501 SW Davie Road
Davie, FL 33314
p. (954) 201-6771
jmuza@broward.edu
http://www.broward.edu/academics/programs/Pages/science-technology-math-engineering-STEM
Administrative Assistant: Nicki Pickett
*Enrollment (2005): A: 10 (10)*

Chair:
Valerio Bartolucci, (M), Bologna, Ge

Professor:
Lewis Fox, (D), Delaware, 1981, Oc
Jay P. Muza, (D), Florida State, 1996, OuPme
Laura Precedo, (D), Emory, 1993, Oc

Lab Manager:
Lynn A. Curtis, (B), Florida Atlantic Univ, 1997, GgPg

## Chipola College

**Dept of Natural Science** (A) (2010)
3094 Indian Circle
Mariana, FL 32446
p. (850) 526-2761
hiltond@chipola.edu
http://www.chipola.edu/instruct/science/index.htm

Professor:
Allan Tidwell, (M), Troy State, Og

## Daytona State College

**School of Biological and Physical Sciences** (A) (2017)
1200 W. International Speedway Blvd.
Daytona Beach, FL 32114
horikas@daytonastate.edu
http://www.daytonastate.edu/CampusDirectory/deptInfo.jsp?dept=SCI

Instructor:
Debra W. Woodall, GgOg

## Eckerd College

**Dept of Geosciences** (B) (2017)
Galbraith Marine Science Laboratory
4200 54th Avenue South
St. Petersburg, FL 33711
p. (727) 864-8200
wetzellr@eckerd.edu
https://www.eckerd.edu/marinescience/
*Enrollment (2016): B: 6 (1)*

Professor:
Gregg R. Brooks, (D), S Florida, 1986, Gu
Joel B. Thompson, (D), Syracuse, 1989, Py
Laura R. Wetzel, (D), Washington Univ (St. Louis), 1997, YrOuGg

## Florida Atlantic University ❐

**Dept of Geosciences** (B,M,D) O (2017)
777 Glades Road
Boca Raton, FL 33431
p. (561) 297-3250
warburto@fau.edu
http://www.geosciences.fau.edu
*Enrollment (2016): B: 156 (38) M: 28 (13) D: 33 (3)*

Chair:
Zhixiao Xie, (D), SUNY (Buffalo), 2002, OirOy

Associate Provost for Programs and Assessment :
Russell L. Ivy, (D), Florida, 1992, On

Professor:
Leonard Berry, (D), Bristol, 1969, Og

Assistant Chair:
David L. Warburton, (D), Chicago, 1978, CgGe

Associate Professor:
Xavier Comas, (D), Rutgers, 2005, Yg
Maria Fadiman, (D), Texas, 2003, On
Scott H. Markwith, (D), Georgia, 2007, Oy
Anton Oleinik, (D), Purdue, 1998, Pe
Charles E. Roberts, (D), Penn State, 1991, Ori
Tara L. Root, (D), Wisconsin, 2005, Hg
Caiyun Zhang, (D), Texas (Dallas), 2010, Or

Associate Scientist:
Tobin Hindle, (D), Florida Atlantic, 2006, HwOi

Assistant Professor:
Tiffany Roberts M. Briggs, (D), South Florida, 2012, On
Weibo Liu, (D), Kansas, 2016, On

Instructor:
James Gammack-Clark, (M), Florida Atlantic, 2001, Ori

Emeritus:
Howard Hanson, (D), Miami, 1979, Oag
Edward J. Petuch, (D), Miami, 1980, Pg
Jorge I. Restrepo, (D), Colorado State, 1987, Hq

## Florida Gateway College

**Mathematics/Science Div** (A,B) (2016)
149 SE College Place
Lake City, FL 32025
p. (386) 752-1822
mustapha.kane@fgc.edu
https://www.fgc.edu/academics/liberal-arts--sciences/math-and-science.aspx

Professor:
Mustapha Kane, (D), Og

Instructor:
Avo Oymayan, Og

## Florida Institute of Technology

**Dept of Marine & Env Systems, Environmental Science Program** (B,M,D) (2014)
150 West University Boulevard
Melbourne, FL 32901
p. (321) 674-8096
dmes@fit.edu
http://www.fit.edu/dmes
Staff Assistant: Carmen Serrano

Head:
George A. Maul, (D), Miami, 1974, Op

Chair:
John G. Windsor, Jr., (D), William & Mary, 1976, Oc

Professor:
Thomas V. Belanger, (D), Florida, 1979, On
John H. Trefry, (D), Texas A&M, 1977, Cm

Associate Professor:
Charles R. Bostater, (D), Delaware, 1990, Op

Assistant Professor:
Steven Lazarus, (D), Oklahoma, Oa

Adjunct Professor:
Joseph A. Angelo, (D), Arizona, 1976, Og
Diane Barile, (M), Florida Inst of Tech, Ou
Michael F. Helmletter, (D), Old Dominion, Oc
J. J. Keaffaber, (D), Florida, Oc
Carlton R. Parks, (M), Oklahoma, 1981, Ow
D. R. Resio, (D), Virginia, Op
Ned P. Smith, (D), Wisconsin, 1972, Op

**Dept of Marine & Env Systems, Ocean Engineering Program** (B,M,D) (2015)
150 West University Boulevard
Melbourne, FL 32901
p. (321) 674-8096
dmes@fit.edu

http://www.fit.edu/dmes
f: https://www.facebook.com/FloridaTechOceanEngineering/?ref=bookmarks
Professor:
Ronnal Reichard, (D), Oo
Geoffry Swain, (D), Southampton, 1981, Oo
Associate Professor:
Prasanta Sahoo, (D), Oo
Robert Weaver, (D), Oo

Florida

**Dept of Marine & Env Systems, Oceanography Program** (B,M,D) (2014)
150 West University Boulevard
Melbourne, FL 32901
p. (321) 674-8096
jwindsor@marine.fit.edu
http://www.fit.edu/dmes
Staff Assistant: Carmen Serrano
Head:
George A. Maul, (D), Miami, 1974, Op
Acting Chair:
John G. Windsor, (D), William & Mary, 1976, Oc
Professor:
Thomas V. Belanger, (D), Florida, Hw
Iver W. Duedall, (D), Dalhousie, 1973, Oc
Geoffrey W.J Swain, (D), Southampton, 1981, Ob
Gary A. Zarillo, (D), Georgia, 1979, Ou
Associate Professor:
Charles R. Bostater, (D), Delaware, 1990, Op
Lee E. Harris, (D), Florida Atlantic, 1995, On
Assistant Professor:
Elizabeth A. Irlandi, (D), North Carolina, 1993, Ob
Eric D. Thosteson, (D), Florida, On
Research Associate:
Simone Metz, (D), Florida Inst of Tech, 1986, Cm
Chih-Shin Shieh, (D), Florida Inst of Tech, 1988, Oc
Robert P. Trocine, (M), Florida Inst of Tech, 1979, Cm
Adjunct Professor:
Diane D. Barile, (M), Florida Inst of Tech, 1975, Ou
Michael F. Helmstetter, (D), Old Dominion, 1992, Oc
Ned P. Smith, (D), Wisconsin, 1972, Op
Emeritus:
Dean R. Norris, (D), Texas A&M, 1969, Ob

**Meteorology Program** (B,M) (2014)
150 West University Boulevard
Melbourne, FL 32901
p. (321) 674-8096
dmes@fit.edu
http://coe.fit.edu/dmes/
*Enrollment (2014): B: 100 (25) M: 60 (20)*
Head:
George A. Maul, (D), Miami, 1974, Op
Professor:
Thomas V. Belanger, (D), Florida, Hw
Joseph R. Dwyer, (D), Chicago, 1994, Oa
Steven M. Lazarus, (D), Oklahoma, Oa
John G. Windsor, Jr., (D), William & Mary, Oc
Associate Professor:
Tom Utley, (D), Florida Inst of Tech, 1996, Ow
Assistant Professor:
Pallav Ray, (D), Miami, 2009, Oa

## Florida International University

**Center for the Study of Matter at Extreme Conditions** (M,D) (2016)
11200 SW 8th ST
VH 150, CeSMEC
Miami, FL 33199
p. (305) 348-3030
chenj@fiu.edu
http://cesmec.fiu.edu
*Enrollment (2012): D: 2 (2)*
Director:
Jiuhua Chen, (D), Japan Grad Univ Adv Studies, 1994, Gyz
Emeritus:
Surendra K. Saxena, (D), Uppsala, 1967, Gy

**Dept of Earth & Environment** (B,M,D) (2014)
Miami, FL 33199
p. (305) 348-2365
geology@fiu.edu
http://www.fiu.edu/orgs/geology
*Enrollment (2007): B: 30 (7) M: 11 (5) D: 20 (5)*
Professor:
Grenville Draper, (D), West Indies, 1979, GctGh
Rosemary Hickey-Vargas, (D), MIT, 1983, Cg
Jose F. Longoria, (D), Texas (Dallas), 1972, Ps
Florentin J-M.R Maurrasse, (D), Columbia, 1973, PsGsn
Gautam Sen, (D), Texas (Dallas), 1981, Gi
Michael C. Sukop, (D), Kentucky, 2001, HwqSp
Dean Whitman, (D), Cornell, 1993, YgGet
Associate Professor:
William T. Anderson, (D), ETH (Switzerland), 2000, Cs
Laurel S. Collins, (D), Yale, 1988, Po
Michael Gross, (D), Penn State, 1993, Gc
Andrew W. Macfarlane, (D), Harvard, 1989, Em
Assistant Professor:
Rene Price, (D), Miami, 2001, Hw
Ping Zhu, (D), Miami, 2002, Oa
Distinguished Research Professor:
Stephen E. Haggerty, (D), London, 1968, GiEgGz
Hugh E. Willoughby, (D), Miami, 1977, Oa
Lecturer:
Neptune Srimal, (D), Rochester, 1986, Gt
Adjunct Professor:
Jose Antonio Barros, (D), Miami, 1995, Or
Michael Wacker, (M), Florida Intl, Gg
Research Scientist:
Edward Robinson, (D), London, 1969, Pm
Research Lab Manager:
Diane H. Pirie, (B), Florida Intl, 1980, Gg
Cooperating Faculty:
Gabriel Guitierrez-Alonso, (D), Oviedo, 1992, Gc

## Florida State University

**Dept of Earth, Ocean, and Atmospheric Science** (B,M,D) (2005)
108 Carraway Building
Tallahassee, FL 32306-4100
p. (850) 644-5861
odom@gly.fsu.edu
http://www.gly.fsu.edu
Program Assistant: Tami S. Karl
Chair:
LeRoy A. Odom, (D), North Carolina, 1971, Cg
James F. Tull, (D), Rice, 1973, Gct
Professor:
James B. Cowart, (D), Florida State, 1974, Cc
Lynn M. Dudley, (D), Washington State, 1983, Sc
Philip Froelich, (D), Rhode Island, 1979, Og
Bill X. Hu, (D), Purdue, 1996, Hw
Yang Wang, (D), Utah, 1992, CgsPe
Associate Professor:
Anthony J. Arnold, (D), Harvard, 1983, Pm
Joseph F. Donoghue, (D), S California, 1981, GueGs
Munir Humayan, (D), Chicago, 1994, XcCg
Stephen A. Kish, (D), North Carolina, 1983, Eg
William C. Parker, (D), Chicago, 1980, Pi
Vincent J. Salters, (D), MIT, 1989, Cg
Assistant Professor:
Jennifer Georgen, (D), MIT/WHOI, 2001, Gu
Ming Ye, (D), Arizona, 2002, Hg
Emeritus:
George W. Devore, (D), Chicago, 1952, Gp
John K. Osmond, (D), Wisconsin, 1954, Cc
Paul C. Ragland, (D), Rice, 1962, Ca
Sherwood W. Wise, Jr, (D), Illinois, 1970, PmGu

## Hillsborough Community College

**Earth Science** (A) (2011)
2112 North 15th St.
Tampa, FL 33605
p. (813) 253-7647
jolney2@hccfl.edu

http://www.hccfl.edu
Instructor:
Marianne O. Caldwell, Og
James W. Fatherree, Og
Thomas M. Klee, (M), S Illinois, 1986, Gg
Jessica L. Olney, (D), N Illinois, 2006, OgGg
Matthew J. Werhner, (M), Adelphi, 1974, Gg
James F. Wysong, Jr., (M), S Florida, 1989, Ow
Adjunct Professor:
Joseph D. Brod, Og
Kyle M. Champion, Og
James R. Douthat, Jr., Og
Van E. Hayes, Og
James E. MacNeil, Og
Brian W. Marlowe, Og
Matthew P. Olney, (D), N Illinois, 2006, OgGg
Leonard T. Roth, Og
Norman E. Soash, Og
Poetchanaporn Tongdee, Og
Kevan A. Van Cleave, Og
William V. Wills, Og

## Jacksonville University

**Dept of Biology & Marine Science** (B,M) (2016)
2800 University Boulevard North
Jacksonville, FL 32211
p. (904) 256-7302
ngoldbe@ju.edu
*Enrollment (2016): M: 3 (2)*
Professor:
Daniel McCarthy, (D), Kings College, Og
Associate Professor:
Nisse Goldberg, (D)

## Miami Dade College, Kendall Campus

**Chemistry-Physics & Earth Science Dept** (A) (2014)
11011 S.W. 104 St.
Room 3291
Miami, FL 33176
p. (305) 237-2492
sorbon@mdc.edu
http://www.mdc.edu/kendall/chmphy/
*Enrollment (2010): A: 100 (0)*
Assistant Professor:
Michael G. McGauley, (M), Miami, 2003, Owp
Emeritus:
John M. Steger, (D), Naval Postgrad Sch, 1997, OgwOg

## Miami-Dade College, Wolfson Campus

**Dept of Natural Sciences, Health & Wellness** (A) (2017)
300 NE 2nd Avenue
Miami, FL 33132
p. (305) 237-3658
mkraus@mdc.edu
Administrative Assistant: Ileana Baldizon
Professor:
Tony Barros, (D), Miami, 1995, Og
Michael Kaldor, (M), SUNY (Buffalo), 1969, Gg

## Palm Beach Community College

**Environmental Science** (1999)
MS #56 4200 Congress Avenue
Lake Worth, FL 33467
p. (561) 868-3475
milesj@palmbeachstate.edu
http://www.pbcc.edu

## Pensacola Junior College

**Dept of Physical Sciences** (2014)
1000 Collge Blvd
Pensacola, FL 32504-8998
p. (850) 484-1189
estout@pensacolastate.edu
Administrative Assistant: Kimberly LaFlamme
District Academic Department Head:
Edwin Stout
Professor:
Lois Dixon
Brooke L. Towery, (M), Ball State, 1969, Gg
Wayne Wooten
Joseph Zayas
Assistant Professor:
Thor Garber
Timothy Hathaway
Instructor:
Bobby Roberson
Michael Stumpe
Emeritus:
Thomas Gee
Science Lab Specialist:
Darrell Kelly

## Saint Petersburg College, Clearwater

**Dept of Natural Sciences** (A) (2013)
2465 Drew Street
Clearwater, FL 33765
p. (727) 791-2534
williams.john@spcollege.edu
http://www.spcollege.edu/clw/science/
Professor:
Carl Opper, (M), Florida, 1982, Hw
Adjunct Professor:
Neva Duncan Tabb, Ow
Hilary Flower, (M), California (Santa Barbara), Gg
Joseph C. Gould, (D), Nova, Gg
Heather L. Judkins, (D), South Florida, Og

## Sante Fe Community College

**Physical Science Dept** (1999)
3000 NW 83rd Street
Gainesville, FL 32606
p. (505) 428-1307
david.johnson@sfcc.edu
http://www.sfcollege.edu

## University of Florida

**Dept of Geological Sciences** (B,M,D) O (2016)
P.O. Box 112120
241 Williamson Hall
1843 Stadium Road
Gainesville, FL 32611-2120
p. (352) 392-2231
info@geology.ufl.edu
http://web.geology.ufl.edu/
*Enrollment (2015): B: 100 (20) M: 16 (6) D: 30 (1)*
Director:
Douglas S. Jones, (D), Princeton, 1980, YgPi
Chair:
David A. Foster, (D), SUNY (Albany), 1989, GtCc
Affiliate Faculty:
Joann Mossa, (D), Louisiana State, 1990, GmHsOy
Professor:
Thomas S. Bianchi, (D), Maryland, 1987, Co
Mark Brenner, (D), Florida, 1983, Gn
James E T Channell, (D), Newcastle upon Tyne, 1975, YmGt
Ellen E. Martin, (D), California (San Diego), 1993, Cl
Jonathan B. Martin, (D), California (San Diego), 1993, ClHg
Joseph G. Meert, (D), 1993, YgGt
Paul A. Mueller, (D), Rice, 1971, Cc
Michael R. Perfit, (D), Columbia, 1977, Giu
Elizabeth J. Screaton, (D), Lehigh, 1995, Hg
Associate Professor:
Peter N. Adams, (D), California (Santa Cruz), 2004, Gm
Paul F. Ciesielski, (D), Florida State, 1978, Pm
John M. Jaeger, (D), SUNY (Stony Brook), 1998, Gs
Raymond Russo, (D), Northwestern, 1990, Yg
Andrew Zimmerman, (D), William & Mary, 2000, Co
Associate Scientist:
Kyoungwon Min, (D), California (Berkeley), 2002, Cc

Florida

Assistant Professor:
Andrea Dutton, (D), Michigan, 2003, CgGsPe
Mark Panning, (D), California (Berkeley), 2004, Ys
Research Associate:
Ray G. Thomas, (B), Florida, 1990, Og
Lecturer:
Matthew C. Smith, (D), Gi
Jim Vogl, (D), California (Santa Barbara), 2000, Gct
Emeritus:
Frank N. Blanchard, (D), Michigan, 1960, Gz
Guerry H. McClellan, (D), Illinois, 1964, En
Neil D. Opdyke, (D), Durham (UK), 1958, Ym
Anthony F. Randazzo, (D), North Carolina, 1968, Gd
Douglas L. Smith, (D), Minnesota, 1972, Yh
Daniel P. Spangler, (D), Arizona, 1969, Hw
Laboratory Director:
Jason H. Curtis, (D), Florida, 1997, Pe
Ann L. Heatherington, (D), Washington (St. Louis), 1988, Cc
George D. Kamenov, (D), Florida, 2004, CgaEm
Assistant Curator:
Jonathan Bloch, (D), Michigan, 2001, Pv
Affiliate Faculty:
Micheal W. Binford, (D), Indiana, 1980, Or
Cooperating Faculty:
David L. Dilcher, (D), Yale, 1964, Pb
Steven R. Manchester, (D), Indiana, 1981, Pb
Bruce J. McFadden, (D), Columbia, 1976, Pv
Claire L. Schelske, (D), Michigan, 1961, Pe

**Florida Museum of Natural History** (2014)
PO Box 117800
Gainesville, FL 32611-7800
p. (352) 846-2000
bmacfadd@flmnh.ufl.edu
http://www.flmnh.ufl.edu
Administrative Assistant: Pam Dennis
Director:
Douglas S. Jones, (D), Princeton, 1980, Pg
Associate Director:
Graig D. Shaak, (D), Pittsburgh, 1972, Pe
Associate Curator:
David W. Steadman, (D), Arizona, 1982, PvGaPs
Associate Curator:
Bruce J. MacFadden, (D), Columbia, 1976, Pv
Steven R. Manchester, (D), Indiana, 1981, Pb
Graduate Research Professor:
David L. Dilcher, (D), Yale, 1964, Pb
Research Associate:
Ann S. Cordell, (M), Florida, 1983, Gx
David M. Jarzen, (D), Toronto, 1973, Pb
Roger W. Portell, (B), Florida, 1985, Pi
Emeritus:
Sylvia J. Scudder, (M), Florida, 1993, Pg

**Soil & Water Science Dept** (B,M,D) (2014)
PO Box 110290
2181 McCarty Hall A
Gainesville, FL 32611-0290
p. (352) 294-3151
krr@ufl.edu
http://soils.ifas.ufl.edu
*Enrollment (2001): B: 34 (6) M: 28 (5) D: 26 (1)*
Professor and Center Director:
Charley Wesley Wood, (D), Os
Chair:
K Ramesh Reddy, (D), Louisiana State, 1976, Sbc
Professor and Center Director:
Nicholas B. Comerford, (D), SUNY (Syracuse), 1980, Sf
John E. Rechcigl, (D), Virginia Tech, 1986, So
Professor:
Teri Balser, (D), Oss
Mary E. Collins, (D), Iowa State, 1980, Sd
James H. Graham, Jr., (D), Oregon State, 1980, Sb
Sabine Grunwald, (D), Giessen (Germany), 1996, OisOr
Willie G. Harris, Jr., (D), Virginia Tech, 1984, Scd
George J. Hochmuth, (D), Wisconsin, 1980, So
Yuncong Li, (D), Maryland, 1993, ScOs
Lena Q. Ma, (D), Colorado State, 1991, Sc
Peter Nkedi-Kizza, (D), California, 1979, Sp
George A. O'Connor, (D), Colorado State, 1970, Sc
Thomas A. Obreza, (D), Florida, 1983, So
Andrew V. Ogram, (D), Tennessee, 1988, Sb
Jerry B. Sartain, (D), North Carolina, 1973, So
Associate Professor:
Zhenli He, (D), Zhejang Ag, 1988, Sc
James W. Jawitz, (D), Florida, 1999, SpHsw
Marc Kramer, (D), Os
S Rao Mylavarapu, (D), Clemson, 1996, Sc
Arnold W. Schumann, (D), Georgia, 1997, So
P. Christopher Wilson, (D), Clemson, 1999, So
Assistant Professor:
Mark W. Clark, (D), Florida, 2000, Sf
Samira H. Daroub, (D), Michigan State, 1994, So
Stefan Gerber, (D), Os
Patrick Inglett, (D), Florida, 2005, Sf
Cheryl Mackowiak, (D), Utah State, 2001, So
Kelly Morgan, (D), So
Maria Silveira, (D), Sao Paulo, 2003, So
Max Teplitski, (D), Ohio State, 2002, On
Gurpal Toor, (D), Lincoln, 2002, Sc
Alan Wright, (D), Sc
Research Associate Professor:
Vimala D. Nair, (D), Gottingen, 1978, Sc
Ann C. Wilkie, (D), Univ Coll (Ireland), 1984, Sb
Lecturer:
James Bonczek, (D), OsHsg
Susan Curry, (M), Os
Research Assistant Professor:
Todd Osborne, (D), Sf
John Thomas, (D), Sp
Assistant In:
Mengsheng Gao, (D), On

## University of Miami

**Dept of Geological Sciences** (B,M,D) (2017)
1301 Memorial Drive
43 Cox Science Building
Coral Gables, FL 33124
p. (305) 284-4253
ruthgoodin@miami.edu
http://www.as.miami.edu/geology/
Administrative Assistant: Ruth Goodin
*Enrollment (2016): B: 37 (7)*
Chair:
Harold R. Wanless, (D), Johns Hopkins, 1971, Gse
Professor:
Larry C. Peterson, (D), Brown, 1984, Gr
Peter Swart
Senior Lecturer:
Teresa A. Hood, (D), Miami, 1991, Gg
Scientist:
Donald F. McNeill, (D), Miami, 1989, GsrGu
Assistant Professor:
James S. Klaus, (D), Illinois, 2005, Gg
Lecturer:
Peter J. Leech, (D), Georgia Tech, 2013
Ta-Shana A. Taylor, (M), Arizona, 2006, OeGePv
Emeritus:
David E. Fisher, (D), Florida, 1958, Cs
John R. Southam, (D), Illinois, 1974, Op

**Dept of Marine Geosciences** (M,D) (2013)
RSMAS
4600 Rickenbacker Causeway
Miami, FL 33149-1098
p. (305) 421-4663
mgg@rsmas.miami.edu
http://rsmas.miami.edu/divs/mgg/
*Enrollment (2013): M: 2 (0) D: 6 (0)*
Professor:
Peter K. Swart, (D), London, 1980, ClPsCs
Professor:
Falk Amelung, (D), Lous Pasteur, 1996, On
Keir Becker, (D), Scripps, 1981, YrGu

Gregor P. Eberli, (D), ETH (Switzerland), 1985, GusYx
Christopher G.A. Harrison, (D), 1964, Yr
James Natland, (D), Scripps, 1975, Gi
Larry C. Peterson, (D), Brown, 1984, Gu
Ruth P. Reid, (D), Miami, 1979, Gs
Associate Professor:
Mark Grasmueck, (D), ETH, 1995, Yr
Shimon Wdowinski, (D), Harvard, 1990, YdGt
Assistant Professor:
Guochin Lin, (D), Scripps, 2005, Yg
Ali Pourmand, (D), Tulane, 2005, Cga
Emeritus:
Tim H. Dixon, (D), Scripps, 1979, OrGte
Robert N. Ginsburg, (D), Chicago, 1953, Gs
Cooperating Faculty:
Patricia L. Blackwelder, (D), South Carolina, 1976, Po

## University of South Florida

**College of Marine Science** (M,D) (2017)
140 7th Avenue South
St. Petersburg, FL 33701
p. (727) 553-1130
Mitchum@usf.edu
http://www.marine.usf.edu
*Enrollment (2009): M: 30 (13) D: 50 (6)*

Distinguished Research Professor:
Robert H. Byrne, (D), Rhode Island, 1974, Cm
John H. Paul, (D), Miami, 1980, Ob
John J. Walsh, (D), Miami, 1969, Ob
Robert H. Weisberg, (D), Rhode Island, 1975, Op
Professor:
Kent A. Fanning, (D), Rhode Island, 1973, Oc
Luis H. Garcia-Rubio, (D), McMaster, 1981, Oc
Pamela Hallock-Muller, (D), Hawaii, 1977, PmGse
Albert C. Hine, (D), South Carolina, 1975, Gs
Gary T. Mitchum, (D), Florida State, 1984, Op
Frank E. Muller-Karger, (D), Maryland, 1988, Or
Joseph J. Torres, (D), California, 1980, Ob
Associate Professor:
Paula G. Coble, (D), MIT/WHOI, 1992, Cm
Kendra Lee Daly, (D), Tennessee, 1995, Ob
Boris Galperin, (D), Israel Inst of Tech, 1982, Op
David J. Hollander, (D), Swiss Fed Inst Tech, 1989, Oc
Mark E. Luther, (D), North Carolina, 1982, Op
David F. Naar, (D), Scripps, 1990, Ou
Ernst B. Peebles, (D), South Florida, 1996, ObCs
Assistant Professor:
Mya Breitbart, (D), California (San Diego), 2006, Ob
Ashanti J. Pyrtle, (D), Texas A&M, 1999, Oc
Research Associate:
Rick Cole, (B), Florida Inst of Tech, 1983, Op
Dwight A. Dieterle, (M), San Jose, 1979, Ob
Jeff C. Donovan, (B), Florida, 1985, Op
Debra E. Huffman, (D), S Florida, 1994, On
Stanley D. Locker, (D), Rhode Island, 1989, Ou
Wensheng Yao, (D), Miami, 1995, Oc
Research Assistant:
David C. English, (M), Washington, 1983, Op
Adjunct Professor:
Serge Andrefouet, (D), Polynesie Francaise, 1998, Opr
Bruce Barber, (D), Ob
Leonard Ciaccio, (D)
Thomas Cuba, (D), USF, Ob
Christopher D'Elia, (D), Georgia, 1974, Ob
George Denton, (D), Og
Cynthia Heil, (D), USF, Ob
Brian Keller, (D)
John Lisle, (D)
Anne Meylan, (D), Florida, 1984, Ob
Terrence Quinn, (D), Ou
Harunur Rashid, (D)
Eugene Shinn, (B), Miami, 1957, Gg
Randy Wells, (D), California (Santa Cruz), 1986, Ob
Emeritus:
Peter R. Betzer, (D), Rhode Island, 1970, Cm
Norman J. Blake, (D), Rhode Island, 1972, Ob
John C. Briggs, (D), Stanford, 1951, Ob
Kendall L. Carder, (D), Oregon State, 1970, Op
Thomas L. Hopkins, (D), Florida State, 1964, Ob
Harold J. Humm, (D), Duke, 1945, Ob
Edward S. Van Vleet, (D), Rhode Island, 1978, Co
Gabriel A. Vargo, (D), Rhode Island, 1976, Ob

## University of South Florida, Tampa

**Dept of Geology** (B,M,D) (2014)
SCA 528
4202 East Fowler Avenue
Tampa, FL 33620-5201
p. (813) 974-2236
mhaney@chuma1.cas.usf.edu
Administrative Assistant: Mary Haney
*Enrollment (2003): B: 54 (15) M: 31 (9) D: 11 (2)*

Chair:
Charles B. Connor, (D), Dartmouth, 1987, Gv
Professor:
Mark T. Stewart, (D), Wisconsin, 1976, Hw
H. Leonard Vacher, (D), Northwestern, 1971, Hw
Associate Professor:
Peter J. Harries, (D), Colorado, 1993, Pg
Jeffrey G. Ryan, (D), Columbia, 1989, Ct
Assistant Professor:
Sarah E. Kruse, (D), MIT, 1989, Yg
Thomas Pichler, (D), Ottawa, 1998, Cl
Mark C. Rains, (D), California (Davis), 2002, Hw
Ping Wang, (D), S Florida, 1995, Gs
Instructor:
Thomas C. Juster, (D), S Florida, 1995, Hw
Eleanour Snow, (D), Brown, 1987, Gz
Emeritus:
Richard A. Davis, Jr., (D), Illinois, 1964, Gs
Assistant Curator:
Raymond Denicourt, Gz
Cooperating Faculty:
Jeffrey Heikoop, Cl
Lisa Robbins, (D), Miami, 1987, Py
James E. Sorauf, (D), Kansas, 1962, Pg
Barbara W. Leyden, (D), Indiana, 1982, Pl
Robert Morton, (D), West Virginia, 1972, Gs
Thomas M. Scott, (D), Florida State, Gr
Sam B. Upchurch, (D), Northwestern, 1970, Hw

## University of West Florida

**Dept of Earth and Environmental Sciences** (B,M) (2016)
11000 University Parkway
Bldg 13
Pensacola, FL 32514
p. (850) 474-3377
environmental@uwf.edu
http://uwf.edu/cse/departments/earth-and-environmental-sciences/
f: https://www.facebook.com/pages/UWF-Department-of-Environmental-Studies/205988446096964
t: @UWF_EES
*Enrollment (2016): B: 133 (39) M: 40 (6)*

Professor:
Johan Liebens, (D), Michigan State, 1995, So
Chairperson:
Matthew C. Schwartz, (D), Delaware, 2002, OcCl
Associate Professor:
Zhiyong Hu, (D), Georgia, 2003, Or
Assistant Professor:
John D. Morgan, (D), Florida State, 2010, Oi
Jason Ortegren, (D), North Carolina (Greensboro), 2008, PeOu
Phillip P. Schmutz, (D), Louisiana State, 2014, Gm
Instructor:
Chasidy Hobbs, (M), West Florida, 2005, OgeOy
Taylor Kirschenfeld, (M), West Florida, 1988, Ob
Adjunct Professor:
Wilbur G. Hugli, (D), West Florida, 2001, Oa
Hilde Snoeckx, (D), Michigan, 1995, Oug
Online GIS Program Coordinator:
Amber Bloechle, (M), West Florida, 2007, Oi

GeoData Center Coordinator:
Nathan McKinney, (M), West Florida, 2011, Oi

## Valencia Community College

**Valencia Community College** (2014)
P.O. Box 3028
Orlando, FL 32802
p. (407) 299-5000
abosley@valenciacollege.edu
http://www.valenciacc.edu



# Georgia

## Columbus State University

**Earth & Space Sciences** (B,M) (2016)
4225 University Avenue
Columbus, GA 31907-5645
p. (706) 507-8091
barineau_clinton@columbusstate.edu
http://ess.columbusstate.edu/
f: https://www.facebook.com/Earth-and-Space-Sciences-at-Columbus-State-University-GA-347767372234/
*Enrollment (2016): B: 39 (9) M: 3 (0)*

Professor:
David R. Schwimmer, (D), SUNY (Stony Brook), 1973, Pv
Chair:
Clinton I. Barineau, (D), Florida State, 2009, GcgGt
Assistant Professor:
Diana Ortega-Ariza, (D), Kansas, 2016, GdgGr
Emeritus:
William J. Frazier, (D), North Carolina, 1973, GsrGd
Thomas B. Hanley, (D), Indiana, 1975, Gc

## Dalton State Community College

**Dept of Natural Sciences** (B) (2015)
650 College Drive
Dalton, GA 30720
p. (706) 272-4440
jmjohnson@daltonstate.edu
http://www.daltonstate.edu/natural-sciences/index.html

Professor:
Jean M. Johnson, (D), Michigan, 1995, Og

## East Georgia State College

**Science & Mathematics** (A,B) (2017)
131 College Circle
Swainsboro, GA 30401
p. (478) 289-2073
stracher@ega.edu
http://faculty.ega.edu/facweb/stracher/stracher.html
*Enrollment (2015): A: 3 (1)*

Professor Emeritus:
Glenn B. Stracher, (D), Nebraska, 1989, GycEc

## Emory University

**Dept of Environmental Studies** (B) (2014)
Mathematics & Science Center
400 Dowman Drive
Atlanta, GA 30322
p. (404) 727-4216
jwegner@emory.edu
http://www.envs.emory.edu/

Head:
Joy Budensiek, (D)
Chair:
Lance Gunderson, (D), Florida, 1992, Sf
Professor:
William B. Size, (D), Illinois, 1971, Gi
Assistant Professor:
Thomas Gillespie, (D), Florida, On
Tracy Yandle, (D), Indiana, 2001, Ob
Instructor:
Anne M. Hall, (M), Georgia Tech, 1985, Gz
Campus Env Officer:
John Wegner, (D), Carleton, 1995, Sf
Lecturer:
Anthony J. Martin, (D), Georgia, 1991, Pe
Adjunct Professor:
Pamela J. W. Gore, (D), George Washington, 1983, Gs
Senior Lecturer:
Charles W. Hickcox, (D), Rice, 1971, Gg
Emeritus:
Howard R. Cramer, (D), Northwestern, 1954, Pg
Willard H. Grant, (D), Johns Hopkins, 1955, Cg
Lore Ruttan, (D), California (Davis), 1999, Ob

## Fort Valley State University

**Cooperative Developmental Energy Program** (2014)
Box 5800 FVSU
1005 State University Drive
Fort Valley, GA 31030
p. (912) 825-6454
crumblyi@fvsu.edu
http://www.fvsu.edu/academics/cdep

Director:
Isaac J. Crumbly, (D), North Dakota State, 1970, On
Assistant Director:
Jackie Hodges, (B), Georgia Southern, 1983, On
Associate Professor:
Aditya Kar, (D), Oklahoma, 1997, Ct

## Georgia Highlands College

**Div of Science and Physical Education** (A,B) (2010)
Main Campus 3175 Hwy 27 South
P.O. Box 1864
Rome, GA 30162
p. (706) 368-7528
bmorris@highlands.edu
http://www.highlands.edu/site/geology

Associate Professor:
Billy Morris, Gg
Instructor:
Tracy Hall, Gg

## Georgia Institute of Technology

**School of Earth & Atmospheric Sciences** (B,M,D) (2014)
311 Ferst Drive
Atlanta, GA 30332-0340
p. (404) 894-3893
rita.anderson@eas.gatech.edu
http://www.eas.gatech.edu
*Enrollment (2011): B: 63 (21) M: 34 (19) D: 76 (13)*

Chair:
Judith A. Curry, (D), Chicago, 1992, OarOn
Professor:
Annalisa Bracco, (D), Genova (Italy), 2000, OpnOn
Kim M. Cobb, (D), Scripps, 2002, PeCsOg
Gregory L. Huey, (D), Wisconsin, 1992, Oa
Ellery D. Ingall, (D), Yale, 1991, CmlOc
Jean Lynch-Stieglitz, (D), Columbia, 1995, PeCsOg
Edward Michael Perdue, (D), Georgia Tech, 1973, Co
Irina Sokolik, (D), Russian Acad of Sci, 1989, Oar
Martial Taillefert, (D), Northwestern, 1997, ClmCa
Rodney J. Weber, (D), Minnesota, 1995, Oa
Peter J. Webster, (D), MIT, 1972, Oap
Senior Scientist:
Hai-ru Chang, (D), Oa
Robert Stickel, (D), Rice, 1979, Oa
Viatcheslav Tatarskii, (D), Oa
Hsiang-Jui Wang, (D), Georgia Tech, 1995, Oa
Associate Professor:
Michael H. Bergin, (D), Carnegie Mellon, 1995, Oa
Robert X. Black, (D), MIT, 1990, Oa
Emanuele Di Lorenzo, (D), Scripps, 2003, OpnOn
Christian Huber, (D), California (Berkeley), 2009
Takamitsu Ito, (D), MIT, 2005
Athanasios Nenes, (D), Caltech, 2002, Oa
Andrew V. Newman, (D), Northwestern, 2000, YdsYg
Marc Stieglitz, (D), Columbia, 1995, Hsg
Yuhang Wang, (D), Harvard, 1997, Oa
James Wray, (D), Cornell, 2010

Assistant Professor:
Yi Deng, (D), Illinois, 2005, OanOn
Josef Dufek, (D), Washington, 2006, GvOnn
Carol M. Paty, (D), Washington, 2006, YmgOn
Zhigang Peng, (D), S California, 2004, YsgOn
Andrew Stack, (D), Wyoming, 2002, Cg
Research Scientist:
Carlos Hoyos, (D), Georgia Tech, 2008, Oa
Hyemi Kim, (D), Oa
Jiping Liu, (D), Columbia, Oa
Chao Luo, (D), Peking, 1990, Oa
James C. St. John, (D), Georgia Tech, 1997, Oa
Henian Zhang, (D), Oa
Adjunct Associate Professor:
Jay Brandes, (D), Washington, Og
Carmen Nappo, (D), Georgia Tech, 1989, Oa
Valerie Thomas, (D), Cornell, 1987, Ou
Adjunct Assistant Professor:
Karim Sabra, (D), Michigan, 2003, Yg
Adjunct Professor:
Clark R. Alexander, (D), North Carolina State, 1990, Cl
Dominic Assimaki, (D)
Jackson O. Blanton, (D), Oregon State, 1968, Cl
Thomas D. Christina, (D), Caltech, 1989, Cl
James Crawford, (D), Georgia Tech, Oa
Heidi Cullen, (D), Columbia, Ow
Rong Fu, (D)
Leonid Germanovich, (D), Moscow State Mining (Russia), 1982, NrYg
Gary G. Gimmestad, (D), Colorado, 1978, Ora
Richard Jahnke, (D), Washington, 1976, Cl
Yongqiang Liu, (D), Ow
Joseph Montoya, (D), Harvard, 1990, Cm
Armistead G. Russell, (D), Caltech, 1985, Oa
Stuart G. Wakeham, (D), Washington, 1976, Co
Herbert L. Windom, (D), California (San Diego), 1968, Cm
Emeritus:
Paul H. Wine, (D), Florida State, 1974, Oa

## Georgia Southern University

**Applied Coastal Research Laboratory** (2015)
10 Ocean Science Circle
Savannah, GA 31411
p. (912) 598-2329
clarkalexander@georgiasouthern.edu
http://cosm.georgiasouthern.edu/icps/acrl/
Director:
Clark R. Alexander, (D), North Carolina State, 1990, OuGsOn
Assistant Professor:
Chester M. Jackson, (D), Georgia, 2010, OnGs
Research Associate:
Michael Robinson, (B), Georgia Southern, 2002, Oin

**Dept of Geology and Geography** (B) (2016)
68 Georgia Avenue
Building 201
PO Box 8149
Statesboro, GA 30460-8149
p. (912) 478-5361
sjunderwood@georgiasouthern.edu
http://cost.georgiasouthern.edu/geo/
*Enrollment (2016): B: 77 (11)*
Chair:
Stephen J. Underwood, (D), Georgia, 1999, Ow
Professor:
James S. Reichard, (D), Purdue, 1995, Hw
Fredrick J. Rich, (D), Penn State, 1979, Pl
Wei Tu, (D), Texas A&M, 2004, Oi
Robert Kelly Vance, (D), New Mexico Tech, 1989, GgEgGi
Mark R. Welford, (D), Illinois, 1993, Oy
Associate Professor:
Robert A. Yarbrough, (D), Georgia, 2006, On
Assistant Professor:
Christine Hladik, (D), Georgia, 2012, SbOr
Chester W. Jackson, (D), Georgia, 2010, GsmGu
Jacque L. Kelly, (D), Hawaii, 2012, Cl
Kathlyn M. Smith, (D), Michigan, 2010, Pv
John Van Stan, (D), Delaware, 2012, HgClSb
Xiaolu Zhou, (D), Illinois, 2014, Or
Emeritus:
Gale A. Bishop, (D), Texas, 1971, Pi
James H. Darrell, (D), Louisiana State, 1973, Pl
Daniel B. Good, (D), Tennessee, 1973, On
Dallas D. Rhodes, (D), Syracuse, 1973, Gm

## Georgia Southwestern State University

**Dept of Geology & Physics** (B) (2014)
800 GSW State University Drive
Americus, GA 31709
p. (229) 931-2353
Deborah.Standridge@gsw.edu
http://www.gsw.edu/%7Egeology/
Administrative Assistant: Debbie Standridge
*Enrollment (2004): B: 15 (2)*
Professor:
Burchard D. Carter, (D), West Virginia, 1981, Po
Thomas J. Weiland, (D), North Carolina, 1988, Gi
Associate Professor:
Samuel T. Peavy, (D), Virginia Tech, 1997, Yg
Assistant Professor:
Svilen Kostov, (D), CUNY, 1992, On
Emeritus:
Daniel D. Arden, (D), California (Berkeley), 1961, Ps
Harland E. Cofer, (D), Illinois, 1957, Gz
John P. Manker, (D), Rice, 1975, Gs

## Georgia State University

**Dept of Geosciences** (B,M,D) O (2017)
340 Kell Hall
PO Box 3965
24 Peachtree Center Avenue
Atlanta, GA 30302-3965
p. (404) 413-5750
deocampo@gsu.edu
geosciences.gsu.edu
f: https://www.facebook.com/gsugeosciences
t: @FromWater2Rocks
Administrative Assistant: Basirat Lawal
*Enrollment (2016): B: 119 (39) M: 49 (18) D: 4 (0)*
VPAA-Provost:
Risa I. Palm, (D), Minnesota, 1972, On
Department Chair:
Daniel M. Deocampo, (D), Rutgers, 2001, Gsn
Associate Professor:
Hassan A. Babaie, (D), Northwestern, 1984, Gc
Dajun Dai, (D), S Illinois, 2007, Oi
Jeremy E. Diem, (D), Arizona, 2000, Oa
W. Crawford Elliott, (D), Case Western, 1988, Cl
Katherine B. Hankins, (D), Georgia, 2004, On
Lawrence M. Kiage, (D), Louisiana State, 2007, OyPlOr
Assistant Professor:
Nadine Kabengi, (D), Florida, ScClGe
Luke Pangle, (D), Oregon State, 2013, HqCs
Katie Price, (D), Georgia, HsGm
Lecturer:
Paulo Hidalgo, (D), Michigan State, 2011, GivGz
Brian Meyer, (D), Georgia State, 2013, GesHw
Ricardo Nogueira, (D), Louisiana State, 2009, Oaw
Christy Visaggi, (D), North Carolina (Wilmington), 2011, PiOe
Adjunct Professor:
J. Marion Wampler, (D), Columbia, 1963, Cc
Emeritus:
Sanford H. Bederman, (D), Minnesota, 1973, On
Timothy E. La Tour, (D), W Ontario, 1979, Gp
W. Robert Power, (D), Johns Hopkins, 1959, En
Seth E. Rose, (D), Arizona, 1987, Hw
Related Staff:
Atieh Tajik, (M), Georgia State, 2005, Gg

## Georgia State University, Perimeter College, Alpharetta Campus

**Dept of Life & Earth Science** (A) (2016)

3705 Brookside Parkway
Alpharetta, GA 30022
p. (678) 240-6227
dstewart29@gsu.edu
Professor:
Dion C. Stewart, (D), The Pennsylvania State Univ, 1980, Gzi

Georgia

## Georgia State University, Perimeter College, Clarkston Campus

**Dept of Life & Earth Science** (A) (2016)
555 North Indian Creek Drive
Clarkston, GA 30021
p. (678) 891-3754
pgore@gsu.edu
*Enrollment (2007): A: 3 (0)*
Professor:
Pamela J. W. Gore, (D), George Washington, 1983, GsOeGn
Associate Professor:
E. Lynn Zeigler, (M), Emory, 1989, Gd
Lecturer:
Stephan Fitzpatrick, (M), Georgia, HwCgOs

## Georgia State University, Perimeter College, Decatur Campus

**Dept of Life & Earth Science** (A) (2016)
3251 Panthersville Road
Decatur, GA 30034-3897
p. (678) 891-2600
dbulger1@gsu.edu
Assistant Professor:
Daniel E. Bulger, (M), NE Illinois, GgOn

## Georgia State University, Perimeter College, Dunwoody Campus

**Dept of Life & Earth Science** (A) (2016)
2101 Womack Road
Dunwoody, GA 30338-4497
p. (770) 274-5050
*Enrollment (2011): A: 4 (0)*
Lecturer:
Rob J. McDowell, (D), Kentucky, 1992, Gts
Kimberly Schulte, (M), Tulsa, 1997, Ge

## Georgia State University, Perimeter College, Newton Campus

**Dept of Geology** (A) (2017)
239 Cedar Lane
Covington, GA 30014
p. (770) 278-1263
Polly.bouker@gpc.edu
http://www.gpc.edu/~newsci/
Associate Professor:
Polly A. Bouker, (M), Georgia, 2006, Gg

## Georgia State University, Perimeter College, Online

**Dept of Life & Earth Science** (A) (2016)
555 North Indian Creek Drive
Clarkston, GA 30021
p. (678) 212-7523
dballero@gsu.edu
Assistant Professor:
Deniz Z. Ballero, (D), Georgia, Gg
Adjunct Professor:
Edward Albin, (D), Georgia, Xm

## Mercer University

**Dept of Environmental Engineering** (B,M) (2004)
1400 Coleman Avenue
Macon, GA 31207
p. (912) 744-2597
lackey_l@mercer.edu
Department Secretary: Brenda Walraven
Professor:
Bruce D. Dod, (D), S Mississippi, 1973, Og
Dan R. Quisenberry, (D), World Open, 1980, Or
Assistant Professor:
Geoffrey W. Hayden, (D), Pennsylvania, 1987, Om
Robert L. Huffman, (D), Massachusetts, 1985, On
Research Associate:
Paul P. Sipiera, (D), Otago (NZ), 1985, Xm
Instructor:
Barbara D. Henley, (M), Mercer, 1978, Gg

## Middle Georgia College

**Geology Dept** (A) (2010)
1100 Second Street
Cochran, GA 31014
tmahaffee@mgc.edu
http://www.mgc.edu
Associate Professor:
Tina Mahaffee, (M), Georgia State, Gg
Assistant Professor:
Daniel Snyder, (D), Iowa, Gg

## University of Georgia

**Dept of Crop & Soil Science** (B,M,D) (2011)
311 Plant Sciences Building
Athens, GA 30602-7272
p. (706) 542-2461
dgs@uga.edu
http://www.cropsoil.uga.edu/
Chair:
Don Shilling
Professor:
Domy C. Adriano, (D), Kansas State, 1970, Sc
Paul M. Bertsch, (D), Kentucky, 1983, Sc
Gary J. Gascho, (D), Michigan State, 1968, Sc
James E. Hook, (D), Penn State, 1975, Sp
Edward T. Kanemasu, (D), Wisconsin, 1969, Sp
David E. Kissel, (D), Kentucky, 1969, Sc
William P. Miller, (D), Virginia Tech, 1981, Sc
David E. Radcliffe, (D), Kentucky, 1984, Sp
William P. Segars, (D), Clemson, 1972, Sc
Larry M. Shuman, (D), Penn State, 1970, Sc
Associate Professor:
Miguel L. Cabrera, (D), Kansas State, 1986, So
Peter G. Hartel, (D), Oregon State, 1984, Sb
Owen C. Plank, (D), Virginia Tech, 1973, Sc
Larry T. West, (D), Texas A&M, 1986, Sd
Assistant Professor:
Glendon H. Harris, (D), Michigan State, 1993, Sb

**Dept of Geology** (B,M,D) ● (2017)
210 Field Street
Room 308
Athens, GA 30602
p. (706) 542-2652
geology@uga.edu
http://www.gly.uga.edu/
*Enrollment (2011): B: 69 (9) M: 32 (11) D: 6 (0)*
Head:
Douglas E. Crowe, (D), Wisconsin, 1990, Em
Grad Coordinator:
Susan T. Goldstein, (D), California (Berkeley), 1984, Pm
Professor:
Ervan G. Garrison, (D), Missouri, 1979, GamGn
Robert B. Hawman, (D), Princeton, 1988, Ys
Steven M. Holland, (D), Chicago, 1990, Gr
John E. Noakes, (D), Texas A&M, 1962, Cc
Valentine A. Nzengung, (D), Georgia Tech, 1993, ClGe
Alberto E. Patino-Douce, (D), Oregon, 1990, OnnOn
L. Bruce Railsback, (D), Illinois, 1989, PeGdCl
Michael F. Roden, (D), MIT, 1982, Gi
Paul A. Schroeder, (D), Yale, 1992, GzEnCl
Sally E. Walker, (D), California (Berkeley), 1988, Pm
James (Jim) E. Wright, (D), California (Santa Barbara), 1981, Gt
Associate Head:
Raymond Freeman-Lynde, (D), Columbia, 1980, Gu

Assistant Professor:
John F. Dowd, (D), Yale, 1984, Hy
Christian Klimczak, (D), GcmGt
Adam Milewski, (D), W Michigan, 2008, Hw
Senior Lecturer:
Marta Patino-Douce, (D), Buenos Aires, 1990, Gi
Adjunct Assistant Professor:
Roger A. Burke, (D), S Florida, 1985, Cl
Adjunct Professor:
Elizabeth J. Reitz, (D), Florida, 1979, Ga
Emeritus:
Gilles O. Allard, (D), Johns Hopkins, 1956, Eg
R. David Dallmeyer, (D), SUNY (Stony Brook), 1972, Gt
Sam Swanson , (D)
David B. Wenner, (D), Caltech, 1971, Cs
James A. Whitney, (D), Stanford, 1972, Giv

## University of West Georgia

**Dept of Geosciences** (B) (2016)
1601 Maple Street
Callaway Building, Room 148
Carrollton, GA 30118
p. (678) 839-6479
jmayer@westga.edu
http://www.westga.edu/geosci/
Administrative Assistant: Anita M. Bryant
*Enrollment (2016): B: 122 (22)*
Chair:
James R. Mayer, (D), Texas, 1995, Hw
Professor:
David M. Bush, (D), Duke, 1991, Ou
Curtis L. Hollabaugh, (D), Washington State, 1980, GzCg
Randal L. Kath, (D), SD Mines, 1990, GcNgHg
Jeong C. Seong, (D), Georgia, 1999, Oir
Associate Professor:
Brad Deline, (D), Cincinnati, 2009, Pi
Georgina G. DeWeese, (D), Tennessee, 2007, Oy
Hannes Gerhardt, (D), Arizona, 2007, On
I. Shea Rose, (D), Florida State, 2008, OyaOw
Karen S. Tefend, (D), Michigan State, 2005, CgGe
Nathan A. Walter, (D), Florida State, 2005, Og
Assistant Professor:
Christopher A. Berg, (D), Texas, 2007, Gpi
Jessie Hong, (D), Colorado, 2012, OieOy
Research Associate:
Randa R. Harris, (M), Tennessee, 2001, Hs
Emeritus:
Timothy M. Chowns, (D), Newcastle, 1968, GsrGg
Thomas J. Crawford, (M), Emory, 1957, Eg
Lab Coordinator:
John D. Congleton, (M), S Methodist, 1990, GeOrPo

## Valdosta State University

**Dept of Physics, Astronomy & Geosciences** (B) (2013)
1500 North Patterson Street
Valdosta, GA 31698-0055
p. (229) 333-5752
echatela@valdosta.edu
http://www.valdosta.edu/phy/
*Enrollment (2010): B: 77 (10)*
Head:
Edward E. Chatelain, (D), Iowa, 1984, PiGlPv
Professor:
Cecilia S. Barnbaum, (D), California (Los Angeles), 1992, On
Frank A. Flaherty, (D), Fordham, 1983, On
Martha A. Leake, (D), Arizona, 1982, Xg
Kenneth S. Rumstay, (D), Ohio State, 1984, Onn
Associate Professor:
Can Denizman, (D), Florida, 1998, Hw
Mary A. Fares, (D), Tennessee Tech, 1992, On
Judy Grable, (D), Tennessee, 2001, Hs
Mark S. Groszos, (D), Florida State, 1996, GctEg
Michael G. Noll, (D), Kansas, 2000, On
Paul Vincent, (D), Texas A&M, 2004, Oig
Pre-Engineering Director:
Barry Hojjatie, (D), Florida, 1990, Om
Assistant Professor:
Jason Allard, (D), Penn State, 2006, Oag
Instructor:
Perry A. Baskin, (M), Valdosta State, 1972, On
Donald Thieme, (D), Georgia, 2003, GaSdGm
Emeritus:
Dennis W. Marks, (D), Michigan, 1970, On
Arnold E. Somers, Jr., (D), Virginia Tech, 1975, On

# Guam

## University of Guam

**College of Natural and Applied Sciences** (B,M) (2013)
Mohammad H. Golabi, PhD
Mangilao, GU 96923
p. (671) 734-9305
mgolabi@uguam.uog.edu
Associate Professor:
Mohammad H. Golabi, (M), OsSpc

**Water & Environmental Research Institute of the Western Pacific** (M) (2014)
UOG Station
Mangilao, GU 96923
p. (671) 735-2685
jjenson@uguamlive.uog.edu
http://www.weriguam.org/
*Enrollment (2014): M: 4 (1)*
Director:
Shahram Khosrowpanah, (D), Colorado State, 1984, Hs
Professor:
Gary R. W. Denton, (D), London, 1974, Ob
John W. Jenson, (D), Oregon State, 1993, HwGel
Associate Professor:
Joseph D. Rouse, (D), Hg
Yuming Wen, (D), Rhode Island, 2004, Oir
Assistant Professor:
Mark A. Lander, (D), Hawaii, 1990, Ow
Research Associate:
Nathan C. Habana, (M), Guam, 2009, Hw
Professor of Hydrology:
Leroy F. Heitz, (D), Idaho, Hg
Professor of Geology:
Henry Galt Siegrist, jr, (D), Penn State, 1961, GdzHy

# Hawaii

## Honolulu Community College

**Natural Sciences** (A) (2017)
874 Dillingham Blvd.
Honolulu, HI 96817
p. (808) 845-9488
brill@hawaii.edu
http://libart.honolulu.hawaii.edu/natsci/geology.php
Professor:
Richard C. Brill, Jr., (M), Hawaii, Gg

## University of Hawai'i, Hilo

**Dept of Geology** (B) (2014)
200 W. Kawili Street
Hilo, HI 96720-4091
p. (808) 933-3383
kenhon@hawaii.edu
Chair:
Ken Hon, (D), Colorado, 1987, Gvi
Professor:
Jene D. Michaud, (D), Arizona, 1992, HqwGm
Associate Professor:
James L. Anderson, (D), S California, 1987, GctYd
Assistant Professor:
Steven P. Lundblad, (D), North Carolina, 1994, GraGe
Affiliate Faculty:
Cheryl A. Gansecki, (D), Stanford, 1998, GviOe
Affiliate Faculty:
John P. Lockwood, (D), Princeton, 1966, Gvg

Emeritus:
Joseph B. Halbig, (D), Penn State, 1969, Ca
Cooperating Faculty:
Christina C. Heliker, (M), W Washington, 1984, Gv

Hawaii

## University of Hawai'i, Manoa

**Dept of Atmospheric Sciences** (B,M,D) (2015)
2525 Correa Road, HIG 350
Honolulu, HI 96822
p. (808) 956-8775
metdept@hawaii.edu
http://www.soest.hawaii.edu/MET
*Enrollment (2013): B: 18 (5) M: 12 (2) D: 18 (4)*

Professor:
Steven Businger, (D), Washington, 1986, Ow
Yi-Leng Chen, (D), Illinois, 1980, Oaw
Pao-Shin Chu, (D), Wisconsin, 1981, Oa
Tim Li, (D), Hawaii, 1993, Oa
Duane E. Stevens, (D), Harvard, 1977, Ow
Bin Wang, (D), Florida State, 1984, Ow
Yuqing Wang, (D), Monash, 1995, Oa
Associate Professor:
Michael Bell, (D), Naval Postgrad Sch, Owr
Jennifer Griswold, (D), California (Santa Cruz), Owr
Emeritus:
Gary M. Barnes, (D), Virginia, 1980, OwGvOe
Cooperating Faculty:
Antony D. Clarke, (D), Washington, 1983, Oc

**Dept of Geology & Geophysics** (B,M,D) (2016)
1680 East-West Road, POST 701
Honolulu, HI 96822
p. (808) 956-7640
krubin@hawaii.edu
http://www.soest.hawaii.edu/gg
*Enrollment (2015): B: 47 (8) M: 20 (10) D: 31 (3)*

Chair:
Kenneth H. Rubin, (D), California (San Diego), 1991, CcGvCg
Dean:
Brian Taylor, (D), Columbia, 1982, Gt
Assoc. Dean:
Charles H. Fletcher, (D), Delaware, 1986, GsOnu
Professor:
Robert A. Dunn, (D), Oregon, 1999, Yrs
Aly I. El-Kadi, (D), Cornell, 1983, Hw
L. Neil Frazer, (D), Princeton, 1978, Ys
Eric J. Gaidos, (D), MIT, 1996, Po
Michael O. Garcia, (D), California (Los Angeles), 1976, Gi
Craig R. Glenn, (D), Rhode Island, 1987, GseCm
Julia E. Hammer, (D), Oregon, 1998, GiCpGv
Bruce F. Houghton, (D), Otago (NZ), 1977, Gv
Garrett T. Ito, (D), MIT/WHOI, 1996, YgrGt
Kevin T. M. Johnson, (D), MIT, 1990, Gi
Stephen J. Martel, (D), Stanford, 1987, Ng
Gregory F. Moore, (D), Cornell, 1977, Gt
Brian N. Popp, (D), Illinois, 1986, Cs
Scott K. Rowland, (D), Hawaii, 1987, On
Paul Wessel, (D), Columbia, 1989, Yr
Associate Professor:
Janet M. Becker, (D), California (San Diego), 1989, Oo
Henrietta Dulai, (D), Florida State, 2005, CcHw
Gregory Ravizza, (D), Yale, 1991, Cm
Kathleen Ruttenberg, (D), Yale, 1990, Cg
Assistant Researcher:
Thomas Shea, (D), Hawaii, 2010, Gv
Assistant Professor:
Jasper G. Konter, (D), California (San Diego), 2007, CcGiCt
Christian A. Miller, (D), MIT/WHOI, 2009, CslCm
Bridget R. Smith-Konter, (D), California (San Diego), 2005, YdXy
Emeritus:
Frederick K. Duennebier, (D), Hawaii, 1972, Yr
Loren W. Kroenke, (D), Hawaii, 1972, Gu
Ralph Moberly, (D), Princeton, 1956, Gu
John M. Sinton, (D), Otago (NZ), 1976, Gi
Jr. Specialist:
Jennifer Engels, (D), Hawaii, 2001, Oe

Director of Student Services:
Leona M. Anthony, (B), 1986, On
Cooperating Faculty:
David A. Clague, (D), California (San Diego), 1974, Gi
Eric H. De Carlo, (D), Hawaii, 1982, Ca
Margo H. Edwards, (D), Columbia, 1992, Gu
Sarah Fagents, (D), Lancaster (UK), 1994, Gv
Luke P. Flynn, (D), Hawaii, 1992, Or
Patricia B. Fryer, (D), Hawaii, 1981, Gu
Milton A. Garces, (D), California (San Diego), 1995, Ys
Alexander N. Krot, (D), Moscow State (Russia), 1989, Xm
Paul G. Lucey, (D), Hawaii, 1986, Xy
Fred T. Mackenzie, (D), Lehigh, 1962, Cm
Murli H. Manghnani, (D), Montana State, 1962, Yx
Fernando Martinez, (D), Columbia, 1988, Yr
Floyd W. McCoy, Jr., (D), Harvard, 1974, Gu
Peter J. Mouginis-Mark, (D), Lancaster (UK), 1977, Xg
Jane E. Schoonmaker, (D), Northwestern, 1981, Cl
Edward R. Scott, (D), Cambridge, 1971, Xm
Shiv K. Sharma, (D), Indian Inst of Tech, 1973, Gx
G. Jeffrey Taylor, (D), Rice, 1970, Xm
Donald M. Thomas, (D), Hawaii, 1977, Ca
Affiliate Graduate Faculty:
James P. Kauahikaua, (D), Hawaii, 1982, Gv
Donald A. Swanson, (D), Johns Hopkins, 1964, Gv

**Dept of Oceanography** (B,M,D) (2015)
1000 Pope Road
Honolulu, HI 96822
p. (808) 956-7633
ocean@soest.hawaii.edu
http://www.soest.hawaii.edu/oceanography/
Administrative Assistant: Kristin Momohara
*Enrollment (2001): B: 56 (0) M: 41 (3) D: 26 (0)*

Chair:
Michael J. Mottl, (D), Harvard, 1976, Ou
Professor:
Barbara Bruno, (D), Hawaii, 1994, Ob
Eric H. De Carlo, (D), Hawaii, 1982, Cm
Jeffrey C. Drazen, (D), San Diego, 2000, Ob
Eric Firing, (D), MIT, 1978, Op
Pierre J. Flament, (D), California (San Diego), 1986, Op
David Ho, (D), Columbia, 2001, Ou
David M. Karl, (D), California (San Diego), 1978, Ob
Paul Kemp, (D), Oregon State, 1985, Ob
Rudolf C. Kloosterziel, (D), Utrecht (Neth), 1990, Op
Douglas S. Luther, (D), MIT, 1980, Op
Julian P. McCreary, (D), California (San Diego), 1977, Op
Margaret Anne McManus, (D), Old Dominion, 1996, Op
Christopher Measures, (D), Southampton, 1978, Oc
Mark A. Merrifield, (D), California (San Diego), 1989, Op
Bo Qiu, (D), Kyoto, 1990, Op
Kelvin J. Richards, (D), Southampton, 1978, Op
Kathleen C. Ruttenberg, (D), Yale, Cg
Francis J. Sansone, (D), North Carolina, 1980, Cm
Niklas Schneider, (D), Hawaii, 1992, Og
Craig R. Smith, (D), California (San Diego), 1983, Ob
Grieg F. Steward, (D), California (San Diego), 1996, Ob
Axel Timmermann, (D), Hamburg (Germany), 1999, Op
Richard E. Zeebe, (D), Bremen, 1998, Ou
Associate Professor:
Glenn Carter, (D), Washington, 2005, Op
Matthew Church, (D), William & Mary, 2003, Ob
Brian Glazer, (D), Cm
Erica Goetze, (D), California (San Diego), 2004, Ob
Christopher Kelley, (D), Hawaii, 1995, Ob
Gary M. McMurtry, (D), Hawaii, 1979, CcGvu
Brian Powell, (D), Colorado, 2005, Op
Karen E. Selph, (D), Hawaii, 1999, Ob
Affiliate Graduate Faculty:
Allen H. Andrews, (D), Rhodes, 2009, Oc
Emeritus:
Paul K. Bienfang, (D), Hawaii, 1977, Ob
Antony D. Clarke, (D), Washington, 1983, Oa
Richard W. Grigg, (D), California (San Diego), 1969, Ob
Barry J. Huebert, (D), Northwestern, 1970, OcCg
Yuan-Hui Li, (D), Columbia, 1967, Cm

Roger Lukas, (D), Hawaii, 1981, Op
Fred T. Mackenzie, (D), Lehigh, 1962, Cm
Lorenz Magaard, (D), Kiel, 1963, Op
Alexander Malahoff, (D), Hawaii, 1965, Cm
Peter Muller, (D), Hamburg, 1974, Op
Jane E. Schoonmaker, (D), Northwestern, 1981, Cm
Stephen V. Smith, (D), Hawaii, 1970, Ou
Richard E. Young, (D), Miami, 1968, Ob

Affiliate Graduate Faculty:
Russell E. Brainard, (D), Naval Postgrad Sch, 1994
Paul G. Falkowski, (D), British Columbia, 1975, Og
Carolyn Jones, (D), Georgia Tech, 2004
Dennis Moore, (D), Harvard, 1968, Og
Jaromir Ruzicka, (D), Tech (Czech), 1963, On
John R. Sibert, (D), Columbia, 1968, Ob
Kevin Weng, (D), Stanford, 2007, Ob

Cooperating Faculty:
Marlin J. Atkinson, (D), Hawaii, 1981, Ob
Whitlow W L. Au, (D), Washington State, 1970, On
Janet M. Becker, (D), California (San Diego), 1989, Og
Michael Cooney, (D), California (Davis), 1992, Og
Walter Dudley, (D), Hawaii, 1976, Ou
Eric J. Gaidos, (D), MIT, 1996, Yg
Ruth D. Gates, (D), Newcastle (UK), 1990, Ob
Petra H. Lenz, (D), California (Santa Barbara), 1983, Og
Jeffrey J. Polovina, (D), California (Berkeley), 1974, On
Brian N. Popp, (D), Illinois, 1986, Cm
Michael S. Rappe, (D), Oregon State, 1997, Og
Florence Thomas, (D), California (Berkeley), 1992, Ob
Robert Toonen, (D), California (Davis), 2001, Ob
John C. Wiltshire, (D), Hawaii, 1983, Og

## Windward Community College

**Natural Sciences** (A) (2014)
45-720 Kea'ahala Rd.
Kane'ohe, HI 96744
p. (808) 236-9115
fmccoy@hawaii.edu
http://windward.hawaii.edu
*Enrollment (2014): A: 6 (1)*

Professor:
Floyd W. McCoy, (D), Harvard, 1974, GgaGu

# Idaho

## Boise State University

**Dept of Geosciences** (B,M,D) ● (2017)
1910 University Drive
Mail Stop 1535
Boise, ID 83725
p. (208) 426-1631
geosciences@boisestate.edu
http://earth.boisestate.edu/
Administrative Assistant: Liz Johansen
*Enrollment (2014): B: 143 (11) M: 29 (6) D: 20 (7)*

University Distinguished Professor:
Matthew Kohn, (D), Rensselaer, 1991, CsGpPv

Dean Graduate College:
John R. Pelton, (D), Utah, 1979, YgeYs

Chair:
James P. McNamara, (D), Alaska (Fairbanks), 1997, Hg

Associate Dean/College of Arts & Sciences:
Clyde J. Northrup, (D), MIT, 1996, Gc

Professor:
Shawn Benner, (D), Waterloo, 2000, Clt
John Bradford, (D), Rice, 1999, Yeg
Nancy Glenn, (D), Nevada (Reno), 2000, OrNgOi
Paul Michaels, (D), Utah, 1993, Ne
Mark D. Schmitz, (D), MIT, 2001, CcaCg

Senior Scientist:
Jim Crowley, (D), Carleton, 1997, Ccg
Samantha Evans, (D), Florida Intl, 2009, Cs

Associate Professor:
Alejandro N. Flores, (D), MIT, 2009, HqOa
Jeffrey Johnson, (D), Washington, 2000, YsGv
Hans-Peter Marshall, (D), Colorado, YxOl
Jennifer L. Pierce, (D), New Mexico, 2004, GmPe
David E. Wilkins, (D), Utah, 1997, OyGm

Research Professor:
Vladimir I. Davydov, (D), St. Petersburg, 1982, Pg

Program Director:
Karen Viskupic, (D), MIT, 2002, OeGc

Clinical Assistant Professor:
Sam Matson, (D), Minnesota, 2010, GrPv

Assistant Professor:
Brittany Brand, (D), Arizona State, 2008, Gsv
Dylan Mikesell, (D), Boies State Univ, 2012, Yg
Dorsey Wanless, (D), Florida, 2010, Gi

Research Professor:
Lee M. Liberty, (M), Wyoming, 1992, Ys

Adjunct Professor:
Virginia S. Gillerman, (D), California (Berkeley), 1982, Eg

Emeritus Research Professor:
Warren Barrash, (D), Idaho, 1986, HwYxGe

Emeritus:
Paul R. Donaldson, (D), Colorado Mines, 1974, Ye
Kenneth M. Hollenbaugh, (D), Idaho, 1968, Eg
Walter S. Snyder, (D), Stanford, 1977, GrcGt
Claude Spinosa, (D), Iowa, 1968, Pi
Charles J. Waag, (D), Arizona, 1968, Gc
Craig M. White, (D), Oregon, 1980, Giv
Monte D. Wilson, (D), Idaho, 1969, Gm
Spencer H. Wood, (D), Caltech, 1975, Gmm

## Brigham Young University, Idaho

**Dept of Geology** (B) ● (2017)
525 South Center Street
Rexburg, ID 83460-0510
p. (208) 496-7670
willisj@byui.edu
http://www.byui.edu/Geology
*Enrollment (2016): B: 170 (37)*

Chair:
Julie B. Willis, (D), Utah, 2009, GtcOi

Professor:
Daniel K. Moore, (D), Rensselaer, 1997, GxiCp
Robert W. Clayton, (D), S California, 1993, GcYgNg
Forest J. Gahn, (D), Michigan, 2004, PogGs
William W. Little, (D), Colorado, 1995, GsdGm
Mark D. Lovell, (M), Idaho, 1998, HyOiGo
Megan Pickard, (D), Penn State, 2015, GiOeGe
Gregory T. Roselle, (D), Wisconsin, 1997, GpCaHg

Instructor:
Coreen Hurst, (M), BYU, Og

Visiting Professor:
Greg Melton, (D), Alberta, 2012, OgEgGp

## College of Southern Idaho

**Dept of Physical Science** (A) (2013)
Box 1238
315 Falls Avenue
Twin Falls, ID 83301
p. (208) 732-6400
swillsey@csi.edu
http://physsci.csi.edu/geology/
*Enrollment (2013): A: 8 (0)*

Professor:
Shawn P. Willsey, (M), N Arizona, 2000, GgcGv

## College of Western Idaho

**Physical Sciences** (A) (2016)
5500 E Opportunity Drive
Nampa, ID 83687
kaejensen@cwidaho.cc
http://cwidaho.cc
*Enrollment (2016): A: 20 (2)*

Assistant Professor:
Ander Sundell, (M), Boise State Univ, Ggc

Adjunct Professor:
Amanda Laib, (M), Boise State, 2016, GeCcGv
Melissa Schlegel, (D), Arizona State, GeHw

## Idaho State University

**Dept of Geosciences** (B,M,D) O (2016)
921 S. 8th Ave STOP 8072
Pocatello, ID 83209-8072
p. (208) 282-3365
geology@isu.edu
http://geology.isu.edu/
f: https://www.facebook.com/idahostategeosciences/
t: @ISUGeoscience
*Enrollment (2016): B: 74 (25) M: 33 (12) D: 3 (1)*

Professor:
Michael O. McCurry, (D), California (Los Angeles), 1985, Gz
Professor:
Paul K. Link, (D), California (Santa Barbara), 1982, Gs
David W. Rodgers, (D), Stanford, 1987, Gct
Glenn D. Thackray, (D), Oregon State, 1989, Gl
Associate Professor:
Benjamin T. Crosby, (D), MIT, 2006, GmHsOy
Leif Tapanila, (D), Utah, 2005, Pi
Assistant Professor:
Donna M. Delparte, (D), Calgary, 2008, Oi
Sarah E. Godsey, (D), California (Berkeley), 2009, HwGm
Shannon E. Kobs-Nawotniak, (D), Buffalo, 2007, Gv
David M. Pearson, (D), Arizona, 2012, Gc
Research Associate:
Diana L. Boyack, (M), Idaho State, 2012, Oi
Lecturer:
H. Carrie Bottenberg, (D), Missouri S&T, 2012, OiGtc
Lori Tapanila, (M), Utah, 2006, Gg
Emeritus:
Scott S. Hughes, (D), Oregon State, 1983, GvXgGi
GIS Director:
Keith Weber, (M), Montana, Oi
Financial Tech:
Melissa Neiers
Related Staff:
Kate Anthony-Zajanc, (M), Idaho State, 2013, Oi
Cooperating Faculty:
John A. Welhan, (D), California, 1981, Hw

## Lewis-Clark State College

**Earth Sciences** (B) (2016)
Div of Natural Science & Mathematics
500 8th Ave
Lewiston, ID 83501
p. (208) 792-2283
klschmidt@lcsc.edu
http://www.lcsc.edu/science/degree-programs/earth-science/
*Enrollment (2016): B: 12 (2)*

Professor:
Keegan L. Schmidt, (D), S California, 2000, Gc

## North Idaho College

**Dept of Geology and Geography** (1999)
1000 West Garden Avenue
Coeur d'Alene, ID 83814
p. (208) 769-3477
Bill_Richards@nic.edu
http://www.nic.edu/

## University of Idaho

**Geological Sciences** (B,M,D) O (2016)
875 Perimeter Drive
MS 3022
Moscow, ID 83844-3022
p. (208) 885-6192
geology@uidaho.edu
http://www.uidaho.edu/sci/geology
*Enrollment (2015): B: 72 (17) M: 8 (9) D: 16 (1)*

Chair:
Leslie L. Baker, (D), Brown Univ, 1996, ClXgGe
Professor:
Jerry P. Fairley, (D), California (Berkeley), 2000, HwGv
Mickey E. Gunter, (D), Virginia Tech, 1987, Gz
Peter E. Isaacson, (D), Oregon State, 1974, Pi
Kenneth F. Sprenke, (D), Alberta, 1983, Ye
Associate Professor:
Thomas R. Wood, (D), Idaho, 2005, Hw
Assistant Professor:
Elizabeth J. Cassel, (D), Stanford, 2010, Gs
Jeffrey Langman, (D), Texas, 2008, ClHws
Eric L. Mittelstaedt, (D), Hawaii, 2008, Yr
Emeritus:
Dennis J. Geist, (D), Oregon, 1985, Giv

**Soil Science Div** (B,M,D) (2010)
Dept of Plant, Soil, & Entomological Sciences
Moscow, ID 83844-2339
p. (208) 885-7012
pses@uidaho.edu

Professor:
Guy Knudsen, (D), Cornell, 1984, Sb
Robert L. Mahler, (D), North Carolina State, 1980, On
Associate Professor:
Bradford D. Brown, (D), Utah State, 1985, On
Matthew J. Morra, (D), Ohio State, 1986, Sb
Assistant Professor:
Jodi Johnson-Maynard, (D), California (Riverside), 1999, Sd
Paul A. McDaniel, (D), North Carolina State, 1989, Sd
Daniel G. Strawn, (D), Delaware, 1999, Sc

# Illinois

## Augustana College

**Department of Geology** (B) (2017)
639 38th St.
Rock Island, IL 61201
p. (309) 794-7318
jeffreystrasser@augustana.edu
http://www.augustana.edu/geology
f: Fryxell Geology Museum at Augustana College
Administrative Assistant: Gail Parsons
*Enrollment (2016): B: 29 (5)*

Professor:
William R. Hammer, (D), Wayne State, 1979, Pv
Jeffrey C. Strasser, (D), Lehigh, 1996, Gm
Michael B. Wolf, (D), Caltech, 1992, Gi
Geology Lab Technician:
Sallie Heine, (M), W Illinois, 2008, Oe
Assistant Curator:
John Oostenryk
Museum Educational Programs Coordinator:
Susan Kornreich Wolf, (B), Hamilton College, 1987, Oe

## Black Hawk College

**Dept of Natural Science & Engineering** (A) (2013)
6600 34th Avenue
Moline, IL 61265
p. (309) 796-5000
harwoodr@bhc.edu
http://www.bhc.edu/academics/departments/natural-science-and-engineering/
Administrative Assistant: Tara Carey

Chair:
Brian Glaser, (D), N Iowa, 1996, On
Professor:
Richard D. Harwood, (M), N Arizona, 1989, Gv

## College of Lake County

**Earth Science Dept** (A) (2017)
19351 W. Washington Steet
Grayslake, IL 60030-1198
p. (847) 543-2504
eng499@clcillinois.edu
http://www.clcillinois.edu/programs/esc/

Professor:
Eric Priest, (M), Creighton, 1986, Owe
Xiaoming Zhai, (D), California (Davis), 1997, Gp

## Columbia College Chicago
**Dept of Science & Mathematics** (2013)
600 South Michigan Avenue
Chicago, IL 60605
p. (312) 369-7396
crasinariu@colum.edu
http://www.colum.edu
Professor:
Gerald E. Adams, (D), Northwestern, 1985, Gi
Robin L. Whatley, (D), California (Santa Barbara), 2004, Gg

## Eastern Illinois University
**Dept of Geology/Geography** (B) (2014)
600 Lincoln Avenue
Charleston, IL 61920-3099
p. (217) 581-2626
geoscience@eiu.edu
http://www.eiu.edu/~geoscience/
Administrative Assistant: Susan Kile
*Enrollment (2014): B: 74 (23)*
Chair:
Michael W. Cornebise, (D), Tennessee, 2003, Og
Professor:
Betty E. Smith, (D), SUNY (Buffalo), 1994, Oi
Associate Professor:
Diane M. Burns, (D), Wyoming, 2006, Gs
James A. Davis, (D), Kansas State, 2001, Ogw
Belayet H. Khan, (D), Pittsburgh, 1985, Oyw
Barry J. Kronenfeld, (D), SUNY (Buffalo), 2004, Oiy
John P. Stimac, (D), Oregon, 1996, Gct
Dave Viertel, (D), Texas State, 2008, Or
Assistant Professor:
Katherine Johnson, (D), Ohio State, 2008, Pmi
Christopher R. Laingen, (D), Kansas State, 2009, SfOn
Instructor:
Robert Cataneo, (M), E Illinois, 2003, Oa
Emeritus:
Alan Baharlou, (D), Tulsa, 1973, Cg
Kathleen M. Bower, (D), New Mexico, 1993, Ng
Craig A. Chesner, (D), Michigan Tech, 1988, GivGz
Vincent P. Gutowski, (D), Pittsburgh, 1986, GmOy
Godson C. Obia, (D), Oklahoma, 1986, On
Raymond N. Pheifer, (D), Indiana, 1979, EcPg
James F. Stratton, (D), Indiana, 1975, Pg
Related Staff:
Steven DiNaso, (M), Indiana State, 2009, Oi

## Elgin Community College
**Dept of Geology** (2010)
1700 Spartan Drive
Elgin, IL 60123
p. (847) 214-7359
agoyal@elgin.edu
http://www.elgin.edu/employeelisting/faculty/default.aspx
Adjunct Professor:
Mark R. Kuntz, Gg
Timothy M. Millen, (M), N Illinois, 1982, Gg
Joseph E. Peterson, Gg

## Field Museum of Natural History
**Dept of Geology** (2014)
1400 S. Lake Shore Drive
Chicago, IL 60605-2496
p. (312) 665-7621
klawson@fieldmuseum.org
http://www.fieldmuseum.org
Administrative Assistant: Karsten L.. Lawson
Administrative Assistant: Elaine Zeiger
Chair:
Olivier C. Rieppel, (D), Basel, 1978, Pv
Curator, Meteoritics:
Meenakshi Wadhwa, (D), Washington (St. Louis), 1994, Xm
Curator, Fossil Fishes:
Lance Grande, (D), CUNY, 1983, Pv
Curator, Fossil Amphibians & Reptiles:
John R. Bolt, (D), Chicago, 1968, Pv
Associate Curator, Fossil Invertebrates:
Scott H. Lidgard, (D), Johns Hopkins, 1984, Pi
Emeritus:
Matthew H. Nitecki, (D), Chicago, 1968, Pi
William D. Turnbull, (D), Chicago, 1967, Pv
Bertram G. Woodland, (D), Chicago, 1962, Gp
Associate Curator, Fossil Invertebrates:
Peter J. Wagner, (D), Chicago, 1995, Pi

## Harold Washington College
**Physcial Science Dept** (A) (2014)
30 E. Lake Street
Room 903
Chicago, IL 60601
p. (312) 553-5791
pvargas21@ccc.edu
http://www.ccc.edu/colleges/washington/departments/Pages/Physical-Sciences.aspx
*Enrollment (2012): A: 17 (0)*

## Heartland Community College
**Math And Science Div** (A) (2010)
2400 Instructional Commons Building
1500 W Raab Road
Normal, IL 61761
p. (309) 268-8640
mark.finley@heartland.edu
http://www.heartland.edu/ms/easc/
Professor:
Robert Dennison, (M), S Mississippi, Ogn
Mark Finley, (M), Iowa State, Gg
Adjunct Professor:
Janet Beach Davis, (B), Illinois (Springfield), Og
Paul Ritter, (M), E Illinois, Og
Steven Travers, (M), Illinois State, Og
Mark Yacucci, (B), Youngstown State, Og

## Illinois Central College
**Math, Science, and Engineering Dept** (A) (2010)
One College Drive
East Peoria, IL 61611
p. (309) 694-5364
sdunham@icc.edu
Administrative Assistant: Diane Weber
Professor:
Martin A. Petit, (M), Illinois State, 1967, Oe
Associate Professor:
Cheryl R. Emerson, (M), N Arizona, 1989, Oe
Assistant Professor:
Ed Stermer, (M), Iowa, Og
Adjunct Professor:
Linda Aylward, (B), Bradley, 1980, Oe
Sean Mulkey, (M), Iowa, 1994, Oe

## Illinois State University
**Dept of Geography-Geology** (B,M) O (2017)
101 South School Street
Campus Box 4400
206 Felmley Hall of Science
Normal, IL 61790-4400
p. (309) 438-7649
geo@ilstu.edu
http://www.geo.ilstu.edu/contactus/
f: https://www.facebook.com/groups/21051078663/
Administrative Assistant: Karen K. Dunton
*Enrollment (2016): B: 86 (16) M: 18 (6)*
Chair:
Dagmar Budikova, (D), Calgary, 2001, OayOi
Professor:
James E. Day, (D), Iowa, 1988, Pi
David H. Malone, (D), Wisconsin, 1994, GcsEg
Eric W. Peterson, (D), Missouri, 2002, Hws
Associate Professor:
John Kostelnick, (D), Kansas, 2006, Oi
Catherine M. O'Reilly, (D), Arizona, 2001, HgCgGg

Illinois

Rex J. Rowley, (D), Kansas, 2009, Oi
Jonathan B. Thayn, (D), Kansas, 2009, Oir
Assistant Professor:
Tenley Banik, (D), Vanderbilt, 2015, GvzGl
Wondwosen M. Seyoum, (D), Georgia, 2016, HwqHs
Lisa M. Tranel, (D), Virginia Tech, 2010, Gmt
Adjunct Professor:
Toby J. Dogwiler, (D), Missouri (Colombia), 2002, Gm
Walton R. Kelly, (D), Virginia, 1993, HgCg
Edward Mehnert, (D), Illinois, 1998, Hg
Andrew Stumpf, (D), New Brunswick, 2001, GgsCg
Steve Van Der Hoven, (D), Utah, 2000, HwSpYg
Emeritus:
Robert S. Nelson, (D), Iowa, 1970, Gm
Coordinator of Academic Services :
Paul A. Meister, (M), Illinois State, 2016, GgHgOn

## Illinois Valley Community College

**Dept of Geology** (A) (2016)
Natural Sciences & Business Division
815 North Orlando Smith Road
Oglesby, IL 61348-9692
p. (815) 224-0394
Mike_Phillips@ivcc.edu
http://www.ivcc.edu/geology.aspx?id=534
*Enrollment (2016): A: 4 (4)*
Professor:
Michael Phillips, (M), S Illinois Univ, 1990, GgeGm

## Kaskaskia College

**Life and Physical Sciences Dept** (A) (2013)
27210 College Road
Centralia, IL 62801
pvig@kaskaskia.edu
http://www.kaskaskia.edu/LPDept/Default.aspx
Instructor:
Pradeep K. Vig, Gg

## Lincoln Land Community College

**Math and Sciences** (A) (2010)
Springfield, IL 62794-9256
p. (217) 786-4923
dean.butzow@llcc.edu
http://www.llcc.edu/mtsc/Sciences/tabid/3280/Default.aspx
*Enrollment (2006): A: 7 (0)*
Professor:
Dean G. Butzow, (M), W Michigan, Oy
Instructor:
Samantha Reif, (M), Michigan Tech, Gg

## McHenry County College

**Geography, Geology and Earth Science** (A) (2015)
8900 US Hwy 14
Crystal Lake, IL 60012
p. (815) 455-3700
pstahmann@mchenry.edu
http://www.mchenry.edu/EarthScience/index.asp
Instructor:
Theodore Erski, (M), Akron, Og
Paul Hamill, (M), Wisconsin, Owg
Kate Kramer, (M), Indiana-Purdue, 2008, GgOg
Paul Stahmann, (M), Brigham Young, 2000, Oyg

## Moraine Valley Community College

**Physical Sciences Dept** (A) (2015)
9000 W. College Pwky
Palos Hills, IL 60465-2478
p. 708-974-5615
syrup@morainevalley.edu
Professor:
Krista A. Syrup, (M), W Michigan, 2002, Cs

## Northeastern Illinois University

**Dept of Earth Science** (B) ● (2016)
5500 N. St. Louis Avenue
Chicago, IL 60625
p. (773) 442-6050
K-Voglesonger@neiu.edu
http://www.neiu.edu/academics/college-of-arts-and-sciences/departments/earth-science
f: https://www.facebook.com/NEIU-Earth-Science-178754214181/
Administrative Assistant: Jeff Wade
*Enrollment (2016): B: 26 (11)*
Professor:
Laura L. Sanders, (D), Kent State, 1986, HwGe
Associate Professor:
Kenneth M. Voglesonger, (D), Arizona State, 2004, ClGe
Assistant Professor:
Elisabet M. Head, (D), GvOrCg
Nadja Insel, (D), Michigan, 2011, Gt
Instructor:
Mohammad Fariduddin, (D), N Illinois, 1999, Pe
Rebekah Fitchett, (M), Illinois (Chicago), 2006, GgCg
Jean M. Hemzacek Laukant, (M), N Illinois, 1986, Sc

## Northern Illinois University

**Dept of Geology and Environmental Geosciences** (B,M,D) ○ (2017)
Davis Hall 312, Normal Rd.
De Kalb, IL 60115-2854
p. (815) 753-1943
askgeology@niu.edu
http://www.niu.edu/geology/
*Enrollment (2016): B: 60 (13) M: 27 (7) D: 7 (0)*
Chair:
Mark P. Fischer, (D), Penn State, 1994, Gco
Professor:
Philip J. Carpenter, (D), New Mexico Tech, 1984, YgHwNg
Ross D. Powell, (D), Ohio State, 1980, Gs
Reed P. Scherer, (D), Ohio State, 1991, Pm
James A. Walker, (D), Rutgers, 1982, Giv
Graduate Program Director:
Mark R. Frank, (D), Maryland, 2001, Cp
Associate Professor:
Melissa E. Lenczewski, (D), Tennessee, 2001, Ge
Paul R. Stoddard, (D), Northwestern, 1989, Gt
Assistant Professor:
Justin P. Dodd, (D), New Mexico, 2011, Cs
Nicole D. LaDue, (D), Michigan State, 2013, Oe
Nathan D. Stansell, (D), Pittsburgh, 2009, Gl
Research Associate:
Anna Buczynska, (D), Antwerp (Belgium), 2014, Csa
Adjunct Professor:
B. Brandon Curry, (D), Illinois, 1995, GelPs
Virginia Naples, (D), Massachusetts, 1980, Pv
Karen Samonds, (D), Stony Brook, 2006, Pv
Emeritus:
Jonathan H. Berg, (D), Massachusetts, 1976, Gp
Colin J. Booth, (D), Penn State, 1984, Hw
C. Patrick Ervin, (D), Wisconsin, 1972, Yv
Hsin-Yi Ling, (D), Washington (St. Louis), 1963, Pm
Carla W. Montgomery, (D), MIT, 1977, Cc
Eugene C. Perry, (D), MIT, 1963, Cs
Office Manager:
Nina Slack, (B)

## Northwestern University

**Earth and Planetary Sciences** (B,D) (2016)
2145 Sheridan Rd
Technological Institute
Office F374
Evanston, IL 60208-3130
p. (847) 491-3238
earth@northwestern.edu
http://www.earth.northwestern.edu/
t: @DepEarthandPlan
Administrative Assistant: Gina Allen
Administrative Assistant: Lisa Collins
Administrative Assistant: Benjamin Rice
*Enrollment (2016): B: 26 (11) D: 28 (4)*

Chair:
Bradley B. Sageman, (D), Colorado, 1991, PsGsPe
Professor:
Craig R. Bina, (D), Northwestern, 1987, YgOmGy
Neal E. Blair, (D), Stanford, 1980, CoSbOc
Steven D. Jacobsen, (D), Colorado, 2001, GyzOm
Andrew D. Jacobson, (D), Michigan, 2001, ClPeCa
Donna M. Jurdy, (D), Michigan, 1974, YgXyg
Seth Stein, (D), Caltech, 1978, YsdYg
Suzan van der Lee, (D), Princeton, 1996, YsxYg
Associate Professor:
Matthew T. Hurtgen, (D), Penn State, 2003, GsPoGr
Assistant Professor:
Daniel Horton, (D), Michigan, 2011, OaPeOn
Magdalena Osburn, (D), Caltech, 2013, Py
Assistant Chair:
Patricia A. Beddows, (D), Bristol (UK), 2004, HyGma
Adjunct Professor:
Gilbert Klapper, (D), Iowa, 1962, Pm
Visiting Professor:
Gilbert Klapper, (D), Iowa, 1962, PgiPo
Wayne Marko, (D), Texas Tech, 2012, Gp
Adjunct Professor Emeritus:
Johannes Weertman, (D), Carnegie Inst, 1951, Om
Emeritus:
Abraham Lerman, (D), Harvard, 1964, CgGnCm
Emile A. Okal, (D), Caltech, 1978, YsgYr

## Oakton Community College
**Earth Science** (A) (2013)
1600 E. Golf Road
Des Plaines, IL 60016
p. (847) 376-7042
cumpston@oakton.edu
http://www.oakton.edu/academics/academic_departments/earth_science/
Chair:
John Carzoli, (D), Oklahoma, 2000, On
Professor:
Thomas R. Brehman, (M), NE Illinois, 1974, Po
Assistant Professor:
Jennifer Cumpston, (M), Og

## Olivet Nazarene University
**Dept of Chemistry and Geosciences** (B) (2016)
One University Avenue
Bourbonnais, IL 60914
p. (815) 939-5394
ccarriga@olivet.edu
http://geology.olivet.edu/
f: https://www.facebook.com/pages/Olivet-Nazarene-University-Geosciences/50458842809
*Enrollment (2016): B: 19 (9)*
Acting Director:
Charles W. Carrigan, (D), Michigan, 2005, GzCgGc
Professor:
Max W. Reams, (D), Washington Univ (St. Louis), 1968, GsmPg
Director, Strickler Planetarium:
Stephen Case, (D), Notre Dame, 2014, GhXg

## Prairie State College
**Dept of Physical Science, Earth Science, and Geology** (1999)
202 S. Halsted St
Chicago Heights, IL 60411
p. (708) 709-3674
lburrough@prairiestate.edu
http://www.prairie.cc.il.us

## Principia College
**Dept of Geology** (2014)
1 Maybeck Place
Elsah, IL 62028
p. (618) 374-5294
Chrissy.mcallister@principia.edu
http://principia.edu/mammoth

Chair:
Janis D. Treworgy, (D), Illinois (Urbana), 1985, GgPvGe

## Richard J. Daley Community College
**Physical Sciences** (1999)
7500 South Pulaski Rd
Chciago, IL 60652
p. (773) 838-7636
gguerra@ccc.edu
http://daley.ccc.edu/

## Richland Community College
**Richland Community College** (1999)
One College Park
Decatur, IL 62521
p. (217) 875-7211
jsmith@richland.edu
http://www.richland.edu/

## Rock Valley College
**Rock Valley College** (1999)
3301 North Mulford Road
Rockford, IL 61114
J.Hankins@RockValleyCollege.edu
http://ednet.rockvalleycollege.edu/

## Sauk Valley Community College
**Earth Science** (2014)
173 IL Rt. 2
Dixon, IL 61021
michael.j.bates@svcc.edu
http://www.svcc.edu

## South Suburban College
**Dept of Physical Science** (1999)
15800 South State Street
South Holland, IL 60473
p. (708) 596-2000 x2417
ahelwig@ssc.edu
http://www.ssc.edu

## Southern Illinois University, Carbondale
**Dept of Geology** (B,M,D) O (2016)
1259 Lincoln Drive, Mailcode 4324
Parkinson 102
Carbondale, IL 62901
p. (618) 453-3351
geology@geo.siu.edu
http://www.geology.siu.edu/
f: Geology at Southern Illinois University
Administrative Assistant: Mona Martin
*Enrollment (2016): B: 34 (14) M: 26 (8) D: 6 (0)*
Chair:
Steven P. Esling, (D), Iowa, 1983, HwGl
Professor:
Ken B. Anderson, (D), Melbourne, 1989, Co
Eric C. Ferre, (D), Toulouse, 1989, GcYmGt
Scott E. Ishman, (D), Ohio, 1990, Pe
Susan M. Rimmer, (D), Penn State, 1985, EcCgGo
John L. Sexton, (D), Indiana, 1974, Ye
Associate Professor:
James A. Conder, (D), Brown, 2001, Ysr
Justin Filiberto, (D), SUNY (Stony Brook), 2006, GiCpXc
Liliana Lefticariu, (D), N Illinois, 2004, CslGe
Assistant Professor:
Harvey Henson, (M), S Illinois, 1989, Yg
Daniel Hummer, (D), Penn State, 2010, Gz
Sally Potter-McIntyre, (D), Utah, 2013, Gs
Research Associate:
William Huggett, (M), S Illinois, 1981, Ec
Leader, Coal Characterization Lab:
John C. Crelling, (D), Penn State, 1973, Ec
Emeritus:
Richard Fifarek, (D), Oregon State, 1985, Em

Charles O. Frank, (D), Syracuse, 1973, Gx
Stanley E. Harris, Jr., (D), Iowa, 1947, Ge
John E. Marzolf, (D), California (Los Angeles), 1970, Gs
Paul D. Robinson, (M), S Illinois, 1963, Gz
Jay Zimmerman, Jr., (D), Princeton, 1968, Gc



## Southwestern Illinois College, Belleville Campus

**Dept of Earth Science** (1999)
2500 Carlyle Avenue
Belleville, IL 62221
p. (618) 235-2700
http://www.swic.edu

## Southwestern Illinois College, Sam Wolf Granite City Campus

**Physical Science** (A) (2013)
4950 Maryville Road
Granite City, IL 62040
p. (618) 235-2700
http://www.swic.edu

Professor:
Joy Branlund, (D), Washington Univ (St. Louis), 2008, Og

## Triton College

**Triton College** (2014)
2000 Fifth Ave.
River Grove, IL 60171
p. (708) 456-0300
austinweinstock@triton.edu
http://www.triton.edu

## University of Chicago

**Dept of the Geophysical Sciences** (B,M,D) (2015)
5734 S. Ellis Avenue
Chicago, IL 60637
p. (773) 702-8101
info@geosci.uchicago.edu
http://geosci.uchicago.edu
Administrative Assistant: David J. Taylor
*Enrollment (2009): B: 20 (9) D: 31 (4)*

Professor:
David Archer, (D), Washington, 1990, Yr
Peter R. Crane, (D), Reading, 1981, Po
Andrew Davis, (D), Yale, 1977, Xc
Michael J. Foote, (D), Chicago, 1989, Pg
John E. Frederick, (D), Colorado, 1975, Oa
Lawrence Grossman, (D), Yale, 1972, Xc
David Jablonski, (D), Yale, 1979, Pi
Douglas R. MacAyeal, (D), Princeton, 1983, Yr
Raymond T. Pierrehumbert, (D), MIT, 1980, Oa
Frank M. Richter, (D), Chicago, 1972, Gt
David B. Rowley, (D), SUNY (Albany), 1983, GtgGc

Associate Professor:
Dion L. Heinz, (D), California (Berkeley), 1986, Yx
Noboru Nakamura, (D), Princeton, 1989, Oa

Assistant Professor:
Charles Kevin Boyce, (D), Harvard, 2001, Pb
Fred Ciesla, (D), Arizona, 2003, Xcg
Albert Colman, (D), Yale, 2002, PyCs
Nicolas Dauphas, (D), Centre Pétro et Géochimiques, 2002, Xc
Pamela Martin, (D), California (Santa Barbara), 2000, Pe
Elisabeth Moyer, (D), Caltech, 2001, Oa
Mark Webster, (D), California (Riverside), 2003, Pi

Visiting Professor:
Ho-kwang (David) Mao, (D), Rochester, 1968, Yg

Emeritus:
Alfred T. Anderson, Jr., (D), Princeton, 1963, Gv
Robert N. Clayton, (D), Caltech, 1955, Cs
Dave Fultz, (D), Chicago, 1947, Oa
Susan M. Kidwell, (D), Yale, 1982, Gs
Michael C. LaBarbera, (D), Duke, 1976, Po
David M. Raup, (D), Harvard, 1957, Po
Ramesh C. Srivastava, (D), McGill, 1964, OarOw

## University of Illinois at Chicago

**Dept of Earth and Environmental Sciences** (B,M,D) (2016)
845 W. Taylor Street
2440 SES
MC 186
Chicago, IL 60607-7059
p. (312) 996-3155
klnagy@uic.edu
http://eaes.uic.edu/
*Enrollment (2016): B: 82 (17) M: 3 (3) D: 11 (2)*

Head:
Kathryn L. Nagy, (D), Texas A&M, 1988, Cg

Professor:
Peter T. Doran, (D), Nevada (Reno), 1996, HswOw
Fabien Kenig, (D), Orleans (France), 1991, Co
Roy E. Plotnick, (D), Chicago, 1983, PoiGq
Carol A. Stein, (D), Columbia, 1984, Yg

Associate Professor:
Andrew J. Dombard, (D), Washington (St. Louis), 2000, Xy
DArcy Meyer Dombard, (D), Washington (St. Louis), 2004, Py

Assistant Professor:
Max Berkelhammer, (D), S California, 2010, Cg

Adjunct Assistant Professor:
Andrew King, (D), California (San Diego), 2008, Og

Adjunct Professor:
Paul Fenter, (D), Pennsylvania, 1990, Gy
Barry Lesht, (D), Chicago, 1977, Op

Visiting Assistant Professor:
Sarah Cadieux, (D), Indiana, 2015, Cg

Emeritus:
Jean E. Bogner, (D), N Illinois Univ, 1996, ClSbHy
Martin F. J. Flower, (D), Manchester, 1971, Gi
Stephen J. Guggenheim, (D), Wisconsin, 1976, Gz
August F. Koster Van Groos, (D), Leiden (Neth), 1966, Cp
Kelvin S. Rodolfo, (D), S California, 1967, GueGs

Clinical Assistant Professor:
Stefany Sit, (D), Miami, 2013, Ygs

Research Specialist:
Kenneth Kearney, (M), Illinois (Chicago), 2013, Cg

Office Support Specialist:
Minnie O. Jones, On

Asst to the Head/Business Manager:
Edna L. Rivera, (B), Illinois (Chicago), 2004, On

## University of Illinois, Urbana-Champaign

**Dept of Atmospheric Sciences** (B,M,D) ● (2016)
105 S. Gregory Street
Urbana, IL 61801-3070
p. (217) 333-2046
atmos-sci@illinois.edu
http://www.atmos.illinois.edu
f: https://www.facebook.com/dasuiuc/?hc_ref=SEARCH&fref=nf
Administrative Assistant: Tammy R. Warf
*Enrollment (2016): B: 74 (23) M: 19 (9) D: 27 (3)*

Head:
Robert M. Rauber, (D), Colorado State, 1985, Oa

Professor:
Larry Di Girolamo, (D), McGill, 1996, Or
Atul K. Jain, (D), India, 1988, Oa
Sonia Lasher-Trapp, (D), Oklahoma, 1998, Oa
Greg M. McFarquhar, (D), Toronto, 1993, Oaw
Michael E. Schlesinger, (D), California (Los Angeles), 1976, Oa
Robert J. Trapp, (D), Oklahoma, 1994, Oa
Donald J. Wuebbles, (D), California (Davis), 1983, Oa

Associate Professor:
Stephen W. Nesbitt, (D), Utah, 2003, Oa
Nicole Riemer, (D), Karlsruhe, 2002, Oa
Zhuo Wang, (D), Hawaii, 2004, Oa

Assistant Professor:
Francina Dominguez, (D), Illinois, 2006, Oa
Deanna Hence, (D), Washington, 2011, Oa
Ryan Sriver, (D), Purdue, 2008, Oa

Research Scientist:
Brian Jewett, (D), Illinois, 1996, Oa
Jun shik Um, (D), Illinois, 2009

Research Associate:
Guangyu Zhao, (D), Illinois, 2006
Lecturer:
Donna J. Charlevoix, (M), California (Davis), 1996, Oa
Emeritus:
Kenneth V. Beard, (D), California (Los Angeles), 1970, Oa
Mankin Mak, (D), MIT, 1968, Oa
Walter A. Robinson, (D), Columbia, 1985, Oa
John E. Walsh, (D), MIT, 1974, Oa
Robert B. Wilhelmson, (D), Illinois, 1972, Oa
Director of Undergraduate Study:
Eric R. Snodgrass, (M), Illinois, 2006, Oa
Clinical Assistant Professor:
Jeffrey Frame, (D), Penn State, 2008, Oaw
Chief Clerk:
Karen Eichelberger

**Dept of Geology** (B,M,D) ● (2016)
605 E. Springfield Avenue
152 Computing Applications Bldg.
Champaign, IL 61820
p. (217) 333-3540
geology@illinois.edu
http://www.geology.illinois.edu
*Enrollment (2016): B: 80 (32) M: 14 (7) D: 26 (0)*
Threet Professor:
James L. Best, (D), London, 1985, Gs
Johnson Professor:
Gary Parker, (D), Minnesota, 1974, Gm
Head:
Thomas M. Johnson, (D), California (Berkeley), 1995, HwCls
Grim Professor:
Jay D. Bass, (D), SUNY (Stony Brook), 1982, Gy
Feng-Sheng Hu, (D), Washington, 1994, Pe
Affiliate Faculty:
Stanley Ambrose, (D), California (Santa Barbara)
Kenneth T. Christensen, (D), Illinois
Marcelo Garcia, (D), Minnesota, 1989, Gm
Bruce Rhoads, (D), Arizona, Gm
Charles J. Werth, (D), Stanford
Professor:
Bruce W. Fouke, (D), SUNY (Stony Brook), 1993, GsdGb
Craig C. Lundstrom, (D), California (Santa Cruz), 1996, Cp
Stephen Marshak, (D), Columbia, 1983, Gc
Xiaodong Song, (D), Caltech, 1994, Ys
Associate Head:
Stephen P. Altaner, (D), Illinois, 1985, Gz
Assistant Clinical Professor:
Michael A. Stewart, (D), Duke, 2000, Gi
Affiliate Faculty:
Scott Olson, (D), Illinois
Associate Professor:
Alison M. Anders, (D), Washington, 2005, Gm
Affiliate Faculty:
Surangi W. Punyasena, (D), Chicago, 2007, PblPe
Assistant Professor:
Jessica Conroy, (D), Arizona, 2011, Cs
Jennifer Druhan, (D), California (Berkeley), 2012, HwCs
Patricia Gregg, (D), MIT/WHOI, 2008, YgCg
Lijun Liu, (D), Caltech, 2010, YgGtm
Wendy Yang, (D), California (Berkeley), 2010
Lecturer:
Jacalyn Wittmer Malinowski, (D), Virginia Tech, 2014, PgGsPy
Adjunct Professor:
Ercan Alp, (D), S Illinois, 1984
Kurtis Burmeister, (D), Illinois, 2005
Brandon Curry, Illinois, 1995
Przemyslaw Dera, (D), Mickiewicz (Poland), 2000
Robert J. Finley, (D), South Carolina, 1975, Go
Leon R. Follmer, (D), Illinois, 1970, Sa
Hannes E. Leetaru, (D), Illinois, 1997, GoYe
Morris W. Leighton, (D), Chicago, 1951, Eo
George Roadcap, (D), Illinois, 2004
William W. Shilts, (D), Syracuse, 1970, Gs
M. Scott Wilkerson, (D), Illinois, 1991, Gc
Walgreen Chair Emeritus Professor:
Susan W. Kieffer, (D), Caltech, 1971, Gv
Grim Emeritus Professor:
Craig M. Bethke, (D), Illinois, 1984, Hw
Emeritus:
Thomas F. Anderson, (D), Columbia, 1967, CslCm
Daniel B. Blake, (D), California (Berkeley), 1966, Pi
Albert V. Carozzi, (D), Geneva, 1948, Gd
Chu-Yung Chen, (D), MIT, 1983, Cg
Wang-Ping Chen, (D), MIT, 1979, YsGt
Donald L. Graf, (D), Columbia, 1950, Cl
Ralph L. Langenheim, Jr., (D), Minnesota, 1951, Gr
Alberto S. Nieto, (D), Illinois, 1974, Ng
Philip A. Sandberg, (D), Stockholm, 1965, Gd
Teaching Specialist:
Ann D. Long, (M), Leeds, 1981, Gm
Research Scientist:
Jonathan H. Tomkin, (D), Australian National, 2002, Gm
Research Associate Professor:
Robert A. Sanford, (D), Michigan State, 1996, Py
Lecturer:
Eileen A. Herrstrom, (D), Iowa, 1995, Gi
Assistant Research Professor:
J. Cory Pettijohn, (D), Boston, 2008, Hg

**Dept of Natural Resources & Environmental Sciences**
(B,M,D) (2014)
W-503 Turner Hall
1102 South Goodwin Avenue
Urbana, IL 61801-4798
p. (217) 333-2770
g-rolfe@uiuc.edu
http://nres.illinois.edu/
Chair:
Jeffrey Brawn
Professor:
Charles W. Boast, (D), Iowa State, 1970, Sp
Mark B. David, (D), New York, 1983, Sf
John J. Hassett, (D), Utah State, 1970, Sc
Robert L. Jones, (D), Illinois, 1962, Sc
Richard L. Mulvaney, (D), Illinois, 1983, Sc
Theodore R. Peck, (D), Wisconsin, 1962, Sc
Joseph W. Stucki, (D), Purdue, 1975, Sc
Associate Professor:
Robert G. Darmody, (D), Maryland, 1980, Sd
Timothy R. Ellsworth, (D), California, 1989, Sp
Kenneth R. Olson, (D), Cornell, 1983, Sc
F. William Simmons, (D), North Carolina State, 1987, Sp
Assistant Professor:
Robert J. M. Hudson, (D), MIT, 1989, Cg
Gregory F. McIsaac, (D), Illinois, 1994, Hs

## Waubonsee Community College

**Earth Sciences** (A) (2016)
Rt. 47 at Waubonsee Dr.
Sugar Grove, IL 60554
p. (630) 466-2783
dvoorhees@waubonsee.edu
https://www.waubonsee.edu/learning/academics/disciplines/science/earth/
*Enrollment (2016): A: 5 (0)*
Associate Professor:
David H. Voorhees, (M), Rensselaer, 1982, OgGg
Assistant Professor:
Karl Schulze, (M), Texas A&M, 2003, OgwOa
Alfred W. Weiss, (M), S Illinois, 2000, Oiy

## Western Illinois University

**Dept of Geology** (B) ○ (2016)
1 University Circle
Macomb, IL 61455
p. (309) 298-1151
geology@wiu.edu
http://www.wiu.edu/cas/geology/
Administrative Assistant: Cerese Wright
*Enrollment (2016): B: 30 (11)*
Professor:
Kyle R. Mayborn, (D), California (Davis), 2000, Gic

Leslie A. Melim, (D), S Methodist, 1991, Gd
Associate Professor:
Steven W. Bennett, (D), Indiana, 1994, Hw
Assistant Professor:
Thomas A. Hegna, (D), Yale, PioPg
Instructor:
Sara Bennett, (M), Indiana, Gg
Emeritus:
Jack B. Bailey, (D), Illinois, 1975, Pg
Peter L. Calengas, (D), Indiana, 1977, En
Cooperating Faculty:
Robert E. Johnson, (B), W Illinois, 1978, On



## Wheaton College

**Dept of Geology & Environmental Science** (B) (2015)
501 E. College Ave.
Wheaton, IL 60187
p. (630) 752-5063
geology@wheaton.edu
http://www.wheaton.edu/geology/
Administrative Assistant: Jamie L. Fearon
*Enrollment (2015): B: 50 (12)*
Chair:
Stephen O. Moshier, (D), Louisiana State, 1986, Gd
Professor:
James A. Clark, (D), Colorado, 1977, GmHy
Jeffrey K. Greenberg, (D), North Carolina, 1978, Gt
Charles Keil, (D), Illinois (Chicago), 1994, On
Instructor:
Lisa Heidlauf, (M), Illinois (Urbana), 1986, Gg
Emeritus:
Gerald H. Haddock, (D), Oregon, 1967, Gi

# Indiana

## Ball State University

**Dept of Geology** (B,M,D) O (2016)
2000 University Avenue
AR117
Muncie, IN 47306
p. (765) 285-8270
geology@bsu.edu
http://www.bsu.edu/geology
Administrative Assistant: Brenda J. Rathel
*Enrollment (2016): B: 47 (8) M: 14 (4) D: 4 (0)*
Chair:
Richard H. Fluegeman, (D), Cincinnati, 1987, Pm
Associate Dean:
Jeffry D. Grigsby, (D), Cincinnati, 1989, GdClGo
Professor:
Kirsten N. Nicholson, (D), Joseph Fourier, 1999, Gi
R. Scott Rice-Snow, (D), Penn State, 1983, Gm
Associate Professor:
Carolyn B. Dowling, (D), Rochester, 2002, Hg
Klaus Neumann, (D), Alabama, 1999, Cl
Assistant Professor:
Shawn J. Malone, (D), Iowa, 2012, GtxGm
Emeritus:
Alan C. Samuelson, (D), Penn State, 1972, Hw
Technician, Environmental Science:
Eric Lange, (M), Ball State Univ, 2014, Cg
Geological Tech:
Michael Kutis, (B), Wisconsin (Platteville), 1989, Gg

## College of the Holy Cross

**Dept of Geosciences** (A,B) (2010)
P.O. Box 308
Notre Dame, IN 46556
p. (219) 239-8417
wcaponigri@hcc-nd.edu
Professor:
Winifred Caponigri Farquhar, (M), Notre Dame, 1971, GrcGe

## DePauw University

**Dept of Geosciences** (B) (2017)
602 South College Avenue
Julian Science Center
Greencastle, IN 46135
p. (765) 658-4654
jmills@depauw.edu
http://www.depauw.edu/acad/geosciences/
f: https://www.facebook.com/DePauwGeosciences/
Administrative Assistant: Mary M.. Donohue
*Enrollment (2015): B: 32 (17)*
Professor:
James G. Mills, (D), Michigan State, 1991, Gi
Frederick M. Soster, (D), Case Western, 1984, Gs
M. Scott Wilkerson, (D), Illinois, 1991, Gct
Associate Professor:
Tim D. Cope, (D), Stanford, 2003, GsOiGr
Jeanette K. Pope, (D), Virginia Tech, 2002, Ge
Emeritus:
James A. Madison, (D), Washington (St. Louis), 1968, Gz

## Earlham College

**Geology Dept** (B) (2016)
801 National Road West
Richmond, IN 47374-4095
p. (765) 983-1429
streeme@earlham.edu
http://www.earlham.edu/geology/
*Enrollment (2006): B: 18 (4)*
Associate Professor:
Cynthia M. Fadem, (D), Washington Univ (St. Louis), 2009, GaOsCs
Andrew Moore, (D), Washington, 1999, Gm
Meg Streepey Smith, (D), Michigan, 2001, Gt
Emeritus:
Jon W. Branstrator, (D), Cincinnati, 1975, Pg
Charles W. Martin, (D), Wisconsin, 1962, Gg

## Hanover College

**Dept of Geology** (B) (2011)
PO Box 890
Hanover, IN 47243-0890
p. (812) 866-7306
worcestr@hanover.edu
*Enrollment (2007): B: 12 (5)*
Chair:
Peter A. Worcester, (D), Miami (Ohio), 1976, Gi
Professor:
Kenneth A. Bevis, (D), Oregon State, 1995, GlmGe
Heyo Van Iten, (D), Michigan, 1989, Pg
Emeritus:
Stanley M. Totten, (D), Illinois, 1962, Gl

## Indiana State University

**Dept of Earth and Environmental Systems** (B,M,D) (2014)
159 Science Building
Terre Haute, IN 47809
p. (812) 237-2444
isu-ees@mail.indstate.edu
http://www.indstate.edu/ess/
f: www.facebook.com/isu.ees
*Enrollment (2009): B: 87 (13) M: 16 (7) D: 6 (2)*
Professor:
Gregory Bierly, (D), Michigan State, Oaw
Sandra S. Brake, (D), Colorado Mines, 1989, GeEm
Anthony Rathburn, (D), Duke, 1992, PyOu
James Speer, (D), Tennessee, 2001, PeOyGe
C. Russell Stafford, (D), Arizona State, 1981, Gam
Qihao Weng, (D), Georgia, Oru
Associate Professor:
Susan Berta, (D), Oklahoma, 1986, GmOr
Kathleen M. Heath, (D), Utah, 2001, Pb
Jennifer C. Latimer, (D), Indiana, 2004, CmGeb
Nancy J. Obermeyer, (D), Chicago, 1987, Oi
Shawn Phillips, (D), New York, 2001
Emeritus:
William Dando, (D)
Prodip K. Dutta, (D), Indiana, 1983, Gs

## Indiana University/Purdue University, Fort Wayne

**Dept of Geosciences** (B) (2015)
2101 East Coliseum Boulevard
Fort Wayne, IN 46805-1499
p. (260) 481-6249
argast@ipfw.edu
http://www.geosci.ipfw.edu/
*Enrollment (2011): B: 27 (3)*

Vice Chancellor for Academic Affairs:
Carl N. Drummond, (D), Michigan, 1994, Gs
Chair:
Anne S. Argast, (D), SUNY (Binghamton), 1986, Gdz
Professor:
Solomon A. Isiorho, (D), Case Western, 1987, HwGeOy
Associate Professor:
Benjamin F. Dattilo, (D), Cincinnati, 1994, GrPiGs
Aranzazu Pinan-Llamas, (D), Boston, 2007, GcdGp
LTL:
Mick Cseri, (M), W Michigan, 1990, Oiy
Lecturer:
Raynond F. Gildner, (D), Cornell, 1990, GgPq
Emeritus:
Dipak K. Chowdhury, (D), Texas A&M, 1961, YsGtYg
James O. Farlow, (D), Yale, 1980, PvoPe
Department Secretary:
Diana Weber
Dept. Technician:
Clarence Tennis

## Indiana University/Purdue University, Indianapolis

**Dept of Earth Sciences** (B,M) (2014)
723 West Michigan Street
Indianapolis, IN 46202-5132
p. (317) 274-7484
ibsz100@iupui.edu
http://www.geology.iupui.edu

Professor:
Andrew P. Barth, (D), S California, 1989, Gi
Gabriel M. Filippelli, (D), California (Santa Cruz), 1994, CmGb
Lin Li, (D), Brown, 2002, XgOrg
Associate Professor:
Kathy J. Licht, (D), Colorado, 1999, Gl
Joseph F. Pachut, Jr., (D), Michigan State, 1977, Po
Lenore P. Tedesco, (D), Miami, 1991, Gs
Assistant Professor:
Pierre-Andre Jacinthe, (D), Ohio State, 1995, Cl
Lecturer:
R. Jeffrey Swope, (D), Colorado, 1997, Gz
Adjunct Professor:
Timothy S. Brothers, (D), California (Los Angeles), 1985, Oy
Timothy G. Fisher, (D), Calgary, 1993, Gl
Swapan K. Ghosh, (D), Syracuse, 1975, Cl
Hendrik M. Haitjema, (D), Minnesota, 1982, Hy
Frederick W. Kleinhans, (D), Ohio State, 1971, Xg

## Indiana University Northwest

**Dept of Geosciences** (A,B) (2016)
3400 Broadway
Gary, IN 46408
p. (219) 980-6738
zkilibar@iun.edu
http://www.iun.edu/~geos/
*Enrollment (2016): B: 24 (5)*

Chair:
Zoran Kilibarda, (D), Nebraska, 1994, GsmGd
Associate Professor:
Erin Argyilan, (D), Illinois (Chicago), 2004, GeHwOw
Kristin Huysken, (D), Michigan State, 1996, GizGc
Emeritus:
Robert Votaw, (D), PiGrs

## Indiana University, Bloomington

**Dept of Geological Sciences** (B,M,D) O (2015)
1001 E. Tenth Street
Bloomington, IN 47405
p. (812) 855-5582
brophy@indiana.edu
http://geology.indiana.edu/
*Enrollment (2014): B: 67 (20) M: 33 (0) D: 39 (0)*

Chair:
Lisa M. Pratt, (D), Princeton, 1981, Co
Professor:
Abhijit Basu, (D), Indiana, 1975, Gd
David L. Bish, (D), Penn State, 1977, Gz
Simon C. Brassell, (D), Bristol, 1980, Co
James G. Brophy, (D), Johns Hopkins, 1984, Gi
Michael W. Hamburger, (D), Cornell, 1986, YsGtv
Gary L. Pavlis, (D), Washington, 1982, YsGtYe
P David Polly, (D), California (Berkeley), 1993, PvgPq
Edward M. Ripley, (D), Penn State, 1976, Em
Juergen Schieber, (D), Oregon, 1985, Gs
Robert P. Wintsch, (D), Brown, 1975, Gp
Chen Zhu, (D), Johns Hopkins, 1992
Academic Director Judson Mead Geologic Field Station:
Bruce Douglas, (D), Princeton, 1983, Gc
Senior Scientist:
Chusi Li, (D), Toronto, 1993, GiEm
Arndt Schimmelmann, (D), California (Los Angeles), 1985, CsGsPe
Associate Professor:
Claudia C. Johnson, (D), Colorado, 1993, Pe
Kaj Johnson, (D), Stanford, 2004, Yg
Associate Scientist:
Erika R. Elswick, (D), Cincinnati, 1998, Cl
Edward W. Herrmann, (D), Indiana, 2013, Sa
Peter Sauer, (D), Colorado, 1997, Pe
Assistant Professor:
Douglas A. Edmonds, (D), Penn State, 2009, Gsm
Julie C. Fosdick, (D), Stanford, 2011, Gt
Chanh Q. Kieu, (D), Maryland, 2008, Oa
Jackson K. Njau, (D), Rutgers, 2006, Pv
Paul W. Staten, (D), Utah, 2013, Oa
Laura E. Wasylenki, (D), Caltech, 1999, Cl
Lab Manager - SIRF:
Ben Underwood, (B), Indiana, 2009, Ca
Lab Manager - SESAME:
Alice Hui, (M), Indiana, 2013, Cl
Lecturer:
Cody Kirkpatrick, (D), Alabama (Huntsville), 2010, Oa
Adjunct Professor:
Maria D. Mastalerz, (D), Poland, 1988, Ec
Jeffrey White, (D), Syracuse, 1984, Hs
Emeritus:
Robert F. Blakely, (D), Indiana, 1973, Ys
J. Robert Dodd, (D), Caltech, 1961, Po
John B. Droste, (D), Illinois, 1956, Gs
Jeremy D. Dunning, (D), North Carolina, 1978, Gc
Donald E. Hattin, (D), Kansas, 1954, Gr
John M. Hayes, (D), MIT, 1966, Co
Erle G. Kauffman, (D), Michigan, 1961, Po
Enrique Merino, (D), California (Berkeley), 1973, Cl
Greg A. Olyphant, (D), Iowa, 1979, Hw
Lee J. Suttner, (D), Wisconsin, 1966, Gd
Senior Lecturer:
Bruce Douglas, (D), Princeton, 1983, Gc
Facilities Administrator:
John L. Hettle Jr.

**School of Public & Environmental Affairs** (B,M,D) (2015)
SPEA Building
Bloomington, IN 47408
p. (812) 855-7485
spea@indiana.edu
http://www.spea.indiana.edu/home/

Geology Librarian:
Christina Sheley, On
Chair:
Philip S. Stevens, (D), Harvard, 1990, Oa
Professor:
Ronald A. Hites, (D), MIT, 1968, Co
Jeffrey R. White, (D), Syracuse, 1984, Hg

Associate Professor:
Diane S. Henshel, (D), Washington, 1987, On
Todd Royer, (D), Idaho State, 1999, HsOg
Assistant Professor:
Vicky J. Meretsky, (D), Arizona, 1995, Sf
Flynn W. Picardal, (D), Arizona, 1992, So
Lecturer:
Melissa Clark, (M), Indiana, 1999, SfOg

## Indiana University, Indianapolis

**Dept of Geography** (A,B) (2011)
213 Cavanaugh Hall
Indianapolis, IN 46202
p. (317) 274-8877
tbrother@iupui.edu
http://www.iupui.edu/~geogdept/
Chair:
Jeffey S. Wilson, (D), Indiana State, 1998, Oy
Professor:
F. L. Bein, (D), Florida, 1974, Oy
Associate Professor:
Timothy S. Brothers, (D), California (Los Angeles), 1985, Oy
Catherine J. Souch, (D), British Columbia, 1990, Gm

## Purdue University

**Dept of Earth, Atmospheric, and Planetary Sciences** (B,M,D) (2017)
550 Stadium Mall Drive
West Lafayette, IN 47907-2051
p. (765) 494-3258
eas-info@purdue.edu
http://www.eaps.purdue.edu
*Enrollment (2014): M: 13 (13) D: 58 (6)*
Department Head:
Indrajeet Chaubey, (D), Oklahoma State, 1997, Hg
University Distinguished Professor:
H. Jay Melosh, (D), Caltech, 1972, XmgXy
Distinguished Professor:
John Cushman, (D), Iowa State, 1978, Sp
Paul B. Shepson, (D), Penn State, 1982, Oa
Associate Department Head:
Darryl E. Granger, (D), California (Berkeley), 1996, Oa
Harshvardhan, (D), SUNY (Stony Brook), 1976, Oa
Professor:
Ernest M. Agee, (D), Missouri, 1968, Oa
Lawrence W. Braile, (D), Utah, 1973, Ys
Maarten de Hoop, (D), Delft (Neth), 1992, Gq
Timothy R. Filley, (D), Penn State, 1997, Co
Andrew M. Freed, (D), Arizona, 1998, Gg
Andrei Gabrielov, (D), Moscow State (Russia), 1973, Ys
Alexander Gluhovsky, (D), USSR Acad of Sci, 1973, Oa
Jon M. Harbor, (D), Washington, 1990, HwGlm
Robert L. Nowack, (D), MIT, 1985, Ys
James G. Ogg, (D), Scripps, 1981, PsGsYm
Kenneth D. Ridgway, (D), Rochester, 1992, Gs
Daniel P. Shepardson, (D), Iowa, 1990, Oe
Yuch-Ning Shieh, (D), Caltech, 1968, CsGpv
Wen-Yih Sun, (D), Chicago, 1975, Oa
Terry R. West, (D), Purdue, 1966, Ng
Qianlai Zhuang, (D), Alaska (Fairbanks), Oa
William J. Zinsmeister, (D), California (Riverside), 1974, Ps
Indiana State Climatologist:
Dev Niyogi, (D), North Carolina State, 2000, Oa
Associate Professor:
Chris Andronicos, (D), Princeton, 1999, Gg
Michael Baldwin, (D), Oklahoma, 2003, Oa
Lucy M. Flesch, (D), Carnegie Inst, 2005, Yg
Hersh Gilbert, (D), Colorado, 2001, Ys
Nathaniel A. Lifton, (D), Arizona, 1997, GmCc
Greg Michalski, (D), California (San Diego), 2003, CsGeOa
Wen-wen Tung, (D), California (Los Angeles), 2002, Oa
Assistant Professor:
Julie Elliott, (D), Alaska (Fairbanks), 2011, Yd
Marty Frisbee, (D), New Mexico Tech, 2010, Hws
Saad Haq, (D), SUNY (Stony Brook), 2004, Gt
Briony Horgan, (D), Cornell, 2010, Xg
David Minton, (D), Arizona, 2009, Xg
Lisa Welp, (D), Caltech, 2006, Cs
Yutian Wu, (D), Columbia, 2011, Oa
Emeritus:
William J. Hinze, (D), Wisconsin, 1957, Ye
Arvid M. Johnson, (D), Penn State, 1965, Gc
Gerald H. Krockover, (D), Iowa, 1970, Oe
Darrell I. Leap, (D), Penn State, 1974, Hy
Phillip J. Smith, (D), Wisconsin, 1967, Oa
Thomas M. Tharp, (D), Wisconsin, 1978, Nm
Dayton G. Vincent, (D), MIT, 1969, Oa

## University of Notre Dame

**Dept of Civil & Environmental Engineering & Earth Sciences** (B,M,D) (2015)
156 Fitzpatrick Hall
Notre Dame, IN 46556
p. (219) 631-5380
ceees@nd.edu
http://www.nd.edu/~ceees
*Enrollment (2001): B: 12 (11) M: 1 (1) D: 16 (5)*
Chair:
Joannes J. Westerink, (D), MIT, 1984, On
Professor:
Peter C. Burns, (D), Manitoba, 1994, Gz
Jeremy B. Fein, (D), Northwestern, 1989, Cl
Joe Fernando, (D), Johns Hopkins, 1983, Oa
Patricia Maurice, (D), Stanford, 1994, Cl
Clive R. Neal, (D), Leeds (UK), 1985, Gi
Associate Professor:
Andrew Kennedy, (D), Monash Univ (Australia), 1998, On
Tony Simonetti, (D), Carleton, 1994, Ct
Assistant Professor:
Melissa Berke, (D), Minnesota, 2011, Cs
Diogo Bolster, (D), Calfornia (San Diego), 2007, Hw
Alan Hamlet, (D), Washington, 2006, Hs
Amy E. Hixon, (D), Clemson, 2013, Cl
David Richter, (D), Stanford, 2011, Oa
Instructor:
Stephanie Simonetti, (D), McGill, 2002, On

## University of Southern Indiana

**Dept of Geology & Physics** (B) (2016)
8600 University Boulevard
Evansville, IN 47712
p. (812) 464-1701
wselliott@usi.edu
http://www.usi.edu/science/geology-and-physics/
Administrative Assistant: Kim E. Schauss
*Enrollment (2016): B: 34 (10)*
Professor:
Joseph A. DiPietro, (D), Oregon State, 1990, Gc
Paul K. Doss, (D), N Illinois, 1991, HwGeOu
Chair:
William S. Elliott, Jr., (D), Indiana, 2002, GsrCl
Associate Professor:
James Durbin, (D), Nebraska, 1999, Gm
Tony Maria, (D), Rhode Island, 2000, Gv
Instructor:
Carrie L. Wright, (M), Wright State, 2006, Oe
Emeritus:
Norman R. King, (D), Indiana, 1973, Gr

# Iowa

## Cornell College

**Dept of Geology** (B) ● (2016)
600 First Street SW
Mount Vernon, IA 52314
p. (319) 895-4306
rdenniston@cornellcollege.edu
http://www.cornellcollege.edu/geology
*Enrollment (2016): B: 18 (8)*
Chair:
Emily O. Walsh, (D), California (Santa Barbara), 2003, GptCc
Professor:
Rhawn F. Denniston, (D), Iowa, 2000, GeCs

Benjamin J. Greenstein, (D), Cincinnati, 1990, GuPie

## Drake University

**Dept of Environmental Science and Policy** (B) (2005)
Des Moines, IA 50311
p. (515) 271-2803
thomas.rosburg@drake.edu

## Iowa State University of Science & Technology

**Dept of Agronomy** (B,M,D) (2014)
2101 Agronomy Hall
Ames, IA 50011-1010
p. (515) 294-1360
slf@iastate.edu
http://www.agron.iastate.edu/
Department Secretary: Pam Hinderaker

Director:
Dennis R. Keeney, (D), Iowa, 1965, So
Chair:
Kendall Lamkey, (D), Iowa State, 1985
Professor:
Raymond W. Arritt, (D), Colorado State, 1985, Oa
Richard E. Carlson, (D), Iowa State, 1971, Ow
Richard M. Cruse, (D), Minnesota, 1978, So
V. P. (Bill) Evangelou, (D), California (Davis), 1981, Sc
William J. Gutowski, (D), MIT, 1984, Ow
Robert Horton, (D), New Mexico State, 1981, Sp
Douglas L. Karlen, (D), Kansas State, 1978, So
Randy J. Killorn, (D), Idaho, 1983, So
Thomas E. Loynachan, (D), North Carolina State, 1975, So
Thomas B. Mooreman, (D), Washington State, 1983, So
Jonathan A. Sandor, (D), California (Berkeley), 1983, Sd
John W. Schafer, Jr., (D), Michigan State, 1968, So
Eugene S. Takle, (D), Iowa State, 1971, Oaw
M. Ali Tatabatai, (D), Iowa State, 1965, Sc
Elwynn Taylor, (D), Washington (St. Louis), 1970, Oa
Michael L. Thompson, (D), Ohio State, 1980, Sc
Regis D. Voss, (D), Iowa State, 1962, So
Douglas N. Yarger, (D), Arizona, 1967, Oa
Associate Professor:
Cynthia Cambardella, (D), Colorado, 1991, So
Thomas A. Kaspar, (D), Iowa State, 1982, So
David A. Laird, (D), Iowa State, 1987, Sc
Antonio W. Mallarino, (D), Iowa State, 1991, So
Assistant Professor:
Lee Burras, (D), Ohio State, 1992, Sd
Larry Halverson, (D), Wisconsin, 1991, So
Stanley J. Henning, (D), Oregon State, 1975, So
Thomas A. Polito, (D), Iowa State, 1987, So
Associate Dean, College of Agriculture:
Gerald A. Miller, (D), Iowa State, 1974, SdaSo
Geology Librarian:
Peter A. Peterson, (D), Illinois, 1953, On

**Dept of Geological & Atmospheric Sciences** (B,M,D) ○ (2017)
253 Science I
2237 Osborn Drive
Ames, IA 50011-3212
p. (515) 294-4477
geology@iastate.edu
http://www.ge-at.iastate.edu/
f: https://www.facebook.com/ISUgeology
Administrative Assistant: DeAnn Frisk
*Enrollment (2015): B: 166 (30) M: 23 (5) D: 20 (0)*

Director:
Aaron R. Wood, (D), Michigan, 2009, PveGr
Chair:
William W. Simpkins, (D), Wisconsin, 1989, HwGlCs
Professor:
Igor A. Beresnev, (D), USSR Acad of Sci, 1986, YgsYe
Cinzia C. Cervato, (D), ETH (Switzerland), 1990, GgOe
Tsing-Chang Chen, (D), Michigan, 1975, Oa
William A. Gallus, (D), Colorado State, 1993, Oa
William J. Gutowski, (D), MIT, 1984, Oa
Neal R. Iverson, (D), Minnesota, 1989, Gl
Paul G. Spry, (D), Toronto, 1984, Em
Eugene S. Takle, (D), Iowa State, 1971, Oa
Xiaoqing Wu, (D), California (Los Angeles), 1992, Oa
Associate Professor:
Kristie Franz, (D), California (Irvine), 2006, Hs
Chris Harding, (D), Houston, 2001, Gq
Alan D. Wanamaker, (D), Maine, 2007, CsPeOg
Assistant Professor:
Beth E. Caissie, (D), Massachusetts, 2012, Pe
Franciszek J. Hasiuk, (D), Michigan, 2008, GdoCl
Jacqueline Reber, (D), Oslo, 2012, Gct
Yuyu Zhou, (D), Rhode Island, 2008, Oyi
Senior Lecturer:
James V. Aanstoos, (D), Purdue, 1996, Or
Jane P. Dawson, (D), New Mexico, 1995, Gp
David M. Flory, (M), Iowa State, 2003, OwaGt
Lecturer:
Rachindra Mawalagedara, (D), Nebraska (Lincoln), 2013, Oa
Adjunct Professor:
Michael R. Burkart, (D), Iowa, 1976, Hy
Emeritus:
Robert Cody, (D), Colorado, 1968, Gz
Carl E. Jacobson, (D), California (Los Angeles), 1980, Gc
Karl E. Seifert, (D), Wisconsin, 1963, Ct
Carl F. Vondra, (D), Nebraska, 1963, Gr
Kenneth E. Windom, (D), Penn State, 1976, Cp
Douglas N. Yarger, (D), Arizona, 1967, Oa
Teaching Lab Coordinator:
Mark E. Mathison, (M), Iowa State, 2000, Oe

## Scott Community College

**Environmental Science** (2014)
500 Belmont Road
Bettendorf, IA 52722
p. 563-441-4001
eiccinfo@eicc.edu
http://www.eicc.edu

## University of Iowa

**Earth & Environmental Sciences** (B,M,D) ● (2017)
115 Trowbridge Hall
123 N. Capitol Street
Iowa City, IA 52242
p. (319) 335-1818
geology@uiowa.edu
http://clas.uiowa.edu/ees/
f: https://www.facebook.com/UIowaGeoscience/?fref=nf
Administrative Assistant: Angela Bellew
*Enrollment (2016): B: 56 (14) M: 19 (6) D: 17 (4)*

Chair:
Charles T. Foster Jr., (D), Johns Hopkins, 1975, GptGg
Professor:
Jonathan M. Adrain, (D), Alberta, 1993, Pi
E. Arthur Bettis III, (D), Iowa, 1995, SdGmSa
Christopher A. Brochu, (D), Texas, 1997, Pv
Jane A. Gilotti, (D), Johns Hopkins, 1987, GctGp
William C. McClelland, (D), Arizona, 1990, GtCcGc
David W. Peate, (D), Open Univ (UK), 1989, CgGi
Mark K. Reagan, (D), California (Santa Cruz), 1987, Gi
Associate Professor:
Jeffrey A. Dorale, (D), Minnesota, 2001, Cs
Ingrid Ukstins Peate, (D), Royal Holloway, 2003, GviXg
Frank H. Weirich, (D), Toronto, 1982, Gm
Assistant Professor:
William Barnhart, (D), Cornell, 2013, Gt
Bradley D. Cramer, (D), Ohio State, 2009, GrPsGs
Emily Finzel, (D), Purdue, 2010, GsrGt
Adjunct Assistant Professor:
Caroline A. Davis, (D), Missouri-Rolla, 2009, Hw
Lecturer:
Mary Kosloski, (D), Cornell, 2012, Pio
Kate Tierney, (D), Ohio State, 2010, GrCsGg
Adjunct Associate Professor:
Brian J. Witzke, (D), Iowa, 1981, Ps
Adjunct Assistant Professor:
Raymond R. Anderson, (D), Iowa, 1992, Gr
Rhawn F. Denniston, (D), Iowa, 2000, Cs

Keith E. Schilling, (D), Iowa, 2009, Hgs
Douglas J. Schnoebelen, (D), Indiana, 1999, Hg
Emily O. Walsh, (D), California (Santa Barbara), 2003, GptCc
Adjunct Professor:
David L. Campbell, (D), California (Berkeley), 1969, Yg
Emeritus:
Richard G. Baker, (D), Colorado, 1969, Pl
Ann F. Budd, (D), Johns Hopkins, 1978, Poi
Robert S. Carmichael, (D), Pittsburgh, 1967, YexGg
Lon D. Drake, (D), Ohio State, 1968, Hy
Philip H. Heckel, (D), Rice, 1966, GrPsGd
George R. McCormick, (D), Ohio State, 1964, Gz
Holmes A. Semken, Jr., (D), Michigan, 1965, Pv
Keene Swett, (D), Edinburgh, 1965, Gs
You-Kuan Zhang, (D), Arizona, 1990, Hw
Rock Lab Manager:
Matthew J. Wortel, (M), Iowa, 2007, Gx
Collections Manager, Paleontology:
Tiffany S. Adrain, (M), Iowa, 2003, Pg
Department Secretary:
Christine K. Harms, (N)



## University of Northern Iowa

**Dept of Earth and Environmental Sciences** (B) (2016)
121 Latham Hall
Cedar Falls, IA 50614-0335
p. (319) 273-2759
siobahn.morgan@uni.edu
http://www.earth.uni.edu/
f: https://www.facebook.com/pages/University-of-Northern-Iowa-Earth-and-Environmental-Science/56860203682
t: @uni_earthsci
Administrative Assistant: Nora Janssen
*Enrollment (2016): B: 95 (16)*
Head:
Siobahn M. Morgan, (D), Washington, 1991, On
Professor:
Alan C. Czarnetzki, (D), Wisconsin, 1992, Ow
Thomas A. Hockey, (D), New Mexico State, 1988, On
Mohammad Z. Iqbal, (D), Indiana, 1994, Hw
Associate Professor:
Kyle R. Gray, (D), Akron, 2009, OeGg
Chad E. Heinzel, (D), N Illinois, 2005, GmaSa
Assistant Professor:
Alexa Sedlacek, (D), Ohio State, 2013, Cs
Xinhua Shen, (D), Colorado State, 2011, Oaw
Instructor:
Paula Even, (M), N Iowa, 2005, On
Lee S. Potter, (D), Texas, 1996, Gz
Aaron Spurr, (M), N Iowa, 1997, Oeg
Michael Stevens, (M), N Iowa, 1992, Oe
Emeritus:
Wayne I. Anderson, (D), Iowa, 1964, Pg
Lynn A. Brant, (D), Penn State, 1980, Gn
Timothy M. Cooney, (D), N Colorado, 1976, Oe
Walter E. De Kock, (D), Ohio State, 1972, Oe
Kenneth J. De Nault, (D), Stanford, 1974, Gz
James C. Walters, (D), Rutgers, 1975, Gm
Lab Tech:
Steven J. Smith, (M), N Iowa, 2000, Og

# Kansas

## Emporia State University

**Earth Science Program** (B,M) ○ (2016)
1 Kellogg Circle
Emporia, KS 66801-5087
p. (620) 341-5330
mschulme@emporia.edu
http://www.emporia.edu/earthsci/
*Enrollment (2016): B: 24 (7) M: 18 (4)*
Head:
Marcia K. Schulmeister, (D), Kansas, 2000, HwClGe
Professor:
James S. Aber, (D), Kansas, 1978, Gl
Kenneth W. Thompson, (D), Iowa State, 1991, Oe
Associate Professor:
Michael A. Morales, (D), California (Berkeley), 1987, PvGrg
Richard O. Sleezer, (D), Kansas, 2001, Os
Assistant Professor:
Alivia J. Allison, (D), Missouri (Kansas City), 2013, GgaGm
Emeritus:
Paul J. Johnston, (M), Kansas, Gg

## Fort Hays State University

**Dept of Geosciences** (B,M) ● (2016)
600 Park Street
Tomanek Hall
Hays, KS 67601-4099
p. (785) 628-5389
geosciences@fhsu.edu
http://www.fhsu.edu/geo
f: https://www.facebook.com/GeoFHSU/
t: @GeoFHSU
Administrative Assistant: Patricia Duffey
*Enrollment (2015): B: 154 (18) M: 43 (3)*
Emeritus:
Paul Phillips, (D), Kansas, 1977, Oe
Professor:
Kenneth R. Neuhauser, (D), South Carolina, 1973, Gc
Chair:
P. Grady Dixon, (D), Arizona State, 2005, OwyOa
Associate Professor:
Richard Lisichenko, (D), Kansas State, 2000, Oi
Tom Schafer, (D), Kansas State, 2000, Oyi
Assistant Professor:
Hendratta N. Ali, (D), Oklahoma State, 2010, GogYs
Keith A. Bremer, (D), Texas State Univ, 2011, OuiOn
Laura E. Wilson, (D), Colorado (Boulder), 2012, PveGs
Chunfu Zhang, (D), Florida State, 2011, Ggz
Instructor:
Eamonn Coveney, (M), Fort Hays State, 2006, Oyn
Amelia A. Fox, (D), Mississippi State, 2015, Or
William H. Heimann, (M), Fort Hays State, 1987, HgwGm
Kara Kuntz, (M), Kansas State, 1990, On
Emeritus:
Richard J. Zakrzewski, (D), Michigan, 1968, PvsPe
Cooperating Faculty:
Joseph R. Thomasson, (D), Iowa State, 1976, Pb

## Johnson County Community College

**Science Division** (A) (2017)
12345 College Blvd
Overland Park, KS 66210
p. (913) 469-3826
csilla@jccc.edu
http://www.jccc.edu/academics/math-science/index.html
*Enrollment (2015): A: 7 (2)*
Professor:
Lynne Beatty, (M), S Illinois Univ, 1985, GgOy
Assistant Professor:
John P. Harty, (D), Oy
Adjunct Professor:
John Maher, (D), Maryland (College Park), 1999, Oy

## Kansas State University

**Dept of Agronomy** (B,M,D) (2016)
2004 Throckmorton
Manhattan, KS 66506
p. (785) 532-6101
agronomy@ksu.edu
http://www.agronomy.k-state.edu/
f: https://www.facebook.com/kstate.agronomy
t: @KStateAgron
*Enrollment (2014): B: 70 (33) M: 41 (9) D: 43 (10)*
Depament Head:
Gary M. Pierzynski, (D), Ohio State, 1989, Sc
Professor:
Stewart Duncan, (D), Kansas State, 1991, So
Walter H. Fick, (D), Texas Tech, 1978, Sf
Dale Fjell, (D), Kansas State, 1982, So
Allan Fritz, (D), Kansas State, 1994, So

Mary Beth Kirkham, (D), Wisconsin, Sp
Gerard J. Kluitenberg, (D), Iowa State, 1989, Sp
David Mengel, (D), Purdue, 1975, So
Clenton E. Owensby, (D), Kansas State, 1969, Sf
Dallas Peterson, (D), North Dakota State, 1987, So
Vara Prasad, (D), Reading (United Kingdom), 1999, So
Michel D. Ransom, (D), Ohio State, 1984, SdcGm
Chuck W. Rice, (D), Kentucky, 1983, Sb
Bill T. Schapaugh, (D), Purdue, 1979, So
Alan Schlegel, (D), Purdue, 1985, So
Phillip Stahlman, (D), Wyoming, 1989, So
Daniel Sweeny, (D), Florida, So
Curtis Thompson, (D), Idaho, 1993, So
Steve M. Welch, (D), Michigan State, 1977, Sp

Associate Professor:
Robert Aiken, (D), Michigan State, 1992, SoYhHq
Ignacio Ciampitti, (D), Purdue, 2012, So
Gary Cramer, (D), Nebraska, 1998, So
Ganga Hettiarachchi, (D), Kansas State, 2000, Sc
John Holman, (D), Idaho, 2005, So
Doo-Hong Min, (D), Maryland, 1998, So
Nathan Nelson, (D), North Carolina State, 2004, So
DeAnn Presley, (D), Kansas State, 2007, So
Kraig Roozeboom, (D), Kansas State, 2006, So
Dorivar Ruiz-Diaz, (D), Iowa State, 2007, So
Gretchen Sassenrath, (D), Illinois, 1988, So
Tesfaye Tesso, (D), Kansas State, 2002, So

Assistant Professor:
Eric Adee, (D), Wisconsin, 1993, So
Lucas Haag, (D), Kansas State, 2013, So
Mithila Jugulam, (D), Guelph, 2004, So
Xiaomao Lin, (D), Nebraska, 1999, On
Colby Moorberg, (D), North Carolina, 2014, So
Geoffrey Morris, (D), Chicago, 2007, So
Augustine Obour, (D), Florida, 2010, So
Ram Perumal, (D), Tamil Nadu Ag Univ, 1993, So
Eduardo Santos, (D), Guelph, 2011, So
Peter Tomlinson, (D), Arkansas, 2006, So
Guorong Zhang, (D), North Dakota State, 2007, So

Emeritus:
Mark Claassen, (D), Iowa State, 1971, So
Bill Eberle, (D), Illinois, 1973, SoOu
Stan Ehler, (D), Missouri, 1975, So
Barney Gordon, (D), South Dakota, 1990, So
Keith Janssen, (D), Michigan State, 1973, So
George Liang, (D), Wisconsin, 1965, So
Gerry L. Posler, (D), Iowa State, 1969, Sf
Kevin Price, (D), Utah, 1987, Or
David Regehr, (D), Illinois, 1975, So
Jim Shroyer, (D), Iowa State, 1980, So
Loyd Stone, (D), South Dakota, 1973, Sp
Steve J. Thien, (D), Purdue, 1971, Sc
Richard Vanderlip, (D), So
D.A. Whitney, (D), Iowa State, 1966, So

Agronomist:
Doug Shoup, (D), Kansas State, 2006, So

Librarian:
Nancy William, (N), On

**Dept of Geology** (B,M) (2013)
108 Thompson Hall
Manhattan, KS 66506-3201
p. (785) 532-6724
rocknrat@ksu.edu
http://www.k-state.edu/geology/
Administrative Assistant: Lori Page-Willyard
*Enrollment (2013): B: 69 (15) M: 27 (5)*

Department Head:
Pamela D. Kempton, (D), S Methodist, 1984, GiCt

Professor:
Sambhudas Chaudhuri, (D), Ohio State, 1966, CcGoe
George R. Clark, II, (D), Caltech, 1969, Poy

Associate Professor:
Allen W. Archer, (D), Indiana, 1983, Gr
Matthew E. Brueseke, (D), Miami Univ, 2006, GitGv
Abdelmoneam E. Raef, (D), AGH (Poland), 2001, YesOr
Matthew W. Totten, (D), Oklahoma, 1992, Gso

Assistant Professor:
Saugata Datta, (D), W Ontario, 2001, ClHwGe
Keith B. Miller, (D), Rochester, 1988, Pe
Joel Q.G. Spencer, (D), Glasgow (UK), 1996, CcGs

Emeritus:
Robert L. Cullers, (D), Wisconsin, 1971, Ct
Charles G. Oviatt, (D), Utah, 1984, Gm
Ronald R. West, (D), Oklahoma, 1970, PoeGs

## University of Kansas

**Dept of Geology** (B,M,D) (2017)
1475 Jayhawk Boulevard
Room 120
Lindley Hall
Lawrence, KS 66045-7613
p. (785) 864-4974
geology@ku.edu
http://geo.ku.edu/
f: https://www.facebook.com/KUGeology?ref=hl
*Enrollment (2013): B: 72 (14) M: 50 (17) D: 34 (9)*

Chair:
Luis A. Gonzalez, (D), Michigan, 1989, CsGd

Union Pacific Distinguished Professor:
J. Douglas Walker, (D), MIT, 1985, Gc

Ritchie Distinguished Professor:
Michael D. Blum, (D), Texas (Austin), 1997, GsrGm

Gulf-Hedberg Distinguished Professor:
Paul A. Selden, (D), Cambridge (UK), 1979, Pio

Courtesy Professor:
Rolfe Mandel, (D), Kansas, 1991, Gam

Associate Dean of Mathematics and Natural Sciences:
Robert H. Goldstein, (D), Wisconsin, 1986, Gs

Professor:
J. Rick Devlin, (D), Waterloo, 1994, Hw
Evan K. Franseen, (D), Wisconsin, 1989, Gd
Stephen T. Hasiotis, (D), Colorado, 1997, Pe
Mary C. Hill, (D), Princeton, 1985, Hgq
Jennifer A. Roberts, (D), Texas (Austin), 2000, PyCl

Hubert H. and Kathleen M. Hall Professor of Geology:
Gene Rankey, (D), Kansas, 1996, GsrGu

Associate Professor:
Ross A. Black, (D), Wyoming, 1989, Ye
David A. Fowle, (D), Notre Dame, 2000, Py
Diane Kamola, (D), Georgia, 1989, Gs
Gwen L. Macpherson, (D), Texas, 1989, Hw
Craig Marshall, (D), Technology (Australia), 2001, GzOm
Leigh Stearns, (D), Maine, 2008, Ol
Michael H. Taylor, (D), California (Los Angeles), 2004, Gt
George P. Tsoflias, (D), Texas (Austin), 1999, YgeGe
Anthony W. Walton, (D), Texas (Austin), 1972, GsEoGv

Assistant Professor:
Noah McLean, (D), MIT, 2012, CcGq
Andreas Möller, (D), Christian-Albrechts (Germany), 1996, Gt
Alison Olcott Marshall, (D), S California, 2006, Py
Randy Stotler, (D), Waterloo, 2008, Hw
Chi Zhang, (D), Rutgers, 2012, Ygx

Courtesy Professor:
John H. Doveton, (D), Edinburgh, 1969, Gq
William C. Johnson, (D), Wisconsin, 1976, Grm
Leonard Krishtalka, (D), Kansas, 1976, Pv
Bruce S. Lieberman, (D), Columbia, 1994, Poi
Gregory A. Ludvigson, (D), Iowa, 1988, Gs
Richard D. Miller, (M), Kansas, 1983, Ye
Edith Taylor, (D), Ohio State, 1983, Pb
Thomas N. Taylor, (D), Illinois, 1964, Pb
W. Lynn Watney, (D), Kansas, 1985, Gr
Donald O. Whittemore, (D), Penn State, 1973, Cl

Courtesy Associate Professor:
Geoff Bohling, (D), Kansas, 1999, Hw
Gaisheng Liu, (D), Alabama, 2004, Hg

Courtesy Assistant Professor:
Andrea Brookfield, (D), Waterloo, 2009, Hw
Jon Smith, (D), Kansas, 2008, GsPe

Adjunct Professor:
Timothy R. Carr, (D), Wisconsin, 1981, Go
John Gosse, (D), Lehigh, 1994, Gm
Daniel F. Stockli, (D), Stanford, 1999, Gc

Visiting Professor:
Kelsey S. Bitting, (D), Rutgers, 2013, Oe
Emeritus:
Ernest E. Angino, (D), Kansas, 1961, Cl
Wakefield Dort, Jr., (D), Stanford, 1955, Gm
Gisela Dreschoff, (D), Tech (Braunschweig), 1972, Ce
Paul Enos, (D), Yale, 1965, GsdGu
Lee C. Gerhard, (D), Kansas, 1964, GsoGd
Carl D. McElwee, (D), Kansas, 1970, Hg
Richard A. Robison, (D), Texas (Austin), 1962, PiGr
Albert J. Rowell, (D), Leeds, 1953, Pi
Don W. Steeples, (D), Stanford, 1975, YseGe
W. Randall Van Schmus, (D), California (Los Angeles), 1964, Cc

## Wichita State University
**Dept of Geology** (B,M) (2011)
1845 Fairmount
Wichita, KS 67260-0027
p. (316) 978-3140
john.gries@wichita.edu
Administrative Assistant: K. L. Smith
Chair:
William C. Parcell, (D), Alabama, 2000, Gr
Professor:
William D. Bischoff, (D), Northwestern, 1985, Cl
John C. Gries, (D), Texas, 1970, Gt
Salvatore J. Mazzullo, (D), Rensselaer, 1974, Go
Associate Professor:
Collette D. Burke, (D), Wisconsin (Milwaukee), 1983, Pm
Assistant Professor:
Hongsheng Cao, (D), Florida State, 2001, Cl
Wan Yang, (D), Texas, 1999, Gs
Lecturer:
Toni K. Jackman, (M), Wichita State, 1984, Ge
David L. Schaffer, (B), Utah, 1982, Ow
Emeritus:
James N. Gundersen, (D), Minnesota, 1958, Ga
Daniel F. Merriam, (D), Kansas, 1961, Gr
Peter G. Sutterlin, (D), Northwestern, 1958, Gd

# Kentucky

## Alice Lloyd College
**Div of Natural Sciences & Mathematics** (2004)
Pippa Passes, KY 41844-9701
p. (606) 368-2101 x5405

## Bluegrass Community and Technical College
**Environmental Science Technology Program** (2015)
470 Cooper Dr
Lexington, KY 40506
p. (859) 246-6448
jean.watts@kctcs.edu
http://www.bluegrass.kctcs.edu/Natural_Sciences/Environmental_Science_Technology.aspx

## Eastern Kentucky University
**Dept of Geosciences** (B) (2015)
521 Lancaster Avenue
Roark 103
Richmond, KY 40475-3102
p. (859) 622-1273
melissa.dieckmann@eku.edu
http://www.geoscience.eku.edu
*Enrollment (2013): B: 102 (22)*
Chair:
Melissa S. Dieckmann, (D), Notre Dame, 1995, OeCg
Professor:
Walter S. Borowski, (D), North Carolina, 1998, CgGo
Stewart S. Farrar, (D), SUNY (Binghamton), 1976, Gpt
John C. White, (D), Baylor, 2002, GiCgt
Donald M. Yow, (D), South Carolina, 2003, OwyOg
David Zurick, (D), Hawaii, 1986, On
Associate Professor:
French T. Huffman, (D), Connecticut, 2006, Oi
Assistant Professor:
Robert T. Lierman, (D), George Washington, 1995, Gsr
Kelly Watson, (D), Florida State, 2012, Or
Lecturer:
Glenn A. Campbell, (M), Marshall, 1995, Oy
Sonja H. Yow, (D), Kentucky, 2008, On
Emeritus:
Gary L. Kuhnhenn, (D), Illinois, 1976, Gd

## Morehead State University
**Dept of Earth and Space Sciences** (B) (2014)
235 Martindale Drive
Morehead, KY 40351
p. (606) 783-2381
c.mason@moreheadstate.edu
http://www.moreheadstate.edu/physsci/
Administrative Assistant: Amanda Holbrook
*Enrollment (2014): B: 40 (4)*
Department Chair:
Benjamin K. Malphrus, (D), West Virginia, 1990
Professor:
Charles E. Mason, (M), George Washington, 1981, PiGrs
Associate Professor:
Marshall Chapman, (D), Massachusetts, 1996, Gv
Eric Jerde, (D), California (Los Angeles), 1991, Cg
Jennifer O'Keefe, (D), Kentucky, 2008, PlEcOe
Steven K. Reid, (D), Texas A&M, 1991, Gs

## Murray State University
**Dept of Geosciences** (B,M) O (2014)
334 Blackburn Hall
Murray, KY 42071
p. (270) 809-2591
george.kipphut@murraystate.edu
http://www.murraystate.edu/geosciences
*Enrollment (2012): B: 44 (9) M: 8 (4)*
Chair:
George W. Kipphut, (D), Columbia, 1978, GeOg
Qiaofeng (Robin) Zhang, (D), W Ontario, 2002, OirOu
Professor:
Haluk Cetin, (D), Purdue, 1993, OiGe
Kit W. Wesler, (D), North Carolina, 1981, Oni
Assistant Professor:
Sung-ho Hong, (D), New Mexico Tech, 2008, HyGe
Research Associate:
Jane L. Benson, (M), Murray State, 1986, Oi
Instructor:
Anthony L. Ortmann, (D), Tulane, 2007, OnYg
Adjunct Professor:
Michael R. Busby, (M), Murray State, 1996, Oi

## Northern Kentucky University
**Dept of Geology** (B) O (2016)
204H Natural Sciences Center
Highland Heights, KY 41099
p. (859) 572-5309
rockawayj@nku.edu
http://nku.edu/
*Enrollment (2016): B: 50 (9)*
Associate Professor:
Janet Bertog, (D), Cincinnati, 2002, Gd
Samuel Boateng, (D), Missouri (Rolla), 1996, Hw
John D. Rockaway, (D), Purdue, 1968, Ng
Lecturer:
Reuben G. Bullard, (M), Cincinnati, 2000, Gd
Sarah E. Johnson, (M), Purdue, 1997, Ng

## Owensboro Community and Technical College
**Owensboro Community and Technical College** (1999)
4800 New Hartford Road
Owensboro, KY 42303
http://www.owensboro.kctcs.edu/

## University of Kentucky
**Dept of Earth and Environmental Sciences** (B,M,D) O (2017)

101 Slone Research Building
121 Washington St.
Lexington, KY 40506-0053
p. (859) 257-3758
moker@uky.edu
http://ees.as.uky.edu
*Enrollment (2016): B: 52 (9) M: 29 (7) D: 12 (2)*

Chair:
David P. Moecher, (D), Michigan, 1988, GxtCg

Professor:
Frank R. Ettensohn, (D), Illinois, 1975, PsGdPg
James C. Hower, (D), Penn State, 1978, Ec
Dhananjay Ravat, (D), Purdue, 1989, YgmYv
Edward W. Woolery, (D), Kentucky, 1998, Ys

Associate Professor:
Alan E. Fryar, (D), Alberta, 1992, HwGeCl
Kevin M. Yeager, (D), Texas A&M, 2002, GsCmc

Assistant Professor:
Sean Bemis, (D), Oregon, 2010, GctYs
Andrea M. Erhardt, (D), Stanford Univ, 2013, CgsCm
Rebecca Freeman, (D), Tulane, 2011, GrPi
Sora L. Kim, (D), California (Santa Cruz), 2010, CsPeo
Michael M. McGlue, (D), Arizona, 2011, Gso
Keely A. O'Farrell, Toronto, 2013, Ygd
Ryan Thigpen, (D), Virginia Tech, 2009, Gtc

Lecturer:
Kent Ratajeski, (D), North Carolina, 1999, Gi

Adjunct Professor:
J. Richard Bowersox, (D), South Florida, 2006, GoPq
Cortland F. Eble, (D), West Virginia, 1988, Pl
Stephen F. Greb, (D), Kentucky, 1992, GsEcGr
Christopher Groves, (D), Virginia, 1993, Hg
Hickman B. John, (D), Kentucky, 2011, Gto
Thomas M. Parris, (D), California (Santa Barbara), 1998, Cg
Thomas Robl, (D), Kentucky, 1977, EcCo
Zhenming Wang, (D), Kentucky, 1998, Yse
Junfeng Zhu, (D), Arizona, 2005

Emeritus:
William H. Blackburn, (D), MIT, 1967, GxCg
Bruce R. Moore, (D), Melbourne, 1967, Gd
Kieran D. O'Hara, (D), Brown, 1984, Gc
Lyle V. A. Sendlein, (D), Iowa State, 1964, Hw
Ronald L. Street, (D), St. Louis, 1975, Ys
William A. Thomas, (D), Virginia Tech, 1960, GtcGr

Laboratory Director:
Peter J. Idstein, (M), Eastern Kentucky, 1992, Gg

**Dept of Mining Engineering** (B,M,D) (2015)
230 Mining & Mineral Resources Building
Lexington, KY 40506-0107
p. (859) 257-8026
rick.honaker@uky.edu
http://www.engr.uky.edu/mng/
t: @UK_Mining
*Enrollment (2015): B: 125 (36) M: 14 (7) D: 17 (3)*

Chair:
Rick Honkaer, (D), Virginia Tech, 1992, Nm

Professor:
Zach Agioutantis, (D), Virginia Tech, 1987, NrmEc
Braden Lusk, (D), Missouri (Rolla), 2006, Nm
Thomas Novak, (D), Penn State, 1984, Nm
Joseph Sottile, Jr., (D), Penn State, 1991, Nm

Assistant Professor:
Kyle Perry , (D), Kentucky, 2010, Nr
Jhon Silva-Castro, (D), Kentucky, 2012, Nm
William Chad Wedding, (D), Kentucky, 2014, Nm

Emeritus:
Kot F. Unrug, (D), Academy of Mining-Metallurgy, 1966, Nr
Andrew M. Wala, (D), Academy of Mining-Metallurgy, 1972, Nm

## Western Kentucky University

**Dept of Geography & Geology** (B,M) ○ (2016)
1906 College Heights Blvd
#31066
Bowling Green, KY 42101-1066
p. (270) 745-4555
david.keeling@wku.edu
http://www.wku.edu/geoweb
*Enrollment (2016): B: 110 (33) M: 25 (9)*

Head:
David J. Keeling, (D), Oregon, 1992, Og

Professor:
Catherine Algeo, (D), Louisiana State, 1997, Oi
Stuart Foster, (D), Ohio State, 1988, Oa
Christopher Groves, (D), Virginia, 1993, Hg
Rezaul Mahmood, (D), Oklahoma, 2000, Oa
Michael May, (D), Indiana, 1993, Ge

Associate Professor:
Josh Durkee, (D), Georgia, Oa
Xingang Fan, (D), Lanzhou, Oa
M. Royhan Gani, (D), Texas (Dallas), 2005, GsrGg
Greg Goodrich, (D), Arizona, Oa
Margaret "Peggy" Gripshover, (D), Tennessee, On
Jason Polk, (D), South Florida, Gm
Fredrick D. Siewers, (D), Illinois, 1995, Pg
Andrew Wulff, (D), Massachusetts, 1993, Gv
Jun Yan, (D), Buffalo, 2004, Oi

Assistant Professor:
Nahid Gani, (D), Texas, Gc
Leslie North, (D), South Florida, Og

Instructor:
William Blackburn, (M), W Kentucky, 2002, Oy
Kevin Cary, (M), W Kentucky, 2000, Oi
Margaret Crowder, (D), WKU, 2012, Oe
Scott Dobler, (M), Bowling Green, 1996, Oa
Patricia Kambesis, (D), Mississippi State, 2014, OigHw

Instructor:
Amy Nemon, (M), WKU, 2005, Oig

Emeritus:
Nicholas Crawford, (D), Clark, 1977, Hg

# Louisiana

## Centenary College of Louisiana

**Dept of Geology** (B) (2016)
2911 Centenary Boulevard
Shreveport, LA 71104
p. (318) 869-5234
dbieler@centenary.edu
*Enrollment (2016): B: 12 (5)*

Professor:
Scott K. Vetter, (D), South Carolina, 1989, Gi

Associate Professor:
David B. Bieler, (D), Illinois, 1983, GtrYe

## Delgado Community College

**Science & Math Div** (A) (2014)
615 City Park Avenue
New Orleans, LA 70119
p. (504) 671-6480
jwood@dcc.edu
http://www.dcc.edu/divisions/sciencemath/

Professor:
Jacqueline Wood, (M), New Orleans, Gg

## Louisiana State University

**Dept of Geography & Anthropology** (B,M,D) (2011)
227 Howe-Russell Geoscience Complex
Baton Rouge, LA 70803
p. (225) 578-5942
gachair@lsu.edu
http://www.ga.lsu.edu
*Enrollment (2006): B: 54 (23) M: 12 (6) D: 22 (7)*

Chair:
Kevin Robbins

Professor:
Patrick A. Hesp, (D), Sydney, 1981, Gm
Richard H. Kesel, (D), Maryland, 1971, Hg

Associate Professor:
Steven Namikas, (D), S California, 1999, Gm

**Dept of Geology & Geophysics** (B,M,D) ○ (2017)

Louisiana

E235 Howe Russell Kniffen Geoscience Complex
Baton Rouge, LA 70803-4101
p. (225) 578-3353
geology@lsu.edu
www.lsu.edu/science/geology/
f: http://www.facebook.com/LSUGeology
t: @LSUGeology
*Enrollment (2015): B: 158 (26) M: 38 (7) D: 29 (4)*

Chair:
Carol M. Wicks, (D), Virginia, 1992, HwClHy
Professor:
Huiming Bao, (D), Princeton, 1998, Cs
Samuel J. Bentley, (D), SUNY (Stony Brook), 1998, Gu
Peter Clift, (D), Edinburgh, 1993, GsuGo
Peter Doran, (D), Hgs
Barbara L. Dutrow, (D), S Methodist, 1985, Gz
Brooks B. Ellwood, (D), Rhode Island, 1977, Ym
Darrell J. Henry, (D), Wisconsin, 1981, Gp
Associate Professor:
Philip J. Bart, (D), Rice, 1998, Gr
Juan M. Lorenzo, (D), Columbia, 1991, Ys
Sophie Warny, (D), Catholic (Belgium), 1999, Pl
Field Camp Director:
Amy Luther, (D), New Mexico Tech, Gc
Assistant Professor of Research:
Yongbo Peng, (D), LSU, Cg
Assistant Professor:
Achim Herrmann, (D), Penn State, 2005, PeGd
Suniti Karuntillake, (D), Cornell, 2008, Xg
Karen Luttrell, (D), Scripps, Yg
Patricia Persaud, (D), Caltech, Ys
Jianwei Wang, (D), Illinois (Urbana-Champaign), ClOmGy
Carol A. Wilson, (D), Boston Univ, 2013, GsOnCc
Guangsheng Zhuang, (D), California (Santa Cruz), 2011, CocGs
Instructor:
Yanxia Ma, (D), Og
Associate Curator, LSU Museum of Natural Science:
Judith A. Schiebout, (D), Texas, 1973, Pv
Emeritus:
Ajoy K. Baksi, (D), Toronto, 1970, Cc
Gary R. Byerly, (D), Michigan State, 1974, Gi
Ray E. Ferrell, Jr., (D), Illinois, 1966, Cl
Jeffrey S. Hanor, (D), Harvard, 1967, Cl
George Hart, (D), Sheffield, 1961, Gg
Clyde Moore, On
Jeffrey A. Nunn, (D), Northwestern, 1981, Yg
James E. Roche, (D), Illinois, 1969, Gg
Barun K. Sen Gupta, (D), Indian Inst of Tech, 1963, Pm

**Dept of Oceanography & Coastal Sciences** (M,D) (2014)
1002 Energy Coast and Environment Building
Baton Rouge, LA 70803
p. (225) 578-6308
ocean@lsu.edu
http://www.ocean.lsu.edu
Administrative Assistant: Gaynell Gibbs
*Enrollment (2010): M: 63 (11) D: 92 (11)*

Professor:
Robert P. Gambrell, (D), North Carolina State, 1974, Cg
Paul A. LaRock, (D), Rensselaer, 1968, Ob
Irving A. Mendelssohn, (D), North Carolina State, 1974, Ob
Richard F. Shaw, (D), Maine, 1981, Ob
Robert E. Turner, (D), Georgia, 1974, Ob
Associate Professor:
Donald M. Baltz, (D), California (Davis), 1980, Ob
Robert S. Carney, (D), Oregon State, 1977, Ob
Lawrence J. Rouse, Jr., (D), Louisiana State, 1969, Op
Assistant Professor:
Mark C. Benfield, (D), Texas A&M, 1991, Ob
Adjunct Professor:
Dubravko Justic, (D), Zagreb, 1989, Op
Nancy N. Rabalais, (D), Texas, 1983, Ob
Harry H. Roberts, (D), Louisiana State, 1969, Gs
Paul W. Sammarco, (D), SUNY, 1978, On
Nan D. Walker, (D), Cape Town, 1989, Og

**School of Plant, Environmental and Soil Sciences** (B,M,D) (2016)
104 M. B. Sturgis Hall
Baton Rouge, LA 70803
p. (225) 578-2110
spess@lsu.edu
http://www.spess.lsu.edu
f: https://www.facebook.com/LsuSchoolOfPlantEnvironmental-SoilSciences/

Professor:
Hussein M. Selim, (D), Iowa State, 1971, Sp
Maud Walsh, (D), Louisiana State, 1989, GeOg
Jim Wang, (D), Iowa State, 1990, Sc
Assistant Professor:
Lewis A. Gaston, (D), Florida, 1987, Sc

## Louisiana Tech University

**Geosciences Program** (B,M) (2011)
600 W. Arizona Street
Ruston, LA 71272
p. (318) 257-3972
mm@engr.latech.edu
http://www.coes.latech.edu/geo/
Department Secretary: Connie McKenzie

Chair:
Gary S. Zumwalt, (D), California (Davis), 1976, Pi
Associate Professor:
Maureen McCurdy, (D), Wisconsin, 1990, Hw
Emeritus:
Leo A. Herrmann, (D), Johns Hopkins, 1951, Go

## Nicholls State University

**Dept of Physical Sciences** (2016)
P.O. Box 2022
Thibodaux, LA 70310
p. (985) 448-4502
marguerite.moloney@nicholls.edu
http://www.nicholls.edu/phsc

Assistant Professor:
Marguerite M. Moloney, (M), S Illinois, 2004, GeOnn
Instructor:
Adam Beyer, (M), S Illinois, Gg

## Northwestern State University

**Dept of Chemistry and Physics** (B) (2015)
Natchitoches, LA 71497
p. (318) 357-5501
chinc@nsula.edu

Director:
Paul Withey
Professor:
Carol S. Chin, (D)
Kelly Knowlton, (D), Texas A&M, 1991, Gg

## South Louisiana Community College

**South Louisiana Community College** (1999)
320 Devalcourt
Lafayette, LA 70506
p. (337) 521-8983
http://www.slcc.cc.la.us

## Tulane University

**Dept of Earth and Environmental Sciences** (B,M,D) ● (2016)
6823 St. Charles Ave.
101 Blessey Hall
New Orleans, LA 70118
p. (504) 865-5198
mreine@tulane.edu
http://www.tulane.edu/sse/eens
*Enrollment (2015): B: 6 (0) M: 0 (2) D: 0 (2)*

Chair:
Torbjörn E. Törnqvist, (D), Utrecht (Neth), 1993, Gs
Professor:
Mead A. Allison, (D), SUNY (Stony Brook), 1993, Gs
Karen Haley Johannesson, (D), Nevada, 1993, HgCg
Associate Professor:
Nancye H. Dawers, (D), Columbia, 1997, Gc

George C. Flowers, (D), California (Berkeley), 1979, Hw
Stephen A. Nelson, (D), California (Berkeley), 1979, Gi
Assistant Professor:
Nicole Gasparini, (D), MIT, 2003, Ngg
Brent M. Goehring, (D), Columbia, 2010, CcGlm
Kyle Martin Straub, (D), MIT, 2007, GgCg
Visiting Professor:
Jeffrey G. Agnew, (D), Louisiana State, 2008, Poe
Reda Amer, (D), St. Louis, 2011, OirGe
Jeffrey M. Sigler, (D), Yale, 2006, GgOwa

## University of Louisiana at Lafayette

**School of Geosciences** (B,M) O (2017)
BOX 43705
Lafayette, LA 70504-4530
p. (337) 482-6468
geology@louisiana.edu
http://geology.louisiana.edu
f: UL Lafayette School of Geosciences
Administrative Assistant: Nadean S. Bienvenu
Administrative Assistant: Pauline R. Greene
*Enrollment (2012): B: 135 (33) M: 49 (17)*
Director:
David M. Borrok, (D), Notre Dame, 2005, ClGeo
Assistant Director:
Durga Poudel, (D), Georgia (Athens), 1998, Ge
Carl Richter, (D), Tubingen, 1990, Ym
Professor:
Gary L. Kinsland, (D), Rochester, 1974, YgeGt
Brian E. Lock, (D), Cambridge, 1969, Gs
Jenneke M. Visser, (D), Louisiana State, 1989, Ge
Senior Scientist:
James E. Martin, (D), Washington, 1979, PvGg
Resource Facilitator:
Jim Foret, (M), Iowa State, 1971, Ge
Associate Professor:
Timothy W. Duex, (D), Texas (Austin), 1983, HgGex
Assistant Professor:
Katie H. Costigan, (D)
Aubrey Hillman, (D), Pittsburgh, 2015, GneGa
Brian Schubert, (D), Binghamton, 2008, Cg
Instructor:
Elisabeth L. Boudreaux, (M), Louisiana (Lafayette), 2013, GgOge
Kristie Cornell, (M), Louisiana (Lafayette), 2003, Gg
Jennifer E. Hargrave, (D), Oklahoma, 2009, GgPg
Adjunct Professor:
F. Clayton Breland, Jr., (D)
Emeritus:
Walter P. Kessinger, (D), Louisiana State, 1974, Pm

## University of Louisiana, Monroe

**School of Science, Atmospheric Science Program** (B) (2017)
700 University Avenue
Monroe, LA 71209-0550
p. (318) 342-1822
casehanks@ulm.edu
http://www.ulm.edu/atmos/
*Enrollment (2012): B: 43 (9)*
Associate Dean of Arts and Sciences:
Michael A. Camille, (D), Texas A&M, 1991, Oy
Professor:
Eric A. Pani, (D), Texas Tech, 1987, Oa
Associate Professor:
Sean Chenoweth, (D), Wisconsin (Milwaukee), 2003, GmOir
Department Head:
Anne T. Case Hanks, (D), Georgia Tech, 2008, Oae
Assistant Professor:
Ken Leppert, (D), Alabama (Huntsville), Oa
Todd Murphy, (D), Alabama (Huntsville), Oa
Station Archeologist/Poverty Point:
Diana M. Greenlee, (D), Washington, 2002, GaCo

## University of New Orleans

**Dept of Earth and Environmental Sciences** (B,M,D) (2014)
2000 Lakeshore Drive
New Orleans, LA 70148
p. (504) 280-6325
djreed@uno.edu
http://www.uno.edu/geology/
*Enrollment (2005): B: 106 (11) M: 35 (6) D: 4 (0)*
Professor:
William H. Busch, (D), Oregon State, 1981, Ou
Terry L. Pavlis, (D), Utah, 1982, Gt
Denise J. Reed, (D), Cambridge, 1986, Gm
A. K. Mostofa Sarwar, (D), Indiana, 1983, Ye
Laura F. Serpa, (D), Cornell, 1986, Ys
William B. Simmons, (D), Michigan, 1973, Gz
Ronald K. Stoessell, (D), California (Berkeley), 1977, Cl
Associate Professor:
Kraig L. Derstler, (D), California (Davis), 1985, PoiPv
Frank R. Hall, (D), Rhode Island, 1991, Ym
Mark A. Kulp, (D), Kentucky, 2000, GrmGs
Assistant Professor:
Christopher D. Parkinson, (D), London (UK), 1991, Gp
Adjunct Professor:
Miles O. Hayes, (D), Texas, 1965, Gs
Karen L. Webber, (D), Rice, 1988, Gv
Michael A. Wise, (D), Manitoba, 1987, Gz
Emeritus:
Gary C. Allen, (D), North Carolina, 1968, Gp
Jacqueline Michel, (D), South Carolina, 1980, Cg



# Maine

## Bates College

**Geology** (B) (2016)
Carnegie Science Center
44 Campus Avenue
Lewiston, ME 04240-6084
p. (207) 786-6490
http://www.bates.edu/geology/
f: https://www.facebook.com/groups/243958415633596/
Administrative Assistant: Sylvia Deschaine
*Enrollment (2016): B: 21 (0)*
Chair:
Michael J. Retelle, (D), Massachusetts, 1985, Gl
Professor:
J. Dykstra Eusden, Jr., (D), Dartmouth, 1988, Gc
Beverly J. Johnson, (D), Colorado, 1998, Cl
Assistant Professor:
Genevieve Robert, (D), Missouri (Columbia), 2014, CpGv
Assistant Instructor:
Marita Bryant, (M), Free (Berlin), 1984, Gg
Emeritus:
Gene A. Clough, (D), Caltech, 1978, Ym
John W. Creasy, (D), Harvard, 1974, Gx

## Bowdoin College

**Dept of Earth and Oceanographic Science** (B) (2014)
6800 College Station
Brunswick, ME 04011
p. (207) 725-3628
mparker@bowdoin.edu
http://www.bowdoin.edu/earth-oceanographic-science/
Administrative Assistant: Marjorie Parker
*Enrollment (2014): B: 60 (27)*
Rusack Professor of Environmental Studies:
Philip Camill III, (D), Duke, 1999, PebSb
Professor:
Rachel J. Beane, (D), Stanford, 1997, GziGp
Chair:
Collin Roesler, (D), Washington, 1992, OgpOr
Associate Professor:
Peter D. Lea, (D), Colorado, 1989, Gl
Assistant Professor:
Michéle LaVigne, (D), Rutgers, 2010, OcPe
Emily M. Peterman, (D), California (Santa Barbara), 2009, Gtp
Service Learning Coord/Lab Instr:
Cathryn K. Field, (M), Smith, 2000, Ge
Lab Instructor:
Joanne Urquhart, (M), Dartmouth, 1987, Gge

Associate Professor:
Edward P. Laine, (D), MIT, 1977, Gu

## Colby College

**Department of Geology** (B) (2016)
5800 Mayflower Hill
Waterville, ME 04901-8858
p. (207) 859-5800
amridky@colby.edu
http://www.colby.edu/geologydept/
Administrative Assistant: Alice M. Ridky
*Enrollment (2015): B: 29 (7)*

Chair:
Bill Sullivan, (D), Wyoming, 2007, Gct
Professor:
Robert A. Gastaldo, (D), S Illinois, 1978, PbGs
Robert E. Nelson, (D), Washington, 1982, Pe
Assistant Professor:
Tasha L. Dunn, (D), Tennessee, 2008, XmGiz
Visiting Professor:
Bruce F. Rueger, (D), Colorado, 2002, Pl
Emeritus:
Donald B. Allen, (D), Illinois, 1972, Eg
Harold R. Pestana, (D), Iowa, 1965, Pi

## University of Maine

**School of Earth and Climate Sciences** (B,M,D) (2016)
5790 Bryand Global Sciences Center
Orono, ME 04469-5790
p. (207) 581-2152
linda.cappuccio@umit.maine.edu
http://www.umaine.edu/earthclimate/
f: https://www.facebook.com/pages/UMaine-School-of-Earth-and-Climate-Sciences/238244500701
*Enrollment (2015): B: 42 (6) M: 13 (4) D: 14 (1)*

Research Professor:
Edward S. Grew, (D), Harvard, 1971, Gpz
Roger L. Hooke, (D), Caltech, 1965, Gm
Director:
Scott E. Johnson, (D), James Cook, 1989, Gc
Professor:
Daniel F. Belknap, (D), Delaware, 1979, Gu
George H. Denton, (D), Yale, 1965, Gl
Brenda L. Hall, (D), Maine, 1997, Gl
Gordon S. Hamilton, (D), Cambridge, 1992, Ol
Joseph T. Kelley, (D), Lehigh, 1980, Gua
Peter O. Koons, (D), ETH (Switzerland), 1982, Gt
Karl J. Kreutz, (D), New Hampshire, 1998, Cs
Daniel R. Lux, (D), Ohio State, 1981, Gi
Kirk A. Maasch, (D), Yale, 1989, Oa
Paul A. Mayewski, (D), Ohio State, 1973, Pe
Aaron E. Putnam, (D), Maine, 2011, Gll
Andrew S. Reeve, (D), Syracuse, 1996, Hw
Associate Professor:
Christopher C. Gerbi, (D), Maine, 2005, Gt
Research Professor:
Ellyn M. Enderlin, (D), Ohio State, 2013, OlrOg
Assistant Professor:
Katherine A. Allen, (D), Columbia, 2013, Gu
Alicia M. Cruz-Uribe, (D), Penn State, 2014, GpCpt
Amanda A. Olsen, (D), Virginia Tech, 2007, Cl
Sean Smith, (D), Johns Hopkins, 2010, Hs
Research Assistant Professor:
Seth W. Campbell, (D), Maine, 2010, Gl
Researach Assistant Professor:
Gordon R. M. Bromley, (D), Maine, 2010, Gl
Instructor:
Alice R. Kelley, (D), Maine, 2006, GalGm
Martin G. Yates, (D), Indiana, 1988, EgGzx
Emeritus:
Harold W. Borns, Jr., (D), Boston Univ, 1959, Gl
Joseph V. Chernosky, Jr., (D), MIT, 1973, Gg
Terence J. Hughes, (D), Northwestern, 1968, Gl
Stephen A. Norton, (D), Harvard, 1967, Cl
Research Assistant Professor:
Sean Birkel, (D), Maine, 2010, Gl

**School of Marine Sciences** (B,M,D) O (2014)
5706 Aubert Hall, Rm 360
Orono, ME 04469-5706
p. (207) 581-4381
fchai@maine.edu
http://www.umaine.edu/marine/
*Enrollment (2010): B: 145 (38) M: 56 (0) D: 9 (0)*

## University of Maine, Farmington

**Dept of Geology** (B) (2017)
173 High Street
Farmington, ME 04938
p. (207) 778-7402
dgibson@maine.edu
http://sciences.umf.maine.edu
*Enrollment (2016): B: 20 (7)*

Professor:
David Gibson, (D), Queen's (Ireland), 1984, GigGz
Associate Professor:
Julia F. Daly, (D), Maine, 2002, GmlGu
Douglas N. Reusch, (D), Maine, 1998, GtCgOu
Emeritus:
Thomas E. Eastler, (D), Columbia, 1970, GeOrGg

## University of Maine, Presque Isle

**Div of Mathematics & Science** (B) (2014)
181 Main Street
Presque Isle, ME 04769
p. (207) 768-9482
kevin.mccartney@umpi.edu
http://www.umpi.edu/
Department Secretary: Connie Leveque

Chair:
Michael Knopp
Professor:
Kevin McCartney, (D), Florida State, 1988, Pm

# Maryland

## College of Southern Maryland

**Biological and Physical Sciences** (A) (2010)
8730 Mitchell Rd
La Plata, MD 20646
p. (301) 934-7841
jenniferh@csmd.edu
http://www.csmd.edu/bio/index.html

Professor:
Tom Russ, (M), Kutztown, GgSo

## Community College of Baltimore County, Catonsville

**School of Mathematics & Science** (A) (2017)
800 S. Rolling Road
MASH 010
Catonsville, MD 21228
p. (443) 840-5935
DLudwikoski@ccbcmd.edu
http://www.ccbcmd.edu/math_science/geology.html
Administrative Assistant: Annjeannette Black
*Enrollment (2015): A: 3 (0)*

Associate Professor:
David J. Ludwikoski, (M), Toledo, 1993, OegOn

## Frederick Community College

**Science Dept** (A) (2014)
7932 Opossumtown Pike
Frederick, MD 21702
p. (301) 846-2510
shsmith@frederick.edu
http://www.frederick.edu/courses_and_programs/dept_science.aspx

Professor:
Richard Gottfried, (M), U of Penn, 1977, GgdGz
Associate Professor:
Natasha Cleveland, (M), Oge

## Frostburg State University

**Dept of Geography** (B) (2010)
101 Braddock Rd.
Frostburg, MD 21532
p. (301) 687-4369
jsaku@frostburg.edu
http://www.frostburg.edu/dept/geog/
Administrative Assistant: Gale Yutzy
*Enrollment (2010): B: 65 (16)*

Chair:
Craig L. Caupp, (D), Utah State, 1986, OnHs
Professor:
Henry W. Bullamore, (D), Iowa, 1978, On
Francis L. Precht, (D), Georgia, 1989, Oy
James C. Saku, (D), Saskatchewan, 1995, On
Associate Professor:
Fritz Kessler, (D), Kansas, 1999, Or
George W. White, (D), Oregon, 1994, On
Assistant Professor:
Phillip Allen, (D), Coventry (UK), 2005, GmePe
David L. Arnold, (D), Indiana, 1994, Oa
Matthew E. Ramspott, (D), Kansas, 2006, Or

## Howard Community College

**Science, Engineering, and Technology Division** (1999)
10901 Little Patuxent Pkwy
Columbia, MD 21044
p. (443) 518-1000
pturner@howardcc.edu
http://www.howardcc.edu/programs-courses/academics/academic-divisions/science-engineering-technology/sciences/

## Johns Hopkins University

**Dept of Environmental Health & Engineering** (B,M,D) (2014)
313 Ames Hall
34th & Charles Streets
Baltimore, MD 21218-2681
p. (410) 516-7092
dogee@jhu.edu
http://www.jhu.edu/dogeeChair:
Edward John Bouwer, (D), Stanford, 1982, On
Professor:
William P. Ball, (D), Stanford, 1989, On
Grace S. Brush, (D), Harvard, 1956, Pl
Hugh Ellis, (D), Waterloo, 1984, On
Steve H. Hanke, (D), Colorado, 1969, On
Benjamin F. Hobbs, (D), Cornell, 1983, On
A. Lynn Roberts, (D), MIT, 1991, On
Erica Schoenberger, (D), California (Berkeley), 1984, On
Alan T. Stone, (D), Caltech, 1983, ClSc
Peter W. Wilcock, (D), MIT, 1987, Gm
Associate Professor:
Markus Hilpert, (D), On
Assistant Professor:
Kai Loon Chen, (D), Yale, 2008, On
Seth Guikema, (D), Stanford, 2003, On
Catherine Norman, (D), California (Santa Barbara), 2005, On
Lecturer:
Hedy Alavi, (D), Ohio State, 1983, On
Emeritus:
John J. Boland, (D), Johns Hopkins, 1973, On
Charles R. O'Melia, (D), Michigan, 1963, On
Eugene D. Shchukin, (D), Moscow State, 1958, On
Senior Academic Program Coordinator:
Adena Rojas, (M), On

**The Morton K. Blaustein Dept of Earth & Planetary Sciences** (B,M,D) (2015)
3400 N Charles Street
301 Olin Hall
Baltimore, MD 21218
p. (410) 516-7135
kgaines@jhu.edu
http://eps.jhu.edu/
*Enrollment (2013): D: 33 (5)*

Chair:
Thomas W.N. Haine, (D), Southampton, 1993, Op
Professor:
Anand Gnanadesikan, (D), MIT/WHOI, 1994, OpcOa
Peter L. Olson, (D), California (Berkeley), 1977, Yg
Darrell F. Strobel, (D), Harvard, 1969, OanOn
Dimitri A. Sverjensky, (D), Yale, 1980, Cl
Darryn W. Waugh, (D), Cambridge, 1991, Oa
Assistant Professor:
Sarah Horst, (D), Arizona, 2011
Naomi Levin, (D), Utah, 2008, Gs
Kevin Lewis, (D), Caltech, 2009
Benjamin Zaitchik, (D), Yale, 2006, Oa
Research Professor:
Katalin Szlavecz, (D), Eotvos (Hungary), 1981, Pi

## Maryland Department of Natural Resources

**Maryland Geological Survey** (2017)
2300 St. Paul Street
Baltimore, MD 21218-5210
p. (410) 554-5500
dale.shelton@maryland.gov
http://www.mgs.md.gov/

Director:
Richard A. Ortt, Jr., (B), Johns Hopkins, 1991, Op
Program Chief:
David W. Bolton, Hw
Hydrogeologist:
Jon Achmad, Hw
David C. Andreasen, (B), Maryland, Hw
Geologist/Program Chief:
Stephen Van Ryswick, (B), Maryland, 2002, Ge
Senior Scientist:
David K. Brezinski, (D), Pittsburgh, 1984, Gr
Hydrogeologist:
Andrew Staley, Hw
Tiffany J. VanDerwerker, (M), Virginia Tech, Hw
Geologist Lead/Adv:
Katherine A. Knippler, (B), Susquehanna Univ, 2001, On
Geologist:
Christopher B. Connallon, (M), Michigan State Univ, 2015, GmOi
Rebecca H. Kavage-Adams, (M), Virginia Tech, 2002, GmOiGe
Heather A. Quinn, (M), Florida, GgHg
Elizabeth R. Sylvia, (M), Towson Univ, 2013, Ge
Education Specialist:
Dale W. Shelton, (B), Towson Univ, 1986, OeGg

## Montgomery College

**Dept of Physics, Engineering & Geosciences** (A) (2011)
51 Mannakee Street
Rockville, MD 20850
p. (301) 279-5230
Muhammad.Kehnemouyi@montgomerycollege.edu
http://www.montgomerycollege.edu/Departments/phengrv/
Administrative Assistant: Mary (Deep) McGregor

Professor:
Alan Cutler, (D), Geology, PgGh
Instructional Laboratory Coordinator for Geosciences:
Kimberly Kelly, (B), Pittsburgh, On

## St. Charles Community College

(2014)
4601 Mid Rivers Mall Drive
Cottleville, MD 63376
p. (636) 922-8000
nfo_desk@stchas.edu
http://www.stchas.edu/

## Towson University

**Dept of Physics, Astronomy & Geosciences** (B) (2017)
8000 York Road
Towson, MD 21252-0001
p. (410) 704-3020
dschaefer@towson.edu
http://wwwnew.towson.edu/physics/geosciences/

*Enrollment (2016): B: 40 (8)*
Professor:
Rachel J. Burks, (D), Texas, 1985, Gc
David A. Vanko, (D), Northwestern, 1982, Gxz
Assistant Professor:
Joel Moore, (D), Penn State, 2008, ClsSc
Wendy Nelson, (D), Penn State, 2009, GiCc
Amy Williams, (D), California (Davis), 2014, Ca
Lecturer:
Tsigabu A. Gebrehiwet, (D), W Michigan, 2007, Hw
Gregory A. Shofner, (D), Maryland, 2011, Cg



## United States Naval Academy

**Dept of Oceanography** (B) (2014)
572C Holloway Road
Annapolis, MD 21402-5026
p. (410) 293-6550
petrunci@usna.edu
http://www.nadn.navy.mil/Oceanography/
Administrative Assistant: Cynthia A.. Ervin
*Enrollment (2014): B: 216 (117)*
Professor:
Peter L. Guth, (D), MIT, 1980, GcOiy
Chair:
David R. Smith, (D), Texas A&M, 1979, Ow
Associate Professor:
Andrew C. Muller, (D), Old Dominion, 1999, OpnOu
Cecily N. Steppe, (D), Delaware, 2001, Ob
Assistant Professor:
Bradford S. Barrett, (D), Oklahoma, 2007, Oa
Gina R. Henderson, (D), Delaware, 2010, Oa
Emil T. Petruncio, (D), Naval Postgrad Sch, 1996, Opr
Elizabeth R. Sanabia, (D), Naval Postgrad Sch, 2010, Ow
William J. Schulz, (D), Old Dominion, 1999, Opw
Joseph P. Smith, (D), Massachusetts (Boston), 2007, CmOg
Instructor:
Dwight E. Smith, (M), Naval Postgrad Sch, 2009, Owg
Megan D. Thomas, (M), Naval Postgrad Sch, 2005, Og
Oceanography Reference Librarian:
Barbara Yoakum, (M), South Carolina, 1985, On

## University of Maryland

**Dept of Geology** (B,M,D) (2016)
Geology Building (#237), Room 1118
8000 Regents Drive
College Park, MD 20742-4211
p. (301) 405-4082
geology@umd.edu
http://www.geol.umd.edu
f: https://www.facebook.com/UMDGeology
Administrative Assistant: Dorothy Brown
*Enrollment (2014): B: 48 (10) M: 4 (5) D: 24 (1)*
Chair:
Richard J. Walker, (D), SUNY (Stony Brook), 1984, CcXcCg
Research Professor:
Robert Tucker, (D), Yale, 1985
Affiliate Professor:
Antonio Busalacchi, (D), Florida State, 1982, Og
Bruce James, (D), Vermont, 1981, Sc
Michael Kearney, (D), W Ontario, 1981, On
Raghuram G. Murtugudde, (D), Columbia, 1994, Obr
Jessica Sunshine, (D), Brown, 1993
Ning Zeng, (D), Arizona, 1994, OaCgGl
Professor:
Michael Brown, (D), Keele, 1975, GxpGt
Philip A. Candela, (D), Harvard, 1982, CpgEg
James Farquhar, (D), Alberta, 1995, Cs
Alan J. Kaufman, (D), Indiana, 1990, Csg
Daniel Lathrop, (D), Texas, 1991, Yg
William F. McDonough, (D), Australian National, 1988, Cg
Roberta L. Rudnick, (D), Australian National, 1988, CgsCt
Senior Research Scientist:
Philip M. Piccoli, (D), Maryland, 1992, Cg
Igor Puchtel, (D), Russian Acad of Sci, 1992, CcgGi
Associate Professor:
Michael N. Evans, (D), Columbia, 1999, PeOaCs
Sujay Kaushal, (D), Colorado, 2003, OuGeHs
Laurent G.J. Montesi, (D), MIT, 2002, YgXyGt
Sarah Penniston-Dorland, (D), Johns Hopkins, 2005, Gp
Karen L. Prestegaard, (D), California (Berkeley), 1982, Hgw
Wenlu Zhu, (D), SUNY (Stony Brook), 1996, YrNrHg
Associate Research Scientist:
Richard Ash, (D), Open, 1990, Ca
Assistant Professor:
Derrick Lampkin, (D)
Vedran Lekic, (D), California (Berkeley), 2009, Ysg
Nicholas Schmerr, (D), Arizona State, 2008, XyYs
Visiting Research Associate:
Timothy Johnson, (D), Derby, 1999
Zoltan Zajacz, (D), Swiss Fed Inst Tech, 2007
Research Associate:
Jabrane Labidi, (D), Inst Physique du Globe de Paris, 2012, Cg
NSF Post-doc:
Scott Burdick, (D), MIT, 2014, Ys
Research Associate:
Katherine Bermingham, (D), Westfalische Wilhelms, 2011, XcCcg
Shuiwang Duan, (D), Tulane, 2005, HsOu
Melodie French, (D), Texas A&M, 2014, Ygx
Joost Hoek, (D), Pennsylvania, 2004, CsgCa
Tolulope M. Olugboji, (D), Yale, 2014, Ysg
Senior Lecturer:
Thomas R. Holtz, Jr., (D), Yale, 1992, PvgPo
John W. Merck, Jr., (D), Texas, 1997, Pv
Lecturer:
Tracey Centorbi, (B), Maryland, 2003, Cg
Adjunct Associate Professor:
Elizabeth Cottrell, (D), Columbia, 2004, CpgGx
Adjunct Assistant Professor:
Anat Shahar, (D), California (Los Angeles), 2008, CspCa
Adjunct Professor:
John Bohlke, (D), California (Berkeley), 1986, Cag
Yingwei Fei, (D), CUNY, 1989, CpGxCg
Steven Shirey, (D), SUNY (Stony Brook), 1984, CgaCc
Deborah Smith, (D), California (San Diego), 1985
Sorena Sorensen, (D), California (Los Angeles), 1984, Gp
Visiting Professor:
Saswata Hier-Majumder, (D), Minnesota, 2004, YgGq
Emeritus Professor:
Ann G. Wylie, (D), Columbia, 1972, Gz
Emeritus Affiliate Research Professor:
George Helz, (D), Penn State, 1970, HsGeCg
Associate Professor Emeritus:
Peter B. Stifel, (D), Utah, 1964, Pg
Faculty Research Assistant:
Todd Karwoski, (B), Maryland, 2005, Gg
Rebecca Plummer, (M), Michigan, 2000, Csa
Valentina Puchtel, (M), Moscow Geological Prospecting Academy, 1983

**Dept of Plant Science & Landscape Architecture** (B,M,D) (2014)
2102 Plant Sciences Building
College Park, MD 20742-4432
p. (301) 405-4356
asmurphy@umd.edu
http://www.psla.umd.edu/
Professor:
Christopher Walsh, (D), Cornell, 1980, On
Associate Professor:
Gary D. Coleman, (D), Nebraska, 1989, On
Jack B. Sullivan, (M), Virginia, 1980, Ou
Program Management Specialist:
Kathy Hunt, On
Coordinator:
Sue Burk, On

**Marine-Estuarine-Environmental Sciences Graduate Program** (M,D) (2014)
1213 HJ Patterson Hall
University of Maryland
College Park, MD 20742
p. (301) 405-6938
mees@umd.edu
http://www.mees.umd.edu

*Enrollment (2013): M: 6 (0) D: 12 (2)*
Professor:
Shenn-Yu Chao, (D), North Carolina State, 1979, Op
Keith N. Eshleman, (D), MIT, 1985, Hg
Thomas R. Fisher, Jr., (D), Duke, 1975, ObiCl
Patricia M. Gilbert, (D), Harvard, 1982, Ob
Lawrence P. Sanford, (D), MIT, 1984, Op
Diane Stoecker, (D), SUNY (Stony Brook), 1979, Ob
Associate Professor:
William Boicourt, (D), Johns Hopkins, 1973, Op
James Carton, (D), Princeton, 1983, Op
Micheal S. Kearney, (D), Ontario, 1981, Gu
Karen L. Prestegaard, (D), California (Berkeley), 1982, Hw
Research Associate Professor:
Jeffery C. Cornwell, (D), Alaska, 1983, Oc
Assistant Professor:
Mark S. Castro, (D), Virginia, 1991, Oa
Raleigh Hood, (D), California (San Diego), 1990, Ob
Alba Torrents, (D), Johns Hopkins, 1992, Sb
Research Associate Professor:
Todd M. Kana, (D), Harvard, 1982, Ob

# Massachusetts

## Amherst College
**Dept of Geology** (B) (2016)
P.O. Box 2238
Amherst, MA 01002-5000
p. (413) 542-2233
dbhutton@amherst.edu
https://www.amherst.edu/academiclife/departments/geology
*Enrollment (2016): B: 112 (13)*
Professor:
John T. Cheney, (D), Wisconsin, 1975, Gi
Peter D. Crowley, (D), MIT, 1985, Gc
Tekla A. Harms, (D), Arizona, 1986, Gt
Anna M. Martini, (D), Michigan, 1997, Cl
Assistant Professor:
David S. Jones, (D), Harvard, 2009, GsClGr
Emeritus:
Edward S. Belt, (D), Yale, 1963, Gs
Margery C. Coombs, (D), Columbia, 1971, Pv

## Bard College at Simon's Rock
**Bard College at Simon's Rock** (2014)
84 Alford Rd.
Great Barrington, MA 01230
p. 413 644-4400 .
admin@simons-rock.edu
http://www.simons-rock.edu/

## Bentley University
**Dept of Natural and Applied Sciences** (B) (2016)
175 Forest Street
Waltham, MA 02452-4705
p. (781) 891-2980
pdavis@bentley.edu
http://www.bentley.edu/academics/departments/natural-and-applied-sciences
Administrative Assistant: Martha E.. Keating
*Enrollment (2016): B: 1 (0)*
Professor:
P. Thompson Davis, (D), Colorado, 1980, Gl
Rick Oches, (D), Massachusetts, 1994, GeOePe
Associate Professor:
David Szymanski, (D), Michigan State, 2007, GvfCg
Adjunct Professor:
Robert P. Ackert, (D), MIT/WHOI, 2000, CcGlm
Mark J. Benotti, (D), Stony Brook, 2006, OgGgOg
Janette Gartner, (M), Massachusetts, 2000, Hy
Laboratory Director:
Anna K. Tary, (M), Boston, 1999, Gl

## Berkshire Community College
**Environmental and Life Science Dept** (A) (2014)
Pittsfield, MA 01201
p. (413) 236-4601
saleksa@berkshirecc.edu
Chair:
Clifford D. Myers, (D), Maine, On
Professor:
Timothy Flanagan, (M), Antioch Univ, 1983, Oei
Thomas F. Tyning, (M), On
Charles E. Weinstein, (M), Wisconsin, On
Emeritus:
Richard L. Ferren, (M), Louisiana State, On
George Hamilton, (M), North Adams State Coll, On
Mary R. Mercuri, (M), Catholic

## Boston College
**Dept of Earth & Environmental Sciences** (B,M) (2017)
140 Commonwealth Avenue
213 Devlin Hall
Chestnut Hill, MA 02467-3809
p. (617) 552-3640
http://www.bc.edu/geology
*Enrollment (2013): B: 23 (23) M: 14 (14)*
Professor:
John E. Ebel, (D), Caltech, 1981, Ys
Gail C. Kineke, (D), Washington, 1993, On
Chair:
Ethan Baxter, (D), California (Berkeley), 2000, Ccg
Associate Professor:
Rudolph Hon, (D), MIT, 1976, Gi
Alan L. Kafka, (D), SUNY (Stony Brook), 1980, Yg
Noah Snyder, (D), MIT, 2001, GgHg
Assistant Professor:
Seth C. Kruckenberg, (D), Minnesota (Twin Cities), 2009, Gct
Jeremy D. Shakun, (D), Oregon State, 2010, Gg
Corinne I. Wong, (D), Texas (Austin), 2013, PeCsHw
Lab Coordinator:
Kenneth G. Galli, (D), Massachusetts, 2003, GsdGg
Adjunct Professor:
Paul K. Strother, (D), Harvard, 1980, PlbPy
Emeritus:
J. Christopher Hepburn, (D), Harvard, 1972, Gpt
James W. Skehan, S.J., (D), Harvard, 1953, Gc

**Weston Observatory** (2015)
381 Concord Road
Weston, MA 02493
p. (617) 552-8300
weston.observatory@bc.edu
http://www.bc.edu/westonobservatory
Director:
Alan L. Kafka, (D), Stony Brook, 1980, Ys
Science Education:
Michael Barnett, (D), Indiana, 2003, Oe
Senior Scientist:
John E. Ebel, (D), Caltech, 1981, Yg
Seismology, Seismic Network Development:
Michael Hagerty, (D), California (Santa Cruz), 1998, Ys
Seismic Analyst and Educational Seismologist:
Anastasia Moulis, (M), Boston, 2003, Ys
Research Scientist:
Seth Kruckenberg, (D), Minnesota (Twin Cities), 2009, Gc
Research Scientist:
John J. Cipar, (D), Caltech, 1981, Ys
Research Associate:
John H. Beck, (D), Boston, 1998, Pb
Adjunct Professor:
Paul Strother, (D), Harvard, 1980, Pb
Alfredo Urzua, (D), MIT, 1981, Ng
Visiting Professor:
Vincent Murphy, (M), Boston, 1957, Yg
Emeritus:
J. Christopher Hepburn, (D), Harvard, 1972, Gg
James W. Skehan, (D), Harvard, 1953, Gc

Massachusetts

## Boston University

**Center for Remote Sensing** (2014)
685 Commonwealth Avenue
Room 433
Boston, MA 02215
p. (617) 353-9709
crsadmin@bu.edu
Administrative Assistant: Emily P. Johnson

Director:
Farouk El-Baz, (D), Missouri (Rolla), 1964, Or
Professor:
Sucharita Gopal, (D), California (Santa Barbara), 1988, Oi
Alan Strahler, (D), Johns Hopkins, 1969, Or
Curtis E. Woodcock, (D), California (Santa Barbara), 1986, Or
Research Associate Professor:
Magaly Koch, (D), Boston Univ, 1993, HyOri
Associate Professor:
Mark A. Friedl, (D), California (Santa Barbara), 1994, Or
Kenneth L. Kvamme, (D), California (Santa Barbara), 1983, Ga
Guido D. Salvucci, (D), MIT, 1994, Hq
Research Associate Professor:
Cordula Robinson, (D), Univ Coll (UK), 1991, Gm
Crystal Schaaf, (D), Boston, 1994, Or
Research Associate:
Eman Ghoneim, (D), Southampton, 2002, Gm
Geology Librarian:
Nasim Momen, On

**Dept of Earth & Environment** (B,M,D) (2015)
675 Commonwealth Avenue
Boston, MA 02215
p. (617) 353-2525
earth@bu.edu
http://www.bu.edu/earth/
*Enrollment (2015): B: 70 (15) M: 15 (10) D: 50 (8)*

Chair:
David R. Marchant, (D), Edinburgh, 1994, Gm
Professor:
Bruce Anderson, (D), Scripps, 1998, Oap
James Lawford Anderson, (D), Gi
Duncan M. FitzGerald, (D), South Carolina, 1977, On
Mark Friedl, (D), California (Santa Barbara), 1993, Or
Sucharita Gopal, California (Santa Barbara), 1988, Oi
Tony Janetos, (D), 1980, On
Richard Murray, (D), California (Berkeley), 1991, Cm
Ranga Myneni, (D), Antwerp, 1985, Or
Nathan Phillips, (D), Duke, 1997, Oy
Guido D. Salvucci, (D), MIT, 1994, Hq
Curtis Woodcock, (D), California (Santa Barbara), 1986, Or
Associate Professor:
Rachel Abercrombie, (D), Reading, 1991, Gt
Michael Dietze, (D), Duke, 2006, SfOr
Sergio Fagherazzi, (D), Padua, 1999, GmOnGs
Robinson Fulweiler, (D), Rhode Island, 2007, Cm
Lucy Hutyra, (D), Harvard, 2007, Oar
Andrew Kurtz, (D), Cornell, 2000, Clg
Assistant Professor:
Dan Li, (D), Princeton, 2013, HqOa
Christine Regalla, (D), Penn State, 2013, GtcGm
Diane Thompson, Arizona, 2013, ObPe
Research Associate:
Farouk El-Baz, (D), Missouri, 1964, Or

## Bridgewater State University

**Dept of Geological Sciences** (B) (2017)
Conant Science Building
Bridgewater, MA 02325
p. (508) 531-1390
Brenda.Flint@bridgew.edu
http://www.bridgew.edu/academics/colleges-departments/department-geological-sciences
Administrative Assistant: Brenda Flint
*Enrollment (2010): B: 70 (0)*

Chairperson:
Robert D. Cicerone, (D), MIT, 1991, YsgXy
Professor:
Richard L. Enright, (D), Rutgers, 1969, HgOrEg
Michael A. Krol, (D), Lehigh, 1996, GzxGt
Peter J. Saccocia, (D), Minnesota, 1991, Cm
Lecturer:
Joseph Doyle, (M), New Hampshire, Gg
Suzanne R. O'Brien, (M), New Hampshire, 1995, Og
Michael A. Penzo, (M), SUNY (Binghamton), 1981, Ge
Visiting Professor:
Christine M. Brandon, (D), Massachusetts, 2015, Gs
Emeritus:
Robert F. Boutilier, (D), Boston, 1963, Gz
Ira E. Furlong, (D), Boston, 1960, Gm

## Bristol Community College

**Div of Mathematics, Science and Engineering** (A) (2010)
777 Elsbree Street
Fall River, MA 02720
p. (508) 678-2811
John.Ahola@bristolcc.edu

Instructor:
John Ahola, Gg

## Cape Cod Community College

**Environmental Technology Program** (2015)
2240 Iyannough Rd
West Barnstable, MA 02668
p. (508) 362-2131 x4468
junderwood@capecod.edu
http://www.capecod.edu/web/natsci/env

## Fitchburg State University

**Earth and Geographic Sciences** (B) (2016)
160 Pearl Street
Fitchburg, MA 01420-2697
p. (978) 665-4636
egordon3@fitchburgstate.edu
http://www.fitchburgstate.edu/academics/academic-departments/earth-and-geographic-sciences-dept/
f: https://www.facebook.com/FitchburgStateUniversityEGS
Administrative Assistant: Melissa Barrette
*Enrollment (2016): B: 38 (0)*

Associate Professor:
Elizabeth S. Gordon, (D), OgaOg
Lawrence R. Guth, (D), Rice, 1991, Gc
Jane Huang, (D), Oi
Assistant Professor:
Reid A. Parsons, (D), XgHs

## Hampshire College

**Dept of Geology** (B) (2010)
Amherst, MA 01002
p. (413) 582-5373
sroof@hampshire.edu
Administrative Assistant: Joan Barrett

Associate Professor:
Steven Roof, (D), Massachusetts, 1995, OiPe
Assistant Professor:
Christina Cianfrani, (D), Vermont, Hw
Cooperating Faculty:
Helaine Selin, (M), SUNY (Albany), On

## Harvard University

**Dept of Earth and Planetary Sciences** (B,D) ● (2016)
Hoffman Laboratory
20 Oxford Street
Cambridge, MA 02138-2902
p. (617) 495-2351
moffatt@eps.harvard.edu
http://www.eps.harvard.edu
*Enrollment (2016): B: 35 (18) D: 58 (7)*

Chair:
John Shaw, (D), Princeton, 1993, Gt
Professor:
James G. Anderson, (D), Colorado, 1970, Oa

Jeremy Bloxham, (D), Cambridge, 1985, Yg
Brian F. Farrell, (D), Harvard, 1981, Ow
Peter Huybers, (D), MIT, 2004, Pe
Miaki Ishii, (D), Harvard, 2003, Yg
Daniel J. Jacob, (D), Caltech, 1985, Oa
Stein B. Jacobsen, (D), Caltech, 1980, Cc
David T. Johnston, (D), Maryland, 2007, CgPy
Andrew H. Knoll, (D), Harvard, 1977, Pb
Zhiming Kuang, (D), CalTech, 2003, Oa
Charles H. Langmuir, (D), SUNY (Stony Brook), 1980, Cg
Scot T. Martin, (D), Caltech, 1995, Cg
James T. McCarthy, (D), Scripps, 1971, Ob
Michael B. McElroy, (D), Queen's, 1962, Oa
Brendan Meade, (D), MIT, 2004, Yg
Jerry X. Mitrovica, (D), Toronto, 1991, Yg
Ann Pearson, (D), MIT/WHOI, 2000, Cm
James R. Rice, (D), Lehigh, 1964, Yg
Daniel P. Schrag, (D), California (Berkeley), 1993, Cg
Eli Tziperman, (D), MIT/WHOI, 1987, Op
Steven C. Wofsy, (D), Harvard, 1971, Oa

Associate Professor:
Francis Macdonald, (D), Harvard, 2009, Pe

Assistant Professor:
Marine Denolle, (D), Stanford, 2014, Ys

Visiting Professor:
Carl Wunsch, (D), MIT, 1966, Op

Emeritus:
Charles W. Burnham, (D), MIT, 1961, GzyGe
Paul F. Hoffman, (D), Johns Hopkins, 1970, Gc
Ulrich Petersen, (D), Harvard, 1963, Eg

## Massachusetts Institute of Technology

**Dept of Earth, Atmospheric, & Planetary Sciences** (B,M,D) ● (2016)
77 Massachusetts Avenue, 54-918
Cambridge, MA 02139
p. (617) 253-2127
eapsinfo@mit.edu
http://eapsweb.mit.edu/
f: https://www.facebook.com/EAPS.MIT
t: https://twitter.com/eapsmit
*Enrollment (2016): B: 20 (9) M: 6 (6) D: 149 (22)*

Head:
Robert van der Hilst, (D), Utrecht (Neth), 1990, Ys

Vice-President for Research:
Maria T. Zuber, (D), Brown, 1986, Xy

Professor:
Richard P. Binzel, (D), Texas, 1986, Xm
Samuel A. Bowring, (D), Kansas, 1985, Gg
Edward A. Boyle, (D), MIT, 1976, Oc
Kerry A. Emanuel, (D), MIT, 1978, Ow
Dara Entekhabi, (D), MIT, 1990, Hg
Raffaele Ferrari, (D), Scripps, 2001, OpaOb
Glenn R. Flierl, (D), Harvard, 1975, Op
Timothy L. Grove, (D), Harvard, 1976, Gi
Bradford H. Hager, (D), Harvard, 1978, Ys
Thomas A. Herring, (D), MIT, 1983, Yd
Paola M. Malanotte-Rizzoli, (D), California (San Diego), 1978, Op
John C. Marshall, (D), Imperial Coll (UK), 1980, Op
F Dale Morgan, (D), MIT, 1981, Yg
Ronald G. Prinn, (D), MIT, 1971, Oa
Daniel H. Rothman, (D), Stanford, 1986, Yg
Leigh H. Royden, (D), MIT, 1982, Gt
Sara Seager, (D), Harvard, 1999, Xy
Susan Solomon, (D), California (Berkeley), 1981, Oa
Roger Summons, (D), New South Wales, 1972, PyCo
Benjamin Weiss, (D), Caltech, 2003, Ym
Jack Wisdom, (D), Caltech, 1981, Og

Senior Research Scientist:
William Durham, (D), MIT, 1975, Cg
Michael Fehler, MIT, 1979, Yg
Chien Wang, (D), SUNY (Albany), 1992, Oa

Principal Research Scientist:
Nilanjan Chatterjee, (D), CUNY, 1989, Cg
Stephanie Dutkiewicz, (D), Rhode Island, 1997, Cg
Robert W. King, (D), MIT, 1975, Yg
Eduardo Andrade Lima, (D), Catholic Univ. of Rio de Janeiro, Ym
Sai Ravela, (D), Massachusetts, 2002
Robert E. Reilinger, (D), Cornell, 1979, Yg
William Rodi, (D), 1989, Yg
C. Adam Schlosser, (D), Maryland, 1995, Oa

Associate Professor:
Tanja Bosak, (D), Caltech, 2004, Py
Kerri Cahoy, (D), Stanford, 2008
Dan Cziczo, (D), Chicago, 1999, Oa
Michael Follows, (D), 1990, Op
Colette Heald, (D), Harvard, 2005, Oa
Oliver Jagoutz, (D), ETH (Switzerland), 2004, GitGc
Paul O'Gorman, (D), Caltech, 2004, Oa
Shuhei Ono, (D), Penn State, 2001, PyCs
Taylor Perron, (D), California (Berkeley), 2006, Gm
Noelle Selin, (D), Harvard, 2007, Oa

Assistant Professor:
Andrew Babbin, (D), Princeton Univ, 2014, Py
Kristin Bergmann, (D), Caltech, 2013, GsCl
Timothy W. Cronin, (D), MIT, 2014, Oa
Gregory Fournier, (D), Connecticut, 2009, Py
David McGee, (D), Columbia, 2009, Cl

Senior Lecturer:
Lodovica Illari, (D), Imperial Coll (UK), 1982, Ow

Lecturer:
Amanda Bosh, (D), MIT, 1994

Emeritus:
B. Clark Burchfiel, (D), Yale, 1961, Gc
Charles C. Counselman, III, (D)
J. Brian Evans, (D), MIT, 1978, Yx
Frederick A. Frey, (D), Wisconsin, 1967, CtGvi
Richard S. Lindzen, (D), Harvard, 1964, Ow
Gordon H. Pettengill, (D), California (Berkeley), 1955, Og
Raymond A. Plumb, (D), Manchester, 1972, Ow
M. Gene Simmons, (D), Harvard, 1962, Yg
John B. Southard, (D), Harvard, 1966, Gs
Peter H. Stone, (D), Harvard, 1964, Oa
M Nafi Toksoz, (D), Caltech, 1963, Ys
Carl I. Wunsch, (D), MIT, 1967, Op

Principal Research Engineer:
Christopher Hill, (D)

## Mount Holyoke College

**Dept of Geology** (B) (2016)
50 College Street
Clapp Laboratory #304
South Hadley, MA 01075-6419
p. (413) 538-2278
rforjwuo@mtholyoke.edu
http://www.mtholyoke.edu/acad/geology
f: https://www.facebook.com/groups/mhcgeoalums/
Administrative Assistant: Rhodaline Forjwuor
*Enrollment (2016): B: 22 (14)*

Professor:
Steven R. Dunn, (D), Wisconsin, 1989, Gp
Girma Kebbede, (D), Syracuse, 1981, Oy
Mark McMenamin, (D), California (Santa Barbara), 1984, PgGst
Thomas L. Millette, (D), Clark, 1989, OriGm
Alan Werner, (D), Colorado, 1988, GlOuGm

Associate Professor:
Michelle J. Markley, (D), Minnesota, 1998, GcgGt

Assistant Professor:
Serin D. Houston, (D), Syracuse, 2012, Oyn

Geoprocessing Lab Manager:
Eugenio J. Marcano, (D), Cornell, 1994, OiSoOn

Visiting Assistant Professor:
Samuel Tuttle, (D), Boston Univ, 2015, Gq

Emeritus:
Martha M. Godchaux, (D), Oregon, 1969, Gv

Laboratory Director:
Penny M. Taylor, (M), SUNY (Oneonta), 2000, Gg

Geology Technician:
Gerard Marchand, (B), Westfield State, 1990, Gga

## Northeastern University

**Dept of Marine and Environmental Sciences** (B,M,D) ○ (2017)
14 Holmes Hall
360 Huntington Ave

Boston, MA 02115
p. (617) 373-3176
environment@neu.edu
http://www.northeastern.edu/mes
f: https://www.facebook.com/northeastern.mes
Administrative Assistant: Danielle Walquist. Lynch
*Enrollment (2016): B: 277 (52) M: 32 (7) D: 34 (3)*

Professor:
Joseph Ayers, (D), California, Santa Cruz, 1975, On
Richard H. Bailey, (D), North Carolina, 1973, Pi
William Detrich, (D), Yale, On
Brian Helmuth, (D), Washington, 1997, On
Mark Patterson, (D), Harvard, 1995, On
Hanumant Singh, (D), MIT, 1995, On

Chair:
Geoffrey Trussell, (D), William & Mary, 1998, On

Associate Chair:
Rebeca Rosengaus, (D), Boston Univ, 1993, On

Associate Professor:
Jonathan Grabowski, (D), North Carolina (Chapel Hill), 2012, GuEg
Malcolm D. Hill, (D), California (Santa Cruz), 1979, OiGze
Justin Ries, (D), Johns Hopkins, 2005, Cm
Martin E. Ross, (D), Idaho, 1978, Gie
Steven Scyphers, (D), South Alabama, 2012, On

Assistant Professor:
Jennifer Bowen, (D), Boston Univ, On
Loretta Fernandez, (D), MIT, 2010, Co
Tarik Gouhier, (D), McGill, On
Randall Hughes, (D), California (Davis), 2006, On
David Kimbro, (D), California, On
Kathleen Lotterhos, (D), Florida State, 2011, On
Amy Mueller, (D), MIT, 2012, On
Steve Vollmer, (D), Harvard, On

Lecturer:
Daniel C. Douglass, (D), Wisconsin (Madison), GlSdOa
Tara Duffy, (D), Stony Brook, Ob
Stephanie Eby, (D), Syracuse Univ, 2010, On

Emeritus:
Donald Cheney, (D), South Florida, On
Gwilym Jones, (D), Indiana State, On
Peter S. Rosen, (D), William & Mary, 1976, Onu

Business and Operations Manager:
Heather Sears, (D), Massachusetts Inst of Technology, 2004, On

Co-op Coordinator:
Sarah Klionsky, (M), Wisconsin (Madison), 2009

## Salem State University

**Geological Sciences Dept** (B) ○ (2017)
352 Lafayette Street
Salem, MA 01970
p. (978) 542-6282
dallen@salemstate.edu
*Enrollment (2009): B: 67 (13)*

Chair:
Douglas Allen, (D), Minnesota, 2003, CgHsCt

Professor:
James L. Cullen, (D), Brown, 1984, GsPmOu
Lindley S. Hanson, (D), Boston Univ, 1988, GmlOn

Associate Professor:
J Bradford Hubeny, (D), Rhode Island, 2005, GeOnCs

Assistant Professor:
Sara Mana, (D), Gci

Visiting Professor:
Patricia Nadeau, (D), Gv

Laboratory Director:
Renee Knudstrup, (B), Ca

## Smith College

**Dept of Geosciences** (B) (2016)
Clark Science Center
44 College Lane
Northampton, MA 01063
p. (413) 585-3805
dkortes@smith.edu
http://www.science.smith.edu/departments/Geology/
Administrative Assistant: Donna M. Kortes
*Enrollment (2010): B: 27 (6)*

Chair:
Bosiljka Glumac, (D), Tennessee, 1997, Gs

Professor:
John B. Brady, (D), Harvard, 1975, Gx
Robert M. Newton, (D), Massachusetts, 1978, Gm
Amy L. Rhodes, (D), Dartmouth, 1996, ClGe

Associate Professor:
Sara B. Pruss, (D), S California, 2004, Pg

Assistant Professor:
Jack Loveless, (D), Cornell, 2008, Gtc

Lecturer:
Mark E. Brandriss, (D), Stanford, 1994, Gx

Emeritus:
H. Robert Burger, (D), Indiana, 1966, GcYeOi
H. Allen Curran, (D), North Carolina, 1968, Pg

Geoscience Technician:
Michael Vollinger, (B), Massachusetts, 1993

## Tufts University

**Dept of Earth and Ocean Sciences** (B) (2016)
Lane Hall
Medford, MA 02155
p. (617) 627-3494
jack.ridge@tufts.edu
http://eos.tufts.edu/
f: https://www.facebook.com/TuftsEOS
Administrative Assistant: Janet Silvano
*Enrollment (2016): B: 20 (5)*

Professor:
John C. Ridge, (D), Syracuse, 1985, GlnGg

Professor:
G Garven, (D), British Columbia, 1982, HwqHy

Associate Professor:
Anne F. Gardulski, (D), Syracuse, 1987, Grs

Assistant Professor:
Andrew Kemp, (D), Pennsylvania, 2009, OnPmOg

Lecturer:
Jacob Benner, (M), Utah, 2002, Pge

Emeritus:
Robert L. Reuss, (D), Michigan, 1970, Gzi

## University of Massachusetts Lowell

**Dept of Environmental, Earth, & Atmospheric Sciences** (B,M) (2016)
1 University Avenue
University of Massachusetts
Lowell, MA 01854
p. (978) 934-3900
Erica_Gavin@uml.edu
https://www.uml.edu/sciences/eeas/
f: https://www.facebook.com/Earth.Sciences.UMass.Lowell?ref=hl
*Enrollment (2016): B: 103 (14) M: 13 (9)*

Chair:
Daniel Obrist, (D), Nevada, 2002, Oa

Professor:
Frank P. Colby, (D), MIT, 1983, Ow
G. Nelson Eby, (D), Boston, 1971, Cg

Associate Professor:
Mathew Barlow, (D), Maryland, 1999, Oa
Jian-Hua Chen, (D), North Carolina State, 1996, Oa

Assistant Professor:
Richard M. Gaschnig, (D), Washington State, 2010, CctCa
Kate Swanger, (D), Boston, 2009, Gl

Lecturer:
Lori Weeden, (M), Boston Univ, 2002, Gge

Emeritus:
Arnold L. O'Brien, (D), Boston, 1973, Hw

## University of Massachusetts, Amherst

**Dept of Geosciences** (B,M,D) ● (2017)
233 Morrill Science Center
611 North Pleasant St.
Amherst, MA 01003-9297
p. (413) 545-2286
juliebg@geo.umass.edu

http://www.geo.umass.edu/
f: https://www.facebook.com/UMass-Geosciences
*Enrollment (2015): B: 99 (35) M: 25 (10) D: 35 (3)*

Department Head:
Julie Brigham-Grette, (D), Colorado, 1985, GlPe
Distinguished University Professor:
Raymond S. Bradley, (D), Colorado, 1974, Oa
Professor:
Stephen J. Burns, (D), Duke, 1987, Cs
Michele L. Cooke, (D), Stanford, 1996, Gc
Robert DeConto, (D), Colorado, 1996, Oa
Piper Gaubatz, (D), California (Berkeley), 1968, Oy
R. Mark Leckie, (D), Colorado, 1984, PmGru
J. Michael Rhodes, (D), Australian National, 1970, GviCa
Sheila J. Seaman, (D), New Mexico, 1988, Gi
Michael L. Williams, (D), New Mexico, 1987, Gc
Associate Extension Professor:
William P. Clement, (D), Wyoming, 1995, Yge
Associate Professor:
David Boutt, (D), New Mexico Tech, 2004, HwqHy
Steven Petsch, (D), Yale, 2000, Col
Stan Stevens, (D), California (Berkeley), 1983, Oy
Eve Vogel, (D), Oregon, 2007, Oy
Jonathan D. Woodruff, (D), MIT, 2008, Gs
Qian Yu, (D), California (Berkeley), 2005, Ori
Extension Assistant Professor:
Christine Hatch, (D), California (Santa Cruz), 2007, Hw
Michael Rawlins, (D), New Hampshire, 2006, Oa
Assistant Professor:
Isla Castaneda, (D), Minnesota, 2007, Ca
Haiying Gao, (D), Rhode Island, 2012, Ys
Isaac J. Larsen, (D), Washington, 2013, Gm
Assistant Professor :
Forrest J. Bowlick, (D), Texas A&M, 2016, Oig
Lecturer:
Christopher D. Condit, (D), New Mexico, 1984, Gi
Michael J. Jercinovic, (D), New Mexico, 1988, Gx
Adjunct Professor:
Douglas R. Hardy, (D), Massachusetts, 1995, Hy
Thomas C. Johnson, (D), California (San Diego), 1975, PeCoOu
Douglas Kowaleski, (D), Boston, 2009, Gm
Eileen McGowan, (D), Massachusetts, 2010, Xg
Thomas L. Millette, (D), Clark, 1989, Oy
Stephen Nathan, (D), Massachusetts, 2005, PmOg
Liang Ning, (M), Nanjing, 2007, Ow
Peter T. Panish, (D), Massachusetts, 1989, Gp
Nicholas Venti, (D), Delaware, 2012, Og
Emeritus:
Laurie Brown, (D), Oregon State, 1974, Ym
James A. Hafner, (D), Michigan, 1970, On
John Hubert, (D), Penn State, 1958, Gs
William D. McCoy, (D), Colorado, 1981, Oy
George E. McGill, (D), Princeton, 1958, Gc
Stearns A. Morse, (D), McGill, 1962, Gi
Rutherford H. Platt, (D), Chicago, 1971, Ou
Peter Robinson, (D), Harvard, 1963, Gc
Richard W. Wilkie, (D), Washington, 1968, OynOn
Richard F. Yuretich, (D), Princeton, 1976, Cl
Massachusetts State Geologist:
Stephen B. Mabee, (D), Massachusetts, 1992, GgHw

## University of Massachusetts, Boston

**School for the Environment** (B,M,D) (2014)
100 Morrissey Boulevard
Boston, MA 02125
p. (617) 287-7440
sfe@umb.edu
http://www.umb.edu/environment
*Enrollment (2010): B: 153 (27) M: 22 (12) D: 18 (4)*

Director:
Jack Wiggin, (M), Boston State Coll, 1981, Ou
Dean:
Robyn Hannigan, (D), Rochester, ClmCt
Chair, Biology:
Rick Kesseli, (D), California (Davis), 1985, Obn
Associate Dean, College of Management:
David Levy, (D), Harvard, On
Professor:
Bob Chen, (D), California (San Diego), 1992, CoOc
Ron Etter, (D), Harvard, 1987, ObPoOn
Zhongping Lee, (D), South Florida, Org
William Robinson, (D), Northeastern, 1981, On
Crystal Schaaf, (D), Boston, 1994, Orw
Michael Shiaris, (D), Tennessee, 1979, On
David Terkla, (D), California (Berkeley), 1979, Onn
Roberta Wollons, (D), Chicago, On
Wei Zhang, (D), Pittsburgh, On
Meng Zhou, (D), SUNY (Stony Brook), Op
VP for Research, New England Aquarium:
Scott Krauss, (D), New Hampshire, 2002, Ob
Director for Research, New England Aquarium:
Michael Tlusty, (D), Syracuse, 1996, Ob
Senior Scientist:
Moira Brown, (D), Guelph, 1995, Ob
Julie Cavin, North Carolina, 2007, Ob
Phillip Hamilton, (M), Massachusetts (Boston), 2002, Ob
Kathleen Hunt, (D), Washington, 1997, Ob
Charles Innis, (D), Penn, 1994, Ob
Meghan Jeans, (D), Vermont Law School, 2002, On
Amy Knowlton, (M), Rhode Island, 1997, Ob
John Mandelman, (D), Northeastern, 2006, Ob
Daniel Pendleton, (D), Cornell, 2010, Ob
Rosalind Rolland, (D), Tufts, 1984, Ob
Heather Tausig, (M), Boston, 1993, Og
Timothy Werner, (M), Stanford, Ob
Director, Environmental Studies Program:
Alan D. Christian, (D), Miami, 2002, HsGmCs
Co-Director, Center for Governance and Sustainability:
Maria Ivanova, (D), Yale, On
Chair, English:
Cheryl Nixon, (D), Harvard, On
Associate Professor:
Conevery Bolton Valencius, (D), Harvard, 1998, Gh
Robert Bowen, (D), S California, 1981, OnnOn
Amy Den Ouden, (D), Connecticut, OnGa
Ellen Douglas, (D), Tufts, 2002, HwqHy
John Duff, (D), Washington, 1995, OnnOn
Eugene Gallagher, (D), Washington, 1983, Ob
Allen Gontz, (D), Maine, 2005, GmlYr
Jose Martinez-Reyes, (D), Massachusetts, GaOn
Deyang Qu, (D), Ottawa, Om
Joshua Reid, (D), California (Davis), Ga
Karen Ricciardi, (D), Vermont, Gq
Juanita Urban-Rich, (D), Memorial, Ob
Associate Scientist:
Andrew Rhyne, (D), Florida Inst of Tech, 2006, Ob
Randi Rotjan, (D), Tufts, 2007, Ob
Assistant Professor:
Jennifer Bowen, (D), Boston, 2005, ObPyOn
Jarrett Byrnes, (D), California (Davis), 2008, ObnOn
Steven Gray, (D), Rutgers, 2010, On
Nardia Haigh, (D), Queensland, On
Helen Poynton, (D), California (Berkeley), On
David Timmons, (D), Massachusetts, On
Research Associate:
Brooke Wikgren, (M), Miami, 2010, Oi
Lecturer:
Deborah Metzel, (D), Maryland, 1998, EgOuy
Michael Trust, Or
Adjunct Professor:
Michael E. Brookfield, (D), Reading (UK), 1973, Gst
Director, Nantucket Field Station:
Sarah Oktay, (D), Texas A&M, Oc
Research Engineer:
Francesco Peri, (M), Massachusetts (Boston), On
Program Manager, Nantucket Semester Program:
Elizabeth Boyle, (D), Massachusetts (Boston), Obn
Manager, GIS Lab:
Helenmary Hotz, (M), Massachusetts (Boston), Oi
Director, Green Harbors Project:
Anamarija Frankic, (D), VIMS, Oe

## Wellesley College

**Dept of Geosciences** (B) (2011)
106 Central Street
Wellesley, MA 02481-8203
p. (781) 283-3151
dbraband@wellesley.edu
http://www.wellesley.edu/Geosciences/
Administrative Assistant: Rita Purcell
*Enrollment (2011): B: 25 (9)*

Chair:
Daniel J. Brabander, (D), Brown, 1997, Cl
Associate Professor:
James Besancon, (D), MIT, 1975, Hww
David Hawkins, (D), MIT, 1996, Gzi
Emeritus:
Margaret D. Thompson, (D), Harvard, 1976, Gc
Instructor:
Kathleen W. Gilbert, (M), Miami, 1995, Cs

## Williams College

**Dept of Geosciences** (B) (2017)
947 Main Street
Williamstown, MA 01267
p. (413) 597-2221
patricia.e.acosta@williams.edu
http://www.williams.edu/Geoscience
f: https://www.facebook.com/groups/williamsgeosciences/?fref=ts
Administrative Assistant: Patricia E. Acosta
*Enrollment (2016): B: 37 (16)*

Chair:
Ronadh Cox, (D), Stanford, 1993, GsXg
Professor:
David P. Dethier, (D), Washington, 1977, Gm
Paul Karabinos, (D), Johns Hopkins, 1981, Gc
Reinhard A. Wobus, (D), Stanford, 1966, Giz
Associate Professor:
Mea S. Cook, (D), MIT/WHOI, 2006, OuGe
Assistant Professor:
Phoebe A. Cohen, (D), Harvard, 2010, Poy
Jose A. Constantine, (D), California (Santa Barbara), 2008, GmOi
Research Associate:
B. Gudveig Baarli, (D), Oslo, 1989, PssPi
Mark E. Brandriss, (D), Stanford, 1993, Gi
Lecturer:
Alex Apotsos, (D), MIT, 2007, Ge
Emeritus:
William T. Fox, (D), Northwestern, 1962, Gs
Markes E. Johnson, (D), Chicago, 1977, Ps
Cooperating Faculty:
Helena F. Warburg, (M), Indiana Sch Lib Sci, 1987, On

## Woods Hole Oceanographic Institution

**Dept of Geology & Geophysics** (D) (2016)
Woods Hole, MA 02543-1541
p. (508) 289-2388
etulka@whoi.edu
http://www.whoi.edu/page.do?pid=7145
Administrative Assistant: Maryanne F. Ferreira
*Enrollment (2014): D: 26 (0)*

Chair:
Daniel C. McCorkle, (D), Washington, 1987, Cm
Senior Scientist:
Mark D. Behn, (D), MIT/WHOI, 2002, Yr
Joan M. Bernhard, (D), California (San Diego), 1990, Cg
Henry J B. Dick, (D), Yale, 1975, GitGc
Jeffrey Donnelly, (D), Brown, 2000, Gu
Robert L. Evans, (D), Cambridge (UK), 1991, Yr
Daniel J. Fornari, (D), Columbia, 1978, Gu
Chris German, (D), Cambridge, 1988, Gt
Susan E. Humphris, (D), MIT/WHOI, 1977, GuCm
Lloyd D. Keigwin, (D), Rhode Island, 1979, CsGu
Jian Lin, (D), Brown, 1988, Gt
Olivier Marchal, (D), Paris, 1996, Pe
Delia W. Oppo, (D), Columbia, 1989, Pm
Deborah K. Smith, (D), California, 1985, Yr
Ralph A. Stephen, (D), Cambridge, 1978, Ys
Maurice A. Tivey, (D), Washington, 1988, Gt
Tenured Associate Scientist:
Andrew Ashton, (D), Duke, 2005, On
Juan Pablo Canales Cisneros, (D), Barcelona, 1997, Yr
Anne L. Cohen, (D), Cape Town, 1993, Pe
Sarah B. Das, (D), Penn State, 2003, Pm
Virginia Edgcomb, (D), Delaware, 1997, Cg
Glenn A. Gaetani, (D), MIT, 1996, Gi
Liviu Giosan, (D), SUNY (Stony Brook), 2001, Gu
Daniel Lizarralde, (D), MIT/WHOI, 1997, Ys
Jeffrey J. McGuire, (D), MIT, 2000, Ys
Robert A. Sohn, (D), California (San Diego), 1996, Yr
S. Adam Soule, (D), Oregon, 2003, Gvu
Associate Scientist:
Weifu Guo, (D), Caltech, 2008, CslPe
Sune G. Nielsen, (D), ETH (Switzerland), 2005, Cs
Assistant Scientist:
Veronique Le Roux, (D), Macquarie, 2008, Gi
Senior Research Specialist:
James E. Broda, (B), Penn State, 1970, Gs
John A. Collins, (D), MIT/WHOI, 1989, Ys
Ann P. McNichol, (D), MIT, 1986, Oc
Mark L. Roberts, (D), Duke, 1988, Yg
Research Specialist:
Jurek Blusztajn, (D), Polish Acad of Sci, 1985, CcGi
Alan R. Gagnon, (B), New Hampshire, 1983, Cs
Li Xu, (D), Xiamen, 1992, Yr
Engineer:
Peter B. Landry, (B), NEIT, 1990, Ca
Research Associate:
Kathryn L. Elder, (B), Massachusetts, 1985, Cs
James R. Elsenbeck, (M), MIT/WHOI, 2007, Yr
Kalina D. Gospodinova, (M), MIT/WHOI, 2012, Cm
Joshua D. Hlavenka, (B), North Texas
Peter C. Lemmond, (B), Lehigh, 1978, On
Brett Longworth, (M), Massachusetts, 2005, Ou
Brian D. Monteleone, (D), Syracuse, 2000, Cc
Kathryn R. Pietro, (M), California (Davis), 2007, Gu
Adjunct Scientist:
Peter D. Bromirski, (D), Hawaii, 1993
Johnson R. Cann, (D), Cambridge, 1963, Ys
Colin Devey, (D), Oxford, 1986
Javier Escartin, (D), MIT, 1996, Gg
Andrea D. Hawkes, (D), Pennsylvania, 2008
Gregory Hirth, (D), Brown, 1991, Gc
Kuo-Fang (Denner) Huang, (D), National Cheng Kung, 2007
Peter B. Kelemen, (D), Washington, 1987, Gi
Yajing Liu, (D), Harvard, 2007
John Maclennan, (D), Cambridge, 2000
Larry Mayer, (D), Scripps, 1979, Gg
Andrew M. McCaig, (D), Cambridge, 1983
Jerry F. McManus, (D), Columbia, 1997, Pe
Uri S. ten Brink, (D), Columbia, 1986, Yg
David Thornalley, (D), Churchill, 2008
Masako Tominaga, (D), Texas A&M, 2009
Oceanographer Emeritus:
Graham S. Giese, (D), Chicago, 1966, On
Steven J. Manganini, (B), Nasson, 1974, Ou
Robert J. Schneider, (D), Oberlin, 1968, Yr
Stephen A. Swift, (D), MIT/WHOI, 1986, Yr
Karl F. Von Reden, (D), Hamburg, 1983, Ct
Emeritus:
William A. Berggren, (D), Stockholm, 1962, Ps
Carl O. Bowin, (D), Princeton, 1960, YrGtg
William B. Curry, (D), Brown, 1980, Pm
Stanley R. Hart, (D), MIT, 1960, CgGvCp
John M. Hayes, (D), MIT, 1966, Ct
Susumu Honjo, (D), Hokkaido, 1961, Ou
George P. Lohmann, (D), Brown, 1972, Pe
David A. Ross, (D), California (San Diego), 1965, GuOg
Hans Schouten, (D), Utrecht (Neth), 1970, Yr
Nobumichi Shimizu, (D), Tokyo, 1968, Ca
William G. Thompson, (D), Columbia, 2005, PeCc
Brian E. Tucholke, (D), MIT/WHOI, 1973, Gut
Elazar Uchupi, (D), S California, 1962, Ou
Frank B. Wooding, (B), Harvard, 1965, Yr

**Dept of Marine Chemistry & Geochemistry** (M,D) (2017)
360 Woods Hole Road
MS 25
Woods Hole, MA 02543-1541
p. (508) 289-2328
mcg@whoi.edu
http://www.whoi.edu/page.do?pid=7146
Administrative Assistant: Linda Cannata
Administrative Assistant: Sheila A. Clifford
Administrative Assistant: Donna Mortimer
Administrative Assistant: Mary Zawoysky
*Enrollment (2015): D: 28 (5)*

Chair:
Scott C. Doney, (D), MIT, 1991, Cm
Scientist Emeritus:
Werner G. Deuser, (D), Penn State, 1963, CsGsu
Associate Dean:
Margaret K. Tivey, (D), Washington, 1989, Cm
Senior Scientist:
Ken O. Buesseler, (D), MIT/WHOI, 1986, Oc
Matthew A. Charette, (D), Rhode Island, 1998, Cm
Konrad A. Hughen, (D), Colorado, 1997, Cc
Mark D. Kurz, (D), MIT/WHOI, 1982, Cc
Bernhard Peucker-Ehrenbrink, (D), Max Planck Inst, 1994, CmlHs
Christopher M. Reddy, (D), Rhode Island, 1997, Co
Daniel J. Repeta, (D), MIT/WHOI, 1982, Co
Jeffrey S. Seewald, (D), Minnesota, 1990, Cp
James A. Yoder, (D), Rhode Island, 1979, Cm
Associate Scientist:
Elizabeth B. Kujawinski, (D), MIT/WHOI, 2000, Oc
Valier Galy, (D), Institut National Polytechnique de Lorrain, 2007, Cm
Colleen Hansel, (D), Stanford, 2004, Cm
Frieder Klein, (D), Bremen, 2009, Cm
Tracy Mincer, (D), California (San Diego), 2004, Cm
Mak A. Saito, (D), MIT/WHOI, 2001, Oc
Amanda C. Spivak, (D), William & Mary, 2008, CmoOn
Benjamin Van Mooy, (D), Washington, 2003, Oc
Z. Aleck Wang, (D), Georgia, 2003, Cm
Assistant Scientist:
Amy Apprill, (D), Hawaii, 2009, Ob
Research Associate:
Joshua M. Curtice, (B), Massachusetts, 1992, Cc
Helen Fredricks, (D), Plymouth (UK), 2000, Oc
Matt McIlvin, (D), Massachusetts, 2004, Oc
Steven M. Pike, (M), Rhode Island, 1998, Ca
Sean Sylva, Oc
Information Systems Associate:
Cyndy Chandler, (B), SUNY (Geneseo), 1975, Og
Information Systems Assoc:
Ivan D. Lima, (D), Miami, 1999, Oc
Research Associate:
Dawn Moran, Ob
Jennie Rheuban, (M), Virginia, 2013, Oc
Melissa Soule, Oc
Gretchen Swarr, Oc
Kristen Whalen, (D), MIT, 2008, Ob
Adjunct Professor:
Minhan Dai, Cm
Thomas Trull, Cm
Senior Research Specialist:
Nelson M. Frew, (D), Washington, 1971, Ca
Dempsey E. Lott, (M), Florida State, 1973, Oc
Scientist Emeritus:
Michael P. Bacon, (D), MIT/WHOI, 1976, Oc
John W. Farrington, (D), Rhode Island, 1972, ComOc
Frederick L. Sayles, (D), Manchester, 1968, Oc
Geoffrey Thompson, (D), Manchester, 1965, Cm
Oliver C. Zafiriou, (D), Johns Hopkins, 1966, Oc
Oceanographer Emeritus:
Jean K. Whelan, (D), MIT, 1965, Co
Emeritus:
William J. Jenkins, (D), McMaster, 1974, Oc
William R. Martin, (D), MIT/WHOI, 1985, Oc
Edward R. Sholkovitz, (D), California (San Diego), 1972, Oc
Senior Research Specialist:
David M. Glover, (D), Alaska, 1985, Oc
Research Specialist:
Carl G. Johnson, (M), Maine, 1983, Ca
Robert K. Nelson, (B), C Connecticut State, 1980, Co
Senior Research Assistant:
Justin Ossolinski, Oc
Research Specialist:
Heather Benway, (D), Oregon State, 2005, Oc
Krista Longnecker, (D), Oregon State, 2004, Oc
Research Assistant:
Jessica Drysdale, Oc
Michaela Fendrock, Oc
Kelsey Gosselin, Oc
William Oestreich, Oc
Zoe Sandwith, Oc
Postdoc:
Tristan Horner, Oxford, 2012, Cm
Dept. Administrator:
Mary Murphy
Assistant Scientist:
David Nicholson, (D), Washington, 2009, Gu
Scott Wankel, (D), Stanford, 2007, Cm
Senior Research Assistant (ret.):
Margaret Sulanowska, Cm
Senior Research Assistant:
Joanne Goudreau, Cm
Research Associate:
Paul Henderson, (D), Cm
Research Assistant:
Kevin Cahill, Cm

## Worcester State University

**Earth, Environment and Physics** (B) (2016)
486 Chandler Street
Worcester, MA 01602-2597
p. (508) 929-8583
whansen@worcester.edu
http://www.worcester.edu/Earth-Environment-and-Physics/
f: https://www.facebook.com/worcesterstatedeep
*Enrollment (2016): B: 93 (11)*

Associate Professor:
Patricia A. Benjamin, (D), Clark, 2002, Og
Allison L. Dunn, (D), Harvard, 2006, OayHg
William J. Hansen, (D), CUNY, 2002, OiyOr
Assistant Professor:
Timothy L. Cook, (D), Massachusetts, 2009, GnsOg
Douglas E. Kowalewski, (D), Boston, 2009, GmlOy
Alexander Tarr, (D), California, Berkley, 2014, Og
Instructor:
Adam M. Davis, (D), Indiana, 2011, GgPyOy
Mark O. Johnson, (D), Clark, 1993, Oy
Ryan A. Portner, (D), Macquarie, 2010, GuvXg

# Michigan

## Adrian College

**Geology Dept** (A,B) (2015)
110 S. Madison St.
Adrian, MI 49221
p. (517) 265-5161
tmuntean@adrian.edu
http://adrian.edu/academics/academic-departments/geology/
f: https://www.facebook.com/groups/55153346895/
*Enrollment (2015): A: 1 (1) B: 14 (4)*

Chair:
Thomas Muntean, (D), Nevada, 2012, Gde
Professor:
Sarah L. Hanson, (D), Utah, 1995, Giz

## Albion College

**Dept of Geological Sciences** (B) O (2016)
611 E. Porter St.
Albion, MI 49224
p. (517) 629-0759
blincoln@albion.edu
http://www.albion.edu/geology/
*Enrollment (2016): B: 18 (9)*

co-chair:
Beth Z. Lincoln, (D), California (Los Angeles), 1985, Gc
Timothy N. Lincoln, (D), California (Los Angeles), 1978, ClHw
Professor:
William S. Bartels, (D), Michigan, 1986, Pv
Thomas I. Wilch, (D), New Mexico Tech, 1997, GlvGm
Associate Professor:
Carrie A. Menold, (D), California (Los Angeles), 2006, Gpz
Assistant Professor:
Michael McRivette, (D), California (Los Angeles), 2011, GtOir
Emeritus:
Russell G. Clark, (D), Dartmouth, 1972, GiOi
Lawrence D. Taylor, (D), Ohio State, 1962, Gl



## Calvin College

**Dept of Geology, Geography, & Environmental Studies** (B) (2014)
3201 Burton SE
Grand Rapids, MI 49546
p. (616) 526-8415
jbascom@calvin.edu
http://www.calvin.edu/academic/geology/
*Enrollment (2012): B: 60 (14)*
Chair:
Johnathan Bascom, (D), Iowa, 1989, Oy
Professor:
Henry Aay, (D), Clark, 1978, Oyu
Janel M. Curry, (D), Minnesota, 1985, Oy
Ralph F. Stearley, (D), Michigan, 1990, PvOgGh
Associate Professor:
Deanna van Dijk, (D), Waterloo, 1998, GmOny
Assistant Professor:
Kenneth A. Bergwerff, (M), Grand Valley State, 1988, Oe
James Skillen, (D), Cornell, 2006, Ou
Jason VanHorn, (D), Ohio State, 2007, Oi
Laboratory Manager/Instructor:
Margene Brewer, (M), W Michigan, 1991, OnGgOn
Emeritus:
Clarence Menninga, (D), Purdue, 1966, CcGg
Gerald K. Van Kooten, (D), California (Santa Barbara), 1980, GoCePi
Davis A. Young, (D), Brown, 1969, GihGz

## Central Michigan University

**Dept of Earth and Atmospheric Sciences** (B) (2014)
314 Brooks Hall
Mount Pleasant, MI 48859
p. (989) 774-3179
sven.morgan@cmich.edu
https://www.cmich.edu/colleges/cst/earth_atmos/Pages/default.aspx
*Enrollment (2013): B: 49 (14)*
Director:
James J. Student, (D), Virginia Tech, 2002, CaGiz
Professor:
Lawrence D. Lemke, (D), Michigan, 2003, HwGes
Sven S. Morgan, (D), Virginia Tech, 1998, Gc
Richard N. Mower, (D), Ow
Mona Sirbescu, (D), Missouri, 2002, Gi
Reed Wicander, (D), California (Los Angeles), 1973, Pl
Department Chair:
Leigh Orf, (D), Wisconsin, 1997, Ow
Associate Professor:
Martin Baxter, (D), St. Louis, 2006, Ow
Patrick Kinnicutt, (D), MIT, 1995, Gq
Lecturer:
Maria Mercedes Gonzalez, (D), Nacional del Sur (Argentina), 1997, Gd

## Charles Stewart Mott Community College

**Divison of Science and Mathematics** (A) (2016)
1401 East Court St.
Flint, MI 48503
p. (810) 762-0279
sheila.swyrtek@mcc.edu
http://www.mcc.edu/science_math/index.shtml
Professor:
Sheila Swyrtek, (M), Minnesota (Duluth), 1996, Ga

## Concordia University

**Div of Natural Sciences** (A,B,M) (2013)
4090 Geddes Road
Ann Arbor, MI 48105-2750
p. (734) 995-7300
nskov@cuaa.edu
Assistant Professor:
James L. Refenes, (M), E Michigan, 2009, Oe

## Delta College

**Dept of Geology** (A) (2014)
1961 Delta Road
University Center, MI 48710-0002
p. (989) 686-9252
andreabair@delta.edu
Department Secretary: Barb Jurmanovich
*Enrollment (2011): A: 8 (4)*
Department Chair:
Timothy L. Clarey, (D), W Michigan, 1996, HwGcPv
Lecturer:
Mary C. Gorte, (M), Rensselaer, 1983, Gs
Emeritus:
Barry A. Carlson, (D), Michigan State, 1974, Yg
Paul A. Catacosinos, (D), Michigan State, 1972, Gr
Cooperating Faculty:
Kevin T. Dehne, (M), E Michigan, 1992, Oe

## Eastern Michigan University

**Dept of Geography & Geology** (B) (2014)
Ypsilanti, MI 48197
p. (743) 487-8589
yulanda.woods@EMICH.EDU
*Enrollment (2001): B: 135 (0)*
Head:
Michael Kasenow, (D), W Michigan, 1994, Hg
Richard A. Sambrook, (D)
Professor:
Eugene Jaworski, (D), Louisiana State, 1971, Or
Carl F. Ojala, (D), Georgia, 1972, Oa
Constantine N. Raphael, (D), Louisiana State, 1967, Gu
Associate Professor:
Steven T. LoDuca, (D), Rochester, 1990, Pi
Assistant Professor:
Kevin Blake, (D), Wisconsin, 1998, Gl
Michael Bradley, (D), Utah, 1988, Gc
Maria-Serena Poli, (D), Padova, 1995, Pe

## Ferris State University

**Dept of Physical Sciences** (2016)
ASC-3021
Big Rapids, MI 49307
p. (616) 592-2580
heckf@ferris.edu
http://www.ferris.edu/htmls/colleges/artsands/Physical-Sciences/HOME-Physical-Sciences.htm
Head:
David Frank, (D), Purdue, 1985
Professor:
Frederick R. Heck, (D), Northwestern, 1987, Gg

## Gogebic Community College

**Math-Science Div** (A) (2013)
E4946 Jackson Rd.
Ironwood, MI 49938
p. (906) 932-4231
admissions@gogebic.edu
http://www.gogebic.edu/academics/Math_Science/
Professor:
Bill Perkis, Gg

## Grand Rapids Community College

**Physical Sciences Dept** (A) (2016)
143 Bostwick Avenue, NE
Grand Rapids, MI 49503
p. (616) 234-4248

jqualls@grcc.edu
http://www.grcc.edu/physci
*Enrollment (2015): A: 2 (1)*
Assistant Professor:
Tari Mattox, (M), N Illinois, 1984, GgvCg

## Grand Valley State University

**Dept of Geology** (B) (2016)
1 Campus Drive
118 Padnos Hall
Allendale, MI 49401
p. (616) 331-3728
geodept@gvsu.edu
http://www.gvsu.edu/geology
Administrative Assistant: Janet H. Potgeter
*Enrollment (2016): B: 155 (24)*
Head:
Figen A. Mekik, (D), Massachusetts, 1992, Gxc
Professor:
Patrick M. Colgan, (D), Wisconsin, 1996, GmlPe
Stephen R. Mattox, (D), N Illinois, 1992, OeGv
Virginia L. Peterson, (D), Massachusetts, 1992, Gxc
Patricia E. Videtich, (D), Brown, 1982, Gd
John C. Weber, (D), Northwestern, 1995, Gc
Associate Professor:
Kevin C. Cole, (D), Arizona, 1990, Gz
Peter E. Riemersma, (D), Wisconsin, 1997, Hw
Peter J. Wampler, (D), Oregon State, 2004, HsOiGm
Assistant Professor:
Caitlin N. Callahan, (D), W Michigan, 2013, GgOg
Tara A. Kneeshaw, (D), Texas A&M, 2008, ClGe
Instructor:
Kelly L. Heid, (M), Mississippi State, 2011, Oe
Ryan G. Vannier, (D), Michigan State, 2014, CgGn
Visiting Professor:
Jeremy C. Gouldey, (D), Northwestern Univ, 2015, CgPeCg
Kevin G. Thaisen, (D), Tennessee, 2012, XgGxOi
Emeritus:
Thomas E. Hendrix, (D), Wisconsin, 1960, Gc
William J. Neal, (D), Missouri, 1968, GsdGr
Norman W. Ten Brink, (D), Washington, 1971, Gm

## Henry Ford Community College

**Science Div** (2004)
Dearborn, MI 48128
p. (313) 845-9632
cjacobs@hfcc.edu

## Hope College

**Dept of Geological & Environmental Sciences** (B) (2017)
35 E 12th Street
P.O. Box 9000
Holland, MI 49422-9000
p. (616) 395-7540
hansen@hope.edu
http://www.hope.edu/academic/geology/
*Enrollment (2015): B: 24 (8)*
Acting Chair:
Edward C. Hansen, (D), Chicago, 1983, GmpGc
Professor:
Brian E. Bodenbender, (D), Michigan, 1994, PiGs
Jon W. Peterson, (D), Chicago, 1989, Ge
Adjunct Professor:
Suzanne DeVries-Zimmerman, (M), Princeton, 1989, Gge

## Jackson College

**Dept of Geology & Geography** (A,B) (2017)
2111 Emmons Road
Jackson, MI 49201
AlbeeScSteven@jccmi.edu
http://www.jccmi.edu/academics/science/geo
*Enrollment (2012): A: 200 (20)*
Wilbur L. Dungy Endowed Chair in the Sciences:
Steven R. Albee-Scott, (D), Michigan, 2005, GePeSf

## Lake Michigan College

**Lake Michigan College** (A) (2014)
Napier Avenue Campus
2755 E. Napier Avenue
Benton Harbor, MI 49022
lovett@lakemichigancollege.edu
http://www.lakemichigancollege.edu/index.php?option=com_content&task=view&id=509&Itemid=157
Instructor:
Cole Lovett, (D), W Michigan, 1995, Gg

## Lake Superior State University

**Geology and Physics** (B) ○ (2015)
650 W. Easterday Avenue
Sault Ste. Marie, MI 49783
p. (906) 635-2267
pkelso@lssu.edu
http://geology.lssu.edu/
Administrative Assistant: Donna White
*Enrollment (2014): B: 37 (7)*
Professor:
Paul R. Kelso, (D), Minnesota, 1993, YmGct
Assistant Professor:
Anna Lindquist, (D), Minnesota, 2013, YmGz
Robin Mattheus, (D), North Carolina, 2009, GsOn
Matt Spencer, (D), Penn State, 2005, Gl
Emeritus:
Lewis M. Brown, (D), New Mexico, 1973, PiOe
Science Lab Manager:
Benjamin Southwell, (M), Central Michigan, 2012, On

## Macomb Community College, Center Campus

**Dept of Science (Geology)** (A) (2014)
44575 Garfield Road
Clinton Township, MI 48038-1139
p. (586) 286-2154
schaferc@macomb.edu
*Enrollment (2014): A: 9 (0)*
Professor:
Carl M. Schafer, (M), Montana, 1998, Gg

## Michigan State University

**Dept of Earth and Environmental Sciences** (B,M,D) (2015)
288 Farm Lane
East Lansing, MI 48824-1115
p. (517) 355-4626
geosci@msu.edu
https://glg.natsci.msu.edu/
f: Geological Sciences
*Enrollment (2014): M: 4 (0) D: 20 (0)*
Chair:
Ralph E. Taggart, (D), Michigan State, 1971, Pb
Professor:
jiquan Chen, (D), Washington, 1991, Og
Kazuya Fujita, (D), Northwestern, 1979, Ys
Julie C. Libarkin, (D), Arizona, 1999, Oen
David T. Long, (D), Kansas, 1977, ClGeb
Michael A. Velbel, (D), Yale, 1984, ClGdSc
Associate Professor:
Bruno Basso, (D), Michigan State, 2002, On
Danita S. Brandt, (D), Yale, 1985, PiGr
Michael D. Gottfried, (D), Kansas, 1991, Pv
David W. Hyndman, (D), Stanford, 1995, Hy
Assistant Professor:
Tyrone Rooney, (D), Penn State, 2006, Gi
Matt Schrenk, (D), Washington, 2005, Py
Jay Zarnetske, (D), Oregon State, 2011, Hwg

## Michigan Technological University

**A. E. Seaman Mineral Museum** (2015)
1404 E. Sharon Avenue
Houghton, MI 49931
p. (906) 487-2572
tjb@mtu.edu
http://www.museum.mtu.edu

Director:
Theodore J. Bornhorst, (D), New Mexico, 1980, EgCgGx
Associate Curator:
Christopher J. Stefano, (D), Michigan, 2010, GzxCg



**Dept of Geological & Mining Engineering & Sciences** (B,M,D) O (2016)
1400 Townsend Drive
Dow Building, Room 630
Houghton, MI 49931-1295
p. (906) 487-2531
geo@mtu.edu
http://www.mtu.edu/geo
Administrative Assistant: Kelly M. McLean
*Enrollment (2016): B: 115 (13) M: 39 (19) D: 17 (3)*
Chair & Professor:
John S. Gierke, (D), Michigan Tech, 1990, Hy
Provost and Acting Dean, Graduate School:
Jacqueline E. Huntoon, (D), Penn State, 1990, Gs
Professor:
Alex S. Mayer, (D), North Carolina, 1992, Hy
James R. Wood, (D), Johns Hopkins, 1973, Cl
Associate Professor:
Simon Carn, (D), Cambridge, 1999, OrGv
Thomas Oommen, (D), Tufts, 2009, NeOiNg
Aleksey K. Smirnov, (D), Rochester, 2002, Yv
Gregory P. Waite, (D), Utah, 2004, YsGv
Shiliang Wu, (D), Harvard, 2007, OaCg
Assistant Professor:
Roohollah Askari, (D), Calgary, 2013, YxxNp
Snehamoy Chatterjee, (D), Indian Inst. of Technology, Nm
Ebrahim Tarshizi, (D), Nevada (Reno), 2014, Nm
Department Facilities Manager:
Robert Barron, (B), Michigan Tech, 1979, OgGzi
Instructor:
James M. Gillis, (B), Michigan Tech, 1983, Nmr
Adjunct Professor and Director, Great Lakes Research Ctr.:
Guy A. Meadows, (D), Purdue, 1977, Gu
Senior Lecturer:
Jeremy Shannon, (D), Michigan Tech, 2006, Gg
Postdoctoral Research Fellow and Temporary Faculty:
Rudiger Escobar-Wolf, (D), Michigan Tech, 2013, Gv
Research Scientist:
Carol Asiala, (B), Michigan Tech, 1985
Cooperating Faculty:
William I. Rose, (D), Dartmouth, 1970, Gv
Roger M. Turpening, (D), Michigan, 1966, Ye

## Mott Community College

**Science and Mathematics** (A) (2016)
1401 East Court St.
Flint, MI 48503
p. (810) 232-9312
sheila.swyrtek@mcc.edu
http://www.mcc.edu
*Enrollment (2015): A: 4 (0)*
Professor:
Sheila M. Swyrtek, (M), Minnesota (Duluth), 1997, GgOg

## Muskegon Community College

**Dept of Mathematics & Physical Sciences** (A) (2016)
221 S. Quarterline Road
Muskegon, MI 49442
p. (231) 777-0289
amber.kumpf@muskegoncc.edu
http://www.muskegoncc.edu/pages/652.asp
Administrative Assistant: Tamera Owens
Instructor:
Amber C. Kumpf, (M), Rhode Island, 2010, GgYrOu

## Northwestern Michigan College

**Northwestern Michigan College** (2014)
1701 East Front Street
Traverse City, MI 49686
p. (231) 995-1000
information@nmc.edu
http://www.nmc.edu

## Oakland Community College

**Natural Sciences Dept** (A) (2015)
2900 Featherstone Road
Auburn Hills, MI 48326
p. (248) 232-4538
lgkodosk@oaklandcc.edu

## Schoolcraft College

**Dept of Geology** (A) (2010)
18600 Haggerty Road
Livonia, MI 48151
p. (734) 462-4400
jrexius@schoolcraft.edu
*Enrollment: No data reported since 1999*
Instructor:
James E. Rexius, (M), E Michigan, 1978, Gl

## University of Michigan

**Dept of Climate and Space Sciences and Engineering** (B,M,D) (2015)
2455 Hayward
Ann Arbor, MI 48109-2143
p. (734) 615-3583
aoss-um@umich.edu
http://aoss.engin.umich.edu/
f: www.facebook.com/umclasp
t: @umclasp
*Enrollment (2015): B: 31 (14) M: 44 (45) D: 53 (6)*
Chair:
James Slavin, (D), California (Los Angeles)
Professor:
Sushil Atreya, (D), Michigan, Oy
John Boyd, (D), Harvard, 1976, Opa
R. Paul Drake, (D), Johns Hopkins
Lennard Fisk, (D), California (San Diego)
Brian Gilchrist, (D), Stanford
Tamas Gombosi, (D), Lóránd Eötvös
Michael Liemohn, (D), Michigan
Mark B. Moldwin, (D), Boston Univ, 1993, On
Joyce Penner, (D), Harvard
Nilton Renno, (D), MIT
Aaron Ridley, (D), Michigan
Richard B. Rood, (D), Florida State, 1982, Owa
Christopher Ruf, (D), Massachusetts (Amherst)
Perry Samson, (D), Wisconsin (Madison)
Thomas Zurbuchen, (D), Bern
Associate Professor:
Jeremy Bassis, (D), Scripps, 2007, OlgOg
Xianglei Huang, (D), Caltech
Christiane Jablonowski, (D), Michigan, 2004, Oaw
Xianzhe Jia, (D), California (Los Angeles)
Justin C. Kasper, (D), MIT, 2003
Susan Lepri, (D), Michigan
Derek Posselt, (D), Colorado State
Allison Steiner, (D), Georgia Tech
Shasha Zou, (D), California (Los Angeles)
Associate Research Scientist:
Natalia Ganjushkina, (D), Moscow State (Russia), 1997
Assistant Professor:
Mark Flanner, (D), California (Irvine)
Gretchen Keppel-Aleks, (D), Caltech
Eric A. Kort, (D), Harvard
Assistant Research Scientist:
Orenthal Tucker, (D)

**Dept of Earth and Environmental Sciences** (B,M,D) (2014)
2534 C.C. Little Building
1100 North University Avenue
Ann Arbor, MI 48109-1005
p. (734) 764-1435
mukasa@umich.edu
http://www.lsa.umich.edu/geo/
Administrative Assistant: Robert J. Patterer

Professor:
Tomasz K. Baumiller, (D), Chicago, 1990, Pg
Joel D. Blum, (D), Caltech, 1990, CsaCl
Maria Clara Castro, (D), Paris, 1995, Hy
Daniel C. Fisher, (D), Harvard, 1975, Pi
Stephen E. Kesler, (D), Stanford, 1966, Eg
Rebecca A. Lange, (D), California (Berkeley), 1989, Gi
Kyger C. Lohmann, (D), SUNY (Stony Brook), 1977, CsGsCl
Samuel B. Mukasa, (D), California (Santa Barbara), 1984, Cc
Henry N. Pollack, (D), Michigan, 1963, Yg
Jeroen Ritsema, (D), California (Santa Cruz), 1995, Ys
Larry J. Ruff, (D), Caltech, 1981, Ys
Gerald R. Smith, (D), Michigan, 1965, Ps
Ben van der Pluijm, (D), New Brunswick, 1984, GctOe
Rob van der Voo, (D), Utrecht (Neth), 1969, Ym
Lynn M. Walter, (D), Miami, 1983, Cl
Youxue Zhang, (D), Columbia, 1989, Cp
Associate Professor:
Udo Becker, (D), Virginia Tech, 1995, On
Robyn J. Burnham, (D), Washington, 1987, Pb
Todd A. Ehlers, (D), Utah, 2001, Yg
Benjamin Passey, (D), Utah, 2007, CsaCl
Christopher J. Poulsen, (D), Penn State, 1999, Pe
Peter J. van Keken, (D), Utrecht (Neth), 1989, Yg
Assistant Professor:
Marin Clark, (D), MIT, 2003, Gm
Ingrid Hendy, (D), California (Santa Barbara), 2000, Py
Nathan Niemi, (D), Caltech, 2001, Gc
Jeffrey A. Wilson, (D), Chicago, 1999, Gg
Research Scientist:
Jeffrey C. Alt, (D), Miami, 1984, Ou
Catherine E. Badgley, (D), Yale, 1982, Pv
Associate Research Scientist:
Chris Hall, (D), Toronto, 1982, Cc
Shaopeng Huang, (D), Acad Sinica, 1990, Yh
Josep M. Pares, (D), Barcelona, 1988, Ym
Assistant Research Scientist:
James D. Gleason, (D), Arizona, 1994, Cg
Jie Lian, (D), Michigan, 2003, Gz
Mirjam Schaller, (D), Bern, 2001, Gm
Adjunct Assistant Research Scientist:
Roland C. Rouse, (D), Michigan, 1972, Gz
Adjunct Assistant Professor:
Karen L. Webber, (D), Rice, 1988, Gv
Adjunct Professor:
John W. Geissman, (D), Michigan, 1980, Ym
William B. Simmons, (D), Michigan, 1973, Gz
Emeritus:
Charles Beck, (D), Cornell, 1955, Pg
William Farrand, (D), Michigan, 1960, Ga
William C. Kelly, (D), Columbia, 1954, Eg
Philip A. Meyers, (D), Rhode Island, 1972, Co
Theodore C. Moore, (D), California (San Diego), 1968, Ou
James O'Neil, (D), Chicago, 1963, Cs
Robert M. Owen, (D), Wisconsin (Madison), 1975, OcGu
Donald R. Peacor, (D), MIT, 1962, Gz
David K. Rea, (D), Oregon State, 1974, Gu
Bruce H. Wilkinson, (D), Texas (Austin), 1973, GseGt

## University of Michigan, Dearborn

**Dept of Natural Sciences** (B,M) (2013)
4901 Evergreen Road
Dearborn, MI 48128
p. (313) 593-5277
kmurray@umd.umich.edu
http://www.umd.umich.edu/?id=570101
*Enrollment (2013): B: 30 (4) M: 14 (3)*
Acting Chair:
Kent S. Murray, (D), California (Davis), 1981, Hw
Professor:
Don Bord, (D), Dartmouth, 1976, Xy
Associate Professor:
Jacob Napieralski, (D), Purdue, 2004, GmlOi
John Riebesell, (D), Chicago, 1975, Ou
Lecturer:
David Matzke, (M), Michigan, 1975, Xy

Adjunct Professor:
Michael Favor, (B), Michigan, 1985, On

## Washtenaw Community College

**Dept of Geology** (A) (2010)
4800 E. Huron River Drive
Ann Arbor, MI 48105
p. (734) 677-5111
salbach@wccnet.edu
Head:
Suzanne M. Albach, (M), Mississippi State, 2005, OegGe

## Wayne State University

**Geology Dept** (B,M) (2015)
0224 Old Main Building
4831 Cass Avenue
Detroit, MI 48201
p. (313) 577-2506
baskaran@wayne.edu
http://sun2.science.wayne.edu/~geology/
*Enrollment (2011): B: 101 (15) M: 9 (1)*
Acting Head:
David Njus
Professor:
Mark M. Baskaran, (D), Phy Res Lab (India), 1985, CcGg
Associate Professor:
Jeffrey L. Howard, (D), California (Santa Barbara), 1987, Gs
Assistant Professor:
Edmond H. van Hees, (D), Michigan, 2000, GzxEg
Lecturer:
Charles F. Barker, (M), Boston, 1988, Gg
Ann Purdy, (D), Michigan State, 1995, Gg
Grazyna Sledzinski, (M), Wayne State, 1994, Gg
John M. Zawiskie, (M), Wayne State, 1979, Gg
Academic Services Officer:
David J. Lowrie, (B), Wayne State, 1964, Gg

## Western Michigan University

**Dept of Geosciences** (B,M,D) O (2016)
1903 W Michigan Ave
Kalamazoo, MI 49008-5241
p. (269) 387-5486
kathryn.wright@wmich.edu
http://www.wmich.edu/geology/
f: https://www.facebook.com/wmugeosciences/?fref=nf
*Enrollment (2015): B: 108 (18) M: 51 (17) D: 12 (4)*
Chair:
Mohamed Sultan, (D), Washington Univ, 1984, OriGe
Dean of the College of Arts and Sciences:
Carla Koretsky, (D), Johns Hopkins, 1998, CglHs
Professor:
Alan E. Kehew, (D), Idaho, 1977, GmlHw
Michelle A. Kominz, (D), Columbia, 1986, OuYgGu
R. V. Krishnamurthy, (D), Physical Research Lab-Dept of Space (India), 1984, Cs
Geosciences Specialist and Co-Director of the Hydrogeology Field Course:
Thomas R. Howe III, (B), W Michigan, 2010, HwGem
Co-Director of the Hydrogeology Field Course:
Donald Matthew M. Reeves, (D), Nevada (Reno), 2006, HwqNr
Associate Professor:
Daniel P. Cassidy, (D), Notre Dame, 1995, Hw
Johnson R. Haas, (D), Washington Univ, 1993, Cl
Duane R. Hampton, (D), Colorado State, 1989, HwSpHg
Heather L. Petcovic, (D), Oregon State, 2004, Oe
William A. Sauck, (D), Arizona, 1972, YgeYv
Assistant Professor:
Robb Gillespie, (D), SUNY, GsoGm
Stephen E. Kaczmarek, (D), Michigan State, 2005, GdCgGz
Joyashish Thakurta, (D), Indiana, 2008, GiEg
Director of CoreKids and Part-time Instructor:
Peter J. Voice, (D), Virginia Tech, 2010, GsOn
Director of Michigan Geological Repository for Research and Education and Professor Emeritus:
William B. Harrison, III, (D), Cincinnati, 1974, GsrPg

# Minnesota

## Bemidji State University

**Center for Environmental, Economic, Earth, & Space Studies** (A,B,M) (2016)
#27, 1500 Birchmont Drive NE
Bemidji, MN 56601
p. (218) 755-2783
tkroeger@bemidjistate.edu
http://www.bemidjistate.edu/academics/departments/ceeess/
*Enrollment (2012): B: 109 (17) M: 23 (3)*

Professor of Geology:
Timothy J. Kroeger, (D), North Dakota, 1995, PlGsHw
Professor:
Dragoljub D. Bilanovic, (D), Technion Israel Inst Tech, 1990, GePyCa
Assistant Professor:
Carl Isaacson, (D), Oregon State Univ, 2007, Ge
Miriam Rios-Sanchez, (D), Michigan Tech, 2012, HwOr

## Carleton College

**Dept of Geology** (B) (2015)
One North College Street
Northfield, MN 55057
p. (507) 222-4407
ehaberot@carleton.edu
http://www.carleton.edu/departments/geol/
Administrative Assistant: Ellen T. Haberoth
*Enrollment (2015): B: 38 (22)*

Charles L. Denison Professor of Geology:
Mary E. Savina, (D), California (Berkeley), 1982, Gm
Professor:
Clinton A. Cowan, (D), Queen's, 1992, GsPe
Cameron Davidson, (D), Princeton, 1991, GptCc
Bereket Haileab, (D), Utah, 1994, GxzGi
Associate Professor:
Sarah J. Titus, (D), Wisconsin, 2006, Gc
Director, Science Ed Resource Center:
Cathryn A. Manduca, (D), Caltech, 1988, Gi
Emeritus:
Caryl E. Buchwald, (D), Kansas, 1966, Gg
Technical Director:
Jonathon L. Cooper, (B), W Washington, Gg

## Century College

**Earth Science** (A) (2016)
3300 Century Avenue North
White Bear Lake, MN 55110
p. (651) 779-3242
joe.osborn@century.edu
http://www.century.edu/futurestudents/programs/pnd.aspx?id=66

Head:
Joe Osborn, (M), Og
Instructor:
Jill Bries-Korpik, (M), Og
John Oughton, Og

## Fond du Lac Tribal and Community College

**Fond du Lac Tribal and Community College** (A) (2010)
2101 14th Street
Cloquet, MN 55720
glang@fdltcc.edu
http://www.fdltcc.edu/

Instructor:
Glenn Langhorst, (M), Minnesota (Duluth), Gg

## Gustavus Adolphus College

**Dept of Geology** (B) (2017)
800 West College Avenue
St Peter, MN 56082
p. (507) 933-7333
welsh@gustavus.edu
http://www.gustavus.edu/geology
Administrative Assistant: Jennifer Kruse
*Enrollment (2016): B: 24 (5)*

Chair:
James L. Welsh, (D), Wisconsin, 1982, Gxc
Associate Provost and Dean of Sciences and Education:
Julie K. Bartley, (D), California (Los Angeles), 1994, GsPgGd
Associate Professor:
Laura Triplett, (D), Minnesota, 2008, GemCl
Visiting Professor:
Andrew Elstad-Haveles, (D), Minnesota, 2015, PogGs
Emeritus:
Keith J. Carlson, (D), Chicago, 1966, Pv
Cooperating Faculty:
Daniel Mollner, On

## Inver Hills Community College

**Geology Dept** (A) (2015)
2500 East 80th Street
Inver Grove Heights, MN 55076
jkorpik@inverhills.mnscu.edu
http://www.inverhills.edu/Departments/Geology/

Instructor:
Jill Bries Korpik, Gg

## Itasca Community College

**Geography/Geographic Info Systems (GIS)** (A) (2016)
1851 East Highway 169
Grand Rapids, MN 55744
p. (218) 322-2364
timothy.fox@itascacc.edu

Instructor:
Mike LeClair
Kenneth Tapp

## Lake Superior College

**Liberal Arts & Sciences Dept** (A) (2010)
2101 Trinity Rd.
Duluth, MN 55811
m.whitehill@lsc.edu
http://www.lsc.edu

Instructor:
Matthew Whitehill, Gg

## Macalester College

**Geology Dept** (B) (2015)
1600 Grand Avenue
St Paul, MN 55105
p. (651) 696-6000
macgregor@macalester.edu
http://www.macalester.edu/geology/
*Enrollment (2015): B: 45 (11)*

Chair:
Kelly MacGregor, (D), California (Santa Cruz), 2002, GmOl
Professor:
John P. Craddock, (D), Michigan, 1988, Gc
Raymond R. Rogers, (D), Chicago, 1995, Gs
Associate Professor:
Kristina A. Curry Rogers, (D), SUNY (Stony Brook), 2001, Pv
Karl R. Wirth, (D), Cornell, 1991, GiOn
Assistant Professor:
Alan Chapman, (D), Caltech, 2011, Gct
Geology Lab Supervisor:
Jeffrey T. Thole, (M), Washington State, 1991, GgCaHw

## Minneapolis Community and Technical College

**Div of Arts and Sciences** (1999)
1501 Hennepin Ave S
Minneapolis, MN 55403
chuck.paulson@minneapolis.edu
http://www.minneapolis.edu/academics/artsandsciences.cfm

## Minnesota State University

**Geology Program** (B,M) (2016)
Department of Geography
206 Morris Hall
Mankato, MN 56001

p. (507) 389-2617
phillip.larson@mnsu.edu
http://sbs.mnsu.edu/earthscience/
*Enrollment (2004): B: 50 (10)*

Director:
Phillip H. Larson, (D), Arizona State, 2013, GmOyHs

Professor:
Donald A. Friend, (D), Arizona State, 1997, OyGm
Bryce W. Hoppie, (D), California (Santa Cruz), 1996, GeHw
Steven Losh, (D), Yale, 1985, GzxGo
Martin D. Mitchell, (D), Illinois, 1993, Ou
Ron Schirmer, (D), Minnesota, 2002, GaPb
Fei Yuan, (D), Minnesota, 2003, Ori

Associate Professor:
Rama P. Mohapatra, (D), Wisconsin (Milwaukee), 2012, OirOg
Ginger L. Schmid, (D), Texas State Univ, 2004, SdGmSf
Forrest Wilkerson, (D), Texas State, 2004, OawGm
Chad Wittkop, (D), Minnesota, GsCgGl

Assistant Professor:
Jonathan H. Anderson, (D), Penn State, 2009, Ga

## Normandale Community College

**Dept of Geography and Geology** (A) (2017)
9700 France Avenue South
Bloomington, MN 55431
p. (952) 358-7032
paul.sabourin@normandale.edu
http://www.normandale.edu/departments/stem-and-education/geography-and-geology

Instructor:
David J. Berner, (M), Colorado
Douglas J. Claycomb, (D), Texas A&M
Richard P. Dunning, (D), Wisconsin
Carolyn Dykoski, (M), Minnesota
Annia Fayon, (D), Arizona State
Lindsay Iredale, (M), Minnesota
Paul D. Sabourin, (D), Minnesota
Ronald D. Ward, (D), Georgia

## North Hennepin Community College

**Geology Dept** (2016)
7411 85th Avenue N.
Brooklyn Park, MN 55445
p. (763) 424-0869
megan.jones@nhcc.edu
http://www.nhcc.edu/academic-programs/academic-departments/geology

## Northland Community & Technical College

**Liberal Arts Program** (1999)
2022 Central Avenue NE
East Grand Forks, MN 56721
p. (218) 683-8694
http://www.northlandcollege.edu

## Rochester Community & Technical College

**Department of Earth Sciences** (2011)
851 30th Ave SE
Rochester, MN 55904
p. (507) 285-7220
Cory.Rubin@roch.edu
http://www.rctc.edu

Instructor:
John C. Tacinelli, (D), Minnesota, 2000, Gig

## Saint Cloud State University

**Dept of Atmospheric and Hydrologic Sciences** (B) (2014)
720 4th Avenue South
St Cloud, MN 56301-4498
p. (320) 308-3260
arhansen@stcloudstate.edu
http://www.stcloudstate.edu/eas/
Administrative Assistant: Debbie Schlumpberger
*Enrollment (2012): B: 119 (31)*

Professor:
Anthony R. Hansen, (D), Iowa State, 1981, Ow

Professor:
Kate S. Pound, (D), Otago (NZ), 1993, GeOe
Robert A. Weisman, (D), SUNY (Albany), 1988, Ow

Associate Professor:
Juan J. Fedele, (D), Illinois, 2003, Hg
Jean L. Hoff, (D), North Dakota, 1989, HwOeGg
Rodney Kubesh, (D), Illinois, 1991, Ow

Assistant Professor:
Brian J. Billings, (D), Nevada (Reno), 2009, Ow

## University of Minnesota, Duluth

**Dept of Earth & Environmental Science** (B,M,D) ● (2016)
229 Heller Hall
1114 Kirby Drive
Duluth, MN 55812
p. (218) 726-8385
dees@d.umn.edu
http://www.d.umn.edu/dees/
Administrative Assistant: Laura L. Chapin
Administrative Assistant: Claudia J. Rock
*Enrollment (2016): B: 54 (19) M: 14 (13) D: 1 (0)*

Professor:
Erik T. Brown, (D), MIT, 1990, Og
John W. Goodge, (D), California (Los Angeles), 1987, Gp
Vicki L. Hansen, (D), California (Los Angeles), 1987, GtXgGc
Howard Mooers, (D), Minnesota, 1988, GleHw

Associate Professor:
Christina D. Gallup, (D), Minnesota, 1997, Cc
Karen B. Gran, (D), Washington, 2005, Gms
John B. Swenson, (D), Minnesota, 2000, Gr
Nigel J. Wattrus, (D), Minnesota, 1984, Yr

Assistant Professor:
Latisha A. Brengman, (D), Tennessee, 2015, Gd
Fred A. Davis, (D), Minnesota (Twin Cities), 2012, CpGi
Salli F. Dymond, (D), Minnesota, 2014, Ge
Christian Schardt, (D), Tasmania, EgCg
Byron A. Steinman, (D), Pittsburgh, 2011, Gn

Emeritus:
Steve Colman, (D), Colorado, 1977, GuPe
James A. Grant, (D), Caltech, 1964, Gp
John C. Green, (D), Harvard, 1960, Gi
Timothy Holst, (D), Minnesota, Gct
James D. Miller, (D), Minnesota, 1986, Gi
Ronald Morton, (D), Carleton, 1976, EgGv
Richard W. Ojakangas, (D), Stanford, 1964, Gd
George R. Rapp, (D), Penn State, 1960, GaeCg

## University of Minnesota, Morris

**Div of Science & Mathematics** (B) (2014)
Geology Discipline
Science Building
600 East 4th Street
Morris, MN 56267
p. (320) 589-6300
geol@mrs.umn.edu
http://www.mrs.umn.edu
*Enrollment (2002): B: 26 (9)*

Coordinator:
James F. Cotter, (D), Lehigh, 1984, Gl

Professor:
Keith A. Brugger, (D), Minnesota, 1992, OlGlm
James B. Van Alstine, (D), North Dakota, 1980, Pg

Emeritus:
Peter M. Whelan, (D), California (Santa Cruz), 1988, Gx

## University of Minnesota, Twin Cities

**Dept of Civil, Environmental and Geo-Engineering** (B,M,D) (2015)
500 Pillsbury Drive SE
Minneapolis, MN 55455
p. (612) 625-5522
cege@umn.edu
http://www.ce.umn.edu/
*Enrollment (2010): B: 26 (8) M: 11 (4) D: 49 (2)*

Minnesota

Head:
John S. Gulliver, (D), Minnesota, 1980, Hs
Professor:
Emmanuel M. Detournay, (D), Minnesota, 1983, NrpNm
Andrew Drescher, (D), Inst of Fund Tech Res (Poland), 1968, So
Efi Foufoula-Georgiou, (D), Florida, 1985, Hg
Joseph F. Labuz, (D), Northwestern, 1985, Nr
Otto D. Strack, (D), Delft (Neth), 1973, Hw
Vaughan R. Voller, (D), Sunderland, 1980, Om
Associate Professor:
Randal J. Barnes, (D), Colorado Mines, 1985, On
Bojan B. Guzina, (D), Colorado, 1996, Nr
Karl A. Smith, (D), Minnesota, 1980, Nx
Research Associate:
Sonia Mogilevskaya, (D), Russian Acad of Sci, 1987, Ng
Adjunct Professor:
Peter A. Cundall, (D), Nr

**Dept of Earth Sciences** (B,M,D) O (2017)
108 Pillsbury Hall
310 Pillsbury Drive SE
Minneapolis, MN 55455-0219
p. (612) 624-1333
esci@umn.edu
http://www.esci.umn.edu
*Enrollment (2016): B: 48 (15) M: 15 (2) D: 41 (7)*
Head:
Donna L. Whitney, (D), Washington, 1991, Gpt
Teaching Professor:
Kent C. Kirkby, (D), Wisconsin, 1994, Gg
Regents Professor:
R. Lawrence Edwards, (D), Caltech, 1988, Cc
Director, Minnesota Geological Survey:
Harvey Thorleifson, (D), Colorado, 1989, Gg
Director of IRM:
Bruce M. Moskowitz, (D), Minnesota, 1980, Ym
Professor:
Hai Cheng, (D), Nanjing Univ (China), 1988, CcPe
David L. Fox, (D), Michigan, 1999, Pg
Marc M. Hirschmann, (D), Washington, 1992, Gi
Peter J. Hudleston, (D), Imperial Coll (UK), 1969, Gc
Emi Ito, (D), Chicago, 1979, Cs
David L. Kohlstedt, (D), Illinois, 1970, YxGzi
Sally G. Kohlstedt, (D), Illinois, 1972, On
Katsumi Matsumoto, (D), Columbia, 2000, Oc
Christopher Paola, (D), MIT/WHOI, 1983, Gs
Justin Revenaugh, (D), MIT, 1989, Ys
William E. Seyfried, Jr., (D), S California, 1977, Cm
Christian P. Teyssier, (D), Monash Univ (Australia), 1986, GctGg
David A. Yuen, (D), California (Los Angeles), 1978, Yg
Assoc Director of IRM:
Joshua Feinberg, (D), California (Berkeley), 2005, YmGze
Associate Professor:
Jake Bailey, (D), S California, 2008, Py
Karen L. Kleinspehn, (D), Princeton, 1982, Gt
Assistant Professor:
Max Bezada, (D), Rice, 2010, Ys
Crystal Ng, (D), MIT, 2008, Hy
Cara M. Santelli, (D), MIT/WHOI, 2007, Py
Ikuko Wada, (D), Victoria, 2009, OnYgGt
Andrew D. Wickert, (D), Colorado (Boulder), 2014, Gml
Researcher 5:
Alexander Morrison, (M), Gn
Research Fellow:
Brad Herried, (M), Minnesota, 2010, Oi
Microprobe Manager:
Anette von der Handt, (D), GiCt
Research Associate:
Dario Bilardello, (D), Lehigh, 2009, Ym
Randy Calcote, (D), Minnesota, 2000, Py
Kang Ding, (D), Acad Sinica, 1987, Cg
Beverley Flood, (D), Py
Stefan Liess, (D), Hg
Jed Mosenfelder, (D), Cp
Amy Myrbo, (D), Minnesota, 2006, PeOe
Nick Seaton, (D)
Mark Shapley, Minnesota, 2005, Gn
Ivanka Stefanova, (D), Sofia, 1991, Pl
Chunyang Tan, (D), Zhejiang Univ, 2011, Cm
Mark Zimmerman, (D), Minnesota, 1999, Yx
Adjunct Professor:
James E. Almendinger, (D), Minnesota, 1988, Pe
Mike Berndt, (D), Minnesota, 1987, Cg
Val W. Chandler, (D), Purdue, 1977, Ye
Mark B. Edlund, (D), Michigan, 1998, Pe
Daniel R. Engstrom, (D), Minnesota, 1983, Pe
Annia K. Fayon, (D), Arizona State, 1997, Gc
Carrie E. Jennings, (D), Minnesota, 1996, Gl
Robert G. Johnson, (D), Iowa State, 1952, Pe
Shenghua Mei, (D), Yx
James D. Miller, (D), Gx
Kristina Curry Rogers, (D), Pg
Raymond Rogers, (D), Pg
Anthony Runkel, (D), Texas, 1988, Gs
Emeritus:
E. Calvin Alexander, Jr., (D), Missouri (Rolla), 1970, HwCcl
Subir K. Banerjee, (D), Cambridge, 1963, Ym
Roger LeB Hooke, (D), Gl
Hans Olaf Pfannkuch, (D), Paris, 1962, HwGeh
James H. Stout, (D), Harvard, 1970, GpzGt
Paul W. Weiblen, (D), Minnesota, 1965, Gi
XRCT Lab Manager:
Brian Bagley, (D), Minnesota, 2011
Student Services:
Jennifer Petrie
Research Specialist:
Betty Wheeler, (M), Minnesota, Hw
Remote Sensing Specialist:
Cathleen Torres Parisian, (M), Or
Post]:
Pu Zhang, (D), Ct
Postdoc:
Joseph Byrnes, (D), Oregon, 2016, Ys
Nestor Cerpa Gilvonio, (D), Yg
Matej Pec, (D), Yx
PostDoc:
Alejandra Quintanilla Terminal, (D), Yx
NSF Postdoc:
Carla Rosenfeld, (D), Py
Geochemical Analyst:
Elizabeth Lundstrom, (M), Ca
Director, CSDCO:
Anders Noren, Gn
Dept Administrator:
Sharon J. Kressler, (B), On
Department Safety Officer:
Scott Alexander, (B), Hw
Curator, LacCore:
Kristina Brady, (M), Minnesota, 2006, Gn
Agouron Institute Fellow:
Dan Jones, (D), Py
Research Fellow:
Thomas Juntunen, (M), Minnesota, 2011, Oi
Geology Librarian:
Carolyn Bishoff, (M), On
Cooperating Faculty:
Martin O. Saar, (D), California (Berkeley), 2003, Hg

**Dept of Soil, Water, & Climate** (B,M,D) ● (2016)
1991 Upper Buford Circle
St. Paul, MN 55108-6028
p. (612) 625-8114
crosen@umn.edu
http://www.swac.umn.edu
Administrative Assistant: Marjorie J. Bonse
*Enrollment (2016): B: 269 (67) M: 11 (4) D: 22 (1)*
Head:
Carl J. Rosen, (D), California (Davis), 1983, Sc
Professor:
James C. Bell, (D), Penn State, 1990, SdfOe
Timothy J. Griffis, (D), McMaster, 2000, Oa
Satish C. Gupta, (D), Utah State, 1972, Sp
Dylan B. Millet, (D), California (Berkeley), 2003, Oa
David J. Mulla, (D), Purdue, 1983, Sp

Edward A. Nater, (D), California (Davis), 1987, Sd
Michael J. Sadowsky, (D), Hawaii, 1983, Sb
Michael A. Schmitt, (D), Illinois, 1985, Sc
Mark W. Seeley, (D), Nebraska, 1977, Oa
Jeffrey S. Strock, (D), North Carolina State, 1999, So
Associate Professor:
Daniel E. Kaiser, (D), Iowa State, 2007, Sc
Albert L. Sims, (D), North Carolina State, 1992, Sc
Peter K. Snyder, (D), Wisconsin, 2003, Oa
Brandy M. Toner, (D), California (Berkeley), 2004, ClmSc
Tracy E. Twine, (D), Wisconsin, 2003, Oa
Kyungsoo Yoo, (D), California (Berkeley), 2003, Sd
Assistant Professor:
Fabian G. Fernandez, (D), Purdue, 2006, ScbSo
Jessica L. M. Gutknecht, (D), Wisconsin, 2007, Sb
Satoshi Ishii, (D), Minnesota, 2007, SbPy
Nicolas Jelinski, (D), Minnesota, 2014, SdCs
Paulo H. Pagliari, (D), Wisconsin, 2012, Sc
Adjunct Professor:
John M. Baker, (D), Texas A&M, 1987, Sp
Gary Feyereisen, (D), Minnesota, 2005, Sp
Jane M F Johnson, (D), Minnesota, 1995, So
Randall K. Kolka, (D), Minnesota, 1996, So
Pamela J. Rice, (D), Iowa State, 1996, So
Kurt Spokas, (D), Minnesota, 2005, Spo
Rodney T. Venterea, (D), California (Davis), 2000, Sp
Emeritus:
Deborah L. Allan, (D), California (Riverside), 1987, Sb
James L. Anderson, (D), Wisconsin, 1976, Sd
Paul R. Bloom, (D), Cornell, 1978, Sc
H. H. Cheng, (D), Illinois, 1961, Sb
Terence H. Cooper, (D), Michigan State, 1975, Sd
Robert H. Dowdy, (D), Michigan State, 1966, Sc
William C. Koskinen, (D), Washington State, 1980, Sc
John A. Lamb, (D), Nebraska, 1984, Sc
Gary L. Malzer, (D), Purdue, 1973, Sc
Jean-Alex E. Molina, (D), Cornell, 1967, SbCg
John F. Moncrief, (D), Wisconsin, 1981, Sp
Gyles W. Randall, (D), Wisconsin, 1972, So
George W. Rehm, (D), Minnesota, 1969, Sc
Michael P. Russelle, (D), Nebraska, 1982, Sc
Executive Secretary:
Kari A. Jarcho, (B), On

## University of Saint Thomas

**Dept of Geology** (B) (2010)
OWS 153
2115 Summit Avenue
St Paul, MN 55105
p. (651) 962-5241
kmtheissen@stthomas.edu
http://www.stthomas.edu/geology/
*Enrollment (2010): B: 16 (0)*
Chair:
Melissa A. Lamb, (D), Stanford, 1998, GtcGs
Associate Professor:
Thomas A. Hickson, (D), Stanford, 1999, GsmGe
Kevin Theissen, (D), Stanford, GnOg
Assistant Professor:
Jennifer McGuire, (D), Michigan State, CgHsGb
Laboratory Director:
Erik Smith, (B), On

## Vermilion Community College

**Vermilion Community College** (1999)
1900 East Camp Street
Ely, MN 55731
p. (218) 235-2173
http://www.vcc.edu/

## Winona State University

**Dept of Geoscience** (B) (2014)
P.O. Box 5838
Winona, MN 55987
p. (507) 457-5260
geoscience@winona.edu
http://www.winona.edu/geology
Administrative Assistant: Abigail Kugel
*Enrollment (2009): B: 61 (20)*
Professor:
Stephen T. Allard, (D), Wyoming, 2003, GcpGz
Candace L. Kairies-Beatty, (D), Pittsburgh, 2003, GeCl
Jamie Ann Meyers, (D), Indiana, 1971, Gsd
Associate Professor & Director Southeastern Minnesota Water Resources Center:
Toby Dogwiler, (D), Missouri, 2002, GmHqOw
Assistant Professor:
Jennifer LB Anderson, (D), Brown, 2004, XgYgOe
William L. Beatty, (D), Pittsburgh, 2003, Pg
Emeritus:
John F. Donovan, (D), Cornell, 1963, Eg
Dennis N. Nielsen, (D), North Dakota, 1973, Gm
Associate VP, Academic Affairs:
Nancy O. Jannik, (D), New Mexico Tech, 1989, Hw
College Laboratory Services Specialist:
Luke Zwiefelhofer, (B), Wyoming, 2004, On

# Mississippi

## Jackson State University

**Physics, Atmospheric Sciences and Geoscience** (B) (2014)
1400 J. R. Lynch St.
JSU Box 17660
Jackson, MS 39217
p. (601) 979-7012
mfadavi@jsums.edu
http://www.jsums.edu/cset/phyat.htm
*Enrollment (2014): B: 9 (4)*

## Millsaps College

**Dept of Geosciences** (B) O (2017)
Box 150648
1701 North State Street
Jackson, MS 39210
p. (601) 974-1340
musseza@millsaps.edu
http://www.millsaps.edu/geology/
*Enrollment (2016): B: 17 (3)*
Professor:
Stan Galicki, (D), Mississippi, 2002, Ged
James B. Harris, (D), Kentucky, 1992, Ye
Chair:
Zachary A. Musselman, (D), Kentucky, 2006, Gm
Emeritus:
Delbert E. Gann, (D), Missouri Sch of Mines, 1976, Gz

## Mississippi State University

**Dept of Geosciences** (B,M,D) (2015)
P. O. Box 5448
108 Hilbun Hall
East Lee Blvd.
Mississippi State, MS 39762
p. (662) 325-3915
whc5@geosci.msstate.edu
http://www.geosciences.msstate.edu/
f: https://www.facebook.com/pages/Department-of-Geosciences-Mississippi-State-University/275657317621
*Enrollment (2014): B: 505 (86) M: 224 (103) D: 24 (4)*
Head:
William H. Cooke, III, (D), Mississippi State, 1997, OriOg
Professor:
Michael E. Brown, (D), North Carolina, 1999, Ow
Darrel W. Schmitz, (D), Texas A&M, 1991, Hw
Associate Professor:
Shrinidhi Ambinakudige, (D), Florida State, 2006, OiyOr
Renee M. Clary, (D), Louisiana State, 2003, OePgGe
Jamie Dyer, (D), Georgia, 2005, Ow
Brenda L. Kirkland, (D), Louisiana State, 1992, Gd
John C. Rodgers, (D), Georgia, 1999, Oy
Kathleen M. Sherman-Morris, (D), Florida State, 2006, Oa
Assistant Professor:
Padmanava Dash, (D), Louisiana State, 2011, OrHg

Christopher Fuhrmann, (D), 2011, Oa
Rinat Gabitov, (D), Rensselaer Polytechnic Inst, 2005, Cas
Qingmin Meng, (D), Georgia, 2006, Oig
Andrew Mercer, (D), Oklahoma, 2008, Owa
Adam Skarke, (D), Delaware, 2013, GuYrOn
Kim Wood, (D), Arizona, 2012, OawOr
Research Associate:
Katarzynz Grala, (M), Iowa State, 2004, Oi
Instructor:
Christa M. Haney, (M), Mississippi State, 1999, Ow
Amy P. Moe-Hoffman, (M), Colorado, 2002, On
John A. Morris, (M), Mississippi State, 2007, OirOw
Lindsey Morschauser, (M), Mississippi State, Ow
Athena Nagel, (D), Mississippi State, 2014, OgiOr
Greg Nordstrom, (M), Mississippi State, 2007, Ow
Tim Wallace, (M), Mississippi State, 1994, Ow
Adjunct Professor:
Paul J. Croft, (D), Rutgers, 1991, Ow
Patrick J. Fitzpatrick, (D), Colorado State, 1995, Ow
James May, (D), Texas A&M, 1988, Hw
Jack C. Pashin, (D), Kentucky, 1990, GoEcGs
Janet E. Simms, (D), Texas A&M, 1991, Yg
Jayaram Veeramony, (D), Delaware, 1999, Og
Emeritus:
John M. Kaye, (D), Louisiana State, 1974, Gg
John E. Mylroie, (D), Rensselaer, 1977, Gm
Charles L. Wax, (D), Louisiana State, 1977, Oa
Distance Academic Coordinator:
Mary A. Dean
Business Manager:
Jerri Wright, (B), Mississippi Univ for Women, 1996
Academic Coordinator:
Tina Davis, (M), Mississippi Univ for Women
Related Staff:
Cynthia Bell

## University of Mississippi

**Dept of Geology & Geological Engineering** (B,M,D) ● (2017)
School of Engineering
P.O. Box 1848
120A Carrier Hall
University, MS 38677-1848
p. (662) 915-7498
geology@olemiss.edu
http://www.engineering.olemiss.edu/gge/
*Enrollment (2016): B: 301 (45) M: 21 (6) D: 5 (0)*
Professor:
Adnan Aydin, (D), Memorial, 1994, NgrYe
Gregg R. Davidson, (D), Arizona, 1995, Hw
Gregory L. Easson, (D), Missouri (Rolla), 1996, Or
Robert M. Holt, (D), New Mexico Tech, 2000, Hq
Associate Professor:
Louis Zachos, (D), Texas (Austin), 2008, Ger
Assistant Professor:
Jennifer N. Gifford, (D), Florida, 2013, GtxCg
Andrew M. O'Reilly, (D), Florida, 2012, Hwq
Brian F. Platt, (D), Kansas, 2009, GsPe
Lance D. Yarbrough, (D), Mississippi, 2006, Ng
Instructor:
Inoka Widanagamage, (D), Kent State, 2015, ClGep
Lecturer:
Cathy A. Grace, (M), Mississippi, 1996, Gg

## University of Southern Mississippi

**Dept of Geography and Geology** (B,M,D) (2015)
118 College Drive, Box 5051
Walker Science Building, Room 127
Hattiesburg, MS 39406
p. (601) 266-4729
franklin.heitmuller@usm.edu
http://www.usm.edu/geography-geology
*Enrollment (2014): B: 114 (19) M: 30 (5) D: 4 (1)*
Professor:
Andy Reese, (D), Louisiana State, 2003, Oy
Professor:
Greg Carter, (D), Wyoming, 1985, Oyi
Clifton Dixon, (D), Texas A&M, 1988, On
Maurice A. Meylan, (D), Hawaii, 1978, Gu
Mark Miller, (D), Arizona, 1988, On
David M. Patrick, (D), Oklahoma, 1972, Ng
Associate Professor:
Jerry Bass, (D), Texas, 2003, On
David Cochran, (D), Kansas, 2005, On
David Holt, (D), Arkansas, 2002, Oyi
Bandana Kar, (D), South Carolina, 2008, Oi
George Raber, (D), South Carolina, 2003, Oi
Assistant Professor:
Grant Harley, (D), Tennessee, 2012, Oy
Omar Harvey, (D), Texas A&M, 2010, GeHwg
Frank Heitmuller, (D), Texas, 2009, Gms
Instructor:
Lin F. Pope, (M), S Mississippi, 1983, Gz
Vicki Tinnon, (D), Kansas State, 2010, On

**Division of Marine Science** (B,M,D) (2017)
1020 Balch Boulevard
National Aeronautics and Space Administration
Stennis Space Center, MS 39529
p. (228) 688-3177
marine.science@usm.edu
http://www.usm.edu/marine
f: https://www.facebook.com/SouthernMissDivisionOfMarine-Science/
t: https://twitter.com/USMMarineSci
*Enrollment (2016): B: 33 (2) M: 31 (14) D: 13 (1)*
Director of the School of Ocean Science and Technology:
William (Monty) Graham, (D), California (Santa Cruz), 1994, Ob
Interim Chair of the Division of Marine Science:
Jerry Wiggert, (D), S California, 1995, Op
Director of the Center for Gulf Studies:
Denis A. Wiesenburg, (D), Oceanography, 1980, Oc
Professor:
Vernon L. Asper, (D), MIT/WHOI, 1986, Ou
Donald G. Redalje, (D), Hawaii, 1980, Ob
Alan M. Shiller, (D), California (San Diego), 1982, OcCtHs
Associate Professor:
Stephan Howden, (D), Rhode Island, 1996, Op
Scott Milroy, (D), South Florida, 2007, Ob
Dmitri Nechaev, (D), Shirshov Inst, 1986, Op
Assistant Professor:
Maarten Buijsman, (D), Utrecht, 2007, Op
Christopher T. Hayes, (D), 2013, Oc
Jessica E. Pilarczyk, (D), Geology, 2011, Ou
Davin Wallace, (D), Rice, 2010, Ou
Director of the Hydrographic Science Research Center:
RADM (Retired) Ken E. Barbor, (M), Naval Postgrad Sch, 1978, Own
Director of the Hydrographic Science Academic Graduate Degree Program:
Maxim F. van Norden, (M), Naval Postgrad Sch, 1979, Op
Instructor:
Danielle Greenhow, (D), USF St. Petersburg, 2013, Og

**Gulf Coast Research Laboratory** (2006)
Department of Coastal Sciences
Geology Section
Ocean Springs, MS 39566-7000
p. (601) 872-4200
joe.griffitt@usm.edu
http://www.coms.usm.edu
Administrative Assistant: Angelia Bone
Head:
Ervin G. Otvos, (D), Massachusetts, 1964, On

# Missouri

## Metropolitan Community College, Blue River

**Geology and Geography Program** (A) (2015)
20301 E. 78 Highway
Independence, MO 64057-2053
p. (816) 220-6622
benjamin.wolfe@mcckc.edu
http://www.mcckc.edu/progs/geol/geology/overview.asp

## Metropolitan Community College, Kansas City

**Geology Dept** (A) (2013)
3200 Broadway
Kansas City, MO 64111
p. (816) 604-3335
melissa.renfrow@mcckc.edu
http://www.mcckc.edu/programs/geology/

Instructor:
Alice Fuerst, Gg
John Horn, (D), Nebraska, GgOy
Carl Priesendorf, (M), Central Missouri, 1987, GgOy
Laura Veverka, (M), Missouri (Kansas City), Oy
Ben Wolfe, (M), Alaska (Fairbanks), 2001, GgOyg

Adjunct Professor:
Janet Raymer, (M), Missouri S&T, GgEo

## Mineral Area College

**Science Dept** (A) (2017)
5270 Flat River Road
P.O. Box 1000
Park Hills, MO 63601
p. (573) 518-2314
bscheidt@MineralArea.edu
http://www.mineralarea.edu/faculty/academicDepartments/science.aspx
*Enrollment (2016): A: 2 (2)*

Assistant Professor:
Brian Scheidt, (M), S Illinois, Hw

Adjunct Professor:
Katherine Perkins, (M)
Kelly Webb, (M)

Emeritus:
James G. Hrouda, (M), Illinois, 1963, Og

## Missouri State University

**Dept of Geography, Geology & Planning** (B,M) (2016)
901 S. National
Springfield, MO 65897
p. (417) 836-5800
ggp@missouristate.edu
http://geosciences.missouristate.edu
f: https://www.facebook.com/MSUggp/
t: @MSUggp
Administrative Assistant: Deana Gibson
*Enrollment (2016): B: 185 (56) M: 38 (8)*

Department Head:
Toby Dogwiler, (D), Missouri, 2002, OiGmOy

Professor:
Kevin R. Evans, (D), Kansas, 1997, GrsGt
Douglas R. Gouzie, (D), Kentucky, 1986, HwGeCl
Melida Gutierrez, (D), Texas (El Paso), 1992, Cg
Rajinder S. Jutla, (D), Virginia Tech, 1995, On
Kevin L. Mickus, (D), Texas (El Paso), 1989, Yg
Robert T. Pavlowsky, (D), Wisconsin, 1995, Gm
Charles W. Rovey, (D), Wisconsin (Milwaukee), 1990, HyGl

Associate Professor:
Alice (Jill) Black, (D), Missouri, 2003, OeaHw
Jun Luo, (D), Wisconsin (Milwaukee), Oi
Judith Meyer, (D), Wisconsin, 1994, On
Xin Miao, (D), California (Berkeley), 2005, Or
Xiaomin Qiu, (D), Texas State, Oi

Assistant Professor:
Mario Daoust, (D), McGill, Oa
Ron Malega, (D), Georgia, On
Matthew P. McKay, (D), West Virginia Univ, 2015, Gct
Gary Michelfelder, (D), Montana State, 2015, GiCgGv

Senior Instructor:
Deborah Corcoran, (M), Michigan State, 1980, On

Associate Provost:
John C. Catau, (D), Michigan State, 1973, On

Emeritus:
David A. Castillon, (D), Michigan State, 1972, Gm
William H. Cheek, (D), Michigan State, 1976, On
William Corcoran, (D), Michigan State, 1981, Oa
Stanley C. Fagerlin, (D), Missouri, 1980, Pg
Russel L. Gerlach, (D), Nebraska, 1974, On
Dimitri Ioannides, (D), Rutgers Univ, 1994, On
Elias Johnson, (D), Oklahoma, 1977, Or
Vincent E. Kurtz, (D), Oklahoma, 1960, Ps
Erwin J. Mantei, (D), Missouri (Rolla), 1965, Ct
Diane M. May, (M), S Illinois (Edwardsville), 1974, On
James F. Miller, (D), Wisconsin, 1970, Ps
Thomas D. Moeglin, (D), Nebraska, 1978, Ng
Thomas G. Plymate, (D), Minnesota, 1986, Gx
Milton D. Rafferty, (D), Nebraska, 1970, On
Paul A. Rollinson, (D), Illinois, 1988, On

Senior Instructor:
Damon Bassett, (M), Missouri, 2003, Pg
Linnea Iantria, (M), George Washington, On

## Missouri University of Science and Technology

**Dept of Geology & Geophysics** (B,M,D) (2004)
125 McNutt Hall
Rolla, MO 65409-0410
p. (573) 341-4616
rocks@umr.edu
Administrative Assistant: Katherine W. Mattison
*Enrollment (2001): B: 76 (16) M: 18 (3) D: 11 (1)*

Lecturer:
Patrick S. Mulvany, (D), Missouri (Rolla), 1996, Gg

Adjunct Professor:
John F. Burst, (D), Missouri, 1950, Eo
Waldemar M. Dressel, (B), Missouri (Rolla), 1943, Gg
Charles E. Robertson, (M), Maryland, 1960, Gc
James E. Vandike, (M), SD Mines, 1979, Hg
James H. Williams, (D), Missouri (Rolla), 1975, Gg

Emeritus:
Shelton K. Grant, (D), Utah, 1966, Gz
Richard D. Hagni, (D), Missouri, 1962, Em
Geza K. Kisvarsanyi, (D), Missouri, 1966, Em
Richard D. Rechtien, (D), Washington, 1964, Ye
Gerald B. Rupert, (D), Missouri, 1964, Ye
Alfred C. Spreng, (D), Wisconsin, 1950, Gr

## Moberly Area Community College, Columbia Campus

**Moberly Area Community College, Columbia Campus** (1999)
601 Business Loop 70 West
Columbia, MO 65203
p. (573) 234-1067
SandyAnderson@macc.edu
http://www.macc.edu/

## Northwest Missouri State University

**Dept of Geology-Geography** (B,M) (2014)
800 University Drive
Maryville, MO 64468
p. (660) 562-1723
geosci@nwmissouri.edu
*Enrollment (2003): B: 176 (53)*

Chair:
C. Renee Rohs, (D), Kansas, 2000, Cg

Professor:
Gregory D. Haddock, (D), Idaho, 1996, Oi

Associate Professor:
Theodore L. Goudge, (D), Oklahoma State, 1984, Oy

Assistant Professor:
James Hickey, (D), Dartmouth, 2006, Ge
Ming-Chih Hung, (D), Utah, 2003, Or
Yanfen Le, (D), Georgia, 2005, Oi
Leah D. Manos, (M), Tennessee, 1997, Oy

Instructor:
Jeffrey Bradley, (M), Oklahoma State, 1991, Og

Emeritus:
Richard M. Felton, (M), Missouri, 1979, Pg

## Ozarks Technical Community College

**Physical Sciences** (1999)
1001 E. Chestnut Expressway
Springfield, MO 65802
p. (417) 447-8238
ehrichp@otc.edu

Missouri

## Saint Louis University
**Earth & Atmospheric Sciences** (B,M,D) (2016)
3642 Lindell Blvd
O'Neil Hall
Room 205
St. Louis, MO 63108
p. (314) 977-3116
herrmarb@slu.edu
http://www.slu.edu/department-of-earth-and-atmospheric-sciences-home
*Enrollment (2016): B: 30 (0) M: 7 (5) D: 7 (0)*

Chair:
William P. Dannevik, (D), St. Louis, 1984, Oa
Benjamin de Foy, (D), Cambridge (UK), 1998, Oa

Professor:
David J. Crossley, (D), British Columbia, 1973, Yg
Jack Fishman, (D), Saint Louis Univ, 1977, Oar
Jack Fishman, (D), Oa
Daniel M. Hanes, (D), Scripps, 1983, GesGm
Robert B. Herrmann, (D), St. Louis, 1974, Ys
Zaitao Pan, (D), Iowa State, 1996, Oa
Lupei Zhu, (D), Caltech, 1998, YsGt

Associate Professor:
Karl Chauff, (D), Pg
John Encarnacion, (D), Michigan, 1994, Gx
Charles E. Graves, (D), Iowa State, 1988, Oa
Robert W. Pasken, (D), St. Louis, 1981, Ow

Assistant Professor:
Lisa Chambers, (D)
Cathy Finlley, (D), Ow
Elizabeth Hasenmueller, (D)
Linda Warren, (D), Yg
Valorie Wilmoth, (D), Ow
Wasit Wulamu, (D), Oi

## University of Missouri
**Dept of Geological Sciences** (B,M,D) (2014)
101 Geology Building
Columbia, MO 65211
p. (573) 882-6785
HuckabeyM@missouri.edu
http://geology.missouri.edu/
Administrative Assistant: Marsha Huckabey
*Enrollment (2013): B: 60 (7) M: 13 (11) D: 21 (0)*

Director of Geology Field Studies:
Miriam Barquero-Molina, (D), Texas (Austin), 2009, Gct

Chair:
Kevin L. Shelton, (D), Yale, 1982, EmCs
Alan Whittington, (D), Open Univ (UK), 1997, GivCp

Professor:
Cheryl A. Kelley, (D), North Carolina, 1993, Co
Mian Liu, (D), Arizona, 1989, Yg
Kenneth A. MacLeod, (D), Washington, 1992, Pg
Peter I. Nabelek, (D), SUNY (Stony Brook), 1983, Ct
Eric A. Sandvol, (D), New Mexico State, 1995, Ys

Associate Professor:
Martin Appold, (D), Johns Hopkins, 1998, HwEm
Francisco Gomez, (D), Cornell, 1999, Gt

Assistant Professor:
John Huntley, (D), Virginia Tech, 2007, PqoPe
James Schiffbauer, (D), Virginia Tech, 2009, Poi

Adjunct Professor:
G. Randy Keller, (D), Texas Tech, 1973, Yg
Timothy McHargue, Gs
James Ni, (D), Cornell, 1984, Ygs
Angela Speck, (D), Univ Coll (London), 1998, Xc
Samson Tesfaye, (D), Colorado, 1999, GtcOr

Emeritus:
Robert L. Bauer, (D), Minnesota, 1981, GcpGt
Raymond L. Ethington, (D), Iowa, 1958, Pm
Thomas J. Freeman, (D), Texas, 1962, Gd
Glen R. Himmelberg, (D), Minnesota, 1965, Gp
Michael B. Underwood, (D), Cornell, 1984, GsuGt

Cooperating Faculty:
Stephen Stanton, On

## University of Missouri, Columbia
**School of Natural Resources** (B,M,D) (2014)
302 Anheuser-Busch Natural Resources Building
Columbia, MO 65211-7250
p. (573) 882-6301
MarketP@missouri.edu
http://www.snr.missouri.edu/seas/
*Enrollment (2012): B: 165 (15) M: 24 (3) D: 13 (0)*

Chair:
Patrick S. Market, (D), St. Louis, 1999, Oa

Professor:
Stephen H. Anderson, (D), North Carolina State, 1985, Sp
Clark J. Gantzer, (D), Minnesota, 1980, Os
Anthony R. Lupo, (D), Purdue, 1995, Oa
Peter P. Motavalli, (D), Cornell, 1989, Os

Extension:
Patrick E. Guinan, (D), Missouri, 2004, Oa

Associate Professor:
Neil I. Fox, (D), Salford (UK), 1998, Oa
Keith W. Goyne, (D), Penn State, 2003, Sc
Randall J. Miles, (D), Texas A&M, 1981, Sd

Assistant Professor:
Bohumil Svoma, (D), Arizona State, 2013, Oa

Instructor:
Eric A. Aldrich, (M), Missouri, 2011, Oa

Adjunct Professor:
Robert J. Kremer, (D), Mississippi State, 1981, Os
Christopher K. Wikle, (D), Iowa State, Oa

Emeritus:
Robert W. Blanchar, (D), Minnesota, 1964, Sc
James R. Brown, (D), Iowa State, 1963, Os
Ernest C. Kung, (D), Wisconsin, 1963, Oa
Stephen E. Mudrick, (D), MIT, 1973, Oa

Other:
E Eugene Alberts, (D), Purdue, 1979, Hw
J Glenn Davis, (D), Iowa State, Os
Frieda Eivazi, (D), Iowa State, 1980, Sc
Newell R. Kitchen, (D), Colorado State, 1990, Sp
Robert N. Lerch, (D), Colorado State, 1990, Os
W. Gene Stevens, (D), Mississippi State, 1992, Os

## University of Missouri, Kansas City
**Dept of Geosciences** (B,M,D) ● (2016)
5100 Rockhill Road
Room 420, Robert H. Flarsheim Hall
Kansas City, MO 64110-2499
p. (816) 235-1334
geosciences@umkc.edu
http://cas.umkc.edu/Geosciences/default.asp
t: UMKC_geosci
*Enrollment (2007): B: 85 (19) M: 13 (2) D: 11 (1)*

Chair:
Wei Ji, (D), Connecticut, 1991, OriOy

Professor:
James B. Murowchick, (D), Penn State, 1984, CgGz
Tina M. Niemi, (D), Stanford, 1992, Gt

Associate Professor:
Jimmy Adegoke, (D), Penn State, 2000, OayOr
Caroline P. Davies, (D), Arizona State, 2000, Pl
Jejung Lee, (D), Northwestern, 2001, GqHw

Emeritus:
Raymond M. Coveney Jr, (D), Michigan, 1972, EmGg
Steven L. Driever, (D), Georgia, 1977, OnnOn
Richard J. Gentile, (D), Missouri, 1965, Gr
Syed E. Hasan, (D), Purdue, 1978, Ng

## Washington University in St. Louis
**Dept of Earth & Planetary Sciences** (B,M,D) (2016)
Campus Box 1169
Rudolph Hall
1 Brookings Drive
St Louis, MO 63130-4899
p. (314) 935-5610
slava@wustl.edu
http://eps.wustl.edu

f: https://www.facebook.com/WashU.EPSci/
t: @WUSTL_EPS
*Enrollment (2016): B: 41 (19) D: 28 (4)*

Chair:
Viatcheslav S. Solomatov, (D), Moscow Inst of Physics & Tech, 1990, Yg
Rudolph Professor of Earth & Planeary Sciences:
Bradley L. Jolliff, (D), SD Mines, 1987, Gx
Research Professor:
Randy L. Korotev, (D), Wisconsin (Madison), 1976, XgCtXm
James S. McDonnell Distinguished University Professor:
Raymond E. Arvidson, (D), Brown, 1974, Xg
Professor:
Robert E. Criss, (D), Caltech, 1981, Cs
Robert F. Dymek, (D), Caltech, 1977, GpiGa
M. Bruce Fegley, (D), MIT, 1980, Xc
William B. McKinnon, (D), Caltech, 1981, Xg
Jill D. Pasteris, (D), Yale, 1980, GbzGe
Jennifer R. Smith, (D), Pennsylvania, 2001, Ga
William H. Smith, (D), Princeton, 1966, Xc
Douglas A. Wiens, (D), Northwestern, 1985, YsrYg
Michael E. Wysession, (D), Northwestern, 1991, Ys
Research Professor:
Katharina Lodders-Fegley, (D), Max Planck, 1991, Xc
Alian Wang, (D), Sci/Tech (France), 1987, Ca
Associate Professor:
Jeffrey G. Catalano, (D), Stanford, 2004, Cl
David A. Fike, (D), MIT, 2007, CsmPy
David Fike, (D), MIT, 2007, Cs
Associate Scientist:
Philip Skemer, (D), Yale, 2007, GcCp
Assistant Professor:
Alexander S. Bradley, (D), MIT, 2008, Py
Michael J. Krawczynski, (D), MIT, 2011, Cg
Rita Parai, (D), Harvard, 2014, Cs
Kun Wang, (D), Washington Univ, 2013, Xc
Administrative Officer:
Robert Gemignani, (M), Notre Dame de Namur Univ, 1998, On
Cooperating Faculty:
Anne M. Hofmeister, (D), Caltech, 1984, Yg

**Environmental Studies Program** (B) (2010)
Box 1169
One Brookings Drive
St. Louis, MO 63130-4899
p. (314) 935-7047
enstadmin@levee.wustl.edu
http://enst.wustl.edu
Administrative Assistant: Barbara Winston
*Enrollment (2010): B: 161 (59)*

Director:
Jan P. Amend, (D), California (Berkeley), 1995, Co
Professor:
Raymond E. Arvidson, (D), Brown, 1974, On
Richard Axelbaum, (D), California (Davis), 1988, Ng
Pratim Biswas, (D), Caltech, 1985, Ng
Robert Blankenship, (D), California (Berkeley), 1975, On
Robert E. Criss, (D), Caltech, 1981, Cs
Willem H. Dickhoff, (D), Free (Amsterdam), On
Mike Dudukovic, (D), Illinois Inst Tech, 1971, On
Robert F. Dymek, (D), Caltech, 2006, Gi
Claude Evans, (D), SUNY (Stony Brook), On
Bruce Fegley, (D), MIT, 1980, Xc
T.R. Kidder, (D), Harvard, 1988, On
Maxine I. Lipeles, (D), Harvard, 1979, On
William R. Lowry, (D), Stanford, 1988, On
Jill D. Pasteris, (D), Yale, 1980, On
Bruce Petersen, (D), Harvard, On
Robert Pollak, (D), MIT, 1964, On
Tab Rasmussen, (D), Duke, On
Barbara Schaal, (D), Yale, 1974, On
Glenn D. Stone, (D), Arizona, 1988, On
Robert W. Sussman, (D), Duke, 1972, On
Alan R. Templeton, (D), Michigan, 1972, On
Associate Professor:
Jon M. Chase, (D), Chicago, On
Clare Palmer, (D), On
Jen R. Smith, (D), Pennsylvania, 2001, Ga
Jay Turner, (D), Washington (St. Louis), 1993, On
Assistant Professor:
Jeff Catalano, (D), Stanford, 2004, Ca
Geoff Childs, (D), Indiana, On
Ellen Damschen, (D), North Carolina State, On
Daniel Giammar, (D), Caltech, 2001, On
Young-Shin Jun, Harvard, 2005, On
Tiffany Knight, (D), Pittsburgh, 2003, Pe
John Orrock, (D), Iowa State, 2004, On
Engineering & Science Director:
Beth Martin, (M), Washington (St. Louis), 1996, Ng
Geology Librarian:
Clara McLeod, On

## William Jewell College

**Dept of Biology** (2011)
Liberty, MO 64068
p. (816) 781-3806 x230
allent@william.jewell.edu
http://www.jewell.edu

Chair:
Tara Allen
Associate Professor:
Charles F. J. Newlon, (M), Missouri, 1962, Og

# Montana

## Flathead Valley Community College

**Geology and Geography** (A) (2016)
777 Grandview Drive
Kalispell, MT 59901
p. (406) 756-3873
aho@fvcc.edu

Associate Professor:
Anita Ho, (D), Oregon, Gg

## Montana State University

**Dept of Earth Sciences** (B,M,D) (2014)
P.O. Box 173480
226 Traphagen Hall
Bozeman, MT 59717-3480
p. (406) 994-3331
earth@montana.edu
http://www.montana.edu/wwwes/
*Enrollment (2013): B: 219 (48) M: 37 (8) D: 17 (3)*

Professor:
David R. Lageson, (D), Wyoming, 1980, Gc
David W. Mogk, (D), Washington, 1984, Gp
James G. Schmitt, (D), Wyoming, 1982, Gr
Cathy Whitlock, (D), Washington, 1983, Pey
William K. Wyckoff, (D), Syracuse, 1982, On
Associate Professor:
Todd C. Feeley, (D), California (Los Angeles), 1993, Gi
Michael Gardner, (D), Colorado Mines, 1993, Gsr
Jordy Hendrikx, (D), Canterbury (NZ), 2005, OnHsOa
Jian-yi Liu, (D), Minnesota, 1992, On
Mark L. Skidmore, (D), Alberta, 2001, ClPyOl
David J. Varricchio, (D), Montana State, 1995, Pv
Assistant Professor:
Jean Dixon, (D), Dartmouth, 2009, GmScOg
Julia H. Haggerty, (D), Colorado, 2004, Ou
Jamie McEvoy, (D), Arizona, 2013, Oe
Research Assistant Professor:
Frankie Jackson, (D), Montana State, 2007, Pv
David B. McWethy, (D), Montana State, 2007, PeSb
Colin Shaw, (D), New Mexico, 2001, GcEmGp
Kaj Williams, Colorado, Oa
Assistant Teaching Professor:
Stuart Challender, (M), Utah State, 1986, Oi
Research Assistant Professor:
David W. Bowen, (D), Colorado, 2001, GroGs
Regents Professor:
John R. Horner, Penn State, 2006, Pv

**Dept of Land Resources & Environmental Sciences** (B,M,D) (2014)

334 Leon Johnson Hall
P.O. Box 173120
Bozeman, MT 59717-3120
p. (406) 994-7060
jefj@montana.edu
http://landresources.montana.edu/
Department Secretary: Peggy Humphrey
*Enrollment (2000): B: 93 (13) M: 25 (2) D: 13 (0)*

Chair:
Tracy M. Sterling, (D)
Professor:
James W. Bauder, (D), Utah State, 1974, Sf
Lisa J. Graumlich, (D), Washington, 1985, Ou
William P. Inskeep, (D), Minnesota, 1985, Sc
Jeffrey S. Jacobsen, (D), Oklahoma State, 1985, So
Gerald A. Nielsen, (D), Wisconsin, 1963, Sd
David M. Ward, (D), Wisconsin, 1975, On
Senior Scientist:
Dennis R. Neuman, (M), Montana State, 1972, On
Associate Professor:
Richard E. Engel, (D), Minnesota, 1983, Sc
Jon M. Wraith, (D), Utah State, 1989, Sp
Assistant Professor:
Paul B. Hook, (D), Colorado State, 1992, Sf
Rick L. Lawrence, (D), Oregon State, 1998, Or
Adjunct Professor:
Douglas J. Dollhopf, (D), Montana State, 1975, On
Emeritus:
Cliff Montagne, (D), Montana State, 1976, Sd

Montana

## Montana State University, Billings

**Dept of Biological & Physical Sciences** (2011)
1500 N. 30th Street
Billings, MT 59101
p. (406) 657-2341 x2028
nsuits@msubillings.edu

Head:
Stanley Wiatr
Dean:
Tasneem Khaleel, (D), Bangalore, 1970, Po
Professor:
Matt Benacquista, (D), Montana State, 1989, Xy
Thomas T. Zwick, (D), N Colorado, 1977, Oe

## Montana Tech

**Geophysical Engineering** (B,M) O (2016)
1300 West Park Street
Butte, MT 59701
p. (406) 496-4401
mspeece@mtech.edu
http://www.mtech.edu/academics/mines/geophysical/
*Enrollment (2016): B: 16 (4) M: 8 (5)*

Head:
Marvin A. Speece, (D), Wyoming, 1992, Ye
Associate Professor:
Xiaobing Zhou, (D), Alaska (Fairbanks), 2002, Or
Assistant Professor:
Mohamed Khalil, (D), Giessen, 2002, YegGe
Khalid Miah, (D), Texas (Austin), 2008, YexYs
Emeritus:
Curtis A. Link, (D), Houston, 1993, Ye

## Montana Tech of the University of Montana

**Dept of Chemistry & Geochemistry** (B,M) (2014)
1300 West Park Street
Butte, MT 59701-8997
p. (406) 496-4207
dhobbs@mtech.edu
Department Secretary: Wilma Immonen

Professor:
Douglas Cameron, (D), Purdue, 1979, Ca
Douglas A. Coe, (D), Oregon State, 1974, Cg
Douglas A. Drew, (D), Wyoming, 1971, Cg
Donald Stierle, (D), California (Riverside), 1979, Co
Assistant Professor:
John D. Hobbs, (D), New Mexico, 1991, On
Research Associate:
Wayne Olmsted, (B), Montana State, 1962, Ca
Andrea Stierle, (D), Montana State, On
Instructor:
Stephen R. Parker, (M), Indiana, 1972, On
Emeritus:
Frank E. Diebold, (D), Colorado Mines, 1967, Cl
Alexis Volborth, (D), Helsinki, 1954, Ca

**Dept of Geological Engineering** (B,M) (2015)
1300 West Park Street
Butte, MT 59701
p. (406) 496-4262
dconrad@mtech.edu
http://www.mtech.edu/geo_eng
*Enrollment (2015): B: 34 (5) M: 24 (12)*

Chair:
Christopher H. Gammons, (D), Penn State, 1988, Cg
Professor:
Mary M. MacLaughlin, (D), California (Berkeley), 1997, Nr
Diane Wolfgram, (D), California (Berkeley), 1977, EgoNm
Associate Professor:
Glenn D. Shaw, (D), California (Merced), 2009, HwCsHs
Larry N. Smith, (D), New Mexico, 1988, GsoGm

## Rocky Mountain College

**Dept of Geology** (B) (2016)
1511 Poly Drive
Billings, MT 59102
p. (406) 657-1101
kalakayt@rocky.edu
http://www.rocky.edu/academics/academic-programs/undergraduate-majors/geology/
*Enrollment (2016): B: 20 (6)*

Professor:
Thomas J. Kalakay, (D), Wyoming, 2001, GcpGi
Associate Professor:
Derek Sjostrom, (D), Dartmouth, 2002, ClGsHs
Assistant Professor:
Emily Geraghty Ward, (D), Montana, 2007, OeGc

## Salish Kootenai College

**Dept of environmental Science** (1999)
58138 US Hwy 93
Ronan, MT 59855
bill_swaney@skc.edu
http://www.skc.edu

## University of Montana

**Dept. of Geosciences** (B,M,D) (2017)
32 Campus Drive #1296
Missoula, MT 59812-1296
p. (406) 243-2341
james.staub@umontana.edu
http://www.umt.edu/geosciences
Administrative Assistant: Christine Foster
Administrative Assistant: Loreene Skeel
*Enrollment (2016): B: 70 (32) M: 14 (7) D: 5 (4)*

Chair:
James R. Staub, (D), South Carolina, 1985, GsrEo
Professor:
Joel Harper, (D), Wyoming, 1998, Ol
Marc S. Hendrix, (D), Stanford, 1992, Gs
Nancy W. Hinman, (D), California (San Diego), 1987, Cgl
James W. Sears, (D), Queens, 1979, Gc
George D. Stanley, Jr., (D), Kansas, 1977, Pi
Associate Professor:
Julia Baldwin, (D), MIT, 2003, Gp
Rebecca Bendick, (D), Colorado, 2000, Yg
Marco Maneta, (D), Extremadura (Spain), 2006, Hsg
Andrew Wilcox, (D), Colorado State, 2005, Gm
Assistant Professor:
Payton Gardner, (D), Utah, 2009, Hw
Hilary R. Martens, (D), Caltech, 2016, YgsYd
Research Associate:
Carrine E. Blank, (D), Berkley, 2002, On

Michael Hofmann, (D), Montana, 2005, GosGr
Lecturer:
Kathleen Harper, (D), Wyoming, 1997
IT/GIS:
Aaron M. Deskins, (B), Montana, 2005, Oi

## University of Montana Western

**Environmental Sciences Dept** (B) (2016)
710 South Atlantic Street
Dillon, MT 59725-3598
p. (406) 683-7615
rob.thomas@umwestern.edu
https://www.umwestern.edu
f: https://www.facebook.com/UMWenvirosciences/
*Enrollment (2016): B: 95 (9)*

Chair:
R. Stephen Mock, (D), Montana State, 1989, Ca
Professor:
Eric G. Dyreson, (D), Arizona, 1997, Gq
Linda M. Lyon, (D), Washington State, 2003, Ou
Delena Norris-Tull, (D), Texas, 1990, Oe
Robert C. Thomas, (D), Washington, 1993, GseOe
Eric S. Wright, (D), Colorado, 2002, Gq
Craig E. Zaspel, (D), Montana State, 1975, Yg
Associate Professor:
Michelle Anderson, (D), Montana, 2008, Hs
Tyler Seacrest, (D), Nebraska, 2011, Gq
Assistant Professor:
Rebekah Levine, (D), New Mexico, 2016, HgGmOa
Wendy M. Ridenour, (D), Montana, 2006, Pe
Spruce W. Schoenemann, (D), Washington, 2015, GeCgs
Adjunct Professor:
Heidi Anderson-Folnagy, (D), Idaho, 2011, Gs
Brenda J. Buck, (D), New Mexico State, 1996, Sdo
Emeritus:
Sheila M. Roberts, (D), Calgary, 1996, Ge

# Nebraska

## Central Community College

**Physical Science** (A) (2015)
4500 63rd Street
PO Box 1027
Columbus, NE 68602
p. (402) 562-1216
dcondreay@cccneb.edu
http://www.cccneb.edu/Science-and-Math/

Instructor:
Denise Condreay, Og

## Chadron State College

**Dept of Geosciences** (B) O (2015)
1000 Main Street
Chadron, NE 69337
p. (308) 432-6377
mleite@csc.edu
http://www.csc.edu/geoscience
Administrative Assistant: Stacy Mittleider
*Enrollment (2013): B: 7 (4)*

Chair:
Wendy Jamison, (D)
Professor:
Michael B. Leite, (D), Wyoming, 1992, Gg
Instructor:
Jennifer L. Balmat, (M), Chadron State Coll, 2008, GgOe

## Creighton University

**Dept of Atmospheric Sciences** (B,M) (2014)
2500 California Plaza
Omaha, NE 68178
p. (402) 280-2641
zehnder@creighton.edu
*Enrollment (2012): B: 10 (0) M: 3 (0)*

Professor:
Joseph A. Zehnder, (D), Chicago, 1986, Oa
Associate Professor:
Jon M. Schrage, (D), Purdue, 1998, Ow
Assistant Professor:
Timothy J. Wagner, (D), Wisconsin, 2011, Oa
Adjunct Professor:
Richard Ritz, (M), Texas A&M, Ow
Emeritus:
Arthur V. Douglas, (D), Arizona, 1976, Og

## University of Nebraska at Omaha

**Dept of Geography and Geology** (B,M) (2016)
6001 Dodge Street
DSC 260
Omaha, NE 68182-0199
p. (402) 554-2662
rshuster@unomaha.edu
http://www.unomaha.edu/college-of-arts-and-sciences/geology/
Administrative Assistant: Brenda Todd
*Enrollment (2016): B: 85 (12)*

Professor:
George F. Engelmann, (D), Columbia, 1978, PvGdr
Harmon D. Maher, Jr., (D), Wisconsin, 1984, GcsGt
Associate Professor:
Robert D. Shuster, (D), Kansas, 1985, GiaOe
Assistant Professor:
Bradley Bereitschaft, (D), North Carolina (Greensboro), 2012, Oyu
Ashlee LD Dere, (D), Penn State, 2014, SdGmCl
James J. Hayes, (D), Indiana, 2008, OruOi
Emeritus:
Jeffrey S. Peake, (D), Louisiana State, 1977, OyaOa
John F. Shroder, Jr., (D), Utah, 1967, GmlOy

## University of Nebraska, Lincoln

**Dept of Earth & Atmospheric Sciences** (B,M,D) (2016)
214 Bessey Hall
Lincoln, NE 68588-0340
p. (402) 472-2663
tfrank2@unl.edu
http://eas.unl.edu/
f: https://www.facebook.com/UNLEarthAtmosSci
Administrative Assistant: Janelle Gerry
Administrative Assistant: Tina M. Schinstock
*Enrollment (2016): B: 105 (21) M: 36 (13) D: 22 (5)*

Chair:
Tracy D. Frank, (D), Michigan, 1996, GssCs
Professor:
Christopher R. Fielding, (D), Durham (UK), 1982, Gs
Sherilyn C. Fritz, (D), Minnesota, 1985, Pem
David M. Harwood, (D), Ohio State, 1986, PmGul
Qi S. Hu, (D), Colorado State, 1992, Oa
Robert M. Joeckel, (D), Iowa, 1993, GrSd
David B. Loope, (D), Wyoming, 1981, Gsg
Robert J. Oglesby, (D), Yale, 1990, Oa
Clinton M. Rowe, (D), Delaware, 1988, Oa
David K. Watkins, (D), Florida State, 1984, PmGu
Vitaly A. Zlotnik, (D), Natl Inst Hydrogeology and Engineering Geology (Russia), 1979, HwyHq
Associate Professor:
Mark R. Anderson, (D), Colorado, 1985, Oa
Adam L. Houston, (D), Illinois, 2004, Oaw
Richard M. Kettler, (D), Michigan, 1990, Cl
Ross Secord, (D), Michigan, 2004, PvCs
Peter J. Wagner, (D), Chicago, 1995, PioPq
Karrie A. Weber, (D), Alabama, 2002, PyCl
Assistant Professor:
Leilani A. Arthurs, (D), Notre Dame, 2007, OeCl
Caroline M. Burberry, (D), Imperial Coll (UK), 2008, Gct
Lynne J. Elkins, (D), MIT/WHOI, 2009, GivCt
Irina Filina, (D), Texas (Austin), 2007, YgGtYe
Matthew S. Van Den Broeke, (D), Oklahoma, 2011, Oar
Research Assistant Professor:
Mindi L. Searls, (D), Washington, 2007, YgGg
Professor of Practice:
Mary Anne Holmes, (D), Florida State, 1989, GsOu
Emeritus:
Ronald G. Goble, (D), Toronto, GxEg

Priscilla C. Grew, (D), California (Berkeley), 1967, GpeOe
Robert M. Hunt, (D), Columbia, 1971, Pv
Merlin P. Lawson, (D), Clark, 1973, OwaOr
Nancy Lindsley-Griffin, (D), California (Davis), 1982, Gt
Darryl T. Pederson, (D), North Dakota, 1971, Hw
Norman D. Smith, (D), Brown, 1967, Gsm
Michael R. Voorhies, (D), Wyoming, 1966, Pv
William J. Wayne, (D), Indiana, 1952, Gm



**School of Natural Resource Sciences** (B,M,D) (2016)
Hardin Hall
3310 Holdrege Street
Lincoln, NE 68583-0961
p. (402) 472-3471
jcarroll2@unl.edu
http://snr.unl.edu/
*Enrollment (2014): M: 4 (0) D: 9 (2)*

Professor:
Ken F. Dewey, (D), Toronto, 1973, Ow
Michael J. Hayes, (D), Missouri, 1994, Oa
Qi (Steve) Hu, (D), Colorado State, 1992, Oa
Kenneth G. Hubbard, (D), Utah State, 1981, Oa
Robert Oglesby, (D), Yale, 1990, Oa
Elizabeth A. Walter-Shea, (D), Nebraska, 1987, Oa
Research Assistant Professor:
Tsegaye Tadesse, (D), Nebraska, 2002, Oar
Associate Professor:
Martha D. Shulski, (D), Minnesota, 2002, Oa
Andrew E. Suyker, (D), Nebraska, 2000, Oa
Assistant Professor:
Guillermo Baigorria, (D), Wageningen Univ, 2005, Ow
Emeritus:
Donald A. Wilhite, (D), Nebraska, 1975, OayOn
Climate Scientist:
Deborah J. Bathke, (D), Ohio State, 2004, Oae

## University of Nebraska, Kearney

**Dept of Geography** (B) (2015)
203 Copeland Hall
Kearney, NE 68849
p. (308) 865-8355
combshj@unk.edu
*Enrollment (2002): B: 165 (0)*

Chair:
Jason Combs, (D), Nebraska, 2000
Professor:
Vijay Boken, (D), Manitoba, 1999
Paul Burger, (D), Oklahoma State, 1997
Jeremy Dillon, (D), Kansas, 2001
Associate Professor:
John Bauer, (D), Kansas, 2006
Instructor:
Nate Eidem, (D), Oregon State, 2011
Matt Engel, (D), Nebraska, 2007

# Nevada

## College of Southern Nevada, West Charleston Campus

**Dept of Physical Sciences** (A) (2015)
6375 W. Charleston Blvd.
Las Vegas, NV 89146
p. (702) 651-7475
physic@csn.edu
http://www.csn.edu/pages/2497.asp

Lead Faculty:
Barbara Graham, (M), Oyw
John E. Keller, (D), S Illinois, 2009, GeHw
Cynthia S. Shroba, (D), Illinois, Gg
Professor:
Patrick D. Clennan, (M), OyGe
Gale D. Martin, (M), Gg
Instructor:
Douglas Sims, (D), Kingston, 2011, GeSp

## Desert Research Institute

**Earth & Ecosystems Sciences** (2015)
2215 Raggio Parkway
Reno, NV 89512-1095
p. (775) 673-7300
bj@dri.edu

Professor:
John Arnone, (D), Yale, 1988, Py
Colleen M. Beck, (D), California (Berkeley), 1979, Sa
Christian H. Fritsen, (D), S California, 1996, Py
Nicholas Lancaster, (D), Cambridge (UK), 1977, Gm
Eric McDonald, (D), New Mexico, 1994, Sd
Alison E. Murray, (D), California (Santa Barbara), 1998, Py
David E. Rhode, (D), Washington, 1987, Pe
Staff Geomorphologist:
Sophie Baker, (M), Dalhousie, 2005, Gm
Associate Research Geomorphologist:
Steven N. Bacon, (M), Humboldt State, 2003, Gm
Associate Professor:
Kenneth D. Adams, (D), Nevada (Reno), 1997, Gm
Thomas F. Bullard, (D), New Mexico, 1995, Gm
Mary Cablk, (D), Oregon State, 1997, Or
Lynn Fenstermaker, (D), Nevada, 2003, Or
Giles Marion, (D), California (Berkeley), 1974, Sc
Kenneth C. McGwire, (D), California (Santa Barbara), 1992, Oy
David A. Mouat, (D), Oregon State, 1974, Ge
Assistant Professor:
JoseLuis Antinao, (D), Dalhousie, 2009, Gm
Amanda Keen-Zebert, (D), Texas State, 2007, Oy
Donald E. Sabol, Jr, (D), Washington, 1991, Or
GIS/Remote Sensing Scientist:
Timothy B. Minor, (M), California (Santa Barbara), 1982, Or
Assistant Research Ecologist:
Richard Jasoni, (D), Texas A&M, 1998, So
Archaeological Technician:
David Page, (M), Nevada (Reno), 2008, Ga

## Great Basin College

**Science Dept** (A) (2016)
1500 College Parkway
Elko, NV 89801
p. (775) 753-2120
caroline.meisner@gbcnv.edu
http://www2.gbcnv.edu/departments/SCI.html

Professor:
Caroline Bruno Meisner, (M), Oregon State, 2003, OgSo
Adjunct Professor:
Mira Kurka, Gg

## Truckee Meadows Community College

**Dept of Physical Sciences** (2014)
7000 Dandini Boulevard
Reno, NV 89512
p. (775) 673-7183
loanderson@tmcc.edu
http://www.tmcc.edu/physicalsci/

## University of Nevada, Las Vegas

**Geoscience Dept** (B,M,D) ● (2017)
4505 S. Maryland Parkway
Box 454010
Las Vegas, NV 89154-4010
p. (702) 895-3262
geodept@unlv.edu
http://geoscience.unlv.edu
Administrative Assistant: Maria I. Rojas
Administrative Assistant: Elizabeth Y. Smith
*Enrollment (2016): B: 18 (18) M: 14 (14) D: 0 (3)*

Professor:
Brenda J. Buck, (D), New Mexico State, 1996, GbSd
Andrew D. Hanson, (D), Stanford, 1998, Co
Ganqing Jiang, (D), Columbia, 2002, GsClGr
David K. Kreamer, (D), Arizona, 1982, Hw
Matthew S. Lachniet, (D), Syracuse, 2001, PeCs
Rodney V. Metcalf, (D), New Mexico, 1990, Gp
Margaret N. Rees, (D), Kansas, 1984, Gs

Stephen Rowland, (D), California (Santa Cruz), 1978, Pi
Wanda J. Taylor, (D), Utah, 1989, Gc
Michael L. Wells, (D), Cornell, 1991, Gc
Water Resourse Director/ Co-Chair:
Michael J. Nicholl, (D), Nevada (Reno), 1993
Department Chair:
Terry L. Spell, (D), SUNY (Albany), 1991, CcGvi
Associate Scientist:
Kathleen Zanetti, (M), Idaho, 1997, Gg
Assistant Professor:
Elisabeth M. Hausrath, (D), Penn State, 2007, ScCl
Associate Chair:
Eugene I. Smith, (D), New Mexico, 1970, Gi
Emeritus:
Frederick W. Bachhuber, (D), New Mexico, 1971, Pl
Jean S. Cline, (D), Virginia Tech, 1990, EmCe
Associate Reseach Professor:
Pamela C. Burnley, (D), California (Davis), 1990, GpyGg

## University of Nevada, Reno

**Center for Neotectonic Studies** (2010)
MS 0169
Reno, NV 89557-0169
p. (775) 784-6067
wesnousky@unr.edu
http://neotectonics.seismo.unr.edu/CNSHome.html
Director:
Steven Wesnousky, (D), Columbia, 1982, Ys

**Center for Research in Economic Geology** (M,D) (2017)
Mail Stop 1169
Reno, NV 89557-1169
p. (775) 784-1382
dawnsnell@unr.edu
Administrative Assistant: Dawn Lee Snell
*Enrollment (2015): M: 3 (3)*
Director:
John Muntean, (D), Stanford, Eg

**Dept of Geography** (B,M,D) (2014)
Mail Stop 0154
Reno, NV 89557-0154
p. (775) 784-6995
kberry@unr.edu
http://www.unr.edu/geography/
Administrative Assistant: Shari Baughman
Chair:
Kate Berry, (D), Colorado, 1993, Oy
Professor:
Scott A. Mensing, (D), California (Berkeley), 1993, Oy
Paul F. Starrs, (D), California (Berkeley), 1989, Oy
Assistant Professor:
Scott Bassett, (D), Harvard, 2001, Ou
P. Anthony Brinkman, (D), California (Berkeley), 2003, Ou
Jill Heaton, (D), Oregon State, 2001, Oi
Research Associate:
Abbey Grimmer, (B), Nevada (Reno), 2009, Oi
Scotty Strachan, (M), Nevada (Reno), 2001, Pe
Instructor:
Mella Harmon, (M), Ou
Adjunct Professor:
Douglas Boyle, (D), Arizona, Hg
Jake Haughland, (D), Colorado (Boulder), 2003, Gm
Kenneth McGwire, (D), California (Santa Barbara), 1992, Or
Ken Nussear, (D), Nevada (Reno), 2004, On
Victoria Randlett, (D), California (Berkeley), 1999, Ou
Christopher Ryan, (M), Nevada (Reno), 1998, Oy
Peter Wigand, (D), Washington State, 1985, Pe

**Dept of Geological Sciences and Engineering** (B,M,D) O (2017)
1664 N. Virginia St., MS 0172
Reno, NV 89557-0172
p. (775) 784-6050
geology@mines.unr.edu
http://www.unr.edu/geology
Research Professor:
Patricia H. Cashman, (D), S California, 1979, Gc
Simon R. Poulson, (D), Penn State, 1990, Cs
Director, Great Basin Center for Geothermal Energy:
Wendy M. Calvin, (D), Colorado, 1991, OrYg
Professor:
John G. Anderson, (D), Columbia, 1976, YsNeYg
Greg B. Arehart, (D), Michigan, 1992, Eg
James R. Carr, (D), Arizona, 1983, NgOuGg
John N. Louie, (D), Caltech, 1987, Ye
Paula J. Noble, (D), Texas, 1993, PmGnPs
James H. Trexler, (D), Washington, 1984, Gs
Scott W. Tyler, (D), Nevada (Reno), 1990, Hw
Robert J. Watters, (D), Imperial Coll (UK), 1972, Nrg
Steven G. Wesnousky, (D), Columbia, 1982, Ys
Associate Dean, College of Science:
Regina Tempel, (D), Colorado Mines, 1993, Cl
Assistant Professor:
Ronald J. Breitmeyer, (D), Wisconsin, 2011, GeHwSp
Wenrong Cao, (D), Univ. of Southern California, 2015, Gtc
Stacia Gordon, (D), Minnesota, 2009, Gi
Scott W. McCoy, (D), Colorado, 2012, GmNg
Philipp P. Ruprecht, (D), Washington, 2009, Giv
Lecturer:
John K. McCormack, (D), Nevada (Reno), 1997, Gz
Adjunct Professor:
Lisa Stillings, (D), Penn State, 1994, Cl
Emeritus:
Robert Karlin, (D), Oregon State, 1984, Ou
Tommy B. Thompson, (D), New Mexico, 1966, Eg
Cooperating Faculty:
John W. Bell, (M), Arizona State, 1974, Ng
Geoffrey Blewitt, (D), Caltech, 1986, Yd
Tom Bullard, (D), New Mexico, 1995, Gm
James E. Faulds, (D), New Mexico, 1989, Gc
Nick Lancaster, (D), Chamberlain, 1977, Gm
Greg Pohll, (D), Nevada (Reno), 1996, Hq
Lisa A. Shevenell, (D), Nevada (Reno), 1990, Hw

**Dept of Mining Engineering** (B,M,D) (2014)
Mail Stop 0173
Reno, NV 89557-0173
p. (775) 784-6961
jameshendrix@unr.edu
http://www.unr.edu/cos/mining/
Administrative Assistant: Carla Scott
*Enrollment (2011): B: 66 (5) M: 8 (0) D: 5 (0)*
Chair:
Danny L. Taylor, (D), Colorado Mines, 1980, Nm
Professor:
Jaak Daemen, (D), Minnesota, 1975, Nr
George Danko, (D), Budapest, 1985, Nm
Pierre Mousset-Jones, (D), London, 1988, Nm
Associate Professor:
Carl Nesbitt, (D), Nevada, 1990, Nx
Thom Seal, (D), Idaho, 2004, Nx
Emeritus:
Maurice Feunstenau, (D), Nx
Development Technician:
John D. Leland, (M), Stanford, 1983

**Graduate Program of Hydrologic Sciences** (M,D) (2016)
1664 N. Virginia Street
MS 0186
Reno, NV 89557-0175
p. (775) 784-6221
hydro@unr.edu
http://www.hydro.unr.edu
*Enrollment (2015): M: 32 (10) D: 8 (1)*
Research Professor:
Kumud Acharya, (D), Saitama, HsGeHq
Kenneth Adams, (D), Nevada (Reno), 1997, Gam
Braimah Apambire, (D)
John J. Arnone, (D), Yale, 1988, Sp
Gayle L. Dana, (D), Nevada (Reno), 1997, Or
Joseph Grzymski, (D)
Roger Jacobson, (D), Penn State, 1973, Cg

Nick Lancaster, (D), Cambridge, 1977, Gm
Joseph McConnell, (D), Arizona, 1997, On
Eric McDonald, (D), New Mexico, 1994, Sp
Alison Murray, (D)
Daniel Obrist, (D), Nevada (Reno), 2002
Simon Poulson, (D), Penn State, 1990, Cls
Ken Taylor, (D)

Associate Director:
Rina Schumer, (D), Nevada (Reno), 2002, HqwGm

Research Professor:
Christian H. Fritsen, (D), S California, 1995, Og

Professor:
Franco Biondi, (D), Arizona, 1994
Wendy Calvin, (D), Colorado (Boulder), 1991, YgOr
George Danko, (D), Hungarian Acad of Sci, 1985, Hq
Mae Gustin, (D), Arizona, 1988, Cg
David Kreamer, (D), Arizona, 1982, HwGn
John Louie, (D), Caltech, 1987, Yg
Maureen McCarthy, (D)
Paula Noble, (D), Texas (Austin), 1993, Gn
Anna Panorska, (D), California (Santa Barbara), 1992
Mark Pinsky, (D)
Greg Pohll, (D), Nevada (Reno), 1996, Hwq
Robert G. Qualls, (D), Georgia, 1989, OsHsSf
Loretta Singletary, (D)
Scott Tyler, (D), Nevada (Reno), 1990, Hg
Mark Walker, (D), Cornell, 1998, On

Research Hydrologist:
Brian Andraski, (D), Nevada (Reno)

Senior Scientist:
Richard Niswonger, (D)

Research Hydrogeologist and Civil Engineer:
Dave Decker, (D), Nevada (Reno), Hw

Associate Research Professor:
Marcus Berli, (D), Swiss Fed Inst Tech, Sp
Li Chen, (D), Chinese Acad of Sci
Clay A. Cooper, (D), Nevada (Reno), 1999, HySpHq
Alan Heyvaert, (D), California (Davis), 1998, GnHs
Justin Huntington, (D), Nevada (Reno), 2011, HgOr
Richard Jasoni, (D), Texas A&M
Alexandra Lutz, (D), Nevada (Reno), Hw
Kenneth McGwire, (D), California (Santa Barbara), 1992, Or
Don Sada, (D)
Rick Susfalk, (D)
Julian Zhu, (D), Dalhousie

Associate Professor:
Sudeep Chandra, (D), California (Davis), 2003, Hs
Keith E. Dennett, (D), Georgia Tech, 1995, Hs
Eric Marchand, (D), Colorado (Boulder), 2000
Sherman Swanson, (D), Oregon State, 1983, Hs
Aleksey Telyakovskiy, (D), Wyoming, 2002, Hq
Gina Tempel, (D), Colorado Mines, 1993, Cl

Associate Research Professor:
Ronald L. Hershey, (D), Nevada (Reno), 2010, HwCls

Associate Scientist:
Lisa Shevenell, (D), Nevada (Reno), 1990, Hw

Assistant Research Professor:
Rishi Parashar, (D), Purdue, 2008, Hq
Seshadri Rajagopal, (D)
Casey Schmidt, (D)

Assistant Research Hydrogeologist:
Rosemary Carroll, (D), Nevada (Reno), 2010, Hq

Assistant Professor:
Ronald Breitmeyer, (D), Wisconsin, 2011, HwGe
Adrian Harpold, (D), Cornell, 2010, HqsOr
Scott McCoy, (D)
Ben Sullivan, (D)
Paul Verburg, (D), Wageningen Ag Univ, 1998, Sp
Steve G. Wells, (D), Cincinnati, 1976, On
Yu (Frank) Yang, (D)

Research Associate:
Ramon Naranjo, (D), Nevada (Reno), 2012, CgHw
Lisa Stillings, (D), Penn State, 1994, Cm

Assistant Research Professor:
Tom Bullard, (D), New Mexico, 1995, Gm

Adjunct Professor:
Jonathan Price, (D)
Jim Thomas, (D), Nevada (Reno), 1996, Cg

Emeritus:
Dale Johnson, (D)
Wally Miller, (D)
Steve Wheatcraft, (D), Hawaii, 1979, Hw

Cooperating Faculty:
Chris Benedict
Jeanne Chambers, (D), Sf
David Prudic, (D), Nevada (Reno)
Michael Rosen, (D), Texas (Austin)
Keirith Snyder, (D)
Mark Weltz, (D)

**Great Basin Center for Geothermal Energy** (2015)
Mail Stop 0172
1664 N. Virginia St
Reno, NV 89557-0172
p. (775) 784-7018
geothermal@unr.edu
www.gbcge.org

Director:
Wendy Calvin, (D), Colorado, 1991, OrYgXg

Professor:
John Louie, (D), Caltech, 1987, Yse

Research Professor:
James Faulds, (D), New Mexico, 1989, Gc

Research Associate:
Nick Hinz, (M), Nevada (Reno), 2004, Gcg

Staff Research Associate:
Chris Sladek, (B), Colorado State, 1994, Gg

**Mackay School of Earth Sciences and Engineering (Director's Office)** (2016)
Mail Stop 0168
Reno, NV 89557-0168
p. (775) 784-6987
juliehill@unr.edu
http://www.mines.unr.edu/Mackay/
Administrative Assistant: Julie Hill

Director:
Russell Fields, (M), Nevada (Reno), 1985, Gg
Thom Seal, (D), Idaho, 2004, NxEmNm

**Nevada Seismological Lab** (2014)
Mail Stop 0174
Reno, NV 89557-0174
p. (775) 784-4975
mainofc@seismo.unr.edu
http://www.seismo.unr.edu
Administrative Assistant: Lori McClelland
Administrative Assistant: Erik Williams

Director:
Graham Kent, (D), California (San Diego), 1992, Ys

Professor:
John G. Anderson, (D), Columbia, 1976, Ys
John N. Louie, (D), Caltech, 1987, Ye
Steve Wesnousky, (D), Columbia, 1982, Ys

Research Associate Professor:
Glenn P. Biasi, (D), Oregon, 1994, Ys
Ileana Tibuleac, (D), S Methodist, 1999, Ys

Assistant Director, Seismic Network Manager & Development Director:
Ken D. Smith, (D), Nevada (Reno), 1991, YsGt

Adjunct Research Assistant Professor:
Satish Pullammanappallil, (D), Nevada (Reno), 1994, Ys

Adjunct Research Associate:
Bill Honjas, (M), Nevada (Reno), Ys

Emeritus Seismic Network Manager:
David von Seggern, (D), Penn State, 1982, Ys

Emeritus:
James N. Brune, (D), Columbia, 1961, Ys

Development Technician:
Ryan Presser, (A)

Volunteer Adjunct Faculty:
Aasha Pancha, (D), Nevada (Reno), 2007, Ys

Seismic Systems Analyst:
David Slater, (B), Calgary, 1990, Ys

Seismic Records Technician:
Tom Rennie, (D), Nevada (Reno), 2007, Ys
Programmer/Analyst, Seismic Network:
Gabriel Plank, (B), Cornell, 1994, Ys
Network Seismologist:
Diane dePolo, (M), Nevada (Reno), 1989, Ys
Development Technician:
Kent Straley
Associate Engineer:
John Torrisi, (A)

## Wassuk College

**Dept of Natural History** (B) (2015)
Box 236
East Ely, NV 89315
p. (702) 289-2168
Associate Professor:
Patrick J. Landon, (M), Montana, 1980, Oy
Adjunct Professor:
Jacob Rajala, (M), E Washington, 1976, On

## Western Nevada College

**Western Nevada College** (2014)
2201 West College Parkway
Carson City, NV 89703
p. (775) 445-4442
Winnie.Kortemeier@wnc.edu
http://www.wnc.edu/academics/division/sme/

# New Hampshire

## Dartmouth College

**Dept of Earth Sciences** (B,M,D) (2016)
228 Fairchild Hall, HB 6105
Hanover, NH 03755
p. (603) 646-2373
earth.sciences@dartmouth.edu
http://www.dartmouth.edu/~earthsci
f: www.facebook.com/DartmouthEarthSciences
*Enrollment (2015): B: 35 (9) M: 0 (5) D: 0 (3)*
Chair:
William Brian Dade, (D), Washington, GsHgEo
Professor:
Xiahong Feng, (D), Case Western, 1991, Cs
Carl E. Renshaw, (D), Stanford, 1993, HgGc
Associate Professor:
Robert L. Hawley, (D), Washington, 2005, Gl
Meredith Kelly, (D), Bern, 2003, Gl
Mukul Sharma, (D), Rochester, Ge
Leslie J. Sonder, (D), Harvard, 1986, Gq
Assistant Professor:
Erich C. Osterberg, (D), Maine, 2007, OaCl
Devon Renock, (D), Michigan, GzCa
Justin V. Strauss, (D), Harvard, 2015, GsCgGt
Emeritus:
Gary D. Johnson, (D), Iowa State, 1971, GrdYm
Research Professor:
Brian P. Jackson, (D), Georgia, 1998, CaGeCt

## Keene State College

**Dept of Geology** (B) (2015)
Mail Stop 2001
229 Main Street
Keene, NH 03435-2001
p. (603) 358-2553
pnielsen@keene.edu
http://www.keene.edu/academics/programs/geol/
*Enrollment (2014): B: 17 (4)*
Chair:
Peter A. Nielsen, (D), Alberta, 1977, GxzGc
Associate Professor:
Steven D. Bill, (D), Case Western, 1982, Pg
Adjunct:
Carol Leger, (B), Keene State Coll, 2005, Og
Dave Obolewicz, (M), Montana Tech, 1978, GgEmOw
Adjunct Professor:
Charles M. Kerwin, (D), New Hampshire, 2006, Gg
Edward M. Pokras, (D), Columbia, 1985, GgPmOb
Cooperating Faculty:
Timothy T. Allen, (D), Dartmouth, 1992, Hw

## Plymouth State University

**Environmental Science and Policy Dept** (B,M,D) (2014)
17 High Street
Plymouth, NH 03264
p. (603) 536-2573
warrent@plymouth.edu
http://oz.plymouth.edu/esp
Chair:
Warren Tomkiewicz, (D), Boston, 1987, OggOe
Professor:
James P. Koermer, (D), Utah, 1980, Ow
Larry T. Spencer, (D), Colorado State, 1968, Gg
Associate Professor:
Mark P. Turski, (D), Texas, 1994, Og
Assistant Professor:
Eric Hoffman, (D), SUNY (Albany), 2000, Ow

## University of New Hampshire

**Dept of Earth Sciences** (B,M,D) (2016)
214 James Hall
56 College Road
Durham, NH 03824
p. (603) 862-1718
earth.sciences@unh.edu
http://ceps.unh.edu/earth-sciences
Administrative Assistant: Susan E. Clark
*Enrollment (2016): B: 54 (10) M: 33 (9) D: 23 (1)*
Chair:
Julia G. Bryce, (D), California (Santa Barbara), 1998, Cg
Research Professor:
Stephen E. Frolking, (D), New Hampshire, 1993, Oa
Cameron P. Wake, (D), New Hampshire, 1993, Gl
Affiliate Professor:
Christopher E. Parrish, (D), Wisconsin (Madison), Or
Professor:
Julia G. Bryce, (D), California (Santa Barbara), 1998, CgcCt
William C. Clyde, (D), Michigan, 1997, Pg
John E. Hughes-Clarke, (D), Dalhousie, 1988, Og
Joseph M. Licciardi, (D), Oregon State, 2000, Gl
Larry A. Mayer, (D), California (San Diego), 1979, Gu
David C. Mosher, (D), Dalhousie, 1993, Og
Ruth K. Varner, (D), New Hampshire, 2000, Cl
Research Associate Professor:
Jack E. Dibb, (D), SUNY (Binghamton), 1988, Oa
Larry G. Ward, (D), South Carolina, 1978, Gu
Associate Professor:
Michael W. Palace, (D), New Hampshire, On
Affiliate Research Associate Professor:
Mark A. Fahnestock, (D), Caltech, 1991, Ol
Affiliate Associate Professor:
Joseph Salisbury, On
Mary D. Stampone, (D), Delaware, 2009, On
Associate Professor:
Margaret S. Boettcher, (D), MIT/WHOI, 2005, Ygs
Rosemarie E. Came, (D), MIT/WHOI, 2005, PeOnn
J. Matthew Davis, (D), New Mexico Tech, 1994, HwYh
Joel E. Johnson, (D), Oregon State, 2004, Gus
Jo Laird, (D), Caltech, 1977, Gp
Anne F. Lightbody, (D), MIT, 2007, HgsOn
Thomas C. Lippmann, (M), Oregon State, 1992, On
James M. Pringle, (D), MIT/WHOI, 1998, Op
Affiliate Professor:
Andrew Armstrong, (M), Johns Hopkins, 1991, On
Affiliate Professor:
Douglas C. Vandemark, (M), New Hampshire, 2005, Or
Emeritus:
Franz E. Anderson, (D), Washington, 1967, Yr
Francis S. Birch, (D), Princeton, 1969, Yg
Wallace A. Bothner, (D), Wyoming, 1967, Gc
Janet W. Campbell, (D), Virginia Tech, 1973, Oa

S. Lawrence Dingman, (D), Harvard, 1970, Hq
Henri E. Gaudette, (D), Illinois, 1963, Cc
Francis R. Hall, (D), Stanford, 1961, Hw
Theodore C. Loder, (D), Alaska, 1971, Oc
Cecil J. Schneer, (D), Cornell, 1954, Gz
Herbert Tischler, (D), Michigan, 1961, Ps
Affiliate Faculty:
Rochelle Wigley, (D), Cape Town (South Africa), 2005, CgGu

**Dept of Natural Resources and the Environment** (B,M,D) (2016)
215 James Hall
56 College Rd
Durham, NH 03824-3589
p. (603) 862-1020
john.halstead@unh.edu
Administrative Assistant: Linda Scogin
Chair:
Theodore E. Howard, (D), Oregon State, 1982, On
Professor:
Robert D. Harter, (D), Purdue, 1966, Sc
Assistant Professor:
Elizabeth A. Rochette, (D), Washington State, 1994, Sc

# New Jersey

## Bergen Community College
**Physical Sciences Depart** (1999)
400 County Rd 62
Paramus, NJ 07652
p. (201) 447-7100
mnotholt@bergen.edu
http://bergen.edu/academics/academic-divisions-departments/physical-science/

## College of New Jersey
**Physics Dept** (2017)
2000 Pennington Rd.
Ewing , NJ 08628
p. 609-771-2569
physics@tcnj.edu
Associate Professor:
Margaret Benoit, Penn State, 2005, Ys
Nathan Magee, (D), Penn State, 2006, Oa
Adjunct Professor:
Thomas Gillespie, (M), Rutgers, 1986, Gc

## Kean University
**School of Environmental and Sustainability Sciences** (B) (2014)
1000 Morris Avenue
Union, NJ 07083-0411
p. (908) 737-3737
fqi@kean.edu
http://www.kean.edu/KU/College-of-Natural-Applied-Health-Sciences
*Enrollment (2006): B: 160 (0)*
Professor:
Robert Metz, (D), Rensselaer, 1967, Gr
Shing Yoh, (D), Drexel, 1989, Ow
Constantine S. Zois, (D), Rutgers, 1980, Ow
Executive Director:
Paul J. Croft, (D), Rutgers, 1991, OwaOe
Associate Professor:
Carrie M. Manfrino, (D), Miami, 1995, GuOb
Feng Qi, (D), Wisconsin (Madison), 2005, OiyOg
Assistant Professor:
Kikombo Ngoy, (D), Oregon, 1996, Oy
Lecturer:
William C. Heyniger, (B), Montclair State, 2011, Ow
Secretary:
Christina Pacia, (N), On

## Montclair State University
**Dept of Earth & Environmental Studies** (B,M) (2014)
1 Normal Avenue
Upper Montclair, NJ 07043
p. (973) 655-4448
ophorid@mail.montclair.edu
http://www.csam.montclair.edu/earth/eesweb
*Enrollment (2002): B: 48 (12) M: 40 (19)*
Chair:
Jonathan M. Lincoln, (D), Northwestern, 1990, Gr
Associate Dean, Science & Mathematics:
Michael A. Kruge, (D), California (Berkeley), 1985, Co
Professor:
Huan E. Feng, (D), SUNY (Stony Brook), 1997, Cm
Gregory A. Pope, (D), Arizona State, 1994, GmaOy
Harbans Singh, (D), Rutgers, 1973, On
William Solecki, (D), Rutgers, 1990, Ou
Rolf Sternberg, (D), Syracuse, 1971, On
Robert W. Taylor, (D), St. Louis, 1971, On
John V. Thiruvathukal, (D), Oregon State, 1968, Yg
Associate Dean:
Duke U. Ophori, (D), Alberta, 1986, Hw
Associate Professor:
Matthew L. Gorring, (D), Cornell, 1997, Gi
Assistant Professor:
Mark J. Chopping, (D), Nottingham, 1998, Or
Adjunct Professor:
Kathryn Black, (M), Oklahoma, 1966, Oy
Matthew S. Tomaso, (M), Texas, 1995, Ga
Christine Valenti, (M), Montclair State, 1997, Gg
Emeritus:
Barbara De Beus, On
Laboratory Director:
Yoko Sato, (M), Montclair State, 2000, Gg

## New Jersey City University
**Dept of Earth and Environmental Sciences** (B) (2014)
Rossey Hall - Room 608
2039 Kennedy Boulevard
Jersey City, NJ 07305-1597
p. (201) 200-3161
lengland@njcu.edu
http://www.njcu.edu/dept/geoscience%5Fgeography/
*Enrollment (2006): B: 37 (19)*
Chair:
Deborah Freile, (D), Boston Univ, 1992, GseGg
Professor:
Martin Abend, (D), Syracuse, 1955, Oy
Research Associate:
John M. O'Brien, (D), California (Santa Barbara), 1973, Gs
Lecturer:
William W. Montgomery, (D), W Michigan, 1998, Hw
Adjunct Professor:
George Papcun, (M), Oe
Howard Zlotkin, (M), Oe
Emeritus:
John Marchisin, (M), Montclair State, 1965, Gg

## Princeton University
**Dept of Civil and Environmental Engineering** (B,M,D) (2014)
E-Quad, Olden Street
Princeton, NJ 08544
p. (609) 258-3598
jsmith@Princeton.EDU
http://cee.princeton.edu/
Professor:
Michael A. Celia, (D), Princeton, 1983, Hw
Francois M.M. Morel, (D), Caltech, 1971, Cm
Jean-Herve Prevost, (D), Stanford, 1974, Ng
James A. Smith, (D), Johns Hopkins, Hg
Associate Professor:
Peter R. Jaffe, (D), Vanderbilt, 1981, Hg

**Dept of Geosciences** (B,D) (2016)
113 Guyot Hall
Princeton, NJ 08544-1003
p. (609) 258-4101
mrusso@Princeton.EDU
http://www.geoweb.princeton.edu
Administrative Assistant: Mary Rose Russo
*Enrollment (2016): B: 24 (17) D: 39 (8)*

Chair:
Bess B. Ward, (D), Washington, 1982, Ob
Associate Chair:
Thomas S. Duffy, (D), Caltech, 1992, GyYgOm
Professor:
Gerta Keller, (D), Stanford, 1978, Pm
Francois M M. Morel, (D), Caltech, 1971, Cl
Satish C B. Myneni, (D), Ohio State, 1995, Sc
Tullis C. Onstott, (D), Princeton, 1980, Py
Michael Oppenheimer, (D), Chicago, 1970, OaGe
S. George H. Philander, (D), Harvard, 1970, Op
Allan M. Rubin, (D), Stanford, 1988, Yg
Jorge L. Sarmiento, (D), Columbia, 1978, Oc
Daniel M. Sigman, (D), MIT, 1997, Cg
Jeroen Tromp, (D), Princeton, 1992, Yss
Professional Specialist:
Amal Jayakumar, (D), Goa, 1999, Cm
Associate Professor:
Stephan A. Fueglistaler, (D), ETH (Switzerland), 2002, Oa
Adam C. Maloof, (D), Harvard, 2004, Ym
R Blair Schoene, (D), MIT, 2006, Cc
Frederik J. Simons, (D), MIT, 2002, YsdYm
Research Scholar:
Anne Morel-Kraepiel, (D), Princeton, 2001, Em
Xinning Zhang, (D), CalTech, 2010, Ou
Assistant Professor:
John A. Higgins, (D), Harvard, 2009, Ob
Jessica Irving, (D), Trinity Coll (Cambridge), 2009, Yg
Associate Professional Specialist:
Sergey Oleynik, (D), Moscow State (Russia), 1999, Cs
Research Associate:
Oliver Baars, (D), IFM-GEOMAR, 2011, Co
Dmitry Borisov, (D), IPGP, 2014, Ysg
Tra Dinh, (D), Washington, 2012, Oa
Michael P. Eddy, (D), MIT, 2016, Cc
Hom Nath Gharti, (D), Oslo, 2011, YsdYg
Alya Pamukcu, (D), Vanderbilt, 2014, NgGe
Youyi Ruan, (D), Virginia Tech, 2012, Ysg
Daniel Stolper, (D), Caltech, 2014, Pe
Sally June Tracy, (D), Caltech, 2015, Gy
Nicolas Van Oostende, (D), Ghent, 2011, Ob
Umair Waheed, (D), King Abdullah Univ of Sci & Tech, 2015, Ys
June Wicks, (D), Caltech, 2013, Cg
Emeritus:
Michael L. Bender, (D), Columbia, 1970, Cg
Lincoln S. Hollister, (D), Caltech, 1966, GptGz
Robert A. Phinney, (D), Caltech, 1961, Ys
Post Doctoral Research Fellow:
Clara Blattler, (D), Oxford, 2012, Gg
Associate Research Scholar:
Paul Gauthier, (D), Paris, 2010, Sb
Chui Yim Maggie Lau, (D), Hong Kong, 2007, Og
Sarah Jane White, (D), MIT, 2012, Ge
Undergraduate/Graduate Coordinator:
Sheryl A. Robas, (A), 1975, On
Academic Lab Manager:
Laurel P. Goodell, (M), Princeton, 1983, Oe
Danielle M. Schmitt, (M), W Michigan, 1999, Oe

**Environmental Engineering & Water Resources Program** (B,M,D) (2014)
E-220 Engineering Quad
Princeton, NJ 08544
p. (609) 258-4655
celia@princeton.edu
Department Secretary: Maryann Rothberg
Director:
Michael A. Celia, (D), Princeton, 1983, Hw
Professor:
Peter R. Jaffe, (D), Vanderbilt, 1981, SbHgw
James A. Smith, (D), Johns Hopkins, 1981, Hg
Eric F. Wood, (D), MIT, 1974, HgOrHq
Assistant Professor:
Catherine A. Peters, (D), Carnegie Mellon, 1992, Hw

**Program in Atmospheric & Oceanic Sciences** (D) (2015)
300 Forrestal Road, Sayre Hall
Princeton, NJ 08540-6654
p. (609) 258-6677
stf@princeton.edu
http://www.princeton.edu/aos/
*Enrollment (2012): D: 28 (6)*
Professor:
Michael Bender, (D), Columbia
Denise L. Mauzerall, (D), Harvard, 1996, Oa
Michael Oppenheimer, (D), Chicago
Stephen Pacala, (D), Stanford
George Philander, (D), Harvard
Jorge L. Sarmiento, (D), Columbia, 1978, On
James Smith, (D), Johns Hopkins
Senior Scientist:
Kirk Bryan, (D), MIT
Syukuro Manabe, (D), Toyko, 1958, Oa
Isidoro Orlanski, (D), MIT
Assistant Professor:
Stephan Fueglistaler, (D), ETH (Switzerland)
David Medvigy, (D), Harvard
Mark Zondlo, (D), Colorado
Lecturer:
Thomas Delworth, (D), Wisconsin
Leo Donner, (D), Chicago
Stephen Garner, (D), MIT
Stephen Griffies, (D), Pennsylvania
Robert Hallberg, (D), Washington
Isaac Held, (D), Princeton
Larry Horowitz, (D), Harvard
Sonya Legg, (D), Imperial Coll (UK)
Yi Ming, (D), Princeton
V. Ramaswamy, (D), SUNY (Albany)
Gabriel Vecchi, (D), Washington
Rong Zhang, (D), MIT, Op
Emeritus:
George Mellor, (D), MIT, 1957, Op

## Raritan Valley Community College
**Dept of Biology** (1999)
118 Lamington Road
Branchburg, NJ 08878
p. (908) 526-1200
dtrybuls@raritanval.edu

## Richard Stockton College of New Jersey
**Dept of Environmental Sciences** (B,M) (2013)
Division of Natural Science and Mathematics
101 Vera King Farris Drive
Galloway, NJ 08205
p. (609) 652-4620
george.zimmermann@stockton.edu
http://intraweb.stockton.edu/eyos/page.cfm?siteID=183&pageID=23
*Enrollment (2005): B: 122 (31)*
Director, Coastal Research Center:
Stewart Farrell, (D), Massachusetts, 1972, On
Professor:
Lynn F. Stiles, (D), Cornell, 1970, On
George Zimmermann, (D), Rutgers, 1982, On
Program Coordinator:
Michael D. Geller, (D), SUNY (Binghamton), 1979, On
Michael J. Hozik, (D), Massachusetts, 1981, Ym
Associate Professor:
Tait Chirenje, (D), Florida, Sc
William J. Cromartie, (D), Cornell, 1974, On
Weihong Fan, (D), Colorado State, 1993, Oi
Assistant Professor:
Tracy Baker, (D), Wyoming, HgOr
Judith Turk, (D), 2012, Sfo
Emeritus:
Raymond G. Mueller, (D), Kansas, 1981, SoGam

## Rider University
**Geological, Environmental, & Marine Sciences (GEMS)** (B) (2016)
2083 Lawrenceville Road
Lawrenceville, NJ 08648-3099

p. (609) 896-5092
husch@rider.edu
www.rider.edu/gems
f: www.facebook.com/riderclas/
t: www.rider.twitter.com/RiderCLAS
*Enrollment (2016): B: 7 (2)*

Chair:
Jonathan M. Husch, (D), Princeton, 1982, GieXg
Professor:
Hongbing Sun, (D), Florida State, 1995, HwCgSc
Associate Professor:
Kathleen M. Browne, (D), Miami, 1993, GuOe
Daniel L. Druckebrod, (D), Virginia, 2003, Onn
Reed A. Schwimmer, (D), Delaware, 1999, GsOgn
Gabriela W. Smalley, (D), Maryland, 2002, ObcOp
Adjunct Professor:
William B. Gallagher, (D), Pennsylvania, 1990, PvoGr
Emeritus:
Mary Jo Hall, (D), Lehigh, 1981, Gs

## Rutgers, The State University of New Jersey

**Earth and Planetary Sciences** (B,M,D) (2016)
Wright Lab
610 Taylor Road
Piscataway, NJ 08854-8066
p. (848) 445-2044
cswish@eps.rutgers.edu
http://geology.rutgers.edu/
*Enrollment (2016): B: 40 (12) M: 0 (3) D: 0 (1)*

Chair:
Carl C. Swisher III, (D), California (Berkeley), 1992, Cc
Undergraduate Program Director, Distinguished Professor:
Gail M. Ashley, (D), British Columbia, 1977, GsmSa
Graduate Program Director:
James D. Wright, (D), Columbia, 1991, CsOuPm
Distinguished Professor:
Paul G. Falkowski, (D), British Columbia, 1975, ObCmPe
Dennis V. Kent, (D), Columbia, 1974, Ym
George R. McGhee, Jr., (D), Rochester, 1978, PoqPi
Kenneth G. Miller, (D), MIT/WHOI, 1982, GurPm
Yair Rosenthal, (D), WHOI, CmlOc
Professor:
Craig S. Feibel, (D), Utah, 1988, GrsGa
Mark D. Feigenson, (D), Princeton, 1982, CgGiv
Claude T. Herzberg, (D), Edinburgh, 1975, CpGi
Vadim Levin, (D), Columbia, 1996, Ysg
Gregory S. Mountain, (D), Columbia, 1981, GuYrGr
Ying Fan Reinfelder, (D), Utah State, 1992, HwyOw
Roy W. Schlische, (D), Columbia, 1990, Gct
Robert M. Sherrell, (D), MIT/WHOI, 1991, CmOu
Martha O. Withjack, (D), Brown, 1977, Gct
Nathan Yee, (D), Notre Dame, 2001, PyCo
Associate Professor:
Juliane Gross, (D), Ruhr- Bochum, Germany, 2009, Xcm
Robert E. Kopp, (D), Caltech, 2007, PeyOg
Assistant Professor:
Sonia Tikoo-Schanz, (D), MIT, 2014, YmXy
Jill A. Van Tongeren, (D), Columbia, 2010, GpcGt
Research Associate:
Richard Mortlock, (D), Rutgers, 2017, CgmCs
Brent D. Turrin, (D), California (Berkeley), 1996, CcGvYm
Assistant Research Professor:
James V. Browning, (D), Rutgers, 1996, GgsGr
Adjunct Professor:
Lauren N. Adamo, (D), Rutgers, 2015, GuPm
Peter P. Sugarman, (D), Rutgers, 1995, Gr
Distinguished Visiting Professor:
William A. Berggren, (D), Stockholm, 1962, PmsPe
Emeritus:
Michael J. Carr, (D), Dartmouth, 1974, Gv
Richard K. Olsson, (D), Princeton, 1958, PmGr
Robert E. Sheridan, (D), Columbia, 1968, YrGu
Cooperating Faculty:
Jeremy S. Delaney, (D), Belfast, 1978, Xg
John R. Reinfelder, (D), SUNY (Stony Brook), 1993, Cl

## Rutgers, The State University of New Jersey, Newark

**Dept of Earth & Environmental Sciences** (B,M,D) (2015)
101 Warren Street
Smith Hall, room 135
Newark, NJ 07102
p. (973) 353-5100
morrin@andromeda.rutgers.edu
http://www.ncas.rutgers.edu/ees
Administrative Assistant: M. Elizabeth Morrin
*Enrollment (2014): B: 25 (6) M: 12 (3) D: 19 (1)*

Chair:
Lee S. Slater, (D), Lancaster (UK), 1997, Yg
Professor:
Yuan Gao, (D), Rhode Island, 1994, Oa
Alexander E. Gates, (D), Virginia Tech, 1986, Gc
Associate Professor:
Evert J. Elzinga, (D), Delaware, 2000, Sc
Andrew E. Kasper, (D), Connecticut, 1970, Pb
Adam B. Kustka, (D), Stony Brook, 2002, OgCmHs
Assistant Professor:
Mihaela Glamoclija, (D), D'Annunzio Univ (Italy), 2005, PyXgGe
Kristina M. Keating, (D), Stanford, 2009, Yg
Ashaki Rouff, (D), Stony Brook, 2004, Cl
Research Associate:
Dimitrios Ntarlagiannis, (D), Rutgers, 2006, Yg
Judith Robinson, (D), Rutgers, 2015, Yg
Emeritus:
Warren Manspeizer, (D), Rutgers, 1964, Gr
John H. Puffer, (D), Stanford, 1969, GivGe
Andreas H. Vassiliou, (D), Columbia, 1969, Gz

## Stockton University

**Geology** (B) ● (2016)
101 Vera King Farris Dr
Galloway, NJ 08205
p. 609-626-6857
matthew.severs@stockton.edu
http://intraweb.stockton.edu/eyos/page.cfm?siteID=183&pageID=33
*Enrollment (2016): B: 35 (7)*

Professor:
Michael J. Hozik, (D), Massachusetts, 1976, GcYmg
Associate Professor:
Matthew Rocky Severs, (D), Virginia Tech, 2007, GxCgEg
Assistant Professor:
Susanne Moskalski, (D), Delaware, GsOn
Judy Turk, (D), SpGmSo
Jeffrey R. Webber, (D), Massachusetts (Amherst), 2016, GcpGt
Emma Witt, (D), Hgs
Other:
Stewart C. Farrell, (D), Massachusetts, 1972, On

## Union County College

**Dept of Biology** (A) (2014)
1033 Springfield Ave.
Cranford, NJ 07016
p. (201) 709-7196
daly@ucc.edu
Administrative Assistant: Helen Gmitro

Associate Professor:
Raymond J. Daly, (M), Rutgers, 1975, Pv

## William Paterson University

**Dept of Environmental Science** (B) (2013)
Science Hall
Wayne, NJ 07470
p. (201) 595-2721
beckerm2@wpunj.edu
http://www.wpunj.edu/cosh/departments/environmental-science/
*Enrollment (2011): B: 90 (3)*

Chairman:
Martin A. Becker, (D), Brooklyn Coll, 1997, PgOw
Professor:
Richard R. Pardi, (D), Pennsylvania, 1983, Cc

Assistant Professor:
Jennifer R. Callanan, (D), Montclair State, 2008, ScGm
Karen Swanson, (D), Penn State, 1989, Cc

# New Mexico

## Eastern New Mexico University

**Dept of Physical Sciences** (B) (2016)
1500 S Ave K
STA 33
Portales, NM 88130
p. (575) 562-2174
jim.constantopoulos@enmu.edu
http://www.enmu.edu/physical/department-physical-sciences
*Enrollment (2016): B: 25 (3)*

Professor:
James T. Constantopoulos, (D), Idaho, 1989, GezGx

## Mesalands Community College

**Mesalands Community College** (A) (2010)
911 South Tenth Street
Pikes Peak community College
Tucumcari, NM 88401
p. (575) 461-4413
axelh@mesalands.edu
http://www.mesalands.edu/

Instructor:
Axel Hungerbuehler, (D), Bristol, Gg

## New Mexico Community College

**New Mexico Community College** (2014)
525 Buena Vista Dr. SE
Albuquerque, NM 87106
p. (505) 224-3000
contactcenter@cnm.edu
http://www.cnm.edu

## New Mexico Highlands University

**Natural Resources Management Dept** (B,M) (2013)
P.O. Box 9000
Las Vegas, NM 87701
p. (505) 454-3000
lindlinej@nmhu.edu
http://www.nmhu.edu/academics/undergraduate/arts_science/natural_resources/
*Enrollment (2012): B: 16 (4) M: 6 (1)*

Professor:
Jennifer Lindline, (D), Bryn Mawr, 1997, GizOe

Associate Professor:
Michael S. Petronis, (D), New Mexico, 2005, YmGcv

## New Mexico Institute of Mining and Technology

**Dept of Earth & Environmental Science** (B,M,D) O (2015)
801 Leroy Place
Socorro, NM 87801
p. (575) 835-5634
geos@nmt.edu
http://www.ees.nmt.edu/
*Enrollment (2014): B: 54 (9) M: 36 (18) D: 19 (2)*

Chair:
Penelope Boston, (D), Colorado (Boulder), 1985, On

Professor:
Susan L. Bilek, (D), California (Santa Cruz), 2001, Ys
Jan M. H. Hendrickx, (D), New Mexico State, 1984, Hw
Philip R. Kyle, (D), Victoria Univ of Wellington (NZ), 1976, Giv
Peter S. Mozley, (D), California (Santa Barbara), 1988, Gs
Mark A. Person, (D), Johns Hopkins, 1990, HyYhg
Fred M. Phillips, (D), Arizona, 1981, Hw

Emeritus Senior Environmental Geologist:
John W. Hawley, (D), Illinois (Urbana Champaign), 1962, GeHwGr

Senior Volcanologist, NMBG:
William C. McIntosh, (D), New Mexico Tech, 1990, Cc

Associate Professor:
Gary Axen, (D), Harvard, 1991, Gct
Bruce I. Harrison, (D), New Mexico, 1992, GmOsSf
Glenn Spinelli, (D), California (Santa Cruz), 2002, Hw

Assistant Professor:
Daniel Cadol, (D), Colorado State, 2010, HsGe
Ronni Grapenthin, (D), Alaska (Fairbanks), 2012, Gv
Kierran Maher, Washington State, Eg
Jolante Van Wijk, (D), Vrije Universiteit (Amsterdam), 2002, GtoYe

Associate Research Professor:
Mark Murray, (D), MIT, Yds
David B. Reusch, (D), Penn State, 2003, On
Dana S. Ulmer-Scholle, (D), S Methodist, 1992, GsdCl

Map Production Coordinator, NMBG:
Phillip Miller

Technical Staff Member, Seimologist:
Charlotte A. Rowe, (D), New Mexico Tech, 2000, Ys

Sr. Geochronologist/Co-Director NM Geochronology Research:
Matthew T. Heizler, (D), California (Los Angeles), 1993, Cc

Senior Scientist:
Robert S. Balch, (D), New Mexico Tech, 1997, Yse

Senior Mineralogist/Economic Geologist/ Director XRD Lab/Curator Mineral Museum:
Virgil L. Lueth, (D), Texas (El Paso), 1988, GzEg

Senior Field Geologist:
Steven M. Cather, (D), Texas (Austin), 1986, Gc
Knning Daniel
Daniel Koning, (M), New Mexico, 1999

Senior Economic Geologist:
Virginia T. McLemore, (D), Texas (El Paso), 1993, Eg

Senior Associate Hydrogeologist/Water Resources Engineer :
James T. McCord, (D), New Mexico Tech, 1989, Hq

Research Scientist:
Thomas Dewers, (D), Indiana, 1990, Yx

Research Associate, NMBG:
Matthew Zimmerer, New Mexico Tech

Professor of Biology:
Thomas L. Kieft, (D), New Mexico, 1983

Principal Senior Petroleum Geologist:
Ronald F. Broadhead, (M), Cincinnati, 1979, GorGs

Principal Senior Environmental Geologist:
David W. Love, (D), New Mexico, 1980, Ge

Principal Geologist:
Paul W. Bauer, (D), New Mexico Tech, 1988, Gc

Postdoc Fellow, USGS:
Jesus Gomez, New Mexico Tech, 2014

Planetary Protection Officer, NASA:
Catharine A. Conley, (D), Cornell, 1994, Py

Geophysicist/Field Geologist/Web Information Specialist:
Shari A. Kelley, (D), S Methodist, 1984, Gt

Geologic Mapping Program Manager:
J Michael Timmons, (D), New Mexico, 2004, GctGs

Geochemist & Deputy Director Manger of Electron Microprobe Lab:
Nelia W. Dunbar, (D), New Mexico Tech, 1989, Cg

Emeritus Senior Principal Geophysicist:
Marshall A. Reiter, (D), Virginia Tech, 1969, Yh

Emeritus Senior Field Geologist:
Richard Chamberlin, (D), Colorado Mines, 1980, GcvGg

Emeritus Director and State Geologist:
Charles E. Chapin, (D), Colorado Mines, 1965, Gv

Distinguished Member of Technical Staff:
Vincent C. Tidwell, (D), New Mexico Tech, 1999, Hwq

Chair/Associate Professor:
Michelle Creech-Eakman, (D), Denver, 1997, XgPy

Cave & Karst Hydrologist:
Lewis Land, (D), North Carolina (Chapel Hill), 1999, Hy

Associate Dean of Science and Wellness at Broward College:
Michael J. Pullin, (D), Kent State, 1999

Assistant Professor, Boise State:
Jeff Johnson, (D), Washington, 2000

Assistant Professor of Hydrogeology and Applied Geology, Purdue:
Marty Frisbee, New Mexico Tech, 2010

Assistant Professor:
Nigel J.F. Blamey, (D), New Mexico Tech, 2000, CaeCl

Adjunct Professor:
Denis Cohen
Derek Ford
Charles (Jack) Oviatt
Michael Underwood
Patrizia Walder

Emeritus:
Antonius J. Budding, (D), Amsterdam
Andrew R. Campbell, (D), Harvard, 1984, Cs
Kent C. Condie, (D), California (San Diego), 1965, Ct
Gerardo W. Gross, (D), Penn State, 1959, Yx
David B. Johnson, (D), Iowa, 1978, Ps
Allan R. Sanford, (D), Caltech, 1958, Ys
John W. Schlue, (D), California (Los Angeles), 1975, YsgOg
John L. Wilson, (D), MIT, 1974, Hw
Visiting Professor of Geochemistry:
Ingar Walder, (D), New Mexico Tech
Seismic Lab Associate:
Shane Ingate

**Dept of Mineral Engineering** (B,M,D) (2010)
Campus Station
Socorro, NM 87801-9990
p. (505) 835-5345
mojtabai@nmt.edu
Department Secretary: Lucero Joanna
Chair:
Navid Mojtabai, (D), Arizona, 1990, Nr
Professor:
William X. Chavez, Jr., (D), California (Berkeley), 1984, Em
Associate Professor:
Cathrine T. Aimone-Martin, (D), Northwestern, 1982, Ng
Assistant Professor:
Baolin Deng, (D), Johns Hopkins, 1995, Cg
Randal S. Martin, (D), Washington State, 1992, Oa
Adjunct Professor:
William Haneberg, (D), Cincinnati, 1989, Ng
Per-Anders Persson, (D), Cambridge, 1960, Nr
Ingar F. Walder, (D), New Mexico Tech, 1991, Cl
Emeritus:
George B. Griswold, (D), Arizona, Nx
Kalman I. Oravecz, (D), Witwatersrand, 1967, Nr

## New Mexico State University, Alamogordo

**New Mexico State University, Alamogordo** (2014)
2400 N. Scenic Drive
Alamogordo, NM 88310
hrnmsua@nmsu.edu
http://www.nmsua.edu

## New Mexico State University, Grants

**New Mexico State University - Grants** (2014)
1500 N Third St
Grants, NM 87020
p. (505) 287-6678
ssgrants@nmsu.edu
http://www.grants.nmsu.edu

## New Mexico State University, Las Cruces

**Department of Physics** (M,D) (2016)
MSC 3D
P.O. Box 30001
Las Cruces, NM 88003-8001
p. (505) 646-3831
thearn@nmsu.edu
http://physics.nmsu.edu
*Enrollment (2016): D: 2 (2)*
Head:
Stefan Zollner, (D), Arizona State, 1991, Om
Associate Professor:
Thomas M. Hearn, (D), Caltech, 1985, Ys
Boris Kiefer, (D), Michigan, GyYgOm
Assistant Professor:
Lauren Waszek, (D), Cambridge, 2012, Ys
Emeritus:
James F. Ni, (D), Cornell, 1984, YsGtYg

**Dept of Geological Sciences** (B,M) ● (2016)
MSC 3AB, Box 30001
1255 N. Horseshoe
Gardiner Hall, Room 171
Las Cruces, NM 88003
p. (575) 646-2708
geology@nmsu.edu
http://geology.nmsu.edu/
Administrative Assistant: Lee Hubbard
*Enrollment (2016): B: 49 (11) M: 18 (4)*
Head:
Nancy J. McMillan, (D), S Methodist, 1986, Gi
Professor:
Jeffrey M. Amato, (D), Stanford, 1995, GcCc
Associate Professor:
Frank C. Ramos, (D), California (Los Angeles), 2000, CgGg
Assistant Professor:
Reed J. Burgette, (D), Oregon, 2008, Gtc
Brian A. Hampton, (D), Purdue, 2006, Gst
Adjunct Professor:
Emily R. Johnson, (D), Oregon, 2008, Gvi

**Dept of Plant & Environmental Sciences** (A,B,M,D) (2011)
Box 30003
Dept. 3Q
Las Cruces, NM 88003-0003
p. (505) 646-3405
lmeyer@nmsu.edu
http://aces.nmsu.edu/academics/pes/
Department Secretary: Paula Ross
Head:
LeRoy A. Daugherty, (D), Cornell, 1975, Sd
Richard Pratt, (D)
Professor:
William C. Lindemann, (D), Minnesota, 1978, Sb
Bobby D. McCaslin, (D), Minnesota, 1974, Sc
Theodore W. Sammis, (D), Arizona, 1974, Sp
Assistant Professor:
Dean Heil, (D), California (Berkeley), 1991, Sc
Tim L. Jones, (D), Washington State, 1989, Sp
H. C. Monger, (D), New Mexico State, 1990, Sa

## San Juan College

**San Juan College** (A) (2016)
4601 College Blvd.
Farmington, NM 87402
p. (505) 566-3325
burrisj@sanjuancollege.edu
http://www.sanjuancollege.edu/geology
*Enrollment (2016): A: 27 (5)*
Professor:
John H. Burris, (D), Michigan State, 2004, GghGz

## University of New Mexico

**Dept of Earth & Planetary Sciences** (B,M,D) ○ (2015)
221 Yale Blvd NE
Northrop Hall, Room 141
MSC03 2040
Albuquerque, NM 87131-0001
p. (505) 277-4204
epsdept@unm.edu
http://epswww.unm.edu/
*Enrollment (2009): B: 184 (36) M: 34 (9) D: 27 (4)*
Chair:
Laura J. Crossey, (D), Wyoming, 1985, Cl
Professor:
Carl A. Agee, (D), Columbia, 1988, Cp
Yemane Asmerom, (D), Arizona, 1988, CcPeCg
Adrian J. Brearley, (D), Manchester (UK), 1984, Gz
Maya Elrick, (D), Virginia Tech, 1990, Gs
Peter J. Fawcett, (D), Penn State, 1994, Pe
Tobias Fischer, (D), Arizona State, 1999, Gv
David J. Gutzler, (D), MIT, 1986, Oa
Karl E. Karlstrom, (D), Wyoming, 1980, Gt
Leslie M. McFadden, (D), Arizona, 1982, Sd
Grant A. Meyer, (D), New Mexico, 1993, Gm
James J. Papike, (D), Minnesota, 1964, Ca
Louis A. Scuderi, (D), California (Los Angeles), 1984, PeGms
Zachary D. Sharp, (D), Michigan, 1987, Cs
Gary Weissmann, (D), California (Davis), Hw

Senior Scientist:
Nieu-Viorel Atudorei, (D), Lausanne, 1998, Cs
Victor J. Polyak, (D), Texas Tech, 1998, CcGzm
Associate Professor:
Joseph Galewsky, (D), California (Santa Cruz), 1996, Ow
Chester J. Weiss, (D), Texas A&M, 1998, Ye
Assistant Professor:
Corinne E. Myers, (D), Kansas, 2013, Po
Brandon Schmandt, (D), Oregon, 2011, Yg
Lindsay Lowe Worthington, (D), Texas (Austin), 2010, Yg
Research Associate:
Frans J.M. Rietmeijer, (D), Utrecht (Neth), 1979, Gp
Lecturer:
Aurora Pun, (D), New Mexico, 1996, Ca
Adjunct Professor:
Fraser Goff, (D), California (Santa Cruz), 1977, Cg
Sean McKenna, (D), Colorado Mines
Duane M. Moore, (D), Illinois (Urbana), 1963, Sc
Thomas E. Williamson, (D), New Mexico, 1993, Pv
Kenneth H. Wohletz, (D), Arizona State, 1980, Gv
Emeritus:
Roger Y. Anderson, (D), Stanford, 1960, Gn
Wolfgang E. Elston, (D), Columbia, 1953, Eg
Rodney E. Ewing, (D), Stanford, 1974, Gz
John W. Geissman, (D), Michigan, 1980, Ym
Rhian Jones, (D), Manchester (UK), 1986, Gz
Cornelis Klein, (D), Harvard, 1965, Gz
Barry S. Kues, (D), Indiana, 1974, Pi
Jane E. Selverstone, (D), MIT, 1985, Gpt
Lee A. Woodward, (D), Washington, 1962, Gc

**Inst of Meteoritics** (B,M,D) (2016)
MSC03 2050
1 University of New Mexico
Albuquerque, NM 87131
p. (505) 277-2747
iom@unm.edu
meteorite.unm.edu
*Enrollment (2016): M: 4 (0) D: 2 (1)*
Professor:
Carl B. Agee, (D), Columbia, 1988, XcGyCp
Senior Scientist III:
Horton E. Newsom, (D), Arizona, 1981, XgcXm
Research Professor:
James J. Papike, (D), Minnesota, 1964, Gz
Robert C. Reedy, (D), Columbia, 1969, Xcy
Senior Scientist III:
Charles K. Shearer, Jr., (D), Massachusetts, 1983, Gi
Assistant Professor:
Jin Zhang, (D), Illinois at Urbana-Champaign, 2014, Gy
Senior Scientist:
Karen Ziegler, (D), Reading/UK, 1993, CsXc
Research Specialist:
Paul V. Burger, (M), New Mexico, 2005, Gz
Research Scientist III:
Michael N. Spilde, (M), SD Mines, 1987, Gz
Program Manager:
Shannon Clark, On

**Water Resources Program** (2004)
1915 Roma NE, Room 1044
Albuquerque, NM 87131-1217
p. (505) 277-5249
fleckj@unm.edu
http://www.unm.edu/~wrp/
Director:
Michael E. Campana, On

## University of New Mexico, Taos
**University of New Mexico, Taos** (A,B) ● (2014)
1157 County Road 110
Ranchos de Taos, NM 87557
colnic@unm.edu
http://www.taos.unm.edu
Adjunct Professor:
Deborah Ragland, (D), Gg

## University of New Mexico, Gallup
**Div of Arts and Sciences** (2015)
200 College Road
Gallup, NM 87301
p. (505) 863-7500
pwatt@unm.edu
http://www.gallup.unm.edu/

## Western New Mexico University
**Dept of Natural Sciences** (2016)
P.O. Box 680
1000 West College Avenue
Silver City, NM 88062
p. (575) 538-6352
corrie.neighbors@wnmu.edu
http://natsci.wnmu.edu/
Assistant Professor:
Corrie Neighbors, (D), Univ of California, Riverside, 2016, Ysr

# New York



## Adelphi University
**Environmental Studies Program** (B,M) (2015)
South Avenue
Garden City, NY 11530
p. (516) 877-4170
schlosse@adelphi.edu
http://environmental-studies.adelphi.edu
*Enrollment (2010): B: 35 (6) M: 22 (8)*
Professor:
Anthony E. Cok, (D), Dalhousie, 1970, OuGeu
Assistant Professor:
Beth A. Christensen, (D), South Carolina, 1997, PmeGs

## Adirondack Community College
**Science Div** (A) (2014)
640 Bay Road
Queensbury, NY 12804
p. (518) 743-2325
minkeld@sunyacc.edu
http://www.sunyacc.edu

## Alfred University
**Dept of Geology** (B) (2017)
Saxon Drive
Alfred, NY 14802
p. (607) 871-2208
fmuller@alfred.edu
http://ottohmuller.com/ENSweb2008/
*Enrollment (2012): B: 15 (10)*
Professor:
Michele M. Hluchy, (D), Dartmouth, 1988, Cl
Otto H. Muller, (D), Rochester, 1974, Gc

## American Museum of Natural History
**Dept of Earth & Planetary Sciences** (2016)
Central Park West at 79th Street
New York, NY 10024-5192
p. (212)769-5100
http://www.amnh.org/our-research/physical-sciences/earth-and-planetary-sciences
Administrative Assistant: Nanette Nicholson
Chair and Curator:
Denton S. Ebel, (D), Purdue, 1993, XmCgOe
Curator:
George E. Harlow, (D), Princeton, 1977, Gz
James D. Webster, (D), Arizona State, 1987, Eg
Senior Scientific Assistant:
Jamie Newman, (M), Brooklyn Coll, Gg
Scientific Assistant:
Sam Alpert, (B), Case Western Reserve Univ., 2012
Saebyul Choe, (B), Bates, 2014
Curator Emeritus:
Edmond A. Mathez, (D), Washington, 1981, Gx

Specialist:
Kim V. Fendrich, (M), Arizona, 2016, Gz
Postdoctoral Fellow:
Shuo Ding, (D)
Nicholas Tailby, (D)
N. Alex Zirakparvar, (D), Syracuse, 2012

**Div of Paleontology** (D) (2010)
Central Park West at 79th Street
New York, NY 10024
p. (212) 769-5815
norell@amnh.org
http://paleo.amnh.org/
Administrative Assistant: Judy Galkin
*Enrollment (2009): D: 8 (0)*
Chair, Professor and Curator:
Mark A. Norell, (D), Yale, 1989, Pv
Provost, Professor and Curator:
Michael J. Novacek, (D), California (Berkeley), 1978, Pv
Professor and Curator:
Niles Eldredge, (D), Columbia, 1969, Pi
Neil H. Landman, (D), Yale, 1982, Pi
John G. Maisey, (D), London, 1974, Pv
Jin Meng, (D), Columbia, 1991, Pv
Dean of the Richard Gilder Graduate School, Professor, and Frick Curator:
John J. Flynn, (D), Columbia, 1983, Pv
Frick Curator Emeritus:
Richard H. Tedford, (D), California (Berkeley), 1960, Pv
Curator Emeritus:
Roger L. Batten, (D), Columbia, 1956, Pi
Eugene S. Gaffney, (D), Columbia, 1969, Pv

## Binghamton University

**Dept of Geological Sciences and Environmental Studies** (B,M,D) (2017)
PO Box 6000
Binghamton, NY 13902-6000
p. (607) 777-2264
demicco@binghamton.edu
http://geology.binghamton.edu
Administrative Assistant: Carol Slavetskas
*Enrollment (2012): B: 30 (0) M: 4 (0) D: 12 (0)*
Chair:
Robert V. Demicco, (D), Johns Hopkins, 1981, Gs
Professor:
Steven R. Dickman, (D), California (Berkeley), 1977, Yg
Joseph R. Graney, (D), Michigan, 1994, ClGeHs
David M. Jenkins, (D), Chicago, 1980, CpGz
Tim K. Lowenstein, (D), Johns Hopkins, 1982, Cl
H. Richard Naslund, (D), Oregon, 1980, Gi
Associate Professor:
Richard E. Andrus, (D), SUNY (Syracuse), 1974, Sf
Jeffrey S. Barker, (D), Penn State, 1984, Ys
Peter L. K. Knuepfer, (D), Arizona, 1984, Gt
Assistant Professor:
Thomas Kulp, (D), Indiana, 2002, Py
Alex Nikulin, (D), Rutgers, 2011, Ys
Molly Patterson, (D), Victoria Univ - New Zealand, 2015, CmOl
Jeff Pietras, (D), Wisconsin, 2003, GsEo
Research Associate:
Alan Jones, (D), Purdue, 1964, Ys
Emeritus:
Donald R. Coates, (D), Columbia, 1956, Gm
Thomas W. Donnelly, (D), Princeton, 1959, Gg
William D. MacDonald, (D), Princeton, 1965, Gc
Karen M. Salvage, (D), Penn State, 1998, Hw
James E. Sorauf, (D), Kansas, 1962, Pi
Francis T. Wu, (D), Caltech, 1966, Ys
Related Staff:
Michael Hubenthal, (M), Binghamton, 2010, OeYg

## Brooklyn College (CUNY)

**Dept of Earth and Environmental Science** (B,M) ● (2016)
2900 Bedford Avenue
Brooklyn, NY 11210
p. (718) 951-5416
wpowell@brooklyn.cuny.edu
http://depthome.brooklyn.cuny.edu/geology/
*Enrollment (2009): B: 24 (5) M: 10 (4) D: 3 (1)*
Chair:
Wayne G. Powell, (D), Queen's, 1994, GaCsOe
Professor:
John A. Chamberlain, (D), Rochester, 1971, Po
Constantin Cranganu, (D), Oklahoma, 1997, GoYhHw
Peter Groffman, SbCl
John Marra, (D), Dalhousie, 1977, Ob
David E. Seidemann, (D), Yale, 1975, Cc
Chair:
Jennifer Cherrier, OcCm
Associate Professor:
Stephen U. Aja, (D), Washington State, 1989, Cl
Zhongqi Cheng, (D), Ohio State, 2001, Ca
Assistant Professor:
Rebecca Boger, (D), William & Mary, 2002, OiHs
Brett Branco, (D), Connecticut, 2007, OnHsOg
Kennet Flores, (D), Gpc
Brianne Smith, (D), Hs
Lecturer:
Matt Garb, (M), Brooklyn Coll, Pg
Laboratory Director:
Guillermo Rocha, (M), CUNY, 1994, GgeCg

## Broome Community College

**Dept of Physical Sciences** (A) (2015)
Upper Front Street
Box 1017
Binghamton, NY 13902
p. (607) 778-5000
smithjj@sunybroome.edu
Professor:
Bruce K. Oldfield, (M), SUNY (Binghamton), 1988, Gg
Assistant Professor:
Jason J. Smith, (M), Binghamton, 2009, Ggs

## Buffalo State College

**Dept of Earth Sciences** (B) ● (2016)
1300 Elmwood Avenue
Buffalo, NY 14222
p. (716) 878-6731
solargs@buffalostate.edu
http://www.buffalostate.edu/earthsciences
Administrative Assistant: Cindy Wong
*Enrollment (2016): B: 80 (20)*
Chair:
Elisa T. Bergslien, (D), SUNY (Buffalo), 2002, ClHw
Professor:
Jill K. Singer, (D), Rice, 1986, GsOp
Planetarium Director:
Kevin K. Williams, (D), Johns Hopkins, 2002, Xg
Associate Professor:
Gary S. Solar, (D), Maryland, 1999, GcpGt
Kevin K. Williams, (D), Johns Hopkins, 2002, GmXg
Assistant Professor:
Bettina Martinez-Hackert, (D), SUNY (Buffalo), 2006, Gvm
Emeritus:
John E. Mack, (D), Fordham, 1971, Xg
Irving Tesmer, (D), Syracuse, Pg

## Cayuga Community College

**Math and Science** (A) (2013)
197 Franklin Street
Auburn, NY 13021
p. (315) 255-1743
waters@cayuga-cc.edu
http://www.cayuga-cc.edu/academics/programs_of_study/math_and_science.php
Professor:
Abu Z. Badruddin, (D), SUNY Coll Env Sci, Oi
Raymond F. Leszczynski, (M), SUNY (Albany), 1965, GglGm

## City College (CUNY)

**Dept of Earth & Atmospheric Sciences** (B,M) (2010)
New York, NY 10031
p. (212) 650-6984
kmcdonald2@ccny.cuny.edu

Chair:
Jeffrey Steiner, (D), Stanford, 1970, Cp
Professor:
Stanley Gedzelman, (D), MIT, 1970, Oa
Edward E. Hindman, (D), Washington, 1975, Oa
Margaret A. Winslow, (D), Columbia, 1979, Gc
Associate Professor:
Patricia M. Kenyon, (D), Cornell, 1986, Yg
Federica Raia, (D), Naples, 1997, Giv
Pengfei Zhang, (D), Utah, 2000, Hw

## Colgate University

**Dept of Geology** (B) ● (2016)
13 Oak Drive
Hamilton, NY 13346
p. (315) 228-7201
jmcnamara@colgate.edu
http://www.colgate.edu/academics/departments-and-programs/geology
Administrative Assistant: Jodi McNamara
*Enrollment (2016): B: 32 (17)*

Associate Professor:
Martin Wong, (D), California (Santa Barbara), 2005, Gtc
Professor:
Richard April, (D), Massachusetts (Amherst), 1978, CgGze
Karen Harpp, (D), Cornell, 1994, GvCgGi
Amy Leventer, (D), Rice, 1988, Ou
William H. Peck, (D), Wisconsin, 2000, GpiCs
Bruce Selleck, (D), Rochester, 1975, Gs
Constance M. Soja, (D), Oregon, 1985, Pi
Assistant Professor:
Aubreya Adams, (D), Penn State, 2010, Yg
Senior Lecturer:
Dianne M. Keller, (M), Colgate, 1988, Gz
Emeritus:
James McLelland, (D), Chicago, 1961, Gp
Paul Pinet, (D), Rhode Island, 1972, Ou

## College of Staten Island

**Engineering Science & Physics** (A,B) (2017)
2800 Victory Boulevard
Staten Island, NY 10314
p. (718) 982-2827
alan.benimoff@csi.cuny.edu
http://www.library.csi.cuny.edu/dept/as/geo/geo.html

Professor:
William J. Fritz, (D), Montana, 1980, Grv
Athanasios Koutavas, (D), Columbia, 2003, PeOg
Assistant Professor:
David Lindo Atichati, (D), the Canary Islands, 2012, Gue
Lecturer:
Jane L. Alexander, (D), Univ Coll (London), 1998, Gse
Alan I. Benimoff, (D), Lehigh, 1984, GizGp
College Laboratory Technician:
Samantha Gebauer, (M)
Adjunct Lecturer:
Imad Harone, (M), CUNY (Staten Island), Gg
Edward Johnson, (M), CUNY (Staten Island), Gg
Vladimir Jovanovic, (M), CUNY (Staten Island), Ge
Adjunct Associate Professor:
Mosbah Kolkas, (D), CUNY, Gg
Adjunct Assistant Professor:
Noureddin Amaach, (D), CUNY, Gg
Rosemary McCall, (D), Gg
Caitlyn Nichols, (D), CUNY, Ge

## Columbia University

**Dept of Earth & Environmental Engineering** (2010)
Henry Krumb School of Mines
500 West 120 Street
918 Mudd Bldg
New York, NY 10027
p. (212) 894-2905
schlosser@ldeo.columbia.edu
http://www.eee.columbia.edu/
Administrative Assistant: Co'Quesie Gilbert
Department Administrator: Barbara Algin

Acting Chair:
Nickolas J. Themelis, (D), McGill, 1961, Nx
Professor:
Paul F. Duby, (D), Columbia, 1962, Nx
Peter Schlosser, (D), Heidelberg, 1985, Cg
Ponisseril Somasundaran, (D), California (Berkeley), 1964, Nx
Tuncel M. Yegulalp, (D), Columbia, 1968, Nm
Associate Professor:
Ross Bagtzoglou, (D), California (Berkeley), 1990, Hw
Senior Research Scientist:
Roelof Versteeg, (D), Paris VII, 1991, Yg
Adjunct Professor:
Vasilis M. Fthenakis, (D), New York, 1991, Oa
Emeritus:
Stefan H. Boshkov, (M), Columbia, 1942, Nm
John T. Kuo, (D), Stanford, 1958, Yg
Malcolm T. Wane, (M), Columbia, 1954, Nm

**Dept of Earth & Environmental Sciences** (B,M,D) (2014)
P.O. Box 1000
61 Route 9W
Palisades, NY 10964
p. (845) 365-8550
carolm@ldeo.columbia.edu
http://eesc.columbia.edu

Vice Chair:
Peter B. de Menocal, (D), Columbia, 1991, PeOuCg
Director, Lamont Doherty Earth Observatory:
Sean Solomon, (D), MIT, 1971, Xy
Dir. Graduate Studies:
Goran Ekstrom, (D), Harvard, 1987, Ys
Chair:
Peter B. Kelemen, (D), Washington, 1988, GiCg
Professor:
Wallace S. Broecker, (D), Columbia, 1958, CmOcPe
Mark A. Cane, (D), MIT, 1975, OapPe
Nicholas Christie-Blick, (D), California (Santa Barbara), 1979, Gst
Joel E. Cohen, (D), Harvard, 1970, On
Hugh Ducklow, (D), Harvard, 1977, Ob
Peter M. Eisenberger, (D), Harvard, 1967, On
Steven L. Goldstein, (D), Columbia, 1986, Csg
Arnold L. Gordon, (D), Columbia, 1965, Op
Kevin L. Griffin, (D), Duke, 1994, PbeOn
Sidney R. Hemming, (D), SUNY (Stony Brook), 1994, CscPe
Jerry F. McManus, (D), Columbia, 1989, PeOu
William H. Menke, (D), Columbia, 1981, YsGvq
John C. Mutter, (D), Columbia, 1982, YrGtYs
Paul E. Olsen, (D), Yale, 1983, PvoGr
Stephanie L. Pfirman, (D), MIT, 1985, OpGe
Terry A. Plank, (D), Columbia, 1993, GxCg
Lorenzo M. Polvani, (D), MIT, 1988, OwaGq
G. Michael Purdy, (D), Cambridge, 1974, Yr
Peter Schlosser, (D), Heidelberg, 1985, Hw
Christopher H. Scholz, (D), MIT, 1967, YxNr
Adam H. Sobel, (D), MIT, 1998, Oa
Marc W. Spiegelman, (D), Cambridge, 1989, GqxCg
Martin Stute, (D), Heidelberg, 1989, CsHgGe
David Walker, (D), Harvard, 1972, CpGzCg
Associate Professor:
Sonya Dyhrman, Scripps, 1999, Ob
Arlene M. Fiore, (D), Harvard, 2003, Oa
Baerbel Hoenisch, (D), Bremen, 2002, PeObCm
Meredith Nettles, (D), Harvard, 2005, YsGl
Maria Tolstoy, (D), California (San Diego), 1994, Yr
Assistant Professor:
Ryan P. Abernathey, (D), MIT, 2012, Op
Tiffany A. Shaw, (D), Toronto, 2009, Oa
Lecturer:
Roger N. Anderson, (D), California (San Diego), 1973, Yh
Anthony G. Barnston, (M), Illinois (Urbana), 1976, On
Alberto Malinverno, (D), Columbia, 1989, Gug

Benjamin S. Orlove, (D), California (Berkeley), 1975, Oa
Andreas M. Thurnherr, (D), Southampton, 2000, Op
Christopher J. Zappa, (D), Washington, 1999, Op

Adjunct Assistant Professor:
Natalie T. Boelman, (D), Columbia, 2004, PbeOr



Adjunct Professor:
Robert F. Anderson, (D), MIT, 1981, OcCmPe
W. Roger Buck, IV, (D), MIT, 1984, YgGt
John J. Flynn, (D), Columbia, 1983, PvyYm
Alessandra Giannini, (D), Columbia, 2001, Oap
Lisa M. Goddard, (D), Princeton, 1995, Oa
Andrew Juhl, (D), California (San Diego), 2000, Ob
Arthur L. Lerner-Lam, (D), California (San Diego), 1982, Ys
Douglas G. Martinson, (D), Columbia, 1982, OpGq
Ronald L. Miller, (D), MIT, 1990, Oa
Mark A. Norell, (D), Yale, 1988, Pv
Dorothy M. Peteet, (D), NYU, 1983, PleOn
Andrew W. Robertson, (D), Reading (UK), 1984, Oa
Joerg Schaefer, (D), ETH Zurich, 2000, Cg
Christopher Small, (D), California (San Diego), 1993, OrYr
Taro Takahashi, (D), Columbia, 1957, OcCg
Mingfang Ting, (D), Princeton, 1990, Oa
Felix Waldhauser, (D), ETH (Switzerland), 1996, YsgGt
Spahr C. Webb, (D), California (San Diego), 1984, Yrg
Gisela Winckler, (D), Heidelberg, 1998, Cm

Emeritus:
Dennis E. Hayes, (D), Columbia, 1966, Yr
James D. Hays, (D), Columbia, 1964, PemOu
Paul G. Richards, (D), Caltech, 1970, YsOn
David H. Rind, (D), Columbia, 1976, Oar
H. James Simpson, Jr., (D), Columbia, 1970, Cm
Lynn R. Sykes, (D), Columbia, 1965, YsGtOn

Senior Administrative Manager:
Carol S. Mountain, (B), Columbia, 1974, On

Business Manager:
Sarah K. Odland, (M), Colorado, 1981, On

Asst. Director Climate and Society Program:
Cynthia Thomson, (M), Columbia, 2009, On

**Lamont-Doherty Earth Observatory** (M,D) (2017)
P.O. Box 1000
61 Route 9W
Palisades, NY 10964
p. (845) 359-2900
director@ldeo.columbia.edu
http://www.ldeo.columbia.edu
*Enrollment (2006): M: 40 (0) D: 83 (6)*

Professor:
Sean C. Solomon, (D), MIT, 1971, Xgy

Chair of DEES:
Peter Kelemen, (D), Washington, 1987, Gi

Professor:
Wallace S. Broecker, (D), Columbia, 1957, Cg
Nicholas Christie-Blick, (D), California (Santa Barbara), 1979, Gs
Peter B. deMenocal, (D), Columbia, 1991, Pe
Hugh W. Ducklow, (D), Harvard, 1977
Peter Eisenberger, (D), Harvard, 1967, On
Goran Ekstrom, (D), Harvard, 1987, Yg
Steven Goldstein, (D), Columbia, 1986, Cg
Arnold L. Gordon, (D), Columbia, 1965, Op
Kevin Griffin, (D), Duke, 1994, On
Sidney Hemming, (D), SUNY (Stony Brook), 1994, Cg
Jerry McManus, (D), Cm
William H. Menke, (D), Columbia, 1981, Ys
John C. Mutter, (D), Columbia, 1982, Yr
Paul E. Olsen, (D), Yale, 1983, Gm
Terry Plank, (D), Columbia, 1993, Cg
G. Michael Purdy, (D), Cambridge, 1974, Yr
Peter Schlosser, (D), Heidelberg, 1985, Hw
Christopher H. Scholz, (D), MIT, 1967, Ys
Adam Sobel, (D), MIT, 1998, Oa
Marc Spiegelman, (D), Cambridge, 1989, Ys

Sr. PGI Research Scientist:
Robin E. Bell, (D), Columbia, 1989, Yr
Richard Seager, (D), Columbia, 1990, Ow

Senior Research Scientist:
Sean M. Higgins, (D), Columbia, 2002, Gs
Kerstin Lehnert, (D), Albert-Ludwigs Freiburg, 1989, Gi

Research Scientist:
Robert Newton, (D), Columbia, 2001, Ct

Paros Sr. Research Scientist:
Spahr Webb, (D), California (San Diego), 1984, Yr

Lamont Research Professor:
Roger W. Buck, (D), MIT, 1984, Yr
Brendon Buckley, (D), Tasmania, 1997, Pe
Steven Chillrud, (D), Columbia, 1995, Cg
Rosanne D'Arrigo, (D), Columbia, 1989, On
James Davis, (D), MIT, 1986, Yd
Suzana de Camargo, (D), Tech (Munich), 1992, Oa
James Gaherty, (D), MIT, 1995, Ys
Joaquim Goes, (D), Nagoya, 1996, Gu
David S. Goldberg, (D), Columbia, 1985, Yr
Won-Young Kim, (D), Uppsala, 1986, Ys
Yochanan Kushnir, (D), Oregon State, 1985, Oa
Braddock Linsley, (D), New Mexico, 1990, Pe
Alberto Malinverno, (D), Columbia, 1989, Go
Douglas G. Martinson, (D), Columbia, 1982, Op
Joerg Schaefer, (D), Swiss Fed Inst, 2000, Pe
Bruce Shaw, (D), Chicago, 1989, Ys
Christopher Small, (D), California (San Diego), 1993, Or
Michael Steckler, (D), Columbia, 1980, Yv
Ajit Subramaniam, (D), SUNY (Stony Brook), 1995, Obr
Marco Tedasco, (D), Italian National Research Council, 2003, Olr
Andreas Thurnherr, (D), Southampton, 1999, Op
Mingfang Ting, (D), Princeton, 1990, Oa
Alexander Van Geen, (D), MIT/WHOI, 1989, Cg
Felix Waldhauser, (D), ETH (Switzerland), 1996, Ys
Gisela Winckler, (D), Heidelberg, 1998, Cg
Xiaojun Yuan, (D), California (San Diego), 1994, Op

Heezen Senior Research Scientist:
Suzanne Carbotte, (D), California (Santa Barbara), 1992, Yr

Ewing LDEO Rsrch Professor:
Robert F. Anderson, (D), MIT, 1981, Cg
Edward R. Cook, (D), Arizona, 1985, Hw
Taro Takahashi, (D), Columbia, 1957, Cm

Dir.Core Repository:
Maureen Raymo, (D), Columbia, 1989, Pe

Deputy Director:
Arthur L. Lerner-Lam, (D), California (San Diego), 1982, Ys

Associate Professor:
Baerbel Hoenisch, (D), Alfred Wegener Inst (Germany), 2002, Cm
Arlene Fiore, (D), Harvard, 2003, Oa
Meredith Nettles, (D), Harvard, 2005, YsOl
Maria Tolstoy, (D), California (San Diego), 1994, Yr

Lamont Associate Research Professor:
Michela Biasutti, (D), Washington, 2003, Yr
Benjamin C. Bostick, (D), Stanford, 2002, Sc
Connie Class, (D), Karlsruhe, 1994, Cg
Benjamin Holtzman, (D), Minnesota, 2003, Ys
Andrew Juhl, (D), Scripps, 2000, Ob
Alexey Kaplan, (D), Gubkin Inst of Tech (Moscow), 1990, Op
Michael Kaplan, (D), Colorado, 1999, Gl
Mikhail Kogan, (D), Inst of Physics (Moscow), 1977, Yd
Wade McGillis, (D), California (Berkeley), 1993, Cm
Raymond N. Sambrotto, (D), Alaska, 1983, Ob
David Schaff, (D), Stanford, 2001, Ys
Donna Shillington, (D), Wyoming, 2004, Yr
Jason Smerdon, (D), Michigan, 2004, Pe
Colin Stark, (D), Leeds (UK), 1991, Gg
Susanne Straub, (D), Kiel, 1991, Gv
Christopher Zappa, (D), Washington, 1999, Op

Assistant Professor:
Ryan Abernathy, (D), MIT, 2012, Og

Special Research Scientist:
Pierre E. Biscaye, (D), Yale, 1964, Cm
Enrico Bonatti, (D), Pisa, 1967, Yr
Dake Chen, (D), SUNY (Stony Brook), 1989, Op
James R. Cochran, (D), Columbia, 1977, Yr
Klaus H. Jacob, (D), Goethe (Frankfurt), 1968, Ys
Stanley Jacobs, (B), MIT, 1962, Op
Walter Pitman, (D), Columbia, 1967, Yr
Paul G. Richards, (D), Caltech, 1970, Ys
William B. F. Ryan, (D), Columbia, 1961, Yr
Leonardo Seeber, (B), Columbia, 1964, Ys

William Smethie, (D), Washington, 1979, Cg
Emeritus:
Mark A. Cane, (D), MIT, 1975, Op
David Walker, (D), Harvard, 1972, Gx
Research Scientist:
Andrew Barclay, (D), Oregon, 1998, Ys
Victoria Ferrini, (D), SUNY (Stony Brook), 2004, Ou
Helga Gomes, (D), Bombay, 1985, Ob
Gilles Guerin, (D), Columbia, 2000, Gu
Naomi Henderson, (D), Wisconsin, 1987, Yr
Timothy Kenna, (D), MIT/WHOI, 2002, Cg
Frank Nitsche, (D), Alfred Wegener Inst (Germany), 1997, Gu
Lamont Associate Research Professor:
Beizhan Yan, (D), Rensselaer, 2004, Cg
Lamont Associate Jr. Rsrch Professor:
Timothy Crone, Washington, 2007, Gu
William Joseph D'Andrea, (D), Brown, 2008, Pe
Pratigya J. Polissar, (D), Massachusetts, 2005, Gg
Heather M. Savage, (D), Penn State, 2007, Ys
Lamont Assist. Rsrch Professor:
Liala Andru-Hayles, (D), Barcelona, 2007, Pe
Anne Becel, (D), Institut de Physique du Globe de Paris, 2006, Gu
Natalie Boelman, (D), Columbia, 2004, Or
Timothy T. Creyts, (D), British Columbia, 2007, Ol
Solange Duhamel, (D), Aix-Maraseille II, Pe
Einat Lev, (D), MIT, 2009, Gv
Jonathan E. Nichols, (D), Brown, 2009, Pe
Michael Previdi, (M), Rutgers, 2006, Oa
Philipp Ruprecht, (D), Washington, 2009, Gv

## Cornell University

**Dept of Earth & Atmospheric Sciences** (B,M,D) O (2017)
2122 Snee Hall
Ithaca, NY 14853-1504
p. (607) 255-3474
easinfo@cornell.edu
http://www.eas.cornell.edu/
*Enrollment (2016): B: 36 (14) M: 6 (4) D: 28 (2)*
Wold Family Professor in Environmental Balance for Human Sustainability:
John F.H. Thompson, (D), Toronto, 1982, EmgNx
Professor:
Geoffrey A. Abers, (D), MIT, 1989, YsGt
Richard W. Allmendinger, (D), Stanford, 1979, Gc
Warren D. Allmon, (D), Harvard, 1988, PoePi
Larry D. Brown, (D), Cornell, 1976, Ye
Lawrence M. Cathles, (D), Princeton, 1968, Hy
John L. Cisne, (D), Chicago, 1973, Pg
Stephen J. Colucci, (D), SUNY (Albany), 1982, Oa
Arthur DeGaetano, (D), Rutgers, 1989, Oa
Louis A. Derry, (D), Harvard, 1989, Cl
Charles H. Greene, (D), Washington, 1985, Ob
David Hysell, (D), Cornell, 1992, Onr
Teresa E. Jordan, (D), Stanford, 1979, GrtOg
Suzanne M. Kay, (D), Brown, 1975, GiCg
Natalie M. Mahowald, (D), MIT, 1996, Oa
Sara C. Pryor, (D), East Anglia, 1992, Oan
Susan Riha, (D), Washington, 1980, Sf
John F. H. Thompson, (D), Toronto, 1982, EgNx
William M. White, (D), Rhode Island, 1977, CctCa
Daniel Wilks, (D), Oregon, 1986, Oa
Associate Professor:
Rowena B. Lohman, (D), Caltech, 2004, YsOrGt
Matthew E. Pritchard, (D), Caltech, 2003, YdGvOl
Assistant Professor:
Toby R. Ault, (D), Arizona, 2011, Ow
Katie M. Keranen, Stanford Univ, 2008, Ys
Sr. Lecturer:
Bruce Monger, (D), Hawaii (Manoa), 1993, Ob
Senior Lecturer:
Mark Wysocki, (M), Cornell, 1988, Oa
Adjunct Asst. Professor:
Robert M. Ross, (D), Harvard, 1990, PoOeGs
Adjunct Associate Professor:
Gregory P. Dietl, (D), North Carolina State, 2002, Peo
Adjunct Professor:
Martin J. Evans, (D), Wales, 1985, Eo
Paula Mikkelsen, (D), Florida Inst of Technology, 1994, Pi
Jason Phipps Morgan, (D), Brown, 1985, GvCmPe
Manfred Strecker, (D), Cornell, 1987, Gt
Martyn Unsworth, (D), Cambridge, 1991, Ye
Emeritus:
Muawia Barazangi, (D), Columbia, 1971, Ys
William A. Bassett, (D), Columbia, 1959, Gz
John M. Bird, (D), Rensselaer, 1962, Gt
Arthur L. Bloom, (D), Yale, 1959, Gm
Bryan L. Isacks, (D), Columbia, 1965, YsGtm
Daniel E. Karig, (D), California (San Diego), 1970, YrGl
Robert W. Kay, (D), Columbia, 1970, Giz
Warren Knapp, (D), Wisconsin, 1968, Oa
Frank H. T. Rhodes, (D), Birmingham, 1950, Pi
Cooperating Faculty:
Ludmilla Aristilde, (D), California (Berkeley), 2008, Ge
Rebecca J. Barthelmie, On
J. Thomas Brenna, (D), Cornell, 1985, Cg
Oliver H. Gao, (D), California (Davis), 2004, On
Alexander G. Hayes, (D), Caltech, 2011, Xg
Greg C. McLaskey, (D), California (Berkeley), 2011, Ys
Thomas D. O'Rourke, (D), Illinois, 1975, Ng
Andy L. Ruina, (D), Brown, 1981, Yx
Steven W. Squyres, (D), Cornell, 1981, Xg
Tammo S. Steenhuis, (D), Wisconsin, 1977, Hg
Zellman Warhaft, (D), London, 1975, On
Max Zhang, (D), California (Davis), 2004, On

**Institute for the Study of the Continents** (2016)
2122 Snee Hall
Ithaca, NY 14853-1504
p. (607) 255-3474
easinfo@cornell.edu
http://www.eas.cornell.edu/
*Enrollment (2010): M: 9 (0) D: 29 (4)*
Professor:
Geoffrey A. Abers, (D), MIT, 1989, Ys
Richard W. Allmendinger, (D), Stanford, 1979, Gct
Larry D. Brown, (D), Cornell, 1976, Ye
Lawrence M. Cathles, (D), Princeton, 1968, EgYgGe
Louis A. Derry, (D), Harvard, 1989, Cg
David L. Hysell, (D), Cornell, 1992, Onr
Teresa E. Jordan, (D), Stanford, 1979, GrtOg
Suzanne M. Kay, (D), Brown, 1975, Gp
William M. White, (D), Rhode Island, 1977, Ce
Associate Professor:
Rowena B. Lohman, (D), Caltech, 2004, Yg
Matthew E. Pritchard, (D), Caltech, 2003, YdGvOl
Assistant Professor:
Katie M. Keranen, (D), Stanford, 2008, Ys
Greg C. McLaskey, (D), California (Berkeley), 2011, YsNre
Visiting Professor:
Franklin G. Horowitz, (D), Cornell, 1989, YgOnYe
Emeritus:
Muawia Barazangi, (D), Columbia, 1971, Ys
Bryan L. Isacks, (D), Gmt
Robert W. Kay, (D), Columbia, 1970, Gi

## Dowling College

**Dept of Earth & Marine Sciences** (2006)
Oakdale, NY
asmirnov@dowling.edu

## Dutchess Community College

**Physical Science** (A) (2017)
53 Pendell Road
Poughkeepsie, NY 12601
p. (845) 431-8550
rambo@sunydutchess.edu
http://www.sunydutchess.edu/academics/departments/mathematicsphysicalandcomputersciences/
*Enrollment (2016): A: 5 (2)*
Chair:
Tim Welling, (M), GeOs
Professor:
Mark McConnaughhay, OgwEo

Associate Professor:
Susan H. Conrad, GsmHs

## Graduate School of the City University of New York

**PhD Program in Earth & Environmental Sciences** (D) (2015)
365 Fifth Avenue
New York, NY 10016
p. (212) 817-8240
ees@gc.cuny.edu
http://www.gc.cuny.edu/Page-Elements/Academics-Research-Centers-Initiatives/Doctoral-Programs/Earth-and-Environmental-Sciences
Administrative Assistant: Lina C. McClain
*Enrollment (2004): D: 40 (4)*

Distinguished Professor:
Gerald M. Friedman, (D), Columbia, 1952, Gs
Executive Officer:
Yehuda Klein
Professor:
Hannes K. Brueckner, (D), Yale, 1968, Cc
John A. Chamberlain, (D), Rochester, 1971, Po
Nicholas K. Coch, (D), Yale, 1965, Gs
Kathleen Crane, (D), California (San Diego), 1977, Yh
Eric Delson, (D), Columbia, 1973, Pv
Robert M. Finks, (D), Columbia, 1959, Pi
Stanley D. Gedzelman, (D), MIT, 1970, Ow
Victor Goldsmith, (D), Massachusetts, 1972, On
Daniel Habib, (D), Penn State, 1965, Pl
Charles A. Heatwole, (D), Michigan State, 1974, On
Edward E. Hindman, (D), Washington, 1975, Ow
Reza M. Khanbilvardi, (D), Penn State, 1983, Hy
Arthur M. Langer, (D), Columbia, 1965, Gz
Irene S. Leung, (D), California (Berkeley), 1969, Gz
David J. Leveson, (D), Columbia, 1960, On
David C. Locke, (D), Kansas State, 1965, Ca
Allan Ludman, (D), Pennsylvania, 1969, Gg
Cherukupalli E. Nehru, (D), Madras (India), 1963, Gi
Wayne G. Powell, (D), Queen's, 1994, EmGaCs
Robert S. Prezant, (D), Delaware (Lewes), 1981, Py
David E. Seidemann, (D), Yale, 1976, Cc
David H. Speidel, (D), Penn State, 1964, Cg
Dennis Weiss, (D), New York, 1971, Pg
Margaret S. Winslow, (D), Columbia, 1979, Gc
Associate Professor:
Sean C. Ahearn, (D), Wisconsin, 1986, Or
Stephen U. Aja, (D), Washington State, 1989, Cl
Patrick W. G. Brock, (D), Leeds, 1963, Gg
Lin A. Ferrand, (D), Princeton, 1988, Hq
Patricia M. Kenyon, (D), Cornell, 1986, Yg
Cecilia M. McHugh, (D), Columbia, 1993, Ou
Inez Miyares, (D), Arizona State, 1994, On
Jeffrey Steiner, (D), Stanford, 1970, Gx
Assistant Professor:
Charles R. Ehlschlaeger, (D), California (Santa Barbara), 1998, On
Mohamed B. Ibrahim, (D), Alberta, 1985, Oy
Robert P. Nolan, (D), CUNY, 1986, Co
William G. Wallace, (D), SUNY (Stony Brook), 1996, Og
Yan Zheng, (D), Columbia, 1999, Cm
Adjunct Professor:
Niles Eldredge, (D), Columbia, 1969, Pg
George E. Harlow, (D), Princeton, 1977, Gz
Neil H. Landman, (D), Yale, 1982, Ps
Edmond A. Mathez, (D), Washington, 1981, Gi
Jin Meng, (D), Columbia, 1991, Pv
Martin Prinz, (D), Columbia, 1961, Gi
Karl H. Szekielda, (D), Aix (France), 1967, Or
Emeritus:
Somdev Bhattacharji, (D), Chicago, 1959, Gt
Saul B. Cohen, (D), Harvard, 1955, Oy
Otto L. Franke, (D), Karlsruhe, 1962, Gc
William H. Harris, (D), Brown, 1971, Hw
Richard S. Liebling, (D), Columbia, 1963, Gz
Peter H. Mattson, (D), Princeton, 1957, Gc
Andrew McIntyre, (D), Columbia, 1967, Pe
Sara L. McLafferty, (D), Iowa, 1979, On
Joaquin Rodriguez, (D), Indiana, 1960, Pi
Surenda K. Saxena, (D), Uppsala, 1964, Cg
B. Charlotte Schrieber, (D), Rensselaer, 1974, Gd
Frederick C. Shaw, (D), Harvard, 1965, Pi
David L. Thurber, (D), Columbia, 1964, Cl

## Hamilton College

**Geosciences Department** (B) (2016)
198 College Hill Road
Clinton, NY 13323
p. (315) 859-4142
dbailey@hamilton.edu
https://my.hamilton.edu/academics/departments?dept=Geosciences
*Enrollment (2016): B: 23 (15)*

Chair:
David G. Bailey, (D), Washington State, 1990, GizGa
Professor:
Cynthia R. Domack, (D), Rice, 1985, Pg
Todd W. Rayne, (D), Wisconsin, 1993, Hy
Barbara J. Tewksbury, (D), Colorado, 1981, Gc
Associate Professor:
Michael L. McCormick, (D), Michigan, 2002, Py
Assistant Professor:
Catherine C. Beck, (D), Rutgers, 2015, GsnGr

## Hartwick College

**Dept of Geological and Environmental Sciences** (B) (2011)
Johnstone Science Center
1 Hartwick Drive
Oneonta, NY 13820
p. (607) 431-4658
griffingd@hartwick.edu
http://www.hartwick.edu/geology.xml
Administrative Assistant: Nancy Heffernan
*Enrollment (2011): B: 28 (9)*

Chair:
David H. Griffing, (D), Binghamton, 1994, GduGm
Professor Emeritus:
David Hutchison, (D), West Virginia, 1968, GxgGi
Professor:
Eric L. Johnson, (D), SUNY (Binghamton), 1990, GpcGi
Robert C. Titus, (D), Boston, 1974, Ps
Associate Professor:
Zsuzsanna Balogh-Brunstad, (D), Washington State, 2006, ClHgOs

## Hobart & William Smith Colleges

**Dept of Geoscience** (B) ● (2017)
300 Pulteney Street
Geneva, NY 14456
p. (315) 781-3586
geoscience@hws.edu
http://www.hws.edu/academics/geoscience/
*Enrollment (2015): B: 42 (14)*

Professor:
Nan Crystal Arens, (D), Harvard, 1993, Po
John D. Halfman, (D), Duke, 1987, GeHsGn
Neil Laird, (D), Illinois, 2001, Oa
D. Brooks McKinney, (D), Johns Hopkins, 1985, Gx
Associate Professor:
Tara M. Curtin, (D), Arizona, 2001, PeGs
David C. Kendrick, (D), Harvard, 1997, Po
Assistant Professor:
David Finkelstein, (D), Illinois (Urbana-Champaign), Cgs
Nicholas Metz, (D), SUNY (Albany), 2011, Oa
Technician:
Barbara Halfman, On

## Hofstra University

**Dept of Geology, Environment, and Sustainability** (B,M) (2016)
114 Hofstra University
Hempstead, NY 11549
p. (516) 463-5564
j.b.bennington@hofstra.edu
http://www.hofstra.edu/Academics/Colleges/HCLAS/GEOL/
f: https://www.facebook.com/GESatHU/
Administrative Assistant: Lena Hiller

*Enrollment (2016): B: 13 (5)*
Professor and Chair:
J Bret Bennington, (D), Virginia Tech, 1994, PeGs
Vice Provost for Research and Engagement:
Robert Brinkmann, (D), Wisconsin, 1989, Os
Associate Professor:
Emma Christa Farmer, (D), Columbia, 2005, Ou
Director of Sustainability Studies:
Sandra J. Garren, (D), South Florida, 2014, Oi
Assistant Professor:
Jase Bernhardt, Penn State, 2016, Owa
Antonios Marsellos, (D), SUNY (Albany), 2008, GtqOi
Instructor:
Annetta Centrella-Vitale, (M), Stony Brook Univ, 1998
Adina Hakimian, (M), CUNY Queens College, 2012, Gg
Steven C. Okulewicz, (M), CUNY (Brooklyn), 1979, CgGue
Adjunct Professor:
Nehru Cherukupalli, (D), Madras, 1963, Gg
Lillian Hess Tanguay, (D), CUNY, 1993, Gg
Richard Liebling, (D), Columbia, 1963, Gg
Emeritus:
Charles M. Merguerian, (D), Columbia, 1985, Gc
Dennis Radcliffe, Queens Univ, 1966, Gzx

## Hudson Valley Community College

**Biology, Chemistry, Physics Dept** (2010)
80 Vandenburgh Ave.
Troy, NY 12180
p. (518) 629-7453
p.schaefer@hvcc.edu
Assistant Professor:
Ruth H. Major, (M), Syracuse, 1989, Ggh

## Hunter College (CUNY)

**Dept of Geography** (B,M) (2014)
695 Park Avenue
Room 1006 North Building
New York, NY 10021
p. (212) 772-5265
imiyares@hunter.cuny.edu
http://www.geo.hunter.cuny.edu
Administrative Assistant: Dana G. Reimer
*Enrollment (2006): B: 135 (30) M: 40 (8)*
Chair:
Ines Miyares, (D), Arizona State, 1994, On
Professor:
Sean C. Ahearn, (D), Wisconsin, 1986, Oi
Jochen Albrecht, (D), Vechta (Germany), 1995, OiyOu
Philip Gersmehl, (D), Georgia, 1970, Oy
Charles A. Heatwole, (D), Michigan State, 1974, On
Wenge Ni-Meister, (D), Boston Univ, 1997, Or
William Solecki, (D), Rutgers, 1990, Ou
Associate Professor:
Allan Frei, (D), Rutgers, 1997, Hs
Rupal Oza, (D), Rutgers, 1999, On
Marianna Pavlovskaya, (D), Clark, 1998, Oi
Haydee Salmun, (D), Johns Hopkins, 1989, OpgOa
Assistant Professor:
Frank Buonaiuto, (D), SUNY (Stony Brook), 2003, On
Hongmian Gong, (D), Georgia, 1997, On
Mohamed Ibrahim, (D), Alberta, 1985, On
Randye L. Rutberg, (D), Columbia, 2000, Cc
Director, SPARs Lab:
Thomas Walter, (M), Miami Univ, 1984, OyeOw
Research Associate:
Carol Gersmehl, (M), Georgia, 1970, Oe
Adjunct Professor:
Jack Eichenbaum, (D), Michigan, 1972, Oy
Anthony Grande, (M), CUNY (Baruch), 1999, On
Edward Linky, (D), Duquesne Law, 1973, On
Teodosia Manecan, (D), Bucharest, 1985, Gp
Faye Melas, (D), CUNY, 1989, Gs
Karl H. Szekielda, (D), Marseille, 1967, Org
Douglas Williamson, (D), CUNY, 2003, Oi
GeoScience Lab Tech:
Amy Jeu, (M), Minnesota, 2002, Oi

## Lehman College (CUNY)

**Earth, Environmental and Geospatial Sciences** (B) (2015)
250 Bedford Park Boulevard West
Bronx, NY 10468-1589
p. (718) 960-8660
heather.sloan@lehman.cuny.edu
http://www.lehman.edu/academics/eggs/
*Enrollment (2014): B: 24 (5)*
Professor:
Eric Delson, (D), Pv
Irene S. Leung, (D), California (Berkeley), 1969, Gz
Associate Professor:
Juliana Maantay, (D), Rutgers, Oi
Assistant Professor:
Hari Pant, (D), Dalhousie, Cg
Heather Sloan, (D), Paris VI, 1993, Yr
Emeritus:
Frederick C. Shaw, (D), Harvard, 1965, Ps

## Long Island University, Brooklyn Campus

**Dept of Physics** (2014)
1 University Plaza
Brooklyn, NY 11201-8423
p. (718) 488-1011
bkln-admissions@liu.edu
http://www.liu.edu/Home/Brooklyn
Professor:
Richard Macomber, (D), Iowa, 1963, Pg
Adjunct Professor:
Richard A. Jackson, (D), Massachusetts, 1980, Gc
Alan Siegelberg, (M), Brooklyn, 1977, Gg
Emeritus:
Samuel R. Kamhi, (D), Columbia, 1963, Gz

## Long Island University, C.W. Post Campus

**Dept of Earth & Environmental Sciences** (B,M) (2013)
720 Northern Boulevard
Brookville, NY 11548-1300
p. (516) 299-2318
maboorst@liu.edu
Administrative Assistant: Beth Rondot
Chair:
Margaret F. Boorstein, (D), Columbia, 1977, Og
Associate Professor:
Victor DiVenere, (D), Columbia, 1995, Ym
Lillian Hess-Tanguay, (D), CUNY, 1993, Gs
E Mark Pires, (D), Michigan State, 1998, Og
Emeritus:
Robert S. Harrison, (D), Cambridge, 1965, Oy
Heinrich Toots, (D), Wyoming, 1965, Pg

## Monroe Community College

**Geoscience Dept** (A) (2016)
1000 E. Henrietta Road
Rochester, NY 14623
p. (716) 292-2425
drobertson@monroecc.edu
http://www.monroecc.edu/depts/geochem/
Administrative Assistant: Judy Miller
*Enrollment (2013): A: 7 (0)*
Associate Professor:
Jessica Barone, (M), Ball State, Ge
Michael Boester, (M), Oy
Amanda Colosimo, (M), North Carolina, 2004, Gg
Daniel E. Robertson, (M), Arizona State, 1986, Eg
Assistant Professor:
Jonathan Little, (M), OylOi
Jason Szymanski, (M), GlPe
Instructor:
Heather Pierce, (M), Connecticut, Oyi

## Orange County Community College

**Dept of Science, Engineering, and Architecture** (A) (2015)
115 South Street

Middletown, NY 10940
p. (845) 341-4570
lawrenceobrien@sunyorange.edu
*Enrollment (2015): A: 3 (1)*

Professor:
Lawrence E. O'Brien, (M), Michigan, 1972, Gg

New York

## Pace University, New York Campus
**Dept of Chemistry & Physical Sciences** (2011)
1 Pace Plaza
New York, NY 10038
p. (212) 346-1502
mshirigarakani@pace.edu
http://www.pace.edu/dyson/academic-departments-and-programs/chemistry-and-physical-sciences---nyc
Department Secretary: Pat Calegari

Chair:
Nigel Yartlett, (D)

Assistant Professor:
Stephen T. Lofthouse, (M), Hunter (CUNY), 1974, Og

Adjunct Professor:
Anatole Dolgoff, (M), Miami (Ohio), 1960, Og
William Hansen, (M), Hunter, 1991, Oa
John Marchisin, (M), Rutgers, 1965, Og
Nathan Reiss, (D), New York, 1973, Oa

## Paleontological Research Institution
**Paleontological Research Instituion** (2014)
1259 Trumansburg Road
Ithaca, NY 14850
p. (607) 273-6623
allmon@museumoftheearth.org
http://www.priweb.org

Director:
Warren D. Allmon, (D), Harvard, 1988, Po

Education Director:
Robert M. Ross, (D), Harvard, 1990, PoOe

## Plattsburgh State University (SUNY)
**Center for Earth & Environmental Science** (B) (2011)
101 Broad Street
102 Hudson Hall
Plattsburgh, NY 12901
p. (518) 564-2028
cees@plattsburgh.edu
http://www.plattsburgh.edu/cees
*Enrollment (2006): B: 58 (0)*

Chair:
Robert Fuller

Distinguished Service Professor:
James C. Dawson, (D), Wisconsin, 1970, Ge

Professor:
Donald D. Adams, (D), Dalhousie, 1973, Cl
Donald J. Bogucki, (D), Tennessee, 1970, Or
David A. Franzi, (D), Syracuse, 1984, Gl
Thomas H. Wolosz, (D), SUNY (Stony Brook), 1984, Pe

Assistant Professor:
Edwin A. Romanowicz, (D), Syracuse, 1993, Hw

Adjunct Professor:
Carol Treadwell-Steitz, (D), New Mexico, 1996, Gg

## Queens College (CUNY)
**School of Earth & Environmental Sciences** (B,M,D) (2014)
65-30 Kissena Boulevard
Flushing, NY 11367
p. (718) 997-3300
gregory.omullan@qc.cuny.edu
http://www.qc.edu/EES
Administrative Assistant: Gladys Sapigao
*Enrollment (2012): B: 150 (19) M: 19 (6) D: 11 (2)*

Distinguished Professor, Director & Chair:
George Hendrey, (D), Washington, 1973, Og

Professor:
Nicholas K. Coch, (D), Yale, 1965, GseOu
N. Gary Hemming, (D), Stony Brook, 1993, CaOcGg
Allan Ludman, (D), Pennsylvania, 1969, GgtGr
Steven Markowitz, (D), Columbia Coll, 1981, GbOn
Cecilia McHugh, (D), Columbia, 1993, Ou
Alfredo Morabia, (D), Johns Hopkins, 1989, GbOn
Stephen Pekar, (D), Rutgers, 1999, PeGrOu
Gillian Stewart, (D), SUNY (Stony Brook), 2005, OcCm
Yan Zheng, (D), Columbia, 1998, CgHwCm

Associate Professor:
Jeffrey Bird, (D), California (Davis), 2001, Sbc

Assistant Professor:
Timothy Eaton, (D), Wisconsin, 2002, HgwGg
Gregory O'Mullan, (D), Princeton, 2005, PyOb
Ashaki Rouff, (D), Stony Brook, 2004, Cac
Chuixiang Yi, (D), Nanjing, 1991, Oaw

Emeritus:
Eugene A. Alexandrov, (D), Columbia, 1959, Eg
Patrick W. G. Brock, (D), Leeds, 1963, Gg
Hannes K. Brueckner, (D), Yale, 1968, GgtCg
Robert M. Finks, (D), Columbia, 1959, Pi
Daniel Habib, (D), Penn State, 1965, Pl
Peter H. Mattson, (D), Princeton, 1957, Gc
Andrew McIntyre, (D), Columbia, 1967, Pe
B. Charlotte Schreiber, (D), Rensselaer, 1974, Gd
David H. Speidel, (D), Penn State, 1964, Cgp
David L. Thurber, (D), Columbia, 1964, Cl

## Queensborough Community College
**Dept of Biological Sciences and Geology** (1999)
222-05 56th Avenue
Bayside, NY 11364
p. (718) 631-6335
MGorelick@qcc.cuny.edu
http://www.qcc.cuny.edu/biologicalsciences/advisors.asp

## Rensselaer Polytechnic Institute
**Dept of Earth & Environmental Sciences** (B,M,D) (2015)
Science Center 1W19
110 8th Street
Troy, NY 12180-3590
p. (518) 276-6474
ees@rpi.edu
http://www.rpi.edu/dept/ees/
*Enrollment (2015): B: 30 (8) M: 2 (4) D: 13 (3)*

Head:
Frank S. Spear, (D), California (Los Angeles), 1976, GpCpGc

Professor:
Peter A. Fox, (D), Monash Univ (Australia), 1985, GqOig
Steven W. Roecker, (D), MIT, 1981, Yg
E. Bruce Watson, (D), MIT, 1976, CpcCt

Associate Professor:
Richard F. Bopp, (D), Columbia, 1979, Co
Miriam E. Katz, (D), Rutgers, 2001, PmeGu

Assistant Professor:
Karyn L. Rogers, (D), Washington Univ, 2006, PyCl
Morgan F. Schaller, (D), Rutgers, 2011, CsPe

Research Associate Professor:
Daniele J. Cherniak, (D), SUNY (Albany), 1990, Gx

Research Associate:
Nichlos D. Tailby, (D), Australian National, 2010, Cpt

Emeritus:
M. Brian Bayly, (D), Chicago, 1962, Gc
Samuel Katz, (D), Columbia, 1955, Yg
Robert G. La Fleur, (D), Rensselaer, 1961, Hw
Donald S. Miller, (D), Columbia, 1960, Cc

Laboratory Director:
Jared W. Singer, (D), Alfred Univ, 2013, CaOm

## Skidmore College
**Dept of Geosciences** (B) (2015)
815 North Broadway
Saratoga Springs, NY 12866
p. (518) 580-5190
knichols@skidmore.edu
http://www.skidmore.edu/academics/geo/
*Enrollment (2015): B: 15 (11)*

Associate Professor:
Amy Frappier, (D), New Hampshire, 2006, GePe
Kyle K. Nichols, (D), Vermont, 2002, Gm
Visiting Assistant Professor:
Margaret Estapa, (D), Maine, 2011, Oc
Assistant Professor:
Greg Gerbi, (D), MIT/WHOI, OgYg
Instructor:
Jennifer Cholnoky, (M), Rensselaer, 2013, Gg
Emeritus:
Richard H. Lindemann, (D), Rensselaer, 1980, Pi

## St. Lawrence University

**Dept of Geology** (B) (2016)
Brown Hall
Canton, NY 13617-1475
p. (315) 229-5851
skelly@stlawu.edu
http://www.stlawu.edu/academics/programs/geology
f: https://www.facebook.com/SLUGeology
Administrative Assistant: Sherrie Kelly
*Enrollment (2016): B: 35 (17)*

Professor:
Jeffrey R. Chiarenzelli, (D), Kansas, 1989, GzCg
Chair:
Antun Husinec, (D), Zagreb, 2002, GsCsGo
Associate Professor:
Alexander K. Stewart, (D), Cincinnati, 2007, GlmHw
Assistant Professor:
Judith Nagel-Myers, (D), Muenster, 2006, PisPe
Research Associate:
George W. Robinson, (D), Queens, 1978, Gz
Visiting Professor:
Erkan Toraman, (D), Minnesota, 2014, GciGp
Emeritus:
J. Mark Erickson, (D), North Dakota, 1971, Poi
Technician:
Matthew F. Van Brocklin, (M), Akron, 1996, GgOgn

## Suffolk County Community College, Ammerman Campus

**Dept of Physical Science** (A) (2016)
533 College Road
Selden, NY 11784
p. (631) 451-4338
butkosd@sunysuffolk.edu
http://depthome.sunysuffolk.edu/Selden/PhysicalScience/
*Enrollment (2016): A: 27 (0)*

Professor:
Darryl J. Butkos, (M), GgHw
Michael Inglis, (D), Og
Scott Mandia, (M), Penn State, 1990, Ow
Associate Professor:
Matthew Pappas, Og
Assistant Professor:
Sean Tvelia, (M), SUNY (Stony Brook), Ggl
Adjunct Professor:
Jessica Dutton, Og
Michael Flanagan, Og
Philip Harrington, Og
Margaret Lomaga, Og
Brian Vorwald, Og

## Sullivan County Community College

**Mathematics and Natural Sciences** (1999)
112 College Road
Lock Sheldrake, NY 12759
dlewkiewicz@sunysullivan.edu

## SUNY, Fredonia

**Department of Geology and Environmental Sciences** (B) (2016)
280 Central Avenue
Fredonia, NY 14063-1020
p. (716) 673-3303
earth@fredonia.edu
http://www.fredonia.edu/earth
*Enrollment (2016): B: 40 (7)*

Chair:
Gordon C. Baird, (D), Rochester, 1975, Ps
Professor:
Gordon C. Baird, (D), Rochester, 1975, GrPgGs
Gary G. Lash, (D), Lehigh, 1980, Gr
Sherri A. Mason, (D), Montana, 2001, OgGe
Associate Professor:
Ann K. Deakin, (D), SUNY (Buffalo), 1996, OiGgOr
Instructor:
Kimberly Weborg-Benson, (M), Illinois, 1991, GgOaPy

## SUNY, Jefferson

**Science** (1999)
1220 Coffeen Street
Watertown, NY 13601
p. (315) 786-2200
cebeyhoneycutt@sunyjefferson.edu
http://www.sunyjefferson.edu/academics/programs-study/liberal-arts-sciences-mathematics-science

## SUNY, Potsdam

**Department of Geology** (B) (2016)
220 Timerman Hall
44 Pierrepont Avenue
Potsdam, NY 13676
p. (315) 267-2286
rygelmc@potsdam.edu
http://www.potsdam.edu/academics/AAS/Geology/
f: https://www.facebook.com/potsdamgeology/
Administrative Assistant: Roberta Greene
*Enrollment (2016): B: 67 (18)*

Chair:
Michael C. Rygel, (D), Dalhousie, 2005, GsrGo
Assistant Professor:
Sara E. Bier, (D), Penn State, 2010, GctYg
Page C. Quinton, (D), Missouri, 2016, PeCsPm
Emeritus:
Robert L. Badger, (D), Virginia Tech, 1989, Gip
James D. Carl, (D), Illinois, 1962, Gz
William T. Kirchgasser, (D), Cornell, 1967, Ps
Neal R. O'Brien, (D), Illinois, 1963, Gs
Frank A. Revetta, (D), Rochester, 1970, Yg

## SUNY, Albany

**Dept of Atmospheric and Environmental Sciences** (B,M,D) (2014)
1400 Washington Avenue
Albany, NY 12222
p. (518) 442-4466
chair@atmos.albany.edu
http://www.atmos.albany.edu
*Enrollment (2004): B: 81 (27) M: 8 (5) D: 6 (4)*

Chair:
Vincent P. Idone, (D), SUNY (Albany), 1982, Oa
Professor:
Lance F. Bosart, (D), MIT, 1969, Oa
Daniel Keyser, (D), Penn State, 1981, Oa
John E. Molinari, (D), Florida State, 1979, Oa
Senior Research Professor:
David R. Fitzjarrald, (D), Virginia, 1980, Oaw
Richard R. Perez, (D), SUNY (Albany), 1983, Oa
James J. Schwab, (D), Harvard, 1983, Oa
Christopher J. Walcek, (D), California (Los Angeles), 1983, Oa
Wei-Chyung Wang, (D), Columbia, 1973, Oa
Associate Professor:
Robert G. Keesee, (D), Colorado, 1979, Oa
Christopher D. Thorncroft, (D), Reading, 1988, Oa
Research Associate:
Stephen S. Howe, (M), Penn State, 1981, Cs
David Knight, (D), Washington, 1987, Oa

## SUNY, Buffalo

**Dept of Geology** (B,M,D) O (2016)

126 Cooke Hall
Buffalo, NY 14260
p. (716) 645-3489
geology@buffalo.edu
http://www.geology.buffalo.edu
t: @UBGeology
Administrative Assistant: Alison A. Lagowski
*Enrollment (2014): B: 76 (38) M: 39 (13) D: 14 (4)*

New York

Chair:
Marcus I. Bursik, (D), Caltech, 1988, Gv
SUNY Distinguished Teaching Professor:
Charles E. Mitchell, (D), Harvard, 1983, Po
Director of Graduate Studies:
Gregory Valentine, (D), Santa Barbara, 1988, Gv
Professor:
Richelle Allen-King, (D), Waterloo, 1991, Cg
Mary Alice Coffroth, (D), Miami, 1988, Gu
Beata M. Csatho, (D), Miskolc (Hungary), 1993, OlrYg
Rossman F. Giese, (D), Columbia, 1962, Gz
Howard R. Lasker, (D), Chicago, 1978, Gu
Associate Professor:
Jason P. Briner, (D), Colorado, 2003, Gl
Tracy K. P. Gregg, (D), Arizona State, 1995, Gv
Research Assistant Professor:
Gerald J. Smith, (D), SUNY (Buffalo), 1997, Gs
Assistant Professor:
Estelle Chaussard, (D), Miami, 2013, YdGtv
Christopher S. Lowry, (D), Wisconsin, 2008, Hw
Erasmus K. Oware, (D), Clemson, 2014, YgHq
Elizabeth K. Thomas, (D), Brown, 2014, Cos
Research Associate:
Anton Schenk, (D), Switzerland, 1972, Or
Adjunct Professor:
Gary S. Lash, (D), Lehigh, 1980, Gc
Carel J. van Oss, (D), Paris, 1955, Gy
Emeritus:
Parker Calkin, (D), Ohio, 1963, Gl
Robert D. Jacobi, (D), Columbia, 1980, GctGs
Michael F. Sheridan, (D), Stanford, 1965, Gv

## SUNY, Cortland

**Geology Dept** (B,M) (2011)
342 Bowers Hall
PO Box 2000
Cortland, NY 13045
p. (607) 753-2815
GeologyDept@cortland.edu
http://www.cortland.edu/geology/
Administrative Assistant: Susan K. Nevins
*Enrollment (2007): B: 48 (15) M: 8 (6)*

Professor:
David J. Barclay, (D), SUNY (Buffalo), 1998, Glm
Christopher P. Cirmo, (D), Syracuse, 1994, Hg
Robert S. Darling, (D), Syracuse, 1992, GpzCp
Associate Professor:
Christopher A. McRoberts, (D), Syracuse, 1994, Pg
Associate Scientist:
Gayle C. Gleason, (D), Brown, 1993, Gct
Lecturer:
Julie L. Barclay, (M), SUNY (Buffalo), 1997, Gg
Laboratory Director:
John R. Driscoll, (A), Northwest Electronic, 1974, On

## SUNY, Geneseo

**Dept of Geological Sciences** (B) (2015)
1 College Circle
ISC 235
Geneseo, NY 14454
p. (585) 245-5291
lounsbur@geneseo.edu
http://www.geneseo.edu/geology
Administrative Assistant: Diane E. Lounsbury
*Enrollment (2015): B: 130 (27)*

Professor:
Scott D. Giorgis, (D), Wisconsin, 2003, GctYm
D. Jeffrey Over, (D), Texas Tech, 1990, Ps

Chair:
Benjamin J.C. Laabs, (D), Wisconsin, 2004, GmlGe
Associate Professor:
Dori J. Farthing, (D), Johns Hopkins, 2001, Gz
Amy L. Sheldon, (D), Utah, 2002, Hw
Assistant Professor:
Nicholas H. Warner, (D), Arizona State, 2008, XgGrHg
Emeritus:
Phillip D. Boger, (D), Ohio State, 1976, Cg
William J. Brennan, (D), Colorado, 1968, Gc
Richard B. Hatheway, (D), Cornell, 1969, Gx
James W. Scatterday, (D), Ohio State, 1963, Po
Richard A. Young, (D), Washington Univ (St. Louis), 1966, GmXgOr

## SUNY, Maritime College

**Science Dept** (B) (2010)
6 Pennyfield Avenue
Bronx, NY 10465
p. (718) 409-7380
kolszewski@sunymaritime.edu
*Enrollment (2009): B: 68 (22)*

Chair:
Kathy Olszewski, (D), SUNY (Stony Brook), 1994, Cg
Associate Professor:
Marie deAngelis, (D), Washington, 1989, Oc
Assistant Professor:
Anthony Manzi, (M), Montclair State, Ow

## SUNY, New Paltz

**Geology** (B) (2016)
1 Hawk Drive
New Paltz, NY 12561
p. (845) 257-3760
vollmerf@newpaltz.edu
http://www.newpaltz.edu/geology/
*Enrollment (2010): B: 98 (12)*

Chair:
Frederick W. Vollmer, (D), Minnesota, 1985, GctGx
Associate Professor:
Alexander J. Bartholomew, (D), Cincinnati, 2006, GrPi
Shafiul H. Chowdhury, (D), W Michigan, 1999, Hw
Alvin S. Konigsberg, (D), Syracuse, 1969, Oa
John A. Rayburn, (D), Binghamton, 2004, Gml
Assistant Professor:
Gordana Garapić, (D), Boston, 2013, Gz
Lecturer:
Kaustubh Patwardhan, (D), Johns Hopkins, 2009, Ggi
Adjunct Professor:
Laurel Mutti, (M), Johns Hopkins, 2004, Gg
Emeritus:
Gilbert J. Brenner, (D), Penn State, 1962, Pl
Constantine Manos, (D), Illinois, 1963, Gs
Martin S. Rutstein, (D), Brown, 1969, Gz
Related Staff:
Donald R. Hodder, (B), SUNY (Geneseo), 1985, Gg

## SUNY, Oneonta

**Dept of Earth and Atmospheric Sciences** (B) (2016)
108 Ravine Parkway
209 Sci. #1
Oneonta, NY 13820-4015
p. (607) 436-3707
James.Ebert@oneonta.edu
http://www.oneonta.edu/academics/earths/
*Enrollment (2012): B: 114 (39) M: 1 (1)*

Chair:
Jerome B. Blechman, (D), Wisconsin - Madison, 1979, Oa
Distinguished Teaching Professor:
James R. Ebert, (D), SUNY (Binghamton), 1984, GrsGd
Associate Professor:
Melissa Godek, (D), Delaware, 2009, OwaOg
Leslie E. Hasbargen, (D), Minnesota, 2003, GmHgOr
Assistant Professor:
Keith Brunstad, (D), Washington State, 2013, GviEg
Leigh M. Fall, (D), Texas A&M, 2010, Pgq
Christopher Karmosky, (D), Penn State, 2013, OwaOy

Pragnyadipta Sen, (M), Gct
Dr.:
Marta Clepper, (D), Kentucky, 2011, OeGgs
Emeritus:
P. Jay Fleisher, (D), Washington State, 1967, Gl
Arthur N. Palmer, (D), Indiana, 1969, Hg

## SUNY, Oswego

**Dept of Atmospheric and Geological Sciences** (B) O (2017)
394 Shineman Science Center
Oswego, NY 13126
p. (315) 312-3065
christine.dallas@oswego.edu
http://www.oswego.edu/ags
*Enrollment (2015): B: 160 (25)*
Professor:
Alfred J. Stamm, (D), Wisconsin, 1976, Ow
Paul B. Tomascak, (D), Maryland, 1995, CgGzi
David W. Valentino, (D), Virginia Tech, 1993, GtcGp
Associate Professor:
Scott Steiger, (D), Texas A&M, 2005, Ow
Assistant Professor:
J Graham Bradley, (D), College (London), 2012, GeHwNg
Rachel J. Lee, (D), Pittsburgh, 2013, GvOr
Steven T. Skubis, (D), SUNY (Albany), 1994, Ow
Michael Veres, (D), Nebraska Lincoln, 2014, Oa

## SUNY, Purchase

**Environmental Studies Program** (B) (2016)
735 Anderson Hill Road
Purchase, NY 10577
p. (914) 251-6646
naturalsciences@purchase.edu
http://www.purchase.edu/Departments/AcademicPrograms/las/sciences/EnvStudies/
*Enrollment (2016): B: 59 (12)*
Professor:
George P. Kraemer, (D), California (Los Angeles), 1989, Ob
Board of Study Coordinator:
Ryan W. Taylor, (D), Oregon State, 2006, OiyOg
Emeritus:
James M. Utter, (D), Rutgers, 1971, On

## SUNY, Stony Brook

**Dept of Geosciences** (B,M,D) (2014)
Nicolls Road
Stony Brook, NY 11794-2100
p. (631) 632-8200
daniel.davis@stonybrook.edu
http://www.geosciences.stonybrook.edu
*Enrollment (2005): B: 70 (15) M: 3 (4) D: 41 (3)*
Professor:
Daniel M. Davis, (D), MIT, 1983, Yg
Gilbert N. Hanson, (D), Minnesota, 1964, GgOe
William E. Holt, (D), Arizona, 1989, Ys
Robert C. Lieberman, (D), Columbia, 1969, GyYsx
Scott M. McLennan, (D), Australian National, 1981, Cg
Hanna Nekvasil, (D), Penn State, 1986, Cp
Artem Oganov, (D), Univ Coll (London), 2002, Gz
John B. Parise, (D), James Cook, 1980, Gz
Brian L. Phillips, (D), Illinois, 1990, Gz
Richard J. Reeder, (D), California (Berkeley), 1980, ClGz
Martin A. Schoonen, (D), Penn State, 1989, Cl
Donald J. Weidner, (D), MIT, 1972, Gy
Lianxing Wen, (D), Caltech, 1998, Ys
Associate Professor:
Timothy Glotch, (D), Arizona State, 2004, XgGz
E. Troy Rasbury, (D), SUNY (Stony Brook), 1998, Cc
Assistant Professor:
Michael Sperazza, (D), Montana, 2006, On
Curator:
Stephen C. Englebright, (M), SUNY (Stony Brook), 1975, Oe
Lecturer:
Christiane W. Stidham, (D), California (Berkeley), 1999, YgGe
Adjunct Professor:
Robert C. Aller, (D), Yale, 1977, Cm
Henry J. Bokuniewicz, (D), Yale, 1976, Yr
J. Kirk Cochran, (D), Yale, 1979, Oc
Roger D. Flood, (D), MIT, 1978, Gu
Baosheng Li, (D), SUNY (Stony Brook), 1996, Gy
Maureen O'Leary, (D), Johns Hopkins, 1997, Pv
Michael T. Vaughan, (D), SUNY (Stony Brook), 1979, Yx
Emeritus:
Garman Harbottle, (D), Columbia, 1949, Cc
David W. Krause, (D), Michigan, 1982, Pv
Donald H. Lindsley, (D), Johns Hopkins, 1961, Cp
Teng-fong Wong, (D), MIT, 1981, Yx
Laboratory Director:
Owen C. Evans, (D), SUNY (Stony Brook), 1994, Cg

**School of Marine and Atmospheric Sciences** (B,M,D) (2015)
145 Endeavour Hall
Stony Brook, NY 11794-5000
p. (516) 632-8700
minghua.zhang@stonybrook.edu
http://www.somas.stonybrook.edu
Professor:
Josephine Y. Aller, (D), S California, 1975, Oba
Edmund K. M. Chang, (D), Princeton, 1993, Oa
J. Kirk Cochran, (D), Yale, 1979, Oc
Brian Colle, (D), Washington, 1997, Oa
David O. Conover, (D), Massachusetts, 1981, Ob
Roger D. Flood, (D), MIT/WHOI, 1978, Ou
Marvin A. Geller, (D), MIT, 1969, Oa
Christopher Gobler, (D), SUNY (Stony Brook), 1999, Ob
Sultan Hameed, (D), Manchester, 1968, Oa
Darcy J. Lonsdale, (D), Maryland, 1979, Ob
Glenn R. Lopez, (D), SUNY (Stony Brook), 1976, Ob
John E. Mak, (D), California (San Diego), 1992, OaCas
Anne McElroy, (D), MIT/WHOI, 1985, Oc
Ellen K. Pikitch, Indiana, 1983, Ob
Mary I. Scranton, (D), MIT/WHOI, 1977, Oc
R. Lawrence Swanson, (D), Oregon State, 1971, Og
Gordon T. Taylor, (D), S California, 1983, Obc
Minghua Zhang, (D), Inst of Atm Physics, 1987, Oa
Associate Professor:
Bassem Allam, (D), W Brittany (France), 1998, Ob
Robert A. Armstrong, (D), Minnesota, 1975, Ocb
David Black, (D), Ou
Bruce J. Brownawell, (D), MIT/WHOI, 1986, Oc
Robert M. Cerrato, (D), Yale, 1980, Ob
Jackie Collier, (D), Stanford, 1994, Ob
Michael Frisk, (D), Maryland, 2004
Marat Khairoutdinov, (D), Oklahoma, 1997, Oa
Daniel A. Knopf, (D), Swiss Fed Inst Tech, 2003, Oa
Kamazima M. Lwiza, (D), Wales, 1991, Oc
Bradley Peterson, (D), S Alabama, 1998, Ob
Joseph Warren, (D), MIT, 2001, Ob
Robert E. Wilson, (D), Johns Hopkins, 1973, Op
Assistant Science Director, IOCS:
Demian Chapman, (D), Nova Southeastern, 2007
Assistant Professor:
Anthony Dvarskas, (D), Maryland (College Park), 2007
Hyemi Kim, (D), Seoul Nat, 2008
Janet Nye, (D), Maryland, 2008, Ob
Christopher Wolfe, (D), Oregon State, 2006, Op
Qingzhi Zhu, (D), Xiamen, 1997, Oc
Faculty Director Semester By The Sea:
Kurt Bretsch, (D), South Carolina, 2005
Lecturer:
Lesley Thorne, (D), Duke, 2010
Research Scientist:
Wuyin Lin, (D), Stony Brook, 2002, Oa
Research Professor:
Charles Flagg, (D), MIT, 1977, Op
Engineer:
Douglas Hill, (D), Columbia, 1977
Ecologist, Author:
Carl Safina, (D), Rutgers, 1987
Director Riverhead Foundation:
Robert A. DiGiovanni, Jr., (M), Stony Brook, 2002
Adjunct Professor:
James Ammerman, (D), Scripps, 1983

Howard Bluestein, MIT, 1976, Oa
Paul Bowser, (D), Auburn, 1978, Obb
Carl Brenninkmeijer, (D), Groningen, 1983
Michael J. Cahill, (D), DePaul, 1978, On
Andre Y. Chistoserdov, (D), Inst of Genetics & Selection (USSR), 1985
Alistair Dove, Queensland, 1999
Anga Engel, Oc
Emmanuelle pales Espinosa, (D), Nante (France), 1999
Mark Fast, (D), Dalhousie, 2005, Ob
Scott Ferson, (D), Stony Brook, 1988
Scott Fowler, (D), 1969
Roxanne Karimi, (D), Dartmouth, 2007
Kathryn Kavanagh, (D), James Cook, 1998
Yangang Liu, (D), Nevada (Reno), 1998
Stephan Munch, (D), Stony Brook, 2002, Ob
John Rapaglia, (D), Stony Brook, 2007
Frank J. Roethel, (D), SUNY (Stony Brook), 1981, Oc
Jeffrey Tongue, Oa
Andrew Vogelmann, (D), Penn State, 1994, Oa
Duane E. Waliser, (D), Oa
Douglas W. R. Wallace, (D), Dalhousie, 1985, Oc
Jian Wang, (D), Caltech, 2002, Oa

Distinguished Professor:
Cindy Lee, (D), California (San Diego), 1975, OcCo

Emeritus:
Dong-Ping Wang, (D), Miami, 1975

Affiliated and Joint Faculty:
Heather L. Lynch, (D), Harvard, 2006

Cooperating Faculty:
Resit Akcakaya, (D), Stony Brook, 1989
Stephen Baines, (D), Yale, 1993
Lee K. Koppelman, (D), New York, 1968, On
Jeffrey Levinton, (D), Yale, 1971, Ob
Dianna K. Padilla, (D), Alberta, 1987
Sheldon Reaven, (D), California (Berkeley), 1975, On

## SUNY, The College at Brockport

**Dept of the Earth Sciences** (B) (2014)
350 New Campus Drive
Brockport, NY 14420-2936
p. (585) 395-2636
earthsci@esc.brockport.edu
http://www.brockport.edu/esc
*Enrollment (2012): B: 117 (16)*

Associate Professor:
James A. Zollweg, (D), Cornell, 1994, Hs

Professor:
Whitney J. Autin, (D), Louisiana State, 1989, Gs
Mark R. Noll, (D), Delaware, 1989, Cl

Associate Professor:
Paul L. Richards, (D), Penn State, 1999, Hg
Scott M. Rochette, (D), St. Louis, 1998, Ow

Adjunct Professor:
David A. Boehm, (M), SUNY (Buffalo), 2003, Gg
Christine Crafts, (B), SUNY (Brockport), 2000, Ow
Jutta S. Dudley, (D), SUNY (Buffalo), 1998, Gg
William G. Glynn, (M), Texas A&M, 1984, Gg
Linda J. Schaffer, (M), SUNY (Brockport), 1988, Oe

Department Secretary:
Lauri A. Kifer, (A), Monroe Comm Coll, 1997, On

Emeritus:
Robert W. Adams, (D), Johns Hopkins, 1964, Gs
John E. Hubbard, (D), Colorado State, 1968, Hg
Richard M. Liebe, (D), Iowa, 1962, Ps
Judy A. Massare, (D), Johns Hopkins, 1983, Pv
John M. Williams, (B), Goshen, Oa

On Leave:
Robert S. Weinbeck, (D), Iowa State, 1980, Oa

Systems Administrator:
Thomas M. McDermott, On

## SUNY, Ulster County Community College

**STEM** (A) (2015)
Burroughs 105
491 Cottekill Road
Stone Ridge, NY 12484
p. (845) 687-5230
schimmrs@sunyulster.edu
http://people.sunyulster.edu/esc

Professor:
Steven Schimmrich, (M), SUNY (Albany), 1991, Og

Assistant Professor:
Karen Helgers, (M), SUNY, 1987, OeHgGe

## Syracuse University

**Dept of Earth Sciences** (B,M,D) (2017)
204 Heroy Geology Laboratory
Syracuse, NY 13244-1070
p. (315) 443-2672
jofitch@syr.edu
http://earthsciences.syr.edu
*Enrollment (2014): B: 54 (11) M: 13 (4) D: 16 (4)*

Chair:
Donald I. Siegel, (D), Minnesota, 1981, Hw

Professor:
Suzanne L. Baldwin, (D), SUNY (Albany), 1988, Cc
Paul G. Fitzgerald, (D), Melbourne (Australia), 1988, GtCc
Linda C. Ivany, (D), Harvard, 1997, PeoCs
Jeffrey Karson, (D), SUNY (Albany), 1977, GtcGv
Cathryn R. Newton, (D), California (Santa Cruz), 1983, Po
Scott D. Samson, (D), Arizona, 1990, Cc
Christopher A. Scholz, (D), Duke, 1989, Gs

Associate Professor:
Gregory Hoke, (D), Cornell, 2006, OgGmt
Laura Lautz, (D), Syracuse, 2005, Hg

Assistant Professor:
Christopher Junium, (D), Penn State, 2010, CsPeCo
Zunli Lu, (D), Rochester, 2008, Cg
Robert Moucha, (D), Toronto, 2003, GtYeg
Jay Thomas, (D), Virginia Tech, 2003, Gi

Instructor:
Daniel Curewitz, (D), Duke, 1999, OgGct

Emeritus:
M. E. Bickford, (D), Illinois, 1960, Cc
James C. Brower, (D), Wisconsin, 1964, Po

Other:
Bruce Wilkinson, (D), Texas, 1974, Gs

## Union College

**Geology Dept** (B) (2015)
807 Union Street
Schenectady, NY 12308-3107
p. (518) 388-6770
geology@union.edu
http://www.union.edu/academic_depts/geology/
Administrative Assistant: Deborah A. Klein
*Enrollment (2015): B: 40 (15)*

Chair:
Donald T. Rodbell, (D), Colorado, 1991, GlmGe

Professor:
John I. Garver, (D), Washington, 1989, GtCcGr
Kurt T. Hollocher, (D), Massachusetts (Amherst), 1985, GxCga

Associate Professor:
Holli M. Frey, (D), Michigan, 2005, GvCag
David P. Gillikin, (D), Vrije Univ (Brussel), 2005, CsmCl

Lecturer:
Matthew R. Manon, (D), Michigan, 2008, Gp
Anouk Verheyden-Gillikin, (D), Vrije Univ (Brussel), 2004

Emeritus:
George H. Shaw, (D), Washington, 1971, Yx

Related Staff:
William S. Neubeck, (M), SUNY (Binghamton), 1980, Gm

## United States Military Academy

**Dept of Geography & Environmental Engineering** (B) (2016)
West Point, NY 10996
p. (914) 938-2300
Wiley.Thompson@usma.edu
http://www.usma.edu/gene/SitePages/Home.aspx
Department Secretary: Jean Keller
*Enrollment (2010): B: 342 (112)*

Professor:
John A. Brockhaus, (D), Idaho, 1987, Yd
Marie C. Johnson, (D), Brown, 1990, Gi
Jon C. Malinowski, (D), North Carolina, 1995, On
Department Chair:
Wiley C. Thompson, (D), Oregon State, 2008, Oy

## University of Rochester

**Dept of Earth & Environmental Sciences** (B,M,D) (2016)
120 Trustee Road
227 Hutchison Hall
Box 270221
Rochester, NY 14620
p. (585) 275-5713
ees@earth.rochester.edu
http://www.earth.rochester.edu
Administrative Assistant: Marjorie Goodison

Professor:
Asish R. Basu, (D), California (Davis), 1975, Gx
Cynthia J. Ebinger, (D), MIT/WHOI, 1988, Gt
Gautam Mitra, (D), Johns Hopkins, 1977, GctNr
Robert J. Poreda, (D), California (San Diego), 1983, Cg
John A. Tarduno, (D), Stanford, 1987, YmGtXm
Associate Professor:
Carmala N. Garzione, (D), Arizona, 2000, Gs
Associate Scientist:
Rory D. Cottrell, (D), Rochester, 2000, Ym
Pennilyn Higgins, (D), Wyoming, 2000, Cs
Emeritus:
Udo Fehn, (D), Munich, 1973, GeCgOu
Lawrence W. Lundgren, (D), Yale, 1958, Ge

## Utica College

**Dept of Geology** (B) (2015)
Gordon Science Center
1600 Burrstone Road
Utica, NY 13502
p. (315) 792-3134
skanfoush@utica.edu
https://www.utica.edu/academic/as/geoscience/new/bachelors.cfm
*Enrollment (2015): B: 10 (6)*

Chair:
Adam Schoonmaker, (D), SUNY (Albany), 2005, GczGx
Associate Professor:
Sharon L. Kanfoush, (D), Florida, 2002, GsOuGn
Adjunct Professor:
Lindsey Geary, (M), Florida State, 2008, Geg
Tiffany McGivern, (M), Utica College, 2014, Geg
Emeritus:
Herman Muskatt, (D), Syracuse, 1963, GgrPg

## Vassar College

**Dept of Earth Science & Geography** (B) (2015)
Box 735
124 Raymond Avenue
Poughkeepsie, NY 12604-0735
p. (845) 437-5540
geo@vassar.edu
http://earthscienceandgeography.vassar.edu/
Administrative Assistant: Lois Horst
*Enrollment (2015): B: 12 (3)*

Chair:
Mary A. Cunningham, (D), Minnesota, 2001, OiyOu
Professor:
Brian J. Godfrey, (D), California (Berkeley), 1984, Ou
Kirsten M. Menking, (D), California (Santa Cruz), 1995, GmPeGc
Joseph Nevins, (D), California (Los Angeles), Ou
Jill S. Schneiderman, (D), Harvard, 1987, Gs
Jeffrey R. Walker, (D), Dartmouth, 1987, GzvGp
Yu Zhou, (D), Minnesota, 1995, Oun
Lab Technician and Collections Manager:
Richard Jones, (M), California (Santa Cruz), 1996
GIS specialist:
Neil Curri, (B), Oi

## York College (CUNY)

**Dept of Earth and Physical Sciences** (B) ○ (2016)
94-20 Guy R. Brewer Blvd
Jamaica, NY 11451
p. (718)262- 2654
nkhandaker@york.cuny.edu
http://www.york.cuny.edu/academics/departments/earth-and-physical-sciences/
*Enrollment (2016): B: 23 (4)*

Geology Discipline Coordinator & CUNY Doctoral Faculty in Earth and Environmental Sciences:
Nazrul I. Khandaker, (D), Iowa State, 1991, GdeOe
Chair:
Timothy Paglione, (D), Boston, Xm
Professor:
Stanley Schleifer, (D), CUNY, 1996, Ge
Associate Professor:
Ratan K. Dhar, (D), CUNY Graduate Center, 2006, Hw
Emeritus:
Stephen Lakatos, (D), Rensselaer, 1971, Cc
Arthur P. Loring, (D), New York, 1966, Gm

# North Carolina

## Appalachian State University

**Dept of Geography & Planning** (B,M) (2014)
323 Rankin Science West
ASU Box 32066
Boone, NC 28608-2066
p. (828) 262-3000
youngje@appstate.edu
http://www.geo.appstate.edu
*Enrollment (2001): B: 82 (46) M: 15 (7)*

Chair:
James E. Young, (D), Minnesota, 1994, Oe
Professor:
Michael W. Mayfield, (D), Tennessee, 1984, Hs
Peter T. Soule, (D), Georgia, 1989, Oa
Roger A. Winsor, (D), Illinois, 1975, On
Associate Professor:
Jeff Colby, (D), Colorado, 1995, Oi
Richard J. Crepeau, (D), California (Irvine), 1995, On
Kathleen Schroeder, (D), Minnesota, 1995, On
Assistant Professor:
Chris Badurek, (D), Buffalo, 2005, Oi
Rob Brown, (D), Louisiana State, 2001, On
Jana Carp, (D), Illinois (Chicago), 1999, On
Gabrielle Katz, (D), Colorado, 2001, Hg
Baker Perry, (D), North Carolina, 2006, Og
Instructor:
Arthur B. Rex, (M), Appalachian State, 1980, Oi
Lecturer:
Terence Milstead, (D), Florida State, 2008, On
Saskia van de Gevel, (D), Tennessee, 2008, Og
Emeritus:
Neal G. Lineback, (D), Tennessee, 1970, Oy

**Dept of Geology** (B) (2014)
PO Box 32067
033 Rankin Science West
Boone, NC 28608-2067
p. (828) 262-3049
millerlj@appstate.edu
http://www.geology.appstate.edu/
*Enrollment (2010): B: 59 (12)*

Chair:
William P. Anderson, (D), North Carolina State, 1999, HwqHs
Johnny A. Waters, (D), Indiana, 1976, Pi
Professor:
Richard N. Abbott, (D), Harvard, 1977, Gzi
Ellen A. Cowan, (D), N Illinois, 1988, Gma
Andrew B. Heckert, (D), New Mexico, 2001, PvgOe
Roy Sidle, (D), Penn State, 1973, Hy
Associate Professor:
Steven J. Hageman, (D), Illinois, 1992, Pie

Assistant Professor:
Sarah Carmichael, (D), Johns Hopkins, 2006, GxCgGd
Chuanhui Gu, (D), Virginia, 2007, Cl
Cynthia Liutkus, (D), Rutgers, 2005, Gd
Scott Marshall, (D), Massachusetts, 2008, YxGc
Katherine Scharer, (D), Oregon, 2005, Gc
Lecturer:
Laura Mallard, (M), Vermont, 2000, GtOe
Christyanne Melendez, (M), N Arizona, Ggv
Crystal Wilson, (M), Tennessee, 2006, Gt
Brian Zimmer, (M), N Arizona, Gg
Adjunct Professor:
Mark G. Adams, (D), North Carolina, 1995, Gp
Keith C. Seramur, (M), N Illinois, 1988, Hw
Adjunct Faculty:
Marg J. McKinney, (M), North Carolina, 1968, Pg
Emeritus:
John E. Callahan, (D), Queen's, 1973, Eg
Frank K. McKinney, (D), North Carolina, 1970, Pi
Loren A. Raymond, (D), California (Davis), 1973, Gx
Fred Webb, (D), Virginia Tech, 1965, Gr



## Asheville-Buncombe Technical Community College

**Dept of Chemistry and Physics** (A) (2013)
340 Victoria Road
Asheville, NC 28801
p. (828) 254-1921
mfender@abtech.edu
http://www1.abtech.edu/content/arts-and-sciences/chemistryphysics/chemistry-and-physics-overview

Instructor:
John Bultman, Gg
Adjunct Professor:
Dan Murphy, Gg

## Brevard College

**Geology Program** (B) ○ (2015)
1 Brevard College Drive
Brevard, NC 28712
p. (828) 884-8377
reynoljh@brevard.edu
http://www.brevard.edu/reynoljh/
Administrative Assistant: Beth Banks

Geology Minor Coordinator:
James H. Reynolds, (D), Dartmouth, 1987, GrvGe

## Cape Fear Community College

**Dept of Science** (2011)
411 N. Front Street
Wilmington, NC 28401
akolb@cfcc.edu
http://cfcc.edu/programs/science/

Chair:
Joy Smoots, (D)

**Science Dept** (A) (2016)
411 N Front St
Wilmington, NC 28401
p. (910) 362-7674
jsmoots@cfcc.edu
http://cfcc.edu/programs/science/
*Enrollment (2016): A: 5 (0)*

Instructor:
Alvin L. Coleman, (M), Tennessee, 2008, Gg
Phil Garwood, (D), Edith Cowans (Australia), 1978, Gg

## Central Piedmont Community College

**Sciences Div** (A) (2013)
PO Box 35009
Charlotte, NC 28235
p. (704) 330-6750
David.Privette@cpcc.edu
http://www.cpcc.edu/

Chair:
Steppen Murphy, (M), S Illinois, Gue
Division Director:
David Privette, (M), Georgia, 1978, Oy
Instructor:
Alisa Hylton, (M), Wichita State, Gg

## Coastal Carolina University

**Dept of Marine Science** (2004)
Conway, NC 29528

## Duke University

**Div of Earth & Ocean Sciences** (B,M,D) (2014)
Nicholas School of the Environment and Earth Sciences
Box 90227
Durham, NC 27708-0227
p. (919) 684-5847
bill.chameides@duke.edu
http://www.nicholas.duke.edu/eos
*Enrollment (2010): B: 37 (15) D: 21 (0)*

Chair:
M. Susan Lozier, (D), Washington, 1989, Op
Professor:
Paul A. Baker, (D), California (San Diego), 1981, PeCg
Alan E. Boudreau, (D), Washington, 1986, Gi
Bruce H. Corliss, (D), Rhode Island, 1978, Pm
Peter K. Haff, (D), Virginia, 1970, On
Robert B. Jackson, (D), Utah State, 1992, On
Emily M. Klein, (D), Columbia, 1988, Gi
Lincoln F. Pratson, (D), Columbia, 1993, Gs
Avner Vengosh, (D), Australian National, 1990, Hw
Associate Professor:
A. Bradshaw Murray, (D), Minnesota, 1995, Ou
Associate Scientist:
Gary S. Dwyer, (D), Duke, 1996, Gs
Assistant Professor:
Nicolas Cassar, (D), Hawaii, 2003, Ob
Wenhong Li, (D), Georgia Tech, On
Lecturer:
Alexander Glass, (D), Illinois (Urbana Champaign), 2006, PiGgOe
Adjunct Professor:
David J. Erickson, (D), Rhode Island, 1987
Peter E. Malin, (D), Princeton, 1978, Ys
Bruce F. Molnia, (D), South Carolina, 1972, Or
Daniel D. Richter, (D), Duke, 1980, Os
William H. Schlesinger, (D), Cornell, 1976, On
Emeritus:
Richard T. Barber, (D), Stanford, 1967, Ob
Duncan Heron, (D), North Carolina, 1958, Grg
Daniel A. Livingstone, (D), Yale, 1953, Pe
Ronald D. Perkins, (D), Indiana, 1962, Gdo
Orrin H. Pilkey, Jr., (D), Florida State, 1962, Ou
Other:
Fred K. Boadu, (D), Georgia Tech, 1994, Yg
James S. Clark, (D), Minnesota, 1988, On
Mark N. Feinglos, (D), McGill, 1973, Gz
Richard F. Kay, (D), Yale, 1973, On

## East Carolina University

**Dept of Geological Sciences** (B,M) ○ (2016)
101 Graham Building
Greenville, NC 27858-4353
p. (252) 328-6360
culvers@ecu.edu
http://www.geology.ecu.edu
Administrative Assistant: Dare Merritt
*Enrollment (2016): B: 66 (24) M: 31 (12)*

Chair:
Stephen J. Culver, (D), Wales, 1976, Pm
Distinguished Research Professor:
Stanley R. Riggs, (D), Montana, 1967, Gu
Professor:
David Reide Corbett, (D), Florida State, 1999, Oc
David Mallinson, (D), S Florida, 1995, Gu
Richard Miller, (D), North Carolina State, 1984, Or
Catherine A. Rigsby, (D), California (Santa Cruz), 1989, Gs
Teaching Associate Professor:
Stephen B. Harper, (D), Georgia, 1996, GmgGe

Associate Professor:
Adriana Heimann, (D), Iowa, 2006, GzpGi
Eduardo Leorri, (D), Univ of Basque Country (Spain), 2003, GsPm
Alex K. Manda, (D), Massachusetts, 2009, HwgHq
Siddhartha Mitra, (D), William & Mary, 1997, Co
Donald W. Neal, (D), West Virginia, 1979, Gr
Richard K. Spruill, (D), North Carolina, 1981, Hw
Terri L. Woods, (D), South Florida, 1988, ClGz
Assistant Professor:
Eric Horsman, (D), Wisconsin, 2006, Gc

## Guilford College

**Dept of Geology & Earth Science** (B) (2016)
5800 West Friendly Avenue
Greensboro, NC 27410-4173
p. (336) 316-2263
ddobson@guilford.edu
http://www.guilford.edu/academics/departments-and-programs/geology/
f: Guilford College Department of Geology and Earth Sciences
*Enrollment (2016): B: 22 (7)*
Professor:
David M. Dobson, (D), Michigan, 1997, GusPe
Marlene McCauley, (D), California (Los Angeles), 1986, Cp
Assistant Professor:
Holly Peterson, (D), British Columbia, 2014, HwGe
Emeritus:
Cyril H. Harvey, (D), Nebraska, 1960, Gr

## North Carolina Agricultural & Tech State University

**Dept of Natural Resources and Environmental Design** (B,M) (2016)
1601 E. Market St
Carver Hall Room 238
Greensboro, NC 27411
p. (336) 334-7543
ash@ncat.edu
*Enrollment (2016): B: 58 (8) M: 7 (2)*
Professor:
Godfrey A. Uzochukwu, (D), Nebraska, 1983, Gz
Associate Professor:
Charles Raczkowski, (D), North Carolina State, So

## North Carolina Central University

**Department of Environmental, Earth and Geospatial Sciences** (B,M) ● (2016)
1801 Fayetteville Street
Durham, NC 27707-19765
p. (919) 530-5296
gvlahovic@nccu.edu
http://www.nccu.edu/academics/sc/artsandsciences/geo-spatialscience/index.cfm
f: NCCU Geography
t: DEEGS_NCCU
*Enrollment (2016): B: 24 (0) M: 20 (0)*
Associate Professor and Chair:
Gordana Vlahovic, (D), North Carolina at Chapel Hill, 1999, YsGtOe
Associate Professor:
John Bang, (D), Texas at El Paso, Ge
Chris McGinn, (D), North Carolina at Greensboro, Ogi
Harris Williams, (D), Arizona State Univ, Ogg
Assistant Professor:
Carresse Gerald, (D), NC A&T State Univ, GeOg
Timothy Mulrooney, (D), North Carolina at Greensboro, Oir
Zhiming Yang, (D), Oklahoma State Univ, Ory

## North Carolina State University

**Dept of Marine, Earth & Atmospheric Sciences** (B,M,D) O (2017)
P.O. Box 8208
Raleigh, NC 27695-8208
p. (919) 515-3711
webmaster_meas@ncsu.edu
http://www.meas.ncsu.edu
f: https://www.facebook.com/measncsu
t: @ncsumeas
*Enrollment (2016): B: 213 (105) M: 39 (20) D: 47 (12)*
Head:
Jay F. Levine, (D)
Director of Graduate Programs:
Elana L. Leithold, (D), Washington, 1987, Gs
Professor:
Viney P. Aneja, (D), North Carolina State, 1977, Oa
David J. DeMaster, (D), Yale, 1979, Oc
David B. Eggleston, (D), William & Mary, 1991, Ob
Ronald V. Fodor, (D), New Mexico, 1972, Gi
Ruoying He, (D), S Florida, 2002, Op
Gary M. Lackmann, (D), SUNY (Albany), 1995, Oa
David McConnell, (D), Texas A&M, 1987, GgOeGc
Helena Mitasova, (D), Slovak Tech, 1987, Oi
Walter Robinson, (D), Columbia, 1985, Oa
Fred H. M. Semazzi, (D), Nairobi, 1983, Oa
William J. Showers, (D), Hawaii, 1982, Cs
Lian Xie, (D), Miami, 1992, Oa
Sandra Yuter, (D), Washington, 1996, Oa
Yang Zhang, (D), Iowa, 1994, Oa
Associate Professor:
Anantha Aiyyer, (D), SUNY (Albany), 2003, Oa
DelWayne Bohnenstiehl, (D), Columbia, 2002, YrsGt
Jingpu P. Liu, (D), William & Mary, 2001, Gu
Nicholas Meskhidze, (D), Georgia Tech, 2003, Oa
Chris Osburn, (D), Lehigh, 2000, Cms
Matthew Parker, (D), Colorado State, 2002, Ow
Astrid Schnetzer, (D), Vienna, 2001, Ob
Assistant Professor:
Stuart P. Bishop, (D), Rhode Island, 2012, OpGq
Paul K. Byrne, (D), Dublin Trinity College, 2010, XgGcOr
Lisa Falk, (D), Columbia Univ., 2014, Og
Erin Hestir, (D), UC Davis, 2010, OrHs
Ethan Hyland, (D), Michigan, 2014, GrCsPe
Markus Petters, (D), Wyoming, 2004, Oa
Karl Wegmann, (D), Lehigh, 2008, Gm
Emeritus:
Satyapal S. Arya, (D), Colorado State, 1968, Oa
John C. Fountain, (D), California (Santa Barbara), 1975, Hw
James P. Hibbard, (D), Cornell, 1988, GctOn
Gerald S. Janowitz, (D), Johns Hopkins, 1967, Op
Daniel Kamykowski, (D), California (San Diego), 1973, Ob
Charles E. Knowles, (D), Texas A&M, 1970, Op
Leonard J. Pietrafesa, (D), Washington, 1973, Op
Sethu S. Raman, (D), Colorado, 1972, Oa
Allen J. Riordan, (D), Wisconsin, 1977, Oa
Dale A. Russell, (D), Columbia, 1964, Pv
Ping-Tung Shaw, (D), MIT/WHOI, 1982, Op
Edward F. Stoddard, (D), California (Los Angeles), 1976, Gp
Charles W. Welby, (D), MIT, 1952, Hw
Donna L. Wolcott, (D), California (Berkeley), 1972, Ob
Thomas G. Wolcott, (D), California (Berkeley), 1971, Ob

**Dept of Soil Science** (B,M,D) (2010)
Box 7619
Raleigh, NC 27695-7619
p. (919) 515-2655
jeff_mullahey@ncsu.edu
Administrative Assistant: Ashru Shah
Professor:
Aziz Amoozegar, (D), Arizona, 1977, Sp
Stephen W. Broome, (D), North Carolina, 1973, Sf
Donald K. Cassel, (D), California (Davis), 1968, Sp
John L. Havlin, (D), Colorado State, 1983, Sc
Dean L. Hesterberg, (D), California (Riverside), 1988, Sc
Michael T. Hoover, (D), Penn State, 1983, Sd
Greg D. Hoyt, (D), Georgia, 1981, Sb
Daniel W. Israel, (D), Oregon State, 1973, Sb
Harold J. Kleiss, Illinois, 1972, Sd
Deanna L. Osmond, (D), Cornell, 1991, Sb
Wayne P. Robarge, (D), Wisconsin, 1975, Sc
Thomas J. Smyth, (D), North Carolina State, 1981, Sc
Michael J. Vepraskas, (D), Texas A&M, 1980, Sd
Michael G. Wagger, (D), Kansas State, 1983, Sb
Associate Professor:
David A. Crouse, (D), North Carolina State, 1996, Sc
Carl Crozier, (D), North Carolina State, 1992, Sb

David Lindbo, (D), Massachusetts, 1990, Sd
Richard A. McLaughlin, (D), Purdue, 1985, Sc
Assistant Professor:
Alexandria Graves, (D), Virginia Tech, 2003, Sb
Wei Shi, (D), Purdue, Sb
Jeffrey G. White, (D), Cornell, 1988, Or
Emeritus:
James W. Gilliam, (D), Mississippi State, 1965, Sb



## University of North Carolina, Wilmington

**Dept of Earth and Ocean Sciences** (B,M) O (2016)
601 South College Road
Wilmington, NC 28403-5944
p. (910) 962-3490
lynnl@uncw.edu
http://www.uncw.edu/earsci/
Administrative Assistant: Alexis Lee
*Enrollment (2007): B: 78 (21) M: 32 (5)*
Chair:
Lynn A. Leonard, (D), S Florida, 1993, Gu
Professor:
Michael M. Benedetti, (D), Wisconsin, 2000, GmSaOy
Nancy R. Grindlay, (D), Rhode Island, 1991, Yg
W. Burleigh Harris, (D), North Carolina, 1975, GrdGs
Richard A. Laws, (D), California (Berkeley), 1983, Pm
Michael S. Smith, (D), Washington Univ (St. Louis), 1990, GzpGa
Associate Professor:
Lewis J. Abrams, (D), Rhode Island, 1992, Gu
David E. Blake, (D), Washington State, 1991, Gp
Douglas W. Gamble, (D), Georgia, 2000, Oy
Joanne N. Halls, (D), South Carolina, 1996, Oi
Eric J. Henry, (D), Arizona, 2001, Hw
Mary E. Hines, (D), Louisiana State, 1992, Ou
Lecturer:
Roger D. Shew, (M), North Carolina, 1979, Eo
Adjunct Professor:
Patricia H. Kelley, (D), Harvard, 1979, Pgi
Emeritus:
Robert T. Argenbright, (D), California (Berkeley), 1990, On
William J. Cleary, (D), South Carolina, 1972, Gu
James A. Dockal, (D), Iowa, 1980, GdcEm
Paul A. Thayer, (D), North Carolina, 1967, GdoHg
Laboratory Director:
Yvonne Marsan, (B), North Carolina (Wilmington), Oi

## University of North Carolina, Asheville

**Dept of Atmospheric Sciences** (B) O (2016)
CPO 2450
One University Heights
Asheville, NC 28804-3299
p. (828) 251-6149
chennon@unca.edu
http://www.atms.unca.edu
f: https://www.facebook.com/uncaweather/
t: @uncaweather
*Enrollment (2016): B: 36 (9)*
Professor:
Alex Huang, (D), Purdue, 1984, Oa
Doug Miller, (D), Purdue, 1996, OanOn
Chair:
Chris Hennon, (D), Ohio State, 2003, OanOn
Associate Professor:
Chris Godfrey, (D), Oklahoma, 2007, OanOn

**Dept of Environmental Studies** (B) (2014)
CPO 2330
Asheville, NC 28804-8511
p. (828) 251-6441
irossell@unca.edu
http://envr.unca.edu
Administrative Assistant: Debra C. Robbins
*Enrollment (2013): B: 144 (44)*
Professor:
Delores M. Eggers, (D), North Carolina, 1999, On
Kevin K. Moorhead, (D), Florida, 1986, Sb
Barbara C. Reynolds, (D), Georgia, 2000, On
Irene M. Rossell, (D), SUNY (Syracuse), 1995, On
Associate Professor:
David P. Gillette, (D), Oklahoma, 2008, On
Jeffrey D. Wilcox, (D), Wisconsin, 2007, HwClGg
Assistant Professor:
Jackie M. Langille, Tennessee, 2012, GctGe
Brittani D. MdNamee, (D), Idaho, 2013, GzxEg

## University of North Carolina, Chapel Hill

**Dept of Geological Sciences** (B,M,D) (2017)
CB 3315, Mitchell Hall
Chapel Hill, NC 27599-3315
p. (919) 966-4516
jonathan.lees@unc.edu
http://www.geosci.unc.edu
*Enrollment (2016): B: 66 (21) M: 4 (5) D: 20 (2)*
Chair:
Jonathan M. Lees, (D), Washington, 1989, YsGv
Professor:
Larry K. Benninger, (D), Yale, 1976, ClGuCt
Joseph G. Carter, (D), Yale, 1976, Po
Drew S. Coleman, (D), Kansas, 1991, CcGit
Allen F. Glazner, (D), California (Los Angeles), 1981, Git
Jose A. Rial, (D), Caltech, 1979, Ys
Associate Professor:
Laura J. Moore, (D), California (Santa Cruz), 1998, Gme
Tamlin M. Pavelsky, (D), California (Los Angeles), 2008, Hg
Kevin G. Stewart, (D), California (Berkeley), 1987, Gc
Donna M. Surge, (D), Michigan, 2001, Pe
Assistant Professor:
Xiaoming Liu, (D), Maryland, 2013, Cgl
Emeritus:
Paul D. Fullagar, (D), Illinois, 1963, Cc
A. Conrad Neumann, (D), Lehigh, 1963, Gu
Joseph St. Jean, (D), Indiana, 1956, Pm
Daniel A. Textoris, (D), Illinois, 1963, Gd

## University of North Carolina, Charlotte

**Dept of Geography & Earth Sciences** (B,M,D) (2014)
9201 University City Boulevard
Charlotte, NC 28223-0001
p. (704) 687-5973
ges@uncc.edu.
http://www.geoearth.uncc.edu/
*Enrollment (2010): B: 280 (70) M: 86 (16) D: 32 (8)*
Professor:
John F. Bender, (D), SUNY (Stony Brook), 1980, GiCtOu
John A. Diemer, (D), SUNY (Binghamton), 1985, Gs
Associate Professor:
Craig J. Allan, (D), York, 1992, Hg
Andy R. Bobyarchick, (D), SUNY (Albany), 1983, Gc
Scott P. Hippensteel, (D), Delaware, 2000, Gr
Walter Martin, (D), Tennessee, 1984, Oa
Ross Meentemeyer, (D), North Carolina, 2000, Oi
Assistant Professor:
Manda S. Adams, (D), Wisconsin, 2005, Ow
Matt Eastin, (D), Colorado State, 2003, Ow
Martha Cary Eppes, (D), New Mexico, 2002, Os
Lecturer:
Jake Armour, (M), New Mexico, 2002, Gg
Terry Shirley, (M), Penn State, 2003, Oaw
Emeritus:
Anne Jefferson, (D), Oregon State, 2006, HgGm
Laboratory Director:
William Garcia, (M), Cincinnati, Pv

## University of North Carolina, Pembroke

**Geology & Geography Dept** (B) (2016)
PO Box 1510
Pembroke, NC 28372-1510
p. (910) 775-4024
geo@uncp.edu
http://www.uncp.edu/geo/
*Enrollment (2016): B: 13 (4)*
Chair:
Martin B. Farley, (D), Penn State, 1987, PlOe

Associate Professor:
Dennis J. Edgell, (D), Kent State, 1992, Oa
Assistant Professor:
Jeff B. Chaumba, (D), Georgia, 2009, GxEgGz
Daren T. Nelson, (D), Utah, 2012, GmHwOe
Instructor:
Jesse Rouse, (M), West Virginia, 2000, Oni
Lecturer:
Amy Gross, (M), North Carolina (Wilmington), 2006, Gg
Nathan E. Phillippi, (M), South Dakota State, 2004, Oni
Emeritus:
Suellen Cabe, (D), North Carolina, 1984, Gg
Thomas E. Ross, (D), Tennessee, 1977, Oy

### Wake Technical Community College

**Natural Sciences Dept** (1999)
9101 Fayetteville Rd
Raleigh, NC 27603
p. (919) 866-5000
cdtatum@waketech.edu
http://www.waketech.edu/programs-courses/credit/natural-sciences

### Western Carolina University

**Dept of Geosciences & Natural Resources** (B) (2016)
Stillwell Building Room 331
Cullowhee, NC 28723-9047
p. (828) 227-7367
mlord@wcu.edu
http://geology.wcu.edu
*Enrollment (2015): B: 52 (12)*
Whitmire Prof Env Sci:
Jerry R. Miller, (D), S Illinois, 1990, GmeGf
Director Prog Study Dev Shorelines:
Robert S. Young, (D), Duke, 1995, GsOn
Dept. Head:
Mark L. Lord, (D), North Dakota, 1988, HwGmOe
Associate Provost:
Brandon E. Schwab, (D), Oregon, 2000, GizGv
Assoc. Dean, Arts & Sciences:
David A. Kinner, (D), Colorado, 2003, HgOe
Associate Professor:
Benjamin R. Tanner, (D), Tennessee, 2005, Cos
Cheryl Waters-Tormey, (D), Wisconsin, 2004, Gct
Assistant Professor:
Amy Fagan, (D), Notre Dame, 2013, GiXg
Frank Forcino, (D), Alberta, 2013, PgOeGr
John P. Gannon, (D), Virginia Tech, 2014, HgSo
Instructor:
Emily Stafford, (D), Alberta, 2014, PgGg
Emeritus:
Steven P. Yurkovich, (D), Brown, 1972, Gx

## North Dakota

### Dickinson State University

**Dept of Natural Science** (B) (2011)
Dickinson, ND 58601
p. (701) 227-2114
Michael.Hastings@dickinsonstate.edu
http://www.dsu.nodak.edu/
Chair:
Michael Hastings
Associate Professor:
Larry D. League, (M), Kansas, 1971, Oy

### Minot State University

**Dept of Geoscience** (B) (2016)
500 University Avenue West
Minot, ND 58707
p. (701) 858-3873
john.webster@minotstateu.edu
http://www.minotstateu.edu/geology/
*Enrollment (2016): B: 35 (6)*
Head:
John R. Webster, (D), Indiana, 1992, Giz
Assistant Professor:
Joseph Collette, (D), California (Riverside), 2014, PgGsr
Nathan R. Hopkins, (D), Lehigh, 2016, GlmOi
Kathyrn Kilroy, (D), Nevada (Reno), 1992, Hg

### North Dakota State University

**Dept of Geosciences** (B) (2016)
NDSU Dept. 2745
P.O. Box 6050
Fargo, ND 58108-6050
p. (701) 231-8455
peter.oduor@ndsu.edu
http://www.ndsu.edu/geosci
*Enrollment (2016): B: 35 (17)*
Professor:
Kenneth E. Lepper, (D), Oklahoma State, 2001, CcGml
Chair:
Peter Oduor, (D), Missouri (Rolla), 2004, Oi
Associate Professor:
Bernhardt Saini-Eidukat, (D), Minnesota, 1991, GiCg
Assistant Professor:
Stephanie S. Day, (D), Minnesota, 2012, GmOiu
Benjamin J. C. Laabs, (D), Wisconsin (Madison), 2004, GlOl
Lydia S. Tackett, (D), S California, 2014, PiGsPo
Instructor:
Jessie Rock, (M), North Dakota State, 2009, Og
Emeritus:
Allan C. Ashworth, (D), Birmingham, 1969, Pe
Donald P. Schwert, (D), Waterloo, 1978, PeGe

**Dept of Soil Science** (B,M,D) (2014)
Walster Hall
Fargo, ND 58105
p. (701) 231-8690
Thomas.Desutter@ndsu.edu
Administrative Assistant: Jacinda Wollan
*Enrollment (2014): B: 18 (0) M: 10 (0) D: 4 (0)*
Professor:
Francis Casey, (D), Iowa State, 2000, SpHq
Dave Franzen
Robert J. Goos, (D), Colorado, 1980, Sc
Associate Professor:
Larry J. Cihacek, (D), Iowa State, 1976, Og
Assistant Professor:
Amitava Chatterjee, Wyoming
Aaron Daigh, (D), Iowa State
Tom DeSutter, (D), Kansas State, Os
Ann-Marie Fortuna
David G. Hopkins, (D), North Dakota State, 1997, Sd
Abbey Wick, Wyoming

### University of North Dakota

**Dept of Geography and Geographic Information Science** (B,M) (2016)
221 Centennial Drive
Stop 9020
Grand Forks, ND 58202
p. (701) 777-4246
cindy.purpur@und.edu
http://arts-sciences.und.edu/geography/
f: https://www.facebook.com/groups/91074880028/
*Enrollment (2016): B: 27 (7) M: 12 (4)*
Chair:
Gregory S. Vandeberg, (D), Kansas State, 2005, GmOiGl
Professor:
Douglas C. Munski, (D), Illinois, 1978, Oe
Bradley C. Rundquist, (D), Kansas State, 2000, OriOy
Paul Todhunter, (D), California (Los Angeles), 1986, Oy
Associate Professor:
Enru Wang, (D), Washington, 2005, Ogi
Assistant Professor:
Christopher Atkinson, (D), Kansas, 2010, Oai
Michael A. Niedzielski, (D), Ohio State, 2009, Oi

Instructor:
Mbongowo J. Mbuh, (D), George Mason Univ, 2015, OrHs
Emeritus:
Devon A. Hansen, (D), Utah, 1999, On

Ohio

**Harold Hamm School of Geology & Geological Engineering**
(B,M,D) (2014)
81 Cornell Street
Stop 8358
Grand Forks, ND 58202
p. (701) 777-2248
jolene.marsh@engr.und.edu
http://engineering.und.edu/geology-and-geological-engineering/
*Enrollment (2014): B: 78 (0) M: 17 (0) D: 10 (0)*
Director:
Joseph H. Hartman, (D), Minnesota, 1984, Pi
Professor:
William D. Gosnold, (D), S Methodist, 1976, YhCt
Stephan Nordeng, (D), Michigan, Go
Dexter Perkins, III, (D), Michigan, 1979, Gp
Associate Professor:
Philip J. Gerla, (D), Arizona, 1983, Hw
Ronald K. Matheney, (D), Arizona State, 1989, Cs
Jaakko Putkonen, (D), Washington, 1997, Gml
Assistant Professor:
Nels F. Forsman, (D), North Dakota, 1985, GdXgGh
I-Hsuan Ho, (D), Iowa, Ng
Dongmei Wang, (D), China, Ng
Emeritus:
Richard D. LeFever, (D), California (Los Angeles), 1979, GsrGo

# Ohio

## Ashland University
**Dept of Chemistry/Geology/Physics** (B) O (2017)
401 College Avenue
Ashland, OH 44805
p. (419) 289-5268
rcorbin@ashland.edu
http://www.ashland.edu/departments/geology
*Enrollment (2016): B: 12 (0)*
Professor:
Nigel Brush, (D), California (Los Angeles), 1992, GmdGa
Instructor:
Clifford P. Ambers, (D), Indiana Univ, 1993, GggGg
Elizabeth Mazzocco, (M), Ohio State Univ, Oa

## Bowling Green State University
**Dept of Geology** (B,M) (2013)
190 Overman Hall
Bowling Green, OH 43403
p. (419) 372-2886
geology@bgsu.edu
http://www.bgsu.edu/departments/geology/
Administrative Assistant: Pat A. Wilhelm
*Enrollment (2010): B: 29 (13) M: 23 (2)*
Chair:
Sheila J. Roberts, (D), Arizona, 1992, CgHw
Professor:
James E. Evans, (D), Washington, 1988, GsHsGe
Charles M. Onasch, (D), Penn State, 1977, Gc
Robert K. Vincent, (D), Michigan, 1973, Yg
Peg M. Yacobucci, (D), Harvard, 1999, Po
Associate Professor:
John R. Farver, (D), Brown, 1988, Gy
Joseph P. Frizado, (D), Northwestern, 1980, Oi
Enrique Gomezdelcampo, (D), Tennessee, 2003, Ge
Peter Gorsevski, (D), Idaho, 2002, OirGm
Kurt S. Panter, (D), New Mexico Tech, 1995, Gv
Jeffrey A. Snyder, (D), Ohio State, 1996, Gm
Lecturer:
Nichole Elkins, (M), Georgia, 2002, GaOe
Christopher Pepple, (M), Oe
Paula J. Steinker, (D), Bowling Green, 1982, Pg
Emeritus:
Don C. Steinker, (D), California (Berkeley), 1969, Po

Department IT:
William Butcher, (B), Rochester, 1972, Gg

## Case Western Reserve University
**Dept of Earth, Environmental and Planetary Sciences**
(B,M,D) (2016)
10900 Euclid Avenue
A.W. Smith #112
Cleveland, OH 44106-7216
p. (216) 368-3690
lmd3@case.edu
http://eeps.case.edu/
*Enrollment (2016): B: 15 (0) M: 2 (0) D: 5 (0)*
Chair:
James A. Van Orman, (D), MIT, 2000, CgGyCp
Professor:
Steven A. Hauck, II, (D), Washington Univ (St. Louis), 2001, XyYgXg
Gerald Matisoff, (D), Johns Hopkins, 1978, Cl
Peter L. McCall, (D), Yale, 1975, Po
Peter J. Whiting, (D), California (Berkeley), 1990, Gm
Associate Professor:
Mulugeta Alene Araya, (D), Univ Turin
Ralph P. Harvey, (D), Pittsburgh, 1990, Xm
Beverly Z. Saylor, (D), MIT, 1996, Gs
Assistant Professor:
Carlo DeMarchi, (D), Georgia Inst Tech
Zhicheng Jing, (D), Yale Univ, 2010
Adjunct Professor:
Andrew Dombard, (D), Washington, 2000, Xy
Joseph T. Hannibal, (D), Kent State, 1990, Pi
David Saja, (D), Pennsylvania, 1999, Gc
Emeritus:
Samuel M. Savin, (D), Caltech, 1967, Cs

## Cedarville University
**Dept of Science and Mathematics** (B) (2016)
251 North Main Street
Cedarville, OH 45314
p. (937) 766-7940
trice@cedarville.edu
http://www.cedarville.edu/Academics/Science-and-Mathematics/Geology.aspx
f: https://www.facebook.com/cedarvillegeology
*Enrollment (2016): B: 29 (7)*
Professor:
Steven Gollmer, (D), Purdue, GzOwg
John H. Whitmore, (D), Loma Linda, 2003, PgGsg
Associate Professor:
Mark Gathany, (D), Colorado State, OiuSb
Assistant Professor:
Thomas L. Rice, (M), Colorado Mines, 1987, GemEo
Adjunct Professor:
Steve A. Austin, (D), Penn State, 1979, GsEcGd

## Central State University
**Intl Center for Water Resources Management** (B) (2016)
C.J. McLin Bldg
Wilberforce, OH 45384
p. (937) 376-6212
knedunuri@centralstate.edu
*Enrollment (2015): B: 45 (3)*
Professor and Chair:
Krishna Kumar Nedunuri, (D), Purdue, 1999, HwCaSo
Dean of College of Science and Engineering:
Subramania I. Sritharan, (D), Colorado State, 1984, Hg
Professor:
Sam Laki, (D), Michigan State, 1992, EgSoHs
Associate Professor:
Ramanitharan Kandiah, (D), Tulane, 2004, HqSoOi
Xiaofang Wei, (D), Indiana State, 2008, OrgOe
De Bonne N. Wishart, (D), Rutgers Univ (Newark), 2008, YgCa
Assistant Professor:
Ning Zhang, (D), West Virginia, 2012, Emo
Emeritus:
Samuel Okunade, (D), Kent State, 1986, Gm

## Cincinnati Museum Center
**Geier Collections and Research Center** (2015)
1301 Western Avenue
Cincinnati, OH 45203
p. (513) 287-7000
information@cincymuseum.org
http://www.cincymuseum.org
Associate Vice President:
Glenn W. Storrs, (D), Yale, 1986, Pv
Curator:
Brenda Hunda, (D), California (Riverside), 2004, Pi
Research Associate:
Nigel C. Hughes, (D), Bristol, 1990, Pi
Arnold I. Miller, (D), Chicago, 1986, Pq
Joshua H. Miller, (D), Chicago, 2009, Pe
Andrew Webber, (D), Cincinnati, 2007, Pi

## Cleveland Museum of Natural History
**Dept of Mineralogy** (2016)
1 Wade Oval Drive
University Circle
Cleveland, OH 44106-1767
p. (216) 231-4600 x3229
dsaja@cmnh.org
http://www.cmnh.org
t: @cmnhmineralogy

**Dept of Paleobotany** (2010)
1 Wade Oval Drive
University Circle
Cleveland, OH 44106-1767
p. (216) 231-4600
dsu@cmnh.org
http://www.cmnh.org/site/ResearchandCollections/Paleobotany.aspx
Head:
Shyamala Chitaley, (D), Reading (UK), 1955, Pb

**Dept of Paleobotany & Paleoecology** (2015)
1 Wade Oval Drive
Cleveland, OH 44106-1767
p. (216) 231-4600 x3240
dsu@cmnh.org
http://www.cmnh.org/site/ResearchandCollections/InvertebratePaleontology.aspx
Head:
Joseph T. Hannibal, (D), Kent State, 1990, Pi

**Dept of Vertebrate Paleontology** (2016)
1 Wade Oval Drive
University Circle
Cleveland, OH 44106-1767
p. (216) 231-4600
mryan@cmnh.org
http://www.cmnh.org/site/ResearchandCollections/VertebratePaleontology.aspx
Head Technician:
Lee E. Hall, (B)
Curator:
Michael J. Ryan, (D)
Collections Manager:
Amanda R. McGee, (M)

## College of Wooster
**Dept of Geology** (B) (2016)
Scovel Hall
944 College Mall
Wooster, OH 44691-2363
p. (330) 263-2380
preeder@wooster.edu
http://www.wooster.edu/academics/areas/geology
f: https://www.facebook.com/pages/College-of-Wooster-Geology-Department/143144126437
Administrative Assistant: Patrice Reeder
*Enrollment (2016): B: 53 (14)*
Chair:
Gregory C. Wiles, (D), SUNY (Buffalo), 1992, Gl
Professor:
Mark A. Wilson, (D), California (Berkeley), 1982, Pi
Associate Professor:
Shelley Judge, (D), Ohio State, 2007, GtsGc
Associate Scientist:
Meagen Pollock, (D), Duke, 2007, Gi

## Cuyahoga Community College - Western Campus
**Earth Science** (A) (2011)
11000 Pleasant Valley Road
Parma, OH 44130
p. (216) 987-5278
enroll@tri-c.edu
Assistant Professor:
Robert Zaleha, Og
Instructor:
John L. Ezerskis, Og
Carol Fondran, Og
Joseph M. Lane, Og
Abby N. Norton-Krane, Og
Kathryn Sasowsky, Og
Adjunct Professor:
Gloria CC Britton, (D), OgGmOn
Jennifer Deka, Og

## Denison University
**Dept of Geosciences** (B) (2016)
F.W. Olin Science Hall
100 W. College Street
Granville, OH 43023
p. (740) 587-6217
hall@denison.edu
http://www.denison.edu/academics/departments/geosciences/
Administrative Assistant: Jude Hall
*Enrollment (2013): B: 9 (9)*
Professor:
Tod A. Frolking, (D), Wisconsin, 1985, Oy
David C. Greene, (D), Nevada (Reno), 1995, GctGe
Associate Professor:
David H. Goodwin, (D), Arizona, 2003, PiCs
Assistant Professor:
Erik Klemetti, (D), Oregon State, 2005, GvxGg
Kate E. Tierney, (D), Ohio State, 2010, CgGrOg
Emeritus:
Kennard B. Bork, (D), Indiana, 1967, PiGhs
Robert J. Malcuit, (D), Michigan State, 1973, Gx

## Hocking College
**GeoEnvironmental Science Program** (2015)
3301 Hocking Parkway
Nelsonville, OH 45764-9704
p. (740) 753-6277
caudill_m@hocking.edu
http://www.hocking.edu/programs/geoenvironmental
Professor:
Michael R. Caudill, (D), Tennessee, 1996, SdHwNg
Instructor:
Kimberly S. Caudill, (B), Ohio Univ, 1984, GeOuHw

## Kent State University
**Dept of Geology** (B,M,D) O (2015)
221 McGilvrey Hall
Kent, OH 44242
p. (330) 672-2680
geology@kent.edu
http://www.kent.edu/geology/
*Enrollment (2015): B: 154 (23) M: 22 (10) D: 9 (3)*
Chair:
Daniel K. Holm, (D), Harvard, 1992, Gc
Professor:
Joseph D. Ortiz, (D), Oregon State, 1995, Gs
Carrie E. Schweitzer, (D), Kent State, 2000, Pi

Alison J. Smith, (D), Brown, 1991, Gn
Neil A. Wells, (D), Michigan, 1984, Gs
Associate Professor:
David B. Hacker, (D), Kent State, 1998, GcHw
Anne J. Jefferson, (D), Oregon State, 2006, HgGm
Assistant Professor:
Tathagata Dasgupta, (D), Syracuse, 2010, Cg
Elizabeth M. Herndon, (D), Penn State, 2012, ClSc
Christopher J. Rowan, (D), Southhampton, 2006, Gt
David M. Singer, (D), Stanford, 2008, Gze
Jeremy C. Williams, (D), Massachusetts (Boston), 2014, Cg
Emeritus:
Rodney M. Feldmann, (D), North Dakota, 1967, Pi
Donald F. Palmer, (D), Princeton, 1968, Yg
Abdul Shakoor, (D), Purdue, 1982, Ng
Related Staff:
Merida Keatts, (M), Kent State, 2000, On



## Kent State University at Stark

**Dept of Geology** (B) O (2015)
6000 Frank Avenue NW
North Canton, OH 44720
p. 330-244-3303
cschweit@kent.edu
http://www.personal.kent.edu/~cschweit/Stark/
Administrative Assistant: Debra Stimer
*Enrollment (2015): B: 10 (0)*
Chair:
Carrie E. Schweitzer, (D), Kent State, 2000, PiGg

## Lakeland Community College

**Geoscience Dept** (A) (2015)
7700 Clocktower Drive
Kirtland, OH 44094
p. (440) 525-7341
dpierce@lakelandcc.edu
http://www.lakelandcc.edu/academic/sh/geol/index.asp
Professor:
David Pierce, (D), GgHsOw

## Marietta College

**Dept of Petroleum Engineering & Geology** (B) (2016)
215 Fifth Street
Marietta, OH 45750
p. (740) 376-4775
sch030@marietta.edu
http://w3.marietta.edu/departments/Petroleum_Engineering/
*Enrollment (2016): B: 35 (9)*
Assistant Professor:
Tej Gautam, (D), Kent State, 2012, NgGeOi
David L. Jeffery, (D), Texas A&M, 2003, Go
Instructor:
Wendy Bartlett, (M), Texas A&M, Geo
Veronica Freeman, (M), Texas, 1993, Pg
Administrative Coordinator:
Susan Hiser, (M), Springfield College, 1991, On

## Miami University

**Dept of Geology and Environmental Earth Sciences** (B,M,D) O (2017)
250 S,. Patterson Avenue
118 Shideler Hall
Oxford, OH 45056
p. (513) 529-3216
edwardca@MiamiOh.edu
http://www.units.muohio.edu/geology/
f: https://www.facebook.com/MiamiGeology/?ref=aymt_homepage_panel
t: geologymiamioh
Administrative Assistant: Cathy Edwards
*Enrollment (2016): B: 155 (28) M: 12 (8) D: 15 (6)*
Janet & Elliot Baines Bicentenntial Professor & Chair:
Elisabeth Widom, (D), California (Santa Cruz), 1991, CcGve
IUGS Vice President:
Yildirim Dilek, (D), California, 1989, GtcGi
Professor:
Michael Brudzinski, (D), Illinois, 2002, Ys
Brian S. Currie, (D), Arizona, 1998, GstGc
Yildirm Dilek, (D), California (Davis), 1989, Gt
Hailiang Dong, (D), Michigan, 1997, Cg
John F. Rakovan, (D), SUNY (Stony Brook), 1996, GzCl
Jason Rech, (D), Arizona, 2001, Gm
Associate Professor (Hamilton Campus:
Mark Krekeler, (D), Illinois (Chicago), 2003, Ge
Associate Professor & Director of IES:
Jonathan Levy, (D), Wisconsin, 1993, Hw
Assistant Professor:
Claire McLeod, (D), Durham, 2012, Gx
Carrie Tyler, (D), Virginia Tech, 2012, Pb
Instructor:
Jill Mignery, (M), Miami, 2004
Lecturer (Middletown Campus:
Tammie Gerke, (D), Cincinnati, 1995
Lecturer:
Todd Dupont, (D), Penn State, 2004, Gl
Euan Mitchell, (D), Univ. of New Mexico, 2013
Adjunct Professor:
Patri Larrea, (D), Zaragoza, 2014
Visiting Assistant Professor:
Trent Garrison, (D), Kentucky, 2015, GeHgEc
Visiting Professor:
Hassan Mirnejad, (M), Carleton, 2001, GxCg
Emeritus:
A. Dwight Baldwin, Jr., (D), Stanford, 1966, Hw
Mark R. Boardman, (D), North Carolina, 1978, On
William K. Hart, (D), Case Western, 1982, GivCc
John M. Hughes, (D), Dartmouth, 1981, Gz
Robert G. McWilliams, (D), Washington, 1968, Ps
John K. Pope, (D), Cincinnati, 1966, Pi
David M. Scotford, (D), Johns Hopkins, 1950, Gp
Geochemistry Lab Manager:
Amy Wolfe, (D), Pittsburgh, 2010, Cs
Laboratory Director:
David C. Kuentz, (M), Texas (Arlington), 1986, Ca
Postdoctoral Research Scholar:
Fara Rasoazanamparany, (D), Miami, 2015
Accounting Technician:
Gail Burger
Director Limper Geology Museum:
Kendall Hauer, (D), Miami Univ, 1995, GgOeCg
Adjunct:
John P. Morton, (D), Texas, 1983
Cooperating Faculty:
R. Hays Cummins, (D), Texas A&M, 1984, Ge

## Mount Union College

**Dept of Geology** (B) (2014)
Alliance, OH 44601
p. (330) 823-3672
graylm@mountunion.edu
http://www.mountunion.edu/gy
*Enrollment (2010): B: 8 (2)*
Professor:
Lee M. Gray, (D), Rochester, 1985, PoGg
Department Chair:
Mark A. McNaught, (D), Rochester, 1991, Gcg
Adjunct Professor:
Leonard G. Epp, (D), Penn State, 1970, Ob

## Muskingum University

**Dept of Geology** (B) (2016)
163 Stormont Street
New Concord, OH 43762
p. (740) 826-8306
svanhorn@muskingum.edu
http://muskingum.edu/dept/geology/index.html
*Enrollment (2016): B: 33 (6)*
Associate Professor:
Stephen R. Van Horn, (D), Connecticut, 1996, GeOiEo

Associate Professor:
Eric W. Law, (D), Case Western, 1982, Gp
David L. Rodland, (D), Virginia Tech, 2003, PgiPe

## Northwest State Community College

**Div of Arts and Sciences** (1999)
22600 State Route 34
Archbold, OH 43502
levans@northweststate.edu
http://www.northweststate.edu

## Oberlin College

**Dept of Geology** (B) (2016)
52 West Lorain Street
Oberlin, OH 44074-1044
p. (440) 775-8350
geology@oberlin.edu
http://new.oberlin.edu/arts-and-sciences/departments/geology/
*Enrollment (2016): B: 46 (18)*

Chair:
Dennis K. Hubbard, (D), South Carolina, 1977, Gsu
Professor:
Karla M. Parsons-Hubbard, (D), Rochester, 1993, PoGu
Bruce M. Simonson, (D), Johns Hopkins, 1982, GsXm
Steven F. Wojtal, (D), Johns Hopkins, 1982, Gc
Associate Professor:
F Zeb Page, (D), Michigan, 2005, GpCsg
Assistant Professor:
Amanda H. Schmidt, (D), Washington, 2010, Gm
Visiting Professor:
Andrew J. Horst, (D), Stanford, 2013, Gcg
Cooperating Faculty:
Alison Ricker, (M), Rhode Island, 1977, On

## Ohio State University

**Atmospheric Sciences Program** (B,M,D) (2016)
103 Bricker Hall
190 North Oval Mall
Columbus, OH 43210-1361
p. (614) 292-2514
mark.9@osu.edu
http://www.geography.ohio-state.edu/atmospheric-and-climatic-studies

Director:
Jeffery C. Rogers, (D), Colorado, 1979, Oa
Professor:
John N. Rayner, (D), Canterbury (NZ), 1965, Oa

**Dept of Civil, Environmental & Geodetic Engineering** (B,M,D) (2016)
470 Hitchcock Hall
2070 Neil Avenue
Columbus, OH 43210
p. (612) 292-2771
Grejner-Brzezinska.1@osu.edu
http://ceg.osu.edu
*Enrollment (2016): M: 11 (4) D: 8 (2)*

Professor and Chair:
Dorota A. Grejner-Brzezinska, (D), Ohio State, 1995, Yd
Research Professor:
Charles K. Toth, (D), OrYd
Professor:
Harvey J. Miller, (D), Ohio State, 1991, Oi
Associate Professor:
Alper Yilmaz, (D), Central Florida, 2004, Or

**Dept of Geography** (B,M,D) (2015)
1036 Derby Hall
154 North Oval Mall
Columbus, OH 43210-1361
p. (614) 292-2514
sui.10@osu.edu
http://www.geography.osu.edu/

Chair:
Daniel Sui
Professor:
David H. Bromwich, (D), Wisconsin, 1979, Oa
Ellen E. Mosley-Thompson, (D), Ohio State, 1979, Oa
Jeffrey C. Rogers, (D), Colorado, 1979, Oa
Associate Professor:
Jay S. Holgood, (D), Ohio State, 1984, Oa
Jialin Lin, (D), SUNY (Stony Brook), 2001, Oa
Bryan G. Mark, (D), Syracuse, 2001, Oa
Assistant Professor:
Alvaro Montenegro, (D), Florida State, 2003, OgaOy
Emeritus:
A. John Arnfield, (D), McMaster, 1973, Oa
John N. Rayner, (D), Canterbury (NZ), 1965, Oa

**School of Earth Sciences** (B,M,D) (2016)
275 Mendenhall Lab
125 South Oval Mall
Columbus, OH 43210-1398
p. (614) 292-2721
earthsciences@osu.edu
http://www.earthsciences.osu.edu
Administrative Assistant: Jill Bryant
Administrative Assistant: Theresa Mooney
Administrative Assistant: Angeletha M. Rogers
*Enrollment (2011): M: 36 (16) D: 40 (2)*

Director, Byrd Polar Research Center:
William B. Lyons, (D), Connecticut, 1979, Cl
Ohio Research Scholar:
David R. Cole, (D), Pennsylvania, 1980, Cg
Ohio Eminent Scholar:
Michael G. Bevis, (D), Cornell, 1982, Yx
Frank W. Schwartz, (D), Illinois, 1972, Hg
Associate Director, Administration:
Lawrence A. Krissek, (D), Oregon State, 1982, Gsu
Associate Dean:
Anne E. Carey, (D), Nevada (Reno), 1995, Hq
Professor:
Loren E. Babcock, (D), Kansas, 1990, Po
Edwin S. Bair, (D), Penn State, 1980, Hw
Michael Barton, (D), Manchester, 1975, GiCpt
Yu-Ping Chin, (D), Michigan, 1988, Hg
Jeffrey J. Daniels, (D), Colorado Mines, 1974, Yg
Christopher Jekeli, (D), Ohio State, 1981, Yd
Mark A. Kleffner, (D), Ohio State, 1988, PmsGr
Matthew R. Saltzman, (D), California (Los Angeles), 1996, Gs
CK Shum, (D), Texas, 1982, Yd
Lonnie G. Thompson, (D), Ohio State, 1976, Gl
Ralph R. B. von Frese, (D), Purdue, 1980, Ye
Terry J. Wilson, (D), Columbia, 1983, Gc
Design Engineer:
Dana Caccamise, (M)
Senior Scientist:
John W. Olesik, (D), Wisconsin, 1982, Ca
Yuchan Yi, (D), Ohio State, 1995, Ydv
Associate Professor:
Douglas E. Alsdorf, (D), Cornell, 1996, Hg
Ozeas S. Costa, Jr, (D), Plymouth (UK), 2002, HsClOu
Andrea G. Grottoli, (D), Houston, 1998, OcPe
Ian M. Howat, (D), California (Santa Cruz), 2006, OlYd
Motomu Ibaraki, (D), Waterloo, 1994, Hg
Daniel N. Leavell, (D), Massachusetts, 1983, Ge
Steven K. Lower, (D), Virginia Tech, 2001, Py
Wendy R. Panero, (D), California (Berkeley), 2001, Yg
Alan J. Saalfeld, (D), Maryland, 1993, Yd
Burkhard A. Schaffrin, (D), Bonn, 1983, Yd
Senior Research Associate:
Christopher Gardner, (M)
Research Scientist:
Paolo Gabrielli, (D), LGGE Grenoble, 2004, CtPeOl
Susan A. Welch, (D), Delaware, 1997, ClGeCs
Assistant Professor:
Joel D. Barker, (D), Alberta, 2007, OlClPe
Ann Cook, (D), Columbia, 2010, Gu
Thomas Darrah, (D), Rochester, 2009, Ge
Michael T. Durand, (D), California, 2007, HgYd
Joachim Moortgat, (D), Radboud, 2006
Audrey Sawyer

Ohio

Derek Sawyer
Michael Wilkins, (D), Manchester, 2006, PyCl
Research Associate:
Junyi Guo
Eric Kendrick
How-wai (Peter) Luk
Anthony Lutton, (D)
Julie Sheets
Stefanie Sherman, (D), Ohio State
Lecturer:
Christena Cox, (D), Ohio State, 1994, GgoGc
Christina Millan, (D), Ggc
Museum Curator :
Dale Gnidovec, (M), Fort Hays State, 1978, Ggr
Adjunct Professor:
Thomas G. Naymik, (D), Ohio State, 1978, Hw
Visiting Professor:
Samuel Haines
David Young
Emeritus:
William I. Ausich, (D), Indiana, 1978, PoGs
Stig M. Bergstrom, (D), Lund (Sweden), 1961, PsGg
James Bradley
James W. Collinson, (D), Stanford, 1966, Ps
Charles E. Corbato, (D), California (Los Angeles), 1960, Ye
James W. Downs, (D), Virginia Tech, 1983, Gz
David H. Elliot, (D), Birmingham, 1965, GitGv
Gunter Faure, (D), MIT, 1961, Cc
Kenneth A. Foland, (D), Brown, 1972, Cc
Charles Herdendorf
Garry D. McKenzie, (D), Ohio State, 1968, Gl
Ivan Mueller
David Nickey
Hallan C. Noltimier, (D), Newcastle upon Tyne, 1965, Ym
Douglas E. Pride, (D), Illinois, 1969, Em
Richard Rapp
Thomas N. Taylor, (D), Illinois, 1964, Pb
Rodney T. Tettenhorst, (D), Illinois, 1960, Gz
Russell O. Utgard, (D), Indiana, 1969, Ge
Peter N. Webb, (D), Utrecht (Neth), 1966, Pm
Laboratory Director:
Michael Johnston
Courtesy Appt.:
Bryan Mark
Other:
Mike Kositzke
Jeff Melarango
Undergraduate Advisor:
Karen Royce, (D), Ohio State
Systems Manager:
Dan Dunlap
Systems Developer:
Michael Seufer
Related Staff:
Christopher Hadad

**School of Environment and Natural Resources** (B,M,D) (2014)
2021 Coffey Road
Room 210 Koffman Hall
Columbus, OH 43210
p. (614) 292-2265
hendrick.15@osu.edu
http://senr.osu.edu/
Administrative Assistant: Mary Capoccia
Eminent Scholar:
Richard P. Dick, (D), Iowa State, 1986, Sb
Associate Director:
Donald J. Eckert, (D), Ohio State, 1978, Sc
Professor:
Jerry M. Bigham, (D), North Carolina State, 1977, Sd
Frank G. Calhoun, (D), Florida, 1971, Sd
Warren A. Dick, (D), Iowa State, Sb
Rattan Lal, (D), Ohio State, 1968, Sp
Associate Professor:
Nicholas T. Basta, (D), Iowa State, 1989, Sc
Edward L. McCoy, (D), Oregon State, 1984, Sp

Assistant Professor:
Dawn Ferris, (D), Minnesota, 1997, Sf
Brian K. Slater, (D), Wisconsin, 1994, Sd

## Ohio University

**Geological Sciences** (B,M) ● (2016)
316 Clippinger Lab
Athens, OH 45701
p. (740) 593-1101
geological_sciences@ohio.edu
http://www.ohio.edu/geology/
*Enrollment (2016): B: 75 (15) M: 19 (13)*
Chair:
Dina L. Lopez, (D), Louisiana State, 1992, CgGeq
Professor:
R. Damian Nance, (D), Cambridge, 1978, Gtc
Alycia L. Stigall, (D), Kansas, 2004, Po
Associate Professor:
Douglas H. Green, (D), Wisconsin, 1989, Yg
Daniel Hembree, (D), Kansas, 2005, Pi
David L. Kidder, (D), California (Santa Barbara), 1987, Gs
Eung Seok Lee, (D), Indiana, 1999, Hw
Keith A. Milam, (D), Tennessee, 2006, Xg
Gregory C. Nadon, (D), Toronto, 1991, Gs
Gregory S. Springer, (D), Colorado State, 2002, Gm
Assistant Professor:
Craig Grimes, Wyoming, 2008, GipCg
Technician & Information Tech:
Timothy A. Grubb, (A), Washington State Comm Coll, 2002
Administrative Associate:
Cheri Sheets, (B), Ohio

## Ohio Wesleyan University

**Dept of Geology & Geography** (B) (2016)
61 S. Sandusky Street
Delaware, OH 43015
p. (740) 368-3615
bsmartin@owu.edu
http://geo.owu.edu
Administrative Assistant: Kathryn M. Boger
*Enrollment (2016): B: 22 (7)*
Director, Environmental Studies Program:
John B. Krygier, (D), Penn State, 1995, Oi
Chair:
Barton S. Martin, (D), Massachusetts, 1991, GxvCg
Professor:
Karen H. Fryer, (D), Illinois, 1986, GcpGt
Keith O. Mann, (D), Iowa, 1987, Pi
Assistant Professor:
Nathanael S. Amador, (D), Penn State, 2015, OylOr
Visiting Professor:
Jennifer T. Mokos, (D), Vanderbilt Univ, 2016, On
Emeritus:
Richard D. Fusch, (D), Oregon, 1972, Ou
David H. Hickcox, (D), Oregon, 1979, Oyw

## Shawnee State University

**Dept of Natural Sciences** (A,B) ● (2016)
940 Second Street
Portsmouth, OH 45662
p. (740) 351-3456
kshoemaker@shawnee.edu
http://www.shawnee.edu
Administrative Assistant: Sharon Messer
*Enrollment (2016): B: 21 (3)*
Associate Professor:
Kurt A. Shoemaker, (D), Miami Univ, 2004, GxzGm
Visiting Professor:
Richard Bayless, (M), Ohio Univ, 1984, GgHg
Other:
Jeffrey A. Bauer, (D), Ohio State, 1987, Ps

## University of Akron

**Dept of Geosciences** (B,M) O (2017)
Akron, OH 44325-4101

p. (330) 972-7630
butch@uakron.edu
http://www.uakron.edu/geology/
Administrative Assistant: Elaine L. Butcher
*Enrollment (2014): B: 150 (21) M: 41 (12)*

Acting Head:
Stephen C. Weeks, (D), Rutgers, 1991
Professor of Instruction:
Meera Chatterjee, (D), Oy
Professor:
John A. Peck, (D), Rhode Island, 1995, GseGn
Ira D. Sasowsky, (D), Penn State, 1992, HwGmo
David N. Steer, (D), Cornell, 1996, Yse
Associate Professor:
Linda R. Barrett, (D), Michigan State, 1995, SdOri
Assistant Professor of Instruction:
Meagan E. Ankney, (D), Wisconsin (Madison), 2014, Gzg
John F. Beltz, (M), Akron, 1992, Ggh
Thomas J. Quick, (M), Akron, 1983, GgCaYe
Jeremy M. Spencer, (D), Kent State, 2015, Oya
Assistant Professor:
Shanon P. Donnelly, (D), Indiana, 2009, Oi
Caleb Holyoke, (D), Brown, 2005, GcNrYx
John M. Senko, (D), Oklahoma, 2004, Cg
James R. Thomka, (D), Cincinnati, 2015, PogGg
Adjunct Professor:
Timothy Matney, (D), Pennsylvania, 1993, Ga

## University of Akron, Wayne College

**University of Akron - Wayne College** (2014)
1901 Smucker Road
Orrville, OH 44667
p. (330) 972-8934
WayneCommunityRelations@uakron.edu
http://www.wayne.uakron.edu/

## University of Cincinnati

**Dept of Geology** (B,M,D) (2013)
500 Geology/Physics Building
P. O. Box 210013
Cincinnati, OH 45221-0013
p. (513) 556-3732
krista.smilek@uc.edu
http://www.artsci.uc.edu/departments/geology.html
*Enrollment (2013): B: 86 (0) M: 10 (0) D: 11 (0)*

Head:
Lewis Owen, (D), Leicester (UK), 1988, GmlGt
Professor:
Thomas J. Algeo, (D), Michigan, 1989, Gs
Carlton E. Brett, (D), Michigan, 1978, Pe
Warren D. Huff, (D), Cincinnati, 1963, Gz
Attila I. Kilinc, (D), Penn State, 1969, CpGvCg
Thomas V. Lowell, (D), SUNY (Buffalo), 1986, Gl
J. Barry Maynard, (D), Harvard, 1972, Cl
David L. Meyer, (D), Yale, 1971, Pi
Arnold I. Miller, (D), Chicago, 1986, Pi
David Nash, (D), Michigan, 1977, Gm
Associate Professor:
Craig Dietsch, (D), Yale, 1985, Gp
Assistant Professor:
Brooke E. Crowley, (D), California (Santa Cruz), 2009, CsPe
Andrew D. Czaja, (D), California (Los Angeles), 2006, PoCg
Aaron Diefendorf, (D), Penn State, 2010, Co
Eva Enkelmann, (D), CcGmt
Amy Townsend-Small, (D), Texas, 2006, Co
Dylan Ward, (D), Colorado, 2010, GmqOi
Research Associate:
Carlton E. Brett, (D), Michigan, 1978, PeGrPi
Adjunct Professor:
Brenda R. Hanke, (D), California (Riverside), 2004, Pi
Glenn W. Storrs, (D), Yale, 1986, Pv
Emeritus:
Madeleine Briskin, (D), Brown, 1973, Pe
John E. Grover, (D), Yale, 1972, Gz
David L. Meyer, (D), Yale, 1971, Pi
Paul E. Potter, (D), Chicago, 1952, Gs

Research Professor:
Joshua H. Miller, (D), Chicago, 2009, PgvPe
Yurena Yanes, PgeCs

## University of Dayton

**Dept of Geology** (B) ● (2016)
300 College Park
SC 179
Dayton, OH 45469-2364
p. (937) 229-3432
dgoldman1@udayton.edu
http://www.udayton.edu/artssciences/geology/
Administrative Assistant: Darla Titus
*Enrollment (2016): B: 40 (7)*

Chair:
Daniel Goldman, (D), SUNY (Buffalo), 1993, PoqGr
Professor:
Donald Pair, (D), Syracuse, 1991, Gl
Michael R. Sandy, (D), London, 1984, Pi
Associate Professor:
Umesh Haritashya, (D), Indian Inst of Tech, 2005, OrGm
Andrea M. Koziol, (D), Chicago, 1988, Cp
Allen J. McGrew, (D), Wyoming, 1992, GtcGp
Shuang-Ye Wu, (D), Cambridge, 2000, Oy
Visiting Assistant Professor:
Zelalem Bedaso, (D), South Florida, 2011, CslGs
Adjunct Professor:
Sue Klosterman, (M), Wright State, 2005, Gg
Andrew Rettig, (D), Cincinnati, 2014, Oy
Emeritus:
Charles J. Ritter, (D), Michigan, 1971, Ct

## University of Toledo

**Dept of Environmental Sciences** (B,M,D) (2014)
2801 W. Bancroft Street
Toledo, OH 43606-3390
p. (419) 530-2009
eccarson@wisc.edu
http://www.utoledo.edu/nsm/envsciences/
Administrative Assistant: Patricia Hacker
*Enrollment (2011): B: 13 (3) M: 10 (5)*

Director of the Lake Erie Research Center:
Carol A. Stepien, (D), S California, 1985, On
Chair:
Timothy G. Fisher, (D), Calgary, 1993, GlmGn
Associate Professor:
Richard H. Becker, (D), W Michigan, 2008, OrGe
Mark J. Camp, (D), Ohio State, 1974, Pi
Daryl F. Dwyer, (D), Michigan State, 1986, On
Johan F. Gottgens, (D), Florida, 1992, On
Scott Heckathorn, (D), Illinois, 1995, On
David E. Krantz, (D), South Carolina, 1988, Gs
James Martin-Hayden, (D), Connecticut, 1994, Hq
Daryl L. Moorhead, (D), Tennessee, 1985, On
Alison L. Spongberg, (D), Texas A&M, 1994, Co
Donald J. Stierman, (D), Stanford, 1977, YgGeYs
Assistant Professor:
Jonathon Bossenbroek, (D), Colorado State (Ft. Collins), 2004, On
Thomas Bridgeman, (D), Michigan, 2001, On
Christine M. Mayer, (D), Illinois, 1998, On
William V. Sigler, (D), Purdue, 1999, On
Michael N. Weintraub, California, 2004, On

## Wittenberg University

**Dept of Geology** (B) (2015)
P.O. Box 720
Springfield, OH 45501-0720
p. (937) 327-7335
jritter@wittenberg.edu
http://www4.wittenberg.edu/academics/geol/
f: https://www.facebook.com/WittenbergGeology
*Enrollment (2015): B: 14 (11)*

Professor:
Kenneth W. Bladh, (D), Arizona, 1978, Gz
John B. Ritter, (D), Penn State, 1990, Gm

Associate Professor:
Michael J. Zaleha, (D), SUNY (Binghamton), 1994, Gs
Assistant Professor:
Sarah K. Fortner, (D), Ohio State, 2008, ClOeHg
Emeritus:
Katherine L. Bladh, (D), Arizona, 1976, Gi
Robert W. Morris, (D), Columbia, 1969, PiGg



## Wright State University

**Dept of Earth and Environmental Science** (B,M) (2014)
3640 Colonel Glenn Highway
260 Brehm Lab
Dayton, OH 45435
p. (937) 775-2201
david.schmidt@wright.edu
http://www.wright.edu/ees/
*Enrollment (2010): B: 66 (8) M: 48 (15)*
Director of Undergraduate Programs:
David Schmidt, (D), Ohio State, PiGd
Acting Chair:
David F. Dominic, (D), West Virginia, 1988, Gsr
Professor:
Christopher C. Barton, (D), Yale, 1983, GqYgHq
C. B. Gregor, (D), Utrecht (Neth), 1967, Gs
Allen G. Hunt, (D), California (Riverside), 1983, GqSpCl
Robert W. Ritzi, Jr., (D), Arizona, 1989, Hw
Associate Professor:
Abinash Agrawal, (D), North Carolina, 1990, ClHwPy
Songlin Cheng, (D), Arizona, 1984, Hw
Ernest C. Hauser, (D), Wisconsin, 1982, Ye
William Slattery, (D), CUNY, 1994, OeGr
Doyle Watts, (D), Michigan, 1979, YeOr
Assistant Professor:
Chad Hammerschmidt, (D), Connecticut, 2005, CmOc
Rebecca Teed, (D), Minnesota, 1999, OePe
Director of Sustainability:
Huntting (Hunt) Brown, (D), Ge
Emeritus:
Byron Kulander, (D), West Virginia, 1966, Gc
Benjamin H. Richard, (D), Indiana, 1966, Ye
Paul J. Wolfe, (D), Case Western, 1966, Ye

## Youngstown State University

**Dept of Geological & Environmental Sciences** (B,M) (2014)
One University Plaza
2120 Moser Hall
Youngstown, OH 44555
p. (330) 941-3612
amjacobs@ysu.edu
http://www.as.ysu.edu/~geology/
*Enrollment (2007): B: 48 (4) M: 21 (1)*
Chair:
Jeffrey C. C., (D), Kent State, 1992, EoNgHy
Director, Env Studies Program:
Isam E. Amin, (D), Nevada (Reno), 1987, Hw
Professor:
Raymond E. Beiersdorfer, (D), California (Davis), 1992, Cg
Alan M. Jacobs, (D), Indiana, 1967, Ge
Assistant Professor:
Joseph E. Andrew, (D), Kansas, 2002, Gc
Felicia P. Armstrong, (D), Oklahoma State, 2003, Sb
Shane V. Smith, (D), Washington State, 2005, Gs
Instructor:
Harry Bircher, (M), Youngstown State, 1995, Ge
Brian M. Greene, (D), Kent State, 2001, Ng
Lawrence P. Gurlea, (M), Pennsylvania, 1971, Cg
Engineering Adjunct:
Scott C. Martin, (D), Clarkson, 1984, Ge
Douglas M. Price, (D), Notre Dame, 1988, Cg
Emeritus:
Ann G. Harris, (M), Miami (Ohio), 1958, Ge
Ikram U. Khawaja, (D), Indiana, 1969, Ec
Charles R. Singler, (D), Nebraska, 1969, Gs
Cooperating Faculty:
Thomas P. Diggins, (D), SUNY, 1997, Py
Carl G. Johnston, (D), Cincinnati, 1992, Py

## Zane State College

**Oil and Gas Engineering Technology** (A) (2015)
1555 Newark Rd
Zanesville, OH 43701
p. (740) 588-1282
nwelch@zanestate.edu
http://www.zanestate.edu
f: https://www.facebook.com/ZaneStateCollege
t: @ZaneStateC

# Oklahoma

## Oklahoma City Community College

**Physical Sciences** (1999)
7777 South May Avenue
Oklahoma City, OK 73159
p. (405) 682-1611
dgregory@occc.edu
http://www.occc.edu

## Oklahoma State University

**Boone Pickens School of Geology** (B,M,D) O (2015)
105 Noble Research Center
Stillwater, OK 74078-3031
p. (405) 744-6358
sandy.earls@okstate.edu
http://geology.okstate.edu
Administrative Assistant: Sandy Earls
*Enrollment (2015): B: 157 (23) M: 64 (13) D: 24 (0)*
Endowed Sun Chair:
Estella Atekwana, (D), Dalhousie, 1990, YgGt
Professor:
Mohamed Abdelsalam, (D), Gct
Michael Grammer, (D), GsEo
Jay M. Gregg, (D), Michigan State, 1982, Gs
Todd Halihan, (D), Texas, 2000, Hw
Jack Pashin, (D)
Associate Professor:
Daniel A. La, (D), Pittsburgh, 2008, GctGg
James Puckette, (D), Oklahoma State, 1996, Go
Assistant Professor:
Eliot Atekwana, (D), W Michigan, 1996, Cg
Jeffrey Byrnes, (D), Pittsburgh, 2002, GgvOn
Priyank Jaiswal, (D), Yg
Tracy Quan, (D), MIT/WHOI, 2005, Og
Natascha Riedinger, (D), Bremen, 2005, CgGu
Javier Vilcaez, (D), Tohoku Univ, 2009, HwNpCl
Visiting Professor:
Mary E. Hileman, (D), Michigan, 1973, GorGd
Emeritus:
Arthur Hounslow, (D), Carleton, 1968, Cl
Douglas Kent, (D), Iowa State, 1969, Hw
Wayne Pettyjohn, (D), Boston, 1964, Hw
Vernon Scott, (D), Utah, 1975, Oe
Gary Stewart, (D), Kansas, 1973, Go
John D. Vitek, (D), Iowa, 1973, Gm

## Rogers State University

**Dept of Mathematics and Physical Sciences** (1999)
1701 W. Will Rogers Blvd.
Claremore, OK 74107
p. (918) 343-6812
vwood@rsu.edu
http://www.rsu.edu

## University of Tulsa

**Dept of Geosciences** (B,M,D) (2016)
800 S. Tucker Drive
Tulsa, OK 74104-9700
p. (918) 631-2517
pjm@utulsa.edu
https://engineering.utulsa.edu/academics/geosciences/
Administrative Assistant: Beverly A. Phelps
*Enrollment (2016): B: 61 (9) M: 20 (15) D: 6 (1)*

Professor:
Peter J. Michael, (D), Columbia, 1983, GivGz
Vice President for Research & Dean of Graduate School:
Janet Haggerty, (D), Hawaii, 1982, Gu
Professor:
Kerry Sublette, (D), Tulsa, 1985, Ge
Decker Dawson Associate Professor of Applied Geophysics:
Jingyi Chen, (D), Chinese Acad of Sci, 2005, Yes
Associate Professor:
Dennis R. Kerr, (D), Wisconsin, 1989, Gs
J. Bryan Tapp, (D), Oklahoma, 1983, Gc
Assistant Professor:
Junran Li, (D), Virginia, 2008, Gme
Bethany P. Theiling, (D), New Mexico, 2012, Cs
Research Associate:
Robert W. Scott, (D), Kansas, 1967, GsrPi
Applied Associate Professor:
Winton Cornell, (D), Rhode Island, 1987, GivCa
Emeritus:
Colin Barker, (D), Oxford, 1965, CgoGo

## Tulsa Community College

**Science and Math** (A) (2016)
909 S. Boston Avenue
Tulsa, OK 74119
p. (918) 595-7246
claude.bolze@tulsacc.edu
http://www.tulsacc.edu
*Enrollment (2016): A: 32 (4)*

Professor:
Claude E. Bolze, (M), Wright State, 1974, GgoPg
Adjunct Professor:
Martin Bregman, (D), New Mexico, 1971, YgGcg

## University of Oklahoma ❐

**ConocoPhillips School of Geology & Geophysics** (B,M,D) O (2015)
100 East Boyd
710 Energy Center
Norman, OK 73019-0628
p. (405) 325-3253
geology@ou.edu
http://geology.ou.edu
*Enrollment (2006): B: 87 (0) M: 40 (0) D: 24 (0)*

Director:
R. Douglas Elmore, (D), Michigan, 1981, YmGs
Professor:
Younane N. Abousleiman, (D), Delaware, 1991, Gg
Michael H. Engel, (D), Arizona, 1980, Co
G. Randy Keller, (D), Texas Tech, 1973, Yg
David London, (D), Arizona State, 1981, CpGiz
Kurt J. Marfurt, (D), Columbia, 1978, Ys
Shankar Mitra, (D), Johns Hopkins, 1977, Gc
R. Paul Philp, (D), Sydney, 1972, Co
Matthew J. Pranter, (D), Colorado Mines, Go
Matthew J. Pranter, Gos
Roger M. Slatt, (D), Alaska, 1970, Go
Gerilyn S. Soreghan, (D), Arizona, 1992, Gs
Stephen R. Westrop, (D), Toronto, 1984, Pi
Associate Professor:
Andrew S. Elwood Madden, (D), Virginia Tech, 2005, ClScGz
Megan E. Elwood Madden, (D), Virginia Tech, 2005, ClXgCg
Richard Lupia, (D), Chicago, 1997, Pm
John D. Pigott, (D), Northwestern, 1981, GoYeCg
Barry L. Weaver, (D), Birmingham (UK), 1980, Ct
Assistant Professor:
Xiaowei Chen, (D), California (San Diego), 2013, YsGt
Shannon Dulin, (D), Oklahoma, 2014, YmGs
Michael J. Soreghan, (D), Arizona, 1994, Gs
Adjunct Professor:
Richard L. Cifelli, (D), Columbia, 1983, Pv
Emeritus:
Judson L. Ahern, (D), Cornell, 1980, Yg
James M. Forgotson, Jr., (D), Northwestern, 1956, Go
M. Charles Gilbert, (D), California (Los Angeles), 1965, Cp
Charles W. Harper, Jr., (D), Caltech, 1964, Pi
G. Randy Keller, (D), Texas Tech, 1973, YgGtEo
David W. Stearns, (D), Texas A&M, 1969, Gc
Cooperating Faculty:
Neil Suneson, (D), California (Santa Barbara), 1980, Gr
Electron Microprobe Operator:
George B. Morgan, (D), Oklahoma, 1988, Gi

**Mewbourne School of Petroleum & Geological Engineering** (B,M,D) (2015)
100 East Boyd Street
Sarkeys Energy Center 1210
Norman, OK 73019-1001
p. (405) 325-2921
mpge@ou.edu
http://mpge.ou.edu/
f: https://www.facebook.com/OUMPGE
t: @ou_mpge
*Enrollment (2015): B: 975 (133) M: 101 (25) D: 37 (4)*

Director:
Chandra S. Rai, (D), Hawaii, 1977, Np
Graduate Liaison:
Deepak Devegowda, (D), Texas A&M, 2008, Np
Director Natural Gas Engineering & Management:
Suresh Sharma, (D), Oklahoma, 1968, Np
Professor:
Younane Abousleiman, (D), Delaware, 1991, Nr
Ramadan Ahmed, (D), Norwegian Univ of Sci & Tech, 2001, Np
Jeff Callard, (D), Louisiana State, 1994, Np
Ahmed Ghassemi, (D), Oklahoma, 1996, Nr
Ben Shiau, (D), Oklahoma, 1995
Carl H. Sondergeld, (D), Cornell, 1977, NpYex
Musharraf Zaman, (D), Bangladesh Univ of Eng and Tech, 1975, Np
Associate Professor:
Mashhad Fahes, (D), Imperial Coll (London), 2006, Np
Ahmad Jamili, (D), Kansas, 2004, Np
Rouzbeh Moghanloo, (D), Texas, 2012, Np
Maysam Pournik, (D), Texas A&M, 2008, Np
Catalin Teodoriu, (D), Ploiesti, Np
Xingru Wu, (D), Texas, 2006, Np
Assistant Professor:
Siddharth Misra, (D), Texas (Austin), 2015, Np
Ahmad Sakhaee-Pour, (D), Texas (Austin), 2012, Np
Instructor:
Ilham El-Monier, (D), Texas A&M, 2012, Np
Emeritus:
Faruk Civan, (D), Oklahoma, 1978, Np
Roy M. Knapp, (D), Kansas, 1973, Np
Jean-Claude Roegiers, (D), Minnesota, 1974, Nr
Subhash N. Shah, (D), New Mexico, 1974, Np

# Oregon

## Central Oregon Community College

**Dept of Science** (A) (2014)
2600 NW College Way
Bend, OR 97701
p. (541) 383-7557
breynolds@cocc.edu
http://science.cocc.edu/Programs_Classes/Geology/default.aspx

Associate Professor:
Robert W. Reynolds, (D), Idaho, 1994, Gv

## Oregon State University ❐

**College of Earth, Ocean, and Atmospheric Sciences** (B,M,D) ● (2017)
104 CEOAS Administration Building
Corvallis, OR 97331-5503
p. (541) 737-1201
contact@coas.oregonstate.edu
http://ceoas.oregonstate.edu
*Enrollment (2010): B: 68 (9) M: 18 (5) D: 18 (1)*

Exec Dir-Marine Studies Initiative:
Jack A. Barth, (D), MIT, 1987, On
Director-Budget & Fiscal Planning:
Sherman H. Bloomer, (D), California (San Diego), 1982, Gi

Director of Ocean Science Program:
Rob Wheatcroft, (D), Washington, 1990, OgGg
Director of OCCRI:
Philip Mote, (D), Washington, 1994, Oag
Director of Marine Resource Management Program:
Flaxen D. Conway, (M), Oregon State, 1986, Gu
Director of Geology Program:
Edward J. Brook, (D), MIT/WHOI, 1993, ClPeOl
Director of Geography Program:
Julia A. Jones, (D), Johns Hopkins, 1983, So
Director of Environmental Science Program:
Laurence Becker, (D), California (Berkeley), 1989, On
Professor:
Jeffrey R. Barnes, (D), Washington, 1983, Oa
Hal Batchelder, (D), Oregon State, 1986, Op
Kelly Benoit-Bird, (D), Hawaii, 2003, Gu
Michael E. Campana, (D), Arizona, 1975, Hg
Lorenzo Ciannelli, (D), Washington, 2002, Og
Peter U. Clark, (D), Colorado, 1984, Gl
Frederick (Rick) Colwell, (D), Virginia Tech, 1986, On
Byron Crump, (D), Washington, 1999
Shanika de Silva, (D), Open (UK), 1987, Gv
John H. Dilles, (D), Stanford, 1984, Em
Gary D. Egbert, (D), Washington, 1987, Yr
Chris Goldfinger, (D), Oregon State, 1994, Yr
Miguel A. Goni, (D), Washington, 1992, Gu
David W. Graham, (D), MIT, 1987, Cm
Burke R. Hales, Washington, 1995, Oc
Merrick C. Haller, (D), Delaware, 1999, OnrOp
Robert N. Harris, (D), Utah, 1996, Yhr
Michael Harte, (D), Victoria, 1994, Gu
Adam J. R Kent, (D), Australian National, 1995, GiCga
Michael Kosro, (D), Scripps, 1985, OprOn
Ricardo Letelier, (D), Hawaii, 1994, Obr
Ricardo Matano, (D), Princeton, 1991, Oap
Andrew J. Meigs, (D), S California, 1995, Gc
Robert N. Miller, (D), California (Berkeley), 1976, OpnOa
Alan C. Mix, (D), Columbia, 1986, Cs
James N. Moum, (D), British Columbia, 1984, Op
Roger L. Nielsen, (D), S Methodist, 1983, GizGv
Anne Nolin, (D), California (Santa Barbara), 1993, Or
David Noone, (D), Melbourne, 2001, Owg
Tuba Ozkan-Haller, (D), Delaware, 1998, On
Clare Reimers, (D), Oregon State, 1982, Oc
Roger Samelson, (D), Oregon State, 1987, Op
Adam Schultz, (D), Washington, 1986, Yg
Eric Skyllingstad, (D), Wisconsin, 1986, Ow
William D. Smyth, (D), Toronto, 1990, Op
Yvette H. Spitz, (D), Old Dominion, 1995, Yr
Marta E. Torres, (D), Oregon State, 1988, Cm
Anne M. Trehu, (D), MIT/WHOI, 1982, Yr
Aaron T. Wolf, (D), Wisconsin, 1992, Hg
Senior Research:
Brian Haley, (D), Oregon State, 2004, Cs
Director, Water Resources Graduate Program:
Mary V. Santelmann, (D), Minnesota, 1988, Oy
Director of Atmospheric Science Program:
Karen M. Shell, (D), Scripps, 2004, Oa
Associate Professor:
Kim S. Bernard, (D), Rhodes, 2007
Anders Carlson, (D), Oregon State, 2006, GlCt
Simon P. de Szoeke, (D), Washington, 2004, Opa
Edward P. Dever, (D), MIT/WHOI, 1995, Op
Theodore Durland, (D), Hawaii, 2006, Yg
Hannah Gosnell, (D), Colorado, 2000, Ou
Randall A. Keller, (D), Oregon State, 1996, Gu
Eric Kirby, (D)
Anthony Koppers, (D), Free (Amsterdam), 1988, CgGv
Alexander Kurapov, (D), St Petersburg, 1994, Yr
Stephen Lancaster, (D), MIT, 1999, Hs
Jim Lerczak, (D), Scripps, 2000, Op
John L. Nabelek, (D), MIT/WHOI, 1984, GtYs
Jonathan Nash, (D), Oregon State, 2000, Op
Peter Ruggiero, (D), Oregon State, 1997, On
Andreas Schmittner, (D), Bern (Switzerland), 1999, OaYr
Kipp Shearman, (D), Oregon State, 1999, Opn
Joseph Stoner, (D), Québec (Montréal), 1995, Gsr
Frank J. Tepley, III, (D), California (Los Angeles), 1999, GiCs
George Waldbusser, (D), Maryland (Baltimore), 2008, Ob
Angelicque White, (D), Oregon State, 2006, Ob
Associate Director of the Institute for Water and Watersheds :
W. Todd Jarvis, (D), Oregon State, 2006, Hw
Assistant Professor:
Christo Buizert, (D), Copenhagen, 2012, Yr
Louise A. Copeman, (D), Memorial Univ (Newfoundland), 2011, On
Jessica Creveling, (D), Harvard, 2012, GgsGr
Jennifer Fehrenbacher, (D), N Illinois, 1997, Cm
Jonathan Fram, (D), California (Berkeley), 2005, Op
Jennifer Hutchings, Ol
Lauren W. Juranek, Washington, 2007, Cm
Robert Kennedy, (D), Oregon State, 2004, Ori
Jennifer L. McKay, (D), British Columbia, 2004, Cm
Larry O'Neill, (D), Oregon State, 2007, Owg
David Rupp, (D), Oregon State Univ, 2005, Oap
Alyssa Shiel, (D), British Columbia, 2010, Cgs
Emily L. Shroyer, (D), Oregon State, 2009, On
Jenna Tilt, (D), Washington, 2007, Ou
Jamon Van Den Hoek, (D), Wisconsin (Madison), 2012, Our
Justin Wettstein, (D), Washington, 2007, Oa
Greg Wilson, (D), Oregon State Univ, 2013, Onp
Bo Zhao, (D), Ohio State Univ, 2015, Oi
Senior Instructor/Program Coordinator:
Kaplan Yalcin, (D), New Hampshire, 2005, GgeOe
Senior Instructor:
Lorene Yokoyama Becker, (M), Wisconsin, 1999, Oi
Steve Cook, (D), Florida, 1995, On
Demian Hommel, (D), Oregon, 2009, On
Instructor:
Shireen Hyrapiet, (D), Oklahoma State Univ, 2000, On
Rebecca Yalcin, (M), Maine, 2001, Gg
Distinguished Emeritus Professor:
Dudley B. Chelton, (D), California (San Diego), 1980, Op
Patricia Wheeler, (D), California (Irvine), 1976, Cm
Associate Dean for Academic Programs:
Anita L. Grunder, (D), Stanford, 1986, Gi
Emeritus:
John S. Allen, Jr., (D), Princeton, 1968, Op
Andrew Bennett, (D), Harvard, 1971, Op
John V. Byrne, (D), S California, 1957, Gu
Douglas Caldwell, (D)
Robert W. Collier, (D), MIT, 1981, Cm
Timothy J. Cowles, (D), Duke, 1977, Ob
Robert A. Duncan, (D), Yr
Martin R. Fisk, (D), Rhode Island, 1978, Yr
Robert Holman, (D), Dalhousie, 1979, On
Philip L. Jackson, (D), Kansas, 1977, Oy
George H. Keller, (D), Illinois, 1966, YgGus
A. Jon Kimerling, (D), Wisconsin, 1976, Oir
Gary Klinkhammer, (D), Rhode Island, 1979, CaEmCg
Robert J. Lillie, (D), Cornell, 1984, Ye
Gordon E. Matzke, (D), Syracuse, 1975, Og
Alan R. Niem, (D), Wisconsin, 1971, GdsGr
Nicklas Pisias, (D), Rhode Island, 1978, Gu
Fredrick G. Prahl, (D), Washington, 1982, Oc
Barry Sherr, (D), Georgia, 1977, Ob
Evelyn Sherr, (D), Duke, 1977, Ob
Lawrence Small, (D), Ob
Robert Lloyd Smith, (D), Oregon State, 1964, OpgOn
Ted P. Strub, (D), California (Davis), 1983, Oa
Richard J. Vong, (D), Washington, 1985, Oa
Robert S. Yeats, (D), Washington, 1958, GtYrNg
J. Ronald Zaneveld, (D), Oregon State, 1971, Yr
Dean:
Roberta Marinelli, (D), S Carolina-Columbia Univ, 1991, GuOcGe
Associate Vice President for Research:
Roy D. Haggerty, (D), Stanford, 1995, Hw

## Portland Community College, Sylvania Campus

**Physical Science Dept** (A) O (2017)
12000 SW 49th Ave.
Portland, OR 97219
p. 971-722-8209
patty.maazouz@pcc.edu
http://www.pcc.edu/programs/geology/

Instructor:
Talal Abdulkareem, (D)
Kali Abel, (M)
Sharon Delcambre, (D), Oag
Gretchen Gebhardt, (M)
Melinda Hutson, (D), Xm
Hollie Oakes-Miller, (M), Gg
Kristy Schepker, (M)
Steve Todd, Ow
Jonathan Weatherford, (M), Gg

## Portland State University

**Dept of Geology** (B,M,D) (2016)
P.O. Box 751
Portland, OR 97207
p. (503) 725-3022
streckm@pdx.edu
http://www.pdx.edu/geology
*Enrollment (2015): B: 156 (9) M: 20 (12) D: 1 (0)*

Acting Chair:
Martin J. Streck, (D), Oregon State, 1994, GivCt
Professor:
Andrew G. Fountain, (D), Washington, 1992, Gl
Associate Professor:
Kenneth M. Cruikshank, (D), Purdue, 1991, Gc
Robert Benjamin Perkins, (D), Portland State, 2000, Hq
Alexander (Alex) M. Ruzicka, (D), Arizona, 1996, Xm
Assistant Professor:
John T. Bershaw, (D), Rochester, 2011, GsCsHy
Adam M. Booth, (D), Oregon, 2012, GmOr
Nancy A. Price, (D), Maine, 2012, GcOeGt
Maxwell L. Rudolph, (D), California (Berkeley), 2012, Gq
Ashley Streig, (D), Oregon, 2014, YsGt
Research Associate:
Richard Hugo, (D), Washington State, 1998, Gy
Adjunct Professor:
Sheila Alfsen
Matthew Brunengo, (D), Portland State, 2012, GgNgGm
Elizabeth Carter, (D), Lausanne, 1993, Pm
Frank D. Granshaw, (D), Portland State Univ, 2011, OnGl
Melinda Hutson, (D), Arizona, 1996, Xmc
William Orr, (D), Michigan State, 1968, GgOg
Dick Pugh
Arron Steiner
Barry Walker, (D), Oregon State, 2011, GiCgGv
Emeritus:
Scott F. Burns, (D), Colorado, 1980, Ng
Michael L. Cummings, (D), Wisconsin, 1978, GgHg
Paul E. Hammond, (D), Washington, 1963, Gi
Ansel G. Johnson, (D), Stanford, 1973, Ye
Richard E. Thoms, (D), California, 1965, Ps
Instructor:
David Percy, (B), Portland State, 1999, Oi
Cooperating Faculty:
Christina L. Hulbe, (D), Chicago, 1998, Gl

## Rogue Community College

**Science Dept-Physical Sciences-Geology** (1999)
3345 Redwood Highway
Grants Pass, OR 97527
p. (541) 245-7527
jvanbrunt@roguecc.edu
http://learn.roguecc.edu/science/physical.htm

## Southern Oregon University

**Dept of Geology** (B) (2014)
1250 Siskiyou Boulevard
Ashland, OR 97520
p. (541) 552-6479
lane@sou.edu
Administrative Assistant: Susan Koralek
*Enrollment (2006): B: 45 (5)*

Chair:
Charles L. Lane, (D), California (Los Angeles), 1987, Hg
Dean, Science:
Joseph L. Graf, Jr., (D), Yale, 1975, Em
Professor:
Jad A. D'Allura, (D), California (Davis), 1977, Gc
Associate Professor:
Eric Dittmer, (M), San Jose State, 1972, Gg
Adjunct Professor:
Vernon J. Crawford, (M), Oregon, 1970, Gg
Harry W. Smedes, (D), Washington, 1959, Gi
Richard Ugland, (M), Utah, 1974, Gg
Emeritus:
Monty A. Elliott, (D), Oregon State, 1971, Gr
William B. Purdom, (D), Arizona, 1960, Gx

## Southwestern Oregon Community College

**Dept of Geology** (A) (2015)
1988 Newmark
Coos Bay, OR 97420-2912
p. (541) 888-7216
rmetzger@socc.edu
*Enrollment (2013): A: 2 (0)*

Professor:
Ronald A. Metzger, (D), Iowa, 1991, PmOePs

## Tillamook Bay Community College

**Associate of Science** (1999)
2510 First Street
Tillamook, OR 97141
bannan@tillamookbay.cc

## Treasure Valley Community College

**Career and Technical Education** ● (2015)
650 College Blvd.
Ontario, OR 97914
p. (541) 881-8866
dtinkler@tvcc.cc
http://www.tvcc.cc.or.us/science/Index.htm
f: https://www.facebook.com/profile.php?id=100005312898088

Professor:
Dorothy Tinkler, (D), Texas Tech, 2003, Oi

## Umpqua Community College

**Dept of Geology Transfer Program** (1999)
P.O. Box 967
1140 Umpqua College Rd.
Roseburg, OR 97470
p. (541) 440-4654
Jason.Aase@umpqua.edu
http://www.umpqua.edu/degree-programs/55

## University of Oregon

**Dept of Earth Sciences** (B,M,D) ○ (2017)
1272 University of Oregon
Eugene, OR 97403-1272
p. (541) 346-4573
sthoms@uoregon.edu
http://earthsciences.uoregon.edu/
*Enrollment (2015): B: 122 (23) M: 8 (7) D: 23 (4)*

Department Head:
Paul Wallace, (D), California (Berkeley), 1991, GviCg
Professor:
Rebecca J. Dorsey, (D), Princeton, 1989, Gr
Eugene D. Humphreys, (D), Caltech, 1985, YsGt
Mark H. Reed, (D), California (Berkeley), 1977, CgEm
Alan W. Rempel, (D), Cambridge (UK), 2001, YgNrOl
Gregory J. Retallack, (D), New England, 1978, PbOsGa
Joshua J. Roering, (D), California (Berkeley), 2000, Gm
Douglas R. Toomey, (D), MIT/WHOI, 1987, YsGt
Ray J. Weldon, (D), Caltech, 1986, Gc
Associate Professor:
Ilya N. Bindeman, (D), Chicago, 1998, Cs
Emilie E. Hooft, (D), MIT, 1996, Yr
Samantha Hopkins, (D), California (Berkeley), 2005, Pg
Qusheng Jin, (D), Illinois (Urbana-Champaign), 2003, Cg

Assistant Professor:
Edward Davis, (D), California (Berkeley), 2005, Pg
Thomas Giachetti, (D), Clermont-Ferrand, 2010, Gv
Dave Sutherland, (D), MIT, 2008, Op
Amanda Thomas, (D), California (Berkeley), 2012, Ys
James Watkins, (D), California (Berkeley), 2010, Cg
Instructor:
Marli G. Miller, (D), Washington, 1997, Gc
Assoc Dean Natural Science:
A. Dana Johnston, (D), Minnesota, 1983, Cp
Emeritus:
Sam Boggs, Jr., (D), Colorado, 1964, Gd
Marvin A. Kays, (D), Washington Univ (St. Louis), 1960, GpcGt
Alexander R. McBirney, (D), California (Berkeley), 1961, Gv
William N. Orr, (D), Michigan State, 1967, Pm
Norman M. Savage, (D), Sydney, 1968, Pi
Harve S. Waff, (D), Oregon, 1970, Yg
Office & Business Manager:
Sandy K. Thoms, (B), Portland State Univ, 1993
Related Staff:
Elise Mezger Weldon, (M), S California, 1986, Gmt
Cooperating Faculty:
John M. Logan, (D), Oklahoma, 1965, Nr

**Dept of Geography** (B,M,D) (2016)
107 Condon Hall
1251 University of Oregon
Eugene, OR 97403-1251
p. (541) 346-4555
uogeog@uoregon.edu
http://geography.uoregon.edu
f: https://www.facebook.com/universityoforegongeography/
*Enrollment (2009): B: 167 (55) M: 22 (9) D: 19 (5)*
Head:
Amy Lobben, (D), Michigan State, 1999, OinOn
Dean:
W. Andrew Marcus, (D), Colorado, 1987
Professor:
Patrick J. Bartlein, (D), Wisconsin, 1978, Pe
Patricia F. McDowell, (D), Wisconsin, 1980, Gm
Alexander B. Murphy, (D), Chicago, 1987, On
Laura Pulido, (D), California, Los Angeles, 1991
Peter A. Walker, (M), Oregon, 1966, On
Associate Professor:
Daniel Buck, (D), California (Berkeley), 2002, On
Shaul E. Cohen, (D), Chicago, 1991, On
Mark Fonstad, (D), Arizona State, 2000, Oy
Dan Gavin, (D), Washington, 2000, OyPle
Xiaobo Su, (D), Singapore, 2007, On
Assistant Professor:
Christopher Bone, (D), Simon Fraser, 2009, Oi
Leigh Johnson, (D), California, Berkeley, 2011
Katharine Meehan, (D), Arizona, 2010, On
Hedda R. Schmidtke, (D), Hamburg, 2004
Lucas Silva, (D), Guelph, 2011
Instructor:
Nicholas Kohler, (D), Oregon, 2004, Oi
Emeritus:
Stanton A. Cook, (D), California (Berkeley), 1960, On
Carl L. Johannessen, (D), California (Berkeley), 1959, On
Alvin W. Urquhart, California (Berkeley), 1962, On
Ronald Wixman, (D), Columbia, 1978, On

## Western Oregon University
**Earth and Physical Science Dept.** (B) (2017)
345 N. Monmouth Ave.
Monmouth, OR 97361
p. (503) 838-8398
taylors@wou.edu
http://www.wou.edu/earthscience
*Enrollment (2016): B: 40 (8)*
Professor:
Jeffrey A. Myers, (D), Santa Barbara, 1998, Gs
Stephen B. Taylor, (D), West Virginia, 1999, Gm
Jeffrey H. Templeton, (D), Oregon State, 1998, GviOe
Assistant Professor:
Melinda Shimizu, (D), Arizona State, 2014, OirOg
Instructor:
Don Ellingson, (M), W Oregon, 1988, OwgXg
Jeremiah Oxford, (M), Oregon State, 2006, Og
Grant Smith, (D), Oregon State, 2012, Og
Phillip Wade, (M), San Diego State, 1991, Oge

## Willamette University
**Environmental and Earth Sciences Dept** (B) (2016)
900 State Street
Salem, OR 97301
p. (503) 370-6587
spike@willamette.edu
http://www.willamette.edu/cla/ees/
f: https://www.facebook.com/groups/186102921735448/
*Enrollment (2016): B: 79 (0)*
Professor:
Karen Arabas, (D), Penn State, 1997, On
Endowed Dempsey Chair:
Joe Bowersox, (D), Wisconsin, 1995, On
Assistant Professor:
Melinda Butterworth, (D), Arizona
Katja Meyer, (D), Penn State Univ
Scott Pike, (D), Georgia, 2000, Gag

# Pennsylvania

## Allegheny College
**Dept of Geology** (B) (2015)
520 North Main Street
Meadville, PA 16335
p. (814) 332-2350
robrien@allegheny.edu
http://www.allegheny.edu/academics/geo/
*Enrollment (2015): B: 20 (0)*
Associate Professor:
Rachel O'Brien, (D), Washington State, 2000, HwGe
Assistant Professor:
Theresa M. Schwartz, (D), Stanford, 2015, Gs
Visiting Professor:
Erin M. Birsic, (M), Wisconsin (Madison), 2015, Gi
Currently Provost and Dean of the College:
Ron B. Cole, (D), Rochester, 1993, Gt

## Bloomsburg University
**Dept of Environmental, Geographical, and Geological Sciences** (B) (2014)
400 East Second Street
Bloomsburg, PA 17815-1301
p. (570) 389-4108
dspringe@bloomu.edu
http://departments.bloomu.edu/geo/
*Enrollment (2007): B: 45 (12)*
Professor:
Shahalam M.N. Amin, (D), Kent State, 1991, GeScOn
John E. Bodenman, (D), Penn State, 1995, Ou
Sandra J. Kehoe-Forutan, (D), Queensland, 1991, Onu
Michael K. Shepard, (D), Washington Univ (St. Louis), 1994, XgYgOr
Dale A. Springer, (D), Virginia Tech, 1982, PivOe
Karen M. Trifonoff, (D), Kansas, 1994, Oi
Associate Professor:
Patricia J. Beyer, (D), Arizona State, 1997, HsOyGm
Cynthia Venn, (D), Pittsburgh, 1996, OgClOc
Assistant Professor:
Jeffrey C. Brunskill, (D), SUNY (Buffalo), 2005, OiwOy
John G. Hintz, (D), Kentucky, 2005, Oin
Brett T. McLaurin, (D), Wyoming, 2000, GrsGd
Jennifer B. Whisner, (D), Tennesse, 2010, GmHw
S. Christopher Whisner, (D), Tennesse, 2005, GciGz
Department Secretary:
Jade L. Swartwood, On

## Bryn Mawr College
**Dept of Geology** (B) (2016)
101 North Merion Avenue

Bryn Mawr, PA 19010-2899
p. (610) 526-5115
aweil@brynmawr.edu
http://www.brynmawr.edu/geology/
*Enrollment (2015): B: 32 (14)*

Professor:
Arlo B. Weil, (D), Michigan, 2001, GctYm
Associate Professor:
Donald C. Barber, (D), Colorado, 2001, Gs
Pedro J. Marenco, (D), S California, 2007, PgCsPo
Assistant Professor:
Selby Cull, (D), Washington, 2011, XgGzx
Research Associate:
Frank S. Welsh
Lecturer:
Katherine N. Marenco, (D), S California, 2008, PiePo
Emeritus:
Maria Luisa B. Crawford, (D), California (Berkeley), 1965, Gxz
William A. Crawford, (D), California (Berkeley), 1965, Cg
Lucian B. Platt, (D), Yale, 1960, Gc
W. Bruce Saunders, (D), Iowa, 1970, Pi

## Bucknell University

**Geology and Environmental Geosciences** (B) (2016)
231 O'Leary Center
Lewisburg, PA 17837
p. (570) 577-1382
cdaniel@bucknell.edu
http://www.bucknell.edu/Geology
f: http://www.facebook.com/BucknellGeology
Administrative Assistant: Carilee Dill
*Enrollment (2016): B: 36 (7)*

Chair:
Christopher G. Daniel, (D), Rensselaer, 1998, Gp
Professor:
Mary Beth Gray, (D), Rochester, 1991, Gc
Carl S. Kirby, (D), Virginia Tech, 1993, Cl
R. Craig Kochel, (D), Texas, 1980, Gm
Jeffrey M. Trop, (D), Purdue, 2000, Gs
Associate Professor:
Ellen K. Herman, (D), Penn State, 2006, Hw
Robert W. Jacob, (D), Brown, 2006, Yg
Emeritus:
Jack C. Allen, (D), Princeton, 1962, GizGe
Edward Cotter, (D), Princeton, 1963, GmPg
Laboratory Director:
Bradley C. Jordan, (M), Rhode Island, 1983, Gg

## California University of Pennsylvania

**Dept of Earth Sciences** (B) (2015)
250 University Avenue
California, PA 15419
p. (724) 938-4180
wickham@calu.edu
http://www.cup.edu/eberly/earthscience
Administrative Assistant: Pamela Higinbotham
*Enrollment (2015): B: 195 (25)*

Chair:
Thomas Wickham, (D), Penn State, 2000, Ou
Professor:
Kyle Fredrick, (D), SUNY (Buffalo), 2008, GgHgGm
Associate Professor:
Thomas Mueller, (D), Illinois, 1999, Oi
Assistant Professor:
John Confer, (D), Penn State, 1997, On
Swarndeep S. Gill, (D), Wyoming, 2002, Oa
Chad Kauffman, (D), Nebraska, 2000, Oa
Susan Ryan, (D), Calgary, 2005, Og

## Carnegie Museum of Natural History

**Section of Vertebrate Paleontology** (2014)
4400 Forbes Avenue
Pittsburgh, PA 15213
p. (412) 622-5782
beardc@CarnegieMNH.org
http://www.carnegiemnh.org/vp/

Curator:
K. Christopher Beard, (D), Johns Hopkins, 1989, Pv
Curator:
David S. Berman, (D), California (Los Angeles), 1969, PgoPv
Mary R. Dawson, (D), Kansas, 1957, Pv
Preparator:
Dan Pickering, (B), Carnegie Mellon, 1983, On
Alan R. Tabrum, (M), SD Mines, 1981, Pv
Norman Wuerthele, (B), Pittsburgh, 1966, Pv
Curator & Associate Director:
Zhexi Luo, (D), California (Berkeley), 1987, Pv
Collections Manager:
Amy C. Henrici, (M), Pittsburgh, 1990, Pv
Assistant Curator:
Matthew C. Lamanna, (D), Pennsylvania, 2004, PvoPe

## Clarion University

**Dept of Biology and Geosciences** (B) O (2016)
840 Wood Street
Clarion, PA 16214
p. (814) 393-2317
cking@clarion.edu
http://www.clarion.edu/BIGS
*Enrollment (2015): B: 97 (24)*

Professor:
Yasser M. Ayad, (D), Montreal, 2000, Oi
Valentine U. James, (D), Texas A&M, OunOn
Anthony J. Vega, (D), Louisiana State, 1994, OawOp
Craig E. Zamzow, (D), Texas, 1983, Gi
Assistant Professor:
Shawn Collins, (D), McMaster Univ, 2004, GsuGc

## Delaware County Community College

**STEM** (A) (2013)
901 Media Line Road
Media, PA 19063
p. (610) 359-5082
jsnyder2@dccc.edu
http://www.dccc.edu/

Professor:
Daniel Childers, (D), Delaware, 2014, OguOi
Associate Professor:
Jennifer L. Snyder, (D), W Michigan, 1998, Og

## Dickinson College

**Dept of Earth Sciences** (B) (2016)
P.O. Box 1773
Carlisle, PA 17013-2896
p. (717) 245-1355
key@dickinson.edu
http://www.dickinson.edu/homepage/96/earth_sciences
Administrative Assistant: Debra Peters
*Enrollment (2016): B: 24 (5)*

Chair:
Marcus M. Key, Jr., (D), Yale, 1988, PiGsa
Professor:
Benjamin R. Edwards, (D), British Columbia (Canada), 1997, GivCg
Associate Professor:
Peter B. Sak, (D), Penn State, 2002, Gcm
Assistant Professor:
Jorden L. Hayes, (D), Wyoming, 2016, Ygs
Alyson M. Thibodeau, (D), Arizona, 2012, CcGaCs
Emeritus:
Jeffery W. Niemitz, (D), S California, 1978, ClOuGn
Noel Potter, Jr., (D), Minnesota, 1969, Gmc
William W. Vernon, (D), Lehigh, 1964, Gz
Technician:
Robert Dean, (M), Texas (El Paso), 2004, Gi

**Environmental Studies & Environmental Science** (2004)
P.O. Box 1773
Carlisle, PA 17013-2896
p. (717) 245-1355
arnoldt@dickinson.edu
Administrative Assistant: Patricia Braught

Chair:
Michael Heiman, (D), California (Berkeley), 1983, Ou
Associate Professor:
Candie Wilderman, (D), Johns Hopkins, 1984, Hs
Visiting Professor:
Kirsten Hural, (D), Cornell, 1997, Gg



## Drexel University

**Dept of Biodiversity, Earth & Environmental Science** (B,M,D) (2016)
3201 Arch St
Suite 240
Philadelphia, PA 19104
p. (215) 571-4639
bees@drexel.edu
http://drexel.edu/bees/
f: https://www.facebook.com/groups/DrexelBEES/
*Enrollment (2016): B: 22 (1) D: 4 (3)*
Pilsbry Chair of Malacology:
Gary Rosenberg, (D), Harvard, 1989, Pi
Professor:
David Velinsky, (D), Old Dominion, 1987, OcCms
Associate Professor:
Ted Daeschler, (D), Pennsylvania, 1998, PvGgPy
Assistant Professor:
Amanda Lough, (D), Washington Univ (St. Louis), 2014, Ysg
Loyc Vanderkluysen, (D), Hawaii, 2008, GviGz
Adjunct Professor:
Mitch Cron, (M), Pennsylvania, 2013, Gge
Jerry V. Mead, (D), SUNY-Syracuse, 2007, OiHgGm

## Edinboro University of Pennsylvania

**Dept of Geosciences** (B) (2015)
126 Cooper Hall
230 Scotland Road
Edinboro, PA 16444-0001
p. (814) 732-2529
geosciences@edinboro.edu
http://www.edinboro.edu/academics/schools-and-departments/cshp/departments/geosciences/
f: https://www.facebook.com/pages/Edinboro-University-Geosciences-Department/426407214146067
*Enrollment (2015): B: 141 (24)*
Chair:
Laurie A. Parendes, (D), Oregon State, 1997, On
Brian S. Zimmerman, (D), Washington State, 1991, EgGzi
Professor:
Baher A. Ghosheh, (D), SUNY (Buffalo), 1988, On
David Hurd, (D), Cleveland State, 1997, Og
Henry Lawrence, (D), Oregon, 1985, Ou
Kerry A. Moyer, (D), Penn State, 1993, Oa
Joseph F. Reese, (D), Texas (Austin), 1995, GctOe
Eric Straffin, (D), Nebraska, 2000, GmsOs
Dale Tshudy, (D), Kent State, 1993, Pi
Associate Professor:
Karen Eisenhart, (D), Colorado, 2004, Oy
Wook Lee, (D), Ohio State, Oi
Assistant Professor:
Richard Deal, (D), South Carolina, 2000, Oi
Tamara Misner, (D), Pittsburgh, 2013, GmHs

## Elizabethtown College

**Dept of Engineering and Physics** (B) (2016)
One Alpha Drive
Esbenshade Room 160
Elizabethtown, PA 17022-2298
p. (717) 361-1392
mcfaddenj@etown.edu
http://www.etown.edu/PhysicsEngineering.aspx
Administrative Assistant: Jennifer McFadden
Associate Professor:
Michael A. Scanlin, (D), Penn State, Ye
Emeritus:
David -- Ferruzza, (M), MIT, 1967, Oan

## Franklin and Marshall College

**Dept of Earth and Environment** (B) (2015)
PO Box 3003
Lancaster, PA 17604-3003
p. (717) 291-4133
diane.kadyk@fandm.edu
http://www.fandm.edu/earthandenvironment.xml
f: https://www.facebook.com/pages/FM-Department-of-Earth-Environment/153052838071694
t: @FandMENE
Administrative Assistant: Diane L. Kadyk
*Enrollment (2014): B: 99 (28)*
Chair:
Dorothy J. Merritts, (D), Arizona, 1987, Gm
Professor:
Carol B. de Wet, (D), Cambridge, 1989, Gs
Stanley A. Mertzman, (D), Case Western, 1971, GizOm
Associate Professor:
Andrew P. deWet, (D), Cambridge, 1989, Ge
Zeshan Ismat, (D), Rochester, 2002, Gc
James E. Strick, (D), Princeton, 1997, Ghe
Robert C. Walter, (D), Case Western, 1981, Cc
Christopher J. Williams, (D), Pennsylvania, 2002, PeSb
Assistant Professor:
Eilzabeth De Santo, (D), Univ Coll (London), On
Paul Harnik, (D), Chicago, 2009, Pg
Director of Public Policy:
Richard V. Pepino, (M), Villanova, 1970, Ou
Adjunct Professor:
Suzanna L. Richter, (D), Pennsylvania, 2006, Ge
Visiting Professor:
Timothy D. Bechtel, (D), Brown, 1989, GeYg
Emeritus:
Robert S. Sternberg, (D), Arizona, 1982, YmGa
Roger D. K. Thomas, (D), Harvard, 1970, PoGh
Laboratory Director:
Steven Sylvester, (M), Franklin & Marshall, 1971, Ca

## Gannon University

**Earth Science Program** (B) (2011)
109 University Square
PMB 3183
Erie, PA 16541
p. (814) 871-7453
olanrewa001@gannon.edu
http://www.gannon.edu/
Assistant Professor:
Johnson Olanrewaju, (D), Penn State, 2002, Cg

## Harrisburg Area Community College

**Science** (A) (2017)
One HACC Drive
Harrisburg, PA 17110
p. 800-222-4222
http://www.hacc.edu/AboutHACC/ContactUs/index.cfm
http://www.hacc.edu/index.cfm
f: https://www.facebook.com/HACC64
t: @hacc_info
*Enrollment (2016): A: 2 (0)*
Professor:
James E. Baxter, P.G., (M), Penn State, 1983, GgHwGm

## Indiana University of Pennsylvania

**Dept of Geoscience** (B) (2017)
111 Walsh Hall
Indiana, PA 15705
p. (724) 357-2379
geoscience-info@iup.edu
http://www.iup.edu/geoscience
*Enrollment (2016): B: 87 (17)*
Chair:
Steven A. Hovan, (D), Michigan, 1993, Ou
Professor:
Karen Rose Cercone, (D), Michigan, 1984, Oeg
John F. Taylor, (D), Missouri, 1984, PsiGs
Associate Professor:
Kenneth S. Coles, (D), Columbia, 1988, XgOeYg

Katie Farnsworth, (D), Virginia Inst Marine Sci, Ong
Jon C. Lewis, (D), Connecticut, Gc
Assistant Professor:
Yvonne K. Branan, (D), Michigan Tech, 2007, Gv
Nicholas Deardorff, (D), Oregon, Gv
Gregory Mount, (D), Florida Atlantic, 2014, Hy
Jonathan P. Warnock, (D), N Illinois, 2013, PemPv
Adjunct Professor:
Thomas R. Moore, (M), Univ Missouri, Eo
Emeritus:
Joseph C. Clark, (D), Stanford, 1966, Gr
Frank W. Hall, (D), Montana, 1969, Gc
Darlene S. Richardson, (D), Columbia, 1974, Gd
Connie J. Sutton, (M), Indiana (Penn), 1968, Oe

## Juniata College

**Dept of Environmental Science & Studies** (2004)
1700 Moore Street
Huntingdon, PA 16652
johnson@juniata.edu
http://www.juniata.edu/departments/environmental/

**Dept of Geology** (B) ● (2016)
1700 Moore St
Huntingdon, PA 16652
p. (814) 641-3601
johanesen@juniata.edu
http://departments.juniata.edu/geology
*Enrollment (2016): B: 26 (0)*
Dept. Chair:
Ryan Mathur, (D), Arizona, 2000, CeYgHw
Associate Professor:
Matthew G. Powell, (D), Johns Hopkins, 2005, PieGs
Assistant Professor:
Katharine Johanesen, (D), Southern California, 2011, GzxGc
Emeritus:
Laurence J. Mutti, (D), Harvard, 1978, GxzSc
J. Peter Trexler, (D), Michigan, 1964, Ps
Robert H. Washburn, (D), Columbia, 1966, Gs

## Kutztown University of Pennsylvania

**Dept of Physical Science** (B) (2017)
Boehm Science Building Room 135
Kutztown, PA 19530
p. (610) 683-4447
simpson@kutztown.edu
http://www.kutztown.edu/acad/geology
Department Secretary: Donna Moore
*Enrollment (2016): B: 14 (40)*
Chair:
Edward L. Simpson, (D), Virginia Tech, 1987, Gs
Professor:
Kurt Friehauf, (D), Stanford, 1998, EmCgGx
Sarah E. Tindall, (D), Arizona, 2000, Gc
Associate Professor:
Laura Sherrod, (D), W Michigan, 2007, YgHw
Assistant Professor:
Erin Kraal, (D), California (Santa Cruz), XgGm
Adrienne Oakley, (D), Hawaii, 2009, YrOun
Jacob Sewall, (D), California (Santa Cruz), 2004, GeOa

## La Salle University

**Dept of Geology & Environmental Science** (B) (2016)
20th and Olney Avenue
Philadelphia, PA 19141
p. (215) 951-1269
hoersch@lasalle.edu
http://www.lasalle.edu/geology
*Enrollment (2015): B: 30 (9)*
Chair:
Alice L. Hoersch, (D), Johns Hopkins, 1977, GpiOi
Adjunct Professor:
Natalie Flynn, (D), Temple, 1999, Gz
Emeritus:
Henry A. Bart, (D), Nebraska, 1974, Gs

## Lafayette College

**Dept of Geology & Environmental Geosciences** (B) (2014)
116 Van Wickle Hall
4 South College Drive
Easton, PA 18042
p. (610) 330-5193
geology@lafayette.edu
http://geology.lafayette.edu
Department Secretary: Rohana Meyerson
*Enrollment (2013): B: 57 (13)*
Professor:
Dru Germanoski, (D), Colorado State, 1989, Gm
Associate Professor:
Kira Lawrence, (D), Brown, 2006, Gg
Lawrence L. Malinconico, (D), Dartmouth, 1982, Yg
David Sunderlin, (D), Chicago, 2004, Gg
Research Associate:
Mary Ann Malinconico, (D), Columbia, 2002, Eo
Emeritus:
Richard W. Faas, (D), Iowa State, 1964, Gs
Guy L. Hovis, (D), Harvard, 1971, GzxCg
Laboratory Director:
John R. Wilson, (M), Virginia Tech, 2001, Oi

## Lehigh University

**Dept of Earth & Environmental Sciences** (B,M,D) (2014)
1 W. Packer Ave.
Bethlehem, PA 18015-3001
p. (610) 758-3660
ljc0@lehigh.edu
http://www.ees.lehigh.edu
Administrative Assistant: Laura J. Cambiotti
*Enrollment (2007): M: 15 (0) D: 12 (0)*
Chair:
Frank J. Pazzaglia, (D), Penn State, 1993, Gm
Professor:
David J. Anastasio, (D), Johns Hopkins, 1988, Gc
Gray E. Bebout, (D), California (Los Angeles), 1989, Gp
Claudio Berti, (D), Chieti, 2009, GmtOn
Edward B. Evenson, (D), Michigan, 1972, Gl
Kenneth P. Kodama, (D), Stanford, 1978, Ym
Anne S. Meltzer, (D), Rice, 1988, Ys
Dork Sahagian, (D), Chicago, 1987, PeGvr
Peter K. Zeitler, (D), Dartmouth, 1983, Cc
Associate Professor:
Robert K. Booth, (D), Wyoming, 2003, On
Benjamin S. Felzer, (D), Brown, 1995, Oa
Bruce R. Hargreaves, (D), California (Berkeley), 1977, Ob
Donald P. Morris, (D), Colorado, 1990, On
Stephen C. Peters, (D), Michigan, 2001, Cl
Joan Ramage, (D), Cornell, 2001, OrGlm
Zicheng Yu, (D), Toronto, 1997, Pe
Research Associate:
Bruce D. Idleman, (D), SUNY (Albany), 1990, Cc
Joshua Stachnik, (D), Wyoming, 2010, Ys
Emeritus:
Bobb Carson, (D), Washington, 1971, Gus
Paul B. Myers, (D), Lehigh, 1960, Hw

## Lock Haven University

**Dept of Geology & Physics** (B) (2013)
301 West Church Street
Lock Haven, PA 17745-2390
p. (570) 484-2048
mkhalequ@lhup.edu
Administrative Assistant: Barbara Greene
*Enrollment (2007): B: 30 (5)*
Professor:
Md. Khalequzzaman, (D), 1998, HwOni
Associate Scientist:
Loretta D. Dickson, (D), Connecticut, 2006, GicGe
Thomas C. Wynn, (D), Virginia Tech, 2004, GsPiEo

## Mansfield University

**Dept of Geosciences** (A,B) O (2016)
Belknap Hall

Mansfield, PA 16933
p. (570) 662-4613
jdemchak@mansfield.edu
http://geoggeol.mansfield.edu/
*Enrollment (2016): A: 10 (6) B: 124 (42)*

Chair:
Jennifer Demchak, (D), West Virginia, 2005, Hs
Associate Professor:
Christopher F. Kopf, (D), Massachusetts (Amherst), 1999, Gcp
Associate Scientist:
Lee Stocks, (D), Kent State, 2010, OyGgm
Assistant Professor:
Linda Kennedy, (D), UNC Greensboro, 2012, Oy

## Mercyhurst University

**Dept of Geology** (B,M,D) (2012)
501 East 38th Street
Erie, PA 16546
p. (814) 824-2581
rbuyce@mercyhurst.edu
http://mai.mercyhurst.edu
*Enrollment (2012): B: 16 (3)*

Director:
James M. Adovasio, (D), Utah, 1970, GafGs
Professor:
M. Raymond Buyce, (D), Rensselaer, 1975, GsaOn
Assistant Professor:
Nicholas Lang, (D), Minnesota, 2006, GvXgGc
Scott McKenzie, (B), Edinboro, 1976, PgXmOe
Lyman Perscio, (D), New Mexico, 2012, GsSpHg
Adjunct Professor:
Frank Vento, (D), Pittsburgh, 1985, GaOsGm

## Millersville University

**Dept of Earth Sciences** (B,M) (2013)
PO Box 1002
Millersville, PA 17551
p. (717) 872-3289
esci@millersville.edu
http://www.millersville.edu/esci
*Enrollment (2010): B: 23 (4)*

Chair:
Richard D. Clark, (D), Wyoming, 1987, Oa
Professor:
Alex J. DeCaria, (D), Maryland, 2000, Oa
L. Lynn Marquez, (D), Northwestern, 1998, Cg
Sepideh Yalda, (D), St. Louis, 1997, Oa
Associate Professor:
Sam Earman, (D), New Mexico Tech, 2004, HwCsl
Ajoy Kumar, (D), Old Dominion, 1996, Opr
Jason R. Price, (D), Michigan State, 2003, Gs
Todd D. Sikora, (D), Penn State, 1996, Oa
Assistant Professor:
Robert Vaillancourt, (D), Rhode Island, 1996, Obc
Instructor:
Joseph Calhoun, (B), Penn State, Oa
Mary Ann Schlegel, (M), MIT/WHOI, 1998, Og
Professor:
Robert S. Ross, (D), Florida State, 1977, Oa
Emeritus:
William M. Jordan, (D), Wisconsin, 1965, Gs
Bernard L. Oostdam, (D), Delaware, 1971, Ou
Charles K. Scharnberger, (D), Washington (St. Louis), 1971, Gc

## Montgomery County Community College

**Dept of Science, Technology, Engineering, and Math** (A) (2010)
Blue Bell, PA 19422
p. (215) 641-6446
rkuhlman@mc3.edu

Professor:
Robert Kuhlman, (M), Bryn Mawr, 1975, Gg
Instructor:
George Buchanan, (M), Drexel, 1992, Ng
Adjunct Professor:
Laurie Martin-Vermilyea, (D), South Carolina, 1992, Oe
Frank Roberts, (D), Bryn Mawr, 1969, Gp
Kelly C. Spangler, (M), Drexel, 2003
Anthony Stevens, (M), Florida, 1981, Ge

## Moravian College

**Dept of Physics & Earth Science** (B) (2016)
1200 Main Street
Bethlehem, PA 18018-6650
p. (610) 861-1437
krieblek@moravian.edu
http://www.physics.moravian.edu
Department Secretary: Lou Ann Vlahovic
*Enrollment (2016): B: 1 (0)*

Chair:
Kelly Krieble, (D), Lehigh, 1993, OnnOn
Assistant Professor:
Erik Larson, (D), Mississippi State, 2010, GmsGg

## Pennsylvania State University, Erie

**Geoscience Dept** (2011)
Erie, PA 16510
p. (814) 898-6277
amf11@psu.edu
http://www.personal.psu.edu/faculty/a/m/amf11/

Chair:
Anthony M. Foyle
Assistant Professor:
Eva Tucker, (M), Cincinnati, 1962, Gg

## Pennsylvania State University, Monaca

**Dept of Geosciences** (B) (2011)
100 University Drive
Monaca, PA 15061
p. (412) 773-3867
jac7@psu.edu
http://www.br.psu.edu/default.htm
*Enrollment (1998): B: 15 (0)*

Assistant Professor:
John A. Ciciarelli, (D), Penn State, 1971, Gg

## Pennsylvania State University, University Park

**Dept of Geosciences** (B,M,D) ● (2016)
503 Deike Building
University Park, PA 16802-2714
p. (814) 865-6711
lkump@psu.edu
http://www.geosc.psu.edu/
Administrative Assistant: Tina Vancas
*Enrollment (2016): B: 122 (16) M: 31 (10) D: 54 (5)*

Head:
Lee R. Kump, (D), S Florida, 1986, ClPeCm
Evan Pugh Professor:
Richard B. Alley, (D), Wisconsin, 1987, Gl
James F. Kasting, (D), Michigan, 1979, Oa
Distinguished Professor:
Katherine H. Freeman, (D), Indiana, 1991, Cos
Director, Earth & Mineral Science Museum:
Russell W. Graham, (D), Texas, 1976, Pv
Director, Earth & Env Systems Inst:
Susan L. Brantley, (D), Princeton, 1987, Cg
Director, Astrobiology Research Center:
Christopher H. House, (D), California (Los Angeles), 1999, Py
Associate Head, Undergraduate Program:
Peter J. Heaney, (D), Johns Hopkins, 1989, Gz
Associate Head, Graduate Programs:
Demian M. Saffer, (D), California (Santa Cruz), 1999, Hw
Professor:
Charles J. Ammon, (D), Penn State, 1991, Ys
Sridhar Anandakrishnan, (D), Wisconsin, 1990, Ys
Michael A. Arthur, (D), Princeton, 1979, OuClGs
David M. Bice, (D), California (Berkeley), 1989, Gg
Timothy J. Bralower, (D), California (San Diego), 1986, Pe
Terry Engelder, (D), Texas A&M, 1973, Nr
Donald M. Fisher, (D), Brown, 1988, GctOm
Kevin P. Furlong, (D), Utah, 1981, Gt
Tanya Furman, (D), MIT, 1989, Cg

Klaus Keller, (D), Princeton, 2000, Og
Michael E. Mann, (D), Yale, 1998, Oa
Chris Marone, (D), Columbia, 1988, Yx
Andrew A. Nyblade, (D), Michigan, 1992, Yg
Mark E. Patzkowsky, (D), Chicago, 1992, Po
Peter D. Wilf, (D), Penn, 1998, Py
Senior Scientist:
Todd Sowers, (D), Rhode Island, 1991, Cs
Associate Professor:
Matthew S. Fantle, (D), California (Berkeley), 2005, Cs
Peter C. LaFemina, (D), Miami, 2005, Yd
Jennifer L. Macalady, (D), California (Davis), 2000, PyClSb
Eliza Richardson, (D), MIT, 2002, YsOe
Assistant Professor:
Roman DiBiase, (D), Arizona State, 2011, Gmt
Maureen D. Feineman, (D), California (Berkeley), 2004, Cp
Bradford Foley, (D), Yale, 2014, YgXyGt
Elizabeth Hajek, (D), Wyoming, 2009, Gs
Tess A. Russo, (D), California (Santa Cruz), 2012, Hw
Christelle Wauthier, (D), Leige, 2011, On
Emeritus:
Shelton S. Alexander, (D), Caltech, 1963, Ys
Hubert L. Barnes, (D), Columbia, 1958, CgEmCe
Roger J. Cuffey, (D), Indiana, 1966, PiePv
David H. Eggler, (D), Colorado, 1967, CpGiv
David (Duff) P. Gold, (D), McGill, 1963, GcgGx
Earl K. Graham, Jr., (D), Penn State, 1969, Yx
Roy J. Greenfield, (D), MIT, 1965, Yg
Albert L. Guber, (D), Illinois, 1962, Pe
Benjamin F. Howell, Jr., (D), Caltech, 1949, Ys
Derrill M. Kerrick, (D), California (Berkeley), 1968, Gp
Hiroshi Ohmoto, (D), Princeton, 1969, Cs
Richard R. Parizek, (D), Illinois, 1961, Hw
Arthur W. Rose, (D), Caltech, 1958, Cge
Rudy L. Slingerland, (D), Penn State, 1977, Gs
Barry Voight, (D), Columbia, 1965, GveGg
William B. White, (D), Penn State, 1962, CgGzy

**Dept of Meteorology** (B,M,D) O (2016)
503 Walker Building
University Park, PA 16802-5013
p. (814) 865-0478
meteodept@meteo.psu.edu
http://www.met.psu.edu
Professor and Head:
David J. Stensrud, (D), Penn State, 1992, Ow
Professor:
Peter R. Bannon, (D), Colorado, 1979, Ow
William H. Brune, (D), Johns Hopkins, 1978, Ow
Eugene E. Clothiaux, (D), Brown, Ow
Kenneth J. Davis, (D), Colorado, 1992, Ow
Jenni L. Evans, (D), Monash, 1990, Ow
Jerry Y. Harrington, (D), Colorado State, 1997, Ow
Gregory S. Jenkins, (D), Ow
James F. Kasting, (D), Michigan, 1979, Ow
Johannes Verlinde, (D), Colorado State, 1992, Ow
George S. Young, (D), Colorado State, Ow
Associate Head:
Hampton N. Shirer, (D), Penn State, 1978, Ow
Associate Professor:
Raymond G. Najjar, (D), Princeton, 1990, Ow
Assistant Professor:
Steven J. Greybush, (D), Maryland, 2011, Ow
Matthew R. Kumjian, (D), Oklahoma, 2012, Ow
Sukyoung Lee, (D), Princeton, 1991, Ow
Research Assistant:
William F. Ryan, (M), Maryland, 1990, Ow
Director, Meteorology Computing:
Charles Pavloski, (D), Penn State, 2004, Ow
Research Associate:
Aijun Deng, (D), Penn State, 1999, Ow
Arthur Person, (M), Penn State, 1983, Ow
William Syrett, (M), Penn State, 1987, Ow
Instructor:
Frederick J. Gadomski, (M), Penn State, 1983, Ow
Paul Knight, (M), Penn State, 1977, Ow

Distinguished Professor:
J. Michael Fritsch, (D), Colorado State, 1978, Ow
Associate Research Professor:
David R. Stauffer, (D), Penn State, 1990, Ow
Emeritus:
Craig F. Bohren, (D), Arizona, 1975, Ow
John J. Cahir, (D), Penn State, 1971, Ow
Toby N. Carlson, (D), Imperial Coll (UK), 1965, Ow
John H. E. Clark, (D), Florida State, 1969, Ow
John A. Dutton, (D), Wisconsin, Ow
William M. Frank, (D), Colorado State, 1976, Ow
Alistair B. Fraser, (D), Imperial Coll (UK), 1968, Ow
Charles L. Hosler, (D), Penn State, 1951, Ow
Dennis Lamb, (D), Washington, 1970, Ow
Nelson L. Seaman, (D), Penn State, 1977, Ow
Dennis W. Thomson, (D), Wisconsin, 1968, Ow
John C. Wyngaard, (D), Penn State, 1967, Ow

**Dept of Plant Science** (A,B,M,D) (2014)
119 Tyson Building
University Park, PA 16802
p. (814) 865-6541
rpm12@psu.edu
http://plantscience.psu.edu/
Professor:
Douglas B. Beegle, (D), Penn State, 1983, ScOs
Jean-Marc Bollag, (D), Basel, 1959, Sb
Edward J. Ciolkosz, (D), Wisconsin, 1967, Sd
Daniel D. Fritton, (D), Iowa State, 1968, Sp
Sridhar Komarneni, (D), Wisconsin, 1973, Sc
Gary W. Petersen, (D), Wisconsin, 1965, Sd
Associate Professor:
Peter J. Landschoot, (D), Rhode Island, 1988, So
Gregory W. Roth, (D), Penn State, 1987, So
Assistant Professor:
Rick L. Day, (D), Penn State, 1991, Ou
Research Associate:
Barry M. Evans, (M), Penn State, 1977, Sd
Adjunct Professor:
Andrew S. Rogowski, (D), Iowa State, 1964, Sp
Lawrence A. Schardt, (M), Penn State, 1996, Sd

**Earth and Mineral Sciences Museum & Art Gallery** (2015)
116 Deike Building
University Park, PA 16802
p. (814) 865-6336
museum@ems.psu.edu
http://www.ems.psu.edu/outreach/museum
Director:
Russell W. Graham, (D), Texas (Austin), 1976, PveOe

**John and Willie Leone Family Dept of Energy and Mineral Engineering** (B,M,D) (2014)
110 Hosler Building
University Park, PA 16802
p. (814) 865-3437
eme@ems.psu.edu
http://www.eme.psu.edu/mnge
*Enrollment (2000): B: 16 (6) M: 6 (2)*
Program Chair, and Deike Endowed Chair in Mining Engineering:
Jeffery L. Kohler, (D), Penn State, 1982, Nm
Associate Professor:
Antonio Nieto, (D), Colorado Mines, 2001, NmOeGq
Jamal Rostami, (D), Colorado Mines, 1997, Nm
Assistant Professor:
Shimin Liu, (D), S Illinois, 2013, Nm

## Point Park University

**Dept of Environmental Studies** (A,B) (2015)
201 Wood Street
Pittsburgh, PA 15222-1994
p. (412) 392-3900
jkudlac@pointpark.edu
http://www.pointpark.edu/Academics/Schools/SchoolofArtsandSciences/Departments/NaturalSciencesandEngineeringTechnology
Administrative Assistant: Roberta T. Gallick

Head:
Mark O. Farrell, (D), Carnegie Mellon, 1978, On
Professor:
John J. Kudlac, (D), Pittsburgh, Ng

Pennsylvania

## Shippensburg University

**Geography-Earth Science Dept** (B,M) (2016)
1871 Old Main Drive
Shippensburg, PA 17257
p. (717) 477-1685
tlmyers@ship.edu
http://www.ship.edu/~geog/
Administrative Assistant: Tammy Myers

Chair:
William L. Blewett, (D), Michigan State, 1991, Oy
Professor:
Scott A. Drzyzga, (D), Michigan State, 2007, OiuOl
Thomas P. Feeney, (D), Georgia, 1997, Gm
Kurtis G. Fuellhart, (D), Penn State, 1998, On
Timothy W. Hawkins, (D), Arizona State, 2004, Oa
Claire A. Jantz, (D), Maryland, 2005, Oui
Paul G. Marr, (D), Denver, 1996, Oi
George M. Pomeroy, (D), Akron, 1999, Ou
Janet S. Smith, (D), Georgia, 1999, Oi
Christopher J. Woltemade, (D), Wisconsin, 1993, HgGmHs
Associate Professor:
Michael T. Applegarth, (D), Arizona State, 2001, Or
Sean Cornell, (D), Cincinnati, 2008, Gg
Alison E. Feeney, (D), Michigan State, 1998, On
Kay R. Williams, (D), Georgia, 1995, Oa
Joseph T. Zume, (D), Oklahoma, 2007, HgYg

## Slippery Rock University

**Dept of Geography, Geology, and the Environment** (B) (2011)
Slippery Rock, PA 16057
p. (724) 738-2048
jack.livingston@sru.edu
http://academics.sru.edu/gge/
Administrative Assistant: Bonita L. Vinton
*Enrollment (2006): B: 50 (9)*

Chair:
Jack Livingston
Professor:
Tamra A. Schiappa, (D), Idaho, 1999, PiGrOe
Michael J. Zieg, (D), Johns Hopkins, 2001, Gi
Associate Professor:
Patrick A. Burkhart, (D), Lehigh, 1994, Hg
Assistant Professor:
Patricia A. Campbell, (D), Pittsburgh, 1994, Gc
Xianfeng Chen, (D), West Virginia, 2005, Or
Julie A. Snow, (D), Rhode Island, 2002, Oa
Michael G. Stapleton, (D), Delaware, 1995, So

## State Museum of Pennsylvania

**Section of Paleontology & Geology** (2014)
300 North Street
Harrisburg, PA 17120-0024
p. (717) 783-9897
c-sjasinsk@pa.gov
http://www.statemuseumpa.org/geologyc.html
f: https://www.facebook.com/StateMuseumofPA

Acting Curator:
Steven E. Jasinski, (D), Pennsylvania, pend, PvgPy

## Susquehanna University

**Dept of Earth & Environmental Sciences** (B) (2010)
514 University Ave
Selinsgrove, PA 17870
p. (570) 372-4216
straubk@susqu.edu
http://www.susqu.edu/ees
*Enrollment (2009): B: 29 (7)*

Chair:
Jennifer M. Elick, (D), Tennessee, 1999, Pe
Katherine H. Straub, (D), Colorado State, 2002, Ow
Associate Professor:
Daniel E. Ressler, (D), Iowa State, 1998, Sp
Derek J. Straub, (D), Oa
Assistant Professor:
Ahmed Lachhab, (D), Iowa, 2006, Hw

## Temple University

**Earth & Environmental Science** (B,M,D) (2014)
1901 N. 13th Street
Beury Hall, Rm. 326
Philadelphia, PA 19122-6081
p. (215) 204-8227
scox@temple.edu
http://www.temple.edu/geology
Administrative Assistant: Shelah Cox
*Enrollment (2005): M: 11 (0)*

Chair:
David E. Grandstaff, (D), Princeton, 1974, Cl
(Emeritus 2014):
George H. Myer, (D), Yale, 1965, Gz
Professor:
Jonathan Nyquist, (D), Wisconsin, 1986, Yg
Laura Toran, (D), Wisconsin, 1986, Hw
Associate Professor:
Ilya Buynevich, (D), Boston Univ, 2001, GsOuGm
Alexandra Davatzes, (D), Stanford, 2007, GsXg
Nicholas Davatzes, (D), Stanford, 2003, GcNrOn
Dennis O. Terry, (D), Nebraska, 1998, Gr
Assistant Professor:
Bojeong Kim, (D)
Sujith Ravi, (D)
Emeritus:
Gene C. Ulmer, (D), Penn State, 1964, Cp
Asst. Lab Manager/Bldg. Coordinator:
Donald Deigh-Kai, (M), Temple, 2008

## Thiel College

**Dept of Environmental Science** (B) (2010)
Greenville, PA 16125
p. (412) 589-2821
areinsel@thiel.edu
*Enrollment (2004): B: 23 (3)*

Professor:
James H. Barton, (D), N Colorado, 1977, Oy

## University of Pennsylvania

**Dept of Earth & Environmental Science** (B,M,D) (2015)
240 S. 33rd Street
Philadelphia, PA 19104-6316
p. (215) 898-5724
earth@sas.upenn.edu
http://www.sas.upenn.edu/earth/

Chair:
Reto Giere, (D), ETH (Switzerland), 1990, GzCgGe
Professor:
Peter Dodson, (D), Yale, 1974, Pv
Hermann W. Pfefferkorn, (D), Muenster, 1968, PbePs
Associate Professor:
David Goldsby, (D)
Douglas Jerolmack, (D), MIT, 2006, GmYxHq
Stephen P. Phipps, (D), Princeton, 1984, Gc
Alain F. Plante, (D), Alberta, 2001, SbOsCo
Assistant Professor:
Irina Marinov
Lauren Sallan, (D)
Jane Willenbring, (D), Dalhousie, 2006, GmCcOl
Lecturer:
Edward L. Doheny, (D), Indiana, 1967, Ng
Stanley L. Laskowski, (M), Drexel, 1973, Ge
Willig B. Sarah, (D), Pennsylvania, 1988, Ge
Emeritus:
Robert F. Giegengack, Jr., (D), Yale, 1968, Gg
Arthur H. Johnson, (D), Cornell, 1975, So
Labratory Manager:
David R. Vann, (D), Pennsylvania, 1993, GeSfXm

Teaching Faculty:
Jane Dmochowski, (D), Caltech, 2004, Yg
Gomaa I. Omar, (D), Pennsylvania, 1985, Cc
Graduate Group Coordinator:
Joan Buccilli
Director, Professional Masters Programs:
Yvette Bordeaux, (D), Pennsylvania, 2000, Po
Department Administrator:
Arlene Mand, (B), 1971, On
Associate Director:
Maria Andrews, (M)

## University of Pittsburgh

**Dept of Geology & Environmental Science** (B,M,D) ● (2017)
200 SRCC Building
4107 O'Hara Street
Pittsburgh, PA 15260-3332
p. (412) 624-8780
gpsgrad@pitt.edu
http://www.geology.pitt.edu
*Enrollment (2015): M: 15 (0) D: 24 (0)*
Chair:
Mark B. Abbott, (D), Minnesota, 1995, Gs
Associate Professor:
Rosemary C. Capo, (D), California (Los Angeles), 1990, Cl
William P. Harbert, (D), Stanford, 1987, Ym
Michael S. Ramsey, (D), Arizona State, 1996, Or
Brian W. Stewart, (D), California (Los Angeles), 1990, CclGe
Josef Werne, (D), Northwestern, 2000, Co
Assistant Professor:
Daniel J. Bain, (D), Johns Hopkins, 2004, HgGmOu
Sergio Contreras, (D), Universidad de Concepcion, 2008, GnCos
Emily M. Elliott, (D), Johns Hopkins, 2003, Hg
Nadine McQuarrie, (D), Arizona, 2001, Gct
Environmental Reporter, Pittsburgh Post-Gazette:
S. Don Hopey
Instructor:
Emily Collins
Marion Divers, (D), Pittsburgh, 2013
Lecturer:
R. Ward Allebach
Mark Collins, (M), Pittsburgh, 1985, On
Charles E. Jones, (D), Oxford, 1992, Gg
Adjunct Professor:
Robert S. Hedin, (D), Rutgers, 1987, Ge
Matthew C. Lamanna, (D), Pennsylvania, 2004, Pv
Visiting Professor:
J. Brian Balta
Emeritus:
Thomas Anderson
William Cassidy, (D), 1969, Xg
Jack Donahue
Bruce W. Hapke, (D), Cornell, 1962, Xy
Edward G. Lidiak, (D), Rice, 1963, Gi
Harold B. Rollins, (D), Columbia, 1967, Pe
Lecturer:
Steven C. Latta
Other:
John S. Pallister, (D), California (Santa Barbara), 1980, GviGg
Matthew Watson

## University of Pittsburgh, Bradford

**Dept of Petroleum Technology** (A) (2012)
300 Campus Drive
Bradford, PA 16701-2898
p. (814) 362-7569
aap@pitt.edu
http://www.upb.pitt.edu/academics/petroleumtechnology.aspx
Administrative Assistant: Janet Shade
*Enrollment (2010): A: 28 (8)*
Program Director:
Assad I. Panah, (D), Oklahoma, 1966, GorGc

## West Chester University

**Dept of Earth and Space Sciences** (B,M) (2014)
750 South Church Street
Merion Science Center
West Chester, PA 19383
p. (610) 436-2727
mhelmke@wcupa.edu
http://geology.wcupa.edu
*Enrollment (2014): B: 106 (23) M: 29 (9)*
Professor:
Richard M. Busch, (D), Pittsburgh, 1984, Oe
Marc R. Gagne, (D), Georgia, 1994, On
Steven C. Good, (D), Colorado, 1993, Pe
Timothy M. Lutz, (D), Pennsylvania, 1979, Gq
LeeAnn Srogi, (D), Pennsylvania, 1988, Gp
Chair:
Martin F. Helmke, (D), Iowa State, 2003, HwSdOg
Associate Professor:
Cynthia G. Fisher, (D), Colorado, 1991, Ou
Joby Hilliker, (D), Penn State, 2001, Ow
Karen M. Schwarz, (D), Arizona State, 1997, On
Arthur Smith, (D), Pennsylvania, 1975, Oe
Assistant Professor:
Howell Bosbyshell, (D), Bryn Mawr, 2000, Gc
Cynthia V. Hall, (D), Georgia Tech, 2008, Cg
Daria L. Nikitina, (D), Delaware, 2000, Gm

## Wilkes University

**Dept of Environmental Engineering & Earth Sciences** (B) (2016)
84 West South Street
Wilkes-Barre, PA 18766
p. (570) 408-4610
sid.halsor@wilkes.edu
http://wilkes.edu/academics/colleges/science-and-engineering/environmental-engineering-earth-sciences/
*Enrollment (2016): B: 25 (6)*
Chair:
Sid P. Halsor, (D), Michigan Tech, 1989, Giv
Professor:
Dale A. Bruns, (D), Idaho State, 1981, OirSf
Kenneth M. Klemow, (D), SUNY (Syracuse), 1982, Py
Prahlad N. Murthy, (D), Texas A&M, 1993, OanOn
Brian T. Redmond, (D), Rensselaer, 1982, Gs
Michael A. Steele, (D), Wake Forest, 1988, Py
Marleen Troy, (D), Drexel, 1989, Hw
Brian E. Whitman, (D), Michigan Tech, 1998, HqwSb
Associate Professor:
Holly Frederick, (D), Penn State, 1999, Sdb
Assistant Professor:
Matthew S. Finkenbinder, (D), Pittsburgh, 2015, GsmGl
Lecturer:
Mark A. Kaster, (M), Saint Louis, 1993, Ow
Julie McMonagle, (M), Lehigh, 1991, Gge
Emeritus:
James M. Case, (D), Dalhousie, 1979, Ob

## York College of Pennsylvania

**Dept of Physical Science** (2010)
York, PA 17405
p. (717) 846-7788 x333
jforesma@ycp.edu
Assistant Professor:
William (Bill) Kreiger, (D), Penn State, 1976, GisOg
Adjunct Professor:
Ralph Eisenhart, (M), Penn State, 1994, Gg
Jeri L. Jones, (B), Catawba, 1977, Ga

# Puerto Rico

## University of Puerto Rico

**Dept of Geology** (B,M) ○ (2017)
PO Box 9000
Mayaguez, PR 00681-9000
p. (787) 265-3845
lizzette.rodriguez1@upr.edu
http://geology.uprm.edu/
*Enrollment (2016): B: 150 (0) M: 24 (0)*
Professor:
Fernando Gilbes, (D), S Florida, 1996, OrGu

James Joyce, (D), Northwestern, 1985, NgGcp
Wilson R. Ramirez, (D), Tulane, 2000, GudOu
Hernan Santos, (D), Colorado, 1999, PiGdr

Director:
Lizzette A. Rodriguez, (D), Michigan Tech, 2007, GviOr

Associate Professor:
Lysa Chizmadia, (D), New Mexico, 2004, GzXm
Alberto Lopez, (D), Northwestern Univ, 2006, YdGtYs

Assistant Professor:
Eugenio Asencio, (D), South Carolina, 2002, YseGt
Thomas Hudgins, (D), Michigan, 2014, GiCpGv
Kenneth Stephen Hughes, (D), North Carolina State, 2014, Gcm
Elizabeth Vanacore, (D), Rice Univ, 2008, Yxs



# Rhode Island

## Brown University

**Dept of Earth, Environmental and Planetary Sciences** (B,M,D) (2016)
Box 1846, 324 Brook Street
Providence, RI 02912
p. (401) 863-3339
DEEPS@brown.edu
http://www.brown.edu/academics/earth-environmental-planetary-sciences/
f: https://www.facebook.com/BrownGeologicalSciences
t: @BrownGeoSci
*Enrollment (2015): B: 44 (23) M: 0 (6) D: 51 (12)*

Chair:
Greg Hirth, (D), Brown, 1991, GcyYg

Professor:
Reid F. Cooper, (D), Cornell, 1983, Gy
Karen M. Fischer, (D), MIT, 1988, Ys
Donald W. Forsyth, (D), MIT/WHOI, 1974, YrsGt
James W. Head, III, (D), Brown, 1969, XgGvt
Timothy D. Herbert, (D), Princeton, 1987, Pe
Yongsong Huang, (D), Bristol (UK), 1997, Co
Yan Liang, (D), Chicago, 1994, Cp
Amanda H. Lynch, (D), Melbourne, 1993, OaGe
John F. Mustard, (D), Brown, 1990, Or
E. Marc Parmentier, (D), Cornell, 1975, Yg
Alberto E. Saal, (D), MIT/WHOI, 2000, Cg

Associate Professor of Research:
Steven C. Clemens, (D), Brown, 1990, Ou

Associate Professor:
Baylor Fox-Kemper, (D), MIT, 2003, Op
Meredith Hastings, (D), Princeton, 2004, Oa
Stephen Parman, (D), MIT, 2001, CpgCt
James M. Russell, (D), Minnesota, 2004, Gn

Assistant Professor:
Colleen Dalton, (D), Harvard, 2007, Ys
Christian Huber, (D), Berkeley, 2009, Gv
Brandon C. Johnson, (D), Purdue, 2013, Xg
Jung-Eun Lee, (D), California (Berkeley), 2005
Ralph E. Milliken, (D), Brown, 2006

Research Associate:
David Murray, (D), Oregon State, 1987, Ou

Adjunct Professor:
Mark Altabet, (D), Oc
Maureen Conte, (D), Columbia, 1989, CgPe

Emeritus:
L. Peter Gromet, (D), Caltech, 1979, Cc
John F. Hermance, (D), Toronto, 1967, HqOrHw
Paul C. Hess, (D), Harvard, 1968, Gi
Carle M. Pieters, (D), MIT, 1977, Or
Warren L. Prell, (D), Columbia, 1974, Ou
Malcolm J. Rutherford, (D), Johns Hopkins, 1968, Cp
Peter H. Schultz, (D), Texas, 1972, Xg
Jan A. Tullis, (D), California (Los Angeles), 1971, Gcy
Terry E. Tullis, (D), California (Los Angeles), 1971, Yx
Thompson Webb, III, (D), Wisconsin, 1971, PelOa

Department Manager:
Nancy Fjeldheim, (B), Ohio State, 1972, On

Academic Program Manager:
Patricia M. Davey, (B), Rhode Island Coll, 1986, On

## Community College of Rhode Island

**Dept of Physics (Geology & Oceanography Div)** (A) (2016)
400 East Avenue
Warwick, RI 02886
p. (401) 333-7443
pbanerjee@ccri.edu
http://www.ccri.edu/physics/

Professor:
Karen Kortz, (D), Rhode Island, 2009, OeGg

Associate Professor:
Emily Burns, (D), Rhode Island, GgOgi
Paul White, (D), Gg

Assistant Professor:
Duayne Rieger, (D), Ys

## Providence College

**Biology Dept** (A) (2004)
Providence, RI 02918
p. (401) 865-2250
chwood@providence.edu

Professor:
Craig B. Wood, (D), Harvard, 1992, Pv

Associate Professor:
Michael S. Zavada, (D), Connecticut, 1982, On

## Roger Williams University

**College of Arts & Sciences** (B) (2011)
Bristol, RI 02809
p. (401) 254-3087
jborden@rwu.edu
Department Secretary: Valerie Catalano

Head:
Mark D. Gould, (D), Rhode Island, 1973, Ob

Chair:
Paul Webb

Professor:
Thomas Doty, (D), Rhode Island, 1977, Ob
Richard Heavers, (D), Rhode Island, 1977, Op
Thomas J. Holstein, (D), Brown, 1969, On
Martine Villalard-Bohnsack, (D), Rhode Island, 1971, Ob

Assistant Professor:
Tim Scott, (D), SUNY (Stony Brook), 1993, Ob

## University of Rhode Island

**Dept of Geosciences** (B,M,D) (2015)
9 East Alumni Ave.,
Kingston, RI 02881
p. (401) 874-2265
http://www.uri.edu/cels/geo/
Administrative Assistant: Lorraine Bailey
*Enrollment (2015): B: 75 (10) M: 14 (9) D: 1 (2)*

Professor & Chair:
David E. Fastovsky, (D), Wisconsin, 1986, PvGsPs

Associate Dean:
Anne I. Veeger, (D), Arizona, 1991, HwCl

Professor:
Thomas B. Boving, (D), Arizona, 1999, Hw

Associate Professor:
Brian K. Savage, (D), Caltech, 2004, Ys

Assistant Professor:
Dawn Cardace, (D), Washington Univ, GpPy
Simon E. Engelhart, (D), Pennsylvania, OnGmYs
Soni M. Pradhanang, (D), HsqHq

Emeritus:
J. Allan Cain, (D), Northwestern, 1962, Gx
O Don Hermes, (D), North Carolina, 1967, Gi
Daniel P. Murray, (D), Brown, 1976, GpOe

**Graduate School of Oceanography** (M,D) (2014)
215 South Ferry Road
Narragansett, RI 02882
p. (401) 874-6222
TheDean@gso.uri.edu
http://www.gso.uri.edu
*Enrollment (2010): M: 41 (12) D: 31 (8)*

Research Professor:
Theodore J. Smayda, (D), Oslo, 1967, Ob
Associate Dean:
Mark Wimbush, (D), California (San Diego), 1969, Op
Professor:
Robert D. Ballard, (D), Rhode Island, 1974, Ga
Steven N. Carey, (D), Rhode Island, 1982, Ou
Jeremy S. Collie, (D), MIT/WHOI, 1985, Ob
Peter Cornillon, (D), Cornell, 1973, Op
Steven L. D'Hondt, (D), Princeton, 1989, Ou
Edward G. Durbin, (D), Rhode Island, 1976, Ob
Isaac Ginis, (D), Inst Exp Meteor, 1986, Op
Tetsu Hara, (D), MIT, 1990, Op
Paul E. Hargraves, (D), William & Mary, 1968, Ob
David L. Hebert, (D), Dalhousie, 1988, Op
Christopher Kincaid, (D), Johns Hopkins, 1990, Ou
John King, (D), Minnesota, 1983, Ou
Roger Larson, (D), California (San Diego), 1970, Ou
Margaret Leinen, (D), Rhode Island, 1980, Ou
John T. Merrill, (D), Colorado, 1976, Oc
S. Bradley Moran, (D), Dalhousie, 1991, Oc
Scott W. Nixon, (D), North Carolina, 1970, Ob
Candace Oviatt, (D), Rhode Island, 1967, Ob
Hans Thomas Rossby, (D), MIT, 1966, Op
Lewis Rothstein, (D), Hawaii, 1983, Op
Haraldur Sigurdsson, (D), Durham (UK), 1970, Ou
Jennifer Specker, (D), Oregon State, 1980, Ob
Robert Tyce, (D), California (San Diego), 1976, Ou
D. Randolph Watts, (D), Cornell, 1973, Op
Karen Wishner, (D), California (San Diego), 1979, Ob
Associate Professor:
Brian G. Heikes, (D), Michigan, 1984, Oa
Yang Shen, (D), Brown, 1994, Ou
David C. Smith, (D), California (San Diego), 1994, Ob
Marine Research Scientist:
Percy Donaghay, (D), Oregon State, 1980, Ob
Kathleen Donohue, (D), Rhode Island, 1996, Op
Alfred K. Hanson, Jr., (D), Rhode Island, 1981, Oc
Robert D. Kenney, (D), Rhode Island, 1984, Ob
Barbara K. Sullivan-Watts, (D), Oregon State, 1977, Og
Associate Dean:
John Farrell, (D), Brown, 1991, Ou
Adjunct Professor:
Lawrence J. Buckley, (D), New Hampshire, 1975, Oc
Richard J. Pruell, (D), Rhode Island, 1984, Co
Charles T. Roman, (D), Delaware, 1981, Ob
Marine Research Scientist:
Dian J. Gifford, (D), Dalhousie, 1986, Ob
Emeritus:
H. Perry Jeffries, (D), Rutgers, 1959, Ob
John A. Knauss, (D), California, 1959, Op
Theodore A. Napora, (D), Yale, 1964, Ob
Michael E. Pilson, (D), California (San Diego), 1964, Oc
James G. Quinn, (D), Connecticut, 1967, Oc
Kenneth A. Rahn, (D), Michigan, 1971, Oc
Saul B. Saila, (D), Cornell, 1952, Ob
Jean-Guy Schilling, (D), MIT, 1966, Ou
John M. Sieburth, (D), Minnesota, 1954, Ob
Elijah V. Swift, (D), Johns Hopkins, 1967, Ob

# South Carolina

## Clemson University

**Bob Campbell Geology Museum** (A,B,M,D) (2014)
140 Discovery Lane
Clemson, SC 29634-0130
p. (864) 656-4602
tsteadm@clemson.edu
http://www.clemson.edu/geomuseum
Director:
Todd A. Steadman, (M), Louisiana State, 1987, OnnOn
Curator:
David J. Cicimurri, (M), SD Mines, 1998, Pv
Curator:
Christian M. Cicimurri, (M), SD Mines, 1999, Pv

**Environmental Engineering and Earth Sciences** (B,M) (2014)
321 Calhoun Drive
Room 445 Brackett Hall
Clemson, SC 29634-0919
p. (864) 656-3438
bcowans@clemson.edu
http://www.clemson.edu/ces/departments/eees/
Administrative Assistant: Cynthia Rae Gravely
*Enrollment (2013): B: 40 (9) M: 8 (6)*
Chair:
Tanju Karanfil, (D), Michigan, 1995, NgOnn
Professor:
James W. Castle, (D), Illinois, 1978, GseHw
Ronald W. Falta, (D), California (Berkeley), 1990, Hq
Cindy M. Lee, (D), Colorado Mines, 1990, GeClOe
Lawrence C. Murdoch, (D), Cincinnati, 1991, Hw
Mark Schlautman, (D), Caltech, 1992, ClHgSc
Assistant Professor:
Stephen M.J Moysey, (D), Stanford, 2005, Hw
Brian A. Powell, (D), Clemson, 2004, Cg
Lindsay C. Shuller-Nickles, (D), Michigan, 2010, Gz
Research Associate:
Scott E. Brame, (M), Clemson, 1993, Hw
Lecturer:
Alan B. Coulson, (D), South Carolina, 2009, PoGgCs
Adjunct Professor:
C. Brannon Andersen, (D), Syracuse, 1994, Cl
Christian M. Cicimurri, (M), South Dakota, 1999, Pv
Brian Looney, (D), Minnesota, 1984, Hq
Vaneaton Price, (D), North Carolina, 1969, Ce
Tommy Temples, (D), South Carolina, 1996, GoYg
Emeritus:
Lois B. Krause, (D), Clemson, 1996, Oe
Fred Molz, (D), Stanford, 1970, Sp
John R. Wagner, (D), South Carolina, 1993, Oe
Richard D. Warner, (D), Stanford, 1971, Gz

## College of Charleston

**Dept of Geology & Environmental Geosciences** (B,M) (2017)
66 George Street
Charleston, SC 29424
p. (843) 953-5589
callahant@cofc.edu
http://geology.cofc.edu/
f: https://www.facebook.com/Geology.CofC/
Administrative Assistant: Stacey K. Hassard
*Enrollment (2016): B: 130 (32)*
Chair:
Timothy J. Callahan, (D), New Mexico Tech, 2001, Hw
Associate Professor:
Erin K. Beutel, (D), Northwestern, 2000, Gct
Mitchell W. Colgan, (D), California (Santa Cruz), 1990, PeGe
Scott Harris, (D), Delaware, 1990, GaOn
Steven C. Jaumé, (D), Columbia, 1994, Ys
Norman S. Levine, (D), Purdue, 1995, OiGeNg
Cassandra R. Runyon, (D), Hawaii, 1988, XgOse
Leslie R. Sautter, (D), South Carolina, 1990, Ob
Vijay M. Vulava, (D), Swiss Fed Inst Tech, 1998, CgHwSc
Assistant Professor:
K. Adem Ali, (D), Kent State, 2011, OrHwNg
Barbara Beckingham, (D), Maryland, 2011, Cl
John Chadwick, (D), Florida, 2002, GiOrGt
Emeritus:
James L. Carew, (D), Texas (Austin), 1978, PoGd
Michael P. Katuna, (D), North Carolina, 1974, Gus
Robert L. Nusbaum, (D), Missouri (Rolla), 1984, Gz
Alexander W. Ritchie, (D), Texas, 1975, Gc
Laboratory Director:
Robin Humphreys, (M), Charleston (South Carolina), 2000, Ge

## Furman University

**Earth and Environmental Sciences** (B) (2017)
3300 Poinsett Highway
Greenville, SC 29613
p. (864) 294-2052
nina.anthony@furman.edu

http://ees.furman.edu
Administrative Assistant: Nina Anthony
*Enrollment (2016): B: 74 (15)*

Chair:
C. Brannon Andersen, (D), Syracuse, 1994, ClGe
Professor:
John M. Garihan, (D), Penn State, 1973, Gct
William A. Ranson, (D), Massachusetts, 1979, GxzGp
Associate Professor:
Weston R. Dripps, (D), Wisconsin, 2003, HwsGe
Suresh Muthukrishnan, (D), Purdue, 2002, OiGmOr
Assistant Professor:
Karen Allen, (D), Georgia, 2016, On
Ruth F. Aronoff, (D), Purdue Univ, 2016, Gc
Matt Cohen, (D), Arizona State, 2015, On
Adjunct Professor:
Courtney Quinn, (D), Nebraska-Lincoln, 2012, On
Melissa Ranhofer, (D), South Carolina, 2009, Gg
Visiting Research Professor:
Christopher Romanek, (D), Texas A&M Univ, 1991, Cls
Emeritus:
Kenneth A. Sargent, (D), Oklahoma, 1973, Hy
Laboratory Director:
Lori Nelsen, (M), Furman Univ, 2000, Ca



## University of South Carolina

**School of the Earth, Ocean & Environment** (B,M,D) ● (2017)
701 Sumter St
EWS 617
Columbia, SC 29208
p. (803) 777-4535
khamilton@geol.sc.edu
http://seoe.sc.edu
f: https://www.facebook.com/MarineScienceSC/
*Enrollment (2016): B: 489 (90) M: 55 (16) D: 40 (8)*

Senior Associate Dean:
Robert C. Thunell, (D), Rhode Island, 1978, Pe
Director of the School of the Earth, Ocean & Environment:
Carol Boggs, (D), Texas, Austin, 1979, On
Director of the Belle Baruch Marine Institute:
James Pinckney, (D), South Carolina, 1992, Ob
Direcctor of ESRI:
Camelia Knapp, (D), Cornell, 2000, YeGt
Professor:
Claudia R. Benitez-Nelson, (D), MIT/WHOI, 1999, Oc
Ron Benner, (D), Georgia, 1984, Ob
Subrahmanyam Bulusu, (D), Southampton (U.K.), 1998, Op
James N. Kellogg, (D), Princeton, 1981, Yg
James H. Knapp, (D), MIT, 1989, Ye
Venkataraman Lakshmi, (D), Princeton, 1995, Hg
Thomas J. Owens, (D), Utah, 1984, Ys
Joseph Quattro, (D), Rutgers, 1991, Ob
Tammi Richardson, (D), Dalhousie, 1996, Ob
Raymond Torres, (D), California (Berkeley), 1997, Hy
George Voulgaris, (D), Southampton (U.K.), 1992, On
Research Professor, Director of Undergraduate Studies:
Gwendelyn Geidel, (D), South Carolina, 1982, HwGeCg
Associate Professor:
David Barbeau, Jr., (D), Arizona, 2003, Gs
Michael Bizimis, (D), Florida State, 2001, GiCg
Jean Ellis, (D), Texas A&M, 2006, Gm
Blaine Griffin, (D), New Hampshire, 2007, Ob
Andrew L. Leier, (D), Arizona, 2005, Gs
Thomas Lekan, (D), On
Howie Scher, (D), Florida, 2005, GuCl
Scott M. White, (D), California (Santa Barbara), 2001, Yr
Alicia M. Wilson, (D), Johns Hopkins, 1999, Hw
Neal Woods, (D), On
Sasha Yankovsky, (D), Marine Hydrophysical (Ukraine), 1991, OpnOw
Gene M. Yogodzinkski, (D), Cornell, 1993, Gi
Assistant Professor:
Jessica Barnes, (D), On
Conor Harrison, (D), On
David Kneas, (D), On
Susan Q. Lang, (D), Washington, 2006, Oc
Ryan Rykaczewski, (D), Scripps, 2009, Ob
Lori A. Ziolkowski, (D), California (Irvine), 2009, CoOcCs
Research Associate:
Dennis Allen, (D), Lehigh, 1978, Ob
Duke Brantley, (D), South Carolina, Ye
Dianne Greenfield, (D), Stony Brook, 2002, Ob
Matthew Kimball, (D), Rutgers, 2008, Ob
Dwayne Porter, (D), OgHs
Jennifer Pournelle, (D), California (San Diego), On
Erik Smith, (D), Maryland, 2000, HsOb
Emeritus:
Philip Barnes, (D), On
John R. Carpenter, (D), Florida State, 1964, Oe
Arthur D. Cohen, (D), Penn State, 1968, Ec
Bruce Coull, (D), Lehigh, 1968, Ob
John Mark Dean, (D), Purdue, 1962, Ob
Robert Ehrlich, (D), Louisiana State, 1965, Gs
Robert Feller, (D), Washington, 1977, Ob
Madilyn Fletcher, (D), Univ.Coll. (N. Wales), 1975, Ob
Christopher G. Kendall, (D), Imperial Coll (UK), 1966, Gs
Ian Lerche, (D), Manchester, 1965, Yg
Willard S. Moore, (D), SUNY (Stony Brook), 1969, Cc
Donald Secor, (D), Stanford, 1962, Gc
W. Edwin Sharp, (D), California (Los Angeles), 1964, Gz
Stephen Stancyk, (D)
Pradeep Talwani, (D), Stanford, 1973, Ys
Dan Tufford, (D), On
Douglas F. Williams, (D), Rhode Island, 1976, Cs
Sarah Woodin, (D), Washington, 1972, Ob
Richard Zingmark, (D)

## University of South Carolina - Lancaster

**University of South Carolina - Lancaster** (A) (2016)
P.O. Box 889
Lancaster, SC 29721
p. (803) 313-7129
martek@mailbox.sc.edu
http://usclancaster.sc.edu

Instructor:
Lynnette Martek, (M), Emporia State Univ, 1994, Ogw

## Winthrop University

**Dept of Chemistry, Physics, & Geology** (2015)
Sims Science Building
Winthrop University
Rock Hill, SC 29733
p. (803) 323-4949
bolandi@winthrop.edu
http://chem.winthrop.edu

Chair, Environmental Sciences and Studies Program:
Marsha S. Bollinger, (D), South Carolina, 1986, CmOgCc
Professor:
Irene B. Boland, (D), South Carolina, 1996, GtgOe
Associate Professor:
Gwen M. Daley, (D), Virginia Tech, 1999, PqgGs
Scott P. Werts, (D), Johns Hopkins, 2006, ScbPy

## Wofford College

**Dept of Geology** (2011)
Wofford College
429 North Church Street
Spartanburg, SC 29303-3663
p. (864) 597-4527
fergusonta@wofford.edu
http://www.wofford.edu/geology/

Director:
Terry A. Ferguson, (D), Tennessee, 1988, Ga

# South Dakota

## Black Hills State University

**School of Natural Sciences** (B) (2015)
1200 University Street, Unit 9008
Spearfish, SD 57799-9008
p. (605) 642-6506
Abigail.Domagall@bhsu.edu

http://www.bhsu.edu/Academics/ProgramsMajors/NaturalSciences/EnvironmentalPhysicalScience/tabid/888/Default.aspx
f: https://www.facebook.com/EnvPhysSciBHSU/?ref=hl
*Enrollment (2015): B: 29 (6)*
Professor:
Mark Gabel, (D), Iowa State, 1982, Pb
Associate Professor:
Abigail M S Domagall, (D), SUNY (Buffalo), 2008, GveOe

## Oglala Lakota College
**Dept of Math, Science, & Technology** (1999)
P.O. Box 490
Kyle, SD 57755
p. (605) 455-6124
hlagarry@olc.edu
http://www.olc.edu/local_links/smet/

## South Dakota School of Mines & Technology
**Dept of Atmospheric and Environmental Sciences** (B,M,D) (2014)
501 E. St. Joseph Street
Rapid City, SD 57701-3995
p. (605) 394-2291
Pamela.Cox@sdsmt.edu
http://www.ias.sdsmt.edu/
*Enrollment (2011): B: 15 (0) M: 14 (3)*
Professor:
Andrew G. Detwiler, (D), SUNY (Albany), 1980, Oa
Associate Professor:
William J. Capehart, (D), Penn State, 1997, Oa
Donna V. Kliche, (D), SD Mines, 2007, OaaOa
P. V. Sundareshwar, (D), South Carolina, 2002, OeCgSb
Assistant Professor:
Adam French, (D), North Carolina State, 2011, Oaw
Lisa Kunza, (D), Wyoming, 2012, HsOe
Instructor:
Darren R. Clabo, (M), SD Mines, 2009, OaaOa
Emeritus:
John H. Helsdon, (D), SUNY (Albany), 1979, Oa
Mark R. Hjelmfelt, (D), Chicago, 1980, Oa
Paul L. Smith, (D), Carnegie Inst, 1960, Oa

**Dept of Geology & Geological Engineering** (B,M,D) O (2016)
501 E. Saint Joseph St.
Rapid City, SD 57701-3901
p. (605) 394-2461
geologyinfo@sdsmt.edu
http://geology.sdsmt.edu
f: https://www.facebook.com/SDSMTGeologyGeologicalEngineering
Administrative Assistant: Cleo J. Heenan
*Enrollment (2016): B: 180 (30) M: 25 (11) D: 16 (2)*
Field Station Director:
Nuri Uzunlar, (D), SD Mines, 1993, EmGcEo
Department Head; Director Museum of Geology:
Laurie C. Anderson, (D), Wisconsin, 1991, Pio
Professor:
Edward F. Duke, (D), Dartmouth, 1984, GipOr
Timothy L. Masterlark, (D), Wisconsin, 2000, Yg
Maribeth H. Price, (D), Princeton, 1995, Oir
Larry D. Stetler, (D), Washington State, 1993, Ng
Senior Scientist:
William M. Roggenthen, (D), Princeton, 1980, NgYg
Associate Professor:
Darrin C. Pagnac, (D), California (Riverside), 2005, Pv
J. Foster Sawyer, (D), SD Mines, 2006, NgGo
Assistant Professor:
Zeynep O. Baran, (D), Miami, 2012, Gco
Christina L. Belanger, (D), Chicago, 2011, PimPe
Kurt W. Katzenstein, (D), Nevada (Reno), 2008, NgrOr
Liangping Li, (D), Polytechnic Univ of Valencia (Spain), 2011, Hqw
Gokce K. Ustunisik, (D), Cincinnati, 2009, Cp
Coordinator and Instructor:
Christopher J. Pellowski, (D), SD Mines, 2012, Gg
Associate Director Museum of Geology:
Sally Y. Shelton, (M), Texas Tech, 1984, Pg
Emeritus:
Arden D. Davis, (D), SD Mines, 1983, Hw
James E. Fox, (D), Wyoming, 1972, Gs
Alvis L. Lisenbee, (D), Penn State, 1972, Gc
James E. Martin, (D), Washington, 1979, Ps
Colin J. Paterson, (D), Otago (NZ), 1978, Eg
Perry H. Rahn, (D), Penn State, 1965, Ng
Jack A. Redden, (D), Harvard, 1956, Gp

## University of South Dakota
**Dept of Earth Sciences & Physics** (B) (2011)
414 East Clark Street
Vermillion, SD 57069-2390
p. (605) 677-5649
esci@usd.edu
http://www.usd.edu/earthsciences/
*Enrollment (2011): B: 25 (9)*
Chair:
Timothy H. Heaton, (D), Harvard, 1988, PvOg
Associate Professor:
Brennan T. Jordan, (D), Oregon State, 2002, GiOw
Mark R. Sweeney, (D), Washington State, 2004, Gms
Instructor:
Jeanne M. Fromm, (M), Idaho State, 1995, GgHs

# Tennessee

## Austin Peay State University
**Geosciences Dept** (B) (2017)
601 College St
Clarksville, TN 37044
p. (931) 221-7454
deibertj@apsu.edu
http://www.apsu.edu/geosciences
*Enrollment (2009): B: 72 (9)*
Chair:
Jack Deibert, (D), Wyoming, GsrGo
Interim Director:
Robert A. Sirk, (D), Kent State, 1991, GeOyu
Professor:
Phyllis A. Camilleri, (D), Wyoming, 1994, Gct
Daniel L. Frederick, (D), Tennessee, 1994, Gsr
Phillip R. Kemmerly, (D), Oklahoma State, 1973, GmeHg
Gregory D. Ridenour, (D), Texas A&M, 1993, HsOgn
Assistant Professor:
Christopher Gentry, (D), Indiana State, 2008, Oyi
Christine Mathenge, (D), Indiana, 2008, Oy
Emeritus:
D. M. S. Bhatia, (D), Missouri (Rolla), 1976, Ca
James X. Corgan, (D), Louisiana State, 1967, Pg
Byron J. Webb, (M), Memphis, Oy
R. Kenton Wibking, (D), Nebraska, Oy
Laboratory Director:
Richard F. Wheeler, (M), Brigham Young Univ, 1980, Ggo

## Middle Tennessee State University
**Dept of Geosciences** (B,M) (2016)
Box 9
Davis Science Building
Room 241
Murfreesboro, TN 37132
p. (615) 898-2726
karen.wolfe@mtsu.edu
http://mtsu.edu/geosciences
f: MTSU Geosciences
t: @MTSUGeosciences
Administrative Assistant: Karen M. Wolfe
*Enrollment (2016): B: 95 (27) M: 19 (8)*
Director:
Zada Law, (M), Wisconsin, 1980, Oin
Chair:
Warner Cribb, (D), Ohio State, 1993, GipGz
Professor:
Mark J. Abolins, (D), Caltech, 1999, Gc
James A. Henry, (D), Kansas, 1978, Oa
Ronald L. Zawislak, (D), Wyoming, 1980, Ye

Associate Professor:
Clay D. Harris, (D), Indiana, 1992, Gs
Melissa Lobegeier, (D), James Cook, 2001, Pmg
Assistant Professor:
Jeremy Aber, (D), Kansas State, 2011, Oyi
Patricia Boda, (D), Minnesota, 2007, OinOn
Racha El Kadiri, (D), Western Michigan Univ, 2014, Hg
Henrique G. Momm, (D), Mississippi, 2008, OiHsOr
Lecturer:
Alan Brown, (M), Illinois State, 2005, Gg
Laura Collins, (M), Mississippi State, 2005, Gg
Michael W. Hiett, (M), Kentucky, 1995, Gg



## Motlow State Community College

**Dept of Natural Sciences** (A) (2014)
PO Box 8500
Lynchburg, TN 37352-8500
p. (931) 393-1810
lmayo@mscc.edu
http://www.mscc.edu/natural_science/
Instructor:
Lisa L Herring Mayo, (M), Mississippi State, 2000, GgOge

## Pellissippi State Community College

**Natural and Behavioral Sciences** (1999)
10915 Hardin Valley Road
P.O. Box 22990
Knoxville, TN 37801
p. (865) 694-6685
jkelley@pstcc.edu
http://www.pstcc.edu
Adjunct Professor:
Peter J. Lemiszki, (D), Tennessee, 1992, GcgOi

## Roane State Community College, Oak Ridge

**Mathematics and Sciences (Geology)** (A) (2011)
276 Patton Lane
Harriman , TN 37748
p. (865) 481-2000
leea@roanestate.edu
http://aclee1234.fortunecity.com
Professor:
Arthur C. Lee, (D), S California, 1994, Ges

## Sewanee: University of the South

**Dept of Earth and Environmental Systems** (B) (2010)
735 University Avenue
Sewanee, TN 37383-1000
p. (931) 598-1271
Sherwood@sewanee.edu
http://www.sewanee.edu/EnvStudies
Chair:
Scott Torreano, (D), Georgia, 1991, Sf
Professor:
Martin A. Knoll, (D), Texas (El Paso), 1988, Hw
Donald B. Potter, Jr., (D), Massachusetts, 1985, Gc
Stephen A. Shaver, (D), Stanford, 1984, Eg
Associate Professor:
C. Ken Smith, (D), Florida, 1996, Sf
Adjunct Professor:
Glendon W. Smalley, (D), Tennessee, 1975, Sf

## Tennessee Tech University

**Dept of Earth Sciences** (B) (2016)
PO Box 5062
Cookeville, TN 38505
p. (931) 372-3121
MHarrison@tntech.edu
http://www.tntech.edu/earth/home/
Administrative Assistant: Peggy Medlin
*Enrollment (2016): B: 69 (11)*
Chair:
Michael J. Harrison, (D), Illinois (Urbana-Champaign), 2002, Gc
Professor:
Evan A. Hart, (D), Tennessee, 2000, Oy
H. Wayne Leimer, (D), Missouri, 1969, Gz
Ping-Chi Li, (D), Iowa, 1992, Oi
Assistant Professor:
Joseph Asante, (D), Nevada (Las Vegas), 2012, HwOrGe
Lauren Michel, (D), Baylor Univ, 2014, PeCs
Jeannette Wolak, (D), Montana State, 2011, Gsr
Adjunct Professor:
Jason E. Duke, (M), Tennessee Tech, 1995, Oi
Emeritus:
Larry W. Knox, (D), Indiana, 1974, Pmg

## University of Memphis

**Center for Earthquake Research & Information (CERI)** (M,D) (2014)
3876 Central Avenue, Suite 1
Memphis, TN 38152-3050
p. (901) 678-2007
clangstn@memphis.edu
http://www.ceri.memphis.edu/
*Enrollment (2011): M: 7 (3) D: 16 (1)*
Chair, Department of Earth Sciences:
Mervin J. Bartholomew, (D), Virginia Tech, 1971, Gt
CERI Founding Director:
Archibald C. Johnston, (D), Colorado, 1979, Ys
CERI Director of Academic Programs:
Christine A. Powell, (D), Princeton, 1976, Ys
CERI Director:
Charles A. Langston, (D), Caltech, 1976, Ys
Professor:
Jer-Ming Chiu, (D), Cornell, 1982, Ys
Associate Professor:
Randel Tom Cox, (D), Missouri, 1995, GtcGm
Jose M. Pujol, (D), Wyoming, 1985, Ye
DES Graduate Coordinator:
Arleen Alice Hill, (D), Og
USGS:
Oliver Boyd, (D), Colorado, 2004, YgGtg
Research Scientist:
Stephen P. Horton, (D), Nevada (Reno), 1992, Ys
Assoc. Research Professor :
Chris Cramer, (D), Stanford, 1976, Ys
Assoc. Research Professor:
Maria Beatrice Magnani, (D), Studi di Perugia, 2000, YesYg
Assoc. Research Professor :
Robert Smalley, Jr., (D), Cornell, 1988, Yd
Assoc. Research Professor:
Mitchell M. Withers, (D), New Mexico Tech, 1997, Ys
Assistant Research Professor:
Heather DeShon, (D), California (Santa Cruz), 2004, Ysg
Research Scientist:
Shu-Choiung Chiu, (D), Memphis, 2010, Ys

**Dept of Earth Sciences** (B,M,D) O (2016)
111 Johnson Hall
488 Patterson Street
Memphis, TN 38152-3550
p. (901) 678-4571 or 678-4358
dlarsen@memphis.edu
http://memphis.edu/earthsciences/
*Enrollment (2016): B: 154 (21) M: 67 (20) D: 53 (6)*
Chair:
Daniel Larsen, (D), New Mexico, 1994, Cl
Director, Confucius Institute:
Hsiang-Te Kung, (D), Tennessee, 1980, Oy
Professor:
Mervin J. Bartholomew, (D), Virginia Tech, 1971, Gtc
Randel T. Cox, (D), Missouri, 1995, GcmGt
David H. Dye, (D), Washington, 1980, Ga
Jose Pujol, (D), Wyoming, 1985, Ye
Roy B. Van Arsdale, (D), Utah, 1979, Gcm
Roy B. Van Arsdale, (D), Utah, 1979, Gc
Associate Professor:
Arleen A. Hill, (D), South Carolina, 2002, On
Andrew M. Mickelson, (D), Ohio State, 2002, Ga
Esra Ozdenerol, (D), Louisiana State, 2000, Oi
Assistant Professor:
Anzhelika Antipova, (D), Louisiana State, 2010, On
Youngsang Kwon, (D), SUNY (Buffalo), 2012, Oi

Ryan M. Parish, (D), Memphis, 2013, Ga
Instructor:
Julie Johnson, (D), Florida International, 2012, Gx
Emeritus:
Phili B. Deboo, (D), Louisiana State, 1963, Pg
Robert W. Deininger, (D), Rice, 1964, Gx
James Dorman, (D), Columbia, 1961, Ys
Archibald C. Johnston, (D), Colorado, 1979, Ys
David N. Lumsden, (D), Illinois (Urbana), 1965, Gd

## University of Tennessee at Chattanooga

**Biology, Geology, and Environmental Science** (B) (2016)
615 McCallie Ave., Dept. 2653
Chattanooga, TN 37403
p. (423) 425-4341
http://www.utc.edu/biology-geology-environmental-science/division-geology/
f: https://www.facebook.com/GeologyatUTC
*Enrollment (2016): B: 53 (4)*

Professor:
Habte G. Churnet, (D), Tennessee, 1979, GxgOg
Jonathan W. Mies, (D), North Carolina, 1990, GctHg
Associate Professor:
Amy Brock-Hon, (D), Nevada (Las Vegas), 2007, GmSc
Assistant Professor:
Ann E. Holmes, (D), Columbia, 1996, GsrPi
A K M Azad Hossain, (D), Mississippi, 2008, Org
Adjunct Professor:
Gregory Brodie, (M), Purdue, 1979, Ge
Geology Laboratory Coordinator:
Wayne K. Williams, (M), Memphis State Univ, 1980, Gdg

## University of Tennessee, Knoxville

**Dept of Earth & Planetary Sciences** (B,M,D) (2017)
306 Earth & Planetary Sciences Building
Knoxville, TN 37996-1410
p. (865) 974-2366
eps@utk.edu
http://web.eps.utk.edu/
f: https://www.facebook.com/UTEPS
*Enrollment (2016): B: 139 (41) M: 17 (12) D: 26 (4)*

Head:
Larry D. McKay, (D), Waterloo, 1991, HyGe
Distinguished Scientist:
Robert D. Hatcher, Jr., (D), Tennessee, 1965, Gc
Professor:
Thomas W. Broadhead, (D), Iowa, 1978, Pi
William M. Dunne, (D), Bristol (UK), 1980, Gc
Annette S. Engel, (D), Texas, 2004, Cl
Christopher Fedo, (D), Virginia Tech, 1994, GsXg
Linda Kah, (D), Harvard, 1997, Gs
Michael L. McKinney, (D), Yale, 1985, GePo
Jeffery E. Moersch, (D), Cornell, 1997, XgOr
Edmund Perfect, (D), Cornell, 1986, SpHwGq
Lawrence A. Taylor, (D), Lehigh, 1968, GiCpXg
Associate Professor:
Devon Burr, (D), Arizona, 2003, XgGm
Micah Jessup, (D), Virginia Tech, 2007, Gc
Associate Scientist:
Joshua Emery, (D), Arizona, 2003, Xg
Assistant Professor:
Andrew Steen, (D), North Carolina (Chapel Hill), 2009, CoOb
Colin Sumrall, (D), Texas, 1997, Pi
Anna Szynkiewicz, (D), Wroclaw, Poland, 2004, Cs
Lecturer:
William Deane, (M), Tennessee, 1998, Gg
Adjunct Professor:
Hassina Z. Bilheux, (D), 2003, On
David R. Cole, (D), Penn State, 1980, Ca
Steven Driese, (D), Wisconsin, SaPeGs
Alice Layton, (D), Purdue, 1987, On
Timothy McCoy, (D), Xm
David Mittlefehldt, (D), Xm
Scott Murchie, (D), Xg
Ryan Otter, (D), On
Tommy Phelps, (D), On
Johnny Waters, (D), Pg
Richard T. Williams, II, (D), Virginia Tech, 1979, Yg
Emeritus:
Don W. Byerly, (D), Tennessee, 1966, GeNg
G. Michael Clark, (D), Penn State, 1966, Gm
Theodore C. Labotka, (D), Caltech, 1978, GpCg
Harry Y. McSween, Jr., (D), Harvard, 1977, XcGi
Kula C. Misra, (D), W Ontario, 1973, Eg
Kenneth R. Walker, (D), Yale, 1969, Pe
Other:
Melanie A. Mayes, (D), Tennessee, 2006, ClGeHw
Cooperating Faculty:
Janet L. Hopson, (D), Tennessee, 1994, On
Robert Riding, (D), UK, Gs

## University of Tennessee, Martin

**Dept of Agriculture, Geosciences, and Natural Resources** (B) (2014)
256 Brehm Hall
Martin, TN 38238
p. (731) 881-7260
mehlhorn@utm.edu
http://www.utm.edu/departments/caas/agnr/geosciences/
*Enrollment (2013): B: 29 (13)*

Professor:
Paula M. Gale, (D), Arkansas, 1988, OsSdc
Michael A. Gibson, (D), Tennessee, 1988, PgiOe
Jefferson S. Rogers, (D), Illinois, 1995, On
Robert M. Simpson, (D), Indiana State, 2000, OawOi
Associate Professor:
Stan P. Dunagan, (D), Tennessee, 1998, GsSa
Assistant Professor:
Thomas A. DePriest, (D), Union (Jackson), 2009, Oe
Benjamin P. Hooks, (D), Maine, 2009, GciNr
Instructor:
Eleanor E. Gardner, (M), Georgia, GgPg
Emeritus:
William T. McCutchen, (M), Berea, 1967, Gg
Robert P. Self, (D), Rice, 1971, Gs
Helmut C. Wenz, (M), W Michigan, 1968, On

## Vanderbilt University

**Earth and Environmental Sciences** (B,M,D) (2016)
SC Science & Engineering Bldg
2301 Vanderbilt Place, SC5726
VU Station B 351805
Nashville, TN 37235
p. (615) 322-2976
jewell.beasleystanley@vanderbilt.edu
http://www.vanderbilt.edu/ees/
f: https://www.facebook.com/groups/393646734010753/
*Enrollment (2016): B: 37 (10) M: 7 (6) D: 12 (2)*

Chair:
Steven L. Goodbred, Jr., (D), William & Mary, 1999, GsOn
Professor:
John C. Ayers, (D), Rensselaer, 1991, Cg
Ralf Bennartz, (D), Free Univ of Berlin, 1997, Oar
David J. Furbish, (D), Colorado, 1985, GmHg
George Hornberger, (D), Stanford, 1970, HgwHs
Calvin F. Miller, (D), California (Los Angeles), 1977, Gi
Director of Graduate Studies:
Guilherme Gualda, (D), Chicago, 2010, GivGz
Associate Professor:
Jonathan M. Gilligan, (D), Yale, 1991, OnaOn
Assistant Professor:
Simon Darroch, (D), Yale, 2014, Poi
Larisa R.G DeSantis, (D), Florida, 2009, Pve
Maria Luisa Jorge, (D), Illinois (Chicago), 2007, Py
Jessica L. Oster, (D), California (Davis), 2010, Cls
Senior Lecturer:
Lily L. Claiborne, (D), Vanderbilt, 2011, GivCg
Neil P. Kelley, (D), California (Davis), 2012, GgOgPg
Garrett W. Tate, (D), Princeton Univ, 2014, Gc
Emeritus:
Leonard P. Alberstadt, (D), Oklahoma, 1967, Gd
Molly F. Miller, (D), California (Los Angeles), 1977, Pe

Arthur L. Reesman, (D), Missouri, 1966, Gg
William G. Siesser, (D), Cape Town, 1971, Pm
Richard G. Stearns, (D), Northwestern, 1953, Yg
Senior Lecturer:
Daniel J. Morgan, (D), Washington, 2009, GmCc
Admin. Asst. II:
Jewell D. Beasley-Stanley, (D), On



## Volunteer State Community College

**Volunteer State Community College** (2014)
1800 Nashville Pike
Gallatin, TN 37006
p. (615) 230-3294
Clark.Cropper@volstate.edu
http://www.volstate.edu

## Walters State Community College

**Walters State Community College** (1999)
500 South Davy Crockett Parkway
Morristown, TN 37813
p. (423) 585-6764
http://www.ws.edu

# Texas

## Alamo Colleges, San Antonio College

**Natural Sciences** (2016)
1819 N Main Ave
San Antonio, TX 78212
p. (210) 486-0045
dlambert@alamo.edu
http://alamo.edu/sac/earthsci/
Professor:
Dean Lambert, (D), UT-Austin, Oy
Associate Professor:
Anne D. Dietz, (M), Tx A&M - Kingsville, OggGg
Full-Time Adjunct:
Dwight Jurena, (M), RPI, GgOg
Adjunct Professor:
Thomas S. Girhard, (M), Texas State, Oyw
Robert Janusz, (M), UT-San Antonio, Og
Ryan E. Rudnicki, (D), Penn State, Oy
Charles K. Smith, (M), Texas State, Oy

## Alamo Colleges, Palo Alto College

**Dept of Geology** (A) (2015)
1400 W. Villaret Blvd
San Antonio, TX 78224
p. (210) 486-3000
ghagen@alamo.edu
http://alamo.edu/pac/geology/

## Alvin Community College

**Dept of Geology** (A) (2013)
3110 Mustang Rd
Alvin, TX 77511
p. (281) 756-5670
ddevery@alvincollege.edu
*Enrollment (2012): A: 2 (1)*
Dept. Chair of Physical Sciences:
Dora Devery, (M), Texas Christian Univ, 1979, Gg

## Amarillo College

**Dept of Physical Science** (2016)
P.O. Box 447
Amarillo, TX 79178
p. (806) 371-5333
rdhobbs@actx.edu
https://www.actx.edu/pscience/
*Enrollment (2009): A: 3 (0)*
Professor:
Richard D. Hobbs, (D), Wyoming, 1998, OrGc
Adjunct Professor:
David Pertl, (M), West Texas A&M, 1984, Go

## Angelo State University

**Dept of Physics and Geosciences** (B) ○ (2015)
ASU Station #10904
San Angelo, TX 76909
p. (325) 942-2242
joseph.satterfield@angelo.edu
http://www.angelo.edu/dept/physics/Geosciences/geoscience.php
*Enrollment (2015): B: 63 (14)*
Planetarium Director:
Mark S. Sonntag, (D), Indiana, 1971, Xg
Professor:
Joseph I. Satterfield, (D), Rice, 1995, Gc
Visiting Assistant Professor:
Fred L. Wilson, (D), Kansas, 1964, Gm
Assistant Professor:
Fawn M. Last, (D), Manitoba, 2013, Gsn
Heather L. Lehto, (D), South Florida, 2012, GvYsOe
James W. Ward, (D), Kentucky, 2008, HwClOn
Adjunct Professor:
Cary D. Carman, (B), Angelo State, 1999, Hg
Steven Lyons, (D), Hawaii, 1981, Ow
Robert Purkiss, (M), Texas Tech, 1991, Gg

## Austin Community College District

**Dept of Earth and Environmental Sciences** (A) (2016)
11928 Stonehollow Drive
Austin, TX 78758-3190
p. (512) 223-4276
rblodget@austincc.edu
http://sites.austincc.edu/ees/
Administrative Assistant: Oralia Guerra
*Enrollment (2015): A: 478 (0)*
Chair:
Robert H. Blodgett, (D), Texas (Austin), 1990, GseOe
Professor:
Ronald A. Johns, (D), Texas (Austin), 1993, Pi
Assistant Professor:
Peter J. Wehner, (M), Vanderbilt, 1992, Gv
Adjunct Professor:
Heather L. Beatty, (M), Texas Tech, 1992, Ge
Peter A. Boone, (D), Texas A&M, 1972, Gro
Thomas W. Brown, (M), Indiana, 1987, Ges
M. Jennifer Cooke, (D), Texas (Austin), 2005, GmCl
Amy J. Cunningham, (M), Cincinnati, 1992, OeGg
Leslie M. Davis, (M), Florida State, 1986, Op
Meredith Y. Denton-Hedrick, (M), Texas A&M, 1992, OeYeEo
Kusali R. Gamage, (D), Florida, 2005, HyOuPm
Khaled W. Hasan, (D), Texas A&M, 1995, HsOri
Ian C. Jones, (D), Texas (Austin), 2002, Hw
Richard V. McGehee, (D), Texas (Austin), 1963, GgOn
Ata U. Rahman, (D), Texas Tech, 1987, GgdGe
Fabienne M. Rambaud, (M), Texas (Austin), 2005, Eg
Carolyn M. Riess, (M), Texas (El Paso), 1984, GoeGg
Mark A. Shepherd, (D), Nebraska (Medical Center), 2015, Oa
Raymond M. Slade, Jr., (B), Southwest Texas, 1971, HgsHq
Jason H. Stephens, (D), Texas (Austin), 2014, YrGuYe
Wenxian Tan, (D), Texas (Austin), 2013, Obg
Anne Turner, (M), Texas (Austin), 1986, Hg
Science Laboratory Technician:
John S. Conners, Oe
Shannon M. Grace, (B), Texas State, 2010, On
Deanna M. Sharp, (B), Angelo State, 2002, One
Cooperating Faculty:
David J. Froehlich, (D), Texas (Austin), 1996, Pv

## Baylor University

**Dept of Geosciences** (B,M,D) ● (2016)
One Bear Place #97354
101 Bagby Ave.
BSB, 4th Floor, Rm. D409
Waco, TX 76798-7354
p. (254) 710-2361
paulette_penney@baylor.edu
http://www.baylor.edu/geology/
Administrative Assistant: Janelle Atchley

Administrative Assistant: Jamie J.. Ruth
*Enrollment (2016): B: 62 (22) M: 15 (9) D: 25 (6)*
Chair:
Stacy C. Atchley, (D), Nebraska, 1990, GroGo
W.M. Keck Foundation Professor of Geophysics:
Robert Jay Pulliam, (D), California (Berkeley), 1991, YsgYe
Associate Graduate Dean for Research:
Steven G. Driese, (D), Wisconsin, 1982, SaGe
Professor:
Peter M. Allen, (D), S Methodist, 1977, HgNg
Rena M. Bonem, (D), Oklahoma, 1975, PieOb
Vincent S. Cronin, (D), Texas A&M, 1988, GctNg
Stephen I. Dworkin, (D), Texas, 1991, ClGd
Stephen Forman, (D), Colorado, Cc
Don M. Greene, (D), Oklahoma, 1980, Oyw
Lee C. Nordt, (D), Texas A&M, 1996, SdGa
Kenneth Wilkins, (D), Florida, 1982, Pv
Joe C. Yelderman, Jr., (D), Wisconsin, 1983, HwgOu
Graduate Program Director:
Daniel J. Peppe, (D), Yale, 2009, PbYmPe
Associate Professor:
John A. Dunbar, (D), Texas, 1989, YgHg
William C. Hockaday, (D), Ohio State, 2006, CoaSb
Joseph D. White, (D), Montana, 1998, Or
Assistant Professor:
Kenneth S. Befus, (D), Texas (Austin), 2014, GviGv
Scott C. James, (D), California (Irvine), 2001, HwGeEo
Emeritus:
Harold H. Beaver, (D), Wisconsin, 1954, Go
William G. Brown, (D), Alaska, 1987, Gc
Thomas T. Goforth, (D), S Methodist, 1973, Yg
Don F. Parker, (D), Texas, 1976, Giv
Instrumentation Specialist:
Timothy Meredith
Ren Zhang, (D), McMaster, 2007, Cs
Laboratory Director:
Sharon Browning, Gg
Office Manager:
Paulette Penney, On

## Blinn College
**Agricultural and Natural Science Programs** (A) (2013)
902 College Avenue
Brenham, TX 77833
p. (979) 830-4200
cl.metz@blinn.edu
http://www.blinn.edu/natscience/index.htm
Instructor:
Michael Dalman, Gg
Cynthia Lawry, (D), Gg

## Blinn College - Brazos Campus
**Geology Dept** (1999)
902 College Ave
Brenham, TX 77833
p. (979) 830-4000
cl.metz@blinn.edu
http://www.blinn.edu/STEM/Geology/index.html

## Brookhaven College
**Science/Math Div - Geology Dept** (2017)
3939 Valley View Lane
Farmers Branch, TX 75244
p. (972) 860-4758
LannaBradshaw@dcccd.edu
http://www.brookhavencollege.edu/instruction/math-science/science/geology/
Chair:
Lanna K. Bradshaw, (M), TAMU-C, 2000, GgOg

## Coastal Bend College
**Science Div** (A) (2013)
3800 Charco Road
Beeville, TX 78102
p. (361) 354-2423
amgarza@coastalbend.edu
http://www.coastalbend.edu/acdem/science/
Administrative Assistant:
*Enrollment (2011): A: 3 (0)*
Instructor:
Danny Burns, (M), Ball State, 1984, Ogn
Richard Cowart, (D), Texas A&M (Corpus Christi), 2008, GeEo

## Collin College - Central Park Campus
**Deptartment of Geology** (A) (2016)
2200 W. University Drive
McKinney, TX 75071
p. (972) 548-6790
bburkett@collin.edu
http://www.collin.edu/geology/
*Enrollment (2016): A: 30 (0)*
Professor:
Brett Burkett, (M), SUNY (Buffalo), 2008, Gvg
Related Staff:
Shannon Burkett, (M), SUNY (Buffalo), 2005

## Collin College - Preston Ridge Campus
**Dept of Geology and Environmental Science** (A) (2017)
9700 Wade Boulevard
Frisco, TX 75035
p. (972) 377-1635
smay@collin.edu
Professor:
Heinrich Goetz, (M), Texas A&M, 1997, Ge
Paul Manganelli, (M), Boston Coll, 1998
S Judson May, (D), New Mexico, 1980, GgoGc

## Collin College - Spring Creek Campus
**Dept of Geology and Environmental Science** (A) (2015)
2800 E. Spring Creek Parkway
Plano, TX 75074
p. (972) 578-5518
dbabcock@collin.edu
http://www.collin.edu/academics/programs/geology.html
*Enrollment (2012): A: 10 (0)*
Chair:
Daphne H. Babcock, (M), Memphis, 1989, Geg
Geology Lab Instructor:
Mark Turner, Gg

## Del Mar College
**Dept of Natural Sciences** (A) (2013)
Corpus Christi, TX 78404
p. (512) 886-1240
jhalcomb@delmar.edu
Professor:
Roger T. Steinberg, (M), Tennessee, 1981, GgoPg
Associate Professor:
Walter V. Kramer, (M), Texas (El Paso), 1970, GgoGx
Emeritus:
Mary S. Thorpe, (M), Baylor, 1966, Ge

## El Centro College, Dallas Community College District
**Geology Program** (A) (2016)
801 Main Street
Dallas, TX 75202
p. (214) 860-2429
nfields@dcccd.edu
http://elcentrocollege.edu/programs/geology
Coordinator:
Nancy Fields, (M), Baylor, Og
Professor:
Steven McCauley, (M), Mississippi State Univ, GgOw
Bethan Salle, (M), Northern Colorado, Gg
Adjunct Professor:
Anna F. Banda, (M), Baylor, 2007, GgeGv
David Coffman, (M), Baylor, Gg
Stephanie Coffman, (M), Baylor, Gg
Alice Ruffel, (M), Oklahoma, Gg

Texas

## El Paso Community College
**Dept of Geological Sciences** (A) (2015)
P.O. Box 20500
El Paso, TX 79998
p. (915) 831-5161
jvillal6@epcc.edu
http://epcc.edu/InstructionalPrograms/geologicalsciences
Chair:
Joshua Villalobos, (M), Texas (El Paso), 200, Gg
Coordinator - Valle Verde:
Russell Smith, (M), Florida, Gg
Coordinator - Transmountain Campus:
Kathleen Devaney, (D), California (Los Angeles), 1992, Gg
Coordinator - Northwest Campus:
Deborah Caskey, (M), Texas, Gg
Professor:
Sulaiman Abushagur, (D), Texas (El Paso), Gg
Instructor:
Robert Rohbaugh, (M), UTEP
Adjunct Professor:
Brenda Barnes, (D), Texas (El Paso), Gg
Lawrence Bothern, (M), New Mexico State, Gg
Sabrina Canalda, (M), Texas (El Paso), Gg
Emile Couroux, Gg
Alexandra Falcon, Gg
Musa Hussein, (D), Texas (El Paso), Gg
Adriana Perez, (M), Texas (El Paso), Gg
Kirk Rothemund, Gg

## Hardin-Simmons University
**Dept of Geology & Environmental Science** (B) (2015)
Box 16164
2200 Hickory Street
Abilene, TX 79698-6164
p. (325) 670-1383
ouimette@hsutx.edu
http://www.hsutx.edu/academics/undergraduate/holland/geology
*Enrollment (2015): B: 20 (5)*
Head:
Mark A. Ouimette, (D), Texas (El Paso), 1994, GieGt
Associate Professor:
Marla Potess, (D), Texas Tech, 2011, Ogu
Steven Rosscoe, (D), Texas Tech, 2008, PmGsPs

## Hill College
**Div of Mathematics and Sciences** (1999)
112 Lamar
Hillsboro, TX 76645
p. 254-659-7500
rroberts@hillcollege.edu
http://www.hillcollege.edu/academics/Traditional/Math_Science_EdServices/Geology.html

## Houston Community College System
**Geology Dept** (A) (2016)
1010 W. Sam Houston Pkwy.N.
Houston, TX 77043
p. (713) 718-5641
dwight.kranz@hccs.edu
http://learningwebsys.hccs.edu/discipline/geology/
Professor:
Dwight S. Kranz, (M), Texas A&M, 1980, Gg

## Kilgore College
**Dept of Chemistry and Geology** (A) (2016)
1100 Broadway
Kilgore, TX 75662
p. (903) 983-8253
pbuchanan@kilgore.edu
*Enrollment (2016): A: 5 (2)*
Professor:
Paul C. Buchanan, (D), Houston, 1995, GgiXm

## Lamar University
**Dept of Earth and Space Sciences** (B) (2010)
P.O. Box 10031
Beaumont, TX 77710
p. (409) 880-8236
jim.jordan@lamar.edu
http://ess.lamar.edu
Professor:
Jim L. Jordan, (D), Rice, 1975, XcgXm
University Professor:
James W. Westgate, (D), Texas (Austin), 1988, Pev
Professor:
Roger W. Cooper, (D), Minnesota, 1978, Gi
Donald E. Owen, (D), Kansas, 1963, Gr
Associate Professor:
Joseph M. Kruger, (D), Arizona, 1991, YgGgOi
Instructor:
Bennetta Schmidt, (D), Gg
Captain:
Mark Adams, (M), Houston (Clear Lake), Xg
Adjunct Professor:
Cynthia L. Parish, (M), Lamar, 2004, Gg
Carla M. Tucker, (M), Texas (Austin), 1990, Hw
Laboratory Coordinator:
Karen M. Woods, (B), Lamar, On

## Laredo Community College
**Dept of Natural Sciences & Kinesiology** (2014)
Cigarroa Science Building Room 217
West End Washington Street
Laredo, TX 78040
p. (956) 721-5195
glenn.blaylock@laredo.edu
http://www.laredo.edu/cms/LCC/Instruction/Divisions/Sciences/Natural_Sciences/Science/
Professor:
Glenn W. Blaylock, (M), Brigham Young, 1998, Gg
Instructor:
Mark T. Childre, (M), Texas (San Antonio)

## Lee College
**Dept of Physical Sciences** (A) (2017)
P. O.Box 818
Baytown, TX 77522
p. (281) 425-6552
jdobberstine@lee.edu
*Enrollment (2016): A: 2 (0)*
Professor:
Sharon Gabel, (D), SUNY, 1991, GseGg

## Lonestar College, CyFair
**Geology Dept** (2013)
9191 Barker Cypress Road
Cypress, TX 77433
p. (281) 290-3919
michael.r.konvicka@lonestar.edu
http://www.lonestar.edu/geology-dept-cyfair.htm

## Lonestar College, Kingwood
**Geology Dept** (1999)
20000 Kingwood Drive
Kingwood, TX 77339
p. (281) 312-1629
Jean.Whileyman@lonestar.edu
http://www.lonestar.edu/geology-dept-kingwood.htm

## Lonestar College, Montgomery
**Geology Dept** (2013)
3200 College Park Drive
Conroe, TX 77384
p. (936) 273-7077
Michael.J.Sundermann@lonestar.edu
http://www.lonestar.edu/geology-dept-montgomery.htm
Professor:
Nathalie N. Brandes, (M), New Mexico Tech

John R. Kleist, (D), Texas

## Lonestar College, North Harris

**Geology Dept** (2014)
2700 W.W. Thorne Drive
Houston, TX 77073-3499
p. (281) 618-5685
tom.hobbs@lonestar.edu
*Enrollment (2007): A: 20 (0)*

Head:
Thomas M. C. Hobbs, (M), Texas (El Paso), 1979, Gg
Professor:
Peter E. Price, (M), Kentucky, 1979, Oi
Adjunct Professor:
Penni Major, (M), Gg
Michelle Mc Mahon, (D), Aberdeen, 1993, Go
Victor S. Resnic, (D), Nat Pet Inst (Russia), 1971, Go
Linda C. Tran, (B), Texas A&M, 1999, Oi

## Lonestar College, Tomball

**Geology Dept** (1999)
30555 Tomball Parkway
Tomball, TX 77375
p. (281) 351-3324
David.O.Bary@lonestar.edu
http://www.lonestar.edu/geology-dept-tomball.htm

## McLennan Community College

**Geology Dept** (A) (2010)
1400 College Drive
Waco, TX 76708
p. (254) 299-8442
efagner@mclennan.edu
http://www.mclennan.edu/departments/geol/

Instructor:
Elaine Alexander, Gg

## Midland College

**Math and Science Div** (A) (2015)
3600 N. Garfield
Midland, TX 79705
p. (432) 685-4612
kwaggoner@midland.edu
http://www.midland.edu/~msd/
*Enrollment (2011): A: 14 (6)*

Associate Professor:
Joan Gawloski, (M), Baylor, GgzGe
Antony Giles, (M), Sul Ross State Univ, 2006, GgvGi
Assistant Professor:
Keonho Kim, (D), GgOwPi
Adjunct Professor:
Karen Waggoner, (D), Texas Tech, GghGc

## Midwestern State University

**Kimbell School of Geosciences** (B,M) (2016)
3410 Taft
Wichita Falls, TX 76308
p. (940) 397-4250
geology.program@mwsu.edu
http://www.mwsu.edu/academics/scienceandmath/geosciences/index
f: https://www.facebook.com/MSUGeosProgram/
*Enrollment (2016): B: 66 (18) M: 16 (0)*

Chair & Prothro Distinguished Associate Professor of Geological Science:
Jonathan D. Price, (D), Oklahoma, 1998, GiCgGz
Robert L. Bolin Distinguished Professor of Petroleum Geology:
W. Scott Meddaugh, (D), Harvard, 1982, GoEgCg
Associate Professor and Graduate Advisor:
Rebecca L. Dodge, (D), Colorado Mines, 1982, Ore
Assistant Professor:
Jesse R. Carlucci, (D), Oklahoma, 2012, PisGs
Adjunct Professor:
Black L. Lisa, (M), Texas Christian, 2008, EoGg
Jay D. Murray, (D), Califiornia Inst of Technology, 1978, Gi

Emeritus:
John Kocurko, (D), Texas Tech, 1972, Gs

## North Lake College, Dallas Community College District

**Math and Natural Sciences** (A) (2010)
5001 North MacArthur Boulevard
Irving, TX 75038
p. (972) 273-3500
memays@dcccd.edu
http://www.northlakecollege.edu/academics/mathscience/index.html

Instructor:
Leonard Kubicek, Gg

## Odessa College

**Dept of Geology, Anthropology & Geography** (A) (2013)
201 W. University
Odessa, TX 79762
p. (915) 335-6558
dedwards@odessa.edu
*Enrollment (2010): A: 1 (1)*

Associate Professor:
Gerald B. McAfee, (M), Sul Ross State, 1966, Pi

## Paris Junior College

**Dept of Science** (2014)
2400 Clarksville Street
Paris, TX 75460
p. 903-782-0481
mbarnett@parisjc.edu
http://www.parisjc.edu/index.php/pjc2/directory-index/C208

## Rice University

**Center for Computational Earth Science** (M,D) (2016)
MS-126
PO Box 1892
Houston, TX 77251-1892
p. (713) 348-3574
dmberry@rice.edu
http://earthscience.rice.edu/centers/ccg/
*Enrollment (2016): M: 7 (2) D: 335 (4)*

Associate Director:
Alan Levander, (D), Stanford, 1984, Ys
William S. Symes, (D), Harvard, 1975, On
Professor:
Richard G. Gordon, (D), Stanford, 1979, Gt
Adrian Lenardic, (D), California (Los Angeles), 1995, Gc
Julia K. Morgan, (D), Cornell, 1993, Gc
Fenglin Niu, (D), Tokyo, 1997, Ysg
Dale S. Sawyer, (D), MIT, 1982, Ys
Colin A. Zelt, (D), British Columbia, 1989, Ys
Assistant Professor:
Helge Gonnermann, (D), California (Berkeley), 2004, Gv

**Dept of Civil and Environmental Engineering** (M,D) (2014)
P.O. Box 1892
MS 317
Houston, TX 77251-1892
p. (713) 348-4951
alvarez@rice.edu
http://ceve.rice.edu/

Chair:
Joseph B. Hughes, (D), Iowa, 1992, On
Professor:
Philip B. Bedient, (D), Florida, 1975, Hw
Arthur A. Few, Jr., (D), Rice, 1969, Oa
Mason B. Tomson, (D), Oklahoma State, 1972, Cg
Calvin H. Ward, (D), Cornell, 1960, Og
Mark R. Wiesner, (D), Johns Hopkins, 1985, Og
Lecturer:
James B. Blackburn, (D), Texas, 1972, Ou
Adjunct Associate Professor:
Stanley M. Pier, (D), Purdue, 1952, Co

Adjunct Assistant Professor:
Charles J. Newell, (D), Rice, 1989, Hw
Adjunct Professor:
Jean-Yves Bottero, (D), Nancy, 1979, On
Cooperating Faculty:
John Hunter, (M), Indiana Sch Lib Sci, 1974, On



**Dept of Earth Science** (B,M,D) (2016)
MS 126
PO Box 1892
Houston, TX 77251-1892
p. (713) 348-4880
geol@rice.edu
http://earthscience.rice.edu/
*Enrollment (2015): B: 28 (8) M: 45 (18) D: 40 (5)*
Chair:
Cin-Ty A. Lee, (D), Harvard, 2001, Cg
Professor:
John B. Anderson, (D), Florida State, 1972, Gu
Rajdeep Dasgupta, (D), Minnesota, 2006, GxCg
Gerald R. Dickens, (D), Michigan, 1996, Yr
Andre W. Droxler, (D), Miami, 1984, Gs
Richard G. Gordon, (D), Stanford, 1979, Ym
Adrian Lenardic, (D), California, 1995, Yg
Alan R. Levander, (D), Stanford, 1984, Ye
Caroline A. Masiello, (D), California (Irvine), 1999, On
Julia K. Morgan, (D), Cornell, 1993, Gt
Fenglin Niu, (D), Tokyo, 1997, Ys
Dale S. Sawyer, (D), MIT, 1982, Yr
Colin A. Zelt, (D), British Columbia, 1989, Yg
Associate Professor:
Helge Gonnermann, (D), California (Berkeley), 2004, Gv
Assistant Professor:
Melodie E. French, (D), Texas A&M Univ, 2014, Gc
Jeffrey A. Nittrouer, (D), Texas, 2010, Gms
Laurence Y. Yeung, (D), Caltech, 2010, CsOaCl
Adjunct Professor:
Vitor Abreu, (D), Rice, 1998, Gg
K. K. Bissada, (D), Washington (St. Louis), 1967, Cg
Stephen H. Danbom, (D), Connecticut, 1975, Gg
Jeffrey J. Dravis, (D), Rice, 1980, Gs
Paul M. Harris, (D), Miami, 1977, Yr
N Ross Hill, (D), Virginia, 1978
Thomas A. Jones, (D), Northwestern, 1969, Gq
Stephen J. Mackwell, (D), Australian National, 1985, Yg
Patrick J. McGovern, (D), MIT, 1996, Xy
David L. Olgaard, (D), MIT, 1985, Yg
W C. Riese, (D), New Mexico, 1980, Eg
Stephanie S. Shipp, (D), Rice, 1999, Oe
Emeritus:
Albert W. Bally, (D), Zurich, 1953, Gg
H. C. Clark, (D), Stanford, 1966, Ng
Gerald H. F. Gardner, (D), Princeton, 1953, Yg
Dieter Heymann, (D), Amsterdam, 1958, Xm
William P. Leeman, (D), Oregon, 1974, Gi
Andreas Luttge, (D), Tubingen, 1990, Cg
John C. Stormer, Jr., (D), California (Berkeley), 1971, Gi
Manik Talwani, (D), Columbia, 1959, Yr
Peter R. Vail, (D), Northwestern, 1959, Gr
Department Coordinator:
Sandra Flechsig, On
Department Administrator:
Lee Willson, On

## Saint Mary's University

**Dept of Physics and Earth Sciences** (2011)
One Camino Santa Maria
San Antonio, TX 78228-8569
p. (210) 436-3235
dfitzgerald@stmarytx.edu
http://www.stmarytx.edu/acad/physicsandearthscience
Chair:
Paul Nienaber
Professor:
David Fitzgerald, (M), Iowa, 1977, Gd
Associate Professor:
Gene W. Lene, (D), Texas, 1981, Ge

## Sam Houston State University

**Geology Program** (B) (2011)
Box 2148
Huntsville, TX 77341
p. (409) 294-1566
bio_bjc@shsu.edu
*Enrollment (2009): B: 53 (7)*
Chair:
Brian J. Cooper, (D), Virginia Tech, 1988, Gz
Professor:
Dennis I. Netoff, (D), Colorado, 1977, Gm
Associate Professor:
Joseph C. Hill, (D), Missouri, 2006, Gcp
Emeritus:
Christopher T. Baldwin, (D), Liverpool, 1976, GsPoGu

## San Antonio Community College

**Dept of Chemistry/Earth Sciences/Astronomy** (A) (2013)
1300 San Pedro Avenue
CG Rm 207
San Antonio, TX 78212
p. (210) 486-0045
tstaggs@alamo.edu
http://www.alamo.edu/sac/earthsci
Professor:
Dean P. Lambert, (D), Texas, 1992, Oyi
Associate Professor:
Anne D. Dietz, (M), Texas A&M (Kingsville), 1989, PgOg
George R. Stanley, (M), Texas (San Antonio), 2006, Og
David A. Wood, (D), Arizona, 2000, Onn
Adjunct Professor:
Thomas Adams, (D), 2011, PvGg
T Scott Girhard, (M), SW Texas State, Oyw
Robert Janusz, (M), Texas (San Antonio), Gg
Dwight Jurena, (M), Rensselaer, GgOg
Ryan E. Rudnicki, (D), Penn State, 1979, Oyr
C. Keith Smith, (M), Texas State, 2005, Oy
Cooperating Faculty:
Steve Dingman, On

## San Jacinto Community College, Central

**Geology Dept** (1999)
8060 Spencer Hwy.
Pasadena, TX 77089
p. (281) 998-6150 x1882
Karen.Purpera@sjcd.edu
http://www.sanjac.edu/

## San Jacinto Community College, North

**Geology Dept** (2013)
5800 Uvalde
Houston, TX 77049
p. (281) 998-6150 x7210
Kevin.Davis@sjcd.edu
http://www.sanjac.edu/

## San Jacinto Community College, South

**Geology Dept** (1999)
13735 Beamer Rd.
Houston, TX 77089
p. (281) 998-6150 x4662
Joe.Granata@sjcd.edu
http://www.sanjac.edu/

## Southern Methodist University

**Roy M. Huffington Dept of Earth Sciences** (B,M,D) (2013)
Post Office Box 750395
Dallas, TX 75275-0395
p. (214) 768-2750
geol@smu.edu
http://www.smu.edu/earthsciences
Administrative Assistant: Stephanie L. Schwob
*Enrollment (2013): B: 60 (15) M: 13 (3) D: 17 (3)*

Chair:
Robert T. Gregory, (D), Caltech, 1981, Cs
Professor:
Bonnie F. Jacobs, (D), Arizona, 1983, Pl
Louis L. Jacobs, (D), Arizona, 1977, Pv
Zhong Lu, (D), Alaska (Fairbanks), 1996, Or
Brian W. Stump, (D), California (Berkeley), 1979, Ys
John V. Walther, (D), California (Berkeley), 1978, Cg
Crayton J. Yapp, (D), Caltech, 1980, Csl
Associate Professor:
Heather R. DeShon, (D), California (Santa Cruz), 2004, Ysr
Matthew J. Hornbach, (D), Wyoming, 2005, YrhYe
M. Beatrice Magnani, (D), Perugia, 2000, YsGtYe
Neil J. Tabor, (D), California (Davis), 2002, Sd
Associate Scientist:
Kurt M. Ferguson, (D), S Methodist, 1990, Cs
Christopher T. Hayward, (D), S Methodist, 1997, Ys
Ian J. Richards, (D), Tennessee, 1994, Cs
Adjunct Professor:
Anthony R. Fiorillo, (D), Pennsylvania, 1989, Pv
John B. Wagner, (D), Texas (Dallas), 2000, Gs
Alisa Winkler, (D), S Methodist, 1990, Pv
Dale A. Winkler, (D), Texas, 1985, Pv
Emeritus:
David D. Blackwell, (D), Harvard, 1968, Yh
James E. Brooks, (D), Washington, 1954, Gr
Michael J. Holdaway, (D), California (Berkeley), 1963, Gp
Robert L. Laury, (D), Wisconsin, 1966, Gd
A. Lee McAlester, (D), Yale, 1960, Pe

## Stephen F. Austin State University

**Dept of Geology** (B,M) O (2016)
PO Box 13011 SFA Station
Nacogdoches, TX 75962
p. (936) 468-3701
geology@sfasu.edu
http://www.geology.sfasu.edu/
Administrative Assistant: Shana Scott
*Enrollment (2010): B: 71 (13) M: 14 (3)*
Chair:
Wesley A. Brown, (D), Texas (El Paso), 2004, YgsGt
Professor:
R. LaRell Nielson, (D), Utah, 1981, Grs
Associate Professor:
Chris A. Barker, (D), South Carolina, 1998, Gc
Kevin W. Stafford, (D), New Mexico Tech, 2008, Hw
Assistant Professor:
Melinda Shaw Faulkner, (D), Stephen F. Austin, 2016, GeClEg
Liane M. Stevens, (D), Montana, 2015, Gpc
Instructor:
Patricia S. Sharp, (M), Stephen F. Austin, 1978, Gg
Lab Coordinator:
Wesley L. Turner, (M), Stephen F. Austin, 2016, Gg

## Sul Ross State University

**Dept of Biology, Geology and Physical Sciences** (B,M) O (2016)
Box C-64
Alpine, TX 79832
p. (432) 837-8112
measures@sulross.edu
http://www.sulross.edu/BGPS
*Enrollment (2016): B: 31 (0) M: 13 (3)*
Professor:
Elizabeth A. Measures, (D), Idaho, 1992, Gdq
David M. Rohr, (D), Oregon State, 1978, PiGd
Kevin Urbanczyk, (D), Washington State, 1993, GiOi
Lecturer:
Jesse Kelsch, (M), New Mexico, Gct

## Tarleton State University

**Chemistry, Geosciences and Physics** (B) (2017)
Box T-540
Stephenville, TX 76402
p. (254) 968-9143
srinivasan@tarleton.edu
http://www.tarleton.edu/CHGP/index.html
Administrative Assistant: Kate Caballero
*Enrollment (2010): B: 42 (7)*
Professor:
Stephen W. Field, (D), Massachusetts, 1988, Gz
Carol A. Thompson, (D), Iowa, 1993, HwGem
Assistant Professor:
Ryan Morgan, (D), Baylor, 2016, Pei
Catherine Ronck, (D), Go

## Tarrant County College, Southeast Campus

**Physical Sciences Dept** (1999)
2100 Southeast Pkwy
Arlington, TX 76018
p. (817) 515-8223
thomas.awtry@tccd.edu
http://www.tccd.edu/academics/tcc-catalog/courses-and-programs/geology/

## Tarrant County College, Northeast Campus

**Natural Science Dept** (A) (2015)
828 Harwood Road
Hurst, TX 76054
p. (817) 515-6565
marles.mccurdy@tccd.edu
https://www.tccd.edu/academics/tcc-catalog/courses-and-programs/geology/
*Enrollment (2011): A: 650 (30)*
Associate Professor:
Meena Balakrishnan, (D), GggGg
Kevin M. Barrett, (D), Texas State, 2012, OagOr
Professor:
Hayden R. Chasteen, (M), Northeast Louisiana, 1981, GgeGg

## Temple College

**Temple College** (2014)
2600 South First Street
Temple, TX 76504
p. (254) 298-8472
john.mcclain@templejc.edu
http://www.templejc.edu/

## Texas A&M University

**Center For Tectonophysics** (M,D) (2017)
3115 TAMU
Department Geology & Geophysics
College of Geosciences
College Station, TX 77843-3115
p. (979) 845-3296
chesterf@tamu.edu
http://tectono.tamu.edu/
*Enrollment (2016): M: 7 (1) D: 13 (1)*
Director:
Frederick Chester, (D), Texas A&M, 1988, GcNrYx
Assistant Director:
Andreas Kronenberg, (D), Brown, 1983, GtyGc
Professor:
Judith Chester, (D), Texas A&M, 1992, GcNrGo
Associate Professor:
Benchuan Duan, (D), California (Riverside), 2006, YsgGq
Julie Newman, (D), Rochester, 1982, Gct
Marcelo Sanchez, (D), Univ Politecnica de Catalunya, 2004, SpNr
David Sparks, (D), Brown, 1992, YdGq
Assistant Professor:
Patrick M. Fulton, (D), Penn State, 2008, YhHyOu
Hiroko Kitajima, (D), Texas A&M, 2010, YxSpGc
Julia S. Reece, (D), Texas (Austin), 2011, SpGso

**Dept of Atmospheric Sciences** (B,M,D) (2016)
3150 TAMU
College Station, TX 77843-3150
p. (979) 845-7671
brady-dennis@tamu.edu
http://atmo.tamu.edu/
*Enrollment (2014): B: 117 (36) M: 48 (5) D: 20 (4)*

Instructional Professor:
Don T. Conlee, (D), Texas A&M, 1994
Distinguished Professor:
Gerald North, (D), 1966, Oa
Distinguished Professor :
Renyi Zhang, (D), MIT, 1993, Oa
Department Head:
Ping Yang, (D), Utah, 1995, Oa
Professor:
Kenneth P. Bowman, (D), Princeton, 1984, Ow
Ping Chang, (D), Princeton, 1988, Op
Donald R. Collins, (D), Caltech, 1999, Oa
Andrew Dessler, (D), Harvard, 1994, On
John Nielsen-Gammon, (D), MIT, 1990, Oa
Richard L. Panetta, (D), Wisconsin, 1978, Ow
R. Saravanan, (D), Princeton, 1990, Oa
Courtney Schumacher, (D), Washington, 2003, Oa
Istvan Szunyogh, (D), Hungarian Academy of Sciences, 1994, Oa
Associate Professor:
Sarah D. Brooks, (D), Colorado, 2002, Oa
Craig Epifanio, (D), Washington, Oa
Robert Korty, (D), MIT, 2005, Oa
Mark Lemmon, (D), Arizona, 1994, Oa
Gunnar Schade, (D), Johannes-Gutenberg (Germany), 1997, Oa
Instructional Assistant Professor:
Tim Logan
Assistant Professor:
Christopher J. Nowotarski, (D), Penn State, 2013, Owa
Anita Rapp, (D), Colorado State, 2008, Or
Yuxuan Wang, (D), Harvard, 2005, Oa
Adjunct Professor:
Larry Carey, (D), Colorado State, 1999, Oa
Christopher A. Davis, (D), MIT, 1990, Ow
Alex Dessler, (D), Duke, 1956, Oa
Hung-Lung Allen Huang, (D), Wisconsin, 1989, Or
Christian D. Kummerow, (D), Minnesota, 1987, Oa
Steve Lyons, (D), Hawaii, 1981, Ow
Chris Snyder, (D), MIT, 1989, Oa
Wei-Kuo Tao, (D), Illinois, 1982, Oa
Emeritus:
Richard E. Orville, (D), Arizona, 1966, Oa

**Dept of Geography** (B,M,D) (2014)
810 Eller O&M Building
3147 TAMU
College Station, TX 77843-3147
p. (979) 845-7141
cbruton@geog.tamu.edu
http://geog.tamu.edu/
*Enrollment (2009): B: 156 (40) M: 16 (8) D: 18 (3)*
Head:
David M. Cairns, (D), Iowa, 1995, Oy
Professor:
Robert S. Bednarz, (D), Chicago, 1975, Oe
Sarah W. Bednarz, (D), Texas A&M, 1992, Oe
John R. Giardino, (D), Nebraska, 1979, Gm
Andrew G. Klein, (D), Cornell, 1997, OriOl
Michael R. Waters, (D), Arizona, 1983, Ga
Associate Professor:
Daniel Z. Sui, (D), Goergia, 1993, Or
Vatche P. Tchakerian, (D), California (Los Angeles), 1989, Gm
Assistant Professor:
Anne Chin, (D), Arizona State, 1994, Gm
Research Associate:
Jean A. Bowman, (M), Rutgers, 1984, Hg
Emeritus:
Clarissa T. Kimber, (D), Wisconsin, 1969, Oy

**Dept of Geology & Geophysics** (B,M,D) (2015)
3115 TAMU
College Station, TX 77843-3115
p. (979) 845-2451
bruton@geo.tamu.edu
http://geoweb.tamu.edu
*Enrollment (2012): B: 278 (41) M: 81 (21) D: 66 (5)*
Head:
John R. Giardino, (D), Nebraska, 1979, GmNg

Regents Professor:
Mary J. Richardson, (D), MIT, 1980, Op
Professor:
Tom Blasingame, (D), Texas A&M, 1989, Np
Richard L. Carlson, (D), Washington, 1976, GtYd
Frederick M. Chester, (D), Texas A&M, 1988, GcNr
Judith Chester, (D), Texas A&M, 1992, GcNr
Mark Everett, (D), Toronto, 1991, Ym
Richard L. Gibson, Jr., (D), MIT, 1991, Yse
Ethan L. Grossman, (D), S California, 1982, Csl
Andrew Hajash, (D), Texas A&M, 1975, Cp
Bruce Herbert, (D), California (Riverside), 1992, Ge
Andreas Kronenberg, (D), Brown, 1983, GyNr
Franco Marcantonio, (D), Columbia, 1994, Ct
Anne Raymond, (D), Chicago, 1983, Pb
William W. Sager, (D), Hawaii, 1979, Ou
Yuefeng Sun, (D), Columbia, 1994, GoYe
Thomas E. Yancey, (D), California (Berkeley), 1971, Pg
Associate Research Professor:
Renald Guillemette, (D), Stanford, 1983, Gz
Associate Professor:
Benchun Duan, (D), California (Riverside), 2006, Ys
Will Lamb, (D), Wisconsin (Madison), 1987, GpCpGz
Brent Miller, (D), Dalhousie, 1997, Cc
Julie Newman, (D), Rochester, 1993, Gc
Thomas Olszewski, (D), Penn State, 2000, Pe
Michael Pope, (D), Virginia Tech, 1995, GrCc
David Sparks, (D), Brown, 1992, Yg
Debbie Thomas, (D), North Carolina, 2002, Ou
Hongbin Zhan, (D), Nevada (Reno), 1996, Hw
Assistant Professor:
Mike Tice, (D), Stanford, 2006, Py
Assistant Dean :
Eric Riggs, (D), California (Riverside), 2000, Gy
Lecturer:
Alfonso Benavides-Iglesias, (D), Texas A&M, 2007, Ys
Emeritus:
Christopher C. Mathewson, (D), Arizona, 1971, NgHwEg
John D. Vitek, (D), Iowa, 1973, GmOe
Technical Laboratory Director:
Michael Heaney, (D), Texas A&M, 1998, Pg
Dean of College of Geosciences:
Kate Miller, (D), Stanford, 1991, Ys

**Dept of Oceanography** (M,D) (2017)
1204 Eller O&M Blg
MS 3146
College Station, TX 77843-3146
p. (979) 845-7211
dthomas@ocean.tamu.edu
http://ocean.tamu.edu
*Enrollment (2015): M: 28 (7) D: 40 (8)*
Director, Texas Sea Grant:
Pamela Plotkin, (D), Texas A&M, 1994, Ob
Head:
Debbie Thomas, (D), North Carolina, 2002, Ou
Executive Associate Dean for Research:
Jack G. Baldauf, (D), California (Berkeley), 1984, Ou
Director, Geochemical and Environmental Research Group:
Anthony Knap, (D), Southampton, Oc
Professor:
Douglas C. Biggs, (D), MIT/WHOI, 1976, Ob
David A. Brooks, (D), Miami, 1975, Op
Lisa Campbell, (D), SUNY (Stony Brook), 1985, Ob
Ping Chang, (D), Princeton, 1988, Op
Piers Chapman, (D), Univ Coll of North Wales, 1982, OcCm
Steven F. DiMarco, (D), Texas (Dallas), 1991, OpnOg
Wilford D. Gardner, (D), MIT/WHOI, 1978, OugOr
Benjamin S. Giese, (D), Washington, 1989, Op
Gerardo Gold-Bouchot, (D), Center for Research & Advan Studies (Mexico), 1991, OcCm
Robert D. Hetland, (D), Florida State, 1999, Op
Alejandro H. Orsi, (D), Texas A&M, 1993, Op
Mary Jo Richardson, (D), MIT/WHOI, 1980, Gu
Gilbert T. Rowe, (D), Duke, 1968, Ob
Peter H. Santschi, (D), Switzerland (Berne), 1975, Oc
Niall C. Slowey, (D), MIT, 1991, Ou

Shari A. Yvon-Lewis, (D), Miami, 1994, Oc
Senior Scientist:
Norman Guinasso, (D), Texas A&M, 1975, Op
Troy Holcombe, (D), Columbia, OuGcg
Matthew K. Howard, (D), Texas A&M, 1992, Op
Ann E. Jochens, (D), Texas A&M, 1977, Op
Adam Klaus, (D), Hawaii, 1991, Gu
Associate Professor:
Rainier Amon, (D), Texas, 1995, Ob
Ayal Anis, (D), Oregon State, 1993, Op
Timothy M. Dellapenna, (D), William & Mary, 1999, Ou
Anja Schulze, (D), Victoria, 2001, Ob
Achim St, (D), Hamburg, 1990, OpwOl
Daniel C.O Thornton, (D), Queen Mary (London), 1995, Ob
Associate Scientist:
Steven K. Baum, (D), Texas A&M, 1996, Op
Jose L. Sericano, (D), Texas A&M, 1993, Cm
Assistant Professor:
Jessica N. Fitzsimmons, (D), MIT/WHOI, 2013, OcCta
Kathryn E. F. Shamberger, (D), Washington, 2011, Oc
Jason B. Sylvan, (D), Rutgers, 2008, ObPy
Yige Zhang, (D), Yale, 2014, Co
Research Associate:
Shinichi Kobara, (D), Texas A&M, Oi
Marion Stoessel, (M), Hamburg, 1984, Op
Zhankun Wang, (D), Massachusetts (Dartmouth), 2009, Op
Instructor:
Chrissy Wiederwohl, (D), Texas A&M, 2012, Op
Distinguished Professor Emeritus:
Robert A. Duce, (D), MIT, 1964, Oc
Worth D. Nowlin, Jr., (D), Texas A&M, 1966, Opn
Distinguished Professor:
Gerald R. North, (D), Wisconsin, 1966, Ow
Director, Texas Sea Grant College Program:
Robert R. Stickney, (D), Florida State, 1971, Ob
Emeritus:
George A. Jackson, (D), Caltech, 1976, Op
Bobby J. Presley, (D), California, 1969, Oc
Robert H. Stewart, (D), California (San Diego), 1969, Op
Laboratory Director:
Terry L. Wade, (D), Rhode Island, 1978, OcCao

**Dept of Soil & Crop Sciences** (B,M,D) (2016)
TAMU 2474
College Station, TX 77843-2474
p. (979) 845-4678
cmorgan@tamu.edu
http://soilcrop.tamu.edu
Administrative Assistant: Carol J. Rhodes
*Enrollment (2014): M: 12 (2) D: 12 (3)*
Professor:
Terry Gentry, (D), Arizona, Sb
Kevin J. McInnes, (D), Kansas State, 1985, Sp
Cristine L. S. Morgan, (D), Wisconsin, 2003, Spd
Tony L. Provin, (D), Purdue, 1995, Sc
Edward C. A. Runge, (D), Iowa State, 1963, Sd
Paul Schwab, (D), Colorado State, 1981, Sc
Associate Professor:
Jacqueline A. Aitkenhead-Peterson, (D), New Hampshire, 2000, HsOsGf
Paul DeLaune, (D), Univ of Arkansas, 2002, Os
Youjun Deng, (D), Texas A&M, 2001, GzSc
Fugen Dou, (D), Texas A&M, 2005, Os
Julie Howe, (D), Wisconsin - Madison, 2004, Sc
Donald McGahan, (D), California (Davis), 2007, Sdc
Assistant Professor:
Jourdan Bell, Texas A&M Univ, 2014
Katie Lewis, (D), Texas A&M Univ, 2014, Sco
Jake Mowrer, (D), Georgia, 2014, Sc
Haly L. Neely, (D), Texas A&M, 2014, Spd
Anil Somenhally, (D), Texas A&M, 2010, Sb

## Texas A&M University, Commerce

**Dept of Biological & Environmental Sciences** (B,M) (2011)
Commerce, TX 75429
p. (903) 886-5378
Haydn_Fox@tamu-commerce.edu
http://www.tamu-commerce.edu/biology/
Assistant Professor:
Haydn A "Chip" Fox, (D), South Carolina, 1994, HgGe

## Texas A&M University, Corpus Christi

**Dept of Physical and Environmental Sciences** (B) (2017)
6300 Ocean Drive
Corpus Christi, TX 78412
p. (361) 825-6000
Valeriu.Murgulet@tamucc.edu
http://geology.tamucc.edu/
*Enrollment (2016): B: 84 (13)*
Chair, Dept, Physical and Environmental Sciences:
Richard Coffin, (D), Delaware, 1986, Ouc
Professor:
Jennifer M. Smith-Engle, (D), Georgia, 1983, Gs
Associate Professor:
Thomas H. Naehr, (D), GEOMAR (Kiel), 1996, GuCm
Director, Center for Water Supply Studies:
Dorina Murgulet, (D), Alabama, 2009, Hw
Endowed Associate Research Professor:
James Gibeaut, (D), South Florida, 1991, Oni
Adjunct Professor:
Clinton Randall (R Bissell, (M), Oklahoma State, 1984, Gg
Erika Locke, (M), California, Los Angeles, 1998, Ggo
Coordinator, Geology Program:
Valeriu Murgulet, (D), Alabama, 2010, Cg

**Environmental Science Program** (B,M) (2017)
6300 Ocean Drive
Corpus Christi, TX 78412
p. (361) 825-2814
jennifer.smith-engle@tamucc.edu
http://pens.tamucc.edu/
*Enrollment (2016): B: 194 (27) M: 36 (12)*
Director of National Spill Control School:
Howard Wood, (M), American Military Univ, 2011, OeHsOn
Program Coordinator:
Jennifer M. Smith-Engle, (D), Georgia, 1983, Gs
Endowed Chair for Socioeconomics:
David Yoskowitz, (D), Texas Tech, 1997, On
Endowed Chair for Fisheries and Ocean Health:
Greg Stunz, (D), Texas A&M, 1999, On
Endowed Chair for Ecosystems Studies and Modeling:
Paul Montagna, (D), South Carolina, 1983, On
Chair, Physical and Environmental Sciences Dept.:
Richard Coffin, (D), Delaware, 1986, Oc
Professor:
Richard McLaughlin, (D), California (Berkeley), On
Associate Professor:
Fereshteh Billiot, (D), Louisiana State, 2000, On
Darek Bogucki, (D), S California, 1996, Op
Gregory Buck, (D), Georgia State, 1999, On
Kirk Cammerata, (D), Kentucky, 1987, On
Patrick Larkin, (D), Texas A&M, 1999, On
Cherie McCollough, (D), Texas (Austin), 2005, On
Riccardo Mozzachiodi, (D), Pisa, 1999, On
Thomas Naehr, (D), Christian-Albrechts (Germany), 1996, Gu
Toshiaki Shinoda, (D), Hawaii, Oa
James Silliman, (D), Michigan, 1998, Co
Delbert Lee Smee, (D), Georgia Tech, 2006, On
Michael Wetz, (D), Oregon State, 2006, Ob
Chair, Center for Water Supply Studies:
Dorina Murgulet, (D), Alabama, 2009, Hw
Assistant Professor:
Hussain Abdulla, (D), Old Dominion Univ, 2009, Co
Jeremy Conkle, (D), Louisiana State, 2010, On
Joseph David Felix, (D), Pittsburgh, 2012, On
Xinping Hu, (D), Old Dominion, 2007, Oc
Chuntao Liu, (D), Wyoming, 2003, Oa
Jennifer Pollack, (D), South Carolina, 2006, Ob
Brandi Reese, (D), Texas A&M, 2011, Ob
Michael Starek, (D), Florida, 2008, Oi
Kim Withers, (D), Texas A&M, 1994, On
Feiqin Xie, (D), Arizona, 2006, OarOw
Lin Zhang, (D), Cs
Research Associate:
Philippe Tissot, (D), Texas A&M, 1994, On

Texas

## Texas A&M University, Kingsville

**Dept of Geosciences** (B) O (2017)
Campus Box 164
Kingsville, TX 78363
p. (512) 595-3310
kftlm00@tamuk.edu
*Enrollment (2016): B: 52 (6)*

Professor:
Thomas L. McGehee, (D), Texas (Dallas), 1987, Cl
Assistant Professor:
Mark T. Ford, (D), Oregon State, GxzCg
Brent Hedquist, (D), Arizona State Univ, Oiw
Veronica I. Sanchez, (D), Houston, GctGm
Robert V. Schneider, (D), Texas El Paso, GoYes
Haibin Su, (D), Cinncinati, Oir
Subbarao Yelisetti, (D), Victoria B.C., 2014, Ygs
Lecturer:
Richard M. Parker, (M), Texas A&M, 2000, GgPg

## Texas Christian University

**School of Geology, Energy, and the Environment** (B,M) (2016)
TCU Box 298830
2950 West Bowie
Fort Worth, TX 76129
p. (817) 257-7270
a.busbey@tcu.edu
https://sgee.tcu.edu/
Administrative Assistant: Krista Scapelli
*Enrollment (2016): B: 76 (13) M: 30 (6)*

TCU Provost and Academic Vice Chancellor:
R. Nowell Donovan, (D), Newcastle upon Tyne, 1972, GdcGt
Professor of Professional Practice:
Becky Johnson, (M), Texas Christian Univ, 1995, GeHwOu
Director of the TCU Institute for Environmental Studies:
Michael C. Slattery, (D), Oxford, 1994, HgSpOy
Director of the TCU Energy Institute:
Ken M. Morgan, (D), Wisconsin, 1978, OrGeOi
Professor:
Richard E. Hanson, (D), Columbia, 1983, GxvGt
John M. Holbrook, (D), Indiana, 1992, Gsr
Geology Program Coordinator:
Arthur B. Busbey, (D), Chicago, 1982, PgvGr
Chair:
Helge Alsleben, (D), S California, 2005, GctGg
Associate Professor:
Rhiannon G. Mayne, (D), Tennessee, 2008, XmgXc
Assistant Professor:
Victoria J. Bennett, (D), Leeds, 2004, On
Omar Harvey, (D), Texas A&M Univ, 2010, OsGe
Xiangyang Xie, (D), Wyoming, 2007, GoEog
Lecturer:
Kristi Argenbright, (M), Texas Christian, GgOn
Adjunct Professor:
Floyd Henk, Jr., (M), Texas Christian, 1981, Go
Emeritus:
John Breyer, (D), Nebraska, 1977, GsdEo
Arthur J. Ehlmann, (D), Utah, 1953, GzXm
Leo Newland, (D), Wisconsin, CaOs
Professor of Professional Practice:
Richard Denne, (D), Louisiana State Univ, 1990, GoEo
Milton Enderlin, (M), Texas Christian, 2010, Nr
Tamie Morgan, (M), Texas Christian, 1984, Oi

## Texas Tech University

**Dept of Geosciences** (B,M,D) O (2016)
Box 41053
2500 Broadway
Science 125
Lubbock, TX 79409-1053
p. (806) 834-0497
alison.winton@ttu.edu
www.geosciences.ttu.edu
*Enrollment (2015): B: 390 (33) M: 48 (21) D: 23 (0)*

Chair:
Jeffrey A. Lee, (D), Arizona State, 1990, Oy
Horn Professor:
Sankar Chatterjee, (D), Calcutta, 1970, Pv
Professor:
George B. Asquith, (D), Wisconsin, 1966, Go
Calvin G. Barnes, (D), Oregon, 1982, Gi
James E. Barrick, (D), Iowa, 1978, Psm
Gary S. Elbow, (D), Pittsburgh, 1972
Juske Horita, (D), Texas A&M, 1997
Thomas M. Lehman, (D), Texas, 1985, Gs
Moira K. Ridley, (D), Nebraska, 1997, Cl
John L. Schroeder, (D), Texas Tech, 1999, Oa
Paul J. Sylvester, (D)
Aaron S. Yoshinobu, (D), S California, 1999, GctGu
Senior Scientist:
Melanie A. Barnes, (D), Texas Tech, 2001, CaGie
Associate Professor:
Eric C. Bruning, (D), Oklahoma, 2008, Oa
Perry L. Carter, (D), Ohio State, 1998
Harold Gurrola, (D), California (San Diego), 1995, Ys
Callum J. Hetherington, (D), Basel (Switzerland), 2001, GxzCg
Haraldur R. Karlsson, (D), Chicago, 1988, Cl
David W. Leverington, (D), Manitoba, 2001, Oiy
Kevin R. Mulligan, (D), Texas A&M, 1997, OiGmSo
Seiichi Nagihara, (D), Texas, 1992, Oi
Christopher C. Weiss, (D), Oklahoma, 2004, OaaOa
Assistant Professor:
Brian C. Ancell, (D), Washington, 2006, Oa
Guofeng Cao, (D), California (Santa Barbara), 2011, Oi
Johannes M L Dahl, (D), Ludwig-Maximilians (Germany), 2010, Owa
Song-Lak Kang, (D), Penn State, 2007, OaHg
Dustin E. Sweet, (D), Oklahoma, 2009, Gsd
Instructor:
Steven R. Cobb, Oa
Linda L. Jones, (M), California (Los Angeles), 1986
Justin E. Weaver, (M), Texas Tech, 1992, Oa
Professor:
Richard E. Peterson, (D), Missouri, 1971, Oa
Unit Coordinator:
Alisan C. Sweet, (M), Oklahoma, 2011, Gs
Debra J. Walker
Senior Technician:
James M. Browning
Senior Business Assistant:
Alison Winton, (B), Texas Tech, 1992, OnnOn
Computer Technician:
Darren W. Hedrick
Academic Advisor:
Celeste N. Yoshinobu, (M), San Diego State, 1994

## Trinity University

**Dept of Geosciences** (B) (2017)
One Trinity Place, #45
San Antonio, TX 78212-7200
p. (210) 999-7092
gkroeger@trinity.edu
https://new.trinity.edu/academics/departments/geosciences
*Enrollment (2016): B: 29 (15)*

Acting Chair:
Glenn C. Kroeger, (D), Stanford, 1987, Yg
Professor:
Thomas W. Gardner, (D), Cincinnati, 1978, GmtHg
Daniel J. Lehrmann, (D), Kansas, 1993, PiGs
Diane R. Smith, (D), Rice, 1984, Gi
Kathleen D. Surpless, (D), Stanford, 2001, Gs
Associate Professor:
Benjamin E. Surpless, (D), Stanford, 1999, GctGi
Adjunct Professor:
Leslie F. Bleamaster III, (D), S Methodist, 2003, Xg
Emeritus:
Walter Coppinger, (D), Miami Univ, 1974, Gcg
Robert L. Freed, (D), Michigan, 1966, Gz

## Tyler Junior College

**Dept of Geology** (1999)
1327 South Baxter Avenue

Tyler, TX 75701
p. (903) 510-2232
gbra@tjc.edu
http://www.tjc.edu/

## University of Houston

**Allied Geophysical Lab** (2014)
Houston, TX 77204-4231
p. (713) 743-9150
rrstewart@uh.edu
http://www.agl.uh.edu

Director:
Robert R. Stewart, (D), MIT, 1983, Ye

**Dept of Earth and Atmospheric Sciences** (B,M,D) O (2017)
SR1, Rm 312
3507 Cullen Blvd
Houston, TX 77204-5007
p. (713) 743-3399
hzhou@uh.edu
http://www.eas.uh.edu
Administrative Assistant: Hannah Dahdouh
Administrative Assistant: Jim Parker
Administrative Assistant: Anja Wells
*Enrollment (2016): B: 413 (69) M: 129 (41) D: 128 (10)*

Department Chair:
Hua-Wei Zhou, (D), Caltech, 1989, YseYg
Professor:
Alan Brandon, (D), Alberta, 1992, GiCac
Kevin Burke, (D), London, 1953, Gt
John F. Casey, (D), SUNY (Albany), 1980, GtiGu
John P. Castagna, (D), Texas, 1983, Ye
Henry S. Chafetz, (D), Texas, 1970, Gds
Evgeny Chesnokov, (D), Russian Acad of Sci, 1987, YseYx
Stuart A. Hall, (D), Newcastle, 1976, Ym
Shuhab D. Khan, (D), Texas (Dallas), 2001, OrGtYg
Thomas Lapen, (D), Wisconsin, 2005, Gz
Aibing Li, (D), Brown, 2000, Ys
Rosalie F. Maddocks, (D), Kansas, 1965, Pm
Paul Mann, (D), SUNY (Albany), 1983
Michael Murphy, (D), California (Los Angeles), 2000, Gc
Bernhard Rappenglueck, (D), Munich, 1996, Oa
Arch M. Reid, (D), Pittsburgh, 1964, GizXm
William W. Sager, (D), Hawaii, 1983, GtYrm
Jonathan Snow, (D), MIT/WHOI, 1992, Ca
Robert Stewart, (D), MIT, 1983, Ye
John Suppe, (D), Yale Univ, 1969, Gtc
Robert Talbot, (D), Wisconsin (Madison), 1981
Arthur B. Weglein, (D), CUNY, 1980, Ye
University Distinguished Research Professor:
Fred Hilterman, (D), Colorado Mines, 1970, Ye
Research Professor:
Adry Bissada, (D), Washington, 1967, Co
Gennady Goloshubin, (D), Inst Physics of Earth (Moscow), 1991, Ye
De-hua Han, (D), Stanford, 1987, Ye
Leon Thomsen, (D), Columbia, 1969, Ye
Associate Professor:
Regina M. Capuano, (D), Arizona, 1988, Hw
Peter Copeland, (D), SUNY (Albany), 1990, CcGt
William R. Dupre, (D), Stanford, 1975, On
Xun Jiang, (D), Caltech, 2006
Alexander Robinson, (D), California (Los Angeles), 2005, Gc
Guoquan Wang, (D), Inst of Geology (China), 2001
Research Associate Professor:
Yongjun Gao, (D), Goettingen, Germany, 2004, CstCa
Virginia Sisson, (D), Princeton Univ, 1985, Gi
Donald Van Niewenhuise, (D), South Carolina, 1978, Ps
Instructional Assistant Professor:
Heather Bedle, (D), Northwestern, 2008, Yes
Daniel Hauptvogel, (D), CUNY, 2015
Jennifer N. Lytwyn, (D), Houston, 1993, Gi
Assistant Professor:
Yunsoo Choi, (D), Georgia Tech, 2007
Qi Fu, (D), Minnesota, 2006, CosCa
Margarete Jadamec, (D), California (Davis), 2009
Joel Saylor, (D), Arizona, 2008
Juan Carlos Silva-Tamayo, (D), Univ Bern, 2009
Yuxuan Wang, (D), Harward Univ, 2005, Oa
Julia Wellner, (D), Rice, 2001, Gs
Jonny Wu, (D), Royal Holloway Univ U(K), 2010, GtcGo
Yingcai Zheng, (D), California (Santa Cruz), 2007
Research Scientist:
Tom Bjorklund, (D), Houston, 2002, Gco
Martin Cassidy, (D), Houston, 2005, GoCo
Nikolay Dyaur, (D), Russian Acad of Sci, 1986, Yxs
Xiangshan Li, (D), Tulane, 2000, Oaw
Research Professor:
James Lawrence, (D), Caltech, 1970, Cl
Peter Percell, (D), California (Berkeley), 1973, Oa
Research Associate Professor:
Dale Bird, (D), Houston, 2004
Robert Wiley, (D), Colorado Mines, 1980
Research Assistant Professor:
James Flynn, (M), Houston, 1991
Charlotte Sjunneskog, (D), Uppsala, 2002
Lecturer:
Peter Bartok, (M), SUNY (Buffalo), 1972, Go
Senior Researcher:
Mike Darnell, (B), Texas A&M, 1974
Postdoctoral Fellow:
Hao Hu, (D), CAS Inst of Geology and Geophysics, 2015, Ye
Wonbae Jeon, (D), Pusan National Univ, 2014
Adjunct:
Amy Kelly, (D), MIT, 2009
Gary Morris, (D), Rice, 1995
Researcher:
Min Sun
Ewa Szymczyk
Fuyong Yan
Laboratory Supervisor:
Minako Righter, (D), Graduate Univ for Advanced Studies, 2006
IT Staff:
Jay Krishnan, (B), Houston, 2001, On

## University of Houston Downtown

**Dept of Natural Sciences** (2014)
1 Main Street
Houston, TX 77002
p. (713) 221-8015
merrillg@uhd.edu

Professor:
Glen K. Merrill, (D), Louisiana State, 1968, Pi
Penny A. Morris-Smith, (D), California (Berkeley), 1975
Associate Professor:
Kenneth S. Johnson, (D), Texas Tech, 1995, GiCac
Lecturer:
Donald S. Musselwhite, (D), 1995

## University of North Texas

**Dept of Geography** (B,M) (2015)
1155 Union Circle #305279
Denton, TX 76203
p. (940) 565-2091
geog@unt.edu
http://www.geography.unt.edu
Administrative Assistant: Tami Deaton

Professor:
Paul F. Hudak, (D), California (Santa Barbara), 1991, Hw
Professor:
Pinliang Dong, (D), New Brunswick, 2003, Oir
C. Reid Ferring, (D), Texas (Dallas), 1993, Ga
Joseph R. Oppong, (D), Alberta, 1992, On
Harry F. L. Williams, (D), Simon Fraser, 1989, Gm
Associate Professor:
Kent M. McGregor, (D), Kansas, 1982, Oag
Lisa A. Nagaoka, (D), Washington, 2000, Ga
Feifei Pan, (D), Georgia Tech, 2002, Hw
Murray Rice, (D), Saskatchewan, 1995, On
Chetan Tiwari, (D), Iowa, 2008, Oi
Steve Wolverton, (D), North Texas, 2001, Ga
Assistant Professor:
Waquar Ahmed, (D), Clark, 2007, On

Ipsita Chatterjee, (D), Clark, 2007, On
Matthew Fry, (D), Texas (Austin), 2008, On
Alexandra Ponette-Gonzalez, (D), Yale, 2011, On

## University of Texas at Austin

Texas

**Dept of Marine Science** (M,D) ● (2015)
750 Channel View Drive
Port Aransas, TX 78373-5015
p. (361) 749-6730
facsearch@utlists.utexas.edu
http://www.utmsi.utexas.edu
f: www.facebook.com/utmsi
*Enrollment (2015): M: 16 (6) D: 16 (1)*

Associate Chair:
Edward J. Buskey, (D), Rhode Island, 1983, Ob
Professor:
Kenneth H. Dunton, (D), Alaska, 1985, Ob
Lee A. Fuiman, (D), Michigan, 1983, Ob
Peter Thomas, (D), Leicester (UK), 1977, Ob
Tracy A. Villareal, (D), Rhode Island, 1989, Ob
Associate Professor:
Bryan A. Black, (D), Penn State, 2003, Sf
Deana L. Erdner, (D), MIT/WHOI, 1997, Ob
Zhanfei Liu, (D), Stony Brook, 2006, Og
James W. McClelland, (D), Boston Univ, 1998, CgHsCs
Assistant Professor:
Brett Baker, (D), Michigan, 2014, Oe
Brad Erisman, (D), California (San Diego), 2008, Ob
Andrew J. Esbaugh, (D), Queens, 2005, Ob
Amber K. Hardison, (D), William & Mary, 2010, Cm
Lecturer:
Gerard C. Shank, (D), North Carolina, 2003, Ob
Emeritus:
Wayne S. Gardner, (D), Wisconsin, 1971, Ob
Gloria J. Holt, (D), Texas A&M, 1976, Ob

**Inst for Geophysics** (2014)
JJ Pickle Research Campus
10100 Burnet Road, Bldg. 196 (ROC)
Austin, TX 78758
p. (512) 471-6156
utig@ig.utexas.edu
http://www.ig.utexas.edu/

Associate Director:
Ian W. D. Dalziel, (D), Edinburgh, 1963, Gt
Cliff Frohlich, (D), Cornell, 1976, Ys
Research Professor:
Stephen P. Grand, (D), Caltech, 1986, Ys
Yosio Nakamura, (D), Penn State, 1963, Ys
Mrinal K. Sen, (D), Hawaii, 1987, Ye
Professor:
Paul L. Stoffa, (D), Columbia, 1974, Ye
Research Scientist:
Gail L. Christeson, (D), MIT/WHOI, 1994, YrsGt
Craig S. Fulthorpe, (D), Northwestern, 1988, Gu
Senior Scientist:
James A. Austin, Jr., (D), MIT/WHOI, 1978, Gu
Nathan L. Bangs, (D), Columbia, 1990, Gu
John A. Goff, (D), MIT/WHOI, 1990, Yr
Lawrence A. Lawver, (D), California (San Diego), 1976, YrGt
Paul Mann, (D), SUNY (Albany), 1983, Gt
Thomas H. Shipley, (D), Rice, 1975, Yr
Frederick W. Talyor, (D), Cornell, 1979, Pe
Research Scientist:
Donald D. Blankenship, (D), Wisconsin, 1989, Yg
Kirk D. McIntosh, (D), California (Santa Cruz), 1992, Yr
Robert J. Pulliam, (D), California (Berkeley), 1991, Ys
Associate Professor:
Luc L. Lavier, (D), Columbia, 1999, YgGt
Research Associate:
Sean S. Gulick, (D), Lehigh, 1999, Yr
John W. Holt, (D), Caltech, 1997, Ye
Charles Jackson, (D), Chicago, 1998, Pe
David L. Morse, (D), Washington, 1997, Yg
Hillary C. Olson, (D), Stanford, 1988, Ps
Robert B. Scott, (D), McGill, 1999, Og
Roustam K. Seifoullaev, (D), Baku State, 1979, Ye
Harm Van Avendonk, (D), Scripps, 1998, Yg
Research Professor:
William E. Galloway, (D), Texas (Austin), 1971, GsoGr
Emeritus:
Milo M. Backus, (D), MIT, 1956, Ye
Arthur E. Maxwell, (D), Scripps, 1959, Og
Postdoc:
Christina Holland, (D), S Florida, 2003, Og
Matthew Hornbach, (D), Wyoming, 2004, Yg
Timothy Whiteaker, (D), Texas, 2004, Gs
Project Coordinator:
Patricia E. Ganey-Curry, (B), Texas A&M, 1978, On
Program Manager:
Katherine K. Ellins, (D), Columbia, 1988, On
Related Staff:
Mark Wiederspahn, (B), Bucknell, 1975, On

**Jackson School of Geosciences** (B,M,D) (2017)
Jackson School of Geosciences
2225 Speedway, Stop C1160
Austin, TX 78712-1692
p. (512) 471-5172
ckerans@jsg.utexas.edu
http://www.jsg.utexas.edu
*Enrollment (2012): B: 315 (66) M: 142 (56) D: 157 (12)*

Dean, Jackson School of Geosciences:
Sharon Mosher, (D), Illinois, 1978, Gc
Chair, Department of Geological Sciences:
Ronald J. Steel, (D), Glasgow (UK), 1970, Gs
Director, Institute for Geophysics:
Terry Quinn, (D), Brown, 1989, Pe
Director, Bureau of Economic Geology:
Scott W. Tinker, (D), Colorado, 1996, Go
Acting Director, Energy & Earth Resources Graduate Program:
William L. Fisher, (D), Kansas, 1961, GsrGo
Professor:
Jay L. Banner, (D), SUNY (Stony Brook), 1986, Cl
Christopher J. Bell, (D), California (Berkeley), 1997, Pv
Philip Bennett, (D), Syracuse, 1988, Hg
Julia A. Clarke, (D), Yale, 2002, PvoPq
Mark P. Cloos, (D), California (Los Angeles), 1981, Gcx
Kerry H. Cook, (D), North Carolina State, 1984, OaPe
Ian W. D. Dalziel, (D), Edinburgh, 1963, Gt
Robert E. Dickinson, (D), MIT, 1966, Oa
Peter B. Flemings, (D), Cornell, 1990, Gr
Sergey B. Fomel, (D), Stanford, 2001, YesEo
Rong Fu, (D), Columbia, 1991, Oa
James E. Gardner, (D), Rhode Island, 1993, Gv
Omar Ghattas, (D), Duke, 1988, GqYg
Stephen P. Grand, (D), McGill, 1986, Yg
Brian Horton, (D), Arizona, 1998, Gs
Charles Kerans, (D), Carleton, 1982, GsrGo
Richard A. Ketcham, (D), Texas (Austin), 1995, Ggq
Gary A. Kocurek, (D), Wisconsin, 1980, Gs
J. Richard Kyle, (D), W Ontario, 1977, EmnCe
David Mohrig, (D), Washington, 1994, GsmGr
Timothy B. Rowe, (D), California (Berkeley), 1987, Pv
Mrinal K. Sen, (D), Hawaii, 1987, Ye
John M. Sharp, Jr., (D), Illinois, 1974, HqwGe
Daniel Stockli, (D), Stanford, 1999, Gc
Paul L. Stoffa, (D), Columbia, 1974, Ye
Robert H. Tatham, (D), Columbia, 1975, Yg
Clark R. Wilson, (D), California (San Diego), 1975, Yg
Zong-Liang Yang, (D), Macquarie, 1992, OwaHq
Senior Research Scientist:
James A. Austin, Jr., (D), MIT/WHOI, 1979, YrGr
Nathan L. Bangs, (D), Columbia, 1990, Gc
Donald D. Blankenship, (D), Wisconsin, 1989, Gl
Gail S. Christeson, (D), MIT, 1994, Yr
Shirley P. Dutton, (D), Texas, 1986, Gs
Clifford A. Frohlich, (D), Cornell, 1976, YsGt
Craig S. Fulthorpe, (D), Northwestern, 1988, GusGr
John A. Goff, (D), MIT, 1990, Gu
Bob A. Hardage, (D), Oklahoma State, 1967, Yes
Susan D. Hovorka, (D), Texas, 1990, Gs
Michael R. Hudec, (D), Wyoming, 1990, Gc
Stephen E. Laubach, (D), Illinois, 1986, Gs

Lawrence A. Lawver, (D), Scripps, 1976, Yr
Robert G. Loucks, (D), Texas (Austin), 1976, Gso
F. Jerry Lucia, (M), Minnesota, 1954, Eo
Kitty L. Milliken, (D), Texas, 1985, Gd
Jeffrey G. Paine, (D), Texas, 1991, On
Stephen C. Ruppel, (D), Tennessee, 1979, Gs
Bridget R. Scanlon, Kentucky, Hg
Thomas H. Shipley, (D), Rice, 1975, Ys
Frederick W. Taylor, (D), Cornell, 1978, GtmGe
Lesli J. Wood, (D), Colorado State, 1992, Eo
Michael H. Young, (D), Arizona, 1995, GeHw
Hongliu Zeng, (D), Texas, 1994, YsGs

Research Scientist:
Peter Eichhubl, (D), California (Santa Barbara), 1997, GcCgNr

Principal Research Scientist:
Patrick Heimbach, (D), Max Planck Inst, 1998, OplOa

Associate Professor:
Bayani Cardenas, (D), New Mexico Tech, 2006, Hw
Elizabeth J. Catlos, (D), California (Los Angeles), 2000, GzCg
Marc A. Hesse, (D), Stanford, 2008, GqoYg
John Lassiter, (D), California (Berkeley), 1995, Cc
Randall A. Marrett, (D), Cornell, 1990, Gc

Associate Scientist:
Ginny A. Catania, (D), Washington, 2004, Ol

Assistant Professor:
Jaime D. Barnes, (D), New Mexico, 2006, CgsCc
Whitney Behr, (D), S California, 2011, Gc
Daniel O. Breecker, (D), New Mexico, 2008, ScCs
Joel P. Johnson, (D), MIT, 2007, Gs
Wonsuck Kim, (D), Minnesota, 2007, Gsr
Luc L. Lavier, (D), Columbia, 1999, Gt
Jung-Fu Lin, (D), Chicago, 2002, GyYm
Timothy M. Shanahan, (D), Arizona, 2007, PeGsCg
Kyle T. Spikes, (D), Stanford, 2008, Ye

Senior Energy Economist:
Gurcan Gulen, (D), Boston Coll, 1996, Ego

Energy Economist:
Svetlana Ikonnikova, (D), Humboldt (Berlin), 2007, Ego

Research Associate:
Todd Caldwell, (D), Nevada (Reno), 2011, Sp
Sigrid Clift, (B), Texas, 1989, Og
Brent Elliott, (D), Helsinki, 2001, GzcGt
Andras Fall, (D), Virginia Tech, 2008, CgGcEg
Peter P. Flaig, (D), Alaska (Fairbanks), 2010, GsrEo
Qilong Fu, (D), Regina, 2005, Gsr
Nicholas W. Hayman, (D), Washington, 2003, Ou
Seyyed Abolfazi Hosseini, (D), Tulsa, 2008, Eo
Farzam Javadpour, (D), Calgary, 2006, NpEo
Carey King, (D), Texas, 2004, Eog
Gang Luo, (D), Missouri, Gtc
Lorena G. Moscardelli, (D), Texas, 2007, GmYs
Hardie S. Nance, (M), Texas, 1978, Grc
Maria-Aikaterini Nikolinakou, (D), MIT, 2008, Np
Yuko Okumura, (D), Hawaii, 2005, OaPe
Cornel Olariu, (D), Texas (Dallas), 2005, Gr
Mariana Olariu, (D), Texas (Dallas), 2007, Gr
Christopher Omelon, Cg
Diana C. Sava, (D), Stanford, 2004, Yg
Timothy L. Whiteaker, (D), Texas, 2004, Oi
Brad Wolaver, (D), Texas (Austin), 2008, HwGe
Changbing Yang, Hw
Christopher K. Zahm, (D), Colorado Mines, 2002, Yg
Mehdi Zeidouni, (D), Calgary, 2011, Eo
Tongwei Zhang, (D), Chinese Acad of Sci, 1999, Cgs

Senior Lecturer:
Mark A. Helper, (D), Texas (Austin), 1985, GctXg

Lecturer:
Mary F. Poteet, (D), California (Berkeley), 2001, Pg

Adjunct Professor:
Laurie S. Duncan, (D), Texas
Marcus Gary, (D), Texas, 2009, Hgs

Emeritus:
Daniel Barker, (D), Princeton, 1961, Gi
Robert E. Boyer, (D), Michigan, 1959, Gt
Richard T. Buffler, (D), California (Berkeley), 1967, Gu
William D. Carlson, (D), California (Los Angeles), 1980, Gp
Peter T. Flawn, (D), Yale, 1951, Eg
Robert L. Folk, (D), Penn State, 1952, Gd
William E. Galloway, (D), Texas, 1971, Gs
Edward C. Jonas, (D), Illinois, 1954, Gz
Wann Langston, Jr., (D), California (Berkeley), 1952, Pv
Leon E. Long, (D), Columbia, 1959, Cc
Ernest L. Lundelius, (D), Chicago, 1954, PvePo
Earle F. McBride, (D), Johns Hopkins, 1960, Gd
Yosio none Nakamura, (D), Penn State, 1963, YsGt
Douglas Smith, (D), Caltech, 1969, Gi
James T. Sprinkle, (D), Harvard, 1971, PioGr

Research Associate Professor:
Sean S. Gulick, (M), Lehigh, 1999, Gc
John W. Holt, (D), Caltech, 1997, OlXyg

## University of Texas, Arlington

**Dept of Earth & Environmental Sciences** (B,M,D) O (2017)
Box 19049
500 Yates Street
Arlington, TX 76019
p. (817) 272-2987
geology@uta.edu
http://www.uta.edu/ees/
*Enrollment (2014): B: 189 (29) M: 70 (21) D: 28 (3)*

Professor:
Asish Basu, (D), California (Davis), 1975, GxCcs
Glen Mattioli, (D), Northwestern, 1987, GitYd
Merlynd K. Nestell, (D), Oregon State, 1966, PmiPs
John S. Wickham, (D), Johns Hopkins, 1969, Gc

Associate Professor:
Qinhong (Max) Hu, (D), Arizona, 1995, GoHwNg
Andrew Hunt, (D), Liverpool, 1988, GbCg
Arne M. Winguth, (D), Hamburg, 1997, Opa

Assistant Professor:
Majie Fan, (D), Arizona, 2009, GstCs
Ashley Griffith, (D), Stanford, 2008, Gct
Liz Griffith, (D), Stanford, 2008, CslCm
Ashanti Johnson, (D), OgGe

Adjunct Professor:
John Damuth, (D), Columbia, GsuYr
Galina P. Nestell, (D), VSEGEI (Russia), 1990, Pms
Cornelia Winguth, (D), Hamburg, 1998, GugOm

Emeritus:
Brooks Ellwood, (D), Rhode Island, 1977, Yg
Christopher R. Scotese, (D), Chicago, 1985, Gt

## University of Texas, Dallas

**Dept of Geosciences** (B,M,D) (2015)
Mail Stop ROC21
800 W Campbell Rd
Richardson, TX 75083-3021
p. (972) 883-2401
geosciences@utdallas.edu
Administrative Assistant: Gloria J. Eby
*Enrollment (2006): B: 40 (2) M: 18 (6) D: 30 (7)*

Professor and Program :
John Dr. Geissman, (D), Gt

Professor :
John S. Oldow, (D), Northwestern, 1978, Gc

Director, Center for Lithospheric Studies:
George A. McMechan, (M), Toronto, 1972, Ys

Professor:
Carlos L. V. Aiken, (D), Arizona, 1976, Yv
William I. Manton, (D), Witwatersrand, 1968, Cc
Robert J. Stern, (D), California (San Diego), 1979, Gt

Research Professor:
Robert B. Finkelman, (D), Maryland, 1980, GbEcCa

Associate Professor:
Tom H. Brikowski, (D), Arizona, 1987, Hw
John F. Ferguson, (D), S Methodist, 1981, Yg

Sr. Lecturer :
Prabin . Shilpakar, (D), Texas (Dallas), 2014, GcYdGc

Senior Lecturer:
William R. Griffin, (D), Texas (Dallas), 2008, Gg
Ignacio Pujana, (D), Texas (Dallas), 1997, Pm

Retired Professor:
Richard M. Mitterer, (D), Florida State, 1966, Col

Emile A. Pessagno, Jr., (D), Princeton, 1960, Pm
Dean C. Presnall, (D), Penn State, 1963, Cp
Retired President and Professor:
Robert H. Rutford, (D), Minnesota, 1969, Gl
Retired Associate Professor:
James L. Carter, (D), Rice, 1965, Eg

**Science/Mathematics Education Program** (M) (2011)
P. O. Box 830688 FN33
Richardson, TX 75080-9688
p. (972) 883-2496
mont@utdallas.edu
http://www.utdallas.edu/dept/SciMathEd
Professor:
Thomas R. Butts, (D), Michigan State, 1973, On
Fred L. Fifer, (D), Vanderbilt, 1973, On
Associate Professor:
Cynthia E. Ledbetter, (D), Texas A&M, 1987, On
Assistant Professor:
Homer A. Montgomery, (D), Texas (Dallas), 1988, Pg
Mary Urquhart, (D), Colorado, 1999, Xy
Instructor:
Barbara Curry, (M), Texas (Dallas), 1998, On

## University of Texas, El Paso

**Dept of Geological Sciences** (B,M,D) O (2016)
500 W. University Avenue
101 Geological Sciences
El Paso, TX 79968-0555
p. (915) 747-5501
jdkubicki@utep.edu
http://science.utep.edu/geology/
*Enrollment (2014): B: 178 (32) M: 55 (15) D: 29 (5)*
Professor:
James D. Kubicki, (D), Yale, 1989, ClGe
Professor:
Elizabeth Y. Anthony, (D), Arizona, 1986, Gi
Diane I. Doser, (D), Utah, 1984, Ys
Katherine A. Giles, (D), Go
Thomas E. Gill, (D), California (Davis), 1995, GmOaCl
Philip C. Goodell, (D), Harvard, 1970, Ce
Steven H. Harder, (D), Texas (El Paso), 1986, Yg
Jose M. Hurtado, (D), MIT, 2002, Gt
Richard S. Jarvis, (D), Cambridge (UK), 1975, OyGmOa
Richard P. Langford, (D), Utah, 1989, Gs
Terry L. Pavlis, (D), Utah, 1982, Gc
Nicholas E. Pingitore, Jr., (D), Brown, 1973, Cl
Laura F. Serpa, (D), Cornell, 1986, Ye
Aaron A. Velasco, (D), California (Santa Cruz), 1993, Yx
Associate Professor:
Marianne Karplus, (D), Stanford, 2012, Ysg
Deana Pennington, (D), Oregon State, 2002, Oi
Assistant Professor:
Benjamin Brunner, (D), ETH, 2003, Cs
Tina Carrick, (D), Texas (El Paso), 2014, OnGg
Lixin Jin, (D), Michigan, 2007, Ge
Lin Ma, (D), Hg
Jie Xu
Science/Eng Research Tech:
Galen M. Kaip, (M), Texas (El Paso), 1998, Yx
Research Associate:
Gail Arnold, (D), Rochester, 2004, Cs
Tina Carrick
Lecturer:
Hector Gonzalez-Huizar, (D), Texas (El Paso), Ys
Vicki Harder
Musa Hussein, (D), Texas (El Paso), 2007, YveYm
Stanley Mubako
Adriana Perez
Jason Ricketts, (D), New Mexico, 2014, GctGi
Maryam Zarei Chaleshtori, (D), Texas (El Paso), 2010
Emeritus:
Kenneth F. Clark, (D), New Mexico, 1966, Eg
David V. Le Mone, (D), Michigan State, 1964, Ps
Network Manager:
Carlos J. Montana, (M), Texas (El Paso), 1992, Yg

## University of Texas, Pan American

**Dept of Physics & Geology** (B) (2011)
1201 W. University Drive
Edinburg, TX 78539
p. (512) 381-3523
bhatti@panam.edu
http://www.utpa.edu/dept/physci/
Chair:
Steven Tidrow
Assistant Professor:
Ruben A. Mazariegos, (D), Texas A&M, 1993, Ye
Eric R. Rieken, (D), Washington State, 1993, Gt

## University of Texas, Permian Basin

**Dept of Geology** (B,M) (2011)
4901 E. University Boulevard
Odessa, TX 79762-0001
p. (432) 552-2243
stoudt_e@utpb.edu
*Enrollment (2005): B: 18 (4) M: 12 (0)*
Chair:
Emilio Mutis-Duplat, (D), Texas (Austin), 1972, Gp
Assistant Professor:
Emily L. Stoudt, (D), Ohio State, 1975, GdPsi
Lecturer:
William L. Basham, (D), Oklahoma, 1978, Ye
Lori L. Manship, (D), Texas Tech, 2008, PioOi
Robert C. Trentham, (D), Texas (El Paso), 1981, Go

## University of Texas, San Antonio

**Dept of Geological Sciences** (B,M) ● (2017)
One UTSA Circle
San Antonio, TX 78249-0663
p. (210) 458-4455
geosciences@utsa.edu
http://www.utsa.edu/geosci
*Enrollment (2015): B: 188 (0) M: 52 (5)*
Professor:
Lance L. Lambert, (D), Iowa, 1992, Pg

## Victoria College

**Science, Mathematics, & Physical Education** (A) (2015)
2200 E. Red River
Victoria, TX 77901
p. (361)573-3291 (Ext.3432)
Alisha.Stearman@VictoriaCollege.edu
https://www.victoriacollege.edu/sciencemathematics
physicaleducation

## Wayland Baptist University

**Geology Department** (B) ● (2016)
1900 W. 7th Street
Plainview, TX 79072
p. (806) 291-1115
walsht@wbu.edu
http://www.wbu.edu/academics/schools/math_and_science/
geology/default.htm
f: https://www.facebook.com/Wayland-Baptist-University-
Geology-Department-198757676854342/
*Enrollment (2016): B: 7 (1)*
Professor:
Don Parker, (D), GivCg
Tim R. Walsh, (D), Texas Tech, 2002, GsPmGo
Assistant Professor:
Mark Bryan, (M), Oklahoma State, 2001, GeoPs

## Weatherford College

**Weatherford College** (1999)
225 College Park Drive
Weatherford, TX 76086
p. (817) 598-6277
http://www.wc.edu

## West Texas A&M University

**Dept of Life, Earth & Environmental Sciences** (B,M) (2015)
P. O. Box 60808, WT Station
Canyon, TX 79016-0001
p. (806) 651-2570
dsissom@wtamu.edu
http://www.wtamu.edu/academics/life-earth-environmental-sciences.aspx
Administrative Assistant: Debi Adams
*Enrollment (2013): B: 13 (2)*

Chair:
David Sissom, (D), On

Professor:
Joseph C. Cepeda, (D), Texas, 1977, GiHw
David B. Parker, (D), Nebraska, 1996, Sbf
William J. Rogers, (D), Texas A&M, 1999, On
Gerald E. Schultz, (D), Michigan, 1966, PvGzOg

Associate Professor:
Gary C. Barbee, (D), Texas A&M, 2004, HwOsi

Instructor:
Cindy D. Meador, (M), West Texas A&M, 1989, Og
Joe D. Rogers, (M), WTSU (WTAMU), 1987, Ga
William C. Rogers, (M), WTAMU, 2005, Ge
Lynn C. Rosa, (M), Oklahoma, 1984, Oge

## Western Texas College

**Math and Science Dept** (2011)
6200 College Avenue
Snyder, TX 79549
p. (325) 573-8511
http://wtc.edu/science/index.html

Assistant Professor:
Troy Lilly, (M), Texas Tech, 1981, GgOg

## Wharton County Junior College, Sugarland Campus

**Wharton County Junior College - Sugarland Campus** (2014)
14004 University Blvd.
Sugarland, TX 77479
p. (281) 239-1559
dannyg@wcjc.edu
http://www.wcjc.edu

## Wharton County Junior College, Wharton Campus

**Geology** (2014)
911 Boling Highway
Wharton, TX 77488
p. (979) 532-6506
dannyg@wcjc.edu
https://www.wcjc.edu/Programs/math-and-science/geology/

# Utah

## Brigham Young University

**Dept of Geography** (B) (2017)
690 SWKT
Provo, UT 84602
p. (801) 378-3851
geography@byu.edu
http://www.geography.byu.edu/
Department Secretary: Karen R. Bryce

Chair:
Matthew J. Shumway, (D), Indiana, 1991, On

Professor:
Richard H. Jackson, (D), Clark, 1970, Ou
Samuel M. Otterstrom, (D), Louisiana State, 1997, Ou

Associate Professor:
Matthew F. Bekker, (D), Iowa, 2002, Oy
James A. Davis, (D), Arizona State, 1992, On
Chad Emmett, (D), Chicago, 1991, On
Perry J. Hardin, (D), Utah, 1989, Or

Assistant Professor:
Jeffrey O. Durrant, (D), Hawaii, 2001, Ge
Mark W. Jackson, (D), South Carolina, 2001, Or
Brandon Plewe, (D), Buffalo, 1997, Oi

Instructor:
Jeffry S. Bird, (M), Brigham Young, 1990, Oy

Emeritus:
Lloyd E. Hudman, (D), Kansas, 1968, On

**Dept of Geological Sciences** (B,M) (2013)
S 389 ESC
Provo, UT 84602
p. (801) 422-3918
geology.office@byu.edu
http://www.geology.byu.edu
Administrative Assistant: Kristine B.. Mortenson
*Enrollment (2013): B: 152 (30) M: 38 (7)*

Professor:
Barry R. Bickmore, (D), Virginia Tech, 2000, ClGz
Eric H. Christiansen, (D), Arizona State, 1981, GiXgGv
Michael J. Dorais, (D), Georgia, 1987, Gi
Ron Harris, (D), London (UK), 1989, Gc
Jeffrey D. Keith, (D), Wisconsin, 1982, Eg
Bart J. Kowallis, (D), Wisconsin (Madison), 1981, GgCcGz
John H. McBride, (D), Cornell, 1987, YeGoe
Thomas H. Morris, (D), Wisconsin, 1986, Gr
Stephen T. Nelson, (D), California (Los Angeles), 1991, Cs
Scott M. Ritter, (D), Wisconsin, 1986, Ps
David G. Tingey, (M), Brigham Young, 1989, On

Associate Professor:
Brooks B. Britt, (D), Calgary, 1993, Pg
Jani Radebaugh, (D), Arizona, 2005, Xg
Summer Rupper, (D), Washington, 2004, OlPe
Randall Skinner, (M), Brigham Young, 1996, Gg

Assistant Professor:
Gregory T. Carling, (D), Utah, 2012, HwCts

Adjunct Professor:
Thomas C. Anderson, Gg

Visiting Professor:
R. William Keach II, (M), Cornell, 1986, YeEoGg

Emeritus:
James L. Baer, (D), Brigham Young, 1968, Go
Myron G. Best, (D), California (Berkeley), Gg
Dana T. Griffen, (D), Virginia Tech, 1976, Gz
Lehi F. Hintze, (D), Columbia, 1951, Pg
Alan L. Mayo, (D), Idaho, 1982, Hw
Wade E. Miller, (D), California (Berkeley), 1968, Pv
R. Paul Nixon, (D), Brigham Young, 1972, Go
Morris S. Petersen, (D), Iowa, 1962, Gso
William R. Phillips, (D), Utah, 1954, Gz

## Salt Lake City Community College, Jordan Campus

**Dept of Geosciences** (1999)
4600 South Redwood Road
Salt Lake City, UT 84123
p. (801)957-4150
adam.dastrup@slcc.edu
http://www.slcc.edu

## Snow College

**Dept of Geology** (A) (2016)
150 College Ave E
Ephraim, UT 84627
p. (435) 283-7519
renee.faatz@snow.edu
https://www.snow.edu/academics/science_math/geology/index.html
f: Snow College Geology
*Enrollment (2016): A: 7 (1)*

Chair:
Renee M. Faatz, (M), Ohio State, 1985, Gg

Assistant Professor:
Ted L. Olson, (M), Utah, 1976, Ys

## Southern Utah University

**Dept of Physical Science** (B) O (2015)
351 West University Blvd.
Cedar City, UT 84720
p. (435) 586-7900
jenniferhargrave@suu.edu

http://www.suu.edu/geology
f: https://www.facebook.com/SUUGeologyClub
Administrative Assistant: Rhonda Riley
*Enrollment (2015): B: 30 (8)*

Professor:
Robert L. Eves, (D), Washington State, 1991, CgOeCa
Assistant Professor:
Jennifer E. Hargrave, (D), Oklahoma, 2009, PvGrs
Jason Kaiser, (D), Oregon State, 2014, GzvCg
John S. MacLean, (D), Montana, 2009, GctOe
Emeritus:
C. Frederick Lohrengel, II, (D), Brigham Young, 1968, GsrPs

## University of Utah

**Dept of Atmospheric Sciences** (B,M,D) (2016)
135 S 1460 E, Rm 819
Salt Lake City, UT 84112-0110
p. (801) 581-6136
atmos-advising@lists.utah.edu
http://www.atmos.utah.edu
*Enrollment (2016): B: 49 (10) M: 17 (4) D: 26 (1)*

Professor:
Timothy J. Garrett, (D), Washington, 2000, Oa
John Horel, (D), Washington, 1982, Oa
Steven Krueger, (D), California (Los Angeles), 1985, Oa
Gerald Mace, (D), Penn State, 1994, Oa
Zhaoxia Pu, (D), Lanzhou, 1997, Oa
Jim Steenburgh, (D), Washington, 1995, Oa
Edward Zipser, (D), Florida State, 1965, Oa
Chair:
Kevin D. Perry, (D), Washington, 1995, Oa
Associate Professor:
Anna Gannet Hallar, (D), Colorado (Boulder), 2003, Oa
John Chun-Han Lin, (D), Harvard, 2003, Oa
Thomas Reichler, (D), Calfornia (San Diego), 2003, Oa
Courtenay Strong, (D), Virginia, 2005, Oa

**Dept of Geography** (B,M,D) (2014)
260 S. Central Campus Drive
Rm 270
Salt Lake City, UT 84112-9155
p. (801) 581-8218
thomas.kontuly@geog.utah.edu
http://www.geog.utah.edu
Administrative Assistant: Susan Van Roosendaal

Chair:
George F. Hepner, (D), Arizona State, 1979, Oy
Professor:
Donald R. Currey, (D), Kansas, 1969, Gm
Thomas M. Kontuly, (D), Pennsylvania, 1978, On
Chung M. Lee, (D), Michigan, 1961, Ou
Assistant Professor:
Thomas J. Cova, (D), California (Santa Barbara), 1999, Oi
Richard R. Forster, (D), Cornell, 1997, Or
Adjunct Professor:
Jeffrey R. Keaton, (D), Texas A&M, 1988, On
Elliott W. Lips, (M), Colorado State, 1990, Gm
Emeritus:
Philip C. Emmi, (D), North Carolina, 1979, Ou
Roger M. Mccoy, (D), Kansas, 1967, Or
Merrill K. Ridd, (D), Northwestern, 1963, Or
Professor-Lecturer:
Arthur Hampson, (D), Hawaii, 1980, Oy
Assistant Professor:
Trevor J. Davis, (D), British Columbia, 1999, Oi
Other:
Fred E. May, (D), Virginia Tech, 1976, On

**Dept of Geology & Geophysics** (B,M,D) ● (2015)
Frederick Albert Sutton Building, Rm 383
115 South 1460 East
Salt Lake City, UT 84112-0102
p. (801) 581-7062
gg@utah.edu
http://www.earth.utah.edu/
Administrative Assistant: Judy Martinez
Administrative Assistant: Dustin Porlas
*Enrollment (2014): B: 55 (0) M: 50 (0) D: 25 (0)*

Research Professor:
Walter J. Arabasz, (D), Caltech, 1971, Ys
Departmental Chair:
John M. Bartley, (D), MIT, 1981, GciCc
Dean, College of Mines & Earth Sciences:
Francis H. Brown, (D), California (Berkeley), 1971, Gg
Professor:
John R. Bowman, (D), Michigan, 1976, Cs
Paul D. Brooks, (D), Colorado, 1995, HgOg
Thure E. Cerling, (D), California (Berkeley), 1977, Cl
Marjorie A. Chan, (D), Wisconsin, 1982, Gs
David A. Dinter, (D), MIT, 1994, GtcGu
Allan A. Ekdale, (D), Rice, 1974, Pe
Cari Johnson, (D), Stanford, 2003, Gs
William P. Johnson, (D), Colorado, 1993, Ng
Barbara P. Nash, (D), California (Berkeley), 1971, Gi
Erich U. Petersen, (D), Michigan, 1983, Em
Peter H. Roth, (D), ETH (Switzerland), 1970, Pm
Douglas K. Solomon, (D), Waterloo, 1992, Hw
Michael S. Zhdanov, (D), Moscow State, 1968, Ye
Research Associate Professor:
Kristine L. Pankow, (D), California, 1999, Ys
Director, Seismograph Stations:
Keith Koper, (D), Washington, 1998
Associate Director, Global Change and Sustainability Center:
Brenda Bowen, (D), Gse
Associate Professor:
Gabriel Bowen, (D), California (Santa Cruz), 2003, Cs
Paul W. Jewell, (D), Princeton, 1989, Hg
David L. Naftz, (D), Colorado Mines, 1993, Cg
Michael Thorne, (D), Arizona State, 2005, YsxYv
Research Associate Professor:
James C. Pechmann, (D), Caltech, 1983, Ys
Research Assistant Professor:
Diego Fernandez, (D), Buenos Aires, 1991, Cg
Research:
Alex Gribenko, (D), Yg
Lecturer:
Holly Godsey, (D), Oe
Assistant Professor:
Lauren Birgenheier, (D), Nebraska (Lincoln), 2007, Gso
Randall B. Irmis, (D), California (Berkeley), 2008, Sa
Fan-Chi Lin, (D), Colorado (Boulder), 2009, Ygs
Peter C. Lippert, (D), California (Santa Cruz), 2010, YmGtPe
Lowell Miyagi, (D), California (Berkeley), 2009, Gz
Jeffrey Moore, (D), California (Berkeley), 2007, NgrGm
Lisa Stright, (D), Stanford, 2011, Ggo
Research Assistant Professor:
Desmond E. Moser, (D), Queen's, 1993, Cc
Adjunct Associate Professor:
Robert N. Harris, (D), Utah, 1996, Gu
Victor Heilweil, (D), Utah, 2003, Hw
James I. Kirkland, (D), Colorado, 1990, PvsPi
Virginia B. Sisson, (D), Princeton, 1985, Gp
Adjunct Assistant Professor:
Kathleen Nicoll, (D), Arizona, 1998, Gs
Adjunct Professor:
Richard Allis, (N)
David Applegate, (D), MIT, 1994, Gt
Lukas Baumgartner, (N)
Harley M. Benz, (D), Utah, 1986, Ys
John M. Harris, (D), Texas, Pv
William G. Pariseau, (D), Minnesota, 1966, Nr
Gerard T. Schuster, (D), Columbia, 1984, Ye
Aurel Trandafir, (D), Kyoto, 2004, Nr
Phillip E. Wannamaker, (D), Utah, 1983, Ye
Emeritus:
Ronald L. Bruhn, (D), Columbia, 1976, Gc
David S. Chapman, (D), Michigan, 1976, Yh
Susan L. Halgedahl, (D), California (Santa Barbara), 1981, Ym
Richard D. Jarrard, (D), California (San Diego), 1975, Gu
William T. Parry, (D), Utah, 1961, CgGz
M. Dane Picard, (D), Princeton, 1963, Gd
Robert B. Smith, (D), Utah, 1967, Ys

**Dept of Mining Engineering** (B,M,D) (2011)
135 South 1460 East
Room 313
Salt Lake City, UT 84112-0113
p. (801) 581-7198
mineeng@mines.utah.edu
http://www.mines.utah.edu/mining
*Enrollment (2009): B: 48 (11) M: 6 (3) D: 1 (0)*

Chair:
Michael G. Nelson, (D), West Virginia, 1989, Nm
Professor:
Michael K. McCarter, (D), Utah, 1972, Nm
William G. Pariseau, (D), Minnesota, 1966, Nr
Adjunct Associate Professor:
Stephen Bessinger, (D), Nm
Associate Professor:
Felipe Calizaya, (D), Colorado Mines, 1985, Nm
Adjunct Associate Professor:
Helmut H. Doelling, (D), Utah, 1964, Gg
Duane L. Whiting, (B), Utah, 1959, Hw
Adjunct Professor:
James Donovan, (D), Virginia Tech, 2003, NmrOr
Krishna P. Sinha, (D), Minnesota, 1979, Nr
Jeffrey Whyatt, (D), Nm
Zavis Zavodni, (D), Nm

**Energy & Geoscience Institute** (M) ● (2016)
423 Wakara Way
Suite 300
Salt Lake City, UT 84108
p. (801) 581-5126
egidirector@egi.utah.edu
http://www.egi.utah.edu

Director:
Raymond A. Levey, (D), South Carolina, 1981, Eo
Professor:
Richardson B. Allen, (D), Columbia, 1983, GctOi
Alastair Fraser, (D), Edinburgh, Gco
Brian McPherson, (D), Utah, 1996, Ye
Joseph N. Moore, (D), Penn State, 1975, Gv
Michal Nemcok, (D), Comenius Univ (Slovakia), 1991, GctGo
Peter E. Rose, (D), Utah, 1993, Np
Rasoul Sorkhabi, (D), Japan, 1991, GctGc
Phillip E. Wannamaker, (D), Utah, 1983, Ye
Scientific Staff :
Julia Kotulova, (D), Komenius Univ Bratislava (Slovakia), CoGo
Senior Scientist:
Bryony Richards-McClung, (D), Go
Stuart Simmons, (D), Minnesota, 1986, Gz
Lansing Taylor, (D), Stanford, 1999, Gc
David Thul, (D), Colorado Mines, 2014, Gg
Associate Professor:
Glenn W. Johnson, (D), South Carolina, 1997, Gq
John D. McLennan, (D), Toronto, 1980, Nrp
Greg Nash, (D), Utah, Gt
Marylin Segall, (D), 1991, GeOu
Assistant Professor:
Shu Jiang , (D), China Univ of Geosciences (Wuhan), GoYsNp
Sudeep Kanungo, (D), Univ Coll (London), Pms
Research Associate:
Kenneth L. Shaw, (B), British Columbia, 1965, Ye
Instructor:
William Keach, (M), Cornell, 1986, GoYes
Adjunct Professor:
Ian Walton, (D), Manchester (UK), 1972, Gq

## Utah State University

**Dept of Geology** (B,M,D) ● (2016)
4505 Old Main Hill
Logan, UT 84322-4505
p. (435) 797-1273
geology@usu.edu
http://geology.usu.edu/
*Enrollment (2016): B: 70 (10) M: 28 (10) D: 4 (0)*

Head:
Joel L. Pederson, (D), New Mexico, 1999, Gm
Professor:
James P. Evans, (D), Texas A&M, 1987, Gc
Susanne U. Janecke, (D), Utah, 1991, Gt
W. David Liddell, (D), Michigan, 1979, GsPo
John W. Shervais, (D), California (Santa Barbara), 1979, Gi
Associate Professor:
Carol M. Dehler, (D), New Mexico, 2001, Gs
Michelle Fleck, (D), Wyoming, 2001, Oe
Thomas E. Lachmar, (D), Idaho, 1989, Hw
Anthony R. Lowry, (D), Utah, 1994, Yds
Tammy M. Rittenour, (D), Nebraska, 2004, GmCcGa
Assistant Professor:
Alexis K. Ault, (D), Colorado, 2012, GtCc
Dennis L. Newell, (D), New Mexico, 2007, Cls
Research Associate:
Kelly K. Bradbury, (D), Utah State Univ, 2012, Gc
Curator of Paleontology, CEU Prehistoric Museum:
Kenneth Carpenter, (D), Colorado, 1996, PvoPe
Emeritus:
Donald W. Fiesinger, (D), Calgary, 1975, Gi
Peter T. Kolesar, (D), California (Riverside), 1973, Cg
Robert Q. Oaks, Jr., (D), Yale, 1965, GsHwYv

## Utah State University Eastern

**Dept of Geology** (A) (2016)
451 East 400 North
Price, UT 84501
p. (435) 613-5232
michelle.fleck@usu.edu
*Enrollment (2016): A: 6 (0)*

Associate Professor:
Michelle Cooper Fleck, (D), Wyoming, 2001, GgOy

## Utah Valley University

**Dept. of Earth Science** (B) (2016)
800 West University Parkway
Orem, UT 84058
p. (801) 863-8582
HORNSDA@uvu.edu
http://www.uvu.edu/
*Enrollment (2016): B: 179 (16)*

Professor:
Daniel Horns, (D), NgGet
Associate Professor:
Joel Bradford, (M), Utah, Gae
Michael Bunds, (D), GctGe
Eddy Cadet, (D), Ge
James Callison, (D), GeSfOs
Steven H. Emerman, (D), Cornell, 1984, HgYgGe
Daniel Stephen, (D), PiGsPe
Assistant Professor:
Hilary Hungerford, (D), Ou
S. McKenzie Skiles, (D), California (Los Angeles), 2014, HsOri
Nathan Toke', (D), Arizona State, 2011, GtmOi
Weihong Wang, (D), CmGeOi
Alessandro Zanazzi, (D), CgsPe

## Weber State University

**Dept of Geosciences** (B) ○ (2016)
1415 Edvalson St. Dept 2507
Ogden, UT 84408-2507
p. (801) 626-7139
rford@weber.edu
http://weber.edu/geosciences/
f: https://www.facebook.com/WSUGeosciences?ref=hl
Administrative Assistant: Marianne Bischoff
*Enrollment (2016): B: 97 (16)*

Dean, College of Science:
David J. Matty, (D), Rice, 1984, Gig
Chair:
Richard L. Ford, (D), California (Los Angeles), 1997, GmOe
Professor:
Michael W. Hernandez, (D), Utah, 2004, Ori
Marek Matyjasik, (D), Kent State, 1997, HwGe
W. Adolph Yonkee, (D), Utah, 1990, Gct

Assistant Professor:
Elizabeth A. Balgord, (D), Arizona, 2015, Grs
Carie M. Frantz, (D), S California, 2013, PyCgOn
Instructor:
Amanda L. Gentry, (M), Nevada (Las Vegas), 2016, GsCcGg
Stephen C. Hallin, (M), Colorado State, 1991, Ow
David Larsen, (M), Brigham Young, 1987, Ge
Gregory B. Nielsen, (D), Utah, 2010, Gg
Sara Summers, (M), Notre Dame, 2012, Gzg
Emeritus:
Sydney R. Ash, (D), Reading (UK), 1966, Pb
Jeffrey G. Eaton, (D), Colorado, 1987, PsGs
Richard W Moyle, (D), Iowa, 1963, Ps
E. Fred Pashley, (D), Arizona, 1966, Ge
James R. Wilson, (D), Utah, 1976, Gz



# Vermont

## Castleton University
**Dept of Natural Sciences (Geology)** (B) (2016)
Castleton University
233 South Street
Castleton, VT 05735
p. (802) 468-1238
tim.grover@castleton.edu
http://www.castleton.edu/academics/undergraduate-programs/geology/
*Enrollment (2015): B: 12 (3)*
Professor:
Timothy W. Grover, (D), Oregon, 1988, Gp
Helen N. Mango, (D), Dartmouth, 1992, CgGe

## Johnson State College
**Dept of Environmental & Health Sciences** (A,B) (2013)
337 College Hill
Johnson, VT 05656-9464
p. (802) 635-1325
Tania.Bacchus@jsc.edu
*Enrollment (2010): B: 37 (11)*
Professor:
Tania S. Bacchus, (D), Maine, 1993, GuOw
Leslie H. Kanat, (D), Cambridge (UK), 1986, Gce

## Lyndon State College
**Dept of Atmospheric Sciences** (B) (2016)
PO Box 919
Lyndonville, VT 05851
p. (802) 626-6254
lsc-metchair@lyndonstate.edu
http://meteorology.lyndonstate.edu
Department Secretary: Brenda Sweet
*Enrollment (2012): B: 96 (9)*
Professor:
Nolan T. Atkins, (D), California (Los Angeles), 1995, Oa
Bruce F. Berryman, (D), Wisconsin, 1974, Oa
Assistant Professor:
Shafer Jason, (D), Utah, 2005, Oa
System Administrator:
Tucker Mark, (B), Lyndon State, Oa

## Middlebury College
**Geology Dept** (B) (2016)
276 Bicentennial Hall
Middlebury, VT 05753
p. (802) 443-5029
geology_chair@middlebury.edu
http://www.middlebury.edu/academics/geol
f: https://www.facebook.com/MiddleburyGeology
Administrative Assistant: Eileen Brunetto
*Enrollment (2016): B: 30 (15)*
Chair:
David P. West, (D), Maine (Orono), 1993, Gc
Professor:
Patricia L. Manley, (D), Columbia, 1989, YrGu
Jeffrey S. Munroe, (D), Wisconsin, 2001, Gln
Peter C. Ryan, (D), Dartmouth, 1994, Cl
Assistant Professor:
Will Amidon, (D), Caltech, 2010, Gm
Kristina J. Walowski, (D), Oregon, 2015, Gi
Visiting Professor:
Thomas O. Manley, (D), Columbia, 1981, Op

## Norwich University
**Dept of Earth and Environmental Sciences** (B) (2016)
158 Harmon Drive
Northfield, VT 05663
p. (802) 485-2304
rdunn@norwich.edu
http://scimath.norwich.edu/geology-environmental-science/
*Enrollment (2016): B: 15 (3)*
Professor:
Richard K. Dunn, (D), Delaware, 1998, GslGa
Professor:
David S. Westerman, (D), Lehigh, 1972, Gxi
Assistant Professor:
G. Christopher Koteas, (D), Massachusetts, 2010, GitYh
Research Associate:
George E. Springston, (M), Massachusetts (Amherst), 1990, GmOiYg
Lecturer:
Laurie Grigg, (D), Oregon, 2000, PeOiGe

## University of Vermont
**Dept of Geology** (B,M) (2016)
Delehanty Hall
Burlington, VT 05405-0122
p. (802) 656-3396
geology@uvm.edu
Administrative Assistant: Robin Hopps
Administrative Assistant: Srebrenka Mrsic
*Enrollment (2015): B: 48 (0) M: 13 (0)*
Professor:
Paul R. Bierman, (D), Washington, 1993, GmeCc
John M. Hughes, (D), Dartmouth, 1981, Gz
Keith A. Klepeis, (D), Texas (Austin), 1993, Gct
Charlotte J. Mehrtens, (D), Chicago, 1979, Gs
Chair:
Andrea Lini, (D), ETH (Switzerland), 1994, CsGn
Associate Professor:
Laura E. Webb, (D), Stanford, 1999, GtCcGc
Research Assistant Professor:
Nicolas Perdrial, (D), Strasbourg (France), 2007, GezCl
Andrew W. Schroth, (D), Dartmouth, 2007, ClGz
Assistant Professor:
Julia Perdrial, (D), Strasbourg, 2008, ClSc
Lecturer:
Stephen F. Wright, (D), Minnesota, 1988, Gc
Emeritus:
David P. Bucke, (D), Oklahoma, 1969, GgdGs
Barry L. Doolan, (D), SUNY (Binghamton), 1970, Gx
John C. Drake, (D), Harvard, 1967, Cl
Senior Research Technician:
Gabriela Mora-Klepeis, (M), Texas (Austin), 1992, CacGi

**Dept of Plant & Soil Science** (B,M,D) (2014)
Jeffords Hall
63 Carrigan Drive
Burlington, VT 05405-1737
p. (802) 656-2630
pss@uvm.edu
http://www.uvm.edu/~pss/

# Virginia

## Central Virginia Community College
**Science, Math, and Engineering** (A) (2013)
3506 Wards Road
Lynchburg, VA 24502

p. (434) 832-7707
laubj@cvcc.vccs.edu
http://www.cvcc.vccs.edu/Academics/SME/default.asp
Instructor:
Mark Tinsley, Og

## College of William & Mary

**Dept of Geology** (B) (2016)
PO Box 8795
McGlothlin-Street Hall
Williamsburg, VA 23187-8795
p. (757) 221-2440
crroex@wm.edu
www.wm.edu/as/geology
Administrative Assistant: Carol Roe
*Enrollment (2016): B: 62 (35)*
Chair:
Christopher M. Bailey, (D), Johns Hopkins, 1994, Gc
Professor:
Gregory S. Hancock, (D), California (Santa Cruz), 1998, Gm
Rowan Lockwood, (D), Chicago, 2001, Po
R. Heather Macdonald, (D), Wisconsin, 1984, GsOe
Brent E. Owens, (D), Washington Univ (St. Louis), 1992, Gx
Associate Professor:
James Kaste, (D), Dartmouth, 2003, ClcSc
Assistant Professor:
Nicholas Balascio, (D), Massachusetts (Amherst), 2011, OyGnl
Research Associate:
Carl R. Berquist, Jr., (D), William & Mary, 1986, Ou
Lecturer:
Rebecca Jiron, (D), California (Santa Barbara), 2015, GmtOg
Emeritus:
Stephen C. Clement, (D), Cornell, 1964, Gz
P. Geoffrey Feiss, (D), Harvard
Gerald H. Johnson, (D), Indiana, 1965, Pg
Director of Laboratories & Technical Support:
Linda D. Morse, (B), Virginia Tech, 1983, Ge

**School of Marine Science** (M,D) (2015)
Virginia Institute of Marine Science
P. O. Box 1346
Gloucester Point, VA 23062-1246
p. (804) 684-7105
ad-as@vims.edu
http://www.vims.edu
*Enrollment (2015): M: 6 (1) D: 11 (1)*
Chair:
Deborah A. Bronk, (D), Maryland, 1992, Oc
Research Professor:
Jian Shen, (D), William & Mary, 1996, Op
Department Chair:
Carl T. Friedrichs, (D), MIT/WHOI, 1993, On
Professor:
Elizabeth A. Canuel, (D), North Carolina, 1992, Oc
Courtney K. Harris, (D), Virginia, 1999, On
Steven A. Kuehl, (D), North Carolina State, 1985, Ou
Jerome P-Y. Maa, (D), Florida, 1986, On
Harry Wang, (D), Johns Hopkins, 1983, Op
Research Associate Professor:
William G. Reay, (D), Virginia Tech, 1992, Hg
Y. Joseph Zhang, (D), Wollongong (Australia), 1996, Op
Associate Professor:
John M. Brubaker, (D), Oregon State, 1979, Op
Assistant Professor:
Donglai Gong, (D), Rutgers, 2010, Op
Christopher J. Hein, (D), Boston, 2012, Ou
Matthew L. Kirwan, (D), Duke, 2007, Gm
Elizabeth H. Shadwick, (D), Dalhousie, 2010, Oc

## George Mason University

**Dept of Atmospheric, Oceanic, and Earth Sciences** (B,M,D) O (2016)
Research Hall 206
4400 University Drive
Fairfax, VA 22030-4444
p. (703) 993-9587
eschnei1@gmu.edu
https://cos.gmu.edu/aoes
Administrative Assistant: Maria D'Souza
Administrative Assistant: Stephanie O'Neill
*Enrollment (2016): B: 87 (74) D: 19 (5)*
University Professor:
Edwin K. Schneider, (D), Harvard, 1976, Oa
University Professor:
Jagadish Shukla, (D), MIT, 1976, Oa
Professor:
Timothy DelSole, (D), Harvard, 1993, Oa
Paul A. Dirmeyer, (D), Maryland, 1992, Oa
Robert M. Hazen, (D), Harvard, 1975, Om
Linda Hinnov, (D), Johns Hopkins, 1994, GrYg
Bohua Huang, (D), Maryland, 1992, Op
Jim Kinter, (D), Princeton, 1984, Oa
David M. Straus, (D), Cornell, 1977, Oa
Stacey Verardo, (D), CUNY, 1995, Pei
Associate Professor:
Mark H. Anders, (D), California Berkeley, 1989, Gc
Zafer Boybeyi, (D), North Carolina State, 1993, Oa
Long S. Chiu, (D), MIT, 1980, OarOw
Barry Klinger, (D), MIT/WHOI, 1992, Op
Randolph McBride, (D), Louisiana State, 1997, On
Julia Nord, (D), CUNY (Brooklyn), 1989, Gzg
Cristiana Stan, (D), Colorado State, 2005, Oa
Assistant Professor:
Natalie Burls, (D), Cape Town, 2010, Oap
Andrew J. Hutsky, (D), Nebraska (Lincoln), 2015, GsrEo
Giuseppina Kysar Mattietti, (D), George Washington, 2001, GiOeGa
Kathy Pegion, (D), George Mason, 2007, Oap
Mark Uhen, (D), Michigan, 1996, Pv
Emeritus:
Richard J. Diecchio, (D), North Carolina, 1980, Gr
Paul S. Schopf, (D), Princeton, 1978, Opa

## Hampton University

**Center for Marine & Coastal Environmental Studies** (B) (2014)
100 Queen Street
Hampton, VA 23668
p. (757) 727-5783
george.burbanck@hamptonu.edu
*Enrollment (2010): B: 32 (9)*
Chair:
George P. Burbanck, (D), Delaware, 1981, On
Professor:
Benjamin E. Cuker, (D), North Carolina State, 1981, Ob
Associate Professor:
Robert A. Jordan, (D), Michigan, 1970, Ob
Assistant Professor:
Deidre M. Gibson, (D), Georgia, 2000, Ob
Adjunct Faculty:
Emory Morgan, (M), Oe
Related Staff:
Gary Morgan, (B), North Carolina State, 2007, OnnOn

## James Madison University

**Dept of Geology & Environmental Science** (B) (2014)
MSC 6903
Memorial Hall
Harrisonburg, VA 22807
p. (703) 568-6130
ulansksl@jmu.edu
http://www.jmu.edu/geology/
Department Secretary: Sandra Delawder
Head:
Stanley L. Ulanski, (D), Virginia, 1977, Og
Professor:
Roddy V. Amenta, (D), Bryn Mawr, 1971, Gc
Lynn S. Fichter, (D), Michigan, 1972, Gr
Lance E. Kearns, (D), Delaware, 1977, Gz
William C. Sherwood, (D), Lehigh, 1961, So
Steven J. Whitmeyer, (D), Boston Univ, 2004, GctOi
Associate Professor:
Steven J. Baedke, (D), Indiana, 1998, Hy
Lewis S. Eaton, (D), Virginia, 1999, Gm
Eric J. Pyle, (D), Georgia, 1995, Oe

Kristen E. St. John, (D), Ohio State, 1998, Ou

## Lynchburg College

**Environmental Sciences, Studies, and Sustainability** (B,M) O (2016)
1501 Lakeside Dr.
Lynchburg, VA 24501
p. (434) 544-8415
haiar@lynchburg.edu
http://www.lynchburg.edu/envsci.xml
*Enrollment (2013): B: 40 (10)*

Professor:
David R. Perault, (D), Oklahoma, 1998, Ge

Associate Professor:
Brooke Haiar, (D), Oklahoma, 2008, PoGeg

## Mary Washington College

**Dept of Geography** (B) (2014)
1301 College Avenue
Fredericksburg, VA 22401-5358
p. (540) 654-1470
shanna@umw.edu
http://www.mwc.edu/geog
*Enrollment (2006): B: 80 (0)*

Chair:
Joseph W. Nicholas, (D), Georgia, 1991, Oy

Associate Professor:
Donald N. Rallis, (D), Penn State, 1992, On

Assistant Professor:
Dawn S. Bowen, (D), Queen's, 1998, On
Stephen P. Hanna, (D), Kentucky, 1997, Or
Farhang Rouhani, (D), Arizona, 2001, On

## Mountain Empire Community College

**Mountain Empire Community College** (1999)
3441 Mountain Empire Road
Big Stone Gap, VA 24219
p. (276) 523-7460
creynolds@mecc.edu
http://www.me.vccs.edu/

## Northern Virginia Community College, Alexandria

**Geology Program** (A) (2017)
5000 Dawes Avenue
Alexandria, VA 22311-5097
p. 703.845.6507
vzabielski@nvcc.edu
http://www.nvcc.edu/campuses-and-centers/alexandria/academic-divisions/science/geology.html

Assistant Dean of Geology:
Victor Zabielski, (D), Brown Univ, 2001, GgOg

## Northern Virginia Community College, Annandale

**Geology Program** (A) (2015)
8333 Little River Turnpike
Annandale, VA 22003
p. (703) 323-3276
cbentley@nvcc.edu
http://www.nvcc.edu/campuses-and-centers/annandale/academic-divisions/math-science--engineering/gol.html
*Enrollment (2011): A: 10 (10)*

Professor:
Kenneth Rasmussen, (D), North Carolina, 1989, Gus

Assistant Professor:
Callan Bentley, (M), Maryland, 2004, Gg
Shelley Jaye, (M), Wayne State, 1984, Giy

## Northern Virginia Community College, Loudoun Campus

**Natural and Applied Science Div - Dept of Geology** (A) (2017)
21200 Campus Drive
Sterling, VA 20164
p. (703) 450-2612
wbour@nvcc.edu
http://www.nvcc.edu/campuses-and-centers/loudoun/academic-divisions/natural/geo.html

Associate Professor:
William Bour, (M), George Washington, 1993, Gge
William Straight, (D), North Carolina State, GgPv

Assistant Professor:
Okia Ikwuazorm, (M), Gg

## Northern Virginia Community College, Woodbridge

**Geology Program** (A) (2015)
2645 College Drive
Woodbridge, VA 22191
p. (703) 878-5614
eburtis@nvcc.edu
http://www.nvcc.edu/woodbridge/divisions/natural.html
*Enrollment (2014): A: 6 (1)*

Instructor:
Erik Burtis, (M), Montana, GgcGi

## Old Dominion University

**Dept of Ocean, Earth & Atmospheric Sciences** (B,M,D) (2016)
4600 Elkhorn Avenue
Norfolk, VA 23529-0496
p. (757) 683-4285
rharvey@odu.edu
http://www.ocean.odu.edu
*Enrollment (2015): B: 113 (13) M: 22 (14) D: 16 (1)*

Professor and Chair:
H. Rodger Harvey, (D), 1985, CoOc

Professor:
Larry P. Atkinson, (D), Dalhousie, 1972, Op
David J. Burdige, (D), California (San Diego), 1983, OcCmo
Gregory A. Cutter, (D), California (Santa Cruz), 1982, OcCtHs
Fred C. Dobbs, (D), Florida State, 1987, Ob
Eileen E. Hofmann, (D), North Carolina State, 1980, Op
John M. Klinck, (D), Iowa, 1980, Op
Margaret Mulholland, (D), Maryland, 1998, Ob

Associate Professor:
Alexander Bochdansky, (D), Memorial, 1997, Ob
Jennifer Georgen, (D), MIT/WHOI, 2001, Gu
John R. McConaugha, (D), S California, 1977, Ob
Nora K. Noffke, (D), Oldenburg, 1997, GsPyGu
Matthew Schmidt, (D), California, 2005, Ou
G. Richard Whittecar, Jr., (D), Wisconsin, 1979, GmHwOe

Assistant Professor:
P. Dreux Chappell, (D), MIT/WHOI, 2009, Oc
Benjamin Hamlington, (D), Colorado, 2011

Emeritus:
Dennis A. Darby, (D), Wisconsin, 1971, Ou
Ann Gargett, (D), British Columbia, 1970, Op
Chester E. Grosch, (D), Stevens Inst of Tech, 1967, Opp
Thomas Royer, (D), Texas A&M, 1969, Op
Joseph H. Rule, (D), Missouri, 1972, ClOsSc
Donald J. P. Swift, (D), North Carolina, 1964, Ou
George T. F. Wong, (D), MIT/WHOI, 1976, Oc

## Patrick Henry Community College

**Art, Science, Business, and Technology Dept** (A) (2011)
645 Patriot Avenue
Martinsville, VA 24112
bdooley@ph.vccs.edu
http://www.ph.vccs.edu/

Assistant Professor:
Brett Dooley, (M), Virginia Tech, 2005, Gg

## Piedmont Virginia Community College

**Associate of Science** (1999)
501 College Drive
Charlottesville, VA 22902
p. (434) 961-5446
khudson@pvcc.edu
http://www.pvcc.edu/

## Radford University

**Dept of Geology** (B) (2014)
Box 6939

Radford University
Radford, VA 24142
p. (540) 831-5652
geology@radford.edu
http://www.radford.edu/~geol-web
Administrative Assistant: Theresa Gawthrop
*Enrollment (2013): B: 62 (12)*
Associate Professor:
Jonathan L. Tso, (D), Virginia Tech, 1987, Gcp
Professor:
Rhett B. Herman, (D), Montana State, 1996, Yg
Parvinder S. Sethi, (D), North Carolina State, 1994, Gz
Chester F. Watts, (D), Purdue, 1983, NgrHw
Director, Museum of the Earth Sciences:
Stephen W. Lenhart, (D), Kentucky, 1985
Associate Professor:
Elizabeth McClellan, (D), Tennessee, GipGd
Research Associate:
Judy Ehlen, (D), Birmingham, 1990, Gm
Emeritus:
Robert C. Whisonant, (D), Florida State, 1967, Gsa

## Randolph-Macon College

**Environmental Studies Program** (B) (2016)
P.O. Box 5005
Ashland, VA 23005
p. (804) 752-3745
mfenster@rmc.edu
http://www.rmc.edu/academics/environmental-studies.aspx
*Enrollment (2015): B: 4 (7)*
Watts Professor of Science:
Michael S. Fenster, (D), Boston, 1995, OnGue
Adjunct Professor:
Charles Saunders, (M), East Carolina, 1990, Ghx

## Rappa Hannock Community College

**Rappa Hannock Community College** (2014)
12745 College Drive
Glenns, VA 23149
p. (804) 435-8970
babdul-malik@rappahannock.edu
http://www.rappahannock.edu

## Tidewater Community College

**Geophysical Sciences Dept** (A) (2010)
Virginia Beach, VA 23456
p. (757) 822-7264
tcclayr@tcc.edu
Instructor:
Rodney Clayton, Gg
Jim Coble, Gg
Mike Lyle, Gg
Azam Tabrizi, (M), London, 1978, Pm
John Waugh, Gg

## University of Mary Washington

**Dept of Earth and Environmental Sciences** (B) ● (2016)
1301 College Avenue
Fredericksburg, VA 22401-5358
p. (540) 654-1016
jhayob@umw.edu
http://cas.umw.edu/ees/
*Enrollment (2016): B: 45 (16)*
Chair:
Jodie Hayob, (D), Michigan, 1994, GxzGg
Professor:
Michael L. Bass, (D), Virginia Tech, 1976, Sf
Grant R. Woodwell, (D), Yale, 1985, Gc
Associate Professor:
Ben O. Kisila, (D), Arkansas, 2002, Hs
Melanie Szulczewski, (D), Wisconsin (Madison), 1999, ScClOg
Charles Whipkey, (D), Pittsburgh, 1999, Cg
Instructor:
Sarah A. Morealli, (M), Pittsburgh, 2010, Gg
Emeritus:
Robert L. McConnell, (D), California (Santa Barbara), 1972, Ge

## University of Virginia

**Blandy Experimental Farm** (2014)
400 Blandy Farm Lane
Boyce, VA 22620
p. (540) 837-1758
Blandy@virginia.edu
http://blandy.virginia.edu/
Director:
David E. Carr, (D), Maryland, 1990, On
Associate Director:
Kyle J. Haynes, (D), Louisiana State, 2004, On

**Dept of Environmental Sciences** (B,M,D) (2016)
Clark Hall
291 McCormick Road
Box 400123
Charlottesville, VA 22904
p. (804) 924-7761
cba4a@virginia.edu
http://www.evsc.virginia.edu
Research Full Professor:
Peter Berg, (D), Tech (Denmark), 1988, So
Jack Cosby, (D), Virginia, 1982, Hs
William Keene, (M), Virginia, 1981, Oa
G. Carleton Ray, (D), Columbia, 1960, Ob
Robert J. Swap, (D), Virginia, 1996, On
Chair:
Michael L. Pace, (D), Georgia, 1981, On
Professor:
Paolo D'Odorico, (D), Padova, 1998, Hg
Robert Davis, (D), Delaware, 1988, OawOy
Howard E. Epstein, (D), Colorado State, 1997, On
James N. Galloway, (D), California (San Diego), 1972, Oa
Janet S. Herman, (D), Penn State, 1982, ClHw
Alan D. Howard, (D), Johns Hopkins, 1970, GmXg
Deborah Lawrence, (D), Duke, 1998, On
Manuel Lerdau, (D), Stanford, 1994, On
Stephen A. Macko, (D), Texas (Austin), 1981, CosOc
Karen McGlathery, (D), Cornell, 1992, Ob
Aaron L. Mills, (D), Cornell, 1975, So
Herman H. Shugart, Jr., (D), Georgia, 1971, Og
David E. Smith, (D), Texas A&M, 1982, Ob
Vivian E. Thomson, (D), Virginia, 1997, On
Patricia L. Wiberg, (D), Washington, 1987, Og
Research Associate Professor:
Linda K. Blum, (D), Cornell, 1980, Sb
David E. Carr, (D), Maryland, 1990, On
Kyle J. Haynes, (D), Louisiana State, 2004, On
Jennie L. Moody, (D), Michigan, 1986, Oa
John H. Porter, (D), Virginia, 1988, Or
Associate Professor:
Stephan F. J DeWekker, (D), British Columbia, 2002, Oa
Matthew A. Reidenbach, (D), Stanford, 2004, Hg
Todd M. Scanlon, (D), Virginia, 2002, Hg
Thomas M. Smith, (D), Tennessee, 1982, Og
Research Assistant Professor:
Karen C. Rice, (D), Virginia, 2001, HgCg
Arthur C. Schwarzschild, (D), Virginia, 2004, On
Assistant Professor:
Kevin M. Grise, (D), Colorado State, 2011, Oa
Sally Pusede, (D), Oa
Lecturer:
Thomas H. Biggs, (D), Arizona, 1997, OyGxe
Emeritus:
Robert Dolan, (D), Louisiana State, 1965, Gm
Bruce P. Hayden, (D), Chicago, 1968, Oa
Bruce W. Nelson, (D), Illinois, 1955, Gs
Wallace E. Reed, (D), Chicago, 1967, Or
William F. Ruddiman, (D), Columbia, 1969, Gu

**Shenandoah Watershed Study** (B,M,D) (2015)
Department of Environmental Sciences
Clark Hall
Charlottesville, VA 22904

p. (434) 924-3382
tms2v@virginia.edu
http://people.virginia.edu/~swas/POST/scripts/overview.php

Associate Professor:
Todd M. Scanlon, (D), Virginia, 2002, Hgs

Research Scientist:
Ami L. Riscassi, (D), Virginia, 2009

**Virginia Coast Reserve Long Term Ecological Research** (2015)
291 McCormick Road
P.O. Box 400123
Charlottesville, VA 22904-4123
p. 434-924-0558
kjm4k@virginia.edu
http://www.vcrlter.virginia.edu

Professor:
Karen J. McGlathery, (D), Cornell, 1992, Obn

## University of Virginia College, Wise

**Dept of Natural Science** (B) (2016)
1 College Avenue
Wise, VA 24293
p. (276) 328-0203
blw@uvawise.edu
Administrative Assistant: Brenda Whitaker

Instructor:
Robert D. VanGundy, (M), North Carolina, 1983, GeHgOg

## Virginia Highlands Community College

**Virginia Highlands Community College** (2014)
100 VHCC Drive
Abingdon, VA 24212
p. (276)739-2433
jsurber@vhcc.edu
http://www.vhcc.edu

## Virginia Polytechnic Institute & State University

**Dept of Crop & Soil Environmental Sciences** (B,M,D) (2010)
240 Smyth Hall
Blacksburg, VA 24061-0404
p. (540) 231-6305
tlthomps@vt.edu
http://www.cses.vt.edu
Department Secretary: Nancy Shields
*Enrollment (2001): B: 98 (0) M: 14 (0) D: 7 (0)*

Head:
John R. Hall, III, (D), Ohio State, 1971, So

Professor:
Marcus M. Alley, (D), Virginia Tech, 1975, Sc
James C. Baker, (D), Virginia Tech, 1978, Sd
Walter L. Daniels, (D), Virginia Tech, 1985, SdGeCg
Stephen J. Donohue, (D), Purdue, 1974, Sc
Gregory K. Evanylo, (D), Georgia, 1982, Sc
Charles Hagedorn, (D), Iowa, 1974, Sb
Gregory L. Mullins, (D), Purdue, 1985, Sc
Raymond B. Reneau, (D), Florida, 1969, Sc
Lucian W. Zelazny, (D), Virginia Tech, 1970, Sc

Associate Professor:
Duane F. Berry, (D), Michigan State, 1984, Sb
Matthew J. Eick, (D), Delaware, 1995, Sc
Naraine Persaud, (D), Florida, 1978, Sp

Assistant Professor:
John M. Galbraith, (D), Cornell, 1997, Sp
Carl E. Zipper, (D), Virginia Tech, 1986, Oe

Adjunct Professor:
Domy C. Adriano, (D), Kansas State, 1970, Sc
V. C. Baligar, (D), Mississippi State, 1975, Sc
Pamela J. Thomas, (D), Virginia Tech, 1998, Sd

**Dept of Geography** (B,M,D) (2010)
115 Major Williams Hall
Blacksburg, VA 24061-0115
p. (540) 231-7557
carstens@vt.edu
*Enrollment (2010): B: 170 (49) M: 17 (9) D: 8 (0)*

Head:
Laurence W. Carstensen, (D), North Carolina, 1981, Oi

Professor:
James B. Campbell, (D), Kansas, 1976, Or

Associate Professor:
Lawrence S. Grossman, (D), Australian National, 1979, On
Lisa M. Kennedy, (D), Tennessee, Pe
Resler M. Lynn, (D), Texas State, Oy

Instructor:
David Carroll, (M), Mississippi State, Ow

**Dept of Geosciences** (B,M,D) (2014)
4044 Derring Hall
Blacksburg, VA 24061
p. (540) 231-6521
wilcar@vt.edu
http://www.geos.vt.edu
Administrative Assistant: Carolyn S. Williams
*Enrollment (2010): B: 92 (16) M: 14 (10) D: 42 (6)*

Research Professor:
Ross J. Angel, (D), Cambridge, 1986, Gz
Robert P. Lowell, (D), Oregon State, 1972, Yrh

Professor:
Robert J. Bodnar, (D), Penn State, 1985, Cg
Thomas J. Burbey, (D), Nevada (Reno), 1994, HwyOr
Patricia M. Dove, (D), Princeton, 1991, Cl
Kenneth A. Eriksson, (D), Witwatersrand, 1977, Gs
Michael F. Hochella, Jr., (D), Stanford, 1981, ClGze
Scott D. King, (D), Caltech, 1990, Yg
Michal J. Kowalewski, (D), Arizona, 1995, Py
Richard D. Law, (D), London, 1981, Gc
J. Fred Read, (D), W Australia, 1971, Gs
Nancy L. Ross, (D), Arizona State, 1985, Gz
Robert J. Tracy, (D), Massachusetts, 1975, Gp
Shuhai Xiao, (D), Harvard, 1998, Pi

Research Associate Professor:
Martin C. Chapman, (D), Virginia Tech, 1998, Ys

Associate Professor:
John A. Hole, (D), British Columbia, 1993, Ye
Madeline E. Schreiber, (D), Wisconsin, 1999, Hw
James A. Spotila, (D), Caltech, 1998, Gt

Assistant Professor:
Barbara M. Bekken, (D), Stanford, 1990, Eg
Ying Zhou, (D), Princeton, 2004, Ys

Sr. Research Associate:
Jing Zhao, (D), Chinese Acad of Sci, 1997

Research Associate:
Nizhou Han, (D), Iowa State, 1996, Sc

Instructor:
Neil E. Johnson, (D), Virginia Tech, 1986, GzEg

Adjunct Professor:
James S. Beard, (D), California (Davis), 1985, GiClg
John A. Chermak, (D), Virginia Tech, 1989, Cg
Benedetto DeVivo, (D)
Alton C. Dooley, (D), Louisiana State, 1998
Nicholas C. Fraser, (D), Aberdeen, 1984, Pv
William S. Henika, (M), Virginia, 1969, Ng
David W. Houseknecht, (D), Penn State, 1978, Gd
Jerry L. Hunter, (D), North Carolina, 1991, Ca
Matthew J. Mikulich, (D), Utah, 1971, Ye
Csaba Szabo, (D), Virginia Tech, 1994, Gi
Chester F. Watts, (D), Purdue, 1983, Ngg

Emeritus:
Richard K. Bambach, (D), Yale, 1969, Po
G. A. Bollinger, (D), St. Louis, 1967, Ys
Cahit Coruh, (D), Istanbul, 1970, YesGt
James R. Craig, (D), Lehigh, 1965, Eg
Gordon C. Grender, (D), Penn State, 1960, Go
David A. Hewitt, (D), Yale, 1970, Cp
Wallace D. Lowry, (D), Rochester, 1943, Gc
Dewey M. McLean, (D), Stanford, 1969, Pe
Paul H. Ribbe, (D), Cambridge, 1963, Gz
J. Donald Rimstidt, (D), Penn State, 1979, Clg
Edwin S. Robinson, (D), Wisconsin, 1964, Yg
A. Krishna Sinha, (D), California (Santa Barbara), 1969, Gt
J. Arthur Snoke, (D), Yale, 1969, Ys

Educational Administrator for Outreach:
S. Llyn Sharp, (M), Virginia Tech, 1990, Oe
Cooperating Faculty:
Edward Lener, (M), Virginia Tech, 1997, On

## Virginia Tech

**Dept of Civil & Environmental Engineering** (B,M,D) (2016)
750 Drillfield Drive
200 Patton Hall
Blacksburg, VA 24061-0105
p. (703) 231-6635
mwiddows@vt.edu
http://www.cee.vt.edu/
Administrative Assistant: Beth Lucus
Graduate Chair:
Mark A. Widdowson, (D), Auburn, HwSpHq
Professor:
Jennifer Irish, (D), Delaware, Oog
Associate Professor:
Randel Dymond, (D), Penn State, HqsOi
Erich Hester, (D), North Carolina, Hsw
Kyle Strom, (D), Iowa, HsGs

## Virginia Wesleyan College

**Dept of Earth and Environmental Sciences** (B) (2015)
1584 Wesleyan Dr
Norfolk, VA 23502
p. 757.455.3200
jchaley@vwc.edu
http://www.vwc.edu/earth-and-envirmental-sciences/
f: https://www.facebook.com/vwc.ees?fref=ts
*Enrollment (2015): B: 22 (0)*
Professor:
John C. Haley, (D), Johns Hopkins, 1986, GgeOi
Associate Professor:
Elizabeth Malcolm, (D), Michigan, 2002, OaCgOg
Garry Noe, (D), California (Riverside), 1982, HsOin

## Virginia Western Community College

**Geology Dept** (A) (2010)
3095 Colonial Avenue
Roanoke, VA 24038
p. (540) 857-7273
abalog-szabo@virginiawestern.edu
Professor:
Anna Balog-Szabo, (D), Virginia Tech, 1996, Gs

## Washington & Lee University

**Dept of Geology** (B) (2017)
204 West Washington Street
Lexington, VA 24450
p. (540) 458-8800
geology@wlu.edu
http://geology.wlu.edu
Administrative Assistant: Sarah Wilson
*Enrollment (2016): B: 28 (23)*
Professor:
Christopher Connors, (D), Princeton, 1999, GoYe
Lisa Greer, (D), Miami, 2001, GsPe
David J. Harbor, (D), Colorado State, 1990, Gm
Associate Professor:
Elizabeth P. Knapp, (D), Virginia, 1997, Cl
Jeffrey Rahl, (D), Yale, 2005, Gt
Emeritus:
Frederick Schwab, (D), Harvard, 1968, Gdg
Edgar W. Spencer, (D), Columbia, 1957, GcOge

# Washington

## Bellevue College

**Earth and Space Sciences Program** (A) (2014)
3000 Landerholm Circle SE
Bellevue, WA 98007
p. (425) 564-3158
kshort@bellevuecollege.edu
http://scidiv.bellevuecollege.edu/
Instructor:
Cary Easterday, (M), GgPg
Gwyn Jones, (M), Gg
Deborah Minium, (M), Gg
Rob Viens, (D), Washington, GgOg

## Central Washington University

**Dept of Geological Sciences** (B,M) (2013)
400 East University Way
MS 7418
Ellensburg, WA 98926-7418
p. (509) 963-2701
chair@geology.cwu.edu
http://www.geology.cwu.edu
*Enrollment (2010): B: 97 (17) M: 20 (8)*
Chair:
Carey A. Gazis, (D), Caltech, 1994, Cs
Director, PANGA Laboratory:
Timothy I. Melbourne, (D), Caltech, 1998, Yd
Professor:
Wendy A. Bohrson, (D), California (Los Angeles), 1993, GivCc
Lisa L. Ely, (D), Arizona, 1992, Gm
Jeffrey Lee, (D), Stanford, 1990, Gct
GPS Data Analyst:
Marcelo Santillan, (M), Memphis, 2003, Yd
Associate Scientist:
Marie A. Ferland, (D), Sydney, 1991, Gu
Assistant Professor:
Susan Kaspari, (D), Maine, 2007, Pe
Audrey Huerta, (D), MIT, 1998, Ygd
Breanyn MacInnes, (D), Washington, 2010, GseGu
Chris Mattinson, (D), Stanford, 2006, Gzp
Walter Szeliga, (D), Colorado, 2010, YgsYd
Research Associate:
Paul Winberry, Penn State, 2008, Yg
Scientific Instructional Tech Supervisor:
Nick Zentner, (M), Idaho State, 1989, Gg
Lecturer:
Keegan Fengler, (M), C Washington, Gg
Winston Norrish, (D), Cincinnati, 1990, GgeGo
Emeritus:
Robert Bentley, (D), Columbia, 1969, Gi
Steven Farkas, (D), New Mexico, 1965, Gs
James R. Hinthorne, (D), California (Santa Barbara), 1974, Ca
M. Meghan Miller, (D), Stanford, 1987, Yd
L. Don Ringe, (D), Washington State, 1968, Gml
Charles M. Rubin, (D), Caltech, 1990, Gt
Systems Administrator:
Craig Scrivner, (D), Caltech, 1998
PANGA Network Engineer:
Andy Miner, (M), C Washington, Yd

## Centralia College

**Earth Sciences Program** (A) (2016)
600 Centralia College Blvd
Centralia, WA 98531
p. (360) 736-9391
ppringle@centralia.edu
http://www.centralia.edu/academics/earthscience/index.html
*Enrollment (2015): A: 3 (1)*
Professor:
Patrick Pringle, (M), Akron, 1982, OgGvOn

## Eastern Washington University

**Dept of Geology** (B) (2017)
130 Science Building
Cheney, WA 99004-2439
p. (509) 359-2286
jthomson@ewu.edu
http://cshe.cslabs.ewu.edu/deptGEO/x55894.html
*Enrollment (2009): B: 52 (12)*
Chair:
Jennifer A. Thomson, (D), Massachusetts, 1992, Gp
Professor:
John P. Buchanan, (D), Colorado State, 1985, Hw

Linda B. McCollum, (D), SUNY (Binghamton), 1980, Po
Associate Professor:
Carmen A. Nezat, (D), Michigan, 2006, ClGeSc
Richard L. Orndorff, (D), Kent State, 1994, Ge
Assistant Professor:
Chad Pritchard, (D), Washington State Univ, Gct
Lecturer:
Jeanne Case, (M), California, Riverside, Gg
Sharen Keattch, (M), Kent State Univ, Kent, OH, 1993, Gg
Emeritus:
Ernest H. Gilmour, (D), Montana, 1967, Pi
Eugene P. Kiver, (D), Wyoming, 1968, Gl



## Edmonds Community College

**Geology** (A) (2013)
20000 68th Ave W
Lynnwood, WA 98036
p. (425) 640-1918
mkelly@edcc.edu
http://www.edcc.edu/stem/geology/
*Enrollment (2013): A: 4 (0)*
Instructor:
Maria Kelly, Og
Adjunct Professor:
Dylan Ahearn, (D), California (Davis)
Thomas Hamilton, Og

## Everett Community College

**Dept of Physical Sciences** (A) (2014)
2000 Tower Street
Everett, WA 98201
p. (425) 388-9429
sdamp@everettcc.edu
http://www.everettcc.edu/programs/mathsci/physical/
Instructor:
Steve Grupp, (M), Colorado Mines, GgOg
Adjunct Professor:
Alecia Spooner, Og

## Grays Harbor College

**Dept of Geology & Oceanography** (A) (2010)
Aberdeen, WA 98520
p. (206) 538-4299
john.hillier@ghc.edu
Professor:
John Hillier, (D), Cornell, OgCa
Emeritus:
James B. Phipps, (D), Oregon State, 1974, Gu

## Green River Community College

**Dept of Geology** (A) (2015)
12401 S.E. 320th
Auburn, WA 98092-3699
p. (253) 833-9111 (Ext. 4248)
kclay@greenriver.edu
http://www.greenriver.edu/academics/areas-of-study/details/earth-science.htm
Instructor:
Kathryn A. Hoppe, (D), Princeton, 1999, OggPg
Katy Shaw, (M), Washington, OggGg

## Highline College

**Physical Sciences Dept/Geology Program** (A) (2016)
MS 29-3
PO Box 98000
2400 S. 240th St.
Des Moines, WA 98198-9800
p. (206) 878-3710
ebaer@highline.edu
http://geology.highline.edu/
*Enrollment (2016): A: 8 (0)*
Instructor:
Eric M. Baer, (D), California (Santa Barbara), 1995, GgvOg
Stephaney Puchalski, (D), Indiana, 2011, PgGg
Michael Valentine, (D), Massachusetts (Amherst), 1990, Gt
Carla Whittington, (M), Indiana, GgvOg

## Lower Columbia College

**Earth Sciences (Natural Sciences Dept.)** (A) (2010)
P.O. Box 3010
Longview, WA 98632
p. (360) 442-2883
dcordero@lowercolumbia.edu
http://lowercolumbia.edu/nr/exeres/394D4C4D-1A8A-4F33-AAA9-F85C71386644
*Enrollment (2001): A: 8 (4)*
Professor:
David I. Cordero, (M), Portland State, 1997, Og

## North Seattle Community College, North Campus

**Earth Science** (1999)
9600 College Way North
Seattle, WA 98103
p. (206) 934-4509
tfurutani@sccd.ctc.edu
http://www.northseattle.edu/

## Northwest Indian College

**Native Environmental Studies** (2014)
2522 Kwina Road
Bellingham, WA 98226
p. (360) 392-4256
toreiro@nwic.edu
http://www.nwic.edu/degrees-and-certificates/bsnes-bachelors-degree

## Olympic College

**Mathematics, Engineering, Sciences, and Health Div** (A) (2014)
Bremerton, WA 98310
p. (360) 475-7777
SMacias@olympic.edu
http://www.olympic.edu/Students/AcadDivDept/MESH/Sciences/Geology/
Instructor:
Steve E. Macias, (M), Washington, 1996, Gg
Adjunct Professor:
Katie Howard, Gg

## Pacific Lutheran University

**Dept of Geosciences** (B) (2013)
12180 S. Park
Tacoma, WA 98447
p. (253) 535-7378
geos@plu.edu
http://www.plu.edu/geosciences/
*Enrollment (2013): B: 52 (11)*
Chair:
Jill M. Whitman, (D), California (San Diego), 1989, OuGuOe
Professor:
Duncan Foley, (D), Ohio State, 1978, GeHyOe
Associate Professor:
Rosemary McKenney, (D), Penn State, 1997, Gm
Claire E. Todd, (D), Washington, 2007, Glm
Assistant Professor:
Peter B. Davis, (D), Minnesota, 2008, GcpGz

## Peninsula College

**Environmental Science** (1999)
1502 East Lauridsen Boulevard
Port Angeles, WA 98362
p. (360) 452-9277
jganzhorn@pencol.edu
http://www.pc.ctc.edu

## Pierce College

**Pierce College** (2014)
9401 Farwest Drive SW
Lakewood, WA 98498
p. (253) 964-6676

zayacjm@piercecollege.edu
http://www.piercecollege.edu/departments/physics_planetary_sciences/geology.asp

## Seattle Central Community College

**Div of Science & Mathematics** (2010)
1701 Broadway
Seattle, WA 98122
p. (206) 587-3858
jhull@sccd.ctc.edu
http://seattlecentral.edu/sci-math/

Instructor:
Katie Gagnon, (D), California (San Diego), 2007, Oug
Joseph M. Hull, (D), Rochester, 1988, Gc
Adjunct Professor:
Michael Harrell, Gg

## Shoreline Community College

**Geology & Earth Sciences** (A) (2010)
16101 Greenwood Avenue North
Seattle, WA 98133
p. (206) 546-4659
eagosta@shoreline.edu
http://www.shoreline.edu/science/geology.htm

## Spokane Community College

**Dept of Earth Science** (1999)
1810 N. Greene Street
Spokane, WA 99217
Andy.Buddington@scc.spokane.edu
http://www.scc.spokane.edu/ArtsSciences/Science/

## Tacoma Community College

**Dept of Earth Sciences** (A) (2015)
6501 South 19th Street
Tacoma, WA 98466
p. (253) 566-5060
rhitz@tacomacc.edu
http://www.tacomacc.edu/academics/mathematicssciencesandengineeringdivision/science/programs/earthscience/
*Enrollment (2015): A: 8 (2)*

Professor:
Ralph B. Hitz, (D), California (Santa Barbara), 1997, PvSaOi
Adjunct Professor:
Jim McDougall, (D), Og
James Peet, (D), Washington
Michael Valentine, (D), Og

## University of Puget Sound

**Geology Dept** (B) (2016)
1500 N. Warner Street
Tacoma, WA 98416-1048
p. (253) 879-3814
mvalentine@pugetsound.edu
http://www.pugetsound.edu/academics/departments-and-programs/undergraduate/geology/
f: https://www.facebook.com/upsgeology
Administrative Assistant: Leslie Levenson
*Enrollment (2015): B: 27 (10)*

Chair:
Michael J. Valentine, (D), Massachusetts, 1990, YmGc
Professor:
Barry Goldstein, (D), Minnesota, 1985, GlmGs
Jeffrey H. Tepper, (D), Washington, 1991, GiCgGv
Associate Professor:
Kena L. Fox-Dobbs, (D), California (Santa Cruz), 2006, PgCsGe
Instructor:
Ken Clark, (M), W Washington, 1989, GgcGi
Emeritus:
Z. Frank Danes, (D), Charles (Prague), 1949, Yv
Albert A. Eggers, (D), Dartmouth, 1972, Gv

## University of Washington

**Dept of Atmospheric Sciences** (B,M,D) (2014)
Box 351640
Seattle, WA 98195-1640
p. (206) 543-4250
chair@atmos.washington.edu
http://www.atmos.washington.edu
*Enrollment (2012): B: 60 (17) M: 30 (9) D: 35 (9)*

Affiliate Associate Professor:
Philip W. Mote, (D), Washington, 1994, Oa
Professor and Chair:
Gregory J. Hakim, (D), SUNY (Albany), 1997, Oa
Professor:
Thomas P. Ackerman, (D), Washington, 1976, Oa
David S. Battisti, (D), Washington, 1988, Oa
Christopher S. Bretherton, (D), MIT, 1984, Oa
Dale R. Durran, (D), MIT, 1981, Oa
Qiang Fu, (D), Utah, 1991, Oa
Dennis L. Hartmann, (D), Princeton, 1975, Oa
Robert A. Houze, (D), MIT, 1972, Oa
Lyatt Jaegle, (D), Caltech, 1996, Oa
Daniel A. Jaffe, (D), Washington, 1987, Oa
Clifford F. Mass, (D), Washington, 1978, Oa
Peter B. Rhines, (D), Trinity Coll (Cambridge), 1967, Op
Stephen G. Warren, (D), Harvard, 1973, Oal
Research Associate Professor:
Roger T. Marchand, (D), Virginia Tech, 1997, Oaa
Associate Professor:
Becky Alexander, (D), California (San Diego), 2002, OaCsPe
Cecilia M. Bitz, (D), Washington, 1997, Oa
Dargan M. Frierson, (D), Princeton, 2005, Oa
Joel A. Thornton, (D), California (Berkeley), 2002, Oa
Robert Wood, (D), Manchester, 1997, Oa
Research Assistant Professor:
Jerome Patoux, (D), Washington, 2003, Oa
Assistant Professor:
Abigail L. S. Swann, (D), California (Berkeley), 2010, Oan
Sr. Lecturer PT:
Lynn A. McMurdie, (D), Washington, 1989, Oa
Adjunct Associate Professor:
Jessica D. Lundquist, (D), Scripps, 2004, Oa
Nathan J. Mantua, (D), Washington, 1994, Oa
Adjunct Professor:
David C. Catling, (D), Oxford, 1994, Oa
Gerard H. Roe, (D), MIT, 1999, Oa
Eric J. Steig, (D), Washington, 1996, CsOla
LuAnne Thompson, (D), MIT, 1990, Og
Ka-Kit Tung, (D), Harvard, 1977, Oa
Research Professor Emeritus:
James E. Tillman, (M), MIT, Oa
Emeritus:
Marcia B. Baker, (D), Washington, Oa
Robert A. Brown, (D), Washington, 1969, Oa
Joost A. Businger, (D), State (Utrecht), Oa
Robert J. Charlson, (D), Washington, Oa
David S. Covert, (D), Washington, 1974, Oa
Robert G. Fleagle, (D), New York, Oa
Thomas C. Grenfell, (D), Washington, 1972, Oa
Halstead Harrison, (D), Stanford, 1960, Oa
Dean A. Hegg, (D), Washington, 1979, Oa
Gary A. Maykut, (D), Washington, Oa
Edward S. Sarachik, (D), Brandeis, Oa
John M. Wallace, (D), MIT, 1966, Oa
Affiliate Professor:
Shuyi S. Chen, (D), Penn State, 1990, Oa
D. Edmunds Harrison, (D), Harvard, 1977, Oa
James E. Overland, (D), New York, 1973, Oa
Lawrence F. Radke, (D), Washington, 1968, Oa
Affiliate Associate Professor:
Timothy S. Bates, (D), Washington, 1992, Oa
Nicholas A. Bond, (D), Washington, 1986, Oa
Bradley R. Colman, (D), MIT, 1984, Oa
F. Anthony Eckel, (D), Washington, 2003, Oa
Mark T. Stoelinga, (D), Washington, 1993, Oa
Sandra E. Yuter, (D), Washington, 1996, Oa
Affiliate Assistant Professor:
Bonnie Light, (D), Washington, 2000, Oa

**Dept of Earth & Space Sciences** (B,M,D) ● (2017)
070 Johnson Hall
Box 351310
Seattle, WA 98195-1310
p. (206) 543-1190
essadv@uw.edu
http://www.ess.washington.edu
f: https://www.facebook.com/pages/University-of-Washington-Earth-and-Space-Sciences/132505186787510
t: https://twitter.com/UW_ESS
*Enrollment (2016): B: 236 (61) M: 0 (19) D: 0 (3)*

Research Professor:
Howard B. Conway, (D), Canterbury (NZ), 1986, Ol
Dale P. Winebrenner, (D), Washington, 1985, Olr

Affiliate Professor:
Brian Atwater, (D), Ys
Arthur Frankel, (D), Ys
Joan Gomberg, (D), YsGt
Frank Gonzalez, (D), Or
Tony Irving, (D), CgXmGi
Richard Sack, (D), CgOn

Affiliate Assistant Professor:
Ralph Haugerud, (D), Oi
Brian L. Sherrod, (D), YsGt

Professor:
George W. Bergantz, (D), Johns Hopkins, 1988, Gi
J. Michael Brown, (D), Minnesota, 1980, Gy
Roger Buick, (D), W Australia, 1986, PeoPm
David C. Catling, (D), Oxford, 1994, On
Darrel S. Cowan, (D), Stanford, 1972, Gct
Kenneth C. Creager, (D), California (San Diego), 1984, Ys
Bernard Hallet, (D), California (Los Angeles), 1975, GmOl
Robert H. Holzworth, (D), California (Berkeley), 1977, XyOag
Heidi B. Houston, (D), Caltech, 1987, YsGt
David R. Montgomery, (D), California (Berkeley), 1991, Gm
Bruce K. Nelson, (D), California (Los Angeles), 1985, CcGie
Charles A. Nittrouer, (D), Washington, 1978, GuYr
Eric J. Steig, (D), Washington, 1996, CsOla
John E. Vidale, (D), Caltech, 1986, YsGtYg
Edwin D. Waddington, (D), British Columbia, 1981, Ol
Peter D. Ward, (D), McMaster, 1976, Po
Stephen G. Warren, (D), Harvard, 1973, Oal
Robert M. Winglee, (D), Sydney, 1984, XyYx

Research Associate Professor:
Evan H. Abramson, (D), MIT, 1985, Gy
Paul A. Bodin, (D), Colorado, 1992, Ys
Michael P. McCarthy, (D), Washington, 1988, XyYx
Ronald S. Sletten, (D), Washington, 1995, OsSb

Research Assistant Professor:
Erika M. Harnett, (D), Washington, 2003, XyGev

Affiliate Associate Professor:
Olivier Bachmann, (D), Geneva, 2000, Giv
Amit Mushkin, (D), OrCcGm

Adjunct Associate Professor:
Katherine Sian Davies-Vollum, (D), Oxford (UK), 1994, GsrPe

Associate Professor:
Juliet Crider, (D), Stanford, 1998, GcmNr
Drew J. Gorman-Lewis, (D), Notre Dame, 2006, PyCl
Katharine W. Huntington, (D), MIT, 2006, Gt
Elizabeth (Liz) Nesbitt, (D), California (Berkeley), 1982, Pim
Gerard H. Roe, (D), MIT, 1999, OaGmOl
David Schmidt, (D), California, Berkeley, 2002, Gt
John O.H Stone, (D), Cambridge, 1986, Cca
Fangzhen Teng, (D), Maryland, 2005, Cg

Research Assistant Professor:
Michelle Koutnik, (D), Washington, 2009, GlOl
Stephen E. Wood, (D), California (Los Angeles), 1999, On

Affiliate Assistant Professor:
Scott Bennett, (D), Geology, California, Davis, 2013, Gnt

Adjunct Assistant Professor:
Christian A. Sidor, (D), Chicago, 2000, Pv

Assistant Professor:
Alison Duvall, (D), Michigan, 2011, GmcGt
Alexis Licht, (D), Gsn

Senior Lecturer:
Terry W. Swanson, (D), Washington, 1994, CcGe
Kathy G. Troost, (D), Oun

Lecturer:
Brian Collins, Gm

Adjunct Professor:
David S. Battisti, (D), Washington, 1988, Oa
Donald E. Brownlee, (D), Washington, 1971, Xgy
John R. Delaney, (D), Arizona, 1977, GuYr
Harlan Paul Johnson, (D), Washington, 1972, Gu
John D. Sahr, (D), Cornell, 1990, OrXym
William S.D Wilcock, (D), MIT, 1992, Gu

Research Professor Emeritus:
Stephen D. Malone, (D), Nevada (Reno), 1972, YsGt
Gary A. Maykut, (D), Washington, 1969, OalYr

Research Professor:
Alan R. Gillespie, (D), Caltech, 1982, GmOlr

Research Associate Professor:
Robert I. Odom, (D), Washington, 1980, YrGu

Affiliate Professor:
Charlotte Schreiber, (D), Gsd

Emeritus:
John B. Adams, (D), Washington, 1961, GcXgOr
Patricia M. Anderson, (D), Brown, 1982, Ple
Marcia B. Baker, (D), Washington, 1971, OaYg
Marcia Baker, (D), Oa
John R. Booker, (D), California (San Diego), 1968, YmGt
Joanne (Jody) Bourgeois, (D), Wisconsin, 1980, GsrGh
Eric S. Cheney, (D), Yale, 1964, Eg
Robert S. Crosson, (D), Stanford, 1966, YsGct
Bernard W. Evans, (D), Oxford (UK), 1959, Gp
I. Stewart (Stu) McCallum, (D), Chicago, 1968, GivGz
Ronald T. Merrill, (D), California (Berkeley), 1967, Ygm
George K. Parks, (D), California (Berkeley), 1966, YxXy
Stephen C. Porter, (D), Yale, 1962, Gl
Charles F. Raymond, (D), Caltech, 1969, Ol
John M. Rensberger, (D), California (Berkeley), 1967, Pvo
Stewart W. Smith, (D), Caltech, 1961, Ys
Minze Stuiver, (D), Groningen (Neth), 1958, Cc
Joseph A. Vance, (D), Washington, 1957, Gi

Senior Computer Specialist:
Harvey Greenberg, (M), Cal State (Chico), 1977, Gml

Postdoctoral Researcher:
Jon Toner, (D)

**School of Oceanography** (B,M,D) (2014)
Box 357940
Seattle, WA 98195-7940
p. (206) 543-5060
admin@ocean.washington.edu
http://www.ocean.washington.edu/
*Enrollment (2003): B: 62 (13) M: 45 (14) D: 47 (10)*

Director:
Russell E. McDuff, (D), California (San Diego), 1978, Oc

Research Professor:
Mark L. Holmes, (D), Washington, 1975, Ou
Ronald L. Shreve, (D), Caltech, 1959, Yr

Professor:
Knut Aagaard, (D), Washington, 1966, Op
John A. Baross, (D), Washington, 1972, Ob
Roy Carpenter, (D), California (San Diego), 1968, Oo
William O. Criminale, Jr., (D), Johns Hopkins, 1960, Op
Eric A. D'Asaro, (D), MIT/WHOI, 1980, Op
John R. Delaney, (D), Arizona, 1977, Yr
Jody W. Deming, (D), Maryland, 1981, Ob
Allan H. Devol, (D), Washington, 1975, Oc
Steven R. Emerson, (D), Columbia, 1974, Oc
Charles C. Eriksen, (D), MIT, 1977, Op
Bruce W. Frost, (D), California (San Diego), 1969, Ob
Michael C. Gregg, (D), California (San Diego), 1971, Op
G. Ross Heath, (D), Scripps, 1968, Cm
Barbara M. Hickey, (D), California (San Diego), 1975, Op
H. Paul Johnson, (D), Washington, 1972, Yr
Marvin D. Lilley, (D), Oregon State, 1983, Ob
Seelye Martin, (D), Johns Hopkins, 1967, Op
James W. Murray, (D), MIT/WHOI, 1973, Oc
Charles A. Nittrouer, (D), Washington, 1978, Ou
Arthur R. M. Nowell, (D), British Columbia, 1975, Ou
Paul D. Quay, (D), Columbia, 1977, Oc

Peter B. Rhines, (D), Cambridge, 1967, Op
Jeffrey E. Richey, (D), California (Davis), 1973, Oc
Associate Professor:
Virginia E. Armbrust, (D), MIT/WHOI, 1990, Ob
Susan L. Hautala, (D), Washington, 1992, Op
Bruce M. Howe, (D), California (San Diego), 1986, Op
Mitsuhiro Kawase, (D), Princeton, 1986, Op
Richard G. Keil, (D), Delaware, 1991, Oc
Deborah S. Kelley, (D), Dalhousie, 1990, Ou
Evelyn J. Lessard, (D), Rhode Island, 1984, Ob
Parker MacCready, (D), Washington, Op
Stephen C. Riser, (D), Rhode Island, 1980, Op
LuAnne Thompson, (D), MIT/WHOI, 1990, Op
Mark J. Warner, (D), California (San Diego), 1988, Op
William S. D. Wilcock, (D), MIT/WHOI, 1992, Ou
Kevin L. Williams, (D), Washington State, 1985, Op
Research Assistant Professor:
Miles G. Logsdon, (D), Washington, 1997, Oi
Andrea S. Ogston, (D), Washington, 1997, Ou
Assistant Professor:
Andrew Barclay, (D), Oregon, 1998, Gu
Daniel Grunbaum, (D), Cornell, 1992, Ob
Anitra E. Ingalls, (D), SUNY (Stony Brook), 2002, Oc
Jeffrey D. Parsons, (D), Illinois, 1998, Ou
Gabrielle L. Rocap, (D), MIT/WHOI, 2000, Ob
Senior Lecturer:
Christina M. Emerick, (D), Oregon State, 1985, Ou
Lecturer:
Richard M. Strickland, (M), Washington, 1975, ObgOg
Adjunct Professor:
Rose Ann Cattolico, (D), SUNY (Stony Brook), 1973, On
Robert Francis, (D), Washington, 1970, Ob
Barbara B. Krieger-Brockett, (D), Wayne State, 1976, Ob
Ronald T. Merrill, (D), California (Berkeley), 1967, Ym
Bruce K. Nelson, (D), California (Los Angeles), 1985, Cc
Edward Sarachik, (D), Brandeis, 1966, Oa
Robert C. Spindel, (D), Yale, 1971, Op
Emeritus:
George C. Anderson, (D), Washington, 1954, Ob
Karl Banse, (D), Kiel, 1955, Ob
Joe S. Creager, (D), Texas A&M, 1958, Ou
Alyn C. Duxbury, (D), Texas A&M, 1963, On
Terry E. Ewart, (D), Washington, 1965, Op
Richard H. Gammon, (D), Harvard, 1970, Oc
Eric L. Kunze, (D), Washington, 1985, Op
Joyce C. Lewin, (D), Yale, 1953, Ob
Brian T. R. Lewis, (D), Wisconsin, 1970, Yr
Dean A. McManus, (D), Kansas, 1959, Ou
Gunnar I. Roden, (M), California (Los Angeles), 1956, Op
David A. Rothrock, (D), Cambridge, 1968, Op
Thomas B. Sanford, (D), MIT, 1967, Op
Richard W. Sternberg, (D), Washington, 1965, Ou
Cooperating Faculty:
Matthew H. Alford, (D), California (San Diego), 1998, Op
Edward T. Baker, (D), Washington, 1973, Ou
Laurie Balistrieri, (M), Washington, 1977, Oc
John Bullister, (D), California (San Diego), 1984, Oc
David A. Butterfield, (D), Washington, 1990, Yr
Glenn A. Cannon, (D), Johns Hopkins, 1969, Op
Meghan F. Cronin, (D), Rhode Island, 1993, Op
Brian D. Dushaw, (D), California (San Diego), 1992, Op
Richard A. Feely, (D), Texas A&M, 1974, Oc
Don E. Harrison, (D), Harvard, 1977, Op
Albert Hermann, (N), Op
Robin T. Holcomb, (D), Stanford, 1981, Ou
Gregory C. Johnson, (D), MIT/WHOI, 1990, Op
Peter A. Jumars, (D), California (San Diego), 1974, Ob
Kathryn A. Kelly, (D), California (San Diego), 1983, Op
William S. Kessler, (D), Washington, 1989, Op
Craig M. Lee, (D), Washington, 1995, Op
Michael J. McPhaden, (D), California (San Diego), 1980, Og
Curtis D. Mobley, (D), Maryland, 1977, Ob
Harold O. Mofjeld, (D), Washington, 1970, Op
Dennis W. Moore, (D), Harvard, 1968, Op
James H. Morison, (D), Washington, 1980, Op
Jeffrey Napp, (D), California (San Diego), 1986, Ob
Jan Newton, (D), Washington, 1989, Ob
Jeffrey A. Nystuen, (D), California (San Diego), 1985, Op
Joan M. Oltman-Shay, (D), California, 1986, Op
Mary Jane Perry, (D), California (San Diego), 1974, Ob
Thomas Pratt, (N), Gu
Joseph A. Resing, (D), Hawaii, 1997, Oc
Christopher L. Sabine, (D), Hawaii, 1992, Oc
Randy Shuman, (D), Washington, 1978, Ob
Laurenz A. Thomsen, (D), Kiel, 1992, Ob
Cynthia T. Tynan, (D), California (San Diego), 1993, Ob
Rebecca A. Woodgate, (D), Oxford, 1994, Op

## Walla Walla Community College

**Walla Walla Community College** (2014)
500 Tausick Way
Walla Walla, WA 99362
p. (509) 527-4278
steven.may@wwcc.edu
http://www.wwcc.edu

## Washington State University

**Dept of Crop & Soil Sciences** (B,M,D) (2014)
201 Johnson Hall
P.O. Box 646420
Pullman, WA 99164-6420
p. (509) 335-3475
bc.johnson@wsu.edu
http://css.wsu.edu/
Chair:
Richard Koenig, (D)
Professor:
David F. Bezdicek, (D), Minnesota, 1967, Sb
James B. Harsh, (D), California (Berkeley), 1983, Sc
Shiou Kuo, (D), Maine, 1973, Sc
Thomas A. Lumpkin, (D), Hawaii, 1978, On
William L. Pan, (D), North Carolina, 1983, Sb
John P. Reganold, (D), California (Davis), 1980, SoOs
Senior Scientist:
Robert G. Stevens, (D), Colorado State, 1971, Sb
Associate Professor:
Craig G. Cogger, (D), Cornell, 1979, Ou
Joan R. Davenport, (D), Guelph, 1985, Sb
Frank J. Peryea, (D), California, 1983, Sc
Assistant Professor:
Markus Flury, (D), Swiss Fed Inst Tech, 1994, Sp
Adjunct Professor:
Alan J. Busacca, (D), California (Davis), 1982, Sa
Ann C. Kennedy, (D), North Carolina State, 1985, Sb
Robert I. Papendick, (D), South Dakota State, 1962, Sp
Jeffery L. Smith, (D), Washington State, 1983, Sb

**Peter Hooper GeoAnalytical Lab** (B,M,D) (2016)
School of the Environment
1228 Webster Physical Sciences Building
Washington State University
Pullman, WA 99164-2812
p. (509) 335-1626
geolab@mail.wsu.edu
http://environment.wsu.edu/facilities/geolab/
Professor:
John A. Wolff, (D), London (UK), 1983, Gi
Research Associate:
Owen K. Neill, (D), Alaska (Fairbanks), 2012, Ca
Emeritus:
Franklin F. Foit, Jr., (D), Michigan, 1968, Gz
Laboratory Manager:
Scott Boroughs, (D), Washington State, 2010, GiCa
Research Technologist:
Charles Knaack, (M), Washington State, 1991, Ca

**School of the Environment** (B,M,D) ● (2016)
Webster 1228
P.O. Box 642812
Pullman, WA 99164-2812
p. (509) 335-3009
soe@wsu.edu
http://environment.wsu.edu/

*Enrollment (2009): B: 34 (11) M: 30 (9) D: 11 (5)*
Professor:
Keith Blatner , (D), 1983, On
Stephen Bollens, (D), Washington, 1990, Ob
Matt Carroll, (D), Washington, 1984, On
David R. Gaylord, (D), Wyoming, 1983, Gs
C. Kent Keller, (D), Waterloo, 1987, Hw
Peter B. Larson, (D), Caltech, 1983, Cs
Dirk Schulze-Makuch, (D), Wisconsin (Milwaukee), 1996, Hw
Jeffrey D. Vervoort, (D), Cornell, 1994, CcaGt
John A. Wolff, (D), Imperial Coll (UK), 1983, Giv
Associate Professor:
Sean P. Long , (D), Princeton, 2010, Gc
Clinical Assistant Professor:
Allyson Beall King, (D)
Clinical Assistant Professor :
Mike Berger , (D), Oregon, 2004
Assistant Professor:
Catherine Cooper, (D), Rice, 2005, Yg
Research Associate:
Scott Boroughs, (D), Washington State, 2010, Gv
Victor A. Valencia, (D), Arizona, 2005, OgGit
Instructor:
Kurt Wilkie, (D)
Adjunct Professor:
Cailin Huyck Orr
Regan L. Patton, (D), Washington State, 1997, Gc
Stephen P. Reidel, (D), Washington State, 1978, GivGt
Emeritus:
Franklin F. Foit, Jr., (D), Michigan, 1968, Gza
Andrew Ford, (D), Dartmouth, 1975, Ou
Eldon H. Franz, (D), Illinois, 1971, Sf
George Hinman
Philip E. Rosenberg, (D), Penn State, 1960, GzCg
A. John Watkinson, (D), Imperial Coll (UK), 1972, Gc
Gary D. Webster, (D), California, 1966, Pis

## Wenatchee Valley College

**Earth Sciences** (A) (2012)
1300 Fifth Street
Wenatchee, WA 98801
p. (509) 682-6754
rdawes@wvc.edu
https://www.wvc.edu/directory/departments/geology/default.asp
Professor:
Ralph Dawes, (D), Washington, 1993, GgeOw
Adjunct Professor:
Kelsay Stanton, (M), W Washington, 2005, GgOe

## Western Washington University

**Dept of Geology** (B,M) O (2017)
516 High Street
Bellingham, WA 98225-9080
p. (360) 650-3582
kate.blizzard@wwu.edu
http://geology.wwu.edu/
Administrative Assistant: Kate Blizzard
*Enrollment (2016): B: 187 (76) M: 30 (9)*
Chair:
Bernard A. Housen, (D), Michigan, 1994, Ym
Professor:
Susan M. DeBari, (D), Stanford, 1991, GxOe
Thor A. Hansen, (D), Yale, 1978, Pg
Scott R. Linneman, (D), Wyoming, 1990, GmOe
Robert J. Mitchell, (D), Michigan Tech, 1996, Hw
Elizabeth R. Schermer, (D), MIT, 1989, Gt
Associate Professor:
Colin B. Amos, (D), California (Santa Barbara), 2007, Gc
Jackie Caplan-Auerbach, (D), Hawaii, Ys
Douglas H. Clark, (D), Washington, 1995, Gl
Assistant Professor:
Robyn Dahl, (D), California, Riverside, 2015, PiOe
Brady Z. Foreman, (D), Wyoming, 2012, Grs
Sean R. Mulcahy, (D), California (Davis), 2009, GpzGp
Melissa S. Rice, (D), Cornell, 2012, XgGsm
Pete Stelling, (D), Alaska (Fairbanks), 2003, GxEgGv
Research Associate:
M. Clark Blake, (D), Stanford, 1965, Gt
Russell F. Burmester, (D), Princeton, 1974, Ym
George Mustoe, (M), Western Washington State College, 1972
Charles A. Ross, (D), Yale, 1964, Pm
Brian Rusk, (D), Oregon, 2003
Emeritus:
R. Scott Babcock, (D), Washington, 1970, Cg
Edwin H. Brown, (D), California (Berkeley), 1966, Gp
Don J. Easterbrook, (D), Washington, 1962, GlmGe
David C. Engebretson, (D), Stanford, 1983, Gt
James L. Talbot, (D), Adelaide, 1962, Gc

## Whatcom Community College

**Sciences Dept** (A) (2016)
237 W. Kellogg Road
Bellingham, WA 98226
p. (360) 383-3539
kkraft@whatcom.ctc.edu
http://www.whatcom.ctc.edu/home
Assistant Professor:
Kaatje Kraft, (D), Arizona State, 2014, Gg
Adjunct Professor:
Bernie Dougan, (M), W Washington, 1990, Gg

## Whitman College

**Dept of Geology** (B) (2016)
345 Boyer
Walla Walla, WA 99362
p. (509) 527-5225
baderne@whitman.edu
http://www.whitman.edu/geology
Administrative Assistant: Patti Moss
*Enrollment (2016): B: 48 (13)*
Chair, Assistant Professor:
Nicholas E. Bader, (D), California (Santa Cruz), 2006, SoHgOi
Professor:
Kevin R. Pogue, (D), Oregon State, 1993, Gc
Patrick K. Spencer, (D), Washington, 1984, Pg
Associate Professor:
Kirsten P. Nicolaysen, (D), MIT, 2001, GiCgGa
Assistant Professor:
Lyman P. Persico, (D), New Mexico, 2012, Gm
Visiting Assistant Professor:
Bryn Kimball, (D), Penn State, 2009, ClEgGz
Emeritus:
Robert J. Carson, (D), Washington, 1970, GmeGl
Geology Technologist:
Angela McGuire, (M), W Washington, 2014, Gi

## Yakima Valley College

**Dept of Geology** (A) (2010)
P. O. Box 22520
16th Ave. & Nob Hill Blvd.
Yakima, WA 98907
p. (509) 575-2350 x2366
dhuycke@yvcc.edu
http://www.yvcc.edu/FutureStudents/AcademicOptions/Programs/PhysicalSciences/Geology/Pages/default.aspx
*Enrollment (2005): A: 4 (0)*
Instructor:
David Huycke, (M), Wyoming, 1979, Gg

# West Virginia

## Concord University

**Dept of Physical Sciences** (B) O (2016)
1000 Vermillion St.
Athens, WV 24712-1000
p. (304) 384-5327
allenj@concord.edu
http://hub.concord.edu/physci
t: @CUGeology
*Enrollment (2016): B: 31 (7)*

Chair:
Joseph L. Allen, (D), Kentucky, 1994, Gc
Associate Professor:
Stephen C. Kuehn, (D), Washington State, 2002, CaGv
Research Associate:
Lewis Cook, (D), West Virginia, 2010, GbPg
Robert Peck, (D), Pennsylvania, 1969, Pg
Instructor:
Jennifer Phillippe, (M), California Davis, 2005, GegOe
Adjunct Professor:
Alyce Lee, (D), Texas A&M, 2009, Og

## Fairmont State University

**College of Science and Technology** (B) (2015)
1201 Locust Avenue
Fairmont, WV 26554
p. (304) 367-4393
dhemler@fairmontstate.edu
Coordinator of Geoscience Program:
Deb Hemler, (D), West Virginia, 1997, Oe
Instructor:
Pamela Casto, (M), Oe
Todd Ensign, (M), N Arizona, Oe
Jaime L. Ford, (M), Fairmont State, 2014, Oe
Visiting Professor:
Marcie Raol, (M), West Virginia, Oe

## Marshall University

**Dept of Geology** (B,M) (2015)
One John Marshall Drive
Huntington, WV 25755
p. (304) 696-6720/(304) 696-6756
geology@marshall.edu
http://marshall.edu/geology
*Enrollment (2015): B: 28 (5) M: 3 (0)*
Professor:
Ronald L. Martino, (D), Rutgers, 1981, Gs
Chair:
William L. Niemann, (D), Missouri (Rolla), 1999, Ng
Associate Professor:
Aley El Shazly, (D), Stanford, 1991, GpiCg
Assistant Professor:
Mitchell Scharman, (D), Texas (El Paso), 2011, GctOg

## Potomac State College

**Dept of Geology & Geography** (A) (2010)
101 Fort Avenue
Keyser, WV 26726
p. (304) 788-6956
JJNinesteel@mail.wvu.edu
Professor:
Judy J. Ninesteel, (M), West Virginia, 1995, Hg

## West Virginia University

**Dept of Geology & Geography** (B,M,D) (2017)
330 Brooks Hall
P.O. Box 6300
98 Beechurst Ave.
Morgantown, WV 26506-6300
p. (304) 293-5603
tim.carr@mail.wvu.edu
http://www.geo.wvu.edu
*Enrollment (2011): B: 165 (21) M: 45 (10) D: 13 (1)*
Professor:
Robert E. Behling, (D), Ohio State, 1971, Gm
Timothy R. Carr, (D), Wisconsin, 1981, GoYe
Joseph J. Donovan, (D), Penn State, 1992, Hq
Jason A. Hubbart, (D), Idaho, 2007, HqgHs
John J. Renton, (D), West Virginia, 1965, Ec
Timothy A. Warner, (D), Purdue, 1992, Or
Thomas H. Wilson, (D), West Virginia, 1980, YeGcYg
Chair:
J. Steven Kite, (D), Wisconsin, 1983, GmHsGa
Associate Professor:
Kathleen Benison, (D), GrsCm
Dengliang Gao, (D), Duke, 1997, Ye
Jaime Toro, (D), Stanford, 1998, Gc
Dorothy Vesper, (D), Penn State, 2002, Hw
Teaching Assistant Professor:
Joe Lebold, (D), West Virginia, 2005, PeGrg
Assistant Professor:
Graham D. M. Andrews, (D), Leicester, 2006, GvtGx
Shikha Sharma, (D), Lucknow Univ (India), 1998, Csa
Amy L. Weislogel, (D), Stanford, 2006, Gs
Adjunct Professor:
Katherine Lee Avary, (M), North Carolina, 1978, Go
Bascombe (Mitch) Blake, (D), West Virginia, 2009, GsEcPb
Alan Brown, (D), Louisiana State, 2002, EoYe
Katherine R. Bruner, (D), West Virginia, 1991, Gd
David Campagna, (D), Purdue, 1990, Gc
Blaine Cecil, (D), West Virginia, 1965, Ec
Phillip Dinterman
Harry Edenborn, (D), Rutgers, 1982, Scb
Evan J. Fedorko, (M), West Virginia, Oi
Nick Fedorko, (M), West Virginia, 1998, Ec
William Grady, (M), West Virginia, 1978, Ec
Michael Ed Hohn, (D), Indiana, 1976, Gq
Ronald McDowell, (D), Colorado Mines, 1987, GsPiCe
Zhihong Zheng
Paul F. Ziemkiewicz,
Associate Chair:
Helen M. Lang, (D), Oregon, 1983, Gpz
Emeritus:
Alan C. Donaldson, (D), Penn State, 1959, Gs
Thomas W. Kammer, (D), Indiana, 1982, Pi
Henry W. Rauch, (D), Penn State, 1972, HwGe
Robert C. Shumaker, (D), Cornell, 1960, Go
Richard A. Smosna, (D), Illinois, 1973, GrEo

# Wisconsin

## Beloit College

**R.D. Salisbury Dept of Geology** (B) (2016)
700 College Street
Beloit, WI 53511
p. (608) 363-2223
rougviej@beloit.edu
http://www.beloit.edu/geology/
f: https://www.facebook.com/groups/47867916990/
*Enrollment (2015): B: 20 (11)*
Chair:
James R. Rougvie, (D), Texas, 1999, GxpGz
Professor:
Carl V. Mendelson, (D), California (Los Angeles), 1981, Pm
Susan K. Swanson, (D), Wisconsin, 2001, Hw
Tech & Safety Officer:
Stephen M. Ballou, (B), Beloit, 1998, Gg
Cooperating Faculty:
Carol Mankiewicz, (D), Wisconsin, 1987, Gs

## Lawrence University

**Dept of Geology** (B) (2017)
711 E Boldt Way
Appleton, WI 54911
p. (920) 832-6731
knudsena@lawrence.edu
http://www.lawrence.edu/dept/geology/
Administrative Assistant: Ellen c. Walsh
*Enrollment (2016): B: 24 (8)*
Professor:
Marcia Bjornerud, (D), Wisconsin, 1987, Gc
Associate Professor:
Jeffrey J. Clark, (D), Johns Hopkins, 1997, Gm
Assistant Professor:
Andrew Knudsen, (D), Idaho, 2002, Cg
Visiting Professor:
Andrew Malone, (D), Chicago, 2016, Ol
Emeritus:
John C. Palmquist, (D), Iowa, 1961, GcpGg
Theodore W. Ross, (D), Washington State, 1969, Gg
Ronald W. Tank, (D), Indiana, 1962, Ge

## Milwaukee Public Museum

**Dept of Geology** (2011)
800 W. Wells Street
Milwaukee, WI 53233
p. (414) 278-2741
sheehan@mpm.edu
http://www.mpm.edu/collect/geology/geosec-noframes.html

Chair:
Peter M. Sheehan, (D), California (Berkeley), 1971, Pi
Adjunct Professor:
Gary D. Rosenberg, (D), California (Los Angeles), 1972, GhPg

## Northland College

**Dept of Geosciences** (B) ● (2017)
1411 Ellis Avenue
Ashland, WI 54806
p. (715) 682-1852
tfitz@northland.edu
*Enrollment (2016): B: 39 (8)*

Professor:
Thomas J. Fitz, (D), Delaware, 1999, GgxGc
Assistant Professor:
Cynthia May, (M), Penn State, 2007, OirOy
David J. Ullman, (D), Wisconsin (Madison), 2013, GlPeGm

## Saint Norbert College

**Geology Dept** (B) (2016)
100 Grant Street
De Pere, WI 54115
p. (920) 403-3987
tim.flood@snc.edu
http://www.snc.edu
*Enrollment (2016): B: 27 (6)*

Professor:
Tim P. Flood, (D), Michigan State, 1987, Gv
Nelson R. Ham, (D), Wisconsin, 1994, Gl
Associate Professor:
Rebecca McKean , (D), Nebraska, 2009, GdPgGh

## University of Wisconsin, Fox Valley

**Dept of Geography and Geology** (1999)
1478 Midway Rd
Menasha, WI 54952
p. (920) 832-2600
joanne.kluessendorf@uwc.edu
http://uwc.edu/depts/geography-geology/geography-geology

## University of Wisconsin Colleges

**Dept of Geography & Geology** (A) (2016)
UW Sheboygan
1 University Dr
Sheboygan , WI 53081
p. (920) 459-6619
jim.mccluskey@uwc.edu
http://www.uwc.edu/depts/geography-geology

Chair:
Karl Byrand, (D), Maryland, 1999, Ony
Chair/CEO UW Marathon County:
Keith Montgomery, (D), Waterloo, 1986, OyGg
Professor:
Norlene Emerson, (D), Wisconsin (Madison), 2002, GgPiGs
Michael Jurmu, (D), Indiana State, 1999, OyGgOn
Diann Kiesel, (D), Wisconsin, 1998, GgOyGg
Robert McCallister, (D), Wisconsin, 1996, OyGgOn
Dean/CEO UW Washington County:
Alan Paul Price, (D), California (Los Angeles), 1998, OyGmOg
Associate Professor:
Iddrisu Adam, (D), Wilfrid Laurier, 2001, OynOn
Assistant Professor:
Beth Johnson, (D), N Illinois, 2009, Glh
James McCluskey, (D), Rutgers, 1987, OinOn
Miclelle Palma, (D), Georgia, 2012, OnnOn
Andrew Shears, (D), Kent State, 2011, Oin
Keith West, (D), Wisconsin (Milwaukee), 2007, OynOn
Instructor:
Sanborn Robert, (M), 1991, GgOyn
Mengist Teklay Berhe
Lecturer:
Jane Fairchild, (M), Wisconsin, 2003, Ow
Seth Rankin, (M), Wisconsin (Milwaukee), 1974, OnnOn
Emeritus:
Thomas Bitner, (M), OynOn
James Brey, (D), Wisconsin, OywOn
Richard Cleek, (M), OnyOn
Garret Deckert, (M), OynOn
Edwin Dommissee, (M), OyGgOn
James Heidt, (M), On
Cary Komoto, (D), Minnesota, 1994, OnnOn
Kenneth Korb, (D)
Gene E. Musolf, (D), Wisconsin, 1970, Oy
Shamim Naim, (D)
Randall Rohe, (D), Colorado, 1978, OyGgOn
Leonard Weis, (D)
Barbara Williams, (D)

## University of Wisconsin, Eau Claire

**Dept of Geology** (B) (2016)
157 Phillips Hall
Eau Claire, WI 54702-4004
p. (715) 836-3732
steinklm@uwec.edu
http://www.uwec.edu/academic/geology
Administrative Assistant: Lorilie M. Steinke
*Enrollment (2011): B: 88 (0)*

Chair:
Kent M. Syverson, (D), Wisconsin, 1992, Gl
Professor:
Karen G. Havholm, (D), Texas, 1991, Oe
Robert L. Hooper, (D), Washington State, 1983, Gz
Phillip D. Ihinger, (D), Caltech, 1991, Gx
J. Brian Mahoney, (D), British Columbia, 1994, GstEg
Associate Professor:
Scott K. Clark, (D), Illinois (Urbana-Champaign), 2007, Oe
Assistant Professor:
Robert Lodge, (D), Gc
Senior Lecturer:
Lori D. Snyder, (M), British Columbia, 1994, Gx

## University of Wisconsin, Extension

**Dept of Environmental Sciences** (2009)
3817 Mineral Point Road
Madison, WI 53705-5100
p. (608) 262-1705
kmzwettl@wisc.edu

Professor:
John W. Attig, (D), Wisconsin, 1984, Gl
Thomas J. Evans, (D), Wisconsin, 1994, Gg
James M. Robertson, (D), Michigan, 1972, Eg
Associate Professor:
Madeline B. Gotkowitz, (M), New Mexico, 1993, Hw
David J. Hart, (D), Wisconsin, 2000, Hw
Assistant Professor:
Eric C. Carson, (D), Wisconsin, 2003, Gl
Patrick I. McLaughlin, (D), Cincinnati, 2006, Gr
Emeritus:
Lee Clayton, (D), Illinois, 1965, Gl

## University of Wisconsin, Green Bay

**Dept of Natural and Applied Science - Geoscience Program** (B,M) (2016)
LS-465
2420 Nicolet Drive
Green Bay, WI 54311
p. (920) 465-2371
luczajj@uwgb.edu
http://www.uwgb.edu/geoscience/
*Enrollment (2016): B: 15 (3) M: 3 (0)*

Interim Chair:
Kevin J. Fermanich, (D), Wisconsin, 1995, SpHwOu

Professor:
John A. Luczaj, (D), Johns Hopkins, 2000, HwGsCl
Associate Professor:
Steven J. Meyer, (D), Nebraska, 1990, OawOg
Assistant Professor:
Ryan Currier, (D), Johns Hopkins, 2011, GizGg
Emeritus:
Steven I. Dutch, (D), Columbia, 1976, GcgGe
Ronald D. Stieglitz, (D), Illinois, 1972, Gr

## University of Wisconsin, Madison

**Center for Limnology** (M,D) (2016)
680 North Park Street
Madison, WI 53706
p. (608)262-3014
fms@ls.wisc.edu
http://limnology.wisc.edu/
Chair:
Kenneth Potter, (D), Johns Hopkins, Ng
Professor:
Michael S. Adams, (D), California (Riverside), 1968, Ob
Anders W. Andren, (D), Florida State, 1972, Oc
David E. Armstrong, (D), Wisconsin, 1966, Oc
Steve Carpenter, (D), Wisconsin, 1979, Ob
Calvin B. DeWitt, (D), Michigan, 1963, Ob
Stanley I. Dodson, (D), Washington, 1970, Ob
Linda K. Graham, (D), Michigan, 1975, Ob
John A. Hoopes, (D), MIT, 1965, Oo
James F. Kitchell, (D), Colorado, 1970, Ob
John E. Kutzbach, (D), Wisconsin, 1966, Pe
William C. Sonzogni, (D), Wisconsin, 1974, Oc
Joy Zedler, (D), Wisconsin, 1968, Ob
Associate Professor:
Emily Stanley, (D), Arizona State, 1993, Ob
Assistant Professor:
Sarah Hotchkiss, (D), Minnesota, 1998, Ob
Carol Lee, (D), Washington, 1998, Ob
Zheng-yu Liu, (D), MIT, 1991, Op
Katherine McMahon, (D), California (Berkeley), 2002, Ng
Jake Vander Zanden, (D), McGill, 1999, On
Chin Wu, (D), MIT, Ng

**Department of Geography** (B,M,D) (2016)
550 N. Park St.
Room 160 Science Hall
Madison, WI 53706
p. (608) 262-2138
smkahn@geography.wisc.edu
http://www.geography.wisc.edu
*Enrollment (2010): B: 157 (30) M: 59 (4) D: 0 (1)*
Affiliate :
John W. Williams, (D), Brown, 1999, PeOyPl

**Department of Geoscience** (B,M,D) ● (2016)
1215 West Dayton St
Madison, WI 53706
p. (608) 262-8960
geodept@geology.wisc.edu
http://geoscience.wisc.edu/geoscience/
*Enrollment (2009): B: 58 (17) M: 20 (6) D: 42 (4)*
Chair:
Harold Tobin, (D), California (Santa Cruz), 1995, Yg
Professor:
Jean M. Bahr, (D), Stanford, 1986, Hw
Philip E. Brown, (D), Michigan, 1980, EmGzp
Alan R. Carroll, (D), Stanford, 1991, Gs
D. Charles DeMets, (D), Northwestern, 1988, Ym
Kurt Feigl, (D), MIT, 1991, Yd
Laurel Goodwin, (D), California (Berkeley), 1988, Gc
Laurel B. Goodwin, (D), California (Berkeley), 1989, Gc
Clark M. Johnson, (D), Stanford, 1986, CgcCs
D. Clay Kelly, (D), North Carolina, 1999, Pm
Eric E. Roden, (D), Maryland, 1990, PySbCl
Bradley S. Singer, (D), Wyoming, 1990, Cc
Clifford H. Thurber, (D), MIT, 1981, YsGv
Basil Tikoff, (D), Minnesota, 1994, Gc
John W. Valley, (D), Michigan, 1980, CsGxz
Herbert F. Wang, (D), MIT, 1971, YgHwYx
Senior Scientist:
Brian L. Beard, (D), Wisconsin, 1992, Cc
John Fournelle, (D), Johns Hopkins, 1989, GiOn
Associate Professor:
Shanan Peters, (D), Chicago, 2003, Gs
Assistant Professor:
Michael A. Cardiff, (D), Stanford, 2010, HwYg
Shaun Marcott, (D), Oregon State, 2011, GlPe
Huifang Xu, (D), Johns Hopkins, 1993, Gz
Lucas Zoet, (D), Penn State, 2012, OlGlm
Adjunct Professor:
Randy Hunt, (D), Wisconsin, 1993, HwCls
Emeritus:
Mary P. Anderson, (D), Stanford, 1973, Hw
Charles R. Bentley, (D), Columbia, 1959, Yg
Carl J. Bowser, (D), California (Los Angeles), 1965, Cl
Charles W. Byers, (D), Yale, 1973, Pe
Nikolas I. Christensen, (D), Wisconsin, 1963, Yx
David L. Clark, (D), Iowa, 1957, Pm
Robert H. Dott, Jr., (D), Columbia, 1955, Gs
Dana H. Geary, (D), Harvard, 1986, Po
Louis J. Maher, Jr., (D), Minnesota, 1961, Pl
Levi Gordon Medaris, Jr., (D), California, 1966, GipSa
David M. Mickelson, (D), Ohio State, 1971, Gl
Museum Director:
Richard Slaughter, (D), Iowa, 2001, Pg
Geology Librarian:
Marie Dvorzak, (M), On
Related Staff:
Kenneth R. Bradbury, (D), Wisconsin, 1982, Hw
Dante Fratta, (D), Georgia Tech, 1999, Ng
Madeline B. Gotkowitz, (M), New Mexico Tech, 1993, Hw
David J. Hart, (D), Wisconsin, 2000, Hg
Thomas S. Hooyer, (D), Iowa State, 1999, Hg
David Krabbenhoft, (D), Wisconsin, 1988, ClGeHw
James M. Robertson, (D), Michigan, 1972, Eg

**Dept of Atmospheric & Oceanic Sciences** (B,M,D) (2016)
1225 W. Dayton Street
Atmospheric, Oceanic & Space Science Building
Madison, WI 53706-1695
p. (608)262-2828
aos@aos.wisc.edu
http://www.aos.wisc.edu
*Enrollment (2011): B: 24 (14) M: 32 (9) D: 34 (5)*
Chair:
Gregory J. Tripoli, (D), Colorado State, 1986, Ow
Professor:
Steven A. Ackerman, (D), Colorado State, 1987, Or
Ankur R. Desai, (D), Penn State, 2006, Oaw
Matthew H. Hitchman, (D), Washington, 1985, Oa
Zhengyu Liu, (D), MIT, 1991, Op
Jonathan E. Martin, (D), Washington, 1992, Ow
Galen McKinley, (D), MIT, 2002, OcpOg
Michael C. Morgan, (D), MIT, 1994, Ow
Grant W. Petty, (D), Washington, 1990, OarOw
Senior Scientist:
Edwin W. Eloranta, (D), Wisconsin, 1972, Yr
Affiliate:
Tracey Holloway, (D), Princeton, 2001, Oa
Affiliate :
Chris Kucharik, (D), Wisconsin, 1997, Oa
Associate Professor:
Daniel J. Vimont, (D), Washington, 2001, Oa
Affiliate:
Samuel Stechmann, (D), Courant Inst, 2008, Ona
Assistant Professor:
Larissa E. Back, (D), Washington, 2007, Oa
Tristan S. L'Ecuyer, (D), Colorado State, 2001, Oa
Adjunct Associate Professor:
Jeffrey R. Key, (D), Colorado, 1988, Ora
Adjunct :
Andrew Heidinger, (D), Colorado State, 1998, Or
Adjunct Professor:
James Kossin, (D), Colorado State, 2000, Oa
Steven Platnick, (D), Arizona, Or

Emeritus:
Linda M. Keller, (M), Wisconsin, 1971, Oa
Francis P. Bretherton, (D), Cambridge, 1961, Og
Stefan L. Hastenrath, (D), Bonn, 1959, Oa
Donald R. Johnson, (D), Wisconsin, 1965, Oa
John E. Kutzbach, (D), Wisconsin, 1966, Oa
John M. Norman, (D), Wisconsin, 1971, So
Robert A. Ragotzkie, (D), Wisconsin, 1953, Ob
Pao-Kuan Wang, (D), California (Los Angeles), 1978, Oa
John A. Young, (D), MIT, 1966, Oa
Research Specialist:
Dierk T. Polzin, (B), Wisconsin, Oa



**Dept of Soil Science** (B,M,D) (2011)
1525 Observatory Drive
Madison, WI 53706-1299
p. (608) 262-2633
slspeth@wisc.edu
http://www.soils.wisc.edu
*Enrollment (2011): B: 34 (4) M: 16 (2) D: 8 (2)*
Chair:
William L. Bland, (D), Wisconsin, 1984, Sp
Professor:
Phillip W. Barak, (D), Hebrew (Israel), 1988, ScGz
William F. Bleam, (D), Cornell, 1984, Sc
James G. Bockheim, (D), Washington, 1972, Sf
William J. Hickey, (D), California (Riverside), 1990, Sb
King-Jau S. Kung, (D), Cornell, 1984, Sp
Birl Lowery, (D), Oregon, 1980, Sp
Frederick W. Madison, (D), Wisconsin, 1972, Sd
J. Mark Powell, (D), Texas A&M, 1989, So
Stephen J. Ventura, (D), Wisconsin, 1989, OrSo
Associate Professor:
Teri C. Balser, (D), California (Berkeley), 2000, Sb
Nick J. Balster, (D), Idaho, 1999, Sf
Carrie A.M Laboski, (D), Minnesota, 2001, So
Joel A. Pedersen, (D), California (Los Angeles), 2001, So
Assistant Professor:
Matthew D. Ruark, (D), Purdue, 2006, So
Douglas J. Soldat, (D), Cornell, 2007, So
Emeritus:
Larry G. Bundy, (D), Iowa, 1973, So
Robin F. Harris, (D), Wisconsin, 1972, Sb
Philip A. Helmke, (D), Wisconsin, 1971, Sc
Keith A. Kelling, (D), Wisconsin, 1974, So
Wayne R. Kusssow, (D), Wisconsin, 1966, So
Kevin McSweeney, (D), Illinois, 1984, SdaSf
John M. Norman, (D), Wisconsin, 1971, Sp
E. Jerry Tyler, (D), North Carolina, 1975, Sd

**Geological Engineering** (B,M,D) ● (2016)
1415 Engineering Drive
2205 Engineering Hall
Madison, WI 53706-1691
p. (608) 890-2662
likos@wisc.edu
http://www.gle.wisc.edu
*Enrollment (2016): B: 144 (0) M: 10 (0) D: 3 (0)*
Chair:
William J. Likos, (D), Colorado Mines, 2000, NgSp
Professor:
Jean M. Bahr, (D), Stanford, 1987, Hw
Kurt L. Feigl, (D), Gt
Tracey Holloway, (D), Princeton, Oa
Kenneth W. Potter, (D), Johns Hopkins, 1976, Hg
Clifford Thurber, (D), MIT, 1981, Ysx
Basil Tikoff, (D), Minnesota, 1997, Gc
Harold J. Tobin, Yr
Herbert F. Wang, (D), MIT, 1971, Yx
Chin Wu, (D), MIT, Onp
Associate Professor:
Dante Fratta, (D), Georgia Tech, 1999, Yx
Steven P. Loheide, (D), Stanford, 2006, HwSp
James M. Tinjum, (D), Wisconsin, 2003, Ng
Assistant Professor:
Matt Ginder-Vogel, (D), Stanford, 2006, Sc
Andrea Hicks, (D), Illinois (Chicago), On
Hiroki Sone, (D), Stanford, 2012, NrYx
Lucas Zoet, (D), Penn State, Gl
Emeritus:
Mary P. Anderson, NAE, (D), Stanford, 1973, Hw
Tuncer B. Edil, (D), Northwestern, 1973, Ng
Bezalel C. Haimson, (D), Minnesota, 1968, Nr

## University of Wisconsin, Milwaukee

**Dept of Geosciences** (B,M,D) (2017)
P.O. Box 413
Milwaukee, WI 53201-0413
p. (414) 229-4561
geosci-office@uwm.edu
http://www4.uwm.edu/letsci/geosciences/
f: https://www.facebook.com/UWMgeosci
*Enrollment (2014): B: 75 (17) M: 19 (8) D: 10 (1)*
Professor:
Timothy J. Grundl, (D), Colorado, 1988, HwCl
Mark T. Harris, (D), Johns Hopkins, 1988, GsrGd
John L. Isbell, (D), Ohio State, 1990, Gs
Keith A. Sverdrup, (D), Scripps, 1981, YsGt
Associate Professor:
Barry I. Cameron, (D), N Illinois, 1998, Gvi
Dyanna M. Czeck, (D), Minnesota, 2001, Gc
Stephen Q. Dornbos, (D), S California, 2003, PiePy
Margaret L. Fraiser, (D), S California, 2005, Pi
Thomas S. Hooyer, (D), Iowa State, 1999, Gl
Lindsay J. McHenry, (D), Rutgers, 2004, GzaXg
Shangping Xu, (D), Princeton, 2005, Hg
Assistant Professor:
Julie Bowles, (D), Calfornia (San Diego), 2005, Ym
Erik L. Gulbranson, (D), California (Davis), 2011, CgPe
Weon Shik Han, (D), New Mexico Tech, 2008, Hw
Adjunct Professor:
Daniel T. Feinstein, (M), Wisconsin, 1986, Hw
Peter M. Sheehan, (D), California (Berkeley), 1971, Pie
Emeritus:
Douglas S. Cherkauer, (D), Princeton, 1972, Hw
William F. Kean, Jr., (D), Pittsburgh, 1972, Ym
Norman P. Lasca, (D), Michigan, 1965, Gml

## University of Wisconsin, Oshkosh

**Geography and Urban Planning Department** (B) (2017)
800 Algoma Avenue
Oshkosh, WI 54901
p. (414) 424-4105
longco@uwosh.edu
http://www.uwosh.edu/geography/
*Enrollment (2016): B: 36 (12)*

**Geology Dept** (B) O (2016)
645 Dempsey Trail
Oshkosh, WI 54901-8649
p. (920) 424-4460
mode@uwosh.edu
http://www.uwosh.edu/geology/
Administrative Assistant: Courtney Maron
*Enrollment (2016): B: 53 (12)*
Chair:
William N. Mode, (D), Colorado, 1980, Gl
Professor:
Eric E. Hiatt, (D), Colorado, 1997, GsOcGu
Maureen A. Muldoon, (D), Wisconsin, 1999, Hw
Timothy S. Paulsen, (D), Illinois, 1997, Gc
Jennifer M. Wenner, (D), Boston, 2001, Gi
Assistant Professor:
Benjamin W. Hallett, (D), Rensselaer Polytechnic Inst, 2012, Gpx
Joseph E. Peterson, (D), N Illinois, 2010, PgGr
Lecturer:
Christie M. Demosthenous, (M), Illinois, 1996, Gz
Emeritus:
Norris W. Jones, (D), Virginia Tech, 1968, Gi
Gene L. La Berge, (D), Wisconsin, 1963, Eg
Thomas S. Laudon, (D), Wisconsin, 1963, Yg
James W. McKee, (D), Louisiana State, 1967, Gr
Brian K. McKnight, (D), Oregon State, 1970, Gdu

Instrumentation Specialist:
Thomas J. Suszek, (M), Minnesota (Duluth), 1991, Gg

## University of Wisconsin, Parkside
**Dept of Geosciences** (B) (2016)
Box 2000
900 Wood Road
Kenosha, WI 53141
p. (262) 595-2744
li@uwp.edu
*Enrollment (2016): B: 30 (8)*
Chair:
Li Zhaohui, (D), SUNY (Buffalo), 1994, ClGzHw
Professor:
John D. Skalbeck, (D), Nevada (Reno), HwYgGg
Assistant Professor:
Rachel Headley, (D), Washington, 2011, Gl
Research Associate:
Julie Kindelman, (D), Hs
Emeritus:
Gerald A. Fowler, PgGg
Allan F. Schneider, (D), Gl
James H. Shea, (D), Gs

## University of Wisconsin, Platteville
**Dept of Geography & Geology** (B) (2011)
Platteville, WI 53818
p. (608) 342-1791
rawlingj@uwplatt.edu
http://www.uwplatt.edu/geography/
Professor:
Charles W. Collins, (D), Nova, 1985, Gm
Associate Professor:
Richard A. Waugh, (D), Wisconsin, 1995, Gc
Assistant Professor:
Mari A. Vice, (D), S Illinois, 1993, Gd
Emeritus:
William A. Broughton, (B), Wisconsin, 1938, Em
Kenneth A. Shubak, (M), Michigan, 1966, Pi

## University of Wisconsin, River Falls
**Dept of Plant & Earth Science** (B) (2016)
410 S. Third Street
River Falls, WI 54022
p. (715) 425-3345
holly.dolliver@uwrf.edu
http://www.uwrf.edu/pes/geol/
Program Assistant: Sue Freiermuth
*Enrollment (2016): B: 50 (0)*
Professor:
Robert W. Baker, (D), Minnesota, 1976, GlmOl
Eric M. Sanden, (D), Texas Tech, 1993, Or
Donavon Taylor, (D), Minnesota, 1981, Sp
Ian S. Williams, (D), California (Santa Barbara), 1981, Yg
Associate Professor:
Holly A.S. Dolliver, (D), Minnesota, 2007, GmSd
Kerry L. Keen, (D), Minnesota, 1992, Hw

## University of Wisconsin, Stevens Point
**Dept of Geography and Geology** (B) (2015)
2001 Fourth Avenue
Stevens Point, WI 54481
p. (715) 346-2629
Dozsvath@uwsp.edu
http://www.uwsp.edu/geo/
Administrative Assistant: Mary Clare Sorenson
*Enrollment (2010): B: 49 (8)*
Professor:
Kevin P. Hefferan, (D), Duke, 1992, GctGe
Neil C. Heywood, (D), Colorado, 1988, Oy
Eric J. Larson, Oregon State, 2001, Or
Karen A. Lemke, (D), Iowa, 1988, GmOy
David L. Ozsvath, (D), SUNY (Binghamton), 1985, Hw
Keith W. Rice, (D), Kansas, 1989, Oir
Michael E. Ritter, (D), Indiana, 1986, Ow
Associate Professor:
Samantha W. Kaplan, (D), Wisconsin, 2003, GnPlGd
Ismaila Odogba, (D), Louisville, 2009, Oui
Instructor:
Timothy T. Kennedy, (M), Wisconsin, 2009, Oir

## University of Wisconsin, Superior
**Dept of Natural Sciences** (B,M) (2014)
PO Box 2000
Belknap & Catlin
Superior, WI 54880
p. (715) 394-8322
natsci@uwsuper.edu
Professor:
William Bajjali, (D), Ottawa, 1994, HwCsg
Associate Professor:
Andrew Breckenridge, (D), Minnesota, 2005, GlnGs
Assistant Professor:
Kristin E. Riker-Coleman, (D), Minnesota, 2008, Gg

## University of Wisconsin, Whitewater
**Geography, Geology, and Environmental Science** (B) (2016)
800 W Main Street
Whitewater, WI 53190
p. (262) 472-1071
geography@uww.edu
http://www.uww.edu/cls/departments/geography-geology-env-sci
*Enrollment (2016): B: 57 (20)*
Professor:
Peter Jacobs, (D), UW-Madison, 1994, SdaGm
Associate Professor:
Prajukti Bhattacharyya, (D), Minnesota, 2000, Gce
Rex Hanger, (D), California (Berkeley), 1992, PigGs
Assistant Professor:
Stephen J. Levas, (D), Ohio State Univ, 2015, CmGe
Science Outreach Director:
Anna Courtier, (D), Minnesota, 2009, Ysg

# Wyoming

## Casper College
**Department of Earth Sciences** (A) (2017)
School of Science
125 College Drive
Casper, WY 82601
p. (307) 268-2513
mconnely@caspercollege.edu
http://www.caspercollege.edu/physcience/geology/
Administrative Assistant: Shereen Mosier
*Enrollment (2012): A: 16 (4)*
Head:
Kent A. Sundell, (D), California (Santa Barbara), 1985, Gg
Instructor:
Melissa Connely, (M), Utah State, 2000, Gs
Kenneth Kreckel, (B), Michigan Tech, 1970, EoYeGg
Karen Sue McCutcheon, (M), Oeg
Beth Wisely, (D), Univ. Oregon, 2005, YgHwGc

## Central Wyoming College
**Earth, Energy, Environment** (A) (2015)
2660 Peck Avenue
Riverton, WY 82501
p. (307) 855-2000
ssmaglik@cwc.edu
http://www.cwc.edu/academics/programs/earthenviron
f: https://www.facebook.com/cwc.edu/
*Enrollment (2015): A: 5 (0)*
Professor:
Suzanne M. Smaglik, (M), Colorado Mines, 1987, GgOeCg
Assistant Professor:
Jacki Klancher, (M), Oi

## Eastern Wyoming College
**Eastern Wyoming College** (A) (2010)

3200 West C. Street
Torrington, WY 82240
p. (307) 532-8330
stuart.nelson@ewc.wy.edu
http://www.ewc.wy.edu/
Instructor:
Christopher Wenzel, (D), Wyoming, 2019, SbfHg

Algeria

## Northwest College

**Northwest College** (2014)
231 West 6th Street
Powell, WY 82435
p. 307.754.6405
Mark.Kitchen@northwestcollege.edu
http://www.nwc.cc.wy.us/

## University of Wyoming

**Dept of Geology and Geophysics** (B,M,D) ● (2015)
Dept. 3006
1000 E. University Ave.
Laramie, WY 82071
p. (307) 766-3386
geol-geophys@uwyo.edu
http://www.uwyo.edu/geolgeophys/
*Enrollment (2015): B: 157 (26) M: 28 (7) D: 31 (2)*

Head:
Paul L. Heller, (D), Arizona, 1983, GrsGt

Professor:
Carrick M. Eggleston, (D), Stanford, 1991, Cl
B. Ronald Frost, (D), Washington, 1973, Gp
Carol D. Frost, (D), Cambridge (UK), 1984, CcGit
W. Steven Holbrook, (D), Stanford, 1989, Ys
Neil F. Humphrey, (D), Washington, 1987, OlGmYh
Barbara E. John, (D), California (Santa Barbara), 1987, Gci
Subhashis Mallick, (D), Hawaii, 1987, YseYg
James D. Myers, (D), Johns Hopkins, 1979, Gi
Kenneth W.W Sims, (D), California (Berkeley), 1995, CgGvCa

Senior Scientist:
Susan M. Swapp, (D), Yale, 1982, Gp

Associate Professor:
Michael J. Cheadle, (D), Cambridge, 1989, Yg
Po Chen, (D), S California, 2005, YseYg
Mark T. Clementz, (D), California (Santa Cruz), 2002, Py
Kenneth G. Dueker, (D), Oregon, 1994, Ys
Robert R. Howell, (D), Arizona, 1980, XgOr
John P. Kaszuba, (D), Colorado Mines, 1997, CgGze
Clifford S. Riebe, (D), California (Berkeley), 2000, ClGme
Bryan N. Shuman, (D), Brown, 2001, PeGne
Ye Zhang, (D), Indiana, 2005, Hy

Associate Scientist:
Janet Dewey, (M), Auburn, 1993, Ca
Laura Vietti, (D), Minnesota, 2014, Pv

Assistant Professor:
Ellen Currano, (D), Penn State, 2008, Pbe
Dario Grana, (D), Stanford, 2013, Ye
Brandon McElroy, (D), Texas (Austin), 2009, Gsm
Andrew D. Parsekian, (D), Rutgers, 2011, Yg

Research Associate:
Kevin R. Chamberlain, (D), Washington (St. Louis), 1990, Cc

Lecturer:
Erin A. Campbell-Stone, (D), Wyoming, 1997, GcEo

Adjunct Professor:
Vladimir Alvarado, (D), Minnesota, 1996, Np
Eric A. Erslev, (D), Harvard, 1981, GctGg
Warren B. Hamilton, (D), California (Los Angeles), 1951, Gt
Peter H. Hennings, (D), Texas, 1991, Gc
Ranie Lynds, (D), Wyoming, 2005, Gsr

Department Secretary:
Deborah Prusia

Academic Coordinator:
Lexi N. Edwards

## Western Wyoming Community College

**Western Wyoming Community College** (2014)
2500 College Drive
Rock Springs, WY 82901
p. (307) 382-1662
webmaster@wwcc.wy.edu
http://www.wwcc.wy.edu/academics/geology/default.htm

# Algeria

## Algerian Petroleum Institute

**Algerian Petroleum Institute** (B) (2010)
Avenue 1er Novembre
Boumerdes, Skikda, Oran 35000
p. +213 (0) 24 81 90 56
iap@iap.dz
http://www.iap.dz/

## Centre Universitaire de Khemis Miliana

**Institut de Sciences de la Nature et de la Terre** (B) (2010)
Route de Theniet El-Had
Khemis Miliana, Khemis Miliana 44225
p. +213-27-66-42-32
kouben55@hotmail.fr
http://www.cukm.org/

## Universite Abou Bekr Belkaid de Tlemcen

**Faculté des sciences de la nature et de la vie et sciences de la terre et de l'univers** (B) (2014)
BP 119
Tlemcen 13000
p. (+213) 040.91.59.09
webcri@mail.univ-tlemcen.dz
http://snv.univ-tlemcen.dz/

## Universite Badji Mokhtar

**Faculte des Sciences de la Terre: Amenagement du territoire, Geologie, Hydraulique, Mines** (B) (2010)
BP 12
Annaba 23000
d.fst@univ-annaba.dz
http://www.univ-annaba.org/

## Universite D'Oran

**Faculté des Sciences de la Terre, de la Géographie et de l'Amenagement du Territoire** (B) (2015)
Rue du Colonel Lofti
Es-Senia, El-Mnouar, Oran 31000
p. +213 (0) 41 58 19 47
contact@univ-oran.dz
http://www.univ-oran.dz/facultes/F_Terre/index.html

## Universite de Annaba

**Inst of Natural Sciences** (2013)
BP 12
El Hadjar
Annaba
http://www.univ-annaba.org/

## Universite de Batna

**Dept de Sciences de la Terre** (B) (2014)
1, Rue Chahid Boukhlouf Mohamed
El Hadi, Batna 5000
p. (+213) 033.86.06.02
recteur@univ-batna.dz
http://www.univ-batna.dz/

## Universite de Jijel

**Dept de Geologie** (B,M,D) (2012)
Ouled Aissa, Jijel
p. +213 (34) 498016
webmaster@univ-jijel.dz
http://www.univ-jijel.dz/

## Universite de Mentouri

**Dept de l'Amenagement du territoire** (B) (2014)
Campus Ahmed Zouaghi

B.P. 325 Route d Ain El Bey
, Constantine 25000
p. (+213) 31.90.02.07
univ-constantine@fr.fm
http://www.umc.edu.dz/vf/index.php/recherche-scientifique/annuaire-des-laboratoires/110-faculte-des-sciences-de-la-terre-/398-de

**Dept des Sciences de la Terre/Geologie** (B) (2011)
Campus Ahmed Zouaghi
B.P. 325 Route d Ain El Bey
Constantine 25000
p. 213 031 90 38 52
univ-constantine@fr.fm
http://www.umc.edu.dz/fst/

### Universite des Sciences et de la Technologie
**Inst of Earth Sciences** (2011)
Houari Boumediene
B.P. 139 Dar El Beida
Eldjazair

### Universite des Sciences et de la Technologie d'Oran
**Dept de Geologie** (B) (2011)
Oran
http://www.univ-usto.dz/

### Universite des Sciences et de la Technologie Houari Boumediene
**Faculte de Sciences de la Tere, Geographie et Amenagement du Territoire** (B) (2014)
USTHB-IST BP 32 El Alia
Bab-Ezzouar Alger 16123
p. (+213) 21247904
aouabadi@usthb.dz
http://www.usthb.dz/fst/

### Universite Djillali Liabes Sidi Bel Abbes
**Faculte des Sciences de la Nature et de la vie** (B) (2010)
BP 89
Sidi Bel Abbes 22000
p. +213 (48) 543018
mbouziani@univ-sba.dz
http://www.univ-sba.dz/

### Universite Kasdi Merbah Ouargla
**Dept de Geologie** (B) (2010)
Route de Ghardaia
Ouargla
p. +213 (29) 712468
info@ouargla-univ.dz
http://193.194.92.30/spip/dept/filiere-hydrocarbures-chimie.htm

**Dept des Hydrocarbures** (B) (2010)
Route de Ghardaia
Ouargia
p. +213 (29) 712468
info@ouargla-univ.dz
http://www.ouargla-univ.dz/index-FR.htm

### Universite M'hamed Bouguerra de Boumerdes
**Faculte des Hydrocarbures et de la chimie** (B) (2010)
Avenue de Ildependance
Boumerdes 35000
p. +213 (24) 816420
doyen_fhc@umbb.dz
http://www.umbb.dz/

### Universite Saad Dahlab, Blida
**Dept des Sciences de l'Eau et de l'Environnement** (B) (2010)
BP 270
Blida 9000
p. +213 (25) 433625
contact@univ-blida.dz
http://www.univ-blida.dz/fac_ingenieur/index.html

## Angola

### Universidade Agostinho Neto
**Faculdade de Geologia** (B) (2014)
Av. 4 de Fevereiro
No. 7
2 andar
Luanda C. P. 815
p. +244-222-333-816
info@geologia-uan.com
http://www.geologia-uan.com/

### Universidade Independente de Angola
**Engenharia dos Recursos Naturais e Ambiente** (B) (2016)
Rua da Missao
Bairro Morro Bento II - Corimba
Luanda
p. (+244) 222 33 89 70
unia@unia.ao
http://www.unia.ao/curso.php?cr=6
Vice-Rector:
Nuno Nascimento Gomes, (M), Lisbon, 1992, GesGu

### Universite de Angola
**Dept of Geology** (2004)
Avenida 4 de Fevereiro 7
Caixa postal 815-C
Luanda

## Argentina

### Servicio Geologico Minero Argentino
**Servicio Geologico Minero Argentino** (2011)
Av. Julio Argentino Roca 651. P.B
Capital Federal
mjanit@mecon.gov.ar
http://www.segemar.gov.ar/

### Universidad Nacional de Buenos Aires
**Dept Ciencias Geologicas** (2011)
University City Pavilion II
Mayor Braque 2160 - CP: C1428EHA
Buenos Aires 1653
p. (011) 4576-3329
graciela@gl.fcen.uba.ar
http://www.gl.fcen.uba.ar/

### Universidad Nacional de Catamarca
**Facultad de Tecnologia Y Ciencias Aplicadas** (2011)
Maximio Victoria No. 55
San Fernando del Valle de Catamarca (470, Catamarca CP 4700
p. (54) 0383-4435 112 int. 112
daa@tecno.unca.edu.ar
http://tecno.unca.edu.ar

### Universidad Nacional de Cordoba
**Facultad de Ciencias Exactas, Fisicas Y Naturales** (2015)
Apartado 35-2060 UCR
San Pedro de Montes de Oca
San Jose, Cordoba
p. +03514332098
geologia@ucr.ac.cr
http://www.geologia.ucr.ac.cr/

### Universidad Nacional de Jujuy
**Instituto de Geologia y Mineria** (2011)
San Salvador de Jujuy , Jujuy
narce@idgym.unju.edu.ar
http://www.idgym.unju.edu.ar/

## Universidad Nacional de la Pampa
**Dept of Geologia** (2011)
Uruguay 151, 6300 Santa Rosa
La Pampa
susanapaccapelo@exactas.unlpam.edu.ar
http://www.exactas.unlpam.edu.ar/

## Universidad Nacional de la Patagonia San Juan Bosco
**Facultad de Ciencias Naturales** (2011)
Comodoro Rivadavia , Chubut
secacademica@unp.edu.ar
http://www.unp.edu.ar/

## Universidad Nacional de La Plata
**School of Astronomy and Geophysics** (2015)
Paseo del Bosque s/n
B1900FWA
p. (0221)-423-6593
academic@fcaglp.unlp.edu.ar
http://www.fcaglp.unlp.edu.ar/

## Universidad Nacional de Rio Cuarto
**Dept. of Geology** (B,M,D) O (2016)
ocampanella@gmail.com
mvillegas@exa.unrc.edu.ar
Rio Cuarto , Cordoba C.P. X5804BYA
p. (0358) 467-6198
webgeo@exa.unrc.edu.ar
http://geo.exa.unrc.edu.ar/
*Enrollment (2016): B: 32 (0) M: 5 (0) D: 2 (0)*

Associate Professor:
Monica B. Villegas, (M), UNRC, 1995, Gsc
Associate Scientist:
Juan Enrique Otamendi, (D), Universidad Nacional de Río Cuarto, 2000, CgaCp
Adjunct Professor:
Hector Daniel Origlia, (M), Arizona State, 1999, NgOn

## Universidad Nacional de Rio Negro
**Dept. of Geology** (2011)
Building Perito Moreno - Pacheco 460
General Roca , Rio Negro
sedevallemedio@unrn.edu.ar
http://www.unrn.edu.ar

## Universidad Nacional de Salta
**Facultad de Ciencias Naturales** (2014)
Salta
p. (0387) 425-5413
decnat@unsa.edu.ar
http://naturales.unsa.edu.ar/

## Universidad Nacional de San Juan
**Facultad de Ciencias Exactas, Físicas y Naturales** (2015)
Parque de Mayo
5400 San Juan
comunicacionfcefyn1@gmail.com
http://exactas.unsj.edu.ar/

## Universidad Nacional de San Luis
**Facultad de Ciencias Exactas, Fisico, Matematicas Y Naturales** (2011)
Ejército de los Andes 950
San Luis D5700HHW
p. +54 (2652) 424027
sacadfmn@unsl.edu.ar
http://webfmn.unsl.edu.ar/index.php

## Universidad Nacional de Tucuman
**Facultad de Ciencias Naturales E Instituto Miguel Lillo** (D) (2013)
Miguel Lillo 205
S.M. de Tucuman
San Miguel de Tucuman, Tucuman CP4000
p. 0381-4239456
info@csnat.unt.edu.ar
http://www.csnat.unt.edu.ar/

## Universidad Nacional del Comahue
**Dept of Geology and Petroleum** (2015)
School of Engineering
Buenos Aires 1400
Neuquen 18300 , Patagonia
p. +54-299-4490300
sacadfi@uncoma.edu.ar
http://www.uncoma.edu.ar/

## Universidad Nacional del Sur
**Departamento de Geologia** (2011)
San Juan 670
Primer Piso
Bahia Blanca
Buenos Aires 8000
p. 54-(0291)-4595147
secgeo@uns.edu.ar
http://www.uns.edu.ar/

# Australia

## Australian National University
**Research School of Earth Sciences** (M,D) (2015)
Mills Rd.
Canberra, ACT 0200
p. +61 2 6125 3406
Director.rses@anu.edu.au
http://rses.anu.edu.au/

Emeritus:
Patrick De Deckker, (D), Adelaide, 1981, GuPmGn

**Dept of Earth & Marine Sciences** (B,M,D) (2005)
47 Daley Road
Canberra, ACT 0200
p. 02 6125 2056
ems@anu.edu.au
http://ems.anu.edu.au/

Reader:
D C "Bear" McPhail, (D), Princeton, 1991, Cl

## Curtin University
**Dept of Mining Engineering** (B,M,D) (2012)
P. O. Box 597
Kalgoorlie
e.topal@curtin.edu.au

**Dept of Applied Geology** (B,M,D) (2012)
GPO Box U1987
Perth, WA 6845
p. +618 92667968
P.Kinny@curtin.edu.au
http://geology.curtin.edu.au/

**Dept of Exploration Geophysics** (B,M,D) (2017)
GPO Box U1987
Perth, WA 6845
p. +61 8 9266-3565
Geophysics-GeneralEnquiries@curtin.edu.au
http://www.geophysics.curtin.edu.au/
f: https://www.facebook.com/ExplorationGeophysicsCurtinUniversity
t: @CurtinGeophys
*Enrollment (2014): B: 27 (13) M: 26 (4) D: 28 (6)*

Professor:
Boris Gurevich, (D), Moscow, 1988, Ye
Anton W. Kepic, (D), British Columbia, 1990, YexYg
Senior Lecturer:
Vassily Mikhaltsevitch, (D), Kaliningrad State, 1997, Ye
Andrew Peter Squelch, (D), Nottingham, 1998, NmYe

Research Fellow:
Michael Carson, (D), Univ Coll (Dublin), 2000, Yx
Aleksandar Dzunic, (M), Belgrade, 1989, YgsYe
Head of Department:
Andrej Bona, (D), Calgary, 2002, Yx
Associate Professor:
Brett D. Harris, (D), Curtin, 2002, HwYve
Maxim Lebedev, (D), Mowcow Inst Phys & Tech, 1990, Yx
Roman Pevzner, (D), Moscow State, 2004, Yxx
Milovan Urosevic, (D), Curtin, 1998, YerYx
Lecturer:
Robert Galvin, (D), Curtin, 2006, Ye
Stanislav Glubokovskikh, (D), Lomonosov Moscow State, 2012, Yxe
Mahyar Madadi, (D), Inst Adv Studies Basic Sci (Zanjan), 1997, Ye
Andrew Pethick, (D), Curtin, 2013, YevYg
Konstantin Tertyshnikov, (D), Curtin, 2014, Ye
Stephanie Vialle, (D), Paris 7 Univ, 2008, Yx
Sasha Ziramov, (M), Belgrade, 2005, Ye
Senior Technical Officer:
Dominic J. Howman, (B), Curtin, 1994, Ye

## Federation University Australia

**Dept of Geology** (B,M,D) ● (2016)
Federation University
P. O. Box 663
Ballarat, VIC 3353
p. 613 53279354
k.dowling@federation.edu.au

## Flinders University

**School of the Environment** (B,M,D) ● (2017)
GPO Box 2100
Adelaide, SA 5001
p. +61 8 8201 7577
dean.sote@flinders.edu.au
http://www.flinders.edu.au/science_engineering/environment/
*Enrollment (2012): B: 44 (14) M: 15 (5) D: 37 (4)*
Professor of Hydrogeology:
Okke Batelaan, (D), Free Univ Brussels, Hw
Research Fellow:
Peter Cook, (D), Flinders Univ, Hw
ARC Future Fellow:
Adrian Werner, (D), Queensland (Australia), Hw
Academic Status:
Nancy Cromar, (D), Napier (UK), On
Professor:
Howard Fallowfield, (D), Dundee (UK), On
Iain Hay, (D), Washington, 1989, Oyy
Patrick Hesp, (D), Sydney, GmOy
Andrew Millington, (D), Sussex (UK), OuyOr
Craig Simmons, (D), Adelaide (Australia), Hw
Director ARA:
Jorg Hacker, (D), Bonn Germany, Ora
Academic Status:
John Edwards, (D), Adelaide (Australia), On
Andew Love, (D), Hw
Associate Professor:
Erick Bestland, (D), Oregon, GsSa
Beverley Clarke, (D), Adelaide (Australia), Oyn
Huade Guan, (D), New Mexico Tech, 2005, HgsHg
Jochen Kaempf, (D), Hamburg, Op
Udoy Saikia, (D), Flinders Univ, On
Research Fellow:
Juliette Woods, (D)
DECRA Research Fellow:
Margaret Shanafield, (D), Nevada, Hw
Sessional Lecturer:
Harpinder Sandhu, (D), Lincoln (NZ), On
Maria Zotti, On
Senior Lecturer:
David Bass, (D), New England, Oy
Kirstin Ross, (D), South Australia, On
Adjunct Senior Lecturer:
Simon Benger, (D), Australian National Univ, Oiy
John Hutson, (D), Natal, Spc
Vincent Post, (D), Amsterdam, Hw
Adjunct Lecturer:
Stephen Fildes, (M), Adelaide (Australia), Ori
Ben van den Akker, (D), Flinders (Australia), On
Lecturer:
Kathryn Bellette, (M), On
Dylan Irvine, (D), Flinders Univ, Hw
Mark Lethbridge, (D), South Australia, Oiy
Graziela Miot da Silva, (D), Fed Rio Grande do Sul (Brazil), Gug
Michael Taylor, (D), Flinders Univ, On
Ilka Wallis, (D), Flinders Univ, HqCg
Harriet Whiley, (D), Flinders Univ, On
Academic Status:
Gour Dasvarma, (D), Australian National Univ, On
Adjunct Professor:
Jim Smith, On
Research Fellow:
Eddie Banks, (D), Flinders Univ, Hw
Adjunct Academic Status:
Samantha de Ritter, (M), Flinders (Australia), Oe
Academic Status:
Glenn Harrington, (D), Flinders Univ, Hw
Andrew McGrath, (D), Ora

## Geoscience Australia

**Geoscience Australia** (2016)
GPO Box 378
Canberra, ACT 2601
p. +61 2 6249 9111
ref.library@ga.gov.au
http://www.ga.gov.au/
t: @GeoAusLibrary

## Government of South Australia, Department for State Development

**Geological Survey of South Australia** (2017)
GPO Box 320
Adelaide, SA 5001
p. +61 8 8463 3000
DSD.minerals@sa.gov.au
http://www.minerals.statedevelopment.sa.gov.au
t: @PACE_sagov

## James Cook University

**Geosciences** (B,M,D) ○ (2017)
Building 034
College of Science & Engineering
James Cook University
Townsville, QLD 4811
p. 07 4781 6947
eric.roberts@jcu.edu.au
https://www.jcu.edu.au/college-of-science-and-engineering
*Enrollment (2016): B: 85 (54) M: 6 (2) D: 25 (7)*
Associate Professor:
Eric M. Roberts, (D), Utah, 2005, GsPvCc
Professor:
Michael I. Bird, (D), Australian National Univ, 1988, ClsSa
Senior Scientist:
Karen E. Joyce, (D), UQ, 2005
Associate Professor:
Zhaoshan Chang, (D), Washington State, 2003, EmCce
Carl Spandler, (D), Univ, 2005, GxEmCg
Postdoctoral Research Fellow:
Arianne Ford, (D), James Cook Univ, 2008, GqOiEg
Dr.:
Christa Placzek, (D), Arizona, 2006, CclGe
Lecturer:
James J. Daniell, (D), Sydney, 2010, GumYr
Jan Marten Huizenga, (D), Vrije Universiteit Amsterdam, 1995, GpCgGc
Ioan Sanislav, (D), James Cook Univ, 2009, GctGp
Peter W. Whitehead, (B), La Trobe Univ, 1986, Ggv
Adjunct Professor:
Robert Henderson, (D), Victoria Welington, 1967, GtPiGg

Postdoctoral Research Fellow & Lecturer:
Hannah L. Hilbert-Wolf, (D), James Cook Univ, 2016, GsCc

## James Cook University of North Queensland

**Economic Geology Research Centre** (A,B,M,D) ○ (2014)
EGRU
James Cook University
Townsville, QLD 4811
p. +61 7 4781 4726
egru@jcu.edu.au
http://www.jcu.edu.au/egru

Australia

## La Trobe University

**Environmental Geoscience** (B,M,D) (2016)
Environmental Geoscience
La Trobe University
Victoria 3086
Melbourne, VIC 3086
p. +61 3 9479 1273
john.webb@latrobe.edu.au
http://www.latrobe.edu.au/environmental-geoscience
*Enrollment (2016): B: 12 (10) M: 3 (4) D: 8 (4)*

Associate Professor:
John A. Webb, (D), Queensland, 1982, GeHwGa
Lecturer:
David Steart, (D), Victoria Univ, 2003, GgPb
Susan Q. White, (D), La Trobe, 2005, GmgGe
Other:
Vincent J. Morand, (D), Sydney, 1987, GgcGt

## Macquarie University

**Dept of Earth and Planetary Sciences** (B,M,D) ○ (2016)
North Ryde
Sydney, NSW 2109
p. 61-2-98508426
eps-admin@mq.edu.au
http://www.eps.mq.edu.au/
f: https://www.facebook.com/MQeps/?fref=ts

**Department of Environmental Sciences** (B,M,D) (2016)
North Ryde, NSW 2109
neil.saintilan@mq.edu.au
http://www.mqu.edu.au/

## Monash University

**School of Earth, Atmosphere and Environment** (B,M,D) (2016)
PO Box 28E
Clayton, VIC 3800
p. +61 3 99054884
earth-atmosphere-environment@monash.edu
http://www.earth.monash.edu.au/

Professor:
Peter A. Cawood, (D), Sydney, 1980, GtcGg

## New South Wales Resources & Energy

**New South Wales Resources & Energy** (2017)
NSW Resource & Energy, PO Box 344
Hunter Regional Mail Centre
, NSW 2310
p. 1300 736 122
geologicalsurvey.info@trade.nsw.gov.au
http://www.resourcesandenergy.nsw.gov.au/

## Queensland University of Technology

**School of Earth, Environmental & Biological Sciences** (B,M,D) ● (2016)
GPO Box 2434
2 George Street
Brisbane, QLD 4001
p. 61 7 3138 2324
nm.davis@qut.edu.au
https://www.qut.edu.au/science-engineering/our-schools/school-of-earth-environmental-and-biological-sciences/earth-sciences
f: https://www.facebook.com/QUTEarthScienceUndergraduatePage/
Administrative Assistant: Noelene Davis
*Enrollment (2016): B: 144 (18) M: 11 (4) D: 12 (2)*

Professor:
Peter R. Grace, (D), Queensland, 1989, Sb
David A. Gust, (D), Australian National, 1982, GiCpGt
Senior Research Fellow :
Charlotte M. Allen, (D), Virginia Tech, CcaGi
Research Fellow:
Henrietta Cathey, (D), Utah, 2006, Giv
Associate Professor:
Scott E. Bryan, (D), Monash Univ (Australia), 1999, GitGs
Senior Lecturer:
Craig Sloss, (D), Wollongong (Australia), 2005, GsrGm
Jessica Trofimovs, (D), Monash Univ (Australia), 2003, GsvGu
Research Associate:
Coralie Siegel, (D), QUT, 2015, GiCc
Lecturer:
Oliver M. Gaede, (D), W Australia, YeNrYs
Patrick Hayman, (D), Monash, 2009, EgGvp
David T. Murphy, (D), Queensland, 2002, CgEgCc
Luke Nothdurft, (D), Queensland Tech, 2009, GdCmGg
Clemens Scheer, (D), Bonn, 2008, SbGe
Christoph Schrank, (D), Toronto, 2009, GcqGt
Adjunct Professor:
Malcolm E. Cox, (D), Auckland Univ, 1986, HwGeEg
Christopher Fielding, (D), Durham, Gsr
Emeritus:
John Rigby, Pb
Analytical Laboratory Coordinator :
Wan-Ping (Sunny) Hu, (D), Canterbury, Ca
Other:
Shane Russell, (B), Queensland Tech, Ca
Technician - Geology:
Alex Hepple, (M), QUT, 2012, GpiGg
Senior Technician - Geology :
Will Stearman, (B), Queensland Tech, 2011, Geg

## Southern Cross University

**School of Environment, Science & Engineering** (A,B,M,D) ○ (2017)
P. O. Box 157
Lismore, NSW 2480
p. +61 2 6620 3766
ese@scu.edu.au
http://scu.edu.au/environment-science-engineering/
f: https://www.facebook.com/scu.sese

Professor:
Bill Boyd, (D), Glasgow, 1982, Pe
Bradley Eyre, (D), Queensland Tech, Gs
Peter Harrison, (D), James Cook
Associate Professor:
Danny Bucher, (D), S Cross, 2001
Malcolm William Clark, (D), Ca
Simon Hartley, (B), New England
Graham Jones, (D), James Cook, Ca
John Doland Nichols, (D), N Arizona
Amanda Reichelt-Brushett, (D), S Cross
Lecturer:
Graeme Palmer, (B), Australian National
Sumith Pathirana, (D), Kent State, Or
Kathryn Taffs, (D), Adelaide, Pe
Michael Whelan

## University of Adelaide

**School of Earth and Environmental Sciences** (2013)
Adelaide, SA 5000
p. +61 8 8303 3999
graham.heinson@adelaide.edu.au
http://www.ees.adelaide.edu.au/

**Australian School of Petroleum** (B,M,D) ● (2016)
Santos Petroleum Engineering Building
North Terrace

Adelaide, SA 5005
p. 61-8-8313-8000
admin@asp.adelaide.edu.au
http://www.asp.adelaide.edu.au/

## University of Canberra

**Institute for Applied Ecology** (2011)
sarre@aerg.canberra.edu.au
http://enterprise.canberra.edu.au/WWW/www-crcfe.nsf

**School of Education, Science, Tecnology and Maths** (2004)
P. O. Box 1
Belconnen, ACT
Geoffrey.Riordan@canberra.edu.au
http://scides.canberra.edu.au/rehs/

## University of Melbourne

**School of Earth Sciences** (B,M,D) (2016)
Melbourne, VIC 3010
p. +61 3 8344 9866
head@earthsci.unimelb.edu.au
http://www.earthsci.unimelb.edu.au

## University of New England

**Earth Sciences** (A,B,M,D) O (2015)
Armidale
NSW
2351
Armidale, NSW 2351
p. +61-2-67732101
geology@une.edu.au
http://www.une.edu.au/about-une/academic-schools/school-of-environmental-and-rural-science/research/life-earth-and-environment/e

Associate Professor:
John R. Paterson, (D), Macquarie, 2005, PioPs

Lecturer:
Phil R. Bell, (D), Alberta, 2011, PgvPo
Luke Milan, (D), GtcGp
Nancy Vickery, (D), New England, GgeYg

Adjunct Professor:
Ian Metcalfe, (D), Leeds (UK), 1976, PmGtg

Emeritus:
Paul Ashley, (D), Macquarie, EgGeEm
Peter Flood, (D), EcGt

## University of New South Wales

**School of Biological, Earth & Environmental Sciences** (B,M,D) (2015)
UNSW
Sydney, NSW 2052
p. 61 2 93852961
bees@unsw.edu.au
http://www.bees.unsw.edu.au/index.html
*Enrollment (2013): B: 80 (24) M: 5 (0) D: 14 (4)*

Associate Professor:
David R. Cohen, (D), New South Wales, 1990, Cet

Professor:
Michael Archer, (D), Pv
Andy Baker, (D), HwCs
James Goff, (D), Gsm
Suzanne J. Hand, (D), Pv
Martin van Kranendonk, (D), Queens (Canada), GxPeGd

Associate Professor:
Bryce Kelly, (D), New South Wales, HwYg
Shawn Laffan, (D), Oiy

Dr:
Ian Graham, (D), UTS Sydney, 1995, GxEg

Lecturer:
Catherine Chague-Goff, (D), CgGs

Adjunct Professor:
Derecke Palmer, (D), New South Wales, 2001, Ye

Dr:
Paul G. Lennox, (D), Monash Univ (Australia), 1985, GcgOe

Visiting Professor:
Colin R. Ward, (D), New South Wales, 1971, EcGd

## University of Newcastle

**Discipline of Earth Sciences** (B,M,D) O (2014)
School of Environmental & Life Sciences
University Drive
Callaghan, NSW 2308
p. +61 2 4921 8976
name.surname@newcastle.edu.au
http://www.newcastle.edu.au/school/environ-life-science/index.html
*Enrollment (2014): B: 132 (121) D: 5 (3)*

Associate Professor:
Silvia Frisia, (D), Milan, Italy, 1991, ClPeGd
Phil Geary, (D), UWS, HwOsHw
Gregory Hancock, (D), OiGmm

Lecturer:
Judy Bailey, (D), Newcastle, 1993, Ec
David Boutelier, (D), Gtc
Alistair Hack, (D), GcCgEm
Anthony Kiem, (D), HqOai
Bill Landenberger, (D), Newcastle, 1997, GiCg
Danielle Verdon-Kidd, (D), HqOai

## University of Queensland

**School of Earth Sciences** (B,M,D) O (2015)
Faculty of Science
Steele Building
Brisbane, QLD 4072
p. +61 7 3365 1180
enquiries@earth.uq.edu.au
http://www.earth.uq.edu.au/
*Enrollment (2015): B: 106 (57) M: 38 (5) D: 60 (6)*

## University of South Australia

**School of Natural and Built Environments** (B) O (2016)
GPO Box 2471
Adelaide, SA 5001
Julie.Mills@unisa.edu.au
http://www.unisa.edu.au/it-engineering-and-the-environment/natural-and-built-environments/

## University of Sydney

**School of Geosciences** (A,B,M,D) (2014)
Madsen Building F09
Sydney, NSW 2006
p. +61 2 9351 2912
sue.taylor@sydney.edu.au
http://www.geosci.usyd.edu.au/index.shtml

Professor:
Jonathan Aitchison, (D), UNE, 1989, GrtPm

Professor:
Geoffrey L. Clarke, (D), Melbourne, 1987, Gp
Dietmar Muller, (D), Scripps, 1989, Yrg

## University of Tasmania

**ARC Center of Excellence in Ore Deposits** (M,D) O (2017)
Private Bag 79
Hobart, TAS 7001
steve.calladine@utas.edu.au
www.utas.edu.au/codes

**School of Earth Sciences / ARC Centre for Excellence in Ore Deposits (CODES)** (B,M,D) (2013)
Private Bag 79
Hobart, TAS 7001
p. 61 3 6226 2476
Rose.Pongratz@utas.edu.au
http://www.utas.edu.au/earth-sciences/home

Director:
J Bruce Gemmell, (D), Dartmouth, 1987, Eg
Ross R. Large, (D), New England, Eg

Professor:
David R. Cooke, (D), Monash, Eg
Anthony J. Crawford, (D), Melbourne, Gi
Jocelyn McPhie, (D), New England, Gv
Associate Professor:
Ron F. Berry, (D), Flinders (Australia), Gc
Lecturer:
Garry J. Davidson, (D), Tasmania, Cs
Peter J. McGoldrick, (D), Melbourne, Ce
Anya Reading, (D), Ysg
Michael Roach, (N), Tasmania, Yg



### University of Western Australia
**School of Earth & Environment** (B,M,D) (2013)
M004
35 Stirling Highway
Crawley, WA 6009
p. 6488 1921
enquiry-see@uwa.edu.au
http://www.see.uwa.edu.au/

### University of Wollongong
**School of Earth & Environmental Sciences** (B,M,D) (2017)
Northfields Avenue
Wollongong, NSW 2522
p. 61 2 4221 4419
anutman@uow.edu.au
http://www.uow.edu.au/science/eesc/

## Austria

### Geological Survey of Austria
**Geologische Bundesanstalt von Osterreich** (2011)
Neulinggasse 38, A 1030
Vienna
p. 43-1-712 56 74
office@geologie.ac.at
http://www.geolba.ac.at/

### Karl-Franzens-Universitaet Graz
**Inst for Earth Science** (A,B,M,D) ○ (2017)
Heinrichstraße 26
Graz 8010
p. +43 316 380-5587
erdwissenschaften@uni-graz.at
http://erdwissenschaften.uni-graz.at/index_en.php
*Enrollment (2016): A: 10 (1) B: 40 (35) M: 20 (14) D: 12 (2)*

### Leopold-Franzens-Universitaet Innsbruck
**Inst fuer Geologie** (2016)
Innrain 52-f
6020 Innsbruck
p. +43 (0) 512 / 507 - 96125
dekanat-geowiss@uibk.ac.at
http://www.uibk.ac.at/fakultaeten/geo_und_atmosphaeren-wissenschaften/

### Montan Universitaet
**Dept of Applied Geological Sciences and Geophysics** (2011)
Inst. for Prospecting
A-8700 Leoben
geologie@mu-leoben.at

### Technische Universitaet Graz
**Earth Sciences** (2011)
Rechbauerstrasse 12
A-8010 Graz
p. +43(0)316 873 6360
martin.dietzel@tugraz.at

### University Leoben
**Chair of Geology and Economic Geology** (B,M,D) (2016)
Montauniversitat Leoben
Peter-Tunnerstrasse 5
Leoben A-8700
p. +43 3842 402 6101
geologie@unileoben.ac.at
http://www.unileoben.ac.at/~buero62/geologie/geologie.html
*Enrollment (2016): B: 336 (17) M: 30 (13) D: 17 (2)*
Professor:
Frank Melcher, (D), Leoben, 1993, GgEg
Senior Scientist:
Heinrich Mali, (D), GcEgNm
Associate Professor:
Walter Prochaska, (D), Eg
Gerd Rantitsch, (D), Ggq
Research Associate:
Peter Onuk, (M), Gg

### University of Innsbruck
**Institute of Geology** (A,B,M,D) (2015)
Innrain 52f
A-6020
Innsbruck, Austria 6020
p. +43 512 507-54300
regina.gratzl@uibk.ac.at
http://www.uibk.ac.at/geologie

### University of Salzburg
**Inst fuer Geowissenschaften** (2011)
Hellbrunner Strasse 34
A-5020 Salzburg
p. +43-662-8044-5200 oder 5400
geo@sbg.ac.at
http://www.uni-salzburg.at/geo

### University of Vienna
**Dept of Meteorology and Geophysics** (B,M,D) ● (2016)
Althanstrasse 14
A-1090 Vienna
p. 0043 1 4277 53701
img-wien@univie.ac.at
http://imgw.univie.ac.at/en/imgw/
*Enrollment (2016): B: 40 (30) M: 20 (16) D: 10 (4)*
Univ.-Prof.:
Goetz Bokelmann, (D), Princeton, 1992, Ygs
o.Univ.-Prof.:
Reinhold Steinacker, (D), Innsbruck, 1975, Owa

**Dept of Geography** (2017)
Althanstrasse 14
Vienna A-1090
p. 00431427753401
elisabeth.aufhauser@univie.ac.at
http://www.univie.ac.at/Geologie/
Head:
Thilo Hofmann, (D)

## Azerbaijan

### Baku State University
**The Faculty of Geology** (2005)
23 Z. Khalilov Street
370145 Baku
m.mahluga@rambler.ru
http://www.ceebd.co.uk/ceeed/un/az/az003.htm

## Bangladesh

### Jahangirnagar University
**Dept of Geological Sciences** (2011)
University Campus, Savar, Dhaka
Dhaka
p. 0088 2 7791045-51 (Ext. 1402)
rabiulju@gmail.com
http://www.juniv.edu/home.php?pg=faculty_science

### Rajshahi University
**Dept of Geology & Mining** (2016)

Motihar, Rajshahi
Rajshahi
p. +880 721-750041
chair.geology@ru.ac.bd
http://www.ru.ac.bd/geol/index.htm

## University of Dhaka

**Dept of Geology** (B,M,D) ● (2017)
Curzon Hall Campus
Dhaka 1000
p. (008) 2- 966 1920-73 (Ext. 730
geology@du.ac.bd
http://www.univdhaka.edu/academic/department_item/GLG

Professor:
Kazi Matin Ahmed, (D), UK, 1994, HwGgHg

# Belgium

## Faculte Polytechnique de Mons

**Dept of Geology** (2004)
Fue de Houdain 9
7000 Mons
yves.quinif@fpms.ac.be

## Faculte Univ Catholique de Mons

**Dept of Geology** (2004)
Chaussee de Binche 151
B-7000 Mons

## Facultes Univ Notre Dame de la Paix

**Dept Geologie** (2004)
Rue de Bruxelles 61
B-5000 Namur
vincent.hallet@fundp.ac.be

## Geological Survey of Belgium

**Service geologique de Belgique** (2011)
Department VII
of the Royal Belgian Institute of Natural Sciences (RBINS)
13, Rue Jenner
1000 Brussels
p. +32 (0)2.788.76.61
bgd@natuurwetenschappen.be
http://www.naturalsciences.be/geology/

## Ghent University

**Dept of Geology** (B,M,D) ● (2016)
Krijgslaan 281-S8
Gent B-9000
p. +32 9 264 45 94
we13@ugent.be
http://www.earthweb.ugent.be/index.php/
f: https://www.facebook.com/GeologieUGent

Head:
Marc De Batist, (D), Ghent Univ (Belgium), 1989, GnuGs

Professor:
Stephen Louwye, (D), Pm

Associate Professor:
Veerle Cnudde, (D), Om
David Van Rooij, (D), Yr
Kristine Walraevens, (D), Hw

Assistant Professor:
Sebastien Bertrand, (D), 2001, GsClPe
Johan De Grave, (D), Cc
Stijn Dewaele, (D), Cg
Thijs Vandenbroucke, (D), Ps

## Katholieke Universiteit Leuven

**Dept of Earth and Environmental Science** (B,M,D) (2014)
Department of Earth & Environmental Sciences
Redingenstraat 16
B-3000 Leuven-Heverlee
p. (+32) 016 321450
erik.mathjis@ees.kuleaven.be
http://ees.kuleuven.be/

Professor:
Patrick Degryse, (D), GaCeGf
Jan Elsen, (D), Gz
Philippe Muchez, (D), EmCe
Manuel Sintubin, (D), GtcGa
Robert Speijer, (D), Pym
Rudy Swennen, Gso

Associate Professor:
Okke Batelaan, (D), Hw
Sarah Fowler, (D), Gip

Assistant Professor:
Marijke Huysmans, (D), Hw

**Dept of Geology - Geography** (2004)
Redingenstraat 16
B-3000 Leuven
erik.mathijs@kuleuven.be

## Universite Catholique de Louvain

**Dept of Geology** (2014)
Bâtiment Mercator
place Louis Pasteur 3
B-1348 Louvain-la-Neuve
p. (32) 10 47 32 97
monique.descampsl@uclouvain.be
http://www.uclouvain.be/geo.html

## Universite de Gembloux

**Les ressources vivantes et l'environnement** (2004)
B-5800 Gembloux

## Universite de Liege

**Dept of Geology** (B,M,D) (2007)
Batiment B18 (secretariat)
Sart Tilman
Liege B-4000
p. 32 4 366 22 51
TH.Billen@ulg.ac.be
http://www.ulg.ac.be/geolsed/geologie

chargé de cours:
Nathalie Fagel, (D), GsuGe
Hans-Balder Havenith, (D), Liège, Ge

Professor:
Frederic P. Boulvain, (D), Brussels, 1990, GdgGs
Andre-Mathieu Fransolet, (D), Liege, Gz
Emmanuelle Javaux, (D), Dalhousie, 1999, Pl
Jacqueline Vander Auwera, (D), Louvain-la-Neuve, 1988, Giv

Emeritus:
Edouard Poty, (D), Liege, 1981, PiGsPe

## Universite de Mons

**Dept of Geology** (2004)
Place du Pare 20
B-7000 Mons

## Universite Libre de Bruxelles (ULB)

**Département des Sciences de la Terre et de l'Environnement** (2004)
50, Ave. F. Roosevelt
Brussels 1050
Pierre.Regnier@ulb.ac.be

## University of Mons

**Dept of Geology and Applied Geology** (M,D) ○ (2016)
rue de Houdain, 9
Mons
gfa@umons.ac.be

## Vrije University Brussel

**Earth System Sciences** (D) (2015)
ESSC-WE-VUB
Faculty of Sciences
Pleinlaan 2

Brussels 1050
p. 0113226293394
phclaeys@vub.ac.be
http://we.vub.ac.be/~essc
*Enrollment (2015): D: 2 (0)*
Head:
Philippe Claeys, (D), California (Davis), 1993, CgPyXc
Professor:
Edward Keppens, (D), Vrije (Brussel), 1983, CsGeg

Bolivia

# Bolivia

## Universidad Mayor de San Andres
**Dept of Geology** (2014)
P. O. Box 12198
Campus Universitario Cota Cota, Calle 27
La Paz
p. (591-2) 2441983
webmaster@umsa.bo
http://www.geologia.umsa.bo/

# Botswana

## Geological Survey of Botswana
**Geological Survey of Botswana** (2014)
Khama One Avenue
Plot 1734
Lobatse
p. +267 5330327
http://www.gov.bw/en/Ministries--Authorities/Ministries/Ministry-of-Minerals-Energy-and-Water-Resources-MMWER/Departments1/Depar

## University of Botswana
**Dept of Geology** (B) (2014)
4775 Notwane Rd.
Private Bag UB00704
Gaborone
p. (267)355 2529
geology@mopipi.ub.bw
http://www.ub.bw/home/ac/1/fac/1/dep/79/Geology/

**Dept of Environmental Science** (B) (2010)
4775 Notwane Road
Private Bag UB 0022
Gaborone
p. (267) 355-0000
Sedilamoya.Pansiri@mopipi.ub.bw
http://www.ub.bw/learning_faculties.cfm?pid=588

# Brazil

## Federal University of Bahia Geophysics
**CPGG** (2004)
Salvador
BA
geofisic@ufba.br
http://www/pppg.ufba.br/

## Geological Survey of Brazil
**Servico Geologico do Brasil** (2011)
Av. SGAN- Quadra 603 - conjunto J
Parte A - 1º andar
, Brasilia - DF 70830-030
cprmsede@df.cprm.gov.br
http://www.cprm.gov.br/

## Universidad Federal de Rio Grande do Sul
**Institute de Geociencia** (2004)
Avenida Bento Gonçalves, 9500
Porto Alegre, RS - 91.501-970
p. +55 51 3308-6337
igeo@ufrgs.br
http://www.ufrgs.br/english/the-university/institutes-faculties-and-schools/institute-of-geoscience

## Universidade de Brasília
**Instituto de Geociências** (2011)
Campus Universitário Darcy Ribeiro ICC - Ala Central
CEP 70.910-900 - Brasilia DF
Caixa Postal 04465.
CEP 70919-970
p. +61 3307-2433
igd@unb.br
http://www.igd.unb.br/

## Universidade de Sao Paulo
**Inst de Geociencias** (2014)
Rua do Lago
562 Cidade Universitaria
05508-080 São Paulo
p. 63.025.530/0007-08
gmgigc@usp.br
http://www.igc.usp.br/

**Inst Oceanografico** (2004)
Cidade Universitaria, Av. Prof. Luciano Gualberto, Travessa 3 no 380, 05508-900 Sao Paulo - SP

## Universidade do Vale do Rio Dos Sinos
**Inst de Geociencias** (2004)
Av. Unisinos, 950 - Cristo Rei, 93 000 Sao Leopoldo - RS

## Universidade Federal da Bahia
**Inst de Geociencias** (2004)
Rua Augusto Viana s/n - Canela, 40 410 Salvador - BA
olivia@ufba.br

## Universidade Federal de Minas Gerais
**Inst de Geociencias** (2014)
Av. Antonio Carlos 6.627
Pampulha, , 31 270 Belo Horizont
p. 55(31) 3409-5420
dir@igc.ufmg.br
http://www.igc.ufmg.br/

## Universidade Federal de Ouro Preto
**Dept de Geologia** (2014)
Campus Morro do Cruzeiro
35 400 Ouro Preto
p. (31) 3559-1600
web@degeo.ufop.br
http://www.degeo.ufop.br/

## Universidade Federal de Pernambuco
**Centro de Tecnologia - Geociencias** (2014)
Av. Agamenon Magalhaes s/n - Santo Amaro, 50 000 Recife - PE
p. (81) 2126.8105
secretaria.proacad@ufpe.br
http://www.ufpe.br/proacad/index.php?option=com_content&view=article&id=150&Itemid=138

## Universidade Federal do Ceara
**Inst de Geociencias** (2004)
Av. Da Universidade, 2853 - Benfica, 60 000 Fortaleza - CE

## Universidade Federal do Para
**Inst. de Geociencias** (2014)
Campus Guama
Caixa postal 1611
Belem
p. (91) 3201-7107
dirig@ufpa.br
http://www.ig.ufpa.br/site/

## Universidade Federal Fluminense
**Dept de Geociencias** (2014)
Av. Gal. Milton Tavares de Souza
p. +55 (21) 2629-5951

gge@vm.uff.br
http://www.uff.br/degeografia/

### Universidade Federal Rural do Rio de Janeiro
**Inst de Geociencias** (2004)
Km 47 da Antigua Rodovia Rio/Sao Paulo - Seropedico, 23 460
Itaguai - RJ

### University of Campinas Geoscience Institute
**Inst de Geosciencias** (2004)
Caixa Postal: 6152
13083-970
Campinas
secretaria.prpg@reitoria.unicamp.br
http://www.ige.unicamp.br/

## Bulgaria

### Bulgarian Academy of Sciences
**Geological Institute** (D) (2013)
Acad.G.Bonchev st. bl.24
Sofia 1113
p. +0359 2 8723 563
geolinst@geology.bas.bg
http://www.geology.bas.bg/
Professor Dr, DSc:
Kristalina Christova Stoykova, (D), Bulgarian Academy of Sciences, 2008, PemPi
Associate Professor:
Iliana Boncheva, (D)
Thomas Noubar Kerestedjian, (D), Bulgaria Acad Sci, 1990, GzeEg

### Mining and Geology University
**Geology Dept** (2013)
Studentski Grad
Sofia 1756
p. 8060221
dekangpf@mgu.bg

### Ministry of Environment and Water
**Ministry of Environment and Water** (2011)
22 Maria Louiza Blvd.
Sofia, 1000
minister@moew.government.bg
http://www.moew.government.bg/

### Sofia University St. Kliment Ohridski
**Dept of Geology and Paleontology** (2016)
1504 Sofia
15 Tsar Osvobodtel Blvd.
Geology of Fossil, Fuel, Organic Petrology, Organic Geochemistry
room 276
gigeor@gea.uni-sofia.bg
http://www.uni-sofia.bg/newweb/faculties/geo/departments/geo

**Dept of Cartography & GIS** (B,M,D) (2012)
1504 Sofia
15 Tsar Osvobodtel Blvd.
GIS, General Physical Geography, Landscape Ecology, Agro-ecology
Sofia, Sofia 1504
p. 9308261
popov@gea.uni-sofia.bg
http://www.gis.gea.uni-sofia.bg/

### University of Mining and Geology, St. Ivan Rilski
**Faculty of Geology** (2004)
Studentski grad
1100 Sofia
dekangpf@mgu.bg
http://www.mgu.bg/frame/html

## Burkina Faso

### Universite de Ouagadougou
**L'Unité de Formation et de Recherche en Sciences de la Vie et de la Terre (UFR/SVT)** (B) (2014)
03 BP 7021, Ougadougou 03
p. +226 50-30-70-64/65
webmaster@univ-ouaga.bf
http://www.univ-ouaga.bf.html/formations/ufr_SVT/frFormtnsSVTdip1.html

## Burundi

### Universite du Burundi
**Dept of Earth Sciences** (B) O (2015)
B.P. 2700
Bujumbura, Burundi
p. 00257 22 22 55 56
webmaster@ub.edu.bi
http://www.ub.edu.bi/

**Faculté des Sciences** (2014)
B.P. 2700
Bjumbura
p. 0025722222059
http://www.ub.edu.bi/ub-fac6.php

## Cameroon

### Bamenda University of Science and Technology
**Dept of Geology** (B) (2014)
PO Box 277
Bamenda, NW Province
p. +237 7726 1789
http://www.bamendauniversity.com/

### Universite de Buea
**Dept of Geology and Environmental Science** (B) (2014)
PO Box 63
Buea, South West Province
p. 237-332-2134
vpktitanji@yahoo.co.uk
http://ubuea.cm/

### Universite de Douala
**Faculty of Sciences** (B,M,D) (2014)
BP 2701
Douala, Cameroun
p. (237) 33 40 75 69
infos.fs@univ-douala.com
http://www.facsciences-univ-douala.cm/index.php/contact

### Universite de Dschang
**Faculte D'Agronomie et des Sciences Agricoles** (B,M) (2017)
POB 96
Dschang
p. (237) 33 45 15 66
agro.50tenair@gmail.com

### Universite de Ngaoundere
**School of Geology and Mining** (B) (2014)
BP 454
Ngaoundere
p. +(237) 225 2767
http://www.univ-ndere.cm/index.php?LANG=EN&ETS=0433A030F35FE61&RUB=E2C0BE24560D78C

### Universite de Yaounde 1
**Dept des Sciences de la Terre** (B) (2014)
BP 812
Yanounde
p. (237) 222 56 60

facsciences@uy1.uninet.cm
http://www.facsciences.uninet.cm/act_aca_dst.html

# Canada

Canada

## Aboriginal Affairs and Northern Development Canada

**Mineral Resources Division** (2017)
Mineral Resources Division
PO Box 100
Iqaluit, NU X0A 0H0
p. (867) 975-4293
nunavutarchives@aandc.gc.ca
http://nunavutgeoscience.ca/

## Acadia University

**Dept of Earth and Environmental Science** (B,M) ● (2016)
Huggins Science Hall
12 University Avenue
Wolfville, NS B4P 2R6
p. (902) 585-1208
ees@acadiau.ca
http://ees.acadiau.ca/
*Enrollment (2016): B: 51 (25) M: 4 (4)*

Head:
Ian S. Spooner, (D), Calgary, 1994, Ge

Professor:
Sandra M. Barr, (D), British Columbia, 1973, Gi
Nelson O'Driscoll, (D), Ottawa, 2003, CgGgOn
Peir K. Pufahl, (D), British Columbia, 2001, Gs
Robert P. Raeside, (D), Calgary, 1982, GptGc
Clifford R. Stanley, (D), British Columbia, 1988, Cg

Assistant Professor:
Alice Cohen, (D), British Columbia, 2011, OnnOn

Instructor:
David W.A. McMullin, (D), British Columbia, 1991, Oe

Geology Tech:
Pam Frail, Gg

## Brandon University

**Dept of Geology** (B,M) ● (2016)
270 18th Street
Brandon, MB R7A 6A9
p. (204) 727-9677
somarina@brandonu.ca
http://www.brandonu.ca/geology
*Enrollment (2016): B: 66 (17) M: 1 (1)*

Professor:
Rong-Yu Li, (D), Alberta, 2002, PgmGg
A. Hamid Mumin, (D), W Ontario, 1994, EgGzt
Simon A. J Pattison, (D), McMaster, 1992, GsoGd
Alireza Somarin, (D), New England, 1999, Gig

Emeritus:
Robert K. Springer, (D), California (Davis), 1971, Gi
Harvey R. Young, (D), Queen's, 1973, Gd

Instructional Associate:
Peter J. Adamo, (B), Brandon, 2000, Gg

Laboratory Director:
Michelle Huminicki, (D), Memorial, 2005, EgGz

Other:
Paul Alexandre

## Brock University

**Dept of Earth Sciences** (B,M) ● (2017)
1812 Sir Isaac Brock Way
St. Catharines, ON L2S 3A1
p. 905 688-5550
earth@brocku.ca
http://www.brocku.ca/mathematics-science/departments-and-centres/earth-sciences
*Enrollment (2016): B: 74 (22) M: 17 (5)*

Professor:
Uwe Brand, (D), Ottawa, 1979, ClsOa
Richard J. Cheel, (D), McMaster, 1984, Gs
Frank Fueten, (D), Toronto, 1989, Gc
Martin J. Head, (D), Aberdeen, 1990, Pl
Francine G. McCarthy, (D), Dalhousie, 1992, Pm
John Menzies, (D), Edinburgh (UK), 1976, GlmGs

Associate Professor:
Gregory C. Finn, (D), Memorial, 1989, Gx
Daniel P. McCarthy, (D), Saskatchewan, 1993, Gl
Mariek Schmidt, (D), Oregon State, 2005, GivXg

Assistant Professor:
Nigel Blamey, (D), New Mexico Tech, 2000, EgCg

Adjunct Professor:
Paul Budkewitsch, (M), Toronto, 1990, GeOrHq
Andrew W. Panko, (D), McMaster, 1985, Ge

## Cape Breton University

**Math, Physics, Geology** (B) (2013)
P.O. Box 5300
Sydney, NS B1P 6L2
p. (902) 539-5300
fenton_isenor@cbu.ca

Instructor:
Fenton M. Isenor, (M), Acadia, 2000, GgNgGm

Emeritus:
Erwin L. Zodrow, (D), Pi

## Capilano University

**Geology Dept** (A,B) (2010)
2055 Purcell Way
North Vancouver, BC V7J 3H5
p. (604) 986-1911
sciences@capilanou.ca
http://www.capilanou.ca/programs/geology.html

Head:
Dileep J A Athaide, (M), British Columbia, 1974, GgOge

Professor:
Jennifer Getsinger, (D), British Columbia

## Carleton University

**Dept of Earth Sciences** (B,M,D) ● (2015)
1125 Colonel By Drive
Ottawa, ON K1S 5B6
p. (613) 520-5633
earth.sciences@carleton.ca
http://earthsci.carleton.ca/
f: https://www.facebook.com/pages/Department-of-Earth-Sciences-Carleton-University/510369329037382
t: @ErthSciCarleton

Professor:
Keith Bell, (D), Oxford, 1964, Cc
George R. Dix, (D), Syracuse, 1988, Gsd
R. Timothy Patterson, (D), California (Los Angeles), 1986, Pm
Giorgio Ranalli, (D), Illinois, 1970, Yg
Claudia Schroder-Adams, (D), Dalhousie, 1986, Pm
George B. Skippen, (D), Johns Hopkins, 1966, Cg
Richard P. Taylor, (D), Leicester, 1980, Cg
David H. Watkinson, (D), Penn State, 1965, Em

Associate Professor:
Gail M. Atkinson, (D), W Ontario, 1993, Ne
John Blenkinsop, (D), British Columbia, 1972, Cc
Sharon D. Carr, (D), Carleton, 1990, Gt
Brian L. Cousens, (D), California (Santa Barbara), 1990, GiCgc
Fred A. Michel, (D), Waterloo, 1982, Hw

Lecturer:
Ildi Munro, (M), Waterloo, 1975, Pg

Adjunct Professor:
Robert Berman, (D), British Columbia, 1983, Gp
Steve L. Cumbaa, (D), Florida, 1975, Pv
J. Allan Donaldson, (D), Johns Hopkins, 1960, Gs
T. Scott Ercit, (D), Manitoba, 1986, Gz
Harold Gibson, (D), Carleton, 1990, Em
Simon Hanmer, (D), Chelsea (UK), 1977, Gc
Mark D. Hannington, (D), Toronto, 1989, Em
Jarmila Kukalova-Peck, (D), Charles (Prague), 1962, Pi
Dale A. Leckie, (D), McMaster, 1983, Gs
R. R. Rainbird, (D), Western, 1991, Gs

Emeritus:
F. K. North, (D), Oxford, 1951, Go

## Dalhousie University

**Dept of Earth Sciences** (B,M,D) O (2017)
Halifax, NS B3H 3J5
p. (902) 494-2358
earth.sciences@dal.ca
http://earthsciences.dal.ca/index2.html

Chair:
James Brenan, (D), Rensselaer, 1990, Gi
James M. Brenan, (D), Rensselaer, 1990, CptEm

Retired:
D Barrie Clarke, (D), Ediburgh, Gi

Professor:
John C. Gosse, (D), Lehigh, 1994, Cg
Djordje Grujic, (D), ETH (Switzerland), 1992, Gt
Grant D. Wach, (D), Oxford, 1993, GorGs

Associate Professor:
Isabelle Coutand, (D), Rennes, Gg
Nicholas Culshaw, (D), Ottawa, 1983, Gc

Assistant Professor:
Lawrence Plug, (D), Alaska (Fairbanks), 2000, Gm

Research Associate:
Lubomir F. Jansa, (D), Charles (Prague), 1967, Gr
Robert Raeside, (D), Calgary, 1982, Gx
Alan Ruffman, (M), Dalhousie, 1966, Ys

Senior Instructor:
Charles C. Walls, (M), Dalhousie, 1996, Oi

Honorary Research Associate:
Prasanta Mukhopadhyay, (D), Jadavpur, 1971, Ec
Graham Williams, (D), Sheffield, 1964, Pl

Adjunct Professor:
Juergen Adam, (D), Tech (Berlin)
Alan J. Anderson, (D), Queen's, Cg
Sandra Barr, (D), British Columbia, 1973, Gx
Hugo Beltrami, (D), Quebec (Montreal), Gg
John Calder, (D), Dalhousie, 1991, Ec
Sonya Dehler, (D), British Columbia, Gg
Jarda Dostal, (D), McMaster, 1974, Cg
Don Fox, (D), Dalhousie, Gg
Paul T. Gayes, (D), SUNY (Stony Brook), 1986, Gs
Peter E. Jones, (D), British Columbia, 1963, Op
Lisa M. Kellman, (D), Quebec (Montreal), 1998, Cg
Daniel Kontak, (D), Queen's, Gg
Joel Kronfeld, (D), Rice, 1972, Cg
Michael Melchin, (D), Western, 1987, Pg
Peta J. Mudie, (D), Dalhousie, 1980, Pl
J. Brendan Murphy, (D), McGill, 1982, Gx
Michael Parsons, (D), Stanford, Gg
Georgia Pe-Piper, (D), Cambridge, 1971, Gv
David Piper, (D), Cantab, 1969, Gs
Paul T. Robinson, (D), California (Berkeley), 1964, Gv
Andre Rochon, (D), Quebec, Gg
Matthew Salisbury, (D), Washington, 1974, Yr
Ralph Stea, (D), Dalhousie, 1995, Gu
Hans J. Wielens, (D), Utrecht (Neth), 1979, Go

Emeritus:
H.B. S. Cooke, (D), Witwatersrand, 1947, Pv
Martin R. Gibling, (D), Ottawa, 1978, Gs
Franco Medioli, (D), Parma, 1959, Pm
G. Clinton Milligan, (D), Harvard, 1961, Gc
Patrick J.C. Ryall, (D), Dalhousie, 1974, Yg
David B. Scott, (D), Dalhousie, 1977, Pm
Marcos Zentilli, (D), Queen's, 1974, Eg

Administrator:
Ann Bannon, (B), Dalhousie, 2001, On

Cooperating Faculty:
Christopher Beaumont, (D), Dalhousie, 1973, Ys

**Dept of Oceanography** (B,M,D) (2017)
Life Sciences Centre
1355 Oxford Street
PO Box 15000
Halifax, NS B3H 4R2
p. (902) 494-3557
oceanography@dal.ca
http://oceanography.dal.ca/
*Enrollment (2015): M: 22 (6) D: 31 (2)*

Chair:
Paul S. Hill, (D), Washington, 1992, Gs

Professor:
Christopher Beaumont, (D), Dalhousie, 1973, Yg
Bernard P. Boudreau, (D), Yale, 1985, CmOcCl
Katja Fennel, (D), Rostock, 1998, Ob
Jonathan Grant, (D), South Carolina, 1981, Ob
Alex E. Hay, (D), British Columbia, 1981, OpnGu
Dan Kelley, (D), Dalhousie, 1986, Op
Markus Kienast, (D), British Columbia, 2002, Cs
Hugh MacIntyre, (D), Deleware, 1996, Ob
Anna Metaxas, (D), Dalhousie, 1994, Ob
Jinyu Sheng, (D), Memorial, 1991, Op
Christopher T. Taggart, (D), McGill, 1986, Ob
Helmuth Thomas, (D), Rostock, 1997, Oc
Keith R. Thompson, (D), Liverpool, 1979, Op
Douglas Wallace, (D), Dalhousie, 1985, Oc

Associate Professor:
Tetjana Ross, (D), Manitoba, 2003, Op

Assistant Professor:
Christopher Algar, (D), Dalhousie, 2009, Obc
David R. Barclay, (D), California (San Diego), 2011, Opo
Carolyn Buchwald, (D), MIT/WHOI, 2013, Oc
Stephanie Kienast, (D), British Columbia, 2002, Gu
Eric Oliver, (D), Dalhousie, 2011, Op

Adjunct Professor:
Mark Baumgartner, (D), Oregon, 2002, Ob
Susanne Craig, (D), Strathclyde, 2000, Op
Peter Cranford, (D), Dalhousie, 1998, Ob
Claudio DiBacco, (D), California (San Diego), 1999, Ob
Dale Ellis, (D), McMaster, 1976, Op
Kenneth Frank, (D), Toledo, 1978, Ob
Richard Greatbatch, (D), Cambridge, 1981, Op
David Greenberg, (D), Liverpool, 1975, Op
David Hebert, (D), Dalhousie, 1988, Op
Paul Hines, (D), Bath, 1989, Op
Bruce D. Johnson, (D), Dalhousie, 1979, Oc
Sebastian Krastel, (D), Kiel, 1999, Yr
William K. W. Li, (D), Dalhousie, 1978, Ob
Keith E. Louden, (D), MIT, 1976, Yr
Youyu Lu, (D), Victoria, 1997, Op
Timothy Milligan, (M), Dalhousie, 1997, Gs
David C. Mosher, (D), Dalhousie, 1993, Ou
Andreas Oschlies, (D), Kiel, 1994, Ob
William Perrie, (D), MIT, 1979, Opw
David J.W. Piper, (D), Cambridge (UK), 1969, GusGo
Harold C. Ritchie, (D), McGill, 1982, Ow
Barry R. Ruddick, (D), MIT, 1977, Op
Peter C. Smith, (D), MIT/WHOI, 1973, Op
Ulrich Sommer, (D), Vienna, 1977, Ob
Toste Tanhua, (D), Goteborg (Sweden), 1997, Oc

Emeritus:
Anthony J. Bowen, (D), California (San Diego), 1967, On
John J. Cullen, (D), California (San Diego), 1980, Ob
Robert O. Fournier, (D), Rhode Island, 1967, Ob
Marlon R. Lewis, (D), Dalhousie, 1984, Ob
Eric L. Mills, (D), Yale, 1964, ObnOn
Robert M. Moore, (D), Southampton, 1977, Oc

## Douglas College

**Dept of Earth and Environmental Sciences, Faculty of Science and Technology** (A) O (2016)
P.O. Box 2503
New Westminster, BC V3L 5B2
p. (604) 527-5400
waddingtond@douglascollege.ca
http://www.douglascollege.ca/programs-courses/faculties/science-technology/earth-and-environmental-sciences

Chair:
David C. Waddington, (M), Queen's (Canada), 1991, OgGuc

Instructor:
Randal Mindell, (D), Alberta, 2008, PbGsg
Derek Turner, (D), Simon Fraser Univ, 2014, Gem
Nathalie Vigouroux-Caillibot, (D), Simon Fraser, 2013, GvzGg

Other:
Michael C. Wilson, (D), Calgary, 1981, GaPgGs

Related Staff:
Denis Beausoleil, (M), Victoria, 2010, Og
Tiffany Johnson, (M), Toronto, 2009, Og

## Lakehead University

**Geology** (B,M) ● (2016)
955 Oliver Road
Thunder Bay, ON P7B 5E1
p. (807) 343-8461
kristine.carey@lakeheadu.ca
https://www.lakeheadu.ca/academics/departments/geology
f: https://www.facebook.com/lakeheaduniversity
t: https://twitter.com/mylakehead
Administrative Assistant: Kristine M. Carey
*Enrollment (2016): B: 68 (16) M: 14 (5)*

Chair:
Peter N. Hollings, (D), Saskatchewan, 1998, Eg
Professor:
Philip W. Fralick, (D), Toronto, 1985, Gs
Mary Louise Hill, (D), Princeton, 1985, Gc
Associate Professor:
Andrew G. Conly, (D), Toronto, 2003, EgCgGz
Amanda Diochon, (D), Dalhousie, 2009, Ge
Assistant Professor:
Shannon Zurevinski, (D), Alberta, 2009, Gz
Emeritus:
Graham J. Borradaile, (D), Liverpool, 1971, Ym
Manfred M. Kehlenbeck, (D), Queen's, 1971, Gc
Stephen A. Kissin, (D), Toronto, 1974, Em
Edward L. Mercy, (D), Imperial Coll (UK), 1955, Cg
Roger H. Mitchell, (D), McMaster, 1969, Gi
Geology Technician:
Kristi Tavener, On
Geology Technician:
Anne Hammond, (B), Lakehead, 1999, On

## Laurentian University, Sudbury

**Harquail School of Earth Sciences** (B,M,D) ● (2016)
Ramsey Lake Road
Sudbury, ON P3E 2C6
p. (705) 675-1151 x 6575
hes@laurentian.ca
http://hes.laurentian.ca/
Administrative Assistant: Roxane J. Mehes
*Enrollment (2016): B: 90 (0) M: 66 (0) D: 21 (0)*

Director:
Doug Tinkham, (D), Alabama, 2002, Gp
Professor:
Harold L. Gibson, (D), Carleton, 1990, EmGv
Daniel J. Kontak, (D), Queens, 1985, Eg
Bruno Lafrance, (D), New Brunswick, 1990, Gc
Michael Lesher, (D), W Australia, 1984, Em
Andrew M. McDonald, (D), Carleton, 1992, Gz
Richard S. Smith, (D), Toronto, Ye
Elizabeth C. Turner, (D), Queen's (Canada), 1999, GrdPo
Associate Professor:
Pedro J. Jugo, (D), Alberta, 2003, CpGi
Matthew I. Leybourne, (D), Ottawa, Cg
Michael Schindler, (D), Frankfurt, Gz
Assistant Professor:
Alessandro Ielpi, (D), Siena, 2013, GsmGr
Emeritus:
Anthony E. Beswick, (D), London, 1965, Gi
Paul Copper, (D), Imperial Coll (UK), 1965, Pi
Richard James, (D), Manchester, 1967, GpiEm
Reid R. Keays, (D), McMaster, 1968, Em
Darrel Long, (D), W Ontario, 1976, GsaGl
Don H. Rousell, (D), Manitoba, 1965, Gc
Robert E. Whitehead, (D), New Brunswick, 1973, Ce
Other:
Bruce C. Jago, (D), Toronto, 1990, EgCe
Cooperating Faculty:
David A. Pearson, (D), London, 1967, Oe
Phillips C. Thurston, (D), Western, Og

## Manitoba Museum

**Dept. of Geology & Paleontology** (2016)
190 Rupert Avenue
Winnipeg, MB R3B 0N2
p. (204) 956-2830
gyoung@manitobamuseum.ca
http://www.manitobamuseum.ca

Curator:
Graham A. Young, (D), New Brunswick, 1988, Pie

## McGill University

**Dept of Earth & Planetary Sciences** (B,M,D) ● (2016)
3450 University Street
Room 238
Montreal, QC H3A 0E8
p. (514) 398-6767
kristy.thornton@mcgill.ca
http://www.mcgill.ca/eps
*Enrollment (2016): B: 31 (15) M: 16 (5) D: 31 (6)*

Chair:
Jeffrey M. McKenzie, (D), Syracuse, 2005, Hw
Retired:
Reinhard Hesse, (D), Tech (Munich), 1964, Gs
Professor:
Don Baker, (D), Penn State, 1985, Cp
Olivia G. Jensen, (D), British Columbia, 1971, Yg
Alfonso Mucci, (D), Miami, 1981, CmlOc
John Stix, (D), Toronto, 1989, Gv
Anthony E. Williams-Jones, (D), Queen's, 1973, Ce
Associate Professor:
Galen Halverson, (D), Harvard, 2003, GsCsGt
Jeanne Paquette, (D), SUNY (Stony Brook), 1991, Gz
Assistant Professor:
Kim Berlo, (D), Bristol (UK), 2006, Gv
Nicolas B. Cowan, (D), Washington, 2009, OaXyg
Peter Douglas, (D), Yale, 2014, Cs
Natalya Gomez, (D), Harvard, 2013, Og
Rebecca Harrington, (D), California (Los Angeles), 2008, Ys
James Kirkpatrick, (D), Glasgow, 2008, Gct
Yajing Liu, (D), Harvard, 2007, Ygs
Christie Rowe, (D), California (Santa Cruz), 2007, GctEm
Vincent van Hinsberg, (D), Bristol, 2006, EgCg
Earth System Science Faculty Lecturer:
William G. Minarik, (D), Rensselaer, 1993, Cp
Adjunct Professor:
Eric Galbraith, (D), British Columbia, 2006, OgPe
Heather Short, (D), Maine, 2006, Gc
Bjorn Sundby, (D), Bergen (Norway), 1966, Cm
Emeritus:
Jafar Arkani-Hamed, (D), MIT, 1969, Xy
Don Francis, (D), MIT, 1974, Gi
Andrew J. Hynes, (D), Cambridge, 1972, Gc
Wallace H. MacLean, (D), McGill, 1968, Em
Robert F. Martin, (D), Stanford, 1969, GziEm
Colin Stearn, (D), Yale, Ps

**Dept of Mining & Materials Engineering** (B,M,D) (2013)
Rm 125 FDA Building
3450 University Street
Montreal, QC H3A 2A7
p. (514) 398-4986
roussos.dimitrakopoulos@mcgill.ca
http://www.mcgill.ca/minmat/mining

Chair:
Stephen Yue, (D), Leeds, 1979, Nx
Professor:
George Demopoulos, (D), McGill, 1982, Nx
Roussos Dimitrakopoulos, (D), Ecole Polytechnique, 1989, Nm
James A. Finch, (D), McGill, 1973, Nx
Raynald Gauvin, (D)
Roderick I. Guthrie, (D), London, 1967, Nx
Ferri Hassani, (D), Nottingham, 1981, Nm
Hani Mitri, (D), Nottingham, 1981, Nm
Hani Mitri, (D), Nottingham
Frank Mucciardi, (D), McGill, 1980, Nx

Post-Retirement:
Michel L. Bilodeau, (D), McGill, 1975, Nm
Associate Professor:
Mathieu Brochu, (D), McGill
Mainul Hasan, (D), McGill, 1987, Nx
Showan Nazhat, (D)
Mihriban Pekguleryuz, (D), McGill
Assistant Professor:
Kirk Bevan, (D), Purdue
Marta Cerruti, (D)
Richard Chromik, (D), SUNY
In-Ho Jung, (D), Ecole Poly (Montreal)
Nathaniel Quitoriano, (D), MIT
Jun Song, (D), Princeton
Kristian Waters, (D)
Lecturer:
John W. Mossop, (B), McGill, 1955, Nm
Forence Paray, (D), McGill
Adjunct Professor:
Robin A.L. Drew, (D), Newcastle, 1980, Nx
Ahmad Hemani, (D), Salford (UK), Nm
Emeritus:
John E. Gruzleski, (D), Toronto, 1967, Nx
John J. Jonas, (D), Cambridge, 1960, Nx
Gordon W. Smith, (D), McGill, 1967, Nx

**Dept of Atmospheric & Oceanic Sciences** (B,M,D) (2016)
805 Sherbrooke Street West
Room 945
Montreal, QC H3A 0B9
p. (514) 398-3764
john.gyakum@mcgill.ca
http://www.mcgill.ca/meteo
Administrative Assistant: Lucy Nunez
*Enrollment (2015): B: 22 (7) M: 18 (12) D: 33 (4)*
Professor:
Parisa A. Ariya, (D), York, 1996, Oa
Peter Bartello, (D), McGill, 1988, Oa
John R. Gyakum, (D), MIT, 1981, Ow
Man Kong Yau, (D), MIT, 1977, Oa
Associate Professor:
Frederic Fabry, (D), McGill, 1994, Oa
Daniel Kirshbaum, (D), Washington, 2005, Oaw
David Straub, (D), Washington, 1990, Op
Bruno Tremblay, (D), McGill, Op
Assistant Professor:
Yi Huang, (D), Princeton, 2007, OrYgOa
Timothy Merlis, (D), Caltech, 2011, Oaw
Thomas Colin Preston, (D), Duke, 2011, OanOn
Andreas Zuend, (D), ETH (Switzerland), 2008, Oaw
Adjunct Professor:
Gilbert Brunet, (D), McGill, 1989, Ow
Ashu Dastoor, (D), IIT
Luc Fillion, (D), McGill, 1991, Owa
Pierre Gauthier, (D), McGill, 1988, Oa
Pavlos Kollias, (D), Miami, 2000, Ora
Hai Lin, (D), McGill, 1995, Oa
Jaime Palter, (D), Duke, 2007, OagOe
Emeritus:
Jacques F. Derome, (D), Michigan, 1968, Oa
Henry G. Leighton, (D), Alberta, 1968, Oa
Lawrence A. Mysak, (D), Harvard, 1966, Op
Isztar I. Zawadzki, (D), McGill, 1972, Oar

## McMaster University

**School of Geography & Earth Sciences** (B,M,D) (2016)
1280 Main Street West
General Science Building
Room 206
Hamilton, ON L8S 4K1
p. (905) 525-9140 (Ext. 24535)
geograd@mcmaster.ca
http://www.science.mcmaster.ca/geo/geomain.html
*Enrollment (2016): B: 282 (85) M: 41 (19) D: 46 (6)*
Director:
K. Bruce Newbold, (D), McMaster, 1994, Ge
Professor:
M. Altaf Arain, (D), Arizona, 1997, Hg
Janok Bhattacharya, (D), McMaster, 1989, Gr
Sean Carey, (D), McMaster, 2000, Hg
Vera Chouinard, (D), McMaster, 1986, On
Paulin Coulibaly, (D), Laval, 2000, Hg
Alan P. Dickin, (D), Oxford, 1981, Cc
Carolyn H. Eyles, (D), Toronto, 1986, Gs
Richard S. Harris, (D), Queen's, 1981, Ou
H. Antonio Paez, (D), Tohoku, 2000, Eg
Edward G. Reinhardt, (D), Carleton, 1996, Pm
Darren M. Scott, (D), McMaster, 2000, Oi
Gregory F. Slater, (D), Toronto, 2001, Cg
James E. Smith, (D), Waterloo, 1995, Hw
J. Michael Waddington, (D), York, 1995, Pe
Allison M. Williams, (D), York, 1997, On
Robert D. Wilton, (D), S California, 1999, On
Associate Professor:
Joseph I. Boyce, (D), Toronto, 1997, Og
Sang-Tae Kim, (D), McGill, 2006, Cg
Maureen Padden, (D), Zurich, 2001, Cg
Niko Yiannakoulias, (D), Alberta, 2006, Ge
Adjunct:
Matthias Peichl, (D), McMaster, 2009
Assistant Professor:
Luc Bernier, (D), McMaster, 2007, Cg
Michael Mercier, (D), McMaster, 2004, Ou
Suzanne Mills, (D), Saskatchewan, 2007, On
Matthias Sweet, (D), Pennsylvania, 2012
Adjunct:
Stacey Mater, (M), McMaster, 2011
Lecturer:
John MacLachlan, (D), McMaster, 2011, Gs
Adjunct Professor:
Howard Barker, (D), McMaster, 1991
Jing Chen, (D), Reading (UK), 1986, OwrOy
Ian G. Droppo, (D), Exeter (UK), 2000
Susan J. Elliott, (D), McMaster, 1992, Ge
Tim Lotimer, (M), Waterloo, 1977
Hanna Maoh, (D), McMaster, 2005
Dan McKenny, (D), Australian National
Nidhi Nagabhatla, (D), ISRO, 2005
Michael Pisaric, (D), Queens
Ulrich Riller, (D), Toronto, 1996, Og
Dominique Rissolo, (D), California (Riverside), 2001
Corinne Schuster-Wallace, (D), Wilfred Laurier, 2001
Amanjot Singh, (D), Braunschweig (Germany), 2001
Spencer Snowling, (D), McMaster, 2000
S Martin Taylor, (D), Victoria (Canada), 1974
Ross Upshur, (D), McMaster, 1986
Lesley A. Warren, (D), Toronto, 1994, CgHg
Christopher L. Werner, (D), Florida State, 2007, GsoHq
University Professor:
John D. Eyles, (D), London, 1983, Ge
Henry P. Schwarcz, (D), Caltech, 1960, Cs
Emeritus:
Brian T. Bunting, (D), London, 1970
Andrew F. Burghardt, (D), Wisconsin, 1958
Paul Clifford, (D), London, 1956
James H. Crocket, (D), MIT
John J. Drake, (D), McMaster, 1973
Derek C. Ford, (D), Oxford, 1963, GmHwCc
Doug Grundy, (D), Manchester (UK), 1966
Fred L. Hall, (D), MIT, 1975
Leslie J. King, (D), Iowa, 1960
James R. Kramer, (D), Michigan, 1958, Cl
Kao Lee Liaw, (D), Clark, 1972
Robert McNutt, (D), MIT, 1965
Gerry V. Middleton, (D), London (UK), 1954
William A. Morris, (D), Open, 1974, Yg
Yorgos Papageorgiou, (D), Ohio State, 1970
Walter G. Peace, (D), McMaster, 1996, Ou
W. Jack Rink, (D), Florida State, 1990, Cc
Michael J. Risk, (D), S California, 1971, Po
Wayne R. Rouse, (D), McGill, 1968
Roger G. Walker, (D), Oxford, 1964
Ming-Ko Woo, (D), British Columbia, 1972, Hg

## Memorial University of Newfoundland
**Dept of Earth Sciences** (B,M,D) (2015)
Centre for Earth Resources Research, Room ER 4063
Alexander Murray Building
St. John's, NL A1B 3X5
p. (709) 864-8142
earthsci@mun.ca
http://www.mun.ca/earthsciences
Professor:
Ali E. Aksu, (D), Dalhousie, 1980, Gu
Elliott T. Burden, (D), Calgary, 1982, Pl
Gregory R. Dunning, (D), Memorial, 1984, CcGit
George A. Jenner, (D), Tasmania, 1982, Gi
Toby C. J. S. Rivers, (D), Ottawa, 1976, Gp
Derek H.C. Wilton, (D), Memorial, 1984, EgCae
Associate Professor:
Tomas J. Calon, (D), Leiden (Neth), 1978, Gc
Charles A. Hurich, (D), Wyoming, 1988, Ys
Aphrodite D. Indares, (D), Montreal, 1989, Gp
Roger A. Mason, (D), Aberdeen, 1978, Gz
Michael A. Slawinski, (D), Calgary, 1996, Ys
Paul J. Sylvester, (D), Washington (St. Louis), 1984, Ca
Assistant Professor:
Alison Leitch, (D), Australian National, 1986, Yg
First Year Lab Instructor:
Roberta (Robbie) Hicks, (M), Gg
Emeritus:
Jeremy Hall, (D), Glasgow, 1971, Ys
Richard N. Hiscott, (D), McMaster, 1977, Gs
Joseph P. Hodych, (D), Toronto, 1971, Ym
Henry Longerich, (D), Indiana, 1967, Cg
Michael G. Rochester, (D), Utah, 1959, Yg

## Mount Allison University
**Dept of Geography** (B,M) (2004)
144 Main Street
Sackville, NB E4L 1A7
p. (506) 364-2326
dmossman@mta.ca
http://www.mta.ca/departments/geography
Associate Professor:
Jeffery W. Ollerhead, (D), Guelph, 1994, Yr
Assistant Professor:
James Xinxia Jiang, (D), Southampton, Gg
Research Associate:
Thomas A. Clair, (D), McMaster, 1991, Hy
Research Professor:
David J. Mossman, (D), Otago (NZ), 1970, EmOn
Emeritus:
Laing Ferguson, (D), Edinburgh, 1960, Pi

## Mount Royal University
**Dept of Earth and Environmental Sciences** (A,B) (2013)
4825 Mount Royal Gate SW
Calgary, AB T3E 6K6
p. (403) 440-6165
lstadnyk@mtroyal.ca
http://www.mtroyal.ab.ca/scitech/earth.shtml
Administrative Assistant: Leona Stadnyk
Chair:
Paul Johnston, (D), W Australia, 1986, Pi
Professor:
John Cox, (D), Aberdeen Univ, 1994, EoGs
Barbara McNicol, (D), Calgary, 1997, Og
Associate Professor:
Katherine Boggs, (D), Calgary, 2004, Gt
Pamela MacQuarrie, (M), Calgary, 1988, Oy
Related Staff:
Michael Clark, (B), Adams, 1976, Og

## Natural Resources Canada
**Ressources Naturelles Canada** (2011)
580 Booth Street , 21st Floor
Ottawa, ON K1A 0E4
debra.tompkinscaron@canada.ca
http://www.nrcan.gc.ca/
Senior Scientist:
Sergey V. Samsonov, (D), Western, 2007, YdOrYg

## New Brunswick Dept of Energy and Mines
**New Brunswick Dept of Energy and Mines** (2017)
Hugh John Flemming Forestry Centre
P. O. Box 6000
Fredericton, NB E3B 5H1
p. (506) 453-3826
geoscience@gnb.ca
http://www.gnb.ca/energy

## Queen's University
**Dept of Geological Sciences and Geological Engineering** (B,M,D) ● (2017)
#240, Bruce Wing, Miller Hall
36 Union St.
Kingston, ON K7L 3N6
p. (613) 533-2597
geolundergradassistant@queensu.ca
www.queensu.ca/geol
*Enrollment (2011): B: 232 (47) M: 46 (20) D: 18 (5)*
Head:
Jean Hutchinson, (D), Toronto, 1992
Professor:
Mark Diederichs, (D), Toronto, Nr
Georgia Fotopoulos, (D), Calgary, 2003, Yd
Laurent Godin, (D), Carleton Univ, 1999, Gct
Noel P. James, (D), McGill, 1972, Gd
Kurt Kyser, (D), California (Berkeley), CeaCs
Guy M. Narbonne, (D), Ottawa, 1981, PiyGs
Gema Olivo, (D), Québec (Montréal), 1995, Eg
Ronald C. Peterson, (D), Virginia Tech, 1980, Gz
Victoria H. Remenda, (D), Waterloo, 1993, Hw
Associate Professor:
Alexander Braun, (D), Goethe Univ (Germany), 1999, YgdYv
John A. Hanes, (D), Toronto, 1979, Cc
Heather E. Jamieson, (D), Queen's, 1982, Ge
Daniel Layton-Matthews, (D), Toronto, 2006, EgCte
Research Associate:
Doug A. Archibald, (D), Queen's, 1982, Cc
Adjunct Professor:
Rob Harrap, (M), Carleton, 1990, Gc
Emeritus:
Alan H. Clark, (D), Manchester, 1964, Em
Robert W. Dalrymple, (D), McMaster, 1977, Gs
John M. Dixon, (D), Connecticut, 1974, Gc
Herwart Helmstaedt, (D), New Brunswick, 1968, Gc
Raymond A. Price, (D), Princeton, 1958, GtcGe

## Royal Ontario Museum
**Dept of Palaeobiology** (2013)
100 Queen's Park
Toronto, ON M5S 2C6
p. (416) 586-5591
davidru@rom.on.ca
Curator Of Vertebrate Palaeontology:
David C. Evans, (D), Toronto, 2007, PvoPq
Assistant Curator:
Kevin L. Seymour, (D), Toronto, 1999, PvoPg

**Dept of Mineralogy** (2015)
100 Queen's Park
Toronto, ON M5S 2C6
p. (416) 586-5820
naturalhistory@rom.on.ca
t: @ROMEarthSci
Emeritus:
Robert I. Gait, (D), Manitoba, 1967, Gz
Frederick J. Wicks, (D), Oxford, 1969, Gz
Curator:
Kimberly T. Tait, (D), Arizona, 2007, GzyXm

**Dept of Earth Sciences** (2010)

100 Queen's Park
Toronto, ON M5S 2C6
p. (416) 856-5811
annetteb@rom.on.ca
Curatorial Assistant:
Vincent Vertolli, (M), Toronto, 1974, Gx

## Royal Tyrrell Museum of Palaeontology

**Royal Tyrrell Museum of Palaeontology** (2013)
P.O. Box 7500
Drumheller, AB T0J 0Y0
p. (403) 823-7707
tyrrell.info@gov.ab.ca
http://www.tyrrellmuseum.com/
Executive Director:
Andrew G. Neuman, (M), Alberta, 1986, Pv
Director, Preservation and Research:
Donald B. Brinkman, (D), McGill, 1979, Pv
Curator:
Dennis R. Braman, (D), Calgary, 1981, Pl
David A. Eberth, (D), Toronto, 1987, Gs
James D. Gardner, (D), Alberta, 2000, Pv
Donald Henderson, (D), Bristol, 2000, Pv
Craig Scott, (D), Alberta, 2008, Pv
Curator:
Francois Therrien, (D), Johns Hopkins, 2004, Pv
Post-doctoral fellow:
Caleb Brown, (D), Toronto, 2013, Pv

## Saint Francis Xavier University

**Dept of Earth Sciences** (B,M) (2015)
P.O Box 5000
Antigonish, NS B2G 2W5
p. (902) 867-5109
igreen@stfx.ca
http://sites.stfx.ca/earth_sciences/
*Enrollment (2015): B: 11 (15) M: 13 (3)*
Professor:
Alan J. Anderson, (D), Queen's, 1990, Gx
Hugo Beltrami, (D), UQAM, 1993, YhOa
Lisa M. Kellman, (D), Quebec (Montreal), 1997, CsOs
Michael J. Melchin, (D), Western, 1987, Pi
J. Brendan Murphy, (D), McGill, 1982, Gt
Associate Professor:
Dave A. Risk, (D), Dalhousie, 2006, Osr
Instructor:
Cindy Murphy, (M), McGill, 1986, Gg
Colette Rennie, (M), Queen's, 1987, Gg
Matthew Schumacher, (M), Waterloo, 2006, Og
Sid Taylor, (M), Memorial, 1977, Gg

## Simon Fraser University

**Dept of Earth Sciences** (B,M) (2010)
8888 University Drive
Burnaby, BC V5A 1S6
p. (604) 291-5387
bcward@sfu.ca
http://www.sfu.ca/earth-sciences
*Enrollment (2003): B: 546 (0) M: 25 (0)*
Chair:
Diana M. Allen, (D), Carleton, 1996, Hy
Professor:
John J. Clague, (D), British Columbia, 1973, Gl
Douglas Stead, (D), Nottingham, 1984, Ng
Associate Professor:
Andrew J. Calvert, (D), Cambridge, 1985, Ys
H. Daniel (Dan) Gibson, (D), Carleton Univ, 2003, GcCcGp
James A. Mac Eachern, (D), Alberta, 1994, Gs
Dan D. Marshall, (D), Lausanne, 1995, Ca
Peter S. Mustard, (D), Carleton, 1990, Gs
Derek J. Thorkelson, (D), Carleton, 1992, Gt
Brent C. Ward, (D), Alberta, 1992, Ge
Glyn Williams-Jones, (D), Open Univ (UK), 2001, GvCeYe
Assistant Professor:
Gwenn Flowers, (D), British Columbia, 2000, Gl
Dirk Kirste, (D), Calgary, 2001, Cg

Lecturer:
Kevin Cameron, (M), Memorial, 1986, Gg
Roberta Donald, (M), British Columbia, 1984, GsOe

## Sir Wilfred Grenfell College

**Environmental Science Unit** (B) (2010)
Division of Science
University Drive
Corner Brook, NF A2H 6P9
p. (709) 637-6289
dparkins@grenfell.mun.ca
http://www.swgc.mun.ca
Department Secretary: Phyllis Langdon
Chair:
William J. Iams, (D), Memorial, 1977, Og
Assistant Professor:
Pierre M. Rouleau, (D), Alberta, 1994, Yg
Michael P. Rutherford, (D), Alberta, 1991, Sb
Adjunct Professor:
Antony R. Berger, (D), Liverpool (UK), 1967, Ge

## Universite du Quebec

**INRS, centre Eau Terre Environnement (Quebec Geoscience Center)** (M,D) (2016)
490 de la Couronne
Quebec, QC G1K 9A9
p. + (418) 654-4677
info@ete.inrs.ca
http://www.ete.inrs.ca
Administrative Assistant: Pascale Cote
Professor:
Normand Bergeron, (D), SUNY (Buffalo), 1994, Gm
Monique Bernier, (D), Or
Fateh Chebana, Hq
Karem Chokmani, Oi
Pierre Francus, (D), 1997, PeGse
Bernard Giroux, Yeg
Erwan Gloaguen, Ye
Yves Gratton, Op
Lyal B. Harris, GcYg
Marc R. LaFleche, (D), Montpellier II, 1991, Ct
Rene Lefebvre, (D), Laval, 1994, Hw
Michel Malo, (D), Montreal Univ (Canada), 1986, GctGs
Richard Martel, (D), Laval, 1996, Hw
Claudio Paniconi, Hw
Pierre-Simon Ross, GvEg
Senior Scientist:
Jean H. Bedard, (D), Montreal, 1985, Gi
Christian Begin, (D), Laval, 1991, PeGmOs
Louise Corriveau, (D), McGill, 1989, Gp
Benoit Dube, (D), Quebec (Chicoutimi), 1990, Em
Denis Lavoie, (D), Laval, 1988, Gs
Yves Michaud, (D), Laval, 1991, Gm
Michel Parent, (D), W Ontario, 1987, Gl
Didier Henri Perret, (D), Laval, 1995, NgeOu
Associate Scientist:
Esther Asselin, (M), Laval, 1988, PlGuh
Research Associate:
Eric Boisvert, (M), Quebec (Montreal), 1994, On
Pierre Brouillette, (B), Laval, 1982, Gg
Kathleen Lauziere, (M), Quebec (Chicoutimi), 1989, Gg
Marc R. Luzincourt, (B), Quebec (Montreal), 1982, Cs
Anna Smirnov, (M), Memorial, 1997, Cs

## Universite du Quebec a Chicoutimi

**Sciences de la Terre** (B,M,D) ● (2016)
555, Boulevard de l'Universite
Chicoutimi, QC G7H 2B1
p. (418) -545- 5011 (Ext. 5202)
sue_sc-terre@uqac.ca
*Enrollment (2016): B: 64 (8) M: 40 (15) D: 16 (1)*
Director:
Michael D. Higgins, (D), McGill, 1980, Gi
Professor:
Sarah- Jane Barnes, (D), Toronto, 1983, EgGiCt
Paul Bedard, (D), UQAC, 1992, CaGzEg

Romain Chesnaux, (D), École Polytechnique de Montréal, Québec, Canada, 2005, Hw
Pierre Cousineau, (D), Laval, 1986, Gd
Real Daigneault, (D), Laval, 1991, GcEgOi
Damien Gaboury, (D), UQAC, 1999, EmCe
Ali Saeidi, (D), Lorraine INP (France), 2010, Nrm
Edward W. Sawyer, (D), Toronto, 1983, GptGi

Instructor:
Denis Cote, (M), Quebec (Chicoutimi), 1986, Gi

Lecturer:
Philippe Page, (D), INRS-ETE, 2006, GiCgEm

Adjunct Professor:
Sylvain Raffini, (D), UQAM, 2008, GcHw

Emeritus:
Guy Archambault, (D), Nr
Edward H. Chown, (D), Johns Hopkins, 1963, Gx
Jayanta Guha, (D), Jadavpur, 1967, Eg
Alain Rouleau, (D), Waterloo, 1984, Hw
Denis W. Roy, (D), Princeton, 1976, Gc

Laboratory Director:
Dany Savard, (M), UQAC, 2010, Ca

## Universite du Quebec a Montreal

**Département des sciences de la Terre et de l'atmosphere** (B,M,D) (2015)
C.P. 8888, succursale Centre-ville
Montreal, QC H3C 3P8
p. +1 514-987-3000
dept.sct@uqam.ca
http://scta.uqam.ca
Administrative Assistant: France Beauchemin

Professor:
Florent Barbecot, (D), Hw
Jean-Pierre Blanchet, (D), Toronto, 1984, Pe
Gilles Couture, (D), British Columbia, 1987, Yg
Fiona Ann Darbyshire, (D), Cambridge, 2000, Ys
Anne de Vernal, (D), UdeM, Montreal, 1986, Pm
Alessandro Marco Forte, (D), Toronto, 1989, Yg
Pierre Gauthier, (D), McGill, 1988, Ow
Eric Girard, (D), McGill, 1999, Ow
Normand Goulet, (D), Queen's, 1976, Gc
Cherif Hamzaoui, (D), Alziers, 1980, Yg
Alfred Jaouich, (D), Minnesota, 1975, Sc
Michel Jebrak, (D), Orleans, 1984, Eg
Michel Lamothe, (D), W Ontario, 1985, Gl
Rene Laprise, (D), Toronto, 1988, Oa
Marie Larocque, (D), de Poitiers, France, 1998, Hw
Marc Michel Lucotte, (D), McGill, 1987, Oc
Daniele Luigi Pinti, (D), France, 1993, Cs
Martin Roy, (D), Oregon, 1998, Gl
Ross Stevenson, (D), Arizona, 1989, Cg
Laxmi Sushama, (D), Melbourne, 1999, Hy
Julie Mireille Thériault, (D), McGill, 2009, On
Enrico Torlaschi, (D), Ecole Polytechnique (Milan), 1976, Ow
Alain Tremblay, (D), Laval, 1989, Gct
David Widory, (D), IPGP-Université, 1999, Cs

Associate Professor:
Sanda Balescu, (D), Libre de Bruxelles, 1988, Pe
Jean Côté, Oa
Bernard Dugas, Oa
Stephane Faure, (D), INRS-Georessources, 1995, Gc
Philippe Gachon, (D), UQAM, 1999, Oa
Michel Gauthier, (D), Polytechnique (Montreal), 1982, Eg
Jean-François Helie, (D), UQAC, 2004, Cs
Claude Hillaire-Marcel, (D), Paris VI, 1979, Cs
Jean-Claude Mareschal, (D), Texas A&M, 1975, Yg
Andre Poirier, (D), UQAM, 2005, Cs
Gilbert P. Prichonnet, (D), Bordeaux, 1967, Pi
William W. Shilts
Gabriel-Constantin Voicu, (D), UQAM, 1995, GcCg

## Universite du Quebec a Rimouski

**Inst des sciences de la mer de Rimouski** (M,D) (2014)
310, Allee des Ursulines
Rimouski, QC G5L 3A1
p. (418) 723- 8617
andre_rochon@qc.ca
http://www.ismer.ca

Director:
Serge Demers, (D), Laval, 1981, Ob

Professor:
Celine Audet, (D), Laval, 1985, Ob
Jean-Claude Brethes, (D), Aix-Marseille, 1978, Ob
Jean-Pierre Gagne, (D), Montreal, 1993, Oc
Michel Gosselin, (D), Laval, 1990, Ob
Vladimir G. Koutitonsky, (D), SUNY (Stony Brook), 1985, Op
Jocelyne Pellerin, (D), Laval, 1982, Ob
Emilien Pelletier, (D), McGill, 1983, Oc
Suzanne Roy, (D), Dalhousie, 1986, Ob
Bjorn Sundby, (D), Bergen (Norway), 1966, Oc
Bruno Zakardjian, (D), Paris, 1994, Ob

## Universite Laval

**Dept de geologie** (B,M,D) (2010)
Pavillon Pouliot
Faculte des Sciences et de genie
Ste-Foy, QC G1K 7P4
p. (418) 656-2193
marc.constantin@ggl.ulaval.ca
http://www.ggl.ulaval.ca
*Enrollment (2002): B: 127 (27) M: 30 (6) D: 20 (3)*

Head:
Marc Constantin, (D), Brest, 1995, Gi
Josee Duchesne, (D), Laval, 1993, Gz

Professor:
Georges Beaudoin, (D), Ottawa, 1991, EgGzCa
Richard Fortier, (D), Montreal, 1994, Yg
Paul W. Glover, (D), East Anglia (UK), 1989, Yx
Rejean J. Hebert, (D), Brest, 1985, Gi
Jacques E. Locat, (D), Sherbrooke, 1982, Ng
Fritz Neuweiler, (D), Berlin, 1995, GsdPo
Rene Therrien, (D), Waterloo, 1992, HwGe

Eletron Microprobe Specialist:
Marc Choquette, (D), Laval, 1988, On

Associate Scientist:
Pauline Dansereau, (M), Laval, 1988, Gs
Andre Levesque, (B), Laval, 1979, Gz
Pierre Therrien, (M), Laval, 1986, Gq

Research Associate:
Danielle Cloutier, (D), Laval, GeOg

Adjunct Professor:
Benoit Fournier, (D), Laval, 1993, NgOmGg

Technician:
Jean Frenette, (B), Laval, 1985, Gz
Martin Plante, Cg

## University of Alberta

**Dept of Earth & Atmospheric Sciences** (B,M,D) ● (2016)
126 Earth Sciences Building
Edmonton, AB T6G 2E3
p. (780) 492-3265
eas@ualberta.ca
http://www.ualberta.ca/EAS/
f: https://www.facebook.com/UofAEarthandAtmosphericSciencesDepartment
t: @UofA_EAS
*Enrollment (2016): B: 449 (129) M: 0 (17) D: 0 (18)*

Chair:
Stephen T. Johnston, (D), Alberta, 1993, GctGg

Professor & Inaugural Director, Planning Program:
Sandeep Agrawal, (D), Ou

Professor:
Robert W. Luth, (D), California (Los Angeles), 1985, Gi
Martin J. Sharp, (D), Aberdeen, 1982, Ol

Distinguished University Professor:
S. George Pemberton, (D), McMaster, 1979, Gr

Canada Research Chair:
Thomas Stachel, (D), Wurzburg, 1991, GiCs

Associate Dean (Research):
Larry M. Heaman, (D), McMaster, 1986, Cc

Associate Chair:
Thomas Chacko, (D), North Carolina, 1987, Gp
Murray Gingras, (D), North Carolina, 1987, Go
Professor:
Andrew B.G Bush, (D), Toronto, 1995, OapPe
Michael W. Caldwell, (D), McGill, 1995, Pv
Octavian Catuneanu, (D), Toronto, 1996, Gs
Robert A. Creaser, (D), La Trobe, 1990, Cc
Philippe Erdmer, (D), Queen's, 1982, Gc
Duane Froese, (D), Calgary, 2001, GlmPe
John Gamon, (D), California (Davis), 1989, Or
Christopher Herd, (D), New Mexico, 2001, XcGiCp
Brian Jones, (D), Ottawa, 1974, Ps
Kurt Konhauser, (D), W Ontario, 1993, Py
Lindsey Leighton, (D), Michigan, 1999, Pi
Hans G. Machel, (D), McGill, 1985, Go
Carl A. Mendoza, (D), Waterloo, 1993, Hw
Karlis Muehlenbachs, (D), Chicago, 1971, Cs
Graham Pearson, (D), Leeds, Cec
David Potter, (D), Newcastle upon Tyne, Ye
Gerhard Reuter, (D), McGill, 1985, Ow
Jeremy P. Richards, (D), Australian National, 1990, Em
Benoit Rivard, (D), Washington (St. Louis), 1990, Or
Benjamin J. Rostron, (D), Alberta, 1995, HwGoCg
G. Arturo Sanchez-Azofeifa, (D), New Hampshire, 1996, OirOg
Bruce Sutherland, (D), Toronto, 1994, Cm
Martyn Unsworth, (D), Cambridge, GtvYe
John W.F Waldron, (D), Edinburgh, 1981, Gc
John D. Wilson, (D), Guelph, 1980, Ow
John-Paul Zonneveld, (D), Alberta, 1999, Go
Associate Professor:
Damian Collins, (D), Simon Fraser, 2004, On
Theresa D. Garvin, (D), McMaster, 1999, On
Nicholas Harris, (D), Stanford, Gso
Jeffrey Kavanaugh, (D), British Columbia, 2001, Ng
G. Peter Kershaw, (D), Alberta, 1983, Gm
Tara McGee, (D), Australian National, 1996, On
Paul Myers, (D), Victoria, 1996, Og
Assistant Professor, Campus Alberta Innovates Program Chair in Watershed Science:
Monireh Faramarzi, (D), ETH Zurich, 2010, HgwHq
Assistant Professor:
Daniel S. Alessi, (D), Notre Dame, 2009, ClGeHw
Jeff Birchall, (D), Canterbury, 2013, OuyOg
Leith Deacon, (D), Ou
Long Li, (D), Lehigh, 2006, CsGpv
Manish Shirgaokar, (D), Ou
Amrita Singh, (D), Irvine, 2015, Ou
Distinguished University Professor:
Nathaniel W. Rutter, (D), Alberta, 1966, Ge
Emeritus:
Halfdan Baadsgaard, (D), ETH (Switzerland), 1955, Cc
Ronald A. Burwash, (D), Minnesota, 1955, Gp
Ian A. Campbell, (D), Colorado, 1968, Gm
Brian D. Chatterton, (D), Australian National, 1970, Pi
David M. Cruden, (D), London, 1969, Nr
John England, (D), Colorado, 1974, Gml
Kenneth J. Fairbairn, (D), Melbourne, 1968, On
Richard C. Fox, (D), Kansas, 1965, Pv
Keith D. Hage, (D), Chicago, 1957, Ow
M. John Hodgson, (D), Toronto, 1973, On
R. Geoffrey Ironside, (D), Durham, 1965, On
Edgar L. Jackson, (D), Toronto, 1974, On
Leszek A. Kosinski, (D), Warsaw, 1958, On
Arleigh H. Laycock, (D), Minnesota, 1957, Hg
Edward P. Lozowski, (D), Toronto, 1970, Oa
Harry J. McPherson, (D), McGill, 1967, Gm
Roger D. Morton, (D), Nottingham, 1959, Em
O.F. George Sitwell, (D), Toronto, 1968, On
Peter J. Smith, (D), Edinburgh, 1964, On
Charles R. Stelck, (D), Stanford, 1951, Ps
József Tóth, (D), Utrecht (Neth), 1965, Hw
Related Staff:
Robert Summers, (D), Guelph, 2005, On

**Inst of Geophysical Research** (B,M,D) (2010)
Mailstop #615
CEB/Physics
Edmonton, AB T6G 2G7
p. (780) 492-3521
msacchi@ualberta.ca
http://www-geo.phys.ualberta.ca/institute/
Administrative Assistant: Lee Grimard
Director:
Douglas R. Schmitt, (D), Caltech, 1987, YxNrYe
Professor:
Robert A. Creaser, (D), La Trobe, 1992, Cc
T. Bryant Moodie, (D), Toronto, 1972, On
Robert Rankin, (D), North Wales, 1984, Xy
Gerhard W. Reuter, (D), McGill, 1985, Oa
Wojciech Rozmus, (D), Inst Nuc Res (Poland), On
John C. Samson, (D), Alberta, 1971, Xy
Martin J. Sharp, (D), Aberdeen, Gl
John Shaw, (D), Reading, 1969, Gl
Samuel S. Shen, (D), Wisconsin, Oa
Bruce R. Sutherland, (D), Toronto, 1994, On
Gordon E. Swaters, (D), British Columbia, 1985, Op
Richard D. Sydora, (D), Texas, 1985, Xy
Martyn Unsworth, (D), Cambridge, On
John D. Wilson, (D), Guelph, 1980, Oa
Associate Professor:
Andrew B. G. Bush, (D), Toronto, Oa
Carl Mendoza, (D), Waterloo, 1993, Hw
Benoit Rivard, (D), Washington, 1990, Or
Ben Rostron, (D), Alberta, On
Mauricio D. Sacchi, (D), British Columbia, Ys
Assistant Professor:
Francis Fenrich, (D), Alberta, 1997, Xy
Jeff Gu, (D), Harvard, 2001, Ys
Moritz Heimpel, (D), Johns Hopkins, 1995, On
Vadim Kravchinsky, (D), Irkutsk, 1996, Ym
Paul Myers, (D), Victoria, 1992, Op
Emeritus:
Michael E. Evans, (D), Australian National, 1969, Ym
Keith D. Hage, (D), Chicago, 1957, Oa
F. Walter Jones, (D), McGill, 1968, Ym
Edward P. Lozowski, (D), Toronto, 1970, Oa
Roger D. Morton, (D), Nottingham, 1959, Em
Edo Nyland, (D), California (Los Angeles), 1967, Ys
David Rankin, (D), Alberta, 1960, Ym
Gordon Rostoker, (D), British Columbia, 1966, Xy
T.J.T (Tim) Spanos, (D), Alberta, 1977, YgsEo

## University of British Columbia

**Dept of Earth, Ocean, and Atmospheric Sciences** (B,M,D) O (2015)
2020-2207 Main Mall
Vancouver, BC V6T 1Z4
p. (604) 827-5284
acairns@eos.ubc.ca
http://www.eoas.ubc.ca
*Enrollment (2010): B: 361 (105) M: 111 (20) D: 93 (15)*
Director, Geological Engineering Program:
Roger D. Beckie, (D), Princeton, 1992, HwCl
Honorary Professor:
Mati Raudsepp, (D), Manitoba, 1984, GzeOm
Dean, Science:
Simon M. Peacock, (D), California (Los Angeles), 1985, GtpYs
Canada Research Chair:
Roger Francois, (D), British Columbia, 1987, Cm
Dominique Weis, (D), Libre de Bruxelles, 1982, Cc
Professor:
Susan E. Allen, (D), Cambridge, 1988, Op
Raymond J. Andersen, (D), California (San Diego), 1975, Oc
Neil Balmforth, (D), Cambridge, 1990, Oa
Michael G. Bostock, (D), Australian National, 1991, YsGt
R. Marc Bustin, (D), British Columbia, 1979, Ec
Gregory M. Dipple, (D), Johns Hopkins, 1991, Gp
Erik Eberhardt, (D), Saskatchewan, 1998, Ng
Lee A. Groat, (D), Manitoba, 1988, Gz
Felix J. Herrmann, (D), Delft (Neth), 1997, Yx
Oldrich Hungr, (D), Alberta, 1981, Gm
Mark Jellinek, (D), Australian National, 1999, Gg
Catherine L. Johnson, (D), San Diego, 1994

Canada

Ulrich Mayer, (D), Waterloo, 1999, Nm
James K. Mortensen, (D), California (Santa Barbara), 1983, Cc
Douglas W. Oldenburg, (D), California (Santa Barbara), 1974, Yg
Evgeny Pakhomov, (D), Russian Acad of Sci, 1992, Ob
James Kelly Russell, (D), Calgary, 1984, GviCg
James S. Scoates, (D), Wyoming, 1994, Gi
Paul L. Smith, (D), McMaster, 1981, Pg
Douw G. Steyn, (D), British Columbia, 1980, Oaw
Roland B. Stull, (D), Washington, 1975, Oa
Curtis Suttle, (D), British Columbia, 1987, Ob

NSERC Industrial Research Chair in Computational Geoscience:
Eldad Haber, (D), British Columbia, 1997, YgOn

Director, MDRU:
Craig J.R. Hart, (D), W Australia, 2004, EmGiCc

Canada Research Chair:
Maria Maldonado, (D), McGill, 1999, Ob
Christian Schoof, (D), Oxford, 2002, Ol

Associate Professor:
Philip Austin, (D), Washington, 1987, Oa
Kurt A. Grimm, (D), California (Santa Cruz), 1992, Gs
Mark S. Johnson, (D), Cornell, 2005, SpHsq
Lori Kennedy, (D), Texas A&M, 1996, Gc
Maya G. Kopylova, (D), Moscow, 1990, Gi
Kristin J. Orians, (D), California (Santa Cruz), 1988, Cm
Richard A. Pawlowicz, (D), MIT/WHOI, 1994, Op
Philippe Tortell, (D), Princeton, 2001, Ob

Assistant Professor:
Kenneth A. Hickey, (D), James Cook Univ, 1995, EgGcg
Valentina Radic, (D), Alaska (Fairbanks), 2008, OlaOg

Honorary Research Associate:
Kevin Kingdon, (M), British Columbia, 1998, Yg
Michael Maxwell, (D), British Columbia, 1986, Yg

Research Associate:
Thomas Bissig, (D), Queens, 2001, EmCg
Farhad Bouzari, (D), Queens, 2003, Eg
Amanda Bustin, (D), Victoria, 2006, YgNg
Philip T. Hammer, (D), California (San Diego), 1991, Ys
Brian Hunt, (D), Tasmania, 2005, Ob
Bruno Kieffer, (D), Grenoble, 2002, Cs
Henryk Modzelewski, (D), British Columbia, 2004, Oa
Roger Pieters, (D), California (Santa Barbara), Op
Lin-Ping Song, (D), Sichuan (China), 1996, Ng
Dave E. Williams, (D), British Columbia, 1987, Oc

Instructor:
Mary Lou Bevier, (D), California (Santa Barbara), 1982, Cc
Sara Harris, (D), Oregon State, 1998, Ou
Tara Ivanochko, (D), British Columbia, 2004, Ou
Stuart Sutherland, (D), Leicester (UK), 1992, Pg

Lecturer:
Brett Gilley, (M), Simon Fraser, 2003, Gg
Francis H. M. Jones, (M), British Columbia, 1987, YeOe

Honorary Lecturer:
Dileep Athaide, (M), British Columbia, 1975, Gg

Adjunct Professor:
Robert G. Anderson, (D), Carleton, 1983, Gt
Stephen Billings, (D), Sydney, 1998, Yg
Alex Cannon, (D), British Columbia, 2009, Oa
Edward Carmack, (D), Washington, 1972, Op
Michael G. G. Foreman, (D), British Columbia, 1984, Op
James Haggart, (D), California (Davis), 1984, PgGr
Catherine J. Hickson, (D), British Columbia, 1987, GvEmOe
Mark Holzer, (D), Simon Fraser, 1990, Oa
R. Lynn Kirlin, (D), Utah State, 1968, Oa
Doug McCollor, (D), British Columbia, 2008, Oaw
Barry Narod, (D), British Columbia, 1979, Yx
Michael Orchard, (D), Hull, 1975, PgGg
K. Wayne Savigny, (D), Alberta, 1980, Ng
Barbara H. Scott Smith, (D), Edinburgh, 1977, Gz
John F. H. Thompson, (D), Toronto, 1984, Em
Richard E. Thomson, (D), British Columbia, 1971, Op
Richard Tosdal, (D), California (Santa Barbara), 1988, Em
Knut von Salzen, (D), Hamburg (Germany), 1997, Oa
Chi S. Wong, (D), California (San Diego), 1968, Cm

Emeritus:
Peter M. Bradshaw, (D), Durham (UK), 1965, EgCeGe
Stephen E. Calvert, (D), California (San Diego), 1964, Cm
Richard L. Chase, (D), Princeton, 1963, Gu
Garry K. C. Clarke, (D), Toronto, 1967, Yg
Ronald M. Clowes, (D), Alberta, 1969, Ys
Robert M. Ellis, (D), Alberta, 1964, Ys
William K. Fletcher, (D), Imperial Coll (UK), 1968, Ce
Michael Healey, (D), Aberdeen, 1969, Ob
William W. Hsieh, (D), British Columbia, 1981, Opa
Alan G. Lewis, (D), Hawaii, 1961, Ob
Stephen G. Pond, (D), British Columbia, 1965, Op
R. Doncaster Russell, (D), Toronto, 1954, Yg
Alastair J. Sinclair, (D), British Columbia, 1964, EmGg
J. Leslie Smith, (D), British Columbia, 1978, Hw
Frank J. R. Taylor, (D), Cape Town, 1965, Ob

Undergraduate Program Coordinator:
Teresa Woodley, On

Secretary to the Head:
Selene Chan, (B), British Columbia, 1995, On

Office Support:
Alicia Warkentin, (N), On

Human Resources Manager:
Cary Thomson, On

Graduate Coordinator:
Audrey Van Slyck

Finance Clerk:
Anita Lam, On
Kathy Scott, On

Director of Resources and Operations:
Renee Haggart, On

Computer Department Manager:
John Amor, On

**Soil Science** (B,M,D) O (2016)
2357 Main Mall
Vancouver, BC V6T 1Z4
p. (604) 822-0252
maja.krzic@ubc.ca
http://www.landfood.ubc.ca/academics/graduate/soil-science-msc-phd/
*Enrollment (2016): B: 4 (4) M: 15 (7) D: 7 (2)*

Professor:
T. Andrew Black, (D), Wisconsin, 1969, Sp
Christopher Chanway, (D), British Columbia, 1987, Sb
Sue Grayston, (D), Sheffield, 1988, Sb
Cindy Prescott, (D), Calgary, 1988, Sb
Suzanne Simard, (D), Oregon State, 1995, Sb

Associate Professor:
Maja Krzic, (D), British Columbia, 1997, Sf

Assistant Professor:
Sean Smukler, (D), California (Davis), 2008, Os

Lecturer:
Sandra Brown, (D), British Columbia, 1997, Sp

Emeritus:
Arthur A. Bomke, (D), Illinois, 1972, Sc
Leslie M. Lavkulich, (D), Cornell, 1969, Sd
Hans D. Schreier, (D), British Columbia, 1976, Og

## University of Calgary

**Dept of Geoscience** (B,M,D) (2016)
2500 University Drive NW
Calgary, AB T2N 1N4
p. (403) 220-5841
bmayer@ucalgary.ca
http://www.geo.ucalgary.ca/
*Enrollment (2003): B: 268 (73) M: 61 (19) D: 27 (4)*

Head:
Larry R. Lines, (D), British Columbia, 1976, Ye

Professor:
Laurence R. Bentley, (D), Princeton, 1990, Hw
James R. Brown, (D), Uppsala, 1972, Ys
Kenneth Duckworth, (D), Leeds, 1964, Ye
Edward D. Ghent, (D), California, 1964, Gp
Terence M. Gordon, (D), Princeton, 1969, Gq
Masaki Hayashi, (D), Waterloo, 1997, HwSp
Charles M. Henderson, (D), Calgary, 1988, Pm
Ian E. Hutcheon, (D), Carleton, 1977, Cl
Edward S. Krebes, (D), Alberta, 1980, Ys
Donald C. Lawton, (D), Auckland, 1979, Ye
James W. Nicholls, (D), California, 1969, Gi

Gerald D. Osborn, (D), California, 1972, Gm
David R. Pattison, (D), Edinburgh, 1985, Gp
Ronald J. Spencer, (D), Johns Hopkins, 1981, Cg
Deborah A. Spratt, (D), Johns Hopkins, 1980, Gc
Robert R. Stewart, (D), MIT, 1983, Ye
Patrick Wu, (D), Toronto, 1982, Ys
Associate Professor:
Russell L. Hall, (D), McMaster, 1976, Pi
Alan R. Hildebrand, (D), Arizona, 1992, Xm
John C. Hopkins, (D), McGill, 1972, Gs
Gary F. Margrave, (D), Alberta, 1981, Ye
Bernhard Mayer, (D), Ludwig-Maximilians, 1993, Cs
Brian J. Moorman, (D), Carlton, 1997, Gm
Cynthia L. Riediger, (D), Waterloo, 1991, Co
Cathy Ryan, (D), Waterloo, 1994, Hw
Assistant Professor:
Jean-Michel Maillol, (D), Alberta, 1992, Ym
Senior Instructor:
Jon W. Jones, (D), Calgary, 1972, Gp
Honorary Professor:
Brian Norford, (D), Yale, 1959, Gr
Adjunct Professor:
John C. Bancroft, (D), Brigham Young, 1975, Ng
Benoit Beauchamp, (D), Calgary, 1987, Pm
Philip J. Currie, (D), McGill, 1981, Pv
Susan L. Gordon, (D), Waterloo, 1999, Hw
Stephen E. Grasby, (D), Calgary, 1997, Hw
William D. Gunter, (D), Johns Hopkins, 1974, Cl
J. Bruce Jamieson, (D), Calgary, 1996, Ne
Thomas F. Moslow, (D), South Carolina, 1980, Co
Peter E. Putnam, (D), Calgary, 1985, Gs
Gerald M. Ross, (D), Carleton, 1983, Gt
Selim Sayegh, (D), McGill, 1980, Gs
John-Paul Zonneveld, (D), Alberta, 1999, Pm
Emeritus:
Peter Bayliss, (D), New South Wales, 1967, Gz
Finley A. Campbell, (D), Princeton, 1958, Em
Frederick A. Cook, (D), Cornell, 1980, YeGtYs
Peter E. Gretener, (D), ETH (Switzerland), 1953, Ng
Leonard V. Hills, (D), Alberta, 1965, Pl
Federico F. Krause, (D), Calgary, 1979, GsdGo
Alfred A. Levinson, (D), Michigan, 1952, Eg
Alan E. Oldershaw, (D), Liverpool, 1967, Gs
Philip S. Simony, (D), London, 1963, Gc
Norman C. Wardlaw, (D), Glasgow, 1960, Go

## University of Guelph

**School of Environmental Sciences** (B,M,D) (2015)
Guelph, ON N1G 2W1
p. (519) 824-2052
earnaud@uoguelph.ca
http://www.ses.uoguelph.ca
*Enrollment (2005): B: 89 (31) M: 48 (11) D: 20 (3)*
Director:
Jonathan Newman, (D), Albany, 1990, On
Director, Controlled Environment Systems Research Facility:
Michael Dixon, (D), Edinburgh, On
Canada Research Chair in Recombinant Antibody Technology:
Christopher C. Hall, (D), Alberta, On
Associate Dean, Research, OAC:
Beverley A. Hale, (D), Guelph, 1989, Ct
Associate Dean, Academics, OAC:
Jonathan Schmidt, (D), Toronto, On
Professor:
Paul Goodwin, (D), California (Davis), On
Andrew Gordon, (D), Alaska (Fairbanks), 1984, On
Ernesto Guzman, (D), California (Davis), On
Tom Hsiang, (D), Washington, On
Hung Lee, (D), McGill, On
Stephen Marshall, (D), Guelph, On
Gard Otis, (D), Kansas, On
Cynthia Scott-Dupree, (D), Simon Fraser, On
Jack Trevors, (D), Waterloo, On
R. Paul Voroney, (D), Saskatchewan, 1983, Sb
Claudia Wagner Riddle, (D), Guelph, 1992, Ow
Director of the Arboretum:
Shelley Hunt, (D), Guelph, On
Dean, OAC:
Rob J. Gordon, (D), Guelph, Oa
Canada Research Chair in Microbial Ecology:
Kari Dunfield, (D), Saskatchewan, 2002, Sb
Associate Professor:
Madhur Anand, (D), W Ontario, On
Emmanuelle Arnaud, (D), McMaster, 2002, Gls
Susan Glasauer, (D), Tech (Munich), 1995, Py
Marc Habash, (D), Guelph, On
Rebecca Hallett, (D), Simon Fraser, On
Richard Heck, (D), Saskatchewan, Sd
John Lauzon, (D), Guelph, So
Ivan O'Halloran, (D), Saskatchewan, 1986, So
Gary Parkin, (D), Guelph, 1994, Hw
Paul Sibley, (D), Waterloo, On
Jon Warland, (D), Guelph, 1999, Oa
Assistant Professor:
Tim Rennie, On
Laura Van Eerd, (D), Guelph, On
Alan Watson, (M), Guelph, On
Instructor:
Neil Rooney, (D), McGill, On
Emeritus:
George Barron, (D), Iowa State, On
Greg J. Boland, (D), Guelph, On
Ward Chesworth, (D), McMaster, 1967, Cl
Les J. Evans, (D), Wales, 1974, Cl
Austin Fletcher, (D), Alberta, On
Terry J. Gillespie, (D), Guelph, 1968, Ow
Michael J. Goss, (D), Reading, So
Pieter H. Groenevelt, (D), Wageningen (Neth), 1969, Sp
Robert Hall, (D), Melbourne, On
Stewart G. Hilts, (D), Toronto, 1981, On
Peter Kevan, (D), Alberta, On
Kenneth M. King, (D), Wisconsin, 1956, Oa
I. Peter Martini, (D), McMaster, 1966, Gsl
Raymond G. McBride, (D), Guelph, 1982, Os
Murray H. Miller, (D), Purdue, 1957, So
Kaushik Narinder, On
Leonard Ritter, On
Keith Solomon, (D), Illinois, On
Gerry Stephenson, On
George W. Thurtell, (D), Wisconsin, 1965, Oa
H. Peter Van Straaten, (D), Goettingen (Germany), 1974, En

## University of Lethbridge

**Dept of Geography** (B,M) (2004)
4401 University Drive
Lethbridge, AB T1K 3M4
p. (403) 329-2225
geography.chair@uleth.ca
http://home.uleth.ca/geo
Administrative Assistant: Margaret Cook
*Enrollment (2002): B: 94 (24) M: 9 (1)*
Professor:
Walter E. Aufrecht, (D), Toronto, On
Rene W. Barendregt, (D), Queen's (Canada), 1977, Gm
Ian R. MacLachlan, (D), Toronto, 1990, Oe
Robert J. Rogerson, (D), Macquarie, 1979, Gm
Associate Professor:
James M. Byrne, (D), Alberta, 1990, Hg
Hester Jiskoot, (D), Leeds, 2000, OlGmOy
Thomas Johnston, (D), Waterloo, 1988, Oe
Derek R. Peddle, (D), Waterloo, 1997, Or
Assistant Professor:
Craig Coburn, (D), Simon Fraser, 2002, Oy
Susan Dakin, (D), Waterloo, 2000, Oe
Stefan Kienzle, (D), Heidelberg, 1993, Oi
Ivan Townshend, (D), Calgary, 1997, On
Wei Xu, (D), Guelph, 1998, Oi
Adjunct Professor:
John Dormaar, (D), Alberta, 1961, So
Ron Hall, On
Larry Herr, (D), On
Daniel L. Johnson, (D), British Columbia, 1983, On
Pano George Karkanis, (D), Uppsala, 1966, So
Ross McKenzie, (D), On

Anne Smith, (D)
Derald Smith, (D), On
Emeritus:
Roy J. Fletcher, (D), Clark, 1968, Oa



## University of Manitoba

**Geological Sciences** (B,M,D) (2014)
240 Wallace Building
125 Dysart Road
Winnipeg, MB R3T 2N2
p. (204) 474-9371
Mostafa.Fayek@ad.umanitoba.ca
http://www.umanitoba.ca/geoscience
Administrative Assistant: Brenda Miller
*Enrollment (2014): M: 12 (3) D: 8 (2)*

Professor:
Anton Chakhmouridian, (D), St. Petersburg, 1997, Gz
Nancy Chow, (D), Memorial, 1986, Gs
Robert J. Elias, (D), Cincinnati, 1979, Pi
Mostafa Fayek, (D), Saskatchewan, 1996, Cs
Ian J. Ferguson, (D), Australian National, 1988, Ye
Andrew Frederiksen, (D), British Columbia, 2001, YsGt
Norman M. Halden, (D), Glasgow, 1983, Cg
Frank C. Hawthorne, (D), McMaster, 1973, Gz
William M. Last, (D), Manitoba, 1980, GsnGo
Soren Rysgaard, (D), Aarhus (Denmark), 1995, Gl
Elena Sokolova, (D), Moscow, 1980, Gz
Senior Scholar:
George S. Clark, (D), Columbia, 1967, Cc
Barbara L. Sherriff, (D), McMaster, 1988, Gz
Allan C. Turnock, (D), Johns Hopkins, 1960, Gp
Associate Professor:
Alfredo Camacho, (D), Australian National, 1998, Gt
Assistant Professor:
Genevieve Ali, (D), Montreal, 2010, Hw
Zou Zou Kuzyk, (D), Manitoba, 2009, Cg
Research Associate:
Yassir Abdu, (D), Uppsala, 2004, Gz
Instructor:
Karen Ferreira, (M), Manitoba, 1984, Gg
William S. Mandziuk, (M), Manitoba, 1989, Gg
Jeffrey Young, (M), Manitoba, 1992, GcOe
Adjunct Professor:
Scott Anderson, (D), Dalhousie, 1998
Christian Bohm, (D), ETH (Switzerland), 1996, Cg
William Buhay
David Corrigan, (D), Carleton, Gt
Jody Deming, (D), Maryland, 1981, On
Michel Houle, (D), Laurentian, 2008, Gv
Brooke Milne, (D), Ga
Vince Palace, (D), Manitoba, 1996
James Reist, (D), Toronto, 1983
Graham A. Young, (D), New Brunswick, 1988, Pi
Emeritus:
William C. Brisbin, (D), California (Los Angeles), 1970, Gc
Robert B. Ferguson, (D), Toronto, 1948, Gz
Wooil Moon, (D), British Columbia, 1976, Ys
James T. Teller, (D), Cincinnati, 1970, Gs
Related Staff:
Neil Ball, (B), Manitoba, 1982, Gz
Laura Bergen, (M), Manitoba, 2013
Mark Cooper, (M), Manitoba, 1997, Gz
Mulu Serzu, (D), Manitoba, 1990, Yg
Ryan Sharpe, (M), Manitoba, 2012
Ravinder Sidhu
Panseok Yang, (D), Memorial, 2002, CaGp
Misuk Yun, (M), Manitoba, 1986, Cs

## University of New Brunswick

**Dept of Geology** (B,M,D) (2014)
Fredericton, NB E3B 5A3
p. (506) 453-4803
geology@unb.ca
Administrative Secretary: Merrill Ann Beatty

Chair:
Joseph C. White, (D), W Ontario, 1979, Gc
Professor:
Bruce E. Broster, (D), W Ontario, 1982, Gl
John Todd Dunn, (D), Alberta, 1983, Gi
John G. Spray, (D), Cambridge, 1980, Gp
Paul F. Williams, (D), Sydney, 1969, Gc
Associate Professor:
Nicholas J. Susak, (D), Princeton, 1981, Cg
Assistant Professor:
Karl Butler, (D), British Columbia, 1996, Yg
Cliff S.J. Shaw, (D), W Ontario, 1994, Gi
Research Associate:
James Whitehead, (D), New Brunswick, 1998, Gg
Honorary Research Professor:
Henk W. Van De Poll, (D), Swansea, 1970, Gs
Adjunct Professor:
Richard A. F. Grieve, (D), Toronto, 1970, Xg
David R. Lentz, (D), Ottawa, 1992, Cg
Randall F. Miller, (D), Waterloo, 1984, Pg
Emeritus:
Arnold L. McAllister, (D), McGill, 1950, Em
Geology Librarian:
Eszter Schwenke, On

## University of New Brunswick Saint John

**Dept of Biological Sciences** (B,M,D) (2016)
PO Box 5050
100 Tucker Park Road
Saint John, NB E2L 4L5
p. (506) 648-5607
lwilson@unbsj.ca
http://www.unbsj.ca/sase/biology/
*Enrollment (2007): B: 2 (0) M: 1 (0)*

Professor:
Lucy A. Wilson, (D), Paris VI, 1986, Ga
Emeritus:
Alan Logan, (D), Durham (UK), 1962, Ob

## University of Ottawa

**Dept of Earth and Environmental Sciences** (B,M,D) (2017)
15025, 120 University
PO Box 450, Station A
Ottawa, ON K1N 6N5
p. (613) 562-5800 x6870
cpoirie5@uottawa.ca
http://www.science.uottawa.ca/est/eng/welcome.html
*Enrollment (2012): B: 78 (18) M: 29 (11) D: 12 (2)*

Chair:
Andre Desrochers, (D), Memorial, 1986, Gs
Professor:
R. William C. Arnott, (D), Alberta, 1987, GsSo
Ian D. Clark, (D), Paris - Sud, 1988, Hw
Jack Robert Cornett, (D), McGill, 1982, CcaGe
Danielle Fortin, (D), Quebec, 1992, Py
Mark Hannington, (D), Toronto, 1989, Eg
Keiko Hattori, (D), Tokyo, 1977, Cg
Michel R. Robin, (D), Waterloo, 1991, Hw
Associate Professor:
Glenn A. Milne, (D), Toronto, 1998, YgOag
David Schneider, (D), Lehigh, 2000, GtCcGg
Replacement Professor:
Olivier Nadeau, (D), McGill, 2011, Gx
Assistant Professor:
Tom A. Al, (D), Waterloo, 1996, HwCgOg
Pascal Audet, (D), British Columbia, 2008, YgGc
Sarah A.S Dare, (D), Cardiff, 2008, GiCtEg
Jonathan O'Neil, (D), McGill, 2009, Cgc
Lecturer:
Simone Dumas, (D), Ottawa, 2004, GsOg
Adjunct Professor:
Frits P. Agterberg, (D), Utrecht (Neth), 1961, Gq
Eric De Kemp, (D), Québec (Chicoutimi), 2000, OirOg
David Andrew Fisher, (D), Copenhagen, 1978, Ol
William K. Fyson, (D), Reading (UK), 1960, Gc
Quentin Gall, (D), Carleton Univ, 1994, GsEmGe
Richard Goulet, (D), Ottawa, 2001, CgPyGe
Jeffrey Wayne Hedenquist, (D), Auckland, 1983, Eg

Donald D. Hogarth, (D), McGill, 1959, Gz
Dogan Paktunc, (D), Ottawa, 1983, Gz
Pat E. Rasmussen, (D), Waterloo, 1993, GeOaGb
Emeritus:
Jan Veizer, (D), Australian National, 1971, Cl

## University of Regina

**Dept of Geology** (B,M,D) ● (2017)
3737
Wascana Parkway
Regina, SK S4S 0A2
p. (306) 585-4147
geology.office@uregina.ca
http://www.uregina.ca/science/geology/
Administrative Assistant: S. May Ngakham
Professor:
Stephen L. Bend, (D), Newcastle, Gox
Kathryn Bethune, (D), Queens, GcpGt
Guoxiang Chi, (D), Queens, Eg
Ian Coulson, (D), Birmingham (UK), 1996, Gvi
Hairuo Qing, (D), McGill, 1991, GsCg
Associate Professor:
Janis Dale, (D), Queens, Gml
Osman Salad Hersi
Assistant Professor:
Tsilavo Raharimahefa
Maria Velez, (D), Amsterdam, Pg
Adjunct Professor:
Kenneth E. Ashton, (D), Gp
Pier L. Binda, (D), Alberta, 1970, Ge
Donald M. Kent, (D), Alberta, 1968, Gs
Emeritus:
Laurence W. Vigrass, (D), Stanford, 1961, Eo

## University of Saskatchewan

**Dept of Geological Sciences** (B,M,D) ○ (2016)
114 Science Place
Saskatoon, SK S7N 5E2
p. (306) 966-5683
sam.butler@usask.ca
artsandscience.usask.ca/geology/
*Enrollment (2016): B: 189 (56) M: 40 (6) D: 25 (3)*
Head:
Sam Butler, (D), Toronto, 2000, Yg
Professor:
Kevin M. Ansdell, (D), Saskatchewan, 1992, EgCgGt
James F. Basinger, (D), Alberta, 1979, Pb
Luis Buatois, (D), Buenos Aires, 1992, Ps
Graham George, (D), Sussex, 1983, Cg
Jim Hendry, (D), Waterloo, 1984, HwClHy
Chris Holmden, (D), Alberta, 1995, Csc
Gabriela Mangano, (D), Buenos Aires, 1992, PsGs
James B. Merriam, (D), York, 1976, Yg
Igor B. Morozov, (D), Moscow State (Russia), 1985, YsGy
Yuanming Pan, (D), W Ontario, 1990, Gx
William P. Patterson, (D), Michigan, 1995, CsPeGn
Ingrid J. Pickering, (D), Imperial Coll (UK), 1990, CgtOn
Brian R. Pratt, (D), Toronto, 1989, PgGsr
Assistant Professor:
Matthew B. Lindsay, (D), Waterloo, 2009, ClGg
Joyce M. McBeth, (D), Manchester, 2007, PyGeSb
Camille Partin, (D), Gtc
Adjunct Professor:
Irvine Annesley, (D), Ottawa, 1989, Gi
Ning Chen, (D), Saskatchewan, 2001, Gz
David Greenwood, (D), Pg
Tom Kotzer, (D), Saskatchewan, 1993, CgEg
Kyle Larson, (D)
Brett Moldovan, (D), Cg
Len Wassenaar, (D), Waterloo, 1990, Hw
Derek A. Wyman, (D), Saskatchewan, 1990, Cg
Emeritus:
Willi K. Braun, (D), Tubingen, 1958, Ps
William G. E. Caldwell, (D), Glasgow, 1957, Ps
Leslie C. Coleman, (D), Princeton, 1955, Gz
Donald J. Gendzwill, (D), Saskatchewan, 1969, Ye
Zoltan Hajnal, (D), Manitoba, 1970, YsGtc
Robin W. Renaut, (D), London, 1982, Gs
Mel R. Stauffer, (D), Australian National, 1964, Gc

## University of Toronto

**Dept of Physics, Geophysics Div** (B,M,D) (2013)
60 St. George Street
Toronto, ON M5S 1A7
p. (416) 978-5175
cliao@physics.utoronto.ca
http://www.physics.utoronto.ca
*Enrollment (2004): M: 4 (3) D: 10 (0)*
Professor:
Bernd Milkereit, (B), Kiel, 1984, Ye
Emeritus:
Richard C. Bailey, (D), Cambridge, 1970, Yg
Richard N. Edwards, (D), Cambridge, 1970, Yr

**Dept of Earth Sciences** (B,M,D) ● (2016)
Earth Sciences Centre
22 Russell Street
Toronto, ON M5S 3B1
p. (416) 978-3022
welcome@es.utoronto.ca
http://www.es.utoronto.ca
Business Officer: Silvanna Papaleo
Chair:
Alexander Cruden, (D), Uppsala, 1989, Gt
Professor:
Richard C. Bailey, (D), Cambridge, 1970, Yg
Nicholas Eyles, (D), Leicester (UK), 1978, Gl
Grant F. Ferris, (D), Guelph, 1985, Gn
Henry C. Halls, (D), Toronto, 1970, Ym
Martin J. Head, (D), Aberdeen, 1991, Ple
Kenneth W.F. Howard, (D), Birmingham, 1979, Hw
Andrew D. Miall, (D), Ottawa, 1969, Gso
James E. Mungall, (D), McGill, 1993, GiEgCg
Barbara Sherwood Lollar, (D), Waterloo, 1990, Cs
Edward T. C. Spooner, (D), Manchester (UK), 1976, En
Peter H. von Bitter, (D), Kansas, 1971, Pm
Director, Jack Satterly Geochronology Lab:
Michael Andrew Hamilton, (D), Massachusetts, 1993, Cc
Associate Professor:
Donald W. Davis, (D), Alberta, 1978, Cc
Grant S. Henderson, (D), W Ontario, 1983, Gz
Russell N. Pysklywec, (D), Toronto, 1998, Yg
Daniel J. Schulze, (D), Texas, 1982, Gi
Assistant Professor:
Jörg Bollmann, (D), Swiss Fed Inst Tech, 1995, Pm
Rebecca Ghent, (D), S Methodist, 2002, Or
Jochen Halfar, (D), Stanford, 1999, Pe
Gopalan Srinivasan, (D), Maharaja SayajiRao, 1995, Ow
Ulrich B. Wortmann, (D), Tech (Munich), Gg
Research Associate:
Colin Bray, (D), Oxford, 1980, Cs
Lecturer:
Carl-Georg Bank, (D), British Columbia, 2002, Og
Adjunct Professor:
Jean-Bernard Caron, (D), Toronto, 2005, Pi
Marianne Douglas, (D), Queens, 1993, Gn
John H. McAndrews, (D), Minnesota, 1964, Pl
Myrna Joyce S. Simpson, (D), Alberta, 1999, Ge
Emeritus:
G M. Anderson, (D), Toronto, 1961, Cp
J. J. Fawcett, (D), Manchester, 1961, Gp
John Gittins, (D), Cambridge (UK), 1959, GipGz
Alan M. Goodwin, (D), Wisconsin, 1953, Gg
Thomas E. Krogh, (D), MIT, 1964, Cc
Anthony J. Naldrett, (D), Queen's, 1964, EmGiCp
Geoffrey Norris, (D), Cambridge, 1964, Pl
Pierre-Yves F. Robin, (D), MIT, 1974, Gc
John C. Rucklidge, (D), Manchester, 1962, Ge
Walfried M. Schwerdtner, (D), Free Univ Berlin (Germany), 1961, GctGq
Steven D. Scott, (D), Penn State, 1968, GuEmCm
John A. Westgate, (D), Alberta, 1964, Gl
Frederick J. Wicks, (D), Oxford, 1969, Gz

Chief Librarian:
Bruce Garrod, On

## University of Victoria

**School of Earth & Ocean Sciences** (B,M,D) O (2015)
P.O. Box 1700
3800 Finnerty Road
Victoria, BC V8W 2Y2
p. (250) 721-6120
seos@uvic.ca
http://www.uvic.ca/science/seos/

Director:
Stephen T. Johnston, (D), Alberta, 1993, Gc
Associate Director (SCIENCE) NEPTUNE Canada:
Kim Juniper, (D), Canterbury, 1982, Ob
Associate Dean of Science:
Kathryn Gillis, (D), Dalhousie, 1987, Gp
Professor:
Dante Canil, (D), Alberta, 1989, Gi
Laurence Coogan, (D), Leicester, UK, Gi
Stanley E. Dosso, (D), British Columbia, 1990, Op
Adam Monahan, (D), British Columbia, 2000, On
Thomas F. Pedersen, (D), Edinburgh (UK), 1979, CmOcCs
George D. Spence, (D), British Columbia, 1984, Ys
Verena Tunnicliffe, (D), Yale, 1980, Ob
Michael J. Whiticar, (D), Kiel, 1978, CosGo
Associate Professor:
Jay Cullen, (D), Rutgers, 2001, Oc
John Dower, (D), Victoria, 1994, Ob
Roberta C. Hamme, (D), Washington, 2003, Oc
Thomas S. James, (D), Princeton, 1991, YdvYg
Jody Klymak, (D), Washington, 2001, Op
Vera Pospelova, (D), McGill, 2003, PelOb
Eileen Van Der Flier-Keller, (D), W Ontario, 1985, Cl
Diana Varela, (D), British Columbia, 1998, Ob
Assistant Professor:
Colin Goldblatt, (D), East Anglia, 2008, Oa
Lucinda Leonard, (D), Victoria, Yg
Kristin Morell, (D), Penn State, Gc
Adjunct Professor:
Vivek Arora, (D), Melbourne, Oa
Vaughn Barrie, (D), Wales, 1986, Gu
Melvyn E. Best, (D), MIT, 1970, Ye
John Cassidy, (D), UBC, Ys
James R. Christian, (D), Hawaii, 1995, Obc
Kenneth L. Denman, (D), British Columbia, 1972, Op
Richard K. Dewey, (D), British Columbia (Canada), 1987, Opg
Gregory M. Flato, (D), Dartmouth, 1991, Oa
John C. Fyfe, (D), McGill, 1987, Oa
Richard J. Hebda, (D), British Columbia, 1977, Pl
Roy D. Hyndman, (D), Australian National, 1967, Ys
Dave Lefebure, (D), Carleton, 1986, Eg
Victor M. Levson, (D), Alberta, 1995, Gl
David L. Mackas, (D), Dalhousie, 1977, Ob
Norman McFarlane, (D), Michigan, 1974, Oa
Garry C. Rogers, (D), British Columbia, 1983, Ys
George J. Simandl, (D), Ecole Polytechnique, 1992, En
Richard Thomson, (D), British Columbia, 1971, Og
Kelin Wang, (D), W Ontario, 1989, Yg
Michael Wilmut, (D), Queen's, 1971, On
Emeritus:
Christopher R. Barnes, (D), Ottawa, 1964, PmOgPe
Ross N. Chapman, (D), British Columbia, 1975, Op
Christopher J. R. Garrett, (D), Cambridge, 1968, Op
John T. Weaver, (D), Saskatchewan, 1959, Ym
On Leave:
Andrew J. Weaver, (D), British Columbia, 1987, Op
Administrative Officer:
Terry P. Russell, (B), Victoria, 2000, On

## University of Waterloo

**Dept of Earth and Environmental Sciences** (B,M,D) ● (2016)
Waterloo, ON N2L 3G1
p. (519) 888-4567, ext. 32069
klalbrec@uwaterloo.ca
https://uwaterloo.ca/earth-environmental-sciences/

Administrative Assistant: Lorraine Albrecht
Chair:
Barry G. Warner, (D)
Professor:
David W. Blowes, (D), Waterloo, 1990, Cg
Mario Coniglio, (D), Memorial, 1985, Gr
Maurice B. Dusseault, (D), Alberta, 1977, Ng
Thomas W. D. Edwards, (D), Waterloo, 1987, Cs
Stephen George Evans, (D), Alberta, 1983, Ng
Shaun K. Frape, (D), Queen's, 1979, Cl
Shoufa Lin, (D), New Brunswick, 1992, On
Carol J. Ptacek, (D), Waterloo, 1992, Cg
David L. Rudolph, (D), Waterloo, 1989, Hq
Sherry L. Schiff, (D), Columbia, 1986, Ng
Edward A. Sudicky, (D), Waterloo, 1983, Hw
Philippe Van Cappellen, (D)
Associate Professor:
Anthony E. Endres, (D), British Columbia, 1991, Pe
Walter Illman, (D)
Martin Ross, (D)
Andre Unger, (D), Waterloo, 1995, Hw
Assistant Professor:
Nandita Basu, (D)
Carl Guilmette, (D)
Brian Kendall, (D)
Lingling Wu, (D)
Lecturer:
Eric C. Grunsky, (D)
John Johnson, (D)
Isabelle McMartin, (D)
Brent Wolfe, (D)
Adjunct Professor:
Edward C. Appleyard, (D), Cambridge, 1963, Gp
Gail Atkinson, (D)
James F. Barker, (D), Waterloo, 1979, Co
Steven Berg, (D)
Alec Blyth, (D)
Thomas Bullen, (D)
Lauren Charlet, (D)
John A. Cherry, (D), Illinois, 1966, Hw
Peter Condon, (D)
Rick Devlin
Michael English, (D)
Nicholas Eyles, (D)
John F. Gartner, On
John J. Gibson, (D), Waterloo, 1996, Cs
Susan Glasuer, (D)
David Good, (D)
John Gosse, (D)
Douglas Gould, (D)
Norman Halden, (D)
Daniel Hammarlund, (D)
Jens Hartman, (D)
Lin Huang, (D)
Daniel Hunkeler
Hyoun-Tae Hwang, (D)
Richard Jackson, (D)
Sun-Wook Jeen, (D)
Michael Krom, (D)
David R. Lee, (D), Virginia Tech, 1976, Hw
Yuri Leonenko, (D)
John Lin, (D)
Robert Linnen, (D), McGill, 1992, Eg
Benoit Made, (D)
Uli Mayer, (D)
Hossein Memarian, (D)
John Molson
Alan V. Morgan, (D), Birmingham, 1970, Gl
Christopher Neville, (D)
Dogan Paktunc, (D)
Sorab Panday, (D)
Young-Jin Park, (D), Hq
Gary Parkin, (N)
Peter Pehme, (D)
Jennifer Pell, (D)
Richard Peltier, (D)
Terry D. Prowse, (D), Canterbury, 1981, Cg

William Quinton, (D)
Eric J. Reardon, (D), Penn State, 1974, Cl
Rashid Rehan, (D)
Iain Samson, (D)
Houston C. Saunderson, (N)
James Sloan, (D)
James Smith
John Spoelstra, (D)
Andrew Stumpf, (D)
William Taylor, (D)
Rene Therrien, (D)
Harvey Thorleifson, (D)
Martin Thullner, (D)
Benoit Valley, (D)
Garth Van der Kamp, (D), Amsterdam, 1973, Hw
Jan van der Kruk, (D)
Cees van Staal
Andrea Vander Woude, (D)
Owen L. White, (D), Illinois, 1970, Ng
C. Wolkersdorfer, (D)
Wenjiao Xiao, (D)
Leiming Zhang, (D)

Emeritus:
Emil O. Frind, (D), Toronto, 1971, Hq
Robert W. Gillham, (D), Illinois, 1973, Hq
Paul Karrow, (D), Illinois, 1957, Gl

Research Professor:
Ramon Aravena, (D), Waterloo, 1993, Co

Research Associate Professor:
Will Robertson, (D), Waterloo, 1992, Hw

Other:
Brewster Conant, Hw

## University of Windsor

**Dept of Earth and Environmental Sciences** (B,M,D) (2011)
401 Sunset Avenue
Windsor, ON N9B 3P4
p. (519) 253-3000 x2486
earth@uwindsor.ca
http://www.uwindsor.ca/ees
Administrative Assistant: Sharon Horne
*Enrollment (2001): B: 52 (0) M: 8 (0)*

Head:
Iain M. Samson, (D), Strathclyde, 1983, EgCg

Professor:
Ihsan S. Al-Aasm, (D), Ottawa, 1985, GdClGo
Aaron Fisk, (D), Manitoba, 1998, ObCs
V. Chris Lakhan, (D), Toronto, 1982, OriOn
Ali Polat, (D), Saskatchewan, 1998, GixCc
Frank Simpson, (D), Jagellonian (Krakow), 1968, Gs
Alan S. Trenhaile, (D), Wales, 1969, Gm

Associate Professor:
Maria T. Cioppa, (D), Lehigh, 1997, Ym
Joel E. Gagnon, (D), McGill, 2006, CalEg
Phil A. Graniero, (D), Toronto, 2001, OinOy
Cyril G. Rodrigues, (D), Carleton, 1980, Pm
Christopher Weisener, (D), South Australia, 2003, CoGePy
Jianwen Yang, (D), Toronto, 1997, HwYg

Instructor:
Denis Tetrault, (D), Western, 2002, Gg

Emeritus:
Brian J. Fryer, (D), MIT, 1971, Cla
Peter P. Hudec, (D), Rensselaer, 1965, NgEmGe
Terence E. Smith, (D), Wales, 1963, Gi
David T. A. Symons, (D), Toronto, 1965, YmGtEm
Andrew Turek, (D), Australian National, 1966, Cc

## Western University

**Dept of Earth Sciences** (B,M,D) O (2016)
Biology & Geology Building
Room 1026
1151 Richmond Street North
London, ON N6A 5B7
p. (519) 661-3187
earth-sc@uwo.ca
http://www.uwo.ca/earth/
f: Earth Sciences at Western University
t: @westernuEarth
Administrative Assistant: Marie Schell
*Enrollment (2016): B: 123 (15) M: 77 (16) D: 54 (8)*

Robert Hodder Chair:
Robert Linnen, (D), McGill, 1992, EmCe

NSERC Industrial Research Chair:
Gail M. Atkinson, (D), W Ontario, 1993, YssYs

Cross-Appt.-Home Dept: Physics & Astronomy, UWO:
Peter Brown, (D), W Ontario, 1999, Xym

Chair, Dept. Earth Sciences (P.Eng.):
R. Gerhard Pratt, (D), Imperial Coll (UK), 1989, Yeg

Canada Research Chair:
Frederick J. Longstaffe, (D), McMaster, 1978, Cs

(P.Eng.,P.Geo.):
Robert A. Schincariol, (D), Ohio State, 1993, HwsNg

Professor:
Stephen R. Hicock, (D), W Ontario, 1980, Gl
Jisuo Jin, (D), Saskatchewan, 1988, Pi
A Guy Plint, (D), Oxford, 1980, Gs
Richard A. Secco, (D), W Ontario, 1988, Gy
Kristy F. Tiampo, (D), Colorado, 2000, Yd

Industrial Research Chair, Joint Appt Physics & Astronomy, U.W.O:
Gordon R. Osinski, (D), New Brunswick, 2004, XgcGp

Industrial Research Chair:
Neil Banerjee, (D), Victoria, 2001, EgCgGx

Cross-Appt.UWO: Physics&Astronomy:
Robert Shcherbakov, (D), Cornell, 2002, YgsYd
Sean R. Shieh, (D), Hawaii, 1998, YxGzOm

Canada Research Chair, Cross Appt. Biology&Geography, UWO:
Brian Branfireun, (D), McGill U, Ge

Associate Professor:
Burns A. Cheadle, (D), W Ontario, 1986, Go
Patricia Corcoran, (D), Dalhousie, 2001, Gd
Roberta L. Flemming, (D), Queen's (Canada), 1997, GzXmOm
Dazhi Jiang, (D), New Brunswick, 1996, GcgGg
Elizabeth A. Webb, (D), W Ontario, 2000, CslSa

Research Adjunct Professor:
Natalie Pietrzak-Renaud, (D), W Ontario, 2011, Eg

Joint appointment with Geography, U.W.O.:
Desmond Moser, (D), Queens, 1993, GtCcXg

Canada Research Chair:
Audrey Bouvier, (D), École Normale Supérieure de Lyon, 2005, Xc

Assistant Professor:
Rob Carpenter, (D), W Ontario, 2004, Em
Phil JA McCausland, (D), Western, 2002, YmXm
Sheri Molnar, (D), U Victoria, 2011, Ys
Catherine Neish, (D), U Arizona, 2008, Xg
Cameron J. Tsujita, (D), McMaster, 1995, Pi
Tony Withers, (D), Bristol, 1997, Cp

Research Scientist:
Li Huang, (M), Concordia, 1992, Yr

Research Engineer:
Matthew Bourassa, (D), Xg

Research Associate:
Stephane Perrouty, (D), Toulouse, 2012, EgGcYe
Po-Yu (Paul) Shen, (D), W Ontario, 1975, Yh

W.S. Fyfe Visiting Scientist- in- Residence:
David Good, (D), McMaster, 1992, Em

Research Associate:
Brian R. Hart, (D), W Ontario, 1995, Ca

Brandon University:
Rong-Yu Li, (D), U Alberta, 2002, Pie

Associate Curator of Mineralogy, Royal Ontario Museum/Professor, University of Toronto:
Kim Tait, (D), Arizona, 2007, Xm

Adjunct Professor:
Ed Cloutis, (D), U Alberta, 1989, Oi
Claudia Cochrane, (M), W Ontario, Go
Yunpeng Dong, (D), Northwest Univ (China), 1997, GtcCg
Richard Grieve, (D), U Toronto, 1970, Xg
Matt Izawa, (D), W Ontario, 2012, Gz
Gero Michel, (D), Harvard Business School, Ys
Sobhi Nasir, (D), Wuerzburg, 1989
Brenden Smithyman, (D), Ye
Gordon Southam, (D), Guelph, 1990, Py
Yves Thibault, (D), W Ontario, 1990, Gz

Livio Tornabene, (D), Tennessee, 2007, XgOr
Lisa Van Loon, (D), Ohio State, 2007, Ca
Emeritus:
Alan Beck, (D), Australian National, 1964, On
W. Glen E. Caldwell, (D), Glasgow, 1957, Pm
William R. Church, (D), Wales, 1961, Gt
Norman A. Duke, (D), Manitoba, 1983, Eg
Michael E. Fleet, (D), Manchester, 1963, Gz
Akio Hayatsu, (D), U Toronto, 1965, Cc
Robert W. Hodder, (D), California (Berkeley), 1959, Eg
Alfred C. Lenz, (D), Princeton, 1959, PiePi
Robert F. Mereu, (D), W Ontario, 1962, Ys
H Wayne Nesbitt, (D), Johns Hopkins, 1975, Cl
H. Currie Palmer, (D), Princeton, 1963, YmGv
Grant M. Young, (D), Glasgow (UK), 1967, GrnGt
Research Scientist:
Charles T. Wu, (D), W Ontario, 1984, Ca
Laboratory Director:
Kim Law, Cs
Wenjun Yong, (D), Yx
Research Technologist:
Jon Jacobs, Gg
Research Technician:
Shamus Duff, (M), Ottawa, 2015, OiXg
Lab Technician:
Marc Beauchamp, Ca
Blair Gibson, Ca
Lab Assistant:
Ivan Barker, Cc
Computer Technician:
Bernie Dunn, Ye

## York University

**Earth and Space Science and Engineering** (B,M,D) ● (2015)
4700 Keele Street
Toronto, ON M3J 1P3
p. (416) 736-5245
esse@yorku.ca
http://www.yorku.ca/esse
Administrative Assistant: Paola Panaro
*Enrollment (2010): B: 100 (38)*
Professor:
Qiuming Cheng, (D), Ottawa, 1994, Oi
Christian Haas, (D), Bremen, 1996, Yg
Gary T. Jarvis, (D), Cambridge, 1978, Yg
Ian C. McDade, (D), Belfast, 1979, Or
Tom McElroy, (D), York, 1985, Oa
Spiros Pagiatakis, (D), New Brunswick, 1988, Yd
Peter A. Taylor, (D), Bristol, 1967, Oa
Chair:
Regina Lee, (D), Toronto, 2000
Associate Professor:
Costas Armenakis, (D), New Brunswick, 1988, Or
Sunil Bisnath, (D), New Brunswick, 2004
Yongsheng Chen, Oa
Michael Daly, (D)
Baoxin Hu, (D), Boston, 1998, Or
Mary Ann Jenkins, (D), Toronto, 1986, Oa
Gary P. Klaassen, (D), Toronto, 1983, Oa
Brendan Quine, (D)
Jinjun Shan, (D)
Gunho Sohn, (D)
George Vukovich, (D), Oa
James Whiteway, (D), York, 1994, Oa
Zheng Hong (George) Zhu, (D)
Assistant Professor:
William Colgan, (D), Yg
Mark Gordon, (D), Oa
John E. Moores, (D), Arizona, 2008
Lecturer:
Hugh Chesser, (M), Toronto, 1987
Franz Newjland, (D)
Jian-Guo Wang, (D)
Emeritus:
Keith D. Aldridge, (D), MIT, 1967, YmdYh
John Miller, (D), Saskatchewan, 1969, Or
Gordon G. Shepherd, (D), Toronto, 1956, OarOn

Douglas E. Smylie, (D), Toronto, 1963, Yg
Anthony M. K. Szeto, (D), Australian National, 1982, Yg

# Cape Verde

## Universidade de Cabo Verde

**Dept de Ciencia e Tecnologia** (M,D) (2015)
Campus de Palmarejo - CP 279
Palmarejo, Praia - Cabo Verde
Praia
Praca Antonio Lereno, Praia CP 379C
p. +238 261 99 01
joao.semedo@docente.unicv.edu.cv
http://www.unicv.edu.cv/dct

**Departamento de Engenharias e Ciencias do Mar** (B,M) (2013)
Ribeira de Julião
CP 163
Mindelo, São Vicente CP 163
p. +238 2326561/62
alexandra.delgado@docente.unicv.edu.cv
http://www.unicv.edu.cv

## Universidade Jean Piaget de Cabo Verde

**Ecologia e Desenvolvimento** (B) (2014)
Campus Universitario da Cidade da Praia
Praia
p. +238 629085
info@unipiaget.cv
http://www.unipiaget.cv/pdf/cursos/ecdm.pdf

# Central African Republic

## Universite de Bangui

**Dept de Chimie-Biologie-Geologie** (B) (2014)
BP 1450
Avenue des Martyrs
Bangui
p. (236) 61 20 05
info@univ-bangui.info
http://www.univ-bangui.org/

# Chad

## Universite de N'Djamena

**Dept de Geologie** (B) (2010)
BP 1117
Avenue Mobutu
N'Djamena
sg@undt.info
http://www.undt.info/

# Chile

## Servicio National de Geologia y Mineria de Chile

**Servicio National de Geologia y Mineria de Chile** (2013)
Avda Sta Maria Nº 0104
Providence
p. 56-2-7375050
oirs@sernageomin.cl
http://www.sernageomin.cl/

## Universidad Austral de Chile

**Inst de Geociencias** (2004)
Campus "Isla Teja", Casilla 567, Valdivia
p. +56 63 2293861
cienciasdelatierra@uach.cl

## Universidad Catolica del Norte

**Dept de Ciencias Geologicas** (2014)
Avenida Angamos 610, ,
Casilla 1280, Antofagasta
p. (55) 2355968

mbembow@ucn.cl
http://www.ucn.cl/facultades/SitioDeInteres/?cod=2&codItem=110&codPrincipal=1124

## Universidad de Chile

**Dept de Geologia** (2004)
Plaza Ercilla 803, Casilla 13148, Correo 21, Santiago
colegios@ing.uchile.cl

**Dept de Geofisica** (2005)
Blanco Encalada 2002
Casilla 2777
Santiago
p. (56 2)696 6563
rmunoz@dgf.uchile.cl
http://www.dgf.uchile.cl/index.html

## Universidad de Concepcion

**Inst de Geologia Economica Aplicada** (2014)
Barrio Universitario - Victor Lamas 1290
Casilla 4107
Concepcion, Region del Bio-Bio
p. (56) 41 220 44 88
maravenah@udec.cl
http://www.udec.cl/postgrado/?q=node/39&codigo=4161

**Dept de Geociencia** (2004)
Cabina 13 - Barrio Universitario, Casilla 3-C, 4250-1 Concepcion, Region del Bio-Bio

# China

## Capital Normal University

**College of Resources, Environment and Tourism** (2011)
West 3rd Ring Road North 105#
Beijing 100048
p. 86 10 68903321
info@mail.cnu.edu.cn
http://www.cnu.edu.cn

## China Geological Survey

**China Geological Survey** (2014)
24 Huangsi Dajie,
Xicheng District
Beijing 100011
p. +86 10 51632963 51632906
enwebmaster@mail.cgs.gov.cn
http://www.cgs.gov.cn/

## China University of Geosciences, Beijing

**Geosciences** (2011)
No.29, Xueyuan Road,
Haidian District
Beijing 100083
http://www.cugb.edu.cn/EnglishWeb/index.html

## China University of Geosciences, Wuhan

**Faculty of Geosciences** (2011)
No. 388, Lumo Road, Hongshan District
Wuhan 430074
p. 86-27-87481030
cugxb@cug.edu.cn
http://en.cug.edu.cn/cug/index.asp

## China University of Mining and Technology, Beijing

**College of Geoscience and Surveying Engineering** (2011)
Beijing
dcxy@cumtb.edu.cn
http://dcxy.cumtb.edu.cn/

## Nanjing University

**Dept of Earth Sciences** (2011)
22 Hankou Road
Nanjing , Jiangsu
wsj@nju.edu.cn
http://www.nju.edu.cn/cps/site/njueweb/fg/index.php

## University of Hong Kong

**Dept of Earth Sciences** (B,M,D) (2016)
James Lee Building
Pokfulam Road
Hong Kong
p. (852) 2859 1084
earthsci@hku.hk
http://www.earthsciences.hku.hk

# Colombia

## EAFIT University

**Dept of Geology** (2014)
Carrera 49 No. 7 Sur - 50
Medellin
p. 01 8000515 900
contacto@eafit.edu.co

## Escuela de Ingenieria de Antioquia

**Dept of Geologic Engineering** (2004)
Calle 25 Sur
#42-63 Envigado
Medellin
http://www.eia.edu.co/site/index.php/pregrados/programas/ing-geologica.html

## Universidad de Santander

**Environmental Engineering** (B,M) (2014)
Facultad Ingenierias
Cll 70 No. 55-210 Campus Universitario Lagos del Cacique
Bucaramanga, Santander
p. +57 7 6516500
nmantilla@udes.edu.co
http://www.udes.edu.co/programas-profesionales/facultad-ingenierias/ingenieria-ambiental.html

## Universidad Industrial de Santander

**School of Geology** (B,M) ○ (2016)
Cra 27 Calle 9 Ciudad Universitaria
Bucaramanga, Santander 1
p. 57-7-6343457
escgeo@uis.edu.co
http://geologia.uis.edu.co/eisi/

Director:
Juan Diego Colegial Gutierrez, (D), Universidad Politecnica de Madrid (Spain), 2004, GeNgOr

Professor:
Luis Carlos Mantilla Figueroa, (D), EgCec

## Universidad Nacional de Colombia

**Dept de Geociencias** (2004)
Carrera 30 No. 45-03, Edificio 224
Calle 45, Cr. 30
Bogota
jjsancheza@unal.edu.co

# Costa Rica

## Universidad de Costa Rica

**Escuela Centroamericana de Geologia** (B,M) ● (2017)
Apartado 35-2060 UCR, San Pedro de Montes de Oca, San Jose
San José
p. 506 25118128
geologia@ucr.ac.cr
f: https://www.facebook.com/ECG.UCR/

Director:
Rolando Mora Chinchilla, (M)

Professor:
Mario Enrique Arias Salguero, (M)
Allan Astorga Gattgens, (D)
Elena Badilla Coto, (M)

Marco Barahona Palomo, (D)
Lolita Campos Bejarano, (D)
Percy Denyer Chavarría, (D)
Lepolt Linkimer Abarca, (D)
Oscar Lucke Castro, (D)
Raúl Mora Amador, (M)
Mauricio Manuel Mora Fernández, (D), Université de Savoie, France, 2003
Stephanie Murillo Maikut, (D)
Luis Guillermo Obando Acuña, (M)
Giovanni Marino Peraldo Huertas, (M)

Associate Professor:
Guaria Cárdenes Sandí, (D)
Maximiliano Garnier Villarreal, (D)
María Isabel Sandoval Gutiérrez, (D)

Instructor:
Patrick Durán Leiva, (B)
María del Pilar Madrigal Quesada, (D)
Vanessa Rojas Herrera, (A)
Luis Guillermo Salazar Mondragón, (M)
Ingrid Vargas Azofeifa, (M)



# Croatia

## Croatian Institute of Geology

**Croatian Geological Survey** (2011)
Sachsova 2
P.O.Box 268
Zagreb HR-10000
p. +38516160888
ured@hgi-cgs.hr
http://www.hgi-cgs.hr/

## University of Zagreb

**Dept of Geology, Faculty of Science** (B,M,D) (2010)
Horvatovac 102a
Zagreb HR-10000
p. +38514605960
godsjek@geol.pmf.hr
http://www.geol.pmf.hr
*Enrollment (2010): B: 121 (27) M: 28 (11) D: 30 (6)*

full professor:
Mladen Juraèiæ, (D), Zagreb, 1987, GueCm
Darko Tibljaš, (D), Zagreb, 1996, GzScGx
Vlasta Æosoviæ, (D), Zagreb, 1996, PemPs

associate professor:
Dražen Balen, (D), Zagreb, 1999, GxpGi

emeritus:
Ivan Gušiæ, (D), Zagreb, 1974, GgPsm

# Cyprus

## Ministry of Agriculture, Rural Development and Environment

**Cyprus Geological Survey Department** (2016)
1 Lefkonos Street
Lefkosia 2064
p. +357 22409213
director@gsd.moa.gov.cy
http://www.moa.gov.cy/gsd
t: https://twitter.com/CY_earthquakes

# Czech Republic

## Charles University

**Inst of Hydrogeology, Engineering Geology and Applied Geophysics** (2005)
Albertov 6
128 43 Praha 2
fischer@natur.cuni.cz
http://prfdec.natur.cuni.cz/~geophys/ustav/katedra.htm

**Institute of Petrology and Structural Geology** (B,M,D) (2014)
Albertov 6
Praha 2 128 43
p. 00420-221951524
petrol@natur.cuni.cz
http://petrol.natur.cuni.cz
*Enrollment (2012): B: 3 (5) M: 15 (4) D: 15 (0)*

Assoc. Prof.:
David Dolejs, (D), 2004, GiCpg

Professor:
Shah Wali Faryad, (D), 1990, GpzGi

**Dept of Geophysics** (B,M,D) (2013)
KG MFF UK
V Holesovickach 2
180 00 Prague 8, Czech Republic
Prague, Czech Republic 180 00
geo@mff.cuni.cz
http://geo.mff.cuni.cz

## Czech Geological Survey

**Cesky Geologicky Ustav** (2013)
Klarov 3
118 21 Praha 1
p. +420 257 089 500
secretar@geology.cz
http://www.geology.cz

**Geofond** (2011)
Kostelni 26
Prague PSC 170 06
p. +420 234 742 111
vstrupl@geofond.cz
http://www.geofond.cz/cz/domu

## Masaryk University

**Dept of Geological Sciences** (B,M,D) O (2017)
Faculty of Science
Kotlarska 2
Brno 611 37
p. +420 549 49 4322
geologie@sci.muni.cz
http://ugv.cz/
f: https://www.facebook.com/chci.byt.geolog/?fref=ts
*Enrollment (2016): B: 128 (43) M: 88 (39) D: 76 (3)*

Head:
Milan Novak, (A), GzpGi

Professor:
Jiri Kalvoda, (A), PsePm
Antonín Prichystal, (A), GgaGv

Associate Professor:
Ondrej Babek, (D), Gds
Martin Ivanov, (A), PgiPe
Jaromir Leichmann, (A), GxpGi
Zdenek Losos, (D), Gzy
Rostislav Melichar, (A), GtcGr
Slavomír Nehyba, (A), GsdGg
Marek Slobodnik, (A), GeEg
Josef Zeman, (D), Comenius Univ (Slovakia), 1987, CgpCe

Assistant Professor:
Martin Knizek, (D), NgmNr
Tomas Kuchovsky, (D), Hgy
Adam Ricka, (D), Hgy

Research Associate:
Renata Copjakova, (D), Gz
Radek Skoda, (D), Masaryk Univ, GzOmGi

Lecturer:
Nela Dolakova, (A), Plg

Emeritus:
Rostislav Brzobohaty, (A), PgsPe
Rudolf Musil, (D), Charles Univ (Prague), 1968, PgvGe

**Dept of Mineralogy, Petrology and Geochemistry** (2004)
Kotlarska 2
611 37 Brno
sona@sci.muni.cz
http://www.muni.cz/sci/structure/315020.html

**Dept of Geography** (2004)

Kotlarska 2
611 37 Brno
dobro@sci.muni.cz
http://www.muni.cz/sci/structure/315030.html

## Technical University of Ostrava
**Faculty of Mining & Geology** (2004)
Vysoka Skola Banska v Ostrava
17.listopadu 15/2172
708-33 Ostrava
p. +420 597 325 456
dekan.hgf@vsb.cz
http://www.hgf.vsb.cz/cs

# D.R. of Congo

## Marien Ngouabi University
**Faculty of Science and Technology** (B,M,D) O (2014)
Face Square Général Charles De Gaulle
Bacongo
BP:69
Brazzaville, Congo
p. (+242) 06 623 61 22
jm_ouamba@yahoo.fr
http://www.univ-mngb.net/fs/

## Universite de Kinshasa
**Dept des Sciences de la Terre** (B) (2014)
PO Box 190
Kinshasa XI
p. (+243) 82 333 96 93
rectorat@unikin.cd
www.unikin.cd

**Départements et filières Sciences Agronomiques** (B) (2010)
PO Box 190
Kinshas XI
p. +243 89 89 20 507
fbkapuku@hotmail.com
http://www.unikin.cd/ogec/

## Universite de Lubumbashi
**Dept de Geologie** (B) (2011)
B.P. 1825
Lubumbashi, Katanga
http://www.unilu.ac.cd/En/Pages/default.aspx

## Universite de Lumumbashi
**Faculté des sciences** (B) (2014)
BP 1825
Lubumbashi
p. +243 263-22-5403
unilu@unilu.net
http://www.unilu.ac.cd/En/Pages/default.aspx

# Denmark

## Aarhus University
**Department of Geoscience** (B,M,D) (2016)
Hoegh-Guldbergs Gade 2
DK-8000 Arhus C
geologi@au.dk
http://geo.au.dk

## Geological Survey of Denmark and Greenland (GEUS)
**Geological Survey of Denmark and Greenland (GEUS)** (2016)
Oester Voldgade 10
Copenhagen DK-1350
p. +45 38 20 00
geus@geus.dk
http://www.geus.dk/

## Roskilde University
**QuadLab, ENSPAC** (2014)
Universitetsvej 1
Roskilde DK-4000
p. 45 46743097
storey@ruc.dk
http://www.quadlab.dk/bcms-ui-base/

## Technical University of Denmark
**Dept of Earth Sciences** (2004)
Anker Engelunsved
DK-2800 Lyngby
thho@env.dtu.dk
http://www.igg.dtu.dk/index_z.htm

## University of Copenhagen
**Inst for Geography and Geology** (2011)
Oster Voldgage 5-7
DK-1350 Copenhagen K
p. +4535322400
pab@ign.ku.dk
http://geo.ku.dk/

**Dept of Geophysics** (2015)
Juliane Maries Vej 30
Copenhagen 2100 Copenhagen
p. +45 353 20605
losos@sci.muni.cz
http://www.nbi.ku.dk/theinstitute/page52794.htm

**Geological Museum** (2011)
Oster Voldgade 5-7
Copenhagen K DK-1350
p. +45 35322345
rcp@snm.ku.dk
http://geologi.snm.ku.dk/english/

# Djibouti

## Insitut de Physique du Globe de Paris, (IPGP)
**Dept of Volcanic Systems** (B) (2010)
Observatoire Géophysique d'Arta, BP 1888
p. (253) 42 21 92
komorow@ipgp.fr
http://volcano.ipgp.jussieu.fr/djibouti/stationdj.html

## Universite de Djibouti
**Dept de Geologie/Biologie** (B) (2010)
BP 1904
p. +253-250459
webmaster@univ.edu.dj
http://www.univ.edu.dj/facultes.fsti.bg.html

# Dominican Republic

## Mineterio de Energie y Minas Republica Dominica
**Servicio Geologico Nacional Republica Dominica** (2016)
Ave Winston Churchill No 75
Edificio J.F. Martinez 3er Piso
Santo Domingo
p. (809) 732 0363
smunoz@sgn.gov.do
http://www.sgn.gov.do

# Ecuador

## Escuela Politecnica Nacional
**Facultad de Geologia, Minas y Petroleos** (2014)
Ladron de Guevara E11
253 Quito
p. (593-02) 2-507 - 126
deparamento.geologia@epn.edu.ec
http://www.epn.edu.ec/index.php?option=com_content
&view=article&id=1202%3Adepartamento-de-geologia-

dg&catid=163&Itemid=342

## Escuela Superior Politecnica Del Litoral
**Facultad de Ciencias de la Tierra** (2014)
Km 30.5, Via Perimentral
Guayaquil EC090150

## Universidad de Guayaquil
**Facultad de Ciencias Naturales** (2004)
Av Raul Gomez Lynx s / n Av Juan Tanca Marengo
Guayaquil
p. 3080777 to 3080758
carmenbonifaz@hotmail.com
http://fccnnugye.com/

# Egypt

## Ain Shams University
**Dept of Geology** (B) (2010)
Abbassia 11566
Cairo 11566
p. +20(2)6830963
deans@asunet.shams.edu.eg
http://sci.shams.edu.eg/Departments_Geology_Index.ASP

**Dept of Geophysics** (B) (2010)
Abbassia 11566
Cairo 11566
p. +20(2)6830963
deans@asunet.shams.edu.eg
http://sci.shams.edu.eg/Departments_Geophysics_Index.ASP

## Al Azhar University
**Dept of Geology** (B) (2010)
Yosief Abbas Street
Cairo 11787
p. +20 (2) 262 3274
k.ubeid@alazhar.edu.ps
http://www.uazhar.edu.eg/bfac/sci/jiolojy.htm

## Alexandria University
**Dept of Environmental Sciences**
(B) (2010)
22 Al-Guish Avenue
Alexandria
p. +20 (3) 591 1152
anwar_elfiky@hotmail.com
http://www.alex.edu.eg/dept.jsp?FC=4&CODE=09

**Dept of Geology** (A,B,M,D) O (2015)
Baghdad Street
Qism Moharram Bek
Alexandria 21511
p. (+203) 3921595
sc-dean@alexu.edu.eg
http://www.sci.alexu.edu.eg/en/Departments/Default.aspx?Dept=4
f: https://www.facebook.com/groups/GSAU.alex/
*Enrollment (2014): B: 26 (0)*
Professor:
Galal Mohamed I Galal, (D), Germany, 1994, GgPe
Khalil I. Khalil Ebeid, (D), Germany, 1995, GgEg
Kadry Nasser Sediek, (D), USSR, 1991, GgsGd
Mohamad Nasser Shaaban, (D), U.S.A., 1992, GgsGd
Assistant Lecturer:
Ahmed Ibrahim Dyab, (M), Egypt, 2013, Gg
Assistant Professor:
Hossam EL-Din Ahmed EL-S Helba, (D), Germany, 1994, GgOm
Ahmed Sadek Mansour, (D), Egypt, 1999, Ggs
Instructor:
Sara Akram M. Mahmoud, (B), Egypt, 2006, Gg
Emeritus Professor:
Mohamed Waguih El Dakkak, (D), Egypt, 1971, GgsGd
Galal Abd EL-Ham Ewas, (D), Norway, 1969, GgNr
Hanafy M. Holail, (D), 1988, Ggs
Rousine Tanios Toni, (D), Egypt, 1967, GgPg
Emeritus Lecturer:
Mohamed M. Tamish, (D), West Germany, 1988, GgsCg
Emeritus Professor:
Mohamed Ahmed Rashed, (D), Moscow - USSR, 1984, GgOs

**Oceanography Dept** (B) (2010)
22 Al-Guish Avenue
Alexandria
p. +20 (3) 591 1152
h_mitwally@sci.alex.edu.eg
http://www.alex.edu.eg/dept.jsp?FC=4&CODE=07

## American University of Cairo
**Dept of Petroleum and Energy Engineering** (B) (2010)
New Cairo Campus: AUC Avenue
PO Box 74
New Cairo 11835
p. +20.2.2615.1000
amrserag@aucegypt.edu
http://www.aucegypt.edu/academics/dept/peng/Pages/default.aspx

## Assiut University
**Dept of Geology** (B) (2010)
Assiut Governorate
Assiut City
PO Box 71515
Assiut 71515
p. +20 (88) 235 7007
sci@aun.edu.eg
http://www.aun.edu.eg/fac_sci/depart/ge1.htm

**Mining & Metallurgical Engineering Dept** (B) (2010)
Assiut Governorate
Assiut City
PO Box 71515
Assiut 71515
p. +20 (88) 235 7007
mamoah@aun.edu.eg
http://www.aun.edu.eg/fac_eng?Dpart/Mining/Mining.htm

## Beni Suef University
**Geology Dept** (B) (2015)
Salah Salem Street
Beni Suef 62511
p. +(20) 822324879, +(20) 2082232
elsherif_zakaria@yahoo.com
http://193.227.1.224/sci/

## British University in Egypt
**Petroleum Engineering** (B) (2010)
Misr Ismalia Road
El Sherouk City 11837
enas.sabry@bue.edu.eg
http://www.bue.edu.eg/

## Cairo University
**Dept of Geophysics** (B) (2010)
University Avenue - Univeristy Square
Giza
p. +20 (2) 572 9584
http://www.freewebtown.com/geophysics2/

**Mining, Petroleum, and Metallurgical Engineering** (B) (2014)
Univerisity Avenue - University Square
Giza
p. +20 (2) 572 9584
nahed_ecae2003@yahoo.com
http://www.eng.cu.edu.eg/dept/en/mpm/index.htm

**Dept of Geology** (B) (2010)
University Avenue - University Square
Giza, Cairo
p. +20 (2) 572 9584

portal@cu.edu.eg
http://www.cu.edu.eg/english/

## El Mansoura University
**Dept of Geology** (B,M,D) (2014)
Faculty of Science
at Mansoura
Department of Geology
Mansoura, Dakahliya governorat 35516
p. +20 050 2242388 Ext. 582
ghazala@mans.edu.eg
http://www.mans.edu.eg/facscim/english/GeologyHome/default.htm
*Enrollment (2012): B: 250 (230) M: 25 (13) D: 15 (7)*
Head:
Hosni H. Ghazala, (D), Mansoura, 1990, Ygx

## El Zagazig University
**Dept of Geology** (B) (2010)
Zagazig
p. +20 (55) 324 577
info@zu.edu.eg
http://www.zu.edu.eg/

## Fayoum University
**Geology Dept** (B,M,D) (2016)
fayoum-Egypt University Zone
ags00@fayoum.edu.eg
Fayoum 63514
p. +20 01005856505
ags00@fayoum.edu.eg
http://www.fayoum.edu.eg/English/Science/Geology/About-Board.aspx
*Enrollment (2016): B: 20 (18) M: 4 (0) D: 2 (0)*
Professor:
ahmad gaber shedied, (D), Cairo, 1995, GgHww

## Minia University
**Geology & Chemistry Dept** (B) (2010)
Minia
p. +(208-6) 324 420 #321 443
minia@frcu.eun.rg
http://www.minia.edu.eg

## Sohag University
**Geology Dept** (B) (2010)
EL Kawaser
PO Box 82524
Sohag 82524
p. +(20) 93 4605745
a_abudeif@yahoo.com
http://www.sohag-univ.edu.eg/

## South Valley University
**Dept of Geology (Aswan Campus)** (B) (2010)
Qena 83523
p. +20 (96) 339 756
mohammedsm2003@yahoo.com
http://www.svu.edu.eg/arabic/aswan/sci/en/depart/Geology/Geology.htm

## Tanta University
**Dept of Geology** (B) (2014)
El-Geish Street
Tanta 31527
p. +20 (40) 337 7929
president@tainta.edu.eg
http://www.tanta.edu.eg/ar/Tanta/3elom/geology.html

## The Egyptian Geological Survey and Mining Authority
(2011)
egov@ad.gov.eg
http://www.egsma.gov.eg/

## Zagazig University
**Dept of Geology** (B,M,D) (2011)
Sharkia Governorate
Zagazig City
geology@zu.edu.eg
http://www.zu.edu.eg/

# Eritrea

## University of Asmara
**Dept of Marine Sciences** (B) (2011)
PO Box 1220
Asmara
p. 291 1 161926 (Ext. 259)
Zekeria@marine.uoa.edu.er
http://www.uoa.edu.er/academics/dmarine/index.html

# Estonia

## Geological Survey of Estonia
**Eesti Geoloogiakeskus OU** (2011)
Juniper tee 82
Tallinn 12618
p. 672 0094
egk@egk.ee
http://www.egk.ee/

## University of Tartu
**Inst of Geology** (B,M,D) ● (2016)
Ravila 14a
Tartu 50411
p. +372 7 375 891
geol@ut.ee
http://www.geoloogia.ut.ee/et
*Enrollment (2016): B: 3 (0) M: 10 (0)*

# Ethiopia

## Addis Ababa University
**Dept of Planetary and Earth Sciences** (2011)
PO Box 1176
Addis Ababa
p. 251-111239462
balem@geo.aau.edu.at
http://www.aau.edu.et/index.php/earth-sciences

**School of Civil & Environmental Engineering** (B) (2010)
PO Box 1176
Addis Ababa
youngkyunkim@aait.edu.et
http://www.aau.edu.et/index.php/earth-sciences

## Arba Minch University
**Dept of Geology** (B) (2011)
PO Box 21
, Arba Minch
p. 251-468814972
yoditayalew@fastmail.fm
http://www.arbaminch-univ.com/

**College of Natural Sciences** (B) (2010)
PO Box 21
Arba Minch
p. +251-46-8810070
alemayehu.hailemicael@amu.edu.et
http://www.arbamich-univ.com/WTIMeteorlogy.html#

**Water & Environmental Engineering** (B) (2011)
PO Box 21
Arba Minch
p. +251-46-8810070

nigussie_tg@yahoo.com
http://www.arbaminch-univ.com/WTIWEE.html

## Mekelle University

**Dept of Earth Science** (B) (2014)
P.O.Box: 231
Management Building, Main Campus
Mekelle
p. (+251) 344 40 40 05
cciad.mu@gmail.com
http://www.mu.edu.et/index.php/department-of-earth-science

**Institute of Geo-information and Earth Observation Sciences** (B) (2014)
P.O.Box: 231
Management Building, Main Campus
Makelle
p. +251 914 720398
meseleagw@yahoo.com
http://www.mu.edu.et/index.php/programs/ethiopia-institute-of-technology-mekelle/institute-of-geo-information-and-earth-observat

## Semera University

**Dept of Geology** (B) (2011)
PO Box 132
Semera
p. 251-336660603
semerauniversity@ethionet.et

## Wollega University

**Dept of Geology** (B) (2011)
PO Box 395 Nekemte
Nekemte
p. 251-576615038
wu@ethionet.et
http://www.wuni.edu.et/

# Fiji

## Ministry of Lands and Mineral Resources

**Mineral Resources Dept of Fiji** (2011)
Private Mail Bag, GPO
Suva
p. (679) 338 1611
director@mrd.gov.fj
http://www.mrd.gov.fj/

## University of the South Pacific

**Div of Earth and Environment Science** (2007)
Private Mail Bag
GPO Suva
Suva, Fiji Islands
kahsai_k@usp.ac.fj
http://www.sidsnet.org/pacific/usp/earth/

# Finland

## Aalto University

**Dept of civil engineering** (B,M,D) (2016)
Rakentajanaukio 4
Espoo
SF-02150 Espoo 00076 AALTO
leena.korkiala-tanttu@aalto.fi
http://civileng.aalto.fi/en/

## Abo Akademi University

**Geology and Mineralogy** (B,M,D) O (2016)
Domkyrkotorget 1
Turku Fi-20500
p. (+)358442956429
geologi@abo.fi
http://www.abo.fi/fakultet/geologi
*Enrollment (2015): B: 26 (8) M: 50 (4) D: 9 (2)*
Professor:
Olav Eklund, (D), Abo Akademi Univ, 1993, GxeGz

## Geological Survey of Finland

**Geologian tutkimuskeskus** (2015)
P.O.Box 96
Espoo FI-02151
p. +358 29 503 0000
gtk@gtk.fi
www.gtk.fi
f: https://www.facebook.com/GTK.FI
t: @GTK_FI
Director, Strategy and Planning:
Jarmo Kohonen, (D)
Director, Projects and Customers:
Petri Lintinen, (D)
Director, Operative Units:
Olli Breilin, (M)
Director, Human Resources, Talent Management and Working Environments:
Helena Tammi, (M)
Director, Communications and Marketing:
Marie-Louise M.Sc. Wiklund, (M), 1987, OnnOn
Director General:
Mika Nykänen, (M)
Digital Innovations and Corporate data:
Mikko Eklund, (M)
Director, Science and Innovation:
Pekka Nurmi, (D)

## University of Helsinki

**Dept of Geosciences and Geography** (B,M,D) (2015)
Gustaf Hällströmink 2
P.O. BOX 68
Helsinki 00014
p. 358-294150827
Mia.Kotilainen@helsinki.fi
http://www.helsinki.fi/geology/
*Enrollment (2007): B: 0 (6) M: 26 (15) D: 6 (0)*
Head:
Juha A. Karhu, (D), Helsinki, 1993, Csg

## University of Oulu

**Dept of Geology** (B,M,D) O (2014)
Linnanmaa
FIN-90014
Oulu
vesa.peuraniemi@oulu.fi
http://cc.oulu.fi/~geolwww/Geology.htm

**Dept of Geophysics** (2007)
Pertti.Kaikkonen@oulu.fi
http://www.gh.oulu.fi/

# France

## Institut de Physique du Globe de Paris

**Institut de Physique du Globe de Paris** (2015)
1 rue Jussieu
75238 Paris Cedex 05
p. (018) 395-7400
secretdir@ipgp.fr
http://www.ipgp.fr

## Bates University

**Quaternary Geology and Sedimentology Laboratory** (2004)
70, avenue Leon Lachamp - Case 907
13288 Marseille Cedex 09
sdescha2@bates.edu

## Bureau de Recherches Geologiques et Minieres

**BRGM-Orleans** (2011)

3 avenue Claude-Guillemin
BP 36009
, Orleans 45060 Cedex 2
p. +33 (0)2 38 64 34 34
http://www.brgm.fr/

## Catholic University of the West

**Environmental Management** (2004)
3, place Andre Leroy
BP 808
49008 Anger

## Centre de Recheches Petrographiques et geochimiques

**Center of Petrographic and Geochemical Research (CRPG du CNRS)** (2004)
15, rue Notre Dame des Pauvres
BP 20
54501 Vandoeuvre Les Nancy Cedex
p. + 33 (0)3 83 59 42 02
info-eureka@lorraine.eu
http://www.crpg.cnrs-nancy.fr

## Ecole des Mines de Paris

**Ecole des Mines de Paris** (2014)
60-62, Boulevard Saint Michel
75272 PARIS cedex 06
Paris, France 75272
isabelle.olzenski@mines-paristech.fr
http://www.ensmp.fr/Eng/ENSMP/aboutENSMP.html

## Ecole Nationale Supérieure de Géologie (ENSG)

**ENSGéologie** (M,D) (2017)
2 Rue du Doyen Marcel Roubault
TSA 70605
VANDOEUVRE-LES-NANCY, Lorraine F-54518
p. 33 (0)3 83 59 64 15
ensg-contact@univ-lorraine.fr
http://www.ensg.univ-lorraine.fr/

Director:
MONTEL Jean-Marc, (D), Eg

## Ecole Nationale Superieure des Mines de Nancy

**Ecole des Mines de Nancy** (M,D) ● (2016)
Campus ARTEM
92 rue du Sergent Blandan
CS14234
NANCY 54042
p. +33 0355662600
mines-nancy-scolarite-ficm@univ-lorraine.fr
http://www.mines-nancy.univ-lorraine.fr/content/g%C3%A9oing%C3%A9nierie
f: https://www.facebook.com/groups/geoingenierie/
t: https://twitter.com/minesnancy

## Ecole Nationale Superieure des Mines de Saint Etiene

(2014)
158 Cours Fauriel
CS 62362
Saint Etienne 42023
p. (+33) (0)4 77 420 278
accueil@ccsti-larotonde.com
http://www.mines-stetienne.fr/fr

## Ecole Normale Superieure de Paris

**Dept of Geosciences** (M,D) ● (2015)
24 rue Lhomond
Paris 75005
p. (+33) (0)1 44 32 22 11
delescluse@geologie.ens.fr
http://www.geosciences.ens.fr
*Enrollment (2015): M: 20 (10) D: 29 (6)*

## ENSPM - Institut Francais du Petrole

**Centre Exploration** (2004)
1-4 avenue de Bois Preau
BP 311
92506 Rueil Malmaison

## French Institute for Research Exploitation of the Sea (IFREMER)

**Dept of Marine Geosciences** (2004)
BP 70
29263 Plouzane

## Higher Natl School of Mines at Saint-Etienne (ENSME)

**Departement Geosciences et environnement** (A,B,M,D) (2013)
158, cours Fauriel
42023 Saint-Etienne Cedex 02
guy@emse.fr

## Higher Natl School of Mines in Paris (ENSMP)

**Department of Earth Sciences and environment** (2004)
35, rue Saint-Honore
77305 Fontainebleau Cedex

## Inst of Interdisciplinary Research de Geologie et de Mecanique (IRIGM)

**Dept of Geosciences and Environment** (2017)
Domaine Univ de Grenoble - BP 53 X
35041 Grenoble Cedex
lise.reymond@unil.ch

## Inst of Physics of the Globe-Paris VII

**Lab de Geochimie, Geomateriaux, Geomag, Sismo, Tech, Obser Volcanologiques** (2004)
Tour 14/24 - 2e etage, 4, place Jussieu
75252 Paris Cedex 05

## Institut de Physique Du Globe De Paris

**Institut de Physique Du Globe De Paris** (M,D) (2016)
1 rue Jussieu
Paris 75005
p. 01 83 95 74 00
accueil@ipgp.fr
http://www.ipgp.fr/

## Institut Polytechnique LaSalle Beauvais (ex-IGAL)

**Dept of Geosciences** (B,M,D) (2014)
19 Rue Pierre WAGUET - BP 30313
Beauvais 60026 cedex
p. +33 (0) 3 44 06 89 91
yannick.vautier@lasalle-beauvais.fr
http://www.lasalle-beauvais.fr/
Administrative Assistant: Nathalie Lermurier

Director of the Geosciences Department:
Yannick Vautier, Inst Géologique Albert-de-Laparent, 1999, GodCo
Director of the Education Program in Geosciences:
Hervé Leyrit, (D), Gv
Director of School/Companies relations:
Pascal Barrier, (D), Inst Géologique Albert-de-Laparent, PmGsPs
Professor:
Olivier Pourret, (D), Rennes I (France), 2006, CgaGe
Lahcen Zouhri, (D), Lille (France), 2000, HwgGe
Geotechnics:
Bassam Barakat, (D), Ecole Centrale Paris (France), 1991, NrgGq
Engineer in Mining & Quarry:
Lucien Corbineau, Inst Géologique Albert-de-Laparent, 2007, EgGgNm
Engineer in Marine Geology:
Olivier Bain, Inst Géologique Albert-de-Laparent, 2000, OuGgOi
Engineer in Geotechnics:
Jean-David Vernhes, Engineer of Polytech Paris UPMC (France), 1998, YgNgr

France

Engineer in Geology:
Benoit Proudhon, Inst Géologique Albert-de-Laparent, 1997, GgcGt
Associate Professor:
Jessica Bonhoure, (D), CREGU-Nancy, 2007, GziCc
Sadek Brahmi, (D), UPMC Paris VI (France), 1991, Nr
Claudia Cherubini, (D), Bari (Italia), 2007, HwGe
Cyril Gagnaison, (D), Sorbonne (France), 2006, PgGsa
Sébastien Laurent-Charvet, (D), Orléans (France), 2001, GctOe
Pascale Lutz, (D), Pau (France), 2002, YgxYd
Mohamed Nasraoui, (D), ENSMP, 1996, EgGzEm
Elsa Ottavi-Pupier, (D), CRPG-CNRS Nancy, 1996, GvzGi
Sébastien Potel, (D), Bâle, 2001, GpzGi
Elodie Saillet, (D), Glasgow (UK), 2009, GctGo
Renaud Toullec, (D), Bordeaux (France), 2006, GsoEo
Ghislain Trullenque, (D), Basel (Switzerland), 2005, GcNrGg

## Laboratoire d'Hydrologie et de Geochimie de Strasbourg

(2014)
1, rue Blessing
67084 Strasbourg Cedex
p. (+33) (0)3 68 85 04 02
marie-claie.pierret@unistra.fr
http://lhyges.unistra.fr/PIERRET-Marie-Claire,250?lang=fr

## National Institute of Applied Science

**Lab de Mineralogie et Geotechnique** (2014)
20, avenue des Buttes de Coesmes
35043 Rennes Cedex 7
p. (+33) (0)2 23 23 82 00
olivier.guillou@insa-rennes.fr
http://www.insa-rennes.fr/insa-rennes.html

## National Polytechnic Inst. of Grenoble

**Ecole Doctorale Terre, Univers, Environnement** (2004)
Domaine Univ de Grnoble - BP 95
46 avenue Félix Viallet
Cedex 1
Grenoble 38031

## Pytheas Institute - Earth Sciences and Astronomy Observatory (PYTHEAS)

**Faculte des Sciences de Luminy** (2004)
Laboratoires de Geologie Marine et Sedimentologie
70 avenue Leon Lachamp
13288 Marseille

## Toulouse University

**Satellite Geophysics and Oceanography Laboratory** (M,D) (2014)
14, Avenue Edouard Belin
31400 Toulouse
p. (+33) (0)5 61 33 29 02
directeur@legos.obs-mip.fr
http://www.legos.obs-mip.fr/Presentation-generale?set_language=en&cl=en

## Universite Blaise Pascal (Clermont Ferrand II)

**Dept des Sciences de la Terra** (2014)
34 Avenue Carnot
63038 Clermont Ferrand Cedex
p. (+33) 04 73 34 67 22
cecile.sergere@univ-bpclermont.fr
http://wwwobs.univ-bpclermont.fr/lmv/cursus/

## Université Claude Bernard Lyon 1

**Laboratoire de Géologie de Lyon: Terre, Planètes, Environnement** (B,M,D) (2016)
Bd du 11 Novembre
Campus de La Doua
Bâtiment Géode
Villeurbanne 69622
p. 0033 (0)472445800
emanuela.mattioli@univ-lyon1.fr
http://lgltpe.ens-lyon.fr/
Administrative Assistant: Marie-Jeanne Barrière
*Enrollment (2012): B: 36 (30) M: 16 (16) D: 10 (10)*

Professor:
Pascal Allemand
Nicolas Coltice
Fabrice Cordey
Gilles Cuny
Isabelle Daniel
Gilles Dromart
philippe Gillet
Stephane Labrosse
Christophe Lécuyer
Emanuela Mattioli
Guillemette Ménot
Cathy Quantin
Pierre Thomas
Senior Scientist:
Thierry Alboussiere
Vincent Balter
janne Blichert toft
Bernard Bourdon
Razvan Caracas
Eric Debayle
Vincent Grossi
Serge Légendre
Philippe Herve Leloup
Bruno Reynard
yanick Ricard
Jean Vannier
Ricard Yanick
Associate Professor:
Muriel Andréani
Anne-Marie Aucourt
Frederic Chambat
Regis Chirat
Claude Colombié
Véronique Daviéro-Gomez
Renaud Deguen
Véronique Gardien
Bernard Gomez
Vincent Langlois
Gweltaz Mahéo
Matthew Makou
Jean-Emmanuel Martelat
Davide Olivero
Vincent Perrier
Jean-Philippe Perrillat
Sylvain Pichat
Bernard Pittet
Frédéric Quillévéré
Stéphane Reboulet
Philippe Sorrel
Guillaume Suan
Benoît Tauzin
Associate Scientist:
Romain Amiot
Thomas Bodin
caroline Fitoussi
Bertrand Lefebvre
Laurence Lemelle
Jérémy Martin
Jan Matas
Emeritus:
francis Albarede
Related Staff:
Emmanuelle Albalat
Ingrid Antheaume
Florent Arnaud-Godet
Brigitte Barchasz
Ghislaine Broillet
Herve Cardon
Fabien Dubuffet
Philippe Fortin
François Fourel
Philippe Grandjean

Naïma Khrouz
Aline Lamboux
Gilles Montagnac
Sophie Passot
Emmanuel Robert
Magali Seris
Philippe Telouk

## Universite d'Orleans

**Dept des Sciences de la Terre** (2004)
BP. 6759
45067 Orleans
scolarite-osuc@univ-orleans.fr

## Universite de Bordeaux I

**Dept des Sciences de la Terra et de la Mer** (2014)
341 Cours de la Liberation
Talence 33400
p. (+33) 05 40 00 88 79
termer@adm.u-bordeaux1.fr
http://www.u-bordeaux1.fr/universite/organisation/composantes-ufr-instituts/ufr-des-sciences-de-la-terre-et-de-la-mer.html

## Universite de Bourgogne

**UFR Sciences Vie, Terre & Environnement** (B,M,D) (2015)
6 boulevard Gabriel
21100 Dijon
direction-ufrsvte@u-bourgogne.fr
http://ufr-svte.u-bourgogne.fr/

## Universite de Bretagne Occidentale

**Departement des Sciences de la Terre** (2014)
3 Rue des Archives
Brest 29287
p. (+33) 02 98 01 61 88
alain.cottignies@univ-brest.fr
http://www.univ-brest.fr/ufr-sciences/menu/Les_departements/Sciences_de_la_Terre

## Universite de Caen

**Dept des Sciences de la Terre** (2014)
Esplanade de la Paix
Caen 14000
p. (+33) 02 31 56 55 87
isabelle.villette@unicagen.fr
http://ufrsciences.unicaen.fr/departements/departement-des-sciences-de-la-terre/

## Universite de Lorraine - Faculte des Sciences et Technologies

**Dept Geosciences** (B,M) ● (2014)
Boulevard des Aiguillettes
BP 239
54506 Vandoeuvre Les Nancy
p. (+33) (0)3 83 68 47 18
bernard.lathuiliere@univ-lorraine.fr
http://www.geologie.uhp-nancy.fr/Php/index.php
*Enrollment (2014): B: 31 (0) M: 51 (0)*

## Universite de Montpellier

**Geosciences Montpellier** (M,D) (2017)
2 place Eugene Bataillon
CC 060
Montpellier cedex 5 34095
p. (+33) (0)4 67 14 36 43
dirgm@gm.univ-montp2.fr
http://www.gm.univ-montp2.fr/?lang=fr

## Universite de Paris Sud (Orsay)

**Dept Sciences de la Terre** (B,M,D) ● (2016)
15 Rue Georges Clemenceau
Orsay 91400
p. (+33) (0)1 69 15 49 09
hermann.zeyen@u-psud.fr
http://geosciences.geol.u-psud.fr/
f: https://www.facebook.com/geosciences.paris.sud/
t: https://twitter.com/GEOPS_Orsay
*Enrollment (2015): B: 91 (36) M: 156 (43) D: 28 (17)*

## Universite de Paris VI

**Lab of Sub-Marine Geodynamics (CEROV)** (2004)
Port de la Darse - BP 48
06230 Villefranche-Sur-Mer

**Center of Geosynamics Research** (2014)
4 Place Jussieu
Paris 75005
p. (+33) 01 44 27 46 98
licence.sciterre@upmc.fr
http://www.upmc.fr/en/education/diplomas/sciences_and_technologies/bachelor_s_degrees/department_of_earth_sciences.html

## Universite de Paris VII

**Dept des Sciences Physiques de la Terre** (2014)
4 place Jussieu
75251 Paris
p. (+33) (0)1 57 27 57 27
zarie.rouas@univ-paris-diderot.fr

## Universite de Pau

**Dept of Geology** (A,B,M,D) (2014)
I.P.R.A. - Geologie
Avenue de L'Universite
Pau 64230
anne-sophie.laloge@univ-pau.fr
http://www.univ-pau.fr/RECHERCHE/GEOPHY/

## Universite de Pau et des Pays de l'Adour

**Faculte des Sciences** (2014)
Avenue de l'Universite
64000 Pau

## Universite de Rennes I

**Dept. de Geosciences** (2015)
2 Rue du Thabot
35065 Rennes Cedex
p. (+33) (0)2 23 23 60 76
florentin.paris@orange.fr
http://www.geosciences.univ-rennes1.fr/?lang=en

## Universite de Strasbourg

**Ecole et Observatoire des Sciences de la Terre** (B,M,D) ● (2014)
5 rue Descartes
67084 Strasbourg cedex F
p. (+33) (0)3 68 85 03 53
eost-contact@unistra.fr
http://eost.unistra.fr/nouveautes-du-site/
Director:
Frederic Masson, (D)
Laboratory Director:
Ulrich Achauer, (D), Karlsruhe, 1990, Ysg

## Université Jean Monnet, Saint-Etienne

**Département de Géologie - Faculte des Sciences et Techiques** (B,M,D) ● (2015)
23 rue du Dr. Paul Michelon
Saint-Etienne Cedex F-42023
p. (+33) (0)4 77 48 15 85
veronique.lavastre@univ-st-etienne.fr
http://portail.univ-st-etienne.fr/bienvenue/presentation/ufr-des-sciences-dpt-geologie-327810.kjsp
Head:
Bertrand N. MOINE, (D), Macquarie Uni., Sydney, & Uni. J. Mon-

net St-Etienne, 2000, CgtGx
Professor:
Jean-Yves COTTIN, (D), Muséum d Histoires Naturelles, Paris, 1978
Damien GUILLAUME, (D)
Jean-François MOYEN, (D), 2000, GiCtGp
Assistant Professor:
Marie-Christine GERBE, (D), GisOe

France

## Universite Paul Sabatier (Toulouse III)

**Dept Sciences de la Terre** (2015)
118 Route de Narbonne
31062 Toulouse
p. (+33) (0)5 82 52 57 21
fsi.sec@univ-tlse3.fr
http://www.univ-tlse3.fr/04628859/0/fiche___pagelibre/&RH=ACCUEIL&RF=1237305837890

## Université de Franche-Comté

**Sciences environnementales** (2011)
Sciences et techniques
Service Scolarité
16, route de Gray
25030 Besançon cedex
p. 03 81 66 62 09 / 62 11
Scolarite.UFR-ST@univ-fcomte.fr
http://sciences.univ-fcomte.fr/formations/listedesformations.htm

**Faculte des Sciences et Techniques** (2014)
16, route de Gray
25030 Besancon Cedex
p. (+33) (0)3 81 66 62 09
scolarite.ufr-2t@univ-fcomte.fr
http://sciences.univ-fcomte.fr/pages/fr/menu3795/formations/licence--sciences-de-la-vie-16862-15340.html

## University of Caen

**Dept des Sciences de la Terre** (2004)
Lab de Geologie structurals
Esplanadde de la Paix
14032 Caen Cedex

## University of Franche - Comte

**Lab de Geol Strucurale et appliquee** (2004)
Faculte des Sci et Techniques
Place Leclerc
25030 Besancon

## University of Francois Rabelais

**Laboratoire de Geologie** (2014)
Parc de Grandmot
37200 Tours

## University of Lille

**Dept of Geology** (B,M,D) ● (2017)
Cite Scientifique
Batiment SN-5
Villeneuve D'Ascq 59655
bruno.vendeville@univ-lille1.fr
http://www.univ-lille1.fr/geosciences/

## University of Louis Pasteur (Strasbourg 1)

**School and Observatory of Earth Sciences** (2004)
1, rue Blessing
67084 Strasbourg Cedex

## University of Maine

**Faculte des Sci - Lab de Geologie** (2014)
Avenue Olivier Messiasen
72085 Le Mans
p. (+33) (0)2 43 83 30 00
biogel@univ-lemans.fr
http://sciences.univ-lemans.fr/Biologie-Geosciences

## University of Montpellier II

**Observatoire des sciences de l'univers OREME** (2004)
2, place Eugene Bataillon
34095 Montpellier Cedex 05
contact@oreme.org

## University of Nantes

**Departement des Sci de la Terre** (2004)
2, rue de la Houssiniere
44072 Nantes Cedex 03

## University of Nice

**Dept of Earth Science** (2004)
Faculte des Sciences
28, avenue Valrose
06034 Nice Cedex
secretariat-master-PPA@unice.fr

## University of Nice-Sopia Antipollis-1

**Laboratoire des Sciences de la Terre** (2017)
Rue A. Einstein
06560 Valbonne
dir.recherche@unice.fr
http://www-geoazur.unice.fr/

## University of Orleans

**Dept des Sciences de la Terre** (2004)
Domaine de la Source
BP 6759
45067 Orleans Cedex 02

## University of Perpignan

**Centre de Sed et Geochimie marines** (2014)
52 Avenue Paul Alduy
66100 Perpignan
p. (+33) (0)4 68 66 21 39
facscien@univ-perp.fr

## University of Picardy

**Dept de Geologie** (2014)
Chemin du Thil
80000 Amiens
p. (+33) (0)4 22 82 76 65
mohamed.benlahsen@u-picardie.fr
http://www.u-picardie.fr/jsp/fiche_structure.jsp?STNAV=US&RUBNAV=&CODE=US&LANGUE=0

## University of Pierre & Marie Curie

**Lab de Paleobotanique et Palnologue evol** (2004)
12, rue Cuvier
75005 Paris

## University of Pierre & Marie Curie (Paris VI)

**Dept de Geotectonique** (2011)
Tour 26 - ler etage (Boite 219)
4, place Jussieu
75252 Paris Cedex 05

**Dept of Living Earth and Environment** (2010)
Tour 15 - 4e etage
4, place Jussieu
75252 Paris Cedex 05
chrystele.sanloup@upmc.fr
http://www.ipgp.jussieu.fr/

## University of Poitiers

**Dept des Sciences de la Terre** (2014)
40, ave du Recteur Pineau
86022 Poitiers Cedex

## University of Provence (Aix-Marseille I)
**Lab de Geol Structurale et appliquee** (2004)
Centre Saint Charles
3, place Victor Hugo - Case 28
13331 Marseille Cedex 3

## University of Reims-Champagne
**Dept des Sciences de la Terre** (2014)
Moulin de la Housse - BP 1039
51687 Reims Cedex
p. (+33) (0)3 26 91 34 19
scolarite.sceiences@univ-reims.fr
http://www.univ-reims.fr/formation/ufr-instituts-et-ecoles/ufr-sciences-exactes-et-naturelles/presentation,8370.html?

## University of Rouen-Upper Normandy
**Dept de Geologie** (2015)
1 Rue Thomas Becket
76821 Mont-Saint-Aignan
p. (+33) (0)2 35 14 68 26
francoise.baillot@univ-rouen.fr

## University of Savoy
**Laborotoire de Geologie** (2004)
Faculte des Sci et Techniques
Campus de Technolac, BP 1104
73011 Chambery Cedex

## University of Science & Technology of Lille
**Inst Des Sciences De La Terre** (2004)
Flandres - Artois, Cite Scientifique
Bat SN5, BP 36
59655 Villeneuve D'Ascq Cedex

## University of West Brittany
**Earth Sciences** (2004)
Groupement de Recherche: GEDO
6, avenue Le Gorgeu
29287 Brest

# Gabon

## Universite des Sciences et Techniques de Masuku
**Dept de Geologie** (B) (2014)
BP 901
Franceville
p. (+241) 01 67 75 78
http://www.labogabon.net/ustm/facscience/index.html

## Universite Omar Bongo
**Dept de Geologie** (B) (2010)
BP 13 131
Boulevard Leon Mba
Libreville
p. +241 73 20 45, (241-72) 69 10
uob@internetgabon.com
http://www.uob.ga/

# Germany

## Aachen University of Technology
**Inst of Structural Geology, Tectonics and Geomechanics** (2005)
Lochnerstr 4-20
D-52056
ged@ged.rwth-aachen.de
http://www.ged.rwth-aachen.de

## Baden-Wuerttemberg - Landesamt fuer Geologie, Rohstoffe und Bergbau (LGRB)
**Regional Government of Freiburg State Office of Geology, Raw Materials and Mining** (2015)
Albertstrasse 5
Freiburg, Baden-Wuerttemberg 79104
p. ++49 9761 208 3000
abteilung9@rpf.bwl.de
http://www.lgrb-bw.de/

## Brandenburg - Landesamt fuer Geowissenschaften und Rohstoffe (LGRB)
**State Office for Mining , Geology and Minerals of Brandenburg** (2011)
Inselstrabe 26
03046 Cottbus
info@lbgr.brandenburg.de
http://www.lbgr.brandenburg.de

## Bundesanstalt fur Geowissenschaften und Rohstoffe (BGR)
**Federal Institute for Geosciences and Natural Resources** (2015)
Stilleweg 2
Hannover 30655
poststelle@bgr.de
http://www.bgr.bund.de

## Christian-Albrechts-Universitaet
**Geology Dept** (2011)
Olshausenstrasse 40/60
2300 Kiel 1

## Ernst-Moritz-Arndt Universitaet
**Institut fur Geographie und Geologie** (2014)
Domstrabe 11
17489 Griesswald
p. (+49) (0) 3834 86-4502
geogra@uni-greifswald.de
http://www.mnf.uni-greifswald.de/institute/geo.html

## Freiburg University College
**Major Earth and Environmental Sciences** (B) (2017)
Bertoldstrasse 17
79085 Freiberg
p. (+49) 761 203-67342
studyinfo@ucf.uni-freiburg.de
http://www.ucf.uni-freiburg.de/

## Freie Universitaet Berlin
**Inst fuer Palaontologie** (2011)
Maltese-Strasse
74-100, Haus D
D-12249 Berlin
palaeont@zedat.fu-berlin.de
http://userpage.fu-berlin.de/~palaeont/WELCOME.HTM

**Inst fur Geologie** (2013)
Malteserstr. 74-100
12249 Berlin
p. (030) 838-70 575
plansec@zedat.fu-berlin.de
http://www.geo.fu-berlin.de/geol/index.html

## Friedrich-Schiller-University Jena
**Institute for Geosciences** (B,M,D) ● (2014)
Burgweg 11
Jena 07749
p. 0049(0)3641/948600
geowissenschaften@uni-jena.de
http://www.geo.uni-jena.de
*Enrollment (2014): B: 17 (43) M: 16 (22) D: 69 (12)*

Univ.-Prof. Dr.:
Sabine Attinger, (D), Hy
Florian Bleibinhaus, (D), Yx
Georg Büchel, (D), Gg
Christoph Heubeck, (D), Stanford, 1994, GsdPy
Nina Kukowski, (D), Ygx
Falko H. Langenhorst, (D), Gz

Juraj Majzlan, (D), Gz
Kai U. Totsche, (D), Bayreuth, 1994, HwClOs
Kamil Ustaszewski, (D), Gc
Lothar Viereck, (D)
Jun.-Prof. Dr.:
Anke Hildebrandt, (D), MIT, 2005, HgSpf



## Garching Technical University
**Inst fur Geologie und Mineralogie** (2004)
Lichtenbergergstrasse 4
0-8046 Garching
rieder@tum.de

## Geologischer Dienst
**Bavarian Environment Agency** (2013)
Buergermeister-Ulrich-Strasse
160
Augsburg 86179
poststelle@lfu.bayern.de
http://www.lfu.bayern.de/geologie/

## Geologischer Dienst Nordrhein-Westfalen
**NRW Geological Survey** (2013)
Postfach 10 07 63
De-Greiff-Strasse 195
Krefeld D-47707
p. +49-2151-897-0
geoinfo@gd.nrw.de
http://www.gd.nrw.de/home.php

## Geologisches Institut der Universitaet
**Dept of Geology** (2011)
Plelcherwael 1
87 Wuerzburg

## Georg-August University of Goettingen
**Dept of Geobiology** (B,M,D) (2016)
Goldschmidstr. 3
Goettingen, Lower Saxony 37077
p. +49 551 397951
jreitne@gwdg.de
http://www.geobiologie.uni-goettingen.de/
*Enrollment (2016): B: 6 (3) M: 4 (3) D: 7 (0)*
Prof.Dr.:
Joachim Reitner, (D), PyyPy
Professor:
Volker Thiel, Dr., (D), CooCo
Dr.:
Gernot Arp, Dr., (D), PyyPy
Andreas Reimer, (D), OcCmm
Dr.:
Jan-Peter Duda, (D), 2014, PyCoGs

## Goethe Universitaet
**Fachbereich Geowissenschaften** (2014)
Altenhoferallee 1
60438 Frankfurt
p. (+49) (0)69 798 40208
dekanat-geowiss@em.uni-frankfurt.de
http://www.uni-frankfurt.de/fb/fb11/index.html

## Hamburg - Geological Survey
**Ministry of Urban Development and Environmental Protection Agency for the Environment** (2014)
Neuenfelder Straße 19
21109 Hamburg
gla@bsu.hamburg.de
http://www.geologie.hamburg.de

## Helmholtz-Zentrum Potsdam Deutsches GeoForschungsZentrum
**GeoForschungs Zentrum GFZ** (2014)
Telegrafenberg 14473
14473 Potsdam
p. (+49) 331 288-1045
presse@gfz-potsdam.de
http://www.gfz-potsdam.de/startseite/

## Hessen - Landesamt fuer Umwelt und Geologie
(2011)
65203 Wiesbaden, Rheingaustraße 186
p. 0611-6939-0
http://www.hlug.de/

## Humboldt Universitaet zu Berlin
**Palaontologisches Museum** (2011)
Invaliden Strasse 43
0-4010 Berlin

## Institut fur Palaontologie
**Dept of Geology** (2004)
O. Weidlich, Fohrenweg 19
W-8504 Stein
palsek@univie.ac.at

## Karlsruhe Institute of Technology
**Institute for Applied Geosciences** (2011)
Hertzstr. 16
Karlsruhe
susanne.winter@kit.edu
http://www.agw.kit.edu/

**Geophysical Institut** (B,M,D) (2010)
Hertzstr. 16
Karlsruhe 76187
p. +49-721-6084431
geophysics@gpi.kit.edu
http://www-gpi.physik.uni-karlsruhe.de/
*Enrollment (2010): B: 15 (0) D: 10 (0)*
Professor:
Thomas Bohlen, (D), Yx
Senior Scientist:
Thomas Forbriger, (D), Ys
Joachim R.R. Ritter, (D), YsGtv
Research Associate:
Rebecca Harrington, (D), Ys

**Institute for Mineralogy and Geochemistry** (2011)
Adenauerring 20b
Karlsruhe D-76131
p. +49 721 608- 43323
img@img.uka.de
http://www.img.kit.edu/

**Dept of Geology** (2014)
Kaiserstrasse 12
76131 Karlsruhe
p. (+49) 721 608 4219 2/43651
geophysik@gpi.kit.edu

http://www.bgu.kit.edu/index.php

## Ludwig-Maximilians-Universitaet Muenchen
**Dept of Earth & Enviromental Sciences** (B,M,D) (2009)
Theresienstr. 41/III
Munich 80333
p. 0049/89/21804250
dingwell@lmu.de
http://min.geo.uni-muenchen.de/
*Enrollment (2007): B: 173 (28) M: 12 (0) D: 24 (0)*
Director:
Donald Bruce Dingwell, Giv
Professor:
Alexander Altenbach, Pg
Wladyslaw Altermann, Gg
Michael Amler, Pg
Valerian Bachtadse, Yg
Hans-Peter Bunge, Yg

Christina De Campos, Gz
Karl Thomas Fehr, Gz
Friedrich Frey, Gz
Anke Friedrich, (D)
Anke Friedrich, (D), MIT, 1998, GgtCc
Helmut Gebrande, (D), LMU Munich (Germany), 1975, YgGyEo
Stuart Gilder, Yg
Peter Gille, (D), Humboldt Universit (Germany), 1984, GzOm
Wolfgang Heckel, Gz
Ernst Hegner, Gg
Soraya Heuss-Aßbichler, Gz
Heiner Igel, Yg
Harald Immel, Pg
Bernd Lammerer, (D), LMU Munich (Germany), 1972, GgtGc
Reinhold Leinfelder, Pg
Rocco Malservisi, Yg
Robert Marschik, Gg
Ludwig Masch, Gz
Wolfgang Moritz, Gz
Bettina Reichenbacher, Pg
Wolfgang Schmahl, Gz
Klaus Weber-Diefenbach, Gg
Christian Wolkersdorfer, Gg

Adjunct Professor:
Frank Trixler, (D), GzOmXc

Other:
Kirill Aldushin, Gz
Greta Barbi, Yg
Robert Barsch, Yg
Johannes Birner, Gg
Florian Bleibinhaus, Yg
Hans Boysen, Gz
Gilbert Britzke, Yg
Benoit Cordonnier, Gz
Alexander Dorfman, Gz
Thomas Dorfner, Gz
Barbara Emmer, Yg
Werner Ertel-Ingrisch, Gz
Andreas Fichtner, Yg
Michaela Frei, Gg
Frantisek Gallovic, Yg
Alexander Gigler, Gz
Stefan Grießl, Gz
Hagen Göttlich, Gz
Marc Hennemeyer, Gz
Katja Henßel, Pg
Kai-Uwe Hess, Gz
Maria Linda Iaccheri, Gg
Giampiero Iaffaldano, Yg
Guntram Jordan, Gz
Ines Kaiser-Bischoff, Gz
Thorsten Kowalke, Pg
Thomas Kunzmann, Gz
Martin Käser, Yg
Markus Lackinger, Gz
Yan Lavallee, Yg
Maike Lübbe, Gz
Götz Meisterernst, Gz
Timo Casjen Merkel, Gz
Marcus Mohr, Yg
Lena Müller, Gz
Dieter Müller-Sohnius, Gz
Malik Naumann, Pg
Jens Oser, Yg
Sohyun Park, Gz
Karin Paschert, Gg
Rossitza Pentcheva, Gz
Nikolai Petersen, Yg
Helen Pfuhl, Yg
Antonio Sebastian Piazzoni, Yg
Christina Plattner, Yg
Josep Puente Alvarez de la
Oliver Riedel, Gz
Alexander Rocholl, Gz
Javier Rubio-Sierra, Gz
Gertrud Rößner, Pg
Dieter Schmid, Pg
Julius Schneider, Gz
Bernhard Schuberth, Yg
Oliver Spieler, Gz
Robert Stark, Gz
Katja Steffens, Gg
Stefan Strasser, Gz
Marco Stupazzini, Yg
Frank Söllner, Gg
Ferdinand Walther, Gz
Joachim Wassermann, Yg
Laura Wehrmann, Pg
Michael Winklhofer, Yg
Ayhan Yurtsever, Gz
Matthias Zeitlhöfler, Gg
Albert Zink, Gz

**Div of Palaeontology & Geobiology, Dept of Earth and Env Sci** (B,M,D) O (2016)
Richard-Wagner-Strasse 10
Muenchen, Bavaria 80333
geobiologie@geo.lmu.de
http://www.palmuc.de

Prof. Dr.:
Gert Woerheide, (D), Pyo

Prof. Dr.:
William Orsi, (D), PyOn
Bettina Reichenbacher, (D), PogPv

## Marburg Chome University

**Geology Dept** (2014)
Deutschhausstrabe 10
35032 Marburg
p. (+49) 06421/28-24257
edda.walz@geo.uni-marburg.de
https://www.uni-marburg.de/fb19_alt

## Martin-Luther-Universitaet Halle-Wittenberg

**Inst for Geosciences and Geography** (B,M,D) (2015)
Von-Seckendorff-Platz 3
D-06120 Halle
p. ++49-045-55 26010
direktor@geo.uni-halle.de
http://www.geo.uni-halle.de

Director:
Herbert Pöllmann, (D), 1984, Gz

## Mecklenburg-Vorpommern - Landesamt fuer Umwelt, Naturschutz und Geologie

**State office for the Environment, Nature Conservation and Geology** (2017)
Goldberger Strasse 12
18273 Gustrow
p. ++49-3843-777 0
poststelle@lung.mv-regierung.de
http://www.lung.mv-regierung.de/

## Niedersachsen - Landesamt fuer Bodenforschung

(2014)
p. +49 (0)511-643-0
poststelle-hannover@lbeg.niedersachsen.de
http://www.lbeg.niedersachsen.de

## Rheinisch-Westfaelische Technische Hochschule Aachen

**Fachgruppe fuer Geowissenschaften und Geographie** (2011)
Lochnerstr. 4-20 (Haus B)
Aachen 52056
p. 0241 80 96219
geowiss@rwth-aachen.de
http://www.fgeo.rwth-aachen.de/

## Rheinland-Pfalz - Landesamt fuer Geologie und Bergbau
**State Office for Geology and Mining** (2011)
Emy-Roeder-Strabe 5
PO Box 100 255
Mainz-Hechtsheim D-55129
office@lgb-rlp.de
http://www.lgb-rlp.de/

## RWTH Aachen University
**Institute of Geology and Palaeontology** (2004)
Facultat fur Bergbau
Huttenwesen und Geowissenschaften/Wulwestr 2
D-5100 Aacen
geo.sek@emr.rwth-aachen.de

## Sachsen - Landesamt fuer Umwelt und Geologie (LfUG)
**Saxon State Agency for Environment, Agriculture and Geology** (2011)
Postfach 54 01 37
Dresden 01 311
lfulg@smul.sachsen.de
http://www.smul.sachsen.de/

## State Survey for Geology and Mining of Sachsen-Anhalt
**State Survey for Geology and Mining of Sachsen-Anhalt** (2016)
Halle, Saxony-Anhalt D-06035
p. +49 345 52 12 0
poststelle@lagb.mw.sachsen-anhalt.de
http://www.lagb.sachsen-anhalt.de

## Technische Hochschule
**Institute of Applied Geoscience** (2014)
Karolinenplatz 5
64289 Darmstadt
p. (+49) 6151 16-2171
herrmann@geo.tu-darmstadt.de
http://www.geo.tu-darmstadt.de/iag/index.de.jsp

## Technische Universitaet Bergakademie Freiberg
**Faculty for Geosciences, Geoengineering and Mining** (2011)
Bergakademis Freiberg
Postfach 47/Bernhard-Cotta Strasse
Freiberg
p. +49 (0)3731 / 39 - 3249
Andrea.Thuemmel@geort.tu-freiberg.de
http://tu-freiberg.de/fakult3/index.en.html

## Technische Universitaet Berlin
**Applied Geosciences** (2011)
Str. Des 17 Juni 135
W-1000 Berlin 12
p. +49 30314-7260 5
http://www.geo.tu-berlin.de/

## Technische Universitaet C.W. Braunschweig
**Inst fuer Geologie und Palaontologie** (2013)
Fachbereich fur Physik und Geowissenschaften
Pockelstrasse 14
D-3300 Braunschweig DEN

**Institute of Environmental Geology** (2011)
Postfach 3329
Braunschweig D-38023
geosecret@tu-bs.de
http://www.tu-braunschweig.de/iug

**Inst fuer Geophysik und Meteorologie** (2011)
Mendelssohnstr 2-3
D-3300 Braunschweig

## Technische Universitaet Clausthal
**Institut fuer Geologie und Palaeontologie** (2013)
LeibnizstraBe 10
D-38678
Clausthal-Zellerfeld 38678
Office@geologie.tu-clausthal.de
http://www.geologie.tu-clausthal.de/

**Inst for Petroleum Engineering** (2015)
Agricolastrasse 10
38678 Clausthal-Zellerfeld
p. +49-5323-72-2239
marion.bischof@tu-clausthal.de
http://www.ite.tu-clausthal.de/

## Technische Universitaet Darmstadt
**Institute of Applied Geosciences** (B,M,D) (2011)
Schnittspahnstr. 9
Darmstadt D-64287
p. +49 6151 16 2171
iag@geo.tu-darmstadt.de
http://www.geo.tu-darmstadt.de/
*Enrollment (2011): B: 280 (3) M: 54 (0) D: 29 (2)*

Acting head:
Hans-Joachim Kleebe, GzOm
Professor:
Rafael Ferreiro Maehlmann, Gxp
Matthias Hinderer, Gs
Andreas Hoppe, GgOui
Stephan Kempe, GgCgHs
Ingo Sass, Ng
Christoph Schueth, Hw
Stephan Weinbruch, GzOwm

## Technische Universitaet Muenchen
**Engineering Geology** (B,M,D) ● (2014)
Arcisstr. 21
Munich, Bavaria 80333
p. +498928925851
geologie@tum.de
http://www.eng.geo.tum.de/
Administrative Assistant: Katja Lokau
*Enrollment (2014): B: 55 (0) M: 66 (0) D: 19 (4)*

Chair-Prof.:
Kurosch Thuro, (D), Technical Munich, 1995, NgrGg
Professor:
Hans Albert Gilg, (D), ETH (Switzerland), EgnEg
Michael Krautblatter, (D), Erlangen, GmNrOi
Senior Scientist:
Gerhard Lehrberger, (D), TUM, EgGxg
Bernhard Lempe, (D), TUM, 2012, GlgNg
Florian M. Menschik, (D), Technical Munich, 2015, NgOiNr
Michael Michael Rieder, (D), Technical Munich, GgxGs
Associate Professor:
Inga Moeck, (D), GcgOn
Research Associate:
Peter Ellecosta, (M), TUM, Ngr
Sibylle Knapp, (M), ETH Zurich, GsgGm
Mathias Köster, (M), TU Dresden, CgGzEg
Philipp Mamot, (M), Bonn, GmOg
Lukas Paysen-Petersen, (M), Technische Univ Muenchen, 2013, NgrGq
Bettina Sellmeier, (M), Technische Univ Muenchen, Ngr
Carola Wieser, (M), Technische Univ Muenchen, Ngr
Lisa Wilfing, (M), Technische Univ Muenchen, Ngr
Lecturer:
Kerry Leith, (D), ETH Zurich, 2012, GmNrOi
Marion Nickmann, (D), Technische Univ Muenchen, 2006, Ng
Laboratory Director:
Heiko Käsling, (D), TUM, 2009, NrmNg

## Technische Universitat Dresden
**Dept of Geosciences** (2015)
Mommsenstr 13
0-8027 Dresden
p. +49 351 463-32863

doris.salomon@tu-dresden.de
http://tu-dresden.de/die_tu_dresden/fakultaeten/fakultaet_forst_geo_und_hydrowissenschaften/fachrichtung_geowissenschaften

## Thuringen - Landesanstalt fuer Umwelt und Geologie
**Thuringen State Institute for Environment and Geology** (2011)
Goeschwitzer Str forty-one
07745 Jena
Poststelle@tlug.thueringen.de
http://www.tlug-jena.de/de/tlug/

## Universitaet Bochum
**Institut fuer Geologie, Mineralogie und Geophysik** (B,M,D) (2013)
Universitaetsstr. 150
Bochum 44780
p. +49 234 32 23233
sabine.sitter@ruhr-uni-bochum.de
http://www.ruhr-uni-bochum.de/gmg/

## Universitaet Bonn
**Steinmann-Institut für Geologie, Mineralogie und Paläontologie** (B,M,D) ● (2014)
Nussallee 8
Bonn 53115
p. +49 228 73 4803
tmartin@uni-bonn.de
http://www.steinmann.uni-bonn.de/

## Universitaet Erlangen-Nuernberg
**Institut fur Geographie** (2011)
Kochstr. 4/4
91054 Erlangen
p. 09131/85-22633
common@geographie.uni-erlangen.de
http://www.geographie.uni-erlangen.de/

## Universitaet Erlangen-Nürnberg
**Geozentrum Nordbayern** (2011)
Schlossgarten 5
91054 Erlangen
p. 09131/85-22615
geologie@geol.uni-erlangen.de
http://www.geol.uni-erlangen.de

## Universitaet Frankfurt
**Institut fur Geowissenschaften** (2011)
Fachbereisch 17
Senckenberganlage 32-34
6000 Frankfurt
geowissenschaften@em.uni-frankfurt.de
http://www.geo.uni-frankfurt.de/ifg/

## Universitaet Freiburg
**Institut fur Geo- und Umweltnaturwissenschaften - Geologie** (B,M,D) (2016)
Albertstr. 23-B
Freiburg i. Br., Baden-Wuerttemberg 79104
ulmer@uni-freiburg.de
https://portal.uni-freiburg.de/geologie

## Universitaet Giessen
**Inst fuer Geowissenschaften und LithosphSrenforshung** (2014)
Senckenbergstrasse 3
35390 Giessen
p. (+49) 641 990
brigitte.becker-lins@geolo.uni-giessen.de
http://www.uni-giessen.de/fbr08/geolith/

## Universitaet Goettingen
**Geowissenschaftliches Zentrum** (2011)
Goldschmidstr. 3
Lower Saxony
Gottingen D-37077
p. +49 551 397951
bhinz@gwdg.de
http://www.uni-goettingen.de/de/125309.html

## Universitaet Greifswald
**Institute for Geography and Geology** (2011)
F.L. Jahn Strasse 17
Greifswald
p. +49 (0)3834 86-4570
geologie@uni-greifswald.de
http://www.mnf.uni-greifswald.de/institute/geo.html

## Universitaet Halle-Wittenberg
**Institute for Geosciences and Geographie** (A,B,M,D) ● (2015)
Von-Seckendorff-Platz 3/4
Halle D-06120
p. +49-0345-55 26055
direktor@geo.uni-halle.de
http://www.geo.uni-halle.de/

## Universitaet Hamburg
**Institute of Geophysics** (B,M,D) ● (2016)
Bundesstrasse 55
Hamburg 20146
p. +4940428382973
dirk.gajewski@uni-hamburg.de
www.geo.uni-hamburg.de/de/geophysik.html
*Enrollment (2016): B: 47 (23) M: 14 (11) D: 12 (8)*
Professor:
Dirk J. Gajewski, (D), Karlsruhe, 1987, YesYg

**Center for Earth System Research and Sustainability** (M,D) (2015)
Bundesstrasse 53
Hamburg 20146
anke.allner@uni-hamburg.de
http://www.cen.uni-hamburg.de/

## Universitaet Hannover
**Institut fur Geologie** (2011)
Fachbeleich Erdwissenschaften
Callinstrasse 30
D-30167 Hannover
p. +49-(0)511-762 2343
sekretariat@geowi.uni-hannover.de
http://www.geologie.uni-hannover.de

## Universitaet Heidelberg
**Facultat fur Chemie und Geowissenschaften** (2014)
Im Neuenheimer Feld 234
69120 Heidelberg
p. (+49) 06221/544 844
dcg@urz.uni-heidelberd.de
http://www.chemgeo.uni-hd.de/

**Insitut fuer Geowissenschaften** (2011)
Im Neuenheimer Feld 236
Heidelberg D-69120
p. 06221-54-8291
Elfriede.Hofmann@geow.uni-heidelberg.de
http://www.geow.uni-heidelberg.de/

## Universitaet Leipzig
**Institut fur Geophysik und Geologie** (2011)
Talstrasse 35
04103 Leipzig
geologie@rz.uni-leipzig.de
http://www.uni-leipzig.de/~geo/

## Universitaet Marburg
**Inst fuer Geologie und Palaontologie** (2011)
Biergenstrasse 12
Lahnberge

D-3550 Marburg-Lahn

## Universitaet Muenster

**Institut fuer Mineralogie** (2011)
Corrensstrabe 24
Muenster 48149
p. +49 251 83-33464
minsek@uni-muenster.de
http://www.uni-muenster.de/Mineralogie/

**Institut fuer Geologie und Palaeontologie** (B,M,D) ● (2016)
Correnssstr. 24
D-48149 Muenster
p. +49 251-83 33974
sklaus@uni-muenster.de
http://www.uni-muenster.de/GeoPalaeontologie/en/index.html
*Enrollment (2016): B: 270 (41) M: 46 (21) D: 25 (3)*

Professor:
Christine Achten, (D), Frankfurt, 2002, CloEg
Heinrich Bahlburg, (D), Technical Univ (Germany), 1986, GsgCc
Ralf Thomas Becker, (D), Bochum, 1990, Pg
Ralf Hetzel, (D), Gcm
Hans Kerp, (D), Utrecht, 1986, PblGg
Harald Strauss, (D), Göttingen, 1985, CsGg

## Universitaet of Kiel

**Institute of Geoscience** (2014)
Christian-Albrechts-Platz 4
24118 Kiel
p. (+49) 431 880 3900
wussow@geophysik.uni-kiel.de
http://www.ifg.uni-kiel.de/6+M52087573ab0.html

## Universitaet Oldenburg

**Institute for Chemistry and Biology of the Marine Environment** (2014)
Carl-von-Ossietzky-Str. 9-11
Box 2503
Oldenburg 26111
p. +49-0441-798-5342
director@icbm.de
http://www.icbm.uni-oldenburg.de/

## Universitaet Potsdam

**Institute for Earth and Environmental Sciences** (B,M,D) (2013)
Karl-Liebknecht-Str. 24-25
Potsdam 14476
p. +49 331 977 2116
sekretariat@geo.uni-potsdam.de
http://www.geo.uni-potsdam.de/
*Enrollment (2013): B: 319 (14) M: 79 (11) D: 122 (16)*

## Universitaet Stuttgart

**Inst fuer Geol und Palaontogisch** (2011)
Fakultaet-7-Geo. und Biowissenschaften
Boblinger Str. 72
D-7000 Stuttgart 10

## Universitaet Trier

**Lehrstuhl fuer Geologie** (B,M,D) (2016)
Geowissenschaften (FB VI)
Trier 54286
p. 0651 201 4647
ensch@uni-trier.de
http://www.uni-trier.de/index.php?id=2632
*Enrollment (2015): B: 67 (12) M: 41 (9) D: 5 (4)*

**Regional and Environmental Science** (2014)
Behringstrabe 21
D-54296 Trier
p. (+49) (0) 651-201-4528
dekanatfb6@uni-trier.de
http://www.uni-trier.de/index.php?id=2220

## Universitaet Tuebingen

**Institute for Geoscience** (2011)
72076 Tuebingen
Sigwartstr. 10
Christophe.Pascal@rub.de
http://www.ifg.uni-tuebingen.de/

## Universitaet Würzburg

**Institut für Geographie und Geologie** (2011)
Am Hubland
97074 Würzburg
p. +49 (0) 931 / 31-85555
http://www.geographie.uni-wuerzburg.de

**Institut für Paläontologie** (2011)
Pleicherwall 1
D-97070 Würzburg
p. 49 0931- 31 25 97
i-palaeontologie@mail.uni-wuerzburg.de
http://www.palaeontologie.uni-wuerzburg.de

## Universitaet zu Koeln

**Institute for Geology and Mineralogy** (2011)
Cologne
p. +49 221 470-5619
ellen.stefan@uni-koeln.de
http://www.geologie.uni-koeln.de/

## University of Bayreuth

**Bayerisches Geoinstitut** (M,D) (2015)
Bayreuth 95440
p. +49(0)921 55-3700
bayerisches.geoinstitut@uni-bayreuth.de
http://www.bgi.uni-bayreuth.de/

## University of Bremen

**Fachbereich 5 Geowissenschaften** (A,B,M,D) ● (2016)
Klagenfurter Strasse
Bremen, Bremen 28359
info@geo.uni-bremen.de
http://www.geo.uni-bremen.de
*Enrollment (2015): A: 249 (0) B: 274 (52) M: 249 (62) D: 111 (29)*

## University of Cologne

**Dept of Geology and Mineralogy** (B,M,D) ● (2014)
Zuelpicherstr. 49a
50674 Koeln
p. +49-221-470-5619
ellen.stefan@uni-koeln.de
http://www.uni-koeln.de/math-nat-fak/geologie/index.html

**Institute of Geophysics and Meteorology** (B,M,D) (2014)
Pohligstraße 3
D-50969 Köln
p. +49 (0)221 470 2552
sekretar@geo.uni-koeln.de
http://www.geomet.uni-koeln.de/en/general/home/

## Universität Mainz

**Institut für Geowissenschaften** (B,M,D) ● (2017)
Joh.-J.-Becher-Weg 21
55128 Mainz
p. +49 6131 39 22857
mertz@uni-mainz.de
http://www.geowiss.uni-mainz.de/

Head:
Bernd R. Schöne, (D), PegPi

Professor:
Jonathan M. Castro, (D), Gv
Boris PJ Kaus, (D), Yg
Michael Kersten, (D), Ca
Cees W. Passchier, (D), Gt

Denis Scholz, (D), Cl
Frank Sirocko, (D), Gs
Richard W. White, (D), Gp
Adjunct Professor:
Dieter Mertz, (D), Cc

## West Wilhelms Universitat
**Dept of Applied Geology** (2004)
Schlossplatz 2
4400 Muenster

## Westfälische Wilhelms-Universität Münster
**Fachbereich Geowissenschaften** (B,M,D) (2016)
Heisenbergstrasse 2
Münster D-48149
p. +49 251 83-30002
dekangeo@uni-muenster.de
http://www.uni-muenster.de/Geowissenschaften/

# Ghana

## Kwame Nkrumah University of Science and Tech
**Petroleum Engineering** (B) (2010)
http://www.knust.edu.gh/ceng/faculties.php

**Dept of Geomatic Engineering** (B,M,D) ○ (2017)
geomaticeng@knust.edu.gh
head.geomatic@knust.edu.gh
Kumasi PMB KNUST
p. +233 (0)3220 60227
geomaticeng@knust.edu.gh
http://www.knust.edu.gh/ceng/faculties.php

**Geological Engineering** (B,M,D) ● (2015)
Geological Engineering Department
College of Engineering
KNUST
Kumasi PMB
geologicaleng@knust.edu.gh
http://archive.knust.edu.gh/pages/index.php?siteid=geoleng
Senior Scientist:
Emmanuel K. Appiah-Adjei, (D), Hohai Univ (China), 2013, HwgGg
Associate Professor:
Simon K. Gawu, (D), Witwatersrand (South Africa), 2004, GgEgOn
Lecturer:
Bukari Ali, (D), Birmingham (England), 1996, HwNgYg
Godfrey C. Amedjoe, (M), Ghana, 2005, GidGc
Samuel Banash, (M)
Gordon Foli, (M), 2004
Solomon S. Gidigasu, (M), Sci & Tech (Ghana), 2013, GeOrNr
Emmanuel Mensah, (M), Sci & Tech (China), 1998, EgGg

## University of Development Studies
**Earth and Environmental Science** (B) (2010)
PO Box 1350
Tamale
p. +233 (71) 22422
edambayi@uds.edu.gh
http://www.uds.edu.gh/

## University of Ghana
**Dept of Earth Science** (B,M,D) ● (2014)
PO Box LG 58
Legon
Accra
p. +233-244-116-879
pmnude@ug.edu.gh
http://www.ug.edu.gh/index1.php?linkid=185&sublinkid=40&subsublinkid=36
*Enrollment (2013): B: 336 (104) M: 105 (33) D: 2 (1)*
Dean of Science:
Daniel K. Asiedu, (D), GsCgGd
Professor:
Bruce K. Banoeng-Yakubo, (D), Ghana, GcHw
Senior Scientist:
Thomas K. Armah, (D), YgeYe
Jacob M. Kutu, (D), GctGc
Patrick Asamoah Sakyi, (D), Cg
Head of Department:
Prosper M. NUDE, (D), Ghana, GipGx
Associate Professor:
Thomas M. Akabzaa, (D), Ghana, 2004, EgGeNm
David Atta-Peters, (D), Ghana, Pl
Johnson Manu, (D), GzEgGz
Frank K. Nyame, (D), Cg
Sandow M. Yidana, (D), Hwq
Lecturer:
Francis Achampong, (D), Ngr
Chris Y. Anani, (D), Gsd
Larry-Pax CHEGBELEH, (D), Ng
Yvonne A.S Loh, (M), Hw
Issac A. Oppong, (D), Np

## University of Mines and Technology
**Dept of Geological Engineering** (B) (2014)
PO Box 237
Tarkwa
p. +233 3123 20935
sps@umat.edu.gh
http://www.umat.edu/gh/UndergraduatePrograms/geology.html

**Dept of Geomatic Engineering** (B) (2014)
PO Box 237
Tarkwa
p. +233(0)362 20324
gm@umat.edu.gh
http://www.umat.edu.gh/UndergraduatePrograms/geomatic.html

**Dept of Mineral Engineering** (B) (2014)
PO Box 237
Tarkwa
p. +233(0)362 21136
mr@umat.edu.gh
http://www.umat.edu.gh/UndergraduatePrograms/mineral.html

**Dept of Mining Engineering** (B) (2014)
PO Box 237
Tarkwa
p. +233 3123 20935
mn@umat.edu.gh
http://www.umat.edu.gh/UndergraduatePrograms/Mining.html

# Greece

## Aristotle University of Thessaloniki
**School of Geology** (2014)
GR-541 24
Thessaloniki 54124
p. (+30) 2310 99.8450
info@geo.auth.gr
http://www.geo.auth.gr/index_en.htm

## Greek Institute of Geology & Mineral Exploration
**Greek Institute of Geology & Mineral Exploration** (2014)
1, Spirou Louis St.,
Olympic Village
Acharnae P.C. 13677
dirgen@igme.gr
http://www.igme.gr

## National Technical University of Athens
**Dept of Mining Engineering and Metallurgy** (2011)
Zographou Campus
15 780 Athens
secretary@metal.ntua.gr
http://www.metal.ntua.gr/div-geology/index.html

### University of Athens
**Faculty of Geology and Geoenvironment** (2004)
National & Capodistran Univ. of Athens
Panepistimioupolis, Ilistra
Athens
secr@geol.uoa.gr
http://www.geol.uoa.gr/engindex.htm

### University of Patras
**Dept of Geology** (2004)
26110 Patras

## Guatemala

### Universidad sde San Carlos de Guatemala
**Centro de Estudios Superiores de Energia y Minas** (2014)
Ciudad Universitaria Zona 12, Guatemala City 01012
p. (502) 2418-9139 ex. 86211
usacesem@ing.usac.edu.gt
http://cesem.ingenieria.usac.edu.gt/

## Guinea

### Univerité Gamal Abdel Nasser de Conakry
**Faculty of Geology and Mining** (M) ● (2014)
BP 1147
Conakry, Guinée
p. +224 631 54 48 38
sekoumoussa@gmail.com
http://www.uganc.org/index.php/2013-01-12-06-22-44/cere

## Guyana

### Guyana Geology and Mines Commission
**Guyana Geology and Mines Commission** (2011)
Upper Brickdam
Georgetown
p. 226-5591, 225-2862
http://www.ggmc.gov.gy/

## Haiti

### Ecole Nacionale de Geologie Appliquee
**Ecole Nacionale de Geologie Appliquee** (2004)
B.P. 1560-Varreux, Port-au-Prince

## Hungary

### Eotvos Lorand University
**Insitute of Geography and Earth Sciences** (B,M,D) ● (2014)
Pazmany Peter setany 1/C
Budapest H-1117
p. (+36 1) 381 2191
ffi@ffi.elte.hu
http://geosci.elte.hu/en_index.htm
*Enrollment (2014): B: 900 (360) M: 300 (120) D: 40 (18)*

Professor:
Judit Bartholy, (D), Eotvos Lorand, 1978, Ow
Professor:
Kristof Petrovay, (D), Eotvos Lorand, 1992, Xy
Associate Professor:
Gábor Timár, (D), Eotvos Lorand, 2003, YgdOi
Research Professor:
János Lichtenberger, (D), Hungarian Academy of Sciences, 1996, OarXy
Professor:
Miklos Kazmer, (D), Eotvos Lorand, 1982, PgoGh
Associate Professor:
Agnes Gorog, (D), Eotvos Lorand, Pmi
László Lenkey, (D), VU Amsterdam, 1999, YhHwGa
Balázs Székely, (D), Uni Tuebingen, 2001, GmOri
Associate Scientist:
Orsolya Ferencz, (D), Budapest, 2001, XyOr
Anikó Kern, (D), Eotvos Lorand, 2012, OrwOa
Gábor Molnár, (D), Eotvos Lorand, 2004, OrYgOi
Assistant Professor:
László Balázs, (D), Eotvos Lorand, 2009, EoYeg
Attila Galsa, (D), Eotvos Lorand Univ (Hungary), 2004, YghYe
Attila Osi, (D), Eotvos Lorand, PviPe
Emoke Toth, (D), Eotvos Lorand, PmsPb
Research Associate:
Istvan Szente, (D), Eotvos Lorand, Pig
Professor:
Ferenc Horváth, (D), Eotvos Lorand, 1971, YgGhc
Péter Márton, (D), Eotvos Lorand, 1971, YmgYs

### Hungarian Geological Survey
**Hungarian Geological Institute** (1999)
Stefánia út 14
Budapest HU-1143
p. 36 1267 1433
hamort@mgsz.hu

### Hungarian Office for Mining and Geology
**Hungarian Office for Mining and Geology** (2014)
PO Box 95
1590 Budapest 1145
p. (+36-1) 301-2900
hivatal@mbfh.hu
http://www.mbfh.hu

### Jozsef Attila University
**Dept of Mineralogy, Geochem & Petrology** (2004)
P O Box 651
H-6701 Szeged

**Dept of Geology & Paleontology** (2004)
Egyetum u. 2-6
H-6722 Szeged

### Univesity of Szeged
**Dept of Physical Geography and Geoinformatics** (B,M,D) (2017)
SZTE Termeszeti Foldrajzi Tanszek
Szeged, 6722 Egyetem utca 2. PF:653
Szeged
p. 0036 62544158
mezosi@geo.u-szeged.hu
http://www.geo.u-szeged.hu/ANG/HPfirst.html
*Enrollment (2012): D: 20 (0)*

Professor:
Janos Rakonczai, (D), 1978, Oe
dean of faculty:
lászló Mucsi, (D), Szeged, 1996, OirOy
Associate Professor:
Andrea Farsang, (D), HAS Hungarian Academy + szeged, 2016, SdOe
Timea Kiss, (D), HAS Hungarian Academy + Debrecen, 2001, GmOy
György Sipos, (D), Szeged, 2007, HqYm
József Szatmári, (D), Szeged, 2007, OriOy

## Iceland

### Landmælingar Íslands
**National Land Survey of Iceland** (2017)
Stillholt 16-18 300 Akranes
Akranes 300
p. 4309000
lmi@lmi.is
http://www.lmi.is/

### University of Iceland
**Faculty of Earth Science** (2014)
Öskju, Sturlugötu 7
101 Reykjavik
p. (+354) 5254600
dagrun@hi.is

http://english.hi.is/sens/faculty_of_earth_sciences/about_faculty

# India

## Aligarh Muslim University

**Dept of Geology** (B,M,D) ● (2014)
Aligarh - U.P.
Aligarh, Uttar Pradesh 202002
p. 0571-2700615
lakrao@yahoo.com
http://www.amu.ac.in/fsc/no4.html

## Andhra University

**Dept of Geology** (2011)
Vishakha-patnam-530 003
College of Science and Technology
, Andhra Pradesh
p. 91-891-2844888
principal_science@andhrauniversity.info
http://www.andhrauniversity.info/

## Anna University

**Dept of Geology** (2011)
Guindy, Chennai-600 025,
Tamil Nadu
p. 044-22358442 / 8444 / 8452
geonag@gmail.com
http://www.annauniv.edu/index.php

## Annamalai University

**Faculty of Marine Sciences** (2011)
Annamalainagar-608 025
Tamil Nadu
p. 91 - 4144 - 238248
info@annamalaiuniversity.ac.in
http://annamalaiuniversity.ac.in/marinesciences.htm

## Banaras Hindu University

**Dept of Geology** (B,M,D) (2016)
Varanasi
U.P. 221005
hbsrivastava@gmail.com
http://www.bhu.ac.in/Geology/index.html

Professor:
HARI BAHADUR SRIVASTAVA, (D), BHU, 1980, GctGp

## Bharathidasan University

**Dept of Geology** (2011)
School of Geosciences
Tiruchirapalli 620 024
p. +91 431 2407034
drmm_bdu@yahoo.co.uk
http://www.bdu.ac.in/schools/geo_sciences/geology/

## Candigarh University

**Dept of Geology** (2016)
National Highway 95
Chandigarh, Punjab 140413
p. (+91) 2724481
chairperson_geology@pu.ac.in
http://geology.puchd.ac.in/

## Cochin University of Science & Technology

**Dept of Geology** (2014)
Kochi
p. +91(484)2577550
akv@cusat.ac.in
http://www.cusat.ac.in/

## Geological Survey of India

**Geological Survey of India** (2011)
27, J.L.Nehru Road
West Bengal Kolkata-700016
dg@gsi.gov.in
http://www.portal.gsi.gov.in

## Goa University

**Dept of Earth Science** (M,D) ● (2017)
Taleigao Plateau
Goa University P.O.
Taleigao Plateau
Panaji, Goa 403 206
p. +91-0832-6519329
mkotha@unigoa.ac.in
http://www.unigoa.ac.in/department.php?adepid=11&mdepid=3
*Enrollment (2016): M: 50 (297) D: 11 (9)*

Head:
Adiveppa G. Chachadi, (D), IIT, Roorkee, 1989, HwYeHs

Professor:
Mahender Kotha, (D), IIT Bombay, 1987, GsoOi

Associate Professor:
Anthony V. Viegas, (D), Goa, 1997, Gig

On Contract:
Poornima Dhavaskar, (M), Geology, 2013, Gge
Raghav Gadgil, (M), Geology, 2012, GiEg

## Guru Nanak Dev University

**Dept of Geology** (2011)
Amritsar
http://www.gndu.ac.in/

## Indian Institute of Science Education and Research, Kolkata

**Dept of Earth Science** (B,M,D) (2013)
Mohanpur Campus
Nadia
Mohanpur (Kolkata) , West Bengal 741252
des.chair@iiserkol.ac.in
http://earth.iiserkol.ac.in/

## Indian Institute of Technology

**Dept of Applied Geology** (2011)
Bombay, Powai
Bombay
ayaz@iitb.ac.in
http://www.geos.iitb.ac.in/

## Indian Institute of Technology, Kharagpur

**Dept of Geology & Geophysics** (B,M,D) (2011)
Kharagpur, Kharagpur
West Bengal 721302
Kharagpur, West Bengal 721302
p. +91-3222-282268
head@gg.iitkgp.ernet.in
http://www.iitkgp.ac.in/departments/home.php?deptcode=MG

## Indian Institute of Technology, Roorkee

**Dept of Earth Sciences** (M,D) (2010)
Department of Earth Sciences
IITRoorkee
Roorkee, Uttrakhand 247667
p. +911332285532
dpkesfes@iitr.ac.in
http://members.tripod.com/rurkiu/acd-earth/index.html

## Indian School of Mines

**Dept of Applied Geology** (2011)
Dhanbad
West Bengal 826 004
p. +91-326-2296616
agl@ismdhanbad.ac.in
http://www.ismdhanbad.ac.in/depart/geology/index.htm

India

## Institute of Science
**Dept of Geology** (2011)
Aurangabad 431004
Aurangabad , Maharashtra
p. +91 (0240) 2400586
director@inosca.org
http://www.inosca.org

## Jadavpur University
**Dept of Applied Geology** (B,M,D) (2013)
Jadavpur
Calcutta
p. +913324572268
hod@geology.jdvu.ac.in
http://www.jaduniv.edu.in/view_department.php?deptid=76

## Jmia Millia Islamia
**Dept of Geography** (2011)
Jamia Nagar
New Delhi-25
rocketibrahim@yahoo.com
http://jmi.ac.in/aboutjamia/departments/geography/introduction

## Lucknow University
**Dept of Geology** (2011)
Lucknow 226007 U.P.
info@lkouniv.ac.in
http://www.lkouniv.ac.in/dept_geology.htm

## Maharaja Sayajirao University of Baroda
**Dept of Geology** (2011)
Baroda - Gujarat
p. (919) 426721547
ischamyal@yahoo.com
http://www.msubaroda.ac.in/science/index.php

## Nagpur University
**Dept of Geology** (2011)
Rao Bahadur D. Laxminarayan Educational Campus
Law College Square
Amravati Road
Nagpur 440 001
p. +91-712-253241
geodeptplacecell@gmail.com
http://www.nagpuruniversity.org/links/FacultyofScience.htm

## Osmania University
**Dept of Geology** (2011)
Hyderabad 500007 A.P.
http://www.osmania.ac.in/

## Panjab University
**Centre for Petroleum and Applied Geology (U.I.E.A.S.T.)** (2011)
Chandigarh 160014
rpatnaik@pu.ac.in
http://pu.ac.in/

## Pondicherry University
**Dept of Earth Science** (B,M,D) (2013)
The Head
Department of Earth Sciences
Pondicherry University
Pondicherry 605 014
p. +914132656741
head.esc@pondiuni.edu.in
http://www.pondiuni.edu.in/department/department-earth-sciences
Professor:
Balakrishnan Srinivasan, (D), Jawaharlal Nehru (India), 1986, CcGiz

## Presidency College
**Dept of Geology** (2011)
College Street
Calcutta
hnb.geol@presidencycollegekolkata.ac.in
http://www.concentric.net/~slahiri/pc/Department/Geology/geo

## Pt. Ravishankar Shukla University
**School of Studies in Geology and Water Resource Management** (M,D) (2014)
SOS Geology
Pt. Ravishankar Shukla UNiversity
Raipur, Chhattisgarh 492010
geology@prsu.com
http://www.prsu.ac.in/
*Enrollment (2013): M: 15 (0) D: 4 (1)*
Professor:
Srikant k. Pande, (D), Nagoya, 1986, GgHyg
Professor:
Ninad Bodhankar, (D), Raipur, 1992, GcNm
Kosiyath Rghavan Na Hari, (D), Vikram Univ Ujjain, 1992, GiiCp
Mohammad Wahdat Yar Khan, (D), AMU, Aligarh, 1979, GdEgCl

## University College of Science, Osmania University
**Centre of Exploration Geophysics** (M,D) ○ (2016)
Department of Geophysics
Osmania University Campus
Hyderabad, Telengana 500007
p. (+91) 40-27097116
head.geophysics@osmania.ac.in
http://www.osmania.ac.in/
*Enrollment (2016): M: 59 (32) D: 15 (0)*
Head:
Dr. Veeraiah B, (D), Osmania, 2005, YexYg
Assistant Professor:
Dr Udaya Laxm Gakka, (D), Osmania, 2009, HwYge
Dr Ram Raj Mathur, (D), Osmania, YemYx

## University of Baroda
**Dept of Geology** (2011)
Baroda , Vadodara 390 002
swanikhil@yahoo.co.in
http://www.msubaroda.ac.in/deptindex.php?ffac_code=2&fdept_code=6

## University of Delhi
**Dept of Geology** (B,M,D) ● (2016)
Department of Geology
University of Delhi
Delhi
New Delhi 110 007
p. 27667073
csdubey@gmail.com
http://www.du.ac.in/
*Enrollment (2014): B: 105 (1200) M: 60 (600) D: 30 (130)*
Director COL:
C S. Dubey, (D), Delhi Univ, 1993, GexGt

## University of Madras
**Dept of Geology** (2011)
Maraimalai Adigalar Campus
Chennai 600 005
p. 044 - 22202790
spmohan50@hotmail.com
http://www.unom.ac.in/departments/geology/geology.html

**Dept of Geology** (2011)
Madras, 600 025
Maraimalai Adigalar Campus
spmohan50@hotmail.com
http://www.unom.ac.in/departments/geology/geology.html

## University of Mysore
**Dept of Geology** (2011)
Mysore 570 005
p. +91 821 2419724
bb@geology.uni-mysore.ac.in
http://www.uni-mysore.ac.in/geology/#A

## University of Pune
**Dept of Geology** (2011)
Ganeshkhind Road
Pune, Maharashtra 411 007
p. +91-020-25601360
geology@unipune.ac.in
http://www.unipune.ac.in/dept/science/geology/default.htm

## Vijayanagara SriKrishnaDevaraya University
**Dept of Applied Geology** (M,D) (2016)
Dr.C.Venkataiah
Professor of Geology
Vijayanagara SriKrishnaDevaraya University
Bellary, Karnataka 583 104
p. (937) 909-0588
venkataiah.c@gmail.com
http://www.vskub.ac.in

# Indonesia

## Direktorat Vulkanologi
**Volcanological Survey of Indonesia** (2011)
Jl. Diponegoro 57
Bandung , West Java
http://portal.vsi.esdm.go.id/

## Gadjah Mada University
**Dept of Geology** (2014)
Jalan Bulaksmur
Yogzakarta 55281
p. (+62) 274 6492340
geografi@geo.ugm.ac.id
http://geo.ugm.ac.id/main/

## Indonesian Directorate General of Geology and Mineral Resources
**Ministry of Mines and Energy** (2011)
pengaduan@esdm.go.id
http://www.esdm.go.id

## Institut Teknologi Bandung
**Dept of Geology** (2014)
Jalan Ganesa, No. 10
Bandung 40132, Jawa Barat
p. 022 2514990
sisfo@fitb.ac.id
http://www.fitb.itb.ac.id/en/

**Dept of Geological Engineering** (2004)
Jl. Ganesha 10
Bandung, 40135
geologi@gc.itb.ac.id

## Trisakti University
**Faculty of Earth and Energy Technologies** (2004)
Jl. Kiai Tapa, Grogel
Jakarta 11440
p. 5663232 Ext. 8510

## Universitas Hasanudin
**Dept of Geological Engineering** (2004)
Jl. Perintis Kemerdekaan
Ujung Pandang 90245

## Universitas Padjadjaran
**Dept of Geological Engineering** (2004)
Jl. Raya Bandung-Sumedang km.21,
Jatinangor

## Universitas Pakuan
**Dept of Geological Engineering** (2014)
Jl. Pakuan, PO Box 452
Bogor 16143, Jawa Barat
p. 0251-8312206
rektorat@unpak.ac.id
http://www.unpak.ac.id/

# Iran

## Bu-Ali University
**Dept of Geology** (A,D) ○ (2015)
Barati@basu.ac.ir
Hamedan , Hamedan 65174
p. 0098-0811-38381460
Barati@basu.ac.ir
http://www.basu.ac.ir
Administrative Assistant: Meisam Gholipoor

## Shahid Bahonar University of Kerman
**Dept of Geology** (B,M,D) ● (2016)
Kerman
contactus@mail.uk.ac.ir
http://www.uk.ac.ir
Associate Professor:
Ahmad Abbasnejad, (D), GemHg
Reza Derakhshani, (D), GtmGc

## University of Tabriz
**Dept of Earth Sciences** (B,M,D) ● (2014)
29 Bahman Blvd.
University of Tabriz
Tabriz , East Azerbaijan 51664
p. +98 (411) 3335 9894
moazzen@tabrizu.ac.ir
http://www.tabrizu.ac.ir
*Enrollment (2014): B: 22 (33) M: 35 (28) D: 12 (7)*
Dean of Faculty:
Asghar Asghari Moghadam, (D), UK, 1992, HwqHy
Professor:
Ali Asghar Calagari, (D), UK, 1997, EgmCg
Ahmad Jahangiri, (D), India, 2000, GivGg
Mohsen -. Moayyed, (D), Iran, 2000, Gig
Mohssen -. Moazzen, (D), UK, 1999, GpCgGz
Associate Professor:
Robab Hajialioghli, (D), Iran, 2007, Gpt
Hosseinzadeh Mohammad Reza, (D), Iran, 2005, Egm
Head of the Department:
Ali Kadkhodai Ilkhchi, (D), Iran, 2008, NpEo
Assistant Professor:
Ebrahim Asghari Ka, (D), Iran, 2001, Ngr
Ghafour Alavi, (D), Eg
Nasir Amel, Iran, 2004, GivGg
Ghodrat Barzegari, (D), Iran, Nr
Mohammad Hassanpour Sedghi, Iran, 2011, YsNeYg
Fatemeh Mesbahi, (D), Iran, Gt
Ata Allah Nadiri, (D), Iran, 2012, HwGqe
Kamal Siahcheshm, (D), Iran, 2011, Egm
Reza Vaezi, (D), Iran, 2010, GeHgs
Behzad Zamani, (D), Iran, 2004, GtYs
Instructor:
Rahim Jomeiri, (M), Iran, 1992, Yg
Maqsood Orouji, (M), Ca
Naser Samimi, (B), Iran, Gg
Lecturer:
Siavosh Sartipzadeh, (M), Iran, Pm

# Ireland

## Geological Survey of Ireland
**Geological Survey of Ireland** (2011)
Beggars Bush

Haddington Road
, Dublin 4
john.butler@gsi.ie
http://www.gsi.ie/

## Geological Survey of Northern Ireland

**Geological Survey of Northern Ireland** (2011)
Colby House
Stranmillis Court
Belfast BT9 5BF
gsni@detini.gov.uk
http://www.bgs.ac.uk/GSNI/

## National University of Ireland Galway

**Earth and Ocean Sciences** (B,D) (2014)
Earth & Ocean Sciences
National University of Ireland
University Road
Galway
p. + 353 (0)91 492 126
lorna.larkin@nuigalway.ie
http://www.nuigalway.ie/eos/

Head:
Martin Feely, (D), National Univ of Ireland, Goz
Professor:
Peter Croot, (D), Otago, Ct
Lecturer:
Rachel R. Cave, (D), South Hampton, Cm
Eve Daly, (D), National Univ of Ireland, Yg
Tiernan Henry, (M), Wisconsin (Madison), Hg
John Murray, (D), Trinity College, PgGs
Robin Raine, (D), National Univ of Ireland, Yg
Tyrrell Shane, (D), Univ Coll (Dublin), Gs
Martin White, (D), Southampton, Op

## Trinity College

**Dept of Geology** (2016)
Department of Geology
Museum Building
Trinity College-Dublin
, Dublin 2
p. +353 01 896 1074
earth@tcd.ie
http://www.tcd.ie/Geology/

Chair:
Balz S. Kamber, (D), CgGgp
Associate Professor:
David Chew, (D), Univ Coll Dublin, 2001, CcGgt
Catherine Coxon, GeOu
Robin Edwards, Ge
Patrick N. Wyse Jackson, (D), Dublin (Ireland), 1992, PiGh
Assistant Professor:
Quentin G. Crowley, Cc
Seán h. mcClenaghan, Eg
Chris Nicholas, Eo
Juan Diego Rodriguez-Blanco, (D), 2006, GzClOm
Catherine V. Rose, Cg
Emma L. Tomlinson, Cc

## University College Cork

**Dept of Geology** (B,M,D) (2005)
Donovans Road
Cork
p. +353 21 4902657
s.culloty@ucc.ie
http://www.ucc.ie/ucc/depts/geology/
Administrative Assistant: Patricia Hegarty

Head:
John Gamble, Gi
Professor:
Ken Higgs, Pl
Research Associate:
Tara Davis, Yg
Jim Smith, Pl
Lecturer:
Alistair Allen, Gp
Bettie Higgs, Yg
Ed Jarvis, Pl
Ivor MacCarthy, (D), 1974, Gs
Pat Meere, (D), NUI (Ireland), 1992, Gc
John Reavy, Gi
Andy J. Wheeler, (D), Cambridge, 1994, Gu
Other:
Mary Lehane, On
Mick O'Callaghan, On
Dan Rose, On

## University College Dublin

**School of Earth Sciences** (2014)
Science Centre West
University College Dublin
Belfield, Dublin 4
p. +353 1 716 2331
geology@ucd.ie
http://www.ucd.ie/geology/

Head:
J. Stephen Daly
Professor:
Peter D. W. Haughton
Frank McDermott, (D), Open Univ (UK), 1987, GgCgOg
Patrick M. Shannon
John J. Walsh
Associate Professor:
Christopher J. Bean
Julian F. Menuge, (D), Cambridge (UK), 1983, EmCcs
Ian D. Somerville
Lecturer:
Conrad Childs
Aggeliki Georgiopopoulou, (D), Southampton (UK), 2007, GusOg
Ivan Lokmer
Patrick J. Orr
Adjunct Professor:
Tom Manzocchi, (D), Goc

# Israel

## Ben Gurion University of the Negev

**Dept of Geological and Env Sciences** (2009)
P.O. Box 653
84105 Beer Sheva
hatzor@bgu.ac.il
http://www.bgu.ac.il/geol/

## Geological Survey of Israel

**Geological Survey of Israel** (2011)
30 Malkhe Israel St.
Jerusalem, 95501
ask_gsi@gsi.gov.il
http://www.gsi.gov.il/

## Hebrew University of Jerusalem

**Faculty of Advanced Environmental Studies** (2004)
Givat Ram
Jerusalem 91904
msfeitel@mscc.huji.ac.il

## Tel Aviv University

**Dept of Geophysical, Atmospheric and Planetary Sciences** (B,M,D) ● (2016)
Ramat Aviv
P O Box 30940
Tel Aviv 69978 69978
p. 972-3-6408633
batshevc@tauex.tau.ac.il
http://geophysics.tau.ac.il/

Full Professor:
Shmuel Marco, (D), Gg

**Geosciences** (B,M,D) ● (2014)

Levanon Road
Ramat Aviv
Tel Aviv 6997801
p. 972-3-6408633
shmulikm@tau.ac.il
http://geophysics.tau.ac.il/

Emeritus:
Akiva Bar-Nun, Hebrew, 1968, Xc

Professor:
Pinhas Alpert, (D), Hebrew (Israel), 1980, OarOw
Zvi Ben-Avraham, (D), WHOI-MIT, 1973, YrGt
Shmulik Marco, (D), Hebrew U, Jerusalem, 1997, Ggc
Morris Podolak, (D), Yeshiva Univ, 1974, OnnOn
Colin G. Price, (D), Columbia, 1993, Oa
Moshe Reshef, (D), Tel-Aviv Univ, 1985, Yeg

Principal Research Assoc. (Assoc. Professor):
Lev Eppelbaum, (D), Inst. of Geophysics of Georgia, 1989, YvmGt

Senior Scientist:
Pavel Kishcha, (D), Russian Academy of Sciences, 1985, Oa

Dr:
Nili Harnik, (D), MIT, 2000, Oa

Assistant Professor:
Ravit Helled, (M), Tel-Aviv Univ, 2008, Xy

Lecturer:
Gilles Hillel Wust-Bloch, (D), 1990, Ys
Alon Ziv, (D), Ys

## The Hebrew University of Jerusalem

**Institute of Earth Sciences** (B,M,D) (2017)
Jerusalem 91904
Jerusalem 91904
p. 97226584686
arimatmon@mail.huji.ac.il
http://earth.es.huji.ac.il/machon/

# Italy

## Alma Mater Studiorum Università di Bologna

**Dipartimento di Scienze Biologiche, Geologiche e Ambientali** (B,M,D) O (2016)
Piazza di Porta San Donato 1
Bologna 40126
p. ++39 - 051 - 2094238
alessandro.gargini@unibo.it
http://www.bigea.unibo.it/it
*Enrollment (2016): B: 120 (30) M: 120 (30) D: 28 (7)*

## Servizio Geologico d'Italia

**Servizio Geologico d'Italia** (2011)
Via Vitaliano Brancati
Roma 48-00144
p. (+39) 0650071
emergenzeambientali@isprambiente.it
http://www.isprambiente.gov.it/it/servizi-per-lambiente/il-servizio-geologico-ditalia

## Università degli Studi di Padova

**Dept of Geosciences** (B,M,D) (2015)
Via G. Gradenigo 6
Padova 35131
p. +390498279110
geoscienze.direzione@unipd.it
http://www.geoscienze.unipd.it/

Full Professor:
Gilberto Artioli, (D), Chicago, 1985, GzyOm
Giuliano Bellieni, (M), Gv
Alessandro Caporali, (D), Ludwig Maximilian (Germany), 1979, YdsOr
Alberto Carton, (M), Gm
Giorgio Cassiani, (D), YgHs
Bernardo Cesare, (D), Gp
Silvana Martin, Gc
Fabrizio Nestola, (D), Modena and Reggio Emilia, 2003, Gz
Giorgio Pennacchioni, Gc
Cristina Stefani, Gd
Massimiliano Zattin, (D), Gs

Associate Professor:
Claudia Agnini, (D), Padova, 2007, PmeGu
Andrea D'Alpaos, (D), Hs
Giulio Di Toro, (D), Gc
Paolo Fabbri, Hw
Elena Fornaciari, (D), Pgm
Massimiliano Ghinassi, (D), Gs
Andrea Marzoli, (D), Cg
Matteo Massironi, (D), XgOr
Claudio Mazzoli, Gx
Stefano Monari, Pg
Paolo Nimis, Eg
Nereo Preto, (D), Gs
Gabriella Salviulo, Gz
Raffaele Sassi, Ggx
Paolo Scotton, Hg
Luciano Secco, Gz
Richard Spiess, Gp
Nicola Surian, Gm
Dario Visona', Gi
Annalisa Zaja, Yg
Dario Zampieri, Gc

Assistant Professor:
Jacopo Boaga, (D), Yg
Aldino Bondesan, (M), Padova, 1986, GmlOi
Anna Breda, (D), Padova, 2003, GsrGg
Luca Capraro, (D), Padova, 2002, PelGg
Maria Chiara Dalconi, (D), Gz
Manuele Faccenda, (D), Gq
Mario Floris, (D), NgOr
Alessandro Fontana, (D), Gm
Antonio Galgaro, Ng
Roberto Gatto, Pg
Luca Giusberti, (D), Pm
Lara Maritan, (D), Gx
Christine Marie Meyzen, (D), Cg
Paolo Mozzi, (D), Gm
Leonardo Piccinini, (D), Hg
Manuel Rigo, (D), Gs
Alberta Silvestri, (D), Gz

## Università degli Studi di Pavia

**Dept of Earth and Environmental Sciences** (A,B,M,D) O (2014)
via Ferrata, 1
Pavia 27100
p. +30 0382 985754
dvagnini@unipv.it
http://sciter.unipv.eu/site/home.html
*Enrollment (2014): B: 45 (25) M: 25 (14) D: 15 (5)*

## Universita Degli Studi di Siena

**Centro di GeoTecnologie** (2011)
Via Vetri Vecchi 34
San Giovanni Valdarno 52027
p. +390559119400
bottacchi@unisi.it
http://www.geotecnologie.unisi.it

## Universita di Bari

**Dept Geomineralogico** (2013)
Via E. Orabona
4-70125 Bari
scandale@geomin.uniba.it
http://www.geomin.uniba.it/frame.htm

## Universita di Cagliari

**Dept di Geoingegneria e Tecnologie Ambientali** (2006)
Via Marengo,3
09124 Cagliari
p. +39 70 675 52/29
mazzella@unica.it
http://geoing.unica.it/digita.htm

Professor:
Prof. Antonio AM MAZZELLA, Gq

## Universita di Calabria
**Dept of Biology Ecology and Earth Science (DIBEST)** (D) (2013)
87036 Arcavacata di Rende
Calabria
crisci@unical.it
http://www.unical.it

Italy

## Universita di Camerino
**Dept Scienze della Terra** (2014)
via E. Betti 1/A
62032 Camerino
p. (+39) 737402126
sst@pec.unicam.it
http://www.sst.unicam.it/SST/en/course-of-degree

## Universita di Catania
**Inst Scienze della Terra** (2014)
Corso Italia 57
95129 Catania
p. (+39) 095-7195730
giovali@unict.it
http://www3.unict.it/idgeg/

## Universita di Firenze
**Dept Scienze della Terra** (B,M,D) ● (2014)
via La Pira 4
50121 Firenze
direttore@geo.unifi.it
http://www.dst.unifi.it/

## Universita di Genova
**Dept Scienze della Terra Ambiente e Vita - DISTAV** (B,D) (2015)
Corso Europa 26
16132 Genoa
p. +39 010 353 8311
direttore@dipteris.unige.it
http://www.distav.unige.it/drupalint/index.php

Professor:
Egidio Armadillo, (D), Ye

Research Associate:
Donato Belmonte, (D), Cg

## Università di Modena e Reggio Emilia
**Dept di Scienze Chimiche e Geologiche** (B,M,D) (2013)
Largo S. Eufemia, 19
Modena, Italy 41121
p. 390592055885
direttore.chimgeo@unimore.it
http://www.dscg.unimore.it/
*Enrollment (2013): B: 23 (13) M: 14 (4) D: 10 (9)*

## Universita di Napoli FEDERICO II
**Dipartimento di Scienze della Terra, dell'Ambiente e delle Risorse** (B,M,D) ● (2016)
L.go S. Marcellino, 10
Naples 80138
p. +390812538112
domenico.calcaterra@unina.it
http://www.distar.unina.it

## Universita di Perrugia
**Dept Scienze della Terra** (2014)
Piazza Università, 1
06100 Perugia
cclgeol@unipg.it
http://cclgeol.unipg.it/cclgeol/

## Università di Torino
**Dept Scienze della Terra** (B,M,D) ● (2016)
via Valperga Caluso 35
10125 Torino
p. 011.6705144
segreteria.dst@unito.it
http://www.dst.unito.it/do/home.pl
*Enrollment (2015): B: 201 (20) M: 77 (6) D: 31 (6)*

Professor:
Fernando Camara Artigas, (D), Gz
Rodolfo Carosi, (D), Gc
Daniele Castelli, (D), Gx

Associate Professor:
Rossella Arletti, (D), Gz
Elena Belluso, (D), Gz
Piera Benna, (M), Gz
Alessandro Borghi, (D), Gx
Marco Bruno, (D), Gz
Paola Cadoppi, (D), Gc
Giorgio Carnevale, (D), Pg
Cesare Comina, (D), Yx
Domenico Antonio De Luca, (D), Hy
Francesco dela Pierre, (D), Gd
Massimo Delfino, (D), Pg
Anna Maria Ferrero, (D), Nr
Andrea Festa, (D), Gc
Maria Gabriella Forno, (M), Gg
Giandomenico Fubelli, (D), Gm
Marco Gattiglio, (D), Gc
Marco Giardino, (D), Gm
Roberto Giustetto, (D), Gz
Giuseppe Mandrone, (D), Ng
Luca Martire, (D), Gd
Michele Motta, (M), Oy
Mauro Prencipe, (D), Gz
Franco Rolfo, (D), Gx
Piergiorgio Rossetti, (M), Eg
Marco Rubbo, (M), Gz

Assistant Professor:
Roberto Ajassa, (M), Oy
Gianni Balestro, (D), Gc
Carlo Bertok, (D), Gd
Sabrina Maria Rita Bonetto, (D), Ng
Silvana Capella, (D), Gz
Corrdao Cigolini, (M), Gv
Emanuele Costa, (D), Gz
Anna d'Atri, (D), Gr
Simona Ferrando, (D), Gx
Simona Fratianni, (D), Oy
Franco Gianotti, (D), Gg
Daniele Giordano, (D), Gv
Chiara Teresa Groppo, (D), Gx
Francesca Lozar, (D), Pg
Edoardo Martinetto, (D), Pb
Luciano Masciocco, (M), Ge
Luigi Motta, (M), Oy
Marco Davide Tonon, (M), Oe
Sergio Carmelo Vinciguerra, (D), Nr
Elena Zanella, (D), Ym

Emeritus:
Emiliano Bruno, (M), Gz
Ezio Callegari, (M), Gx
Roberto Compagnoni, (M), Gx
Giovanni Ferraris, (M), Gz
Giulio Pavia, (M), Pg
Germano Rigault de la Longrais, (M), Gz

## Università di Trieste
**Dipartimento di Matematica e Geoscienze** (D) (2016)
via Weiss 2
Trieste , Italy I-34128
p. +39 040 5582055
dmg@pec.units.it
http://www.geoscienze.units.it/

## Università di Udine
**Dipartimento di Georisorse e Territorio** (2010)
Dipartimento di Georisorse e Territorio
Via Cotonificio 114
33100 Udine
ta.tempoindeterminato@uniud.it

http://udgtls.dgt.uniud.it/

## Universita di Urbino
**Inst di Geologia Applicata** (2014)
Via Muzio Oddi 14
61029 Urbino

## Universita Pisa
**Dept of Geosciences** (D) (2004)
via S. Maria 53
56126 Pisa
p. +39050847260
martinelli@dst.unipi.it
http://www.dst.unipi.it/

## University of Ferrara
**Dept of Geology** (2014)
Via Savonarola, 9
44121 Ferrara
p. (+39) 0532 293493
mob_int@unife.it
http://www.unife.it/international/education/double-degrees-1/geological-sciences-ferrara-cadiz

## University of Milano
**Dept of Geology** (2014)
Via Festa Del Perdono 7
20126 Milano
p. (+39) 02 6448 1
foreign.office@unimib.it
http://www.unimib.it/go/46204/Home/English/Academic-Programs/Mathematics-Physics-and-Natural-Sciences/Geological-Sciences-and-Te

## University of Parma
**Dept of Physics and Earth Sciences** (B,M,D) (2014)
Parco Area delle Scienze 157/A
Parma 43100
p. +39 0521 905326
dipterr@unipr.it
http://www.difest.unipr.it
Professor:
Fulvio Celico, (D), Hgw
Research Associate:
Andrea Artoni, (D), Gtr
Tiziano Boschetti, (D), Cgs

## University of Siena
**Dept of Physical Science, Earth, and Environment** (B,M,D) ● (2017)
Strada Laterina, 8
Siena 53100
p. (+39) 0577 233938
pec.dsfta@pec.unisipec.it
http://www.dsfta.unisi.it/it
f: https://www.facebook.com/dsfta.siena?ref=hl
Professor:
Mauro Coltorti, (M), GmYg
Associate Professor:
Luca Maria Foresi, Gs
Cecilia Viti, Gz

# Jamaica

## University of the West Indies Mona Campus
**Dept of Geography and Geology** (B,M,D) (2014)
University of the West Indies
Mona
Kingston KGN7
p. 876-927-2728
geoggeol@uwimona.edu.jm
http://myspot.mona.uwi.edu/dogg/
Head:
Simon F. Mitchell, (D), Liverpool, 1993, GsgGs

# Japan

## Akita University
**Deparment of Earth Science and Technology** (2004)
1-1 Tagata Gakuen-cho, Akita-shi
Akita

## Ehime University
**Dept of Earth Sciences** (2004)
hori.rie.mm@ehime-u.ac.jp
http://www.ehime-u.ac.jp/~cutie/index.html

## Geological Survey of Japan
**Geological Survey of Japan** (2011)
http://www.gsj.jp/

## Hirosaki University
**Dept of Earth Science** (2004)
wata@cc.hirosaki-u.ac.jp
http://sci.hirosaki-u.ac.jp/~earth/index.html

## Hiroshima University
**Dept of Earth and Planetary Systems Science** (2005)
Kagami-yama 1-3-1
Higashi-Hiroshima
Hiroshima 739
toiawase@geol.sci.hiroshima-u.ac.jp
http://www.geol.sci.hiroshima-u.ac.jp/index_e.html

## Kagoshima University
**Faculty of Science** (2004)
koko@sci.kagoshima-u.ac.jp
http://earth.sci.kagoshima-u.ac.jp/index.html

## Kanazawa University
**Dept of Earth Sciences** (2004)
Kakuma-machi
Kanazawa 920-1192
fsci-pla-director@edu.kobe-u.ac.jp
http://earth.s.kanazawa-u.ac.jp/

## Kobe University
**Earth & Planetary Sciences** (2004)
fsci-pla-director@edu.kobe-u.ac.jp
http://shidahara1.earth.s.kobe-u.ac.jp/index.html

## Kumamoto University
**Dept of Earth and Environment** (B,M,D) (2011)
2-39-1, Kurokami
Kumamoto City 860-8555
p. 81-96-342-3411
tadao@sci.kumamoto-u.ac.jp
http://www.sci.kumamoto-u.ac.jp/earthsci
Professor:
Shiro Hasegawa, (D), Pm
Toshiaki Hasenaka, (D), Gv
Hiroki Matsuda, (D), Gs
Tadao Nishiyama, (D), Gp
Hidetoshi Shibuya, (D), Ymg
Jun Shimada, (D), Hw
Akira Yoshiasa, (D), Gz

## Nagoya University
**Dept of Earth and Planetary Sciences** (2006)
env@post.jimu.nagoya-u.ac.jp
http://www.eps.nagoya-u.ac.jp/

## Shizuoka University
**Inst of Geosciences** (2005)
Shizuoka 422-8529
setmasu@ipc.shizuoka.ac.jp
http://www.sci.shizuoka.ac.jp/~geo/Welcome.html

## Tohoku University
**Dept of Mineralogy, Petrology and Economic Geology** (2015)
Sendai 980-8578
kaiho@m.tohoku.ac.jp
http://www.ganko.tohoku.ac.jp/

## Tsukuba University
**College of Geoscience** (2004)
kankyojoho@ynu.ac.jp
http://www.geo.tsukuba.ac.jp/

## University of Tokyo
**Deparment of Earth and Planetary Science** (2016)
3-1, Hongo 7chome, Bunkyo-ku
Tokyo
Hirata@eri.u-tokyo.ac.jp
http://www.eri.u-tokyo.ac.jp/

## University of Toyama
**Deptt of Earth Sciences** (B,M,D) (2015)
3190 Gofuku
Toyama City, Toyama Prefecture 930-8555
p. 81-76-445-6654
takeuchi@sci.u-toyama.ac.jp
http:/www.sci.u-toyama.ac.jp/earth/index-en.html
*Enrollment (2015): B: 208 (0) M: 36 (0) D: 7 (0)*
Professor:
Akira Takeuchi, (D), Osaka City, 1979, GtcYe
Tohru Watanabe, (D), Tokyo, 1991, YxsGv
Associate Professor:
Shigekazu Kusumoto, (D), Kyoto, 1999, YvdGt

## Yokohama National University
**Dept of Environment and Natural Sciences** (2004)
kankyojoho@ynu.ac.jp
http://chigaku.ed.ynu.ac.jp/geology-e.html

# Kenya

## Jomo Kenyatta University of Agriculture & Technology
**Geomatic Engineering and Geospatial Information Systems** (B,M,D) (2014)
PO Box 62000
-00200 Nairobi
Nairobi
p. +254-67-5352391
gegis@eng.jkuat.ac.ke
http://www.jkuat.ac.ke/departments/gegis/
*Enrollment (2014): B: 175 (35) M: 45 (19) D: 8 (0)*
Chair:
Thomas G. Ngigi, (D), Chiba Univ (Japan), 2007, OriOr
Lecturer:
Nathan O. Agutu, Jomo Kenyatta Univ of Ag and Tech, OgiOr
Mark Boitt, (M), Stuttgart Univ of Applied Sci, 2010, OgrOi
George W. Chege, (M), Jomo Kenyatta Univ of Ag and Tech, 2014, OgrOi
Charles Gaya, (M), Stuttgart Univ of Applied Sci, 2002, OrgOi
Andrew Imwati, (D), Nairobi, OggOi
Benson K. Kenduiywo, (D), Technical Univ Darmstadt, 2016, OriYd
Fridah K. Kirimi, (M), Jomo Kenyatta Univ of Ag and Tech, 2010, OiiOr
Moffat G. Magondu, (M), Jomo Kenyatta Univ of Ag and Tech, 2014, OirOg
Felix N. Mutua, (D), Tokyo, 2012, OigOr
Nancy Mwangi, (M), Jomo Kenyatta Univ of Ag and Tech, 2008, OiiOr
Mercy W. Mwaniki, (M), Jomo Kenyatta Univ of Ag and Tech, 2010, OiiOr
Eunice W. Nduati, (M), Jomo Kenyatta Univ of Ag and Tech, 2014, OrrOi
Patroba A. Odera, (D), Kyoto, 2012, YdOge
Hunja Waithaka, (D), Hokkaido, YdOgi
Charles B. Wasomi, (M), Stuttgart Univ of Applied Sci, OrrOi

**Dept of Environmental Studies** (B) (2011)
P.O. Box 62000 00200
Nairobi
pro@jkuat.ac.ke
http://www.jkuat.ac.ke/

## Kenyatta University
**School of Environmental Sciences** (B,M,D) (2010)
Maseno
p. +254-057-351620/2
dean-envs@ku.ac.ke
http://www.maseno.ac.ke/index3.php?section=schools&page=school-encironment

## University of Nairobi
**Dept of Geology** (B) (2016)
PO Box 30197
Nairobi
p. +254 20 4449856
geology@uonbi.ac.ke
http://geology.uonbi.ac.ke/

**Dept of Geography** (B) (2010)
PO Box 30917
Nairobi
p. +254 (020) 318262 (28400)
samowuor@uonbi.ac.ke
http://uonbi.ac.ke/departments/?dept_code=HF&&face_code=32

**Dept of Meteorology** (B,M,D) (2011)
PO Box 30197
Nairobi 00100
p. 254 020-4449004 (2070)
dept-meteo@uonbi.ac.ke
http://www.uonbi.ac.ke

# Korea, South

## Korea University
**Earth and Environmental Sciences** (B,M,D) (2015)
Anam-dong, Seongbuk-Gu
Seoul 136-713
p. 82-2-3290-3170
sjchoh@korea.ac.kr
http://ees.korea.ac.kr/
*Enrollment (2015): B: 100 (27) M: 30 (13) D: 8 (4)*
Head:
Meehye Lee, (D), Rhode Island, 1999, OagCm
Vice President for Research Affairs:
Seong-Taek Yun, (D), Korea, 1991, ClHyCs
Vice Dean of Office of Planning and Budget:
Young Jae Lee, (D), SUNY Stony Brook, 2005, ScGze
Vice Dean of College of Science:
Ho Young Jo, (D), Wisconsin, 2003, NgCaGe
Professor:
Seon-Gyu Choi, (D), Waseda, 1983, GzEmg
Seong-Jae Doh, (D), Rhode Island, 1987, YmgYe
Jin-Han Ree, (D), SUNY (Albany), 1991, Gct
Associate Professor:
Suk-Joo Choh, (D), Texas (Austin), 2004, GdsPe
Scott A. Whattam, (D), Hong Kong, 2002, GiCtg

# Latvia

## Latvija Valst Geologijas Dienests
**Latvia State Geological Survey** (2011)
Moscow street 165
Riga, LV-1019
lvgma@lvgma.gov.lv
http://mapx.map.vgd.gov.lv/geo3/

## University of Latvia
**Faculty of Geographical and Earth Sciences** (2004)
10 Alberta Str.
202-203 Room
Riga
zeme@lu.lv
http://www.lu.lv/e_strukt/fakult/geog/info/

# Lebanon

## American University of Beirut

**Dept of Geology** (2004)
Faculty of Arts and Sciences
P.O.Box 11-0236
Riad El-Solh
Beirut 1107 2020
p. (961)-1-340460/ext. 4160
arahman@aub.edu.lb

# Lesotho

## National University of Lesotho

**Dept of Geology & Mines** (B) (2010)
P.O. Box 750, Maseru
p. (266) 34 0601
registrar@nul.ls
http://www.nul.ls/

**Dept of Geography and Environmental Science** (B) (2014)
P.O. Roma 180
Maseru 100
p. +266 2234 0601
registrar@nul.ls
http://www.nul.ls/faculties/fost/geography/index.html

# Liberia

## University of Liberia

**Dept of Geology** (B) (2014)
Capitol Hill
PO Box 9020
Monrovia 9020
p. 231-6-422-304
weekso@fiu.edu
http://www.universityliberia.org/ul_course_master_list_biology.htm#geology

**Dept of Mining Engineering** (B) (2010)
Captiol Hill
PO Box 9020
Monrovia 9020
p. +231 6422304
weekso@fiu.edu
http://www.universityliberia.org/ul_course_master_list_mining.htm#mine

# Libya

## Oil Companies School

**Oil Companies School** (B) (2010)
info@ocslibya.com
http://www.geocites.com/chatalaine/OCS.html

## Petroleum Training and Qualifying Institute (PTQI)

(B) (2014)
Gergarish Road 9KM - Asiahia
Tripoli
p. +218 21 4833771-5
info@ptqi.edu.ly
http://www.ptqi.edu.ly/online/en_home.php#

## University of Garyounis

**Dept of Geology** (B) (2014)
PO Box 1308
Benghazi
p. +218-61-86304
uni.office@uob.edu.ly
http://www.garyounis.edu/

# Lithuania

## Lithuanian Geological Survey under the Ministry of Environment

**Lithuanian Geological Survey under the Ministry of Environment** (2015)
S. Konarskio St. 35
LT-03123 Vilnius
indre.virbickiene@lgt.lt
http://www.lgt.lt/

## Vilniaus Pedagoginis Universitetas

**Dept of Geology** (2014)
Studentu 39, 2034 Vilnius
p. (8 5) 275 89 35
gmtf.dekanatas@leu.lt
http://www.leu.lt/lt/gmtf/gmtf_apie_mus/all.html

## Vilniaus Universitetas

**Dept of Hydrogeology & Engin Geol** (2014)
Giurlionio 21/27, 2009, Vilnius
p. 239 8278
robert.mokrik@gf.vu.lt
http://www.vu.lt/en/scientific-report-2012/faculties-and-institutes/faculty-of-natural-sciences#DEPARTMENT_OF_HYDROGEOLOGY_AND_E

**Dept of Geology and Mineralogy** (B,M,D) ● (2015)
Ciurlionio str. 21/27
LT03101
Vilnius LT03101
p. +370 5 2398272
eugenija.rudnickaite@gf.vu.lt
http://www.geol.gf.vu.lt/lt
*Enrollment (2015): B: 8 (8) M: 11 (11) D: 10 (3)*

# Luxembourg

## Service Geologique du Luxembourg

**Service Geologique du Luxembourg** (2011)
23, rue du Chemin de Fer
L-8057 Bertrange
geologie@pch.etat.lu
http://www.pch.public.lu/administration/organigramme/geo/

# Madagascar

## Universite d' Antananarivo

**Dept of Chemistry** (B) (2010)
BP 566
Antananarivo 101
p. +261 20 22 326 39
rtianasoa.manoelson@gmail.com
http://www.univ-antananarivo.mg/

**Dept of Biology** (B) (2010)
BP 566
Antananarivo 101
p. +261 20 22 329 39
arovonjy@yahoo.fr
http://www.univ-antananarivo.mg/

**Dept des Sciences de la Terre** (A,B,M,D) (2017)
BP 906
Antananarivo 101 , Madagascar
p. 00261324205328
saphirzzf@gmail.com
http://www.univ-antananarivo.mg/
*Enrollment (2016): A: 144 (138) B: 127 (111) M: 63 (14) D: 1 (0)*

## Universite de Mahajanga

**Dept de Sciences de la Terre et Sciences de l'Environnement** (B) (2014)
p. +261 32 05 579 44

# Malawi

## Mzuzu University

**Faculty of Environmental Sciences** (B) (2014)
Private Bag 201
Mzuzu
p. +(265) 1 320 722/ 320 575
ur@mzuni.ac.mw
http://www.mzuni.ac.mw/index.php?option=com_content&view=article&id=21&Itemid=15

## University of Malawi

**Dept of Natural Resources Management** (B) (2010)
PO Box 219
Lilongwe
p. (265) 01 277 260
david.mkwambisi@bunda.luanar.mw
http://www.bunda.unima.mw/nrm.thm

**Chancellor College, Natural Resources and Environment Centre** (B) (2014)
PO Box 280
Zomba
p. +(265) 1 524 685
geo@chanco.unima.mw
http://www.chanco.unima.mw/department/department.php?DepartmentID=4&Source=Department_of_Geography_and_Earth_Sciences

# Malaysia

## Minerals and Geoscience Department Malaysia

**Minerals and Geoscience Dept Malaysia** (2011)
20th Floor, Bangunan Tabung Haji
Jalan Tun Razak
50658 Kuala Lumpur
jmgkll@jmg.gov.my
http://www.jmg.gov.my/

## National University of Malaysia

**Geology Dept** (2004)
43600 Bangi
Selangor
dftsm@ftsm.ukm.my

## Universiti Putra Malaysia

**Dept of Environmental Sciences** (2004)
43400 UPM Serdang
puziah@env.upm.edu.my
http://fsas.upm.edu.my/~sas/envpage/Dept.html

## University of Malaya

**Dept of Geology** (B,M,D) (2007)
Dept. of Geology
University of Malaya
50603 Kuala Lumpur, Malaysia.
Kuala Lumpur, Selangor 50603
p. 03-79674203
ketua_geologi@um.edu.my
http://www.um.edu.my

Head:
Wan Hasiah Abdullah, (D), Newcastle upon Tyne, 1994, Ec
Professor:
Teh Guan Hoe, (D), Heidelberg, Ca
Lee Chai Peng, (D), Liverpool, Pg
John Kuna Raj, (D), Malaya, 1983, Ng
Denis N.K Tan, (M), Malaya, Go
Associate Professor:
Abd Rashid Ahmad, (D), Oxford, Sc
Azman Abdul Ghani, (D), Liverpool, Gi
Mohamed Ali Hasan, (M), Ulster, Hg
Azhar Hj Hussin, (D), London, Gs
Tajul Anuar Jamaluddin, (D), Wales, 1997, Ng
Mustaffa Kamal Shuib, (M), London, 1986, Gc
Samsudin Hj. Taib, (D), Durham, Yg
Lecturer:
Nuraiteng Tee Abdullah, (D), London, Gr
Ahmad Tajuddin Hj Ibrahim, (D), Newcastle upon Tyne, Ng
Che Noorliza Lat, (M), Nevada (Reno), 1989, Yg
Mat Ruzlin Maulud, (B), Malaya, On
Nur Iskandar Taib, (D), Indiana, Gi
Ismail Yusoff, (D), Norwich, Hg

# Mali

## Universite de Bamako

**Sciences de la Terre** (B) (2010)
BP 3206
Bamako
p. (223) 222 32 44
phine.romagnoli@unige.ch
http://www.ml.refer.org/u-bamako/spip.php?article.137

# Malta

## University of Malta

**Dept of Geosciences** (M,D) O (2015)
Department of Geosciences
University of Malta
Msida Campus
Msida MSD2080
p. (+356) 2340 2362
geo.sci@um.edu.mt
http://www.um.edu.mt/science/geosciences
*Enrollment (2015): M: 5 (0) D: 4 (0)*

Dr:
Pauline Galea, (D), Victoria Univ of Wellington (NZ), 1993, YsgOg
Associate Professor:
Aldo Drago, (D), Southampton, Opg
Raymond Ellul, (D), Oa
Dr:
Sebastiano D'Amico, (D), YsgOg
Anthony Galea, (D), Trieste, Opg
Aaron Micallef, (D), Southampton, 2008, GumOu
Lecturer:
Noel Aquilina, (D), Birmingham, Oaw
Adam Gauci, (M), Op

# Mauritania

## Universite de Nouakchott

**Dept de Geologie** (B) (2014)
BP 5026
Nouakchott
p. (222) 25 13 82
awa@univ-nkc.mr
http://www.univ-nkc.mr/spip.php?rubrique34

# Mauritius

## University of Mauritius

**Mauritius Radio Telescope** (B) (2014)
Reduit
p. (230) 454 1041
nalini@uom.ac.mu
http://www.uom.ac.mu/mrt/mrt2.html

**Dept of Chemical and Environmental Engineering** (B) (2010)
Reduit
p. (230) 454 1041
d.surroop@uom.ac.mu
http://www.uom.ac.mu/Faculties/FOE/CEE/index.asp

# Mexico

## Centro de Estudios Superiores del Estado Sonora

**Escuela Superior de Geociencias** (B) (2004)
Col. Las Quintas
Hermosillo, SO 83240

p. 621/5-37-78
manuel.valenzuela@ues.mx
Department Secretary: Manuel Valenzuela-Renteria
Director:
Marco A. Gonzalez-Juarez, (B), Sonora, 1977, Gg
Associate Professor:
Jesus E. Cruz-Teran, (B), Sonora, 1979, On
Leopoldo Diaz-Encinas, (B), Chihuahua, 1988, Em
Gustavo E. Durazo-Tapia, (B), Sonora, 1981, Em
Francisco A. Esparza-Yanez, (B), Sonora, 1983, Gg
Raul A. Gongora-Jurado, (B), Autonoma de Chihuahua, 1977, Nm
Maria B. Hurtado-De La Ree, (B), Sonora, 1987, Nx
Leobardo Lopez-Pineda, (B), Nacional Auton, 1980, Yg
Jesus F. Maytorena-Silva, (B), Sonora, 1981, Gr
Rogelio Monreal-Saavedra, (D), Texas (Dallas), 1989, Gr
Gerardo Monteverde-Gutierrez, (B), Sonora, 1980, Nm
Arnulfo Salazar-Avila, (B), Sonora, 1981, Nx
Angel Slistan-Grijalva, (B), Sonora, 1981, Nm
Instructor:
Luis G. Vite-Picazo, (B), Nacional Auton, 1946, Nm

## Centro de Investigación Científica y de Educación Superior de Ensenada

**Earth Sciences Division** (M,D) (2016)
Carretera Tijuana-Ensenada # 3918
Zona Playitas
Ensenada, Baja California 22860
p. (01152-664)1750500
dir-ct@cicese.mx
http://www.cicese.mx
*Enrollment (2010): M: 44 (8) D: 20 (2)*
Titular researcher:
Edgardo Canon-Tapia, (D), Hawaii, 1996, GvYmGt
Research Associate:
Jose G. Acosta, (M), CICESE (Ensenada), 1980, Ne
Jesus M. Brassea, (M), Cinvestav, 1986, Ye
Juan M. Espinosa, (M), CICESE (Ensenada), 1983, Ye
Jose J. Gonzalez, (M), CICESE (Ensenada), 1986, Yd
Alejandro Hinojosa, (M), CICESE (Ensenada), 1988, Or
Luis H. Mendoza, (M), CICESE (Ensenada), 1982, Ne
Alfonso Reyes, (B), UNAM, Ne
Titular researcher:
Raul Castro, (D), Nevada (Reno), 1991, Ys
Juan Contreras, (D), Columbus, 1999, Gt
Luis A. Delgado-Argote, (D), UNAM, 2000, GitGu
Francisco Esparza, (D), CICESE (Ensenada), 1991, Ye
John Fletcher, (D), Utah, 1994, Gpt
Carlos Flores, (D), Toronto, 1986, Ye
Jose Frez, (M), California (Los Angeles), 1980, Ys
Juan Garcia, (D), Oregon State, 1990, Yv
Ewa Glowacka, (D), Polish Acad of Sci, 1991, Ys
Enrique Gomez, (D), Toronto, 1981, Ye
Titular Researcher:
Mario Gonzalez-Escobar, (D), CICESE, 2002, YesYs
Titular researcher:
Antonio Gonzalez-Fernandez, (D), Complutense, 1996, Gu
Javier Helenes, (D), Stanford, 1980, Gr
Thomas Kretzschmar, (D), Tubingen, 1995, Ge
Margarita Lopez, (D), Toronto, 1985, Cc
Arturo Martin, (D), Paris XI Orsay, 1988, Gd
Luis Munguia, (D), California (San Diego), 1982, Ys
Alejandro Nava, (D), California (San Diego), 1980, Ys
Marco A. Perez, (D), CICESE, 1995, Ye
Jose M. Romo, (D), CICESE (Ensenada), 2002, Ye
Pratap Sahay, (D), Alberta, 1986, Ys
Rogelio Vazquez, (D), CICESE (Ensenada), 2002, Ye
Antonio Vidal, (D), CICESE (Ensenada), 2001, Ys
Bodo Weber, (D), Gp

## Ciudad Universitaria

**Facultad de Ingenieria** (2014)
C.P. 04510
p. 56 22 08 66
fainge@servidor.unam.mx
http://www.ingenieria.unam.mx/

## SGM

**Servicio Geológico Mexicano** (2016)
Blvd. F. Angeles 93.50-4 km.
Hidalgo 42080, Pachuca 42083
gciadoctec@sgm.gob.mx
http://www.sgm.gob.mx/

## Universidad Autonoma de Baja California Sur

**Departamento Académico de Geología Marina** (2004)
Carretera Sur Km 5-1/2
Cd. Universitaria
La Paz

**Dept of Marine Geology** (B) (2004)
Carretera al Sur km 5.5, Box 19-B
La Paz, BS 23000
p. 682/2-47-55/2-01-40/2-45-69
oceanologia.fcm@uabc.edu.mx
Chair:
Alejandro Alvarez-Arellano, (M), Nacional Auton, 1984, Gs
Professor:
Rodolfo Cruz-Orozco, (D), Louisiana State, 1974, Gu
Javier Gaitan-Moran, (M), Intl Inst-Aerospace Sur & Earth Sci, 1986, Or
Carlos A. Galli-Olivier, (D), Utah, 1968, Gd
Jose I. Peredo-Jaime, (B), Auto de Baja California, 1978, Og
Luis R. Segura-Vernis, (D), Nacional Autonoma, 1977, Pi
Associate Professor:
Alejandro J. Carillo-Chavez, (M), Cincinnati, 1981, Gp
Efrain Cornejo-Luna, (B), Nacional Auton, 1969, Gq
Genaro Martinez-Gutierrez, (B), Inst Politecnico Nac, 1982, Gr
Cesar Martinez-Noriega, (B), Nacional Auton, Ou
Jaime I. Monroy-Sanchez, (M), 1987, Ge
Miriam Nunez-Velazco, (B), Tech (Madero), 1980, Go
Jose A. Perez-Venzor, (B), Autonoma (SLP), 1978, Gi
Ramon Pimentel-Hernandez, (B), Inst Politecnico Nac, 1980, Yg
Humberto Rojas-Soriano, (B), Nacional Auton, 1983, Gz
Paulino Rojo-Garcia, (B), Inst Politecnico Nac, 1983, Gc
Assistant Professor:
Luis A. Herrera-Gil, (B), Nacional Auton, 1980, Pg
Lecturer:
Cesar A. Lopez-Ferreira, (B), Nacional Auton, 1979, Hw
Cooperating Faculty:
Oscar Rodriguez-Plasencia, (B), Escuela Militar de Met, 1959, Ow

## Universidad Autonoma de Chihuahua

**Facultad de Ingenieria** (B,M) (2004)
Circuito Universitario Campus 2
Chihuahua, CH 31170
p. (614) 442-9500
ancorral@uach.mx
Director:
Arturo Leal-Bejarano, (B), Autonoma de Chihuahua, 1976, On
Head:
Arturo Lujan-Lopez, (M), Essex, 1975, On
Chair:
Socorro I. Aguirre-Moriel, (B), ITR (Switzerland), 1979, Nx
Hector M. Mendoza-Aguilar, (B), Autonoma de Chihuahua, 1980, Nx
Professor:
Rafael Chavez-Aguirre, (M), Autonoma de Chihuahua, 1993, Hw
Adolfo Chavez-Rodriguez, (D), Arizona, 1987, Hy
Miguel Franco-Rubio, (M), Nacional Autonoma, 1978, Eg
Rafael Madrigal-Rubio, (B), Nacional Auton, 1969, Ng
Teodulo Mena-Zambrano, (B), Inst Politecnico Nac, 1964, Ng
Hector Minor-Velazquez, (B), Autonoma de Chihuahua, 1978, Nx
Ignacio A. Reyes-Cortes, (B), Texas (El Paso), 1997, Gx
Manuel Reyes-Cortes, (B), Nacional Auton, 1970, Gi
Miguel Royo-Ochoa, (M), Autonoma de Chihuahua, 1997, Hw
David H. Ruiz-Cisneros, (B), Autonoma de Chihuahua, 1974, Nx
Luis M. Trevizo-Cano, (B), Autonoma de Chihuahua, 1978, Nx

## Universidad Autonoma de Nuevo Leon

**Facultad de Ciencias de la Tierra** (2004)
Hacienda Guadalupe Km. 8 Camino a Cerro Prieto. - Linares, N 67700

p. (81) 8329 4170 Ext. 4170
fmedina@fct.uanl.mx
http://www.fct.uanl.mx/portal/index.php

Mexico

## Universidad Autonoma de San Luis Potosi

**Area de Ciencias de la Terra** (B,M) (2004)
Facultad de Ingenieri#2.a
Av. Dr. Manuel Nava No. 8
San Luis Potosi#2., SL 78290
p. (48) 13-82-22
mlr@fciencias.uaslp.mx
Department Secretary: Juan Manuel Torres Aguilera

Director:
Hector David Atisha-Castillo, (M), Texas, 1978, On

Head:
Joel Milan-Navarro, (M), Nancy (France), 1979, Or

Professor:
Jose Refugio Acevedo-Arroyo, (B), Nacional Auton, 1957, Ng
Luis Garcia-Gutierrez, (M), Stanford, 1951, Eg
Panfilo R. Martinez-Macias, (M), Nacional Auton, 1995, Gc
Francisco Javier Orozco-Villasenor, (M), Colorado State, 1983, Eg
Ramon Ortiz-Aguirre, (M), Madrid, 1979, Hg
Carlos Francisco Puente-Muniz, (B), Autonoma (SLP), 1979, Hw
Delfino C. Ruvalcaba-Ruiz, (D), Colorado State, 1982, Em
Juan Manuel Torres-Agulera, (B), Autonoma (SLP), 1980, Gx

Librarian:
Panfilo R. Martinez-Macias, On

## Universidad Autonoma de Sonora

**Ejecutivo de la Unidad Regional Norte** (2004)
Av. Universidad e Irigoyen
Col. Ortiz
Hermosillo

## Universidad de Guanajuato

**Departamento de Ingeniería en Minas, Metalurgia y Geología** (B) O (2016)
Sede San Matias
Ex Hda. de San Matías s/n
Col. San Javier
Guanajuato, Guanajuato C.P. 36025
p. ((+473) 73 2 2291 (Ext. 5304)
juancho@ugto.mx
http://www.di.ugto.mx/minas/index.php/antecedentes
*Enrollment (2015): B: 7 (7)*

## Universidad Nacional Autonoma de Mexico

**Facultad de Ingeniería** (B) (2014)
Av. Universidad 3000, Ciudad Universitaria
Coyoacán, CP 4510
p. 56 22 08 66
fainge@servidor.unam.mx
http://www.ingenieria.unam.mx/paginas/Carreras/ingenieriaMinas/ingMinas_Desc.php

Director:
Antonio Nieto Antunez, (M), Stanford, 1970, Nm

Professor:
Angelica Casillas, (B), Inst Politecnico Nac, 1976, Hw
Esteban Cedillo, (D), Heidelberg, 1988, Em
Juventino Martinez, (D), Paris VI, 1980, Gt
Edgardo Meave, (B), Guanajuato, 1947, Gg
Francisco Medina, (B), Guanajuato, 1982, Og
Ricardo Navarro, (B), Inst Politecnico Nac, 1973, Nm
Salvador Ulloa, (B), Guanajuato, 1947, Eg
Fernando Vasallo, (D), Lomonosov (USSR), 1982, Em
Carlos Yanez, (B), Inst Politecnico Nac, 1976, Ce

**Div de Estudios de Posgrado-Campus Morelos** (2004)
C.P. 62550
Facultad de Ingeneria
Jiutepec, MR
p. (73) 19 39 57 (x. 135)

Director:
Alvaro Munoz Mendoza, (D), Strathclyde, 1989, On

Acting Head:
Hugo Acosta Borbon, (M), Nacional Auton, 1992, On
Esperanza Ramirez Camperos, (M), Nacional Auton, 1989, On

**Inst de Geofisica** (2004)
losos@sci.muni.cz
http://tlacaelel.igeofcu.unam.mx/index.eng.html

**Inst de Geologia** (M,D) (2010)
Apdo. postal 70-296
Ciudad Universitaria
Delegacion Coyoacan, DF 04510
p. (52) 5622 4314
igl@geologia.unam.Mx
http://132.248.20.1/geol.htm
Administrative Assistant: Ana María Rodriguez Simental

Director:
Dante J. Moran-Zenteno, (D), Nacional Auton, 1992, Cc

Head:
Susana A. Alaniz Alvarez, (D), Nacional Auton, 1995, Gc
Luca Ferrari Pedraglio, (D), Milan, 1992, Gt
Sergio Cevallos Ferriz, (D), Alberta, 1990, Po
Carlos M. Gonzalez Leon, (D), Arizona, 1992, Gr
Klavdia Oleschko Loutkova, (D), Lomosov (Moscow), 1984, So
Maria S. Lozano Garcia, (D), d'Aix (Marseilles), 1979, Pe
Francisco J. Vega Vera, (D), Nacional Auton, 1988, Py

Chair:
Gustavo Tolson Jones, (D), Nacional Auton, 1998, Gc

Senior Scientist:
Shelton Applegate-Pleasants, (D), Chicago, 1961, Pv
Jorge Aranda, (D), Oregon, 1982, Gi
Blanca E. Buitron-Sanchez, (D), Nacional Auton, 1974, Pi
Oscar Carranza-Castaneda, (D), Nacional Auton, 1989, Pv
Gerardo Carrasco-Nunez, (D), Michigan Tech, 1993, Gv
Ana L. Carreno, (D), d'Orsay (Paris), 1979, Pm
Miguel Carrillo Martinez, (D), Paris VI, 1976, Pb
Liberto De Pablo, (D), Ohio State, 1958, Gz
Rodolfo Del Arenal-Capetillo, (B), Nacional Auton, 1960, Hw
Ismael Ferrusquia, (D), Houston, 1971, Gr
David Flores-Roman, (D), Nacional Auton, 1981, So
Celestina Gonzalez-Arreola, (D), Nacional Auton, 1989, Pi
Jose C. Guerrero-Garcia, (D), Texas (Dallas), 1975, Ym
John D. Keppie, (D), Glasgow, 1964, Gt
Victor M. Malpica-Cruz, (D), Bordeaux, 1980, Gs
Enrique Martinez Hernandez, (D), Michigan State, 1979, Pl
Juventino Martinez-Reyes, (D), Paris VI, 1980, Gt
Luis M. Mitre-Salazar, (D), Paris IV, 1978, Ge
Adrian Ortega, (D), Waterloo, 1993, Hw
Fernando Ortega-Gutierrez, (D), Leeds (UK), 1975, Gp
Sergio Palacios-Mayorga, (M), Nacional Auton, 1972, So
Jerjes Pantoja-Alor, (M), Arizona, 1963, Gg
Maria del Carmen Perrilliat-Montoya, (D), Nacional Auton, 1969, Pi
Angel Nieto Samaniego, (D), Nacional Auton, 1994, Gc
Christina D. Siebe Grabach, (D), Hohenheim (Germany), 1993, Sc
Alicia Silva-Pineda, (D), Nacional Auton, 1980, Po
Max Suter-Cargnelutti, (D), Basel, 1978, Gt
Jordi Tritlla, (D), Barcelona, 1994, Eg
Ana B. Villasenor-Martinez, (D), Nacional Auton, 1991, Pi
Reinhard Weber-Gobel, (D), Tubingen, 1967, Po

Associate Scientist:
Gerardo J. Aguirre Diaz, (D), Texas, 1993, Gv
Thierry Calmus, (D), Paris VI, 1983, Gu
Antoni Camprubi, (D), Barcelona, 1998, Eg
Alejandro J. Carillo Chavez, (D), Wyoming, 1996, Hg
Elena Centeno Garcia, (D), Arizona, 1994, Gt
Rodolfo Corona-Esquivel, (M), Nacional Auton, 1985, Em
Mariano Elias-Herrera, (M), Nacional Auton, 1982, Gp
Maria L. Flores-Delgadillo, (M), Nacional Auton, 1987, So
Gilberto Hernandez-Silva, (D), Nacional Auton, 1983, So
Rafael Huizar-Alvarez, (D), Franche Comte, 1989, Hw
Cesar Jacques Ayala, (D), Cincinnati, 1983, Gr
Marisol Montellano-Ballesteros, (D), California (Berkeley), 1986, Pv
Amabel M. Oretega Rivera, (D), Queens, 1997, Cc
Odranoel Quintero, (D), Sorbonne, 1995, Gc
Jose L. Rodriguez-Castaneda, (M), Pittsburgh, 1984, Gg
Jaime Roldan Quintana, (M), Iowa, 1976, Gv
Gerardo Sanchez-Rubio, (M), Imperial Coll (UK), 1984, Gv

Jesu Sole, (D), Barcelona, 1996, Cc
Luis F. Vassallo-Morales, (D), Lomonosov, 1981, Em
Maria G. Villasenor-Cabral, (M), Leeds (UK), 1974, Ca
Research Associate:
Irma Aguilera-Ortiz, (B), Iberoameric, 1971, Ca
Victor M. Davila-Alcocer, (M), Texas (Dallas), 1986, Ps
Jose G. Solorio-Munguia, (B), Nacional Auton, 1958, Gz
Adjunct Professor:
J. Duncan Keppie, (D), Glasgow (Scotland), 1967, Gt
Luis Silva-Mora, (D), d'Aix (Marseilles), 1979, Gi
Emeritus:
Gloria Alencaster-Ybarra, (D), Nacional Auton, 1969, Pi
Zoltan De Cserna, (D), Columbia, 1955, Gt
Geology Librarian:
Teresa Soledad Medina Malagon, (B), On

# Mongolia

## Gazarchin Institute

**Gazarchin Institute** (2011)
Bayanzurkh District
Ulaanbaatar 46
gazarchin_institute@gazarchin.edu.mn
http://www.gazarchin.edu.mn/

## Mineral Resources Authority of Mongolia

**Mineral Resources Authority of Mongolia** (2013)
info@mram.gov.mn
http://www.mram.gov.mn/

## Mongolian University of Science and Technology

**School of Geology and Petroleum Engineering** (2011)
Bagatoiruu-46, P. O. Box-520
Ulaanbaatar 46
p. 976-11-312291
jepces@must.edu.mn
http://www.gs.edu.mn

## National University of Mongolia

**Geology and Geography Faculty** (2011)
Ikh Surguuliin gudamj - 1, Baga Toiruu,
Sukhbaatar district,
Ulaanbaatar
p. 976-11-311890
geo@num.edu.mn
http://geo.num.edu.mn

# Morocco

## Ecole Nationale de l' Industrie Minerale

**Dept de Mines** (B) (2010)
Avenue Hadj Ahmed Cherkaoui
BP 753
Adgal, Rabat
p. (+212) 037 68 02 30
zaydi@enim.ac.ma
http://www.enim.ac.ma/formation/mines

**Dept de Sciences des Materiaux** (B) (2010)
Avenue Hadj Ahmed Cherkaoui
BP 753
Agdal, Rabat
p. (+212) 037 68 02 30
zaydi@enim.ac.ma
http://www.enim.ac.ma/formation/materiaux/

**Dept de Sciences de la Terre** (B) (2010)
Avenue Hadj Ahmed Cherkaoui
BP 753
Agdal, Rabat
p. (+212) 037 68 02 30
zaydi@enim.ac.ma
http://www.enim.ac.ma/formation/sciences_de_la_terre/

## ONAREP - Naional Office of Petroleum Research

**ONAREP - Naional Office of Petroleum Research** (B) (1999)
PO Box 8030
Rabat 10050
p. +212 37 28-1616

## ONHYM

**National Office of Hydrocarbons and Mines** (2011)
rh@onhym.com
http://www.onhym.com/Default.aspx?alias=www.onhym.com/EN

## School of Mines of Marrakech

**School of Mines** (B,M,D) (2011)
Rue Machaar, El Harm Quartier Issil-B.P. 38
Marrakech
p. 212-30-97-79

## Universite Abdelmalek Essaadi

**Faculte de Sciences de Tetouen** (B) (2010)
BP 2121
Tetouan 93002
p. +212 (0) 39 97 24 23
vdap.fs@gmail.com
http://www.fst.ac.ma/stu.html

**Dept de Sciences de la Terre** (B) (2014)
p. +212 539 97 93 16
presidence@uae.ma
http://www.uae.ma/portail/FR/

## Universite Cadi Ayyad

**Dept of Geology** (2014)
Prince Moulay Abdellah
BP 515
Marakech
p. +212 (0)5 24 43 46 49
http://www.fssm.ucam.ac.ma/pages/geologie.php

## Universite Hassan 1er - Settat

**Dept de Geologie Apliquee** (B) (2014)
BP 577
Settat 26000
p. 05.23.40.07.36
http://www.fsts.ac.ma/fsts/index/php?option=com_content&task=view&id=37&Itemid=12

## Universite Hassan II

**Dept of Geology** (B) (2014)
9, Rue Tarik Bnou Zia
Anfa Casablanca
p. 0522 23 06 80
z.hilmi@fsac.ac.ma
http://www.fsac.ac.ma/depart/geo/index.html

## Universite Hassan II - Mohammedia

**Dept de Sciences de la Terre** (B) (2013)
279 Cite Yassmina
Mohammedia, Casablanca
p. +212(33)314635
presidence@univh2m.ac.ma

## Universite Ibn Tofail

**Dept de Geologie** (B,M,D) ● (2014)
Abdelhakboua@Yahoo.com
Kenitra 14 000
p. +212 061984449
abdelhakboua@yahoo.com
http://www.univ-ibntofail.ac.ma/fre/departments.php?esp=4&rub=14&srub=36&srub_=96
*Enrollment (2014): B: 14 (0) M: 23 (0) D: 7 (0)*

## Universite Ibn Zohr, Agadir

**Dept de Sciences de la Terre** (B) (2010)
BP 32/S
Agadir 80000
p. +212 (28) 22 71 25
acma2008@esta.ac.ma
http://www.esta.ac.ma/

## Universite Mohammed 1er (Oujda)

**Ecole Nationale des Sciences Appliquees d'Al Hoceima (ENSAH)** (B) (2010)
BP 724
Oujda 60000
p. +212(56)500612
presidence@ump.ma
http://webserver1.ump.ma/ecoles_facultes/ensah

**Dept des Sciences de la Terre** (B,M,D) (2011)
BP 717
, 60000 Oujda
p. (212)536500601/02
fso@fso.ump.ma
http://sciences1.univ-oujda.ac.ma/index.htm

## Universite Mohammed V

**Dept of Geology** (2014)
BP 554
Rue Michlifen Agdal
Rbat-Chellah
p. +212 05 37 77 54 71
http://www.fsr.ac.ma/ancien/index.php/departement/giolo-gie.html

**Dept of Geology** (B) (2010)
4 Avenue Ibn Battouta
BP 1014 RP
Rabat-Chellah
p. +212 (0) 37 77 18 34/35/38
fs@um5.ac.ma
http://www.fsr.ac.ma/

## Universite Mohammed V (Agdal)

**Dept de Genie Minerale** (B) (2010)
Avenue Ibnsina
BP 765
Adgal Raba
p. (212 - 537) 77.26.47 - 77.19.0
contact@um5a.ac.ma
http://www.emi.ac.ma

## Universite Moulay Ismail, Meknes

**Dept de Sciences de la Terre** (B) (2014)
Marjane II BP 298
Meknes 5003
p. +212 5 35 53 78 96
doyen@fs-umi.ac.ma
http://ww.umi.ac.ma/

## Universite Sidi Mohammed Ben Abdallah

**Dept of Geology** (2014)
Dhar El Mahraz
BP 42
Atlas-Fes
p. 06 61 35 04 81
dept_geo@fsdmfes.ac.ma
http://www.fsdmfes.ac.ma/Presentation/Departements/Geologie.php

## University Mohammed I

**Dept of Geology** (B,M,D) (2013)
Faculté des Sciences Bd. Mohammed VI
BP 717
Oujda, Oujda-Angad 60000
p. (212)667772215
mbouabdellah2002@yahoo.fr
http://webserver1.ump.ma/

# Mozambique

## Universidade Eduardo Mondlane

**Escola Superior de Ciencias Marinhas e Costeiras** (B) (2014)
CP 128
Avenida 1 de Julho
Bairro Chuabo Dembe
Didad de Quelimane, Zambezia
p. +258 24900500/1
valera.dias@uem.mz
http://www.marine.uem.mz/

**Laboratorio de Gemologia** (B,M) (2014)
Av. Julius Nyerere 3453, Departamento de Física
Campus Universitário
Maputo
p. (+258) 21 497003
uthui@zebra.uem.mz
http://www.fisica.uem.mz/index.php?option=com_content&view=article&id=51&Itemid=57

## Universidade Lurio

**Faculdade de Engenharia e Ciencias Naturais** (B) (2014)
Av. 25 de Setembro
N. 958
Pemba, Cabo Delgado
p. (+258) 27 221238
fecn@unilurio.ac.mz
http://www.unilurio.ac.mz/faculdades_pmb_pt.htm

# Namibia

## Geological Survey of Namibia in Windhoek

**Geological Survey of Namibia in Windhoek** (2015)
Private Bag 13297
Windhoek
p. +264-61-2848111
gschneider@mme.gov.na
http://www.mme.gov.na/gsn/

## University of Namibia

**Dept of Geology** (B,M,D) ● (2015)
Private Bag 13301
Windhoek 9000
p. (+264 61) 206 3712
ttjipura@unam.na
http://www.unam.na/faculties/science/departments.html
*Enrollment (2015): B: 16 (12) M: 6 (1)*

Chair:
Ansgar Wanke, Germany, GsCl
Associate Professor:
Frederick Akalemwa Kamona, (D), Germany, 1991, EgGzCe
Benjamin S. Mapani, (D), Melbourne (Australia), 1994, GtcGp
Lecturer:
Martin Harris, (M), Geosciences (Beijing), 2013
Regmi Kamal, (D), Australian National, 2013
Albertina Nakwafila, (M), Stellenbosch, 2014
Ester Shalimba, (M), China Univ of Geosciences, 2014, Ggp
Collen Uahengo, (M), Geosciences (China), 2013, GgoEg
Shoopala Uugulu, (M), France, 2012
Heike Wanke, Germany, Hw

# Nepal

## Tribhuvan University

**Central Dept of Geology** (2004)
Gandhi Bhavan, Kirtipur, Kathmandu
info@geology.edu.np
http://www.geology.edu.np/

# Netherlands

## Delft University of Technology

**Department of Geoscience & Engineering** (B,M,D) ○ (2016)
Stevinweg 1
2628 Delft, the Netherlands
p. +31 15 2781328
J.D.Jansen@tudelft.nl
www.tudelft.nl

## National Geological Survey of the Netherlands

**National Geological Survey of the Netherlands** (2011)
Princetonlaan 6
3584 CB Utrecht
p. (+31)30-2564256
http://www.en.geologicalsurvey.nl/

## Royal NIOZ

**Marine Microbiology and Biogeochemistry** (2016)
Landsdiep 4
't Horntje 1797 SZ
Jaap.Damste@nioz.nl
http://www.nioz.nl

## State University of Groningen

**Faculty of Spatial Sciences** (B,M) (2016)
P.O. Box 800
Groningen, Netherlands 9700 AV
p. +31 50 363 8891
i.l.veen@rug.nl
http://www.rug.nl/frw
f: http://www.facebook.com/FRWRUG
t: @FRW_RUG
*Enrollment (2014): B: 190 (0) M: 60 (0)*

## State University of Utrecht

**Faculty of Geoscience** (B,M,D) (2014)
P.O. Box 80.115
Utrecht 3508 TA
p. (+31) (0)30 253 2044
info@geo.uu.nl
http://www.uu.nl/faculty/geosciences/EN/Contact/Pages/default.aspx

## Technical Univ. of Delft

**Dept of Applied Geology** (A,B,M,D) ● (2014)
P. O. Box 5048
Stevinweg 1
2600-GA Delft
p. +31 15 27 86019
h.h.m.zwiers@tudelft.nl
http://www.citg.tudelft.nl/over-faculteit/afdelingen/geoscience-engineering/

## Universiteit Utrecht

**Faculty of Geosciences** (B,M) (2016)
Budapestlaan 4b
Utrecht 3584 CD
p. +31 30 253 7890
mscinfo.geo@uu.nl
http://www.geo.uu.nl/

## University of Amsterdam

**Dept of Geology** (2014)
Science Park 904
1098 XH Amsterdam
p. +31 (0)20525 8626
info-science@uva.nl

## University of Twente

**Faculty of Geo-Information Sciences and Earth Observation** (M,D) (2017)
P.O. Box 217
7500 AE Enschede
p. +31 (0)53 487 44 44
info-itc@utwente.nl
http://www.itc.nl

## University of Utrecht

**Dept of Earth Sciences** (2013)
Heidelberglaan 2
P.O. Box 80115
Utrecht
C.Marcelis@geo.uu.nl

## VU University Amsterdam

**Faculty of Earth and Life Sciences, Dep. of Earth Sciences** (B,M,D) ● (2017)
De Boelelaan 1085-1087
1081 HV Amsterdam 1081 HV
p. +31 (0)20 59 87310
f.bosse@vu.nl
http://www.falw.vu.nl/en/index.asp#
*Enrollment (2015): B: 248 (56) M: 67 (45)*

Professor:
Gareth R> Davies, (D), GiCcGf
Han J. Dolman, (D), HgOw
Guido R. van der Werf, (D), Sb
Wim van Westrenen, (D), CpGz
Peter Verburg, (D), OyGe
Jan Wijbrans, (D), CcGtv

Associate Professor:
Fraukje M. Brouwer, (D), Utrecht, 2000, GptGc
Gerald M. Ganssen, (D), ObPe
Kees Kasse, (D), Gms
Frank J.C. Peeters, (D), PmOg
Didier Roche, (D), Pe
Wouter Schellart, (D), Gt
Ronald T. van Balen, (D), GgmGt
Ko van Huissteden, (D), CgSb
Jorien Vonk, (D), Hg
Pieter Vroon, (D), GvCcl

Assistant Professor:
Kay J. Beets
Janne Koornneef, (D), GvCa
Klaudia Kuiper, (D), Cc
Kei Ogata, (D), Gt
Maarten A. Prins, (D), GsOg
Monica Sanchez Roman, (D), GdsGg
Nynke Schulp, (D), Oy
Marlies ter Voorde, (D), Gqt
Jeroen van der Lubbe, (D), Gs
Ype van der Velde, (D), Hg
Astrid van Teeffelen, (D), Oy
Jasper van Vliet, (D), Oy
Sander Veraverbeke, (D), Hg
Martijn C. Westhoff, (D), Hg

Research Associate:
Antoon Meesters, (D), OwHg
Emma van der Zande, (D), Oy

Dr:
Bernd Andeweg, (D), 2002, GtcGg
Ron Kaandorp, (D), Gg
Els Ufkes, (D), PmOgGu

# Netherlands Antilles

## Royal NIOZ

**Marine Geology** (2011)
Landsdiep 4
Den Burg 1797 SZ
p. +31(0)222 369 300
Jens.Greinert@nioz.nl
http://www.nioz.nl

**Physical Oceanography** (2011)
Landsdiep 4
Den Burg 1797 SZ
secretary@nioz.nl

http://www.nioz.nl

# New Zealand

## Institute of Geological and Nuclear Sciences Ltd (New Zealand)

**Institute of Geological and Nuclear Sciences Ltd (New Zealand)** (2015)
1 Fairway Drive, Avalon 5010. PO Box 30-368, 5040.
Lower Hutt
p. +64-4-570-1444
G.Alderwick@gns.cri.nz
http://www.gns.cri.nz/
f: https://www.facebook.com/gnsscience

## Massey University

**Soil and Earth Sciences Group** (2004)
Private Bag
Palmerston North
p. 0800 MASSEY (0800 627 739)
contact@massey.ac.nz
http://www.massey.ac.nz/massey/learning/programme-course-paper/programme.cfm?major_code=2044&prog_id=92431

## University of Auckland

**School of Environment**
(2009)
Private Bag 92019
Auckland Mail Center
Auckland 1142
p. 64 9 3737599
p.kench@auckland.ac.nz
http://www.auckland.ac.nz/glg/geology.htm

## University of Canterbury

**Dept of Geological Sciences** (B,M,D) ● (2017)
Private Bag 4800
Christchurch 8140
p. (643) 3694384
geology@canterbury.ac.nz
http://www.geol.canterbury.ac.nz/title.html
*Enrollment (2014): B: 28 (0) M: 0 (35) D: 0 (20)*
Professor Hazard & Disaster Management:
Timothy R H Davies, (D)
Professor:
James W. Cole, (D), Wellington, 1966, Gv
Andy Nicol, (D)
Jarg R. Pettinga, (D), Auckland Univ, 1981, Gtc
Senior Lecturer Hazard & Disaster Management:
Thomas Wilson, (D), Canterbury, Gv
Senior Lecturer:
Kari N. Bassett, (D), Minnesota, 1995, Gs
Catherine Reid, (D), Po
Associate Professor:
Ben Kennedy, (D), Gv
Lecturer:
Darren Gravley, (D), Gv
Samuel Hampton, (D), Canterbury, Gq
Alex Nichols, (D)

## University of Otago

**Dept of Geology** (B,M,D) ○ (2016)
P. O. Box 56
360 Leith Walk
Dunedin 9054
p. +64 3 479-7519
geology@otago.ac.nz
http://www.otago.ac.nz/geology/
*Enrollment (2005): M: 20 (0) D: 22 (0)*
Head of Department:
James D. L. White, (D), California (Santa Barbara), 1989, Gvs
Professor:
Dave Craw, (D), Otago, 1981, Eg
R. Ewan Fordyce, (D), Canterbury (NZ), 1981, Pv
Senior Lecturer:
Candace E. Martin, (D), Yale, 1990, ClcGz
Virginia G. Toy, (D), Otago, 2008, Gct
Senior Lecturer:
J. Michael Palin, (D), Yale, 1992, CcGip
Associate Professor:
Andrew R. Gorman, (D), British Columbia, 2001, YrgYe
Daphne E. Lee, (D), Otago, 1980, Pib
Dr:
Christopher M. Moy, (D), Gus
Adjunct Professor:
Gary S. Wilson, (D), Victoria (NZ), 1993, GuYmGs
Emeritus:
Douglas S. Coombs, (D), Gi
Alan F. Cooper, (D), Otago, 1970, GpiGt
Rick H. Sibson, (D), Imperial Coll (UK), 1977, Gc

## University of Waikato

**Dept of Earth Sciences** (2014)
Private Bag 3105
Hamilton 3240
p. +64 7 838 4625
science@waikato.ac.nz
http://sci.waikato.ac.nz/study/subjects/earth-sciences

## Victoria University of Wellington

**School of Geography, Environment and Earth Sciences** (A,B,M,D) ○ (2014)
P. O. Box 600
Wellington 6040
p. +6444636108
geo-enquiries@vuw.ac.nz
http://www.geo.vuw.ac.nz/
*Enrollment (2012): B: 153 (107) M: 51 (22)*
Professor:
Tim Little, (D), Stanford, 1988, Gct

**Dept of Geology** (2015)
P. O. Box 600
Wellington
p. +64 4 463 5337
geo-enquiries@vuw.ac.nz
http://www.victoria.ac.nz/sgees/study/postgraduate-study/geology/

# Nicaragua

## University of Managua

**Centro de Investigaciones Geocientificas** (2004)
Colonia Miguel Bonilla No. 165, P.O. Box A-131 Cc. Managua, Managua

# Niger

## Institut de Recherche pour le Developpement au Niger

**Institut de Recherche pour le Developpement au Niger** (B) (2010)
Avenue de Maradi
Niamey BP 11 416
p. (227) 20 75 38 27
irdniger@ird.fr
http://www.ird.ne/

## Universite Abdou Moumouni de Niamey

**Dept de Geologie** (2014)
B.P. 10662
Niamey
p. (227) 20 31 50 72
http://www.secheresse.info/article.php3?id_article=1664
Doyen:
Abdoulaye M. Alassane

# Nigeria

## Adekunle Ajasin University

**Dept of Earth Sciences** (B) (2010)
Akungba Akoko, Ikare Akoko 23401
p. +234 (34) 246444
iosazuwa@yahoo.com
http://www.ajasin.edu.ng/academics/index.php

## African University of Science & Technology

**Petroleum Engineering** (B,M) (2010)
PMB 681
Abuja F.C.T.
Km 10 Airport Road
Garki
p. +234 9 7800680
http://aust.edu.ng/

## Ahmadu Bello University

**Dept of Geology** (2014)
Sokoto Road
Samaru-Zaria
Zaria 2222, Kaduna
p. +234 (69) 50 691
ybijimi@yahoo.com
http://www.abu.edu.ng/

## American University of Nigeria

**School of Arts and Sciences** (B) (2010)
Lamido Zubairu Way
PMB 2250, Yola 23455
p. +234 (805) 5485 702
dsas@aun.edu.ng
http://www.abti-american.edu.ng

## Caleb University

**Dept of Environmental Protection and Management** (B) (2010)
PMB 21238
Ikeja GPO, Lagos
p. +234-1-764-7312
info@calebuniversity.edu.ng
http://calebuniversity.edu.ng/courses.php?coll_id=4

**Dept of Surveying and Geoinformatics** (B) (2010)
PMB 21238
Ikeja GPO, Lagos
p. +234-1-764-7312
info@calebuniversity.edu.ng
http://calebuniversity.edu.ng/courses.php?collUid=4

## Crawford University

**Dept of Mathematics and Physical Sciences** (B) (2010)
PMB 2001
Faith City
p. +234 (1) 8134785
ccrawford@rsu.edu
http://www.crawforduniversity.edu.ng

## Delta State University

**Dept of Geology** (B) (2014)
PMB 1
Abraka
p. +234 (54) 66009
delsu03@yahoo.com
htttp://www.delsunigeria.net/

Director:
S. H. O. Egboh, (D)
Chair:
E. Adaikpoh, (D)

## Enugu State University of Science and Technology

**Dept of Geology and Mining** (B) (2010)
MB 01660, Enugu
p. +234 (42) 451319
dean.fans@esut.edu.ng
http://www.esut.edu.ng/

**Dept of Geography and Meteorology** (B) (2010)
MB 01660, Enugu
p. +234 (42) 451319
dean.fans@esut.edu.ng
http://www.esut.edu.ng/

## Federal Polytechnic, Bida

**Dept of Survey and Geoinformatics** (B) (2010)
http://www.fedpolybidaportal.com/

**Dept of Quantitative Surveying** (B) (2010)
http://www.fedpolybidaportal.com/

## Federal Polytechnic, Offa

**School of Environmental Studies** (B) (2010)
http://www.fedpoffa.edu.ng/

## Federal University of Technology

**Dept of Applied Geophysics** (B,M,D) (2014)
PMB 704
Akure, Ondo
p. +234 803-595-9029
joamigun@futa.edu.ng
http://agp.futa.edu.ng/page.php?pageid=179

## Federal University of Technology, Akure

**Earth & Mineral Sciences** (B) (2010)
PMB 704
Akure
p. +234 (34) 243744
sems@futa.edu.ng
http://www.futa.edu.ng/sems/

**Applied Geology** (B) ● (2016)
PMB 704
Akure, Ondo
p. +234 (0) 803 667 2708
agy@futa.edu.ng
http://www.futa.edu.ng/agy/
*Enrollment (2015): B: 85 (70)*

Professor:
Yinusa A. Asiwaju-Bello, (D), Leeds, GgHgNg
Idowu O. Odeyemi, (D), Ibadan, 1976, GgxOr
Doctor:
O A.G Jegede, (D), Akure, GgNgg
C T. Okonkwo, (D), Keele, GgxGc
S A. Opeloye, (D), Bauchi, GgoGs
Mrs:
Oluwaseyi A. Bamisaye, (M), Akure, 2007, GgOr
Mr:
Mohammed O. Adepoju, (M), Akure, 2004, GgEg
Adeshina L. Adisa, (M), Akure, 2012, GgCe
Isaac O. Ajigo, (M), Jos, 2013, GgNm
Timothy I. Asowata, (M), Ibadan, 2010, Gge
Sunday O. Daramola, (M), Akure, 2012, HwGe
Oladimeji Rilwan Egbeyemi, (M), Ibadan, 2004, GgCg
Emmanuel Eghietseme Igonor, (M), Ibadan, 2007, GgCeGi
Osazuwa A. Ogbahon, (M), Akure, 2011, Ggo
Joshua O. Owoseni, (M), Ibadan, GgHw
Doctor:
P S. Ola, (D), Akure, GosGg
Solomon O. Olabode, (D), Akure, GgoGs

## Federal University of Technology, Yola

**Dept of Surveying & Geoinformatics** (B) (2010)
PMB 2076
Yola
p. +234 (803) 6065646
survey@futy.edu.ng
http://www.futy.edu.ng/academic/dsurvey.htm

**Dept of Geography** (B) (2010)

PMB 2076
Yola
p. +234 (803) 6065646
geography@futy.edu.ng
http://www.futy.edu.ng/academic/dgeography.htm

Nigeria

**Dept of Soil Science** (B) (2010)
PMB 2076
Yola
p. +234 805 3372 784
soilscience@futy.edu.ng
http://www.futy.edu.ng/academic/dsoil.htm

**Dept of Geology** (B) (2010)
PMB 2076
Yola
p. +234 (803) 6065646
geology@futy.edu.ng
http://www.futy.edu.ng/academic/dgeology.htm

## Gombe State University

**Dept of Geology** (B) (2010)
contact@gsu.edu.ng
http://www.gomsu.org/

**Dept of Geography** (B) (2010)
contact@gsu.edu.ng
http://www.gomsu.org/

## Ladoke Akintola University of Technology

**Dept of Earth Sciences** (B) (2016)
PMB 4000
Ogbomoso, Oyo 4000
p. +234 8033539911
contact@lautech.edu.ng
http://www.lautech.edu.ng/Academics/undergraduates/FPAS/index.html
*Enrollment (2015): B: 13 (13)*

Dr:
Oyelowo Gabriel Bayowa, (D), Obafemi Awolowo, 2013, GggGg
Dr:
Abosede Olufunmi Adewoye, (D), Ladoke Akintola, 2014, GeeGe
Dr:
Moruffdeen Adedapo Adabanija, (D), Ibadan, 2009, GggGg
Mr:
Olukayode Adegoke Afolabi, (M), Ibadan, 2000, GxxGx
Ismaila Abiodun Akinlabi, (M), Ibadan, 2011, GggGg
Mustapha Taiwo Jimoh, (M), Ibadan, 2006, GxxGx
Lanre Lateef Kolawole, (M), Newcastle, 1985, GeeGe
Olusola Christophe Oduneye, (M), Ibadan, 2009, GooGo
Gbenga Olakunle Ogungbesan, (M), Ibadan, 2006, GooGo

## Lagos State Polytechnic

**School of Environmental Studies (Quantity Surveying)** (B) (2010)
info@mylaspotech.edu.ng
http://mylaspotech.net/

## National Open University of Nigeria

**School of Science & Technology** (B) (2010)
PO Box 1866
Bauchi
p. +234 1-8188849, +234 1-4820720
sst@noun.edu.ng
http://www.nou.edu.ng/noun/indexs.jsp

## Nnamdi Azikiwe University

**Dept of Geology/Meteorology and Env Management** (B) (2010)
PMB 5025
Awka, Nnewi 5025
p. +234 (46) 550018
emmaojukwu@unizik.edu.ng
http://www.unizikeduportal.org/

**Dept of Geological Sciences** (B,M,D) (2016)
Awka Campus
PMB 5025
Awka, Anambra 5025
p. +234 (46) 550018
ne.ajaegwu@unizik.edu.ng
http:wwwww.unizikeduportal.org/

## Obafemi Awolowo University

**Dept of Soil Science** (B) (2010)
Ile-Ife
p. +234 (36) 230 290
midowu@oauife.edu.ng
http://www.oauife.edu.ng/faculties/agric/soil_sci.html

**Dept of Geography** (B,M,D) ○ (2014)
Ile-Ife
p. +234 (70) 37772355
sinaayanlade@yahoo.co.uk
http://www.oauife.edu.ng/faculties/soc_sciences/departments/geo/home.htm

**Dept of Geology** (B,M,D) ● (2015)
Ile-Ife, Osun State 220002
p. 08128181062
geoife@oauife.edu.ng
http://gly.oauife.edu.ng/
*Enrollment (2014): B: 213 (0) M: 24 (0) D: 12 (0)*

Head:
A. A. Adepelumi, (D), YgsNe
Professor:
S. A. Adekola, (D), GdoCg
O. Afolabi, Yg
J. O. Ajayi, Oo
T. R. Ajayi, CgEgGe
O. A. Alao, (D), Yg
U. K. Benjamin, GoCg
S. L. Fadiya, (D), God
A. O. Ige, Cg
A. O. Ige, Cg
E. James-Aworeni, Hg
J. I. Nwachukwu, GoCg
O. O. Ocan, GtzGp
S. B. Ojo, Yg
V. O. Olarewaju, GozCg
A. O. Olorunfemi, Go
M. O. Olorunfemi, Yg
A. A. Oyawale, GeoCg
M. A. Rahaman, GpcGi
B. M. Salami, Hg
I. A. Tubosun
Associate Professor:
A Adetunji, (D), EgGi
Assistant Professor:
Dele E. Falebita, (D), Yg
Research Associate:
M. A. Olayiwola, GugPi
O. M. Oyebanjo

## Rivers State University of Science & Technology

**Inst of Geoscience & Space Technology (IGST)** (2011)
P.M.B. 5080
Nkpolu - Oroworukwo
Port Harcourt, Rivers State
info@ust.edu.ng
http://www.ust.edu.ng/

## Umaru Musa Yar'Adua University

**Dept of Geography** (B) (1999)

## Universite de Niamey

**Dept de Geologie** (2004)
BP 10662, Niamey

## University of Ado - Ekiti

**Geology Dept** (B) (2010)
PMB 5363

Ado-Ekiti
p. +234 (30) 250026
info@eksu.edu.ng
http://www.unadportal.com/

**Dept of Agricultural Sciences** (B) (2010)
PMB 5363
Ado-Ekiti
p. +234 (30) 250026
http://www.unadportal.com/

## University of Benin

**Dept of Petroleum Engineering** (B) (2010)
PMB 115
Benin City, Edo State
p. 234 0802 345 5681
deanengineering@uniben.edu
http://www.uniben.edu/UniversityOfBenin-PetroleumEngineering.html

**Dept of Geology** (B,M,D) ● (2015)
P. M. B. 1155
Benin City, Edo State
p. 234 0802 345 5681
registrar@uniben.edu
http://www.uniben.edu/physicalscielcdfaculty.html
Professor:
Godwin Osarenkhoe Asuen, (D), EcGsg
Professor:
Christopher Nnaemeka Akujieze, (D), HwGeHg
Williams Ogbevire Emofurieta, (D), GzCeGe
Tony Uzozili Onyeobi, (D), NrgGq
Associate Professor:
Isaac Okpeseyi Imasuen, (D), Western Ontario (Canada), Gee
Senior Lecturer:
Ben Obelenwa Ezenabor, (M), HgOrGe
Franklin A. Lucas, (D), PmlGr
Lecturer II:
Aitalokhai Joel Edegbai, (M), GosPi
Efetobore Gladys Maju-Oyovwikowhe, (M), GszGo
Sikiru Adeoye Salami, (D), YgeGg
Lecturer I:
Ovie Odokuma-Alonge, (M), CgGip
Graduate Assistant:
Tomilola Andre-Obayanju, (B), NgHgGe
Assistant Lecturer:
Aiyevbekpen Helen Akenzua-Adamcyzk, (M), GeHgGg
Nosa Samuel Igbinigie, (M), GorGg
Alexander Ogbamikhumi, (M), YsgGg
Sunday Erapkpower Okunuwadje, (M), GsoGd
Edoseghe Edwin Osagiede, (M), GctGx

## University of Calabar

**Dept of Geology** (B) (2014)
PMB 1115 Eta Agbo Road,
Calabar Municipal, Cross River State
p. (+234) 8173740083
general@unical.edu.ng
http://www.unical.edu.ng/pages/programs_courses/sciences.php?nav=departments
Head:
Nse U. Essien

## University of Ibadan

**Dept of Geology** (B,M,D) (2014)
Ibadan, Oyo State
p. 02-8101100-8101104
uigeology@yahoo.com
http://sci.ui.edu.ng/geowelcome
Professor:
I. M. Akaegbobi, Cg
A. A. Elueze, Eg
Ougbenga. Akindeji. Okunlola, (D), Ibadan, 2001, EgCgGe
A. I. Olayinka, Yg
Moshood N. Tijani, (D), Muenster (Germany), 1997, HwCgGe
Associate Professor:
A. S. Olatunji, (D), Ibadan, 2006, CgGeCt
Lecturer:
A. E. Abimbola, Cg
O. C. Adeigbe, (D), Ibadan, 2009, GosGe
G. O. Adeyemi, NgHg
O. A. Boboye, Go
A. T. Bolarinwa, EgGz
O. A. Ehinola, Goe
M. E. Nton, Go
M. A. Oladunjoye, Yg
O. O. Osinowo, Yg
I. A. Oyediran, Ng

**Dept of Petroleum Engineering** (B,M,D) (2017)
Oyo Road
Ibadan
p. +234-2-810-3168
isehunwa@yahoo.com
http://www.ui.edu.ng/?q=petengine
*Enrollment (2016): B: 223 (43) M: 17 (17) D: 10 (2)*

**Dept of Geology** (B,M,D) (2016)
Ibadan
Ibadan, Oyo State 200284
p. +234-8033819066
ehinola01@yahoo.com
http://www.ui.edu.ng/?q=departmentofgeology
Professor:
Olugbenga Ajayi Ehinola, (D), Ibadan, 2002, GoeGs

## University of IFE

**Dept of Geology** (2004)
ILE-IFE

## University of Ilorin

**Dept of Geology and Mineral Sciences** (B) (2014)
PMB 1515
p. +234 (31) 221691
deanfacsci@unilorin.edu.ng
http://www.unilorin.edu.ng/unilorin/index.php/sciences-dept/geology-mineral-sciences
Head:
J.I. D. Adekeye, (D), GoCg
Professor:
S. O. Akande, (D), EgCg
O. Ogunsanwo, Ng
Lecturer:
A. Abdurrahman, (M), CeEg
A. D. Adedcyin, (M), Eg
O. A. Adekeye, (D), Pm
S.M. A. Adelana, (M), HgYgGe
D. A. Alao, NgGz
O. O. Ige, (M), Ng
Kehinde A. Olawuyi, (D), Federal Technology (Nigeria), 2015, YeGge
O. A. Omotoso, (M), Ge
W. O. Raji, (M), Ye

## University of Jos

**Dept of Geosciences and Mining** (B) (2014)
PMB 2084
Jos 930001
p. +234 (73) 559 52
vc@unijos.edu.ng
http://www.unijos.edu.ng/

## University of Lagos

**Dept of Petroleum and Gas Engineering** (B) (2010)
Akoka
Yaba, Lagos
p. +234 (1) 493 2660 3
informationunit@unilag.edu.ng
http://www.unilag.edu.ng/index.php?page=home

Nigeria

**Dept of Metalurgical and Materials Engineering** (B) (2010)
Akoka
Yaba, Lagos
p. +234 (1) 493 2600 3
informationunit@unilag.edu.ng
http://www.unilag.edu.ng/index.php?page=home

**Dept of Civil and Environmental Engineering** (B) (2010)
Akoka
Yaba, Lagos
p. +234 (1) 493 2660 3
informationunit@unilag.edu.ng
http://www.unilag.edu.ng/index.php?page=about_departmentdetail&sno=11

**Dept Surveying and Geoinformatics** (B) (2010)
Akoka
Yaba, Lagos
p. +234 (1) 493 2660 3
informationunit@unilag.edu.ng
http://www.unilag.edu.ng/index.php?page=about_departmentdetail&sno=16

## University of Maiduguri

**Dept of Geology** (2014)
PMB 1069 FEA
Maiduguri
info@unimaid.edu.ng
http://www.unimaid.edu.ng/root/faculty_of_sciences/dept_geology.html

## University of Maiduguri, Borno State

**Dept of Geology** (B,M,D) (2014)
Bama Road
PMB 1069
Maiduguri
p. +234 (76) 231730
info@unimaid.edu.ng
http://www.unimaid.edu.ng/home.php
Professor:
I. B. Goni
Assistant Professor:
Saidu Baba
Lecturer:
Sani Adamu
Kyari M. Ajar
Musa Malana Aji
Fati Bukar
Shettima Bukar
Jalo M. El-Nafaty
Millitus V. Joseph
Samaila C. Kali
Yakubu B. Mohd
Mohammed Poukar
Aishatu Sani
Manaja Mijinyawa Uba
Sidi M. Waru
Asabe Kuku Yahaya
Soloman N. Yusuf
A. Adam Zarma

## University of Nigeria

**Dept of Geology** (B) (2010)
Nsukka
micah.osilike@unn.edu.ng
http://www.unn.edu.ng/physicalsciences/content/view/700/601/

## University of Nigeria Nsukka

**Dept of Geoinformatics and Surveying** (B) (2010)
batho.okolo@unn.edu.ng
http://www.unn.edu.ng/environmentalstudies/content/view/27/44/

**Dept of Geology** (B) (2014)
anthony.okonta@unn.edu.ng
http://unn.edu.ng/department/geology
Lecturer:
Alloysuis Okwudili Anyiam
Luke I. Manah
Smart Chicka Obiora

## University of Port Harcourt

**Dept of Geography & Environmental Management** (B) (2010)
East/ West Road
PMB 5323
Choba, Port Harcourt 500001
p. +234 (84) 230890-99
edulms@chimpgroup.com
http://uniport.edu.ng/

**Dept of Geology** (B) (2014)
East/ West Road
P.M.B. 5323
Choba, Port Harcourt 500001
p. +234 (0)84 817 941
uniport@uniport.edu.ng
http://www.uniport.edu.ng/faculties/science.html

**Dept of Petroleum and Gas Engineering** (B) (2010)
East/ West Road
PMB 5323
Choba, Port Harcourt 500001
p. +234 (84) 230890-99
dean@eng.uniport.edu.ng
http://uniport.edu.ng/

## University of Uyo

**Faculty of Agriculture** (B,M,D) (2015)
PMB 1017
Uyo, Akwa Ibom State 520001
p. +234 (85) 200303
vc@uniuyo.edu.ng
http://www.uniuyo.edu.ng/index.htm

**Faculty of Engineering** (B,M,D) (2016)
PMB 1017
Uyo
p. +234 (85) 200303
vc@uniuyo.edu.ng
http://www.uniuyo.edu.ng/index.htm

**Faculty of Science Education** (B,M,D) (2010)
PMB 1017
Uyo
p. +234 (85) 200303
vc@uniuyo.edu.ng
http://www.uniuyo.edu.ng/index.htm

**Faculty of Social Science** (B,M,D) (2010)
PMB 1017
Uyo
p. +234 (85) 200303
vc@uniuyo.edu.ng
http://www.uniuyo.edu.ng/index.htm

**Faculty of Environmental Studies** (B,M,D) (2010)
PMB 1017
Uyo
p. +234 (85) 200303
vc@uniuyo.edu.ng
http://www.uniuyo.edu.ng/index.htm

## Western Delta University

**Geology and Petroleum Studies** (B) (2010)
PMB 10
Oghara
p. +234 (70) 35400531
info@wdu.edu.ngg

http://www.wduniversity.org/

## Wukari Jubilee University
**Dept of Geography and Environmental Conservation** (B) (2010)
PMB 1019
Wukari, Taraba State
p. +234 (080) 80329138
wukari_jubilee@yahoo.com
http://wukarijubileeuniversity.org/

# Norway

## NLH
**Dept of Soil Sciences** (2015)
P.O. Box 28
1432 AS
p. +47 67 23 00 00
post@nmbu.no
http://www.umb.no/noragric/

## Norges Geologiske Undersokelse
**Geological Survey of Norway** (2017)
Postboks 6315 Sluppen
7491 Trondheim
ngu@ngu.no
http://www.ngu.no/no/

## Norwegian Institute of Science and Technology
**Dept of Geology and Mineral Resources Engineering** (2004)
N-7034 Trondheim
mai.britt.mork@ntnu.no

## Norwegian Institute of Technology
**Faculty of Engineering Science and Technology** (2004)
S. P. Andersens vei 15A
N7034 Trondheim
p. 73594779
ingvald.strommen@ntnu.no

## NTNU Norwegian University of Science and Technology
**Dept of Geology and Mineral Resources Engineering** (B,M,D) (2014)
Sem Sælands vei 1
7491 Trondheim
Trondheim
p. (+47) 73 59 48 00
iigb-info@ivt.ntnu.no
http://www.ntnu.no/igb
f: https://www.facebook.com/geologibergNTNU

## University of Bergen
**Dept of Earth Science** (2008)
Allegaten 41
N-5007 Bergen
post@geo.uib.no
http://www.geo.uib.no/

## University of Oslo
**Deptartment of Geosciences** (B,M,D) (2016)
Post Box 1047, Blindern
N-0316 Oslo
p. +47 22856656
elisabeth.alve@geo.uio.no
http://www.mn.uio.no/geo/
f: https://www.facebook.com/uiogeo/

**Mineralogical-Geological Museum** (2014)
Sars gate 1
P.O. Box 1172
Blindern, Oslo 0318
p. +47 228 55050
postmottak@nhm.uio.no
http://www.nhm.uio.no/english/

## University of Tromso
**Geology Dept** (2004)
N-9037 Tromso
matthias.forwick@uit.no
http://www.ibg.uit.no/geologi/geo_eng_end.html

# Oman

## Sultan Qaboos University
**Dept of Earth Science** (B,M,D) ● (2014)
123 Al-Khod
P.O. Box 36
Muscat
p. +968 2414 6832
khirbash@squ.edu.om
http://www.squ.edu.om/earth-sci/tabid/11718/language/en-US/Default.aspx
*Enrollment (2013): B: 85 (60) M: 5 (25) D: 2 (1)*

Head:
Salah Al-Khirbash, (D)

# Pakistan

## COMSATS
**Institute of Information Technology** (2011)
Park Road, Chak Shahzad
Islamabad
p. +92-51-9247000-2
info@ciit.edu.pk
http://www.comsats.edu.pk/

## Geological Survey of Pakistan
(2014)
p. (+958) 9211032
qta@gsp.gov.pk
http://www.gsp.gov.pk/

## Institute of Space Technology
**Institute of Space Technology** (B,M,D) (2013)
P.O. Box 2750
Islamabad 44000
p. 92.51.9075100
info@ist.edu.pk
http://www.ist.edu.pk/

## National University of Sciences and Technology
**Institute of Geographic Information System** (2011)
RIMMS building, NUST H-12 Campus
Islamabad
p. +92-51-90854400
info@igis.nust.edu.pk
http://igis.nust.edu.pk/

## Sind University Pakistan
**Dept of Geology** (2004)
University of Sindh, Allama I.I. Kazi Campus
Jamshoro
Sindh 76080
p. 92-22-9213172-9213181-90 Ext:
dean.science@usindh.edu.pk

## University of Engineering and Technology
**Geological Engineering** (B,M,D) ● (2016)
Lahore, Punjab 54890
p. 0092-42-99029487
mzubairab1977@gmail.com
http://www.uet.edu.pk/faculties/facultiesinfo/geological/index.html?RID=introduction

Chair:
Muhammad Zubair Abu Bakar, (D), 2012, NrmNg

Assistant Professor:
Muhammad Farooq Ahmed, (D), 2013, Ng

## University of Karachi
**Dept of Geology** (2014)
KARACHI-75270
p. 99261300-06 Ext: 2295
vc@uok.edu.pk
http://uok.edu.pk/faculties/geology/

## University of Sargodha
**Dept of Earth Science** (B,M,D) ● (2014)
Sargodha
earthscience@uos.edu.pk
http://www.uos.edu.pk

**College of Agriculture** (2011)
Sargodha
agri@uos.edu.pk
http://www.uos.edu.pk

## University of the Punjab
**Dept of Geography** (2011)
New Campus
Lahore 54590
chairman@geog.pu.edu.pk
http://pu.edu.pk/home/department/46/Department-of-Geography

# Panama

## Universidad de Panama
**Facultad de Ciencias Naturales, Exactas y Technologia** (2004)
Urbanizacion El Cangrejo, Estafeta Universitaria
ciencias@ancon.up.ac.pa

# Paraguay

## Universidad Nacional se Asuncion
**Dept de Geologia** (2004)
University Campus Km. 11, Asuncion
facen@facen.una.py

# Peru

## Geological Survey of Peru
**Instituto Geologico Minero y Metalurgico - INGEMMET** (2013)
Av. Canada 1470 San Borja
Lima
p. 051-1-6189800
webmaster@ingemmet.gob.pe
http://www.ingemmet.gob.pe/

## San Agustin's Natl University
**Escuela Academico Profesional de Ingenieria Geologica y Geofisica** (2011)
Av. Indraendencia #1015
Arequipa
p. (054) 244498
geologia@unsa.edu.pe
http://www.unsa.edu.pe

## Univ Nacional de Ingeneiria
**Escuela Academico Profesional de Ingenieria Geologica** (2011)
Av. Tupac Amaru S/N
Rimac, Lima
esanchez@uingemmet.gob.pe
http://www.uni.edu.pe/sitio/academico/facultades/geologica/

## Univ Nacional Mayor San Marcos
**Escuela Academico Profesional de Ingenieria Geologica** (2016)
Av. Venezuela S/N
Lima
j_jacay@yahoo.com
http://www.unmsm.edu.pe

# Philippines

## University of the Philippines
**National Institute of Geological Sciences** (B,M,D) ● (2014)
C.P. Garcia corner Velasquez Street
Diliman, Quezon City 1101
p. (+632) 9296046
inquire@nigs.org
http://nigs.org

# Poland

## AGH University of Science and Technology
**Faculty of Geology, Geophysics, and Environmental Protection** (2014)
A. Mickiewicza 30 Ave.
30-059 Krakow
p. +48 12 633 29 36
aglown@geol.agh.edu.pl
http://www.wggios.agh.edu.pl/

**Faculty of Mining and Geoengineering** (2014)
Al. Mickiewicza 30
PL-30059 Krakow
p. +48 12 617 21 15
gorn@agh.edu.pl

## Jagiellonian University
**Institute of Geological Sciences** (2010)
Oleandry 2a Str.
30-063 Krakow
sekretariat.ing@uj.edu.pl
http://www.ing.uj.edu.pl

## Ministry of Environmental Protection, Natural Resources and Forestry
**Dept of Geology** (2016)
st. Wawelska 52/54
Warsaw 00-922
p. +48 22 3692449
Departament.Geologii.i.Koncesji.Geologicznych@mos.gov.pl
http://www.mos.gov.pl/dg/dga1.htm

## Panstwowy Instytut Geologiczny
**Polish Geological Institute** (2011)
Ul. Rakowiecka 4 ,
PL-00-975 Warszawa
sekretariat@pgi.gov.pl
http://www.pgi.gov.pl/

## Polish Academy of Sciences
**Institute of Geological Sciences** (2004)
Al Zwirki i Wigury 93
02-089 Warsaw
p. (+48 22) 620 33 46, 656 60 63
ingpanhard@.pan.pl

## Silesian University of Technology
**Faculty of Mining and Geology** (B,M,D) ● (2015)
Wydzial Gornictwa i Geologii
Politechnika Slaska
ul.Akademicka 2
Gliwice 44-100
p. +48 32 2371283
rg@polsl.pl
http://www.polsl.pl/Wydzialy/RG/Strony/Witamy.aspx
*Enrollment (2014): B: 2556 (615) M: 327 (283) D: 98 (5)*

## Univ of Mining & Metallurgy
**Dept of Mining and Metallurgical Engineering** (2004)

Inst. of Drilling & Oil Exploration
Krakow
jameshendrix@unr.edu

## University of Wroclaw
**Inst of Geological Sciences** (B,M,D) (2017)
Pl. Maksa Borna 9
50-205 Wroclaw
p. (+4871) 321-10-76
sekretariat.ing@uwr.edu.pl
http://www.ing.uni.wroc.pl/home-page
*Enrollment (2016): D: 14 (8)*
Director:
Krystyna Choma-Moryl, (D), 1983, Ng
Professor:
Pawel Aleksandrowski, (D), 1985, Gct
Marek Awdankiewicz, (D), 1997, Gvi
Piotr Gunia, (D), 1983, Gaz
Jacek Gurwin, (D), 1997, Hy
Henryk Marszalek, (D), 1994, Hy
Jacek Puziewicz, (D), 1985, Gix
Jerzy Sobotka, (D), 1992, Yg
Andrzej Solecki, (D), 1987, Eg
Stanislaw Stasko, (D), 1986, Hg
Jacek Szczepanski, (D), 1999, Ggc
Adjunct Professor:
Anna Gorecka-Nowak, (D), 1992, Pg
Maciej Gorka, (D), 2007, OaCa
Jakub Kierczak, (D), 2007, GeSc
Antoni Muszer, (D), 1994, Eg
Anna Pietranik, (D), 2006, CgGi
Marta Rauch, (D), 2002, Gc
Robert Tarka, (D), 1995, Hg
Jurand Wojewoda, (D), 1989, Gs

# Portugal

## Istituto Geologico e Mineiro (IGM)
**Geological and Mining Institute of Portugal** (2011)
Estrada da Portela
Zambujal
atendimento@ineti.pt
http://www.ineti.pt/

## Oporto University
**Dept of Geology** (2004)
Faculdade de Ciencias do Porto
Praca de Gomes Teixeira
4099-002 PORTO
dg.sec@fc.up.pt
http://www.fc.up.pt/depts/geo/indexi.html

## Universidade da Madeira
**Dept de Biologia** (B) (2010)
Campus Universitario da Penteada
9000-390 Funchal
p. +251 291 705 380
mgouveia@uma.pt

http://www.uma.pt/Unidades/Biologia/

## Universidade de Aveiro
**Dept de Geociencias** (B,M,D) (2014)
Campo Universitario Santiago
Aveiro 3810-193
p. +351 234 370 357
tavares.rocha@ua.pt
www.ua.pt/geo
Professor:
Eduardo Ferreira da Silva, (D), CeGbCl

## Universidade de Coimbra
**Dept de Ciencias da Terra** (A,B,M,D) ● (2016)
Rua Silvio Lima, Univ. Coimbra, Pólo II Coimbra
Coimbra 3030-790 Coimbra
p. +351-239 860 514
cgeo@ci.uc.pt
http://www.dct.uc.pt
Associate Professor:
Alcides Castilho Pereira, (D), Coimbra, 1992, GebOr

## Universidade de Lisboa
**Dept of Geology** (A,B,M,D) ● (2015)
Ed. C6, Campo Grande
Campo Grande
Lisboa 1749-016
p. +35121 750 00 66 / + 351 21 75
dgeologia@fc.ul.pt
http://www.fc.ul.pt/en/dg?refer=1
f: https://www.facebook.com/Departamento-de-Geologia-FCUL-192772970763173/?ref=ts
Professor:
Maria da Conceição Pombo Freitas, (D), Lisbon, 1996, GusGe

## Universidade de Trás-os-Montes e Alto Douro
**Departamento de Geologia** (B,M,D) (2013)
Apartado 1013
Vila Real 5001-801
p. +351259350224
geologia@utad.pt
http://www.geologia.utad.pt
Professor:
Maria E.P. Gomes, (D), UTAD, 1996, GiCgGz
Associate Professor:
Ana M.P. Alencoão, (D), UTAD, 1998, HgGgm
Artur A.A. Sá, (D), UTAD, 2005, PgiPs
Assistant Professor:
Maria E.P.S. Abreu, (D), UTAD, 2012, GaaGa
João C.C.V. Baptista, (D), UTAD, 1998, GmmGa
Maria R.M. Costa, (D), UTAD, 2000, Hw
Paulo J.C. Favas, (D), UTAD, 2008, GeCtGg
José M.M. Lourenço, (D), UTAD, 2006, YeOin
Alcino S. Oliveira, (D), uTAD, 2002, Hww
Anabela R.R.C. Oliveira, (D), UTAD, 2011, Gss
Fernando A.L. Pacheco, (D), UTAD, 2001, Hq
Luís M.O. Sousa, (D), 2001, EnGgNg
Rui J.S. Teixeira, (D), UTAD, 2008
Nuno M.O.C.M. Vaz, (D), UTAD, 2010, PmlYg

## Universidade do Minho
**Dept de Ciencias da Terra** (B,M,D) (2004)
Campus de Gualtar
Braga 4710-057
p. 351 253 604 300
sec@dct.uminho.pt
http://www.dct.uminho.pt/index/index.html

## Universidade Nova de Lisboa
**Dept de Ciencias da Terra** (2014)
Qta. da Torre
2829-516 Caparica
p. (351) 212 948 573
dct.secretariado@fct.unl.pt
http://www.dct.fct.unl.pt/

# Qatar

## University of Qatar
**Dept of Geology** (B) (2015)
PO Box 2713
Doha
Doha
hamadsaad@qu.edu.qa
http://www.angelfire.com/ms/GeoQU/
Professor:
Sobhi J. Nasir, (D), Germany, 1986, Gx
Associate Professor:
Hamad A. Al-Saad, (D), Egypt, 1996, PsGsu
Abdulali A. Sadiq, (D), Southampton (UK), 1995, Or

Assistant Professor:
Latifa B. AL-Nouimy, (D), Egypt, 1994, Hgw
Sief A. Alhajari, (D), North Carolina, 1992, Hgw



# Reunion

## Universite de la Reunion

**Faculte de Sciences et Technologie - Sciences de la Terre** (B) (2014)
15, avenue Rene Cassin
CS 92003
Saint-Denis, Cedex 9 97744
p. (262) 0262 938697
jean-lambert.join@univ-reunion.fr
http://sciences.univ-reunion.fr/rubrique.php3?id_rubrique=72
Professor:
Jean-Lambert Join, (D), Hw
Associate Professor:
Fabrice R. Fontaine, (D), Yg

**Laboratoire GeoSciences Reunion** (B,M,D) (2013)
15 avenue Rene Cassin
BP 7151
Saint Denis, Messag Cedex 9 97744
vfamin@univ-reunion.fr
http://www.geosciencesreunion.fr/
Professor:
Laurent Michon, (D), GtvGg
Professor:
Jean-Lambert Join, HwGsHg
Associate Professor:
Vincent Famin, (D), GcCgGg
Fabrice R. Fontaine, (D), Ysg
Claude Smutek, NrOrg

# Romania

## Al. I. Cuza University of Iasi

**Dept of Geology** (B,M,D) (2016)
geology@uaic.ro
B-dul Carol I nr. 20A
Iasi, Iasi 700505-RO, Iasi
geology@uaic.ro
geology.uaic.ro
*Enrollment (2015): B: 59 (26) M: 40 (18) D: 11 (5)*
Head:
Nicolae Buzgar, (D), Al. I. Cuza Iasi, 1998, GziGd
Conferentiar/ Reader:
Crina Miclaus, (D), Al. I. Cuza Iasi, 2001, Gsr
Professor:
Ovidiu Gabriel Iancu, (D), Al. I. Cuza Iasii, 1998, GpXgCg
Dan Stumbea, (D), Univ Babes-Bolya, 1998, Gez
Assistant Professor:
Traian Gavriloaiei, (D), Al. I. Cuza Iasi, 1999, CaOaCg
Lecturer:
Andrei Ionut Apopei, (D), Al. I. Cuza Iasi, 2014, GziGg
Andrei Buzatu, (D), Al. I. Cuza Iasi, 2014, GziGg
Mitica Pintilei, (D), Al. I. Cuza Iasi, 2010, CgeCt
Oana Stan, (D), Al. I. Cuza Iasii, 2009, Ge

## Babes-Bolyai University

**Faculty of Biology and Geology, Dept of Geology** (B,M,D) (2017)
Str. M. Kogalniceanu nr.1
RO - 400084 Cluj-Napoca
Cluj-Napoca 400084
p. +40-264-405371
sorin.filipescu@ubbcluj.ro
http://bioge.ubbcluj.ro/geologie/
Chair:
Sorin Filipescu, (D), Babes-Bolyai, 1996, GrPm
Professor:
Ioan Bucur, (D), Babes-Bolyai, 1991, PgmGs
Vlad Codrea, (D), Babes-Bolyai, 1995, PvGo

## Institul Geologic al Romaniei

**Geological Institute of Romania** (2011)
Caransebes, 1 Sector 1
Bucuresti
office@igr.ro
http://www.igr.ro/

## University of Bucharest

**Department of Geophysics - Faculty of Geology and Geophysics** (B,M,D) ● (2016)
6, Traian Vuia Street
Sector 2
Bucharest, Bucharest 020956
p. +40-21-3125024
bogdan.niculescu@gg.unibuc.ro
http://www.unibuc.ro/e/depts/geologie-geofizica/geofizica/
*Enrollment (2016): B: 23 (20) M: 20 (5)*
Associate Professor:
Bogdan Mihai NICULESCU, (D), Bucharest, 2002, Yge

# Rwanda

## Universite Nationale du Rwanda

**Dept of Geology** (B) (2014)
Faculty of Science
PO Box 56
Butare
p. +250 (0) 252 530 122
info@nur.ac.rw
http://www.nur.ac.rw/

**Dept of Geography** (B) (2010)
PO Box 56
Butare
p. +250 (0) 252 530 122
info@nur.ac.rw
http://ww.nur.ac.rw/

# Saudi Arabia

## King Abdulaziz University

**Dept of Geology** (2015)
P. O. Box 1744
Building No 55
Jeddah 21441
jha25@hotmail.com
http://earthscience.kau.edu.sa/Default.aspx?Site_ID=145&Lng=EN

## King Fahd University of Petroleum and Minerals

**Geosciences Dept** (B,M,D) ● (2015)
P. O. Box 5070
KFUPM
Dhahran 31261
p. +96638602620
c-es@kfupm.edu.sa
http://www.kfupm.edu.sa/departments/es/default.aspx
*Enrollment (2015): B: 52 (8) M: 77 (17) D: 6 (0)*
Professor:
Michael A. Kaminski, (D), MIT/WHOI, 1988, GuPm
Gabor Korvin, (D), Heavy Industry (Hungary), 1979, Yx
Associate Professor:
Osman Abdullatif, (D), Khartoum, 1993, Gso
Abdulwahab Abokhodair, (D), California (Santa Cruz), 1978, Ym
Khalid Al-Ramadan, (D), Uppsala, 2006, GodGs
Abdulaziz Al-Shaibani, (D), Texas A&M, 1999, HwGeg
Abdullatif Al-Shuhail, (D), Texas A&M, 1998, Ygs
Mustafa Hariri, (D), SD Mines, 1995, GcOrGt
Mohammad Makkawi, (D), Colorado State, 1998, HwGoq
Bassam S. Tawabini, (D), King Fahd Petroleum and Minerals, 2002, GeHsCa
Assistant Professor:
Waleed Abdulghani, (D), Go
Abdullah Al-Shuhail, (D), Calgary, 2011, Ye
Ismail Kaka, (D), Carleton (Ottawa), 2006, Ys
Mohammed Qurban, (D)

Lecturer:
Ayman Al-Lehyani, (M), King Fahd, Yg
Mutasim Sami Osman, (M), King Fahd Petroleum and Minerals, 2014, GosGc

### Saudi Geological Survey
**Saudi Geological Survey** (2011)
P.O. Box: 54141
21514
p. 966-2-619-5000
sgs@sgs.org.sa
http://www.sgs.org.sa/Arabic/Pages/default.aspx

## Senegal

### Universite Cheikh Anta Diop
**Dept de Geologie** (B) (2011)
Faculte Des Sciences
BP 5005
Dakar-Fann
p. +221 869.27.66
fst@ucad.edu.sn
http://fst.ucad.sn/index.php?option=com_content&task=view&id=19&Itemid=35

**Institut des Sciences de l' Environnement (ISE)** (B) (2010)
BP 5005
Dakar-Fann
p. +221 869.27.66
diengdiomaye@yahoo.fr
http://196.1.95.4/ise/index_ISE.htm

## Sierra Leone

### Njala University
**Dept of Mining and Metallurgy** (B) (2010)
Freetown
p. +232-22-228788, +232-22-226851
nuc@sierratel.sl
http://www.nu-online.com/

**Dept of Soil Science** (A,B,M,D) (2015)
Freetown
p. +232-22-228788, +232-22-226851
nuc@sierratel.sl
http://ww.nu-online.com/

**Dept of Geography & Rural Development** (B) (2010)
Freetown
p. +232-22-228788, +232-22-226851
nuc@sierratel.sl
http://www.nu-online.com/

**Institute of Environmental Management and Quality Control** (B) (2010)
Freetown
p. +232-22-228788, +232-22-226851
nuc@sierratel.sl
http://www.nu-online.com/

### University of Sierra Leone
**Dept of Geology** (2014)
PO Box 87
Freetown
aiah_gbakima2000@yahoo.com
http://www.tusol.org/programmes

## Slovakia

### Comenius University in Bratislava
**Department of Geology and Palaeontology** (A,B,M,D) O (2017)
Faculty of Natural Sciences
Ilkovičova 6
Mlynska dolina G
Bratislava 842 15
p. 00421260296529
kgp@fns.uniba.sk
http://geopaleo.fns.uniba.sk/
*Enrollment (2016): B: 12 (0) M: 1 (0) D: 5 (0)*
Professor:
Michal Kováč, (D), Gsg
Professor:
Dusan Plasienka, (D), Gt
Daniela Reháková, (D), Pms
Aubrecht Roman, (D), GrXg
Associate Professor:
Jozef Hók, (D), Ggt
Natália Hudáčková, (D), PimPo
Martin Sabol, (D), Pvo
Rastislav Vojtko, (D), GctGm
Associate Scientist:
Matúš Hyžný, Pi
Ján Schlögl, (D), Piy
Assistant Professor:
Peter Joniak, (D), Pv
Marianna Kováčová, (D), PlbPe

### Dionyz Stur Institute of Geology
**Geology Dept** (2015)
Mlynska Dolina 1
81704 Bratislava 11
p. 02 / 59375111
secretary@geology.sk
http://www.geology.sk/new/en

### Geologicka Sluzba Slovenskej Republiky
**Geological Survey of Slovak Republic** (2011)
Mill Valley 1
817 04 Bratislava 11
secretary@geology.sk
http://www.sguds.sk

### Kosice Technical University
**Faculty of Mining, Ecology, Process Control and Geotechnology** (2014)
Fakulta BERG
Park Komenskeho 12
040 01 Kosice
p. +421-55-602 1111
sekrd.fberg@tuke.sk
http://www.fberg.tuke.sk/bergweb/index.php?IdLang=0&Selection=4

## Slovenia

### Geoloski zavod Slovenije
**Geological Survey of Slovenia** (M,D) (2014)
Dimiceva ulica 14
Ljubljana SI - 1000
p. +386 1 2809702
www@geo-zs.si
http://www.geo-zs.si

### University of Ljubljaa
**Oddelek za Geologijo** (2011)
Askerceva 12
Ljubljana
spela.turic@ntf.uni-lj.si
http://www.ntf.uni-lj.si/og

### University of Ljubljana
**Faculty of natural science and engineering** (2014)
Aškerčeva c. 12
Ljubljana 1000
p. 01/470-45-00
tanja.kocevar@ntf.uni-lj.si
http://www.uni-lj.si/academies_and_faculties/faculties/2013071111502957/

# Somalia

## Burao University

**Dept of Community Development** (B) (2010)
p. +252 7126481
info@universityofburao.com
http://www.buraouniversity.com/rural_and_environmental_studies.htm

## Mogadishu University

**Somali Centre for Water and Environment** (B) (2014)
p. (252)-5-932454/ 223433/ 658479
info@mogadishuuniversity.com
http://www.mogadishuuniversity.com/index.html

## Nugaal University

**Institute of Geology** (B,M) ● (2016)
Lasanod
p. +252 275 4063 / 7412619
nugaaluniversity@gmail.com
http://www.nugaaluniversity.com/index.html

# South Africa

## Cape Peninsula University of Technology

**Dept of Environmental and Occupational Studies** (B) (2010)
Cape Town
p. (+27) 021 959 6230
vanderwesthuizenh@cput.ac.za
http://www.cput.ac.za/

## Geological Survey of South Africa

**Council for Geoscience** ● (2015)
Private Bag X112
Pretoria 0001
p. 0027128411911
info@geoscience.org.za
http://www.geoscience.org.za/
Executive Manager:
Fhatuwani L. Ramagwede, (M), 2011, EgGzEg

## Nelson Mandela Metropolitan University

**Dept of Geoscience** (B,M,D) (2014)
PO Box 77000
Summerstrand, Port Elizabeth 6031
p. 041 504 2325
sheila.entress@nmmu.ac.za
http://geosci.nmmu.ac.za/
Administrative Assistant: Sheila Entress
*Enrollment (2002): B: 65 (22) M: 7 (0) D: 1 (0)*
Head:
Nigel Webb, (D)
Professor:
Vincent Kakembo, Sp
Associate Professor:
Moctar Doucoure, (D)
Daniel Mikes, (D)
Post-Doctoral Fellowship:
Bastien Linol, (D), NMMU, 2013, Gs
Lecturer:
Callum Anderson, (M), Port Elizabeth, Gzs
Wilma Britz, (D)
Gideon Brunsdon, (M)
Anton de Wit, (D)
Pakama Syongwana, (D)
Nicolas Tonnelier, (D)
Leizel Williams-Bruinders, (M), On
Emeritus:
Peter Booth, (D)
Anthony Christopher
Related Staff:
Paul Baldwin
Willie Deysel

**Coastal and Marine Research Institute** (B) (2010)
Summerstand Campus (South)
PO Box 77000
Port Elizabeth 6031
p. +27 41 5042877
cmr@nmmu.ac.za
http://www.nmmu.ac.za/default.asp?id=380&bhcp=1

**Centre for African Conservation Ecology** (M,D) ○ (2015)
Summerstand Campus (South)
PO Box 77000
Port Elizabeth 6031
p. +27 41 504 2308
Graham.Kerley@nmmu.ac.za
http://ace.nmmu.ac.za/

## North-West University

**School of Environmental and Health Sciences** (B) (2010)
Private Bag x 1290
Potchefstroom 2520
p. (27) 18 299 2528
monica.mosala@nwu.ac.za
http://www.puk.ac.za/opencms/export/PUK/html/fakulteite/natur/geol/index_e.html

## Rhodes University

**Dept of Geography** (B) (2010)
PO Box 94
Grahamstwon 6140
p. +27 (0)46 603 8111
registrar@ru.ac.za
http://oldwww.ru.ac.za/academic/departments/geography/

**Institute for Water Research** (M,D) (2014)
PO Box 94
Grahamstown 6140
p. +27 46 62224014 / 6222428 / 62
registrar@ru.ac.za
http://www.ru.ac.za/static/institutes/iwr//?request=institutes/iwr/
Professor:
Denis Hughes, (D), Wales, 1978, Hqs

**Dept of Geology** (B,M,D) (2014)
PO Box 94
Grahamstown , Eastern Cape Provinc 6140
p. +27 (0)46 603 8309
geolsec@ru.ac.za
http://www.ru.ac.za/geology/
Administrative Assistant: Ashley Goddard
Administrative Assistant: Vuyokazi Nkayi
Head:
Steve Prevec, Cg
Professor:
Annette Gotz, Gs
Peter Horvath, (D), Gp
Bantubonke Izwe Ntsaluba, Gi
Briony Proctor
Hari Tsikos, CgEg
Yon Yao, Eg
Associate Professor:
Steffen Buttner, (D), Frankfurt (Germany), 1997, Gcp
Associate Scientist:
Billy De Klerk, Pg
Robert Gess, Pg
Rose Prevec
Mike Skinner
Emeritus:
Roger Jacobs
Julian S. Marsh, (D), Cape Town (South Africa), 1973, GiCgGv
Laboratory Director:
Gelu Costin, Gp
Cooperating Faculty:
John Hepple

## Stellenbosch University

**Dept of Process Engineering (Chemical and Mineral Processing)** (B,M,D) (2015)

Private Bag X1
Matieland 7602
p. +27 21 808-4485
chemeng@sun.ac.za
http://www.chemeng.sun.ac.za/

**Centre for Geographical Analysis** (B) (2010)
Private Bag X1
Matieland 7602
p. (27) 21 808 3218
rdonaldson@sun.ac.za
http://academic.sun.ac.za/cga/expertise/user.asp?UserID=11

**Dept of Earth Sciences** (B,M,D) (2015)
Private Bag X1
Matieland, Western Cape 7602
p. +27 (0)21 808 3219
lcon@sun.ac.za
http://www.sun.ac.za/earthSci/
Professor:
Alakendra N. Roychoudhury, (D), Georgia Tech, 1999, ClmOc

## Tshwane University of Technology

**Dept of Environmental, Water & Earth Sciences** (B,M,D) (2015)
Private Bag x680
Pretoria 0001
p. 012 382 6232
gerberme@tut.ac.za
http://www.tut.ac.za/Students/facultiesdepartments/science/departments/environscience/Pages/default.aspx
Administrative Assistant: Sarah Galebies
Administrative Assistant: Retha Gerber
*Enrollment (2015): B: 59 (37) M: 1 (0)*
Lecturer:
Chamunorwa Kambewa, (M), MSc Mineral Exploration, 1998, CgeOe
Thando Majodina, (M), GgzGe
Skhumbuzo Sibeko, (B), GggGg
Other:
Mlindelwa Lupankwa, (D), HywGg

**Dept of Environmental Health** (B) (2010)
Private Bag X680
Staatsartillerie Road
Pretoria West 0001
p. +27 (0)12 382 5911
vanrooyenps@tut.ac.za
http://www.tut.ac.za/Pages/default.aspx

## University of Cape Town

**Dept of Oceanography** (B) (2010)
Private Bag X3
Rhondebosch 7701
p. +27 (0) 21 650-3277
isabelle.ansorge@uct.ac.za
http://www.sea.uct.ac.za/index.php

**Dept of Environmental and Geographical Science** (B) (2010)
Shell Environmental & Geographical Science Building
South Lane, Upper Campus
Private Bad X3
Rondebosch 7701
p. (27) 21 - 6502873 / 4
michael.meadows@uct.ac.za
http://www.egs.uct.ac.za/

**Dept of Geological Sciences** (B,M,D) ● (2016)
Private Bag X3
Rondebosch 7701, Western Cape
p. +27 (0)21-650-2931 
head.geologicalsciences@uct.ac.za
http://www.geology.uct.ac.za/
Administrative Assistant: Lynn Evon
Administrative Assistant: Denise Lesch
*Enrollment (2014): B: 30 (30) M: 30 (8) D: 10 (0)*
Head:
Chris Harris, GiCsGg
Professor:
Steve Richardson, Cg
Senior Scientist:
Nicholas Laidler
Petrus Le Roux
Christel Tinguely
Associate Professor:
John Compton, (D), Harvard, 1986, CmGa
Associate Scientist:
Fayrooza Rawoot
Assistant Professor:
Emese M. Bordy, (D), Rhodes, 2001, GsPes
Lecturer:
Johann Diener, Gp
Lynnette N. Greyling, (D), Witwatersrand, 2009, EgmCe
Phillip Janney, (D), California (San Diego), 1996, GiCaXc
Beth Kahle, YgEoYg
Laboratory Director:
Kerryn Gray, (M)

## University of Fort Hare

**Dept of Geology** (2014)
Sobukwe Walk 8
Livingstone Hall
Alice, Eastern Cape 5700
p. +27 (0)40 602 2011
wkoll@ufh.ac.za
http://wvw.ufh.ac.za/departments/geology/
Head:
Oswald Gwavava
Lecturer:
CJ Gunter
Vuyokazi Mazomba
Related Staff:
Luzuko Sigabi

**Dept of Geographic Information Systems** (B) (2014)
Private Bag X1314
Alice 5700
p. +27 (0)40 602 2011
wkoll@ufh.ac.za
http://www.ufh.ac.za/departments/gis/gishome.html

**Dept of Geography, Land Use, and Environmental Sciences** (B) (2014)
Private Bag X1314
Alice 5700
p. +27 (0)40 602 2011
wkoll@ufh.ac.za
http://www.ufh.ac.za/

## University of Johannesburg

**Dept of Geology** (B) (2010)
PO Box 524
Kingsway & University (APK Campus)
Auckland Park 2006
p. +27 (11) 559-4701
mdekock@uj.ac.za
http://www.uj.ac.za/Default.aspx?alias=www.uj.ac.za/geology

**Dept of Geography, Environmental Management & Energy Studies** (B,M,D) (2014)
PO Box 524
Auckland Park, Gauteng 2006
p. +27 (0)11 559 2433
science@uj.ac.za
http://www.uj.ac.za/Default.aspx?alias=www.uj.ac.za/geography
Dr:
Isaac T. Rampedi, (D), South Africa, 2010, PySdHg

**Dept of Mining Engineering** (B) (2010)
PO Box 17011

Doornfontein 2028
p. +27 (0)11 559-6628
hgrobler@uj.ac.za
http://www.uj.ac.za/Default.aspx?alias=www.uj.ac.za/mining

**Dept of Mine Surveying** (B) (2010)
p. +27 (0)11 559-6186
hgrobler@uj.ac.za
http://www.uj.ac.za/Default.aspx?alias=www.uj.ac.za/minesurv

South Africa

## University of KwaZulu-Natal

**School of Civil Engineering Surveying & Construction** (B) (2014)
Centenary Building
King George V Avenue
Durban 4041
p. (+27) 2603065
troisc@ukzn.ac.za
http://geomatics.ukzn.ac.za/HomePage2247.aspx

**Centre for Water Resources Research** (B,M,D) (2016)
Room 203
Rabie Saunders Building
Pietermaritzburg, KwaZulu-Natal 3201
p. +27-(0)33-260 5490
smithers@ukzn.ac.za
http://cwrr.ukzn.ac.za/

**School of Environmental Sciences** (B,M,D) (2017)
Department of Geography
King Edward Avenue
Pietermaritzburg Campus
Pietermaritzburg, KwaZulu-Natal 3209
p. + 27 031 260 5346
gijsbertsen@ukzn.ac.za
http://saees.ukzn.ac.za/Homepage.aspx
f: https://www.facebook.com/FriendsOfUkznAgriculture

**Discipline of Geological Sciences, School of Agricultural, Earth and Environmental Sciences** (A,B,M,D) ● (2017)
Private Bag X 54001
Durban, KwaZulu-Natal 4000
p. +27 (0)31 260 2516
mccoshp@ukzn.ac.za
http://www.geology.ukzn.ac.za/

## University of Limpopo

**Dept of Soil Sciences** (B) (2014)
Turfloop Campus
Private Bag X1106
Sovenga 0727
p. +27 (0)15 268 9111
Funso.Kutu@ul.ac.za
http://www.ul.ac.za/index.php?Entity=agri_soil_scie

**Geography & Environmental Studies Dept** (B) (2014)
Turfloop Campus
Private Bag X1106
Sovenga 0727
p. (+27) (0)15 268 3756
salphy.ramokolo@ul.ac.za
http://www.ul.ac.za/index.php?Entity=agri_geo_environ

**School of Physical and Mineral Science** (B) (2015)
Turfloop Campus
Private Bag X1106
Sovenga 0727
p. +27(0) 15 268 3492
SPMS@ul.ac.za
http://www.ul.ac.za/index.php?Entity=School%20Main%20Menu&school_id=9
Administrative Assistant: M. D.. Ramusi

Head:
K. E. Rammutia

Professor:
John Dunlevey, (D), Stellenbosch, 1985, GzxEg
M. Khanyi
M. A. Letsoalo
M. A. Mahladisa
R. M. Makwela
T, Mobakazi
T. E. Mosuang
J. M. T. Mphahlele
T. T. Netshisaulu
M. Netsianda
P. Ntoahee
O. O. Nubi
M. Phala
M. J. Ramusi
L. Wilsenach

Other:
J. P. T. Crafford, (N), OnNgOn

## University of Pretoria

**Centre of Environmental Studies** (M,D) (2010)
Pretora 0002
p. +27 (0)12 420-3111
willemferguson@zoology.up.ac.za
http://www.up.ac.za/centre-environmental-studies/index.php

**University of Pretoria Natural Hazards Centre** (M,D) (2016)
Department of Geology
University of Pretoria, Private Bag X20, Hatfield, PRETORIA, 0028
Pretoria 0028
p. +27 (0)12 420 3613
andrzej.kijko@up.ac.za
http://www.up.ac.za/university-of-pretoria-natural-hazard-centre-africa

Professor:
Andrzej Kijko, (D), Ysg

Research Associate:
Ansie Smit, (M), Mathematical Statistics

**Dept of Mining Engineering** (B,M,D) (2014)
Pretoria 0002
p. +27 0 12 420-3111
ssc@up.ac.za
http://web.up.ac.za/default.asp?ipkCategoryID=2520

**Dept of Geography, Geoinformatics and Meteorology** (B,M,D) (2015)
Room 3-7
Geography Building
Cnr Lynnwood and University Roads
Hatfield, Pretoria 0083
p. +27 (0)12 420 3536
lunga.ngcongo@up.ac.za
http://web.up.ac.za/ggm
Administrative Assistant: M. C.. van Aardt

Head:
C.J. deW. Rautenbach

Associate Professor:
Serena M. Coetzee, (D), Pretoria, 2009, Oi
P. D. Sumner

Lecturer:
Daniel Darkey
Nerhene Davis
Liesl Dyson
Joos Esterhuizen
Natalie S. Haussmann
Michael Loubser
P. L. Philemon Tsela
Barend van der Merwe
Fritz van der Merwe

Related Staff:
Ingrid Booysen
Popi Mahlangu
Tebogo Moremi
Erika Pretorius

**Dept of Geology** (B,M,D) ● (2017)
Pretoria 0002
p. +27 (0)12 420 2454
wlady.Altermann@up.ac.za
http://web.up.ac.za/default.asp?ipkCategoryID=2048

## University of the Free State

**Department of Geology** (B,M,D) ● (2016)
PO Box 339
Bloemfontein, Free State Province 9300
p. +27 (0)51 401 2515
roelofsef@ufs.ac.za
http://www.ufs.ac.za/natagri/departments-and-divisions/geology-home
*Enrollment (2016): B: 234 (66) M: 13 (6) D: 6 (0)*

Dr:
Frederick Roelofse, (D), the Witwatersrand, Gip
Prof:
Wayne Colliston, (D), the Free State, GcNr
Marian Tredoux, (D), the Witwatersrand, Cga
Ms:
Justine Magson, (M), Free State, 2016, CgGi
Thendo Mapholi, (B), the Free State, Ggp
Makhadi Rinae, (B), the Free State, Gge
Mr:
Adriaan Odendaal, (B), Free State, 2010, GgsGr
Raimund Rentel, (M), Stellenbosch, Gz
Dr:
Robert Hansen, (D), Stellenbosch, CgGe
Ms:
Megan Purchase, (M), the Free State, Gx

**Institute for Groundwater Studies** (B,M,D) (2016)
PO Box 339
Internal Bus 56
Bloemfontein 9300
p. +27(0)51-4019111
vermeulend@ufs.ac.za
http://www.ufs.ac.za/igs

**Centre for Environmental Management** (B) (2010)
Po Box 338
Bloemfontein 9300
p. +27(0)51-4019111
avenantmf@ufs.ac.za
http://www.ufs.ac.za/faculties/index.php?FCode=04&DCode=106

## University of the Western Cape

**International Ocean Institute of Southern Africa** (B) (2010)
Private Bag X17
Bellville 7535
p. +27 21 959 3088
ioi-sa@uwc.ac.za
http://www.ioisa.org.za/

**Dept of Earth Science** (B) (2014)
Modderdam Road
Private Bag X17
Bellville, Cape Town 7530
p. +27 (0)21 959 2223
wdavids@uwc.ac.za
http://www.uwc.ac.za/Faculties/NS/EarthScience/Pages/default.aspx
Administrative Assistant: Caroline Barnard
Administrative Assistant: Wasielah Davids
Administrative Assistant: Chantal Johannes
Administrative Assistant: Mandy Naidoo

Head:
Charles Okujeni, (D), Berlin, Germany
Chair:
Jacqueline Goldin, (D), UCT
Yongxin Xu
Professor:
Dominic Mazvimavi, (D), Wageningen Univ, 2003
Jan M. van Bever Donker, (D), Cape Town (South Africa), 1979, GcpGt
Associate Professor:
Ebernard Braune
Lecturer:
Marcelene Andrews
James Ayuk Ayuk, (M), UWC
Lewis Jonkey, (M)
Thokozani Kanyerere
Mimonitu Opuwari, (D), UWC
Henok Solomon, (M), UWC
VLIR Coordinator:
Shauib Dustay
HIVE Manager:
Yafah Hoosain, (M), UWC
Deputy Head:
Theo Scheepers, (M), Stellenbosch
Related Staff:
Janine Becorney
Shamiel Davids
Peter Meyer



## University of the Witwatersrand

**School of Geography, Archaeology and Environmental Studies** (B,M,D) (2016)
Private Bag 32050
Wits, Johannesburg
p. +27 11 717 6503
donna.koch@wits.ac.za
http://web.wits.ac.za/Academic/Science/Geography/Home.htm

**School of Mining Engineering** (B,M,D) ● (2015)
Private Bag 3
WITS, Johannesburg 2050
p. +27 11-717-7003
bekir.genc@wits.ac.za
http://web.wits.ac.za/Academic/EBE/MiningEng/

**School of Geosciences** (B,M,D) ● (2016)
Faculty of Science
Private Bag 3
Wits, Johannesburg 2050
p. 27 11 717 6547
sharon.ellis@wits.ac.za
http://web.wits.ac.za/Academic/Science/GeoSciences/Home.htm

Professor:
Lewis D. Ashwal, (D), Princeton, 1979, GxCgGt
Roger L. Gibson, (D), Cambridge, 1990, GcpGg
Kim AA Hein, (D), Tasmania, 1995, GgEmGc
Judith A. Kinnaird, (D), St. Andrews, 1987, EmGi
Associate Professor:
Paul A.M. Nex, (D), Univ Coll (Cork), 1997, Eg
Dr:
Michael QW Jones, (D), the Witwatersrand, 1981, YhGtYg
Electron Microprobe Scientist:
Peter Horvath, (D), Eotvos Lorand, 2002, GpCpGz
Dr:
Grant M. Bybee, (D), the Witwatersrand, 2013, Gi
Katie A. Smart, (D), Alberta, 2011, CsGx
Doctor:
Zubair A. Jinnah, (D), the Witwatersrand, 2011, GsPg
Emeritus:
Carl R. Anhaeusser, (D), Witwatersrand, 1983, GgEgOg

## University of Venda

**Dept of Geography and Geo-Information Sciences** (B) (2014)
University of Venda
Private Bag X5050
Thohoyandou, Limpopo 0950
p. +27 15 962 8593
nthaduleni.nethengwe@univen.ac.za
http://www.univen.ac.za/enviornmental_sciences/dep_geography_geo_sciences.html

Head:
N. S. Nethengwe, (D), West Virginia

Acting Head:
T. M. Nelwamondo, (D), UP, 2010, GeOuy
Lecturer:
E. Kori, (M), Univen
M. J. Mokgoebo, (M), Univen
N. V. Mudau, (M), Univen
A. Muyoki, (D), Howard
M. Nembudani, (M), Stellenbosch
Related Staff:
K. H. Mathivha, (M), Univen

**Mining and Environmental Geology** (A,B,M,D) (2013)
School of Environmental Science
Private Bag X5050
Thohoyandou , Limpopo Province 0950
p. +27 159628580
ogolaj@univen.ac.za
http://www.univen.ac.za/environmental_sciences/dep_mining_environmental.html
Dr.:
Milton Kataka, (D), Witwatersrand, 2003

## University of Venda for Science & Technology

**GIS Resource Centre** (B,M,D) (2017)
GIS Resource Centre
Private Bag x5050
Thohoyandou, Limpopo 0950
p. +27 15 962 8044
farai.dondofema@univen.ac.za
http://www.univen.ac.za/environmental_sciences/gis_centre.html
t: @chinomukutu3
*Enrollment (2015): B: 90 (0) M: 8 (0) D: 1 (0)*
Chief Technician:
Farai Dondofema, (M), Zimbabwe, 2007, OrHgOi

**Institute of Semi-Arid Environment and Disaster Management** (B) (2014)
p. (+27) 015 962 8513
ndidzum@univen.ac.za
http://www.univen.ac.za/index.php?Entity=Institute%20of%20Semi-Arid%20Environment&Sch=3

**Dept of Hydrology and Water Resources** (B) (2014)
University of Venda
Private Bag X5050
Thohoyandou, Limpopo 0950
p. +27 015 962 8513
environmental@univen.ac.za
http://www.univen.ac.za/environmental_sciences/dep_hydrology_water.html
Head:
J. O. Odiyo, (D)
Lecturer:
J. R. Gumbo, (D)
P. M. Kundu, (D)
R. Makungo, (M)
Tinyiko Rivers Nkuna, (M), Venda, 2012, HgwOa

## University of Zululand

**Dept of Geography and Environmental Studies** (B) (2010)
Private Bag X1001
KwaDlangezwa 3886
p. +27 (035) 902 6282
kamwendog@unizulu.ac.za
http://www.uzulu.ac.za/scie_geo_env.php

**Dept of Hydrology** (B,M,D) O (2017)
Private Bag X1001
KwaDlangezwa 3886
p. +27 (035) 902 6282
SimonisJ@unizulu.ac.za
http://www.uzulu.ac.za/scie_hydro.php
*Enrollment (2016): B: 332 (79) M: 15 (2) D: 2 (0)*

# Spain

## Coruna University

**University Institute of Geology** (D) (2016)
Edificio Servicios Centrales de Investigación
Campus de Elviña s/n
La Coruna 15071
p. 0034 981167000
xeoloxia@udc.es
http://www.udc.es/iux
*Enrollment (2015): D: 8 (80)*
Director:
Juan Ramon Vidal Romaní, (D), Complutense (Madrid), 1983, GmlGc
Professor:
Antonio Paz Gonzalez, (D), Santiago de Compostela, 1982, Sd
Paleontologist:
Aurora Grandal d'Anglade, (D), Coruña, 1993, Pv
Associate Professor:
Elena Pilar de Uña Alvarez, (D), Santiago de Compostela, 1986, Oy
Cruz Iglesias, (D), Corunna, 2001, GaOmNr
María Teresa Taboada Castro, (D), Santiago de Compostela, 1990, Sd
Associate Scientist:
Marcos Vaqueiro Rodriguez, (M), Vigo (Spain), GcmYd
Research Associate:
Jorge Sanjurjo, (D), Corunna, 2005, GaCcPy

## Institute of Marine Sciences (CSIC)

**Marine Geosciences Departmen** (2016)
Passeig Maritim de la Barceloneta
37-49
Barcelona 08003
p. (+34) 63 230 95 00
secredir@icm.csic.es
http://gma.icm.csic.es

## Instituto Geologico y Minero de Espana

**Instituto Geologico y Minero de Espana** (2011)
Rios Rosas, 23
28003 Madrid
igme@igme.es
http://www.igme.es

## Univ Complutense de Madrid

**Facultad de Ciencias Geologicas** (B,M) (2010)
C/ Jose Antonio Novais 12
Cuidad Universitaria
Madrid E-28040
p. 34 91394 4819
infoweb@geo.ucm.es
http://www.ucm.es/centros/webs/fgeo
*Enrollment (2010): M: 89 (0)*
Dean:
Eumenio Ancochea, (D), Fac CC Geologicas (Madrid), 1972, Gvv
International Relations Vicedean:
Agustin P. Pieren, (D), Fac CC Geologicas, 1982, GrEgg

## Universidad Autonoma de Madrid

**Departamento de Geologia y Geoquimica** (M,D) O (2016)
Faculdad de ciencias
Campus de Cantoblanco, C/ Francisco Tomás y Valiente, 7
Módulo 06, 6ª planta
Madrid 28049
p. 91 497 48 00
directora.geologia@uam.es
http://www.uam.es/GyG
*Enrollment (2014): M: 15 (5) D: 11 (7)*

## Universidad de Alicante

**Environment and Earth Sciences Dept** (B,M,D) (2013)
Facultad de Ciencia
San Vicente de Raspeig
Alicante E-03080
p. (+34) 96 590 3552

dctma@ua.es
http://dctma.ua.es/en/environment-and-earth-sciences-department.html

## Universidad de Granada
**Dept de Geologia** (2004)
Andalusian Institute of Geophysics
Campus Universitario de la Cartuja
18071 Granada
jaguirre@ugr.es
http://www.ugr.es/iag/iagpds.html

## Universidad de Huelva
**Geologia** (B,M,D) (2011)
Huelva
p. 959219809
secgeo@uhu.es
http://www.uhu.es/dgeo/

## Universidad de Jaen
**Departamento de Geologia** (B,M,D) ○ (2016)
Edificio B-3
Campus Universitario
Jaen 23071
p. +34-953-212295
jmmolina@ujaen.es
http://geologia.ujaen.es/

## Universidad de Las Palmas de Gran Canaria
**Dept de Fisica** (B) (2011)
C/Juan de Quesada, nº 30
, Las Palmas de Gran C 35001
p. (+34) 928 451 000/023
universidad@ulpgc.es
http://www.dfis.ulpgc.es/

**Facultad de Ciencias del Mar** (B,M,D) (2014)
Edificio de Ciencias Básicas
Campus Universitario de Tafira
Las Palmas de Gran Canaria, Las Palmas 35017
p. +34 928 452 900
sec_dec_fcm@ulpgc.es
www.fcm.ulpgc.es
f: https://www.facebook.com/groups/www.fcm.ulpgc.es/?fref=ts

## Universidad de Murcia
**Dept of Geography and Regional Planning** (2004)
Facultad de Sciencia
Calle Santo Cristo 1
Murcia 30100
p. (868) 88-7446
jagb@um.es

## Universidad de Oviedo
**Dept de Geologia** (2005)
Campus de Llamaquique
Jesus Arias de Velasco, s/n
33005 Oviedo E-33005
geodir@geol.uniovi.es
http://www.geol.uniovi.es/

## Universidad de Pais Vascu
**Dept de Geologia** (2004)
Facultad de Sciencia
37008 Salamanca
p. 946015491
sec-centro.fct@ehu.es

## Universidad de Palma de Mallorca
**Dept de Geologia/Geofisica** (2004)
Facultad de Sciencia
07012 Palma

## Universidad de Sevilla
**Applied Geology to Civil Engineering** (D) (2016)
Departamento de Cristalografía y Mineralogía
calle Profesor García González 1
Seville 410012
p. +34954556318
igonza@us.es
http://www.departamento.us.es/dcmqa/
Professor:
Isabel González, (D), GzeSc

## Universidad de Valencia
**Dept de Geologia** (M,D) (2015)
Faculty of Biological Sciences
Building A (2nd and 3rd floor)
C/ Dr. Moliner, 50
Burjassot- (Valencia ) 46100
p. (+34) 96 354 46 02
dep.geologia@uv.es
http://www.uv.es/geologia

## Universidad de Valladolid
**Dept of Geography** (2004)
Facultad de Faculty of Philosophy and Arts
Palacio Santa Cruz/Plaza Santa Cruz B
Valladolid
p. +34 983 423 005
belart@fyl.uva.es

## Universidad de Zaragoza
**Facultad de Ciencias** (2004)
50.009 Zaragoza
seccienz@unizar.es
http://wzar.unizar.es/acad/fac/geolo/adepo.html

**Museo de Ciencias Naturales** (2016)
Museo de Ciencias Naturales
Universidad de Zaragoza
Edificio Paraninfo. Plaza Basilio Paraiso
Zaragoza, Aragón 50005
museonat@unizar.es
http://museonat.unizar.es/
f: https://www.facebook.com/museopaleounizar
t: @museonat

## Universitat Autonoma de Barcelona
**Geologia** (B,M,D) ● (2016)
Facultat de Ciencies, Edifici C
Campus UAB
Bellaterra, Catalunya 08193
p. +34 93 581 3022
d.geologia@uab.es
http://departaments.uab.cat/geologia/
*Enrollment (2016): B: 60 (0) M: 40 (0) D: 20 (0)*
Chair:
Antoni Teixell Cácharo, GtcYx
Professor:
María Luísa Arboleya Cimadevilla, (D), Gc
Joan Bach Plaza, (D), Gg
Esteve Cardellach López, (D), GzeCs
Esmeralda Caus Gracia, (D), Pm
Eugènia Estop Graells, (D), Gz
David Gómez Gras, (D), Gd
Francisco Martínez Fernández, (D), GpCp
Juan Francesc Piniella Febrer, (D), Gz
Eduard Remacha Grau, (D), Universitat Aut, 1983, Gso
Associate Professor:
Lluís Casas Duocastella, (D), GzaYm
Mercè Corbella Cordomí, (D), GzCl
Elena Druguet Tantiña, (D), Gc
Joan Estalrich López, (D), HgGe
Gúmer García Galán, (D), GiCg
Rogelio Linares, (D), NgYx
Ricard Martínez Ribas, (D), Pi

Oriol Oms Llobet, (D), GrsYm
Josep Maria Pons Muñoz, Pv
Joan Reche Estrada, (D), GpCp
Enric Vicens Batet, (D), Pg
Assistant Professor:
Valentí Oliveras Castro, (D), Gvi
Joan Poch Serra, (D), Gr
Lecturer:
Albert Griera Artigas, (D), GcCl
Mario Zarroca Hernández, (D), YgNg

## Universitat de Barcelona

**Estratigrafia, Paleontologia i Geociencies marines** (2006)
c/Marti Franques, s/n
08028
Barcelona
p. +33 934021384
secretaria-geologia@ub.edu
http://www.ub.es/dpep/1welcom3.htm

**Dept de Geoquimica, Petrologia I Prospeccio Geologica** (2004)
Marti Franques, s/n
Barcelona
8028
dept-geoquimica@ub.edu
http://www.ub.es/geoquimi/index.html

**Facultat de Ciències de la Terra (Earth Sciences Faculty)** (B,M,D) O (2016)
Martí Franqués s/n
Barcelona 08028
p. +34 934 021 335
deganat-geologia@ub.edu
http://www.ub.edu/geologia/en/

## Universitat de les Illes Balears

**Dept de Ciencies de la Terra** (B,M,D) (2013)
Carretera de Valldemossa, km 7.5
Palma, Balearic Islands 07122
p. +34 971172362
dct@uib.es
http://www.uib.es/depart/dctweb/home.htm

## University of the Basque Country UPV/EHU

**Departamento de Geodinamica** (B,M,D) O (2016)
Facultad de Ciencia y Tecnología
Barrio Sarriena, s/n
Leioa, Vizcaya 48940 Leioa
p. + 34. 94 601 2563
julia.cuevas@ehu.es
http://www.geodinamica.ehu.es/s0001-home1/es/
Professor:
Julia Cuevas, (D), 1988, GctGg
Professor:
Benito Abalos, (D), Universidad del Pa, 1990, GctGp
Iñaki Antigüedad, (D), Hw
Jose M. Tubía, (D), GctYg
Senior Scientist:
Aitor Aranguren, (D), GcYm
Associate Professor:
Jose M. Badillo, (M), Gmt
Jose Julian Esteban, (D), GtCcYg
Hilario Llanos, (D), Hgy
Tomas Morales, (D), NgHg
Pablo Puelles, (D), Gct
Vicente Santana, (D), GgYm
Jesus Angel Uriarte, (D), NgHg
Assistant Professor:
Arturo Apraiz, (D), Gpt
Vicente Iribar, (D), HwGm
Luis Miguel Martinez Torres, (D), Ggt
Fernando Sarrionandia, (D), Basque Country, 2006, GigGv
Research Associate:
Nestor Vegas, (D), Universidad del Pa, GcYmGt

## Univesidad de Malaga

**Dept de Geography** (2004)
El Ejido
29071 Malaga
ocana@uma.es

# Sri Lanka

## Sabaragamuwa University of Sri Lanka

**Dept of Natural Resources** (2011)
P.O Box 02
Belihuloya , Sabaragamuwa Provinc
p. 0094-45-2280293
head_nr@sab.ac.lk
http://www.sab.ac.lk

## Sri Lanka Geological Survey & Mines Bureau

**Sri Lanka Geological Survey & Mines Bureau** (2011)
Senanayake Building, No 4, Galle Road,
Dehiwala
gsmb@slt.lk
http://www.gsmb.gov.lk/

## University of Moratuwa

**Dept of Earth Resources Engineering** (2011)
Katubedda, Moratuwa
Western Province
Moratuwa
p. 0094-11-2650353
shiromi@earth.mrt.ac.lk
http://www.ere.mrt.ac.lk/

## University of Peradeniya

**Dept of Geology** (2011)
Central Province
Peradeniya 22000
p. 0094-81 2394 200/201
geology@pdn.ac.lk
http://www.pdn.ac.lk/sci/geology/

## Uwa Wellassa University of Sri Lanka

**Dept Mineral Resources & Technology** (2011)
2nd Mile Post Passara Road
Badulla , Uwa province
p. 0094-55-2226400
info@uwu.ac.lk
http://www.uwu.ac.lk/

# Sudan

## Cairo University

**Dept of Geology** (2004)
Khartoum Branch
PO Box 1055
Khartoum

## El-Neelain University

**Dept of Geology** (B) (2010)
Gamhuria Street
Khartoum
p. +249 (183) 77-441
http://www.neelain.edu/sd/english.index.htm

## Gama'at El Khartoum

**Dept of Geology** (2014)
PO Box 321
Khartoum
geology@uofk.edu
http://science.uofk.edu/index.php?option=com_content&view=article&id=101&Itemid=128&lang=en

## International University of Africa

**Dept of Geology** (B) (2010)

PO Box 2469
Khartoum
p. (+249) 183 223211
minerals@iua.edu.sd
http://www.iua.edu.sd/upldep.htm

## Sudan University of Science and Technology

**Surveying Engineering Dept** (B) (2010)
Southern and Northern Campus
Khartoum
p. +249183468622
elhadielnazier@sustech.edu
http://www.sustech.edu/faculty_en/department.php?coll_no=9&chk=a7b751cd18a66f8cd84b301f24267aab

**College of Water & Environmental Engineering** (B) (2010)
p. +24985200512 / 985200511 / 918
m.ginaya@sustech.edu
http://www.sustech.edu/faculty_en/index.php?coll_no=13&chk=cdf8953eb5124fa6f08f046043be8bf

**College of Petroleum Engineering and Technology** (B) (2010)
Southern Campus
Khartoum
zeinabkhaleel@sustech.edu
http://www.sustech.edu/faculty_en/index.php?coll_no=23&chk=1d4f14cee52598202e0fc32a12b2dfc2

## University of Dongola

**Dept of Geology** (B) (2014)
PO Box 47
Northern Province, Dongola
p. +249-241-821-516, +249-241-821
http://www.uofd.edu.sd/index.php/ar/

## University of Gezira

**Dept of Geology** (2014)
Faculty of Science and Technology
PO Box 20
Wad Medani
p. (002) 49511825724
webinfo@uofg.edu.sd
http://uofg.edu.sd/ENV/

## University of Juba

**Dept of Geology and Mining** (B) (2010)
PO Box 321/1
Khartoum Center, Juba
p. +249 (83) 222125
info@juba.edu.sd
http://www.juba.edu.sd/

**Dept of Environmental Studies** (B) (2010)
PO Box 321/1
Khartoum Center, Juba
p. +249 (83) 222125
info@juba.edu.sd
http://www.juba.edu.sd/

## University of Khartoum

**Dept of Geology** (B,M,D) (2014)
PO Box 321
Khartoum 11115
geology@uofk.edu
http://www.uofk.edu/
Head:
Ibahim Abdu Mohamed
Professor:
Saad Eldin Hamad Moha Ali
Abdelhalim Hassan Elnadi, Goz
Associate Professor:
Abdelwahab Yousif Abbas
Salah Bashir Abdalla, Ng
Osman Mahmoud Abdelatif
Samia Abdelrahman
Badr Eldin Khal Agmed
Fath Elrahman ali Birair
Omar Elbadri ali Elmaki
Abdelhafiz Gad Almula
Mohamed Zaid Awad
Insaf Sanhoory Babiker
Almed Sulaiman Daood
Abdalla Guma Farwa
Instructor:
Eltayeb Elasha Abdalia
Sami Osman Ibrahim
Amro Shaikh Idris Ahmed
Waleed Elmahdi Siddig
Saif Eldin Sir Elkhatim
Lecturer:
Amany Ali Badi
Related Staff:
Walaa Elnasir Ibrahim

**Marine Research Laboratory in Suakin** (B) (2010)
PO Box 321
Khartoum 11115
sbabdalla@hotmail.com
https://www.fkm.utm.my/marine/mtl/?Album:Visitor

## University of Kordofan

**Dept of Soil and Water Sciences** (B) (2014)
p. 860008-611-00249
hadiaabdelatif@kordofan.edu.sd
http://science.kordofan.edu.sd/

## University of Neelain

**Dept of Geology** (B) (2011)
science_dean@neelain.edu.sd
http://www.neelain.edu.sd/

## University of the Red Sea

**Dept of Geology** (B) (2014)
PO Box 24
Red Sea, Port Sudan
p. +249-311-219-28
redseauniv44@hotmail.com

# Suriname

## Anton de Kom Universiteit van Suriname

**Environmental Sciences Dept** (B) O (2016)
Universiteitscomplex Leysweg
P.O.B. 9212
Gebouw 17
Paramaribo, n.a. n.a.
p. 597465558 (Ext. 308)
s.carilho@uvs.edu
http://adekus.uvs.edu/section/sectSection.php?secID=5304&lb=Onderwijs%20.%20Technologie%20.%20Milieu
f: Anton de Kom University of Suriname

## Anton de Kom University of Suriname

**Dept of Geology & Mining** (2004)
Leysweg, P.O. Box 9212, Universiteits Complex, Gebouw VI

# Swaziland

## University of Swaziland

**Dept of Geography, Environmental Science and Planning** (B,M,D) (2014)
Private Bag 4
Kwaluseni M201
p. (+268) 2517-0000
kwaluseni@uniswa.sz
http://www.uniswa.sz/academics/science/gep

# Sweden

## Chalmers University of Technology
**Dept of Geology** (2005)
S-412 96 Goteborg
Gothenburg
sofie.hallden@chalmers.se
http://geo.chalmers.se/index.htm

## Karlstad University
**Risk Management** (B,M,D) ● (2014)
Karlstad University
Karlstad 65188
Magnus.Johansson@kau.se
http://www.kau.se/riskhantering

## Lund University
**Dept of Geology, Historical Geology & Paleontology** (2014)
Soluegatan 12
223 62 Lund
p. (+46) 462221424
mikael.calner@geol.lu.se
http://www.science.lu.se/

## Stockholm University
**Dept of Geological Sciences** (B,M,D) ● (2016)
Svante Arrheniusväg 8
106 91 Stockholm
office@geo.su.se
http://www.geo.su.se

## Sveriges geologiska Undersokning
**Geological Survey of Sweden** (2016)
Box 670
751 28 Uppsala
sgu@sgu.se
http://www.sgu.se/

## Umea Universitet
**Dept of Ecology and Environmental Science** (2013)
Natural Sciences, Johan Bures Road 14, Umeå
S-901 87 Umea
p. +46 90 786 50 00
jolina.orrell@emg.umu.se
http://www.emg.umu.se

## University of Gothenberg
**Master of Earth Science** (2004)
Box 100
S-405 30 Gothenburg
p. +46 31-786 0000

## University of Stockholm
**Dept of Geology and Geochemistry** (2006)
106 91 Stockholm
Barbara@geo.su.se
http://www.geo.su.se/

## Uppsala Universitet
**Dept of Earth Sciences** (B,M,D) ● (2014)
Villavagen 16
UPPSALA SE-752 36
p. +46-18-4710000
PREFEKT@geo.uu.se
http://www.geo.uu.se/default.asp?pageid=1&lan=1

# Switzerland

## ETH Hoenggerberg
**Geophysical Inst ETH** (2014)
Sonneggstrasse 5
8092 Zurich
p. +41 44 633 26 05
christoph.baerlocher@erdw.ethz.ch
http://www.geophysics.ethz.ch/

## ETH Zurich
**Dept of Earth Sciences** (2012)
ETH-Zenrum
CH-8092 Zurich
info@erdw.ethz.ch
http://www.erdw.ethz.ch/

## Federal Office for the Environment (FOEN)
**Federal Office for the Environment (FOEN)** (2015)
info@bafu.admin.ch
Bern 3003
p. 0041313229311
info@bafu.admin.ch
http://www.bafu.admin.ch/
t: @bafuCH

## University of Basel
**Geol & Palaeontological Inst** (2012)
Bernouillistrasse 32
CH-4056 Basale
joelle.glanzmann@unibas.ch
http://www.unibas.ch/earth/GPI/paleo/index.htm

**Inst of Earth Sciences** (2004)
Bernoullistrasse 30
Basel 4056
eberhard.parlow@unibas.ch
http://therion.minpet.unibas.ch/minpet/index.html

**Dept of Earth Sciences** (2012)
Bernoullistr.32
CH-4056 Basel
Joelle.Glanzmann@unibas.ch
http://duw.unibas.ch/

## University of Fribourg
**Div of Earth Sciences** (2017)
Chemin du Musee 6
Perolles
CH-1700 Fribourg
nicole.bruegger@unifr.ch
http://www.unifr.ch/geology/

## University of Geneva
**Section of Earth Sciences and Environment** (2009)
13, rue des Maraichers
Geneva CH-1205
p. 0041223796628
elisabeth.lagut@unige.ch
http://www.unige.ch/sciences/terre/

## University of Lausanne
**Faculty of Geosciences and Environment** (B,M,D) ● (2016)
Geopolis
Lausanne CH-1015
p. +41 21 692 35 00
doyen.gse@unil.ch
www.unil.ch/gse/home.html
*Enrollment (2016): B: 118 (0) M: 125 (0) D: 21 (0)*

## University of Neuchatel
**Centre for Hydrogeology and Geothermics** (B,M,D) (2016)
Rue Emile-Argand 11
CH-2000 Neuchatel
secretariat.chyn@unine.ch
http://www.unine.ch/chyn

## Universität Bern
**Institut für Geologie** (B,M,D) (2017)
Baltzerstrasse 1+3
Bern CH-3012

p. +41 (0) 31 631 87 61
info@geo.unibe.ch
http://www.geo.unibe.ch

# Syria

## Damascus University

**Dept of Geology** (2004)
Jammah Dimasq
Damscus

# Taiwan

## Academia Sinica

**Institute of Earth Sciences** (D) (2016)
128, Section 2, Academia Road, Nangang
Taipei 11529
p. 886-2-2783-9910
jhwang@earth.sinica.edu.tw
http://www.earth.sinica.edu.tw/index_e.php

Research Fellow:
Li Zhao, (D), Princeton, 1995, YsgGt
Distinguished Researcg Fellow:
Jeen-Hwa Wang, (D), SUNY, 1982, Ys
Associate Research Fellow:
Wu-Cheng Chi, (D), California (Berkeley), GtYrs
Assistant Research Fellow:
Wen-che Yu, (D), Stony Brook, 2007, Ys

## Central Geological Survey (MOEA) of Taiwan

**Central Geological Survey (MOEA) of Taiwan** (2013)
District No. 2, Lane 109
Taipei 235
cgs@moeacgs.gov.tw
http://www.moeacgs.gov.tw/main.jsp

## Chinese Culture University

**Dept of Geology** (2004)
Hanaoka Yangmingshan
Taipei 1111455
p. (02) 2861-0511 rpm 26105
crssge@staff.pccu.edu.tw

## National Cheng-Kung University

**Dept Earth Sciences** (2004)
Tainan
wong56@mail.ncku.edu.tw

## National Chung Cheng University

**Inst of Seismology** (2004)
168 University Rd
Min-Hsiung Chia-Yi
seismo@ccu.edu.tw
http://www.eq.ccu.edu.tw

## National Taiwan University

**Inst of Geology** (2004)
245 Choushan Road, Taipei 106-17

# Tanzania

## ARDHI University

**School of Geospatial Sciences and Technolgy** (B,M,D) (2012)
P.O.Box 35176
Dar es Salaam
hagai@aru.ac.tz
http://www.aru.ac.tz/page.php?id=63
*Enrollment (2012): B: 120 (112) M: 10 (2) D: 6 (0)*

## Geological Survey of Tanzania

**Geological Survey of Tanzania** (2011)
P .O Box 903
Dodoma
madini-do@gst.go.tz
http://www.gst.go.tz/

## University of Dar es Salaam

**Dept of Geology** (B,M,D) ● (2015)
PO Box 35052
Dar es Salaam
p. +255 22 2410013
geology@udsm.ac.tz
http://www.conas.udsm.ac.tz/geology/
*Enrollment (2015): B: 65 (53) M: 17 (0)*

Head:
Nelson Boniface, (D), Kiel, GcpGz
Professor:
Justian R. Ikingura, (D), Carleton, 1989, EgCgGi
Makenya A. Maboko, (D), Australian National, 1991, GpCgGx
Senior Lecturer:
Charles Z. Kaaya, (D), Cologne, 1993, GrsGu
Isaac Muneji Marobhe, (D), Helsinki Univ of Tech (Finland), 1990, YegYx
Associate Professor:
Shukrani Manya, (D), Dar Es Salaam, 2008, CgGzg
Hudson H. Nkotagu, (D), TU Berlin, 1994, HwgGe
Lecturer:
Kasanzu Charles, (D), 2014, Ggh
Emmanuel O. Kazimoto, (D), Kiel, 2014, EgGpCg
Elisante E. Mshiu, (D), Martin Luther (Germany), 2014, GgCeOr
Gabriel D. Mulibo, (D), Penn State, 2013, Ysg
Ferdinand W. Richard, (D), Uppsala, Sweden, 1999, YsGt

**Institute of Marine Sciences** (B) (2011)
Mizingani Road
PO Box 35091
Zanzibar
p. 255-24-2232128/2230741
director@ims.udsm.ac.tz
http://www.ims.udsm.ac.tz

## University of Dodoma

**Dept of Geology and Petroleum Studies** (2011)
P.O Box 259
Dodoma
p. +255 26 2310000
vc@udom.ac.tz
http://www.udom.ac.tz/

**School of Mines and Petroleum Engineering** (2011)
PO Box 259
Dodoma
p. +255 22 2410013
vc@udom.ac.tz
http://www.udom.ac.tz

# Thailand

## Asian Institute of Technology

**Dept of Geology** (2014)
P.O. Box 4
Klong Luang
Pathumthani 12120, Bangkok 10501
p. +66 (0) 2524 6057
supamas@ait.ac.th
http://www.set.ait.ac.th/page.php?fol=gte&page=gte

## Dept of Mineral Resources

(2011)
75/10 Rama 6 Road, Phayathai
Bangkok 10400
pornthip@dmr.go.th
http://www.dmr.go.th/

# Togo

## Universite de Lome

**Dept de Geologie** (B) (2014)
BP 1515

Lomé , TOGO
p. (+228) 22-25-50-93

# Tunisia

## Birzeit University

**Dept of Geology** (2011)
Faculty of Science
P. O. Box 14
7021 Jarzouna Tunis
Bizerte
p. 216590717
fsb@fsb.rnu.tn
http://www.fsb.rnu.tn/index_fr/contact.html

## Universite de Carthage

**Dept de Geologie** (B) (2010)
Jarzouna 7021
p. +21671841353
fsb@fsb.rnu.tn
http://www.fsb.rnu.tn/fsbindex.htm

## Universite de Gabes

**Dept de Sciences de la Terre** (B) (2014)
Cite Riadh
Zerig, Gabes 6072
p. +216 75 394 800
mail@fsg.rnu.tn
http:www/fsg.rnu.tn/PRESENTATION.htm

**Inst Superieur des Sciences et Techniques des Eaux de Gabes** (B) (2010)
Cite Riadh
Zerig, Gabes 6072
p. +216 75 394 800
samir.kamal@isstegb.rnu.tn
http://www.isstegb.rnu.tn/francais/index.htm

## Universite de Sfax

**Dept de Sciences de la Terre** (B) (2010)
Route de l Aeroport km 0.5
, Sfax 3029
p. 216 74 276 400
fss@www.fss.rnu.tn
http://www.fss.rnu.tn

**Dept de Genie Materiaux** (B) (2010)
B.P:w.3038
Sfax
p. (216) 74 274 088
enis@enis.rnu.tn
http://www.enis.rnu.tn/content/enis00003.htm

**Dept de Genie Georessources et Environnement** (B) (2010)
B.P:w.3038
Sfax
p. (216) 74 274 088
enis@enis.rnu.tn
http://www.enis.rnu.tn/content/enis00003.htm

## Universite de Tunis

**Dept of Geology** (2011)
Faculty of Science
1 Rue de Beja
Tunis 2092
p. 21671872600
http://www.fst.rnu.tn/fr/index.php

# Turkey

## Canakkale Onsekiz Mart University

**Jeoloji Bolumu** (2011)
Terzioglu Campus
Canakkale 17020
sztutkun@comu.edu.tr



http://jeoloji.comu.edu.tr/

## Cukurova Universitesi

**Jeoloji Muhendisligi Bolumu** (B,M,D) ● (2014)
Faculty of Engineering and Architecture, Department of Geological Engineering
01330 Adana
Balcali
p. (+ 90 322) 338 67 15
parlak@cu.edu.tr
http://jeoloji.cu.edu.tr

Professor:
Osman Parlak, (D), Geneva, 1996, GipCc

## Eskisehir Universitesi

**Maden Muhendisligi Bolumu** (2004)
Muhendislik Fakultesi
Yunus Emre Kampusu
Eskisehir

## Firat University

**Dept of Geological Engineering** (A,B,M,D) ○ (2016)
Firat University, Department of Geological Engineering
23119 Elazig Turkey
Elazig 23100
p. 424-2370000-5979
asasmaz@firat.edu.tr
http://jeo.muh.firat.edu.tr/tr/node/104
f: https://www.facebook.com/groups/firatjeoloji/
*Enrollment (2016): A: 7 (1) B: 2 (15) M: 9 (0) D: 9 (2)*

Chair:
Ahmet Sasmaz, (D), EgGeCe

Professor:
Ercan Aksoy, (A), Ggt
Ahmet Feyzi Bingol, (A), GipGv
Bahattin Cetindag, (A), NgYv
Ahmet Sagiroglu, (A), Egm

Associate Professor:
Dicle Bal Akkoca, (D), EnCo
Melahat Beyarslan, (D), GipGx
Hasan Celik, (A), GscOu
Zulfu Gurocak, (A), Nrg
Leyla Kalender, (A), CgcCs
Calibe Koc Tasgin, (A), GsdGg
Sevcan Kurum, (A), GivGx

Assistant Professor:
Bünyamin Akgul, (D), GzyGi
Murat Inceoz, (D), GctGm
Ayse Didem Kic, (D), GxpGy
Esra Ozel Yildirim, (D), Gxi
Ozlem Oztekin Okan, (D), HgwHs
Melek Ural, (D), GivCc

Research Assistant:
Mehmet Kokum, (M), Indiana, 2012, GttGg

Research Assistant:
Hatice Kara, (M), Cg
Nevin Ozturk, (M), Cg

Research Asistant:
Elif Akgun, (M), GgcGt
Onur Alkaç, GgsGu
Gizem Arslan, (M), GxiGp
Yasemin Aslan, Nr
Serap Colak Erol, (D), 2014, GgcGt
Mehmet Ali Erturk, (D), 2016, GiCc
Mustafa Kanik, (D), 2015, NrSo
Sibel Kaygili, (D), 2016, GgPgs
Mahmut Palutoglu, (D), YgsGc
Mustafa Eren Rizeli, (M), Gxi
Abdullah Sar, (M), GviGp
Ismail Yildirim, ScGix

## Hacettepe University

**Geological Engineering Dept** (2005)
Muhendislik Fakultesi
Beytepe Kampusu

tunay@hacettepe.edu.tr
http://www.jeo.hun.edu.tr/800x600.htm

**Hydrogeological Engineering Programme** (B,M,D) ● (2015)
Beytepe Campus
Ankara 06800
p. +90 (312) 297730
ekmekci@hacettepe.edu.tr
http://www.hacettepe.edu.tr/
*Enrollment (2015): B: 235 (42) M: 12 (0) D: 11 (3)*
Professor:
Sakir Simsek, (D), Istanbul, 1982, Hw
Serdar Bayari, (D), Hacettepe, 1991, Hw
Mehmet Ekmekci, (D), HwyCs
Galip Yüce, (D), Hw
Nur Naciye Özyurt, (D), Hacettepe, Hw
Assistant Professor:
Turker Kurttas, (D), Hacettepe, 1997, HwgOr
Levent Tezcan, (D), Hacettepe, 1993, Hw

## Istanbul Universitesi
**Jeoloji Bolumu** (2014)
Muhendislik Fakultesi
Beyazit
p. 0 (212) 473 70 70 ex. 17600
koral@istanbul.edu.tr
http://muhendislik.istanbul.edu.tr/jeoloji/?p=6909

## ITU
**Jeoloji Bolumu** (2014)
Maden Fakultesi
80394, Macka
Istanbul

## Middle East Technical University
**Inst of Marine Science** (2014)
Div. of Marine Geology & Geophysics
P.O. Box 28
33731 Erdemli-Mersin
p. +90-324 521 3434
adminims@metu.edu.tr
http://www.ims.metu.edu.tr/

**Dept of Geological Engineering** (B,M,D) (2010)
Inonu Bulvari
ODTU
Ankara TR-06531
p. +90-312-2102682
gesin@metu.edu.tr
http://www.geoe.metu.edu.tr/

## Mineral Research and Exploration General Directorate of Turkey
(2014)
p. +90 312 201 11 51
mta@mta.gov.tr
http://www.mta.gov.tr/

## Selcuk Universitesi
**Faculty of Sciences** (2004)
Muhendislik Fakultesi
42040 Konya
p. (+)903322412484
fen@selcuk.edu.tr

# Uganda

## Gulu University
**Faculty of Agriculture and Environment** (B,M,D) (2016)
P.O. Box 166
Gulu
p. +256782673491
d.ongeng@gu.ac.ug
http://www.gu.ac.ug

**Dept of Geography** (B) ● (2015)
P.O. Box 166
Gulu
p. +256772517488
charles.okumu52@gmail.com
http://www.gu.ac.ug/index.php?option=com_content&view=category&layout=blog&id=48&Itemid=66
Lecturer:
Expedito Nuwategeka, (D), Gulu, 2015, OyyGv

## Kabale University
**Dept of Environmental Science and Natural Resources** (B) (2010)
PO Box 317
Kabale
p. 256-4864-22803
akampabf@gmail.com
http://www.kabaleuniversity.ac.ag/

## Makerere University
**Dept of Geology and Petroleum Studies** (2014)
P. O. Box 7062
Kampala 041
p. +256 41 532631-4
http://mak.ac.ug/
Instructor:
Kevin Aanyu, (M), Mak
Wycliff Kawule, (M), ITC
Robert Mamgbi, (M), CT-Prague
Lecturer:
Erasmus Barifajio, (D), Mak
John Mary Kiberu, (D), TUB
Agnes Alaba Kuterama, (M), ITC
Andrew Muwanga, (D), Brausnschweig
Immaculate Nakimera Ssemanda, (D), Mak

**Dept of Geography, Geoinformatics & Climatic Sciences** (B) (2010)
PO Box 7062
Kampala
p. +256-41-53126 1
geog@caes.mak.ac.ug
http://www.geography.mak.ac.ug/

**Institute of Environment and Natural Resources** (B) (2014)
PO Box 7062
Kampala 041
p. 256 414 530134
fkansiime@muienr.mak.ac.ug
http://muienr.mak.ac.ug/

## Ndejje University
**Faculty of Forest Science & Environmental Management** (B) (2014)
PO Box 7088
Kampala, Ndejje Hill
p. +256-0392-730326
forest@ndejjeuniversity.ac.ug
http://www.ndejjeuinversity.ac.ug/academics.htm

# Ukraine

## National Mining University (in Ukraine)
**State Mining University of Ukraine** (2005)
19 Karl Marx Avenue
Dnepropetrovsk
320600
dfr@nmuu.dp.ua
http://www.apex.dp.ua/english/uageo/ukraine.html

# United Arab Emirates

## United Arab Emirates University
**Dept of Geology** (B,M,D) ● (2014)
P.O. Box 15551
College of Science, UAE University
Jamma Street

Al-Ain, Abu Dhabi 9713
p. +971-3-7136380
Ahmed.murad@uaeu.ac.ae

# United Kingdom



## Ulster University, Coleraine

**School of Geography and Environmental Sciences** (A,B,M,D) ● (2017)
Cromore Road
Coleraine BT521SA
p. 0044(0)2870124401
a.moore@ulster.ac.uk
https://www.ulster.ac.uk/faculties/life-and-health-sciences/schools/geography-and-environmental-sciences
t: @UlsterUniGES

## Aberystwyth University

**Dept of Geography & Earth Sciences** (B,M,D) (2015)
Llandinam Building
Penglais
Aberystwyth SY23 3DB
p. 01970 622606
gesstaff@aber.ac.uk
http://www.aber.ac.uk/en/iges

Professor:
Ron Fuge, (D), Wales, GbCgGe

Emeritus:
Michael M J Hambrey, (D), Manchester, 1974, GlsGm

## Anglia Ruskin University

**Dept of Life Sciences** (2014)
Cambridge Campus
East Road
Cambridge CB1 1PT
p. +44 845 271 3333
michael.cole@anglia.ac.uk
http://www.anglia-ruskin.ac.uk/ruskin/en/home/faculties/fst/departments/lifesciences.html

## Bangor University

**School of Ocean Sciences** (B,M,D) (2013)
Menai Bridge,
Isle of Anglesey
LL59 5AB
p. +44-1248-382854
oss011@bangor.ac.uk
http://www.sos.bangor.ac.uk/

Associate Professor:
Andrew J. Davies, (D), Queen's Univ (Ireland), 2004, Og

## Birkbeck College

**Dept of Earth and Planetary Sciences** (2014)
Department of Earth and Planetary Sciences
University of London - Birkbeck
Malet Street
Bloomsbury, London WC1E 7HX
p. +44 (0)20 7631 6665
p.gaunt@bbk.ac.uk
http://www.bbk.ac.uk/geology/
Administrative Assistant: Peter Gaunt

Head:
Gerald Roberts, Ne

Professor:
Charlie Bristow, (D), Leeds (UK), 1987, Gs
Andy Carter, Og
Ian Crawford, On
Hilary Downes, Cg

Instructor:
Karen Hudson-Edwards, Gez

Lecturer:
Andy Beard
Simon Drake, Gv
Dominic Fortes, Og
Peter Grinford, Xg
Steve Hirons, Cl
Philip Hopley, Pe
Phillip Pogge von Strandmann, (D), Csl
Vincent C. H. Tong, Yg
Charlie Underwood, c.un, Pg

## British Antarctic Survey

**British Antarctic Survey** (2016)
High Cross, Madingley Road
Cambridge CB3 0ET
p. +44 (0)1223 221400
trr@bas.ac.uk
https://www.bas.ac.uk/team/science-teams/geosciences/
t: @BAS_News

## British Geological Survey

**British Geological Survey** (2012)
Environmental Science Centre
Nicker Hill
Keyworth
Nottingham NG12 5GG
p. 0115 936 3100
enquiries@bgs.ac.uk
http://www.bgs.ac.uk/

## Cardiff University

**School of Earth and Ocean Sciences** (B,M,D) (2014)
Main Building
Park Place
Cardiff, Wales CF10 3YE
p. +44 (0)29 2087 4830
earth-ug@cf.ac.uk
http://www.cardiff.ac.uk/earth/
*Enrollment (2002): B: 174 (0) M: 18 (0) D: 28 (0)*

Head:
R J. Parkes

Professor:
J A. Cartwright, (D), Ys
Dianne Edwards, (D), Pb
I Hall, On
A Harris
C Harris, Ge
A C. Kerr, (D), Gx
Bernard Elgey Leake, (D), Bristol, 1974, GipGz
R J. Lisle, Gc
C MacLeod, Gu
J A. Pearce, Cg
P N. Pearson, Pe
D Rickard, Cg
V P. Wright, Gs

Commander:
N Rodgers, (D), Ow

Lecturer:
Tiago Alves
Rhoda Ballinger, (D), On
Stephen Barker, Pe
C M. Berry, (D), Pb
P J. Brabham, (D), Ye
L C. Cherns, Pg
Jose Constantine, (D), Hy
J H. Davies, Yg
I Fryett, On
TC Hales, (D), Gm
A Hemsley, Pl
T Jones, Ge
C Lear, Ou
Johan Lissenberg
Sergio Lourenco, (D), Sp
R Perkins, On
J Pike, On
H M. Prichard, Gx
H Sass
H D. Smith, On
S J. Wakefield, Cm
C F. Wooldridge, On
Y Yang, Hg

## Durham University

**Dept of Earth Sciences** (B,M,D) (2015)
Science Laboratories
South Road
Durham DH1 3LE
p. +44 0191 3342300
earth.sciences@durham.ac.uk
http://www.dur.ac.uk/earthsciences/

Director:
Richard Davies, Eo
Ed Llewellin, Gv
Head:
Jon Gluyas, Eo
Professor:
Andrew Aplin, Go
Kevin Burton, Cc
Jon Davidson, Cm
Neil R. Goulty, (D), Cambridge, Yg
Chris Greenwell, (D), GzEoCg
David Harper, Pg
Colin G. Macpherson, (D), London, 1994, CgGiCs
Jim McElwaine, Yg
Yaoling Niu, On
Associate Professor:
Jeroen van Hunen, (D), 2001, Gt
Instructor:
Mark Allen, Gs
Darren Grocke, Pe
Simon Mathias, Oi
Dave Selby, Cc
Lecturer:
James Baldini, Hs
Richard James Brown, Gv
Pablo Cubillas, Cg
Chris Dale, Oa
Nicola De Paola, Gc
Richard Hobbs, Ys
Claie Horwell, Gz
Stuart Jones, Gs
Lara Kalnins
Kenneth J. McCaffrey, (D), Durham, Gc
Christine Peirce, (D), Cambridge, Yg
Helen Williams, Cs
Fred Worrall, (D), Cambridge, Cg
Senior Lecturer:
Howard A. Armstrong, (D), Nottingham, 1983, Pg
Research Assistant:
Jonathan Imber, (D), Durham, Gc
Reader:
Gillian R. Foulger, (D), Durham, Yg
Robert E. Holdsworth, (D), Leeds, Gc

## Edinburgh University

**School of Geosciences** (B,M,D) ● (2016)
Kings Buildings
West Mains Road
Edinburgh EH9 3JW
p. +44 (0) 131 651 7068
sarah.mcallister@ed.ac.uk
http://www.geos.ed.ac.uk/
f: https://www.facebook.com/geosciences/
t: @geosciencesed
*Enrollment (2016): B: 60 (60) M: 20 (20) D: 80 (80)*

Director:
Bruce M. Gittings, (M), Edinburgh, Oi
Paul R. van Gardingen, (D), On
Head:
Alexander W. Tudhope, (D), Cs
Chair:
Chris Dibben, On
Mark D. A. Rounsevell, (D), Ou
Charles W. J. Withers, (D)
Professor:
Emily S. Brady, (D), Glas, Ge
Andrew Curtis, Ys
Andrew J. Dugmore, Gae
J. Godfrey Fitton, Gi
Simon L. Harley, Gt
R. Stuart Haszeldine, (D), Strath, Oa
Gabriele C. Hegerl, (D), Oa
Dick Kroon, Gg
Ian G. Main, (D), Ys
Patrick Meir, (D), Edinburgh
Maurizio Mencuccini, (D), Florence
John B. Moncrieff, Oa
Peter W. Neinow, (D), Cantab, Ol
Paul Palmer, Oa
Jamie R. Pearce, (D), Ge
Alastair H. F. Robertson, (D), Leicester, 1975, GtgGs
Hugh D. Sinclair, (D), Os
David Stevenson, (D), Oa
Simon F. B. Tett, Ow
Kathryn A. Whaler, (D), Cambridge (UK), 1981, Ym
Mathew Williams, Oa
Wyn Williams, Ym
Iain H. Woodhouse, (D), H-W, Oi
Rachel A. Woods, (D), Open, Gs
Anton M. Ziolkowski, (D), Cambridge (UK), 1971, YesYg
Senior Scientist:
Nicola Cayzer, (D), Edinburgh, Yg
Richard Hinton, Cg
Research Associate:
Andrew Bell, Gt
Ian B. Butler, (D), Cg
John Craven, Yg
Jan C. de Hogg, Ge
Walter Geibert, Cc
Chris L. Hayward, Ga
Sian F. Henley, (D), Edinburgh (UK), 2013, CmOcCs
Mark Naylor, (D), Edinburgh, Gt
Anthony J. Newton, (D), Edinburgh, On
Laetitia Pichevin, (D), Bordeaux, Cg
John A. Stevenson, Gv
Instructor:
Richard L. H. Essery, Oa
Raja Ganeshram, Pg
Andrew T. Mcleod, Og
Bryne T. Ngwenya, (D), Reed, Co
Hugh C. Pumphrey, (D), Mississippi, Oa
David Reay, (D), Oa
Tom Slater, (D), London, On
Samantha Staddon
Lecturer:
Simon J. Allen
Mikael Attal, (D), Joseph Fourier Univ (France), 2003, GmtOy
Massimo Bollasina, Oa
Geoffrey Bromily, (D), Ge
Eliza Calder, Gv
Mark Chapman, (D), Yg
Gregory L. Cowie, Co
Julie Cupples, On
Kyle Dexter
Ruth Doherty, (D), Edinburgh, Oa
Rowan Ellis, On
Florian Fusseis, Gc
Daniel Goldberg
Noel Gourmelen
Margaret C. Graham, Ge
Kate V. Heal, Co
Nicholas R. J. Hulton, Ol
Antonio Ioris, Ge
Gail D. Jackson, PbGe
Simon Jung, Og
Linda Kirstein, Gt
Eric Laurier, On
Caroline Lehmann, (D)
Fraser MacDonal
William A. Mackaness, Oi
Ondrej Masek, (D), Ng
Christopher I. McDermott, Hg
Marc J. Metzger, (D), Wageningen, Ge
Nina J. Morris

Simon N. Mudd, (D), Hg
Caroline Nichol
Eva Panagiotakopulu, Pe
Genevieve Patenaude, (D), Ge
Jan Penrose, (D), Toronto
Kanchana Ruwanpura
Casey Ryan
Kate Saunders, (D), GviGz
Simon J. Shackley, Eg
Niamh K. Shortt, Eg
Sran P. Sohi, (D), London, Os
Neil Stuart, Oi
Daniel Swanton, (D), Durham
Jenny A. Tait
Alexander Thomas
Dan van der Horst, (D), Eg
Mark Wilkinson, (D)
Merwether Wilson, Ob
Ronald M. Wilson, (D)
Emeritus:
John Grace, Oa
Teaching Fellow:
Thomas Challads, Hg
Other:
Stuart M. V. Gilfillan, (D), Manchester, Oa
Andrew S. Hein, (D), Edinburgh
Tetsuya Komabayashi, (D), Gz
Isla Myers-Smith
Marisa Wilson, (D), Oxcon

## Exeter University

**Camborne School of Mines** (2014)
College of Engineering, Mathematics and Physical Sciences
University of Exeter
Penryn Campus
Penryn, Cornwall TR10 9FE
p. (+44) 1392 661000
cornwall@exeter.ac.uk
http://emps.exeter.ac.uk/geology/contact/

Director:
John Coggan, (D), Newcastle, Nx
Charlie Moon, (D), Imperial Coll, Nx
Neill Wood, (B), Imperial Coll (UK), 1983, YeOe
Professor:
Hylke J. Glass, (D), NxmEm
Stephen Hasselbo, Gs
Kip Jeffrey, Nx
Bernd Lottermoser, Ca
Lecturer:
Jens Andersen, (D), GiEm
Ian Bailey, (D), Univ Coll London, On
Robert Barley
Christopher Bryan, Nx
Patrick Foster, Nm
Sam Hughes, Gt
Gareth Kennedy, Nm
Kate L. Littler, (D), Univ Coll London, 2011, PeCl
John Macadam, Ge
Lewis Meyer, (D), Exeter (UK), 2001, Nmr
Kathryn Moore, (D), Bristol, 1999, Ge
Kim Moreton, (D), Exeter Univ, 2008, GeEmOu
Richard Pascoe, (D), Gz
Duncan Pirrie
Robin Shail, (D), Keele, 1992, Gtc
Ross Stickland
David Watkins, Hg
Andrew Wetherelt, Nm
Paul Wheeler, Nx
Ben Williamson, (D), London, 1991, EmGv

## Heriot-Watt University

**Institute of Petroleum Engineering** (M,D) (2017)
Research Park, Riccarton
Heriot-Watt University
Edinburgh Campus
Edinburgh EH14 4AS
p. +44 (0) 131 451 3543
egis-staff@hw.ac.uk
https://www.hw.ac.uk/schools/energy-geoscience-infrastructure-society/research/ipe.htm
f: https://www.facebook.com/hwu.egis
t: @HWUPetroleum

Director:
John Ford
Andy Gardiner
Eric Mackay
James M. Somerville, Np
Rink van Dijke
Head:
Dorrik Stow
Chair:
Sebastian Geiger, (D), ETH (Switzerland), 2004, Np
John Underhill
Professor:
Patrick Corbett, Np
Gary Couples, Np
David Davies, Ng
Mahmoud Jamiolahmady, (D), Inst of Petroleum Engineering, 2001, On
Colin MacBeth
Ken Sorbie
Associate Professor:
Andreas Busch
Jingsheng Ma, (D), Np
Uisdean Nicholson
Assistant Professor:
Hamed Amini
Elli-Maria (Elma) Charalampidou
Erkal Ersoy, (D), Eo
Morteza Haghighat Sefat
Research Associate:
Achim Ahrens
Ross Anderson, Eo
Lorraine Boak
Matthew Booth
Rachel Brackenridge
Rod Burgass
Antonio Carvalho
Antonin Chapoy
Romain Chassagne
Ilya Fursov
Alexader Graham
Sally Hamilton, Gs
Oleg Ishkov
Rachel Jamieson
Marco Lorusso
Julian Maes
Pedram Mahzari
Maria-Daphne Mangriotis, Ys
Pedro Martinez Garcia
Robin Shields, Cg
Mike Singleton
Oscar Vazquez
Francesca Watson
Jinhai Yang
Zhen Yin
Zhao Zhang
Zhao Zhang
Stephanie Zihms
Instructor:
Saleh Goodarzian
Steve McDougall
Lecturer:
Vasily Demyanov, Oi
Florian Doster, Hg
Ahmed H. Elsheikh
Zeyun Jiang
Helen Lever, (D), James Cook Univ, 2003, Go
Helen Lewis
Khafiz Muradov, (D)
Gillian E. Pickup, Hw
Asghar Shams, Yg
Karl D. Stephen

Laboratory Director:
Jim Buckman
Mike Christie
Other:
Bahman Tohidi

## Imperial College

**Earth Science and Engineering** (2015)
Imperial College London
South Kensington Campus
London SW7 2AZ
p. +44 (0)20 7589 5111
philip.allen@imperial.ac.uk
http://www3.imperial.ac.uk/earthscienceandengineering

Director:
Nigel Brandon, Ge
Professor:
Martin J. Blunt, (D), Cambridge, 1988, NpHw
Head:
Johannes Cilliers, (D), Cape Town, 1994, Gz
Chair:
Philip Allen, Gs
Alastair Fraser, (D), Glasgow, 1995, Go
Matthew Jackson, (D), Liverpool, 1997
Howard Johnson, Go
Peter King, (D), Cambridge, 1982, Np
Ann H. Muggeridge, (D), Oxford, 1986, Np
Yanghua Wang, (D), Imperial Coll, 1997, Yg
Professor:
John Cosgrove, (D), Imperial Coll (London), 1972, Gc
Sevket Durucan, Ng
Sanjeev Gupta, Gs
Joanna Morgan, (D), Cambridge, 1988, Yg
Jane Plant, Ge
Mark Rehkamper, (D), Mainz, 1995, CsaCc
Mark Sephton, (D), Open, 1996, Xm
Velisa Vesovic, (D), Imperial Coll, 1998, Np
Michael Warner, (D), York, 1979, Ys
Dominik Weiss, (D), Berne, 1998, CgsCl
Robert W. Zimmerman, (D), California (Berkeley), 1984, NrHwNp
Associate Professor:
Lidia Lonergan, (D), Oxford, 1991, GtcGs
Research Associate:
Rebecca Bell, (D), Southampton, 2008, Gt
Branko Bijeljic, (D), Imperial Coll, 2000, Np
Raphael Blumenfeld, (D), Cambridge, On
Rhordri Davies, (D), Cardiff, 2007, Yg
Zita Martins, (D), Leiden, 2007, Xm
Christopher Pain, Yg
Randall Perry, Og
Instructor:
Jenny Collier, (D), Cambridge, 1989, Yr
Saskia Goes, (D), California (Santa Cruz), 1995, Yg
Gary Hampson, (D), Liverpool, 1995, Gd
Christopher Jackson, (D), Manchester, 2002, Gt
Anna Korre, Ng
Jian Guo Liu, (D), Imperial Coll, 2002, Or
Stephen Neethling, (D), UMIST, 1999, Gz
Emma Passmore
Matthew Piggot, Bath, 2001, Hg
Jamie Wilkinson, (D), Southampton, 1989, Em
Lecturer:
Ian Bastow, (D), Leeds, 2005, Yg
Gareth Collins, (D), Imperial Coll (London), 2002, Xm
Matthew Genge, (D), Univ Coll (London), 1993, Xm
Gerard Gorman, (D), Imperial Coll (London), 2005, Yg
Kathryn Hadler, (D), Manchester, 2006, Gz
Cedric John, (D), Potsdam, 2003, Gs
Samuel Krevor, Eo
John-Paul Latham, (D), Imperial Coll, 2003
Philippa J. Mason, Or
Adrian Muxworthy, (D), Oxford, 1998, Ym
Julie Prytulak, (D), Bristol, 2008, Gg
Mark Sutton, (D), Wales-Cardiff, 1996, Pg
Tina van de Flierdt, (D), ETH Zuerich, 2003, Cs
Alex Whittaker, (D), Edinburgh, 2007, Gt
Visiting Professor:
Rosalind Coogon, Cs
Emeritus:
Alain Gringarten, (D), Stanford, 1971, Np
John Woods, (D), Imperial Coll, 1964

## Keele University

**Geography, Geology and the Environment** (B,M,D) (2016)
William Smith Building,
Keele University
Keele, Staffordshire ST5 5BG
p. (+44) 01782 733615
gge@keele.ac.uk
http://www.keele.ac.uk/gge/
f: https://www.facebook.com/KeeleGeologyGeoscience
t: @KeeleGeology
*Enrollment (2014): B: 149 (45) M: 29 (15) D: 10 (2)*

Associate Professor:
Stuart Egan, Ggc
Ralf Gertisser, (D), Freiburg (Germany), 2001, GivGz
Jamie Pringle, Yg
Ian G. Stimpson, (D), Wales, 1987, YesYg
Research Associate:
Sam Toon, Ys
Rachel Westwood, Yx
Lecturer:
Stu Clarke, Gdo
Ralf Halama, GpiCa
Michael Montenari, PmgPs
Steven Rogers, Gg

## Kingston University

**School of Geography, Geology and Environment** (2014)
Penrhyn Road Centre
Penrhyn Road
Kingston upon Thames
Surrey KT1 2EE
p. +44 (0)20 8417 9000
G.Gillmore@kingston.ac.uk
http://www.kingston.ac.uk/geolsci/

Director:
Stuart Downward
Head:
Gavin Gillmore, (D), Univ Coll (London), Gb
Professor:
Ian Jarvis, (D), Oxon, 1980, CgGsr
Peter Treloar, (D), Glasgow, 1978, Gc
Nigel Walford, (D), London, 1981, Oi
Martyn Waller, (D), CNAA
Instructor:
Peter Hooda, (D), London, 1992, Em
Annie Hughes, (D), Bristol, 1997, Oi
Mike Smith, (D), Sheffield, 1999, Or
Lecturer:
Andy Adam-Bredford
Alistair Baird
Douglas Brown, Oi
Kerry Brown, (D), Stony Brook, 2004, Ge
Norman Cheun, Ge
Tracey Coates
Hadrian Cook, (D), East Anglia, 1986, Ge
Peter Garside, (D), Liverpool
Paul Grant
Ian Greatbatch, (D), 2008, Oi
Frances Harris, (D), Kingston, 1995
Mary Kelly
David Kidd, (D), St. Andrews, 2005, Oiy
James Lambert-Smith, (D), Kingston, 2014, Gg
Andrew Miles, (D), Edinburgh, 2012, Gz
Stephanie Mills, (D), Witwatersrand, 2006, Ol
Pamela Murphy
Colin Ryall
Christopher Satow, (D), London, 2012, Ge
Neil Thomas
Visiting Professor:
Rosalind Taylor

Emeritus:
Richard Moody
Andy Rankin
Other:
Andrew Swan



## Leicester University

**Geology Dept** (B,M) (2014)
University Road
Leicester LE1 7RH
p. +44 (0)116 252 3933
geology@le.ac.uk
http://www2.le.ac.uk/departments/geology
Head:
Richard England, Yg
Professor:
Sarah Davies, Gs
Mike Lovell, Yg
Randy Parrish, Cs
Mark Purnell, Po
Mark Williams, Po
Lecturer:
Stewart Fishwick, Yg
Sarah Gabbott, Po
Tom Harvey, Po
Gawen Jenkin
Max Moorkamp, (D), Yg
Mike Norry, Pe
Dan Smith, Ge
Richard Walker, Gs
Jan Zalasiewicz, Po
Emeritus:
Andy D. Saunders, (D), Birmingham (UK), 1976, GiCgGe

## Liverpool John Moores University

**School of Natural Sciences and Psychology** (B,M,D) (2016)
Byrom Street
Liverpool L3 3AF
p. (+44) 0151 904 6300
j.r.kirby@ljmu.ac.uk
https://www.ljmu.ac.uk/research/centres-and-institutes/environment-research-group
f: https://www.facebook.com/groups/LJMU.geography.env.sci/
t: @ljmugeog
*Enrollment (2016): B: 72 (0) M: 3 (0)*
Professor:
Andy Tattersall
Dr:
Jason Kirby

## Newcastle University

**Dept of Civil Engineering** (2015)
Drummond Building
Newcastle upon Tyne, NE1 7RU
C.M.Earle-Storey@newcastle.ac.uk
http://www.ncl.ac.uk/~ncivil/index.htm

## Nottingham Trent University

**Dept of animal, rural, and environmental sciences** (2004)
Burton Street
Nottingham NG1 4BU
p. +44 (0)115 941 8418
anne.coules@ntu.ac.uk
http://www.ntu.ac.uk

## Polytechnic Southwest

**Dept of Earth Sciences** (2014)
Drake Circus
Plymouth, Devon PL4 8AA
p. (+44) (0)1752 584584
science.environment@plymouth.ac.uk
http://www5.plymouth.ac.uk/schools/school-of-geography-earth-and-environmental-sciences/earth-sciences

## Queens University Belfast

**School of Natural and Built Environment (Geography)** (B,M,D) ● (2016)
University Road
Belfast BT7 1NN
p.warke@qub.ac.uk
http://www.qub.ac.uk/schools/NBE/
*Enrollment (2016): B: 320 (106) M: 6 (6) D: 40 (5)*
Head:
Patricia Warke, (D), Queen's Univ (Ireland), 1994, GmOym
Professor:
Keith Lilley, (D)
David Livingstone, (D)
Associate Professor:
Merav Amir, (D)
Iestyn Barr, (D)
Oliver Dunnett, (D)
Paul Ell, (D)
Diarmd Finnegan, (D)
Nuala Johnson, (D)
M. Satish Kumar, (D), Jawaharlal Nehru Univ (India), 1991, OnnOn
Jennifer McKinley, (D)
Will Megarry, (D), Univ Coll Dublin, 2013, Gae
Donal Mullan, (D)
Helen Roe, (D)
Alastair Ruffell, (D)
Ian Shuttleworth, (D)
Tristan Sturm, (D)
Lecturer:
Niall Majury, (D), Toronto, 1999

## Staffordshire Polytechnic

**Dept of Applied Sciences** (2014)
College Road, Stroke-on-Trent
Staffordshire ST4 2DE
p. (+44) 1782 294000
r.boast@staffs.ac.uk
http://www.staffs.ac.uk/academic_depts/sciences/subjects/environment/index.jsp

**Geography and the Environment** (2004)
j.w.wheeler@staffs.ac.uk
http://www.staffs.ac.uk/sands/scis/geology/geology.html#

## Swansea University Prifysgol Abertawe

**Dept of Earth Sciences** (2014)
Wallace Building
Swansea University
Singleton Park
Swansea, Wales SA2 8PP
p. +44 (0) 1792 205678 ext 8112
geography@swansea.ac.uk
http://www.swansea.ac.uk/geography/

## Teesside University

**School of Science and Engineering** (B,M) (2016)
Tees Valley
Middlesborough TS1 3BA
p. 01642 738800
sse-admissions@tees.ac.uk
http://www.tees.ac.uk/schools/sse/index.cfm

## The Open University

**Dept of Environment, Earth & Ecosystems** (B,M,D) (2015)
Faculty of Science
The Open University
Walton Hall
Milton Keynes MK7 6AA
p. +44 (0) 1908 652886
Env-Earth-Ecosystems-Enquiries@open.ac.uk
http://www.open.ac.uk/science/environment-earth-ecosystems/
f: https://www.facebook.com/pages/OU-Environment-Earth-Ecosystems/234968859948707
t: @OU_EEE
Professor:
Fabrizio Ferrucci, Yg

David Gowing, (D), Lancaster, 1991, Pb
Nigel B.W. Harris, (D), Cambridge (UK), 1973, GtiGp
Simon Kelley, (D), London, 1985, Cs
Walter Oechel, Oa

Associate Professor:
Mark A. Brandon, (D), Cambridge (UK), 1995, Og
Dave McGarvie, (D), Lancaster, 1985, GviCt

Associate Scientist:
Peter Sheldon

Research Associate:
Marie-Laure Bagard
Gareth Davies, (D), Cranfield, Oe

Reader:
Neil Edwards, (D)

Instructor:
Stephen Blake, (D), Lancaster, 1982, Gv

Lecturer:
Pallavi Anand, (D), Cambridge (UK), 2002, ClPe
Tom Argles, (D), Oxon, Oi
Angela L. Coe, (D), Oxford, Gs
Anthony Cohen, (D), Cambridge, 1988, Cs
Sarah Davies, (D), Sheffield, Oe
Miranda Dyson, (D), Witwatersrand, 1989, Py
Tamsin Edwards, (D)
Richard Holliman, (D), Open, Oe
Philip Sexton, (D), Southampton, 2000, Ge
Carlton Wood

Visiting Professor:
Stephen Self
Edward Youngs

Emeritus:
Phil Potts
Robert Spicer, (D), Imperial, 1975

## The University of Sunderland

**Faculty of Applied Science** (2014)
Edinburgh Building
City Campus
Chester Rd.
Sunderland SR1 3SD
p. (+44) 191 515 2000
john.macintyre@sunderland.ac.uk
http://www.sunderland.ac.uk/faculties/apsc/ourdepartments/cet/

## University College London

**Dept of Earth Sciences** (B,M,D) (2015)
Gower Street
, London WC1E 6BT
p. +44 (0)20 7679 2363
earthsci@ucl.ac.uk
http://www.ucl.ac.uk/es
f: EarthSciences UcL

Professor:
Dario Alfe, (D), International School for Advanced Studies (Italy), 1997, OmGyYh
Tim Atkinson, (D), Bristol, 1971, Hg
Paul Brown, (D), Univ Coll (London), 1986, Pm
David Dobson, (D), Univ Coll (London), 1995, Gz
Adrian P. Jones, (D), Durham, 1980, GiCgGg
Carolina Lithgow-Bertelloni, (D), California (Berkeley), 1994, Yg
John M. McArthur, (D), Imperial Coll (UK), 1974, ClHwPs
Jacqueline McGlade, (D), Guelph, 1980, Ge
Philip Meredith, (D), Imperial Coll, 1983, Nr
Eric Oelkers, (D), California (Berkeley), 1988, ClGze
Kevin Pickering, (D), Oxford, 1979, Gs
David Price, (D), Cambridge (UK), 1980, GzYgPm
Chris Rapley, (D), London, 1976, OgrOl
Graham Shields-Zhou, (D), Eidgenössische Technische Hochschule Zürich, 1997, Cg
Lars Stixrude, (D), California (Berkeley), 1991, Yg
Julienne Stroeve, (D), Colorado, 1996, Olr
Juergen Thurow, (D), Eberhard-Karls-Universität Tübingen, 1987, Gs
Paul Upchurch, (D), Cambridge, 1993, Po
Lidunka Vocadlo, (D), Univ Coll (London), 1993, Gy
Bridget S. Wade, (D), Edinburgh (UK), PmeCl
Ian Wood, (D), Univ Coll (London), 1977, Gz

Lecturer:
William Graham Burgess, (D), Birmingham, 1987, Hwy
Anna Ferreira, (D), Oxford, 1005, Ys
Andrew Fortes, (D), Univ Coll (London), 2004, Xy
Anjali Goswani, (D), Chicago, 2005, Po
Chris Kilburn, (D), Univ Coll (London), 1984, Yx
Wendy Kirk, (D), Univ Coll London, 1986
Tom Mitchell, Gc
Dominic Papineau, Cg
Phillip Pogge von Strandmann, (D), Open Univ (UK), 2006, Csl
Alex Song, (D), Caltech, Ys
Pieter Vermeesch, (D), Stanford, 2005, Cc

Emeritus:
Bill McGuire, (D), Univ Coll (London), 1980, Gv

## University College of Swansea

**Dept of Geography** (2004)
Singleton Park
Swansea SA2 8PP

## University of Aberdeen

**Dept of Geology and Petroleum Geology** (B,M) (2016)
Meston Building
King's College
Aberdeen AB24 3UE
geology@abdn.ac.uk
http://www.abdn.ac.uk/geology/

Head:
David Jolley, (D)

Chair:
Ian Alsop, (D)
Rob Butler
Adrian Hartley
John Howell
Andrew Hurst
Ben Kneller
David MacDonald
Ron Steel
Randell A. Stephenson, (D), Dalhousie, 1981, YgGtOg

Associate Professor:
Clare Bond, (D), Gct

Research Associate:
Steven Andrews
Robert Daly
Jyldz Tabyldy Kyzy
Sam Spinks
Christian Vallejo

Instructor:
David Muirhead

Lecturer:
Stephen Bowden
David Cornwell, (D), Leicester (UK), 2008, YsGtYe
Dave Healy
Malcolm J. Hole
David Iacopini
Joyce Neilson
Colin North
Nick Schofield

## University of Bedfordshire

**Dept of Life Sciences** (2014)
Park Square
Luton, Bedfordshire LU1 3JU
p. +44 1234 400400
international@beds.ac.uk
http://www.beds.ac.uk/howtoapply/departments/science

## University of Birmingham

**School of Geography, Earth & Environmental Sciences** (B,M,D) (2017)
School of Geography, Earth and Environmental Sciences
University of Birmingham
Edgbaston, Birmingham B15 2TT

p. +44 (0)121 414 5531
w.j.bloss@bham.ac.uk
http://www.birmingham.ac.uk/schools/gees/index.aspx

Director:
Rob MacKenzie
Head:
David Hannah, Hg
Chair:
Eugenia Valsami-Jones, (D), ClGeCa
Professor:
Stuart Harrad
Roy M. Harrison, (D), Birmingham (UK), 1989, Oa
Jamie Lead
Alexander Milner
Tim Reston
Jon Sadler
John Tellam, (D), Hw
Associate Professor:
Paul Anderson
Bridin Carroll
Rosie Day, (D), Univ Coll London, On
Arshad Isakjee
Lloyd Jenkins
Stephen M. Jones, (D), Cambridge (UK), 2000, Og
Jonathan Oldfield, (D), Oy
Research Associate:
Mohamed A. Abdallah
Salim Alam
Mohammed Baalousha
David Beddows
Ian Boomer
Leigh R. Crilley, (D), 2013, Oa
Mark Cuthbert
Simon J. Dixon, (D), Gms
Jonathan Eden
Sophie Hadfield-Hill
James Hale
Stephanie Handley-Sidhu
Marie Hutton
Kieran Khamis
Marcus Kohler
James Levine
Yuning Ma
Paul Martin
Sara Martinez-Loriente
Mauro Masiol
Heiko Moossen
Catherine L. Muller
Martin Müller
Irina Nikolova
Irina Nikolova
Matt O'Callaghan
Isabella Romer Roche
Zongbo Shi
Shei Sia Su
Rick Thomas
Amey S. Tilak
Sarah Jane Veevers
Saskia Warren
Sebastian Watt
Jianxin Yin
Deputy Head of School:
William Bloss
Instructor:
James Bendle
Lee Chapman
Jason Hilton
Dominique Moran
Ian Phillips
Jo Southworth
Lecturer:
Lauren Andres
Daniel Arribas-Bel
Austin Barber
Nicholas Barrand
Rebecca Bartlett
Lesley Batty
Mike Beazley
Chris Bradley
Xioaming Cai
Andy Chambers
Julian Clark
Juana Maria Delgado-Saborit
Warren Eastwood
Steven Emery
Sara Fregonese
Guy Harrington
Alan Hastie
Alan Herbert
Phil Jones
Nick Kettridge
Stephen Krause
Mark Ledger
Peter Lee
Agnieszka Leszczynski
Iseult Lynch
Zena Lynch
Vlad Mykhnenko
Patricia Noxolo
Francis Pope
Jessica Pykett
Adam Ramadan
Joanna Renshaw
Michael Riley
Michael Rivett
John Round
Ivan Sansom
Greg Sambrook Smith
Carl Stevenson
Emmanouil Tranos
James Wheeley
Martin Widman
Emeritus:
Ian Fairchild, (D), Nottingham, 1978, GsClGm
Birmingham Fellow & Academic Keeper, Lapworth Museum of Geology:
Richard Butler
Other:
Tom Dunkley Jones
Related Staff:
Melanie Bickerton

## University of Bristol

**School of Earth Sciences** (B,M,D) (2016)
Wills Memorial Building
Queens Road
Bristol BS8 1RJ
p. +44 117 954 5400
m.j.walter@bristol.ac.uk
http://www.bristol.ac.uk/earthsciences/
f: https://www.facebook.com/School-of-Earth-Sciences-Bristol-University-146277648746662/
t: @UOBEarthscience
*Enrollment (2013): B: 25 (9) M: 44 (5)*

Professor:
Michael Walter , (D), 1991, GiCpg

## University of Cambridge

**Dept of Earth Sciences** (B,D) (2016)
Downing Street
Cambridge , Cambridgeshire CB2 3EQ
p. +44 (0)1223 333400
satr@cam.ac.uk
http://www.esc.cam.ac.uk

Head:
Simon Redfern, Gz
Professor:
Michael Bickle, (D), Oxford, 1973, GtYgGo
Nicholas J. Butterfield, Po
Michael Carpenter, Gy
David Hodell, Ge
Marian Holness, Gi
James A. Jackson, (D), Cambridge (UK), 1980, YgGt
Dan P. McKenzie, (D), YgGt

Simon Conway Morris, Po
Keith Priestley, YgGt
Nick Rawlinson, Ygg
Ekhard Salje, Gy
Nicky White, YgGt
Robert White, YgGt
Eric W. Wolff, (D), Cambridge (UK), 1992, OlPe
Andy Woods, YgGt

Lecturer:
David Al-Attar, YdGt
Alex Copley, YgGt
Sanne Cottaar, Ys
Neil Davies, Gs
Marie Edmonds, Gv
Ian Farnan, Gz
Sally Gibson, Gi
Richard Harrison, Gz
Tim Holland, Gp
John Maclennan, Gi
Kenneth McNamara, Po
Jerome Anthony Neufeld, GoYgGt
David Norman, Po
John Rudge, YgGt
Luke C. Skinner, (D), Cambridge (UK), 2005, GeCmOg
Ed Tipper, Cl
Alexandra Turchyn, Ge

Emeritus:
Nigel Woodcock, (D), Imperial Coll (UK), 1973, GcsGt

## University of Derby

**Geographical, Earth and Environmental Sciences** (B,M) (2014)
Kedleston Road
Derby DE22 1GAB
p. +44 (0) 1332 591703
fehs@derby.ac.uk
http://www.derby.ac.uk/science/gees/
*Enrollment (2013): B: 120 (30) M: 9 (10)*

Head:
Hugh Rollinson, Cg

Professor:
Aradhana Mehra, Co

Instructor:
Jacob Adentunji, Ge

Lecturer:
Martin Whiteley, Go

## University of East Anglia

**School of Env Sciences** (2015)
Norwich Research Park
Norwich NR4 7TJ
p. +44 (0)1603 592542
env.enquiries@uea.ac.uk
http://www.uea.ac.uk/environmental-sciences

Director:
Corinne Le Qu, (D), Og

Professor:
Jan Alexander, (D), Leeds, Gt
Julian E. Andrews, (D), Leicester (UK), 1984, CsGsCl
Simon Clegg, (D), East Anglia (UK), 1986, OacCg
Brett Day, (D), UEA, 2004, Ge
Alastair Grant, (D), Wales, 1983, GeObCt
Karen Heywood, Op
Kevin Hiscock, (D), Birmingham, Hg
Tim Jickells, (D), Southampton, Ob
Phillip D. Jones, (D), Newcastle upon Tyne, 1977, OaPeHg
Andy Jordan, Ge
Andrew A. Lovett, (D), Wales, 1990, Oiu
Coling Murrell, Oa
Timothy J. Osborn, (D), East Anglia (UK), 1995, Oa
Carlos Peres, Sf
Ian Renfrew, Ow
Ian Renfrew, Ow
Bill Sturges, Oa
Roland von Glasow, (D), Max-Planck-Inst in Mainz, Yg
Andrew Watkinson, Ge
Sir Robert Watson

Assistant Professor:
Gill Malin, (D), Liverpool (UK), 1983, Obc

Research Associate:
Amy Binner

Instructor:
Alex Baker, (D), Plymouth Marine Lab, On
Jenni Barclay, (D), Bristol, Gv
Paul Burton, (D), Cambridge, Ys
Mark Chapman, (D), East Anglia, Pm
Paul Dolman, Ou
Jan Kaiser, Cs
Adrian Matthews, (D), Reading, Ow
Claire Reeves, (D), UEA, Cg
Carol Robinson, Ob
Parvadha Suntharalingam, Oc
Cock von Oosterhout, (D), Leiden
Rachel Warren, Eg

Lecturer:
Annela Anger-Kraavi, EgGe
Annela Anger-Kraavi, Eg
Victor Bense, Hg
Alan Bond, (D), Lancaster, Ge
Jason Chilvers, (D), Univ Coll (London), Ge
Pietro Cosentino, Yg
Stephen Dorling, Ow
Aldina Franco, (D), UEA, Ge
Robert Hall, (D), Proudman Oceanographic Laboratory, On
Tom Hargreaves, Ge
Richard Herd, (D), Lancaster, Gv
Martin Johnson, Cm
Manoj Joshi, (D), Oxford, Oi
Iain Lake, Ge
Irene Lorenzoni, Eg
Nikolai Pedentchouk, (D), Penn State, 2004, Cs
Jane Powell, Ge
Brian Reid, GeCg
Gill Seyfang, Og
Congxiao Shang, (D), Queen Mary, Hw
Peter Simmons, Eg
Trevor Tolhurst, Sb
Jenni Turner, (D), 2008, Gt
Naomi Vaughan, (D), UEA, Ow
Charlie Wilson, (D), British Columbia, Ge
Xiaoming Zhai, (D), Dalhousie, Og

Professorial Fellow:
Kerry Turner, GeEg

## University of Exeter

**Camborne School of Mines** (2004)
Redruth, Cornwall
TR15 3SE
C.Jeffrey@exeter.ac.uk
http://www.ex.ac.uk/CSM/

## University of Glasgow

**School of Geographical and Earth Sciences** (2015)
The Gregory Building
East Quadrangle
University Avenue
Glasgow G12 8QQ
p. +44 (0) 141 330 5436
GES-General@glasgow.ac.uk
http://www.gla.ac.uk/schools/ges/

Head:
Maggie Cusack, Gz

Professor:
Paul Bishop
John Briggs, Ou
Roderick Brown, Cc
Deborah Dixon
Jim Hansom, (D), Aberdeen, 1979, OnGmOy
Trevor Hoey, Gs
Martin Lee, Gz
Christopher Philo

Senior Scientist:
Nicolas Beaudoin

Associate Professor:
Gordon Curry, (D), Imperial Coll (UK), 1979, GegPg
David Forrest, (D), Glasgow, 1996, Oi
Research Associate:
Enateri Alakpa
Susan Fitzer, Ob
Michael Gallagher
Lydia Hallis
Ulrich Kelka
Angela Last
Paula Lindgren, Oa
Alistair Mcgowan, Pg
Larissa Naylor
Alan W. Owen, (D), Glasgow, 1977, PgsPo
Instructor:
Anne Dunlop
Mhairi Harvey
Hayden Lorimer
Rhian Meara
Hester Parr
Vernon Phoenix
Daisy Rood
Lecturer:
Brian Bell, So
David Brown, Gvs
Seamus Coveney, Oi
Tim Dempster, Gz
Jane Drummond, Oi
Derek Fabel, Xg
David Featherstone
Nick Kamenos
Ozan Karaman
Daniel Koehn, Gt
Hannah Mathers, (D), GmlGg
Cheryl Mcgeachan
Simon Naylor
Cristina Persano

## University of Leeds

**School of Earth and Environment** (B,M,D) (2014)
Maths/Earth and Environment Building
The University of Leeds
Leeds LS2 9JT
p. +44 113 343 2846
enquiries@see.leeds.ac.uk
http://earth.leeds.ac.uk/
Administrative Assistant: Samantha E. Haynes
Director:
Piers M. Forster, (D), Reading (UK), 1994, Oa
Head:
Robert Mortimer, Ge
Chair:
Christopher Collier
Simon Poulton, (D), Cls
Peter Taylor, Ge
Professor:
John Barrett, Ge
Liane G. Benning, Ce
Alan Blyth, Oa
Ian Brooks, (D), UMIST, On
Ken Carslaw, (D), East Anglia (UK), 1994, Oa
Andy Challinor, (D), Leds, 1999, Ow
Martyn Chipperfield, (D), Cambridge, Oa
Surage Dessai, (D), East Anglia
Andy Dougil, (D), Sheffield, Os
Paul Field
Quentin Fishe, Np
Paul W.J. Glover, (D), East Anglia (UK), 1989, YxGoa
Andy Gouldson, Ge
Alan Haywood, Pe
Steve Hencher, Ng
Andy Hooper, Ydg
Greg Houseman, (D), Cambridge, Yg
Michael Krom, Cm
Bill McCaffrey, Gs
Stephen Mobbs, Oa
Jurgen Neuberg, (D), Colorado, Gv
Jouni Paavola, Ge
Doug Parker, (D), Reading, Ow
Jeff Peakall, Gs
Andrew Shepherd
Lindsay C. Stinger, (D), On
Graham Stuart, Ys
Paul Wignall, PgGs
Tim Wright, Yd
Bruce Yardley, Cg
William Young, (D), Ge
Associate Professor:
Doug Angus, Ys
Stephen Arnold, (D), Leeds, Oa
Wolfgang Buermann, (D), Boston, 2012, Ou
Ian Burke, (D), Southampton, Ge
Robert J. Chapman, (D), Leeds (UK), 1990, EmCe
Steven Dobbie, (D), Dalhousie, Oa
Luuk Flesken, (D), Wageningen Univ, Ge
Phil Livermore, Ym
Christian Maerz, (D), Bremen, 2008, CmOuEm
Jim McQuaid, (D), Leeds (UK), 1999, Oa
Daniel Morgan, (D), Open Univ, 2003, Giv
Jon Mound, (D), Toronto, 2001, Gt
Noelle Odling, (D), Queens, Hg
Caroline Peacock, (D), Bristol, Sb
Sally Russell, (D), Queensland, Ge
Dominick Spracklen, (D), Leeds, 1999, On
Julia Steinberger, Ou
Jared West, Hg
Research Associate:
David Banks, Cg
Simon Bottrell
Lauren Gregoire, (D), Bristol
Ruza Ivanovic, (D), Bristol, 2013, OgaPe
John Marsham, (D), Edinburgh, 2003, Oa
Instructor:
Nick Dixon, (D), Leeds, 2002, Ou
Tim Foxon, Ge
Clare Gordon, Oi
Chris Green
Dave Hodgson
Jacqueline Houghton
Damian Howells
Geoff Lloyd, (D), Birmingham, 1984
Sebastian Rost, Ys
Lecturer:
Ralf Barkemeyer
Emma Bramham, Ym
Roger A. Clark, (D), Leeds, 1982, YseYs
Martin Dallimer, (D), Edinburgh, 2001, Ge
Monica Di Gregorio, (D), London School of Economics, Eg
Jen Dyer, (D), Leeds, Ge
Alan Gadian
Fiona Gill, (D), PoCa
Dabo Guan, Eg
Jason Harvey, (D), 2005, Cg
Mark Hildyard
George Holmes, Ge
Andrea Jackson, (D), Lancaster, Oa
Julia Leventon, (D), Cent, Ge
Crispin Little, (D), Bristol, 1995, Pg
Piroska Lorinczi, (D), Leeds, 2006, Np
Vernon Manville
Andrew McCair, (D), Cambridge, 1983, Gc
Lucie Middlemiss, (D), Leeds, 2009, On
Nigel Mountney, Gs
Rob Newton, (D), Leeds, Ge
Alice OWen, (D), Leeds
Douglas Paton, Gc
Richard Phillips, (D), Oxford, Gt
Claire Quinn, (D), Kings College, Ge
Andrew Ross, (D), Cambridge, Ow
Susannah Sallu, Ga
Ivan Savov, (D), South Florida, 2004, Cg
Anne Tallontire
Mark Thomas, (D), Ng
Taija Torvela, (D), Abo Akademi, 2007, GctEg

James van Alstine, Ge
Xianyun Wen, (D), Sichuan Univ (China), 1982, Og
Visiting Professor:
Jane Francis, Pe
Tim Needham, Gc
Emeritus:
Rob Knipe, Gt

## University of Liverpool

**School of Environmental Sciences** (B,M,D) (2014)
Jane Herdman Building
4 Brownlow Street
Liverpool, Merseyside L69 3GP
p. +44 151 794 5146
envsci@liv.ac.uk
http://www.liv.ac.uk/environmental-sciences/index.html

**Dept of Earth, Ocean, and Ecological Sciences** (2014)
Jane Herdman Building
4 Brownlow Street
Liverpool L69 3GP
p. +44 (0)151 795 4642
Apboyle@liverpool.ac.uk
http://www.liv.ac.uk/earth-ocean-and-ecological-sciences/
Head:
Andreas Rietbrock, YsGv
Professor:
Daniel Faulkner, Ys
Chris Frid, Ob
Richard Holme, Ym
Peter Kokelaar, GstGv
Nick Kusznir, Yd
Yan Lavallee, Gv
Rob Marrs, Os
Jim Marshall, Cs
Jonathan Sharples, Ocp
Stan van der Berg, Em
John Wheeler, Em
Richard Worden, Gs
Research Associate:
Katherine Allen, Oi
James Ball
Helen Bloomfield, Ob
Stephen Crowley, Gs
Jonathan Lauderdale, Og
Claire Melet
Andreas Nilsson, Ym
Vassil Roussenov, Og
Anu Thompson, Oc
Xiao Wang
Tsuyuko Yamanaka, Ob
Lecturer:
Charlotte Jeffrey Abt, On
Andy Biggin, Ym
Alan P. Boyle, (D), Univ Coll London, 1982, GczOe
Bryony Caswell, Ob
Silvio De Angelis, Ys
Rob Duller, Gvs
Liz Fisher, Co
Jonathan Green, Ob
Mimi Hill, Ym
Janine Kavanagh, Gv
Helen Kinvig, Gs
Claire Mahaffey, Oc
Elisabeth Mariani, Nr
Kate Parr, Pe
John Piper, YmGt
Leonie Robinson, Ob
Isabelle Ryder, Gt
Matthew Spencer, Og
Neil Suttie
Alessandro Tagliabue, Oc
Jack Thomson, Ob
Other:
Sara Henton De Angelis
Jackie Kendrick
Rachel Salaun, Oc

Felix von Aulock

## University of London, Royal Holloway & Bedford New College

**Dept of Earth Sciences** (B,M,D) (2016)
Department of Earth Sciences
Royal Holloway University of London
Egham Hill
Egham , Surrey TW20 0EX
p. +44 (0) 1784 443 581
info@es.rhul.ac.uk
https://www.royalholloway.ac.uk/earthsciences/home.aspx
t: @RHULEarthSci
Head:
Dave Mattey, Cs
Professor:
Margaret Collinson, Po
Agust Gudmundsson, Gc
Robert Hall, (D), Univ Coll London, 1974, Gt
Martin King, Oa
Daniel P. Le Heron, (D), Aberystwyth Univ, 2004, OgGsl
Ken R. McClay, (D), Imperial Coll (UK), 1978, GctGo
Jason Morgan, Yg
Wolfgang Muller, (D), ETH (Switzerland), 1998, CcPeGa
Euan Nisbet, Og
Matthew Thirlwall, Cs
Dave Waltham, Yg
Associate Professor:
Kevin C. Clemitshaw, (D), East Anglia (UK), 1986, Oag
Research Associate:
Dave Alderton, Gz
Lecturer:
Jurgen Adam, Gc
Anirban Basu, (D), Illinois (Urbana Champaign), CglCs
Domenico Chiarella, (D), Basilicata (Italy), 2011, Gs
Howard Falcon-Lang, Pg
Francisco J. Hernandez-Molina, Gs
Saswata Hier-Majumder, Yg
Christina Manning, Cg
Nicola Scarselli, (D), Go
Steve Smith, Ge
Giulio Solferino, (D), Gg
Paola Vannucchi, Gu
Ian Watkinson, Gt
Emeritus:
Andrew Cunningham Scott, Pb
Laboratory Director:
Nathalie Grassineau, Cs
David Lowry, Cs

## University of Manchester

**School of Earth and Environmental Sciences** (2015)
Williamson Building
Oxford Road
, Manchester M13 9PL
p. +44 (0) 161 306 9360
earth.support@manchester.ac.uk
Administrative Assistant: Steven Olivier
Director:
Stephen Boult, Hg
Chair:
Jonathan Redfern, (D), Bristol, 1990, GsoGl
Professor:
Mike Bowman, (D), Wales, Go
Gregg Butler, (D), Univ Coll Swansea, 1971, On
Andrew Chamberlain, Ga
Thomas Choularton, (D), Manchester, 1987, Oa
Hue Coe, (D), UMIST, 1993, Oa
Stephen Flint, (D), Leeds (UK), 1985, GrsGo
Martin Gallagher, (D), UMIDST, 1986, Oa
Jamie Gilmour, (D), Sheffield
Colin Hughes, (D), Open, 1992, Go
Francis Livens, Cg
Jonathan Lloyd, (D), Canterbury, 1993, Ge
Ian Lyon, (D), 1993, Xc
Gordon McFiggans, (D), UEA, 2000, Oa

Richard Pattrick, (D), 1980, Gz
Carl Percival, (D), Oxford, 1995, Oa
David Poyla, (D), Manchester, 1987, Cm
Ernest Rutter, (D), Imperial Coll (UK), 1970, NrGct
David Schultz, Ow
Kevin Taylor, Go
David Vaughan, (D), Oxford, 1971, GzClGe
Geraint Vaughan, (D), Oxford, 1982, Oa
Roy Wogelius, (D), Northwestern, 1990, Cg

Associate Professor:
Neil C. Mitchell, (D), Oxford, 1989, GumYg

Research Associate:
Karl Beswick, Oa
Alastair Booth, Oa
Pieter Bots, Ge
Gerard Capes, Oa
Deborah Chavrit, (D), Nantes (France), 2010, Gi
Patricia L. Clay, (D), Open Univ (UK), 2010, CgaGg
Filipa Cox
Ian Crawford, Oa
Sarah Crowther, (D), Oxford, 2003, Oa
Christopher Dearden, Oa
Patrick Dowey
Helen Downie, Ge
Nicholas Edwards, Pg
Christopher Emersic, (D)
Torsten Henkel, Cs
Hazel Jones, Oa
Nimisha Joshi
Richard Kift, Oa
Kimberly Leather, Oa
Dantong Liu, Oa
Douglas Lowe, Oa
William Morgan, Oa
Miquel Poyatos-More
Laura Richards, Ge
Hugo Ricketts, Oa
Athanasios Rizoulis, Ge
Andrew Smedley, Oa
Robert Sparkes, (D), Cambridge, 2012
Jonathan Taylor, (D), Manchester
Karen Theis, (D), Manchester, 2008, On
James Whitehead, Oa
Paul Williams, Oa

Lecturer:
Grant Allen, (D), Leicester, 2005, Oa
Simon Brocklehurst, (D), MIT, 2002, Gm
Kate Brodie, (D), Imperial Coll, 1979, Gc
Rufus Brunt, (D), Leeds, Gs
Michael Buckley, (D), York, 2008, Ga
Ray Burgess, (D), Open, Cs
Victoria Coker, (D), Manchester, 2007, Gz
Paul Connolly, (D), Manchester, 2006, Oa
Stephen Covery-Crump, (D), Univ Coll (London), 1992
Giles Droop, (D), Oxford, 1979, Gp
Victoria Egerton, Pg
David Hodgetts, (D), Keele, 1995, Gsc
Cathy Hollis, (D), Aberdeen, 1995, Gs
Merren Jones, (D), Whyoming, 1997, Gst
Julian Mecklenburgh, Gc
John Nudds, (D), Dunhelm, 1975, Pg
Clare Robinson, (D), Lancaster, 1990, Co
Stefan Schroeder, (D), Bern, 2000, Go
Bart van Dongen, Utrecht, 2003, Co

Emeritus:
Christopher Henderson
Peter Jonas
Grenville Turner, (D), Oxford, 1962, Xc

## University of Newcastle Upon Tyne

**School of Civil Engineering and Geosciences** (2016)
Newcastle University
Newcastle upon Tyne
NE1 7RU
p. +44 (0)191 208 6323
ceg@ncl.ac.uk
http://www.ncl.ac.uk/ceg/

Professor of Soil Science / Head of School :
David Manning

Professor:
Margaret Carol Bell, Ge
Andras Bordossy, Og
Peter J. Clarke, (D), Oxford (UK), 1997, YdOri
Chris Kilsby, Hg
Stephen Larter, Gg
Zhenhong Li, Yd
Jon Mills, (D), Newcastle upon Tyne, 1996, OrYdOg
Phillip Moore, Yd

Associate Professor:
Geoffrey D. Abbott, (D), Southampton, 1980, Co

Research Associate:
David Alderson, Oi
Joana Baptista, Ng
Stephen Blenkinson, On
Bernard Bowler, Go
Aidan Burton
Allistair Ford, Gq
Kirill Palamartchouk, Yd

Instructor:
James Barhurst, Gs
Neil Gray, Py
Helen Talbot, Co

Lecturer:
Jamie Amezaga, Ge
Stuart Barr, Oi
Colin Davie, (D), Glasgow, 2002, Ngr
Stuart Edwards, Gq
Gaetano Elia, gaet, Ng
David Fairbairn, Gq
Rachel Gaulton, Or
Jean Hall, Ng
Geoffrey Parkin, Hg
Nigel Penna, Gq
Paul Quinn, Hg
Crees van der Land, Go

Visiting Professor:
Rick Brassington, Hg

Related Staff:
Peter Cunningham, Hg

## University of Nottingham

**Dept of Mineral Resources** (2004)
University Park
Nottingham HG7 2RD
wpadmin@nottingham.ac.uk

## University of Oxford

**Dept of Earth Sciences** (2014)
Department of Earth Sciences
South Parks Road
Oxford, Oxfordshire OX1 3AN
p. +44 1865 272000
reception@earth.ox.ac.uk
http://www.earth.ox.ac.uk/

Head:
Alex Halliday, Cs

Chair:
Christopher Ballentine, Xg
Phillip England, Gt

Professor:
Martin Brasier, Po
Joe Cartwright, Gs
Shamita Das, Gtv
Donalf G. Fraser, Cg
Gideon Henderson, Cg
Hugh C. Jenkyns, (D), Leicester (UK), 1970, GrsCs
Samar Khatiwala
Conall Mac Niocaill, (D), GtgYm
Tamsin A. Mather, (D), Gv
Barry Parsons, Yd
David M. Pyle, (D), Cambridge (UK), 1990, GviGh
Ros Rickaby, Gz
Mike Searle, Gt

Anthony B. Watts, (D), Durham, 1970, GutYv
Bernard Wood, Gz
John Woodhouse, Yg
Associate Professor:
Stuart Robinson, (D), Oxford (UK), 2002, PeGr
Karin Sigloch, (D), Princeton, 2008, YgsGt
Lecturer:
Roger Benson, Po
Heather Bouman, Cg
Matt Friedman, Po
Lars Hansen, Gz
Helen Johnson, Og
Richard Katz, Cg
Graeme Lloyd, Gs
Don Porcelli, Cg
Richard Walker, Or
Dave Waters, Gp

## University of Plymouth

**Dept of Earth Sciences** (2015)
Drake Circus
Plymouth, Devon PL4 8AA
p. +44 (0)1752 584584
science.technology@plymouth.ac.uk
http://www.plymouth.ac.uk/schools/sogees

Head:
Mark Anderson, (D), Wales Cardiff, Gct
Professor:
Antony Morris, (D), Edinburgh (UK), 1990, YmGtu
Greogry Price, (D), Reading, 1994, Gsr
Associate Professor:
Stephen Grimes, (D), Cardiff, 1998, Cs
Martin Stokes, (D), Plymouth, 1997, Gm
Graeme Taylor, Yg
Matthew Watkinson, (D), Open, 1989, Gs
Lecturer:
Sarah Boulton, (D), Edinburgh, 2005, Or
Paul Cole, (D), 1990, Gv
Arjan Dijkstra, (D), Utrecht, 2001, Gi
Meriel E.J. FitzPatrick, (D), Plymouth, 1992, PlgGr
Luca Menegon, (D), Padua, 2006, Gct
Andrew Merritt, (D), Leeds, 2010, Ng
Kevin Page, (D), Univ Coll (London), 1988, Gr
Christopher Smart, (D), Southampton, 1993, Pg
Colin Wilkins, (D), James Cook (North Queensland), 1991, Eg
Emeritus:
Malcolm Hart, (D), London, Pm

## University of Portsmouth

**School of Earth and Environmental Sciences** (2014)
Burnaby Road
Portsmouth, Hampshire PO1 3QL
p. +44 (0)23 9284 2257
sees.enquiries@port.ac.uk
http://www.sci.port.ac.uk/departments/academic/sees

Head:
Rob Strachan, (D), Keele, 1982, Cc
Professor:
Andrew Gale, (D), King's College (UK), 1984, PgGrCl
Jim Smith, (D), Ge
Associate Professor:
Craig D. Storey, (D), Leicester, 2002, GptCc
Research Associate:
Emilie Braund, (D), Graz, 2011, Gz
Fay Couceiro, (D), On
Penny Lancaster, (D), Bristol, 2011, Gz
Robert Loveridge
Darren Naish
David Ray, (D)
Alan Raybould, (D)
Steve Sweetman, (D), Pg
Mark Whitton, (D), Portsmouth, 2008, Pg
Instructor:
Chris Dewdney, (D), Birkbeck, 1983, Yg
Gary Fones, (D), Central Lancashire, Cm
David Loydell, (D), Gr
David Martill, (D), Po
Lecturer:
John Allen, (D), Southampton, 1996, Og
Hooshyar Assadullahi
Philip Benson, (D)
Michelle Bloor, (D), Hg
Dean Bullen, (D), Gt
Anthony Butcher, (D), PlmOe
James Darling, (D), Bristol, 2009, Cs
Mike Fowler, (D), Imperial, Gi
David Franklin, (D), Ym
Martyn Gardiner, (D)
Andy Gibson, (D), Eg
David Giles, Gm
Michelle Hale, (D), Flinders, Og
Nick Koor
Nicholas Minter, (D), Bristol, 2007, Gs
Derek Rust, (D), Gt
Camen Solana, (D), Gv
Richard Teeuw, (D), Stirring (UK), 1985, OaGmOi
Melvin M. Vopson, (D), Central Lancashire, 2002, Yg
Nick Walton, Hw
John Whalley, (D), Gc
Malcolm Whitworth, (D)

## University of Reading

**Soil Research Centre** (B,M,D) (2012)
Department of Geography and Environmental Science
School of Human and Environmental Sciences
Whiteknights
Reading RG6 6AB
p. +44 (0)118 378 8911
shes@reading.ac.uk
http://www.reading.ac.uk/soil-research-centre

Professor:
Chris Collins, (D), Sb
Associate Professor:
Stuart Black, (D), Lancaster, Cc
Chris Collins, (D), Sc
McGoff Hazel, (D), Liverpool, Pg
Steve Robinson, (D), Sf
Liz Shaw, (D), Sb
Dr:
Joanna Clark, (D), Leeds, Sc

## University of South Wales

**Geology Section** (2004)
Department of Science
Pontypridd

## University of Southampton

**Ocean and Earth Science** (B,M,D) ● (2016)
University of Southampton Waterfront Campus
National Oceanography Centre Southampton
European Way
Southampton, Hampshire SO14 3ZH
p. +44 (0)23 8059 2011
soes@soton.ac.uk
http://www.southampton.ac.uk/oes
f: https://www.facebook.com/UoSGeoscience
https://www.facebook.com/UoSOceanography
t: @OceanEarthUoS

Professor:
Stephen Roberts, (D), Open Univ, 1986, Cg
Royal Society Wolfson Research Merit Award holder:
Eelco J. Rohling, (D), Utrecht, 1991, Op
Professorial Fellow, Deputy Director, GAU-Radioanalytical:
Phillip Warwick, (D), Southampton, 1999, CaGe
Head of Physical Oceanography Research Group:
Sybren Drijfhout, (D), Urtecht, 1992, Op
Head of Palaeoceanography and Palaeoclimate Research Group:
Paul A. Wilson, (D), Cambridge, 1995, GsCg
Head of NOCS Graduate School:
Timothy A. Minshull, (D), Cambridge, 1990, Yr
Head of Marine Biology & Ecology Research Group:
Jorg Wiedenmann, (D), Ulm (Germany), 2000, Ob

Head of Marine Biogeochemistry Research Group:
Toby Tyrrell, (D), Edinburgh (UK), 1993, OgPe
Director of Research:
Rachael James, (D), Cambridge, 1995, Oc
Director of Programmes:
Andy Cundy, (D), Southampton, 1994
Dean, Faculty of Natural and Environmental Sciences:
Rachel A. Mills, (D), Cambridge, 1992, Cm
Professor:
Carl L. Amos, (D), Imperial Coll, 1974, Gs
Nicholas R. Bates, (D), Southampton, 1995, Oc
Thomas Bibby, (D), Imperial Coll (London), 2003, Ob
Colin Brownlee, (D), Newcastle upon Tyne, Ob
Jonathan M. Bull, (D), Edinburgh (UK), 1990, YrGct
Ian Croudace, (D), Birmingham, 1980, Ge
Gavin Foster, (D), Open, 2000, Csg
Stephen J. Hawkins, (D), Liverpool, 1979, Nr
Tim Henstock, (D), Cambridge (UK), 1994, YgrYe
Alan Kemp, (D), Edinburgh, 1985, Pe
Maeve Lohan, (D), Southampton, 2003, Ob
Robert Marsh, (D), Southampton, 2000, Op
John E. A. Marshall, (D), Bristol, 1981, PlGro
Lisa McNeill, (D), Oregon State, 1998, GtYr
Christopher Mark Moore, (D), Southampton, 2002, Ob
Alberto Naveira Garabato, (D), Liverpool, 1999, Op
Martin R. Palmer, (D), Leeds, 1984, Cg
Duncan A. Purdie, (D), Wales, 1982, Ob
Martin Solan, (D), National Univ of Ireland (Galway), 2000, Ob
Damon Teagle, (D), Cambridge, 1993, Cg
Senior Lecturer:
Ian Harding, (D), Cambridge (UK), 1986, PemPl
Rex N. Taylor, (D), Southampton, 1987, GvCa
Head of Geology and Geophysics Research Group:
Justin Dix, (D), St. Andrews, 1994, GaYs
Head of Geochemistry Research Group:
Juerg M. Matter, (D), Swiss Fed Inst Tech, 2001, Ng
Associate Professor:
Jonathan Copley, (D), Southampton, 1998, Ob
Eleanor Frajka-Williams, (D), Washington, 2009, Op
Thomas M. Gernon, (D), Bristol, 2007, OgGvi
Ivan D. Haigh, (D), Southampton, 2009, On
Chris Hauton, (D), Southampton, 1995, Ob
Lawrence Hawkins, (D), Southampton, 1985, Ob
Antony Jensen, (D), Southampton, 1982, Ob
Derek Keir, (D), Royal Holloway (London), 2006, Gtv
Cathy Lucas, (D), Southampton, 1993, Ob
Catherine A. Rychert, (D), Brown, 2007, Yg
Florian Sevellec, (D), UBO, 2007, Op
Sven Thatje, (D), Bremen, 2003, Ob
Clive Trueman, (D), Bristol, 1997, Cg
Neil Wells, (D), Reading (UK), 1974, OpwOn
Jessica H. Whiteside, (D), Columbia, 2006, Pe
Teaching Fellow :
Hachem Kassem, (D), Southampton, 2016, OnGsOi
SMMI Lecturer:
Steven Bohaty, (D), California (Santa Cruz), 2006, Og
Senior Tutor, Principle Teaching Fellow:
Simon R. Boxall, (D), Liverpool, 1985, Op
Senior Tutor:
Andy J. Barker, (D), Univ of Wales (Cardiff), 1983, Gg
NERC Advanced Senior Research Fellow:
Tom Ezard, (D), Imperial Coll (London), 2007
Lecturer:
Amanda E. Bates, Victoria, 2006, Ob
Claudie Beaulieu, (D), Quebec, 2009, Op
Phillip Fenberg, (D), California (San Diego), 2008, Ob
Jasmin A. Godbold, (D), Aberdeen, 2009, Ob
Phillip A. Goodwin, (D), Liverpool, 2007, Pe
Laura Grange, (D), Southampton, 2005, Ob
Nicholas Harmon, (D), Brown, 2007, Yg
Eli Lazarus, (D), Duke, GmOny
Kevin Oliver, (D), East Anglia, 2003, Op
Marc Rius, (D), Barcelona, 2008, Onn
Esther Sumner, (D), Bristol, 2009, Gs
Chuang Xuan, (D), Florida, 2010, Pe
Visiting Professor:
Robert B. Whitmarsh, (D), Cambridge (UK), 1967, Yr
Emeritus:
Ian S. Robinson, (D), Warwick, 1973, Or
John G. Shepherd, (D), Cambridge, 1970, Oa
Peter Statham, (D), Southampton, 1983, Ob
Teaching Fellow:
Judith A. Coggon, (D), Durham, 2010, Cg
Senior Research Fellow:
Ken Collins, (D)
Matthew Cooper, (D), Cambridge, 1999, Oc
Research Fellow:
Charlie Thompson, (D), Southampton, 2003

## University of St. Andrews

**School of Geography & Geosciences** (2004)
Division of Geology
St. Andrews
Fife KY16 9ST

**Dept of Earth and Environmental Science** (B,M,D) (2015)
Irvine Building
North St.
Fife, Scotland KY16 9AL
p. (+44) (0)1334 463940
earthsci@st-andrews.ac.uk
http://earthsci.st-andrews.ac.uk/
f: https://www.facebook.com/EarthSciStA/
t: @EarthSciStA
Director:
Richard Bates, (D), Wales, Yx
Ruth Robinson, (D), Penn State, 1997, Gs
Head:
Tony Prave, (D), Penn State, 1986, GgrGs
Research Associate:
Nicky Allison, (D), Edinburgh, 1994, Oc
Mark Claire, (D), Washington, Oa
Catherine Cole, (D), Southampton, 2013, Cm
Ruth Hindshaw, (D), ETH Zürich, 2011, Cs
Gareth Izon, (D), Cg
Coralie Mills
James Rae, (D), Bristol, Cg
Vincent Rinterknecht, (D), CcGma
Lecturer:
Andrea Burke, Cs
Adriam Finch, Gz
Timothy Hill, (D), Edinburgh, 2003, Ge
Tim Raub, (D), Yale, 2008, Py
Michael Singer, (D), California (Santa Barbara), 2003, Hg
John Walden, (D), Wolverhampton Polytechnic, 1990, Ym
Robert Wilson, (D), W Ontario, 2003, On
Aubrey Zerkle, (D), Penn State, 2006, Co
Emeritus:
Colin K. Ballantyne, (D), Edinburgh (UK), 1981, OyGml

## University of Wales

**Geography and Earth Sciences** (2014)
Llandinam Building
Penglais Campus
Aberystwyth SY23 3DB
p. +44(0) 1970 622 606
dges@aber.ac.uk
http://www.aber.ac.uk/en/iges/
Director:
Neil Glasser
Head:
Rhys Jones
Professor:
Paul A. Brewer, (D), Gm
John Grattan
Matthew Hannah
Alun Hubbard
Bryn Hubbard
David Kay
Henry Lamb
Richard Lucas
Mark Macklin
Alex Maltman

Nick Pearce
Andrew D. Thomas, (D), Swansea Univ, 1996, SbGmSf
Mark Whitehead
Michael Woods
Instructor:
Peter Abrahams
Peter Merriman
Andrew Mitchell
Helen Roberts
Stephen Tooth
Lecturer:
Charlie Bendall
Peter Bunting
Rachel Carr
Rhys Dafydd Jones
Sarah Davies
Carina Fearnley
Elizabeth Gagen
Hywel Griffiths
Kevin Grove
Andrew Hardy
Jesse Heley
Tom Holt
Gareth Hoskins
Tristram Irvine-Fynn
Cerys Jones
Bill Perkings
Kimberly Peters
George Petropoulos
Mitch Rose
Joe Williams
Richard Williams
Sophie Wynne-Jones
Teaching Fellow:
Stefania Amici

**School of Ocean Sciences** (2014)
Bangor University
Menai Bridge
Anglesey, Bangor LL59 5AB
p. (01248) 382851
oss011@bangor.ac.uk
http://www.bangor.ac.uk/oceansciences/
Director:
Colin Jago
Professor:
David Bowers
Alan Davies
Michel Kaiser
Hilary Kennedy
Chris Richardson
Tom Rippeth
James Scourse
John Simpson
David Thomas
Associate Professor:
Simon P. Neill, (D), Strathclyde, 2001, OnpOg
Instructor:
Jan Geert Hiddink
Stuart Jenkins
Lecturer:
Martin Austin
Jaco H. Baas
Paul Butler
Lui Gimenez
Mattias Green
Cara Hughes
Dei Huws
Suzanna Jackson
Lewis LeVay
Shelagh Malham
Irene Martins
Ian McCarthy
Gay Mitchelson-Jacob
Anna Pienkowski
Martin Skov
John Russel Turner
Katrien van Landeghem
Stephanie Wilson
Related Staff:
Timothy Whitton

**Dept of Earth and Ocean Sciences** (2014)
College of Cardiff
Main Building
Park Place
, Cardiff CF1 3YE
p. +44 (0)29 208 74830
earth-ug@cf.ac.uk
http://www.cardiff.ac.uk/earth/
Head:
R. John PArkes, Cg
Professor:
Thomas Blenkinspo
Dianne Edwards, Pb
Ian R. Hall
Richard Lisle
Chris MacLeod, Og
Wolfgang Maier
Paul Pearson, PmCg
Instructor:
Stephen Barker
Huw Davies
Andrew Kerr, Og
Jenny Pike, Og
Lecturer:
Tiego Alves, Og
Liz Bagshaw
Rhoda Ballinger, On
Peter Brabham, Ng
David Buchs
Alan Channing, Gd
Jose Constantine
T. C. Hales, Gt
Alan Hemsley, Pl
Tim Jones
C. Johan Lissenberg
Iain MacDonald, Ca
Rupert Perkins
Phil Renforth
David Reynolds
Henrik Sass, Co
Simon Wakefield
Emeritus:
Hazel Prichard, (D), EgGze
Laboratory Director:
Caroline Lear, Og
Tutorial Fellow:
Ian Fryett
Nick Rodgers, Xm

# Uruguay

## Direccion Nacional de Mineria y Geologia de Uruguay

**Direccion Nacional de Mineria y Geologia de Uruguay** (2015)
Hervidero 2861
Montevideo, Montevideo 11800
p. + 5982 2001951
infomiem@miem.gub.uy
http://www.dinamige.gub.uy/

## Universidad de la Republica Montevideo

**Dept de Geologia** (2004)
Avenida 18 de Julio 1968, Montevideo

## Universidad de la Republica Oriental del Uruguay (UDELAR)

**Instituto de Geología y Paleontología** (2011)
Avenida 18 de Julio 1968, Montevideo
CP11 400
Montevideo 10773

http://www.fcien.edu.uy/menu2/estructura2/ingepa2.html

**Deptartamento de Geografia** (B) (2016)
Iguá 4225 Piso 14 Sur C.P: 11400
Montevideo, Montevideo 11400
p. (598-2) 525 15 52
geotecno@fcien.edu.uy
http://geografia.fcien.edu.uy/
Professor:
Virginia Fernández, (M), Girona, 2001, OirOn
Assistant Professor:
Yuri Resnichenko, (M), la República, 2010, OirOn
Related Staff:
Carlos Miguel, (B), Universidad de Valladolid, 4, OirSf
Virginia Pedemonte, (B), Facultad de Arquitectura, 5, OirOu

**Dept de Suelos y Aguas** (2011)
p. (598 2) 359 82 72
suelosyaguas@fagro.edu.uy
http://suelosyaguas.fagro.edu.uy/

# Uzbekistan

## Institute of Geology and Geophysics
(2011)
http://www.ingeo.uz/

## Institute of Mineral Resources
(2013)
gpniimr@evo.uz
http://www.gpniimr.uz/

## National University of Uzbekistan
**Faculty of Geology** (2011)
Tashkent
University City 100095
p. +998712460224
geology@nuu.uz
http://nuu.uz/geolog

## Tashkent State Technical University
**Faculty of Geology** (2011)
Uzbekistan, Tashkent, 10095 Universitetskaya street, 2
p. +998712464600
TFTU_info@mail.ru

## The State Committee for Nature Protection
(2011)
100 159, Tashkent, pl. Independence, 5
info@uznature.uz
http://www.uznature.uz

## The State Committee of the Republic of Uzbekistan on Geology and Mineral Resources
(2011)
11, T.Shevchenko str., Tashkent, Republic of Uzbekistan 100060
p. +998 (71) 256-8653
geolcom@bcc.com.uz
http://www.uzgeolcom.uz

## Uzbekistan National Oil and Gas Company - Uzbekneftegaz (UNG)
(2014)
100047, city Street, Tashkent.
Istiqbol, 21
p. +998 (71) 233-5757
kans@uzneftegaz.uz
http://www.uzneftegaz.uz

# Venezuela

## Universidad Central de Venezuela
**Inst de Ciencias de la Tierra** (2014)
Los Chaguaramos, Apdo. 3895, 1010-A Caracas
p. 0212 6366236
coordinv@ciens.ucv.ve
http://www.coordinv.ciens.ucv.ve/investigacion/genci/sitios/35/index.html

**Escuela de Geologia, Minas y Geofisica** (2004)
Ciudad Universitaria, 47028 Caracas 1041 A

## Universidad de la Este, Cumana
**Dept de Geologia y Minas** (2004)
Apartado Postal 245, Cumana (Estado Sucre)

## Universidad de Los Andes
**Dept de Geologia y Minas** (2014)
Avenida 3, Independencia, La Hechlcera, Merida
ocamacho@ula.ve
http://llama.adm.ula.ve/pingenieria/index.php?option=com_content&view=article&id=313&Itemid=215

**Escuela de Ingenieria Geologica** (2004)
Av. Tulio Febres C., Merida 5101

## Universidad de Oriente, Nucleo Bolivar
**Escuela de Ciencias de la Tierra** (2004)
Av. Universidad, Campus Universitario La Sabanita, Ciudad Bolivar

## Universidad Simon Bolivar
**Coordinacion de Ingenieria Geofisica** (2015)
Valle de Sarteneja, Baruta, Edo. Miranda 80659, Caracas 1080
p. 9063545
coord-geo@usb.ve
http://www.gc.usb.ve/geocoordweb/index.html

# Vietnam

## Hanoi University of Mining & Geology
**Faculty of Geology** (A,B,M,D) O (2015)
Duc Thang Ward - North Tu Liem Distr.
Hanoi
p. +84-4-38387567
diachat@humg.edu.vn
khoadiachat.edu.vn
*Enrollment (2015): B: 2800 (420) M: 205 (50) D: 24 (6)*
Dean of Faculty:
Lam Van Nguyen, (D), Hanoi Univ of Mining and Geology, 2002, HwGeHg
Deputy Dean of Faculty:
Thanh Xuan Ngo, (D), Okayama Uni of Science (Japan), 2009, GtcGi

# Zambia

## Copperbelt University
**Mining Dept** (B) (2010)
Kitwe
deansot@cbu.ac.zm
http://www.cbu.edu.zm/technology

## University of Zambia
**Dept of Metallurgy and Mineral Processing** (B) (2010)
PO Box 32379
, Lusaka
p. +360-21-1-250871
registrar@unza.zm
http://www.unza.zm/index.php?option=com_content&task=view&id=479&Itemid=574

# Zimbabwe

## Africa University

**Faculty of Agriculture and Natural Sciences** (B) (2014)
Fairview Rd (Off- Nyanga Rd)
PO Box 1320
Old Mutare, Mutare
p. +2632060075
info@africau.edu
http://www.africau.edu/academic/default.htm

## University of Zimbabwe

**Dept of Geology** (B,M,D) (2014)
Building B047
University of Zimbabwe
P.O. Box MP167
Mount Pleasant, Harare
p. 303211 Ext. 15032
gchipari@science.uz.ac.zw
http://www.uz.ac.zw/index.php/2013-07-09-08-51-40/the-department-of-geology/226-sci/dept-sci/geology-dpt/826-dr-lrm-nhamo
*Enrollment (2012): B: 27 (17) M: 1 (0) D: 3 (1)*

Chair:
LRM Nhamo
Lecturer:
Trendai Jnila, (D)
Isidro Rafael Vit Manuel, (D)
Maideyi Lydia Meck, (D)

**Dept of Geography & Environmental Science** (B,M,D) (2013)
PO Box MP167
Mount Pleasant, Harare
p. +263-04-303211
geography@arts.uz.ac.zw
http://www.uz.ac.zw/science/geography/

**Institute of Mining Research** (B) (2010)
PO Box MP167
Mount Pleasant, Harare
p. +263-4-336418
speka@science.uz.ac.zw
http://www.uz.ac.zw/

**Dept of Mining Engineering** (B) (2014)
PO Box MP167
Mount Pleasant, Harare
p. (263) 4 -3335 x ext: 17089)
kudzie@eng.uz.ac.zw
http://www.uz.ac.zw/index.php/mining-about

**Dept of Geoinformatics and Surveying** (B) (2014)
PO Box MP167
Mount Pleasant, Harare
p. +263 772 318 473
bukalt@eng.uz.ac.zw
http://www.uz.ac.zw/index.php/fac-of-eng/department-of-geoinformatics-and-surveying

# Theses and Dissertations, 2014

The following section documents all of the 2014 geoscience dissertations and theses from U.S. and Canadian institutions that were reported to GeoRef Information Services. If you have questions about the data or to make sure your institution's data is included in the future, please contact Monika Long at ml@agiweb.org.



## Acadia University

Masters

**Beresford, Vincent Paul,** *Field relationships, petrology, and tectonic setting of Neoproterozoic plutonic rocks in the southern Cobequid Highlands, Nova Scotia*

**Bisrat, Biniam Tesfay,** *The use of partial digestion techniques in pedogeochemical exploration*

**Drummond, Justin Barclay Rogers,** *Sedimentology and stratigraphy of Neoproterozoic peritidal phosphorite, Sete Lagoas Formation, Brazil; implications for the evolution of the Precambrian phosphorus cycle*

Bachelor

**Plante, Melanie,** *Paleoenvironments of the Devonian-Carboniferous Blue Beach Member of the Horton Bluff Formation, Nova Scotia, Canada*

**Reid, Michael G.,** *Petrography and geochemistry of drill core from the Taylors Brook property in the Stirling Belt, southeastern Cape Breton Island, Nova Scotia*

## Arizona State University

Doctorate

**Adams, Byron Allen,** *Tectonic and climatic influence on the evolution of the Bhutan Himalaya*

**Alsop, Eric Bennie,** *Integrating Metagenomics and Geochemistry: Functional Evolution and Taxonomic Classification of Hot Spring Communities*

**Bohon, Wendy,** *Late Cenozoic-recent tectonics of the southwestern margin of the Tibetan Plateau, Ladakh, northwest India*

**Carlin, Maureen Cassin,** *Comparative Analysis of Horizontal Directional Drilling Construction Methods in China*

**Fertelmes, Craig Martin,** *Vesicular Basalt Provisioning Practices Among the Prehistoric Hohokam of the Salt-Gila Basin, Southern Arizona*

**Haddad, David,** *Effects of Fault Segmentation, Mechanical Interaction, and Structural Complexity on Earthquake-Generated Deformation*

**Hamdan, Abeer,** *Damming Ephemeral Streams: Understanding Biogeomorphic Shifts and Implications to Traversed Streams due to the Central Arizona Project (CAP) Canal, Arizona*

**Jungers, Matthew Cross,** *Post-Tectonic Landscape Evolution of Sedimentary Basins in Southeastern Arizona and Northern Chile*

**Kraft, Katrien Van Der Hoeven,** *Determining Persistence of Community College Students in Introductory Geology Classes*

**Mead, Chris,** *Biogeochemistry Science and Education; Part One: Using Non-Traditional Stable Isotopes as Environmental Tracers; Part Two: Identifying and Measuring Undergraduate Misconceptions in Biogeochemistry*

**Mendez-Barroso, Luis Arturo,** *Integration of Remote Sensing, Field Observations and Modelling for Ecohydrological Studies in Sonora, Mexico*

**Pacheco, Heather Anne,** *Choice and Participation of Career by STEM Professionals with Sensory and Orthopedic Disabilities and the Roles of Assistive Technologies*

**Pagano, Michael Doran,** *Variability of Elemental Abundances in the Local Neighborhood and its Effect on Planetary Systems*

**Palmer, Ronald Evan,** *Analysis of the Spatial Thinking of College Students in Traditional and Web-facilitated Introductory Geography Courses using Aerial Photography and Geo-visualization Technology*

**Rossi, Matthew William,** *Hydroclimatic Controls on Erosional Efficiency in Mountain Landscapes*

**Stroik, Laura Kathryn,** *The Dietary Competitive Environment of the Origination and Early Diversification of Euprimates in North America*

**Ulett, Mark Andrew,** *Definitely Directed Evolution (1890-1926): The Importance of Variation in Major Evolutionary Works by Theodor Eimer, Edward Drinker Cope, and Leo Berg*

**Vance, Kirk Erik,** *Early Age Characterization and Microstructural Features of Sustainable Binder Systems for Concrete*

**Vaughn, Michael David,** *Towards Biohybrid Artificial Photosynthesis*

**Williams, Curtis Davis,** *Early Solar System to Deep Mantle: The Geochemistry of Planetary Systems*

**Wyant, Karl Arthur,** *Soil Moisture Availability and Energetic Controls on Belowground Network Complexity and Function in Arid Ecosystems*

**Young, Kelsey,** *The Use of Terrestrial Analogs in Preparing for Planetary Surface Exploration: Sampling and Radioisotopic Dating of Impactites and Deployment of In Situ Analytical Technologies*

Masters

**Ghadage, Prasannakumar,** *Novel Waypoint Generation Method for Increased Mapping Efficiency*

**Guild, Meghan R.,** *Boron Isotopic Composition of the Subcontinental Lithospheric Mantle*

**Heim, Zackary,** *Probabilistic Based Assessment of the Influence of Nonlinear Soil Behavior and Stratification on the Performance of Laterally Loaded Drilled Pier Foundations*

**Jansen, Michael Andrew,** *A Phylogenetic Revision of Minyomerus Horn, 1876 and Piscatopus Sleeper, 1960 (Curculionidae: Entiminae: Tanymecini: Tanymecina)*

**Lai, Jason Chi-Shun,** *Assessing Martian Bedrock Mineralogy Through "Windows" in the Dust Using Near- and Thermal Infrared Remote Sensing*

**Puruhito, Emil,** *Automated Monitoring and Control Systems for an Algae Photobioreactor*

**Robinson, Scott Michael,** *Quantifying the Temporal and Spatial Response of Channel Steepness to Changes in Rift Basin Architecture*

**Shafer, Zachery,** *Use of X-Ray Diffraction to Identify and Quantify Soil Swelling Potential*

**Wertheim, Alex,** *Growth and Characterization of Pyrite Thin Films for Photovoltaic Applications*

**Wilson, Sean,** *Improving our Understanding of Source Zones at Petroleum Impacted Sites through Physical Model Studies*

## Baylor University

Doctorate

**Huang, Rixiang,** *Effects of surface heterogeneity on the colloidal stability, protein adsorption and bacterial interaction of nanoparticles*

**Jennings, Debra S.,** *Paleopedology of paleo-wetland and barite-bearing, hydric paleosols in the Morrison Formation (Upper Jurassic-Lower Cretaceous), north central Wyoming, USA; a multianalytical approach*

**Meier, Holly A.,** *Analysis of deposition, erosion, and landscape stability during the late Quaternary using multi-proxy evidence from Owl Creek, central Texas, USA*

**Meighan, Hallie E.,** *Seismic analysis of a slab tear in the northeast Caribbean*

**Michel, Lauren Ashley,** *Field, micromorphologic and geochemical study of modern and ancient soils from Riesel, Texas and Rusinga Island, Lake Victoria, Kenya*

**Wegert, Daniel James,** *Lithospheric magmatism in southern Colorado and northern New Mexico*

Masters

**Boling, Kenneth S.,** *Controls on the accumulation of organic matter in the Eagle Ford Group, Central Texas, USA*

**Felda, Garrett Robert,** *Depositional character of Late Triassic fluvial and lacustrine strata of the Owl Rock Member (Chinle Formation), Petrified Forest National Park, Arizona*

**Gunnell, Alan R.,** *Mapping the distribution of methane hydrate beneath Woolsy Mound, Mississippi Canyon Block 118, Gulf of Mexico*

**Ju, David H.,** *Aquifer framework restoration (AFR) in an alluvial aquifer, central Texas*

**Kuijper, Kimberley E.,** *The controls on reservoir continuity within the Late Mississippian Elkton Member at Caroline Field, central Alberta, Canada*

**Reed, Tyler H.,** *Spatial correlation of earthquakes with two known and two suspected seismogenic faults, north Tahoe-Truckee area, California*

**Schwed, Martin,** *Seismic site characterization through joint mod-*

*eling of complementary data functionals, with applications to Santo Domingo, Dominican Republic*

Bachelor

**Boggess, Alex,** *A petrographic and stable isotopic analysis of a Quaternary flowstone from Crystal Cave, Kutztown, Pennsylvania, U.S.A.*

**Davis, Rebecca,** *Burn it to the ground; an investigation of charcoal carbon stability in soils using "C NMR spectroscopy and [14]C radiocarbon dating*

**Fenley, C. William, IV,** *Magnetostratigraphy of Late Cretaceous through early Paleocene deposits in the San Juan Basin, New Mexico, USA*

**Keracik, John Charles,** *Assessing the source rock potential using solid-state [13]C NMR organic geochemical approaches*

**Meyer, Creighton,** *Methods for the isolation of suspended particulate matter and implications for the carbon and nitrogen cycling in the Brazos River*

**Raley, Kristina,** *Sauropod trackways in the Glen Rose Limestone, Coryell County, TX*

## Boise State University

Doctorate

**Babcock, Esther,** *Targeted Full-Waveform Inversion for Recovering Thin- and Ultra-Thin-Layer Properties Using Radar and Seismic Reflection Methods*

**Havens, Scott Christopher,** *Development and Application of Tools for Avalanche Forecasting, Avalanche Detection, and Snowpack Characterization*

**Morrison, Michael Wayne,** *In-Situ Viscoelastic Soil Parameter Estimation Using Love Wave Inversion*

**Thoma, Michael James Jr.,** *Estimating Unsaturated Flow Properties in Coarse Conglomeratic Sediment*

Masters

**Bolvardi, Vahab,** *Electromagnetically Induced Remediation of Contaminated Soil*

**Dawson, Blaine C.,** *Developing and Testing a Greenness-Duration Method for Mapping Irrigated Areas: A Case Study in the Snake River Plain*

**Loughridge, Ricci,** *Identifying Topographic Controls of Terrestrial Vegetation Using Remote Sensing Data in a Semiarid Mountain Watershed, Idaho, USA*

**Najafi, Somayeh,** *The Effect of Electromagnetic Waves on Airflow During Air Sparging*

**Portugais, Brian Richard,** *Dual-State Kalman Filter Forecasting and Control Theory Applications for Proactive Ramp Metering*

## Bowling Green State University

Masters

**Garnes, William Thomas,** *Subsurface facies analysis of the Devonian Berea Sandstone in southeastern Ohio*

**Hernandez, Brett M.,** *Physical volcanology, kinematics, paleomagnetism, and anisotropy of magnetic susceptibility of the Nathrop Volcanics, Colorado*

**Ilangakoon, Nayani Thanuja,** *Relationship between leaf area index (LAI) estimated by terrestrial LiDAR and remotely sensed vegetation indices as a proxy to forest carbon sequestration*

**Liu, Xiaohui,** *Web-based multi-criteria evaluation of spatial trade-offs between environmental and economic implications from hydraulic fracturing in a shale gas region in Ohio*

**Mekonnen, Addisu Dereje,** *Wind Farm Site Suitability Analysis in Lake Erie Using Web-Based Participatory GIS (PGIS)*

**Stouten, Craig A.,** *Subsurface facies analysis of the Clinton Sandstone, located in Perry, Fairfield, and Vinton Counties*

## Brigham Young University

Masters

**Anderson, Kathleen RaFawn,** *Compositional analysis of three clay artifact collections from the Southwestern United States*

**Arnold, Karl D.,** *Sand sea extents and sediment volumes on Titan from dune parameters*

**Carpenter, McLean Kent,** *West Antarctic surface mass balance; do synoptic scale modes of climate contribute to observed variability?*

**Decker, Megan Carolee,** *Paterae on Io; geologic mapping of Tupan Patera and experimental models*

**Dossett, Toby S.,** *The first $^{40}Ar/^{39}Ar$ ages and tephrochronologic framework for the Jurassic Entrada Sandstone in central Utah*

**Fisher, Tsz Man Lau,** *Reconstruction of 1852 Banda Arc megathrust earthquake and tsunami*

**Greenhalgh, Scott Royal,** *Along strike variability of thrust-fault vergence*

**Heiner, Brandon D.,** *Multi-scale neotectonic study of the Clear Lake fault zone in the Sevier Desert Basin (central Utah)*

**Jennings, George R., III,** *Facies analysis, sequence stratigraphy and paleogeography of the Middle Jurassic (Callovian) Entrada Sandstone; traps, tectonics, and analog*

**Johnson, Douglas M.,** *The nature and origin of pebble dikes and associated alteration; Tintic mining district (Ag-Pb-Zn), Utah*

**Liu, Zac Yung-Chun,** *The tectonics of Saturn's moon Titan and tsunami modeling of the 1629 mega-thrust earthquake in eastern Indonesia*

**May, Skyler Bart,** *The Bell Springs Formation; characterization and correlation of Upper Triassic strata in northeast Utah*

**McGuire, Kevin Michael,** *Comparative sedimentology of Lake Bonneville and the Great Salt Lake*

**Meyer, Eric R.,** *Normal fault block or giant landslide? Baldy Block, Wasatch Range, Utah*

**Pahnke, P. D.,** *Characterization of Cretaceous chalk microporosity related to depositional texture; based upon study of the Upper Cretaceous Niobrara Formation, Denver-Julesburg Basin, Colorado and Wyoming*

**Reed, Lincoln H.,** *Build-and-fill development of lower Ismay (Middle Pennsylvanian Paradox Formation) phylloid-algal mounds of the Paradox Basin, southeastern Utah*

**Shurtliff, Ryan Andros,** *Wetlands on the Thousand Lake Mountain mega-landslide as paleoclimate proxies*

**Yaede, Johnathan R.,** *A new geophysical strategy for measuring the thickness of the critical zone*

## California Institute of Technology

Doctorate

**Dougherty, Sara Lyn,** *Seismic structure along transitions from flat to normal subduction; central Mexico, southern Peru, and southwest Japan*

**Fokoua Djodom, Landry,** *Optimal scaling in ductile fracture*

**Miller, Madeline Diane,** *The deep ocean density structure at the last glacial maximum; what was it and why?*

**Morton, Timothy Davies,** *Demographic studies of extrasolar planets*

**Olson, Michael James,** *Cloud computing services for seismic networks*

**Rosenburg, Margaret Anne,** *Interpretation of lunar topography; impact cratering and surface roughness*

**Song, Shiyan,** *A new ground motion intensity measure, peak filtered acceleration (PFA), to estimate collapse vulnerability of buildings in earthquakes*

**Thomas, Marion Y.,** *Frictional properties of faults; from observation on the Longitudinal Valley Fault, Taiwan, to dynamic simulations*

**Wu, Stephen,** *Future of earthquake early warning; quantifying uncertainty and making fast automated decisions for applications*

**Zhan, Zhongwen,** *Exploiting seismic waveforms of ambient noise and earthquakes*

## California Polytechnic State University

Bachelor

**Buckley, David James,** *Detection of an Ultra Low Velocity Zone beneath Central Mexico with PcP Waveform Modeling*

**Fridlund, Joshua,** *The Effect of Increasing Rates of Biochar on Corn Grown in Salinas Clay Loam*

**Jason DeMoss,** *Utilizing Indicator of Reduction in Soils Tubes to Affirm a Serpentinitic Hydric Soil on the California Central Coast*

**McMahan, Michael,** *Relationship of Joint Sets to Folded Diatomite Bedding of the Miguelito Member of the Pismo Formation in Montãña de Oro State Park*

**Meyst, William J.,** *Investigating meter scale topographic variation as a factor of Monterey pine (Pinus radiata) growing conditions at Kenneth Norris Rancho Marino Reserve, Cambria, CA*

**Shor, Grayson Michael,** *Water in the 21st Century*

**Zhao, Vincent,** *Detrital Zircon Geochemistry of the Nacimiento Block, Santa Ynez Mountains, California*

## California State University at Hayward

Masters

**de Jong, Menso,** *Analysis of extrinsic tracer data at seawater intrusion barriers in Los Angeles County*

**Holtz, Marianne L.,** *Investigating the source of nitrate in a water supply well in the Salinas Valley with isotopic tracers*

## California State University at Long Beach

Masters

**Ijeoma, Idu Opral C.,** *A test of diagenetic ordering in siliceous lithofacies, Monterey formation, southwestern Casmalia Hills, Santa Maria Basin, California*

**Kenyon, Scott,** *Using offset geomorphic features to estimate paleoearthquake slip distribution on the Claremont Fault, northern San Jacinto fault zone*

**Torn, Daniel.,** *Sedimentology and stratigraphy of diatomaceous sediments in the Casmalia Hills and Orcutt oil fields in the Santa Maria basin, California*

**Wicker, Cary,** *Tectonic geomorphology of the San Timoteo Badlands: new insights from OSL and LiDAR data*

## Carleton University

### Doctorate

**Neville, Lisa Ann,** *High resolution paleolimnology of lakes in the Athabasca Oil Sands mining region, Alberta, Canada*

**Thomson, Danielle,** *Neoproterozoic carbonate ramp development in the Canadian Arctic (Shaler Supergroup); sedimentology, sequence stratigraphy, and chemostratigraphy*

### Masters

**Anderson, Erika,** *Physical volcanology and geochemistry of Upsal Hogback Volcano, Fallon, Nevada, USA*

**Cutts, Jamie Alistair,** *Age and geochemical character of granite and syenite plutons in the Grenville Province of southeastern Ontario; insights into magmatism during the Ottawan Orogeny and evidence of the Frontenac Intrusive Suite in the Sharbot Lake domain*

**Dionne, Danielle,** *Sedimentary and ecosystem response to Late Cretaceous (Cenomanian-Turonian) paleoenvironmental events along the eastern margin of the Western Interior Seaway, Canada*

**Lu, Wenzhe,** *Mapping graben-fissure systems in the Mead Quadrangle, Venus*

**Mackinder, Alana,** *A petrographic, geochemical and isotopic study of the 780 Ma Gunbarrel large igneous province, western North America*

**Melanson, David,** *Identification of VMS ore lens reflections using vertical seismic profiling and 3D finite difference modeling in Flin Flon, Manitoba, Canada*

**Nasser, Nawaf,** *Arcellaceans (testate lobose amoeba) as proxies for arsenic and heavy metal contamination in the Baker Creek watershed region, Northwest Territories, Canada*

**Prince, John,** *Sequence stratigraphic, lithostratigraphic and stable isotope analysis of the Minto Inlet Formation and Killan Formation of the Shaler Supergroup, Northwest Territories*

**Upiter, Lindsay Maia,** *Effects of system movement in flight on helicopter-borne magnetic gradiometry*

## Colorado School of Mines

### Doctorate

**Al Duhailan, Mohammed,** *Petroleum-expulsion fracturing in organic-rich shales; genesis and impact on unconventional pervasive petroleum systems*

**Bearup, Lindsay A.,** *Changing flow, transport, and geochemistry in the mountain pine beetle-killed forests of Rocky Mountain National Park*

**Broughton, David W.,** *Geology and ore deposits of the Central African Copperbelt*

**Doe, Michael Frederick,** *Reassessment of Paleo- and Mesoproterozoic basin sediments of Arizona; implications for tectonic growth of southern Laurentia and global tectonic configurations*

**Dykes, Gregory B., Jr.,** *Cuttings transport implications for drill string design; a study with computational fluid dynamics*

**Emmanuel, Olusanmi Olatunde,** *Geologic characterization and the recognition of cyclicity in the Middle Devonian Marcellus Shale, Appalachian Basin, NE USA*

**Franklin Dykes, Alyssa,** *Deposition, stratigraphy, provenance, and reservoir characterization of carbonate mudstones; the Three Forks Formation, Williston Basin*

**Gartner, Joseph Eugene,** *Empirical models for post-fire hazard assessments; analysis of terrain, burn severity, rainfall, and soil influences on post-fire erosion by debris flow*

**Gordon, Gregory S.,** *Stratigraphic evolution and architectural analysis of structurally confined submarine fans; a tripartite outcrop-based study*

**Jin, Hui,** *Source rock potential of the Bakken Shales in the Williston Basin, North Dakota and Montana*

**Laugier, Fabien J.,** *Three-dimensional facies and process-regime variability in shelf-edge deltas; implications for shelf-margin progradation and deepwater sediment delivery*

**Mikkelson, Kristin,** *The impact of bark beetle infestation on water quality and hydrology in the Rocky Mountain West*

**Moody, Jeremiah D.,** *Variations in the architecture of fluvial deposits within a marginal marine setting, Eocene Sobrarbe and Escanilla Formations, Spain*

**Rydzy, Marisa B.,** *The effect of hydrate formation on the elastic properties of unconsolidated sediment*

**Stammer, Jane G.,** *Hydrodynamic fractionation of minerals and textures in submarine fans; quantitative analysis from outcrop, experimental, and subsurface studies*

**Tedesco, Steven A.,** *Reservoir characterization and geology of the coals and carbonaceous shales of the Cherokee Group in the Cherokee Basin, Kansas, Missouri and Oklahoma, U.S.A.*

**Theloy, Cosima,** *Integration of geological and technological factors influencing production in the Bakken Play, Williston Basin*

**Vargas-Johnson, Javier,** *Development of experimental methods for intermediate scale testing of deep geologic $CO_2$ sequestration trapping processes at ambient laboratory conditions*

### Masters

**Al Ibrahim, Mustafa Ali H.,** *Multi-scale sequence stratigraphy, cyclostratigraphy, and depositional environment of carbonate mudrocks in the Tuwaiq Mountain and Hanifa Formations, Saudi Arabia*

**Bazzell, Aaron,** *Origin of brecciated intervals and petrophysical analyses; the Three Forks Formation, Williston Basin, North Dakota, U.S.A.*

**Beck, Andrew J.,** *Evaluating best management practice scenarios in Ballona Creek watershed using EPA's SUSTAIN model*

**Beisman, James,** *Development of a parallel reactive transport model with spatially variable nitrate reduction in a floodplain aquifer*

**Bennett, Mitchell M.,** *Cathodoluminescence and fluid inclusion characteristics of hydrothermal quartz from porphyry deposits*

**Bircher, Matthew John,** *Depositional environment and reservoir characterization of the Paleocene Fort Union Formation, Washakie Basin, southwest Wyoming, U.S.A.*

**Carver, Franki J.,** *Diagenesis and dolomitization of the Ouray Formation, San Juan Mountains, Colorado*

**Coleman, Camriel A.,** *Refining magnetic amplitude methodology for use in the presence of remanent magnetization*

**Donovan, Ian P.,** *A probabilistic approach to post-wildfire debris-flow volume modeling*

**Dueñas S., Claudia,** *Understanding rock quality heterogeneity of Montney Shale reservoir, Pouce Coupe Field, Alberta, Canada*

**Eidsnes, Henriette Vilde Hitland,** *Structural and stratigraphic factors influencing hydrocarbon accumulations in the Bakken petroleum system in the Elm Coulee Field, Williston Basin, Montana*

**Finley, Elena,** *3-D seismic characterization of the Niobrara Formation, Silo Field, Laramie County, Wyoming*

**Geesaman, Patrick J.,** *Structural observations and stratigraphic variability in Jurassic strata, Upheaval Dome, Canyonlands National Park, Utah, USA*

**Gibson, Alexander T.,** *Paleoenvironmental analysis and reservoir characterization of the Late Cretaceous Eagle Ford Formation in Frio County, Texas, USA*

**Godinez, Lemuel J.,** *Important mineralogical factors for fluid saturation, specific surface area and pore size distributions based on gas adsorption, cation exchange capacity and 2D dielectric microscopy; a case study of quartz phase porcelanites in the Miocene Monterey Formation*

**Gross, Timothy G.,** *Controls and distribution of Cu-Au mineralization that developed the Island Mountain Deposit, Whistler Property, southcentral Alaska*

**Gutierrez, Claudia,** *Stratigraphy and petroleum potential of the upper Three Forks Formation, North Dakota, Williston Basin, USA*

**Hallau, Daniel Griffin,** *Facies and diagenesis of the Park City Formation; Sheep Mountain Anticline, Wind River Basin, Fremont County, Wyoming*

**Hanneman, Harry,** *Mineralogy and geochemistry of carbonaceous mudstone as a vector to ore; a case study at the Lagunas Norte high-sulfidation gold deposit, Peru*

**Herzog, Matthew T.,** *Pore pressure, and the interdependency between lithology, porosity, and acoustic log response targeting the Vaca Muerta Formation*

**Hollon, Zachary Grant,** *Elemental chemostratigraphy and reservoir properties of the Mowry Shale in the Bighorn and Powder River basins, Wyoming, USA*

**Hood, John Calvin,** *Acoustic monitoring of hydraulic stimulation in granites*

**Ittharat, Detchai,** *3D radio reflection imaging of asteroid interiors*
**Kernan, Henry E.,** *Electrofacies, elemental composition, and source rock characteristics along seismic reflectors of the Vaca Muerta Formation in the Loma La Lata area, Neuquen Basin, Argentina*
**Kocman, Katie Beth,** *Interpreting depositional and diagenetic trends in the Bakken Formation based on handheld X-ray fluorescence analysis, McLean, Dunn and Mountrail Counties, North Dakota*
**Kumar, Sanyog,** *Upper and lower Bakken Shale production contribution to the middle Bakken reservoir*
**Long, Jena Marie,** *Structural controls on roll-front mineralization at the Buss Pit Deposit, Gas Hills District, Wyoming*
**MacFarlane, Tyler L.,** *Amplitude inversion of fast and slow converted waves for fracture characterization of the Montney Formation in Pouce Coupe Field, Alberta, Canada*
**Matthies, Nicholas,** *Understanding and mapping variability of the Niobrara Formation across Wattenberg Field, Denver Basin*
**Oseguera, Olivia,** *The significance of magma mingling and mixing during the formation of the host-rock successions of Archean massive sulfide deposits in the Noranda Camp, Abitibi Subprovince, Quebec*
**Palmer, Justin C.,** *Structural geology and geochemistry of sedimentary rock-hosted gold in the eastern Nadaleen Trend, Yukon Territory, Canada*
**Pederson, Chris Andrew,** *Compensational behavior of three debris-flow fans in southern Colorado*
**Penn, Colin Andre Kress,** *Green to grey; numerical experiments to explain multi-scale hydrologic responses to mountain pine beetle tree mortality in a headwater basin*
**Pless, Claire,** *Implications of scaling relationship for relay ramps; Arches and Canyonlands National Parks, SE Utah*
**Poole, Sheven,** *Quantitative mineralogy and distributions of minerals of the Green River Formation, Piceance Creek basin, western Colorado*
**Pratt, Daniel R.,** *A landslide hazard rating system for Colorado highways*
**Prugue, Rodrigo,** *Identification of reducing conditions and correlated hydrological and biogeochemical properties in a heterogeneous floodplain aquifer*
**Rivera, Saul,** *Ultrasonic and low field nuclear magnetic resonance study of lower Monterey Formation; San Joaquin Basin*
**Smith, Colgan B.,** *Determining the source of spatially variable water chemistry in perennial tributaries in the Grand Canyon, Arizona, USA; influences from water-rock interaction and marine evaporite dissolution*
**Swierenga, Michael,** *Depositional history and lateral variability of microbial carbonates, Three Mile Canyon and Evacuation Creek, eastern Uinta Basin, Utah*
**Webber, Robert B.,** *Structural and stratigraphic controls on fracture distribution within sand bodies of the upper Iles and lower Williams Fork Formations, Mamm Creek and Divide Creek Fields, Piceance Basin, Colorado*
**Weigel, Jacob F.,** *Evaluation of thrusting and folding of the Deadman Creek thrust fault, Sangre de Cristo Range, Saguache County, Colorado*
**Wokasch, Travis,** *Elemental chemostratigraphy and depositional environment interpretation of the Eagle Ford Shale, south Texas*

## Columbia University

Doctorate

**Almeida, Rafael,** *Mechanisms and Magnitude of Cenozoic Crustal Extension in the Vicinity of Lake Mead, Nevada and the Beaver Dam Mountains, Utah*
**Foster, Anna,** *Surface-wave propagation and phase-velocity structure from observations on the USArray Transportable Array*
**Kelley, Colin,** *Recent and future drying of the Mediterranean region: anthropogenic forcing, natural variability and social impacts*
**Lloyd, Alexander,** *Timescales of magma ascent during explosive eruptions: Insights from the re-equilibration of magmatic volatiles*
**Nakamura, Jennifer Anne,** *Hydroclimatology of Extreme Precipitation and Floods Originating from the North Atlantic Ocean*
**Perez, Marc,** *A Model for Optimizing the Combination of Solar Electricity Generation, Supply Curtailment, Transmission and Storage*

## Cornell University

Doctorate

**Trostle, Kyle Daniel,** *Weathering and magnesium isotope fractionation in arid Hawaiian soils*

## Dalhousie University

Masters

**Cullen, Janette,** *Geophysical character of the Mohican Channel gas hydrate, Nova Scotia, Eastern Canada*
**Forde, Tanya C.,** *Coastal evolution over the past 3000 years at Conrads Beach, Nova Scotia, Canada; understanding the dynamics of transgressive coasts*

Bachelor

**Borg, Ian,** *The possible role of petroleum system in the metallogenesis of gold in Meguma metaturbidites; geochemical investigation in the Touquoy Deposit, Moose River, Nova Scotia*
**Campbell, Taylor,** *Seismic stratigraphy and attribute analysis of the Mesozoic and Cenozoic of the Penobscot area, offshore Nova Scotia*
**Ma, Svieda,** *Did the Adula Nappe (European basement) endure UHP metamorphism during the Alpine Orogeny? Insight form a (micro)structural and textural study*
**Mallyon, Deirdre,** *Tectonic and climatic controls on the growth and shape of the Himalayan foreland fold-and-thrust belt; a numerical study*
**Milligan, Rachel S.,** *Reaction of iron-titanium oxide minerals with kimberlite magma; a case study for Orapa kimberlite cluster*
**Minichiello, Jeff,** *Increasing concentrations in aluminum in southwest Nova Scotia from 1980 to 2011*
**Price, Colin Arthur,** *Facies of the vertebrate-bearing Scots Bay Member at Wasson Bluff, Nova Scotia*
**Saric, Adi,** *Stress changes in the Shillong Plateau and the cause of seismic gap in the eastern Himalaya*
**van Drecht, Leigh,** *Sedimentology and paleoenvironment of an Early Jurassic dinosaur bone bed, Wasson Bluff, Parrsboro, Nova Scotia*

## Dartmouth College

Doctorate

**Beal, Samual Adams,** *Determining anthropogenic and natural controls on the global mercury cycle using lake sediment and ice cores*
**Levy, Laura B.,** *Late glacial and holocene fluctuations of local glaciers and the Greenland Ice Sheet, eastern and western Greenland*

Masters

**Niu, Danielle,** *A relict sulfate-methane transition zone in the Mid-Devonian Marcellus Shale*
**Voorhis, James,** *A powder microelectrode study of arsenic redox behavior on pyrite in anoxic environments*

## Florida State University

Masters

**Sezen, Tugba,** *Oligocene-lower Miocene calcareous nannofossil biostratigraphy of ODP Leg 154 Hole 929A from the western Equatorial Atlantic at the Ceara Rise*

## George Mason University

Doctorate

**Slate, Robert,** *Environmental Protection Agency Regulation of Asbestos and Carbon Nanotubes Under the Toxic Substances Control Act: Investigating the Role of Politics, Science, and Policy in Administrative Rulemaking and Implementation*
**Wooten, Alexander,** *The Urban Soil Lead Predicament: The Applicaton of In Situ Fixation Technology as an Ecologically Sustainable Method of Lead Abatement in Urban Soils*

Masters

**Char, Chelia,** *How Does An Environmental Educator Address Student Engagement In A Meaningful Watershed Educational Experience (MWEE)?*

## Georgia State University

Masters

**Braun, Erick,** *Enhancing Georgia's Paleohurricane Record: A Comprehensive Analysis of Vibracores from St. Catherines Island*
**Heck, Sarah,** *Space, Politics and Occupy Wall Street*
**Hottenstein, Aaron,** *A Spatiotemporal And Geochemical Evaluation Of Groundwater Quality Adjacent To Natural Gas Drilling And Hydraulic Fracturing In Dimock Township, Susquehanna County, Pennsylvania*
**Jarrett, Robert E.,** *Authigenic Minerals: Locality 80, Bed I Tuffs, Olduvai Gorge, Tanzania*
**Kelly, Austin,** *GIS Least-Cost Route Modeling Of The Proposed Trans-Anatolian Pipeline In Western Turkey*
**Pickering, Rebecca,** *Tri-Octahedral Domains and Crystallinity in Synthetic Clays: Implications for Lacustrine Paleoenvironmental Reconstruction*
**Puckett, Mechelle,** *Mobility in the Neoliberal City: Atlanta's Left Behind Neighborhoods*
**Simpson, Simmone,** *The Paleoenvironment of the Lower Missis-*

*sippi River Delta During the Late Holocene*
**Skelton, Craig,** *The Implementation and Education of Geographic Information Systems in a Local Government for Municipal Planning: A Case Study of Dangriga, Belize*
**Zhang, Yingzhi,** *Understanding The Influence Of Participants' Preferences On The Affiliation Network Of Churches Using Agent-based Modeling*

## Illinois State University

Masters

**Hoffman, Lauren L.,** *Spatial variability of erosion patterns along the eastern margin of the Rio Grande Rift*
**Meyers, Matthew Douglas,** *The relationship between environmental factors and cyanobacteria population in Lake Bloomington and Evergreen Lake in McLean County, Illinois*
**Schroeder, Kathryn E.,** *Determining the source of anomalous segments in a karst stream*
**Seipel, Logan C.,** *Surficial geology and groundwater investigation of the Garden Prairie, IL 7.5 minute quadrangle*
**Theesfeld, Kristen L.,** *Nitrate uptake in central Illinois streams; a comparison along a transient storage gradient*

## Indiana State University

Masters

**Foxx, Heather A.,** *Using the spatial variability of lead in urban soils and demographic variables to predict exposure risks; an environmental justice analysis in Terre Haute, Indiana*
**Willingham, Jake,** *The ecology and morphology of deep-sea benthic Foraminifera; the Australian margin and epifaunal pore characteristics*

## Iowa State University of Science and Tech

Doctorate

**Fornadel, Andrew P.,** *Stable tellurium isotope variability in ore-forming systems; a theoretical and experimental approach*

Masters

**Bristol, Samantha Kay,** *Geochemical and geochronological constraints on the formation of shear-zone hosted Cu-Au-Bi-Te mineralization in the Stanos District, Chalkidiki, northern Greece*
**Day, Sarah Elizabeth,** *Assessing the style of advance and retreat of the Des Moines Lobe using LiDAR topographic data*

## Lakehead University

Masters

**Dasti, Ian Raymond,** *The geochemistry and petrogenesis of the Ni-Cu-PGE Shakespeare Deposit, Ontario, Canada*
**Kuzmich, Benjamin N.,** *Petrogenesis of the ferrogabbroic intrusions and associated Fe-Ti-V-P mineralization within the McFaulds greenstone belt, Superior Province, Canada*
**Reynolds, Dane,** *Progressive freeze concentration of naphthenic acids*
**Trevisan, Brent E.,** *The petrology, mineralization and regional context of the Thunder mafic to ultramafic intrusion, Midcontinent Rift, Thunder Bay, Ontario*

Bachelor

**Bjorkman, Ruth,** *The mapping and petrography of an Archean maar deposit*
**Dolega, Simon,** *Strain analysis on the Max Lake polymictic conglomerates in the Wabigoon Subprovince, Ontario, Canada*
**Goetz, Matthew M.,** *Heating Experiments of Amethyst from Thunder Bay Amethyst Mine*
**Kemper, George,** *Sedimentology of the Outan Island Formation*
**Liimu, Jared J.,** *The Role of Brittle-Ductile Deformation and Competency Contrast in Gold Mineralization in the C-Zone, Hemlo Gold Camp, Ontario*
**McIntyre, Tim,** *Sedimentology and Geochemistry of Mesoarchean Chemical Sediments of the Red Lake and Wallace Lake Greenstone Belts*
**Molloy, Shannon,** *Remote Sensing Techniques for the Mapping of Arc to Rift Transitional Rocks in Central Baja California*
**Nolan, Ainslee,** *Metamorphism and Deformation at the Wabigoon Quetico Subprovince Boundary in the Decourcey Lake Area*
**Pucci, Alexander,** *Microstructure of Steep Rock Carbonate Precipitates*
**Quinn, Jordan,** *Lithogeochemical and Petrological Analysis of a Mafic Metavolcanic Sequence South of Musselwhite Mine, North Caribou Greeenstone Belt*
**Scheffler, Kayla,** *The Effects of Whole-Tree Harvesting and Fire Disturbances on Carbon and Nitrogen Stores in a Dystric Brunisol*

## Loma Linda University

Masters

**Fleming, Monte A.,** *Sedimentology of marine vertebrate burial in the Miocene Pisco Fm., Peru*
**Stanton, Caleb,** *Correlation and paleoenvironments above west T9.3 tuff, Pisco Formation, Peru*

## Louisiana State University

Doctorate

**Killingsworth, Bryan Alan,** *Untangling Earth system responses recorded in sulfate's sulfur and oxygen isotopes at the dawn of multicellular life and today*

Masters

**Akyuz, Isil,** *Palynology of the Upper Cretaceous Ferron-Notom Sandstone, Utah*
**Sorey, Laura Carolyn,** *Paleoclimatology and paleotempestology study of Blue Hole, Lighthouse Reef, Belize through geochemical proxies*

## McMaster University

Doctorate

**Moumblow, Rebecca M.,** *Nd isotope mapping of crustal boundaries within the eastern Grenville and Makkovik Provinces, southern Labrador*
**Slomka, Jessica M.,** *Architectural element analysis of glaciated terrains*

Masters

**Bradford, Lauren,** *Investigation of microbial community response during oil sands reclamation via lipid and carbon isotope analyses*
**Clay, Samantha L.,** *Identifying the fate of petroleum hydrocarbons released into the environment and their potential biodegradation using stable carbon isotopes and microbial lipid analysis*
**Fraser, Stephanie,** *Evaluating the influence of vegetation on evapotranspiration from waste rock surfaces in the Elk Valley, British Columbia*
**Reid, Michelle,** *Biogeochemical zonation in an Athabasca Oil Sands composite tailings deposit undergoing reclamation wetland construction*
**Tang, Weigang,** *An automated toolkit for hyetograph-hydrograph analysis*
**Wyman, Jillian,** *Oxygen and boron isotope effects in synthetic calcite*

## Miami University (Ohio)

Doctorate

**Huang, Qiuyuan,** *Geomicrobial investigations on extreme environments; linking geochemistry to microbial ecology in terrestrial hot springs and saline lakes*
**Zhang, Jing,** *Interaction of methanogens with clay minerals, organic matter, and metals*

Masters

**Abbott, Elizabeth R.,** *Shallow seismicity patterns in the northwestern section of the Mexico subduction zone*
**Al-Qattan, Nasser M.,** *Interpretation of oxygen isotopic values ($\delta^{18}O$) of North American land snails*
**Catlett, Gentry A.,** *Pluvial deposits in Mudawwara, Jordan and their implications for Mediterranean and monsoonal precipitation in the Levant*
**Glasser, Paul Allen,** *Kinetics and mechanisms of $Cr^{6+}$ reduction by structural Fe(II) in clay minerals*
**Rasor, Bart A.,** *Data mining for tectonic tremor in the IRIS preprocessed quality analysis database*
**Skoumal, Robert J.,** *Optimizing multi-station earthquake template matching through re-examination of the Youngstown, Ohio sequence*
**Turk, Sezer,** *Seismic structure and tectonics of the Alasehir-Gediz Graben, western Turkey*

## Michigan Technological University

Doctorate

**Engelmann, Carol A.,** *Investigation of strategies to promote effective teacher professional development experiences in earth Science*
**Islam, Nayyer,** *Time Lapse Seismic Observations and Effects of Reservoir Compressibility at Teal South Oil Field*
**Kulakov, Evgeniy V.,** *Properties of the Proterozoic geomagnetic field and geological implications of paleomagnetic data from rocks of the North American Midcontinent Rift*
**Zheng, Jianqiu,** *Denitrification in Soils: From Genes to Environmental Outcomes*

Masters

**Gratton, Lorenzo,** *Static and dynamic stress change at 27 volcanoes of the Central American volcanic arc after the MW 7.6 Costa Rica earthquake of 5 September 2012*

**Guo, Qiang,** *Tuning, AVO, and flat-spot effects in a seismic analysis of North Sea Block F3*
**Hetherington, Rachel M.,** *Slope stability analysis of Mount Meager, south-western British Columbia, Canada*
**Hetland, Brianna R.,** *A Surface Displacement Analysis for Volcan Pacaya from October 2001 through March 2013 by Means of 3-D Modeling of Precise Position GPS Data*
**Lerner, Geoffrey A.,** *Comparison of non-heating paleointensity techniques using basalts from Lemptégy Volcano, France and synthetic magnetite-bearing samples*
**Lucas, Evan G.,** *Use of an Electrical Impedance Tomography Method to Detect and Track Fractures in a Gelatin Medium*
**McKenney, Erin Leigh,** *Exploring Changes in Detrital Flocculent Layer Dynamics Due to Shifts in Macrophyte Communities in the Northern Everglades*
**Russo, Elena,** *Evaluation of the Evolving Stress Field of the Yellowstone Volcanic Plateau from 1988 to 2010 from Earthquake First Motions Inversion*
**Veverica, Timothy J.,** *Ionic Liquid Extraction Unveils Previously Occluded Humicbound Iron in Peat Porewater*
**Wargelin, Justin B.,** *Fault Morphology Within The Southern Kenyan Portion of the East African Rift Valley*
**Wellik, John J., II,** *Doing more with short period data; determining magnitudes from clipped and over-run seismic data at Mount St. Helens*

## Mississippi State University

Doctorate

**Jones, John Paul,** *An investigation of geochemical evidence for three paleo-environments*
**Kambesis, Patricia Nancy,** *Influence of coastal processes on speleogenesis and landforms in the Caribbean region*
**Larson, Erik Bond,** *Bahamian Quaternary geology and the global carbon budget*
**Lenz, Richard Jason,** *Petroleum releases from underground storage tanks in northwest Indiana; successful remediation techniques and implications of cost effectiveness*
**Nagel, Athena Marie Owen,** *Investigation of macro and micro scale void spaces; preservation, modeling and biofilm interactions*
**Sumrall, Jeanne Lambert,** *Investigation of sense of place effects in an online learning environment*

Masters

**Banks, John Vernon,** *Water supply potential of the Wilcox aquifers in Lafayette County, Mississippi*
**Breland, Benjamin,** *The updip limit of the Smackover Formation in Clarke County, Mississipppi*
**Brooke, James Michael,** *Geologic analysis of the Upper Jurassic Cotton Valley Formation in Jefferson County, Mississippi*
**Geroux, Jonathon Michael,** *Diffusive gradients in thin film (DGT); a proposed method to find geochemical predictors of sediment oxygen demand*
**Lawrence, Orry Patrick,** *Predicting flank margin cave collapse in the Bahamas*
**Martin, Seth M.,** *Geologic controls of sand boil formation at Buck Chute, Mississippi*
**Ridlen, Nicole Marie,** *Speleothem strontium concentrations in eogenetic carbonates*
**Strange, Ryan Charles,** *The influences of geologic depositional environments on sand boil development, Tara Wildlife Lodge area in Mississippi*
**Tetteh, Lucy Korlekwor,** *A multi-decadal remote sensing study on glacial change in the North Patagonia ice field Chile*
**Thompson, David Luke,** *Stratigraphy, environments of deposition, and mineralogical characterization of heavy minerals from selected Cretaceous formations of the eastern Mississippi Embayment*
**Travis, Ryan William,** *Evaluation and quantification of modern karst features as proxies for paleokarst reservoirs*

## Missouri University of Science and Tech

Doctorate

**Almansour, Abdullah Owidah,** *Investigating and modeling tarmats in a Kuwaiti carbonate reservoir and their role in understanding oil reserves and recovery economics*
**Purevsuren, Uranbaigal,** *Crustal thickness and Vp/Vs beneath the Western United States; constraints from stacking of receiver functions*

Masters

**Al-Alwani, Mustafa Adil,** *Data mining and statistical analysis of completions in the Canadian Montney Formation*
**Chauhan, Pratap D.,** *Data analysis and summary for surfactant-polymer flooding based on oil field projects and laboratory data*
**Cheng, Shengyao,** *PN anisotropy tomography of the Colorado Plateau and adjacent areas*
**Duru, Dennis Chieze,** *Effects of grain properties and compaction on single-tool normal indentation of granular materials*
**Kaba, Azupuri Ayerikujei,** *Effects of indentation speed and water saturation level on the behavior of Roubidoux Sandstone*
**Li, Mengke,** *Influence of erosional unloading on the state of stress for low amplitude single-layer bucklefolds; implications for tensile fracture occurrence*
**Ray, Melissa Ann,** *Shear-wave splitting and mantle deformation beneath the western Tibetan Plateau*
**Wu, Si,** *A receiver function study of crust thickness and composition beneath the Southwestern United States*
**Xue, Liang,** *A remote sensing investigation of elevated sub-horizontal topographic surfaces in the Wichita Mountains, Oklahoma*

## Montana Tech of the University of Montana

Masters

**Mitchell, Katie,** *Groundwater and surface water interactions at Georgetown Lake, Montana with emphasis on quantification of groundwater contribution*
**Vignali, Deven,** *Seismic numerical modeling of geothermal resources*

## Northwestern University

Doctorate

**Baczynski, Allison Ann,** *Evaluating Carbon Cycle Dynamics and Hydrologic Change during the Paleocene-Eocene Thermal Maximum, Bighorn Basin, Wyoming*
**Chang, Yun-Yuan,** *Influence of point defects on the elastic properties of mantle minerals and superhard materials*
**Gomes, Maya Lorraine,** *Modern and Ancient Studies of Sulfur Isotope Cycling in Low-Sulfate Systems*
**Li, Dan Darcy,** *Global Biogeochemical Cycle of Silicon And Silicate Weathering during Soil Development*
**Merino, Miguel,** *Tectonics and Seismicity of Rifts Past and Present*

## Ohio State University

Doctorate

**Bancroft, Alyssa Marie,** *Silurian and Ordovician conodont biostratigraphy of the Moose River Basin and Appalachian Basin*
**Edwards, Cole T.,** *Carbon, sulfur, and strontium isotope stratigraphy of the Lower- Middle Ordovician, Great Basin, USA: implications for oxygenation and causes of global biodiversification*
**Huh, Kyung In,** *Glacier volume changes in the Tropical Andes: a multi-scale assessment in the Cordillera Blanca, Peruvian Andes*
**La Frenierre, Jeff David,** *Assessing the hydrologic implications of glacier recession and the potential for water resources vulnerability at Volcan Chimborazo, Ecuador*
**Lazar, Kelly Best,** *Benthic foraminifera as paleo-sea-ice indicators in the western Arctic Ocean*

Masters

**Fair, Alexandria Corinne,** *Elemental cycling in a flow-through lake in the McMurdo Dry Valleys, Antarctica: Lake Miers*
**Holsinger, John F.,** *The impact of SMCRA on select soil properties in reclaimed mine sites determined by geochemical and hydrological analyses*
**Hwang, Bohyun,** *Water-rock interaction in the Coso geothermal area*
**Imbrogno, David F.,** *Analysis of dam failures and development of a dam safety evaluation program*
**Philippoff, Karl Steven,** *An investigation into the causes of d18O variations in the Dasuopu ice core, central Himalayas, using coral composites and instrumental data*
**Umholtz, Nicholas Moehle,** *Middle to late Ordovician $\delta^{13}C$ and $^{87}Sr/^{86}Sr$ stratigraphy in Virginia and West Virginia: implications for the timing of the Knox unconformity*
**Wehmann, Adam,** *A spatial-temporal contextual kernel method for generating high- quality land-cover time series*
**Yencho, Nathan Andrew,** *Investigation of dynamic liquefaction potential of impounded Class F fly ash*

Bachelor

**Barr, Daniel J.,** *A morphological investigation of submarine volcanic cones and their relation to crustal stress, Adare Basin, Antarctica*
**Blau, Evan S.,** *Effects of sub-surface structures and compositions on contamination in groundwater aquifers in the Midwest*
**Cappell, Marc William,** *An examination of changes within the active zone moisture content and soil swell potential of expansive clay*

soils at a site in Denton County, Texas

**Cowan, Zachariah M.,** *Development of ion milling methods for SEM imaging of the Utica Shale*

**Elston, Harold W.,** *Mineralogical and geochemical assessment of the Eagle Ford Shale*

**Enriquez, Daniel A.,** *Modifying the McKenzie stretching theory for sedimentary basins to account for the depth dependence of sediment density*

**Gomes, Christian,** *How do active sedimentary basins dewater as they deepen and compact? A hypothesis in which pervasive crustal fracturing provides persistent channels for vertical fluid flow*

**Hibbard, Shannon Maria,** *Controls on gravel composition in a proglacial environment, Kaunertal, Austria*

**Koons, Rachel Claudia,** *A study of cut marks on the Orleton Mastodon and the potential implications of anthropogenic modification*

**Lewis, Natasha,** *Coupling of carbonate associated sulfate and carbon isotope trends in Middle Ordovician strata from Meiklejohn Peak, Nevada; implications for atmospheric $O_2$*

**Phetteplace, Thomas,** *Examining the geological potential for helium production in the United States*

**Thompson, Andrew,** *Determining the depths of magma chambers beneath Hawaiian volcanoes using petrological methods*

**Whyte, Colin,** *The mid- to late Pleistocene ice rafted debris record at IODP Site 1308, central North Atlantic*

**Young, Julia,** *The role of bedrock geology in the aqueous chemistry of Cantabrian rivers*

## Old Dominion University

Masters

**Ames, Katherine,** *Crustal accretion and mantle geodynamics at microplates; constraints from gravity analysis and numerical modeling*

## Oregon State University

Doctorate

**Burns, Dale H.,** *Crustal architecture and magma dynamics in a large continental magmatic system; a case study of the Purico-Chascon volcanic complex, northern Chile*

**Grocke, Stephanie B.,** *Magma dynamics and evolution in continental arcs; insights from the Central Andes*

**Kaiser, Jason Frederick,** *Understanding large resurgent calderas and associated magma systems; the Pastos Grandes Caldera Complex, southwest Bolivia*

**Monigle, Patrick W.,** *Seismic wave attenuation of the crust and upper mantle in the Himalaya and south-central Tibetan Plateau*

**Praetorius, Summer Kate,** *Abrupt deglacial climate changes in the North Pacific and implications for climate tipping points*

**Rosen, Julia L.,** *Augmenting and interpreting ice core greenhouse gas records*

**Strano, Sarah Elianna,** *Deep-sea sediment paleomagnetism; a case study from the North Atlantic*

Masters

**Black, Bran,** *Stratigraphic correlation of seismoturbidites and the integration of sediment cores with 3.5 kHz CHIRP subbottom data in southern Cascadia*

**Cuzzone, Joshua,** *An interdisciplinary approach towards understanding late Pleistocene ice sheet change*

**Gavillot, Yann G.,** *Active tectonics of the Kashmir Himalaya (NW India) and earthquake potential on folds, out-of-sequence thrusts, and duplexes*

**Krawl, Kyle A.,** *A petrogenetic model for the Caribbean large igneous province*

**Marston, Brooke E.,** *Improving the representation of large landforms in analytical relief shading*

**Preppernau, Charles A.,** *3D vs. conventional volcanic hazard maps; a user study at Mount Hood*

## Pennsylvania State University at University Park

Doctorate

**Cruz-Uribe, Alicia,** *Assessment and implications of (dis)equilibrium in metamorphic rocks*

**Cui, Ying,** *Assessment of climate change and global carbon cycle perturbation during the end-Permian mass extinction*

**Dere, Ashlee,** *Shale weathering across a latitudinal climosequence*

**Fegyveresi, John M.,** *Physical Properites of the West Antarctic Ice Sheet (WAIS) Divide Deep Core: Development, Evolution, and Interpretation*

**Geirsson, Halldor,** *Crustal deformation and volcanism at active plate boundaries*

**Martino, Amanda,** *Assessment and advancement of approaches for the study of subseafloor microbial communities*

**Peterson, Kristina,** *Nucleation, growth, and phase transformation mechanisms of the iron (oxy)hydroxides*

**Ramirez, Ramses M. ,** *Terrestrial planets under extreme radiative forcings: applications to habitable zones, early Mars, and a high-CO? Earth*

**Schueth, Jonathan D.,** *Investigating the response of calcareous nannoplankton to significant environmental disturbance: links between extinction and environmental change and the role of incumbency in recovery*

**Scuderi, Marco,** *Mechanical properties of the seismogenic zone*

**Shi, Xuhua,** *Deformation of lacustrine shorelines in central Tibet: Implications for lake level history, fault kinematics, and crustal rheology*

**Smith, Karen Elizabeth,** *Exogenous and endogenous sources of organic compounds on the early Earth: investigating carbonaceous meteorites and plausibly prebiotic complex mixtures by liquid chromatography-mass spectrometry*

**Tollerud, Heather J. ,** *The role of water in the development of surface roughness and mineralogical variability on playa surface sediments: implications for aeolian erodibility and dust emission*

**West, Nicole,** *Topographic fingerprints of hillslope erosion in the central Appalachians as revealed by meteoric 10be: Toward understanding the evolution of critical zone architecture*

**Yesavage, Tiffany A. ,** *Chemical and physical weathering in regolith: an investigation of three different Fe-rich sites of varying climate and lithology*

Masters

**Cederberg, James,** *A Quantitative Assessment of the Effects of Base Level Fall and basin Depth on River-dominated Deltas*

**Chorney, Andrew,** *mechanisms of sulfur isotope fractionation during thermochemical sulfate reduction by amino acids: implications for the Mif-s record in archean rocks*

**Chung, Angela,** *Evaluation of Variability in Soil Organic Carbon Isotope Record and Implications for Organic Matter Preservation*

**Culp, Brian M. ,** *Impact of CO2 on fracture complexity when used as a fracture fluid in rock*

**Doman, Christine,** *Stable isotope and lipid signatures of plants across a climate gradient: implications for biomarker-based paleoclimate reconstructions*

**Ethier Colon, William,** *In situ pE-pH analysis on the oxidation kinetics of manganese bearing solution*

**Gonzales, Matthew,** *In-situ observation of calcium carbonate phase heterogeneity during mineral precipitation: Implications for the interpretation of calcium carbonate based proxies*

**Grieve, Paul,** *Measuring concentrations of natural gas in three streams in Pennsylvania to estimate methane fluxes from the subsurface*

**Johnston, Thomas,** *The role of poroelasticity during disequilibrium compaction and hydrocarbon generation on horizontal stress in the Devonian section of the Appalachian Basin*

**Kachingwe, Marsella,** *Crustal structure in eastern and southern Africa with implications for secular variation in Precambrian crustal genesis*

**Meghani, Nooreen,** *Uplift and shortening in the Nepalese and Indian sub-Himalaya determined with quantitative geomorphology*

**Merkhofer, Lisa,** *Australian analogs for an Eocene Patagonian Paleorainforest*

**Meyers, Beth,** *Analyzing the state of lithospheric stress in greater thailand through finite element modeling*

**Modlich, Mitchell,** *Thermal Maturity of Gas Shales in the Appalachian Plateau of Pennsylvania*

**Orlando, Joseph,** *The anatomy of weathering profiles on different lithologies in the tropical forest of northeastern Puerto Rico: from bedrock to clouds*

**Tsao, Leah E.,** *Culture-dependent and independent studies of sulfur oxidizing bacteria from the Frasassi Caves*

**Wang, Jiuyuan,** *The origin of a basin-scale thin limestone: the Middle Devonian Cherry Valley Member, Marcellus Formation*

**Wysocki, Nathaniel,** *Fine Sediment Deposition and Storage in Sandy-Fluvial Systems*

## Princeton University

Bachelor

**Beale, Andrea,** *Fluid inclusions in marine halite as a window into the Mg isotopic composition of past oceans*

**Bluher, Sarah,** *An integrated chemostratigraphic approach to understanding the Siluro-Devonian positive carbon isotope excursion*

**Shepherd, Robert,** *Diagenesis in the Great Bahamas Bank; an*

*analysis of Mg isotopes in the sediment cores from ODP Site 1003*
**Sutjiawan, Djohan,** *New constraints on the size of Io's core*

## Purdue University

Doctorate

**Johnson, Alexandria V.,** *The formation of ice in maritime cumuli: Insights from new observations and modeling*
**Zhu, Qing,** *Improving the quantification of terrestrial ecosystem carbon budget with models of biogeochemistry and atmospheric transport and chemistry using in Situ and satellite observational data*
**Zhu, Xudong,** *Quantifying exchanges of methane and carbon dioxide between terrestrial ecosystems and the atmosphere in the northern high latitudes*

Masters

**Pope, Ian C.,** *Deforestation of cloud forest in the central highlands of Guatemala: soil erosion and sustainability implications for Q'Eqchi'Maya communities*
**Robertson, Peter B.,** *Part I: Neoacadian to Alleghanian foreland basin development and provenance in the central Appalachian orogen, pine mountain thrust sheet, Part II: Structural configuration of a modified mesozoic to cenozoic forearc basin system, South-central Alaska*

## Rensselaer Polytechnic Institute

Doctorate

**Borrelli, Chiara,** *Reconstructing past ocean circulation and chemistry using benthic Foraminifera; traditional and novel approaches*

Masters

**Rodzianko, Anastasia,** *Imaging the subsurface of Taiwan using ambient noise tomography and full waveform inversion*

## San Diego State University

Masters

**Alexander, Paul,** *Amplification of long period ground motion by the Los Angeles Basin*
**Bhakta, Shardul Sanjay,** *Rapid decision tool to predict earthquake destruction in Sumatra by using First Motion*
**Boppidi, Ravikanth Reddy,** *A mobile tool about causes and distribution of dramatic natural phenomena*
**Carrasco, Taylor Lang,** *Identifying a reference frame for calculating mass change during weathering: a case study utilizing the C# program assessing element immobility*
**Chatterjee, Moumita,** *GIS learning tool for world's largest earthquakes and their causes*
**Fisch, Gregory Zane,** *Arc-rift transition volcanism in the Volcanic Hills, Jacumba and Coyote Mountains, San Diego and Imperial Counties, California*
**Khaire, Atul Sudhakar,** *Robust CSV to Shapefile utility and DBF file interpreter*
**Taniguchi, Kristine Teru,** *Regional impacts of urbanization on stream channel geometry: importance of watershed area and channel particle size*

## San Jose State University

Masters

**Clay, Pamela Jamie,** *Structure, construction, and emplacement of Jurassic and Cretaceous plutons in the Keiths Dome-Echo Lake area, southwest of Lake Tahoe, California*
**Jensen, Kaai,** *The role of lithology in glacial valley cross-sectional shape in Sierra Nevada, California*
**Johnson, Brent Michael,** *Thermal and compositional evolution of the mid-Miocene Searchlight magmatic system (Nevada, USA) as recorded in zircon*
**Jurek, Anne C.,** *Vulnerability of groundwater to perchloroethylene contamination from dry cleaners in the Niles Cone groundwater basin, southern Alameda County, California*
**Kuoch, Alan,** *Examination of multiparticle transport as a function of slope and sediment volume*
**Lidzbarski, Marsha Izabella,** *U-Pb geochronology of the Miocene Peach Spring Tuff supereruption and precursor Cook Canyon Tuff, western Arizona, USA*
**Qin, Zhengzheng,** *An unsaturated zone flux study in a highly-fractured bedrock area; ground water recharge processes at the Masser recharge site, east-central Pennsylvania*
**Short, Lauren Elizabeth,** *The role of large woody debris in inhibiting the dispersion of a post-fire sediment pulse*

## Simon Fraser University

Doctorate

**Holding, Shannon Tyler,** *Risk to water security on small islands; a numerical modeling approach*
**Turner, Derek Glen,** *Pleistocene stratigraphy, glacial limits and paleoenvironments of White River and Silver Creek, southwest Yukon*

Masters

**Ertolahti, Leila,** *Contemporary subsidence and settlement of the Fraser River delta inferred from SqueeSAR(TM)-type InSAR data*
**Foster, Simon,** *Characterizing groundwater-surface water interactions within a mountain to ocean watershed, Lake Cowichan, British Columbia*
**Ricci, Liam,** *Architecture and facies analysis of Allomember F, Upper Cretaceous Horseshoe Canyon Formation, Drumheller, Alberta*

## Southern Illinois University at Carbondale

Masters

**Chartrand, Zachary,** *Partial melting experiments on an Mg # 80 Martian mantle and their implications for basalt genesis*
**Heij, Gerhard W.,** *Cataclastic flow kinematics inferred from magnetic fabrics at the Heart Mountain Detachment, Wyoming*
**Jarvis, Stephanie K.,** *Thecamoebians as an environmental proxy for the Middle Mississippi River floodplain*
**Payne, Caldwell,** *Alteration, mineralization, and stable isotope geochemistry of the Wind Mountain and Willard-Colado epithermal deposits, Nevada: implications for amagmatic hydrothermal activity*
**Tarlow, Scott,** *Three dimensional modeling of mantle melt underneath the Lau Back-Arc spreading center and Tofua volcanic arc*

## Southern Methodist University

Masters

**Kweik, Ramsey Sharif,** *Thermal and mass history of Fairway Field in East Texas; implication for geothermal energy development in an oil and gas setting*
**Zhu, Lu,** *Early Permian paleosol morphologies and paleoatmospheric $pCO_2$ estimates, north-central Texas, U.S.A.*

## Stanford University

Masters

**Alkawai, Wisam,** *Integrating basin modeling with seismic technology and rock physics*
**Bandyopadhyay, Parag,** *Robust nonlinear regression for parameter estimation in pressure transient analysis*
**Iskhakov, Ruslan,** *High-resolution numerical simulation of $CO_2$ sequestration in saline aquifers*
**Jiang, Rui,** *Pressure preconditioning using proper orthogonal decomposition*
**Lagasca, Joseph Paul,** *Simulation of large-scale $CO_2$ plumes in deep saline aquifer*
**Li, Peipei,** *Sensitivity analysis of pre-stack seismic inversion on facies classification using statistical rock physics*
**Lowry, Brent,** *Direct conditioning of upscaled reservoir models to fine scale well data using direct sampling*
**Mabunda, Ha-andza,** *Utilization of temperature data to determine reservoir thickness*
**Moriarty, Dylan,** *Rapid surface detection of $CO_2$ leaks from geologic sequestration sites*
**Smith, Stuart Sweeney,** *Exergetic life cycle assessment of carbon capture and sequestration technology*
**Tan, Xiaojin,** *Comparing training-image based algorithms using an analysis of distance*
**Tian, Chuan,** *Applying machine learning and data mining techniques to interpret flow rate, pressure and temperature data from permanent downhole gauges*
**Tran Minh Tuan,** *Formation evaluation of an unconventional shale reservoir; application to the North Slope Alaska*
**Vogel, Steven E.,** *Analysis of a mass-based compositional multiscale formulation*
**Wang, Beibei,** *The role of kerogen versus clay in the adsorption mechanisms of $CO_2$ and $CH_4$ in gas shales*
**Yun, Wonjin,** *Micro visual investigation of polymer retention in a micromodel*
**Zahasky, Christopher,** *Quantification and intervention strategies for potential leakage from carbon storage reservoirs*
**Zentner, Danielle Bridgette,** *Stratigraphy, sedimentology and provenance of the ca. 3.26 Ga Mapepe Formation in the Manzimnyama Syncline, Barberton greenstone belt, South Africa*

## Stephen F. Austin State University

Masters

**Diaz-Garcia, Nancy,** *Stratigraphy, sedimentology, geochemical analysis and hydrocarbon potential of the Woodford Shale, Central Basin Platform, West Texas*

**Kaufmann, Emily Luise,** *The paleodepositional environment and stratigraphy of the Upper Cretaceous Ozan Formation in Collin County, Texas U.S.A.*

## SUNY, Binghamton University

Masters

**Terry, Lee Raymond,** *Autotrophic oxidation of antimony (III) by bacteria isolated from contaminated mine sediments*

## SUNY, University at Buffalo

Doctorate

**Kelley, Samuel Edward,** *Holocene ice margin fluctuations of the Greenland Ice Sheet in the Disko Bugt region, West Greenland*

Masters

**Ceperley, Elizabeth Grayce,** *Reconstructing Late Pleistocene and Holocene glacier fluctuations using cosmogenic 10Be exposure dating and lacustrine sediment, Brooks Range, Arctic Alaska*

**Ciruzzi, Dominick Michael,** *Quantifying Seasonal Volume of Groundwater in High Elevation Meadows: Implications for Complex Aquifer Geometry in a Changing Climate*

**Cronauer, Sandra L.,** *Reconstructing ice sheet and alpine glacier margins during the Early Holocene on Nuussuaq in central West Greenland*

**Harp, Andrew Gary,** *Shallow Plumbing Geometry and Eruptive Processes of a Monogenetic Volcano, Lunar Crater Volcanic Field, Nevada*

## Texas Christian University

Masters

**Adams, Ashley Lynn,** *Stratigraphy and paleontology of Upper Cretaceous to Paleocene strata on the western Rosillos Mountain Ranch, Brewster County, Texas*

**Allen, Sarah Dawn,** *Reverse meanders, pseudo point bars, and the enigma of meandering in braided rivers*

**Ashley, Tyler Lee,** *Lithology, diagenesis, and depositional environment of the (Mississippian) Barnett Shale and Limestone in the Dangelmayr A #7 core in the Fort Worth Basin, Cooke County, Texas*

**Dizayee, Rubar,** *Groundwater degradation and sustainability of the Erbil Basin, Erbil, Kurdistan region, Iraq*

**Finical, Rene Gabrielle,** *Examination of selected dolomite horizons in the Lower Ordovician Kindblade Formation, Blue Creek Canyon, Slick Hills, S.W. Oklahoma*

**Hill, Jennifer Anne,** *Complex Eocene-Oligocene hypabyssal intrusive systems associated with basaltic phreatomagmatic vents in the area east of Peña Mountain, Big Bend National Park, West Texas*

**Huling, Galen Alden,** *Evidence for clustering of delta-lobe reservoirs within fluvio-lacustrine systems, Jurassic Kayenta Formation, Utah*

**Lowry, Garrett W.,** *Assessment of hydrocarbon potential in the lower Eagle Ford Shale, Madison County, Texas*

## Université Laval

Doctorate

**Cloutier, Catherine,** *évaluation du comportement cinétique et du risque associé aux glissements de terrain rocheux actifs à l'aide de mesures de surveillance; le cas du glissement de Gascons, Gaspésie, Canada*

**Cochand, Fabien,** *Impact des changements climatiques et du développement urbain sur les ressources en eaux du bassin versant de la rivière Saint-Charles*

Masters

**Fabien-Ouellet, Gabriel,** *Mesures sismiques à faible profondeur; une approche intégrée*

**Lalonde, Erik,** *Alteration and Cu-Zn mineralization of the Turgeon volcanogenic massive sulfide deposit (New Brunswick, Canada)*

**Lesbros-Piat-Desvial, Marion,** *Hydrothermal alteration and uranium mineralization at the Camie River Prospect (Otish Basin, Québec)*

## University of Akron

Masters

**Ampomah, Richard Owusu,** *Sediment harvesting, beneficial use and the impact of climate and land-use/land-cover change on sediment load*

**Anderson, Michael D.,** *Analysis of upper mantle reflections beneath the Trans-Uralian and East-Uralian zones of the Ural Mountains, Russia*

**Filiano, Gina L.,** *Role of joints and rock stresses in the formation of sandstone caves in northeastern Ohio*

**Fretz, Chrystal E.,** *Development of a system to evaluate geochemical gradients in acid min drainage induced sediments in a column format*

**Haake, Zachary J.,** *Biogeochemical gradients within an acid mine drainage-derived iron mound, North Lima, Ohio / Zachary J Haake*

**Pacanovsky, Aaron J.,** *Petrology of gold ore-bearing carbonates of the Helen Zone, Cove deposit, Lander County, Nevada / Aaron J Pacanovsky*

**Scaggs, Laura M.,** *A geophysical study of subsurface paleokarst features and voids at Ohio Caverns, Champaign County, Ohio*

**Schnell, Andrew J.,** *Petrology of hydrothermal zebra dolomite at the Cove mine, McCoy mining district, Northern Fish Creek Mountains, Lander County, Nevada*

**Zoller, Kevin M.,** *Porphyritic intrusions of the Helen Zone in the Cove deposit, Lander County, Nevada*

## University of Alabama at Tuscaloosa

Doctorate

**Khanal, Subodha,** *Structural and kinematic evolution of the Himalayan thrust belt, central Nepal*

**Mandal, Subhadip,** *Structural, kinematic and geochronologic evolution of the Himalayan fold-thrust belt in Kumaun, Uttaranchal, northwest India*

Masters

**Aldridge, David E.,** *Hydroclimate time-series archived in a 4300 year old stalagmite from DeSoto Caverns (Alabama, USA)*

**Alesce, Meghan Elizabeth,** *Geomorphological relationships through the use of 2-D seismic reflection data, Lidar, and aerial imagery*

**Black, Heather Dawn,** *$\delta^{15}N$ in mollusk shells as a potential paleoenvironmental proxy for nitrogen loading in Chesapeake Bay*

**Cato, Craig Lee,** *Seismic interpretation and structural analysis of the Alleghanian fold-thrust belt in central Alabama*

**Gregg, Andrea Christine,** *Tectonic evolution of the west Florida Basin, eastern Gulf of Mexico*

**Hunter, Ian,** *Origin and development of the Apalachicola Basin*

**Kenyon, Lindsey Metcalf,** *Determining crustal thickness beneath the Transantarctic Mountains and the Wilkes subglacial basin using S-wave receiver functions*

**Legg, Joel Arthur,** *Assessment of the petroleum generation potential of the Neal Shale in the Black Warrior Basin, Alabama*

**Li, Xiaping,** *High resolution molecular characterization of photochemical and microbial transformation of dissolved organic matter in temperate streams of different watershed land use*

**Nicosia, Alexandra Rose,** *Effects of the 1982-1983 El Niño mega event on bivalve mollusk biomineralization*

**Slavic, David R.,** *Enhanced-solubilization of multicomponent dense immiscible liquid in homogeneous porous media*

**Wei, Jiexin,** *Sclerochronological and geochemical study of modern and ancient Semele corrugata from north coastal Peru*

**Yildirim, Aycan,** *Subsurface fracture analysis using FMI logs; implications for regional state of stress prediction in the Black Warrior Basin, Alabama*

## University of Alaska at Fairbanks

Doctorate

**Gong, Wenyu,** *Long-term monitoring of geodynamic surface deformation using SAR interferometry*

**Mori, Hirotsugu,** *Osteology, relationships and paleoecology of a new arctic hadrosaurid (Dinosauria, Ornithopoda) from the Prince Creek Formation of northern Alaska*

**Salazar Jaramillo, Susana,** *Paleoclimate and paleoenvironment of the Prince Creek and Cantwell Formations, Alaska; terrestrial evidence of middle Maastrichtian greenhouse event*

**Yoon, Seok J.,** *Environment, health and safety management in mining and other industries*

Masters

**Bolz, Patrizia,** *Development of a methodology for the characterization of mafic rocks with respect to their use for mineral carbonation; the mineralogy, petrology, and geochemistry of the Portage Lake Volcanics in the Keweenaw Peninsula, Michigan*

**Broman, Bonnie Nell,** *Metamorphism and element redistribution; investigations of Ag-bearing and associated minerals in the Arctic volcanogenic massive sulfide deposit, southwest Brooks Range, northwest Alaska*

**Bruton, Christopher Patrick,** *Automatic classification of volcanic earthquakes using multi-station waveforms and dynamic neural networks*

**Frohman, Rachel A.,** *Identification and evolution of tectonic faults in the Greater Fairbanks area, Alaska*

**Hutton, Eric M.,** *Surface to subsurface correlation of the Shublik Formation; implications for Triassic paleoceanography and source rock accumulation*

**Miller, Ulrika,** *The effect of topography on the seismic wavefield*
**Rahilly, Kristen E.,** *Mapping methods and observations of surficial snow/ice cover at Redoubt and Pavlof Volcanoes, Alaska using optical satellite imagery*
**Smith, Jacquelyn R.,** *Patterns and potential solutions to coastal geohazards at Golovin, Alaska*
**Wentz, Raelene,** *Fracture characteristics and distribution in Cretaceous rocks near the Umiat Anticline, North Slope of Alaska*
**Westbrook, Rachel E.,** *Evidence for a glacial refugium in south-central Beringia using modern analogs; a 152.2 kyr palynological record from IODP Expedition 323 sediment*
**Worden, Anna K.,** *Monitoring small scale explosive activity as a precursor to periods of heightened volcanic unrest*

## University of Arizona

Doctorate

**Edge, Russ,** *Rifting of the Guinea margin in the Equatorial Atlantic from 112 to 84 Ma; implications of paleo-reconstructions for structure and sea-surface circulation*
**Fay, Hannah Isabel,** *Studies of copper-cobalt mineralization at Tenke-Fungurume, Central African Copperbelt; and developments in geology between 1550 and 1750 A.D.*
**Girardi, James Daniel,** *Comparison of Mesozoic magmatic evolution and iron oxide (-copper-gold) ("IOCG") mineralization, Central Andes and western North America*
**Holland, Austin Adams,** *Imaging time dependent crustal deformation using GPS geodesy and induced seismicity, stress and optimal fault orientations in the North American Mid-continent*
**Ibanez-Mejia, Mauricio,** *Timing and rates of Precambrian crustal genesis and deformation in northern South America*
**Orem, Caitlin Anne,** *The frequency and magnitude of flood discharges and post-wildfire erosion in the Southwestern U.S.*
**Pepper, Martin Bailey,** *Magmatic history and crustal genesis of South America; constraints from U-Pb ages and Hf isotopes of detrital zircons in modern rivers*
**Routson, Cody Craig,** *The context of megadrought; multiproxy paleoenvironmental perspectives from the south San Juan Mountains, Colorado*
**Spinler, Joshua C.,** *Investigating crustal deformation associated with the North America-Pacific Plate boundary in Southern California with GPS geodesy*

Masters

**Cross, Edward A., III,** *The structure, stratigraphy, and evolution of the Lesser Himalaya of central Nepal*
**Dominguez, Ada R.,** *Significance of climate and magmatism on copper-bearing deposits and implications for ore-forming processes*
**Harris-Park, Erin,** *The micromorphology of Younger Dryas-aged black mats from Nevada, Arizona, Texas and New Mexico*
**Hoge, Aryn Kinley,** *The Jackson-Lawton-Bowman normal fault system and its relationship to carlin-type gold mineralization, Eureka District, Nevada*
**Melgar Pauca, Mauro Joel,** *Epithermal mineralization associated with diatreme breccia and rhyolitic dome, La Miel, Haiti*
**Richardson, Carson A.,** *Reconstruction of normal fault blocks in the Ann-Mason and Blue Hill areas, Yerington District, Nevada*

## University of Arkansas at Fayetteville

Masters

**Abbott, Shailyn,** *Structure and stratigraphy of a complex anticlinal feature, Backbone Anticline, Arkoma Basin, Arkansas*
**Anderson, Paula,** *Comparisons of hydrogeologic modeling methods to define capture zones for public water supply wells in northern Arkansas*
**Benson, Richard Craddock,** *3D seismic mapping of probable tripolitic chert bodies in Osage County, Oklahoma*
**Buratowski, Greg,** *Application of detrital zircon geochronology to determine the sedimentary provenance of the middle Bloyd Sandstone, Arkoma Shelf, northern Arkansas*
**Cahill, Thomas E.,** *Subsurface sequence stratigraphy and reservoir characterization of the Mississippian limestone (Kinderhookian to Meramecian), south central Kansas and north central Oklahoma*
**Jennings, Caleb James,** *Mechanical stratigraphy of the Mississippian in Osage County, Oklahoma*
**Kirk, Clara J.,** *A hazard assessment and proposed risk index for art, architecture, archive and artifact protection; case studies for assorted international museums*
**Marchese, Elizabeth Ann,** *Porosity development within lobes to downslope ramp deposits on a prograding carbonate shelf of the Kinderhookian to Osagean Series in northwest Arkansas*
**Martin, Adam T.,** *Depositional history and stratigraphic framework of Upper Cretaceous (Campanian to Maastrichtian) strata in the Minerva-Rockdale oil field of Milam County and adjacent counties, Texas*
**Studebaker, Elizabeth,** *Structural and stratigraphic transition from the Arkoma Shelf into the Arkoma Basin during basin subsidence; Arkoma Basin, northwest Arkansas*
**Trotter, Christopher,** *The significance of dolomitized Hunton strata in the Kinta and Bonanza Fields of the Arkoma Basin*

## University of British Columbia

Doctorate

**Betrie, Getnet Dubale,** *Risk management of acid rock drainage under uncertainty*
**Blanchette-Guertin, Jean-François,** *Seismic energy propagation in highly scattering environments and constraints on lunar interior structure from the scattered signals of the Apollo passive seismic experiment*
**Bordet, Esther-Jeanne,** *Eocene volcanic response to the tectonic evolution of the Canadian Cordillera*
**Cienciala, Piotr,** *Hydrogeomorphic controls on spatial pattern of fish habitat in a mountain stream*
**Crawford-Flett, Kaley A.,** *An improved hydromechanical understanding of seepage-induced instability phenomena in soil*
**Golding, Martyn Lee,** *Biostratigraphy and sedimentology of Triassic hydrocarbon-bearing rocks in northeastern British Columbia*
**Harrison, Anna Lee,** *Mechanisms of carbon mineralization from the pore to field scale; implications for carbon dioxide sequestration*
**Jacobs, Anthony David,** *Quantifying the mineral carbonation potential of mine waste material; a new parameter for geospatial estimation*
**Leaney, W. Scott,** *Microseismic source inversion in anisotropic media*
**Luzi, David Steven,** *Sediment transport and morphological response of a semi-alluvial channel; insights from a Froude scaled laboratory model*
**Marsh, Diana Elizabeth,** *From "Extinct Monsters" to Deep Time; an ethnography of fossil exhibits production at the Smithsonian's National Museum of Natural History*
**Peterson, Holly Esther,** *Unsaturated hydrology, evaporation, and geochemistry of neutral and acid rock drainage in highly heterogeneous mine waste rock at the Antamina Mine, Peru*
**Royer, Alexandra Amélie,** *Studies of seismic deconvolution and low-frequency earthquakes*
**Shamekhi, Seyedeh Elham,** *Probabilistic assessment of rock slope stability using response surfaces determined from finite element models of geometric realizations*
**Sihota, Natasha Julie Jane,** *Novel approaches for quantifying source zone natural attenuation of fossil and alternative fuels*
**Smith, Evan Mathew,** *Fluid inclusions in fibrous and octahedrally-grown diamonds*
**Wyllie, Duncan C.,** *Rock fall engineering; development and calibration of an improved model for analysis of rock fall hazards on highways and railways*
**Yang, Dikun,** *Geophysical survey decomposition and efficient 3D inversion of time-domain electromagnetic data*

Masters

**Ashwood, Wesley,** *Numerical model for the prediction of total dynamic landslide forces on flexible barriers*
**Baldeon Vera, Geidy Adriana,** *Geo-referenced landslide information system for characterization of landslide hazards at the reservoir scale, Bridge River watershed, southwestern, B.C.*
**Bird, Lawrence,** *Hydrology and thermal regime of a proglacial lake fed by a calving glacier*
**Bluemel, Britt,** *Biogeochemical expressions of buried REE mineralization at the Norra Karr alkaline complex, southern Sweden*
**Chernos, Matthew,** *The relative importance of calving and surface ablation at a lacustrine terminating glacier; a detailed assessment of ice loss at Bridge Glacier, British Columbia*
**Colborne, Jacqueline,** *Stratigraphic, depositional and diagenetic controls on reservoir development, Upper Devonian Big Valley Formation, southern Alberta*
**Elgueta, Maria Alejandra,** *Channel adjustment of a gravel-bed stream under episodic sediment supply regimes*
**Estepho, Mathiew,** *Seepage induced consolidation test; characterization of mature fine tailings*
**Greenlaw, Lauren,** *Surface lithogeochemistry of the Relincho porphyry copper-molybdenum deposit, Atacama region, Chile*
**Hames, Benjamin P.,** *Evolution of the Late Cretaceous Whistler*

Theses & Dissertations

*Au-(Cu) porphyry corridor and magmatic-hydrothermal system, Kahiltna Terrane, southwestern Alaska, USA*

**Hou, Pengfei,** *Sinemurian (Early Jurassic) stratigraphy at Last Creek, British Columbia and Five Card Draw, Nevada; paleontology and environmental implications*

**MacKenzie, Lucy,** *Modelling channel morphodynamics; the effects of large wood and bed grain size distribution*

**Mak, Stephen W.,** *Assessing fracture network connectivity of prefeasibility-level high temperature geothermal projects using discrete fracture network modelling*

**Manor, Matthew John,** *Convergent margin Ni-Cu-PGE deposits; geology, geochronology, and geochemistry of the Giant Mascot magmatic sulphide deposit, Hope, British Columbia*

**Matharu, Gian,** *Crustal anisotropy in a subduction zone forearc; northern Cascadia*

**McKenzie, Greg,** *Geology and mineralization at Independence Creek; Dawson Range, west-central Yukon Territory*

**McNulty, Brian,** *Geology, alteration, lithogeochemistry and hydrothermal fluid characterization of the Neoproterozoic Niblack polymetallic volcanic-hosted massive sulfide camp, southeast Alaska, USA*

**Miao, Lina,** *Efficient seismic imaging with spectral projector and joint sparsity*

**Postlethwaite, Benjamin,** *Seismic velocities and composition of the Canadian crust*

**Rodriguez Madrid, Alfonso Luis,** *Geology, alteration, mineralization and hydrothermal evolution of the La Bodega-La Mascota Deposits, California-Vetas mining district, Eastern Cordillera of Colombia, Northern Andes*

**Ryan, Amy G.,** *Water solubility and bubble growth dynamics in rhyolitic silicate melts at atmospheric pressure*

**Santos, Jair,** *A method for analysis of rock blocks with complex arbitrary geometries*

**Seidalinova, Ainur,** *Monotonic and cyclic shear loading response of fine-grained gold tailings*

**Taguchi, Genki,** *Fault tree analysis of slurry and dewatered tailings management; a framework*

**Velasquez, Alejandro,** *Trace element analysis of native gold by laser ablation ICP-MS; a case study in greenstone-hosted quartz-carbonate vein ore deposits, Timmins, Ontario*

**Ya'acoby, Avee,** *The petrology and petrogenesis of the Ren carbonatite sill and fenites, southeastern British Columbia, Canada*

## University of California at Berkeley

Doctorate

**French, Scott Winfield,** *Global full-waveform tomography using the spectral element method; new constraints on the structure of Earth's interior*

**Huang, Mong-Han,** *Crustal deformation during Co- and post-seismic phases of the earthquake cycle inferred from geodetic and seismic data*

**Kaercher, Pamela Michelle,** *Crystallographic preferred orientation and deformation of deep Earth minerals*

**Kim, Hyojin,** *Water chemistry evolution through the critical zone*

**Palucis, Marisa Christina,** *Using quantitative topographic analysis to understand the role of water on transport and deposition processes on crater walls*

**Zhang, Shuo,** *Mineralization of carbon dioxide sequestered in volcanogenic sandstone reservoir rocks*

**Zheng, Zhao,** *Refining constraints on seismic discontinuities and elastic structure in the Earth's upper mantle*

## University of California at Davis

Doctorate

**Brown, Rocko Anthony,** *The analysis and synthesis of river topography*

**Dumlao, Matthew Ruhland,** *From soil stability to nitrogen management; using a plant development perspective to investigate effects of root and shoot systems on soil stability, soil erosion, and soil nitrate management*

**Elliott, Austin John,** *Control of rupture behavior by a restraining double-bend from slip rates on the Altyn Tagh Fault*

**Gray, Andrew,** *Sediment transport and sedimentation dynamics in small mountainous, dry-summer river systems*

**Joshi, Geetika,** *Regulation and environmental abundance of methyl-tert-butyl ether degradation genes of Methylibium petroleiphilum PM1*

**Matiasek, Sandrine Journet,** *Dissolved organic matter sources and dynamics in an agricultural watershed; contribution from sediment desorption and insights from an amino acids time series*

**Stelten, Mark Evan,** *The post-caldera magmatic system at Yellowstone Plateau; compositional evolution, rhyolite generation, and the physical nature of the magma reservoir*

**Williams, Amy Jo,** *Microbial biosignature preservation at Iron Mountain, California*

Masters

**Alldritt, Katelin,** *Hydropedology and hydrologic connectivity of an oak-woodland hillslope in the northern Sierra foothills of California*

**Anderson, Carolyn G.,** *Effects of biosolids-derived pharmaceuticals on microbial communities and nitrogen processes in soil*

**Applegate, Olin Brice,** *Impact of dairy farming on groundwater salinity in California's Central Valley; a mass balance approach*

**Carlson, Emily Margaret,** *Soils of the watershed of Lake Atitlan, Guatemala; their role in eutrophication with a focus on phosphorus sorption*

**Gonzalez, Robert Lee,** *Re-envisioning cross-sectional hydraulic geometry as spatially explicit hydraulic topography*

**Griffin, Julie McDermott,** *Refining geochemical estimates of Carboniferous glacioeustasy; an integrated stratigraphic and conodont oxygen isotope study*

**Hathaway, Emily Marie,** *Comparing migration pathways of biodegradation products from petroleum hydrocarbon natural attenuation*

**Kirk, Emilie Rachelle,** *Measuring subsidence in peatland soils using a nitrogen budget approach*

**Liu, Yunjie,** *Modeling study of groundwater and surface water interaction using high resolution integrated model*

**Malazian, Armen Isaac,** *Soil moisture regime and associated water balance in the southern Sierra Nevada*

**Newcomb, Nicholas Jordan,** *Evaluating the impact of groundwater pumping on meadow hydrology and streamflow in the Yosemite Valley*

**Saal, Matthew Montgomery Bigole Bovee,** *Repackaging soil survey into a decision-support tool for agricultural groundwater banking in California*

**Starnes, Jesslyn Kathleen,** *Multi-stage metamorphism of amphibolite in Franciscan Melange, Ring Mountain, California*

## University of California at Santa Barbara

Doctorate

**Stern, Joseph Vincent,** *Regional benthic foraminiferal oxygen isotope stacks for the last glacial cycle*

## University of California at Santa Cruz

Doctorate

**Avants, Megan,** *Effects of near-source heterogeneity on wave fields emanating from crustal sources observed at regional and teleseismic distances*

**Reid, Rachel Elizabeth Brown,** *Dietary ecology of coastal coyotes (Canis latrans); marine-terrestrial linkages from the Holocene to present*

**Russell, Nicole L.,** *Sea-level rise, El Niño, and the future of the California coastline*

**Wasem, Christina A. Dwyer,** *The physics and chemistry of terrestrial planet and satellite accretion*

**Yue, Han,** *Toward resolving stable high-resolution kinematic rupture models of large earthquakes by joint inversion of seismic, geodetic and tsunami observations*

Masters

**Beitch, Marci Jillian,** *Greenland ice sheet retreat since the Little Ice Age*

**Bezore, Rhiannon Victoria Ann,** *A comparative study of passive versus dynamic sea-level rise inundation models for the island of Kauai*

**Coggan, Brian,** *Shoreline change in Southern California during the 2009/2010 El Niño Modoki*

**Hernandez, Stephen,** *The magnitude distribution of dynamically triggered earthquakes*

**Perera, Viranga,** *Lunar geophysics; the Moon's fundamental shape and paleomagnetism studies*

**Racz, Andrew,** *Spatial and temporal infiltration dynamics during managed aquifer recharge*

## University of Colorado at Boulder

Doctorate

**Higgins, Stephanie,** *River delta subsidence measured with interferometric synthetic aperture radar (InSAR)*

**Hopkins, Michelle Diane,** *Geochemical signatures in zircons as probes of the impact history of the inner solar system*

**Leckey, Erin H.,** *Changing climates and the evolution of insect* *herbivory in western oaks*

**Levandowski, William Brower,** *Geophysical investigations of the origins and effects of density variations in the crust and upper mantle beneath the Western and Central United States*

**Quillmann, Ursula,** *The influence of subpolar gyre dynamics on centennial to millennial scale Holocene climate variability in the high-latitude North Atlantic*

**Rengers, Francis K.,** *The influence of transient perturbations on landscape evolution; exploring gully and post-wildfire erosion*

**Wickert, Andrew David,** *Impacts of Pleistocene glaciation and its geophysical effects on North American river systems*

Masters

**Michaels, Julian Martin H.,** *Pore systems of the B Chalk and lower A Marl zones of the Niobrara Formation, Denver-Julesburg Basin, Colorado*

**Ring, Jeremy Daniel,** *Petrophysical evaluation of lithology and mineral distribution with an emphasis on feldspars and clays, middle and upper Williams Fork Formation, Grand Valley Field, Piceance Basin, Colorado*

## University of Hawaii at Manoa

Doctorate

**Colman, Alice,** *Effects of variable magma supply on magma reservoirs and eruption characteristics along the Gálapagos spreading center*

**Crites, Sarah Tinney,** *Interactions of galactic cosmic rays with the lunar surface*

**Gasda, Patrick J.,** *The aqueous alteration of carbon-bearing phases in CR carbonaceous chondrules*

**Trang, David,** *A remote analysis of the lunar landscape*

Masters

**Czeck, Benjamin C.,** *Growth, yield, and nutrient changed of sweet potato grown across the spectrum of $CO_2$ concentrations projected in the next 150 years*

**Glancy, Sarah Elizabeth,** *Petrology and geochemistry of boninites and related lavas from the Mata Volcanoes, NE Lau Basin*

**Larin, Penny N.,** *Perception of vulnerability relating to sea level rise and climate change in island communities; insights from Hawai'i*

**Maher, Sarah M.,** *The case for a chron 21 change in Africa absolute plate motion*

**Waters, Christine A.,** *Variability in submarine groundwater discharge composition and the fate of groundwater delivered nutrients at Kiholo Bay and Honokohau Harbor, north Kona District, Hawai'i*

**Zaiss, Jessica,** *Osmium geochemical behavior and global isotopic changed associated with the Cretaceous-Paleogene impact*

## University of Houston

Doctorate

**Bernal-Olaya, Rocio,** *Tectonostratigraphic evolution of the Sinu accretionary prism and lower Magdalena forearc basin formed above a shallow-dipping subduction zone, Caribbean margin of Colombia*

**Epps, Jonathan,** *Subsidence and sea-level rise for the northern Gulf of Mexico; an integrated multi-sensor approach*

**Famubode, Oyebode,** *Paleosol evolution in a sequence-stratigraphic framework, Cretaceous Ferron Notom Delta south central Utah, U.S.A.*

**Lisi, Arianna,** *Shear-wave velocity structure in the crust and upper mantle beneath the central Tien Shan from surface-wave tomography*

**Oyem, Arnold,** *Application of constrained least-squares spectral analysis*

**Yenugu, Malleswara Rao,** *Elastic, microstructural and geochemical characterization of kerogen maturity for shales*

**Zhao, Luanxiao,** *Poroelastic and seismic characterization of heterogeneous reservoir rocks*

Masters

**Akhun, Selin,** *Evolution of late Cenozoic minibasins and growth faulting, Green Canyon area, northwestern Gulf of Mexico*

**Atekpa, Musa,** *Comparison and integration of frequency multiattributes in thin layer visualization; a Stratton Field case study*

**Coskun, Suleyman,** *3-D seismic survey design via modeling and reverse time migration; Pierce Junction salt dome, Texas*

**Khadeeva, Anna,** *Linking elastic and petrophysical properties associated with deformation bands*

**Lamb, Julianne,** *An examination of the Arkoma foreland basin and its petroleum system through burial and thermal history modeling*

**Liang, Chen,** *Spectral bandwidth extension; invention versus harmonic extrapolation*

**Lim, Un Young,** *Investigation of and correction for waveform changes arising from NMO stretch*

**Lombardi, Michael,** *Convex hull method for SRV estimation*

**Myziuk, Nicholas,** *A relation between adiabatic and isothermal moduli*

**O'Keeffe, Kevin,** *Geometry and kinematics of the Canones Fault and the effects of lithology on the distribution of strain within the Canones Fault damage zone*

**Okonkwo, Onochie,** *Improved interpretation of incised valley system-channel delineation and thickness estimation using spectral inversion, Blackfoot Field, Alberta, Canada*

**Piscen, Sercan,** *Reservoir characterization via amplitude versus offset analysis and impedance inversion in the Thrace Basin of northwestern Turkey*

**Sansal, Tuna Altay,** *Contribution of seismic amplitude anomaly information in prospect risk analysis*

**Wang, Zi,** *Target formation pore pressure prediction using 3D seismic data via Fillippone and Eaton approaches in southern Sichuan, China*

**Xia, Keyao,** *Azimuthal P-wave AVA inversion for fracture orientation and density*

**Yazan, Kenan,** *Investigation of the imbricated system along the Muroto transect and its relationship to the subducting Muroto Seamount*

## University of Idaho

Doctorate

**Abplanalp, Jason M.,** *Pennsylvanian conodont biostratigraphy of the Snaky Canyon Formation, east-central Idaho*

**Clevy, June Renee,** *Geospatial analysis and seasonal changes in water-equivalent hydrogen in eastern equatorial Mars*

**Martin, Emily S.,** *The fractured ice shell of Saturn's moon Enceladus; insights into the global stress history and interior structure*

**Rader, Erika,** *Spatter, scoria bombs, and upper mantle xenoliths; cooling histories of rocks in three terrestrial settings*

Masters

**Boylan, Ryan.,** *Monitoring and modeling sediment and organic carbon loads from the dryland cropping region of the Inland Pacific Northwest*

**Browning, David A.,** *Hyperspectral remote sensing in mineral exploration; ammonium-illite as a pathfinder for gold*

**Cook, Casey Ann,** *Global contraction/expansion and polar lithospheric thinning on Titan from patterns of tectonism*

**Lubenow, Brady,** *Best practices for shallow ground temperature measurements*

**Mitchell, Liza R. J.,** *Stable isotope composition of stream biota: partitioning variability and identifying environmental correlations in a wilderness watershed*

**Moody, Alex C.,** *Stochastic and analytical exploration of enhanced geothermal system viability on the Snake River plain, Idaho*

**Morrow, Thomas A.,** *Geochemistry of the Galapagos transform fault*

**Rodriguez, Aaron P.,** *Devonian and Mississippian Sappington Formation in southwest Montana; stratigraphic framework and facies relationships*

**Schwartz, Darin,** *Volcanic, structural, and morphological history of Santa Cruz Island, Galápagos Archipelago*

## University of Illinois at Urbana-Champaign

Doctorate

**Eke, Esther,** *Numerical modeling of river migration incorporating erosional and depositional bank processes*

**Engel, Frank,** *The fluvial dynamics of compound meander bends*

**Kanissery, Ramdas,** *Bioavailability of metolachlor and glyphosate in aerobic and anaerobic soils*

**Moon, Sungwoo,** *Integrated computational-experimental soil behavior characterization from direct simple shear tests on Boston blue clay*

**Oviedo Salcedo, Diego,** *Accounting for parameter uncertainty and temporal variability in coupled groundwater-surface water models using component and systems reliability analysis*

**Verma, Siddhartha,** *Predictability and trends of annual pollutant loads in Midwestern watersheds*

**Wang, Xiangli,** *Chromium and uranium isotopic exchange kinetics and isotope fractionation during oxidation of tetravalent uranium by dissolved oxygen*

Masters

**Gajos, Norbert,** *Spatially controlled Fe isotope variations at Torres del Paine*

**Hermosillo, Armando,** *Investigating flat-slab subduction underneath South America using 4D numerical simulations*

**Kent, Graham,** *Modeling and analysis of management for an agroecosystem using an agent-based model interface for the soil and water assessment tool (SWAT)*

**Neal, Conor,** *Suspended sediment supply dominated by channel processes in a low-gradient agricultural watershed, Wildcat Slough, Fisher, IL, USA*

**Zatwarnicki, Katelyn,** *Modeling effective isotopic fractionation in groundwater aquifers with heterogeneous reaction rates*

## University of Iowa

Doctorate

**Grass, Andy Darrell,** *Inferring lifestyle and locomotor habits of extinct sloths through scapula morphology and implications for convergent evolution in extant sloths*

**Petrie, Meredith Blair,** *Evolution of eclogite facies metamorphism in the St. Cyr Klippe, Yukon-Tanana Terrane, Yukon, Canada*

**Rocheford, Mary Kathryn,** *Determining geomorphological and land use effects through physico-chemical fingerprinting of soils*

Masters

**Even, Matthew James,** *How does groundwater subsidy of vegetation change as a function of landscape position and soil profile characteristics at the Ciha Fen (Johnson County, IA, USA)?*

**Guest, Rachel L.,** *Description and phylogenetic analysis of a new alligatoroid from the Eocene of Laredo, Texas*

**Isard, Sierra Juliane,** *Origin of the Tower Peak Unit in the St. Cyr area, Yukon, Canadian Cordillera*

**Knauss, Georgia Ellen,** *A morphological description of Baptemys wyomingensis and an analysis of its phylogenetic relationship within Kinosternoidea*

**Thieme, Clara,** *Controlled flooding on the Colorado River; using GIS methods to assess sandbar development*

## University of Kansas

Doctorate

**Bidgoli, Tandis S.,** *Low-temperature thermochronometric constraints on Cenozoic intraplate deformation in the central Basin and Range*

**Bonilla-Rodriguez, Alvin J.,** *Depositional and Paleoenvironmental Settings of Cretaceous Limestones in the Greater Antilles*

**Falk, Amanda Renee,** *Foot and hindlimb morphology, soft tissues, and tracemaking behaviors of Early Cretaceous birds from China and the Republic of Korea with a comparison to modern avian morphology and behavior*

**Gapp, Ian Wesley,** *Phylogenetic analyses of trilobites from the Cambrian and Ordovician radiations*

**Hager, Christian,** *Integrated tectonic and quantitative thermochronometric investigation of the Xainza Rift, Tibet*

**Lamsdell, James Christopher,** *Selectivity in the evolution of Palaeozoic arthropod groups, with focus on mass extinctions and radiations; a phylogenetic approach*

**Saupe, Erin E.,** *Integrating ecology and evolution in deep time; using ecological niche modeling to study species' evolutionary responses to climate from the Pliocene to the present-day biodiversity crisis*

Masters

**Bailey, Bevin L.,** *High-resolution shear-wave reflection profiling to image offset in unconsolidated near-surface sediments*

**Baker, Matthew Peter,** *Ground-penetrating radar imaging of fluid flow through a discrete fracture*

**Burris, Natalie Lynn,** *The Effect of Point Velocity Probe Size on Groundwater Velocity Estimation in Noncohesive Sediments*

**Downen, Matthew Ross,** *The taxonomy and taphonomy of fossil spiders from the Crato Formation of Brazil*

**Huff, Breanna L.,** *Microbial and geochemical characterization of Wellington oil field, southcentral Kansas, and potential applications to microbial enhanced oil recovery*

**Katz, Britney S.,** *Analysis of Chemical Storage and Transit Times to Characterize Water Movement Through a Thick Unsaturated Zone Overlying the High Plains Aquifer, Northwestern Kansas*

**Konetchy, Brant Evan,** *High-resolution quantification of groundwater flux using a heat tracer: laboratory sandbox tests*

**Lippert, Peter Gregory,** *Detrital U-Pb geochronology provenance analyses; case studies in the greater Green River basin, Wyoming, and the Book Cliffs, Utah*

**Liu, Huan,** *Inorganic and organic carbon variations in surface water, Konza prairie LTER site, USA, and Maolan karst experimental site, China*

**Long, Molly,** *Characterizing the Groundwater-Surface Water Interactions in Different Subsurface Geologic Environments Using Geochemical and Isotopic Analyses*

**Stolz, Dustin J.,** *Reservoir character of the Avalon Shale (Bone Spring Formation) of the Delaware Basin, West Texas and southeast New Mexico; effect of carbonate-rich sediment gravity flows*

**Thompson, Angela R.,** *Effect of flow rate on clogging processes in small diameter aquifer storage and recovery injection wells*

**Voegerl, Ryan Scott,** *Quantifying the carboxyl group density of microbial cell surfaces as a function of salinity; insights into microbial precipitation of low-temperature dolomite*

**Waynick, Michael Anthony,** *Fluvial to shelfal strata of the Late Cretaceous to Paleogene Dorotea and Tres Pasos Formations, Magallanes Basin, El Calafate, Argentina*

## University of Kentucky

Masters

**Brengman, Clayton M. J.,** *Instrument correction and dynamic site profile validation at the Central United States Seismic Observatory, New Madrid seismic zone*

**Burkett, Corey A.,** *Late Quaternary crustal deformation at the apex of the Mount McKinley restraining bend of the Denali Fault, Alaska*

**Federschmidt, Sara E.,** *Paleoseismic and structural characterization of the Hines Creek Fault; Denali National Park and Preserve, Alaska*

**Johnston, Michelle N.,** *A petrographic characterization of the Leatherwood coal bed in eastern Kentucky*

**Kelly, Evan A.,** *Age of the Walden Creek Group, western Blue Ridge Province; resolving a decades-old controversy via detrital mineral geochronology and sedimentary provenance analysis*

**O'Bryan, Alice C.,** *The nature and origin of cyclicity in the Cleveland Member of the Ohio Shale (Upper Devonian), northeastern Kentucky, U.S.A.*

**Orton, Alice M.,** *Science and public policy of earthquake hazard mitigation in the New Madrid seismic zone*

**Spaulding, Daniel F.,** *Geology of the west half of the Cove Creek Gap Quadrangle and adjacent areas, western North Carolina; insights into eastern Great Smoky Mountains tectonometamorphism*

**Walker, Laurel Anne,** *Determining hillslope diffusion rates in a boreal forest; Quaternary fluvial terraces in the Nenana River valley, central Alaska Range*

**Wolfe, Phillip,** *Holocene sedimentary responses to growth faulting in a back-barrier setting; east Matagorda Peninsula, Texas*

**Woodruff, Olivia P.,** *Temporal and spatial characterization of Macondo 252 signatures in Gulf of Mexico shelf and slope sediments*

## University of Maine

Doctorate

**Campbell, Seth William,** *Determining Basin Geometry, Stability, and Flow Dynamics of Valley Glaciers with Ground-Penetrating Radar*

**Koffman, Tobias Nicholas Buttersworth,** *Paleoclimatic Events of the Past 15,000 Years Inferred from Beryllium-10 Chronologies of Moraines in New Zealand*

**Millette, Patricia M.,** *The Effects of Conducting Authentic Field-Geology Research on High School Students' Understanding of the Nature of Science, and Their Views of Themselves as Research Scientists*

**Weitz, Nora Amelie,** *Modeling Glaciers and Ice Sheets in Greenland and Antarctica withan Emphasis on Embedded Modeling and Subglacial Hydrology*

Masters

**Beers, Thomas Manuel,** *State of the Art High Resolution Glacio-Chemistry from Greenland and Antarctic Ice Cores, Implications to Climate Reconstructions*

**Braddock, Scott,** *Holocene Sea-Surface Temperatures in the McMurdo Sound Region, Antarctica, Reconstructed from Isotope Records of Adamussium colbecki*

**Cronkite, Eliza M.,** *The Late Quaternary Stratigraphy and Paleogeographic Evolution of the New Meadows Embayment, Northeastern Casco Bay, Gulf of Maine*

**Jacobacci, Kara,** *Testing GPR and LiDAR Techniques for Identifying Landslides on the Maine Coast*

**Lennon, Jennifer Renee,** *$^{10}$Be Surface-Exposure Chronology of the Tres Hermanos and Rosa Irene Moraines Near Bahía Inútil, Chilean Patagonia*

**Strand, Peter Douglas,** *A $^{10}$Be Surface Exposure Chronology of Left-Lateral Moraines of the Pukaki Glacier, Southern Alps, New Zealand*

**Taylor, Agnes R.,** *Elemental Release Behavior of Serpentinite in the Presence of Organic Acids: Implications for Studying the Habitability of Europa*

## University of Manitoba

Doctorate

**Adetunji, Ademola Quadri,** *Resistivity structure of the Precambrian Grenville Province, Canada*

## University of Memphis

Masters

**Olson, Kristian,** *Tectonic controls on stratigraphy and sedimentology of Pliocene-Pleistocene Lake Tecopa Beds, southeastern California*

**Plunk, Lindsay,** *Chemical and mineralogical analysis of varney red-filmed ceramics from the lower Mississippi River valley*

## University of Michigan

Doctorate

**Hyland, Ethan Gordon,** *Multiproxy terrestrial records of climatic and ecological change during the early Eocene climatic optimum*

## University of Minnesota at Duluth

Doctorate

**Dyess, Jonathan,** *Multi-scale structural and kinematic analysis of a Neoarchean shear zone in northeastern Minnesota; implications for assembly of the southern Superior Province*

Masters

**Goscinak, Christopher Joseph,** *Quartz fabric analysis of the Kawishiwi shear zone, NE Minnesota*

**Nissen, Chelsea I.,** *New evidence of Proterozoic high P-T metamorphism in East Antarctica from thermobarometry and in-situ U-Pb age dating of monazite in metamorphic glacial clasts, central Transantarctic Mountains, Antarctica*

**Steiner, Ronald Alexander,** *Genesis of sulfide mineralization within the granite footwall of the Maturi Deposit of the South Kawishiwi Intrusion, Duluth complex, NE Minnesota*

**Votava, Jillian Emilia,** *The Holocene history of Lake Kivu (East Africa); new perspectives from new cores*

## University of Missouri at Columbia

Masters

**Brown, Keith Michael,** *Quantifying bottomland hardwood forest and agricultural grassland evapotranspiration in floodplain reaches of a mid Missouri stream*

## University of Nevada at Las Vegas

Masters

**Lacy, Alison C.,** *Garnet dating, pressure-temperature time paths and kinematic analysis of the schist of Upper Narrows, Raft River Mountains, northwestern Utah; tectonic implications of pressure-temperature-time-deformation paths*

## University of New Hampshire

Doctorate

**Corr, Chelsea A.,** *The analysis of in situ and retrieved aerosol properties measured during three airborne field campaigns*

**Figueroa-Nieves, Debora,** *Impacts of treated sewage effluents on stream ecology in Puerto Rico*

**Hunter, Maria O'Healy,** *Carbon stocks and cycling in the Amazon basin: measurement and modeling of natural disturbance and recovery using airborne LIDAR*

**Kelsey, Eric P.,** *Towards understanding North Pacific climate variability with instrumental and ice core records*

**Swarthout, Robert F.,** *Volatile organic compound emissions from unconventional natural gas production: source signatures and air quality impacts*

**Treat, Claire C.,** *Effects of climate change on carbon and nitrogen cycling in permafrost soils of Alaska*

Masters

**Amante, Jacqueline M.,** *The distribution and movement of black carbon and other chemical impurities and the effects on snow albedo in New Hampshire*

**Lawrence, Katherine D.,** *Effects of discharge in residence time distributions in a small headwaters wetland in the Ipswich River watershed*

**Roddy, Samantha,** *Vegetation influences on the ebullition of methane in a temperate wetland*

**Rosengarten, David,** *Spatial and temporal variability of nitrate cycling in a New England headwater wetland and stream*

**Uter, Melika Coleen,** *The impact of ocean acidification on metal contaminated marine sediment*

**Yao, Fang,** *Uncertainty analysis on photogrammetry-derived national shoreline*

## University of North Carolina at Wilmington

Masters

**Albritton, Casey,** *A high-resolution stable carbon and oxygen isotope stratigraphy spanning the Late Triassic (Carnian-Norian) from the Blue Mountains Province of Oregon, U.S.A.*

**Baggarley, Jehan M.,** *Remote sensing and GIS techniques for the assessment of flash flood risk in Quseir, Red Sea Coast, Egypt*

**Hall, Kristen,** *Assessment on decadal and annual beach width and dune height patterns on Masonboro Island, North Carolina*

**Kerr, James,** *Assessing the influence of escalation during the Mesozoic marine revolution; shell breakage and adaptation against enemies in Mesozoic ammonites*

**Pierce, Courtney N.,** *Development of a high resolution sea-level record in the Florida Panhandle*

**Stokes, Kirsten Elise,** *Diatoms as proxies of late-Pleistocene/Holocene climatic events in the Bellingshausen Sea, western Antarctic Peninsula*

## University of North Dakota

Doctorate

**Ashu, Richard Agbor Tabe,** *Stratigraphy, depositional environments, and petroleum potential of the Three Forks Formation -- Williston Basin, North Dakota*

**Bibby, Theodore C.,** *Landscape evolution and preservation of ice over one million years old quantified with cosmogenic nuclides $^{26}Al$, $^{10}Be$, and $^{21}Ne$, Ong Valley, Antarctica*

**Gbolo, Prosper,** *Quantifying nutrient cycling and fate within an abandoned feedlot and adjacent wetlands*

Masters

**Bosshart, Nicholas Wade,** *Characterization, diagenesis, and geocellular modeling of Winnipegosis Formation pinnacle reefs in the Williston Basin, North Dakota*

**Ochsner, Aaron Thomas,** *Patterns and mechanisms of heat transport in the northern Denver Basin Nebraska, South Dakota and Wyoming*

**Sebade, Matthew Joseph,** *The depositional environment and diagenetic effects on sand bodies within the unconventional resource play of the Spearfish Formation (triassic) in north central North Dakota*

**Zimny, Eric Gerald,** *Geoneutrino production of the northern Black Hills, South Dakota, United States of America / by Eric Gerald Zimny*

Bachelor

**Rasanen, Ryan A.,** *WISCO special waste landfill: site location northwest 1/4 of section 26, township 154 north, range 104 west, Williams County, North Dakota*

## University of Oklahoma

Doctorate

**Bejarano, Carlos Luis Cabarcas,** *Integrated interpretation of borehole microseismic data from hydraulic fracturing stimulation treatments*

**Dulin, Shannon Ann,** *Paleomagnetism as a tool for determining diagenetic, depositional, and erosional events as recorded in sedimentary and igneous rocks*

**Guo, Shiguang,** *Seismic solutions for unconventional resource plays*

**Paul, Debapriya,** *A detailed analysis of rift related structures; insights from laser scanned clay models and 3-D seismic interpretation*

**Romero, Andrea Antonietta Miceli,** *Subsurface and outcrop organic geochemistry of the Eagle Ford Shale (Cenomanian-Coniacian) in west, southwest, central and east Texas*

**Wallet, Bradley C.,** *Seismic attribute expression of fluvial-deltaic and turbidite systems*

**Zhang, Bo,** *Long offset seismic data analysis for resource plays*

Masters

**Bailey, William Joseph,** *Fault plane geomorphology and structural analysis of a Middle East giant carbonate oil field*

**Cardona-Valencia, Luis Felipe,** *Integrated characterization of the Woodford Shale in the southern Cherokee Platform, Oklahoma*

**Carrell, Jordan,** *Field-scale hydrogeologic modeling of water injection into the Arbuckle Zone of the Midcontinent*

**Dixon, Emily,** *Flow-through dissolution rates of jarosite in brines*

**Dyer, Steven Michael,** *An investigation into the stability field of zirconium-rich garnets (kimzeyite/kerimasite) at carbonated aqueous conditions*

**Ha, Thang,** *Seismic reprocessing and interpretation of a shallow "buried hill" play; Texas Panhandle*

**Karam, Pierre,** *Analogue modeling of salt structures*

**Kowalczyk, Michael Warren Brock,** *Integrated petrophysical-geophysical carbonate sequence analysis for reservoir characterization of a Middle East field*

**Leavitt, Roger Earl,** *Structural analysis and 3D modeling of* *rotational fault blocks and implications of transfer zones along the Nordland Ridge, offshore mid-Norway*

**Leonet, Richard Jesus Brito,** *Geological characterization and sequence stratigraphic framework of the Brown Shale, Central Sumatra Basin, Indonesia; implications as an unconventional resource*

**Lin, Tengfei,** *Attribute analysis of time- vs. depth-migrated seismic data; application to Renqiu Field, China*

**Liu, Li,** *Source and fate of oils in the Lawton Oilfield, southwestern Oklahoma*

**Long, Sara,** *Fracture height growth characterization from microseismic data in the Granite Wash*

**Mann, Elizabeth,** *Stratigraphic study of organic-rich microfacies of the Woodford Shale, Anadarko Basin, Oklahoma*

**Mattei, Gabriel A.,** *Seismic variability of the Alaskan Peninsula and eastern Aleutian megathrust*

**McCullough, Brenton Joseph,** *Sequence-stratigraphic framework and characterization of the Woodford Shale on the southern Cherokee Platform of central Oklahoma*

**Mutlu, Onur,** *Seismic attribute illumination of tectonic deformation in Chicontepec Basin, Mexico*

**Pfeifer, Lily Springman,** *Rapid Permian exhumation of the Montagne Noire Dome recorded in provenance of Upper Paleozoic clastics in the Graissessac-Lodève Basin, France*

**Prada, Adriana Gomez,** *Integrated geological characterization and distribution of the Salada Member, La Luna Formation, in the central area of the Middle Magdalena Basin, Colombia*

**Riley, Brett R.,** *Subsurface stratigraphy, petrography, structure, and reservoir characteristics of some Mississippian-Pennsylvanian sandstones, McGee Valley area, Atoka and Pushmataha Counties, Oklahoma*

**Salisbury, Megan Nicole,** *Depositional setting of the Wolfcamp Formation, Midland Basin, Texas, using organic geochemistry*

**Sarkar, Debajyoti Basu,** *Rheology and thermal state of Titan's crust; potential role of methane clathrates*

**Sefein, Kirellos John,** *Organic geochemistry and paleoenvironment of the Brown Shale Formation, Kiliran Sub-basin, Central Sumatra Basin, Indonesia*

**Steullet, Alex Kayla,** *An integrated paleomagnetic and diagenetic study of the Marcellus Shale within the plateau province of the Appalachian Basin*

**Swain, Brandon Paul,** *Structure, stratigraphy, and economic viability of pre-middle Eocene deepwater deposits; present day northwestern Gulf of Mexico*

**Toth, Christopher Robert,** *Separation of the earthquake tomography inverse problem to refine hypocenter locations and tomographic models; a case study from central Oklahoma*

**Tréanton, Jessica,** *Outcrop-derived chemostratigraphy of the Woodford Shale, Murray County, Oklahoma*

**Trumbo, Daniel Blaine,** *A production calibrated reservoir characterization of the Mississippi Lime in a mature field utilizing reprocessed legacy 3D seismic data, Kay County, Oklahoma*

**Yalcin, Esra,** *Delaware Basin thermal evolution from constrained vitrinite reflectance; tectonic versus flexural subsidence*

**Yates, Rachel,** *Microseismic event location sensitivity to anisotropy; a Granite Wash case study*

**Zahrai, Shayda,** *Interpreting climatic change during the late Pleistocene/early Holocene in Oklahoma based on the stable isotope and amino acid composition of fossil bison bones*

**Zhou, Yuqi,** *High resolution spectral gamma ray sequence stratigraphy of shelf edge to basin floor upper Capitan Permian carbonates, Guadalupe Mountains, Texas and Delaware Basin, New Mexico*

## University of Rhode Island

Doctorate

**Brounce, Maryjo N.,** *A geochemical investigation of oxygen fugacity in the Marianas subduction factory*

**Covellone, Brian M.,** *Investigation into 3D earth structure and sources using full seismic waveforms*

Masters

**Logan, Luke Andrew,** *Performance analysis of an underwater wire flying profiling vehicle*

**Morissette, Cameron E.,** *Paleoenvironmental and paleolandscape reconstructions of Greenwich Bay region, RI*

## University of Shippensburg

Masters

**Markowitz, Graham D.,** *Alterations in channel geometry, shear stress, and bedload sediments within a reconstructed stream channel: Larry's Creek, Lycoming County, Pennsylvania*

## University of South Florida at Saint Petersburg

Doctorate

**Patten, James,** *Investigations of the physical and analytical chemistry of iron in aqueous solutions*

**Williams, Clare Carlisle,** *A multi-proxy approach to understanding abrupt climate change and Laurentide ice sheet melting history based on Gulf of Mexico sediments*

Masters

**Snow, Tasha,** *Timing of Svalbard/Barents Sea ice sheet decay during the last glacial termination*

## University of South Florida at Tampa

Doctorate

**Boop, Liana M.,** *Characterization of the depositional environment of phreatic overgrowths on speleothems in the littoral caves of Mallorca (Spain); a physical, geochemical and stable isotopic study*

**Brutsche, Katherine E.,** *Evolution and equilibraiton of artifical morphologic perturbations in the form of nearshore berm nourishments along the Florida Gulf Coast*

**Callahan, Michael Kroh,** *Groundwater controls on physical and chemical processes in streamside wetlands and headwater streams in the Kenai Peninsula, Alaska*

**Knorr, Paul Octavius,** *Response of benthic Foraminifera to ocean acidification and impact on Florida's carbonate sediment production*

**Mondal, Subhronil,** *Short- and long-term trends in ecological interactions; form predator-prey interactions to Phanerozoic diversification*

**Stremtan, Ciprian Cosmin,** *Mantle-curst interaction in granite petrogenesis in post-collisional settings; insights form the Danubian Variscan plutons of the Romanian Southern Carpathians*

Masters

**Croft, Lance C.,** *Interpolating beach profile data using linear and non-linear functions*

**Harburger, Aleeza M.,** *Probabilistic modeling of lava flows; a hazard assessment for the San Francisco Volcanic Field, Arizona*

**Xie, Ming,** *Verification and comparison of two commonly used numerical modeling systems in hydrodynamic simulation at a dual-inlet system, west-central Florida*

## University of Southern Mississippi

Masters

**Ayers, Michael Charles,** *Historical channel adjustments in the Pascagoula River Basin and adjacent systems, Southeast Mississippi*

**Funderburk, William Richard,** *Mapping the distribution of barrier island slash pine-oak woodland and determining growth responses of Pinus elliottii to Hurricane Katrina (2005) on Cat Island, Mississippi*

**Jeter, Guy Wilburn, Jr.,** *A vegetation analysis on Horn Island, Mississippi, ca. 1940 using characteristic dimensions derived from historical aerial photography*

**Sutherlin, Timothy Guy,** *Assessing landscape change in highland Peru with an emphasis on tree cover change, 1948-2012*

**Williams, Serena Elizabeth,** *Education and economic development: an untapped alliance*

Bachelor

**McCracken, Stephanie Lane,** *Utilizing Landsat TM and OLI in predicting Oncomelania hupensis habitats around Poyang Lake before and after Three Gorges Dam completion*

## University of Tennessee at Knoxville

Doctorate

**Allen, Melissa Ree,** *Impacts of Climate Change on the Evolution of the Electrical Grid*

**Cropper, Samuel Clark,** *Comparisons of Point and Average Capillary Pressure - Saturation Functions for Porous Media*

**Halty, Isaac Andres Jeldes,** *Stability, Erosion, and Morphology Considerations for Sustainable Slope Design*

**Keenan, Sarah Wheeler,** *Gastrointestinal microbial diversity and diagenetic alteration of bone from the American alligator (Alligator mississippiensis)*

**Udry, Arya Sigrid Waltraud,** *Exploring martian magmas: From the mantle to the regolith*

Masters

**Berg, Ashley René Manning,** *Calcitized Evaporites in the Precambrian: Deposition and Diagenesis in a Low Sulfate Ocean*

**Buongiorno, Joy,** *Mineralized microbialites as archives of environmental evolution of a hypersaline lake basin: Laguna Negra, Catamarca Province, Argentina*
**Ennis, Megan Elizabeth,** *Crystallization Kinetics of Olivine-Phyric Shergottites*
**O'Dell, Debra Blumberg,** *Measuring Carbon Dioxide (CO2) Flux of Agricultural Practices in Sub-Saharan Africa*
**Rehrer, Justin Randolph,** *Characterizing the relationship of part of the Inner Piedmont and Pine Mountain window, Georgia, from detailed geologic mapping, geochemistry, geochronology, and structural analysis at the southwestern end of the Cat Square terrane*
**Reichert, Benjamin Lee,** *Investigating the Effects of Urbanization on Residual Forest Soils in Knox Co., Tennessee*
**Simmons, Paul Vanterpool,** *A Spatial Analysis of Streambank Heterogeneity and its Contribution to Bank Stability*
**Smith, Gina Brianne,** *Metamorphic Evolution of Eastern Blue Ridge Calc-Silicates in Southwestern North Carolina, Northeastern Georgia, and Northwestern South Carolina*
**Street, Derek Lee,** *Transport of Fecal Pollution Indicators: Impacts from the Land Spreading of Liquid Manure on Water Quality*
**Tran, Vi Thi Tuong,** *An Analysis of the Suspended Sediment Rating Curve Parameters in the Upper Mississippi River Basin at the Monthly and Annual Levels*
**Walker, Ann Elizabeth,** *Structural analysis of the Tablerock thrust sheet, Grandfather Mountain window, northwestern North Carolina: Emplacement kinematics of a large horse in a major thrust system*
**White, Kyle Vincent,** *Structural and metamorphic evolution of the Haimanta Group: Insights from P-T-t-D path modeling of rocks near the Leo Pargil dome, NW India*

## University of Texas at Austin

### Doctorate

**Alvarez, Tricia Grier,** *The southeastern Caribbean subduction to strike-slip transition zone; a study of the effects on lithospheric structures and overlying clastic basin evolution and fill*
**Befus, Kenneth Stephen,** *Storage, ascent, and emplacement of rhyolite lavas*
**Bhowmik, Sayantan,** *Particle tracking proxies for prediction of $CO_2$ plume migration within a model selection framework*
**Byerly, Benjamin Lee,** *Constraints from mantle xenoliths on the geodynamic evolution of Earth's upper mantle*
**Cao, Fei,** *Development of a two-phase flow coupled capacitance resistance model*
**Carter, Russell Wirkus,** *Fluid characterization at the Cranfield $CO_2$ injection site; quantitative seismic interpretation from rock-physics modeling and seismic inversion*
**Chen, Peila,** *Enhanced oil recovery in fractured vuggy carbonates*
**Eakin, Daniel Hoyt, Jr.,** *An analysis of subduction related tectonics offshore southern and eastern Taiwan*
**Frooqnia, Amir,** *Numerical simulation and interpretation of borehole fluid-production measurements*
**Ganjdanesh, Reza,** *Integrating carbon capture and storage with energy production from saline aquifers*
**Hwang, Jongsoo,** *Factors affecting injection well performance and fracture growth in waterflooded reservoirs*
**Jiang, Meijuan,** *Seismic reservoir characterization of the Haynesville Shale; rock-physics modeling, prestack seismic inversion and grid searching*
**Kong, Xianhui,** *Petrophysical modeling and simulation study of geological $CO_2$ sequestration*
**Li, Siwei,** *Seismic imaging and velocity model building with the linearized eikonal equation and upwind finite-differences*
**Mehmani, Yashar,** *Modeling single-phase flow and solute transport across scales*
**Moore, Stephanie Jean,** *Crystallization of metamorphic garnet; nucleation mechanisms and yttrium and rare-earth-element uptake*
**Ortega, Edwin Yamid,** *Inversion-based petrophysical interpretation of multi-detector logging-while-drilling Sigma measurements*
**Parker, William Gibson, Jr.,** *Taxonomy and phylogeny of the Aetosauria (Archosauria, Pseudosuchia) including a new species from the Upper Triassic of Arizona*
**Robertson, Wendy Marie,** *Anthropogenic impacts on recharge processes and water quality in basin aquifers of the desert southwest; a coupled field observation and modeling study*
**Salazar, Migdalys Beatriz,** *The impact of shelf margin geometry and tectonics on shelf-to-sink sediment dynamics and resultant basin fill architectures*
**Santillan, Eugenio Felipe Unson,** *Microbial responses to $CO_2$ during carbon sequestration; insights into an unexplored extreme environment*
**Schroeder, Dustin Matthew,** *Characterizing the subglacial hydrology of Thwaites Glacier, West Antarctica using airborne radar sounding*
**Singh, Gurpreet,** *Coupled flow and geomechanics modeling for fractured poroelastic reservoirs*
**Venkatraman, Ashwin,** *Gibbs free energy minimization for flow in porous media*
**Wu, Kan,** *Numerical modeling of complex hydraulic fracture development in unconventional reservoirs*

### Masters

**Al-Waily, Mustafa Badieh,** *Depth-registration of 9-compnent 3-dimensional seismic data in Stephens County, Oklahoma*
**Alabbad, Emad Abbad,** *Experimental investigation of geomechanical aspects of hydraulic fracturing unconventional formations*
**Altubayyeb, Abdulaziz Samir,** *A numerical study of the impact of waterflood pattern size on ultimate recovery in undersaturated oil reservoirs*
**Apiwatcharoenkul, Woravut,** *Uncertainty in proved reserves estimation by decline curve analysis*
**Atakturk, Katelyn Rahsan,** *Deciphering the P-T-t conditions of garnet-bearing metamorphic rocks in the southern Menderes Massif, SW Turkey*
**Aybar, Umut,** *Investigation of analytical models incorporating geomechanical effects on production performance of hydraulically and naturally fractured unconventional reservoirs*
**Basu, Saptaswa,** *Fracture diagnostics using low frequency electromagnetic induction*
**Becker, Lauren Elizabeth,** *The seismic response to fracture clustering; a finite element wave propagation study*
**Betts, William Salter,** *Compressibility and permeability of Gulf of Mexico mudrocks, resedimented and in-situ*
**Browne, Katharine Elizabeth,** *Phylogenetic and ecological significance of variation in the scleral ring in aquatic foraging birds*
**Cardenas, Benjamin Thomas,** *Evidence for changes in coastline-controlled base level from fluvial stratigraphy at Aeolis Dorsa, Mars*
**Chilongo, Owen Chasoba,** *Understanding Kabwe's lead pollution*
**Cronin, Michael Brett,** *Core-scale heterogeneity and dual-permeability pore structure in the Barnett Shale*
**Das, Jasaswee Triyambak,** *Evaluation of the rate of secondary swelling in expansive clays using centrifuge technology*
**Dawes, Colleen Marie,** *Energy transitions on the Hawaiian Islands: water resource implications for Hawaii's electrical power system*
**Decker, Luke Adam,** *Seismic diffraction imaging methods and applications*
**Eftekhari, Behzad,** *A rule based model of creating complex networks of connected fractures*
**Ergene, Suzan Muge,** *Lithologic heterogeneity of the Eagle Ford Formation, south Texas*
**Erturk, Nurtac,** *Comparison of direct-S modes produced by different source types*
**Ferreira, Elton Luiz Diniz,** *Improved estimation of pore connectivity and permeability in deepwater carbonates with the construction of multi-layer static and dynamic petrophysical models*
**Flinker, Raquel Henriques,** *Modeling of soil moisture dynamics of grasslands in response to $CO_2$ and biodiversity manipulations at BioCON*
**Galdeano Alexandres, Carlos,** *Modeling stormwater sewer systems using high resolution data*
**Gips, Jameson Parker,** *Shale characterization using TGA, Py-GC-MS, and NMR*
**Gustie, Patrick John,** *Characterization of VTI media with $PS_v$ AVO attributes*
**Hatch, Rosemary,** *Trace element incorporation in modern speleothem calcite and implications for paleoclimate reconstruction*
**Hester, Stephen Albert, III,** *Engineering and economics of enhanced oil recovery in the Canadian oil sands*
**Liang, Yu,** *Experimental study of convective dissolution of carbon dioxide in porous media*
**Maleski, Jacqueline Patrice,** *Direct shear wave polarization corrections at multiple offsets for anisotropy analysis in multiple layers*
**Mathisen, Maren Gabriella,** *Spatial and temporal evolution of the cave graben fault system, Guadalupe Mountains, New Mexico*
**Naraghi, Morteza Elahi,** *Geostatistical data integration in complex reservoirs*
**O'Neill, Laurie Christine,** *REE-Be-U-F mineralization of the Round Top Laccolith, Sierra Blanca peaks, Trans-Pecos Texas*

**Okafor, Brandon,** *Investigating pedogenic carbonate formation by measuring the stable isotope composition of water in Vertisols*
**Ozkul, Canalp,** *Fracture abundance and strain in folded Cardium Formation, Alberta fold-and-thrust belt, Canada*
**Pierre, Jon Paul,** *Impacts from above-ground activities in the Eagle Ford Shale play on landscapes and hydrologic flows, La Salle County, Texas*
**Pollard, Brittney Maryah,** *Reactivation of fractures as discrete shear zones from fluid enhanced reaction softening, Harquahala metamorphic core complex, west-central Arizona*
**Pommer, Maxwell Elliott,** *Quantitative assessment of pore types and pore size distribution across thermal maturity, Eagle Ford Formation, south Texas*
**Roberts, Julia Nicole,** *Direct in-situ evaluation of liquefaction susceptibility*
**Sadahiro, Makoto,** *Analysis of GPU-based convolution for acoustic wave propagation modeling with finite differences; Fortran to CUDA-C step-by-step*
**Shakiba, Mahmood,** *Modeling and simulation of fluid flow in naturally and hydraulically fractured reservoirs using embedded discrete fracture model (EDFM)*
**Shin, Timothy Andrew,** *Tectonic evolution of Aegean metamorphic core complexes, Andros and Tinos Islands, Greece*
**Simon, Rebekah Elizabeth,** *Syndepositional fault control on dolomitization of a steep-walled carbonate platform margin, Yates Formation, Rattlesnake Canyon, New Mexico*
**Smith, Brittany Claire,** *The effects of vegetation on island geomorphology in the Wax Lake Delta, Louisiana*
**Song, Dong Hee,** *Using simple models to describe oil production from unconventional reservoirs*
**Sun, Yuhao,** *Investigation of buoyant plumes in a quasi-2D domain; characterizing the influence of local capillary trapping and heterogeneity on sequestered $CO_2$; a bench scale experiment*
**Taylor, Jacob Matthew,** *Quantifying thermally driven fracture geometry during $CO_2$ storage*
**Tokan-Lawal, Adenike O.,** *Understanding fluid flow in rough-walled fractures using X-ray microtomography images*
**Tudor, Eugen Petrut,** *Facies variability in deep water channel-to-lobe transition zone; Jurassic Los Molles Formation, Neuquen Basin, Argentina*
**Verma, Rahul,** *Validation of level set contact angle method for multiphase flow in porous media*
**Victor, Rodolfo Araujo,** *Pore scale modeling of rock transport properties*
**Wang, Weiwei,** *Study of natural and hydraulic fracture interaction using semi-circular bending experiments*

Bachelor

**Eljuri, Audrey,** *Role of rain gardens and vegetated retention ponds in redirecting stormwater runoff into evaporation or recharge in downtown Austin, Texas*
**Kocis, Tiffany Noel,** *Understanding surface water - groundwater interactions on the Blanco River, Hays County, TX*
**Kurka, Nicole,** *Tracking the transition from aerial to aquatic flight in birds using quill markings*
**Le, Daniel N.,** *Optical remote sensing of glaciers in the northern Himachal Pradesh*
**Neely, Wesley R.,** *An examination of full tensor gravity gradiometry data; linking tensor components to structural features of the Vinton Salt Dome, Louisiana*
**Salin, Aaron Zachary,** *Quaternary slip transfer between the Banning and Mission Creek Faults of the southern San Andreas Fault, Southern California*
**Zurbuchen, Julie,** *Imaging evidence for Hubbard Glacier advances and retreats since the last glacial maximum in Yakutat and Disenchantment Bays, Alaska*

## University of Texas at Dallas

Doctorate

**Bao Dinh Nguyen,** *Efficient strategies and imaging conditions for elastic prestack reverse-time migration of reflection seismic data*

## University of Toledo

Masters

**Fugate, Joseph,** *Measurements of land subsidence rates on the northwestern portion of the Nile Delta using radar interferometry techniques*
**Maike, Christopher,** *A flood-tidal delta complex, the Holocene/Pleistocene boundary, and seismic stratigraphy in the Quaternary section off the southern Assateague Island coast, Virginia, USA*

## University of Tulsa

Masters

**Akintomide, Akinbobola,** *Structural analysis of Lower Paleozoic and Pennsylvanian rocks in West Velma Field, Stephens County, Oklahoma*
**Al Shawareb, Ahmed,** *Late Ordovician (Katian-Hirnantian) chitinozoans from northwest Saudi Arabia; biostratigraphic and paleoenvironmental implications*
**El-Waraky, Mohamed,** *Organic biomarker classification of oil families and oil source rock correlation for the West Velma and Sholem Alechem Fields in southern Oklahoma*
**Kocyigit, Zeynep,** *Subsurface structural analysis of the northern Sho-Vel-Tum region, Stephens County, Oklahoma*
**Lai, Xin,** *Upper Albian sequence stratigraphy and geochemical events of Comanche Shelf Interior Washita Group, southwest Texas*
**Liu, Xiaobo,** *Application of convolutional perfectly matched layer (C-PML) to seismic wave-field simulation with irregular free surface and graphic processing unit (GPU) implementation*
**Miller, Summer A.,** *Post eruptive source modeling for Okmok Volcano, Alaska using GPS and InSAR*
**Sanders, Cheryl M.,** *Structural geology of the Big Bend Anticline, Brooks Range foothills, Alaska*
**Shang, Peng,** *Tracking paleoredox depositional conditions in the Late Cretaceous ocean; a case study of the Mancos Shale, Smoky Hill Member*
**Tripplehorn, Tyler,** *Structural characterization and elastic flexure modeling of basement geometry in south central Oklahoma*
**Wang, Yulun,** *Chemostratigraphy of upper Albian (Lower Cretaceous) Comanche shelf margin*
**Zhang, Jie,** *Constraining large lava flows on Ontong Java Plateau by chemical correlation of flows and by crystallinity*
**Zhao, Zhencong,** *Simultaneous inversion of velocity and interface geometry using seismic travel times based on finite-frequency signals*

## University of Utah

Doctorate

**Lehane, James Richard Woodson,** *Applications of quantitative methods and chaos theory in ichnology for analysis of invertebrate behavior and evolution*
**Stolp, Bernard Jan,** *Determining mean transit times of groundwater flow systems*
**Wang, Xin,** *Multisource crosstalk reduction and plane-wave least-squares Kirchhoff migration*

Masters

**Bierman, David Frank,** *Comparison study of magnetotelluric inversion using different transfer functions*
**Boswell, Jonathan T.,** *Porphyry system fertility discrimination and mineralization vectoring using igneous apatite substitutions to derive pre-exsolution melt mineralization component concentrations*
**Dame, Brittany Elise,** *Developing a new passive diffusion sampling array to detect helium anomalies associated with volcanic unrest*
**Linville, Lisa Mae,** *Dynamic earthquake triggering potential across EarthScope's transportable array*
**Pierson, Joel Thomas,** *Assessing nutrient loads from in-situ fertilizer amendments in Willard Spur*
**Solder, John Edward Eberly,** *Quantifying groundwater-surface water exchange; development and testing of Shelby tubes and seepage blankets as discharge measurement and sample collection devices*
**Szwarc, Tyler Scott,** *Interactions between axial and transverse drainage systems in the Late Cretaceous Cordilleran foreland basin; evidence from detrital zircons in the Straight Cliffs Formation, southern Utah*
**Turner, Alexandre Marcel,** *Temporal variability in fluvial deposition and the implications for static reservoir connectivity; John Henry Member, Straight Cliffs Formation (Cretaceous, Utah)*
**Whittaker, Stefanie Nicole,** *Seismic array constraints on the D" discontinuity beneath Central America*

## University of Victoria

Doctorate

**Anderson, Brock,** *A modelling study of ridge flank hydrothermal circulation globally, constrained by fluid and rock chemistry, and seafloor heat flow*
**Brant, Casey Ojistoh,** *An investigation of high- and low-temperature mid-ocean ridge hydrothermal systeMaster's using trace*

*element geochemistry and lithium isotopes*

**Gehrmann, Romina,** *Non-linear Bayesian inversion of controlled source electromagnetic data offshore Vancouver Island, Canada, and in the German North Sea*

**MacDougall, Andrew Hugh,** *A modelling study of the permafrost carbon feedback to climate change feedback strength, timing, and carbon cycle consequences*

**Scholz, Nastasja Anaïs,** *Submarine landslides offshore Vancouver Island, British Columbia and the [west coast] possible role of gas hydrates in slope stability*

**Simon, Karen,** *Improved glacial isostatic adjustment models for northern Canada*

**Steininger, Gavin,** *Determination of Seabed Acoustic Scattering Properties by Trans-Dimensional Bayesian Inversion*

**Yelisetti, Subbarao,** *Seismic structure, gas hydrate, and slumping studies on the Northern Cascadia margin using multiple migration and full waveform inversion of OBS and MCS data*

Masters

**Abhar, Kimia,** *Spatial-temporal analysis of blowout dunes in Cape Cod National Seashore using sequential air photos and LiDAR*

**Byrne, Brendan,** *Radiative Forcings of Well-Mixed Greenhouse Gases*

**Hobbs, Tiegan Elizabeth,** *Insights from the Mw 7.8 2012 Haida Gwaii earthquake: static stress Modelling and empirical Green's function analysis*

**Linton, Hayley Christina,** *Spatial and Temporal Variations in Hydroclimatic Variables Affecting Streamflow across Western Canada*

**Murowinski, Emma Christina,** *Parameterizing the breaking and scattering of a mode-1 internal tide on abrupt step topography*

**Nephin, Jessica,** *Benthic Macrofaunal and Megafaunal Distribution on the Canadian Beaufort Shelf and Slope*

**St-Hilaire, Vikki Maria,** *Holocene glacial history of the Bowser River Watershed, Northern Coast Mountains, British Columbia*

**Stacey, Cooper D.,** *Frequency and initiation mechanics of submarine slides on the Fraser Delta front*

**Teeter, Lianna,** *Modelling oxygen and argon to improve estimation of net community productivity in a coastal upwelling zone using $O_2/Ar$*

**Tesdal, Jan-Erik,** *The spatial and temporal distribution of oceanic dimethylsulfide and its effects on atmospheric composition and aerosol forcing*

**Vallarino, Amy,** *Implications of shallow groundwater and surface water connections for nitrogen movement in typical Boreal Plain landscapes*

**Zoeller, Khalhela,** *Insights into the distribution and mobility of metals in the sheeted dike complex formed at fast-spreading ridges (Pito Deep, EPR)*

## University of Western Ontario

Doctorate

**Eshaghi, Attieh,** *Magnitude estimation for earthquake and tsunami early warning systems*

**Fereidoni, Azadeh,** *Seismicity processes in the Charlevoix seismic zone, Eastern Canada*

**Laarman, Jordan E.,** *A detailed metallogenic study of the McFaulds Lake chromite deposits, northern Ontario*

**Renaud, Jim Antonio,** *The Aricheng basement-hosted albitite-type uranium deposit, Roraima Basin, co-operative Republic of Guyana, South America*

**Sanchez, Laura Anabelle,** *Statistical analysis and computer modelling of volcanic eruptions*

Masters

**Buitenhuis, Eric,** *The Latte gold zone, Kaminak's Coffee gold project, Yukon, Canada; geology, geochemistry, and metallogeny*

**Bynoe, Laurisha,** *Shear zone influence on the emplacement of a giant pegmatite; the Whabouchi lithium pegmatite, Quebec, Canada*

**Edey, David R.,** *Micro-computed tomography semi-empirical beam hardening correction; method and application to meteorites*

**Nuhn, Anna M.,** *Morphologic and structural mapping of layered central uplifts on Mars*

**Pickersgill, Annemarie E.,** *Shock metamorphic effects in lunar and terrestrial plagioclase feldspar investigated by optical petrography and micro-X-ray diffraction*

**Skuce, Mitchell E.,** *Isotopic fingerprinting of shallow and deep groundwaters in southwestern Ontario and its applications to abandoned well remediation*

**Ulanowski, Thomas A.,** *Hydrology and biogeochemistry of a bog-fen-tributary complex in the Hudson Bay Lowlands, Ontario, Canada*

## University of Wisconsin at Madison

Doctorate

**Ankney, Meagan Elise,** *Isotopic records of deep and shallow magmatic evolution of Mount Mazama, Crater Lake, Oregon*

**Aswasereelert, Wasinee,** *Astronomical and stochastic influences on lacustrine and marine environments during the Cenozoic; case studies from the Green River Formation (Eocene) and the world's ocean (late Paleogene-Present)*

**Percak-Dennett, Elizabeth Maria,** *Microbial iron redox cycling in terrestrial environments*

**Rook, Deborah L.,** *Unearthing evolution; connecting geological and biological histories in the fossil record of North America*

**Winsor, Kelsey,** *A chronology of late Quaternary southwestern Greenland ice sheet retreat using terrestrial and marine records*

**Wu, Tao,** *A study of microbial reduction of iron oxides and phyllosilicates in natural subsurface sediments*

Masters

**Castongia, Ethan E. E.,** *An experimental investigation of distributed acoustic sensing (DAS) on lake ice*

**Haroldson, Erik L.,** *Fluid inclusion and stable isotope study of Magino; a magmatic related Archean gold deposit*

**Haserodt, Megan Joy,** *Effects of roads on groundwater flow patterns in peatlands and implications for nearby salmon streams on the Kenai Peninsula, AK*

**Jeppson, Tamara,** *Multi-scale analysis of San Andreas fault zone physical properties*

**Jiang, Xintong,** *Crystal growth of calcite and Mg-bearing calcite {104} surfaces; in-situ observation using a flow-through fluid cell*

**Lemon, Sarah A. M.,** *Ambient noise tomography of the Katmai volcanic cluster*

**Rawles, Christopher Jeffrey,** *Microseismicity near the central Alpine Fault, South Island, New Zealand; focal mechanisms and state of stress*

**Schaen, Allen J.,** *Eocene to Pleistocene evolution of the Delarof Islands, Aleutian Arc from $^{40}Ar/^{39}Ar$ geochronology and geochemistry of plutonic and volcanic rocks*

**Wende, Allison Marie,** *Identifying distinct mantle and crustal influences in individual cone-building stages at Mt. Shasta using U-Th and Sr isotopes*

## University of Wisconsin at Milwaukee

Doctorate

**Feriancikova, Lucia,** *The spread of emerging contaminants in the soil-groundwater system*

**Huang, Wei,** *Spatial dimensions of tower karst and cockpit karst; a case study of Guilin, China*

Masters

**Block, Jane,** *The Rice Bay and Northeast Bay gneiss domes; a kinematic study of competent rock bodies in the Rainy Lake region of Ontario, Canada*

**Calhoun, Justin,** *Fabric and microstructural analysis of the Loch Borralan Pluton, Northwest Highlands, Scotland*

**Egan, Alice,** *Simulating recharge in a Wisconsin watershed; the effect of sub annual precipitation patterns*

**Fedorchuk, Nicholas David,** *Evaluating the biogenicity of fluvial-lacustrine stromatolites from the Mesoproterozoic Copper Harbor Conglomerate, Upper Peninsula of Michigan, USA*

**Greenwood, Steven Michael,** *Mineralogy and geochemistry of Pleistocene volcanics at Embagai Caldera and Natron Basin, Tanzania; potential constraints on the stratigraphy of Olduvai Gorge*

**Jung, Na-Hyun,** *Fault-controlled advective, diffusive, and eruptive $CO_2$ leakage from natural reservoirs in the Colorado Plateau, east-central Utah*

**Pauls, Kathryn N.,** *Sedimentology and paleoecology of fossil-bearing, high-latitude marine and glacially influenced deposits in the Tepuel Basin, Patagonia, Argentina*

**Rolle, Jenna,** *Early Triassic echinoids of the Western United States; their implications for paleoecology and the habitable zone hypothesis following the Permo-Triassic mass extinction*

**Watson, Zachary T.,** *An analysis of $CO_2$-driven cold-water geysers in Green River, Utah and Chimayo, New Mexico*

## University of Wyoming

Doctorate

**Churchill, Morgan,** *The evolution and paleoecology of seals and walruses (Carnivora, Pinnipedia)*

**Everson, Erik D.,** *Seismic structure of the Costa Rican subduction system from active-source onshore-offshore seismic data and* *imaging plate boundary processes at the Cascadia subduction zone offshore Washington*

**Li, Ye,** *An uncertainty analysis of modeling geologic carbon sequestration in a naturally fractured reservoir at Teapot Dome, Wyoming*

Masters

**Bagdonas, Davin A.,** *Petrogenesis of the Neoarchean Wyoming Batholith, central Wyoming*

**Chang, Yang,** *$CO_2$ sequestration and EOR cooptimization in a mature field; Teapot Dome, Wyoming*

**Garnier, Bridget,** *Evaluation and interpretation of faulting on the Rock Springs Uplift, SW Wyoming*

**Lightner, Erik,** *Reconstruction of $C_4$ abundance using carbon isotope ratios of coprolites; rabbit coprolites track vegetation change better than rodent coprolites*

**Mandl, Maximilian B.,** *Late-Pleistocene and early-Holocene climate signals in lake-sediment $\delta^{18}O$ records from the Northeastern United States*

**Ratigan, Deirdre,** *Paleosol geochemistry and mineralogy in the Upper Jurassic Morrison Formation, southeast Utah*

**Wang, Dongdong,** *Physically based stochastic inversion parameter uncertainty assessment on a confined aquifer with a highly scalable parallel solver*

**Yu, Meng,** *Modeling and reservoir simulation of $CO_2$ storage in the Tensleep Formation; surrogate models based on the Teapot Dome*

**Zhang, Yifan,** *Physics-based groundwater inversion of fractured aquifers with unknown boundary conditions*

## Utah State University

Doctorate

**Kessler, James Andrew,** *In-situ stress and geology from the MH-2 borehole, Mountain Home, Idaho; implications for geothermal exploration from fractures, rock properties, and geomechanics*

**Petrie, Elizabeth Sandra,** *Rock strength of caprock seal lithologies; evidence for past seal failure, migration of fluids and the analysis of the reservoir seal interface in outcrop and the subsurface*

**Potter, Katherine Elizabeth,** *The Kimama Core; a 6.4 Ma record of volcanism, sedimentation, and magma petrogenesis on the axial volcanic high, Snake River plain, ID*

Masters

**Flores, Santiago L.,** *Mesoscale deformational features near outcrop analogs of a reservoir-seal interface; implications for seal failure*

**Geiger, Faye L.,** *Landscape evolution of the Needles fault zone, Utah, investigated through chronostratigraphic and terrain analysis*

## Vanderbilt University

Doctorate

**Pamukcu, Ayla Susan,** *Understanding the what, when, where, and why of supereruptions*

Masters

**Copeland, Marja Antoinell,** *Stream-channel morphology in a mixed-bedrock valley; the Harpeth River watershed, Middle Tennessee*

**Doane, Tyler Hill,** *Hillslope characteristics and behavior in relation to nonlocal sediment transport*

**Katsiaficas, Nathan James,** *Provenance of modern soils of middle Tennessee assessed using zircon U-Pb geochronology and element mass fluxes*

**Myers, Christopher Glen,** *A high resolution speleothem record from NE India; paleoseismic and modern climate insights through U-Th and d18O analysis*

**Williams, Lauren Alexandra,** *Late Quaternary stratigraphy and infilling of the Meghna River valley along the tectonically active eastern margin of the Ganges-Brahmaputra-Meghna Delta*

**Worland, Scott Campbell,** *Source, transport, and evolution of saline groundwater in a shallow Holocene aquifer on the tidal deltaplain of southwest Bangladesh*

## Washington University

Doctorate

**Fraeman, Abigail Ann,** *Materials and surface processes at Gale Crater and the moons of Mars derived from high spatial and spectral resolution orbital datasets*

**Metzger, J. Garrecht,** *The Late Ordovician biogeochemical carbon cycle*

## Western Michigan University

Doctorate

**Alharbi, Talal Ghazi,** *Integrated (remote sensing, GIS, and modeling) hydrological investigation and landslide susceptibility studies in the Arabian Shield*

**El Kadiri, Racha,** *An integrated approach (remote sensing, GIS, engineering, data mining) for modeling, assessing and mitigating slope stability hazards in mountainous environments*

Masters

**Guzman, Ivan R.,** *Stratigraphic framework and landsystem correlation for deposits of the Saginaw Lobe, Michigan, USA*

## Western Washington University

Masters

**Barber, Alec,** *Sediment and vegetation monitoring during a levee removal project on the Stillaguamish River Delta at Port Susan Bay, WA*

**Clement, Curtis R.,** *Estimating sediment yield from the Swift Creek landslide, Whatcom County, Washington State*

**Cota, Angela C.,** *A geochemical study of the Riddle Peaks Gabbro, north Cascades; evidence for amphibole accumulation in the mid-crust of an arc*

**Ferreira, Benjamin R.,** *High-resolution lidar mapping and analysis to quantify surface movement of Swift Creek landslide, Whatcom County, WA*

**Gouran, Brian D.,** *Variation in state seismic mitigation policies: a comparative analysis of seismic risk and policy development*

**Heard, Kathryn Elizabeth,** *Systematic analysis of terrestrial carbon stocks in a small catchment of the Kolyma watershed*

**Hoffnagle, Eric A.,** *Age, origin, and tectonic evolution of the Yellow Aster Complex; northwest Washington State*

**Mack, Chelsea Joy,** *Quantifying submarine eruptive flux from interpretation of hydroacoustic signals, West Mata Volcano, Lau Basin*

**Messa, Stephanie,** *The question of resilient and effective ecosystem governance: a case study of the Abbotsford-Sumas Aquifer International Task Force*

**Messé Graham Thomas,** *Magnetostratigraphy and block rotation of the Mecca Hills, CA*

**Messe, Graham T.,** *Magnetostratigraphy and block rotation of the Mecca Hills, CA*

**Schierl, Zachary P.,** *Effectiveness of time-lapse videos as a method to teach rates of surface geological processes*

**Winter, Hanna Maria,** *Temperature and moisture effects on respiration in the organic horizon of a Pacific Northwest forest soil*

## Woods Hole Oceanographic Institution

Doctorate

**Gonneea, Meagan E.,** *Temporal variability in chemical cycling of the subterranean estuary and associated chemical loading to the coastal ocean*

**Munson, Kathleen M.,** *Transformations of mercury in the marine water column*

**Ohnemus, Daniel Chester,** *The biogeochemistry of marine particulate trace metals*

**Toomey, Michael,** *Quaternary morphology and paleoenvironmental records of carbonate islands*

Masters

**Amrhein, Daniel Edward,** *An inverse approach to understanding benthic oxygen isotope records from the last deglaciation*

# Geological Surveys of the United States

## Alabama

### Geological Survey of Alabama

**Geological Survey of Alabama** (2016)
420 Hackberry Lane
P.O. Box 869999
Tuscaloosa, AL 35486-6999
p. (205) 349-2852
ntew@gsa.state.al.us
http://www.gsa.state.al.us/

State Geologist:
Berry H (Nick) Tew , (D), Alabama, 1999, GroGs
Division Manager, Groundwater Assessment:
Stephen C. Jones, (M), Alabama, 1996, HgGeCg
Division Manager, Geologic Investigations Division:
Sandy M. Ebersole, (D), Alabama, 2009, Ogi
Division Manager, Energy Investigations Program:
Denise J. Hills, (M), Delaware, 1998, YeEoOn
Division Manager, Ecosystems Investigations :
Stuart W. McGregor, (M), Tennessee Tech, 1987, Hs
Deputy Director, GSA:
Patrick E. O'Neil, (D), Alabama, 1993, Hs
Manager, Petroleum Systems & Technology:
David C. Kopaska-Merkel, (D), Kansas, 1983, GsPiOe
Manager, Geologic Mapping:
Gene Daniel Irvin, (M), Alabama, 1994, Gg
Visiting Professor:
William A. Thomas, (D), Virginia Tech, 1960, GtcGr
Scientific Aid:
Arthur McLin
GIS Specialist:
Elizabeth A. Wynn, (M), Alabama, 2008, HsOyi
Geologist & Paleontology Curator:
T. Lynn Harrell, Jr., (D), Alabama, 2016, PvGg
Geologist:
Mirza A. Beg, (M), Roorkee, 1964, Gz
Richard E. Carroll, (D), Michigan State, 1992, EcPlGo
Craig Cato, (M), Alabama, 2014, Gco
Jamekia Dawson, (B), Alabama, 2015, Hg
William T. Jackson Jr, (M), Memphis, 2012, GcsGt
Guohai Jin, (M), Zhejiang, 1989, Np
Mac McKinney, (B), Alabama, 2007, Gg
Neil E. Moss, (M), Alabama, 1987, Hw
Marcella Redden, (M), Alabama, 2004, Gg
David Tidwell, (M), South Florida, 2005, Gm
Dane S. VanDervoort, (M), Auburn, 2016, GcpOi
Environmental Engineering Specialist:
Amye S. Hinson, (B), Alabama, 2006, HwOn
Chemist:
Rick Wagner, (B), Texas (San Antonio), 1982, Cg
Biologist:
Rebecca A. Bearden, (B), Auburn, 2007, Hs

## Alaska

### Alaska Division of Geological & Geophysical Surveys

**Department of Natural Resources** (2017)
3354 College Road
Fairbanks, AK 99709-3707
p. (907) 451-5010
dggspubs@alaska.gov
http://www.dggs.alaska.gov
f: http://www.facebook.com/pages/Fairbanks-AK/Alaska-DGGS/346699054500
t: @akdggs

Director & State Geologist:
Steven S. Masterman, (M), Alaska (Fairbanks), 1990, EgNg
Petroleum Geologist I:
David Lepain, (D), Alaska (Fairbanks), 1993, EoGos
GeoScientist I:
Melanie B. Werdon, (D), Alaska (Fairbanks), Eg
Geologist V:
Janet R. G. Schaefer, (M), Alaska (Fairbanks), Gv
De Anne S.P. Stevens, (M), Alaska (Fairbanks), Ng
Geologist IV:
Marwan A. Wartes, (M), Wisconsin, GsoGt
Gabriel J. Wolken, (D), Gm
Geohydrologist-Geologist IV:
Ronnie Daanen, (D), Minnesota (St. Paul), Hy
Division Operation Manager:
Kenneth R. Papp, (M), Alaska (Fairbanks), Gg

## Alberta

### Alberta Geological Survey

**Alberta Geological Survey** (2016)
402, Twin Atria Building
4999 - 98 Avenue
Edmonton, AB T6B 2X3
p. (780) 638-4491
AGS-Info@aer.ca
http://ags.aer.ca/

## Arizona

### University of Arizona

**Arizona Geological Survey** (2017)
1955 E 6th St.
PO Box 210184
Tucson, AZ 85721
p. (520) 621-2352
fmconway@email.arizona.edu
http://www.azgs.az.gov
f: https://www.facebook.com/AZ.Geological.Survey
t: @AZGeology

Director & State Geologist:
Philip A. Pearthree, (D), Arizona, Ge
Research Geologist :
Joseph P. Cook, (M), Arizona, Gm
Research Geologist:
Charles A. Ferguson, (D), Calgary, Gcv
Brian F. Gootee, (M), Arizona State Univ, Gm
Brad Johnson, (D), Carleton, 1994, Gc
Jeri J. Young, (D), Arizona State Univ, Ys
Chief, Environmental Geology:
Ann Youberg, (D), Arizona, Ge
Chief, Geologic Extension Service:
Michael F. Conway, (D), Michigan Tech, 1993, OeGvg

## Arkansas

### Arkansas Geological Survey

**Arkansas Geological Survey** (2014)
Vardelle Parham Geology Center
3815 West Roosevelt Road
Little Rock, AR 72204
p. (501) 296-1877
ags@arkansas.gov
http://www.geology.ar.gov/home/
Administrative Assistant: Laure Hinze

Director & State Geologist:
Bekki C. White, (M), Centenary, 1993, GoEoGg
Geologist Supervisor:
William L. Prior, (M), Memphis State, 1979, GggEc
Information Systems Analyst:
James K. Curry, (M), S Methodist, 1978, On
Geologist:
Sandra Chandler, Oe
Andrew Haner, Oi
David Johnston, Gg
Lea Nondorf, Cg
Senior Petroleum Geologist:
Peng Li, (D), Alabama, 2007, Gog
M. Ed Ratchford, (D), Idaho, 1994, GocEo

Professional Geologist:
Richard S. Hutto, (B), Arkansas Tech Univ, 1994, Gg
GIS Analyst:
Nathan H. Taylor, (B), 2007, OiyGe
Geology Supervisor:
Scott Ausbrooks, (B), Arkansas, 2001, YsGeg
Angela Chandler, (M), Arkansas, 1996, GgrGs
Doug Hanson, (M), Memphis State, 1991, Gg
Geologist:
Ty Johnson, (M), Arkansas, 2008, GgOi
Daniel S. Rains, (B), Arkansas, 2002, GgOi
Deputy Director & Asst State Geologist:
Mac B. Woodward, (B), S State, 1957, Gog

# British Columbia

## British Columbia Geological Survey & Development

**British Columbia Geological Survey & Development** (2011)
PO Box 9333 Stn Prov Govt
Victoria, BC V8W 9N3
Geological.Survey@gov.bc.ca
http://www.empr.gov.bc.ca/MINING/GEOSCIENCE/Pages/default.aspx

# California

## California Geological Survey

**California Geological Survey** (2016)
801 K Street
Suite 1200
Sacramento, CA 95814
p. (916) 445-1825
cgshq@consrv.ca.gov
http://www.consrv.ca.gov/CGS/Pages/Index.aspx
State Geologist:
John G. Parrish, (D)
Supervising Engineering Geologist:
John Clinkenbeared, (B), Eg
Anthony F. Shakal, (D), MIT, 1980, Ys
Christopher J. Wills, (M), Wisconsin, Ng
Supervising Engineering Geologist:
Timothy P. McCrink, (M), New Mexico Tech, 1982, Ng
William Short, (M), Ng
Senior Engineering Geologist (Supervisor):
Jennifer Thornburg, (M), California (Santa Cruz), Ng
Senior Engineering Geologist:
Rui Chen, (D), Edmonton, Nr
Ron C. Churchill, (D), Minnesota, 1980, Cg
Cliff Davenport, (B), Ng
Tim Dawson, (M), Ng
Marc Delattre, (M), Ng
Jim Falls, (B), Ng
Pamela J. Irvine, (M), California (Berkeley), 1977, Ng
Pamela Irvine, (M), Ng
Donald Lindsay, (M), Ng
David Longstreth, (M), Ng
Gerald Marshall, (B), Ng
Steven Reynolds, (M), Ng
Anne Rosinski, (M), Ng
Michael A. Silva, (B), California (Davis), 1978, Ng
Jim Thompson, (B), Ng
Rick I. Wilson, (B), Fresno State, 1987, Ge
Senior Engineer:
Moh J. Huang, (D), Caltech, 1983, Ne
Geologist:
Lawrence Busch, (A), Eg
Chris T. Higgins, (M), California (Davis), 1977, Gg
Engineering Geologist:
Patrick Brand, (M), Ng
David Branum, (B), Ng
John Church, (B), Eg
Kevin Doherty, (M), Ng
Michael Fuller, (B), Ng
Carlos Guiterrez, (B), Sacramento State, Ng
Will Harris, (B), Ng
Wayne Haydon, (M), Ng
Cheryl Hayhurst, (B), Ng
Janis Hernandez, (M), Ng
Peter Holland, (M), Ng
Jeremy Lancaster, (M), Ng
Mike Manson, (B), Ng
Maxime Mareschal, (M), Ng
Ante Mlinarevic, (M), Ng
Brian Olson, (M), Ng
John Oswald, (B), Ng
Florante Perez, Ng
Cindy L. Pridmore, (M), San Diego State, 1983, Ng
Pete Roffers, (M), Ng
Ron Rubin, (M), Ng
Gordon Seitz, (D), Ng
Joshua Smith, (B), Eg
Eleanor Spangler, (M), Ng
Brian Swanson, (M), Ng
Mark Weigers, (M), Ng
Chase White, (M), Ng
Civil Engineer:
Badie Rowshandel, (D), Ne
Associate Scientist:
Carla Rosa, (M), Ng

# Colorado

## Colorado School of Mines

**Colorado Geological Survey** (2017)
1801 19th Street
Golden, CO 80401
p. 303-384-2655
cgs_pubs@mines.edu
http://www.coloradogeologicalsurvey.org/
f: https://www.facebook.com/ColoradoGeologicalSurvey
State Geologist:
Karen Berry, (B), Colorado Mines, NgOu
Senior Mapping Geologist:
Matt Morgan, (M), Colorado Mines, 2006, GgmXm
Senior Hydrogeologist:
Peter Barkmann, (M), Montana, 1984, HwGcg
Senior Geothermal Geologist:
Paul Morgan, (D), Imperial Coll, 2003, YhGtYg
Senior Engineering Geologist:
Jonathan Lovekin, (M), Colorado Mines, 2007, NgOuGs
Engineering Geologist:
Jill Carlson, (B), Wesleyan, 1987, NgOu
Senior Engineering Geologist:
Jonathan White, (B), Eastern Illinois Univ, 1983, NgOuGg
Hydrogeologist:
Lesley Sebol, (D), Waterloo, 2005, HwGeOg
GIS Hazard Analyst:
Francis Scot Fitzgerald, (M), Denver, 2011, OirGg
Geologist:
Kassandra Lindsey, (M), Portland State, 2015, NgGmOi
Mike O'Keeffe, (M), New Mexico Tech, 1994, GgeEg
Engineering Geologist:
Kevin McCoy, (D), Colorado Mines, 2015, NgOiHw
Scientific & Technical Graphic Designer:
Larry Scott, (B), the Arts, 1985, Oy

# Connecticut

## Dept of Energy and Environmental Protection

**Connecticut Geological Survey** (2017)
Office of Information Management
79 Elm Street, 6th floor
Hartford, CT 06106-5127
p. (860) 424-3540
deep.ctgeosurvey@ct.gov
http://www.ct.gov/deep/geology
State Geologist:
Margaret A. Thomas, (M), Connecticut, 1983, Gg
Senior Research Associate:
Randolph P. Steinen, (D), Brown, 1973, Gs
Civil Engineer:
Thomas E. Nosal, (M), C Connecticut State, 1992, On

Plant Ecologist:
Nelson DeBarros, (M), Penn State, 2011, On
Research Associate:
Lindsey Belliveau, (M), Connecticut, 2016, Gmt
Teresa K. Gagnon, (M), Boston Coll, 1992, Gg
Resource Assistant:
James M. Bogart, (B), S Connecticut State, 2015, Ge

# Delaware

## University of Delaware

**Delaware Geological Survey** (2016)
Delaware Geological Survey
257 Academy Street
Newark, DE 19716-7501
p. (302) 831-2833
delgeosurvey@udel.edu
http://www.dgs.udel.edu
Administrative Assistant: Karen L. D'Amato
Administrative Assistant: Laura K. Wisk
State Geologist:
David R. Wunsch, (D), Kentucky, 1992, ClHwNg
Hydrogeologist:
A. Scott Andres, (M), Lehigh, 1984, Hg
Senior Scientist:
Peter P. McLaughlin, (D), Louisiana State, 1989, HgPm
Hydrogeologist:
Thomas E. McKenna, (D), Texas (Austin), 1997, Hg
Hydrogeologist:
Changming He, (D), Nevada (Reno), 2004, GqHg
Associate Scientist:
Stefanie J. Baxter, (M), Delaware, 1994, On
Research Associate:
John A. Callahan, (M), Delaware, 2014, Ori
Jaime L. Tomlinson, (M), Delaware, 2006, Hg
Emeritus:
John H. Talley, (M), Franklin and Marshall, 1974, Oa
Scientist:
Kelvin W. Ramsey, (D), Delaware, 1988, Oa
William Schenck, (M), Delaware, 1997, OiGi
GIS Specialist:
Lillian T. Wang, (M), Delaware, 2005, Oi

# Florida

## Florida Geological Survey

**Florida Dept of Environmental Protection** (2016)
Commonwealth Building
3000 Commonwealth Blvd, Suite 1
Tallahassee, FL 32303
p. (850) 617-0300
jonathan.arthur@dep.state.fl.us
http://www.dep.state.fl.us/geology/
State Geologist:
Jonathan D. Arthur, (D), Florida State, 1994, HwCgGe
Assistant State Geologist:
Guy H. Means, (M), Florida State, 2009, PgGe

# Georgia

## Georgia Dept of Natural Resources

**Georgia Environmental Protection Div** (2016)
Environmental Protection Division
2 Martin Luther King Jr. Dr., Suite 1152
East Tower
Atlanta, GA 30334-9004
p. (404) 657-5947
askepd@gaepd.org
http://epd.georgia.gov/
Senior Scientist:
James Kennedy, (D), Texas A&M Univ, 1981

# Hawaii

## Dept of Land & Natural Resources

**Commission on Water Resource Management** (2017)
Kalanimoku Building
P.O. Box 621
1151 Punchbowl Street, #227
Honolulu, HI 96809
p. (808) 587-0214
dlnr.cwrm@hawaii.gov
http://dlnr.hawaii.gov/cwrm

# Idaho

## University of Idaho

**Idaho Geological Survey** (2015)
875 Perimeter Dr. MS 3014
University of Idaho
Moscow, ID 83844-3014
p. (208) 885-7991
igs@uidaho.edu
http://www.idahogeology.org/
Director:
Michael Ratchford, (D), Idaho, 1994, GcoEg
Professor:
John A. Welhan, (D), California (San Diego), 1981, Hw
Associate Professor:
Virginia S. Gillerman, (D), California (Berkeley), 1982, Eg
Associate Scientist:
Reed S. Lewis, (D), Oregon State, 1990, GgiEg
Assistant Professor:
William M. Phillips, (D), Arizona, 1997, Gm
Senior Petroleum Geologist:
Renee L. Breedlovestrout, (D), Idaho, 2011, PbGro
Senior Geologist:
Dennis M. Feeney, (M), W Washington, 2008, GigYg
Emeritus Director :
Roy M. Breckenridge, (D), Wyoming, 1975, Gg
Emeritus Director:
Kurt L. Othberg, (D), Idaho, 1991, GmgGl
Manager, Digital Geological Mapping:
Loudon R. Stanford, (M), Idaho, 1982, GlmGg
Staff:
Jane S. Freed, (B), Idaho, 1995, Oi
Research Assistant:
Glenda K. Bull

# Illinois

## Illinois State Geological Survey

**Energy & Earth Resource Center** (2014)
615 E. Peabody Drive.
Champaign, IL 61820-6964
p. (217) 244-2430
finley@isgs.uiuc.edu
Center Director:
Robert J. Finley, (D), South Carolina, 1975, Eo
Scientist:
Latif A. Khan, (D), Tech, 1971, Nx
Senior Scientist:
Richard A. Cahill, (D), Illinois, 1980, Ca
Scott M. Frailey, (D), Missouri (Rolla), 1989, Ng
Massoud Rostam-Abadi, (D), Wayne State, 1982, On
William R. Roy, (D), Illinois, 1985, Sc
Scientist:
Mei-In (Melissa) Chou, (D), Michigan State, 1977, Co
Sheng-Fu Joseph Chou, (D), Michigan State, 1977, Co
Joseph A. Devera, (M), S Illinois, 1985, Gr
Ivan G. Krapac, (M), Illinois, 1987, Ca
Zakaria Lasemi, (D), Miami, 1990, En
Hannes E. Leetaru, (D), Illinois, 1997, Eo
Donald G. Mikulic, (D), Oregon State, 1979, Ps
Beverly Seyler, (M), SUNY, 1978, Eo
Associate Scientist:
Cheri A. Chenoweth, (B), Illinois, 1979, Ec
Joan E. Crockett, (B), Illinois, 1983, EoOg
John P. Grube, (M), Colorado Mines, 1984, Eo
Bryan G. Huff, (M), Illinois, 1984, Eo
Rex A. Knepp, (M), Go

Geological Surveys

Assistant Scientist:
F. Brett Denny, (B), Missouri (Rolla), 1985, Gr
Christopher P. Korose, (B), Illinois, 1995, Ec
Vinodkumar A. Patel, (B), Inst of Tech, 1973, On
Emeritus:
Pam Cookus, On
Assistant Scientist:
Kathleen M. Henry, (B), Illinois State, 1982, Gg

**Geologic Mapping and Hydrogeology Center** (2017)
615 East Peabody Drive
Champaign, IL 61820-6964
p. (217) 244-2430
keefer@isgs.illinois.edu
Administrative Assistant: Bonnie Renfrew
Head:
Donald Keefer, (M), Illinois (Urbana), 1992, Hw
Senior Scientist:
Robert A. Bauer, (M), Illinois, 1983, Nr
Leon R. Follmer, (D), Illinois, 1970, Sd
Keith C. Hackley, (M), Illinois, 1984, Cs
Ardith K. Hansel, (D), Illinois, 1980, Gl
Samuel V. Panno, (M), S Illinois, 1978, Cg
Scientist:
Michael L. Barnhardt, (D), Illinois, 1979, Gm
Michael J. Chrzastowski, (D), Delaware, 1986, Ou
Brandon B. Curry, (D), Illinois, 1995, Pe
David R. Larson, (M), Nebraska, 1976, Hw
Edward Mehnert, (D), Illinois, 1997, Hw
Christopher J. Stohr, (D), Illinois, 1996, Ng
C. Pius Weibel, (D), Illinois, 1987, Gr
Assistant Scientist:
Edward C. Smith, (B), Illinois State, 1985, Hw
Andrew J. Stumpf, (D), New Brunswick, 2001, Gl
Associate Scientist:
William S. Dey, (M), Illinois, 1983, Hw
David A. Grimley, (D), Illinois, 1996, Ym
Hue-Hwa Hwang, (D), Illinois, 1985, Cs
Richard J. Rice, (B), Illinois State, 1980, Hw
Wen-June Su, (M), Illinois, 1985, Ng
Robert C. Vaiden, (M), Illinois, 1985, Hw
Hong Wang, (D), Illinois, 1996, Cc
Assistant Scientist:
Sallie E. Greenberg, (M), Illinois, 1997, Cs
Andrew C. Phillips, (D), Illinois (Chicago), 1993, Gm

## Illinois State Water Survey

**Analytical Chemistry and Technology** (2004)
2204 Griffith Drive
Champaign, IL 61820
p. (217) 333-9321
smothers@sws.uiuc.edu
http://www.sws.uiuc.edu
Head, Analytical Chemistry & Tech:
Kent W. Smothers, (B), Blackburn, 1980, Cg
Senior Chemist:
Gary R. Peyton, (M), North Texas, 1968, On
Senior Scientist:
Michael E. Caughey, (D), Texas, 1988, Co
Thomas R. Holm, (D), Caltech, 1978, Hw
Shundar Lin, (D), Syracuse, 1967, Hs
Michael L. Machesky, (D), Wisconsin, 1986, Cg
Donald P. Roseboom, (M), Bradley, 1976, Hy
Associate Chemist:
Jane E. Rothert, (M), Washington State, 1977, Og
Lab Director:
Daniel Webb, (M), Illinois, Hg
Laboratory Director:
Loretta M. Skowron, (B), Illinois, 1976, Ca

**Atmospheric Sciences Div** (2004)
2204 Griffith Drive
Champaign, IL 61820
p. (217) 333-2210
dkristo@illinois.edu

**Illinois State Water Survey** (2004)
2204 Griffith Drive
Champaign, IL 61820
p. (217) 333-2210
demissie@illinois.edu
Principal Scientist:
Nani G. Bhowmik, (D), Colorado State, 1968, Hs
Senior Professional Scientist:
H. Vernon Knapp, (M), Kansas, 1980, Hs
Associate Hydrogeologist:
Steven D. Wilson, (M), Illinois, 1988, Hw
Professional Scientist:
William C. Bogner, (M), Illinois, 1983, Hs
Associate Professional Scientist:
Walton R. Kelly, (D), Virginia, 1993, Hw
George S. Roadcap, (M), Ohio State, 1990, Hw
Renjie Xia, (D), Illinois, 1991, Hs
Assistant Professional Scientist:
Deva K. Borah, (D), Mississippi, 1979, Hq
Assistant Hydrogeologist:
Randall A. Locke, III, (M), Iowa, 1994, Hw
Scott C. Meyer, (M), North Carolina, 1987, Hw

**Office of the Director** (2013)
2204 Griffith Drive
Champaign, IL 61820
p. (217) 244-5459
demissie@illinois.edu
http://www.isws.illinois.edu

## University of Illinois, Urbana-Champaign

**Illinois State Geological Survey** (2017)
615 East Peabody Drive
Champaign, IL 61820-6964
p. 217-333-4747
info@isgs.illinois.edu
https://www.isgs.illinois.edu/
f: https://www.facebook.com/ILGeoSurvey/
t: https://twitter.com/ILGeoSurvey
Administrative Assistant: Tamra S. Montgomery
Senior Geologist:
Anne L. Ellison, (M), Ge
Geoscience Information Stewardship:
Mark Yacucci, (M), Illinois State
Associate Geologist:
Scott D. Elrick, (M), GoEc
Chief Scientist:
Richard C. Berg, (D), Illinois, 1979, Gem
Senior Geologist:
Zakaria Lasemi, (D), Illinois, Eg
Senior Geochemist:
Samuel V. Panno, (M), S Illinois Univ, 1978, ClGgHw
Interim Chief Scientist, Head of Quaternary and Engineering Geology:
Steven E. Brown, (M), Wisconsin (Madison), 1990, Glm
Head Hydrogeology and Geophysics:
Donald A. Keefer, (M), Illinois, 1992, HwOn
Head Geochemistry:
Randall Locke, (M), Cl
Senior Scientist:
Edward Mehnert, (D), Illinois (Urbana-Champaign), 1998, Hyw
Associate Quaternary Geologist:
Andrew C. Phillips, (D), Illinois (Chicago), 1993, GsmOi
Assistant Petroleum Geologist:
Nathan D. Webb, (M), Illinois (Urbana Champaign), 2009, EoGo
Senior Bedrock Geologist:
C. Pius Weibel, (D), Gl
Illinois State Geologist:
E. Donald McKay III, (D), Illinois, 1977, GlsGr
Wetlands Geology Specialist:
Colleen Long, (M), North Carolina (Chapel Hill), 2012, Gg
Jessica Monson, (M), Gg
Wetlands Geologist and Head:
James J. Miner, (M), Gg
Wetlands Geologist:
Jessica R. Ackerman, On
Team Leader/Associate Geologist:
Dale R. Schmidt, (M), Ge

Team Leader:
D. Adomaitis, Ge
Senior Petroleum Geologist:
Hannes E. Leetaru, (D), Illinois, 1997, Gor
Senior Paleontologist:
Joseph A. Devera, (M), Pg
Senior Geophysicist:
Timothy H. Larson, (D)
Remote Sensing Data Manager:
Janet Carmarca
Program Manager, Illinois Height Modernization Project:
Sheena K. Beaverson, Or
Principal Engineering Geologist:
R. A. Bauer, (M), Ng
Petroleum Geologist:
Bryan G. Huff, (M), EcGo
Media and Information Technology Administrator:
Daniel Byers
Map Standards Coordinator:
Jennifer Carrell, Og
Hydrogeologist and Assistant Section Head:
Yu-Feng Forrest Lin, (D), Wisconsin (Madison), 2002, Hg
Geospatial Applications Developer/GIS Specialist:
Melony Barrett, Oi
Geologic Specialist:
Zohreh Askari, (M), Azad Univ, 1997, Gsc
Alan R. Myers, (B), EcGo
Jennifer M. Obrad, (M), EcGo
Mary J. Seid, (M), Gz
Enviromental Data Coordinator:
Clint Beccue, Ge
Engineering Geologist:
Christopher J. Stohr, (D), Illinois (Urbana-Champaign), OrGeNg
Associate Wetlands Geologist:
Steven Benton, Gg
Keith W. Carr, (M), Gg
Eric T. Plankell, (M), Hw
Geoff Pociask, (M), Hw
Associate Sedimentologist:
Xiaodong Miao, (D), Wisconsin (Madison), 2005, GmsGl
Associate Quaternary Geologist:
Olivier J. Caron, (D), Univ du Québec à Montréal, Ng
David A. Grimley, (D), Illinois, 1996, Ng
Andrew J. Stumpf, (D), Gl
Associate Petroleum Geologist:
Joan Crockett, (B), GoEc
Associate Geologist:
Curtis C. Albert, (B), HgYg
Cheri Chenoweth, GoEc
Christopher P. Korose, (M), EcGo
Tim Young, (B), Illinois (Urbana Champaign), 1989, Hw
Associate Geohydrologist:
Jason F. Thomason, (D), Hw
Associate Geochemist:
Shari E. Effert-Fanta, (M), Cas
Hue-Hwa Ellen Hwang, (D), CgGs
Associate Engineering Geologist:
Andrew Anderson, Ng
Greg A. Kientop, (M), Ge
Associate Economic Geologist:
F. Brett Denny, (M), GzEg
Associate Director Advanced Energy Technology Initiative:
Sallie Greenberg, (D), Illinois (Urbana Champaign), 2013, ClOg
Assistant Wetlands Geologist:
Kathleen E. Bryant, (M), Gg
Melinda C. Higley, (D)
Assistant Section Head, Environmental Assessments:
Mark Collier, Ge
Assistant Section Head:
B. Brandon Curry, (D), Illinois (Urbana-Champaign), 1995, Gel
Assistant Geologist:
James Damico, (M)
Craig R. Decker, (B)
Scott R. Ellis, (B), Ge
Bradley Ettlie, (B), Ge
Jared Freiburg, (M), GgzGx
James W. Geiger, (B), Ge
Nathan P. Grigsby, (B)
Matthew P. Spaeth, (M), Ge
Assistant Geochemist:
Peter M. Berger, (M), Illinois (Urbana-Champaign), 2008, Cg
Assistant Director:
Mona M. Knight, (M), On
Other:
Hong Wang, Illinois (Urbana-Champaign), 1996, Cg
Related Staff:
Torie L. Strole, (B), On

## Indiana

### Indiana University

**Indiana Geological Survey** (2016)
611 North Walnut Grove Avenue
Bloomington, IN 47405
p. (812) 855-7636
igsinfo@indiana.edu
http://igs.indiana.edu/

State Geologist & Director:
Todd A. Thompson, (D), Bloomington, 1987, GsmGu
Assistant Director, Technical Services:
Richard T. Hill, (B), Oi
State Geologist Emeritus/Senior Scientist:
John C. Steinmetz, (D), Miami, 1978, Pm
Research Geophysicist/Hydrologist:
Kevin M. Ellett, (M), California (Davis), 2002, YeHw
Head, Subsurface Geology:
Charles W. Zuppann, (M), Vanderbilt, 1974, Go
Head, Geologic Mapping:
Nancy R. Hasenmueller, (M), Ohio State, Ge
Senior Scientist:
Tracy D. Branam, (M), Indiana, CasEm
Walter Hasenmueller, (M), Ohio State, 1970, Gr
Sally L. Letsinger, (D), Indiana, 2001, HqOir
Maria Mastalerz, (D), Silesian Tech, 1988, Ec
John A. Rupp, (M), E Washington, 1980, Go
Reservoir Geologist:
Cristian R. Medina, (M), Indiana, 2007, GoeHw
Head, Center for Geospatial Data Analysis:
Shawn Naylor, (M), Indiana, Hw
Associate Scientist:
Christopher Dintaman, (M), Indiana, Hw
Agnieszka Drobniak, (D), Ec
Rebecca A. Meyer, Ec
Research Scientist:
Patrick I. McLaughlin, (D), Cincinnati, 2006, GrClGs
Quaternary Geologist:
Henry Loope, (D), Wisconsin, 2013, Gls

## Iowa

### Iowa Dept of Natural Resources

**Iowa Geological and Water Survey** (2014)
109 Trowbridge Hall
Iowa City, IA 52242-1319
p. (319) 335-1575
MaryPat.Heitman@dnr.iowa.gov
http://www.igsb.uiowa.edu
Administrative Assistant: Mary Pat. Heitman

Research Geologist:
Richard A. Langel, (M), Iowa, 1996, Gg
State Geologist:
Robert D. Libra, Gg
Section Supervisor, Geology and Groundwater Studies:
J. Michael Gannon, (M), Arizona, Gg
Section Supervisor, Geographic Information:
Chris Ensminger
Research Geologist:
Mary R. Howes, Gg
Lynette S. Seigley, (M), Iowa, Gg
Natural Resource Biologist:
Jacklyn Gautsch, (B), Wisconsin, On
GIS Technician:
Chris Kahle, (M), Kansas, Oi

Geologist 3, Research Geologist:
Paul Hiaibao Liu, (D), Nebraska, Gg
Robert M. McKay, (B), Tulane, Gg
Deborah J. Quade, (M), Iowa, Gl
Robert Rowden, (M), Iowa, Gg
Keith Schilling, (M), Iowa State, Gg
Stephanie Tassier-Surine, (M), Massachusetts, Gg
Paul E. VanDorpe, (M), Wayne State, Gg
Geologist 3, Remote Sensing Analyst:
James D. Giglierano, (M), Purdue, GgOri
Pete Kollasch, (M), Iowa, GgOri
Geologist 3, GIS Analyst:
Kathryne Clark, (M), New Mexico, GgOi
Geologist 3, Geographic Information System Analyst:
Calvin Wolter, (B), Arizona, GgOi
Geologist 2, Research Geologist:
Michael Bounk, (M), Iowa, Gg
Chad Fields, (M), N Iowa, Gg
Geologist 3, NRGIS Library Manager and GIS Analyst:
Casey Kohrt, (B), Iowa State, GgOi

## Kansas

### University of Kansas

**Kansas Geological Survey** (2015)
1930 Constant Avenue
West Campus
Lawrence, KS 66047-3724
p. (785) 864-3965
jbogle@kgs.ku.edu
http://www.kgs.ku.edu/
f: www.facebook.com/KansasGeologicalSurvey
t: @ksgeology

Interim Director:
Rex C. Buchanan, (M), Wisconsin, 1982, Oe
Section Chief, Senior Scientist:
James J. Butler, Jr., (D), Stanford, 1987, Hw
Richard D. Miller, (D), Leoben, 2007, Ye
Section Chief, Senior Research Associate:
Robert S. Sawin, (M), Kansas State, 1977, Gg
Manager, Wichita Well Sample Library:
Mike Dealy, (B), Fort Hays State, 1979, Gg
Manager, GIS Section/DASC:
Kenneth A. Nelson, (B), Kansas, 1993, Oy
Manager, Geohydrology Support Services:
Blake Wilson, (M), Kansas State, 1993, Oi
Senior Scientific Fellow:
John H. Doveton, (D), Edinburgh, 1969, Go
Evan K. Franseen, (D), Wisconsin, 1989, Gs
Lynn W. Watney, (D), Kansas, 1985, Gr
Donald O. Whittemore, (D), Penn State, 1973, Hw
Emeritus, Senior Scientist:
Ricardo A. Olea, (D), Kansas, 1982, GqoEc
Senior Scientist:
Greg A. Ludvigson, (D), Iowa, 1988, GrCsPe
Rolfe D. Mandel, (D), Kansas, 1991, Ga
Assistant Scientist:
Andrea Brookfield, (D), Waterloo, 2009, Hy
Gaisheng Liu, (D), Alabama, 2004, Hy
Kerry D. Newell, (D), Kansas, 1996, Go
Jon J. Smith, (D), Kansas, 2007, Gg
Associate Scientist:
Geoffrey C. Bohling, (D), Kansas, 1999, Hw
Senior Research Associate:
Jason Rush, (M), Texas, 2001, Gg
Senior Research Assistant:
Eileen Battles, (B), Kansas, 1993, Oy
Edward Reboulet, (M), Boise State, 2003, Hy
Petroleum Engineer:
Yehven I. Holubnyak, (M), North Dakota, 2008, On
Geologist, Data Resources Library:
Daniel R. Suchy, (D), McGill, 1992, Gg
Assistant Research Professor:
Julian Ivanov, (D), Kansas, 2002, Yg
Courtesy Professor:
James J. Butler, (D), Stanford, 1986, Hw
Emeritus, Senior Scientist:
Pieter Berendsen, (D), California (Riverside), 1971, Cg
Robert W. Buddemeier, (D), Washington, 1969, Hy
Tim R. Carr, (D), Wisconsin, 1981, Gg
John C. Davis, (D), Wyoming, 1967, GqoGd
Lee C. Gerhard, (D), Kansas, 1964, Gr
Daniel F. Merriam, (D), Kansas, 1961, Gg
Emeritus, Senior Scientific Fellow:
Lawrence L. Brady, (D), Kansas, 1971, Gg
Emeritus, Associate Scientist:
Truman Waugh, (B), Washburn, 1963, Ca
Emeritus, Assistant Director:
Lawrence H. Skelton, (M), Wichita State, 1991, Gg
Senior Research Assistant:
Brett Bennet, (B), Kansas, 1982, Ng
Hydrogeochemist:
Jordi Batlle-Aguilar, (D), Liege, 2008, Hy
Geology Extension Coordinator:
Susan G. Stover, (M), Kansas, 1993, Gg
Assistant Scientist:
Tandis Bidgoli, (D), Kansas, 2014, Gc

## Kentucky

### University of Kentucky

**Kentucky Geological Survey** (2017)
228 Mining & Minerals Resources Building
504 Rose Street
Lexington, KY 40506-0107
p. (859) 257-5500
jerryw@uky.edu
http://www.uky.edu/KGS

Interim State Geologist and Director:
Gerald A. Weisenfluh, (D), South Carolina, 1982, Ec
Head, Western Kentucky Office:
David A. Williams, (M), E Kentucky, 1979, Eg
Head, Water Resources Section:
Chuck J. Taylor, (M), Kentucky, 1992, Gg
Head, Geoscience Information Management:
Doug C. Curl, (M), Tennessee, 1998, Gc
Head, Geologic Hazards Section:
Zhenming Wang, (D), Kentucky, 1998, Ys
Head, Energy & Minerals:
David C. Harris, (M), SUNY (Stony Brook), 1982, Go
Manager, Well Sample & Core Library:
Patrick J. Gooding, (M), E Kentucky, 1983, Eo
Geologist:
Rick Bowersox, (D), South Florida, 2006, Go
Geologist:
Stephen F. Greb, (D), Kentucky, 1992, GsEcGs
Adjunct Professor:
Gerald A. Weisenfluh, (D), South Carolina, 1982, EcOi
Technology Transfer Officer:
Michael J. Lynch, (B), E Kentucky, 1975, On
Manager, Administration:
Kathryn E. Ellis, (M), Kentucky, 2012, On
Hydrogeologist:
E. Glynn Beck, (M), East Carolina, 1997, Hw
James C. Currens, (M), E Kentucky, 1978, Hg
Junfeng Zhu, (D), Arizona, 2005, Hw
Geologist:
Matt Crawford, (M), E Kentucky, 2001, On
Bart Davidson, (M), E Kentucky, 1986, Hg
Cortland F. Eble, (D), West Virginia, 1988, Pl
John Hickman, (D), Kentucky, 2011, Eo
Brandon C. Nuttall, (B), E Kentucky, 1971, Eo
Thomas M. Parris, (D), California (Santa Barbara), 1998, Eo
Thomas N. Sparks, (M), Duke, 1979, Gg

## Louisiana

### Louisiana State University

**Basin Research Energy Section** (2004)
Louisiana Geological Survey/LSU
208 Howe Russell Geoscience Complex
Baton Rouge, LA 70803-4101

p. (225) 578-8328
mhorn@lsu.edu
http://www.bri.lsu.edu
Office Coordinator: Cherri B. Webre
Associate Professor:
Ronald K. Zimmerman, (D), Louisiana State, 1966, Go
Assistant Professor:
Clayton F. Breland, (D), Tennessee, 1980, Ye
John B. Echols, (D), Louisiana State, 1966, Go
Research Associate:
Brian J. Harder, (B), Louisiana State, 1981, Eo
Bobby L. Jones, (B), Louisiana State, 1953, Go
Phillip W. Lemay, (B), Centenary, 1999, Gg
Michael B. Miller, (M), North Carolina, 1982, Go
Lloyd R. Milner, (B), Louisiana State, 1985, Gg
Patrick M. O'Neill, (B), Louisiana State, 1985, Sf
Computer Analyst:
Reed J. Bourgeois, (B), Louisiana State, 1983, On
Accountant Technician:
Carla Domingue, On

**Louisiana Geological Survey** (2014)
3079-Energy,
Coast and Environment Bldg.
Baton Rouge, LA 70803
p. (225) 578-5320
hammer@lsu.edu
http://www.lgs.lsu.edu
State Geologist and Professor:
Chacko J. John, (D), Delaware, 1977, GosGe
Assistant Professor:
Douglas A. Carlson, (D), Wisconsin (Milwaukee), 2001, Hw
Marty R. Horn, (D), Texas (Arlington), 1996, EoGgr
GIS Coordinator:
R Hampton Peele, (M), Louisiana State, 2000, OirOg
Computer Analyst:
Reed J. Bourgeois, (B), Nichols State, 1985, On
Research Associate:
Brian J. Harder, (B), Louisiana State, 1981, Eo
Paul V. Heinrich, (M), Illinois, 1982, GsaGm
Bobby Jones
Richard P. McCulloh, (M), Texas (Austin), 1977, Gg
Lloyd R. Milner, (B), Louisiana State, 1985, Gg
Patrick M. O'Neill, (B), Louisiana State, 1985, Sf
Robert L. Paulsell, (B), Louisiana State, 1987, Oy
Lisa G. Pond, (B), Louisiana State, 1987, Og
Arren Schulingkamp, Gg
Cartographic Manager:
John I. Snead, (B), Louisiana State, 1978, Gm
Assistant Director:
John E. Johnston, III, (M), Texas, 1977, EoGe
Office Coordinator:
Melissa H. Esnault, (B), On
Accountant Technician:
Jeanne Johnson, (N), On

## Maine

### Dept of Agriculture, Conservation, and Forestry

**Maine Geological Survey** (2016)
93 State House Station
Augusta, ME 04333-0093
p. (207) 287-2801
mgs@maine.gov
http://www.maine.gov/dacf/mgs/
Administrative Assistant: Aline Smith
State Geologist:
Robert G. Marvinney, (D), Syracuse, 1986, Gg
State Soil Scientist:
David Rocque, Sd
Senior Geologist:
Lindsay Spigel, (D), Wisconsin, 2006, Gm
Physical Geologist:
Henry N. Berry IV, (D), Massachusetts, 1989, Gg
Marine Geologist:
Stephen M. Dickson, (D), Maine, 1999, OnGuOu
Peter Slovinsky, (M), South Carolina, 2001, Gu
Hydrogeologist:
Ryan Gordon, (D), Syracuse, 2014, HwGm
Daniel B. Locke, (B), Maine, 1982, Hw
Thomas K. Weddle, (D), Boston, 1991, GlHw
Director, Earth Resources Information:
Christian Halsted, (B), Maine, 1995, Oi
GIS Coordinator:
Amber Whittaker, (M), New Mexico, 2006, OiGcCg

## Manitoba

### Manitoba Geological Survey

**Manitoba Growth, Enterprise and Trade** (2016)
360-1395 Ellice Avenue
Winnipeg, MB R3G 3P2
p. 1-800-223-5215
christian.bohm@gov.mb.ca
http://www.manitoba.ca/iem/index.html
Senior Scientist:
Scott Anderson, EgGcCg
Christian Bohm, (D), ETH (Switzerland), 1996, GgCgEg
Michelle P.B. Nicolas, (M), Manitoba, 1997, GrsGo

## Maryland

### Maryland Department of Natural Resources

**Hydrogeology & Hydrology Program** (2013)
2300 St. Paul Street
Baltimore, MD 21218-5210
p. (410) 554-5500
JHalka@dnr.state.md.us
http://www.mgs.md.gov
Administrative Assistant: Donajean M. Appel
Hydrogeologist-Sedimentary Geologist:
Andrew W. Staley, (M), Wisconsin, 1992, Hy
Hydrogeologist:
Grufon Achmad, (D), Missouri, 1973, Hw
David W. Bolton, (M), W Michigan, 1988, Hw
David D. Drummond, (M), George Washington, 1988, Hy
Mark T. Duigon, (M), Indiana, 1977, Hy
John M. Wilson, (B), Maryland, 1976, Hw
Environmental Geologist:
Heather Quinn, (M), Florida, 1988, Ge

**Maryland Geological Survey** (2017)
2300 St. Paul Street
Baltimore, MD 21218-5210
p. (410) 554-5500
MGS.info@maryland.gov
http://www.mgs.md.gov/
Director:
Richard A. Ortt, (B), Johns Hopkins, 1991, GeOrYr
Program Chief, Hydrology & Hydrogeology:
David Bolton, (M), W Michigan, 1988, Hgw
Program Chief, Coastal and Environmental Geology:
Stephen Van Ryswick, (B), Maryland, 2002, GeSoYr
Hydrogeologist:
Grufron Achmad, (D), Missouri, 1973, HwGgz
David Andreasen, (B), Maryland, 1985, Hw
Andrew Staley, (M), Wisconsin, 1992, Hw
Geologist (Outreach Coordinator):
Dale W. Shelton, (B), Towson, 1986, GgOe
Geologist:
David K. Brezinski, (D), Pittsburgh, 1984, GrPgGg
Johanna Gemperline, (M), Illinois, 2013, Hw
Heather Quinn, (M), Florida, 1988, GrOiHw

## Massachusetts

### Massachusetts Geological Survey

**Dept of Geosciences** (2016)
Univ of Massachusetts (Amherst)
611 North Pleasant Street
Amherst, MA 01002
p. (413) 545-2286
sbmabee@geo.umass.edu

Geological Surveys

State Geologist:
Stephen B. Mabee, (D), Massachusetts, 1992, Hw

## Minnesota

### University of Minnesota

**Minnesota Geological Survey** (2016)
2609 West Territorial Road
Saint Paul, MN 55114-1009
p. 612-626-2969
mgs@umn.edu
http://www.mngs.umn.edu/
f: https://www.facebook.com/MinnesotaGeologicalSurvey
Director:
Harvey Thorleifson, (D), Colorado, 1989, Gl
Geologist:
Terrence Boerboom, (M), Minnesota (Duluth), 1987, GivGc
Geologist:
Jennifer Horton, (M), Toledo, 2015, Gl
Geologist:
Jacqueline Hamilton, (M), Minnesota, 2007, OiGe
Andrew J. Retzler, (M), Idaho State, 2013, GrsPi

## Mississippi

### Mississippi Office of Geology

**Environmental Geology Div** (2009)
P.O. Box 20307
2380 Highway 80 West
Jackson, MS 39289-1307
p. (601) 961-5500
john_marble@deq.state.ms.us
Division Director:
John C. Marble, (B), Mississippi State, 1974, Ge

**Geospatial Resources Div** (2013)
P.O. Box 2279
700 N. State St
Jackson, MS 39225-2279
p. (601) 961-5500
barbara_yassin@deq.state.ms.us
www.deq.state.ms.us
Geologist:
Steven D. Champlin, (B), Alabama, 1976, Go
GIS Analyst:
Barbara E. Yassin, (B), Illinois State, 1989, OyiGm
Geologist:
Peter S. Hutchins, (B), Millsaps, 1990, On

**Mining & Reclamation Div** (2004)
P.O. Box 20307
2380 Highway 80 West
Jackson, MS 39289-1307
p. (601) 961-5500
Ken_McCarley@deq.state.ms.us
Secretary: Tamara Duckworth
Secretary: Sandra Saik
Division Director:
J.Kendrick McCarley, (M), Mississippi, 1996, Ng
Assistant Division Director:
Stanley C. Thieling, (M), Iowa, 1973, Go
Geologist:
Michael Akin, (B), Mississippi State, 1995, Ge
James L. Matheny, (B), Delta State, 1994, On
Jim F. McMullin, (B), Millsaps, 1959, Gg
Thomas M. Ray, (M), S Mississippi, 1975, Gg
Biologist:
David J. Wickens, (B), Mississippi, 1991, On
Related Staff:
Robert J. Millette, (B), SW Louisiana, 1989, On
James E. Starnes, (B), Millsaps, 1996, Gg

**Mississippi Dept of Environmental Quality** (2016)
P.O. Box 2279
Jackson, MS 39225-2279
p. (601) 961-5500
mbograd@mdeq.ms.gov
http://www.deq.state.ms.us/
State Geologist:
Michael B. E. Bograd, (M), Mississippi, 2002, Gg

**Surface Geology Div** (2010)
P.O. Box 2279
700 North State Street
Jackson, MS 39225
p. (601) 961-5500
david_dockery@deq.state.ms.us
Division Director:
David T. Dockery, (D), Tulane, 1991, PiGgPs
Geologist:
James E. Starnes, (B), Millsaps, 1996, GgaPi
David E. Thompson, (B), Mississippi State, 1986, GrEg
Environmental Scientist:
Kenneth D. Davis, (B), Mississippi State, 1968, Gr
Cooperating Faculty:
Daniel W. Morse, (B), Texas, 1982, On

## Missouri

### Missouri Dept of Agriculture

**Land Survey Program** (2013)
PO Box 937
Land Survey Building/1251A Gale Drive
Rolla, MO 65402-0937
p. (573) 368-2300
darrell.pratte@mda.mo.gov
http://mda.mo.gov/weights/landsurvey/
State Land Surveyor:
Darrell D. Pratte, On

### Missouri Dept of Natural Resources

**Dam & Reservoir Safety** (2015)
PO Box 250
Buehler Bldg/111 Fairgrounds Rd
Rolla, MO 65402
p. (573) 368-2175
bob.clay@dnr.mo.gov
Professional Staff:
Robert Clay, (M), Oklahoma State, 1977, On
Professional Staff:
Glenn D. Lloyd, (B), On
Other:
Paul Simon, (B), MST
Ryan Stack, (B), MST

**Div of Geology and Land Survey** (2014)
PO Box 250
111 Fairgrounds Road
Rolla, MO 65402-0250
p. (573) 368-2100
joe.gillman@dnr.mo.gov
http://www.dnr.state.mo.us/geology.htm
Administrative Assistant: Tami L.. Allison
State Geologist & Division Director:
Mimi R. Garstang, (B), Southwest Missouri State, Ge
Deputy Director & Assistant State Geologist:
James W. Duley, (B), C Missouri, 1975, Hy

**Geological Survey Program** (2016)
PO Box 250
111 Fairgrounds Rd
Rolla, MO 65402-0250
p. (573) 368-2143
gspgeol@dnr.mo.gov
http://www.dnr.mo.gov/geology
Professional Staff:
Joe Gillman, (B), Missouri State, 1992, Gg
Geologist:
Michael A. Siemens, (M), Wichita State, 1985, GgHgOi
Senior Scientist:
Pat Mulvany, (D), Missouri S&T, 1996

Unit Chief:
Justin Davis, (M), Missouri S&T, 2012
Cheryl M. Seeger, (D), Missouri S&T, 2003, Gig
Chris Vierrether, (M), Missouri S&T, 1988, OnEc
Glen Young, (M), SE Missouri State, 1994, Gg
Section Chief:
Larry Pierce, (M), On
Peter Price, (B), Missouri S&T, 1977, GeHy
Kyle Rollins, (B), Missouri State, 1986, OyGg
Program Director:
Carey Bridges, (M), Missouri (Columbia), 1999, Gg
Geologist:
Joey Baughman, (B), Missouri S&T, 2012, Gg
Peter Bachle, (B), Missouri S&T, 1997, Gg
Fletcher Bone, (B), Central Missouri, 2007, Gg
David L. Bridges, (D), Missouri S&T, 2011, GgiPg
John H. Corley, (M), Missouri, 2014, Geg
Jeff Crews, (M), Missouri S&T, 2004, Hy
Trevor Ellis, (B), Missouri S&T, 2010, Gg
Kyle Ganz, (M), Missouri S&T, 2013
Airin Haselwander, (B), Missouri S&T, 2011, Gg
Terry Hawkins, (B), Brigham Young Univ, 1988, Geg
Thomas Herbst, (M), Missouri Sci & Tech, 2014, Gg
Jeremiah Jackson, (M), Missouri State, 2011, Gg
Brenna McDonald, (B), SE Missouri State, 1997, Ge
Brad Mitchell, (M), Missouri S&T, 2010
Matt Parker, (B), Missouri S&T, 1993
John Pate, (B), Tennessee, 2007, Ge
Molly Starkey, (M), Missouri State, 2011
Vicki Voigt, (B), Missouri S&T, 2010, Gg
Geological Tech:
Cecil Boswell, On
Eric Hohl, On
Karen Loveland, On
Dan Nordwald, On
Patrick Scheel, On
Fred Shaw, On
Environmental Specialist:
Andrew Combs
Deputy Director:
Jerry L. Prewett, (B), Missouri State, 1992, GgHw

**Water Resources Program** (2016)
PO Box 250
111 Fairgrounds Rd
Rolla, MO 65401-0250
p. (573) 368-2175
mowaters@dnr.mo.gov
http://www.dnr.state.mo.us/dgls/
Deputy Director:
Andrea Collier
Geologist:
Scott Kaden, Hw
Associate Scientist:
Robert Bacon, Hs
Professional Staff:
Cynthia Brookshire, (B), Hw
Charles Du Charme, (B), Hs

# Montana

## Montana Tech of The University of Montana

**Montana Bureau of Mines & Geology** (2015)
1300 West Park Street
Butte, MT 59701-8997
p. (406) 496-4180
jmetesh@mtech.edu
http://www.mbmg.mtech.edu
f: https://www.facebook.com/MontanaGeology/
Administrative Assistant: Margaret Delaney
Administrative Assistant: Charlotte McKenzie
Administrative Assistant: Bette Wasik
State Geologist:
John J. Metesh, (D), Montana, 2003, HwCa
Assistant Director RET:
Marvin R. Miller, (M), Indiana, 1965, Hw
Director, Contracts & Grants:
Carleen Cassidy, (B), Montana Tech, On
Research Division Chief:
Thomas W. Patton, (M), Montana Tech, 1987, Hw
Associate Research Geologist:
Catherine McDonald, (M), California (Davis), 1992, Grs
Senior Hydrogeologist:
Jon C. Reiten, (M), North Dakota, 1983, Hw
Geologist:
Susan M. Vuke, (M), Montana, 1982, GgrGs
Senior Research Hydrogeologist:
John R. Wheaton, (M), Montana, 1987, HwEc
Hydrogeologist:
Ginette ABDO, (M), Penn State, 1989, Gg
Andrew L. Bobst, (M), Binghamton, 2000, HwCgEo
Groundwater Assessment Program Manager:
John I. La Fave, (M), Texas, 1987, Hw
Geologist:
Michael C. Stickney, (M), Montana, 1980, YsGtm
Assistant Research Geologist:
Jeff Lonn, (M), Montana, 1985, GgtGc
Publications Editor:
Susan A. Barth, (M), Montana Tech, 2009, On
Hydrogeologist:
Gary Icopini, (D), Michigan State, 2000, Hw
Assistant Research Hydrogeologist:
Camela Carstarphen, (M), Oregon State, 1991, Hw
Assistant Research Geologist:
Phyllis Hargrave, (M), Montana Tech, 1990, Gg
Sr. Research Hydrogeologist:
Thomas E. Michalek, (M), Montana, 2001, HgwHs
Sr. Hydrogeologist:
Kirk B. Waren, (M), Wright State, 1988, HwsOe
Hydrogeologist:
Terence E. Duaime, (B), Montana Tech, 1978, Hws
Research Associate:
Colleen G. Elliott, (D), New Brunswick, 1988, GctGg
Senior Research Geologist, Museum Curator:
Richard B. Berg, (D), Montana, 1964, EnGz
Seismic Analyst:
Deborah Smith
Research Assistant III:
Jaqueline R. Timmer, (B), Montana Tech, 1990, Ca
Professional Scientist/Hydro:
Nicholas Tucci
Hydrogeologist:
Daniel D. Blythe, (B), Montana State, 2006, Hws
GIS Specialist:
Ken L. Sandau, (A), Montana Tech, 1999, Oi
Paul R. Thale, (M), Montana State, 1994, Oi
Geologic Cartographer:
Susan M. Smith, (B), Montana State, 1970, On
Chemist:
Ashley Huft, (B), Montana Tech, 2008, Ca
Associate Research Professor:
Steve F. McGrath, (M), Montana Tech, 1992, CaEgCe
Assistant Hydrogeologist:
Mary K. Sutherland, (M), Montana, 2009, Hws
Accounting Associate:
Joanne Lee
Other:
Nancy Favero, Info Sys Tech
Computer Software Eng/Applications:
Luke Buckley, (B), Montana Tech, 1995, On

# Nebraska

## University of Nebraska, Lincoln

**Conservation & Survey Div** (D) (2014)
Conservation & Survey Division
3310 Holdrege Street
616 Hardin Hall
Lincoln, NE 68583-0996
p. (402) 472-3471
mkuzila1@unl.edu
http://snr.unl.edu/csd/

Director:
Mark S. Kuzila, (D), Nebraska, 1988, Sd
Professor:
Xun-Hong Chen, (D), Wyoming, 1994, Hq
David C. Gosselin, (D), SD Mines, 1987, CgHw
James W. Merchant, (D), Kansas, 1984, Oru
Jozsef Szilagyi, (D), California (Davis), 1997, HqOa
Senior Scientist:
Susan Olafsen-Lackey, (B), SD Mines, 1982, Hw
Steven S. Sibray, (M), New Mexico, 1977, HwGg
Associate Professor:
Paul Hanson, (D), Nebraska, 2005, Gm
Matt Joeckel, (D), Iowa, 1993, GsrGg
Research Associate:
Leslie M. Howard, (M), Nebraska, 1989, Oyi
Emeritus:
Marvin P. Carlson, (D), Nebraska, 1969, Gt
Robert F. Diffendal, Jr., (D), Nebraska, 1971, GrPiGh
Duane A. Eversoll, (M), Nebraska, 1977, Ng
Anatoly Gitelson, (D), Inst Radio Technology (Russia), 1972, On
James W. Goeke, (M), Colorado State, 1970, HwGmYe
Donald C. Rundquist, (D), Nebraska, 1977, Ori
James Swinehart, (D), Gg

## Nevada

### University of Nevada

**Nevada Bureau of Mines and Geology** (2017)
Mail Stop 178
University of Nevada
Reno, NV 89557-0088
p. (775) 682-8766
jfaulds@unr.edu
http://www.nbmg.unr.edu/
f: https://www.facebook.com/Nevada-Bureau-of-Mines-and-Geology-106397989390636/
Administrative Assistant: Alex Nesbitt
State Geologist/Professor:
James E. Faulds, (D), New Mexico, 1989, Gc
Professor:
Geoffrey Blewitt, (D), Caltech, 1996, Yd
William Hammond, (D), Oregon, 2000, Yd
Christopher D. Henry, (D), Texas (Austin), 1975, GtvCc
Geologic Mapping Specialist:
Seth Dee, (M), Oregon, 2006, Gg
Nicholas Hinz, (M), Nevada (Reno), 2007, Gg
Senior Scientist:
Alan R. Ramelli, (M), Nevada (Reno), 1988, Ng
Director, Center for Research in Economic Geology:
John Muntean, (D), Stanford, 1998, Eg
Associate Professor:
Bridget Ayling, (D), Australia, 2006, CgYhGo
Craig M. dePolo, (D), Nevada (Reno), 1998, Ng
Corne Kreemer, (D), SUNY (Stony Brook), 2001, Yd
Assistant Professor:
Rich Koehler, (D), Nevada (Reno), 2009, GmtNg
Mike Ressel, (D), Nevada (Reno), 2005, Eg
Andrew V. Zuza, (D), UCLA, 2016, Gc
Geologic Information Specialist:
David A. Davis, (M), Nevada (Reno), 1990, Gg
Emeritus:
John W. Bell, (M), Arizona State, 1974, NgGm
Stephen B. Castor, (D), Nevada (Reno), 1972, Eg
Larry J. Garside, (M), Nevada (Reno), 1968, Gg
Liang-Chi Hsu, (D), California (Los Angeles), 1966, Cp
Daphne D. LaPointe, (M), Montana, 1977, Gg
Paul J. Lechler, (D), Nevada (Reno), 1995, Cg
Jonathan G. Price, (D), California (Berkeley), 1977, Gg
Lisa Shevenell, (D), Nevada, 1990, Cg
Joseph V. Tingley, (M), Nevada (Reno), 1963, Em
Susan L. Tingley, (B), California (Los Angeles), 1966, Oy

## New Hampshire

### New Hampshire Geological Survey

**New Hampshire Dept of Environmental Services** (2014)
29 Hazen Drive
P.O.Box 95
Concord, NH 03302-0095
p. (603) 271-1975
geology@des.nh.gov
http://des.nh.gov/organization/commissioner/gsu/
State Geologist:
Frederick H. Chormann, Jr., (M), New Hampshire, 1985, HyOiGm
Geoscience Program Specialist:
Gregory A. Barker, (B), Rhode Island, 1985, GglOi
Outreach Coordinator:
Lee Wilder, (M), New Hampshire, 1972, GgOee
Hydrologist:
Jeremy D. Nicoletti, (B), James Madison, 2009, GgHg
Neil F. Olson, (M), Idaho State, 2010, GeHy
Fluvial Geomorphology Specialist:
Shane Csiki, (D), Illinois (Urbana Champaign), 2014, OyGm

## New Jersey

### New Jersey Geological and Water Survey

**New Jersey Geological and Water Survey** (2016)
PO Box 420, Mail Code:29-01
Trenton, NJ 08625-0420
p. (609) 292-1185
njgsweb@dep.nj.gov
http://www.state.nj.us/dep/njgs/
Administrative Assistant: Tenika Jacobs
State Geologist:
Jeffrey L. Hoffman, (M), Princeton, 1981, Hq
Bureau Chief:
David L. Pasicznyk, (B), Temple, 1979, Yg
Research Scientist 1:
Steven E. Spayd, (M), UMD New Jersey, 2004, HwGb
Peter J. Sugarman, (D), Rutgers, 1995, Gr
Supervising Geologist:
Eric W. Roman, (M), Temple, 1999, Hw
Supervising Env Specialist:
Helen L. Rancan, (M), Stevens, 1995, Hs
Supervising Env Engineer:
Richard Shim-Chim, (B), Toronto, 1973, Hw
Research Scientist 2:
James T. Boyle, (M), New Mexico Tech, 1984, Hw
Suhas L. Ghatge, (M), W Michigan, 1984, Yve
Donald H. Monteverde, (D), Rutgers, 2008, Grg
Ronald W. Witte, (M), Lehigh, 1988, Gl
Research Scientist 1:
Scott D. Stanford, (D), Rutgers, 2001, Gml
GIS Specialist 1:
Zehdreh Allen-Lafayette, (M), Syracuse, 1994, Oi
Mark A. French, (B), Rutgers, 1989, Hq
Ted J. Pallis, (M), Montclair State, 1994, Oi
Ronald Pristas, (B), Penn State, 1990, Oi
Principal Environmental Specialist:
Raymond T. Bousenberry, (M), New Jersey Inst Tech, 2007, Ge
Steven E. Domber, (M), Wisconsin, 2000, Hw
Gregg M. Steidl, (B), Rutgers, 1995, Ge
Research Scientist 2:
John H. Dooley, (M), New Mexico Tech, 1983, CgGfe
GIS Specialist 2:
Michael W. Girard, (B), Bloomsberg, 1996, Oi
Section Chief:
William P. Graff, (B), Clark, 1979, On
Investigator:
Walter Marzulli, Oy

## New Mexico

### New Mexico Institute of Mining & Technology

**New Mexico Bureau of Geology & Mineral Resources** (2015)
801 Leroy Place
Socorro, NM 87801-4796
p. (575) 835-5420
greer@nmbg.nmt.edu
http://geoinfo.nmt.edu/

Director and State Geologist:
L Greer Price, (M), Washington, 1974, Gg
Senior Volcanologist:
William C. McIntosh, (D), New Mexico Tech, 1990, Cc
Field Geologist:
Bruce Allen, (D), New Mexico, 1993, Gm
Senior Field Geologist:
Steven M. Cather, (D), Texas, 1986, Gs
Principal Senior Geologist:
Paul W. Bauer, (D), New Mexico Tech, 1988, Gp
Adjunct Faculty:
George S. Austin, (D), Iowa, 1971, En
David W. Love, (D), New Mexico, 1980, Ge
Senior Mining Engineer:
Robert W. Eveleth, (B), New Mexico Tech, 1969, Nm
Senior Field Geologist:
Richard M. Chamberlin, (D), Colorado Mines, 1980, Gg
Senior Env Geologist Emeritus:
John W. Hawley, (D), Illinois, 1962, Ge
Senior Chemist:
Lynn A. Brandvold, (M), North Dakota State, 1964, Ca
Principal Senior Geophysicist:
Marshall A. Reiter, (D), Virginia Tech, 1969, Yh
Geologist:
James M. Barker, (M), California (Santa Barbara), 1972, En
Emeritus Director & State Geologist:
Peter Scholle, (D)
Emeritus:
Charles E. Chapin, (D), Colorado Mines, 1966, Gt
Ibrahim H. Gundiler, (D), New Mexico Tech, 1975, Nx
Webmaster/Geologist:
Adam S. Read, (M), New Mexico, 1997, Gg
Senior Lab Associate:
Lisa Peters, (M), Texas (El Paso), 1987, Gg
Senior Geophysicist, Field Geologist:
Shari Kelley, (D), S Methodist, 1984, Yg
Principal Senior Petroleum Geologist:
Ronald F. Broadhead, (M), Cincinnati, 1979, Go
Principal Senior Hydrogeologist:
Peggy Johnson, (M), New Mexico Tech, 1990, Hw
Minerals Outreach Liason:
Virginia McLemore, (D), Texas (El Paso), 1994, Em
Mineralogist/Economic Geologist:
Virgil W. Lueth, (D), Texas, 1988, Gg
Manager, Digital Cartography Lab:
Glen Jones, (B), New Mexico Tech, 1989, On
Hydrologist; Hydrogeology Program Manager:
Stacy Timmons, (M), Oregon State, 2002, Gg
Hydrogeologist:
Lewis A. Land, (D), North Carolina, 1999, Hg
GIS Cartographer:
David J. McCraw, (M), New Mexico, 1985, Oy
Geological Librarian:
Maureen Wilks, (D), New Mexico Tech, 1991, Gp
Geochronologist:
Matthew T. Heizler, (D), California (Los Angeles), 1993, Cc
Field Geologist:
Geoffrey Rawling, (D), New Mexico Tech, 2002, Gg
Economic Geologist:
Douglas Bland, (M), Wyoming, 1982, Eg
Deputy Director; Geochemist:
Nelia W. Dunbar, (D), New Mexico Tech, 1989, Gi
Deputy Director & Manager, Geologic Mapping Program:
Michael Timmons, (D), New Mexico, 2004, Gg
Chemistry Lab Manager:
Bonnie A. Frey, (M), New Mexico Tech, 2002, Cg
Related Staff:
Gretchen K. Hoffman, (M), Arizona, 1979, Ec

## New York

### New York State Geological Survey

**New York State Geological Survey** (2014)
3000 Cultural Education Center
Madison Avenue
Albany, NY 12230
p. (518) 473-6262
djornov@mail.nysed.gov
http://www.nysm.nysed.gov/nysgs/
Administrative Assistant: Donna Jornov
State Paleontologist and Paleontology Curator:
Ed Landing, (D), Michigan, 1979, PseGr
Geoarchaeologist, Quaternary Geologist:
Julieann Van Nest, (D), Iowa, 1997, GamOi
Curator of Sedimentary Geology:
Charles Ver Straeten, (D), Rochester, 1996, Grs
Curator of Geology:
Marian V. Lupulescu, (D), IASI (Romania), 1987, GzxEm
Senior Scientist:
Andrew Kozlowski, (D), W Michigan, 2002, Gl
Langhorne Smith, (D), Virginia Tech, 1996, Go
Museum Scientist 1:
Brian Bird, (D), GlOi
Project Geologist:
James Leone, (B), SUNY (Albany), 2006, EoGoCg
Brian Slater, (M), SUNY (Albany), 2007, GoEoGr
Education Specialist:
Kathleen Bonk, (B), SUNY (Albany), 2009, GoCg
Brandon L. Graham
Paleontology Collections Technician:
Frank Mannolini
Technician:
Michael Pascussi, (A), Schenectady Comm Coll, 1999, Eo
Related Staff:
Michael E. Hawkins, (B), USNY Regents, 1982, Gz
Linda A. VanAller-Hernick, (B), St Rose, 1974, Pgb
Cooperating Faculty:
Barry Floyd, (M), Rensselaer, 1987, On

## North Carolina

### North Carolina Geological Survey

**Dept of Environmental Quality** (2016)
1612 Mail Service Center
Raleigh, NC 27699-1612
p. 919-707-9210
kenneth.b.taylor@ncdenr.gov
http://portal.ncdenr.org/lr/geological_home
Administrative Assistant: Joyce Sanford
Division Director:
Tracy E. Davis, (B), North Carolina State, 1987
State Geologist:
Kenneth B. Taylor, (D), Saint Louis Univ, 1991, YsGgEo
Senior Geologist:
Philip Bradley, (M), North Carolina State
Senior Geologist :
Bart Cattanach, (M), North Carolina State
Senior Geologist:
Kathleen M. Farrell, (D), Louisiana State, 1989, Gs
Jeffrey C. Reid, (D), Georgia, 1981
Engineering Geologist:
Richard M. Wooten, (M), Georgia, 1980
Project Geologist:
Randy Bechtel, (M), Oe
Geologist I:
Nick Bozdog, (B), Gg
Kenny Gay, (M), East Carolina Univ, 1980, GzsPi
Heather Hanna, (M), Duke, 2004, On

## North Dakota

### North Dakota Geological Survey

**North Dakota Geological Survey** (2016)
1016 E. Calgary Ave.
600 East Boulevard Avenue
Bismarck, ND 58505-0840
p. (701) 328-8000
emurphy@nd.gov
https://www.dmr.nd.gov/ndgs/
State Geologist:
Edward C. Murphy, (M), North Dakota, 1983, On
Paleontologist:
Jeff Person, Pg

Geologist:
Ned Kruger, Gg
Julie A. LeFever, Gg
Lorraine A. Manz, (D), London, 1982, GmlGg
Tim Nesheim, (M), Iowa, 2009, Gg

# Northern Territory

## Northern Territory Government Minerals and Energy

**Northern Territory Government Minerals and Energy** (2011)
48-50 Smith St
Paspalis Centrepoint Building
GPO Box 3000
Darwin, NT 0800
p. +61 8 8999 5511
minerals@nt.gov.au
http://www.minerals.nt.gov.au

## Northwest Territories Geological Survey

**Industry Tourism and Investment - Government of the Northwest Territories** (2016)
P.O. Box 1320
Yellowknife, NT X1A 2L9
p. (867) 767-9211
NTGS@gov.nt.ca
http://www.nwtgeoscience.ca/

# Nova Scotia

## Geological Survey of Canada

**Atlantic Div** (2012)
Bedford Institute of Oceanography
1 Challenger Drive
P.O. Box 1006
Dartmouth, NS B2Y 4A2
p. (902) 426-4386
Pat.Dennis@NRCan-RNCan.gc.ca
http://gsc.nrcan.gc.ca/org/atlantic/index_e.php
Senior Scientist:
Kumiko Azetsu-Scott, (D), Dalhousie, 1992, Ocp

## Nova Scotia Natural Resources

**Nova Scotia Natural Resources** (2014)
P.O. Box 698
Founders Square
Halifax, NS B3J 3M8
p. (902) 424-5935
http://www.gov.ns.ca/natr/

# Ohio

## Ohio Dept of Natural Resources

**Div of Geological Survey** (2017)
2045 Morse Road Bldg. C-1
Columbus, OH 43229
p. (614) 265-6576
geo.survey@dnr.state.oh.us
http://www.ohiodnr.com/geosurvey/
State Geologist:
Thomas J. Serenko, (D), Imperial Coll (UK), EgnEm
Senior Scientist:
Frank Fugitt, (B), Ohio Univ, GrHw
Geologic Assistant:
Madge R. Fitak, (B), Mt Union, 1972, Gg
Geologist:
Daniel R. Blake, (M), Wright State, 2013, Ysg
geologist:
Michael P. Solis, (M), Kentucky, 2010, GgcGt

# Oklahoma

## University of Oklahoma

**Oklahoma Geological Survey** (2014)
100 East Boyd
Energy Center
Suite N-131
Norman, OK 73019-0628
p. (405) 325-3031
ogs@ou.edu
http://www.ogs.ou.edu/homepage.php
Geologist:
Thomas M. Stanley, (D), Kansas, 2000, PiGsc
Adjunct Professor:
Kyle E. Murray, (D), Colorado Mines, 2003, HwOiGo
Seismologist:
Amberlee Darold, (M), Oregon, 2012, Ys
Geologist:
Julie M. Chang, (D), Texas (El Paso), 2006, GgOnn
Brittany Pritchett, Go

# Ontario

## Ontario Geological Survey

**Ontario Geological Survey** (2016)
933 Ramsey Lake Road
Sudbury, ON P3E6B5
p. (705) 670-5758
tracy.livingstone@ontario.ca
http://www.mndm.gov.on.ca/en/mines-and-minerals
f: www.facebook.com/OGSgeology
t: @OGSgeology
Director:
Jack R. Parker, (B), Lakehead Univ, 1980, GgEgOe
A/Senior Manager, Earth Resources & Geoscience Mapping Section:
James Schweyer
A/Senior Manager, Resident Geologist Program:
Mark Smyk
A/Senior Manager, Geoservices Section:
Renée-Luce Simard

# Oregon

## Oregon Dept of Geology & Mineral Industries

**Baker City Field Office** (2014)
1995 Third St, Suite D
Baker City, OR 97814
p. (541) 523-3133
mark.ferns@dogami.state.or.us
http://www.oregongeology.org
Field Geologist:
Jason D. McClaughry, (M), Washington State, 2003, GrvGs
Regional Geologist:
Mark L. Ferns, (M), Oregon, 1979, GrtGi

**Coastal Field Office** (2014)
PO Box 1033
Newport, OR 97365
p. (541) 574-6658
Jonathan.Allan@dogami.state.or.us
Regional Geologist:
Rob Witter, (D), Oregon, 1999, Og
Geologist:
George R. Priest, (D), Oregon State, 1980, Gg
Coastal Section Supervisor:
Jonathan C. Allan, (D), Canterbury (NZ), 1998, Gm

**Grants Pass Field Office** (2004)
5375 Monument Drive
Grants Pass, OR 97526-8513
p. (541) 476-2496
tom.wiley@dogami.state.or.us
http://www.oregongeology.org
Department Secretary: Kathleen McGee
Resident Geologist:
Frank Hladky, (M), Idaho State, 1986, Gg
Regional Geologist:
Tom Wiley, (M), Stanford, 1983, Gg

**Oregon Dept of Geology and Mineral Industries** (2014)
800 NE Oregon Street
Suite 965
Portland, OR 97232-2162
p. (971) 673-1555
alyssa.pratt@dogami.state.or.us
http://www.oregongeology.org
State Geologist:
Vicki S. McConnell, (D), Alaska (Fairbanks), 1996, Gv
Regional Geologist:
Thomas J. Wiley, (M), Stanford, 1983, Gg
Chief Scientist:
Ian P. Madin, (M), Oregon State, Gg
Regional Geologist:
Jonathan C. Allan, (D), Canterbury (NZ), 1998, Gm
Jason McClaughry, (M), WSU, 2003, Gg
Reclamationist:
Robert Brinkmann, (B), Colorado State, 1983, HwGoEm
Ben Mundie, Ge
Industrial Minerals Geologist:
Clark Niewendorp, (M), Eg
Geotechnical Earthquake Engineer:
Yumei Wang, (M), California (Berkeley), 1988, Ne
Engineering Gelogist:
Bill Burns, (M), Portland State, 1999, Ng
Natural Resource Specialist, Oil and Gas Program:
Robert Houston, (M), Oregon, Ge
Geologist:
Lina Ma, (B), Oregon, 2002
Assistant Director - MLRR:
Tom Ferrero, (M), Hw

# Pennsylvania

## Pennsylvania Bureau of Topographic & Geologic Survey

**DCNR-Pennsylvania Geological Survey** (2016)
3240 Schoolhouse Road
Middletown, PA 17057-3534
p. (717) 702-2017
ra-askdcnr@pa.gov
http://www.dcnr.state.pa.us/topogeo/
f: https://www.facebook.com/PennsylvaniaGeology
Administrative Assistant: Connie Cross
Administrative Assistant: Elizabeth C. Lyon
State Geologist and Bureau Director:
Gale C. Blackmer, (D), Penn State, 1992, Gc
Geologist Manager:
Michael E. Moore, (B), Penn State, 1975, HwOi
Assistant Bureau Director :
Kristin M. Carter, (M), Lehigh, 1993, GoHw
Senior Geologic Scientist:
Rose-Anna Behr, (M), New Mexico Tech, 1999, Gcr
Helen L. Delano, (M), SUNY (Binghamton), 1979, GgmNg
Clifford H. Dodge, (M), Northwestern, 1976, GrEcGh
Kristen Hand, (B), Nicholls State, HwGg
William E. Kochanov, (M), West Virginia, 1983, GemGg
Antonette K. Markowski, (M), S Illinois, 1990, Eco
Victoria V. Neboga, (M), Kiev State, 1985, Hw
Caron E. O'Neil, (M), Pittsburgh, 1986, GgOn
Katie Schmid, (M), Pittsburgh, 2005, Go
Stephen G. Shank, (D), Penn State, 1993, GziGp
James R. Shaulis, (M), Penn State, 1985, EcOe
Thomas G. Whitfield, (B), West Virginia, 1973, Oi
Geologist Supervisor:
John H. Barnes, (M), SUNY (Buffalo), 1972, Gg
Brian Dunst, (B), Indiana of Pennsylvania, 1982, EgoHw
Gary M. Fleeger, (M), Illinois, 1980, GlrHw
Stuart O. Reese, (M), Tennessee, 1986, Hw
Geologic Scientist:
Robin Anthony, (B), Case Western, Go
Aaron D. Bierly, (B), Pitt-Johnstown, Gg
Leonard J. Lentz, (M), North Carolina State, 1983, Ec
John C. Neubaum, (M), American Military Univ, 2001, EcnGg
Librarian:
Jody L. Smale, (M), Clarion, 2010
IT Technician:
Mark A Dornes
IT Generalist:
David Fletcher
IT Adminstrator:
Sandipkumar P. Patel, (B)
Clerk Typist:
Lynn J. Levino, Go
Jody L. Rebuck, (B), Messiah
Renee Speicher

# Puerto Rico

## Puerto Rico Bureau of Geology

**Dept of Natural & Environmental Resources** (2004)
Apartado 9066600
Puerta de Tierra Station
San Juan, PR 00906-6600
p. (787) 722-2526
Director:
Vanessa del S. Rodriguez, On

# Quebec

## Ministère de l'Énergie et des Ressources naturelles Québec

**Ministère de l'Énergie et des Ressources naturelles Québec** (2017)
5700 4eme avenue ouest
local A-409
Quebec, QC G1H 6R1
p. 1-866-248-6936
services.clientele@mern.gouv.qc.ca
http://www.mern.gouv.qc.ca/

# Queensland

## Queensland Environment and Resource Management

**Queensland Environment and Resource Management** (2011)
GPO Box 2454
Brisbane, QLD 4001
info@derm.qld.gov.au
http://www.derm.qld.gov.au/

# Rhode Island

## University of Rhode Island

**Rhode Island Geological Survey** (D) (2015)
9 East Alumni Ave.
314 Woodward Hall
University of Rhode Island
Kingston, RI 02881
p. 401.874.2191
rigsurv@etal.uri.edu
http://www.uri.edu/cels/geo/GEO_risurvey.html
Administrative Assistant: Cheryl Grasso
Chair:
David Fastovsky, (D), Wisconsin (Madison), 1986, Pv
State Geologist (Research Professor Emeritus):
Jon C. Boothroyd, (D), South Carolina, 1974, GslOn
Professor:
Thomas Boving, (D), Arizona (Tucson), 1999, Hg
Assistant Professor:
Dawn Cardace, (D), Washington Univ (St. Louis), 2006, Py
Simon Engelhard, (D), Pennsylvania, 2010, Gg
Brian Savage, (D), Caltech, 2004, Ys
Lecturer:
Elizabeth Laliberte, (D), Rhode Island, 1997, Og
Research Associate:
Bryan A. Oakley, (D), Rhode Island, 2012, Gsl

# Saskatchewan

## Saskatchewan Energy and Resources

**Saskatchewan Ministry of the Economy** (2016)
1000 - 2103 - 11th Avenue
Regina, SK S4P 3Z8
p. (306) 787-9581
dmoadmin.econ@gov.sk.ca
http://www.saskatchewan.ca/

# South Carolina

## Dept of Natural Resources

**South Carolina Geological Survey** (2016)
5 Geology Road
Columbia, SC 29212
p. 803.896.7931
scgs@dnr.sc.gov
http://www.dnr.sc.gov/geology/

Director:
Charles W. Clendenin, Jr., (D), Witwatersrand, 1989, Eg
Geologist III:
William R. Doar, III, (D), South Carolina, 2014, GrdGp
Chief Geologist:
C. Scott Howard, (D), Delaware
Geologist II:
Katherine E. Luciano, (M), Coll of Charleston, GusGm
Program Manager-Drilling:
Joe Koch, Og
Digitizer II:
Matt Henderson, (B), South Carolina, Oi

# South Dakota

## South Dakota Dept of Environment and Natural Resources

**Geological Survey Program** (2016)
Akeley-Lawrence Science Center
University of South Dakota
414 East Clark Street
Vermillion, SD 57069-2390
p. (605) 677-5227
derric.iles@usd.edu
http://www.sdgs.usd.edu/

State Geologist:
Derric L. Iles, (M), Iowa State, 1977, HwGg
Hydrologist:
Dragan Filipovic, (M), Belgrade, 1985, Hwq

# Tennessee

## Tennessee Geological Survey

**Dept of Environment & Conservation** (2017)
William R. Snodgrass TN Tower
312 Rosa L. Parks Ave., 12th Floor
Nashville, TN 37243
p. (615) 532-1502
Ronald.Zurawski@tn.gov
http://www.tn.gov/environment/tdg

State Geologist:
Ronald P. Zurawski, (M), Vanderbilt, 1973, GgEg
Environmental Consultant:
Albert B. Horton, (M), Vanderbilt, 1981, GgOi
Peter J. Lemiszki, (D), Tennessee, 1992, GcOiGg
Barry W. Miller, (M), Tennessee, 1989, EcOiGg
Environmental Scientist:
Vince Antonacci, (M), Ball State, 1987, GgOi
Ronald J. Clendening, (B), Tennessee Tech, 1986, Gg
Admin. Services Asst. 2 :
Carolyn A. Patton

# Texas

## University of Texas at Austin, Jackson School of Geosciences

**Bureau of Economic Geology** (2014)
University Station, Box X
Austin, TX 78713-8924
p. (512) 471-1534
begmail@beg.utexas.edu
http://www.beg.utexas.edu

Director:
Scott W. Tinker, (D), Colorado, 1996, Gor
Associate Director:
Jay P. Kipper, (M), Trinity (San Antonio), 1983, On
Eric C. Potter, (M), Oregon State, 1975, Eo
Senior Research Scientist:
Shirley P. Dutton, (D), Texas (Austin), 1986, Gd
Bob A. Hardage, (D), Oklahoma State, 1967, Yg
Susan D. Hovorka, (D), Texas (Austin), 1990, Gse
Mike Hudec, (D), Wyoming, 1990, Gg
Martin P. A. Jackson, (D), Cape Town, 1976, Gc
Stephen E. Laubach, (D), Illinois, 1986, GcNrGz
Bob Loucks, (D), Texas, 1976, Go
Bridget R. Scanlon, (D), Kentucky, 1985, Hw
Research Scientist:
William A. Ambrose, (M), Texas, 1983, Gs
Peter Eichhubl, (D), California (Santa Barbara), 1997, GcCgNr
Jean-Philippe Nicot, (D), Texas (Austin), 1998, HwCg
Senior Scientist:
Ian Duncan, (D), British Columbia, 1982, Gg
F. Jerry Lucia, (M), Minnesota, 1954, Go
Stephen C. Ruppel, (D), Tennessee, 1979, Gd
Senior Research Scientist:
Kitty L. Milliken
Research Scientist:
Sergey Fomel, (D), Stanford, 2002, Yx
Jeffrey G. Paine, (D), Washington, 1991, Gr
Julia Stowell Gale, (D), Exeter (UK), 1987, Gc
Hongliu Zeng, (D), Texas, 1994, Gs
Associate Professor:
Charles Kerans, (D), Carleton, 1982, Gd
Research Scientist:
Tim Dooley, (D), London (UK), 1994, Gc
Research Scientist Associate:
John R. Andrews, (B), North Carolina, 1990, Oi
Research Associate:
Bruce Cutright
Qilong Fu, (D), Regina, 2005, Gso
H, Scott Hamlin, (D), Texas, Gr
Ursula Hammes, (D), Colorado, 1992, Gg
Farzam Javadpour, (D)
Timothy A. Meckel, (D)
Osareni C. Ogiesoba, (D)
Katherine D. Romanak, (D)
Diana Sava, (D), Stanford, 2004, Yg
Changbing Yang, (D)
Christopher K. Zahm, (D)
Beverly Blakeney DeJarnett, (M), Penn State, 1986, Gg
Edward W. Collins, (M), Stephen F. Austin, 1978, Ge
Micheal V. DeAngelo, (M), Texas (El Paso), 1988, Ye
Xavier Janson, (D), Miami, 2002, Gr
Katherine D. Romanak, (D), Texas (Austin), 1997, CgScGe
Research Scientist Associate:
Seay Nance, (D), Texas, 1988, Gg
Research Scientist Associate IV:
Robert M. Reed, (D), Texas (Austin), 1999, GxoGc
Robert C. Reedy, (M), New Mexico Tech, 1996, Hg
IT Manager:
Ron Russell, (M), Oklahoma, 1993, On
Associate Director:
Michael H. Young, Arizona, 1995, Ge
512-471-7135:
George Bush, (B), Texas Tech
Research Scientist Associate V:
Tucker F. Hentz, (M), Kansas, 1982, GrsGo
Research Scientist Associate IV:
Thomas A. Tremblay, (M), Texas, 1992, Oy
Research Scientist Associate:
Caroline Breton, (B), Texas, 2001, Oi
Dallas B. Dunlap, (B), Texas, 1997, Ye
Tiffany Hepner, (M), S Florida, 2000, Og

Project Manager:
Rebecca C. Smyth, (M), Texas, 1995, Hg
Ramon H. Trevino, (M), Texas (Arlington), 1988, GsoGe
Related Staff:
Joseph S. Yeh, (B), Fu-Jen Catholic, 1977, Gm

# Utah

## Utah Geological Survey

**Dept of Natural Resources** (2016)
1594 West North Temple, Ste 3110
Box 146100
Salt Lake City, UT 84114-6100
p. (801) 537-3300
rickallis@utah.gov
geology.utah.gov
f: https://www.facebook.com/UTGeologicalSurvey/
t: @utahgeological
State Geologist:
Richard G. Allis, (D), Toronto, 1977, Go
Program Manager:
Steve D. Bowman, (D), Nevada (Reno), 2002, NgGe
Michael D. Hylland, (M), Oregon State, 1990, Gg
Mike Lowe, (M), Utah State, 1985, Hw
Craig D. Morgan, (B), Utah, 1975, Go
Grant C. Willis, (M), Brigham Young, 1983, Gg
Curator Core Research Center:
Peter J. Nielsen, (M), Brigham Young, 1992, Eo
GIS Analyst:
J Buck Ehler, (B), Utah State, 2004, Oi
Financial Manager II:
Jodi T. Patterson, (M), Weber State, 2002, On

**Environmental Sciences Program** (2004)
1594 West North Temple, Ste 3110
PO Box 146100
Salt Lake City, UT 84114-6100
p. (801) 537-3389
rickallis@utah.gov
http://geology.utah.gov
Manager:
Mike Lowe, (M), Utah State, 1985, Hw
Geologist:
Charles E. Bishop, (B), Utah, 1986, Hw
Hugh A. Hurlow, (D), Washington, 1982, Hw
Janae Wallace, (M), N Arizona, 1993, Hw
Senior Geologist:
James I. Kirkland, (D), Colorado, 1990, Pg
Senior Scientist:
David B. Madsen, (D), Missouri, 1973, Ga
Research Associate:
Martha C. Hayden, (B), Utah, 1978, Pv
Geotechnician:
Alison Corey, On

**Geologic Hazards Program** (2014)
1594 West North Temple
P O Box 146100
Salt Lake City, UT 84114-6100
p. (801) 537-3300
rickallis@utah.gov

http://geology.utah.gov/ghp/
Program Manager:
Steve D. Bowman, (D), Nevada (Reno), 2002, NgOn
Senior Scientist:
William R. Lund, (B), Idaho, 1970, Ng
Paleoseismologist:
Christopher B. DuRoss, (M), Utah, 2004, Ng
Landslide Geologist:
Gregg Beukelman, (M), Boise State, 1997, NgOi
Hazards Mapping Geologist:
Jessica Castleton, (B), Weber State, 2005, Ng
Adam McKean, (M), Brigham Young Univ, 2011, Gmg
Hazards Geologist:
Tyler R. Knudsen, (M), Nevada (Las Vegas), 2005, Ng
Gregory N. McDonald, (B), Utah, 1992, Ng
Hazard Mapping Geologist:
Ben A. Erickson, (M), Utah, 2011, Ng
Debris Flow/Landslide Geologist:
Richard E. Giraud, (M), Idaho, 1986, Ng
Geologist:
Mike Hylland, (M), Oregon State, 1990, PgYs
GIS Analyst:
Corey Unger, (B), Weber State, 2001, Oi

**Geologic Information and Outreach Program** (2010)
1594 West North Temple Ste 3110
PO Box 146100
Salt Lake City, UT 84114-6100
p. (801) 537-3325
rebeccamedina@utah.gov
http://geology.utah.gov
Manager:
Sandra N. Eldredge, (B), Skidmore, 1978, Oe
Geologist:
William F. Case, (B), Westminster, 1967, Ng
Mark R. Milligan, (M), Utah, 1995, Gg
Christine M. Wilkerson, (B), Utah, 1985, Gg
Bookstore Manager:
Patricia Stokes, On
Geology Librarian:
Mage Yonetani, On

**Geologic Mapping Program** (2015)
1594 West North Temple, Ste 3110
PO Box 146100
Salt Lake City, UT 84114-6100
p. (801) 537-3355
grantwillis@utah.gov
http://geology.utah.gov
Senior Scientist:
Robert Biek, (M), N Illinois, 1988, Gg
Senior Geologist:
Donald Clark, (M), N Illinois, 1987, Gg
Jonathan K. King, (M), Wyoming, 1984, Gg
Douglas A. Sprinkel, (M), Utah State, 1977, GosGg
Manager:
Grant C. Willis, (M), Brigham Young, 1983, Gg
GIS Analyst:
Basia Matyjasik, (B), Warsaw, 1988, Gg
Senior GIS Analyst:
Kent D. Brown, On

**Utah Dept of Natural Resources** (2014)
1594 West North Temple, Suite 3110
P.O. Box 146100
Salt Lake City, UT 84114-6100
p. (801) 537-3300
rickallis@utah.gov
http://geology.utah.gov
Administrative Assistant: Dianne Davis
Director:
Richard G. Allis, (D), Toronto, 1977, Eg
Deputy Director:
Kimm M. Harty, (M), Alberta, 1984, Gm
Program Manager:
Steve Bowman, (D), U Nev Reno, 2000, Ng

# Vermont

## Agency of Natural Resources, Dept of Environmental Conservation

**Vermont Geological Survey** (2016)
1 National Life Drive
Main 2
Montpelier, VT 05620-3902
p. (802) 522-5210
marjorie.gale@vermont.gov
http://www.anr.state.vt.us/dec/geo/vgs.htm
State Geologist:
Marjorie H. Gale, (M), Vermont, 1980, GgcGe

Geologist -Environmental Scientist:
Jonathan Kim, (D), Buffalo, 1995, GetCg

## Victoria

### Geological Survey of Victoria, Australia

**Victoria - Dept of Economic Development, Jobs, Transport and Resources** (2016)
GPO Box 2392
Melbourne, VIC 3001
p. +61 3 94528906
gsv.info@ecodev.vic.gov.au
http://www.earthresources.vic.gov.au/earth-resources

### Victoria, Dept of Sustainability and Environment

**Dept of Sustainability and Environment** (2014)
8 Nicholson Street
East Melbourne, VIC 3002
p. +61 3 5332 5000
peter.walsh@parliament.vic.gov.au
http://www.dse.vic.gov.au/

## Virginia

### Division of Geology and Mineral Resources

**Southwestern Minerals & Geology Section** (2004)
P.O. Box 144
453 West Main St
Abingdon, VA 24210
p. (540) 676-5577
steve.walz@dmme.virginia.gov
http://www.dmme.virginia.gov/divisionmineralresources.shtml

**Virginia Dept of Mines, Minerals & Energy** (2016)
Fontaine Research Park
900 Natural Resources Drive
Suite 500
Charlottesville, VA 22903
p. (434) 951-6341
david.spears@dmme.virginia.gov
http://dmme.virginia.gov/DGMR/divisiongeology mineralresources.shtml

State Geologist:
David B. Spears, (M), Virginia Tech, 1983, GcEg
Manager, Geologic Mapping:
Matthew J. Heller, (M), North Carolina State, 1996, Gg
Manager, Economic Geology:
William L. Lassetter, (M), Nevada (Reno), 1996, EgHg

### University of Virginia

**Virginia State Climatology Office** (2014)
291 McCormick Road
P.O. Box 400123
Charlottesville, VA 22904-4123
p. 434-924-0548
climate@virginia.edu
http://climate.virginia.edu/home.htm

Director:
Patrick J. Michaels, (D), Wisconsin, 1979, Oa

## Washington

### Geological Survey of Western Australia

**Geological Survey of Western Australia** (2016)
Mineral House
100 Plain Street
East Perth , WA 6004
p. +61 8 9222 3333
geological.survey@dmp.wa.gov.au
http://www.dmp.wa.gov.au/Geological-Survey/Geological-Survey-262.aspx

### Washington Division of Geology & Earth Resources

**Washington Dept of Natural Resources** (2015)
1111 Washington Street, SE, MS 47007
Olympia, WA 98504-7007
p. (360) 902-1450
geology@dnr.wa.gov
http://www.dnr.wa.gov/geology/

Geology Librarian:
Stephanie Earls, (M), Washington, 2010, Gge
State Geologist:
David K. Norman, (M), Utah, 1980, Gg

## West Virginia

### West Virginia Geological & Economic Survey

**West Virginia Geological & Economic Survey** (2016)
Mont Chateau Research Center
1 Mont Chateau Road
Morgantown, WV 26508-8079
p. (304) 594-2331
info@geosrv.wvnet.edu
http://www.wvgs.wvnet.edu
f: https://www.facebook.com/WVGeoSurvey

State Geologist:
Michael Ed. Hohn, (D), Indiana, 1976, Eg
Adjunct Professor:
Paula J. Hunt, (M), Purdue, 1988, GgHw

## Wisconsin

### University of Wisconsin - Extension

**Wisconsin Geological and Natural History Survey** (2016)
3817 Mineral Point Road
Madison, WI 53705-5100
p. (608) 262-1705
jill.pongetti@wgnhs.uwex.edu
http://www.wisconsingeologicalsurvey.org

Hydrogeologist:
Madeline B. Gotkowitz, (D), New Mexico Tech, 1993, Hw
David J. Hart, (D), Wisconsin, 2000, HwYg
Director and State Geologist:
Kenneth R. Bradbury, (D), Wisconsin, 1982, Hw
Professor:
Kenneth R. Bradbury, (D), Wisconsin (Madison), 1982, Hw
Geologist:
Eric C. Carson, (D), Wisconsin, 2003, Gl
J. Elmo Rawling, (D), Gl
Geologist:
James J. Zambito, (D), Cincinnati, 2011, Gsg
State Geologist and Director:
James M. Robertson, (D), Michigan, 1972, Eg
Professor:
Thomas J. Evans, (D), Wisconsin, 1994, Eg
Emeritus:
Bruce A. Brown, (D), Manitoba, 1984, GcEg
Ron G. Hennings, (M), Wisconsin, 1977
Fred W. Madison
Roger M. Peters, (B), Wisconsin, 1969, Gg
Samples and Laboratories Manager:
Val L. Stanley, (M), Minnesota, 2010, GmpOi
Outreach Manager:
M. Carol McCartney, (D), Wisconsin, 1979, HwGle
Hydrogeologist:
Mike J. Parsen, (M), Hw
Geotechnician:
Peter M. Chase, (B), Wisconsin (Milwaukee), 1985, Hw
Geologist:
William G. Batten, (B), Wisconsin, 1973, GgHw
Irene D. Lippelt, (B), Manitoba, 1978, Gg
Esther K. Stewart, (M), Idaho State, 2008, Gg
Related Staff:
Grace E. Graham, (B), Beloit College, 2013, Hw
Jacob J. Krause, (B), Hw

# Wyoming

## Wyoming State Geological Survey

**Wyoming State Geological Survey** (2016)
P.O. Box 1347
Laramie, WY 82073
p. (307) 766-2286
wsgs-info@wyo.gov
http://www.wsgs.wyo.gov/

State Geologist:
Thomas A. Drean, (M), Penn State, 1978, Gg
Head:
Erin Campbell-Stone, (D), Ng
Seth Wittke, Ng
Geologist:
Karl Taboga, (M), Wyoming, 1979, HwGe
Geologist:
Chris J. Carroll, (B), California (Santa Barbara), 1982, EcgGg
Robert Gregory, En
Ranie Lynds, (D), Wyoming, 2005, Eo
Jim Rodgers, Gg
James Stafford, (M), Wyoming, 2013, HgyEg
Wayne Sutherland, En
Rachel Toner, Eo
Other:
Jacob Carnes, (M), Gg
Andrea M. Loveland, (M), Alaska (Fairbanks), Gct

# Yukon

## Yukon Geological Survey

**Yukon Geological Survey** (2011)
suzanne.roy@gov.yk.ca
http://www.geology.gov.yk.ca/

# Agencies and Intl Organizations

## International Union of Geological Sciences

**Executive Secretariat** (2015)
Chinese Academy of Geologic Sciences
No26, Baiwanzhuang Road
Xicheng District
Beijing 100037
p. 86-10-8833-3287
iugs.beijing@gmail.com
http://www.iugs.org

IUGS President:
Roland Oberhansli
IUGS Treasurer:
Dong Shuwen
IUGS Secretary General:
Ian Lambert
IUGS Councillor (2012-2016):
Hassina Mouri
Yujiro Ogawa
IUGS Councillor (2010-2014):
Wesley Hill
Sampat K. Tandon

## United Nations Education, Scientific, and Cultural Organization

**International Geoscience Programme** (2014)
Division of Ecological and Earth Sciences
1 rue Miollis
Paris, Cedex 15 F-75732
p. +33 (0)1 45 68 10 00
m.alaawah@unesco.org
http://www.unesco.org/new/en/natural-sciences/environment/earth-sciences/international-geoscience-programme/

IGCP Chairperson:
Patricia Vickers-Rich, (D), Columbia, 1972
Team Leader, Hydrogeology:
Gil Mahe
Team Leader, Global Change and Evolution of Life:
Guy Narbonne
Team Leader, Geohazards:
Andrej Gosar
Team Leader, Geodynamic:
George Gibson
Team Leader, Earth Resources:
Robert Moritz
IGCP/SIDA Representative:
Vivi Vajda, (D), Lund, 1998

## Ministry of Environmental Protection, Natural Resources and Forestry

**Dept of Geology** (2016)
st. Wawelska 52/54
Warsaw 00-922
p. +48 22 3692449
Departament.Geologii.i.Koncesji.Geologicznych@mos.gov.pl
http://www.mos.gov.pl/dg/dga1.htm

## Sveriges geologiska Undersokning

**Geological Survey of Sweden** (2016)
Box 670
751 28 Uppsala
sgu@sgu.se
http://www.sgu.se/

## Argonne National Laboratory

**Chemical Technology Div** (2004)
9700 South Cass Avenue
Building 205
Argonne, IL 60439
p. (630) 252-4383
ebunel@anl.gov

Senior Scientist:
Milton Blander, (D), Yale, 1953, Xc
Associate Scientist:
Allen J. Bakel, (D), Oklahoma, 1990, Co
Ronald P. Chiarello, (D), Northeastern, 1990, Om
Donald G. Graczyk, (D), Wisconsin, 1975, Cs
Ben D. Holt, (D), Illinois Inst of Tech, 1969, Ca
James J. Mazer, (D), Northwestern, 1987, Cl
David J. Wronkiewicz, (D), New Mexico Tech, 1989, Cg
Adjunct Professor:
Neil C. Sturchio, (D), Washington (St. Louis), 1983, Cg
Related Staff:
Alice M. Essling, (B), St. Xavier, 1956, Ca
Edmund A. Huff, (M), Chicago, 1957, Ca
Francis J. Markun, (B), Lewis, 1961, Cc
Florence P. Smith, (B), Southern, 1968, Ca

**Geosciences & Information Technology Section** (2014)
9700 South Cass Avenue
Argonne, IL 60439
p. (630) 252-6034
djmiller@anl.gov
Department Secretary: Sue Baumann

Manager, Geosciences & Information Tech:
Lisa A. Durham, (M), Purdue, 1989, Hw
Assistant System Engineer:
Cheong-yip R. Yuen, (D), Wisconsin (Milwaukee), 1986, Ge
Associate Scientist:
Robert L. Johnson, (D), Cornell, 1991, On
Research Associate:
John Ditmars, (D), Caltech, 1971, On
Jennifer Herbert, (B), N Illinois, 1997, Gg
Zhenhua Jiang, (D), Duke, 1992, On
David S. Miller, (D), Johns Hopkins, 1995, Ge
Terri L. Patton, (M), NE Illinois, 1989, Cl
John J. Quinn, (M), Minnesota, 1992, Hq
Geology Librarian:
Swati Wagh, On

**Environmental Research Div** (2014)
9700 South Cass Avenue
Argonne, IL 60439
p. (630) 252-3879
bmlesht@anl.gov
http://www.anl.gov/ER/

Senior Scientist:
Jeffrey S. Gaffney, (D), California (Riverside), 1975, Oa
Raymond M. Miller, (D), Illinois State, 1975, Sb
Marvin L. Wesely, (D), Wisconsin, 1970, Oa
Associate Scientist:
Jacqueline C. Burton, (D), Tennessee, 1978, Cg
Richard L. Coulter, (D), Penn State, 1976, Oa
Paul V. Doskey, (D), Wisconsin, 1982, Oa
Paul A. Fenter, (D), Pennsylvania, 1990, Gy
Julie D. Jastrow, (D), Chicago, 1994, Sb
Lorraine M. LaFreniere, (D), Wisconsin, 1980, On
In Young Lee, (D), California (Los Angeles), 1975, Oa
Barry M. Lesht, (D), Chicago, 1977, Op
Nancy A. Marley, (D), Florida State, 1984, Oa
William T. Meyer, (D), Imperial Coll (UK), 1973, Cg
Robert A. Sedivy, (M), Georgia Tech, 1979, Hw
Jack D. Shannon, (D), Oklahoma, 1975, Ow
Douglas L. Sisterson, (M), Wyoming, 1975, Oa
Mohamed Sultan, (D), Washington (St. Louis), 1984, Cg
John L. Walker, (D), Imperial Coll (UK), 1964, Cg
Y Eugene Yan, (D), Ohio State, 1998, Gg
Research Associate:
Richard H. Becker, (M), Washington (St. Louis), Or
Clyde B. Dennis, (M), Florida, 1979, On
Richard L. Hart, (B), Illinois Inst of Tech, 1970, Oa
Timothy J. Martin, (B), Beloit, 1974, Oa
Barney W. Nashold, (M), Illinois (Chicago Circle), 1976, On
Kent A. Orlandini, (B), Illinois, 1957, Cc

Candace M. Rose, (B), Benedictine Coll, 1985, On
Other:
David R. Cook, (M), Penn State, 1977, OwaOs

**Energy Systems Div** (2005)
9700 South Cass Avenue
Bldg 362 E-340
Argonne, IL 60439-4815
p. (630) 252-3392
energy_systems@anl.gov
Administrative Assistant: Barbara Sullivan
Head:
Donald O. Johnson, (D), Illinois, 1972, Gr
Senior Scientist:
Lyle D. McGinnis, (D), Illinois, 1965, Yg
Scientist:
Kenneth L. Brubaker, (D), Wisconsin, 1972, Oa
Dorland E. Edgar, (D), Purdue, 1976, Gm
Steven F. Miller, (B), Knox, 1984, Gg
R. Eric Zimmerman, (D), Northwestern, 1972, Ng
Research Associate:
Paul C. Heigold, (D), Illinois, 1969, Yg
Theresa C. Scholtz, (B), N Illinois, 1985, Ge
Michael D. Thompson, (M), N Illinois, 1989, Yg
Related Staff:
John F. Schneider, (M), N Illinois, 1977, Ca
Linda M. Shem, (M), Northwestern, 1991, Ge
Patrick L. Wilkey, (M), Illinois, 1976, Ng

**Chemistry Div** (2005)
9700 South Cass Avenue
Argonne, IL 60439
p. (630) 972-3570
ebunel@anl.gov
Senior Scientist:
Dieter M. Gruen, (D), Chicago, 1951, Om
Michael J. Pellin, (D), Illinois, 1978, Om
Associate Scientist:
Wallis F. Calaway, (D), Indiana, 1975, Om

## Bureau of Land Management

**Eastern States Office** (1999)
7450 Boston Boulevard
Springfield, VA 22153
p. (703) 440-1600
es_general_web@blm.gov

**Headquarters Directorate** (1999)
1849 C Street NW
Rm 5665
Washington, DC 20240
p. (202) 208-3801
director@blm.gov

**Colorado State Office** (1999)
2850 Youngfield Street
Lakewood, CO 80215
p. (303) 239-3600
hhankins@blm.gov

**California State Office** (1999)
2800 Cottage Way
Suite W-1623
Sacramento, CA 95825
p. (916) 978-4400
mdipinto@blm.gov

**Arizona State Office** (1999)
One North Central Ave
Suite 800
Phoenix, AZ 85004
p. (602) 417-9500
egomez@blm.gov

**Alaska State Office** (1999)
222 West Seventh Ave
#13
Anchorage, AK 99513
p. (907) 271-3212
dlassuy@blm.gov

**National Training Center** (1999)
9828 North 31st Avenue
Phoenix, AZ 85051
p. (602) 906-5500
dwilkins@blm.gov

**Nevada State Office** (1999)
1340 Financial Blvd
Reno, NV 89502
p. (775) 861-6400
jswickard@blm.gov

**Idaho State Office** (1999)
1387 South Vinnell Way
Boise, ID 83709
p. (208) 373-4000
sellis@blm.gov

**Montana State Office** (1999)
5001 Southgate Drive
Billings, MT 59101
p. (406) 896-5012
kiszler@blm.gov

**New Mexico State Office** (1999)
301 Dinosaur Trail
Santa Fe, NM 87502
p. (505) 954-2098
jjuen@blm.gov

**Oregon State Office** (1999)
333 SW 1st Avenue
Portland, OR 97204
p. (503) 808-6001
b2jackso@blm.gov

**Utah State Office** (1999)
440 West 200 South
Suite 500
Salt Lake City, UT 84101
p. (801) 539-4001
utsomail@blm.gov

**Wyoming State Office** (1999)
PO Box 1828
Cheyenne, WY 82003
p. (307) 775-6256
jcamargo@blm.gov

**National Operations Center** (1999)
PO Box 25047
Denver, CO 80225
p. (303) 236-8857
lgraham@blm.gov

## Department of Interior Office of Surface Mining Reclamation and Enforcement

**OSMRE Geospatial Information Services** (2016)
OSMRE Attn: David Carter
1999 Broadway, Suite 3320
Denver, CO 80225
p. (303) 293-5019
dcarter@osmre.gov
www.osmre.gov
Branch Manager:
David Carter, (M), Denver, 1999, Oir
Physical Scientist:
Carrie A. Middleton, (M), Colorado Mines, 2008, OrGfOi

## Dept of Agriculture

**Natural Resources Conservation Service** (2016)
14th and Independence Ave., SW

Washington, DC 20250
p. (202) 720-7246
jason.weller@wdc.usda.gov
http://www.nrcs.usda.gov

## Dept of Commerce

**National Inst of Standards & Technology** (2015)
100 Bureau Drive
Stop 1070
Gaithersburg, MD 20899-3460
p. (301) 975-2758
inquiries@nist.gov
http://www.nist.gov/

**National Oceanic & Atmospheric Administration** (2004)
Silver Spring Metro Center 3
1315 East-West Highway
Silver Spring, MD 20910-3282
p. (202) 482-3436
kathryn.sullivan@noaa.govkathryn.sullivan@noaa.gov
http://www.noaa.gov

## Dept of Defense

**Naval Intelligence Command** (2004)
4251 Suitland Road
Washington, DC 20395-5720
p. (301) 669-4000
http://www.nmic.navy.mil/

**Naval Oceanography Command** (2004)
U.S. Naval Observatory
34th Street & Massachusetts Avenue, NW
Washington, DC 20007
p. (202) 762-1020
http://www.oceanographer.navy.mil

**Space & Naval Warfare Systems Command** (2004)
4301 Pacific Highway
San Diego, CA 92110-3127
p. (619) 524-7053
http://www.spawar.navy.mil/

**U.S. Army Corps of Engineers** (2004)
441 G Street, NW
Washington, DC 20314
p. (202) 761-0001
http://www.usace.army.mil/working.html

**U.S. Special Operations Command (Air Force)** (2004)
100 Bartley Street
Hurlburt Field, FL 32544-5273
p. (850) 884-2323
http://www.af.mil/sites/afsoc.html

**Space Command** (2004)
150 Vandenberg Street
Peterson Air Force Base, CO 80914-4020
p. (719) 554-3001
http://www.af.mil/sites/afspc.shtml

**Air Weather Service** (2004)
106 Peacekeeper Drive
Offutt Air Force Base, NE 68113-4039
p. (402) 294-5749

**U.S. Army Research Laboratory Command** (2004)
The Pentagon
Washington, DC 20310
p. (703) 695-0363
http://www.army.mil/csa/

**U.S. Army Chemical & Biological Defense Command** (2004)
The Pentagon
Washington, DC 20310
p. (703) 695-0363
http://www.army.mil/csa/

**Defense Threat Reduction Agency** (2004)
8725 John T. Kingman Road
MS 6201
Ft. Belvoir, VA 22060-6201
p. (703) 767-4883
dtra.publicaffairs@dtra.mil
http://www.dtra.mil

**Air Force Center for Environmental Excellence** (2004)
3207 North Road
Brooks Air Force Base, TX 78235-5363
p. (210) 536-2162

**Office of Naval Research** (2004)
Ballston Towers #1
800 North Quincy Street
Arlington, VA 22203
p. (703) 696-4767
http://www.onr.navy.mil/

## Dept of Energy

**Office of Oil & Gas** (2004)
Forrestal Building
1000 Independence Avenue, SW
Washington, DC 20585-0640
p. (202) 586-6012

**Assistant Secretary for Environment, Safety & Health** (2004)
Forrestal Building
1000 Independence Avenue, SW
Washington, DC 20585-0119
p. (202) 586-6151
http://www.eh.doe.gov/

**Energy Information Administration** (2004)
Forrestal Building
1000 Independence Avenue, SW
Washington, DC 20585-0601
p. (202) 586-8800
howard.gruenspecht@eia.gov
http://www.eia.doe.gov/

**Office of Resource Management** (2004)
Forrestal Building
1000 Independence Avenue, SW
Washington, DC 20585-0620
p. (202) 586-3521

**Office of Coal, Nuclear, Electric & Alternate Fuels** (2004)
COMSAT Building
950 L'Enfant Plaza, SW
Washington, DC 20024
p. (202) 287-7990

**Office of Energy Projects** (2004)
Federal Energy Regulatory Commission
888 First Street, NE
Washington, DC 20426
p. (202) 219-2700

**Assistant Secretary for Environmental Management** (2014)
Forrestal Building
1000 Independence Avenue, SW
Washington, DC 20585-0113
p. (202) 586-7709
EM.WebContentManager@em.doe.gov
http://www.em.doe.gov/

**Coal & Power Systems** (2004)
Forrestal Building
1000 Independence Avenue, SW
Washington, DC 20585-0320
p. (202) 586-1650

**Office of Nuclear Energy, Science & Technology** (2004)
Forrestal Building
1000 Independence Avenue, SW
Washington, DC 20585-0117
p. (202) 586-6630
http://www.ne.doe.gov/

**Assistant Secretary for Energy Efficiency & Renewable Energy** (2004)
Forrestal Building
1000 Independence Avenue, SW
Washington, DC 20585-0121
p. (202) 586-9220
http://www.eren.doe.gov/

## Dept of Health & Human Services

**Centers for Disease Control & Prevention** (2004)
1600 Clifton Rd.
Atlanta, GA 30333
p. (404) 639-3535
http://www.cdc.gov

**Agency for Toxic Substances & Disease Registry** (2004)
1600 Clifton Road
Atlanta, GA 30333
p. (404) 639-7000
http://www.atsdr.cdc.gov/atsdrhome.html

## Dept of Labor

**Mine Safety & Health Administration** (2004)
Balston Tower #3
4015 Wilson Boulevard
Arlington, VA 22203
p. (703) 235-1385
http://www.msha.gov

## Dept of State

**Intl Boundary & Water Commission, US & Mexico** (2004)
The Commons
4171 North Mesa, Suite C-310
El Paso, TX 79902-1441
p. 1-800-262-8857
john.merino@ibwc.state.gov

## Dept of the Interior

**Office of Surface Mining, Reclamation & Enforcement** (2014)
1951 Constitution Ave. N.W.
Washington, DC 20240
p. (202) 208-2565
osm-getinfo@osmre.gov
http://www.osmre.gov/

**U.S. Fish & Wildlife Service** (2004)
1849 C. Street N.W.
Washington, DC 20240
p. (202) 208-4545
http://www.fws.gov/

**National Park Service** (2004)
1849 C. Street N.W.
Washington, DC 20240
p. (202) 208-4621
http://www.nps.gov

**Office of the Secretary** (2014)
1849 C. Street N.W.
Washington, DC 20240
p. (202) 208-3100
feedback@ios.doi.gov
http://www.doi.gov

**Bureau of Land Management** (2014)
Office of Public Affairs
1849 C Street, Room 406-LS
Washington, DC 20240
p. (202) 208-3801
director@blm.gov
http://www.blm.gov

**Bureau of Reclamation** (2004)
1849 C. Street N.W.
Washington, DC 20240
p. (202) 513-0501
http://www.usbr.gov

**Bureau of Indian Affairs** (2004)
1849 C. Street N.W.
Washington, DC 20240
p. (202) 208-7163
http://www.doi.gov/bureau-indian-affairs.html

## Dept of the Treasury

**Internal Revenue Service** (2004)
1111 Constitution Avenue, NW
Washington, DC 20224
p. (202) 622-9511
http://www.irs.ustreas.gov

## Dept of Transportation

**Federal Aviation Administration** (2004)
800 Independence Avenue, SW
Washington, DC 20591
p. (202) 267-3484
http://www.faa.gov

**U.S. Coast Guard** (2004)
2100 Second Street, SW
Washington, DC 20593
p. (202) 366-4000
http://www.uscg.mil

## Environmental Protection Agency

**Office of Wetlands, Oceans & Watersheds** (2004)
Fairchild Building
499 South Capitol Street, SW
Washington, DC 20003
p. (202) 260-7166
best-wong.benita@epa.gov

**Office of Underground Storage Tanks** (2004)
Crystal Gateway One
1235 Jefferson Davis Highway
Arlington, VA 22202
p. (703) 603-9900
hoskinson.carolyn@epa.gov

**Office of Science Policy** (2016)
1200 Pennsylvania Ave., NW
MC-8104R
Washington, DC 20460
p. (202) 564-6705
hauchman.fred@epa.gov

**Office of Science & Technology** (2004)
1200 Pennsylvania Avenue, NW
Washington, DC 20460
p. (202) 260-5400
Southerland.elizabeth@Epa.gov

**National Center for Environmental Research** (2004)
Ronald Reagan Building
1300 Pennsylvania Avenue, NW
Washington, DC 20004
p. (202) 564-6825
johnson.jim@epa.gov

**National Center for Environmental Assessment** (2016)
USEPA-ORD-NCEA-8623P
1200 Pennsylvania Avenue, NW

Washington, DC 20460
p. (703) 347-8623
frithsen.jeff@epa.gov
http://www.epa.gov/ncea

**Office of Solid Waste** (2014)
1301 Constitutiion Ave. NW
Arlington, VA 22202
p. (202) 566-0200
aastanislaus@epa.gov
http://www2.epa.gov/aboutepa/about-office-solid-waste-and-emergency-response-oswer

**Office of Atmospheric Programs** (2004)
501 Third Street, NW
Washington, DC 20001
p. (202) 564-9140
McCabe.janet@Epa.gov

**Office of Emergency & Remedial Response (Superfund/Oil Programs)** (2004)
Crystal Gateway One
1235 Jefferson Davis Highway
Arlington, VA 22202
p. (703) 603-8960
oilinfo@epamail.epa.gov
http://www.epa.gov/superfund

**Office of Ground Water & Drinking Water** (2004)
1200 Pennsylvania Avenue, NW
Washington, DC 20460
p. (202) 260-5400
grevatt.peter@epa.gov

**Assistant Administrator for Environmental Information** (2004)
1200 Pennsylvania Avenue, NW
Washington, DC 20460
p. (202) 564-6665
Nelson.kimberly@Epa.gov

**Safety, Health & Environmental Management Div** (2014)
1200 Pennsylvania Avenue, NW
Washington, DC 20460
p. (202) 564-1640
dfe@epa.gov

**National Air & Radiation Environmental Laboratory** (2004)
540 South Morris Avenue
Montgomery, AL 36115-2601
p. (334) 270-3404
griggs.john@epa.gov

**Assistant Administrator for Water Programs** (2004)
1200 Pennsylvania Avenue, NW
Washington, DC 20460
p. (202) 260-5700
Beauvais.joel@Epa.gov

## Federal Emergency Management Agency

**Office of the Director** (2004)
500 C Street, SW
Washington, DC 20472
p. (202) 646-4600
http://www.fema.gov/

## Jet Propulsion Laboratory

**Earth & Space Sciences Div** (2014)
Geology & Planetology Section
Pasadena, CA 91109
p. (818) 354-3440
daniel.j.mccleese@jpl.nasa.gov
Department Administrative Manager: Murray Geller

Lead Scientist:
Bruce E. Banerdt, (D), S California, 1983, Xy
Section Manager:
Ronald G. Blom, (D), California (Santa Barbara), 1987, Or
Section Member:
Michael J. Abrams, (M), Caltech, 1973, Or
Ronald E. Alley, (M), Northwestern, 1972, Or
Diana L. Blaney, (D), Hawaii, 1990, Xg
Bonnie J. Buratti, (D), Cornell, 1983, Or
Robert E. Crippen, (D), California (Santa Barbara), 1989, Or
Joy A. Crisp, (D), Princeton, 1984, Gv
Thomas C. Duxbury, (M), Purdue, 1966, On
Diane L. Evans, (D), Washington, 1981, Gm
Tom G. Farr, (D), Washington, 1981, Gm
Ken E. Herkenhoff, (D), Caltech, 1989, Xg
Simon J. Hook, (D), Durham (UK), 1989, Or
Erik R. Ivins, (M), California (Los Angeles), 1976, Gt
Anne B. Kahle, (D), California (Los Angeles), 1975, Or
Harold R. Lang, (D), Calgary, 1983, Gr
Kyle C. McDonald, (D), Michigan, 1991, On
Robert M. Nelson, (D), Pittsburgh, 1978, Xg
Eni G. Njoku, (D), MIT, 1976, Or
Frank D. Palluconi, (M), Penn State, 1963, Or
David C. Pieri, (D), Cornell, 1979, Gt
Jeffrey J. Plant, (D), Washington (St. Louis), 1991, On
Jeffrey Plescia, (D), S California, 1985, On
Carol A. Raymond, (D), Columbia, 1989, Yr
Suzanne E. Smrekar, (D), S Methodist, 1990, Xy
Linda J. Spilker, (D), California (Los Angeles), 1992, Xy
Ellen R. Stofan, (D), Brown, 1989, Xg
Glenn J. Veeder, (D), Caltech, 1974, On
Section Member:
Matthew P. Golombek, (D), Massachusetts, 1981, XgyGt
Senior Scientist:
Ronald S. Sanders, (D), Brown, 1970, Gr

## Lawrence Livermore National Laboratory

**Atmospheric, Earth and Energy Div** (2015)
7000 East Avenue
PO Box 808
Livermore, CA 94551
p. (925) 423-4412
pls-webmaster@mail.llnl.gov
http://www-pls.llnl.gov

**Atmospheric, Earth and Energy** (2013)
7000 East Avenue
L-203
Livermore, CA 94550
p. (925) 423-1848
antoun1@llnl.gov
https://www-pls.llnl.gov/?url=about_pls-atmospheric_earth_and_energy_division
Administrative Assistant: Laura Long

Head:
Kenneth J. Jackson, (D), California (Berkeley), 1983, Cl
Senior Scientist:
Roger D. Aines, (D), Caltech, 1984, Cg
Bill L. Bourcier, (D), Oregon State, 1983, Cg
Thomas A. Buscheck, (D), California (Berkeley), 1984, Ng
Steven Carle, (D), California (Davis), 1996, Hw
Charles R. Carrigan, (D), California (Los Angeles), 1977, Yg
Susan A. Carroll, (D), Northwestern, 1988, Cl
M Lee Davisson, (M), California (Davis), 1992, Cg
Quingyun Duan, (D), Arizona, 1991, Hg
Robert C. Finkel, (D), California (San Diego), 1974, Cg
Samuel (Julio) J. Friedmann, (D), MIT, Gg
Richard B. Knapp, (D), Arizona, 1978, Ng
Gayle A. Pawloski, (B), Cal State (Hayward), 1979, Gg
Abelardo L. Ramirez, (M), Purdue, 1979, Ng
Sarah K. Roberts, (M), Arizona State, 1990, Cl
Andrew F B Tompson, (D), Princeton, 1985, Ng
Jeffrey L. Wagoner, (M), California (Riverside), 1977, Gs
Ananda M. Wijesinghe, (D), MIT, 1978, Nr
Sun Yunwei, (D), Israel Inst of Tech, 1995, Ng
Mavrik Zavarin, Yr

## Los Alamos National Laboratory

**Earth & Environmental Sciences Div** (2016)
P.O. Box 1663
Los Alamos, NM 87545
p. (505) 667-3644
wallacet@lanl.gov
*Enrollment (1996): B: 17 (17) M: 6 (6)*

Technical Staff:
James E. Bossert, (D), Colorado State, 1990, Oa
Director:
Terry C. Wallace, Jr., (D), Caltech, 1983, Ys
Technical Staff:
W. Scott Baldridge, (D), Caltech, 1979, GicGt
Kay H. Birdsell, (M), Colorado, 1985, Hqy
James W. Carey, (D), Harvard, 1990, CpYxNr
Lianjie Huang, (D), Paris, 1994, Yg
Technical Staff:
Paul Aamodt, (M), Nevada, 1991, On
Douglas Alde, (D), Illinois, 1979, On
M. James Aldrich, (D), New Mexico, 1972, Gc
Fairley J. Barnes, (D), New Mexico State, 1986, On
Naomi M. Becker, (D), Wisconsin, 1991, Hg
James D. Blacic, (D), California (Los Angeles), 1971, Nr
Rainer Bleck, (D), Penn State, 1968, Ow
Christopher R. Bradley, (D), MIT, 1993, Yg
Thomas L. Brake, (M), New Mexico State, 1985, Nr
David E. Broxton, (M), New Mexico, 1976, Gx
Wendee M. Brunish, (D), Illinois, 1981, On
Gilles Y. Bussod, (D), California (Los Angeles), 1988, Yx
Katherine Campbell, (D), New Mexico, 1979, Gq
Theodore C. Carney, (M), Colorado State, 1981, Ng
Gregory L. Cole, (D), Arizona, 1990, Og
Steffanie Coonley, (M), New Mexico, 1984, On
Keeley R. Costigan, (D), Colorado State, 1992, Oa
William Cottingame, (D), Texas, 1984, Oa
James L. Craig, (B), Nevada, 1974, Gg
Bruce M. Crowe, (D), California (Santa Barbara), 1974, Gv
Zora V. Dash, (M), New Mexico, 1988, On
Deborah J. Daymon, (M), Idaho, 1994, On
Kalpak Dighe, (D), Clemson, 1994, On
John C. Dinsmoor, (B), Colorado Mines, 1989, Nm
Alison M. Dorries, (D), Harvard, 1986, On
Donald S. Dreesen, (B), New Mexico, 1968, Np
David V. Duchane, (D), Michigan, 1978, On
Michael H. Ebinger, (D), Purdue, 1988, So
C L Edwards, (D), New Mexico Tech, 1975, Yg
C. James Elliott, (D), Yale, On
Scott M. Elliott, (D), California, 1983, Oa
Perry D. Farley, (D), Oklahoma, 1981, On
George T. Farmer, (D), Cincinnati, 1968, Hw
David N. Fogel, (M), Cal State, 1997, On
Carl W. Gable, (D), Harvard, 1989, Gt
Edward S. Gaffney, (D), Caltech, 1973, Yg
Anthony F. Gallegos, (D), Colorado State, 1970, Py
Jamie N. Gardner, (D), California (Davis), 1985, Gg
Fraser Goff, (D), California (Santa Cruz), 1977, Cg
Jeffrey C. Hansen, (B), Weber State, 1971, On
Charles D. Harrington, (D), Indiana, 1970, Gm
Hans E. Hartse, (D), New Mexico Tech, 1991, Ys
Ward L. Hawkins, (B), Nevada (Reno), 1977, Og
Grant Heiken, (D), California (Santa Barbara), 1972, Gv
Donald D. Hickmott, (D), MIT, 1988, Gp
Steve T. Hildebrand, (D), Texas (Dallas), 1993, Ys
Emil F. Homuth, (D), Washington, 1974, Ye
Leigh S. House, (D), Columbia, 1982, Ys
Paul A. Johnson, (M), Arizona, 1984, Yg
Eric M. Jones, (D), Wisconsin, 1970, Oa
Hemendra N. Kalia, (D), Missouri Sch of Mines, 1970, Nm
Jim C. Kao, (D), Illinois, 1985, Oa
Danny Katzman, (M), New Mexico, 1991, Gg
Elizabeth Keating, (D), Wisconsin, 1995, Gg
Sharad Kelkar, (M), Texas, 1979, Np
C. F. Keller, Jr., (D), Indiana, 1969, Oa
Richard G. Kovach, (M), Naval Postgrad Sch, 1979, Nm
Donathon J. Krier, (M), New Mexico, 1980, Gv
Thomas D. Kunkle, (D), Hawaii, 1978, Xy
Edward M. Kwicklis, (M), Colorado, 1987, Gg
Chung Chieng A. Lai, (D), Texas A&M, 1984, Ow
Schon S. Levy, (M), Texas, 1975, Gi
Peter C. Lichtner, (D), Mainz, 1974, On
Rodman Linn, (D), New Mexico State, 1997, Ng
Lynn McDonald, (M), Cal State, On
Maureen A. McGraw, (D), California (Berkeley), 1996, On
Laurie A. McNair, (D), Carnegie Mellon, 1995, On
Wayne R. Meadows, (B), Cal State (Bakersfield), 1974, Yg
Theodore Mockler, (M), Carnegie Mellon, 1986, On
Orrin B. Myers, (D), Colorado State, 1992, Py
Balu Nadiga, (D), Caltech, 1992, On
Brent D. Newman, (D), New Mexico Tech, 1996, Gg
John W. Nyhan, (D), Colorado State, 1972, So
Ronald D. Oliver, (B), Oregon Inst of Tech, 1972, Yg
Howard J. Patton, (D), MIT, 1978, Yg
Frank V. Perry, (D), California, 1988, Gv
William S. Phillips, (D), MIT, 1985, Yg
Eugene W. Pokorny, (B), Missouri, 1979, Nm
William M. Porch, (D), Washington, 1971, Yg
Allyn R. Pratt, (M), Boise State, 1977, Ge
George Randall, (D), SUNY (Binghamton), Yg
Steen Rasmussen, (D), Tech (Denmark), 1985, Yg
Jon M. Reisner, (D), Iowa State, Oa
Steven L. Reneau, (D), California (Berkeley), 1988, Gm
Douglas O. Revelle, (D), Michigan, 1974, Oa
Peter Roberts, (D), MIT, 1989, Ys
Bruce A. Robinson, (D), MIT, 1985, On
R. Roussel-Dupre, (D), Colorado, 1979, On
Thomas J. Shankland, (D), Harvard, 1966, Yx
Catherine H. Smith, (D), New Mexico State, 1995, Ca
Wendy E. Soll, (D), MIT, 1991, Hw
Everett P. Springer, (D), Utah State, 1983, Hq
Lee Steck, (D), California, Gg
Robert P. Swift, (D), Washington, 1969, Nr
E.M.D. Symbalisty, (D), Chicago, 1984, On
Steven R. Taylor, (D), MIT, 1980, Ys
James Tencate, (D), Texas, 1992, On
Bryan J. Travis, (D), Florida State, 1974, Hq
David T. Vaniman, (D), California (Santa Cruz), 1976, Gx
Richard G. Warren, (M), New Mexico, 1972, Gi
Douglas J. Weaver, (M), Nevada, 1995, On
Thomas A. Weaver, (D), Chicago, 1973, Yg
Rodney W. Whitaker, (D), Indiana, 1976, Oa
Earl M. Whitney, (D), Utah, On
Judith L. Winterkamp, (B), Texas, 1977, On
Kenneth H. Wohletz, (D), Arizona State, 1980, Gv
Giday WoldeGabriel, (D), Case Western, 1987, Gx
Andrew V. Wolfsberg, (D), Stanford, 1993, Hw
George A. Zyvoloski, (D), California (Santa Barbara), 1975, Ng
Project Leader:
Mark T. Peters, (D), Chicago, 1992, Yg
Technical Staff:
James N. Albright, (D), Chicago, 1969, Yg
Donald W. Brown, (M), California (Los Angeles), 1961, Nr
Larry Allan Jones, (B), Nebraska, 1972, Ng
Other:
Christina “Tina” Behr-Andres, (D), Michigan Tech, 1992, On

**Chemistry Div** (B,M,D) (2004)
P.O. Box 1663
Los Alamos, NM 87545
p. (505) 667-4457
chemistry@lanl.gov
http://pearl1.lanl.gov/external/default.htm

Division Leader:
Alexander J. Gancarz, (D), Caltech, 1976, Cc
Staff:
Kent D. Abney, (D), Colorado State, 1987, On
Stephen F. Agnew, (D), Washington State, 1981, On
Moses Attrep, (D), Arkansas, 1965, Oe
Timothy M. Benjamin, (D), Caltech, 1979, Ca
Scott M. Bowen, (D), New Mexico, 1983, Ca
James R. Brainard, (D), Indiana, 1979, On
Jeffrey C. Bryan, (D), Washington, 1988, On
Carol J. Burns, (D), California (Berkeley), 1987, On
Timothy P. Burns, (D), Nebraska, 1986, Co
Gilbert W. Butler, (D), California (Berkeley), 1967, Cc

Edwin P. Chamberlain, (D), Texas A&M, 1971, Cc
David L. Clark, (D), Indiana, 1986, On
Dean A. Cole, (D), Iowa, 1985, Co
David B. Curtis, (D), Oregon State, 1974, Cc
Paul R. Dixon, (D), Yale, 1989, On
Robert J. Donahoe, (D), North Carolina State, 1985, On
Stephen K. Doorn, (D), Northwestern, 1989, On
Clarence J. Duffy, (D), British Columbia, 1977, Cg
Deward W. Efurd, (D), Arkansas, 1975, Ct
Phillip G. Eller, (D), Ohio State, 1971, On
June T. Fabryka-Martin, (D), Arizona, 1988, Hg
Bryan L. Fearey, (D), Iowa State, 1986, Cc
David L. Finnegan, (D), Maryland, 1984, Oa
John R. Fitzpatrick, (M), New Mexico, 1983, On
Malcolm M. Fowler, (D), Washington (St. Louis), 1972, Oa
Sammy R. Garcia, (B), New Mexico Highlands, 1972, Ca
Russell E. Gritzo, (B), New Mexico State, 1983, On
Richard C. Heaton, (D), Illinois, 1973, On
Sara B. Helmick, (B), Southwestern, 1955, On
David R. Janecky, (D), Minnesota, 1982, Cl
Kung King-Hsi, (D), Cornell, 1989, On
Scott Kinkead, (D), Idaho, 1983, On
Gregory J. Kubas, (D), Northwestern, 1970, Oa
Pat J. Langston-Unkefer, (D), Texas A&M, 1978, Co
Patrick A. Longmire, (D), New Mexico, 1991, On
Michael MacInnes, (M), Wisconsin, 1969, On
Allen S. Mason, (D), Miami, 1974, Oa
Charles M. Miller, (D), Stanford, 1980, Cc
Geoffrey G. Miller, (D), Rensselaer, 1984, Cc
Terrance L. Morgan, (M), Colorado State, 1984, Ng
David Morris, (D), North Carolina State, 1984, Cc
Eugene J. Mroz, (D), Maryland, 1976, Oa
Michael T. Murrell, (D), California (San Diego), 1980, Cc
Allen E. Ogard, (D), Chicago, 1957, Hw
Jose A. Olivares, (D), Iowa State, 1985, Cs
Hain Oona, (D), Arizona, 1979, On
Kevin C. Ott, (D), Caltech, 1983, On
Edward S. Patera, (D), Arizona State, 1982, Cg
Richard E. Perrin, (B), Denver, 1952, Cc
Eugene J. Peterson, (D), Arizona State, 1976, Ct
Dennis Phillips, (D), Hawaii, 1976, Co
Jane Poths, (D), Chicago, 1982, Cc
Pamela Z. Rogers, (D), California (Berkeley), 1981, Cl
Donald J. Rokop, (D), Lake Forest, 1985, Cc
Robert S. Rundberg, (D), CUNY, 1978, Gq
Nancy N. S. Sauer, (D), Iowa State, 1986, On
Norman C. Schroeder, (D), Iowa State, 1985, Cc
Louis A. Silks, (B), Suffolk, 1978, On
Paul H. Smith, (D), California (Berkeley), 1987, On
Zita V. Svitra, (B), Roosevelt, 1967, On
Basil I. Swanson, (D), Northwestern, 1969, On
C. Drew Tait, (D), North Carolina State, 1984, Cp
Wayne A. Taylor, (B), New Mexico, 1978, Cl
Kimberly W. Thomas, (D), California (Berkeley), 1978, Ct
Joseph L. Thompson, (D), Penn State, 1963, Ct
Ines Triay, (D), Miami, 1985, Hg
Clifford J. Unkefer, (D), Minnesota, 1981, On
David J. Vieira, (D), California (Berkeley), 1978, On
Jerry B. Wilhelmy, (D), California (Berkeley), 1969, On
Kurt Wolfsberg, (D), Washington (St. Louis), 1959, Cc
William H. Woodruff, (D), Purdue, 1972, On
Mary Anne Yates, (D), Carnegie Mellon, 1976, On

Emeritus:
Ernest A. Bryant, (D), Washington (St. Louis), 1956, Cg
Merle E. Bunker, (D), Indiana, 1950, On
William R. Daniels, (D), New Mexico, 1965, Ct
Donald L. Hull, (M), Iowa State, 1974, On

## National Aeronautics & Space Administration

**NASA Headquarters** (2004)
Washington, DC 20546-0001
p. (202) 358-2345
info-center@hq.nasa.gov
http://www.hq.nasa.gov/

**Ames Research Center** (2004)
Building BN200
Moffett Field, CA 94035-1000
p. (650) 604-5111
michael.s.mewhinney@nasa.gov

**Lyndon B. Johnson Space Center** (2004)
Houston, TX 77058-3696
p. (281) 483-5309
ellen.ochoa-1@nasa.gov

**Langley Research Center Office of Communications** (2017)
11 Langley Boulevard
Hampton, VA 23681-2199
p. (757) 864-6120
rob.wyman@nasa.gov
http://www.nasa.gov/langley
f: https://www.facebook.com/nasalarc
t: @NASA_Langley

**George C. Marshall Space Flight Center** (2004)
Marshall Space Flight Center, AL 35812-0001
p. (256) 544-1910
patrick.e.scheuermann@nasa.gov

**Goddard Space Flight Center** (2004)
Greenbelt Road
Greenbelt, MD 20771-0001
p. (301) 286-5121
stephanie.s.keene@nasa.gov

## National Oceanic and Atmospheric Administration

**National Centers for Environmental Information** (2016)
David Skaggs Research Center
325 Broadway
Boulder, CO 80303
p. (303) 497-6826
ncei.info@noaa.gov
http://www.ngdc.noaa.gov/
f: http://www.facebook.com/NOAANCEIoceangeo
t: @NOAANCEIocngeo

## National Science Foundation

**Ocean Sciences Division** (2016)
4201 Wilson Boulevard
Arlington, VA 22230
p. (703) 292-8580
rwmurray@nsf.gov
http://www.nsf.gov/div/index.jsp?div=OCE

Division Director:
Richard W. Murray
Section Head for Integrative Programs Section:
Bauke Houtman
Program Director for Ship Operations Program:
Rose Dufour
Program Director for Physical Oceanography Program:
Eric C. Itsweire
Baris Mete Uz
Program Director for Oceanographic Instrumentation and Technical Service Programs:
James Holik
Program Director for Ocean Observatories Initiative:
Jean M. McGovern
Program Director for Ocean Education Programs:
Elizabeth L. Rom
Program Director for Ocean Drilling Programs:
James F. Allan
Program Director for Marine Geology and Geophysics Program:
Candace O. Major
Barbara Ransom
Program Director for Chemical Oceanography Program:
Donald Rice
Program Director for Biological Oceanography Program:
David L. Garrison
Program Director:
John Walter
Program Directof for Ocean Drilling Programs:
Thomas Janacek

Associate Program Director for Oceanographic Technology and Interdisciplinary Coordination Program:
Kandace S. Binkley

**Atmospheric and Geospace Sciences Div** (2015)
4201 Wilson Boulevard
Arlington, VA 22230
p. (703) 292-8520
pshepson@nsf.gov
http://www.nsf.gov/div/index.jsp?div=AGS
Division Director:
Paul B. Shepson
Section Head for NCAR/Facilities Section:
Stephan P. Nelson
Section Head for Geospace Section:
Richard A. Behnke
Section Head for Atmosphere Section:
David J. Verardo
Program Director for Solar Terrestrial Research Program:
Therese M. Jorgensen
Program Director for Physical and Dynamic Meteorology Program:
A. Gannet Hallar
Chungu Lu
Bradley F. Smull
Program Director for Paleoclimate Program:
Candace O. Major
Program Director for Magnetospheric Physics Program:
Raymond J. Walker
Program Director for Geospace Facilities:
Robert M. Robinson
Program Director for Education and Cross Disciplinary Activities Program:
Linda George
Program Director for Climate and Large Scale Dynamics Program & Carbon and Water in Earth Systems Program:
Eric T. DeWeaver
Program Director for Climate and Large Scale Dynamics Program:
Anjuli Bamzai
Program Director for Atmospheric Chemistry Program:
Peter Milne
Anne-Marie Schmoltner
Program Director for Aeronomy Program:
Anja Stromme
Program Coodinator for NCAR/Facilities Section:
Sarah L. Ruth
Facilities Program Manager for NCAR/Facilities Section:
Linnea M. Avallone
Assistant Program Director for Atmosphere Section:
Nicholas F. Anderson

**Office of the Director** (2015)
4201 Wilson Boulevard
Arlington, VA 22230
p. (703) 292-5111
fcordova@nsf.gov
http://www.nsf.gov
f: https://www.facebook.com/US.NSF
t: @NSF

**Directorate for Geosciences** (2016)
4201 Wilson Boulevard
Room 705
Arlington, VA 22230
p. (703) 292-8500
mcavanau@nsf.gov
https://www.nsf.gov/dir/index.jsp?org=GEO
f: https://www.facebook.com/US.NSF
t: @NSF
Assistant Director:
Roger Wakimoto
Deputy Assistant Director:
Margaret Cavanaugh
Senior Advisor:
Craig R. Robinson
Program Director for International Activities:
Maria Uhle

**Earth Sciences Div** (2017)
4201 Wilson Boulevard
Arlington, VA 22230
p. (703) 292-8550
cfrost@nsf.gov
http://www.nsf.gov/div/index.jsp?div=EAR
t: @NSF_EAR
Program Director for Hydrologic Sciences:
Thomas Torgersen
Program Director for Instrumentation and Facilities:
David Lambert
Section Head:
Gregory J. Anderson
Program Director for Tectonics Program:
David M. Fountain
Stephen S. Harlan
Program Director for Petrology and Geochemistry Program:
Sonia Esperanca
Jennifer Wade
Program Director for Integrated Earth Systems:
Leonard E. Johnson
Program Director for Instrumentation and Facilities:
Russell C. Kelz
Program Director for Geophysics Program:
Robin Reichlin
Program Director for Geobiology and Low Temp Geochemistry:
Enriqueta C. Barrera
Program Director for Education and Human Resources:
Lina Patino
Program Director for Earthscope:
Margaret Benoit
Program Director :
Jonathan Wynn

**Office of Polar Programs** (2016)
4201 Wilson Boulevard
Arlington, VA 22230
p. (703) 292-8030
kfalkner@nsf.gov
http://www.nsf.gov/div/index.jsp?org=PLR
f: https://www.facebook.com/pages/Division-of-Polar-Programs-National-Science-Foundation/139290131761511
Division of Polar Programs:
Kelly K. Falkner, (D)
Section for Arctic Sciences:
Eric Saltzman, (D)
Section for Antarctic Sciences:
Scott G. Borg, (D), Arizona State, 1984, Git
Section for Antarctic Infrastructure and Logistics:
Brian W. Stone, (M)
Antarctic Astrophysics and Geospace Science Program Director:
Vladimir Papitashvili, (D), Russian Acad of Sci, 1981, YgmOa
Specialized Support Manager, Antarctic Infrastructure & Logistics Section:
Arthur J. Brown
Senior Advisor:
Susanne M. LaFratta-Decker
Program Manager, Techology Development, Antarctic Infrastructure & Logistics Section:
Patrick D. Smith
Program Manager, System Operations & Logistics, Antarctic Infrastructure & Logistics Section:
Paul Sheppard
Outreach and Education Program Manager:
Peter West
Operations Manager, Antarctic Infrastructure & Logistics Section:
Margaret A. Knuth
Ocean Project Manager, Antarctic Infrastructure & Logistics Section:
Timothy M. McGovern
Facilities Engineering Projects Manager, Antarctic Infrastructure & Logistics Section:
Ben D. Roth
Environmental Officer:
Polly A. Penhale, (D)
Arctic System Science Program Director:
Neil R. Swanberg, (D)

Arctic Social Sciences Program Director:
Anna M. Kerttula de Echave, (D)
Arctic Research Support & Logistics Program Manager:
Patrick R. Haggerty
Arctic Research Support & Logistics Associate Program Manager:
Renee Crain
Arctic Natural Sciences Program Director:
William J. Wiseman, (D)
Antarctic Research Support Manager:
Jessie L. Crain
Antarctic Research & Logistics Integration Associate Program Director:
Nature McGinn, (D)
Antarctic Ocean and Atmospheric Sciences Program Director:
Peter Milne, (D)
Antarctic Glaciology Program Director:
Julie M. Palais, (D)

## Nuclear Regulatory Commission

**NRC Headquarters** (2004)
One White Flint North Building
11555 Rockville Pike
Rockville, MD 20852
p. (301) 415-8200
http://www.nrc.gov/

## Oak Ridge National Laboratory

**Environmental Sciences Div** (2014)
P.O. Box 2008
Mail Stop 6035
Oak Ridge, TN 37831
p. (865) 574-7374
envsci@ornl.gov
http://www.esd.ornl.gov
Director:
Stephen G. Hildebrand, (D), Michigan, 1973, On
Senior Scientist:
Marshall Adams, (D), North Carolina, 1974, Ob
Jeff Amthor, (D), Yale, 1987, Sf
Tom Ashwood, (M), Murray State, 1975, Ge
Mark Bevelhimer, (D), Tennessee, 1990, Hs
Terence J. Blasing, (D), Wisconsin, 1975, Oa
Thomas Boden, (M), Miami, 1985, Oa
Craig Brandt, (M), Tennessee, 1988, Ge
Scott C. Brooks, (D), Virginia, 1994, Cl
Robert S. Burlage, (D), Tennessee, 1990, Sb
Meng-Dawn Cheng, (D), Illinois, 1986, Oa
Robert B. Cook, (D), Columbia, 1981, Cg
William E. Doll, (D), Wisconsin, 1983, Yg
Thomas O. Early, (D), Washington (St. Louis), 1969, Cl
Baohua Gu, (D), California (Berkeley), 1991, Cl
Philip M. Jardine, (D), Virginia Tech, 1985, Sc
Liyuan Liang, (D), Caltech, 1988, Cl
Steven E. Lindberg, (D), Florida State, 1979, Cl
John F. McCarthy, (D), Rhode Island, 1975, Co
Gerilynn R. Moline, (D), Wisconsin, 1992, Hw
Tony V. Palumbo, (D), North Carolina State, 1980, Sb
Tommy J. Phelps, (D), Wisconsin, 1985, Sb
Ellen D. Smith, (M), Wisconsin, 1979, Hg
Brian P. Spalding, (D), Cornell, 1976, Sc
Robert S. Turner, (D), Pennsylvania, 1983, Cl
David B. Watson, (M), New Mexico Tech, 1983, Hw
Olivia M. West, (D), MIT, 1991, Ng
Associate Scientist:
Tammy Beaty, Oi
Mary Anna Bogle, (M), Miami, 1975, Ge
Norman D. Farrow, (B), Oregon State, 1974, Hg
Deputy Director:
Gary K. Jacobs, (D), Penn State, 1981, Cg

**Energy Div** (2004)
P.O. Box 2008
Oak Ridge, TN 37831-6187
p. (615) 574-5510
khaleelma@ornl.gov
Administrative Assistant: Teresa D. Ferguson
Director:
Robert B. Shelton, (D), S Illinois, 1970, On
Chair:
Donald W. Lee, (D), Michigan, 1977, Hw
Senior Scientist:
Richard H. Ketelle, (M), Tennessee, 1977, Ge
Russell Lee, (D), McMaster, 1978, Oy
William P. Staub, (D), Iowa State, 1969, Ng
Research Associate:
Arthur C. Curtis, (M), Colorado State, 1993, Ge
Robert O. Johnson, (D), Tennessee, 1984, Hg
Richard R. Lee, (M), Temple, 1982, Gr
John D. Tauxe, (D), Texas, 1994, Hw

## Pacific Northwest National Laboratory

**Hydrology** (2014)
Environmental Technology Div
PO Box 999 MSIN K9-36
Richland, WA 99352
p. (509) 372-6045
mike.fayer@pnnl.gov
Head:
Mark D. Freshley, (M), Arizona, 1982, Hw
Senior Research Engineer:
Mark D. White, (D), Colorado State, 1986, Hy
Senior Scientist:
Diana H. Bacon, (D), Washington State, 1997, Hq
Michael J. Fayer, (D), Massachusetts, 1984, Hq
Randy R. Kirkham, (D), Washington, 1993, Hq
Philip D. Meyer, (D), Illinois, 1992, Hw
Martinus Oostrom, (B), Wageningen Ag, 1984, Hq
Marshall C. Richmond, (D), Iowa, 1987, Hs
Timothy D. Scheibe, (D), Stanford, 1993, Hw
Lance W. Vail, (M), Montana State, 1982, Hg
Mark S. Wigmosta, (D), Washington, 1991, Hs
Steven B. Yabusaki, (M), Washington, 1986, Hg
Associate Scientist:
Cynthia L. Rakowski, (M), Utah State, 1996, On
Research Associate:
William A. Perkins, (B), Oregon State, 1985, Hg

**Applied Geology & Geochemistry** (2014)
Environmental Technology Div
Wayne J. Martin, Manager
PO Box 999 MSIN K6-81
Richland, WA 99352
p. (509) 376-5952
Christopher.Brown@pnnl.gov
http://www.pnl.gov/agg
Administrative Assistant: Charissa J. Chou
Head:
George R. Holdren, (D), Johns Hopkins, 1977, Cl
Staff Scientist:
Christopher J. Murray, (D), Stanford, 1992, Gq
Lab Fellow:
Bernard P. McGrail, (D), Columbia Southern, 1996, Cg
Senior Scientist:
Douglas B. Barnett, (M), E Washington, 1985, Em
Bruce N. Bjornstad, (M), E Washington, 1980, Gs
Kirk J. Cantrell, (D), Georgia Tech, 1989, Cl
Amy P. Gamerdinger, (D), Cornell, 1989, Cl
Tyler J. Gilmore, (M), Idaho, 1987, Hg
Floyd N. Hodges, (D), Texas, 1975, Cg
Duane G. Horton, (D), Illinois, 1983, Gz
Kenneth M. Krupka, (D), Penn State, 1984, Cl
George V. Last, (M), Washington State, 1997, Ge
Jonathan W. Lindberg, (M), Washington State, 1995, Ec
Shas V. Mattigod, (D), Washington State, 1976, Cl
Alan C. Rohay, (D), Washington, 1982, Ys
Herbert T. Schaef, (M), Texas Tech, 1991, Cg
R. Jeffrey Serne, (B), Washington, 1969, CgSoCl
Mark D. Sweeney, (B), C Washington, 1985, Yg
Bruce A. Williams, (B), Colorado Mines, 1980, Ng
Associate Scientist:
Yi-Ju Chien, (M), Stanford, 1998, Gq
Jonathan P. Icenhower, (D), Oklahoma, 1995, Cl
David C. Lanigan, (B), Michigan Tech, 1980, Hg

Virginia L. Legore, (B), Oregon State, 1974, Cg
Clark W. Lindenmeier, (M), E Washington, 1995, Cg
Paul F. Martin, (B), Pacific Lutheran, 1982, Cg
Kent E. Parker, (M), Washington State, 1995, Cg
Research Associate:
Alexandra B. Amonette, (M), CUNY, 1976, Cg
Deborah S. Burke, (B), E Washington, 1992, Sc
Elsa A. Camacho, (B), Heritage, 1996, Cg
Matthew J. O'Hara, (B), Montana, 1996, Cg
Robert D. Orr, (B), N State, 1994, Ca

## Sandia National Laboratory

**Geoscience and Environment** (2014)
1515 Eubank SE
P.O. Box 5800
Albuquerque, NM 87185
gobrela@sandia.gov
http://www.sandia.gov/

## U.S. Geological Survey

**Regional Director, Southeast** (2014)
1770 Corporate Drive, Suite 500
Norcross, GA 30093
p. (770) 409-7701
jdweaver@usgs.gov
http://www.usgs.gov

**Regional Director, Northeast** (2014)
12201 Sunrise Valley Drive, MS 953
Reston, VA 20192
p. (703) 648-6660
druss@usgs.gov
http://www.usgs.gov

**Regional Director, Midwest** (2014)
1451 Green Road
Ann Arbor, MI 48105
p. (737) 214-7207
lcarl@usgs.gov
http://www.usgs.gov

**Regional Director, Northwest** (2014)
909 1st Avenue
Seattle, WA 98104
p. (206) 220-4600
ddlynch@usgs.gov
http://www.usgs.gov

**Regional Director, Pacific** (2014)
Modoc Hall
3030 State University Drive East
Suite 3005
Sacramento, CA 95819
p. (916) 278-9551
mark_sogge@usgs.gov
http://www.usgs.gov

**Office of the Director** (2014)
12201 Sunrise Valley Drive, MS 100
Reston, VA 20192
p. (703) 648-7411
abwade@usgs.gov
www.usgs.gov

**Director, Office of Science Quality & Integrity** (2015)
12201 Sunrise Valley Drive, MS 911
Reston, VA 20192
p. (703) 648-6601
athornhill@usgs.gov
http://www.usgs.gov

**Associate Director, Climate and Land-Use Change** (2014)
12201 Sunrise Valley Drive, MD 409
Reston, VA 20192
p. (703) 648-5215
sryker@usgs.gov
http://www.usgs.gov

**Northwest Regional Safety Manager** (2016)
PO Box 596
Brightwood, OR 97011
p. (503) 622-4432
wsimonds@usgs.gov
http://www.usgs.gov

**Associate Director, Water** (2014)
12201 Sunrise Valley Drive, MD 150
Reston, VA 20192
p. (703) 648-4557
whwerkhe@usgs.gov
http://www.usgs.gov

**Natural Hazards Mission Area** (2016)
12201 Sunrise Valley Drive, MS 111
Reston, VA 20192
p. (703) 648-6600
applegate@usgs.gov
http://www.usgs.gov/natural_hazards/

**Associate Director, Human Capital** (2014)
12201 Sunrise Valley Drive, MS 201
Reston, VA 20192
p. (703) 648-7261
dwade@usgs.gov
http://www.usgs.gov

**Associate Director, Office of Communications & Publishing** (2015)
12201 Sunrise Valley Drive, MS 119
Reston, VA 20192
p. (703) 648-5750
bwainman@usgs.gov
http://www.usgs.gov

**Associate Director, Administration & Enterprise Information** (2004)
12201 Sunrise Valley Drive, MS 201
Reston, VA 20192
p. (703) 648-7261
http://www.usgs.gov

**Associate Director, Office of Budget, Planning, & Integration** (2014)
12201 Sunrise Valley Drive, MS 105
Reston, VA 20192
p. (703) 648-4443
cburzyk@usgs.gov
http://www.usgs.gov

# SPECIALTY CODES

Specialty codes are used to indicate the research or teaching specialities of faculty members listed in the directory. Bold numbers are totals for each major category. Numbers in parentheses are individual specialty totals.



**GEOLOGY 6343**

Gg General Geology (1271)
Ga Archaeological Geology (129)
Ge Environmental Geology (700)
Gm Geomorphology (590)
Gl Glacial Geology (239)
Gu Marine Geology (238)
Gz Mineralogy & Crystallography (542)
Gn Paleolimnology (56)
Go Petroleum Geology (293)
Gx General Petrology (210)
Gi Igneous Petrology (529)
Gp Metamorphic Petrology (277)
Gd Sedimentary Petrology (185)
Gs Sedimentology (832)
Gr Physical Stratigraphy (322)
Gc Structural Geology (748)
Gt Tectonics (678)
Gv Volcanology (347)
Gq Mathematical Geology (79)
Gy Mineral Physics (71)
Gh History of Geology (39)
Gb Geomedicine (20)
Gf Forensic Geology (10)

**ECONOMIC GEOLOGY 592**

Eg General Economic Geology (282)
Ec Coal (69)
Em Metals (136)
En Non-Metals (24)
Eo Oil and Gas (110)

**GEOCHEMISTRY 2093**

Cg General Geochemistry (628)
Ca Analytical Geochemistry (183)
Cp Experimental Petrology/Phase Equilibria (99)
Ce Exploration Geochemistry (54)
Cc Geochronology & Radioisotopes (318)
Cl Low-temperature Geochemistry (378)
Cm Marine Geochemistry (188)
Co Organic Geochemistry (134)
Cs Stable Isotopes (327)
Ct Trace Element Distribution (80)

**GEOPHYSICS 1688**

Yg General Geophysics (619)
Yx Experimental Geophysics (101)
Ye Exploration Geophysics (247)
Yd Geodesy (87)
Ym Geomagnetism & Paleomagnetism (146)
Yv Gravity (27)
Yh Heat Flow (33)
Ys Seismology (527)
Yr Marine Geophysics (172)

**PALEONTOLOGY 1587**

Pg General Paleontology (261)
Ps Paleostratigraphy (111)
Pm Micropaleontology (172)
Pb Paleobotany (72)
Pl Palynology (70)
Pq Quantitative Paleontology (16)
Pv Vertebrate Paleontology (240)
Pi Invertebrate Paleontology (253)
Po Paleobiology (193)
Pe Paleoecology & Paleoclimatology (360)
Py Geobiology (128)

**HYDROLOGY 1907**

Hg General Hydrology (327)
Hw Ground Water/Hydrogeology (695)
Hq Quantitative Hydrology (113)
Hs Surface Waters (164)
Hy Geohydrology (102)

**SOIL SCIENCE 1278**

Sp Soil Physics/Hydrology (103)
Sc Soil Chemistry/Mineralogy (180)
Sd Pedology/Classification/Morphology (85)
Sf Forest Soils/Rangelands/Wetlands (58)
Sb Soil Biology/Biochemistry (107)
Sa Paleopedology/Archeology (23)
So Other Soil Science (156)

**ENGINEERING GEOLOGY 606**

Ng General Engineering Geology (309)
Ne Earthquake Engineering (22)
Nx Mining Tech/Extractive Metallurgy (48)
Nm Mining Engineering (94)
Np Petroleum Engineering (62)
Nr Rock Mechanics (120)

**OCEANOGRAPHY 1555**

Og General Oceanography (221)
Ob Biological Oceanography (466)
Oc Chemical Oceanography (255)
Ou Geological Oceanography (153)
Op Physical Oceanography (405)
On Shore and Nearshore Processes (122)

**PLANETOLOGY 335**

Xc Cosmochemistry (62)
Xg Extraterrestrial Geology (148)
Xy Extraterrestrial Geophysics (90)
Xm Meteorites & Tektites (75)

**OTHER 4381**

Og General Earth Sciences (395)
Oa Atmospheric Sciences (900)
Oe Earth Science Education (301)
Oy Physical Geography (329)
Oo Ocean Engineering/Mining (26)
Or Remote Sensing (419)
Os Soil Science (76)
Ow Meteorology (336)
Om Material Science (56)
Ou Land Use/Urban Geology (116)
Oi Geographic Information Systems (479)
Ol Glaciology (71)
On Not Elsewhere Classified (1417)

**TOTAL 22365**

# Faculty Specialty Index

## GEOLOGY

### General Geology

Baughman, Joey, Missouri Dept of Natural Resources
Abbott, Lon, University of Colorado
ABDO, Ginette, Montana Tech of The University of Montana
Abousleiman, Younane N., University of Oklahoma
Abreu, Vitor, Rice University
Abushagur, Sulaiman, El Paso Community College
Adabanija, Moruffdeen A., Ladoke Akintola University of Tech
Adamo, Peter J., Brandon University
Adepoju, Mohammed O., Federal University of Tech, Akure
Adisa, Adeshina L., Federal University of Tech, Akure
Ahola, John, Bristol Community College
Ajigo, Isaac O., Federal University of Tech, Akure
Akgun, Elif , Firat University
Akinlabi, Ismaila A., Ladoke Akintola University of Tech
Aksoy, Ercan, Firat University
Alexander, Dane, Western Michigan University
Alexander, Elaine, McLennan Community College
Alkaç, Onur, Firat University
Allison, Alivia J., Emporia State University
Altermann, Wladyslaw, Ludwig-Maximilians-Universitaet Muenchen
Amaach, Noureddin, College of Staten Island
Ambers, Clifford P., Ashland University
Anderson, Paula, University of Arkansas, Fayetteville
Anderson, Thomas C., Brigham Young University
Andronicos, Chris, Purdue University
Anhaeusser, Carl R., University of the Witwatersrand
Antonacci, Vince, Tennessee Geological Survey
Argenbright, Kristi, Texas Christian University
Armour, Jake, University of North Carolina, Charlotte
Ashwell, Paul A., University of Canterbury
Asiwaju-Bello, Yinusa A., Federal University of Tech, Akure
Asowata, Timothy I., Federal University of Tech, Akure
Athaide, Dileep, University of British Columbia
Athaide, Dileep J., Capilano University
Athey, Jennifer E., Alaska Division of Geological & Geophysical Surveys
Bach Plaza, Joan, Universitat Autonoma de Barcelona
Bachle, Peter, Missouri Dept of Natural Resources
Baer, Eric M., Highline College
Baker, Cathy, Arkansas Tech University
Balakrishnan, Meena, Tarrant County College- Northeast Campus
Ballero, Deniz Z., Georgia State University, Perimeter College, Online
Ballou, Stephen M., Beloit College
Bally, Albert W., Rice University
Balmat, Jennifer L., Chadron State College
Balsley, Christopher, Southern Connecticut State University
Bamisaye, Oluwaseyi A., Federal University of Tech, Akure
Banda, Anna F., El Centro College - Dallas Community College District
Barclay, Julie L., SUNY, Cortland
Barker, Andy J., University of Southampton
Barker, Charles F., Wayne State University
Barker, Gregory A., New Hampshire Geological Survey
Barminski, Robert, Hartnell College
Barnes, Brenda, El Paso Community College
Barnes, John H., Pennsylvania Bureau of Topographic & Geologic Survey
Batten, William G., University of Wisconsin - Extension
Baxter, P.G., James E., Harrisburg Area Community College
Bayless, Richard, Shawnee State University
Bayowa, Oyelowo G., Ladoke Akintola University of Tech
Beatty, Lynne, Johnson County Community College
Beltrami, Hugo, Dalhousie University
Beltz, John F., University of Akron
Bender, Hallock J., Mt. San Antonio College
Bennett, Sara, Western Illinois University
Benson, Donna M., Mesa Community College
Benson, Robert G., Adams State University
Bentley, Callan, Northern Virginia Community College - Annandale
Benton, Steven, University of Illinois, Urbana-Champaign
Berg, Richard, University of Illinois
Berry IV, Henry N., Dept of Agriculture, Conservation, and Forestry
Best, Myron G., Brigham Young University
Beyer, Adam, Nicholls State University
Bice, David M., Pennsylvania State University, University Park
Biek, Robert, Utah Geological Survey
Bierly, Aaron D., Pennsylvania Bureau of Topographic & Geologic Survey
Birner, Johannes, Ludwig-Maximilians-Universitaet Muenchen
Bissell, Clinton R., Texas A&M University, Corpus Christi
Blakeney DeJarnett, Beverly, University of Texas at Austin
Blattler, Clara, Princeton University
Blaylock, Glenn W., Laredo Community College
Boehm, David A., SUNY, The College at Brockport
Bograd, Michael B., Mississippi Office of Geology
Bohm, Christian, Manitoba Geological Survey
Bolze, Claude E., Tulsa Community College
Bone, Fletcher, Missouri Dept of Natural Resources
Boroughs, Terry J., American River College
Bothern, Lawrence, El Paso Community College
Boudreaux, Elisabeth L., University of Louisiana at Lafayette
Bouker, Polly A., Georgia State University, Perimeter College, Newton Campus
Bounk, Michael, Iowa Dept of Natural Resources
Bour, William, Northern Virginia Community College - Loudoun Campus
Bouse, Robin, El Camino College
Bowen, Esther E., Argonne National Laboratory
Bowles, Zack, Mesa Community College
Bowring, Samuel A., Massachusetts Institute of Tech
Bozdog, Nick, North Carolina Geological Survey
Bradshaw, Lanna K., Brookhaven College
Brady, Lawrence L., University of Kansas
Branciforte, Chloe, Mesa Community College
Breckenridge, Roy M., University of Idaho
Bridges, Carey, Missouri Dept of Natural Resources
Bridges, David L., Missouri Dept of Natural Resources
Bries Korpik, Jill, Inver Hills Community College
Brill, Jr., Richard C., Honolulu Community College
Brock, Patrick W. G., Queens College (CUNY)
Brock, Patrick W. G., Graduate School of the City University of New York
Brouillette, Pierre, Universite du Quebec
Brown, Alan, Middle Tennessee State University
Brown, Francis H., University of Utah
Browning, James V., Rutgers, The State University of New Jersey
Browning, Sharon, Baylor University
Brueckner, Hannes K., Queens College (CUNY)
Brunengo, Matthew , Portland State University
Bruschke, Freddi Jo, California State University, Fullerton
Bryant, Kathleen E., University of Illinois, Urbana-Champaign
Bryant, Marita, Bates College
Buchanan, Paul C., Kilgore College
Buchwald, Caryl E., Carleton College
Bucke, David P., University of Vermont
Bulger, Daniel E., Georgia State University, Perimeter College, Decatur Campus
Bultman, John, Asheville-Buncombe Technical Community College
Burd, Aurora, Antelope Valley College
Burns, Emily, Community College of Rhode Island
Burris, John H., San Juan College
Burtis, Erik, Northern Virginia Community College - Woodbridge
Butcher, William, Bowling Green State University
Butkos, Darryl J., Suffolk County Community College, Ammerman Campus
Büchel, Georg, Friedrich-Schiller-University Jena
Byrnes, Jeffrey, Oklahoma State University
Cabe, Suellen, University of North Carolina, Pembroke
Caldwell, Andy, Front Range Community College - Larimer
Callahan, Caitlin N., Grand Valley State University
Cameron, Kevin, Simon Fraser University
Canalda, Sabrina, El Paso Community College
Carnes, Jacob, Wyoming State Geological Survey
Carr, Keith W., University of Illinois, Urbana-Champaign
Carr, Tim R., University of Kansas
Case, Jeanne, Eastern Washington University
Caskey, Deborah, El Paso Community College
Cervato, Cinzia C., Iowa State University of Science & Tech
Chadima, Sarah A., South Dakota Dept of Environment and Natural Resources
Chamberlin, Richard M., New Mexico Institute of Mining & Tech

Chandler, Angela, Arkansas Geological Survey
Chang, Julie M., University of Oklahoma
Charles, Kasanzu, University of Dar es Salaam
Chasteen, Hayden R., Tarrant County College- Northeast Campus
Chernosky, Jr., Joseph V., University of Maine
Cherukupalli, Nehru, Hofstra University
Cholnoky, Jennifer, Skidmore College
Christensen, Wesley P., South Dakota Dept of Environment and Natural Resources
Ciciarelli, John A., Pennsylvania State University, Monaca
Clark, Donald, Utah Geological Survey
Clark, Kathryne, Iowa Dept of Natural Resources
Clark, Ken, University of Puget Sound
Clayton, Rodney, Tidewater Community College
Clendening, Ronald J., Tennessee Geological Survey
Coble, Jim, Tidewater Community College
Coffman, David, El Centro College - Dallas Community College District
Coffman, Stephanie, El Centro College - Dallas Community College District
Colak Erol, Serap, Firat University
Coleman, Alvin L., Cape Fear Community College
Collins, Laura, Middle Tennessee State University
Colosimo, Amanda, Monroe Community College
Conquy, Xenia, Broward College
Cooper, Jennifer, Southern Connecticut State University
Cooper, Jonathon L., Carleton College
Cornell, Kristie, University of Louisiana at Lafayette
Cornell, Sean, Shippensburg University
Cotera, Jr., Augustus S., Northern Arizona University
Couroux, Emile, El Paso Community College
Coutand, Isabelle, Dalhousie University
Cox, Christena, Ohio State University
Craig, James L., Los Alamos National Laboratory
Crain, John R., College of the Sequoias
Crawford, Vernon J., Southern Oregon University
Creveling, Jessica, Oregon State University
Criswell, James, Cape Fear Community College
Cron, Mitch, Drexel University
Cronoble, James M., Metropolitan State College of Denver
Cummings, Michael L., Portland State University
Curtis, Lynn A., Broward College, Central Campus
Cvetko Tešoviæ, Blanka, University of Zagreb
Dalman, Michael, Blinn College
Damir, Buckoviæ, University of Zagreb
Danbom, Stephen H., Rice University
Davis, Adam M., Worcester State University
Davis, David A., University of Nevada
Dawes, Ralph, Wenatchee Valley College
Day, Damon P., Mt. San Antonio College
Deakin, Joann, Cochise College
Dealy, Mike, University of Kansas
Deane, William, University of Tennessee, Knoxville
Dee, Seth, University of Nevada
deGroot, Robert, California State University, Fullerton
Dehler, Sonya, Dalhousie University
Delano, Helen L., Pennsylvania Bureau of Topographic & Geologic Survey
Devaney, Kathleen, El Paso Community College
Devery, Dora, Alvin Community College
Devlahovich, Vincent A., College of the Canyons
DeVries-Zimmerman, Suzanne, Hope College
Dhavaskar, Poornima, Goa University
Di Fiori, Sara, El Camino College
Dinklage, William, Santa Barbara City College
Dittmer, Eric, Southern Oregon University
Doelling, Helmut H., University of Utah
Donnelly, Thomas W., Binghamton University
Dooley, Brett, Patrick Henry Community College
Dougan, Bernie, Whatcom Community College
Doyle, Joseph, Bridgewater State University
Drean, Thomas A., Wyoming State Geological Survey
Dressel, Waldemar M., Missouri University of Science and Tech
Drobny, Gerry, Cosumnes River College
Dudley, Jutta S., SUNY, The College at Brockport
Duncan, Ian, University of Texas at Austin, Jackson School of Geosciences
Dushman, Beth, American River College
Dyab, Ahmed I., Alexandria University
Earls, Stephanie, Washington Division of Geology & Earth Resources
Easterday, Cary, Bellevue College
Edson, Carol, Las Positas College
Egan, Stuart, Keele University
Egbeyemi, Oladimeji R., Federal University of Tech, Akure
Eisenhart, Ralph, York College of Pennsylvania
El Dakkak, Mohamed W., Alexandria University
Ellis, Trevor, Missouri Dept of Natural Resources
Emerson, Norlene, University of Wisconsin Colleges
Engelhard, Simon, University of Rhode Island
Escartin, Javier, Woods Hole Oceanographic Institution
Esparza-Yanez, Francisco A., Centro de Estudios Superiores del Estado Sonora
Evans, Thomas J., University of Wisconsin, Extension
Ewas, Galal A., Alexandria University
Faatz, Renee M., Snow College
Fagnan, Brian A., South Dakota Dept of Environment and Natural Resources
Fahrenbach, Mark D., South Dakota Dept of Environment and Natural Resources
Falcon, Alexandra, El Paso Community College
Fantozzi, Joanna , California State University, Fullerton
Farquharson, Phil, MiraCosta College
Farrell, Michael, Cuyamaca College
Fengler, Keegan , Central Washington University
Ferreira, Karen, University of Manitoba
Fields, Chad, Iowa Dept of Natural Resources
Fields, Russell, University of Nevada, Reno
Figg, Sean, Palomar College
Filkorn, Harry, Los Angeles Pierce College
Finley, Mark, Heartland Community College
Fitak, Madge R., Ohio Dept of Natural Resources
Fitchett, Rebekah, Northeastern Illinois University
Fitz, Thomas J., Northland College
Fleck, Michelle C., Utah State University Eastern
Flower, Hilary, Saint Petersburg College, Clearwater
Flowers Falls, Emily, Washington & Lee University
Forno, Maria Gabriella, Università di Torino
Fox, Don, Dalhousie University
Frail, Pam, Acadia University
Fredrick, Kyle, California University of Pennsylvania
Freed, Andrew M., Purdue University
Frei, Michaela, Ludwig-Maximilians-Universitaet Muenchen
Freiburg, Jared, University of Illinois, Urbana-Champaign
Friedmann, Samuel (Julio) J., Lawrence Livermore National Laboratory
Friedrich, Anke, Ludwig-Maximilians-Universitaet Muenchen
Fromm, Jeanne M., University of South Dakota
Frost, Gina M., San Joaquin Delta College
Fuerst, Alice, Metropolitan Community College-Kansas City
Gagnon, Teresa K., Dept of Energy and Environmental Protection
Galal, Galal M., Alexandria University
Gale, Marjorie H., Agency of Natural Resources, Dept of Environmental Conservation
Gannon, J. Michael, Iowa Dept of Natural Resources
Gardner, Eleanor E., University of Tennessee, Martin
Gardner, Jamie N., Los Alamos National Laboratory
Garrett, Maureen, Arizona Western College
Garside, Larry J., University of Nevada
Garwood, Phil, Cape Fear Community College
Gawloski, Joan, Midland College
Gawu, Simon K., Kwame Nkrumah University of Science and Tech
Gianotti, Franco, Università di Torino
Giegengack, Jr., Robert F., University of Pennsylvania
Giglierano, James D., Iowa Dept of Natural Resources
Gildner, Raynond F., Indiana University / Purdue University, Fort Wayne
Giles, Antony, Midland College
Gilley, Brett, University of British Columbia
Gillman, Joe, Missouri Dept of Natural Resources
Glynn, William G., SUNY, The College at Brockport
Gnidovec, Dale, Ohio State University
Gonzalez-Juarez, Marco A., Centro de Estudios Superiores del Estado Sonora
Goodwin, Alan M., University of Toronto
Gottfried, Richard, Frederick Community College
Gould, Joseph C., Saint Petersburg College, Clearwater
Grace, Cathy A., University of Mississippi

Griffin, William R., University of Texas, Dallas
Gross, Amy, University of North Carolina, Pembroke
Grover, Jeffrey A., Cuesta College
Grupp, Steve, Everett Community College
Gušiæ, Ivan, University of Zagreb
Hagadorn, James W., Denver Museum of Nature & Science
Hakimian, Adina, Hofstra University
Hale, Leslie J., Smithsonian Institution / National Museum of Natural History
Haley, John C., Virginia Wesleyan College
Hall, Tracy, Georgia Highlands College
Hamdan, Abeer, Phoenix College
Hamecher, Emily A., California State University, Fullerton
Hammes, Ursula, University of Texas at Austin, Jackson School of Geosciences
Hams, Jacquelyn E., Los Angeles Valley College
Hand, Linda M., College of San Mateo
Hanna, Ruth L., Las Positas College
Hanson, Doug, Arkansas Geological Survey
Hanson, Gilbert N., SUNY, Stony Brook
Hargrave, Jennifer E., University of Louisiana at Lafayette
Hargrave, Phyllis, Montana Tech of The University of Montana
Harma, Roberta L., Moorpark College
Harone, Imad, College of Staten Island
Harrell, Michael, Seattle Central Community College
Harrison, Linda , Western Michigan University
Hart, George, Louisiana State University
Hartley, Susan, Lake Superior College
Haselwander, Airin, Missouri Dept of Natural Resources
Hauer, Kendall, Miami University
Hayes, Garry F., California State University, Stanislaus
Hayes, Garry F., Modesto Junior College
Heck, Frederick R., Ferris State University
Heerschap, Lauren, Fort Lewis College
Hegner, Ernst, Ludwig-Maximilians-Universitaet Muenchen
Heidlauf, Lisa, Wheaton College
Hein, Kim A., University of the Witwatersrand
Helba, Hossam EL-Din A., Alexandria University
Heller, Matthew J., Division of Geology and Mineral Resources
Henley, Barbara D., Mercer University
Henry, Kathleen M., Illinois State Geological Survey
Hepburn, J. Christopher, Boston College
Herbert, Jennifer, Argonne National Laboratory
Herbst, Thomas, Missouri Dept of Natural Resources
Hernandez, Larry, MiraCosta College
Herring Mayo, Lisa L., Motlow State Community College
Herzig, Chuck, El Camino College
Hess Tanguay, Lillian, Hofstra University
Hetherington, Jean, Diablo Valley College
Hickcox, Charles W., Emory University
Hicks, Roberta (Robbie), Memorial University of Newfoundland
Hiett, Michael W., Middle Tennessee State University
Higgins, Chris T., California Geological Survey
Hill, Catherine B., Arizona Western College
Hill, Chris, Grossmont College
Hinz, Nicholas, University of Nevada
Hirner, Sarah M., University of Northern Colorado
Ho, Anita, Flathead Valley Community College
Hobbs, Thomas M., Lonestar College - North Harris
Hochstaedter, Alfred, Monterey Peninsula College
Hodder, Donald R., SUNY, New Paltz
Hók, Jozef, Comenius University in Bratislava
Holail, Hanafy M., Alexandria University
Holmes, Stevie L., South Dakota Dept of Environment and Natural Resources
Hood, Teresa A., University of Miami
Hood, William C., Colorado Mesa University
Hopkins, David M., University of Alaska, Fairbanks
Hoppe, Andreas, Technische Universitaet Darmstadt
Horn, John, Metropolitan Community College-Kansas City
Horton, Albert B., Tennessee Geological Survey
House, Martha, Pasadena City College
Howard, Katie, Olympic College
Howes, Mary R., Iowa Dept of Natural Resources
Hudec, Mike, University of Texas at Austin, Jackson School of Geosciences
Hungerbuehler, Axel, Mesalands Community College
Hunt, Paula J., West Virginia Geological & Economic Survey
Hural, Kirsten, Dickinson College
Hussein, Musa, El Paso Community College
Hutto, Richard S., Arkansas Geological Survey
Huycke, David, Yakima Valley College
Hylland, Michael D., Utah Geological Survey
Hylton, Alisa, Central Piedmont Community College
Iaccheri, Maria Linda, Ludwig-Maximilians-Universitaet Muenchen
Idstein, Peter J., University of Kentucky
Igonor, Emmanuel E., Federal University of Tech, Akure
Ikwuazorm, Okia, Northern Virginia Community College - Loudoun Campus
Irvin, Gene D., Geological Survey of Alabama
Isenor, Fenton M., Cape Breton University
Jackson, Hiram, Cosumnes River College
Jackson, Jeremiah, Missouri Dept of Natural Resources
Jacobs, Jon, Western University
Jaecks, Glenn, American River College
Janusz, Robert, San Antonio Community College
Jegede, O A., Federal University of Tech, Akure
Jellinek, Mark, University of British Columbia
Jensen, Ann R., South Dakota Dept of Environment and Natural Resources
Jiang, James Xinxia, Mount Allison University
Johnson, Darren J., South Dakota Dept of Environment and Natural Resources
Johnson, Edward, College of Staten Island
Johnson, Kurt, Alaska Division of Geological & Geophysical Surveys
Johnson, Ty, Arkansas Geological Survey
Johnston, David, Arkansas Geological Survey
Johnston, Paul J., Emporia State University
Jones, Charles E., University of Pittsburgh
Jones, Gwyn, Bellevue College
Jordan, Bradley C., Bucknell University
Jurena, Dwight, Alamo Colleges - San Antonio College
Jurena, Dwight, San Antonio Community College
Kaandorp, Ron, VU University Amsterdam
Kaldor, Michael, Miami-Dade College (Wolfson Campus)
Kapp, Jessica, University of Arizona
Karwoski, Todd, University of Maryland
Katzman, Danny, Los Alamos National Laboratory
Kaye, John M., Mississippi State University
Kaygili, Sibel, Firat University
Keating, Elizabeth, Los Alamos National Laboratory
Keattch, Sharen , Eastern Washington University
Keene, Deborah A., University of Alabama
Kelley, Neil P., Vanderbilt University
Kempe, Stephan, Technische Universitaet Darmstadt
Kerwin, Charles M., Keene State College
Kerwin, Michael W., University of Denver
Ketcham, Richard A., University of Texas at Austin
Khalil Ebeid, Khalil I., Alexandria University
Kiesel, Diann, University of Wisconsin Colleges
Kim, Keonho, Midland College
King, Jonathan K., Utah Geological Survey
Kirkby, Kent C., University of Minnesota, Twin Cities
Klaus, James S., University of Miami
Klee, Thomas M., Hillsborough Community College
Klosterman, Sue, University of Dayton
Knowlton, Kelly, Northwestern State University
Kodosky, Larry, Oakland Community College
Koenig, Brian, College of the Desert
Kohrt, Casey, Iowa Dept of Natural Resources
Kolkas, Mosbah, College of Staten Island
Kollasch, Pete, Iowa Dept of Natural Resources
Kontak, Daniel, Dalhousie University
Kowallis, Bart J., Brigham Young University
Kraft, Kaatje, Mesa Community College
Kraft, Kaatje, Whatcom Community College
Kramer, J. Curtis, University of the Pacific
Kramer, Kate, McHenry County College
Kramer, Walter V., Del Mar College
Kranz, Dwight S., Houston Community College System
Kroon, Dick, Edinburgh University
Kruger, Ned , North Dakota Geological Survey
Kubicek, Leonard, North Lake College - Dallas Community College District

Kuhlman, Robert, Montgomery County Community College
Kukoè, Duje, University of Zagreb
Kumpf, Amber C., Muskegon Community College
Kuntz, Mark R., Elgin Community College
Kurka, Mira, Great Basin College
Kutis, Michael, Ball State University
Kwicklis, Edward M., Los Alamos National Laboratory
Lacy, Tor, Cerritos College
Lambert-Smith, James, Kingston University
Lammerer, Bernd, Ludwig-Maximilians-Universitaet Muenchen
Langel, Richard A., Iowa Dept of Natural Resources
Langhorst, Glenn, Fond du Lac Tribal and Community College
LaPointe, Daphne D., University of Nevada
Larter, Stephen, University of Newcastle Upon Tyne
Lauziere, Kathleen, Universite du Quebec
Lawrence, Kira, Lafayette College
Lawry, Cynthia, Blinn College
LeFever, Julie A., North Dakota Geological Survey
Leite, Michael B., Chadron State College
Lemay, Phillip W., Louisiana State University
Leszczynski, Raymond F., Cayuga Community College
Levy, Melissa H., American River College
Lewis, Chris, City College of San Francisco
Lewis, Mary, Contra Costa College
Lewis, Reed S., University of Idaho
Liauw, Henri L., Broward College
Libra, Robert D., Iowa Dept of Natural Resources
Liebling, Richard, Hofstra University
Lilly, Troy, Western Texas College
Lippelt, Irene D., University of Wisconsin - Extension
Liu, Paul Hiaibao, Iowa Dept of Natural Resources
Locke, Erika , Texas A&M University, Corpus Christi
Lombard, Armand J., Mesa Community College
Long, Colleen, University of Illinois, Urbana-Champaign
Lonn, Jeff, Montana Tech of The University of Montana
Lovett, Cole, Lake Michigan College
Lowrie, David J., Wayne State University
Lu, Yuehan, University of Alabama
Ludman, Allan, Queens College (CUNY)
Ludman, Allan, Graduate School of the City University of New York
Lueth, Virgil W., New Mexico Institute of Mining & Tech
Lužar-Oberiter, Borna, University of Zagreb
Lyle, Mike, Tidewater Community College
Lyman, John C., Bakersfield College
Mabee, Stephen B., University of Massachusetts, Amherst
Macias, Steve E., Olympic College
MacLachlan, James, Metropolitan State College of Denver
Madin, Ian P., Oregon Dept of Geology & Mineral Industries
Magee, Robert, Virginia Wesleyan College
Mahaffee, Tina, Middle Georgia College
Mahlen, Nancy J., SUNY, Geneseo
Mahmoud, Sara A., Alexandria University
Majodina, Thando, Tshwane University of Tech
Major, Penni, Lonestar College - North Harris
Major, Ruth H., Hudson Valley Community College
Mandziuk, William S., University of Manitoba
Mansour, Ahmed S., Alexandria University
Mapholi, Thendo, University of the Free State
Marchand, Gerard, Mount Holyoke College
Marchisin, John, New Jersey City University
Marco, Shmuel, Tel Aviv University
Marco, Shmulik, Tel Aviv University
Marschik, Robert, Ludwig-Maximilians-Universitaet Muenchen
Marshall, Thomas R., South Dakota Dept of Environment and Natural Resources
Martin, Charles W., Earlham College
Martin, Gale D., College of Southern Nevada - West Charleston Campus
Martinez Torres, Luis Miguel, University of the Basque Country UPV/EHU
Martinez-Hackert, Bettina, SUNY, Buffalo
Martinuš, Maja, University of Zagreb
Marty, Kevin, Imperial Valley College
Martz, Todd, University of California, San Diego
Marvinney, Robert G., Dept of Agriculture, Conservation, and Forestry
Mata, Scott, California State University, Fullerton
Mattox, Tari, Grand Rapids Community College
Matyjasik, Basia, Utah Geological Survey
May, S J., Collin College - Preston Ridge Campus
Mayer, Larry, Woods Hole Oceanographic Institution
McArthur, Russell , City College of San Francisco
McCall, Rosemary, College of Staten Island
McCauley, Steven, El Centro College - Dallas Community College District
McClaughry, Jason, Oregon Dept of Geology & Mineral Industries
McConnell, David, North Carolina State University
McCoy, Floyd W., Windward Community College
McCraley, Tia, Arizona Western College
McCulloh, Richard P., Louisiana State University
McCutchen, William T., University of Tennessee, Martin
McDermott, Frank, University College Dublin
McGehee, Richard V., Austin Community College District
McKay, Robert M., Iowa Dept of Natural Resources
McKinney, Mac, Geological Survey of Alabama
McMonagle, Julie, Wilkes University
Meave, Edgardo, Universidad Nacional Autonoma de Mexico
Meister, Paul A., Illinois State University
Melcher, Frank, University Leoben
Meldahl, Keith H., MiraCosta College
Melendez, Christyanne, Appalachian State University
Merriam, Daniel F., University of Kansas
Metzler, Christopher V., MiraCosta College
Meyer, Jeffrey W., Santa Barbara City College
Michel, Suzanne, Cuyamaca College
Millan, Christina, Ohio State University
Millen, Timothy M., Elgin Community College
Miller, David, California State University, Bakersfield
Miller, Steven F., Argonne National Laboratory
Miller-Hicks, Bryan, Cuyamaca College
Milligan, Mark R., Utah Geological Survey
Milner, Lloyd R., Louisiana State University
Miner, James J., University of Illinois, Urbana-Champaign
Minium, Deborah, Bellevue College
Moll, Nancy E., College of the Desert
Monet, Julie, California State University, Chico
Monson, Jessica, University of Illinois, Urbana-Champaign
Montayne, Simone, Alaska Division of Geological & Geophysical Surveys
Montwill, Gail F., Santiago Canyon College
Morand, Vincent J., La Trobe University
Morealli, Sarah A., University of Mary Washington
Morgan, Matt, Colorado School of Mines
Morris, Billy, Georgia Highlands College
Mshiu, Elisante E., University of Dar es Salaam
Mulvany, Patrick S., Missouri University of Science and Tech
Munn, Barbara J., California State University, Sacramento
Murphy, Cindy, Saint Francis Xavier University
Murphy, Dan, Asheville-Buncombe Technical Community College
Muskatt, Herman, Utica College
Mutti, Laurel, SUNY, New Paltz
Nagy-Shadman, Elizabeth, Pasadena City College
Nance, Seay, University of Texas at Austin, Jackson School of Geosciences
Nesheim, Tim, North Dakota Geological Survey
Newman, Brent D., Los Alamos National Laboratory
Newman, Jamie, American Museum of Natural History
Nicholl, Amy, University of Northern Colorado
Nicoletti, Jeremy D., New Hampshire Geological Survey
Nielsen, Gregory B., Weber State University
Norman, David K., Washington Division of Geology & Earth Resources
Norrish, Winston, Central Washington University
O'Brien, Lawrence E., Orange County Community College
O'Keeffe, Mike, Colorado School of Mines
O'Neil, Caron E., Pennsylvania Bureau of Topographic & Geologic Survey
Oakes-Miller, Hollie, Portland Community College - Sylvania Campus
Obolewicz, Dave, Keene State College
Odendaal, Adriaan, University of the Free State
Odeyemi, Idowu O., Federal University of Tech, Akure
Ogbahon, Osazuwa A., Federal University of Tech, Akure
Okonkwo, C T., Federal University of Tech, Akure
Olabode, Solomon O., Federal University of Tech, Akure
Oldfield, Bruce K., Broome Community College
Oldham, Richard L., American River College
Onuk, Peter, University Leoben

Opeloye, S A., Federal University of Tech, Akure
Opperman, William, Broward College
Orr, William , Portland State University
Oswald, Peter J., University of Alaska, Anchorage
Owoseni, Joshua O., Federal University of Tech, Akure
Pande, Srikant k., Pt. Ravishankar Shukla University
Pantoja-Alor, Jerjes, Universidad Nacional Autonoma de Mexico
Papp, Kenneth R., Alaska Division of Geological & Geophysical Surveys
Parish, Cynthia L., Lamar University
Parker, Jack R., Ontario Geological Survey
Parker, Richard M., Texas A&M University, Kingsville
Parnella, Bill, University of Delaware
Parrick, Brittany, Ohio Dept of Natural Resources
Parsons, Michael, Dalhousie University
Paschert, Karin, Ludwig-Maximilians-Universitaet Muenchen
Patton, Jason A., Arkansas Tech University
Patton, Terri, Argonne National Laboratory
Patwardhan, Kaustubh, SUNY, New Paltz
Pawloski, Gayle A., Lawrence Livermore National Laboratory
Pellowski, Christopher J., South Dakota School of Mines & Tech
Perez, Adriana, El Paso Community College
Perkis, Bill, Gogebic Community College
Pesavento, Jim, Palomar College
Peters, Lisa, New Mexico Institute of Mining & Tech
Peters, Roger M., University of Wisconsin - Extension
Peterson, Joseph E., Elgin Community College
Phelps, William, Riverside City College
Phillips, Michael, Illinois Valley Community College
Pierce, David, Lakeland Community College
Pirie, Diane H., Florida International University
Pokras, Edward M., Keene State College
Polissar, Pratigya J., Columbia University
Poole, T. Craig, Fresno City College
Posiloviæ, Hrvoje, University of Zagreb
Prave, Tony, University of St. Andrews
Prewett, Jerry L., Missouri Dept of Natural Resources
Price, Jonathan G., University of Nevada
Price, L G., New Mexico Institute of Mining & Tech
Prichystal, Antonín, Masaryk University
Priesendorf, Carl, Metropolitan Community College-Kansas City
Priest, George R., Oregon Dept of Geology & Mineral Industries
Prior, William L., Arkansas Geological Survey
Prothero, Donald, Los Angeles Pierce College
Proudhon, Benoit, Institut Polytechnique LaSalle Beauvais (ex-IGAL)
Prytulak, Julie, Imperial College
Purdy, Ann, Wayne State University
Purkiss, Robert, Angelo State University
Putnam, Roger, California State University, Stanislaus
Quick, Thomas J., University of Akron
Quinn, Heather A., Maryland Department of Natural Resources
Ragland, Deborah, University of New Mexico - Taos
Rahman, Ata U., Austin Community College District
Rains, Daniel S., Arkansas Geological Survey
Ranhofer, Melissa, Furman University
Rantitsch, Gerd, University Leoben
Rashed, Mohamed A., Alexandria University
Rath, Carolyn, California State University, Fullerton
Rathburn, Sara L., Colorado State University
Rawling, Geoffrey, New Mexico Institute of Mining & Tech
Ray, Waverly, Cuyamaca College
Raymer, Janet, Metropolitan Community College-Kansas City
Read, Adam S., New Mexico Institute of Mining & Tech
Redden, Marcella, Geological Survey of Alabama
Reesman, Arthur L., Vanderbilt University
Reif, Samantha, Lincoln Land Community College
Rennie, Colette, Saint Francis Xavier University
Repka, James, Saddleback Community College
Rieder, Michael M., Technische Universitaet Muenchen
Riker-Coleman, Kristin E., University of Wisconsin, Superior
Rinae, Makhadi, University of the Free State
Riordan, Jean, Alaska Division of Geological & Geophysical Surveys
Rivera, Mark, University of Alaska, Anchorage
Robert, Sanborn, University of Wisconsin Colleges
Robinson, Francis J., Yale University
Robinson, Richard, Santa Monica College
Robinson, Sarah, United States Air Force Academy
Rocha, Guillermo, Brooklyn College (CUNY)
Roche, James E., Louisiana State University
Rochon, Andre, Dalhousie University
Rodgers, Jim, Wyoming State Geological Survey
Rodriguez-Castaneda, Jose L., Universidad Nacional Autonoma de Mexico
Rogers, Steven, Keele University
Ross, Theodore W., Lawrence University
Rothemund, Kirk, El Paso Community College
Rounds, Steven W., California State University, Sacramento
Rowden, Robert, Iowa Dept of Natural Resources
Ruetz, Joseph W., Stanford University
Ruffel, Alice, El Centro College - Dallas Community College District
Rumrill, Julie, Southern Connecticut State University
Ruppert, Kelly R., California State University, Fullerton
Rush, Jason, University of Kansas
Russ, Tom, College of Southern Maryland
Sacramentogrilo, Isabelle, San Diego State University
Salle, Bethan, El Centro College - Dallas Community College District
Samimi, Naser, University of Tabriz
Santana, Vicente, University of the Basque Country UPV/EHU
Sassi, Raffaele, Università degli Studi di Padova
Sato, Yoko, Montclair State University
Sawin, Robert S., University of Kansas
Schafer, Carl M., Macomb Community College, Center Campus
Scheffler, Joanna, Mesa Community College
Schenck, William S., University of Delaware
Schilling, Keith, Iowa Dept of Natural Resources
Schmidt, Bennetta, Lamar University
Schulingkamp, Arren, Louisiana State University
Schultz, Jan, Santa Barbara City College
Sediek, Kadry N., Alexandria University
Seigley, Lynette S., Iowa Dept of Natural Resources
Shaaban, Mohamad N., Alexandria University
Shade, Harry, West Valley College
Shakun, Jeremy D., Boston College
Shalimba, Ester, University of Namibia
Shannon, Jeremy, Michigan Technological University
Sharp, Patricia S., Stephen F. Austin State University
shedied, ahmad g., Fayoum University
Sheets, H. David, SUNY, Buffalo
Shimizu, Melinda, Mesa Community College
Shinn, Eugene, University of South Florida
Shorey, Christian V., Colorado School of Mines
Shroba, Cynthia S., College of Southern Nevada - West Charleston Campus
Sibeko, Skhumbuzo, Tshwane University of Tech
Sicard, Karri, Alaska Division of Geological & Geophysical Surveys
Siegelberg, Alan, Long Island University, Brooklyn Campus
Siemens, Michael A., Missouri Dept of Natural Resources
Sigler, Jeffrey M., Tulane University
Skelton, Lawrence H., University of Kansas
Skinner, Randall, Brigham Young University
Sladek, Chris, University of Nevada, Reno
Sledzinski, Grazyna, Wayne State University
Smaglik, Suzanne M., Central Wyoming College
Smith, Jason J., Broome Community College
Smith, Jon J., University of Kansas
Smith, Mike, Front Range Community College - Larimer
Smith, Russell, El Paso Community College
Smithson, Jayne, Contra Costa College
Snyder, Daniel, Middle Georgia College
Snyder, Noah, Boston College
Solferino, Giulio, University of London, Royal Holloway & Bedford New College
Solis, Michael P., Ohio Dept of Natural Resources
Sparks, Thomas N., University of Kentucky
Speed, Don, Phoenix College
Spencer, Larry T., Plymouth State University
Stakes, Debra, Cuesta College
Stanton, Kathryn, Sacramento City College
Stanton, Kelsay, Wenatchee Valley College
Stark, Colin, Columbia University
Steart, David, La Trobe University
Steck, Lee, Los Alamos National Laboratory
Steffens, Katja, Ludwig-Maximilians-Universitaet Muenchen
Steinberg, Roger T., Del Mar College
Stewart, Esther K., University of Wisconsin - Extension

Stine, Alexander, San Francisco State University
Stover, Susan G., University of Kansas
Straight, William, Northern Virginia Community College - Loudoun Campus
Straub, Kyle M., Tulane University
Stright, Lisa, University of Utah
Stumpf, Andrew , Illinois State University
Suchy, Daniel R., University of Kansas
Sundell, Ander, College of Western Idaho
Sundell, Kent A., Casper College
Sunderlin, David, Lafayette College
Suszek, Thomas J., University of Wisconsin, Oshkosh
Swyrtek, Sheila M., Mott Community College
Szczepanski, Jacek, University of Wroclaw
Söllner, Frank, Ludwig-Maximilians-Universitaet Muenchen
Tajik, Atieh, Georgia State University
Tamish, Mohamed M., Alexandria University
Tapanila, Lori, Idaho State University
Tassier-Surine, Stephanie, Iowa Dept of Natural Resources
Taylor, Carolyn, Mesa Community College
Taylor, Chuck J., University of Kentucky
Taylor, Penny M., Mount Holyoke College
Taylor, Sid, Saint Francis Xavier University
Tetrault, Denis, University of Windsor
Thole, Jeffrey T., Macalester College
Thomas, Margaret A., Dept of Energy and Environmental Protection
Thorleifson, Harvey, University of Minnesota, Twin Cities
Thornberry-Ehrlich, Trista L., Colorado State University
Thul, David, University of Utah
Timmons, Michael, New Mexico Institute of Mining & Tech
Timmons, Stacy, New Mexico Institute of Mining & Tech
Tolley, William, Southern Connecticut State University
Tomiæ, Vladimir, University of Zagreb
Toni, Rousine T., Alexandria University
Towery, Brooke L., Pensacola Junior College
Treadwell-Steitz, Carol, Plattsburgh State University (SUNY)
Treworgy, Janis D., Principia College
Tucker, Eva, Pennsylvania State University, Erie
Turbeville, John, MiraCosta College
Turner, Mark, Collin College - Spring Creek Campus
Turner, Wesley L., Stephen F. Austin State University
Turner, III, Henry, University of Arkansas, Fayetteville
Tvelia, Sean, Suffolk County Community College, Ammerman Campus
Uahengo, Collen, University of Namibia
Ugland, Richard, Southern Oregon University
Urquhart, Joanne, Bowdoin College
Valenti, Christine, Montclair State University
van Balen, Ronald T., VU University Amsterdam
Van Brocklin, Matthew F., St. Lawrence University
Vance, Robert K., Georgia Southern University
VanDorpe, Paul E., Iowa Dept of Natural Resources
Vickery, Nancy, University of New England
Vidoviæ, Jelena, University of Zagreb
Viens, Rob, Bellevue College
Vig, Pradeep K., Kaskaskia College
Villalobos, Joshua, El Paso Community College
Vinton, Bonita L., Slippery Rock University
Voigt, Vicki, Missouri Dept of Natural Resources
Vuke, Susan M., Montana Tech of The University of Montana
Wacker, Michael, Florida International University
Wadleigh, Hank, Cypress College
Waggoner, Karen, Midland College
Wakefield, Kelli, Mesa Community College
Waugh, John, Tidewater Community College
Weatherford, Jonathan, Portland Community College - Sylvania Campus
Weber, Diane, Illinois Central College
Weber-Diefenbach, Klaus, Ludwig-Maximilians-Universitaet Muenchen
Weborg-Benson, Kimberly, SUNY Fredonia
Weeden, Lori, University of Massachusetts Lowell
Wein, Agatha, Cuyamaca College
Werhner, Matthew J., Hillsborough Community College
West, Robert, East Los Angeles College
Whatley, Robin L., Columbia College Chicago
Wheeler, Richard F., Austin Peay State University
White, Paul, Community College of Rhode Island
Whitehead, James, University of New Brunswick
Whitehead, Peter W., James Cook University
Whitehill, Matthew, Lake Superior College
Whittier, Michael, California State University, Stanislaus
Whittington, Carla, Highline College
Wilbur, Bryan, Pasadena City College
Wilder, Lee, New Hampshire Geological Survey
Wiley, Thomas J., Oregon Dept of Geology & Mineral Industries
Wilkerson, Christine M., Utah Geological Survey
Willahan, Duane, Gavilan College
Williams, Curtis J., Cypress College
Williams, James H., Missouri University of Science and Tech
Willis, Grant C., Utah Geological Survey
Willis, Marc, Fullerton College
Willsey, Shawn P., College of Southern Idaho
Wilson, Jeffrey A., University of Michigan
Wolfe, Ben, Metropolitan Community College-Kansas City
Wolkersdorfer, Christian, Ludwig-Maximilians-Universitaet Muenchen
Wolter, Calvin, Iowa Dept of Natural Resources
Wood, Jacqueline, Delgado Community College
Woodall, Debra W., Daytona State College
Wortmann, Ulrich B., University of Toronto
Wypych, Alicja, Alaska Division of Geological & Geophysical Surveys
Yalcin, Kaplan, Oregon State University
Yalcin, Rebecca, Oregon State University
Yan, Y E., Argonne National Laboratory
Yoshida, Glenn, Los Angeles Southwest College
Young, Glen, Missouri Dept of Natural Resources
Zabel, Garrett E., Colorado Mountain College
Zabielski, Victor, Northern Virginia Community College - Alexandria
Zanetti, Kathleen, University of Nevada, Las Vegas
Zawiskie, John M., Wayne State University
Zayac, John M., Los Angeles Pierce College
Zeitlhöfler, Matthias, Ludwig-Maximilians-Universitaet Muenchen
Zentner, Nick, Central Washington University
Zhang, Chunfu, Fort Hays State University
Zimmer, Brian, Appalachian State University
Zurawski, Ronald P., Tennessee Geological Survey

## Archaeological Geology

Abreu, Maria E., Universidade de Trás-os-Montes e Alto Douro
Adams, Kenneth, University of Nevada, Reno
Adovasio, James M., Mercyhurst University
Anderson, Jonathan H., Minnesota State University
Ballard, Robert D., University of Rhode Island
Bowers, Peter, University of Alaska, Fairbanks
Bradford, Joel, Utah Valley University
Buckley, Michael, University of Manchester
Chamberlain, Andrew, University of Manchester
Degryse, Patrick, Katholieke Universiteit Leuven
Dix, Justin, University of Southampton
Dugmore, Andrew J., Edinburgh University
Dye, David H., University of Memphis
Elkins, Nichole, Bowling Green State University
Fadem, Cynthia M., Earlham College
Farrand, William, University of Michigan
Ferguson, Terry A., Wofford College
Ferring, C. Reid, University of North Texas
Garrison, Ervan G., University of Georgia
Greenlee, Diana M., University of Louisiana, Monroe
Gundersen, James N., Wichita State University
Gunia, Piotr, University of Wroclaw
Harris, Scott, College of Charleston
Haynes, Jr., C. Vance, University of Arizona
Hayward, Chris L., Edinburgh University
Holliday, Vance, University of Arizona
Iglesias, Cruz, Coruna University
Jones, Jeri L., York College of Pennsylvania
Kelley, Alice R., University of Maine
Kvamme, Kenneth L., Boston University
Lee-Gorishti, Yolanda, Southern Connecticut State University
Madsen, David B., Utah Geological Survey
Mandel, Rolfe D., University of Kansas
Martinez-Reyes, Jose, University of Massachusetts, Boston
Mason, Owen, University of Alaska, Fairbanks
Matney, Timothy, University of Akron
Megarry, Will, Queens University Belfast
Mickelson, Andrew M., University of Memphis
Milne, Brooke, University of Manitoba
Nagaoka, Lisa A., University of North Texas

Page, David, Desert Research Institute
Parish, Ryan M., University of Memphis
Pike, Scott, Willamette University
Pope, Richard, University of Derby
Powell, Wayne G., Brooklyn College (CUNY)
Rapp, George R., University of Minnesota, Duluth
Reid, Joshua, University of Massachusetts, Boston
Reitz, Elizabeth J., University of Georgia
Rogers, Joe D., West Texas A&M University
Sallu, Susannah, University of Leeds
Sanjurjo, Jorge, Coruna University
Schiffman, Robert A., Bakersfield College
Schirmer, Ron, Minnesota State University
Smith, Jen R., Washington University in St. Louis
Smith, Jennifer R., Washington University in St. Louis
Stafford, C. Russell, Indiana State University
Swyrtek, Sheila, Charles Stewart Mott Community College
Thieme, Donald, Valdosta State University
Tomaso, Matthew S., Montclair State University
Towner, Ronald, University of Arizona
Van Nest, Julieann, New York State Geological Survey
Vento, Frank, Mercyhurst University
Waters, Michael R., Texas A&M University
Wilson, Lucy A., University of New Brunswick Saint John
Wilson, Michael C., Douglas College
Wolverton, Steve, University of North Texas

## Environmental Geology

Abbasnejad, Ahmad, Shahid Bahonar University of Kerman
Aden, Douglas, Ohio Dept of Natural Resources
Adentunji, Jacob, University of Derby
Adewoye, Abosede O., Ladoke Akintola University of Tech
Adomaitis, D., University of Illinois, Urbana-Champaign
Akenzua-Adamcyzk, Aiyevbekpen H., University of Benin
Albee-Scott, Steven R., Jackson College
Amezaga, Jamie, University of Newcastle Upon Tyne
Amin, Shahalam M., Bloomsburg University
Apotsos, Alex, Williams College
Argyilan, Erin, Indiana University Northwest
Aristilde, Ludmilla, Cornell University
Ashwood, Tom, Oak Ridge National Laboratory
Babcock, Daphne H., Collin College - Spring Creek Campus
Bang, John, North Carolina Central University
Barone, Jessica, Monroe Community College
Barrett, John, University of Leeds
Bartholemew, Paul, University of New Haven
Bartlett, Wendy, Marietta College
Bartolucci, Valerio, Broward College, Central Campus
Beatty, Heather L., Austin Community College District
Beccue, Clint, University of Illinois, Urbana-Champaign
Bechtel, Timothy D., Franklin and Marshall College
Beebe, Alex, University of South Alabama
Bell, Margaret C., University of Newcastle Upon Tyne
Berg, Richard C., University of Illinois, Urbana-Champaign
Berger, Antony R., Sir Wilfred Grenfell College
Bilanovic, Dragoljub D., Bemidji State University
Binda, Pier L., University of Regina
Bircher, Harry, Youngstown State University
Bogart, James M., Dept of Energy and Environmental Protection
Bogle, Mary Anna, Oak Ridge National Laboratory
Bond, Alan, University of East Anglia
Boss, Stephen K., University of Arkansas, Fayetteville
Bots, Pieter, University of Manchester
Bousenberry, Raymond T., New Jersey Geological and Water Survey
Bradley, J G., SUNY, Oswego
Brady, Emily S., Edinburgh University
Brake, Sandra S., Indiana State University
Brandon, Nigel, Imperial College
Brandt, Craig, Oak Ridge National Laboratory
Branfireun, Brian, Western University
Breitmeyer, Ronald J., University of Nevada, Reno
Brodie, Gregory, University of Tennessee at Chattanooga
Bromily, Geoffrey, Edinburgh University
Brown, Huntting (Hunt), Wright State University
Brown, Kerry, Kingston University
Brown, Thomas W., Austin Community College District
Bryan, Mark, Wayland Baptist University
Budkewitsch, Paul, Brock University
Burke, Ian, University of Leeds
Byerly, Don W., University of Tennessee, Knoxville
Cadet, Eddy, Utah Valley University
Callison, James, Utah Valley University
Caudill, Kimberly S., Hocking College
Chappell, James R., Colorado State University
Chernoff, Barry, Wesleyan University
Cheun, Norman, Kingston University
Chilvers, Jason, University of East Anglia
Cloutier, Danielle, Universite Laval
Colegial Gutierrez, Juan D., Universidad Industrial de Santander
Collier, Mark, University of Illinois, Urbana-Champaign
Collins, Edward W., University of Texas at Austin, Jackson School of Geosciences
Congleton, John D., University of West Georgia
Constantopoulos, James T., Eastern New Mexico University
Cook, Hadrian, Kingston University
Corley, John H., Missouri Dept of Natural Resources
Cowart, Richard, Coastal Bend College
Coxon, Catherine, Trinity College
Croudace, Ian, University of Southampton
Cummins, R. Hays, Miami University
Curry, B. B., University of Illinois, Urbana-Champaign
Curry, B. B., Northern Illinois University
Curry, Gordon, University of Glasgow
Curtis, Arthur C., Oak Ridge National Laboratory
Dallimer, Martin, University of Leeds
Darrah, Thomas, Ohio State University
Davis, R. Laurence, University of New Haven
Davis, Steven J., University of California, Irvine
Dawson, James C., Plattsburgh State University (SUNY)
Day, Brett, University of East Anglia
de Hogg, Jan C., Edinburgh University
Denniston, Rhawn F., Cornell College
deWet, Andrew P., Franklin and Marshall College
Dimmick, Charles W., Central Connecticut State University
Diochon, Amanda, Lakehead University
Downie, Helen, University of Manchester
Dubey, C S., University of Delhi
Duncan, Ian J., University of Texas at Austin
Durrant, Jeffrey O., Brigham Young University
Dyer, Jen, University of Leeds
Dymond, Salli F., University of Minnesota, Duluth
Eastler, Thomas E., University of Maine - Farmington
Edwards, Robin, Trinity College
Elliott, Susan J., McMaster University
Ellis, Scott R., University of Illinois, Urbana-Champaign
Ellison, Anne L., University of Illinois, Urbana-Champaign
Erdmann, Anne L., University of Illinois
Ettlie, Bradley, University of Illinois, Urbana-Champaign
Eyles, John D., McMaster University
Favas, Paulo J., Universidade de Trás-os-Montes e Alto Douro
Fehn, Udo, University of Rochester
Field, Cathryn K., Bowdoin College
Flesken, Luuk, University of Leeds
Foley, Duncan, Pacific Lutheran University
Foret, Jim, University of Louisiana at Lafayette
Foxon, Tim, University of Leeds
Franco, Aldina, University of East Anglia
Frappier, Amy, Skidmore College
Galicki, Stan, Millsaps College
Garrison, Trent, Miami University
Garstang, Mimi R., Missouri Dept of Natural Resources
Geary, Lindsey, Utica College
Geiger, James W., University of Illinois, Urbana-Champaign
Gerald, Carresse, North Carolina Central University
Gidigasu, Solomon S., Kwame Nkrumah University of Science and Tech
Goetz, Heinrich, Collin College - Preston Ridge Campus
Gomes, Nuno N., Universidade Independente de Angola
Gomezdelcampo, Enrique, Bowling Green State University
Gouldson, Andy, University of Leeds
Graham, Margaret C., Edinburgh University
Grant, Alastair, University of East Anglia
Halfman, John D., Hobart & William Smith Colleges
Hanes, Daniel M., Saint Louis University
Hargreaves, Tom, University of East Anglia
Harris, Ann G., Youngstown State University

Harris, C, Cardiff University
Harris, Jr., Stanley E., Southern Illinois University Carbondale
Harvey, Omar, University of Southern Mississippi
Hasenmueller, Nancy R., Indiana University
Havenith, Hans-Balder, Universite de Liege
Hawkins, Terry, Missouri Dept of Natural Resources
Hawley, John W., New Mexico Institute of Mining and Tech
Hedin, Robert S., University of Pittsburgh
Herbert, Bruce, Texas A&M University
Hickey, James, Northwest Missouri State University
Hill, Timothy, University of St. Andrews
Hodell, David, University of Cambridge
Holmes, George, University of Leeds
Hoppie, Bryce W., Minnesota State University
Houston, Robert, Oregon Dept of Geology & Mineral Industries
Howell, Dave, Leicester University
Hubeny, J B., Salem State University
Hudson-Edwards, Karen, Birkbeck College
Humphreys, Robin, College of Charleston
Ioris, Antonio, Edinburgh University
Isaacson, Carl, Bemidji State University
Jackman, Toni K., Wichita State University
Jacobs, Alan M., Youngstown State University
Jamieson, Heather E., Queen's University
Jin, Lixin, University of Texas, El Paso
Johnson, Becky, Texas Christian University
Jones, T, Cardiff University
Jordan, Andy, University of East Anglia
Jovanovic, Vladimir, College of Staten Island
Kairies-Beatty, Candace L., Winona State University
Keller, John E., College of Southern Nevada - West Charleston Campus
Ketelle, Richard H., Oak Ridge National Laboratory
Kientop, Greg A., University of Illinois, Urbana-Champaign
Kierczak, Jakub, University of Wroclaw
Kim, Jonathan, Agency of Natural Resources, Dept of Environmental Conservation
Kipphut, George W., Murray State University
Kirchner, James W., University of California, Berkeley
Kochanov, William E., Pennsylvania Bureau of Topographic & Geologic Survey
Kolawole, Lanre L., Ladoke Akintola University of Tech
Krekeler, Mark, Miami University
Kretzschmar, Thomas, Centro de Investigación Científica y de Educación Superior de Ensenada
Laib, Amanda, College of Western Idaho
Lake, Iain, University of East Anglia
Larsen, David, Weber State University
Laskowski, Stanley L., University of Pennsylvania
Last, George V., Pacific Northwest National Laboratory
Leavell, Daniel N., Ohio State University
Lee, Arthur C., Roane State Community College - Oak Ridge
Lee, Cindy M., Clemson University
Lenczewski, Melissa E., Northern Illinois University
Lene, Gene W., Saint Mary's University
Leventon, Julia, University of Leeds
Lloyd, Jonathan, University of Manchester
Love, David W., New Mexico Institute of Mining & Tech
Lundgren, Lawrence W., University of Rochester
Macadam, John, Exeter University
Martin, Scott C., Youngstown State University
Masciocco, Luciano, Università di Torino
Matthews, Robert A., University of California, Davis
May, Michael, Western Kentucky University
Mayer, Margaret, Dine' College
McConnell, Robert L., University of Mary Washington
McCullough, Jr., Edgar J., University of Arizona
McDonald, Brenna, Missouri Dept of Natural Resources
McGivern, Tiffany, Utica College
McGlade, Jacqueline, University College London
McKinney, Michael L., University of Tennessee, Knoxville
Metzger, Marc J., Edinburgh University
Meyer, Brian, Georgia State University
Meyer, W. Craig, Los Angeles Pierce College
Miller, David S., Argonne National Laboratory
Mitre-Salazar, Luis M., Universidad Nacional Autonoma de Mexico
Moloney, Marguerite M., Nicholls State University
Monroy-Sanchez, Jaime I., Universidad Autonoma de Baja California Sur
Moore, Kathryn, Exeter University
Moreton, Kim, Exeter University
Morse, Linda D., College of William & Mary
Mortimer, Robert, University of Leeds
Mouat, David A., Desert Research Institute
Mundie, Ben, Oregon Dept of Geology & Mineral Industries
Nelwamondo, T. M., University of Venda
Newbold, K. B., McMaster University
Newton, Rob, University of Leeds
Nichols, Caitlyn, College of Staten Island
Null, E. Jan, San Francisco State University
O'Connor, Yuet-Ling, Pasadena City College
Oches, Rick, Bentley University
Olson, Neil F., New Hampshire Geological Survey
Omotoso, O. A., University of Ilorin
Orndorff, Richard L., Eastern Washington University
Oyawale, A. A., Obafemi Awolowo University
Paavola, Jouni, University of Leeds
Panko, Andrew W., Brock University
Pashley, E. F., Weber State University
Pate, John, Missouri Dept of Natural Resources
Patenaude, Genevieve, Edinburgh University
Pearce, Jamie R., Edinburgh University
Pearthree, Philip A., University of Arizona
Penzo, Michael A., Bridgewater State University
Perault, David R., Lynchburg College
Perdrial, Nicolas, University of Vermont
Pereira, Alcides C., Universidade de Coimbra
Peterson, Jon W., Hope College
Phillippe, Jennifer, Concord University
Piotrowski, Alexander, University of Cambridge
Plant, Jane, Imperial College
Pohopien, Kazimierz M., Mt. San Antonio College
Pope, Jeanette K., DePauw University
Poudel, Durga, University of Louisiana at Lafayette
Pound, Kate S., Saint Cloud State University
Powell, Jane, University of East Anglia
Pratt, Allyn R., Los Alamos National Laboratory
Price, Peter, Missouri Dept of Natural Resources
Quinn, Claire, University of Leeds
Rasmussen, Pat E., University of Ottawa
Reeves, Donald Matt, University of Alaska, Anchorage
Reid, Brian, University of East Anglia
Rezaie-Boroon, Mohammad H., California State University, Los Angeles
Rice, Thomas L., Cedarville University
Richards, Laura, University of Manchester
Richter, Suzanna L., Franklin and Marshall College
Rizoulis, Athanasios, University of Manchester
Roberts, Sheila M., University of Montana Western
Rogers, William C., West Texas A&M University
Rucklidge, John C., University of Toronto
Ruhl, Laura S., University of Arkansas at Little Rock
Russell, Sally, University of Leeds
Rutter, Nathaniel W., University of Alberta
Sarah, Willig B., University of Pennsylvania
Satow, Christopher, Kingston University
Schlegel, Melissa, College of Western Idaho
Schleifer, Stanley, York College (CUNY)
Schmidt, Dale R., University of Illinois, Urbana-Champaign
Schneider, Jim, Orange Coast College
Schoenemann, Spruce W., University of Montana Western
Scholtz, Theresa C., Argonne National Laboratory
Schulte, Kimberly, Georgia State University, Perimeter College, Dunwoody Campus
Scott, Robert B., University of Texas at Austin
Segall, Marylin, University of Utah
Sewall, Jacob, Kutztown University of Pennsylvania
Sexton, Philip, The Open University
Sharma, Mukul, Dartmouth College
Shaw Faulkner, Melinda, Stephen F. Austin State University
Shem, Linda M., Argonne National Laboratory
Simpson, Myrna Joyce S., University of Toronto
Sims, Douglas , College of Southern Nevada - West Charleston Campus
Sirk, Robert A., Austin Peay State University
Skinner, Luke C., University of Cambridge
Slobodnik, Marek, Masaryk University
Smith, Dan, Leicester University

Smith, Jim, University of Portsmouth
Smith, Steve, University of London, Royal Holloway & Bedford New College
Snider, Henry I., Eastern Connecticut State University
Spaeth, Matthew P., University of Illinois, Urbana-Champaign
Spahr, Paul, Ohio Dept of Natural Resources
Spooner, Ian S., Acadia University
Stahle, David W., University of Arkansas, Fayetteville
Stan, Oana, Al. I. Cuza University of Iasi
Stearman, Will, Queensland University of Tech
Steidl, Gregg M., New Jersey Geological and Water Survey
Stevens, Anthony, Montgomery County Community College
Stumbea, Dan, Al. I. Cuza University of Iasi
Sturm, Diana, University of South Alabama
Sublette, Kerry, The University of Tulsa
Sylvia, Elizabeth R., Maryland Department of Natural Resources
Tank, Ronald W., Lawrence University
Tawabini, Bassam S., King Fahd University of Petroleum and Minerals
Taylor, Peter, University of Leeds
Thorpe, Mary S., Del Mar College
Triplett, Laura, Gustavus Adolphus College
Turchyn, Alexandra, University of Cambridge
Turner, Derek, Douglas College
Turner, Kerry, University of East Anglia
Utgard, Russell O., Ohio State University
Vaezi, Reza, University of Tabriz
van Alstine, James, University of Leeds
Van Horn, Stephen R., Muskingum University
Van Ryswick, Stephen, Maryland Department of Natural Resources
VanGundy, Robert D., University of Virginia College, Wise
Vann, David R., University of Pennsylvania
Visser, Jenneke M., University of Louisiana at Lafayette
Walsh, Maud, Louisiana State University
Ward, Brent C., Simon Fraser University
Watkinson, Andrew, University of East Anglia
Webb, John A., La Trobe University
Welling, Tim, Dutchess Community College
White, Sarah Jane, Princeton University
Wilson, Charlie, University of East Anglia
Wilson, Rick I., California Geological Survey
Witkowski, Christine, Middlesex Community College
Yiannakoulias, Niko, McMaster University
Youberg, Ann, University of Arizona
Young, Michael H., University of Texas at Austin, Jackson School of Geosciences
Young, William, University of Leeds
Yuen, Cheong-yip R., Argonne National Laboratory
Zachos, Louis, University of Mississippi

## Geomorphology

Adams, Kenneth D., Desert Research Institute
Adams, Peter N., University of Florida
Allan, Jonathan C., Oregon Dept of Geology & Mineral Industries
Allen, Bruce, New Mexico Institute of Mining & Tech
Allen, Phillip, Frostburg State University
Amidon, Will, Middlebury College
Anders, Alison M., University of Illinois, Urbana-Champaign
Anderson, Robert, University of Colorado
Anderson, Robert S., University of California, Santa Cruz
Anderson, Suzanne P., University of Colorado
Antinao, JoseLuis, Desert Research Institute
Aslan, Andres, Colorado Mesa University
Attal, Mikael, Edinburgh University
Bacon, Steven N., Desert Research Institute
Badillo, Jose M., University of the Basque Country UPV/EHU
Baker, Sophie, Desert Research Institute
Baker, Victor R., University of Arizona
Baptista, João C., Universidade de Trás-os-Montes e Alto Douro
Barendregt, Rene W., University of Lethbridge
Beget, James E., University of Alaska, Fairbanks
Behling, Robert E., West Virginia University
Belliveau, Lindsey, Dept of Energy and Environmental Protection
Benedetti, Michael M., University of North Carolina Wilmington
Bergeron, Normand, Universite du Quebec
Berta, Susan, Indiana State University
Berti, Claudio, Lehigh University
Bierman, Paul R., University of Vermont
Blisniuk, Kimberly, San Jose State University
Bloom, Arthur L., Cornell University
Bondesan, Aldino, Università degli Studi di Padova
Bookhagen, Bodo, University of California, Santa Barbara
Booth, Adam M., Portland State University
Bradley, William C., University of Colorado
Brewer, Paul A., University of Wales
Brock-Hon, Amy, University of Tennessee at Chattanooga
Brocklehurst, Simon, University of Manchester
Brush, Nigel, Ashland University
Bull, William B., University of Arizona
Bullard, Thomas F., Desert Research Institute
Bullard, Tom, University of Nevada, Reno
Burbank, Doug, University of California, Santa Barbara
Burbank, Douglas W., University of California, Santa Barbara
Burke, Raymond M., Humboldt State University
Campbell, Ian A., University of Alberta
Carson, Robert J., Whitman College
Carton, Alberto, Università degli Studi di Padova
Carver, Gary A., Humboldt State University
Castillon, David A., Missouri State University
Chenoweth, Sean, University of Louisiana, Monroe
Chin, Anne, Texas A&M University
Clark, G. Michael, University of Tennessee, Knoxville
Clark, James A., Wheaton College
Clark, Jeffrey J., Lawrence University
Clark, Marin, University of Michigan
Coates, Donald R., Binghamton University
Colgan, Patrick M., Grand Valley State University
Collins, Brian, University of Washington
Collins, Charles W., University of Wisconsin, Platteville
Coltorti, Mauro, University of Siena
Connallon, Christopher B., Maryland Department of Natural Resources
Constantine, Jose A., Williams College
Cook, Joseph P., University of Arizona
Cooke, M. J., Austin Community College District
Cornwell, Kevin J., California State University, Sacramento
Cotter, Edward, Bucknell University
Cowan, Ellen A., Appalachian State University
Crosby, Benjamin T., Idaho State University
Currey, Donald R., University of Utah
Dale, Janis, University of Regina
Daly, Julia F., University of Maine - Farmington
Daniels, Michael, University of Denver
Day, Stephanie S., North Dakota State University
Dethier, David P., Williams College
DiBiase, Roman , Pennsylvania State University, University Park
Dietrich, William E., University of California, Berkeley
Dixon, Jean, Montana State University
Dixon, John C., University of Arkansas, Fayetteville
Dixon, Simon J., University of Birmingham
Dogwiler, Toby J., Illinois State University
Dogwiler, Toby, Winona State University
Dolan, Robert, University of Virginia
Dolliver, Holly A., University of Wisconsin, River Falls
Dort, Jr., Wakefield, University of Kansas
Durbin, James, University of Southern Indiana
Duvall, Alison, University of Washington
Eaton, Lewis S., James Madison University
Edgar, Dorland E., Argonne National Laboratory
Ehlen, Judy, Radford University
Ellis, Jean, University of South Carolina
Ely, Lisa L., Central Washington University
England, John, University of Alberta
Evans, Diane L., Jet Propulsion Laboratory
Fagherazzi, Sergio, Boston University
Farr, Tom G., Jet Propulsion Laboratory
Feeney, Thomas P., Shippensburg University
Finnegan, Noah J., University of California, Santa Cruz
Fontana, Alessandro, Università degli Studi di Padova
Ford, Derek C., McMaster University
Ford, Richard L., Weber State University
Fubelli, Giandomenico, Università di Torino
Furbish, David J., Vanderbilt University
Furlong, Ira E., Bridgewater State University
Gabet, Emmanuel, San Jose State University
Garcia, Antonio F., California Polytechnic State University
Garcia, Marcelo, University of Illinois, Urbana-Champaign

Gardner, Thomas W., Trinity University
Germanoski, Dru, Lafayette College
Ghoneim, Eman, Boston University
Giardino, John R., Texas A&M University
Giardino, Marco, Università di Torino
Giles, David, University of Portsmouth
Gill, Thomas E., University of Texas, El Paso
Gillam, Mary L., Fort Lewis College
Gillespie, Alan R., University of Washington
Gontz, Allen, University of Massachusetts, Boston
Gootee, Brian F., University of Arizona
Gordon, Steven J., United States Air Force Academy
Gosse, John, University of Kansas
Gran, Karen B., University of Minnesota, Duluth
Greenberg, Harvey, University of Washington
Guccione, Margaret J., University of Arkansas, Fayetteville
Gutowski, Vincent P., Eastern Illinois University
Hales, TC, Cardiff University
Hallet, Bernard, University of Washington
Hancock, Gregory S., College of William & Mary
Hansen, Edward C., Hope College
Hanson, Lindley S., Salem State University
Hanson, Paul, Unversity of Nebraska - Lincoln
Harbor, David J., Washington & Lee University
Harper, Stephen B., East Carolina University
Harrington, Charles D., Los Alamos National Laboratory
Harrison, Bruce I., New Mexico Institute of Mining and Tech
Hasbargen, Leslie E., SUNY, Oneonta
Haughland, Jake, University of Nevada, Reno
Heinzel, Chad E., University of Northern Iowa
Heitmuller, Frank, University of Southern Mississippi
Hesp, Patrick A., Louisiana State University
Hesp, Patrick, Flinders University
Higgins, Charles G., University of California, Davis
Hooke, Roger L., University of Maine
Hooks, W. Gary, University of Alabama
Hopkins, Kenneth D., University of Northern Colorado
Howard, Alan D., University of Virginia
Hungr, Oldrich, University of British Columbia
Hyatt, James A., Eastern Connecticut State University
Isacks, Bryan L., Cornell University
Jerolmack, Douglas, University of Pennsylvania
Jiron, Rebecca, College of William & Mary
Kasse, Kees, VU University Amsterdam
Kaufman, Darrell S., Northern Arizona University
Kavage-Adams, Rebecca H., Maryland Department of Natural Resources
Kehew, Alan E., Western Michigan University
Keller, Edward A., University of California, Santa Barbara
Kelsey, Harvey M., Humboldt State University
Kemmerly, Phillip R., Austin Peay State University
Kendrick, Katherine J., University of California, Riverside
Kenny, Ray, Fort Lewis College
Kershaw, G. Peter, University of Alberta
Kirwan, Matthew L., College of William & Mary
Kiss, Timea, Univesity of Szeged
Kite, J. Steven, West Virginia University
Knott, Jeffrey R., California State University, Fullerton
Kochel, R. Craig, Bucknell University
Koehler, Rich, University of Nevada
Kowaleski, Douglas, University of Massachusetts, Amherst
Kowalewski, Douglas E., Worcester State University
Krautblatter, Michael, Technische Universitaet Muenchen
Laabs, Benjamin J., SUNY, Geneseo
Lamb, Michael P., California Institute of Tech
Lancaster, Nicholas, Desert Research Institute
Lancaster, Nick, University of Nevada, Reno
Larsen, Isaac J., University of Massachusetts, Amherst
Larson, Erik, Moravian College
Larson, Phillip H., Minnesota State University
Lasca, Norman P., University of Wisconsin, Milwaukee
Lazarus, Eli, University of Southampton
Lehre, Andre K., Humboldt State University
Leith, Kerry, Technische Universitaet Muenchen
Lemke, Karen A., University of Wisconsin, Stevens Point
Li, Junran, The University of Tulsa
Lifton, Nathaniel A., Purdue University
Linneman, Scott R., Western Washington University
Lips, Elliott W., University of Utah
Long, Ann D., University of Illinois, Urbana-Champaign
Loring, Arthur P., York College (CUNY)
MacGregor, Kelly, Macalester College
Madej, Mary Ann, Humboldt State University
Mamot, Philipp, Technische Universitaet Muenchen
Manz, Lorraine A., North Dakota Geological Survey
Marchant, David R., Boston University
Marchetti, David W., Western State Colorado University
Marshall, Jeffrey S., California State Polytechnic University, Pomona
Mathers, Hannah, University of Glasgow
McCoy, Scott W., University of Nevada, Reno
McDowell, Patricia F., University of Oregon
McKean, Adam, Utah Geological Survey
McKenney, Rosemary, Pacific Lutheran University
McMillan, Margaret E., University of Arkansas at Little Rock
McPherson, Harry J., University of Alberta
Menking, Kirsten M., Vassar College
Merritts, Dorothy J., Franklin and Marshall College
Meyer, Grant A., University of New Mexico
Miao, Xiaodong, University of Illinois, Urbana-Champaign
Michaud, Yves, Universite du Quebec
Miller, Jerry R., Western Carolina University
Misner, Tamara, Edinboro University of Pennsylvania
Montgomery, David R., University of Washington
Moore, Andrew, Earlham College
Moore, Laura J., University of North Carolina, Chapel Hill
Moorman, Brian J., University of Calgary
Morgan, Daniel J., Vanderbilt University
Moscardelli, Lorena G., University of Texas at Austin
Mossa, Joann, University of Florida
Mount, Jeffrey F., University of California, Davis
Mozzi, Paolo, Università degli Studi di Padova
Musselman, Zachary A., Millsaps College
Mylroie, John E., Mississippi State University
Namikas, Steven, Louisiana State University
Napieralski, Jacob, University of Michigan, Dearborn
Nash, David, University of Cincinnati
Nelson, Daren T., University of North Carolina, Pembroke
Nelson, Robert S., Illinois State University
Netoff, Dennis I., Sam Houston State University
Neubeck, William S., Union College
Newton, Robert M., Smith College
Nichols, Kyle K., Skidmore College
Nielsen, Dennis N., Winona State University
Nikitina, Daria L., West Chester University
Nittrouer, Jeffrey A., Rice University
Okunade, Samuel, Central State University
Olsen, Paul E., Columbia University
ONeal, Michael, University of Delaware
Orme, Amalie, California State University, Northridge
Osborn, Gerald D., University of Calgary
Oskin, Michael, University of California, Davis
Othberg, Kurt L., University of Idaho
Ouimet, William, University of Connecticut
Oviatt, Charles G., Kansas State University
Owen, Lewis, University of Cincinnati
Parker, Gary, University of Illinois, Urbana-Champaign
Pasternack, Gregory B., University of California, Davis
Patton, Peter C., Wesleyan University
Pavlowsky, Robert T., Missouri State University
Pazzaglia, Frank J., Lehigh University
Pederson, Joel L., Utah State University
Pelletier, Jon D., University of Arizona
Perron, Taylor, Massachusetts Institute of Tech
Persico, Lyman P., Whitman College
Phillips, William M., University of Idaho
Pierce, Jennifer L., Boise State University
Pinter, Nicholas, University of California, Davis
Pitlick, John, University of Colorado
Pizzuto, James E., University of Delaware
Plug, Lawrence, Dalhousie University
Polk, Jason, Western Kentucky University
Pope, Gregory A., Montclair State University
Potter, Jr., Noel, Dickinson College
Putkonen, Jaakko, University of North Dakota
Rayburn, John A., SUNY, New Paltz

Rech, Jason, Miami University
Reed, Denise J., University of New Orleans
Refsnider, Kurt, Prescott College
Reneau, Steven L., Los Alamos National Laboratory
Renwick, William, Miami University
Rhoads, Bruce, University of Illinois, Urbana-Champaign
Rhodes, Dallas D., Georgia Southern University
Rice-Snow, R. Scott, Ball State University
Ringe, L. Don, Central Washington University
Rittenour, Tammy M., Utah State University
Ritter, John B., Wittenberg University
Robinson, Cordula, Boston University
Rockwell, Thomas K., San Diego State University
Roering, Joshua J., University of Oregon
Rogerson, Robert J., University of Lethbridge
Savina, Mary E., Carleton College
Sawyer, Carol F., University of South Alabama
Schaller, Mirjam, University of Michigan
Schiefer, Erik, Northern Arizona University
Schmidt, Amanda H., Oberlin College
Schmutz, Phillip P., University of West Florida
Shepherd, Stephanie L., Auburn University
Shreve, Ronald L., University of California, Los Angeles
Shroder, Jr., John F., University of Nebraska at Omaha
Sklar, Leonard, San Francisco State University
Snead, John I., Louisiana State University
Snyder, Jeffrey A., Bowling Green State University
Souch, Catherine J., Indiana University, Indianapolis
Spigel, Lindsay, Dept of Agriculture, Conservation, and Forestry
Springer, Gregory S., Ohio University
Springston, George E., Norwich University
Stanford, Scott D., New Jersey Geological and Water Survey
Stanley, Val L., University of Wisconsin - Extension
Stine, Scott W., California State University, East Bay
Stokes, Martin, University of Plymouth
Straffin, Eric, Edinboro University of Pennsylvania
Strasser, Jeffrey C., Augustana College
Surian, Nicola, Università degli Studi di Padova
Sweeney, Mark R., University of South Dakota
Székely, Balázs, Eotvos Lorand University
Taylor, Stephen B., Western Oregon University
Tchakerian, Vatche P., Texas A&M University
Ten Brink, Norman W., Grand Valley State University
Thorson, Robert M., University of Connecticut
Tidwell, David, Geological Survey of Alabama
Tomkin, Jonathan H., University of Illinois, Urbana-Champaign
Toy, Terrence J., University of Denver
Tranel, Lisa M., Illinois State University
Trenhaile, Alan S., University of Windsor
Tucker, Gregory E., University of Colorado
van Dijk, Deanna, Calvin College
Vandeberg, Gregory S., University of North Dakota
Vidal Romaní, Juan Ramon, Coruna University
Vitek, John D., Oklahoma State University
Vitek, John D., Texas A&M University
Wakabayashi, John, California State University, Fresno
Walters, James C., University of Northern Iowa
Ward, Dylan, University of Cincinnati
Warke, Patricia, Queens University Belfast
Wayne, William J., University of Nebraska, Lincoln
Webb, Robert H., University of Arizona
Wegmann, Karl, North Carolina State University
Weirich, Frank H., University of Iowa
Weldon, Elise M., University of Oregon
Werner, Bradley T., University of California, San Diego
Whisner, Jennifer B., Bloomsburg University
White, Susan Q., La Trobe University
Whiting, Peter J., Case Western Reserve University
Whittecar, Jr., G. Richard, Old Dominion University
Wickert, Andrew D., University of Minnesota, Twin Cities
Wilcock, Peter W., Johns Hopkins University
Wilcox, Andrew, University of Montana
Willenbring, Jane, University of Pennsylvania
Williams, Harry F. L., University of North Texas
Williams, Kevin K., Buffalo State College
Wilson, Fred L., Angelo State University
Wilson, Greg C., Grand Valley State University
Wilson, Monte D., Boise State University
Wohl, Ellen E., Colorado State University
Wolken, Gabriel J., Alaska Division of Geological & Geophysical Surveys
Wood, Spencer H., Boise State University
Yeh, Joseph S., University of Texas at Austin, Jackson School of Geosciences
Young, Richard A., SUNY, Geneseo

## Glacial Geology

Aber, James S., Emporia State University
Alley, Richard B., Pennsylvania State University, University Park
Andrews, John T., University of Colorado
Angle, Michael, Ohio Dept of Natural Resources
Arnaud, Emmanuelle, University of Guelph
Attig, John W., University of Wisconsin, Extension
Baker, Robert W., University of Wisconsin, River Falls
Barclay, David J., SUNY, Cortland
Berthold, Angela, University of Minnesota
Bevis, Kenneth A., Hanover College
Bird, Brian, New York State Geological Survey
Birkel, Sean, University of Maine
Blake, Kevin, Eastern Michigan University
Blankenship, Donald D., University of Texas at Austin
Borns, Jr., Harold W., University of Maine
Breckenridge, Andrew, University of Wisconsin, Superior
Brigham-Grette, Julie, University of Massachusetts, Amherst
Briner, Jason P., SUNY, Buffalo
Bromley, Gordon R., University of Maine
Broster, Bruce E., University of New Brunswick
Brown, Steven E., University of Illinois, Urbana-Champaign
Calkin, Parker, SUNY, Buffalo
Campbell, Seth W., University of Maine
Carlson, Anders, Oregon State University
Carson, Eric C., University of Wisconsin, Extension
Clague, John J., Simon Fraser University
Clark, Douglas H., Western Washington University
Clark, Peter U., Oregon State University
Clayton, Lee, University of Wisconsin, Extension
Clebnik, Sherman M., Eastern Connecticut State University
Cotter, James F., University of Minnesota, Morris
Crossen, Kristine J., University of Alaska, Anchorage
Davis, P. Thompson, Bentley University
Dengler, Elizabeth, University of Minnesota
Denton, George H., University of Maine
Douglass, Daniel C., Northeastern University
Dupont, Todd, Miami University
Easterbrook, Don J., Western Washington University
Erber, Nathan, Ohio Dept of Natural Resources
Evenson, Edward B., Lehigh University
Eyles, Nicholas, University of Toronto
Fisher, Timothy G., Indiana University / Purdue University, Indianapolis
Fisher, Timothy G., University of Toledo
Fleeger, Gary M., Pennsylvania Bureau of Topographic & Geologic Survey
Fleisher, P. Jay, SUNY, Oneonta
Flowers, Gwenn, Simon Fraser University
Fountain, Andrew G., Portland State University
Franzi, David A., Plattsburgh State University (SUNY)
Froese, Duane, University of Alberta
Goldstein, Barry, University of Puget Sound
Gowan, Angela S., University of Minnesota
Hall, Brenda L., University of Maine
Ham, Nelson R., Saint Norbert College
Hambrey, Michael M., Aberystwyth University
Hawley, Robert L., Dartmouth College
Headley, Rachel, University of Wisconsin, Parkside
Hicock, Stephen R., Western University
Hooke, Roger L., University of Minnesota, Twin Cities
Hooyer, Thomas S., University of Wisconsin, Milwaukee
Hopkins, Nathan R., Minot State University
Horton, Jennifer, University of Minnesota
Hughes, Terence J., University of Maine
Hughes III, Richard O., Crafton Hills College
Hulbe, Christina L., Portland State University
Iverson, Neal R., Iowa State University of Science & Tech
Jennings, Carrie E., University of Minnesota, Twin Cities
Johnson, Beth , University of Wisconsin Colleges
Johnson, James B., Colorado Mesa University

Faculty by Specialty

Kaplan, Michael, Columbia University
Karrow, Paul, University of Waterloo
Kelly, Meredith, Dartmouth College
Kiver, Eugene P., Eastern Washington University
Knaeble, Alan, University of Minnesota
Koutnik, Michelle, University of Washington
Kozlowski, Andrew, New York State Geological Survey
Laabs, Benjamin J., North Dakota State University
Lamothe, Michel, Universite du Quebec a Montreal
Larour, Eric Y., SUNY, Buffalo
Lea, Peter D., Bowdoin College
Lempe, Bernhard, Technische Universitaet Muenchen
Leonard, Eric M., Colorado College
Levson, Victor M., University of Victoria
Licciardi, Joseph M., University of New Hampshire
Licht, Kathy J., Indiana University / Purdue University, Indianapolis
Loope, Henry, Indiana University
Lowell, Thomas V., University of Cincinnati
Lusardi, Barb, University of Minnesota
Marcott, Shaun, University of Wisconsin, Madison
Marshall, Katherine, University of Minnesota
McCarthy, Daniel P., Brock University
McKay III, E. Donald, University of Illinois, Urbana-Champaign
McKenzie, Garry D., Ohio State University
Menzies, John, Brock University
Meyer, Gary, University of Minnesota
Mickelson, David M., University of Wisconsin, Madison
Mode, William N., University of Wisconsin, Oshkosh
Mooers, Howard, University of Minnesota, Duluth
Morgan, Alan V., University of Waterloo
Munroe, Jeffrey S., Middlebury College
Nash, T A., Ohio Dept of Natural Resources
Nguyen, Maurice, University of Minnesota
Pair, Donald, University of Dayton
Parent, Michel, Universite du Quebec
Porter, Stephen C., University of Washington
Putnam, Aaron E., University of Maine
Quade, Deborah J., Iowa Dept of Natural Resources
Rawling, J. E., University of Wisconsin - Extension
Retelle, Michael J., Bates College
Rexius, James E., Schoolcraft College
Ridge, John C., Tufts University
Rodbell, Donald T., Union College
Roy, Martin, Universite du Quebec a Montreal
Rutford, Robert H., University of Texas, Dallas
Rysgaard, Soren, University of Manitoba
Schneider, Allan F., University of Wisconsin, Parkside
Schroeder, Dustin M., Stanford University
Schulz, Layne D., South Dakota Dept of Environment and Natural Resources
Sharp, Martin J., University of Alberta
Shaw, John, University of Alberta
Spencer, Matt, Lake Superior State University
Staley, Amie, University of Minnesota
Stanford, Loudon R., University of Idaho
Stansell, Nathan D., Northern Illinois University
Stewart, Alexander K., St. Lawrence University
Straw, Byron, University of Northern Colorado
Stumpf, Andrew J., University of Illinois, Urbana-Champaign
Swanger, Kate, University of Massachusetts Lowell
Syverson, Kent M., University of Wisconsin, Eau Claire
Szymanski, Jason, Monroe Community College
Tary, Anna K., Bentley University
Taylor, Lawrence D., Albion College
Thackray, Glenn D., Idaho State University
Thompson, Lonnie G., Ohio State University
Thorleifson, Harvey, University of Minnesota
Todd, Claire E., Pacific Lutheran University
Totten, Stanley M., Hanover College
Tulaczyk, Slawek, University of California, Santa Cruz
Ullman, David J., Northland College
Wagner, Kaleb, University of Minnesota
Wake, Cameron P., University of New Hampshire
Weddle, Thomas K., Dept of Agriculture, Conservation, and Forestry
Weibel, C. P., University of Illinois, Urbana-Champaign
Werner, Alan, Mount Holyoke College
Westgate, John A., University of Toronto
Wilch, Thomas I., Albion College
Wiles, Gregory C., College of Wooster
Witte, Ronald W., New Jersey Geological and Water Survey
Zoet, Lucas, University of Wisconsin, Madison

## Marine Geology

Abrams, Lewis J., University of North Carolina Wilmington
Adamo, Lauren N., Rutgers, The State University of New Jersey
Aiello, Ivano, Moss Landing Marine Laboratories
Aksu, Ali E., Memorial University of Newfoundland
Allen, Katherine A., University of Maine
Aluwihare, Lihini I., University of California, San Diego
Anderson, John B., Rice University
Austin, Jr., James A., University of Texas at Austin
Bacchus, Tania S., Johnson State College
Bangs, Nathan L., University of Texas at Austin
Barclay, Andrew, University of Washington
Barrie, Vaughn, University of Victoria
Becel, Anne, Columbia University
Belknap, Daniel F., University of Maine
Benoit-Bird, Kelly, Oregon State University
Bentley, Samuel J., Louisiana State University
Brooks, Gregg R., Eckerd College
Brown, Kevin M., University of California, San Diego
Browne, Kathleen M., Rider University
Buffler, Richard T., University of Texas at Austin
Byrne, John V., Oregon State University
Calmus, Thierry, Universidad Nacional Autonoma de Mexico
Carlin, Joe, California State University, Fullerton
Carson, Bobb, Lehigh University
Chase, Richard L., University of British Columbia
Cleary, William J., University of North Carolina Wilmington
Coffroth, Mary Alice, SUNY, Buffalo
Colman, Steve, University of Minnesota, Duluth
Conway, Flaxen D., Oregon State University
Cook, Ann, Ohio State University
Crone, Timothy, Columbia University
Cruz-Orozco, Rodolfo, Universidad Autonoma de Baja California Sur
Curray, Joseph R., University of California, San Diego
Daniell, James J., James Cook University
De Deckker, Patrick, Australian National University
Delaney, John R., University of Washington
Dobson, David M., Guilford College
Donnelly, Jeffrey, Woods Hole Oceanographic Institution
Donoghue, Joseph F., Florida State University
Driscoll, Neal W., University of California, San Diego
Dugan, Brandon, Colorado School of Mines
Eberli, Gregor P., University of Miami
Edwards, Margo H., University of Hawai'i, Manoa
Ferland, Marie A., Central Washington University
Flood, Roger D., SUNY, Stony Brook
Fornari, Daniel J., Woods Hole Oceanographic Institution
Freeman-Lynde, Raymond, University of Georgia
Freitas, Maria da Conceição P., Universidade de Lisboa
Fryer, Patricia B., University of Hawai'i, Manoa
Fulthorpe, Craig S., University of Texas at Austin
Gee, Jeffrey S., University of California, San Diego
Georgen, Jennifer, Old Dominion University
Georgen, Jennifer, Florida State University
Georgiopopoulou, Aggeliki, University College Dublin
Giosan, Liviu, Woods Hole Oceanographic Institution
Glickson, Deborah, National Academy of Sciences/National Research Council
Goes, Joaquim, Columbia University
Goff, John A., University of Texas at Austin
Goni, Miguel A., Oregon State University
Gonzalez-Fernandez, Antonio, Centro de Investigación Científica y de Educación Superior de Ensenada
Grabowski, Jonathan, Northeastern University
Greene, H. Gary, Moss Landing Marine Laboratories
Greenstein, Benjamin J., Cornell College
Guerin, Gilles, Columbia University
Haggerty, Janet, The University of Tulsa
Harris, Robert N., University of Utah
Harte, Michael, Oregon State University
Humphris, Susan E., Woods Hole Oceanographic Institution
Jacobson, Gary L., Grossmont College
Jarrard, Richard D., University of Utah

Johnson, Harlan P., University of Washington
Johnson, Joel E., University of New Hampshire
Juraèiæ, Mladen, University of Zagreb
Kaminski, Michael A., King Fahd University of Petroleum and Minerals
Katuna, Michael P., College of Charleston
Kearney, Micheal S., University of Maryland
Keller, Randall A., Oregon State University
Kelley, Joseph T., University of Maine
Kienast, Stephanie, Dalhousie University
Klasik, John A., California State Polytechnic University, Pomona
Klaus, Adam, Texas A&M University
Kroenke, Loren W., University of Hawai'i, Manoa
Laine, Edward P., Bowdoin College
Lasker, Howard R., SUNY, Buffalo
Leonard, Lynn A., University of North Carolina Wilmington
Lindo Atichati, David, College of Staten Island
Liu, Jingpu P., North Carolina State University
Lonsdale, Peter F., University of California, San Diego
Luciano, Katherine E., Dept of Natural Resources
MacLeod, C, Cardiff University
Malinverno, Alberto , Columbia University
Mallinson, David, East Carolina University
Manfrino, Carrie M., Kean University
Marinelli, Roberta, Oregon State University
Mayer, Larry A., University of New Hampshire
McCoy, Jr., Floyd W., University of Hawai'i, Manoa
Meadows, Guy A., Michigan Technological University
Meylan, Maurice A., University of Southern Mississippi
Micallef, Aaron, University of Malta
Miller, Kenneth G., Rutgers, The State University of New Jersey
Miot da Silva, Graziela, Flinders University
Mitchell, Neil C., University of Manchester
Moberly, Ralph, University of Hawai'i, Manoa
Mountain, Gregory S., Rutgers, The State University of New Jersey
Moy, Christopher M., University of Otago
Naehr, Thomas H., Texas A&M University, Corpus Christi
Neumann, A. Conrad, University of North Carolina, Chapel Hill
Nicholson, David, Woods Hole Oceanographic Institution
Nitsche, Frank, Columbia University
Nittrouer, Charles A., University of Washington
Norris, Richard D., University of California, San Diego
Olayiwola, M. A., Obafemi Awolowo University
Patton, Jason R., Humboldt State University
Peterson, Larry C., University of Miami
Phipps, James B., Grays Harbor College
Pietro, Kathryn R., Woods Hole Oceanographic Institution
Piper, David J., Dalhousie University
Pisias, Nicklas, Oregon State University
Portner, Ryan A., Worcester State University
Pratt, Thomas, University of Washington
Ramirez, Wilson R., University of Puerto Rico
Raphael, Constantine N., Eastern Michigan University
Rasmussen, Kenneth, Northern Virginia Community College - Annandale
Rea, David K., University of Michigan
Reed, Donald L., San Jose State University
Richardson, Mary Jo, Texas A&M University
Richaud, Mathieu, California State University, Fresno
Riggs, Stanley R., East Carolina University
Rodolfo, Kelvin S., University of Illinois at Chicago
Ross, David A., Woods Hole Oceanographic Institution
Ruddiman, William F., University of Virginia
Scher, Howie, University of South Carolina
Schwartz, David, Cabrillo College
Scott, Steven D., University of Toronto
Skarke, Adam, Mississippi State University
Slovinsky, Peter, Dept of Agriculture, Conservation, and Forestry
Stea, Ralph, Dalhousie University
Tucholke, Brian E., Woods Hole Oceanographic Institution
Van Andel, Tjeerd H., Stanford University
Vannucchi, Paola, University of London, Royal Holloway & Bedford New College
Ward, Larry G., University of New Hampshire
Watts, Anthony B., University of Oxford
Wehmiller, John F., University of Delaware
Wheeler, Andy J., University College Cork
Wilcock, William S., University of Washington
Wilson, Gary S., University of Otago
Winguth, Cornelia, University of Texas, Arlington

## Mineralogy & Crystallography

Abbott, Richard N., Appalachian State University
Abdu, Yassir, University of Manitoba
Ague, Jay J., Yale University
Akgul, Bünyamin, Firat University
Alderton, Dave, University of London, Royal Holloway & Bedford New College
Aldushin, Kirill, Ludwig-Maximilians-Universitaet Muenchen
Altaner, Stephen P., University of Illinois, Urbana-Champaign
Anderson, Callum, Nelson Mandela Metropolitan University
Angel, Ross J., Virginia Polytechnic Institute & State University
Ankney, Meagan E., University of Akron
Apopei, Andrei I., Al. I. Cuza University of Iasi
Arletti, Rossella, Università di Torino
Artioli, Gilberto, Università degli Studi di Padova
Ball, Neil, University of Manitoba
Bassett, William A., Cornell University
Bayliss, Peter, University of Calgary
Beane, Rachel J., Bowdoin College
Beg, Mirza A., Geological Survey of Alabama
Belluso, Elena, Università di Torino
Benna, Piera, Università di Torino
Bermanec, Vladimir, University of Zagreb
Bish, David L., Indiana University, Bloomington
Bladh, Kenneth W., Wittenberg University
Blanchard, Frank N., University of Florida
Bonhoure, Jessica, Institut Polytechnique LaSalle Beauvais (ex-IGAL)
Boutilier, Robert F., Bridgewater State University
Boysen, Hans, Ludwig-Maximilians-Universitaet Muenchen
Braund, Emilie, University of Portsmouth
Brearley, Adrian J., University of New Mexico
Brown, Jr., Gordon E., Stanford University
Bruno, Emiliano, Università di Torino
Bruno, Marco, Università di Torino
Burger, Paul V., University of New Mexico
Burnham, Charles, Fort Lewis College
Burnham, Charles W., Harvard University
Burns, Peter C., University of Notre Dame
Buzatu, Andrei, Al. I. Cuza University of Iasi
Buzgar, Nicolae, Al. I. Cuza University of Iasi
Camara Artigas, Fernando, Università di Torino
Capella, Silvana, Università di Torino
Cardellach López, Esteve, Universitat Autonoma de Barcelona
Carl, James D., SUNY Potsdam
Carrigan, Charles W., Olivet Nazarene University
Casas Duocastella, Lluís, Universitat Autonoma de Barcelona
Catlos, Elizabeth J., University of Texas at Austin
Chakhmouridian, Anton, University of Manitoba
Chen, Ning, University of Saskatchewan
Chiarenzelli, Jeffrey R., St. Lawrence University
Chizmadia, Lysa, University of Puerto Rico
Choi, Seon-Gyu, Korea University
Cilliers, Johannes, Imperial College
Clement, Stephen C., College of William & Mary
Cody, Robert, Iowa State University of Science & Tech
Cofer, Harland E., Georgia Southwestern State University
Coker, Victoria, University of Manchester
Cole, Kevin C., Grand Valley State University
Coleman, Leslie C., University of Saskatchewan
Collins, Lorence G., California State University, Northridge
Cooper, Brian J., Sam Houston State University
Cooper, Mark, University of Manitoba
Copjakova, Renata, Masaryk University
Corbella Cordomí, Mercè, Universitat Autonoma de Barcelona
Cordonnier, Benoit, Ludwig-Maximilians-Universitaet Muenchen
Costa, Emanuele, Università di Torino
Cusack, Maggie, University of Glasgow
Dalconi, Maria C., Università degli Studi di Padova
De Campos, Christina, Ludwig-Maximilians-Universitaet Muenchen
De Nault, Kenneth J., University of Northern Iowa
De Pablo, Liberto, Universidad Nacional Autonoma de Mexico
Demosthenous, Christie M., University of Wisconsin, Oshkosh
Dempster, Tim, University of Glasgow
Deng, Youjun, Texas A&M University
Denicourt, Raymond, University of South Florida, Tampa
Denny, F. B., University of Illinois, Urbana-Champaign

Dobson, David, University College London
Dollase, Wayne A., University of California, Los Angeles
Dorfman, Alexander, Ludwig-Maximilians-Universitaet Muenchen
Dorfner, Thomas, Ludwig-Maximilians-Universitaet Muenchen
Downs, James W., Ohio State University
Duchesne, Josee, Universite Laval
Dunlevey, John, University of Limpopo
Dutrow, Barbara L., Louisiana State University
Ehlmann, Arthur J., Texas Christian University
Elliott, Brent, University of Texas at Austin
Elsen, Jan, Katholieke Universiteit Leuven
Emofurieta, Williams O., University of Benin
Ercit, T. Scott, Carleton University
Ertel-Ingrisch, Werner, Ludwig-Maximilians-Universitaet Muenchen
Estop Graells, Eugènia, Universitat Autonoma de Barcelona
Ewing, Rodney C., Stanford University
Ewing, Rodney E., University of New Mexico
Farnan, Ian, University of Cambridge
Farthing, Dori J., SUNY, Geneseo
Farzaneh, Akbar, University of Tabriz
Feather, Russell, Smithsonian Institution / National Museum of Natural History
Fehr, Karl Thomas, Ludwig-Maximilians-Universitaet Muenchen
Feinglos, Mark N., Duke University
Fendrich, Kim V., American Museum of Natural History
Ferguson, Robert B., University of Manitoba
Ferraris, Giovanni, Università di Torino
Field, Stephen W., Tarleton State University
Finch, Adriam, University of St. Andrews
Fleet, Michael E., Western University
Flemming, Roberta L., Western University
Flynn, Natalie, La Salle University
Foit, Jr., Franklin F., Washington State University
Fransolet, Andre-Mathieu, Universite de Liege
Freed, Robert L., Trinity University
Frenette, Jean, Universite Laval
Frey, Friedrich, Ludwig-Maximilians-Universitaet Muenchen
Frueh, Alfred J., University of Connecticut
Gait, Robert I., Royal Ontario Museum
Gann, Delbert E., Millsaps College
Garapić,, Gordana, SUNY, New Paltz
Gay, Kenny, North Carolina Geological Survey
Giere, Reto, University of Pennsylvania
Giese, Rossman F., SUNY, Buffalo
Gigler, Alexander, Ludwig-Maximilians-Universitaet Muenchen
Gille, Peter, Ludwig-Maximilians-Universitaet Muenchen
Giustetto, Roberto, Università di Torino
Gollmer, Steven, Cedarville University
González, Isabel , Universidad de Sevilla
Grant, Shelton K., Missouri University of Science and Tech
Greenwell, Chris, Durham University
Grießl, Stefan, Ludwig-Maximilians-Universitaet Muenchen
Griffen, Dana T., Brigham Young University
Groat, Lee A., University of British Columbia
Grover, John E., University of Cincinnati
Guggenheim, Stephen J., University of Illinois at Chicago
Guillemette, Renald, Texas A&M University
Gunter, Mickey E., University of Idaho
Göttlich, Hagen, Ludwig-Maximilians-Universitaet Muenchen
Hadler, Kathryn, Imperial College
Hall, Anne M., Emory University
Hansen, Lars, University of Oxford
Harlow, George E., Graduate School of the City University of New York
Harlow, George E., American Museum of Natural History
Harrison, Richard, University of Cambridge
Hawkins, David, Wellesley College
Hawkins, Michael E., New York State Geological Survey
Hawthorne, Frank C., University of Manitoba
Hazen, Robert M., Carnegie Institution of Washington
Heaney, Peter J., Pennsylvania State University, University Park
Heckel, Wolfgang, Ludwig-Maximilians-Universitaet Muenchen
Heimann, Adriana, East Carolina University
Henderson, Grant S., University of Toronto
Hennemeyer, Marc, Ludwig-Maximilians-Universitaet Muenchen
Hess, Kai-Uwe, Ludwig-Maximilians-Universitaet Muenchen
Heuss-Aßbichler, Soraya, Ludwig-Maximilians-Universitaet Muenchen
Hogarth, Donald D., University of Ottawa
Hollabaugh, Curtis L., University of West Georgia
Hooper, Robert L., University of Wisconsin, Eau Claire
Horton, Duane G., Pacific Northwest National Laboratory
Horwell, Claie, Durham University
Hovis, Guy L., Lafayette College
Huff, Warren D., University of Cincinnati
Hughes, John M., Miami University
Hughes, John M., University of Vermont
Hummer, Daniel, Southern Illinois University Carbondale
Izawa, Matt, Western University
Johanesen, Katharine, Juniata College
Johnson, Neil E., Virginia Polytechnic Institute & State University
Jonas, Edward C., University of Texas at Austin
Jones, Rhian, University of New Mexico
Jordan, Guntram, Ludwig-Maximilians-Universitaet Muenchen
Kaiser, Jason, Southern Utah University
Kaiser-Bischoff, Ines, Ludwig-Maximilians-Universitaet Muenchen
Kamhi, Samuel R., Long Island University, Brooklyn Campus
Kampf, Anthony R., Los Angeles County Museum of Natural History
Kearns, Lance E., James Madison University
Keller, Dianne M., Colgate University
Kerestedjian, Thomas N., Bulgarian Academy of Sciences
Kleebe, Hans-Joachim, Technische Universitaet Darmstadt
Klein, Cornelis, University of New Mexico
Komabayashi, Tetsuya, Edinburgh University
Krol, Michael A., Bridgewater State University
Kunzmann, Thomas, Ludwig-Maximilians-Universitaet Muenchen
Lackinger, Markus, Ludwig-Maximilians-Universitaet Muenchen
Lancaster, Penny, University of Portsmouth
Langenhorst, Falko H., Friedrich-Schiller-University Jena
Langer, Arthur M., Graduate School of the City University of New York
Lapen, Thomas, University of Houston
Le Voyer, Marion, Smithsonian Institution / National Museum of Natural History
Leavens, Peter B., University of Delaware
Lee, Martin, University of Glasgow
Leimer, H. Wayne, Tennessee Tech University
Leitz, Robert E., Metropolitan State College of Denver
Leung, Irene S., Lehman College (CUNY)
Leung, Irene S., Graduate School of the City University of New York
Levesque, Andre, Universite Laval
Lian, Jie, University of Michigan
Liebling, Richard S., Graduate School of the City University of New York
Losh, Steven, Minnesota State University
Losos, Zdenek, Masaryk University
Lueth, Virgil L., New Mexico Institute of Mining and Tech
Lupulescu, Marian V., New York State Geological Survey
Lübbe, Maike, Ludwig-Maximilians-Universitaet Muenchen
Madison, James A., DePauw University
Majzlan, Juraj, Friedrich-Schiller-University Jena
Manu, Johnson, University of Ghana
Marshall, Craig, University of Kansas
Martin, Robert F., McGill University
Masch, Ludwig, Ludwig-Maximilians-Universitaet Muenchen
Mason, Roger A., Memorial University of Newfoundland
Mattinson, Chris, Central Washington University
McCormack, John K., University of Nevada, Reno
McCormick, George R., University of Iowa
McCurry, Michael O., Idaho State University
McDonald, Andrew M., Laurentian University, Sudbury
McHenry, Lindsay J., University of Wisconsin, Milwaukee
MdNamee, Brittani D., University of North Carolina, Asheville
Meisterernst, Götz, Ludwig-Maximilians-Universitaet Muenchen
Merkel, Timo Casjen, Ludwig-Maximilians-Universitaet Muenchen
Miles, Andrew, Kingston University
Miyagi, Lowell, University of Utah
Moritz, Wolfgang, Ludwig-Maximilians-Universitaet Muenchen
Myer, George H., Temple University
Müller, Lena, Ludwig-Maximilians-Universitaet Muenchen
Müller-Sohnius, Dieter, Ludwig-Maximilians-Universitaet Muenchen
Neethling, Stephen, Imperial College
Nestola, Fabrizio, Università degli Studi di Padova
Nord, Julia, George Mason University
Novak, Milan , Masaryk University
Nusbaum, Robert L., College of Charleston
Oganov, Artem, SUNY, Stony Brook
Paktunc, Dogan, University of Ottawa

Papike, James J., University of New Mexico
Paquette, Jeanne, McGill University
Parise, John B., SUNY, Stony Brook
Park, Sohyun, Ludwig-Maximilians-Universitaet Muenchen
Pascoe, Richard, Exeter University
Pattrick, Richard, University of Manchester
Peacor, Donald R., University of Michigan
Pentcheva, Rossitza, Ludwig-Maximilians-Universitaet Muenchen
Peterson, Ronald C., Queen's University
Phillips, Brian L., SUNY, Stony Brook
Phillips, William R., Brigham Young University
Piniella Febrer, Juan Francesc, Universitat Autonoma de Barcelona
Pope, Lin F., University of Southern Mississippi
Post, Jeffrey E., Smithsonian Institution / National Museum of Natural History
Potter, Lee S., University of Northern Iowa
Prencipe, Mauro, Università di Torino
Price, David, University College London
Pöllmann, Herbert, Martin-Luther-Universitaet Halle-Wittenberg
Radcliffe, Dennis, Hofstra University
Rakovan, John F., Miami University
Raudsepp, Mati, University of British Columbia
Redfern, Simon, University of Cambridge
Renock, Devon, Dartmouth College
Rentel, Raimund, University of the Free State
Reuss, Robert L., Tufts University
Ribbe, Paul H., Virginia Polytechnic Institute & State University
Rickaby, Ros, University of Oxford
Riedel, Oliver, Ludwig-Maximilians-Universitaet Muenchen
Rigault de la Longrais, Germano, Università di Torino
Robinson, George W., St. Lawrence University
Robinson, Paul D., Southern Illinois University Carbondale
Rocholl, Alexander, Ludwig-Maximilians-Universitaet Muenchen
Rodriguez-Blanco, Juan Diego, Trinity College
Rojas-Soriano, Humberto, Universidad Autonoma de Baja California Sur
Rosenberg, Philip E., Washington State University
Ross, Nancy L., Virginia Polytechnic Institute & State University
Rossman, George R., California Institute of Tech
Rouse, Roland C., University of Michigan
Rubbo, Marco, Università di Torino
Rubio-Sierra, Javier, Ludwig-Maximilians-Universitaet Muenchen
Rutstein, Martin S., SUNY, New Paltz
Salviulo, Gabriella, Università degli Studi di Padova
Schindler, Michael, Laurentian University, Sudbury
Schmahl, Wolfgang, Ludwig-Maximilians-Universitaet Muenchen
Schneer, Cecil J., University of New Hampshire
Schneider, Julius, Ludwig-Maximilians-Universitaet Muenchen
Schroeder, Paul A., University of Georgia
Scott Smith, Barbara H., University of British Columbia
Secco, Luciano, Università degli Studi di Padova
Seid, Mary J., University of Illinois, Urbana-Champaign
Sethi, Parvinder S., Radford University
Shank, Stephen G., Pennsylvania Bureau of Topographic & Geologic Survey
Sharp, Thomas G., Arizona State University
Sharp, W. Edwin, University of South Carolina
Shen, Guoyin, Carnegie Institution of Washington
Sherriff, Barbara L., University of Manitoba
Shuller-Nickles, Lindsay C., Clemson University
Silvestri, Alberta, Università degli Studi di Padova
Simmons, Stuart, University of Utah
Simmons, William B., University of Michigan
Simmons, William B., University of New Orleans
Singer, David M., Kent State University
Skinner, H. Catherine W., Yale University
Skoda, Radek, Masaryk University
Smith, Michael S., University of North Carolina Wilmington
Smyth, Joseph R., University of Colorado
Snow, Eleanour, University of South Florida, Tampa
Sokolova, Elena, University of Manitoba
Solorio-Munguia, Jose G., Universidad Nacional Autonoma de Mexico
Spieler, Oliver, Ludwig-Maximilians-Universitaet Muenchen
Spilde, Michael N., University of New Mexico
Stark, Robert, Ludwig-Maximilians-Universitaet Muenchen
Stefano, Christopher J., Michigan Technological University
Stewart, Dion C., Georgia State University, Perimeter College, Alpharetta Campus
Strasser, Stefan, Ludwig-Maximilians-Universitaet Muenchen
Summers, Sara, Weber State University
Swope, R. J., Indiana University / Purdue University, Indianapolis
Tait, Kimberly T., Royal Ontario Museum
Tettenhorst, Rodney T., Ohio State University
Thibault, Yves, Western University
Tibljaš, Darko, University of Zagreb
Tomašiæ, Nenad, University of Zagreb
Trixler, Frank, Ludwig-Maximilians-Universitaet Muenchen
Uzochukwu, Godfrey A., North Carolina Agricultural & Tech State University
van Hees, Edmond H., Wayne State University
Vassiliou, Andreas H., Rutgers, The State University of New Jersey, Newark
Vaughan, David, University of Manchester
Vernon, William W., Dickinson College
Viti, Cecilia, University of Siena
Walker, Jeffrey R., Vassar College
Walther, Ferdinand, Ludwig-Maximilians-Universitaet Muenchen
Warner, Richard D., Clemson University
Webb, Christine, Smithsonian Institution / National Museum of Natural History
Weinbruch, Stephan, Technische Universitaet Darmstadt
Wenk, Hans-Rudolf, University of California, Berkeley
Wicks, Frederick J., University of Toronto
Wicks, Frederick J., Royal Ontario Museum
Wilson, James R., Weber State University
Winchell, Robert E., California State University, Long Beach
Wise, Michael A., Smithsonian Institution / National Museum of Natural History
Wise, Michael A., University of New Orleans
Wood, Bernard, University of Oxford
Wood, Ian, University College London
Wylie, Ann G., University of Maryland
Xu, Huifang, University of Wisconsin, Madison
Yoshiasa, Akira, Kumamoto University
Yurtsever, Ayhan, Ludwig-Maximilians-Universitaet Muenchen
Zink, Albert, Ludwig-Maximilians-Universitaet Muenchen
Zurevinski, Shannon, Lakehead University
Šæavnièar, Stjepan, University of Zagreb
Žigoveèki Gobac, Željka, University of Zagreb

## Paleolimnology

Anderson, Roger Y., University of New Mexico
Axford, Yarrow, Northwestern University
Bennett, Scott, University of Washington
Brady, Kristina, University of Minnesota, Twin Cities
Brant, Lynn A., University of Northern Iowa
Brenner, Mark, University of Florida
Buchheim, Paul, Loma Linda University
Contreras, Sergio, University of Pittsburgh
Cook, Timothy L., Worcester State University
De Batist, Marc, Ghent University
Douglas, Marianne, University of Toronto
Ferris, Grant F., University of Toronto
Glaser, Paul H., University of Minnesota, Twin Cities
Heyvaert, Alan, University of Nevada, Reno
Hillman, Aubrey, University of Louisiana at Lafayette
Kaplan, Samantha W., University of Wisconsin, Stevens Point
Kirby, Matthew E., California State University, Fullerton
Lisiecki, Lorraine, University of California, Santa Barbara
Morrison, Alexander, University of Minnesota, Twin Cities
Noren, Anders, University of Minnesota, Twin Cities
Porter, Susannah, University of California, Santa Barbara
Russell, James M., Brown University
Shapley, Mark, University of Minnesota, Twin Cities
Smith, Alison J., Kent State University
Steinman, Byron A., University of Minnesota, Duluth
Stevens (Landon), Lora R., California State University, Long Beach
Theissen, Kevin, University of Saint Thomas

## Petroleum Geology

Abdulghani, Waleed, King Fahd University of Petroleum and Minerals
Adeigbe, O. C., University of Ibadan
Adekeye, J.I. D., University of Ilorin
Al-Ramadan, Khalid, King Fahd University of Petroleum and Minerals
Ali, Hendratta N., Fort Hays State University
Allis, Richard G., Utah Geological Survey

Andrews, Richard D., University of Oklahoma
Anthony, Robin, Pennsylvania Bureau of Topographic & Geologic Survey
Aplin, Andrew, Durham University
Arnold, Dan, Heriot-Watt University
Asquith, George B., Texas Tech University
Avary, Katherine L., West Virginia University
Baer, James L., Brigham Young University
Bartok, Peter, University of Houston
Beaver, Harold H., Baylor University
Bend, Stephen L., University of Regina
Benjamin, U. K., Obafemi Awolowo University
Boboye, O. A., University of Ibadan
Bonk, Kathleen, New York State Geological Survey
Bowersox, J. Richard, University of Kentucky
Bowler, Bernard, University of Newcastle Upon Tyne
Bowman, Mike, University of Manchester
Broadhead, Ronald F., New Mexico Institute of Mining and Tech
Burton, Bradford R., Western State Colorado University
Carr, Timothy R., West Virginia University
Carr, Timothy R., University of Kansas
Carter, Kristin M., Pennsylvania Bureau of Topographic & Geologic Survey
Cassidy, Martin, University of Houston
Champlin, Steven D., Mississippi Office of Geology
Cheadle, Burns A., Western University
Chenoweth, Cheri, University of Illinois, Urbana-Champaign
Cochrane, Claudia, Western University
Connors, Christopher, Washington & Lee University
Cranganu, Constantin, Brooklyn College (CUNY)
Crockett, Joan, University of Illinois, Urbana-Champaign
Curtis, John B., Colorado School of Mines
Dallegge, Todd A., University of Northern Colorado
Denne, Richard, Texas Christian University
Doveton, John H., University of Kansas
Echols, John B., Louisiana State University
Edegbai, Aitalokhai J., University of Benin
Ehinola, Olugbenga A., University of Ibadan
Eide, Elizabeth A., National Academy of Sciences/National Research Council
Elnadi, Abdelhalim H., University of Khartoum
Elrick, Scott D., University of Illinois, Urbana-Champaign
Fadiya, S. L., Obafemi Awolowo University
Fakhari, Mohammad D., Ohio Dept of Natural Resources
Feely, Martin, National University of Ireland Galway
Finley, Robert J., University of Illinois, Urbana-Champaign
Folorunso, I. O., University of Ilorin
Forgotson, Jr., James M., University of Oklahoma
Fraser, Alastair, Imperial College
Giles, Katherine A., University of Texas, El Paso
Gillespie, Janice, California State University, Bakersfield
Gingras, Murray, University of Alberta
Graham, Stephan A., Stanford University
Grender, Gordon C., Virginia Polytechnic Institute & State University
Hanks, Catherine L., University of Alaska, Fairbanks
Harris, David C., University of Kentucky
Harun, Nina, Alaska Division of Geological & Geophysical Surveys
Haynes, Samantha E., University of Leeds
Henk, Jr., Floyd, Texas Christian University
Herrmann, Leo A., Louisiana Tech University
Hileman, Mary E., Oklahoma State University
Hofmann, Michael, University of Montana
Hu, Qinhong (Max), University of Texas, Arlington
Hughes, Colin, University of Manchester
Igbinigie, Nosa S., University of Benin
Jeffery, David L., Marietta College
Jiang , Shu, University of Utah
John, Chacko J., Louisiana State University
Johnson, Howard, Imperial College
Jones, Bobby L., Louisiana State University
Keach, William, University of Utah
Knepp, Rex A., Illinois State Geological Survey
Leetaru, Hannes E., University of Illinois, Urbana-Champaign
Leetaru, Hannes E., University of Illinois, Urbana-Champaign
Lever, Helen, Heriot-Watt University
Levino, Lynn J., Pennsylvania Bureau of Topographic & Geologic Survey
Li, Peng, Arkansas Geological Survey
Loucks, Bob, University of Texas at Austin, Jackson School of Geosciences
Lucia, F. Jerry, University of Texas at Austin, Jackson School of Geosciences
Machel, Hans G., University of Alberta
Malinverno, Alberto, Columbia University
Manzocchi, Tom, University College Dublin
Mazzullo, Salvatore J., Wichita State University
Mc Mahon, Michelle, Lonestar College - North Harris
Meddaugh, W. S., Midwestern State University
Medina, Cristian R., Indiana University
Miller, Michael B., Louisiana State University
Minzoni, Marcello, University of Alabama
Morgan, Craig D., Utah Geological Survey
Neufeld, Jerome A., University of Cambridge
Newell, Kerry D., University of Kansas
Nixon, R. Paul, Brigham Young University
Nordeng, Stephan, University of North Dakota
North, F. K., Carleton University
Nton, M. E., University of Ibadan
Nunez-Velazco, Miriam, Universidad Autonoma de Baja California Sur
Nwachukwu, J. I., Obafemi Awolowo University
Oduneye, Olusola C., Ladoke Akintola University of Tech
Ogungbesan, Gbenga O., Ladoke Akintola University of Tech
Ola, P S., Federal University of Tech, Akure
Olarewaju, V. O., Obafemi Awolowo University
Olorunfemi, A. O., Obafemi Awolowo University
Osman, Mutasim S., King Fahd University of Petroleum and Minerals
Panah, Assad I., University of Pittsburgh, Bradford
Pashin, Jack C., Mississippi State University
Pertl, David, Amarillo College
Pigott, John D., University of Oklahoma
Pranter, Matthew J., University of Oklahoma
Pritchett, Brittany, University of Oklahoma
Puckette, James, Oklahoma State University
Ratchford, M. E., Arkansas Geological Survey
Resnic, Victor S., Lonestar College - North Harris
Richards-McClung, Bryony, University of Utah
Riess, Carolyn M., Austin Community College District
Ronck, Catherine, Tarleton State University
Rupp, John A., Indiana University
Scarselli, Nicola, University of London, Royal Holloway & Bedford New College
Schmid, Katie, Pennsylvania Bureau of Topographic & Geologic Survey
Schneider, Robert V., Texas A&M University, Kingsville
Schroeder, Stefan, University of Manchester
Shumaker, Robert C., West Virginia University
Simpson, Altus, Cypress College
Slater, Brian, New York State Geological Survey
Slatt, Roger M., University of Oklahoma
Smith, Langhorne, New York State Geological Survey
Sonnenberg, Stephen A., Colorado School of Mines
Sprinkel, Douglas A., Utah Geological Survey
Stewart, Gary, Oklahoma State University
Stright, Lisa, Colorado State University
Sullivan, Raymond, San Francisco State University
Sun, Yuefeng, Texas A&M University
Tan, Denis N., University of Malaya
Taylor, Kevin, University of Manchester
Temples, Tommy, Clemson University
Tinker, Scott W., University of Texas at Austin, Jackson School of Geosciences
Trentham, Robert C., University of Texas, Permian Basin
van der Land, Crees, University of Newcastle Upon Tyne
Van Kooten, Gerald K., Calvin College
Vautier, Yannick, Institut Polytechnique LaSalle Beauvais (ex-IGAL)
Wach, Grant D., Dalhousie University
Wardlaw, Norman C., University of Calgary
Weimer, Paul, University of Colorado
White, Bekki C., Arkansas Geological Survey
Whiteley, Martin, University of Derby
Wielens, Hans J., Dalhousie University
Woodward, Mac B., Arkansas Geological Survey
Xie, Xiangyang, Texas Christian University
Zimmerman, Ronald K., Louisiana State University
Zonneveld, John-Paul, University of Alberta
Zuppann, Charles W., Indiana University

## General Petrology

Afolabi, Olukayode A., Ladoke Akintola University of Tech
Ague, Jay J., Yale University
Aird, Hannah M., California State University, Chico
Anderson, Alan J., Saint Francis Xavier University
Arslan, Gizem, Firat University
Ashwal, Lewis D., University of the Witwatersrand
Balen, Dražen, University of Zagreb
Barr, Sandra, Dalhousie University
Basu, Asish R., University of Rochester
Basu, Asish, University of Texas, Arlington
Bickel, Charles E., San Francisco State University
Biševac, Vanja, University of Zagreb
Blackburn, William H., University of Kentucky
Borghi, Alessandro, Università di Torino
Brady, John B., Smith College
Brandriss, Mark E., Smith College
Brown, Michael, University of Maryland
Broxton, David E., Los Alamos National Laboratory
Cain, J. Allan, University of Rhode Island
Callegari, Ezio, Università di Torino
Cameron, Kenneth L., University of California, Santa Cruz
Carmichael, Sarah, Appalachian State University
Castelli, Daniele, Università di Torino
Chaumba, Jeff B., University of North Carolina, Pembroke
Cherniak, Daniele J., Rensselaer Polytechnic Institute
Chown, Edward H., Universite du Quebec a Chicoutimi
Churnet, Habte G., University of Tennessee at Chattanooga
Compagnoni, Roberto, Università di Torino
Compton, Robert R., Stanford University
Cordell, Ann S., University of Florida
Cortes, Joaquin A., SUNY, Buffalo
Crawford, Maria Luisa B., Bryn Mawr College
Creasy, John W., Bates College
Dasgupta, Rajdeep, Rice University
DeBari, Susan M., Western Washington University
Deininger, Robert W., University of Memphis
Dobrzhinetskaya, Larissa F., University of California, Riverside
Doolan, Barry L., University of Vermont
Eklund, Olav, Abo Akademi University
Encarnacion, John, Saint Louis University
Erickson, Rolfe C., Sonoma State University
Ferrando, Simona, Università di Torino
Ferreiro Maehlmann, Rafael, Technische Universitaet Darmstadt
Finn, Gregory C., Brock University
Ford, Mark T., Texas A&M University, Kingsville
Frank, Charles O., Southern Illinois University Carbondale
Ganguly, Jiba, University of Arizona
Goble, Ronald G., University of Nebraska, Lincoln
Gonzales, David A., Fort Lewis College
Graham, Ian, University of New South Wales
Gray, Norman H., University of Connecticut
Groppo, Chiara Teresa, Università di Torino
Hacker, Bradley R., University of California, Santa Barbara
Haileab, Bereket, Carleton College
Hanson, Richard E., Texas Christian University
Hatheway, Richard B., SUNY, Geneseo
Hayob, Jodie, University of Mary Washington
Hetherington, Callum J., Texas Tech University
Hollocher, Kurt T., Union College
Hutchison, David, Hartwick College
Ihinger, Phillip D., University of Wisconsin, Eau Claire
Jercinovic, Michael J., University of Massachusetts, Amherst
Jimoh, Mustapha T., Ladoke Akintola University of Tech
Johnson, Julie, University of Memphis
Jolliff, Bradley L., Washington University in St. Louis
Kılıc, Ayse Didem, Firat University
Kerr, A C., Cardiff University
Keskinen, Mary J., University of Alaska, Fairbanks
Leichmann, Jaromir, Masaryk University
Lewy, Robert, Bakersfield College
Malcuit, Robert J., Denison University
Manning, Craig E., University of California, Los Angeles
Maritan, Lara, Università degli Studi di Padova
Martin, Barton S., Ohio Wesleyan University
Mathez, Edmond A., American Museum of Natural History
Mazzoli, Claudio, Università degli Studi di Padova
McKinney, D. Brooks, Hobart & William Smith Colleges
McLeod, Claire, Miami University
Mekik, Figen A., Grand Valley State University
Miller, James D., University of Minnesota, Twin Cities
Mirnejad, Hassan , Miami University
Moecher, David P., University of Kentucky
Moore, Daniel K., Brigham Young University - Idaho
Murphy, J. Brendan, Dalhousie University
Mutti, Laurence J., Juniata College
Nadeau, Olivier, University of Ottawa
Nasir, Sobhi J., University of Qatar
Nesse, William D., University of Northern Colorado
Nielsen, Peter A., Keene State College
Noblett, Jeffrey B., Colorado College
Owens, Brent E., College of William & Mary
Ozel Yıldırım, Esra , Firat University
Pan, Yuanming, University of Saskatchewan
Peterson, Virginia L., Grand Valley State University
Plank, Terry A., Columbia University
Plymate, Thomas G., Missouri State University
Prichard, H M., Cardiff University
Ptacek, Anton D., San Diego State University
Purchase, Megan, University of the Free State
Purdom, William B., Southern Oregon University
Raeside, Robert, Dalhousie University
Ranson, William A., Furman University
Raymond, Loren A., Appalachian State University
Reed, Robert M., University of Texas at Austin, Jackson School of Geosciences
Reyes-Cortes, Ignacio A., Universidad Autonoma de Chihuahua
Rizeli, Mustafa Eren, Firat University
Roadifer, Jack E., Colorado Mesa University
Rolfo, Franco, Università di Torino
Rougvie, James R., Beloit College
Severs, Matthew R., Stockton University
Sharma, Shiv K., University of Hawai'i, Manoa
Shoemaker, Kurt A., Shawnee State University
Snyder, Lori D., University of Wisconsin, Eau Claire
Spandler, Carl, James Cook University
Steiner, Jeffrey, Graduate School of the City University of New York
Stelling, Pete, Western Washington University
Thompson, Allan M., University of Delaware
Thy, Peter, University of California, Davis
Torres-Agulera, Juan Manuel, Universidad Autonoma de San Luis Potosi
Van Buer, Nicholas J., California State Polytechnic University, Pomona
van Kranendonk, Martin, University of New South Wales
Vaniman, David T., Los Alamos National Laboratory
Vanko, David A., Towson University
Vertolli, Vincent, Royal Ontario Museum
Walker, David, Columbia University
Warren, Jessica, University of Delaware
Weaver, Stephen G., Colorado College
Welsh, James L., Gustavus Adolphus College
Westerman, David S., Norwich University
Whelan, Peter M., University of Minnesota, Morris
WoldeGabriel, Giday, Los Alamos National Laboratory
Wortel, Matthew J., University of Iowa
Yurkovich, Steven P., Western Carolina University

## Igneous Petrology

Adams, Gerald E., Columbia College Chicago
Albee, Arden L., California Institute of Tech
Allen, Jack C., Bucknell University
Amedjoe, Godfrey C., Kwame Nkrumah University of Science and Tech
Amel, Nasir, University of Tabriz
Andersen, Jens, Exeter University
Anderson, James Lawford, Boston University
Annesley, Irvine, University of Saskatchewan
Anthony, Elizabeth Y., University of Texas, El Paso
Aranda, Jorge, Universidad Nacional Autonoma de Mexico
Bachmann, Olivier, University of Washington
Badger, Robert L., SUNY Potsdam
Bailey, David G., Hamilton College
Baldridge, W. Scott, Los Alamos National Laboratory
Barker, Daniel, University of Texas at Austin
Barnes, Calvin G., Texas Tech University
Barr, Sandra M., Acadia University
Barth, Andrew P., Indiana University / Purdue University, Indianapolis
Barton, Michael, Ohio State University

Beard, James S., Virginia Polytechnic Institute & State University
Bedard, Jean H., Universite du Quebec
Bender, E. E., Orange Coast College
Bender, John F., University of North Carolina, Charlotte
Benimoff, Alan I., College of Staten Island
Bentley, Robert, Central Washington University
Bergantz, George W., University of Washington
Beswick, Anthony E., Laurentian University, Sudbury
Beyarslan, Melahat, Firat University
Bingol, Ahmet Feyzi, Firat University
Birsic, Erin M., Allegheny College
Bizimis, Michael, University of South Carolina
Blackerby, Bruce A., California State University, Fresno
Bladh, Katherine L., Wittenberg University
Bloomer, Sherman H., Oregon State University
Boerboom, Terrence, University of Minnesota
Bohrson, Wendy A., Central Washington University
Borg, Scott G., National Science Foundation
Boroughs, Scott, Washington State University
Boudreau, Alan E., Duke University
Brandon, Alan, University of Houston
Brandriss, Mark E., Williams College
Brenan, James, Dalhousie University
Brophy, James G., Indiana University, Bloomington
Browne, Brandon L., Humboldt State University
Brueseke, Matthew E., Kansas State University
Bryan, Scott E., Queensland University of Tech
Bybee, Grant M., University of the Witwatersrand
Byerly, Gary R., Louisiana State University
Canil, Dante, University of Victoria
Carmichael, Ian S., University of California, Berkeley
Castillo, Paterno R., University of California, San Diego
Cathey, Henrietta, Queensland University of Tech
Cecil, M. R., California State University, Northridge
Cepeda, Joseph C., West Texas A&M University
Chadwick, John, College of Charleston
Chavrit, Deborah, University of Manchester
Cheney, John T., Amherst College
Chesner, Craig A., Eastern Illinois University
Christiansen, Eric H., Brigham Young University
Clague, David A., University of Hawai'i, Manoa
Claiborne, Lily L., Vanderbilt University
Clark, Russell G., Albion College
Clarke, D B., Dalhousie University
Clausen, Benjamin L., Loma Linda University
Clemens-Knott, Diane, California State University, Fullerton
Coleman, Robert G., Stanford University
Condit, Christopher D., University of Massachusetts, Amherst
Constantin, Marc, Universite Laval
Coogan, Laurence, University of Victoria
Coombs, Douglas S., University of Otago
Cooper, Kari M., University of California, Davis
Cooper, Roger W., Lamar University
Cornell, Winton, The University of Tulsa
Cote, Denis, Universite du Quebec a Chicoutimi
Cottle, John, University of California, Santa Barbara
Cousens, Brian L., Carleton University
Crawford, Anthony J., University of Tasmania
Cribb, Warner, Middle Tennessee State University
Currier, Ryan, University of Wisconsin, Green Bay
Dare, Sarah A., University of Ottawa
Davies, Gareth R., VU University Amsterdam
Dean, Robert, Dickinson College
Delgado-Argote, Luis A., Centro de Investigación Científica y de Educación Superior de Ensenada
Dick, Henry J B., Woods Hole Oceanographic Institution
Dickson, Loretta D., Lock Haven University
Dijkstra, Arjan, University of Plymouth
Dingwell, Donald Bruce, Ludwig-Maximilians-Universitaet Muenchen
Dolejs, David, Charles University
Dorais, Michael J., Brigham Young University
Downs, Robert T., University of Arizona
Duke, Edward F., South Dakota School of Mines & Tech
Dunbar, Nelia W., New Mexico Institute of Mining & Tech
Dunn, John Todd, University of New Brunswick
Dymek, Robert F., Washington University in St. Louis
Edwards, Benjamin R., Dickinson College
Elders, Wilfred A., University of California, Riverside
Elkins, Lynne J., University of Nebraska, Lincoln
Elliot, David H., Ohio State University
Erturk, Mehmet Ali, Firat University
Fagan, Amy, Western Carolina University
Feeley, Todd C., Montana State University
Feeney, Dennis M., University of Idaho
Fernandez, Louis A., California State University, San Bernardino
Fiesinger, Donald W., Utah State University
Filiberto, Justin, Southern Illinois University Carbondale
Fitton, J. G., Edinburgh University
Fleming, Thomas H., Southern Connecticut State University
Flower, Martin F. J., University of Illinois at Chicago
Fodor, Ronald V., North Carolina State University
Fournelle, John, University of Wisconsin, Madison
Fowler, Mike, University of Portsmouth
Fowler, Sarah, Katholieke Universiteit Leuven
Fox, Lydia K., University of the Pacific
Francis, Don, McGill University
Gadgil, Raghav, Goa University
Gaetani, Glenn A., Woods Hole Oceanographic Institution
Gamble, John, University College Cork
Garcia, Michael O., University of Hawai'i, Manoa
García Galán, Gúmer, Universitat Autonoma de Barcelona
Garrison, Jennifer M., California State University, Los Angeles
Geist, Dennis J., University of Idaho
GERBE, Marie-Christine, Université Jean Monnet, Saint-Etienne
Gertisser, Ralf, Keele University
Ghani, Azman Abdul, University of Malaya
Gibson, David, University of Maine - Farmington
Gibson, Sally, University of Cambridge
Gill, James B., University of California, Santa Cruz
Gittins, John, University of Toronto
Glazner, Allen F., University of North Carolina, Chapel Hill
Gomes, Maria E., Universidade de Trás-os-Montes e Alto Douro
Gordon, Stacia , University of Nevada, Reno
Gorring, Matthew L., Montclair State University
Green, John C., University of Minnesota, Duluth
Grimes, Craig, Ohio University
Grove, Timothy L., Massachusetts Institute of Tech
Grunder, Anita L., Oregon State University
Gualda, Guilherme, Vanderbilt University
Gust, David A., Queensland University of Tech
Haddock, Gerald H., Wheaton College
Haggerty, Stephen E., Florida International University
Halsor, Sid P., Wilkes University
Hammer, Julia E., University of Hawai'i, Manoa
Hammersley, Lisa, California State University, Sacramento
Hammond, Paul E., Portland State University
Hannah, Judith L., Colorado State University
Hanson, Sarah L., Adrian College
Hari, Kosiyath R., Pt. Ravishankar Shukla University
Harris, Chris, University of Cape Town
Hart, William K., Miami University
Hearn Jr, B. Carter, Smithsonian Institution / National Museum of Natural History
Hebert, Rejean J., Universite Laval
Hermes, O D., University of Rhode Island
Herrstrom, Eileen A., University of Illinois, Urbana-Champaign
Hess, Paul C., Brown University
Hidalgo, Paulo, Georgia State University
Higgins, Michael D., Universite du Quebec a Chicoutimi
Hirschmann, Marc M., University of Minnesota, Twin Cities
Hirt, William H., College of the Siskiyous
Holness, Marian, University of Cambridge
Hon, Rudolph, Boston College
Hudgins, Thomas, University of Puerto Rico
Husch, Jonathan M., Rider University
Huysken, Kristin, Indiana University Northwest
Irvine, T. Neil, Carnegie Institution of Washington
Jagoutz, Oliver, Massachusetts Institute of Tech
Jahangiri, Ahmad, University of Tabriz
Janney, Phillip, University of Cape Town
Jaye, Shelley, Northern Virginia Community College - Annandale
Jenner, George A., Memorial University of Newfoundland
Johnson, Kenneth S., University of Houston Downtown
Johnson, Kevin T. M., University of Hawai'i, Manoa

Johnson, Marie C., United States Military Academy
Jones, Adrian P., University College London
Jones, Norris W., University of Wisconsin, Oshkosh
Jordan, Brennan T., University of South Dakota
Kay, Robert W., Cornell University
Kay, Suzanne M., Cornell University
Kelemen, Peter B., Woods Hole Oceanographic Institution
Kelemen, Peter B., Columbia University
Kempton, Pamela D., Kansas State University
Kent, Adam J., Oregon State University
Klein, Emily M., Duke University
Kopylova, Maya G., University of British Columbia
Koteas, G. Christopher, Norwich University
Kreiger, William (Bill), York College of Pennsylvania
Kurum, Sevcan, Firat University
Kyle, Philip R., New Mexico Institute of Mining and Tech
Kysar Mattietti, Giuseppina, George Mason University
Lackey, Jade Star, Pomona College
Landenberger, Bill, University of Newcastle
Lange, Rebecca A., University of Michigan
Le Roux, Veronique, Woods Hole Oceanographic Institution
Leake, Bernard E., Cardiff University
Leeman, William P., Rice University
Lehnert, Kerstin, Columbia University
Lesher, Charles E., University of California, Davis
Levy, Schon S., Los Alamos National Laboratory
Lidiak, Edward G., University of Pittsburgh
Lindline, Jennifer, New Mexico Highlands University
Luth, Robert W., University of Alberta
Lux, Daniel R., University of Maine
Lytwyn, Jennifer N., University of Houston
Maclennan, John, University of Cambridge
Mahood, Gail A., Stanford University
Manduca, Cathryn A., Carleton College
Marsh, Julian S., Rhodes University
Mathez, Edmond A., Graduate School of the City University of New York
Mattioli, Glen, University of Texas, Arlington
Matty, David J., Weber State University
Mayborn, Kyle R., Western Illinois University
McCallum, I. Stewart (Stu), University of Washington
McClellan, Elizabeth, Radford University
McGuire, Angela, Whitman College
McMillan, Nancy J., New Mexico State University, Las Cruces
Medaris, Jr., Levi G., University of Wisconsin, Madison
Memeti, Valbone, California State University, Fullerton
Mertzman, Stanley A., Franklin and Marshall College
Michael, Peter J., The University of Tulsa
Michelfelder, Gary, Missouri State University
Miller, Calvin F., Vanderbilt University
Miller, James D., University of Minnesota, Duluth
Miller, Jonathan S., San Jose State University
Mills, James G., DePauw University
Mitchell, Roger H., Lakehead University
Moayyed, Mohsen -., University of Tabriz
Morgan, Daniel, University of Leeds
Morgan, George B., University of Oklahoma
Morse, Stearns A., University of Massachusetts, Amherst
MOYEN, Jean-François, Université Jean Monnet, Saint-Etienne
Mungall, James E., University of Toronto
Murray, Jay D., Midwestern State University
Mustart, David A., San Francisco State University
Myers, James D., University of Wyoming
Nash, Barbara P., University of Utah
Naslund, H. Richard, Binghamton University
Natland, James, University of Miami
Naumann, Terry R., University of Alaska, Anchorage
Neal, Clive R., University of Notre Dame
Nehru, Cherukupalli E., Graduate School of the City University of New York
Nelson, Stephen A., Tulane University
Nelson, Wendy, Towson University
Nicholls, James W., University of Calgary
Nicholson, Kirsten N., Ball State University
Nicolaysen, Kirsten P., Whitman College
Nielsen, Roger L., Oregon State University
Ntsaluba, Bantubonke I., Rhodes University
NUDE, Prosper M., University of Ghana
Ouimette, Mark A., Hardin-Simmons University
Page, Philippe, Universite du Quebec a Chicoutimi
Parker, Don, Wayland Baptist University
Parker, Don F., Baylor University
Parlak, Osman, Cukurova Universitesi
Patino-Douce, Marta, University of Georgia
Perez-Venzor, Jose A., Universidad Autonoma de Baja California Sur
Perfit, Michael R., University of Florida
Petrinec, Zorica, University of Zagreb
Philpotts, Anthony R., University of Connecticut
Pickard, Megan, Brigham Young University - Idaho
Polat, Ali, University of Windsor
Pollock, Meagen, College of Wooster
Price, Jonathan D., Midwestern State University
Prinz, Martin, Graduate School of the City University of New York
Puffer, John H., Rutgers, The State University of New Jersey, Newark
Puziewicz, Jacek, University of Wroclaw
Raia, Federica, City College (CUNY)
Ratajeski, Kent, University of Kentucky
Reagan, Mark K., University of Iowa
Reavy, John, University College Cork
Reid, Arch M., University of Houston
Reid, Mary R., Northern Arizona University
Reidel, Stephen P., Washington State University
Reyes-Cortes, Manuel, Universidad Autonoma de Chihuahua
Roden, Michael F., University of Georgia
Roelofse, Frederick, University of the Free State
Rooney, Tyrone, Michigan State University
Ross, Martin E., Northeastern University
Ruprecht, Philipp P., University of Nevada, Reno
Saini-Eidukat, Bernhardt, North Dakota State University
Sarrionandia, Fernando, University of the Basque Country UPV/EHU
Saunders, Andy D., Leicester University
Schmidt, Mariek, Brock University
Schulze, Daniel J., University of Toronto
Schwab, Brandon E., Western Carolina University
Schwab, Brandon E., Humboldt State University
Schwartz, Joshua J., California State University, Northridge
Scoates, James S., University of British Columbia
Seaman, Sheila J., University of Massachusetts, Amherst
Seeger, Cheryl M., Missouri Dept of Natural Resources
Sen, Gautam, Florida International University
Shaw, Cliff S., University of New Brunswick
Shearer, Jr., Charles K., University of New Mexico
Shervais, John W., Utah State University
Shuster, Robert D., University of Nebraska at Omaha
Siegel, Coralie, Queensland University of Tech
Sigurdson, David, California State University, Dominguez Hills
Silva-Mora, Luis, Universidad Nacional Autonoma de Mexico
Sinton, John M., University of Hawai'i, Manoa
Sirbescu, Mona, Central Michigan University
Sisson, Virginia, University of Houston
Size, William B., Emory University
Smedes, Harry W., Southern Oregon University
Smith, Diane R., Trinity University
Smith, Douglas, University of Texas at Austin
Smith, Eugene I., University of Nevada, Las Vegas
Smith, Matthew C., University of Florida
Smith, Terence E., University of Windsor
Somarin, Alireza, Brandon University
Spera, Frank J., University of California, Santa Barbara
Springer, Robert K., Brandon University
Stachel, Thomas, University of Alberta
Stern, Charles R., University of Colorado
Stewart, Michael A., University of Illinois, Urbana-Champaign
Stork, Allen L., Western State Colorado University
Stormer, Jr., John C., Rice University
Streck, Martin J., Portland State University
Stull, Robert J., California State University, Los Angeles
Szabo, Csaba, Virginia Polytechnic Institute & State University
Tacinelli, John C., Rochester Community & Technical College
Taib, Nur Iskandar, University of Malaya
Taylor, Lawrence A., University of Tennessee, Knoxville
Tepley, III, Frank J., Oregon State University
Tepper, Jeffrey H., University of Puget Sound
Thakurta, Joyashish, Western Michigan University
Thomas, Jay, Syracuse University

Ural, Melek, Firat University
Urbanczyk, Kevin, Sul Ross State University
Vance, Joseph A., University of Washington
Vander Auwera, Jacqueline, Universite de Liege
Vetter, Scott K., Centenary College of Louisiana
Viegas, Anthony V., Goa University
Visona', Dario, Università degli Studi di Padova
von der Handt, Anette, University of Minnesota, Twin Cities
Wadsworth, William B., Whittier College
Walker, Barry , Portland State University
Walker, James A., Northern Illinois University
Walowski, Kristina J., Middlebury College
Walter , Michael , University of Bristol
Wanless, Dorsey, Boise State University
Warren, Richard G., Los Alamos National Laboratory
Webster, John R., Minot State University
Weiblen, Paul W., University of Minnesota, Twin Cities
Weiland, Thomas J., Georgia Southwestern State University
Wendlandt, Richard F., Colorado School of Mines
Wenner, Jennifer M., University of Wisconsin, Oshkosh
Whattam, Scott A., Korea University
White, Craig M., Boise State University
White, John C., Eastern Kentucky University
Whitney, James A., University of Georgia
Whittington, Alan, University of Missouri
Wiese, Katryn, City College of San Francisco
Wirth, Karl R., Macalester College
Wittke, James, Northern Arizona University
Wobus, Reinhard A., Williams College
Wolf, Michael B., Augustana College
Wolff, John A., Washington State University
Wolff, John A., Washington State University
Worcester, Peter A., Hanover College
Woyski, Margaret S., California State University, Fullerton
Wulff, Andrew H., Whittier College
Yogodzinkski, Gene M., University of South Carolina
Young, Davis A., Calvin College
Zamzow, Craig E., Clarion University
Zieg, Michael J., Slippery Rock University

## Metamorphic Petrology

Adams, Mark G., Appalachian State University
Allen, Alistair, University College Cork
Allen, Gary C., University of New Orleans
Appleyard, Edward C., University of Waterloo
Apraiz, Arturo, University of the Basque Country UPV/EHU
Ashton, Kenneth E., University of Regina
Baldwin, Julia, University of Montana
Bauer, Paul W., New Mexico Institute of Mining & Tech
Bebout, Gray E., Lehigh University
Berg, Christopher A., University of West Georgia
Berg, Jonathan H., Northern Illinois University
Berman, Robert, Carleton University
Blake, David E., University of North Carolina Wilmington
Brouwer, Fraukje M., VU University Amsterdam
Brown, Edwin H., Western Washington University
Burnley, Pamela C., University of Nevada, Las Vegas
Burwash, Ronald A., University of Alberta
Cardace, Dawn, University of Rhode Island
Carillo-Chavez, Alejandro J., Universidad Autonoma de Baja California Sur
Carlson, William D., University of Texas at Austin
Cesare, Bernardo, Università degli Studi di Padova
Chacko, Thomas, University of Alberta
Clarke, Geoffrey L., University of Sydney
Cooper, Alan F., University of Otago
Corriveau, Louise, Universite du Quebec
Costin, Gelu, Rhodes University
Cruz-Uribe, Alicia M., University of Maine
Daniel, Christopher G., Bucknell University
Darling, Robert S., SUNY, Cortland
Davidson, Cameron, Carleton College
Dawson, Jane P., Iowa State University of Science & Tech
Day, Howard W., University of California, Davis
DeAngelis, Michael T., University of Arkansas at Little Rock
Devore, George W., Florida State University
Diener, Johann, University of Cape Town
Dietsch, Craig, University of Cincinnati
Dipple, Gregory M., University of British Columbia
Droop, Giles, University of Manchester
Dunn, Steven R., Mount Holyoke College
Dymek, Robert F., Washington University in St. Louis
Eckert, James O., Yale University
El Shazly, Aley, Marshall University
Elias-Herrera, Mariano, Universidad Nacional Autonoma de Mexico
Evans, Bernard W., University of Washington
Farrar, Stewart S., Eastern Kentucky University
Faryad, Shah Wali, Charles University
Fawcett, J. J., University of Toronto
Fletcher, John, Centro de Investigación Científica y de Educación Superior de Ensenada
Flores, Kennet, Brooklyn College (CUNY)
Foster Jr., Charles T., University of Iowa
Frost, B. Ronald, University of Wyoming
Ghent, Edward D., University of Calgary
Giaramita, Mario J., California State University, Stanislaus
Gillis, Kathryn, University of Victoria
Goodge, John W., University of Minnesota, Duluth
Grant, James A., University of Minnesota, Duluth
Grew, Edward S., University of Maine
Grew, Priscilla C., University of Nebraska, Lincoln
Grover, Timothy W., Castleton University
Hacker, Bradley R., University of California, Santa Barbara
Hajialioghli, Robab, University of Tabriz
Halama, Ralf, Keele University
Hallett, Benjamin W., University of Wisconsin, Oshkosh
Henry, Darrell J., Louisiana State University
Hepburn, J. Christopher, Boston College
Hepple, Alex, Queensland University of Tech
Hickmott, Donald D., Los Alamos National Laboratory
Himmelberg, Glen R., University of Missouri
Hoersch, Alice L., La Salle University
Hoisch, Thomas D., Northern Arizona University
Holdaway, Michael J., Southern Methodist University
Holland, Tim, University of Cambridge
Hollister, Lincoln S., Princeton University
Horvath, Peter, Rhodes University
Horvath, Peter, University of the Witwatersrand
Huizenga, Jan Marten, James Cook University
Iancu, Ovidiu G., Al. I. Cuza University of Iasi
Indares, Aphrodite D., Memorial University of Newfoundland
James, Richard, Laurentian University, Sudbury
Joesten, Raymond, University of Connecticut
Johnson, Eric L., Hartwick College
Jones, Jon W., University of Calgary
Kay, Suzanne M., Cornell University
Kays, Marvin A., University of Oregon
Kerrick, Derrill M., Pennsylvania State University, University Park
La Tour, Timothy E., Georgia State University
Labotka, Theodore C., University of Tennessee, Knoxville
Laird, Jo, University of New Hampshire
Lamb, Will, Texas A&M University
Lang, Helen M., West Virginia University
Law, Eric W., Muskingum University
Leech, Mary L., San Francisco State University
Liou, Juhn G., Stanford University
Maboko, Makenya A., University of Dar es Salaam
Magloughlin, Jerry F., Colorado State University
Mahan, Kevin H., University of Colorado
Manecan, Teodosia, Hunter College (CUNY)
Manon, Matthew R., Union College
Marko, Wayne, Northwestern University
Martínez Fernández, Francisco, Universitat Autonoma de Barcelona
McLelland, James, Colgate University
Menold, Carrie A., Albion College
Metcalf, Rodney V., University of Nevada, Las Vegas
Metzger, Ellen P., San Jose State University
Moazzen, Mohssen -., University of Tabriz
Mogk, David W., Montana State University
Morton, Douglas M., University of California, Riverside
Mulcahy, Sean R., Western Washington University
Murray, Daniel P., University of Rhode Island
Mutis-Duplat, Emilio, University of Texas, Permian Basin
Nishiyama, Tadao, Kumamoto University
Ortega-Gutierrez, Fernando, Universidad Nacional Autonoma de Mexico

Page, F Zeb, Oberlin College
Panish, Peter T., University of Massachusetts, Amherst
Parkinson, Christopher D., University of New Orleans
Pattison, David R., University of Calgary
Peck, William H., Colgate University
Penniston-Dorland, Sarah, University of Maryland
Perkins, III, Dexter, University of North Dakota
Pervunina, Aelita, Institute of Geology (Karelia, Russia)
Plummer, Charles C., California State University, Sacramento
Potel, Sébastien, Institut Polytechnique LaSalle Beauvais (ex-IGAL)
Radakovich, Amy, University of Minnesota
Raeside, Robert P., Acadia University
Rahaman, M. A., Obafemi Awolowo University
Reche Estrada, Joan, Universitat Autonoma de Barcelona
Redden, Jack A., South Dakota School of Mines & Tech
Rietmeijer, Frans J., University of New Mexico
Rivers, Toby C. J. S., Memorial University of Newfoundland
Roberts, Frank, Montgomery County Community College
Roselle, Gregory T., Brigham Young University - Idaho
Rosenfeld, John L., University of California, Los Angeles
Rumble, III, Douglas, Carnegie Institution of Washington
Sawyer, Edward W., Universite du Quebec a Chicoutimi
Schiffman, Peter, University of California, Davis
Scotford, David M., Miami University
Selverstone, Jane E., University of New Mexico
Sisson, Virginia B., University of Utah
Sorensen, Sorena S., Smithsonian Institution / National Museum of Natural History
Sorensen, Sorena, University of Maryland
Spear, Frank S., Rensselaer Polytechnic Institute
Spiess, Richard, Università degli Studi di Padova
Spray, John G., University of New Brunswick
Srogi, LeeAnn, West Chester University
Stevens, Liane M., Stephen F. Austin State University
Stoddard, Edward F., North Carolina State University
Storey, Craig D., University of Portsmouth
Stout, James H., University of Minnesota, Twin Cities
Stowell, Harold H., University of Alabama
Swapp, Susan M., University of Wyoming
Thomson, Jennifer A., Eastern Washington University
Tinkham, Doug, Laurentian University, Sudbury
Tracy, Robert J., Virginia Polytechnic Institute & State University
Turnock, Allan C., University of Manitoba
Van Tongeren, Jill A., Rutgers, The State University of New Jersey
Walsh, Emily O., Cornell College
Walsh, Emily O., University of Iowa
Waters, Dave, University of Oxford
Weber, Bodo, Centro de Investigación Científica y de Educación Superior de Ensenada
White, Richard W., Universität Mainz
Whitney, Donna L., University of Minnesota, Twin Cities
Wilks, Maureen, New Mexico Institute of Mining & Tech
Wintsch, Robert P., Indiana University, Bloomington
Woodland, Bertram G., Field Museum of Natural History
Zhai, Xiaoming, College of Lake County

## Sedimentary Petrology

Adekola, S. A., Obafemi Awolowo University
Al-Aasm, Ihsan S., University of Windsor
Alberstadt, Leonard P., Vanderbilt University
Argast, Anne S., Indiana University / Purdue University, Fort Wayne
Babek, Ondrej, Masaryk University
Basu, Abhijit, Indiana University, Bloomington
Benson, Donald J., University of Alabama
Bertog, Janet, Northern Kentucky University
Bertok, Carlo, Università di Torino
Boggs, Jr., Sam, University of Oregon
Boulvain, Frederic P., Universite de Liege
Brengman, Latisha A., University of Minnesota, Duluth
Bruner, Katherine R., West Virginia University
Budd, David A., University of Colorado
Bullard, Reuben G., Northern Kentucky University
Carozzi, Albert V., University of Illinois, Urbana-Champaign
Chafetz, Henry S., University of Houston
Channing, Alan, University of Wales
Choh, Suk-Joo, Korea University
Clarke, Stu, Keele University
Corcoran, Patricia, Western University
Cousineau, Pierre, Universite du Quebec a Chicoutimi
DeCelles, Peter G., University of Arizona
dela Pierre, Francesco, Università di Torino
Dockal, James A., University of North Carolina Wilmington
Donovan, R. Nowell, Texas Christian University
Dutton, Shirley P., University of Texas at Austin, Jackson School of Geosciences
Fitzgerald, David, Saint Mary's University
Folk, Robert L., University of Texas at Austin
Forsman, Nels F., University of North Dakota
Franseen, Evan K., University of Kansas
Freeman, Thomas J., University of Missouri
Fritsche, A. E., California State University, Northridge
Galli-Olivier, Carlos A., Universidad Autonoma de Baja California Sur
Gómez Gras, David, Universitat Autonoma de Barcelona
Gonzalez, Maria M., Central Michigan University
Griffing, David H., Hartwick College
Grigsby, Jeffry D., Ball State University
Hampson, Gary , Imperial College
Hasiuk, Franciszek J., Iowa State University of Science & Tech
Horton, Jr., Robert A., California State University, Bakersfield
Houseknecht, David W., Virginia Polytechnic Institute & State University
James, Noel P., Queen's University
Kaczmarek, Stephen E., Western Michigan University
Kerans, Charles, University of Texas at Austin, Jackson School of Geosciences
Khan, Mohammad Wahdat Y., Pt. Ravishankar Shukla University
Khandaker, Nazrul I., York College (CUNY)
Kirkland, Brenda L., Mississippi State University
Kuhnhenn, Gary L., Eastern Kentucky University
Lasemi, Zakaria, University of Illinois
Laury, Robert L., Southern Methodist University
Liutkus, Cynthia, Appalachian State University
Lowe, Donald R., Stanford University
Lumsden, David N., University of Memphis
Martin, Arturo, Centro de Investigación Científica y de Educación Superior de Ensenada
Martire, Luca, Università di Torino
McBride, Earle F., University of Texas at Austin
McKean , Rebecca, Saint Norbert College
McKnight, Brian K., University of Wisconsin, Oshkosh
Measures, Elizabeth A., Sul Ross State University
Melim, Leslie A., Western Illinois University
Milliken, Kitty L., University of Texas at Austin
Montañez, Isabel P., University of California, Davis
Moore, Bruce R., University of Kentucky
Moshier, Stephen O., Wheaton College
Muntean, Thomas, Adrian College
Niem, Alan R., Oregon State University
Nothdurft, Luke, Queensland University of Tech
Ojakangas, Richard W., University of Minnesota, Duluth
Ortega-Ariza, Diana, Columbus State University
Pedone, Vicki A., California State University, Northridge
Perkins, Ronald D., Duke University
Picard, M. Dane, University of Utah
Portner, Ryan, San Jose State University
Randazzo, Anthony F., University of Florida
Richardson, Darlene S., Indiana University of Pennsylvania
Ruppel, Stephen C., University of Texas at Austin, Jackson School of Geosciences
Saja, David B., Cleveland Museum of Natural History
Sanchez Roman, Monica, VU University Amsterdam
Sandberg, Philip A., University of Illinois, Urbana-Champaign
Savrda, Charles E., Auburn University
Schreiber, B. Charlotte, Queens College (CUNY)
Schrieber, B. Charlotte, Graduate School of the City University of New York
Schwab, Frederick, Washington & Lee University
Siegrist, jr, Henry G., University of Guam
Stefani, Cristina, Università degli Studi di Padova
Stoudt, Emily L., University of Texas, Permian Basin
Sutterlin, Peter G., Wichita State University
Suttner, Lee J., Indiana University, Bloomington
Sutton, Sally J., Colorado State University
Textoris, Daniel A., University of North Carolina, Chapel Hill
Thayer, Paul A., University of North Carolina Wilmington
Uddin, Ashraf, Auburn University

Vice, Mari A., University of Wisconsin, Platteville
Videtich, Patricia E., Grand Valley State University
Walker, Theodore R., University of Colorado
Williams, Wayne K., University of Tennessee at Chattanooga
Young, Harvey R., Brandon University
Zeigler, E. Lynn, Georgia State University, Perimeter College, Clarkston Campus

## Sedimentology

Abbott, Mark B., University of Pittsburgh
Abbott, Patrick L., San Diego State University
Abdullatif, Osman, King Fahd University of Petroleum and Minerals
Adams, Robert W., SUNY, The College at Brockport
Alexander, Jane L., College of Staten Island
Algeo, Thomas J., University of Cincinnati
Allen, Mark, Durham University
Allen, Philip, Imperial College
Allison, Mead A., Tulane University
Alvarez-Arellano, Alejandro, Universidad Autonoma de Baja California Sur
Ambrose, William A., University of Texas at Austin, Jackson School of Geosciences
Amos, Carl L., University of Southampton
Anani, Chris Y., University of Ghana
Andersen, David W., San Jose State University
Anderson, Thomas B., Sonoma State University
Anderson-Folnagy, Heidi, University of Montana Western
Anfinson, Owen, Sonoma State University
Anglen, Jeff, California State University, Fresno
Arnott, R. William C., University of Ottawa
Ashley, Gail M., Rutgers, The State University of New Jersey
Asiedu, Daniel K., University of Ghana
Askari, Zohreh, University of Illinois, Urbana-Champaign
Austin, Steve A., Cedarville University
Autin, Whitney J., SUNY, The College at Brockport
Bahlburg, Heinrich, Universitaet Muenster
Baldwin, Christopher T., Sam Houston State University
Balog-Szabo, Anna, Virginia Western Community College
Barbeau, Jr., David, University of South Carolina
Barber, Donald C., Bryn Mawr College
Barhurst, James, University of Newcastle Upon Tyne
Bart, Henry A., La Salle University
Bartley, Julie K., Gustavus Adolphus College
Bassett, Kari N., University of Canterbury
Beck, Catherine C., Hamilton College
Behl, Richard J., California State University, Long Beach
Bekker, Andrey, University of California, Riverside
Belt, Edward S., Amherst College
Berger, Wolfgang H., University of California, San Diego
Bergmann, Kristin, Massachusetts Institute of Tech
Bershaw, John T., Portland State University
Bertrand, Sebastien, Ghent University
Best, James L., University of Illinois, Urbana-Champaign
Bestland, Erick, Flinders University
Bhattacharya, Janok, University of Houston
Birgenheier, Lauren, University of Utah
Bjornstad, Bruce N., Pacific Northwest National Laboratory
Blake, Bascombe (Mitch), West Virginia University
Blodgett, Robert H., Austin Community College District
Blum, Michael D., University of Kansas
Boothroyd, Jon C., University of Rhode Island
Bordy, Emese M., University of Cape Town
Bourgeois, Joanne (Jody), University of Washington
Bowen, Brenda , University of Utah
Brand, Brittany, Boise State University
Brandon, Christine M., Bridgewater State University
Breda, Anna, Università degli Studi di Padova
Breyer, John, Texas Christian University
Bristow, Charlie, Birkbeck College
Broda, James E., Woods Hole Oceanographic Institution
Brunt, Rufus, University of Manchester
Burns, Diane M., Eastern Illinois University
Busby, Cathy, University of California, Santa Barbara
Buyce, M. Raymond, Mercyhurst University
Buynevich, Ilya, Temple University
Caputo, Mario V., San Diego State University
Carrapa, Barbara, University of Arizona
Carroll, Alan R., University of Wisconsin, Madison
Cartwright, Joe, University of Oxford
Caruthers, Andrew, Western Michigan University
Cassel, Elizabeth J., University of Idaho
Castle, James W., Clemson University
Cather, Steven M., New Mexico Institute of Mining & Tech
Catuneanu, Octavian, University of Alberta
Celik, Hasan, Firat University
Chan, Marjorie A., University of Utah
Cheel, Richard J., Brock University
Chiarella, Domenico, University of London, Royal Holloway & Bedford New College
Chow, Nancy, University of Manitoba
Chowns, Timothy M., University of West Georgia
Christie-Blick, Nicholas, Columbia University
Clift, Peter, Louisiana State University
Coch, Nicholas K., Graduate School of the City University of New York
Coch, Nicholas K., Queens College (CUNY)
Coe, Angela L., The Open University
Colburn, Ivan P., California State University, Los Angeles
Cole, Rex D., Colorado Mesa University
Collins, Shawn, Clarion University
Connely, Melissa, Casper College
Conrad, Susan H., Dutchess Community College
Cope, Tim D., DePauw University
Corsetti, Frank A., University of Southern California
Cowan, Clinton A., Carleton College
Cox, Ronadh, Williams College
Crowley, Stephen, University of Liverpool
Cullen, James L., Salem State University
Currie, Brian S., Miami University
Dade, William B., Dartmouth College
Dalrymple, Robert W., Queen's University
Damuth, John, University of Texas, Arlington
Dansereau, Pauline, Universite Laval
Davatzes, Alexandra, Temple University
Davies, Neil, University of Cambridge
Davies, Sarah, Leicester University
Davies-Vollum, Katherine Sian, University of Washington
Davis, Jr., Richard A., University of South Florida, Tampa
de Wet, Carol B., Franklin and Marshall College
Dehler, Carol M., Utah State University
Deibert, Jack, Austin Peay State University
Demicco, Robert V., Binghamton University
Deocampo, Daniel M., Georgia State University
Desrochers, Andre, University of Ottawa
Diemer, John A., University of North Carolina, Charlotte
Dix, George R., Carleton University
Dominic, David F., Wright State University
Donald, Roberta, Simon Fraser University
Donaldson, Alan C., West Virginia University
Donaldson, J. Allan, Carleton University
Dott, Jr., Robert H., University of Wisconsin, Madison
Dravis, Jeffrey J., Rice University
Droste, John B., Indiana University, Bloomington
Droxler, Andre W., Rice University
Drummond, Carl N., Indiana University / Purdue University, Fort Wayne
Dumas, Simone, University of Ottawa
Dunagan, Stan P., University of Tennessee, Martin
Dunn, Richard K., Norwich University
Dutta, Prodip K., Indiana State University
Dutton, Shirley P., University of Texas at Austin
Dwyer, Gary S., Duke University
Eberth, David A., Royal Tyrrell Museum of Palaeontology
Edmonds, Douglas A., Indiana University, Bloomington
Egenhoff, Sven O., Colorado State University
Ehrlich, Robert, University of South Carolina
Elliott, Jr., William S., University of Southern Indiana
Elrick, Maya, University of New Mexico
Enos, Paul, University of Kansas
Eriksson, Kenneth A., Virginia Polytechnic Institute & State University
Ethridge, Frank G., Colorado State University
Evans, James E., Bowling Green State University
Eyles, Carolyn H., McMaster University
Eyre, Bradley, Southern Cross University
Faas, Richard W., Lafayette College
Fagel, Nathalie, Universite de Liege
Fairchild, Ian, University of Birmingham

Fan, Majie, University of Texas, Arlington
Fantozzi, Joanna, Orange Coast College
Farkas, Steven, Central Washington University
Farrell, Kathleen M., North Carolina Geological Survey
Fedo, Christopher, University of Tennessee, Knoxville
Fielding, Christopher, Queensland University of Tech
Fielding, Christopher R., University of Nebraska, Lincoln
Fillmore, Robert P., Western State Colorado University
Finkenbinder, Matthew S., Wilkes University
Finzel, Emily, University of Iowa
Fischer, Alfred G., University of Southern California
Fisher, William L., University of Texas at Austin
Flaig, Peter P., University of Texas at Austin
Fletcher, Charles H., University of Hawai'i, Manoa
Foresi, Luca Maria, University of Siena
Fouke, Bruce W., University of Illinois, Urbana-Champaign
Fox, James E., South Dakota School of Mines & Tech
Fox, William T., Williams College
Fralick, Philip W., Lakehead University
Frank, Tracy D., University of Nebraska, Lincoln
Franseen, Evan K., University of Kansas
Frazier, William J., Columbus State University
Freile, Deborah, New Jersey City University
Friedman, Gerald M., Graduate School of the City University of New York
Fu, Qilong, University of Texas at Austin, Jackson School of Geosciences
Gabel, Sharon, Lee College
Gaines, Robert R., Pomona College
Gall, Quentin, University of Ottawa
Galli, Kenneth G., Boston College
Galloway, William E., University of Texas at Austin
Gani, M. Royhan, Western Kentucky University
Gardner, Michael, Montana State University
Garrison, Robert E., University of California, Santa Cruz
Garzione, Carmala N., University of Rochester
Gayes, Paul T., Dalhousie University
Gaylord, David R., Washington State University
Gentry, Amanda L., Weber State University
Gerhard, Lee C., University of Kansas
Ghinassi, Massimiliano, Università degli Studi di Padova
Gianniny, Gary, Fort Lewis College
Gibling, Martin R., Dalhousie University
Gillespie, Robb, Western Michigan University
Ginsburg, Robert N., University of Miami
Glenn, Craig R., University of Hawai'i, Manoa
Glumac, Bosiljka, Smith College
Goff, James, University of New South Wales
Goldstein, Robert H., University of Kansas
Goodbred, Jr., Steven L., Vanderbilt University
Gore, Pamela J. W., Emory University
Gore, Pamela J. W., Georgia State University, Perimeter College, Clarkston Campus
Gorte, Mary C., Delta College
Gotz, Annette, Rhodes University
Grammer, Michael, Oklahoma State University
Greb, Stephen F., University of Kentucky
Greene, Todd, California State University, Chico
Greer, Lisa, Washington & Lee University
Gregg, Jay M., Oklahoma State University
Gregor, C. B., Wright State University
Grimm, Kurt A., University of British Columbia
Grippo, Alessandro, Santa Monica College
Grotzinger, John P., California Institute of Tech
Grove, Karen, San Francisco State University
Guo, Junhua "., California State University, Bakersfield
Gupta, Sanjeev, Imperial College
Hajek, Elizabeth, Pennsylvania State University, University Park
Hall, Mary Jo, Rider University
Halverson, Galen, McGill University
Hamilton, Sally, Heriot-Watt University
Hampton, Brian A., New Mexico State University, Las Cruces
Harris, Clay D., Middle Tennessee State University
Harris, Mark T., University of Wisconsin, Milwaukee
Harris, Nicholas, University of Alberta
Harrison, III, William B., Western Michigan University
Hasselbo, Stephen, Exeter University
Hayes, Miles O., University of New Orleans
Haywick, Douglas W., University of South Alabama
Hazel, Jr., Joseph E., Northern Arizona University
Heinrich, Paul V., Louisiana State University
Hendrix, Marc S., University of Montana
Hernandez-Molina, Francisco J., University of London, Royal Holloway & Bedford New College
Hess-Tanguay, Lillian, Long Island University, C.W. Post Campus
Hesse, Reinhard, McGill University
Heubeck, Christoph, Friedrich-Schiller-University Jena
Hiatt, Eric E., University of Wisconsin, Oshkosh
Hickson, Thomas A., University of Saint Thomas
Higgins, Sean M., Columbia University
Hilbert-Wolf, Hannah L., James Cook University
Hill, Paul S., Dalhousie University
Hill, Philip R., University of Victoria
Hinderer, Matthias, Technische Universitaet Darmstadt
Hine, Albert C., University of South Florida
Hiscott, Richard N., Memorial University of Newfoundland
Hodgetts, David, University of Manchester
Hoey, Trevor, University of Glasgow
Holbrook, John M., Texas Christian University
Hollis, Cathy, University of Manchester
Holmes, Ann E., University of Tennessee at Chattanooga
Holmes, Mary Anne, University of Nebraska, Lincoln
Hopkins, John C., University of Calgary
Horton, Brian, University of Texas at Austin
Hovorka, Susan D., University of Texas at Austin, Jackson School of Geosciences
Hovorka, Susan D., University of Texas at Austin
Howard, Jeffrey L., Wayne State University
Hubbard, Dennis K., Oberlin College
Hubert, John, University of Massachusetts, Amherst
Huntoon, Jacqueline E., Michigan Technological University
Hurtgen, Matthew T., Northwestern University
Husinec, Antun, St. Lawrence University
Hussin, Azhar Hj, University of Malaya
Hutsky, Andrew J., George Mason University
Ielpi, Alessandro, Laurentian University, Sudbury
Ingersoll, Raymond V., University of California, Los Angeles
Isbell, John L., University of Wisconsin, Milwaukee
Jackson, Chester W., Georgia Southern University
Jaeger, John M., University of Florida
Janson, Xavier, University of Texas at Austin
Jiang, Ganqing, University of Nevada, Las Vegas
Jinnah, Zubair A., University of the Witwatersrand
Joeckel, Matt, Unversity of Nebraska - Lincoln
John, Cedric, Imperial College
Johnson, Cari, University of Utah
Johnson, Joel P., University of Texas at Austin
Jones, David S., Amherst College
Jones, Lawrence S., Colorado Mesa University
Jones, Merren, University of Manchester
Jones, Stuart, Durham University
Jordan, William M., Millersville University
Kah, Linda, University of Tennessee, Knoxville
Kamola, Diane, University of Kansas
Kanfoush, Sharon L., Utica College
Kendall, Christopher G., University of South Carolina
Kent, Donald M., University of Regina
Kerans, Charles, University of Texas at Austin
Kerr, Dennis R., The University of Tulsa
Kidder, David L., Ohio University
Kidwell, Susan M., University of Chicago
Kilibarda, Zoran, Indiana University Northwest
Kim, Wonsuck, University of Texas at Austin
Kinvig, Helen, University of Liverpool
Knapp, Sibylle, Technische Universitaet Muenchen
Koc Tasgin, Calibe, Firat University
Kocurek, Gary A., University of Texas at Austin
Kocurko, John, Midwestern State University
Kokelaar, Peter, University of Liverpool
Kopaska-Merkel, David C., Geological Survey of Alabama
Kotha, Mahender, Goa University
Kováč, Michal, Comenius University in Bratislava
Kovaèiæ, Marijan, University of Zagreb
Kraft, John C., University of Delaware
Krantz, David E., University of Toledo

Kraus, Mary J., University of Colorado
Krause, Federico F., University of Calgary
Krissek, Lawrence A., Ohio State University
Kurtanjek, Dražen, University of Zagreb
Langford, Richard P., University of Texas, El Paso
Last, Fawn M., Angelo State University
Last, William M., University of Manitoba
Laubach, Stephen E., University of Texas at Austin
Lavoie, Denis, Universite du Quebec
Leckie, Dale A., Carleton University
LeFever, Richard D., University of North Dakota
Lehman, Thomas M., Texas Tech University
Leier, Andrew L., University of South Carolina
Leithold, Elana L., North Carolina State University
Leorri, Eduardo, East Carolina University
Levin, Naomi, Johns Hopkins University
Licht, Alexis, University of Washington
Liddell, W. David, Utah State University
Lierman, Robert T., Eastern Kentucky University
Link, Paul K., Idaho State University
Linn, Anne M., National Academy of Sciences/National Research Council
Linol, Bastien, Nelson Mandela Metropolitan University
Little, William W., Brigham Young University - Idaho
Lloyd, Graeme, University of Oxford
Lock, Brian E., University of Louisiana at Lafayette
Lohrengel, II, C. Frederick, Southern Utah University
Long, Darrel, Laurentian University, Sudbury
Loope, David B., University of Nebraska, Lincoln
Loucks, Robert G., University of Texas at Austin
Ludvigson, Gregory A., University of Kansas
Lynds, Ranie, University of Wyoming
Mac Eachern, James A., Simon Fraser University
MacCarthy, Ivor, University College Cork
Macdonald, R. Heather, College of William & Mary
MacInnes, Breanyn, Central Washington University
Macintyre, Ian G., Smithsonian Institution / National Museum of Natural History
MacLachlan, John, McMaster University
Mahoney, J. Brian, University of Wisconsin, Eau Claire
Maju-Oyovwikowhe, Efetobore G., University of Benin
Malpica-Cruz, Victor M., Universidad Nacional Autonoma de Mexico
Manker, John P., Georgia Southwestern State University
Mankiewicz, Carol, Beloit College
Manos, Constantine, SUNY, New Paltz
Marjanac, Tihomir, University of Zagreb
Marsaglia, Kathleen M., California State University, Northridge
Martini, I. Peter, University of Guelph
Martino, Ronald L., Marshall University
Marzolf, John E., Southern Illinois University Carbondale
Matsuda, Hiroki, Kumamoto University
Mattheus, Robin, Lake Superior State University
McCaffrey, Bill, University of Leeds
McCarthy, Paul, University of Alaska, Fairbanks
McCrone, Alistair W., Humboldt State University
McDowell, Ronald, West Virginia University
McElroy, Brandon, University of Wyoming
McGlue, Michael M., University of Kentucky
McHargue, Timothy, University of Missouri
McNeill, Donald F., University of Miami
Mehrtens, Charlotte J., University of Vermont
Melas, Faye, Hunter College (CUNY)
Merrill, Robert D., California State University, Fresno
Meyers, Jamie A., Winona State University
Miall, Andrew D., University of Toronto
Miclaus, Crina, Al. I. Cuza University of Iasi
Middleton, Larry T., Northern Arizona University
Milligan, Timothy, Dalhousie University
Minter, Nicholas, University of Portsmouth
Minzoni, Rebecca T., University of Alabama
Mitchell, Simon F., University of the West Indies Mona Campus
Mohrig, David, University of Texas at Austin
Morton, Robert, University of South Florida, Tampa
Moskalski, Susanne, Stockton University
Mount, Jeffrey F., University of California, Davis
Mountney, Nigel, University of Leeds
Mozley, Peter S., New Mexico Institute of Mining and Tech
Mrinjek, Ervin, University of Zagreb
Mustard, Peter S., Simon Fraser University
Myers, Jeffrey A., Western Oregon University
Myrow, Paul M., Colorado College
Nadon, Gregory C., Ohio University
Nalin, Ronald, Loma Linda University
Neal, William J., Grand Valley State University
Nehyba, Slavomír, Masaryk University
Nelson, Bruce W., University of Virginia
Neuweiler, Fritz, Universite Laval
Nick, Kevin, Loma Linda University
Nicoll, Kathleen, University of Utah
Noffke, Nora K., Old Dominion University
O'Brien, John M., New Jersey City University
O'Brien, Neal R., SUNY Potsdam
Oakley, Bryan A., University of Rhode Island
Oaks, Jr., Robert Q., Utah State University
OConnell, Suzanne B., Wesleyan University
Okunuwadje, Sunday E., University of Benin
Oldershaw, Alan E., University of Calgary
Oliveira, Anabela R., Universidade de Trás-os-Montes e Alto Douro
Ortiz, Joseph D., Kent State University
Paola, Christopher, University of Minnesota, Twin Cities
Pattison, Simon A., Brandon University
Peakall, Jeff, University of Leeds
Pearson, Eugene F., University of the Pacific
Peck, John A., University of Akron
Perscio, Lyman, Mercyhurst University
Peters, Shanan, University of Wisconsin, Madison
Petersen, Morris S., Brigham Young University
Phillips, Andrew C., University of Illinois, Urbana-Champaign
Pickering, Kevin, University College London
Pietras, Jeff, Binghamton University
Pikelj, Kristina, University of Zagreb
Piper, David, Dalhousie University
Platt, Brian F., University of Mississippi
Plink-Bjorklund, Piret, Colorado School of Mines
Plint, A G., Western University
Potter, Paul E., University of Cincinnati
Potter-McIntyre, Sally, Southern Illinois University Carbondale
Powell, Ross D., Northern Illinois University
Pratson, Lincoln F., Duke University
Preto, Nereo, Università degli Studi di Padova
Price, Greogry, University of Plymouth
Price, Jason R., Millersville University
Prins, Maarten A., VU University Amsterdam
Pufahl, Peir K., Acadia University
Putnam, Peter E., University of Calgary
Qing, Hairuo, University of Regina
Rainbird, R. R., Carleton University
Ramirez, Pedro C., California State University, Los Angeles
Ramsey, Kelvin W., University of Delaware
Rankey, Gene, University of Kansas
Read, J. Fred, Virginia Polytechnic Institute & State University
Reams, Max W., Olivet Nazarene University
Redfern, Jonathan, University of Manchester
Redmond, Brian T., Wilkes University
Rees, Margaret N., University of Nevada, Las Vegas
Reid, Ruth P., University of Miami
Reid, Steven K., Morehead State University
Remacha Grau, Eduard, Universitat Autonoma de Barcelona
Renaut, Robin W., University of Saskatchewan
Ridgway, Kenneth D., Purdue University
Riding, Robert, University of Tennessee, Knoxville
Rigo, Manuel, Università degli Studi di Padova
Rigsby, Catherine A., East Carolina University
Roberts, Eric M., James Cook University
Roberts, Harry H., Louisiana State University
Robinson, Ruth, University of St. Andrews
Rogers, Raymond R., Macalester College
Runkel, Anthony, University of Minnesota, Twin Cities
Ruppel, Stephen C., University of Texas at Austin
Rygel, Michael C., SUNY Potsdam
Saltzman, Matthew R., Ohio State University
Sayegh, Selim, University of Calgary
Saylor, Beverly Z., Case Western Reserve University
Schieber, Juergen, Indiana University, Bloomington
Schneiderman, Jill S., Vassar College

Scholz, Christopher A., Syracuse University
Schreiber, Charlotte, University of Washington
Schwartz, Theresa M., Allegheny College
Schwimmer, Reed A., Rider University
Scott, Robert W., The University of Tulsa
Self, Robert P., University of Tennessee, Martin
Selleck, Bruce, Colgate University
Setterholm, Dale, University of Minnesota
Shane, Tyrrell, National University of Ireland Galway
Shaw, John B., University of Arkansas, Fayetteville
Shea, James H., University of Wisconsin, Parkside
Shilts, William W., University of Illinois, Urbana-Champaign
Simms, Alexander, University of California, Santa Barbara
Simonson, Bruce M., Oberlin College
Simpson, Edward L., Kutztown University of Pennsylvania
Simpson, Frank, University of Windsor
Singer, Jill K., Buffalo State College
Singler, Charles R., Youngstown State University
Sirocko, Frank, Universität Mainz
Slingerland, Rudy L., Pennsylvania State University, University Park
Sloss, Craig, Queensland University of Tech
Smith, Gerald J., SUNY, Buffalo
Smith, Jon, University of Kansas
Smith, Larry N., Montana Tech of the University of Montana
Smith, Norman D., University of Nebraska, Lincoln
Smith, Shane V., Youngstown State University
Smith-Engle, Jennifer M., Texas A&M University, Corpus Christi
Soreghan, Gerilyn S., University of Oklahoma
Soreghan, Michael J., University of Oklahoma
Soster, Frederick M., DePauw University
Southard, John B., Massachusetts Institute of Tech
Staub, James R., University of Montana
Steel, Ronald J., University of Texas at Austin
Steenberg, Julia, University of Minnesota
Steinen, Randolph P., Dept of Energy and Environmental Protection
Stoner, Joseph, Oregon State University
Strauss, Justin V., Dartmouth College
Sumner, Dawn Y., University of California, Davis
Sumner, Esther, University of Southampton
Surpless, Kathleen D., Trinity University
Sweet, Alisan C., Texas Tech University
Sweet, Dustin E., Texas Tech University
Swennen, Rudy, Katholieke Universiteit Leuven
Swett, Keene, University of Iowa
Syvitski, Jai P., University of Colorado
Tedesco, Lenore P., Indiana University / Purdue University, Indianapolis
Teller, James T., University of Manitoba
Thomas, Robert C., University of Montana Western
Thompson, Todd A., Indiana University
Thurow, Juergen, University College London
Totten, Matthew W., Kansas State University
Toullec, Renaud, Institut Polytechnique LaSalle Beauvais (ex-IGAL)
Trevino, Ramon H., University of Texas at Austin, Jackson School of Geosciences
Trexler, James H., University of Nevada, Reno
Triplehorn, Donald M., University of Alaska, Fairbanks
Trofimovs, Jessica, Queensland University of Tech
Trop, Jeffrey M., Bucknell University
Törnqvist, Torbjörn E., Tulane University
Ulmer-Scholle, Dana S., New Mexico Institute of Mining and Tech
Underwood, Michael B., University of Missouri
Van De Poll, Henk W., University of New Brunswick
van der Lubbe, Jeroen, VU University Amsterdam
Villegas, Monica B., Universidad Nacional de Rio Cuarto
Voice, Peter J., Western Michigan University
Wagner, John B., Southern Methodist University
Wagoner, Jeffrey L., Lawrence Livermore National Laboratory
Walker, Richard, Leicester University
Walsh, J.P., East Carolina University
Walsh, Tim R., Wayland Baptist University
Walton, Anthony W., University of Kansas
Wang, Ping, University of South Florida, Tampa
Wanke, Ansgar, University of Namibia
Wanless, Harold R., University of Miami
Ward, Wesley, Northern Arizona University
Wartes, Marwan A., Alaska Division of Geological & Geophysical Surveys
Washburn, Robert H., Juniata College
Watkinson, Matthew, University of Plymouth
Weislogel, Amy L., West Virginia University
Wellner, Julia, University of Houston
Wells, Neil A., Kent State University
Werner, Christopher L., McMaster University
Whalen, Michael T., University of Alaska, Fairbanks
Whisonant, Robert C., Radford University
Whiteaker, Timothy, University of Texas at Austin
Wilkinson, Bruce, Syracuse University
Wilkinson, Bruce H., University of Michigan
Wilson, Carol A., Louisiana State University
Wilson, Paul A., University of Southampton
Wittkop, Chad, Minnesota State University
Wojewoda, Jurand, University of Wroclaw
Wolak, Jeannette, Tennessee Tech University
Wood, Lesli, Colorado School of Mines
Woodruff, Jonathan D., University of Massachusetts, Amherst
Woods, Adam D., Santiago Canyon College
Woods, Adam D., California State University, Fullerton
Woods, Rachel A., Edinburgh University
Worden, Richard, University of Liverpool
Wright, V P., Cardiff University
Wynn, Thomas C., Lock Haven University
Yang, Wan, Wichita State University
Yeager, Kevin M., University of Kentucky
Young, Robert S., Western Carolina University
Zaleha, Michael J., Wittenberg University
Zambito, James J., University of Wisconsin - Extension
Zattin, Massimiliano, Università degli Studi di Padova
Zeng, Hongliu, University of Texas at Austin, Jackson School of Geosciences

## Physical Stratigraphy

Aalto, Kenneth R., Humboldt State University
Abdullah, Nuraiteng Tee, University of Malaya
Aitchison, Jonathan, University of Sydney
Alvarez, Walter, University of California, Berkeley
Anderson, Raymond R., University of Iowa
Archer, Allen W., Kansas State University
Aschoff, Jennifer, University of Alaska, Anchorage
Atchley, Stacy C., Baylor University
Baird, Gordon C., SUNY Fredonia
Balgord, Elizabeth A., Weber State University
Bart, Philip J., Louisiana State University
Bartholomew, Alexander J., SUNY, New Paltz
Benison, Kathleen , West Virginia University
Bhattacharya, Janok, McMaster University
Blakeman, Audrey, Ohio Dept of Natural Resources
Blakey, Ronald C., Northern Arizona University
Boone, Peter A., Austin Community College District
Bowen, David W., Montana State University
Brezinski, David K., Maryland Department of Natural Resources
Brooks, James E., Southern Methodist University
Catacosinos, Paul A., Delta College
Chipping, David H., California Polytechnic State University
Clark, Joseph C., Indiana University of Pennsylvania
Coniglio, Mario, University of Waterloo
Cramer, Bradley D., University of Iowa
d'Atri, Anna, Università di Torino
Dalton, Richard F., New Jersey Geological and Water Survey
Dattilo, Benjamin F., Indiana University / Purdue University, Fort Wayne
Denny, F. Brett, Illinois State Geological Survey
Devera, Joseph A., Illinois State Geological Survey
Diecchio, Richard J., George Mason University
Diffendal, Jr., Robert F., Unversity of Nebraska - Lincoln
Doar, III, William R., Dept of Natural Resources
Dodge, Clifford H., Pennsylvania Bureau of Topographic & Geologic Survey
Dorsey, Rebecca J., University of Oregon
Drzewiecki, Peter A., Eastern Connecticut State University
Ebert, James R., SUNY, Oneonta
Elliott, Monty A., Southern Oregon University
Evanoff, Emmett, University of Northern Colorado
Evans, Kevin R., Missouri State University
Farquhar, Winifred C., College of the Holy Cross
Feibel, Craig S., Rutgers, The State University of New Jersey
Ferns, Mark L., Oregon Dept of Geology & Mineral Industries

Ferrusquia, Ismael, Universidad Nacional Autonoma de Mexico
Fichter, Lynn S., James Madison University
Filipescu, Sorin, Babes-Bolyai University
Fisher, Victor A., California State University, Chico
Flemings, Peter B., University of Texas at Austin
Flint, Stephen, University of Manchester
Foreman, Brady Z., Western Washington University
Freeman, Rebecca, University of Kentucky
Fritz, William J., College of Staten Island
Fugitt, Frank, Ohio Dept of Natural Resources
Gardulski, Anne F., Tufts University
Gentile, Richard J., University of Missouri-Kansas City
Gerhard, Lee C., University of Kansas
Gonzalez Leon, Carlos M., Universidad Nacional Autonoma de Mexico
Hamlin, H, Scott, University of Texas at Austin, Jackson School of Geosciences
Harris, W. Burleigh, University of North Carolina Wilmington
Harvey, Cyril H., Guilford College
Hasenmueller, Walter, Indiana University
Hattin, Donald E., Indiana University, Bloomington
Heckel, Philip H., University of Iowa
Heermance, Richard V., California State University, Northridge
Helenes, Javier, Centro de Investigación Científica y de Educación Superior de Ensenada
Heller, Paul L., University of Wyoming
Hentz, Tucker F., University of Texas at Austin, Jackson School of Geosciences
Heron, Duncan, Duke University
Hinnov, Linda, George Mason University
Hippensteel, Scott P., University of North Carolina, Charlotte
Holland, Steven M., University of Georgia
Hyland, Ethan, North Carolina State University
Jacques Ayala, Cesar, Universidad Nacional Autonoma de Mexico
Jansa, Lubomir F., Dalhousie University
Janson, Xavier, University of Texas at Austin, Jackson School of Geosciences
Jenkyns, Hugh C., University of Oxford
Joeckel, Robert M., University of Nebraska, Lincoln
Johnson, Donald O., Argonne National Laboratory
Johnson, Gary D., Dartmouth College
Johnson, William C., University of Kansas
Jordan, Robert R., University of Delaware
Jordan, Teresa E., Cornell University
Kaaya, Charles Z., University of Dar es Salaam
Kehler, Philip L., University of Arkansas at Little Rock
King, Norman R., University of Southern Indiana
King, Jr., David T., Auburn University
Kulp, Mark A., University of New Orleans
Lang, Harold R., Jet Propulsion Laboratory
Langenheim, Jr., Ralph L., University of Illinois, Urbana-Champaign
Lash, Gary G., SUNY Fredonia
Lee, Richard R., Oak Ridge National Laboratory
Lincoln, Jonathan M., Montclair State University
Loydell, David, University of Portsmouth
Lozinsky, Richard P., Fullerton College
Ludvigson, Greg A., University of Kansas
Lundblad, Steven P., University of Hawai'i, Hilo
Mancini, Ernest A., University of Alabama
Manger, Walter, University of Arkansas, Fayetteville
Manspeizer, Warren, Rutgers, The State University of New Jersey, Newark
Martinez-Gutierrez, Genaro, Universidad Autonoma de Baja California Sur
Matson, Sam, Boise State University
Maytorena-Silva, Jesus F., Centro de Estudios Superiores del Estado Sonora
McClaughry, Jason D., Oregon Dept of Geology & Mineral Industries
McDonald, Catherine, Montana Tech of The University of Montana
McKee, James W., University of Wisconsin, Oshkosh
McLaughlin, Patrick I., Indiana University
McLaughlin, Patrick I., University of Wisconsin, Extension
McLaughlin, Jr., Peter P., University of Delaware
McLaurin, Brett T., Bloomsburg University
Meckel, Timothy A., University of Texas at Austin
Merriam, Daniel F., Wichita State University
Metz, Robert, Kean University
Monreal-Saavedra, Rogelio, Centro de Estudios Superiores del Estado Sonora
Monteverde, Donald H., New Jersey Geological and Water Survey
Morris, Thomas H., Brigham Young University
Nance, Hardie S., University of Texas at Austin
Neal, Donald W., East Carolina University
Nicolas, Michelle P., Manitoba Geological Survey
Nielson, R. LaRell, Stephen F. Austin State University
Norford, Brian, University of Calgary
Olariu, Cornel, University of Texas at Austin
Olariu, Mariana, University of Texas at Austin
Oms Llobet, Oriol, Universitat Autonoma de Barcelona
Osleger, David A., University of California, Davis
Owen, Donald E., Lamar University
Page, Kevin, University of Plymouth
Paine, Jeffrey G., University of Texas at Austin, Jackson School of Geosciences
Parcell, William C., Wichita State University
Pemberton, S. George, University of Alberta
Peterson, Gary L., San Diego State University
Peterson, Larry C., University of Miami
Pieren, Agustin P., Univ Complutense de Madrid
Poch Serra, Joan, Universitat Autonoma de Barcelona
Pope, Michael , Texas A&M University
Retzler, Andrew J., University of Minnesota
Reynolds, James H., Brevard College
Roman, Aubrecht, Comenius University in Bratislava
Sadler, Peter M., University of California, Riverside
Sanders, Ronald S., Jet Propulsion Laboratory
Schmitt, James G., Montana State University
Scott, Thomas M., University of South Florida, Tampa
Simmons, Lizanne V., Santiago Canyon College
Smosna, Richard A., West Virginia University
Snyder, Walter S., Boise State University
Spreng, Alfred C., Missouri University of Science and Tech
Stieglitz, Ronald D., University of Wisconsin, Green Bay
Sugarman, Peter J., New Jersey Geological and Water Survey
Sugarman, Peter P., Rutgers, The State University of New Jersey
Suneson, Neil , University of Oklahoma
Swenson, John B., University of Minnesota, Duluth
Terry, Dennis O., Temple University
Tew , Berry H., Geological Survey of Alabama
Tierney, Kate, University of Iowa
Turner, Elizabeth C., Laurentian University, Sudbury
Vail, Peter R., Rice University
Valenzuela-Renteria, Manuel, Centro de Estudios Superiores del Estado Sonora
Van de Water, Peter, California State University, Fresno
Ver Straeten, Charles, New York State Geological Survey
Vondra, Carl F., Iowa State University of Science & Tech
Watney, Lynn W., University of Kansas
Watney, W. Lynn, University of Kansas
Webb, Fred, Appalachian State University
Weimer, Robert J., Colorado School of Mines
Young, Grant M., Western University
Zachry, Doy L., University of Arkansas, Fayetteville

## Structural Geology

Bodhankar, Ninad, Pt. Ravishankar Shukla University
Abalos, Benito, University of the Basque Country UPV/EHU
Abdelsalam, Mohamed, Oklahoma State University
Abolins, Mark J., Middle Tennessee State University
Adam, Jurgen, University of London, Royal Holloway & Bedford New College
Adams, John B., University of Washington
Alaniz Alvarez, Susana A., Universidad Nacional Autonoma de Mexico
Aldrich, M. James, Los Alamos National Laboratory
Aleksandrowski, Pawel, University of Wroclaw
Allard, Stephen T., Winona State University
Allen, Joseph L., Concord University
Allen, Richardson B., University of Utah
Allison, David T., University of South Alabama
Allmendinger, Richard W., Cornell University
Alsleben, Helge, Texas Christian University
Amato, Jeffrey M., New Mexico State University, Las Cruces
Amenta, Roddy V., James Madison University
Amos, Colin B., Western Washington University
Anastasio, David J., Lehigh University
Anders, Mark H., George Mason University
Anderson, James L., University of Hawai'i, Hilo

Anderson, Mark, University of Plymouth
Andrew, Joseph E., Youngstown State University
Aranguren, Aitor, University of the Basque Country UPV/EHU
Arboleya Cimadevilla, María Luísa, Universitat Autonoma de Barcelona
Armstrong, Phillip A., California State University, Fullerton
Aronoff, Ruth F., Furman University
Arrowsmith, Ramon, Arizona State University
Axen, Gary, New Mexico Institute of Mining and Tech
Aydin, Atilla, Stanford University
Babaie, Hassan A., Georgia State University
Bailey, Christopher M., College of William & Mary
Baird, Graham, University of Northern Colorado
Balestro, Gianni, Università di Torino
Bangs, Nathan L., University of Texas at Austin
Banoeng-Yakubo, Bruce K., University of Ghana
Baran, Zeynep O., South Dakota School of Mines & Tech
Barineau, Clinton I., Columbus State University
Barker, Chris A., Stephen F. Austin State University
Barnes, Charles W., Northern Arizona University
Barquero-Molina, Miriam, University of Missouri
Bartley, John M., University of Utah
Bauer, Paul W., New Mexico Institute of Mining and Tech
Bauer, Robert L., University of Missouri
Bayly, M. Brian, Rensselaer Polytechnic Institute
Behr, Rose-Anna, Pennsylvania Bureau of Topographic & Geologic Survey
Behr, Whitney, University of Texas at Austin
Bemis, Sean, University of Kentucky
Berry, Ron F., University of Tasmania
Bethune, Kathryn, University of Regina
Beutel, Erin K., College of Charleston
Bhattacharyya, Prajukti, University of Wisconsin, Whitewater
Bidgoli, Tandis, University of Kansas
Bier, Sara E., SUNY Potsdam
Bilodeau, William L., California Lutheran University
Bishop, Kim, California State University, Los Angeles
Bjorklund, Tom, University of Houston
Bjornerud, Marcia, Lawrence University
Blackmer, Gale C., Pennsylvania Bureau of Topographic & Geologic Survey
Bobyarchick, Andy R., University of North Carolina, Charlotte
Bond, Clare, University of Aberdeen
Boniface, Nelson, University of Dar es Salaam
Bosbyshell, Howell, West Chester University
Bothner, Wallace A., University of New Hampshire
Boyle, Alan P., University of Liverpool
Bradbury, Kelly K., Utah State University
Bradley, Michael, Eastern Michigan University
Brandon, Mark T., Yale University
Brennan, William J., SUNY, Geneseo
Brisbin, William C., University of Manitoba
Brodie, Kate, University of Manchester
Brogan, George, Irvine Valley College
Brown, Bruce A., University of Wisconsin - Extension
Brown, William G., Baylor University
Bruhn, Ronald L., University of Utah
Bunds, Michael, Utah Valley University
Burberry, Caroline M., University of Nebraska, Lincoln
Burchfiel, B. C., Massachusetts Institute of Tech
Burger, H. Robert, Smith College
Burks, Rachel J., Towson University
Burmeister, Kurtis C., University of the Pacific
Buttner, Steffen, Rhodes University
Bykerk-Kauffman, Ann, California State University, Chico
Byrne, Timothy, University of Connecticut
Cadoppi, Paola, Università di Torino
Calon, Tomas J., Memorial University of Newfoundland
Camilleri, Phyllis A., Austin Peay State University
Campagna, David, West Virginia University
Campbell, Patricia A., Slippery Rock University
Campbell-Stone, Erin A., University of Wyoming
Carlson, Diane H., California State University, Sacramento
Carosi, Rodolfo, Università di Torino
Cashman, Patricia H., University of Nevada, Reno
Cashman, Susan M., Humboldt State University
Caskey, John, San Francisco State University
Cather, Steven M., New Mexico Institute of Mining and Tech
Cato, Craig, Geological Survey of Alabama
Çemen, Ibrahim, University of Alabama
Chamberlin, Richard, New Mexico Institute of Mining and Tech
Chapman, Alan, Macalester College
Chester, Frederick, Texas A&M University
Chester, Judith, Texas A&M University
Clayton, Robert W., Brigham Young University - Idaho
Cloos, Mark P., University of Texas at Austin
Coleman, Jr., Paul J., University of California, Los Angeles
Colliston, Wayne, University of the Free State
Connelly, Jeffrey B., University of Arkansas at Little Rock
Coogan, James, Western State Colorado University
Cooke, Michele L., University of Massachusetts, Amherst
Coppinger, Walter, Trinity University
Cosgrove, John, Imperial College
Cowan, Darrel S., University of Washington
Cox, Randel T., University of Memphis
Craddock, John P., Macalester College
Crespi, Jean M., University of Connecticut
Crider, Juliet, University of Washington
Cronin, Vincent S., Baylor University
Crowley, Peter D., Amherst College
Cruikshank, Kenneth M., Portland State University
Cuevas, Julia, University of the Basque Country UPV/EHU
Culshaw, Nicholas, Dalhousie University
Curl, Doug C., University of Kentucky
Czeck, Dyanna M., University of Wisconsin, Milwaukee
D'Allura, Jad A., Southern Oregon University
Daigneault, Real, Universite du Quebec a Chicoutimi
Davatzes, Nicholas, Temple University
Davis, George H., University of Arizona
Davis, Gregory A., University of Southern California
Davis, Peter B., Pacific Lutheran University
Dawers, Nancye H., Tulane University
De Paola, Nicola, Durham University
Di Toro, Giulio, Università degli Studi di Padova
DiPietro, Joseph A., University of Southern Indiana
Dixon, John M., Queen's University
Dooley, Tim, University of Texas at Austin, Jackson School of Geosciences
Douglas, Bruce, Indiana University, Bloomington
Draper, Grenville, Florida International University
Druguet Tantiña, Elena, Universitat Autonoma de Barcelona
Duebendorfer, Ernest M., Northern Arizona University
Dumitru, Trevor, Stanford University
Dunne, George C., California State University, Northridge
Dunne, William M., University of Tennessee, Knoxville
Dunning, Jeremy D., Indiana University, Bloomington
Dutch, Steven I., University of Wisconsin, Green Bay
Eichhubl, Peter, University of Texas at Austin
Eichhubl, Peter, University of Texas at Austin, Jackson School of Geosciences
Elliott, Colleen G., Montana Tech of The University of Montana
Erdmer, Philippe, University of Alberta
Erslev, Eric A., University of Wyoming
Erslev, Eric A., Colorado State University
Eusden, Jr., J. Dykstra, Bates College
Evans, James P., Utah State University
Evans, Mark, Central Connecticut State University
Famin, Vincent, Universite de la Reunion
Faulds, James E., University of Nevada, Reno
Faure, Stephane, Universite du Quebec a Montreal
Fayon, Annia K., University of Minnesota, Twin Cities
Ferguson, Charles A., University of Arizona
Ferre, Eric C., Southern Illinois University Carbondale
Festa, Andrea, Università di Torino
Fischer, Mark P., Northern Illinois University
Fisher, Donald M., Pennsylvania State University, University Park
Franke, Otto L., Graduate School of the City University of New York
Fraser, Alastair, University of Utah
French, Melodie E., Rice University
Fryer, Karen H., Ohio Wesleyan University
Fryxell, Joan E., California State University, San Bernardino
Fueten, Frank, Brock University
Fusseis, Florian, Edinburgh University
Fyson, William K., University of Ottawa
Gale, Julia F., University of Texas at Austin

Gani, Nahid, Western Kentucky University
Garihan, John M., Furman University
Gates, Alexander E., Rutgers, The State University of New Jersey, Newark
Gattiglio, Marco, Università di Torino
Gibson, H. Daniel (Dan), Simon Fraser University
Gibson, Roger L., University of the Witwatersrand
Gibson, Ronald C., Golden West College
Gillespie, Thomas, College of New Jersey
Gilotti, Jane A., University of Iowa
Giorgis, Scott D., SUNY, Geneseo
Girty, Gary H., San Diego State University
Gleason, Gayle C., SUNY, Cortland
Godin, Laurent, Queen's University
Gold, David (Duff) P., Pennsylvania State University, University Park
Goodwin, Laurel B., University of Wisconsin, Madison
Goulet, Normand, Universite du Quebec a Montreal
Gray, Mary Beth, Bucknell University
Greene, David C., Denison University
Griera Artigas, Albert, Universitat Autonoma de Barcelona
Griffith, Ashley, University of Texas, Arlington
Groshong, Jr., Richard H., University of Alabama
Gross, Michael, Florida International University
Groszos, Mark S., Valdosta State University
Gudmundsson, Agust, University of London, Royal Holloway & Bedford New College
Guitierrez-Alonso, Gabriel, Florida International University
Gulick, Sean S., University of Texas at Austin
Guth, Lawrence R., Fitchburg State University
Guth, Peter L., United States Naval Academy
Hack, Alistair, University of Newcastle
Hacker, David B., Kent State University
Hall, Frank W., Indiana University of Pennsylvania
Hanley, Thomas B., Columbus State University
Hanmer, Simon, Carleton University
Hannula, Kimberly, Fort Lewis College
Hariri, Mustafa, King Fahd University of Petroleum and Minerals
Harrap, Rob, Queen's University
Harris, Lyal B., Universite du Quebec
Harris, Ron, Brigham Young University
Harrison, Michael J., Tennessee Tech University
Hatcher, Jr., Robert D., University of Tennessee, Knoxville
Hawkins, John F., Auburn University
Hefferan, Kevin P., University of Wisconsin, Stevens Point
Heimsath, Arjun, Arizona State University
Helmstaedt, Herwart, Queen's University
Helper, Mark A., University of Texas at Austin
Hendrix, Thomas E., Grand Valley State University
Hennings, Peter H., University of Wyoming
Hetherington, Eric D., College of the Sequoias
Hetzel, Ralf, Universitaet Muenster
Hibbard, James P., North Carolina State University
Hill, Joseph C., Sam Houston State University
Hill, Mary Louise, Lakehead University
Hinz, Nick, University of Nevada, Reno
Hirth, Greg, Brown University
Hirth, Gregory, Woods Hole Oceanographic Institution
Hodges, Kip V., Arizona State University
Hoffman, Paul F., Harvard University
Holdsworth, Robert E., Durham University
Holm, Daniel K., Kent State University
Holst, Timothy, University of Minnesota, Duluth
Holyoke, Caleb, University of Akron
Hooks, Benjamin P., University of Tennessee, Martin
Horsman, Eric, East Carolina University
Horst, Andrew J., Oberlin College
Hozik, Michael J., Stockton University
Hudec, Michael R., University of Texas at Austin
Hudleston, Peter J., University of Minnesota, Twin Cities
Hughes, Kenneth S., University of Puerto Rico
Hull, Joseph M., Seattle Central Community College
Hynes, Andrew J., McGill University
Imber, Jonathan, Durham University
Inceoz, Murat, Firat University
Ismat, Zeshan, Franklin and Marshall College
Jackson, Martin P. A., University of Texas at Austin, Jackson School of Geosciences
Jackson, Richard A., Long Island University, Brooklyn Campus
Jackson Jr, William T., Geological Survey of Alabama
Jacobi, Robert D., SUNY, Buffalo
Jacobson, Carl E., Iowa State University of Science & Tech
Jessup, Micah, University of Tennessee, Knoxville
Jiang, Dazhi, Western University
Jirsa, Mark, University of Minnesota
John, Barbara E., University of Wyoming
Johnson, Arvid M., Purdue University
Johnson, Brad, University of Arizona
Johnson, Scott E., University of Maine
Johnston, Scott, California Polytechnic State University
Johnston, Stephen T., University of Alberta
Johnston, Stephen T., University of Victoria
Jones, Gustavo Tolson, Universidad Nacional Autonoma de Mexico
Kalakay, Thomas J., Rocky Mountain College
Kanat, Leslie H., Johnson State College
Karabinos, Paul, Williams College
Kath, Randal L., University of West Georgia
Kehlenbeck, Manfred M., Lakehead University
Kelsch, Jesse, Sul Ross State University
Kelty, Thomas, California State University, Long Beach
Kennedy, Lori, University of British Columbia
Kirkpatrick, James, McGill University
Klepeis, Keith A., University of Vermont
Klimczak, Christian, University of Georgia
Kopf, Christopher F., Mansfield University
Kruckenberg, Seth C., Boston College
Krugh, W C., California State University, Bakersfield
Kuiper, Yvette D., Colorado School of Mines
Kulander, Byron, Wright State University
Kutu, Jacob M., University of Ghana
La, Daniel A., Oklahoma State University
Lafrance, Bruno, Laurentian University, Sudbury
Lageson, David R., Montana State University
Langille, Jackie M., University of North Carolina, Asheville
Lapusta, Nadia, California Institute of Tech
Lash, Gary S., SUNY, Buffalo
Laubach, Stephen E., University of Texas at Austin, Jackson School of Geosciences
Laurent-Charvet, Sébastien, Institut Polytechnique LaSalle Beauvais (ex-IGAL)
Law, Richard D., Virginia Polytechnic Institute & State University
Lee, Jeffrey, Central Washington University
Lemiszki, Peter J., Tennessee Geological Survey
Lemiszki, Peter J., Pellissippi State Community College
Lenardic, Adrian, Rice University
Lennox, Paul G., University of New South Wales
Lewis, Jon C., Indiana University of Pennsylvania
Lincoln, Beth Z., Albion College
Lisenbee, Alvis L., South Dakota School of Mines & Tech
Lisle, R J., Cardiff University
Little, Tim, Victoria University of Wellington
Livaccari, Richard F., Colorado Mesa University
Lodge, Robert, University of Wisconsin, Eau Claire
Long , Sean P., Washington State University
Loveland, Andrea M., Wyoming State Geological Survey
Lowry, Wallace D., Virginia Polytechnic Institute & State University
Luther, Amy, Louisiana State University
Ma, Chong, Auburn University
MacDonald, William D., Binghamton University
MacLean, John S., Southern Utah University
Maher, Jr., Harmon D., University of Nebraska at Omaha
Mali, Heinrich, University Leoben
Malo, Michel, Universite du Quebec
Malone, David H., Illinois State University
Mana, Sara , Salem State University
Markley, Michelle J., Mount Holyoke College
Marrett, Randall A., University of Texas at Austin
Marshak, Stephen, University of Illinois, Urbana-Champaign
Martin, Silvana, Università degli Studi di Padova
Martinez-Macias, Panfilo R., Universidad Autonoma de San Luis Potosi
Marty, Kevin G., Imperial Valley College
Mattson, Peter H., Graduate School of the City University of New York
Mattson, Peter H., Queens College (CUNY)
McCaffrey, Kenneth J., Durham University
McCair, Andrew, University of Leeds
McClay, Ken R., University of London, Royal Holloway & Bedford New

College
McGill, George E., University of Massachusetts, Amherst
McIntosh, Kirk D., University of Texas at Austin
McKay, Matthew P., Missouri State University
McNaught, Mark A., Mount Union College
McNulty, Brendan A., California State University, Dominguez Hills
McQuarrie, Nadine, University of Pittsburgh
Mecklenburgh, Julian, University of Manchester
Meere, Pat, University College Cork
Meigs, Andrew J., Oregon State University
Menegon, Luca, University of Plymouth
Merguerian, Charles M., Hofstra University
Mezger, Jochen E., University of Alaska, Fairbanks
Mies, Jonathan W., University of Tennessee at Chattanooga
Miller, Elizabeth L., Stanford University
Miller, Marli G., University of Oregon
Miller, Robert B., San Jose State University
Milligan, G. Clinton, Dalhousie University
Miranda, Elena A., California State University, Northridge
Mitchell, Tom, University College London
Mitra, Gautam, University of Rochester
Mitra, Shankar, University of Oklahoma
Moeck, Inga, Technische Universitaet Muenchen
Mookerjee, Matty, Sonoma State University
Moore, J. Casey, University of California, Santa Cruz
Morell, Kristin, University of Victoria
Morgan, Julia K., Rice University
Morgan, Sven S., Central Michigan University
Mosher, Sharon, University of Texas at Austin
Muller, Otto H., Alfred University
Murphy, Michael, University of Houston
Needham, Tim, University of Leeds
Nemcok, Michal, University of Utah
Neuhauser, Kenneth R., Fort Hays State University
Newman, Julie, Texas A&M University
Niemi, Nathan, University of Michigan
Northrup, Clyde J., Boise State University
Nourse, Jonathan A., California State Polytechnic University, Pomona
O'Hara, Kieran D., University of Kentucky
O'Meara, Stephanie, Colorado State University
Oldow, John S., University of Texas, Dallas
Onasch, Charles M., Bowling Green State University
Osagiede, Edoseghe E., University of Benin
Palmquist, John C., Lawrence University
Paterson, Scott R., University of Southern California
Paton, Douglas, University of Leeds
Patton, Regan L., Washington State University
Paulsen, Timothy S., University of Wisconsin, Oshkosh
Pavlis, Terry L., University of Texas, El Paso
Pawley, Alison, University of Manchester
Pearson, David M., Idaho State University
Pennacchioni, Giorgio, Università degli Studi di Padova
Perry, Bruce, California State University, Long Beach
Petrie, Elizabeth S., Western State Colorado University
Phipps, Stephen P., University of Pennsylvania
Pinan-Llamas, Aranzazu, Indiana University / Purdue University, Fort Wayne
Platt, John P., University of Southern California
Platt, Lucian B., Bryn Mawr College
Pogue, Kevin R., Whitman College
Pollard, David D., Stanford University
Potter, Jr., Donald B., Sewanee: University of the South
Price, Nancy A., Portland State University
Pritchard, Chad, Eastern Washington University
Puelles, Pablo, University of the Basque Country UPV/EHU
Quintero, Odranoel, Universidad Nacional Autonoma de Mexico
Raffini, Sylvain, Universite du Quebec a Chicoutimi
Ratchford, Michael , University of Idaho
Rauch, Marta, University of Wroclaw
Reber, Jacqueline, Iowa State University of Science & Tech
Ree, Jin-Han, Korea University
Reese, Joseph F., Edinboro University of Pennsylvania
Reinen, Linda A., Pomona College
Resor, Phillip G., Wesleyan University
Reynolds, Stephen J., Arizona State University
Ricketts, Jason, University of Texas, El Paso
Ritchie, Alexander W., College of Charleston
Robertson, Charles E., Missouri University of Science and Tech
Robin, Pierre-Yves F., University of Toronto
Robinson, Alexander, University of Houston
Robinson, Delores, University of Alabama
Robinson, Kevin, San Diego State University
Robinson, Peter, University of Massachusetts, Amherst
Rodgers, David W., Idaho State University
Rogers, Robert D., California State University, Stanislaus
Rojo-Garcia, Paulino, Universidad Autonoma de Baja California Sur
Rousell, Don H., Laurentian University, Sudbury
Rowe, Christie, McGill University
Roy, Denis W., Universite du Quebec a Chicoutimi
Rusmore, Margaret E., Occidental College
Saillet, Elodie, Institut Polytechnique LaSalle Beauvais (ex-IGAL)
Saja, David, Case Western Reserve University
Sak, Peter B., Dickinson College
Saleeby, Jason B., California Institute of Tech
Samaniego, Angel Nieto, Universidad Nacional Autonoma de Mexico
Sanchez, Veronica I., Texas A&M University, Kingsville
Sanislav, Ioan, James Cook University
Satterfield, Joseph I., Angelo State University
Scharer, Katherine, Appalachian State University
Scharman, Mitchell, Marshall University
Scharnberger, Charles K., Millersville University
Schlische, Roy W., Rutgers, The State University of New Jersey
Schmidt, Keegan L., Lewis-Clark State College
Schoonmaker, Adam, Utica College
Schrank, Christoph, Queensland University of Tech
Schwerdtner, Walfried M., University of Toronto
Sears, James W., University of Montana
Secor, Donald, University of South Carolina
Sen, Pragnyadipta, SUNY, Oneonta
Shaw, Colin, Montana State University
Shilpakar, Prabin ., University of Texas, Dallas
Short, Heather, McGill University
Shuib, Mustaffa Kamal, University of Malaya
Sibson, Rick H., University of Otago
Siddoway, Christine S., Colorado College
Simony, Philip S., University of Calgary
Singleton, John, Colorado State University
Skehan, James W., Boston College
Skehan, S.J., James W., Boston College
Skemer, Philip, Washington University in St. Louis
Solar, Gary S., Buffalo State College
Sorkhabi, Rasoul, University of Utah
Spencer, Edgar W., Washington & Lee University
Spratt, Deborah A., University of Calgary
SRIVASTAVA, HARI B., Banaras Hindu University
Stauffer, Mel R., University of Saskatchewan
Stearns, David W., University of Oklahoma
Stewart, Kevin G., University of North Carolina, Chapel Hill
Stimac, John P., Eastern Illinois University
Stinson, Amy L., Santiago Canyon College
Stinson, Amy L., Irvine Valley College
Stockli, Daniel, University of Texas at Austin
Stockli, Daniel F., University of Kansas
Stowell Gale, Julia, University of Texas at Austin, Jackson School of Geosciences
Strayer, Luther M., California State University, East Bay
Sullivan, Bill, Colby College
Suneson, Neil H., University of Oklahoma
Surpless, Benjamin E., Trinity University
Talbot, James L., Western Washington University
Tapp, J. B., The University of Tulsa
Tate, Garrett W., Vanderbilt University
Taylor, Lansing, University of Utah
Taylor, Wanda J., University of Nevada, Las Vegas
Tewksbury, Barbara J., Hamilton College
Teyssier, Christian P., University of Minnesota, Twin Cities
Thompson, Margaret D., Wellesley College
Tikoff, Basil, University of Wisconsin, Madison
Timmons, J M., New Mexico Institute of Mining and Tech
Tindall, Sarah E., Kutztown University of Pennsylvania
Titus, Sarah J., Carleton College
Tobisch, Othmar T., University of California, Santa Cruz
Toraman, Erkan, St. Lawrence University
Toro, Jaime, West Virginia University

Torvela, Taija, University of Leeds
Toy, Virginia G., University of Otago
Treloar, Peter, Kingston University
Tremblay, Alain, Universite du Quebec a Montreal
Trudgill, Bruce D., Colorado School of Mines
Trullenque, Ghislain, Institut Polytechnique LaSalle Beauvais (ex-IGAL)
Tso, Jonathan L., Radford University
Tubía, Jose M., University of the Basque Country UPV/EHU
Tull, James F., Florida State University
Tullis, Jan A., Brown University
Twiss, Robert J., University of California, Davis
Ustaszewski, Kamil, Friedrich-Schiller-University Jena
Van Arsdale, Roy B., University of Memphis
van Bever Donker, Jan M., University of the Western Cape
van der Pluijm, Ben, University of Michigan
VanDervoort, Dane S., Geological Survey of Alabama
Vaqueiro Rodriguez, Marcos, Coruna University
Vegas, Nestor, University of the Basque Country UPV/EHU
Vogl, Jim, University of Florida
Voicu, Gabriel-Constantin, Universite du Quebec a Montreal
Vojtko, Rastislav , Comenius University in Bratislava
Vollmer, Frederick W., SUNY, New Paltz
Waag, Charles J., Boise State University
Waldron, John W., University of Alberta
Walker, J. Douglas, University of Kansas
Waters-Tormey, Cheryl, Western Carolina University
Watkinson, A. John, Washington State University
Waugh, Richard A., University of Wisconsin, Platteville
Webber, Jeffrey R., Stockton University
Weber, John C., Grand Valley State University
Weil, Arlo B., Bryn Mawr College
Weldon, Ray J., University of Oregon
Wells, Michael L., University of Nevada, Las Vegas
West, David P., Middlebury College
Whalley, John, University of Portsmouth
Whisner, S. Christopher, Bloomsburg University
White, Joseph C., University of New Brunswick
Whitmeyer, Steven J., James Madison University
Wickham, John S., University of Texas, Arlington
Wilkerson, M. S., University of Illinois, Urbana-Champaign
Wilkerson, M. Scott, DePauw University
Williams, Michael L., University of Massachusetts, Amherst
Williams, Paul F., University of New Brunswick
Wilson, Terry J., Ohio State University
Winslow, Margaret S., Graduate School of the City University of New York
Winslow, Margaret A., City College (CUNY)
Withjack, Martha O., Rutgers, The State University of New Jersey
Wojtal, Steven F., Oberlin College
Woodcock, Nigel, University of Cambridge
Woodward, Lee A., University of New Mexico
Woodwell, Grant R., University of Mary Washington
Workman-Ford, Kerry, California State University, Fresno
Wright, Stephen F., University of Vermont
Yikilmaz, Burak, University of California, Davis
Yonkee, W. A., Weber State University
Yoshinobu, Aaron S., Texas Tech University
Young, Jeffrey, University of Manitoba
Zampieri, Dario, Università degli Studi di Padova
Zimmerman, Jr., Jay, Southern Illinois University Carbondale
Zuza, Andrew V., University of Nevada

## Tectonics

Abercrombie, Rachel, Boston University
Alexander, Jan, University of East Anglia
Anderson, Robert G., University of British Columbia
Andeweg, Bernd , VU University Amsterdam
Applegate, David, University of Utah
Artoni, Andrea, University of Parma
Ault, Alexis K., Utah State University
Avouac, Jean-Philippe, California Institute of Tech
Barnhart, William, University of Iowa
Barth, Nicolas, University of California, Riverside
Bartholomew, Mervin J., University of Memphis
Bell, Andrew, Edinburgh University
Bell, Rebecca, Imperial College
Bhattacharji, Somdev, Graduate School of the City University of New York
Bickle, Michael, University of Cambridge
Bieler, David B., Centenary College of Louisiana
Bird, John M., Cornell University
Blake, M. Clark, Western Washington University
Boggs, Katherine, Mount Royal University
Boland, Irene B., Winthrop University
Boutelier, David, University of Newcastle
Boyer, Robert E., University of Texas at Austin
Bradshaw, John, University of Canterbury
Brown, Sarah, California State University, Bakersfield
Bullen, Dean, University of Portsmouth
Burgette, Reed J., New Mexico State University, Las Cruces
Burgmann, Roland, University of California, Berkeley
Burke, Kevin, University of Houston
Busby, Cathy J., University of California, Davis
Camacho, Alfredo, University of Manitoba
Cao, Wenrong, University of Nevada, Reno
Carlson, Marvin P., Unversity of Nebraska - Lincoln
Carlson, Richard L., Texas A&M University
Carr, Sharon D., Carleton University
Casey, John F., University of Houston
Cawood, Peter A., Monash University
Centeno Garcia, Elena, Universidad Nacional Autonoma de Mexico
Chapin, Charles E., New Mexico Institute of Mining & Tech
Chi, Wu-Cheng, Academia Sinica
Church, William R., Western University
Cole, Ron B., Allegheny College
Contreras, Juan, Centro de Investigación Científica y de Educación Superior de Ensenada
Corrigan, David, University of Manitoba
Cowgill, Eric S., University of California, Davis
Cox, Randel T., University of Memphis
Cruden, Alexander, University of Toronto
Cunningham, William D., Eastern Connecticut State University
Dallmeyer, R. David, University of Georgia
Dalziel, Ian W. D., University of Texas at Austin
Das, Shamita, University of Oxford
De Cserna, Zoltan, Universidad Nacional Autonoma de Mexico
Derakhshani, Reza, Shahid Bahonar University of Kerman
Dewey, John, University of California, Davis
Dickinson, William R., University of Arizona
Dilek, Yildirim, Miami University
Dinter, David A., University of Utah
Dolan, James F., University of Southern California
Dong, Yunpeng, Western University
Dooley, Tim P., University of Texas at Austin
Driscoll, Peter E., Carnegie Institution of Washington
Ducea, Mihai N., University of Arizona
Dumond, Gregory, University of Arkansas, Fayetteville
Ebinger, Cynthia J., University of Rochester
Engebretson, David C., Western Washington University
England, Phillip, University of Oxford
Esteban, Jose Julian, University of the Basque Country UPV/EHU
Feigl, Kurt L., University of Wisconsin, Madison
Ferrari Pedraglio, Luca, Universidad Nacional Autonoma de Mexico
Fitzgerald, Paul G., Syracuse University
Fosdick, Julie C., Indiana University, Bloomington
Foster, David A., University of Florida
Furlong, Kevin P., Pennsylvania State University, University Park
Gable, Carl W., Los Alamos National Laboratory
Gans, Phillip B., University of California, Santa Barbara
Garver, John I., Union College
Gehrels, George E., University of Arizona
Geissman, John D., University of Texas, Dallas
Gerbi, Christopher C., University of Maine
German, Chris, Woods Hole Oceanographic Institution
Gifford, Jennifer N., University of Mississippi
Gomez, Francisco, University of Missouri
Gordon, Richard G., Rice University
Greenberg, Jeffrey K., Wheaton College
Gries, John C., Wichita State University
Grujic, Djordje, Dalhousie University
Guenthner, Willy, University of Illinois, Urbana-Champaign
Hales, T. C., University of Wales
Hall, Clarence A., University of California, Los Angeles
Hall, Robert, University of London, Royal Holloway & Bedford New College

Hamilton, Warren B., University of Wyoming
Hansen, Vicki L., University of Minnesota, Duluth
Haq, Saad, Purdue University
Harley, Simon L., Edinburgh University
Harms, Tekla A., Amherst College
Harris, Nigel B., The Open University
Hemphill-Haley, Mark, Humboldt State University
Henderson, Robert, James Cook University
Henry, Christopher D., University of Nevada
Hughes, Sam, Exeter University
Huntington, Katharine W., University of Washington
Hurtado, Jose M., University of Texas, El Paso
Insel, Nadja, Northeastern Illinois University
Ivins, Erik R., Jet Propulsion Laboratory
Jackson, Christopher, Imperial College
Janecke, Susanne U., Utah State University
John, Hickman B., University of Kentucky
Judge, Shelley, College of Wooster
Kapp, Paul, University of Arizona
Karlstrom, Karl E., University of New Mexico
Karson, Jeffrey, Syracuse University
Kato, Terence T., California State University, Chico
Keir, Derek, University of Southampton
Keller, Edward A., University of California, Santa Barbara
Kelley, Shari A., New Mexico Institute of Mining and Tech
Keppie, J. Duncan, Universidad Nacional Autonoma de Mexico
Keppie, John D., Universidad Nacional Autonoma de Mexico
Kirstein, Linda, Edinburgh University
Kleinspehn, Karen L., University of Minnesota, Twin Cities
Klemperer, Simon L., Stanford University
Knipe, Rob, University of Leeds
Knuepfer, Peter L. K., Binghamton University
Koehn, Daniel, University of Glasgow
Kokum, Mehmet, Firat University
Koons, Peter O., University of Maine
Kronenberg, Andreas, Texas A&M University
Lamb, Melissa A., University of Saint Thomas
Lavier, Luc L., University of Texas at Austin
Lin, Jian, Woods Hole Oceanographic Institution
Lindsley-Griffin, Nancy, University of Nebraska, Lincoln
Lonergan, Lidia, Imperial College
Loveless, Jack, Smith College
Luo, Gang, University of Texas at Austin
Mac Niocaill, Conall, University of Oxford
Mallard, Laura, Appalachian State University
Malone, Shawn J., Ball State University
Mann, Paul, University of Texas at Austin
Mapani, Benjamin S., University of Namibia
Marsellos, Antonios, Hofstra University
Martinez, Juventino, Universidad Nacional Autonoma de Mexico
Martinez-Reyes, Juventino, Universidad Nacional Autonoma de Mexico
May, Daniel J., University of New Haven
McClelland, William C., University of Iowa
McDowell, Rob J., Georgia State University, Perimeter College, Dunwoody Campus
McGill, Sally F., California State University, San Bernardino
McGrew, Allen J., University of Dayton
McNeill, Lisa, University of Southampton
McPherson, Robert C., Humboldt State University
McRivette, Michael, Albion College
Melichar, Rostislav, Masaryk University
Melosh, Henry J., University of Arizona
Mesbahi, Fatemeh, University of Tabriz
Michon, Laurent, Universite de la Reunion
Milan, Luke, University of New England
Molnar, Peter, University of Colorado
Moore, Gregory F., University of Hawai'i, Manoa
Moores, Eldridge M., University of California, Davis
Morgan, Julia K., Rice University
Morrow, Robert H., Dept of Natural Resources
Moser, Desmond, Western University
Moucha, Robert, Syracuse University
Mound, Jon, University of Leeds
Mueller, Karl J., University of Colorado
Murphy, J. Brendan, Saint Francis Xavier University
Möller, Andreas, University of Kansas
Nabelek, John L., Oregon State University
Nadin, Elisabeth S., University of Alaska, Fairbanks
Nance, R. Damian, Ohio University
Nash, Greg, University of Utah
Naylor, Mark, Edinburgh University
Ngo, Thanh X., Hanoi University of Mining & Geology
Niemi, Tina M., University of Missouri-Kansas City
Ocan, O. O., Obafemi Awolowo University
Ogata, Kei, VU University Amsterdam
Onderdonk, Nate, California State University, Long Beach
Partin, Camille, University of Saskatchewan
Passchier, Cees W., Universität Mainz
Pavlis, Terry L., University of New Orleans
Peacock, Simon M., University of British Columbia
Peterman, Emily M., Bowdoin College
Pettinga, Jarg R., University of Canterbury
Phillips, Richard, University of Leeds
Pieri, David C., Jet Propulsion Laboratory
Plasienka, Dusan, Comenius University in Bratislava
Pluhar, Chris, California State University, Fresno
Price, Raymond A., Queen's University
Rahl, Jeffrey, Washington & Lee University
Regalla, Christine, Boston University
Reusch, Douglas N., University of Maine - Farmington
Richter, Frank M., University of Chicago
Rieken, Eric R., University of Texas, Pan American
Robertson, Alastair H., Edinburgh University
Roeske, Sarah M., University of California, Davis
Ross, Gerald M., University of Calgary
Rowan, Christopher J., Kent State University
Rowley, David B., University of Chicago
Royden, Leigh H., Massachusetts Institute of Tech
Rubin, Charles M., Central Washington University
Rust, Derek, University of Portsmouth
Ryder, Isabelle, University of Liverpool
Sager, William W., University of Houston
Schellart, Wouter , VU University Amsterdam
Schermer, Elizabeth R., Western Washington University
Schmidt, David , University of Washington
Schneider, David, University of Ottawa
Scotese, Christopher R., University of Texas, Arlington
Searle, Mike, University of Oxford
Shail, Robin, Exeter University
Shaw, John, Harvard University
Sinha, A. Krishna, Virginia Polytechnic Institute & State University
Sintubin, Manuel, Katholieke Universiteit Leuven
Spotila, James A., Virginia Polytechnic Institute & State University
Srimal, Neptune, Florida International University
Steltenpohl, Mark G., Auburn University
Stern, Robert J., University of Texas, Dallas
Stock, Joann M., California Institute of Tech
Stoddard, Paul R., Northern Illinois University
Strecker, Manfred, Cornell University
Streepey Smith, Meg, Earlham College
Suppe, John, University of Houston
Suter-Cargnelutti, Max, Universidad Nacional Autonoma de Mexico
Takeuchi, Akira, University of Toyama
Taylor, Brian, University of Hawai'i, Manoa
Taylor, Frederick W., University of Texas at Austin
Taylor, Michael H., University of Kansas
Teixell Cácharo, Antoni, Universitat Autonoma de Barcelona
Tesfaye, Samson, University of Missouri
Thigpen, Ryan, University of Kentucky
Thomas, William A., Geological Survey of Alabama
Thomas, William A., University of Kentucky
Thorkelson, Derek J., Simon Fraser University
Tivey, Maurice A., Woods Hole Oceanographic Institution
Toke', Nathan, Utah Valley University
Turner, Jenni, University of East Anglia
Umhoefer, Paul J., Northern Arizona University
Unsworth, Martyn, University of Alberta
Valentine, Michael , Highline College
Valentino, David W., SUNY, Oswego
van Hunen, Jeroen, Durham University
Van Wijk, Jolante, New Mexico Institute of Mining and Tech
vanKeken, Peter E., Carnegie Institution of Washington
Wallace, Laura, University of Texas at Austin
Watkinson, Ian, University of London, Royal Holloway & Bedford New

College
Webb, Laura E., University of Vermont
Wernicke, Brian P., California Institute of Tech
Whipple, Kelin, Arizona State University
Whittaker, Alex, Imperial College
Willis, Julie B., Brigham Young University - Idaho
Wilson, Crystal, Appalachian State University
Wong, Martin, Colgate University
Wright, James (Jim) E., University of Georgia
Wu, Jonny, University of Houston
Yeats, Robert S., Oregon State University
Yin, An, University of California, Los Angeles
Yule, J. Douglas, California State University, Northridge
Zamani, Behzad, University of Tabriz

## Volcanology

, Anthony F., Universite de la Reunion
Aguirre Diaz, Gerardo J., Universidad Nacional Autonoma de Mexico
Anderson, Steve W., University of Northern Colorado
Anderson, Jr., Alfred T., University of Chicago
Andrews, Benjamin, Smithsonian Institution / National Museum of Natural History
Andrews, Graham D., West Virginia University
Awdankiewicz, Marek, University of Wroclaw
Banik, Tenley, Illinois State University
Barclay, Jenni, University of East Anglia
Befus, Kenneth S., Baylor University
Bellieni, Giuliano, Università degli Studi di Padova
Blake, Stephen, The Open University
Boroughs, Scott, Washington State University
Branan, Yvonne K., Indiana University of Pennsylvania
Branney, Mike, Leicester University
Brown, David, University of Glasgow
Brown, Richard J., Durham University
Brunstad, Keith, SUNY, Oneonta
Burkett, Brett, Collin College - Central Park Campus
Bursik, Marcus I., SUNY, Buffalo
Calder, Eliza, Edinburgh University
Cameron, Barry I., University of Wisconsin, Milwaukee
Cameron, Cheryl, Alaska Division of Geological & Geophysical Surveys
Camp, Victor E., San Diego State University
Canon-Tapia, Edgardo, Centro de Investigación Científica y de Educación Superior de Ensenada
Carr, Michael J., Rutgers, The State University of New Jersey
Carrasco-Nunez, Gerardo, Universidad Nacional Autonoma de Mexico
Castro, Jonathan M., Universität Mainz
Chapin, Charles E., New Mexico Institute of Mining and Tech
Chapman, Marshall, Morehead State University
Cigolini, Corrdao, Università di Torino
Clarke, Amanda, Arizona State University
Cole, James W., University of Canterbury
Cole, Paul, University of Plymouth
Connor, Charles B., University of South Florida, Tampa
Coulson, Ian, University of Regina
Crisp, Joy A., Jet Propulsion Laboratory
Crowe, Bruce M., Los Alamos National Laboratory
Davis, Fred, Smithsonian Institution / National Museum of Natural History
de Silva, Shanika, Oregon State University
Deardorff, Nicholas, Indiana University of Pennsylvania
Dennen, Rob, Smithsonian Institution / National Museum of Natural History
Domagall, Abigail M., Black Hills State University
Drake, Simon, Birkbeck College
Dufek, Josef, Georgia Institute of Tech
Duffield, Wendell A., Northern Arizona University
Duller, Rob, University of Liverpool
Edmonds, Marie, University of Cambridge
Eggers, Albert A., University of Puget Sound
Escobar-Wolf, Rudiger, Michigan Technological University
Fagents, Sarah, University of Hawai'i, Manoa
Fischer, Tobias, University of New Mexico
Fiske, Richard S., Smithsonian Institution / National Museum of Natural History
Flood, Tim P., Saint Norbert College
Frey, Holli M., Union College
Gansecki, Cheryl A., University of Hawai'i, Hilo
Gardner, James E., University of Texas at Austin
Genareau, Kimberly, University of Alabama
Giachetti, Thomas, University of Oregon
Giordano, Daniele, Università di Torino
Godchaux, Martha M., Mount Holyoke College
Gonnermann, Helge, Rice University
Graettinger, Alison H., SUNY, Buffalo
Grapenthin, Ronni, New Mexico Institute of Mining and Tech
Gravley, Darren, University of Canterbury
Green, Jack, California State University, Long Beach
Gregg, Tracy K. P., SUNY, Buffalo
Gutmann, James T., Wesleyan University
Harpp, Karen, Colgate University
Harwood, Richard D., Black Hawk College
Hasenaka, Toshiaki, Kumamoto University
Hausback, Brian, California State University, Sacramento
Hazlett, Richard W., Pomona College
Head, Elisabet M., Northeastern Illinois University
Heiken, Grant, Los Alamos National Laboratory
Heliker, Christina C., University of Hawai'i, Hilo
Herd, Richard, University of East Anglia
Herrick, Julie, Smithsonian Institution / National Museum of Natural History
Hickson, Catherine J., University of British Columbia
Holm, Richard F., Northern Arizona University
Hon, Ken, University of Hawai'i, Hilo
Houghton, Bruce F., University of Hawai'i, Manoa
Houle, Michel, University of Manitoba
Huber, Christian, Brown University
Hughes, Scott S., Idaho State University
Johnson, Emily R., New Mexico State University, Las Cruces
Kauahikaua, James P., University of Hawai'i, Manoa
Kavanagh, Janine, University of Liverpool
Kennedy, Ben, University of Canterbury
Kieffer, Susan W., University of Illinois, Urbana-Champaign
Klemetti, Erik, Denison University
Kobs-Nawotniak, Shannon E., Idaho State University
Kohut, Ed, University of Delaware
Koornneef, Janne, VU University Amsterdam
Krier, Donathon J., Los Alamos National Laboratory
Lang, Nicholas, Mercyhurst University
Larsen, Jessica F., University of Alaska, Fairbanks
Lavallee, Yan, University of Liverpool
Lee, Rachel J., SUNY, Oswego
Lehto, Heather L., Angelo State University
LeMasurier, Wesley E., University of Colorado, Denver
Lev, Einat , Columbia University
Leyrit, Hervé, Institut Polytechnique LaSalle Beauvais (ex-IGAL)
Llewellin, Ed, Durham University
Lockwood, John P., University of Hawai'i, Hilo
Magsino, Sammantha L., National Academy of Sciences/National Research Council
Manga, Michael, University of California, Berkeley
Maria, Tony, University of Southern Indiana
Martinez-Hackert, Bettina, Buffalo State College
Mather, Tamsin A., University of Oxford
McBirney, Alexander R., University of Oregon
McConnell, Vicki S., Oregon Dept of Geology & Mineral Industries
McGarvie, Dave, The Open University
McGuire, Bill, University College London
McPhie, Jocelyn, University of Tasmania
Melson, William G., Smithsonian Institution / National Museum of Natural History
Moore, Joseph N., University of Utah
Nadeau, Patricia, Salem State University
Neuberg, Jurgen, University of Leeds
Oliveras Castro, Valentí, Universitat Autonoma de Barcelona
Ort, Michael H., Northern Arizona University
Ottavi-Pupier, Elsa, Institut Polytechnique LaSalle Beauvais (ex-IGAL)
Pallister, John S., University of Pittsburgh
Panter, Kurt S., Bowling Green State University
Pe-Piper, Georgia, Dalhousie University
Perry, Frank V., Los Alamos National Laboratory
Phipps Morgan, Jason, Cornell University
Popp, Christoph, Smithsonian Institution / National Museum of Natural History
Putirka, Keith D., California State University, Fresno
Pyle, David M., University of Oxford

Reynolds, Robert W., Central Oregon Community College
Rhodes, J. Michael, University of Massachusetts, Amherst
Riggs, Nancy, Northern Arizona University
Robinson, Paul T., Dalhousie University
Rodriguez, Lizzette A., University of Puerto Rico
Roldan Quintana, Jaime, Universidad Nacional Autonoma de Mexico
Roman, Diana C., Carnegie Institution of Washington
Rose, William I., Michigan Technological University
Ross, Pierre-Simon, Universite du Quebec
Ruprecht, Philipp, Columbia University
Russell, James K., University of British Columbia
Sanchez-Rubio, Gerardo, Universidad Nacional Autonoma de Mexico
Sar, Abdullah, Firat University
Saunders, Kate, Edinburgh University
Schaefer, Janet R. G., Alaska Division of Geological & Geophysical Surveys
Sennert, Sally K., Smithsonian Institution / National Museum of Natural History
Shea, Thomas, University of Hawai'i, Manoa
Sheridan, Michael F., SUNY, Buffalo
Smith, Alan L., California State University, San Bernardino
Solana, Camen, University of Portsmouth
Soule, S. Adam, Woods Hole Oceanographic Institution
Stevenson, John A., Edinburgh University
Stix, John, McGill University
Straub, Susanne, Columbia University
Swanson, Donald A., University of Hawai'i, Manoa
Szymanski, David, Bentley University
Taylor, Rex N., University of Southampton
Teasdale, Rachel, California State University, Chico
Templeton, Jeffrey H., Western Oregon University
Ukstins Peate, Ingrid, University of Iowa
Valentine, Gregory, SUNY, Buffalo
Van Ry, Michael, Orange Coast College
Vanderkluysen, Loyc, Drexel University
Venzke, Edward, Smithsonian Institution / National Museum of Natural History
Vigouroux-Caillibot, Nathalie, Douglas College
Voight, Barry, Pennsylvania State University, University Park
Vroon, Pieter, VU University Amsterdam
Wallace, Paul, University of Oregon
Webber, Karen L., University of Michigan
Webber, Karen L., University of New Orleans
Wehner, Peter J., Austin Community College District
White, James D., University of Otago
Williams, Stanley N., Arizona State University
Williams-Jones, Glyn, Simon Fraser University
Wilson, Thomas, University of Canterbury
Wohletz, Kenneth H., Los Alamos National Laboratory
Wohletz, Kenneth H., University of New Mexico
Wulff, Andrew, Western Kentucky University
Wunderman, Richard, Smithsonian Institution / National Museum of Natural History

## Mathematical Geology

Agterberg, Frits P., University of Ottawa
Barton, Christopher C., Wright State University
Bolton, Edward W., Yale University
Campbell, Katherine, Los Alamos National Laboratory
Chien, Yi-Ju, Pacific Northwest National Laboratory
Chou, Charissa J., Pacific Northwest National Laboratory
Cornejo-Luna, Efrain, Universidad Autonoma de Baja California Sur
Davis, John C., University of Kansas
de Hoop, Maarten, Purdue University
Doveton, John H., University of Kansas
Dyreson, Eric G., University of Montana Western
Edwards, Stuart, University of Newcastle Upon Tyne
Faccenda, Manuele, Università degli Studi di Padova
Fairbairn, David, University of Newcastle Upon Tyne
Fialko, Yuri A., University of California, San Diego
Ford, Allistair, University of Newcastle Upon Tyne
Ford, Arianne, James Cook University
Fox, Peter A., Rensselaer Polytechnic Institute
Ghattas, Omar, University of Texas at Austin
Gordon, Terence M., University of Calgary
Hampton, Samuel, University of Canterbury
Harbaugh, John W., Stanford University
Harding, Chris, Iowa State University of Science & Tech
He, Changming, University of Delaware
Hesse, Marc A., University of Texas at Austin
Hohn, Michael E., West Virginia University
Hunt, Allen G., Wright State University
Johnson, Glenn W., University of Utah
Jones, Thomas A., Rice University
Journel, Andre G., Stanford University
Kinnicutt, Patrick, Central Michigan University
Lee, Jejung, University of Missouri-Kansas City
Lutz, Timothy M., West Chester University
MAZZELLA, Prof. Antonio A., Universita di Cagliari
Murray, Christopher J., Pacific Northwest National Laboratory
Olea, Ricardo A., University of Kansas
Penna, Nigel, University of Newcastle Upon Tyne
Ricciardi, Karen, University of Massachusetts, Boston
Rogova, Galina L., SUNY, Buffalo
Rudolph, Maxwell L., Portland State University
Rundberg, Robert S., Los Alamos National Laboratory
Seacrest, Tyler, University of Montana Western
Sonder, Leslie J., Dartmouth College
Spiegelman, Marc W., Columbia University
Sun, Alexander, University of Texas at Austin
ter Voorde, Marlies, VU University Amsterdam
Therrien, Pierre, Universite Laval
Tuttle, Samuel , Mount Holyoke College
Walton, Ian, University of Utah
Wright, Eric S., University of Montana Western

## Mineral Physics

Abramson, Evan H., University of Washington
Bass, Jay D., University of Illinois, Urbana-Champaign
Brown, J. Michael, University of Washington
Bukowinski, Mark S., University of California, Berkeley
Carpenter, Michael, University of Cambridge
Chen, Jiuhua, Florida International University
Cohen, Ronald E., Carnegie Institution of Washington
Cooper, Reid F., Brown University
Duffy, Thomas S., Princeton University
Farver, John R., Bowling Green State University
Fenter, Paul A., Argonne National Laboratory
Fenter, Paul, University of Illinois at Chicago
Goncharov, Alexander F., Carnegie Institution of Washington
Hugo, Richard, Portland State University
Jackson, Jennifer M., California Institute of Tech
Jacobsen, Steven D., Northwestern University
Jeanloz, Raymond, University of California, Berkeley
Karato, Shun-ichiro, Yale University
Kavner, Abby, University of California, Los Angeles
Kiefer, Boris, New Mexico State University, Las Cruces
Knittle, Elise, University of California, Santa Cruz
Kronenberg, Andreas, Texas A&M University
Lee, Kanani K., Yale University
Li, Baosheng, SUNY, Stony Brook
Lieberman, Robert C., SUNY, Stony Brook
Lin, Jung-Fu, University of Texas at Austin
Liu, Zhenxian, Carnegie Institution of Washington
Militzer, Burkhard, University of California, Berkeley
Riggs, Eric, Texas A&M University
Salje, Ekhard, University of Cambridge
Saxena, Surendra K., Florida International University
Secco, Richard A., Western University
Stracher, Glenn B., East Georgia State College
Struzhkin, Viktor V., Carnegie Institution of Washington
Tracy, Sally J., Princeton University
Tschauner, Oliver, University of Nevada, Las Vegas
Tyburczy, James A., Arizona State University
van Oss, Carel J., SUNY, Buffalo
Vocadlo, Lidunka, University College London
Weidner, Donald J., SUNY, Stony Brook
Williams, Quentin, University of California, Santa Cruz
Zhang, Jin, University of New Mexico

## History of Geology

Bolton Valencius, Conevery, University of Massachusetts, Boston
Case, Stephen, Olivet Nazarene University
Chyba, Christopher F., Stanford University
Drobnyk, John W., Southern Connecticut State University
Rosenberg, Gary D., Milwaukee Public Museum

Saunders, Charles, Randolph-Macon College
Strick, James E., Franklin and Marshall College

### Geomedicine

Buck, Brenda J., University of Nevada, Las Vegas
Cook, Lewis, Concord University
Finkelman, Robert B., University of Texas, Dallas
Fuge, Ron, Aberystwyth University
Gillmore, Gavin, Kingston University
Hunt, Andrew, University of Texas, Arlington
Markowitz, Steven, Queens College (CUNY)
Morabia, Alfredo, Queens College (CUNY)
Pasteris, Jill D., Washington University in St. Louis

## ECONOMIC GEOLOGY

### General Economic Geology

Adedcyin, A. D., University of Ilorin
Adetunji, A, Obafemi Awolowo University
Akabzaa, Thomas M., University of Ghana
Akande, S. O., University of Ilorin
Alavi, Ghafour, University of Tabriz
Alexandrov, Eugene A., Queens College (CUNY)
Allard, Gilles O., University of Georgia
Allen, Donald B., Colby College
Anderson, Scott, Manitoba Geological Survey
Anger-Kraavi, Annela, University of East Anglia
Ansdell, Kevin M., University of Saskatchewan
Arehart, Greg B., University of Nevada, Reno
Ashley, Paul, University of New England
Banerjee, Neil, Western University
Barnes, Sarah- J., Universite du Quebec a Chicoutimi
Barton, Mark D., University of Arizona
Beaudoin, Georges, Universite Laval
Bekken, Barbara M., Virginia Polytechnic Institute & State University
Blamey, Nigel, Brock University
Bland, Douglas, New Mexico Institute of Mining & Tech
Bolarinwa, A. T., University of Ibadan
Bornhorst, Theodore J., Michigan Technological University
Bouzari, Farhad, University of British Columbia
Bradshaw, Peter M., University of British Columbia
Brimhall, George H., University of California, Berkeley
Burt, Donald M., Arizona State University
Busch, Lawrence, California Geological Survey
Calagari, Ali Asghar, University of Tabriz
Callahan, John E., Appalachian State University
Camprubi, Antoni, Universidad Nacional Autonoma de Mexico
Carlisle, Donald, University of California, Los Angeles
Carter, James L., University of Texas, Dallas
Castor, Stephen B., University of Nevada
Cathles, Lawrence M., Cornell University
Cheney, Eric S., University of Washington
Cheng, Yanbo, James Cook University
Chi, Guoxiang, University of Regina
Church, John, California Geological Survey
Clark, Kenneth F., University of Texas, El Paso
Clendenin, Jr., Charles W., Dept of Natural Resources
Clinkenbeared, John, California Geological Survey
Conly, Andrew G., Lakehead University
Cook, Robert B., Auburn University
Cooke, David R., University of Tasmania
Corbineau, Lucien, Institut Polytechnique LaSalle Beauvais (ex-IGAL)
Coron, Cynthia R., Southern Connecticut State University
Corral, Isaac, James Cook University
Craig, James R., Virginia Polytechnic Institute & State University
Craw, Dave, University of Otago
Crawford, Thomas J., University of West Georgia
Di Gregorio, Monica, University of Leeds
Donovan, John F., Winona State University
Duke, Norman A., Western University
Dunst, Brian, Pennsylvania Bureau of Topographic & Geologic Survey
Elston, Wolfgang E., University of New Mexico
Elueze, A. A., University of Ibadan
Evans, Thomas J., University of Wisconsin - Extension
Feltrin, Leo, Western University
Flawn, Peter T., University of Texas at Austin
Franco-Rubio, Miguel, Universidad Autonoma de Chihuahua
Garcia-Gutierrez, Luis, Universidad Autonoma de San Luis Potosi
Gauthier, Michel, Universite du Quebec a Montreal
Gemmell, J B., University of Tasmania
Gibson, Andy, University of Portsmouth
Gilg, Hans Albert, Technische Universitaet Muenchen
Gillerman, Virginia S., Boise State University
Gillerman, Virginia S., University of Idaho
Greyling, Lynnette N., University of Cape Town
Guan, Dabo, University of Leeds
Guha, Jayanta, Universite du Quebec a Chicoutimi
Gulen, Gurcan, University of Texas at Austin
Hannington, Mark, University of Ottawa
Harris, DeVerle P., University of Arizona
Hayman, Patrick, Queensland University of Tech
Hedenquist, Jeffrey W., University of Ottawa
Hickey, Kenneth A., University of British Columbia
Hodder, Robert W., Western University
Hohn, Michael E., West Virginia Geological & Economic Survey
Hollenbaugh, Kenneth M., Boise State University
Holley, Elizabeth, Colorado School of Mines
Hollings, Peter N., Lakehead University
Huminicki, Michelle, Brandon University
Ikingura, Justian R., University of Dar es Salaam
Ikonnikova, Svetlana, University of Texas at Austin
Jago, Bruce C., Laurentian University, Sudbury
Jean-Marc, MONTEL, Ecole Nationale Supérieure de Géologie (ENSG)
Jebrak, Michel, Universite du Quebec a Montreal
Jessey, David R., California State Polytechnic University, Pomona
Kamilli, Robert J., University of Arizona
Kamona, Frederick A., University of Namibia
Karginoglu, Yusuf , Firat University
Kazimoto, Emmanuel O., University of Dar es Salaam
Keith, Jeffrey D., Brigham Young University
Kelly, William C., University of Michigan
Kesler, Stephen E., University of Michigan
Kish, Stephen A., Florida State University
Kontak, Daniel J., Laurentian University, Sudbury
La Berge, Gene L., University of Wisconsin, Oshkosh
Laki, Sam, Central State University
Large, Ross R., University of Tasmania
Lasemi, Zakaria, University of Illinois, Urbana-Champaign
Lassetter, William L., Division of Geology and Mineral Resources
Layton-Matthews, Daniel, Queen's University
Lefebure, Dave, University of Victoria
Lehrberger, Gerhard, Technische Universitaet Muenchen
Levinson, Alfred A., University of Calgary
Linnen, Robert, University of Waterloo
Lorenzoni, Irene, University of East Anglia
Maher, Kierran, New Mexico Institute of Mining and Tech
Mantilla Figueroa, Luis C., Universidad Industrial de Santander
Masterman, Steven S., Alaska Division of Geological & Geophysical Surveys
mcClenaghan, Seán h., Trinity College
McLemore, Virginia T., New Mexico Institute of Mining and Tech
Mensah, Emmanuel, Kwame Nkrumah University of Science and Tech
Metzel, Deborah, University of Massachusetts, Boston
Misra, Kula C., University of Tennessee, Knoxville
Mohammad Reza, Hosseinzadeh, University of Tabriz
Morton, Ronald, University of Minnesota, Duluth
Mumin, A. Hamid, Brandon University
Muntean, John, University of Nevada, Reno
Muszer, Antoni, University of Wroclaw
Nasraoui, Mohamed, Institut Polytechnique LaSalle Beauvais (ex-IGAL)
Nex, Paul A., University of the Witwatersrand
Niewendorp, Clark, Oregon Dept of Geology & Mineral Industries
Nimis, Paolo, Università degli Studi di Padova
Okunlola, Ougbenga. A., University of Ibadan
Olivo, Gema, Queen's University
Orozco-Villasenor, Francisco Javier, Universidad Autonoma de San Luis Potosi
Paez, H. A., McMaster University
Paterson, Colin J., South Dakota School of Mines & Tech
Perrouty, Stephane, Western University
Petersen, Ulrich, Harvard University
Pietrzak-Renaud, Natalie, Western University
Potra, Adriana, University of Arkansas, Fayetteville
Prichard, Hazel, University of Wales

Prochaska, Walter, University Leoben
Ramagwede, Fhatuwani L., Geological Survey of South Africa
Rambaud, Fabienne M., Austin Community College District
Ressel, Mike, University of Nevada
Ridley, John R., Colorado State University
Riese, W C., Rice University
Robertson, Daniel E., Monroe Community College
Robertson, James M., University of Wisconsin, Extension
Robertson, James M., University of Wisconsin, Madison
Rossetti, Piergiorgio, Università di Torino
Sagiroglu, Ahmet, Firat University
Samson, Iain M., University of Windsor
Sasmaz, Ahmet, Firat University
Schardt, Christian, University of Minnesota, Duluth
Seedorff, Eric, University of Arizona
Serenko, Thomas J., Ohio Dept of Natural Resources
Shackley, Simon J., Edinburgh University
Shaver, Stephen A., Sewanee: University of the South
Shortt, Niamh K., Edinburgh University
Siahcheshm, Kamal, University of Tabriz
Simmons, Peter, University of East Anglia
Smith, Joshua, California Geological Survey
Solecki, Andrzej, University of Wroclaw
Steele-MacInnes, Matthew, University of Arizona
Stucker, James D., Ohio Dept of Natural Resources
Talley, John H., University of Delaware
Thompson, John F., Cornell University
Thompson, Tommy B., University of Nevada, Reno
Titley, Spencer R., University of Arizona
Tritlla, Jordi, Universidad Nacional Autonoma de Mexico
Twelker, Evan, Alaska Division of Geological & Geophysical Surveys
Ulloa, Salvador, Universidad Nacional Autonoma de Mexico
van der Horst, Dan, Edinburgh University
van Hinsberg, Vincent, McGill University
Warren, Rachel, University of East Anglia
Webster, James D., American Museum of Natural History
Werdon, Melanie B., Alaska Division of Geological & Geophysical Surveys
Wilkins, Colin, University of Plymouth
Williams, David A., University of Kentucky
Wilton, Derek H., Memorial University of Newfoundland
Wolfgram, Diane, Montana Tech of the University of Montana
Yao, Yon, Rhodes University
Yates, Martin G., University of Maine
Yellich, John, Western Michigan University
Zentilli, Marcos, Dalhousie University
Zimmerman, Brian S., Edinboro University of Pennsylvania

## Coal

Abdullah, Wan Hasiah, University of Malaya
Asuen, Godwin O., University of Benin
Bailey, Judy, University of Newcastle
Bustin, R. Marc, University of British Columbia
Calder, John, Dalhousie University
Cardott, Brian J., University of Oklahoma
Carroll, Chris J., Wyoming State Geological Survey
Carroll, Richard E., Geological Survey of Alabama
Cecil, Blaine, West Virginia University
Chenoweth, Cheri A., Illinois State Geological Survey
Cohen, Arthur D., University of South Carolina
Crelling, John C., Southern Illinois University Carbondale
Drobniak, Agnieszka, Indiana University
Elrick, Scott D., University of Illinois
Fedorko, Nick, West Virginia University
Flood, Peter, University of New England
Grady, William, West Virginia University
Hoffman, Gretchen K., New Mexico Institute of Mining & Tech
Hower, James C., University of Kentucky
Huff, Bryan G., University of Illinois, Urbana-Champaign
Huggett, William, Southern Illinois University Carbondale
Khawaja, Ikram U., Youngstown State University
Korose, Christopher P., Illinois State Geological Survey
Korose, Christopher P., University of Illinois, Urbana-Champaign
Lentz, Leonard J., Pennsylvania Bureau of Topographic & Geologic Survey
Lindberg, Jonathan W., Pacific Northwest National Laboratory
Markowski, Antonette K., Pennsylvania Bureau of Topographic & Geologic Survey
Mastalerz, Maria D., Indiana University, Bloomington
Meyer, Rebecca A., Indiana University
Miller, Barry W., Tennessee Geological Survey
Mukhopadhyay, Prasanta, Dalhousie University
Myers, Alan R., University of Illinois, Urbana-Champaign
Neubaum, John C., Pennsylvania Bureau of Topographic & Geologic Survey
Obrad, Jennifer M., University of Illinois, Urbana-Champaign
Pheifer, Raymond N., Eastern Illinois University
Renton, John J., West Virginia University
Rimmer, Susan M., Southern Illinois University Carbondale
Robl, Thomas , University of Kentucky
Shaulis, James R., Pennsylvania Bureau of Topographic & Geologic Survey
Smallwood, Shane, Ohio Dept of Natural Resources
Ward, Colin R., University of New South Wales
Weisenfluh, Gerald A., University of Kentucky

## Metals

Barnett, Douglas B., Pacific Northwest National Laboratory
Bissig, Thomas, University of British Columbia
Broughton, William A., University of Wisconsin, Platteville
Brown, Philip E., University of Wisconsin, Madison
Brueckner, Stefanie M., Auburn University
Campbell, Finley A., University of Calgary
Carpenter, Rob, Western University
Cedillo, Esteban, Universidad Nacional Autonoma de Mexico
Chang, Zhaoshan, James Cook University
Chapman, Robert J., University of Leeds
Chavez, Jr., William X., New Mexico Institute of Mining and Tech
Clark, Alan H., Queen's University
Cline, Jean S., University of Nevada, Las Vegas
Corona-Esquivel, Rodolfo, Universidad Nacional Autonoma de Mexico
Coveney Jr, Raymond M., University of Missouri-Kansas City
Crowe, Douglas E., University of Georgia
Diaz-Encinas, Leopoldo, Centro de Estudios Superiores del Estado Sonora
Dilles, John H., Oregon State University
Dube, Benoit, Universite du Quebec
Durazo-Tapia, Gustavo E., Centro de Estudios Superiores del Estado Sonora
Einaudi, Marco T., Stanford University
Fifarek, Richard, Southern Illinois University Carbondale
Friehauf, Kurt, Kutztown University of Pennsylvania
Gaboury, Damien, Universite du Quebec a Chicoutimi
Gibson, Harold, Carleton University
Gibson, Harold L., Laurentian University, Sudbury
Good, David, Western University
Graf, Jr., Joseph L., Southern Oregon University
Guilbert, John M., University of Arizona
Hagni, Richard D., Missouri University of Science and Tech
Hannington, Mark D., Carleton University
Hart, Craig J., University of British Columbia
Hooda, Peter, Kingston University
Keays, Reid R., Laurentian University, Sudbury
Kinnaird, Judith A., University of the Witwatersrand
Kissin, Stephen A., Lakehead University
Kisvarsanyi, Geza K., Missouri University of Science and Tech
Koestel, Mark, Mt. San Antonio College
Kyle, J. Richard, University of Texas at Austin
Lesher, Michael, Laurentian University, Sudbury
Linnen, Robert, Western University
Macfarlane, Andrew W., Florida International University
MacLean, Wallace H., McGill University
McAllister, Arnold L., University of New Brunswick
McLemore, Virginia, New Mexico Institute of Mining & Tech
Melchiorre, Erik, California State University, San Bernardino
Menuge, Julian F., University College Dublin
Monecke, Thomas, Colorado School of Mines
Morel-Kraepiel, Anne, Princeton University
Morton, Roger D., University of Alberta
Mossman, David J., Mount Allison University
Muchez, Philippe, Katholieke Universiteit Leuven
Naldrett, Anthony J., University of Toronto
Newberry, Rainer J., University of Alaska, Fairbanks
Petersen, Erich U., University of Utah
Powell, Wayne G., Graduate School of the City University of New York
Pride, Douglas E., Ohio State University
Richards, Jeremy P., University of Alberta

Ripley, Edward M., Indiana University, Bloomington
Romberger, Samuel B., Colorado School of Mines
Ruvalcaba-Ruiz, Delfino C., Universidad Autonoma de San Luis Potosi
Salaun, Pascal, University of Liverpool
Shelton, Kevin L., University of Missouri
Sinclair, Alastair J., University of British Columbia
Spry, Paul G., Iowa State University of Science & Tech
Stensrud, Howard L., California State University, Chico
Thompson, John F., Cornell University
Thompson, John F. H., University of British Columbia
Tingley, Joseph V., University of Nevada
Tosdal, Richard, University of British Columbia
Uzunlar, Nuri, South Dakota School of Mines & Tech
van der Berg, Stan, University of Liverpool
Vasallo, Fernando, Universidad Nacional Autonoma de Mexico
Vassallo-Morales, Luis F., Universidad Nacional Autonoma de Mexico
Watkinson, David H., Carleton University
Wheeler, Greg, California State University, Sacramento
Wheeler, John, University of Liverpool
Wilkinson, Jamie, Imperial College
Williamson, Ben, Exeter University
Zhang, Ning, Central State University

### Non-Metals
Austin, George S., New Mexico Institute of Mining & Tech
Bal Akkoca, Dicle, Firat University
Barker, James M., New Mexico Institute of Mining & Tech
Berg, Richard B., Montana Tech of The University of Montana
Calengas, Peter L., Western Illinois University
Gregory, Robert, Wyoming State Geological Survey
Krukowski, Stanley T., University of Oklahoma
Lasemi, Zakaria, Illinois State Geological Survey
McClellan, Guerry H., University of Florida
Power, W. Robert, Georgia State University
Simandl, George J., University of Victoria
Sousa, Luís M., Universidade de Trás-os-Montes e Alto Douro
Spooner, Edward T. C., University of Toronto
Sutherland, Wayne, Wyoming State Geological Survey
Van Straaten, H. Peter, University of Guelph

### Oil and Gas
Anderson, Ross, Heriot-Watt University
Balázs, László, Eotvos Lorand University
Brown, Alan, West Virginia University
Burst, John F., Missouri University of Science and Tech
C., Jeffrey C., Youngstown State University
Cox, John, Mount Royal University
Crockett, Joan E., Illinois State Geological Survey
Davies, Richard, Durham University
Deisher, Jeffrey , Ohio Dept of Natural Resources
Ersoy, Erkal, Heriot-Watt University
Evans, Martin J., Cornell University
Finley, Robert J., Illinois State Geological Survey
Gillis, Robert, Alaska Division of Geological & Geophysical Surveys
Gluyas, Jon, Durham University
Gooding, Patrick J., University of Kentucky
Grube, John P., Illinois State Geological Survey
Harder, Brian J., Louisiana State University
Herriott, Trystan, Alaska Division of Geological & Geophysical Surveys
Hickman, John, University of Kentucky
Hooks, Chris H., Geological Survey of Alabama
Horn, Marty R., Louisiana State University
Hosseini, Seyyed Abolfazi, University of Texas at Austin
Huff, Bryan G., Illinois State Geological Survey
Johnston, III, John E., Louisiana State University
King, Carey, University of Texas at Austin
Kreckel, Kenneth, Casper College
Krevor, Samuel, Imperial College
Leetaru, Hannes E., Illinois State Geological Survey
Leighton, Morris W., University of Illinois, Urbana-Champaign
Leone, James, New York State Geological Survey
Lepain, David, Alaska Division of Geological & Geophysical Surveys
Levey, Raymond A., University of Utah
Lisa, Black L., Midwestern State University
Lucia, F. J., University of Texas at Austin
Lynds, Ranie, Wyoming State Geological Survey
Malinconico, Mary Ann, Lafayette College
Moore, Thomas R., Indiana University of Pennsylvania
Nicholas, Chris, Trinity College
Nielsen, Peter J., Utah Geological Survey
Nuttall, Brandon C., University of Kentucky
Oldershaw, Michael, Bakersfield College
Parris, Thomas M., University of Kentucky
Pascussi, Michael, New York State Geological Survey
Potter, Eric C., University of Texas at Austin, Jackson School of Geosciences
Seyler, Beverly, Illinois State Geological Survey
Shew, Roger D., University of North Carolina Wilmington
Tew, Nick, University of Alabama
Toner, Rachel, Wyoming State Geological Survey
Vigrass, Laurence W., University of Regina
Waid, Christopher, Ohio Dept of Natural Resources
Webb, Nathan D., University of Illinois, Urbana-Champaign
Wood, Lesli J., University of Texas at Austin
Zeidouni, Mehdi, University of Texas at Austin

## GEOCHEMISTRY

### General Geochemistry
Abimbola, A. E., University of Ibadan
Aines, Roger D., Lawrence Livermore National Laboratory
Ajayi, T. R., Obafemi Awolowo University
Akaegbobi, I. M., University of Ibadan
Allen, Douglas, Salem State University
Allen-King, Richelle, SUNY, Buffalo
Amonette, Alexandra B., Pacific Northwest National Laboratory
Anbar, Ariel, Arizona State University
Anderson, Alan J., Dalhousie University
Anderson, Robert F., Columbia University
April, Richard, Colgate University
Atekwana, Eliot, Oklahoma State University
Ayers, John C., Vanderbilt University
Ayling, Bridget, University of Nevada
Babcock, R. S., Western Washington University
Baharlou, Alan, Eastern Illinois University
Bales, Roger C., University of Arizona
Bank, Tracy L., SUNY, Buffalo
Banks, David, University of Leeds
Barker, Colin, The University of Tulsa
Barnes, Hubert L., Pennsylvania State University, University Park
Barnes, Jaime D., University of Texas at Austin
Basu, Anirban, University of London, Royal Holloway & Bedford New College
Beiersdorfer, Raymond E., Youngstown State University
Belmonte, Donato, Universita di Genova
Bender, Michael L., Princeton University
Berendsen, Pieter, University of Kansas
Berger, Peter M., University of Illinois, Urbana-Champaign
Berkelhammer, Max, University of Illinois at Chicago
Berndt, Mike, University of Minnesota, Twin Cities
Bernhard, Joan M., Woods Hole Oceanographic Institution
Bernier, Luc, McMaster University
Bird, Dennis K., Stanford University
Bissada, K. K., Rice University
Blake, Ruth E., Yale University
Blowes, David W., University of Waterloo
Bodnar, Robert J., Virginia Polytechnic Institute & State University
Boger, Phillip D., SUNY, Geneseo
Bohm, Christian, University of Manitoba
Borowski, Walter S., Eastern Kentucky University
Boschetti, Tiziano, University of Parma
Bouman, Heather, University of Oxford
Bourcier, Bill L., Lawrence Livermore National Laboratory
Brantley, Susan L., Pennsylvania State University, University Park
Brenna, J. T., Cornell University
Broecker, Wallace S., Columbia University
Brooks, Paul D., University of Arizona
Bryant, Ernest A., Los Alamos National Laboratory
Bryce, Julia G., University of New Hampshire
Burton, Jacqueline C., Argonne National Laboratory
Buseck, Peter R., Arizona State University
Butler, Ian B., Edinburgh University
Cadieux, Sarah, University of Illinois at Chicago
Camacho, Elsa A., Pacific Northwest National Laboratory
Centorbi, Tracey, University of Maryland

Chague-Goff, Catherine, University of New South Wales
Chatterjee, Nilanjan, Massachusetts Institute of Tech
Chen, Chu-Yung, University of Illinois, Urbana-Champaign
Chermak, John A., Virginia Polytechnic Institute & State University
Chillrud, Steven, Columbia University
Chorover, Jonathan D., University of Arizona
Churchill, Ron C., California Geological Survey
Claeys, Philippe, Vrije University Brussel
Clark, Jordan, University of California, Santa Barbara
Class, Connie, Columbia University
Clay, Patricia L., University of Manchester
Coe, Douglas A., Montana Tech of the University of Montana
Coggon, Judith A., University of Southampton
Cole, David R., Ohio State University
Collier, James D., Fort Lewis College
Conte, Maureen, Brown University
Cook, Robert B., Oak Ridge National Laboratory
Cowman, Tim C., South Dakota Dept of Environment and Natural Resources
Crawford, William A., Bryn Mawr College
Cubillas, Pablo, Durham University
Dasgupta, Tathagata, Kent State University
Davisson, M L., Lawrence Livermore National Laboratory
Dawson, M. Robert, Iowa State University of Science & Tech
Day, James, University of California, San Diego
Deng, Baolin, New Mexico Institute of Mining and Tech
Derry, Louis A., Cornell University
Dewaele, Stijn, Ghent University
Ding, Kang, University of Minnesota, Twin Cities
Dong, Hailiang, Miami University
Dooley, John H., New Jersey Geological and Water Survey
Dostal, Jarda, Dalhousie University
Downes, Hilary, Birkbeck College
Drew, Douglas A., Montana Tech of the University of Montana
Duffy, Clarence J., Los Alamos National Laboratory
Dunbar, Nelia W., New Mexico Institute of Mining and Tech
Durham, William, Massachusetts Institute of Tech
Dutkiewicz, Stephanie, Massachusetts Institute of Tech
Dutton, Andrea, University of Florida
Eby, G. Nelson, University of Massachusetts Lowell
Edgcomb, Virginia, Woods Hole Oceanographic Institution
Eiler, John M., California Institute of Tech
Erhardt, Andrea M., University of Kentucky
Evans, Owen C., SUNY, Stony Brook
Eves, Robert L., Southern Utah University
Fajkoviæ, Hana, University of Zagreb
Fall, Andras, University of Texas at Austin
Farley, Kenneth A., California Institute of Tech
Farmer, G. Lang, University of Colorado
Feigenson, Mark D., Rutgers, The State University of New Jersey
Fernandez, Diego, University of Utah
Finkel, Robert C., Lawrence Livermore National Laboratory
Finkelstein, David, Hobart & William Smith Colleges
Flegal, Russell, University of California, Santa Cruz
Flowers, Rebecca M., University of Colorado
Foustoukos, Dionysis, Carnegie Institution of Washington
Fraser, Donalf G., University of Oxford
Frey, Bonnie A., New Mexico Institute of Mining & Tech
Furman, Tanya, Pennsylvania State University, University Park
Gambrell, Robert P., Louisiana State University
Gammons, Christopher H., Montana Tech of the University of Montana
Garbesi, Karina, California State University, East Bay
George, Graham, University of Saskatchewan
Gleason, James D., University of Michigan
Goff, Fraser, University of New Mexico
Goff, Fraser, Los Alamos National Laboratory
Goldstein, Steven, Columbia University
Gosse, John C., Dalhousie University
Gosselin, David C., Unversity of Nebraska - Lincoln
Gouldey, Jeremy C., Grand Valley State University
Goulet, Richard, University of Ottawa
Grant, Willard H., Emory University
Gulbranson, Erik L., University of Wisconsin, Milwaukee
Gurlea, Lawrence P., Youngstown State University
Gustin, Mae, University of Nevada, Reno
Gutierrez, Melida, Missouri State University
Gysi, Alex, Colorado School of Mines
Halden, Norman M., University of Manitoba
Hall, Cynthia V., West Chester University
Hansen, Robert, University of the Free State
Hart, Stanley R., Woods Hole Oceanographic Institution
Harvey, Jason, University of Leeds
Hattori, Keiko, University of Ottawa
Hemming, Sidney, Columbia University
Henderson, Gideon, University of Oxford
Hervig, Richard, Arizona State University
Hickey-Vargas, Rosemary, Florida International University
Hinman, Nancy W., University of Montana
Hinton, Richard, Edinburgh University
Hodges, Floyd N., Pacific Northwest National Laboratory
Hudson, Robert J. M., University of Illinois, Urbana-Champaign
Hwang, Hue-Hwa E., University of Illinois, Urbana-Champaign
Ige, A. O., Obafemi Awolowo University
Irving, Tony, University of Washington
Izon, Gareth, University of St. Andrews
Jacobs, Gary K., Oak Ridge National Laboratory
Jacobson, Roger, University of Nevada, Reno
Jarvis, Ian, Kingston University
Jerde, Eric, Morehead State University
Jin, Qusheng, University of Oregon
Johnson, Clark M., University of Wisconsin, Madison
Johnston, David T., Harvard University
Kalender, Leyla, Firat University
Kallemeyn, Gregory, University of California, Los Angeles
Kamber, Balz S., Trinity College
Kambewa, Chamunorwa , Tshwane University of Tech
Kamenov, George D., University of Florida
Kaplan, Isaac R., University of California, Los Angeles
Kara, Hatice, Firat University
Kaszuba, John P., University of Wyoming
Katz, Richard, University of Oxford
Kearney, Kenneth, University of Illinois at Chicago
Kellman, Lisa M., Dalhousie University
Kenna, Timothy, Columbia University
Kim, Sang-Tae, McMaster University
Kirste, Dirk, Simon Fraser University
Knudsen, Andrew, Lawrence University
Kolesar, Peter T., Utah State University
Koppers, Anthony, Oregon State University
Koretsky, Carla, Western Michigan University
Kotzer, Tom, University of Saskatchewan
Krawczynski, Michael J., Washington University in St. Louis
Kronfeld, Joel, Dalhousie University
Ku, Teh-Lung, University of Southern California
Kuzyk, Zou Zou, University of Manitoba
Köster, Mathias, Technische Universitaet Muenchen
Labidi, Jabrane, University of Maryland
Lange, Eric, Ball State University
Langmuir, Charles H., Harvard University
Le Roex, Anton, University of Cape Town
Lechler, Paul J., University of Nevada
Lee, Cin-Ty A., Rice University
Legore, Virginia L., Pacific Northwest National Laboratory
Lentz, David R., University of New Brunswick
Lerman, Abraham, Northwestern University
Leybourne, Matthew I., Laurentian University, Sudbury
Lindenmeier, Clark W., Pacific Northwest National Laboratory
Liu, Xiaoming, University of North Carolina, Chapel Hill
Livens, Francis, University of Manchester
Locke II, Randall A., University of Illinois
Longerich, Henry, Memorial University of Newfoundland
Lopez, Dina L., Ohio University
Loyd, Sean, California State University, Fullerton
Lu, Zunli, Syracuse University
Luttge, Andreas, Rice University
Lyons, Timothy W., University of California, Riverside
Macpherson, Colin G., Durham University
Magson, Justine, University of the Free State
Mango, Helen N., Castleton University
Manning, Christina, University of London, Royal Holloway & Bedford New College
Manya, Shukrani, University of Dar es Salaam
Marquez, L. Lynn, Millersville University
Martin, Paul F., Pacific Northwest National Laboratory

Martin, Scot T., Harvard University
Marzoli, Andrea, Università degli Studi di Padova
McClelland, James W., University of Texas at Austin
McDonough, William F., University of Maryland
McGrail, Bernard P., Pacific Northwest National Laboratory
McGuire, Jennifer, University of Saint Thomas
McIntosh, Jennifer C., University of Arizona
McKibben, Michael A., University of California, Riverside
McLennan, Scott M., SUNY, Stony Brook
Meduniæ, Gordana, University of Zagreb
Meixner, Thomas, University of Arizona
Mercy, Edward L., Lakehead University
Meyer, William T., Argonne National Laboratory
Meyzen, Christine M., Università degli Studi di Padova
Michel, Jacqueline, University of New Orleans
MOINE, Bertrand N., Université Jean Monnet, Saint-Etienne
Moldovan, Brett, University of Saskatchewan
Mortlock, Richard , Rutgers, The State University of New Jersey
Mukhopadhyay, Sujoy, University of California, Davis
Murgulet, Valeriu, Texas A&M University, Corpus Christi
Murowchick, James B., University of Missouri-Kansas City
Murphy, David T., Queensland University of Tech
Naftz, David L., University of Utah
Nagy, Kathryn L., University of Illinois at Chicago
Naranjo, Ramon, University of Nevada, Reno
Nondorf, Lea, Arkansas Geological Survey
Nyame, Frank K., University of Ghana
O'Driscoll, Nelson, Acadia University
O'Hara, Matthew J., Pacific Northwest National Laboratory
O'Neil, Jonathan, University of Ottawa
Odokuma-Alonge, Ovie, University of Benin
Odom, LeRoy A., Florida State University
Okulewicz, Steven C., Hofstra University
Olanrewaju, Johnson, Gannon University
Olatunji, A. S., University of Ibadan
Olsen, Khris B., Pacific Northwest National Laboratory
Olszewski, Kathy, SUNY, Maritime College
Omelon, Christopher, University of Texas at Austin
Otamendi, Juan E., Universidad Nacional de Rio Cuarto
Ozturk, Nevin, Firat University
Padden, Maureen, McMaster University
Palinkaš, Ladislav, University of Zagreb
Palmer, Martin R., University of Southampton
Pant, Hari, Lehman College (CUNY)
Papineau, Dominic, University College London
Parker, Kent E., Pacific Northwest National Laboratory
PArkes, R. J., University of Wales
Parris, Thomas M., University of Kentucky
Parry, William T., University of Utah
Patera, Edward S., Los Alamos National Laboratory
Pearce, J A., Cardiff University
Peate, David W., University of Iowa
Peng, Yongbo, Louisiana State University
Piccoli, Philip M., University of Maryland
Pichevin, Laetitia, Edinburgh University
Pickering, Ingrid J., University of Saskatchewan
Pietranik, Anna, University of Wroclaw
Pintilei, Mitica, Al. I. Cuza University of Iasi
Plank, Terry, Columbia University
Plante, Martin, Universite Laval
Pommier, Anne, University of California, San Diego
Porcelli, Don, University of Oxford
Poreda, Robert J., University of Rochester
Pourmand, Ali, University of Miami
Pourret, Olivier, Institut Polytechnique LaSalle Beauvais (ex-IGAL)
Powell, Brian A., Clemson University
Prevec, Steve, Rhodes University
Price, Douglas M., Youngstown State University
Prohiæ, Esad, University of Zagreb
Prowse, Terry D., University of Waterloo
Ptacek, Carol J., University of Waterloo
Rae, James, University of St. Andrews
Ramos, Frank C., New Mexico State University, Las Cruces
Reed, Mark H., University of Oregon
Reeves, Claire, University of East Anglia
Reiners, Peter, University of Arizona
Richardson, Steve, University of Cape Town
Rickard, D, Cardiff University
Riedinger, Natascha, Oklahoma State University
Roberts, Sheila J., Bowling Green State University
Roberts, Stephen, University of Southampton
Rohs, C. Renee, Northwest Missouri State University
Rollinson, Hugh, University of Derby
Romanak, Katherine D., University of Texas at Austin
Rose, Arthur W., Pennsylvania State University, University Park
Rose, Catherine V., Trinity College
Rudnick, Roberta L., University of Maryland
Ruiz, Joaquin, University of Arizona
Ruttenberg, Kathleen, University of Hawai'i, Manoa
Ruttenberg, Kathleen C., University of Hawai'i, Manoa
Saal, Alberto E., Brown University
Sack, Richard, University of Washington
Sakyi, Patrick A., University of Ghana
Salters, Vincent J., Florida State University
Savov, Ivan, University of Leeds
Saxena, Surenda K., Graduate School of the City University of New York
Schaef, Herbert T., Pacific Northwest National Laboratory
Schaefer, Joerg, Columbia University
Schauble, Edwin A., University of California, Los Angeles
Schlosser, Peter, Columbia University
Schrag, Daniel P., Harvard University
Schubert, Brian, University of Louisiana at Lafayette
Seitz, Jeffery C., California State University, East Bay
Senko, John M., University of Akron
Serne, R. Jeffrey, Pacific Northwest National Laboratory
Shahar, Anat, Carnegie Institution of Washington
Sheldon, Amy L., SUNY, Buffalo
Shevenell, Lisa, University of Nevada
Shiel, Alyssa, Oregon State University
Shields, Robin, Heriot-Watt University
Shields-Zhou, Graham, University College London
Shirey, Steven, University of Maryland
Shofner, Gregory A., Towson University
Sigman, Daniel M., Princeton University
Sims, Kenneth W., University of Wyoming
Skinner, Brian J., Yale University
Skippen, George B., Carleton University
Slater, Gregory F., McMaster University
Smethie, William, Columbia University
Speidel, David H., Queens College (CUNY)
Speidel, David H., Graduate School of the City University of New York
Spencer, Ronald J., University of Calgary
Stack, Andrew, Georgia Institute of Tech
Stanley, Clifford R., Acadia University
Stebbins, Jonathan F., Stanford University
Steele, Kenneth F., University of Arkansas, Fayetteville
Stevenson, Ross, Universite du Quebec a Montreal
Strmiæ Palinkaš, Sabina, University of Zagreb
Sturchio, Neil C., Argonne National Laboratory
Sultan, Mohamed, Argonne National Laboratory
Susak, Nicholas J., University of New Brunswick
Taylor, Richard P., Carleton University
Teagle, Damon, University of Southampton
Tefend, Karen S., University of West Georgia
Teng, Fangzhen, University of Washington
Thomas, Jim, University of Nevada, Reno
Tierney, Kate E., Denison University
Tomascak, Paul B., SUNY, Oswego
Tomson, Mason B., Rice University
Tredoux, Marian, University of the Free State
Trueman, Clive, University of Southampton
Tsikos, Hari, Rhodes University
Van Geen, Alexander, Columbia University
van Huissteden, Ko, VU University Amsterdam
Van Orman, James A., Case Western Reserve University
Vannier, Ryan G., Grand Valley State University
Varekamp, Johan C., Wesleyan University
Vulava, Vijay M., College of Charleston
Wagner, Rick, Geological Survey of Alabama
Walker, John L., Argonne National Laboratory
Walther, John V., Southern Methodist University
Wang, Hong, University of Illinois, Urbana-Champaign
Wang, Yang, Florida State University
Warburton, David L., Florida Atlantic University

Warren, Lesley A., McMaster University
Watkins, James, University of Oregon
Weiss, Dominik, Imperial College
Whipkey, Charles, University of Mary Washington
White, William B., Pennsylvania State University, University Park
Wicks, June, Princeton University
Wigley, Rochelle, University of New Hampshire
Williams, Jeremy C., Kent State University
Winckler, Gisela, Columbia University
Wogelius, Roy, University of Manchester
Wooden, Joseph, Stanford University
Worrall, Fred, Durham University
Wronkiewicz, David J., Argonne National Laboratory
Wyman, Derek A., University of Saskatchewan
Yan, Beizhan, Columbia University
Yardley, Bruce, University of Leeds
Zanazzi, Alessandro, Utah Valley University
Zeman, Josef, Masaryk University
Zhang, Tongwei, University of Texas at Austin
Zheng, Yan, Queens College (CUNY)
Zolotov, Mikhail Y., Arizona State University

## Analytical Geochemistry

Aguilera-Ortiz, Irma, Universidad Nacional Autonoma de Mexico
Ash, Richard, University of Maryland
Barnes, Melanie A., Texas Tech University
Beauchamp, Marc, Western University
Bedard, Paul, Universite du Quebec a Chicoutimi
Benjamin, Timothy M., Los Alamos National Laboratory
Bhatia, D. M. S., Austin Peay State University
Bhattacharyya, Sidhartha, University of Alabama
Blamey, Nigel J., New Mexico Institute of Mining and Tech
Bohlke, John, University of Maryland
Bowen, Scott M., Los Alamos National Laboratory
Branam, Tracy D., Indiana University
Brandvold, Lynn A., New Mexico Institute of Mining & Tech
Cahill, Richard A., Illinois State Geological Survey
Cameron, Douglas, Montana Tech of the University of Montana
Castaneda, Isla, University of Massachusetts, Amherst
Catalano, Jeff, Washington University in St. Louis
Cheng, Zhongqi, Brooklyn College (CUNY)
Clark, Malcolm W., Southern Cross University
Cole, David R., University of Tennessee, Knoxville
De Carlo, Eric H., University of Hawai'i, Manoa
Dewey, Janet, University of Wyoming
Effert-Fanta, Shari E., University of Illinois, Urbana-Champaign
Essling, Alice M., Argonne National Laboratory
Frew, Nelson M., Woods Hole Oceanographic Institution
Gabitov, Rinat, Mississippi State University
Gagnon, Joel E., University of Windsor
Garcia, Sammy R., Los Alamos National Laboratory
Gavriloaiei, Traian, Al. I. Cuza University of Iasi
Gibson, Blair, Western University
Halbig, Joseph B., University of Hawai'i, Hilo
Hart, Brian R., Western University
Hemming, N. G., Queens College (CUNY)
Hinthorne, James R., Central Washington University
Hoe, Teh Guan, University of Malaya
Holt, Ben D., Argonne National Laboratory
Hu, Wan-Ping (Sunny), Queensland University of Tech
Huff, Edmund A., Argonne National Laboratory
Huft, Ashley, Montana Tech of The University of Montana
Hunter, Jerry L., Virginia Polytechnic Institute & State University
Jackson, Brian P., Dartmouth College
Jedrysek, Mariusz O., University of Wroclaw
Johnson, Carl G., Woods Hole Oceanographic Institution
Jones, Graham, Southern Cross University
Kersten, Michael, Universität Mainz
Klinkhammer, Gary, Oregon State University
Knaack, Charles, Washington State University
Knudstrup, Renee, Salem State University
Krapac, Ivan G., Illinois State Geological Survey
Kuehn, Stephen C., Concord University
Kuentz, David C., Miami University
Landry, Peter B., Woods Hole Oceanographic Institution
Locke, David C., Graduate School of the City University of New York
Lottermoser, Bernd, Exeter University
Lundstrom, Elizabeth, University of Minnesota, Twin Cities
MacDonald, Iain, University of Wales
Marshall, Dan D., Simon Fraser University
McGrath, Steve F., Montana Tech of The University of Montana
Messo, Charles W., University of Dar es Salaam
Mock, R. Stephen, University of Montana Western
Mora-Klepeis, Gabriela, University of Vermont
Mujumba, Jean K., University of Dar es Salaam
Navarre-Sitchler, Alexis, Colorado School of Mines
Neill, Owen K., Washington State University
Nelsen, Lori, Furman University
Newland, Leo, Texas Christian University
Olesik, John W., Ohio State University
Olmsted, Wayne, Montana Tech of the University of Montana
Orouji, Maqsood, University of Tabriz
Orr, Robert D., Pacific Northwest National Laboratory
Papike, James J., University of New Mexico
Pike, Steven M., Woods Hole Oceanographic Institution
Pun, Aurora, University of New Mexico
Ragland, Paul C., Florida State University
Ranville, James, Colorado School of Mines
Rose, Timothy, Smithsonian Institution / National Museum of Natural History
Rouff, Ashaki, Queens College (CUNY)
Russell, Shane, Queensland University of Tech
Savard, Dany, Universite du Quebec a Chicoutimi
Schneider, John F., Argonne National Laboratory
Shimizu, Nobumichi, Woods Hole Oceanographic Institution
Singer, Jared W., Rensselaer Polytechnic Institute
Smith, Catherine H., Los Alamos National Laboratory
Smith, Florence P., Argonne National Laboratory
Snow, Jonathan, University of Houston
Student, James J., Central Michigan University
Sylvester, Paul J., Memorial University of Newfoundland
Sylvester, Steven, Franklin and Marshall College
Thomas, Donald M., University of Hawai'i, Manoa
Thompson, Christopher J., Pacific Northwest National Laboratory
Timmer, Jaqueline R., Montana Tech of The University of Montana
Underwood, Ben, Indiana University, Bloomington
Van Loon, Lisa, Western University
Villasenor-Cabral, Maria G., Universidad Nacional Autonoma de Mexico
Voelker, Bettina, Colorado School of Mines
Volborth, Alexis, Montana Tech of the University of Montana
Wang, Alian, Washington University in St. Louis
Warwick, Phillip, University of Southampton
Waugh, Truman, University of Kansas
Williams, Amy, Towson University
Williams, Kim R., Colorado School of Mines
Wu, Charles T., Western University
Yang, Panseok, University of Manitoba

## Experimental Petrology/Phase Equilibria

Agee, Carl A., University of New Mexico
Anderson, G M., University of Toronto
Asimow, Paul D., California Institute of Tech
Baker, Don, McGill University
Brenan, James M., Dalhousie University
Candela, Philip A., University of Maryland
Carey, James W., Los Alamos National Laboratory
Cottrell, Elizabeth, University of Maryland
Cottrell, Elizabeth, Smithsonian Institution / National Museum of Natural History
Davis, Fred A., University of Minnesota, Duluth
Eggler, David H., Pennsylvania State University, University Park
Ernst, W. Gary, Stanford University
Fei, Yingwei, Carnegie Institution of Washington
Fei, Yingwei, University of Maryland
Feineman, Maureen D., Pennsylvania State University, University Park
Frank, Mark R., Northern Illinois University
Gilbert, M. Charles, University of Oklahoma
Hajash, Andrew, Texas A&M University
Herzberg, Claude T., Rutgers, The State University of New Jersey
Hewitt, David A., Virginia Polytechnic Institute & State University
Hsu, Liang-Chi, University of Nevada
Jenkins, David M., Binghamton University
Johnston, A. Dana, University of Oregon
Kilinc, Attila I., University of Cincinnati
Koster Van Groos, August F., University of Illinois at Chicago
Koziol, Andrea M., University of Dayton

Lazar, Codi, California State University, San Bernardino
Liang, Yan, Brown University
Lindsley, Donald H., SUNY, Stony Brook
London, David, University of Oklahoma
Lundstrom, Craig C., University of Illinois, Urbana-Champaign
McCauley, Marlene, Guilford College
Minarik, William G., McGill University
Mosenfelder, Jed, University of Minnesota, Twin Cities
Munoz, James L., University of Colorado
Mysen, Bjorn O., Carnegie Institution of Washington
Nekvasil, Hanna, SUNY, Stony Brook
Newton, Robert C., University of California, Los Angeles
Parman, Stephen, Brown University
Presnall, Dean C., University of Texas, Dallas
Robert, Genevieve, Bates College
Rutherford, Malcolm J., Brown University
Seewald, Jeffrey S., Woods Hole Oceanographic Institution
Simmons, Craig, Colorado School of Mines
Spera, Frank J., University of California, Santa Barbara
Steiner, Jeffrey, City College (CUNY)
Stolper, Edward M., California Institute of Tech
Tailby, Nichlos D., Rensselaer Polytechnic Institute
Tait, C. Drew, Los Alamos National Laboratory
Ulmer, Gene C., Temple University
Ustunisik, Gokce K., South Dakota School of Mines & Tech
van Westrenen, Wim, VU University Amsterdam
Walker, David, Columbia University
Watson, E. Bruce, Rensselaer Polytechnic Institute
Windom, Kenneth E., Iowa State University of Science & Tech
Withers, Tony, Western University
Wyllie, Peter J., California Institute of Tech
Zhang, Youxue, University of Michigan

## Exploration Geochemistry

Abdurrahman, A., University of Ilorin
Benning, Liane G., University of Leeds
Borojeviæ Šoštariæ, Sibila, University of Zagreb
Closs, L. Graham, Colorado School of Mines
Cohen, David R., University of New South Wales
da Silva, Eduardo F., Universidade de Aveiro
Dreschoff, Gisela, University of Kansas
Fletcher, William K., University of British Columbia
Goodell, Philip C., University of Texas, El Paso
Kyser, Kurt, Queen's University
Mathur, Ryan, Juniata College
McGoldrick, Peter J., University of Tasmania
Pearson, Graham, University of Alberta
Price, Vaneaton, Clemson University
White, William M., Cornell University
Whitehead, Robert E., Laurentian University, Sudbury
Williams-Jones, Anthony E., McGill University
Yanez, Carlos, Universidad Nacional Autonoma de Mexico

## Geochronology & Radioisotopes

Ackert, Robert P., Bentley University
Allen, Charlotte M., Queensland University of Tech
Archibald, Doug A., Queen's University
Asmerom, Yemane, University of New Mexico
Baadsgaard, Halfdan, University of Alberta
Baksi, Ajoy K., Louisiana State University
Baldwin, Suzanne L., Syracuse University
Barker, Ivan, Western University
Baskaran, Mark M., Wayne State University
Baxter, Ethan, Boston College
Beard, Brian L., University of Wisconsin, Madison
Bell, Keith, Carleton University
Bevier, Mary Lou, University of British Columbia
Bickford, M. E., Syracuse University
Black, Stuart, University of Reading
Blackburn, Terrence, University of California, Santa Cruz
Blenkinsop, John, Carleton University
Blusztajn, Jurek, Woods Hole Oceanographic Institution
Blythe, Ann, Occidental College
Brown, Roderick, University of Glasgow
Brueckner, Hannes K., Graduate School of the City University of New York
Burton, Kevin, Durham University
Butler, Gilbert W., Los Alamos National Laboratory
Carlson, Richard W., Carnegie Institution of Washington
Chamberlain, Edwin P., Los Alamos National Laboratory
Chamberlain, Kevin R., University of Wyoming
Chaudhuri, Sambhudas, Kansas State University
Cheng, Hai, University of Minnesota, Twin Cities
Chew, David, Trinity College
Clark, George S., University of Manitoba
Coleman, Drew S., University of North Carolina, Chapel Hill
Copeland, Peter, University of Houston
Cornett, Jack R., University of Ottawa
Cowart, James B., Florida State University
Creaser, Robert A., University of Alberta
Crowley, Jim, Boise State University
Crowley, Quentin G., Trinity College
Curtice, Joshua M., Woods Hole Oceanographic Institution
Curtis, David B., Los Alamos National Laboratory
Curtis, Garniss H., University of California, Berkeley
Davis, Donald W., University of Toronto
De Grave, Johan, Ghent University
DePaolo, Donald J., University of California, Berkeley
Dickin, Alan P., McMaster University
Dulai, Henrietta, University of Hawai'i, Manoa
Dunning, Gregory R., Memorial University of Newfoundland
Eddy, Michael P., Princeton University
Edwards, R. Lawrence, University of Minnesota, Twin Cities
Enkelmann, Eva, University of Cincinnati
Faure, Gunter, Ohio State University
Fearey, Bryan L., Los Alamos National Laboratory
Fenton, Cassandra, Colorado Mesa University
Foland, Kenneth A., Ohio State University
Forman, Stephen, Baylor University
Frost, Carol D., University of Wyoming
Fullagar, Paul D., University of North Carolina, Chapel Hill
Gallup, Christina D., University of Minnesota, Duluth
Gancarz, Alexander J., Los Alamos National Laboratory
Gaschnig, Richard M., University of Massachusetts Lowell
Gaudette, Henri E., University of New Hampshire
Geibert, Walter, Edinburgh University
Georgiev, Svetoslav, Colorado State University
Goehring, Brent M., Tulane University
Gromet, L. P., Brown University
Hall, Chris, University of Michigan
Hames, Willis E., Auburn University
Hamilton, Michael Andrew, University of Toronto
Hanan, Barry B., San Diego State University
Hanes, John A., Queen's University
Harbottle, Garman, SUNY, Stony Brook
Harrison, T. Mark, University of California, Los Angeles
Hauri, Erik H., Carnegie Institution of Washington
Hayatsu, Akio, Western University
Heaman, Larry M., University of Alberta
Heatherington, Ann L., University of Florida
Heizler, Matthew T., New Mexico Institute of Mining and Tech
Horan, Mary F., Carnegie Institution of Washington
Hourigan, Jeremy, University of California, Santa Cruz
Hughen, Konrad A., Woods Hole Oceanographic Institution
Idleman, Bruce D., Lehigh University
Jacobsen, Stein B., Harvard University
Jull, A. J. Timothy, University of Arizona
Karner, Daniel B., Sonoma State University
Kimbrough, David L., San Diego State University
Konter, Jasper G., University of Hawai'i, Manoa
Krogh, Thomas E., University of Toronto
Krummenacher, Daniel, San Diego State University
Kuiper, Klaudia, VU University Amsterdam
Kurz, Mark D., Woods Hole Oceanographic Institution
Lakatos, Stephen, York College (CUNY)
Lassiter, John, University of Texas at Austin
Lepper, Kenneth E., North Dakota State University
Lively, Rich, University of Minnesota
Long, Leon E., University of Texas at Austin
Lopez, Margarita, Centro de Investigación Científica y de Educación Superior de Ensenada
Manton, William I., University of Texas, Dallas
Markey, Richard, Colorado State University
Markun, Francis J., Argonne National Laboratory
McDowell, Fred W., University of Texas at Austin

McIntosh, William C., New Mexico Institute of Mining and Tech
McLean, Noah, University of Kansas
McMurtry, Gary M., University of Hawai'i, Manoa
McWilliams, Michael O., Stanford University
Menninga, Clarence, Calvin College
Mertz, Dieter, Universität Mainz
Miller, Brent, Texas A&M University
Miller, Charles M., Los Alamos National Laboratory
Miller, Donald S., Rensselaer Polytechnic Institute
Miller, Geoffrey G., Los Alamos National Laboratory
Miller, Gifford H., University of Colorado
Min, Kyoungwon, University of Florida
Mock, Timothy D., Carnegie Institution of Washington
Monteleone, Brian D., Woods Hole Oceanographic Institution
Montgomery, Carla W., Northern Illinois University
Moore, Willard S., University of South Carolina
Moran-Zenteno, Dante J., Universidad Nacional Autonoma de Mexico
Morris, David, Los Alamos National Laboratory
Mortensen, James K., University of British Columbia
Moser, Desmond E., University of Utah
Mueller, Paul A., University of Florida
Mukasa, Samuel B., University of Michigan
Muller, Wolfgang, University of London, Royal Holloway & Bedford New College
Murrell, Michael T., Los Alamos National Laboratory
Nelson, Bruce K., University of Washington
Noakes, John E., University of Georgia
Omar, Gomaa I., University of Pennsylvania
Oretega Rivera, Amabel M., Universidad Nacional Autonoma de Mexico
Orlandini, Kent A., Argonne National Laboratory
Osmond, John K., Florida State University
Palin, J. Michael, University of Otago
Papanastassiou, D.A. (Dimitri), California Institute of Tech
Pardi, Richard R., William Paterson University
Perrin, Richard E., Los Alamos National Laboratory
Placzek, Christa, James Cook University
Polyak, Victor J., University of New Mexico
Poths, Jane, Los Alamos National Laboratory
Puchtel, Igor, University of Maryland
Rasbury, E. Troy, SUNY, Stony Brook
Renne, Paul, University of California, Berkeley
Rhodes, Edward J., University of California, Los Angeles
Rink, W. J., McMaster University
Rinterknecht, Vincent, University of St. Andrews
Rokop, Donald J., Los Alamos National Laboratory
Rubin, Kenneth H., University of Hawai'i, Manoa
Rutberg, Randye L., Hunter College (CUNY)
Samson, Scott D., Syracuse University
Schmitz, Mark D., Boise State University
Schoene, R B., Princeton University
Schroeder, Norman C., Los Alamos National Laboratory
Seidemann, David E., Graduate School of the City University of New York
Seidemann, David E., Brooklyn College (CUNY)
Selby, Dave, Durham University
Shea, Erin, University of Alaska, Anchorage
Shirey, Steven B., Carnegie Institution of Washington
Silver, Leon T., California Institute of Tech
Singer, Bradley S., University of Wisconsin, Madison
Sole, Jesu, Universidad Nacional Autonoma de Mexico
Southon, John, University of California, Irvine
Spell, Terry L., University of Nevada, Las Vegas
Spencer, Joel Q., Kansas State University
Srinivasan, Balakrishnan, Pondicherry University
Stewart, Brian W., University of Pittsburgh
Stone, John O., University of Washington
Strachan, Rob, University of Portsmouth
Stuiver, Minze, University of Washington
Swanson, Karen, William Paterson University
Swanson, Terry W., University of Washington
Swisher III, Carl C., Rutgers, The State University of New Jersey
Tera, Fouad, Carnegie Institution of Washington
Thibodeau, Alyson M., Dickinson College
Tomlinson, Emma L., Trinity College
Turek, Andrew, University of Windsor
Turrin, Brent D., Rutgers, The State University of New Jersey
Van Schmus, W. Randall, University of Kansas
Vermeesch, Pieter, University College London
Vervoort, Jeffrey D., Washington State University
Walker, Richard J., University of Maryland
Walter, Robert C., Franklin and Marshall College
Wampler, J. Marion, Georgia State University
Wasserburg, Gerald J., California Institute of Tech
Weis, Dominique, University of British Columbia
White, William M., Cornell University
Widom, Elisabeth, Miami University
Wielicki, Matthew, University of Alabama
Wijbrans, Jan, VU University Amsterdam
Wolfsberg, Kurt, Los Alamos National Laboratory
Yang, Gang, Colorado State University
Yin, Qing-zhu, University of California, Davis
Zeitler, Peter K., Lehigh University
Zimmerman, Aaron, Colorado State University
Zou, Haibo, Auburn University
Zreda, Marek, University of Arizona

## Low-temperature Geochemistry

Achten, Christine, Universitaet Muenster
Adams, Donald D., Plattsburgh State University (SUNY)
Agrawal, Abinash, Wright State University
Aharon, Paul, University of Alabama
Aja, Stephen U., Graduate School of the City University of New York
Aja, Stephen U., Brooklyn College (CUNY)
Alessi, Daniel S., University of Alberta
Alexander, Clark R., Georgia Institute of Tech
Amend, Jan, University of Southern California
Anand, Pallavi, The Open University
Andersen, C. B., Clemson University
Andersen, C. Brannon, Furman University
Angino, Ernest E., University of Kansas
Arrhenius, Gustaf, University of California, San Diego
Baker, Leslie L., University of Idaho
Balogh-Brunstad, Zsuzsanna, Hartwick College
Banner, Jay L., University of Texas at Austin
Baron, Dirk, California State University, Bakersfield
Beckingham, Barbara, College of Charleston
Benner, Shawn, Boise State University
Benninger, Larry K., University of North Carolina, Chapel Hill
Bergslien, Elisa T., Buffalo State College
Bertine, Kathe K., San Diego State University
Bickmore, Barry R., Brigham Young University
Bird, MIchael I., James Cook University
Bischoff, William D., Wichita State University
Blanton, Jackson O., Georgia Institute of Tech
Bogner, Jean E., University of Illinois at Chicago
Borrok, David M., University of Louisiana at Lafayette
Bowser, Carl J., University of Wisconsin, Madison
Brabander, Daniel J., Wellesley College
Brand, Uwe, Brock University
Brook, Edward J., Oregon State University
Brooks, Scott C., Oak Ridge National Laboratory
Burke, Roger A., University of Georgia
Cantrell, Kirk J., Pacific Northwest National Laboratory
Cao, Hongsheng, Wichita State University
Capo, Rosemary C., University of Pittsburgh
Carroll, Susan A., Lawrence Livermore National Laboratory
Casey, William H., University of California, Davis
Catalano, Jeffrey G., Washington University in St. Louis
Cerling, Thure E., University of Utah
Chesworth, Ward, University of Guelph
Christina, Thomas D., Georgia Institute of Tech
Clark, Jordan F., University of California, Santa Barbara
Cody, Anita M., Iowa State University of Science & Tech
Crossey, Laura J., University of New Mexico
Datta, Saugata, Kansas State University
Derry, Louis A., Cornell University
Diebold, Frank E., Montana Tech of the University of Montana
Donahoe, Rona J., University of Alabama
Dostie, Philip, Bates College
Dove, Patricia M., Virginia Polytechnic Institute & State University
Drake, John C., University of Vermont
Dworkin, Stephen I., Baylor University
Early, Thomas O., Oak Ridge National Laboratory
Eggleston, Carrick M., University of Wyoming
Elliott, W. Crawford, Georgia State University

Elswick, Erika R., Indiana University, Bloomington
Elwood Madden, Andrew S., University of Oklahoma
Elwood Madden, Megan E., University of Oklahoma
Engel, Annette S., University of Tennessee, Knoxville
Evans, Les J., University of Guelph
Fein, Jeremy B., University of Notre Dame
Ferrell, Jr., Ray E., Louisiana State University
Fortner, Sarah K., Wittenberg University
Frape, Shaun K., University of Waterloo
Frisia, Silvia, University of Newcastle
Fryer, Brian J., University of Windsor
Gamerdinger, Amy P., Pacific Northwest National Laboratory
Ghosh, Swapan K., Indiana University / Purdue University, Indianapolis
Graf, Donald L., University of Illinois, Urbana-Champaign
Grandstaff, David E., Temple University
Graney, Joseph R., Binghamton University
Graustein, William C., Yale University
Greenberg, Sallie, University of Illinois, Urbana-Champaign
Gu, Baohua, Oak Ridge National Laboratory
Gu, Chuanhui, Appalachian State University
Gunter, William D., University of Calgary
Haas, Johnson R., Western Michigan University
Hannigan, Robyn, University of Massachusetts, Boston
Hanor, Jeffrey S., Louisiana State University
Harrison, Wendy J., Colorado School of Mines
Heikoop, Jeffrey, University of South Florida, Tampa
Herman, Janet S., University of Virginia
Herndon, Elizabeth M., Kent State University
Hirons, Steve, Birkbeck College
Hixon, Amy E., University of Notre Dame
Hluchy, Michele M., Alfred University
Hochella, Jr., Michael F., Virginia Polytechnic Institute & State University
Holdren, George R., Pacific Northwest National Laboratory
Hounslow, Arthur, Oklahoma State University
Hui, Alice, Indiana University, Bloomington
Hutcheon, Ian E., University of Calgary
Icenhower, Jonathan P., Pacific Northwest National Laboratory
Jacinthe, Pierre-Andre, Indiana University / Purdue University, Indianapolis
Jackson, Kenneth J., Lawrence Livermore National Laboratory
Jacobson, Andrew D., Northwestern University
Jahnke, Richard, Georgia Institute of Tech
Janecky, David R., Los Alamos National Laboratory
Johnson, Beverly J., Bates College
Karlsson, Haraldur R., Texas Tech University
Kaste, James, College of William & Mary
Kelly, Jacque L., Georgia Southern University
Kettler, Richard M., University of Nebraska, Lincoln
Kimball, Bryn, Whitman College
Kirby, Carl S., Bucknell University
Knapp, Elizabeth P., Washington & Lee University
Kneeshaw, Tara A., Grand Valley State University
Krabbenhoft, David, University of Wisconsin, Madison
Kramer, James R., McMaster University
Krupka, Kenneth M., Pacific Northwest National Laboratory
Ku, Timothy C., Wesleyan University
Kubicki, James D., University of Texas, El Paso
Kump, Lee R., Pennsylvania State University, University Park
Kurtz, Andrew, Boston University
Langman, Jeffrey, University of Idaho
Larsen, Daniel, University of Memphis
Lawrence, James, University of Houston
Leckie, James O., Stanford University
Liang, Liyuan, Oak Ridge National Laboratory
Lincoln, Timothy N., Albion College
Lindberg, Steven E., Oak Ridge National Laboratory
Lindsay, Matthew B., University of Saskatchewan
Locke, Randall, University of Illinois, Urbana-Champaign
Long, Austin, University of Arizona
Long, David T., Michigan State University
Lowenstein, Tim K., Binghamton University
Lyons, William B., Ohio State University
Macalady, Donald L., Colorado School of Mines
Martin, Candace E., University of Otago
Martin, Ellen E., University of Florida
Martin, Jonathan B., University of Florida
Martini, Anna M., Amherst College
Matisoff, Gerald, Case Western Reserve University
Mattigod, Shas V., Pacific Northwest National Laboratory
Maurice, Patricia, University of Notre Dame
Mayes, Melanie A., University of Tennessee, Knoxville
Maynard, J. Barry, University of Cincinnati
Mazer, James J., Argonne National Laboratory
McArthur, John M., University College London
McGee, David, Massachusetts Institute of Tech
McGehee, Thomas L., Texas A&M University, Kingsville
McPhail, D C "Bear", Australian National University
Merino, Enrique, Indiana University, Bloomington
Moore, Joel, Towson University
Morel, Francois M M., Princeton University
Munk, LeeAnn, University of Alaska, Anchorage
Murphy, William M., California State University, Chico
Nelson, Marc A., University of Arkansas, Fayetteville
Nesbitt, H W., Western University
Neumann, Klaus, Ball State University
Newell, Dennis L., Utah State University
Nezat, Carmen A., Eastern Washington University
Niemitz, Jeffery W., Dickinson College
Noll, Mark R., SUNY, The College at Brockport
Norton, Stephen A., University of Maine
Nzengung, Valentine A., University of Georgia
Oelkers, Eric, University College London
Olsen, Amanda A., University of Maine
Oster, Jessica L., Vanderbilt University
Panno, Samuel V., University of Illinois, Urbana-Champaign
Parnell, Roderic A., Northern Arizona University
Patton, Terri L., Argonne National Laboratory
Perdrial, Julia, University of Vermont
Peters, Stephen C., Lehigh University
Pichler, Thomas, University of South Florida, Tampa
Pingitore, Jr., Nicholas E., University of Texas, El Paso
Poulson, Simon, University of Nevada, Reno
Poulton, Simon, University of Leeds
Rademacher, Laura K., University of the Pacific
Reardon, Eric J., University of Waterloo
Reeder, Richard J., SUNY, Stony Brook
Reinfelder, John R., Rutgers, The State University of New Jersey
Rhodes, Amy L., Smith College
Ridley, Moira K., Texas Tech University
Riebe, Clifford S., University of Wyoming
Rimstidt, J. Donald, Virginia Polytechnic Institute & State University
Roberts, Sarah K., Lawrence Livermore National Laboratory
Rogers, Pamela Z., Los Alamos National Laboratory
Romanek, Christopher, Furman University
Rouff, Ashaki, Rutgers, The State University of New Jersey, Newark
Roychoudhury, Alakendra N., Stellenbosch University
Rule, Joseph H., Old Dominion University
Runnells, Donald D., University of Colorado
Rustad, James R., University of California, Davis
Ryan, Peter C., Middlebury College
Sample, James C., Northern Arizona University
Saunders, James A., Auburn University
Schlautman, Mark, Clemson University
Scholz, Denis, Universität Mainz
Schoonen, Martin A., SUNY, Stony Brook
Schoonmaker, Jane E., University of Hawai'i, Manoa
Schroth, Andrew W., University of Vermont
Sjostrom, Derek, Rocky Mountain College
Skidmore, Mark L., Montana State University
Stillings, Lisa, University of Nevada, Reno
Stoessell, Ronald K., University of New Orleans
Stone, Alan T., Johns Hopkins University
Sturchio, Neil C., University of Delaware
Suarez, Celina, University of Arkansas, Fayetteville
Sverjensky, Dimitri A., Johns Hopkins University
Swart, Peter K., University of Miami
Taillefert, Martial, Georgia Institute of Tech
Taylor, Wayne A., Los Alamos National Laboratory
Telmer, Kevin, University of Victoria
Tempel, Gina, University of Nevada, Reno
Tempel, Regina, University of Nevada, Reno
Templeton, Alexis, University of Colorado
Thurber, David L., Queens College (CUNY)
Thurber, David L., Graduate School of the City University of New York

Tipper, Ed, University of Cambridge
Toner, Brandy M., University of Minnesota, Twin Cities
Turner, Robert S., Oak Ridge National Laboratory
Valsami-Jones, Eugenia, University of Birmingham
Van Der Flier-Keller, Eileen, University of Victoria
Varner, Ruth K., University of New Hampshire
Veizer, Jan, University of Ottawa
Velbel, Michael A., Michigan State University
Voglesonger, Kenneth M., Northeastern Illinois University
Walder, Ingar F., New Mexico Institute of Mining and Tech
Walker, Charles T., California State University, Long Beach
Wallace, Adam F., University of Delaware
Walter, Lynn M., University of Michigan
Wang, Jianwei, Louisiana State University
Wasylenki, Laura E., Indiana University, Bloomington
Wehmiller, John F., University of Delaware
Welch, Susan A., Ohio State University
West, A. Joshua (Josh), University of Southern California
Whittemore, Donald O., University of Kansas
Widanagamage, Inoka, University of Mississippi
Wood, James R., Michigan Technological University
Woods, Terri L., East Carolina University
Wunsch, David R., University of Delaware
Yun, Seong-Taek, Korea University
Yuretich, Richard F., University of Massachusetts, Amherst
Zhaohui, Li, University of Wisconsin, Parkside

## Marine Geochemistry

Hoenisch, Baerbel, Columbia University
Adkins, Jess F., California Institute of Tech
Aller, Robert C., SUNY, Stony Brook
Berelson, William M., University of Southern California
Betzer, Peter R., University of South Florida
Biscaye, Pierre E., Columbia University
Bollinger, Marsha S., Winthrop University
Boudreau, Bernard P., Dalhousie University
Broecker, Wallace S., Columbia University
Byrne, Robert H., University of South Florida
Cahill, Kevin, Woods Hole Oceanographic Institution
Calvert, Stephen E., University of British Columbia
Cave, Rachel R., National University of Ireland Galway
Charette, Matthew A., Woods Hole Oceanographic Institution
Church, Thomas M., University of Delaware
Coble, Paula G., University of South Florida
Cole, Catherine, University of St. Andrews
Collier, Robert W., Oregon State University
Compton, John, University of Cape Town
Cuomo, Carmela, University of New Haven
Dai, Minhan, Woods Hole Oceanographic Institution
Davidson, Jon, Durham University
De Carlo, Eric H., University of Hawai'i, Manoa
Dickson, Andrew G., University of California, San Diego
Doney, Scott C., Woods Hole Oceanographic Institution
Fehrenbacher, Jennifer, Oregon State University
Feng, Huan E., Montclair State University
Filippelli, Gabriel M., Indiana University / Purdue University, Indianapolis
Fones, Gary, University of Portsmouth
Francois, Roger, University of British Columbia
Fulweiler, Robinson, Boston University
Galy, Valier, Woods Hole Oceanographic Institution
Gerwick, William H., University of California, San Diego
Glazer, Brian, University of Hawai'i, Manoa
Gospodinova, Kalina D., Woods Hole Oceanographic Institution
Goudreau, Joanne, Woods Hole Oceanographic Institution
Graham, David W., Oregon State University
Grassian, Vicki, University of California, San Diego
Hammerschmidt, Chad, Wright State University
Hammond, Douglas E., University of Southern California
Hansel, Colleen, Woods Hole Oceanographic Institution
Hardison, Amber K., University of Texas at Austin
Heath, G. Ross, University of Washington
Henderson, Paul, Woods Hole Oceanographic Institution
Henley, Sian F., Edinburgh University
Horner, Tristan, Woods Hole Oceanographic Institution
Ingall, Ellery D., Georgia Institute of Tech
Jayakumar, Amal, Princeton University
Johnson, Martin, University of East Anglia
Juranek, Lauren W., Oregon State University
Kastner, Miriam, University of California, San Diego
Klein, Frieder, Woods Hole Oceanographic Institution
Krom, Michael, University of Leeds
Latimer, Jennifer C., Indiana State University
Levas, Stephen J., University of Wisconsin, Whitewater
Li, Yuan-Hui, University of Hawai'i, Manoa
Luther, III, George W., University of Delaware
Mackenzie, Fred T., University of Hawai'i, Manoa
Maerz, Christian, University of Leeds
Malahoff, Alexander, University of Hawai'i, Manoa
McCorkle, Daniel C., Woods Hole Oceanographic Institution
McGillis, Wade, Columbia University
McKay, Jennifer L., Oregon State University
McManus, Jerry, Columbia University
Metz, Simone, Florida Institute of Tech
Mills, Rachel A., University of Southampton
Mincer, Tracy, Woods Hole Oceanographic Institution
Montoya, Joseph, Georgia Institute of Tech
Moore, Bradley S., University of California, San Diego
Morel, Francois M., Princeton University
Mucci, Alfonso, McGill University
Murray, Richard, Boston University
Nuester, Jochen, California State University, Chico
Orians, Kristin J., University of British Columbia
Osburn, Chris, North Carolina State University
Patterson, Molly , Binghamton University
Pearson, Ann, Harvard University
Pedersen, Thomas F., University of Victoria
Peucker-Ehrenbrink, Bernhard, Woods Hole Oceanographic Institution
Popp, Brian N., University of Hawai'i, Manoa
Poyla, David, University of Manchester
Ravizza, Gregory, University of Hawai'i, Manoa
Ries, Justin, Northeastern University
Rosenthal, Yair, Rutgers, The State University of New Jersey
Russell, Ann D., University of California, Davis
Saccocia, Peter J., Bridgewater State University
Sansone, Francis J., University of Hawai'i, Manoa
Schoonmaker, Jane E., University of Hawai'i, Manoa
Sericano, Jose L., Texas A&M University
Seyfried, Jr., William E., University of Minnesota, Twin Cities
Sherrell, Robert M., Rutgers, The State University of New Jersey
Simpson, Jr., H. James, Columbia University
Smith, Joseph P., United States Naval Academy
Spivak, Amanda C., Woods Hole Oceanographic Institution
Stillings, Lisa, University of Nevada, Reno
Sturz, Anne A., University of San Diego
Sulanowska, Margaret, Woods Hole Oceanographic Institution
Sundby, Bjorn, McGill University
Sutherland, Bruce, University of Alberta
Takahashi, Taro, Columbia University
Tan, Chunyang, University of Minnesota, Twin Cities
Thompson, Geoffrey, Woods Hole Oceanographic Institution
Tivey, Margaret K., Woods Hole Oceanographic Institution
Torres, Marta E., Oregon State University
Trefry, John H., Florida Institute of Tech
Tripati, Aradhna, University of California, Los Angeles
Trocine, Robert P., Florida Institute of Tech
Trull, Thomas, Woods Hole Oceanographic Institution
Ullman, William J., University of Delaware
Valentine, David, University of California, Santa Barbara
Wakefield, S J., Cardiff University
Wang, Weihong, Utah Valley University
Wang, Z. Aleck, Woods Hole Oceanographic Institution
Wankel, Scott, Woods Hole Oceanographic Institution
Weiss, Ray F., University of California, San Diego
Wheeler, Patricia, Oregon State University
Winckler, Gisela, Columbia University
Windom, Herbert L., Georgia Institute of Tech
Wong, Chi S., University of British Columbia
Yoder, James A., Woods Hole Oceanographic Institution
Zachos, James C., University of California, Santa Cruz
Zheng, Yan, Graduate School of the City University of New York

## Organic Geochemistry

Abbott, Geoffrey D., University of Newcastle Upon Tyne
Abdulla, Hussain, Texas A&M University, Corpus Christi
Amend, Jan P., Washington University in St. Louis
Anderson, Ken B., Southern Illinois University Carbondale

Aravena, Ramon, University of Waterloo
Baars, Oliver, Princeton University
Bada, Jeffrey L., University of California, San Diego
Bakel, Allen J., Argonne National Laboratory
Banfield, Jillian, University of California, Berkeley
Barker, James F., University of Waterloo
Bianchi, Thomas S., University of Florida
Bissada, Adry, University of Houston
Blair, Neal E., Northwestern University
Bopp, Richard F., Rensselaer Polytechnic Institute
Brassell, Simon C., Indiana University, Bloomington
Burns, Timothy P., Los Alamos National Laboratory
Chen, Bob, University of Massachusetts, Boston
Chou, Mei-In (Melissa), Illinois State Geological Survey
Chou, Sheng-Fu Joseph, Illinois State Geological Survey
Cody, George D., Carnegie Institution of Washington
Cole, Dean A., Los Alamos National Laboratory
Cowie, Gregory L., Edinburgh University
Cowley, Scott W., Colorado School of Mines
Diefendorf, Aaron, University of Cincinnati
Engel, Michael H., University of Oklahoma
Farrington, John W., Woods Hole Oceanographic Institution
Feakins, Sarah, University of Southern California
Fernandez, Loretta, Northeastern University
Filley, Timothy R., Purdue University
Fisher, Liz, University of Liverpool
Francis, Robert D., California State University, Long Beach
Freeman, Katherine H., Pennsylvania State University, University Park
Fu, Qi, University of Houston
Hanson, Andrew D., University of Nevada, Las Vegas
Hartnett, Hilairy E., Arizona State University
Harvey, H. Rodger, Old Dominion University
Hayes, John M., Indiana University, Bloomington
Heal, Kate V., Edinburgh University
Hites, Ronald A., Indiana University, Bloomington
Hockaday, William C., Baylor University
Huang, Yongsong, Brown University
Kelley, Cheryl A., University of Missouri
Kenig, Fabien, University of Illinois at Chicago
Kotulova, Julia, University of Utah
Kruge, Michael A., Montclair State University
Langston-Unkefer, Pat J., Los Alamos National Laboratory
Liss, Peter, University of East Anglia
Love, Gordon, University of California, Riverside
MacCarthy, Patrick, Colorado School of Mines
Macko, Stephen A., University of Virginia
McCarthy, John F., Oak Ridge National Laboratory
Mehra, Aradhana, University of Derby
Meyers, Philip A., University of Michigan
Mitra, Siddhartha, East Carolina University
Mitterer, Richard M., University of Texas, Dallas
Moldowan, J. Michael, Stanford University
Moslow, Thomas F., University of Calgary
Nelson, Robert K., Woods Hole Oceanographic Institution
Ngwenya, Bryne T., Edinburgh University
Nolan, Robert P., Graduate School of the City University of New York
Perdue, Edward Michael, Georgia Institute of Tech
Petsch, Steven, University of Massachusetts, Amherst
Phillips, Dennis, Los Alamos National Laboratory
Philp, R. Paul, University of Oklahoma
Pier, Stanley M., Rice University
Pratt, Lisa M., Indiana University, Bloomington
Pruell, Richard J., University of Rhode Island
Reddy, Christopher M., Woods Hole Oceanographic Institution
Repeta, Daniel J., Woods Hole Oceanographic Institution
Riediger, Cynthia L., University of Calgary
Robinson, Clare, University of Manchester
Sass, Henrik, University of Wales
Sepúlveda, Julio C., University of Colorado
Sessions, Alex L., California Institute of Tech
Silliman, James, Texas A&M University, Corpus Christi
Spongberg, Alison L., University of Toledo
Steen, Andrew, University of Tennessee, Knoxville
Stierle, Donald, Montana Tech of the University of Montana
Talbot, Helen, University of Newcastle Upon Tyne
Tanner, Benjamin R., Western Carolina University
Thiel, Dr., Volker, Georg-August University of Goettingen
Thomas, Elizabeth K., SUNY, Buffalo
Tierney, Jessica, University of Arizona
Townsend-Small, Amy, University of Cincinnati
van Dongen, Bart, University of Manchester
Van Vleet, Edward S., University of South Florida
Venkatesan, M I., University of California, Los Angeles
Visscher, Pieter T., University of Connecticut
Voorhees, Kent J., Colorado School of Mines
Wakeham, Stuart G., Georgia Institute of Tech
Weisener, Christopher, University of Windsor
Werne, Josef, University of Pittsburgh
Whelan, Jean K., Woods Hole Oceanographic Institution
Whiticar, Michael J., University of Victoria
Zerkle, Aubrey, University of St. Andrews
Zhang, Yige, Texas A&M University
Zhuang, Guangsheng, Louisiana State University
Zimmerman, Andrew, University of Florida
Ziolkowski, Lori A., University of South Carolina

## Stable Isotopes

Anderson, Thomas F., University of Illinois, Urbana-Champaign
Anderson, William T., Florida International University
Andrews, Julian E., University of East Anglia
Arnold, Gail, University of Texas, El Paso
Atudorei, Nieu-Viorel, University of New Mexico
Bao, Huiming, Louisiana State University
Bedaso, Zelalem, University of Dayton
Berke, Melissa, University of Notre Dame
Bindeman, Ilya N., University of Oregon
Blum, Joel D., University of Michigan
Bowen, Gabriel, University of Utah
Bowman, John R., University of Utah
Bray, Colin, University of Toronto
Brunner, Benjamin, University of Texas, El Paso
Buczynska, Anna, Northern Illinois University
Burgess, Ray, University of Manchester
Burke, Andrea, University of St. Andrews
Burns, Stephen J., University of Massachusetts, Amherst
Campbell, Andrew R., New Mexico Institute of Mining and Tech
Clayton, Robert N., University of Chicago
Cohen, Anthony, The Open University
Conroy, Jessica, University of Illinois, Urbana-Champaign
Coogon, Rosalind, Imperial College
Criss, Robert E., Washington University in St. Louis
Crowley, Brooke E., University of Cincinnati
Darling, James, University of Portsmouth
Davidson, Garry J., University of Tasmania
Denniston, Rhawn F., University of Iowa
Dettman, David, University of Arizona
Deuser, Werner G., Woods Hole Oceanographic Institution
Dodd, Justin P., Northern Illinois University
Dorale, Jeffrey A., University of Iowa
Douglas, Peter, McGill University
Eastoe, Chris J., University of Arizona
Edwards, Thomas W. D., University of Waterloo
Elder, Kathryn L., Woods Hole Oceanographic Institution
Evans, Samantha, Boise State University
Fantle, Matthew S., Pennsylvania State University, University Park
Farquhar, James, University of Maryland
Fayek, Mostafa, University of Manitoba
Feng, Xiahong, Dartmouth College
Ferguson, Kurt M., Southern Methodist University
Fike, David, Washington University in St. Louis
Fisher, David E., University of Miami
Foster, Gavin, University of Southampton
Fricke, Henry C., Colorado College
Gagnon, Alan R., Woods Hole Oceanographic Institution
Gao, Yongjun, University of Houston
Gazis, Carey A., Central Washington University
Gibson, John J., University of Waterloo
Gilbert, Kathleen W., Wellesley College
Gillikin, David P., Union College
Goldstein, Steven L., Columbia University
Gonzalez, Luis A., University of Kansas
Graczyk, Donald G., Argonne National Laboratory
Grassineau, Nathalie, University of London, Royal Holloway & Bedford New College
Gregory, Robert T., Southern Methodist University

Griffith, Liz, University of Texas, Arlington
Grimes, Stephen, University of Plymouth
Grossman, Ethan L., Texas A&M University
Guo, Weifu, Woods Hole Oceanographic Institution
Haley, Brian, Oregon State University
Halliday, Alex, University of Oxford
Helie, Jean-François, Universite du Quebec a Montreal
Hemming, Sidney R., Columbia University
Henkel, Torsten, University of Manchester
Higgins, Pennilyn, University of Rochester
Hillaire-Marcel, Claude, Universite du Quebec a Montreal
Hilton, David R., University of California, San Diego
Hindshaw, Ruth, University of St. Andrews
Hoek, Joost, University of Maryland
Holk, Gregory J., California State University, Long Beach
Holmden, Chris, University of Saskatchewan
Horton, Travis, University of Canterbury
Howe, Stephen S., SUNY, Albany
Hren, Michael, University of Connecticut
Ingram, B. Lynn, University of California, Berkeley
Ito, Emi, University of Minnesota, Twin Cities
Junium, Christopher, Syracuse University
Kaiser, Jan, University of East Anglia
Kaplan, Isaac R., University of California, Los Angeles
Karhu, Juha A., University of Helsinki
Kaufman, Alan J., University of Maryland
Keigwin, Lloyd D., Woods Hole Oceanographic Institution
Kelley, Simon, The Open University
Kellman, Lisa M., Saint Francis Xavier University
Keppens, Edward, Vrije University Brussel
Kieffer, Bruno, University of British Columbia
Kienast, Markus, Dalhousie University
Kim, Sora L., University of Kentucky
Kirtland Turner, Sandra, University of California, Riverside
Knauth, L. Paul, Arizona State University
Kohn, Matthew, Boise State University
Krantz, David E., University of Delaware
Kreutz, Karl J., University of Maine
Krishnamurthy, R. V., Western Michigan University
Lambert, W. J., University of Alabama
Larson, Peter B., Washington State University
Law, Kim, Western University
Leavitt, Steven W., University of Arizona
Lefticariu, Liliana, Southern Illinois University Carbondale
Li, Long, University of Alberta
Lini, Andrea, University of Vermont
Lohmann, Kyger C., University of Michigan
Longstaffe, Frederick J., Western University
Lowry, David, University of London, Royal Holloway & Bedford New College
Luzincourt, Marc R., Universite du Quebec
Marshall, Jim, University of Liverpool
Matheney, Ronald K., University of North Dakota
Mattey, Dave, University of London, Royal Holloway & Bedford New College
Mayer, Bernhard, University of Calgary
Michalski, Greg, Purdue University
Miller, Christian A., University of Hawai'i, Manoa
Mix, Alan C., Oregon State University
Muehlenbachs, Karlis, University of Alberta
Nelson, Stephen T., Brigham Young University
Nielsen, Sune G., Woods Hole Oceanographic Institution
O'Neil, James, University of Michigan
Ohmoto, Hiroshi, Pennsylvania State University, University Park
Oleynik, Sergey, Princeton University
Olivares, Jose A., Los Alamos National Laboratory
Parai, Rita, Washington University in St. Louis
Parrish, Randy, Leicester University
Passey, Benjamin, University of Michigan
Patterson, William P., University of Saskatchewan
Pedentchouk, Nikolai, University of East Anglia
Perry, Eugene C., Northern Illinois University
Pinti, Daniele Luigi, Universite du Quebec a Montreal
Plummer, Rebecca, University of Maryland
Pogge von Strandmann, Phillip, Birkbeck College
Pogge von Strandmann, Phillip, University College London
Poirier, Andre, Universite du Quebec a Montreal
Popp, Brian N., University of Hawai'i, Manoa
Poulson, Simon R., University of Nevada, Reno
Ravelo, Christina, University of California, Santa Cruz
Rehkamper, Mark, Imperial College
Richards, Ian J., Southern Methodist University
Rye, Danny M., Yale University
Savin, Samuel M., Case Western Reserve University
Schaller, Morgan F., Rensselaer Polytechnic Institute
Schimmelmann, Arndt, Indiana University, Bloomington
Schwarcz, Henry P., McMaster University
Sedlacek, Alexa, University of Northern Iowa
Shahar, Anat, University of Maryland
Sharma, Shikha, West Virginia University
Sharp, Zachary D., University of New Mexico
Sherwood Lollar, Barbara, University of Toronto
Shieh, Yuch-Ning, Purdue University
Showers, William J., North Carolina State University
Smart, Katie A., University of the Witwatersrand
Smirnov, Anna, Universite du Quebec
Sowers, Todd, Pennsylvania State University, University Park
St. Amour, Natalie , Western University
Steig, Eric J., University of Washington
Strauss, Harald, Universitaet Muenster
Stute, Martin, Columbia University
Syrup, Krista A., Moraine Valley Community College
Szynkiewicz, Anna, University of Tennessee, Knoxville
Taylor, Hugh P., California Institute of Tech
Theiling, Bethany P., The University of Tulsa
Thirlwall, Matthew, University of London, Royal Holloway & Bedford New College
Tudhope, Alexander W., Edinburgh University
Valley, John W., University of Wisconsin, Madison
van de Flierdt, Tina, Imperial College
Wanamaker, Alan D., Iowa State University of Science & Tech
Wang, Jianhua, Carnegie Institution of Washington
Webb, Elizabeth A., Western University
Welp, Lisa, Purdue University
Wenner, David B., University of Georgia
White, James W. C., University of Colorado
Widory, David, Universite du Quebec a Montreal
Williams, Douglas F., University of South Carolina
Williams, Helen, Durham University
Wolfe, Amy, Miami University
Wright, James D., Rutgers, The State University of New Jersey
Yapp, Crayton J., Southern Methodist University
Yeung, Laurence Y., Rice University
Young, Edward D., University of California, Los Angeles
Yun, Misuk, University of Manitoba
Zhang, Lin, Texas A&M University, Corpus Christi
Zhang, Ren, Baylor University
Ziegler, Karen, University of New Mexico
Zierenberg, Robert A., University of California, Davis

## Trace Element Distribution

Coale, Kenneth H., Moss Landing Marine Laboratories
Condie, Kent C., New Mexico Institute of Mining and Tech
Croot, Peter, National University of Ireland Galway
Cullers, Robert L., Kansas State University
Daniels, William R., Los Alamos National Laboratory
Efurd, Deward W., Los Alamos National Laboratory
Frey, Frederick A., Massachusetts Institute of Tech
Gabrielli, Paolo, Ohio State University
Hale, Beverley A., University of Guelph
Hayes, John M., Woods Hole Oceanographic Institution
Kar, Aditya, Fort Valley State University
Kyte, Frank T., University of California, Los Angeles
LaFleche, Marc R., Universite du Quebec
Liese, Homer C., University of Connecticut
Mantei, Erwin J., Missouri State University
Marcantonio, Franco, Texas A&M University
Nabelek, Peter I., University of Missouri
Newton, Robert, Columbia University
Nicolescu, Stefan, Yale University
Peterson, Eugene J., Los Alamos National Laboratory
Ritter, Charles J., University of Dayton
Ryan, Jeffrey G., University of South Florida, Tampa
Seifert, Karl E., Iowa State University of Science & Tech
Simonetti, Tony, University of Notre Dame

Thomas, Kimberly W., Los Alamos National Laboratory
Thompson, Joseph L., Los Alamos National Laboratory
Von Reden, Karl F., Woods Hole Oceanographic Institution
Weaver, Barry L., University of Oklahoma
Wildeman, Thomas R., Colorado School of Mines
Zhang, Pu, University of Minnesota, Twin Cities

## GEOPHYSICS

### General Geophysics

Adams, Aubreya, Colgate University
Adepelumi, A. A., Obafemi Awolowo University
Afolabi, O., Obafemi Awolowo University
Ahern, Judson L., University of Oklahoma
Aihaiti, Muhetaer, Carnegie Institution of Washington
Al-Lehyani, Ayman, King Fahd University of Petroleum and Minerals
Al-Shuhail, Abdullatif, King Fahd University of Petroleum and Minerals
Alao, O. A., Obafemi Awolowo University
Albright, James N., Los Alamos National Laboratory
Anderson, Megan L., Colorado College
Angelopoulos, Vassilis, University of California, Los Angeles
Armah, Thomas K., University of Ghana
Atekwana, Estella, Oklahoma State University
Audet, Pascal, University of Ottawa
Bachtadse, Valerian, Ludwig-Maximilians-Universitaet Muenchen
Bailey, Richard C., University of Toronto
Barbi, Greta, Ludwig-Maximilians-Universitaet Muenchen
Barsch, Robert, Ludwig-Maximilians-Universitaet Muenchen
Bastow, Ian, Imperial College
Batzle, Michael L., Colorado School of Mines
Beaumont, Christopher, Dalhousie University
Becker, Alex, University of California, Berkeley
Bendick, Rebecca, University of Montana
Bentley, Charles R., University of Wisconsin, Madison
Bercovici, David, Yale University
Beresnev, Igor A., Iowa State University of Science & Tech
Billen, Magali I., University of California, Davis
Billings, Stephen, University of British Columbia
Bina, Craig R., Northwestern University
Birch, Francis S., University of New Hampshire
Bird, G. Peter, University of California, Los Angeles
Blankenship, Donald D., University of Texas at Austin
Bleibinhaus, Florian, Ludwig-Maximilians-Universitaet Muenchen
Bleistein, Norman, Colorado School of Mines
Bloxham, Jeremy, Harvard University
Boadu, Fred K., Duke University
Boaga, Jacopo, Università degli Studi di Padova
Boehler, Reinhard, Carnegie Institution of Washington
Boettcher, Margaret S., University of New Hampshire
Bokelmann, Goetz, University of Vienna
Boyd, Oliver, University of Memphis
Bradley, Christopher R., Los Alamos National Laboratory
Braun, Alexander, Queen's University
Bregman, Martin, Tulsa Community College
Briggs, Derek E., Yale University
Britzke, Gilbert, Ludwig-Maximilians-Universitaet Muenchen
Brodholt, John, University College London
Brooks, Debra A., Santiago Canyon College
Brown, Wesley A., Stephen F. Austin State University
Buck, IV, W. R., Columbia University
Bunge, Hans-Peter, Ludwig-Maximilians-Universitaet Muenchen
Bustin, Amanda, University of British Columbia
Butler, Karl, University of New Brunswick
Butler, Sam, University of Saskatchewan
Calvin, Wendy, University of Nevada, Reno
Campbell, David L., University of Iowa
Carlson, Barry A., Delta College
Carpenter, Philip J., Northern Illinois University
Carrigan, Charles R., Lawrence Livermore National Laboratory
Cassiani, Giorgio, Università degli Studi di Padova
Cayzer, Nicola, Edinburgh University
Cerpa Gilvonio, Nestor, University of Minnesota, Twin Cities
Chapman, Mark, Edinburgh University
Chase, Clement G., University of Arizona
Cheadle, Michael J., University of Wyoming
Chow, Paul, Carnegie Institution of Washington
Clarke, Garry K. C., University of British Columbia
Clement, William P., University of Massachusetts, Amherst
Coakley, Bernard, University of Alaska, Fairbanks
Colgan, William, York University
Comas, Xavier, Florida Atlantic University
Cooper, Catherine , Washington State University
Copley, Alex, University of Cambridge
Cosentino, Pietro, University of East Anglia
Couture, Gilles, Universite du Quebec a Montreal
Craig, Mitchell S., California State University, East Bay
Craven, John, Edinburgh University
Crossley, David J., Saint Louis University
Daly, Eve, National University of Ireland Galway
Daniels, Jeffrey J., Ohio State University
Davies, J H., Cardiff University
Davies, Rhordri, Imperial College
Davis, Daniel M., SUNY, Stony Brook
Davis, Tara, University College Cork
Dengler, Lorinda, Humboldt State University
Dewdney, Chris, University of Portsmouth
Dickman, Steven R., Binghamton University
Dieterich, James H., University of California, Riverside
Dmochowski, Jane, University of Pennsylvania
Doll, William E., Oak Ridge National Laboratory
Dorman, Leroy M., University of California, San Diego
Dunbar, John A., Baylor University
Durland, Theodore, Oregon State University
Dzunic, Aleksandar, Curtin University
Ebel, John E., Boston College
Edwards, C L, Los Alamos National Laboratory
Ehlers, Todd A., University of Michigan
Ekstrom, Goran, Columbia University
Ellwood, Brooks, University of Texas, Arlington
Emmer, Barbara, Ludwig-Maximilians-Universitaet Muenchen
England, Richard, Leicester University
Falebita, Dele E., Obafemi Awolowo University
Fehler, Michael, Massachusetts Institute of Tech
Ferguson, John F., University of Texas, Dallas
Ferrucci, Fabrizio, The Open University
Fichtner, Andreas, Ludwig-Maximilians-Universitaet Muenchen
Filina, Irina, University of Nebraska, Lincoln
Fishwick, Stewart, Leicester University
Flesch, Lucy M., Purdue University
Foley, Bradford, Pennsylvania State University, University Park
Fontaine, Fabrice R., Universite de la Reunion
Forte, Alessandro Marco, Universite du Quebec a Montreal
Fortier, Richard, Universite Laval
Foulger, Gillian R., Durham University
French, Melodie, University of Maryland
Gaffney, Edward S., Los Alamos National Laboratory
Gaidos, Eric J., University of Hawai'i, Manoa
Gallovic, Frantisek, Ludwig-Maximilians-Universitaet Muenchen
Galsa, Attila, Eotvos Lorand University
Gardner, Gerald H. F., Rice University
Gebrande, Helmut, Ludwig-Maximilians-Universitaet Muenchen
Ghazala, Hosni H., El Mansoura University
Gilder, Stuart, Ludwig-Maximilians-Universitaet Muenchen
Goes, Saskia, Imperial College
Goforth, Thomas T., Baylor University
Goodliffe, Andrew, University of Alabama
Gorman, Gerard, Imperial College
Goulty, Neil R., Durham University
Graham, Gina R., Alaska Division of Geological & Geophysical Surveys
Gramsch, Stephen, Carnegie Institution of Washington
Grand, Stephen P., University of Texas at Austin
Green, Douglas H., Ohio University
Greenfield, Roy J., Pennsylvania State University, University Park
Gregg, Patricia, University of Illinois, Urbana-Champaign
Gribenko, Alex, University of Utah
Grindlay, Nancy R., University of North Carolina Wilmington
Haas, Christian, York University
Haber, Eldad, University of British Columbia
Hamzaoui, Cherif, Universite du Quebec a Montreal
Hardage, Bob A., University of Texas at Austin, Jackson School of Geosciences
Harder, Steven H., University of Texas, El Paso
Harmon, Nicholas, University of Southampton
Harry, Dennis L., Colorado State University

Hayes, Jorden L., Dickinson College
Heigold, Paul C., Argonne National Laboratory
Henson, Harvey, Southern Illinois University Carbondale
Henstock, Tim, University of Southampton
Henyey, Thomas L., University of Southern California
Herman, Rhett B., Radford University
Hier-Majumder, Saswata, University of Maryland
Hier-Majumder, Saswata, University of London, Royal Holloway & Bedford New College
Higgs, Bettie, University College Cork
Hofmeister, Anne M., Washington University in St. Louis
Hornbach, Matthew, University of Texas at Austin
Horowitz, Franklin G., Cornell University
Horváth, Ferenc, Eotvos Lorand University
Houseman, Greg, University of Leeds
Huang, Lianjie, Los Alamos National Laboratory
Huerta, Audrey, Central Washington University
Iaffaldano, Giampiero, Ludwig-Maximilians-Universitaet Muenchen
Igel, Heiner, Ludwig-Maximilians-Universitaet Muenchen
Irving, Jessica, Princeton University
Ishii, Miaki, Harvard University
Ito, Garrett T., University of Hawai'i, Manoa
Ivanov, Julian, University of Kansas
Jackson, James A., University of Cambridge
Jacob, Robert W., Bucknell University
Jaiswal, Priyank, Oklahoma State University
Jarvis, Gary T., York University
Jensen, Olivia G., McGill University
Johnson, Kaj, Indiana University, Bloomington
Johnson, Paul A., Los Alamos National Laboratory
Johnson, Verner C., Colorado Mesa University
Jomeiri, Rahim, University of Tabriz
Jones, Douglas S., University of Florida
Jurdy, Donna M., Northwestern University
Kafka, Alan L., Boston College
Kahle, Beth, University of Cape Town
Kaplinski, Matthew A., Northern Arizona University
Katz, Samuel, Rensselaer Polytechnic Institute
Kaus, Boris P., Universität Mainz
Kavner, Abby, University of California, Los Angeles
Keating, Kristina M., Rutgers, The State University of New Jersey, Newark
Keller, G. Randy, University of Oklahoma
Keller, G. Randy, University of Missouri
Keller, George H., Oregon State University
Kelley, Shari, New Mexico Institute of Mining & Tech
Kellogg, James N., University of South Carolina
Kellogg, Louise H., University of California, Davis
Kenyon, Patricia M., City College (CUNY)
Kenyon, Patricia M., Graduate School of the City University of New York
King, Robert W., Massachusetts Institute of Tech
King, Scott D., Virginia Polytechnic Institute & State University
Kingdon, Kevin, University of British Columbia
Kinsland, Gary L., University of Louisiana at Lafayette
Korenaga, Jun, Yale University
Kroeger, Glenn C., Trinity University
Kruger, Joseph M., Lamar University
Kruse, Sarah E., University of South Florida, Tampa
Kukowski, Nina, Friedrich-Schiller-University Jena
Kuo, John T., Columbia University
Käser, Martin, Ludwig-Maximilians-Universitaet Muenchen
Lat, Che Noorliza, University of Malaya
Lathrop, Daniel, University of Maryland
Laudon, Thomas S., University of Wisconsin, Oshkosh
Lavallee, Yan, Ludwig-Maximilians-Universitaet Muenchen
Lavier, Luc L., University of Texas at Austin
Leitch, Alison, Memorial University of Newfoundland
Lenardic, Adrian, Rice University
Leonard, Lucinda, University of Victoria
Lerche, Ian, University of South Carolina
Lewis, Stephen D., California State University, Fresno
Lin, Fan-Chi, University of Utah
Lin, Guochin, University of Miami
Lithgow-Bertelloni, Carolina, University College London
Liu, Lanbo, University of Connecticut
Liu, Lijun, University of Illinois, Urbana-Champaign
Liu, Mian, University of Missouri
Liu, Yajing, McGill University
Lohman, Rowena B., Cornell University
Lopez-Pineda, Leobardo, Centro de Estudios Superiores del Estado Sonora
Louie, John, University of Nevada, Reno
Lovell, Mike, Leicester University
Luttrell, Karen, Louisiana State University
Lutz, Pascale, Institut Polytechnique LaSalle Beauvais (ex-IGAL)
Mackwell, Stephen J., Rice University
Mahatsente, Rezene, University of Alabama
Malinconico, Lawrence L., Lafayette College
Malservisi, Rocco, Ludwig-Maximilians-Universitaet Muenchen
Mao, Ho-kwang (David), University of Chicago
Mareschal, Jean-Claude, Universite du Quebec a Montreal
Martens, Hilary R., University of Montana
Masterlark, Timothy L., South Dakota School of Mines & Tech
Mavko, Gerald M., Stanford University
Maxwell, Michael, University of British Columbia
McElwaine, Jim, Durham University
McGinnis, Lyle D., Argonne National Laboratory
McKenzie, Dan P., University of Cambridge
Meade, Brendan, Harvard University
Meadows, Wayne R., Los Alamos National Laboratory
Meert, Joseph G., University of Florida
Meng, Yue, Carnegie Institution of Washington
Merriam, James B., University of Saskatchewan
Merrill, Ronald T., University of Washington
Mickus, Kevin L., Missouri State University
Mikesell, Dylan, Boise State University
Milne, Glenn A., University of Ottawa
Mitrovica, Jerry X., Harvard University
Mohr, Marcus, Ludwig-Maximilians-Universitaet Muenchen
Montana, Carlos J., University of Texas, El Paso
Montesi, Laurent G., University of Maryland
Moorkamp, Max, Leicester University
Morgan, F D., Massachusetts Institute of Tech
Morgan, Jason, University of London, Royal Holloway & Bedford New College
Morgan, Joanna, Imperial College
Morris, William A., McMaster University
Morrison, H. Frank, University of California, Berkeley
Morse, David L., University of Texas at Austin
Mukerji, Tapan, Stanford University
Murphy, Vincent, Boston College
Ni, James, University of Missouri
NICULESCU, Bogdan M., University of Bucharest
Ntarlagiannis, Dimitrios, Rutgers, The State University of New Jersey, Newark
Nunn, Jeffrey A., Louisiana State University
Nur, Amos M., Stanford University
Nyblade, Andrew A., Pennsylvania State University, University Park
Nyquist, Jonathan, Temple University
O'Farrell, Keely A., University of Kentucky
Ogiesoba, Osareni C., University of Texas at Austin
Ojo, S. B., Obafemi Awolowo University
Oladunjoye, M. A., University of Ibadan
Olayinka, A. I., University of Ibadan
Oldenburg, Douglas W., University of British Columbia
Olgaard, David L., Rice University
Olhoeft, Gary R., Colorado School of Mines
Oliver, Ronald D., Los Alamos National Laboratory
Olorunfemi, M. O., Obafemi Awolowo University
Olson, Peter L., Johns Hopkins University
Oser, Jens, Ludwig-Maximilians-Universitaet Muenchen
Osinowo, O. O., University of Ibadan
Oware, Erasmus K., SUNY, Buffalo
Pain, Christopher, Imperial College
Palmer, Donald F., Kent State University
Palutoglu, Mahmut, Firat University
Panero, Wendy R., Ohio State University
Papitashvili, Vladimir, National Science Foundation
Park, Changyong, Carnegie Institution of Washington
Parmentier, E. Marc, Brown University
Parsekian, Andrew D., University of Wyoming
Pasicznyk, David L., New Jersey Geological and Water Survey
Patton, Howard J., Los Alamos National Laboratory
Peavy, Samuel T., Georgia Southwestern State University

Peirce, Christine, Durham University
Pelton, John R., Boise State University
Peters, Mark T., Los Alamos National Laboratory
Petersen, Nikolai, Ludwig-Maximilians-Universitaet Muenchen
Pfuhl, Helen, Ludwig-Maximilians-Universitaet Muenchen
Phillips, William S., Los Alamos National Laboratory
Piazzoni, Antonio Sebastian, Ludwig-Maximilians-Universitaet Muenchen
Pimentel-Hernandez, Ramon, Universidad Autonoma de Baja California Sur
Plattner, Christina, Ludwig-Maximilians-Universitaet Muenchen
Pollack, Henry N., University of Michigan
Porch, William M., Los Alamos National Laboratory
Priestley, Keith, University of Cambridge
Pringle, Jamie, Keele University
Pulliam, Jay, University of Texas at Austin
Pysklywec, Russell N., University of Toronto
Raine, Robin, National University of Ireland Galway
Ranalli, Giorgio, Carleton University
Randall, George, Los Alamos National Laboratory
Rasmussen, Steen, Los Alamos National Laboratory
Ravat, Dhananjay, University of Kentucky
Rawlinson, Nick, University of Cambridge
Reilinger, Robert E., Massachusetts Institute of Tech
Rempel, Alan W., University of Oregon
Revetta, Frank A., SUNY Potsdam
Rice, James R., Harvard University
Richard, Robert, University of California, Los Angeles
Richards, Mark A., University of California, Berkeley
Richardson, Randall M., University of Arizona
Roach, Michael, University of Tasmania
Roberts, Mark L., Woods Hole Oceanographic Institution
Robinson, Edwin S., Virginia Polytechnic Institute & State University
Robinson, Judith, Rutgers, The State University of New Jersey, Newark
Rochester, Michael G., Memorial University of Newfoundland
Rodi, William, Massachusetts Institute of Tech
Roecker, Steven W., Rensselaer Polytechnic Institute
Romanovsky, Vladimir, University of Alaska, Fairbanks
Rothman, Daniel H., Massachusetts Institute of Tech
Rouleau, Pierre M., Sir Wilfred Grenfell College
Rubin, Allan M., Princeton University
Rudge, John, University of Cambridge
Russell, Christopher T., University of California, Los Angeles
Russell, R. Doncaster, University of British Columbia
Russo, Raymond, University of Florida
Ryall, Patrick J., Dalhousie University
Rychert, Catherine A., University of Southampton
Sabra, Karim, Georgia Institute of Tech
Salami, Sikiru A., University of Benin
Sammis, Charles G., University of Southern California
Sass, John H., Northern Arizona University
Sauck, William A., Western Michigan University
Sava, Diana, University of Texas at Austin, Jackson School of Geosciences
Schmandt, Brandon, University of New Mexico
Schriver, David, University of California, Los Angeles
Schubert, Gerald, University of California, Los Angeles
Schuberth, Bernhard, Ludwig-Maximilians-Universitaet Muenchen
Schultz, Adam, Oregon State University
Searls, Mindi L., University of Nebraska, Lincoln
Segall, Paul, Stanford University
Serzu, Mulu, University of Manitoba
Shams, Asghar, Heriot-Watt University
Shcherbakov, Robert, Western University
Sherrod, Laura, Kutztown University of Pennsylvania
Sigloch, Karin, University of Oxford
Simila, Gerald W., California State University, Northridge
Simmons, M. G., Massachusetts Institute of Tech
Simms, Janet E., Mississippi State University
Sinogeikin, Stanislav, Carnegie Institution of Washington
Sit, Stefany , University of Illinois at Chicago
Slater, Lee S., Rutgers, The State University of New Jersey, Newark
Sleep, Norman H., Stanford University
Smylie, Douglas E., York University
Sobotka, Jerzy, University of Wroclaw
Solomatov, Viatcheslav S., Washington University in St. Louis
Solomon, Sean C., Carnegie Institution of Washington
Sonett, Charles P., University of Arizona
Spanos, T.J.T (Tim), University of Alberta
Sparks, David, Texas A&M University
Stearns, Richard G., Vanderbilt University
Stein, Carol A., University of Illinois at Chicago
Stephenson, Randell A., University of Aberdeen
Stidham, Christiane W., SUNY, Stony Brook
Stierman, Donald J., University of Toledo
Stixrude, Lars, University College London
Stupazzini, Marco, Ludwig-Maximilians-Universitaet Muenchen
Suckale, Jenny, Stanford University
Sweeney, Mark D., Pacific Northwest National Laboratory
Szeliga, Walter, Central Washington University
Szeto, Anthony M. K., York University
Taib, Samsudin Hj., University of Malaya
Tatham, Robert H., University of Texas at Austin
Taylor, Graeme, University of Plymouth
ten Brink, Uri S., Woods Hole Oceanographic Institution
Thiruvathukal, John V., Montclair State University
Thompson, George A., Stanford University
Thompson, Michael D., Argonne National Laboratory
Timár, Gábor, Eotvos Lorand University
Tobin, Harold, University of Wisconsin, Madison
Tong, Vincent C., Birkbeck College
Tsai, Victor, California Institute of Tech
Tsoflias, George P., University of Kansas
Turcotte, Donald L., University of California, Davis
Van Avendonk, Harm, University of Texas at Austin
van Keken, Peter J., University of Michigan
Velli, Marco, University of California, Los Angeles
Vernhes, Jean-David, Institut Polytechnique LaSalle Beauvais (ex-IGAL)
Versteeg, Roelof, Columbia University
Vincent, Robert K., Bowling Green State University
von Glasow, Roland, University of East Anglia
Vopson, Melvin M., University of Portsmouth
Waff, Harve S., University of Oregon
Waltham, Dave, University of London, Royal Holloway & Bedford New College
Wang, Chi-Yuen, University of California, Berkeley
Wang, Herbert F., University of Wisconsin, Madison
Wang, Kelin, University of Victoria
Wang, Yanghua, Imperial College
Warren, Linda, Saint Louis University
Wassermann, Joachim, Ludwig-Maximilians-Universitaet Muenchen
Weaver, Thomas A., Los Alamos National Laboratory
Weeraratne, Dayanthie, California State University, Northridge
Wettlaufer, John S., Yale University
White, Nicky, University of Cambridge
White, Robert, University of Cambridge
Whitman, Dean, Florida International University
Williams, Ian S., University of Wisconsin, River Falls
Williams, II, Richard T., University of Tennessee, Knoxville
Wilson, Clark R., University of Texas at Austin
Winberry, Paul, Central Washington University
Winklhofer, Michael, Ludwig-Maximilians-Universitaet Muenchen
Wisely, Beth , Casper College
Wishart, De Bonne N., Central State University
Woodhouse, John, University of Oxford
Woods, Andy, University of Cambridge
Worthington, Lindsay L., University of New Mexico
Xiao, Yuming, Carnegie Institution of Washington
Yelisetti, Subbarao, Texas A&M University, Kingsville
Yuen, David A., University of Minnesota, Twin Cities
Zahm, Christopher K., University of Texas at Austin
Zaja, Annalisa, Università degli Studi di Padova
Zarroca Hernández, Mario, Universitat Autonoma de Barcelona
Zaspel, Craig E., University of Montana Western
Zelt, Colin A., Rice University
Zha, Changsheng, Carnegie Institution of Washington
Zhang, Chi, University of Kansas
Zoback, Mark D., Stanford University

## Experimental Geophysics

Anderson, Orson L., University of California, Los Angeles
Askari, Roohollah, Michigan Technological University
Bates, Richard, University of St. Andrews
Bevis, Michael G., Ohio State University
Bleibinhaus, Florian, Friedrich-Schiller-University Jena

Bohlen, Thomas, Karlsruhe Institute of Tech
Bona, Andrej, Curtin University
Bridges, Frank, University of California, Santa Cruz
Bussod, Gilles Y., Los Alamos National Laboratory
Carson, Michael, Curtin University
Christensen, Nikolas I., University of Wisconsin, Madison
Comina, Cesare, Università di Torino
Dewers, Thomas, New Mexico Institute of Mining and Tech
Dyaur, Nikolay, University of Houston
Evans, J. B., Massachusetts Institute of Tech
Fomel, Sergey, University of Texas at Austin, Jackson School of Geosciences
Fratta, Dante, University of Wisconsin, Madison
Glover, Paul W., University of Leeds
Glover, Paul W., Universite Laval
Glubokovskikh, Stanislav, Curtin University
Graham, Jr., Earl K., Pennsylvania State University, University Park
Green, II, Harry W., University of California, Riverside
Gross, Gerardo W., New Mexico Institute of Mining and Tech
Heinz, Dion L., University of Chicago
Herrmann, Felix J., University of British Columbia
Kaip, Galen M., University of Texas, El Paso
Key, Kerry, University of California, San Diego
Kilburn, Chris, University College London
Kitajima, Hiroko, Texas A&M University
Kohlstedt, David L., University of Minnesota, Twin Cities
Korvin, Gabor, King Fahd University of Petroleum and Minerals
Lebedev, Maxim, Curtin University
Manghnani, Murli H., University of Hawai'i, Manoa
Mao, Ho-kwang, Carnegie Institution of Washington
Marone, Chris, Pennsylvania State University, University Park
Marshall, Hans-Peter, Boise State University
Marshall, Scott, Appalachian State University
McNamara, Allen, Arizona State University
Mei, Shenghua, University of Minnesota, Twin Cities
Narod, Barry, University of British Columbia
Parks, George K., University of Washington
Pec, Matej, University of Minnesota, Twin Cities
Pevzner, Roman, Curtin University
Quintanilla Terminal, Alejandra, University of Minnesota, Twin Cities
Ruina, Andy L., Cornell University
Schmitt, Douglas R., University of Alberta
Scholz, Christopher H., Columbia University
Shankland, Thomas J., Los Alamos National Laboratory
Shaw, George H., Union College
Shieh, Sean R., Western University
Shu, Jinfu, Carnegie Institution of Washington
Somayazulu, Maddury, Carnegie Institution of Washington
Tullis, Terry E., Brown University
Vanacore, Elizabeth, University of Puerto Rico
Vanorio, Tiziana, Stanford University
Vaughan, Michael T., SUNY, Stony Brook
Velasco, Aaron A., University of Texas, El Paso
Vialle, Stephanie , Curtin University
Wang, Herbert F., University of Wisconsin, Madison
Watanabe, Tohru, University of Toyama
Westwood, Rachel, Keele University
Wong, Teng-fong, SUNY, Stony Brook
Yong, Wenjun, Western University
Zebker, Howard A., Stanford University
Zimmerman, Mark, University of Minnesota, Twin Cities

## Exploration Geophysics

Al-Shuhail, Abdullah, King Fahd University of Petroleum and Minerals
Alumbaugh, David L., University of California, Berkeley
Armadillo, Egidio, Universita di Genova
B, Dr. V., University College of Science, Osmania University
Backus, Milo M., University of Texas at Austin
Basham, William L., University of Texas, Permian Basin
Becker, Alex, University of California, Berkeley
Bedle, Heather, University of Houston
Benson, Robert D., Colorado School of Mines
Best, Melvyn E., University of Victoria
Biondi, Biondo L., Stanford University
Black, Ross A., University of Kansas
Brabham, P J., Cardiff University
Bradford, John, Boise State University
Brantley, Duke, University of South Carolina
Brassea, Jesus M., Centro de Investigación Científica y de Educación Superior de Ensenada
Breland, Clayton F., Louisiana State University
Brown, Larry D., Cornell University
Carmichael, Robert S., University of Iowa
Castagna, John P., University of Houston
Chandler, Val W., University of Minnesota, Twin Cities
Chen, Jingyi, The University of Tulsa
Claerbout, Jon F., Stanford University
Clayton, Robert W., California Institute of Tech
Cook, Frederick A., University of Calgary
Corbato, Charles E., Ohio State University
Coruh, Cahit, Virginia Polytechnic Institute & State University
Davis, Thomas L., Colorado School of Mines
DeAngelo, Micheal V., University of Texas at Austin, Jackson School of Geosciences
Donaldson, Paul R., Boise State University
Duckworth, Kenneth, University of Calgary
Dunlap, Dallas B., University of Texas at Austin, Jackson School of Geosciences
Dunn, Bernie, Western University
Ellett, Kevin M., Indiana University
Esparza, Francisco, Centro de Investigación Científica y de Educación Superior de Ensenada
Espinosa, Juan M., Centro de Investigación Científica y de Educación Superior de Ensenada
Ferguson, Ian J., University of Manitoba
Flores, Carlos, Centro de Investigación Científica y de Educación Superior de Ensenada
Fomel, Sergey B., University of Texas at Austin
Gaede, Oliver M., Queensland University of Tech
Gajewski, Dirk J., Universitaet Hamburg
Galvin, Robert, Curtin University
Gao, Dengliang, West Virginia University
Gendzwill, Donald J., University of Saskatchewan
Giroux, Bernard, Universite du Quebec
Gloaguen, Erwan, Universite du Quebec
Goloshubin, Gennady, University of Houston
Gomez, Enrique, Centro de Investigación Científica y de Educación Superior de Ensenada
Gonzalez-Escobar, Mario, Centro de Investigación Científica y de Educación Superior de Ensenada
Grana, Dario, University of Wyoming
Gurevich, Boris, Curtin University
Hale, Dave, Colorado School of Mines
Han, De-hua, University of Houston
Hardage, Bob A., University of Texas at Austin
Harris, James B., Millsaps College
Hauser, Ernest C., Wright State University
Hills, Denise J., Geological Survey of Alabama
Hilterman, Fred, University of Houston
Hinze, William J., Purdue University
Hole, John A., Virginia Polytechnic Institute & State University
Holt, John W., University of Texas at Austin
Homuth, Emil F., Los Alamos National Laboratory
Howman, Dominic J., Curtin University
Hu, Hao, University of Houston
Jiracek, George R., San Diego State University
Johnson, Ansel G., Portland State University
Jones, Francis H., University of British Columbia
Keach II, R. William, Brigham Young University
Kepic, Anton W., Curtin University
Khalil, Mohamed, Montana Tech
Knapp, Camelia, University of South Carolina
Knapp, James H., University of South Carolina
Larner, Kenneth L., Colorado School of Mines
Lawton, Donald C., University of Calgary
Lee, Ki Ha, University of California, Berkeley
Levander, Alan R., Rice University
Li, Yaoguo, Colorado School of Mines
Lillie, Robert J., Oregon State University
Lines, Larry R., University of Calgary
Link, Curtis A., Montana Tech
Louie, John N., University of Nevada, Reno
Lourenço, José M., Universidade de Trás-os-Montes e Alto Douro
Madadi, Mahyar, Curtin University
Magnani, Maria B., University of Memphis

Margrave, Gary F., University of Calgary
Marobhe, Isaac M., University of Dar es Salaam
Mathur, Dr R., University College of Science, Osmania University
Mazariegos, Ruben A., University of Texas, Pan American
McBride, John H., Brigham Young University
McPherson, Brian, University of Utah
Miah, Khalid, Montana Tech
Mikhaltsevitch, Vassily, Curtin University
Mikulich, Matthew J., Virginia Polytechnic Institute & State University
Milkereit, Bernd, University of Toronto
Miller, Richard D., University of Kansas
Morrison, Huntly Frank, University of California, Berkeley
Nabighian, Misac, Colorado School of Mines
Olawuyi, Kehinde A., University of Ilorin
Olsen, Kim B., San Diego State University
Palmer, Derecke, University of New South Wales
Perez, Marco A., Centro de Investigación Científica y de Educación Superior de Ensenada
Pethick, Andrew, Curtin University
Potter, David, University of Alberta
Pratt, R. Gerhard, Western University
Pujol, Jose M., University of Memphis
Raef, Abdelmoneam E., Kansas State University
Raji, W. O., University of Ilorin
Rechtien, Richard D., Missouri University of Science and Tech
Rector, James W., University of California, Berkeley
Reshef, Moshe, Tel Aviv University
Richard, Benjamin H., Wright State University
Romo, Jose M., Centro de Investigación Científica y de Educación Superior de Ensenada
Rupert, Gerald B., Missouri University of Science and Tech
Sarwar, A. K. Mostofa, University of New Orleans
Sava, Paul, Colorado School of Mines
Sbar, Marc, University of Arizona
Scanlin, Michael A., Elizabethtown College
Schuster, Gerard T., University of Utah
Seifoullaev, Roustam K., University of Texas at Austin
Sen, Mrinal K., University of Texas at Austin
Serpa, Laura F., University of Texas, El Paso
Sexton, John L., Southern Illinois University Carbondale
Shaw, Kenneth L., University of Utah
Smith, Richard S., Laurentian University, Sudbury
Smithyman, Brenden, Western University
Snieder, Roel, Colorado School of Mines
Speece, Marvin A., Montana Tech
Spikes, Kyle T., University of Texas at Austin
Sprenke, Kenneth F., University of Idaho
Sternberg, Ben K., University of Arizona
Stewart, Robert R., University of Houston
Stewart, Robert R., University of Calgary
Stimpson, Ian G., Keele University
Stoffa, Paul L., University of Texas at Austin
Tertyshnikov, Konstantin, Curtin University
Thomsen, Leon, University of Houston
Tsvankin, Ilya D., Colorado School of Mines
Turpening, Roger M., Michigan Technological University
Unsworth, Martyn, Cornell University
Urosevic, Milovan, Curtin University
Vazquez, Rogelio, Centro de Investigación Científica y de Educación Superior de Ensenada
von Frese, Ralph R., Ohio State University
Wannamaker, Phillip E., University of Utah
Watts, Doyle, Wright State University
Weglein, Arthur B., University of Houston
Weiss, Chester J., University of New Mexico
Wilson, Thomas H., West Virginia University
Wolfe, Paul J., Wright State University
Wood, Neill, Exeter University
Zawislak, Ronald L., Middle Tennessee State University
Zhang, Bo, University of Alabama
Zhdanov, Michael S., University of Utah
Ziolkowski, Anton M., Edinburgh University
Ziramov, Sasha, Curtin University

## Geodesy

Al-Attar, David, University of Cambridge
Bennett, Richard, University of Arizona
Blewitt, Geoffrey, University of Nevada, Reno
Brockhaus, John A., United States Military Academy
Caporali, Alessandro, Università degli Studi di Padova
Chaussard, Estelle, SUNY, Buffalo
Clarke, Peter J., University of Newcastle Upon Tyne
Davis, James , Columbia University
Elliott, Julie, Purdue University
Feigl, Kurt, University of Wisconsin, Madison
Flake, Rex , Central Washington University
Fotopoulos, Georgia, Queen's University
Freymueller, Jeffrey T., University of Alaska, Fairbanks
Funning, Gareth J., University of California, Riverside
Gonzalez, Jose J., Centro de Investigación Científica y de Educación Superior de Ensenada
Grejner-Brzezinska, Dorota A., Ohio State University
Hammond, William, University of Nevada
Herring, Thomas A., Massachusetts Institute of Tech
Hooper, Andy, University of Leeds
James, Thomas S., University of Victoria
Jekeli, Christopher, Ohio State University
Kogan, Mikhail, Columbia University
Kreemer, Corne, University of Nevada
Kusznir, Nick, University of Liverpool
LaFemina, Peter C., Pennsylvania State University, University Park
Li, Zhenhong, University of Newcastle Upon Tyne
Lopez, Alberto, University of Puerto Rico
Lowry, Anthony R., Utah State University
Melbourne, Timothy I., Central Washington University
Miller, M. Meghan, Central Washington University
Miner, Andy, Central Washington University
Moore, Phillip, University of Newcastle Upon Tyne
Murray, Mark, New Mexico Institute of Mining and Tech
Newman, Andrew V., Georgia Institute of Tech
Odera, Patroba A., Jomo Kenyatta University of Agriculture & Tech
Pagiatakis, Spiros, York University
Palamartchouk, Kirill, University of Newcastle Upon Tyne
Parsons, Barry, University of Oxford
Pritchard, Matthew E., Cornell University
Rundle, John, University of California, Davis
Saalfeld, Alan J., Ohio State University
Samsonov, Sergey V., Natural Resources Canada
Santillan, Marcelo, Central Washington University
Schaffrin, Burkhard A., Ohio State University
Shum, CK, Ohio State University
Smalley, Jr., Robert, University of Memphis
Smith-Konter, Bridget R., University of Hawai'i, Manoa
Sparks, David, Texas A&M University
Spinler, Joshua C., University of Arkansas at Little Rock
Stegman, Dave, University of California, San Diego
Tiampo, Kristy F., Western University
Waithaka, Hunja, Jomo Kenyatta University of Agriculture & Tech
Wdowinski, Shimon, University of Miami
Wells, David E., University of Southern Mississippi
Wright, Tim, University of Leeds
Yi, Yuchan, Ohio State University

## Geomagnetism & Paleomagnetism

Abokhodair, Abdulwahab, King Fahd University of Petroleum and Minerals
Aldridge, Keith D., York University
Backus, George E., University of California, San Diego
Banerjee, Subir K., University of Minnesota, Twin Cities
Biggin, Andy, University of Liverpool
Bilardello, Dario, University of Minnesota, Twin Cities
Bogue, Scott W., Occidental College
Booker, John R., University of Washington
Borradaile, Graham J., Lakehead University
Bowles, Julie, University of Wisconsin, Milwaukee
Braginsky, Stanislav I., University of California, Los Angeles
Bramham, Emma , University of Leeds
Brown, Laurie, University of Massachusetts, Amherst
Burmester, Russell F., Western Washington University
Channell, James E., University of Florida
Chen, Xiaojia, Carnegie Institution of Washington
Cioppa, Maria T., University of Windsor
Clough, Gene A., Bates College
Coe, Robert S., University of California, Santa Cruz
Constable, Catherine G., University of California, San Diego
Cottrell, Rory D., University of Rochester

De Boer, Jelle Z., Wesleyan University
DeMets, D. Charles, University of Wisconsin, Madison
DiVenere, Victor, Long Island University, C.W. Post Campus
Doh, Seong-Jae, Korea University
Douglass, David N., Pasadena City College
Dulin, Shannon, University of Oklahoma
Ellwood, Brooks B., Louisiana State University
Elmore, R. Douglas, University of Oklahoma
Evans, David A., Yale University
Evans, Michael E., University of Alberta
Everett, Mark, Texas A&M University
Feinberg, Joshua, University of Minnesota, Twin Cities
Franklin, David, University of Portsmouth
Fraser-Smith, Antony C., Stanford University
Geissman, John W., University of New Mexico
Geissman, John W., University of Michigan
Glatzmaier, Gary A., University of California, Santa Cruz
Gordon, Richard G., Rice University
Guerrero-Garcia, Jose C., Universidad Nacional Autonoma de Mexico
Halgedahl, Susan L., University of Utah
Hall, Frank R., University of New Orleans
Hall, Stuart A., University of Houston
Halls, Henry C., University of Toronto
Harbert, William P., University of Pittsburgh
Hill, Mimi, University of Liverpool
Hodych, Joseph P., Memorial University of Newfoundland
Holme, Richard, University of Liverpool
Housen, Bernard A., Western Washington University
Hozik, Michael J., Richard Stockton College of New Jersey
Jackson, Michael, University of Minnesota, Twin Cities
Jones, F. Walter, University of Alberta
Kean, Jr., William F., University of Wisconsin, Milwaukee
Kelso, Paul R., Lake Superior State University
Kent, Dennis V., Rutgers, The State University of New Jersey
Kodama, Kenneth P., Lehigh University
Kravchinsky, Vadim, University of Alberta
Larson, Edwin E., University of Colorado
Layer, Paul, University of Alaska, Fairbanks
Lima, Eduardo A., Massachusetts Institute of Tech
Lindquist, Anna, Lake Superior State University
Lippert, Peter C., University of Utah
Livermore, Phil, University of Leeds
Lund, Steven P., University of Southern California
Maillol, Jean-Michel, University of Calgary
Maloof, Adam C., Princeton University
Marshall, Claude Monte, San Diego State University
Márton, Péter, Eotvos Lorand University
McCausland, Phil J., Western University
Merrill, Ronald T., University of Washington
Morris, Antony, University of Plymouth
Moskowitz, Bruce M., University of Minnesota, Twin Cities
Muxworthy, Adrian, Imperial College
Negrini, Robert M., California State University, Bakersfield
Nilsson, Andreas, University of Liverpool
Noltimier, Hallan C., Ohio State University
Opdyke, Neil D., University of Florida
Palmer, H. C., Western University
Pares, Josep M., University of Michigan
Park, Stephen K., University of California, Riverside
Parker, Robert L., University of California, San Diego
Paty, Carol M., Georgia Institute of Tech
Petronis, Michael S., New Mexico Highlands University
Piper, John, University of Liverpool
Rankin, David, University of Alberta
Richter, Carl, University of Louisiana at Lafayette
Roberts, Paul H., University of California, Los Angeles
Shibuya, Hidetoshi, Kumamoto University
Srnka, Len, University of California, San Diego
Sternberg, Robert S., Franklin and Marshall College
Stone, David B., University of Alaska, Fairbanks
Symons, David T., University of Windsor
Tarduno, John A., University of Rochester
Tauxe, Lisa, University of California, San Diego
Tikoo-Schanz, Sonia, Rutgers, The State University of New Jersey
Valentine, Michael J., University of Puget Sound
van der Voo, Rob, University of Michigan
Verosub, Kenneth L., University of California, Davis
Walden, John, University of St. Andrews
Weaver, John T., University of Victoria
Weiss, Benjamin, Massachusetts Institute of Tech
Whaler, Kathryn A., Edinburgh University
Williams, Wyn, Edinburgh University
Zanella, Elena, Università di Torino
Zhao, Xixi, University of California, Santa Cruz

## Gravity

Aiken, Carlos L., University of Texas, Dallas
Danes, Z. Frank, University of Puget Sound
Eppelbaum, Lev, Tel Aviv University
Ervin, C. Patrick, Northern Illinois University
Garcia, Juan, Centro de Investigación Científica y de Educación Superior de Ensenada
Ghatge, Suhas L., New Jersey Geological and Water Survey
Grannell, Roswitha B., California State University, Long Beach
Hussein, Musa, University of Texas, El Paso
Kusumoto, Shigekazu, University of Toyama
Smirnov, Aleksey K., Michigan Technological University
Steckler, Michael, Columbia University

## Heat Flow

Anderson, Roger N., Columbia University
Beltrami, Hugo, Saint Francis Xavier University
Blackwell, David D., Southern Methodist University
Chapman, David S., University of Utah
Crane, Kathleen, Graduate School of the City University of New York
Fulton, Patrick M., Texas A&M University
Harris, Robert N., Oregon State University
Hsieh, Wen-Pin, Academia Sinica
Huang, Shaopeng, University of Michigan
Jones, Michael Q., University of the Witwatersrand
Lenkey, László, Eotvos Lorand University
Morgan, Paul, Northern Arizona University
Morgan, Paul, Colorado School of Mines
Reiter, Marshall A., New Mexico Institute of Mining and Tech
Sclater, John G., University of California, San Diego
Shen, Po-Yu (Paul), Western University
Smith, Douglas L., University of Florida
Starrfield, Sumner, Arizona State University

## Seismology

Abers, Geoffrey A., Cornell University
Achauer, Ulrich, Universite de Strasbourg
Agnew, Duncan C., University of California, San Diego
Alexander, Shelton S., Pennsylvania State University, University Park
Allen, Clarence R., California Institute of Tech
Allen, Richard M., University of California, Berkeley
Ammon, Charles J., Pennsylvania State University, University Park
Ampuero, Jean-Paul, California Institute of Tech
Anandakrishnan, Sridhar, Pennsylvania State University, University Park
Anderson, John G., University of Nevada, Reno
Angus, Doug, University of Leeds
Arabasz, Walter J., University of Utah
Archuleta, Ralph J., University of California, Santa Barbara
Asencio, Eugenio, University of Puerto Rico
Assatourians, Karen, Western University
Aster, Rhicard C., Colorado State University
Aster, Richard, Colorado State University
Atkinson, Gail M., Western University
Atwater, Brian, University of Washington
Ausbrooks, Scott, Arkansas Geological Survey
Balch, Robert S., New Mexico Institute of Mining and Tech
Barazangi, Muawia, Cornell University
Barclay, Andrew, Columbia University
Barker, Jeffrey S., Binghamton University
Beaumont, Christopher, Dalhousie University
Beck, Susan L., University of Arizona
Beghein, Caroline, University of California, Los Angeles
Ben-Zion, Yehuda, University of Southern California
Benavides-Iglesias, Alfonso, Texas A&M University
Benoit, Margaret, College of New Jersey
Benz, Harley M., University of Utah
Beroza, Gregory C., Stanford University
Bezada, Max, University of Minnesota, Twin Cities
Biasi, Glenn P., University of Nevada, Reno
Bilek, Susan L., New Mexico Institute of Mining and Tech
Blake, Daniel R., Ohio Dept of Natural Resources

Blakely, Robert F., Indiana University, Bloomington
Bodin, Paul A., University of Washington
Bollinger, G. A., Virginia Polytechnic Institute & State University
Borisov, Dmitry, Princeton University
Bostock, Michael G., University of British Columbia
Braile, Lawrence W., Purdue University
Brodsky, Emily, University of California, Santa Cruz
Brown, James R., University of Calgary
Brudzinski, Michael, Miami University
Brumbaugh, David S., Northern Arizona University
Brune, James N., University of Nevada, Reno
Burdick, Scott, University of Maryland
Burton, Paul, University of East Anglia
Byrnes, Joseph, University of Minnesota, Twin Cities
Calvert, Andrew J., Simon Fraser University
Cann, Johnson R., Woods Hole Oceanographic Institution
Caplan-Auerbach, Jackie, Western Washington University
Cardimona, Steve, Mendocino College
Cartwright, J A., Cardiff University
Cassidy, John, University of Victoria
Castro, Raul, Centro de Investigación Científica y de Educación Superior de Ensenada
Chapman, Martin C., Virginia Polytechnic Institute & State University
Chen, Po, University of Wyoming
Chen, Wang-Ping, University of Illinois, Urbana-Champaign
Chen, Xiaowei, University of Oklahoma
Chesnokov, Evgeny, University of Houston
Chiu, Jer-Ming, University of Memphis
Chiu, Shu-Choiung, University of Memphis
Chowdhury, Dipak K., Indiana University / Purdue University, Fort Wayne
Christensen, Douglas, University of Alaska, Fairbanks
Cicerone, Robert D., Bridgewater State University
Cipar, John J., Boston College
Clark, Roger A., University of Leeds
Clowes, Ronald M., University of British Columbia
Cochran, Elizabeth, University of California, Riverside
Collins, John A., Woods Hole Oceanographic Institution
Conder, James A., Southern Illinois University Carbondale
Cormier, Vernon F., University of Connecticut
Cornwell, David, University of Aberdeen
Cottaar, Sanne , University of Cambridge
Courtier, Anna, University of Wisconsin, Whitewater
Cramer, Chris, University of Memphis
Creager, Kenneth C., University of Washington
Crosson, Robert S., University of Washington
Curtis, Andrew, Edinburgh University
D'Amico, Sebastiano, University of Malta
Dalton, Colleen, Brown University
Darbyshire, Fiona Ann, Universite du Quebec a Montreal
Darold, Amberlee, University of Oklahoma
Davis, Paul M., University of California, Los Angeles
Day, Steven M., San Diego State University
De Angelis, Silvio, University of Liverpool
Denolle, Marine, Harvard University
dePolo, Diane, University of Nevada, Reno
DeShon, Heather R., Southern Methodist University
DeShon, Heather, University of Memphis
Dorman, James, University of Memphis
Doser, Diane I., University of Texas, El Paso
Dreger, Douglas S., University of California, Berkeley
Duan, Benchuan, Texas A&M University
Dueker, Kenneth G., University of Wyoming
Dunham, Eric M., Stanford University
Ebel, John E., Boston College
Ekstrom, Goran, Columbia University
Ellis, Robert M., University of British Columbia
Ellsworth, Bill, Stanford University
Faulkner, Daniel, University of Liverpool
Ferreira, Anna, University College London
Fischer, Karen M., Brown University
Fontaine, Fabrice R., Universite de la Reunion
Forbriger, Thomas, Karlsruhe Institute of Tech
Ford, Heather, University of California, Riverside
Fox, Jeff, Ohio Dept of Natural Resources
Frankel, Arthur, University of Washington
Frazer, L. N., University of Hawai'i, Manoa
Frederiksen, Andrew, University of Manitoba
Frez, Jose, Centro de Investigación Científica y de Educación Superior de Ensenada
Frohlich, Cliff, University of Texas at Austin
Frohlich, Clifford A., University of Texas at Austin
Fujita, Kazuya, Michigan State University
Gabrielov, Andrei, Purdue University
Gaherty, James, Columbia University
Galea, Pauline, University of Malta
Gao, Haiying, University of Massachusetts, Amherst
Garces, Milton A., University of Hawai'i, Manoa
Garnero, Edward, Arizona State University
Gharti, Hom Nath, Princeton University
Ghosh, Abhijit, University of California, Riverside
Gibson, Jr., Richard L., Texas A&M University
Gilbert, Hersh, Purdue University
Gilpin, Bernard J., Golden West College
Glowacka, Ewa, Centro de Investigación Científica y de Educación Superior de Ensenada
Gomberg, Joan, University of Washington
Gonzalez-Huizar, Hector, University of Texas, El Paso
Grand, Stephen P., University of Texas at Austin
Gu, Jeff, University of Alberta
Gurnis, Michael C., California Institute of Tech
Gurrola, Harold, Texas Tech University
Hager, Bradford H., Massachusetts Institute of Tech
Hagerty, Michael, Boston College
Hajnal, Zoltan, University of Saskatchewan
Hall, Jeremy, Memorial University of Newfoundland
Hamburger, Michael W., Indiana University, Bloomington
Hamilton, Warren, Colorado School of Mines
Hammer, Philip T., University of British Columbia
Hansen, Samantha E., University of Alabama
Harkrider, David G., California Institute of Tech
Harrington, Rebecca, Karlsruhe Institute of Tech
Harrington, Rebecca, McGill University
Harris, Jerry M., Stanford University
Hartse, Hans E., Los Alamos National Laboratory
Hassanpour Sedghi, Mohammad, University of Tabriz
Hauksson, Egill, California Institute of Tech
Hawman, Robert B., University of Georgia
Hayward, Christopher T., Southern Methodist University
Hearn, Thomas M., New Mexico State University, Las Cruces
Helmberger, Donald V., California Institute of Tech
Herrmann, Robert B., Saint Louis University
Hildebrand, Steve T., Los Alamos National Laboratory
Hobbs, Richard, Durham University
Holbrook, W. S., University of Wyoming
Holland, Austin A., University of Oklahoma
Holt, William E., SUNY, Stony Brook
Holtzman, Benjamin, Columbia University
Honjas, Bill, University of Nevada, Reno
Horton, Stephen P., University of Memphis
House, Leigh S., Los Alamos National Laboratory
Houston, Heidi B., University of Washington
Howell, Jr., Benjamin F., Pennsylvania State University, University Park
Huerfano, Victor, University of Puerto Rico
Humphreys, Eugene D., University of Oregon
Hurich, Charles A., Memorial University of Newfoundland
Hyndman, Roy D., University of Victoria
Isacks, Bryan L., Cornell University
Jackson, David D., University of California, Los Angeles
Jacob, Klaus H., Columbia University
James, David E., Carnegie Institution of Washington
Jasbinsek, John J., California Polytechnic State University
Jaumé, Steven C., College of Charleston
Ji, Chen, University of California, Santa Barbara
Johnson, Jeffrey , Boise State University
Johnson, Lane R., University of California, Berkeley
Johnson, Roy A., University of Arizona
Johnston, Archibald C., University of Memphis
Jones, Alan, Binghamton University
Jones, Craig H., University of Colorado
Jordan, Thomas H., University of Southern California
Kafka, Alan L., Boston College
Kaka, Ismail, King Fahd University of Petroleum and Minerals
Kanamori, Hiroo, California Institute of Tech
Karplus, Marianne, University of Texas, El Paso

Kent, Graham, University of Nevada, Reno
Keranen, Katie M., Cornell University
Kijko, Andrzej, University of Pretoria
Kim, Won-Young, Columbia University
Knopoff, Leon, University of California, Los Angeles
Kovach, Robert L., Stanford University
Kowalke, Sara, Ohio Dept of Natural Resources
Krebes, Edward S., University of Calgary
Langston, Charles A., University of Memphis
Laske, Gabi, University of California, San Diego
Lay, Thorne, University of California, Santa Cruz
Lees, Jonathan M., University of North Carolina, Chapel Hill
Lekic, Vedran, University of Maryland
Lerner-Lam, Arthur L., Columbia University
Levander, Alan, Rice University
Levin, Vadim, Rutgers, The State University of New Jersey
Li, Aibing, University of Houston
Li, Yong-Gang, University of Southern California
Liberty, Lee M., Boise State University
Linde, Alan T., Carnegie Institution of Washington
Liner, Christopher , University of Arkansas, Fayetteville
Lizarralde, Daniel, Woods Hole Oceanographic Institution
Lohman, Rowena B., Cornell University
Long, Maureen D., Yale University
Lorenzo, Juan M., Louisiana State University
Lough, Amanda, Drexel University
Louie, John, University of Nevada, Reno
Ma, Shuo, San Diego State University
Magnani, M. Beatrice, Southern Methodist University
Main, Ian G., Edinburgh University
Malin, Peter E., Duke University
Mallick, Subhashis, University of Wyoming
Malone, Stephen D., University of Washington
Mangriotis, Maria-Daphne , Heriot-Watt University
Marfurt, Kurt J., University of Oklahoma
Masters, T. Guy, University of California, San Diego
McClelland, Lori, University of Nevada, Reno
McGuire, Jeffrey J., Woods Hole Oceanographic Institution
McLaskey, Greg C., Cornell University
McLaskey, Greg C., Cornell University
McMechan, George A., University of Texas, Dallas
McNally, Karen C., University of California, Santa Cruz
Meltzer, Anne S., Lehigh University
Meng, Lingsen, University of California, Los Angeles
Menke, William H., Columbia University
Mereu, Robert F., Western University
Michel, Gero, Western University
Miller, Kate, Texas A&M University
Minster, J. Bernard H., University of California, San Diego
Molnar, Sheri, Western University
Moon, Wooil, University of Manitoba
Morozov, Igor B., University of Saskatchewan
Moulis, Anastasia, Boston College
Mulibo, Gabriel D., University of Dar es Salaam
Munguia, Luis, Centro de Investigación Científica y de Educación Superior de Ensenada
Nakamura, Yosio, University of Texas at Austin
Nakamura, Yosio n., University of Texas at Austin
Nava, Alejandro, Centro de Investigación Científica y de Educación Superior de Ensenada
Neighbors, Corrie, Western New Mexico University
Nettles, Meredith, Columbia University
Ni, James F., New Mexico State University, Las Cruces
Nikulin, Alex, Binghamton University
Nissen-Meyer, Targe, University of Oxford
Niu, Fenglin, Rice University
Nowack, Robert L., Purdue University
Nyland, Edo, University of Alberta
Ogbamikhumi, Alexander, University of Benin
Oglesby, David D., University of California, Riverside
Okal, Emile A., Northwestern University
Okaya, David A., University of Southern California
Oliver, Adolph A., Chabot College
Olson, Ted L., Snow College
Olugboji, Tolulope M., University of Maryland
Owens, Thomas J., University of South Carolina
Pancha, Aasha, University of Nevada, Reno
Pankow, Kristine L., University of Utah
Panning, Mark, University of Florida
Park, Jeffrey J., Yale University
Pavlis, Gary L., Indiana University, Bloomington
Pechmann, James C., University of Utah
Peng, Zhigang, Georgia Institute of Tech
Pennington, Wayne D., Michigan Technological University
Persaud, Patricia, Louisiana State University
Phinney, Robert A., Princeton University
Plank, Gabriel, University of Nevada, Reno
Polet, Jascha, California State Polytechnic University, Pomona
Powell, Christine A., University of Memphis
Pride, Steven, University of California, Berkeley
Pullammanappallil, Satish, University of Nevada, Reno
Pulliam, Robert J., Baylor University
Pulliam, Robert J., University of Texas at Austin
Reading, Anya, University of Tasmania
Rector, III, James W., University of California, Berkeley
Rennie, Tom, University of Nevada, Reno
Revenaugh, Justin, University of Minnesota, Twin Cities
Rial, Jose A., University of North Carolina, Chapel Hill
Richard, Ferdinand W., University of Dar es Salaam
Richards, Paul G., Columbia University
Richardson, Eliza, Pennsylvania State University, University Park
Rieger, Duayne, Community College of Rhode Island
Rietbrock, Andreas, University of Liverpool
Ritsema, Jeroen, University of Michigan
Ritter, Joachim R., Karlsruhe Institute of Tech
Roberts, Peter, Los Alamos National Laboratory
Rogers, Garry C., University of Victoria
Rohay, Alan C., Pacific Northwest National Laboratory
Romanowicz, Barbara A., University of California, Berkeley
Rost, Sebastian, University of Leeds
Rowe, Charlotte A., New Mexico Institute of Mining and Tech
Ruan, Youyi, Princeton University
Ruff, Larry J., University of Michigan
Ruffman, Alan, Dalhousie University
Sacchi, Mauricio D., University of Alberta
Sacks, I. Selwyn, Carnegie Institution of Washington
Sahay, Pratap, Centro de Investigación Científica y de Educación Superior de Ensenada
Salyards, Stephen L., University of California, Los Angeles
Sandvol, Eric A., University of Missouri
Sanford, Allan R., New Mexico Institute of Mining and Tech
Savage, Brian K., University of Rhode Island
Savage, Heather M., Columbia University
Sawyer, Dale S., Rice University
Schaff, David, Columbia University
Schlue, John W., New Mexico Institute of Mining and Tech
Scholz, Christopher H., Columbia University
Schutt, Derek L., Colorado State University
Schwab, Fred, University of California, Los Angeles
Schwartz, Susan Y., University of California, Santa Cruz
Seeber, Leonardo, Columbia University
Serpa, Laura F., University of New Orleans
Shakal, Anthony F., California Geological Survey
Shaw, Bruce, Columbia University
Shearer, Peter M., University of California, San Diego
Sheehan, Anne F., University of Colorado
Sherrod, Brian L., University of Washington
Shipley, Thomas H., University of Texas at Austin
Simons, Frederik J., Princeton University
Simons, Mark, California Institute of Tech
Slater, David, University of Nevada, Reno
Slawinski, Michael A., Memorial University of Newfoundland
Smith, Ken D., University of Nevada, Reno
Smith, Robert B., University of Utah
Smith, Stewart W., University of Washington
Smythe, William, University of California, Los Angeles
Snoke, J. Arthur, Virginia Polytechnic Institute & State University
Song, Alex, University College London
Song, Xiaodong, University of Illinois, Urbana-Champaign
Spence, George D., University of Victoria
Spetzler, Hartmut A., University of Colorado
Spiegelman, Marc, Columbia University
Stachnik, Joshua, Lehigh University
Steeples, Don W., University of Kansas

Steer, David N., University of Akron
Steidl, Jamison H., University of California, Santa Barbara
Stein, Seth, Northwestern University
Stephen, Ralph A., Woods Hole Oceanographic Institution
Stickney, Michael C., Montana Tech of The University of Montana
Street, Ronald L., University of Kentucky
Streig, Ashley, Portland State University
Stuart, Graham, University of Leeds
Stump, Brian W., Southern Methodist University
Sverdrup, Keith A., University of Wisconsin, Milwaukee
Sykes, Lynn R., Columbia University
Talwani, Pradeep, University of South Carolina
Tanimoto, Toshiro, University of California, Santa Barbara
Tape, Carl, University of Alaska, Fairbanks
Taylor, Kenneth B., North Carolina Geological Survey
Taylor, Steven R., Los Alamos National Laboratory
Teng, Ta-liang, University of Southern California
Thomas, Amanda, University of Oregon
Thorne, Michael, University of Utah
Thurber, Clifford H., University of Wisconsin, Madison
Tibuleac, Ileana, University of Nevada, Reno
Toksoz, M N., Massachusetts Institute of Tech
Toomey, Douglas R., University of Oregon
Toon, Sam, Keele University
Tromp, Jeroen, Princeton University
Van Avendonk, Harm J., University of Texas at Austin
van der Hilst, Robert, Massachusetts Institute of Tech
van der Lee, Suzan, Northwestern University
Vidal, Antonio, Centro de Investigación Científica y de Educación Superior de Ensenada
Vidale, John E., University of Washington
Vlahovic, Gordana, North Carolina Central University
von Seggern, David, University of Nevada, Reno
Wagner, Lara S., Carnegie Institution of Washington
Waheed, Umair, Princeton University
Waite, Gregory P., Michigan Technological University
Waldhauser, Felix, Columbia University
Wallace, Jr., Terry C., Los Alamos National Laboratory
Wang, Jeen-Hwa, Academia Sinica
Wang, Zhenming, University of Kentucky
Ward, Steven N., University of California, Santa Cruz
Warner, Michael, Imperial College
Waszek, Lauren, New Mexico State University, Las Cruces
Wen, Lianxing, SUNY, Stony Brook
Wesnousky, Steven G., University of Nevada, Reno
Wiens, Douglas A., Washington University in St. Louis
Williams, Erik, University of Nevada, Reno
Withers, Mitchell M., University of Memphis
Wolf, Lorraine W., Auburn University
Wolny, Dave, Colorado Mesa University
Woolery, Edward W., University of Kentucky
Wu, Francis T., Binghamton University
Wu, Patrick, University of Calgary
Wu, Ru-shan, University of California, Santa Cruz
Wust-Bloch, Gilles H., Tel Aviv University
Wysession, Michael E., Washington University in St. Louis
Xie, Xiao-bi, University of California, Santa Cruz
Young, Jeri J., University of Arizona
Yu, Wen-che, Academia Sinica
Zandt, George, University of Arizona
Zelt, Colin A., Rice University
Zeng, Hongliu, University of Texas at Austin
Zhan, Zhongwen, California Institute of Tech
Zhao, Li, Academia Sinica
Zhou, Hua-Wei, University of Houston
Zhou, Ying, Virginia Polytechnic Institute & State University
Zhu, Lupei, Saint Louis University
Ziv, Alon, Tel Aviv University

## Marine Geophysics

Anderson, Franz E., University of New Hampshire
Antonellini, Marco, Stanford University
Archer, David, University of Chicago
Austin, Jr., James A., University of Texas at Austin
Barry, James P., Stanford University
Becker, Keir, University of Miami
Behn, Mark D., Woods Hole Oceanographic Institution
Bell, Robin E., Columbia University
Ben-Avraham, Zvi, Tel Aviv University
Biasutti, Michela, Columbia University
Bohnenstiehl, DelWayne, North Carolina State University
Bokuniewicz, Henry J., SUNY, Stony Brook
Bonatti, Enrico, Columbia University
Borja, Ronaldo I., Stanford University
Bowin, Carl O., Woods Hole Oceanographic Institution
Buck, Roger W., Columbia University
Buizert, Christo, Oregon State University
Bull, Jonathan M., University of Southampton
Butterfield, David A., University of Washington
Canales Cisneros, Juan Pablo, Woods Hole Oceanographic Institution
Cande, Steven C., University of California, San Diego
Carbotte, Suzanne, Columbia University
Chamberlain, C. Page, Stanford University
Christeson, Gail L., University of Texas at Austin
Cochran, James R., Columbia University
Collier, Jenny, Imperial College
Constable, Steven C., University of California, San Diego
Cooper, Alan K., Stanford University
Delaney, John R., University of Washington
Dickens, Gerald R., Rice University
Duennebier, Frederick K., University of Hawai'i, Manoa
Duncan, Robert A., Oregon State University
Dunn, Robert A., University of Hawai'i, Manoa
Edwards, Richard N., University of Toronto
Egbert, Gary D., Oregon State University
Eloranta, Edwin W., University of Wisconsin, Madison
Elsenbeck, James R., Woods Hole Oceanographic Institution
Evans, Robert L., Woods Hole Oceanographic Institution
Farges, Francois, Stanford University
Fisk, Martin R., Oregon State University
Forsyth, Donald W., Brown University
Freyberg, David L., Stanford University
Fricker, Helen A., University of California, San Diego
Goff, John A., University of Texas at Austin
Goldberg, David S., Columbia University
Goldfinger, Chris, Oregon State University
Gorelick, Steven, Stanford University
Gorman, Andrew R., University of Otago
Grasmueck, Mark, University of Miami
Gulick, Sean S., University of Texas at Austin
Harris, Paul M., Rice University
Harrison, Christopher G., University of Miami
Hayes, Dennis E., Columbia University
Henderson, Naomi, Columbia University
Hildebrand, John A., University of California, San Diego
Hooft, Emilie E., University of Oregon
Hornbach, Matthew J., Southern Methodist University
Huang, Li, Western University
Johnson, H. Paul, University of Washington
Karig, Daniel E., Cornell University
Kitanidis, Peter K., Stanford University
Krastel, Sebastian , Dalhousie University
Kummerow, Christian D., Colorado State University
Kurapov, Alexander, Oregon State University
Lawver, Lawrence A., University of Texas at Austin
Lewis, Brian T. R., University of Washington
Louden, Keith E., Dalhousie University
Lowell, Robert P., Virginia Polytechnic Institute & State University
Luyendyk, Bruce P., University of California, Santa Barbara
MacAyeal, Douglas R., University of Chicago
Mack, Seymour, California State University, Fresno
Madsen, John A., University of Delaware
Manley, Patricia L., Middlebury College
Martinez, Fernando, University of Hawai'i, Manoa
McClain, James S., University of California, Davis
McIntosh, Kirk D., University of Texas at Austin
Minshull, Timothy A., University of Southampton
Mittelstaedt, Eric L., University of Idaho
Monismith, Stephen G., Stanford University
Muller, Dietmar, University of Sydney
Mutter, John C., Columbia University
Nilsson, Anders, Stanford University
Oakley, Adrienne, Kutztown University of Pennsylvania
Odom, Robert I., University of Washington
Ollerhead, Jeffery W., Mount Allison University

Orcutt, John A., University of California, San Diego
Pitman, Walter, Columbia University
Purdy, G. Michael, Columbia University
Raymond, Carol A., Jet Propulsion Laboratory
Rodriguez Simental, Ana María, Universidad Nacional Autonoma de Mexico
Ryan, William B. F., Columbia University
Salisbury, Matthew, Dalhousie University
Sandwell, David T., University of California, San Diego
Sawyer, Dale S., Rice University
Schneider, Robert J., Woods Hole Oceanographic Institution
Schouten, Hans, Woods Hole Oceanographic Institution
Sheridan, Robert E., Rutgers, The State University of New Jersey
Shillington, Donna, Columbia University
Shipley, Thomas H., University of Texas at Austin
Shreve, Ronald L., University of Washington
Silver, Eli A., University of California, Santa Cruz
Sloan, Heather, Lehman College (CUNY)
Smith, Deborah K., Woods Hole Oceanographic Institution
Sohn, Robert A., Woods Hole Oceanographic Institution
Sorlien, Christopher C., University of California, Santa Barbara
Spitz, Yvette H., Oregon State University
Spormann, Alfred, Stanford University
Stephens, Jason H., Austin Community College District
Swift, Stephen A., Woods Hole Oceanographic Institution
Talwani, Manik, Rice University
Thompson, David W. J., Colorado State University
Tobin, Harold J., University of Wisconsin, Madison
Tolstoy, Maria, Columbia University
Trehu, Anne M., Oregon State University
Van Rooij, David, Ghent University
Wattrus, Nigel J., University of Minnesota, Duluth
Webb, Spahr C., Columbia University
Wessel, Paul, University of Hawai'i, Manoa
Wetzel, Laura R., Eckerd College
White, Scott M., University of South Carolina
Whitmarsh, Robert B., University of Southampton
Wooding, Frank B., Woods Hole Oceanographic Institution
Xu, Li, Woods Hole Oceanographic Institution
Zaneveld, J. R., Oregon State University
Zavarin, Mavrik, Lawrence Livermore National Laboratory
Zhu, Wenlu, University of Maryland

## PALEONTOLOGY

### General Paleontology

Adrain, Tiffany S., University of Iowa
Altenbach, Alexander, Ludwig-Maximilians-Universitaet Muenchen
Amler, Michael, Ludwig-Maximilians-Universitaet Muenchen
Anderson, Wayne I., University of Northern Iowa
Armstrong, Howard A., Durham University
Bailey, Jack B., Western Illinois University
Bajraktareviæ, Zlatan, University of Zagreb
Bassett, Damon, Missouri State University
Baumiller, Tomasz K., University of Michigan
Beatty, William L., Winona State University
Beck, Charles, University of Michigan
Becker, Ralf Thomas, Universitaet Muenster
Bell, Phil R., University of New England
Benner, Jacob, Tufts University
Berman, David S., Carnegie Museum of Natural History
Berry, David R., California State Polytechnic University, Pomona
Bill, Steven D., Keene State College
Boyer, Diana L., SUNY, Buffalo
Branstrator, Jon W., Earlham College
Bray, Emily, University of Colorado
Britt, Brooks B., Brigham Young University
Brzobohaty, Rostislav, Masaryk University
Bucur, Ioan, Babes-Bolyai University
Busbey, Arthur B., Texas Christian University
Carnevale, Giorgio, Università di Torino
Chauff, Karl, Saint Louis University
Cherns, L C., Cardiff University
Cisne, John L., Cornell University
Clyde, William C., University of New Hampshire
Collette, Joseph, Minot State University
Coorough, Patricia J., Milwaukee Public Museum
Corgan, James X., Austin Peay State University
Cramer, Howard R., Emory University
Curran, H. Allen, Smith College
Davis, Edward, University of Oregon
Davydov, Vladimir I., Boise State University
De Klerk, Billy, Rhodes University
Deboo, Phili B., University of Memphis
Delfino, Massimo, Università di Torino
Devera, Joseph A., University of Illinois, Urbana-Champaign
Dietz, Anne D., San Antonio Community College
Domack, Cynthia R., Hamilton College
Edwards, Nicholas, University of Manchester
Egerton, Victoria, University of Manchester
Eldredge, Niles, Graduate School of the City University of New York
Fagerlin, Stanley C., Missouri State University
Falcon-Lang, Howard, University of London, Royal Holloway & Bedford New College
Fall, Leigh M., SUNY, Oneonta
Felton, Richard M., Northwest Missouri State University
Fio, Karmen, University of Zagreb
Fisher, Frank, University of Colorado
Foote, Michael J., University of Chicago
Forcino, Frank, Western Carolina University
Fornaciari, Elena, Università degli Studi di Padova
Fowler, Gerald A., University of Wisconsin, Parkside
Fox, David L., University of Minnesota, Twin Cities
Fox-Dobbs, Kena L., University of Puget Sound
Freeman, Veronica, Marietta College
Gagnaison, Cyril, Institut Polytechnique LaSalle Beauvais (ex-IGAL)
Gale, Andrew, University of Portsmouth
Ganeshram, Raja, Edinburgh University
Garb, Matt, Brooklyn College (CUNY)
Gatto, Roberto, Università degli Studi di Padova
Gess, Robert, Rhodes University
Gibson, Michael A., University of Tennessee, Martin
Gorecka-Nowak, Anna, University of Wroclaw
Green, Keith E., Cypress College
Greenwood, David , University of Saskatchewan
Haggart, James, University of British Columbia
Hansen, Thor A., Western Washington University
Harnik, Paul, Franklin and Marshall College
Harper, David, Durham University
Harries, Peter J., University of South Florida, Tampa
Hazel, McGoff, University of Reading
Heaney, Michael , Texas A&M University
Henderson, Wayne G., California State University, Fullerton
Henßel, Katja, Ludwig-Maximilians-Universitaet Muenchen
Herrera-Gil, Luis A., Universidad Autonoma de Baja California Sur
Hintze, Lehi F., Brigham Young University
Hopkins, Samantha, University of Oregon
Hylland, Mike, Utah Geological Survey
Immel, Harald, Ludwig-Maximilians-Universitaet Muenchen
Ivanov, Martin, Masaryk University
Johnson, Gerald H., College of William & Mary
Jones, Douglas S., University of Florida
Kazmer, Miklos, Eotvos Lorand University
Kelley, Patricia H., University of North Carolina Wilmington
Kern, J. Philip, San Diego State University
Kirkland, James I., Utah Geological Survey
Kirschvink, Joseph L., California Institute of Tech
Klapper, Gilbert , Northwestern University
Kowalke, Thorsten, Ludwig-Maximilians-Universitaet Muenchen
Lambert, Lance L., University of Texas, San Antonio
Laporte, Leo F., University of California, Santa Cruz
Leinfelder, Reinhold, Ludwig-Maximilians-Universitaet Muenchen
Li, Rong-Yu, Brandon University
Linsley, David, Colgate University
Little, Crispin, University of Leeds
Lozar, Francesca, Università di Torino
Lucas-Clark, Joyce, City College of San Francisco
MacLeod, Kenneth A., University of Missouri
Macomber, Richard, Long Island University, Brooklyn Campus
Malinowski, Jacalyn W., University of Illinois, Urbana-Champaign
Mandra, York T., San Francisco State University
Marenco, Pedro J., Bryn Mawr College
Mcgowan, Alistair, University of Glasgow
McHugh, Julia, Colorado Mesa University

McKenzie, Scott, Mercyhurst University
McKinney, Marg J., Appalachian State University
McMenamin, Mark, Mount Holyoke College
McRoberts, Christopher A., SUNY, Cortland
Means, Guy H., Florida Geological Survey
Melchin, Michael, Dalhousie University
Mezga, Aleksandar, University of Zagreb
Miller, Joshua H., University of Cincinnati
Miller, Randall F., University of New Brunswick
Moe-Hoffman, Amy P., University of Colorado
Monaco, Pat, University of Colorado
Monari, Stefano, Università degli Studi di Padova
Montgomery, Homer A., University of Texas, Dallas
Moro, Alan, University of Zagreb
Munro, Ildi, Carleton University
Murray, John, National University of Ireland Galway
Musil, Rudolf, Masaryk University
Naumann, Malik, Ludwig-Maximilians-Universitaet Muenchen
Nudds, John, University of Manchester
Orchard, Michael, University of British Columbia
Owen, Alan W., University of Glasgow
Pavia, Giulio, Università di Torino
Peck, Robert, Concord University
Peng, Lee Chai, University of Malaya
Perez-Huerta, Alberto, University of Alabama
Person, Jeff, North Dakota Geological Survey
Peterson, Joseph E., University of Wisconsin, Oshkosh
Petuch, Edward J., Florida Atlantic University
Pezelj, Đurðica, University of Zagreb
Poteet, Mary F., University of Texas at Austin
Pratt, Brian R., University of Saskatchewan
Pruss, Sara B., Smith College
Puchalski, Stephaney, Highline College
Reichenbacher, Bettina, Ludwig-Maximilians-Universitaet Muenchen
Rodland, David L., Muskingum University
Rogers, Kristina C., University of Minnesota, Twin Cities
Rogers, Raymond, University of Minnesota, Twin Cities
Rößner, Gertrud, Ludwig-Maximilians-Universitaet Muenchen
Sá, Artur A., Universidade de Trás-os-Montes e Alto Douro
Schmid, Dieter, Ludwig-Maximilians-Universitaet Muenchen
Scudder, Sylvia J., University of Florida
Shelton, Sally Y., South Dakota School of Mines & Tech
Shropshire, K. Lee, University of Northern Colorado
Siewers, Fredrick D., Western Kentucky University
Slaughter, Richard, University of Wisconsin, Madison
Smart, Christopher, University of Plymouth
Smith, Dena M., University of Colorado
Smith, Paul L., University of British Columbia
Sorauf, James E., University of South Florida, Tampa
Spencer, Patrick K., Whitman College
Stafford, Emily, Western Carolina University
Steinker, Paula J., Bowling Green State University
Stifel, Peter B., University of Maryland
Stratton, James F., Eastern Illinois University
Superchi-Culver, Tonia, University of Colorado
Sutherland, Stuart, University of British Columbia
Sutton, Mark, Imperial College
Sweetman, Steve, University of Portsmouth
Tesmer, Irving, Buffalo State College
Thompson, Jann W. M., Smithsonian Institution / National Museum of Natural History
Toots, Heinrich, Long Island University, C.W. Post Campus
Underwood, Charlie, Birkbeck College
Van Alstine, James B., University of Minnesota, Morris
Van Iten, Heyo, Hanover College
VanAller-Hernick, Linda A., New York State Geological Survey
Velez, Maria, University of Regina
Vicens Batet, Enric, Universitat Autonoma de Barcelona
Wallace, Steve, University of Colorado
Waters, Johnny, University of Tennessee, Knoxville
Wehrmann, Laura, Ludwig-Maximilians-Universitaet Muenchen
Weiss, Dennis, Graduate School of the City University of New York
White, Lisa D., San Francisco State University
Whitmore, John H., Cedarville University
Whitton, Mark, University of Portsmouth
Wignall, Paul, University of Leeds
Yancey, Thomas E., Texas A&M University
Yanes, Yurena, University of Cincinnati

## Paleostratigraphy

Al-Saad, Hamad A., University of Qatar
Arden, Daniel D., Georgia Southwestern State University
Baarli, B. Gudveig, Williams College
Baird, Gordon C., SUNY Fredonia
Barrick, James E., Texas Tech University
Bauer, Jeffrey A., Shawnee State University
Berggren, William A., Woods Hole Oceanographic Institution
Bergstrom, Stig M., Ohio State University
Braun, Willi K., University of Saskatchewan
Buatois, Luis, University of Saskatchewan
Caldwell, William G. E., University of Saskatchewan
Cohen, Andrew S., University of Arizona
Collinson, James W., Ohio State University
Davila-Alcocer, Victor M., Universidad Nacional Autonoma de Mexico
Eaton, Jeffrey G., Weber State University
Ettensohn, Frank R., University of Kentucky
Finney, Stanley C., California State University, Long Beach
Flory, Richard A., California State University, Chico
Grubb, Barbara, Mt. San Antonio College
Johnson, David B., New Mexico Institute of Mining and Tech
Johnson, Markes E., Williams College
Jones, Brian, University of Alberta
Kalvoda, Jiri, Masaryk University
Kirchgasser, William T., SUNY Potsdam
Kurtz, Vincent E., Missouri State University
Landing, Ed, New York State Geological Survey
Landman, Neil H., Graduate School of the City University of New York
Le Mone, David V., University of Texas, El Paso
Leatham, W. Britt, California State University, San Bernardino
Lewis, Gerald L., Pasadena City College
Liebe, Richard M., SUNY, The College at Brockport
Longoria, Jose F., Florida International University
Mangano, Gabriela, University of Saskatchewan
Martin, James E., South Dakota School of Mines & Tech
Maurrasse, Florentin J., Florida International University
McWilliams, Robert G., Miami University
Mikulic, Donald G., Illinois State Geological Survey
Miller, James F., Missouri State University
Miller, Richard H., San Diego State University
Moyle, Richard W., Weber State University
Murphy, Michael A., University of California, Riverside
Ogg, James G., Purdue University
Olson, Hillary C., University of Texas at Austin
Over, D. Jeffrey, SUNY, Geneseo
Ritter, Scott M., Brigham Young University
Sageman, Bradley B., Northwestern University
Shaw, Frederick C., Lehman College (CUNY)
Smith, Gerald R., University of Michigan
Stearn, Colin, McGill University
Stelck, Charles R., University of Alberta
Stevens, Calvin H., San Jose State University
Taylor, John F., Indiana University of Pennsylvania
Thoms, Richard E., Portland State University
Tischler, Herbert, University of New Hampshire
Titus, Robert C., Hartwick College
Tobin, Tom S., University of Alabama
Trexler, J. Peter, Juniata College
Van Niewenhuise, Donald, University of Houston
Vandenbroucke, Thijs, Ghent University
Witzke, Brian J., University of Iowa
Zinsmeister, William J., Purdue University

## Micropaleontology

Adekeye, O. A., University of Ilorin
Agnini, Claudia, Università degli Studi di Padova
Arnold, Anthony J., Florida State University
Barnes, Christopher R., University of Victoria
Barrier, Pascal, Institut Polytechnique LaSalle Beauvais (ex-IGAL)
Beauchamp, Benoit, University of Calgary
Berggren, William A., Rutgers, The State University of New Jersey
Bollmann, Jörg, University of Toronto
Brown, Paul, University College London
Burke, Collette D., Wichita State University
Buzas, Martin A., Smithsonian Institution / National Museum of Natural History

Caldwell, W. Glen E., Western University
Carreno, Ana L., Universidad Nacional Autonoma de Mexico
Carter, Elizabeth, Portland State University
Caus Gracia, Esmeralda, Universitat Autonoma de Barcelona
Chapman, Mark, University of East Anglia
Christensen, Beth A., Adelphi University
Ciesielski, Paul F., University of Florida
Clark, David L., University of Wisconsin, Madison
Corliss, Bruce H., Duke University
Culver, Stephen J., East Carolina University
Curry, William B., Woods Hole Oceanographic Institution
Das, Sarah B., Woods Hole Oceanographic Institution
de Vernal, Anne, Universite du Quebec a Montreal
Eicher, Donald, University of Colorado
Ethington, Raymond L., University of Missouri
Finger, Ken, University of California, Berkeley
Fluegeman, Richard H., Ball State University
Frankel, Larry, University of Connecticut
Fritts, Paul J., California State University, Long Beach
Giusberti, Luca, Università degli Studi di Padova
Goldstein, Susan T., University of Georgia
Gorog, Agnes, Eotvos Lorand University
Hallock-Muller, Pamela, University of South Florida
Hart, Malcolm, University of Plymouth
Harwood, David M., University of Nebraska, Lincoln
Hasegawa, Shiro, Kumamoto University
Hemphill-Haley, Eileen, Humboldt State University
Henderson, Charles M., University of Calgary
Huber, Brian T., Smithsonian Institution / National Museum of Natural History
Hunt, Gene, Smithsonian Institution / National Museum of Natural History
Ingle, Jr., James C., Stanford University
Johnson, Katherine, Eastern Illinois University
Kanungo, Sudeep, University of Utah
Katz, Miriam E., Rensselaer Polytechnic Institute
Keller, Gerta, Princeton University
Kelly, D. Clay, University of Wisconsin, Madison
Kessinger, Walter P., University of Louisiana at Lafayette
Klapper, Gilbert, Northwestern University
Kleffner, Mark A., Ohio State University
Knox, Larry W., Tennessee Tech University
Laws, Richard A., University of North Carolina Wilmington
Leadbetter, Jared R., California Institute of Tech
Leckie, R. Mark, University of Massachusetts, Amherst
Ling, Hsin-Yi, Northern Illinois University
Lobegeier, Melissa, Middle Tennessee State University
Louwye, Stephen, Ghent University
Lucas, Franklin A., University of Benin
Lundin, Robert F., Arizona State University
Lupia, Richard, University of Oklahoma
Maddocks, Rosalie F., University of Houston
Martin, Ronald E., University of Delaware
McCarthy, Francine G., Brock University
McCartney, Kevin, University of Maine, Presque Isle
Medioli, Franco, Dalhousie University
Mendelson, Carl V., Beloit College
Metcalfe, Ian, University of New England
Metzger, Ronald A., Southwestern Oregon Community College
Montenari, Michael, Keele University
Nathan, Stephen, University of Massachusetts, Amherst
Nathan , Stephen, Eastern Connecticut State University
Nations, Jack D., Northern Arizona University
Nestell, Galina P., University of Texas, Arlington
Nestell, Merlynd K., University of Texas, Arlington
Noble, Paula J., University of Nevada, Reno
Olsson, Richard K., Rutgers, The State University of New Jersey
Oppo, Delia W., Woods Hole Oceanographic Institution
Orr, William N., University of Oregon
Patterson, R. Timothy, Carleton University
Pearson, Paul, University of Wales
Peeters, Frank J., VU University Amsterdam
Pessagno, Jr., Emile A., University of Texas, Dallas
Pujana, Ignacio, University of Texas, Dallas
Reháková, Daniela, Comenius University in Bratislava
Reinhardt, Edward G., McMaster University
Roberts, Charles, Mt. San Antonio College
Robinson, Edward, Florida International University
Rodrigues, Cyril G., University of Windsor
Ross, Charles A., Western Washington University
Rosscoe, Steven, Hardin-Simmons University
Roth, Peter H., University of Utah
Sartipzadeh, Siavosh, University of Tabriz
Scherer, Reed P., Northern Illinois University
Schroder-Adams, Claudia, Carleton University
Scott, David B., Dalhousie University
Sen Gupta, Barun K., Louisiana State University
Siesser, William G., Vanderbilt University
Sloan, Doris, University of California, Berkeley
Sloan, Jon R., California State University, Northridge
St. Jean, Joseph, University of North Carolina, Chapel Hill
Steinmetz, John C., Indiana University
Stott, Lowell D., University of Southern California
Tabrizi, Azam, Tidewater Community College
Thomas, Ellen, Wesleyan University
Thomas, Ellen, Yale University
Toth, Emoke, Eotvos Lorand University
Ufkes, Els , VU University Amsterdam
Vaz, Nuno M., Universidade de Trás-os-Montes e Alto Douro
von Bitter, Peter H., University of Toronto
Wade, Bridget S., University College London
Walker, Sally E., University of Georgia
Watkins, David K., University of Nebraska, Lincoln
Webb, Peter N., Ohio State University
Wise, Jr, Sherwood W., Florida State University
Zonneveld, John-Paul, University of Calgary

## Paleobotany

Ash, Sydney R., Weber State University
Basinger, James F., University of Saskatchewan
Beck, John H., Boston College
Berry, C M., Cardiff University
Boelman, Natalie T., Columbia University
Boyce, Charles K., University of Chicago
Breedlovestrout, Renee L., University of Idaho
Burnham, Robyn J., University of Michigan
Carrillo Martinez, Miguel, Universidad Nacional Autonoma de Mexico
Chitaley, Shyamala, Cleveland Museum of Natural History
Currano, Ellen, University of Wyoming
Dilcher, David L., University of Florida
DiMichele, William A., Smithsonian Institution / National Museum of Natural History
Donoghue, Michael J., Yale University
Doyle, James A., University of California, Davis
Edwards, Dianne, Cardiff University
Edwards, Dianne, University of Wales
Erwin, Diane, University of California, Berkeley
Gabel, Mark, Black Hills State University
Gastaldo, Robert A., Colby College
Gowing, David, The Open University
Griffin, Kevin L., Columbia University
Heath, Kathleen M., Indiana State University
Hu, Shusheng, Yale University
Jackson, Gail D., Edinburgh University
Jarzen, David M., University of Florida
Kasper, Andrew E., Rutgers, The State University of New Jersey, Newark
Kerp, Hans, Universitaet Muenster
Knoll, Andrew H., Harvard University
Looy, Cynthia, University of California, Berkeley
Manchester, Steven R., University of Florida
Martinetto, Edoardo, Università di Torino
McElwain, Jenny, Field Museum of Natural History
Miller, Ian, Denver Museum of Nature & Science
Mindell, Randal, Douglas College
Peppe, Daniel J., Baylor University
Pfefferkorn, Hermann W., University of Pennsylvania
Punyasena, Surangi W., University of Illinois, Urbana-Champaign
Raymond, Anne, Texas A&M University
Retallack, Gregory J., University of Oregon
Rigby, John, Queensland University of Tech
Scott, Andrew C., University of London, Royal Holloway & Bedford New College
Strother, Paul, Boston College
Taggart, Ralph E., Michigan State University
Taylor, Edith, University of Kansas

Taylor, Thomas N., Ohio State University
Taylor, Thomas N., University of Kansas
Thomasson, Joseph R., Fort Hays State University
Tiffney, Bruce H., University of California, Santa Barbara
Tyler, Carrie, Miami University
Wing, Scott L., Smithsonian Institution / National Museum of Natural History

## Palynology

Anderson, Patricia M., University of Washington
Asselin, Esther, Universite du Quebec
Atta-Peters, David, University of Ghana
Bachhuber, Frederick W., University of Nevada, Las Vegas
Baker, Richard G., University of Iowa
Braman, Dennis R., Royal Tyrrell Museum of Palaeontology
Brenner, Gilbert J., SUNY, New Paltz
Brush, Grace S., Johns Hopkins University
Burden, Elliott T., Memorial University of Newfoundland
Butcher, Anthony, University of Portsmouth
Darrell, James H., Georgia Southern University
Davies, Caroline P., University of Missouri-Kansas City
Davis, Owen K., University of Arizona
Delusina, Irina, University of California, Davis
Dolakova, Nela, Masaryk University
Eble, Cortland F., University of Kentucky
Farley, Martin B., University of North Carolina, Pembroke
FitzPatrick, Meriel E., University of Plymouth
Fowell, Sarah, University of Alaska, Fairbanks
Habib, Daniel, Queens College (CUNY)
Habib, Daniel, Graduate School of the City University of New York
Head, Martin J., University of Toronto
Head, Martin J., Brock University
Hebda, Richard J., University of Victoria
Hemsley, A, Cardiff University
Hemsley, Alan, University of Wales
Higgs, Ken, University College Cork
Hills, Leonard V., University of Calgary
Jacobs, Bonnie F., Southern Methodist University
Jarvis, Ed, University College Cork
Javaux, Emmanuelle, Universite de Liege
Kováčová, Marianna, Comenius University in Bratislava
Kroeger, Timothy J., Bemidji State University
Leyden, Barbara W., University of South Florida, Tampa
Maher, Jr., Louis J., University of Wisconsin, Madison
Marshall, John E., University of Southampton
Martinez Hernandez, Enrique, Universidad Nacional Autonoma de Mexico
McAndrews, John H., University of Toronto
Mudie, Peta J., Dalhousie University
Norris, Geoffrey, University of Toronto
O'Keefe, Jennifer, Morehead State University
Peteet, Dorothy M., Columbia University
Rich, Fredrick J., Georgia Southern University
Rueger, Bruce F., Colby College
Smith, Jim, University College Cork
Stefanova, Ivanka, University of Minnesota, Twin Cities
Strother, Paul K., Boston College
Warny, Sophie, Louisiana State University
Wicander, Reed, Central Michigan University
Williams, Graham, Dalhousie University

## Quantitative Paleontology

Belasky, Paul, Ohlone College
Daley, Gwen M., Winthrop University
Huntley, John, University of Missouri
Marshall, Charles R., University of California, Berkeley
Miller, Arnold I., Cincinnati Museum Center

## Vertebrate Paleontology

Adams, Thomas, San Antonio Community College
Applegate-Pleasants, Shelton, Universidad Nacional Autonoma de Mexico
Archer, Michael, University of New South Wales
Badgley, Catherine E., University of Michigan
Barnosky, Anthony D., University of California, Berkeley
Bartels, William S., Albion College
Beard, K. Christopher, Carnegie Museum of Natural History
Behrensmeyer, Anna K., Smithsonian Institution / National Museum of Natural History
Bell, Christopher J., University of Texas at Austin
Bloch, Jonathan, University of Florida
Bolt, John R., Field Museum of Natural History
Brand, Leonard R., Loma Linda University
Brinkman, Donald B., Royal Tyrrell Museum of Palaeontology
Brochu, Christopher A., University of Iowa
Brown, Caleb, Royal Tyrrell Museum of Palaeontology
Caldwell, Michael W., University of Alberta
Campbell, Kenneth E., Los Angeles County Museum of Natural History
Carlson, Keith J., Gustavus Adolphus College
Carpenter, Kenneth, University of Colorado
Carpenter, Kenneth, Utah State University
Carrano, Matthew T., Smithsonian Institution / National Museum of Natural History
Carranza-Castaneda, Oscar, Universidad Nacional Autonoma de Mexico
Chatterjee, Sankar, Texas Tech University
Chiappe, Luis M., Los Angeles County Museum of Natural History
Cicimurri, Christian M., Clemson University
Cicimurri, David J., Clemson University
Cifelli, Richard L., University of Oklahoma
Clarke, Julia A., University of Texas at Austin
Clemens, William A., University of Alaska, Fairbanks
Clemens, William A., University of California, Berkeley
Codrea, Vlad, Babes-Bolyai University
Cooke, H.B. S., Dalhousie University
Coombs, Margery C., Amherst College
Covert, Herbert, University of Colorado
Cumbaa, Steve L., Carleton University
Currie, Philip J., University of Calgary
Curry Rogers, Kristina A., Macalester College
Daeschler, Ted, Drexel University
Daly, Raymond J., Union County College
Dawson, Mary, University of Colorado
Dawson, Mary R., Carnegie Museum of Natural History
Delson, Eric, Graduate School of the City University of New York
Delson, Eric, Lehman College (CUNY)
DeSantis, Larisa R., Vanderbilt University
Dodson, Peter, University of Pennsylvania
Druckenmiller, Patrick S., University of Alaska, Fairbanks
Dundas, Robert G., California State University, Fresno
Dzanh, Trihn, University of Colorado
Eberle, Jaelyn J., University of Colorado
Elliott, David K., Northern Arizona University
Emry, Robert, Smithsonian Institution / National Museum of Natural History
Engelmann, George F., University of Nebraska at Omaha
Esperante, Raul, Loma Linda University
Evans, David C., Royal Ontario Museum
Farlow, James O., Indiana University / Purdue University, Fort Wayne
Fastovsky, David E., University of Rhode Island
Fearon, Jamie L., Wheaton College
Fiorillo, Anthony R., Southern Methodist University
Flynn, John J., Columbia University
Flynn, John J., American Museum of Natural History
Fordyce, R. Ewan, University of Otago
Fox, Richard C., University of Alberta
Fraser, Nicholas C., Virginia Polytechnic Institute & State University
Froehlich, David J., Austin Community College District
Gaffney, Eugene S., American Museum of Natural History
Gallagher, William B., Rider University
Garcia, William, University of North Carolina, Charlotte
Gardner, James D., Royal Tyrrell Museum of Palaeontology
Gauthier, Jacques A., Yale University
Getty, Mike, Denver Museum of Nature & Science
Gillette, David D., Northern Arizona University
Goodwin, Mark B., University of California, Berkeley
Gottfried, Michael D., Michigan State University
Graham, Russell W., Pennsylvania State University, University Park
Grandal d'Anglade, Aurora, Coruna University
Grande, Lance, Field Museum of Natural History
Gray, Robert S., Santa Barbara City College
Green, Harry W., University of California, Berkeley
Guthrie, Dale, University of Alaska, Fairbanks
Hammer, William R., Augustana College
Hand, Suzanne J., University of New South Wales
Hargrave, Jennifer E., Southern Utah University
Harrell, Jr., T. L., Geological Survey of Alabama
Harris, John M., University of Utah

Harris, Judith A., University of Colorado
Hayden, Martha C., Utah Geological Survey
Heaton, Timothy H., University of South Dakota
Heckert, Andrew B., Appalachian State University
Henderson, Donald, Royal Tyrrell Museum of Palaeontology
Henrici, Amy C., Carnegie Museum of Natural History
Hitz, Ralph B., Tacoma Community College
Holroyd, Pat, University of California, Berkeley
Holtz, Jr., Thomas R., University of Maryland
Horner, John R., Montana State University
Hunt, Robert M., University of Nebraska, Lincoln
Indeck, Jeff, University of Colorado
Ivy, Logan D., Denver Museum of Nature & Science
Jackson, Frankie, Montana State University
Jacobs, Louis L., Southern Methodist University
Jasinski, Steven E., State Museum of Pennsylvania
Johnson, Ned K., University of California, Berkeley
Joniak, Peter , Comenius University in Bratislava
Kirkland, James I., University of Utah
Knell, Michael J., Southern Connecticut State University
Koch, Paul L., University of California, Santa Cruz
Krause, David W., SUNY, Stony Brook
Krishtalka, Leonard, University of Kansas
Lamanna, Matthew C., University of Pittsburgh
Lamanna, Matthew C., Carnegie Museum of Natural History
Lamb, James P., University of West Alabama
Langston, Jr., Wann, University of Texas at Austin
Lidicker, Jr., William Z., University of California, Berkeley
Lindsay, Everett H., University of Arizona
Lundelius, Ernest L., University of Texas at Austin
Luo, Zhexi, Carnegie Museum of Natural History
Lyson, Tyler, Denver Museum of Nature & Science
MacFadden, Bruce J., University of Florida
Maisey, John G., American Museum of Natural History
Makovicky, Peter J., Field Museum of Natural History
Marcot, Jonathan, University of Colorado
Martin, James E., University of Louisiana at Lafayette
Massare, Judy A., SUNY, The College at Brockport
McDonald, Greg, University of Colorado
McFadden, Bruce J., University of Florida
McLeod, Samuel A., Los Angeles County Museum of Natural History
Meng, Jin, Graduate School of the City University of New York
Meng, Jin, American Museum of Natural History
Merck, Jr., John W., University of Maryland
Miller, Wade E., Brigham Young University
Montellano-Ballesteros, Marisol, Universidad Nacional Autonoma de Mexico
Morales, Michael A., Emporia State University
Motani, Ryosuke, University of California, Davis
Naples, Virginia, Northern Illinois University
Neuman, Andrew G., Royal Tyrrell Museum of Palaeontology
Njau, Jackson K., Indiana University, Bloomington
Norell, Mark A., American Museum of Natural History
Norell, Mark A., Columbia University
Norris, Christopher A., Yale University
Novacek, Michael J., American Museum of Natural History
O'Leary, Maureen, SUNY, Stony Brook
Olsen, Paul E., Columbia University
Osi, Attila, Eotvos Lorand University
Padian, Kevin, University of California, Berkeley
Pagnac, Darrin C., South Dakota School of Mines & Tech
Parham, James, California State University, Fullerton
Patton, James L., University of California, Berkeley
Polly, P D., Indiana University, Bloomington
Pomeroy, Diana, Orange Coast College
Pons Muñoz, Josep Maria, Universitat Autonoma de Barcelona
Ray, Clayton E., Smithsonian Institution / National Museum of Natural History
Rensberger, John M., University of Washington
Rieppel, Olivier C., Field Museum of Natural History
Robinson, Peter, University of Colorado
Rowe, Timothy B., University of Texas at Austin
Russell, Dale A., North Carolina State University
Sabol, Martin, Comenius University in Bratislava
Sagebiel, J. C., San Bernardino County Museum
Samonds, Karen, Northern Illinois University
Sankey, Julia, California State University, Stanislaus
Schiebout, Judith A., Louisiana State University
Schultz, Gerald E., West Texas A&M University
Schwartz, Hilde, University of California, Santa Cruz
Schwimmer, David R., Columbus State University
Scott, Craig, Royal Tyrrell Museum of Palaeontology
Sears, Karen, University of Colorado
Secord, Ross, University of Nebraska, Lincoln
Semken, Jr., Holmes A., University of Iowa
Sertich, Joseph, Denver Museum of Nature & Science
Seymour, Kevin L., Royal Ontario Museum
Shaw, Christopher A., Los Angeles County Museum of Natural History
Sidor, Christian A., University of Washington
Smith, Kathlyn M., Georgia Southern University
Steadman, David W., University of Florida
Stearley, Ralph F., Calvin College
Storrs, Glenn W., University of Cincinnati
Storrs, Glenn W., Cincinnati Museum Center
Sumida, Stuart S., California State University, San Bernardino
Tabrum, Alan R., Carnegie Museum of Natural History
Tedford, Richard H., American Museum of Natural History
Therrien, Francois, Royal Tyrrell Museum of Palaeontology
Turnbull, William D., Field Museum of Natural History
Uhen, Mark, George Mason University
Varricchio, David J., Montana State University
Vietti, Laura, University of Wyoming
Voorhies, Michael R., University of Nebraska, Lincoln
Vrba, Elisabeth S., Yale University
Wilkins, Kenneth, Baylor University
Williamson, Thomas E., University of New Mexico
Wilson, Laura E., Fort Hays State University
Winkler, Alisa, Southern Methodist University
Winkler, Dale A., Southern Methodist University
Wood, Aaron R., Iowa State University of Science & Tech
Wood, Craig B., Providence College
Woodburne, Michael O., University of California, Riverside
Wuerthele, Norman, Carnegie Museum of Natural History
Wyss, Andre R., University of California, Santa Barbara
Zakrzewski, Richard J., Fort Hays State University

## Invertebrate Paleontology

Adrain, Jonathan M., University of Iowa
Alencaster-Ybarra, Gloria, Universidad Nacional Autonoma de Mexico
Anderson, Laurie C., South Dakota School of Mines & Tech
Bailey, Richard H., Northeastern University
Batten, Roger L., American Museum of Natural History
Belanger, Christina L., South Dakota School of Mines & Tech
Beus, Stanley S., Northern Arizona University
Bishop, Gale A., Georgia Southern University
Blake, Daniel B., University of Illinois, Urbana-Champaign
Bodenbender, Brian E., Hope College
Bonem, Rena M., Baylor University
Bonuso, Nicole, California State University, Fullerton
Bork, Kennard B., Denison University
Brandt, Danita S., Michigan State University
Briggs, Derek E., Yale University
Broadhead, Thomas W., University of Tennessee, Knoxville
Brown, Lewis M., Lake Superior State University
Buitron-Sanchez, Blanca E., Universidad Nacional Autonoma de Mexico
Buss, Leo W., Yale University
Butts, Susan H., Yale University
Cairns, Stephen, Smithsonian Institution / National Museum of Natural History
Caldwell, Roy L., University of California, Berkeley
Camp, Mark J., University of Toledo
Carlucci, Jesse R., Midwestern State University
Caron, Jean-Bernard, University of Toronto
Chatelain, Edward E., Valdosta State University
Chatterton, Brian D., University of Alberta
Clark, Murlene W., University of South Alabama
Copper, Paul, Laurentian University, Sudbury
Cuffey, Roger J., Pennsylvania State University, University Park
Dahl, Robyn, Western Washington University
Day, James E., Illinois State University
Deline, Brad, University of West Georgia
Dornbos, Stephen Q., University of Wisconsin, Milwaukee
Dutro, Thomas, Smithsonian Institution / National Museum of Natural History
Eldredge, Niles, American Museum of Natural History

Elias, Robert J., University of Manitoba
Erwin, Douglas H., Smithsonian Institution / National Museum of Natural History
Evanoff, Emmett, University of Colorado
Feldmann, Rodney M., Kent State University
Ferguson, Laing, Mount Allison University
Finks, Robert M., Graduate School of the City University of New York
Finks, Robert M., Queens College (CUNY)
Finnegan, Seth, University of California, Berkeley
Fisher, Daniel C., University of Michigan
Fraiser, Margaret L., University of Wisconsin, Milwaukee
Full, Robert J., University of California, Berkeley
Gangloff, Roland A., University of Alaska, Fairbanks
Gilmour, Ernest H., Eastern Washington University
Glass, Alexander, Duke University
Gonzalez-Arreola, Celestina, Universidad Nacional Autonoma de Mexico
Goodwin, David H., Denison University
Hageman, Steven J., Appalachian State University
Hall, Russell L., University of Calgary
Hanger, Rex, University of Wisconsin, Whitewater
Hanke, Brenda R., University of Cincinnati
Hannibal, Joseph T., Case Western Reserve University
Hannibal, Joseph T., Cleveland Museum of Natural History
Harper, Jr., Charles W., University of Oklahoma
Hartman, Joseph H., University of North Dakota
Hegna, Thomas A., Western Illinois University
Hembree, Daniel, Ohio University
Hudáčková, Natália , Comenius University in Bratislava
Hughes, Nigel C., Cincinnati Museum Center
Hunda, Brenda, Cincinnati Museum Center
Hyžný, Matúš , Comenius University in Bratislava
Isaacson, Peter E., University of Idaho
Jablonski, David, University of Chicago
James, Matthew J., Sonoma State University
Jin, Jisuo, Western University
Johns, Ronald A., Austin Community College District
Johnston, Paul, Mount Royal University
Kammer, Thomas W., West Virginia University
Key, Jr., Marcus M., Dickinson College
Kosloski, Mary, University of Iowa
Kues, Barry S., University of New Mexico
Kukalova-Peck, Jarmila, Carleton University
Labandeira, Conrad C., Smithsonian Institution / National Museum of Natural History
Landman, Neil H., American Museum of Natural History
Lee, Daphne E., University of Otago
Leggitt, Leroy, Loma Linda University
Lehrmann, Daniel J., Trinity University
Leighton, Lindsey, University of Alberta
Lenz, Alfred C., Western University
Lewis, Ronald D., Auburn University
Li, Rong-Yu, Western University
Lidgard, Scott H., Field Museum of Natural History
Lindberg, David R., University of California, Berkeley
Lindemann, Richard H., Skidmore College
Lockley, Martin G., University of Colorado, Denver
LoDuca, Steven T., Eastern Michigan University
Mann, Keith O., Ohio Wesleyan University
Manship, Lori L., University of Texas, Permian Basin
Marenco, Katherine N., Bryn Mawr College
Martínez Ribas, Ricard, Universitat Autonoma de Barcelona
Mason, Charles E., Morehead State University
McAfee, Gerald B., Odessa College
McKinney, Frank K., Appalachian State University
Melchin, Michael J., Saint Francis Xavier University
Merrill, Glen K., University of Houston Downtown
Meyer, David L., University of Cincinnati
Mikkelsen, Paula, Cornell University
Mikkelsen, Paula, Paleontological Research Institution
Miller, Arnold I., University of Cincinnati
Miller, William C., Humboldt State University
Morris, Robert W., Wittenberg University
Nagel-Myers, Judith, St. Lawrence University
Narbonne, Guy M., Queen's University
Nesbitt, Elizabeth (Liz), University of Washington
Nitecki, Matthew H., Field Museum of Natural History
Parker, William C., Florida State University
Paterson, John R., University of New England
Perrilliat-Montoya, Maria del Carmen, Universidad Nacional Autonoma de Mexico
Pestana, Harold R., Colby College
Pojeta, John, Smithsonian Institution / National Museum of Natural History
Pope, John K., Miami University
Portell, Roger W., University of Florida
Poty, Edouard, Universite de Liege
Powell, Matthew G., Juniata College
Prichonnet, Gilbert P., Universite du Quebec a Montreal
Rhodes, Frank H. T., Cornell University
Ritterbush, Linda A., California Lutheran University
Robison, Richard A., University of Kansas
Rodriguez, Joaquin, Graduate School of the City University of New York
Rohr, David M., Sul Ross State University
Rosenberg, Gary, Drexel University
Rowell, Albert J., University of Kansas
Rowland, Stephen, University of Nevada, Las Vegas
Sandy, Michael R., University of Dayton
Santos, Hernan, University of Puerto Rico
Saunders, W. Bruce, Bryn Mawr College
Savage, Norman M., University of Oregon
Schiappa, Tamra A., Slippery Rock University
Schlögl, Ján , Comenius University in Bratislava
Schmidt, David, Wright State University
Schweitzer, Carrie E., Kent State University at Stark
Segura-Vernis, Luis R., Universidad Autonoma de Baja California Sur
Selden, Paul A., University of Kansas
Shaw, Frederick C., Graduate School of the City University of New York
Sheehan, Peter M., University of Wisconsin, Milwaukee
Sheehan, Peter M., Milwaukee Public Museum
Shubak, Kenneth A., University of Wisconsin, Platteville
Soja, Constance M., Colgate University
Sorauf, James E., Binghamton University
Spinosa, Claude, Boise State University
Springer, Dale A., Bloomsburg University
Sprinkle, James T., University of Texas at Austin
Squires, Richard L., California State University, Northridge
Stanley, Thomas M., University of Oklahoma
Stanley, Jr., George D., University of Montana
Stephen, Daniel, Utah Valley University
Stock, Carl W., University of Alabama
Sumrall, Colin, University of Tennessee, Knoxville
Szente, Istvan, Eotvos Lorand University
Szlavecz, Katalin, Johns Hopkins University
Tackett, Lydia S., North Dakota State University
Tapanila, Leif, Idaho State University
Tshudy, Dale, Edinboro University of Pennsylvania
Tsujita, Cameron J., Western University
Villasenor-Martinez, Ana B., Universidad Nacional Autonoma de Mexico
Visaggi, Christy, Georgia State University
Votaw, Robert, Indiana University Northwest
Wagner, Peter J., University of Nebraska, Lincoln
Wagner, Peter J., Field Museum of Natural History
Waller, Thomas R., Smithsonian Institution / National Museum of Natural History
Warme, John E., Colorado School of Mines
Waters, Johnny A., Appalachian State University
Webber, Andrew, Cincinnati Museum Center
Webster, Gary D., Washington State University
Webster, Mark, University of Chicago
Westrop, Stephen R., University of Oklahoma
Wilson, Mark A., College of Wooster
Wyse Jackson, Patrick N., Trinity College
Xiao, Shuhai, Virginia Polytechnic Institute & State University
Young, Graham A., Manitoba Museum
Young, Graham A., University of Manitoba
Zodrow, Erwin L., Cape Breton University
Zumwalt, Gary S., Louisiana Tech University

## Paleobiology

Allmon, Warren D., Paleontological Research Institution
Allmon, Warren D., Cornell University
Arens, Nan Crystal, Hobart & William Smith Colleges
Ausich, William I., Ohio State University
Awramik, Stanley M., University of California, Santa Barbara
Babcock, Loren E., Ohio State University

Bambach, Richard K., Virginia Polytechnic Institute & State University
Benson, Roger, University of Oxford
Blackwelder, Patricia L., University of Miami
Bordeaux, Yvette, University of Pennsylvania
Brasier, Martin, University of Oxford
Brehman, Thomas R., Oakton Community College
Brower, James C., Syracuse University
Budd, Ann F., University of Iowa
Bush, Andrew M., University of Connecticut
Butterfield, Nicholas J., University of Cambridge
Caldwell, Roy, University of California, Berkeley
Carew, James L., College of Charleston
Carlson, Sandra J., University of California, Davis
Caron, Jean-Bernard, Royal Ontario Museum
Carter, Burchard D., Georgia Southwestern State University
Carter, Joseph G., University of North Carolina, Chapel Hill
Chamberlain, John A., Brooklyn College (CUNY)
Chamberlain, John A., Graduate School of the City University of New York
Cheetham, Alan H., Smithsonian Institution / National Museum of Natural History
Chin, Karen, University of Colorado
Clark, II, George R., Kansas State University
Cohen, Phoebe A., Williams College
Collins, Laurel S., Florida International University
Collinson, Margaret, University of London, Royal Holloway & Bedford New College
Coulson, Alan B., Clemson University
Cowen, Richard, University of California, Davis
Crane, Peter R., University of Chicago
Czaja, Andrew D., University of Cincinnati
Darroch, Simon, Vanderbilt University
Derstler, Kraig L., University of New Orleans
Dietl, Gregory, Paleontological Research Institution
Dodd, J. Robert, Indiana University, Bloomington
Droser, Mary L., University of California, Riverside
Elstad-Haveles, Andrew, Gustavus Adolphus College
Erickson, J. Mark, St. Lawrence University
Ferriz, Sergio Cevallos, Universidad Nacional Autonoma de Mexico
Flessa, Karl W., University of Arizona
Friedman, Matt, University of Oxford
Gabbott, Sarah, Leicester University
Gahn, Forest J., Brigham Young University - Idaho
Gaidos, Eric J., University of Hawai'i, Manoa
Geary, Dana H., University of Wisconsin, Madison
Gill, Fiona, University of Leeds
Goldman, Daniel, University of Dayton
Goswani, Anjali, University College London
Gray, Lee M., Mount Union College
Haiar, Brooke, Lynchburg College
Harvey, Tom, Leicester University
Hickman, Carole S., University of California, Berkeley
Hughes, Nigel C., University of California, Riverside
Jackson, Jeremy B. C., University of California, San Diego
Kauffman, Erle G., Indiana University, Bloomington
Kendrick, David C., Hobart & William Smith Colleges
Khaleel, Tasneem, Montana State University, Billings
Koehl, Mimi A. R., University of California, Berkeley
LaBarbera, Michael C., University of Chicago
Lieberman, Bruce S., University of Kansas
Lindberg, David R., University of California, Berkeley
Lockwood, Rowan, College of William & Mary
Martill, David, University of Portsmouth
Mattheus, Paul E., University of Alaska, Fairbanks
McCall, Peter L., Case Western Reserve University
McCollum, Linda B., Eastern Washington University
McGhee, Jr., George R., Rutgers, The State University of New Jersey
McNamara, Kenneth, University of Cambridge
Mead, James I., Northern Arizona University
Mitchell, Charles E., SUNY, Buffalo
Morris, Simon C., University of Cambridge
Myers, Corinne E., University of New Mexico
Newton, Cathryn R., Syracuse University
Norman, David, University of Cambridge
Pachut, Jr., Joseph F., Indiana University / Purdue University, Indianapolis
Parsons-Hubbard, Karla M., Oberlin College
Patzkowsky, Mark E., Pennsylvania State University, University Park
Plotnick, Roy E., University of Illinois at Chicago
Purnell, Mark, Leicester University
Raup, David M., University of Chicago
Reichenbacher, Bettina, Ludwig-Maximilians-Universitaet Muenchen
Reid, Catherine, University of Canterbury
Risk, Michael J., McMaster University
Ross, Robert M., Cornell University
Ross, Robert M., Paleontological Research Institution
Runnegar, Bruce, University of California, Los Angeles
Scatterday, James W., SUNY, Geneseo
Schiffbauer, James, University of Missouri
Schopf, J. William, University of California, Los Angeles
Silva-Pineda, Alicia, Universidad Nacional Autonoma de Mexico
Slatkin, Montgomery, University of California, Berkeley
Smith, Dena M., University of Colorado
Steinker, Don C., Bowling Green State University
Stetter, Karl O., University of California, Los Angeles
Stigall, Alycia L., Ohio University
Thomas, Roger D. K., Franklin and Marshall College
Thomka, James R., University of Akron
Upchurch, Paul, University College London
Valentine, James W., University of California, Berkeley
Wake, David B., University of California, Berkeley
Wake, Marvalee H., University of California, Berkeley
Ward, Peter D., University of Washington
Weber-Gobel, Reinhard, Universidad Nacional Autonoma de Mexico
West, Ronald R., Kansas State University
Williams, Mark, Leicester University
Yacobucci, Peg M., Bowling Green State University
Zalasiewicz, Jan, Leicester University

## Paleoecology & Paleoclimatology

, Susan K., Central Washington University
Almendinger, James E., University of Minnesota, Twin Cities
Andru-Hayles, Liala , Columbia University
Andrus, C. Fred T., University of Alabama
Ashworth, Allan C., North Dakota State University
Baker, Paul A., Duke University
Balescu, Sanda, Universite du Quebec a Montreal
Barker, Stephen, Cardiff University
Bartlein, Patrick J., University of Oregon
Begin, Christian, Universite du Quebec
Bennington, J Bret, Hofstra University
Billups, Katharina, University of Delaware
Blanchet, Jean-Pierre, Universite du Quebec a Montreal
Bottjer, David J., University of Southern California
Boyd, Bill, Southern Cross University
Bralower, Timothy J., Pennsylvania State University, University Park
Brett, Carlton E., University of Cincinnati
Briskin, Madeleine, University of Cincinnati
Buckley, Brendon, Columbia University
Buick, Roger, University of Washington
Byers, Charles W., University of Wisconsin, Madison
Caissie, Beth E., Iowa State University of Science & Tech
Came, Rosemarie E., University of New Hampshire
Camill III, Philip, Bowdoin College
Capraro, Luca, Università degli Studi di Padova
Chapin, F. Stuart, University of California, Berkeley
Charles, Christopher D., University of California, San Diego
Clapham, Matthew E., University of California, Santa Cruz
Cleaveland, Malcolm, University of Arkansas, Fayetteville
Cobb, Kim M., Georgia Institute of Tech
Cohen, Anne L., Woods Hole Oceanographic Institution
Cole, Julia E., University of Arizona
Colgan, Mitchell W., College of Charleston
Curtin, Tara M., Hobart & William Smith Colleges
Curtis, Jason H., University of Florida
D'Andrea, William J., Columbia University
de Menocal, Peter B., Columbia University
Dekens, Petra, San Francisco State University
deMenocal, Peter B., Columbia University
Dietl, Gregory P., Cornell University
Duhamel, Solange, Columbia University
Edlund, Mark B., University of Minnesota, Twin Cities
Ekdale, Allan A., University of Utah
Elick, Jennifer M., Susquehanna University
Endres, Anthony E., University of Waterloo
Engstrom, Daniel R., University of Minnesota, Twin Cities

Evans, Michael N., University of Maryland
Fariduddin, Mohammad, Northeastern Illinois University
Fawcett, Peter J., University of New Mexico
Francis, Jane, University of Leeds
Francus, Pierre, Universite du Quebec
Fritts, Harold C., University of Arizona
Fritz, Sherilyn C., University of Nebraska, Lincoln
Good, Steven C., West Chester University
Goodwin, Phillip A., University of Southampton
Grigg, Laurie, Norwich University
Grocke, Darren, Durham University
Guber, Albert L., Pennsylvania State University, University Park
Halfar, Jochen, University of Toronto
Harding, Ian, University of Southampton
Hasiotis, Stephen T., University of Kansas
Hays, James D., Columbia University
Haywood, Alan, University of Leeds
He, Helen, University of East Anglia
Herbert, Timothy D., Brown University
Herrmann, Achim, Louisiana State University
Hill, Tessa M., University of California, Davis
Hirschboeck, Katherine K., University of Arizona
Hoenisch, Baerbel, Columbia University
Hopley, Philip, Birkbeck College
Hu, Feng-Sheng, University of Illinois, Urbana-Champaign
Hughes, Malcolm K., University of Arizona
Huybers, Peter, Harvard University
Ishman, Scott E., Southern Illinois University Carbondale
Ivany, Linda C., Syracuse University
Jackson, Charles S., University of Texas at Austin
Johnson, Claudia C., Indiana University, Bloomington
Johnson, Robert G., University of Minnesota, Twin Cities
Johnson, Thomas C., University of Massachusetts, Amherst
Kemp, Alan, University of Southampton
Kennedy, Lisa M., Virginia Polytechnic Institute & State University
Knight, Tiffany, Washington University in St. Louis
Kopp, Robert E., Rutgers, The State University of New Jersey
Koutavas, Athanasios, College of Staten Island
Kutzbach, John E., University of Wisconsin, Madison
Lachniet, Matthew S., University of Nevada, Las Vegas
Lea, David W., University of California, Santa Barbara
Lebold, Joe, West Virginia University
Linsley, Braddock, Columbia University
Littler, Kate L., Exeter University
Livingstone, Daniel A., Duke University
Lohmann, George P., Woods Hole Oceanographic Institution
Lozano Garcia, Maria S., Universidad Nacional Autonoma de Mexico
Lynch-Stieglitz, Jean, Georgia Institute of Tech
Macdonald, Francis, Harvard University
Mann, Daniel H., University of Alaska, Fairbanks
Marchal, Olivier, Woods Hole Oceanographic Institution
Martin, Anthony J., Emory University
Martin, Pamela, University of Chicago
Mayewski, Paul A., University of Maine
McAlester, A. Lee, Southern Methodist University
McIntyre, Andrew, Queens College (CUNY)
McIntyre, Andrew, Graduate School of the City University of New York
McLean, Dewey M., Virginia Polytechnic Institute & State University
McManus, Jerry F., Columbia University
McManus, Jerry F., Woods Hole Oceanographic Institution
McWethy, David B., Montana State University
Medina Elizalde, Martin A., Auburn University
Michel, Lauren, Tennessee Tech University
Miller, Joshua H., Cincinnati Museum Center
Miller, Keith B., Kansas State University
Miller, Molly F., Vanderbilt University
Mishler, Brent, University of California, Berkeley
Morgan, Ryan, Tarleton State University
Myrbo, Amy, University of Minnesota, Twin Cities
Nelson, Robert E., Colby College
Nichols, Jonathan E., Columbia University
Norry, Mike, Leicester University
Oleinik, Anton, Florida Atlantic University
Olszewski, Thomas, Texas A&M University
Ortegren, Jason, University of West Florida
Panagiotakopulu, Eva, Edinburgh University
Parr, Kate, University of Liverpool
Pearson, P N., Cardiff University
Pekar, Stephen, Queens College (CUNY)
Poli, Maria-Serena, Eastern Michigan University
Pospelova, Vera, University of Victoria
Poulsen, Christopher J., University of Michigan
Power, Mary E., University of California, Berkeley
Powers, Elizabeth, California State University, Bakersfield
Quinn, Terry, University of Texas at Austin
Quinton, Page C., SUNY Potsdam
Railsback, L. Bruce, University of Georgia
Raymo, Maureen, Columbia University
Rhode, David E., Desert Research Institute
Ridenour, Wendy M., University of Montana Western
Rindsberg, Andrew K., University of West Alabama
Robinson, Stuart, University of Oxford
Roche, Didier, VU University Amsterdam
Rollins, Harold B., University of Pittsburgh
Royer, Dana, Wesleyan University
Sahagian, Dork, Lehigh University
Salzer, Matthew W., University of Arizona
Sauer, Peter, Indiana University, Bloomington
Schaefer, Joerg, Columbia University
Schellenberg, Stephen A., San Diego State University
Schelske, Claire L., University of Florida
Schwert, Donald P., North Dakota State University
Schöne, Bernd R., Universität Mainz
Scuderi, Louis A., University of New Mexico
Severinghaus, Jeffrey P., University of California, San Diego
Shaak, Graig D., University of Florida
Shanahan, Timothy M., University of Texas at Austin
Shroat-Lewis, Rene A., University of Arkansas at Little Rock
Shuman, Bryan N., University of Wyoming
Sinha, Ashish, California State University, Dominguez Hills
Sloan, Lisa C., University of California, Santa Cruz
Smerdon, Jason, Columbia University
Speer, James, Indiana State University
Spero, Howard J., University of California, Davis
Sremac, Jasenka, University of Zagreb
Stolper, Daniel, Princeton University
Stoykova, Kristalina C., Bulgarian Academy of Sciences
Strachan, Scotty, University of Nevada, Reno
Surge, Donna M., University of North Carolina, Chapel Hill
Swetnam, Thomas W., University of Arizona
Taffs, Kathryn, Southern Cross University
Talyor, Frederick W., University of Texas at Austin
Thompson, William G., Woods Hole Oceanographic Institution
Thunell, Robert C., University of South Carolina
Trouet, Valerie, University of Arizona
Verardo, Stacey, George Mason University
Vermeij, Geerat J., University of California, Davis
Waddington, J. M., McMaster University
Walker, Kenneth R., University of Tennessee, Knoxville
Warnock, Jonathan P., Indiana University of Pennsylvania
Webb, III, Thompson, Brown University
Weldeab, Syee, University of California, Santa Barbara
Westgate, James W., Lamar University
Whiteside, Jessica H., University of Southampton
Whitlock, Cathy, Montana State University
Wigand, Peter, University of Nevada, Reno
Williams, Christopher J., Franklin and Marshall College
Williams, John W., University of Wisconsin, Madison
Wolosz, Thomas H., Plattsburgh State University (SUNY)
Wong, Corinne I., Boston College
Woodhouse, Connie, University of Arizona
Xuan, Chuang, University of Southampton
Yu, Zicheng, Lehigh University
Æosoviæ, Vlasta, University of Zagreb

## Geobiology

Arnone, John, Desert Research Institute
Arp, Dr., Gernot, Georg-August University of Goettingen
Babbin, Andrew, Massachusetts Institute of Tech
Bailey, Jake, University of Minnesota, Twin Cities
Bosak, Tanja, Massachusetts Institute of Tech
Bradley, Alexander S., Washington University in St. Louis
Calcote, Randy, University of Minnesota, Twin Cities
Cardace, Dawn, University of Rhode Island
Chan, Clara, University of Delaware

Chaput, Dominique, Smithsonian Institution / National Museum of Natural History
Clementz, Mark T., University of Wyoming
Colman, Albert, University of Chicago
Conley, Catharine A., New Mexico Institute of Mining and Tech
Diggins, Thomas P., Youngstown State University
Duda, Jan-Peter, Georg-August University of Goettingen
Dupraz, Christophe, University of Connecticut
Dyson, Miranda, The Open University
Farmer, Jack D., Arizona State University
Flood, Beverley, University of Minnesota, Twin Cities
Fortin, Danielle, University of Ottawa
Fournier, Gregory, Massachusetts Institute of Tech
Fowle, David A., University of Kansas
Frantz, Carie M., Weber State University
Fritsen, Christian H., Desert Research Institute
Gallegos, Anthony F., Los Alamos National Laboratory
Glamoclija, Mihaela, Rutgers, The State University of New Jersey, Newark
Glasauer, Susan, University of Guelph
Gorman-Lewis, Drew J., University of Washington
Gray, Neil, University of Newcastle Upon Tyne
Hendy, Ingrid, University of Michigan
House, Christopher H., Pennsylvania State University, University Park
Johnston, Carl G., Youngstown State University
Jones, Dan, University of Minnesota, Twin Cities
Jorge, Maria Luisa, Vanderbilt University
Klemow, Kenneth M., Wilkes University
Konhauser, Kurt, University of Alberta
Kowalewski, Michal J., Virginia Polytechnic Institute & State University
Kulp, Thomas, Binghamton University
Lower, Steven K., Ohio State University
Macalady, Jennifer L., Pennsylvania State University, University Park
McBeth, Joyce M., University of Saskatchewan
McCormick, Michael L., Hamilton College
Meyer Dombard, DArcy, University of Illinois at Chicago
Mock, Thomas, University of East Anglia
Murray, Alison E., Desert Research Institute
Myers, Orrin B., Los Alamos National Laboratory
Nealson, Kenneth, University of Southern California
Newman, Dianne K., California Institute of Tech
O'Mullan, Gregory, Queens College (CUNY)
Olcott Marshall, Alison, University of Kansas
Ono, Shuhei, Massachusetts Institute of Tech
Onstott, Tullis C., Princeton University
Orphan, Victoria, California Institute of Tech
Orsi, William, Ludwig-Maximilians-Universitaet Muenchen
Osburn, Magdalena , Northwestern University
Prezant, Robert S., Graduate School of the City University of New York
Rampedi, Isaac T., University of Johannesburg
Rathburn, Anthony, Indiana State University
Raub, Tim, University of St. Andrews
Reitner, Joachim, Georg-August University of Goettingen
Robbins, Lisa, University of South Florida, Tampa
Roberts, Jennifer A., University of Kansas
Roden, Eric E., University of Wisconsin, Madison
Rogers, Karyn L., Rensselaer Polytechnic Institute
Rosenfeld, Carla, University of Minnesota, Twin Cities
Sanford, Robert A., University of Illinois, Urbana-Champaign
Santelli, Cara, Smithsonian Institution / National Museum of Natural History
Santelli, Cara M., University of Minnesota, Twin Cities
Sanudo, Sergio, University of Southern California
Schrenk, Matt, Michigan State University
Seibt, Ulrike, University of California, Los Angeles
Shapiro, Russell, California State University, Chico
Southam, Gordon, Western University
Speijer, Robert, Katholieke Universiteit Leuven
Steele, Michael A., Wilkes University
Summons, Roger, Massachusetts Institute of Tech
Thompson, Joel B., Eckerd College
Tice, Mike, Texas A&M University
Vega Vera, Francisco J., Universidad Nacional Autonoma de Mexico
Weber, Karrie A., University of Nebraska, Lincoln
Wilf, Peter D., Pennsylvania State University, University Park
Wilkins, Michael, Ohio State University
Woerheide, Gert, Ludwig-Maximilians-Universitaet Muenchen
Yee, Nathan, Rutgers, The State University of New Jersey

## HYDROLOGY

### General Hydrology

Adelana, S.M. A., University of Ilorin
AL-Nouimy, Latifa B., University of Qatar
Albert, Curtis C., University of Illinois, Urbana-Champaign
Alencoão, Ana M., Universidade de Trás-os-Montes e Alto Douro
Alhajari, Sief A., University of Qatar
Allan, Craig J., University of North Carolina, Charlotte
Allen, Peter M., Baylor University
Alsdorf, Douglas E., Ohio State University
Andres, A. S., University of Delaware
Arain, M. A., McMaster University
Atkinson, Tim, University College London
Bain, Daniel J., University of Pittsburgh
Baker, Tracy, Richard Stockton College of New Jersey
Barnard, Holly R., University of Colorado
Becker, Naomi M., Los Alamos National Laboratory
Bennett, Philip, University of Texas at Austin
Bense, Victor, University of East Anglia
Bloor, Michelle, University of Portsmouth
Bork, Karrigan, University of the Pacific
Borsa, Adrian, University of California, San Diego
Boult, Stephen, University of Manchester
Boving, Thomas, University of Rhode Island
Bowman, Jean A., Texas A&M University
Boyle, Douglas, University of Nevada, Reno
Brassington, Rick, University of Newcastle Upon Tyne
Brooks, Paul D., University of Utah
Brown, David L., California State University, Chico
Burkhart, Patrick A., Slippery Rock University
Byrne, James M., University of Lethbridge
Campana, Michael E., Oregon State University
Carey, Sean, McMaster University
Carillo Chavez, Alejandro J., Universidad Nacional Autonoma de Mexico
Carman, Cary D., Angelo State University
Celico, Fulvio, University of Parma
Challads, Thomas, Edinburgh University
Chaubey, Indrajeet, Purdue University
Chin, Yu-Ping, Ohio State University
Cirmo, Christopher P., SUNY, Cortland
Coulibaly, Paulin, McMaster University
Crawford, Nicholas, Western Kentucky University
Cunningham, Peter, University of Newcastle Upon Tyne
Currens, James C., University of Kentucky
D'Odorico, Paolo, University of Virginia
Dahlke, Helen E., University of California, Davis
Darby, Jeannie, University of California, Davis
Davidson, Bart, University of Kentucky
Dawson, Jamekia, Geological Survey of Alabama
Dolman, Han J., VU University Amsterdam
Doran, Peter, Louisiana State University
Doster, Florian, Heriot-Watt University
Dowling, Carolyn B., Ball State University
Duan, Quingyun, Lawrence Livermore National Laboratory
Duex, Timothy W., University of Louisiana at Lafayette
Durand, Michael T., Ohio State University
Eaton, Timothy, Queens College (CUNY)
El Kadiri, Racha, Middle Tennessee State University
Elliott, Emily M., University of Pittsburgh
Emerman, Steven H., Utah Valley University
Enright, Richard L., Bridgewater State University
Entekhabi, Dara, Massachusetts Institute of Tech
Eshleman, Keith N., University of Maryland
Estalrich López, Joan, Universitat Autonoma de Barcelona
Ezenabor, Ben O., University of Benin
Fabryka-Martin, June T., Los Alamos National Laboratory
Faramarzi, Monireh, University of Alberta
Farrow, Norman D., Oak Ridge National Laboratory
Fedele, Juan J., Saint Cloud State University
Foufoula-Georgiou, Efi, University of Minnesota, Twin Cities
Fox, Haydn A., Texas A&M University, Commerce
Gannon, John P., Western Carolina University
Gary, Marcus, University of Texas at Austin
Gilmore, Tyler J., Pacific Northwest National Laboratory

Grieneisen, Michael L., University of California, Davis
Groves, Christopher, University of Kentucky
Groves, Christopher, Western Kentucky University
Guan, Huade, Flinders University
Hagan, Robert M., University of California, Davis
Hamann, Hillary, University of Denver
Hannah, David, University of Birmingham
Hanson, Blaine R., University of California, Davis
Hart, David J., University of Wisconsin, Madison
Harter, Thomas, University of California, Davis
Hasan, Mohamed Ali, University of Malaya
Heimann, William H., Fort Hays State University
Heitz, Leroy F., University of Guam
Henry, Tiernan, National University of Ireland Galway
Hernes, Peter J., University of California, Davis
Hildebrandt, Anke, Friedrich-Schiller-University Jena
Hill, Mary C., University of Kansas
Hiscock, Kevin, University of East Anglia
Hooyer, Thomas S., University of Wisconsin, Madison
Hornberger, George, Vanderbilt University
Hubbard, John E., SUNY, The College at Brockport
Huntington, Justin, University of Nevada, Reno
Ibaraki, Motomu, Ohio State University
Inamdar, Shreeram, University of Delaware
Jacobs, Katharine L., University of Arizona
Jaffe, Peter R., Princeton University
James-Aworeni, E., Obafemi Awolowo University
Jefferson, Anne, University of North Carolina, Charlotte
Jefferson, Anne J., Kent State University
Jewell, Paul W., University of Utah
Johannesson, Karen H., Tulane University
Johnson, Robert O., Oak Ridge National Laboratory
Johnson, Robert L., Argonne National Laboratory
Jones, Stephen C., Geological Survey of Alabama
Kasenow, Michael, Eastern Michigan University
Katz, Gabrielle, Appalachian State University
Kelly, Walton R., Illinois State University
Kesel, Richard H., Louisiana State University
Keyantash, John, California State University, Dominguez Hills
Kilroy, Kathyrn, Minot State University
Kilsby, Chris, University of Newcastle Upon Tyne
Kinner, David A., Western Carolina University
Kuchovsky, Tomas, Masaryk University
Lakshmi, Venkataraman, University of South Carolina
Land, Lewis A., New Mexico Institute of Mining & Tech
Lane, Charles L., Southern Oregon University
Lanigan, David C., Pacific Northwest National Laboratory
Lautz, Laura, Syracuse University
Laycock, Arleigh H., University of Alberta
Lee, Michael D., California State University, East Bay
Levine, Rebekah, University of Montana Western
Liess, Stefan, University of Minnesota, Twin Cities
Lightbody, Anne F., University of New Hampshire
Lin, Yu-Feng F., University of Illinois, Urbana-Champaign
Liu, Gaisheng, University of Kansas
Llanos, Hilario, University of the Basque Country UPV/EHU
Ma, Lin, University of Texas, El Paso
McDermott, Christopher I., Edinburgh University
McElwee, Carl D., University of Kansas
McKenna, Thomas E., University of Delaware
McLaughlin, Peter P., University of Delaware
McNamara, James P., Boise State University
Megdal, Sharon B., University of Arizona
Mehnert, Edward, Illinois State University
Michalek, Thomas E., Montana Tech of The University of Montana
Molotch, Noah, University of Colorado
Monohan, Carrie, California State University, Chico
Mudd, Simon N., Edinburgh University
Ninesteel, Judy J., Potomac State College
Nkuna, Tinyiko R., University of Venda for Science & Tech
O'Driscoll, Michael A., East Carolina University
O'Reilly, Catherine M., Illinois State University
Odling, Noelle, University of Leeds
Ortiz-Aguirre, Ramon, Universidad Autonoma de San Luis Potosi
Oztekin Okan, Ozlem, Firat University
Palmer, Arthur N., SUNY, Oneonta
Parkin, Geoffrey, University of Newcastle Upon Tyne
Pavelsky, Tamlin M., University of North Carolina, Chapel Hill
Perkins, William A., Pacific Northwest National Laboratory
Pettijohn, J. C., University of Illinois, Urbana-Champaign
Piccinini, Leonardo, Università degli Studi di Padova
Piggot, Matthew, Imperial College
Potter, Kenneth W., University of Wisconsin, Madison
Prestegaard, Karen L., University of Maryland
Puente, Carlos E., University of California, Davis
Quinn, Paul, University of Newcastle Upon Tyne
Reay, William G., College of William & Mary
Reedy, Robert C., University of Texas at Austin, Jackson School of Geosciences
Reidenbach, Matthew A., University of Virginia
Renshaw, Carl E., Dartmouth College
Rice, Karen C., University of Virginia
Richards, Paul L., SUNY, The College at Brockport
Ricka, Adam, Masaryk University
Root, Tara L., Florida Atlantic University
Rouse, Joseph D., University of Guam
Saar, Martin O., University of Minnesota, Twin Cities
Salami, B. M., Obafemi Awolowo University
Sandoval, Samuel, University of California, Davis
Sandoval Solis, Samuel, University of California, Davis
Scanlon, Bridget R., University of Texas at Austin
Scanlon, Todd M., University of Virginia
Schilling, Keith E., University of Iowa
Schnoebelen, Douglas J., University of Iowa
Schwartz, Frank W., Ohio State University
Scotton, Paolo, Università degli Studi di Padova
Screaton, Elizabeth J., University of Florida
Singer, Michael, University of St. Andrews
Slade, Jr., Raymond M., Austin Community College District
Slattery, Michael C., Texas Christian University
Smith, Ellen D., Oak Ridge National Laboratory
Smith, James A., Princeton University
Smith, Michael, Geological Survey of Alabama
Smith, Ronald M., Pacific Northwest National Laboratory
Smyth, Rebecca C., University of Texas at Austin, Jackson School of Geosciences
Sorooshian, Soroosh, University of Arizona
Sorooshian, Soroosh, University of California, Irvine
Spane, Frank A., Pacific Northwest National Laboratory
Sritharan, Subramania I., Central State University
Stafford, James, Wyoming State Geological Survey
Stasko, Stanislaw, University of Wroclaw
Steele, Ken, University of Arkansas, Fayetteville
Steenhuis, Tammo S., Cornell University
Tarka, Robert, University of Wroclaw
Thorne, Paul D., Pacific Northwest National Laboratory
Tomlinson, Jaime L., University of Delaware
Triay, Ines, Los Alamos National Laboratory
Turner, Anne, Austin Community College District
Tyler, Scott, University of Nevada, Reno
Vail, Lance W., Pacific Northwest National Laboratory
van der Velde, Ype, VU University Amsterdam
Van Stan, John, Georgia Southern University
Vandike, James E., Missouri University of Science and Tech
Veraverbeke, Sander, VU University Amsterdam
Vermeul, Vincent R., Pacific Northwest National Laboratory
Vonk, Jorien, VU University Amsterdam
Wallender, Wesley W., University of California, Davis
Watkins, David, Exeter University
West, Jared, University of Leeds
Westhoff, Martijn C., VU University Amsterdam
White, Jeffrey R., Indiana University, Bloomington
Witt, Emma, Stockton University
Wolf, Aaron T., Oregon State University
Woltemade, Christopher J., Shippensburg University
Woo, Ming-Ko, McMaster University
Wood, Eric F., Princeton University
Xu, Shangping, University of Wisconsin, Milwaukee
Yabusaki, Steven B., Pacific Northwest National Laboratory
Yang, Y, Cardiff University
Ye, Ming, Florida State University
You, Jinsheng, University of Nebraska, Lincoln
Yusoff, Ismail, University of Malaya
Zaccaria, Daniele, University of California, Davis

Zhang, Minghua , University of California, Davis
Ziemer, Robert R., Humboldt State University
Zume, Joseph T., Shippensburg University

## Ground Water/Hydrogeology

Simsek, Sakir, Hacettepe University
Ahmed, Kazi M., University of Dhaka
Akujieze, Christopher N., University of Benin
Al, Tom A., University of Ottawa
Al-Shaibani, Abdulaziz, King Fahd University of Petroleum and Minerals
Alberts, E E., University of Missouri, Columbia
Alexander, Scott, University of Minnesota, Twin Cities
Alexander, Jr., E. Calvin, University of Minnesota, Twin Cities
Ali, Bukari, Kwame Nkrumah University of Science and Tech
Ali, Genevieve, University of Manitoba
Allen, Timothy T., Keene State College
Amin, Isam E., Youngstown State University
Anderson, Mary P., University of Wisconsin, Madison
Anderson, William P., Appalachian State University
Anderson, NAE, Mary P., University of Wisconsin, Madison
Andreasen, David C., Maryland Department of Natural Resources
Andres, A. S., University of Delaware
Antigüedad, Iñaki, University of the Basque Country UPV/EHU
Appiah-Adjei, Emmanuel K., Kwame Nkrumah University of Science and Tech
Appold, Martin, University of Missouri
Arthur, Jonathan D., Florida Geological Survey
Asante, Joseph, Tennessee Tech University
Asghari Moghadam, Asghar, University of Tabriz
Bagtzoglou, Ross, Columbia University
Bahr, Jean M., University of Wisconsin, Madison
Bair, Edwin S., Ohio State University
Bajjali, William, University of Wisconsin, Superior
Baker, Andy, University of New South Wales
Baldwin, Jr., A. Dwight, Miami University
Banks, Eddie, Flinders University
Barbecot, Florent, Universite du Quebec a Montreal
Barbee, Gary C., West Texas A&M University
Barkmann, Peter, Colorado School of Mines
Barrash, Warren, Boise State University
Batelaan, Okke, Flinders University
Batelaan, Okke, Katholieke Universiteit Leuven
Bayari, Serdar, Hacettepe University
Beck, E. G., University of Kentucky
Becker, Matthew, California State University, Long Beach
Beckie, Roger D., University of British Columbia
Bedient, Philip B., Rice University
Behnke, Jerold J., California State University, Chico
Belanger, Thomas V., Florida Institute of Tech
Bennett, Steven W., Western Illinois University
Bentley, Laurence R., University of Calgary
Bergamaschi, Brian, California State University, Sacramento
Besancon, James, Wellesley College
Bethke, Craig M., University of Illinois, Urbana-Champaign
Bishop, Charles E., Utah Geological Survey
Blythe, Daniel D., Montana Tech of The University of Montana
Boateng, Samuel, Northern Kentucky University
Bobst, Andrew L., Montana Tech of The University of Montana
Bohling, Geoff, University of Kansas
Bohling, Geoffrey C., University of Kansas
Bolster, Diogo, University of Notre Dame
Bolton, David W., Maryland Department of Natural Resources
Booth, Colin J., Northern Illinois University
Boutt, David, University of Massachusetts, Amherst
Boving, Thomas B., University of Rhode Island
Boyle, James T., New Jersey Geological and Water Survey
Bradbury, Kenneth R., University of Wisconsin, Madison
Bradbury, Kenneth R., University of Wisconsin - Extension
Brahana, J. Van, University of Arkansas, Fayetteville
Brame, Scott E., Clemson University
Bratcher, Susan, California State University, Fresno
Breitmeyer, Ronald, University of Nevada, Reno
Brikowski, Tom H., University of Texas, Dallas
Brinkmann, Robert, Oregon Dept of Geology & Mineral Industries
Brookfield, Andrea, University of Kansas
Brookshire, Cynthia, Missouri Dept of Natural Resources
Brusseau, Mark L., University of Arizona
Buchanan, John P., Eastern Washington University
Burbey, Thomas J., Virginia Polytechnic Institute & State University
Burgess, William G., University College London
Burnett, Earl E., Arizona Western College
Butler, James J., University of Kansas
Butler, Jr., James J., University of Kansas
Callahan, Timothy J., College of Charleston
Capuano, Regina M., University of Houston
Cardenas, Bayani, University of Texas at Austin
Cardiff, Michael A., University of Wisconsin, Madison
Carle, Steven, Lawrence Livermore National Laboratory
Carling, Gregory T., Brigham Young University
Carlson, Catherine A., Eastern Connecticut State University
Carlson, Douglas A., Louisiana State University
Carstarphen, Camela, Montana Tech of The University of Montana
Casillas, Angelica, Universidad Nacional Autonoma de Mexico
Cassidy, Daniel P., Western Michigan University
Celia, Michael A., Princeton University
Chachadi, Adiveppa G., Goa University
Chase, Peter M., University of Wisconsin - Extension
Chavez-Aguirre, Rafael, Universidad Autonoma de Chihuahua
Cheng, Songlin, Wright State University
Cherkauer, Douglas S., University of Wisconsin, Milwaukee
Cherry, John A., University of Waterloo
Cherubini, Claudia, Institut Polytechnique LaSalle Beauvais (ex-IGAL)
Chesnaux, Romain, Universite du Quebec a Chicoutimi
Chowdhury, Shafiul H., SUNY, New Paltz
Cianfrani, Christina, Hampshire College
Clarey, Timothy L., Delta College
Clark, Ian D., University of Ottawa
Conant, Brewster, University of Waterloo
Cook, Edward R., Columbia University
Cook, Peter, Flinders University
Costa, Maria R., Universidade de Trás-os-Montes e Alto Douro
Cox, Malcolm E., Queensland University of Tech
Croxen III, Fred W., Arizona Western College
Davidson, Gregg R., University of Mississippi
Davis, Arden D., South Dakota School of Mines & Tech
Davis, Caroline A., University of Iowa
Davis, J. Matthew, University of New Hampshire
Davis, Ralph K., University of Arkansas, Fayetteville
Decker, Dave, University of Nevada, Reno
Del Arenal-Capetillo, Rodolfo, Universidad Nacional Autonoma de Mexico
Denizman, Can, Valdosta State University
Devlin, J. R., University of Kansas
Dhar, Ratan K., York College (CUNY)
Dimova, Natasha T., University of Alabama
Dintaman, Christopher, Indiana University
Domber, Steven E., New Jersey Geological and Water Survey
Doss, Paul K., University of Southern Indiana
Douglas, Ellen, University of Massachusetts, Boston
Dripps, Weston R., Furman University
Druhan, Jennifer, University of Illinois, Urbana-Champaign
Duaime, Terence E., Montana Tech of The University of Montana
Durham, Lisa, Argonne National Laboratory
Earman, Sam, Millersville University
Ekmekci, Mehmet, Hacettepe University
El-Kadi, Aly I., University of Hawai'i, Manoa
Ellis, Andre, California State University, Los Angeles
Esling, Steven P., Southern Illinois University Carbondale
Fabbri, Paolo, Università degli Studi di Padova
Fairley, Jerry P., University of Idaho
Famiglietti, James, University of California, Irvine
Farmer, George T., Los Alamos National Laboratory
Feinstein, Daniel T., University of Wisconsin, Milwaukee
Ferre, Paul A., University of Arizona
Ferrero, Tom, Oregon Dept of Geology & Mineral Industries
Filipovic, Dragan, South Dakota Dept of Environment and Natural Resources
Fisher, Andrew T., University of California, Santa Cruz
Fitzpatrick, Stephan, Georgia State University, Perimeter College, Clarkston Campus
Flowers, George C., Tulane University
Fogg, Graham E., University of California, Davis
Fountain, John C., North Carolina State University
Freshley, Mark D., Pacific Northwest National Laboratory
Frisbee, Marty, Purdue University

Fryar, Alan E., University of Kentucky
Gakka, Dr U., University College of Science, Osmania University
Gardner, Payton, University of Montana
Garven, G, Tufts University
Ge, Shemin, University of Colorado
Geary, Phil, University of Newcastle
Gebrehiwet, Tsigabu A., Towson University
Geidel, Gwendelyn, University of South Carolina
Gemperline, Johanna M., Maryland Department of Natural Resources
Gentry, Randall, Argonne National Laboratory
Gerla, Philip J., University of North Dakota
Godsey, Sarah E., Idaho State University
Goeke, James W., Unversity of Nebraska - Lincoln
Gordon, Ryan, Dept of Agriculture, Conservation, and Forestry
Gordon, Susan L., University of Calgary
Gorelick, Steven M., Stanford University
Gotkowitz, Madeline B., University of Wisconsin, Madison
Gotkowitz, Madeline B., University of Wisconsin, Extension
Gouzie, Douglas R., Missouri State University
Graham, Grace E., University of Wisconsin - Extension
Grasby, Stephen E., University of Calgary
Grismer, Mark E., University of California, Davis
Grundl, Timothy J., University of Wisconsin, Milwaukee
Gurdak, Jason, San Francisco State University
Habana, Nathan C., University of Guam
Haggerty, Roy D., Oregon State University
Halihan, Todd, Oklahoma State University
Hall, Francis R., University of New Hampshire
Hampton, Duane R., Western Michigan University
Han, Weon Shik, University of Wisconsin, Milwaukee
Hand, Kristen, Pennsylvania Bureau of Topographic & Geologic Survey
Harbor, Jon M., Purdue University
Hargis, David, University of Arizona
Harrington, Glenn, Flinders University
Harris, Brett D., Curtin University
Harris, William H., Graduate School of the City University of New York
Hart, David J., University of Wisconsin, Extension
Harter, Thomas L., University of California, Davis
Harter, Thomas, University of California, Davis
Hatch, Christine, University of Massachusetts, Amherst
Hayashi, Masaki, University of Calgary
Hays, Phillip D., University of Arkansas, Fayetteville
Heilweil, Victor, University of Utah
Helmke, Martin F., West Chester University
Hendrickx, Jan M., New Mexico Institute of Mining and Tech
Hendry, Jim, University of Saskatchewan
Henry, Eric J., University of North Carolina Wilmington
Herman, Ellen K., Bucknell University
Hershey, Ronald L., University of Nevada, Reno
Hibbs, Barry, California State University, Los Angeles
Hindle, Tobin, Florida Atlantic University
Hinson, Amye S., Geological Survey of Alabama
Hoff, Jean L., Saint Cloud State University
Hoover, Karin A., California State University, Chico
Horner, Tim C., California State University, Sacramento
Howard, Kenneth W., University of Toronto
Howe III, Thomas R., Western Michigan University
Hu, Bill X., Florida State University
Hudak, Paul F., University of North Texas
Huizar-Alvarez, Rafael, Universidad Nacional Autonoma de Mexico
Hunt, Randy, University of Wisconsin, Madison
Hurlow, Hugh A., Utah Geological Survey
Huysmans, Marijke, Katholieke Universiteit Leuven
Icopini, Gary, Montana Tech of The University of Montana
Iles, Derric L., South Dakota Dept of Environment and Natural Resources
Iqbal, Mohammad Z., University of Northern Iowa
Iribar, Vicente , University of the Basque Country UPV/EHU
Irvine, Dylan, Flinders University
Isiorho, Solomon A., Indiana University / Purdue University, Fort Wayne
James, Scott C., Baylor University
Jannik, Nancy O., Winona State University
Jarvis, W. T., Oregon State University
Jenson, John W., University of Guam
Johnson, Peggy, New Mexico Institute of Mining & Tech
Johnson, Thomas M., University of Illinois, Urbana-Champaign
Join, Jean-Lambert, Universite de la Reunion
Jones, Ian C., Austin Community College District
Juster, Thomas C., University of South Florida, Tampa
Kaden, Scott, Missouri Dept of Natural Resources
Keefer, Donald A., University of Illinois, Urbana-Champaign
Keen, Kerry L., University of Wisconsin, River Falls
Keller, C. Kent, Washington State University
Kelly, Bryce, University of New South Wales
Kelly, Walton R., Illinois State Water Survey
Kent, Douglas, Oklahoma State University
Khalequzzaman, Md., Lock Haven University
Knoll, Martin A., Sewanee: University of the South
Krause, Jacob J., University of Wisconsin - Extension
Kreamer, David, University of Nevada, Reno
Kreamer, David K., University of Nevada, Las Vegas
Kurttas, Turker, Hacettepe University
La Fave, John I., Montana Tech of The University of Montana
La Fleur, Robert G., Rensselaer Polytechnic Institute
Lachhab, Ahmed, Susquehanna University
Lachmar, Thomas E., Utah State University
LaFreniere, Lorraine, Argonne National Laboratory
Lambert, Carolyn D., University of Northern Colorado
Larocque, Marie, Universite du Quebec a Montreal
Laton, W. R., California State University, Fullerton
Lee, David R., University of Waterloo
Lee, Donald W., Oak Ridge National Laboratory
Lee, Eung Seok, Ohio University
Lee, Ming-Kuo, Auburn University
Lefebvre, Rene, Universite du Quebec
Lemke, Lawrence D., Central Michigan University
Leonhart, Leo S., University of Arizona
Levy, Jonathan, Miami University
Locke, Daniel B., Dept of Agriculture, Conservation, and Forestry
Locke, III, Randall A., Illinois State Water Survey
Loh, Yvonne A., University of Ghana
Loheide, Steven P., University of Wisconsin, Madison
Lopez-Ferreira, Cesar A., Universidad Autonoma de Baja California Sur
Lord, Mark L., Western Carolina University
Love, Andew, Flinders University
Lowe, Mike, Utah Geological Survey
Lowry, Christopher S., SUNY, Buffalo
Luczaj, John A., University of Wisconsin, Green Bay
Lutz, Alexandra, University of Nevada, Reno
Mabee, Stephen B., Massachusetts Geological Survey
MacNish, Robert, University of Arizona
Macpherson, Gwen L., University of Kansas
Makkawi, Mohammad, King Fahd University of Petroleum and Minerals
Manda, Alex K., East Carolina University
Marino, M. A., University of California, Davis
Marino, Miguel A., University of California, Davis
Martel, Richard, Universite du Quebec
Matyjasik, Marek, Weber State University
May, James, Mississippi State University
Mayer, James R., University of West Georgia
Mayo, Alan L., Brigham Young University
McCartney, M. C., University of Wisconsin - Extension
McCurdy, Maureen, Louisiana Tech University
McKenzie, Jeffrey M., McGill University
Mendoza, Carl A., University of Alberta
Metesh, John J., Montana Tech of The University of Montana
Meyer, Philip D., Pacific Northwest National Laboratory
Meyer, Scott C., Illinois State Water Survey
Michael, Holly A., University of Delaware
Michel, Fred A., Carleton University
Milewski, Adam, University of Georgia
Miller, Marvin R., Montana Tech of The University of Montana
Misra, Debsmita, University of Alaska, Fairbanks
Mitchell, Robert J., Western Washington University
Moline, Gerilynn R., Oak Ridge National Laboratory
Montgomery, William W., New Jersey City University
Moore, Michael E., Pennsylvania Bureau of Topographic & Geologic Survey
Moran, Jean E., California State University, East Bay
Moss, Neil E., Geological Survey of Alabama
Moysey, Stephen M., Clemson University
Muldoon, Maureen A., University of Wisconsin, Oshkosh
Murdoch, Lawrence C., Clemson University
Murgulet, Dorina, Texas A&M University, Corpus Christi
Murray, Kent S., University of Michigan, Dearborn

Murray, Kyle E., University of Oklahoma
Myers, Paul B., Lehigh University
Nadiri, Ata Allah, University of Tabriz
Narasimhan, T. N., University of California, Berkeley
Naylor, Shawn, Indiana University
Naymik, Thomas G., Ohio State University
Neboga, Victoria V., Pennsylvania Bureau of Topographic & Geologic Survey
Nedunuri, Krishna K., Central State University
Neuman, Shlomo P., University of Arizona
Newcomer, Darrell R., Pacific Northwest National Laboratory
Newell, Charles J., Rice University
Nguyen, Lam V., Hanoi University of Mining & Geology
Nicot, Jean-Philippe, University of Texas at Austin, Jackson School of Geosciences
Nkotagu, Hudson H., University of Dar es Salaam
Noonan, Mathew T., South Dakota Dept of Environment and Natural Resources
Norman, Ralph R., Geological Survey of Alabama
Noyes, Joanne M., South Dakota Dept of Environment and Natural Resources
O'Brien, Arnold L., University of Massachusetts Lowell
O'Brien, Rachel, Allegheny College
O'Reilly, Andrew M., University of Mississippi
Oberdorfer, June A., San Jose State University
Ogard, Allen E., Los Alamos National Laboratory
Olafsen-Lackey, Susan, Unversity of Nebraska - Lincoln
Oliveira, Alcino S., Universidade de Trás-os-Montes e Alto Douro
Olyphant, Greg A., Indiana University, Bloomington
Ophori, Duke U., Montclair State University
Opper, Carl, Saint Petersburg College, Clearwater
Ortega, Adrian, Universidad Nacional Autonoma de Mexico
Osborn, Stephen G., California State Polytechnic University, Pomona
Ozsvath, David L., University of Wisconsin, Stevens Point
Pan, Feifei, University of North Texas
Paniconi, Claudio, Universite du Quebec
Parizek, Richard R., Pennsylvania State University, University Park
Parkin, Gary, University of Guelph
Parsen, Mike J., University of Wisconsin - Extension
Patton, Thomas W., Montana Tech of The University of Montana
Pederson, Darryll T., University of Nebraska, Lincoln
Peters, Catherine A., Princeton University
Peterson, Eric W., Illinois State University
Peterson, Holly, Guilford College
Pettyjohn, Wayne, Oklahoma State University
Pfannkuch, Hans O., University of Minnesota, Twin Cities
Phillips, Fred M., New Mexico Institute of Mining and Tech
Pickup, Gillian E., Heriot-Watt University
Plankell, Eric T., University of Illinois, Urbana-Champaign
Pociask, Geoff, University of Illinois, Urbana-Champaign
Pohll, Greg, University of Nevada, Reno
Post, Vincent , Flinders University
Prestegaard, Karen L., University of Maryland
Price, Rene, Florida International University
Puente-Muniz, Carlos Francisco, Universidad Autonoma de San Luis Potosi
Quinn, John, Argonne National Laboratory
Rains, Mark C., University of South Florida, Tampa
Rauch, Henry W., West Virginia University
Reese, Stuart O., Pennsylvania Bureau of Topographic & Geologic Survey
Reeve, Andrew S., University of Maine
Reeves, Donald Matthew M., Western Michigan University
Reichard, James S., Georgia Southern University
Reinfelder, Ying Fan, Rutgers, The State University of New Jersey
Reiten, Jon C., Montana Tech of The University of Montana
Remenda, Victoria H., Queen's University
Remson, Irwin, Stanford University
Rich, Thomas B., South Dakota Dept of Environment and Natural Resources
Riemersma, Peter E., Grand Valley State University
Rios-Sanchez, Miriam, Bemidji State University
Ritzi, Jr., Robert W., Wright State University
Roadcap, George S., Illinois State Water Survey
Robbins, Gary A., University of Connecticut
Robertson, Will, University of Waterloo
Robin, Michel R., University of Ottawa
Roman, Eric W., New Jersey Geological and Water Survey
Romanowicz, Edwin A., Plattsburgh State University (SUNY)
Ronayne, Michael J., Colorado State University
Rose, Seth E., Georgia State University
Rostron, Benjamin J., University of Alberta
Rouleau, Alain, Universite du Quebec a Chicoutimi
Royo-Ochoa, Miguel, Universidad Autonoma de Chihuahua
Russo, Tess A., Pennsylvania State University, University Park
Ryan, Cathy, University of Calgary
Saffer, Demian M., Pennsylvania State University, University Park
Saint, Prem K., California State University, Fullerton
Salvage, Karen M., Binghamton University
Samuelson, Alan C., Ball State University
Sanders, Laura L., Northeastern Illinois University
Sanford, William E., Colorado State University
Sasowsky, Ira D., University of Akron
Scanlon, Bridget R., University of Texas at Austin, Jackson School of Geosciences
Scheibe, Timothy D., Pacific Northwest National Laboratory
Scheidt, Brian, Mineral Area College
Schincariol, Robert A., Western University
Schlosser, Peter, Columbia University
Schmitz, Darrel W., Mississippi State University
Schreiber, Madeline E., Virginia Polytechnic Institute & State University
Schueth, Christoph, Technische Universitaet Darmstadt
Schulmeister, Marcia K., Emporia State University
Schulze-Makuch, Dirk, Washington State University
Scott, Christopher A., University of Arizona
Sebol, Lesley, Colorado School of Mines
Sedivy, Robert A., Argonne National Laboratory
Sendlein, Lyle V. A., University of Kentucky
Seramur, Keith C., Appalachian State University
Seyoum, Wondwosen M., Illinois State University
Shanafield, Margaret, Flinders University
Shang, Congxiao, University of East Anglia
Shaw, Glenn D., Montana Tech of the University of Montana
Sheldon, Amy L., SUNY, Geneseo
Shevenell, Lisa A., University of Nevada, Reno
Shim-Chim, Richard, New Jersey Geological and Water Survey
Shimada, Jun, Kumamoto University
Sibray, Steven S., Unversity of Nebraska - Lincoln
Siegel, Donald I., Syracuse University
Simmons, Craig, Flinders University
Simpkins, William W., Iowa State University of Science & Tech
Singha, Kamini, Colorado School of Mines
Skalbeck, John D., University of Wisconsin, Parkside
Smith, J. Leslie, University of British Columbia
Smith, James E., University of Arizona
Smith, James E., McMaster University
Smith, Karen, Argonne National Laboratory
Soll, Wendy E., Los Alamos National Laboratory
Solomon, Douglas K., University of Utah
Spangler, Daniel P., University of Florida
Spayd, Steven E., New Jersey Geological and Water Survey
Spinelli, Glenn, New Mexico Institute of Mining and Tech
Springer, Abraham E., Northern Arizona University
Spruill, Richard K., East Carolina University
Stafford, Kevin W., Stephen F. Austin State University
Staley, Andrew, Maryland Department of Natural Resources
Stewart, Mark T., University of South Florida, Tampa
Stotler, Randy, University of Kansas
Strack, Otto D., University of Minnesota, Twin Cities
Sudicky, Edward A., University of Waterloo
Suen, C. J., California State University, Fresno
Sukop, Michael C., Florida International University
Sun, Hongbing, Rider University
Sutherland, Mary K., Montana Tech of The University of Montana
Swanson, Susan K., Beloit College
Tabidian, M. Ali, California State University, Northridge
Taboga, Karl, Wyoming State Geological Survey
Tauxe, John D., Oak Ridge National Laboratory
Tellam, John, University of Birmingham
Tezcan, Levent, Hacettepe University
Therrien, Rene, Universite Laval
Thomason, Jason F., University of Illinois, Urbana-Champaign
Thompson, Carol A., Tarleton State University
Tick, Geoffrey, University of Alabama
Tidwell, Vincent C., New Mexico Institute of Mining and Tech

Tijani, Moshood N., University of Ibadan
Tilton, Eric E., University of Colorado
Tipping, Robert, University of Minnesota
Toran, Laura, Temple University
Tóth, József, University of Alberta
Totsche, Kai U., Friedrich-Schiller-University Jena
Troy, Marleen, Wilkes University
Tucker, Carla M., Lamar University
Tyler, Scott W., University of Nevada, Reno
Unger, Andre, University of Waterloo
Upchurch, Sam B., University of South Florida, Tampa
Vacher, H. Leonard, University of South Florida, Tampa
Van Der Hoven, Steve, Illinois State University
Van der Kamp, Garth, University of Waterloo
VanDerwerker, Tiffany J., Maryland Department of Natural Resources
Veeger, Anne I., University of Rhode Island
Vengosh, Avner, Duke University
Vesper, Dorothy, West Virginia University
Vilcaez, Javier, Oklahoma State University
Wallace, Janae, Utah Geological Survey
Walraevens, Kristine, Ghent University
Walton, Nick, University of Portsmouth
Wang, Dong, California State University, Fresno
Wanke, Heike, University of Namibia
Ward, James W., Angelo State University
Waren, Kirk B., Montana Tech of The University of Montana
Wassenaar, Len, University of Saskatchewan
Watson, David B., Oak Ridge National Laboratory
Weissmann, Gary, University of New Mexico
Welby, Charles W., North Carolina State University
Welhan, John A., University of Idaho
Welhan, John A., Idaho State University
Werner, Adrian , Flinders University
Wheatcraft, Steve, University of Nevada, Reno
Wheaton, John R., Montana Tech of The University of Montana
Wheeler, Betty, University of Minnesota, Twin Cities
Whiting, Duane L., University of Utah
Whittemore, Donald O., University of Kansas
Wicks, Carol M., Louisiana State University
Widdowson, Mark A., Virginia Tech
Wilcox, Jeffrey D., University of North Carolina, Asheville
Wilson, Alicia M., University of South Carolina
Wilson, John L., New Mexico Institute of Mining and Tech
Wilson, Lorne G., University of Arizona
Wilson, Steven D., Illinois State Water Survey
Wolaver, Brad, University of Texas at Austin
Wolfsberg, Andrew V., Los Alamos National Laboratory
Wood, Thomas R., University of Idaho
Yan, Eugene, Argonne National Laboratory
Yang, Changbing, University of Texas at Austin
Yang, Jianwen, University of Windsor
Yelderman, Jr., Joe C., Baylor University
Yidana, Sandow M., University of Ghana
Young, Tim, University of Illinois, Urbana-Champaign
Yüce, Galip, Hacettepe University
Zarnetske, Jay, Michigan State University
Zhan, Hongbin, Texas A&M University
Zhang, Pengfei, City College (CUNY)
Zhang, You-Kuan, University of Iowa
Zheng, Chunmiao, University of Alabama
Zhu, Junfeng, University of Kentucky
Zlotnik, Vitaly A., University of Nebraska, Lincoln
Zouhri, Lahcen, Institut Polytechnique LaSalle Beauvais (ex-IGAL)
Zreda, Marek G., University of Arizona
Özyurt, Nur N., Hacettepe University

## Quantitative Hydrology

Bacon, Diana H., Pacific Northwest National Laboratory
Benson, David A., Colorado School of Mines
Birdsell, Kay H., Los Alamos National Laboratory
Borah, Deva K., Illinois State Water Survey
Carey, Anne E., Ohio State University
Carroll, Rosemary, University of Nevada, Reno
Chebana, Fateh, Universite du Quebec
Chen, Xun-Hong, Unversity of Nebraska - Lincoln
Covington, Matthew, University of Arkansas, Fayetteville
Danko, George, University of Nevada, Reno
Dingman, S. Lawrence, University of New Hampshire
Donovan, Joseph J., West Virginia University
Duckstein, Lucien, University of Arizona
Dymond, Randel, Virginia Tech
Evans, David G., California State University, Sacramento
Falta, Ronald W., Clemson University
Fayer, Michael J., Pacific Northwest National Laboratory
Ferrand, Lin A., Graduate School of the City University of New York
Flores, Alejandro N., Boise State University
French, Mark A., New Jersey Geological and Water Survey
Frind, Emil O., University of Waterloo
Gillham, Robert W., University of Waterloo
Gupta, Hoshin V., University of Arizona
Harpold, Adrian, University of Nevada, Reno
Hermance, John F., Brown University
Hoffman, Jeffrey L., New Jersey Geological and Water Survey
Holt, Robert M., University of Mississippi
Hubbart, Jason A., West Virginia University
Hughes, Denis, Rhodes University
Kandiah, Ramanitharan, Central State University
Kiem, Anthony, University of Newcastle
Kirkham, Randy R., Pacific Northwest National Laboratory
Letsinger, Sally L., Indiana University
Li, Dan, Boston University
Li, Liangping, South Dakota School of Mines & Tech
Loague, Keith, Stanford University
Looney, Brian, Clemson University
Maddock, III, Thomas, University of Arizona
Martin-Hayden, James, University of Toledo
McCord, James T., New Mexico Institute of Mining and Tech
Meko, David M., University of Arizona
Michaud, Jene D., University of Hawai'i, Hilo
Oostrom, Martinus, Pacific Northwest National Laboratory
Pacheco, Fernando A., Universidade de Trás-os-Montes e Alto Douro
Pangle, Luke, Georgia State University
Parashar, Rishi, University of Nevada, Reno
Park, Young-Jin, University of Waterloo
Perkins, Robert B., Portland State University
Pohll, Greg, University of Nevada, Reno
Puente, Carlos E., University of California, Davis
Quinn, John J., Argonne National Laboratory
Restrepo, Jorge I., Florida Atlantic University
Rudolph, David L., University of Waterloo
Salvucci, Guido D., Boston University
Schumer, Rina, University of Nevada, Reno
Sharp, Jr., John M., University of Texas at Austin
Sipos, György, Univesity of Szeged
Springer, Everett P., Los Alamos National Laboratory
Stockton, Charles W., University of Arizona
Szilagyi, Jozsef, Unversity of Nebraska - Lincoln
Telyakovskiy, Aleksey, University of Nevada, Reno
Thorbjarnarson, Kathryn W., San Diego State University
Travis, Bryan J., Los Alamos National Laboratory
Verdon-Kidd, Danielle, University of Newcastle
Wallis, Ilka, Flinders University
Whitman, Brian E., Wilkes University
Yeh, Tian-Chyi J., University of Arizona
Zhang, Yong, University of Alabama

## Surface Waters

Acharya, Kumud, University of Nevada, Reno
Aitkenhead-Peterson, Jacqueline A., Texas A&M University
Anderson, Michelle, University of Montana Western
Bacon, Robert, Missouri Dept of Natural Resources
Baldini, James, Durham University
Bauer, Carl J., University of Arizona
Bearden, Rebecca A., Geological Survey of Alabama
Bevelhimer, Mark, Oak Ridge National Laboratory
Beyer, Patricia J., Bloomsburg University
Bhowmik, Nani G., Illinois State Water Survey
Blanken, Peter D., University of Colorado
Bogner, William C., Illinois State Water Survey
Cadol, Daniel, New Mexico Institute of Mining and Tech
Chandra, Sudeep, University of Nevada, Reno
Christian, Alan D., University of Massachusetts, Boston
Clark, Robert A., University of Arizona
Cosby, Jack, University of Virginia
Costa, Jr, Ozeas S., Ohio State University
D'Alpaos, Andrea, Università degli Studi di Padova

Demchak, Jennifer, Mansfield University
Dennett, Keith E., University of Nevada, Reno
Doran, Peter T., University of Illinois at Chicago
Du Charme, Charles, Missouri Dept of Natural Resources
Duan, Shuiwang, University of Maryland
Franz, Kristie, Iowa State University of Science & Tech
Frei, Allan, Hunter College (CUNY)
Goodrich, David C., University of Arizona
Grable, Judy, Valdosta State University
Gulliver, John S., University of Minnesota, Twin Cities
Hamlet, Alan, University of Notre Dame
Harris, Randa R., University of West Georgia
Hasan, Khaled W., Austin Community College District
Hawkins, R. B., University of Arizona
Helz, George, University of Maryland
Hester, Erich, Virginia Tech
Ince, Simon, University of Arizona
Johnston, K. R. Gina, California State University, Chico
Khosrowpanah, Shahram, University of Guam
Kindelman, Julie, University of Wisconsin, Parkside
Kisila, Ben O., University of Mary Washington
Knapp, H. Vernon, Illinois State Water Survey
Knight, Allen W., University of California, Davis
Kunza, Lisa, South Dakota School of Mines & Tech
Lancaster, Stephen, Oregon State University
Lansey, Kevin E., University of Arizona
Lisle, Thomas, Humboldt State University
Lund, Jay R., University of California, Davis
Mahat, Vinod, Argonne National Laboratory
Maneta, Marco, University of Montana
Maxwell, Reed M., Colorado School of Mines
Mayfield, Michael W., Appalachian State University
McGregor, Stuart W., Geological Survey of Alabama
McIsaac, Gregory F., University of Illinois, Urbana-Champaign
Noe, Garry, Virginia Wesleyan College
O'Neil, Patrick E., Geological Survey of Alabama
Pradhanang, Soni M., University of Rhode Island
Price, Katie, Georgia State University
Rancan, Helen L., New Jersey Geological and Water Survey
Richard, Gigi, Colorado Mesa University
Richmond, Marshall C., Pacific Northwest National Laboratory
Ridenour, Gregory D., Austin Peay State University
Royer, Todd, Indiana University, Bloomington
Scott, Verne H., University of California, Davis
Shuttleworth, W. James, University of Arizona
Skiles, S. McKenzie, Utah Valley University
Smerdon, Ernest T., University of Arizona
Smith, Brianne, Brooklyn College (CUNY)
Smith, Erik, University of South Carolina
Smith, Laurence C., University of California, Los Angeles
Smith, Sean, University of Maine
Stieglitz, Marc, Georgia Institute of Tech
Strom, Kyle, Virginia Tech
Swanson, Sherman, University of Nevada, Reno
Troch, Peter A., University of Arizona
Valdes, Juan B., University of Arizona
Wampler, Peter J., Grand Valley State University
Washburne, James C., University of Arizona
White, Jeffrey, Indiana University, Bloomington
Wigmosta, Mark S., Pacific Northwest National Laboratory
Wilderman, Candie, Dickinson College
Wynn, Elizabeth A., Geological Survey of Alabama
Xia, Renjie, Illinois State Water Survey
Zollweg, James A., SUNY, The College at Brockport

## Geohydrology

Allen, Diana M., Simon Fraser University
Attinger, Sabine, Friedrich-Schiller-University Jena
Baedke, Steven J., James Madison University
Batlle-Aguilar, Jordi, University of Kansas
Beddows, Patricia A., Northwestern University
Brookfield, Andrea, University of Kansas
Buddemeier, Robert W., University of Kansas
Burkart, Michael R., Iowa State University of Science & Tech
Castro, Maria Clara, University of Michigan
Cathles, Lawrence M., Cornell University
Chavez-Rodriguez, Adolfo, Universidad Autonoma de Chihuahua
Chormann, Jr., Frederick H., New Hampshire Geological Survey
Clair, Thomas A., Mount Allison University
Constantine, Jose, Cardiff University
Cooper, Clay A., University of Nevada, Reno
Crews, Jeff, Missouri Dept of Natural Resources
Daanen, Ronnie, Alaska Division of Geological & Geophysical Surveys
De Luca, Domenico Antonio, Università di Torino
Dowd, John F., University of Georgia
Drake, Lon D., University of Iowa
Duley, James W., Missouri Dept of Natural Resources
Fogg, Graham E., University of California, Davis
Gamage, Kusali R., Austin Community College District
Gartner, Janette, Bentley University
Gierke, John S., Michigan Technological University
Gurwin, Jacek, University of Wroclaw
Haitjema, Hendrik M., Indiana University / Purdue University, Indianapolis
Hardy, Douglas R., University of Massachusetts, Amherst
Hong, Sung-ho, Murray State University
Hyndman, David W., Michigan State University
Khanbilvardi, Reza M., Graduate School of the City University of New York
Knight, Rosemary J., Stanford University
Koch, Magaly, Boston University
Land, Lewis, New Mexico Institute of Mining and Tech
Leap, Darrell I., Purdue University
Liu, Gaisheng, University of Kansas
Lovell, Mark D., Brigham Young University - Idaho
Lupankwa, Mlindelwa , Tshwane University of Tech
Marszalek, Henryk, University of Wroclaw
Mayer, Alex S., Michigan Technological University
McIntosh, Jennifer, University of Arizona
McKay, Larry D., University of Tennessee, Knoxville
McKenna, Thomas E., University of Delaware
Mehnert, Edward, University of Illinois, Urbana-Champaign
Mount, Gregory, Indiana University of Pennsylvania
Ng, Crystal, University of Minnesota, Twin Cities
Oshun, Jasper, Humboldt State University
Person, Mark A., New Mexico Institute of Mining and Tech
Rayne, Todd W., Hamilton College
Reboulet, Edward, University of Kansas
Rovey, Charles W., Missouri State University
Sargent, Kenneth A., Furman University
Sidle, Roy, Appalachian State University
Sushama, Laxmi, Universite du Quebec a Montreal
Torres, Raymond, University of South Carolina
White, Mark D., Pacific Northwest National Laboratory
Zhang, Ye, University of Wyoming

## SOIL SCIENCE

## Soil Physics/Hydrology

Amoozegar, Aziz, North Carolina State University
Anderson, Stephen H., University of Missouri, Columbia
Arnone, John J., University of Nevada, Reno
Baker, John M., University of Minnesota, Twin Cities
Berli, Marcus, University of Nevada, Reno
Black, T. Andrew, University of British Columbia
Bland, William L., University of Wisconsin, Madison
Boast, Charles W., University of Illinois, Urbana-Champaign
Brown, Sandra, University of British Columbia
Burger, Martin, University of California, Davis
Caldwell, Todd, University of Texas at Austin
Casey, Francis, North Dakota State University
Cassel, Donald K., North Carolina State University
Cushman, John, Purdue University
Ellsworth, Timothy R., University of Illinois, Urbana-Champaign
Fermanich, Kevin J., University of Wisconsin, Green Bay
Feyereisen, Gary, University of Minnesota, Twin Cities
Flury, Markus, Washington State University
Fritton, Daniel D., Pennsylvania State University, University Park
Galbraith, John M., Virginia Polytechnic Institute & State University
Grattan, Stephen R., University of California, Davis
Grismer, Mark E., University of California, Davis
Groenevelt, Pieter H., University of Guelph
Gupta, Satish C., University of Minnesota, Twin Cities
Hook, James E., University of Georgia
Hopmans, Jan W., University of California, Davis

Horton, Robert, Iowa State University of Science & Tech
Hsiao, Theodore C., University of California, Davis
Hutson, John , Flinders University
Jawitz, James W., University of Florida
Johnson, Mark S., University of British Columbia
Jones, Tim L., New Mexico State University, Las Cruces
Kakembo, Vincent, Nelson Mandela Metropolitan University
Kanemasu, Edward T., University of Georgia
Kirkham, Mary Beth, Kansas State University
Kitchen, Newell R., University of Missouri, Columbia
Kluitenberg, Gerard J., Kansas State University
Kung, King-Jau S., University of Wisconsin, Madison
Lal, Rattan, Ohio State University
Lourenco, Sergio, Cardiff University
Lowery, Birl, University of Wisconsin, Madison
McCoy, Edward L., Ohio State University
McDonald, Eric, University of Nevada, Reno
McInnes, Kevin J., Texas A&M University
Molz, Fred, Clemson University
Moncrief, John F., University of Minnesota, Twin Cities
Morgan, Cristine L., Texas A&M University
Mulla, David J., University of Minnesota, Twin Cities
Neely, Haly L., Texas A&M University
Nielsen, Donald R., University of California, Davis
Niu, Guo-Yue, University of Texas at Austin
Nkedi-Kizza, Peter, University of Florida
Norman, John M., University of Wisconsin, Madison
Papendick, Robert I., Washington State University
Perfect, Edmund, University of Tennessee, Knoxville
Persaud, Naraine, Virginia Polytechnic Institute & State University
Radcliffe, David E., University of Georgia
Reece, Julia S., Texas A&M University
Ressler, Daniel E., Susquehanna University
Rogowski, Andrew S., Pennsylvania State University, University Park
Sammis, Theodore W., New Mexico State University, Las Cruces
Sanchez, Marcelo, Texas A&M University
Selim, Hussein M., Louisiana State University
Simmons, F. William, University of Illinois, Urbana-Champaign
Spokas, Kurt, University of Minnesota, Twin Cities
Stone, Loyd, Kansas State University
Taylor, Donavon, University of Wisconsin, River Falls
Thomas, John, University of Florida
Turk, Judy, Stockton University
Venterea, Rodney T., University of Minnesota, Twin Cities
Verburg, Paul, University of Nevada, Reno
Wang, Zhi (Luke), California State University, Fresno
Warrick, Arthur W., University of Arizona
Welch, Steve M., Kansas State University
Wierenga, Peter J., University of Arizona
Wraith, Jon M., Montana State University
Yeh, Jim, University of Arizona

## Soil Chemistry/Mineralogy

Adriano, Domy C., University of Georgia
Adriano, Domy C., Virginia Polytechnic Institute & State University
Ahmad, Abd Rashid, University of Malaya
Alley, Marcus M., Virginia Polytechnic Institute & State University
Appel, Christopher (Chip) S., California Polytechnic State University
Artiola, Janick F., University of Arizona
Baligar, V. C., Virginia Polytechnic Institute & State University
Barak, Phillip W., University of Wisconsin, Madison
Basta, Nicholas T., Ohio State University
Beegle, Douglas B., Pennsylvania State University, University Park
Bertsch, Paul M., University of Georgia
Blanchar, Robert W., University of Missouri, Columbia
Bleam, William F., University of Wisconsin, Madison
Bloom, Paul R., University of Minnesota, Twin Cities
Bomke, Arthur A., University of British Columbia
Bostick, Benjamin C., Columbia University
Breecker, Daniel O., University of Texas at Austin
Burke, Deborah S., Pacific Northwest National Laboratory
Callanan, Jennifer R., William Paterson University
Casey, William H., University of California, Davis
Chirenje, Tait, Richard Stockton College of New Jersey
Chorover, Jonathan, University of Arizona
Clark, Joanna, University of Reading
Collins, Chris, University of Reading
Crouse, David A., North Carolina State University
Curry, Joan E., University of Arizona
Dahlgran, Randy, University of California, Davis
Donohue, Stephen J., Virginia Polytechnic Institute & State University
Dowdy, Robert H., University of Minnesota, Twin Cities
Dudley, Lynn M., Florida State University
Eckert, Donald J., Ohio State University
Edenborn, Harry, West Virginia University
Eick, Matthew J., Virginia Polytechnic Institute & State University
Eivazi, Frieda, University of Missouri, Columbia
Elzinga, Evert J., Rutgers, The State University of New Jersey, Newark
Engel, Richard E., Montana State University
Evangelou, V. P., Iowa State University of Science & Tech
Evanylo, Gregory K., Virginia Polytechnic Institute & State University
Fendorf, Scott E., Stanford University
Fernandez, Fabian G., University of Minnesota, Twin Cities
Gascho, Gary J., University of Georgia
Gaston, Lewis A., Louisiana State University
Ginder-Vogel, Matt, University of Wisconsin, Madison
Goos, Robert J., North Dakota State University
Goyne, Keith W., University of Missouri, Columbia
Guertal, Elizabeth A., Auburn University
Han, Nizhou, Virginia Polytechnic Institute & State University
Harris, Jr., Willie G., University of Florida
Harsh, James B., Washington State University
Harter, Robert D., University of New Hampshire
Hassett, John J., University of Illinois, Urbana-Champaign
Havlin, John L., North Carolina State University
He, Zhenli, University of Florida
Heil, Dean, New Mexico State University, Las Cruces
Helmke, Philip A., University of Wisconsin, Madison
Hemzacek Laukant, Jean M., Northeastern Illinois University
Hesterberg, Dean L., North Carolina State University
Hettiarachchi, Ganga, Kansas State University
Howe, Julie, Auburn University
Howe, Julie, Texas A&M University
Inskeep, William P., Montana State University
James, Bruce, University of Maryland
Jaouich, Alfred, Universite du Quebec a Montreal
Jardine, Philip M., Oak Ridge National Laboratory
Jones, Robert L., University of Illinois, Urbana-Champaign
Kabengi, Nadine, Georgia State University
Kaiser, Daniel E., University of Minnesota, Twin Cities
Kissel, David E., University of Georgia
Komarneni, Sridhar, Pennsylvania State University, University Park
Koskinen, William C., University of Minnesota, Twin Cities
Kuo, Shiou, Washington State University
Laird, David A., Iowa State University of Science & Tech
Lamb, John A., University of Minnesota, Twin Cities
Lee, Young Jae, Korea University
Lewis, Katie, Texas A&M University
Li, Yuncong, University of Florida
Ma, Lena Q., University of Florida
Malzer, Gary L., University of Minnesota, Twin Cities
Marion, Giles, Desert Research Institute
McCaslin, Bobby D., New Mexico State University, Las Cruces
McLaughlin, Richard A., North Carolina State University
Miller, William P., University of Georgia
Mitchell, Charles C., Auburn University
Moore, Duane M., University of New Mexico
Mowrer, Jake, Texas A&M University
Mullins, Gregory L., Virginia Polytechnic Institute & State University
Mulvaney, Richard L., University of Illinois, Urbana-Champaign
Mylavarapu, S R., University of Florida
Myneni, Satish C B., Princeton University
Nair, Vimala D., University of Florida
O'Connor, George A., University of Florida
Odom, John W., Auburn University
Olson, Kenneth R., University of Illinois, Urbana-Champaign
Pagliari, Paulo H., University of Minnesota, Twin Cities
Peck, Theodore R., University of Illinois, Urbana-Champaign
Peryea, Frank J., Washington State University
Pierzynski, Gary M., Kansas State University
Plank, Owen C., University of Georgia
Provin, Tony L., Texas A&M University
Quade, Jay, University of Arizona
Rehm, George W., University of Minnesota, Twin Cities
Reneau, Raymond B., Virginia Polytechnic Institute & State University

Robarge, Wayne P., North Carolina State University
Rochette, Elizabeth A., University of New Hampshire
Rosen, Carl J., University of Minnesota, Twin Cities
Roy, William R., Illinois State Geological Survey
Russelle, Michael P., University of Minnesota, Twin Cities
Schmitt, Michael A., University of Minnesota, Twin Cities
Schwab, Paul, Texas A&M University
Segars, William P., University of Georgia
Shuman, Larry M., University of Georgia
Siebe Grabach, Christina D., Universidad Nacional Autonoma de Mexico
Sims, Albert L., University of Minnesota, Twin Cities
Smyth, Thomas J., North Carolina State University
Spalding, Brian P., Oak Ridge National Laboratory
Strawn, Daniel G., University of Idaho
Stucki, Joseph W., University of Illinois, Urbana-Champaign
Szulczewski, Melanie, University of Mary Washington
Tatabatai, M. A., Iowa State University of Science & Tech
Thien, Steve J., Kansas State University
Thompson, Michael L., Iowa State University of Science & Tech
Toor, Gurpal, University of Florida
Touchton, Joseph T., Auburn University
Trumbore, Susan E., University of California, Irvine
Werts, Scott P., Winthrop University
Wood, Charles W., Auburn University
Wright, Alan, University of Florida
Zelazny, Lucian W., Virginia Polytechnic Institute & State University

## Pedology/Classification/Morphology

Anderson, James L., University of Minnesota, Twin Cities
Baker, James C., Virginia Polytechnic Institute & State University
Barrett, Linda R., University of Akron
Bell, James C., University of Minnesota, Twin Cities
Bettis III, E. A., University of Iowa
Bigham, Jerry M., Ohio State University
Birkeland, Peter W., University of Colorado
Buck, Brenda J., University of Montana Western
Burras, Lee, Iowa State University of Science & Tech
Calhoun, Frank G., Ohio State University
Caudill, Michael R., Hocking College
Ciolkosz, Edward J., Pennsylvania State University, University Park
Collins, Mary E., University of Florida
Cooper, Terence H., University of Minnesota, Twin Cities
Daniels, Walter L., Virginia Polytechnic Institute & State University
Darmody, Robert G., University of Illinois, Urbana-Champaign
Daugherty, LeRoy A., New Mexico State University, Las Cruces
Dere, Ashlee L., University of Nebraska at Omaha
Evans, Barry M., Pennsylvania State University, University Park
Farsang, Andrea, Univesity of Szeged
Frederick, Holly, Wilkes University
Heck, Richard, University of Guelph
Hoover, Michael T., North Carolina State University
Hopkins, David G., North Dakota State University
Jacobs, Peter, University of Wisconsin, Whitewater
Jelinski, Nicolas, University of Minnesota, Twin Cities
Johnson-Maynard, Jodi, University of Idaho
Kleiss, Harold J., North Carolina State University
Kuzila, Mark S., Unversity of Nebraska - Lincoln
Lavkulich, Leslie M., University of British Columbia
Lindbo, David, North Carolina State University
Madison, Frederick W., University of Wisconsin, Madison
Mbila, Monday O., Alabama A&M University
McDaniel, Paul A., University of Idaho
McDonald, Eric, Desert Research Institute
McFadden, Leslie M., University of New Mexico
McGahan, Donald, Texas A&M University
McSweeney, Kevin, University of Wisconsin, Madison
Miles, Randall J., University of Missouri, Columbia
Miller, Gerald A., Iowa State University of Science & Tech
Montagne, Cliff, Montana State University
Nater, Edward A., University of Minnesota, Twin Cities
Nielsen, Gerald A., Montana State University
Nordt, Lee C., Baylor University
Paz Gonzalez, Antonio, Coruna University
Petersen, Gary W., Pennsylvania State University, University Park
Post, Donald F., University of Arizona
Ransom, Michel D., Kansas State University
Rice, Thomas J., California Polytechnic State University
Rocque, David, Dept of Agriculture, Conservation, and Forestry
Runge, Edward C. A., Texas A&M University
Ryder, Roy, University of South Alabama
Sandor, Jonathan A., Iowa State University of Science & Tech
Schardt, Lawrence A., Pennsylvania State University, University Park
Schmid, Ginger L., Minnesota State University
Shaw, Joey N., Auburn University
Slater, Brian K., Ohio State University
Taboada Castro, María T., Coruna University
Tabor, Neil J., Southern Methodist University
Thomas, Pamela J., Virginia Polytechnic Institute & State University
Tyler, E. Jerry, University of Wisconsin, Madison
Vepraskas, Michael J., North Carolina State University
West, Larry T., University of Georgia
Yoo, Kyungsoo, University of Minnesota, Twin Cities

## Forest Soils/Rangelands/Wetlands

Amthor, Jeff, Oak Ridge National Laboratory
Andrus, Richard E., Binghamton University
Balster, Nick J., University of Wisconsin, Madison
Bass, Michael L., University of Mary Washington
Bauder, James W., Montana State University
Black, Bryan A., University of Texas at Austin
Bockheim, James G., University of Wisconsin, Madison
Broome, Stephen W., North Carolina State University
Chambers, Jeanne, University of Nevada, Reno
Clark, Mark W., University of Florida
Clark, Melissa, Indiana University, Bloomington
Comerford, Nicholas B., University of Florida
David, Mark B., University of Illinois, Urbana-Champaign
Dietze, Michael, Boston University
Ferris, Dawn, Ohio State University
Fick, Walter H., Kansas State University
Franz, Eldon H., Washington State University
Gunderson, Lance, Emory University
Hallock, Brent G., California Polytechnic State University
Hook, Paul B., Montana State University
Inglett, Patrick, University of Florida
Krzic, Maja, University of British Columbia
Laingen, Christopher R., Eastern Illinois University
Meretsky, Vicky J., Indiana University, Bloomington
Miner, James J., University of Illinois
O'Neill, Patrick M., Louisiana State University
Osborne, Todd, University of Florida
Owensby, Clenton E., Kansas State University
Peres, Carlos, University of East Anglia
Posler, Gerry L., Kansas State University
Riha, Susan, Cornell University
Robinson, Steve, University of Reading
Smalley, Glendon W., Sewanee: University of the South
Smith, C. Ken, Sewanee: University of the South
Taskey, Ronald D., California Polytechnic State University
Torreano, Scott, Sewanee: University of the South
Turk, Judith , Richard Stockton College of New Jersey
Wegner, John, Emory University

## Soil Biology/Biochemistry

Allan, Deborah L., University of Minnesota, Twin Cities
Armstrong, Felicia P., Youngstown State University
Balser, Teri C., University of Wisconsin, Madison
Berry, Duane F., Virginia Polytechnic Institute & State University
Bezdicek, David F., Washington State University
Bird, Jeffrey, Queens College (CUNY)
Blum, Linda K., University of Virginia
Bollag, Jean-Marc, Pennsylvania State University, University Park
Burlage, Robert S., Oak Ridge National Laboratory
Chanway, Christopher, University of British Columbia
Cheng, H. H., University of Minnesota, Twin Cities
Collins, Chris, University of Reading
Crozier, Carl, North Carolina State University
Davenport, Joan R., Washington State University
Dick, Richard P., Ohio State University
Dick, Warren A., Ohio State University
Dunfield, Kari, University of Guelph
Feng, Yucheng, Auburn University
Gauthier, Paul, Princeton University
Gentry, Terry, Texas A&M University
Gerba, Charles P., University of Arizona
Gilliam, James W., North Carolina State University

Grace, Peter R., Queensland University of Tech
Graham, Jr., James H., University of Florida
Graves, Alexandria, North Carolina State University
Grayston, Sue, University of British Columbia
Groffman, Peter, Brooklyn College (CUNY)
Gutknecht, Jessica L., University of Minnesota, Twin Cities
Hagedorn, Charles, Virginia Polytechnic Institute & State University
Harris, Glendon H., University of Georgia
Harris, Robin F., University of Wisconsin, Madison
Hartel, Peter G., University of Georgia
Hickey, William J., University of Wisconsin, Madison
Hladik, Christine, Georgia Southern University
Hoyt, Greg D., North Carolina State University
Ishii, Satoshi, University of Minnesota, Twin Cities
Israel, Daniel W., North Carolina State University
Jaffe, Peter R., Princeton University
Jastrow, Julie D., Argonne National Laboratory
Kennedy, Ann C., Washington State University
Knudsen, Guy, University of Idaho
Lindemann, William C., New Mexico State University, Las Cruces
Matson, Pamela A., Stanford University
Miller, Raymond M., Argonne National Laboratory
Molina, Jean-Alex E., University of Minnesota, Twin Cities
Moorhead, Kevin K., University of North Carolina, Asheville
Morra, Matthew J., University of Idaho
Ogram, Andrew V., University of Florida
Oikawa, Patricia, California State University, East Bay
Osmond, Deanna L., North Carolina State University
Palumbo, Tony V., Oak Ridge National Laboratory
Pan, William L., Washington State University
Parker, David B., West Texas A&M University
Peacock, Caroline, University of Leeds
Phelps, Tommy J., Oak Ridge National Laboratory
Plante, Alain F., University of Pennsylvania
Prescott, Cindy, University of British Columbia
Reddy, K R., University of Florida
Rice, Chuck W., Kansas State University
Ruehr, Thomas A., California Polytechnic State University
Rutherford, Michael P., Sir Wilfred Grenfell College
Sadowsky, Michael J., University of Minnesota, Twin Cities
Scheer, Clemens, Queensland University of Tech
Shaw, Liz, University of Reading
Shi, Wei, North Carolina State University
Silva, Lucas C., University of California, Davis
Simard, Suzanne, University of British Columbia
Smith, Jeffery L., Washington State University
Somenhally, Anil, Texas A&M University
Stevens, Robert G., Washington State University
Thomas, Andrew D., University of Wales
Tolhurst, Trevor, University of East Anglia
Torrents, Alba, University of Maryland
van der Werf, Guido R., VU University Amsterdam
Voroney, R. Paul, University of Guelph
Wagger, Michael G., North Carolina State University
Wenzel, Christopher, Eastern Wyoming College
Wilkie, Ann C., University of Florida

## Paleopedology/Archeology

Beck, Colleen M., Desert Research Institute
Busacca, Alan J., Washington State University
Driese, Steven G., Baylor University
Driese, Steven, University of Tennessee, Knoxville
Follmer, Leon R., University of Illinois, Urbana-Champaign
Herrmann, Edward W., Indiana University, Bloomington
Irmis, Randall B., University of Utah
Monger, H. C., New Mexico State University, Las Cruces

## Other Soil Science

Adamsen, Floyd, University of Arizona
Adee, Eric, Kansas State University
Aiken, Robert, Kansas State University
Bader, Nicholas E., Whitman College
Bell, Brian, University of Glasgow
Berg, Peter, University of Virginia
Bundy, Larry G., University of Wisconsin, Madison
Cabrera, Miguel L., University of Georgia
Cambardella, Cynthia, Iowa State University of Science & Tech
Ciampitti, Ignacio, Kansas State University
Claassen, Mark, Kansas State University
Coleman, Tommy L., Alabama A&M University
Cramer, Gary, Kansas State University
Cruse, Richard M., Iowa State University of Science & Tech
Daroub, Samira H., University of Florida
Dingus, Delmar D., California Polytechnic State University
Dormaar, John, University of Lethbridge
Drescher, Andrew, University of Minnesota, Twin Cities
Duncan, Stewart, Kansas State University
Eberle, Bill, Kansas State University
Ebinger, Michael H., Los Alamos National Laboratory
Ehler, Stan, Kansas State University
Ferre, Paul (Ty), University of Arizona
Fjell, Dale, Kansas State University
Flores-Delgadillo, Maria L., Universidad Nacional Autonoma de Mexico
Flores-Roman, David, Universidad Nacional Autonoma de Mexico
Fritz, Allan, Kansas State University
Gordon, Barney, Kansas State University
Goss, Michael J., University of Guelph
Haag, Lucas, Kansas State University
Hall, III, John R., Virginia Polytechnic Institute & State University
Halverson, Larry, Iowa State University of Science & Tech
Henning, Stanley J., Iowa State University of Science & Tech
Hernandez-Silva, Gilberto, Universidad Nacional Autonoma de Mexico
Hochmuth, George J., University of Florida
Holman, John, Kansas State University
Hopmans, Jan W., University of California, Davis
Houlton, Ben, University of California, Davis
Jacobsen, Jeffrey S., Montana State University
Janssen, Keith, Kansas State University
Jasoni, Richard, Desert Research Institute
Johnson, Arthur H., University of Pennsylvania
Johnson, Jane M., University of Minnesota, Twin Cities
Jones, Julia A., Oregon State University
Jugulam, Mithila, Kansas State University
Karkanis, Pano G., University of Lethbridge
Karlen, Douglas L., Iowa State University of Science & Tech
Kaspar, Thomas A., Iowa State University of Science & Tech
Keeney, Dennis R., Iowa State University of Science & Tech
Kelling, Keith A., University of Wisconsin, Madison
Killorn, Randy J., Iowa State University of Science & Tech
Kolka, Randall K., University of Minnesota, Twin Cities
Kusssow, Wayne R., University of Wisconsin, Madison
Laboski, Carrie A., University of Wisconsin, Madison
Landschoot, Peter J., Pennsylvania State University, University Park
Lauzon, John, University of Guelph
Liang, George, Kansas State University
Liebens, Johan, University of West Florida
Loutkova, Klavdia Oleschko, Universidad Nacional Autonoma de Mexico
Loynachan, Thomas E., Iowa State University of Science & Tech
Mackowiak, Cheryl, University of Florida
Mallarino, Antonio W., Iowa State University of Science & Tech
Mengel, David, Kansas State University
Miller, Murray H., University of Guelph
Mills, Aaron L., University of Virginia
Min, Doo-Hong, Kansas State University
Moorberg, Colby, Kansas State University
Mooreman, Thomas B., Iowa State University of Science & Tech
Moran, Susan, University of Arizona
Morgan, Kelly, University of Florida
Morris, Geoffrey, Kansas State University
Mueller, Raymond G., Richard Stockton College of New Jersey
Nelson, Nathan, Kansas State University
Norman, John M., University of Wisconsin, Madison
Nyhan, John W., Los Alamos National Laboratory
O'Halloran, Ivan, University of Guelph
Obour, Augustine, Kansas State University
Obreza, Thomas A., University of Florida
Palacios-Mayorga, Sergio, Universidad Nacional Autonoma de Mexico
Pedersen, Joel A., University of Wisconsin, Madison
Perumal, Ram, Kansas State University
Peterson, Dallas, Kansas State University
Picardal, Flynn W., Indiana University, Bloomington
Polito, Thomas A., Iowa State University of Science & Tech
Powell, J. Mark, University of Wisconsin, Madison
Prasad, Vara , Kansas State University
Presley, DeAnn, Kansas State University

Raczkowski, Charles, North Carolina Agricultural & Tech State University
Randall, Gyles W., University of Minnesota, Twin Cities
Rechcigl, John E., University of Florida
Reganold, John P., Washington State University
Regehr, David, Kansas State University
Rice, Pamela J., University of Minnesota, Twin Cities
Rolston, Dennis E., University of California, Davis
Roozeboom, Kraig, Kansas State University
Roth, Gregory W., Pennsylvania State University, University Park
Ruark, Matthew D., University of Wisconsin, Madison
Ruiz-Diaz, Dorivar, Kansas State University
Santos, Eduardo, Kansas State University
Sartain, Jerry B., University of Florida
Sassenrath, Gretchen, Kansas State University
Schaap, Marcel, University of Arizona
Schafer, Jr., John W., Iowa State University of Science & Tech
Schapaugh, Bill T., Kansas State University
Schlegel, Alan, Kansas State University
Schumann, Arnold W., University of Florida
Sherwood, William C., James Madison University
Shoup, Doug, Kansas State University
Shroyer, Jim, Kansas State University
Shuford, James W., Alabama A&M University
Silveira, Maria, University of Florida
Silvertooth, Jeffrey C., University of Arizona
Smith, Terry L., California Polytechnic State University
Soldat, Douglas J., University of Wisconsin, Madison
Stahlman, Phillip, Kansas State University
Stapleton, Michael G., Slippery Rock University
Strock, Jeffrey S., University of Minnesota, Twin Cities
Stubler, Craig, California Polytechnic State University
Sweeny, Daniel, Kansas State University
Szecsody, James E., Pacific Northwest National Laboratory
Tesso, Tesfaye, Kansas State University
Thompson, Curtis, Kansas State University
Tomlinson, Peter, Kansas State University
Vanderlip, Richard, Kansas State University
Voss, Regis D., Iowa State University of Science & Tech
Walworth, James, University of Arizona
Whitney, D.A., Kansas State University
Wilson, P. Christopher, University of Florida
Zhang, Guorong, Kansas State University

## ENGINEERING GEOLOGY

### General Engineering Geology

, Ebrahim A., University of Tabriz
Abdalla, Salah B., University of Khartoum
Acevedo-Arroyo, Jose Refugio, Universidad Autonoma de San Luis Potosi
Achampong, Francis, University of Ghana
Adams, Herbert G., California State University, Northridge
Adeyemi, G. O., University of Ibadan
Ahmed, Muhammad F., University of Engineering and Tech
Aimone-Martin, Cathrine T., New Mexico Institute of Mining and Tech
Alao, D. A., University of Ilorin
Anderson, Andrew, University of Illinois, Urbana-Champaign
Andre-Obayanju, Tomilola, University of Benin
Axelbaum, Richard, Washington University in St. Louis
Aydin, Adnan, University of Mississippi
Bancroft, John C., University of Calgary
Baptista, Joana, University of Newcastle Upon Tyne
Bauer, R. A., University of Illinois, Urbana-Champaign
Bell, David, University of Canterbury
Bell, John W., University of Nevada, Reno
Bennet, Brett, University of Kansas
Berry, Karen, Colorado School of Mines
Beukelman, Gregg, Utah Geological Survey
Biswas, Pratim, Washington University in St. Louis
Bonetto, Sabrina Maria Rita, Università di Torino
Bower, Kathleen M., Eastern Illinois University
Bowman, Steve D., Utah Geological Survey
Brabham, Peter, University of Wales
Brady, Roland H., California State University, Fresno
Brand, Patrick, California Geological Survey
Branum, David, California Geological Survey
Brunkal, Holly, Western State Colorado University
Buchanan, George, Montgomery County Community College
Burns, Bill, Oregon Dept of Geology & Mineral Industries
Burns, Scott F., Portland State University
Buscheck, Thomas A., Lawrence Livermore National Laboratory
Campbell-Stone, Erin, Wyoming State Geological Survey
Carlson, Jill, Colorado School of Mines
Carney, Theodore C., Los Alamos National Laboratory
Caron, Olivier J., University of Illinois, Urbana-Champaign
Carr, James R., University of Nevada, Reno
Case, William F., Utah Geological Survey
Castleton, Jessica, Utah Geological Survey
Cato, Kerry, California State University, San Bernardino
Cetindag, Bahattin, Firat University
CHEGBELEH, Larry-Pax, University of Ghana
Choma-Moryl, Krystyna, University of Wroclaw
Clark, H. C., Rice University
Davenport, Cliff, California Geological Survey
Davie, Colin, University of Newcastle Upon Tyne
Davies, David, Heriot-Watt University
Dawson, Tim, California Geological Survey
Delattre, Marc, California Geological Survey
dePolo, Craig M., University of Nevada
Doheny, Edward L., University of Pennsylvania
Doherty, Kevin, California Geological Survey
DuRoss, Christopher B., Utah Geological Survey
Durucan, Sevket, Imperial College
Dusseault, Maurice B., University of Waterloo
Eberhardt, Erik, University of British Columbia
Edil, Tuncer B., University of Wisconsin, Madison
Elia, Gaetano, University of Newcastle Upon Tyne
Ellecosta, Peter , Technische Universitaet Muenchen
Erickson, Ben A., Utah Geological Survey
Evans, Stephen G., University of Waterloo
Eversoll, Duane A., Unversity of Nebraska - Lincoln
Falls, Jim, California Geological Survey
Ferriz, Horacio, California State University, Stanislaus
Floris, Mario, Università degli Studi di Padova
Fournier, Benoit, Universite Laval
Frailey, Scott M., Illinois State Geological Survey
Fratta, Dante, University of Wisconsin, Madison
Fuller, Michael, California Geological Survey
Galgaro, Antonio, Università degli Studi di Padova
Gasparini, Nicole, Tulane University
Gautam, Tej, Marietta College
Giraud, Richard E., Utah Geological Survey
Greene, Brian M., Youngstown State University
Gretener, Peter E., University of Calgary
Grimley, David A., University of Illinois, Urbana-Champaign
Guiterrez, Carlos, California Geological Survey
Hall, Jean, University of Newcastle Upon Tyne
Haneberg, William, New Mexico Institute of Mining and Tech
Harris, Will, California Geological Survey
Hasan, Syed E., University of Missouri-Kansas City
Haydon, Wayne, California Geological Survey
Hayhurst, Cheryl, California Geological Survey
Hencher, Steve, University of Leeds
Henika, William S., Virginia Polytechnic Institute & State University
Hernandez, Janis, California Geological Survey
Higgins, Jerry D., Colorado School of Mines
Ho, I-Hsuan, University of North Dakota
Holland, Peter, California Geological Survey
Holzer, Thomas, Stanford University
Horns, Daniel, Utah Valley University
Hubbard, Trent, Alaska Division of Geological & Geophysical Surveys
Hudec, Peter P., University of Windsor
Ibrahim, Ahmad Tajuddin Hj, University of Malaya
Ige, O. O., University of Ilorin
Irvine, Pamela J., California Geological Survey
Jamaluddin, Tajul Anuar, University of Malaya
Jo, Ho Young, Korea University
Johnpeer, Gary D., Cerritos College
Johnson, Sarah E., Northern Kentucky University
Johnson, William P., University of Utah
Jones, Larry Allan, Los Alamos National Laboratory
Joyce, James, University of Puerto Rico
Karanfil, Tanju, Clemson University
Katzenstein, Kurt W., South Dakota School of Mines & Tech

Kavanaugh, Jeffrey, University of Alberta
Knapp, Richard B., Lawrence Livermore National Laboratory
Knizek, Martin, Masaryk University
Knudsen, Tyler R., Utah Geological Survey
Korre, Anna, Imperial College
Kreylos, Oliver, University of California, Davis
Krohn, James P., Los Angeles Pierce College
Kudlac, John J., Point Park University
Lancaster, Jeremy, California Geological Survey
Likos, William J., University of Wisconsin, Madison
Linares, Rogelio, Universitat Autonoma de Barcelona
Lindsay, Donald, California Geological Survey
Lindsey, Kassandra, Colorado School of Mines
Linn, Rodman, Los Alamos National Laboratory
Locat, Jacques E., Universite Laval
Lokau, Katja, Technische Universitaet Muenchen
Longstreth, David, California Geological Survey
Lovekin, Jonathan, Colorado School of Mines
Lund, William R., Utah Geological Survey
Lupogo, Keneth, University of Dar es Salaam
Madrigal-Rubio, Rafael, Universidad Autonoma de Chihuahua
Mandrone, Giuseppe, Università di Torino
Manson, Mike, California Geological Survey
Mareschal, Maxime, California Geological Survey
Marshall, Gerald, California Geological Survey
Martel, Stephen J., University of Hawai'i, Manoa
Martin, Beth, Washington University in St. Louis
Masek, Ondrej, Edinburgh University
Mathewson, Christopher C., Texas A&M University
Matter, Juerg M., University of Southampton
McCoy, Kevin, Colorado School of Mines
McCrink, Timothy P., California Geological Survey
McDonald, Gregory N., Utah Geological Survey
McMahon, Katherine, University of Wisconsin, Madison
Mena-Zambrano, Teodulo, Universidad Autonoma de Chihuahua
Menschik, Florian M., Technische Universitaet Muenchen
Merifield, Paul M., University of California, Los Angeles
Merritt, Andrew, University of Plymouth
Mlinarevic, Ante, California Geological Survey
Moeglin, Thomas D., Missouri State University
Mogilevskaya, Sonia, University of Minnesota, Twin Cities
Moore, Jeffrey, University of Utah
Morales, Tomas, University of the Basque Country UPV/EHU
Morgan, Terrance L., Los Alamos National Laboratory
Nickmann, Marion, Technische Universitaet Muenchen
Niemann, William L., Marshall University
Nieto, Alberto S., University of Illinois, Urbana-Champaign
O'Rourke, Thomas D., Cornell University
Ogunsanwo, O., University of Ilorin
Olson, Brian , California Geological Survey
Origlia, Hector D., Universidad Nacional de Rio Cuarto
Oswald, John, California Geological Survey
Oyediran, I. A., University of Ibadan
Pamukcu, Alya, Princeton University
Patrick, David M., University of Southern Mississippi
Paysen-Petersen, Lukas, Technische Universitaet Muenchen
Pennuto, Christopher, SUNY, Buffalo
Perez, Florante, California Geological Survey
Perret, Didier H., Universite du Quebec
Pestrong, Raymond, San Francisco State University
Pipkin, Bernard W., University of Southern California
Poeter, Eileen P., Colorado School of Mines
Potter, Kenneth, University of Wisconsin, Madison
Poulton, Mary M., University of Arizona
Poulton, Mary, University of Arizona
Prevost, Jean-Herve, Princeton University
Pridmore, Cindy L., California Geological Survey
Rahn, Perry H., South Dakota School of Mines & Tech
Raj, John K., University of Malaya
Ramelli, Alan R., University of Nevada
Ramirez, Abelardo L., Lawrence Livermore National Laboratory
Reynolds, Steven, California Geological Survey
Rockaway, John D., Northern Kentucky University
Roffers, Pete, California Geological Survey
Roggenthen, William M., South Dakota School of Mines & Tech
Rosa, Carla, California Geological Survey
Rosinski, Anne, California Geological Survey
Rubin, Ron, California Geological Survey
Santi, Paul M., Colorado School of Mines
Sass, Ingo, Technische Universitaet Darmstadt
Savigny, K. Wayne, University of British Columbia
Sawyer, J. F., South Dakota School of Mines & Tech
Schiff, Sherry L., University of Waterloo
Seitz, Gordon, California Geological Survey
Sellmeier, Bettina, Technische Universitaet Muenchen
Shakoor, Abdul, Kent State University
Short, William, California Geological Survey
Silva, Michael A., California Geological Survey
Song, Lin-Ping, University of British Columbia
Spangler, Eleanor, California Geological Survey
Staub, William P., Oak Ridge National Laboratory
Stead, Douglas, Simon Fraser University
Stetler, Larry D., South Dakota School of Mines & Tech
Stevens, De Anne S., Alaska Division of Geological & Geophysical Surveys
Swanson, Brian, California Geological Survey
Thomas, Mark, University of Leeds
Thompson, Jim, California Geological Survey
Thornburg, Jennifer, California Geological Survey
Thuro, Kurosch, Technische Universitaet Muenchen
Tinjum, James M., University of Wisconsin, Madison
Tompson, Andrew F., Lawrence Livermore National Laboratory
Turner, A. Keith, Colorado School of Mines
Uriarte, Jesus Angel, University of the Basque Country UPV/EHU
Urzua, Alfredo, Boston College
Villeneuve, Marlene C., University of Canterbury
Walton, Gabriel, Colorado School of Mines
Wang, Dongmei, University of North Dakota
Watts, Chester F., Virginia Polytechnic Institute & State University
Watts, Chester F., Radford University
Weber, Gerald E., University of California, Santa Cruz
Weigers, Mark, California Geological Survey
West, Olivia M., Oak Ridge National Laboratory
West, Terry R., Purdue University
White, Chase, California Geological Survey
White, Jonathan, Colorado School of Mines
White, Owen L., University of Waterloo
Wieser, Carola, Technische Universitaet Muenchen
Wilfing, Lisa, Technische Universitaet Muenchen
Wilkey, Patrick L., Argonne National Laboratory
Williams, Bruce A., Pacific Northwest National Laboratory
Williams, John W., San Jose State University
Wills, Christopher J., California Geological Survey
Wittke, Seth, Wyoming State Geological Survey
Wu, Chin, University of Wisconsin, Madison
Yarbrough, Lance D., University of Mississippi
Yunwei, Sun, Lawrence Livermore National Laboratory
Zhou, Wendy W., Colorado School of Mines
Zimmerman, R. Eric, Argonne National Laboratory
Zyvoloski, George A., Los Alamos National Laboratory

## Earthquake Engineering

Acosta, Jose G., Centro de Investigación Científica y de Educación Superior de Ensenada
Atkinson, Gail M., Carleton University
Heaton, Thomas H., California Institute of Tech
Huang, Moh J., California Geological Survey
Jamieson, J. Bruce, University of Calgary
Kutter, Bruce, University of California, Davis
Mendoza, Luis H., Centro de Investigación Científica y de Educación Superior de Ensenada
Michaels, Paul, Boise State University
Oommen, Thomas, Michigan Technological University
Reyes, Alfonso, Centro de Investigación Científica y de Educación Superior de Ensenada
Roberts, Gerald, Birkbeck College
Rowshandel, Badie, California Geological Survey
Wang, Yumei, Oregon Dept of Geology & Mineral Industries

## Mining Tech/Extractive Metallurgy

Aguirre-Moriel, Socorro I., Universidad Autonoma de Chihuahua
Bryan, Christopher, Exeter University
Coggan, John, Exeter University
Demopoulos, George, McGill University
Dessureault, Sean, University of Arizona

Doyle, Fiona M., University of California, Berkeley
Drew, Robin A., McGill University
Duby, Paul F., Columbia University
Evans, James W., University of California, Berkeley
Feunstenau, Maurice, University of Nevada, Reno
Finch, James A., McGill University
Fuerstenau, Douglas W., University of California, Berkeley
Glass, Hylke J., Exeter University
Griswold, George B., New Mexico Institute of Mining and Tech
Gruzleski, John E., McGill University
Gundiler, Ibrahim H., New Mexico Institute of Mining & Tech
Guthrie, Roderick I., McGill University
Hasan, Mainul, McGill University
Hiskey, J. Brent, University of Arizona
Hurtado-De La Ree, Maria B., Centro de Estudios Superiores del Estado Sonora
Jeffrey, Kip, Exeter University
Jonas, John J., McGill University
Khan, Latif A., Illinois State Geological Survey
Kinabo, Crispin P., University of Dar es Salaam
Lee, Jaeheon, University of Arizona
Mendoza-Aguilar, Hector M., Universidad Autonoma de Chihuahua
Minor-Velazquez, Hector, Universidad Autonoma de Chihuahua
Moon, Charlie, Exeter University
Mucciardi, Frank, McGill University
Nesbitt, Carl, University of Nevada, Reno
Ruiz-Cisneros, David H., Universidad Autonoma de Chihuahua
Salazar-Avila, Arnulfo, Centro de Estudios Superiores del Estado Sonora
Sastry, Kalanadh V. S., University of California, Berkeley
Seal, Thom, University of Nevada, Reno
Smith, Gordon W., McGill University
Smith, Karl A., University of Minnesota, Twin Cities
Somasundaran, Ponisseril, Columbia University
Themelis, Nickolas J., Columbia University
Trevizo-Cano, Luis M., Universidad Autonoma de Chihuahua
Wheeler, Paul, Exeter University
Yue, Stephen, McGill University

## Mining Engineering

MASSACCI, Prof. Giorgio, Universita di Cagliari
Bandopadhyay, Sukumar, University of Alaska, Fairbanks
Bessinger, Stephen, University of Utah
Bilodeau, Michel L., McGill University
Boshkov, Stefan H., Columbia University
Brune, Jürgen F., Colorado School of Mines
Calizaya, Felipe, University of Utah
Chatterjee, Snehamoy, Michigan Technological University
Chen, Gang, University of Alaska, Fairbanks
Dagdelen, Kadri, Colorado School of Mines
Danko, George, University of Nevada, Reno
Dimitrakopoulos, Roussos, McGill University
Dinsmoor, John C., Los Alamos National Laboratory
Donovan, James, University of Utah
Eveleth, Robert W., New Mexico Institute of Mining & Tech
Foster, Patrick, Exeter University
Ganguli, Rajive, University of Alaska, Fairbanks
Gillis, James M., Michigan Technological University
Gongora-Jurado, Raul A., Centro de Estudios Superiores del Estado Sonora
Hafez, Sabry ., University of Alaska, Fairbanks
Hassani, Ferri, McGill University
Hemani, Ahmad, McGill University
Honkaer, Rick, University of Kentucky
Kalia, Hemendra N., Los Alamos National Laboratory
Kaunda, Rennie, Colorado School of Mines
Kennedy, Gareth, Exeter University
Kim, Eunhye, Colorado School of Mines
Kim, Kwangmin, University of Arizona
Kohler, Jeffery L., Pennsylvania State University, University Park
Kovach, Richard G., Los Alamos National Laboratory
Kuchta, Mark, Colorado School of Mines
Liu, Shimin, Pennsylvania State University, University Park
Lusk, Braden , University of Kentucky
MANCA, Prof. Pierpaolo P., Universita di Cagliari
Mayer, Ulrich, University of British Columbia
McCarter, Michael K., University of Utah
Meyer, Lewis, Exeter University
Miller, Hugh, Colorado School of Mines
Mitri, Hani, McGill University
Momayez, Moe, University of Arizona
Monteverde-Gutierrez, Gerardo, Centro de Estudios Superiores del Estado Sonora
Mossop, John W., McGill University
Mousset-Jones, Pierre, University of Nevada, Reno
Navarro, Ricardo, Universidad Nacional Autonoma de Mexico
Nelson, Michael G., University of Utah
Nieto, Antonio, Pennsylvania State University, University Park
Nieto Antunez, Antonio, Universidad Nacional Autonoma de Mexico
Novak, Thomas, University of Kentucky
Petr, Vilem, Colorado School of Mines
Pokorny, Eugene W., Los Alamos National Laboratory
Rostami, Jamal, Pennsylvania State University, University Park
Silva-Castro, Jhon, University of Kentucky
Slistan-Grijalva, Angel, Centro de Estudios Superiores del Estado Sonora
Sottile, Jr., Joseph, University of Kentucky
Squelch, Andrew P., Curtin University
Tarshizi, Ebrahim, Michigan Technological University
Taylor, Danny L., University of Nevada, Reno
Tenorio, Victor O., University of Arizona
Tharp, Thomas M., Purdue University
Vite-Picazo, Luis G., Centro de Estudios Superiores del Estado Sonora
Wala, Andrew M., University of Kentucky
Wane, Malcolm T., Columbia University
Wedding, William C., University of Kentucky
Wetherelt, Andrew, Exeter University
Whyatt, Jeffrey, University of Utah
Yegulalp, Tuncel M., Columbia University
Zavodni, Zavis, University of Utah
Zhang, Jinhong, University of Arizona
Zipf, Karl, Colorado School of Mines

## Petroleum Engineering

Ahmed, Ramadan, University of Oklahoma
Alvarado, Vladimir, University of Wyoming
Bijeljic, Branko, Imperial College
Blasingame, Tom, Texas A&M University
Blunt, Martin J., Imperial College
Callard, Jeff, University of Oklahoma
Civan, Faruk, University of Oklahoma
Cooper, George A., University of California, Berkeley
Corbett, Patrick, Heriot-Watt University
Couples, Gary, Heriot-Watt University
Devegowda, Deepak, University of Oklahoma
Dreesen, Donald S., Los Alamos National Laboratory
El-Monier, Ilham, University of Oklahoma
Fahes, Mashhad, University of Oklahoma
Fishe, Quentin , University of Leeds
Geiger, Sebastian, Heriot-Watt University
Gringarten, Alain, Imperial College
Jamili, Ahmad, University of Oklahoma
Javadpour, Farzam, University of Texas at Austin
Jin, Guohai, Geological Survey of Alabama
Kadkhodai Ilkhchi, Ali, University of Tabriz
Kelkar, Sharad, Los Alamos National Laboratory
King, Peter, Imperial College
Knapp, Roy M., University of Oklahoma
Lorinczi, Piroska, University of Leeds
Ma, Jingsheng, Heriot-Watt University
Misra, Siddharth, University of Oklahoma
Moghanloo, Rouzbeh, University of Oklahoma
Muggeridge, Ann H., Imperial College
Nikolinakou, Maria-Aikaterini, University of Texas at Austin
Oppong, Issac A., University of Ghana
Patzek, Tad W., University of California, Berkeley
Pournik, Maysam, University of Oklahoma
Rai, Chandra S., University of Oklahoma
Rose, Peter E., University of Utah
Sakhaee-Pour, Ahmad, University of Oklahoma
Shah, Subhash N., University of Oklahoma
Sharma, Suresh, University of Oklahoma
Somerville, James M., Heriot-Watt University
Sondergeld, Carl H., University of Oklahoma
Teodoriu, Catalin, University of Oklahoma
Vesovic, Velisa, Imperial College
Wu, Xingru, University of Oklahoma
Zaman, Musharraf, University of Oklahoma

## Rock Mechanics

Abousleiman, Younane, University of Oklahoma
Abu Bakar, Muhammad Z., University of Engineering and Tech
Agioutantis, Zach, University of Kentucky
Archambault, Guy, Universite du Quebec a Chicoutimi
Asbury, Brian, Colorado School of Mines
Aslan, Yasemin, Firat University
Barakat, Bassam, Institut Polytechnique LaSalle Beauvais (ex-IGAL)
Barzegari, Ghodrat, University of Tabriz
Blacic, James D., Los Alamos National Laboratory
Brahmi, Sadek, Institut Polytechnique LaSalle Beauvais (ex-IGAL)
Brake, Thomas L., Los Alamos National Laboratory
Brown, Donald W., Los Alamos National Laboratory
Chen, Rui, California Geological Survey
Cruden, David M., University of Alberta
Cundall, Peter A., University of Minnesota, Twin Cities
Daemen, Jaak, University of Nevada, Reno
Detournay, Emmanuel M., University of Minnesota, Twin Cities
Diederichs, Mark, Queen's University
Enderlin, Milton, Texas Christian University
Engelder, Terry, Pennsylvania State University, University Park
Ferrero, Anna Maria, Università di Torino
Germanovich, Leonid, Georgia Institute of Tech
Ghassemi, Ahmed, University of Oklahoma
Gordon, Robert B., Yale University
Gurocak, Zulfu, Firat University
Guzina, Bojan B., University of Minnesota, Twin Cities
Haimson, Bezalel C., University of Wisconsin, Madison
Hawkins, Stephen J., University of Southampton
Kanik, Mustafa, Firat University
Kemeny, John M., University of Arizona
Kulatilake, Pinnaduwa H. S. W., University of Arizona
Käsling, Heiko, Technische Universitaet Muenchen
Labuz, Joseph F., University of Minnesota, Twin Cities
Logan, John M., University of Oregon
MacLaughlin, Mary M., Montana Tech of the University of Montana
Mariani, Elisabeth, University of Liverpool
McLennan, John D., University of Utah
Meredith, Philip, University College London
Mojtabai, Navid, New Mexico Institute of Mining and Tech
Nelson, Priscilla P., Colorado School of Mines
Onyeobi, Tony U., University of Benin
Oravecz, Kalman I., New Mexico Institute of Mining and Tech
Ozbay, M. Ugur, Colorado School of Mines
Pariseau, William G., University of Utah
Perry , Kyle, University of Kentucky
Persson, Per-Anders, New Mexico Institute of Mining and Tech
Roegiers, Jean-Claude, University of Oklahoma
Rutter, Ernest, University of Manchester
Saeidi, Ali, Universite du Quebec a Chicoutimi
Sinha, Krishna P., University of Utah
Smutek, Claude, Universite de la Reunion
Sone, Hiroki, University of Wisconsin, Madison
Swift, Robert P., Los Alamos National Laboratory
Trandafir, Aurel, University of Utah
Unrug, Kot F., University of Kentucky
Vinciguerra, Sergio, Leicester University
Vinciguerra, Sergio Carmelo, Università di Torino
Watters, Robert J., University of Nevada, Reno
Wijesinghe, Ananda M., Lawrence Livermore National Laboratory
Zimmerman, Robert W., Imperial College

# OCEANOGRAPHY

## General Oceanography

Abernathy, Ryan, Columbia University
Allen, John, University of Portsmouth
Alves, Tiego, University of Wales
Arrigo, Kevin, Stanford University
Barros, Tony, Miami-Dade College (Wolfson Campus)
Becker, Janet M., University of Hawai'i, Manoa
Benotti, Mark J., Bentley University
Bohaty, Steven, University of Southampton
Brandes, Jay, Georgia Institute of Tech
Brandon, Mark A., The Open University
Brown, Erik T., University of Minnesota, Duluth
Buchanan, Donald G., San Bernardino Valley College
Busalacchi, Antonio, University of Maryland
Chamberlin, William S., Fullerton College
Chandler, Cyndy, Woods Hole Oceanographic Institution
Ciannelli, Lorenzo, Oregon State University
Cooney, Michael, University of Hawai'i, Manoa
Curewitz, Daniel, Syracuse University
Davies, Andrew J., Bangor University
Deen, Patricia A., Palomar College
DeJesus, Roman, Fullerton College
Denton, George, University of South Florida
Dong, Charles, El Camino College
Eicken, Hajo, University of Alaska, Fairbanks
Falkowski, Paul G., University of Hawai'i, Manoa
Field, Richard T., University of Delaware
Fritsen, Christian H., University of Nevada, Reno
Froelich, Philip, Florida State University
Galbraith, Eric, McGill University
Gerbi, Greg, Skidmore College
Gomez, Natalya, McGill University
Gordon, Elizabeth S., Fitchburg State University
Greenhow, Danielle, University of Southern Mississippi
Grossman, Walter, San Bernardino Valley College
Hale, Michelle, University of Portsmouth
Hepner, Tiffany, University of Texas at Austin, Jackson School of Geosciences
Hernes, Peter, University of California, Davis
Hickman, Anna, University of Southampton
Holland, Christina, University of Texas at Austin
Holliday, Joseph W., El Camino College
Hoyt, William H., University of Northern Colorado
Hughes-Clarke, John E., University of New Hampshire
Iams, William J., Sir Wilfred Grenfell College
Ivanovic, Ruza, University of Leeds
Johnson, Ashanti, University of Texas, Arlington
Johnson, Helen, University of Oxford
Judkins, Heather L., Saint Petersburg College, Clearwater
Jung, Simon, Edinburgh University
Keller, Klaus, Pennsylvania State University, University Park
Kerr, Andrew, University of Wales
King, Andrew, University of Illinois at Chicago
Kustka, Adam B., Rutgers, The State University of New Jersey, Newark
Kuwabara, James, City College of San Francisco
Laliberte, Elizabeth, University of Rhode Island
Lauderdale, Jonathan, University of Liverpool
Leach, Harry, University of Liverpool
Lear, Caroline, University of Wales
Ledbetter, Michael T., University of Arkansas at Little Rock
Lee, Alyce, Concord University
Lee, Stephen C., Los Angeles Pierce College
Leinen, Margaret, University of California, San Diego
Lenz, Petra H., University of Hawai'i, Manoa
Liu, Zhanfei, University of Texas at Austin
Ma, Yanxia, Louisiana State University
MacLeod, Chris, University of Wales
Maxwell, Arthur E., University of Texas at Austin
McCarthy, Daniel, Jacksonville University
McPhaden, Michael J., University of Washington
McWilliams, James, University of California, Los Angeles
Montenegro, Alvaro, Ohio State University
Moore, Dennis, University of Hawai'i, Manoa
Morris, Kalon, Saddleback Community College
Mosher, David C., University of New Hampshire
Muir, William, San Bernardino Valley College
Myers, Paul, University of Alberta
Noyes, Jim, El Camino College
Peredo-Jaime, Jose I., Universidad Autonoma de Baja California Sur
Pike, Jenny, University of Wales
Porter, Dwayne, University of South Carolina
Quan, Tracy, Oklahoma State University
Radulski, Robert, Southern Connecticut State University
Rappe, Michael S., University of Hawai'i, Manoa
Roesler, Collin, Bowdoin College
Roussenov, Vassil, University of Liverpool
Schlegel, Mary A., Millersville University
Schneider, Niklas, University of Hawai'i, Manoa
Schroeder, William W., Dauphin Island Sea Lab
Scott, Robert B., University of Texas at Austin

Seyfang, Gill, University of East Anglia
Spencer, Matthew, University of Liverpool
Steger, John M., Miami Dade College (Kendall Campus)
Sullivan-Watts, Barbara K., University of Rhode Island
Swanson, R. L., SUNY, Stony Brook
Tausig, Heather, University of Massachusetts, Boston
Thomas, Megan D., United States Naval Academy
Thompson, LuAnne, University of Washington
Thomson, Richard, University of Victoria
Thurman, Harold V., Mt. San Antonio College
Tomkiewicz, Warren, Plymouth State University
Trujillo, Alan P., Palomar College
Tucker, Stevens P., Naval Postgraduate School
Tyrrell, Toby, University of Southampton
Ulanski, Stanley L., James Madison University
Veeramony, Jayaram, Mississippi State University
Venn, Cynthia, Bloomsburg University
Venti, Nicholas, University of Massachusetts, Amherst
Walker, Nan D., Louisiana State University
Wallace, William G., Graduate School of the City University of New York
Wen, Xianyun, University of Leeds
Wheatcroft, Rob, Oregon State University
Wiberg, Patricia L., University of Virginia
Williams, Harris, North Carolina Central University
Williams, Ric, University of Liverpool
Wiltshire, John C., University of Hawai'i, Manoa
Wolcott, Ray, Cuyamaca College
Wolff, George, University of Liverpool
Yin, Jianjun, University of Arizona
Yon, Lisa, Palomar College
Zhai, Xiaoming, University of East Anglia

## Biological Oceanography

Adams, Marshall, Oak Ridge National Laboratory
Adams, Michael S., University of Wisconsin, Madison
Algar, Christopher, Dalhousie University
Allam, Bassem, SUNY, Stony Brook
Allen, Dennis, University of South Carolina
Allen, Eric E., University of California, San Diego
Aller, Josephine Y., SUNY, Stony Brook
Amon, Rainier, Texas A&M University
Anderson, George C., University of Washington
Apprill, Amy, Woods Hole Oceanographic Institution
Armbrust, Virginia E., University of Washington
Atkinson, Marlin J., University of Hawai'i, Manoa
Audet, Celine, Universite du Quebec a Rimouski
Auster, Peter, University of Connecticut
Azam, Farooq, University of California, San Diego
Baltz, Donald M., Louisiana State University
Banse, Karl, University of Washington
Barber, Bruce, University of South Florida
Barber, Richard T., Duke University
Barlow, Jay P., University of California, San Diego
Baross, John A., University of Washington
Bartlett, Douglas H., University of California, San Diego
Bates, Amanda E., University of Southampton
Baumann, Hannes, University of Connecticut
Baumann, Zofia, University of Connecticut
Baumgartner, Mark, Dalhousie University
Benfield, Mark C., Louisiana State University
Benner, Ron, University of South Carolina
Bibby, Thomas, University of Southampton
Bienfang, Paul K., University of Hawai'i, Manoa
Biggs, Douglas C., Texas A&M University
Blake, Norman J., University of South Florida
Bloomfield, Helen, University of Liverpool
Boar, Rosalind, University of East Anglia
Bochdansky, Alexander, Old Dominion University
Bollens, Stephen, Washington State University
Boudrias, Michel A., University of San Diego
Bowen, Jennifer, University of Massachusetts, Boston
Bowser, Paul, SUNY, Stony Brook
Boyle, Elizabeth , University of Massachusetts, Boston
Breitbart, Mya, University of South Florida
Brethes, Jean-Claude, Universite du Quebec a Rimouski
Briggs, John C., University of South Florida
Brown, Moira, University of Massachusetts, Boston
Brownlee, Colin, University of Southampton
Bruno, Barbara, University of Hawai'i, Manoa
Bucklin, Ann, University of Connecticut
Burton, Ronald S., University of California, San Diego
Buskey, Edward J., University of Texas at Austin
Byrnes, Jarrett, University of Massachusetts, Boston
Campbell, Lisa, Texas A&M University
Carlile, Amy L., University of New Haven
Carney, Robert S., Louisiana State University
Carpenter, Steve, University of Wisconsin, Madison
Carr, Mark, University of California, Santa Cruz
Case, James M., Wilkes University
Cassar, Nicolas, Duke University
Caswell, Bryony, University of Liverpool
Cavin, Julie, University of Massachusetts, Boston
Cerrato, Robert M., SUNY, Stony Brook
Checkley, David M., University of California, San Diego
Christian, James R., University of Victoria
Church, Matthew, University of Hawai'i, Manoa
Clague, David A., Stanford University
Collie, Jeremy S., University of Rhode Island
Collier, Jackie, SUNY, Stony Brook
Conover, David O., SUNY, Stony Brook
Copley, Jonathan, University of Southampton
Coull, Bruce, University of South Carolina
Cowles, Timothy J., Oregon State University
Cranford, Peter, Dalhousie University
Croll, Don, University of California, Santa Cruz
Cuba, Thomas, University of South Florida
Cuker, Benjamin E., Hampton University
Cullen, John J., Dalhousie University
D'Elia, Christopher, University of South Florida
Daly, Kendra L., University of South Florida
Dam, Hans G., University of Connecticut
Dayton, Paul K., University of California, San Diego
Dean, John Mark, University of South Carolina
Demers, Serge, Universite du Quebec a Rimouski
Deming, Jody W., University of Washington
Denton, Gary R. W., University of Guam
DeWitt, Calvin B., University of Wisconsin, Madison
DiBacco, Claudio, Dalhousie University
Dieterle, Dwight A., University of South Florida
Dobbs, Fred C., Old Dominion University
Dodson, Stanley I., University of Wisconsin, Madison
Donaghay, Percy, University of Rhode Island
Doty, Thomas, Roger Williams University
Dower, John, University of Victoria
Drazen, Jeffrey C., University of Hawai'i, Manoa
Ducklow, Hugh, Columbia University
Duffy, Tara , Northeastern University
Dunton, Kenneth H., University of Texas at Austin
Durbin, Edward G., University of Rhode Island
Dyhrman, Sonya, Columbia University
Eggleston, David B., North Carolina State University
Ellis, Hugh I., University of San Diego
Epp, Leonard G., Mount Union College
Erdner, Deana L., University of Texas at Austin
Erisman, Brad, University of Texas at Austin
Esbaugh, Andrew J., University of Texas at Austin
Etter, Ron , University of Massachusetts, Boston
Falkowski, Paul G., Rutgers, The State University of New Jersey
Fast, Mark, SUNY, Stony Brook
Felbeck, Horst, University of California, San Diego
Feller, Robert, University of South Carolina
Fenberg, Phillip, University of Southampton
Fennel, Katja, Dalhousie University
Fisher, Jr., Thomas R., University of Maryland
Fisk, Aaron, University of Windsor
Fitzer, Susan, University of Glasgow
Fletcher, Madilyn, University of South Carolina
Fournier, Robert O., Dalhousie University
Francis, Robert, University of Washington
Frank, Kenneth, Dalhousie University
Franks, Peter J. S., University of California, San Diego
Frid, Chris, University of Liverpool
Frost, Bruce W., University of Washington
Fuiman, Lee A., University of Texas at Austin
Gaasterland, Terry, University of California, San Diego

Gallagher, Eugene, University of Massachusetts, Boston
Ganssen, Gerald M., VU University Amsterdam
Gardner, Wayne S., University of Texas at Austin
Gates, Ruth D., University of Hawai'i, Manoa
Gibson, Deidre M., Hampton University
Gifford, Dian J., University of Rhode Island
Gilbert, Patricia M., University of Maryland
Gobler, Christopher, SUNY, Stony Brook
Godbold, Jasmin A., University of Southampton
Goetze, Erica, University of Hawai'i, Manoa
Gomes, Helga, Columbia University
Gosselin, Michel, Universite du Quebec a Rimouski
Gould, Mark D., Roger Williams University
Graham, Linda K., University of Wisconsin, Madison
Graham, Michael, Moss Landing Marine Laboratories
Graham, William (., University of Southern Mississippi
Grange, Laura, University of Southampton
Grant, Jonathan, Dalhousie University
Green, Jonathan, University of Liverpool
Greene, Charles H., Cornell University
Greenfield, Dianne, University of South Carolina
Griffin, Blaine, University of South Carolina
Grigg, Richard W., University of Hawai'i, Manoa
Grunbaum, Daniel, University of Washington
Haderlie, Eugene C., Naval Postgraduate School
Hamilton, Phillip, University of Massachusetts, Boston
Hargraves, Paul E., University of Rhode Island
Hargreaves, Bruce R., Lehigh University
Harvey, James T., Moss Landing Marine Laboratories
Hastings, Philip A., University of California, San Diego
Hauton, Chris, University of Southampton
Hawkins, Lawrence, University of Southampton
Healey, Michael, University of British Columbia
Heath, Carolyn, Fullerton College
Heil, Cynthia, University of South Florida
Hessler, Robert R., University of California, San Diego
Higgins, John A., Princeton University
Holland, Nicholas D., University of California, San Diego
Holt, Gloria J., University of Texas at Austin
Hood, Raleigh, University of Maryland
Hopkins, Thomas S., Dauphin Island Sea Lab
Hopkins, Thomas L., University of South Florida
Hotchkiss, Sarah, University of Wisconsin, Madison
Humm, Harold J., University of South Florida
Hunt, Brian, University of British Columbia
Hunt, Kathleen, University of Massachusetts, Boston
Innis, Charles, University of Massachusetts, Boston
Irlandi, Elizabeth A., Florida Institute of Tech
Jeffries, H. Perry, University of Rhode Island
Jehl, Joseph R., University of San Diego
Jensen, Antony, University of Southampton
Jickells, Tim, University of East Anglia
Jordan, Robert A., Hampton University
Juhl, Andrew, Columbia University
Jumars, Peter A., University of Washington
Juniper, Kim, University of Victoria
Kamykowski, Daniel, North Carolina State University
Kana, Todd M., University of Maryland
Karl, David M., University of Hawai'i, Manoa
Kaufmann, Ronald S., University of San Diego
Kelley, Christopher, University of Hawai'i, Manoa
Kelly, John, University of New Haven
Kemp, Paul, University of Hawai'i, Manoa
Kenney, Robert D., University of Rhode Island
Kent, Donald B., University of San Diego
Kesseli, Rick, University of Massachusetts, Boston
Kimball, Matthew, University of South Carolina
Kirschenfeld, Taylor, University of West Florida
Kitchell, James F., University of Wisconsin, Madison
Knowlton, Amy , University of Massachusetts, Boston
Kooyman, Gerald L., University of California, San Diego
Kraemer, George P., SUNY, Purchase
Krauss, Scott , University of Massachusetts, Boston
Krieger-Brockett, Barbara B., University of Washington
Landry, Michael R., University of California, San Diego
LaRock, Paul A., Louisiana State University
Lee, Carol, University of Wisconsin, Madison
Leichter, James J., University of California, San Diego
Lessard, Evelyn J., University of Washington
Letelier, Ricardo, Oregon State University
Levin, Lisa A., University of California, San Diego
Levinton, Jeffrey, SUNY, Stony Brook
Lewin, Joyce C., University of Washington
Lewis, Alan G., University of British Columbia
Lewis, Marlon R., Dalhousie University
Li, William K., Dalhousie University
Lilley, Marvin D., University of Washington
Lin, Senjie, University of Connecticut
Logan, Alan, University of New Brunswick Saint John
Lohan, Maeve, University of Southampton
Lonsdale, Darcy J., SUNY, Stony Brook
Lopez, Glenn R., SUNY, Stony Brook
Lowery, Mary Sue, University of San Diego
Lucas, Cathy, University of Southampton
MacIntyre, Hugh, Dalhousie University
Mackas, David L., University of Victoria
Maldonado, Maria, University of British Columbia
Malin, Gill, University of East Anglia
Mandelman, John, University of Massachusetts, Boston
Marra, John, Brooklyn College (CUNY)
Martiny, Adam, University of California, Irvine
McCarthy, James T., Harvard University
McConaugha, John R., Old Dominion University
McGlathery, Karen J., University of Virginia
McGowan, John A., University of California, San Diego
McManus, George B., University of Connecticut
Mendelssohn, Irving A., Louisiana State University
Metaxas, Anna, Dalhousie University
Meylan, Anne, University of South Florida
Miller, Douglas C., University of Delaware
Mills, Eric L., Dalhousie University
Milroy, Scott, University of Southern Mississippi
Mobley, Curtis D., University of Washington
Monger, Bruce, Cornell University
Moore, Christopher M., University of Southampton
Moran, Dawn, Woods Hole Oceanographic Institution
Mueller, Erich M., University of South Alabama
Mulholland, Margaret, Old Dominion University
Munch, Stephan, SUNY, Stony Brook
Murtugudde, Raghuram G., University of Maryland
Napora, Theodore A., University of Rhode Island
Napp, Jeffrey, University of Washington
Newman, William A., University of California, San Diego
Newton, Jan, University of Washington
Nixon, Scott W., University of Rhode Island
Norris, Dean R., Florida Institute of Tech
Nye, Janet, SUNY, Stony Brook
Ohman, Mark D., University of California, San Diego
Oliver, John S., Moss Landing Marine Laboratories
Oliver, Matthew J., University of Delaware
Oschlies, Andreas, Dalhousie University
Oviatt, Candace, University of Rhode Island
Pakhomov, Evgeny, University of British Columbia
Palenik, Brian, University of California, San Diego
Paul, John H., University of South Florida
Peebles, Ernst B., University of South Florida
Pellerin, Jocelyne, Universite du Quebec a Rimouski
Pendleton, Daniel, University of Massachusetts, Boston
Perry, Mary Jane, University of Washington
Peterson, Bradley, SUNY, Stony Brook
Pikitch, Ellen K., SUNY, Stony Brook
Pinckney, James, University of South Carolina
Plotkin, Pamela, Texas A&M University
Pollack, Jennifer, Texas A&M University, Corpus Christi
Potts, Donald C., University of California, Santa Cruz
Powers, Sean, University of South Alabama
Purdie, Duncan A., University of Southampton
Quattro, Joseph, University of South Carolina
Rabalais, Nancy N., Louisiana State University
Ragotzkie, Robert A., University of Wisconsin, Madison
Ray, G. Carleton, University of Virginia
Redalje, Donald G., University of Southern Mississippi
Reese, Brandi, Texas A&M University, Corpus Christi
Rhyne, Andrew, University of Massachusetts, Boston

Richardson, Tammi, University of South Carolina
Robinson, Carol, University of East Anglia
Robinson, Leonie, University of Liverpool
Rocap, Gabrielle L., University of Washington
Rolland, Rosalind, University of Massachusetts, Boston
Roman, Charles T., University of Rhode Island
Rotjan, Randi, University of Massachusetts, Boston
Roughgarden, Joan, Stanford University
Rouse, Gregory W., University of California, San Diego
Rowe, Gilbert T., Texas A&M University
Roy, Suzanne, Universite du Quebec a Rimouski
Ruttan, Lore, Emory University
Rykaczewski, Ryan, University of South Carolina
Saila, Saul B., University of Rhode Island
Sambrotto, Raymond N., Columbia University
Sautter, Leslie R., College of Charleston
Schnetzer, Astrid, North Carolina State University
Schulze, Anja, Texas A&M University
Scott, Tim, Roger Williams University
Selph, Karen E., University of Hawai'i, Manoa
Shank, Gerard C., University of Texas at Austin
Shaw, Richard F., Louisiana State University
Sherr, Barry, Oregon State University
Sherr, Evelyn, Oregon State University
Shuman, Randy, University of Washington
Shumway, Sandra, University of Connecticut
Sibert, John R., University of Hawai'i, Manoa
Sieburth, John M., University of Rhode Island
Silver, Mary W., University of California, Santa Cruz
Small, Lawrence, Oregon State University
Smalley, Gabriela W., Rider University
Smayda, Theodore J., University of Rhode Island
Smith, Craig R., University of Hawai'i, Manoa
Smith, David C., University of Rhode Island
Smith, David E., University of Virginia
Smith, Jennifer E., University of California, San Diego
Solan, Martin, University of Southampton
Sommer, Ulrich , Dalhousie University
Sousa, Wayne P., University of California, Berkeley
Specker, Jennifer, University of Rhode Island
Stakes, Debra S., Stanford University
Stanley, Emily, University of Wisconsin, Madison
Statham, Peter, University of Southampton
Steppe, Cecily N., United States Naval Academy
Steward, Grieg F., University of Hawai'i, Manoa
Stewart, Brent S., University of San Diego
Stickney, Robert R., Texas A&M University
Stoecker, Diane, University of Maryland
Strickland, Richard M., University of Washington
Subramaniam, Ajit, Columbia University
Sugihara, George, University of California, San Diego
Suttle, Curtis, University of British Columbia
Swain, Geoffrey W., Florida Institute of Tech
Swift, Elijah V., University of Rhode Island
Sylvan, Jason B., Texas A&M University
Taggart, Christopher T., Dalhousie University
Tan, Wenxian, Austin Community College District
Taylor, Frank J. R., University of British Columbia
Taylor, Gordon T., SUNY, Stony Brook
Thatje, Sven, University of Southampton
Thomas, Florence, University of Hawai'i, Manoa
Thomas, Peter, University of Texas at Austin
Thompson, Diane, Boston University
Thomsen, Laurenz A., University of Washington
Thomson, Jack, University of Liverpool
Thornton, Daniel C., Texas A&M University
Tlusty, Michael, University of Massachusetts, Boston
Toonen, Robert, University of Hawai'i, Manoa
Torres, Joseph J., University of South Florida
Tortell, Philippe, University of British Columbia
Treude, Tina, University of California, Los Angeles
Tunnicliffe, Verena, University of Victoria
Turner, Robert E., Louisiana State University
Tynan, Cynthia T., University of Washington
Urban-Rich, Juanita, University of Massachusetts, Boston
Vacquier, Victor D., University of California, San Diego
Vaillancourt, Robert, Millersville University
Van Oostende, Nicolas, Princeton University
Varela, Diana, University of Victoria
Vargo, Gabriel A., University of South Florida
Vaudrey, Jamie, University of Connecticut
Villalard-Bohnsack, Martine, Roger Williams University
Villareal, Tracy A., University of Texas at Austin
Waldbusser, George, Oregon State University
Walsh, John J., University of South Florida
Ward, Bess B., Princeton University
Ward, J. Evan, University of Connecticut
Warren, Joseph, SUNY, Stony Brook
Wells, Randy, University of South Florida
Welschmeyer, Nicholas A., Moss Landing Marine Laboratories
Weng, Kevin, University of Hawai'i, Manoa
Werner, Timothy, University of Massachusetts, Boston
Wetz, Michael, Texas A&M University, Corpus Christi
Whalen, Kristen, Woods Hole Oceanographic Institution
White, Angelicque, Oregon State University
Wiedenmann, Jorg, University of Southampton
Wilson, Merwether, Edinburgh University
Wishner, Karen, University of Rhode Island
Wolcott, Donna L., North Carolina State University
Wolcott, Thomas G., North Carolina State University
Woodin, Sarah, University of South Carolina
Yamanaka, Tsuyuko, University of Liverpool
Yandle, Tracy, Emory University
Young, Richard E., University of Hawai'i, Manoa
Zakardjian, Bruno, Universite du Quebec a Rimouski
Zedler, Joy, University of Wisconsin, Madison
Zehr, Jonathan P., University of California, Santa Cruz
Zhang, Huan, University of Connecticut

## Chemical Oceanography

Allison, Nicky, University of St. Andrews
Altabet, Mark, Brown University
Andersen, Raymond J., University of British Columbia
Anderson, Robert F., Columbia University
Andersson, Andreas, University of California, San Diego
Andren, Anders W., University of Wisconsin, Madison
Andrews, Allen H., University of Hawai'i, Manoa
Armstrong, David E., University of Wisconsin, Madison
Armstrong, Robert A., SUNY, Stony Brook
Azetsu-Scott, Kumiko, Geological Survey of Canada
Bacon, Michael P., Woods Hole Oceanographic Institution
Balistrieri, Laurie, University of Washington
Barbeau, Katherine A., University of California, San Diego
Bates, Nicholas R., University of Southampton
Benitez-Nelson, Claudia R., University of South Carolina
Benway, Heather, Woods Hole Oceanographic Institution
Bishop, James K., University of California, Berkeley
Boyle, Edward A., Massachusetts Institute of Tech
Brewer, Peter G., Stanford University
Bronk, Deborah A., College of William & Mary
Brownawell, Bruce J., SUNY, Stony Brook
Bruland, Kenneth W., University of California, Santa Cruz
Buchwald, Carolyn, Dalhousie University
Buckley, Lawrence J., University of Rhode Island
Buesseler, Ken O., Woods Hole Oceanographic Institution
Bullister, John, University of Washington
Burdige, David J., Old Dominion University
Canuel, Elizabeth A., College of William & Mary
Chan, Kwan M., California State University, Long Beach
Chapman, Piers, Texas A&M University
Chappell, P. D., Old Dominion University
Cherrier, Jennifer, Brooklyn College (CUNY)
Clarke, Antony D., University of Hawai'i, Manoa
Cochran, J. Kirk, SUNY, Stony Brook
Coffin, Richard, Texas A&M University, Corpus Christi
Cooper, Matthew, University of Southampton
Corbett, David R., East Carolina University
Cornwell, Jeffery C., University of Maryland
Crutzen, Paul, University of California, San Diego
Culberson, Charles H., University of Delaware
Cullen, Jay, University of Victoria
Cutter, Gregory A., Old Dominion University
deAngelis, Marie, SUNY, Maritime College
Delaney, Margaret L., University of California, Santa Cruz
DeMaster, David J., North Carolina State University

Devol, Allan H., University of Washington
Druffel, Ellen R. M., University of California, Irvine
Drysdale, Jessica, Woods Hole Oceanographic Institution
Duce, Robert A., Texas A&M University
Duedall, Iver W., Florida Institute of Tech
Emerson, Steven R., University of Washington
Emile-Geay, Julien, University of Southern California
Engel, Anga, SUNY, Stony Brook
Estapa, Margaret, Skidmore College
Fanning, Kent A., University of South Florida
Feely, Richard A., University of Washington
Fendrock, Michaela, Woods Hole Oceanographic Institution
Fenical, William H., University of California, San Diego
Fisher, Nicholas S., SUNY, Stony Brook
Fitzgerald, William F., University of Connecticut
Fitzsimmons, Jessica N., Texas A&M University
Fogel, Marilyn L., University of Delaware
Fox, Lewis, Broward College, Central Campus
Fredricks, Helen, Woods Hole Oceanographic Institution
Gagne, Jean-Pierre, Universite du Quebec a Rimouski
Gammon, Richard H., University of Washington
Garcia-Rubio, Luis H., University of South Florida
Gieskes, Joris M., University of California, San Diego
Glover, David M., Woods Hole Oceanographic Institution
Gold-Bouchot, Gerardo, Texas A&M University
Gosselin, Kelsey, Woods Hole Oceanographic Institution
Granger, Julie, University of Connecticut
Grottoli, Andrea G., Ohio State University
Hales, Burke R., Oregon State University
Hamme, Roberta C., University of Victoria
Hanson, Jr., Alfred K., University of Rhode Island
Hayes, Christopher T., University of Southern Mississippi
Haymet, Anthony D., University of California, San Diego
Helmletter, Michael F., Florida Institute of Tech
Helmstetter, Michael F., Florida Institute of Tech
Hollander, David J., University of South Florida
Hu, Xinping, Texas A&M University, Corpus Christi
Huebert, Barry J., University of Hawai'i, Manoa
Hughes, Chambers, University of California, San Diego
Ingalls, Anitra E., University of Washington
James, Rachael, University of Southampton
Jenkins, William J., Woods Hole Oceanographic Institution
Johnson, Bruce D., Dalhousie University
Keaffaber, J. J., Florida Institute of Tech
Keil, Richard G., University of Washington
Kiene, Ronald P., University of South Alabama
Knap, Anthony, Texas A&M University
Kujawinski, Elizabeth B., Woods Hole Oceanographic Institution
Lang, Susan Q., University of South Carolina
LaVigne, Michéle, Bowdoin College
Lee, Cindy, SUNY, Stony Brook
Lima, Ivan D., Woods Hole Oceanographic Institution
Loder, Theodore C., University of New Hampshire
Longnecker, Krista, Woods Hole Oceanographic Institution
Lott, Dempsey E., Woods Hole Oceanographic Institution
Lucotte, Marc Michel, Universite du Quebec a Montreal
Lund, David, University of Connecticut
Lwiza, Kamazima M., SUNY, Stony Brook
Mahaffey, Claire, University of Liverpool
Marchitto, Thomas M., University of Colorado
Martin, William R., Woods Hole Oceanographic Institution
Mason, Robert, University of Connecticut
Matsumoto, Katsumi, University of Minnesota, Twin Cities
McDuff, Russell E., University of Washington
McElroy, Anne, SUNY, Stony Brook
McIlvin, Matt, Woods Hole Oceanographic Institution
McKinley, Galen, University of Wisconsin, Madison
McNichol, Ann P., Woods Hole Oceanographic Institution
Measures, Christopher, University of Hawai'i, Manoa
Merrill, John T., University of Rhode Island
Moffett, James, University of Southern California
Moore, Robert M., Dalhousie University
Moran, S. Bradley, University of Rhode Island
Murray, James W., University of Washington
Nuzzio, Donald B., University of Delaware
Oestreich, William, Woods Hole Oceanographic Institution
Oktay, Sarah, University of Massachusetts, Boston
Ossolinski, Justin, Woods Hole Oceanographic Institution
Owen, Robert M., University of Michigan
Paytan, Adina, Stanford University
Pellenbarg, Robert, College of the Desert
Pelletier, Emilien, Universite du Quebec a Rimouski
Pennock, Jonathan R., Dauphin Island Sea Lab
Pilson, Michael E., University of Rhode Island
Prahl, Fredrick G., Oregon State University
Precedo, Laura, Broward College, Central Campus
Presley, Bobby J., Texas A&M University
Pyrtle, Ashanti J., University of South Florida
Quay, Paul D., University of Washington
Quinn, James G., University of Rhode Island
Rahn, Kenneth A., University of Rhode Island
Ravelo, Ana C., University of California, Santa Cruz
Reimer, Andreas, Georg-August University of Goettingen
Reimers, Clare, Oregon State University
Resing, Joseph A., University of Washington
Rheuban, Jennie, Woods Hole Oceanographic Institution
Richey, Jeffrey E., University of Washington
Roethel, Frank J., SUNY, Stony Brook
Russell, Joellen, University of Arizona
Sabine, Christopher L., University of Washington
Saito, Mak A., Woods Hole Oceanographic Institution
Salaun, Rachel, University of Liverpool
Sandwith, Zoe, Woods Hole Oceanographic Institution
Santschi, Peter H., Texas A&M University
Sarmiento, Jorge L., Princeton University
Sayles, Frederick L., Woods Hole Oceanographic Institution
Schwartz, Matthew C., University of West Florida
Scranton, Mary I., SUNY, Stony Brook
Shadwick, Elizabeth H., College of William & Mary
Shamberger, Kathryn E., Texas A&M University
Sharp, Jonathan H., University of Delaware
Sharples, Jonathan, University of Liverpool
Shieh, Chih-Shin, Florida Institute of Tech
Shiller, Alan M., University of Southern Mississippi
Sholkovitz, Edward R., Woods Hole Oceanographic Institution
Simjouw, Jean-Paul, University of New Haven
Skoog, Annelie, University of Connecticut
Sonzogni, William C., University of Wisconsin, Madison
Soule, Melissa, Woods Hole Oceanographic Institution
Stewart, Gillian, Queens College (CUNY)
Sundby, Bjorn, Universite du Quebec a Rimouski
Suntharalingam, Parvadha, University of East Anglia
Swarr, Gretchen, Woods Hole Oceanographic Institution
Sylva, Sean, Woods Hole Oceanographic Institution
Tagliabue, Alessandro, University of Liverpool
Takahashi, Taro, Columbia University
Tanhua, Toste, Dalhousie University
Thomas, Helmuth, Dalhousie University
Thompson, Anu, University of Liverpool
Timmermans, Mary-Louise, Yale University
Tobias, Craig, University of Connecticut
Traganza, Eugene D., Naval Postgraduate School
Ullman, William J., University of Delaware
Van Mooy, Benjamin, Woods Hole Oceanographic Institution
Velinsky, David, Drexel University
Veron, Alain J., University of Delaware
Vlahos, Penny, University of Connecticut
Wade, Terry L., Texas A&M University
Wallace, Douglas, Dalhousie University
Wallace, Douglas W. R., SUNY, Stony Brook
Wiesenburg, Denis A., University of Southern Mississippi
Williams, Dave E., University of British Columbia
Windsor, John G., Florida Institute of Tech
Windsor, Jr., John G., Florida Institute of Tech
Wong, George T. F., Old Dominion University
Yao, Wensheng, University of South Florida
Yvon-Lewis, Shari A., Texas A&M University
Zafiriou, Oliver C., Woods Hole Oceanographic Institution
Zhu, Qingzhi, SUNY, Stony Brook

## Geological Oceanography

Alexander, Clark R., Georgia Southern University
Alt, Jeffrey C., University of Michigan
Arthur, Michael A., Pennsylvania State University, University Park
Asper, Vernon L., University of Southern Mississippi

Bain, Olivier, Institut Polytechnique LaSalle Beauvais (ex-IGAL)
Baker, Edward T., University of Washington
Baldauf, Jack G., Texas A&M University
Berquist, Jr., Carl R., College of William & Mary
Billups, Katharina, University of Delaware
Black, David, SUNY, Stony Brook
Busch, William H., University of New Orleans
Bush, David M., University of West Georgia
Byrne, Timothy, University of Connecticut
Carey, Steven N., University of Rhode Island
Clemens, Steven C., Brown University
Coffin, Richard, Texas A&M University, Corpus Christi
Cok, Anthony E., Adelphi University
Cook, Mea S., Williams College
Creager, Joe S., University of Washington
D'Hondt, Steven L., University of Rhode Island
Darby, Dennis A., Old Dominion University
Dellapenna, Timothy M., Texas A&M University
Dudley, Walter, University of Hawai'i, Manoa
Emerick, Christina M., University of Washington
Farmer, Emma C., Hofstra University
Farrell, John, University of Rhode Island
Ferrini, Victoria, Columbia University
Finney, Bruce P., University of Alaska, Fairbanks
Fisher, Cynthia G., West Chester University
Flood, Roger D., SUNY, Stony Brook
Gagnon, Katie, Seattle Central Community College
Gardner, Wilford D., Texas A&M University
Harris, Sara, University of British Columbia
Hawkins, James W., University of California, San Diego
Hayman, Nicholas W., University of Texas at Austin
Hein, Christopher J., College of William & Mary
Ho, David, University of Hawai'i, Manoa
Holcomb, Robin T., University of Washington
Holcombe, Troy, Texas A&M University
Holmes, Mark L., University of Washington
Honjo, Susumu, Woods Hole Oceanographic Institution
Hovan, Steven A., Indiana University of Pennsylvania
Ivanochko, Tara, University of British Columbia
Karlin, Robert, University of Nevada, Reno
Kelley, Deborah S., University of Washington
Kincaid, Christopher, University of Rhode Island
King, John, University of Rhode Island
Kominz, Michelle A., Western Michigan University
Kuehl, Steven A., College of William & Mary
Larson, Roger, University of Rhode Island
Lear, C, Cardiff University
Leinen, Margaret, University of Rhode Island
Leventer, Amy, Colgate University
Locker, Stanley D., University of South Florida
Longworth, Brett, Woods Hole Oceanographic Institution
Manganini, Steven J., Woods Hole Oceanographic Institution
Martinez-Noriega, Cesar, Universidad Autonoma de Baja California Sur
McHugh, Cecilia M., Graduate School of the City University of New York
McHugh, Cecilia, Queens College (CUNY)
McManus, Dean A., University of Washington
Moore, Theodore C., University of Michigan
Mosher, David C., Dalhousie University
Mottl, Michael J., University of Hawai'i, Manoa
Murray, A. Bradshaw, Duke University
Murray, David, Brown University
Muza, Jay P., Broward College, Central Campus
Naar, David F., University of South Florida
Nittrouer, Charles A., University of Washington
Nowell, Arthur R. M., University of Washington
Ogston, Andrea S., University of Washington
Oostdam, Bernard L., Millersville University
Parsons, Jeffrey D., University of Washington
Pilarczyk, Jessica E., University of Southern Mississippi
Pilkey, Jr., Orrin H., Duke University
Pinet, Paul, Colgate University
Prell, Warren L., Brown University
Quinn, Terrence, University of South Florida
Sager, William W., Texas A&M University
Schilling, Jean-Guy, University of Rhode Island
Schmidt, Matthew, Old Dominion University
Shen, Yang, University of Rhode Island
Sigurdsson, Haraldur, University of Rhode Island
Slowey, Niall C., Texas A&M University
Smith, Stephen V., University of Hawai'i, Manoa
Snoeckx, Hilde, University of West Florida
Sommerfield, Christopher K., University of Delaware
Spaziani, Amy, Ohio Dept of Natural Resources
St. John, Kristen E., James Madison University
Stanley, Daniel J., Smithsonian Institution / National Museum of Natural History
Sternberg, Richard W., University of Washington
Swift, Donald J. P., Old Dominion University
Thomas, Debbie, Texas A&M University
Tyce, Robert, University of Rhode Island
Uchupi, Elazar, Woods Hole Oceanographic Institution
Wallace, Davin, University of Southern Mississippi
Whitman, Jill M., Pacific Lutheran University
Wilcock, William S., University of Washington
Zarillo, Gary A., Florida Institute of Tech
Zeebe, Richard E., University of Hawai'i, Manoa
Zhang, Xinning, Princeton University

## Physical Oceanography

Aagaard, Knut, University of Washington
Abernathey, Ryan P., Columbia University
Alford, Matthew H., University of Washington
Allen, Susan E., University of British Columbia
Allen, Jr., John S., Oregon State University
Andrefouet, Serge, University of South Florida
Anis, Ayal, Texas A&M University
Armi, Laurence, University of California, San Diego
Atkinson, Larry P., Old Dominion University
Barclay, David R., Dalhousie University
Batchelder, Hal, Oregon State University
Baum, Steven K., Texas A&M University
Beaulieu, Claudie, University of Southampton
Bennett, Andrew, Oregon State University
Bishop, Stuart P., North Carolina State University
Bogucki, Darek, Texas A&M University, Corpus Christi
Bohlen, Walter F., University of Connecticut
Boicourt, William, University of Maryland
Bordoni, Simona, California Institute of Tech
Bostater, Charles R., Florida Institute of Tech
Bourke, Robert H., Naval Postgraduate School
Bowman, Malcolm J., SUNY, Stony Brook
Boxall, Simon R., University of Southampton
Boyd, John, University of Michigan
Bracco, Annalisa, Georgia Institute of Tech
Breaker, Laurence, Moss Landing Marine Laboratories
Brooks, David A., Texas A&M University
Brubaker, John M., College of William & Mary
Buckingham, Michael J., University of California, San Diego
Buijsman, Maarten, University of Southern Mississippi
Bulusu, Subrahmanyam, University of South Carolina
Cane, Mark A., Columbia University
Cannon, Glenn A., University of Washington
Carder, Kendall L., University of South Florida
Carmack, Edward, University of British Columbia
Carter, Glenn, University of Hawai'i, Manoa
Carton, James, University of Maryland
Cessi, Paola, University of California, San Diego
Chang, Ping, Texas A&M University
Chao, Shenn-Yu, University of Maryland
Chapman, Ross N., University of Victoria
Chelton, Dudley B., Oregon State University
Chen, Dake, Columbia University
Chiu, Ching-Sang, Naval Postgraduate School
Chu, Peter C., Naval Postgraduate School
Cole, Rick, University of South Florida
Collins, Curtis A., Naval Postgraduate School
Cornillon, Peter, University of Rhode Island
Craig, Susanne , Dalhousie University
Crease, James, University of Delaware
Criminale, Jr., William O., University of Washington
Cronin, Meghan F., University of Washington
D'Asaro, Eric A., University of Washington
Davis, Leslie M., Austin Community College District
de Szoeke, Simon P., Oregon State University
Denman, Kenneth L., University of Victoria

Dever, Edward P., Oregon State University
Dewey, Richard K., University of Victoria
Di Lorenzo, Emanuele, Georgia Institute of Tech
Dierssen, Heidi, University of Connecticut
DiMarco, Steven F., Texas A&M University
Donohue, Kathleen, University of Rhode Island
Donovan, Jeff C., University of South Florida
Dorman, Clive E., San Diego State University
Dosso, Stanley E., University of Victoria
Drago, Aldo, University of Malta
Drijfhout, Sybren, University of Southampton
Dushaw, Brian D., University of Washington
Ellis, Dale, Dalhousie University
English, David C., University of South Florida
Eriksen, Charles C., University of Washington
Ewart, Terry E., University of Washington
Fedorov, Alexey V., Yale University
Ferrari, Raffaele, Massachusetts Institute of Tech
Fewings, Melanie, University of Connecticut
Firing, Eric, University of Hawai'i, Manoa
Flagg, Charles, SUNY, Stony Brook
Flament, Pierre J., University of Hawai'i, Manoa
Flierl, Glenn R., Massachusetts Institute of Tech
Flynn, Russell L., Cypress College
Follows, Michael, Massachusetts Institute of Tech
Foreman, Michael G. G., University of British Columbia
Fox-Kemper, Baylor, Brown University
Frajka-Williams, Eleanor, University of Southampton
Fram, Jonathan, Oregon State University
Galea, Anthony, University of Malta
Galperin, Boris, University of South Florida
Garfield, Newell (Toby), San Francisco State University
Gargett, Ann, Old Dominion University
Garrett, Christopher J. R., University of Victoria
Garwood, Roland W., Naval Postgraduate School
Gauci, Adam, University of Malta
Giese, Benjamin S., Texas A&M University
Gille, Sarah T., University of California, San Diego
Ginis, Isaac, University of Rhode Island
Gnanadesikan, Anand, Johns Hopkins University
Gong, Donglai, College of William & Mary
Gordon, Arnold L., Columbia University
Gratton, Yves, Universite du Quebec
Greatbatch, Richard , Dalhousie University
Greenberg, David, Dalhousie University
Gregg, Michael C., University of Washington
Grosch, Chester E., Old Dominion University
Guinasso, Norman, Texas A&M University
Haine, Thomas W., Johns Hopkins University
Hara, Tetsu, University of Rhode Island
Harrison, Don E., University of Washington
Hautala, Susan L., University of Washington
Hay, Alex E., Dalhousie University
He, Ruoying, North Carolina State University
Heavers, Richard, Roger Williams University
Hebert, David L., University of Rhode Island
Hebert, David, Dalhousie University
Heimbach, Patrick, University of Texas at Austin
Hendershott, Myrl C., University of California, San Diego
Herbers, Thomas H., Naval Postgraduate School
Hermann, Albert, University of Washington
Hetland, Robert D., Texas A&M University
Heywood, Karen, University of East Anglia
Hickey, Barbara M., University of Washington
Hines, Paul, Dalhousie University
Hofmann, Eileen E., Old Dominion University
Howard, Matthew K., Texas A&M University
Howden, Stephan, University of Southern Mississippi
Howe, Bruce M., University of Washington
Hsieh, William W., University of British Columbia
Huang, Bohua, George Mason University
Ierley, Glenn R., University of California, San Diego
Jackson, George A., Texas A&M University
Jacobs, Stanley, Columbia University
Janowitz, Gerald S., North Carolina State University
Jochens, Ann E., Texas A&M University
Johnson, Gregory C., University of Washington
Jones, Peter E., Dalhousie University
Justic, Dubravko, Louisiana State University
Kaempf, Jochen, Flinders University
Kaplan, Alexey, Columbia University
Kawase, Mitsuhiro, University of Washington
Kelley, Dan, Dalhousie University
Kelly, Kathryn A., University of Washington
Kessler, William S., University of Washington
Klinck, John M., Old Dominion University
Klinger, Barry, George Mason University
Kloosterziel, Rudolf C., University of Hawai'i, Manoa
Klymak, Jody, University of Victoria
Knauss, John A., University of Rhode Island
Knowles, Charles E., North Carolina State University
Kosro, Michael, Oregon State University
Koutitonsky, Vladimir G., Universite du Quebec a Rimouski
Kumar, Ajoy, Millersville University
Kunze, Eric L., University of Washington
Kuperman, William A., University of California, San Diego
Lebow, Ruth Y., Los Angeles Pierce College
Lee, Craig M., University of Washington
Lerczak, Jim, Oregon State University
Lesht, Barry, University of Illinois at Chicago
Lesht, Barry M., Argonne National Laboratory
Liu, Zheng-yu, University of Wisconsin, Madison
Lombardo, Kelly, University of Connecticut
Lozier, M. Susan, Duke University
Lu, Youyu, Dalhousie University
Lukas, Roger, University of Hawai'i, Manoa
Luther, Douglas S., University of Hawai'i, Manoa
Luther, Mark E., University of South Florida
MacCready, Parker, University of Washington
Magaard, Lorenz, University of Hawai'i, Manoa
Malanotte-Rizzoli, Paola M., Massachusetts Institute of Tech
Manley, Thomas O., Middlebury College
Marsh, Robert, University of Southampton
Marshall, John C., Massachusetts Institute of Tech
Martin, Seelye, University of Washington
Martinson, Douglas G., Columbia University
Maslowski, Wieslaw, Naval Postgraduate School
Maul, George A., Florida Institute of Tech
McCreary, Julian P., University of Hawai'i, Manoa
McManus, Margaret Anne, University of Hawai'i, Manoa
Mellor, George, Princeton University
Melville, W. Kendall, University of California, San Diego
Merrifield, Mark A., University of Hawai'i, Manoa
Miller, Robert N., Oregon State University
Mitchum, Gary T., University of South Florida
Mofjeld, Harold O., University of Washington
Moore, Dennis W., University of Washington
Moore, Jefferson K., University of California, Irvine
Morison, James H., University of Washington
Moum, James N., Oregon State University
Muller, Andrew C., United States Naval Academy
Muller, Peter, University of Hawai'i, Manoa
Munk, Walter H., University of California, San Diego
Myers, Paul, University of Alberta
Mysak, Lawrence A., McGill University
Nash, Jonathan, Oregon State University
Naveira Garabato, Alberto, University of Southampton
Nechaev, Dmitri, University of Southern Mississippi
Nowlin, Jr., Worth D., Texas A&M University
Nystuen, Jeffrey A., University of Washington
O'Donnell, James, University of Connecticut
Oliver, Eric, Dalhousie University
Oliver, Kevin, University of Southampton
Oltman-Shay, Joan M., University of Washington
Orsi, Alejandro H., Texas A&M University
Ortt, Jr., Richard A., Maryland Department of Natural Resources
Paduan, Jeffrey D., Naval Postgraduate School
Pawlowicz, Richard A., University of British Columbia
Perrie, William, Dalhousie University
Perry, David J., Chabot College
Petruncio, Emil T., United States Naval Academy
Pfirman, Stephanie L., Columbia University
Philander, S. George H., Princeton University
Pieters, Roger, University of British Columbia

Pietrafesa, Leonard J., North Carolina State University
Pinkel, Robert, University of California, San Diego
Pond, Stephen G., University of British Columbia
Powell, Brian, University of Hawai'i, Manoa
Powell, Thomas M., University of California, Berkeley
Primeau, Francois, University of California, Irvine
Pringle, James M., University of New Hampshire
Qiu, Bo, University of Hawai'i, Manoa
Radko, Timour, Naval Postgraduate School
Resio, D. R., Florida Institute of Tech
Rhines, Peter B., University of Washington
Richards, Kelvin J., University of Hawai'i, Manoa
Richardson, Mary J., Texas A&M University
Riser, Stephen C., University of Washington
Roden, Gunnar I., University of Washington
Roemmich, Dean H., University of California, San Diego
Rohling, Eelco J., University of Southampton
Ross, Tetjana, Dalhousie University
Rossby, Hans T., University of Rhode Island
Rothrock, David A., University of Washington
Rothstein, Lewis, University of Rhode Island
Rouse, Jr., Lawrence J., Louisiana State University
Royer, Thomas, Old Dominion University
Ruddick, Barry R., Dalhousie University
Rudnick, Daniel L., University of California, San Diego
Salmon, Richard L., University of California, San Diego
Salmun, Haydee, Hunter College (CUNY)
Samelson, Roger, Oregon State University
Sanford, Lawrence P., University of Maryland
Sanford, Thomas B., University of Washington
Schopf, Paul S., George Mason University
Schulz, William J., United States Naval Academy
Semtner, Albert J., Naval Postgraduate School
Send, Uwe, University of California, San Diego
Sevellec, Florian, University of Southampton
Shaw, Ping-Tung, North Carolina State University
Shearman, Kipp, Oregon State University
Shen, Jian, College of William & Mary
Sheng, Jinyu, Dalhousie University
Smith, Ned P., Florida Institute of Tech
Smith, Peter C., Dalhousie University
Smith, Robert L., Oregon State University
Smyth, William D., Oregon State University
Southam, John R., University of Miami
Spindel, Robert C., University of Washington
St, Achim, Texas A&M University
Stanton, Timothy P., Naval Postgraduate School
Stewart, Robert H., Texas A&M University
Stoessel, Marion, Texas A&M University
Stramski, Dariusz, University of California, San Diego
Straub, David, McGill University
Sutherland, Dave, University of Oregon
Swaters, Gordon E., University of Alberta
Talley, Lynne D., University of California, San Diego
Thompson, Andrew F., California Institute of Tech
Thompson, Keith R., Dalhousie University
Thompson, LuAnne, University of Washington
Thomson, Richard E., University of British Columbia
Thurnherr, Andreas M., Columbia University
Timmermann, Axel, University of Hawai'i, Manoa
Tokmakian, Robin T., Naval Postgraduate School
Tremblay, Bruno, McGill University
Tziperman, Eli, Harvard University
van Norden, Maxim F., University of Southern Mississippi
Veronis, George, Yale University
Von Schwind, Joseph J., Naval Postgraduate School
Wang, Harry, College of William & Mary
Wang, Zhankun, Texas A&M University
Warner, Mark J., University of Washington
Watts, D. Randolph, University of Rhode Island
Weaver, Andrew J., University of Victoria
Webster, Ferris, University of Delaware
Weisberg, Robert H., University of South Florida
Wells, Neil, University of Southampton
White, Martin, National University of Ireland Galway
Whitney, Michael, University of Connecticut
Wickham, Jacob B., Naval Postgraduate School
Wiederwohl, Chrissy, Texas A&M University
Wiggert, Jerry, University of Southern Mississippi
Williams, Kevin L., University of Washington
Wilson, Robert E., SUNY, Stony Brook
Wimbush, Mark, University of Rhode Island
Winguth, Arne M., University of Texas, Arlington
Wolfe, Christopher, SUNY, Stony Brook
Woodgate, Rebecca A., University of Washington
Wu, Jin, University of Delaware
Wunsch, Carl, Harvard University
Wunsch, Carl I., Massachusetts Institute of Tech
Yankovsky, Sasha, University of South Carolina
Young, William R., University of California, San Diego
Yuan, Xiaojun, Columbia University
Zappa, Christopher J., Columbia University
Zhang, Rong, Princeton University
Zhang, Y. J., College of William & Mary
Zhou, Meng, University of Massachusetts, Boston

## Shore and Nearshore Processes

Ashton, Andrew, Woods Hole Oceanographic Institution
Baker, Alex, University of East Anglia
Ballinger, Rhoda, University of Wales
Barth, Jack A., Oregon State University
Baxter, Stefanie J., University of Delaware
Boardman, Mark R., Miami University
Bowen, Anthony J., Dalhousie University
Branco, Brett, Brooklyn College (CUNY)
Briggs, Tiffany Roberts M., Florida Atlantic University
Buonaiuto, Frank, Hunter College (CUNY)
Burbanck, George P., Hampton University
Camann, Eleanor J., Red Rocks Community College
Couceiro, Fay, University of Portsmouth
Dickson, Stephen M., Dept of Agriculture, Conservation, and Forestry
Dunbar, Robert B., Stanford University
Dupre, William R., University of Houston
Duxbury, Alyn C., University of Washington
Engelhart, Simon E., University of Rhode Island
Farnsworth, Katie, Indiana University of Pennsylvania
Farrell, Stewart, Richard Stockton College of New Jersey
Fenster, Michael S., Randolph-Macon College
FitzGerald, Duncan M., Boston University
Friedrichs, Carl T., College of William & Mary
Gibeaut, James, Texas A&M University, Corpus Christi
Giese, Graham S., Woods Hole Oceanographic Institution
Goldsmith, Victor, Graduate School of the City University of New York
Griggs, Gary B., University of California, Santa Cruz
Guza, Robert T., University of California, San Diego
Haigh, Ivan D., University of Southampton
Hall, Robert, University of East Anglia
Haller, Merrick C., Oregon State University
Hansom, Jim, University of Glasgow
Harris, Courtney K., College of William & Mary
Harris, Lee E., Florida Institute of Tech
Helmuth, Brian, Northeastern University
Holman, Robert, Oregon State University
Jackson, Chester M., Georgia Southern University
Jones, D. M., Ohio Dept of Natural Resources
Kassem, Hachem, University of Southampton
Kearney, Michael, University of Maryland
Kemp, Andrew, Tufts University
Kennedy, Andrew, University of Notre Dame
Kineke, Gail C., Boston College
Koppelman, Lee K., SUNY, Stony Brook
Lippmann, Thomas C., University of New Hampshire
Maa, Jerome P-Y., College of William & Mary
MacMahan, Jamie, Naval Postgraduate School
McBride, Randolph, George Mason University
McManus, Margaret A., University of California, Santa Cruz
Neill, Simon P., University of Wales
Niu, Yaoling, Durham University
Oakley, Bryan, Eastern Connecticut State University
Otvos, Ervin G., University of Southern Mississippi
Ozkan-Haller, Tuba, Oregon State University
Paine, Jeffrey G., University of Texas at Austin
Rosen, Peter S., Northeastern University
Ruggiero, Peter, Oregon State University
Seibel, Erwin, San Francisco State University

Shroyer, Emily L., Oregon State University
Sommerfield, Christopher K., University of Delaware
Thornton, Edward B., Naval Postgraduate School
Thosteson, Eric D., Florida Institute of Tech
Trembanis, Arthur C., University of Delaware
Vander Zanden, Jake, University of Wisconsin, Madison
Voulgaris, George, University of South Carolina
Westerink, Joannes J., University of Notre Dame
Wilson, Greg, Oregon State University
Winant, Clinton D., University of California, San Diego
Wu, Chin, University of Wisconsin, Madison

## PLANETOLOGY

### Cosmochemistry

Agee, Carl B., University of New Mexico
Alexander, Conel M., Carnegie Institution of Washington
Bar-Nun, Akiva, Tel Aviv University
Bermingham, Katherine, University of Maryland
Blake, Geoffrey A., California Institute of Tech
Blander, Milton, Argonne National Laboratory
Bodenheimer, Peter, University of California, Santa Cruz
Bouvier, Audrey, Western University
Boynton, William V., University of Arizona
Brown, Robert H., University of Arizona
Burnett, Donald S., California Institute of Tech
Ciesla, Fred, University of Chicago
Dauphas, Nicolas, University of Chicago
Davis, Andrew, University of Chicago
Drake, Michael J., University of Arizona
Fegley, Bruce, Washington University in St. Louis
Fegley, M. Bruce, Washington University in St. Louis
Greenwood, James P., Wesleyan University
Gross, Juliane, Rutgers, The State University of New Jersey
Grossman, Lawrence, University of Chicago
Herd, Christopher, University of Alberta
Humayan, Munir, Florida State University
Huntress, Jr., Wesley T., Carnegie Institution of Washington
Jordan, Jim L., Lamar University
Korycansky, Don, University of California, Santa Cruz
Kring, David, University of Arizona
Lauretta, Dante, University of Arizona
Lewis, John S., University of Arizona
Lin, Douglas, University of California, Santa Cruz
Lodders-Fegley, Katharina, Washington University in St. Louis
Lunine, Jonathan I., University of Arizona
Lyon, Ian, University of Manchester
MacPherson, Glenn J., Smithsonian Institution / National Museum of Natural History
McKeegan, Kevin D., University of California, Los Angeles
McSween, Jr., Harry Y., University of Tennessee, Knoxville
Mojzsis, Stephen J., University of Colorado
Nittler, Larry R., Carnegie Institution of Washington
Reedy, Robert C., University of New Mexico
Smith, William H., Washington University in St. Louis
Speck, Angela, University of Missouri
Swindle, Timothy, University of Arizona
Turner, Grenville, University of Manchester
Vogt, Steven, University of California, Santa Cruz
Wang, Kun, Washington University in St. Louis

### Extraterrestrial Geology

Adams, Mark, Lamar University
Anderson, Jennifer L., Winona State University
Arvidson, Raymond E., Washington University in St. Louis
Asphaug, Erik, Arizona State University
Asphaug, Erik, University of California, Santa Cruz
Baker, Victor R., University of Arizona
Ballentine, Christopher, University of Oxford
Batygin, Konstantin, California Institute of Tech
Blaney, Diana L., Jet Propulsion Laboratory
Bleamaster III, Leslie F., Trinity University
Bourassa, Matthew, Western University
Brownlee, Donald E., University of Washington
Burr, Devon, University of Tennessee, Knoxville
Byrne, Paul K., North Carolina State University
Cahill, Karen R., Smithsonian Institution / National Museum of Natural History
Cassidy, William, University of Pittsburgh
Christensen, Philip R., Arizona State University
Coles, Kenneth S., Indiana University of Pennsylvania
Craddock, Robert A., Smithsonian Institution / National Air & Space Museum
Creech-Eakman, Michelle, New Mexico Institute of Mining and Tech
Cull, Selby, Bryn Mawr College
Delaney, Jeremy S., Rutgers, The State University of New Jersey
Dietz, Richard D., University of Northern Colorado
Ehlmann, Bethany, California Institute of Tech
Elachi, Charles, California Institute of Tech
Emery, Joshua, University of Tennessee, Knoxville
Fabel, Derek, University of Glasgow
Gilmore, Martha S., Wesleyan University
Glotch, Timothy, SUNY, Stony Brook
Golombek, Matthew P., Jet Propulsion Laboratory
Grant, John A., Smithsonian Institution / National Air & Space Museum
Grieve, Richard, Western University
Grieve, Richard A. F., University of New Brunswick
Grinford, Peter, Birkbeck College
Grosfils, Eric B., Pomona College
Hayes, Alexander G., Cornell University
Head, III, James W., Brown University
Herkenhoff, Ken E., Jet Propulsion Laboratory
Horgan, Briony, Purdue University
Howell, Robert R., University of Wyoming
Hynek, Brian M., University of Colorado
Jakosky, Bruce M., University of Colorado
Johnson, Brandon C., Brown University
Karuntillake, Suniti, Louisiana State University
Kleinhans, Frederick W., Indiana University / Purdue University, Indianapolis
Knutson, Heather, California Institute of Tech
Korotev, Randy L., Washington University in St. Louis
Kraal, Erin, Kutztown University of Pennsylvania
Leake, Martha A., Valdosta State University
Li, Lin, Indiana University / Purdue University, Indianapolis
Lorenz, Ralph, University of Arizona
Mack, John E., Buffalo State College
Massironi, Matteo, Università degli Studi di Padova
Maxwell, Ted A., Smithsonian Institution / National Air & Space Museum
McEwen, Alfred S., University of Arizona
McGowan, Eileen, University of Massachusetts, Amherst
McKinnon, William B., Washington University in St. Louis
Milam, Keith A., Ohio University
Minton, David, Purdue University
Moersch, Jeffery E., University of Tennessee, Knoxville
Mouginis-Mark, Peter J., University of Hawai'i, Manoa
Murchie, Scott, University of Tennessee, Knoxville
Neish, Catherine, Western University
Nelson, Robert M., Jet Propulsion Laboratory
Newsom, Horton E., University of New Mexico
Osinski, Gordon R., Western University
Parsons, Reid A., Fitchburg State University
Piatex, Jennifer L., Central Connecticut State University
Radebaugh, Jani, Brigham Young University
Regas, James L., California State University, Chico
Rice, Melissa S., Western Washington University
Runyon, Cassandra R., College of Charleston
Schultz, Peter H., Brown University
Shepard, Michael K., Bloomsburg University
Sheppard, Scott S., Carnegie Institution of Washington
Solomon, Sean C., Columbia University
Sonntag, Mark S., Angelo State University
Squyres, Steven W., Cornell University
Stofan, Ellen R., Jet Propulsion Laboratory
Strom, Robert G., University of Arizona
Thaisen, Kevin G., Grand Valley State University
Tornabene, Livio, Western University
Turtle, Elizabeth, University of Arizona
Warner, Nicholas H., SUNY, Geneseo
Watters, Thomas R., Smithsonian Institution / National Air & Space Museum
Weinberger, Alycia J., Carnegie Institution of Washington
Williams, Kevin K., Buffalo State College
Zimbelman, James R., Smithsonian Institution / National Air & Space Museum

## Extraterrestrial Geophysics

Angelopoulos, Vassilis, University of California, Los Angeles
Arkani-Hamed, Jafar, McGill University
Asphaug, Erik, University of California, Santa Cruz
Aurnou, Jonathan M., University of California, Los Angeles
Banerdt, Bruce E., Jet Propulsion Laboratory
Benacquista, Matt, Montana State University, Billings
Bord, Don, University of Michigan, Dearborn
Brown, Michael E., California Institute of Tech
Brown, Peter, Western University
Coleman, Paul J., University of California, Los Angeles
Dombard, Andrew, Case Western Reserve University
Dombard, Andrew J., University of Illinois at Chicago
Fenrich, Francis, University of Alberta
Ferencz, Orsolya, Eotvos Lorand University
Fortes, Andrew, University College London
Garrick-Bethell, Ian, University of California, Santa Cruz
Giacalone, Joe, University of Arizona
Goldreich, Peter M., California Institute of Tech
Greenberg, Richard J., University of Arizona
Hapke, Bruce W., University of Pittsburgh
Harnett, Erika M., University of Washington
Hauck, II, Steven A., Case Western Reserve University
Helled, Ravit , Tel Aviv University
Holzworth, Robert H., University of Washington
Hood, Lonnie ("Lon") L., University of Arizona
Hubbard, William B., University of Arizona
Jewitt, David, University of California, Los Angeles
Jokipii, Jack R., University of Arizona
Kivelson, Margaret G., University of California, Los Angeles
Kulkarni, Shrinivas R., California Institute of Tech
Kunkle, Thomas D., Los Alamos National Laboratory
Larsen, Kristine, Central Connecticut State University
Lucey, Paul G., University of Hawai'i, Manoa
Margot, Jean-Luc, University of California, Los Angeles
Matzke, David, University of Michigan, Dearborn
McCarthy, Michael P., University of Washington
McGovern, Patrick J., Rice University
McPherron, Robert L., University of California, Los Angeles
Muhleman, Duane O., California Institute of Tech
Newman, William I., University of California, Los Angeles
Paige, David A., University of California, Los Angeles
Petrovay, Kristof, Eotvos Lorand University
Rankin, Robert, University of Alberta
Robinson, R. Craig, Central Connecticut State University
Rostoker, Gordon, University of Alberta
Russell, Christopher T., University of California, Los Angeles
Samson, John C., University of Alberta
Schlichting, Hilke, University of California, Los Angeles
Schmerr, Nicholas, University of Maryland
Seager, Sara, Massachusetts Institute of Tech
Smrekar, Suzanne E., Jet Propulsion Laboratory
Solomon, Sean , Columbia University
Spilker, Linda J., Jet Propulsion Laboratory
Steele, Andrew, Carnegie Institution of Washington
Stevenson, David J., California Institute of Tech
Sydora, Richard D., University of Alberta
Urquhart, Mary, University of Texas, Dallas
Walker, Raymond J., University of California, Los Angeles
Winglee, Robert M., University of Washington
Zuber, Maria T., Massachusetts Institute of Tech

## Meteorites & Tektites

Albin, Edward, Georgia State University, Perimeter College, Online
Beck, Andrew, Smithsonian Institution / National Museum of Natural History
Binzel, Richard P., Massachusetts Institute of Tech
Bullock, Emma, Smithsonian Institution / National Museum of Natural History
Cartwright, Julia, University of Alabama
Clarke, Jr., Roy S., Smithsonian Institution / National Museum of Natural History
Collins, Gareth, Imperial College
Corrigan, Catherine, Smithsonian Institution / National Museum of Natural History
Dunn, Tasha L., Colby College
Ebel, Denton S., American Museum of Natural History
French, Bevan, Smithsonian Institution / National Museum of Natural History
Gardner-Vandy, Kathryn, Smithsonian Institution / National Museum of Natural History
Genge, Matthew, Imperial College
Glass, Billy P., University of Delaware
Goreva, Yulia, Smithsonian Institution / National Museum of Natural History
Grocholski, Brent, Smithsonian Institution / National Museum of Natural History
Hartman, Ron N., Mt. San Antonio College
Harvey, Ralph P., Case Western Reserve University
Heymann, Dieter, Rice University
Hildebrand, Alan R., University of Calgary
Hutson, Melinda, Portland State University
Hutson, Melinda, Portland Community College - Sylvania Campus
Krot, Alexander N., University of Hawai'i, Manoa
Martins, Zita, Imperial College
Mayne, Rhiannon G., Texas Christian University
McCoy, Timothy, University of Tennessee, Knoxville
McCoy, Timothy J., Smithsonian Institution / National Museum of Natural History
Melosh, H. J., Purdue University
Mittlefehldt, David, University of Tennessee, Knoxville
Paglione, Timothy, York College (CUNY)
Rodgers, Nick, University of Wales
Rubin, Alan E., University of California, Los Angeles
Ruzicka, Alexander (Alex) M., Portland State University
Scott, Edward R., University of Hawai'i, Manoa
Sephton, Mark, Imperial College
Singerling, Sheryl, Smithsonian Institution / National Museum of Natural History
Sipiera, Paul P., Mercer University
Swindle, Timothy D., University of Arizona
Tait, Kim, Western University
Taylor, G. Jeffrey, University of Hawai'i, Manoa
Wadhwa, Meenakshi, Field Museum of Natural History
Warren, Paul, University of California, Los Angeles
Wasson, John T., University of California, Los Angeles
Welzenbach, Linda, Smithsonian Institution / National Museum of Natural History

# OTHER

## General Earth Sciences

Agutu, Nathan O., Jomo Kenyatta University of Agriculture & Tech
Admire, Amanda R., Humboldt State University
Ahrens, Donald C., Modesto Junior College
Angelo, Joseph A., Florida Institute of Tech
Balch, Jennifer K., University of Colorado
Bank, Carl-Georg, University of Toronto
Barron, Robert, Michigan Technological University
Beach Davis, Janet, Heartland Community College
Beausoleil, Denis, Douglas College
Benjamin, Patricia A., Worcester State University
Berry, Leonard, Florida Atlantic University
Boitt, Mark , Jomo Kenyatta University of Agriculture & Tech
Booher, Gary, El Camino College
Boorstein, Margaret F., Long Island University, C.W. Post Campus
Bordossy, Andras, University of Newcastle Upon Tyne
Boyce, Joseph I., McMaster University
Bradley, Jeffrey, Northwest Missouri State University
Branlund, Joy, Southwestern Illinois College - Sam Wolf Granite City Campus
Bretherton, Francis P., University of Wisconsin, Madison
Bries-Korpik, Jill, Century College
Britton, Gloria C., Cuyahoga Community College - Western Campus
Brod, Joseph D., Hillsborough Community College
Bruening, Susan, Eastern Connecticut State University
Burlakova, Lyubov, SUNY, Buffalo
Burns, Danny, Coastal Bend College
Busa, Mark, Middlesex Community College
Caldwell, Marianne O., Hillsborough Community College
Carlson, Galen R., California State University, Fullerton
Carlson, Heath, Eastern Connecticut State University
Carrell, Jennifer, University of Illinois, Urbana-Champaign
Carter, Andy, Birkbeck College

Champion, Kyle M., Hillsborough Community College
Chege, George W., Jomo Kenyatta University of Agriculture & Tech
Chen, jiquan, Michigan State University
Childers, Daniel, Delaware County Community College
Cihacek, Larry J., North Dakota State University
Clark, Michael, Mount Royal University
Cleveland, Natasha, Frederick Community College
Clift, Sigrid, University of Texas at Austin
Cole, Gregory L., Los Alamos National Laboratory
Condreay, Denise, Central Community College
Cordero, David I., Lower Columbia College
Cornebise, Michael W., Eastern Illinois University
Cumpston, Jennifer, Oakton Community College
Davis, James A., Eastern Illinois University
de Wit, Cary, University of Alaska, Fairbanks
Deka, Jennifer, Cuyahoga Community College - Western Campus
DeLima, Lynn-Ann, Eastern Connecticut State University
Dennison, Robert, Heartland Community College
Dietz, Anne D., Alamo Colleges - San Antonio College
Dod, Bruce D., Mercer University
Dolgoff, Anatole, Pace University, New York Campus
Douglas, Arthur V., Creighton University
Douthat, Jr., James R., Hillsborough Community College
Dutton, Jessica, Suffolk County Community College, Ammerman Campus
Ebersole, Sandy M., Geological Survey of Alabama
Erski, Theodore, McHenry County College
Ezerskis, John L., Cuyahoga Community College - Western Campus
Falk, Lisa, North Carolina State University
Fatherree, James W., Hillsborough Community College
Fielding, Lynn, El Camino College
Fields, Nancy, El Centro College - Dallas Community College District
Flanagan, Michael, Suffolk County Community College, Ammerman Campus
Fondran, Carol, Cuyahoga Community College - Western Campus
Fortes, Dominic, Birkbeck College
Foss, Donald J., College of Marin
Gernon, Thomas M., University of Southampton
Gibbs, Samantha J., University of Southampton
Goetz, Andrew R., University of Denver
Goulden, Michael L., University of California, Irvine
Hamilton, Thomas, Edmonds Community College
Harrington, Philip, Suffolk County Community College, Ammerman Campus
Haverluk, Terry W., United States Air Force Academy
Hawkins, Ward L., Los Alamos National Laboratory
Hayes, Van E., Hillsborough Community College
Hendrey, George, Queens College (CUNY)
Hill, Arleen A., University of Memphis
Hillier, John, Grays Harbor College
Hobbs, Chasidy, University of West Florida
Hoke, Gregory, Syracuse University
Hoppe, Kathryn A., Green River Community College
Hrouda, James G., Mineral Area College
Hurd, David, Edinboro University of Pennsylvania
Hurst, Coreen, Brigham Young University - Idaho
Ide, Kayo, University of California, Los Angeles
Imwati, Andrew, Jomo Kenyatta University of Agriculture & Tech
Inglis, Michael, Suffolk County Community College, Ammerman Campus
Janusz, Robert, Alamo Colleges - San Antonio College
Johnson, Jean M., Dalton State Community College
Johnson, Kathleen, University of California, Irvine
Johnson, Tiffany, Douglas College
Jones, Stephen M., University of Birmingham
Kane, Mustapha, Florida Gateway College
Keeling, David J., Western Kentucky University
Kellner, Patricia, Los Angeles Harbor College
Kelly, Maria, Edmonds Community College
Khade, Vishnu R., Eastern Connecticut State University
Koch, Joe, Dept of Natural Resources
Lane, Joseph M., Cuyahoga Community College - Western Campus
Lau, Chui Yim Maggie, Princeton University
Le Heron, Daniel P., University of London, Royal Holloway & Bedford New College
Le Qu, Corinne, University of East Anglia
Leger, Carol, Keene State College
Leighty, Robert S., Mesa Community College
Lerch, Derek, Feather River College
Levasseur, Emile , Eastern Connecticut State University
Locke, James L., College of Marin
Lofthouse, Stephen T., Pace University, New York Campus
Lomaga, Margaret, Suffolk County Community College, Ammerman Campus
Lowe, John C., George Washington University
Mack, John, Los Angeles Harbor College
MacNeil, James E., Hillsborough Community College
Marchisin, John, Pace University, New York Campus
Marlowe, Brian W., Hillsborough Community College
Martek, Lynnette, University of South Carolina - Lancaster
Martin, John W., San Jose City College
Mason, Sherri A., SUNY Fredonia
Matzke, Gordon E., Oregon State University
McConnaughhay, Mark, Dutchess Community College
McDougall, Jim, Tacoma Community College
McGinn, Chris, North Carolina Central University
Mcleod, Andrew T., Edinburgh University
McNicol, Barbara, Mount Royal University
Meador, Cindy D., West Texas A&M University
Medina, Francisco, Universidad Nacional Autonoma de Mexico
Meisner, Caroline B., Great Basin College
Melton, Greg, Brigham Young University - Idaho
Mitra, Chandana, Auburn University
Moreno, Rafael, Metropolitan State College of Denver
Morton, Bruce, Eastern Connecticut State University
Motyka, James, Eastern Connecticut State University
Munasinghe, Tissa, Los Angeles Harbor College
Nagel, Athena, Mississippi State University
Narey, Martha A., University of Denver
Neumann, Patricia, El Camino College
Neves, Douglas, El Camino College
Newlon, Charles F. J., William Jewell College
Nisbet, Euan, University of London, Royal Holloway & Bedford New College
North, Leslie, Western Kentucky University
Norton-Krane, Abby N., Cuyahoga Community College - Western Campus
O'Brien, Suzanne R., Bridgewater State University
Olney, Jessica L., Hillsborough Community College
Olney, Matthew P., Hillsborough Community College
Osborn, Joe, Century College
Oughton, John, Century College
Oxford, Jeremiah, Western Oregon University
Oymayan, Avo, Florida Gateway College
Pappas, Matthew, Suffolk County Community College, Ammerman Campus
Paradise, Thomas R., University of Arkansas, Fayetteville
Pedley, Kate, University of Canterbury
Peprah, Ebenezer, El Camino College
Perry, Baker, Appalachian State University
Perry, Randall, Imperial College
Pettengill, Gordon H., Massachusetts Institute of Tech
Pires, E M., Long Island University, C.W. Post Campus
Pittman, Jason, Folsom Lake College
Pond, Lisa G., Louisiana State University
Potess, Marla, Hardin-Simmons University
Pringle, Patrick, Centralia College
Rapley, Chris, University College London
Renfrew, Melanie, Los Angeles Harbor College
Riller, Ulrich, McMaster University
Ritter, Paul, Heartland Community College
Rock, Jessie, North Dakota State University
Rosa, Lynn C., West Texas A&M University
Roth, Leonard T., Hillsborough Community College
Ryan, Susan, California University of Pennsylvania
Sasowsky, Kathryn, Cuyahoga Community College - Western Campus
Saunders, Ralph, California State University, Dominguez Hills
Schimmrich, Steven, SUNY, Ulster County Community College
Schreier, Hans D., University of British Columbia
Schulze, Karl, Waubonsee Community College
Schumacher, Matthew, Saint Francis Xavier University
Shaw, Katy, Green River Community College
Shugart, Jr., Herman H., University of Virginia
Smilnak, Roberta A., Metropolitan State College of Denver

Smith, Grant, Western Oregon University
Smith, H. Dixon, Metropolitan State College of Denver
Smith, Steven J., University of Northern Iowa
Smith, Thomas M., University of Virginia
Snyder, Jennifer L., Delaware County Community College
Soash, Norman E., Hillsborough Community College
Spooner, Alecia, Everett Community College
Springer, Kathleen B., San Bernardino County Museum
Stanley, George R., San Antonio Community College
Stermer, Ed, Illinois Central College
Stone, Jim, Aims Community College
Sullivan, Donald G., University of Denver
Tarr, Alexander, Worcester State University
Taylor, Matthew, University of Denver
Thomas, Ray G., University of Florida
Thurston, Phillips C., Laurentian University, Sudbury
Tidwell, Allan, Chipola College
Tinsley, Mark, Central Virginia Community College
Tongdee, Poetchanaporn, Hillsborough Community College
Travers, Steven, Heartland Community College
Turski, Mark P., Plymouth State University
Valencia, Victor A., Washington State University
Valentine, Michael, Tacoma Community College
Van Cleave, Kevan A., Hillsborough Community College
van de Gevel, Saskia, Appalachian State University
Veblen, Thomas T., University of Colorado
Voorhees, David H., Waubonsee Community College
Vorwald, Brian, Suffolk County Community College, Ammerman Campus
Waddington, David C., Douglas College
Wade, Phillip, Western Oregon University
Walter, Nathan A., University of West Georgia
Wang, Enru, University of North Dakota
Ward, Calvin H., Rice University
Warger, Jane, Chaffey College
Webb, Craig A., Mt. San Antonio College
White, Susan, Los Angeles Harbor College
Wiesner, Mark R., Rice University
Williams, Mark W., University of Colorado
Wills, William V., Hillsborough Community College
Winterbottom, Wesley, Eastern Connecticut State University
Wisdom, Jack, Massachusetts Institute of Tech
Witter, Rob, Oregon Dept of Geology & Mineral Industries
Yacucci, Mark, Heartland Community College
Zaleha, Robert, Cuyahoga Community College - Western Campus

## Atmospheric Sciences

Ackerman, Thomas P., University of Washington
Adegoke, Jimmy, University of Missouri-Kansas City
Agee, Ernest M., Purdue University
Aiyyer, Anantha, North Carolina State University
Aldrich, Eric A., University of Missouri, Columbia
Alexander, Becky, University of Washington
Allard, Jason, Valdosta State University
Allen, Grant, University of Manchester
Allen, Robert J., University of California, Riverside
Alpert, Pinhas, Tel Aviv University
Anastasio, Cort, University of California, Davis
Ancell, Brian C., Texas Tech University
Anderson, Bruce, Boston University
Anderson, James G., Harvard University
Anderson, Mark R., University of Nebraska, Lincoln
Aneja, Viney P., North Carolina State University
Antonescu, Adrian, University of Manchester
Aquilina, Noel , University of Malta
Arakawa, Akio, University of California, Los Angeles
Ariya, Parisa A., McGill University
Arnfield, A. John, Ohio State University
Arnold, David L., Frostburg State University
Arnold, Stephen, University of Leeds
Arora, Vivek, University of Victoria
Arritt, Raymond W., Iowa State University of Science & Tech
Arya, Satyapal S., North Carolina State University
Atkins, Nolan T., Lyndon State College
Atkinson, Christopher, University of North Dakota
Austin, Philip, University of British Columbia
Back, Larissa E., University of Wisconsin, Madison
Baker, Marcia B., University of Washington
Baker, Marcia, University of Washington
Baldwin, Michael, Purdue University
Balmforth, Neil, University of British Columbia
Barlow, Mathew, University of Massachusetts Lowell
Barnes, Elizabeth A., Colorado State University
Barnes, Jeffrey R., Oregon State University
Barrett, Bradford S., United States Naval Academy
Barrett, Kevin M., Tarrant County College- Northeast Campus
Bartello, Peter, McGill University
Bates, Timothy S., University of Washington
Bathke, Deborah J., University of Nebraska, Lincoln
Battisti, David S., University of Washington
Beard, Kenneth V., University of Illinois, Urbana-Champaign
Bell, Michael M., Colorado State University
Bennartz, Ralf, Vanderbilt University
Bergin, Michael H., Georgia Institute of Tech
Berryman, Bruce F., Lyndon State College
Beswick, Karl, University of Manchester
Bierly, Gregory, Indiana State University
Birner, Thomas, Colorado State University
Bitz, Cecilia M., University of Washington
Black, Robert X., Georgia Institute of Tech
Blasing, Terence J., Oak Ridge National Laboratory
Blechman, Jerome B., SUNY, Oneonta
Bluestein, Howard, SUNY, Stony Brook
Blyth, Alan, University of Leeds
Boden, Thomas, Oak Ridge National Laboratory
Boering, Kristie, University of California, Berkeley
Bollasina, Massimo, Edinburgh University
Bond, Nicholas A., University of Washington
Bonneau, Laurent, Yale University
Booth, Alastair, University of Manchester
Bosart, Lance F., SUNY, Albany
Bossert, James E., Los Alamos National Laboratory
Boybeyi, Zafer, George Mason University
Bradley, Raymond S., University of Massachusetts, Amherst
Bretherton, Christopher S., University of Washington
Bromwich, David H., Ohio State University
Brooks, Sarah D., Texas A&M University
Brown, Robert A., University of Washington
Brubaker, Kenneth L., Argonne National Laboratory
Bruning, Eric C., Texas Tech University
Budikova, Dagmar, Illinois State University
Burls, Natalie, George Mason University
Bush, Andrew B. G., University of Alberta
Businger, Joost A., University of Washington
Calhoun, Joseph, Millersville University
Campbell, Janet W., University of New Hampshire
Cane, Mark A., Columbia University
Cannon, Alex, University of British Columbia
Capehart, William J., South Dakota School of Mines & Tech
Capes, Gerard, University of Manchester
Carey, Larry, Texas A&M University
Carslaw, Ken, University of Leeds
Case Hanks, Anne T., University of Louisiana, Monroe
Castro, Mark S., University of Maryland
Cataneo, Robert, Eastern Illinois University
Catling, David C., University of Washington
Cess, Robert D., SUNY, Stony Brook
Chameides, William L., Duke University
Chang, Edmund K., SUNY, Stony Brook
Chang, Hai-ru, Georgia Institute of Tech
Chang, Young-Soo, Argonne National Laboratory
Charlevoix, Donna J., University of Illinois, Urbana-Champaign
Charlson, Robert J., University of Washington
Chen, Jian-Hua, University of Massachusetts Lowell
Chen, Shuyi S., University of Washington
Chen, Tsing-Chang, Iowa State University of Science & Tech
Chen, Yi-Leng, University of Hawai'i, Manoa
Chen, Yongsheng, York University
Cheng, Meng-Dawn, Oak Ridge National Laboratory
Chipperfield, Martyn, University of Leeds
Chiu, Long S., George Mason University
Choularton, Thomas, University of Manchester
Chu, Pao-Shin, University of Hawai'i, Manoa
Chuang, Patrick Y., University of California, Santa Cruz
Clabo, Darren R., South Dakota School of Mines & Tech

Claire, Mark, University of St. Andrews
Clark, Richard D., Millersville University
Clarke, Antony D., University of Hawai'i, Manoa
Clegg, Simon, University of East Anglia
Clemitshaw, Kevin C., University of London, Royal Holloway & Bedford New College
Cobb, Steven R., Texas Tech University
Coe, Hue, University of Manchester
Colle, Brian, SUNY, Stony Brook
Collett, Jr., Jeffrey L., Colorado State University
Collins, Donald R., Texas A&M University
Collins, William D., University of California, Berkeley
Collis, Scott, Argonne National Laboratory
Colman, Bradley R., University of Washington
Colucci, Stephen J., Cornell University
Connolly, Paul, University of Manchester
Cook, Kerry H., University of Texas at Austin
Corcoran, William, Missouri State University
Costigan, Keeley R., Los Alamos National Laboratory
Côté, Jean, Universite du Quebec a Montreal
Cottingame, William, Los Alamos National Laboratory
Coulter, Richard L., Argonne National Laboratory
Covert, David S., University of Washington
Cowan, Nicolas B., McGill University
Crawford, Ian, University of Manchester
Crawford, James, Georgia Institute of Tech
Crilley, Leigh R., University of Birmingham
Cronin, Timothy W., Massachusetts Institute of Tech
Crowther, Sarah, University of Manchester
Curry, Judith A., Georgia Institute of Tech
Cziczo, Dan, Massachusetts Institute of Tech
Dale, Chris, Durham University
Dannevik, William P., Saint Louis University
Daoust, Mario, Missouri State University
Davis, Robert, University of Virginia
de Camargo, Suzana, Columbia University
de Foy, Benjamin, Saint Louis University
Dearden, Christopher, University of Manchester
DeCaria, Alex J., Millersville University
DeConto, Robert, University of Massachusetts, Amherst
DeGaetano, Arthur, Cornell University
Delcambre, Sharon, Portland Community College - Sylvania Campus
DelSole, Timothy, George Mason University
Dempsey, David P., San Francisco State University
Deng, Yi, Georgia Institute of Tech
Denning, A. Scott, Colorado State University
Derome, Jacques F., McGill University
Desai, Ankur R., University of Wisconsin, Madison
Dessler, Alex, Texas A&M University
Detwiler, Andrew G., South Dakota School of Mines & Tech
DeWekker, Stephan F., University of Virginia
Dibb, Jack E., University of New Hampshire
Dickinson, Robert E., University of Texas at Austin
Diem, Jeremy E., Georgia State University
Dinh, Tra, Princeton University
Dirmeyer, Paul A., George Mason University
Dobbie, Steven, University of Leeds
Dobler, Scott, Western Kentucky University
Doherty, Ruth, Edinburgh University
Dominguez, Francina, University of Illinois, Urbana-Champaign
Doskey, Paul V., Argonne National Laboratory
Dugas, Bernard, Universite du Quebec a Montreal
Dunn, Allison L., Worcester State University
Durkee, Josh, Western Kentucky University
Durran, Dale R., University of Washington
Dutcher, Allen, University of Nebraska, Lincoln
Dwyer, Joseph R., Florida Institute of Tech
Eckel, F. Anthony, University of Washington
Edgell, Dennis J., University of North Carolina, Pembroke
Elliott, Scott M., Los Alamos National Laboratory
Ellul, Raymond , University of Malta
Epifanio, Craig, Texas A&M University
Essery, Richard L., Edinburgh University
Esslinger, Kelly L., Arizona Western College
Fabry, Frederic, McGill University
Fan, Xingang, Western Kentucky University
Felzer, Benjamin S., Lehigh University
Feng, Song, University of Arkansas, Fayetteville
Feng, Yan, Argonne National Laboratory
Fernando, Joe, University of Notre Dame
Ferruzza, David -., Elizabethtown College
Few, Jr., Arthur A., Rice University
Fink, Uwe, University of Arizona
Finnegan, David L., Los Alamos National Laboratory
Fiore, Arlene, Columbia University
Fiore, Arlene M., Columbia University
Fischer, Emily, Colorado State University
Fishman, Jack, Saint Louis University
Fitzjarrald, David R., SUNY, Albany
Flato, Gregory M., University of Victoria
Fleagle, Robert G., University of Washington
Fletcher, Roy J., University of Lethbridge
Forster, Piers M., University of Leeds
Foster, Stuart, Western Kentucky University
Fovell, Robert, University of California, Los Angeles
Fowler, Malcolm M., Los Alamos National Laboratory
Fox, Neil I., University of Missouri, Columbia
Frame, Jeffrey, University of Illinois, Urbana-Champaign
Frederick, John E., University of Chicago
French, Adam, South Dakota School of Mines & Tech
Frierson, Dargan M., University of Washington
Frolking, Stephen E., University of New Hampshire
Fthenakis, Vasilis M., Columbia University
Fu, Qiang, University of Washington
Fu, Rong, University of Texas at Austin
Fueglistaler, Stephan A., Princeton University
Fuhrmann, Christopher, Mississippi State University
Fultz, Dave, University of Chicago
Fung, Inez, University of California, Berkeley
Fyfe, John C., University of Victoria
Gachon, Philippe, Universite du Quebec a Montreal
Gaffney, Jeffrey S., Argonne National Laboratory
Gallagher, Martin, University of Manchester
Galloway, James N., University of Virginia
Gallus, William A., Iowa State University of Science & Tech
Gao, Yuan, Rutgers, The State University of New Jersey, Newark
Garcia, Oswaldo, San Francisco State University
Garrett, Timothy J., University of Utah
Gauthier, Pierre, McGill University
Gedzelman, Stanley, City College (CUNY)
Geller, Marvin A., SUNY, Stony Brook
Ghate, Virendra P., Argonne National Laboratory
Ghil, Michael, University of California, Los Angeles
Giannini, Alessandra, Columbia University
Gilfillan, Stuart M., Edinburgh University
Gill, Swarndeep S., California University of Pennsylvania
Gluhovsky, Alexander, Purdue University
Goddard, Lisa M., Columbia University
Godfrey, Chris, University of North Carolina at Asheville
Goldblatt, Colin, University of Victoria
Goodman, Paul, University of Arizona
Goodrich, Greg, Western Kentucky University
Gordon, Mark, York University
Gordon, Rob J., University of Guelph
Gorka, Maciej, University of Wroclaw
Grace, John, Edinburgh University
Granger, Darryl E., Purdue University
Graves, Charles E., Saint Louis University
Grenfell, Thomas C., University of Washington
Griffis, Timothy J., University of Minnesota, Twin Cities
Grise, Kevin M., University of Virginia
Grotjahn, Richard , University of California, Davis
Guinan, Patrick E., University of Missouri, Columbia
Gutowski, William J., Iowa State University of Science & Tech
Gutzler, David J., University of New Mexico
Hage, Keith D., University of Alberta
Hakim, Gregory J., University of Washington
Hallar, Anna G., University of Utah
Hameed, Sultan, SUNY, Stony Brook
Hansen, William, Pace University, New York Campus
Hanson, Howard, Florida Atlantic University
Harnik, Nili, Tel Aviv University
Harrison, D. Edmunds, University of Washington
Harrison, Halstead, University of Washington

Harrison, Roy M., University of Birmingham
Harshvardhan, Purdue University
Hart, Richard L., Argonne National Laboratory
Hartmann, Dennis L., University of Washington
Hastenrath, Stefan L., University of Wisconsin, Madison
Hastings, Meredith, Brown University
Haszeldine, R. S., Edinburgh University
Hauser, Rolland K., California State University, Chico
Hawkins, Timothy W., Shippensburg University
Hayden, Bruce P., University of Virginia
Hayes, Michael J., University of Nebraska, Lincoln
Heald, Colette, Massachusetts Institute of Tech
Hegerl, Gabriele C., Edinburgh University
Hegg, Dean A., University of Washington
Heikes, Brian G., University of Rhode Island
Helsdon, John H., South Dakota School of Mines & Tech
Hence, Deanna, University of Illinois, Urbana-Champaign
Henderson, Gina R., United States Naval Academy
Hennon, Chris, University of North Carolina at Asheville
Henry, James A., Middle Tennessee State University
Hickmon, Nicki, Argonne National Laboratory
Hindman, Edward E., City College (CUNY)
Hirschboeck, Katherine K., University of Arizona
Hitchman, Matthew H., University of Wisconsin, Madison
Hjelmfelt, Mark R., South Dakota School of Mines & Tech
Holberg, Jay B., University of Arizona
Holgood, Jay S., Ohio State University
Holloway, Tracey, University of Wisconsin, Madison
Holzer, Mark, University of British Columbia
Horel, John, University of Utah
Horton, Daniel, Northwestern University
Houston, Adam L., University of Nebraska, Lincoln
Houze, Robert A., University of Washington
Hoyos, Carlos, Georgia Institute of Tech
Hu, Qi S., University of Nebraska, Lincoln
Hu, Qi (Steve), University of Nebraska, Lincoln
Huang, Alex, University of North Carolina at Asheville
Hubbard, Kenneth G., University of Nebraska, Lincoln
Huey, Gregory L., Georgia Institute of Tech
Hugli, Wilbur G., University of West Florida
Hunten, Donald M., University of Arizona
Hutyra, Lucy, Boston University
Idone, Vincent P., SUNY, Albany
Ingersoll, Andrew P., California Institute of Tech
Jablonowski, Christiane, University of Michigan
Jackson, Andrea, University of Leeds
Jacob, Daniel J., Harvard University
Jaegle, Lyatt, University of Washington
Jaffe, Daniel A., University of Washington
Jain, Atul K., University of Illinois, Urbana-Champaign
Jason, Shafer, Lyndon State College
Jenkins, Mary Ann, York University
Jewett, Brian, University of Illinois, Urbana-Champaign
Johnson, Donald R., University of Wisconsin, Madison
Jones, Eric M., Los Alamos National Laboratory
Jones, Hazel, University of Manchester
Jones, Phillip D., University of East Anglia
Kang, Song-Lak, Texas Tech University
Kao, Jim C., Los Alamos National Laboratory
Kasting, James F., Pennsylvania State University, University Park
Kauffman, Chad, California University of Pennsylvania
Keables, Michael J., University of Denver
Keeling, Ralph F., University of California, San Diego
Keene, William, University of Virginia
Keesee, Robert G., SUNY, Albany
Keller, Linda M., University of Wisconsin, Madison
Keller, Jr., C. F., Los Alamos National Laboratory
Kennel, Charles F., University of California, San Diego
Keyser, Daniel, SUNY, Albany
Khairoutdinov, Marat, SUNY, Stony Brook
Kieu, Chanh Q., Indiana University, Bloomington
Kift, Richard, University of Manchester
Kim, Hyemi, Georgia Institute of Tech
Kim, Saewung, University of California, Irvine
King, Kenneth M., University of Guelph
King, Martin, University of London, Royal Holloway & Bedford New College
Kinter, Jim, George Mason University
Kirkpatrick, Cody, Indiana University, Bloomington
Kirlin, R. Lynn, University of British Columbia
Kirshbaum, Daniel, McGill University
Kishcha, Pavel, Tel Aviv University
Klaassen, Gary P., York University
Kliche, Donna V., South Dakota School of Mines & Tech
Knapp, Warren, Cornell University
Knight, David, SUNY, Albany
Knopf, Daniel A., SUNY, Stony Brook
Koehler, Thomas , United States Air Force Academy
Konigsberg, Alvin S., SUNY, New Paltz
Korty, Robert, Texas A&M University
Kossin, James, University of Wisconsin, Madison
Kota, Jozsef, University of Arizona
Kotamarthi, Rao, Argonne National Laboratory
Kreidenweis, Sonia M., Colorado State University
Krueger, Steven, University of Utah
Kuang, Zhiming, Harvard University
Kubas, Gregory J., Los Alamos National Laboratory
Kucharik, Chris, University of Wisconsin, Madison
Kummerow, Christian D., Texas A&M University
Kung, Ernest C., University of Missouri, Columbia
Kursinski, Robert, University of Arizona
Kushnir, Yochanan, Columbia University
Kutzbach, John E., University of Wisconsin, Madison
L'Ecuyer, Tristan S., University of Wisconsin, Madison
Lackmann, Gary M., North Carolina State University
Laird, Neil, Hobart & William Smith Colleges
Laprise, Rene, Universite du Quebec a Montreal
Lasher-Trapp, Sonia, University of Illinois, Urbana-Champaign
Lazarus, Steven, Florida Institute of Tech
Lazarus, Steven M., Florida Institute of Tech
Leather, Kimberly, University of Manchester
Lee, In Young, Argonne National Laboratory
Lee, Meehye, Korea University
Leighton, Henry G., McGill University
Lemmon, Mark, Texas A&M University
Leppert, Ken, University of Louisiana, Monroe
Lew, Jeffrey, University of California, Los Angeles
Li, Tim, University of Hawai'i, Manoa
Li, Xiangshan, University of Houston
Lichtenberger, János, Eotvos Lorand University
Light, Bonnie, University of Washington
Lin, Gong-yuh, California State University, Northridge
Lin, Hai, McGill University
Lin, Jialin, Ohio State University
Lin, John C., University of Utah
Lin, Wuyin, SUNY, Stony Brook
Lindgren, Paula, University of Glasgow
Liou, Kuo Nan, University of California, Los Angeles
Liu, Chuntao, Texas A&M University, Corpus Christi
Liu, Dantong, University of Manchester
Liu, Jiping, Georgia Institute of Tech
Liu, Ping, SUNY, Stony Brook
Lowe, Douglas, University of Manchester
Lozowski, Edward P., University of Alberta
Lundquist, Jessica D., University of Washington
Luo, Chao, Georgia Institute of Tech
Lupo, Anthony R., University of Missouri, Columbia
Lynch, Amanda H., Brown University
Lyons, Lawrence, University of California, Los Angeles
Lyons, Walter A., University of Northern Colorado
Maasch, Kirk A., University of Maine
Mace, Gerald, University of Utah
Magee, Nathan, College of New Jersey
Magnusdottir, Gudrun, University of California, Irvine
Mahmood, Rezaul, Western Kentucky University
Mahowald, Natalie M., Cornell University
Mak, John E., SUNY, Stony Brook
Mak, Mankin, University of Illinois, Urbana-Champaign
Malcolm, Elizabeth, Virginia Wesleyan College
Maloney, Eric, Colorado State University
Manabe, Syukuro, Princeton University
Mann, Michael E., Pennsylvania State University, University Park
Manning, Andrew, University of East Anglia
Mantua, Nathan J., University of Washington

Marchand, Roger T., University of Washington
Mark, Bryan G., Ohio State University
Mark, Tucker, Lyndon State College
Market, Patrick S., University of Missouri, Columbia
Marley, Nancy A., Argonne National Laboratory
Marsham, John, University of Leeds
Martin, Randal S., New Mexico Institute of Mining and Tech
Martin, Timothy J., Argonne National Laboratory
Martin, Walter, University of North Carolina, Charlotte
Mason, Allen S., Los Alamos National Laboratory
Mass, Clifford F., University of Washington
Matano, Ricardo, Oregon State University
Mauzerall, Denise L., Princeton University
Mawalagedara, Rachindra, Iowa State University of Science & Tech
Maykut, Gary A., University of Washington
Mayor, Shane D., California State University, Chico
Mazzocco, Elizabeth, Ashland University
McCollor, Doug, University of British Columbia
McElroy, Michael B., Harvard University
McElroy, Tom, York University
McFarlane, Norman, University of Victoria
McFarquhar, Greg M., University of Illinois, Urbana-Champaign
McFiggans, Gordon, University of Manchester
McGregor, Kent M., University of North Texas
McMurdie, Lynn A., University of Washington
McQuaid, Jim, University of Leeds
McWilliams, James C., University of California, Los Angeles
Mechoso, Carlos R., University of California, Los Angeles
Merlis, Timothy, McGill University
Meskhidze, Nicholas, North Carolina State University
Metz, Nicholas, Hobart & William Smith Colleges
Meyer, Steven J., University of Wisconsin, Green Bay
Michaels, Patrick J., University of Virginia
Miller, Doug, University of North Carolina at Asheville
Miller, Ronald L., Columbia University
Millet, Dylan B., University of Minnesota, Twin Cities
Mitchell, Jonathan, University of California, Los Angeles
Mobbs, Stephen, University of Leeds
Modzelewski, Henryk, University of British Columbia
Molina, Mario J., University of California, San Diego
Molinari, John E., SUNY, Albany
Monahan, Edward C., University of Connecticut
Moncrieff, John B., Edinburgh University
Monteverdi, John P., San Francisco State University
Moody, Jennie L., University of Virginia
Morgan, William, University of Manchester
Mosley-Thompson, Ellen E., Ohio State University
Mote, Philip, Oregon State University
Mote, Philip W., University of Washington
Moyer, Elisabeth, University of Chicago
Moyer, Kerry A., Edinboro University of Pennsylvania
Mroz, Eugene J., Los Alamos National Laboratory
Mudrick, Stephen E., University of Missouri, Columbia
Mullen, Steven L., University of Arizona
Murphy, Todd, University of Louisiana, Monroe
Murrell, Coling, University of East Anglia
Murthy, Prahlad N., Wilkes University
Nakamura, Noboru, University of Chicago
Nappo, Carmen, Georgia Institute of Tech
Nathan, Terrence R., University of California, Davis
Neelin, J. David, University of California, Los Angeles
Nenes, Athanasios, Georgia Institute of Tech
Nesbitt, Stephen W., University of Illinois, Urbana-Champaign
Newman, Steven B., Central Connecticut State University
Nielsen-Gammon, John, Texas A&M University
Niyogi, Dev, Purdue University
Nogueira, Ricardo, Georgia State University
Norris, Joel R., University of California, San Diego
North, Gerald, Texas A&M University
O'Gorman, Paul, Massachusetts Institute of Tech
Obrist, Daniel, University of Massachusetts Lowell
Oechel, Walter, The Open University
Oglesby, Robert, University of Nebraska, Lincoln
Oglesby, Robert J., University of Nebraska, Lincoln
Ojala, Carl F., Eastern Michigan University
Okalebo, Jane, University of Nebraska, Lincoln
Okumura, Yuko, University of Texas at Austin
Oppenheimer, Michael, Princeton University
Orlove, Benjamin S., Columbia University
Orville, Richard E., Texas A&M University
Osborn, Timothy J., University of East Anglia
Osterberg, Erich C., Dartmouth College
Overland, James E., University of Washington
Overpeck, Jonathan T., University of Arizona
Palmer, Paul, Edinburgh University
Palter, Jaime, McGill University
Pan, Zaitao, Saint Louis University
Pani, Eric A., University of Louisiana, Monroe
Patoux, Jerome, University of Washington
Paulson, Suzanne, University of California, Los Angeles
Pegion, Kathy, George Mason University
Percell, Peter, University of Houston
Percival, Carl, University of Manchester
Perez, Richard R., SUNY, Albany
Perry, Kevin D., University of Utah
Peterson, Richard E., Texas Tech University
Petters, Markus, North Carolina State University
Petty, Grant W., University of Wisconsin, Madison
Pierce, Jeff, Colorado State University
Pierrehumbert, Raymond T., University of Chicago
Polzin, Dierk T., University of Wisconsin, Madison
Powell, Mark L., Los Angeles Pierce College
Prather, Kimberly A., University of California, San Diego
Prather, Michael J., University of California, Irvine
Preston, Thomas C., McGill University
Previdi, Michael, Columbia University
Price, Colin G., Tel Aviv University
Prinn, Ronald G., Massachusetts Institute of Tech
Pryor, Sara C., Cornell University
Pu, Zhaoxia, University of Utah
Pumphrey, Hugh C., Edinburgh University
Pusede, Sally, University of Virginia
Radke, Lawrence F., University of Washington
Raman, Sethu S., North Carolina State University
Ramanathan, V., University of California, San Diego
Ramsey, Kelvin W., University of Delaware
Randall, David A., Colorado State University
Rappenglueck, Bernhard, University of Houston
Rasmussen, Kristen L., Colorado State University
Rauber, Robert M., University of Illinois, Urbana-Champaign
Ravishankara, A.R., Colorado State University
Rawlins, Michael, University of Massachusetts, Amherst
Ray, Pallav, Florida Institute of Tech
Rayner, John N., Ohio State University
Reay, David, Edinburgh University
Reichler, Thomas, University of Utah
Reisner, Jon M., Los Alamos National Laboratory
Reiss, Nathan, Pace University, New York Campus
Reuter, Gerhard W., University of Alberta
Revelle, Douglas O., Los Alamos National Laboratory
Richter, David, University of Notre Dame
Ricketts, Hugo, University of Manchester
Riemer, Nicole, University of Illinois, Urbana-Champaign
Rind, David H., Columbia University
Riordan, Allen J., North Carolina State University
Ritsche, Michael, Argonne National Laboratory
Robertson, Andrew W., Columbia University
Robinson, Walter, North Carolina State University
Robinson, Walter A., University of Illinois, Urbana-Champaign
Roe, Gerard H., University of Washington
Rogers, Jeffery C., Ohio State University
Ross, Robert S., Millersville University
Rowe, Clinton M., University of Nebraska, Lincoln
Rupp, David, Oregon State University
Russell, Armistead G., Georgia Institute of Tech
Russell, Lynn M., University of California, San Diego
Rutledge, Steven A., Colorado State University
Saltzman, Eric S., University of California, Irvine
Sandel, Bill R., University of Arizona
Sarachik, Edward S., University of Washington
Saravanan, R., Texas A&M University
Schade, Gunnar, Texas A&M University
Schlesinger, Michael E., University of Illinois, Urbana-Champaign
Schlosser, C. Adam, Massachusetts Institute of Tech

Schmittner, Andreas, Oregon State University
Schneider, Edwin K., George Mason University
Schneider, Tapio, California Institute of Tech
Schroeder, John L., Texas Tech University
Schumacher, Courtney, Texas A&M University
Schumacher, Russ, Colorado State University
Schwab, James J., SUNY, Albany
Seeley, Mark W., University of Minnesota, Twin Cities
Selin, Noelle, Massachusetts Institute of Tech
Semazzi, Fred H. M., North Carolina State University
Serreze, Mark, University of Colorado
Shaw, Tiffany A., Columbia University
Shell, Karen M., Oregon State University
Shellito, Lucinda, University of Northern Colorado
Shen, Samuel S., University of Alberta
Shen, Xinhua, University of Northern Iowa
Shepherd, Gordon G., York University
Shepherd, John G., University of Southampton
Shepherd, Mark A., Austin Community College District
Shepson, Paul B., Purdue University
Sherman-Morris, Kathleen M., Mississippi State University
Shinoda, Toshiaki, Texas A&M University, Corpus Christi
Shirley, Terry, University of North Carolina, Charlotte
Shukla, Jagadish, George Mason University
Shulski, Martha D., University of Nebraska, Lincoln
Sikora, Todd D., Millersville University
Simpson, Robert M., University of Tennessee, Martin
Sisterson, Doug, Argonne National Laboratory
Sisterson, Douglas L., Argonne National Laboratory
Smedley, Andrew, University of Manchester
Smith, Paul L., South Dakota School of Mines & Tech
Smith, Phillip J., Purdue University
Smith, Ronald B., Yale University
Snodgrass, Eric R., University of Illinois, Urbana-Champaign
Snow, Julie A., Slippery Rock University
Snyder, Chris, Texas A&M University
Snyder, Peter K., University of Minnesota, Twin Cities
Snyder, Richard L., University of California, Davis
Sobel, Adam, Columbia University
Sobel, Adam H., Columbia University
Sokolik, Irina, Georgia Institute of Tech
Solomon, Susan, Massachusetts Institute of Tech
Somerville, Richard C., University of California, San Diego
Soule, Peter T., Appalachian State University
Srivastava, Ramesh C., University of Chicago
Sriver, Ryan, University of Illinois, Urbana-Champaign
St. John, James C., Georgia Institute of Tech
Stan, Cristiana, George Mason University
Staten, Paul W., Indiana University, Bloomington
Steenburgh, Jim, University of Utah
Stevens, Philip S., Indiana University, Bloomington
Stevenson, David, Edinburgh University
Steyn, Douw G., University of British Columbia
Stickel, Robert, Georgia Institute of Tech
Stoelinga, Mark T., University of Washington
Stone, Peter H., Massachusetts Institute of Tech
Stramler, Kirstie L., City College of San Francisco
Straub, Derek J., Susquehanna University
Straus, David M., George Mason University
Strobel, Darrell F., Johns Hopkins University
Strong, Courtenay, University of Utah
Strub, Ted P., Oregon State University
Stull, Roland B., University of British Columbia
Sturges, Bill, University of East Anglia
Stutz, Jochen P., University of California, Los Angeles
Sun, Wen-Yih, Purdue University
Suyker, Andrew E., University of Nebraska, Lincoln
Svoma, Bohumil, University of Missouri, Columbia
Swann, Abigail L., University of Washington
Szunyogh, Istvan, Texas A&M University
Tadesse, Tsegaye, University of Nebraska, Lincoln
Takle, Eugene S., Iowa State University of Science & Tech
Talley, John H., University of Delaware
Tao, Wei-Kuo, Texas A&M University
Tatarskii, Viatcheslav, Georgia Institute of Tech
Taylor, Elwynn, Iowa State University of Science & Tech
Taylor, Gregory R., California State University, Chico
Taylor, Peter A., York University
Teeuw, Richard, University of Portsmouth
Thorncroft, Christopher D., SUNY, Albany
Thorne, Richard M., University of California, Los Angeles
Thornton, Joel A., University of Washington
Thurtell, George W., University of Guelph
Tillman, James E., University of Washington
Ting, Mingfang, Columbia University
Tomasko, Martin G., University of Arizona
Tongue, Jeffrey, SUNY, Stony Brook
Trapp, Robert J., University of Illinois, Urbana-Champaign
Tung, Ka-Kit, University of Washington
Tung, Wen-wen, Purdue University
Turco, Richard, University of California, Los Angeles
Twine, Tracy E., University of Minnesota, Twin Cities
Van Den Broeke, Matthew S., University of Nebraska, Lincoln
van den Heever, Sue, Colorado State University
Vanos, Jennifer K., University of California, San Diego
Vaughan, Geraint, University of Manchester
Vega, Anthony J., Clarion University
Veres, Michael, SUNY, Oswego
Vimont, Daniel J., University of Wisconsin, Madison
Vincent, Dayton G., Purdue University
Vogelmann, Andrew, SUNY, Stony Brook
von Salzen, Knut, University of Victoria
von Salzen, Knut, University of British Columbia
Vong, Richard J., Oregon State University
Vukovich, George, York University
Wagner, Timothy J., Creighton University
Walcek, Christopher J., SUNY, Albany
Waliser, Duane E., SUNY, Stony Brook
Wallace, John M., University of Washington
Walsh, John E., University of Illinois, Urbana-Champaign
Walter-Shea, Elizabeth A., University of Nebraska, Lincoln
Wang, Chien, Massachusetts Institute of Tech
Wang, Hsiang-Jui, Georgia Institute of Tech
Wang, Jian, SUNY, Stony Brook
Wang, Pao-Kuan, University of Wisconsin, Madison
Wang, Wei-Chyung, SUNY, Albany
Wang, Yuhang, Georgia Institute of Tech
Wang, Yuqing, University of Hawai'i, Manoa
Wang, Yuxuan, Texas A&M University
Wang, Yuxuan, University of Houston
Wang, Zhuo, University of Illinois, Urbana-Champaign
Warland, Jon, University of Guelph
Warren, Stephen G., University of Washington
Waugh, Darryn W., Johns Hopkins University
Wax, Charles L., Mississippi State University
Weare, B. C., University of California, Davis
Weaver, Justin E., Texas Tech University
Weber, Rodney J., Georgia Institute of Tech
Webster, Peter J., Georgia Institute of Tech
Weinbeck, Robert S., SUNY, The College at Brockport
Weiss, Christopher C., Texas Tech University
Wennberg, Paul O., California Institute of Tech
Wesely, Marvin L., Argonne National Laboratory
Wettstein, Justin, Oregon State University
Whitaker, Rodney W., Los Alamos National Laboratory
Whitehead, James, University of Manchester
Whiteway, James, York University
Wikle, Christopher K., University of Missouri, Columbia
Wilhelmson, Robert B., University of Illinois, Urbana-Champaign
Wilhite, Donald A., University of Nebraska, Lincoln
Wilkerson, Forrest, Minnesota State University
Wilks, Daniel, Cornell University
Williams, John M., SUNY, The College at Brockport
Williams, Kaj, Montana State University
Williams, Kay R., Shippensburg University
Williams, Mathew, Edinburgh University
Williams, Paul, University of Manchester
Willoughby, Hugh E., Florida International University
Wilson, John D., University of Alberta
Wine, Paul H., Georgia Institute of Tech
Wofsy, Steven C., Harvard University
Wood, Kim, Mississippi State University
Wood, Robert, University of Washington
Wu, Shiliang, Michigan Technological University

Wu, Xiaoqing, Iowa State University of Science & Tech
Wu, Yutian, Purdue University
Wuebbles, Donald J., University of Illinois, Urbana-Champaign
Wysocki, Mark, Cornell University
Xie, Feiqin, Texas A&M University, Corpus Christi
Xie, Lian, North Carolina State University
Yalda, Sepideh, Millersville University
Yang, Ping, Texas A&M University
Yarger, Douglas N., Iowa State University of Science & Tech
Yau, Man Kong, McGill University
Yi, Chuixiang, Queens College (CUNY)
Young, John A., University of Wisconsin, Madison
Yu, Jin-Yi, University of California, Irvine
Yung, Yuk L., California Institute of Tech
Yuter, Sandra, North Carolina State University
Yuter, Sandra E., University of Washington
Zaitchik, Benjamin, Johns Hopkins University
Zawadzki, Isztar I., McGill University
Zehnder, Joseph A., Creighton University
Zender, Charles, University of California, Irvine
Zeng, Ning, University of Maryland
Zhang, Henian, Georgia Institute of Tech
Zhang, Minghua, SUNY, Stony Brook
Zhang, Renyi, Texas A&M University
Zhang, Xi, University of California, Santa Cruz
Zhang, Yang, North Carolina State University
Zhu, Ping, Florida International University
Zhuang, Qianlai, Purdue University
Zipser, Edward, University of Utah
Zuend, Andreas, McGill University

## Earth Science Education

Albach, Suzanne M., Washtenaw Community College
Arthurs, Leilani A., University of Nebraska, Lincoln
Attrep, Moses, Los Alamos National Laboratory
Aylward, Linda, Illinois Central College
Baker, Brett, University of Texas at Austin
Barnett, Michael, Boston College
Bechtel, Randy, North Carolina Geological Survey
Bednarski, Marsha, Central Connecticut State University
Bednarz, Robert S., Texas A&M University
Bednarz, Sarah W., Texas A&M University
Bergwerff, Kenneth A., Calvin College
Bitting, Kelsey S., University of Kansas
Black, Alice (Jill), Missouri State University
Blankenbicker, Adam, Smithsonian Institution / National Museum of Natural History
Buchanan, Rex C., University of Kansas
Burns, Sandra, Central Connecticut State University
Busch, Richard M., West Chester University
Butcher, Patricia M., California State University, Fullerton
Carpenter, John R., University of South Carolina
Casto, Pamela, Fairmont State University
Cercone, Karen Rose, Indiana University of Pennsylvania
Chandler, Sandra, Arkansas Geological Survey
Clark, Scott K., University of Wisconsin, Eau Claire
Clary, Renee M., Mississippi State University
Clepper, Marta, SUNY, Oneonta
Coburn, Daniel, Southern Connecticut State University
Conners, John S., Austin Community College District
Conway, Michael F., University of Arizona
Cooney, Timothy M., University of Northern Iowa
Crowder, Margaret, Western Kentucky University
Cunningham, Amy J., Austin Community College District
d'Alessio, Matthew, California State University, Northridge
Dakin, Susan, University of Lethbridge
Davies, Gareth, The Open University
Davies, Sarah, The Open University
De Kock, Walter E., University of Northern Iowa
de Ritter, Samantha, Flinders University
Dehne, Kevin T., Delta College
DeKraker, Dan, Cerritos College
Denton-Hedrick, Meredith Y., Austin Community College District
DePriest, Thomas A., University of Tennessee, Martin
Dieckmann, Melissa S., Eastern Kentucky University
Eldredge, Sandra N., Utah Geological Survey
Elkins, Joe T., University of Northern Colorado
Emerson, Cheryl R., Illinois Central College
Engels, Jennifer, University of Hawai'i, Manoa
Englebright, Stephen C., SUNY, Stony Brook
Ensign, Todd, Fairmont State University
Ferguson, Julie , University of California, Irvine
Flanagan, Timothy, Berkshire Community College
Fleck, Michelle, Utah State University
Ford, Jaime L., Fairmont State University
Frankic, Anamarija, University of Massachusetts, Boston
Geraghty Ward, Emily, Rocky Mountain College
Gersmehl, Carol, Hunter College (CUNY)
Godsey, Holly, University of Utah
Goodell, Laurel P., Princeton University
Gray, Kyle R., University of Northern Iowa
Hamane, Angel, California State University, Los Angeles
Havholm, Karen G., University of Wisconsin, Eau Claire
Heid, Kelly L., Grand Valley State University
Heine, Sallie, Augustana College
Helgers, Karen, SUNY, Ulster County Community College
Hemler, Deb, Fairmont State University
Herman, Theodore C., West Valley College
Holliman, Richard, The Open University
Hubenthal, Michael, Binghamton University
Hunter, Arlene , The Open University
Jahanyar, Alireza, University of Tabriz
Johnston, Thomas, University of Lethbridge
King-Rundel, Judith A., California State University, Dominguez Hills
Kornreich Wolf, Susan, Augustana College
Kortz, Karen, Community College of Rhode Island
Krause, Lois B., Clemson University
Krockover, Gerald H., Purdue University
Kusnick, Judith E., California State University, Sacramento
LaDue, Nicole D., Northern Illinois University
Libarkin, Julie C., Michigan State University
Ludwikoski, David J., Community College of Baltimore County, Catonsville
MacLachlan, Ian R., University of Lethbridge
Martin-Vermilyea, Laurie, Montgomery County Community College
Mathison, Mark E., Iowa State University of Science & Tech
Mattox, Stephen R., Grand Valley State University
McCutcheon, Karen Sue, Casper College
McEvoy, Jamie , Montana State University
McGeary, Susan, University of Delaware
McMullin, David W., Acadia University
McNeal, Karen S., Auburn University
Messina, Paula, San Jose State University
Morgan, Emory, Hampton University
Morton, Allan E., Central Arizona College
Mulkey, Sean, Illinois Central College
Munski, Douglas C., University of North Dakota
Nakagawa, Masami, Colorado School of Mines
Norris-Tull, Delena, University of Montana Western
Papcun, George, New Jersey City University
Pearson, David A., Laurentian University, Sudbury
Pepple, Christopher, Bowling Green State University
Petcovic, Heather L., Western Michigan University
Petit, Martin A., Illinois Central College
Phillips, Paul, Fort Hays State University
Preston, William L., California Polytechnic State University
Pyle, Eric J., James Madison University
Rakonczai, Janos, Univesity of Szeged
Raol, Marcie, Fairmont State University
Refenes, James L., Concordia University
Schaffer, Linda J., SUNY, The College at Brockport
Schmitt, Danielle M., Princeton University
Scott, Vernon, Oklahoma State University
Semken, Steven, Arizona State University
Sharp, S. L., Virginia Polytechnic Institute & State University
Shelton, Dale W., Maryland Department of Natural Resources
Shepardson, Daniel P., Purdue University
Shipp, Stephanie S., Rice University
Slattery, William, Wright State University
Smay, Jessica J., San Jose City College
Smith, Arthur, West Chester University
Spurr, Aaron, University of Northern Iowa
Stevens, Michael, University of Northern Iowa
Sundareshwar, P. V., South Dakota School of Mines & Tech
Sutton, Connie J., Indiana University of Pennsylvania

Taylor, Ta-Shana A., University of Miami
Teed, Rebecca, Wright State University
Thompson, Kenneth W., Emporia State University
Tonon, Marco Davide, Università di Torino
Viskupic, Karen, Boise State University
Wagner, John R., Clemson University
Weaver Bowman, Kristin, California State University, Fullerton
White, Lisa D., University of California, Berkeley
Wood, Howard, Texas A&M University, Corpus Christi
Wright, Carrie L., University of Southern Indiana
Young, James E., Appalachian State University
Zipper, Carl E., Virginia Polytechnic Institute & State University
Zlotkin, Howard, New Jersey City University
Zwick, Thomas T., Montana State University, Billings

## Physical Geography

Aay, Henry, Calvin College
Abend, Martin, New Jersey City University
Aber, Jeremy, Middle Tennessee State University
Adam, Iddrisu, University of Wisconsin Colleges
Ajassa, Roberto, Università di Torino
Amador, Nathanael S., Ohio Wesleyan University
Aubert, John E., American River College
Balascio, Nicholas, College of William & Mary
Ballantyne, Colin K., University of St. Andrews
Barnett, Roger T., University of the Pacific
Barton, James H., Thiel College
Bascom, Johnathan, Calvin College
Bass, David, Flinders University
Bates, Mary, College of the Canyons
Battles, Eileen, University of Kansas
Beeton, Jared M., Adams State University
Bein, F. L., Indiana University, Indianapolis
Bekker, Matthew F., Brigham Young University
Bereitschaft, Bradley, University of Nebraska at Omaha
Berry, Kate, University of Nevada, Reno
Biggs, Thomas H., University of Virginia
Bird, Jeffry S., Brigham Young University
Bitner, Thomas, University of Wisconsin Colleges
Black, Kathryn, Montclair State University
Blackburn, William, Western Kentucky University
Blewett, William L., Shippensburg University
Boester, Michael, Monroe Community College
Brey, James, University of Wisconsin Colleges
Brothen, Jerry, El Camino College
Brothers, Timothy S., Indiana University / Purdue University, Indianapolis
Brothers, Timothy S., Indiana University, Indianapolis
Butzow, Dean G., Lincoln Land Community College
Cairns, David M., Texas A&M University
Camille, Michael A., University of Louisiana, Monroe
Campbell, Glenn A., Eastern Kentucky University
Carter, Greg, University of Southern Mississippi
Chacko, Elizabeth, George Washington University
Chaddock, Lisa, Cuyamaca College
Chaney, Phil L., Auburn University
Chatterjee, Meera, University of Akron
Clarke, Beverley, Flinders University
Clennan, Patrick D., College of Southern Nevada - West Charleston Campus
Coburn, Craig, University of Lethbridge
Cohen, Saul B., Graduate School of the City University of New York
Comrie, Andrew, University of Arizona
Coveney, Eamonn, Fort Hays State University
Croft, Gary M., San Bernardino Valley College
Csiki, Shane, New Hampshire Geological Survey
Curry, Janel M., Calvin College
Dawsey, Cyrus B., Auburn University
de Uña Alvarez, Elena P., Coruna University
Deckert, Garret, University of Wisconsin Colleges
DeWeese, Georgina G., University of West Georgia
Dexter, Leland R., Northern Arizona University
Dommissee, Edwin, University of Wisconsin Colleges
Dubsky, Scott , United States Air Force Academy
Duncan, Ian, City College of San Francisco
Ebiner, Matt, El Camino College
Eichenbaum, Jack, Hunter College (CUNY)
Eisenhart, Karen, Edinboro University of Pennsylvania
Engstrom, Vanessa, San Bernardino Valley College
Foltltin, Thomas, George Washington University
Fonstad, Mark, University of Oregon
Fratianni, Simona, Università di Torino
Friend, Donald A., Minnesota State University
Frolking, Tod A., Denison University
Fuller, Doult O., George Washington University
Gamble, Douglas W., University of North Carolina Wilmington
Gaubatz, Piper, University of Massachusetts, Amherst
Gavin, Dan, University of Oregon
Gentry, Christopher, Austin Peay State University
Gersmehl, Philip, Hunter College (CUNY)
Girhard, T S., San Antonio Community College
Girhard, Thomas S., Alamo Colleges - San Antonio College
Goudge, Theodore L., Northwest Missouri State University
Graham, Barbara, College of Southern Nevada - West Charleston Campus
Greene, Don M., Baylor University
Hall, Luke D., Ventura College
Hampson, Arthur, University of Utah
Harley, Grant, University of Southern Mississippi
Harrison, Robert S., Long Island University, C.W. Post Campus
Hart, Evan A., Tennessee Tech University
Harty, John P., Johnson County Community College
Hay, Iain, Flinders University
Heibel, Todd, San Bernardino Valley College
Hepner, George F., University of Utah
Hess, Darrel E., City College of San Francisco
Heywood, Neil C., University of Wisconsin, Stevens Point
Hickcox, David H., Ohio Wesleyan University
Holt, David, University of Southern Mississippi
Houston, Serin D., Mount Holyoke College
Howard, Leslie M., Unversity of Nebraska - Lincoln
Ibrahim, Mohamed B., Graduate School of the City University of New York
Jackson, Philip L., Oregon State University
Jain, Cathy, Palomar College
Jarvis, Richard S., University of Texas, El Paso
Johnson, Mark O., Worcester State University
Jordan, Karen J., University of South Alabama
Jurmu, Michael, University of Wisconsin Colleges
Karpilo, Jr, Ronald J., Colorado State University
Kebbede, Girma, Mount Holyoke College
Keen-Zebert, Amanda, Desert Research Institute
Kennedy, Linda, Mansfield University
Key, Doug, Palomar College
Khan, Belayet H., Eastern Illinois University
Kiage, Lawrence M., Georgia State University
Kilcoyne, John R., Metropolitan State College of Denver
Kimber, Clarissa T., Texas A&M University
Kung, Hsiang-Te, University of Memphis
Laity, Julie E., California State University, Northridge
Lambert, Dean, Alamo Colleges - San Antonio College
Landon, Patrick J., Wassuk College
League, Larry D., Dickinson State University
Lee, Jeffrey A., Texas Tech University
Lee, Russell, Oak Ridge National Laboratory
Lineback, Neal G., Appalachian State University
Little, Jonathan, Monroe Community College
LoVetere, Crystal, Cerritos College
Lynn, Resler M., Virginia Polytechnic Institute & State University
MacQuarrie, Pamela, Mount Royal University
Maher, John, Johnson County Community College
Maio, Chris, University of Alaska, Fairbanks
Mann, Dan, University of Alaska, Fairbanks
Manos, Leah D., Northwest Missouri State University
Markwith, Scott H., Florida Atlantic University
Marzulli, Walter, New Jersey Geological and Water Survey
Mathenge, Christine, Austin Peay State University
McCallister, Robert, University of Wisconsin Colleges
McCoy, William D., University of Massachusetts, Amherst
McCraw, David J., New Mexico Institute of Mining & Tech
McGrath, Jr., Dorn C., George Washington University
McGwire, Kenneth C., Desert Research Institute
Mensing, Scott A., University of Nevada, Reno
Millette, Thomas L., University of Massachusetts, Amherst
Minnich, Richard A., University of California, Riverside
Montgomery, Keith, University of Wisconsin Colleges
Montoya, Judith, Northern Arizona University

Motta, Luigi, Università di Torino
Motta, Michele, Università di Torino
Musolf, Gene E., University of Wisconsin Colleges
Nelson, Kenneth A., University of Kansas
Ngoy, Kikombo, Kean University
Nicholas, Joseph W., Mary Washington College
Norwood, James, Auburn University
Nuwategeka, Expedito, Gulu University
Ogbuchiekwe, Edmund Jekwu, San Bernardino Valley College
Oldfield, Jonathan, University of Birmingham
Olson, Kimberly, American River College
Paulsell, Robert L., Louisiana State University
Peake, Jeffrey S., University of Nebraska at Omaha
Pesses, Michael, Antelope Valley College
Phillips, Nathan, Boston University
Pierce, Heather, Monroe Community College
Pinnt, Todd, Arizona Western College
Precht, Francis L., Frostburg State University
Price, Alan P., University of Wisconsin Colleges
Price, Marie D., George Washington University
Privette, David, Central Piedmont Community College
Rahman, Abu, Antelope Valley College
Reese, Andy, University of Southern Mississippi
Rettig, Andrew, University of Dayton
Robinson, Michael A., Santa Barbara City College
Rodgers, John C., Mississippi State University
Rohe, Randall, University of Wisconsin Colleges
Rollins, Kyle, Missouri Dept of Natural Resources
Rose, I. S., University of West Georgia
Ross, Thomas E., University of North Carolina, Pembroke
Rudnicki, Ryan E., Alamo Colleges - San Antonio College
Ryan, Christopher, University of Nevada, Reno
Sandlin, Stephen H., San Bernardino Valley College
Santelmann, Mary V., Oregon State University
Schafer, Tom, Fort Hays State University
Schmidt, Lisa, San Bernardino Valley College
Schulp, Nynke , VU University Amsterdam
Scott, Larry, Colorado School of Mines
Scott, Thomas A., University of California, Riverside
Sharkey, Debra, Cosumnes River College
Smith, C. K., San Antonio Community College
Smith, Charles K., Alamo Colleges - San Antonio College
Smith, Steven C., American River College
Spencer, Jeremy M., University of Akron
Stahmann, Paul, McHenry County College
Starrs, Paul F., University of Nevada, Reno
Stern, Herschel I., MiraCosta College
Stevens, Stan, University of Massachusetts, Amherst
Stocks, Lee, Mansfield University
Swarts, Stanley W., Northern Arizona University
Thompson, Wiley C., United States Military Academy
Thomsen, Charles E., American River College
Tingley, Susan L., University of Nevada
Todhunter, Paul, University of North Dakota
Tremblay, Thomas A., University of Texas at Austin, Jackson School of Geosciences
van der Zande, Emma, VU University Amsterdam
van Leeuwen, Willem, University of Arizona
van Teeffelen, Astrid, VU University Amsterdam
van Vliet, Jasper, VU University Amsterdam
Verburg, Peter, VU University Amsterdam
Veverka, Laura, Metropolitan Community College-Kansas City
Vogel, Eve, University of Massachusetts, Amherst
Walter, Thomas, Hunter College (CUNY)
Webb, Byron J., Austin Peay State University
Welford, Mark R., Georgia Southern University
West, Keith, University of Wisconsin Colleges
White, Scott, Fort Lewis College
Wibking, R. Kenton, Austin Peay State University
Wilkie, Richard W., University of Massachusetts, Amherst
Wilkins, David E., Boise State University
Wilson, Jeffey S., Indiana University, Indianapolis
Wu, Shuang-Ye, University of Dayton
Wyckoff, John W., University of Colorado, Denver
Yassin, Barbara E., Mississippi Office of Geology
Ye, Hengchun, California State University, Los Angeles
Zhou, Yuyu, Iowa State University of Science & Tech
Zorba, Molly, Antelope Valley College

## Ocean Engineering/Mining

Ajayi, J. O., Obafemi Awolowo University
Becker, Janet M., University of Hawai'i, Manoa
Carpenter, Roy, University of Washington
Gibson, Carl H., University of California, San Diego
Hodgkiss, Jr., William S., University of California, San Diego
Hoopes, John A., University of Wisconsin, Madison
Irish, Jennifer, Virginia Tech
Reichard, Ronnal, Florida Institute of Tech
Sahoo, Prasanta, Florida Institute of Tech
Swain, Geoffry, Florida Institute of Tech
Weaver, Robert, Florida Institute of Tech

## Remote Sensing

Aanstoos, James V., Iowa State University of Science & Tech
Abdalati, Waleed, University of Colorado
Abrams, Michael J., Jet Propulsion Laboratory
Ackerman, Steven A., University of Wisconsin, Madison
Ahearn, Sean C., Graduate School of the City University of New York
Ali, K. Adem, College of Charleston
Alley, Ronald E., Jet Propulsion Laboratory
Aly, Mohamed H., University of Arkansas, Fayetteville
Applegarth, Michael T., Shippensburg University
Barros, Jose Antonio, Florida International University
Beaverson, Sheena K., University of Illinois, Urbana-Champaign
Becker, Richard H., Argonne National Laboratory
Becker, Richard H., University of Toledo
Berlin, Graydon L., Northern Arizona University
Bernier, Monique, Universite du Quebec
Binford, Micheal W., University of Florida
Blom, Ronald G., Jet Propulsion Laboratory
Boelman, Natalie, Columbia University
Bogucki, Donald J., Plattsburgh State University (SUNY)
Boulton, Sarah, University of Plymouth
Brewer, Micheal, George Washington University
Broadfoot, Lyle A., University of Arizona
Buratti, Bonnie J., Jet Propulsion Laboratory
Cablk, Mary, Desert Research Institute
Callahan, John A., University of Delaware
Calvin, Wendy, University of Nevada, Reno
Campbell, Bruce A., Smithsonian Institution / National Air & Space Museum
Campbell, James B., Virginia Polytechnic Institute & State University
Carn, Simon, Michigan Technological University
Chen, Shu-Hua, University of California, Davis
Chen, Xianfeng, Slippery Rock University
Chopping, Mark J., Montclair State University
Cooke, III, William H., Mississippi State University
Crippen, Robert E., Jet Propulsion Laboratory
Dana, Gayle L., University of Nevada, Reno
Dash, Padmanava, Mississippi State University
Di Girolamo, Larry, University of Illinois, Urbana-Champaign
Dixon, Tim H., University of Miami
Dodge, Rebecca L., Midwestern State University
Dondofema, Farai, University of Venda for Science & Tech
Easson, Gregory L., University of Mississippi
El-Baz, Farouk, Boston University
Fenstermaker, Lynn, Desert Research Institute
Ferris, Michael H., California State University, Dominguez Hills
Fildes, Stephen , Flinders University
Flynn, Luke P., University of Hawai'i, Manoa
Forster, Richard R., University of Utah
Fox, Amelia A., Fort Hays State University
Frankenberg, Christian, California Institute of Tech
Frazin, Richard, University of Michigan
Friedl, Mark, Boston University
Friedl, Mark A., Boston University
Frost, Eric G., San Diego State University
Gaitan-Moran, Javier, Universidad Autonoma de Baja California Sur
Gammack-Clark, James, Florida Atlantic University
Gamon, John, University of Alberta
Gaulton, Rachel, University of Newcastle Upon Tyne
Gaya, Charles , Jomo Kenyatta University of Agriculture & Tech
Ghent, Rebecca, University of Toronto
Gibson, Glen, United States Air Force Academy
Gilbes, Fernando, University of Puerto Rico

Gimmestad, Gary G., Georgia Institute of Tech
Giraldo, Mario A., California State University, Northridge
Glenn, Nancy, Boise State University
Goetz, Alexander, University of Colorado
Gonzalez, Frank, University of Washington
Hacker, Jorg, Flinders University
Hanna, Stephen P., Mary Washington College
Hardin, Perry J., Brigham Young University
Haritashya, Umesh, University of Dayton
Hay, Rodrick, California State University, Dominguez Hills
Hayes, James J., University of Nebraska at Omaha
Heidinger, Andrew, University of Wisconsin, Madison
Hernandez, Michael W., Weber State University
Hestir, Erin, North Carolina State University
Hinojosa, Alejandro, Centro de Investigación Científica y de Educación Superior de Ensenada
Hobbs, Richard D., Amarillo College
Hook, Simon J., Jet Propulsion Laboratory
Hossain, A K M Azad, University of Tennessee at Chattanooga
Hu, Baoxin, York University
Hu, Zhiyong, University of West Florida
Huang, Hung-Lung Allen, Texas A&M University
Huang, Yi, McGill University
Hung, Ming-Chih, Northwest Missouri State University
Hutchinson, Charles F., University of Arizona
Jackson, Mark W., Brigham Young University
Jaworski, Eugene, Eastern Michigan University
Ji, Wei, University of Missouri-Kansas City
Jin, Yufang, University of California, Davis
Johnson, Elias, Missouri State University
Kahle, Anne B., Jet Propulsion Laboratory
Kenduiywo, Benson K., Jomo Kenyatta University of Agriculture & Tech
Kennedy, Robert, Oregon State University
Kern, Anikó, Eotvos Lorand University
Kessler, Fritz, Frostburg State University
Key, Jeffrey R., University of Wisconsin, Madison
Khan, Shuhab D., University of Houston
Klein, Andrew G., Texas A&M University
Klemas, Victor, University of Delaware
Kollias, Pavlos, McGill University
Kudela, Raphael M., University of California, Santa Cruz
Lakhan, V. Chris, University of Windsor
Larson, Eric J., University of Wisconsin, Stevens Point
Lawrence, Rick L., Montana State University
Lee, Keenan, Colorado School of Mines
Lee, Zhongping , University of Massachusetts, Boston
Liu, Jian G., Imperial College
Lu, Zhong, Southern Methodist University
Lulla, Kamlesh, University of Delaware
Lyon, Ronald J. P., Stanford University
Mason, Philippa J., Imperial College
Mbuh, Mbongowo J., University of North Dakota
Mccoy, Roger M., University of Utah
McDade, Ian C., York University
McGrath, Andrew, Flinders University
McGwire, Kenneth, University of Nevada, Reno
Merchant, James W., Unversity of Nebraska - Lincoln
Meyer, Franz J., University of Alaska, Fairbanks
Miao, Xin, Missouri State University
Middleton, Carrie A., Department of Interior Office of Surface Mining Reclamation and Enforcement
Milan-Navarro, Joel, Universidad Autonoma de San Luis Potosi
Miller, John, York University
Miller, Richard, East Carolina University
Millette, Thomas L., Mount Holyoke College
Mills, Jon, University of Newcastle Upon Tyne
Minor, Timothy B., Desert Research Institute
Molnár, Gábor, Eotvos Lorand University
Molnia, Bruce F., Duke University
Morgan, Ken M., Texas Christian University
Muller-Karger, Frank E., University of South Florida
Mushkin, Amit, University of Washington
Mustard, John F., Brown University
Myneni, Ranga , Boston University
Nduati, Eunice W., Jomo Kenyatta University of Agriculture & Tech
Ngigi, Thomas G., Jomo Kenyatta University of Agriculture & Tech
Ni-Meister, Wenge, Hunter College (CUNY)
Njoku, Eni G., Jet Propulsion Laboratory
Nolin, Anne, Oregon State University
Palluconi, Frank D., Jet Propulsion Laboratory
Parrish, Christopher E., University of New Hampshire
Pathirana, Sumith, Southern Cross University
Peddle, Derek R., University of Lethbridge
Peltzer, Gilles, University of California, Los Angeles
Pieters, Carle M., Brown University
Platnick, Steven, University of Wisconsin, Madison
Porter, John H., University of Virginia
Prakash, Anupma, University of Alaska, Fairbanks
Price, Kevin, Kansas State University
Quisenberry, Dan R., Mercer University
Ramage, Joan, Lehigh University
Ramsey, Michael S., University of Pittsburgh
Ramspott, Matthew E., Frostburg State University
Rapp, Anita, Texas A&M University
Reed, Wallace E., University of Virginia
Ridd, Merrill K., University of Utah
Rivard, Benoit, University of Alberta
Roberts, Charles E., Florida Atlantic University
Robinson, Ian S., University of Southampton
Rundquist, Bradley C., University of North Dakota
Rundquist, Donald C., Unversity of Nebraska - Lincoln
Sadiq, Abdulali A., University of Qatar
Sahr, John D., University of Washington
Sanden, Eric M., University of Wisconsin, River Falls
Schaaf, Crystal, University of Massachusetts, Boston
Schaaf, Crystal, Boston University
Schenk, Anton, SUNY, Buffalo
Shi, Xuan, University of Arkansas, Fayetteville
Small, Christopher, Columbia University
Smith, Mike, Kingston University
Smith, Peter, University of Arizona
Stohr, Christopher J., University of Illinois, Urbana-Champaign
Strahler, Alan, Boston University
Strain, Priscilla L., Smithsonian Institution / National Air & Space Museum
Sui, Daniel Z., Texas A&M University
Sultan, Mohamed, Western Michigan University
Szatmári, József, Univesity of Szeged
Szekielda, Karl H., Graduate School of the City University of New York
Torres Parisian, Cathleen, University of Minnesota, Twin Cities
Toth, Charles K., Ohio State University
Trust, Michael, University of Massachusetts, Boston
Tullis, Jason A., University of Arkansas, Fayetteville
Turner, Eugene, California State University, Northridge
Ustin, Susan L., University of California, Davis
Vandemark, Douglas C., University of New Hampshire
Velicogna, Isabella, University of California, Irvine
Ventura, Stephen J., University of Wisconsin, Madison
Viertel, Dave, Eastern Illinois University
Walker, Richard, University of Oxford
Warner, Timothy A., West Virginia University
Wasomi, Charles B., Jomo Kenyatta University of Agriculture & Tech
Watson, Kelly, Eastern Kentucky University
Wei, Xiaofang, Central State University
Weng, Qihao, Indiana State University
White, Jeffrey G., North Carolina State University
White, Joseph D., Baylor University
Woo, David, California State University, East Bay
Woodcock, Curtis E., Boston University
Yan, Xiao-Hai, University of Delaware
Yang, Zhiming, North Carolina Central University
Yilmaz, Alper, Ohio State University
Yool, Stephen R., University of Arizona
Yu, Qian, University of Massachusetts, Amherst
Yuan, Fei, Minnesota State University
Zhang, Caiyun, Florida Atlantic University
Zhou, Xiaobing, Montana Tech
Zhou, Xiaolu, Georgia Southern University

## Soil Science

Balser, Teri , University of Florida
Bonczek, James, University of Florida
Brinkmann, Robert, Hofstra University
Brown, James R., University of Missouri, Columbia
Budke, William, Ventura College

Curry, Susan, University of Florida
Dahlgren, Randy A., University of California, Davis
Davis, J G., University of Missouri, Columbia
DeLaune, Paul, Texas A&M University
DeSutter, Tom, North Dakota State University
Dou, Fugen, Texas A&M University
Dougil, Andy, University of Leeds
Eppes, Martha C., University of North Carolina, Charlotte
Gale, Paula M., University of Tennessee, Martin
Gantzer, Clark J., University of Missouri, Columbia
Geisseler, Daniel, University of California, Davis
Gerber, Stefan, University of Florida
Golabi, Mohammad H., University of Guam
Harvey, Omar, Texas Christian University
Hopmans, Jan W., University of California, Davis
Horwath, William R., University of California, Davis
Inglett, Kanika Sharma, University of Florida
Jackson, Louise E., University of California, Davis
Kramer, Marc, University of Florida
Kremer, Robert J., University of Missouri, Columbia
Lerch, Robert N., University of Missouri, Columbia
Marrs, Rob, University of Liverpool
McBride, Raymond G., University of Guelph
Mishra, Umakant, Argonne National Laboratory
Motavalli, Peter P., University of Missouri, Columbia
O'Geen, Toby, University of California, Davis
Parikh, Sanjai, University of California, Davis
Qualls, Robert G., University of Nevada, Reno
Richards, James H., University of California, Davis
Richter, Daniel D., Duke University
Risk, Dave A., Saint Francis Xavier University
Scow, Kate, University of California, Davis
Sinclair, Hugh D., Edinburgh University
Sleezer, Richard O., Emporia State University
Sletten, Ronald S., University of Washington
Smukler, Sean, University of British Columbia
Sohi, Sran P., Edinburgh University
Southard, Randal J., University of California, Davis
Stevens, W. G., University of Missouri, Columbia
Wesley Wood, Charley, University of Florida

## Meteorology

Adams, Manda S., University of North Carolina, Charlotte
Ault, Toby R., Cornell University
Babb, David M., Pennsylvania State University, University Park
Bahrmann, Chad, Pennsylvania State University, University Park
Baigorria, Guillermo, University of Nebraska, Lincoln
Bannon, Peter R., Pennsylvania State University, University Park
Barbor, RADM (Retired) Ken E., University of Southern Mississippi
Barnes, Gary M., University of Hawai'i, Manoa
Bartholy, Judit, Eotvos Lorand University
Baxter, Martin, Central Michigan University
Bell, Michael, University of Hawai'i, Manoa
Bernhardt, Jase, Hofstra University
Billings, Brian J., Saint Cloud State University
Blackwell, Keith G., University of South Alabama
Bleck, Rainer, Los Alamos National Laboratory
Bohren, Craig F., Pennsylvania State University, University Park
Bowman, Kenneth P., Texas A&M University
Brown, Michael E., Mississippi State University
Brown, Paul W., University of Arizona
Brune, William H., Pennsylvania State University, University Park
Brunet, Gilbert, McGill University
Businger, Steven, University of Hawai'i, Manoa
Cahir, John J., Pennsylvania State University, University Park
Carlson, Richard E., Iowa State University of Science & Tech
Carlson, Toby N., Pennsylvania State University, University Park
Carroll, David, Virginia Polytechnic Institute & State University
Challinor, Andy, University of Leeds
Chang, Chih-Pei, Naval Postgraduate School
Chen, Hway-Jen, Naval Postgraduate School
Chen, Jing, McMaster University
Clark, John H., Pennsylvania State University, University Park
Clothiaux, Eugene E., Pennsylvania State University, University Park
Colby, Frank P., University of Massachusetts Lowell
Cook, David R., Argonne National Laboratory
Corona, Thomas J., Metropolitan State College of Denver
Coulter, Richard L., Argonne National Laboratory
Cox, Helen M., California State University, Northridge
Crafts, Christine, SUNY, The College at Brockport
Creasey, Robert L., Naval Postgraduate School
Croft, Paul J., Kean University
Croft, Paul J., Mississippi State University
Cullen, Heidi, Georgia Institute of Tech
Czarnetzki, Alan C., University of Northern Iowa
Dahl, Johannes M., Texas Tech University
Davidson, Kenneth L., Naval Postgraduate School
Davis, Christopher A., Texas A&M University
Davis, Kenneth J., Pennsylvania State University, University Park
Deng, Aijun, Pennsylvania State University, University Park
Dewey, Ken F., University of Nebraska, Lincoln
Dixon, P. Grady, Fort Hays State University
Dorling, Stephen, University of East Anglia
Duncan Tabb, Neva, Saint Petersburg College, Clearwater
Durkee, Philip A., Naval Postgraduate School
Dutton, John A., Pennsylvania State University, University Park
Dyer, Jamie, Mississippi State University
Eastin, Matt, University of North Carolina, Charlotte
Edson, James B., University of Connecticut
Ellingson, Don, Western Oregon University
Elsberry, Russell L., Naval Postgraduate School
Emanuel, Kerry A., Massachusetts Institute of Tech
Estberg, Gerald N., University of San Diego
Evans, Jenni L., Pennsylvania State University, University Park
Fairchild, Jane, University of Wisconsin Colleges
Farrell, Brian F., Harvard University
Ferger, Marisa, Pennsylvania State University, University Park
Fillion, Luc, McGill University
Finley, Jason P., Los Angeles Pierce College
Finlley, Cathy, Saint Louis University
Fitzpatrick, Patrick J., Mississippi State University
Flory, David M., Iowa State University of Science & Tech
Flynn, Wendilyn, University of Northern Colorado
Frank, William M., Pennsylvania State University, University Park
Fraser, Alistair B., Pennsylvania State University, University Park
Frederickson, Paul A., Naval Postgraduate School
Fritsch, J. Michael, Pennsylvania State University, University Park
Fullmer, James W., Southern Connecticut State University
Gadomski, Frederick J., Pennsylvania State University, University Park
Galewsky, Joseph, University of New Mexico
Gauthier, Pierre, Universite du Quebec a Montreal
Gedzelman, Stanley D., Graduate School of the City University of New York
Gillespie, Terry J., University of Guelph
Girard, Eric, Universite du Quebec a Montreal
Godek, Melissa, SUNY, Oneonta
Greybush, Steven J., Pennsylvania State University, University Park
Griswold, Jennifer, University of Hawai'i, Manoa
Guest, Peter S., Naval Postgraduate School
Gutowski, William J., Iowa State University of Science & Tech
Gyakum, John R., McGill University
Hacker, Joshua P., Naval Postgraduate School
Hage, Keith D., University of Alberta
Hallin, Stephen C., Weber State University
Hamill, Paul, McHenry County College
Haney, Christa M., Mississippi State University
Haney, Robert, Naval Postgraduate School
Hansen, Anthony R., Saint Cloud State University
Harr, Patrick A., Naval Postgraduate School
Harrington, Jerry Y., Pennsylvania State University, University Park
Hehr, John G., University of Arkansas, Fayetteville
Heyniger, William C., Kean University
Hilliker, Joby, West Chester University
Hindman, Edward E., Graduate School of the City University of New York
Hoffman, Eric, Plymouth State University
Hosler, Charles L., Pennsylvania State University, University Park
Illari, Lodovica, Massachusetts Institute of Tech
Jenkins, Gregory S., Pennsylvania State University, University Park
Jordan, Mary S., Naval Postgraduate School
Karmosky, Christopher, SUNY, Oneonta
Kaster, Mark A., Wilkes University
Kasting, James F., Pennsylvania State University, University Park
Kimball, Sytske K., University of South Alabama
Knight, Paul, Pennsylvania State University, University Park

Koermer, James P., Plymouth State University
Kubesh, Rodney, Saint Cloud State University
Kumjian, Matthew R., Pennsylvania State University, University Park
Lai, Chung Chieng A., Los Alamos National Laboratory
Lamb, Dennis, Pennsylvania State University, University Park
Lander, Mark A., University of Guam
Lawson, Merlin P., University of Nebraska, Lincoln
Lee, Sukyoung, Pennsylvania State University, University Park
Leonard, Meredith L., Los Angeles Valley College
Lerach, David G., University of Northern Colorado
Lindzen, Richard S., Massachusetts Institute of Tech
Liu, Yongqiang, Georgia Institute of Tech
Lyons, Steve, Texas A&M University
Lyons, Steven, Angelo State University
Mandia, Scott, Suffolk County Community College, Ammerman Campus
Manzi, Anthony, SUNY, Maritime College
Martin, Jonathan E., University of Wisconsin, Madison
Matthews, Adrian, University of East Anglia
McGauley, Michael G., Miami Dade College (Kendall Campus)
Meesters, Antoon, VU University Amsterdam
Mendenhall, Larry, Mt. San Antonio College
Mercer, Andrew, Mississippi State University
Montgomery, Michael T., Naval Postgraduate School
Moore, Richard W., Naval Postgraduate School
Morgan, Michael C., University of Wisconsin, Madison
Morschauser, Lindsey, Mississippi State University
Mower, Richard N., Central Michigan University
Murphree, James Thomas, Naval Postgraduate School
Murray, Andrew, University of South Alabama
Najjar, Raymond G., Pennsylvania State University, University Park
Nielsen, Kurt E., Naval Postgraduate School
Nietfeld, Daniel, Creighton University
Ning, Liang, University of Massachusetts, Amherst
Noone, David, Oregon State University
Nordstrom, Greg, Mississippi State University
North, Gerald R., Texas A&M University
Nowotarski, Christopher J., Texas A&M University
Nuss, Wendell A., Naval Postgraduate School
O'Neill, Larry, Oregon State University
Orf, Leigh, Central Michigan University
Panetta, Richard L., Texas A&M University
Park, Myung-Sook, Naval Postgraduate School
Parker, Doug, University of Leeds
Parker, Matthew, North Carolina State University
Parks, Carlton R., Florida Institute of Tech
Pasken, Robert W., Saint Louis University
Pavloski, Charles, Pennsylvania State University, University Park
Penny, Andrew, Naval Postgraduate School
Person, Arthur, Pennsylvania State University, University Park
Plumb, Raymond A., Massachusetts Institute of Tech
Polvani, Lorenzo M., Columbia University
Priest, Eric, College of Lake County
Renard, Robert J., Naval Postgraduate School
Renfrew, Ian, University of East Anglia
Reuter, Gerhard, University of Alberta
Ritchie, Harold C., Dalhousie University
Ritter, Michael E., University of Wisconsin, Stevens Point
Ritz, Richard, Creighton University
Rochette, Scott M., SUNY, The College at Brockport
Rockwood, Anthony A., Metropolitan State College of Denver
Rodgers, N, Cardiff University
Rodriguez-Plasencia, Oscar, Universidad Autonoma de Baja California Sur
Rood, Richard B., University of Michigan
Ross, Andrew, University of Leeds
Russell, William H., Los Angeles Pierce College
Ryan, William F., Pennsylvania State University, University Park
Sanabia, Elizabeth R., United States Naval Academy
Scarnato, Barbara V., Naval Postgraduate School
Schaffer, David L., Wichita State University
Schrage, Jon M., Creighton University
Schultz, David, University of Manchester
Seager, Richard, Columbia University
Seaman, Nelson L., Pennsylvania State University, University Park
Shannon, Jack D., Argonne National Laboratory
Shirer, Hampton N., Pennsylvania State University, University Park
Skubis, Steven T., SUNY, Oswego
Skyllingstad, Eric, Oregon State University
Smith, David R., United States Naval Academy
Smith, Dwight E., United States Naval Academy
Srinivasan, Gopalan, University of Toronto
Stamm, Alfred J., SUNY, Oswego
Stauffer, David R., Pennsylvania State University, University Park
Steiger, Scott, SUNY, Oswego
Steinacker, Reinhold, University of Vienna
Stensrud, David J., Pennsylvania State University, University Park
Stevens, Duane E., University of Hawai'i, Manoa
Straub, Katherine H., Susquehanna University
Syrett, William, Pennsylvania State University, University Park
Terwey, Wes, University of South Alabama
Tett, Simon F., Edinburgh University
Thomson, Dennis W., Pennsylvania State University, University Park
Todd, Steve, Portland Community College - Sylvania Campus
Torlaschi, Enrico, Universite du Quebec a Montreal
Tripoli, Gregory J., University of Wisconsin, Madison
Underwood, Stephen J., Georgia Southern University
Utley, Tom, Florida Institute of Tech
Vaughan, Naomi, University of East Anglia
Verlinde, Johannes, Pennsylvania State University, University Park
Vernon, James Y., Los Angeles Pierce College
Wagner Riddle, Claudia, University of Guelph
Wallace, Tim, Mississippi State University
Wang, Bin, University of Hawai'i, Manoa
Wang, Qing, Naval Postgraduate School
Wash, Carlyle H., Naval Postgraduate School
Weisman, Robert A., Saint Cloud State University
Williams, Aaron, University of South Alabama
Williams, Forrest, Naval Postgraduate School
Williams, Roger T., Naval Postgraduate School
Wilmoth, Valorie, Saint Louis University
Wilson, John D., University of Alberta
Wyngaard, John C., Pennsylvania State University, University Park
Wysong, Jr., James F., Hillsborough Community College
Yang, Zong-Liang, University of Texas at Austin
Yoh, Shing, Kean University
Young, George S., Pennsylvania State University, University Park
Yow, Donald M., Eastern Kentucky University
Zois, Constantine S., Kean University

## Material Science

Alfe, Dario, University College London
Calaway, Wallis F., Argonne National Laboratory
Chiarello, Ronald P., Argonne National Laboratory
Chrzan, Daryl, University of California, Berkeley
Cnudde, Veerle, Ghent University
deFontaine, Didier, University of California, Berkeley
DeJonghe, Lutgard, University of California, Berkeley
Devine, Thomas M., University of California, Berkeley
Eberhart, Mark E., Colorado School of Mines
Ferrari, Mauro, University of California, Berkeley
Glaeser, Andreas, University of California, Berkeley
Gronsky, Ronald, University of California, Berkeley
Gruen, Dieter M., Argonne National Laboratory
Haller, Eugene, University of California, Berkeley
Hayden, Geoffrey W., Mercer University
Hazen, Robert M., George Mason University
Hojjatie, Barry, Valdosta State University
Morris, Jr., J. W., University of California, Berkeley
Navrotsky, Alexandra, University of California, Davis
Pellin, Michael J., Argonne National Laboratory
Pylypenko, Svitlana, Colorado School of Mines
Qu, Deyang, University of Massachusetts, Boston
Ritchie, Robert O., University of California, Berkeley
Sands, Timothy, University of California, Berkeley
Strobel, Timothy A., Carnegie Institution of Washington
Voller, Vaughan R., University of Minnesota, Twin Cities
Weber, Eicke, University of California, Berkeley
Weertman, Johannes, Northwestern University
Zollner, Stefan, New Mexico State University, Las Cruces

## Land Use/Urban Geology

Agrawal, Sandeep , University of Alberta
Barile, Diane D., Florida Institute of Tech
Bassett, Scott, University of Nevada, Reno
Birchall, Jeff, University of Alberta

Blackburn, James B., Rice University
Bodenman, John E., Bloomsburg University
Bonine, Michael E., University of Arizona
Bremer, Keith A., Fort Hays State University
Briggs, John, University of Glasgow
Brinkman, P. Anthony, University of Nevada, Reno
Buermann, Wolfgang, University of Leeds
Cogger, Craig G., Washington State University
Davidson, Fiona M., University of Arkansas, Fayetteville
Day, Rick L., Pennsylvania State University, University Park
Deacon, Leith, University of Alberta
Dixon, Nick, University of Leeds
Dolman, Paul, University of East Anglia
Emmi, Philip C., University of Utah
Ford, Andrew, Washington State University
Fusch, Richard D., Ohio Wesleyan University
Godfrey, Brian J., Vassar College
Gosnell, Hannah, Oregon State University
Graff, Thomas O., University of Arkansas, Fayetteville
Graumlich, Lisa J., Montana State University
Haggerty, Julia H., Montana State University
Harmon, Mella, University of Nevada, Reno
Harris, Richard S., McMaster University
Heiman, Michael, Dickinson College
Hines, Mary E., University of North Carolina Wilmington
Hintz, Rashauna, University of Arkansas, Fayetteville
Holl, Karen D., University of California, Santa Cruz
Holland, Edward C., University of Arkansas, Fayetteville
Hungerford, Hilary, Utah Valley University
Jackson, Richard H., Brigham Young University
James, Valentine U., Clarion University
Jantz, Claire A., Shippensburg University
Kaushal, Sujay, University of Maryland
Larson, David J., California State University, East Bay
Lawrence, Henry, Edinboro University of Pennsylvania
Lee, Chung M., University of Utah
Lyon, Linda M., University of Montana Western
Marston, Sallie A., University of Arizona
Mercier, Michael , McMaster University
Millington, Andrew, Flinders University
Mitchell, Martin D., Minnesota State University
Nevins, Joseph, Vassar College
Odogba, Ismaila, University of Wisconsin, Stevens Point
Otterstrom, Samuel M., Brigham Young University
Peace, Walter G., McMaster University
Pepino, Richard V., Franklin and Marshall College
Platt, Rutherford H., University of Massachusetts, Amherst
Pomeroy, George M., Shippensburg University
Randlett, Victoria, University of Nevada, Reno
Riebesell, John, University of Michigan, Dearborn
Rounsevell, Mark D., Edinburgh University
Shirgaokar, Manish, University of Alberta
Singh, Amrita, University of Alberta
Skillen, James, Calvin College
Solecki, William, Montclair State University
Solecki, William, Hunter College (CUNY)
Steinberger, Julia, University of Leeds
Sullivan, Jack B., University of Maryland
Thomas, Valerie, Georgia Institute of Tech
Tilt, Jenna, Oregon State University
Troost, Kathy G., University of Washington
Van Den Hoek, Jamon, Oregon State University
Wickham, Thomas, California University of Pennsylvania
Wiggin, Jack, University of Massachusetts, Boston
Zhou, Yu, Vassar College

## Geographic Information Systems

Adhikari, Sanchayeeta, California State University, Northridge
Ahearn, Sean C., Hunter College (CUNY)
Albrecht, Jochen, Hunter College (CUNY)
Alderson, David, University of Newcastle Upon Tyne
Algeo, Catherine, Western Kentucky University
Allen, Katherine, University of Liverpool
Allen-Lafayette, Zehdreh, New Jersey Geological and Water Survey
Ambinakudige, Shrinidhi, Mississippi State University
Amer, Reda, Tulane University
Andrews, John R., University of Texas at Austin, Jackson School of Geosciences
Anthony-Zajanc, Kate, Idaho State University
Argles, Tom, The Open University
Ayad, Yasser M., Clarion University
Badruddin, Abu Z., Cayuga Community College
Badurek, Chris, Appalachian State University
Bailey, Keiron D., University of Arizona
Barr, Stuart, University of Newcastle Upon Tyne
Barrett, Melony, University of Illinois, Urbana-Champaign
Bauer, Emily, University of Minnesota
Beaty, Tammy, Oak Ridge National Laboratory
Becker, Lorene Y., Oregon State University
Benger, Simon, Flinders University
Benson, Jane L., Murray State University
Betchwars, Corey, University of Minnesota
Bloechle, Amber, University of West Florida
Bloomgren, Bruce, University of Minnesota
Boda, Patricia, Middle Tennessee State University
Boger, Rebecca, Brooklyn College (CUNY)
Bone, Christopher, University of Oregon
Boroushaki, Soheil, California State University, Northridge
Bottenberg, H. Carrie, Idaho State University
Bowlick, Forrest J., University of Massachusetts, Amherst
Boyack, Diana L., Idaho State University
Breton, Caroline, University of Texas at Austin, Jackson School of Geosciences
Brown, Douglas, Kingston University
Bruns, Dale A., Wilkes University
Brunskill, Jeffrey C., Bloomsburg University
Burton, Christopher G., Auburn University
Busby, Michael R., Murray State University
Cao, Guofeng, Texas Tech University
Carstensen, Laurence W., Virginia Polytechnic Institute & State University
Carter, David, Department of Interior Office of Surface Mining Reclamation and Enforcement
Cary, Kevin, Western Kentucky University
Cetin, Haluk, Murray State University
Challender, Stuart, Montana State University
Cheng, Qiuming, York University
Cheung, Wing, Palomar College
Chokmani, Karem, Universite du Quebec
Chopra, Prame N., Australian National University
Christopherson, Gary, University of Arizona
Cloutis, Ed, Western University
Coetzee, Serena M., University of Pretoria
Colby, Jeff, Appalachian State University
Cothren, Jackson D., University of Arkansas, Fayetteville
Cova, Thomas J., University of Utah
Coveney, Seamus, University of Glasgow
Cseri, Mick, Indiana University / Purdue University, Fort Wayne
Cunningham, Mary A., Vassar College
Curri, Neil, Vassar College
Dai, Dajun, Georgia State University
Dark, Shawna J., California State University, Northridge
Davis, Trevor J., University of Utah
De Kemp, Eric, University of Ottawa
Deakin, Ann K., SUNY Fredonia
Deal, Richard, Edinboro University of Pennsylvania
Delparte, Donna M., Idaho State University
Demyanov, Vasily, Heriot-Watt University
Deskins, Aaron M., University of Montana
DiNaso, Steven, Eastern Illinois University
Diver, Kim, Wesleyan University
Dogwiler, Toby, Missouri State University
Dong, Pinliang, University of North Texas
Donnelly, Shanon P., University of Akron
Drummond, Jane, University of Glasgow
Drzyzga, Scott A., Shippensburg University
Duff, Shamus, Western University
Duke, Jason E., Tennessee Tech University
Earle, Robert, American River College
Ehler, J B., Utah Geological Survey
Fan, Weihong, Richard Stockton College of New Jersey
Fedorko, Evan J., West Virginia University
Fernández, Virginia, Universidad de la Republica Oriental del Uruguay (UDELAR)
Fitzgerald, Francis S., Colorado School of Mines
Forrest, David, University of Glasgow

Freed, Jane S., University of Idaho
Frizado, Joseph P., Bowling Green State University
Garren, Sandra J., Hofstra University
Gathany, Mark, Cedarville University
Gauthier, Donald J., Los Angeles Valley College
Girard, Michael W., New Jersey Geological and Water Survey
Gittings, Bruce M., Edinburgh University
Gopal, Sucharita, Boston University
Gordon, Clare, University of Leeds
Gorsevski, Peter, Bowling Green State University
Grala, Katarzynz, Mississippi State University
Graniero, Phil A., University of Windsor
Greatbatch, Ian, Kingston University
Grimmer, Abbey, University of Nevada, Reno
Grunwald, Sabine, University of Florida
Haddock, Gregory D., Northwest Missouri State University
Halls, Joanne N., University of North Carolina Wilmington
Halsted, Christian, Dept of Agriculture, Conservation, and Forestry
Hamilton, Jacqueline, University of Minnesota
Hancock, Gregory, University of Newcastle
Haner, Andrew, Arkansas Geological Survey
Hansen, William J., Worcester State University
Haugerud, Ralph, University of Washington
Heaton, Jill, University of Nevada, Reno
Hedquist, Brent, Texas A&M University, Kingsville
Henderson, Matt, Dept of Natural Resources
Herried, Brad, University of Minnesota, Twin Cities
Hick, Steven, University of Denver
Hill, Malcolm D., Northeastern University
Hill, Richard T., Indiana University
Hintz, John G., Bloomsburg University
Hong, Jessie, University of West Georgia
Hotz, Helenmary, University of Massachusetts, Boston
Howard, Hugh H., American River College
Huang, Jane, Fitchburg State University
Huffman, French T., Eastern Kentucky University
Hughes, Annie, Kingston University
Jennings, Nathan, American River College
Jeu, Amy, Hunter College (CUNY)
Joshi, Manoj, University of East Anglia
Juntunen, Thomas, University of Minnesota, Twin Cities
Kahle, Chris, Iowa Dept of Natural Resources
Kambesis, Patricia, Western Kentucky University
Kar, Bandana, University of Southern Mississippi
Kelleher, Cole, University of Minnesota, Twin Cities
Kennedy, Timothy T., University of Wisconsin, Stevens Point
Kidd, David, Kingston University
Kienzle, Stefan, University of Lethbridge
Kimerling, A. Jon, Oregon State University
Kirimi, Fridah K., Jomo Kenyatta University of Agriculture & Tech
Klancher, Jacki, Central Wyoming College
Kobara, Shinichi, Texas A&M University
Kohler, Nicholas, University of Oregon
Kostelnick, John, Illinois State University
Krizek, Jeffrey, San Bernardino Valley College
Kronenfeld, Barry J., Eastern Illinois University
Krygier, John B., Ohio Wesleyan University
Kwon, Youngsang, University of Memphis
Laffan, Shawn, University of New South Wales
Law, Zada, Middle Tennessee State University
Le, Yanfen, Northwest Missouri State University
Lee, Wook, Edinboro University of Pennsylvania
Lethbridge, Mark, Flinders University
Leverington, David W., Texas Tech University
Levine, Norman S., College of Charleston
Li, Gary, California State University, East Bay
Li, Ping-Chi, Tennessee Tech University
Limp, W. F., University of Arkansas, Fayetteville
Lisichenko, Richard, Fort Hays State University
Lobben, Amy, University of Oregon
Logsdon, Miles G., University of Washington
Lovett, Andrew A., University of East Anglia
Lukinbeal, Christopher, University of Arizona
Luo, Jun, Missouri State University
Lupo, Tom, American River College
Maantay, Juliana, Lehman College (CUNY)
Maas, Regan, California State University, Northridge
Machovina, Brett , United States Air Force Academy
Mackaness, William A., Edinburgh University
Magondu, Moffat G., Jomo Kenyatta University of Agriculture & Tech
Marcano, Eugenio J., Mount Holyoke College
Marr, Paul G., Shippensburg University
Marsan, Yvonne, University of North Carolina Wilmington
Marzen, Luke J., Auburn University
Mathias, Simon, Durham University
May, Cynthia, Northland College
McCluskey, James, University of Wisconsin Colleges
McKinney, Nathan, University of West Florida
Mead, Jerry V., Drexel University
Meentemeyer, Ross, University of North Carolina, Charlotte
Meng, Qingmin, Mississippi State University
Metcalf, Meredith, Eastern Connecticut State University
Miguel, Carlos, Universidad de la Republica Oriental del Uruguay (UDELAR)
Miller, Harvey J., Ohio State University
Miller, Max, Front Range Community College - Westminster
Mitasova, Helena, North Carolina State University
Mohapatra, Rama P., Minnesota State University
Momm, Henrique G., Middle Tennessee State University
Morgan, John D., University of West Florida
Morgan, Tamie, Texas Christian University
Morris, John A., Mississippi State University
Mucsi, lászló, Univesity of Szeged
Mueller, Thomas, California University of Pennsylvania
Mulligan, Kevin R., Texas Tech University
Mulrooney, Timothy, North Carolina Central University
Muthukrishnan, Suresh, Furman University
Mutua, Felix N., Jomo Kenyatta University of Agriculture & Tech
Mwangi, Nancy , Jomo Kenyatta University of Agriculture & Tech
Mwaniki, Mercy W., Jomo Kenyatta University of Agriculture & Tech
Nagihara, Seiichi, Texas Tech University
Nemon, Amy, Western Kentucky University
Niedzielski, Michael A., University of North Dakota
Nimako, Solomon Nana Kwaku, San Bernardino Valley College
Obermeyer, Nancy J., Indiana State University
Oduor, Peter, North Dakota State University
Ozdenerol, Esra, University of Memphis
Palladino, Steve D., Ventura College
Pallis, Ted J., New Jersey Geological and Water Survey
Palmer, Evan, United States Air Force Academy
Pavlovskaya, Marianna, Hunter College (CUNY)
Pedemonte, Virginia, Universidad de la Republica Oriental del Uruguay (UDELAR)
Peele, R Hampton, Louisiana State University
Pennington, Deana, University of Texas, El Paso
Percy, David, Portland State University
Plewe, Brandon, Brigham Young University
Porter, Claire, University of Minnesota, Twin Cities
Portillo, Danny, United States Air Force Academy
Price, Maribeth H., South Dakota School of Mines & Tech
Price, Peter E., Lonestar College - North Harris
Pristas, Ronald, New Jersey Geological and Water Survey
Proctor, Elizabeth, City College of San Francisco
Qi, Feng, Kean University
Qiu, Xiaomin, Missouri State University
Raber, George, University of Southern Mississippi
Resnichenko, Yuri, Universidad de la Republica Oriental del Uruguay (UDELAR)
Rex, Arthur B., Appalachian State University
Rice, Keith W., University of Wisconsin, Stevens Point
Robinson, Lori, University of Minnesota
Robinson, Michael, Georgia Southern University
Roof, Steven, Hampshire College
Rowley, Rex J., Illinois State University
Sadd, James L., Occidental College
Sanchez-Azofeifa, G. Arturo, University of Alberta
Sandau, Ken L., Montana Tech of The University of Montana
Schenck, William, University of Delaware
Scott, Darren M., McMaster University
Seong, Jeong C., University of West Georgia
Shears, Andrew, University of Wisconsin Colleges
Shimizu, Melinda, Western Oregon University
Smith, Betty E., Eastern Illinois University
Smith, Janet S., Shippensburg University

Starek, Michael , Texas A&M University, Corpus Christi
Stuart, Neil, Edinburgh University
Stutsman, Sam, University of South Alabama
Su, Haibin, Texas A&M University, Kingsville
Sun, Yifei, California State University, Northridge
Sutton, Paul C., University of Denver
Taylor, Nathan H., Arkansas Geological Survey
Taylor, Ryan W., SUNY, Purchase
Thale, Paul R., Montana Tech of The University of Montana
Thayn, Jonathan B., Illinois State University
Tinkler, Dorothy, Treasure Valley Community College
Tiwari, Chetan, University of North Texas
Tong, Daoqin, University of Arizona
Tran, Linda C., Lonestar College - North Harris
Trifonoff, Karen M., Bloomsburg University
Tu, Wei, Georgia Southern University
Unger, Corey, Utah Geological Survey
VanHorn, Jason, Calvin College
Veisze, Paul M., American River College
Vincent, Paul, Valdosta State University
Wahl, Tim, University of Minnesota
Walford, Nigel, Kingston University
Walls, Charles C., Dalhousie University
Wang, Lillian T., University of Delaware
Weber, Keith, Idaho State University
Weiss, Alfred W., Waubonsee Community College
Wen, Yuming, University of Guam
Whiteaker, Timothy L., University of Texas at Austin
Whitfield, Thomas G., Pennsylvania Bureau of Topographic & Geologic Survey
Whittaker, Amber, Dept of Agriculture, Conservation, and Forestry
Wikgren, Brooke, University of Massachusetts, Boston
Williamson, Douglas, Hunter College (CUNY)
Wilson, Blake, University of Kansas
Wilson, John R., Lafayette College
Wilson, Roy R., Eastern Connecticut State University
Woodhouse, Iain H., Edinburgh University
Wulamu, Wasit, Saint Louis University
Xie, Zhixiao, Florida Atlantic University
Xu, Wei, University of Lethbridge
Yacucci, Mark A., University of Illinois
Yan, Jun, Western Kentucky University
Ye, Gordon, City College of San Francisco
Zhang, Qiaofeng (Robin), Murray State University
Zhao, Bo, Oregon State University

## Glaciology

Barker, Joel D., Ohio State University
Bassis, Jeremy, University of Michigan
Booth, Adam, Imperial College
Brugger, Keith A., University of Minnesota, Morris
Catania, Ginny A., University of Texas at Austin
Conway, Howard B., University of Washington
Creyts, Timothy T., Columbia University
Csatho, Beata M., SUNY, Buffalo
Dupont, Todd, University of California, Irvine
Enderlin, Ellyn M., University of Maine
Fahnestock, Mark A., University of New Hampshire
Fisher, David A., University of Ottawa
Hamilton, Gordon S., University of Maine
Harper, Joel, University of Montana
Hock, Regine M., University of Alaska, Fairbanks
Holt, John W., University of Texas at Austin
Howat, Ian M., Ohio State University
Hulton, Nicholas R., Edinburgh University
Humphrey, Neil F., University of Wyoming
Hutchings, Jennifer, Oregon State University
Jiskoot, Hester, University of Lethbridge
Malone, Andrew, Lawrence University
Mills, Stephanie, Kingston University
Neinow, Peter W., Edinburgh University
Pettit, Erin C., University of Alaska, Fairbanks
Radic, Valentina, University of British Columbia
Raymond, Charles F., University of Washington
Rignot, Eric, University of California, Irvine
Rupper, Summer, Brigham Young University
Schoof, Christian, University of British Columbia
Sharp, Martin J., University of Alberta
Stearns, Leigh, University of Kansas
Stroeve, Julienne, University College London
Tedasco, Marco, Columbia University
Waddington, Edwin D., University of Washington
Winebrenner, Dale P., University of Washington
Wolff, Eric W., University of Cambridge
Zoet, Lucas, University of Wisconsin, Madison

## Not Elsewhere Classified

(Deep) McGregor, Mary, Montgomery College
Aamodt, Paul, Los Alamos National Laboratory
Abney, Kent D., Los Alamos National Laboratory
Abt, Charlotte J., University of Liverpool
Ackerman, Jessica R., University of Illinois, Urbana-Champaign
Acosta, Patricia E., Williams College
Acosta Borbon, Hugo, Universidad Nacional Autonoma de Mexico
Adams, Debi, West Texas A&M University
Agnew, Stephen F., Los Alamos National Laboratory
Aguilera, Juan Manuel Torres, Universidad Autonoma de San Luis Potosi
Ahmed, Waquar, University of North Texas
Alavi, Hedy, Johns Hopkins University
Albrecht, Lorraine, University of Waterloo
Albright, Katia, University of Nevada, Reno
Alde, Douglas, Los Alamos National Laboratory
Algin, Barbara, Columbia University
Ali-Bray, Julie, Mt. San Antonio College
Allen, Gina, Northwestern University
Allen, Karen, Furman University
Allison, Tami L., Missouri Dept of Natural Resources
Ambruster, W. Scott, University of Alaska, Fairbanks
Amelung, Falk, University of Miami
Amor, John, University of British Columbia
Anand, Madhur, University of Guelph
Andersen, Elaine, Stanford University
Antar, Ali A., Central Connecticut State University
Anthony, Leona M., University of Hawai'i, Manoa
Anthony, Nina, Furman University
Antipova, Anzhelika, University of Memphis
Arabas, Karen, Willamette University
Argenbright, Robert T., University of North Carolina Wilmington
Armstrong, Andrew, University of New Hampshire
Arundale, Wendy H., University of Alaska, Fairbanks
Arvidson, Raymond E., Washington University in St. Louis
Ashour-Abdalla, Maha, University of California, Los Angeles
Atisha-Castillo, Hector David, Universidad Autonoma de San Luis Potosi
Au, Whitlow W L., University of Hawai'i, Manoa
Aufrecht, Walter E., University of Lethbridge
Ayers, Joseph , Northeastern University
Bailey, Ian, Exeter University
Bailey, Lorraine, University of Rhode Island
Baldizon, Ileana, Miami-Dade College (Wolfson Campus)
Ball, Elizabeth, University of Nevada, Reno
Ball, William P., Johns Hopkins University
Ballinger, Rhoda, Cardiff University
Banks, Beth, Brevard College
Bannon, Ann, Dalhousie University
Barnmore, Garrett, University of Nevada, Reno
Barnbaum, Cecilia S., Valdosta State University
Barnes, Fairley J., Los Alamos National Laboratory
Barnes, Jessica, University of South Carolina
Barnes, Philip, University of South Carolina
Barnes, Randal J., University of Minnesota, Twin Cities
Barnston, Anthony G., Columbia University
Barrett, Joan, Hampshire College
Barron, George, University of Guelph
Barth, Susan A., Montana Tech of The University of Montana
Barthelmie, Rebecca J., Cornell University
Bartl, Simona, Moss Landing Marine Laboratories
Baskin, Perry A., Valdosta State University
Bass, Jerry, University of Southern Mississippi
Basso, Bruno, Michigan State University
Baumann, Sue, Argonne National Laboratory
Beasley-Stanley, Jewell D., Vanderbilt University
Beatty, Merrill Ann, University of New Brunswick
Beck, Alan, Western University
Becker, Laurence, Oregon State University
Becker, Udo, University of Michigan
Bederman, Sanford H., Georgia State University

Behr-Andres, Christina "Tina", Los Alamos National Laboratory
Belanger, Thomas V., Florida Institute of Tech
Bellette, Kathryn, Flinders University
Bellew, Angela, University of Iowa
Bennett, Victoria J., Texas Christian University
Berchem, Jean, University of California, Los Angeles
Bernknopf, Richard, Stanford University
Betancourt, Julio, University of Arizona
Bezdecny, Kris, California State University, Los Angeles
Bienvenu, Nadean S., University of Louisiana at Lafayette
Bilheux, Hassina Z., University of Tennessee, Knoxville
Billiot, Fereshteh, Texas A&M University, Corpus Christi
Bischoff, Marianne, Weber State University
Bishoff, Carolyn, University of Minnesota, Twin Cities
Black, Annjeannette , Community College of Baltimore County, Catonsville
Blank, Carrine E., University of Montana
Blankenship, Robert, Washington University in St. Louis
Blatner , Keith , Washington State University
Blenkinson, Stephen, University of Newcastle Upon Tyne
Blizzard, Kate, Western Washington University
Blumenfeld, Raphael, Imperial College
Boger, Kathryn M., Ohio Wesleyan University
Boggs, Carol, University of South Carolina
Boisvert, Eric, Universite du Quebec
Boland, Greg J., University of Guelph
Boland, John J., Johns Hopkins University
Booth, Robert K., Lehigh University
Boss, Alan P., Carnegie Institution of Washington
Bossenbroek, Jonathon , University of Toledo
Boston, Penelope, New Mexico Institute of Mining and Tech
Boswell, Cecil, Missouri Dept of Natural Resources
Bottero, Jean-Yves, Rice University
Bourgeois, Reed J., Louisiana State University
Bouwer, Edward John, Johns Hopkins University
Bowen, Dawn S., Mary Washington College
Bowen, Jennifer, Northeastern University
Bowen, Robert, University of Massachusetts, Boston
Bowersox, Joe, Willamette University
Bowles, Ann B., University of San Diego
Bozzato, Edda, Laurentian University, Sudbury
Bradley, Michael D., University of Arizona
Brainard, James R., Los Alamos National Laboratory
Braught, Patricia, Dickinson College
Brewer, Margene, Calvin College
Bridgeman, Thomas, University of Toledo
Brooks, Ian, University of Leeds
Brown, Bradford D., University of Idaho
Brown, Kent D., Utah Geological Survey
Brown, Rob, Appalachian State University
Brunetto, Eileen, Middlebury College
Brunish, Wendee M., Los Alamos National Laboratory
Bryan, Jeffrey C., Los Alamos National Laboratory
Bryant, Anita M., University of West Georgia
Bryce, Karen R., Brigham Young University
Buck, Daniel, University of Oregon
Buck, Gregory, Texas A&M University, Corpus Christi
Buckley, Luke, Montana Tech of The University of Montana
Bullamore, Henry W., Frostburg State University
Bunker, Merle E., Los Alamos National Laboratory
Burk, Sue, University of Maryland
Burns, Carol J., Los Alamos National Laboratory
Burns, John W., Mt. San Antonio College
Busse, Friedrich H., University of California, Los Angeles
Butler, Gregg, University of Manchester
Butler, R. Paul, Carnegie Institution of Washington
Butts, Thomas R., University of Texas, Dallas
Byrand, Karl, University of Wisconsin Colleges
Caballero, Kate, Tarleton State University
Cahill, Michael J., SUNY, Stony Brook
Cailliet, Gregor M., Moss Landing Marine Laboratories
Calegari, Pat, Pace University, New York Campus
Cambiotti, Laura J., Lehigh University
Cammerata, Kirk , Texas A&M University, Corpus Christi
Campana, Michael E., University of New Mexico
Cantarero, Debra A., Pasadena City College
Capoccia, Mary, Ohio State University
Carey, Kristine M., Lakehead University
Carey, Tara, Black Hawk College
Carp, Jana, Appalachian State University
Carr, David E., University of Virginia
Carrick, Tina, University of Texas, El Paso
Carroll, Matt, Washington State University
Carzoli, John, Oakton Community College
Cassidy, Carleen, Montana Tech of The University of Montana
Catalano, Valerie, Roger Williams University
Catau, John C., Missouri State University
Catling, David C., University of Washington
Cattolico, Rose Ann, University of Washington
Caupp, Craig L., Frostburg State University
Chambers, John E., Carnegie Institution of Washington
Chan, Selene, University of British Columbia
Chase, Anne, University of Arizona
Chase, Jon M., Washington University in St. Louis
Chatterjee, Ipsita, University of North Texas
Cheek, William H., Missouri State University
Chen, Kai Loon, Johns Hopkins University
Cheney, Donald, Northeastern University
Chief, Karletta, University of Arizona
Childs, Geoff, Washington University in St. Louis
Choquette, Marc, Universite Laval
Chouinard, Vera, McMaster University
Christopher, Micol, Mt. San Antonio College
Clark, David L., Los Alamos National Laboratory
Clark, James S., Duke University
Clark, Robert O., Northern Arizona University
Clark, Shannon, University of New Mexico
Clay, Robert, Missouri Dept of Natural Resources
Cleek, Richard, University of Wisconsin Colleges
Cochran, David, University of Southern Mississippi
Coffin, Richard B., University of Delaware
Cohen, Alice, Acadia University
Cohen, Joel E., Columbia University
Cohen, Matt, Furman University
Cohen, Shaul E., University of Oregon
Colby, Bonnie C., University of Arizona
Coleman, Gary D., University of Maryland
Collins, Damian, University of Alberta
Collins, Lisa, Northwestern University
Collins, Mark, University of Pittsburgh
Colwell, Frederick (Rick), Oregon State University
Confer, John, California University of Pennsylvania
Conkle, Jeremy, Texas A&M University, Corpus Christi
Cook, Stanton A., University of Oregon
Cook, Steve, Oregon State University
Cookus, Pam, Illinois State Geological Survey
Coonley, Steffanie, Los Alamos National Laboratory
Copeman, Louise A., Oregon State University
Corcoran, Deborah, Missouri State University
Corey, Alison, Utah Geological Survey
Costello, Margaret, California State University, Long Beach
Cote, Pascale, Universite du Quebec
Cox, Shelah, Temple University
Crafford, J. P., University of Limpopo
Crawford, Ian, Birkbeck College
Crawford, Matt, University of Kentucky
Crepeau, Richard J., Appalachian State University
Crimmins, Michael, University of Arizona
Cromar, Nancy, Flinders University
Cromartie, William J., Richard Stockton College of New Jersey
Crozier, George F., Dauphin Island Sea Lab
Crumbly, Isaac J., Fort Valley State University
Cruz-Teran, Jesus E., Centro de Estudios Superiores del Estado Sonora
Csaplar, Csilla, Stanford University
Cupples, Julie, Edinburgh University
Curry, Barbara, University of Texas, Dallas
Curry, James K., Arkansas Geological Survey
D'Arrigo, Rosanne, Columbia University
Damschen, Ellen, Washington University in St. Louis
Dash, Zora V., Los Alamos National Laboratory
Dasvarma, Gour, Flinders University
Daugherty, Carolyn M., Northern Arizona University
Davey, Patricia M., Brown University
Davis, James A., Brigham Young University

Day, Rosie, University of Birmingham
Daymon, Deborah J., Los Alamos National Laboratory
De Beus, Barbara, Montclair State University
De Santo, Eilzabeth, Franklin and Marshall College
Dean, Jeffrey S., University of Arizona
Deaton, Tami, University of North Texas
DeBarros, Nelson, Dept of Energy and Environmental Protection
DeGraff, Jerry, California State University, Fresno
Delawder, Sandra, James Madison University
Deming, Jody, University of Manitoba
Den Ouden, Amy, University of Massachusetts, Boston
Dennis, Clyde B., Argonne National Laboratory
Dennis, Pam, University of Florida
Desch, Steven, Arizona State University
Deschaine, Sylvia, Bates College
Dessler, Alexander, University of Arizona
Dessler, Andrew, Texas A&M University
Detrich, William, Northeastern University
Dibben, Chris, Edinburgh University
Dickerson, Richard E., University of California, Los Angeles
Dickhoff, Willem H., Washington University in St. Louis
Dighe, Kalpak, Los Alamos National Laboratory
Dill, Carilee, Bucknell University
Dingman, Steve, San Antonio Community College
Ditmars, John, Argonne National Laboratory
Divine, Aaron, Northern Arizona University
Dixon, Clifton, University of Southern Mississippi
Dixon, Michael, University of Guelph
Dixon, Paul R., Los Alamos National Laboratory
Doherty, Cheryl, University of Delaware
Dollhopf, Douglas J., Montana State University
Domingue, Carla, Louisiana State University
Donahoe, Robert J., Los Alamos National Laboratory
Donohue, Mary M., DePauw University
Doorn, Stephen K., Los Alamos National Laboratory
Dorries, Alison M., Los Alamos National Laboratory
Driever, Steven L., University of Missouri-Kansas City
Driscoll, John R., SUNY, Cortland
Druckebrod, Daniel L., Rider University
Duchane, David V., Los Alamos National Laboratory
Dudukovic, Mike, Washington University in St. Louis
Duff, John, University of Massachusetts, Boston
Duffey, Patricia, Fort Hays State University
Dunton, Karen K., Illinois State University
Duxbury, Thomas C., Jet Propulsion Laboratory
Dvorzak, Marie, University of Wisconsin, Madison
Dwyer, Daryl F., University of Toledo
Earls, Sandy, Oklahoma State University
Ebert, David, Moss Landing Marine Laboratories
Eby, Gloria J., University of Texas, Dallas
Eby, Stephanie, Northeastern University
Edwards, John , Flinders University
Eggers, Delores M., University of North Carolina, Asheville
Ehlschlaeger, Charles R., Graduate School of the City University of New York
Eisenberger, Peter M., Columbia University
Eller, Phillip G., Los Alamos National Laboratory
Ellins, Katherine K., University of Texas at Austin
Elliott, C. James, Los Alamos National Laboratory
Ellis, Hugh, Johns Hopkins University
Ellis, Kathryn E., University of Kentucky
Ellis, Rowan, Edinburgh University
Emmett, Chad, Brigham Young University
Epstein, Howard E., University of Virginia
Esnault, Melissa H., Louisiana State University
Esser, Corinne, University of California, Davis
Evans, Claude, Washington University in St. Louis
Even, Paula, University of Northern Iowa
Fadiman, Maria, Florida Atlantic University
Fairbairn, Kenneth J., University of Alberta
Fallowfield, Howard, Flinders University
Fares, Mary A., Valdosta State University
Farley, Perry D., Los Alamos National Laboratory
Farrell, Mark O., Point Park University
Favor, Michael, University of Michigan, Dearborn
Feeney, Alison E., Shippensburg University
Felix, Joseph D., Texas A&M University, Corpus Christi
Ferguson, Teresa D., Oak Ridge National Laboratory
Ferreira, Maryanne F., Woods Hole Oceanographic Institution
Ferren, Richard L., Berkshire Community College
Fifer, Fred L., University of Texas, Dallas
Fish, Jennifer M., University of California, Santa Cruz
Fitzpatrick, John R., Los Alamos National Laboratory
Fitzsimmons, Kevin, University of Arizona
Fjeldheim, Nancy, Brown University
Flaherty, Frank A., Valdosta State University
Flechsig, Sandra, Rice University
Fletcher, Austin, University of Guelph
Flores-Garcia, Mari C., California State University, Northridge
Floyd, Barry, New York State Geological Survey
Fogel, David N., Los Alamos National Laboratory
Foley, John P., Montana Tech of The University of Montana
Foster, Christine, University of Montana
Foster, Michael, Moss Landing Marine Laboratories
Foti, Pamela, Northern Arizona University
Fox, Laurel R., University of California, Santa Cruz
Freiermuth, Sue, University of Wisconsin, River Falls
Frisk, DeAnn, Iowa State University of Science & Tech
Fry, Matthew, University of North Texas
Fryett, I, Cardiff University
Fuellhart, Kurtis G., Shippensburg University
Gagne, Marc R., West Chester University
Gallick, Roberta T., Point Park University
Ganey-Curry, Patricia E., University of Texas at Austin
Gao, Mengsheng, University of Florida
Gao, Oliver H., Cornell University
Garcia, Jan, Occidental College
Garrod, Bruce, University of Toronto
Gartner, John F., University of Waterloo
Garvin, Theresa D., University of Alberta
Gautsch, Jacklyn, Iowa Dept of Natural Resources
Gehrels, Tom, University of Arizona
Geller, Jonathan, Moss Landing Marine Laboratories
Geller, Michael D., Richard Stockton College of New Jersey
Geller, Murray, Jet Propulsion Laboratory
Gemignani, Robert, Washington University in St. Louis
Gerhardt, Hannes, University of West Georgia
Gerlach, Russel L., Missouri State University
Gerry, Janelle, University of Nebraska, Lincoln
Ghosheh, Baher A., Edinboro University of Pennsylvania
Giammar, Daniel, Washington University in St. Louis
Gibbs, Gaynell, Louisiana State University
Gibson, Deana, Missouri State University
Gifford-Gonzalez, Dianne, University of California, Santa Cruz
Gilbert, Co'Quesie, Columbia University
Gillespie, Thomas, Emory University
Gillette, David P., University of North Carolina, Asheville
Gilligan, Jonathan M., Vanderbilt University
Gitelson, Anatoly, Unversity of Nebraska - Lincoln
Glaser, Brian, Black Hawk College
Glenn, Ed, University of Arizona
Gmitro, Helen, Union County College
Goldhamer, David A., University of California, Davis
Gong, Hongmian, Hunter College (CUNY)
Gonzalez, Richard, University of San Diego
Good, Daniel B., Georgia Southern University
Goodin, Ruth, University of Miami
Goodison, Marjorie, University of Rochester
Goodwin, Paul, University of Guelph
Gordon, Andrew, University of Guelph
Gottgens, Johan F., University of Toledo
Gouhier, Tarik, Northeastern University
Grace, Shannon M., Austin Community College District
Graff, William P., New Jersey Geological and Water Survey
Grande, Anthony, Hunter College (CUNY)
Granshaw, Frank D., Portland State University
Grasso, Cheryl, University of Rhode Island
Grattan, Stephen R., University of California, Davis
Gravely, Cynthia Rae, Clemson University
Gray, Sarah, University of San Diego
Gray, Steven, University of Massachusetts, Boston
Green, Brittany, Kansas State University
Greene, Barbara, Lock Haven University
Greene, Pauline R., University of Louisiana at Lafayette

Greene, Roberta, SUNY Potsdam
Griffin, Kevin, Columbia University
Griffith, Caitlin, University of Arizona
Gripshover, Margaret "Peggy", Western Kentucky University
Gritzo, Russell E., Los Alamos National Laboratory
Groppi, Christopher, Arizona State University
Grossman, Lawrence S., Virginia Polytechnic Institute & State University
Guerra, Oralia, Austin Community College District
Guerrieri, Mary, University of Arizona
Guikema, Seth, Johns Hopkins University
Guzman, Ernesto, University of Guelph
Habash, Marc, University of Guelph
Hacker, Patricia, University of Toledo
Haddad, Brent, University of California, Santa Cruz
Haff, Peter K., Duke University
Hafner, James A., University of Massachusetts, Amherst
Haggart, Renee, University of British Columbia
Haigh, Nardia, University of Massachusetts, Boston
Halfman, Barbara, Hobart & William Smith Colleges
Hall, Christopher C., University of Guelph
Hall, I, Cardiff University
Hall, Jude, Denison University
Hall, Robert, University of Guelph
Hall, Ron, University of Lethbridge
Hallett, Rebecca, University of Guelph
Hamilton, George, Berkshire Community College
Hammersley, Charles, Northern Arizona University
Hammond, Anne, Lakehead University
Haney, Mary, University of South Florida, Tampa
Hanke, Steve H., Johns Hopkins University
Hankins, Katherine B., Georgia State University
Hanna, Heather, North Carolina Geological Survey
Hansen, Devon A., University of North Dakota
Hansen, Jeffrey C., Los Alamos National Laboratory
Hardy, Shaun J., Carnegie Institution of Washington
Harris, Virginia M., Wesleyan University
Harrison, Conor, University of South Carolina
Hassard, Stacey K., College of Charleston
Hawley, Rebecca D., Northern Arizona University
Haynes, Kyle J., University of Virginia
Heaton, Richard C., Los Alamos National Laboratory
Heatwole, Charles A., Graduate School of the City University of New York
Heatwole, Charles A., Hunter College (CUNY)
Heckathorn, Scott, University of Toledo
Heenan, Cleo J., South Dakota School of Mines & Tech
Hegarty, Patricia, University College Cork
Heidt, James, University of Wisconsin Colleges
Heimpel, Moritz, University of Alberta
Heitman, Mary P., Iowa Dept of Natural Resources
Helmick, Sara B., Los Alamos National Laboratory
Hendrikx, Jordy, Montana State University
Henshel, Diane S., Indiana University, Bloomington
Herr, Larry, University of Lethbridge
Hicks, Andrea, University of Wisconsin, Madison
Higinbotham, Pamela, California University of Pennsylvania
Hildebrand, Stephen G., Oak Ridge National Laboratory
Hill, Arleen A., University of Memphis
Hill, Julie, University of Nevada, Reno
Hiller, Lena, Hofstra University
Hilpert, Markus, Johns Hopkins University
Hilts, Stewart G., University of Guelph
Hinderaker, Pam, Iowa State University of Science & Tech
Hiser, Susan, Marietta College
Hobbs, Benjamin F., Johns Hopkins University
Hobbs, John D., Montana Tech of the University of Montana
Hockey, Thomas A., University of Northern Iowa
Hodges, Jackie, Fort Valley State University
Hodgson, M. John, University of Alberta
Hohl, Eric, Missouri Dept of Natural Resources
Holbrook, Amanda, Morehead State University
Hollenbeck, Diane, Metropolitan State College of Denver
Holstein, Thomas J., Roger Williams University
Holubnyak, Yehven I., University of Kansas
Hommel, Demian, Oregon State University
Hopson, Janet L., University of Tennessee, Knoxville
Horne, Sharon, University of Windsor
Howard, Theodore E., University of New Hampshire
Hsiang, Tom, University of Guelph
Hsiao, Theodore C., University of California, Davis
Huang, Norden E., University of Delaware
Huckabey, Marsha, University of Missouri
Hudman, Lloyd E., Brigham Young University
Huffman, Debra E., University of South Florida
Huffman, Robert L., Mercer University
Hughes, Joseph B., Rice University
Hughes, Randall, Northeastern University
Hull, Donald L., Los Alamos National Laboratory
Humphrey, Peggy, Montana State University
Hunt, Kathy, University of Maryland
Hunt, Shelley, University of Guelph
Hunter, John, Rice University
Huntoon, Laura, University of Arizona
Hutchins, Peter S., Mississippi Office of Geology
Hyrapiet, Shireen, Oregon State University
Hysell, David L., Cornell University
Iantria, Linnea, Missouri State University
Ibrahim, Mohamed, Hunter College (CUNY)
Immonen, Wilma, Montana Tech of the University of Montana
Ioannides, Dimitri, Missouri State University
Ironside, R. Geoffrey, University of Alberta
Ivanova, Maria, University of Massachusetts, Boston
Ivy, Russell L., Florida Atlantic University
Jackson, Edgar L., University of Alberta
Jackson, Robert B., Duke University
Jacobs, Tenika, New Jersey Geological and Water Survey
Jamiolahmady, Mahmoud, Heriot-Watt University
Janetos, Tony, Boston University
Jarcho, Kari A., University of Minnesota, Twin Cities
Jeans, Meghan, University of Massachusetts, Boston
Jensen, Scott W., South Dakota Dept of Environment and Natural Resources
Jiang, Zhenhua, Argonne National Laboratory
Joanna, Lucero, New Mexico Institute of Mining and Tech
Johannessen, Carl L., University of Oregon
Johnson, Daniel L., University of Lethbridge
Johnson, Emily P., Boston University
Johnson, Jeanne, Louisiana State University
Johnson, Judy L., University of Alaska, Fairbanks
Johnson, Robert E., Western Illinois University
Johnson, Robert L., Argonne National Laboratory
Johnston, Karin, George Washington University
Jones, Glen, New Mexico Institute of Mining & Tech
Jones, Gwilym, Northeastern University
Jones, Minnie O., University of Illinois at Chicago
Jones, III, John P., University of Arizona
Jornov, Donna, New York State Geological Survey
Joseph, Miranda, University of Arizona
Jun, Young-Shin, Washington University in St. Louis
Jurmanovich, Barb, Delta College
Juszczyk , Carmen , University of Colorado
Jutla, Rajinder S., Missouri State University
Karl, Tami S., Florida State University
Kay, Richard F., Duke University
Keala, Lori, Pomona College
Keating, Martha E., Bentley University
Keaton, Jeffrey R., University of Utah
Keatts, Merida, Kent State University
Keeling, Robyn, Utah Geological Survey
Kehoe-Forutan, Sandra J., Bloomsburg University
Keil, Charles, Wheaton College
Keller, Jean, United States Military Academy
Kelly, Kimberly, Montgomery College
Kelly, Sherrie, St. Lawrence University
Kennedy, Christina B., Northern Arizona University
Kennel, Charles F., University of California, Los Angeles
Kettmann, Elizabeth, Sonoma State University
Kevan, Peter, University of Guelph
Keyser, Jan, University of South Alabama
Khurana, Krishan, University of California, Los Angeles
Kidder, T.R., Washington University in St. Louis
Kifer, Lauri A., SUNY, The College at Brockport
Kile, Susan, Eastern Illinois University
Kim, Stacy, Moss Landing Marine Laboratories

Kimbro, David, Northeastern University
King-Hsi, Kung, Los Alamos National Laboratory
Kinkead, Scott, Los Alamos National Laboratory
Kipper, Jay P., University of Texas at Austin, Jackson School of Geosciences
Knauss, Virginia L., California State University, Dominguez Hills
Kneas, David, University of South Carolina
Knight, Mona M., University of Illinois, Urbana-Champaign
Knippler, Katherine A., Maryland Department of Natural Resources
Kohlstedt, Sally G., University of Minnesota, Twin Cities
Komoto, Cary, University of Wisconsin Colleges
Kontuly, Thomas M., University of Utah
Koralek, Susan, Southern Oregon University
Kornberg, Amy, University of California, Santa Cruz
Kosinski, Leszek A., University of Alberta
Kostov, Svilen, Georgia Southwestern State University
Krauss, Lawrence, Arizona State University
Kressler, Sharon J., University of Minnesota, Twin Cities
Krieble, Kelly, Moravian College
Krishnan, Jay, University of Houston
Kruse, Jennifer, Gustavus Adolphus College
Kugel, Abigail, Winona State University
Kumar, M. Satish, Queens University Belfast
Kuntz, Kara, Fort Hays State University
LaBella, Joel, Wesleyan University
LaDochy, Steve, California State University, Los Angeles
LaFreniere, Lorraine M., Argonne National Laboratory
Lagowski, Alison A., SUNY, Buffalo
Lam, Anita, University of British Columbia
Lane, Mark, Palomar College
Langdon, Phyllis, Sir Wilfred Grenfell College
Lanoue, Christopher A., South Dakota Dept of Environment and Natural Resources
Larkin, Patrick, Texas A&M University, Corpus Christi
Larock, B. E., University of California, Davis
Larson, Harold P., University of Arizona
Laurier, Eric, Edinburgh University
Lavallee, Daniel, University of California, Santa Barbara
Lawrence, Deborah, University of Virginia
Layton, Alice, University of Tennessee, Knoxville
Leal-Bejarano, Arturo, Universidad Autonoma de Chihuahua
Lebofsky, Larry A., University of Arizona
Ledbetter, Cynthia E., University of Texas, Dallas
Lee, Alexis, University of North Carolina Wilmington
Lee, Hung, University of Guelph
Lehane, Mary, University College Cork
Lekan, Thomas, University of South Carolina
Leland, John, University of Nevada, Reno
Lemmond, Peter C., Woods Hole Oceanographic Institution
Lener, Edward, Virginia Polytechnic Institute & State University
Lerdau, Manuel, University of Virginia
Lemurier, Nathalie, Institut Polytechnique LaSalle Beauvais (ex-IGAL)
Leveque, Connie, University of Maine, Presque Isle
Leveson, David J., Graduate School of the City University of New York
Levy, David, University of Massachusetts, Boston
Li, Wenhong, Duke University
Lichtner, Peter C., Los Alamos National Laboratory
Lin, Hsing K., University of Alaska, Fairbanks
Lin, Shoufa, University of Waterloo
Lin, Xiaomao, Kansas State University
Lindley, Stacy, California State University, Sacramento
Linky, Edward, Hunter College (CUNY)
Lipeles, Maxine I., Washington University in St. Louis
Liu, Jian-yi, Montana State University
Liu, Weibo, Florida Atlantic University
Lloyd, Glenn D., Missouri Dept of Natural Resources
Loeb, Valerie, Moss Landing Marine Laboratories
Long, Laura, Lawrence Livermore National Laboratory
Long, Lisa, Ohio Dept of Natural Resources
Longmire, Patrick A., Los Alamos National Laboratory
Lotterhos, Kathleen, Northeastern University
Lounsbury, Diane E., SUNY, Geneseo
Loveland, Karen, Missouri Dept of Natural Resources
Lowry, William R., Washington University in St. Louis
Loxsom, Fred, Eastern Connecticut State University
Lucus, Beth, Virginia Tech
Lujan-Lopez, Arturo, Universidad Autonoma de Chihuahua
Lumpkin, Thomas A., Washington State University
Lynch, Michael J., University of Kentucky
Lyon, Elizabeth C., Pennsylvania Bureau of Topographic & Geologic Survey
Maciha, Mark J., Northern Arizona University
MacInnes, Michael, Los Alamos National Laboratory
Mahler, Robert L., University of Idaho
Maier, Raina M., University of Arizona
Malagon, Teresa Soledad Medina, Universidad Nacional Autonoma de Mexico
Malega, Ron, Missouri State University
Malinowski, Jon C., United States Military Academy
Mand, Arlene, University of Pennsylvania
Mangel, Mark S., University of California, Santa Cruz
Marks, Dennis W., Valdosta State University
Marshall, Stephen, University of Guelph
Martin, Mona, Southern Illinois University Carbondale
Martinez, Judy, University of Utah
Martinez-Macias, Panfilo R., Universidad Autonoma de San Luis Potosi
Masiello, Caroline A., Rice University
Massey, Michael, California State University, East Bay
Mattison, Katherine W., Missouri University of Science and Tech
Maulud, Mat Ruzlin, University of Malaya
May, Diane M., Missouri State University
May, Fred E., University of Utah
Mayer, Christine M., University of Toledo
McArthur, Russell, San Francisco State University
McAtee, Mike, California State Polytechnic University, Pomona
McClain, Lina C., Graduate School of the City University of New York
McCollough, Cherie, Texas A&M University, Corpus Christi
McConnell, Joseph, University of Nevada, Reno
McCoy, Sue, Cosumnes River College
McCulley, Dawn, California State University, Stanislaus
McDermott, Thomas M., SUNY, The College at Brockport
McDonald, Kyle C., Jet Propulsion Laboratory
McDonald, Lynn, Los Alamos National Laboratory
McFadden, Jennifer, Elizabethtown College
McGee, Tara, University of Alberta
McGraw, Maureen A., Los Alamos National Laboratory
McKenzie, Charlotte, Montana Tech of The University of Montana
McKenzie, Connie, Louisiana Tech University
McKenzie, Phyllis, Smithsonian Institution / National Museum of Natural History
McKenzie, Ross, University of Lethbridge
McLafferty, Sara L., Graduate School of the City University of New York
McLaughlin, Richard, Texas A&M University, Corpus Christi
McLeod, Clara, Washington University in St. Louis
McMillan, Robert S., University of Arizona
McNair, Laurie A., Los Alamos National Laboratory
McNamara, Jodi, Colgate University
Medlin, Peggy, Tennessee Tech University
Meehan, Katharine, University of Oregon
Mehes, Roxane J., Laurentian University, Sudbury
Merritt, Dare, East Carolina University
Messer, Sharon, Shawnee State University
Meyer, Christopher, Smithsonian Institution / National Museum of Natural History
Meyer, Judith, Missouri State University
Meyerson, Rohana, Lafayette College
Middlemiss, Lucie, University of Leeds
Miller, Judy, Monroe Community College
Miller, Mark, University of Southern Mississippi
Miller, Ted R., South Dakota Dept of Environment and Natural Resources
Mills, Suzanne, McMaster University
Milstead, Terence, Appalachian State University
Mitchneck, Beth A., University of Arizona
Mittleider, Stacy, Chadron State College
Miyares, Ines, Hunter College (CUNY)
Miyares, Inez, Graduate School of the City University of New York
Mockler, Theodore, Los Alamos National Laboratory
Moe-Hoffman, Amy P., Mississippi State University
Mokos, Jennifer T., Ohio Wesleyan University
Moldwin, Mark B., University of Michigan
Mollner, Daniel, Gustavus Adolphus College
Momen, Nasim, Boston University
Momohara, Kristin, University of Hawai'i, Manoa
Monahan, Adam, University of Victoria

Montagna, Paul, Texas A&M University, Corpus Christi
Montgomery, Tamra S., University of Illinois, Urbana-Champaign
Moodie, T. Bryant, University of Alberta
Mooney, Phillip, Sonoma State University
Moore, Clyde, Louisiana State University
Moore, Donna, Kutztown University of Pennsylvania
Moorhead, Daryl L., University of Toledo
Morehouse, Barbara, University of Arizona
Morgan, Gary, Hampton University
Morgan, Siobahn M., University of Northern Iowa
Morin, Paul, University of Minnesota, Twin Cities
Morrin, M. Elizabeth, Rutgers, The State University of New Jersey, Newark
Morris, Brenda, Palomar College
Morris, Donald P., Lehigh University
Mortenson, Kristine B., Brigham Young University
Moss, Patti, Whitman College
Mountain, Carol S., Columbia University
Mozzachiodi, Riccardo, Texas A&M University, Corpus Christi
Mueller, Amy, Northeastern University
Munoz Mendoza, Alvaro, Universidad Nacional Autonoma de Mexico
Murphy, Alexander B., University of Oregon
Murphy, Edward C., North Dakota Geological Survey
Myers, Clifford D., Berkshire Community College
Myers, Tammy, Shippensburg University
Nadiga, Balu, Los Alamos National Laboratory
Nakatsuka, James, University of California, Los Angeles
Narinder, Kaushik, University of Guelph
Nashold, Barney W., Argonne National Laboratory
Nesbitt, Alex, University of Nevada
Neuman, Dennis R., Montana State University
Nevins, Susan K., SUNY, Cortland
Newman, Jonathan, University of Guelph
Newton, Anthony J., Edinburgh University
Newton, Seth A., Geological Survey of Alabama
Nichols, Liz, Irvine Valley College
Nichols, Terry E., University of Arkansas, Fayetteville
Nicholson, Nanette, American Museum of Natural History
Nielsen, Mary, University of South Dakota
Nixon, Cheryl, University of Massachusetts, Boston
Noll, Michael G., Valdosta State University
Nordwald, Dan, Missouri Dept of Natural Resources
Norman, Catherine, Johns Hopkins University
Nosal, Thomas E., Dept of Energy and Environmental Protection
Nussear, Ken, University of Nevada, Reno
O'Callaghan, Mick, University College Cork
O'Day, Sandra, Central Connecticut State University
O'Melia, Charles R., Johns Hopkins University
O'Neil, Jennifer, University of Nevada, Reno
Obia, Godson C., Eastern Illinois University
Occhiuzzi, Tony, Mesa Community College
Odland, Sarah K., Columbia University
Oglesby, Elizabeth, University of Arizona
Okafor, Florence A., Alabama A&M University
Olson, James R., South Dakota Dept of Environment and Natural Resources
Oona, Hain, Los Alamos National Laboratory
Oppong, Joseph R., University of North Texas
Orrock, John, Washington University in St. Louis
Ortmann, Anthony L., Murray State University
Oswalt, Ginny L., University of Arkansas at Little Rock
Otis, Gard, University of Guelph
Ott, Kevin C., Los Alamos National Laboratory
Otter, Ryan, University of Tennessee, Knoxville
Otto, LeeAnn, University of San Diego
Ouellette, Vicki L., University of Northern Colorado
Owens, Tamera, Muskegon Community College
Oza, Rupal, Hunter College (CUNY)
Pace, Michael L., University of Virginia
Pacia, Christina, Kean University
Palace, Michael W., University of New Hampshire
Palm, Risa I., Georgia State University
Palma, Miclelle, University of Wisconsin Colleges
Palmer, Christina, California State University, San Bernardino
Palmer, Clare, Washington University in St. Louis
Papaleo, Silvanna, University of Toronto
Parendes, Laurie A., Edinboro University of Pennsylvania
Parker, Jim, University of Houston
Parker, Joan, Moss Landing Marine Laboratories
Parker, Marjorie, Bowdoin College
Parker, Stephen R., Montana Tech of the University of Montana
Parrish, Pia, San Diego State University
Parsons, Gail, Augustana College
Pasteris, Jill D., Washington University in St. Louis
Patel, Vinodkumar A., Illinois State Geological Survey
Patino-Douce, Alberto E., University of Georgia
Patronas, Dennis, Dauphin Island Sea Lab
Patterson, Jodi T., Utah Geological Survey
Patterson, Mark, Northeastern University
Penney, Paulette, Baylor University
Pepper, Ian L., University of Arizona
Peri, Francesco, University of Massachusetts, Boston
Perkins, R, Cardiff University
Petersen, Bruce, Washington University in St. Louis
Peterson, Peter A., Iowa State University of Science & Tech
Phelps, Tommy, University of Tennessee, Knoxville
Phillippi, Nathan E., University of North Carolina, Pembroke
Pickering, Dan, Carnegie Museum of Natural History
Pickett, Nicki, Broward College, Central Campus
Pierce, Larry, Missouri Dept of Natural Resources
Pike, J, Cardiff University
Plane, David A., University of Arizona
Plant, Jeffrey J., Jet Propulsion Laboratory
Plescia, Jeffrey, Jet Propulsion Laboratory
Podolak, Morris, Tel Aviv University
Pollak, Robert, Washington University in St. Louis
Polovina, Jeffrey J., University of Hawai'i, Manoa
Ponette-Gonzalez, Alexandra, University of North Texas
Porlas, Dustin, University of Utah
Pournelle, Jennifer, University of South Carolina
Powers, Roger W., University of Alaska, Fairbanks
Poynton, Helen, University of Massachusetts, Boston
Pratte, Darrell D., Missouri Dept of Agriculture
Prewitt, Charles, University of Arizona
Prichard, Terry L., University of California, Davis
Pugh, Teresa, Vanderbilt University
Pundsack, Jonathan, University of Minnesota, Twin Cities
Purcell, Rita, Wellesley College
Purtle, Jennifer M., University of Arkansas, Fayetteville
Quinn, Courtney, Furman University
Quintero, Sylvia, University of Arizona
Rafferty, Milton D., Missouri State University
Rajala, Jacob, Wassuk College
Rakowski, Cynthia L., Pacific Northwest National Laboratory
Rallis, Donald N., Mary Washington College
Ramirez Camperos, Esperanza, Universidad Nacional Autonoma de Mexico
Rankin, Seth, University of Wisconsin Colleges
Rasmussen, Tab, Washington University in St. Louis
Reaven, Sheldon, SUNY, Stony Brook
Rennie, Tim, University of Guelph
Reusch, David B., New Mexico Institute of Mining and Tech
Reynolds, Barbara C., University of North Carolina, Asheville
Rhoads, James, Arizona State University
Rhodes, Carol J., Texas A&M University
Rice, Benjamin, Northwestern University
Rice, Murray, University of North Texas
Ricker, Alison, Oberlin College
Ridgwell, Andy, University of California, Riverside
Ridky, Alice M., Colby College
Rieke, George H., University of Arizona
Riley, James, University of Arizona
Riley, Rhonda, Southern Utah University
Ritter, Leonard, University of Guelph
Rius, Marc, University of Southampton
Rivera, Edna L., University of Illinois at Chicago
Robas, Sheryl A., Princeton University
Robbins, Debra C., University of North Carolina, Asheville
Roberts, A. Lynn, Johns Hopkins University
Robinson, Bruce A., Los Alamos National Laboratory
Robinson, David, The Open University
Robinson, Mark, Arizona State University
Robinson, William, University of Massachusetts, Boston
Rodriguez, Vanessa del S., Puerto Rico Bureau of Geology

Roe, Carol, College of William & Mary
Roemer, Elizabeth, University of Arizona
Rogers, Jefferson S., University of Tennessee, Martin
Rogers, William J., West Texas A&M University
Roinstad, Lori L., South Dakota Dept of Environment and Natural Resources
Rojas, Adena, Johns Hopkins University
Rollinson, Paul A., Missouri State University
Rondot, Beth, Long Island University, C.W. Post Campus
Rooney, Neil, University of Guelph
Rose, Candace M., Argonne National Laboratory
Rose, Dan, University College Cork
Rosengaus, Rebeca, Northeastern University
Ross, Kirstin, Flinders University
Ross, Paula, New Mexico State University, Las Cruces
Rossell, Irene M., University of North Carolina, Asheville
Rostam-Abadi, Massoud, Illinois State Geological Survey
Rostron, Ben, University of Alberta
Rothberg, Maryann, Princeton University
Rouhani, Farhang, Mary Washington College
Rouse, Jesse, University of North Carolina, Pembroke
Roussel-Dupre, R., Los Alamos National Laboratory
Rowland, Scott K., University of Hawai'i, Manoa
Rozmus, Wojciech, University of Alberta
Rumstay, Kenneth S., Valdosta State University
Russell, Ron, University of Texas at Austin, Jackson School of Geosciences
Russell, Terry P., University of Victoria
Russell, Theresa J., University of Arkansas, Fayetteville
Russo, Mary Rose, Princeton University
Ruzicka, Jaromir, University of Hawai'i, Manoa
Sabala-Foreman, Susan M., Northern Arizona University
Saikia, Udoy, Flinders University
Saku, James C., Frostburg State University
Salisbury, Joseph, University of New Hampshire
Sammarco, Paul W., Louisiana State University
Sanchez, Charles, University of Arizona
Sandhu, Harpinder, Flinders University
Santander, Erma, University of Arizona
Sapigao, Gladys, Queens College (CUNY)
Sarin, Manmohan, University of Delaware
Saripalli, Srikanth, Arizona State University
Sarmiento, Jorge L., Princeton University
Sauer, Nancy N. S., Los Alamos National Laboratory
Scannapieco, Evan, Arizona State University
Scapelli, Krista, Texas Christian University
Schaal, Barbara, Washington University in St. Louis
Schauss, Kim E., University of Southern Indiana
Scheel, Patrick, Missouri Dept of Natural Resources
Scheidemen, Kathy J., University of California, Santa Barbara
Schell, Marie, Western University
Schlesinger, William H., Duke University
Schlumpberger, Debbie, Saint Cloud State University
Schmidt, Jonathan, University of Guelph
Schoenberger, Erica, Johns Hopkins University
Schroeder, Kathleen, Appalachian State University
Schwankl, Larry J., University of California, Davis
Schwarz, Karen M., West Chester University
Schwarzschild, Arthur C., University of Virginia
Schwenke, Eszter, University of New Brunswick
Schwob, Stephanie L., Southern Methodist University
Scogin, Linda, University of New Hampshire
Scott, Kathy, University of British Columbia
Scott-Dupree, Cynthia, University of Guelph
Scyphers, Steven, Northeastern University
Sears, Heather, Northeastern University
Selin, Helaine, Hampshire College
Serrano, Carmen, Florida Institute of Tech
Shade, Janet, University of Pittsburgh, Bradford
Shah, Ashru, North Carolina State University
Sharma, Govind, Alabama A&M University
Sharp, Deanna M., Austin Community College District
Shaw, Fred, Missouri Dept of Natural Resources
Shchukin, Eugene D., Johns Hopkins University
Sheley, Christina, Indiana University, Bloomington
Shelton, Robert B., Oak Ridge National Laboratory
Shiaris, Michael, University of Massachusetts, Boston
Shields, Nancy, Virginia Polytechnic Institute & State University
Shock, Everett, Arizona State University
Showman, Adam, University of Arizona
Shumway, Matthew J., Brigham Young University
Sibley, Paul, University of Guelph
Sigler, William V., University of Toledo
Silks, Louis A., Los Alamos National Laboratory
Silvano, Janet, Tufts University
Simon, Kathy, California Polytechnic State University
Simonetti, Stephanie, University of Notre Dame
Simovich, Marie, University of San Diego
Singh, Hanumant, Northeastern University
Singh, Harbans, Montclair State University
Sissom, David , West Texas A&M University
Sitwell, O.F. George, University of Alberta
Skeel, Loreene, University of Montana
Slater, Tom, Edinburgh University
Slavetskas, Carol, Binghamton University
Smee, Delbert L., Texas A&M University, Corpus Christi
Smith, Derald, University of Lethbridge
Smith, Donald R., University of California, Santa Cruz
Smith, Elizabeth Y., University of Nevada, Las Vegas
Smith, Erik, University of Saint Thomas
Smith, H D., Cardiff University
Smith, Jim, Flinders University
Smith, K. L., Wichita State University
Smith, Paul H., Los Alamos National Laboratory
Smith, Peter J., University of Alberta
Smith, Susan M., Montana Tech of The University of Montana
Solomon, Keith, University of Guelph
Somers, Jr., Arnold E., Valdosta State University
Sorenson, Mary Clare, University of Wisconsin, Stevens Point
Southwell, Benjamin, Lake Superior State University
Spears, Ellen, University of Alabama
Spencer, Mary R., University of Kentucky
Sperazza, Michael, SUNY, Stony Brook
Spracklen, Dominick, University of Leeds
Stadnyk, Leona, Mount Royal University
Stahl, Terry L., Carnegie Institution of Washington
Stampone, Mary D., University of New Hampshire
Standridge, Debbie, Georgia Southwestern State University
Stanton, Stephen, University of Missouri
Starr, Richard, Moss Landing Marine Laboratories
Steadman, Todd A., Clemson University
Stechmann, Samuel, University of Wisconsin, Madison
Steller, Diana, Moss Landing Marine Laboratories
Stephenson, Gerry, University of Guelph
Stepien, Carol A., University of Toledo
Sternberg, Rolf, Montclair State University
Stevens, Joan M., California Polytechnic State University
Stierle, Andrea, Montana Tech of the University of Montana
Stiles, Lynn F., Richard Stockton College of New Jersey
Stimer, Debra, Kent State University at Stark
Stinger, Lindsay C., University of Leeds
Stokes, Patricia, Utah Geological Survey
Stone, Glenn D., Washington University in St. Louis
Strangeway, Robert J., University of California, Los Angeles
Strole, Torie L., University of Illinois, Urbana-Champaign
Strong, Ellen, Smithsonian Institution / National Museum of Natural History
Stunz, Greg, Texas A&M University, Corpus Christi
Su, Xiaobo, University of Oregon
Sullivan, Barbara, Argonne National Laboratory
Summers, Robert, University of Alberta
Sussman, Robert W., Washington University in St. Louis
Sutherland, Bruce R., University of Alberta
Svitra, Zita V., Los Alamos National Laboratory
Swanson, Basil I., Los Alamos National Laboratory
Swap, Robert J., University of Virginia
Swartwood, Jade L., Bloomsburg University
Sweet, Brenda, Lyndon State College
Swetnam, Thomas W., University of Arizona
Swindall, Diane, University of California, Davis
Symbalisty, E.M.D., Los Alamos National Laboratory
Symes, William S., Rice University
Taney, R. Marieke, Northern Arizona University
Tanji, K. K., University of California, Davis

Tavener, Kristi, Lakehead University
Taylor, David J., University of Chicago
Taylor, Michael, Flinders University
Taylor, Robert W., Montclair State University
Templeton, Alan R., Washington University in St. Louis
Tencate, James, Los Alamos National Laboratory
Teplitski, Max, University of Florida
Ter-Simonian, Vardui, University of Southern California
Terkla, David, University of Massachusetts, Boston
Theis, Karen, University of Manchester
Thériault, Julie Mireille, Universite du Quebec a Montreal
Thompson, Glennis, University of California, Berkeley
Thomson, Cary, University of British Columbia
Thomson, Cynthia, Columbia University
Thomson, Vivian E., University of Virginia
Timmons, David, University of Massachusetts, Boston
Tingey, David G., Brigham Young University
Tinnon, Vicki, University of Southern Mississippi
Tissot, Philippe , Texas A&M University, Corpus Christi
Todd, Brenda, University of Nebraska at Omaha
Torcellini, Paul, Eastern Connecticut State University
Townshend, Ivan, University of Lethbridge
Tracy, Matthew, United States Air Force Academy
Trevors, Jack, University of Guelph
Triplehorn, Judy, University of Alaska, Fairbanks
Trost, G K., University of Arkansas, Fayetteville
Trout, Jennifer, Western Michigan University
Trussell, Geoffrey, Northeastern University
Tufford, Dan, University of South Carolina
Tuller, Markus, University of Arizona
Turner, Jay, Washington University in St. Louis
Tyning, Thomas F., Berkshire Community College
Unkefer, Clifford J., Los Alamos National Laboratory
Unsworth, Martyn, University of Alberta
Urquhart, Alvin W., University of Oregon
Utter, James M., SUNY, Purchase
van den Akker, Ben, Flinders University
Van Eerd, Laura, University of Guelph
van Gardingen, Paul R., Edinburgh University
Van Otten, George A., Northern Arizona University
Van Roosendaal, Susan, University of Utah
Vancas, Tina, Pennsylvania State University, University Park
Varadi, Ferenc D., University of California, Los Angeles
Varady, Robert G., University of Arizona
Veeder, Glenn J., Jet Propulsion Laboratory
Vieira, David J., Los Alamos National Laboratory
Vierrether, Chris, Missouri Dept of Natural Resources
Vlahovic, Lou Ann, Moravian College
Vollmer, Steve, Northeastern University
Wada, Ikuko, University of Minnesota, Twin Cities
Wade, Jeff, Northeastern Illinois University
Wadhwa, Meenakshi, Arizona State University
Wagh, Swati, Argonne National Laboratory
Waite, Cynthia, University of Southern California
Walker, Mark, University of Nevada, Reno
Walker, Peter A., University of Oregon
Walker, Raymond J., University of California, Los Angeles
Walraven, Brenda, Mercer University
Walrod, Amanda G., University of Arkansas, Fayetteville
Walsh, Christopher, University of Maryland
Walsh, Daniel E., University of Alaska, Fairbanks
Walsh, Ellen c., Lawrence University
Warburg, Helena F., Williams College
Ward, David M., Montana State University
Ward, Marie D., University of Alaska, Fairbanks
Warhaft, Zellman, Cornell University
Warkentin, Alicia, University of British Columbia
Warren, Richard L., University of Arizona
Waterstone, Marvin, University of Arizona
Watson, Alan, University of Guelph
Wauthier, Christelle, Pennsylvania State University, University Park
Weaver, Douglas J., Los Alamos National Laboratory
Webre, Cherri B., Louisiana State University
Weinstein, Charles E., Berkshire Community College
Weintraub, Michael N., University of Toledo
Wells, Steve G., University of Nevada, Reno
Welton, Leicha, University of Alaska, Fairbanks
Wenz, Helmut C., University of Tennessee, Martin
Wesler, Kit W., Murray State University
Wetzel, Dan L., University of Alaska, Fairbanks
Whiley, Harriet, Flinders University
Whitaker, Brenda, University of Virginia College, Wise
White, George W., Frostburg State University
Whitney, Earl M., Los Alamos National Laboratory
Wiederspahn, Mark, University of Texas at Austin
Wiklund, Marie-Louise M., Geological Survey of Finland
Wilder, Margaret, University of Arizona
Wilhelmy, Jerry B., Los Alamos National Laboratory
William, Nancy, Kansas State University
Williams, Allison M., McMaster University
Williams, Carolyn S., Virginia Polytechnic Institute & State University
Williams-Bruinders, Leizel, Nelson Mandela Metropolitan University
Willson, Lee, Rice University
Wilmut, Michael, University of Victoria
Wilson, Robert, University of St. Andrews
Wilson, Sarah, Washington & Lee University
Wilson, Thomas B., University of Arizona
Wilton, Robert D., McMaster University
Winsor, Roger A., Appalachian State University
Winston, Barbara, Washington University in St. Louis
Winterkamp, Judith L., Los Alamos National Laboratory
Winton, Alison, Texas Tech University
Withers, Kim, Texas A&M University, Corpus Christi
Wixman, Ronald, University of Oregon
Wolfe, Karen M., Middle Tennessee State University
Wollan, Jacinda, North Dakota State University
Wollons, Roberta, University of Massachusetts, Boston
Wong, Cindy, Buffalo State College
Wood, Carolyn F., Dauphin Island Sea Lab
Wood, Stephen E., University of Washington
Woodard, Gary C., University of Arizona
Woodley, Teresa, University of British Columbia
Woodruff, William H., Los Alamos National Laboratory
Woods, Karen M., Lamar University
Woods, Neal, University of South Carolina
Wooldridge, C F., Cardiff University
Wright, Kathym, Western Michigan University
Wu, David T., Colorado School of Mines
Wyckoff, William K., Montana State University
Yarbrough, Robert A., Georgia Southern University
Yates, Mary Anne, Los Alamos National Laboratory
Yelle, Roger, University of Arizona
Yoakum, Barbara, United States Naval Academy
Yochem, Pam, University of San Diego
Yonetani, Mage, Utah Geological Survey
Yoskowitz, David , Texas A&M University, Corpus Christi
Young, Donald W., University of Arizona
Young, Patrick, Arizona State University
Young, Priscilla E., South Dakota Dept of Environment and Natural Resources
Yow, Sonja H., Eastern Kentucky University
Yutzy, Gale, Frostburg State University
Zajac, Roman N., University of New Haven
Zavada, Michael S., Providence College
Zeiger, Elaine, Field Museum of Natural History
Zelizer, Nora, Princeton University
Zhang, Max, Cornell University
Zhang, Wei, University of Massachusetts, Boston
Zimmermann, George, Richard Stockton College of New Jersey
Zotti, Maria, Flinders University
Zurick, David, Eastern Kentucky University
Zwiefelhofer, Luke, Winona State University

# Faculty Index

## A

Aagaard, Knut, (206) 543-8942 aagaard@apl.washington.edu, Univ of Washington – Op
Aalto, Kenneth R., (707) 826-4978 kra1@humboldt.edu, Humboldt State Univ – Gr
Aamodt, Paul, (505) 665-1331 plaamod@lanl.gov, Los Alamos National Laboratory – On
Aanstoos, James V., 515-294-1032 aanstoos@iastate.edu, Iowa State Univ of Science & Tech – Or
Aay, Henry, (616) 526-7033 aay@calvin.edu, Calvin Coll – Oyu
Abalos, Benito, benito.abalos@ehu.eus, Univ of the Basque Country UPV/EHU – GctGp
Abbott, Geoffrey D., +44 (0) 191 208 6608 geoff.abbott@ncl.ac.uk, Univ of Newcastle Upon Tyne – Co
Abbott, Lon, (303) 492-6172 lon.abbott@colorado.edu, Univ of Colorado – GgtGm
Abbott, Mark B., (412) 624-1408 mabbott1@pitt.edu, Univ of Pittsburgh – Gs
Abbott, Patrick L., (619) 594-5591 pabbott@mail.sdsu.edu, San Diego State Univ – Gs
Abbott, Richard N., (828) 262-3055 abbottrn@appstate.edu, Appalachian State Univ – Gzi
Abdalati, Waleed, WaleednAbdalati@colorado.edu, Univ of Colorado – Or
Abdelsalam, Mohamed, 405-744-6358 mohamed.abdel_salam@okstate.edu, Oklahoma State Univ – Gct
ABDO, Ginette, 406-496-4152 gabdo@mtech.edu, Montana Tech of The Univ of Montana – Gg
Abdu, Yassir, (204) 474-7356 abdu@cc.umanitoba.ca, Univ of Manitoba – Gz
Abdulghani, Waleed, +96638602848 wmaghani@kfupm.edu.sa, King Fahd Univ of Petroleum and Minerals – Go
Abdulla, Hussain, 361-825-6050 Hussain.Abdulla@tamucc.edu, Texas A&M Univ, Corpus Christi – Co
Abdullah, Nuraiteng Tee, 03-79674229 naiteng@um.edu.my, Univ of Malaya – Gr
Abdullah, Wan Hasiah, 03-79674232 wanhasia@um.edu.my, Univ of Malaya – Ec
Abdullatif, Osman, +96638601479 osmanabd@kfupm.edu.sa, King Fahd Univ of Petroleum and Minerals – Gso
Abdurrahman, A., aabdrrahman@unilorin.edu.ng, Univ of Ilorin – CeEg
Abend, Martin, (201) 200-3161 mabend@njcu.edu, New Jersey City Univ – Oy
Aber, James S., jaber@emporia.edu, Emporia State Univ – Gl
Abernathey, Ryan P., 845-365-8185 rpa@ldeo.columbia.edu, Columbia Univ – Op
Abney, Kent D., (505) 665-3894 Los Alamos National Laboratory – On
Abokhodair, Abdulwahab, +96638602625 akwahab@kfupm.edu.sa, King Fahd Univ of Petroleum and Minerals – Ym
Abolins, Mark J., (615)904-8372 Mark.Abolins@mtsu.edu, Middle Tennessee State Univ – Gc
Abousleiman, Younane, 405-325-2900 yabousle@ou.edu, Univ of Oklahoma – Nr
Abrams, Lewis J., 910-962-2350 abramsl@uncw.edu, Univ of North Carolina Wilmington – Gu
Abrams, Michael J., (818) 354-0937 Jet Propulsion Laboratory – Or
Abramson, Evan H., 206-616-4388 evan@ess.washington.edu, Univ of Washington – Gy
Abreu, Maria E., msabreu@utad.pt, Universidade de Trás-os-Montes e Alto Douro – GaaGa
Abt, Charlotte J., +440151 794 5178 Chj@liverpool.ac.uk, Univ of Liverpool – On
Abu Bakar, Muhammad Z., 0092-42-99029487 mzubairab1977@gmail.com,Univ of Engineering and Tech – NrmNg
Abushagur, Sulaiman, 915-831-2539 sulaiman@epcc.edu, El Paso Community Coll – Gg
Acharya, Kumud, (702) 862-5371 kumud.acharya@dri.edu, Univ of Nevada, Reno – HsGeHq
Achauer, Ulrich, ulrich.achauer@unistra.fr, Universite de Strasbourg – Ysg
Achten, Christine, +49-251-83-33941 achten@uni-muenster.de, Universitaet Muenster – CloEg
Ackerman, Jessica R., 217-333-4258 jracker@illinois.edu, Univ of Illinois, Urbana-Champaign – On
Ackerman, Steven A., (608)263-3647 stevea@ssec.wisc.edu, Univ of Wisconsin, Madison – Or
Ackerman, Thomas P., 206-221-2767 ackerman@atmos.washington.edu, Univ of Washington – Oa
Ackert, Robert P., 617 496-6449 rackert@fas.harvard.edu, Bentley Univ – CcGlm
Acosta, Jose G., jchang@cicese.mx, Centro de Investigación Científica y de Educación Superior de Ensenada – Ne
Acosta, Patricia E., (413) 597-2221 patricia.e.acosta@williams.edu, Williams Coll – On
Adabanija, Moruffdeen A., +2348037786592 maadabanija@lautech.edu.ng, Ladoke Akintola Univ of Tech – GggGg
Adam, Iddrisu, iddrisu.adam@uwc.edu, Univ of Wisconsin Colls – OynOn
Adam, Jurgen, +44 1784 414258 Jurgen.Adam@rhul.ac.uk, Univ of London, Royal Holloway & Bedford New Coll – Gc
Adamo, Lauren N., lauren.adamo@rutgers.edu, Rutgers, The State Univ of New Jersey – GuPm
Adamo, Peter J., 204-727-9683 adamop@brandonu.ca, Brandon Univ – Gg
Adams, Aubreya, 315-228-7202 aadams@colgate.edu, Colgate Univ – Yg
Adams, Debi, 806-651-2570 dadams@wtamu.edu, West Texas A&M Univ – On
Adams, Donald D., (518) 564-4037 adamsdd@plattsburgh.edu, Plattsburgh State Univ (SUNY) – Cl
Adams, Gerald E., 312-344-7540 gadams@colum.edu, Columbia Coll Chicago – Gi
Adams, Herbert G., (818) 677-2575 herb.adams@csun.edu, California State Univ, Northridge – Ng
Adams, John B., adams@ess.washington.edu, Univ of Washington – GcXgOr
Adams, Kenneth, 775.673.7345 ken.adams@dri.edu, Univ of Nevada, Reno – Gam
Adams, Manda S., 704-687-5984 Manda.Adams@uncc.edu, Univ of North Carolina, Charlotte – Ow
Adams, Mark, capt.mark.adams@gmail.com, Lamar Univ – Xg
Adams, Mark G., (828) 262-3049 adamsmg@appstate.edu, Appalachian State Univ – Gp
Adams, Marshall, (865) 574-7335 sma@ornl.gov, Oak Ridge National Laboratory – Ob
Adams, Michael S., (608) 263-5994 Univ of Wisconsin, Madison – Ob
Adams, Peter N., (352) 846-0825 adamsp@ufl.edu, Univ of Florida – Gm
Adams, Thomas, 210-486-0045 tadams67@alamo.edu, San Antonio Community Coll – PvGg
Adamsen, Floyd, (602) 437-1702 f.j.adamsen@gmail.com, Univ of Arizona – So
Adedcyin, A. D., deloadedoyin@yahoo.com, Univ of Ilorin – Eg
Adee, Eric, (785) 354-7236 eadee@ksu.edu, Kansas State Univ – So
Adegoke, Jimmy, (816) 235-2978 adegokej@umkc.edu, Univ of Missouri-Kansas City – OayOr
Adeigbe, O. C., oc.adeigbe@mail.ui.edu.ng, Univ of Ibadan – GosGe
Adekola, S. A., 234-708-928-5170 adekoladsolo@gmail.com, Obafemi Awolowo Univ – GdoCg
Adelana, S.M. A., adelana@gmx.net, Univ of Ilorin – HgYgGe
Aden, Douglas, (614) 265-6579 doug.aden@dnr.state.oh.us, Ohio Dept of Natural Resources – GeOnGl
Adentunji, Jacob, j.adetunji@derby.ac.uk, Univ of Derby – Ge
Adepelumi, A. A., 234-8128181062 aadepelu@oauife.edu.ng, Obafemi Awolowo Univ – YgsNe
Adepoju, Mohammed O., 08034722855 moadepoju@futa.edu.ng, Federal Univ of Tech, Akure – GgEg
Adetunji, A, 234-8137433622 Obafemi Awolowo Univ – EgGi
Adewoye, Abosede O., +2347032045714, aoadewoye55@lautech.edu.ng, Ladoke Akintola Univ of Tech – GeeGe
Adeyemi, G. O., go.adeyemi@mail.ui.edu.ng, Univ of Ibadan – NgHg
Adhikari, Sanchayeeta, 818-677-5630 sadhikari@csun.edu, California State Univ, Northridge – Oi
Adisa, Adeshina L., 08029458886 adeshina4me2000@gmail.com, Federal Univ of Tech, Akure – GgCe
Adkins, Jess F., (626) 395-8550 jess@gps.caltech.edu, California Inst of Tech – Cm
Admire, Amanda R., 707-826-3111 ara11@humboldt.edu, Humboldt State Univ – Og
Adomaitis, D., 217-244-8872 Dadomait@illinois.edu, Univ of Illinois, Urbana-Champaign – Ge

Adovasio, James M., 814-824-2581 adovasio@mercyhurst.edu, Mercyhurst Univ – GafGs

Adrain, Jonathan M., (319) 335-1539 jonathan-adrain@uiowa.edu, Univ of Iowa – Pi

Adrain, Tiffany S., (319) 335-1822 tiffany-adrain@uiowa.edu, Univ of Iowa – Pg

Adriano, Domy C., (803) 725-2752 Univ of Georgia – Sc

Afolabi, O., 234-803-512-4838 toltobi@yahoo.com, Obafemi Awolowo Univ – Yg

Afolabi, Olukayode A., +2348052158722 oaafolabi@lautech.edu.ng, Ladoke Akintola Univ of Tech – GxxGx

Agee, Carl A., (505) 277-1644 agee@unm.edu, Univ of New Mexico – Cp

Agee, Ernest M., (765) 494-3282 eagee@purdue.edu, Purdue Univ – Oa

Agioutantis, Zach, 859-257-953 zach.agioutantis@uky.edu, Univ of Kentucky – NrmEc

Agnew, Duncan C., (858) 534-2590 dagnew@ucsd.edu, Univ of California, San Diego – Ys

Agnew, Stephen F., (505) 665-1764 Los Alamos National Laboratory – On

Agnini, Claudia, 39-049-827918 claudia.agnini@unipd.it, Università degli Studi di Padova – PmeGu

Agrawal, Abinash, (937) 775-3455 abinash.agrawal@wright.edu, Wright State Univ – ClHwPy

Agrawal, Sandeep , 780-492-1230 sagrawal@ualberta.ca, Univ of Alberta – Ou

Agterberg, Frits P., 613-996-2374 Frits.Agterberg@nrcan-rncan.gc.ca, Univ of Ottawa – Gq

Ague, Jay J., (203) 432-3171 jay.ague@yale.edu, Yale Univ – Gx

Aharon, Paul, (205) 348-2528 paharon@ua.edu, Univ of Alabama – ClmPy

Ahearn, Sean C., (212) 772-5327 Graduate School of the City Univ of New York – Or

Ahern, Judson L., (405) 325-4480 jahern@ou.edu, Univ of Oklahoma – Yg

Ahmad, Abd Rashid, 03-79674156 abrashid@um.edu.my, Univ of Malaya – Sc

Ahmed, Kazi M., +880 1711846840 kmahmed@du.ac.bd, Univ of Dhaka – HwGgHg

Ahmed, Muhammad F., mfageo@hotmail.com, Univ of Engineering and Tech – Ng

Ahmed, Waquar, 940-565-2721 waquar.ahmed@unt.edu, Univ of North Texas – On

Ahola, John, John.Ahola@bristolcc.edu, Bristol Community Coll – Gg

Ahrens, Donald C., (209) 575-6300 Modesto Junior Coll – Og

Aiello, Ivano, (831) 771-4400 iaiello@mlml.calstate.edu, Moss Landing Marine Laboratories – Gu

Aiken, Carlos L., (972) 883-2450 aiken@utdallas.edu, Univ of Texas, Dallas – Yv

Aiken, Robert, (785) 462-6281 raiken@ksu.edu, Kansas State Univ – SoYhHq

Aines, Roger D., (925) 423-7184 aines1@llnl.gov, Lawrence Livermore National Laboratory – Cg

Aird, Hannah M., haird@csuchico.edu, California State Univ, Chico – GxEg

Aitkenhead-Peterson, Jacqueline A., (979) 845-3682 jacqui_a-p@tamu.edu, Texas A&M Univ – HsOsGf

Aiyyer, Anantha, 919-515-7973 aaiyyer@ncsu.edu, North Carolina State Univ – Oa

Aja, Stephen U., (718) 951-5405 Graduate School of the City Univ of New York – Cl

Aja, Stephen U., 718-951-5000 x2881 suaja@brooklyn.cuny.edu, Brooklyn Coll (CUNY) – Cl

Ajassa, Roberto, roberto.ajassa@unito.it, Università di Torino – Oy

Ajayi, J. O., 234-803-401-4357 owoajayi@oauife.edu.ng, Obafemi Awolowo Univ – Oo

Ajayi, T. R., 234-803-725-8924 traajayi@oauife.edu.ng, Obafemi Awolowo Univ – CgEgGe

Ajigo, Isaac O., 08032107038 ioajigo@futa.edu.ng, Federal Univ of Tech, Akure – GgNm

Akabzaa, Thomas M., 233-246325685 akabzaa@ug.edu.gh, Univ of Ghana – EgGeNm

Akaegbobi, I. M., izumike2002@yahoo.com, Univ of Ibadan – Cg

Akgul, Bünyamin, 00904242370000-5988 bakgul@firat.edu.tr, Firat Univ – GzyGi

Akgun, Elif , 00904242370000-5972 efiratligil@firat.edu.tr, Firat Univ – GgcGt

Akinlabi, Ismaila A., +2348050225113 iaakinlabi@lautech.edu.ng, Ladoke Akintola Univ of Tech – GggGg

Aksoy, Ercan, 00904242370000-5974 Firat Univ – Ggt

Aksu, Ali E., (709) 737-8385 aaksu@sparky2.esd.mun.ca, Memorial Univ of Newfoundland – Gu

Akujieze, Christopher N., krisjiaku@yahoo.com, Univ of Benin – Hw-GeHg

Al, Tom A., 613-562-5800 6966 tom.al@uottawa.ca, Univ of Ottawa – HwCgOg

Al-Aasm, Ihsan S., (519) 253-3000 ext 2494 alaasm@uwindsor.ca, Univ of Windsor – GdClGo

Al-Attar, David, +44 (0) 1223 348935 da380@cam.ac.uk, Univ of Cambridge – YdGt

Al-Lehyani, Ayman, +96638601661 allehyani@kfupm.edu.sa, King Fahd Univ of Petroleum and Minerals – Yg

Al-Ramadan, Khalid, +96638607175 ramadank@kfupm.edu.sa, King Fahd Univ of Petroleum and Minerals – GodGs

Al-Saad, Hamad A., hamadsaad@qu.edu.qa, Univ of Qatar – PsGsu

Al-Shaibani, Abdulaziz, +96638604002 shaibani@kfupm.edu.sa, King Fahd Univ of Petroleum and Minerals – HwGeg

Al-Shuhail, Abdullah, +966138602538 shuhail@kfupm.edu.sa, King Fahd Univ of Petroleum and Minerals – Ye

Alao, O. A., 234-805-466-7314 olade77@yahoo.com, Obafemi Awolowo Univ – Yg

Alavi, Hedy, (410) 516-7091 alavi@jhu.edu, Johns Hopkins Univ – On

Albach, Suzanne M., 734-677-5111 salbach@wccnet.edu, Washtenaw Community Coll – OegGe

Albee, Arden L., 626.395.6260 aalbee@caltech.edu, California Inst of Tech – GizXg

Albee-Scott, Steven R., (517) 796-8526 albeescsteven@jccmi.edu, Jackson Coll – GePeSf

Alberstadt, Leonard P., 615-322-2160 leonard.p.alberstadt@vanderbilt.edu, Vanderbilt Univ – Gd

Albert, Curtis C., 217-244-2188 abert@illinois.edu, Univ of Illinois, Urbana-Champaign – HgYg

Alberts, E E., 573-882-1144 Univ of Missouri, Columbia – Hw

Albin, Edward, ealbin@gsu.edu, Georgia State Univ, Perimeter Coll, Online – Xm

Albrecht, Jochen, (212) 772-5221 jochen@hunter.cuny.edu, Hunter Coll (CUNY) – OiyOu

Albrecht, Lorraine, klalbrec@uwaterloo.ca, Univ of Waterloo – On

Albright, James N., (505) 667-4318 j_albright@lanl.gov, Los Alamos National Laboratory – Yg

Albright, Katia, (775) 682-8370 kalbright@unr.edu, Univ of Nevada, Reno – On

Alde, Douglas, (505) 667-0488 dxa@lanl.gov, Los Alamos National Laboratory – On

Alderson, David, +44 (0) 191 208 7121 david.alderson@ncl.ac.uk, Univ of Newcastle Upon Tyne – Oi

Alderton, Dave, +44 1784 443585 D.Alderton@rhul.ac.uk, Univ of London, Royal Holloway & Bedford New Coll – Gz

Aldrich, Eric A., 573-882-6301 aldriche@missouri.edu, Univ of Missouri, Columbia – Oa

Aldrich, M. James, (505) 667-1495 jaldrich@lanl.gov, Los Alamos National Laboratory – Gc

Aldridge, Keith D., (416) 399-0124 keith@yorku.ca, York Univ – YmdYh

Aldushin, Kirill, 089/2180 4337 kirill.aldushin@lrz.uni-muenchen.de, Ludwig-Maximilians-Universitaet Muenchen – Gz

Aleksandrowski, Pawel, pawel.aleksandrowski@uwr.edu.pl, Univ of Wroclaw – Gct

Alessi, Daniel S., (780) 492-3265 alessi@ualberta.ca, Univ of Alberta – ClGeHw

Alexander, Becky, (206) 543-0164 beckya@atmos.washington.edu, Univ of Washington – OaCsPe

Alexander, Clark R., 912-598-2329 clark.alexander@skio.usg.edu, Georgia Southern Univ – OuGsOn

Alexander, Conel M., (202) 478-8478 calexander@carnegiescience.edu, Carnegie Institution of Washington – Xc

Alexander, Dane, (269) 387-5486 dane.alexander@wmich.edu, Western Michigan Univ – GgOnn

Alexander, Elaine, (254) 299-8442 ealexander@mclennan.edu, McLennan Community Coll – Gg

Alexander, Jan, +44 (0)1603 59 3759 j.alexander@uea.ac.uk, Univ of East Anglia – Gt

Alexander, Jane L., 718-982-3013 jane.alexander@csi.cuny.edu, Coll of Staten Island – Gse

Alexander, Scott, 612-626-4164 alexa107@umn.edu, Univ of Minnesota, Twin Cities – Hw

Alexander, Shelton S., (814) 863-7246 shel@geosc.psu.edu, Pennsylvania State Univ, Univ Park – Ys
Alexander, Jr., E. Calvin, (612) 624-3517 alexa001@umn.edu, Univ of Minnesota, Twin Cities – HwCcl
Alfe, Dario, +44 202 7679 32361 d.alfe@ucl.ac.uk, Univ Coll London – OmGyYh
Alford, Matthew H., 206-221-3257 malford@apl.washington.edu, Univ of Washington – Op
Algar, Christopher, (902) 494-7192 calgar@dal.ca, Dalhousie Univ – Obc
Algeo, Catherine, (270) 745-5922 katie.algeo@wku.edu, Western Kentucky Univ – Oi
Algeo, Thomas J., (513) 556-4195 thomas.algeo@uc.edu, Univ of Cincinnati – Gs
Algin, Barbara, (212) 854-2905 ba110@columbia.edu, Columbia Univ – On
Ali, Bukari, +233 20 330 7976 bukariali@yahoo.co.uk, Kwame Nkrumah Univ of Science and Tech – HwNgYg
Ali, Genevieve, 204-474-7266 Genevieve.Ali@umanitoba.ca, Univ of Manitoba – Hw
Ali, Hendratta N., (785) 628-4608 hnali@fhsu.edu, Fort Hays State Univ – GogYs
Ali, K. Adem, (843) 953-0877 alika@cofc.edu, Coll of Charleston – OrHwNg
Alkaç, Onur, 00904242370000-5970 oalkac@firat.edu.tr, Firat Univ – GgsGu
Allam, Bassem, (632) 632-8745 bassem.allam@stonybrook.edu, SUNY, Stony Brook – Ob
Allan, Craig J., 704-687-5999 cjallan@email.uncc.edu, Univ of North Carolina, Charlotte – Hg
Allan, Deborah L., (612) 625-3158 dallan@soils.umn.edu, Univ of Minnesota, Twin Cities – Sb
Allan, Jonathan C., (541) 574-6658 jonathan.allan@dogami.state.or.us, Oregon Dept of Geology & Mineral Industries – Gm
Allard, Gilles O., (706) 542-2420 Univ of Georgia – Eg
Allard, Jason, (229) 333-5752 jmallard@valdosta.edu, Valdosta State Univ – Oag
Allard, Stephen T., (507) 457-2739 sallard@winona.edu, Winona State Univ – GcpGz
Allen, Alistair, +353 21 902769 a.allen@ucc.ie, Univ Coll Cork – Gp
Allen, Bruce, 505-366-2531 allenb@gis.nmt.edu, New Mexico Inst of Mining & Tech – Gm
Allen, Charlotte M., +61 7 3138 0177 cm.allen@qut.edu.au, Queensland Univ of Tech – CcaGi
Allen, Clarence R., (626) 395-6904 allen@gps.caltech.edu, California Inst of Tech – Ys
Allen, Dennis, dallen@belle.baruch.sc.edu, Univ of South Carolina – Ob
Allen, Diana M., (604) 291-3967 dallen@sfu.ca, Simon Fraser Univ – Hy
Allen, Donald B., dballen@colby.edu, Colby Coll – Eg
Allen, Eric E., (858) 534-2570 eallen@ucsd.edu, Univ of California, San Diego – Ob
Allen, Gary C., gallen@uno.edu, Univ of New Orleans – Gp
Allen, Gina, (847) 491-3238 gina@earth.northwestern.edu, Northwestern Univ – On
Allen, Grant, +44 0161 306-6851 Grant.Allen@manchester.ac.uk, Univ of Manchester – Oa
Allen, John, +44 023 9284 2257 john.allen@port.ac.uk, Univ of Portsmouth – Og
Allen, Joseph L., 304-384-5238 allenj@concord.edu, Concord Univ – Gc
Allen, Karen, (864)294-2504 karen.allen@furman.edu, Furman Univ – On
Allen, Katherine, +44 0151 795 4646 K.A.Allen@liverpool.ac.uk, Univ of Liverpool – Oi
Allen, Katherine A., Katherine.Allen@maine.edu, Univ of Maine – Gu
Allen, Mark, +44 191 33 42344 m.b.allen@durham.ac.uk, Durham Univ – Gs
Allen, Peter M., (254)710-2189 peter_allen@baylor.edu, Baylor Univ – HgNg
Allen, Philip, +44 20 759 47363 philip.allen@imperial.ac.uk, Imperial Coll – Gs
Allen, Phillip, 301/687-4891 ppallen@frostburg.edu, Frostburg State Univ – GmePe
Allen, Richard M., (510) 642-1275 rallen@seismo.berkeley.edu, Univ of California, Berkeley – Ys
Allen, Richardson B., (801) 581-7574 pallen@egi.utah.edu, Univ of Utah – GctOi
Allen, Robert J., (951) 827-4870 robert.allen@ucr.edu, Univ of California, Riverside – Oa
Allen, Susan E., (604) 822-2828 sallen@eos.ubc.ca, Univ of British Columbia – Op
Allen, Timothy T., (603) 358-2571 tallen@keene.edu, Keene State Coll – HwGe
Allen, Jr., John S., jallen@coas.oregonstate.edu, Oregon State Univ – Op
Allen-King, Richelle, (716) 645-4287 richelle@geology.buffalo.edu, SUNY, Buffalo – Cg
Allen-Lafayette, Zehdreh, (609) 292-2576 zehdreh.allen-lafayette@dep.nj.gov, New Jersey Geological and Water Survey – Oi
Aller, Josephine Y., (631) 632-8655 josephine.aller@stonybrook.edu, SUNY, Stony Brook – Oba
Aller, Robert C., (631) 632-8746 robert.aller@stonybrook.edu, SUNY, Stony Brook – Cm
Alley, Marcus M., (540) 231-9777 malley@vt.edu, Virginia Polytechnic Inst & State Univ – Sc
Alley, Richard B., (814) 863-1700 ralley@essc.psu.edu, Pennsylvania State Univ, Univ Park – Gl
Alley, Ronald E., (818) 354-0751 ron@lithos.jpl.nasa.gov, Jet Propulsion Laboratory – Or
Allis, Richard G., (801) 537-3301 rickallis@utah.gov, Utah Geological Survey – Go
Allison, Alivia J., 620-341-5984 aalliso2@emporia.edu, Emporia State Univ – GgaGm
Allison, David T., 251-460-6381 dallison@southalabama.edu, Univ of South Alabama – Gc
Allison, Mead A., (504) 862-3270 meadallison@tulane.edu, Tulane Univ – Gs
Allison, Nicky, +44 01334 463952 na9@st-andrews.ac.uk, Univ of St. Andrews – Oc
Allison, Tami L., tami.allison@dnr.mo.gov, Missouri Dept of Natural Resources – On
Allmon, Warren D., (607) 273-6623 wda1@cornell.edu, Cornell Univ – PoePi
Allmon, Warren D., (607) 273-6623 (Ext. 14) wda1@cornell.edu, Paleontological Research Institution – Po
Almendinger, James E., 651-433-5953 dinger@smm.org, Univ of Minnesota, Twin Cities – Pe
Alpert, Pinhas, pinhas@post.tau.ac.il, Tel Aviv Univ – OarOw
Alsdorf, Douglas E., 614 247-6908 alsdorf.1@osu.edu, Ohio State Univ – Hg
Alsleben, Helge, (817) 257-5455 h.alsleben@tcu.edu, Texas Christian Univ – GctGg
Alt, Jeffrey C., (734) 764-8380 jalt@umich.edu, Univ of Michigan – Ou
Altabet, Mark, maltabet@umassd.edu, Brown Univ – Oc
Altaner, Stephen P., (217) 244-1244 altaner@illinois.edu, Univ of Illinois, Urbana-Champaign – Gz
Altenbach, Alexander, 089/2180 6598 a.altenbach@lrz.uni-muenchen.de, Ludwig-Maximilians-Universitaet Muenchen – Pg
Altermann, Wladyslaw, 089/2180 6552 wlady.altermann@iaag.geo.uni-muenchen.de, Ludwig-Maximilians-Universitaet Muenchen – Gg
Alumbaugh, David L., (510) 215-4227 dalumbaugh@berkeley.edu, Univ of California, Berkeley – Ye
Aluwihare, Lihini I., (858) 822-4886 laluwihare@ucsd.edu, Univ of California, San Diego – Gu
Alvarado, Vladimir, 307-766-6464 valvarad@uwyo.edu, Univ of Wyoming – Np
Alvarez, Walter, (510) 642-2602 platetec@berkeley.edu, Univ of California, Berkeley – Gr
Alves, Tiego, +44(0)29 208 76754 AlvesT@cf.ac.uk, Univ of Wales – Og
Aly, Mohamed H., 479-575-4524 aly@uark.edu, Univ of Arkansas, Fayetteville – OriGg
Amador, Nathanael S., 740-368-3619 nsamador@owu.edu, Ohio Wesleyan Univ – OylOr
Amato, Jeffrey M., (575) 646-3017 amato@nmsu.edu, New Mexico State Univ, Las Cruces – GcCc
Ambers, Clifford P., cambers2@ashland.edu , Ashland Univ – GggGg
Ambinakudige, Shrinidhi, (662) 268-1032 x210 shrinidhi@geosci.msstate.edu,
Mississippi State Univ – OiyOr
Ambrose, William, 512-471-0258 william.ambrose@beg.utexas.edu, Univ of Texas at Austin – GsrEo
Ambruster, W. Scott, (907) 474-7161 ffwsa@aurora.alaska.edu, Univ of Alaska, Fairbanks – On
Amedjoe, Godfrey C., +233 24 5961 073 chiri.amedjoe@gmail.com, Kwame Nkrumah Univ of Science and Tech – GidGc
Amel, Nasir, +98 (411) 339 2696 amel@tabrizu.ac.ir, Univ of Tabriz – GivGg

Amelung, Falk, 305 421-4949 famelung@rsmas.miami.edu, Univ of Miami – On
Amend, Jan, (213) 740-0652 janamend@usc.edu, Univ of Southern California – ClmCo
Amend, Jan P., (314) 935-8651 amend@wustl.edu, Washington Univ in St. Louis – Co
Amenta, Roddy V., (540) 568-6674 amentarv@jmu.edu, James Madison Univ – Gc
Amer, Reda, 504-862-3220 ramer1@tulane.edu, Tulane Univ – OirGe
Amezaga, Jamie, +44 (0) 191 208 4876 jaime.amezaga@ncl.ac.uk, Univ of Newcastle Upon Tyne – Ge
Amidon, Will, 802-443-5980 wamidon@middlebury.edu, Middlebury Coll – Gm
Amin, Isam E., (330) 941-2293 ieamin@ysu.edu, Youngstown State Univ – Hw
Amler, Michael, 089/2180 6602 amler@lrz.uni-muenchen.de, Ludwig-Maximilians-Universitaet Muenchen – Pg
Ammon, Charles J., (814) 865-2310 cammon@geosc.psu.edu, Pennsylvania State Univ, Univ Park – Ys
Amon, Rainier, 409 740 4719 amonr@tamug.edu, Texas A&M Univ – Ob
Amonette, Alexandra B., (509) 376-5019 ieer@ieer.org, Pacific Northwest National Laboratory – Cg
Amoozegar, Aziz, (919) 515-3967 North Carolina State Univ – Sp
Amor, John, (604) 822-6933 jamor@eos.ubc.ca, Univ of British Columbia – On
Amos, Carl L., +44 (0)23 8059 6068 cla8@noc.soton.ac.uk, Univ of Southampton – Gs
Amos, Colin B., 360-650-3587 colin.amos@wwu.edu, Western Washington Univ – Gc
Ampuero, Jean-Paul, 626.395.6958 ampuero@gps.caltech.edu, California Inst of Tech – Ys
Amthor, Jeff, (865) 576-2773 Oak Ridge National Laboratory – Sf
Anand, Madhur, (519) 824-4120 Ext.56254 manand@uoguelph.ca, Univ of Guelph – On
Anand, Pallavi, +44 (0) 1908 652225 x 52225 pallavi.anand@open.ac.uk, The Open Univ – ClPe
Anandakrishnan, Sridhar, (814) 863-6742 sak@essc.psu.edu, Pennsylvania State Univ, Univ Park – Ys
Anastasio, Cort, 530-754-6095 canastasio@ucdavis.edu, Univ of California, Davis – Oa
Anastasio, David J., (610) 758-5117 dja2@lehigh.edu, Lehigh Univ – Gc
Anbar, Ariel, (480) 965-0767 anbar@asu.edu, Arizona State Univ – Cgs
Ancell, Brian C., 806-834-3143 brian.ancell@ttu.edu, Texas Tech Univ – Oa
Anders, Alison M., 217-244-3917 amanders@illinois.edu, Univ of Illinois, Urbana-Champaign – Gm
Anders, Mark H., mander44@gmu.edu, George Mason Univ – Gc
Andersen, C. Brannon, (864) 294-3366 brannon.andersen@furman.edu, Furman Univ – ClGe
Andersen, David W., (408) 924-5014 david.andersen@sjsu.edu, San Jose State Univ – GsCl
Andersen, Jens, +44 01326 371836 j.c.andersen@exeter.ac.uk, Exeter Univ – GiEm
Andersen, Raymond J., (604) 822-4511 randersn@eos.ubc.ca, Univ of British Columbia – Oc
Anderson, Alan J., (902) 867-2309 aanderso@stfx.ca, Saint Francis Xavier Univ – Gx
Anderson, Alan J., 902-863-5413 aanderso@stfx.ca, Dalhousie Univ – Cg
Anderson, Andrew, 217-244-0995 acandrsn@illinois.edu, Univ of Illinois, Urbana-Champaign – Ng
Anderson, Bruce, brucea@bu.edu, Boston Univ – Oap
Anderson, Callum, 27 41 504 2811 callum.anderson@nmmu.ac.za, Nelson Mandela Metropolitan Univ – Gzs
Anderson, G M., (416) 978-2062 Univ of Toronto – Cp
Anderson, James G., (617) 495-5922 anderson@huarp.harvard.edu, Harvard Univ – Oa
Anderson, James L., 808-974-7640 jamesa@hawaii.edu, Univ of Hawai'i, Hilo – GctYd
Anderson, James L., (612) 625-0279 jandersn@soils.umn.edu, Univ of Minnesota, Twin Cities – Sd
Anderson, James Lawford, lawford@bu.edu, Boston Univ – Gi
Anderson, Jennifer L., 507-457-2457 jlanderson@winona.edu, Winona State Univ – XgYgOe
Anderson, John B., (713) 348-4652 johna@rice.edu, Rice Univ – Gu
Anderson, John G., (775) 784-1954 jga@unr.edu, Univ of Nevada, Reno – YsNeYg
Anderson, Jonathan H., (507) 389-1301 jonathan.anderson@mnsu.edu, Minnesota State Univ – Ga
Anderson, Ken B., 618-453-7389 kanderson@geo.siu.edu, Southern Illinois Univ Carbondale – Co
Anderson, Laurie C., (605) 394-1290 Laurie.Anderson@sdsmt.edu, South Dakota School of Mines & Tech – Pio
Anderson, Mark, +44 1752 584768 M.Anderson@plymouth.ac.uk, Univ of Plymouth – Gct
Anderson, Mark R., (402) 472-6656 manderson4@unl.edu, Univ of Nebraska, Lincoln – Oa
Anderson, Mary P., (914) 365-8335 andy@geology.wisc.edu, Univ of Wisconsin, Madison – Hw
Anderson, Megan L., 719-389-6512 mlanderson@coloradoColl.edu, Colorado Coll – YgGt
Anderson, Michelle, (406) 683-7076 michelle.anderson@umwestern.edu, Univ of Montana Western – Hs
Anderson, Orson L., (310) 825-2386 Univ of California, Los Angeles – Yx
Anderson, Patricia M., pata@uw.edu, Univ of Washington – Ple
Anderson, Paula, 479-575-3355 pea001@uark.edu, Univ of Arkansas, Fayetteville – Gg
Anderson, Raymond R., (319) 335-1589 Raymond.Anderson@dnr.iowa.gov, Univ of Iowa – Gr
Anderson, Robert, 303-735-4684 Robert.S.Anderson@colorado.edu, Univ of Colorado – Gm
Anderson, Robert F., 845-365-8508 Columbia Univ – Cg
Anderson, Robert G., (604) 822-2449 Univ of British Columbia – Gt
Anderson, Robert S., (831) 459-3342 randerson@es.ucsc.edu, Univ of California, Santa Cruz – Gm
Anderson, Roger N., (845) 365-8335 anderson@ldeo.columbia.edu, Columbia Univ – Yh
Anderson, Ross, +44 (0) 131 451 3798 r.anderson@hw.ac.uk, Heriot-Watt Univ – Eo
Anderson, Scott, scott.anderson@gov.mb.ca, Manitoba Geological Survey – EgGcCg
Anderson, Stephen H., 573-882-6303 andersons@missouri.edu, Univ of Missouri, Columbia – Sp
Anderson, Steve W., (970) 351-2973 steven.anderson@unco.edu, Univ of Northern Colorado – GvOeXg
Anderson, Suzanne P., (303) 492-7071 suzanne.anderson@colorado.edu, Univ of Colorado – GmCl
Anderson, Thomas B., (775) 747-1438 tom.anderson@sonoma.edu, Sonoma State Univ – GsrGd
Anderson, Wayne I., (319) 273-3125 wayne.anderson@uni.edu, Univ of Northern Iowa – Pg
Anderson, William P., andersonwp@appstate.edu, Appalachian State Univ – HwqHs
Anderson, William T., (305) 348-2693 andersow@fiu.edu, Florida International Univ – Cs
Anderson, Jr., Alfred T., (773) 702-8138 Univ of Chicago – Gv
Anderson, NAE, Mary P., (608) 262-2396 andy@geology.wisc.edu, Univ of Wisconsin, Madison – Hw
Anderson-Folnagy, Heidi, (406) 683-7134 heidi.anderson@umwestern.edu, Univ of Montana Western – Gs
Andersson, Andreas, (858) 822-2486 aandersson@ucsd.edu, Univ of California, San Diego – Oc
Andeweg, Bernd , 31 20 5987339 bernd.andeweg@vu.nl, VU Univ Amsterdam – GtcGg
Andre-Obayanju, Tomilola, tomilola.obayanju@uniben.edu, Univ of Benin – NgHgGe
Andreasen, David C., (410) 260-8814 david.andreasen@maryland.gov, Maryland Department of Natural Resources – Hw
Andren, Anders W., (608) 262-2470 Univ of Wisconsin, Madison – Oc
Andres, A. S., (302) 831-2833 asandres@udel.edu, Univ of Delaware – HwOi
Andrews, Benjamin, (202) 633-1818 andrewsb@si.edu, Smithsonian Institution / National Museum of Natural History – Gv
Andrews, Graham D., (304) 293-2192 graham.andrews@mail.wvu.edu, West Virginia Univ – GvtGx
Andrews, John R., 512-471-1534 john.andrews@beg.utexas.edu, Univ of Texas at Austin, Jackson School of Geosciences – Oi
Andrews, John T., (303) 492-5183 john.t.andrews@colorado.edu, Univ of Colorado – GluGs
Andrews, Julian E., +44 (0)1603 59 2536 j.andrews@uea.ac.uk, Univ of East Anglia – CsGsCl
Andrews, Richard D., 405-325-3991 rdandrews@ou.edu, Univ of Oklahoma – Go
Andronicos, Chris, (765) 494-5982 candroni@purdue.edu,

Purdue Univ – Gg
Andrus, C. Fred T., (205) 348-5177 fandrus@ua.edu, Univ of Alabama – PeGaCs
Andrus, Richard E., (607) 777-2453 Binghamton Univ – Sf
Aneja, Viney P., 91951557808 viney_aneja@ncsu.edu, North Carolina State Univ – Oa
Anfinson, Owen, anfinson@sonoma.edu, Sonoma State Univ – GsdGg
Angel, Ross J., (540) 231-7974 rangel@vt.edu, Virginia Polytechnic Inst & State Univ – Gz
Angelo, Joseph A., (407) 768-8000 Florida Inst of Tech – Og
Angelopoulos, Vassilis, (310) 794-7090 vassilis@ucla.edu, Univ of California, Los Angeles – XyOaYm
Anger-Kraavi, Annela, +44 (0)1603 59 2633 a.anger-kraavi@uea.ac.uk, Univ of East Anglia – Eg
Angino, Ernest E., (785) 843-7503 Univ of Kansas – Cl
Angle, Michael, 614 265 6602 mike.angle@dnr.state.oh.us, Ohio Dept of Natural Resources – GlgHw
Angus, Doug, +4401133431326 d.angus@leeds.ac.uk, Univ of Leeds – Ys
Anhaeusser, Carl R., 011-7176581 carl.anhaeusser@wits.ac.za, Univ of the Witwatersrand – GgEgOg
Anis, Ayal, (409) 740-4987 anisa@tamug.tamu.edu, Texas A&M Univ – Op
Ankney, Meagan E., 330 972-7633 mankney@uakron.edu, Univ of Akron – Gzg
Ansdell, Kevin M., (306) 966-5698 kevin.ansdell@usask.ca, Univ of Saskatchewan – EgCgGt
Antar, Ali A., (860) 832-2931 antar@ccsu.edu, Central Connecticut State Univ – On
Anthony, Elizabeth Y., (915) 747-5483 eanthony@utep.edu, Univ of Texas, El Paso – Gi
Anthony, Leona M., (808) 956-8763 leonaa@hawaii.edu, Univ of Hawai'i, Manoa – On
Anthony, Nina, 864-294-2052 nina.anthony@furman.edu, Furman Univ – On
Anthony, Robin, 412-442-4295 robanthony@pa.gov, Pennsylvania Bureau of Topographic & Geologic Survey – Go
Anthony-Zajanc, Kate, anthcath@isu.edu, Idaho State Univ – Oi
Antinao, JoseLuis, 775-673-7450 JoseLuis.antinao@dri.edu, Desert Research Inst – Gm
Antipova, Anzhelika, antipova@memphis.edu, Univ of Memphis – On
Antonacci, Vince, (615) 532-1507 Vince.Antonacci@tn.gov, Tennessee Geological Survey – GgOi
Antonellini, Marco, marcoa@pangea.stanford.edu, Stanford Univ – Yr
Antonescu, Adrian, +44 0161 306-3911 bogdan.antonescu@manchester.ac.uk, Univ of Manchester – Oa
Aplin, Andrew, +44 191 33 42332 Durham Univ – Go
Apopei, Andrei I., +40232201463 andrei.apopei@uaic.ro, Al. I. Cuza Univ of Iasi – GziGg
Apotsos, Alex, 413-597-5082 alex.apotsos@williams.edu, Williams Coll – Ge
Appel, Christopher (Chip) S., (805) 756-1691 cappel@calpoly.edu, California Polytechnic State Univ – Sc
Appiah-Adjei, Emmanuel K., +233 20 7934 556 ekappiah-adjei.soe@knust.edu.gh, Kwame Nkrumah Univ of Science and Tech – HwgGg
Applegarth, Michael T., (717) 477-1712 mtappl@ship.edu, Shippensburg Univ – Or
Applegate, David, (703) 648-6600 applegate@usgs.gov, Univ of Utah – Gt
Appold, Martin, (573) 882-0701 appoldm@missouri.edu, Univ of Missouri – HwEm
Apprill, Amy, (505) 289-2649 aapprill@whoi.edu, Woods Hole Oceanographic Institution – Ob
Apraiz, Arturo, arturo.apraiz@ehu.eus, Univ of the Basque Country UPV/EHU – Gpt
April, Richard, (315) 228-7212 rapril@colgate.edu, Colgate Univ – CgGze
Aquilina, Noel , noel.aquilina@um.edu.mt, Univ of Malta – Oaw
Arabas, Karen, 503-370-6666 karabas@willamette.edu, Willamette Univ – On
Arabasz, Walter J., (801) 581-7410 arabasz@seis.utah.edu, Univ of Utah – Ys
Arain, M. A., (905) 525-9140 (Ext. 27941) arainm@mcmaster.ca, McMaster Univ – Hg
Arakawa, Akio, (310) 825-9874 aar@atmos.ucla.edu, Univ of California, Los Angeles – Oa
Arboleya Cimadevilla, María Luísa, ++935811951 marialuisa.arboleya@uab.cat, Universitat Autonoma de Barcelona – Gc
Archambault, Guy, garchambault@uqac.ca, Universite du Quebec a Chicoutimi – Nr
Archer, Allen W., (785) 532-2244 aarcher@ksu.edu, Kansas State Univ – Gr
Archer, David, (773) 702-0823 Univ of Chicago – Yr
Archer, Michael, m.archer@unsw.edu.au, Univ of New South Wales – Pv
Archibald, Doug A., (613) 545-6594 Queen's Univ – Cc
Archuleta, Ralph J., (805) 893-8441 archuleta@geol.ucsb.edu, Univ of California, Santa Barbara – Ys
Arehart, Greg B., (775) 784-6470 arehart@unr.edu, Univ of Nevada, Reno – Eg
Arens, Nan Crystal, 315-781-3930 arens@hws.edu, Hobart & William Smith Colls – Po
Argast, Anne S., 260-481-6252 argast@ipfw.edu, Indiana Univ / Purdue Univ, Fort Wayne – Gdz
Argenbright, Kristi, (817)2577273 k.a.argenbright2@tcu.edu, Texas Christian Univ – GgOn
Argenbright, Robert T., (910) 962-3498 argenbrightr@uncw.edu, Univ of North Carolina Wilmington – On
Argles, Tom, tom.argles@open.ac.uk, The Open Univ – Oi
Argyilan, Erin, 219-980-7124 eargyila@iun.edu, Indiana Univ Northwest – GeHwOw
Aristilde, Ludmilla, la31@cornell.edu, Cornell Univ – GeSb
Ariya, Parisa A., (514) 398-3615 parisa.ariya@mcgill.ca, McGill Univ – Oa
Arkani-Hamed, Jafar, 514-398-6767 jafar@physics.utoronto.ca, McGill Univ – Xy
Arletti, Rossella, rossella.arletti@unito.it, Università di Torino – Gz
Armadillo, Egidio, +390103538085 egidio@dipteris.unige.it, Universita di Genova – Ye
Armbrust, Virginia E., (206) 616-1783 armbrust@ocean.washington.edu, Univ of Washington – Ob
Armi, Laurence, (858) 534-6843 larmi@ucsd.edu, Univ of California, San Diego – Op
Armour, Jake, 704-687-5968 jarmour@uncc.edu, Univ of North Carolina, Charlotte – Gg
Armstrong, Andrew, 603-862-4559 andya@ccom.unh.edu, Univ of New Hampshire – On
Armstrong, David E., (608) 262-2470 Univ of Wisconsin, Madison – Oc
Armstrong, Felicia P., 330-941-1385 fparmstrong@ysu.edu, Youngstown State Univ – Sb
Armstrong, Howard A., 0191 3744780 h.a.armstrong@durham.ac.uk, Durham Univ – Pg
Armstrong, Phillip A., (657) 278-3169 parmstrong@fullerton.edu, California State Univ, Fullerton – GcYg
Armstrong, Robert A., (609) 258-5260 robert.armstrong@stonybrook.edu, SUNY, Stony Brook – Ocb
Arnaud, Emmanuelle, 519-824-4120 x58087 earnaud@uoguelph.ca, Univ of Guelph – Gls
Arnold, Anthony J., (850) 644-4228 arnold@gly.fsu.edu, Florida State Univ – Pm
Arnold, Dan, +44 (0) 131 451 8298 d.arnold@hw.ac.uk, Heriot-Watt Univ – Go
Arnold, David L., 301/687-4053 dlarnold@frostburg.edu, Frostburg State Univ – Oa
Arnold, Gail, 915-747-8373 glarnold@utep.edu, Univ of Texas, El Paso – Cs
Arnold, Stephen, +44(0) 113 34 37245 s.arnold@leeds.ac.uk, Univ of Leeds – Oa
Arnone, John, (775) 673-7445 jarnone@dri.edu, Desert Research Inst – Py
Arnott, R. William C., (613)562-5800 6854 warnott@uottawa.ca, Univ of Ottawa – GsSo
Aronoff, Ruth F., (864)294-3363 ruth.aronoff@furman.edu, Furman Univ – Gc
Arora, Vivek, Vivek.Arora@ec.gc.ca, Univ of Victoria – Oa
Arp, Dr., Gernot, +49 (0)551 397986 garp@gwdg.de, Georg-August Univ of Goettingen – PyyPy
Arrhenius, Gustaf, (858) 534-2961 arrhenius@ucsd.edu, Univ of California, San Diego – ClGzCm
Arrigo, Kevin, (650) 723-3599 arrigo@stanford.edu, Stanford Univ – Og
Arritt, Raymond W., (515) 294-9870 rwarritt@iastate.edu, Iowa State Univ of Science & Tech – Oa
Arrowsmith, Ramon, (480) 965-3541 ramon.arrowsmith@asu.edu, Arizona State Univ – GcmGt
Arslan, Gizem, 00904242370000-5972 gturus@firat.edu.tr,

Firat Univ – GxiGp
Arthur, Jonathan D., (850) 617-0320 jonathan.arthur@dep.state.fl.us, Florida Geological Survey – HwCgGe
Arthur, Michael A., (814) 863-6054 maa6@psu.edu, Pennsylvania State Univ, Univ Park – OuClGs
Arthurs, Leilani A., 402-472-2663 Univ of Nebraska, Lincoln – OeCl
Artiola, Janick F., (520) 621-3516 jartiola@email.arizona.edu, Univ of Arizona – ScHsCa
Artioli, Gilberto, 39-049-8279162 gilberto.artioli@unipd.it, Università degli Studi di Padova – GzyOm
Arundale, Wendy H., (907) 474-7039 Univ of Alaska, Fairbanks – On
Arvidson, Raymond E., (314) 935-5679 arvidson@wunder.wustl.edu, Washington Univ in St. Louis – Xg
Arya, Satyapal S., (919) 515-7002 pal_arya@ncsu.edu, North Carolina State Univ – Oa
Asante, Joseph, (931) 372-3576 jasante@tntech.edu, Tennessee Tech Univ – HwOrGe
Asbury, Brian, 303-273-3123 basbury@mines.edu, Colorado School of Mines – Nr
Aschoff, Jennifer, (907) 786-1442 jaschoff@uaa.alaska.edu, Univ of Alaska, Anchorage – GrdGo
Asencio, Eugenio, (787) 265-3845 eugenio.asencio@upr.edu, Univ of Puerto Rico – YseGt
Asghari Moghadam, Asghar, +98 (411) 339 2703 moghadam@tabrizu.ac.ir, Univ of Tabriz – HwqHy
Ash, Richard, (301) 405-7504 rdash@umd.edu, Univ of Maryland – Ca
Ashley, Gail M., (848) 445-2221 gmashley@rci.rutgers.edu, Rutgers, The State Univ of New Jersey – GsmSa
Ashley, Paul, +61-2-67732348 pashley@une.edu.au, Univ of New England – EgGeEm
Ashour-Abdalla, Maha, (310) 825-8881 mabdalla@igpp.ucla.edu, Univ of California, Los Angeles – On
Ashton, Andrew, 508-289-3751 aashton@whoi.edu, Woods Hole Oceanographic Institution – On
Ashwal, Lewis D., +27 11 717 6652 lewis.ashwal@wits.ac.za, Univ of the Witwatersrand – GxCgGt
Ashwell, Paul A., +64 3 364 2987 X4456 paul.ashwell@canterbury.ac.nz, Univ of Canterbury – Gg
Ashwood, Tom, (865) 574-7542 Oak Ridge National Laboratory – Ge
Ashworth, Allan C., 701-231-7919 allan.ashworth@ndsu.edu, North Dakota State Univ – Pe
Asimow, Paul D., (626) 395-4133 asimow@gps.caltech.edu, California Inst of Tech – CpGzy
Asiwaju-Bello, Yinusa A., 08036672708 Federal Univ of Tech, Akure – GgHgNg
Askari, Roohollah, (906) 487-2029 raskari@mtu.edu, Michigan Technological Univ – YxxNp
Askari, Zohreh, (217) 300-1819 askari@illinois.edu, Univ of Illinois, Urbana-Champaign – Gsc
Aslan, Andres, (970) 248-1614 aaslan@coloradomesa.edu, Colorado Mesa Univ – Gm
Aslan, Yasemin, 00904242370000-5959 yaslan@hotmail.com, Firat Univ – Nr
Asmerom, Yemane, (505) 277-4204 asmerom@unm.edu, Univ of New Mexico – CcPeCg
Asowata, Timothy I., 08065348350 Federal Univ of Tech, Akure – Gge
Asper, Vernon L., (228) 688-3178 vernon.asper@usm.edu, Univ of Southern Mississippi – Ou
Asphaug, Erik, (480) 727-2219 easphaug@asu.edu, Arizona State Univ – Xgy
Asphaug, Erik, (831) 459-2260 easphaug@pmc.ucsc.edu, Univ of California, Santa Cruz – Xg
Asquith, George B., 806-834-0497 george.asquith@ttu.edu, Texas Tech Univ – Go
Assatourians, Karen, 519-661-2111, ext. 84715 kassatou@uwo.ca, Western Univ – Ys
Asselin, Esther, (418) 654-2612 esther.asselin@canada.ca, Universite du Quebec – PlGuh
Aster, Richard C., (970) 491-7606 rick.aster@colostate.edu, Colorado State Univ – YsgGv
Atchley, Stacy C., (254) 710-2196 stacy_atchley@baylor.edu, Baylor Univ – GroGo
Atekwana, Eliot, 405-744-6358 eliot.atekwana@okstate.edu, Oklahoma State Univ – Cg
Athaide, Dileep, (604) 984-1771 dathaide@eos.ubc.ca, Univ of British Columbia – Gg
Athaide, Dileep J., 604-986-1911, loc 7552 dathaide@capilanou.ca, Capilano Univ – GgOge
Athey, Jennifer E., (907) 451-5028 jennifer.athey@alaska.gov, Alaska Division of Geological & Geophysical Surveys – Gg
Atkins, Nolan T., 802-626-6238 nolan.atkins@lyndonstate.edu, Lyndon State Coll – Oa
Atkinson, Christopher, christopher.atkinson@und.edu, Univ of North Dakota – Oai
Atkinson, Gail M., 519-661-2111 x.84207 gatkins6@uwo.ca, Western Univ – YssYs
Atkinson, Gail M., gma@ccs.carleton.ca, Carleton Univ – Ne
Atkinson, Larry P., (757) 683-4926 atkinson@ccpo.odu.edu, Old Dominion Univ – Op
Atkinson, Marlin J., 235-2224 mja@hawaii.edu, Univ of Hawai'i, Manoa – Ob
Atkinson, Tim, +44 020 7679 37711 t.atkinson@ucl.ac.uk, Univ Coll London – Hg
Atkinson, Jr., William W., (303) 492-6103 william.atkinson@colorado.edu, Univ of Colorado – EmCg
Atreya, Sushil, (734) 936-0489 atreya@umich.edu, Univ of Michigan – Oy
Attal, Mikael, +44 (0) 131 650 8533 mikael.attal@ed.ac.uk, Edinburgh Univ – GmtOy
Attig, John W., (608) 262-6131 jwattig@wisc.edu, Univ of Wisconsin, Extension – Gl
Attinger, Sabine, 0049(0)3641/948651 sabine.attinger@ufz.de, Friedrich-Schiller-Univ Jena – Hy
Attrep, Moses, (505) 667-0088 Los Alamos National Laboratory – Oe
Atudorei, Nieu-Viorel, atudorei@unm.edu, Univ of New Mexico – Cs
Atwater, Brian, 206-553-2927 atwater@ess.washington.edu, Univ of Washington – Ys
Au, Whitlow W L., 808-247-5026 wau@hawaii.edu, Univ of Hawai'i, Manoa – On
Aubert, John E., (916) 484-8637 aubertj@arc.losrios.edu, American River Coll – Oy
Audet, Celine, (418) 723-1986 (Ext. 1744) celine_audet@uqar.qc.ca, Universite du Quebec a Rimouski – Ob
Audet, Pascal, (613)562-5800 2344 pascal.audet@uottawa.ca, Univ of Ottawa – YgGc
Aufrecht, Walter E., (403) 329-2485 aufrecht@uleth.ca, Univ of Lethbridge – On
Ault, Alexis K., aault@email.arizona.edu, Utah State Univ – GtCc
Ault, Toby R., 607-255-1509 tra38@cornell.edu, Cornell Univ – Ow
Aurnou, Jonathan M., (310) 825-2054 aurnou@epss.ucla.edu, Univ of California, Los Angeles – XyYmx
Ausbrooks, Scott, 501 683-0119 scott.ausbrooks@arkansas.gov, Arkansas Geological Survey – YsGeg
Ausich, William I., (614) 292-3353 ausich.1@osu.edu, Ohio State Univ – PoGs
Auster, Peter, 860-405-9118 peter.auster@uconn.edu, Univ of Connecticut – Ob
Austin, George S., (505) 835-5230 george@gis.nmt.edu, New Mexico Inst of Mining & Tech – En
Austin, Philip, (604) 822-2175 paustin@eos.ubc.ca, Univ of British Columbia – Oa
Austin, Jr., James A., (512) 471-0450 jamie@ig.utexas.edu, Univ of Texas at Austin – Gu
Autin, Whitney J., 585-395-5738 dirtguy@esc.brockport.edu, SUNY, The Coll at Brockport – Gs
Avary, Katherine L., 304 594 2331 avary@geosrv.wvnet.edu, West Virginia Univ – Go
Avouac, Jean-Philippe, 626.395.2350 avouac@gps.caltech.edu, California Inst of Tech – GtYs
Awdankiewicz, Marek, marek.awdankiewicz@uwr.edu.pl, Univ of Wroclaw – Gvi
Awramik, Stanley M., (805) 893-3830 Awramik@geol.ucsb.edu, Univ of California, Santa Barbara – Po
Axelbaum, Richard, rla@me.wustl.edu, Washington Univ in St. Louis – Ng
Axen, Gary, 575.835.5178 gaxen@ees.nmt.edu, New Mexico Inst of Mining and Tech – Gct
Axford, Yarrow, 847.467.2268 yarrow@earth.northwestern.edu, Northwestern Univ – GnPmi
Ayad, Yasser M., yayad@clarion.edu, Clarion Univ – Oi
Aydin, Adnan, (662) 915-1342 aaydin@olemiss.edu, Univ of Mississippi – NgrYe
Aydin, Atilla, (650) 725-8708 aydin@pangea.stanford.edu, Stanford Univ – Gc

Ayers, John C., 615-322-2158 john.c.ayers@vanderbilt.edu, Vanderbilt Univ – Cg
Ayers, Joseph , 7815817370 x309 lobster@neu.edu, Northeastern Univ – On
Ayling, Bridget, (775) 682-8768 bayling@unr.edu, Univ of Nevada – CgYhGo
Aylward, Linda, (309) 694-5256 Illinois Central Coll – Oe
Azam, Farooq, (858) 534-6850 fazam@ucsd.edu, Univ of California, San Diego – Ob
Azetsu-Scott, Kumiko, (902) 426-8572 kumiko.azetsu-scott@mar.dfo-mpo.gc.ca, Geological Survey of Canada – Ocp

## B

Baadsgaard, Halfdan, budb@powersurfr.com, Univ of Alberta – Cc
Baarli, B. Gudveig, (413) 597-2329 gudveig.baarli@williams.edu, Williams Coll – PssPi
Baars, Oliver, 609-258-2489 obaars@princeton.edu, Princeton Univ – Co
Babaie, Hassan A., (404) 413-5766 hbabaie@gsu.edu, Georgia State Univ – Gc
Babb, David M., (814) 863-3918 dmb16@psu.edu, Pennsylvania State Univ, Univ Park – Ow
Babbin, Andrew, babbin@mit.edu, Massachusetts Inst of Tech – Py
Babcock, Daphne H., (972) 578-5518 dbabcock@collin.edu, Collin Coll - Spring Creek Campus – Geg
Babcock, Loren E., (614) 292-0358 babcock.5@osu.edu, Ohio State Univ – Po
Babcock, R. S., (360) 650-3592 babcock@wwu.edu, Western Washington Univ – Cg
Babek, Ondrej, +420 549 49 3163 Masaryk Univ – Gds
Bacchus, Tania S., (802) 635-1329 Tania.Bacchus@jsc.edu, Johnson State Coll – GuOw
Bach Plaza, Joan, ++935811272 joan.bach@uab.cat, Universitat Autonoma de Barcelona – Gg
Bachhuber, Frederick W., bachhubf@nevada.edu, Univ of Nevada, Las Vegas – Pl
Bachle, Peter, (573) 368-2472 peter.bachle@dnr.mo.gov, Missouri Dept of Natural Resources – Gg
Bachmann, Olivier, baolivie@ethz.ch, Univ of Washington – Giv
Bachtadse, Valerian, 089/2180 4237 valerian@geophysik.uni-muenchen.de, Ludwig-Maximilians-Universitaet Muenchen – Yg
Back, Larissa E., 608-262-0776 lback@wisc.edu, Univ of Wisconsin, Madison – Oa
Backus, George E., (858) 534-2468 gebackus@ucsd.edu, Univ of California, San Diego – YmsOp
Bacon, Diana H., (509) 372-6132 diana.bacon@pnl.gov, Pacific Northwest National Laboratory – Hq
Bacon, Michael P., (508) 289-2559 mbacon@whoi.edu, Woods Hole Oceanographic Institution – Oc
Bacon, Robert, (573) 526-0807 Missouri Dept of Natural Resources – Hs
Bacon, Steven N., 775-673-7473 Steve.Bacon@dri.edu, Desert Research Inst – Gm
Bada, Jeffrey L., (858) 534-4258 jbada@ucsd.edu, Univ of California, San Diego – Co
Bader, Nicholas E., (509) 527-5113 baderne@whitman.edu, Whitman Coll – SoHgOi
Badger, Robert L., (315) 267-2624 badgerrl@potsdam.edu, SUNY Potsdam – Gip
Badgley, Catherine E., (734) 763-6448 cbadgley@umich.edu, Univ of Michigan – Pv
Badillo, Jose M., josemaria.badillo@ehu.eus, Univ of the Basque Country UPV/EHU – Gmt
Badruddin, Abu Z., 315 2948610 ext.231 badruddin@cayuga-cc.edu, Cayuga Community Coll – Oi
Badurek, Chris, (828) 262-7054 badurekca@appstate.edu, Appalachian State Univ – Oi
Baedke, Steven J., (540) 568-6156 baedkesj@jmu.edu, James Madison Univ – Hy
Baer, Eric M., (206) 878-3710 ebaer@highline.edu, Highline Coll – GgvOg
Baer, James L., james_baer@byu.edu, Brigham Young Univ – Go
Bagtzoglou, Ross, (212) 854-3154 Columbia Univ – Hw
Baharlou, Alan, 217-581-2626 abaharlou@eiu.edu, Eastern Illinois Univ – Cg
Bahlburg, Heinrich, +49-251-83-33935 bahlbur@uni-muenster.de, Universitaet Muenster – GsgCc
Bahr, Jean M., (608) 262-2396 jmbahr@geology.wisc.edu, Univ of Wisconsin, Madison – Hw
Bahr, Jean M., (608) 262-5513 jmbahr@geology.wisc.edu, Univ of Wisconsin, Madison – Hw
Bahrmann, Chad, 814-865-9500 cbahrmann@psu.edu, Pennsylvania State Univ, Univ Park – Ow
Baigorria, Guillermo, gbaigorria@unl.edu, Univ of Nebraska, Lincoln – Ow
Bailey, Christopher M., (757) 221-2445 cmbail@wm.edu, Coll of William & Mary – Gc
Bailey, David G., (315) 859-4142 dbailey@hamilton.edu, Hamilton Coll – GizGa
Bailey, Ian, +4401326 259322 i.bailey@exeter.ac.uk, Exeter Univ – On
Bailey, Jack B., (309) 298-1481 JB-Bailey@wiu.edu, Western Illinois Univ – Pg
Bailey, Jake, (612) 624-1603 baileyj@umn.edu, Univ of Minnesota, Twin Cities – Py
Bailey, Judy, 61 02 4921 5415 Judy.Bailey@newcastle.edu.au, Univ of Newcastle – Ec
Bailey, Keiron D., kbailey@email.arizona.edu, Univ of Arizona – Oi
Bailey, Lorraine, (401) 874-2265 lbailey@mail.uri.edu, Univ of Rhode Island – On
Bailey, Richard C., (416) 978-3231 bailey@physics.utoronto.ca, Univ of Toronto – Yg
Bailey, Richard H., (617) 373-3181 r.bailey@neu.edu, Northeastern Univ – Pi
Bain, Daniel J., (412) 624-8780 dbain@pitt.edu, Univ of Pittsburgh – HgGmOu
Bain, Olivier, +33(0)3 44069304 olivier.bain@lasalle-beauvais.fr, Institut Polytechnique LaSalle Beauvais (ex-IGAL) – OuGgOi
Bair, Edwin S., (614) 292-6197 bair.1@osu.edu, Ohio State Univ – Hw
Baird, Gordon C., (716) 673-3840 baird@fredonia.edu, SUNY Fredonia – GrPgGs
Baird, Graham, 970-351-2830 Graham.Baird@unco.edu, Univ of Northern Colorado – GcgGz
Bajjali, William, (715) 394 8056 wbajjali@uwsuper.edu, Univ of Wisconsin, Superior – HwCsg
Bajraktareviæ, Zlatan, +38514606098 zbajrak@geol.pmf.hr, Univ of Zagreb – PgvPm
Bakel, Allen J., (630) 252-5486 Argonne National Laboratory – Co
Baker, Alex, +44 (0)1603 59 1529 alex.baker@uea.ac.uk, Univ of East Anglia – On
Baker, Andy, a.baker@unsw.edu.au, Univ of New South Wales – HwCs
Baker, Brett, 361-749-6774 brett_baker@utexas.edu, Univ of Texas at Austin – Oe
Baker, Cathy, (479) 968-0661 cbaker@atu.edu, Arkansas Tech Univ – Gg
Baker, Don, 514-398-7485 don.baker@mcgill.ca, McGill Univ – Cp
Baker, Edward T., (206) 526-6251 baker@pmel.noaa.gov, Univ of Washington – Ou
Baker, James C., (540) 231-9785 suebrown@vt.edu, Virginia Polytechnic Inst & State Univ – Sd
Baker, John M., (612) 625-4249 jbaker@soils.umn.edu, Univ of Minnesota, Twin Cities – Sp
Baker, Leslie L., 885-6192 lbaker@uidaho.edu, Univ of Idaho – ClXgGe
Baker, Marcia B., 206-685-9697 mbbaker@uw.edu, Univ of Washington – OaYg
Baker, Paul A., (919) 684-6450 pbaker@duke.edu, Duke Univ – PeCg
Baker, Richard G., (319) 335-1827 dick-baker@uiowa.edu, Univ of Iowa – Pl
Baker, Robert W., (715) 425-3345 robert.w.baker@uwrf.edu, Univ of Wisconsin, River Falls – GlmOl
Baker, Sophie, 775-673-7434 Sophie.baker@dri.edu, Desert Research Inst – Gm
Baker, Tracy, 609 652-4611 tracy.baker@stockton.edu, Richard Stockton Coll of New Jersey – HgOr
Baker, Victor R., (520) 621-7875 baker@email.arizona.edu, Univ of Arizona – XgGmh
Baksi, Ajoy K., (225) 388-3422 abaksi@geol.lsu.edu, Louisiana State Univ – Cc
Bal Akkoca, Dicle, 00904242370000-5982 dbal@firat.edu.tr, Firat Univ – EnCo
Balakrishnan, Meena, 817-515-6360 meena.balakrishnan@tccd.edu, Tarrant County Coll- Northeast Campus – GggGg
Balascio, Nicholas, 757-221-2880 nbalascio@wm.edu, Coll of William & Mary – OyGnl
Balázs, László, balazslaszlo@windowslive.com, Eotvos Lorand Univ – EoYeg
Balch, Robert S., 575-835-5305 balch@prrc.nmt.edu, New Mexico Inst of Mining and Tech – Yse

Baldauf, Jack G., (979) 845-3651 jbaldauf@ocean.tamu.edu, Texas A&M Univ – Ou
Baldini, James, +44 191 33 42334 james.baldini@durham.ac.uk, Durham Univ – Hs
Baldizon, Ileana, (305) 237-3658 ileana.baldizo@mdc.edu, Miami-Dade Coll (Wolfson Campus) – On
Baldridge, W. Scott, (505) 667-4338 sbaldridge@lanl.gov, Los Alamos National Laboratory – GicGt
Baldwin, Christopher T., (936) 294-1593 baldwin@shsu.edu, Sam Houston State Univ – GsPoGu
Baldwin, Julia, 406-243-5778 jbaldwin@mso.umt.edu, Univ of Montana – Gp
Baldwin, Michael, baldwin@purdue.edu, Purdue Univ – Oa
Baldwin, Suzanne L., (315) 443-2672 sbaldwin@syr.edu, Syracuse Univ – Cc
Baldwin, Jr., A. Dwight, dbbaldwin@comcast.net, Miami Univ – Hw
Balen, Dražen, +38514605967 drbalen@geol.pmf.hr, Univ of Zagreb – GxpGi
Bales, Roger C., rbales@eng.ucmerced.edu, Univ of Arizona – Cg
Balescu, Sanda, 514-987-3000 #3126 sanda.balescu@club-internet.fr, Universite du Quebec a Montreal – Pe
Balestro, Gianni, gianni.balestro@unito.it, Università di Torino – Gc
Balgord, Elizabeth A., 801-626-6225 balgord@email.arizona.edu, Weber State Univ – Grs
Balistrieri, Laurie, (206) 543-8966 balistri@ocean.washington.edu, Univ of Washington – Oc
Ball, Elizabeth, (775) 682-7351 ball@mines.unr.edu, Univ of Nevada, Reno – On
Ball, Neil, (204) 474-8075 na_ball@umanitoba.ca, Univ of Manitoba – Gz
Ball, William P., (410) 516-5434 bball@jhu.edu, Johns Hopkins Univ – On
Ballantyne, Colin K., ckb@st-andrews.ac.uk, Univ of St. Andrews – OyGml
Ballard, Robert D., (401) 874-6115 bballard@gso.uri.edu, Univ of Rhode Island – Ga
Ballentine, Christopher, +44 (1865) 272000 (72938) chrisb@earth.ox.ac.uk, Univ of Oxford – Xg
Ballero, Deniz Z., 678-212-7567 dballero@gsu.edu, Georgia State Univ, Perimeter Coll, Online – Gg
Ballinger, Rhoda, ballingerRC@cf.ac.uk, Cardiff Univ – On
Ballinger, Rhoda, +44(0)29 208 76671 BallingerRC@cf.ac.uk, Univ of Wales – On
Ballou, Stephen M., (608) 363-2388 ballous@beloit.edu, Beloit Coll – Gg
Bally, Albert W., (713) 348-6063 geol@rice.edu, Rice Univ – Gg
Balmat, Jennifer L., (308) 432-6483 jbalmat@csc.edu, Chadron State Coll – GgOe
Balmforth, Neil, (604) 822-9835 njb@math.ubc.ca, Univ of British Columbia – Oa
Balog-Szabo, Anna, 540-857-7222 abalog-szabo@virginiawestern.edu, Virginia Western Community Coll – Gs
Balogh-Brunstad, Zsuzsanna, (607) 431-4734 balogh_brunz@hartwick.edu, Hartwick Coll – ClHgOs
Balser, Teri , 352-294-3157 tcbalser@ufl.edu, Univ of Florida – Oss
Balser, Teri C., (608) 262-0132 tcbalser@wisc.edu, Univ of Wisconsin, Madison – Sb
Balsley, Christopher, 203-392-6647 balsleyc1@southernct.edu, Southern Connecticut State Univ – Gg
Balster, Nick J., (608) 263-5719 njbalster@wisc.edu, Univ of Wisconsin, Madison – Sf
Baltz, Donald M., (225) 388-6512 Louisiana State Univ – Ob
Bambach, Richard K., (617) 484-7457 rbambach@oeb.harvard.edu, Virginia Polytechnic Inst & State Univ – Po
Bamisaye, Oluwaseyi A., 07031533939 adunseyi@gmail.com, Federal Univ of Tech, Akure – GgOr
Bancroft, John C., (403) 220-5026 bancroft@geo.ucalgary.ca, Univ of Calgary – Ng
Banda, Anna F., (903) 802-0186 aperry@dcccd.edu, El Centro Coll - Dallas Community Coll District – GgeGv
Bandopadhyay, Sukumar, (907) 474-6876 sbandopadhyay@alaska.edu, Univ of Alaska, Fairbanks – Nm
Banerdt, Bruce E., (818) 354-5413 Jet Propulsion Laboratory – Xy
Banerjee, Neil, (519) 661-2111 x.83727 nbanerj3@uwo.ca, Western Univ – EgCgGx
Banerjee, Subir K., (612) 624-5722 banerjee@tc.umn.edu, Univ of Minnesota, Twin Cities – Ym
Banfield, Jillian, 510-642-9488 jill@seismo.berkeley.edu, Univ of California, Berkeley – CoGz
Bang, John, (919) 530-6569 jjbang@nccu.edu, North Carolina Central Univ – Ge
Bangs, Nathan L., (512) 471-0424 nathan@ig.utexas.edu, Univ of Texas at Austin – Gu
Banik, Tenley, 309-438-8922 tjbanik@ilstu.edu, Illinois State Univ – GvzGl
Bank, Carl-Georg, (416) 978-4381 bank@geology.utoronto.ca, Univ of Toronto – Og
Bank, Tracy L., tlbank@buffalo.edu, SUNY, Buffalo – Cg
Banks, Beth, 828-884-8164 Beth.Banks@brevard.edu, Brevard Coll – On
Banks, David, +44(0) 113 34 35244 d.banks@see.leeds.ac.uk, Univ of Leeds – Cg
Banks, Eddie, eddie.banks@flinders.edu.au, Flinders Univ – Hw
Banner, Jay L., (512) 471-5016 banner@mail.utexas.edu, Univ of Texas at Austin – Cl
Bannon, Ann, ann.bannon@dal.ca, Dalhousie Univ – On
Bannon, Peter R., (814) 863-1309 bannon@ems.psu.edu, Pennsylvania State Univ, Univ Park – Ow
Banse, Karl, (206) 543-5079 banse@ocean.washington.edu, Univ of Washington – Ob
Bao, Huiming, (225) 578-3419 bao@lsu.edu, Louisiana State Univ – Cs
Baptista, Joana, +44 (0) 191 208 5899 joana.baptista@ncl.ac.uk, Univ of Newcastle Upon Tyne – Ng
Baptista, João C., jbaptist@utad.pt, Universidade de Trás-os-Montes e Alto Douro – GmmGa
Bar-Nun, Akiva, xx 97236406972 akivab@post.tau.ac.il, Tel Aviv Univ – Xc
Barak, Phillip W., (608) 263-5450 pwbarak@wisc.edu, Univ of Wisconsin, Madison – ScGz
Barakat, Bassam, +33(0)3 44068973 bassam.barakat@lasalle-beauvais.fr, Institut Polytechnique LaSalle Beauvais (ex-IGAL) – NrgGq
Baran, Zeynep O., (605) 394-2461 Zeynep.Baran@sdsmt.edu, South Dakota School of Mines & Tech – Gco
Barazangi, Muawia, (607) 255-6411 mb44@cornell.edu, Cornell Univ – Ys
Barazangi, Muawia, mb44@cornell.edu, Cornell Univ – Ys
Barbeau, Katherine A., (858) 822-4339 kbarbeau@ucsd.edu, Univ of California, San Diego – Oc
Barbeau, Jr., David, dbarbeau@geol.sc.edu, Univ of South Carolina – Gs
Barbecot, Florent, 514-987-3000 #7786 barbecot.florent@uqam.ca, Universite du Quebec a Montreal – Hw
Barbee, Gary C., (806) 651-2294 gbarbee@wtamu.edu, West Texas A&M Univ – HwOsi
Barber, Donald C., (610) 526-5110 dbarber@brynmawr.edu, Bryn Mawr Coll – Gs
Barber, Richard T., (919) 728-2111 rbarber@duke.edu, Duke Univ – Ob
Barbi, Greta, 089/2180 4234 barbi@gophysik.uni-muenchen.de, Ludwig-Maximilians-Universitaet Muenchen – Yg
Barbor, RADM (Retired) Ken E., (228) 688-3720 ken.barbor@usm.edu, Univ of Southern Mississippi – Own
Barclay, Andrew, 206-543-8956 barclay@ldeo.columbia.edu, Univ of Washington – Gu
Barclay, David J., (607) 753-2921 david.barclay@cortland.edu, SUNY, Cortland – Glm
Barclay, David R., (902) 494-4164 dbarclay@dal.ca, Dalhousie Univ – Opo
Barclay, Jenni, +44 (0)1603 59 3887 j.barclay@uea.ac.uk, Univ of East Anglia – Gv
Barclay, Julie L., julie.barclay@cortland.edu, SUNY, Cortland – Gg
Barendregt, Rene W., (403) 329-2530 barendregt@uleth.ca, Univ of Lethbridge – Gm
Barhurst, James, +44 (0) 191 208 5431 james.bathurst@ncl.ac.uk, Univ of Newcastle Upon Tyne – Gs
Barile, Diane, (321) 674-8096 dmes@fit.edu, Florida Inst of Tech – Ou
Barineau, Clinton I., 706 569-3026 barineau_clinton@columbusstate.edu, Columbus State Univ – GcgGt
Barker, Andy J., +44 (0)23 80593641 A.J.Barker@soton.ac.uk, Univ of Southampton – Gg
Barker, Charles F., (313) 831-7279 cfbarker@earthlink.net, Wayne State Univ – Gg
Barker, Chris A., 936 468-2340 cbarker@sfasu.edu, Stephen F. Austin State Univ – Gc
Barker, Colin, (918) 631-3014 colin-barker@utulsa.edu, The Univ of Tulsa – CgoGo
Barker, Daniel, (512) 471-5502 danbarker@mail.utexas.edu, Univ of Texas at Austin – Gi

Barker, Gregory A., (603) 271-7332 gbarker@des.state.nh.us, New Hampshire Geological Survey – GglOi
Barker, Ivan, 519-661-2111 ext 88397 ibarke2@uwo.ca, Western Univ – Cc
Barker, James M., (505) 825-5114 jbark@nmt.edu, New Mexico Inst of Mining & Tech – En
Barker, Jeffrey S., (607) 777-2522 jbarker@binghamton.edu, Binghamton Univ – Ys
Barker, Joel D., (740) 725-6097 barker.246@osu.edu, Ohio State Univ – OlClPe
Barker, Stephen, barkers3@cardiff.ac.uk, Cardiff Univ – Pe
Barkmann, Peter, 303-384-2642 barkmann@mines.edu, Colorado School of Mines – HwGcg
Barlow, Jay P., (858) 546-7178 jbarlow@ucsd.edu, Univ of California, San Diego – Ob
Barlow, Mathew, (978) 934-3908 mathew_barlow@uml.edu, Univ of Massachusetts Lowell – Oa
Barminski, Robert, 831-770-7056 rbarminski@aol.com, Hartnell Coll – GgOg
Barmore, Garrett, 775-784-4528 gbarmore@unr.edu, Univ of Nevada, Reno – On
Barnard, Holly R., Holly.Barnard@colorado.edu, Univ of Colorado – Hg
Barnbaum, Cecilia S., (229) 249-2645 cbarnbau@valdosta.edu, Valdosta State Univ – On
Barnes, Calvin G., 806-834-7389 cal.barnes@ttu.edu, Texas Tech Univ – Gi
Barnes, Charles W., (928) 774-6079 chuck.barnes@nau.edu, Northern Arizona Univ – GcXg
Barnes, Christopher R., (250) 721-8847 crbarnes@uvic.ca, Univ of Victoria – PmOgPe
Barnes, Elizabeth A., eabarnes@atmos.colostate.edu, Colorado State Univ – Oa
Barnes, Fairley J., (505) 667-4933 fyb@lanl.gov, Los Alamos National Laboratory – On
Barnes, Gary M., gbarnes@hawaii.edu, Univ of Hawai'i, Manoa – OwGvOe
Barnes, Hubert L., (814) 865-7573 barnes@psu.edu, Pennsylvania State Univ, Univ Park – CgEmCe
Barnes, Jaime D., jdbarnes@jsg.utexas.edu, Univ of Texas at Austin – CgsCc
Barnes, Jeffrey R., (541) 737-5685 barnes@coas.oregonstate.edu, Oregon State Univ – Oa
Barnes, Jessica, jebarnes@mailbox.sc.edu, Univ of South Carolina – On
Barnes, John H., 717.702.2025 jbarnes@pa.gov, Pennsylvania Bureau of Topographic & Geologic Survey – Gg
Barnes, Melanie A., (806) 834-7965 melanie.barnes@ttu.edu, Texas Tech Univ – CaGie
Barnes, Philip, pbarnes@environ.sc.edu, Univ of South Carolina – On
Barnes, Randal J., (612) 625-5828 Univ of Minnesota, Twin Cities – On
Barnes, Sarah- J., 418 545 5011 sjbarnes@uqac.ca, Universite du Quebec a Chicoutimi – EgGiCt
Barnett, Douglas B., (509) 376-3416 brent.barnett@pnl.gov, Pacific Northwest National Laboratory – Em
Barnett, Michael, 617-552-8300 barnetge@bc.edu, Boston Coll – Oe
Barnett, Roger T., rbarnett@pacific.edu, Univ of the Pacific – Oy
Barnhart, William, (319) 384-4732 william-barnhart-1@uiowa.edu, Univ of Iowa – Gt
Barnosky, Anthony D., barnosky@berkeley.edu, Univ of California, Berkeley – Pv
Barnston, Anthony G., (845) 680-4447 tonyb@iri.columbia.edu, Columbia Univ – On
Baron, Dirk, (661) 654-3044 dbaron@csub.edu, California State Univ, Bakersfield – Cl
Barone, Jessica, 585-292-2448 jbarone@monroecc.edu, Monroe Community Coll – Ge
Baross, John A., (206) 543-0833 Univ of Washington – Ob
Barquero-Molina, Miriam, (573) 882-9557 barqueromolinam@missouri.edu, Univ of Missouri – Gct
Barr, Sandra M., (902) 585-1340 sandra.barr@acadiau.ca, Acadia Univ – Gi
Barr, Stuart, +44 (0) 191 208 6449 stuart.barr@ncl.ac.uk, Univ of Newcastle Upon Tyne – Oi
Barrash, Warren, (208) 426-1229 wbarrash@boisestate.edu, Boise State Univ – HwYxGe
Barrett, Bradford S., 410-293-6567 bbarrett@usna.edu, United States Naval Academy – Oa
Barrett, John, +44(0) 113 34 32394 j.r.barrett@leeds.ac.uk, Univ of Leeds – Ge
Barrett, Kevin M., (817) 515-6352 kevin.barrett@tccd.edu, Tarrant County Coll- Northeast Campus – OagOr
Barrett, Linda R., 330 972-7620 barrett@uakron.edu, Univ of Akron – SdOri
Barrett, Melony, 217-333-7917 mebarret@illinois.edu, Univ of Illinois, Urbana-Champaign – Oi
Barrick, James E., (806) 834-2717 jim.barrick@ttu.edu, Texas Tech Univ – Psm
Barrie, Vaughn, (250) 363-6424 Univ of Victoria – Gu
Barrier, Pascal, +33(0)3 44068975 pascal.barrier@lasalle-beauvais.fr, Institut Polytechnique LaSalle Beauvais (ex-IGAL) – PmGsPs
Barron, George, gbarron@uoguelph.ca, Univ of Guelph – On
Barron, Robert, (906) 487-2096 rjbarron@mtu.edu, Michigan Technological Univ – OgGzi
Barros, Tony, (305) 237-3754 tbarros@mdc.edu, Miami-Dade Coll (Wolfson Campus) – Og
Barsch, Robert, 089/2180 4201 barsch@geophysik.uni-muenchen.de, Ludwig-Maximilians-Universitaet Muenchen – Yg
Bart, Henry A., (215) 951-1268 bart@lasalle.edu, La Salle Univ – Gs
Bart, Philip J., (225) 388-3109 pbart@geol.lsu.edu, Louisiana State Univ – Gr
Bartello, Peter, (514) 398-8075 peter.bartello@mcgill.ca, McGill Univ – Oa
Bartels, William S., (517) 629-0313 wbartels@albion.edu, Albion Coll – Pv
Barth, Andrew P., (317) 274-1243 ibsz100@iupui.edu, Indiana Univ / Purdue Univ, Indianapolis – Gi
Barth, Jack A., (541) 737-1607 barth@coas.oregonstate.edu, Oregon State Univ – On
Barth, Nicolas, (951) 827-3138 nic.barth@ucr.edu, Univ of California, Riverside – Gt
Barth, Susan A., (406) 496-4687 sbarth@mtech.edu, Montana Tech of The Univ of Montana – On
Barthelmie, Rebecca J., rb737@cornell.edu, Cornell Univ – Ona
Bartholomew, Alexander J., 845-257-3765 barthola@newpaltz.edu, SUNY, New Paltz – GrPi
Bartholomew, Mervin J., (901) 678-1613 jbrthlm1@memphis.edu, Univ of Memphis – Gt
Bartholy, Judit, +36 20 3722945 bartholy@caesar.elte.hu, Eotvos Lorand Univ – Ow
Bartl, Simona, 831-771-4400 sbartl@mlml.calstate.edu, Moss Landing Marine Laboratories – On
Bartlein, Patrick J., (541) 346-4967 bartlein@uoregon.edu, Univ of Oregon – Pe
Bartlett, Douglas H., (858) 534-5233 dbartlett@ucsd.edu, Univ of California, San Diego – Ob
Bartlett, Wendy, 740-376-4775 bartletw@marietta.edu, Marietta Coll – Geo
Bartley, John M., (801) 58--7162 john.bartley@utah.edu, Univ of Utah – GciCc
Bartley, Julie K., 507-933-7307 jbartley@gustavus.edu, Gustavus Adolphus Coll – GsPgGd
Bartok, Peter, peter@bartokinc.com, Univ of Houston – Go
Bartolucci, Valerio, (954) 201-6678 vbartolu@broward.edu, Broward Coll, Central Campus – Ge
Barton, Christopher C., (727) 215-5538 chris.barton@wright.edu, Wright State Univ – GqYgHq
Barton, James H., (412) 589-2821 Thiel Coll – Oy
Barton, Mark D., (520) 621-8529 mdbarton@email.arizona.edu, Univ of Arizona – Eg
Barton, Michael, (614) 292-3132 barton.2@osu.edu, Ohio State Univ – GiCpt
Barzegari, Ghodrat, Gbarzegari@gmail.com, Univ of Tabriz – Nr
Bascom, Johnathan, (616) 526-7053 jbascom@calvin.edu, Calvin Coll – Oy
Basham, William L., 432 552-2057 basham_w@utpb.edu, Univ of Texas, Permian Basin – Ye
Basinger, James F., (306) 966-5684 jim.basinger@usask.ca, Univ of Saskatchewan – Pb
Baskaran, Mark M., (313) 577-3262 Baskaran@wayne.edu, Wayne State Univ – CcGg
Baskin, Perry A., (229) 259-5052 pbaskin@valdosta.edu, Valdosta State Univ – On
Bass, David, david.bass@flinders.edu.au, Flinders Univ – Oy
Bass, Jay D., (217) 333-1018 jaybass@illinois.edu, Univ of Illinois, Urbana-Champaign – Gy
Bass, Jerry, 601 266-4732 joby@usm.edu, Univ of Southern Mississippi – On

Bass, Michael L., (540) 654-1424 mbass@umw.edu, Univ of Mary Washington – Sf
Bassett, Damon, (417) 836-4897 dbassett@missouristate.edu, Missouri State Univ – Pg
Bassett, Kari N., (03) 366-7001 ext 7732 kari.bassett@canterbury.ac.nz, Univ of Canterbury – Gs
Bassett, Scott, (775) 784-1434 sbassett@unr.edu, Univ of Nevada, Reno – Ou
Bassett, William A., wab7@cornell.edu, Cornell Univ – Gz
Bassis, Jeremy, (734) 615-3606 jbassis@umich.edu, Univ of Michigan – OlgOg
Basta, Nicholas T., 614-292-6282 basta.4@osu.edu, Ohio State Univ – Sc
Bastow, Ian, +44 20 759 42974 i.bastow@imperial.ac.uk, Imperial Coll – Yg
Basu, Abhijit, 8128556654/5581 basu@indiana.edu, Indiana Univ, Bloomington – Gd
Basu, Anirban, +44 1784 414083 anirban.basu@rhul.ac.uk, Univ of London, Royal Holloway & Bedford New Coll – CglCs
Basu, Asish, (817) 272-2987 abasu@uta.edu, Univ of Texas, Arlington – GxCcs
Basu, Asish R., (585) 275-2413 abasu@earth.rochester.edu, Univ of Rochester – Gx
Batchelder, Hal, (541) 737-4500 hbatch@pices.int, Oregon State Univ – Op
Batelaan, Okke, okke.batelaan@ees.kuleuven.be, Katholieke Universiteit Leuven – Hw
Bates, Amanda E., +44(0)23 8059 8046 A.E.Bates@soton.ac.uk, Univ of Southampton – Ob
Bates, Mary, (661) 362-5054 Mary.Bates@canyons.edu, Coll of the Canyons – Oyg
Bates, Nicholas R., N.R.Bates@soton.ac.uk, Univ of Southampton – Oc
Bates, Richard, +44 01334 463997 crb@st-andrews.ac.uk, Univ of St. Andrews – Yx
Bates, Timothy S., 206-526-6248 Tim.Bates@noaa.gov, Univ of Washington – Oa
Bathke, Deborah J., dbathke2@unl.edu, Univ of Nebraska, Lincoln – Oae
Batlle-Aguilar, Jordi, 785-864-2113 jba@kgs.ku.edu, Univ of Kansas – Hy
Batten, William G., (608) 262-9903 bill.batten@wgnhs.uwex.edu, Univ of Wisconsin - Extension – GgHw
Battisti, David S., (206) 543-2019 battisti@uw.edu, Univ of Washington – Oa
Battles, Eileen, (785) 864-2129 battles@kgs.ku.edu, Univ of Kansas – Oy
Batygin, Konstantin, 626-395-2920 kbatygin@gps.caltech.edu, California Inst of Tech – Xg
Batzle, Michael L., (303) 384-2067 mbatzle@mines.edu, Colorado School of Mines – Yg
Bauder, James W., (406) 994-5685 jbauder@montana.edu, Montana State Univ – Sf
Bauer, Carl J., cjbauer@email.arizona.edu, Univ of Arizona – Hs
Bauer, Emily, (612) 626-0909 bauer010@umn.edu, Univ of Minnesota – Oi
Bauer, Jeffrey A., (740) 351-3421 jbauer@shawnee.edu, Shawnee State Univ – Ps
Bauer, Paul W., (505) 835-5106 bauer@nmbg.nmt.edu, New Mexico Inst of Mining & Tech – Gp
Bauer, R. A., 217-244-2394 rabauer@illinois.edu, Univ of Illinois, Urbana-Champaign – Ng
Bauer, Robert L., (573) 882-3759 bauerr@missouri.edu, Univ of Missouri – GcpGt
Baum, Steven K., (979) 845-0793 sbaum@ocean.tamu.edu, Texas A&M Univ – Op
Baumann, Hannes, 860-405-9297 hannes.baumann@uconn.edu, Univ of Connecticut – Ob
Baumann, Zofia, (860) 405-9281 zofia.baumann@uconn.edu, Univ of Connecticut – ObcCm
Baumgartner, Mark, mbaumgartner@whoi.edu, Dalhousie Univ – Ob
Baumiller, Tomasz K., (734) 764-7543 tomaszb@umich.edu, Univ of Michigan – Pg
Baxter, Ethan, 617-552-3640 ethan.baxter@bc.edu, Boston Coll – Ccg
Baxter, Martin, 989-774-2055 baxte1ma@cmich.edu, Central Michigan Univ – Ow
Baxter, Stefanie J., 302-831-1576 steff@udel.edu, Univ of Delaware – On
Baxter, P.G., James E., 717-780-2377 jebaxter@hacc.edu, Harrisburg Area Community Coll – GgHwGm
Bayari, Serdar, +90 (312) 2977740 serdar@hacettepe.edu.tr, Hacettepe Univ – Hw
Bayliss, Peter, (403) 220-5026 Univ of Calgary – Gz
Bayly, M. Brian, (518) 283-1721 bayly.brian@gmail.com, Rensselaer Polytechnic Inst – Gc
Bayowa, Oyelowo G., +2348030820291 obayowa@lautech.edu.ng, Ladoke Akintola Univ of Tech – GggGg
Beach Davis, Janet, (309) 268-8513 janet.beach-davis@heartland.edu, Heartland Community Coll – Og
Beane, Rachel J., (207) 725-3160 rbeane@bowdoin.edu, Bowdoin Coll – GziGp
Beard, Brian L., 608-262-1806 beardb@geology.wisc.edu, Univ of Wisconsin, Madison – Cc
Beard, James S., (276) 634-4170 jim.beard@vmnh.virginia.gov, Virginia Polytechnic Inst & State Univ – GiClg
Beard, K. Christopher, (412) 622-5782 beardc@CarnegieMNH.org, Carnegie Museum of Natural History – Pv
Beard, Kenneth V., (217) 333-1676 k-beard@uiuc.edu, Univ of Illinois, Urbana-Champaign – Oa
Bearden, Rebecca A., 205-247-3623 rbearden@gsa.state.al.us, Geological Survey of Alabama – Hs
Beasley-Stanley, Jewell D., (615) 322-2976 jewell.beasleystanley@vanderbilt.edu, Vanderbilt Univ – On
Beatty, Heather L., heather.beatty@austincc.edu, Austin Community Coll District – Ge
Beatty, Lynne, 9134698500 x3785 lbeatty@jccc.edu, Johnson County Community Coll – GgOy
Beatty, Merrill Ann, (506) 453-4803 mbeatty@unb.ca, Univ of New Brunswick – On
Beatty, William L., 507-474-5789 wbeatty@winona.edu, Winona State Univ – Pg
Beaty, Tammy, (865) 574-0119 taw@ornl.gov, Oak Ridge National Laboratory – Oi
Beauchamp, Marc, 519-661-2111 ext 88104 mbeauch6@uwo.ca, Western Univ – Ca
Beaudoin, Georges, (418) 656-3141 beaudoin@ggl.ulaval.ca, Universite Laval – EgGzCa
Beaulieu, Claudie, +44 (0)23 8059 6412 C.Beaulieu@soton.ac.uk, Univ of Southampton – Op
Beaumont, Christopher, (902) 494-3779 chris.beaumont@dal.ca, Dalhousie Univ – Yg
Beausoleil, Denis, 604-777-6117 beausoleild@douglasColl.ca, Douglas Coll – Og
Beaver, Harold H., (254) 710-2184 harold_beaver@baylor.edu, Baylor Univ – Go
Beaverson, Sheena K., 217-244-9306 sbeavers@illinois.edu, Univ of Illinois, Urbana-Champaign – Or
Bebout, Gray E., (610) 758-5831 geb0@lehigh.edu, Lehigh Univ – Gp
Beccue, Clint, 217-265-5161 cbeccue@illinois.edu, Univ of Illinois, Urbana-Champaign – Ge
Bechtel, Randy, 919-707-9204 randy.bechtel@ncdenr.gov, North Carolina Geological Survey – Oe
Bechtel, Timothy D., 717 291-4133 timothy.bechtel@fandm.edu, Franklin and Marshall Coll – GeYg
Beck, Catherine C., 315-859-4847 ccbeck@hamilton.edu, Hamilton Coll – GsnGr
Beck, Charles, (734) 763-5089 chbeck@umich.edu, Univ of Michigan – Pg
Beck, Colleen M., (702) 795-8077 colleen@dri.edu, Desert Research Inst – Sa
Beck, E. G., 270.827.3414 x23 ebeck@uky.edu, Univ of Kentucky – Hw
Beck, John H., 617-552-8300 john.beck@bc.edu, Boston Coll – Pb
Beck, Susan L., (520) 621-8628 slbeck@email.arizona.edu, Univ of Arizona – Ys
Becker, Alex, (510) 643-9181 Univ of California, Berkeley – Yg
Becker, Janet M., (808) 956-6514 jbecker@soest.hawaii.edu, Univ of Hawai'i, Manoa – Og
Becker, Keir, (305) 421-4661 kbecker@rsmas.miami.edu, Univ of Miami – YrGu
Becker, Laurence, (541) 737-9504 beckerla@geo.oregonstate.edu, Oregon State Univ – On
Becker, Lorene Y., 541-737-6993 beckelo@geo.oregonstate.edu, Oregon State Univ – Oi
Becker, Matthew, 562-985-8983 mbecker3@csulb.edu, California State Univ, Long Beach – Hwg
Becker, Naomi M., (505) 667-2165 nmb@lanl.gov,

Los Alamos National Laboratory – Hg
Becker, Ralf Thomas, +49-251-83-33951 rbecker@uni-muenster.de, Universitaet Muenster – Pg
Becker, Richard H., (630) 252-7595 Argonne National Laboratory – Or
Becker, Richard H., (419) 530-4571 richard.becker@utoledo.edu, Univ of Toledo – OrGe
Becker, Udo, (734) 615-6894 ubecker@umich.edu, Univ of Michigan – On
Beckie, Roger D., (604) 822-6462 rbeckie@eos.ubc.ca, Univ of British Columbia – HwCl
Beckingham, Barbara, (843) 953-0483 beckinghamba@cofc.edu, Coll of Charleston – Cl
Bedard, Jean H., (418) 654-2671 jbedard@nrcan.gc.ca, Universite du Quebec – Gi
Bedard, Paul, 4185455011, 2276 pbedard@uqac.ca, Universite du Quebec a Chicoutimi – CaGzEg
Bedaso, Zelalem, (937) 229-2393 zbedaso1@udayton.edu, Univ of Dayton – CslGs
Beddows, Patricia A., (847) 491-7460 patricia@earth.northwestern.edu, Northwestern Univ – HyGma
Bedient, Philip B., (713) 348-4953 bedient@rice.edu, Rice Univ – Hw
Bedle, Heather, (713) 743-6111 hbedle@uh.edu, Univ of Houston – Yes
Bednarski, Marsha, (860) 832-2943 bednarskim@ccsu.edu, Central Connecticut State Univ – Oe
Bednarz, Robert S., (979) 845-7187 r-bednarz@tamu.edu, Texas A&M Univ – Oe
Bednarz, Sarah W., (979) 845-1579 s-bednarz@tamu.edu, Texas A&M Univ – Oe
Beebe, Alex, dbeebe@southalabama.edu, Univ of South Alabama – Ge
Beegle, Douglas B., (814) 863-1016 dbb@psu.edu, Pennsylvania State Univ, Univ Park – ScOs
Beeton, Jared M., (719) 587-7357 jmbeeton@adams.edu, Adams State Univ – OySd
Befus, Kenneth S., Kenneth_Befus@baylor.edu, Baylor Univ – GviGv
Beg, Mirza A., 205-247-3624 mbeg@gsa.state.al.us, Geological Survey of Alabama – Gz
Beget, James E., ffjeb1@uaf.edu, Univ of Alaska, Fairbanks – Gm
Beghein, Caroline, (310) 825-0742 cbeghein@ucla.edu, Univ of California, Los Angeles – Ysg
Begin, Christian, (418) 654-2648 christian.begin@canada.ca, Universite du Quebec – PeGmOs
Behl, Richard J., (562) 985-5850 behl@csulb.edu, California State Univ, Long Beach – GsPeGd
Behling, Robert E., (304) 293-5603 rbehling@wvu.edu, West Virginia Univ – Gm
Behn, Mark D., (508) 289-3637 mbehn@whoi.edu, Woods Hole Oceanographic Institution – Yr
Behr, Rose-Anna, (717) 702-2035 rosbehr@pa.gov, Pennsylvania Bureau of Topographic & Geologic Survey – Gcr
Behr, Whitney, 512-232-1941 behr@utexas.edu, Univ of Texas at Austin – Gc
Behr-Andres, Christina "Tina", (505) 667-3644 behr-andres@lanl.gov, Los Alamos National Laboratory – On
Behrensmeyer, Anna K., (202) 633-1307 Smithsonian Institution / National Museum of Natural History – Pv
Beiersdorfer, Raymond E., (330) 941-1753 rebeiersdorfer@ysu.edu, Youngstown State Univ – Cg
Bein, F. L., (317) 274-1100 Indiana Univ, Indianapolis – Oy
Bekken, Barbara M., (540) 231-4466 bekken@vt.edu, Virginia Polytechnic Inst & State Univ – Eg
Bekker, Andrey, (951) 827-4611 andrey.bekker@ucr.edu, Univ of California, Riverside – GsrCg
Bekker, Matthew F., (801) 422-1961 matthew_bekker@byu.edu, Brigham Young Univ – Oy
Belanger, Christina L., (605) 394-2461 Christina.Belanger@sdsmt.edu, South Dakota School of Mines & Tech – PimPe
Belanger, Thomas V., (321) 674-8096 belanger@fit.edu, Florida Inst of Tech – Hw
Belasky, Paul, 510-979-7938 pbelasky@ohlone.edu, Ohlone Coll – PqiGs
Belknap, Daniel F., (207) 581-2159 belknap@maine.edu, Univ of Maine – Gu
Bell, Andrew, +44 (0) 131 650 4918 a.bell@ed.ac.uk, Edinburgh Univ – Gt
Bell, Brian, +44 01413306898 Brian.Bell@glasgow.ac.uk, Univ of Glasgow – So
Bell, Christopher J., (512) 471-7301 cjbell@mail.utexas.edu, Univ of Texas at Austin – Pv
Bell, David, (03) 3642-717 david.bell@canterbury.ac.nz, Univ of Canterbury – Ng
Bell, James C., (651) 328-1359 bellx007@umn.edu, Univ of Minnesota, Twin Cities – SdfOe
Bell, John W., (775) 784-1939 jbell@unr.edu, Univ of Nevada – NgGm
Bell, Keith, keith_bell@carleton.ca, Carleton Univ – Cc
Bell, Margaret C., +44 (0) 191 208 7936 margaret.bell@ncl.ac.uk, Univ of Newcastle Upon Tyne – Ge
Bell, Michael, mmbell@hawaii.edu, Univ of Hawai'i, Manoa – Owr
Bell, Rebecca, +44 20 759 40903 rebecca.bell@imperial.ac.uk, Imperial Coll – Gt
Bell, Robin E., 845-365-8827 Columbia Univ – Yr
Bellette, Kathryn, kathryn.bellette@flinders.edu.au, Flinders Univ – On
Bellew, Angela, (319) 335-1819 angela-bellew@uiowa.edu, Univ of Iowa – On
Bellieni, Giuliano, +390498279155 giuliano.bellieni@unipd.it, Università degli Studi di Padova – Gv
Belliveau, Lindsey, 860-424-3581 lindsey.belliveau@ct.gov, Dept of Energy and Environmental Protection – Gmt
Belluso, Elena, elena.belluso@unito.it, Università di Torino – Gz
Belmonte, Donato, +39 010 353 8136 Universita di Genova – Cg
Belt, Edward S., (413) 542-2712 esbelt@amherst.edu, Amherst Coll – Gs
Beltrami, Hugo, 902-867-2326 hbeltram@stfx.ca, Saint Francis Xavier Univ – YhOa
Beltz, John F., 330 972-6687 jfb4@uakron.edu, Univ of Akron – Ggh
Bemis, Sean, sean.bemis@uky.edu, Univ of Kentucky – GctYs
Ben-Zion, Yehuda, (213) 740-6734 ybz@earth.usc.edu, Univ of Southern California – Ys
Benacquista, Matt, (406) 657-2341 Montana State Univ, Billings – Xy
Benavides-Iglesias, Alfonso, a.benavides@geos.tamu.edu, Texas A&M Univ – Ys
Bend, Stephen L., (306) 585-4021 Univ of Regina – Gox
Bender, E. E., (714) 432-5681 ebender@occ.cccd.edu, Orange Coast Coll – Giz
Bender, John F., 704-687- 5956 jfbender@email.uncc.edu, Univ of North Carolina, Charlotte – GiCtOu
Bender, Michael L., (609) 258-5807 bender@princeton.edu, Princeton Univ – Cg
Bendick, Rebecca, 406-243-5774 bendick@mso.umt.edu, Univ of Montana – Yg
Benedetti, Michael M., (910) 962-7650 benedettim@uncw.edu, Univ of North Carolina Wilmington – GmSaOy
Benfield, Mark C., (225) 388-6372 mbenfie@lsu.edu, Louisiana State Univ – Ob
Benger, Simon, simon.benger@flinders.edu.au, Flinders Univ – Oiy
Benimoff, Alan I., 718-982-2835 alan.benimoff@csi.cuny.edu, Coll of Staten Island – GizGp
Benison, Kathleen , 304-293-5603 West Virginia Univ – GrsCm
Benitez-Nelson, Claudia R., 777-0018 cbnelson@geol.sc.edu, Univ of South Carolina – Oc
Benjamin, Patricia A., 508-929-8606 pbenjamin@worcester.edu, Worcester State Univ – Og
Benjamin, Timothy M., (505) 667-5154 Los Alamos National Laboratory – Ca
Benjamin, U. K., 234-8038312732 uzochukwu.benjamin@oauife.edu.ng, Obafemi Awolowo Univ – GoCg
Benna, Piera, piera.benna@unito.it, Università di Torino – Gz
Benner, Jacob, 617-627-2207 jacob.benner@tufts.edu, Tufts Univ – Pge
Benner, Ron, benner@mailbox.sc.edu, Univ of South Carolina – Ob
Bennet, Brett, 785-864-2117 bb@kgs.ku.edu, Univ of Kansas – Ng
Bennett, Philip, (512) 471-3587 pbennett@mail.utexas.edu, Univ of Texas at Austin – Hg
Bennett, Richard, (520) 621-2324 rab@geo.arizona.edu, Univ of Arizona – YdGt
Bennett, Sara, 309/298-1905 SC-Bennett@wiu.edu, Western Illinois Univ – Gg
Bennett, Scott, scottekb@uw.edu, Univ of Washington – Gnt
Bennett, Steven W., (309) 298-1256 Sc-Bennett@wiu.edu, Western Illinois Univ – Hw
Bennett, Victoria J., (817) 257-6603 v.bennett@tcu.edu, Texas Christian Univ – On
Benning, Liane G., +44(0) 113 34 35220 l.g.benning@leeds.ac.uk, Univ of Leeds – Ce
Benninger, Larry K., (919) 962-0699 lbenning@email.unc.edu, Univ of North Carolina, Chapel Hill – ClGuCt
Bennington, J Bret, 516 463-5568 geojbb@hofstra.edu,

Hofstra Univ – PeGs
Benoit-Bird, Kelly, 541-737-2063 kbenoit@coas.oregonstate.edu, Oregon State Univ – Gu
Benotti, Mark J., 781.891.2980 mbenotti@bentley.edu, Bentley Univ – OgGgOg
Bense, Victor, +44 (0)1603 59 1297 v.bense@uea.ac.uk, Univ of East Anglia – Hg
Benson, David A., 303-273-3806 dbenson@mines.edu, Colorado School of Mines – Hq
Benson, Donna M., (480) 461-7247 donnabenson@mesacc.edu, Mesa Community Coll – Gg
Benson, Jane L., (270) 809-3106 jane.benson@murraystate.edu, Murray State Univ – Oi
Benson, Robert D., (303) 273-3455 Colorado School of Mines – Ye
Benson, Robert G., (719) 587-7921 rgbenson@adams.edu, Adams State Univ – Gg
Benson, Roger, +44 (1865) 272000 roger.benson@earth.ox.ac.uk, Univ of Oxford – Po
Bentley, Callan, (703) 323-3276 cbentley@nvcc.edu, Northern Virginia Community Coll - Annandale – Gg
Bentley, Charles R., (608) 238-8873 bentley@geology.wisc.edu, Univ of Wisconsin, Madison – Yg
Bentley, Laurence R., (403) 220-4512 Univ of Calgary – Hw
Bentley, Samuel J., 225-578-5735 sjb@lsu.edu, Louisiana State Univ – Gu
Benton, Steven, 217-244-0082 s-benton@illinois.edu, Univ of Illinois, Urbana-Champaign – Gg
Benway, Heather, (508) 289-2838 hbenway@whoi.edu, Woods Hole Oceanographic Institution – Oc
Benz, Harley M., benz@gldfs.cr.usgs.gov, Univ of Utah – Ys
Berchem, Jean, (310) 206-6484 jberchem@igpp.ucla.edu, Univ of California, Los Angeles – On
Bercovici, David, (203) 432-3168 david.bercovici@yale.edu, Yale Univ – Yg
Bereitschaft, Bradley, 402-554-2674 bbereitschaft@unomaha.edu, Univ of Nebraska at Omaha – Oyu
Berelson, William M., (213) 740-5828 berelson@usc.edu, Univ of Southern California – Cm
Berendsen, Pieter, 785-864-2141 pieterb@kgs.ku.edu, Univ of Kansas – Cg
Beresnev, Igor A., (515) 294-7529 beresnev@iastate.edu, Iowa State Univ of Science & Tech – YgsYe
Berg, Christopher A., (678) 839-4059 cberg@westga.edu, Univ of West Georgia – Gpi
Berg, Jonathan H., (815)753-1943 jongeol@yahoo.com, Northern Illinois Univ – Gp
Berg, Peter, (804) 924-1318 pb8n@virginia.edu, Univ of Virginia – So
Berg, Richard, 217-244-2776 rberg@illinois.edu, Univ of Illinois – Ggg
Berg, Richard B., (406) 496-4172 dberg@mtech.edu, Montana Tech of The Univ of Montana – EnGz
Berg, Richard C., (217) 244-2776 rberg@illinois.edu, Univ of Illinois, Urbana-Champaign – Gem
Bergantz, George W., 206-685-4972 bergantz@uw.edu, Univ of Washington – Gi
Berger, Antony R., (604) 480-0840 Sir Wilfred Grenfell Coll – Ge
Berger, Peter M., 217-333-7078 pmberger@illinois.edu, Univ of Illinois, Urbana-Champaign – Cg
Berger, Wolfgang H., (858) 822-2545 wberger@ucsd.edu, Univ of California, San Diego – Gs
Bergeron, Normand, (418) 654-3703 normand.bergeron@ete.inrs.ca, Universite du Quebec – Gm
Berggren, William A., (508) 289-2593 wberggren@whoi.edu, Woods Hole Oceanographic Institution – Ps
Berggren, William A., 848-445-8523 wberggren@whoi.edu, Rutgers, The State Univ of New Jersey – PmsPe
Bergin, Michael H., (404) 894-9723 mike.bergin@ce.gatech.edu, Georgia Inst of Tech – Oa
Bergmann, Kristin, (617) 253-9852 kdberg@mit.edu, Massachusetts Inst of Tech – GsCl
Bergslien, Elisa T., 716-878-3793 bergslet@buffalostate.edu, Buffalo State Coll – ClHw
Bergstrom, Stig M., (614) 292-4473 bergstrom.1@osu.edu, Ohio State Univ – PsGg
Bergwerff, Kenneth A., (616) 526-6371 kbergwer@calvin.edu, Calvin Coll – Oe
Berke, Melissa, Melissa.Berke.1@nd.edu, Univ of Notre Dame – Cs
Berkelhammer, Max, 312-413-8271 berkelha@uic.edu, Univ of Illinois at Chicago – Cg
Berli, Marcus, 702.862.5452 markus.berli@dri.edu, Univ of Nevada, Reno – Sp
Berlin, Graydon L., lenn.berlin@nau.edu, Northern Arizona Univ – Or
Berlo, Kim, (514) 398-5884 kim.berlo@mcgill.ca, McGill Univ – Gv
Berman, David S., (412) 622-3248 bermand@carnegiemnh.org, Carnegie Museum of Natural History – PgoPv
Bermanec, Vladimir, +38514605972 vberman@public.carnet.hr, Univ of Zagreb – GzEnGi
Bermingham, Katherine, (301) 405-2707 kberming@umd.edu, Univ of Maryland – XcCcg
Bernhard, Joan M., 508-289-3480 jbernhard@whoi.edu, Woods Hole Oceanographic Institution – Cg
Bernhardt, Jase, 516-463-5731 jase.e.bernhardt@hofstra.edu, Hofstra Univ – Owa
Bernier, Luc, (905) 525-9140 (Ext. 26364) berniejm@mcmaster.ca, McMaster Univ – Cg
Bernier, Monique, 418 654-2585 monique.bernier@ete.inrs.ca, Universite du Quebec – Or
Bernknopf, Richard, (650) 725-9692 Stanford Univ – On
Beroza, Gregory C., (650) 723-4958 beroza@stanford.edu, Stanford Univ – Ys
Berquist, Jr., Carl R., (757) 221-2448 crberq@wm.edu, Coll of William & Mary – Ou
Berry, C M., Chris.Berry@earth.cf.ac.uk, Cardiff Univ – Pb
Berry, David R., 909-869-3455 drberry@csupomona.edu, California State Polytechnic Univ, Pomona – PgGo
Berry, Duane F., (540) 231-9792 duberry@vt.edu, Virginia Polytechnic Inst & State Univ – Sb
Berry, Karen, 303-384-2640 kaberry@mines.edu, Colorado School of Mines – NgOu
Berry, Kate, (775) 784-6344 kberry@unr.edu, Univ of Nevada, Reno – Oy
Berry, Leonard, 561 297-2935 berry@fau.edu, Florida Atlantic Univ – Og
Berry, Ron F., 61 3 6226 2456 Ron.Berry@utas.edu.au, Univ of Tasmania – Gc
Berry IV, Henry N., 207-287-7179 henry.n.berry@maine.gov, Dept of Agriculture, Conservation, and Forestry – Gg
Berryman, Bruce F., (802) 626-6478 Lyndon State Coll – Oa
Bershaw, John T., (503) 725-3778 bershaw@pdx.edu, Portland State Univ – GsCsHy
Berta, Susan, 812-237-2261 sberta@indstate.edu, Indiana State Univ – GmOr
Berthold, Angela, (612) 626-6744 berth084@d.umn.edu, Univ of Minnesota – Gl
Berti, Claudio, (610) 758-2581 clb208@lehigh.edu, Lehigh Univ – GmtOn
Bertine, Kathe K., (619) 594-6369 kbertine@geology.sdsu.edu, San Diego State Univ – Cl
Bertog, Janet, 859-572-1523 bertogj@nku.edu, Northern Kentucky Univ – Gd
Bertok, Carlo, carlo.bertok@unito.it, Università di Torino – Gd
Bertsch, Paul M., (803) 725-2752 Univ of Georgia – Sc
Besancon, James, (781) 283-3030 jbesancon@wellesley.edu, Wellesley Coll – Hww
Best, James L., 217-244-1839 jimbest@illinois.edu, Univ of Illinois, Urbana-Champaign – Gs
Best, Melvyn E., (250) 658-0791 best@islandnet.com, Univ of Victoria – Ye
Best, Myron G., best_myron_g@byu.edu, Brigham Young Univ – Gg
Beswick, Anthony E., (705) 675-1151 Laurentian Univ, Sudbury – Gi
Beswick, Karl, karl.beswick@manchester.ac.uk, Univ of Manchester – Oa
Betchwars, Corey, (612) 625-3507 betch038@umn.edu, Univ of Minnesota – Oi
Bettis III, E. A., (319) 335-1831 art-bettis@uiowa.edu, Univ of Iowa – SdGmSa
Betzer, Peter R., (727) 553-1130 pbetzer@marine.usf.edu, Univ of South Florida – Cm
Beukelman, Gregg, (801) 537-3380 greggbeukelman@utah.gov, Utah Geological Survey – NgOi
Beus, Stanley S., beuss3@msn.com, Northern Arizona Univ – Pi
Beutel, Erin K., (843) 953-5591 beutele@cofc.edu, Coll of Charleston – Gct
Bevelhimer, Mark, (865) 576-0266 mb2@ornl.gov, Oak Ridge National Laboratory – Hs
Bevier, Mary Lou, (604) 822-4655 mbevier@eos.ubc.ca, Univ of British Columbia – Cc
Bevis, Kenneth A., (812) 866-7307 bevis@hanover.edu, Hanover Coll – GlmGe
Bevis, Michael G., 614 247 5071 bevis.6@osu.edu, Ohio State Univ – Yx

Beyarslan, Melahat, 00904242370000-5985 mbeyarslan@firat.edu.tr, Firat Univ – GipGx
Beyer, Adam, adam.beyer@nicholls.edu, Nicholls State Univ – Gg
Beyer, Patricia J., (570) 389-4570 pbeyer@bloomu.edu, Bloomsburg Univ – HsOyGm
Bezada, Max, 614-624-3280 mbezada@umn.edu, Univ of Minnesota, Twin Cities – Ys
Bezdecny, Kris, (323) 343-2400 kbezdec@calstatela.edu, California State Univ, Los Angeles – OniOn
Bezdicek, David F., (509) 335-3644 bezdicek@wsu.edu, Washington State Univ – Sb
Bhattacharji, Somdev, (718) 951-5500 Graduate School of the City Univ of New York – Gt
Bhattacharya, Janok, (905) 525-9140 (Ext. 23528) bhattaj@mcmaster.ca, McMaster Univ – Gr
Bhattacharya, Janok, bhattaj@mcmaster.ca, Univ of Houston – Gsr
Bhattacharyya, Prajukti, (262) 472-5257 bhattacj@uww.edu, Univ of Wisconsin, Whitewater – Gce
Bhattacharyya, Sidhartha, 205-348-6427 bhatt001@crimson.ua.edu, Univ of Alabama – CatGe
Bhowmik, Nani G., (217) 333-6775 nbhowmik@uiuc.edu, Illinois State Water Survey – Hs
Bianchi, Thomas S., 352-392-6138 tbianchi@ufl.edu, Univ of Florida – Co
Biasi, Glenn P., (775) 784-4576 glenn@seismo.unr.edu, Univ of Nevada, Reno – Ys
Bibby, Thomas, +44 (0)23 8059 6446 tsb@noc.soton.ac.uk, Univ of Southampton – Ob
Bice, David M., 814-865-4477 dbice@geosc.psu.edu, Pennsylvania State Univ, Univ Park – Gg
Bickel, Charles E., (415) 338-1963 bickel@sfsu.edu, San Francisco State Univ – Gx
Bickford, M. E., (315) 443-2672 mebickfo@syr.edu, Syracuse Univ – Cc
Bickle, Michael, +44 (0) 1223 333484 mb72@esc.cam.ac.uk, Univ of Cambridge – GtYgGo
Bickmore, Barry R., (801) 422-4680 barry_bickmore@byu.edu, Brigham Young Univ – ClGz
Bidgoli, Tandis, (785) 864-3315 bidgoli@kgs.ku.edu, Univ of Kansas – GcoCc
Biek, Robert, (801) 537-3356 bobbiek@utah.gov, Utah Geological Survey – Gg
Bieler, David B., (318) 869-5234 dbieler@centenary.edu, Centenary Coll of Louisiana – GtrYe
Bienfang, Paul K., 808-956-7402 bienfang@soest.hawaii.edu, Univ of Hawai'i, Manoa – Ob
Bienvenu, Nadean S., (337) 482-6647 nadean@louisiana.edu, Univ of Louisiana at Lafayette – On
Bier, Sara E., (315) 267-3482 bierse@potsdam.edu, SUNY Potsdam – GctYg
Bierly, Aaron D., 717-702-2034 aabierly@pa.gov, Pennsylvania Bureau of Topographic & Geologic Survey – Gg
Bierly, Gregory, 812-237-3225 Gregory.Bierly@indstate.edu, Indiana State Univ – Oaw
Bierman, Paul R., (802) 656-4411 pbierman@uvm.edu, Univ of Vermont – GmeCc
Biggin, Andy, +440151 794 3460 A.Biggin@liverpool.ac.uk, Univ of Liverpool – Ym
Biggs, Douglas C., (979) 845-3423 d-biggs@tamu.edu, Texas A&M Univ – Ob
Biggs, Thomas H., (434) 924-0580 thb3k@virginia.edu, Univ of Virginia – OyGxe
Bigham, Jerry M., (614) 292-9066 bigham.1@osu.edu, Ohio State Univ – Sd
Bijeljic, Branko, +44 20 759 46420 b.bijeljic@imperial.ac.uk, Imperial Coll – Np
Bilanovic, Dragoljub D., 218-755-2801 dbilanovic@bemidjistate.edu, Bemidji State Univ – GePyCa
Bilardello, Dario, (612) 624-5049 dario@umn.edu, Univ of Minnesota, Twin Cities – Ym
Bilek, Susan L., 575.835.6510 sbilek@nmt.edu, New Mexico Inst of Mining and Tech – Ys
Bilheux, Hassina Z., (865) 384-9630 bilheuxhn@ornl.gov, Univ of Tennessee, Knoxville – On
Bill, Steven D., (603) 358-2552 sbill@keene.edu, Keene State Coll – Pg
Billen, Magali I., 530-754-5696 mibillen@ucdavis.edu, Univ of California, Davis – Yg
Billings, Brian J., 320-308-3298 bjbillings@stcloudstate.edu, Saint Cloud State Univ – Ow
Billings, Stephen, sbillings@skyri.com, Univ of British Columbia – Yg
Billiot, Fereshteh, 361-825-6067 Fereshteh.Billiot@tamucc.edu, Texas A&M Univ, Corpus Christi – On
Billups, Katharina, 302-645-4249 kbillups@udel.edu, Univ of Delaware – Pe
Bilodeau, Michel L., (514) 398-4755 McGill Univ – Nm
Bilodeau, William L., (805) 493-3264 bilodeau@callutheran.edu, California Lutheran Univ – GcsGg
Bina, Craig R., (847) 491-5097 craig@earth.northwestern.edu, Northwestern Univ – YgOmGy
Bindeman, Ilya N., 541-346-3817 bindeman@uoregon.edu, Univ of Oregon – Cs
Binford, Micheal W., (352) 392-0494 mbinford@geog.ufl.edu, Univ of Florida – Or
Bingol, Ahmet Feyzi, 00904242370000-5973 fbingol@firat.edu.tr, Firat Univ – GipGv
Binzel, Richard P., (617) 253-6486 rpb@mit.edu, Massachusetts Inst of Tech – Xm
Biondi, Biondo L., (650) 723-9831 biondo@sep.stanford.edu, Stanford Univ – Ye
Birch, Francis S., (603) 862-1718 fsb@cisunix.unh.edu, Univ of New Hampshire – Yg
Birchall, Jeff, (780) 248-5758 jeff.birchall@ualberta.ca, Univ of Alberta – OuyOg
Bircher, Harry, 330-549-9051 harrybircher@buckeyecivildesign.com, Youngstown State Univ – Ge
Bird, Brian, bbird@mail.nysed.gov, New York State Geological Survey – GlOi
Bird, Dennis K., (650) 723-1664 bird@pangea.stanford.edu, Stanford Univ – Cg
Bird, G. Peter, (310) 825-1126 pbird@epss.ucla.edu, Univ of California, Los Angeles – YgGtg
Bird, Jeffrey, 718-997-3332 jeffrey.bird@qc.cuny.edu, Queens Coll (CUNY) – Sbc
Bird, Jeffry S., (801) 378-2929 jeff_bird@byu.edu, Brigham Young Univ – Oy
Bird, John M., jmb21@cornell.edu, Cornell Univ – Gt
Bird, MIchael I., 61742421137 michael.bird@jcu.edu.au, James Cook Univ – ClsSa
Birdsell, Kay H., (505) 665-0260 khb@lanl.gov, Los Alamos National Laboratory – Hqy
Birgenheier, Lauren, (801) 585-3158 lauren.birgenheier@utah.edu, Univ of Utah – Gso
Birkel, Sean, sean.birkel@maine.edu, Univ of Maine – Gl
Birkeland, Peter W., (303) 492-6985 birkelap@stripe.colorado.edu, Univ of Colorado – Sd
Birner, Johannes, 089/2180 6565 johannes.birner@web.de, Ludwig-Maximilians-Universitaet Muenchen – Gg
Birner, Thomas, thomas.birner@colostate.edu, Colorado State Univ – Oa
Birsic, Erin M., (814) 332-2872 ebirsic@allegheny.edu, Allegheny Coll – Gi
Biscaye, Pierre E., 845-365-8429 Columbia Univ – Cm
Bischoff, Marianne, 801-626-7139 mariannebischoff@weber.edu, Weber State Univ – On
Bischoff, William D., (316) 978-6659 Wichita State Univ – Cl
Bish, David L., (812) 855-2039 bish@indiana.edu, Indiana Univ, Bloomington – Gz
Bishoff, Carolyn, 612-625-0317 cbishoff@umn.edu, Univ of Minnesota, Twin Cities – On
Bishop, Charles E., (801) 537-3361 nrugs.cbishop@state.ut.us, Utah Geological Survey – Hw
Bishop, Gale A., (912) 681-5361 Georgia Southern Univ – Pi
Bishop, James K., (510) 495-2457 bishop@atmos.berkeley.edu, Univ of California, Berkeley – Oc
Bishop, Kim, 323 343-2409 Kbishop@calstatela.edu, California State Univ, Los Angeles – Gc
Bishop, Stuart P., 919-515-7894 spbishop@ncsu.edu, North Carolina State Univ – OpGq
Bissada, Adry, 713-743-4026 kbissada@uh.edu, Univ of Houston – Co
Bissig, Thomas, 604-822-5503 tbissig@eos.ubc.ca, Univ of British Columbia – EmCg
Biswas, Pratim, pratim.biswas@seas.wustl.edu, Washington Univ in St. Louis – Ng
Bitting, Kelsey S., 785-864-1041 kelsey.bitting@ku.edu, Univ of Kansas – Oe

Bitz, Cecilia M., 206-543-1339 bitz@atmos.washington.edu, Univ of Washington – Oa
Bizimis, Michael, 803-777-5565 mbizimis@geol.sc.edu, Univ of South Carolina – GiCg
Biševac, Vanja, +38514605969 vabisevac@geol.pmf.hr, Univ of Zagreb – GxpGz
Bjorklund, Tom, 281-589-6846 tbjorklund@uh.edu, Univ of Houston – Gco
Bjornerud, Marcia, (920) 832-7015 marcia.bjornerud@lawrence.edu, Lawrence Univ – Gc
Bjornstad, Bruce N., (509) 373-6948 bruce.bjornstad@pnl.gov, Pacific Northwest National Laboratory – Gs
Blacic, James D., (505) 667-6815 jblacic@lanl.gov, Los Alamos National Laboratory – Nr
Black, Alice (Jill), (417) 836-5300 ablack@missouristate.edu, Missouri State Univ – OeaHw
Black, Annjeannette , 443-840-4560 ablack@ccbcmd.edu, Community Coll of Baltimore County, Catonsville – On
Black, Bryan A., 361-749-6789 bryan.black@utexas.edu, Univ of Texas at Austin – Sf
Black, David, (631) 632-8676 david.black@stonybrook.edu, SUNY, Stony Brook – Ou
Black, Kathryn, (973) 655-4448 Montclair State Univ – Oy
Black, Robert X., (404) 894-1756 rob.black@eas.gatech.edu, Georgia Inst of Tech – Oa
Black, Ross A., (785) 864-2740 black@ku.edu, Univ of Kansas – Ye
Black, Stuart, s.black@rdg.ac.uk, Univ of Reading – Cc
Black, T. Andrew, (604) 822-2730 Univ of British Columbia – Sp
Blackburn, James B., (713) 524-1012 blackbur@rice.edu, Rice Univ – Ou
Blackburn, Terrence, terryb@ucsc.edu, Univ of California, Santa Cruz – Cc
Blackburn, William, (270) 745-8849 will.blackburn@wku.edu, Western Kentucky Univ – Oy
Blackmer, Gale C., (717) 702-2032 gblackmer@pa.gov, Pennsylvania Bureau of Topographic & Geologic Survey – Gc
Blackwelder, Patricia L., 305 421-4677 pblackwelder@rsmas.miami.edu, Univ of Miami – Po
Blackwell, Keith G., 251-460-6302 kblackwell@southalabama.edu, Univ of South Alabama – Ow
Bladh, Katherine L., klbladh@wittenberg.edu, Wittenberg Univ – Gi
Bladh, Kenneth W., (937) 327-7334 kbladh@wittenberg.edu, Wittenberg Univ – Gz
Blair, Neal E., (847) 491-8790 n-blair@northwestern.edu, Northwestern Univ – CoSbOc
Blake, Bascombe (Mitch), (304) 594-2331 blake@geosrv.wvnet.edu, West Virginia Univ – GsEcPb
Blake, Daniel B., (217) 333-3833 dblake@illinois.edu, Univ of Illinois, Urbana-Champaign – Pi
Blake, Daniel R., (740) 548-7348 daniel.blake@dnr.state.oh.us, Ohio Dept of Natural Resources – Ysg
Blake, David E., (910) 962-3387 blaked@uncw.edu, Univ of North Carolina Wilmington – Gp
Blake, Geoffrey A., (626) 395-6296 gab@gps.caltech.edu, California Inst of Tech – Xc
Blake, M. Clark, (360) 650-3595 mcblake@nas.com, Western Washington Univ – Gt
Blake, Norman J., (727) 553-1521 nblake@marine.usf.edu, Univ of South Florida – Ob
Blake, Ruth E., (203) 432-3191 ruth.blake@yale.edu, Yale Univ – Cg
Blake, Stephen, stephen.blake@open.ac.uk, The Open Univ – Gv
Blakely, Robert F., (812) 855-1339 Indiana Univ, Bloomington – Ys
Blakeman, Audrey, 614 265 6591 audrey.blakeman@dnr.state.oh.us, Ohio Dept of Natural Resources – GrEc
Blakeney DeJarnett, Beverly, 713-896-6740 bev.dejarnett@beg.utexas.edu, Univ of Texas at Austin, Jackson School of Geosciences – Gg
Blakey, Ronald C., (928) 523-2740 ronald.blakey@nau.edu, Northern Arizona Univ – Gr
Blamey, Nigel, 905 688-5550 3365 nblamey@brocku.ca, Brock Univ – EgCg
Blanchard, Frank N., (352) 392-2296 Univ of Florida – Gz
Blanchet, Jean-Pierre, 514-987-3000 #3316 blanchet.jean-pierre@uqam.ca, Universite du Quebec a Montreal – Pe
Bland, Douglas, (505) 466-6696 dmbland@comcast.net, New Mexico Inst of Mining & Tech – Eg
Bland, William L., (608) 262-0221 wlbland@wisc.edu, Univ of Wisconsin, Madison – Sp
Blander, Milton, (630) 252-4548 Argonne National Laboratory – Xc
Blaney, Diana L., (818) 354-5419 Jet Propulsion Laboratory – Xg
Blank, Carrine E., (406) 241-2038 carrine.blank@umontana.edu, Univ of Montana – On
Blanken, Peter D., blanken@colorado.edu, Univ of Colorado – HsOaw
Blankenbicker, Adam, (202) 633-1123 blankenbickera@si.edu, Smithsonian Institution / National Museum of Natural History – Oe
Blankenship, Donald D., 512-471-0489 blank@ig.utexas.edu, Univ of Texas at Austin – Gl
Blankenship, Robert, blankship@wustl.edu, Washington Univ in St. Louis – On
Blasing, Terence J., (865) 574-7368 blasingtj@ornl.gov, Oak Ridge National Laboratory – Oa
Blasingame, Tom, (979) 845-2292 t-blasingame@spindletop.tamu.edu, Texas A&M Univ – Np
Blatner , Keith , 509-335-4499 blatner@wsu.edu, Washington State Univ – On
Blattler, Clara, 609 258-4101 blattler@princeton.edu, Princeton Univ – Gg
Blaylock, Glenn W., 956-764-5715 glenn.blaylock@laredo.edu, Laredo Community Coll – Gg
Bleam, William F., (608) 262-9956 wfbleam@wisc.edu, Univ of Wisconsin, Madison – Sc
Bleamaster III, Leslie F., (210) 999-7740 lbleamas@trinity.edu, Trinity Univ – Xg
Blechman, Jerome B., (607) 436-3322 Jerome.Blechman@oneonta.edu, SUNY, Oneonta – Oa
Bleck, Rainer, (505) 665-9150 bleck@lanl.gov, Los Alamos National Laboratory – Ow
Bleibinhaus, Florian, 089/2180 4202 bleibi@geophysik.uni-muenchen.de, Ludwig-Maximilians-Universitaet Muenchen – Yg
Bleibinhaus, Florian, 0049(0)3641/948661 Janet.Kressler@uni-jena.de, Friedrich-Schiller-Univ Jena – Yx
Bleistein, Norman, (303) 273-3557 norm3980@aol.com, Colorado School of Mines – Yg
Blenkinson, Stephen, +44 (0)191 208 7933 stephen.blenkinsop@ncl.ac.uk, Univ of Newcastle Upon Tyne – On
Blenkinsop, John, jblenkin@ccs.carleton.ca, Carleton Univ – Cc
Blewett, William L., (717) 477-1513 wlblew@ship.edu, Shippensburg Univ – Oy
Blewitt, Geoffrey, (775) 682-8778 gblewitt@unr.edu, Univ of Nevada, Reno – Yd
Blisniuk, Kimberly, 408 924-5045 kimberly.blisniuk@sjsu.edu, San Jose State Univ – Gmt
Blizzard, Kate, (360) 650-3581 kate.blizzard@wwu.edu, Western Washington Univ – On
Bloch, Jonathan, jbloch@flmnh.ufl.edu, Univ of Florida – Pv
Blodgett, Robert H., (512) 223-4376 rblodget@austincc.edu, Austin Community Coll District – GseOe
Bloechle, Amber, (850) 857-6121 abloechle@uwf.edu, Univ of West Florida – Oi
Blom, Ronald G., (818) 354-4681 Jet Propulsion Laboratory – Or
Bloom, Arthur L., alb11@cornell.edu, Cornell Univ – Gm
Bloom, Paul R., (612) 625-4711 pbloom@soils.umn.edu, Univ of Minnesota, Twin Cities – Sc
Bloomer, Sherman H., (541) 737-4811 sherman.bloomer@oregonstate.edu, Oregon State Univ – Gi
Bloomfield, Helen, +440151 795 4652 H.J.Bloomfield@liverpool.ac.uk, Univ of Liverpool – Ob
Bloomgren, Bruce, (612) 626-4108 bloom005@umn.edu, Univ of Minnesota – Oi
Bloor, Michelle, +44 023 92 842295 michelle.bloor@port.ac.uk, Univ of Portsmouth – Hg
Bloxham, Jeremy, (617) 495-9517 jeremy_bloxham@harvard.edu, Harvard Univ – Yg
Bluestein, Howard, hblue@ou.edu, SUNY, Stony Brook – Oa
Blum, Joel D., (734) 764-1435 jdblum@umich.edu, Univ of Michigan – CsaCl
Blum, Linda K., (434) 924-0560 lkb2e@virginia.edu, Univ of Virginia – Sb
Blum, Michael D., 785-864-4974 Univ of Kansas – GsrGm
Blumenfeld, Raphael, r.blumenfeld@imperial.ac.uk, Imperial Coll – On
Blusztajn, Jurek, (508) 289-2692 jblusztajn@whoi.edu, Woods Hole Oceanographic Institution – CcGi
Blyth, Alan, +44(0) 113 34 31632 a.m.blyth@leeds.ac.uk, Univ of Leeds – Oa
Blythe, Ann, 323 259-2553 ablythe@oxy.edu, Occidental Coll – Cc
Blythe, Daniel D., 406.496.4379 dblythe@mtech.edu, Montana Tech of The Univ of Montana – Hws
Boadu, Fred K., (919) 660-5432 boadu@duke.edu, Duke Univ – Yg

Boaga, Jacopo, +390498279189 jacopo.boaga@unipd.it, Università degli Studi di Padova – Yg
Boar, Rosalind, +44 (0)1603 59 3103 r.boar@uea.ac.uk, Univ of East Anglia – Ob
Boardman, Mark R., (513) 529-3230 boardman@MiamiOh.edu, Miami Univ – On
Boast, Charles W., (217) 333-1278 Univ of Illinois, Urbana-Champaign – Sp
Boateng, Samuel, 859-572-6683 boatengs@nku.edu, Northern Kentucky Univ – Hw
Boboye, O. A., oa.boboye@mail.ui.edu.ng, Univ of Ibadan – Go
Bobst, Andrew L., 406.496.4409 abobst@mtech.edu, Montana Tech of The Univ of Montana – HwCgEo
Bobyarchick, Andy R., 704-687-5998 arbobyar@email.uncc.edu, Univ of North Carolina, Charlotte – Gc
Bochdansky, Alexander, 757-6834933 abochdan@odu.edu, Old Dominion Univ – Ob
Bockheim, James G., (608) 263-5903 bockheim@wisc.edu, Univ of Wisconsin, Madison – Sf
Boda, Patricia, 615-904-8098 Pat.Boda@mtsu.edu, Middle Tennessee State Univ – OinOn
Boden, Thomas, (865) 241-4842 tab@ornl.gov, Oak Ridge National Laboratory – Oa
Bodenbender, Brian E., (616) 395-7541 bodenbender@hope.edu, Hope Coll – PiGs
Bodenheimer, Peter, (831) 459-2064 peter@ucolick.org, Univ of California, Santa Cruz – Xc
Bodenman, John E., (570) 389-4697 boden@bloomu.edu, Bloomsburg Univ – Ou
Bodin, Paul A., 206-616-7315 bodin@uw.edu, Univ of Washington – Ys
Bodnar, Robert J., (540) 231-7455 rjb@vt.edu, Virginia Polytechnic Inst & State Univ – Cg
Boelman, Natalie T., (845) 365-8480 nboelman@ldeo.columbia.edu, Columbia Univ – PbeOr
Boerboom, Terrence, (612) 626-3369 boerb001@umn.edu, Univ of Minnesota – GivGc
Boering, Kristie, (510) 642-3472 boering@cchem.berkeley.edu, Univ of California, Berkeley – Oa
Boester, Michael, mboester@monroecc.edu, Monroe Community Coll – Oy
Boettcher, Margaret S., 603-862-0580 margaret.boettcher@unh.edu, Univ of New Hampshire – Ygs
Bogart, James M., james.bogart@ct.gov, Dept of Energy and Environmental Protection – Ge
Boger, Kathryn M., (740) 368-3615 kmboger@owu.edu, Ohio Wesleyan Univ – On
Boger, Rebecca, 718-951-5000 x2159 rboger@brooklyn.cuny.edu, Brooklyn Coll (CUNY) – OiHs
Boggs, Carol, boggscl@mailbox.sc.edu, Univ of South Carolina – On
Boggs, Katherine, (403) 440-6645 kboggs@mtroyal.ca, Mount Royal Univ – Gt
Boggs, Jr., Sam, (541) 346-4553 sboggs@uoregon.edu, Univ of Oregon – Gd
Bogle, Mary Anna, (865) 574-7824 amb@ornl.gov, Oak Ridge National Laboratory – Ge
Bogner, Jean E., jbogner@uic.edu, Univ of Illinois at Chicago – ClSbHy
Bogner, William C., (217) 333-9546 Illinois State Water Survey – Hs
Bograd, Michael B., (601) 961-5528 mbograd@mdeq.ms.gov, Mississippi Office of Geology – Gg
Bogucki, Darek, 361-825-825-2836 Darek.Bogucki@tamucc.edu, Texas A&M Univ, Corpus Christi – Op
Bogucki, Donald J., (518) 564-4030 Plattsburgh State Univ (SUNY) – Or
Bogue, Scott W., (323) 259-2563 bogue@oxy.edu, Occidental Coll – Ym
Bohaty, Steven, +44 (0)23 8059 3040 S.Bohaty@noc.soton.ac.uk, Univ of Southampton – Og
Bohlen, Thomas, +49-721-6084416 thomas.bohlen@kit.edu, Karlsruhe Inst of Tech – Yx
Bohlen, Walter F., 860-405-9176 walter.bohlen@uconn.edu, Univ of Connecticut – Op
Bohling, Geoff, 785-864-2093 geoff@kgs.ku.edu, Univ of Kansas – Hw
Bohm, Christian, (204) 945-6549 christian.bohm@gov.mb.ca, Manitoba Geological Survey – GgCgEg
Bohnenstiehl, DelWayne, (919) 515-7449 drbohnen@ncsu.edu, North Carolina State Univ – YrsGt
Bohren, Craig F., (814) 865-2951 bohren@ems.psu.edu, Pennsylvania State Univ, Univ Park – Ow
Bohrson, Wendy A., (509) 963-2835 bohrson@geology.cwu.edu, Central Washington Univ – GivCc
Boicourt, William, (410) 221-8426 boicourt@hpl.umces.edu, Univ of Maryland – Op
Boisvert, Eric, (418) 654-3705 eboisvert@nrcan.gc.ca, Universite du Quebec – On
Boitt, Mark , mboitt@jkuat.ac.ke, Jomo Kenyatta Univ of Agriculture & Tech – OgrOi
Bokuniewicz, Henry J., (631) 632-8674 henry.bokuniewicz@stonybrook.edu, SUNY, Stony Brook – Yr
Boland, Greg J., (519) 824-4120 x52755 gboland@uoguelph.ca, Univ of Guelph – On
Boland, Irene B., 803-323-4949 bolandi@winthrop.edu, Winthrop Univ – GtgOe
Boland, John J., (410) 516-7103 jboland@jhu.edu, Johns Hopkins Univ – On
Bolarinwa, A. T., at.bolarinwa@mail.ui.edu.ng, Univ of Ibadan – EgGz
Bollag, Jean-Marc, (814) 863-0843 jmbollag@psu.edu, Pennsylvania State Univ, Univ Park – Sb
Bollasina, Massimo, +44 (0) 131 650 4915 mbollasi@staffmail.ed.ac.uk, Edinburgh Univ – Oa
Bollens, Stephen, 360-546-9116 sbollens@vancouver.wsu.edu, Washington State Univ – Ob
Bollinger, G. A., (540) 231-6521 Virginia Polytechnic Inst & State Univ – Ys
Bollinger, Marsha S., 803-323-4944 bollingerm@winthrop.edu, Winthrop Univ – CmOgCc
Bollmann, Jörg, (416) 978-2061 bollmann@geology.utoronto.ca, Univ of Toronto – Pm
Bolster, Diogo, Diogo.Bolster.5@nd.edu, Univ of Notre Dame – Hw
Bolt, John R., (312) 665-7629 jbolt@fieldmuseum.org, Field Museum of Natural History – Pv
Bolton, David W., (410) 554-5561 david.bolton@maryland.gov, Maryland Department of Natural Resources – Hw
Bolton, Edward W., (203) 432-3149 edward.bolton@yale.edu, Yale Univ – Gq
Bolton Valencius, Conevery, conevery.valencius@umb.edu, Univ of Massachusetts, Boston – Gh
Bolze, Claude E., 918-595-7246 claude.bolze@tulsacc.edu, Tulsa Community Coll – GgoPg
Bomke, Arthur A., (604) 822-6534 Univ of British Columbia – Sc
Bona, Andrej, +61 8 9266 7194 A.Bona@curtin.edu.au, Curtin Univ – Yx
Bonatti, Enrico, 845-365-8699 Columbia Univ – Yr
Bonczek, James, 352-294-3112 bonczek@ufl.edu, Univ of Florida – OsHsg
Bond, Alan, +44 (0)1603 59 3402 alan.bond@uea.ac.uk, Univ of East Anglia – Ge
Bond, Clare, +44 (0)1224 273492 clare.bond@abdn.ac.uk, Univ of Aberdeen – Gct
Bond, Nicholas A., Nicholas.Bond@noaa.gov, Univ of Washington – Oa
Bondesan, Aldino, 39-335-5473369 aldino.bondesan@gmail.com, Università degli Studi di Padova – GmlOi
Bone, Christopher, cbone@uoregon.edu, Univ of Oregon – Oi
Bone, Fletcher, (573) 368-2183 fletcher.bone@dnr.mo.gov, Missouri Dept of Natural Resources – Gg
Bonem, Rena M., (254) 710-2187 rena_bonem@baylor.edu, Baylor Univ – PieOb
Bonetto, Sabrina Maria Rita, sabrina.bonetto@unito.it, Università di Torino – Ng
Bonhoure, Jessica, +33(0)3 44068994 jessica.bonhoure@lasalle-beauvais.fr, Institut Polytechnique LaSalle Beauvais (ex-IGAL) – GziCc
Boniface, Nelson, +255222410013 nelson.boniface@udsm.ac.tz, Univ of Dar es Salaam – GcpGz
Bonine, Michael E., kebonine@u.arizona.edu, Univ of Arizona – Ou
Bonk, Kathleen, 518-473-9988 kbonk@mail.nysed.gov, New York State Geological Survey – GoCg
Bonneau, Laurent, (203) 432-3142 laurent.bonneau@yale.edu, Yale Univ – Oa
Bonuso, Nicole, (657) 278-8451 nbonuso@fullerton.edu, California State Univ, Fullerton – Pie
Booher, Gary, (310) 660-3593 gbooher@elcamino.edu, El Camino Coll – Og
Booker, John R., 206-543-1190 booker@ess.washington.edu, Univ of Washington – YmGt
Bookhagen, Bodo, (805) 893-3568 bodo@geog.ucsb.edu, Univ of California, Santa Barbara – Gm

Faculty Index -B

Boone, Peter A., pboone@austincc.edu, Austin Community Coll District – Gro
Boorstein, Margaret F., (516) 299-2318 maboorst@liu.edu, Long Island Univ, C.W. Post Campus – Og
Booth, Adam, +44 20 759 46528 a.booth@imperial.ac.uk, Imperial Coll – Ol
Booth, Adam M., 503-725-3320 boothad@pdx.edu, Portland State Univ – GmOr
Booth, Alastair, murray.booth@manchester.ac.uk, Univ of Manchester – Oa
Booth, Colin J., (815) 753-1943 cjbooth@niu.edu, Northern Illinois Univ – Hw
Booth, Robert K., rkb205@lehigh.edu, Lehigh Univ – On
Boothroyd, Jon C., (401) 874-2265 jon_boothroyd@uri.edu, Univ of Rhode Island – GslOn
Bopp, Richard F., (518) 276-3075 boppr@rpi.edu, Rensselaer Polytechnic Inst – Co
Borah, Deva K., (217) 244-8856 Illinois State Water Survey – Hq
Bord, Don, (313) 593-5483 dbord@umich.edu, Univ of Michigan, Dearborn – Xy
Bordeaux, Yvette, (215) 898-9191 bordeaux@sas.upenn.edu, Univ of Pennsylvania – Po
Bordoni, Simona, 626.395.2672 bordoni@gps.caltech.edu, California Inst of Tech – Op
Bordossy, Andras, +44 (0) 191 208 6319 andras.bardossy@ncl.ac.uk, Univ of Newcastle Upon Tyne – Og
Bordy, Emese M., 021-650-2901 emese.bordy@uct.ac.za, Univ of Cape Town – GsPes
Borg, Scott G., sborg@nsf.gov, National Science Foundation – Git
Borisov, Dmitry, 609-258-4101 dborisov@princeton.edu, Princeton Univ – Ysg
Bork, Karrigan, 916-733-2818 kbork@PACIFIC.EDU, Univ of the Pacific – Hg
Bork, Kennard B., (928) 554-4942 bork@denison.edu, Denison Univ – PiGhs
Bornhorst, Theodore J., 906-487-2721 tjb@mtu.edu, Michigan Technological Univ – EgCgGx
Borns, Jr., Harold W., (207) 581-2196 borns@maine.edu, Univ of Maine – Gl
Borojeviæ Šoštariæ, Sibila, +38514605961 sborojsost@geol.pmf.hr, Univ of Zagreb – CegCc
Boroughs, Scott, 509-335-1626 scott.boroughs@wsu.edu, Washington State Univ – GiCa
Boroughs, Terry J., (916) 331-8596 BorougT@arc.losrios.edu, American River Coll – GgXgCg
Boroushaki, Soheil, 818-677-4715 soheil.boroushaki@csun.edu, California State Univ, Northridge – Oi
Borowski, Walter S., (859) 622-1277 w.borowski@eku.edu, Eastern Kentucky Univ – CgGo
Borradaile, Graham J., gjborrad@lakeheadu.ca, Lakehead Univ – Ym
Borrok, David M., (337) 482-2888 dmb5953@louisiana.edu, Univ of Louisiana at Lafayette – ClGeo
Borsa, Adrian, (858) 534-6895 aborsa@ucsd.edu, Univ of California, San Diego – Hg
Bosak, Tanja, 617-324-3959 tbosak@mit.edu, Massachusetts Inst of Tech – Py
Bosart, Lance F., (518) 442-4564 bosart@atmos.albany.edu, SUNY, Albany – Oa
Bosbyshell, Howell, (610) 436-2805 hbosbyshell@wcupa.edu, West Chester Univ – Gc
Boss, Alan P., (202) 478-8858 aboss@carnegiescience.edu, Carnegie Institution of Washington – On
Boss, Stephen K., (479) 575-7134 sboss@uark.edu, Univ of Arkansas, Fayetteville – GeYgGg
Bossenbroek, Jonathon , 419-530-8376 jonathon.bossenbroek@utoledo.edu, Univ of Toledo – On
Bossert, James E., (505) 667-6268 bossert@lanl.gov, Los Alamos National Laboratory – Oa
Bostater, Charles R., (321) 674-8096 bostater@fit.edu, Florida Inst of Tech – Op
Bostock, Michael G., (604) 822-2082 bostock@eos.ubc.ca, Univ of British Columbia – YsGt
Boston, Penelope, 575.835.5657 pboston@nmt.edu, New Mexico Inst of Mining and Tech – On
Boswell, Cecil, 573-368-2146 cecil.boswell@dnr.mo.gov, Missouri Dept of Natural Resources – On
Bothern, Lawrence, 915-831-5088 lbothern@epcc.edu, El Paso Community Coll – Gg
Bothner, Wallace A., (603) 862-1718 wally.bothner@unh.edu, Univ of New Hampshire – Gc
Bots, Pieter, pieter.bots@manchester.ac.uk, Univ of Manchester – Ge
Bottenberg, H. Carrie, 208-282-3538 bottcarr@isu.edu, Idaho State Univ – OiGtc
Bottero, Jean-Yves, bottero@arbois.cerege.fr, Rice Univ – On
Bottjer, David J., (213) 740-6100 dbottjer@usc.edu, Univ of Southern California – Pe
Boudreau, Alan E., (919) 684-5646 boudreau@duke.edu, Duke Univ – Gi
Boudreau, Bernard P., (902) 494-8895 bernie.boudreau@dal.ca, Dalhousie Univ – CmOcCl
Boudreaux, Elisabeth L., (337) 482-5349 eboudreaux@louisiana.edu, Univ of Louisiana at Lafayette – GgOge
Boudrias, Michel A., (619) 260-4600 boum@sandiego.edu, Univ of San Diego – Ob
Bouker, Polly A., (770) 278-1320 pbouker@gsu.edu, Georgia State Univ, Perimeter Coll, Newton Campus – Gg
Boult, Stephen, +44 0161 275-3867 s.boult@manchester.ac.uk, Univ of Manchester – Hg
Boulton, Sarah, +44 1752 584762 sarah.boulton@plymouth.ac.uk, Univ of Plymouth – Or
Boulvain, Frederic P., 32 4 366 22 52 fboulvain@ulg.ac.be, Universite de Liege – GdgGs
Bouman, Heather, +44 (1865) 272019 Heather.Bouman@earth.ox.ac.uk, Univ of Oxford – Cg
Bounk, Michael, Michael.Bounk@dnr.iowa.gov, Iowa Dept of Natural Resources – Gg
Bour, William, (703) 450-2612 wbour@nvcc.edu, Northern Virginia Community Coll - Loudoun Campus – Gge
Bourassa, Matthew, 519-661-2111 ext 80370 mbouras@uwo.ca, Western Univ – Xg
Bourcier, Bill L., (925) 423-3745 bourcier1@llnl.gov, Lawrence Livermore National Laboratory – Cg
Bourgeois, Joanne (Jody), (206) 685-2443 jbourgeo@uw.edu, Univ of Washington – GsrGh
Bourgeois, Reed J., (225) 388-8879 reed@lgs.bri.lsu.edu, Louisiana State Univ – On
Bourke, Robert H., (831) 656-2673 rbourke@nps.edu, Naval Postgraduate School – Op
Bouse, Robin, (310) 660-3593 rbouse@elcamino.edu, El Camino Coll – Gg
Bousenberry, Raymond T., (609) 984-6587 raymond.bousenberry@dep.nj.gov, New Jersey Geological and Water Survey – Ge
Boutt, David, (413) 545-2724 dboutt@geo.umass.edu, Univ of Massachusetts, Amherst – HwqHy
Bouvier, Audrey, 519-661-2111 x.88516 audrey.bouvier@uwo.ca, Western Univ – Xc
Bouwer, Edward John, (410) 516-7437 bouwer@jhu.edu, Johns Hopkins Univ – On
Bouzari, Farhad, (604) 822-1874 fbouzari@eos.ubc.ca, Univ of British Columbia – Eg
Boving, Thomas, (401) 874-7053 boving@uri.edu, Univ of Rhode Island – Hg
Bowen, Anthony J., (902) 494-7082 tony.bowen@dal.ca, Dalhousie Univ – On
Bowen, Brenda , (801) 585-5326 brenda.bowen@utah.edu, Univ of Utah – Gse
Bowen, David W., dbowen@montana.edu, Montana State Univ – GroGs
Bowen, Dawn S., (540) 654-1491 dbowen@mwc.edu, Mary Washington Coll – On
Bowen, Esther E., (630) 252-7553 ebowen@anl.gov, Argonne National Laboratory – Gg
Bowen, Gabriel, (801) 585-7925 gabe.bowen@utah.edu, Univ of Utah – Cs
Bowen, Jennifer, 7815817370 x346 je.bowen@northeastern.edu, Northeastern Univ – On
Bowen, Jennifer, jennifer.bowen@umb.edu, Univ of Massachusetts, Boston – ObPyOn
Bowen, Scott M., (505) 667-4313 Los Alamos National Laboratory – Ca
Bower, Kathleen M., 217-581-2626 kmbower@eiu.edu, Eastern Illinois Univ – Ng
Bowersox, J. Richard, (859) 323-0536 j.r.bowersox@uky.edu, Univ of Kentucky – GoPq
Bowersox, Joe, 503-370-6220 jbowerso@willamette.edu, Willamette Univ – On
Bowin, Carl O., (508) 289-2572 cbowin@whoi.edu,

Woods Hole Oceanographic Institution – YrGtg
Bowler, Bernard, +44 (0) 191 208 5931 bernard.bowler@ncl.ac.uk, Univ of Newcastle Upon Tyne – Go
Bowles, Julie, 414-229-6110 bowlesj@uwm.edu, Univ of Wisconsin, Milwaukee – Ym
Bowles, Zack, zbowles@mesacc.edu, Mesa Community Coll – Gg
Bowlick, Forrest J., (413)-577-3816 fbowlick@umass.edu, Univ of Massachusetts, Amherst – Oig
Bowman, Jean A., (409) 862-6544 jbowman@tamu.edu, Texas A&M Univ – Hg
Bowman, John R., (801) 581-7250 john.bowman@utah.edu, Univ of Utah – Cs
Bowman, Kenneth P., (409) 845-4060 jbowman@tamu.edu, Texas A&M Univ – Ow
Bowman, Malcolm J., (631) 632-8669 malcolm.bowman@stonybrook.edu, SUNY, Stony Brook – Op
Bowman, Mike, michael.bowman@manchester.ac.uk, Univ of Manchester – Go
Bowman, Steve D., (801) 537-3304 stevebowman@utah.gov, Utah Geological Survey – NgGe
Bowring, Samuel A., (617) 253-3775 sbowring@mit.edu, Massachusetts Inst of Tech – Gg
Bowser, Carl J., (608) 262-8955 bowser@geology.wisc.edu, Univ of Wisconsin, Madison – Cl
Bowser, Paul, prb4@cornell.edu, SUNY, Stony Brook – Obb
Boxall, Simon R., +44 (0)23 80592744 simon.boxall@soton.ac.uk, Univ of Southampton – Op
Boyack, Diana L., (208) 282-3137 boyadian@isu.edu, Idaho State Univ – Oi
Boybeyi, Zafer, (703) 993-1560 zboybeyi@gmu.edu, George Mason Univ – Oa
Boyce, Charles K., (773) 834-7640 Univ of Chicago – Pb
Boyce, Joseph I., (905) 525-9140 (Ext. 24188) boycej@mcmaster.ca, McMaster Univ – Og
Boyd, Bill, william.boyd@scu.edu.au, Southern Cross Univ – Pe
Boyd, John, (734) 764-3338 jpboyd@umich.edu, Univ of Michigan – Opa
Boyd, Oliver, (901) 678-2007 olboyd@memphis.edu, Univ of Memphis – YgGtg
Boyer, Diana L., dboyer@oswego.edu, SUNY, Buffalo – PgGs
Boyer, Robert E., (512) 471-7228 reboyer@mail.utexas.edu, Univ of Texas at Austin – Gt
Boyle, Alan P., +44 0151 794 5154 apboyle@liverpool.ac.uk, Univ of Liverpool – GczOe
Boyle, Douglas, 775-784-6995 Univ of Nevada, Reno – Hg
Boyle, Edward A., (617) 253-3388 eaboyle@mit.edu, Massachusetts Inst of Tech – Oc
Boyle, Elizabeth , elizabeth.boyle@umb.edu, Univ of Massachusetts, Boston – Obn
Boyle, James T., (609) 984-6587 jim.boyle@dep.nj.gov, New Jersey Geological and Water Survey – Hw
Boynton, William V., (520) 621-6941 wboynton@lpl.arizona.edu, Univ of Arizona – XcOr
Boysen, Hans, 089/2180 4333 boysen@lmu.de, Ludwig-Maximilians-Universitaet Muenchen – Gz
Bozdog, Nick, 828-296-4635 nick.bozdog@ncdenr.gov , North Carolina Geological Survey – Gg
Bozzato, Edda, 7056751151, ext. 2272 ebozzato@laurentian.ca, Laurentian Univ, Sudbury – OnnOn
Brabander, Daniel J., 781-283-3056 dbraband@wellesley.edu, Wellesley Coll – Cl
Brabham, P J., brabham@cf.ac.uk, Cardiff Univ – Ye
Brabham, Peter, +44(0)29 208 74334 Brabham@cardiff.ac.uk, Univ of Wales – Ng
Bracco, Annalisa, (404) 894-1749 annalisa@eas.gatech.edu, Georgia Inst of Tech – OpnOn
Bradbury, Kelly K., kellykbradbury@gmail.com, Utah State Univ – Gc
Bradbury, Kenneth R., (608) 263-7921 ken.bradbury@uwex.edu, Univ of Wisconsin, Madison – Hw
Bradford, Joel, (801) 863-7263 bradfojo@uvu.edu, Utah Valley Univ – Gae
Bradford, John, jbradfor@boisestate.edu, Boise State Univ – Yeg
Bradley, Alexander S., (314) 935-6333 abradley@levee.wustl.edu, Washington Univ in St. Louis – Py
Bradley, Christopher R., (505) 665-6713 cbradley@lanl.gov, Los Alamos National Laboratory – Yg
Bradley, J G., (315) 312-3081 graham.bradley@oswego.edu, SUNY, Oswego – GeHwNg
Bradley, Jeffrey, (660) 562-1818 jbradle@nwmissouri.edu, Northwest Missouri State Univ – Og
Bradley, Michael, (313) 487-0218 Eastern Michigan Univ – Gc
Bradley, Michael D., (520) 621-3865 mdb@hwr.arizona.edu, Univ of Arizona – On
Bradley, Raymond S., (413) 545-2120 rbradley@geo.umass.edu, Univ of Massachusetts, Amherst – Oa
Bradshaw, John, (03) 3667-001 x7487 john.bradshaw@canterbury.ac.nz, Univ of Canterbury – Gt
Bradshaw, Lanna K., 972/860-4713 LannaBradshaw@dcccd.edu, Brookhaven Coll – GgOg
Bradshaw, Peter M., (604) 681-8600 pbradshaw@firstpointminerals.com, Univ of British Columbia – EgCeGe
Brady, Emily S., +44 (0) 131 650 9137 Emily.Brady@ed.ac.uk, Edinburgh Univ – Ge
Brady, John B., (413) 585-3953 jbrady@smith.edu, Smith Coll – Gx
Brady, Kristina, 612-626-7889 brad0311@umn.edu, Univ of Minnesota, Twin Cities – Gn
Brady, Lawrence L., (785) 864-2159 lbrady@kgs.ku.edu, Univ of Kansas – Gg
Brady, Roland H., (559) 278-2391 roland_brady@csufresno.edu, California State Univ, Fresno – NgOuGa
Braginsky, Stanislav I., (310) 794-1331 jbragin@ucla.edu, Univ of California, Los Angeles – Ym
Brahana, J. Van, (479) 575-3355 brahana@uark.edu, Univ of Arkansas, Fayetteville – HwGe
Brahmi, Sadek, +33(0)3 44068978 sadek.brahmi@lasalle-beauvais.fr, Institut Polytechnique LaSalle Beauvais (ex-IGAL) – Nr
Braile, Lawrence W., (765) 494-5979 braile@purdue.edu, Purdue Univ – Ys
Brainard, James R., (505) 667-0150 Los Alamos National Laboratory – On
Brake, Sandra S., (812) 237-2270 sandra.brake@indstate.edu, Indiana State Univ – GeEm
Brake, Thomas L., (702) 794-7828 brake_thomas_l@lanl.gov, Los Alamos National Laboratory – Nr
Bralower, Timothy J., 814-863-1240 bralower@psu.edu, Pennsylvania State Univ, Univ Park – Pe
Braman, Dennis R., 403-823-7707 Royal Tyrrell Museum of Palaeontology – Pl
Brame, Scott E., (864) 656-7167 scott@clemson.edu, Clemson Univ – Hw
Bramham, Emma , +44(0) 113 34 35595 E.K.Bramham@leeds.ac.uk, Univ of Leeds – Ym
Branam, Tracy D., (812) 855-8390 tbranam@indiana.edu, Indiana Univ – CasEm
Branan, Yvonne K., ykbranan@iup.edu, Indiana Univ of Pennsylvania – Gv
Brand, Brittany, 208-426-4154 brittanybrand@boisestate.edu, Boise State Univ – Gsv
Brand, Leonard R., (909) 558-4530 lbrand@llu.edu, Loma Linda Univ – Pv
Brand, Uwe, (905) 688-5550 x3529 uwe.brand@brocku.ca, Brock Univ – ClsOa
Brandes, Jay, (912) 598-2361 jay.brandes@skio.usg.edu, Georgia Inst of Tech – Og
Brandon, Alan, 713-743-2359 abrandon@uh.edu, Univ of Houston – GiCac
Brandon, Christine M., (508) 531-2125 christine.brandon@bridgew.edu, Bridgewater State Univ – Gs
Brandon, Mark A., 01908 655172 mark.brandon@open.ac.uk, The Open Univ – Og
Brandon, Mark T., (203) 432-3135 mark.brandon@yale.edu, Yale Univ – Gc
Brandon, Nigel, +44 20 759 45704 n.brandon@imperial.ac.uk, Imperial Coll – Ge
Brandriss, Mark E., (413) 585-3585 mbrandri@smith.edu, Smith Coll – Gx
Brandt, Craig, (865) 574-1921 fcb@ornl.gov, Oak Ridge National Laboratory – Ge
Brandt, Danita S., (517) 355-6595 brandt@msu.edu, Michigan State Univ – PiGr
Brandvold, Lynn A., (505) 835-5517 lynnb@nmt.edu, New Mexico Inst of Mining & Tech – Ca
Branfireun, Brian, bbranfir@uwo.ca, Western Univ – Ge
Branlund, Joy, (618) 797-7451 joy.branlund@swic.edu, Southwestern Illinois Coll - Sam Wolf Granite City Campus – Og
Branney, Mike, +440116 252 3647 mjb26@le.ac.uk, Leicester Univ – Gv
Branstrator, Jon W., (765) 983-1339 jonb@earlham.edu, Earlham Coll – Pg
Brant, Lynn A., (319) 273-6160 lynn.brant@uni.edu,

Univ of Northern Iowa – Gn
Brantley, Duke, brantley@mailbox.sc.edu, Univ of South Carolina – Ye
Brantley, Susan L., (814) 863-1739 brantley@eesi.psu.edu, Pennsylvania State Univ, Univ Park – Cg
Brasier, Martin, +44 (1865) 272074 martin.brasier@earth.ox.ac.uk, Univ of Oxford – Po
Brassea, Jesus M., jbrassea@cicese.mx, Centro de Investigación Científica y de Educación Superior de Ensenada – Ye
Brassell, Simon C., (812) 855-3786 simon@indiana.edu, Indiana Univ, Bloomington – Co
Brassington, Rick, +44 (0) 192 576 6754 rick.brassington@ncl.ac.uk, Univ of Newcastle Upon Tyne – Hg
Braught, Patricia, (717) 245-1355 braught@dickinson.edu, Dickinson Coll – On
Braun, Alexander, (613) 533-6621 braun@queensu.ca, Queen's Univ – YgdYv
Braund, Emilie, +44 023 9284 2257 emilie.bruand@port.ac.uk, Univ of Portsmouth – Gz
Bray, Colin, (416) 978-6516 cjbray@quartz.geology.utoronto.ca, Univ of Toronto – Cs
Breaker, Laurence, 831-771-4400 lbreaker@mlml.calstate.edu, Moss Landing Marine Laboratories – Op
Brearley, Adrian J., (505) 277-4163 brearley@unm.edu, Univ of New Mexico – Gz
Breckenridge, Andrew, abrecken@uwsuper.edu, Univ of Wisconsin, Superior – GlnGs
Breecker, Daniel O., 512-471-6166 breecker@jsg.utexas.edu, Univ of Texas at Austin – ScCs
Breedlovestrout, Renee L., (208) 885-7560 reneeb@uidaho.edu, Univ of Idaho – PbGro
Brehman, Thomas R., (847) 376-7036 granite@oakton.edu, Oakton Community Coll – Po
Breitbart, Mya, (727) 553-3520 mya@marine.usf.edu, Univ of South Florida – Ob
Breitmeyer, Ronald, 775.682.6049 rbreitmeyer@unr.edu, Univ of Nevada, Reno – HwGe
Breland, Clayton F., (225) 388-8300 clayton@lgs.bri.lsu.edu, Louisiana State Univ – Ye
Bremer, Keith A., (785) 628-4644 kabremer@fhsu.edu, Fort Hays State Univ – OuiOn
Brenan, James, (902) 494-2355 jbrenan@dal.ca, Dalhousie Univ – Gi
Brengman, Latisha A., 218-726-7586 lbrengma@d.umn.edu, Univ of Minnesota, Duluth – Gd
Brenna, J. T., (607) 255-9182 jtb4@cornell.edu, Cornell Univ – Cg
Brenner, Mark, (352) 392-9617 brenner@ufl.edu, Univ of Florida – Gn
Bretherton, Christopher S., 206-685-7414 breth@washington.edu, Univ of Washington – Oa
Bretherton, Francis P., (608)262-7497 fbretherton@charter.net, Univ of Wisconsin, Madison – Og
Brethes, Jean-Claude, (418) 724-1779 cjmichau@globetrotter.net, Universite du Quebec a Rimouski – Ob
Breton, Caroline, 512-471-0322 cari.breton@beg.utexas.edu, Univ of Texas at Austin, Jackson School of Geosciences – Oi
Brett, Carlton E., (513) 556-4556 carlton.brett@uc.edu, Univ of Cincinnati – Pe
Brewer, Margene, (616) 526-8415 mkb25@calvin.edu, Calvin Coll – OnGgOn
Brewer, Micheal, (202) 994-6185 mbrewer@gwu.edu, George Washington Univ – Or
Brewer, Paul A., +44 (0)1970 622 586 pqb@aber.ac.uk, Univ of Wales – Gm
Brezinski, David K., (410) 554-5526 david.brezinski@maryland.gov, Maryland Department of Natural Resources – Gr
Bridgeman, Thomas, 419-530-8373 Thomas.Bridgeman@utoledo.edu, Univ of Toledo – On
Bridges, Carey, 573-368-2143 carey.bridges@dnr.mo.gov, Missouri Dept of Natural Resources – Gg
Bridges, David L., (573) 368-2656 david.bridges@dnr.mo.gov, Missouri Dept of Natural Resources – GgiPg
Bridges, Frank, (831) 459-2893 bridges@cats.ucsc.edu, Univ of California, Santa Cruz – Yx
Bries Korpik, Jill, (651) 450-3726 jkorpik@inverhills.mnscu.edu, Inver Hills Community Coll – Gg
Bries-Korpik, Jill, (651) 779-3434 jill.brieskorpik@century.edu, Century Coll – Og
Briggs, Derek E., (203) 432-8590 derek.briggs@yale.edu, Yale Univ – Yg
Briggs, John, +44 0141 330 8744 John.Briggs@glasgow.ac.uk, Univ of Glasgow – Ou
Briggs, Tiffany Roberts M., (561) 297-4669 briggst@fau.edu, Florida Atlantic Univ – On
Brigham-Grette, Julie, (413) 577-2270 juliebg@geo.umass.edu, Univ of Massachusetts, Amherst – GlPe
Brikowski, Tom H., (972) 883-6242 brikowi@utdallas.edu, Univ of Texas, Dallas – Hw
Brill, Jr., Richard C., (808) 845-9488 brill@hawaii.edu, Honolulu Community Coll – Gg
Brimhall, George H., 510.642.5868 brimhall@eps.berkeley.edu, Univ of California, Berkeley – Eg
Briner, Jason P., 716-645-4326 jbriner@buffalo.edu, SUNY, Buffalo – Gl
Brinkman, Donald B., (403) 823-7707 Royal Tyrrell Museum of Palaeontology – Pv
Brinkman, P. Anthony, 775-784-6995 brinkman@unr.edu, Univ of Nevada, Reno – Ou
Brinkmann, Robert, 516-463-7348 robert.brinkmann@hofstra.edu, Hofstra Univ – Os
Brinkmann, Robert, (541) 967-2039 robert.d.brinkmann@mlrr.oregongeology.com, Oregon Dept of Geology & Mineral Industries – HwGoEm
Brisbin, William C., (204) 474-9454 w_brisbin@umanitoba.ca, Univ of Manitoba – Gc
Briskin, Madeleine, (513) 556-3009 madeleine.briskin@uc.edu, Univ of Cincinnati – Pe
Bristow, Charlie, +44 020 3073 8025 c.bristow@ucl.ac.uk, Birkbeck Coll – Gs
Britt, Brooks B., 801-422-7316 brooks_britt@byu.edu, Brigham Young Univ – Pg
Britton, Gloria C., (216) 987-5228 gloria.britton@tri-c.edu, Cuyahoga Community Coll - Western Campus – OgGmOn
Britzke, Gilbert, 089/2180 4209 gilbert.brietzke@geophysik.uni-muenchen.de, Ludwig-Maximilians-Universitaet Muenchen – Yg
Broadfoot, Lyle A., (520) 621-4303 broadfoot@vega.lpl.arizona.edu, Univ of Arizona – Or
Broadhead, Ronald F., (505) 835-5202 ron@gis.nmt.edu, New Mexico Inst of Mining & Tech – Go
Broadhead, Thomas W., (865) 974-1151 twbroadhead@utk.edu, Univ of Tennessee, Knoxville – Pi
Brochu, Christopher A., (319) 353-1808 chris-brochu@uiowa.edu, Univ of Iowa – Pv
Brock, Patrick W. G., (718) 997-3328 Graduate School of the City Univ of New York – Gg
Brock-Hon, Amy, (423) 425-4409 amy-brock-hon@utc.edu, Univ of Tennessee at Chattanooga – GmSc
Brockhaus, John A., (845) 938-2063 United States Military Academy – Yd
Brocklehurst, Simon, +44 0161 275-3037 Simon.H.Brocklehurst@manchester.ac.uk, Univ of Manchester – Gm
Brod, Joseph D., 813-744-8018 jbrod@hccfl.edu, Hillsborough Community Coll – Og
Broda, James E., (508) 289-2466 jbroda@whoi.edu, Woods Hole Oceanographic Institution – Gs
Brodholt, John, +44 020 7679 32622 j.brodholt@ucl.ac.uk, Univ Coll London – Yg
Brodie, Gregory, (423) 425-5915 gregory-brodie@utc.edu, Univ of Tennessee at Chattanooga – Ge
Brodie, Kate, +44 0161 275-3948 Kate.Brodie@manchester.ac.uk, Univ of Manchester – Gc
Brodsky, Emily, (831) 459-1854 brodsky@ucsc.edu, Univ of California, Santa Cruz – YsgHw
Broecker, Wallace S., (845) 365-8413 broecker@ldeo.columbia.edu, Columbia Univ – CmOcPe
Brogan, George, 949-451-5687 gbrogan@ivc.edu, Irvine Valley Coll – Gc
Bromily, Geoffrey, +44 (0) 131 650 8519 geoffrey.bromiley@ed.ac.uk, Edinburgh Univ – Ge
Bromley, Gordon R., gordon.bromley@maine.edu, Univ of Maine – Gl
Bromwich, David H., (614) 688-5314 bromwich.1@osu.edu, Ohio State Univ – Oa
Bronk, Deborah A., (804) 684-7779 bronk@vims.edu, Coll of William & Mary – Oc
Brook, Edward J., (541) 737-8197 brooke@geo.oregonstate.edu, Oregon State Univ – ClPeOl
Brookfield, Andrea, 785-864-2199 andrea@kgs.ku.edu, Univ of Kansas – Hw

Brooks, David A., (979) 845-5527 dbrooks@ocean.tamu.edu, Texas A&M Univ – Op
Brooks, Debra A., (714) 564-4788 Santiago Canyon Coll – Yg
Brooks, Gregg R., (727) 864-8992 brooksgr@eckerd.edu, Eckerd Coll – Gu
Brooks, Ian, +44(0) 113 34 36743 i.m.brooks@leeds.ac.uk, Univ of Leeds – On
Brooks, Paul D., (520) 621-3424 brooks@hwr.arizona.edu, Univ of Arizona – Cg
Brooks, Paul D., (801) 585-2858 paul.brooks@utah.edu, Univ of Utah – HgOg
Brooks, Sarah D., (979) 845-5632 sbrooks@tamu.edu, Texas A&M Univ – Oa
Brooks, Scott C., (865) 574-6398 3sb@ornl.gov, Oak Ridge National Laboratory – Cl
Broome, Stephen W., (919) 515-2643 North Carolina State Univ – Sf
Brophy, James G., (812) 855-6417 brophy@indiana.edu, Indiana Univ, Bloomington – Gi
Broster, Bruce E., (506) 453-4804 broster@unb.ca, Univ of New Brunswick – Gl
Brothen, Jerry, (310) 660-3593 jbrothen@elcamino.edu, El Camino Coll – Oy
Brothers, Timothy S., (317) 274-1101 Indiana Univ / Purdue Univ, Indianapolis – Oy
Brouillette, Pierre, (418) 654-2567 pbrouill@nrcan.gc.ca, Universite du Quebec – Gg
Brouwer, Fraukje M., +31-20-598 7335 fraukje.brouwer@vu.nl, VU Univ Amsterdam – GptGc
Brower, James C., (315) 443-2672 Syracuse Univ – Po
Brown, Alan, abrown11@houston.oilfield.slb.com, West Virginia Univ – EoYe
Brown, Bradford D., (208) 722-6701 Univ of Idaho – On
Brown, Bruce A., (608) 263-3201 babrown1@wisc.edu, Univ of Wisconsin - Extension – GcEg
Brown, Caleb, (403) 823-7707 Royal Tyrrell Museum of Palaeontology – Pv
Brown, Cathe, (202) 633-1788 brownc@si.edu, Smithsonian Institution / National Museum of Natural History – Gzi
Brown, David, +44 01413307410 David.Brown@glasgow.ac.uk, Univ of Glasgow – Gvs
Brown, David L., (530) 898-4035 dlbrown@csuchico.edu, California State Univ, Chico – Hg
Brown, Donald W., (505) 667-1926 dwb@lanl.gov, Los Alamos National Laboratory – Nr
Brown, Douglas, +44 208 417 2245 Doug.Brown@kingston.ac.uk, Kingston Univ – Oi
Brown, Edwin H., (360) 650-3597 ehbrown@wwu.edu, Western Washington Univ – Gp
Brown, Erik T., (218) 726-8891 etbrown@d.umn.edu, Univ of Minnesota, Duluth – Og
Brown, Francis H., (801) 581-8767 frank.brown@utah.edu, Univ of Utah – Gg
Brown, Huntting (Hunt), 937 775-4996 hunt.brown@wright.edu, Wright State Univ – Ge
Brown, J. Michael, 206-616-6058 brown@ess.washington.edu, Univ of Washington – Gy
Brown, James R., (403) 220-7484 jbrown@geo.ucalgary.ca, Univ of Calgary – Ys
Brown, Kent D., 801-537-3350 kentbrown@utah.gov, Utah Geological Survey – On
Brown, Kerry, K.Brown@kingston.ac.uk, Kingston Univ – Ge
Brown, Kevin M., (858) 534-5368 kmbrown@ucsd.edu, Univ of California, San Diego – Gu
Brown, Larry D., (607) 255-6346 ldb7@cornell.edu, Cornell Univ – Ye
Brown, Laurie, (413) 545-0245 lbrown@geo.umass.edu, Univ of Massachusetts, Amherst – Ym
Brown, Lewis M., (906) 635-2155 lbrown@lssu.edu, Lake Superior State Univ – PiOe
Brown, Michael, (301) 405-4080 mbrown@umd.edu, Univ of Maryland – GxpGt
Brown, Michael E., (626) 395-8423 mbrown@gps.caltech.edu, California Inst of Tech – Xy
Brown, Michael E., (662) 325-2906 mebrown@ra.msstate.edu, Mississippi State Univ – Ow
Brown, Moira, mwbrown@neaq.org, Univ of Massachusetts, Boston – Ob
Brown, Paul, +44 020 7679 32431 p.bown@ucl.ac.uk, Univ Coll London – Pm
Brown, Paul W., (520) 621-1319 pbrown@ag.arizona.edu, Univ of Arizona – Ow
Brown, Peter, 5196612111 ext.86458 pbrown@uwo.ca, Western Univ – Xym
Brown, Philip E., (608) 262-5954 pebrown@wisc.edu, Univ of Wisconsin, Madison – EmGzp
Brown, Richard J., +44 (0) 191 33 42303 richard.brown3@durham.ac.uk, Durham Univ – Gv
Brown, Rob, (828) 262-7222 brownrn@appstate.edu, Appalachian State Univ – On
Brown, Robert A., rabrown@atmos.washington.edu, Univ of Washington – Oa
Brown, Robert H., (520) 626-9045 rhb@lpl.arizona.edu, Univ of Arizona – Xc
Brown, Roderick, +44 01413305460 Roderick.Brown@glasgow.ac.uk, Univ of Glasgow – Cc
Brown, Sandra, 604-822-5965 Univ of British Columbia – Sp
Brown, Steven E., 217-333-5143 steebrow@illinois.edu, Univ of Illinois – Gl
Brown, Thomas W., tbrown1@austincc.edu, Austin Community Coll District – Ges
Brown, Wesley A., (936) 468-2422 brownwa1@sfasu.edu, Stephen F. Austin State Univ – YgsGt
Brown, William G., (254) 710-2075 william_brown@baylor.edu, Baylor Univ – Gc
Brown, Jr., Gordon E., (650) 723-9168 gordon@pangea.stanford.edu, Stanford Univ – Gz
Brownawell, Bruce J., (631) 632-8658 bruce.brownawell@stonybrook.edu, SUNY, Stony Brook – Oc
Browne, Brandon L., (707) 826-3950 blb519@humboldt.edu, Humboldt State Univ – GivGz
Browne, Kathleen M., (609) 896-5408 browne@rider.edu, Rider Univ – GuOe
Browning, James V., 848-445-3368 jvb@rci.rutgers.edu, Rutgers, The State Univ of New Jersey – GgsGr
Brownlee, Colin, +44 (0)1752 633347 Univ of Southampton – Ob
Brownlee, Donald E., 206-543-2888 brownlee@ess.washington.edu, Univ of Washington – Xgy
Broxton, David E., 505-667-2492 broxton@lanl.gov, Los Alamos National Laboratory – Gx
Brubaker, John M., (804) 684-7222 brubaker@vims.edu, Coll of William & Mary – Op
Brubaker, Kenneth L., (630) 252-7630 Argonne National Laboratory – Oa
Brudzinski, Michael, 513-529-9758 brudzimr@MiamiOh.edu, Miami Univ – Ys
Brueckner, Hannes K., (718) 997-3300 Graduate School of the City Univ of New York – Cc
Brueckner, Hannes K., hannes@ldeo.columbia.edu, Queens Coll (CUNY) – GgtCg
Brueckner, Stefanie M., (334) 844-4988 smb0115@auburn.edu, Auburn Univ – EmCag
Bruening, Susan, bruenings@easternct.edu, Eastern Connecticut State Univ – Og
Brueseke, Matthew E., (785) 532-6724 brueseke@ksu.edu, Kansas State Univ – GitGv
Brugger, Keith A., (320) 589-6310 bruggeka@mrs.umn.edu, Univ of Minnesota, Morris – OlGlm
Bruhn, Ronald L., (801) 581-7162 ron.bruhn@utah.edu, Univ of Utah – Gc
Bruland, Kenneth W., (831) 459-4587 bruland@cats.ucsc.edu, Univ of California, Santa Cruz – Oc
Brumbaugh, David S., (928) 523-7191 david.brumbaugh@nau.edu, Northern Arizona Univ – Ys
Brune, James N., (775) 784-4974 brune@seismo.unr.edu, Univ of Nevada, Reno – Ys
Brune, Jürgen F., 303.273.3704 jbrune@mines.edu, Colorado School of Mines – Nm
Brune, William H., (814) 865-3286 whb2@psu.edu, Pennsylvania State Univ, Univ Park – Ow
Brunengo, Matthew , 503-725-3391 mbruneng@pdx.edu, Portland State Univ – GgNgGm
Bruner, Katherine R., (304) 293-5603 bruner@geo.wvu.edu, West Virginia Univ – Gd
Brunet, Gilbert, (514) 421-4617 gilbert.brunet@canada.ca, McGill Univ – Ow
Brunetto, Eileen, (802) 443-5970 efahey@middlebury.edu, Middlebury Coll – On

Bruning, Eric C., 806-834-3120 eric.bruning@ttu.edu, Texas Tech Univ – Oa

Brunish, Wendee M., (505) 667-5724 wb@lanl.gov, Los Alamos National Laboratory – On

Brunkal, Holly, 970.943.2180 hbrunkal@western.edu, Western State Colorado Univ – NgGm

Brunner, Benjamin, 915-747-5507 bbrunner@utep.edu, Univ of Texas, El Paso – Cs

Bruno, Barbara, (808) 956-0901 barb@hawaii.edu, Univ of Hawai'i, Manoa – Ob

Bruno, Emiliano, emiliano.bruno@unito.it, Università di Torino – Gz

Bruns, Dale A., 570-408-4603 dale.bruns@wilkes.edu, Wilkes Univ – OirSf

Brunskill, Jeffrey C., 570-389-4355 jbrunski@bloomu.edu, Bloomsburg Univ – OiwOy

Brunstad, Keith, (607) 436-3066 keith.brunstad@oneonta.edu, SUNY, Oneonta – GviEg

Brunt, Rufus, +44 0161 306-6816 rufus.brunt@manchester.ac.uk, Univ of Manchester – Gs

Bruschke, Freddi Jo, 657-278-3551 fbruschke@fullerton.edu, California State Univ, Fullerton – Gg

Brush, Grace S., (410) 516-7107 gbrush@jhu.edu, Johns Hopkins Univ – Pl

Brush, Nigel, (419) 289-5271 nbrush@ashland.edu, Ashland Univ – GmdGa

Brusseau, Mark L., (520) 621-3244 brusseau@ag.arizona.edu, Univ of Arizona – Hw

Bryan, Christopher, +44 01326 259482 C.G.Bryan@exeter.ac.uk, Exeter Univ – Nx

Bryan, Scott E., +61 7 3138 4827 scott.bryan@qut.edu.au, Queensland Univ of Tech – GitGs

Bryant, Anita M., (678) 839-4051 abryant@westga.edu, Univ of West Georgia – On

Bryant, Ernest A., (505) 667-2422 Los Alamos National Laboratory – Cg

Bryant, Kathleen E., 217-244-9045 kebryant@illinois.edu, Univ of Illinois, Urbana-Champaign – Gg

Bryant, Marita, (207) 786-6452 mbryant@bates.edu, Bates Coll – Gg

Bryce, Julia G., 603-862-3139 julie.bryce@unh.edu, Univ of New Hampshire – Cg

Bryce, Karen R., (801) 378-5470 karen_bryce@byu.edu, Brigham Young Univ – On

Brzobohaty, Rostislav, +420 549 49 3326 rosta@sci.muni.cz, Masaryk Univ – PgsPe

Buatois, Luis, luis.buatois@usask.ca, Univ of Saskatchewan – Ps

Buchanan, Donald G., 9093844399 ext. 5467 dbuchanan@sbccd.cc.ca.us, San Bernardino Valley Coll – Og

Buchanan, George, gbuchana@mc3.edu, Montgomery County Community Coll – Ng

Buchanan, John P., (509) 359-7493 jbuchanan@ewu.edu, Eastern Washington Univ – Hw

Buchanan, Paul C., (903) 983-8253 pbuchanan@kilgore.edu, Kilgore Coll – GgiXm

Buchanan, Rex C., (785) 864-2106 rex@kgs.ku.edu, Univ of Kansas – Oe

Buchheim, Paul, (909) 558-4530 pbuchheim@llu.edu, Loma Linda Univ – GnsPe

Buchwald, Carolyn, (902) 494-3666 cbuchwald@dal.ca, Dalhousie Univ – Oc

Buck, Brenda J., (702) 895-3583 buckb@unlv.nevada.edu, Univ of Montana Western – Sdo

Buck, Brenda J., (702) 895-1694 buckb@unlv.nevada.edu, Univ of Nevada, Las Vegas – GbSd

Buck, Daniel, (541) 346-2353 danielb@uoregon.edu, Univ of Oregon – On

Buck, Gregory, 361-825-3717 Gregory.Buck@tamucc.edu, Texas A&M Univ, Corpus Christi – On

Buck, Roger W., 845-365-8592 Columbia Univ – Yr

Bucke, David P., (802) 899-3584 david.bucke@uvm.edu, Univ of Vermont – GgdGs

Buckingham, Michael J., (858) 534-7977 mbuckingham@ucsd.edu, Univ of California, San Diego – Op

Buckley, Brendon, 845-365-8782 Columbia Univ – Pe

Buckley, Lawrence J., (401) 874-6671 lbuckley@gso.uri.edu, Univ of Rhode Island – Oc

Buckley, Luke, (406) 496-4677 lbuckley@mtech.edu, Montana Tech of The Univ of Montana – On

Buckley, Michael, +44(0)161 306 5175 M.Buckley@manchester.ac.uk, Univ of Manchester – Ga

Bucklin, Ann, 860-405-9260 ann.bucklin@uconn.edu, Univ of Connecticut – Ob

Bucur, Ioan, +40-264-405371 ibucur@bioge.ubbcluj.ro, Babes-Bolyai Univ – PgmGs

Buczynska, Anna, 815-753-7945 abuczynska@niu.edu, Northern Illinois Univ – Csa

Budd, Ann F., (319) 335-1818 ann-budd@uiowa.edu, Univ of Iowa – Poi

Budd, David A., (303) 492-3988 david.budd@colorado.edu, Univ of Colorado – Gd

Buddemeier, Robert W., (785) 864-2112 buddrw@kgs.ku.edu, Univ of Kansas – Hy

Budikova, Dagmar, (309) 438-2546 dbudiko@ilstu.edu, Illinois State Univ – OayOi

Buermann, Wolfgang, +44(0) 113 34 34958 w.buermann@leeds.ac.uk, Univ of Leeds – Ou

Buesseler, Ken O., (508) 289-2309 kbuesseler@whoi.edu, Woods Hole Oceanographic Institution – Oc

Buick, Roger, 206-543-1913 buick@ess.washington.edu, Univ of Washington – PeoPm

Buijsman, Maarten, (228) 688-2385 maarten.buijsman@usm.edu, Univ of Southern Mississippi – Op

Buizert, Christo, 541-737-1572 buizertc@science.oregonstate.edu, Oregon State Univ – Yr

Bukowinski, Mark S., (510) 642-0977 Univ of California, Berkeley – Gy

Bulger, Daniel E., 678-891-2415 dbulger1@gsu.edu, Georgia State Univ, Perimeter Coll, Decatur Campus – GgOn

Bull, Jonathan M., +44 (0)23 8059 3078 bull@noc.soton.ac.uk, Univ of Southampton – YrGct

Bull, William B., (520) 621-6024 Univ of Arizona – Gm

Bullamore, Henry W., (301) 687-4413 hbullamore@frostburg.edu, Frostburg State Univ – On

Bullard, Reuben G., 859-572-6907 bullardr@nku.edu, Northern Kentucky Univ – Gd

Bullard, Thomas F., (775) 673-7420 tbullard@dri.edu, Desert Research Inst – Gm

Bullen, Dean, +44 023 92 842289 dean.bullen@port.ac.uk, Univ of Portsmouth – Gt

Bullister, John, (206) 526-6741 bullister@pmel.noaa.gov, Univ of Washington – Oc

Bullock, Emma, bullocke@si.edu, Smithsonian Institution / National Museum of Natural History – Xm

Bultman, John, (828) 254-1921 ext 319 jbultman@abtech.edu, Asheville-Buncombe Technical Community Coll – Gg

Bulusu, Subrahmanyam, 803-777-2572 sbulusu@geol.sc.edu, Univ of South Carolina – Op

Bunds, Michael, (801) 863-6306 Michael.Bunds@uvu.edu, Utah Valley Univ – GctGe

Bundy, Larry G., (608) 263-2889 lgbundy@wisc.edu, Univ of Wisconsin, Madison – So

Bunge, Hans-Peter, 089/2180 4225 bunge@lmu.de, Ludwig-Maximilians-Universitaet Muenchen – Yg

Buonaiuto, Frank, 212-650-3092 frank.buonaiuto@hunter.cuny.edu, Hunter Coll (CUNY) – On

Buratti, Bonnie J., (818) 354-7427 Jet Propulsion Laboratory – Or

Burbanck, George P., (757) 727-5783 george.burbanck@hamptonu.edu, Hampton Univ – On

Burbank, Doug, (805) 893-7858 burbank@geol.ucsb.edu, Univ of California, Santa Barbara – Gmt

Burberry, Caroline M., (402) 472-7157 cburberry2@unl.edu, Univ of Nebraska, Lincoln – Gct

Burbey, Thomas J., (540) 231-6696 tjburbey@vt.edu, Virginia Polytechnic Inst & State Univ – HwyOr

Burchfiel, B. C., (617) 253-7919 bcburch@mit.edu, Massachusetts Inst of Tech – Gc

Burd, Aurora, aburd@avc.edu, Antelope Valley Coll – GgOg

Burden, Elliott T., (709) 737-8388 etburden@mun.ca, Memorial Univ of Newfoundland – Pl

Burdick, Scott, sburdick@umd.edu, Univ of Maryland – Ys

Burdige, David J., (757) 683-4930 dburdige@odu.edu, Old Dominion Univ – OcCmo

Burger, H. Robert, rburger@smith.edu, Smith Coll – GcYeOi

Burger, Martin, 530-754-6497 mburger@ucdavis.edu, Univ of California, Davis – Spb

Burger, Paul V., (505) 277-3827 pvburger@unm.edu, Univ of New Mexico – Gz

Burgess, Ray, +44 0161 275-3958 ray.burgess@manchester.ac.uk, Univ of Manchester – Cs
Burgess, William G., +44 020 7679 37820 william.burgess@ucl.ac.uk, Univ Coll London – Hwy
Burgette, Reed J., 575-646-3782 burgette@nmsu.edu, New Mexico State Univ, Las Cruces – Gtc
Burgmann, Roland, (510) 643-9545 burgmann@seismo.berkeley.edu, Univ of California, Berkeley – Gt
Burk, Sue, 301-405-6244 Univ of Maryland – On
Burkart, Michael R., mburkart@iastate.edu, Iowa State Univ of Science & Tech – Hy
Burke, Andrea, +44 01334 463910 ab276@st-andrews.ac.uk, Univ of St. Andrews – Cs
Burke, Collette D., (316) 978-3140 Wichita State Univ – Pm
Burke, Deborah S., (509) 372-2483 deborah.burke@pnl.gov, Pacific Northwest National Laboratory – Sc
Burke, Ian, +44(0) 113 34 37532 lab 33965 i.t.burke@leeds.ac.uk, Univ of Leeds – Ge
Burke, Kevin, 713-743-3399 kburke@uh.edu, Univ of Houston – Gt
Burke, Raymond M., (707) 826-4292 rmb2@humboldt.edu, Humboldt State Univ – Gm
Burke, Roger A., (706) 542-2652 Univ of Georgia – Cl
Burkett, Brett, (976) 548-6510 bburkett@collin.edu, Collin Coll - Central Park Campus – Gvg
Burkhart, Patrick A., (724) 738-2502 patrick.burkhart@sru.edu, Slippery Rock Univ – Hg
Burks, Rachel J., (410) 704-3005 rburks@towson.edu, Towson Univ – Gc
Burlage, Robert S., (865) 574-7321 Oak Ridge National Laboratory – Sb
Burlakova, Lyubov, burlakle@buffalostate.edu, SUNY, Buffalo – Og
Burls, Natalie, 703-993-5756 nburls@gmu.edu, George Mason Univ – Oap
Burmeister, Kurtis C., (209) 946-2398 kburmeister@pacific.edu, Univ of the Pacific – GctEg
Burmester, Russell F., (360) 650-3654 russ.burmester@wwu.edu, Western Washington Univ – Ym
Burnett, Donald S., (626) 395-6117 burnett@gps.caltech.edu, California Inst of Tech – Xc
Burnett, Earl E., (928) 329-1030 Arizona Western Coll – HwYgGe
Burnham, Charles, burnham_c@fortlewis.edu, Fort Lewis Coll – Gz
Burnham, Robyn J., (734) 764-0489 rburnham@umich.edu, Univ of Michigan – Pb
Burnley, Pamela C., 702-895-2536 pamela.burnley@unlv.edu, Univ of Nevada, Las Vegas – GpyGg
Burns, Bill, (971) 673-1555 Oregon Dept of Geology & Mineral Industries – Ng
Burns, Carol J., (505) 665-1765 Los Alamos National Laboratory – On
Burns, Danny, 361-354-2405 deburns@coastalbend.edu, Coastal Bend Coll – Ogn
Burns, Diane M., (217) 581-2827 dmburns@eiu.edu, Eastern Illinois Univ – Gs
Burns, Emily, (401) 825-1009 eburns@ccri.edu, Community Coll of Rhode Island – GgOgi
Burns, Peter C., (574) 631-5380 peter.burns.50@nd.edu, Univ of Notre Dame – Gz
Burns, Sandra, (860) 832-2934 burns@ccsu.edu, Central Connecticut State Univ – Oe
Burns, Scott F., (503) 725-3389 burnss@pdx.edu, Portland State Univ – Ng
Burns, Stephen J., (413) 545-0142 sburns@geo.umass.edu, Univ of Massachusetts, Amherst – Cs
Burns, Timothy P., (505) 667-4600 Los Alamos National Laboratory – Co
Burr, Devon, (865) 974-2366 dburr1@utk.edu, Univ of Tennessee, Knoxville – XgGm
Burras, Lee, (515) 294-0559 lburras@iastate.edu, Iowa State Univ of Science & Tech – Sd
Burris, John H., (505) 566-3325 burrisj@sanjuanColl.edu, San Juan Coll – GghGz
Bursik, Marcus I., (716) 645-4265 mib@geology.buffalo.edu, SUNY, Buffalo – Gv
Burst, John F., (573) 341-4616 Missouri Univ of Science and Tech – Eo
Burt, Donald M., (480) 965-6180 donald.burt@asu.edu, Arizona State Univ – EgCgGz
Burtis, Erik, (703) 878-5614 eburtis@nvcc.edu, Northern Virginia Community Coll - Woodbridge – GgcGi
Burton, Bradford R., 970.943.2252 bburton@western.edu, Western State Colorado Univ – GocGt
Burton, Christopher G., 334-844-3418 cgb0038@abuurn.edu, Auburn Univ – Oi
Burton, Jacqueline C., (630) 252-8795 jcburton@anl.gov, Argonne National Laboratory – Cg
Burton, Kevin, +44 (0) 191 33 44298 kevin.burton@durham.ac.uk, Durham Univ – Cc
Burton, Paul, +44 (0)1603 59 2982 p.burton@uea.ac.uk, Univ of East Anglia – Ys
Burton, Ronald S., (858) 822-5784 rburton@ucsd.edu, Univ of California, San Diego – Ob
Burwash, Ronald A., (780) 492-3085 ronald.burwash@telus.net, Univ of Alberta – Gp
Busa, Mark, 860-343-5779 MBusa@mxcc.commnet.edu, Middlesex Community Coll – OgEgYg
Busacca, Alan J., (509) 335-1859 busacca@wsu.edu, Washington State Univ – Sa
Busalacchi, Antonio, (301) 405-5599 tonyb@essic.umd.edu, Univ of Maryland – Og
Busbey, Arthur B., (817) 257-7301 a.busbey@tcu.edu, Texas Christian Univ – PgvGr
Busby, Cathy, (805) 893-4068 Univ of California, Santa Barbara – Gs
Busby, Cathy J., cjbusby@ucdavis.edu, Univ of California, Davis – Gt
Busby, Michael R., (270) 809-3370 michael.busby@murraystate.edu, Murray State Univ – Oi
Busch, Richard M., (610) 436-2716 rbusch@wcupa.edu, West Chester Univ – Oe
Busch, William H., (503) 392-3341 whbusch@gmail.com, Univ of New Orleans – Ou
Buscheck, Thomas A., (925) 423-9390 buscheck1@llnl.gov, Lawrence Livermore National Laboratory – Ng
Buseck, Peter R., (480) 965-3945 pbuseck@asu.edu, Arizona State Univ – Cg
Bush, Andrew B., (780) 492-0351 andrew.bush@ualberta.ca, Univ of Alberta – OapPe
Bush, Andrew M., (860) 486-9358 andrew.bush@uconn.edu, Univ of Connecticut – PoqPi
Bush, Andrew B. G., (780) 492-0351 andrew.bush@ualberta.ca, Univ of Alberta – Oa
Bush, David M., (678) 839-4057 dbush@westga.edu, Univ of West Georgia – Ou
Buskey, Edward J., (361) 749-3102 ed.buskey@utexas.edu, Univ of Texas at Austin – Ob
Buss, Leo W., 203-432-3869 leo.buss@yale.edu, Yale Univ – Pi
Busse, Friedrich H., (310) 825-7698 Univ of California, Los Angeles – On
Bussod, Gilles Y., (505) 667-7220 gbussod@lanl.gov, Los Alamos National Laboratory – Yx
Bustin, Amanda, abustin@eos.ubc.ca, Univ of British Columbia – YgNg
Bustin, R. Marc, (604) 822-6179 mbustin@eos.ubc.ca, Univ of British Columbia – Ec
Butcher, Anthony, +44 023 92 842486 anthony.butcher@port.ac.uk, Univ of Portsmouth – PlmOe
Butcher, Patricia M., (657) 278-3561 pbutcher@fullerton.edu, California State Univ, Fullerton – Oeg
Butkos, Darryl J., (631) 451-4354 butkosd@sunysuffolk.edu, Suffolk County Community Coll, Ammerman Campus – GgHw
Butler, Gilbert W., (505) 667-6005 Los Alamos National Laboratory – Cc
Butler, Gregg, [44]1772 635135 gregg.butler@manchester.ac.uk, Univ of Manchester – On
Butler, Ian B., +44 (0) 131 650 5885 ian.butler@ed.ac.uk, Edinburgh Univ – Cg
Butler, James J., (785) 864-3965 jbutler@ku.edu, Univ of Kansas – Hw
Butler, Karl, (506) 453-4804 Univ of New Brunswick – Yg
Butler, R. Paul, (202) 478-8866 pbutler@carnegiescience.edu, Carnegie Institution of Washington – On
Butler, Sam, (306) 966-5702 sam.butler@usask.ca, Univ of Saskatchewan – Yg
Butler, Jr., James J., (785) 864-2116 jbutler@kgs.ku.edu, Univ of Kansas – Hw
Butterfield, David A., (206) 526-6722 dab3@u.washington.edu, Univ of Washington – Yr
Butterfield, Nicholas J., +44 (0) 1223 333379 njb1005@esc.cam.ac.uk, Univ of Cambridge – Po
Buttner, Steffen, +27 (0)46-603-8775 s.buettner@ru.ac.za, Rhodes Univ – Gcp
Butts, Susan H., (203) 432-3037 susan.butts@yale.edu, Yale Univ – PiGsPe
Butzow, Dean G., 217-786-4923 dean.butzow@llcc.edu, Lincoln Land Community Coll – Oy
Buyce, M. Raymond, rbuyce@mercyhurst.edu, Mercyhurst Univ – GsaOn
Buynevich, Ilya, (215) 204-3635 coast@temple.edu,

Temple Univ – GsOuGm
Buzas, Martin A., (202) 633-1313 Smithsonian Institution / National Museum of Natural History – Pm
Buzatu, Andrei, +40232201463 andrei.buzatu@uaic.ro, Al. I. Cuza Univ of Iasi – GziGg
Buzgar, Nicolae, +40232201462 nicolae.buzgar@uaic.ro, Al. I. Cuza Univ of Iasi – GziGd
Bybee, Grant M., 011 717 6633 grant.bybee@wits.ac.za, Univ of the Witwatersrand – Gi
Büchel, Georg, +49(0)3641 948640 georg.buechel@uni-jena.de, Friedrich-Schiller-Univ Jena – Gg
Byerly, Don W., donbyerly@comcast.net, Univ of Tennessee, Knoxville – GeNg
Byerly, Gary R., (225) 578-5318 glbyer@lsu.edu, Louisiana State Univ – Gi
Byers, Charles W., (608) 262-2361 cwbyers@geology.wisc.edu, Univ of Wisconsin, Madison – Pe
Bykerk-Kauffman, Ann, 530-898-6305 abykerk-kauffman@csuchico.edu, California State Univ, Chico – GcOe
Byrand, Karl, (920) 459-6619 karl.byrand@uwc.edu, Univ of Wisconsin Colls – Ony
Byrne, James M., (403) 329-2002 byrne@uleth.ca, Univ of Lethbridge – Hg
Byrne, John V., john.byrne@coas.oregonstate.edu, Oregon State Univ – Gu
Byrne, Paul K., 919-513-2578 paul.byrne@ncsu.edu, North Carolina State Univ – XgGcOr
Byrne, Robert H., (727) 553-1508 byrne@marine.usf.edu, Univ of South Florida – Cm
Byrne, Timothy, (860) 486-3142 tim.byrne@uconn.edu, Univ of Connecticut – Gc
Byrnes, Jarrett, jarrett.byrnes@umb.edu, Univ of Massachusetts, Boston – ObnOn
Byrnes, Jeffrey, 405-744-6358 jeffrey.byrnes@okstate.edu, Oklahoma State Univ – GgvOn
Byrnes, Joseph, jsbyrnes@umn.edu, Univ of Minnesota, Twin Cities – Ys

## C

C., Jeffrey C., (330) 941-3612 jcdick@ysu.edu, Youngstown State Univ – EoNgHy
Caballero, Kate, (254) 968-9143 caballero@tarleton.edu, Tarleton State Univ – On
Cablk, Mary, (775) 673-7371 mcablk@dri.edu, Desert Research Inst – Or
Cabrera, Miguel L., (706) 542-1242 Univ of Georgia – So
Cadet, Eddy, (801) 863-8881 cadeted@uvu.edu, Utah Valley Univ – Ge
Cadieux, Sarah, 312-355-1182 scadieux@uic.edu, Univ of Illinois at Chicago – Cg
Cadol, Daniel, 575-835-5645 dcadol@ees.nmt.edu, New Mexico Inst of Mining and Tech – HsGe
Cadoppi, Paola, paola.cadoppi@unito.it, Università di Torino – Gc
Cahill, Kevin, (508) 289-2925 kcahill@whoi.edu, Woods Hole Oceanographic Institution – Cm
Cahill, Michael J., (631) 588-8778 mjc@germanocahill.com, SUNY, Stony Brook – On
Cahill, Richard A., 217-244-2532 cahill@isgs.uiuc.edu, Illinois State Geological Survey – Ca
Cahir, John J., (814) 863-8358 cahir@ems.psu.edu, Pennsylvania State Univ, Univ Park – Ow
Cairns, David M., (979) 845-2783 cairns@geog.tamu.edu, Texas A&M Univ – Oy
Cairns, Stephen, cairnss@si.edu, Smithsonian Institution / National Museum of Natural History – Pi
Caissie, Beth E., 515-294-7528 bethc@iastate.edu, Iowa State Univ of Science & Tech – Pe
Calagari, Ali Asghar, +98 (411) 339 2699 calagari@tabrizu.ac.ir, Univ of Tabriz – EgmCg
Calaway, Wallis F., (630) 972-3586 Argonne National Laboratory – Om
Calcote, Randy, 612-624-8526 calco001@umn.edu, Univ of Minnesota, Twin Cities – Py
Calder, Eliza, +44 (0) 131 650 4910 ecalder@staffmail.ed.ac.uk, Edinburgh Univ – Gv
Calder, John, (902) 424-2778 Dalhousie Univ – Ec
Caldwell, Andy, 970-204-8228 andrew.caldwell@frontrange.edu, Front Range Community Coll - Larimer – Gg
Caldwell, Marianne O., mcaldwell@hccfl.edu, Hillsborough Community Coll – Og
Caldwell, Michael W., 780-492-3458 mw.caldwell@ualberta.ca, Univ of Alberta – Pv
Caldwell, Roy, (510) 642-1391 rlcaldwell@berkeley.edu, Univ of California, Berkeley – Po
Caldwell, Todd, 512-471-2003 todd.caldwell@beg.utexas.edu, Univ of Texas at Austin – Sp
Caldwell, W. Glen E., 519-661-3187 gcaldwel@uwo.ca, Western Univ – Pm
Caldwell, William G. E., (519) 661-3857 Univ of Saskatchewan – Ps
Calegari, Pat, (212) 346-1502 Pace Univ, New York Campus – On
Calengas, Peter L., (309) 298-1151 PL-Calengas@wiu.edu, Western Illinois Univ – En
Calhoun, Frank G., (330) 263-3722 scalhoun@coas.oregonstate.edu, Ohio State Univ – Sd
Calhoun, Joseph, 717-872-3289 jcalhoun@hearst.com, Millersville Univ – Oa
Calizaya, Felipe, (801) 581-5422 felipe.calizaya@utah.edu, Univ of Utah – Nm
Callahan, Caitlin N., 616-331-3601 callahac@gvsu.edu, Grand Valley State Univ – GgOg
Callahan, John A., 302-831-3584 john.callahan@udel.edu, Univ of Delaware – Ori
Callahan, John E., (828) 262-2746 callahnje@appstate.edu, Appalachian State Univ – Eg
Callahan, Timothy J., (843) 953-8278 callahant@cofc.edu, Coll of Charleston – Hw
Callanan, Jennifer R., (973) 720-3979 callananj@wpunj.edu, William Paterson Univ – ScGm
Callard, Jeff, callard@ou.edu, Univ of Oklahoma – Np
Callison, James, (801) 863-8679 JCallison@uvu.edu, Utah Valley Univ – GeSfOs
Calon, Tomas J., (709) 737-8398 tcalon@sparky2.esd.mun.ca, Memorial Univ of Newfoundland – Gc
Calvert, Andrew J., (604) 291-5387 acalvert@sfu.ca, Simon Fraser Univ – Ys
Calvert, Stephen E., (604) 822-5210 calvert@eos.ubc.ca, Univ of British Columbia – Cm
Calvin, Wendy, 775.784.1785 wcalvin@unr.edu, Univ of Nevada, Reno – YgOr
Camacho, Alfredo, (204) 474-7413 camacho@cc.umanitoba.ca, Univ of Manitoba – Gt
Camacho, Elsa A., (509) 376-5473 elsa.camacho@pnl.gov, Pacific Northwest National Laboratory – Cg
Camann, Eleanor J., 303-914-6290 eleanor.camann@rrcc.edu, Red Rocks Community Coll – OneGs
Camara Artigas, Fernando, fernando.camaraartigas@unito.it, Università di Torino – Gz
Cambardella, Cynthia, (515) 294-2921 cindy.cambardella@ars.usda.gov, Iowa State Univ of Science & Tech – So
Cambiotti, Laura J., ljc0@lehigh.edu, Lehigh Univ – On
Came, Rosemarie E., 603-862-1720 rosemarie.came@unh.edu, Univ of New Hampshire – PeOnn
Cameron, Barry I., 414-229-3136 bcameron@uwm.edu, Univ of Wisconsin, Milwaukee – Gvi
Cameron, Cheryl, (907) 451-5012 cheryl.cameron@alaska.gov, Alaska Division of Geological & Geophysical Surveys – Gv
Cameron, Douglas, (406) 496-4247 dcameron@mtech.edu, Montana Tech of the Univ of Montana – Ca
Cameron, Kevin, (604) 291-4703 kjc@sfu.ca, Simon Fraser Univ – Gg
Camill III, Philip, 207-721-5149 pcamill@bowdoin.edu, Bowdoin Coll – PebSb
Camille, Michael A., (318) 342-1750 Univ of Louisiana, Monroe – Oy
Camilleri, Phyllis A., (931) 221-7317 camillerip@apsu.edu, Austin Peay State Univ – Gct
Cammerata, Kirk , 361-825-2468 Kirk.Cammerata@tamucc.edu, Texas A&M Univ, Corpus Christi – On
Camp, Mark J., (419) 530-2398 mark.camp@utoledo.edu, Univ of Toledo – Pi
Camp, Victor E., (619) 594-7170 vcamp@mail.sdsu.edu, San Diego State Univ – Gv
Campagna, David, 757 229 1661 david@campagna-associates.com, West Virginia Univ – Gc
Campana, Michael E., (541) 737-2413 michael.campana@oregonstate.edu, Oregon State Univ – Hg
Campbell, Andrew R., (575) 835-5327 campbell@nmt.edu, New Mexico Inst of Mining and Tech – Cs
Campbell, Bruce A., 202-633-2472 Smithsonian Institution / National Air

& Space Museum – Or
Campbell, David L., (319) 335-1314 david-l-campbell@uiowa.edu, Univ of Iowa – Yg
Campbell, Finley A., (403) 220-6801 Univ of Calgary – Em
Campbell, Glenn A., (859) 622-6474 glenn.campbell@eku.edu, Eastern Kentucky Univ – Oy
Campbell, Ian A., ian.campbell@ualberta.ca, Univ of Alberta – Gm
Campbell, James B., (540) 231-5841 Virginia Polytechnic Inst & State Univ – Or
Campbell, Katherine, (505) 667-2799 ksc@lanl.gov, Los Alamos National Laboratory – Gq
Campbell, Kenneth E., (213) 763-3425 kcampbel@nhm.org, Los Angeles County Museum of Natural History – Pv
Campbell, Lisa, (979) 845-5706 lisacampbell@tamu.edu, Texas A&M Univ – Ob
Campbell, Patricia A., (724) 738-4405 patricia.campbell@sru.edu, Slippery Rock Univ – Gc
Campbell, Seth W., seth.campbell@maine.edu, Univ of Maine – Gl
Campbell-Stone, Erin, 307-766-2286 Ext. 230 erin.campbell-stone@wyo.gov, Wyoming State Geological Survey – Ng
Campbell-Stone, Erin A., (307) 766-2053 erincs@uwyo.edu, Univ of Wyoming – GcEo
Canalda, Sabrina, 915-831- 2617 mren@epcc.edu, El Paso Community Coll – Gg
Canales Cisneros, Juan Pablo, (508) 289-2893 jpcanales@whoi.edu, Woods Hole Oceanographic Institution – Yr
Cande, Steven C., (858) 534-1552 scande@ucsd.edu, Univ of California, San Diego – Yr
Candela, Philip A., (301) 405-2783 candela@geol.umd.edu, Univ of Maryland – CpgEg
Cane, Mark A., 845-365-8344 Columbia Univ – Op
Canil, Dante, (250) 472-4180 dcanil@uvic.ca, Univ of Victoria – Gi
Cannon, Alex, (604) 325-7830 acannon@eos.ubc.ca, Univ of British Columbia – Oa
Cantarero, Debra A., (626) 585-7138 dacantarero@pasadena.edu, Pasadena City Coll – On
Cantrell, Kirk J., (509) 376-2136 kirk.cantrell@pnl.gov, Pacific Northwest National Laboratory – Cl
Canuel, Elizabeth A., (804) 684-7134 ecanuel@vims.edu, Coll of William & Mary – Oc
Cao, Guofeng, 806-834-8920 guofeng.cao@ttu.edu, Texas Tech Univ – Oi
Cao, Hongsheng, 316-978-3140 Wichita State Univ – Cl
Cao, Wenrong, 775-784-1770 wenrongc@unr.edu, Univ of Nevada, Reno – Gtc
Capehart, William J., (605) 394-2291 William.Capehart@sdsmt.edu, South Dakota School of Mines & Tech – Oa
Capella, Silvana, silvana.capella@unito.it, Università di Torino – Gz
Caplan-Auerbach, Jackie, 360-650-4153 caplanj@wwu.edu, Western Washington Univ – Ys
Capo, Rosemary C., (412) 624-8873 rcapo@pitt.edu, Univ of Pittsburgh – Cl
Capoccia, Mary, (614) 292-8522 capoccia.6@osu.edu, Ohio State Univ – On
Caporali, Alessandro, 39-049-8279122 alessandro.caporali@unipd.it, Università degli Studi di Padova – YdsOr
Capraro, Luca, 39-049-8279182 luca.capraro@unipd.it, Università degli Studi di Padova – PelGg
Capuano, Regina M., 713-743-2957 capuano@uh.edu, Univ of Houston – Hw
Caputo, Mario V., (909) 214-7742 mvcaputo@earthlink.net, San Diego State Univ – GsdOn
Carbotte, Suzanne, 845-365-8895 Columbia Univ – Yr
Cardace, Dawn, 401-874-9384 cardace@uri.edu, Univ of Rhode Island – GpPy
Cardellach López, Esteve, ++34935813091 esteve.cardellach@uab.cat, Universitat Autonoma de Barcelona – GzeCs
Cardenas, Bayani, 512-471-9425 cardenas@mail.utexas.edu, Univ of Texas at Austin – Hw
Cardiff, Michael A., (608) 262-8960 cardiff@wisc.edu, Univ of Wisconsin, Madison – HwYg
Cardimona, Steve, (707) 468-3219 scardimo@mendocino.edu, Mendocino Coll – Ys
Cardott, Brian J., (405) 325-8065 bcardott@ou.edu, Univ of Oklahoma – EcoGg
Carew, James L., (843) 953-5592 carewj@cofc.edu, Coll of Charleston – PoGd
Carey, Anne E., (614) 292-2375 carey.145@osu.edu, Ohio State Univ – Hq
Carey, James W., (505) 667-5540 bcarey@lanl.gov, Los Alamos National Laboratory – CpYxNr
Carey, Kristine M., (807) 343-8461 kristine.carey@lakeheadu.ca, Lakehead Univ – On
Carey, Larry, 256-961-7909 larry.carey@nsstc.uah.edu, Texas A&M Univ – Oa
Carey, Sean, 905-525-9140 (Ext. 20134) careysk@mcmaster.ca, McMaster Univ – Hg
Carey, Steven N., (401) 874-6209 scarey@gso.uri.edu, Univ of Rhode Island – Ou
Carey, Tara, 309-796-5274 careyt@bhc.edu, Black Hawk Coll – On
Carle, Steven, (925) 423-5039 carle1@llnl.gov, Lawrence Livermore National Laboratory – Hw
Carlile, Amy L., 203-479-4257 acarlile@newhaven.edu, Univ of New Haven – Obn
Carlin, Joe, (657) 278-3054 jcarlin@fullerton.edu, California State Univ, Fullerton – GuOuGs
Carling, Gregory T., (801) 422-2622 greg.carling@byu.edu, Brigham Young Univ – HwCts
Carlisle, Donald, (310) 825-1934 carlisle@epss.ucla.edu, Univ of California, Los Angeles – Eg
Carlson, Anders, 541-737-3625 acarlson@coas.oregonstate.edu, Oregon State Univ – GlCt
Carlson, Barry A., bacarlso@delta.edu, Delta Coll – Yg
Carlson, Catherine A., 860-465-5218 carlsonc@easternct.edu, Eastern Connecticut State Univ – Hwg
Carlson, Diane H., carlsondh@csus.edu, California State Univ, Sacramento – Gc
Carlson, Douglas A., 225/578-3671 dcarlson@lsu.edu, Louisiana State Univ – Hw
Carlson, Galen R., 657-278-3882 gcarlson@Exchange.fullerton.edu, California State Univ, Fullerton – Og
Carlson, Heath, carlsonh@easternct.edu, Eastern Connecticut State Univ – Og
Carlson, Jill, 303-384-2643 carlson@mines.edu, Colorado School of Mines – NgOu
Carlson, Keith J., (507) 933-7307 jcarlson@gac.edu, Gustavus Adolphus Coll – Pv
Carlson, Marvin P., (402) 472-3471 rockdrmpc@hotmail.com, Unversity of Nebraska - Lincoln – Gt
Carlson, Richard E., (515) 294-9868 richard@iastate.edu, Iowa State Univ of Science & Tech – Ow
Carlson, Richard L., 9798451398 carlson@geo.tamu.edu, Texas A&M Univ – GtYd
Carlson, Richard W., (202) 478-8474 carlson@dtm.ciw.edu, Carnegie Institution of Washington – Cc
Carlson, Sandra J., (530) 752-2834 sjcarlson@ucdavis.edu, Univ of California, Davis – Po
Carlson, Toby N., (814) 863-1582 tnc@psu.edu, Pennsylvania State Univ, Univ Park – Ow
Carlson, William D., (512) 471-4770 wcarlson@jsg.utexas.edu, Univ of Texas at Austin – Gp
Carlucci, Jesse R., (940) 397-4448 jesse.carlucci@mwsu.edu, Midwestern State Univ – PisGs
Carmack, Edward, (250) 363-6585 carmack@dfo-mpo.gc.ca, Univ of British Columbia – Op
Carman, Cary D., 325-944-4600 cdcarman@usgs.gov, Angelo State Univ – Hg
Carmichael, Ian S., (510) 642-2577 ian@eps.berkeley.edu, Univ of California, Berkeley – Gi
Carmichael, Robert S., (319) 337-6499 robert-carmichael@uiowa.edu, Univ of Iowa – YexGg
Carmichael, Sarah, (828) 262-8471 carmichaelsk@appstate.edu, Appalachian State Univ – GxCgGd
Carn, Simon, 906/487-1756 scarn@mtu.edu, Michigan Technological Univ – OrGv
Carnes, Jacob, jacob.carnes1@wyo.gov, Wyoming State Geological Survey – Gg
Carney, Robert S., (225) 388-6511 Louisiana State Univ – Ob
Carney, Theodore C., (505) 667-3415 tedc@lanl.gov, Los Alamos National Laboratory – Ng
Caron, Jean-Bernard, (416) 586-5753 jcaron@rom.on.ca, Univ of Toronto – Pi
Caron, Jean-Bernard, jcaron@rom.on.ca, Royal Ontario Museum – PoiPg
Caron, Olivier J., 217-300-0198 caron@illinois.edu,

Univ of Illinois, Urbana-Champaign – Ng
Carosi, Rodolfo, rodolfo.carosi@unito.it, Università di Torino – Gc
Carp, Jana, (828) 262-7091 carpje@appstate.edu, Appalachian State Univ – On
Carpenter, Kenneth, (435) 613-5752 ken.carpenter@usu.edu, Utah State Univ – PvoPe
Carpenter, Michael, +44 (0) 1223 333483 mc43@esc.cam.ac.uk, Univ of Cambridge – Gy
Carpenter, Philip J., (815) 753-1523 pjcarpenter@niu.edu, Northern Illinois Univ – YgHwNg
Carpenter, Rob, rcarpen5@uwo.ca, Western Univ – Em
Carpenter, Roy, (206) 543-8535 rcarp@u.washington.edu, Univ of Washington – Oo
Carr, David E., 540-837-1758 blandy@virginia.edu, Univ of Virginia – On
Carr, James R., (775) 784-4244 carr@unr.edu, Univ of Nevada, Reno – NgOuGg
Carr, Keith W., 217-265-0267 kw-carr@illinois.edu, Univ of Illinois, Urbana-Champaign – Gg
Carr, Mark, (831) 459-3958 carr@biology.ucsc.edu, Univ of California, Santa Cruz – Ob
Carr, Michael J., carr@rutgers.edu, Rutgers, The State Univ of New Jersey – Gv
Carr, Timothy R., (785) 864-2135 tcarr@kgs.ku.edu, Univ of Kansas – Go
Carr, Timothy R., 304-293-9660 tim.carr@mail.wvu.edu, West Virginia Univ – GoYe
Carrano, Matthew T., (202) 633-1506 Smithsonian Institution / National Museum of Natural History – Pv
Carrapa, Barbara, 520-621-4910 bcarrapa@email.arizona.edu, Univ of Arizona – Gs
Carrell, Jennifer, 217-244-2764 jcarrell@illinois.edu, Univ of Illinois, Urbana-Champaign – Og
Carrick, Tina, (915) 831-8875 tcarrick@utep.edu, Univ of Texas, El Paso – OnGg
Carrigan, Charles R., (925) 422-3941 carrigan1@llnl.gov, Lawrence Livermore National Laboratory – Yg
Carrigan, Charles W., (815) 939-5346 CCarriga@olivet.edu, Olivet Nazarene Univ – GzCgGc
Carroll, Alan R., (608) 262-2368 acarroll@geology.wisc.edu, Univ of Wisconsin, Madison – Gs
Carroll, Chris J., (307) 766-2286 Ext. 243 chris.carroll@wyo.gov, Wyoming State Geological Survey – EcgGg
Carroll, David, (540) 231-5469 carrolld@vt.edu, Virginia Polytechnic Inst & State Univ – Ow
Carroll, Matt, 509-335-2235 carroll@wsu.edu, Washington State Univ – On
Carroll, Richard E., 205-247-3551 rcarroll@gsa.state.al.us, Geological Survey of Alabama – EcPlGo
Carroll, Rosemary, rosemary.carroll@dri.edu, Univ of Nevada, Reno – Hq
Carroll, Susan A., (925) 423-7552 Lawrence Livermore National Laboratory – Cl
Carslaw, Ken, +44(0) 113 34 31597 k.s.carslaw@leeds.ac.uk, Univ of Leeds – Oa
Carson, Bobb, bc00@lehigh.edu, Lehigh Univ – Gus
Carson, Eric C., 608-890-1998 eccarson@wisc.edu, Univ of Wisconsin, Extension – Gl
Carson, Michael, +618 9266-4973 Michael.carson@curtin.edu.au, Curtin Univ – Yx
Carson, Robert J., (509) 527-5224 carsonrj@whitman.edu, Whitman Coll – GmeGl
Carstarphen, Camela, 406-496-4633 ccarstarphen@mtech.edu, Montana Tech of The Univ of Montana – Hw
Carstensen, Laurence W., (540) 231-2600 Virginia Polytechnic Inst & State Univ – Oi
Carter, Andy, +44 020 7679 2418 a.carter@ucl.ac.uk, Birkbeck Coll – Og
Carter, Burchard D., (229) 931-2325 burchard.carter@gsw.edu, Georgia Southwestern State Univ – Po
Carter, David, (303) 293-5014 dcarter@osmre.gov, Department of Interior Office of Surface Mining Reclamation and Enforcement – Oir
Carter, Glenn, (808) 956-9267 gscarter@hawaii.edu, Univ of Hawai'i, Manoa – Op
Carter, Greg, (228) 214-3305 greg.carter@usm.edu, Univ of Southern Mississippi – Oyi
Carter, James L., 972 883-2455 jcarter@utdallas.edu, Univ of Texas, Dallas – Eg
Carter, Joseph G., (919) 962-0685 clams@email.unc.edu, Univ of North Carolina, Chapel Hill – Po
Carter, Kristin M., 412.442.4234 krcarter@pa.gov, Pennsylvania Bureau of Topographic & Geologic Survey – GoHw
Carton, Alberto, +390498279141 alberto.carton@unipd.it, Università degli Studi di Padova – Gm
Carton, James, (301) 405-5365 Univ of Maryland – Op
Cartwright, J A., cartwrightJA@cf.ac.uk, Cardiff Univ – Ys
Cartwright, Joe, +44 (1865) 272000 joe.cartwright@earth.ox.ac.uk, Univ of Oxford – Gs
Caruthers, Andrew, (269) 387-8633 andrew.caruthers@wmich.edu, Western Michigan Univ – GsCsGr
Carver, Gary A., (907) 487-4551 Humboldt State Univ – Gmt
Cary, Kevin, (270) 745-2981 kevin.cary@wku.edu, Western Kentucky Univ – Oi
Carzoli, John, (847) 376-7042 jcarzoli@oakton.edu, Oakton Community Coll – On
Casas Duocastella, Lluís, ++34935868365 Lluis.Casas@uab.cat, Universitat Autonoma de Barcelona – GzaYm
Case, James M., (570) 408-4618 james.case@wilkes.edu, Wilkes Univ – Ob
Case, Jeanne, 509-359-4288 jdcase@ewu.edu, Eastern Washington Univ – Gg
Case, Stephen, (815) 939-5681 scase@olivet.edu, Olivet Nazarene Univ – GhXg
Case, William F., (801) 537-3340 nrugs.bcase@state.ut.us, Utah Geological Survey – Ng
Case Hanks, Anne T., (318) 342-1822 casehanks@ulm.edu, Univ of Louisiana, Monroe – Oae
Casey, Francis, francis.casey@ndsu.edu, North Dakota State Univ – SpHq
Casey, John F., 713-743-2824 jfcasey@uh.edu, Univ of Houston – GtiGu
Casey, William H., (916) 752-3211 Univ of California, Davis – Cl
Cashman, Patricia H., (775) 784-6924 pcashman@mines.unr.edu, Univ of Nevada, Reno – Gc
Cashman, Susan M., (707) 826-3114 smc1@humboldt.edu, Humboldt State Univ – Gc
Caskey, Deborah, 915-831-8905 dcaskey1@epcc.edu, El Paso Community Coll – Gg
Caskey, John, (415) 405-0353 caskey@sfsu.edu, San Francisco State Univ – Gc
Cassar, Nicolas, 919 681-8865 nicolas.cassar@duke.edu, Duke Univ – Ob
Cassel, Donald K., (919) 515-1457 North Carolina State Univ – Sp
Cassel, Elizabeth J., (208) 885-4289 ecassel@uidaho.edu, Univ of Idaho – Gs
Cassiani, Giorgio, +390498279189 giorgio.cassiani@unipd.it, Università degli Studi di Padova – YgHs
Cassidy, Carleen, 406-496-4769 ccassidy@mtech.edu, Montana Tech of The Univ of Montana – On
Cassidy, Daniel P., (269) 387-5324 daniel.cassidy@wmich.edu, Western Michigan Univ – Hw
Cassidy, John, cassidy@pgc.nrcan.gc.ca, Univ of Victoria – Ys
Cassidy, Martin, 713-616-5853 mcassidy@uh.edu, Univ of Houston – GoCo
Cassidy, William, (412) 624-8886 Univ of Pittsburgh – Xg
Castagna, John P., 713-743-8699 jpcastagna@uh.edu, Univ of Houston – Ye
Castaneda, Isla, 413 577-1124 isla@geo.umass.edu, Univ of Massachusetts, Amherst – Ca
Castelli, Daniele, daniele.castelli@unito.it, Università di Torino – Gx
Castillo, Paterno R., (858) 534-0383 pcastillo@ucsd.edu, Univ of California, San Diego – Giu
Castillon, David A., (417) 836-5800 DavidCastillon@MissouriState.edu, Missouri State Univ – Gm
Castle, James W., (864) 656-5015 jcastle@clemson.edu, Clemson Univ – GseHw
Castleton, Jessica, (801) 537-3381 jessicacastleton@utah.gov, Utah Geological Survey – Ng
Casto, Pamela, 304-367-8436 Pamela.M.Casto@ivv.nasa.gov, Fairmont State Univ – Oe
Castor, Stephen B., 775-682-8766 scastor@unr.edu, Univ of Nevada – Eg
Castro, Maria Clara, 734-615-3812 mccastro@umich.edu, Univ of Michigan – Hy
Castro, Mark S., (301) 689-3115 castro@al.umces.edu, Univ of Maryland – Oa
Castro, Raul, raul@cicese.mx, Centro de Investigación Científica y de Educación Superior de

Faculty Index -C

Ensenada – Ys
Caswell, Bryony, +440151 795 4390 B.A.Caswell@liverpool.ac.uk, Univ of Liverpool – Ob
Catalano, Jeff, 314-935-6015 catalano@levee.wustl.edu, Washington Univ in St. Louis – Ca
Cataneo, Robert, 217-581-2626 rcataneo@eiu.edu, Eastern Illinois Univ – Oa
Catania, Ginny A., (512) 471-0403 gcatania@utig.ig.utexas.edu, Univ of Texas at Austin – Ol
Catau, John C., (417) 836-4589 johncatau@missouristate.edu, Missouri State Univ – On
Cather, Steven M., (505) 835-5153 steve@gis.nmt.edu, New Mexico Inst of Mining & Tech – Gs
Cathey, Henrietta, +61 7 3138 0416 henrietta.cathey@qut.edu.au, Queensland Univ of Tech – Giv
Cathles, Lawrence M., (607) 255-2844 lmc19@cornell.edu, Cornell Univ – Hy
Catling, David C., dcatling@u.washington.edu, Univ of Washington – Oa
Catlos, Elizabeth J., 512-471-4762 ejcatlos@gmail.com, Univ of Texas at Austin – GzCg
Cato, Craig, 205-247-3693 ccato@gsa.state.al.us, Geological Survey of Alabama – Gco
Cato, Kerry, (909) 537-5409 kerry.cato@csusb.edu, California State Univ, San Bernardino – Ng
Cattolico, Rose Ann, (206) 543-9363 racat@u.washington.edu, Univ of Washington – On
Catuneanu, Octavian, (780) 492-6569 octavian.catuneanu@ualberta.ca, Univ of Alberta – Gs
Caudill, Kimberly S., 740-753-6289 caudillk7007@hocking.edu, Hocking Coll – GeOuHw
Caudill, Michael R., 740-753-6277 caudillm@hocking.edu, Hocking Coll – SdHwNg
Caupp, Craig L., (301) 687-4755 ccaupp@frostburg.edu, Frostburg State Univ – OnHs
Caus Gracia, Esmeralda, ++935812031 esmeralda.caus@uab.cat, Universitat Autonoma de Barcelona – Pm
Cave, Rachel R., +353 (0)91 492 351 rachel.cave@nuigalway.ie, National Univ of Ireland Galway – Cm
Cavin, Julie, j.cavin@hotmail.com, Univ of Massachusetts, Boston – Ob
Cawood, Peter A., (+44) (0)1334 463911 peter.cawood@monash.edu, Monash Univ – GtcGg
Cayzer, Nicola, +44 (0) 131 650 8527 Nicola.Cayzer@ed.ac.uk, Edinburgh Univ – Yg
Cecil, Blaine, 703 648 6415 bcecil@usgs.gov, West Virginia Univ – Ec
Cecil, M. R., (818) 677-7009 robinson.cecil@csun.edu, California State Univ, Northridge – GiCcGz
Celia, Michael A., (609) 258-5425 Princeton Univ – Hw
Celik, Hasan, hasancelik@firat.edu.tr, Firat Univ – GscOu
Çemen, Ibrahim, 205-348-8019 icemen@ua.edu, Univ of Alabama – Gct
Centorbi, Tracey, (301) 405-6965 tlcento@umd.edu, Univ of Maryland – Cg
Cepeda, Joseph C., 806-651-2584 jcepeda@wtamu.edu, West Texas A&M Univ – GiHw
Cercone, Karen Rose, (724) 357-5623 kcercone@iup.edu, Indiana Univ of Pennsylvania – Oeg
Cerling, Thure E., (801) 581-5558 thure.cerling@utah.edu, Univ of Utah – Cl
Cerpa Gilvonio, Nestor, ncerpagi@umn.edu, Univ of Minnesota, Twin Cities – Yg
Cerrato, Robert M., (631) 632-8666 robert.cerrato@stonybrook.edu, SUNY, Stony Brook – Ob
Cervato, Cinzia C., (515) 294-7583 cinzia@iastate.edu, Iowa State Univ of Science & Tech – GgOe
Cesare, Bernardo, +390498279148 bernardo.cesare@unipd.it, Università degli Studi di Padova – Gp
Cessi, Paola, (858) 534-0622 pcessi@ucsd.edu, Univ of California, San Diego – Op
Cetin, Haluk, (270) 809-2085 hcetin@murraystate.edu, Murray State Univ – OiGe
Cetindag, Bahattin, bcetindag@firat.edu.tr, Firat Univ – NgYv
Chachadi, Adiveppa G., +91-832-6519330 chachadi@unigoa.ac.in, Goa Univ – HwYeHs
Chacko, Elizabeth, (202) 994-6185 echack@gwu.edu, George Washington Univ – Oy
Chacko, Thomas, (780) 492-5395 tom.chacko@ualberta.ca, Univ of Alberta – Gp
Chaddock, Lisa, 6196604000 x3033 Lisa.Chaddock@gcccd.edu, Cuyamaca Coll – Oy
Chadima, Sarah A., 605-677-6166 sarah.chadima@usd.edu, South Dakota Dept of Environment and Natural Resources – Gg
Chadwick, John, (843) 953-5950 chadwickj@cofc.edu, Coll of Charleston – GiOrGt
Chafetz, Henry S., (713) 743-3227 hchafetz@uh.edu, Univ of Houston – Gds
Chague-Goff, Catherine, c.chague-goff@unsw.edu.au, Univ of New South Wales – CgGs
Chakhmouridian, Anton, (204) 474-7278 chakhmou@ms.umanitoba.ca, Univ of Manitoba – Gz
Challads, Thomas, +44 (0) 131 650 8543 tchallan@staffmail.ed.ac.uk, Edinburgh Univ – Hg
Challender, Stuart, (406) 994-7566 schallender@montana.edu, Montana State Univ – Oi
Challinor, Andy, +44(0) 113 34 33194 a.j.challinor@leeds.ac.uk, Univ of Leeds – Ow
Chamberlain, Andrew, +44 (0)161 306 4176 andrew.chamberlain@manchester.ac.uk, Univ of Manchester – Ga
Chamberlain, Edwin P., (718) 951-5416 Los Alamos National Laboratory – Cc
Chamberlain, John A., (718) 951-5926 Graduate School of the City Univ of New York – Po
Chamberlain, John A., 7189515000 x2885 johnc@brooklyn.cuny.edu, Brooklyn Coll (CUNY) – Po
Chamberlain, Kevin R., (307) 766-2914 kchamber@uwyo.edu, Univ of Wyoming – Cc
Chamberlin, Richard, (575) 835-5310 richard@nmbg.nmt.edu, New Mexico Inst of Mining and Tech – GcvGg
Chamberlin, William S., (714) 992-7443 schamberlin@fullcoll.edu, Fullerton Coll – OgbOe
Chambers, Jeanne, 775-784-5329 chambers@unr.edu, Univ of Nevada, Reno – SfHg
Chambers, John E., (202) 478-8855 jchambers@carnegiescience.edu, Carnegie Institution of Washington – On
Chameides, William L., 613-8004 bill.chameides@duke.edu, Duke Univ – OaGe
Champion, Kyle M., 813-253-7326 Hillsborough Community Coll – Og
Champlin, Steven D., (601) 961-5506 stephen_champlin@deq.state.ms.us, Mississippi Office of Geology – Go
Chan, Clara, 302-831-1819 cschan@udel.edu, Univ of Delaware – PyCmGz
Chan, Kwan M., (562) 985-4817 kmchan@csulb.edu, California State Univ, Long Beach – Oc
Chan, Marjorie A., (801) 581-6551 marjorie.chan@utah.edu, Univ of Utah – Gs
Chan, Selene, (604) 822-2034 schan@eos.ubc.ca, Univ of British Columbia – On
Chandler, Angela, (501) 683-0111 angela.chandler@arkansas.gov, Arkansas Geological Survey – GgrGs
Chandler, Cyndy, (508) 289-2765 cchandler@whoi.edu, Woods Hole Oceanographic Institution – Og
Chandler, Sandra, (501) 683-0125 sandra.chandler@arkansas.gov, Arkansas Geological Survey – Oe
Chandler, Val, (612) 626-4976 chand004@umn.edu, Univ of Minnesota – Ye
Chandler, Val W., 612-627-4780 Ext. 203 chand004@umn.edu, Univ of Minnesota, Twin Cities – Ye
Chandra, Sudeep, 775.784.6221 sudeep@unr.edu, Univ of Nevada, Reno – Hs
Chaney, Phil L., (334) 844-3420 chanepl@auburn.edu, Auburn Univ – OynOn
Chang, Chih-Pei, 831-656-2840 cpchang@nps.edu, Naval Postgraduate School – Ow
Chang, Edmund K., (631) 632-6170 kar.chang@stonybrook.edu, SUNY, Stony Brook – Oa
Chang, Hai-ru, hrc@eas.gatech.edu, Georgia Inst of Tech – Oa
Chang, Julie M., 405-325-7055 jmchang@ou.edu, Univ of Oklahoma – GgOnn
Chang, Ping, (979) 845-8196 pchang@ocean.tamu.edu, Texas A&M Univ – Op
Chang, Young-Soo, (630) 252-4076 changy@anl.gov, Argonne National Laboratory – Oa
Chang, Zhaoshan, +61 7 47816434 zhaoshan.chang@jcu.edu.au, James Cook Univ – EmCce
Channell, James E., 352-392-3658 jetc@ufl.edu,

Faculty Index -C

Univ of Florida – YmGt
Channing, Alan, +44(0)29 208 76213 ChanningA@cardiff.ac.uk, Univ of Wales – Gd
Chanway, Christopher, christopher.chanway@ubc.ca, Univ of British Columbia – Sb
Chao, Shenn-Yu, (410) 221-8427 chao@hpl.umces.edu, Univ of Maryland – Op
Chapin, Charles E., (505) 835-5613 chapin@gis.nmt.edu, New Mexico Inst of Mining & Tech – Gt
Chapin, F. Stuart, (510) 642-1003 Univ of California, Berkeley – Pe
Chapman, Alan, chapman@macalester.edu, Macalester Coll – Gct
Chapman, David S., (801) 581-7642 david.chapman@utah.edu, Univ of Utah – Yh
Chapman, Mark, +44 (0)1603 59 3114 mark.chapman@uea.ac.uk, Univ of East Anglia – Pm
Chapman, Mark, +44 (0) 131 650 8521 mchapman@staffmail.ed.ac.uk, Edinburgh Univ – Yg
Chapman, Marshall, (606) 783-5397 m.chapman@moreheadstate.edu, Morehead State Univ – Gv
Chapman, Martin C., (540) 231-5036 mcc@vt.edu, Virginia Polytechnic Inst & State Univ – Ys
Chapman, Piers, (979) 845-9399 piers.chapman@tamu.edu, Texas A&M Univ – OcCm
Chapman, Robert J., +44(0) 113 34 33190 r.j.chapman@leeds.ac.uk, Univ of Leeds – EmCe
Chapman, Ross N., (250) 472-4340 chapman@uvic.ca, Univ of Victoria – Op
Chappell, James R., (970) 491-5147 Jim_Chappell@partner.nps.gov, Colorado State Univ – Ge
Chappell, P. D., 757-683-4937 pdchappe@odu.edu, Old Dominion Univ – Oc
Chaput, Dominique, chaputdl@si.edu, Smithsonian Institution / National Museum of Natural History – Py
Charette, Matthew A., (508) 289-3205 mcharette@whoi.edu, Woods Hole Oceanographic Institution – Cm
Charles, Christopher D., (858) 534-5911 ccharles@ucsd.edu, Univ of California, San Diego – Pe
Charles, Kasanzu, Kcharls16@yahoo.com, Univ of Dar es Salaam – Ggh
Charlevoix, Donna J., (217) 244-9575 charlevo@atmos.uiuc.edu, Univ of Illinois, Urbana-Champaign – Oa
Charlson, Robert J., charlson@chem.washington.edu, Univ of Washington – Oa
Chase, Anne, (520) 621-6004 achase@email.arizona.edu, Univ of Arizona – On
Chase, Clement G., (520) 621-2417 cgchase@email.arizona.edu, Univ of Arizona – Yg
Chase, Jon M., jchase@biology2.wustl.edu, Washington Univ in St. Louis – On
Chase, Peter M., (608) 265.6003 peter.chase@wgnhs.uwex.edu, Univ of Wisconsin - Extension – Hw
Chase, Richard L., (604) 822-3086 rchase@eos.ubc.ca, Univ of British Columbia – Gu
Chasteen, Hayden R., 817-515-6694 hayden.chasteen@tccd.edu, Tarrant County Coll- Northeast Campus – GgeGg
Chatelain, Edward E., (229) 333-5758 echatela@valdosta.edu, Valdosta State Univ – PiGlPv
Chatterjee, Ipsita, 940-565-2372 Ipsita.Chatterjee@unt.edu, Univ of North Texas – On
Chatterjee, Meera, 330 972-2394 meera@uakron.edu, Univ of Akron – Oy
Chatterjee, Nilanjan, (617) 253-1995 nchat@mit.edu, Massachusetts Inst of Tech – Cg
Chatterjee, Sankar, 806-834-4590 sankar.chatterjee@ttu.edu, Texas Tech Univ – Pv
Chatterjee, Snehamoy, 906/487-2516 schatte1@mtu.edu, Michigan Technological Univ – Nm
Chatterton, Brian D., (780) 492-3085 brian.chatterton@ualberta.ca, Univ of Alberta – Pi
Chaubey, Indrajeet, (765) 494-3258 ichaubey@purdue.edu, Purdue Univ – Hg
Chaudhuri, Sambhudas, (785) 532-2246 ksuncsc@ksu.edu, Kansas State Univ – CcGoe
Chauff, Karl, (314) 977-3143 chauff@eas.slu.edu, Saint Louis Univ – Pg
Chaumba, Jeff B., (910) 522-5787 jeff.chaumba@uncp.edu, Univ of North Carolina, Pembroke – GxEgGz
Chaussard, Estelle, (716) 645-4291 estellec@buffalo.edu, SUNY, Buffalo – YdGtv
Chavrit, Deborah, +44 0161 275-0760 deborah.chavrit@manchester.ac.uk, Univ of Manchester – Gi
Cheadle, Burns A., 519-661-2111 x.89009 bcheadle@uwo.ca, Western Univ – Go
Cheadle, Michael J., (307) 766-3206 cheadle@uwyo.edu, Univ of Wyoming – Yg
Chebana, Fateh, fateh.chebana@ete.inrs.ca, Universite du Quebec – Hq
Checkley, David M., (858) 534-4228 dcheckley@ucsd.edu, Univ of California, San Diego – Ob
Cheek, William H., (417) 836-4589 billcheek@missouristate.edu, Missouri State Univ – On
Cheel, Richard J., (905) 688-5550 (Ext. 3512) RCheel@brocku.ca, Brock Univ – Gs
Chelton, Dudley B., (541) 737-4017 chelton@coas.oregonstate.edu, Oregon State Univ – Op
Chen, Bob, bob.chen@umb.edu, Univ of Massachusetts, Boston – CoOc
Chen, Dake, (845) 365-8496 Columbia Univ – Op
Chen, Gang, (907)474-6875 gchen@alaska.edu, Univ of Alaska, Fairbanks – Nm
Chen, Hway-Jen, 831-656-3788 hjchen@nps.edu, Naval Postgraduate School – Ow
Chen, Jian-Hua, (978) 934-4861 jianhua_chen@uml.edu, Univ of Massachusetts Lowell – Oa
Chen, Jing, chenj@geog.utoronto.ca, McMaster Univ – OwrOy
Chen, Jingyi, 918-631-2517 jingyi-chen@utulsa.edu, The Univ of Tulsa – Yes
Chen, jiquan, (419) 932-1517 jqchen@msu.edu, Michigan State Univ – Og
Chen, Jiuhua, 305-348-3030 chenj@fiu.edu, Florida International Univ – Gyz
Chen, Po, 307-766-3086 pchen@uwyo.edu, Univ of Wyoming – YseYg
Chen, Shu-Hua, 530-752-1822 shachen@ucdavis.edu, Univ of California, Davis – OraOw
Chen, Shuyi S., schen@rsmas.miami.edu, Univ of Washington – Oa
Chen, Tsing-Chang, (515) 294-9874 tmchen@iastate.edu, Iowa State Univ of Science & Tech – Oa
Chen, Wang-Ping, (217) 333-2744 wpchen@illinois.edu, Univ of Illinois, Urbana-Champaign – YsGt
Chen, Xianfeng, (724) 738-2385 xianfeng.chen@sru.edu, Slippery Rock Univ – Or
Chen, Xiaowei, 508-289-3820 xiaowei.fengr@gmail.com, Univ of Oklahoma – YsGt
Chen, Xun-Hong, (402) 472-0772 xchen2@unl.edu, Unversity of Nebraska - Lincoln – Hq
Chen, Yi-Leng, (808) 956-2570 yileng@hawaii.edu, Univ of Hawai'i, Manoa – Oaw
Chen, Yongsheng, (416)736-2100 #40124 yochen@yorku.ca, York Univ – Oa
Cheney, Donald, d.cheney@neu.edu, Northeastern Univ – On
Cheney, Eric S., 206-543-1163 vaalbara@uw.edu, Univ of Washington – Eg
Cheney, John T., (413) 542-2311 Amherst Coll – Gi
Cheng, H. H., (612) 625-1793 hcheng@soils.umn.edu, Univ of Minnesota, Twin Cities – Sb
Cheng, Hai, 86-29-83395119 cheng021@xjtu.edu.cn, Univ of Minnesota, Twin Cities – CcPe
Cheng, Meng-Dawn, (423) 241-5918 ucn@ornl.gov, Oak Ridge National Laboratory – Oa
Cheng, Qiuming, (416)736-2100 #22842 qiuming@yorku.ca, York Univ – Oi
Cheng, Songlin, 937 775-3455 songlin.cheng@wright.edu, Wright State Univ – Hw
Cheng, Yanbo, 07 47816808 yanbo.cheng1@jcu.edu.au, James Cook Univ – EgGgEm
Cheng, Zhongqi, 7189515000 x2647 zcheng@brooklyn.cuny.edu, Brooklyn Coll (CUNY) – Ca
Chenoweth, Cheri, 217-244-4610 cchenowe@illinois.edu, Univ of Illinois, Urbana-Champaign – GoEc
Chenoweth, Sean, (318) 342-1887 chenoweth@ulm.edu, Univ of Louisiana, Monroe – GmOir
Cherkauer, Douglas S., (262) 628-3672 aquadoc@uwm.edu, Univ of Wisconsin, Milwaukee – Hw
Chermak, John A., 540-231-1785 jchermak@vt.edu, Virginia Polytechnic Inst & State Univ – Cg
Cherniak, Daniele J., (518) 276-3358 chernd@rpi.edu, Rensselaer Polytechnic Inst – Gx
Chernoff, Barry, 860 6852452 bchernoff@wesleyan.edu,

Faculty Index -C

Wesleyan Univ – Ge
Cherrier, Jennifer, 718-951-5000 x2927 Jennifer.Cherrier18@brooklyn.cuny.edu,
Brooklyn Coll (CUNY) – OcCm
Cherubini, Claudia, +33(0)3 44068977 claudia.cherubini@lasalle-beauvais.fr, Institut Polytechnique LaSalle Beauvais (ex-IGAL) – HwGe
Cherukupalli, Nehru, 516 463-6545 geonec@hofstra.edu,
Hofstra Univ – Gg
Chesnaux, Romain, 418-545-5011 ext: 5426 rchesnaux@uqac.ca,
Universite du Quebec a Chicoutimi – Hw
Chesner, Craig A., (217) 581-6323 cachesner@eiu.edu,
Eastern Illinois Univ – GivGz
Chesnokov, Evgeny, 713-743-2579 emchesnokov@uh.edu,
Univ of Houston – YseYx
Chester, Frederick, (979) 845-3296 chesterf@tamu.edu,
Texas A&M Univ – GcNrYx
Chester, Judith, (979) 845-1380 chesterj@geo.tamu.edu,
Texas A&M Univ – GcNr
Chesworth, Ward, (519) 824-4120 (Ext. 52457) wcheswor@uoguelph.ca,
Univ of Guelph – Cl
Cheun, Norman, +44 020 8417 2811 K.W.Cheung@kingston.ac.uk,
Kingston Univ – Ge
Cheung, Wing, 7607441150 x3652 wcheung@palomar.edu,
Palomar Coll – OirOy
Chew, David, + 353 1 8963481 chewd@tcd.ie, Trinity Coll – CcGgt
Chi, Wu-Cheng, 886-2-2783-9910 ext 510 wchi@sinica.edu.tw,
Academia Sinica – GtYrs
Chiappe, Luis M., (213) 863-3323 lchiappe@nhm.org,
Los Angeles County Museum of Natural History – Pv
Chiarella, Domenico, +44 1784 443890 domenico.chiarella@rhul.ac.uk,
Univ of London, Royal Holloway & Bedford New Coll – Gs
Chiarello, Ronald P., (630) 252-9327 Argonne National Laboratory – Om
Chiarenzelli, Jeffrey R., 315-229-5202 jchiarenzelli@stlawu.edu,
St. Lawrence Univ – GzCg
Chief, Karletta, (520) 626-5598 kchief@email.arizona.edu,
Univ of Arizona – On
Chien, Yi-Ju, (509) 373-4822 Yi-Hu.chien@pnl.gov,
Pacific Northwest National Laboratory – Gq
Childers, Daniel, (610) 359-5242 dchilder@dccc.edu,
Delaware County Community Coll – OguOi
Childs, Geoff, gchilds@wustl.edu, Washington Univ in St. Louis – On
Chillrud, Steven, (845) 365-8893 Columbia Univ – Cg
Chilvers, Jason, +44 (0)1603 59 3130 jason.chilvers@uea.ac.uk,
Univ of East Anglia – Ge
Chin, Anne, (409) 845-7141 Texas A&M Univ – Gm
Chin, Karen, 303-735-3074 karen.chin@colorado.edu,
Univ of Colorado – Po
Chin, Yu-Ping, (614) 292-6953 chin.15@osu.edu, Ohio State Univ – Hg
Chipperfield, Martyn, +44(0) 113 34 36459 m.chipperfield@leeds.ac.uk,
Univ of Leeds – Oa
Chipping, David H., (805) 756-1695 dchippin@calpoly.edu,
California Polytechnic State Univ – Grt
Chirenje, Tait, 609-652-4588 tait.chirenje@stockton.edu,
Richard Stockton Coll of New Jersey – Sc
Chiu, Ching-Sang, (831) 656-3239 chiu@nps.edu,
Naval Postgraduate School – Op
Chiu, Jer-Ming, (901) 678-2007 jerchiu@memphis.edu,
Univ of Memphis – Ys
Chiu, Long S., (703) 993-1984 lchiu@gmu.edu,
George Mason Univ – OarOw
Chiu, Shu-Choiung, (901) 678-2007 sachi@memphis.edu,
Univ of Memphis – Ys
Chizmadia, Lysa, (787) 265-3845 Univ of Puerto Rico – GzXm
Choh, Suk-Joo, 82-2-3290-3180 sjchoh@korea.ac.kr, Korea Univ – GdsPe
Choi, Seon-Gyu, 82-2-3290-3174 seongyu@korea.ac.kr,
Korea Univ – GzEmg
Chokmani, Karem, karem.chokmani@ete.inrs.ca,
Universite du Quebec – Oi
Cholnoky, Jennifer, (518) 580-8127 jcholnok@skidmore.edu,
Skidmore Coll – Gg
Choma-Moryl, Krystyna, krystyna.choma-moryl@uwr.edu.pl,
Univ of Wroclaw – Ng
Chopping, Mark J., 973-655-4448 choppingm@mail.montclair.edu,
Montclair State Univ – Or
Chormann, Jr., Frederick H., (603) 271-1975 frederick.chormann@des.nh.gov,
New Hampshire Geological Survey – HyOiGm
Chorover, Jonathan, (520) 621-7228 chorover@cals.arizona.edu,
Univ of Arizona – ScClo
Chorover, Jonathan D., (520) 626-5635 chorover@cals.arizona.edu,
Univ of Arizona – Cg
Chou, Charissa J., (509) 372-3804 Charissa.chou@pnl.gov,
Pacific Northwest National Laboratory – Gq
Chou, Mei-In (Melissa), 217-244-0312 chou@isgs.uiuc.edu,
Illinois State Geological Survey – Co
Chou, Sheng-Fu Joseph, 217-244-2744 jchou@isgs.uiuc.edu,
Illinois State Geological Survey – Co
Chouinard, Vera, (905) 525-9140 (Ext. 23518) chouinar@mcmaster.ca,
McMaster Univ – On
Choularton, Thomas, +44 0161 306-3950 choularton@manchester.ac.uk,
Univ of Manchester – Oa
Chow, Nancy, (204) 474-6451 n_chow@umanitoba.ca,
Univ of Manitoba – Gs
Chowdhury, Dipak K., (260) 481-6249 chowdhur@ipfw.edu,
Indiana Univ / Purdue Univ, Fort Wayne – YsGtYg
Chowdhury, Shafiul H., chowdhus@newpaltz.edu,
SUNY, New Paltz – Hw
Chown, Edward H., (418) 545-5011 Universite du Quebec a Chicoutimi – Gx
Chowns, Timothy M., (678) 839-4052 tchowns@westga.edu,
Univ of West Georgia – GsrGg
Christensen, Beth A., 516-877-4174 christensen@adelphi.edu,
Adelphi Univ – PmeGs
Christensen, Douglas, doug@giseis.alaska.edu,
Univ of Alaska, Fairbanks – Ys
Christensen, Nikolas I., (608) 265-4469 chris@geology.wisc.edu,
Univ of Wisconsin, Madison – Yx
Christensen, Philip R., (480) 965-7105 phil.christensen@asu.edu,
Arizona State Univ – Xg
Christensen, Wesley P., 605-677-6149 wes.christensen@usd.edu,
South Dakota Dept of Environment and Natural Resources – Gg
Christeson, Gail L., (512) 471-0463 gail@ig.utexas.edu,
Univ of Texas at Austin – YrsGt
Christian, Alan D., alan.christian@umb.edu,
Univ of Massachusetts, Boston – HsGmCs
Christiansen, Eric H., 801422 2113 eric_christiansen@byu.edu,
Brigham Young Univ – GiXgGv
Christie-Blick, Nicholas, (845) 365-8180 ncb@ldeo.columbia.edu,
Columbia Univ – Gst
Chrzan, Daryl, (510) 643-1624 dcchrzan@berkeley.edu ,
Univ of California, Berkeley – Om
Chu, Pao-Shin, chu@hawaii.edu, Univ of Hawai'i, Manoa – Oa
Chu, Peter C., (831) 656-3688 pcchu@nps.edu,
Naval Postgraduate School – Opr
Chuang, Patrick Y., (831) 459-1501 pchuang@pmc.ucsc.edu,
Univ of California, Santa Cruz – Oa
Church, Matthew, (808) 956-8779 mjchurch@hawaii.edu,
Univ of Hawai'i, Manoa – Ob
Church, Thomas M., (302) 831-2558 tchurch@udel.edu,
Univ of Delaware – Cm
Church, William R., (519) 661-3192 wrchurch@uwo.ca,
Western Univ – Gt
Churchill, Ron C., (916) 327-0745 California Geological Survey – Cg
Churnet, Habte G., (423) 425-4407 habte-churnet@utc.edu,
Univ of Tennessee at Chattanooga – GxgOg
Chyba, Christopher F., (650) 725-6468 cchyba@princeton.edu,
Stanford Univ – Gh
Ciampitti, Ignacio, (785) 532-6940 ciampitti@ksu.edu,
Kansas State Univ – So
Cianfrani, Christina, ccNS@hampshire.edu, Hampshire Coll – Hw
Ciannelli, Lorenzo, 541-737-3142 lciannelli@coas.oregonstate.edu,
Oregon State Univ – Og
Cicerone, Robert D., (508) 531-2713 rcicerone@bridgew.edu,
Bridgewater State Univ – YsgXy
Ciciarelli, John A., (412) 773-3867 jac7@psu.edu,
Pennsylvania State Univ, Monaca – Gg
Cicimurri, Christian M., (864) 650-8456 cmcici@clemson.edu,
Clemson Univ – Pv
Cicimurri, David J., (864) 656-4601 dcheech@clemson.edu,
Clemson Univ – Pv
Ciesielski, Paul F., (352) 392-2231 pciesiel@ufl.edu,
Univ of Florida – Pm
Ciesla, Fred, (773) 702-8169 Univ of Chicago – Xcg
Cifelli, Richard L., (405) 325-4712 rlc@ou.edu,

Univ of Oklahoma – Pv
Cigolini, Corrdao, corrado.cigolini@unito.it, Università di Torino – Gv
Cihacek, Larry J., (701) 231-8572 North Dakota State Univ – Og
Cilliers, Johannes, +44 20 759 47360 j.j.cilliers@imperial.ac.uk, Imperial Coll – Gz
Ciolkosz, Edward J., (814) 865-1530 f8i@psu.edu, Pennsylvania State Univ, Univ Park – Sd
Cioppa, Maria T., 519-253-3000 ext. 2502 mcioppa@uwindsor.ca, Univ of Windsor – Ym
Cipar, John J., 617-552-8300 cipar@bc.edu, Boston Coll – Ys
Cirmo, Christopher P., (607) 753-2924 cirmoc@cortland.edu, SUNY, Cortland – Hg
Cisne, John L., john.cisne@cornell.edu, Cornell Univ – Pg
Civan, Faruk, (405) 325-6778 fcivan@ou.edu, Univ of Oklahoma – Np
Claassen, Mark, (785) 532-6101 mclaasse@ksu.edu, Kansas State Univ – So
Clabo, Darren R., 605-394-1996 Darren.Clabo@sdsmt.edu, South Dakota School of Mines & Tech – OaaOa
Claerbout, Jon F., (650) 723-3717 Stanford Univ – Ye
Claeys, Philippe, (322) 629-3391 phclaeys@vub.ac.be, Vrije Univ Brussel – CgPyXc
Clague, David A., clague@mbari.org, Stanford Univ – Ob
Clague, David A., (808) 967-8819 clague@mbari.org, Univ of Hawai'i, Manoa – Giv
Clague, John J., (604) 291-5387 Simon Fraser Univ – Gl
Claiborne, Lily L., (615) 343-4515 lily.claiborne@vanderbilt.edu, Vanderbilt Univ – GivCg
Claire, Mark, +44 01334 463688 mc229@st-andrews.ac.uk, Univ of St. Andrews – Oa
Clapham, Matthew E., 831-459-1276 mclapham@pmc.ucsc.edu, Univ of California, Santa Cruz – Pe
Clarey, Timothy L., (989) 686-9252 tlclarey@delta.edu, Delta Coll – HwGcPv
Clark, Alan H., (613) 533-6187 Queen's Univ – Em
Clark, David L., (505) 665-0005 Los Alamos National Laboratory – On
Clark, David L., (608) 262-4972 dlc@geology.wisc.edu, Univ of Wisconsin, Madison – Pm
Clark, Donald, (801) 537-3344 donclark@utah.gov, Utah Geological Survey – Gg
Clark, Douglas H., (360) 650-7939 doug.clark@wwu.edu, Western Washington Univ – Gl
Clark, G. Michael, (865) 974-6006 clarkgmorph@utk.edu, Univ of Tennessee, Knoxville – Gm
Clark, George S., (204) 474-7343 gs_clark@umanitoba.ca, Univ of Manitoba – Cc
Clark, H. C., (713) 527-4887 hcclark@owlnet.rice.edu, Rice Univ – Ng
Clark, Ian D., 613 562-5800 Ext 6834 idclark@uottawa.ca, Univ of Ottawa – Hw
Clark, James A., (630) 752-5163 james.clark@wheaton.edu, Wheaton Coll – GmHy
Clark, James S., (919) 660-7402 jimclark@duke.edu, Duke Univ – On
Clark, Jeffrey J., (920) 832-6733 Lawrence Univ – Gm
Clark, Joanna, j.m.clark@reading.ac.uk, Univ of Reading – Sc
Clark, John H., 814-863-1581 clark@ems.psu.edu, Pennsylvania State Univ, Univ Park – Ow
Clark, Jordan, (805) 893-7838 Univ of California, Santa Barbara – Cg
Clark, Joseph C., (724) 357 5622 Indiana Univ of Pennsylvania – Gr
Clark, Kathryne, Kathryne.Clark@dnr.iowa.gov, Iowa Dept of Natural Resources – GgOi
Clark, Ken, (253) 879-3138 kclark@pugetsound.edu, Univ of Puget Sound – GgcGi
Clark, Kenneth F., 915-581-8371 clark@utep.edu, Univ of Texas, El Paso – Eg
Clark, Malcolm W., malcolm.clark@scu.edu.au, Southern Cross Univ – Ca
Clark, Marin, 734-615-0484 marinkc@umich.edu, Univ of Michigan – Gm
Clark, Mark W., (352) 392-1803 (Ext. 316) clarkmw@ufl.edu, Univ of Florida – Sf
Clark, Melissa, (812) 855-4556 Indiana Univ, Bloomington – SfOg
Clark, Michael, (403) 440-8944 mdclark@mtroyal.ca, Mount Royal Univ – Og
Clark, Murlene W., 251-460-6381 mclark@southalabama.edu, Univ of South Alabama – PiGrg
Clark, Peter U., (541) 737-1247 clarkp@geo.oregonstate.edu, Oregon State Univ – Gl
Clark, Richard D., 717-872-3930 Richard.clark@millersville.edu, Millersville Univ – Oa
Clark, Robert A., (520) 621-3842 clark@hwr.arizona.edu, Univ of Arizona – Hs
Clark, Robert O., (928) 523-1321 robert.clark@nau.edu, Northern Arizona Univ – On
Clark, Roger A., +44(0) 113 34 35221 r.a.clark@leeds.ac.uk, Univ of Leeds – YseYs
Clark, Russell G., (517) 629-0312 rgclark@albion.edu, Albion Coll – GiOi
Clark, Scott K., (715) 836-2958 clarksco@uwec.edu, Univ of Wisconsin, Eau Claire – Oe
Clark, Shannon, (505) 277-1644 skclark@unm.edu, Univ of New Mexico – On
Clark, II, George R., (785) 532-2242 grc@ksu.edu, Kansas State Univ – Poy
Clarke, Amanda, 480-965-6590 amanda.clarke@asu.edu, Arizona State Univ – Gv
Clarke, Antony D., (808) 956-6215 tclarke@soest.hawaii.edu, Univ of Hawai'i, Manoa – Oa
Clarke, Beverley, beverley.clarke@flinders.edu.au, Flinders Univ – Oyn
Clarke, Garry K. C., (604) 822-3602 gclarke@eos.ubc.ca, Univ of British Columbia – Yg
Clarke, Geoffrey L., +61293512919 geoffrey.clarke@usyd.edu.au, Univ of Sydney – Gp
Clarke, Julia A., (512) 232-7563 julia_clarke@jsg.utexas.edu, Univ of Texas at Austin – PvoPq
Clarke, Peter J., +44 (0) 191 208 6351 peter.clarke@ncl.ac.uk, Univ of Newcastle Upon Tyne – YdOri
Clarke, Stu, (+44) 01782 733171 s.m.clarke@keele.ac.uk, Keele Univ – Gdo
Clarke, Jr., Roy S., (202) 633-1789 clarker@si.edu, Smithsonian Institution / National Museum of Natural History – Xm
Clary, Renee M., (662) 268-1032 x215 rclary@geosci.msstate.edu, Mississippi State Univ – OePgGe
Class, Connie, (845) 365-8712 Columbia Univ – Cg
Clausen, Benjamin L., (909) 558-4548 bclausen@llu.edu, Loma Linda Univ – GiCtYg
Clay, Patricia L., +44 0161 275-0407 patricia.clay@manchester.ac.uk, Univ of Manchester – CgaGg
Clay, Robert, 573/368-2177 bob.clay@dnr.mo.gov, Missouri Dept of Natural Resources – On
Clayton, Lee, (608) 263-6839 lclayton@wisc.edu, Univ of Wisconsin, Extension – Gl
Clayton, Robert N., (773) 702-7777 Univ of Chicago – Cs
Clayton, Robert W., (626) 395-6909 clay@gps.caltech.edu, California Inst of Tech – Ye
Clayton, Robert W., 208-496-1906 claytonr@byui.edu, Brigham Young Univ - Idaho – GcYgNg
Clayton, Rodney, 757-822-7264 tcclayr@tcc.edu, Tidewater Community Coll – Gg
Cleary, William J., 910-962-2320 clearyw@uncw.edu, Univ of North Carolina Wilmington – Gu
Cleaveland, Malcolm, (479) 575-6790 mcleavel@uark.edu, Univ of Arkansas, Fayetteville – Pe
Clebnik, Sherman M., (860) 465-4323 clebniks@easternct.edu, Eastern Connecticut State Univ – Gl
Clegg, Simon, +44 (0)1603 59 3185 s.clegg@uea.ac.uk, Univ of East Anglia – OacCg
Clemens, Steven C., (401) 863-1964 steven_clemens@brown.edu, Brown Univ – Ou
Clemens, William A., (510) 642-6675 bclemens@berkeley.edu, Univ of California, Berkeley – Pv
Clemens-Knott, Diane, (657) 278-2369 dclemensknott@fullerton.edu, California State Univ, Fullerton – Gig
Clement, William P., (413) 545-5910 wclement@geo.umass.edu, Univ of Massachusetts, Amherst – Yge
Clementz, Mark T., (307) 766-6048 mclement1@uwyo.edu, Univ of Wyoming – Py
Clemitshaw, Kevin C., +44 1784 414026 k.clemitshaw@rhul.ac.uk, Univ of London, Royal Holloway & Bedford New Coll – Oag
Clendenin, Jr., Charles W., (803) 896-7702 ClendeninB@dnr.sc.gov, Dept of Natural Resources – Eg
Clendening, Ronald J., (615) 532-1504 Ron.Clendening@tn.gov, Tennessee Geological Survey – Gg
Clennan, Patrick D., 702-651-7501 patrick.clennan@csn.edu, Coll of Southern Nevada - West Charleston Campus – OyGe
Clepper, Marta, Marta.Clepper@oneonta.edu,

Faculty Index -C

SUNY, Oneonta – OeGgs
Cleveland, Natasha, 301-846-2605 ncleveland@frederick.edu, Frederick Community Coll – Oge
Clift, Peter, 225-578-2153 pclift@lsu.edu, Louisiana State Univ – GsuGo
Clift, Sigrid, 512-471-0320 sigrid.clift@beg.utexas.edu, Univ of Texas at Austin – Og
Cline, Jean S., (702) 575-9968 jean.cline@unlv.edu, Univ of Nevada, Las Vegas – EmCe
Cloos, Mark P., (512) 471-4170 cloos@jsg.utexas.edu, Univ of Texas at Austin – Gcx
Closs, L. Graham, (303) 273-3856 lcloss@mines.edu, Colorado School of Mines – Ce
Clothiaux, Eugene E., 814-865-2915 cloth@meteo.psu.edu, Pennsylvania State Univ, Univ Park – Ow
Clough, Gene A., (207) 786-6396 gclough@bates.edu, Bates Coll – Ym
Cloutier, Danielle, (418) 656-7679 danielle.cloutier@ggl.ulaval.ca, Universite Laval – GeOg
Cloutis, Ed, 204-786-9386 e.cloutis@uwinnipeg.ca, Western Univ – Oi
Clowes, Ronald M., (604) 822-4138 clowes@eos.ubc.ca, Univ of British Columbia – Ys
Clyde, William C., 603-862-3148 will.clyde@unh.edu, Univ of New Hampshire – Pg
Cnudde, Veerle, Veerle.cnudde@ugent.be, Ghent Univ – Om
Coakley, Bernard, 907-474-5385 bernard.coakley@gi.alaska.edu, Univ of Alaska, Fairbanks – Yg
Coale, Kenneth H., (831) 771-4400 coale@mlml.calstate.edu, Moss Landing Marine Laboratories – CtcOc
Cobb, Kim M., (404) 894-3895 kcobb@eas.gatech.edu, Georgia Inst of Tech – PeCsOg
Cobb, Steven R., 806-834-1395 steve.cobb@ttu.edu, Texas Tech Univ – Oa
Coble, Jim, 757-822-7090 tccoblj@tcc.edu, Tidewater Community Coll – Gg
Coble, Paula G., (727) 553-1631 pcoble@marine.usf.edu, Univ of South Florida – Cm
Coburn, Craig, (403) 317-2818 craig.coburn@uleth.ca, Univ of Lethbridge – Oy
Coburn, Daniel, 203-392-5835 coburnd1@southernct.edu, Southern Connecticut State Univ – Oe
Coch, Nicholas K., (718) 997-3326 Graduate School of the City Univ of New York – Gs
Cochran, David, 601-266-6014 david.cochran@usm.edu, Univ of Southern Mississippi – On
Cochran, Elizabeth, cochran@ucr.edu, Univ of California, Riverside – Ys
Cochran, J. K., (631) 632-8746 kcochran@notes.cc.sunysb.edu, SUNY, Stony Brook – Oc
Cochran, J. Kirk, (631) 632-8733 kirk.cochran@stonybrook.edu, SUNY, Stony Brook – Oc
Codrea, Vlad, +40-264-405300 ext 5205 vcodrea@bioge.ubbcluj.ro, Babes-Bolyai Univ – PvGo
Cody, Anita M., amcody@iastate.edu, Iowa State Univ of Science & Tech – Cl
Cody, George D., 202-478-8980 gcody@carnegiescience.edu, Carnegie Institution of Washington – Co
Cody, Robert, rdcody@iastate.edu, Iowa State Univ of Science & Tech – Gz
Coe, Douglas A., (406) 496-4207 dcoe@mtech.edu, Montana Tech of the Univ of Montana – Cg
Coe, Hue, +44 0161 306-9362 hugh.coe@manchester.ac.uk, Univ of Manchester – Oa
Coe, Robert S., (831) 459-2393 rcoe@pmc.ucsc.edu, Univ of California, Santa Cruz – Ym
Coetzee, Serena M., 27 (0)12 420 3823 serena.coetzee@up.ac.za, Univ of Pretoria – Oi
Coffin, Richard, 361-825-2814 Richard.Coffin@tamucc.edu, Texas A&M Univ, Corpus Christi – Ouc
Coffroth, Mary Alice, 716-645-4871 coffroth@acsu.buffalo.edu, SUNY, Buffalo – Gu
Coggan, John, +44 01326 371824 J.Coggan@exeter.ac.uk, Exeter Univ – Nx
Cogger, Craig G., (206) 840-4512 cogger@wsu.edu, Washington State Univ – Ou
Coggon, Judith A., +44 (0)23 80596539 jude.coggon@soton.ac.uk, Univ of Southampton – Cg
Cohen, Alice, 902-585-1126 alice.cohen@acadiau.ca, Acadia Univ – OnnOn
Cohen, Andrew S., (520) 621-4691 cohen@email.arizona.edu, Univ of Arizona – Ps
Cohen, Anne L., (508) 289-2958 acohen@whoi.edu, Woods Hole Oceanographic Institution – Pe
Cohen, Anthony, anthony.cohen@open.ac.uk, The Open Univ – Cs
Cohen, David R., 612 9385 8084 d.cohen@unsw.edu.au, Univ of New South Wales – Cet
Cohen, Joel E., (212) 327-8883 cohen@rockvax.rockefeller.edu, Columbia Univ – On
Cohen, Matt, (864) 294 - 2505 matthew.cohen@furman.edu, Furman Univ – On
Cohen, Phoebe A., 413-597-2358 phoebe.a.cohen@williams.edu, Williams Coll – Poy
Cohen, Ronald E., 202-478-8937 rcohen@carnegiescience.edu, Carnegie Institution of Washington – Gy
Cohen, Saul B., (212) 772-5412 Graduate School of the City Univ of New York – Oy
Cohen, Shaul E., (541) 346-4500 scohen@uoregon.edu, Univ of Oregon – On
Cok, Anthony E., 516 877 4171 cok@adelphi.edu, Adelphi Univ – OuGeu
Coker, Victoria, +44 0161 275-3803 vicky.coker@manchester.ac.uk, Univ of Manchester – Gz
Colak Erol, Serap, 00904242370000-5995 serapcolak@firat.edu.tr, Firat Univ – GgcGt
Colburn, Ivan P., (323) 343-2413 California State Univ, Los Angeles – Gs
Colby, Bonnie C., (520) 621-4775 bcolby@email.arizona.edu, Univ of Arizona – On
Colby, Frank P., (978) 934-3906 frank_colby@uml.edu, Univ of Massachusetts Lowell – Ow
Colby, Jeff, (828) 262-7126 colbyj@appstate.edu, Appalachian State Univ – Oi
Cole, Catherine, +44 01334 464018 csc5@st-andrews.ac.uk, Univ of St. Andrews – Cm
Cole, David R., (865) 574-5473 cole.618@osu.edu, Univ of Tennessee, Knoxville – Ca
Cole, David R., (614) 688-7407 cole.618@osu.edu, Ohio State Univ – Cg
Cole, Dean A., (505) 665-0832 Los Alamos National Laboratory – Co
Cole, Gregory L., (505) 667-1858 gcole@lanl.gov, Los Alamos National Laboratory – Og
Cole, James W., (03) 3642-766 jim.cole@canterbury.ac.nz, Univ of Canterbury – Gv
Cole, Julia E., (520) 626-2341 jecole@email.arizona.edu, Univ of Arizona – PeCsOa
Cole, Kevin C., 616-331-3791 colek@gvsu.edu, Grand Valley State Univ – Gz
Cole, Paul, +44 1752 585985 paul.cole@plymouth.ac.uk, Univ of Plymouth – Gv
Cole, Rex D., (970) 248-1599 rcole@coloradomesa.edu, Colorado Mesa Univ – GsrGo
Cole, Rick, (727) 553-1522 rickcole@rdsea.com, Univ of South Florida – Op
Cole, Ron B., (814) 332-3393 rcole@allegheny.edu, Allegheny Coll – Gt
Colegial Gutierrez, Juan D., (316) 237-2685 colegial@uis.edu.co, Universidad Industrial de Santander – GeNgOr
Coleman, Alvin L., (910) 362-7365 acoleman@cfcc.edu, Cape Fear Community Coll – Gg
Coleman, Drew S., 919-962-0705 dcoleman@email.unc.edu, Univ of North Carolina, Chapel Hill – CcGit
Coleman, Paul J., (310) 825-1776 pcoleman@igpp.ucla.edu, Univ of California, Los Angeles – Xy
Coleman, Robert G., (650) 723-9205 coleman@pangea.stanford.edu, Stanford Univ – Gi
Coleman, Tommy L., (205) 851-5462 tcoleman@aamu.edu, Alabama A&M Univ – So
Coleman, Jr., Paul J., (310) 825-1776 Univ of California, Los Angeles – Gc
Coles, Kenneth S., (724) 357-5626 kcoles@iup.edu, Indiana Univ of Pennsylvania – XgOeYg
Colgan, Mitchell W., (843) 953-7171 colganm@cofc.edu, Coll of Charleston – PeGe
Colgan, Patrick M., (616) 313-3201 colganp@gvsu.edu, Grand Valley State Univ – GmlPe
Colgan, William, (416)736-2100 #77703 colgan@yorku.ca, York Univ – Yg
Colle, Brian, (631) 632-3174 brian.colle@stonybrook.edu, SUNY, Stony Brook – Oa
Collett, Jr., Jeffrey L., collett@atmos.colostate.edu, Colorado State Univ – Oa
Collette, Joseph, 701-858-4142 joseph.collette@minotstateu.edu,

Minot State Univ – PgGsr
Collie, Jeremy S., (401) 874-6859 jcollie@gso.uri.edu, Univ of Rhode Island – Ob
Collier, Jackie, (631) 632-8696 jackie.collier@stonybrook.edu, SUNY, Stony Brook – Ob
Collier, James D., (970) 247-7129 collier_j@fortlewis.edu, Fort Lewis Coll – Cg
Collier, Jenny, +44 20 759 46547 r.coggon@imperial.ac.uk, Imperial Coll – Yr
Collier, Mark, 217-300-1171 mcollier@illinois.edu, Univ of Illinois, Urbana-Champaign – Ge
Collier, Robert W., (541) 737-4367 rcollier@coas.oregonstate.edu, Oregon State Univ – Cm
Collins, Brian, 206-685-1910 bcollins@uw.edu, Univ of Washington – Gm
Collins, Curtis A., (831) 656-3271 collins@nps.edu, Naval Postgraduate School – Op
Collins, Damian, 780-492-3197 damian.collins@ualberta.ca, Univ of Alberta – On
Collins, Donald R., dcollins@tamu.edu, Texas A&M Univ – Oa
Collins, Edward W., (512) 471-6247 eddie.collins@beg.utexas.edu, Univ of Texas at Austin, Jackson School of Geosciences – Ge
Collins, Gareth, +44 20 759 41518 g.collins@imperial.ac.uk, Imperial Coll – Xm
Collins, John A., (508) 289-2733 jcollins@whoi.edu, Woods Hole Oceanographic Institution – Ys
Collins, Laura, laura.collins@mtsu.edu, Middle Tennessee State Univ – Gg
Collins, Laurel S., (305) 348-1732 collinsl@fiu.edu, Florida International Univ – Po
Collins, Lisa, 847-467-1002 lisa.collins@northwestern.edu, Northwestern Univ – On
Collins, Lorence G., lorencec@sysmatrix.net, California State Univ, Northridge – Gzx
Collins, Mark, (412) 624-6615 mookie@pitt.edu, Univ of Pittsburgh – On
Collins, Mary E., (352) 392-1951 mec@ufl.edu, Univ of Florida – Sd
Collins, Shawn, scollins@clarion.edu, Clarion Univ – GsuGc
Collins, William D., (510) 495-2407 wdcollins@berkeley.edu, Univ of California, Berkeley – Oa
Collinson, James W., collinson.1@osu.edu, Ohio State Univ – Ps
Collinson, Margaret, +44 1784 443607 M.Collinson@rhul.ac.uk, Univ of London, Royal Holloway & Bedford New Coll – Po
Collis, Scott, 630-252-0550 scollis@anl.gov, Argonne National Laboratory – Oa
Colliston, Wayne, +27 (0)51 401 2318 colliswp@ufs.ac.za, Univ of the Free State – GcNr
Colman, Albert, (773) 824-1278 Univ of Chicago – PyCs
Colman, Bradley R., Brad.Colman@noaa.gov, Univ of Washington – Oa
Colosimo, Amanda, (585) 292-2421 acolosimo@monroecc.edu, Monroe Community Coll – Gg
Coltorti, Mauro, +390577233814 mauro.coltorti@unisi.it, Univ of Siena – GmYg
Colucci, Stephen J., (607) 255-1752 sjc25@cornell.edu, Cornell Univ – Oa
Colwell, Frederick (Rick), (541) 737-5220 rcolwell@coas.oregonstate.edu, Oregon State Univ – On
Comas, Xavier, 561 297-3256 xcomas@fau.edu, Florida Atlantic Univ – Yg
Comerford, Nicholas B., (850) 875-7100 nbc@ufl.edu, Univ of Florida – Sf
Comina, Cesare, cesare.comina@unito.it, Università di Torino – Yx
Compton, John, (021) 650-2927 john.compton@uct.ac.za, Univ of Cape Town – CmGa
Comrie, Andrew, (520) 621-1585 comrie@email.arizona.edu, Univ of Arizona – Oy
Conder, James A., (618) 453-7352 conder@geo.siu.edu, Southern Illinois Univ Carbondale – Ysr
Condie, Kent C., (575) 835-5531 kcondie@nmt.edu, New Mexico Inst of Mining and Tech – Ct
Condit, Christopher D., (413) 545-0272 ccondit@geo.umass.edu, Univ of Massachusetts, Amherst – Gi
Condreay, Denise, 402-562-1216 dcondreay@cccneb.edu, Central Community Coll – Og
Confer, John, 724-938-4211 confer@calu.edu, California Univ of Pennsylvania – On
Congleton, John D., (678) 839-4066 jconglet@westga.edu, Univ of West Georgia – GeOrPo
Conkle, Jeremy, 361-825-2682 Jeremy.Conkle@tamucc.edu, Texas A&M Univ, Corpus Christi – On
Conley, Catharine A., 202-358-3912 cassie.conley@nasa.gov, New Mexico Inst of Mining and Tech – Py
Conly, Andrew G., (807) 343-8463 andrew.conly@lakeheadu.ca, Lakehead Univ – EgCgGz
Connallon, Christopher B., (410) 554 5545 christopher.connallon@maryland.gov, Maryland Department of Natural Resources – GmOi
Connelly, Jeffrey B., (501) 569-3546 jbconnelly@ualr.edu, Univ of Arkansas at Little Rock – GcNg
Connely, Melissa, 307-268-2017 mconnely@casperColl.edu, Casper Coll – Gs
Conners, John S., (512) 223-3040 jconners@austincc.edu, Austin Community Coll District – Oe
Connolly, Paul, paul.connolly@manchester.ac.uk, Univ of Manchester – Oa
Connor, Charles B., (813) 974-0325 cconnor@cas.usf.edu, Univ of South Florida, Tampa – Gv
Connors, Christopher, (540) 458-8170 connorsc@wlu.edu, Washington & Lee Univ – GoYe
Conover, David O., David.Conover@stonybrook.edu, SUNY, Stony Brook – Ob
Conrad, Susan H., 845-431- 8534 conrad@sunydutchess.edu, Dutchess Community Coll – GsmHs
Conroy, Jessica, 217-244-4855 jconro@illinois.edu, Univ of Illinois, Urbana-Champaign – Cs
Constable, Catherine G., (858) 534-3183 cconstable@ucsd.edu, Univ of California, San Diego – Ym
Constable, Steven C., (858) 534-2409 sconstable@ucsd.edu, Univ of California, San Diego – Yr
Constantin, Marc, (418) 656-3139 marc.constantin@ggl.ulaval.ca, Universite Laval – Gi
Constantine, Jose, constantineja@cf.ac.uk, Cardiff Univ – Hy
Constantine, Jose A., 413-597-3298 jconstantine@williams.edu, Williams Coll – GmOi
Constantopoulos, James T., (575) 562-2651 jim.constantopoulos@enmu.edu, Eastern New Mexico Univ – GezGx
Conte, Maureen, mconte@mbl.edu, Brown Univ – CgPe
Conway, Flaxen D., (541) 737-1339 fconway@coas.oregonstate.edu, Oregon State Univ – Gu
Conway, Howard B., (206) 685-8085 hcon@uw.edu, Univ of Washington – Ol
Conway, Michael F., 520.621.2352 fmconway@email.arizona.edu, Univ of Arizona – OeGvg
Coogan, James, (970) 943-3425 jcoogan@western.edu, Western State Colorado Univ – Gc
Coogan, Laurence, lacoogan@uvic.ca, Univ of Victoria – Gi
Coogon, Rosalind, +44 20 759 46547 r.coggon@imperial.ac.uk, Imperial Coll – Cs
Cook, Ann, (614) 247-6085 cook.1129@osu.edu, Ohio State Univ – Gu
Cook, David R., (630) 252-5840 drcook@anl.gov, Argonne National Laboratory – OwaOs
Cook, Edward R., (845) 365-8618 Columbia Univ – Hw
Cook, Frederick A., (250) 537-8892 fcook@ucalgary.ca, Univ of Calgary – YeGtYs
Cook, Hadrian, +44 020 8417 67756 H.Cook@kingston.ac.uk, Kingston Univ – Ge
Cook, Joseph P., 520-621-2470 joecook@email.arizona.edu, Univ of Arizona – Gm
Cook, Kerry H., 512-232-7931 kc@jsg.utexas.edu, Univ of Texas at Austin – OaPe
Cook, Lewis, 304-384-5327 lcook@concord.edu, Concord Univ – GbPg
Cook, Mea S., 413-597-4541 mea.cook@williams.edu, Williams Coll – OuGe
Cook, Peter, peter.cook@flinders.edu.au, Flinders Univ – Hw
Cook, Robert B., (334) 844-4282 cookrob@auburn.edu, Auburn Univ – Eg
Cook, Steve, (541) 737-0962 cooks@geo.oregonstate.edu, Oregon State Univ – On
Cook, Timothy L., 508-929-8574 timothy.cook@worcester.edu , Worcester State Univ – GnsOg
Cooke, David R., 61 3 6226 7605 d.cooke@utas.edu.au, Univ of Tasmania – Eg

Faculty Index -C

Cooke, Michele L., (413) 547-3142 cooke@geo.umass.edu, Univ of Massachusetts, Amherst – Gc
Cooke, III, William H., (662) 325-1393 whc5@geosci.msstate.edu, Mississippi State Univ – OriOg
Cookus, Pam, 217-244-2486 cookus@isgs.uiuc.edu, Illinois State Geological Survey – On
Coombs, Douglas S., +64 3 479-7505 doug.coombs@otago.ac.nz, Univ of Otago – Gi
Cooney, Michael, (808) 956-7337 mcooney@hawaii.edu, Univ of Hawai'i, Manoa – Og
Cooney, Timothy M., (319) 273-2918 timothy.cooney@uni.edu, Univ of Northern Iowa – Oe
Coonley, Steffanie, (505) 665-2330 scoonley@lanl.gov, Los Alamos National Laboratory – On
Cooper, Alan F., +64 3 479-7515 alan.cooper@stonebow.otago.ac.nz, Univ of Otago – GpiGt
Cooper, Brian J., 936 294-1566 bjcooper@shsu.edu, Sam Houston State Univ – Gz
Cooper, Catherine , 509-335-1501 cmcooper@wsu.edu, Washington State Univ – Yg
Cooper, Clay A., (775) 673-7372 clay.cooper@dri.edu, Univ of Nevada, Reno – HySpHq
Cooper, Jennifer, 203-392-5842 cooperj1@southernct.edu, Southern Connecticut State Univ – Ggi
Cooper, Jonathon L., (507) 222-4401 jlcooper@carleton.edu, Carleton Coll – Gg
Cooper, Kari M., (530) 754-8826 kmcooper@ucdavis.edu, Univ of California, Davis – Gi
Cooper, Mark, (204) 474-8075 mark_cooper@umanitoba.ca, Univ of Manitoba – Gz
Cooper, Matthew, +44 (0)23 80592062 matthew.cooper@noc.soton.ac.uk, Univ of Southampton – Oc
Cooper, Reid F., (401) 863-2160 Reid_Cooper@Brown.edu, Brown Univ – Gy
Cooper, Roger W., (409) 880-8239 roger.cooper@lamar.edu, Lamar Univ – Gi
Cooper, Terence H., (612) 625-7747 tcooper@soils.umn.edu, Univ of Minnesota, Twin Cities – Sd
Coorough, Patricia J., (414) 278-6155 coorough@mpm.edu, Milwaukee Public Museum – PgiPo
Cope, Tim D., (765) 658-6443 tcope@depauw.edu, DePauw Univ – GsOiGr
Copeland, Peter, 713-743-3649 copeland@uh.edu, Univ of Houston – CcGt
Copeman, Louise A., 541-737-6840 Oregon State Univ – On
Copjakova, Renata, +549 49 3073 copjakova@sci.muni.cz, Masaryk Univ – Gz
Copley, Alex, +44 (0) 1223 748937 acc41@cam.ac.uk, Univ of Cambridge – YgGt
Copley, Jonathan, +44 (0)23 8059 6621 jtc@noc.soton.ac.uk, Univ of Southampton – Ob
Copper, Paul, pcopper@laurentian.ca, Laurentian Univ, Sudbury – Pi
Coppinger, Walter, (406) 287-2288 wcopping@trinity.edu, Trinity Univ – Gcg
Corbato, Charles E., 614-242-0138 corbato.1@osu.edu, Ohio State Univ – Ye
Corbella Cordomí, Mercè, ++34935811973 merce.corbella@uab.cat, Universitat Autonoma de Barcelona – GzCl
Corbett, David R., (252) 328-1367 corbettd@ecu.edu, East Carolina Univ – Oc
Corbett, Patrick, +44 (0) 131 451 3171 p.corbett@hw.ac.uk, Heriot-Watt Univ – Np
Corbineau, Lucien, +33(0)3 44068962 lucien.corbineau@lasalle-beauvais.fr, Institut Polytechnique LaSalle Beauvais (ex-IGAL) – EgGgNm
Corcoran, Deborah, 417-836-6889 debcorcoran@missouristate.edu, Missouri State Univ – On
Corcoran, Patricia, 519-661-2111, ext. 86836 pcorcor@uwo.ca, Western Univ – Gd
Corcoran, William, (417) 836-5781 williamcorcoran@missouristate.edu, Missouri State Univ – Oa
Cordell, Ann S., 3523921721 x491 cordell@flmnh.ufl.edu, Univ of Florida – Gx
Cordero, David I., 360-442-2883 dcordero@lowercolumbia.edu, Lower Columbia Coll – Og
Cordonnier, Benoit, 089/2180 4271 cordonnier@min.uni-muenchen.de, Ludwig-Maximilians-Universitaet Muenchen – Gz
Corey, Alison, (801) 537-3122 nrugs.acorey@state.ut.us, Utah Geological Survey – On
Corley, John H., (573) 368-2132 john.corley@dnr.mo.gov, Missouri Dept of Natural Resources – Geg
Corliss, Bruce H., (919) 684-2951 bruce.corliss@duke.edu, Duke Univ – Pm
Cormier, Vernon F., (860) 486-3547 vernon.cormier@uconn.edu, Univ of Connecticut – YsGt
Cornebise, Michael W., (217) 581-7486 mwcornebise@eiu.edu, Eastern Illinois Univ – Og
Cornell, Kristie, 337-482-1455 kcornell@louisiana.edu, Univ of Louisiana at Lafayette – Gg
Cornell, Sean, 717-477-1310 srcornell@ship.edu, Shippensburg Univ – Gg
Cornell, Winton, (918) 631-3248 winton-cornell@utulsa.edu, The Univ of Tulsa – GivCa
Cornett, Jack R., (613) 562-5800 x7177 rcornett@uottawa.ca, Univ of Ottawa – CcaGe
Cornillon, Peter, (401) 874-6283 pcornillon@gso.uri.edu, Univ of Rhode Island – Op
Cornwell, David, +44 (0)1224 273448 d.cornwell@abdn.ac.uk, Univ of Aberdeen – YsGtYe
Cornwell, Jeffery C., (410) 221-8445 cornwell@hpl.umces.edu, Univ of Maryland – Oc
Cornwell, Kevin J., (916) 278-6667 cornwell@csus.edu, California State Univ, Sacramento – Gm
Coron, Cynthia R., (203) 392-5840 coroncl@southernct.edu, Southern Connecticut State Univ – Eg
Corona, Thomas J., (303) 556-8525 Metropolitan State Coll of Denver – Ow
Corral, Isaac, +61 7 47815681 isaac.corralcalleja@jcu.edu.au, James Cook Univ – EgGvCg
Corrigan, Catherine, (202) 633-1855 corriganc@si.edu, Smithsonian Institution / National Museum of Natural History – XmcXg
Corriveau, Louise, (418) 654-2672 lcorrive@nrcan.gc.ca, Universite du Quebec – Gp
Corsetti, Frank A., (213) 740-6123 fcorsett@earth.usc.edu, Univ of Southern California – Gs
Cortes, Joaquin A., 716-645-4291 caco@buffalo.edu, SUNY, Buffalo – Gx
Coruh, Cahit, (386) 447-1289 coruh@vt.edu, Virginia Polytechnic Inst & State Univ – YesGt
Cosby, Jack, (804) 324-7787 bjc4a@virginia.edu, Univ of Virginia – Hs
Cosentino, Pietro, p.cosentino@uea.ac.uk, Univ of East Anglia – Yg
Cosgrove, John, +44 20 759 46466 j.cosgrove@imperial.ac.uk, Imperial Coll – Gc
Costa, Emanuele, emanuele.costa@unito.it, Università di Torino – Gz
Costa, Jr, Ozeas S., (419) 755-4128 costa.47@osu.edu, Ohio State Univ – HsClOu
Costello, Margaret, (562) 985-4809 mcostel2@csulb.edu, California State Univ, Long Beach – On
Costigan, Keeley R., (505) 665-4788 krc@lanl.gov, Los Alamos National Laboratory – Oa
Costin, Gelu, +27 (0)46-603-8316 g.costin@ru.ac.za, Rhodes Univ – Gp
Cote, Denis, (418) 545-5011 dcote@uqac.ca, Universite du Quebec a Chicoutimi – Gi
Côté, Jean, 514 987-3000 #2351 cote.jean@uqam.ca, Universite du Quebec a Montreal – Oa
Cote, Pascale, (418) 654-2601 pacote@nrcan.gc.ca, Universite du Quebec – On
Cothren, Jackson D., (479) 575-6790 jcothre@uark.edu, Univ of Arkansas, Fayetteville – Oi
Cottaar, Sanne , sc845@cam.ac.uk, Univ of Cambridge – Ys
Cotter, James F., (320) 589-6312 cotterjf@mrs.umn.edu, Univ of Minnesota, Morris – Gl
Cottingame, William, (505) 667-8339 wcottingame@lanl.gov, Los Alamos National Laboratory – Oa
Cottle, John, (805) 893-3471 cottle@geol.ucsb.edu, Univ of California, Santa Barbara – Gi
Cottrell, Elizabeth, (202) 633-1859 cottrelle@si.edu, Smithsonian Institution / National Museum of Natural History – Cp
Cottrell, Rory D., rory@earth.rochester.edu, Univ of Rochester – Ym
Couceiro, Fay, +44 023 92 842294 fay.couceiro@port.ac.uk, Univ of Portsmouth – On
Coulibaly, Paulin, (905) 525-9140 (Ext. 23354) couliba@mcmaster.ca,

McMaster Univ – Hg
Coull, Bruce, bccoull@sc.edu, Univ of South Carolina – Ob
Coulson, Alan B., (864) 656-1897 acoulso@clemson.edu, Clemson Univ – PoGgCs
Coulter, Richard L., (631) 252-5833 rcoulter@anl.gov, Argonne National Laboratory – Ow
Couples, Gary, +44 (0) 131 451 3123 g.couples@hw.ac.uk, Heriot-Watt Univ – Np
Cousens, Brian L., 001-613-520-2600 x4436 brian.cousens@carleton.ca, Carleton Univ – GiCgc
Cousineau, Pierre, (418) 545-5011 pcousine@uqac.ca, Universite du Quebec a Chicoutimi – Gd
Couture, Gilles, 514-987-3000 #8905 couture.gilles@uqam.ca, Universite du Quebec a Montreal – Yg
Cova, Thomas J., (801) 581-7930 tom.cova@geog.utah.edu, Univ of Utah – Oi
Coveney, Seamus, +44 0141 330 7750 Seamus.Coveney@glasgow.ac.uk, Univ of Glasgow – Oi
Coveney Jr, Raymond M., (816) 235-2980 coveneyr@umkc.edu, Univ of Missouri-Kansas City – EmGg
Covert, David S., 206-685-7461 dcovert@atmos.washington.edu, Univ of Washington – Oa
Covington, Matthew, 479-575-3876 mcoving@uark.edu, Univ of Arkansas, Fayetteville – HqGmg
Cowan, Clinton A., (507) 222-7021 ccowan@carleton.edu, Carleton Coll – GsPe
Cowan, Darrel S., (206) 543-4033 darrel@uw.edu, Univ of Washington – Gct
Cowan, Ellen A., (828) 262-2260 cowanea@appstate.edu, Appalachian State Univ – Gma
Cowan, Nicolas B., (514) 398-1967 nicolas.cowan@mcgill.ca, McGill Univ – OaXyg
Cowart, James B., (850) 644-5784 jcowart@mailer.fsu.edu, Florida State Univ – Cc
Cowart, Richard, (361) 644-3049 recowart@coastalbend.edu, Coastal Bend Coll – GeEo
Cowen, Richard, richard@blueaokfarm.com, Univ of California, Davis – Po
Cowgill, Eric S., (530) 754-6574 escowgill@ucdavis.edu, Univ of California, Davis – Gtc
Cowie, Gregory L., +44 (0) 131 650 8502 Dr.Greg.Cowie@ed.ac.uk, Edinburgh Univ – Co
Cowles, Timothy J., tjc@coas.oregonstate.edu, Oregon State Univ – Ob
Cowley, Scott W., (303) 273-3638 scowley@mines.edu, Colorado School of Mines – Cog
Cowman, Tim C., 605-677-6151 tim.cowman@usd.edu, South Dakota Dept of Environment and Natural Resources – CgGg
Cox, Christena, 614-688-2355 cox.1@osu.edu, Ohio State Univ – GgoGc
Cox, Helen M., 818-677-3512 helen.m.cox@csun.edu, California State Univ, Northridge – OwgOr
Cox, John, (403) 440-6160 jcox@mtroyal.ca, Mount Royal Univ – EoGs
Cox, Malcolm E., 61 7 3138 1649 m.cox@qut.edu.au, Queensland Univ of Tech – HwGeEg
Cox, Randel T., (901) 678-4361 randycox@memphis.edu, Univ of Memphis – GcmGt
Cox, Ronadh, (413) 597-2297 ronadh.cox@williams.edu, Williams Coll – GsXg
Cox, Shelah, 215 204-8227 scox@temple.edu, Temple Univ – On
Coxon, Catherine, + 353 1 8962235 cecoxon@tcd.ie, Trinity Coll – GeOu
Craddock, John P., (651) 696-6620 craddock@macalester.edu, Macalester Coll – Gc
Craddock, Robert A., 202-633-2473 Smithsonian Institution / National Air & Space Museum – Xg
Crafford, J. P., +27 (0)15 268 2217 krappie.crafford@ul.ac.za, Univ of Limpopo – OnNgOn
Craig, James L., (505) 665-7996 jlcraig@lanl.gov, Los Alamos National Laboratory – Gg
Craig, James R., (540) 231-5222 jrcraig@vt.edu, Virginia Polytechnic Inst & State Univ – Eg
Craig, Mitchell S., (510) 885-3425 mitchell.craig@csueastbay.edu, California State Univ, East Bay – Yg
Craig, Susanne , (902) 494-4381 susanne.craig@dal.ca, Dalhousie Univ – Op
Crain, John R., (209) 730-3812 Coll of the Sequoias – Gg
Cramer, Bradley D., (319) 335-0704 bradley-cramer@uiowa.edu, Univ of Iowa – GrPsGs
Cramer, Chris, (901) 678-2007 ccramer@memphis.edu, Univ of Memphis – Ys
Cramer, Gary, (620) 662-9021 gcramer@ksu.edu, Kansas State Univ – So
Crane, Kathleen, (212) 772-5265 Graduate School of the City Univ of New York – Yh
Crane, Peter R., (773) 922-9410 Univ of Chicago – Po
Cranford, Peter, (902) 426-3277 cranfordp@mar.dfo-mpo.gc.ca, Dalhousie Univ – Ob
Cranganu, Constantin, 7189515000 x2878 cranganu@brooklyn.cuny.edu, Brooklyn Coll (CUNY) – GoYhHw
Craven, John, +44 (0) 131 650 7887 John.Craven@ed.ac.uk, Edinburgh Univ – Yg
Craw, Dave, +64 3 479-7529 dave.craw@otago.ac.nz, Univ of Otago – Eg
Crawford, Anthony J., 61 3 6226 2490 Tony.Crawford@utas.edu.au, Univ of Tasmania – Gi
Crawford, Ian, +44 020 3073 8026 i.crawford@bbk.ac.uk, Birkbeck Coll – On
Crawford, Ian, +44 0161 306-6850 I.Crawford@manchester.ac.uk, Univ of Manchester – Oa
Crawford, James, 757-864-7231 j.h.crawford@larc.nasa.gov, Georgia Inst of Tech – Oa
Crawford, Maria Luisa B., (610) 526-5111 mcrawfor@brynmawr.edu, Bryn Mawr Coll – Gxz
Crawford, Matt, 859.323.0510 mcrawford@uky.edu, Univ of Kentucky – On
Crawford, Nicholas, (270) 745-5889 nicholas.crawford@wku.edu, Western Kentucky Univ – Hg
Crawford, Thomas J., (678) 839-4062 Univ of West Georgia – Eg
Crawford, Vernon J., (541) 552-6479 crawford@sou.edu, Southern Oregon Univ – Gg
Crawford, William A., (610) 526-5112 Bryn Mawr Coll – Cg
Creager, Kenneth C., 206-685-2803 kcc@ess.washington.edu, Univ of Washington – Ys
Crease, James, (302) 645-4240 Univ of Delaware – Op
Creaser, Robert A., (780) 492-2942 robert.creaser@ualberta.ca, Univ of Alberta – Cc
Creasey, Robert L., 831-656-3178 creasey@nps.edu, Naval Postgraduate School – Ow
Creasy, John W., (207) 786-6153 jcreasy@bates.edu, Bates Coll – Gx
Creech-Eakman, Michelle, 575-835-6756 mce@kestrel.nmt.edu, New Mexico Inst of Mining and Tech – XgPy
Crelling, John C., (618) 453-7361 jcrelling@geo.siu.edu, Southern Illinois Univ Carbondale – Ec
Crepeau, Richard J., (828) 262-7052 crepeaurj@appstate.edu, Appalachian State Univ – On
Crespi, Jean M., (860) 486-0601 crespi@geol.uconn.edu, Univ of Connecticut – Gc
Creveling, Jessica, 541-737-2112 crevelij@oregonstate.edu, Oregon State Univ – GgsGr
Crews, Jeff, (573) 368-2356 jeff.crews@dnr.mo.gov, Missouri Dept of Natural Resources – Hy
Cribb, Warner, (615) 898-2379 warner.cribb@mtsu.edu, Middle Tennessee State Univ – GipGz
Crider, Juliet, (206) 543-8715 juliet.crider@ess.washington.edu, Univ of Washington – GcmNr
Crilley, Leigh R., +44 (0)121 414 5523 l.crilley@bham.ac.uk, Univ of Birmingham – Oa
Criminale, Jr., William O., (206) 543-9506 lascala@amath.washington.edu, Univ of Washington – Op
Crimmins, Michael, (520) 626-4244 crimmins@u.arizona.edu, Univ of Arizona – On
Crippen, Robert E., (818) 354-2475 robert.e.crippen@jpl.nasa.gov, Jet Propulsion Laboratory – Or
Crisp, Joy A., (818) 354-9036 Jet Propulsion Laboratory – Gv
Criss, Robert E., (314) 935-7441 criss@levee.wustl.edu, Washington Univ in St. Louis – Cs
Criss, Robert E., criss@wustl.edu, Washington Univ in St. Louis – Cs
Criswell, James, (910) 392-7536 jcriswell@cfcc.edu, Cape Fear Community Coll – Gg
Crockett, Joan, 217-333-6630 jcrocket@illinois.edu, Univ of Illinois, Urbana-Champaign – GoEc
Crockett, Joan E., (217) 244-2388 crockett@illinois.edu, Illinois State Geological Survey – EoOg
Croft, Gary M., gcroft@valleyColl.edu, San Bernardino Valley Coll – Oy
Croft, Paul J., (908) 737-3737 pcroft@kean.edu, Kean Univ – OwaOe
Croll, Don, (831) 459-3610 dcroll@cats.ucsc.edu,

Faculty Index -C

Univ of California, Santa Cruz – Ob
Cromar, Nancy, nancy.cromar@flinders.edu.au, Flinders Univ – On
Cromartie, William J., (609) 652-4413 iaprod193@pollux.stockton.edu, Richard Stockton Coll of New Jersey – On
Cron, Mitch, mitch.cron@gmail.com, Drexel Univ – Gge
Cronin, Meghan F., (206) 526-6449 Meghan.F.Cronin@noaa.gov, Univ of Washington – Op
Cronin, Timothy W., twcronin@mit.edu, Massachusetts Inst of Tech – Oa
Cronin, Vincent S., (254) 710-2174 vince_cronin@baylor.edu, Baylor Univ – GctNg
Cronoble, James M., (303) 556-3070 Metropolitan State Coll of Denver – Gg
Croot, Peter, + 353 (0)91 492 194 peter.croot@nuigalway.ie, National Univ of Ireland Galway – Ct
Crosby, Benjamin T., (208) 282-2949 crosby@isu.edu, Idaho State Univ – GmHsOy
Crossen, Kristine J., (907) 786-6838 kjcrossen@uaa.alaska.edu, Univ of Alaska, Anchorage – Gl
Crossey, Laura J., (505) 277-4204 lcrossey@unm.edu, Univ of New Mexico – Cl
Crossley, David J., (314) 977-3153 crossley@eas.slu.edu, Saint Louis Univ – Yg
Crosson, Robert S., crosson@ess.washington.edu, Univ of Washington – YsGct
Croudace, Ian, +44 (0)23 8059 2780 iwc@noc.soton.ac.uk, Univ of Southampton – Ge
Crouse, David A., (919) 515-7302 North Carolina State Univ – Sc
Crowder, Margaret, (270) 745-5973 margaret.crowder@wku.edu, Western Kentucky Univ – Oe
Crowe, Bruce M., (702) 794-7206 bmc@lanl.gov, Los Alamos National Laboratory – Gv
Crowe, Douglas E., 706-542-2382 crowe@gly.uga.edu, Univ of Georgia – Em
Crowley, Brooke E., brooke.crowley@uc.edu, Univ of Cincinnati – CsPe
Crowley, Peter D., (413) 542-2715 pdcrowley@amherst.edu, Amherst Coll – Gc
Crowley, Quentin G., + 353 1 8962403 crowleyq@tcd.ie, Trinity Coll – Cc
Crowley, Stephen, +44 0151 794 5163 Sfcrow@liverpool.ac.uk, Univ of Liverpool – Gs
Crowther, Sarah, +44 0161 275-0407 Sarah.Crowther@manchester.ac.uk, Univ of Manchester – Oa
Croxen III, Fred W., 928-344-7586 fred.croxen@azwestern.edu, Arizona Western Coll – HwPvOi
Crozier, Carl, (919) 793-4428 North Carolina State Univ – Sb
Crozier, George F., (334) 861-7557 gcrozier@disl.org, Dauphin Island Sea Lab – On
Cruden, Alexander, (416) 978-3021 chair@geology.utoronto.ca, Univ of Toronto – Gt
Cruden, David M., (780) 492-3085 dave.cruden@ualberta.ca, Univ of Alberta – Nr
Cruikshank, Kenneth M., (503) 725-3383 cruikshankk@pdx.edu, Portland State Univ – Gc
Crumbly, Isaac J., (912) 825-6454 Fort Valley State Univ – On
Cruse, Richard M., (515) 294-7850 rmc@iastate.edu, Iowa State Univ of Science & Tech – So
Crutzen, Paul, pcrutzen@ucsd.edu, Univ of California, San Diego – Oc
Cruz-Uribe, Alicia M., (207) 581-4494 alicia.cruzuribe@maine.edu, Univ of Maine – GpCpt
Csaplar, Csilla, (650) 498-6877 csaplar@stanford.edu, Stanford Univ – On
Csatho, Beata M., (716) 645-4325 bcsatho@buffalo.edu, SUNY, Buffalo – OlrYg
Csiki, Shane, (603) 271-2876 shane.csiki@des.nh.gov, New Hampshire Geological Survey – OyGm
Cubillas, Pablo, +44 (0) 191 33 41710 pablo.cubillas@durham.ac.uk, Durham Univ – Cg
Cuevas, Julia, julia.cuevas@ehu.eus, Univ of the Basque Country UPV/EHU – GctGg
Cuffey, Roger J., (814) 865-1293 rcuffey@psu.edu, Pennsylvania State Univ, Univ Park – PiePv
Cuker, Benjamin E., (757) 727-5783 benjamin.cuker@hamptonu.edu, Hampton Univ – Ob
Cullen, Jay, 250-721-6120 jcullen@uvic.ca, Univ of Victoria – Oc
Cullen, John J., (902) 494-6667 john.cullen@dal.ca, Dalhousie Univ – Ob
Cullers, Robert L., (785) 532-2240 rcullers@ksu.edu, Kansas State Univ – Ct
Culshaw, Nicholas, (902) 494-3501 Dalhousie Univ – Gc
Culver, Stephen J., (252) 328-6360 culvers@ecu.edu, East Carolina Univ – Pm
Cummings, Michael L., (503) 725-3395 cummingsm@pdx.edu, Portland State Univ – GgHg
Cummins, R. Hays, (513) 529-1338 cumminrh@MiamiOh.edu, Miami Univ – Ge
Cumpston, Jennifer, cumpston@oakton.edu, Oakton Community Coll – Og
Cundall, Peter A., (612) 626-0369 pacundall@aol.com, Univ of Minnesota, Twin Cities – Nr
Cunningham, Amy J., acunnin2@austincc.edu, Austin Community Coll District – OeGg
Cunningham, Mary A., 845-437-5547 macunningham@vassar.edu, Vassar Coll – OiyOu
Cunningham, Peter, +44 (0) 191 208 7932 peter.cunningham@ncl.ac.uk, Univ of Newcastle Upon Tyne – Hg
Cunningham, William D., 860-465-4321 cunninghamw@easternct.edu, Eastern Connecticut State Univ – GtcGp
Cuomo, Carmela, ccuomo@newhaven.edu, Univ of New Haven – CmmGu
Cupples, Julie, +44 (0) 131 651 4315 Julie.Cupples@ed.ac.uk, Edinburgh Univ – On
Curewitz, Daniel, 315-443-2672 dcurewit@syr.edu, Syracuse Univ – Og-Gct
Curl, Doug C., 859-323-0519 doug@uky.edu, Univ of Kentucky – Gc
Curran, H. Allen, (413) 585-3943 acurran@smith.edu, Smith Coll – Pg
Currano, Ellen, (307) 766-2819 ecurrano@uwyo.edu, Univ of Wyoming – Pbe
Curray, Joseph R., (858) 459-2130 jcurray@ucsd.edu, Univ of California, San Diego – Gu
Currens, James C., 859-323-0526 currens@uky.edu, Univ of Kentucky – Hg
Currey, Donald R., (801) 581-8218 Univ of Utah – Gm
Curri, Neil, 845-437-7708 necurri@vassar.edu, Vassar Coll – Oi
Currie, Brian S., (513) 529-7578 curriebs@miamioh.edu, Miami Univ – GstGc
Currie, Philip J., (403) 823-7707 Univ of Calgary – Pv
Currier, Ryan, (920) 465-2582 currierr@uwgb.edu, Univ of Wisconsin, Green Bay – GizGg
Curry, B. B., 217-244-5787 b-curry@illinois.edu, Univ of Illinois, Urbana-Champaign – Gel
Curry, Gordon, +44 01413305444 gordon.curry@glasgow.ac.uk, Univ of Glasgow – GegPg
Curry, James K., (501) 682-6992 james.curry@arkansas.gov, Arkansas Geological Survey – On
Curry, Janel M., (616) 526-6869 Calvin Coll – Oy
Curry, Joan E., (520) 626-5081 curry@ag.arizona.edu, Univ of Arizona – Sc
Curry, Judith A., (404) 894-3948 judith.curry@eas.gatech.edu, Georgia Inst of Tech – OarOn
Curry, Susan, (352) 294-3147 scurry@ufl.edu, Univ of Florida – Os
Curry, William B., (508) 289-2591 wcurry@whoi.edu, Woods Hole Oceanographic Institution – Pm
Curry Rogers, Kristina A., 651-696-6288 krogers@macalester.edu, Macalester Coll – Pv
Curtice, Joshua M., (508) 289-2618 jcurtice@whoi.edu, Woods Hole Oceanographic Institution – Cc
Curtin, Tara M., (315) 781-3928 curtin@hws.edu, Hobart & William Smith Colls – PeGs
Curtis, Andrew, +44 (0) 131 650 8515 Andrew.Curtis@ed.ac.uk, Edinburgh Univ – Ys
Curtis, Arthur C., (615) 576-7642 Oak Ridge National Laboratory – Ge
Curtis, David B., (505) 667-4845 Los Alamos National Laboratory – Cc
Curtis, Garniss H., (510) 845-0333 Univ of California, Berkeley – Cc
Curtis, Jason H., (352) 392-2296 jcurtis@che.ufl.edu, Univ of Florida – Pe
Curtis, John B., jbcurtis@mines.edu, Colorado School of Mines – GoCo
Curtis, Lynn A., (954) 201-6942 lcurtis@broward.edu, Broward Coll, Central Campus – GgPg
Cusack, Maggie, +44 01413305491 Maggie.Cusack@glasgow.ac.uk, Univ of Glasgow – Gz
Cushman, John, (765) 494-8040 jcushman@purdue.edu, Purdue Univ – Sp
Cutter, Gregory A., (757) 683-4929 gcutter@odu.edu, Old Dominion Univ – OcCtHs
Cvetko Tešoviæ, Blanka, +38514606107 blankacvetko@yahoo.com, Univ of Zagreb – GgPsm
Czaja, Andrew D., andrew.czaja@uc.edu, Univ of Cincinnati – PoCg
Czarnetzki, Alan C., (319) 273-2152 alan.czarnetzki@uni.edu, Univ of Northern Iowa – Ow

Czeck, Dyanna M., 414-229-3948 dyanna@uwm.edu, Univ of Wisconsin, Milwaukee – Gc
Cziczo, Dan, (617) 324-4882 djcziczo@mit.edu, Massachusetts Inst of Tech – Oa

## D

d'Alessio, Matthew, (818) 677-3647 matthew.dalessio@csun.edu, California State Univ, Northridge – OeGt
D'Allura, Jad A., (541) 552-6480 dallura@sou.edu, Southern Oregon Univ – Gc
D'Alpaos, Andrea, +390498279117 andrea.dalpaos@unipd.it, Università degli Studi di Padova – Hs
D'Amico, Sebastiano, +356 2340 3101 sebdamico@gmail.com, Univ of Malta – YsgOg
D'Asaro, Eric A., (206) 545-2982 dasaro@apl.washington.edu, Univ of Washington – Op
d'Atri, Anna, anna.datri@unito.it, Università di Torino – Gr
D'Hondt, Steven L., (401) 792-6808 dhondt@gso.uri.edu, Univ of Rhode Island – Ou
D'Odorico, Paolo, (434) 924-7241 pd6v@virginia.edu, Univ of Virginia – Hg
da Silva, Eduardo F., eafsilva@ua.pt, Universidade de Aveiro – CeGbCl
Daanen, Ronnie, 907-451-5965 ronald.daanen@alaska.gov, Alaska Division of Geological & Geophysical Surveys – Hy
Dade, William B., 603-646-0286 William.B.Dade@Dartmouth.edu, Dartmouth Coll – GsHgEo
Daemen, Jaak, (775) 7844309 daemen@mines.unr.edu, Univ of Nevada, Reno – Nr
Daeschler, Ted, 215 299 1133 ebd29@drexel.edu, Drexel Univ – PvGgPy
Dagdelen, Kadri, (303) 273-3711 kdagdele@mines.edu, Colorado School of Mines – Nm
Dahl, Johannes M., 806-834-6197 johannes.dahl@ttu.edu, Texas Tech Univ – Owa
Dahl, Robyn, 360-650-7207 Robyn.Dahl@wwu.edu, Western Washington Univ – PiOe
Dahlgran, Randy, 530-752-2814 radahlgren@ucdavis.edu, Univ of California, Davis – Sc
Dahlke, Helen E., (530) 302-5358 hdahlke@ucdavis.edu, Univ of California, Davis – Hgs
Dai, Dajun, 404 413 5750 geoddd@langate.gsu.edu, Georgia State Univ – Oi
Daigneault, Real, (418) 545-5011 rdaignea@uqac.ca, Universite du Quebec a Chicoutimi – GcEgOi
Dakin, Susan, (403) 329-2279 susan.dakin@uleth.ca, Univ of Lethbridge – Oe
Dalconi, Maria C., +390498279163 mariachiara.dalconi@unipd.it, Università degli Studi di Padova – Gz
Dale, Chris, +44 (0) 191 33 42342 christopher.dale@durham.ac.uk, Durham Univ – Oa
Dale, Janis, (306) 585-4840 Univ of Regina – Gml
Daley, Gwen M., 803-323-4973 daleyg@winthrop.edu, Winthrop Univ – PqgGs
Dallegge, Todd A., 970-351-1086 Todd.Dallegge@unco.edu, Univ of Northern Colorado – GosGr
Dallimer, Martin, +44(0) 113 34 35279 M.Dallimer@leeds.ac.uk, Univ of Leeds – Ge
Dallmeyer, R. David, (706) 542-7448 dallmeyr@uga.edu, Univ of Georgia – Gt
Dalman, Michael, (979) 830-4206 mdalman@blinn.edu, Blinn Coll – Gg
Dalrymple, Robert W., (613) 533-6186 dalrympl@queensu.ca, Queen's Univ – Gs
Dalton, Colleen, 401-863-5875 Colleen_Dalton@brown.edu, Brown Univ – Ys
Dalton, Richard F., (609) 292-2576 richard.dalton@dep.nj.gov, New Jersey Geological and Water Survey – Gr
Daly, Eve, +353 (0)91 492 183 eve.daly@nuigalway.ie, National Univ of Ireland Galway – Yg
Daly, Julia F., (207) 778-7403 dalyj@maine.edu, Univ of Maine - Farmington – GmlGu
Daly, Kendra L., (727) 553-1041 kdaly@marine.usf.edu, Univ of South Florida – Ob
Daly, Raymond J., daly@ucc.edu, Union County Coll – Pv
Dalziel, Ian W. D., (512) 471-0431 ian@ig.utexas.edu, Univ of Texas at Austin – Gt
Dam, Hans G., (860) 405-9098 hans.dam@uconn.edu, Univ of Connecticut – Ob
Damir, Buckoviæ, +38514606106 buckovic@geol.pmf.hr, Univ of Zagreb – GgPms
Damschen, Ellen, damschen@wustl.edu, Washington Univ in St. Louis – On
Damuth, John, (817) 272-2987 damuth@uta.edu, Univ of Texas, Arlington – GsuYr
Dana, Gayle L., (775) 674-7538 gdana@dri.edu, Univ of Nevada, Reno – Or
Daniel, Christopher G., (570) 577-1133 cdaniel@bucknell.edu, Bucknell Univ – Gp
Daniell, James J., james.daniell@jcu.edu.au, James Cook Univ – GumYr
Daniels, Jeffrey J., 614-292-6704 daniels.9@osu.edu, Ohio State Univ – Yg
Daniels, Michael, (303) 871-7531 j.michael.daniels@du.edu, Univ of Denver – Gm
Daniels, Walter L., (540) 231-7175 wdaniels@vt.edu, Virginia Polytechnic Inst & State Univ – SdGeCg
Daniels, William R., (505) 667-4546 Los Alamos National Laboratory – Ct
Danko, George, (775) 784-4284 danko@mines.unr.edu, Univ of Nevada, Reno – Hq
Dannevik, William P., (314) 977-3115 dannevik@eas.slu.edu, Saint Louis Univ – Oa
Dansereau, Pauline, dansero@ggl.ulaval.ca, Universite Laval – Gs
Daoust, Mario, 417-836-5301 mariodaoust@missouristate.edu, Missouri State Univ – Oa
Daramola, Sunday O., 08060256588 sunday.daramola@gmail.com, Federal Univ of Tech, Akure – HwGe
Darby, Dennis A., (757) 683-4701 ddarby@odu.edu, Old Dominion Univ – Ou
Darby, Jeannie, (916) 752-5670 Univ of California, Davis – Hg
Darbyshire, Fiona Ann, 514-987-3000 #5054 darbyshire.fiona_ann@uqam.ca, Universite du Quebec a Montreal – Ys
Dare, Sarah A., (613) 562-5800 x6859 sdare@uottawa.ca, Univ of Ottawa – GiCtEg
Dark, Shawna J., 818-677-6937 shawna.dark@csun.edu, California State Univ, Northridge – Oi
Darling, James, +44 023 92 842247 james.darling@port.ac.uk, Univ of Portsmouth – Cs
Darling, Robert S., (607) 753-2923 darlingr@cortland.edu, SUNY, Cortland – GpzCp
Darmody, Robert G., (217) 333-9489 Univ of Illinois, Urbana-Champaign – Sd
Darold, Amberlee, 405/325-8611 Amberlee.P.Darold-1@ou.edu, Univ of Oklahoma – Ys
Daroub, Samira H., (561) 993-1500 sdaroub@ufl.edu, Univ of Florida – So
Darrah, Thomas, (614) 688-2132 darrah.24@osu.edu, Ohio State Univ – Ge
Darrell, James H., (912) 478-5361 JDarrell@GeorgiaSouthern.edu, Georgia Southern Univ – Pl
Darroch, Simon, simon.a.darroch@vanderbilt.edu, Vanderbilt Univ – Poi
Das, Sarah B., 508-289-2464 sdas@whoi.edu, Woods Hole Oceanographic Institution – Pm
Dasgupta, Rajdeep, 713.348.2664 Rajdeep.Dasgupta@rice.edu, Rice Univ – GxCg
Dasgupta, Tathagata, 330-672-4104 tdasgupt@kent.edu, Kent State Univ – Cg
Dash, Padmanava, 662-325-3915 pd175@msstate.edu, Mississippi State Univ – OrHg
Dash, Zora V., (505) 667-1923 zvd@lanl.gov, Los Alamos National Laboratory – On
Dasvarma, Gour, gour.dasvarma@flinders.edu.au, Flinders Univ – On
Datta, Saugata, (785) 532-2241 sdatta@ksu.edu, Kansas State Univ – ClHwGe
Dattilo, Benjamin F., (260) 481-6250 dattilob@ipfw.edu, Indiana Univ / Purdue Univ, Fort Wayne – GrPiGs
Daugherty, Carolyn M., carolyn.daugherty@nau.edu, Northern Arizona Univ – On
Daugherty, LeRoy A., (505) 646-3406 New Mexico State Univ, Las Cruces – Sd
Dauphas, Nicolas, (773) 702-2930 Univ of Chicago – Xc
Davatzes, Alexandra, (215) 204-3907 alix@temple.edu, Temple Univ – GsXg
Davatzes, Nicholas, (215) 204-2837 davatzes@temple.edu, Temple Univ – GcNrOn
Davenport, Joan R., (509) 786-9384 jdavenp@tricity.wsu.edu, Washington State Univ – Sb
Davey, Patricia M., 863-2449 patricia_davey@brown.edu, Brown Univ

– On
David, Mark B., (217) 333-4308 dmnicol@illinois.edu, Univ of Illinois, Urbana-Champaign – Sf
Davidson, Bart, 859-323-0524 bdavidson@uky.edu, Univ of Kentucky – Hg
Davidson, Cameron, (507) 222-7144 cdavidso@carleton.edu, Carleton Coll – GptCc
Davidson, Fiona M., (479) 575-3879 fdavidso@comp.uark.edu, Univ of Arkansas, Fayetteville – Ou
Davidson, Garry J., 61 3 6226 2815 Garry.Davidson@utas.edu.au, Univ of Tasmania – Cs
Davidson, Gregg R., (662) 915-5824 davidson@olemiss.edu, Univ of Mississippi – Hw
Davidson, Jon, +44 (0) 191 33 42328 j.p.davidson@durham.ac.uk, Durham Univ – Cm
Davidson, Kenneth L., 831-656-2309 kldavids@nps.edu, Naval Postgraduate School – Ow
Davie, Colin, +44 (0) 191 208 6458 colin.davie@ncl.ac.uk, Univ of Newcastle Upon Tyne – Ngr
Davies, Andrew J., andrew.j.davies@bangor.ac.uk, Bangor Univ – Og
Davies, David, +44 (0) 131 451 3569 d.davies@hw.ac.uk, Heriot-Watt Univ – Ng
Davies, Gareth, +44(0)7780864555 gareth.davies@open.ac.uk, The Open Univ – Oe
Davies, Gareth R., g.r.davies@vu.nl, VU Univ Amsterdam – GiCcGf
Davies, Neil, +44 (0) 1223 333453 nsd27@cam.ac.uk, Univ of Cambridge – Gs
Davies, Rhordri, +44 20 759 45722 rhodri.davies@imperial.ac.uk, Imperial Coll – Yg
Davies, Richard, +44 (0) 191 33 49308 richard.davies@durham.ac.uk, Durham Univ – Eo
Davies, Sarah, sarah.davies@open.ac.uk, The Open Univ – Oe
Davies, Sarah, +440116 252 3624 sjd27@le.ac.uk, Leicester Univ – Gs
Davies-Vollum, Katherine Sian, 206-692-4626 ksdavies@u.washington.edu, Univ of Washington – GsrPe
Davis, Andrew, (773) 702-8164 Univ of Chicago – Xc
Davis, Arden D., (605) 394-2527 Arden.Davis@sdsmt.edu, South Dakota School of Mines & Tech – Hw
Davis, Caroline A., (563) 288-2886 caroline-davis@uiowa.edu, Univ of Iowa – Hw
Davis, Daniel M., 631-632-8217 daniel.davis@sunysb.edu, SUNY, Stony Brook – Yg
Davis, David A., (775) 682-8767 ddavis@unr.edu, Univ of Nevada – Gg
Davis, Donald W., (416) 946-0365 dond@geology.utoronto.ca, Univ of Toronto – Cc
Davis, Edward, 541-346-3461 edavis@uoregon.edu, Univ of Oregon – Pg
Davis, Fred, (202) 633-1837 davisf@si.edu, Smithsonian Institution / National Museum of Natural History – Gv
Davis, Fred A., (218) 726-8331 fdavis@d.umn.edu, Univ of Minnesota, Duluth – CpGi
Davis, George H., (520) 621-1856 gdavis@u.arizona.edu, Univ of Arizona – Gc
Davis, Gregory A., (213) 740-6726 gdavis@usc.edu, Univ of Southern California – Gc
Davis, J. Matthew, (603) 862-1718 matt.davis@unh.edu, Univ of New Hampshire – HwYh
Davis, James A., (217) 581-5528 jadavis2@eiu.edu, Eastern Illinois Univ – Ogw
Davis, James A., (801) 378-3852 james_davis@byu.edu, Brigham Young Univ – On
Davis, John C., john.davis5@mchsi.com, Univ of Kansas – GqoGd
Davis, Kenneth J., (814) 863-8601 davis@met.psu.edu, Pennsylvania State Univ, Univ Park – Ow
Davis, Leslie M., (512) 223-1790 (Ext. 22763) ldavis1@austincc.edu, Austin Community Coll District – Op
Davis, Owen K., (520) 621-7953 odavis@email.arizona.edu, Univ of Arizona – Pl
Davis, P. Thompson, (781) 891-3479 pdavis@bentley.edu, Bentley Univ – Gl
Davis, Paul M., (310) 825-1343 pdavis@epss.ucla.edu, Univ of California, Los Angeles – YsmYg
Davis, Peter B., 253-538-5770 davispb@plu.edu, Pacific Lutheran Univ – GcpGz
Davis, R. Laurence, (203) 932-7108 rldavis@newhaven.edu, Univ of New Haven – GemHg
Davis, Ralph K., (479) 575-4515 ralphd@comp.uark.edu, Univ of Arkansas, Fayetteville – HwGgg
Davis, Robert, (804) 924-0579 red3u@virginia.edu, Univ of Virginia – OawOy
Davis, Steven J., (949) 824-1821 sjdavis@uci.edu, Univ of California, Irvine – GeCsOu
Davis, Tara, +353 21 4903696 t.davis@ucc.ie, Univ Coll Cork – Yg
Davis, Thomas L., (303) 273-3938 tdavis@mines.edu, Colorado School of Mines – Ye
Davis, Trevor J., (801) 587-9019 Univ of Utah – Oi
Davis, Jr., Richard A., (813) 974-2773 rdavis@chuma.cas.usf.edu, Univ of South Florida, Tampa – Gs
Davisson, M L., (925) 423-5993 davisson2@llnl.gov, Lawrence Livermore National Laboratory – Cg
Davydov, Vladimir I., (208) 426-1119 vdavydov@boisestate.edu, Boise State Univ – Pg
Dawers, Nancye H., (504) 865-5198 ndawers@tulane.edu, Tulane Univ – Gc
Dawes, Ralph, 509-682-6754 rdawes@wvc.edu, Wenatchee Valley Coll – GgeOw
Dawsey, Cyrus B., (334) 844-3418 dawsecb@auburn.edu, Auburn Univ – Oy
Dawson, Jamekia, 205-247-3611 jdawson@gsa.state.al.us, Geological Survey of Alabama – Hg
Dawson, James C., (518) 564-4035 james.dawson@plattsburgh.edu, Plattsburgh State Univ (SUNY) – Ge
Dawson, Jane P., (515) 294-6302 jpdawson@iastate.edu, Iowa State Univ of Science & Tech – Gp
Dawson, Mary R., (412) 622-3246 dawsonm@carnegiemnh.org, Carnegie Museum of Natural History – Pv
Day, Brett, +44 (0)1603 59 1413 brett.day@uea.ac.uk, Univ of East Anglia – Ge
Day, Howard W., (530) 752-2882 hwday@ucdavis.edu, Univ of California, Davis – Gp
Day, James, (858) 534-5431 jmdday@ucsd.edu, Univ of California, San Diego – Cg
Day, James E., (309) 438-8678 jeday@ilstu.edu, Illinois State Univ – Pi
Day, Rick L., (814) 863-1615 r4d@psu.edu, Pennsylvania State Univ, Univ Park – Ou
Day, Rosie, +44 (0)121 41 48096 r.j.day@bham.ac.uk, Univ of Birmingham – On
Day, Stephanie S., 701-231-8837 stephanie.day@ndsu.edu, North Dakota State Univ – GmOiu
Day, Steven M., (619) 594-2663 sday@mail.sdsu.edu, San Diego State Univ – Ys
Daymon, Deborah J., (505) 667-9021 deba@lanl.gov, Los Alamos National Laboratory – On
Dayton, Paul K., (858) 534-6740 pdayton@ucsd.edu, Univ of California, San Diego – Ob
De Angelis, Silvio, +44-151-794-5161 S.De-Angelis@liverpool.ac.uk, Univ of Liverpool – Ys
De Batist, Marc, marc.debatist@ugent.be, Ghent Univ – GnuGs
De Beus, Barbara, (973) 655-4448 debeusb@mail.montclair.edu, Montclair State Univ – On
De Boer, Jelle Z., (860) 685-2254 jdeboer@wesleyan.edu, Wesleyan Univ – YmGta
De Campos, Christina, 089/2180 4264 campos@min.uni-muenchen.de, Ludwig-Maximilians-Universitaet Muenchen – Gz
De Carlo, Eric H., (808) 956-6473 edecarlo@soest.hawaii.edu, Univ of Hawai'i, Manoa – Ca
De Carlo, Eric H., (808) 956-5924 edecarlo@soest.hawaii.edu, Univ of Hawai'i, Manoa – Cm
De Deckker, Patrick, +61-2-6125 2070 patrick.dedeckker@anu.edu.au, Australian National Univ – GuPmGn
de Foy, Benjamin, (314) 977-3122 bdefoy@slu.edu, Saint Louis Univ – Oa
De Grave, Johan, johan.degrave@ugent.be, Ghent Univ – Cc
de Hogg, Jan C., +44 (0) 131 650 8525 cdehoog@staffmail.ed.ac.uk, Edinburgh Univ – Ge
de Hoop, Maarten, (765) 496-6439 mdehoop@math.purdue.edu, Purdue Univ – Gq
De Kemp, Eric, 613-947-3738 edekemp@nrcan.gc.ca, Univ of Ottawa – OirOg
De Klerk, Billy, +27 (0)46-622-2312 b.deklerk@ru.ac.za, Rhodes Univ – Pg
De Kock, Walter E., (319) 266-6577 Univ of Northern Iowa – Oe
De Luca, Domenico Antonio, domenico.deluca@unito.it, Università di Torino – Hy
de Menocal, Peter B., (845) 365-8483 peter@ldeo.columbia.edu, Columbia Univ – PeOuCg

De Nault, Kenneth J., (319) 273-2033 kenneth.denault@uni.edu, Univ of Northern Iowa – Gz
De Santo, Eilzabeth, (717) 358-4555 edesanto@fandm.edu, Franklin and Marshall Coll – On
de Silva, Shanika, (541) 737-1212 desilvsh@geo.oregonstate.edu, Oregon State Univ – Gv
de Szoeke, Simon P., 541-737-8391 sdeszoek@coas.oregonstate.edu, Oregon State Univ – Opa
de Uña Alvarez, Elena P., 00 34 988387137 edeuna@uvigo.es, Coruna Univ – Oy
de Vernal, Anne, 514-987-3000 #8599 devernal.anne@uqam.ca, Universite du Quebec a Montreal – Pm
de Wet, Carol B., (717) 291-4388 Carol.deWet@FandM.edu, Franklin and Marshall Coll – Gs
de Wit, Cary, (907) 474-7141 cwdewit@alaska.edu, Univ of Alaska, Fairbanks – Og
Deacon, Leith, 780-248-5761 deacon1@ualberta.ca, Univ of Alberta – Ou
Deakin, Ann K., (716) 673-3303 deakin@fredonia.edu, SUNY Fredonia – OiGgOr
Deakin, Joann, 520-249-9042 deakinj@cochise.edu, Cochise Coll – Ggg
Deal, Richard, (814) 732-1733 rdeal@edinboro.edu, Edinboro Univ of Pennsylvania – Oi
Dealy, Mike, (316) 943-2343 mdealy@kgs.ku.edu, Univ of Kansas – Gg
Dean, Jeffrey S., (520) 621-2320 jdean@ltrr.arizona.edu, Univ of Arizona – On
Dean, John Mark, jmdean36@gmail.com, Univ of South Carolina – Ob
Dean, Robert, 717-245-1109 deanr@dickinson.edu, Dickinson Coll – Gi
Deane, William, (865) 974-2366 wdeane@utk.edu, Univ of Tennessee, Knoxville – Gg
deAngelis, Marie, 718-409-7380 mdeangelis@sunymaritime.edu, SUNY, Maritime Coll – Oc
DeAngelis, Michael T., 569-3542 mtdeangelis@ualr.edu, Univ of Arkansas at Little Rock – GpCp
DeAngelo, Micheal V., (512) 232-3373 michael.deangelo@beg.utexas.edu, Univ of Texas at Austin, Jackson School of Geosciences – Ye
Dearden, Christopher, +44 0161 306-3911 christopher.dearden@manchester.ac.uk, Univ of Manchester – Oa
Deardorff, Nicholas, 724-357-2379 n.deardorff@iup.edu, Indiana Univ of Pennsylvania – Gv
Deaton, Tami, (940) 565-2091 tami.deaton@unt.edu, Univ of North Texas – On
DeBari, Susan M., (360) 650-3588 debari@wwu.edu, Western Washington Univ – GxOe
DeBarros, Nelson, 860-424-3585 nelson.debarros@ct.gov, Dept of Energy and Environmental Protection – On
Deboo, Phili B., (901) 678-4424 pdeboo@memphis.edu, Univ of Memphis – Pg
DeCaria, Alex J., (717) 871-4739 alex.decaria@millersville.edu, Millersville Univ – Oa
DeCelles, Peter G., (520) 621-4910 decelles@email.arizona.edu, Univ of Arizona – Gd
Decker, Dave, 775.673.7353 dave.decker@dri.edu, Univ of Nevada, Reno – Hw
DeConto, Robert, (413) 545-3426 deconto@geo.umass.edu, Univ of Massachusetts, Amherst – Oa
Dee, Seth, (775) 682-7704 sdee@unr.edu, Univ of Nevada – Gg
Deen, Patricia A., 7607441150 ext. 2512 pdeen@palomar.edu, Palomar Coll – Ogg
DeGaetano, Arthur, (607) 255-0385 atd2@cornell.edu, Cornell Univ – Oa
deGroot, Robert, 657-278-8275 rdegroot@fullerton.edu, California State Univ, Fullerton – Gg
Degryse, Patrick, patrick.degryse@ees.kuleuven.be, Katholieke Universiteit Leuven – GaCeGf
Dehler, Carol M., (435) 797-0764 carol.dehler@usu.edu, Utah State Univ – Gs
Dehne, Kevin T., (989) 686-9326 ktdehne@delta.edu, Delta Coll – Oe
Deibert, Jack, (931) 221-6318 deibertj@apsu.edu, Austin Peay State Univ – GsrGo
Deininger, Robert W., (901) 678-2177 Univ of Memphis – Gx
Deisher, Jeffrey , 740 548 7348 jeffrey.deischer@dnr.state.oh.us, Ohio Dept of Natural Resources – Eo
DeJesus, Roman, 714-992-7462 rdejesus@fullcoll.edu, Fullerton Coll – Og
Deka, Jennifer, (216) 987-5827 Jennifer.Deka@tri-c.edu, Cuyahoga Community Coll - Western Campus – Og
Dekens, Petra, (415) 338-6015 dekens@sfsu.edu, San Francisco State Univ – Pe
DeKraker, Dan, (562) 860-2451 x2668 ddekraker@cerritos.edu, Cerritos Coll – OepGg
Delaney, John R., (206) 543-4830 jdelaney@u.washington.edu, Univ of Washington – Yr
Delaney, Margaret L., (831) 459-4736 delaney@cats.ucsc.edu, Univ of California, Santa Cruz – Oc
Delano, Helen L., 717.702.2031 hdelano@pa.gov, Pennsylvania Bureau of Topographic & Geologic Survey – GgmNg
DeLaune, Paul, 940-552-9941x207 pbdelaune@ag.tamu.edu, Texas A&M Univ – Os
Delawder, Sandra, delawdsa@jmu.edu, James Madison Univ – On
Deline, Brad, (678) 839-4061 bdeline@westga.edu, Univ of West Georgia – Pi
Dellapenna, Timothy M., (409) 740-4952 dellapet@tamug.tamu.edu, Texas A&M Univ – Ou
Delparte, Donna M., 208-282-4419 delparte@isu.edu, Idaho State Univ – Oi
DelSole, Timothy, (703) 993-5715 tdelsole@gmu.edu, George Mason Univ – Oa
Delson, Eric, (718) 960-8405 delson@amnh.org, Graduate School of the City Univ of New York – Pv
Delusina, Irina, (530) 752-1861 idelusina@ucdavis.edu, Univ of California, Davis – Pl
DeMaster, David J., (919) 515-7026 david_demaster@ncsu.edu, North Carolina State Univ – Oc
Demchak, Jennifer, 570-662-4613 jdemchak@mansfield.edu, Mansfield Univ – Hs
Demers, Serge, (418) 723-1986 (Ext. 1483) Universite du Quebec a Rimouski – Ob
DeMets, D. Charles, (608) 262-8598 chuck@geology.wisc.edu, Univ of Wisconsin, Madison – Ym
Demicco, Robert V., (607) 777-2604 demicco@binghamton.edu, Binghamton Univ – Gs
Deming, Jody, jdeming@u.washington.edu, Univ of Manitoba – On
Deming, Jody W., (206) 543-0845 jdeming@u.washington.edu, Univ of Washington – Ob
Demopoulos, George, (514) 398-4755 McGill Univ – Nx
Demosthenous, Christie M., (920) 424-7167 demosthe@uwosh.edu, Univ of Wisconsin, Oshkosh – Gz
Dempsey, David P., (415) 338-7716 ddempsey@norte.sfsu.edu, San Francisco State Univ – Oa
Dempster, Tim, +4401413305445 Tim.Dempster@glasgow.ac.uk, Univ of Glasgow – Gz
Demyanov, Vasily, +44 (0) 131 451 8298 v.demyanov@hw.ac.uk, Heriot-Watt Univ – Oi
Den Ouden, Amy, amy.denouden@umb.edu, Univ of Massachusetts, Boston – OnGa
Deng, Aijun, (814) 863-8253 axd157@psu.edu, Pennsylvania State Univ, Univ Park – Ow
Deng, Baolin, (505) 835-5505 New Mexico Inst of Mining and Tech – Cg
Deng, Yi, 404-385-1821 yi.deng@eas.gatech.edu, Georgia Inst of Tech – OanOn
Deng, Youjun, 979-862-8476 yjd@tamu.edu, Texas A&M Univ – GzSc
Dengler, Elizabeth, (612) 626-3379 edengler@umn.edu, Univ of Minnesota – Gl
Dengler, Lorinda, (707) 826-3115 lad1@humboldt.edu, Humboldt State Univ – Yg
Denicourt, Raymond, (813) 974-2236 Univ of South Florida, Tampa – Gz
Denizman, Can, (229) 333-5752 Valdosta State Univ – Hw
Denman, Kenneth L., (250) 363-8230 denmank@ec.gc.ca, Univ of Victoria – Op
Denne, Richard, 817-257-4423 r.denne@tcu.edu, Texas Christian Univ – GoEo
Dennen, Rob, dennenr@si.edu, Smithsonian Institution / National Museum of Natural History – Gv
Dennett, Keith E., (775) 784-4056 kdennett@unr.edu, Univ of Nevada, Reno – Hs
Denning, A. Scott, (970) 491-8359 scott.denning@colostate.edu, Colorado State Univ – Oag
Dennis, Clyde B., (630) 252-5999 cbdennis@anl.gov, Argonne National Laboratory – On
Dennison, Robert, (309) 268-8646 robert.dennison@heartland.edu, Heartland Community Coll – Ogn
Denniston, Rhawn F., (319) 895-4306 RDenniston@cornellColl.edu, Univ of Iowa – Cs
Denny, F. B., 618-985-3394 x240 fdenny@illinois.edu,

Univ of Illinois, Urbana-Champaign – GzEg
Denny, F. Brett, 618-985-3394 denny@isgs.uiuc.edu, Illinois State Geological Survey – Gr
Denolle, Marine, (617) 496-6441 mdenolle@fas.harvard.edu, Harvard Univ – Ys
Denton, Gary R. W., (671) 735-2690 gdenton@uguam.uog.edu, Univ of Guam – Ob
Denton, George H., (207) 581-2193 debbies@maine.edu, Univ of Maine – Gl
Denton-Hedrick, Meredith Y., mdentonh@austincc.edu, Austin Community Coll District – OeYeEo
Deocampo, Daniel M., 404413-5750 deocampo@gsu.edu, Georgia State Univ – Gsn
DePaolo, Donald J., (510) 642-7686 DJDePaolo@lbl.gov, Univ of California, Berkeley – Cc
dePolo, Craig M., (775) 682-8770 eq_dude@sbcglobal.net, Univ of Nevada – Ng
dePolo, Diane, (775) 784-4976 diane@seismo.unr.edu, Univ of Nevada, Reno – Ys
DePriest, Thomas A., (731) 881-7441 tdepriest@utm.edu, Univ of Tennessee, Martin – Oe
Derakhshani, Reza, derakhshani@uk.ac.ir, Shahid Bahonar Univ of Kerman – GtmGc
Dere, Ashlee L., 402 554 3317 adere@unomaha.edu, Univ of Nebraska at Omaha – SdGmCl
Derome, Jacques F., (514) 398-6079 jacques.derome@mcgill.ca, McGill Univ – Oa
Derry, Louis A., lad9@cornell.edu, Cornell Univ – Cg
Derstler, Kraig L., (504) 280-6799 kderstle@uno.edu, Univ of New Orleans – PoiPv
Desai, Ankur R., (608) 265-9201 desai@aos.wisc.edu, Univ of Wisconsin, Madison – Oaw
DeSantis, Larisa R., 615-343-7831 larisa.desantis@vanderbilt.edu, Vanderbilt Univ – Pve
Desch, Steven, (480) 965-7742 steve.desch@asu.edu, Arizona State Univ – On
Deschaine, Sylvia, (207) 786-6490 sdescha2@bates.edu, Bates Coll – On
DeShon, Heather, 901-678-2007 hdeshon@memphis.edu, Univ of Memphis – Ysg
DeShon, Heather R., 214-768-2916 Southern Methodist Univ – Ysr
Deskins, Aaron M., (406) 243-5853 aaron.deskins@umontana.edu, Univ of Montana – Oi
Desrochers, Andre, 613-562-5838 adesro@uottawa.ca, Univ of Ottawa – Gs
Dessler, Alexander, (520) 621-4589 dessler@vega.lpl.arizona.edu, Univ of Arizona – On
Dessler, Andrew, 979-862-1427 adessler@tamu.edu, Texas A&M Univ – On
Dessureault, Sean, (520) 621-2359 sdessure@email.arizona.edu, Univ of Arizona – Nx
DeSutter, Tom, 701-231-8690 Thomas.Desutter@ndsu.edu, North Dakota State Univ – Os
Dethier, David P., (413) 597-2078 david.p.dethier@williams.edu, Williams Coll – Gm
Detournay, Emmanuel M., (612) 625-5522 detou001@tc.umn.edu, Univ of Minnesota, Twin Cities – NrpNm
Detrich, William, (617) 373-4495 iceman@neu.edu, Northeastern Univ – On
Dettman, David, (520) 621-4618 dettman@email.arizona.edu, Univ of Arizona – Cs
Detwiler, Andrew G., (605) 394-1995 Andrew.Detwiler@sdsmt.edu, South Dakota School of Mines & Tech – Oa
Deuser, Werner G., (508) 289-2551 wdeuser@whoi.edu, Woods Hole Oceanographic Institution – CsGsu
Devaney, Kathleen, 915-831-5161 kdevaney@epcc.edu, El Paso Community Coll – Gg
Devegowda, Deepak, deepak.devegowda@ou.edu, Univ of Oklahoma – Np
Dever, Edward P., 541-737-2749 edever@coas.oregonstate.edu, Oregon State Univ – Op
Devera, Joseph A., 618-985-3394 devera@isgs.uiuc.edu, Illinois State Geological Survey – Gr
Devera, Joseph A., 618-985-3394 x241 j-devera@illinois.edu, Univ of Illinois, Urbana-Champaign – Pg
Devery, Dora, (281) 756-5670 ddevery@alvinColl.edu, Alvin Community Coll – Gg
Devlahovich, Vincent A., (661) 362-3658 Vincent.Devlahovich@canyons.edu, Coll of the Canyons – GgOye
Devlin, J. R., 785-864-4994 jfrickdevlin@gmail.com, Univ of Kansas – Hw
Devol, Allan H., (206) 543-1292 devol@u.washington.edu, Univ of Washington – Oc
DeVries-Zimmerman, Suzanne, (616) 395-7297 zimmerman@hope.edu, Hope Coll – Gge
Dewaele, Stijn, stijndg.dewaele@ugent.be, Ghent Univ – Cg
Dewdney, Chris, +44 023 92 842417 chris.dewdney@port.ac.uk, Univ of Portsmouth – Yg
DeWeese, Georgina G., (678) 839-4065 gdeweese@westga.edu, Univ of West Georgia – Oy
DeWekker, Stephan F., (434) 924-3324 sfd3d@virginia.edu, Univ of Virginia – Oa
Dewers, Thomas, 505-845-0631 tdewers@sandia.gov, New Mexico Inst of Mining and Tech – Yx
deWet, Andrew P., (717) 291-3815 andy.dewet@fandm.edu, Franklin and Marshall Coll – Ge
Dewey, Janet, (307) 223-2265 jdewey2@uwyo.edu, Univ of Wyoming – Ca
Dewey, John, jfdewey@ucdavis.edu, Univ of California, Davis – Gt
Dewey, Richard K., (250) 472-4009 rdewey@uvic.ca, Univ of Victoria – Opg
DeWitt, Calvin B., (608) 263-7771 Univ of Wisconsin, Madison – Ob
Dexter, Leland R., lee.dexter@nau.edu, Northern Arizona Univ – Oy
Dhar, Ratan K., 718-262-2889 rdhar@york.cuny.edu, York Coll (CUNY) – Hw
Di Fiori, Sara, (310) 660-3593 sdifiori@elcamino.edu, El Camino Coll – GgOg
Di Girolamo, Larry, (217) 333-3080 gdi@illinois.edu, Univ of Illinois, Urbana-Champaign – Or
Di Gregorio, Monica, +44(0) 113 34 31592 m.digregorio@leeds.ac.uk, Univ of Leeds – Eg
Di Lorenzo, Emanuele, (404) 894-3994 edl@gatech.edu, Georgia Inst of Tech – OpnOn
Di Toro, Giulio, +390498279105 giulio.ditoro@unipd.it, Università degli Studi di Padova – Gc
DiBacco, Claudio, (902) 426-9778 Claudio.DiBacco@dfo-mpo.gc.ca, Dalhousie Univ – Ob
Dibb, Jack E., 603-862-3063 jack.dibb@unh.edu, Univ of New Hampshire – Oa
Dibben, Chris, +44 (0) 131 650 2552 chris.dibben@ed.ac.uk, Edinburgh Univ – On
DiBiase, Roman , 814-865-7388 rdibiase@psu.edu, Pennsylvania State Univ, Univ Park – Gmt
Dick, Henry J B., (508) 289-2590 hdick@whoi.edu, Woods Hole Oceanographic Institution – GitGc
Dick, Richard P., 614-247-7605 dick.78@osu.edu, Ohio State Univ – Sb
Dick, Warren A., (330) 263-3877 dick.5@osu.edu, Ohio State Univ – Sb
Dickens, Gerald R., 713.348.5130 jerry@rice.edu, Rice Univ – Yr
Dickerson, Richard E., (310) 825-5864 Univ of California, Los Angeles – On
Dickhoff, Willem H., (314) 935-4169 wimd@howdy.wustl.edu, Washington Univ in St. Louis – On
Dickin, Alan P., (905) 525-9140 (Ext. 24365) dickin@mcmaster.ca, McMaster Univ – Cc
Dickinson, Robert E., 512-232-7933 robted@jsg.utexas.edu, Univ of Texas at Austin – Oa
Dickinson, William R., (520) 299-5220 wrdickin@dakotacom.net, Univ of Arizona – Gt
Dickman, Steven R., (607) 777-2857 dickman@binghamton.edu, Binghamton Univ – Yg
Dickson, Andrew G., (858) 822-2990 adickson@ucsd.edu, Univ of California, San Diego – Cm
Dickson, Loretta D., (570) 484-2068 ldickson@lhup.edu, Lock Haven Univ – GicGe
Dickson, Stephen M., (207) 287-7174 stephen.m.dickson@maine.gov, Dept of Agriculture, Conservation, and Forestry – OnGuOu
Diebold, Frank E., (406) 496-4215 Montana Tech of the Univ of Montana – Cl
Diecchio, Richard J., (703) 993-5394 rdiecchi@gmu.edu, George Mason Univ – Gr
Dieckmann, Melissa S., (859) 622-1274 melissa.dieckmann@eku.edu, Eastern Kentucky Univ – OeCg
Diederichs, Mark, 613-533-6504 diederim@queensu.ca, Queen's Univ – Nr

Diefendorf, Aaron, aaron.diefendorf@uc.edu, Univ of Cincinnati – Co
Diem, Jeremy E., (404) 413-5770 jdiem@gsu.edu, Georgia State Univ – Oa
Diemer, John A., 704-687-5994 jadiemer@email.uncc.edu, Univ of North Carolina, Charlotte – Gs
Diener, Johann, 021-650-2925 johann.diener@uct.ac.za, Univ of Cape Town – Gp
Dierssen, Heidi, 860-405-9239 heidi.dierssen@uconn.edu, Univ of Connecticut – Op
Dieterich, James H., (951) 827-2976 james.dieterich@ucr.edu, Univ of California, Riverside – Yg
Dieterle, Dwight A., (727) 553-1114 ddieterle@marine.usf.edu, Univ of South Florida – Ob
Dietl, Gregory, 607-273-6623 x17 gpd3@cornell.edu, Paleontological Research Institution – Po
Dietl, Gregory P., (607) 273-6623 gpd3@cornell.edu, Cornell Univ – Peo
Dietrich, William E., (510) 642-2633 bill@geomorph.berkeley.edu, Univ of California, Berkeley – Gm
Dietsch, Craig, craig.dietsch@uc.edu, Univ of Cincinnati – Gp
Dietz, Anne D., adietz@alamo.edu, Alamo Colls - San Antonio Coll – OggGg
Dietz, Richard D., (970) 351-2950 richard.dietz@unco.edu, Univ of Northern Colorado – Xg
Dietze, Michael, dietze@bu.edu, Boston Univ – SfOr
Diffendal, Jr., Robert F., (402) 472-7546 rfd@unl.edu, Unversity of Nebraska - Lincoln – GrPiGh
Diggins, Thomas P., 330-941-3605 tpdiggins@ysu.edu, Youngstown State Univ – Py
Dighe, Kalpak, (505) 665-1701 kdighe@lanl.gov, Los Alamos National Laboratory – On
Dijkstra, Arjan, +44 1752 584774 arjan.dijkstra@plymouth.ac.uk, Univ of Plymouth – Gi
Dilcher, David L., (352) 392-1721 dilcher@flmnh.ufl.edu, Univ of Florida – Pb
Dilek, Yildirim, +1 513 529 2212 dileky@miamioh.edu, Miami Univ – GtcGi
Dilles, John H., (541) 737-1245 dillesj@geo.oregonstate.edu, Oregon State Univ – Em
DiMarco, Steven F., (979) 862-4168 sdimarco@tamu.edu, Texas A&M Univ – OpnOg
DiMichele, William A., (202) 633-1319 Smithsonian Institution / National Museum of Natural History – Pb
Dimmick, Charles W., (860) 832-2936 dimmick@ccsu.edu, Central Connecticut State Univ – Ge
Dimova, Natasha T., 205-348-0256 ntdimova@ua.edu, Univ of Alabama – HwCcm
DiNaso, Steven, 217-581-6248 sdinaso@eiu.edu, Eastern Illinois Univ – Oi
Ding, Kang, 612-626-1860 mlcd@umn.edu, Univ of Minnesota, Twin Cities – Cg
Dingman, S. Lawrence, (603) 862-1718 Univ of New Hampshire – Hq
Dingus, Delmar D., (805) 756-2753 ddingus@calpoly.edu, California Polytechnic State Univ – So
Dingwell, Donald Bruce, 089/21804136 dingwell@min.uni-muenchen.de, Ludwig-Maximilians-Universitaet Muenchen – Giv
Dinh, Tra, 609-258-4183 tdinh@princeton.edu, Princeton Univ – Oa
Dinklage, William, 805-730-4114 wsdinklage@sbcc.edu, Santa Barbara City Coll – Gge
Dinsmoor, John C., (702) 295-6189 john_dinsmoor@lanl.gov, Los Alamos National Laboratory – Nm
Dintaman, Christopher, (812) 855-7428 cdintama@indiana.edu, Indiana Univ – Hw
Dinter, David A., (801) 581-7937 david.dinter@utah.edu, Univ of Utah – GtcGu
Diochon, Amanda, (807) 343-8444 adiochon@lakeheadu.ca, Lakehead Univ – Ge
DiPietro, Joseph A., (812) 465-7041 dipietro@usi.edu, Univ of Southern Indiana – Gc
Dipple, Gregory M., (604) 827-0653 gdipple@eoas.ubc.ca, Univ of British Columbia – Gp
Dirmeyer, Paul A., (703) 993-5363 pdirmeye@gmu.edu, George Mason Univ – Oa
Ditmars, John, (630) 252-5953 Argonne National Laboratory – On
Dittmer, Eric, (541) 552-6496 dittmer@sou.edu, Southern Oregon Univ – Gg
DiVenere, Victor, (516) 299-2318 divenere@liu.edu, Long Island Univ, C.W. Post Campus – Ym
Diver, Kim, 860-685-2610 kdiver@wesleyan.edu, Wesleyan Univ – OiyOg
Divine, Aaron, (928) 523-7835 Aaron.Divine@nau.edu, Northern Arizona Univ – OnnOn
Dix, George R., (613) 520-2600 george.dix@carleton.ca, Carleton Univ – Gsd
Dix, Justin, +44 (0)23 8059 3057 J.K.Dix@soton.ac.uk, Univ of Southampton – GaYs
Dixon, Clifton, 601-266-4731 c.dixon@usm.edu, Univ of Southern Mississippi – On
Dixon, Jean, (406 ) 994-3342 jean.dixon@montana.edu, Montana State Univ – GmScOg
Dixon, John C., (479) 575-5808 jcdixon@uark.edu, Univ of Arkansas, Fayetteville – GmSdGg
Dixon, John M., 613-533-6172 john.dixon@queensu.ca, Queen's Univ – Gc
Dixon, Michael, (519) 824-4120 Ext.52555 dixon@ces.uoguelph.ca, Univ of Guelph – On
Dixon, Nick, +44(0) 113 34 34931 fuensd@leeds.ac.uk, Univ of Leeds – Ou
Dixon, P. Grady, (785) 628-4536 pgdixon@fhsu.edu, Fort Hays State Univ – OwyOa
Dixon, Tim H., (813) 974-0152 thd@usf.edu, Univ of Miami – OrGte
Dmochowski, Jane, 215-5735388 janeed@sas.upenn.edu, Univ of Pennsylvania – Yg
Doar, III, William R., (803) 609-7065 doarw@dnr.sc.gov, Dept of Natural Resources – GrdGp
Dobbie, Steven, +44(0) 113 34 36725 j.s.e.dobbie@leeds.ac.uk, Univ of Leeds – Oa
Dobbs, Fred C., (757) 683-4301 fdobbs@odu.edu, Old Dominion Univ – Ob
Dobler, Scott, (270) 745-7078 scott.dobler@wku.edu, Western Kentucky Univ – Oa
Dobrzhinetskaya, Larissa F., (951) 827-2028 larissa@ucrac1.ucr.edu, Univ of California, Riverside – Gx
Dobson, David, +44 020 7679 32398 d.dobson@ucl.ac.uk, Univ Coll London – Gz
Dobson, David M., (336) 316-2278 ddobson@guilford.edu, Guilford Coll – GusPe
Dockal, James A., (563) 845-1034 dockal@uncw.edu, Univ of North Carolina Wilmington – GdcEm
Dodd, J. Robert, (516) 632-8204 dodd@indiana.edu, Indiana Univ, Bloomington – Po
Dodd, Justin P., 815-753-7949 jdodd@niu.edu, Northern Illinois Univ – Cs
Dodge, Clifford H., 717.702.2036 cdodge@pa.gov, Pennsylvania Bureau of Topographic & Geologic Survey – GrEcGh
Dodge, Rebecca L., (940) 397-4475 rebecca.dodge@mwsu.edu, Midwestern State Univ – Ore
Dodson, Peter, (215) 898-8784 dodsonp@vet.upenn.edu, Univ of Pennsylvania – Pv
Dodson, Stanley I., (608) 262-6395 Univ of Wisconsin, Madison – Ob
Dogwiler, Toby, 507-457-5267 tdogwiler@winona.edu, Winona State Univ – GmHqOw
Doh, Seong-Jae, 82-2-3290-3173 sjdoh@korea.ac.kr, Korea Univ – YmgYe
Doheny, Edward L., (215) 898-6085 Univ of Pennsylvania – Ng
Doherty, Cheryl, (302) 831-2569 sparrow@udel.edu, Univ of Delaware – On
Doherty, Ruth, +44 (0) 131 650 6759 ruth.doherty@ed.ac.uk, Edinburgh Univ – Oa
Dolakova, Nela, +420 549 49 3542 nela@sci.muni.cz, Masaryk Univ – Plg
Dolan, James F., (213) 740-8599 dolan@usc.edu, Univ of Southern California – Gt
Dolan, Robert, rd5q@virginia.edu, Univ of Virginia – Gm
Dolejs, David, +420-221951525 Charles Univ – GiCpg
Dolgoff, Anatole, (212) 346-1502 Pace Univ, New York Campus – Og
Doll, William E., d8e@ornl.gov, Oak Ridge National Laboratory – Yg
Dollase, Wayne A., (310) 825-3823 dollase@ucla.edu, Univ of California, Los Angeles – Gz
Dollhopf, Douglas J., (406) 944-5594 dollhopf@montana.edu, Montana State Univ – On
Dolliver, Holly A., holly.dolliver@uwrf.edu, Univ of Wisconsin, River Falls – GmSd
Dolman, Han J., +31 20 5987358 han.dolman@vu.nl, VU Univ Amsterdam – HgOw
Dolman, Paul, +44 (0)1603 59 3175 p.dolman@uea.ac.uk, Univ of East Anglia – Ou

Domack, Cynthia R., (315) 859-4710 cdomack@hamilton.edu, Hamilton Coll – Pg
Domagall, Abigail M., (605) 642-6506 abigail.domagall@bhsu.edu, Black Hills State Univ – GveOe
Dombard, Andrew J., (312) 996-9206 adombard@uic.edu, Univ of Illinois at Chicago – Xy
Domber, Steven E., (609) 984-6587 steven.domber@dep.nj.gov, New Jersey Geological and Water Survey – Hw
Domingue, Carla, (225) 388-8407 carla@lgs.bri.lsu.edu, Louisiana State Univ – On
Dominguez, Francina, 217-265-5483 francina@illinois.edu, Univ of Illinois, Urbana-Champaign – Oa
Dominic, David F., 937 775-3455 david.dominic@wright.edu, Wright State Univ – Gsr
Donaghay, Percy, (401) 874-6944 donaghay@gso.uri.edu, Univ of Rhode Island – Ob
Donahoe, Robert J., (505) 667-7603 Los Alamos National Laboratory – On
Donahoe, Rona J., 205-348-1879 rdonahoe@geo.ua.edu, Univ of Alabama – Cl
Donald, Roberta, (778) 782-4925 robbie_donald@sfu.ca, Simon Fraser Univ – GsOe
Donaldson, Alan C., (304) 293-5603 adonalds@wvu.edu, West Virginia Univ – Gs
Donaldson, Paul R., (208) 426-3639 pdonalds@boisestate.edu, Boise State Univ – Ye
Dondofema, Farai, +27 15 962 80044 farai.dondofema@univen.ac.za, Univ of Venda for Science & Tech – OrHgOi
Doney, Scott C., 508-289-3776 sdoney@whoi.edu, Woods Hole Oceanographic Institution – Cm
Dong, Charles, (310) 660-3593 El Camino Coll – Og
Dong, Hailiang, (513) 529-2517 dongh@MiamiOh.edu, Miami Univ – Cg
Dong, Pinliang, (940) 565-2377 pinliang.dong@unt.edu, Univ of North Texas – Oir
Dong, Yunpeng, ydong265@uwo.ca, Western Univ – GtcCg
Donnelly, Jeffrey, 508-289-2994 jdonnelly@whoi.edu, Woods Hole Oceanographic Institution – Gu
Donnelly, Shanon P., 330 972-7630 sd51@uakron.edu, Univ of Akron – Oi
Donoghue, Joseph F., (407) 823-0631 joseph.donoghue@ucf.edu, Florida State Univ – GueGs
Donoghue, Michael J., 203-432-1935 michael.donoghue@yale.edu, Yale Univ – Pbo
Donohue, Kathleen, (401) 874-6615 kdonohue@gso.uri.edu, Univ of Rhode Island – Op
Donohue, Mary M., 765-658-4654 marydonohue@depauw.edu, DePauw Univ – On
Donohue, Stephen J., (540) 231-9740 donohue@vt.edu, Virginia Polytechnic Inst & State Univ – Sc
Donovan, James, (801) 585-3029 james.donovan@utah.edu, Univ of Utah – NmrOr
Donovan, Jeff C., (727) 553-1116 jdonovan@marine.usf.edu, Univ of South Florida – Op
Donovan, Joseph J., (304) 293-5603 donovan@geo.wvu.edu, West Virginia Univ – Hq
Donovan, R. Nowell, (817) 257-7214 r.donovan@tcu.edu, Texas Christian Univ – GdcGt
Doolan, Barry L., (802) 656-0248 bdoolan@uvm.edu, Univ of Vermont – Gx
Dooley, Alton C., 276-634-4173 alton.dooley@vmnh.virginia.gov, Virginia Polytechnic Inst & State Univ –
Dooley, Brett, bdooley@ph.vccs.edu, Patrick Henry Community Coll – Gg
Dooley, John H., (609) 292-2576 john.dooley@dep.state.nj.us, New Jersey Geological and Water Survey – CgGfe
Dooley, Tim, 512-471-8261 tim.dooley@beg.utexas.edu, Univ of Texas at Austin, Jackson School of Geosciences – Gc
Dorais, Michael J., (801) 422-1347 dorais@byu.edu, Brigham Young Univ – Gi
Dorale, Jeffrey A., (319) 335-0822 jeffrey-dorale@uiowa.edu, Univ of Iowa – Cs
Doran, Peter, pdoran@lsu.edu, Louisiana State Univ – Hgs
Doran, Peter T., (225) 578-3955 pdoran@lsu.edu, Univ of Illinois at Chicago – HswOw
Dorfman, Alexander, 089/2180 4275 Ludwig-Maximilians-Universitaet Muenchen – Gz
Dorfner, Thomas, 089/2180 4278 dorfner@min.uni-muenchen.de, Ludwig-Maximilians-Universitaet Muenchen – Gz
Dorling, Stephen, +44 (0)1603 59 2533 s.dorling@uea.ac.uk, Univ of East Anglia – Ow
Dormaar, John, (403) 327-4561 Univ of Lethbridge – So
Dorman, Clive E., (619) 594-5707 cdorman@mail.sdsu.edu, San Diego State Univ – Op
Dorman, James, (901) 678-4753 dorman@comcast.net, Univ of Memphis – Ys
Dorman, Leroy M., (858) 534-2406 ldorman@ucsd.edu, Univ of California, San Diego – Yg
Dornbos, Stephen Q., (414) 229-6630 sdornbos@uwm.edu, Univ of Wisconsin, Milwaukee – PiePy
Dorries, Alison M., (505) 665-6952 adorries@lanl.gov, Los Alamos National Laboratory – On
Dorsey, Rebecca J., (541) 346-4431 rdorsey@uoregon.edu, Univ of Oregon – Gr
Dort, Jr., Wakefield, (785) 864-4974 Univ of Kansas – Gm
Doser, Diane I., 915-747-5851 doser@utep.edu, Univ of Texas, El Paso – Ys
Doskey, Paul V., (630) 252-7662 pvdoskey@anl.gov, Argonne National Laboratory – Oa
Doss, Paul K., (812) 465-7132 pdoss@usi.edu, Univ of Southern Indiana – HwGeOu
Dosso, Stanley E., (250) 472-4341 sdosso@uvic.ca, Univ of Victoria – Op
Dostal, Jarda, (902) 420-5747 Dalhousie Univ – Cg
Doster, Florian, +44 (0)131 451 3171 f.doster@hw.ac.uk, Heriot-Watt Univ – Hg
Dostie, Philip, (207) 786-6485 pdostie@bates.edu, Bates Coll – Cl
Dott, Jr., Robert H., (608) 262-1856 rdott@geology.wisc.edu, Univ of Wisconsin, Madison – Gs
Doty, Thomas, (401) 254-3066 Roger Williams Univ – Ob
Dou, Fugen, (409) 752-2741 ext. 2223 f-dou@aesrg.tamu.edu, Texas A&M Univ – Os
Dougan, Bernie, (360) 383-3877 bdougan@whatcom.ctc.edu, Whatcom Community Coll – Gg
Dougil, Andy, +44(0) 113 34 36782 a.j.dougill@leeds.ac.uk, Univ of Leeds – Os
Douglas, Arthur V., (402) 280-2464 sonora@creighton.edu, Creighton Univ – Og
Douglas, Bruce, 812-855-3848 douglasb@indiana.edu, Indiana Univ, Bloomington – Gc
Douglas, Ellen, ellen.douglas@umb.edu, Univ of Massachusetts, Boston – HwqHy
Douglas, Marianne, (780) 492-0055 marianne.douglas@ualberta.ca, Univ of Toronto – Gn
Douglas, Peter, 514-398-3677 peter.douglas@mcgill.ca, McGill Univ – Cs
Douglass, Daniel C., (617) 373-4381 d.douglass@northeastern.edu, Northeastern Univ – GlSdOa
Douglass, David N., (626) 585-7036 dndouglass@pasadena.edu, Pasadena City Coll – Ym
Dove, Patricia M., (540) 231-2444 pdove@vt.edu, Virginia Polytechnic Inst & State Univ – Cl
Doveton, John H., (785) 864-2100 doveton@kgs.ku.edu, Univ of Kansas – Go
Dowd, John F., (706) 542-2383 Jdowd@uga.edu, Univ of Georgia – Hy
Dowdy, Robert H., (612) 625-7058 bdowdy@soils.umn.edu, Univ of Minnesota, Twin Cities – Sc
Dower, John, 250-721-6120 dower@uvic.ca, Univ of Victoria – Ob
Dowling, Carolyn B., 765-285-8274 cbdowling@bsu.edu, Ball State Univ – Hg
Downes, Hilary, +44 020 3073 8027 h.downes@ucl.ac.uk, Birkbeck Coll – Cg
Downie, Helen, helen.downie@manchester.ac.uk, Univ of Manchester – Ge
Downs, Robert T., (520) 626-8092 rdowns@email.arizona.edu, Univ of Arizona – Gi
Doyle, James A., (530) 752-7591 jadoyle@ucdavis.edu, Univ of California, Davis – PblPs
Doyle, Joseph, j6doyle@bridgew.edu, Bridgewater State Univ – Gg
Drake, John C., (802) 656-0244 jdrake@uvm.edu, Univ of Vermont – Cl
Drake, Lon D., (319) 335-1826 Univ of Iowa – Hy
Drake, Michael J., (520) 621-6962 drake@lpl.arizona.edu, Univ of Arizona – Xc
Drake, Simon, +440203 073 8024 drakesimon1@gmail.com, Birkbeck Coll – Gv
Draper, Grenville, (305) 348-3087 draper@fiu.edu, Florida International Univ – GctGh
Drazen, Jeffrey C., (808) 956-6567 jdrazen@hawaii.edu, Univ of Hawai'i, Manoa – Ob

Drean, Thomas A., (307) 766-2286 x223 tom.drean@wyo.gov, Wyoming State Geological Survey – Gg
Dreesen, Donald S., (505) 667-1913 dreesen@lanl.gov, Los Alamos National Laboratory – Np
Dreger, Douglas S., (510) 643-1719 dreger@seismo.berkeley.edu, Univ of California, Berkeley – Ys
Drescher, Andrew, (612) 625-2374 Univ of Minnesota, Twin Cities – So
Dreschoff, Gisela, (785) 864-4517 Univ of Kansas – Ce
Dressel, Waldemar M., (573) 341-4616 Missouri Univ of Science and Tech – Gg
Drew, Douglas A., (404) 496-4202 ddrew@mtech.edu, Montana Tech of the Univ of Montana – Cg
Drew, Robin A., (514) 398-4755 McGill Univ – Nx
Driese, Steven, 254-710-2361 steven_driese@baylor.edu, Univ of Tennessee, Knoxville – SaPeGs
Driese, Steven G., (254) 710-2177 Steven_Driese@baylor.edu, Baylor Univ – SaGe
Driever, Steven L., (816) 235-2971 drievers@umkc.edu, Univ of Missouri-Kansas City – OnnOn
Drijfhout, Sybren, +44 (0)23 8059 6202 S.S.Drijfhout@soton.ac.uk, Univ of Southampton – Op
Dripps, Weston R., (864) 294-3392 weston.dripps@furman.edu, Furman Univ – HwsGe
Driscoll, John R., (607) 753-2926 driscollj@cortland.edu, SUNY, Cortland – On
Driscoll, Neal W., (858) 822-5026 ndriscoll@ucsd.edu, Univ of California, San Diego – Gu
Driscoll, Peter E., 202-478-8827 pdriscoll@carnegiescience.edu, Carnegie Institution of Washington – GtYm
Drobniak, Agnieszka, (812) 855-2687 agdrobni@indiana.edu, Indiana Univ – Ec
Drobny, Gerry, (916) 691-7204 Cosumnes River Coll – Gg
Droop, Giles, +44 0161 275-3809 giles.droop@manchester.ac.uk, Univ of Manchester – Gp
Droser, Mary L., 951-827-3797 mary.droser@ucr.edu, Univ of California, Riverside – Po
Droxler, Andre W., 713.348.4885 andre@rice.edu, Rice Univ – Gs
Druckebrod, Daniel L., 609-895-5422 ddruckenbrod@rider.edu, Rider Univ – Onn
Druckenmiller, Patrick S., 907-474-6954 ffpsd@uaf.edu, Univ of Alaska, Fairbanks – Pv
Druffel, Ellen R. M., (949) 824-2116 edruffel@uci.edu, Univ of California, Irvine – Oc
Druguet Tantiña, Elena, ++935813730 elena.druguet@uab.cat, Universitat Autonoma de Barcelona – Gc
Druhan, Jennifer, (217) 300-0142 jdruhan@illinois.edu, Univ of Illinois, Urbana-Champaign – HwCs
Drummond, Carl N., (260) 481-5750 drummond@ipfw.edu, Indiana Univ / Purdue Univ, Fort Wayne – Gs
Drummond, Jane, +4401413304208 Jane.Drummond@glasgow.ac.uk, Univ of Glasgow – Oi
Drysdale, Jessica, (508) 289-3739 jdrysdale@whoi.edu, Woods Hole Oceanographic Institution – Oc
Drzewiecki, Peter A., (860) 465-4322 drzewieckip@easternct.edu, Eastern Connecticut State Univ – GrsGd
Drzyzga, Scott A., (717) 477-1307 sadrzy@ship.edu, Shippensburg Univ – OiuOl
Duaime, Terence E., (406) 496-4157 tduaime@mtech.edu, Montana Tech of The Univ of Montana – Hws
Duan, Benchuan, (979) 845-3297 bduan@tamu.edu, Texas A&M Univ – YsgGq
Duan, Quingyun, (925) 422-7704 duan2@llnl.gov, Lawrence Livermore National Laboratory – Hg
Duan, Shuiwang, (301) 405-2407 sgao@fresno.ars.usda.gov, Univ of Maryland – HsOu
Dube, Benoit, (418) 654-2669 bdube@nrcan.gc.ca, Universite du Quebec – Em
Dubey, C S., (981) 107-4867 csdubey@gmail.com, Univ of Delhi – GexGt
Dubsky, Scott , 719-333-3080 scott.dubsky@usafa.edu, United States Air Force Academy – OyGm
Duby, Paul F., (212) 854-2928 pfd1@columbia.edu, Columbia Univ – Nx
Duce, Robert A., (979) 845-5756 rduce@ocean.tamu.edu, Texas A&M Univ – Oc
Ducea, Mihai N., (520) 621-5171 ducea@email.arizona.edu, Univ of Arizona – Gt
Duchane, David V., (505) 667-9893 duchane@lanl.gov, Los Alamos National Laboratory – On
Duchesne, Josee, (418) 656-2177 duchesne@ggl.ulaval.ca, Universite Laval – Gz
Ducklow, Hugh, 845-365-8167 hducklow@ldeo.columbia.edu, Columbia Univ – Ob
Duckstein, Lucien, (520) 621-2274 duckstein@engref.fr, Univ of Arizona – Hq
Duckworth, Kenneth, (403) 220-7375 Univ of Calgary – Ye
Duda, Jan-Peter, +49 551 397960 jan-peter.duda@geo.uni-goettingen.de, Georg-August Univ of Goettingen – PyCoGs
Dudley, Lynn M., 850-644-4214 dudley@gly.fsu.edu, Florida State Univ – Sc
Dudley, Walter, 808-933-3905 dudley@hawaii.edu, Univ of Hawai'i, Manoa – Ou
Dudukovic, Mike, dudu@che.wustl.edu, Washington Univ in St. Louis – On
Duebendorfer, Ernest M., (928) 523-7510 ernie.d@nau.edu, Northern Arizona Univ – Gct
Duedall, Iver W., (321) 674-8096 duedall@fit.edu, Florida Inst of Tech – Oc
Dueker, Kenneth G., (307) 766-2657 dueker@uwyo.edu, Univ of Wyoming – Ys
Duennebier, Frederick K., (808) 956-4779 fred@soest.hawaii.edu, Univ of Hawai'i, Manoa – Yr
Duex, Timothy W., (337) 482-6222 twd7693@louisiana.edu, Univ of Louisiana at Lafayette – HgGex
Dufek, Josef, (404) 984-9472 josef.dufek@eas.gatech.edu, Georgia Inst of Tech – GvOnn
Duff, John, john.duff@umb.edu, Univ of Massachusetts, Boston – OnnOn
Duff, Shamus, (519) 661-2111 x80370 shamus.duff@uwo.ca, Western Univ – OiXg
Duffey, Patricia, 785-628-5389 pduffey@fhsu.edu, Fort Hays State Univ – On
Duffield, Wendell A., (928) 523-4852 Northern Arizona Univ – Gv
Duffy, Clarence J., (505) 667-5154 Los Alamos National Laboratory – Cg
Duffy, Tara , (617) 373-6323 t.duffy@neu.edu, Northeastern Univ – Ob
Duffy, Thomas S., (609) 258-6769 duffy@princeton.edu, Princeton Univ – GyYgOm
Dugan, Brandon, (713) 348-5088 dugan@mines.edu, Colorado School of Mines – GuYrHw
Dugas, Bernard, 514 987-3000 #5319 dugas.bernard@uqam.ca, Universite du Quebec a Montreal – Oa
Dugmore, Andrew J., +44 (0) 131 650 8156 Andrew.Dugmore@ed.ac.uk, Edinburgh Univ – Gae
Duke, Edward F., (605) 394-2388 edward.duke@sdsmt.edu, South Dakota School of Mines & Tech – GipOr
Duke, Jason E., (931) 372-3121 jduke@tntech.edu, Tennessee Tech Univ – Oi
Duke, Norman A., (519) 661-3199 nduke@uwo.ca, Western Univ – Eg
Dulai, Henrietta, (808) 956-0720 hdulaiov@hawaii.edu, Univ of Hawai'i, Manoa – CcHw
Duley, James W., 573-368-2105 bill.duley@dnr.mo.state, Missouri Dept of Natural Resources – Hy
Dulin, Shannon, (405) 325-5488 sdulin@ou.edu, Univ of Oklahoma – YmGs
Duller, Rob, +44-151-794-5179 Robert.Duller@liverpool.ac.uk, Univ of Liverpool – Gvs
Dumas, Simone, (613)562-5800 6230 dumas@uottawa.ca, Univ of Ottawa – GsOg
Dumitru, Trevor, (650) 723-1328 trevor@pangea.stanford.edu, Stanford Univ – Gc
Dumond, Gregory, 479-575-3411 gdumond@uark.edu, Univ of Arkansas, Fayetteville – GtcGg
Dunagan, Stan P., (731) 881-7437 sdunagan@utm.edu, Univ of Tennessee, Martin – GsSa
Dunbar, John A., (254) 710-2191 john_dunbar@baylor.edu, Baylor Univ – YgHg
Dunbar, Nelia W., (575) 835-5783 nelia@nmbg.nmt.edu, New Mexico Inst of Mining & Tech – Gi
Dunbar, Robert B., (650) 725-6830 dunbar@pangea.stanford.edu, Stanford Univ – On
Duncan, Ian, 512-471-5117 ian.duncan@beg.utexas.edu, Univ of Texas at Austin, Jackson School of Geosciences – Gg
Duncan, Ian, rduncan@ccsf.edu, City Coll of San Francisco – Oy
Duncan, Robert A., rduncan@coas.oregonstate.edu, Oregon State Univ – Yr
Duncan, Stewart, 785532277 sduncan@ksu.edu, Kansas State Univ – So

Duncan Tabb, Neva, 721-791-2758 Duncantabb.Neva@spColl.edu, Saint Petersburg Coll, Clearwater – Ow
Dundas, Robert G., (559) 278-6984 rdundas@csufresno.edu, California State Univ, Fresno – PvGr
Dunfield, Kari, (519) 824-4120 x58088 dunfield@uoguelph.ca, Univ of Guelph – Sb
Dunlap, Dallas B., (512) 471-4858 dallas.dunlap@beg.utexas.edu, Univ of Texas at Austin, Jackson School of Geosciences – Ye
Dunlevey, John, +27 (0)15 268 3483 john.dunlevey@ul.ac.za, Univ of Limpopo – GzxEg
Dunn, Allison L., (508) 929-8641 adunn@worcester.edu, Worcester State Univ – OayHg
Dunn, Bernie, 519-850-2362 bdunn@uwo.ca, Western Univ – Ye
Dunn, John Todd, (506) 453-4804 Univ of New Brunswick – Gi
Dunn, Richard K., (802) 485-2304 rdunn@norwich.edu, Norwich Univ – GslGa
Dunn, Robert A., (808) 956-3728 dunnr@hawaii.edu, Univ of Hawai'i, Manoa – Yrs
Dunn, Steven R., 413-538-2531 sdunn@mtholyoke.edu, Mount Holyoke Coll – Gp
Dunn, Tasha L., (207) 859-5808 tldunn@colby.edu, Colby Coll – XmGiz
Dunne, George C., (818) 677-2511 george.dunne@csun.edu, California State Univ, Northridge – Gc
Dunne, William M., (423) 974-4161 wdunne@utk.edu, Univ of Tennessee, Knoxville – Gc
Dunning, Gregory R., (709) 864-8481 gdunning@mun.ca, Memorial Univ of Newfoundland – CcGit
Dunning, Jeremy D., (812) 856-4448 dunning@indiana.edu, Indiana Univ, Bloomington – Gc
Dunst, Brian, 412-442-4230 bdunst@pa.gov, Pennsylvania Bureau of Topographic & Geologic Survey – EgoHw
Dunton, Karen K., (309) 438-7640 kkedunt@ilstu.edu, Illinois State Univ – On
Dunton, Kenneth H., (361) 749-6744 ken.dunton@utexas.edu, Univ of Texas at Austin – Ob
Dupont, Todd, (513) 529-9734 dupontt@miamioh.edu, Miami Univ – Gl
Dupont, Todd, 949-824-1133 tdupont@uci.edu, Univ of California, Irvine – Ol
Dupraz, Christophe, 860-486-1394 christophe.dupraz@uconn.edu, Univ of Connecticut – Py
Dupre, William R., 713-743-4987 wdupre@uh.edu, Univ of Houston – On
Durand, Michael T., (614)247-4835 durand.8@osu.edu, Ohio State Univ – HgYd
Durbin, Edward G., (401) 874-6850 edurbin@gso.uri.edu, Univ of Rhode Island – Ob
Durbin, James, 812-465-1208 jdurbin@usi.edu, Univ of Southern Indiana – Gm
Durham, Lisa, 630-252-3170 ladurham@anl.gov, Argonne National Laboratory – Hw
Durham, William, wbdurham@mit.edu, Massachusetts Inst of Tech – Cg
Durkee, Philip A., 831-656-2517 durkee@nps.edu, Naval Postgraduate School – Owr
Durland, Theodore, 541-737-5058 tdurland@coas.oregonstate.edu, Oregon State Univ – Yg
DuRoss, Christopher B., (801) 537-3348 christopherduross@utah.gov, Utah Geological Survey – Ng
Durran, Dale R., 206-543-7440 durrand@atmos.washington.edu, Univ of Washington – Oa
Durrant, Jeffrey O., (801) 422-4116 jodurrant@byu.edu, Brigham Young Univ – Ge
Durucan, Sevket, +44 20 759 47354 s.durucan@imperial.ac.uk, Imperial Coll – Ng
Dushaw, Brian D., (206) 685-4198 dushaw@apl.washington.edu, Univ of Washington – Op
Dutch, Steven I., (920) 465-2371 dutchs@uwgb.edu, Univ of Wisconsin, Green Bay – GcgGe
Dutcher, Allen, adutcher1@unl.edu, Univ of Nebraska, Lincoln – Oa
Dutkiewicz, Stephanie, 617-253-2454 stephd@ocean.mit.edu, Massachusetts Inst of Tech – Cg
Dutro, Thomas, (202) 633-1322 Smithsonian Institution / National Museum of Natural History – Pi
Dutrow, Barbara L., (225) 388-2525 dutrow@geol.lsu.edu, Louisiana State Univ – Gz
Dutta, Prodip K., (812) 237-2268 Prodip.Dutta@indstate.edu, Indiana State Univ – Gs
Dutton, Andrea, 352-846-2413 adutton@ufl.edu, Univ of Florida – CgGsPe
Dutton, Jessica, duttonj@sunysuffolk.edu, Suffolk County Community Coll, Ammerman Campus – Og
Dutton, John A., 814-865-1534 dutton@ems.psu.edu, Pennsylvania State Univ, Univ Park – Ow
Dutton, Shirley P., (512) 471-0329 shirley.dutton@beg.utexas.edu, Univ of Texas at Austin, Jackson School of Geosciences – Gd
Duvall, Alison, 206-211-8311 aduvall@uw.edu, Univ of Washington – GmcGt
Duxbury, Thomas C., (818) 354-4301 Jet Propulsion Laboratory – On
Dvorzak, Marie, (608) 262-8956 mdvorzak@geology.wisc.edu, Univ of Wisconsin, Madison – On
Dworkin, Stephen I., (254) 710-2186 steve_dworkin@baylor.edu, Baylor Univ – ClGd
Dwyer, Daryl F., 419-5302661 daryl.dwyer@utoledo.edu, Univ of Toledo – On
Dwyer, Gary S., (919) 681-8164 gsd3@duke.edu, Duke Univ – Gs
Dwyer, Joseph R., 321-674-7208 jdwyer@fit.edu, Florida Inst of Tech – Oa
Dyab, Ahmed I., 002-03-3921595 seawolf_5000@yahoo.com, Alexandria Univ – Gg
Dyaur, Nikolay, 713-743-6539 ndyaur@uh.edu, Univ of Houston – Yxs
Dye, David H., 901-678-3330 daviddye@memphis.edu, Univ of Memphis – Ga
Dyer, Jamie, (662) 268-1032 Ext 220 jamie.dyer@msstate.edu, Mississippi State Univ – Ow
Dyer, Jen, +44(0) 113 34 39086 j.dyer@leeds.ac.uk, Univ of Leeds – Ge
Dyhrman, Sonya, 845-365-8165 sdyhrman@ldeo.columbia.edu, Columbia Univ – Ob
Dymek, Robert F., (314) 935-5344 bob_d@levee.wustl.edu, Washington Univ in St. Louis – GpiGa
Dymond, Randel, (540) 231-9962 dymond@vt.edu, Virginia Tech – HqsOi
Dymond, Salli F., (218) 726-6582 dymon003@umn.edu, Univ of Minnesota, Duluth – Ge
Dyreson, Eric G., (406) 683-7275 eric.dyreson@umwestern.edu, Univ of Montana Western – Gq
Dyson, Miranda, +44 (0)1908 653398 x 53398 miranda.dyson@open.ac.uk, The Open Univ – Py
Dzunic, Aleksandar, +61 8 9266 2297 A.Dzunic@curtin.edu.au, Curtin Univ – YgsYe


## E


Earls, Sandy, 405-744-6358 sandy.earls@okstate.edu, Oklahoma State Univ – On
Earls, Stephanie, (360) 902-1473 stephanie.earls@dnr.wa.gov, Washington Division of Geology & Earth Resources – Gge
Early, Thomas O., eot@ornl.gov, Oak Ridge National Laboratory – Cl
Earman, Sam, (717) 871-4336 searman@millersville.edu, Millersville Univ – HwCsl
Easson, Gregory L., (662) 915-5995 geasson@olemiss.edu, Univ of Mississippi – Or
Easterbrook, Don J., (360) 650-3583 don.easterbrook@wwu.edu, Western Washington Univ – GlmGe
Easterday, Cary, cary.easterday@bellevueColl.edu, Bellevue Coll – GgPg
Eastin, Matt, 704-687-5914 mdeastin@uncc.edu, Univ of North Carolina, Charlotte – Ow
Eastler, Thomas E., 207-778-7401 eastler@maine.edu, Univ of Maine - Farmington – GeOrGg
Eastoe, Chris J., (520) 791-7430 eastoe@email.arizona.edu, Univ of Arizona – CsGg
Eaton, Jeffrey G., jeaton@weber.edu, Weber State Univ – PsGs
Eaton, Lewis S., (540) 568-3339 eatonls@jmu.edu, James Madison Univ – Gm
Eaton, Timothy, 718-997-3327 timothy.eaton@qc.cuny.edu, Queens Coll (CUNY) – HgwGg
Ebel, Denton S., (212) 769-5381 debel@amnh.org, American Museum of Natural History – XmCgOe
Ebel, John E., (617) 552-3399 ebel@bc.edu, Boston Coll – Ys
Eberhardt, Erik, (604) 827-5573 eeberhardt@eos.ubc.ca, Univ of British Columbia – Ng
Eberhart, Mark E., (303) 273-3726 meberhar@mines.edu, Colorado School of Mines – Om

Eberle, Bill, (785) 532-6101 weberle@ksu.edu, Kansas State Univ – SoOu
Eberle, Jaelyn J., 303-492-8069 jaelyn.eberle@colorado.edu, Univ of Colorado – Pv
Eberli, Gregor P., (305) 421-4678 geberli@rsmas.miami.edu, Univ of Miami – GusYx
Ebersole, Sandy M., (205) 247-3613 sebersole@gsa.state.al.us, Geological Survey of Alabama – Ogi
Ebert, David, 831-771-4400 debert@mlml.calstate.edu, Moss Landing Marine Laboratories – On
Ebert, James R., (607) 436-3065 James.Ebert@oneonta.edu, SUNY, Oneonta – GrsGd
Eberth, David A., 403-823-7707 Royal Tyrrell Museum of Palaeontology – Gs
Ebiner, Matt, (310) 660-3593 mebiner@elcamino.edu, El Camino Coll – Oy
Ebinger, Cynthia J., 585-276-3364 ebinger@earth.rochester.edu, Univ of Rochester – Gt
Ebinger, Michael H., (505) 667-3147 mhe@lanl.gov, Los Alamos National Laboratory – So
Eble, Cortland F., 8592575500 x 149 eble@uky.edu, Univ of Kentucky – Pl
Eby, G. Nelson, (978) 934-3907 nelson_eby@uml.edu, Univ of Massachusetts Lowell – Cg
Eby, Gloria J., (972) 883-24004 gloria.eby@utdallas.edu, Univ of Texas, Dallas – On
Eby, Stephanie, (617) 373-4380 s.eby@neu.edu, Northeastern Univ – On
Echols, John B., (225) 388-8573 john@lgs.bri.lsu.edu, Louisiana State Univ – Go
Eckel, F. Anthony, 206-685-8537 teckel@atmos.washington.edu, Univ of Washington – Oa
Eckert, Donald J., (614) 292-9048 eckert.1@osu.edu, Ohio State Univ – Sc
Eckert, James O., (203) 432-3181, Yale Univ – Gp
Eddy, Michael P., (609) 258-4101 meddy@princeton.edu, Princeton Univ – Cc
Edegbai, Aitalokhai J., aitalokhai.edegbai@uniben.edu, Univ of Benin – GosPi
Edgar, Dorland E., (630) 252-7596 Argonne National Laboratory – Gm
Edgcomb, Virginia, 508-289-3734 vedgcomb@whoi.edu, Woods Hole Oceanographic Institution – Cg
Edgell, Dennis J., (910) 521-6479 dennis.edgell@uncp.edu, Univ of North Carolina, Pembroke – Oa
Edil, Tuncer B., (608) 262-3225 edil@engr.wisc.edu, Univ of Wisconsin, Madison – Ng
Edlund, Mark B., mbedlund@smm.org, Univ of Minnesota, Twin Cities – Pe
Edmonds, Douglas A., (812) 855-4512 edmondsd@indiana.edu, Indiana Univ, Bloomington – Gsm
Edmonds, Marie, +44 (0) 1223 333463 medm06@esc.cam.ac.uk, Univ of Cambridge – Gv
Edson, Carol, 925-424-1336 cedson@laspositasColl.edu, Las Positas Coll – Gg
Edson, James B., (860) 405-9165 james.edson@uconn.edu, Univ of Connecticut – OwaOp
Edwards, Benjamin R., (717) 254-8934 edwardsb@dickinson.edu, Dickinson Coll – GivCg
Edwards, C L, (505) 667-8464 cledwards@lanl.gov, Los Alamos National Laboratory – Yg
Edwards, Dianne, +44(0)29 208 74264 EdwardsD2@cardiff.ac.uk, Univ of Wales – Pb
Edwards, Dianne, edwardsd2@cf.ac.uk, Cardiff Univ – Pb
Edwards, Margo H., (808) 956-5232 margo@soest.hawaii.edu, Univ of Hawai'i, Manoa – Gu
Edwards, Nicholas, +44 0161 275-3810 nicholas.edwards@manchester.ac.uk, Univ of Manchester – Pg
Edwards, R. Lawrence, 612-626-0207 edwar001@umn.edu, Univ of Minnesota, Twin Cities – Cc
Edwards, Richard N., (416) 978-2267 edwards@core.physics.utoronto.ca, Univ of Toronto – Yr
Edwards, Robin, robin.edwards@tcd.ie, Trinity Coll – Ge
Edwards, Stuart, +44 (0) 191 208 3986 stuart.edwards@ncl.ac.uk, Univ of Newcastle Upon Tyne – Gq
Effert-Fanta, Shari E., 217-244-2192 sfanta@illinois.edu, Univ of Illinois, Urbana-Champaign – Cas
Efurd, Deward W., (505) 667-2437 Los Alamos National Laboratory – Ct
Egan, Stuart, (+44) 01782 733174 s.s.egan@keele.ac.uk, Keele Univ – Ggc
Egbert, Gary D., (541) 737-2947 egbert@coas.oregonstate.edu, Oregon State Univ – Yr
Egbeyemi, Oladimeji R., 08167781509 oregbeyemi@futa.edu.ng, Federal Univ of Tech, Akure – GgCg
Egenhoff, Sven O., (970) 491-0749 sven@warnercnr.colostate.edu, Colorado State Univ – Gs
Egerton, Victoria, victoria.egerton@manchester.ac.uk, Univ of Manchester – Pg
Eggers, Delores M., (828) 251-6654 eggers@unca.edu, Univ of North Carolina, Asheville – On
Eggler, David H., (814) 571-1960 dhe1@psu.eu, Pennsylvania State Univ, Univ Park – CpGiv
Eggleston, Carrick M., (307) 766-6769 carrick@uwyo.edu, Univ of Wyoming – Cl
Eggleston, David B., (919) 515-7840 eggleston@ncsu.edu, North Carolina State Univ – Ob
Ehinola, Olugbenga A., +2348033819066 oa.ehinola, Univ of Ibadan – GoeGs
Ehlen, Judy, judyehlen@hotmail.com, Radford Univ – Gm
Ehler, J B., (801) 537-3343 buckehler@utah.gov, Utah Geological Survey – Oi
Ehler, Stan, (785) 532-6101 swehler@ksu.edu, Kansas State Univ – So
Ehlers, Todd A., (734) 763-5112 tehlers@umich.edu, Univ of Michigan – Yg
Ehlmann, Arthur J., (817) 257-6278 a.ehlmann@tcu.edu, Texas Christian Univ – GzXm
Ehlmann, Bethany, 626.395.6720 ehlmann@caltech.edu, California Inst of Tech – Xg
Ehlschlaeger, Charles R., (212) 772-5321 Graduate School of the City Univ of New York – On
Eichenbaum, Jack, (212) 772-5265 jaconet@aol.com, Hunter Coll (CUNY) – Oy
Eicher, Don L., (303) 492-6187 Univ of Colorado – Pm
Eichhubl, Peter, (512) 475-8829 peter.eichhubl@beg.utexas.edu, Univ of Texas at Austin, Jackson School of Geosciences – GcCgNr
Eick, Matthew J., (540) 231-8943 eick@vt.edu, Virginia Polytechnic Inst & State Univ – Sc
Eicken, Hajo, (907) 474-7280 hajo.eicken@gi.alaska.edu, Univ of Alaska, Fairbanks – Og
Eide, Elizabeth A., (202) 334-2392 eeide@nas.edu, National Academy of Sciences/National Research Council – GogGt
Eiler, John M., (626) 395-6239 eiler@gps.caltech.edu, California Inst of Tech – Cg
Einaudi, Marco T., (650) 723-0575 marco@pangea.stanford.edu, Stanford Univ – Em
Eisenberger, Peter M., (917) 568-6915 petere@ldeo.columbia.edu, Columbia Univ – On
Eisenhart, Karen, (814) 732-2398 keisenhart@edinboro.edu, Edinboro Univ of Pennsylvania – Oy
Eisenhart, Ralph, (717) 846-7788 York Coll of Pennsylvania – Gg
Eivazi, Frieda, 573-681-5459 Univ of Missouri, Columbia – Sc
Ekdale, Allan A., (801) 581-7266 a.ekdale@utah.edu, Univ of Utah – Pe
Eklund, Olav, +358442956429 olav.eklund@abo.fi, Abo Akademi Univ – GxeGz
Ekmekci, Mehmet, +90 312 2977730 ekmekci@hacettepe.edu.tr, Hacettepe Univ – HwyCs
Ekstrom, Goran, (845) 365-8427 ekstrom@ldeo.columbia.edu, Columbia Univ – Ys
El Dakkak, Mohamed W., 002-03-3921595 waguihm@yahoo.com, Alexandria Univ – GgsGd
El Kadiri, Racha, 615-484-7641 racha.elkadiri@mtsu.edu, Middle Tennessee State Univ – Hg
El Shazly, Aley, (304) 696-6756 elshazly@marshall.edu, Marshall Univ – GpiCg
El-Baz, Farouk, (617) 353-5081 farouk@bu.edu, Boston Univ – Or
El-Kadi, Aly I., (808) 956-6331 elkadi@hawaii.edu, Univ of Hawai'i, Manoa – Hw
El-Monier, Ilham, ilham.el-monier@ou.edu, Univ of Oklahoma – Np
Elachi, Charles, 626.395.8858 charles.elachi@jpl.nasa.gov, California Inst of Tech – Xg
Elder, Kathryn L., (508) 289-2513 kelder@whoi.edu, Woods Hole Oceanographic Institution – Cs
Elders, Wilfred A., wilfred.elders@ucr.edu, Univ of California, Riverside – Gi
Eldredge, Niles, (212) 769-5723 American Museum of Natural History – Pi
Eldredge, Sandra N., (801) 537-3328 nrugs.selredg@state.ut.us, Utah Geological Survey – Oe
Elia, Gaetano, +44 (0) 191 208 7934 Univ of Newcastle Upon Tyne – Ng

Elias, Robert J., (204) 474-8862 eliasrj@ms.umanitoba.ca, Univ of Manitoba – Pi
Elick, Jennifer M., 570-372-4214 elick@susqu.edu, Susquehanna Univ – Pe
Elkins, Joe T., 970-351-3060 joe.elkins@unco.edu, Univ of Northern Colorado – OeGgCl
Elkins, Lynne J., (402) 472-7563 lelkins@unl.edu, Univ of Nebraska, Lincoln – GivCt
Ellecosta, Peter , +49 (89) 289 25885 p.ellecosta@tum.de, Technische Universitaet Muenchen – Ngr
Eller, Phillip G., (505) 667-5830 Los Alamos National Laboratory – On
Ellett, Kevin M., 812-856-3671 kmellett@indiana.edu, Indiana Univ – YeHw
Ellingson, Don, 503-838-8865 ellingd@wou.edu, Western Oregon Univ – OwgXg
Ellins, Katherine K., (512) 232-3251 kellins@ig.utexas.edu, Univ of Texas at Austin – On
Elliot, David H., (614) 292-5076 elliot.1@osu.edu, Ohio State Univ – GitGv
Elliott, Brent, 512-471-1812 brent.elliott@beg.utexas.edu, Univ of Texas at Austin – GzcGt
Elliott, C. James, (505) 667-9517 cje@lanl.gov, Los Alamos National Laboratory – On
Elliott, Colleen G., (406) 496-4143 celliott@mtech.edu, Montana Tech of The Univ of Montana – GctGg
Elliott, David K., (928) 523-7188 david.elliott@nau.edu, Northern Arizona Univ – Pv
Elliott, Emily M., (412) 624-8780 eelliott@pitt.edu, Univ of Pittsburgh – Hg
Elliott, Julie, (765) 496-7206 julieelliott@purdue.edu, Purdue Univ – Yd
Elliott, Monty A., (541) 552-6482 Southern Oregon Univ – Gr
Elliott, Scott M., (505) 667-0949 sme@lanl.gov, Los Alamos National Laboratory – Oa
Elliott, Susan J., (519) 888-4567 (Ext. 33923) elliotts@uwaterloo.ca, McMaster Univ – Ge
Elliott, W. Crawford, (404) 413-5756 wcelliott@gsu.edu, Georgia State Univ – Cl
Elliott, Jr., William S., (812) 228-5053 wselliott@usi.edu, Univ of Southern Indiana – GsrCl
Ellis, Andre, aellis3@calstatela.edu, California State Univ, Los Angeles – HwCg
Ellis, Hugh, (901) 678-2007 hugh.ellis@jhu.edu, Johns Hopkins Univ – On
Ellis, Hugh I., ellis@sandiego.edu, Univ of San Diego – Ob
Ellis, Kathryn E., (859) 323-0504 kathryn.ellis@uky.edu, Univ of Kentucky – On
Ellis, Robert M., (604) 822-6574 rellis@eos.ubc.ca, Univ of British Columbia – Ys
Ellis, Rowan, +44 (0) 131 651 4447 Rowan.Ellis@ed.ac.uk., Edinburgh Univ – On
Ellis, Scott R., 217-265-5105 ellis3@illinois.edu, Univ of Illinois, Urbana-Champaign – Ge
Ellis, Trevor, 573-368-2153 Trevor.Ellis@dnr.mo.gov, Missouri Dept of Natural Resources – Gg
Ellison, Anne L., 217-244-2502 aerdmann@illinois.edu, Univ of Illinois, Urbana-Champaign – Ge
Ellsworth, Bill, (650) 723-9390 wellsworth@stanford.edu, Stanford Univ – Ys
Ellsworth, Timothy R., (217) 333-2055 Univ of Illinois, Urbana-Champaign – Sp
Ellwood, Brooks B., (225) 578-3416 ellwood@lsu.edu, Louisiana State Univ – Ym
Elmore, R. Douglas, (405) 325-3253 delmore@ou.edu, Univ of Oklahoma – YmGs
Elnadi, Abdelhalim H., geology@uofk.edu, Univ of Khartoum – Goz
Eloranta, Edwin W., (608)262-7327 eloranta@lidar.ssec.wisc.edu, Univ of Wisconsin, Madison – Yr
Elrick, Maya, (505) 277-5077 dolomite@unm.edu, Univ of New Mexico – Gs
Elrick, Scott D., (217) 333-3222 elrick@illinois.edu, Univ of Illinois – EcGgs
Elsberry, Russell L., 831-656-2373 elsberry@nps.edu, Naval Postgraduate School – Ow
Elsen, Jan, jan.elsen@ees.kuleuven.be, Katholieke Universiteit Leuven – Gz
Elsenbeck, James R., (508) 289-3913 jelsenbeck@whoi.edu, Woods Hole Oceanographic Institution – Yr
Elstad-Haveles, Andrew, 507-933-6475 ahaveles@gustavus.edu, Gustavus Adolphus Coll – PogGs
Elswick, Erika R., (812) 855-2493 eelswick@indiana.edu, Indiana Univ, Bloomington – Cl
Elueze, A. A., aa.elueze@mail.ui.edu.ng, Univ of Ibadan – Eg
Elwood Madden, Andrew S., (405) 325-3253 amadden@ou.edu, Univ of Oklahoma – ClScGz
Ely, Lisa L., (509) 963-2177 ely@cwu.edu, Central Washington Univ – Gm
Elzinga, Evert J., 973-353-5238 elzinga@andromeda.rutgers.edu, Rutgers, The State Univ of New Jersey, Newark – Sc
Emanuel, Kerry A., (617) 253-2462 emanuel@mit.edu, Massachusetts Inst of Tech – Ow
Emerick, Christina M., (206) 543-2491 tina@ocean.washington.edu, Univ of Washington – Ou
Emerman, Steven H., (801) 863-6864 stevene@uvu.edu, Utah Valley Univ – HgYgGe
Emerson, Cheryl R., (309) 694-5373 Illinois Central Coll – Oe
Emerson, Norlene, (608) 647-6186 x109 norlene.emerson@uwc.edu, Univ of Wisconsin Colls – GgPiGs
Emerson, Steven R., (206) 543-0428 emerson@u.washington.edu, Univ of Washington – Oc
Emery, Joshua, (865) 974-2366 jemery2@utk.edu, Univ of Tennessee, Knoxville – Xg
Emile-Geay, Julien, 213-740-2945 julieneg@usc.edu, Univ of Southern California – Oc
Emmer, Barbara, 089/2180 4231 Ludwig-Maximilians-Universitaet Muenchen – Yg
Emmett, Chad, (801) 422-7886 chad_emmett@byu.edu, Brigham Young Univ – On
Emmi, Philip C., (801) 581-5562 pcemmi@geog.utah.edu, Univ of Utah – Ou
Emofurieta, Williams O., wemofu@uniben.edu, Univ of Benin – GzCeGe
Emry, Robert, (202) 633-1323 Smithsonian Institution / National Museum of Natural History – Pv
Encarnacion, John, (314) 977-3119 encarnjp@eas.slu.edu, Saint Louis Univ – Gx
Enderlin, Ellyn M., ellyn.enderlin@maine.edu, Univ of Maine – OlrOg
Enderlin, Milton, 817 257 5318 m.enderlin@tcu.edu, Texas Christian Univ – Nr
Endres, Anthony E., (519) 888-4567 (Ext. 3552) Univ of Waterloo – Pe
Engebretson, David C., (360) 650-3595 david.engebretson@wwu.edu, Western Washington Univ – Gt
Engel, Anga, mpeschke@geomar.de, SUNY, Stony Brook – Oc
Engel, Annette S., 865-974-2366 aengel1@utk.edu, Univ of Tennessee, Knoxville – Cl
Engel, Michael H., (405) 325-4435 ab1635@ou.edu, Univ of Oklahoma – Co
Engel, Richard E., (406) 994-5295 rengel@montana.edu, Montana State Univ – Sc
Engelder, Terry, (814) 865-3620 engelder@geosc.psu.edu, Pennsylvania State Univ, Univ Park – Nr
Engelhard, Simon, (401) 874-2187 engelhart@uri.edu, Univ of Rhode Island – Gg
Engelmann, George F., 402-554-4804 gengelmann@unomaha.edu, Univ of Nebraska at Omaha – PvGdr
Engels, Jennifer, 808-956-2562 engels@hawaii.edu, Univ of Hawai'i, Manoa – Oe
England, John, (780) 492-5673 john.england@ualberta.ca, Univ of Alberta – Gml
England, Phillip, +44 (1865) 272000 philip@earth.ox.ac.uk, Univ of Oxford – Gt
England, Richard, +440116 252 3522 hodgeology@le.ac.uk, Leicester Univ – Yg
Englebright, Stephen C., (631) 632-8230 SUNY, Stony Brook – Oe
English, David C., (727) 553-1503 denglish@marine.usf.edu, Univ of South Florida – Op
Engstrom, Daniel R., (612) 433-5953 (Ext. 18) dre@umn.edu, Univ of Minnesota, Twin Cities – Pe
Engstrom, Vanessa, vengstrom@valleyColl.edu, San Bernardino Valley Coll – Oy
Enos, Paul, (785) 864-9714 enos@ku.edu, Univ of Kansas – GsdGu
Enright, Richard L., (508) 531-1390 enright@bridgew.edu, Bridgewater State Univ – HgOrEg
Ensign, Todd, 304-367-8438 tensign@fairmontstate.edu, Fairmont State Univ – Oe
Entekhabi, Dara, (617) 253-9698 darae@mit.edu, Massachusetts Inst of Tech – Hg

Faculty Index -E

Epifanio, Craig, cepi@tamu.edu, Texas A&M Univ – Oa
Eppelbaum, Lev, +97236405086 levap@post.tau.ac.il, Tel Aviv Univ – YvmGt
Eppes, Martha C., 704-687-5993 meppes@uncc.edu, Univ of North Carolina, Charlotte – Os
Epstein, Howard E., (434) 924-4308 hee2b@virginia.edu, Univ of Virginia – On
Erber, Nathan, 614 265 6579 nathan.erber@dnr.state.oh.us, Ohio Dept of Natural Resources – Gl
Erdmann, Anne L., 217-244-2502 aerdmann@illinois.edu, Univ of Illinois – Ge
Erdmer, Philippe, (403) 492-2676 philippe.erdmer@ualberta.ca, Univ of Alberta – Gc
Erdner, Deana L., (361) 749-6719 derdner@utexas.edu, Univ of Texas at Austin – Ob
Erhardt, Andrea M., 859-257-6931 andrea.erhardt@uky.edu, Univ of Kentucky – CgsCm
Erickson, Ben A., (801) 537-3379 benerickson@utah.gov, Utah Geological Survey – Ng
Erickson, J. Mark, (315) 379-5198 St. Lawrence Univ – Poi
Erickson, Rolfe C., (707) 664-2334 rolfe.erickson@sonoma.edu, Sonoma State Univ – Gx
Eriksen, Charles C., (206) 543-6528 eriksen@u.washington.edu, Univ of Washington – Op
Eriksson, Kenneth A., (540) 231-4680 kaeson@vt.edu, Virginia Polytechnic Inst & State Univ – Gs
Erisman, Brad, 361-749-6833 berisman@utexas.edu, Univ of Texas at Austin – Ob
Ernst, W. Gary, (650) 723-2750 ernst@pangea.stanford.edu, Stanford Univ – Cp
Erski, Theodore, (815) 455-8992 terski@mchenry.edu, McHenry County Coll – Og
Erslev, Eric A., (970) 231-2654 eric.erslev@colostate.edu, Univ of Wyoming – GctGg
Ersoy, Erkal, +44 (0)131 451 4777 erkal.ersoy@pet.hw.ac.uk, Heriot-Watt Univ – Eo
Ertel-Ingrisch, Werner, 089/2180 4275 ertel@min.uni-muenchen.de, Ludwig-Maximilians-Universitaet Muenchen – Gz
Erturk, Mehmet Ali, 00904242370000-5981 maerturk@firat.edu.tr, Firat Univ – GiCc
Ervin, C. Patrick, (815) 753-1943 pervin@niu.edu, Northern Illinois Univ – Yv
Erwin, Diane, (510) 642-3921 dmerwin@berkeley.edu, Univ of California, Berkeley – Pb
Erwin, Douglas H., (202) 633-1324 Smithsonian Institution / National Museum of Natural History – Pi
Esbaugh, Andrew J., (361) 749-6835 a.esbaugh@austin.utexas.edu, Univ of Texas at Austin – Ob
Escobar-Wolf, Rudiger, 906/487-2128 rpescoba@mtu.edu, Michigan Technological Univ – Gv
Eshleman, Keith N., (301) 689-7170 eshleman@al.umces.edu, Univ of Maryland – Hg
Esling, Steven P., (618) 453-7376 esling@siu.edu, Southern Illinois Univ Carbondale – HwGl
Esnault, Melissa H., (225) 578-5320 mesnaul@lsu.edu, Louisiana State Univ – On
Esparza, Francisco, fesparz@cicese.mx, Centro de Investigación Científica y de Educación Superior de Ensenada – Ye
Esser, Corinne, (530) 752-3668 caesser@ucdavis.edu, Univ of California, Davis – On
Essery, Richard L., +44 (0) 131 651 9093 Richard.Essery@ed.ac.uk, Edinburgh Univ – Oa
Essling, Alice M., (630) 252-3493 Argonne National Laboratory – Ca
Esslinger, Kelly L., kelly.esslinger@azwestern.edu, Arizona Western Coll – OapGg
Estalrich López, Joan, ++935811270 joan.estalrich@uab.cat, Universitat Autonoma de Barcelona – HgGe
Estapa, Margaret, 518-589-5477 mestapa@skidmore.edu, Skidmore Coll – Oc
Estop Graells, Eugènia, ++34935813089 eugenia.estop@uab.cat, Universitat Autonoma de Barcelona – Gz
Ethington, Raymond L., (573) 882-6470 ethingtonr@missouri.edu, Univ of Missouri – Pm
Ethridge, Frank G., (970) 491-6195 fredpet@cnr.colostate.edu, Colorado State Univ – Gs
Ettensohn, Frank R., (859) 257-1401 f.ettensohn@uky.edu, Univ of Kentucky – PsGdPg
Etter, Ron , ron.etter@umb.edu, Univ of Massachusetts, Boston – ObPoOn
Ettlie, Bradley, 217-265-6543 ettlie76@illinois.edu, Univ of Illinois, Urbana-Champaign – Ge
Eusden, Jr., J. Dykstra, (207) 786-6152 deusden@bates.edu, Bates Coll – Gc
Evangelou, V. P., (515) 294-9237 Iowa State Univ of Science & Tech – Sc
Evanoff, Emmett, 970-351-2647 Emmett.Evanoff@unco.edu, Univ of Northern Colorado – GrPgGs
Evans, Barry M., (814) 863-3531 bmel@psu.edu, Pennsylvania State Univ, Univ Park – Sd
Evans, Bernard W., 206-543-1163 bwevans@uw.edu, Univ of Washington – Gp
Evans, Claude, (314) 935-6684 Washington Univ in St. Louis – On
Evans, David A., (203) 432-3127 david.evans@yale.edu, Yale Univ – YmGt
Evans, David C., davide@rom.on.ca, Royal Ontario Museum – PvoPq
Evans, David G., (916) 278-7840 dave_evans@csus.edu, California State Univ, Sacramento – HqYg
Evans, Diane L., (818) 354-2418 Jet Propulsion Laboratory – Gm
Evans, J. B., (617) 253-2856 brievans@mit.edu, Massachusetts Inst of Tech – Yx
Evans, James E., (419) 372-2414 evansje@bgnet.bgsu.edu, Bowling Green State Univ – GsHsGe
Evans, James P., (435) 797-1267 james.evans@usu.edu, Utah State Univ – Gc
Evans, Jenni L., (814) 865-3240 evans@meteo.psu.edu, Pennsylvania State Univ, Univ Park – Ow
Evans, Kevin R., (417) 836-5590 kevinevans@missouristate.edu, Missouri State Univ – GrsGt
Evans, Les J., (519) 824-4120 (Ext. 53017) levans@lrs.uoguelph.ca, Univ of Guelph – Cl
Evans, Mark, evansmaa@ccsu.edu, Central Connecticut State Univ – GciGz
Evans, Michael E., (780) 492-5517 evans@phys.ualberta.ca, Univ of Alberta – Ym
Evans, Michael N., (301) 405-8763 mnevans@umd.edu, Univ of Maryland – PeOaCs
Evans, Owen C., (631) 632-8061 owen.evans@sunysb.edu, SUNY, Stony Brook – Cg
Evans, Robert L., (508) 289-2673 revans@whoi.edu, Woods Hole Oceanographic Institution – Yr
Evans, Samantha, 208-426-1121 samevans@boisestate.edu, Boise State Univ – Cs
Evans, Thomas J., (608) 263-4125 tevans@wisc.edu, Univ of Wisconsin, Extension – Gg
Evanylo, Gregory K., (540) 231-9739 gevanylo@vt.edu, Virginia Polytechnic Inst & State Univ – Sc
Eveleth, Robert W., (505) 835-5325 beveleth@gis.nmt.edu, New Mexico Inst of Mining & Tech – Nm
Even, Paula, 319-273-3818 paula.even@uni.edu, Univ of Northern Iowa – On
Evenson, Edward B., (610) 758-3659 ebe0@lehigh.edu, Lehigh Univ – Gl
Everett, Mark, (979) 862-2129 everett@geo.tamu.edu, Texas A&M Univ – Ym
Eversoll, Duane A., (402) 472-7524 deversoll2@unl.edu, University of Nebraska - Lincoln – Ng
Eves, Robert L., (435) 586-1934 eves@suu.edu, Southern Utah Univ – CgOeCa
Ewart, Terry E., (206) 543-1327 ewart@apl.washington.edu, Univ of Washington – Op
Ewas, Galal A., 002-03-3921595 goueiss@gmail.com, Alexandria Univ – GgNr
Ewing, Rodney C., (650) 497-6203 rodewing@umich.edu, Stanford Univ – GzeCl
Eyles, Carolyn H., (905) 525-9140 (Ext. 23524) eylesc@mcmaster.ca, McMaster Univ – Gs
Eyles, John D., (905) 525-9140 (Ext. 23152) eyles@mcmaster.ca, McMaster Univ – Ge
Eyles, Nicholas, 416-287-7195 eyles@utsc.utoronto.ca, Univ of Toronto – Gl
Eyre, Bradley, bradley.eyre@scu.edu.au, Southern Cross Univ – Gs
Ezerskis, John L., (216) 987-5227 John.Ezerskis@tri-c.edu, Cuyahoga Community Coll - Western Campus – Og

## F

Faatz, Renee M., (435) 283-7519 renee.faatz@snow.edu, Snow Coll – Gg
Fabbri, Paolo, +390498279124 paolo.fabbri@unipd.it, Università degli Studi di Padova – Hw
Fabel, Derek, +4401413305473 Derek.Fabel@glasgow.ac.uk, Univ of Glasgow – Xg
Fabry, Frederic, (514) 398-7733 frederic.fabry@staff.mcgill.ca, McGill Univ – Oa
Fabryka-Martin, June T., (505) 665-2300 Los Alamos National Laboratory – Hg
Faccenda, Manuele, +390498279159 manuele.faccenda@unipd.it, Università degli Studi di Padova – Gq
Fadem, Cynthia M., (765) 983-1231 fademcy@earlham.edu, Earlham Coll – GaOsCs
Fadiman, Maria, 561 297-3314 mfadiman@fau.edu, Florida Atlantic Univ – On
Fadiya, S. L., 234-803-332-0230 fadiyalawrence@yahoo.co.uk, Obafemi Awolowo Univ – God
Fagan, Amy, 828 227 3820 alfagan@wcu.edu, Western Carolina Univ – GiXg
Fagel, Nathalie, +32 4 3662209 Nathalie.Fagel@ulg.ac.be, Universite de Liege – GsuGe
Fagents, Sarah, 808-956-3163 fagents@higp.hawaii.edu, Univ of Hawai'i, Manoa – Gv
Fagerlin, Stanley C., (417) 836-5800 Missouri State Univ – Pg
Fagherazzi, Sergio, (617) 353-2092 sergio@bu.edu, Boston Univ – GmOnGs
Fagnan, Brian A., 605-394-6652 brian.fagnan@state.sd.us, South Dakota Dept of Environment and Natural Resources – Gg
Fahes, Mashhad, mashhad.fahes@ou.edu, Univ of Oklahoma – Np
Fahnestock, Mark A., 603-862-0322 mark.fahnestock@unh.edu, Univ of New Hampshire – Ol
Fahrenbach, Mark D., 605-394-6830 mark.fahrenbach@state.sd.us, South Dakota Dept of Environment and Natural Resources – Gg
Fairbairn, David, +44 (0) 191 208 6353 david.fairbairn@ncl.ac.uk, Univ of Newcastle Upon Tyne – Gq
Fairchild, Ian, +44 (0)121 414 4181 i.j.fairchild@bham.ac.uk, Univ of Birmingham – GsClGm
Fairchild, Jane, jane.fairchild@uwc.edu, Univ of Wisconsin Colls – Ow
Fairley, Jerry P., (208) 885-9259 jfairley@uidaho.edu, Univ of Idaho – HwGv
Fajkoviæ, Hana, +38514605969 hanaf@geol.pmf.hr, Univ of Zagreb – CgGeCm
Fakhari, Mohammad D., (614) 265-6584 mohammad.fakhari@dnr.state.oh.us, Ohio Dept of Natural Resources – GocNg
Falcon-Lang, Howard, +44 1784 414039 Howard.Falcon-Lang@rhul.ac.uk, Univ of London, Royal Holloway & Bedford New Coll – Pg
Falebita, Dele E., 234-703-298-7836 delefale@oauife.edu.ng, Obafemi Awolowo Univ – Yg
Falk, Lisa, 919-515-8458 esfalk@ncsu.edu, North Carolina State Univ – Og
Falkowski, Paul G., falco@imcs.rutgers.edu, Univ of Hawai'i, Manoa – Og
Falkowski, Paul G., (848) 932-6555 (Ext. 370) falko@rci.rutgers.edu, Rutgers, The State Univ of New Jersey – ObCmPe
Fall, Andras, (512) 471-8334 andras.fall@beg.utexas.edu, Univ of Texas at Austin – CgGcEg
Fall, Leigh M., 607 436-2615 Leigh.Fall@oneonta.edu, SUNY, Oneonta – Pgq
Fallowfield, Howard, howard.fallowfield@flinders.edu.au, Flinders Univ – On
Falta, Ronald W., (864) 656-0125 faltar@clemson.edu, Clemson Univ – Hq
Famiglietti, James, (949)824-9434 jfamigli@uci.edu, Univ of California, Irvine – Hw
Fan, Majie, (817) 272-2987 mfan@uta.edu, Univ of Texas, Arlington – GstCs
Fan, Weihong, fanw@pollux.stockton.edu, Richard Stockton Coll of New Jersey – Oi
Fanning, Kent A., (727) 553-1594 kaf@marine.usf.edu, Univ of South Florida – Oc
Fantle, Matthew S., 814-863-9968 mfantle@geosc.psu.edu, Pennsylvania State Univ, Univ Park – Cs
Fantozzi, Joanna , (657) 278-5464 jfantozzi@fullerton.edu, California State Univ, Fullerton – Gg
Fantozzi, Joanna, jfantozzi@occ.cccd.edu, Orange Coast Coll – GsPeHg
Fares, Mary A., (229) 333-5755 mfares@valdosta.edu, Valdosta State Univ – On
Fariduddin, Mohammad, 773/442-6059 M-Fariduddin@neiu.edu, Northeastern Illinois Univ – Pe
Farley, Kenneth A., (626) 395-6005 farley@gps.caltech.edu, California Inst of Tech – Cg
Farley, Martin B., 910-521-6478 martin.farley@uncp.edu, Univ of North Carolina, Pembroke – PlOe
Farley, Perry D., (505) 667-2415 dfarley@lanl.gov, Los Alamos National Laboratory – On
Farlow, James O., 260-481-6251 farlow@ipfw.edu, Indiana Univ / Purdue Univ, Fort Wayne – PvoPe
Farmer, Emma C., 516 463-5568 geoecf@hofstra.edu, Hofstra Univ – Ou
Farmer, G. Lang, 303-492-6534 farmer@colorado.edu, Univ of Colorado – Cg
Farmer, George T., (505) 665-0225 gfarmer@lanl.gov, Los Alamos National Laboratory – Hw
Farmer, Jack D., (480) 965-6748 jfarmer@asu.edu, Arizona State Univ – Py
Farnan, Ian, +44 (0) 1223 333431 ifarnan@esc.cam.ac.uk, Univ of Cambridge – Gz
Farnsworth, Katie, katie.farnsworth@iup.edu, Indiana Univ of Pennsylvania – Ong
Farquhar, James, (301) 405-1434 jfarquha@essic.umd.edu, Univ of Maryland – Cs
Farquhar, Winifred C., 574-2338855 wcaponigri@hcc-nd.edu, Coll of the Holy Cross – GrcGe
Farquharson, Phil, pfarquharson@miracosta.edu, MiraCosta Coll – Gg
Farr, Tom G., (818) 354-9057 tom.farr@jpl.nasa.gov, Jet Propulsion Laboratory – Gm
Farrand, William, (734) 764-1435 wfarrand@umich.edu, Univ of Michigan – Ga
Farrar, Stewart S., (859) 622-1279 stewart.farrar@eku.edu, Eastern Kentucky Univ – Gpt
Farrell, Brian F., (617) 495-2998 farrell@seas.harvard.edu, Harvard Univ – Ow
Farrell, John, (401) 874-6561 jfarrell@gso.uri.edu, Univ of Rhode Island – Ou
Farrell, Kathleen M., (919) 733-7353 kathleen.farrell@ncdenr.gov, North Carolina Geological Survey – Gs
Farrell, Mark O., (412) 392-3879 Point Park Univ – On
Farrell, Michael, michaelfarrell.myefolio.com, Cuyamaca Coll – Gg
Farrell, Stewart, (609) 652-4245 farrells@stockton.edu, Richard Stockton Coll of New Jersey – On
Farrington, John W., (508) 289-3911 jfarrington@whoi.edu, Woods Hole Oceanographic Institution – ComOc
Farrow, Norman D., ndf@ornl.gov, Oak Ridge National Laboratory – Hg
Farsang, Andrea, +3662544156 farsang@geo.u-szeged.hu, Univesity of Szeged – SdOe
Farthing, Dori J., 585-245-5298 farthing@geneseo.edu, SUNY, Geneseo – Gz
Farver, John R., (419) 372-7203 jfarver@bgnet.bgsu.edu, Bowling Green State Univ – Gy
Faryad, Shah Wali, +420221951521 faryad@natur.cuni.cz, Charles Univ – GpzGi
Fastovsky, David, (401) 874-2185 defastov@uri.edu, Univ of Rhode Island – Pv
Fatherree, James W., 813-253-7906 jfatherree@hccfl.edu, Hillsborough Community Coll – Og
Faulds, James, (775) 682-8751 jfaulds@unr.edu, Univ of Nevada, Reno – Gc
Faulds, James E., (775) 682-6650 jfaulds@unr.edu, Univ of Nevada – Gc
Faulkner, Daniel, +440151 794 5169 Faulkner@liverpool.ac.uk, Univ of Liverpool – Ys
Faure, Gunter, faure.1@osu.edu, Ohio State Univ – Cc
Faure, Stephane, 514-987-3000 #2369 faure.stephane@uqam.ca, Universite du Quebec a Montreal – Gc
Favas, Paulo J., 351259350220 pjcf@utad.pt, Universidade de Trás-os-Montes e Alto Douro – GeCtGg
Favor, Michael, (313) 593-5235 Univ of Michigan, Dearborn – On
Fawcett, J. J., (416) 978-3027 fawcett@quartz.geology.utoronto.ca, Univ of Toronto – Gp
Fawcett, Peter J., (505) 277-3867 fawcett@unm.edu, Univ of New Mexico – Pe
Fayek, Mostafa, (204) 474-7982 fayek@cc.umanitoba.ca, Univ of Manitoba – Cs

Fayer, Michael J., (509) 372-6045 mike.fayer@pnl.gov, Pacific Northwest National Laboratory – Hq
Fayon, Annia K., 612-626-9805 fayon001@umn.edu, Univ of Minnesota, Twin Cities – Gc
Feakins, Sarah, (213) 740-7168 feakins@usc.edu, Univ of Southern California – CoGuCs
Fearey, Bryan L., (505) 665-2423 Los Alamos National Laboratory – Cc
Fearon, Jamie L., (630) 752-5063 jamie.fearon@wheaton.edu, Wheaton Coll – Pv
Feather, Russell, (202) 633-1793 featherr@si.edu, Smithsonian Institution / National Museum of Natural History – Gz
Fedele, Juan J., 320-308-1049 jjfedele@stcloudstate.edu, Saint Cloud State Univ – Hg
Fedo, Christopher, (865) 974-2366 cfedo@utk.edu, Univ of Tennessee, Knoxville – GsXg
Fedorov, Alexey V., (203) 432-3153 alexey.fedorov@yale.edu, Yale Univ – Op
Feeley, Todd C., (406) 994-6917 tfeely@umontana.edu, Montana State Univ – Gi
Feely, Martin, +353 (0)91 492 129 martin.feely@nuigalway.ie, National Univ of Ireland Galway – Goz
Feely, Richard A., (206) 526-6214 feely@pmel.noaa.gov, Univ of Washington – Oc
Feeney, Alison E., (717) 477-1319 aefeen@ark.ship.edu, Shippensburg Univ – On
Feeney, Dennis M., (208) 885-5203 dmfeeney@uidaho.edu, Univ of Idaho – GigYg
Feeney, Thomas P., (717) 477-1297 tpfeen@ark.ship.edu, Shippensburg Univ – Gm
Fegley, Bruce, (314) 935-4852 bfegley@levee.wustl.edu, Washington Univ in St. Louis – Xc
Fehler, Michael, 617-253-3589 fehler@mit.edu, Massachusetts Inst of Tech – Yg
Fehn, Udo, (585) 244-4868 udo.fehn@rochester.edu, Univ of Rochester – GeCgOu
Fehr, Karl Thomas, 089/2180 4256 fehr@min.uni-muenchen.de, Ludwig-Maximilians-Universitaet Muenchen – Gz
Fehrenbacher, Jennifer, 541-737-6285 fehrenje@coas.oregonstate.edu, Oregon State Univ – Cm
Fei, Yingwei, fei@gl.ciw.edu, Univ of Maryland – CpGxCg
Fei, Yingwei, 202-478-8936 yfel@carnegiescience.edu, Carnegie Institution of Washington – Cp
Feibel, Craig S., (848) 445-2721 feibel@rci.rutgers.edu, Rutgers, The State Univ of New Jersey – GrsGa
Feigenson, Mark D., (848) 445-3149 feigy@rci.rutgers.edu, Rutgers, The State Univ of New Jersey – CgGiv
Feigl, Kurt, 608-262-0176 feigl@wisc.edu, Univ of Wisconsin, Madison – Yd
Feigl, Kurt L., (608) 262-0176 feigl@wisc.edu, Univ of Wisconsin, Madison – Gt
Fein, Jeremy B., (574) 631-6101 fein.1@nd.edu, Univ of Notre Dame – Cl
Feinberg, Joshua, (612) 624-8429 feinberg@umn.edu, Univ of Minnesota, Twin Cities – YmGze
Feineman, Maureen D., 814) 863-6649 mdf12@psu.edu, Pennsylvania State Univ, Univ Park – Cp
Feinglos, Mark N., (919) 668-1367 mark.feinglos@duke.edu, Duke Univ – Gz
Feinstein, Daniel T., (414) 962-2582 dtfeinst@usgs.gov, Univ of Wisconsin, Milwaukee – Hw
Felbeck, Horst, (858) 534-6647 hfelbeck@ucsd.edu, Univ of California, San Diego – Ob
Feldmann, Rodney M., (330) 672-2506 rfeldman@kent.edu, Kent State Univ – Pi
Felix, Joseph D., 361-825-4180 Joseph.Felix@tamucc.edu, Texas A&M Univ, Corpus Christi – On
Feller, Robert, feller@biol.sc.edu, Univ of South Carolina – Ob
Felton, Richard M., (660) 562-1569 rfelton@nwmissouri.edu, Northwest Missouri State Univ – Pg
Feltrin, Leo, lfeltrin@uwo.ca, Western Univ – Eg
Felzer, Benjamin S., (610) 758-3536 bsf208@lehigh.edu, Lehigh Univ – Oa
Fenberg, Phillip, +44 (0)23 80592729 P.B.Fenberg@soton.ac.uk, Univ of Southampton – Ob
Fendorf, Scott E., (650) 723-5238 fendorf@pangea.stanford.edu, Stanford Univ – Sc
Fendrich, Kim V., 212 769- kfendrich@amnh.org, American Museum of Natural History – Gz
Fendrock, Michaela, (508) 289-2209 mfendrock@whoi.edu, Woods Hole Oceanographic Institution – Oc
Feng, Huan E., (973) 655-7549 fengh@mail.montclair.edu, Montclair State Univ – Cm
Feng, Song, (479) 575-4748 songfeng@uark.edu, Univ of Arkansas, Fayetteville – OayOw
Feng, Xiahong, 603-646-1712 xiahong.feng@dartmouth.edu, Dartmouth Coll – Cs
Feng, Yan, 630-252-2550 yfeng@anl.gov, Argonne National Laboratory – Oa
Feng, Yucheng, (334) 844-3967 yfeng@acesag.auburn.edu, Auburn Univ – Sb
Fengler, Keegan , 509.963.2719 keegan@geology.cwu.edu, Central Washington Univ – Gg
Fenical, William H., (858) 534-2133 wfenical@ucsd.edu, Univ of California, San Diego – Oc
Fennel, Katja, (902) 494-4526 katja.fennel@dal.ca, Dalhousie Univ – Ob
Fenrich, Francis, (780) 492-2149 frances@space.ualberta.ca, Univ of Alberta – Xy
Fenster, Michael S., (804)752-3745 mfenster@rmc.edu, Randolph-Macon Coll – OnGue
Fenstermaker, Lynn, (702) 862-5412 lynn@dri.edu, Desert Research Inst – Or
Fenter, Paul, (630) 252-7053 fenter@anl.gov, Univ of Illinois at Chicago – Gy
Fenton, Cassandra, 970-248-1077 cfenton@coloradomesa.edu, Colorado Mesa Univ – CcGmCl
Ferencz, Orsolya, orsi@sas.elte.hu, Eotvos Lorand Univ – XyOr
Ferger, Marisa, (814) 863-4229 mferger@psu.edu, Pennsylvania State Univ, Univ Park – Ow
Ferguson, Charles A., 520-770-3500 caf@email.arizona.edu, Univ of Arizona – Gcv
Ferguson, Ian J., (204) 474-9154 ij_ferguson@umanitoba.ca, Univ of Manitoba – Ye
Ferguson, John F., (972) 883-2410 ferguson@utdallas.edu, Univ of Texas, Dallas – Yg
Ferguson, Julie , (949)824-9411 julie.ferguson@uci.edu, Univ of California, Irvine – Oe
Ferguson, Terry A., (864) 597-4527 fergusonta@wofford.edu, Wofford Coll – Ga
Ferland, Marie A., 509-963-2829 ferlandm@cwu.edu, Central Washington Univ – Gu
Fermanich, Kevin J., (920) 465-2240 fermanik@uwgb.edu, Univ of Wisconsin, Green Bay – SpHwOu
Fernandez, Diego, (801) 587-9366 diego.fernandez@utah.edu, Univ of Utah – Cg
Fernandez, Fabian G., 612-625-7460 fabiangf@umn.edu, Univ of Minnesota, Twin Cities – ScbSo
Fernandez, Loretta, 617.373.5461 l.fernandez@neu.edu, Northeastern Univ – Co
Fernandez, Louis A., (909) 537-5024 California State Univ, San Bernardino – Gi
Fernández, Virginia, (598) 2525 1552 vivi@fcien.edu.uy, Universidad de la Republica Oriental del Uruguay (UDELAR) – OirOn
Fernando, Joe, Harindra.J.Fernando.10@nd.edu, Univ of Notre Dame – Oa
Ferns, Mark L., (541) 523-3133 mark.ferns@state.or.us, Oregon Dept of Geology & Mineral Industries – GrtGi
Ferrand, Lin A., (212) 650-8017 Graduate School of the City Univ of New York – Hq
Ferrando, Simona, simona.ferrando@unito.it, Università di Torino – Gx
Ferrari, Raffaele, (617) 253-1291 raffaele@mit.edu, Massachusetts Inst of Tech – OpaOb
Ferre, Eric C., (618) 453-7368 eferre@geo.siu.edu, Southern Illinois Univ Carbondale – GcYmGt
Ferre, Paul (Ty), (520) 621-6082 ty@hwr.arizona.edu, Univ of Arizona – So
Ferreira, Anna, +44 020 7679 37704 a.ferreira@ucl.ac.uk, Univ Coll London – Ys
Ferreira, Karen, (204) 474-9374 karen.ferreira@umanitoba.ca, Univ of Manitoba – Gg
Ferreira, Maryanne F., (508) 289-2266 mferreira@whoi.edu, Woods Hole Oceanographic Institution – On
Ferrell, Jr., Ray E., (225) 388-5306 rferrell@lsu.edu, Louisiana State Univ – Cl
Ferren, Richard L., (413) 236-4553 rferren@berkshirecc.edu,

Berkshire Community Coll – On
Ferrero, Anna Maria, anna.ferrero@unito.it, Università di Torino – Nr
Ferrero, Tom, (541) 967-2053 Oregon Dept of Geology & Mineral Industries – Hw
Ferring, C. Reid, (940) 565-2993 reid.ferring@unt.edu, Univ of North Texas – Ga
Ferris, Dawn, 419-755-3909 dferris@sciencesocieties.org, Ohio State Univ – Sf
Ferris, Grant F., 416-978-0526 ferris@quartz.geology.utoronto.ca, Univ of Toronto – Gn
Ferris, Michael H., 310 243-3405 mferris@csudh.edu, California State Univ, Dominguez Hills – Or
Ferriz, Horacio, (209) 667-3874 hferriz@csustan.edu, California State Univ, Stanislaus – NgHwYe
Ferruzza, David -., (not) li-sted ferruzzad@etown.edu, Elizabethtown Coll – Oan
Festa, Andrea, andrea.festa@unito.it, Università di Torino – Gc
Few, Jr., Arthur A., (713) 527-4003 (Ext. 3601) few@rice.edu, Rice Univ – Oa
Fewings, Melanie, 860-405-9080 melanie.fewings@uconn.edu, Univ of Connecticut – Op
Feyereisen, Gary, 612-625-0968 gfeyer@umn.edu, Univ of Minnesota, Twin Cities – Sp
Fialko, Yuri A., (858) 822-5028 yfialko@ucsd.edu, Univ of California, San Diego – GqtGv
Fichter, Lynn S., (540) 568-6531 fichtels@jmu.edu, James Madison Univ – Gr
Fichtner, Andreas, 089/2180 4230 andreas.fichtner@geophysik.uni-muenchen.de, Ludwig-Maximilians-Universitaet Muenchen – Yg
Fick, Walter H., (785) 532-7223 whfick@ksu.edu, Kansas State Univ – Sf
Field, Cathryn K., cfield@bowdoin.edu, Bowdoin Coll – Ge
Field, Richard T., 302-831-2695 Univ of Delaware – Og
Field, Stephen W., (254) 968-9887 field@tarleton.edu, Tarleton State Univ – Gz
Fielding, Christopher, +61 7 3138 2261 cfielding2@unl.edu, Queensland Univ of Tech – Gsr
Fielding, Christopher R., (402) 472-9801 cfielding2@unl.edu, Univ of Nebraska, Lincoln – Gs
Fielding, Lynn, (310) 660-3593 lfielding@elcamino.edu, El Camino Coll – Og
Fields, Chad, Chad.Fields@dnr.iowa.gov, Iowa Dept of Natural Resources – Gg
Fields, Nancy, 214-860-2429 nfields@dcccd.edu, El Centro Coll - Dallas Community Coll District – Og
Fields, Russell, (775) 682-8735 rfields@unr.edu, Univ of Nevada, Reno – Gg
Fiesinger, Donald W., don.fiesinger@usu.edu, Utah State Univ – Gi
Figg, Sean, (760)744-1150 ext 2513 sfigg@palomar.edu, Palomar Coll – Gg
Fike, David, 314-935-6607 dfike@levee.wustl.edu, Washington Univ in St. Louis – Cs
Fildes, Stephen , stephen.fildes@flinders.edu.au, Flinders Univ – Ori
Filiberto, Justin, (618) 453-4849 filiberto@siu.edu, Southern Illinois Univ Carbondale – GiCpXc
Filina, Irina, (402) 472-2077 ifilina2@unl.edu, Univ of Nebraska, Lincoln – YgGtYe
Filipescu, Sorin, +40-264-405300 ext 5206 sorin@bioge.ubbcluj.ro, Babes-Bolyai Univ – GrPm
Filipovic, Dragan, (605) 677-6160 dragan.filipovic@usd.edu, South Dakota Dept of Environment and Natural Resources – Hwq
Filippelli, Gabriel M., (317) 274-3795 gfilippe@iupui.edu, Indiana Univ / Purdue Univ, Indianapolis – CmGb
Filkorn, Harry, filkornh@pierceColl.edu, Los Angeles Pierce Coll – GgPi
Filley, Timothy R., (765) 494-6581 filley@purdue.edu, Purdue Univ – Co
Fillion, Luc, 514-421-4770 luc.fillion@ec.gc.ca, McGill Univ – Owa
Fillmore, Robert P., (970) 943-2650 rfillmore@western.edu, Western State Colorado Univ – Gs
Finch, Adriam, +44 (0) 1334 462384 aaf1@st-and.ac.uk, Univ of St. Andrews – Gz
Finch, James A., (514) 398-4755 McGill Univ – Nx
Finger, Ken, 510-6432559 kfinger@berkeley.edu, Univ of California, Berkeley – PmOnn
Fink, Uwe, (520) 2736 uwefink@lpl.arizona.edu, Univ of Arizona – Oa
Finkel, Robert C., (925) 422-2044 finkel1@llnl.gov, Lawrence Livermore National Laboratory – Cg
Finkelman, Robert B., (972) 883-2459 bobf@utdallas.edu, Univ of Texas, Dallas – GbEcCa
Finkelstein, David, (315) 781-4443 finkelstein@hws.edu, Hobart & William Smith Colls – Cgs
Finkenbinder, Matthew S., (570) 408-3871 matthew.finkenbinder@wilkes.edu, Wilkes Univ – GsmGl
Finks, Robert M., (718) 997-3305 Queens Coll (CUNY) – Pi
Finley, Jason P., finleyjp@pierceColl.edu, Los Angeles Pierce Coll – Ow
Finley, Mark, (309) 268-8642 mark.finley@heartland.edu, Heartland Community Coll – Gg
Finley, Robert J., 217-244-8389 finley@isgs.uiuc.edu, Univ of Illinois, Urbana-Champaign – Go
Finlley, Cathy, (314) 977-3126 finleuca@slu.edu, Saint Louis Univ – Ow
Finn, Gregory C., (905) 688-5550 x3528 greg.finn@brocku.ca, Brock Univ – Gx
Finnegan, David L., (505) 667-6548 Los Alamos National Laboratory – Oa
Finnegan, Noah J., 831-459-5110 nfinnegan@pmc.ucsc.edu, Univ of California, Santa Cruz – Gm
Finnegan, Seth, sethf@berkeley.edu, Univ of California, Berkeley – Pio
Finney, Bruce P., (907) 474-7724 Univ of Alaska, Fairbanks – Ou
Finney, Stanley C., (562) 985-8637 scfinney@csulb.edu, California State Univ, Long Beach – Ps
Finzel, Emily, (319) 335-0405 emily-finzel@uiowa.edu, Univ of Iowa – GsrGt
Fio, Karmen, +38514606088 karmen.fio@gmail.com, Univ of Zagreb – PgGg
Fiore, Arlene M., (845) 365-8550 amfiore@ldeo.columbia.edu, Columbia Univ – Oa
Fiorillo, Anthony R., (214) 421-3466 Southern Methodist Univ – Pv
Firing, Eric, (808) 956-7894 efiring@hawaii.edu, Univ of Hawai'i, Manoa – Op
Fischer, Alfred G., (213) 740-5821 fisher@usc.edu, Univ of Southern California – Gs
Fischer, Emily, evf@rams.colostate.edu, Colorado State Univ – Oa
Fischer, Karen M., (401) 863-1360 Karen_Fischer@Brown.edu, Brown Univ – Ys
Fischer, Mark P., (815) 753-0523 mfischer@niu.edu, Northern Illinois Univ – Gco
Fischer, Tobias, (505) 277-0683 fischer@unm.edu, Univ of New Mexico – Gv
Fish, Jennifer M., 831-459-1235 jmsfish@ucsc.edu, Univ of California, Santa Cruz – On
Fishe, Quentin , +44(0) 113 34 31920 q.j.fisher@leeds.ac.uk, Univ of Leeds – Np
Fisher, Andrew T., (831) 459-5598 afisher@ucsc.edu, Univ of California, Santa Cruz – Hw
Fisher, Cynthia G., (610) 436-2108 cfisher@wcupa.edu, West Chester Univ – Ou
Fisher, Daniel C., (734) 764-0488 dcfisher@umich.edu, Univ of Michigan – Pi
Fisher, David A., 613-996-7623 fisher@nrcan-rncan.gc.ca, Univ of Ottawa – Ol
Fisher, David E., (305) 284-3254 dfisher@miami.edu, Univ of Miami – Cs
Fisher, Donald M., (814) 865-3206 fisher@geosc.psu.edu, Pennsylvania State Univ, Univ Park – GctOm
Fisher, Liz, +440151 795 4390 E.H.Fisher@liverpool.ac.uk, Univ of Liverpool – Co
Fisher, Nicholas S., (631) 632-8649 nicholas.fisher@stonybrook.edu, SUNY, Stony Brook – Oc
Fisher, Timothy G., (219) 980-7122 tfisher@indiana.edu, Indiana Univ / Purdue Univ, Indianapolis – Gl
Fisher, Timothy G., (419) 530-2009 timothy.fisher@utoledo.edu, Univ of Toledo – GlmGn
Fisher, William L., (512) 471-5600 wfisher@mail.utexas.edu, Univ of Texas at Austin – GsrGo
Fisher, Jr., Thomas R., (410) 221-8432 fisher@hpl.umces.edu, Univ of Maryland – ObiCl
Fishman, Jack, (314) 977-3132 jfishma2@slu.edu, Saint Louis Univ – Oa
Fishwick, Stewart, +44 0116 252 3810 sf130@le.ac.uk, Leicester Univ – Yg
Fisk, Aaron, 519-253-3000 x4740 afisk@uwindsor.ca, Univ of Windsor – ObCs
Fisk, Martin R., (541) 737-5208 mfisk@coas.oregonstate.edu, Oregon State Univ – Yr
Fiske, Richard S., (202) 633-1794 fisker@si.edu, Smithsonian Institution / National Museum of Natural History – Gv

Fitak, Madge R., 614-265-6585 madge.fitak@dnr.state.oh.us, Ohio Dept of Natural Resources – Gg
Fitchett, Rebekah, 773-442-6052 R-Fitchett@neiu.edu, Northeastern Illinois Univ – GgCg
Fitton, J. G., +44 (0) 131 650 8529 Godfrey.Fitton@ed.ac.uk, Edinburgh Univ – Gi
Fitz, Thomas J., (715) 682-1852 tfitz@northland.edu, Northland Coll – GgxGc
Fitzer, Susan, +44 01413305442 Susan.Fitzer@glasgow.ac.uk, Univ of Glasgow – Ob
Fitzgerald, David, (210) 436-3235 Saint Mary's Univ – Gd
FitzGerald, Duncan M., (617) 353-2530 dunc@bu.edu, Boston Univ – On
Fitzgerald, Francis S., 303-384-2644 ffitzger@mines.edu, Colorado School of Mines – OirGg
Fitzgerald, Paul G., (315) 443-2672 pgfitzge@syr.edu, Syracuse Univ – GtCc
Fitzgerald, William F., 860-405-9158 william.fitzgerald@uconn.edu, Univ of Connecticut – Oc
Fitzjarrald, David R., (518) 437-8735 dfitzjarrald@albany.edu, SUNY, Albany – Oaw
Fitzpatrick, John R., (505) 667-4761 Los Alamos National Laboratory – On
FitzPatrick, Meriel E., +44 1752 584769 m.e.fitzpatrick@plymouth.ac.uk, Univ of Plymouth – PlgGr
Fitzpatrick, Patrick J., (228) 688-1157 fitz@gri.msstate.edu, Mississippi State Univ – Ow
Fitzpatrick, Stephan, (678) 891-3773 sfitzpatrick1@gsu.edu, Georgia State Univ, Perimeter Coll, Clarkston Campus – HwCgOs
Fitzsimmons, Jessica N., jessfitz@tamu.edu, Texas A&M Univ – OcCta
Fitzsimmons, Kevin, (520) 626-3324 kevfitz@ag.arizona.edu, Univ of Arizona – On
Fjeldheim, Nancy, njf@brown.edu, Brown Univ – On
Fjell, Dale, (785) 532-5833 dfjell@ksu.edu, Kansas State Univ – So
Flagg, Charles, (631) 632-3184 Charles.Flagg@stonybrook.edu, SUNY, Stony Brook – Op
Flaherty, Frank A., (229) 333-5665 flaherty@valdosta.edu, Valdosta State Univ – On
Flaig, Peter P., peter.flaig@beg.utexas.edu, Univ of Texas at Austin – GsrEo
Flake, Rex , 509.963.1114 rex@geology.cwu.edu, Central Washington Univ – Yd
Flament, Pierre J., (808) 956-6663 pflament@hawaii.edu, Univ of Hawai'i, Manoa – Op
Flanagan, Michael, flanagam@sunysuffolk.edu, Suffolk County Community Coll, Ammerman Campus – Og
Flanagan, Timothy, (413) 236-4503 tflanaga@berkshirecc.edu, Berkshire Community Coll – Oei
Flato, Gregory M., (250) 363-8223 greg.flato@ec.gc.ca, Univ of Victoria – Oa
Flawn, Peter T., (512) 471-1825 pflawn@po.utexas.edu, Univ of Texas at Austin – Eg
Flechsig, Sandra, 713.348.3326 geol@rice.edu, Rice Univ – On
Fleck, Michelle, (435) 613-5232 michelle.fleck@usu.edu, Utah State Univ – Oe
Fleeger, Gary M., (717) 702-2045 gfleeger@pa.gov, Pennsylvania Bureau of Topographic & Geologic Survey – GlrHw
Fleet, Michael E., (519) 661-3184 mfleet@uwo.ca, Western Univ – Gz
Flegal, Russell, (831) 459-2093 rflegal@es.ucsc.edu, Univ of California, Santa Cruz – Cg
Fleisher, P. Jay, (607) 286-7541 fleishpj@oneonta.edu, SUNY, Oneonta – Gl
Fleming, Thomas H., 203-392-5837 flemingt1@southernct.edu, Southern Connecticut State Univ – GiCgc
Flemings, Peter B., 512-471-6156 flemings@ig.utexas.edu, Univ of Texas at Austin – Gr
Flemming, Roberta L., (519) 661-3143 rflemmin@uwo.ca, Western Univ – GzXmOm
Flesch, Lucy M., (765) 494-0263 lmflesch@purdue.edu, Purdue Univ – Yg
Flesken, Luuk, +31(0) 317 485467 l.fleskens@leeds.ac.uk, Univ of Leeds – Ge
Flessa, Karl W., (520) 621-7336 kflessa@email.arizona.edu, Univ of Arizona – Po
Fletcher, Austin, rfletche@uoguelph.ca, Univ of Guelph – On
Fletcher, Charles H., (808) 956-2582 fletcher@soest.hawaii.edu, Univ of Hawai'i, Manoa – GsOnu
Fletcher, John, jfletche@cicese.mx, Centro de Investigación Científica y de Educación Superior de Ensenada – Gpt
Fletcher, William K., (604) 822-2392 wfletcher@eos.ubc.ca, Univ of British Columbia – Ce
Flierl, Glenn R., (617) 253-4692 glenn@pimms.mit.edu, Massachusetts Inst of Tech – Op
Flint, Stephen, +44 0161 306-6971 stephen.flint@manchester.ac.uk, Univ of Manchester – GrsGo
Flood, Beverley, 612-624-1603 floo0017@umn.edu, Univ of Minnesota, Twin Cities – Py
Flood, Peter, +61-2-67732329 pflood@une.edu.au, Univ of New England – EcGt
Flood, Roger D., (631) 632-6971 roger.flood@stonybrook.edu, SUNY, Stony Brook – Ou
Flood, Tim P., (920) 403-1356 tim.flood@snc.edu, Saint Norbert Coll – Gv
Flores, Alejandro N., lejoflores@boisestate.edu, Boise State Univ – HqOa
Flores, Kennet, 718-951-5000 x3253 KEFlores@brooklyn.cuny.edu, Brooklyn Coll (CUNY) – Gpc
Flores-Garcia, Mari C., (818) 677-3541 mari.flores@csun.edu, California State Univ, Northridge – On
Floris, Mario, +390498279121 mario.floris@unipd.it, Università degli Studi di Padova – NgOr
Flory, David M., (515) 294-0264 flory@iastate.edu, Iowa State Univ of Science & Tech – OwaGt
Flory, Richard A., rflory@csuchico.edu, California State Univ, Chico – Ps
Flowers, George C., (504) 862-3192 flowers@tulane.edu, Tulane Univ – Hw
Flowers, Gwenn, (604) 291-6638 gflowers@sfu.ca, Simon Fraser Univ – Gl
Flowers, Rebecca M., (303) 492-5135 rebecca.flowers@colorado.edu, Univ of Colorado – CgcGt
Flowers Falls, Emily, (540) 458-8868 fallse@wlu.edu, Washington & Lee Univ – Gg
Floyd, Barry, 518-473-9988 esogis@mail.nysed.gov, New York State Geological Survey – On
Fluegeman, Richard H., (765) 285-8267 rfluegem@bsu.edu, Ball State Univ – Pm
Flynn, John J., (212) 769-5806 American Museum of Natural History – Pv
Flynn, Luke P., (808) 956-3154 Univ of Hawai'i, Manoa – Or
Flynn, Natalie, 215.951.1269 La Salle Univ – Gz
Flynn, Wendilyn, 970-351-1071 wendilyn.flynn@unco.edu, Univ of Northern Colorado – Owa
Fodor, Ronald V., (919) 515-7177 ron_fodor@ncsu.edu, North Carolina State Univ – Gi
Fogel, David N., (505) 665-3305 fogel@lanl.gov, Los Alamos National Laboratory – On
Fogg, Graham E., (916) 752-6810 Univ of California, Davis – Hy
Foit, Jr., Franklin F., (509) 335-3093 Washington State Univ – Gz
Foland, Kenneth A., foland.1@osu.edu, Ohio State Univ – Cc
Foley, Bradford, (814) 863-3591 bjf5382@psu.edu, Pennsylvania State Univ, Univ Park – YgXyGt
Foley, Duncan, (253) 535-7568 foleyd@plu.edu, Pacific Lutheran Univ – GeHyOe
Foley, John P., (406) 496-4414 jfoley@mtech.edu, Montana Tech of The Univ of Montana – On
Folk, Robert L., (512) 471-5294 rlfolk@mail.utexas.edu, Univ of Texas at Austin – Gd
Follmer, Leon R., (217) 244-6945 follmer@isgs.uiuc.edu, Univ of Illinois, Urbana-Champaign – Sa
Follows, Michael, (617) 253-5939 mick@plume.mit.edu, Massachusetts Inst of Tech – Op
Folorunso, I. O., foloisgood@yahoo.com, Univ of Ilorin – GoCg
Foltltin, Thomas, (202) 994-7156 George Washington Univ – Oy
Fomel, Sergey, 512-475-9573 sergey.fomel@beg.utexas.edu, Univ of Texas at Austin, Jackson School of Geosciences – Yx
Fondran, Carol, (216) 987-5227 Carol.Fondran@tri-c.edu, Cuyahoga Community Coll - Western Campus – Og
Fones, Gary, +44 023 92 842252 gary.fones@port.ac.uk, Univ of Portsmouth – Cm
Fonstad, Mark, fonstad@uoregon.edu, Univ of Oregon – Oy
Fontaine, Fabrice R., +262262938207 fabrice.fontaine@univ-reunion.fr, Universite de la Reunion – Yg
Fontana, Alessandro, +390498279118 alessandro.fontana@unipd.it, Università degli Studi di Padova – Gm
Foote, Michael J., (773) 702-4320 mfoote@midway.uchicago.edu, Univ of Chicago – Pg
Forbriger, Thomas, +49-721-6084593 thomas.forbriger@kit.edu, Karlsruhe Inst of Tech – Ys
Forcino, Frank, (828) 227-2888 flforcino@wcu.edu,

Western Carolina Univ – PgOeGr
Ford, Allistair, +44 (0) 191 208 7121 alistair.ford@ncl.ac.uk, Univ of Newcastle Upon Tyne – Gq
Ford, Andrew, 509-335-7846 forda@wsu.edu, Washington State Univ – Ou
Ford, Arianne, Arianne.Ford@jcu.edu.au, James Cook Univ – GqOiEg
Ford, Derek C., (905) 525-9140 x20132 dford@mcmaster.ca, McMaster Univ – GmHwCc
Ford, Heather, heather.ford@yale.edu, Univ of California, Riverside – Ys
Ford, Jaime L., 304-367-8379 Jaime.L.Ford@ivv.nasa.gov, Fairmont State Univ – Oe
Ford, Mark T., 361-593-3590 mark.ford@tamuk.edu, Texas A&M Univ, Kingsville – GxzCg
Ford, Richard L., (801) 626-6942 rford@weber.edu, Weber State Univ – GmOe
Fordyce, R. Ewan, +64 3 479-7510/7535 ewan.fordyce@otago.ac.nz, Univ of Otago – Pv
Foreman, Brady Z., 360-650-2546 brady.foreman@wwu.edu, Western Washington Univ – Grs
Foreman, Michael G. G., (250) 363-6306 ForemanM@pac.dfo-mpo.gc.ca, Univ of British Columbia – Op
Foresi, Luca Maria, luca.foresi@unisi.it, Univ of Siena – Gs
Foret, Jim, 337-482-6064 jaf3663@louisiana.edu, Univ of Louisiana at Lafayette – Ge
Forgotson, Jr., James M., (405) 325-4451 jforgot@ou.edu, Univ of Oklahoma – Go
Fornaciari, Elena, +390498279185 eliana.fornaciari@unipd.it, Università degli Studi di Padova – Pgm
Fornari, Daniel J., (508) 289-2857 dfornari@whoi.edu, Woods Hole Oceanographic Institution – Gu
Forno, Maria Gabriella, gabriella.forno@unito.it, Università di Torino – Gg
Forrest, David, +4401413305401 david.forrest@glasgow.ac.uk, Univ of Glasgow – Oi
Forsman, Nels F., (701) 777-4349 nels.forsman@engr.und.edu, Univ of North Dakota – GdXgGh
Forster, Piers M., +44(0) 113 34 36476 p.m.forster@leeds.ac.uk, Univ of Leeds – Oa
Forster, Richard R., (801) 581-3611 rick.forster@geog.utah.edu, Univ of Utah – Or
Forsyth, Donald W., (401) 863-1699 donald_forsyth@brown.edu, Brown Univ – YrsGt
Forte, Alessandro Marco, 514-987-3000 #5607 forte.alessandro@uqam.ca, Universite du Quebec a Montreal – Yg
Fortes, Andrew, +44 020 7679 32383 andrew.fortes@ucl.ac.uk, Univ Coll London – Xy
Fortes, Dominic, andrew.fortes@ucl.ac.uk, Birkbeck Coll – Og
Fortier, Richard, (418) 656-2746 richard.fortier@ggl.ulaval.ca, Universite Laval – Yg
Fortin, Danielle, 613-562-5800 6423 dfortin@uottawa.ca, Univ of Ottawa – Py
Fortner, Sarah K., (937) 327-7328 sfortner@wittenberg.edu, Wittenberg Univ – ClOeHg
Fosdick, Julie C., (812) 855-6109 jfosdick@indiana.edu, Indiana Univ, Bloomington – Gt
Foss, Donald J., (415) 485-9523 jcprice@metro.net, Coll of Marin – Og
Foster, David A., (352) 392-2231 dafoster@ufl.edu, Univ of Florida – GtCc
Foster, Gavin, +44 (0)23 8059 3786 Gavin.Foster@noc.soton.ac.uk, Univ of Southampton – Csg
Foster, Patrick, +44 01326 371828 P.J.Foster@exeter.ac.uk, Exeter Univ – Nm
Foster, Stuart, (270) 745-5976 stuart.foster@wku.edu, Western Kentucky Univ – Oa
Foster Jr., Charles T., (319) 335-1801 tom-foster@uiowa.edu, Univ of Iowa – GptGg
Foti, Pamela, (928) 523-6196 Pam.foti@nau.edu, Northern Arizona Univ – On
Fotopoulos, Georgia, (613) 533-6639 georgia.fotopoulos@queensu.ca, Queen's Univ – Yd
Foufoula-Georgiou, Efi, (612) 626-0369 Univ of Minnesota, Twin Cities – Hg
Fouke, Bruce W., (217) 244-5431 fouke@illinois.edu, Univ of Illinois, Urbana-Champaign – GsdGb
Foulger, Gillian R., 0191 3742514 g.r.foulger@durham.ac.uk, Durham Univ – Yg
Fountain, Andrew G., (503) 725-3386 andrew@pdx.edu, Portland State Univ – Gl
Fountain, John C., 919-515-3717 fountain@ncsu.edu, North Carolina State Univ – Hw
Fournelle, John, (608) 262-7964 johnf@geology.wisc.edu, Univ of Wisconsin, Madison – GiOn
Fournier, Benoit, (418) 656-3930 benoit.fournier@ggl.ulaval.ca, Universite Laval – NgOmGg
Fournier, Gregory, 617-324-6164 g4nier@mit.edu, Massachusetts Inst of Tech – Py
Fournier, Robert O., (902) 494-3666 robert.fournier@dal.ca, Dalhousie Univ – Ob
Fovell, Robert, (310) 206-9956 fovell@atmos.ucla.edu, Univ of California, Los Angeles – Oa
Fowell, Sarah, (907) 474-7810 ffsjf@uaf.edu, Univ of Alaska, Fairbanks – Pl
Fowle, David A., (785) 864-1955 fowle@ku.edu, Univ of Kansas – Py
Fowler, Gerald A., fowler@uwp.edu, Univ of Wisconsin, Parkside – PgGg
Fowler, Malcolm M., (505) 667-5439 Los Alamos National Laboratory – Oa
Fowler, Mike, +44 023 92 842293 mike.fowler@port.ac.uk, Univ of Portsmouth – Gi
Fowler, Sarah, sarah.fowler@ees.kuleuven.be, Katholieke Universiteit Leuven – Gip
Fox, David L., 612-624-6361 dlfox@umn.edu, Univ of Minnesota, Twin Cities – Pg
Fox, James E., (605) 394-2461 James.fox@sdsmt.edu, South Dakota School of Mines & Tech – Gs
Fox, Jeff, 740 548 7348 jeff.fox@dnr.state.oh.us, Ohio Dept of Natural Resources – Ys
Fox, Laurel R., (831) 459-2533 fox@biology.ucsc.edu, Univ of California, Santa Cruz – On
Fox, Lewis, 954-201-6674 lfox@broward.edu, Broward Coll, Central Campus – Oc
Fox, Lydia K., (209) 946-2481 lkfox@pacific.edu, Univ of the Pacific – GizOe
Fox, Neil I., 573-882-2144 foxn@missouri.edu, Univ of Missouri, Columbia – Oa
Fox, Peter A., (518) 727-4862 pfox@cs.rpi.edu, Rensselaer Polytechnic Inst – GqOig
Fox, Richard C., (780) 492-5491 richard.fox@ualberta.ca, Univ of Alberta – Pv
Fox, William T., william.t.fox@williams.edu, Williams Coll – Gs
Fox-Dobbs, Kena L., 253-879-2458 kena@pugetsound.edu, Univ of Puget Sound – PgCsGe
Fox-Kemper, Baylor, 401-863-3979 Baylor_Fox-Kemper@brown.edu, Brown Univ – Op
Foxon, Tim, +44(0) 113 34 37910 t.j.foxon@leeds.ac.uk, Univ of Leeds – Ge
Frail, Pam, (902) 585-1513 pam.frail@acadiau.ca, Acadia Univ – Gg
Frailey, Scott M., 217-244-2430 frailey@isgs.uiuc.edu, Illinois State Geological Survey – Ng
Fraiser, Margaret L., 414-229-3827 mfraiser@uwm.edu, Univ of Wisconsin, Milwaukee – Pi
Frajka-Williams, Eleanor, +44 (0)23 80596044 e.frajka-williams@noc.soton.ac.uk, Univ of Southampton – Op
Fralick, Philip W., (807) 343-8288 philip.fralick@lakeheadu.ca, Lakehead Univ – Gs
Fram, Jonathan, 541-737-3966 jfram@coas.oregonstate.edu, Oregon State Univ – Op
Frame, Jeffrey, 217-244-9575 frame@illinois.edu, Univ of Illinois, Urbana-Champaign – Oaw
Francis, Don, donald.francis@mcgill.ca, McGill Univ – Gi
Francis, Robert, (206) 543-7345 Univ of Washington – Ob
Francis, Robert D., (562) 985-4929 jfrancis@csulb.edu, California State Univ, Long Beach – Co
Franco, Aldina, +44 (0)1603 59 2721 a.franco@uea.ac.uk, Univ of East Anglia – Ge
Frank, Charles O., (618) 453-7365 frank@geo.siu.edu, Southern Illinois Univ Carbondale – Gx
Frank, Kenneth, (902) 426-3498 frankk@mar.dfo-mpo.gc.ca, Dalhousie Univ – Ob
Frank, Mark R., (815) 753-8395 mfrank@niu.edu, Northern Illinois Univ – Cp
Frank, Tracy D., (402) 472-9799 tfrank2@unl.edu, Univ of Nebraska, Lincoln – GssCs
Frank, William M., frank@ems.psu.edu, Pennsylvania State Univ, Univ Park – Ow

Franke, Otto L., (212) 650-6984 Graduate School of the City Univ of New York - Gc
Frankel, Arthur, 206-553-0626 afrankel@uw.edu, Univ of Washington - Ys
Frankenberg, Christian, 626-395-6143 cfranken@caltech.edu, California Inst of Tech - Or
Frankic, Anamarija, anamarija.frankic@umb.edu, Univ of Massachusetts, Boston - Oe
Franklin, David, +44 023 92 843540 david.franklin@port.ac.uk, Univ of Portsmouth - Ym
Franks, Peter J. S., (858) 534-7528 pfranks@ucsd.edu, Univ of California, San Diego - Ob
Franseen, Evan K., (785) 864-2072 evanf@kgs.ku.edu, Univ of Kansas - Gs
Fransolet, Andre-Mathieu, 32 4 366 22 06 amfransolet@ulg.ac.be, Universite de Liege - Gz
Frantz, Carie M., (801) 626-6181 cariefrantz@weber.edu, Weber State Univ - PyCgOn
Franz, Eldon H., (206) 300-9259 franz@wsu.edu, Washington State Univ - Sf
Franz, Kristie, (515) 294-7454 kfranz@iastate.edu, Iowa State Univ of Science & Tech - Hs
Franzi, David A., (518) 564-4033 franzida@plattsburgh.edu, Plattsburgh State Univ (SUNY) - Gl
Frappier, Amy, 518-580-8371 afrappie@skidmore.edu, Skidmore Coll - GePe
Fraser, Alastair, 44-20-7594-6530 afraser@egi.utah.edu, Univ of Utah - Gco
Fraser, Alistair B., abf1@psu.edu, Pennsylvania State Univ, Univ Park - Ow
Fraser, Donalf G., +44 (1865) 272033 don@earth.ox.ac.uk, Univ of Oxford - Cg
Fraser, Nicholas C., (540) 666-8611 Virginia Polytechnic Inst & State Univ - Pv
Fraser-Smith, Antony C., (650) 723-3684 Stanford Univ - Ym
Fratianni, Simona, simona.fratianni@unito.it, Università di Torino - Oy
Fratta, Dante, 608-265-5644 fratta@wisc.edu, Univ of Wisconsin, Madison - Ng
Frazer, L. N., (808) 956-3724 neil@soest.hawaii.edu, Univ of Hawai'i, Manoa - Ys
Frazier, William J., (706) 575-7142 frazier_bill@columbusstate.edu, Columbus State Univ - GsrGd
Frazin, Richard, (734) 647-9689 rfrazin@umich.edu, Univ of Michigan - Or
Frederick, Daniel L., (931) 221-7455 frederickd@apsu.edu, Austin Peay State Univ - Gsr
Frederick, Holly, 570-408-4880 holly.frederick@wilkes.edi, Wilkes Univ - Sdb
Frederick, John E., (773) 702-3237 Univ of Chicago - Oa
Frederickson, Paul A., 831-595-5212 pafreder@nps.edu, Naval Postgraduate School - Ow
Frederiksen, Andrew, (204) 474-9460 andrew_frederiksen@umanitoba.ca, Univ of Manitoba - YsGt
Fredrick, Kyle, (724) 938-4180 fredrick@calu.edu, California Univ of Pennsylvania - GgHgGm
Fredricks, Helen, 508-289-3678 hfredricks@whoi.edu, Woods Hole Oceanographic Institution - Oc
Freed, Andrew M., (765) 496-3738 freed@purdue.edu, Purdue Univ - Gg
Freed, Jane S., (208) 885-7479 jfreed@uidaho.edu, Univ of Idaho - Oi
Freed, Robert L., (210) 999-7092 bfreed@trinity.edu, Trinity Univ - Gz
Freeman, Katherine H., (814) 863-8177 khf4@psu.edu, Pennsylvania State Univ, Univ Park - Cos
Freeman, Rebecca, (859) 257-3758 rebecca.freeman@uky.edu, Univ of Kentucky - GrPi
Freeman, Thomas J., (573) 882-6673 FreemanT@missouri.edu, Univ of Missouri - Gd
Freeman, Veronica, 740-376-4775 freemanv@marietta.edu, Marietta Coll - Pg
Freeman-Lynde, Raymond, (706) 542-2391 rfreeman@uga.edu, Univ of Georgia - Gu
Frei, Allan, (212) 772-5322 afrei@hunter.cuny.edu, Hunter Coll (CUNY) - Hs
Frei, Michaela, 089/2180 6590 michaela.frei@iaag.geo.uni-muenchen.de, Ludwig-Maximilians-Universitaet Muenchen - Gg
Freiburg, Jared, 217-244-2495 freiburg@illinois.edu, Univ of Illinois, Urbana-Champaign - GgzGx
Freiermuth, Sue, (715) 425-3345 susan.m.freiermuth@uwrf.edu, Univ of Wisconsin, River Falls - On
Freile, Deborah, (201) 200-3161 dfreile@njcu.edu, New Jersey City Univ - GseGg
Freitas, Maria da Conceição P., +351217500352 cfreitas@fc.ul.pt, Universidade de Lisboa - GusGe
French, Adam, (605) 394-1649 adam.french@sdsmt.edu, South Dakota School of Mines & Tech - Oaw
French, Bevan, (202) 633-1326 Smithsonian Institution / National Museum of Natural History - Xm
French, Mark A., (609) 984-6587 mark.french@dep.state.nj.us, New Jersey Geological and Water Survey - Hq
French, Melodie, mefrench@umd.edu, Univ of Maryland - Ygx
French, Melodie E., 713.348.5088 mefrench@rice.edu, Rice Univ - Gc
Frenette, Jean, 418-656-8123 frenette@ggl.ulaval.ca, Universite Laval - Gz
Freshley, Mark D., (509) 372-6094 mark.freshley@pnl.gov, Pacific Northwest National Laboratory - Hw
Frew, Nelson M., (508) 289-2489 nfrew@whoi.edu, Woods Hole Oceanographic Institution - Ca
Frey, Bonnie A., (505) 835-5160 bfrey@nmt.edu, New Mexico Inst of Mining & Tech - Cg
Frey, Frederick A., (617) 253-2818 fafrey@mit.edu, Massachusetts Inst of Tech - CtGvi
Frey, Friedrich, 089/21804332 f.frey@lmu.de, Ludwig-Maximilians-Universitaet Muenchen - Gz
Frey, Holli M., 518-388-6418 freyh@union.edu, Union Coll - GvCag
Freymueller, Jeffrey T., (907) 474-7286 jeff@giseis.alaska.edu, Univ of Alaska, Fairbanks - Yd
Frez, Jose, jofrez@cicese.mx, Centro de Investigación Científica y de Educación Superior de Ensenada - Ys
Fricke, Henry C., (719) 389-6514 hfricke@coloradoColl.edu, Colorado Coll - Cs
Fricker, Helen A., (858) 534-6145 hafricker@ucsd.edu, Univ of California, San Diego - Yr
Frid, Chris, +44 0151 795 4382 C.L.J.Frid@liverpool.ac.uk, Univ of Liverpool - Ob
Friedl, Mark, friell@bu.edu, Boston Univ - Or
Friedman, Gerald M., (718) 951-5840 Graduate School of the City Univ of New York - Gs
Friedman, Matt, +44 (1865) 272035 matt.friedman@earth.ox.ac.uk, Univ of Oxford - Po
Friedmann, Samuel (Julio) J., (925) 423-0585 friedmann2@llnl.gov, Lawrence Livermore National Laboratory - Gg
Friedrich, Anke, friedrich@iaag.geo.uni-muenchen.de, Ludwig-Maximilians-Universitaet Muenchen - GgtCc
Friedrichs, Carl T., (804) 684-7303 cfried@vims.edu, Coll of William & Mary - On
Friehauf, Kurt, (610) 683-4446 friehauf@kutztown.edu, Kutztown Univ of Pennsylvania - EmCgGx
Friend, Donald A., (507) 389-2617 donald.friend@mnsu.edu, Minnesota State Univ - OyGm
Frierson, Dargan M., 206-685-7364 dargan@atmos.washington.edu, Univ of Washington - Oa
Frind, Emil O., (519) 888-4567 (Ext. 3959) Univ of Waterloo - Hq
Frisbee, Marty, (765) 494-8678 mfrisbee@purdue.edu, Purdue Univ - Hws
Frisia, Silvia, +61 2 4921 5402 Silvia.Frisia@newcastle.edu.au, Univ of Newcastle - ClPeGd
Frisk, DeAnn, (515) 294-4477 dfrisk@iastate.edu, Iowa State Univ of Science & Tech - On
Fritsch, J. Michael, (814) 863-1842 fritsch@ems.psu.edu, Pennsylvania State Univ, Univ Park - Ow
Fritsche, A. E., a.eugene.fritsche@csun.edu, California State Univ, Northridge - Gd
Fritsen, Christian H., (775) 673-7487 cfritsen@dri.edu, Univ of Nevada, Reno - Og
Fritton, Daniel D., (814) 865-1143 ddf@psu.edu, Pennsylvania State Univ, Univ Park - Sp
Fritts, Harold C., (520) 621-1608 hfritts@ltrr.arizona.edu, Univ of Arizona - Pe
Fritz, Allan, (785) 532-7245 akf@ksu.edu, Kansas State Univ - So
Fritz, Sherilyn C., (402) 472-6431 sfritz2@unl.edu, Univ of Nebraska, Lincoln - Pem
Fritz, William J., 718-982-2400 william.fritz@csi.cuny.edu, Coll of Staten Island - Grv
Frizado, Joseph P., (419) 372-7202 frizado@bgnet.bgsu.edu, Bowling Green State Univ - Oi
Froehlich, David J., (512) 223-4894 eohippus@austincc.edu,

Austin Community Coll District – PvOn
Froelich, Philip, 850-644-4331 froelich@ocean.fsu.edu, Florida State Univ – Og
Froese, Duane, (780) 492-1968 duane.froese@ualberta.ca, Univ of Alberta – GlmPe
Frohlich, Cliff, (512) 471-0460 cliff@ig.utexas.edu, Univ of Texas at Austin – Ys
Frolking, Stephen E., (603) 862-0244 steve.frolking@unh.edu, Univ of New Hampshire – Oa
Frolking, Tod A., (740) 587-6222 frolking@denison.edu, Denison Univ – Oy
Fromm, Jeanne M., (605) 582-3416 Jeanne.Fromm@usd.edu, Univ of South Dakota – GgHs
Frost, B. Ronald, (307) 766-4290 rfrost@uwyo.edu, Univ of Wyoming – Gp
Frost, Bruce W., 206-543-7186 frost@ocean.washington.edu, Univ of Washington – Ob
Frost, Carol D., frost@uwyo.edu, Univ of Wyoming – CcGit
Frost, Eric G., (619) 594-5003 eric.frost@sdsu.edu, San Diego State Univ – Or
Frost, Gina M., 209 954 5380 gfrost@deltaColl.edu, San Joaquin Delta Coll – Gg
Fry, Matthew, 940-369-7576 mfry@unt.edu, Univ of North Texas – On
Fryar, Alan E., (859) 257-4392 alan.fryar@uky.edu, Univ of Kentucky – HwGeCl
Fryer, Brian J., 519-253-3000 ext, 3750 bfryer@uwindsor.ca, Univ of Windsor – Cla
Fryer, Karen H., (740) 368-3618 khfryer@owu.edu, Ohio Wesleyan Univ – GcpGt
Fryer, Patricia B., (808) 956-3146 pfryer@soest.hawaii.edu, Univ of Hawai'i, Manoa – Gu
Fryett, I, fryettI@cf.ac.uk, Cardiff Univ – On
Fryxell, Joan E., (909) 537-5311 jfryxell@csusb.edu, California State Univ, San Bernardino – Gct
Fthenakis, Vasilis M., (516) 282-2830 Columbia Univ – Oa
Fu, Qi, (713) 743-3660 qfu3@uh.edu, Univ of Houston – CosCa
Fu, Qiang, 206-685-2070 qfu@atmos.washington.edu, Univ of Washington – Oa
Fu, Qilong, 512-232-9372 qilong.fu@beg.utexas.edu, Univ of Texas at Austin – Gsr
Fu, Rong, 512-232-7932 rongfu@jsg.utexas.edu, Univ of Texas at Austin – Oa
Fubelli, Giandomenico, giandomenico.fubelli@unito.it, Università di Torino – Gm
Fueglistaler, Stephan A., (609) 258-8238 stf@princeton.edu, Princeton Univ – Oa
Fuellhart, Kurtis G., (717) 477-1309 kgfuel@ship.edu, Shippensburg Univ – On
Fueten, Frank, (905) 688-5550 (Ext. 3856) FFueten@brocku.ca, Brock Univ – Gc
Fuge, Ron, +44 (0)1970 622 642 rrf@aber.ac.uk, Aberystwyth Univ – GbCgGe
Fugitt, Frank, (614) 265-6759 frank.fugitt@dnr.state.oh.us, Ohio Dept of Natural Resources – GrHw
Fuhrmann, Christopher, (662) 268-1032 Ext 219 cmf396@msstate.edu, Mississippi State Univ – Oa
Fuiman, Lee A., (361) 749-6775 lee.fuiman@utexas.edu, Univ of Texas at Austin – Ob
Fujita, Kazuya, 517-355-0142 fujita@msu.edu, Michigan State Univ – Ys
Full, Robert J., (510) 642-9896 rjfull@garnet.berkeley.edu, Univ of California, Berkeley – Pi
Fullagar, Paul D., (919) 962-0677 fullagar@unc.edu, Univ of North Carolina, Chapel Hill – Cc
Fuller, Doult O., (202) 994-8073 dfuller@gwu.edu, George Washington Univ – Oy
Fullmer, James W., (203) 392-5841 fullmerj1@southernct.edu, Southern Connecticut State Univ – Ow
Fulthorpe, Craig S., (512) 471-0459 craig@ig.utexas.edu, Univ of Texas at Austin – Gu
Fulton, Patrick M., (979) 862-2493 pfulton@tamu.edu, Texas A&M Univ – YhHyOu
Fulweiler, Robinson, 617-358-5466 rwf@bu.edu, Boston Univ – Cm
Fung, Inez, (510) 643-9367 inez@atmos.berkeley.edu, Univ of California, Berkeley – Oa
Funning, Gareth J., gareth.funning@ucr.edu, Univ of California, Riverside – YdgYs
Furbish, David J., 615-322-2137 david.j.furbish@vanderbilt.edu, Vanderbilt Univ – GmHg
Furlong, Kevin P., (814) 863-0567 kevin@geodyn.psu.edu, Pennsylvania State Univ, Univ Park – Gt
Furman, Tanya, (814) 865-5782 furman@psu.edu, Pennsylvania State Univ, Univ Park – Cg
Fusch, Richard D., (740) 368-3616 rdfusch@owu.edu, Ohio Wesleyan Univ – Ou
Fusseis, Florian, +44 (0) 131 650 6755 ffusseis@staffmail.ed.ac.uk, Edinburgh Univ – Gc
Fyfe, John C., (250) 363-8236 john.fyfe@ec.gc.ca, Univ of Victoria – Oa
Fyson, William K., 613-745-6645 wfyson@uottawa.ca, Univ of Ottawa – Gc

## G

Gaasterland, Terry, (858) 822-4600 tgaasterland@ucsd.edu, Univ of California, San Diego – Ob
Gabbott, Sarah, +440116 252 3636 sg21@le.ac.uk, Leicester Univ – Po
Gabel, Mark, (605) 642-9035 Black Hills State Univ – Pb
Gabel, Sharon, 281-425-6335 sgabel@lee.edu, Lee Coll – GseGg
Gabet, Emmanuel, 408 924-5035 manny.gabet@sjsu.edu, San Jose State Univ – Gm
Gabitov, Rinat, (662) 268-1032 Ext 218 rg850@msstate.edu, Mississippi State Univ – Cas
Gable, Carl W., (505) 665-3533 gable@lanl.gov, Los Alamos National Laboratory – Gt
Gaboury, Damien, 418-545-5011 dgaboury@uqac.ca, Universite du Quebec a Chicoutimi – EmCe
Gabrielli, Paolo, 614 292 6664 gabrielli.1@osu.edu, Ohio State Univ – CtPeOl
Gabrielov, Andrei, (765) 496-2868 gabriea@purdue.edu, Purdue Univ – Ys
Gachon, Philippe, 514-987-3000 #2601 gachon.philippe@uqam.ca, Universite du Quebec a Montreal – Oa
Gadomski, Frederick J., (814) 863-4229 gadomski@mail.meteo.psu.edu, Pennsylvania State Univ, Univ Park – Ow
Gaede, Oliver M., +61 7 3138 2535 oliver.gaede@qut.edu.au, Queensland Univ of Tech – YeNrYs
Gaetani, Glenn A., (508) 289-3724 ggaetani@whoi.edu, Woods Hole Oceanographic Institution – Gi
Gaffney, Edward S., (505) 665-6387 gaffney@lanl.gov, Los Alamos National Laboratory – Yg
Gaffney, Eugene S., (212) 769-5801 American Museum of Natural History – Pv
Gaffney, Jeffrey S., (630) 252-5178 gaffney@anl.gov, Argonne National Laboratory – Oa
Gagnaison, Cyril, +33(0)3 44068997 cyril.gagnaison@lasalle-beauvais.fr, Institut Polytechnique LaSalle Beauvais (ex-IGAL) – PgGsa
Gagne, Jean-Pierre, (418) 724-1870 jean-pierre_gagne@uqar.ca, Universite du Quebec a Rimouski – Oc
Gagne, Marc R., (610) 436-3014 mgagne@wcupa.edu, West Chester Univ – On
Gagnon, Alan R., (508) 289-2961 agagnon@whoi.edu, Woods Hole Oceanographic Institution – Cs
Gagnon, Joel E., (519) 253-3000 x2496 jgagnon@uwindsor.ca, Univ of Windsor – CalEg
Gagnon, Katie, 206-516-3161 KGagnon@sccd.ctc.edu, Seattle Central Community Coll – Oug
Gagnon, Teresa K., (860) 424-3680 teresa.gagnon@ct.gov, Dept of Energy and Environmental Protection – Gg
Gahn, Forest J., (208) 496-1900 gahnf@byui.edu, Brigham Young Univ - Idaho – PogGs
Gaidos, Eric J., (808) 956-7897 gaidos@hawaii.edu, Univ of Hawai'i, Manoa – Yg
Gaines, Robert R., (909) 621-8674 robert.gaines@pomona.edu, Pomona Coll – GsClPi
Gajewski, Dirk J., +4940428382975 dirk.gajewski@uni-hamburg.de, Universitaet Hamburg – YesYg
Gakka, Dr U., +914027097116 udayalaxmi.g@gmail.com, Univ Coll of Science, Osmania Univ – HwYge
Galal, Galal M., 002-03-3921595 galalgalal2004@yahoo.com, Alexandria Univ – GgPe
Galbraith, Eric, 514-398-6767 eric.galbraith@mcgill.ca, McGill Univ – OgPe
Galbraith, John M., (540) 231-9784 ttcf@vt.edu, Virginia Polytechnic Inst & State Univ – Sp
Gale, Andrew, +44 023 92 846127 andy.gale@port.ac.uk, Univ of Portsmouth – PgGrCl

Gale, Julia F., (512) 232-7957 julia.gale@beg.utexas.edu, Univ of Texas at Austin – GctGo
Gale, Marjorie H., (802) 522-5210 marjorie.gale@vermont.gov, Agency of Natural Resources, Dept of Environmental Conservation – GgcGe
Gale, Paula M., (731) 881-7326 pgale@utm.edu, Univ of Tennessee, Martin – OsSdc
Galea, Anthony, anthony.galea@um.edu.mt, Univ of Malta – Opg
Galea, Pauline, +356 2340 3034 pauline.galea@um.edu.mt, Univ of Malta – YsgOg
Galewsky, Joseph, 505-277-4204 galewsky@unm.edu, Univ of New Mexico – Ow
Galgaro, Antonio, +390498279123 antonio.galgaro@unipd.it, Università degli Studi di Padova – Ng
Galicki, Stan, (601) 974-1340 galics@millsaps.edu, Millsaps Coll – Ged
Gall, Quentin, (613) 234-0188 qgall@rogers.com, Univ of Ottawa – GsEmGe
Gallagher, Eugene, eugene.gallagher@umb.edu, Univ of Massachusetts, Boston – Ob
Gallagher, Martin, +44 0161 306-3937 martin.gallagher@manchester.ac.uk, Univ of Manchester – Oa
Gallagher, William B., 609-896-5000 ext. 7784 wgallagher@rider.edu, Rider Univ – PvoGr
Gallegos, Anthony F., (505) 665-0862 agallegos@lanl.gov, Los Alamos National Laboratory – Py
Galli, Kenneth G., (617) 552-4504 kenneth.galli@bc.edu, Boston Coll – GsdGg
Gallovic, Frantisek, 089/2180 4209 frantisek.gallovic@geophysik.uni-muenchen.de, Ludwig-Maximilians-Universitaet Muenchen – Yg
Galloway, James N., (804) 924-1303 jng@virginia.edu, Univ of Virginia – Oa
Galloway, William E., (512) 471-5673 galloway@austin.utexas.edu, Univ of Texas at Austin – GsoGr
Gallup, Christina D., (218) 726-8984 cgallup@d.umn.edu, Univ of Minnesota, Duluth – Cc
Gallus, William A., (515) 294-2270 wgallus@iastate.edu, Iowa State Univ of Science & Tech – Oa
Galperin, Boris, (727) 553-1101 bgalperin@marine.usf.edu, Univ of South Florida – Op
Galsa, Attila, gali@pangea.elte.hu, Eotvos Lorand Univ – YghYe
Galy, Valier, (508) 289-2340 vgaly@whoi.edu, Woods Hole Oceanographic Institution – Cm
Gamage, Kusali R., kgamage@austincc.edu, Austin Community Coll District – HyOuPm
Gamble, Douglas W., (910) 962-3778 gambled@uncw.edu, Univ of North Carolina Wilmington – Oy
Gamble, John, +353 21 4903955 j.gamble@ucc.ie, Univ Coll Cork – Gi
Gambrell, Robert P., (225) 388-6426 Louisiana State Univ – Cg
Gamerdinger, Amy P., (509) 373-3077 amy.gamerdinger@pnl.gov, Pacific Northwest National Laboratory – Cl
Gammack-Clark, James, 561 297-0314 jgammack@fau.edu, Florida Atlantic Univ – Ori
Gammon, Richard H., (206) 543-1609 gammon@macmail.chem.washington.edu, Univ of Washington – Oc
Gammons, Christopher H., (406) 496-4763 cgammons@mtech.edu, Montana Tech of the Univ of Montana – Cg
Gamon, John, 780-492-0345 gamon@ualberta.ca, Univ of Alberta – Or
Gancarz, Alexander J., (505) 667-2606 Los Alamos National Laboratory – Cc
Ganeshram, Raja, +44 (0) 131 650 7364 R.Ganeshram@ed.ac.uk, Edinburgh Univ – Pg
Ganey-Curry, Patricia E., (512) 471-0408 patty@ig.utexas.edu, Univ of Texas at Austin – On
Ganguli, Rajive, (907) 474-7212 rganguli@alaska.edu, Univ of Alaska, Fairbanks – Nm
Ganguly, Jiba, 520-621-6006 ganguly@email.arizona.edu, Univ of Arizona – Gx
Gani, M. Royhan, (270) 745-5977 royhan.gani@wku.edu, Western Kentucky Univ – GsrGg
Gani, Nahid, nahid.gani@wku.edu, Western Kentucky Univ – Gc
Gann, Delbert E., 601-974-1341 gannde@millsaps.edu, Millsaps Coll – Gz
Gannon, J. Michael, Mike.Gannon@dnr.iowa.gov, Iowa Dept of Natural Resources – Gg
Gannon, John P., (828) 227-3813 jpgannon@wcu.edu, Western Carolina Univ – HgSo
Gans, Phillip B., (805) 893-2642 Univ of California, Santa Barbara – Gt
Gansecki, Cheryl A., (808) 932-7549 gansecki@hawaii.edu, Univ of Hawai'i, Hilo – GviOe
Ganssen, Gerald M., +31 20 59 87369 g.m.ganssen@vu.nl, VU Univ Amsterdam – ObPe
Gantzer, Clark J., 573-882-0611 Univ of Missouri, Columbia – Os
Gao, Dengliang, 304-293-3310 Dengliang.Gao@mail.wvu.edu, West Virginia Univ – Ye
Gao, Haiying, 413 577-1250 haiyinggao@geo.umass.edu, Univ of Massachusetts, Amherst – Ys
Gao, Mengsheng, 352-392-1951 msgao@ufl.edu, Univ of Florida – On
Gao, Oliver H., hg55@cornell.edu, Cornell Univ – On
Gao, Yongjun, 713-743-4382 yongjungao@uh.edu, Univ of Houston – CstCa
Gao, Yuan, 973-353-1139 yuangaoh@andromeda.rutgers.edu, Rutgers, The State Univ of New Jersey, Newark – Oa
Garbesi, Karina, 510-885-3172 karina.garbesi@csueastbay.edu, California State Univ, East Bay – Cg
Garces, Milton A., (808) 325-1558 Univ of Hawai'i, Manoa – Ys
Garcia, Antonio F., (805) 756-2430 afgarcia@calpoly.edu, California Polytechnic State Univ – Gmg
Garcia, Jan, (323) 259-2823 jangarcia@oxy.edu, Occidental Coll – On
Garcia, Juan, jgarcia@cicese.mx, Centro de Investigación Científica y de Educación Superior de Ensenada – Yv
Garcia, Marcelo, 217-244-4484 mhgarcia@illinois.edu, Univ of Illinois, Urbana-Champaign – Gm
Garcia, Michael O., (808) 956-6641 garcia@soest.hawaii.edu, Univ of Hawai'i, Manoa – Gi
Garcia, Oswaldo, (415) 338-2061 ogarcia@sfsu.edu, San Francisco State Univ – Oa
Garcia, Sammy R., (505) 667-4151 Los Alamos National Laboratory – Ca
Garcia, William, 704-687-5982 wjgarcia@uncc.edu, Univ of North Carolina, Charlotte – Pv
García Galán, Gúmer, ++935812835 gumer.galan@uab.cat, Universitat Autonoma de Barcelona – GiCg
Garcia-Rubio, Luis H., (727) 553-1246 garcia@marine.usf.edu, Univ of South Florida – Oc
Gardner, Eleanor E., 731-881-7444 egardne3@utm.edu, Univ of Tennessee, Martin – GgPg
Gardner, Gerald H. F., (412) 421-5514 Rice Univ – Yg
Gardner, James D., (403) 823-7707 Royal Tyrrell Museum of Palaeontology – Pv
Gardner, James E., (512) 471-0953 gardner@mail.utexas.edu, Univ of Texas at Austin – Gv
Gardner, Jamie N., (505) 667-1799 jgardner@lanl.gov, Los Alamos National Laboratory – Gg
Gardner, Michael, (406) 994-6658 mgardner@montana.edu, Montana State Univ – Gsr
Gardner, Payton, (406) 243-2458 payton.gardner@umontana.edu, Univ of Montana – Hw
Gardner, Thomas W., (210) 999-7655 tgardner@trinity.edu, Trinity Univ – GmtHg
Gardner, Wayne S., (361) 749-6823 wayne.gardner@utexas.edu, Univ of Texas at Austin – Ob
Gardner, Wilford D., (979) 845-7211 wgardner@ocean.tamu.edu, Texas A&M Univ – OugOr
Gardner-Vandy, Kathryn, gardner-vandyk@si.edu, Smithsonian Institution / National Museum of Natural History – Xm
Gardulski, Anne F., 617-627-2891 anne.gardulski@tufts.edu, Tufts Univ – Grs
Garfield, Newell (Toby), (415) 338-3713 garfield@sfsu.edu, San Francisco State Univ – Op
Gargett, Ann, (757) 683-6009 gargett@ccpo.odu.edu, Old Dominion Univ – Op
Garihan, John M., (864) 294-3363 jack.garihan@furman.edu, Furman Univ – Gct
Garnero, Edward, (480) 965-7653 garnero@asu.edu, Arizona State Univ – Ys
Garren, Sandra J., 516-463-5565 sandra.j.garren@hofstra.edu, Hofstra Univ – Oi
Garrett, Christopher J. R., (250) 721-7702 garrett@uvphys.phys.uvic.ca, Univ of Victoria – Op
Garrett, Timothy J., 801-581-5768 tim.garrett@utah.edu, Univ of Utah – Oa
Garrick-Bethell, Ian, 831-459-1277 igarrick@ucsc.edu,

Univ of California, Santa Cruz – Xy
Garrison, Ervan G., (706) 542-1097 egarriso@uga.edu, Univ of Georgia – GamGn
Garrison, Jennifer M., (323) 343-2412 jgarris@calstatela.edu, California State Univ, Los Angeles – GiCcGv
Garrison, Robert E., (831) 459-5563 Univ of California, Santa Cruz – Gs
Garrison, Trent, (513) 529-3806 garristm@miamioh.edu, Miami Univ – GeHgEc
Garrod, Bruce, (416) 978-3538 bruce.garrod@utoronto.ca, Univ of Toronto – On
Garside, Larry J., 775-784-6693 lgarside@unr.edu, Univ of Nevada – Gg
Garstang, Mimi R., 573-368-2101 mimi.garstang@dnr.mo.gov, Missouri Dept of Natural Resources – Ge
Gartner, Janette, (781) 891-2901 jgartner@bentley.edu, Bentley Univ – Hy
Garven, G, (617) 627-3795 grant.garven@tufts.edu, Tufts Univ – HwqHy
Garver, John I., (518) 388-6770 garverj@union.edu, Union Coll – GtCcGr
Garvin, Theresa D., (780) 492-4593 theresa.garvin@ualberta.ca, Univ of Alberta – On
Garwood, Phil, 910 392 7111 pgarwood@cfcc.edu, Cape Fear Community Coll – Gg
Garwood, Roland W., garwood@nps.edu, Naval Postgraduate School – Op
Garzione, Carmala N., (585) 273-4572 garzione@earth.rochester.edu, Univ of Rochester – Gs
Gaschnig, Richard M., (978) 934-3706 richard_gaschnig@uml.edu, Univ of Massachusetts Lowell – CctCa
Gascho, Gary J., (912) 386-3329 Univ of Georgia – Sc
Gasparini, Nicole, 504-862-3197 ngaspari@tulane.edu, Tulane Univ – Ngg
Gastaldo, Robert A., (207) 859-5807 ragastal@colby.edu, Colby Coll – PbGs
Gaston, Lewis A., (504) 388-1323 lagaston@agcenter.lsu.edu, Louisiana State Univ – Sc
Gates, Alexander E., 973-353-5034 agates@andromeda.rutgers.edu, Rutgers, The State Univ of New Jersey, Newark – Gc
Gates, Ruth D., (808) 236-7420 rgates@hawaii.edu, Univ of Hawai'i, Manoa – Ob
Gathany, Mark, 937-766-3823 mgathany@cedarville.edu, Cedarville Univ – OiuSb
Gattiglio, Marco, marco.gattiglio@unito.it, Università di Torino – Gc
Gatto, Roberto, +390498279172 roberto.gatto@unipd.it, Università degli Studi di Padova – Pg
Gaubatz, Piper, (413) 545-0768 gaubatz@geo.umass.edu, Univ of Massachusetts, Amherst – Oy
Gauci, Adam, adam.gauci@um.edu.mt, Univ of Malta – Op
Gaudette, Henri E., (603) 862-1718 Univ of New Hampshire – Cc
Gaulton, Rachel, +44 (0) 191 208 6577 rachel.gaulton@ncl.ac.uk, Univ of Newcastle Upon Tyne – Or
Gautam, Tej, (740) 376-4371 tej.gautam@marietta.edu, Marietta Coll – NgGeOi
Gauthier, Donald J., (818) 778-5514 gauthidj@lavc.edu, Los Angeles Valley Coll – OiyOw
Gauthier, Jacques, (203) 432-3150 jacques.gauthier@yale.edu, Yale Univ – Pv
Gauthier, Michel, 514-987-3000 #4560 gauthier.michel@uqam.ca, Universite du Quebec a Montreal – Eg
Gauthier, Paul, (609) 258-7442 ppg@princeton.edu, Princeton Univ – Sb
Gauthier, Pierre, 514-987-3000 #3304 gauthier.pierre@uqam.ca, Universite du Quebec a Montreal – Ow
Gauthier, Pierre, pierre.gauthier@ec.gc.ca, McGill Univ – Oa
Gautsch, Jacklyn, 319-335-1761 jackie.gautsch@dnr.iowa.gov, Iowa Dept of Natural Resources – On
Gavin, Dan, (541) 346-5787 dgavin@uoregon.edu, Univ of Oregon – OyPle
Gavriloaiei, Traian, +40-232-201462 tgavrilo@uaic.ro, Al. I. Cuza Univ of Iasi – CaOaCg
Gawloski, Joan, 432-685-4630 jgawloski@midland.edu, Midland Coll – GgzGe
Gawu, Simon K., +233 244 067804 skygawu@yahoo.com, Kwame Nkrumah Univ of Science and Tech – GgEgOn
Gay, Kenny, 9197337353 x28 kenny.gay@ncdenr.gov, North Carolina Geological Survey – GzsPi
Gaya, Charles , cogaya@jkuat.ac.ke, Jomo Kenyatta Univ of Agriculture & Tech – OrgOi
Gayes, Paul T., (803) 349-2213 Dalhousie Univ – Gs
Gaylord, David R., (509) 335-8127 gaylordd@wsu.edu, Washington State Univ – Gs
Gazis, Carey A., 509.963-2820 cgazis@geology.cwu.edu, Central Washington Univ – Cs
Ge, Shemin, (303) 492-8323 ges@colorado.edu, Univ of Colorado – HwEoYg
Geary, Dana H., dana@geology.wisc.edu, Univ of Wisconsin, Madison – Po
Geary, Lindsey, (315)792-3134 Utica Coll – Geg
Geary, Phil, 61 02 4921 6726 phil.geary@newcastle.edu.au, Univ of Newcastle – HwOsHw
Gebrande, Helmut, 089/2180 4325 gebrande@geophysik.uni-muenchen.de, Ludwig-Maximilians-Universitaet Muenchen – YgGyEo
Gebrehiwet, Tsigabu A., 410-704-2220 Tsigab@gmail.com, Towson Univ – Hw
Gedzelman, Stanley, (212) 650-6470 City Coll (CUNY) – Oa
Gee, Jeffrey S., (858) 534-4707 jsgee@ucsd.edu, Univ of California, San Diego – Gu
Gehrels, George E., (520) 349-4702 ggehrels@email.arizona.edu, Univ of Arizona – Gt
Gehrels, Tom, (520) 621-6970 tgehrels@lpl.arizona.edu, Univ of Arizona – On
Geibert, Walter, +44 (0) 131 651 7704 Walter.Geibert@ed.ac.uk, Edinburgh Univ – Cc
Geidel, Gwendelyn, (803) 777-7171 geidel@geol.sc.edu, Univ of South Carolina – HwGeCg
Geiger, James W., 217-265-8989 jgeiger@illinois.edu, Univ of Illinois, Urbana-Champaign – Ge
Geiger, Sebastian, s.geiger@hw.ac.uk, Heriot-Watt Univ – Np
Geissman, John D., 972-883-2403 geissman@utdallas.edu, Univ of Texas, Dallas – Gt
Geissman, John W., (505) 277-3433 jgeiss@unm.edu, Univ of New Mexico – Ym
Geist, Dennis J., (208) 885-6491 dgeist@uidaho.edu, Univ of Idaho – Giv
Geller, Jonathan, 831-771-4400 geller@mlml.calstate.edu, Moss Landing Marine Laboratories – On
Geller, Marvin A., (631) 632-8701 marvin.geller@stonybrook.edu, SUNY, Stony Brook – Oa
Geller, Michael D., (609) 652-4620 Richard Stockton Coll of New Jersey – On
Gemignani, Robert, (314) 935-4614 rgemigna@levee.wustl.edu, Washington Univ in St. Louis – On
Gemmell, J B., 61 3 6226 2893 bruce.gemmell@utas.edu.au, Univ of Tasmania – Eg
Gemperline, Johanna M., (410) 554-5552 johanna.gemperline@maryland.gov, Maryland Department of Natural Resources – Hw
Genareau, Kimberly, 205-348-1878 kdg@ua.edu, Univ of Alabama – Gvi
Gendzwill, Donald J., don.gendzwill@usask.ca, Univ of Saskatchewan – Ye
Genge, Matthew, +44 20 759 46499 m.genge@imperial.ac.uk, Imperial Coll – Xm
Gentile, Richard J., (816) 235-2974 gentiler@umkc.edu, Univ of Missouri-Kansas City – Gr
Gentry, Amanda L., (303) 319-0695 amandagentry@weber.edu, Weber State Univ – GsCcGg
Gentry, Christopher, 931-221-7478 gentryc@apsu.edu, Austin Peay State Univ – Oyi
Gentry, Randall, 630-252-4440 rgentry@anl.gov, Argonne National Laboratory – Hw
Gentry, Terry, 979-845-5323 tjgentry@tamu.edu, Texas A&M Univ – Sb
George, Graham, 306-966-5722 g.george@usask.ca, Univ of Saskatchewan – Cg
Georgen, Jennifer, 850-645-4987 georgen@gly.fsu.edu, Florida State Univ – Gu
Georgen, Jennifer, 757-683-5198 jgeorgen@odu.edu, Old Dominion Univ – Gu
Georgiev, Svetoslav, (970)-491-3789 svetoslav.georgiev@colostate.edu, Colorado State Univ – CcGo
Georgiopopoulou, Aggeliki, (+353) 1 716 2062 aggie.georg@ucd.ie, Univ Coll Dublin – GusOg
Geraghty Ward, Emily, emily.ward@rocky.edu, Rocky Mountain Coll – OeGc
Gerald, Carresse, (919) 530-7117 cgerald6@nccu.edu, North Carolina Central Univ – GeOg
Gerba, Charles P., (520) 621-6906 gerba@ag.arizona.edu, Univ of Arizona – Sb

Faculty Index -G

GERBE, Marie-Christine, 33-477485123 gerbe@univ-st-etienne.fr, Université Jean Monnet, Saint-Etienne – GisOe
Gerber, Stefan, 352294-3174 sgerber@ufl.edu, Univ of Florida – Os
Gerbi, Christopher C., 207 581-2153 Univ of Maine – Gt
Gerbi, Greg, 518-580-5127 ggerbi@skidmore.edu, Skidmore Coll – OgYg
Gerhard, Lee C., 78538643965 leeg@sunflower.com, Univ of Kansas – GsoGd
Gerhardt, Hannes, (678) 839-4064 hgerhard@westga.edu, Univ of West Georgia – On
Gerla, Philip J., (701) 777-3305 phil.gerla@engr.und.edu , Univ of North Dakota – Hw
Gerlach, Russel L., (417) 836-5800 Missouri State Univ – On
German, Chris, 508-289-2853 cgerman@whoi.edu, Woods Hole Oceanographic Institution – Gt
Germanoski, Dru, (610) 330-5196 germanod@lafayette.edu, Lafayette Coll – Gm
Germanovich, Leonid, (404) 894-2284 leonid@ce.gatech.edu, Georgia Inst of Tech – NrYg
Gernon, Thomas M., +44 (0)23 8059 2670 thomas.gernon@noc.soton.ac.uk, Univ of Southampton – OgGvi
Gerry, Janelle, 402-472-2663 Univ of Nebraska, Lincoln – On
Gersmehl, Carol, 212-772-3534 carol.gersmehl@hunter.cuny.edu, Hunter Coll (CUNY) – Oe
Gertisser, Ralf, (+44) 01782 733181 r.gertisser@keele.ac.uk, Keele Univ – GivGz
Gerwick, William H., (858) 534-0578 wgerwick@ucsd.edu, Univ of California, San Diego – Cm
Gess, Robert, +27 (0)82-7595848 robg@imaginet.co.za, Rhodes Univ – Pg
Getty, Mike, 303-370-6406 Mike.Getty@dmns.org, Denver Museum of Nature & Science – Pv
Ghani, Azman Abdul, 03-79674234 azmangeo@um.edu.my, Univ of Malaya – Gi
Gharti, Hom Nath, (609) 258-2605 hgharti@princeton.edu, Princeton Univ – YsdYg
Ghassemi, Ahmed, ahmad.ghassemi@ou.edu, Univ of Oklahoma – Nr
Ghate, Virendra P., (630) 252-1609 vghate@anl.gov, Argonne National Laboratory – OawOr
Ghatge, Suhas L., (609) 984-6587 suhas.ghatge@dep.nj.gov, New Jersey Geological and Water Survey – Yve
Ghattas, Omar, (512) 232-4304 omar@ices.utexas.edu, Univ of Texas at Austin – GqYg
Ghazala, Hosni H., (109) 688-7904 ghazala@mans.edu.eg, El Mansoura Univ – Ygx
Ghent, Edward D., (403) 220-5847 ghent@geo.ucalgary.ca, Univ of Calgary – Gp
Ghent, Rebecca, (416) 978-0597 ghentr@geology.utoronto.ca, Univ of Toronto – Or
Ghil, Michael, (310) 206-2285 ghil@atmos.ucla.edu, Univ of California, Los Angeles – OaYdOp
Ghinassi, Massimiliano, +390498279181 massimiliano.ghinassi@unipd.it, Università degli Studi di Padova – Gs
Ghoneim, Eman, ghoneime@uncw.edu, Boston Univ – Gm
Ghosh, Abhijit, (951) 827-4493 aghosh@ucr.edu, Univ of California, Riverside – YsGt
Ghosheh, Baher A., (814) 732-2207 ghosheh@edinboro.edu, Edinboro Univ of Pennsylvania – On
Giacalone, Joe, 520-626-8365 giacalone@lpl.arizona.edu, Univ of Arizona – Xy
Giachetti, Thomas, tgiachet@uoregon.edu, Univ of Oregon – Gv
Giannini, Alessandra, (845) 680-4473 alesall@iri.columbia.edu, Columbia Univ – Oap
Gianniny, Gary, (970) 247-7254 gianniny_g@fortlewis.edu, Fort Lewis Coll – Gs
Gianotti, Franco, franco.gianotti@unito.it, Università di Torino – Gg
Giaramita, Mario J., (209) 667-3558 mgiaramita@csustan.edu, California State Univ, Stanislaus – Gpz
Giardino, John R., (979) 845-3224 giardino@geo.tamu.edu, Texas A&M Univ – GmNg
Giardino, John R., (409) 867-9067 rickg@tamu.edu, Texas A&M Univ – Gm
Giardino, Marco, marco.giardino@unito.it, Università di Torino – Gm
Gibbs, Samantha J., +44 (0)23 80592003 Samantha.Gibbs@noc.soton.ac.uk, Univ of Southampton – Og
Gibeaut, James, 361-825-2060 James.Gibeaut@tamucc.edu, Texas A&M Univ, Corpus Christi – Oni
Gibling, Martin R., mgibling@is.dal.ca, Dalhousie Univ – Gs
Gibson, Andy, +44 023 92 842654 andy.gibson@port.ac.uk, Univ of Portsmouth – Eg
Gibson, Blair, bgibso5@uwo.ca, Western Univ – Ca
Gibson, Carl H., (858) 534-3184 cgibson@ucsd.edu, Univ of California, San Diego – Oo
Gibson, David, (207) 778-7402 dgibson@maine.edu, Univ of Maine - Farmington – GigGz
Gibson, Deana, (417) 836-5801 deanagibson@missouristate.edu, Missouri State Univ – On
Gibson, Deidre M., 757.727.5883 deidre.gibson@hamptonu.edu, Hampton Univ – Ob
Gibson, Glen, 719-333-3080 glen.gibson@usafa.edu, United States Air Force Academy – Ori
Gibson, H. Daniel (Dan), (778) 782-7057 hdgibson@sfu.ca, Simon Fraser Univ – GcCcGp
Gibson, Harold L., 7056751151 x2371 hgibson@laurentian.ca, Laurentian Univ, Sudbury – EmGv
Gibson, Michael A., (731) 881-7435 mgibson@utm.edu, Univ of Tennessee, Martin – PgiOe
Gibson, Roger L., 011 717 6553 roger.gibson@wits.ac.za, Univ of the Witwatersrand – GcpGg
Gibson, Ronald C., (714) 895-8194 rgibson@gwc.cccd.edu, Golden West Coll – Gc
Gibson, Sally, +44 (0) 1223 333401 sally@esc.cam.ac.uk, Univ of Cambridge – Gi
Gibson, Jr., Richard L., (979) 862-8653 gibson@tamu.edu, Texas A&M Univ – Yse
Gidigasu, Solomon S., +233 27 7807 707 ssrgidigasu.soe@knust.edu.gh, Kwame Nkrumah Univ of Science and Tech – GeOrNr
Giegengack, Jr., Robert F., (215) 898-5191 gieg@sas.upenn.edu, Univ of Pennsylvania – Gg
Giere, Reto, giere@sas.upenn.edu, Univ of Pennsylvania – GzCgGe
Gierke, John S., (906) 487-2535 jsgierke@mtu.edu, Michigan Technological Univ – Hy
Giese, Benjamin S., (979) 845-2306 bgiese@ocean.tamu.edu, Texas A&M Univ – Op
Giese, Graham S., (508) 289-2297 ggiese@whoi.edu, Woods Hole Oceanographic Institution – On
Giese, Rossman F., (716)645-4263 glgclay@acsu.buffalo.edu, SUNY, Buffalo – Gz
Gieskes, Joris M., (858) 534-4257 jgieskes@ucsd.edu, Univ of California, San Diego – Oc
Gifford, Dian J., (401) 874-6690 dgifford@uri.edu, Univ of Rhode Island – Ob
Gifford, Jennifer N., 662-915-2079 jngiffor@olemiss.edu, Univ of Mississippi – GtxCg
Gifford-Gonzalez, Dianne, (831) 459-2633 Univ of California, Santa Cruz – On
Gigler, Alexander, 089/2180 4185 Ludwig-Maximilians-Universitaet Muenchen – Gz
Giglierano, James D., James.Giglierano@dnr.iowa.gov, Iowa Dept of Natural Resources – GgOri
Gilbert, Co'Quesie, (619) 534-2470 cag65@columbia.edu, Columbia Univ – On
Gilbert, Hersh, (765) 496-9518 hersh@purdue.edu, Purdue Univ – Ys
Gilbert, Kathleen W., 781-283-3086 kgilbert@wellesley.edu, Wellesley Coll – Cs
Gilbert, M. Charles, (405) 325-3253 mcgilbert@ou.edu, Univ of Oklahoma – Cp
Gilbert, Patricia M., (410) 221-8422 Univ of Maryland – Ob
Gilbes, Fernando, 787 2653845 gilbes@cacique.uprm.edu, Univ of Puerto Rico – OrGu
Gilder, Stuart, 089/2180 4239 stuart.gilder@geophysik.uni-muenchen.de, Ludwig-Maximilians-Universitaet Muenchen – Yg
Gildner, Raynond F., 260-481-0249 gildnerr@ipfw.edu, Indiana Univ / Purdue Univ, Fort Wayne – GgPq
Giles, Antony, (432) 685-5580 agiles@midland.edu, Midland Coll – GgvGi
Giles, David, +44 023 92 842248 david.giles@port.ac.uk, Univ of Portsmouth – Gm
Giles, Katherine A., 915-747-7075 kagiles@utep.edu, Univ of Texas, El Paso – Go
Gilfillan, Stuart M., +44 (0) 131 651 3462 Stuart.Gilfillan@ed.ac.uk, Edinburgh Univ – Oa
Gilg, Hans Albert, +49 89 289 25855 agilg@tum.de, Technische Universitaet Muenchen – EgnEg
Gill, Fiona, +44(0) 113 34 35190 f.gill@leeds.ac.uk,

Univ of Leeds – PoCa
Gill, James B., (831) 459-3842 jgill@pmc.ucsc.edu, Univ of California, Santa Cruz – Gi
Gill, Swarndeep S., (724) 938-1677 gill@calu.edu, California Univ of Pennsylvania – Oa
Gill, Thomas E., (915) 747-5168 tegill@utep.edu, Univ of Texas, El Paso – GmOaCl
Gillam, Mary L., gillam@rmi.net, Fort Lewis Coll – Gm
Gille, Peter, 089/2180 4355 peter.gille@lrz.uni-muenchen.de, Ludwig-Maximilians-Universitaet Muenchen – GzOm
Gille, Sarah T., (858) 822-4425 sgille@ucsd.edu, Univ of California, San Diego – Op
Gillerman, Virginia S., (208) 332-4420 vgillerm@uidaho.edu, Univ of Idaho – Eg
Gillerman, Virginia S., (208) 426-4002 vgillerm@boisestate.edu, Boise State Univ – Eg
Gillespie, Alan R., (206) 685-8265 arg3@uw.edu, Univ of Washington – GmOlr
Gillespie, Janice, 661-654-3040 jgillespie@csub.edu, California State Univ, Bakersfield – Go
Gillespie, Robb, (269) 387-5364 robb.gillespie@wmich.edu, Western Michigan Univ – GsoGm
Gillespie, Terry J., (519) 824-4120 (Ext. 54276) tgillesp@uoguelph.ca, Univ of Guelph – Ow
Gillespie, Thomas, (609) 771-2569 Coll of New Jersey – Gc
Gillespie, Thomas, thomas.gillespie@emory.edu, Emory Univ – On
Gillette, David P., (828) 251-6366 dgillett@unca.edu, Univ of North Carolina, Asheville – On
Gilley, Brett, bgilley@eos.ubc.ca, Univ of British Columbia – Gg
Gilliam, James W., (919) 515-2040 North Carolina State Univ – Sb
Gilligan, Jonathan M., (615) 322-2420 jonathan.gilligan@vanderbilt.edu, Vanderbilt Univ – OnaOn
Gillikin, David P., (518) 388-6679 gillikid@union.edu, Union Coll – CsmCl
Gillis, James M., (906) 487-1820 jmgillis@mtu.edu, Michigan Technological Univ – Nmr
Gillis, Kathryn, 250-721-6120 kgillis@uvic.ca, Univ of Victoria – Gp
Gillis, Robert, (907) 451-5024 robert.gillis@alaska.gov, Alaska Division of Geological & Geophysical Surveys – Eo
Gillman, Joe, (573) 368-2101 joe.gillman@dnr.mo.gov, Missouri Dept of Natural Resources – Gg
Gillmore, Gavin, +44 020 8417 2518 G.Gillmore@kingston.ac.uk, Kingston Univ – Gb
Gilmore, Martha S., (860) 685-3129 mgilmore@wesleyan.edu, Wesleyan Univ – XgGmOr
Gilmore, Tyler J., (509) 376-2370 tyler.gilmore@pnl.gov, Pacific Northwest National Laboratory – Hg
Gilmour, Ernest H., (509) 359-7480 egilmour@ewu.edu, Eastern Washington Univ – Pi
Gilotti, Jane A., (319) 335-1097 jane-gilotti@uiowa.edu, Univ of Iowa – GctGp
Gilpin, Bernard J., (714) 895-8233 bgilpin@gwc.cccd.edu, Golden West Coll – Ys
Gimmestad, Gary G., (404) 493-1331 gary.gimmestad@gmail.com, Georgia Inst of Tech – Ora
Ginder-Vogel, Matt, 608-262-0768 mgindervogel@wisc.edu, Univ of Wisconsin, Madison – Sc
Gingras, Murray, 780-492-1963 mgringras@ualberta.ca, Univ of Alberta – Go
Ginis, Isaac, (401) 874-6484 iginis@gso.uri.edu, Univ of Rhode Island – Op
Ginsburg, Robert N., 305 421-4875 rginsburg@rsmas.miami.edu, Univ of Miami – Gs
Giordano, Daniele, daniele.giordano@unito.it, Università di Torino – Gv
Giorgis, Scott D., (585) 245-5293 giorgis@geneseo.edu, SUNY, Geneseo – GctYm
Giosan, Liviu, 508-289-2257 lgiosan@whoi.edu, Woods Hole Oceanographic Institution – Gu
Giraldo, Mario A., (818) 677-4431 mario.giraldo@csun.edu, California State Univ, Northridge – OriOg
Girard, Eric, 514-987-3000 #3325 girard.eric@uqam.ca, Universite du Quebec a Montreal – Ow
Girard, Michael W., (609) 292-2576 mike.girard@dep.state.nj.us, New Jersey Geological and Water Survey – Oi
Giraud, Richard E., (801) 537-3351 richardgiraud@utah.gov, Utah Geological Survey – Ng
Girhard, T S., 486-0045 tgirhard@alamo.edu, San Antonio Community Coll – Oyw
Girty, Gary H., (619) 594-2552 ggirty@mail.sdsu.edu, San Diego State Univ – Gc
Gitelson, Anatoly, gitelson@calmit.unl.edu, Unversity of Nebraska - Lincoln – On
Gittings, Bruce M., +44 (0) 131 650 2558 bruce@ed.ac.uk, Edinburgh Univ – Oi
Gittins, John, (416) 483-9345 j.gittins@utoronto.ca, Univ of Toronto – GipGz
Giusberti, Luca, +390498279183 luca.giusberti, Università degli Studi di Padova – Pm
Giustetto, Roberto, roberto.giustetto@unito.it, Università di Torino – Gz
Glamoclija, Mihaela, (973) 353-2509 m.glamoclija@rutgers.edu, Rutgers, The State Univ of New Jersey, Newark – PyXgGe
Glasauer, Susan, (519) 824-4120 xt52453 glasauer@uoguelph.ca, Univ of Guelph – Py
Glaser, Brian, 309-796-5238 glaserb@bhc.edu, Black Hawk Coll – On
Glaser, Paul H., 612-624-8395 glase001@umn.edu, Univ of Minnesota, Twin Cities – Gn
Glass, Alexander, (919) 684-6167 alex.glass@duke.edu, Duke Univ – PiGgOe
Glass, Billy P., (302) 449-2464 bglass@udel.edu, Univ of Delaware – XmGsz
Glass, Hylke J., +44 01326 371823 h.j.glass@exeter.ac.uk, Exeter Univ – NxmEm
Glatzmaier, Gary A., (541) 214-5882 glatz@es.ucsc.edu, Univ of California, Santa Cruz – YmXy
Glazer, Brian, 808-956-6658 glazer@hawaii.edu, Univ of Hawai'i, Manoa – Cm
Glazner, Allen F., (919) 962-0689 afg@unc.edu, Univ of North Carolina, Chapel Hill – Git
Gleason, Gayle C., (607) 753-2816 gleasong@cortland.edu, SUNY, Cortland – Gct
Gleason, James D., 734-764-9523 jdgleaso@umich.edu, Univ of Michigan – Cg
Glenn, Craig R., (808) 956-2200 glenn@soest.hawaii.edu, Univ of Hawai'i, Manoa – GseCm
Glenn, Ed, (520) 626-2664 eglenn@ag.arizona.edu, Univ of Arizona – On
Glenn, Nancy, 208.221.1245 nancyglenn@boisestate.edu, Boise State Univ – OrNgOi
Glickson, Deborah, (202) 334-2024 dglickson@nas.edu, National Academy of Sciences/National Research Council – GutGg
Gloaguen, Erwan, erwan.gloaguen@ete.inrs.ca, Universite du Quebec – Ye
Glotch, Timothy, (631) 632-1168 timothy.glotch@stonybrook.edu, SUNY, Stony Brook – XgGz
Glover, David M., (508) 289-2656 dglover@whoi.edu, Woods Hole Oceanographic Institution – Oc
Glover, Paul W., 418-656-5180 paul.glover@ggl.ulaval.ca, Universite Laval – Yx
Glover, Paul W., +44(0) 113 34 35213 p.w.j.glover@leeds.ac.uk, Univ of Leeds – YxGoa
Glowacka, Ewa, glowacka@cicese.mx, Centro de Investigación Científica y de Educación Superior de Ensenada – Ys
Glubokovskikh, Stanislav, +618 9266-7190 stanislav.glubokovskikh@curtin.edu.au, Curtin Univ – Yxe
Gluhovsky, Alexander, (765) 494-0670 aglu@purdue.edu, Purdue Univ – Oa
Glumac, Bosiljka, (413) 585-3680 bglumac@smith.edu, Smith Coll – Gs
Gluyas, Jon, +44 (0) 191 33 42302 j.g.gluyas@durham.ac.uk, Durham Univ – Eo
Glynn, William G., wglynn@brockport.edu, SUNY, The Coll at Brockport – Gg
Gnanadesikan, Anand, (410) 516-0722 gnanades@jhu.edu, Johns Hopkins Univ – OpcOa
Gnidovec, Dale, (614) 292-6896 gnidovec.1@osu.edu, Ohio State Univ – Ggr
Gobler, Christopher, (631) 632-5043 Christopher.Gobler@stonybrook.edu, SUNY, Stony Brook – Ob
Godbold, Jasmin A., +44 (0)23 80593639 J.A.Goldbold@soton.ac.uk, Univ of Southampton – Ob
Godchaux, Martha M., (208) 882-9062 Mount Holyoke Coll – Gv
Goddard, Lisa M., (845) 680-4430 goddard@iri.columbia.edu, Columbia Univ – Oa
Godek, Melissa, (607) 436-3375 melissa.godek@oneonta.edu,

SUNY, Oneonta – OwaOg
Godfrey, Brian J., 845-437-5544 godfrey@vassar.edu, Vassar Coll – Ou
Godfrey, Chris, 828-232-5160 cgodfrey@unca.edu, Univ of North Carolina at Asheville – OanOn
Godin, Laurent, (613) 533-3223 godinl@queensu.ca, Queen's Univ – Gct
Godsey, Holly, (801) 587-7865 Univ of Utah – Oe
Godsey, Sarah E., 208-282-3170 godsey@isu.edu, Idaho State Univ – HwGm
Goehring, Brent M., (504) 862-3196 bgoehrin@tulane.edu, Tulane Univ – CcGlm
Goeke, James W., (308) 530-4437 jgoeke@unl.edu, Unversity of Nebraska - Lincoln – HwGmYe
Goes, Saskia, +44 20 759 46434 s.goes@imperial.ac.uk, Imperial Coll – Yg
Goetz, Alexander, 303-492-5086 goetz@cses.colorado.edu, Univ of Colorado – Or
Goetz, Andrew R., (303) 871-2674 agoetz@du.edu, Univ of Denver – Ogu
Goetz, Heinrich, (972) 377-1079 hgoetz@collin.edu, Collin Coll - Preston Ridge Campus – Ge
Goetze, Erica, (808) 956-7156 egoetze@hawaii.edu, Univ of Hawai'i, Manoa – Ob
Goff, Fraser, (505) 667-8060 fraser@lanl.gov, Los Alamos National Laboratory – Cg
Goff, James, j.goff@unsw.edu.au, Univ of New South Wales – Gsm
Goff, John A., 512-471-0476 goff@ig.utexas.edu, Univ of Texas at Austin – Gu
Goforth, Thomas T., (254) 710-2183 tom_goforth@baylor.edu, Baylor Univ – Yg
Golabi, Mohammad H., 671-735-2143 mgolabi@uguam.uog.edu, Univ of Guam – OsSpc
Gold, David (Duff) P., (814) 865-9993 gold@ems.psu.edu, Pennsylvania State Univ, Univ Park – GcgGx
Gold-Bouchot, Gerardo, (979) 845-9826 ggold@tamu.edu, Texas A&M Univ – OcCm
Goldberg, David S., (845) 365-8674 Columbia Univ – Yr
Goldblatt, Colin, 250-721-6120 czg@uvic.ca, Univ of Victoria – Oa
Goldfinger, Chris, (541) 737-9622 gold@coas.oregonstate.edu, Oregon State Univ – Yr
Goldhamer, David A., (559) 646-6500 dagoldhamer@ucdavis.edu, Univ of California, Davis – On
Goldman, Daniel, (937) 229-5637 dgoldman1@udayton.edu, Univ of Dayton – PoqGr
Goldreich, Peter M., (626) 395-6193 pmg@gps.caltech.edu, California Inst of Tech – Xy
Goldsmith, Victor, (212) 772-5450 Graduate School of the City Univ of New York – On
Goldstein, Barry, (253) 879-3822 goldstein@pugetsound.edu, Univ of Puget Sound – GlmGs
Goldstein, Robert H., (785) 864-2738 gold@ku.edu, Univ of Kansas – Gs
Goldstein, Steven, (845) 365-8787 Columbia Univ – Cg
Goldstein, Susan T., (706) 542-2652 sgoldst@gly.uga.edu, Univ of Georgia – Pm
Gollmer, Steven, 937-766-7764 gollmers@cedarville.edu, Cedarville Univ – GzOwg
Golombek, Matthew P., (818) 393-7948 mgolombek@jpl.nasa.gov, Jet Propulsion Laboratory – XgyGt
Goloshubin, Gennady, 713-743-2796 ggoloshubin@uh.edu, Univ of Houston – Ye
Gomberg, Joan, 206-616-5581 gomberg@usgs.gov, Univ of Washington – YsGt
Gomes, Maria E., 351259350261 mgomes@utad.pt, Universidade de Trás-os-Montes e Alto Douro – GiCgGz
Gomes, Nuno N., +244924987900 ngomes999@gmail.com, Universidade Independente de Angola – GesGu
Gomez, Enrique, egomez@cicese.mx, Centro de Investigación Científica y de Educación Superior de Ensenada – Ye
Gomez, Francisco, (573) 882-9744 fgomez@missouri.edu, Univ of Missouri – Gt
Gomez, Natalya, 514-398-4885 natalya.gomez@mcgill.ca, McGill Univ – Og
Gómez Gras, David, ++935813093 david.gomez@uab.cat, Universitat Autonoma de Barcelona – Gd
Gomezdelcampo, Enrique, 419 372 2886 egomezd@bgsu.edu, Bowling Green State Univ – Ge
Goncharov, Alexander F., 202-478-8947 agoncharov@carnegiescience.edu, Carnegie Institution of Washington – Gy
Gong, Donglai, 804-684-7529 gong@vims.edu, Coll of William & Mary – Op
Gong, Hongmian, (212) 772-4658 gong@hunter.cuny.edu, Hunter Coll (CUNY) – On
Goni, Miguel A., (541) 737-0578 mgoni@coas.oregonstate.edu, Oregon State Univ – Gu
Gonnermann, Helge, (713) 348-6263 Helge.M.Gonnermann@rice.edu, Rice Univ – Gv
Gontz, Allen, allen.gontz@umb.edu, Univ of Massachusetts, Boston – GmlYr
Gonzales, David A., (970) 247-7378 gonzales_d@fortlewis.edu, Fort Lewis Coll – Gx
Gonzalez, Frank, 206-290-0903 figonzal@uw.edu, Univ of Washington – Or
González, Isabel , +34954556317 igonza@us.es, Universidad de Sevilla – GzeSc
Gonzalez, Jose J., javier@cicese.mx, Centro de Investigación Científica y de Educación Superior de Ensenada – Yd
Gonzalez, Luis A., 785-864-2743 lgonzlez@ku.edu, Univ of Kansas – CsGd
Gonzalez, Maria M., (989) 774-3179 gonza1mm@cmich.edu, Central Michigan Univ – Gd
Gonzalez, Richard, (619) 260-4600 gonzalez@sandiego.edu, Univ of San Diego – On
Gonzalez Leon, Carlos M., cmgleon@servidor.unam.mx, Universidad Nacional Autonoma de Mexico – Gr
Gonzalez-Huizar, Hector, 915-747-5305 hectorg@utep.edu, Univ of Texas, El Paso – Ys
Good, Daniel B., (912) 478-5361 DanGood@GeorgiaSouthern.edu, Georgia Southern Univ – On
Goodbred, Jr., Steven L., (615) 343-6424 steven.goodbred@vanderbilt.edu, Vanderbilt Univ – GsOn
Goodell, Laurel P., (609) 258-1043 laurel@princeton.edu, Princeton Univ – Oe
Goodell, Philip C., 915-747-5593 goodell@utep.edu, Univ of Texas, El Paso – Ce
Goodge, John W., (218) 726-7491 jgoodge@d.umn.edu, Univ of Minnesota, Duluth – Gp
Goodin, Ruth, 305-284-4253 ruthgoodin@miami.edu, Univ of Miami – On
Gooding, Patrick J., (859) 389-8810 gooding@uky.edu, Univ of Kentucky – Eo
Goodison, Marjorie, 585-275-5713 margie@earth.rochester.edu, Univ of Rochester – On
Goodliffe, Andrew, amg@ua.edu, Univ of Alabama – YgeGt
Goodman, Paul, 520-621-8484 pgoodman@email.arizona.edu, Univ of Arizona – Oa
Goodrich, David C., (520) 670-6380 goodrich@tucson.ars.ag.gov, Univ of Arizona – Hs
Goodrich, Greg, gregory.goodrich@wku.edu, Western Kentucky Univ – Oa
Goodwin, Alan M., (416) 978-5613 Univ of Toronto – Gg
Goodwin, David H., (740) 587-5621 goodwind@denison.edu, Denison Univ – PiCs
Goodwin, Laurel, 608-262-8960 laurel@geology.wisc.edu, Univ of Wisconsin, Madison – Gc
Goodwin, Laurel B., (608) 265-4234 laurel@geology.wisc.edu, Univ of Wisconsin, Madison – Gc
Goodwin, Mark B., (505) 835-5178 mark@berkeley.edu, Univ of California, Berkeley – Pv
Goodwin, Paul, (519) 824-4120 Ext.52754 pgoodwin@uoguelph.ca, Univ of Guelph – On
Goodwin, Phillip A., +44 (0)23 80596161 P.A.Goodwin@soton.ac.uk, Univ of Southampton – Pe
Goos, Robert J., (701) 231-8581 North Dakota State Univ – Sc
Gootee, Brian F., 602.708.8846 bgootee@email.arizona.edu, Univ of Arizona – Gm
Gopal, Sucharita, suchi@bu.edu, Boston Univ – Oi
Gordon, Andrew, (519) 824-4120 ext 52415 agordon@uoguelph.ca, Univ of Guelph – On
Gordon, Arnold L., (845) 365-8325 Columbia Univ – Op
Gordon, Barney, (785) 532-6101 bgordon@ksu.edu, Kansas State Univ – So
Gordon, Clare, +44(0) 113 34 35210 c.e.gordon@leeds.ac.uk, Univ of Leeds – Oi
Gordon, Mark, (416)736-2100 mgordon@yorku.ca, York Univ – Oa

Gordon, Richard G., (713) 348-5279 rgg@rice.edu, Rice Univ – Gt
Gordon, Rob J., (519) 824-4120 Ext.52285 rjgordon@uoguelph.ca, Univ of Guelph – Oa
Gordon, Robert B., (203) 432-3125 robert.gordon@yale.edu, Yale Univ – Nr
Gordon, Ryan, 207-287-7178 ryan.gordon@maine.gov, Dept of Agriculture, Conservation, and Forestry – HwGm
Gordon, Stacia , 775-784-6476 staciag@unr.edu, Univ of Nevada, Reno – Gi
Gordon, Steven J., (719) 333-3067 usafa.dfeg@usafa.edu, United States Air Force Academy – Gm
Gordon, Terence M., (403) 220-8301 tmg@geo.ucalgary.ca, Univ of Calgary – Gq
Gore, Pamela J. W., 678-891-3754 pgore@gsu.edu, Georgia State Univ, Perimeter Coll, Clarkston Campus – GsOeGn
Gorecka-Nowak, Anna, anna.gorecka-nowak@uwr.edu.pl, Univ of Wroclaw – Pg
Gorelick, Steven, (650) 725-2950 gorelick@geo.stanford.edu, Stanford Univ – Yr
Goreva, Yulia, gorevay@si.edu, Smithsonian Institution / National Museum of Natural History – Xm
Gorman, Andrew R., +64 3 479-7516 andrew.gorman@otago.ac.nz, Univ of Otago – YrgYe
Gorman, Gerard, +44 20 759 49985 g.gorman@imperial.ac.uk, Imperial Coll – Yg
Gorman-Lewis, Drew J., 206-543-3541 dgormanl@uw.edu, Univ of Washington – PyCl
Gorog, Agnes, gorog@ludens.elte.hu, Eotvos Lorand Univ – Pmi
Gorring, Matthew L., (973) 655-5409 gorringm@mail.montclair.edu, Montclair State Univ – Gi
Gorsevski, Peter, (419) 372-7201 peterg@bgsu.edu, Bowling Green State Univ – OirGm
Gosnell, Hannah, (541) 737-1222 gosnell@colorado.edu, Oregon State Univ – Ou
Gosnold, William D., (701) 777-2631 will.gosnold@engr.und.edu, Univ of North Dakota – YhCt
Gospodinova, Kalina D., 508-289-3212 kgospodinova@whoi.edu, Woods Hole Oceanographic Institution – Cm
Goss, Michael J., (519) 824-4120 (Ext. 2491) mgoss@uoguelph.ca, Univ of Guelph – So
Gosse, John, 902-494-6632 jcgosse@is.dal.CA, Univ of Kansas – Gm
Gosselin, David C., (402) 472-8919 dgosselin2@unl.edu, University of Nebraska - Lincoln – CgHw
Gosselin, Kelsey, kgosselin@whoi.edu, Woods Hole Oceanographic Institution – Oc
Gosselin, Michel, (418) 724-1761 michel_gosselin@uqar.ca, Universite du Quebec a Rimouski – Ob
Goswani, Anjali, +44 020 7679 32190 a.goswami@ucl.ac.uk, Univ Coll London – Po
Gotkowitz, Madeline B., (608) 262-1580 mbgotkow@wisc.edu, Univ of Wisconsin - Extension – Hw
Gottfried, Michael D., 517-432-5480 gottfrie@msu.edu, Michigan State Univ – Pv
Gottfried, Richard, rGottfried@frederick.edu, Frederick Community Coll – GgdGz
Gottgens, Johan F., (419) 530-8451 johan.gottgens@utoledo.edu, Univ of Toledo – On
Gotz, Annette, +27 (0)46-603-8313 a.gotz@ru.ac.za, Rhodes Univ – Gs
Goudge, Theodore L., (660) 562-1798 tgoudge@nwmissouri.edu, Northwest Missouri State Univ – Oy
Goudreau, Joanne, (508) 289-2560 jgoudreau@whoi.edu, Woods Hole Oceanographic Institution – Cm
Gouhier, Tarik, 7815817370 x302 t.gouhier@neu.edu, Northeastern Univ – On
Gould, Joseph C., gould.joe@spColl.edu, Saint Petersburg Coll, Clearwater – Gg
Gould, Mark D., (401) 254-3087 Roger Williams Univ – Ob
Goulden, Michael L., (949) 824-1983 mgoulden@uci.edu, Univ of California, Irvine – Og
Gouldey, Jeremy C., 616-331-2995 gouldjer@gvsu.edu, Grand Valley State Univ – CgPeCg
Gouldson, Andy, +44(0) 113 34 36417 a.gouldson@leeds.ac.uk, Univ of Leeds – Ge
Goulet, Normand, (514) 987-3375 r27254@er.uqam.ca, Universite du Quebec a Montreal – Gc
Goulet, Richard, 613-943-9922 Richard.Goulet@cnsc-ccsn.gc.ca, Univ of Ottawa – CgPyGe
Goulty, Neil R., 0191 3742513 n.r.goulty@durham.ac.uk, Durham Univ – Yg
Gouzie, Douglas R., (417) 836-5228 douglasgouzie@missouristate.edu, Missouri State Univ – HwGeCl
Gowan, Angela S., (612) 626-6451 gowa0001@umn.edu, Univ of Minnesota – Gl
Gowing, David, +44 (0)1908 659468 x 59468 david.gowing@open.ac.uk, The Open Univ – Pb
Goyne, Keith W., 573-882-0090 Univ of Missouri, Columbia – Sc
Grable, Judy, (229) 333-5752 Valdosta State Univ – Hs
Grabowski, Jonathan, 7815817370 x337 j.grabowski@neu.edu, Northeastern Univ – GuEg
Grace, Cathy A., (662)915-1799 cag@olemiss.edu, Univ of Mississippi – Gg
Grace, John, +44 (0) 131 650 5400 jgrace@ed.ac.uk, Edinburgh Univ – Oa
Grace, Peter R., 61 7 3138 2610 pr.grace@qut.edu.au, Queensland Univ of Tech – Sb
Grace, Shannon M., (512) 223-4891 sgrace@austincc.edu, Austin Community Coll District – On
Graczyk, Donald G., (630) 252-3489 Argonne National Laboratory – Cs
Grady, William, 304 594 2331 grady@geosrv.wvnet.edu, West Virginia Univ – Ec
Graf, Jr., Joseph L., (541) 552-6861 graf@sou.edu, Southern Oregon Univ – Em
Graff, Thomas O., (479) 575-3878 tgraff@comp.uark.edu, Univ of Arkansas, Fayetteville – Ou
Graff, William P., (609) 292-2576 bill.graff@dep.nj.gov, New Jersey Geological and Water Survey – On
Graham, Barbara, 702-651-4173 barbara.graham@csn.edu, Coll of Southern Nevada - West Charleston Campus – Oyw
Graham, David W., (541) 737-4140 dgraham@coas.oregonstate.edu, Oregon State Univ – Cm
Graham, Gina R., 907-451-5031 gina.graham@alaska.gov, Alaska Division of Geological & Geophysical Surveys – Yg
Graham, Grace E., (608) 263-4125 grace.graham@wgnhs.uwex.edu, Univ of Wisconsin - Extension – Hw
Graham, Ian, i.graham@unsw.edu.au, Univ of New South Wales – GxEg
Graham, Linda K., (608) 262-2640 Univ of Wisconsin, Madison – Ob
Graham, Margaret C., +44 (0) 131 650 4767 Margaret.Graham@ed.ac.uk, Edinburgh Univ – Ge
Graham, Michael, (831) 771-4400 mgraham@mlml.calstate.edu, Moss Landing Marine Laboratories – Ob
Graham, Russell W., (814) 865-6336 graham@ems.psu.edu, Pennsylvania State Univ, Univ Park – Pv
Graham, Stephan A., (650) 723-0507 graham@pangea.stanford.edu, Stanford Univ – Go
Graham, William (., (228) 688-3177 monty.graham@usm.edu, Univ of Southern Mississippi – Ob
Graham, Jr., Earl K., (814) 865-2273 graham@ems.psu.edu, Pennsylvania State Univ, Univ Park – Yx
Graham, Jr., James H., (863) 956-1151 jhgraham@ufl.edu, Univ of Florida – Sb
Grala, Katarzynz, (662) 268-1032 Ext 222 kg160@msstate.edu, Mississippi State Univ – Oi
Grammer, Michael, 405-744-6358 michael.grammer@okstate.edu, Oklahoma State Univ – GsEo
Gran, Karen B., 218-726-7406 kgran@d.umn.edu, Univ of Minnesota, Duluth – Gms
Grana, Dario, 307-766-3449 dgrana@uwyo.edu, Univ of Wyoming – Ye
Grand, Stephen P., (512) 471-3005 steveg@maestro.geo.utexas.edu, Univ of Texas at Austin – Yg
Grandal d'Anglade, Aurora, 00 34 981 167000 xeaurora@udc.es, Coruna Univ – Pv
Grande, Anthony, 212-772-5265 tony.grande@hunter.cuny.edu, Hunter Coll (CUNY) – On
Grande, Lance, (312) 665-7632 lgrande@fieldmuseum.org, Field Museum of Natural History – Pv
Grandstaff, David E., (215) 204-8228 grand@temple.edu, Temple Univ – Cl
Graney, Joseph R., (607) 777-6347 jgraney@binghamton.edu, Binghamton Univ – ClGeHs
Grange, Laura, +44 (0)23 80592786 L.J.Grange@noc.soton.a.cuk, Univ of Southampton – Ob
Granger, Darryl E., (765) 494-0043 dgranger@purdue.edu, Purdue Univ – Oa
Granger, Julie, 860-405-9094 julie.granger@uconn.edu,

Univ of Connecticut – Oc
Graniero, Phil A., (519) 253-3000 x2485 graniero@uwindsor.ca, Univ of Windsor – OinOy
Grannell, Roswitha B., (562) 985-4927 grannell@csulb.edu, California State Univ, Long Beach – Yv
Granshaw, Frank D., (503) 725-3391 fgransha@pdx.edu, Portland State Univ – OnGl
Grant, Alastair, +44 (0)1603 59 2537 a.grant@uea.ac.uk, Univ of East Anglia – GeObCt
Grant, James A., (218) 726-7237 jgrant@d.umn.edu, Univ of Minnesota, Duluth – Gp
Grant, John A., 202-633-2474 Smithsonian Institution / National Air & Space Museum – Xg
Grant, Jonathan, (902) 494-2021 jon.grant@dal.ca, Dalhousie Univ – Ob
Grant, Shelton K., (573) 341-4616 Missouri Univ of Science and Tech – Gz
Grapenthin, Ronni, 575-835-5924 rg@nmt.edu, New Mexico Inst of Mining and Tech – Gv
Grasmueck, Mark, 305 421-4858 mgrasmueck@rsmas.miami.edu, Univ of Miami – Yr
Grassian, Vicki, (858) 534-2499 vhgrassian@ucsd.edu, Univ of California, San Diego – Cm
Grassineau, Nathalie, +44 1784 443810 Nathalie.Grassineau@rhul.ac.uk, Univ of London, Royal Holloway & Bedford New Coll – Cs
Grasso, Cheryl, 401-874-2265 cgrasso@uri.edu, Univ of Rhode Island – On
Grattan, Stephen R., (530) 752-1130 srgrattan@ucdavis.edu, Univ of California, Davis – On
Grattan, Stephen R., 530-752-4618 srgrattan@ucdavis.edu, Univ of California, Davis – Spp
Gratton, Yves, yves.gratton@ete.inrs.ca, Universite du Quebec – Op
Graumlich, Lisa J., (406) 994-5320 dalylisa@montana.edu, Montana State Univ – Ou
Graustein, William C., (203) 287-2853 william.graustein@yale.edu, Yale Univ – Cl
Gravely, Cynthia Rae, (864) 656-3438 gravelc@clemson.edu, Clemson Univ – On
Graves, Alexandria, 919-513-0635 alexandria_graves@ncsu.edu, North Carolina State Univ – Sb
Graves, Charles E., (314) 977-3121 gravesce@slu.edu, Saint Louis Univ – Oa
Gravley, Darren, +64 3 3667001 Ext 45683 darren.gravley@canterbury.ac.nz, Univ of Canterbury – Gv
Gray, Kyle R., 319-273-2809 kyle.gray@uni.edu, Univ of Northern Iowa – OeGg
Gray, Lee M., (330) 823-3605 graylm@mountunion.edu, Mount Union Coll – PoGg
Gray, Mary Beth, (570) 577-1146 mbgray@bucknell.edu, Bucknell Univ – Gc
Gray, Neil, +44 (0) 191 208 4887 neil.gray@ncl.ac.uk, Univ of Newcastle Upon Tyne – Py
Gray, Norman H., dlfox@umn.edu, Univ of Connecticut – Gx
Gray, Robert S., (805) 965-0581 gray@sbcc.net, Santa Barbara City Coll – PvGzx
Gray, Sarah, (860) 486-1386 sgray@sandiego.edu, Univ of San Diego – On
Gray, Steven, steven.gray@umb.edu, Univ of Massachusetts, Boston – On
Greatbatch, Ian, +44 020 8417 2879 I.Greatbatch@kingston.ac.uk, Kingston Univ – Oi
Greatbatch, Richard , rgreatbatch@geomar.de, Dalhousie Univ – Op
Greb, Stephen F., (859) 323-0542 greb@uky.edu, Univ of Kentucky – GsEcGs
Green, Brittany, (785) 532-6101 bdgreen@ksu.edu, Kansas State Univ – On
Green, Douglas H., (740) 593-1843 green@ohio.edu, Ohio Univ – Yg
Green, Harry W., (510) 642-3059 Univ of California, Berkeley – Pv
Green, Jack, (562) 985-4198 jgreen3@csulb.edu, California State Univ, Long Beach – Gv
Green, John C., (218) 726-7208 jgreen@d.umn.edu, Univ of Minnesota, Duluth – Gi
Green, Jonathan, +44 0151 795 4385 Jonathan.Green@liverpool.ac.uk, Univ of Liverpool – Ob
Green, II, Harry W., (951) 827-4505 harry.green@ucr.edu, Univ of California, Riverside – Yx
Greenberg, David, (902) 426-2431 greenbergd@mar.dfo-mpo.gc.ca, Dalhousie Univ – Op
Greenberg, Harvey, (206) 685-7981 hgreen@uw.edu, Univ of Washington – Gml
Greenberg, Jeffrey K., (630) 752-5866 jeffrey.greenberg@wheaton.edu, Wheaton Coll – Gt
Greenberg, Richard J., (520) 621-6940 greenberg@lpl.arizona.edu, Univ of Arizona – Xy
Greenberg, Sallie, (217) 244-4068 sallieg@illinois.edu, Univ of Illinois, Urbana-Champaign – ClOg
Greene, Barbara, (570) 484-2048 bgreene@lhup.edu, Lock Haven Univ – On
Greene, Brian M., 412-395-7323 Brian.Greene@usace.army.mil, Youngstown State Univ – Ng
Greene, Charles H., chg2@cornell.edu, Cornell Univ – Ob
Greene, David C., (740) 587-6476 greened@denison.edu, Denison Univ – GctGe
Greene, Don M., (254) 710-2193 don_greene@baylor.edu, Baylor Univ – Oyw
Greene, Pauline R., (337) 482-6468 geology@louisiana.edu, Univ of Louisiana at Lafayette – On
Greene, Roberta, (315) 267-2286 greenera@potsdam.edu, SUNY Potsdam – On
Greene, Todd, 530-898-5546 tjgreene@csuchico.edu, California State Univ, Chico – Gsr
Greenfield, Dianne, dgreenfield@belle.baruch.sc.edu, Univ of South Carolina – Ob
Greenfield, Roy J., (814) 237-1810 roy@geosc.psu.edu, Pennsylvania State Univ, Univ Park – Yg
Greenhow, Danielle, (228) 688-2309 danielle.greenhow@usm.edu, Univ of Southern Mississippi – Og
Greenlee, Diana M., (318) 926-3314 greenlee@ulm.edu, Univ of Louisiana, Monroe – GaCo
Greenstein, Benjamin J., (319) 895-4307 bgreenstein@cornellColl.edu, Cornell Coll – GuPie
Greenwell, Chris, +44 (0) 191 33 42324 chris.greenwell@durham.ac.uk, Durham Univ – GzEoCg
Greenwood, James P., 860 685-2545 jgreenwood@wesleyan.edu, Wesleyan Univ – Xc
Greer, Lisa, 540-458-8871 greerl@wlu.edu, Washington & Lee Univ – GsPe
Gregg, Jay M., 405-744-6358 jay.gregg@okstate.edu, Oklahoma State Univ – Gs
Gregg, Michael C., (206) 543-1353 gregg@apl.washington.edu, Univ of Washington – Op
Gregg, Patricia, 217-333-3540 pgregg@illinois.edu, Univ of Illinois, Urbana-Champaign – YgCg
Gregg, Tracy K. P., (716) 645-4328 tgregg@geology.buffalo.edu, SUNY, Buffalo – Gv
Gregor, C. B., 937 775-3455 Wright State Univ – Gs
Gregory, Robert, (307) 766-2286 Ext. 237 robert.gregory@wyo.gov, Wyoming State Geological Survey – En
Gregory, Robert T., (214) 768-3075 Southern Methodist Univ – Cs
Grejner-Brzezinska, Dorota A., 614-292-3455 grejner-brzezinska.1@osu.edu, Ohio State Univ – Yd
Grender, Gordon C., (540) 231-6521 Virginia Polytechnic Inst & State Univ – Go
Grenfell, Thomas C., tcg@atmos.washington.edu, Univ of Washington – Oa
Gretener, Peter E., (403) 220-5849 Univ of Calgary – Ng
Grew, Edward S., (207) 581-2169 esgrew@maine.edu, Univ of Maine – Gpz
Grew, Priscilla C., (402) 472-2095 pgrew1@unl.edu, Univ of Nebraska, Lincoln – GpeOe
Greybush, Steven J., 814-867-4926 sjg213@psu.edu, Pennsylvania State Univ, Univ Park – Ow
Greyling, Lynnette N., 021-650-294886 l.greyling@uct.ac.za, Univ of Cape Town – EgmCe
Gribenko, Alex, (801) 585-6484 alex.gribenko@utah.edu, Univ of Utah – Yg
Grieneisen, Michael L., mgrien@ucdavis.edu, Univ of California, Davis – Hg
Griera Artigas, Albert, ++935811035 albert.griera@uab.cat, Universitat Autonoma de Barcelona – GcCl
Gries, John C., (316) 978-3140 Wichita State Univ – Gt
Grieve, Richard A. F., (613) 995-5372 geology@unb.ca, Univ of New Brunswick – Xg
Grießl, Stefan, 089/2180 4188 stefan.griessl@uni.muenchen.de, Ludwig-Maximilians-Universitaet Muenchen – Gz
Griffen, Dana T., 801-422-2305 dana_griffen@byu.edu,

Brigham Young Univ – Gz
Griffin, Blaine, bgriffen@biol.sc.edu, Univ of South Carolina – Ob
Griffin, Kevin, (845) 365-8371 Columbia Univ – On
Griffin, William R., (972) 883-2430 griffin@utdallas.edu, Univ of Texas, Dallas – Gg
Griffing, David H., 607-431-4629 griffingd@hartwick.edu, Hartwick Coll – GduGm
Griffis, Timothy J., (612) 625-3117 tgriffis@soils.umn.edu, Univ of Minnesota, Twin Cities – Oa
Griffith, Ashley, (817) 272-2987 wagriff@uta.edu, Univ of Texas, Arlington – Gct
Griffith, Caitlin, (520) 626-3806 griffith@lpl.arizona.edu, Univ of Arizona – On
Griffith, Liz, (817) 272-2987 lgriff@uta.edu, Univ of Texas, Arlington – CslCm
Grigg, Laurie, 802 485 3323 lgrigg@norwich.edu, Norwich Univ – PeOiGe
Grigg, Richard W., (808) 956-7186 rgrigg@soest.hawaii.edu, Univ of Hawai'i, Manoa – Ob
Griggs, Gary B., (831) 459-5006 ggriggs@pmc.ucsc.edu, Univ of California, Santa Cruz – On
Grigsby, Jeffry D., (765) 285-8270 jgrigsby@bsu.edu, Ball State Univ – GdClGo
Grimes, Craig, 740-593-1104 grimesc1@ohio.edu, Ohio Univ – GipCg
Grimes, Stephen, +44 1752 584759 stephen.grimes@plymouth.ac.uk, Univ of Plymouth – Cs
Grimley, David A., 217-244-7324 dgrimley@illinois.edu, Univ of Illinois, Urbana-Champaign – Ng
Grimm, Kurt A., (604) 822-9258 kgrimm@eos.ubc.ca, Univ of British Columbia – Gs
Grimmer, Abbey, 775-784-6869 abbeygrimmer@gmail.com, Univ of Nevada, Reno – Oi
Grindlay, Nancy R., (910) 962-2352 grindlayn@uncw.edu, Univ of North Carolina Wilmington – Yg
Grinford, Peter, +44 020 7679 7986 p.grindrod@ucl.ac.uk, Birkbeck Coll – Xg
Gringarten, Alain, +44 20 759 47440 a.gringarten@imperial.ac.uk, Imperial Coll – Np
Grippo, Alessandro, grippo_alessandro@smc.edu, Santa Monica Coll – Gs
Grise, Kevin M., (434) 924-0433 kmg3r@virginia.edu, Univ of Virginia – Oa
Grismer, Mark E., (530) 304-5797 megrismer@ucdavis.edu, Univ of California, Davis – Spf
Griswold, George B., (505) 299-6192 New Mexico Inst of Mining and Tech – Nx
Griswold, Jennifer, smalljen@hawaii.edu, Univ of Hawai'i, Manoa – Owr
Gritzo, Russell E., (505) 667-0481 Los Alamos National Laboratory – On
Groat, Lee A., (604) 822-4525 lgroat@eos.ubc.ca, Univ of British Columbia – Gz
Grocholski, Brent, grocholskib@si.edu, Smithsonian Institution / National Museum of Natural History – Xm
Grocke, Darren, +44 (0) 191 33 42282 Durham Univ – Pe
Groenevelt, Pieter H., (519) 824-4120 (Ext. 53585) pgroenev@lrs.uoguelph.ca, Univ of Guelph – Sp
Groffman, Peter, 212-413-3143 Peter.Groffman@brooklyn.cuny.edu, Brooklyn Coll (CUNY) – SbCl
Gromet, L. P., 401-863-1920 Peter_Gromet@brown.edu, Brown Univ – Cc
Groppi, Christopher, (480) 965-6436 cgroppi@as.arizona.edu, Arizona State Univ – On
Groppo, Chiara Teresa, chiara.groppo@unito.it, Università di Torino – Gx
Grosch, Chester E., (757) 683-4931 enright@ccpo.odu.edu, Old Dominion Univ – Opp
Grosfils, Eric B., (909) 621-8673 egrosfils@pomona.edu, Pomona Coll – XgGvq
Groshong, Jr., Richard H., 205-348-1882 rhgroshon@cs.com, Univ of Alabama – Gc
Gross, Amy, 910-521-6588 amy.gross@uncp.edu, Univ of North Carolina, Pembroke – Gg
Gross, Gerardo W., grossgw@nmt.edu, New Mexico Inst of Mining and Tech – Yx
Gross, Juliane, (848) 445-3619 jgross@eps.rutgers.edu, Rutgers, The State Univ of New Jersey – Xcm
Gross, Michael, (305) 348-3932 grossm@fiu.edu, Florida International Univ – Gc
Grossman, Ethan L., (979) 845-0637 e-grossman@tamu.edu, Texas A&M Univ – Csl
Grossman, Lawrence, (773) 702-8153 Univ of Chicago – Xc
Grossman, Lawrence S., (540) 231-5116 Virginia Polytechnic Inst & State Univ – On
Grossman, Walter, wgrossman@yahoo.com, San Bernardino Valley Coll – Og
Groszos, Mark S., (229) 333-5664 msgroszo@valdosta.edu, Valdosta State Univ – GctEg
Grotjahn, Richard , 530-752-2246 Grotjahn@ucdavis.edu, Univ of California, Davis – Oa
Grottoli, Andrea G., 614 292 5782 grottoli.1@osu.edu, Ohio State Univ – OcPe
Grotzinger, John P., 626.395.6785 grotz@gps.caltech.edu, California Inst of Tech – GsPeXg
Grove, Karen, (415) 338-2617 kgrove@sfsu.edu, San Francisco State Univ – Gs
Grove, Timothy L., (617) 253-2878 tlgrove@mit.edu, Massachusetts Inst of Tech – Gi
Grover, Jeffrey A., 8055463100 x2759 jgrover@cuesta.edu, Cuesta Coll – Gg
Grover, John E., (513) 556-6674 john.grover@uc.edu, Univ of Cincinnati – Gz
Grover, Timothy W., (802) 468-1289 tim.grover@castleton.edu, Castleton Univ – Gp
Groves, Christopher, (270) 745-5974 chris.groves@wku.edu, Western Kentucky Univ – Hg
Grube, John P., 217-244-1716 grube@isgs.uiuc.edu, Illinois State Geological Survey – Eo
Gruen, Dieter M., (630) 972-3513 Argonne National Laboratory – Om
Grujic, Djordje, (902) 494-2208 dgrujic@is.dal.ca, Dalhousie Univ – Gt
Grunbaum, Daniel, (206) 221-6594 grunbaum@ocean.washington.edu, Univ of Washington – Ob
Grunder, Anita L., (541) 737-5189 grundera@geo.oregonstate.edu, Oregon State Univ – Gi
Grundl, Timothy J., 414-229-4765 grundl@uwm.edu, Univ of Wisconsin, Milwaukee – HwCl
Grunwald, Sabine, (352) 392-1951 sabgru@ufl.edu, Univ of Florida – OisOr
Grupp, Steve, 425-388-9450 sgrupp@everettcc.edu, Everett Community Coll – GgOg
Gruzleski, John E., (514) 398-4755 McGill Univ – Nx
Gu, Baohua, 865-574-7286 gub1@ornl.gov, Oak Ridge National Laboratory – Cl
Gu, Chuanhui, 828-262-7859 guc@appstate.edu, Appalachian State Univ – Cl
Gu, Jeff, (780) 492-2292 jgu@phys.ualberta.ca, Univ of Alberta – Ys
Gualda, Guilherme, 615-322-2976 g.gualda@vanderbilt.edu, Vanderbilt Univ – GivGz
Guan, Dabo, +44(0) 113 34 37432 d.guan@leeds.ac.uk, Univ of Leeds – Eg
Guan, Huade, huade.guan@flinders.edu.au, Flinders Univ – HgsHg
Guccione, Margaret J., (479) 575-3354 guccione@comp.uark.edu, Univ of Arkansas, Fayetteville – GmaGg
Gudmundsson, Agust, +44 1784 276345 Agust.Gudmundsson@rhul.ac.uk, Univ of London, Royal Holloway & Bedford New Coll – Gc
Guenthner, Willy, wrg@illinois.edu, Univ of Illinois, Urbana-Champaign – GtzCg
Guerin, Gilles, (845)-365-8671 Columbia Univ – Gu
Guerra, Oralia, (512) 223-6052 oguerra1@austincc.edu, Austin Community Coll District – On
Guerrieri, Mary, (520) 621-2828 mary@lpl.arizona.edu, Univ of Arizona – On
Guertal, Elizabeth A., eguertal@acesag.auburn.edu, Auburn Univ – Sc
Guest, Peter S., 831-656-2451 pguest@nps.edu, Naval Postgraduate School – Ow
Guggenheim, Stephen J., (312) 996-3263 xtal@uic.edu, Univ of Illinois at Chicago – Gz
Guha, Jayanta, 418 545 5222 jguha@uqac.c, Universite du Quebec a Chicoutimi – Eg
Guilbert, John M., (520) 621-6024 j.guilbert@comcast.net, Univ of Arizona – Em
Guillemette, Renald, (979) 845-6301 guillemette@geo.tamu.edu, Texas A&M Univ – Gz
Guinan, Patrick E., 573-882-5909 guinanp@missouri.edu, Univ of Missouri, Columbia – Oa
Guinasso, Norman, (979) 862-2323 norman@geos.tamu.edu,

Texas A&M Univ – Op
Gulbranson, Erik L., 414-229-1153 gulbrans@uwm.edu, Univ of Wisconsin, Milwaukee – CgPe
Gulen, Gurcan, (713) 654-5404 gurcan.gulen@beg.utexas.edu, Univ of Texas at Austin – Ego
Gulick, Sean S., (512) 471-3262 sean@ig.utexas.edu, Univ of Texas at Austin – Yr
Gulliver, John S., (612) 625-4080 gulli003@tc.umn.edu, Univ of Minnesota, Twin Cities – Hs
Gundersen, James N., (316) 978-3140 Wichita State Univ – Ga
Gunderson, Lance, 404 727 8108 lgunder@emory.edu, Emory Univ – Sf
Gundiler, Ibrahim H., (505) 835-5730 gundiler@gis.nmt.edu, New Mexico Inst of Mining & Tech – Nx
Gunia, Piotr, piotr.gunia@uwr.edu.pl, Univ of Wroclaw – Gaz
Gunter, Mickey E., (208) 885-6015 mgunter@uidaho.edu, Univ of Idaho – Gz
Gunter, William D., (403) 472-4406 Univ of Calgary – Cl
Guo, Weifu, (508) 289-3380 wguo@whoi.edu, Woods Hole Oceanographic Institution – CslPe
Gupta, Hoshin V., (520) 275-5534 hoshin.gupta@hwr.arizona.edu, Univ of Arizona – HqOga
Gupta, Sanjeev, +44 20 759 46527 s.gupta@imperial.ac.uk, Imperial Coll – Gs
Gupta, Satish C., (612) 625-1241 sgupta@soils.umn.edu, Univ of Minnesota, Twin Cities – Sp
Gurdak, Jason, (415) 338-6869 jgurdak@sfsu.edu, San Francisco State Univ – Hw
Gurevich, Boris, +61 8 9266-7359 B.Gurevich@curtin.edu.au, Curtin Univ – Ye
Gurlea, Lawrence P., lisobar@aol.com, Youngstown State Univ – Cg
Gurnis, Michael C., (626) 395-6979 gurnis@gps.caltech.edu, California Inst of Tech – Ys
Gurocak, Zulfu, 00904242370000-5991 zgurocak@firat.edu.tr, Firat Univ – Nrg
Gurrola, Harold, 806-834-8625 harold.gurrola@ttu.edu, Texas Tech Univ – Ys
Gurwin, Jacek, jacek.gurwin@uwr.edu.pl, Univ of Wroclaw – Hy
Gust, David A., 61 7 3138 2217 d.gust@qut.edu.au, Queensland Univ of Tech – GiCpGt
Gustin, Mae, 775.784.4203 mgustin@cabnr.unr.edu, Univ of Nevada, Reno – Cg
Guth, Lawrence R., 978-665-3082 lguth@fitchburgstate.edu, Fitchburg State Univ – Gc
Guth, Peter L., (410) 293-6560 pguth@usna.edu, United States Naval Academy – GcOiy
Guthrie, Roderick I., (514) 398-4755 McGill Univ – Nx
Gutierrez, Melida, (417) 836-5967 mgutierrez@missouristate.edu, Missouri State Univ – Cg
Gutknecht, Jessica L., 612-626-8435 jgutknec@umn.edu, Univ of Minnesota, Twin Cities – Sb
Gutmann, James T., (860) 685-2258 jgutmann@wesleyan.edu, Wesleyan Univ – Gv
Gutowski, Vincent P., 217-581-3825 vpgutowski@eiu.edu, Eastern Illinois Univ – GmOy
Gutowski, William J., (515) 294-5632 gutowski@iastate.edu, Iowa State Univ of Science & Tech – Oa
Gutzler, David J., (505) 277-3328 gutzler@unm.edu, Univ of New Mexico – Oa
Guza, Robert T., (858) 534-0585 rguza@ucsd.edu, Univ of California, San Diego – On
Guzina, Bojan B., (612) 626-0789 guzina@wave.ce.umn.edu, Univ of Minnesota, Twin Cities – Nr
Guzman, Ernesto, (519) 824-4120 Ext.53609 eguzman@uoguelph.ca, Univ of Guelph – On
Gušiæ, Ivan, +38514606102 ivangusic@yahoo.com, Univ of Zagreb – GgPsm
Gyakum, John R., (514) 398-6076 john.gyakum@mcgill.ca, McGill Univ – Ow
Gysi, Alex, (303) 273-3828 agysi@mines.edu, Colorado School of Mines – Cg
Göttlich, Hagen, 089/2180 5615 goettlich@ennab.de, Ludwig-Maximilians-Universitaet Muenchen – Gz

## H

Haag, Lucas, (785) 462-6281 lhaag@ksu.edu, Kansas State Univ – So
Haas, Christian, (416)736-2100 #77705 haasc@yorku.ca, York Univ – Yg
Haas, Johnson R., (269) 387-2878 johnson.haas@wmich.edu, Western Michigan Univ – Cl
Habana, Nathan C., (671) 735-2693 nhabana@uguam.uog.edu, Univ of Guam – Hw
Habash, Marc, (519) 824-4120 Ext.52748 mhabash@uoguelph.ca, Univ of Guelph – On
Haber, Eldad, (604) 822-4525 haber@eos.ubc.ca, Univ of British Columbia – YgOn
Habib, Daniel, (718) 997-3333 Graduate School of the City Univ of New York – Pl
Hacker, Bradley R., (805) 893-7952 hacker@geol.ucsb.edu, Univ of California, Santa Barbara – Gp
Hacker, David B., 330-672-8831 dhacker@kent.edu, Kent State Univ – GcHw
Hacker, Jorg, jorg.hacker@flinders.edu.au, Flinders Univ – Ora
Hacker, Joshua P., 831-656-2722 jphacker@nps.edu, Naval Postgraduate School – Ow
Hacker, Patricia, 419-530-5058 patricia.hacker@utoledo.edu, Univ of Toledo – On
Haddad, Brent, (831) 459-4149 bhaddad@cats.ucsc.edu, Univ of California, Santa Cruz – On
Haddock, Gerald H., (630) 752-5063 Wheaton Coll – Gi
Haddock, Gregory D., (660) 562-1719 haddock@nwmissouri.edu, Northwest Missouri State Univ – Oi
Hadler, Kathryn, +44 20 759 47198 k.hadler@imperial.ac.uk, Imperial Coll – Gz
Hafez, Sabry, (907) 474- 6917 ssabour@alaska.edu, Univ of Alaska, Fairbanks – Nm
Haff, Peter K., (919) 684-5902 haff@duke.edu, Duke Univ – On
Hafner, James A., (413) 545-0778 hafner@geo.umass.edu, Univ of Massachusetts, Amherst – On
Hagadorn, James W., (303) 370-6058 james.hagadorn@dmns.org, Denver Museum of Nature & Science – GgsPg
Hagan, Robert M., (530) 752-0453 Univ of California, Davis – Hg
Hagedorn, Charles, (540) 231-4895 chagedor@vt.edu, Virginia Polytechnic Inst & State Univ – Sb
Hageman, Steven J., (828) 262-6609 hagemansj@appstate.edu, Appalachian State Univ – Pie
Hager, Bradford H., (617) 253-0126 bhhager@mit.edu, Massachusetts Inst of Tech – Ys
Hagerty, Michael, 617-552-8300 hagertmb@bc.edu, Boston Coll – Ys
Haggart, James, jim.haggart@canada.ca, Univ of British Columbia – PgGr
Haggart, Renee, (604) 822-2789 rhaggart@eos.ubc.ca, Univ of British Columbia – On
Haggerty, Janet, (918) 631-2304 janet-haggerty@utulsa.edu, The Univ of Tulsa – Gu
Haggerty, Julia H., (406) 994-6904 julia.haggerty@montana.edu, Montana State Univ – Ou
Haggerty, Roy D., (541) 737-0663 Roy.Haggerty@oregonstate.edu, Oregon State Univ – Hw
Haggerty, Stephen E., (305) 348-7338 haggerty@fiu.edu, Florida International Univ – GiEgGz
Hagni, Richard D., (573) 341-4657 rhagni@umr.edu, Missouri Univ of Science and Tech – Em
Haiar, Brooke, haiar@lynchburg.edu, Lynchburg Coll – PoGeg
Haigh, Ivan D., +44 (023) 80596501 I.D.Haigh@soton.ac.uk, Univ of Southampton – On
Haigh, Nardia, nardia.haigh@umb.edu, Univ of Massachusetts, Boston – On
Haileab, Bereket, (507) 222-5746 bhaileab@carleton.edu, Carleton Coll – GxzGi
Haimson, Bezalel C., (608) 262-2563 bhaimson@wisc.edu, Univ of Wisconsin, Madison – Nr
Haine, Thomas W., (410) 516-7048 thomas.haine@jhu.edu, Johns Hopkins Univ – Op
Hajash, Andrew, (979) 845-0642 hajash@geo.tamu.edu, Texas A&M Univ – Cp
Hajek, Elizabeth, hajek@psu.edu, Pennsylvania State Univ, Univ Park – Gs
Hajnal, Zoltan, (306) 966-5694 zoltan.hajnal@usask.ca, Univ of Saskatchewan – YsGtc
Hakim, Gregory J., 206-685-2439 hakim@atmos.washington.edu, Univ of Washington – Oa
Hakimian, Adina, 516-463-6545 adina.i.hakimian@hofstra.edu, Hofstra Univ – Gg
Halama, Ralf, +44 (0) 1782 7 34960 r.halama@keele.ac.uk, Keele Univ – GpiCa

Halbig, Joseph B., halbig@wazoo.com, Univ of Hawai'i, Hilo – Ca
Halden, Norman M., (204) 474-6910 nm_halden@umanitoba.ca, Univ of Manitoba – Cg
Hale, Beverley A., (519) 824-4120 (Ext. 53434) bhale@uoguelph.ca, Univ of Guelph – Ct
Hale, Dave, 303-273-3461 dhale@mines.edu, Colorado School of Mines – Ye
Hale, Leslie J., (202) 633-1796 halel@si.edu, Smithsonian Institution / National Museum of Natural History – GgOnGh
Hale, Michelle, +44 023 92 842290 michelle.hale@port.ac.uk, Univ of Portsmouth – Og
Hales, Burke R., (541) 737-8121 bhales@coas.oregonstate.edu, Oregon State Univ – Oc
Hales, T. C., +44(0)29 208 74329 HalesT@cardiff.ac.uk, Univ of Wales – Gt
Hales, TC, halest@cf.ac.uk, Cardiff Univ – Gm
Haley, Brian, 541-737-2649 bhaley@coas.oregonstate.edu, Oregon State Univ – Cs
Haley, John C., 757-455-3407 jchaley@vwc.edu, Virginia Wesleyan Coll – GgeOi
Halfar, Jochen, (905) 828-5419 jochen.halfar@utoronto.ca, Univ of Toronto – Pe
Halfman, John D., (315) 781-3918 halfman@hws.edu, Hobart & William Smith Colls – GeHsGn
Halgedahl, Susan L., (801) 581-7062 s.halgedahl@utah.edu, Univ of Utah – Ym
Halihan, Todd, 405-744-6358 todd.halihan@okstate.edu, Oklahoma State Univ – Hw
Hall, Anne M., (404) 727-2863 ahall04@emory.edu, Emory Univ – Gz
Hall, Brenda L., (207) 581-2191 brendah@maine.edu, Univ of Maine – Gl
Hall, Chris, (734) 764-6391 cmhall@umich.edu, Univ of Michigan – Cc
Hall, Christopher C., (519) 824-4120 Ext.52740 jchall@uoguelph.ca, Univ of Guelph – On
Hall, Clarence A., (310) 825-1010 hall@epss.ucla.edu, Univ of California, Los Angeles – Gt
Hall, Cynthia V., (610) 436-1003 chall@wcupa.edu, West Chester Univ – Cg
Hall, Frank R., (504) 280-1105 frhall@mac.com, Univ of New Orleans – Ym
Hall, Jean, +44 (0) 191 208 8783 jean.hall@ncl.ac.uk, Univ of Newcastle Upon Tyne – Ng
Hall, Jeremy, (709) 737-7569 jeremyh@mun.ca, Memorial Univ of Newfoundland – Ys
Hall, Jude, (740) 587-6217 hall@denison.edu, Denison Univ – On
Hall, Luke D., (805) 642-3211 lhall@vcccd.net, Ventura Coll – Oy
Hall, Robert, +44 1784 443897 robert.hall@rhul.ac.uk, Univ of London, Royal Holloway & Bedford New Coll – Gt
Hall, Robert, +44 (0)1603 59 2550 robert.hall@uea.ac.uk, Univ of East Anglia – On
Hall, Russell L., (403) 220-6678 Univ of Calgary – Pi
Hall, Stuart A., 713-743-3416 sahgeo@uh.edu, Univ of Houston – Ym
Hall, Tracy, 678-872-8415 thall@highlands.edu, Georgia Highlands Coll – Gg
Hall, III, John R., (504) 231-6305 jrhall3@vt.edu, Virginia Polytechnic Inst & State Univ – So
Hallar, Anna G., (801) 587-7238 gannet.hallar@utah.edu, Univ of Utah – Oa
Haller, Merrick C., (541) 737-9141 merrick.haller@oregonstate.edu, Oregon State Univ – OnrOp
Hallet, Bernard, 206-685-2409 hallet@uw.edu, Univ of Washington – GmOl
Hallett, Benjamin W., 920-424-0868 hallettb@uwosh.edu, Univ of Wisconsin, Oshkosh – Gpx
Hallett, Rebecca, (519) 824-4120 Ext.54488 rhallett@uoguelph.ca, Univ of Guelph – On
Halliday, Alex, +44 (1865) 272969 alex.halliday@earth.ox.ac.uk, Univ of Oxford – Cs
Hallin, Stephen C., stephenhallin@weber.edu, Weber State Univ – Ow
Hallock, Brent G., (805)756-2436 bhallock@calpoly.edu, California Polytechnic State Univ – Sf
Hallock-Muller, Pamela, (727) 553-1567 pmuller@usf.edu, Univ of South Florida – PmGse
Halls, Henry C., 905-828-5363 hlhalls@utm.utoronto.ca, Univ of Toronto – Ym
Halls, Joanne N., (910) 962-7614 hallsj@uncw.edu, Univ of North Carolina Wilmington – Oi
Halsor, Sid P., (570) 408-4611 sid.halsor@wilkes.edu, Wilkes Univ – Giv
Halsted, Christian, (207) 287-7175 christian.h.halsted@maine.gov, Dept of Agriculture, Conservation, and Forestry – Oi
Halverson, Galen, (514) 398-4894 galen.halverson@mcgill.ca, McGill Univ – GsCsGt
Halverson, Larry, (515) 294-0495 larryh@iastate.edu, Iowa State Univ of Science & Tech – So
Ham, Nelson R., (920) 403-3977 nelson.ham@snc.edu, Saint Norbert Coll – Gl
Hamane, Angel, ahamane2@calstatela.edu, California State Univ, Los Angeles – Oe
Hamann, Hillary, (303) 871-3977 hillary.hamann@du.edu, Univ of Denver – HgOy
Hambrey, Michael M., +44 (0)1970 621 860 mjh@aber.ac.uk, Aberystwyth Univ – GlsGm
Hamburger, Michael W., (812) 855-2934 hamburg@indiana.edu, Indiana Univ, Bloomington – YsGtv
Hamdan, Abeer, abeer.hamdan@phoenixColl.edu, Phoenix Coll – Gg
Hamecher, Emily A., (657) 278-7096 ehamecher@fullerton.edu, California State Univ, Fullerton – GgzGi
Hameed, Sultan, (631) 632-8319 sultan.hameed@stonybrook.edu, SUNY, Stony Brook – Oa
Hames, Willis E., (334) 844-4881 hameswe@auburn.edu, Auburn Univ – CcGpt
Hamill, Paul, (815) 455-8698 phamill@mchenry.edu, McHenry County Coll – Owg
Hamilton, George, (413) 499-4660 bhamilton@berkshirecc.edu, Berkshire Community Coll – On
Hamilton, Gordon S., (207) 581-3446 gordon.hamilton@maine.edu, Univ of Maine – Ol
Hamilton, Jacqueline, (612) 626-8292 stub0035@umn.edu, Univ of Minnesota – OiGe
Hamilton, Michael Andrew, (416) 946-7424 mahamilton@geology.utoronto.ca, Univ of Toronto – Cc
Hamilton, Phillip, phamiltn@neaq.org, Univ of Massachusetts, Boston – Ob
Hamilton, Sally, +44 (0)131 451 3198 s.hamilton@hw.ac.uk, Heriot-Watt Univ – Gs
Hamilton, Thomas, 4256401339x7067 thomas.hamilton@edcc.edu, Edmonds Community Coll – Og
Hamilton, Warren, 303-384-2047 whamilto@mines.edu, Colorado School of Mines – Ys
Hamlin, H, Scott, 512-4759527 scott.hamlin@beg.utexas.edu, Univ of Texas at Austin, Jackson School of Geosciences – Gr
Hamme, Roberta C., (250) 472-4014 rhamme@uvic.ca, Univ of Victoria – Oc
Hammer, Julia E., 808-956-5996 jhammer@hawaii.edu, Univ of Hawai'i, Manoa – GiCpGv
Hammer, Philip T., (604) 822-5703 phammer@eos.ubc.ca, Univ of British Columbia – Ys
Hammer, William R., (309) 794-7487 williamhammer@augustana.edu, Augustana Coll – Pv
Hammerschmidt, Chad, 937 775-3457 chad.hammerschmidt@wright.edu, Wright State Univ – CmOc
Hammersley, Charles, (928) 523-6655 charles.hammersley@nau.edu, Northern Arizona Univ – On
Hammersley, Lisa, 916-278-7200 hammersley@csus.edu, California State Univ, Sacramento – Gig
Hammes, Ursula, 512-471-1891 ursula.hammes@beg.utexas.edu, Univ of Texas at Austin, Jackson School of Geosciences – Gg
Hammond, Anne, (807) 343-8677 anne.hammond@lakeheadu.ca, Lakehead Univ – On
Hammond, Douglas E., (213) 740-5837 dhammond@usc.edu, Univ of Southern California – Cm
Hammond, Paul E., (503) 725-3387 hammondp@pdx.edu, Portland State Univ – Gi
Hammond, William, 775 784-6436 whammond@unr.edu, Univ of Nevada – Yd
Hampson, Arthur, (801) 585-5698 spike.hampson@geog.utah.edu, Univ of Utah – Oy
Hampson, Gary , +44 20 759 46475 g.j.hampson@imperial.ac.uk, Imperial Coll – Gd
Hampton, Brian A., 575-646-2997 bhampton@nmsu.edu, New Mexico State Univ, Las Cruces – Gst
Hampton, Duane R., (269) 387-5496 duane.hampton@wmich.edu,

Western Michigan Univ – HwSpHg
Hampton, Samuel, +64 3 3667001 Ext 6770 samuel.hampton@canterbury.ac.nz, Univ of Canterbury – Gq
Hams, Jacquelyn E., (818) 778-5566 hamsje@lavc.edu, Los Angeles Valley Coll – GgOue
Hamzaoui, Cherif, 514-987-3000 #6837 hamzaoui.cherif@uqam.ca, Universite du Quebec a Montreal – Yg
Han, De-hua, 713-743-9293 dhan@uh.edu, Univ of Houston – Ye
Han, Nizhou, (540) 231-2403 nhan@vt.edu, Virginia Polytechnic Inst & State Univ – Sc
Han, Weon Shik, 414-229-2493 hanw@uwm.edu, Univ of Wisconsin, Milwaukee – Hw
Hanan, Barry B., (619) 594-6710 bhanan@mail.sdsu.edu, San Diego State Univ – Cc
Hancock, Gregory, 61 02 4921 5090 Greg.Hancock@newcastle.edu.au, Univ of Newcastle – OiGmm
Hancock, Gregory S., (757) 221-2446 gshanc@wm.edu, Coll of William & Mary – Gm
Hand, Kristen, 717–702–2046 khand@pa.gov, Pennsylvania Bureau of Topographic & Geologic Survey – HwGg
Hand, Linda M., 650 574-6633 hand@smccd.edu, Coll of San Mateo – GgPgOg
Hand, Suzanne J., s.hand@unsw.edu.au, Univ of New South Wales – Pv
Haneberg, William, (505) 255-8005 New Mexico Inst of Mining and Tech – Ng
Haner, Andrew, (501) 683-0153 andrew.haner@arkansas.gov, Arkansas Geological Survey – Oi
Hanes, Daniel M., (314) 977-3703 dhanes@slu.edu, Saint Louis Univ – GesGm
Hanes, John A., (613) 533-6188 hanes@queensu.ca, Queen's Univ – Cc
Haney, Christa M., 662-268-1032 Ext 224 meloche@geosci.msstate.edu, Mississippi State Univ – Ow
Hanke, Brenda R., 513-556-3732 Univ of Cincinnati – Pi
Hanke, Steve H., (410) 516-7183 hanke@jhu.edu , Johns Hopkins Univ – On
Hankins, Katherine B., 404 413-5775 geokbh@langate.gsu.edu, Georgia State Univ – On
Hanks, Catherine L., 907-474-5562 chanks@gi.alaska.edu, Univ of Alaska, Fairbanks – Go
Hanna, Heather, 9197337353 x30 heather.hanna@ncdenr.gov , North Carolina Geological Survey – On
Hanna, Ruth L., 925-424-1319 rhanna@laspositasColl.edu, Las Positas Coll – Gg
Hanna, Stephen P., (540) 654-1490 shanna@umw.edu, Mary Washington Coll – Or
Hannah, David, +44 (0)121 41 46925 d.m.hannah@bham.ac.uk, Univ of Birmingham – Hg
Hannah, Judith L., (970) 491-1329 jhannah@warnercnr.colostate.edu, Colorado State Univ – GiCc
Hannibal, Joseph T., (216) 231-4600 jhanniba@cmnh.org, Case Western Reserve Univ – Pi
Hannigan, Robyn, robyn.hannigan@umb.edu, Univ of Massachusetts, Boston – ClmCt
Hannington, Mark, 613-562-5292 Mark.Hannington@uottawa.ca, Univ of Ottawa – Eg
Hannula, Kimberly, (970) 247-7463 hannula_k@fortlewis.edu, Fort Lewis Coll – GctGp
Hanor, Jeffrey S., (225) 388-3418 hanor@geol.lsu.edu, Louisiana State Univ – Cl
Hansel, Colleen, (508) 289-3738 chansel@whoi.edu, Woods Hole Oceanographic Institution – Cm
Hansen, Anthony R., 320-308-2009 arhansen@stcloudstate.edu, Saint Cloud State Univ – Ow
Hansen, Devon A., devon.hansen@und.edu, Univ of North Dakota – On
Hansen, Jeffrey C., (505) 667-5043 jchansen@lanl.gov, Los Alamos National Laboratory – On
Hansen, Lars, +44 (1865) 272000 lars.hansen@earth.ox.ac.uk, Univ of Oxford – Gz
Hansen, Robert, +27 (0)51 401 2712 hansenr@ufs.ac.za, Univ of the Free State – CgGe
Hansen, Samantha E., (205) 348-7089 shansen@ua.edu, Univ of Alabama – YsGt
Hansen, Thor A., (360) 650-3648 thor.hansen@wwu.edu, Western Washington Univ – Pg
Hansen, Vicki L., (218) 726-8628 vhansen@d.umn.edu, Univ of Minnesota, Duluth – GtXgGc
Hansen, William, (212) 346-1502 Pace Univ, New York Campus – Oa
Hansen, William J., (508) 929-8608 whansen@worcester.edu, Worcester State Univ – OiyOr
Hansom, Jim, +4401413305406 jim.hansom@glasgow.ac.uk, Univ of Glasgow – OnGmOy
Hanson, Andrew D., (702) 895-1092 Univ of Nevada, Las Vegas – Co
Hanson, Blaine R., (530) 752-1130 brhanson@ucdavis.edu, Univ of California, Davis – Hg
Hanson, Doug, 501 683-0115 doug.hanson@arkansas.gov, Arkansas Geological Survey – Gg
Hanson, Gilbert N., (631) 632-8210 gilbert.hanson@stonybrook.edu, SUNY, Stony Brook – GgOe
Hanson, Howard, 561 297 2460 hphanson@fau.edu, Florida Atlantic Univ – Oag
Hanson, Lindley S., lhanson@salemstate.edu, Salem State Univ – GmlOn
Hanson, Paul, (402) 472-7762 phanson2@unl.edu, University of Nebraska - Lincoln – Gm
Hanson, Richard E., (817) 257-7996 r.hanson@tcu.edu, Texas Christian Univ – GxvGt
Hanson, Sarah L., 517-264-3944 slhanson@adrian.edu, Adrian Coll – Giz
Hanson, Jr., Alfred K., (401) 874-6899 akhanson@gso.uri.edu, Univ of Rhode Island – Oc
Hapke, Bruce W., (412) 624-8876 hapke@pitt.edu, Univ of Pittsburgh – Xy
Haq, Saad, (765) 496-7206 haq@purdue.edu, Purdue Univ – Gt
Hara, Tetsu, (401) 874-6509 thara@uri.edu, Univ of Rhode Island – Op
Harbaugh, John W., (650) 723-3365 harbaugh@pangea.stanford.edu, Stanford Univ – Gq
Harbert, William P., (412) 624-8874 harbert@pitt.edu, Univ of Pittsburgh – Ym
Harbor, David J., 540 458 8871 harbord@wlu.edu, Washington & Lee Univ – Gm
Harbor, Jon M., (765) 494-4753 jharbor@purdue.edu, Purdue Univ – HwGlm
Harbottle, Garman, garman@bnl.gov, SUNY, Stony Brook – Cc
Hardage, Bob A., 512-471-0300 bob.hardage@beg.utexas.edu, Univ of Texas at Austin – Yes
Harder, Brian J., (225) 578-8533 bharde1@lsu.edu, Louisiana State Univ – Eo
Harder, Steven H., (915) 747-5746 harder@utep.edu, Univ of Texas, El Paso – Yg
Hardin, Perry J., (801) 378-6062 perry_hardin@byu.edu, Brigham Young Univ – Or
Harding, Chris, (515) 294-7521 charding@iastate.edu, Iowa State Univ of Science & Tech – Gq
Harding, Ian, +44 (0)23 80592071 ich@noc.soton.ac.uk, Univ of Southampton – PemPl
Hardison, Amber K., (361) 749-6705 amber.hardison@utexas.edu, Univ of Texas at Austin – Cm
Hardy, Douglas R., (802) 649-1829 dhardy@geo.umass.edu, Univ of Massachusetts, Amherst – Hy
Hardy, Shaun J., (202) 478-7960 shardy@carnegiescience.edu, Carnegie Institution of Washington – On
Hargis, David, 619-521-0165 dhargis@hargis.com, Univ of Arizona – Hwq
Hargrave, Jennifer E., 337-482-0678 jhargrave@louisiana.edu, Univ of Louisiana at Lafayette – GgPg
Hargrave, Jennifer E., 435.865.8429 jenniferhargrave@suu.edu, Southern Utah Univ – PvGrs
Hargrave, Phyllis, 406-496-4606 phargrave@mtech.edu, Montana Tech of The Univ of Montana – Gg
Hargraves, Paul E., (401) 874-6241 pharg@gso.uri.edu, Univ of Rhode Island – Ob
Hargreaves, Bruce R., (610) 758-3683 brh0@lehigh.edu, Lehigh Univ – Ob
Hargreaves, Tom, +44 (0)1603 59 3116 tom.hargreaves@uea.ac.uk, Univ of East Anglia – Ge
Hari, Kosiyath R., krharigeology@gmail.com, Pt. Ravishankar Shukla Univ – GiiCp
Hariri, Mustafa, +96638601601 mmhariri@kfupm.edu.sa, King Fahd Univ of Petroleum and Minerals – GcOrGt
Haritashya, Umesh, 937-229-2939 Umesh.Haritashya@notes.udayton.edu, Univ of Dayton – OrGm
Harkrider, David G., (626) 395-6910 California Inst of Tech – Ys
Harley, Grant, 601-266-5884 grant.harley@usm.edu, Univ of Southern Mississippi – Oy
Harley, Simon L., +44 (0) 131 650 4839 Simon.Harley@ed.ac.uk, Edinburgh Univ – Gt

Harlow, George E., (212) 769-5378 Graduate School of the City Univ of New York – Gz
Harma, Roberta L., (805) 553.4161 rharma@vcccd.edu, Moorpark Coll – Gg
Harmon, Mella, 775-784-4046 mellah@unr.edu, Univ of Nevada, Reno – Ou
Harmon, Nicholas, +44 (0)23 80594783 N.Harmon@noc.soton.ac.uk, Univ of Southampton – Yg
Harms, Tekla A., (413) 542-2711 taharms@amherst.edu, Amherst Coll – Gt
Harnett, Erika M., 206-543-0212 eharnett@ess.washington.edu, Univ of Washington – XyGev
Harnik, Nili, 03-640-6359 harnik@tau.ac.il, Tel Aviv Univ – Oa
Harnik, Paul, (717) 358-5946 paul.harnik@fandm.edu, Franklin and Marshall Coll – Pg
Harper, David, +44 (0) 191 33 47143 david.harper@durham.ac.uk, Durham Univ – Pg
Harper, Joel, (406) 243-2341 joel.harper@umontana.edu, Univ of Montana – Ol
Harper, Stephen B., (252) 328-6773 harpers@ecu.edu, East Carolina Univ – GmgGe
Harper, Jr., Charles W., (405) 325-7725 charper@gcn.ou.edu, Univ of Oklahoma – Pi
Harpold, Adrian, (775) 784-6759 aharpold@cabnr.unr.edu, Univ of Nevada, Reno – HqsOr
Harpp, Karen, (315) 228-7211 kharpp@colgate.edu, Colgate Univ – GvCgGi
Harr, Patrick A., 831-656-3787 paharr@nps.edu, Naval Postgraduate School – Ow
Harrap, Rob, 613-533-2553 harrap@queensu.ca, Queen's Univ – Gc
Harrell, Michael, 206-344-4392 MHarrell@sccd.ctc.edu, Seattle Central Community Coll – Gg
Harrell, Jr., T. L., 205-247-3559 TLHarrell@gsa.sate.al.us, Geological Survey of Alabama – PvGg
Harries, Peter J., (813) 974-4974 harries@chuma.cas.usf.edu, Univ of South Florida, Tampa – Pg
Harrington, Charles D., (505) 667-0078 charrington@lanl.gov, Los Alamos National Laboratory – Gm
Harrington, Glenn, glenn.harrington@flinders.edu.au, Flinders Univ – Hw
Harrington, Rebecca, +49-721-6084625 rebecca.harrington@kit.edu, Karlsruhe Inst of Tech – Ys
Harrington, Rebecca, (514) 398-2722 rebecca.harrington@mcgill.ca, McGill Univ – Ys
Harris, Ann G., 330-941-3613 agharris@cc.ysu.edu, Youngstown State Univ – Ge
Harris, Brett D., +618 9266-3089 B.Harris@curtin.edu.au, Curtin Univ – HwYve
Harris, C, harrisC@cf.ac.uk, Cardiff Univ – Ge
Harris, Chris, 021-650-4886 chris.harris@uct.ac.za, Univ of Cape Town – GiCsGg
Harris, Clay D., 615-904-8019 Clay.Harris@mtsu.edu, Middle Tennessee State Univ – Gs
Harris, Courtney K., (804) 684-7194 ckharris@vims.edu, Coll of William & Mary – On
Harris, David C., (859) 323-0545 dcharris@uky.edu, Univ of Kentucky – Go
Harris, DeVerle P., (520) 621-6024 dharris@geo.arizona.edu, Univ of Arizona – Eg
Harris, Glendon H., (706) 542-2968 Univ of Georgia – Sb
Harris, James B., 601-974-1343 harrijb@millsaps.edu, Millsaps Coll – Ye
Harris, Jerry M., (650) 723-0496 Stanford Univ – Ys
Harris, Lee E., (321) 674-8096 lharris@fit.edu, Florida Inst of Tech – On
Harris, Lyal B., lyal.harris@ete.inrs.ca, Universite du Quebec – GcYg
Harris, Mark T., (414) 229-5483 mtharris@uwm.edu, Univ of Wisconsin, Milwaukee – GsrGd
Harris, Nicholas, (780) 492-0356 nharris@ualberta.ca, Univ of Alberta – Gso
Harris, Nigel B., n.b.w.harris@open.ac.uk, The Open Univ – GtiGp
Harris, Randa R., (678) 839-4056 rharris@westga.edu, Univ of West Georgia – Hs
Harris, Richard S., (905) 525-9140 (Ext. 27216) harrisr@mcmaster.ca, McMaster Univ – Ou
Harris, Robert N., (541) 737-4370 rharris@coas.oregonstate.edu, Oregon State Univ – Yhr
Harris, Robert N., (801) 587-9366 rharris@coas.oregonstate.edu, Univ of Utah – Gu
Harris, Robin F., (608) 263-5691 rfharris@wisc.edu, Univ of Wisconsin, Madison – Sb
Harris, Ron, (801) 422-9264 rharris@byu.edu, Brigham Young Univ – Gc
Harris, Sara, (604) 822-5674 sara@eos.ubc.ca, Univ of British Columbia – Ou
Harris, Scott, (843) 953-0864 harriss@cofc.edu, Coll of Charleston – GaOn
Harris, Virginia M., (860) 685-2244 vharris@wesleyan.edu, Wesleyan Univ – On
Harris, W. Burleigh, (910) 962-3492 harrisw@uncw.edu, Univ of North Carolina Wilmington – GrdGs
Harris, William H., (718) 951-5416 Graduate School of the City Univ of New York – Hw
Harris, Jr., Stanley E., (618) 453-3351 Southern Illinois Univ Carbondale – Ge
Harris, Jr., Willie G., (352) 294-3110 apatite@ufl.edu, Univ of Florida – Scd
Harrison, Bruce I., (575) 835-5864 bruce@nmt.edu, New Mexico Inst of Mining and Tech – GmOsSf
Harrison, Christopher G., 305 421-4610 charrison@rsmas.miami.edu, Univ of Miami – Yr
Harrison, Conor, cmharris@mailbox.sc.edu, Univ of South Carolina – On
Harrison, Don E., (206) 526-6225 harrison@pmel.noaa.gov, Univ of Washington – Op
Harrison, Halstead, harrison@atmos.washington.edu, Univ of Washington – Oa
Harrison, Linda , (269) 387-8642 linda.harrison@wmich.edu, Western Michigan Univ – GgPmOn
Harrison, Michael J., (931) 372-3751 mharrison@tntech.edu, Tennessee Tech Univ – Gc
Harrison, Richard, +44 (0) 1223 333380 rjh40@esc.cam.ac.uk, Univ of Cambridge – Gz
Harrison, Robert S., (516) 299-2318 Long Island Univ, C.W. Post Campus – Oy
Harrison, Roy M., +44 (0)121 41 43494 r.m.harrison@bham.ac.uk, Univ of Birmingham – Oa
Harrison, T. Mark, (310) 825-7970 tmh@argon.ess.ucla.edu, Univ of California, Los Angeles – CcGt
Harrison, Wendy J., (303) 273-3821 wharriso@mines.edu, Colorado School of Mines – Cl
Harrison, III, William B., (269) 387-8691 william.harrison_iii@wmich.edu, Western Michigan Univ – GsrPg
Harry, Dennis L., (970) 491-2714 dharry@warnercnr.colostate.edu, Colorado State Univ – Yg
Harsh, James B., (509) 335-3650 harsh@wsu.edu, Washington State Univ – Sc
Harshvardhan, (765) 494-0693 harsh@purdue.edu, Purdue Univ – Oa
Hart, Brian R., bhart@uwo.ca, Western Univ – Ca
Hart, Craig J., (604) 822-5149 chart@eos.ubc.ca, Univ of British Columbia – EmGiCc
Hart, David J., (608) 262-2307 dave.hart@wgnhs.uwex.edu, Univ of Wisconsin - Extension – HwYg
Hart, Evan A., (931) 372-3121 EHart@TnTech.edu, Tennessee Tech Univ – Oy
Hart, Malcolm, +44 1752 584761 M.Hart@plymouth.ac.uk, Univ of Plymouth – Pm
Hart, Richard L., (630) 252-5839 rlhart@anl.gov, Argonne National Laboratory – Oa
Hart, Stanley R., (520) 625-4543 shart@whoi.edu, Woods Hole Oceanographic Institution – CgGvCp
Hart, William K., (513) 529-3217 hartwk@miamioh.edu, Miami Univ – GivCc
Harte, Michael, mharte@coas.oregonstate.edu, Oregon State Univ – Gu
Hartel, Peter G., (706) 542-0898 Univ of Georgia – Sb
Harter, Robert D., (603) 862-1020 Univ of New Hampshire – Sc
Harter, Thomas, 530-752-2709 thharter@ucdavis.edu, Univ of California, Davis – Hw
Harter, Thomas, 530-400-1784 ThHarter@ucdavis.edu, Univ of California, Davis – Hg
Harter, Thomas L., (530) 752-2709 thharter@ucdavis.edu, Univ of California, Davis – Hw
Hartley, Susan, (218)279-2661 s.hartley@lsc.edu, Lake Superior Coll – Gg
Hartman, Joseph H., (701) 777-5055 joseph.hartman@engr.und.edu, Univ of North Dakota – Pi
Hartmann, Dennis L., 206-543-7460 dennis@atmos.washington.edu, Univ of Washington – Oa
Hartnett, Hilairy E., (480) 965-5593 h.hartnett@asu.edu,

Arizona State Univ – ComOc
Hartse, Hans E., (505) 664-8495 Los Alamos National Laboratory – Ys
Harty, John P., jharty1@jccc.edu, Johnson County Community Coll – Oy
Harun, Nina, 907-451-5085 nina.harun@alaska.gov, Alaska Division of Geological & Geophysical Surveys – Go
Harvey, Cyril H., (336) 316-2238 charvey@guilford.edu, Guilford Coll – Gr
Harvey, H. Rodger, 757-683-6298 rharvey@odu.edu, Old Dominion Univ – CoOc
Harvey, James T., harvey@mlml.calstate.edu, Moss Landing Marine Laboratories – Ob
Harvey, Jason, +44(0) 113 34 34033 feejh@leeds.ac.uk, Univ of Leeds – Cg
Harvey, Omar, 817-257-4272 omar.harvey@tcu.edu, Texas Christian Univ – OsGe
Harvey, Omar, 601-266-4529 Omar.harvey@tcu.edu, Univ of Southern Mississippi – GeHwg
Harvey, Ralph P., (216) 368-0198 rph@case.edu, Case Western Reserve Univ – Xm
Harvey, Tom, +440116 252 3644 thph2@le.ac.uk, Leicester Univ – Po
Harwood, David M., (402) 472-2648 dharwood1@unl.edu, Univ of Nebraska, Lincoln – PmGul
Harwood, Richard D., (309) 796-5271 harwoodr@bhc.edu, Black Hawk Coll – Gv
Hasan, Khaled W., khaled.hasan@austincc.edu, Austin Community Coll District – HsOri
Hasan, Mainul, (514) 398-4755 McGill Univ – Nx
Hasan, Mohamed Ali, 03-79674203/4141 alihasan@um.edu.my, Univ of Malaya – Hg
Hasan, Syed E., (816) 235-2976 hasans@umkc.edu, Univ of Missouri-Kansas City – Ng
Hasbargen, Leslie E., (607) 436-2741 leslie.hasbargen@oneonta.edu, SUNY, Oneonta – GmHgOr
Haselwander, Airin, (573) 368-2196 airin.haselwander@dnr.mo.gov, Missouri Dept of Natural Resources – Gg
Hasenmueller, Nancy R., (812) 855-7428 hasenmue@indiana.edu, Indiana Univ – Ge
Hasenmueller, Walter, (812) 855-2687 whasenmu@indiana.edu, Indiana Univ – Gr
Hasiotis, Stephen T., (785) 864-4941 hasiotis@ku.edu, Univ of Kansas – Pe
Hasiuk, Franciszek J., (515) 294-6610 franek@iastate.edu, Iowa State Univ of Science & Tech – GdoCl
Hassani, Ferri, (514) 398-4755 McGill Univ – Nm
Hassanpour Sedghi, Mohammad, +98 (411) 339 2697 Univ of Tabriz – YsNeYg
Hassard, Stacey K., yanagawask@cofc.edu, Coll of Charleston – On
Hasselbo, Stephen, +44 01326 253651 S.P.Hesselbo@exeter.ac.uk, Exeter Univ – Gs
Hassett, John J., (217) 333-9472 Univ of Illinois, Urbana-Champaign – Sc
Hastenrath, Stefan L., (608)262-3659 slhasten@wisc.edu, Univ of Wisconsin, Madison – Oa
Hastings, Meredith, 401-863-3658 Meredith_Hastings@brown.edu, Brown Univ – Oa
Hastings, Philip A., (858) 822-2913 phastings@ucsd.edu, Univ of California, San Diego – Ob
Haszeldine, R. S., +44 (0) 131 650 8549 Stuart.Haszeldine@ed.ac.uk, Edinburgh Univ – Oa
Hatch, Christine, 413 577-2245 chatch@geo.umass.edu, Univ of Massachusetts, Amherst – Hw
Hatcher, Jr., Robert D., (865) 974-6565 bobmap@utk.edu, Univ of Tennessee, Knoxville – Gc
Hatheway, Richard B., hatheway@geneseo.edu, SUNY, Geneseo – Gx
Hattin, Donald E., (812) 855-8232 hattin@indiana.edu, Indiana Univ, Bloomington – Gr
Hattori, Keiko, 613-562-5800 6866 khattori@uottawa.ca, Univ of Ottawa – Cg
Hauck, II, Steven A., (216) 368-3675 hauck@case.edu, Case Western Reserve Univ – XyYgXg
Hauer, Kendall, (513) 529-3220 hauerkl@miamioh.edu, Miami Univ – GgOeCg
Haugerud, Ralph, 206-713-7453 rah@ess.washington.edu, Univ of Washington – Oi
Haughland, Jake, 775-784-6995 Univ of Nevada, Reno – Gm
Hauksson, Egill, (626) 395-6954 hauksson@gps.caltech.edu, California Inst of Tech – Ys
Hauri, Erik H., (202) 478-8471 ehauri@carnegiescience.edu, Carnegie Institution of Washington – Cc
Hausback, Brian, (916) 278-6521 hausback@csus.edu, California State Univ, Sacramento – Gv
Hauser, Ernest C., 937 775-3455 ernest.hauser@wright.edu, Wright State Univ – Ye
Hausrath, Elisabeth M., 702-895-1134 elisabeth.hausrath@unlv.edu, Univ of Nevada, Las Vegas – ScCl
Hautala, Susan L., (206) 543-0596 susanh@ocean.washington.edu, Univ of Washington – Op
Hauton, Chris, +44 (0)23 80595784 ch10@noc.soton.ac.uk, Univ of Southampton – Ob
Havenith, Hans-Balder, 32 4 3662035 HB.Havenith@ulg.ac.be, Universite de Liege – Ge
Haverluk, Terry W., (719) 333-8746 usafa.dfeg@usafa.edu, United States Air Force Academy – Og
Havholm, Karen G., (715) 836-2945 havholkg@uwec.edu, Univ of Wisconsin, Eau Claire – Oe
Havlin, John L., (919) 515-2655 havlin@ncsu.edu, North Carolina State Univ – Sc
Hawkins, David, 781-283-3554 dhawkins@wellesley.edu, Wellesley Coll – Gzi
Hawkins, James W., (858) 534-2161 jhawkins@ucsd.edu, Univ of California, San Diego – OuGip
Hawkins, John F., (334) 844-4894 jfh0005@auburn.edu, Auburn Univ – GcOe
Hawkins, Lawrence, +44 (0)23 80593426 leh@noc.soton.ac.uk, Univ of Southampton – Ob
Hawkins, Michael E., (518) 486-2011 mhawkins@mail.nysed.gov, New York State Geological Survey – Gz
Hawkins, R. B., (520) 621-7273 rhawkins@ag.arizona.edu, Univ of Arizona – Hs
Hawkins, Stephen J., +44 (0)23 8059 3596 S.J.Hawkins@soton.ac.uk, Univ of Southampton – Nr
Hawkins, Terry, (573) 368-2164 terry.hawkins@dnr.mo.gov, Missouri Dept of Natural Resources – Geg
Hawkins, Timothy W., (717) 477-1662 twhawk@ship.edu, Shippensburg Univ – Oa
Hawkins, Ward L., (505) 667-5835 whawkins@lanl.gov, Los Alamos National Laboratory – Og
Hawley, John W., (505) 255-4847 hgeomatters@qwest.net, New Mexico Inst of Mining & Tech – Ge
Hawley, Rebecca D., (928) 523-1251 d.hawley@nau.edu, Northern Arizona Univ – On
Hawley, Robert L., 603-646-2373 Robert.Hawley@dartmouth.edu, Dartmouth Coll – Gl
Hawman, Robert B., (706) 542-2398 rob@seismo.gly.uga.edu, Univ of Georgia – Ys
Hawthorne, Frank C., (204) 474-8861 frank_hawthorne@umanitoba.ca, Univ of Manitoba – Gz
Hay, Alex E., (902) 494-6657 alex.hay@dal.ca, Dalhousie Univ – OpnGu
Hay, Iain, iain.hay@flinders.edu.au, Flinders Univ – Oyy
Hay, Rodrick, (310) 243-2547 rhay@csudh.edu, California State Univ, Dominguez Hills – Or
Hayashi, Masaki, (403) 220-2794 hayashi@ucalgary.ca, Univ of Calgary – HwSp
Hayden, Bruce P., (804) 924-0545 bph@virginia.edu, Univ of Virginia – Oa
Hayden, Geoffrey W., (912) 752-2597 Mercer Univ – Om
Hayden, Martha C., (801) 537-3311 nrugs.mhayden@state.ut.us, Utah Geological Survey – Pv
Hayes, Alexander G., (607) 255-1712 agh4@cornell.edu, Cornell Univ – XgOrGs
Hayes, Christopher T., (228) 688-3469 christopher.t.hayes@usm.edu, Univ of Southern Mississippi – Oc
Hayes, Dennis E., (845) 365-8470 deph@ldeo.columbia.edu, Columbia Univ – Yr
Hayes, Garry F., (209) 575-6294 Modesto Junior Coll – Gg
Hayes, Garry F., GHayes@csustan.edu, California State Univ, Stanislaus – Gg
Hayes, James J., 402-554-3862 jjhayes@unomaha.edu, Univ of Nebraska at Omaha – OruOi
Hayes, John M., (914) 365-8470 jhayes@whoi.edu, Woods Hole Oceanographic Institution – Ct
Hayes, Jorden L., (717) 245-8303 hayesjo@dickinson.edu, Dickinson Coll – Ygs
Hayes, Michael J., (402) 472-4271 mhayes2@unl.edu, Univ of Nebraska, Lincoln – Oa
Hayes, Miles O., 504-280-6325 mhayes@uno.edu,

Univ of New Orleans – Gs
Hayes, Van E., 813-253-7685 vhayes@hccfl.edu, Hillsborough Community Coll – Og
Hayman, Nicholas W., 512-471-7721 hayman@ig.utexas.edu, Univ of Texas at Austin – Ou
Hayman, Patrick, patrick.hayman@qut.edu.au, Queensland Univ of Tech – EgGvp
Haymet, Anthony D., (858) 534-2827 thaymet@ucsd.edu, Univ of California, San Diego – OcgOn
Haynes, Kyle J., (540) 837-1758 kjh8w@virginia.edu, Univ of Virginia – On
Haynes, Samantha E., +44(0) 113 34 34938 s.e.haynes@leeds.ac.uk, Univ of Leeds – GosGg
Haynes, Jr., C. Vance, (520) 621-6307 Univ of Arizona – Ga
Hayob, Jodie, (540) 654-1425 jhayob@umw.edu, Univ of Mary Washington – GxzGg
Hays, James D., (845) 365-8403 jimhays@ldeo.columbia.edu, Columbia Univ – PemOu
Hays, Phillip D., (479) 575-7343 pdhays@uark.edu, Univ of Arkansas, Fayetteville – HwCsGg
Hayward, Chris L., +44 (0) 131 650 5827 chris.hayward@ed.ac.uk, Edinburgh Univ – Ga
Haywick, Douglas W., (334) 460-6381 dhaywick@jaguar1.usouthal.edu, Univ of South Alabama – Gs
Haywood, Alan, +44(0) 113 34 38657 earamh@leeds.ac.uk, Univ of Leeds – Pe
Hazel, McGoff, H.J.McGoff@rdg.ac.uk, Univ of Reading – Pg
Hazel, Jr., Joseph E., (928) 523-9145 joseph.hazel@nau.edu, Northern Arizona Univ – Gs
Hazen, Robert M., (703) 993-2163 rhazen@gmu.edu, George Mason Univ – Om
Hazen, Robert M., (202) 478-8962 rhazen@carnegiescience.edu, Carnegie Institution of Washington – Gz
Hazlett, Richard W., (909) 621-8675 rhazlett@pomona.edu, Pomona Coll – Gv
He, Changming, (302) 831-4917 hchm@udel.edu, Univ of Delaware – GqHg
He, Helen, +44 (0)1603 59 2091 yi.he@uea.ac.uk, Univ of East Anglia – Pe
He, Ruoying, 919-513-0943 ruoying_he@ncsu.edu, North Carolina State Univ – Op
He, Zhenli, 772-468-3922 zhe@ufl.edu, Univ of Florida – Sc
Head, Elisabet M., 773-442-6055 E-Head@neiu.edu, Northeastern Illinois Univ – GvOrCg
Head, Martin J., (905) 688-5550 x5216 mjhead@brocku.ca, Univ of Toronto – Ple
Head, Martin J., (905) 688-5550 (Ext. 5216) mjhead@brocku.ca, Brock Univ – Pl
Head, III, James W., (401) 863-2526 james_head_iii@brown.edu, Brown Univ – XgGvt
Headley, Rachel, headley@uwp.edu, Univ of Wisconsin, Parkside – Gl
Heal, Kate V., +44 (0) 131 650 5420 K.Heal@ed.ac.uk, Edinburgh Univ – Co
Heald, Colette, (617) 324-5666 heald@mit.edu, Massachusetts Inst of Tech – Oa
Healey, Michael, (604) 822-4705 mhealey@eos.ubc.ca, Univ of British Columbia – Ob
Heaman, Larry M., (780) 492-2778 larry.heaman@ualberta.ca, Univ of Alberta – Cc
Heaney, Michael , (979) 845-7841 heaney@geo.tamu.edu, Texas A&M Univ – Pg
Heaney, Peter J., (814) 865-6821 pjheaney@psu.edu, Pennsylvania State Univ, Univ Park – Gz
Hearn, Thomas M., (505) 646-5076 thearn@nmsu.edu, New Mexico State Univ, Las Cruces – Ys
Hearn Jr, B. Carter, (202) 633-1756 hearnc@si.edu, Smithsonian Institution / National Museum of Natural History – Gic
Heath, Carolyn, 714-992-7444 cheath@fullcoll.edu, Fullerton Coll – Ob
Heath, G. Ross, (206) 543-3153 rheath@u.washington.edu, Univ of Washington – Cm
Heath, Kathleen M., (812) 237-3004 kheath@indstate.edu, Indiana State Univ – Pb
Heatherington, Ann L., (352) 392-6220 aheath@ufl.edu, Univ of Florida – Cc
Heaton, Jill, (775) 784-8056 jheaton@unr.edu, Univ of Nevada, Reno – Oi
Heaton, Richard C., (505) 667-1141 Los Alamos National Laboratory – On
Heaton, Thomas H., (626) 395-6897 heaton@gps.caltech.edu, California Inst of Tech – Ne
Heaton, Timothy H., (605) 677-6122 Timothy.Heaton@usd.edu, Univ of South Dakota – PvOg
Heatwole, Charles A., (212) 772-5323 Graduate School of the City Univ of New York – On
Heavers, Richard, (401) 254-3095 Roger Williams Univ – Op
Hebda, Richard J., (250) 387-5493 Univ of Victoria – Pl
Hebert, David, (902) 426-1216 david.hebert@dfo-mpo.gc.ca, Dalhousie Univ – Op
Hebert, David L., (401) 874-6610 herbert@gso.uri.edu, Univ of Rhode Island – Op
Hebert, Rejean J., 418-656-3137 herbert@ggl.ulaval.ca, Universite Laval – Gi
Heck, Frederick R., (231) 592-2588 heckf@ferris.edu, Ferris State Univ – Gg
Heck, Richard, (519) 824-4120 (Ext. 52450) rheck@uoguelph.ca, Univ of Guelph – Sd
Heckathorn, Scott, 419-530-4328 scott.heckathorn@utoledo.edu, Univ of Toledo – On
Heckel, Philip H., (319) 335-1804 philip-heckel@uiowa.edu, Univ of Iowa – GrPsGd
Heckel, Wolfgang, 089/2180 4331 heckl@lmu.de, Ludwig-Maximilians-Universitaet Muenchen – Gz
Heckert, Andrew B., (828) 262-7609 heckertab@appstate.edu, Appalachian State Univ – PvgOe
Hedenquist, Jeffrey W., jhedenquist@gmail.com, Univ of Ottawa – Eg
Hedin, Robert S., (412) 624-8780 Univ of Pittsburgh – Ge
Hedquist, Brent, 361-593-3586 Brent.Hedquist@tamuk.edu, Texas A&M Univ, Kingsville – Oiw
Heenan, Cleo J., (605) 394-2461 Cleo.Heenan@sdsmt.edu, South Dakota School of Mines & Tech – On
Heermance, Richard V., (818) 677-4357 richard.heermance@csun.edu, California State Univ, Northridge – GrCc
Heerschap, Lauren, 970-247-7620 Heerschap_L@fortlewis.edu, Fort Lewis Coll – GgEgGt
Hefferan, Kevin P., (715) 346-4453 kheffera@uwsp.edu, Univ of Wisconsin, Stevens Point – GctGe
Hegarty, Patricia, +353 21 4902533 p.hegarty@ucc.ie, Univ Coll Cork – On
Hegerl, Gabriele C., +44 (0) 131 651 9092 Gabi.Hegerl@ed.ac.uk, Edinburgh Univ – Oa
Hegg, Dean A., 206-543-1984 deanhegg@atmos.washington.edu, Univ of Washington – Oa
Hegna, Thomas A., (309) 298-1366 ta-hegna@wiu.edu, Western Illinois Univ – PioPg
Hegner, Ernst, 089/2180 4274 hegner@lmu.de, Ludwig-Maximilians-Universitaet Muenchen – Gg
Hehr, John G., (479) 575-7428 jghehr@uark.edu, Univ of Arkansas, Fayetteville – Owy
Heibel, Todd, 909-384-8638 theibel@sbccd.cc.ca.us, San Bernardino Valley Coll – Oyi
Heid, Kelly L., 616-331-3783 heidke@gvsu.edu, Grand Valley State Univ – Oe
Heidinger, Andrew, 608-263-6757 heidinger@ssec.wisc.edu, Univ of Wisconsin, Madison – Or
Heigold, Paul C., (630) 252-7861 Argonne National Laboratory – Yg
Heiken, Grant, (505) 667-8477 heiken@lanl.gov, Los Alamos National Laboratory – Gv
Heikes, Brian G., (401) 874-6810 zagar@notos.gso.uri.edu, Univ of Rhode Island – Oa
Heil, Dean, (505) 646-3405 New Mexico State Univ, Las Cruces – Sc
Heilweil, Victor, heilweil@usgs.gov, Univ of Utah – Hw
Heiman, Michael, (717) 245-1338 heiman@dickinson.edu, Dickinson Coll – Ou
Heimann, Adriana, 252 3288636 heimanna@ecu.edu, East Carolina Univ – GzpGi
Heimann, William H., (785) 625-5663 whheimann@fhsu.edu, Fort Hays State Univ – HgwGm
Heimbach, Patrick, (512) 232-7694 heimbach@utexas.edu, Univ of Texas at Austin – OplOa
Heimpel, Moritz, (780) 492-3519 mheimpel@phys.ualberta.ca, Univ of Alberta – On
Heimsath, Arjun, 480-965-5585 arjun.heimsath@asu.edu, Arizona State Univ – Gc
Hein, Christopher J., 804-684-7533 hein@vims.edu,

Coll of William & Mary – Ou
Hein, Kim A., 011 717 6623 kim.ncube0hein@wits.ac.za, Univ of the Witwatersrand – GgEmGc
Heine, Sallie, (309) 794-8108 sallieheine@augustana.edu, Augustana Coll – Oe
Heinrich, Paul V., (225) 578-4398 heinric@lsu.edu, Louisiana State Univ – GsaGm
Heinz, Dion L., (773) 702-3046 heinz@uchicago.edu, Univ of Chicago – Yx
Heinzel, Chad E., (319) 273-6168 chad.heinzel@uni.edu, Univ of Northern Iowa – GmaSa
Heitman, Mary P., MaryPat.Heitman@dnr.iowa.gov, Iowa Dept of Natural Resources – On
Heitmuller, Frank, 601-266-5423 franklin.heitmuller@usm.edu, Univ of Southern Mississippi – Gms
Heizler, Matthew T., (575) 835-5244 matt@nmt.edu, New Mexico Inst of Mining and Tech – Cc
Helba, Hossam EL-Din A., 002-03-3921595 kahralran2009@Yahoo.com, Alexandria Univ – GgOm
Helenes, Javier, jhelenes@cicese.mx, Centro de Investigación Científica y de Educación Superior de Ensenada – Gr
Helgers, Karen, (845) 687-5224 helgersk@sunyulster.edu, SUNY, Ulster County Community Coll – OeHgGe
Helie, Jean-François, 514-987-3000 #2413 helie.jean-francois@uqam.ca, Universite du Quebec a Montreal – Cs
Heliker, Christina C., (808) 967-8807 cheliker@usgs.gov, Univ of Hawai'i, Hilo – Gv
Helled, Ravit , rhelled@post.tau.ac.il, Tel Aviv Univ – Xy
Heller, Matthew J., (434) 951-6351 matt.heller@dmme.virginia.gov, Division of Geology and Mineral Resources – Gg
Heller, Paul L., (307) 766-2245 heller@uwyo.edu, Univ of Wyoming – GrsGt
Helmberger, Donald V., (626) 395-6998 helm@gps.caltech.edu, California Inst of Tech – Ys
Helmick, Sara B., (505) 667-9583 Los Alamos National Laboratory – On
Helmke, Martin F., 610-436-3565 mhelmke@wcupa.edu, West Chester Univ – HwSdOg
Helmke, Philip A., (608) 263-4947 pahelmke@wisc.edu, Univ of Wisconsin, Madison – Sc
Helmstaedt, Herwart, (613) 533-6175 Queen's Univ – Gc
Helmuth, Brian, 7815817370 x307 b.helmuth@northeastern.edu, Northeastern Univ – On
Helper, Mark A., (512) 471-1009 helper@jsg.utexas.edu, Univ of Texas at Austin – GctXg
Helsdon, John H., (605) 394-2291 John.Helsdon@sdsmt.edu, South Dakota School of Mines & Tech – Oa
Hemani, Ahmad, (514) 398-4755 (Ext. 0517) McGill Univ – Nm
Hembree, Daniel, (740) 597-1495 hembree@ohio.edu, Ohio Univ – Pi
Hemler, Deb, (304) 367-4393 dhemler@fairmontstate.edu, Fairmont State Univ – Oe
Hemming, N. G., 718-997-3335 Queens Coll (CUNY) – CaOcGg
Hemming, Sidney, (845) 365-8417 Columbia Univ – Cg
Hemphill-Haley, Eileen, emh21@humboldt.edu, Humboldt State Univ – Pm
Hemphill-Haley, Mark, (707) 826-3933 mah54@humboldt.edu, Humboldt State Univ – Gt
Hemsley, A, hemsleyAR@cf.ac.uk, Cardiff Univ – Pl
Hemsley, Alan, +44(0)29 208 75367 HemsleyAR@cardiff.ac.uk, Univ of Wales – Pl
Hemzacek Laukant, Jean M., 773/442-6056 J-Hemzacek@neiu.edu, Northeastern Illinois Univ – Sc
Hence, Deanna, 217-244-6014 dhence@illinois.edu, Univ of Illinois, Urbana-Champaign – Oa
Hencher, Steve, S.R.Hencher@leeds.ac.uk, Univ of Leeds – Ng
Hendershott, Myrl C., (858) 534-5705 mhendershott@ucsd.edu, Univ of California, San Diego – Op
Henderson, Charles M., (403) 220-6170 henderson@geo.ucalgary.ca, Univ of Calgary – Pm
Henderson, Donald, (403) 823-7707 Royal Tyrrell Museum of Palaeontology – Pv
Henderson, Gideon, gideon.henderson@earth.ox.ac.uk, Univ of Oxford – Cg
Henderson, Gina R., 410-293-6555 ghenders@usna.edu, United States Naval Academy – Oa
Henderson, Grant S., (416) 978-6041 henders@geology.utoronto.ca, Univ of Toronto – Gz
Henderson, Matt, (803) 896-7931 hendersonm@dnr.sc.gov, Dept of Natural Resources – Oi
Henderson, Naomi, (845)-365-8320 Columbia Univ – Yr
Henderson, Paul, (508) 289-3466 phenderson@whoi.edu, Woods Hole Oceanographic Institution – Cm
Henderson, Robert, 747814746 bob.henderson@jcu.edu.au, James Cook Univ – GtPiGg
Henderson, Wayne G., (657) 278-3561 whenderson@fullerton.edu, California State Univ, Fullerton – PgGg
Hendrey, George, 718-997-3325 george.hendrey@qc.cuny.edu, Queens Coll (CUNY) – Og
Hendrickx, Jan M., (575) 835-5892 hendrick@nmt.edu, New Mexico Inst of Mining and Tech – Hw
Hendrikx, Jordy, (406) 994-6918 jordy.hendrikx@montana.edu, Montana State Univ – OnHsOa
Hendrix, Marc S., 406-243-5867 marc.hendrix@umontana.edu, Univ of Montana – Gs
Hendrix, Thomas E., 616-331-3728 Grand Valley State Univ – Gc
Hendry, Jim, (306) 966-5720 jim.hendry@usask.ca, Univ of Saskatchewan – HwClHy
Hendy, Ingrid, 734615-6892 ihendy@umich.edu, Univ of Michigan – Py
Henika, William S., (540) 231-4298 bhenika@vt.edu, Virginia Polytechnic Inst & State Univ – Ng
Henkel, Torsten, +44 0161 275-3810 Torsten.Henkel@manchester.ac.uk, Univ of Manchester – Cs
Henley, Sian F., +44 (0) 131 650 7010 s.f.henley@ed.ac.uk, Edinburgh Univ – CmOcCs
Hennemeyer, Marc, 089/2180 4340 marc@hennemeyer.de, Ludwig-Maximilians-Universitaet Muenchen – Gz
Henning, Stanley J., (515) 294-7846 sjhennin@iastate.edu, Iowa State Univ of Science & Tech – So
Hennings, Peter H., (307) 766-3386 Univ of Wyoming – Gc
Hennon, Chris, (828) 232-5159 chennon@unca.edu, Univ of North Carolina at Asheville – OanOn
Henrici, Amy C., (412) 622-3265 henricia@carnegiemnh.org, Carnegie Museum of Natural History – Pv
Henry, Christopher D., (775) 682-8753) chenry@unr.edu, Univ of Nevada – GtvCc
Henry, Darrell J., (225) 388-2693 dhenry@geol.lsu.edu, Louisiana State Univ – Gp
Henry, Eric J., 910-962-7622 henrye@uncw.edu, Univ of North Carolina Wilmington – Hw
Henry, James A., (615)904-8452 Jim.Henry@mtsu.edu, Middle Tennessee State Univ – Oa
Henry, Kathleen M., (217) 244-8994 kmhenry@illinois.edu, Illinois State Geological Survey – Gg
Henry, Tiernan, +353 (0)91 495 096 tiernan.henry@nuigalway.ie, National Univ of Ireland Galway – Hg
Henshel, Diane S., (812) 855-4556 dhenshel@indiana.edu, Indiana Univ, Bloomington – On
Henson, Harvey, 618-536-6666 henson@geo.siu.edu, Southern Illinois Univ Carbondale – Yg
Henstock, Tim, +44 (0)23 80596491 then@noc.soton.ac.uk, Univ of Southampton – YgrYe
Hentz, Tucker F., (512) 471-7281 tucker.hentz@beg.utexas.edu, Univ of Texas at Austin, Jackson School of Geosciences – GrsGo
Henyey, Thomas L., (213) 740-5832 henyey@usc.edu, Univ of Southern California – Yg
Henßel, Katja, 089/2180 6624 k.henssel@lrz.uni-muenchen.de, Ludwig-Maximilians-Universitaet Muenchen – Pg
Hepburn, J. Christopher, 617-552-3642 john.hepburn@bc.edu, Boston Coll – Gg
Hepner, George F., (801) 581-6021 george.hepner@geog.utah.edu, Univ of Utah – Oy
Hepner, Tiffany, (512) 475-9572 tiffany.hepner@beg.utexas.edu, Univ of Texas at Austin, Jackson School of Geosciences – Og
Hepple, Alex, 3138 5051 a.hepple@qut.edu.au, Queensland Univ of Tech – GpiGg
Herbers, Thomas H., (831) 656-2917 herbers@nps.edu, Naval Postgraduate School – Op
Herbert, Bruce, (979) 845-2405 herbert@geo.tamu.edu, Texas A&M Univ – Ge
Herbert, Jennifer, (630) 252-0493 Argonne National Laboratory – Gg
Herbert, Timothy D., (401) 863-1207 timothy_herbert@brown.edu, Brown Univ – Pe
Herbst, Thomas, 573-368-2143 thomas.herbst@dnr.mo.gov, Missouri Dept of Natural Resources – Gg

Herd, Christopher, (780) 492-5798 herd@ualberta.ca, Univ of Alberta – XcGiCp
Herd, Richard, +44 (0)1603 59 3667 r.herd@uea.ac.uk, Univ of East Anglia – Gv
Herkenhoff, Ken E., (818) 354-3539 ken.e.herkenhoff@jpl.nasa.gov, Jet Propulsion Laboratory – Xg
Herman, Ellen K., (570) 577-3088 ekh008@bucknell.edu, Bucknell Univ – Hw
Herman, Janet S., (434) 924-0553 jsh5w@virginia.edu, Univ of Virginia – ClHw
Herman, Rhett B., 540 831-5441 rherman@radford.edu, Radford Univ – Yg
Hermance, John F., (508) 252-3116 john_hermance@brown.edu, Brown Univ – HqOrHw
Hermann, Albert, 206-526-6495 Albert.J.Hermann@noaa.gov, Univ of Washington – Op
Hermes, O D., dhermes@uri.edu, Univ of Rhode Island – Gi
Hernandez, Larry, (760) 757-2121, x6329 lhernandez@miracosta.edu, MiraCosta Coll – Gg
Hernandez, Michael W., 801-626-8186 mhernandez@weber.edu, Weber State Univ – Ori
Hernandez-Molina, Francisco J., Javier.Hernandez-Molina@rhul.ac.uk, Univ of London, Royal Holloway & Bedford New Coll – Gs
Herndon, Elizabeth M., (330) 672-2680 eherndo1@kent.edu, Kent State Univ – ClSc
Hernes, Peter J., 530-754-4327 pjhernes@ucdavis.edu, Univ of California, Davis – Hg
Heron, Duncan, (919) 684-5321 duncan.heron@duke.edu, Duke Univ – Grg
Herried, Brad, (612) 625-6501 herri147@umn.edu, Univ of Minnesota, Twin Cities – Oi
Herring, Thomas A., (617) 253-5941 tah@mit.edu, Massachusetts Inst of Tech – Yd
Herring Mayo, Lisa L., 931-393-2136 lmayo@mscc.edu, Motlow State Community Coll – GgOge
Herriott, Trystan, (907) 451-5011 trystan.herriott@alaska.gov, Alaska Division of Geological & Geophysical Surveys – Eo
Herrmann, Achim, 225-578-3016 aherrmann@lsu.edu, Louisiana State Univ – PeGd
Herrmann, Edward W., (812) 856-0587 edherrma@indiana.edu, Indiana Univ, Bloomington – Sa
Herrmann, Felix J., (604) 822-8628 fherrmann@eos.ubc.ca, Univ of British Columbia – Yx
Herrmann, Robert B., (314) 977-3120 rbh@eas.slu.edu, Saint Louis Univ – Ys
Herrstrom, Eileen A., (217) 333-7732 herrstro@illinois.edu, Univ of Illinois, Urbana-Champaign – Gi
Hershey, Ronald L., (775) 673-7393 ron.hershey@dri.edu, Univ of Nevada, Reno – HwCls
Hervig, Richard, (480) 965-8427 hervig@asu.edu, Arizona State Univ – Cg
Herzberg, Claude T., (848) 445-3154 herzberg@rci.rutgers.edu, Rutgers, The State Univ of New Jersey – CpGi
Herzig, Chuck, (310) 660-3593 cherzig@elcamino.edu, El Camino Coll – GgOg
Hesp, Patrick, patrick.hesp@flinders.edu.au, Flinders Univ – GmOy
Hesp, Patrick A., (225) 578-6244 pahesp@lsu.edu, Louisiana State Univ – Gm
Hess, Darrel E., (415) 239-3104 dhess@ccsf.edu, City Coll of San Francisco – Oy
Hess, Kai-Uwe, 089/2180 4275 hess@min.uni-muenchen.de, Ludwig-Maximilians-Universitaet Muenchen – Gz
Hess, Paul C., (401) 863-1929 Paul_Hess@Brown.edu, Brown Univ – Gi
Hess Tanguay, Lillian, 516 463-6545 geolht@hofstra.edu, Hofstra Univ – Gg
Hess-Tanguay, Lillian, (516) 299-2318 lhess@liu.edu, Long Island Univ, C.W. Post Campus – Gs
Hesse, Marc A., (512) 471-0768 mhesse@jsg.utexas.edu, Univ of Texas at Austin – GqoYg
Hesse, Reinhard, 514-398-3627 reinhard.hesse@mcgill.ca, McGill Univ – Gs
Hessler, Robert R., (858) 534-2665 rhessler@ucsd.edu, Univ of California, San Diego – Ob
Hester, Erich, (540) 231-9758 ehester@vt.edu, Virginia Tech – Hsw
Hesterberg, Dean L., (919) 515-2636 North Carolina State Univ – Sc
Hestir, Erin, 919-515-7778 erin_hestir@ncsu.edu, North Carolina State Univ – OrHs
Hetherington, Callum J., (806) 834-3110 callum.hetherington@ttu.edu, Texas Tech Univ – GxzCg
Hetherington, Eric D., (209) 730-3812 Coll of the Sequoias – Gc
Hetherington, Jean, (925) 685-1230 (Ext. 462) jhetheri@dvc.edu, Diablo Valley Coll – Gg
Hetland, Robert D., (979) 458-0096 rhetland@ocean.tamu.edu, Texas A&M Univ – Op
Hettiarachchi, Ganga, (785) 532-7209 ganga@ksu.edu, Kansas State Univ – Sc
Hetzel, Ralf, +49-251-83-33908 rahetzel@uni-muenster.de, Universitaet Muenster – Gcm
Heubeck, Christoph, 0049(0)3641/948620 christoph.heubeck@uni-jena.de, Friedrich-Schiller-Univ Jena – GsdPy
Heuss-Aßbichler, Soraya, 089/21804252 soraya@min.uni-muenchen.de, Ludwig-Maximilians-Universitaet Muenchen – Gz
Hewitt, David A., (540) 231-6521 dhewitt@vt.edu, Virginia Polytechnic Inst & State Univ – Cp
Heymann, Dieter, (713) 348-4890 dieter@owlnet.rice.edu, Rice Univ – Xm
Heyniger, William C., (908) 737-3660 wheynige@kean.edu, Kean Univ – Ow
Heyvaert, Alan, 775.673.7322 alan.heyvaert@dri.edu, Univ of Nevada, Reno – GnHs
Heywood, Karen, +44 (0)1603 59 2555 k.heywood@uea.ac.uk, Univ of East Anglia – Op
Heywood, Neil C., (715) 346-4452 nheywood@uwsp.edu, Univ of Wisconsin, Stevens Point – Oy
Hiatt, Eric E., (920) 424-7001 hiatt@uwosh.edu, Univ of Wisconsin, Oshkosh – GsOcGu
Hibbard, James P., (919) 515-7242 jphibbar@ncsu.edu, North Carolina State Univ – GctOn
Hibbs, Barry, (323) 343-2414 bhibbs@calstatela.edu, California State Univ, Los Angeles – Hw
Hick, Steven, 303-871-2535 shick@du.edu, Univ of Denver – Oi
Hickcox, Charles W., (404) 727-0118 geocwh@emory.edu, Emory Univ – Gg
Hickey, Barbara M., (206) 543-4737 bhickey@u.washington.edu, Univ of Washington – Op
Hickey, James, (660) 562-1817 jhickey@nwmissouri.edu, Northwest Missouri State Univ – Ge
Hickey, Kenneth A., (778) 384-7074 khickey@eos.ubc.ca, Univ of British Columbia – EgGcg
Hickey, William J., (608) 262-9018 wjhickey@wisc.edu, Univ of Wisconsin, Madison – Sb
Hickey-Vargas, Rosemary, (305) 348-3471 hickey@fiu.edu, Florida International Univ – Cg
Hickman, Anna, +44 (0)23 80592132 A.Hickman@noc.soton.ac.uk, Univ of Southampton – Og
Hickman, Carole S., (510) 642-3429 Univ of California, Berkeley – Po
Hickman, John, (859) 323-0541 jhickman@uky.edu, Univ of Kentucky – Eo
Hickmon, Nicki, 630-252-7662 nhickmon@anl.gov, Argonne National Laboratory – Oa
Hickmott, Donald D., (505) 667-8753 dhickmott@lanl.gov, Los Alamos National Laboratory – Gp
Hicks, Andrea, hicks5@wisc.edu, Univ of Wisconsin, Madison – On
Hicks, Roberta (Robbie), (709) 737-8349 rhicks@mun.ca, Memorial Univ of Newfoundland – Gg
Hickson, Catherine J., (604) 761-5573 ttgeo@telus.net, Univ of British Columbia – GvEmOe
Hickson, Thomas A., (651) 962-5241 tahickson@stthomas.edu, Univ of Saint Thomas – GsmGe
Hicock, Stephen R., (519) 661-3189 shicock@uwo.ca, Western Univ – Gl
Hidalgo, Paulo, phidalgo@gsu.edu, Georgia State Univ – GivGz
Hier-Majumder, Saswata, (301) 405-6979 saswata@umd.edu, Univ of Maryland – YgGq
Hiett, Michael W., (615)898-5075 Michael.Hiett@mtsu.edu, Middle Tennessee State Univ – Gg
Higgins, Charles G., paulaH88@hotmail.com, Univ of California, Davis – Gm
Higgins, Chris T., (916) 322-9997 California Geological Survey – Gg
Higgins, Jerry D., (303) 273-3817 jhiggins@mines.edu, Colorado School of Mines – Ng
Higgins, John A., (609) 258-2756 jahiggin@princeton.edu, Princeton Univ – Ob
Higgins, Michael D., 4185455011 x 5052 mhiggins@uqac.ca, Universite du Quebec a Chicoutimi – Gi
Higgins, Pennilyn, pennilyn.higgins@rochester.edu,

Faculty Index -H

Univ of Rochester – Cs
Higgs, Bettie, +353 21 4902117 b.higgs@ucc.ie, Univ Coll Cork – Yg
Higgs, Ken, +353 21 4902290 k.higgs@ucc.ie, Univ Coll Cork – Pl
Higinbotham, Pamela, 724-938-4180 higinbotham@calu.edu, California Univ of Pennsylvania – On
Hilbert-Wolf, Hannah L., hannah.hilbertwolf@my.jcu.edu.au, James Cook Univ – GsCc
Hildebrand, Alan R., 403-220-2291 ahildebr@ucalgary.ca, Univ of Calgary – Xm
Hildebrand, John A., (858) 534-4069 jhildebrand@ucsd.edu, Univ of California, San Diego – Yr
Hildebrand, Stephen G., hildebrandsg@ornl.gov, Oak Ridge National Laboratory – On
Hildebrand, Steve T., (505) 667-4318 hildebrand@lanl.gov, Los Alamos National Laboratory – Ys
Hildebrandt, Anke, 349-3641-948651 hildebrandt.a@uni-jena.de, Friedrich-Schiller-Univ Jena – HgSpf
Hileman, Mary E., (405) 744-4341 mary.hileman@okstate.edu, Oklahoma State Univ – GorGd
Hill, Arleen A., 901-678-2589 aahill@memphis.edu, Univ of Memphis – On
Hill, Catherine B., catherine.hill@azwestern.edu, Arizona Western Coll – GgOag
Hill, Chris, 619 644 7342 chris.hill@gcccd.net, Grossmont Coll – Gg
Hill, Joseph C., (936) 294-1560 geojoe@shsu.edu, Sam Houston State Univ – Gcp
Hill, Julie, (775) 784-6987 juliehill@unr.edu, Univ of Nevada, Reno – On
Hill, Malcolm D., (617) 373-4377 m.hill@neu.edu, Northeastern Univ – OiGze
Hill, Mary C., 785-864-2728 mchill@ku.edu, Univ of Kansas – Hgq
Hill, Mary Louise, (807) 343-8319 mary.louise.hill@lakeheadu.ca, Lakehead Univ – Gc
Hill, Mimi, +44 0151 794 3462 M.Hill@liverpool.ac.uk, Univ of Liverpool – Ym
Hill, Paul S., (902) 494-2266 paul.hill@dal.ca, Dalhousie Univ – Gs
Hill, Philip R., philip.hill@canada.ca, Univ of Victoria – Gsu
Hill, Richard T., (812) 855-9583 hill2@indiana.edu, Indiana Univ – Oi
Hill, Tessa M., (530) 752-0179 tmhill@ucdavis.edu, Univ of California, Davis – Pe
Hill, Timothy, +44 01334 464013 tch2@st-andrews.ac.uk, Univ of St. Andrews – Ge
Hillaire-Marcel, Claude, 514-987-3000 #3376 hillaire-marcel.claude@uqam.ca, Universite du Quebec a Montreal – Cs
Hiller, Lena, 516 463-5564 geolzh@hofstra.edu, Hofstra Univ – On
Hillier, John, jhillier@ghc.edu, Grays Harbor Coll – OgCa
Hilliker, Joby, (610) 436-2213 jhilliker@wcupa.edu, West Chester Univ – Ow
Hillman, Aubrey, (337) 482-5162 Univ of Louisiana at Lafayette – GneGa
Hills, Denise J., (205) 247-3694 dhills@gsa.state.al.us, Geological Survey of Alabama – YeEoOn
Hills, Leonard V., (403) 220-5848 Univ of Calgary – Pl
Hilterman, Fred, 713-850-7600 x3318 fhilterman@uh.edu, Univ of Houston – Ye
Hilton, David R., (858) 822-0639 drhilton@ucsd.edu, Univ of California, San Diego – Cs
Hilts, Stewart G., (519) 824-4120 (Ext. 52448) shilts@uoguelph.ca, Univ of Guelph – On
Himmelberg, Glen R., himmelbergg@missouri.edu, Univ of Missouri – Gp
Hindle, Tobin, 561 297-2846 thindle@fau.edu, Florida Atlantic Univ – HwOi
Hindman, Edward E., (212) 650-6469 City Coll (CUNY) – Oa
Hindshaw, Ruth, +44 01334 463936 rh71@st-andrews.ac.uk, Univ of St. Andrews – Cs
Hine, Albert C., (727) 553-1161 hine@usf.edu, Univ of South Florida – Gs
Hines, Mary E., (910) 962-3012 hinese@uncw.edu, Univ of North Carolina Wilmington – Ou
Hines, Paul, phines50@gmail.com, Dalhousie Univ – Op
Hinman, Nancy W., (406) 243-5277 nancy.hinman@umontana.edu, Univ of Montana – Cgl
Hinnov, Linda, 703-993-3082 lhinnov@gmu.edu, George Mason Univ – GrYg
Hinojosa, Alejandro, alhinc@cicese.mx, Centro de Investigación Científica y de Educación Superior de Ensenada – Or
Hinson, Amye S., (205) 247-3577 ahinson@gsa.state.al.us, Geological Survey of Alabama – HwOn
Hinton, Richard, +44 (0) 131 650 8548 Richard.Hinton@ed.ac.uk, Edinburgh Univ – Cg
Hintz, John G., (570)389-4140 jhintz@bloomu.edu, Bloomsburg Univ – Oin
Hintz, Rashauna, 479-575-3355 rmicken@uark.edu, Univ of Arkansas, Fayetteville – Ou
Hintze, Lehi F., 801 422-6361 lehi.hintze@gmail.com, Brigham Young Univ – Pg
Hinz, Nicholas, (775)784-1446 nhinz@unr.edu, Univ of Nevada – Gg
Hinze, William J., wjh730@comcast.net, Purdue Univ – Ye
Hippensteel, Scott P., 704-687-5992 shippens@email.uncc.edu, Univ of North Carolina, Charlotte – Gr
Hirner, Sarah M., 970-351-2398 Sarah.Hirner@unco.edu, Univ of Northern Colorado – Gg
Hirons, Steve, +44 020 3073 8028 s.hirons@ucl.ac.uk, Birkbeck Coll – Cl
Hirschboeck, Katherine K., (520) 621-6466 katie@ltrr.arizona.edu, Univ of Arizona – Pe
Hirschmann, Marc M., 612-625-6698 hirsc022@umn.edu, Univ of Minnesota, Twin Cities – Gi
Hirt, William H., (530) 938-5255 hirt@siskiyous.edu, Coll of the Siskiyous – Gi
Hirth, Greg, (401) 863-7063 Greg_Hirth@Brown.edu, Brown Univ – GcyYg
Hirth, Gregory, 508-289-2776 Greg_Hirth@brown.edu, Woods Hole Oceanographic Institution – Gc
Hiscock, Kevin, +44 (0)1603 59 3104 k.hiscock@uea.ac.uk, Univ of East Anglia – Hg
Hiscott, Richard N., (709) 737-8394 rhiscott@sparky2.esd.mun.ca, Memorial Univ of Newfoundland – Gs
Hiser, Susan, 740-376-4775 sch030@marietta.edu, Marietta Coll – On
Hiskey, J. Brent, (520) 621-6185 jbh@engr.arizona.edu, Univ of Arizona – Nx
Hitchman, Matthew H., (608)262-4653 matt@aos.wisc.edu, Univ of Wisconsin, Madison – Oa
Hites, Ronald A., (812) 855-0193 hitesr@indiana.edu, Indiana Univ, Bloomington – Co
Hitz, Ralph B., (253) 566-5299 rhitz@tacomacc.edu, Tacoma Community Coll – PvSaOi
Hixon, Amy E., (574) 631-1872 ahixon@nd.edu, Univ of Notre Dame – Cl
Hjelmfelt, Mark R., (605) 394-2291 Mark.Hjelmfelt@sdsmt.edu, South Dakota School of Mines & Tech – Oa
Hladik, Christine, (912) 478-0338 chladik@georgiasouthern.edu, Georgia Southern Univ – SbOr
Hluchy, Michele M., (607) 871-2838 Alfred Univ – Cl
Hlusko, Leslea, (510) 643-8838 hlusko@berkeley.edu, Univ of California, Berkeley – Onn
Ho, Anita, (406) 756-3873 aho@fvcc.edu, Flathead Valley Community Coll – Gg
Ho, David, (808) 956-3311 ho@hawaii.edu, Univ of Hawai'i, Manoa – Ou
Ho, I-Hsuan, (701) 777-6156 ihsuan.ho@engr.und.edu, Univ of North Dakota – Ng
Hobbs, Benjamin F., (410) 516-4681 bhobbs@jhu.edu, Johns Hopkins Univ – On
Hobbs, Chasidy, (850) 474-2735 chobbs@uwf.edu, Univ of West Florida – OgeOy
Hobbs, John D., dhobbs@mtech.edu, Montana Tech of the Univ of Montana – On
Hobbs, Richard, +44 (0) 191 33 44295 r.w.hobbs@durham.ac.uk, Durham Univ – Ys
Hobbs, Richard D., (806) 371-5333 rdhobbs@actx.edu, Amarillo Coll – OrGc
Hobbs, Thomas M., 281-618-5796 Tom.Hobbs@nhmccd.edu, Lonestar Coll - North Harris – Gg
Hochella, Jr., Michael F., (540) 231-6227 hochella@vt.edu, Virginia Polytechnic Inst & State Univ – ClGze
Hochmuth, George J., 352-392-1803 318 hoch@ufl.edu, Univ of Florida – So
Hochstaedter, Alfred, (831) 646-4149 ahochstaedter@mpc.edu, Monterey Peninsula Coll – GgOgg
Hock, Regine M., 907-474-7691 regine.hock@gi.alaska.edu, Univ of Alaska, Fairbanks – Ol
Hockaday, William C., (254)7102639 william_hockaday@baylor.edu, Baylor Univ – CoaSb
Hockey, Thomas A., (319) 273-2065 thomas.hockey@uni.edu, Univ of Northern Iowa – On
Hodder, Donald R., (914) 257-3757 SUNY, New Paltz – Gg

Hodder, Robert W., (519) 433-9550 rhodder@uwo.ca, Western Univ – Eg
Hodell, David, +44 (0) 1223 330270 dhod07@esc.cam.ac.uk, Univ of Cambridge – Ge
Hodges, Floyd N., (509) 376-4627 floyd.hodges@pnl.gov, Pacific Northwest National Laboratory – Cg
Hodges, Kip V., 480-965-5331 kvhodges@asu.edu, Arizona State Univ – Gc
Hodgetts, David, David.Hodgetts@manchester.ac.uk, Univ of Manchester – Gsc
Hodgkiss, Jr., William S., (858) 534-1798 whodgkiss@ucsd.edu, Univ of California, San Diego – Oo
Hodgson, M. John, john.hodgson@ualberta.ca, Univ of Alberta – On
Hodych, Joseph P., (709) 864-7567 jhodych@mun.ca, Memorial Univ of Newfoundland – Ym
Hoe, Teh Guan, 03-79674231 tehgh@um.edu.my, Univ of Malaya – Ca
Hoek, Joost, (301) 405-2407 hoekj@umd.edu, Univ of Maryland – CsgCa
Hoenisch, Baerbel, (845) 365-8828 hoenisch@ldeo.columbia.edu, Columbia Univ – PeObCm
Hoersch, Alice L., (215) 951-1269 hoersch@lasalle.edu, La Salle Univ – GpiOi
Hoey, Trevor, +4401413307736 Trevor.Hoey@glasgow.ac.uk, Univ of Glasgow – Gs
Hoff, Jean L., (320) 308-5914 jhoff@stcloudstate.edu, Saint Cloud State Univ – HwOeGg
Hoffman, Gretchen K., (505) 835-5640 gretchen@gis.nmt.edu, New Mexico Inst of Mining & Tech – Ec
Hoffman, Jeffrey L., (609) 292-1185 jeffrey.l.hoffman@dep.nj.gov, New Jersey Geological and Water Survey – Hq
Hoffman, Paul F., (617) 496-6380 paulfhoffman@gmail.com, Harvard Univ – Gc
Hofmann, Eileen E., (757) 683-5334 ehofmann@odu.edu, Old Dominion Univ – Op
Hofmann, Michael, (406) 243-5855 michael.hofmann@umontana.edu, Univ of Montana – GosGr
Hofmeister, Anne M., (314) 935-7440 hofmeist@levee.wustl.edu, Washington Univ in St. Louis – YghOn
Hogarth, Donald D., 613-731-8090 dhogarth@uottawa.ca, Univ of Ottawa – Gz
Hohl, Eric, 573-368-2168 eric.hohl@dnr.mo.gov, Missouri Dept of Natural Resources – On
Hoisch, Thomas D., (928) 523-1904 thomas.hoisch@nau.edu, Northern Arizona Univ – Gp
Hojjatie, Barry, (229) 333-5753 bhojjati@valdosta.edu, Valdosta State Univ – Om
Hók, Jozef, hok@fns.uniba.sk, Comenius Univ in Bratislava – Ggt
Hoke, Gregory, (315) 443-1903 gdhoke@syr.edu, Syracuse Univ – OgGmt
Holail, Hanafy M., 002-03-3921595 hanafyholail@hotmail.com, Alexandria Univ – Ggs
Holberg, Jay B., (520) 621-4571 holberg@vega.lpl.arizona.edu, Univ of Arizona – Oa
Holbrook, Amanda, (606) 783-2381 a.holbrook@moreheadstate.edu, Morehead State Univ – On
Holbrook, John M., 817-257-6275 john.holbrook@tcu.edu, Texas Christian Univ – Gsr
Holbrook, W. S., (307) 766-2427 steveh@uwyo.edu, Univ of Wyoming – Ys
Holcomb, Robin T., (206) 543-5274 rholcomb@ocean.washington.edu, Univ of Washington – Ou
Holcombe, Troy, (979) 845-3528 tholcombe@ocean.tamu.edu, Texas A&M Univ – OuGcg
Holdaway, Michael J., (214) 692-2750 Southern Methodist Univ – Gp
Holdren, George R., (509) 376-2242 rich.holdren@pnl.gov, Pacific Northwest National Laboratory – Cl
Holdsworth, Robert E., 0191 3742529 r.e.holdsworth@durham.ac.uk, Durham Univ – Gc
Hole, John A., (540) 231-3858 hole@vt.edu, Virginia Polytechnic Inst & State Univ – Ye
Holgood, Jay S., (614) 292-3999 hobgood.1@osu.edu, Ohio State Univ – Oa
Holk, Gregory J., 562-985-5006 gholk@csulb.edu, California State Univ, Long Beach – CsGxEm
Holl, Karen D., (831) 459-3668 kholl@cats.ucsc.edu, Univ of California, Santa Cruz – Ou
Hollabaugh, Curtis L., (678) 839-4050 chollaba@westga.edu, Univ of West Georgia – GzCg
Holland, Austin A., 405-325-8497 austin.holland@ou.edu, Univ of Oklahoma – YsOnn
Holland, Edward C., 479-575-6635 echollan@uark.edu, Univ of Arkansas, Fayetteville – Ou
Holland, Nicholas D., (858) 534-2085 nholland@ucsd.edu, Univ of California, San Diego – Ob
Holland, Steven M., (706) 542-0424 stratum@gly.uga.edu, Univ of Georgia – Gr
Holland, Tim, +44 (0) 1223 333466 tjbh@esc.cam.ac.uk, Univ of Cambridge – Gp
Hollander, David J., (727) 553-1019 davidh@usf.edu, Univ of South Florida – Oc
Hollenbaugh, Kenneth M., (208) 426-3700 khollenb@boisestate.edu, Boise State Univ – Eg
Holley, Elizabeth, 303-273-3409 eholley@mines.edu, Colorado School of Mines – Eg
Holliday, Joseph W., 3106603593 xt. 3371 jholliday@elcamino.edu, El Camino Coll – Og
Holliday, Vance, (520) 621-4734 vthollid@email.arizona.edu, Univ of Arizona – GamSa
Holliday, Vance, vthollid@email.arizona.edu, Univ of Arizona – Ga
Holliman, Richard, +44(0) 1908 654 646 x 54646 richard.holliman@open.ac.uk, The Open Univ – Oe
Hollings, Peter N., (807) 343-8329 peter.hollings@lakeheadu.ca, Lakehead Univ – Eg
Hollis, Cathy, +44 0161 306-6583 Cathy.Hollis@manchester.ac.uk, Univ of Manchester – Gs
Hollister, Lincoln S., (609) 258-4106 linc@princeton.edu, Princeton Univ – GptGz
Hollocher, Kurt T., (518) 388-6518 hollochk@union.edu, Union Coll – GxCga
Holloway, Tracey, (608) 262-5356 taholloway@wisc.edu, Univ of Wisconsin, Madison – Oa
Holm, Daniel K., (330) 672-4094 dholm@kent.edu, Kent State Univ – Gc
Holm, Richard F., richard.holm@nau.edu, Northern Arizona Univ – Gv
Holman, John, (620) 276-8286 jholman@ksu.edu, Kansas State Univ – So
Holman, Robert, (541) 737-2914 holman@coas.oregonstate.edu, Oregon State Univ – On
Holmden, Chris, (306) 966-5697 chris.holmden@usask.ca, Univ of Saskatchewan – Csc
Holme, Richard, +44 0151 794 5254 R.T.Holme@liverpool.ac.uk, Univ of Liverpool – Ym
Holmes, Ann E., (423) 425-1704 ann-holmes@utc.edu, Univ of Tennessee at Chattanooga – GsrPi
Holmes, George, +44(0) 113 34 31163 G.Holmes@leeds.ac.uk, Univ of Leeds – Ge
Holmes, Mark L., (206) 543-7313 mholmes@ocean.washington.edu, Univ of Washington – Ou
Holmes, Mary Anne, 402-472-5211 mholmes2@unl.edu, Univ of Nebraska, Lincoln – GsOu
Holmes, Stevie L., 605-677-6147 stevie.holmes@usd.edu, South Dakota Dept of Environment and Natural Resources – Gg
Holness, Marian, +44 (0) 1223 333434 marian@esc.cam.ac.uk, Univ of Cambridge – Gi
Holroyd, Pat, (510) 642-3733 pholroyd@berkeley.edu, Univ of California, Berkeley – Pv
Holst, Timothy, tholst@d.umn.edu, Univ of Minnesota, Duluth – Gct
Holstein, Thomas J., (401) 254-3097 Roger Williams Univ – On
Holt, Ben D., (630) 252-4347 Argonne National Laboratory – Ca
Holt, David, 228-214-3255 david.h.holt@usm.edu, Univ of Southern Mississippi – Oyi
Holt, Gloria J., (361) 749-6716 joanholt@utexas.edu, Univ of Texas at Austin – Ob
Holt, John W., 512-471-0487 jack@ig.utexas.edu, Univ of Texas at Austin – OlXyg
Holt, Robert M., (662) 915-6687 rmholt@olemiss.edu, Univ of Mississippi – Hq
Holt, William E., 631-632-8215 william.holt@sunysb.edu, SUNY, Stony Brook – Ys
Holtz, Jr., Thomas R., (301) 405-4084 tholtz@umd.edu, Univ of Maryland – PvgPo
Holubnyak, Yehven I., (785) 864-2070 eugene@kgs.ku.edu, Univ of Kansas – On
Holyoke, Caleb, (330) 972-7635 cholyoke@uakron.edu, Univ of Akron – GcNrYx
Holzer, Mark, (604) 822-0531 mholzer@langara.bc.ca, Univ of British Columbia – Oa

Faculty Index -H

Holzworth, Robert H., (206) 685-7410 bobholz@uw.edu, Univ of Washington – XyOag
Hommel, Demian, 541-737-5070 hommeld@geo.oregonstate.edu, Oregon State Univ – On
Homuth, Emil F., (702) 794-7351 fred_homuth@notes.ymp.gov, Los Alamos National Laboratory – Ye
Hon, Ken, (808) 974-7302 kenhon@hawaii.edu, Univ of Hawai'i, Hilo – Gvi
Hon, Rudolph, (617) 552-3656 hon@bc.edu, Boston Coll – Gi
Hong, Jessie, (678) 839-5466 jhong@westga.edu, Univ of West Georgia – OieOy
Hong, Sung-ho, 270-809-2591 shong4@murraystate.edu, Murray State Univ – HyGe
Honjas, Bill, (775) 784-6613 bhonjas@optimsoftware.com , Univ of Nevada, Reno – Ys
Honjo, Susumu, (508) 289-2589 shonjo@whoi.edu, Woods Hole Oceanographic Institution – Ou
Honkaer, Rick, 859-257-1108 rick.honaker@uky.edu, Univ of Kentucky – Nm
Hood, Lonnie ("Lon") L., (520) 621-6936 lon@lpl.arizona.edu, Univ of Arizona – XyOa
Hood, Raleigh, (305) 361-4668 rhood@umces.edu, Univ of Maryland – Ob
Hood, Teresa A., 305-284-8647 t.hood@miami.edu, Univ of Miami – Gg
Hood, William C., (970) 241-8020 Colorado Mesa Univ – Gg
Hooda, Peter, +44 020 8417 2155 p.hooda@kingston.ac.uk, Kingston Univ – Em
Hooft, Emilie E., (541) 346-4762 emilie@uoregon.edu, Univ of Oregon – Yr
Hook, James E., (912) 386-3182 Univ of Georgia – Sp
Hook, Paul B., (406) 944-3724 paul@intermountainaquatics.com, Montana State Univ – Sf
Hook, Simon J., (818) 354-0974 simon@lithos.jpl.nasa.gov, Jet Propulsion Laboratory – Or
Hooke, Roger L., (207) 581-2203 rhooke@acadia.net, Univ of Maine – Gm
Hooks, Benjamin P., (731) 881-7430 bhooks@utm.edu, Univ of Tennessee, Martin – GciNr
Hooks, Chris H., (205) 247-3721 chooks@gsa.state.al.us, Geological Survey of Alabama – EoGxo
Hooks, W. Gary, (205) 348-1877 Univ of Alabama – Gm
Hooper, Andy, +44(0) 113 34 37723 a.hooper@leeds.ac.uk, Univ of Leeds – Ydg
Hooper, Robert L., (715) 836-4932 hooperrl@uwec.edu, Univ of Wisconsin, Eau Claire – Gz
Hoopes, John A., (608) 262-2977 Univ of Wisconsin, Madison – Oo
Hoover, Karin A., 530-898-6269 khoover@csuchico.edu, California State Univ, Chico – Hw
Hoover, Michael T., (919) 515-7305 North Carolina State Univ – Sd
Hooyer, Thomas S., 608-263-4175 hooyer@uwm.edu, Univ of Wisconsin, Madison – Hg
Hooyer, Thomas S., 414-229-5594 hooyer@uwm.edu, Univ of Wisconsin, Milwaukee – Gl
Hopkins, David G., (701) 231-8948 North Dakota State Univ – Sd
Hopkins, David M., (907) 474-7565 Univ of Alaska, Fairbanks – Gg
Hopkins, John C., (403) 220-5842 hopkins@geo.ucalgary.ca, Univ of Calgary – Gs
Hopkins, Kenneth D., (970) 351-2853 kenneth.hopkins@unco.edu, Univ of Northern Colorado – Gml
Hopkins, Nathan R., (701) 858-4205 nathan.hopkins@minotstateu.edu, Minot State Univ – GlmOi
Hopkins, Samantha, 541-346-5976 shopkins@uoregon.edu, Univ of Oregon – Pg
Hopkins, Thomas S., (205) 348-1791 Dauphin Island Sea Lab – Ob
Hopley, Philip, +44 020 3073 8029 p.hopley@ucl.ac.uk, Birkbeck Coll – Pe
Hopmans, Jan W., (916) 752-3060 Univ of California, Davis – So
Hoppe, Kathryn A., 253-833-9111 ext. 4323 khoppe@greenriver.edu, Green River Community Coll – OggPg
Hoppie, Bryce W., (507) 389-2315 bryce.hoppie@mnsu.edu, Minnesota State Univ – GeHw
Hopson, Janet L., 865-946-1460 hopsonj@ornl.gov, Univ of Tennessee, Knoxville – On
Horan, Mary F., 202-478-8481 mhoran@carnegiescience.edu, Carnegie Institution of Washington – Cc
Horel, John, 801-581-7091 john.horel@utah.edu, Univ of Utah – Oa
Horgan, Briony, (765) 496-2290 briony@purdue.edu, Purdue Univ – Xg
Horn, John, (816) 604-3132 john.horn@mcckc.edu, Metropolitan Community Coll-Kansas City – GgOy
Horn, Marty R., 225-578-2681 mhorn@lsu.edu, Louisiana State Univ – EoGgr
Hornbach, Matthew J., 214-768-2389 Southern Methodist Univ – YrhYe
Hornberger, George, (615) 343-1144 george.m.hornberger@vanderbilt.edu, Vanderbilt Univ – HgwHs
Horne, Sharon, (519) 253-3000 ext, 2528 shorne@uwindsor.ca, Univ of Windsor – On
Horner, John R., (406) 994-3982 jhorner@montana.edu, Montana State Univ – Pv
Horner, Tim C., (916) 278-5635 hornertc@csus.edu, California State Univ, Sacramento – HwGs
Horner, Tristan, (508) 289-3825 thorner@whoi.edu, Woods Hole Oceanographic Institution – Cm
Horns, Daniel, (801) 863-8582 hornsda@uvu.edu, Utah Valley Univ – NgGet
Horowitz, Franklin G., +16072754955 frank.horowitz@cornell.edu, Cornell Univ – YgOnYe
Horsman, Eric, 252 3285265 horsmane@ecu.edu, East Carolina Univ – Gc
Horst, Andrew J., 440-775-8714 ahirst@oberlin.edu, Oberlin Coll – Gcg
Horton, Albert B., (615) 532-1509 Albert.Horton@tn.gov, Tennessee Geological Survey – GgOi
Horton, Brian, 512-471-5172 horton@mail.utexas.edu, Univ of Texas at Austin – Gs
Horton, Daniel, 847-467-6185 danethan@earth.northwestern.edu, Northwestern Univ – OaPeOn
Horton, Duane G., (509) 376-6868 duane.horton@pnl.gov, Pacific Northwest National Laboratory – Gz
Horton, Jennifer, (612) 626-4067 jmhorton@umn.edu, Univ of Minnesota – Gl
Horton, Robert, (515) 294-7843 rhorton@iastate.edu, Iowa State Univ of Science & Tech – Sp
Horton, Stephen P., (901) 678-2007 shorton@memphis.edu, Univ of Memphis – Ys
Horton, Travis, +64 3 3667001 Ext 7734 travis.horton@canterbury.ac.nz, Univ of Canterbury – Cs
Horton, Jr., Robert A., 661-654-3059 rhorton@csub.edu, California State Univ, Bakersfield – GdsGo
Horváth, Ferenc, frankh@ludens.elte.hu, Eotvos Lorand Univ – YgGhc
Horvath, Peter, +27 (0)46-603-8312 p.horvath@ru.ac.za, Rhodes Univ – Gp
Horvath, Peter, 011 717 6539 peter.horvath@wits.ac.za, Univ of the Witwatersrand – GpCpGz
Horwath, William R., (530) 754-6029 wrhorwath@ucdavis.edu, Univ of California, Davis – OsSb
Horwell, Claie, +44 (0) 191 33 42253 claire.horwell@durham.ac.uk, Durham Univ – Gz
Hosler, Charles L., (814) 863-8358 hosler@ems.psu.edu, Pennsylvania State Univ, Univ Park – Ow
Hossain, A K M Azad, 423-425-4404 Azad-Hossain@utc.edu, Univ of Tennessee at Chattanooga – Org
Hosseini, Seyyed Abolfazi, 512-471-1534 seyyed.hosseini@beg.utexas.edu, Univ of Texas at Austin – Eo
Houghton, Bruce F., (808) 956-2561 bhought@soest.hawaii.edu, Univ of Hawai'i, Manoa – Gv
Hounslow, Arthur, 405-372-2328 Oklahoma State Univ – Cl
Hourigan, Jeremy, 831-459-2879 hourigan@ucsc.edu, Univ of California, Santa Cruz – Cc
House, Christopher H., (814) 865-8802 chouse@geosc.psu.edu, Pennsylvania State Univ, Univ Park – Py
House, Leigh S., (505) 667-1912 house@lanl.gov, Los Alamos National Laboratory – Ys
House, Martha, mahouse@pasadena.edu, Pasadena City Coll – Gg
Houseknecht, David W., 703-648-6466 dhouse@usgs.gov, Virginia Polytechnic Inst & State Univ – Gd
Houseman, Greg, +44(0) 113 34 35206 g.a.houseman@leeds.ac.uk, Univ of Leeds – Yg
Housen, Bernard A., (360) 650-3588 bernieh@wwu.edu, Western Washington Univ – Ym
Houston, Adam L., ahouston2@unl.edu, Univ of Nebraska, Lincoln – Oaw
Houston, Heidi B., 206-616-7092 hhouston@uw.edu, Univ of Washington – YsGt
Houston, Robert, (541) 967-2039 Robert.A.Houston@mlrr.oregongeology.com, Oregon Dept of Geology & Mineral Industries – Ge
Houston, Serin D., 413-538-2055 shouston@mtholyoke.edu,

Mount Holyoke Coll – Oyn
Houze, Robert A., 206-543-6922 houze@atmos.washington.edu, Univ of Washington – Oa
Hovan, Steven A., (724) 357-5625 hovan@iup.edu, Indiana Univ of Pennsylvania – Ou
Hovis, Guy L., (610) 258-3907 hovisguy@lafayette.edu, Lafayette Coll – GzxCg
Hovorka, Susan D., 512-471-4863 susan.hovorka@beg.utexas.edu, Univ of Texas at Austin – Gs
Howard, Alan D., (804) 924-0563 ah6p@virginia.edu, Univ of Virginia – GmXg
Howard, Hugh H., 916 484-8805 howardh@arc.losrios.edu, American River Coll – Oi
Howard, Jeffrey L., (313) 577-3258 aa2675@wayne.edu, Wayne State Univ – Gs
Howard, Katie, 360-475-7700 khoward@llion.org, Olympic Coll – Gg
Howard, Kenneth W., 416-287-7233 Univ of Toronto – Hw
Howard, Leslie M., (402) 472-9192 thoward@unl.edu, Unversity of Nebraska - Lincoln – Oyi
Howard, Matthew K., (979) 862-4169 mkhoward@tamu.edu, Texas A&M Univ – Op
Howard, Theodore E., (603) 862-1020 Univ of New Hampshire – On
Howat, Ian M., (614) 247-8944 howat.4@osu.edu, Ohio State Univ – OlYd
Howden, Stephan, (228) 688-5284 stephan.howden@usm.edu, Univ of Southern Mississippi – Op
Howe, Bruce M., (206) 543-9141 billhowe@cs.washington.edu, Univ of Washington – Op
Howe, Julie, (334) 844-3972 jah0020@auburn.edu, Auburn Univ – Sc
Howe, Julie, jhowe@tamu.edu, Texas A&M Univ – Sc
Howe, Stephen S., (518) 442-5053 showe@albany.edu, SUNY, Albany – Cs
Howe III, Thomas R., (269) 387-5492 thomas.r.howe@wmich.edu, Western Michigan Univ – HwGem
Howell, Dave, +440116 252 3804 dah29@le.ac.uk, Leicester Univ – Ge
Howell, Robert R., (307) 766-6296 rhowell@uwyo.edu, Univ of Wyoming – XgOr
Howell, Jr., Benjamin F., (814) 863-0886 howellbf@aol.com, Pennsylvania State Univ, Univ Park – Ys
Hower, James C., (859) 257-0261 james.hower@uky.edu, Univ of Kentucky – Ec
Howes, Mary R., mary.howes@dnr.iowa.gov, Iowa Dept of Natural Resources – Gg
Howman, Dominic J., +61 8 9266-2329 D.Howman@curtin.edu.au, Curtin Univ – Ye
Hoyos, Carlos, carlos.hoyos@eas.gatech.edu, Georgia Inst of Tech – Oa
Hoyt, Greg D., (919) 684-3562 North Carolina State Univ – Sb
Hoyt, William H., (970) 351-2487 william.hoyt@unco.edu, Univ of Northern Colorado – OgGsOe
Hozik, Michael J., (609) 652-4277 hozikm@stockton.edu, Richard Stockton Coll of New Jersey – Ym
Hren, Michael, 860-486-9511 michael.hren@uconn.edu, Univ of Connecticut – CslCo
Hrouda, James G., (573) 518-2350 jimh@mineralarea.edu, Mineral Area Coll – Og
Hsiang, Tom, (519) 824-4120 Ext.52753 thsiang@uoguelph.ca, Univ of Guelph – On
Hsiao, Theodore C., (530) 752-0691 tchsiao@ucdavis.edu, Univ of California, Davis – SpOwn
Hsieh, Wen-Pin, 886-227839910-509 wphsieh@earth.sinica.edu.tw, Academia Sinica – YhGy
Hsieh, William W., (250) 388-0508 whsieh@eos.ubc.ca, Univ of British Columbia – Opa
Hsu, Liang-Chi, (775) 682-8746 lihsu@unr.edu, Univ of Nevada – Cp
Hu, Baoxin, (416)736-2100 #20557 baoxin@yorku.ca, York Univ – Or
Hu, Bill X., 850-644-3943 hu@gly.fsu.edu, Florida State Univ – Hw
Hu, Feng-Sheng, (217) 244-2982 fhu@illinois.edu, Univ of Illinois, Urbana-Champaign – Pe
Hu, Hao, hhu5@central.uh.edu, Univ of Houston – Ye
Hu, Qi S., 402-472-6642 QHU2@unl.edu, Univ of Nebraska, Lincoln – Oa
Hu, Qinhong (Max), (817) 272-5398 maxhu@uta.edu, Univ of Texas, Arlington – GoHwNg
Hu, Shusheng, 203-432-3790 shusheng.hu@yale.edu, Yale Univ – PblGs
Hu, Wan-Ping (Sunny), +61 7 3138 7314 sunny.hu@qut.edu.au, Queensland Univ of Tech – Ca
Hu, Xinping, (361) 825-3395 xinping.hu@tamucc.edu, Texas A&M Univ, Corpus Christi – Oc
Hu, Zhiyong, 850-474-3494 zhu@uwf.edu, Univ of West Florida – Or
Huang, Alex, (828) 232-5157 Univ of North Carolina at Asheville – Oa
Huang, Bohua, (703) 993-6084 bhuang@gmu.edu, George Mason Univ – Op
Huang, Li, (519) 661-3188 LHuang3@uwo.ca, Western Univ – Yr
Huang, Lianjie, (505) 665-1108 ljh@lanl.gov, Los Alamos National Laboratory – Yg
Huang, Moh J., (916) 322-9304 California Geological Survey – Ne
Huang, Norden E., (301) 286-8879 Univ of Delaware – On
Huang, Shaopeng, (734) 763-3169 shaopeng@umich.edu, Univ of Michigan – Yh
Huang, Yi, 514-398-8217 yi.huang@mcgill.ca, McGill Univ – OrYgOa
Huang, Yongsong, Yongsong_Huang@Brown.edu, Brown Univ – Co
Hubbard, Dennis K., 440-775-8346 dennis.hubbard@oberlin.edu, Oberlin Coll – Gsu
Hubbard, Kenneth G., (402) 472-8294 Univ of Nebraska, Lincoln – Oa
Hubbard, Trent, (907) 451-5009 trent.hubbard@alaska.gov, Alaska Division of Geological & Geophysical Surveys – Ng
Hubbard, William B., (520) 621-6942 hubbard@lpl.arizona.edu, Univ of Arizona – XyYvOa
Hubbart, Jason A., (304) 293-2472 jason.hubbart@mail.wvu.edu, West Virginia Univ – HqgHs
Hubenthal, Michael, (607) 777-4612 hubenth@iris.edu, Binghamton Univ – OeYg
Hubeny, J B., bhubeny@salemstate.edu, Salem State Univ – GeOnCs
Huber, Brian T., (202) 633-1328 Smithsonian Institution / National Museum of Natural History – Pm
Huber, Christian, christian_huber@brown.edu, Brown Univ – Gv
Huckabey, Marsha, (573) 882-2040 HuckabeyM@missouri.edu, Univ of Missouri – On
Hudáčková, Natália , hudackova@fns.uniba.sk, Comenius Univ in Bratislava – PimPo
Hudak, Paul F., (940) 565-4312 paul.hudak@unt.edu, Univ of North Texas – Hw
Hudec, Michael R., 512-471-1428 michael.hudec@beg.utexas.edu, Univ of Texas at Austin – Gc
Hudec, Peter P., (519) 253-3000 x2491 hudec@uwindsor.ca, Univ of Windsor – NgEmGe
Hudgins, Thomas, (787) 265-3845 thomas.hudgins@upr.edu, Univ of Puerto Rico – GiCpGv
Hudleston, Peter J., 612-625-0046 hudle001@umn.edu, Univ of Minnesota, Twin Cities – Gc
Hudman, Lloyd E., (801) 378-4346 lloyd_hudman@byu.edu, Brigham Young Univ – On
Hudson, Robert J. M., (217) 333-7641 rjhudson@uiuc.edu, Univ of Illinois, Urbana-Champaign – Cg
Hudson-Edwards, Karen, +44 0203 073 8030 k.hudson-edwards@bbk.ac.uk, Birkbeck Coll – Gez
Huebert, Barry J., (808) 258-4673 huebert@hawaii.edu, Univ of Hawai'i, Manoa – OcCg
Huerfano, Victor, (787) 265-3845 victor.huerfano@upr.edu, Univ of Puerto Rico – Ysg
Huerta, Audrey, 509.963.2718 huerta@geology.cwu.edu, Central Washington Univ – Ygd
Huey, Gregory L., (404) 894-5541 greg.huey@eas.gatech.edu, Georgia Inst of Tech – Oa
Huff, Bryan G., 217-244-2509 huff@isgs.uiuc.edu, Illinois State Geological Survey – Eo
Huff, Edmund A., (630) 252-3633 Argonne National Laboratory – Ca
Huff, Warren D., (513) 556-3731 warren.huff@uc.edu, Univ of Cincinnati – Gz
Huffman, Debra E., (727) 553-3930 debrah@usf.edu, Univ of South Florida – On
Huffman, French T., (859) 622-6968 tyler.huffman@eku.edu, Eastern Kentucky Univ – Oi
Huffman, Robert L., (912) 752-2704 Mercer Univ – On
Huft, Ashley, (406) 496-4789 ahuft@mtech.edu, Montana Tech of The Univ of Montana – Ca
Huggett, William, 618-453-7392 huggett@geo.siu.edu, Southern Illinois Univ Carbondale – Ec
Hughen, Konrad A., (508) 289-3353 khughen@whoi.edu, Woods Hole Oceanographic Institution – Cc
Hughes, Annie, +44 020 8417 2603 Ku08925@kingston.ac.uk, Kingston Univ – Oi

Faculty Index -H

Hughes, Chambers, (858) 534-7121 chughes@ucsd.edu, Univ of California, San Diego – Oc
Hughes, Colin, colin.hughes@manchester.ac.uk, Univ of Manchester – Go
Hughes, Denis, +27 46 6224014 d.hughes@ru.ac.za, Rhodes Univ – Hqs
Hughes, John M., (802) 656-9443 jmhughes@uvm.edu, Univ of Vermont – Gz
Hughes, John M., John.M.Hughes@uvm.edu, Miami Univ – Gz
Hughes, Joseph B., (713)348-5603 david.v.hughes@rice.edu, Rice Univ – On
Hughes, Kenneth S., (787) 265-3845 kenneth.hughes@upr.edu, Univ of Puerto Rico – Gcm
Hughes, Malcolm K., (520) 621-6470 mhughes@ltrr.arizona.edu, Univ of Arizona – Pe
Hughes, Nigel C., nigel.hughes@ucr.edu, Cincinnati Museum Center – Pi
Hughes, Nigel C., (951) 827-3098 nigel.hughes@ucr.edu, Univ of California, Riverside – Po
Hughes, Randall, 7815817370 x314 rhughes@neu.edu, Northeastern Univ – On
Hughes, Sam, +44 07727 096492 S.P.Hughes@exeter.ac.uk, Exeter Univ – Gt
Hughes, Scott S., (530) 533-1933 hughscot@isu.edu, Idaho State Univ – GvXgGi
Hughes III, Richard O., 909-389-3237 rihughes@craftonhills.edu, Crafton Hills Coll – GlOe
Hughes-Clarke, John E., 603-862-5505 jhc@ccom.unh.edu, Univ of New Hampshire – Og
Hugli, Wilbur G., 850-474-3470 whugli@uwf.edu, Univ of West Florida – Oa
Hugo, Richard, 503-725-3356 hugo@pdx.edu, Portland State Univ – Gy
Hui, Alice, akhui@umail.iu.edu, Indiana Univ, Bloomington – Cl
Huizenga, Jan Marten, (074) 781-4597 jan.huizenga@jcu.edu.au, James Cook Univ – GpCgGc
Hulbe, Christina L., chulbe@pdx.edu, Portland State Univ – Gl
Hull, Donald L., (505) 667-4151 Los Alamos National Laboratory – On
Hull, Joseph M., jhull@sccd.ctc.edu, Seattle Central Community Coll – Gc
Hulton, Nicholas R., +44 (0) 131 650 7543 Nick.Hulton@ed.ac.uk, Edinburgh Univ – Ol
Humayan, Munir, 850-644-5860 humayun@magnet.fsu.edu, Florida State Univ – XcCg
Huminicki, Michelle, huminickim@brandonu.ca, Brandon Univ – EgGz
Hummer, Daniel, 618-453-7386 daniel.hummer@siu.edu, Southern Illinois Univ Carbondale – Gz
Humphrey, Neil F., (307) 314-2332 neil@uwyo.edu, Univ of Wyoming – OlGmYh
Humphrey, Peggy, (406) 994-5718 peggyh@montana.edu, Montana State Univ – On
Humphreys, Eugene D., (541) 852-0091 genehumphreys@gmail.com, Univ of Oregon – YsGt
Humphreys, Robin, (843) 953-7424 humphreysr@cofc.edu, Coll of Charleston – Ge
Humphris, Susan E., (508) 289-3451 shumphris@whoi.edu, Woods Hole Oceanographic Institution – GuCm
Hunda, Brenda, (513) 455-7160 bhunda@cincymuseum.org, Cincinnati Museum Center – Pi
Hung, Ming-Chih, (660) 562-1797 mhung@nwmissouri.edu, Northwest Missouri State Univ – Or
Hungerbuehler, Axel, (575) 461-3466 axelh@mesalands.edu, Mesalands Community Coll – Gg
Hungerford, Hilary, (801) 863-7160 hilary.hungerford@uvu.edu, Utah Valley Univ – Ou
Hungr, Oldrich, (604) 822-8471 ohungr@eos.ubc.ca, Univ of British Columbia – Gm
Hunt, Allen G., (937) 775-3116 allen.hunt@wright.edu, Wright State Univ – GqSpCl
Hunt, Andrew, 817 272 2987 hunt@uta.edu, Univ of Texas, Arlington – GbCg
Hunt, Brian, (604) 822-9135 bhunt@eos.ubc.ca, Univ of British Columbia – Ob
Hunt, Gene, (202) 633-1331 Smithsonian Institution / National Museum of Natural History – Pm
Hunt, Kathleen, huntk@neaq.org, Univ of Massachusetts, Boston – Ob
Hunt, Paula J., (304) 594-2331 phunt@geosrv.wvnet.edu, West Virginia Geological & Economic Survey – GgHw
Hunt, Randy, (608) 821-3847 rjhunt@usgs.gov, Univ of Wisconsin, Madison – HwCls
Hunt, Robert M., (402) 472-4604 rhunt2@unl.edu, Univ of Nebraska, Lincoln – Pv
Hunt, Shelley, (519) 824-420 Ext.53065 shunt@uoguelph.ca, Univ of Guelph – On
Hunten, Donald M., (520) 621-4002 dhunten@lpl.arizona.edu, Univ of Arizona – Oa
Hunter, Arlene , +44 (0) 1908 655400 x55400 a.g.hunter@open.ac.uk, The Open Univ – Oe
Hunter, Jerry L., 540-392-0540 hunterje@vt.edu, Virginia Polytechnic Inst & State Univ – Ca
Huntington, Justin, 775.673.7670 justin.huntington@dri.edu, Univ of Nevada, Reno – HgOr
Huntington, Katharine W., 206-543-1750 kate1@uw.edu, Univ of Washington – Gt
Huntley, John, (573) 884-8083 Univ of Missouri – PqoPe
Huntoon, Jacqueline E., (906) 487-2440 jeh@mtu.edu, Michigan Technological Univ – Gs
Huntoon, Laura, huntoon@email.arizona.edu, Univ of Arizona – On
Huntress, Jr., Wesley T., (202) 478-8910 whuntress@carnegiescience.edu, Carnegie Institution of Washington – Xc
Hural, Kirsten, (717) 245-1632 Dickinson Coll – Gg
Hurd, David, (814) 732-2493 dhurd@edinboro.edu, Edinboro Univ of Pennsylvania – Og
Hurich, Charles A., (709) 737-2384 churich@mun.ca, Memorial Univ of Newfoundland – Ys
Hurlow, Hugh A., (801) 537-3385 nrugs.hharlow@state.ut.us, Utah Geological Survey – Hw
Hurtado, Jose M., (915) 747-5669 jhurtado@utep.edu, Univ of Texas, El Paso – Gt
Hurtgen, Matthew T., (847) 491-7539 matt@earth.northwestern.edu, Northwestern Univ – GsPoGr
Husch, Jonathan M., (609) 896-5330 husch@rider.edu, Rider Univ – GieXg
Husinec, Antun, (315) 229-5248 ahusinec@stlawu.edu, St. Lawrence Univ – GsCsGo
Hussein, Musa, (915) 747-5424 mjhussein@utep.edu, Univ of Texas, El Paso – YveYm
Hussin, Azhar Hj, 03-79674203 azharhh@um.edu.my, Univ of Malaya – Gs
Hutcheon, Ian E., (403) 220-6744 ian@earth.geo.ucalgary.ca, Univ of Calgary – Cl
Hutchings, Jennifer, (541) 737-4453 jhutchings@coas.oregonstate.edu, Oregon State Univ – Ol
Hutchins, Peter S., (601) 961-5505 peter_hutchins@deq.state.ms.us, Mississippi Office of Geology – On
Hutchinson, Charles F., chuck@ag.arizona.edu, Univ of Arizona – Or
Hutchison, David, 607-431-4730 HutchisonD@hartwick.edu, Hartwick Coll – GxgGi
Hutsky, Andrew J., (703) 993-5394 ahutsky@gmu.edu, George Mason Univ – GsrEo
Hutson, John , john.hutson@flinders.edu.au, Flinders Univ – Spc
Hutson, Melinda, 971-722-4146 mhutson@pcc.edu, Portland Community Coll - Sylvania Campus – Xm
Hutson, Melinda, 503-725-3372 mhutson@pdx.edu, Portland State Univ – Xmc
Hutto, Richard S., (501) 683-0151 richard.hutto@arkansas.gov, Arkansas Geological Survey – Gg
Hutyra, Lucy, lrhutyra@bu.edu, Boston Univ – Oar
Huybers, Peter, 617-495-8391 phuybers@fas.harvard.edu, Harvard Univ – Pe
Huycke, David, 509-574-4817 dhuycke@yvcc.edu, Yakima Valley Coll – Gg
Huysken, Kristin, (219) 980-6739 khuysken@iun.edu, Indiana Univ Northwest – GizGc
Huysmans, Marijke, marijke.huysmans@ees.kuleuven.be, Katholieke Universiteit Leuven – Hw
Hwang, Hue-Hwa E., 217-244-9876 hhhwang@illinois.edu, Univ of Illinois, Urbana-Champaign – CgGs
Hyatt, James A., (860) 465-5789 hyattj@easternct.edu, Eastern Connecticut State Univ – Gml
Hyland, Ethan, 207-999-9047 ehyland@unity.ncsu.edu, North Carolina State Univ – GrCsPe
Hylland, Michael D., (801) 537-3382 mikehylland@utah.gov, Utah Geological Survey – Gg
Hylton, Alisa, (704)330-6297 Alisa.Hylton@cpcc.edu, Central Piedmont Community Coll – Gg
Hyndman, David W., 517-353-4442 hyndman@msu.edu, Michigan State Univ – Hy
Hyndman, Roy D., (250) 363-6428 Univ of Victoria – Ys

Hynek, Brian M., 303-735-4312 brian.hynek@colorado.edu, Univ of Colorado – Xg
Hynes, Andrew J., andrew.hynes@mcgill.ca, McGill Univ – Gc
Hyrapiet, Shireen, 541-737-3407 shireen.hyrapiet@oregonstate.edu, Oregon State Univ – On
Hysell, David, dlh37@cornell.edu, Cornell Univ – Onr

## I

Iaccheri, Maria Linda, 089/2180 4274 Ludwig-Maximilians-Universitaet Muenchen – Gg
Iaffaldano, Giampiero, 089/2180 4220 giampiero@geophysik.uni-muenchen.de, Ludwig-Maximilians-Universitaet Muenchen – Yg
Iancu, Ovidiu G., +40232201455 ogiancu@uaic.ro, Al. I. Cuza Univ of Iasi – GpXgCg
Iantria, Linnea, (417) 836-5318 liantria@missouristate.edu, Missouri State Univ – On
Ibaraki, Motomu, (614) 292-7528 ibaraki.1@osu.edu, Ohio State Univ – Hg
Ibrahim, Ahmad Tajuddin Hj, 03-79674230 ahmadt@um.edu.my, Univ of Malaya – Ng
Ibrahim, Mohamed, 212-772-5267 mibrahim@hunter.cuny.edu, Hunter Coll (CUNY) – On
Icenhower, Jonathan P., (509) 372-0078 jon.icenhower@pnl.gov, Pacific Northwest National Laboratory – Cl
Icopini, Gary, 406-496-4841 gicopini@mtech.edu, Montana Tech of The Univ of Montana – Hw
Ide, Kayo, (310) 206-6484 Univ of California, Los Angeles – Og
Idleman, Bruce D., bdi2@lehigh.edu, Lehigh Univ – Cc
Idone, Vincent P., (518) 442-4577 vpi@atmos.albany.edu, SUNY, Albany – Oa
Idstein, Peter J., 859-257-2770 peter.idstein@uky.edu, Univ of Kentucky – Gg
Ielpi, Alessandro, (705) 675-1151 aielpi@laurentian.ca, Laurentian Univ, Sudbury – GsmGr
Ierley, Glenn R., gierley@ucsd.edu, Univ of California, San Diego – Op
Igel, Heiner, 089/2180 4204 heiner.igel@lmu.de, Ludwig-Maximilians-Universitaet Muenchen – Yg
Iglesias, Cruz, 0034981167000 cruzi@udc.es, Coruna Univ – GaOmNr
Igonor, Emmanuel E., 08066943346 eeigonor@futa.edu.ng, Federal Univ of Tech, Akure – GgCeGi
Ihinger, Phillip D., (715) 836-2158 ihinger@uwec.edu, Univ of Wisconsin, Eau Claire – Gx
Ikingura, Justian R., jikingura@gmail.com, Univ of Dar es Salaam – EgCgGi
Ikonnikova, Svetlana, 512-232-9464 s.ikonnikova@mail.utexas.edu, Univ of Texas at Austin – Ego
Iles, Derric L., (605) 677-6148 derric.iles@usd.edu, South Dakota Dept of Environment and Natural Resources – HwGg
Illari, Lodovica, (617) 253-2286 illari@squall.mit.edu, Massachusetts Inst of Tech – Ow
Imasuen, Isaac O., okpeseyi.imasuen@uniben.edu, Univ of Benin – Gee
Imber, Jonathan, 0191 3742508 jonathan.imber@durham.ac.uk, Durham Univ – Gc
Immel, Harald, 089/2180 6615 h.immel@lrz.uni-muenchen.de, Ludwig-Maximilians-Universitaet Muenchen – Pg
Immonen, Wilma, (406) 496-4182 wimmonen@mtech.edu, Montana Tech of the Univ of Montana – On
Inamdar, Shreeram, 302-831-8877 inamdar@udel.edu, Univ of Delaware – Hg
Ince, Simon, (520) 621-5082 Univ of Arizona – Hs
Inceoz, Murat, 00904242370000-5987 minceoz@firat.edu.tr, Firat Univ – GctGm
Indares, Aphrodite D., (709) 737-2456 afin@sparky2.esd.mun.ca, Memorial Univ of Newfoundland – Gp
Ingall, Ellery D., (404) 894-3883 ellery.ingall@eas.gatech.edu, Georgia Inst of Tech – CmlOc
Ingalls, Anitra E., 206-221-6748 aingalls@u.washington.edu, Univ of Washington – Oc
Ingersoll, Andrew P., 626 395-6167 api@gps.caltech.edu, California Inst of Tech – Oa
Ingersoll, Raymond V., (310) 825-8634 ringer@epss.ucla.edu, Univ of California, Los Angeles – Gst
Ingle, Jr., James C., (650) 723-3366 ingle@pangea.stanford.edu, Stanford Univ – Pm
Inglett, Kanika Sharma, 352-294-3164 Univ of Florida – Os
Inglett, Patrick, 352-392-1803 pinglett@ufl.edu, Univ of Florida – Sf
Inglis, Michael, 631-451-4120 ingism@sunysuffolk.edu, Suffolk County Community Coll, Ammerman Campus – Og
Ingram, B. Lynn, (510) 643-1474 ingram@eps.berkeley.edu, Univ of California, Berkeley – Cs
Innis, Charles, clemmys@aol.com, Univ of Massachusetts, Boston – Ob
Insel, Nadja, 773-442-6058 N-Insel@neiu.edu, Northeastern Illinois Univ – Gt
Inskeep, William P., (406) 994-5077 binskeep@montana.edu, Montana State Univ – Sc
Ioannides, Dimitri, 417-836-5800 DIoannides@MissouriState.edu, Missouri State Univ – On
Ioris, Antonio, +44 (0) 131 651 9090 a.ioris@ed.ac.uk, Edinburgh Univ – Ge
Iqbal, Mohammad Z., (319) 273-2998 m.iqbal@uni.edu, Univ of Northern Iowa – Hw
Iribar, Vicente , vicente.iribar@ehu.eus, Univ of the Basque Country UPV/EHU – HwGm
Irish, Jennifer, (540) 231-2298 jirish@vt.edu, Virginia Tech – Oog
Irlandi, Elizabeth A., (321) 674-8096 irlandi@fit.edu, Florida Inst of Tech – Ob
Irmis, Randall B., (801) 581-6555 irmis@umnh.utah.edu, Univ of Utah – Sa
Ironside, R. Geoffrey, geoff.ironside@ualberta.ca, Univ of Alberta – On
Irvin, Gene D., 205/247-3542 dirvin@gsa.state.al.us, Geological Survey of Alabama – Gg
Irvine, Dylan, dylan.irvine@flinders.edu.au, Flinders Univ – Hw
Irvine, Pamela J., (213) 620-4784 California Geological Survey – Ng
Irvine, T. Neil, 202-478-8950 nirvine@carnegiescience.edu, Carnegie Institution of Washington – Gi
Irving, Jessica, 609 258-4536 jirving@princeton.edu, Princeton Univ – Yg
Irving, Tony, irving@ess.washington.edu, Univ of Washington – CgXmGi
Isaacson, Carl, 218-755-4104 cisaacson@bemidjistate.edu, Bemidji State Univ – Ge
Isaacson, Peter E., (208) 885-7969 isaacson@uidaho.edu, Univ of Idaho – Pi
Isacks, Bryan L., bli1@cornell.edu, Cornell Univ – Gmt
Isbell, John L., 414-229-2877 jisbell@uwm.edu, Univ of Wisconsin, Milwaukee – Gs
Isenor, Fenton M., Fenton_Isenor@cbu.ca, Cape Breton Univ – GgNgGm
Ishii, Miaki, 617-384-8066 ishii@eps.harvard.edu, Harvard Univ – Yg
Ishii, Satoshi, 612-624-7902 ishi0040@umn.edu, Univ of Minnesota, Twin Cities – SbPy
Ishman, Scott E., (618) 453-7377 sishman@siu.edu, Southern Illinois Univ Carbondale – Pe
Isiorho, Solomon A., 260-481-6254 isiorho@ipfw.edu, Indiana Univ / Purdue Univ, Fort Wayne – HwGeOy
Ismat, Zeshan, 717-358-4485 zeshan.ismat@fandm.edu, Franklin and Marshall Coll – Gc
Israel, Daniel W., (919) 515-2388 North Carolina State Univ – Sb
Ito, Emi, 612-624-7881 eito@umn.edu, Univ of Minnesota, Twin Cities – Cs
Ito, Garrett T., (808) 956-9717 gito@hawaii.edu, Univ of Hawai'i, Manoa – YgrGt
Ivanochko, Tara, (604) 827-3179 tivanochko@eos.ubc.ca, Univ of British Columbia – Ou
Ivanov, Julian, (785) 864-2089 jivanov@kgs.ku.edu, Univ of Kansas – Yg
Ivanov, Martin, +420 549 49 4600 mivanov@sci.muni.cz, Masaryk Univ – PgiPe
Ivanova, Maria, maria.ivanova@umb.edu, Univ of Massachusetts, Boston – On
Ivanovic, Ruza, +44(0) 113 34 34945 r.ivanovic@leeds.ac.uk, Univ of Leeds – OgaPe
Ivany, Linda C., (315) 443-3626 lcivany@syr.edu, Syracuse Univ – PeoCs
Iverson, Neal R., (515) 294-8048 niverson@iastate.edu, Iowa State Univ of Science & Tech – Gl
Ivins, Erik R., (818) 354-4785 Jet Propulsion Laboratory – Gt
Ivy, Logan D., (303) 370-6474 Logan.Ivy@dmns.org, Denver Museum of Nature & Science – Pv
Ivy, Russell L., ivy@fau.edu, Florida Atlantic Univ – On
Izon, Gareth, +44 01334 463936 gji3@st-andrews.ac.uk, Univ of St. Andrews – Cg

## J

Jablonowski, Christiane, (734) 763-6238 cjablono@umich.edu, Univ of Michigan – Oaw

Jablonski, David, (773) 702-8163 djablons@midway.uchicago.edu, Univ of Chicago – Pi
Jacinthe, Pierre-Andre, pjacinth@iupui.edu, Indiana Univ / Purdue Univ, Indianapolis – Cl
Jackson, Andrea, +44(0) 113 34 36728 a.v.jackson@leeds.ac.uk, Univ of Leeds – Oa
Jackson, Brian P., (603) 646-1272 brian.jackson@dartmouth.edu, Dartmouth Coll – CaGeCt
Jackson, Charles, (512) 471-0401 charles@ig.utexas.edu, Univ of Texas at Austin – Pe
Jackson, Chester M., (912) 598-2328 cjackson@georgiasouthern.edu, Georgia Southern Univ – OnGs
Jackson, Chester W., (912) 478-0174 cjackson@georgiasouthern.edu, Georgia Southern Univ – GsmGu
Jackson, Christopher, +44 20 759 47450v c.jackson@imperial.ac.uk, Imperial Coll – Gt
Jackson, David D., (310) 825-0421 djackson@ucla.edu, Univ of California, Los Angeles – YsdYg
Jackson, Edgar L., ed.jackson@ualberta.ca, Univ of Alberta – On
Jackson, Frankie, (406) 994-6642 frankiej@montana.edu, Montana State Univ – Pv
Jackson, Gail D., +44 (0) 131 650 5436 G.Jackson@ed.ac.uk, Edinburgh Univ – PbGe
Jackson, George A., (979) 845-0405 gjackson@ocean.tamu.edu, Texas A&M Univ – Op
Jackson, Hiram, (916) 691-7605 Cosumnes River Coll – Gg
Jackson, James A., +44 (0) 1223 337197 jaj2@cam.ac.uk, Univ of Cambridge – YgGt
Jackson, Jennifer M., (626) 395-6780 jackson@gps.caltech.edu, California Inst of Tech – Gy
Jackson, Jeremiah, 573-368-2182 jeremiah.jackson@dnr.mo.gov, Missouri Dept of Natural Resources – Gg
Jackson, Jeremy B. C., (858) 518-7613 jbjackson@ucsd.edu, Univ of California, San Diego – PoePi
Jackson, Kenneth J., (925) 422-6053 jackson8@llnl.gov, Lawrence Livermore National Laboratory – Cl
Jackson, Louise E., 530-754-9116 lejackson@ucdavis.edu, Univ of California, Davis – Os
Jackson, Mark W., (801) 378-9753 mark_jackson@byu.edu, Brigham Young Univ – Or
Jackson, Martin P. A., (512) 475-9548 martin.jackson@beg.utexas.edu, Univ of Texas at Austin, Jackson School of Geosciences – Gc
Jackson, Michael, 612-624-5274 jacks057@umn.edu, Univ of Minnesota, Twin Cities – Ym
Jackson, Philip L., jacksonp@geo.oregonstate.edu, Oregon State Univ – Oy
Jackson, Richard A., (718) 460-1476 Long Island Univ, Brooklyn Campus – Gc
Jackson, Richard H., (801) 378-6063 richard_jackson@byu.edu, Brigham Young Univ – Ou
Jackson, Robert B., 919 660 7408 jackson@duke.edu, Duke Univ – On
Jackson Jr, William T., 205-247-3548 wjackson@gsa.state.al.us, Geological Survey of Alabama – GcsGt
Jacob, Daniel J., (617) 495-1794 djj@io.harvard.edu, Harvard Univ – Oa
Jacob, Klaus H., (845) 365-8440 Columbia Univ – Ys
Jacob, Robert W., (570) 577-1791 rwj003@bucknell.edu, Bucknell Univ – Yg
Jacobi, Robert D., (716) 207-2478 rdjacobi@acsu.buffalo.edu, SUNY, Buffalo – GctGs
Jacobs, Alan M., 330-941-2933 amjacobs@ysu.edu, Youngstown State Univ – Ge
Jacobs, Jon, 519-661-2111 ext 86752 jjacobs@uwo.ca, Western Univ – Gg
Jacobs, Katharine L., (520) 626-0684 jacobsk@email.arizona.edu, Univ of Arizona – Hg
Jacobs, Louis L., (214) 768-2773 Southern Methodist Univ – Pv
Jacobs, Peter, jacobsp@uww.edu, Univ of Wisconsin, Whitewater – SdaGm
Jacobs, Stanley, (845) 365-8326 Columbia Univ – Op
Jacobs, Tenika, (609) 984-6587 tenika.jacobs@dep.state.nj.us, New Jersey Geological and Water Survey – On
Jacobsen, Jeffrey S., (406) 994-4605 jefj@montana.edu, Montana State Univ – So
Jacobsen, Stein B., (617) 495-5233 jacobsen@neodymium.harvard.edu, Harvard Univ – Cc
Jacobsen, Steven D., 847-467-1825 steven@earth.northwestern.edu, Northwestern Univ – GyzOm
Jacobson, Andrew D., (847) 491-3132 adj@earth.northwestern.edu, Northwestern Univ – ClPeCa
Jacobson, Carl E., cejac@iastate.edu, Iowa State Univ of Science & Tech – Gc
Jacobson, Roger, (775) 673-7364 roger@dri.edu, Univ of Nevada, Reno – Cg
Jaeger, John M., (352) 846-1381 jmjaeger@ufl.edu, Univ of Florida – Gs
Jaegle, Lyatt, 206-685-2679 jaegle@atmos.washington.edu, Univ of Washington – Oa
Jaffe, Daniel A., djaffe@u.washington.edu, Univ of Washington – Oa
Jaffe, Peter R., (609) 258-4653 Princeton Univ – Hg
Jago, Bruce C., (705) 675-1151 x7227 bjago@laurentian.ca, Laurentian Univ, Sudbury – EgCe
Jagoutz, Oliver, (617) 324-5514 jagoutz@mit.edu, Massachusetts Inst of Tech – GitGc
Jahangiri, Ahmad, +98 (411) 339 2695 jahangiri@tabriu.ac.ir, Univ of Tabriz – GivGg
Jain, Atul K., (217) 333-2128 jain1@illinois.edu, Univ of Illinois, Urbana-Champaign – Oa
Jain, Cathy, 7607441150 xt. 2952 cjain@palomar.edu, Palomar Coll – Oy
Jaiswal, Priyank, 405-744-6358 priyank.jaiswal@okstate.edu, Oklahoma State Univ – Yg
Jakosky, Bruce M., 303-492-8004 bruce.jakosky@colorado.edu, Univ of Colorado – Xg
Jamaluddin, Tajul Anuar, 03-79674152 taj@um.edu.my, Univ of Malaya – Ng
James, Bruce, (301) 405-8573 brjames@umd.edu, Univ of Maryland – Sc
James, David E., djames@carnegiescience.edu, Carnegie Institution of Washington – YsGtYg
James, Matthew J., (707) 664-2301 james@sonoma.edu, Sonoma State Univ – PiGhPo
James, Noel P., (613) 533-6170 jamesn@queensu.ca, Queen's Univ – Gd
James, Rachael, +44 (0)23 80599005 R.H.James@soton.ac.uk, Univ of Southampton – Oc
James, Richard, 7056751151 x2271 rjames@laurentian.ca, Laurentian Univ, Sudbury – GpiEm
James, Scott C., (254) 710-2534 sc_james@baylor.edu, Baylor Univ – HwGeEo
James, Thomas S., (250) 363-6403 thomas.james@canada.ca, Univ of Victoria – YdvYg
James, Valentine U., (814) 393-1938 vjames@clarion.edu, Clarion Univ – OunOn
James-Aworeni, E., 234-705-326-6216 dawnjames@gmail.com, Obafemi Awolowo Univ – Hg
Jamieson, Heather E., (613) 545-6181 Queen's Univ – Ge
Jamieson, J. Bruce, (403) 288-0803 Univ of Calgary – Ne
Jamili, Ahmad, ahmad.jamili@ou.edu, Univ of Oklahoma – Np
Jamiolahmady, Mahmoud, +44 (0) 131 451 3122 m.jamiolahmady@pet.hw.ac.uk, Heriot-Watt Univ – On
Janecke, Susanne U., (435) 797-3877 susanne.janecke@usu.edu, Utah State Univ – Gt
Janecky, David R., (505) 667-7603 Los Alamos National Laboratory – Cl
Janetos, Tony, ajanetos@bu.edu, Boston Univ – On
Janney, Phillip, +27-21-650-2929 phil.janney@uct.ac.za, Univ of Cape Town – GiCaXc
Jannik, Nancy O., njannik@winona.edu, Winona State Univ – Hw
Janowitz, Gerald S., (919) 515-7837 jerry_janowitz@ncsu.edu, North Carolina State Univ – Op
Janson, Xavier, 512-475-9524 xavier.janson@beg.utexas.edu, Univ of Texas at Austin, Jackson School of Geosciences – Gr
Janssen, Keith, (785) 532-6101 kjanssen@ksu.edu, Kansas State Univ – So
Jantz, Claire A., (717) 477-1399 cajant@ship.edu, Shippensburg Univ – Oui
Janusz, Robert, 210-486-0045 San Antonio Community Coll – Gg
Jaouich, Alfred, 514-987-3000 #3378 jaouich.alfred@uqam.ca, Universite du Quebec a Montreal – Sc
Jarcho, Kari A., (612) 625-5251 kjarcho@umn.edu, Univ of Minnesota, Twin Cities – On
Jardine, Philip M., ipj@ornl.gov, Oak Ridge National Laboratory – Sc
Jarrard, Richard D., (801) 581-7062 r.jarrard@utah.edu, Univ of Utah – Gu
Jarvis, Ed, +353 21 4902698 e.jarvis@ucc.ie, Univ Coll Cork – Pl
Jarvis, Gary T., (416)736-2100 #77710 jarvis@yorku.ca, York Univ – Yg
Jarvis, Ian, +44 020 8417 2526 i.jarvis@kingston.ac.uk, Kingston Univ – CgGsr
Jarvis, Richard S., (915) 747-5263 rsjarvis@utep.edu, Univ of Texas, El Paso – OyGmOa
Jarvis, W. T., 541-737-8052 Todd.Jarvis@oregonstate.edu,

Oregon State Univ – Hw
Jarzen, David M., 3523921721 x 245 dilcher@flmnh.ufl.edu, Univ of Florida – Pb
Jasbinsek, John J., (805) 756-2013 jjasbins@calpoly.edu, California Polytechnic State Univ – Ysg
Jasinski, Steven E., (717) 783-9897 c-sjasinsk@pa.gov, State Museum of Pennsylvania – PvgPy
Jason, Shafer, 802-626-6225 jason.shafer@lyndonstate.edu, Lyndon State Coll – Oa
Jasoni, Richard, 775-673-7472 richard.jasoni@dri.edu, Desert Research Inst – So
Jastrow, Julie D., (630) 252-3226 jdjastrow@anl.gov, Argonne National Laboratory – Sb
Jaumé, Steven C., (843) 953-1802 jaumes@cofc.edu, Coll of Charleston – Ys
Javadpour, Farzam, (512) 232-8068 farzam.javadpour@beg.utexas.edu, Univ of Texas at Austin – NpEo
Javaux, Emmanuelle, 32 4 366 54 22 EJ.Javaux@ulg.ac.be, Universite de Liege – Pl
Jawitz, James W., 352-392-1951 (203) jawitz@ufl.edu, Univ of Florida – SpHsw
Jayakumar, Amal, (609) 258-6294 ajayakum@princeton.edu, Princeton Univ – Cm
Jaye, Shelley, (703) 425-5180 sjaye@nvcc.edu, Northern Virginia Community Coll - Annandale – Giy
Jean-Marc, MONTEL, +33 3 83 59 64 00 jean-marc.montel@univ-lorraine.fr, Ecole Nationale Supérieure de Géologie (ENSG) – Eg
Jeanloz, Raymond, (510) 642-2639 jeanloz@uclink.berkeley.edu, Univ of California, Berkeley – Gy
Jeans, Meghan, mjeans@neaq.org, Univ of Massachusetts, Boston – On
Jebrak, Michel, 514-987-3000 #3986 jebrak.michel@uqam.ca, Universite du Quebec a Montreal – Eg
Jedrysek, Mariusz O., mariusz.jedrysek@uwr.edu.pl, Univ of Wroclaw – CaGe
Jefferson, Anne, 704-687-5977 ajefferson@uncc.edu, Univ of North Carolina, Charlotte – HgGm
Jefferson, Anne J., (330) 672-2746 ajeffer9@kent.edu, Kent State Univ – HgGm
Jeffery, David L., 740-376-4844 jefferyd@marietta.edu, Marietta Coll – Go
Jeffrey, Kip, +4401326259442 C.Jeffrey@exeter.ac.uk, Exeter Univ – Nx
Jeffries, H. Perry, (401) 874-6222 jeffries@uri.edu, Univ of Rhode Island – Ob
Jekeli, Christopher, 614 292 7117 jekeli.1@osu.edu, Ohio State Univ – Yd
Jelinski, Nicolas, 612-626-9936 jeli0026@umn.edu, Univ of Minnesota, Twin Cities – SdCs
Jellinek, Mark, mjellinek@eos.ubc.ca, Univ of British Columbia – Gg
Jenkins, David M., (607) 777-2736 dmjenks@binghamton.edu, Binghamton Univ – CpGz
Jenkins, Gregory S., 814-865-0479 gsj1@psu.edu, Pennsylvania State Univ, Univ Park – Ow
Jenkins, Mary Ann, (416)736-2100 #22992 maj@yorku.ca, York Univ – Oa
Jenkins, William J., (508) 289-2554 wjenkins@whoi.edu, Woods Hole Oceanographic Institution – Oc
Jenkyns, Hugh C., +44 (1865) 272023 hugh.jenkyns@earth.ox.ac.uk, Univ of Oxford – GrsCs
Jenner, George A., (709) 737-8387 gjenner@sparky2.esd.mun.ca, Memorial Univ of Newfoundland – Gi
Jennings, Carrie E., (612) 718-1415 carrie@umn.edu, Univ of Minnesota, Twin Cities – Gl
Jensen, Ann R., 605-677-6159 ann.jensen@usd.edu, South Dakota Dept of Environment and Natural Resources – Gg
Jensen, Antony, +44 (0)23 80593428 acj@noc.soton.ac.uk, Univ of Southampton – Ob
Jensen, Olivia G., 514-398-3587 olivia.jensen@mcgill.ca, McGill Univ – Yg
Jensen, Scott W., 605-677-6869 scott.jensen@usd.edu, South Dakota Dept of Environment and Natural Resources – On
Jenson, John W., (671) 735-2689 jjenson@uguam.uog.edu, Univ of Guam – HwGel
Jercinovic, Michael J., (413) 545-2431 mjj@geo.umass.edu, Univ of Massachusetts, Amherst – Gx
Jerde, Eric, (606) 783-5406 e.jerde@moreheadstate.edu, Morehead State Univ – Cg
Jerolmack, Douglas, 215-746-2823 sediment@sas.upenn.edu, Univ of Pennsylvania – GmYxHq
Jessey, David R., 909-869-3457 drjessey@csupomona.edu, California State Polytechnic Univ, Pomona – Eg
Jessup, Micah, (865) 974-2366 mjessup@utk.edu, Univ of Tennessee, Knoxville – Gc
Jeu, Amy, (212) 772-4019 ajeu@hunter.cuny.edu, Hunter Coll (CUNY) – Oi
Jewell, Paul W., (801) 581-6636 paul.jewell@utah.edu, Univ of Utah – Hg
Jewett, Brian, 217-333-3957 bjewett@illinois.edu, Univ of Illinois, Urbana-Champaign – Oa
Jewitt, David, (310) 825-3880 jewitt@epss.ucla.edu, Univ of California, Los Angeles – XyOnn
Ji, Chen, (805) 893-2923 ji@geol.ucsb.edu, Univ of California, Santa Barbara – Ys
Ji, Wei, (816) 235-1334 jiwei@umkc.edu, Univ of Missouri-Kansas City – OriOy
Jiang, Dazhi, 519-661-3192 djiang3@uwo.ca, Western Univ – GcgGg
Jiang, Ganqing, (702) 895-2708 ganqing.jiang@unlv.edu, Univ of Nevada, Las Vegas – GsClGr
Jiang, James Xinxia, (506) 364-2326 Mount Allison Univ – Gg
Jiang , Shu, 801-585-9816 sjiang@egi.utah.edu, Univ of Utah – GoYsNp
Jiang, Zhenhua, (410) 436-6864 Argonne National Laboratory – On
Jickells, Tim, +44 (0)1603 59 3117 t.jickells@uea.ac.uk, Univ of East Anglia – Ob
Jimoh, Mustapha T., +2348052472942 mtjimoh@lautech.edu.ng, Ladoke Akintola Univ of Tech – GxxGx
Jin, Guohai, 205-247-3560 gjin@gsa.state.al.us, Geological Survey of Alabama – Np
Jin, Jisuo, (519) 661-4061 jjin@uwo.ca, Western Univ – Pi
Jin, Lixin, (915) 747-5559 ljin2@utep.edu, Univ of Texas, El Paso – Ge
Jin, Qusheng, 541-346-4999 qjin@uoregon.edu, Univ of Oregon – Cg
Jin, Yufang, (530) 601 9805 jin.yufang@gmail.com, Univ of California, Davis – Or
Jinnah, Zubair A., (011) 717-6554 zubair.jinnah@wits.ac.za, Univ of the Witwatersrand – GsPg
Jiracek, George R., (619) 594-5160 gjiracek@mail.sdsu.edu, San Diego State Univ – Ye
Jiron, Rebecca, 757-221-2484 rljiron@wm.edu, Coll of William & Mary – GmtOg
Jirsa, Mark, (612) 626-4028 jirsa001@umn.edu, Univ of Minnesota – Gc
Jiskoot, Hester, (403) 329-2739 hester.jiskoot@uleth.ca, Univ of Lethbridge – OlGmOy
Jo, Ho Young, 82-2-3290-3179 hyjo@korea.ac.kr, Korea Univ – NgCaGe
Jochens, Ann E., (979) 845-6714 ajochens@ocean.tamu.edu, Texas A&M Univ – Op
Joeckel, Matt, (402) 472-7520 rjoeckel3@unl.edu, University of Nebraska - Lincoln – GsrGg
Joesten, Raymond, (860) 486-6643 raymond.joesten@uconn.edu, Univ of Connecticut – Gp
Johanesen, Katharine, (814) 641-3601 johanesen@juniata.edu, Juniata Coll – GzxGc
Johannesson, Karen H., 504-862-3193 kjohanne@tulane.edu, Tulane Univ – HgCg
John, Barbara E., (307) 766-4232 bjohn@uwyo.edu, Univ of Wyoming – Gci
John, Cedric, +44 20 759 46461 cedric.john@imperial.ac.uk, Imperial Coll – Gs
John, Chacko J., (225) 578-8681 cjohn@lsu.edu, Louisiana State Univ – GosGe
John, Hickman B., 859-323-0541 jhickman@uky.edu, Univ of Kentucky – Gto
Johnpeer, Gary D., (714) 255-9819 gjohnpeer@earthlink.net, Cerritos Coll – Ng
Johns, Ronald A., (512) 223-6002 rjohns@austincc.edu, Austin Community Coll District – Pi
Johnson, Ansel G., (503) 725-3381 Portland State Univ – Ye
Johnson, Arthur H., ahj@sas.upenn.edu, Univ of Pennsylvania – So
Johnson, Ashanti, (817) 272-2987 ashanti@uta.edu, Univ of Texas, Arlington – OgGe
Johnson, Becky, (817) 257-7271 becky.johnson@tcu.edu, Texas Christian Univ – GeHwOu
Johnson, Beth , beth.johnson@uwc.edu, Univ of Wisconsin Colls – Glh
Johnson, Beverly J., 207 786-6062 bjohnso3@bates.edu, Bates Coll – Cl
Johnson, Brad, 520-621-2470 bradjohnson@email.ari, Univ of Arizona – Gc
Johnson, Brandon C., (401) 863-5163 brandon_johnson@brown.edu, Brown Univ – Xg
Johnson, Bruce D., (902) 494-3259 bruce.johnson@dal.ca, Dalhousie Univ – Oc

Johnson, Cari, (801) 585-3782 cari.johnson@utah.edu, Univ of Utah – Gs
Johnson, Carl G., (508) 289-2304 cjohnson@whoi.edu, Woods Hole Oceanographic Institution – Ca
Johnson, Clark M., (828) 265-8680 clarkj@geology.wisc.edu, Univ of Wisconsin, Madison – CgcCs
Johnson, Claudia C., 812-855-0646 claudia@indiana.edu, Indiana Univ, Bloomington – Pe
Johnson, Daniel L., (403) 327-4561 Univ of Lethbridge – On
Johnson, Darren J., 605-677-6162 darren.j.johnson@usd.edu, South Dakota Dept of Environment and Natural Resources – GgEo
Johnson, David B., (575) 835-5635 djohnson@nmt.edu, New Mexico Inst of Mining and Tech – Ps
Johnson, Donald O., (630) 252-3392 Argonne National Laboratory – Gr
Johnson, Donald R., (608)262-2538 donj@ssec.wisc.edu, Univ of Wisconsin, Madison – Oa
Johnson, Elias, (417) 836-5800 Missouri State Univ – Or
Johnson, Emily P., (617) 353-9709 Boston Univ – On
Johnson, Emily R., 575-646-3795 erj@nsmu.edu, New Mexico State Univ, Las Cruces – Gvi
Johnson, Eric L., 1-607-4314658 johnsone@hartwick.edu, Hartwick Coll – GpcGi
Johnson, Gary D., (603) 636-2371 gary.d.johnson@dartmouth.edu, Dartmouth Coll – GrdYm
Johnson, Glenn W., (801) 581-6151 gjohnson@egi.utah.edu, Univ of Utah – Gq
Johnson, Gregory C., (206) 526-6806 Univ of Washington – Op
Johnson, H. Paul, (206) 543-8474 johnson@ocean.washington.edu, Univ of Washington – Yr
Johnson, Helen, +44 01865 272142 Helen.Johnson@earth.ox.ac.uk, Univ of Oxford – Og
Johnson, Howard, +44 20 759 46461 h.d.johnson@imperial.ac.uk, Imperial Coll – Go
Johnson, Jane M., (320) 589-3411 Univ of Minnesota, Twin Cities – So
Johnson, Jean M., (706) 272-2666 jmjohnson@daltonstate.edu, Dalton State Community Coll – Og
Johnson, Jeanne, (225) 578-8407 jeannej@lsu.edu, Louisiana State Univ – On
Johnson, Jeffrey , 208-426-2959 jeffrey.b.johnson@gmail.com, Boise State Univ – YsGv
Johnson, Joel E., 603-862-1718 joel.johnson@unh.edu, Univ of New Hampshire – Gus
Johnson, Joel P., 512-232-5288 joelj@jsg.utexas.edu, Univ of Texas at Austin – Gs
Johnson, Judy L., (907) 474-7388 jljohnson21@alaska.edu, Univ of Alaska, Fairbanks – OnnOn
Johnson, Julie, 901-678-4217 jjhnsn79@memphis.edu, Univ of Memphis – Gx
Johnson, Kaj, 812-855-3612 kajjohns@indiana.edu, Indiana Univ, Bloomington – Yg
Johnson, Katherine, (217) 581-7270 kjohnson4@eiu.edu, Eastern Illinois Univ – Pmi
Johnson, Kathleen, 949-824-6174 kathleen.johnson@uci.edu, Univ of California, Irvine – Og
Johnson, Kenneth S., johnsonk@uhd.edu, Univ of Houston Downtown – GiCac
Johnson, Kevin T. M., 808-956-3444 kjohnso2@hawaii.edu, Univ of Hawai'i, Manoa – Gi
Johnson, Kurt, (907) 696-0079 kurt.johnson@alaska.gov, Alaska Division of Geological & Geophysical Surveys – Gg
Johnson, Lane R., LRJohnson@lbl.gov, Univ of California, Berkeley – Ys
Johnson, Marie C., (845) 938-4855 marie-johnson@usma.edu, United States Military Academy – Gi
Johnson, Mark O., mjohnson15@worcester.edu, Worcester State Univ – Oy
Johnson, Mark S., (778) 999-2102 mark.johnson.2102@gmail.com, Univ of British Columbia – SpHsq
Johnson, Markes E., 413-597-2329 markes.e.johnson@williams.edu, Williams Coll – Ps
Johnson, Martin, +44 (0)1603 59 1299 martin.johnson@uea.ac.uk, Univ of East Anglia – Cm
Johnson, Ned K., (510) 642-3059 Univ of California, Berkeley – Pv
Johnson, Neil E., 540-231-1785 johnsonne@vt.edu, Virginia Polytechnic Inst & State Univ – GzEg
Johnson, Paul A., (507) 933-7442 Los Alamos National Laboratory – Yg
Johnson, Peggy, (575) 835-5819 peggy@nmbg.nmt.edu, New Mexico Inst of Mining & Tech – Hw
Johnson, Robert E., (309) 298-1368 RE-Johnson2@wiu.edu, Western Illinois Univ – On
Johnson, Robert G., 612-626-0853 johns088@umn.edu, Univ of Minnesota, Twin Cities – Pe
Johnson, Robert L., (630) 252-7004 rljohnson@anl.gov, Argonne National Laboratory – On
Johnson, Robert O., (507) 933-7442 Oak Ridge National Laboratory – Hg
Johnson, Roy A., (520) 621-4890 johnson6@email.arizona.edu, Univ of Arizona – Ys
Johnson, Sarah E., 859-572-6907 johnsonsa@nku.edu, Northern Kentucky Univ – Ng
Johnson, Scott E., (207) 581-2142 johnsons@maine.edu, Univ of Maine – Gc
Johnson, Thomas C., (218) 341-4599 tcj@d.umn.edu, Univ of Massachusetts, Amherst – PeCoOu
Johnson, Thomas M., (217) 244-2002 tmjohnsn@illinois.edu, Univ of Illinois, Urbana-Champaign – HwCls
Johnson, Tiffany, (604) 777-6117 johnsont@douglasColl.ca, Douglas Coll – Og
Johnson, Ty, (501) 683-0153 ty.johnson@arkansas.gov, Arkansas Geological Survey – GgOi
Johnson, Verner C., (970) 248-1672 vjohnson@coloradomesa.edu, Colorado Mesa Univ – Yg
Johnson, William C., 785-864-5548 wcj@ku.edu, Univ of Kansas – Grm
Johnson, William P., (801) 581-5033 william.johnson@utah.edu, Univ of Utah – Ng
Johnston, A. Dana, (541) 346-5588 adjohn@uoregon.edu, Univ of Oregon – Cp
Johnston, Archibald C., (901) 678-2007 ajohnstn@memphis.edu, Univ of Memphis – Ys
Johnston, Carl G., 330-941-7151 cgjohnston@ysu.edu, Youngstown State Univ – Py
Johnston, David, (501) 683-0126 david.johnston@arkansas.gov, Arkansas Geological Survey – Gg
Johnston, David T., 617-496-5024 johnston@eps.harvard.edu, Harvard Univ – CgPy
Johnston, K. R. Gina, gjohnston@csuchico.edu, California State Univ, Chico – Hs
Johnston, Karin, (541) 346-5588 George Washington Univ – On
Johnston, Paul, (403) 440-6174 pajohnston@mtroyal.ca, Mount Royal Univ – Pi
Johnston, Paul J., 620-341-5330 jillpaulj@hotmail.com, Emporia State Univ – Gg
Johnston, Scott, 805-756-1650 scjohnst@calpoly.edu, California Polytechnic State Univ – GcpCc
Johnston, Stephen T., 250 721-6120 stj@uvic.ca, Univ of Victoria – Gc
Johnston, Stephen T., (780) 492-5249 stjohnst@ualberta.ca, Univ of Alberta – GctGg
Johnston, Thomas, (403) 329-2534 johnston@uleth.ca, Univ of Lethbridge – Oe
Johnston, III, John E., (225) 578-8657 hammer@lsu.edu, Louisiana State Univ – EoGe
Join, Jean-Lambert, +262262938697 jean-lambert.join@univ-reunion.fr, Universite de la Reunion – Hw
Jokipii, Jack R., (520) 621-4256 jokipii@lpl.arizona.edu, Univ of Arizona – Xy
Jolliff, Bradley L., (314) 935-5622 blj@levee.wustl.edu, Washington Univ in St. Louis – Gx
Jomeiri, Rahim, +98 (411) 339 2704 Univ of Tabriz – Yg
Jonas, John J., (514) 398-4755 McGill Univ – Nx
Jones, Adrian P., +44 020 7679 32415 adrian.jones@ucl.ac.uk, Univ Coll London – GiCgGg
Jones, Alan, (607) 777-2518 Binghamton Univ – Ys
Jones, Bobby L., (225) 388-8328 Louisiana State Univ – Go
Jones, Brian, (780) 492-3074 brian.jones@ualberta.ca, Univ of Alberta – Ps
Jones, Charles E., (412) 624-6347 cejones@pitt.edu,Univ of Pittsburgh – Gg
Jon es, Craig H., (303) 492-6994 craig.jones@colorado.edu, Univ of Colorado – YsGtYg
Jones, D. M., (740) 548-7348 dalton.jones@dnr.state.oh.us, Ohio Dept of Natural Resources – OnGgOe
Jones, Dan, dsjones@umn.edu, Univ of Minnesota, Twin Cities – Py
Jones, David S., (413) 542-2714 djones@amherst.edu, Amherst Coll – GsClGr
Jones, Douglas S., 3523921721 x485 dsjones@flmnh.ufl.edu, Univ of Florida – Pg
Jones, Douglas S., (352) 273-1902 dsjones@flmnh.ufl.edu,

Univ of Florida – YgPi

Jones, Eric M., (505) 667-6386 honais@lanl.gov, Los Alamos National Laboratory – Oa

Jones, F. Walter, (780) 492-0667 wjones@phys.ualberta.ca, Univ of Alberta – Ym

Jones, Francis H., (604) 822-2138 fjones@eos.ubc.ca, Univ of British Columbia – YeOe

Jones, Glen, (505) 835-5627 glen@gis.nmt.edu, New Mexico Inst of Mining & Tech – On

Jones, Jon W., (403) 220-5024 Univ of Calgary – Gp

Jones, Julia A., (541) 737-1224 jonesj@geo.oregonstate.edu, Oregon State Univ – So

Jones, Larry Allan, (505) 667-0142 ljones@lanl.gov, Los Alamos National Laboratory – Ng

Jones, Lawrence S., (970) 248-1708 lajones@coloradomesa.edu, Colorado Mesa Univ – GsmGr

Jones, Merren, Merren.A.Jones@manchester.ac.uk, Univ of Manchester – Gst

Jones, Michael Q., (011) 717-6628 michael.jones@wits.ac.za, Univ of the Witwatersrand – YhGtYg

Jones, Minnie O., (312) 996-3154 mojones@uic.edu, Univ of Illinois at Chicago – On

Jones, Norris W., (920) 424-4460 jonesnw@uwosh.edu, Univ of Wisconsin, Oshkosh – Gi

Jones, Peter E., 902-426-3869 Dalhousie Univ – Op

Jones, Phillip D., +44 (0)1603 59 2090 p.jones@uea.ac.uk, Univ of East Anglia – OaPeHg

Jones, Rhian, 505-277-4204 rjones@unm.edu, Univ of New Mexico – Gz

Jones, Robert L., (217) 333-9490 Univ of Illinois, Urbana-Champaign – Sc

Jones, Stephen C., (205) 247-3601 sjones@gsa.state.al.us, Geological Survey of Alabama – HgGeCg

Jones, Stephen M., +44 (0)121 41 46155 s.jones.4@bham.ac.uk, Univ of Birmingham – Og

Jones, Stuart, +44 (0) 191 33 42319 stuart.jones@durham.ac.uk, Durham Univ – Gs

Jones, T, jonesTP@cf.ac.uk, Cardiff Univ – Ge

Jones, Tim L., (505) 646-3405 New Mexico State Univ, Las Cruces – Sp

Jones, III, John P., jpjones@email.arizona.edu, Univ of Arizona – On

Jordan, Andy, +44 (0)1603 59 2552 a.jordan@uea.ac.uk, Univ of East Anglia – Ge

Jordan, Bradley C., (570) 577-3024 jordan@bucknell.edu, Bucknell Univ – Gg

Jordan, Brennan T., (605) 677-6143 Brennan.Jordan@usd.edu, Univ of South Dakota – GiOw

Jordan, Guntram, 089/2180 4353 guntram.jodan@lrz.uni-muenchen.de, Ludwig-Maximilians-Universitaet Muenchen – Gz

Jordan, Jim L., (409) 880-8211 jim.jordan@lamar.edu, Lamar Univ – XcgXm

Jordan, Karen J., 251-460-6381 kjordan@southalabama.edu, Univ of South Alabama – Oyr

Jordan, Mary S., 831-656-7571 jordan@nps.edu, Naval Postgraduate School – Ow

Jordan, Robert A., (757) 727-5783 robert.jordan@hamptonu.edu, Hampton Univ – Ob

Jordan, Robert R., (302) 831-6415 rrjordan@udel.edu, Univ of Delaware – Gr

Jordan, Teresa E., (607) 255-3596 tej1@cornell.edu, Cornell Univ – GrtOg

Jordan, Thomas H., (213) 821-1237 tjordan@usc.edu, Univ of Southern California – Ys

Jorge, Maria Luisa, (615) 322-2160 malu.jorge@vanderbilt.edu, Vanderbilt Univ – Py

Jornov, Donna, djornov@mail.nysed.gov, New York State Geological Survey – On

Joshi, Manoj, +44 (0)1603 59 3647 m.joshi@uea.ac.uk, Univ of East Anglia – Oi

Journel, Andre G., (650) 723-1594 journel@pangea.stanford.edu, Stanford Univ – Gq

Judge, Shelley, (330) 263-2297 sjudge@wooster.edu, Coll of Wooster – GtsGc

Judkins, Heather L., Judkins.Heather@spColl.edu, Saint Petersburg Coll, Clearwater – Og

Jugo, Pedro J., 7056751151 x2106 pjugo@laurentian.ca, Laurentian Univ, Sudbury – CpGi

Jugulam, Mithila, (785) 532-2755 mithila@ksu.edu, Kansas State Univ – So

Juhl, Andrew, 845-365-8837 andyjuhl@ldeo.columbia.edu, Columbia Univ – Ob

Jull, A. J. Timothy, (520) 621-6816 jull@u.arizona.edu, Univ of Arizona – CclCg

Jung, Simon, +44 (0) 131 650 4837 Simon.Jung@ed.ac.uk, Edinburgh Univ – Og

Juniper, Kim, 250-721-6120 kjuniper@uvic.ca, Univ of Victoria – Ob

Junium, Christopher, 315-443-8969 ckjunium@syr.edu, Syracuse Univ – CsPeCo

Juntunen, Thomas, (612) 626-0505 junt0015@umn.edu, Univ of Minnesota, Twin Cities – Oi

Juraèiæ, Mladen, +38514606099 mjuracic@geol.pmf.hr, Univ of Zagreb – GueCm

Juranek, Lauren W., 541-737-2368 ljuranek@coas.oregonstate.edu, Oregon State Univ – Cm

Jurdy, Donna M., (847) 491-7163 donna@earth.northwestern.edu, Northwestern Univ – YgXyg

Jurena, Dwight, 210-486-0062 djurena@alamo.edu, San Antonio Community Coll – GgOg

Jurena, Dwight, djurena@alamo.edu, Alamo Colls - San Antonio Coll – GgOg

Jurmanovich, Barb, 989-686-9445 Delta Coll – On

Jurmu, Michael, (920) 929-1163 michael.jurmu@uwc.edu, Univ of Wisconsin Colls – OyGgOn

Juster, Thomas C., (813) 974-9691 Univ of South Florida, Tampa – Hw

Juszczyk , Carmen , (303) 492-2330 CarmenJ@Colorado.EDU, Univ of Colorado – On

Jutla, Rajinder S., (417) 836-5298 rajinderjutla@missouristate.edu, Missouri State Univ – On

## K

Kc, Ayse Didem, 00904242370000-5969 adkilic@firat.edu.tr, Firat Univ – GxpGy

Kaandorp, Ron, ron.kaandorp@falw.vu.nl, VU Univ Amsterdam – Gg

Kabengi, Nadine, 404-413-5207 kabengi@gsu.edu, Georgia State Univ – ScClGe

Kaczmarek, Stephen E., (269) 387-5479 stephen.kaczmarek@wmich.edu, Western Michigan Univ – GdCgGz

Kaden, Scott, 573-368-2194 nrkades@mail.dnr.state.mo.us, Missouri Dept of Natural Resources – Hw

Kadkhodai Ilkhchi, Ali, +98 (411) 339 2724 Univ of Tabriz – NpEo

Kaempf, Jochen, jochen.kaempf@flinders.edu.au, Flinders Univ – Op

Kafka, Alan L., (617) 552-3650 kafka@bc.edu, Boston Coll – Yg

Kah, Linda, (865) 974-2366 lckah@utk.edu, Univ of Tennessee, Knoxville – Gs

Kahle, Anne B., (818) 354-7265 anne@aster.jpl.nasa.gov, Jet Propulsion Laboratory – Or

Kahle, Beth, 021-650-2900 beth.kahle@uct.ac.za, Univ of Cape Town – YgEoYg

Kahle, Chris, Chris.Kahle@dnr.iowa.gov, Iowa Dept of Natural Resources – Oi

Kaip, Galen M., (915) 747-6817 gkaip@utep.edu, Univ of Texas, El Paso – Yx

Kairies-Beatty, Candace L., (507) 474-5789 ckairiesbeatty@winona.edu, Winona State Univ – GeCl

Kaiser, Daniel E., 612-624-3482 dekaiser@umn.edu, Univ of Minnesota, Twin Cities – Sc

Kaiser, Jan, +44 (0)1603 59 3393 j.kaiser@uea.ac.uk, Univ of East Anglia – Cs

Kaiser, Jason, (435) 865-8275 jasonkaiser@suu.edu, Southern Utah Univ – GzvCg

Kaiser-Bischoff, Ines, 089/2180 4314 kaiser-bischoff@lmu.de, Ludwig-Maximilians-Universitaet Muenchen – Gz

Kaka, Ismail, +96638603879 skaka@kfupm.edu.sa, King Fahd Univ of Petroleum and Minerals – Ys

Kakembo, Vincent, 27 41 504 4516 Vincent.Kakembo@nmmu.ac.za, Nelson Mandela Metropolitan Univ – Sp

Kalakay, Thomas J., 406-657-1101 kalakayt@rocky.edu, Rocky Mountain Coll – GcpGi

Kaldor, Michael, (305) 237-3025 michael.kaldor@mdc.edu, Miami-Dade Coll (Wolfson Campus) – Gg

Kalender, Leyla, 00904242370000-5984 leylakalender@firat.edu.tr, Firat Univ – CgcCs

Kalia, Hemendra N., (702) 295-5767 hemendra_kalia@notes.ymp.gov, Los Alamos National Laboratory – Nm

Kallemeyn, Gregory, (310) 825-3202 Univ of California, Los Angeles – Cg

Kalvoda, Jiri, +420 549 49 4756 dino@sci.muni.cz, Masaryk Univ – PsePm

Kamber, Balz S., + 353 1 8962957 kamberbs@tcd.ie, Trinity Coll – CgGgp
Kambesis, Patricia, (270) 745-5984 pat.kambesis@wku.edu, Western Kentucky Univ – OigHw
Kambewa, Chamunorwa , 078 321 6710 kambewac@tut.ac.za, Tshwane Univ of Tech – CgeOe
Kamenov, George D., (352) 846-3599 kamenov@ufl.edu, Univ of Florida – CgaEm
Kamhi, Samuel R., (718) 460-1476 Long Island Univ, Brooklyn Campus – Gz
Kamilli, Robert J., (520) 670-5576 bkamilli@usgs.gov, Univ of Arizona – EgmCg
Kaminski, Michael A., +966138602344 kaminski@kfupm.edu.sa, King Fahd Univ of Petroleum and Minerals – GuPm
Kammer, Thomas W., (304) 293-5603 tkammer@wvu.edu, West Virginia Univ – Pi
Kamola, Diane, (785) 864-2724 dlkamola@ku.edu, Univ of Kansas – Gs
Kamona, Frederick A., afkamona@unam.na, Univ of Namibia – EgGzCe
Kampf, Anthony R., (213) 763-3328 akampf@nhm.org, Los Angeles County Museum of Natural History – Gz
Kamykowski, Daniel, dan_kamykowski@ncsu.edu, North Carolina State Univ – Ob
Kana, Todd M., (410) 221-8481 kana@hpl.umces.edu, Univ of Maryland – Ob
Kanamori, Hiroo, (626) 395-6914 hiroo@gps.caltech.edu, California Inst of Tech – Ysg
Kanat, Leslie H., (802) 635-1327 kanatl@jsc.vsc.edu, Johnson State Coll – Gce
Kandiah, Ramanitharan, (937) 376-6260 rkandiah@centralstate.edu, Central State Univ – HqSoOi
Kane, Mustapha, 386.754. 4452 mustapha.kane@fgc.edu, Florida Gateway Coll – Og
Kanemasu, Edward T., (706) 542-2151 Univ of Georgia – Sp
Kanfoush, Sharon L., (315) 792-3134 skanfoush@utica.edu, Utica Coll – GsOuGn
Kang, Song-Lak, (806) 834-1139 song-lak.kang@ttu.edu, Texas Tech Univ – OaHg
Kanik, Mustafa, 00904242370000-5964 Firat Univ – NrSo
Kanungo, Sudeep, 801-585-7852 skanungo@egi.utah.edu, Univ of Utah – Pms
Kao, Jim C., (505) 667-9226 Los Alamos National Laboratory – Oa
Kaplan, Alexey, (845) 365-8689 Columbia Univ – Op
Kaplan, Isaac R., (310) 825-5706 irkaplan@ucla.edu, Univ of California, Los Angeles – Csg
Kaplan, Samantha W., (715) 346-4149 skaplan@uwsp.edu, Univ of Wisconsin, Stevens Point – GnPlGd
Kaplinski, Matthew A., (928) 523-9145 Northern Arizona Univ – Yg
Kapp, Jessica, 520-626-5701 jkapp@email.arizona.edu, Univ of Arizona – Gg
Kapp, Paul, (520) 626-8763 pkapp@email.arizona.edu, Univ of Arizona – GtcGm
Kar, Aditya, (912) 825-6844 Fort Valley State Univ – Ct
Kar, Bandana, 601-266-5786 bandana.kar@usm.edu, Univ of Southern Mississippi – Oi
Kara, Hatice, 00904242370000-5965 haticekara@firat.edu.tr, Firat Univ – Cg
Karabinos, Paul, (413) 597-2079 paul.m.karabinos@williams.edu, Williams Coll – Gc
Karanfil, Tanju, (864) 656-1005 tkaranf@clemson.edu, Clemson Univ – NgOnn
Karato, Shun-ichiro, (203) 432-3147 shun-ichiro.karato@yale.edu, Yale Univ – Gy
Karginoglu, Yusuf , 00904242370000-5989 Firat Univ – Eg
Karhu, Juha A., 358-294150834 Juha.Karhu@helsinki.fi, Univ of Helsinki – Csg
Karig, Daniel E., dek9@cornell.edu, Cornell Univ – YrGl
Karkanis, Pano G., (403) 317-0156 Res. (403) 330-9445 Cel. karkanis@telusplanet.net, Univ of Lethbridge – So
Karl, David M., (808) 956-8964 dkarl@soest.hawaii.edu, Univ of Hawai'i, Manoa – Ob
Karl, Tami S., (850) 644-5861 karl@gly.fsu.edu, Florida State Univ – On
Karlen, Douglas L., (515) 294-3336 Doug.Karlen@ars.usda.gov, Iowa State Univ of Science & Tech – So
Karlin, Robert, (775) 784-1770 karlin@mines.unr.edu, Univ of Nevada, Reno – Ou
Karlsson, Haraldur R., 806-834-7978 hal.karlsson@ttu.edu, Texas Tech Univ – Cl
Karlstrom, Karl E., (505) 277-4346 kek1@unm.edu, Univ of New Mexico – Gt
Karmosky, Christopher, Christopher.Karmosky@oneonta.edu, SUNY, Oneonta – OwaOy
Karner, Daniel B., (707) 664-2854 karner@sonoma.edu, Sonoma State Univ – Cc
Karpilo, Jr, Ronald J., (970) 225-3500 ron_karpilo@partner.nps.gov, Colorado State Univ – Oy
Karplus, Marianne, (915) 747-5413 mkarplus@utep.edu, Univ of Texas, El Paso – Ysg
Karson, Jeffrey, 315-443-7976 jakarson@syr.edu, Syracuse Univ – GtcGv
Karuntillake, Suniti, sunitiw@lsu.edu, Louisiana State Univ – Xg
Karwoski, Todd, (301) 405-0084 karwoski@geol.umd.edu, Univ of Maryland – Gg
Kaspar, Thomas A., (515) 294-8873 Tom.Kaspar@ars.usda.gov, Iowa State Univ of Science & Tech – So
Kasse, Kees, +31 20 59 87381 c.kasse@vu.nl, VU Univ Amsterdam – Gms
Kassem, Hachem, +44 (0)23 8059 6205 hachem.kassem@soton.ac.uk, Univ of Southampton – OnGsOi
Kaste, James, 757-221-2591 jmkaste@wm.edu, Coll of William & Mary – ClcSc
Kaster, Mark A., 570-408-5046 mark.kaster@wilkes.edu, Wilkes Univ – Ow
Kasting, James F., (814) 865-3207 jfk4@psu.edu, Pennsylvania State Univ, Univ Park – Oa
Kastner, Miriam, (858) 534-2065 mkastner@ucsd.edu, Univ of California, San Diego – CmlCm
Kaszuba, John P., 307-766-3392 jkaszub1@uwyo.edu, Univ of Wyoming – CgGze
Kath, Randal L., (678) 839-4063 rkath@westga.edu, Univ of West Georgia – GcNgHg
Kato, Terence T., 530-898-5262 tkato@csuchico.edu, California State Univ, Chico – Gt
Katuna, Michael P., katunam@cofc.edu, Coll of Charleston – Gus
Katz, Gabrielle, (828) 262-3000 katzgl@appstate.edu, Appalachian State Univ – Hg
Katz, Miriam E., (518) 276-8521 katzm@rpi.edu, Rensselaer Polytechnic Inst – PmeGu
Katz, Richard, +44-1865-282122 richard.katz@earth.ox.ac.uk, Univ of Oxford – Cg
Katzenstein, Kurt W., (605) 394-2461 Kurt.Katzenstein@sdsmt.edu, South Dakota School of Mines & Tech – NgrOr
Katzman, Danny, (505) 667-0599 katzman@lanl.gov, Los Alamos National Laboratory – Gg
Kauahikaua, James P., (808) 967-7320 Univ of Hawai'i, Manoa – Gv
Kauffman, Chad, 724 938-5760 kauffman@cup.edu, California Univ of Pennsylvania – Oa
Kauffman, Erle G., (812) 855-5154 Indiana Univ, Bloomington – Po
Kaufman, Alan J., (301) 405-0395 kaufman@geol.umd.edu, Univ of Maryland – Csg
Kaufman, Darrell S., (928) 523-7192 darrell.kaufman@nau.edu, Northern Arizona Univ – Gm
Kaufmann, Ronald S., (619) 260-5904 kaufmann@sandiego.edu, Univ of San Diego – Ob
Kaunda, Rennie, 303-273-3772 rkaunda@mines.edu, Colorado School of Mines – Nm
Kaushal, Sujay, (301) 405-0454 skaushal@umd.edu, Univ of Maryland – OuGeHs
Kavage-Adams, Rebecca H., (410) 554-5553 rebecca.adams@maryland.gov, Maryland Department of Natural Resources – GmOiGe
Kavanagh, Janine, +44-151-794-5150 Janine.Kavanagh@liverpool.ac.uk, Univ of Liverpool – Gv
Kavanaugh, Jeffrey, 780-492-1740 jeff.kavanaugh@ualberta.ca, Univ of Alberta – Ng
Kavner, Abby, (310) 206-3675 akavner@ucla.edu, Univ of California, Los Angeles – GyYg
Kawase, Mitsuhiro, (206) 543-0766 kawase@ocean.washington.edu, Univ of Washington – Op
Kay, Richard F., (919) 684-2143 richard.kay@duke.edu, Duke Univ – On
Kay, Robert W., (607) 255-3461 rwk6@cornell.edu, Cornell Univ – Giz
Kay, Suzanne M., (607) 255-4701 smk16@cornell.edu, Cornell Univ – Gp
Kaye, John M., (662) 325-3915 Mississippi State Univ – Gg
Kaygili, Sibel, 00904242370000-5962 skaygili@firat.edu.tr, Firat Univ – GgPgs
Kays, Marvin A., (541) 346-4578 makays@oregon.uoregon.edu, Univ of Oregon – GpcGt
Kazimoto, Emmanuel O., +255222410013 ekazimoto@udsm.ac.tz,

Univ of Dar es Salaam – EgGpCg
Kazmer, Miklos, +36-1-372-2500 ext. 8627 mkazmer@gmail.com, Eotvos Lorand Univ – PgoGh
Keables, Michael J., (303) 871-2653 michael.keables@du.edu, Univ of Denver – Oa
Keach, William, 801-585-1717 bkeach@egi.utah.edu, Univ of Utah – GoYes
Keach II, R. William, 801-857-7728 bkeach@byu.edu, Brigham Young Univ – YeEoGg
Keaffaber, J. J., (407) 768-8000 Florida Inst of Tech – Oc
Keala, Lori, (909) 621-8675 lkeala@pomona.edu, Pomona Coll – On
Kean, Jr., William F., 414-229-5231 wkean@uwm.edu, Univ of Wisconsin, Milwaukee – Ym
Kearney, Kenneth, 312-413-3655 kkearn3@uic.edu, Univ of Illinois at Chicago – Cg
Kearney, Micheal S., (301) 405-4057 kearneym@umd.edu, Univ of Maryland – Gu
Kearns, Lance E., 540-568-6421 kearnsle@jmu.edu, James Madison Univ – Gz
Keating, Elizabeth, (505) 665-6714 ekeating@lanl.gov, Los Alamos National Laboratory – Gg
Keating, Kristina M., 973-353-1263 kmkeat@andromeda.rutgers.edu, Rutgers, The State Univ of New Jersey, Newark – Yg
Keating, Martha E., (781) 891-2980 mkeating@bentley.edu, Bentley Univ – On
Keaton, Jeffrey R., (801) 581-8218 Univ of Utah – On
Keattch, Sharen , 509-359-7358 skeattch@ewu.edu, Eastern Washington Univ – Gg
Keatts, Merida, 330-672-2897 mkeatts@kent.edu, Kent State Univ – On
Keays, Reid R., (705) 675-1151 Laurentian Univ, Sudbury – Em
Kebbede, Girma, 413-538-2004 gkebbede@mtholyoke.edu, Mount Holyoke Coll – Oy
Keefer, Donald A., (217) 244-2786 dkeefer@illinois.edu, Univ of Illinois, Urbana-Champaign – HwOn
Keeling, David J., (270) 745-4555 david.keeling@wku.edu, Western Kentucky Univ – Og
Keeling, Ralph F., (858) 534-7582 rkeeling@ucsd.edu, Univ of California, San Diego – Oa
Keeling, Robyn, (801) 537-3333 rkeeling@utah.gov, Utah Geological Survey – On
Keen, Kerry L., (715) 425-3729 kerry.l.keen@uwrf.edu, Univ of Wisconsin, River Falls – Hw
Keen-Zebert, Amanda, 775-673-7434 akz@dri.edu, Desert Research Inst – Oy
Keene, Deborah A., 205-348-3334 dakeene@ua.edu, Univ of Alabama – Gga
Keene, William, (804) 924-0586 wck@virginia.edu, Univ of Virginia – Oa
Keeney, Dennis R., (515) 294-8066 drkeeney@iastate.edu, Iowa State Univ of Science & Tech – So
Keesee, Robert G., (518) 442-4566 rgk@atmos.albany.edu, SUNY, Albany – Oa
Kehew, Alan E., (269) 387-5495 alan.kehew@wmich.edu, Western Michigan Univ – GmlHw
Kehoe-Forutan, Sandra J., (570) 389-4106 kehoe@bloomu.edu, Bloomsburg Univ – Onu
Keigwin, Lloyd D., (508) 289-2784 lkeigwin@whoi.edu, Woods Hole Oceanographic Institution – CsGu
Keil, Charles, (630) 752-7271 chris.keil@wheaton.edu, Wheaton Coll – On
Keil, Richard G., (206) 616-1947 rickkeil@ocean.washington.edu, Univ of Washington – Oc
Keir, Derek, +44 (023) 8059 6614 D.Keir@soton.ac.uk, Univ of Southampton – Gtv
Keith, Jeffrey D., 801-422-2189 jeff_keith@byu.edu, Brigham Young Univ – Eg
Kelemen, Peter B., (845) 365-8728 peterk@ldeo.columbia.edu, Columbia Univ – GiCg
Kelkar, Sharad, (505) 667-4639 kelkar@vega.lanl.gov, Los Alamos National Laboratory – Np
Kelleher, Cole, 612-626-0505 Univ of Minnesota, Twin Cities – Oi
Keller, C. Kent, (509) 335-3040 ckkeller@wsu.edu, Washington State Univ – Hw
Keller, Dianne M., (315) 228-7893 dkeller@colgate.edu, Colgate Univ – Gz
Keller, Edward A., (805) 893-4207 keller@geol.ucsb.edu, Univ of California, Santa Barbara – Gt
Keller, G. Randy, (405) 255-0608 grkeller@ou.edu, Univ of Oklahoma – YgGtEo
Keller, George H., george.keller@oregonstate.edu, Oregon State Univ – YgGus
Keller, Gerta, (609) 258-4117 gkeller@princeton.edu, Princeton Univ – Pm
Keller, Jean, (845) 938-4185 United States Military Academy – On
Keller, John E., 702-651-5887 john.keller@csn.edu, Coll of Southern Nevada - West Charleston Campus – GeHw
Keller, Klaus, (814) 865-6718 kkeller@geosc.psu.edu, Pennsylvania State Univ, Univ Park – Og
Keller, Linda M., (608) 265-2209 lmkeller@wisc.edu, Univ of Wisconsin, Madison – Oa
Keller, Randall A., 541-737-7648 kellerr@geo.oregonstate.edu, Oregon State Univ – Gu
Keller, Jr., C. F., (505) 667-0920 cfk@lanl.gov, Los Alamos National Laboratory – Oa
Kelley, Alice R., (207) 581-2056 akelley@maine.edu, Univ of Maine – GalGm
Kelley, Cheryl A., (573) 882-8813 KelleyC@missouri.edu, Univ of Missouri – Co
Kelley, Christopher, (808) 956-7437 ckelley@hawaii.edu, Univ of Hawai'i, Manoa – Ob
Kelley, Dan, (902) 494-1694 dan.kelley@dal.ca, Dalhousie Univ – Op
Kelley, Deborah S., (206) 543-9279 kelley@ocean.washington.edu, Univ of Washington – Ou
Kelley, Joseph T., (207) 581-2162 jtkelley@maine.edu, Univ of Maine – Gua
Kelley, Neil P., neil.p.kelley@vanderbilt.edu, Vanderbilt Univ – GgOgPg
Kelley, Patricia H., (910) 962-7406 kelleyp@uncw.edu, Univ of North Carolina Wilmington – Pgi
Kelley, Shari, (575) 661-6171 sakelley@nmbg.nmt.edu, New Mexico Inst of Mining & Tech – Yg
Kelley, Shari A., 505-412-9269 sakelley@nmbg.nmt.edu, New Mexico Inst of Mining and Tech – Gt
Kelley, Simon, +44 (0)1908 653009 x 53009 simon.kelley@open.ac.uk, The Open Univ – Cs
Kelling, Keith A., (608) 263-2795 kkelling@wisc.edu, Univ of Wisconsin, Madison – So
Kellman, Lisa M., (902) 867-5086 lkellman@stfx.ca, Saint Francis Xavier Univ – CsOs
Kellner, Patricia, pakellner@socal.rr.com, Los Angeles Harbor Coll – Og
Kellogg, James N., (803) 777-4501 kellogg@geol.sc.edu, Univ of South Carolina – Yg
Kellogg, Louise H., (530) 752-3690 kellogg@ucdavis.edu, Univ of California, Davis – Yg
Kelly, Bryce, bryce.kelly@unsw.edu.au, Univ of New South Wales – HwYg
Kelly, D. Clay, (608) 262-1698 ckelly@geology.wisc.edu, Univ of Wisconsin, Madison – Pm
Kelly, Jacque L., (912) 478-8677 JKelly@GeorgiaSouthern.edu, Georgia Southern Univ – Cl
Kelly, John, 203-479-4822 jkelly@newhaven.edu, Univ of New Haven – Obn
Kelly, Kathryn A., (206) 543-9810 kkelly@apl.washington.edu, Univ of Washington – Op
Kelly, Kimberly, (240) 567-5227 Kimberly.Kelly@montgomeryColl.edu, Montgomery Coll – On
Kelly, Maria, (425) 640-1918 mkelly@edcc.edu, Edmonds Community Coll – Og
Kelly, Meredith, 603-646-9647 Meredith.Kelly@dartmouth.edu, Dartmouth Coll – Gl
Kelly, Sherrie, 315-229-5851 skelly@stlawu.edu, St. Lawrence Univ – On
Kelly, Walton R., (217) 333-3729 Illinois State Water Survey – Hw
Kelly, William C., (734) 764-1435 billkell@umich.edu, Univ of Michigan – Eg
Kelsch, Jesse, (432) 837-8657 jkelsch@sulross.edu, Sul Ross State Univ – Gct
Kelsey, Harvey M., (707) 826-3991 hmk1@humboldt.edu, Humboldt State Univ – Gmt
Kelso, Paul R., (906) 635-2158 pkelso@lssu.edu, Lake Superior State Univ – YmGct
Kelty, Thomas, (562) 985-4589 t.kelty@csulb.edu., California State Univ, Long Beach – Gc
Kemeny, John M., (520) 621-4448 kemeny@email.arizona.edu, Univ of Arizona – Nr
Kemmerly, Phillip R., (931) 221-7471 kemmerlyp@apsu.edu, Austin Peay State Univ – GmeHg
Kemp, Alan, +44 (0)23 80592788 aesk@noc.soton.ac.uk, Univ of Southampton – Pe

Kemp, Andrew, (617) 627-0869 andrew.kemp@tufts.edu, Tufts Univ – OnPmOg
Kemp, Paul, (808) 956-6220 paulkemp@hawaii.edu, Univ of Hawai'i, Manoa – Ob
Kempton, Pamela D., (785) 532-6743 pkempton@ksu.edu, Kansas State Univ – GiCt
Kendrick, David C., 315-781-3929 kendrick@hws.edu, Hobart & William Smith Colls – Po
Kendrick, Katherine J., (951) 276-4418 kendrick@gps.caltech.edu, Univ of California, Riverside – Gm
Kenduiywo, Benson K., bkenduiywo@jkuat.ac.ke, Jomo Kenyatta Univ of Agriculture & Tech – OriYd
Kenig, Fabien, (312) 996-3020 fkenig@uic.edu, Univ of Illinois at Chicago – Co
Kenna, Timothy, (845)-365-8513 Columbia Univ – Cg
Kennedy, Andrew, andrew.b.kennedy.117@nd.edu, Univ of Notre Dame – On
Kennedy, Ann C., (509) 335-1554 akennedy@wsu.edu, Washington State Univ – Sb
Kennedy, Ben, +64 3 3667001 Ext 7775 ben.kennedy@canterbury.ac.nz, Univ of Canterbury – Gv
Kennedy, Christina B., tina.kennedy@nau.edu, Northern Arizona Univ – On
Kennedy, Gareth, +44 01326 371876 G.A.Kennedy@exeter.ac.uk, Exeter Univ – Nm
Kennedy, Linda, 570-662-4609 lkennedy@mansfield.edu, Mansfield Univ – Oy
Kennedy, Lisa M., (540) 231-1422 kennedy1@vt.edu, Virginia Polytechnic Inst & State Univ – Pe
Kennedy, Lori, (604) 822-1811 lkennedy@eos.ubc.ca, Univ of British Columbia – Gc
Kennedy, Robert, (541) 737-6332 rkennedy@coas.oregonstate.edu, Oregon State Univ – Ori
Kennedy, Timothy T., 715-346-4934 tkennedy@uwsp.edu, Univ of Wisconsin, Stevens Point – Oir
Kennel, Charles F., (310) 825-4018 Univ of California, Los Angeles – On
Kennel, Charles F., (858) 822-6424 ckennel@ucsd.edu, Univ of California, San Diego – Oa
Kenney, Robert D., (401) 874-6664 rkenney@gso.uri.edu, Univ of Rhode Island – Ob
Kenny, Ray, (970) 247-7462 kenny_r@fortlewis.edu, Fort Lewis Coll – Gm
Kent, Adam J., (541) 737-1205 adam.kent@geo.oregonstate.edu, Oregon State Univ – GiCga
Kent, Dennis V., (848) 445-7049 dvk@rci.rutgers.edu, Rutgers, The State Univ of New Jersey – Ym
Kent, Douglas, 405-377-0166 Oklahoma State Univ – Hw
Kent, Graham, (775) 784-4977 gkent@seismo.unr.edu, Univ of Nevada, Reno – Ys
Kenyon, Patricia M., (212) 650-6472 Graduate School of the City Univ of New York – Yg
Kepic, Anton W., +61 8 9266-7503 A.Kepic@curtin.edu.au, Curtin Univ – YexYg
Keppens, Edward, 32-2-6293395 ekeppens@vub.ac.be, Vrije Univ Brussel – CsGeg
Keppie, J. Duncan, 56224303 duncan@servidor.unam.mx, Universidad Nacional Autonoma de Mexico – Gt
Kerans, Charles, (512) 471-1368 charles.kerans@beg.utexas.edu, Univ of Texas at Austin, Jackson School of Geosciences – Gd
Kerans, Charles, (512) 471-4282 ckerans@jsg.utexas.edu, Univ of Texas at Austin – GsrGo
Kerestedjian, Thomas N., +359 2 979 2244 thomas@geology.bas.bg, Bulgarian Academy of Sciences – GzeEg
Kern, Anikó, anikoc@nimbus.elte.hu, Eotvos Lorand Univ – OrwOa
Kerp, Hans, +49-251-83-23966 kerp@uni-muenster.de, Universitaet Muenster – PblGg
Kerr, A C., kerra@cf.ac.uk, Cardiff Univ – Gx
Kerr, Andrew, +44(0)29 208 74578 KerrA@cardiff.ac.uk, Univ of Wales – Og
Kerr, Dennis R., 918 631 3020 dennis-kerr@utulsa.edu, The Univ of Tulsa – Gs
Kerrick, Derrill M., (814) 865-7574 kerrick@geosc.psu.edu, Pennsylvania State Univ, Univ Park – Gp
Kershaw, G. Peter, (780) 492-0346 peter.kershaw@ualberta.ca, Univ of Alberta – Gm
Kerwin, Charles M., 603.358.2405 ckerwin@keene.edu, Keene State Coll – Gg
Kerwin, Michael W., 303-871-3998 mkerwin@du.edu, Univ of Denver – Gg
Kesel, Richard H., (225) 578-5880 gakesel@lsu.edu, Louisiana State Univ – Hg
Keskinen, Mary J., (907) 474-7769 mjkeskinen@alaska.edu, Univ of Alaska, Fairbanks – Gxz
Kesler, Stephen E., (734) 763-5057 skesler@umich.edu, Univ of Michigan – Eg
Kesseli, Rick, rick.kesseli@umb.edu, Univ of Massachusetts, Boston – Obn
Kessinger, Walter P., (337) 984-3554 Univ of Louisiana at Lafayette – Pm
Kessler, Fritz, (301) 687-4266 fkessler@frostburg.edu, Frostburg State Univ – Or
Kessler, William S., (206) 526-6221 kessler@pmel.noaa.gov, Univ of Washington – Op
Ketcham, Richard A., (512) 471-6942 ketcham@jsg.utexas.edu, Univ of Texas at Austin – Ggq
Kettler, Richard M., (402) 472-0882 rkettler1@unl.edu, Univ of Nebraska, Lincoln – Cl
Kettmann, Elizabeth, (707) 664-2334 meyerel@sonoma.edu, Sonoma State Univ – On
Kevan, Peter, pkevan@uoguelph.ca, Univ of Guelph – On
Key, Doug, 7607441150 ext.2515 dkey@palomar.edu, Palomar Coll – Oy
Key, Jeffrey R., (608) 263-2605 jkey@ssec.wisc.edu, Univ of Wisconsin, Madison – Ora
Key, Kerry, (858) 822-2975 kkey@ucsd.edu, Univ of California, San Diego – Yx
Key, Jr., Marcus M., (717) 245-1448 key@dickinson.edu, Dickinson Coll – PiGsa
Keyantash, John, (310) 243-2363 jkeyantash@csudh.edu, California State Univ, Dominguez Hills – HgOae
Keyser, Daniel, (518) 442-4559 keyser@atmos.albany.edu, SUNY, Albany – Oa
Khairoutdinov, Marat, (631) 632-6339 marat.khairoutdinov@stonybrook.edu, SUNY, Stony Brook – Oa
Khalequzzaman, Md., (570) 484-2075 mkhalequ@lhup.edu, Lock Haven Univ – HwOni
Khalil, Mohamed, 406-496-4716 mkhalil@mtech.edu, Montana Tech – YegGe
Khalil Ebeid, Khalil I., 002-03-3921595 kebeid@yahoo.com, Alexandria Univ – GgEg
Khan, Belayet H., 217-581-6246 bhkhan@eiu.edu, Eastern Illinois Univ – Oyw
Khan, Latif A., 217-244-2383 info@isgs.illinois.edu, Illinois State Geological Survey – Nx
Khan, Mohammad Wahdat Y., (982) 719-7331 mwykhan@rediffmail.com, Pt. Ravishankar Shukla Univ – GdEgCl
Khan, Shuhab D., (713) 743-5404 sdkhan@uh.edu, Univ of Houston – OrGtYg
Khanbilvardi, Reza M., (215) 650-8009 Graduate School of the City Univ of New York – Hy
Khandaker, Nazrul I., (718) 262-2079 nkhandaker@york.cuny.edu, York Coll (CUNY) – GdeOe
Khawaja, Ikram U., john@cis.ysu.edu, Youngstown State Univ – Ec
Khosrowpanah, Shahram, (671) 735-2694 khosrow@uog.edu, Univ of Guam – Hs
Khurana, Krishan, (310) 825-8240 Univ of California, Los Angeles – On
Kiage, Lawrence M., (404) 413-5777 geolkk@langate.gsu.edu, Georgia State Univ – OyPlOr
Kidd, David, +44 020 8417 62541 david.kidd@kingston.ac.uk, Kingston Univ – Oiy
Kidder, David L., (740) 593-1108 kidder@ohio.edu, Ohio Univ – Gs
Kidder, T.R., trkidder@wustl.edu, Washington Univ in St. Louis – On
Kidwell, Susan M., (773) 702-3008 skidwell@midway.uchicago.edu, Univ of Chicago – Gs
Kiefer, Boris, 575 646 1932 bkiefer@nmsu.edu, New Mexico State Univ, Las Cruces – GyYgOm
Kieffer, Bruno, bkieffer@eos.ubc.ca, Univ of British Columbia – Cs
Kieffer, Susan W., (217) 244-6206 skieffer@illinois.edu, Univ of Illinois, Urbana-Champaign – Gv
Kienast, Markus, (902) 494-8338 markus.kienast@dal.ca, Dalhousie Univ – Cs
Kienast, Stephanie, (902) 494-2203 stephanie.kienast@dal.ca, Dalhousie Univ – Gu
Kiene, Ronald P., (251) 861-7526 rkiene@jaguar1.usouthal.edu, Univ of South Alabama – Oc
Kientop, Greg A., 217-265-6581 gkientop@illinois.edu, Univ of Illinois, Urbana-Champaign – Ge

Kienzle, Stefan, stefan.kienzle@uleth.ca, Univ of Lethbridge – Oi
Kiesel, Diann, (608) 355-5223 diann.kiesel@uwc.edu, Univ of Wisconsin Colls – GgOyGg
Kieu, Chanh Q., (812) 856-5704 ckieu@indiana.edu, Indiana Univ, Bloomington – Oa
Kifer, Lauri A., (585) 395-2636 lmulley@brockport.edu, SUNY, The Coll at Brockport – On
Kift, Richard, +44 0161 306-8770 Richard.Kift@manchester.ac.uk, Univ of Manchester – Oa
Kijko, Andrzej, +27 12 420 3613 andrzej.kijko@up.ac.za, Univ of Pretoria – Ysg
Kilburn, Chris, +44 020 7679 37194 c.kilburn@ucl.ac.uk, Univ Coll London – Yx
Kilcoyne, John R., (303) 556-4258 Metropolitan State Coll of Denver – Oy
Kile, Susan, 217-581-2626 skkile@eiu.edu, Eastern Illinois Univ – On
Kilibarda, Zoran, (219) 980-6753 zkilibar@iun.edu, Indiana Univ Northwest – GsmGd
Kilinc, Attila I., (513) 556-5967 attila.kilinc@uc.edu, Univ of Cincinnati – CpGvCg
Killorn, Randy J., (515) 294-1923 rkillorn@iastate.edu, Iowa State Univ of Science & Tech – So
Kilroy, Kathyrn, 701-858-3114 kathryn.kilroy@minotstateu.edu, Minot State Univ – Hg
Kilsby, Chris, +44 (0) 191 208 5614 chris.kilsby@ncl.ac.uk, Univ of Newcastle Upon Tyne – Hg
Kim, Eunhye, 303-273-3428 ekim1@mines.edu, Colorado School of Mines – Nm
Kim, Hyemi, (404) 894-1738 hyemi.kim@eas.gatech.edu, Georgia Inst of Tech – Oa
Kim, Jonathan, (802) 522-5401 jon.kim@vermont.gov, Agency of Natural Resources, Dept of Environmental Conservation – GetCg
Kim, Keonho, 432-685-4739 kkim@midland.edu, Midland Coll – GgOwPi
Kim, Kwangmin, kimkm@email.arizona.edu, Univ of Arizona – NmrNx
Kim, Saewung, (949)824-4531 saewungk@uci.edu, Univ of California, Irvine – Oa
Kim, Sang-Tae, (905) 525-9140 (Ext. 26494) sangtae@mcmaster.ca, McMaster Univ – Cg
Kim, Sora L., 859-323-4463 sora.kim@uky.edu, Univ of Kentucky – CsPeo
Kim, Stacy, 831-771-4400 skim@mlml.calstate.edu, Moss Landing Marine Laboratories – On
Kim, Won-Young, (845) 365-8387 Columbia Univ – Ys
Kim, Wonsuck, 512-471-4203 delta@jsg.utexas.edu, Univ of Texas at Austin – Gsr
Kimball, Bryn, 509-527-4951 kimballb@whitman.edu, Whitman Coll – ClEgGz
Kimball, Matthew, matt@belle.baruch.sc.edu, Univ of South Carolina – Ob
Kimber, Clarissa T., (409) 845-7141 Texas A&M Univ – Oy
Kimbro, David, 7815817370 x310 d.kimbro@neu.edu, Northeastern Univ – On
Kimbrough, David L., (619) 594-1385 dkimbrough@mail.sdsu.edu, San Diego State Univ – Cc
Kimerling, A. Jon, kimerlia@geo.oregonstate.edu, Oregon State Univ – Oir
Kincaid, Christopher, (401) 874-6571 kincaid@gso.uri.edu, Univ of Rhode Island – Ou
Kineke, Gail C., (617) 552-3655 gail.kineke.1@bc.edu, Boston Coll – On
King, Andrew, alking@uic.edu, Univ of Illinois at Chicago – Og
King, Carey, 512-471-5468 cking@jsg.utexas.edu, Univ of Texas at Austin – Eog
King, John, (401) 874-6594 jking@gso.uri.edu, Univ of Rhode Island – Ou
King, Jonathan K., (801) 537-3354 jonking@utah.gov, Utah Geological Survey – Gg
King, Kenneth M., (519) 824-4120 (Ext. 52453) kenmking@rogers.com, Univ of Guelph – Oa
King, Martin, +44 1784 414038 M.King@rhul.ac.uk, Univ of London, Royal Holloway & Bedford New Coll – Oa
King, Norman R., (812) 464-1794 nking@usi.edu, Univ of Southern Indiana – Gr
King, Peter, +44 20 759 47362 peter.king@imperial.ac.uk, Imperial Coll – Np
King, Robert W., (617) 253-7064 rwk@chandler.mit.edu, Massachusetts Inst of Tech – Yg
King, Scott D., 540-231-6521 Virginia Polytechnic Inst & State Univ – Yg
King, Jr., David T., (334) 844-4882 kingdat@auburn.edu, Auburn Univ – GrOn
King-Rundel, Judith A., 310 243-3205 dendrochick@aol.com, California State Univ, Dominguez Hills – Oe
Kingdon, Kevin, kevin.kingdon@skyresearch.com, Univ of British Columbia – Yg
Kinkead, Scott, (505) 665-1760 Los Alamos National Laboratory – On
Kinnaird, Judith A., 27117176583 judith.kinnaird@wits.ac.za, Univ of the Witwatersrand – EmGi
Kinner, David A., (828) 227-3821 dkinner@wcu.edu, Western Carolina Univ – HgOe
Kinnicutt, Patrick, 989-774-2294 kinni1p@cmich.edu, Central Michigan Univ – Gq
Kinsland, Gary L., (337) 288-6421 glkinsland@louisiana.edu, Univ of Louisiana at Lafayette – YgeGt
Kinter, Jim, (703) 993-5700 ikinter@gmu.edu, George Mason Univ – Oa
Kinvig, Helen, +440151 795 4657 H.Kinvig@liverpool.ac.uk, Univ of Liverpool – Gs
Kipper, Jay P., (512) 475-9505 jay.kipper@beg.utexas.edu, Univ of Texas at Austin, Jackson School of Geosciences – On
Kipphut, George W., (270) 809-2847 gkipphut@murraystate.edu, Murray State Univ – GeOg
Kirby, Carl S., (570) 577-1385 kirby@bucknell.edu, Bucknell Univ – Cl
Kirby, Matthew E., (657) 278-2158 mkirby@fullerton.edu, California State Univ, Fullerton – GnPeGg
Kirchgasser, William T., (315) 267-2296 kirchgwt@potsdam.edu, SUNY Potsdam – Ps
Kirchner, James W., (510) 643-8559 kirchner@seismo.berkeley.edu, Univ of California, Berkeley – Ge
Kirimi, Fridah K., fkirimi@jkuat.ac.ke, Jomo Kenyatta Univ of Agriculture & Tech – OiiOr
Kirkby, Kent C., (612) 624-1392 kirkby@tc.umn.edu, Univ of Minnesota, Twin Cities – Gg
Kirkham, Mary Beth, (785) 532-0422 mbk@ksu.edu, Kansas State Univ – Sp
Kirkham, Randy R., (509) 372-6038 rr_kirkham@pnl.gov, Pacific Northwest National Laboratory – Hq
Kirkland, Brenda L., 662-268-1032 Ext 228 kirkland@geosci.msstate.edu, Mississippi State Univ – Gd
Kirkland, James I., (801) 537-3307 nrugs.jkirklan@state.ut.us, Utah Geological Survey – Pg
Kirkpatrick, Cody, (812) 855-3481 codykirk@indiana.edu, Indiana Univ, Bloomington – Oa
Kirkpatrick, James, (514) 398-7442 james.kirkpatrick@mcgill.ca, McGill Univ – Gct
Kirlin, R. Lynn, (250) 721-8681 lkirlin@comcast.net, Univ of British Columbia – Oa
Kirschenfeld, Taylor, 850-474-2746 Univ of West Florida – Ob
Kirschvink, Joseph L., (626) 395-6136 kirschvink@caltech.edu, California Inst of Tech – Pg
Kirshbaum, Daniel, 514-398-3347 daniel.kirshbaum@mcgill.ca, McGill Univ – Oaw
Kirste, Dirk, 604-291-5365 dkirste@sfu.ca, Simon Fraser Univ – Cg
Kirstein, Linda, +44 (0) 131 650 4838 Linda.Kirstein@ed.ac.uk, Edinburgh Univ – Gt
Kirtland Turner, Sandra, (951) 827-3191 sandra.kirtlandturner@ucr.edu, Univ of California, Riverside – Cs
Kirwan, Matthew L., 804-684-7054 kirwan@vims.edu, Coll of William & Mary – Gm
Kish, Stephen A., (850) 644-2064 kish@gly.fsu.edu, Florida State Univ – Eg
Kishcha, Pavel, 972-3-6407411 pavelk@post.tau.ac.il, Tel Aviv Univ – Oa
Kisila, Ben O., (540) 654-1107 bkisila@umw.edu, Univ of Mary Washington – Hs
Kiss, Timea, +3662544156 kisstimi@gmail.com, Univesity of Szeged – GmOy
Kissel, David E., (706) 542-0900 Univ of Georgia – Sc
Kissin, Stephen A., (807) 343-8220 stephen.kissin@lakeheadu.ca, Lakehead Univ – Em
Kisvarsanyi, Geza K., (573) 341-4616 Missouri Univ of Science and Tech – Em
Kitajima, Hiroko, 979.458.2717 kitaji@tamu.edu, Texas A&M Univ – YxSpGc
Kitchell, James F., (608) 262-9512 Univ of Wisconsin, Madison – Ob
Kitchen, Newell R., 573-882-1138 kitchenn@missouri.edu,

Univ of Missouri, Columbia – Sp
Kite, J. Steven, (304) 293-5603 steve.kite@mail.wvu.edu, West Virginia Univ – GmHsGa
Kivelson, Margaret G., (310) 825-3435 mkivelson@igpp.ucla.edu, Univ of California, Los Angeles – Xy
Kiver, Eugene P., (509) 359-7959 eugene.kiver@ewu.edu, Eastern Washington Univ – Gl
Klaassen, Gary P., (416)736-2100 #77727 gklaass@yorku.ca, York Univ – Oa
Klancher, Jacki, 307-855-2205 jklanche@cwc.edu, Central Wyoming Coll – Oi
Klapper, Gilbert, (847) 732-1859 g-klapper@northwestern.edu, Northwestern Univ – Pm
Klapper, Gilbert, g-klapper@northwestern.edu, Northwestern Univ – PgiPo
Klasik, John A., 909-869-3453 jaklasik@csupomona.edu, California State Polytechnic Univ, Pomona – Gu
Klaus, Adam, (979) 845-3055 aklaus@odpemail.tamu.edu, Texas A&M Univ – Gu
Klaus, James S., 305-284-3426 j.klaus@miami.edu, Univ of Miami – Gg
Klee, Thomas M., (813) 253-7259 tklee@hccfl.edu, Hillsborough Community Coll – Gg
Kleffner, Mark A., (614) 295-8208 kleffner.1@osu.edu, Ohio State Univ – PmsGr
Klein, Andrew G., (979) 845-7179 klein@geog.tamu.edu, Texas A&M Univ – OriOl
Klein, Cornelis, (505) 277-2023 cklein@unm.edu, Univ of New Mexico – Gz
Klein, Emily M., (919) 684-5965 ek4@duke.edu, Duke Univ – Gi
Klein, Frieder, fklein@whoi.edu, Woods Hole Oceanographic Institution – Cm
Kleinhans, Frederick W., 317-290-1689 fkleinha@iupui.edu, Indiana Univ / Purdue Univ, Indianapolis – Xg
Kleinspehn, Karen L., 612-624-0537 klein004@umn.edu, Univ of Minnesota, Twin Cities – Gt
Kleiss, Harold J., (919) 515-2643 North Carolina State Univ – Sd
Klemas, Victor, (302) 831-8256 klemas@udel.edu, Univ of Delaware – Or
Klemetti, Erik, 740-587-5788 klemettie@denison.edu, Denison Univ – GvxGg
Klemow, Kenneth M., (570) 408-4758 kenneth.klemow@wilkes.edu, Wilkes Univ – Py
Klemperer, Simon L., (650) 723-8214 Stanford Univ – Gt
Klepeis, Keith A., (802) 656-0247 kklepeis@uvm.edu, Univ of Vermont – Gct
Kliche, Donna V., 605-394-1957 Donna.Kliche@sdsmt.edu, South Dakota School of Mines & Tech – OaaOa
Klimczak, Christian, klimczak@uga.edu, Univ of Georgia – GcmGt
Klinck, John M., (757) 683-6005 Klinck@ccpo.odu.edu, Old Dominion Univ – Op
Klinger, Barry, (703) 993-9227 bklinger@gmu.edu, George Mason Univ – Op
Klinkhammer, Gary, (541) 737-5209 gklinkhammer@coas.oregonstate.edu, Oregon State Univ – CaEmCg
Kloosterziel, Rudolf C., 808-956-7668 rudolf@soest.hawaii.edu, Univ of Hawai'i, Manoa – Op
Klosterman, Sue, (937) 229-2661 Sue.Klosterman@notes.udayton.edu, Univ of Dayton – Gg
Kluitenberg, Gerard J., (785) 532-7215 gjk@ksu.edu, Kansas State Univ – Sp
Klymak, Jody, 250-721-6120 jklymak@uvic.ca, Univ of Victoria – Op
Knaack, Charles, (509) 335-6742 knaack@wsu.edu, Washington State Univ – Ca
Knaeble, Alan, (612) 626-2495 knaeb001@umn.edu, Univ of Minnesota – Gl
Knap, Anthony, 979-862-2323 ext 111 tknap@tamu.edu, Texas A&M Univ – Oc
Knapp, Camelia, 777-8491 camelia@geol.sc.edu, Univ of South Carolina – YeGt
Knapp, Elizabeth P., (540) 458-8867 knappe@wlu.edu, Washington & Lee Univ – Cl
Knapp, H. Vernon, (217) 333-4423 vknapp@uiuc.edu, Illinois State Water Survey – Hs
Knapp, James H., (803) 777-6886 knapp@geol.sc.edu, Univ of South Carolina – Ye
Knapp, Richard B., (925) 423-3328 knapp4@llnl.gov, Lawrence Livermore National Laboratory – Ng
Knapp, Roy M., 405-325-6829 knapp@ou.edu, Univ of Oklahoma – Np
Knapp, Sibylle, +49 (89) 289 25895 sibylle.knap@)tum.de, Technische Universitaet Muenchen – GsgGm
Knapp, Warren, (607) 255-3034 wwk2@cornell.edu, Cornell Univ – Oa
Knauss, John A., (401) 874-6141 jknauss@gso.uri.edu, Univ of Rhode Island – Op
Knauss, Virginia L., (310) 243-3377 vknauss@csudh.edu, California State Univ, Dominguez Hills – On
Knauth, L. Paul, (480) 965-2867 knauth@asu.edu, Arizona State Univ – Cs
Kneas, David, kneas@mailbox.sc.edu, Univ of South Carolina – On
Kneeshaw, Tara A., 616-331-8996 kneeshta@gvsu.edu, Grand Valley State Univ – ClGe
Knell, Michael J., (203) 392-5836 knellm1@southernct.edu, Southern Connecticut State Univ – PvGsOg
Knepp, Rex A., 217-244-2422 knepp@isgs.uiuc.edu, Illinois State Geological Survey – Go
Knight, Allen W., (530) 752-0453 aknight557@aol.com, Univ of California, Davis – Hs
Knight, David, (518) 442-4204 knight@atmos.albany.edu, SUNY, Albany – Oa
Knight, Mona M., (217) 244-2390 mmknight@illinois.edu, Univ of Illinois, Urbana-Champaign – On
Knight, Paul, (814) 863-4229 knight@mail.meteo.psu.edu, Pennsylvania State Univ, Univ Park – Ow
Knight, Rosemary J., (650) 723-4746 Stanford Univ – Hy
Knight, Tiffany, tknight@biology2.wustl.edu, Washington Univ in St. Louis – Pe
Knipe, Rob, +44(0) 113 34 35208 knipe@rdr.leeds.ac.uk, Univ of Leeds – Gt
Knippler, Katherine A., (410) 554-5543 katherine.knippler@maryland.gov, Maryland Department of Natural Resources – On
Knittle, Elise, (831) 459-4949 eknittle@pmc.ucsc.edu, Univ of California, Santa Cruz – Gy
Knizek, Martin, +420 549 49 6298 kniza@sci.muni.cz, Masaryk Univ – NgmNr
Knoll, Andrew H., (617) 495-9306 aknoll@harvard.edu, Harvard Univ – Pb
Knoll, Martin A., (931) 598-1713 mknoll@sewanee.edu, Sewanee: Univ of the South – Hw
Knopf, Daniel A., (631) 632-3092 Daniel.Knopf@stonybrook.edu, SUNY, Stony Brook – Oa
Knopoff, Leon, (310) 825-1885 knopoff@physics.ucla.edu, Univ of California, Los Angeles – Ys
Knott, Jeffrey R., (657) 278-5547 jknott@fullerton.edu, California State Univ, Fullerton – Gmg
Knowles, Charles E., (919) 515-7943 ernie_knowles@ncsu.edu, North Carolina State Univ – Op
Knowlton, Amy, aknowlton@neaq.org, Univ of Massachusetts, Boston – Ob
Knox, Larry W., (931) 372-3523 lknox@tntech.edu, Tennessee Tech Univ – Pmg
Knudsen, Andrew, (920) 832-6731 knudsena@lawrence.edu, Lawrence Univ – Cg
Knudsen, Guy, gknudsen@uidaho.edu, Univ of Idaho – Sb
Knudsen, Tyler R., (435) 865-9036 tylerknudsen@utah.gov, Utah Geological Survey – Ng
Knudstrup, Renee, rknudstrup@salemstate.edu, Salem State Univ – Ca
Knuepfer, Peter L. K., (607) 777-2389 Binghamton Univ – Gt
Knutson, Heather, 626.395.4268 hknutson@caltech.edu, California Inst of Tech – Xg
Kobara, Shinichi, 979 845 4089 shinichi@tamu.edu, Texas A&M Univ – Oi
Kobs-Nawotniak, Shannon E., (208) 282-3365 kobsshan@isu.edu, Idaho State Univ – Gv
Koc Tasgin, Calibe, 00904242370000-5976 calibekoc@firat.edu.tr, Firat Univ – GsdGg
Koch, Joe, (803) 896-4167 kochj@dnr.sc.gov, Dept of Natural Resources – Og
Koch, Magaly, mkoch@bu.edu, Boston Univ – HyOri
Koch, Paul L., (831) 459-5861 pkoch@pmc.ucsc.edu, Univ of California, Santa Cruz – Pv
Kochanov, William E., (717) 702-2033 wkochanov@pa.gov, Pennsylvania Bureau of Topographic & Geologic Survey – GemGg
Kochel, R. Craig, (570) 577-3032 kochel@bucknell.edu, Bucknell Univ – Gm
Kocurek, Gary A., (512) 471-5855 garyk@mail.utexas.edu, Univ of Texas at Austin – Gs

Kocurko, John, 940-397-4250 Midwestern State Univ – Gs
Kodama, Kenneth P., (610) 758-3663 kpk0@lehigh.edu, Lehigh Univ – Ym
Kodosky, Larry, 248-232-4538 lgkodosk@oaklandcc.edu, Oakland Community Coll – GgvEg
Koehl, Mimi A. R., (510) 642-8103 Univ of California, Berkeley – Po
Koehler, Rich, (775) 682-8763 rkoehler@unr.edu, Univ of Nevada – GmtNg
Koehler, Thomas , (719) 333-8712 thomas.koehler@usafa.edu, United States Air Force Academy – Oa
Koehn, Daniel, Daniel.Koehn@glasgow.ac.uk, Univ of Glasgow – Gt
Kogan, Mikhail, (845) 365-8882 Columbia Univ – Yd
Kohler, Jeffery L., 814-865-9834 JK9@psu.edu, Pennsylvania State Univ, Univ Park – Nm
Kohler, Nicholas, (541) 346-4160 nicholas@uoregon.edu, Univ of Oregon – Oi
Kohlstedt, David L., (612) 626-1544 dlkohl@umn.edu, Univ of Minnesota, Twin Cities – YxGzi
Kohlstedt, Sally G., 612-624-9368 sgk@umn.edu, Univ of Minnesota, Twin Cities – On
Kohn, Matthew, mattkohn@boisestate.edu, Boise State Univ – CsGpPv
Kohut, Ed, 302-831-2569 ekoh@udel.edu, Univ of Delaware – GvxGg
Kokelaar, Peter, +440151 794 5188 P.Kokelaar@liverpool.ac.uk, Univ of Liverpool – GstGv
Kokum, Mehmet, 00904242370000-5963 mkokum@firat.edu.tr, Firat Univ – GttGg
Kolawole, Lanre L., +2348032277598 llkolawole@lautech.edu.ng, Ladoke Akintola Univ of Tech – GeeGe
Kolesar, Peter T., peter.t.kolesar@gmail.com, Utah State Univ – Cg
Kolka, Randall K., (218) 326-7100 Univ of Minnesota, Twin Cities – So
Kolkas, Mosbah, mossbah.kolkas@csi.cuny.edu, Coll of Staten Island – Gg
Kollias, Pavlos, 514-398-1500 pavlos.kollias@stonybrook.edu, McGill Univ – Ora
Komabayashi, Tetsuya, +44 (0) 131 650 8518 tetsuya.komabayashi@ed.ac.uk, Edinburgh Univ – Gz
Komarneni, Sridhar, (814) 865-1542 komarneni@psu.edu, Pennsylvania State Univ, Univ Park – Sc
Kominz, Michelle A., (269) 387-5340 michelle.kominz@wmich.edu, Western Michigan Univ – OuYgGu
Komoto, Cary, cary.komoto@normandale.edu, Univ of Wisconsin Colls – OnnOn
Konhauser, Kurt, 780-492-2571 kurtk@ualberta.ca, Univ of Alberta – Py
Konigsberg, Alvin S., (914) 257-3758 SUNY, New Paltz – Oa
Kontak, Daniel J., 7056751151 x2352 dkontak@laurentian.ca, Laurentian Univ, Sudbury – Eg
Konter, Jasper G., 808-956-8705 jkonter@hawaii.edu, Univ of Hawai'i, Manoa – CcGiCt
Kontuly, Thomas M., (801) 581-8218 thomas.kontuly@geog.utah.edu, Univ of Utah – On
Koons, Peter O., (207)-581-2158 peter.koons@maine.edu, Univ of Maine – Gt
Koornneef, Janne, +31 20 59 81824 j.m.koornneef@vu.nl, VU Univ Amsterdam – GvCa
Kooyman, Gerald L., (858) 534-2091 gkooyman@ucsd.edu, Univ of California, San Diego – Ob
Kopaska-Merkel, David C., (205) 247-3695 davidkm@gsa.state.al.us, Geological Survey of Alabama – GsPiOe
Kopf, Christopher F., (570) 662-4615 ckopf@mansfield.edu, Mansfield Univ – Gcp
Kopp, Robert E., (732) 200-2705 robert.kopp@rutgers.edu, Rutgers, The State Univ of New Jersey – PeyOg
Koppers, Anthony, 541-737-5425 akoppers@coas.oregonstate.edu, Oregon State Univ – CgGv
Kopylova, Maya G., (604) 822-0865 mkopylova@eos.ubc.ca, Univ of British Columbia – Gi
Koralek, Susan, koralek@sou.edu, Southern Oregon Univ – On
Korenaga, Jun, (203) 432-7381 jun.korenaga@yale.edu, Yale Univ – YgsCg
Koretsky, Carla, (269) 387-4372 carla.koretsky@wmich.edu, Western Michigan Univ – CglHs
Kornberg, Amy, (831) 459-4137 amylkorn@ucsc.edu, Univ of California, Santa Cruz – On
Kornreich Wolf, Susan, (309) 794-7369 susanwolf@augustana.edu, Augustana Coll – Oe
Korose, Christopher P., 217-333-7256 korose@illinois.edu, Illinois State Geological Survey – Ec
Korotev, Randy L., (314) 935-5637 korotev@wustl.edu, Washington Univ in St. Louis – XgCtXm
Korre, Anna, +44 20 759 47372 a.korre@imperial.ac.uk, Imperial Coll – Ng
Korty, Robert, 979-847-9090 korty@tamu.edu, Texas A&M Univ – Oa
Kortz, Karen, kkortz@ccri.edu, Community Coll of Rhode Island – OeGg
Korvin, Gabor, +96638603265 gabor@kfupm.edu.sa, King Fahd Univ of Petroleum and Minerals – Yx
Korycansky, Don, (831) 459-5843 Univ of California, Santa Cruz – Xc
Koskinen, William C., (612) 625-4276 koskinen@soils.umn.edu, Univ of Minnesota, Twin Cities – Sc
Kosloski, Mary, (319) 335-0893 mary-kosloski@uiowa.edu, Univ of Iowa – Pio
Kosro, Michael, (541) 737-3079 kosro@coas.oregonstate.edu, Oregon State Univ – OprOn
Kossin, James, 608-265-5356 kossin@ssec.wisc.edu, Univ of Wisconsin, Madison – Oa
Kostelnick, John, 309-438-7679 jckoste@ilstu.edu, Illinois State Univ – Oi
Koster Van Groos, August F., (312) 996-8678 kvg@uic.edu, Univ of Illinois at Chicago – Cp
Kostov, Svilen, 229-931-2321 skostov@gsw.edu, Georgia Southwestern State Univ – On
Kota, Jozsef, (520) 621-4396 kota@lpl.arizona.edu, Univ of Arizona – Oa
Kotamarthi, Rao, 630-252-7164 vrkotamarthi@anl.gov, Argonne National Laboratory – Oa
Koteas, G. Christopher, 802 485 3321 gkoteas@norwich.edu, Norwich Univ – GitYh
Kotha, Mahender, +91-832-6519329 mkotha@unigoa.ac.in, Goa Univ – GsoOi
Kotulova, Julia, jkotulova@egi.utah.edu, Univ of Utah – CoGo
Koutavas, Athanasios, 718-982-2972 tom.koutavas@csi.cuny.edu, Coll of Staten Island – PeOg
Koutitonsky, Vladimir G., (418) 724-1986 (Ext. 1763) vgk@uqar.qc.ca, Universite du Quebec a Rimouski – Op
Koutnik, Michelle, 206-221-5041 mkoutnik@uw.edu, Univ of Washington – GlOl
Ková, Michal, +421260296555 kovacm@fns.uniba.sk, Comenius Univ in Bratislava – Gsg
Kovacová, Marianna, kovacova@fns.uniba.sk, Comenius Univ in Bratislava – PlbPe
Kovach, Richard G., (702) 295-6180 rkovach@lanl.gov, Los Alamos National Laboratory – Nm
Kovach, Robert L., (650) 723-4827 Stanford Univ – Ys
Kovaèiæ, Marijan, +38514605963 mkovacic@geol.pmf.hr, Univ of Zagreb – GsdGx
Kowaleski, Douglas, 857-234-9339 dkowal@geo.umass.edu, Univ of Massachusetts, Amherst – Gm
Kowalewski, Douglas E., 508-929-8646 douglas.kowalewski@worcester.edu, Worcester State Univ – GmlOy
Kowalewski, Michal J., (540) 231-5951 michalk@vt.edu, Virginia Polytechnic Inst & State Univ – Py
Kowalke, Sara, 740 548 7348 sara.kowalke@dnr.state.oh.us, Ohio Dept of Natural Resources – Ys
Kowalke, Thorsten, 089/2180 6733 t.kowalke@lrz.uni-muenchen.de, Ludwig-Maximilians-Universitaet Muenchen – Pg
Kowallis, Bart J., (801) 422-2467 bart_kowallis@byu.edu, Brigham Young Univ – GgCcGz
Koziol, Andrea M., (937) 229-2954 Andrea.Koziol@notes.udayton.edu, Univ of Dayton – Cp
Kozlowski, Andrew, 518 486-2012 akozlows@mail.nysed.gov, New York State Geological Survey – Gl
Kraal, Erin, 484-646-5859 kraal@kutztown.edu, Kutztown Univ of Pennsylvania – XgGm
Krabbenhoft, David, (608) 821-3843 dpkrabbe@usgs.gov, Univ of Wisconsin, Madison – ClGeHw
Kraemer, George P., 914-251-6640 george.kraemer@purchase.edu, SUNY, Purchase – Ob
Kraft, Kaatje, (360) 383-3539 kkraft@whatcom.ctc.edu, Whatcom Community Coll – Gg
Kramer, J. Curtis, (209) 946-2482 ckramer@pacific.edu, Univ of the Pacific – Gg
Kramer, James R., (905) 525-9140 kramer@mcmaster.ca, McMaster Univ – Cl
Kramer, Kate, (815) 479-7877 kkramer@mchenry.edu, McHenry County Coll – GgOg
Kramer, Marc, 352-294-3165 mgkramer@ufl.edu, Univ of Florida – Os
Kramer, Walter V., (361) 698-1385 wkramer@delmar.edu, Del Mar Coll – GgoGx

Krantz, David E., (419) 530-2662 david.krantz@utoledo.edu,
Univ of Toledo – Gs
Kranz, Dwight S., (713) 718-5641 dwight.kranz@hccs.edu,
Houston Community Coll System – Gg
Krapac, Ivan G., 217-333-6442 krapac@isgs.uiuc.edu,
Illinois State Geological Survey – Ca
Krastel, Sebastian , skrastel@geophysik.uni-kiel.de,
Dalhousie Univ – Yr
Kraus, Mary J., 303-492-7251 mary.kraus@colorado.edu,
Univ of Colorado – GsSa
Krause, David W., (303) 370-6379 david.krause@sunysb.edu,
SUNY, Stony Brook – Pv
Krause, Federico F., (403) 220-5845 fkrause@ucalgary.ca,
Univ of Calgary – GsdGo
Krause, Jacob J., jjkrause2@wisc.edu, Univ of Wisconsin - Extension – Hw
Krause, Lois B., (864) 656-7653 Clemson Univ – Oe
Krauss, Lawrence, (480) 965-6378 krauss@asu.edu,
Arizona State Univ – On
Krauss, Scott , skrauss@neaq.org, Univ of Massachusetts, Boston – Ob
Krautblatter, Michael, +49 89 28925866 m.krautblatter@tum.de,
Technische Universitaet Muenchen – GmNrOi
Kravchinsky, Vadim, (780) 492-5591 vkrav@phys.ualberta.ca,
Univ of Alberta – Ym
Krawczynski, Michael J., 314-935-6328 mikekraw@levee.wustl.edu,
Washington Univ in St. Louis – Cg
Kreamer, David, 702.895.3553 dave.kreamer@unlv.edu,
Univ of Nevada, Reno – HwGn
Krebes, Edward S., (403) 220-5028 Univ of Calgary – Ys
Kreckel, Kenneth, 307-268-3457 kkreckel@casperColl.edu,
Casper Coll – EoYeGg
Kreemer, Corne, 775 682-8780 kreemer@unr.edu,
Univ of Nevada – Yd
Kreidenweis, Sonia M., sonia@atmos.colostate.edu,
Colorado State Univ – Oa
Kreiger, William (Bill), (717) 815-1379 wkreiger@ycp.edu,
York Coll of Pennsylvania – GisOg
Krekeler, Mark, 513-785-3106 krekelmp@miamioh.edu, Miami Univ – Ge
Kremer, Robert J., 573-882-6408 Univ of Missouri, Columbia – Os
Kressler, Sharon J., 612-625-5068 kress004@umn.edu,
Univ of Minnesota, Twin Cities – On
Kretzschmar, Thomas, tkretzsc@cicese.mx,
Centro de Investigación Científica y de Educación Superior de Ensenada – Ge
Kreutz, Karl J., (207)-581-3011 karl.kreutz@maine.edu,
Univ of Maine – Cs
Krevor, Samuel, s.krevor@imperial.ac.uk,
Imperial Coll – Eo
Krieble, Kelly, (610) 861-1437 krieblek@moravian.edu,
Moravian Coll – OnnOn
Krieger-Brockett, Barbara B., (206) 543-2216 krieger@cheme.washington.edu, Univ of Washington – Ob
Krier, Donathon J., (505) 665-7834 krier@lanl.gov,
Los Alamos National Laboratory – Gv
Kring, David, (281) 486-2119 kring@lpi.usra.edu,
Univ of Arizona – XcGiXg
Krishnamurthy, R. V., (269) 387-5501 r.v.krishnamurthy@wmich.edu,
Western Michigan Univ – Cs
Krishnan, Jay, 713-743-1385 jkrishnan@uh.edu, Univ of Houston – On
Krishtalka, Leonard, (785) 864-4540 krishtalka@ku.edu,
Univ of Kansas – Pv
Krissek, Lawrence A., (614) 292-1924 krissek.1@osu.edu,
Ohio State Univ – Gsu
Krockover, Gerald H., (765) 494-5795 hawk1@purdue.edu,
Purdue Univ – Oe
Kroeger, Glenn C., (210) 999-7607 gkroeger@trinity.edu,
Trinity Univ – Yg
Kroeger, Timothy J., (218) 755-2783 tkroeger@bemidjistate.edu,
Bemidji State Univ – PlGsHw
Kroenke, Loren W., (808) 956-7845 kroenke@soest.hawaii.edu,
Univ of Hawai'i, Manoa – Gu
Krogh, Thomas E., (416) 586-5811 tomk@rom.on.ca,
Univ of Toronto – Cc
Krohn, James P., krohnjp@pierceColl.edu, Los Angeles Pierce Coll – NgGg
Krom, Michael, +44(0) 113 34 30477 M.D.Krom@leeds.ac.uk,
Univ of Leeds – Cm
Kronenberg, Andreas, (979) 845-0132 a-kronenberg@geos.tamu.edu,
Texas A&M Univ – GtyGc
Kronenfeld, Barry J., (217) 581-7014 bjkronenfeld@eiu.edu,
Eastern Illinois Univ – Oiy
Kroon, Dick, +44 (0) 131 651 7089 D.Kroon@ed.ac.uk,
Edinburgh Univ – Gg
Krot, Alexander N., 808-956-3900 sasha@higp.hawaii.edu,
Univ of Hawai'i, Manoa – Xm
Kruckenberg, Seth C., 617-552-3647 seth.kruckenberg@bc.edu,
Boston Coll – Gct
Krueger, Steven, 801-581-3903 steve.krueger@utah.edu,
Univ of Utah – Oa
Kruge, Michael A., 973-655-7668 krugem@mail.montclair.edu,
Montclair State Univ – Co
Kruger, Joseph M., (409) 880-8233 joseph.kruger@lamar.edu,
Lamar Univ – YgGgOi
Kruger, Ned , ekadrmas@nd.gov, North Dakota Geological Survey – Gg
Krugh, W C., 661-654-3126 wkrugh@csub.edu,
California State Univ, Bakersfield – GcmCc
Krukowski, Stanley T., 405-325-3031 skrukowski@ou.edu,
Univ of Oklahoma – En
Krupka, Kenneth M., (509) 376-4412 ken.krupka@pnl.gov,
Pacific Northwest National Laboratory – Cl
Kruse, Jennifer, (507) 933-7333 jkruse@gustavus.edu,
Gustavus Adolphus Coll – On
Kruse, Sarah E., (813) 974-7341 Univ of South Florida, Tampa – Yg
Krygier, John B., (740) 368-3622 jbkrygie@owu.edu,
Ohio Wesleyan Univ – Oi
Krzic, Maja, 604-822-0252 maja.krzic@ubc.ca,
Univ of British Columbia – Sf
Ku, Teh-Lung, (213) 740-5826 rku@usc.edu,
Univ of Southern California – Cg
Ku, Timothy C., (860)685-2265 tcku@wesleyan.edu,
Wesleyan Univ – Cl
Kuang, Zhiming, 617-495-2354 kuang@eps.harvard.edu,
Harvard Univ – Oa
Kubas, Gregory J., (505) 667-5846 Los Alamos National Laboratory – Oa
Kubesh, Rodney, 320-308-4217 rjkubesh@stcloudstate.edu,
Saint Cloud State Univ – Ow
Kubicek, Leonard, (972) 273-3508 lenkubicek@dcccd.edu,
North Lake Coll - Dallas Community Coll District – Gg
Kubicki, James D., 915-747-5501 jdkubicki,
Univ of Texas, El Paso – ClGe
Kucharik, Chris, 608-890-3021 kucharik@wisc.edu,
Univ of Wisconsin, Madison – Oa
Kuchovsky, Tomas, +420 549 49 5452 tomas@sci.muni.cz,
Masaryk Univ – Hgy
Kuchta, Mark, (303) 273-3306 mkuchta@mines.edu,
Colorado School of Mines – Nm
Kudela, Raphael M., (831) 459-3290 kudela@cats.ucsc.edu,
Univ of California, Santa Cruz – Or
Kudlac, John J., (412) 392-3423 Point Park Univ – Ng
Kuehl, Steven A., (804) 684-7118 kuehl@vims.edu,
Coll of William & Mary – Ou
Kuehn, Stephen C., 304-384-6322 sckuehn@concord.edu,
Concord Univ – CaGv
Kuentz, David C., (513) 529-5992 kuentzdc@MiamiOh.edu,
Miami Univ – Ca
Kues, Barry S., (505) 277-3626 bkues@unm.edu,
Univ of New Mexico – Pi
Kugel, Abigail, 507-457-5260 akugel@winona.edu,
Winona State Univ – On
Kuhlman, Robert, rkuhlman@mc3.edu,
Montgomery County Community Coll – Gg
Kuhnhenn, Gary L., (859) 622-8140 gary.kuhnhenn@eku.edu,
Eastern Kentucky Univ – Gd
Kuiper, Klaudia, k.f.kuiper@vu.nl, VU Univ Amsterdam – Cc
Kuiper, Yvette D., 303-273-3105 ykuiper@mines.edu,
Colorado School of Mines – Gc
Kujawinski, Elizabeth B., (508) 289-3696 ekujawinski@whoi.edu,
Woods Hole Oceanographic Institution – Oc
Kukoè, Duje, +38514606111 duje.kukoc@geol.pmf.hr,
Univ of Zagreb – Gg
Kukowski, Nina, 0049(0)3641/948680 nina.kukowski@uni-jena.de,
Friedrich-Schiller-Univ Jena – Ygx
Kulander, Byron, byron.kulander@wright.edu, Wright State Univ – Gc
Kulatilake, Pinnaduwa H. S. W., (520) 621-6064 kulatila@u.arizona.edu,
Univ of Arizona – Nr

Kulkarni, Shrinivas R., (626) 395-4010 srk@astro.caltech.edu, California Inst of Tech – Xy
Kulp, Mark A., (504) 280-1170 mkulp@uno.edu, Univ of New Orleans – GrmGs
Kulp, Thomas, tkulp@binghamton.edu, Binghamton Univ – Py
Kumar, Ajoy, (717) 871-2432 ajoy.kumar@millersville.edu, Millersville Univ – Opr
Kumar, M. Satish, +442890973479 s.kumar@qub.ac.uk, Queens Univ Belfast – OnnOn
Kumjian, Matthew R., 814-863-1581 kumjian@psu.edu, Pennsylvania State Univ, Univ Park – Ow
Kummerow, Christian D., kummerow@atmos.colostate.edu, Colorado State Univ – Yr
Kump, Lee R., (814) 863-1274 lkump@psu.edu, Pennsylvania State Univ, Univ Park – ClPeCm
Kumpf, Amber C., (231) 777-0289 amber.kumpf@muskegoncc.edu, Muskegon Community Coll – GgYrOu
Kung, Ernest C., 573-882-5909 Univ of Missouri, Columbia – Oa
Kung, Hsiang-Te, (901) 678-4538 hkung@memphis.edu, Univ of Memphis – Oy
Kung, King-Jau S., (608) 262-6530 kskung@wisc.edu, Univ of Wisconsin, Madison – Sp
Kunkle, Thomas D., (505) 667-1259 Los Alamos National Laboratory – Xy
Kuntz, Kara, (785) 628-5804 kkuntz@fhsu.edu, Fort Hays State Univ – On
Kuntz, Mark R., (847) 697-1000 mkuntz@elgin.edu, Elgin Community Coll – Gg
Kunza, Lisa, (605) 394-2449 lisa.kunza@sdsmt.edu, South Dakota School of Mines & Tech – HsOe
Kunze, Eric L., (206) 543-8467 kunze@uvic.ca, Univ of Washington – Op
Kunzmann, Thomas, 089/2180 4292 kunzmann@min.uni-muenchen.de, Ludwig-Maximilians-Universitaet Muenchen – Gz
Kuo, Shiou, (206) 840-4573 skuo@wsu.edu, Washington State Univ – Sc
Kuperman, William A., (858) 534-3158 wkuperman@ucsd.edu, Univ of California, San Diego – Op
Kurapov, Alexander, 541-737-2865 kurapov@coas.oregonstate.edu, Oregon State Univ – Yr
Kurka, Mira, 775-777-1054 mira.kurka@gbcnv.edu, Great Basin Coll – Gg
Kursinski, Robert, (520) 626-3338 kursinsk@atmo.arizona.edu, Univ of Arizona – Oa
Kurtanjek, Dražen, +38514605965 dkurtan@inet.hr, Univ of Zagreb – GsdGg
Kurttas, Turker, +90-312-2977760 kurttast@gmail.com, Hacettepe Univ – HwgOr
Kurtz, Andrew, kurtz@bu.edu, Boston Univ – Clg
Kurtz, Vincent E., (417) 836-5800 Missouri State Univ – Ps
Kurum, Sevcan, 00904242370000-5992 skurum@firat.edu.tr, Firat Univ – GivGx
Kurz, Mark D., (508) 289-2328 mkurz@whoi.edu, Woods Hole Oceanographic Institution – Cc
Kushnir, Yochanan, (845) 365-8669 Columbia Univ – Oa
Kusnick, Judith E., (916) 278-4692 California State Univ, Sacramento – Oe
Kusssow, Wayne R., (608) 263-3631 wrkussow@wisc.edu, Univ of Wisconsin, Madison – So
Kustka, Adam B., 973-353-5509 kustka@andromeda.rutgers.edu, Rutgers, The State Univ of New Jersey, Newark – OgCmHs
Kusumoto, Shigekazu, 81-76-445-6653 kusu@sci.u-toyama.ac.jp, Univ of Toyama – YvdGt
Kusznir, Nick, +44-151-794-5182 N.Kusznir@liverpool.ac.uk, Univ of Liverpool – Yd
Kutis, Michael, 765-285-2487 mkutis@bsu.edu, Ball State Univ – Gg
Kutzbach, John E., (608)262-0392 jek@facstaff.wisc.edu, Univ of Wisconsin, Madison – Oa
Kuwabara, James, 415-452-7776 kuwabara@usgs.gov, City Coll of San Francisco – Og
Kuzila, Mark S., (402) 472-7537 mkuzila@unl.edu, University of Nebraska - Lincoln – Sd
Kuzyk, Zou Zou, 204-272-1535 umkuzyk@cc.umanitoba.ca, Univ of Manitoba – Cg
Kvamme, Kenneth L., (617) 353-3415 Boston Univ – Ga
Kwicklis, Edward M., (505) 665-7408 kwicklis@lanl.gov, Los Alamos National Laboratory – Gg
Kyle, J. Richard, (512) 471-4351 rkyle@jsg.utexas.edu, Univ of Texas at Austin – EmnCe
Kyle, Philip R., (575) 835-5995 kyle@nmt.edu, New Mexico Inst of Mining and Tech – Giv
Kysar Mattietti, Giuseppina, (703) 993-9269 gkysar@gmu.edu, George Mason Univ – GiOeGa
Kyser, Kurt, (613) 533-6179 kyserk@queensu.ca, Queen's Univ – CeaCs
Kyte, Frank T., (310) 825-2015 kyte@igpp.ucla.edu, Univ of California, Los Angeles – Ct
Käser, Martin, 089/2180 4138 martin.kaeser@geophysik.uni-muenchen.de, Ludwig-Maximilians-Universitaet Muenchen – Yg
Käsling, Heiko, +49 89 289 25831 heiko.kaesling@tum.de, Technische Universitaet Muenchen – NrmNg
Köster, Mathias, +49 (89) 289 25893 mathias.koester@tum.de, Technische Universitaet Muenchen – CgGzEg

## L

L'Ecuyer, Tristan S., 608-262-2828 Univ of Wisconsin, Madison – Oa
La, Daniel A., (405) 744-6358 daniel.lao_davila@okstate.edu, Oklahoma State Univ – GctGg
La Berge, Gene L., (920) 424-4460 laberge@uwosh.edu, Univ of Wisconsin, Oshkosh – Eg
La Fave, John I., 406-496-4306 jlafave@mtech.edu, Montana Tech of The Univ of Montana – Hw
La Tour, Timothy E., 404413-5767 tlatour@gsu.edu, Georgia State Univ – Gp
Laabs, Benjamin J., (701) 231-6197 benjamin.laabs@ndsu.edu, North Dakota State Univ – GlOl
Laabs, Benjamin J., 585-245-5305 laabs@geneseo.edu, SUNY, Geneseo – GmlGe
Labandeira, Conrad C., (202) 633-1336 Smithsonian Institution / National Museum of Natural History – Pi
LaBarbera, Michael C., (773) 702-8092 mlabarbe@uchicago.edu, Univ of Chicago – Po
LaBella, Joel, (860)685-2242 jlabella@wesleyan.edu, Wesleyan Univ – On
Laboski, Carrie A., (608) 263-2795 laboski@wisc.edu, Univ of Wisconsin, Madison – So
Labotka, Theodore C., (865) 974-2366 tlabotka@utk.edu, Univ of Tennessee, Knoxville – GpCg
Labuz, Joseph F., (612) 625-9060 jlabuz@umn.edu, Univ of Minnesota, Twin Cities – Nr
Lachhab, Ahmed, 570-374-4215 lachhab@susqu.edu, Susquehanna Univ – Hw
Lachmar, Thomas E., (435) 797-1247 tom.lachmar@gmail.com, Utah State Univ – Hw
Lachniet, Matthew S., 702-895-4388 matthew.lachniet@unlv.edu, Univ of Nevada, Las Vegas – PeCs
Lackey, Jade Star, (909) 621-8677 jadestar.lackey@pomona.edu, Pomona Coll – GipCs
Lackinger, Markus, 089/2180 Ludwig-Maximilians-Universitaet Muenchen – Gz
Lackmann, Gary M., 919-515-1439 gary@ncsu.edu, North Carolina State Univ – Oa
Lacy, Tor, tlacy@cerritos.edu, Cerritos Coll – Gg
LaDochy, Steve, (323) 343-3222 sladoch@calstatela.edu, California State Univ, Los Angeles – On
LaDue, Nicole D., (815) 753-7935 nladue@niu.edu, Northern Illinois Univ – Oe
LaFemina, Peter C., 814-865-7326 pfemina@geosc.psu.edu, Pennsylvania State Univ, Univ Park – Yd
Laffan, Shawn, shawn.laffan@unsw.edu.au, Univ of New South Wales – Oiy
LaFleche, Marc R., (418) 654-2670 marc.richer-lafleche@ete.inrs.ca, Universite du Quebec – Ct
Lafrance, Bruno, 7056751151 x2264 blafrance@laurentian.ca, Laurentian Univ, Sudbury – Gc
LaFreniere, Lorraine, (630) 252-7969 lafreniere@anl.gov, Argonne National Laboratory – HwGgEg
Lageson, David R., (406) 994-6913 lageson@montana.edu, Montana State Univ – Gc
Lagowski, Alison A., (716) 645-4856 aal@buffalo.edu, SUNY, Buffalo – On
Lai, Chung Chieng A., (505) 665-6635 cal@lanl.gov, Los Alamos National Laboratory – Ow
Laib, Amanda, amandalaib@cwidaho.cc, Coll of Western Idaho – GeCcGv
Laingen, Christopher R., (217) 581-2999 crlaingen@eiu.edu, Eastern Illinois Univ – SfOn
Laird, David A., (515) 294-1581 dalaird@iastate.edu, Iowa State Univ of Science & Tech – Sc
Laird, Jo, (603) 862-1718 jl@cisunix.unh.edu, Univ of New Hampshire – Gp

Laird, Neil, 315-781-3603 laird@hws.edu, Hobart & William Smith Colls – Oa
Laity, Julie E., (818) 677-3532 julie.laity@csun.edu, California State Univ, Northridge – Oy
Lakatos, Stephen, (718) 262-2589 York Coll (CUNY) – Cc
Lake, Iain, +44 (0)1603 59 3744 i.lake@uea.ac.uk, Univ of East Anglia – Ge
Lakhan, V. Chris, (519) 253-3000 x2183 lakan@uwindsor.ca, Univ of Windsor – OriOn
Laki, Sam, (937) 376-6272 slaki@centralstate.edu, Central State Univ – EgSoHs
Lakshmi, Venkataraman, (803) 777-3552 vlakshmi@geol.sc.edu, Univ of South Carolina – Hg
Lal, Rattan, 614-292-9069 lal.1@osu.edu, Ohio State Univ – Sp
Laliberte, Elizabeth, (401) 874-5512 elalib@mail.uri.edu, Univ of Rhode Island – Og
Lam, Anita, 604-822-2736 alam@eos.ubc.ca, Univ of British Columbia – On
Lamanna, Matthew C., (412) 624-8780 Univ of Pittsburgh – Pv
Lamanna, Matthew C., (412) 578-2696 lamannam@carnegiemnh.org, Carnegie Museum of Natural History – PvoPe
Lamb, Dennis, (814) 865-0174 lno@psu.edu, Pennsylvania State Univ, Univ Park – Ow
Lamb, James P., 205 652 3725 jlamb@uwa.edu, Univ of West Alabama – PvePg
Lamb, John A., (612) 625-1772 jlamb@soils.umn.edu, Univ of Minnesota, Twin Cities – Sc
Lamb, Melissa A., (651) 962-5242 malamb@stthomas.edu, Univ of Saint Thomas – GtcGs
Lamb, Michael P., 626.395.3612 mpl@gps.caltch.edu, California Inst of Tech – Gm
Lamb, Will, (979) 845-3075 lamb@geo.tamu.edu, Texas A&M Univ – GpCpGz
Lambert, Carolyn D., 970-351-2647 Carolyn.Lambert@unco.edu, Univ of Northern Colorado – Hw
Lambert, Dean P., 210-486-0471 dlambert@alamo.edu, San Antonio Community Coll – Oyi
Lambert, Lance L., (210) 458-4455 Lance.Lambert@utsa.edu, Univ of Texas, San Antonio – Pg
Lambert, W. J., 205-348-4404 jlambert@ua.edu, Univ of Alabama – CsPeCg
Lambert-Smith, James, J.S.Lambert-Smith@kingston.ac.uk, Kingston Univ – Gg
Lammerer, Bernd, 089/2180 6517 lammerer@iaag.geo.uni-muenchen.de, Ludwig-Maximilians-Universitaet Muenchen – GgtGc
Lamothe, Michel, 514-987-3000 #3361 lamothe.michel@uqam.ca, Universite du Quebec a Montreal – Gl
Lancaster, Nicholas, (775) 673-7304 nick@dri.edu, Desert Research Inst – Gm
Lancaster, Penny, +44 023 9284 2272 penny.lancaster@port.ac.uk, Univ of Portsmouth – Gz
Lancaster, Stephen, 541-737-9258 lancasts@geo.oregonstate.edu, Oregon State Univ – Hs
Land, Lewis, 575-887-5508 lland@nmbg.nmt.edu, New Mexico Inst of Mining and Tech – Hy
Land, Lewis A., 505-887-5505 lland@gis.nmt.edu, New Mexico Inst of Mining & Tech – Hg
Landenberger, Bill, 61 02 4921 6366 bill.landenberger@newcastle.edu.au, Univ of Newcastle – GiCg
Lander, Mark A., (671) 735-2685 mlander@uguam.uog.edu, Univ of Guam – Ow
Landing, Ed, (518) 473-8071 elanding@mail.nysed.gov, New York State Geological Survey – PseGr
Landman, Neil H., (212) 769-5723 Graduate School of the City Univ of New York – Ps
Landman, Neil H., (212) 769-5712 American Museum of Natural History – Pi
Landry, Michael R., (858) 534-4702 mlandry@ucsd.edu, Univ of California, San Diego – Ob
Landry, Peter B., (508) 289-3443 plandry@whoi.edu, Woods Hole Oceanographic Institution – Ca
Landschoot, Peter J., (814) 863-1017 pcl11@psu.edu, Pennsylvania State Univ, Univ Park – So
Lane, Charles L., (541) 552-6114 lane@sou.edu, Southern Oregon Univ – Hg
Lane, Joseph M., (216) 987-5227 Joseph.Lane@tri-c.edu, Cuyahoga Community Coll - Western Campus – Og
Lane, Mark, 7607441150 xt. 2951 mlane@palomar.edu, Palomar Coll – On
Lang, Harold R., (304) 293-5603 harold@lithos.jpl.nasa.gov, Jet Propulsion Laboratory – Gr
Lang, Helen M., (304) 293-5469 hlang@wvu.edu, West Virginia Univ – Gpz
Lang, Nicholas, 814-824-3646 nlang@mercyhurst.edu, Mercyhurst Univ – GvXgGc
Lang, Susan Q., 803-777-8832 slang@geol.sc.edu, Univ of South Carolina – Oc
Lange, Eric, (765) 285-8272 eslange@bsu.edu, Ball State Univ – Cg
Lange, Rebecca A., (734) 764-7421 becky@umich.edu, Univ of Michigan – Gi
Langel, Richard A., 319-335-4102 richard.langel@dnr.iowa.gov, Iowa Dept of Natural Resources – Gg
Langenheim, Jr., Ralph L., rlangenh@illinois.edu, Univ of Illinois, Urbana-Champaign – Gr
Langenhorst, Falko H., 0049(0)3641/948730 falko.langenhorst@uni-jena.de, Friedrich-Schiller-Univ Jena – Gz
Langer, Arthur M., (718) 951-4793 Graduate School of the City Univ of New York – Gz
Langford, Richard P., (915) 747-5968 langford@utep.edu, Univ of Texas, El Paso – Gs
Langhorst, Glenn, 218-879-0719 glang@fdltcc.edu, Fond du Lac Tribal and Community Coll – Gg
Langille, Jackie M., (828) 251-6453 jlangill@unca.edu, Univ of North Carolina, Asheville – GctGe
Langman, Jeffrey, (208) 885-0310 jlangman@uidaho.edu, Univ of Idaho – ClHws
Langmuir, Charles H., (617) 384-9948 langmuir@eps.harvard.edu, Harvard Univ – Cg
Langston, Charles A., (901) 678-2007 clangstn@memphis.edu, Univ of Memphis – Ys
Langston, Jr., Wann, (512) 471-7736 wannl@mail.utexas.edu, Univ of Texas at Austin – Pv
Langston-Unkefer, Pat J., (505) 665-2556 Los Alamos National Laboratory – Co
Lanigan, David C., (509) 376-9308 david.lanigan@pnl.gov, Pacific Northwest National Laboratory – Hg
Lanoue, Christopher A., 605-677-6153 chris.lanoue@usd.edu, South Dakota Dept of Environment and Natural Resources – On
Lansey, Kevin E., (520) 621-2512 lansey@engr.arizona.edu, Univ of Arizona – Hs
Lapen, Thomas, 713-743-6122 tjlapen@uh.edu, Univ of Houston – Gz
LaPointe, Daphne D., (775) 682-8772 dlapoint@unr.edu, Univ of Nevada – Gg
Laporte, Leo F., laporte@ucsc.edu, Univ of California, Santa Cruz – Pg
Laprise, Rene, 514-987-3000 #3302 laprise.rene@uqam.ca, Universite du Quebec a Montreal – Oa
Lapusta, Nadia, (626) 395-2277 lapusta@caltech.edu, California Inst of Tech – GcYs
Large, Ross R., 61 3 6226 2819 Ross.Large@utas.edu.au, Univ of Tasmania – Eg
Larkin, Patrick, 361-825-3258 Patrick.Larkin@tamucc.edu, Texas A&M Univ, Corpus Christi – On
Larner, Kenneth L., klarner@mines.edu, Colorado School of Mines – Ye
LaRock, Paul A., (225) 388-6307 Louisiana State Univ – Ob
Larocque, Marie, 514-987-3000 #1515 larocque.marie@uqam.ca, Universite du Quebec a Montreal – Hw
Larsen, Daniel, (901) 678-4358 dlarsen@memphis.edu, Univ of Memphis – Cl
Larsen, Isaac J., 413-545-0538 ilarsen@geo.umass.edu, Univ of Massachusetts, Amherst – Gm
Larsen, Jessica F., 907-474-7992 jflarsen@alaska.edu, Univ of Alaska, Fairbanks – Gv
Larsen, Kristine, (860) 832-2938 larsen@ccsu.edu, Central Connecticut State Univ – Xy
Larson, David J., 510-885-3132 david.larson@csueastbay.edu, California State Univ, East Bay – Ou
Larson, Edwin E., 303-492-6172 Univ of Colorado – Ym
Larson, Eric J., 715-346-4098 Eric.Larsen@uwsp.edu, Univ of Wisconsin, Stevens Point – Or
Larson, Erik, (610) 861-1440 larsone@moravian.edu, Moravian Coll – GmsGg
Larson, Harold P., (520) 621-6943 hplarson@u.arizona.edu, Univ of Arizona – On

Larson, Peter B., (509) 335-3095 plarson@wsu.edu, Washington State Univ – Cs
Larson, Phillip H., (507) 389-2617 phillip.larson@mnsu.edu, Minnesota State Univ – GmOyHs
Larson, Roger, (401) 874-6165 rlar@uri.edu, Univ of Rhode Island – Ou
Larter, Stephen, +44 (0) 191 208 5956 alex.leathard@ncl.ac.uk, Univ of Newcastle Upon Tyne – Gg
Lasca, Norman P., (414) 229-4602 nplasca@uwm.edu, Univ of Wisconsin, Milwaukee – Gml
Lasemi, Zakaria, 217-244-6944 lasemi@isgs.uiuc.edu, Illinois State Geological Survey – En
Lash, Gary G., (716) 673-3842 lash@fredonia.edu, SUNY Fredonia – Gr
Lash, Gary S., lash@fredonia.edu, SUNY, Buffalo – Gc
Lasher-Trapp, Sonia, 217-244-4250 slasher@illinois.edu, Univ of Illinois, Urbana-Champaign – Oa
Laske, Gabi, (858) 534-8774 glaske@ucsd.edu, Univ of California, San Diego – Ys
Lasker, Howard R., 716-645-4870 hlasker@buffalo.edu, SUNY, Buffalo – Gu
Laskowski, Stanley L., (215) 573-3164 slaskows@sas.upenn.edu, Univ of Pennsylvania – Ge
Lassetter, William L., (434) 951-6361 william.lassetter@dmme.virginia.gov, Division of Geology and Mineral Resources – EgHg
Lassiter, John, (512) 471-4002 lassiter1@mail.utexas.edu, Univ of Texas at Austin – Cc
Last, Fawn M., Fawn.last@angelo.edu, Angelo State Univ – Gsn
Last, George V., (509) 376-3961 george.last@pnl.gov, Pacific Northwest National Laboratory – Ge
Last, William M., (204) 474-8361 wm_last@umanitoba.ca, Univ of Manitoba – GsnGo
Lat, Che Noorliza, 03-79674157 noorliza@um.edu.my, Univ of Malaya – Yg
Lathrop, Daniel, (301) 405-1594 lathrop@umd.edu, Univ of Maryland – Yg
Latimer, Jennifer C., (812) 237-2254 jen.latimer@indstate.edu, Indiana State Univ – CmGeb
Laton, W. R., (657) 278-7514 wlaton@fullerton.edu, California State Univ, Fullerton – HwGeHy
Lau, Chui Yim Maggie, 609-258-6899 maglau@princeton.edu, Princeton Univ – Og
Laubach, Stephen E., (512) 471-6303 steve.laubach@beg.utexas.edu, Univ of Texas at Austin, Jackson School of Geosciences – GcNrGz
Lauderdale, Jonathan, J.Lauderdale@liverpool.ac.uk, Univ of Liverpool – Og
Laudon, Thomas S., (414) 424-4464 ectoproct@sbcglobal.net, Univ of Wisconsin, Oshkosh – Yg
Laurent-Charvet, Sébastien, +33(0)3 44068995 sébastien.laurent-charvet@lasalle-beauvais.fr, Institut Polytechnique LaSalle Beauvais (ex-IGAL) – GctOe
Lauretta, Dante, (520) 626-1138 lauretta@lpl.arizona.edu, Univ of Arizona – XcmXg
Laurier, Eric, +44 (0) 131 651 4303 Eric.Laurier@ed.ac.uk, Edinburgh Univ – On
Lautz, Laura, 315-443-1196 lklautz@syr.edu, Syracuse Univ – Hg
Lauziere, Kathleen, (418) 654-2658 klauzier@nrcan.gc.ca, Universite du Quebec – Gg
Lauzon, John, (519) 824-4120 (Ext. 52459) lauzonj@uoguelph.ca, Univ of Guelph – So
Lavallee, Daniel, (805) 893-8446 daniel@crustal.ucsb.edu, Univ of California, Santa Barbara – On
Lavallee, Yan, +440151 794 5183 Yan.Lavallee@liverpool.ac.uk, Univ of Liverpool – Gv
Lavallee, Yan, 089/2180 4221 yanlavallee@hotmail.com, Ludwig-Maximilians-Universitaet Muenchen – Yg
Lavier, Luc L., 512-471-0455 luc@ig.utexas.edu, Univ of Texas at Austin – Gt
LaVigne, Michéle, 207-798-4283 mlavign@bowdoin.edu, Bowdoin Coll – OcPe
Lavkulich, Leslie M., (604) 822-3477 Univ of British Columbia – Sd
Lavoie, Denis, (418) 654-2571 delavoie@nrcan.gc.ca, Universite du Quebec – Gs
Law, Eric W., (740) 826-8242 ericlaw@muskingum.edu, Muskingum Univ – Gp
Law, Kim, 519-661-2111 ext 83881 krlaw@uwo.ca, Western Univ – Cs
Law, Richard D., (540) 231-6685 rdlaw@vt.edu, Virginia Polytechnic Inst & State Univ – Gc
Law, Zada, (615) 494-8805 zada.law@mtsu.edu, Middle Tennessee State Univ – Oin
Lawrence, Deborah, (434) 924-0581 dl3c@virginia.edu, Univ of Virginia – On
Lawrence, Henry, (814) 732-1572 hlawrence@edinboro.edu, Edinboro Univ of Pennsylvania – Ou
Lawrence, James, jimrslawrence@gmail.com, Univ of Houston – Cl
Lawrence, Kira, (610) 330-5194 lawrenck@lafayette.edu, Lafayette Coll – Gg
Lawrence, Rick L., (406) 994-5409 Montana State Univ – Or
Lawry, Cynthia, (979) 830-4406 Cynthia.Lawry@blinn.edu, Blinn Coll – Gg
Laws, Richard A., (910) 962-4125 laws@uncw.edu, Univ of North Carolina Wilmington – Pm
Lawson, Merlin P., (402) 202-5392 mlawson1@unl.edu, Univ of Nebraska, Lincoln – OwaOr
Lawton, Donald C., (403) 220-5718 Univ of Calgary – Ye
Lawver, Lawrence A., (512) 471-0433 lawver@ig.utexas.edu, Univ of Texas at Austin – YrGt
Lay, Thorne, (831) 459-3164 tlay@pmc.ucsc.edu, Univ of California, Santa Cruz – Ys
Layer, Paul, 907-474-5514 pwlayer@alaska.edu, Univ of Alaska, Fairbanks – Ym
Layton, Alice, 865-974-8072 alayton@utk.edu, Univ of Tennessee, Knoxville – On
Layton-Matthews, Daniel, (613) 533-6338 dlayton@queensu.ca, Queen's Univ – EgCte
Lazar, Codi, 909-537-5586 clazar@csusb.edu, California State Univ, San Bernardino – Cp
Lazarus, Eli, +44(0)29 208 75563 e.d.lazarus@soton.ac.uk, Univ of Southampton – GmOny
Lazarus, Steven, (321) 674-8096 Florida Inst of Tech – Oa
Lazarus, Steven M., 321-674-2160 slazarus@fit.edu, Florida Inst of Tech – Oa
Le, Yanfen, 660-562-1525 le@nwmissouri.edu, Northwest Missouri State Univ – Oi
Le Heron, Daniel P., +44 1784 443615 daniel.le-heron@rhul.ac.uk, Univ of London, Royal Holloway & Bedford New Coll – OgGsl
Le Mone, David V., 915-747-5501 lemone@utep.edu, Univ of Texas, El Paso – Ps
Le Qu, Corinne, +44 (0)1603 59 2840 c.lequere@uea.ac.uk, Univ of East Anglia – Og
Le Roex, Anton, 021-650-2902 anton.leroex@uct.ac.za, Univ of Cape Town – Cg
Le Roux, Veronique, (508) 289-3549 vleroux@whoi.edu, Woods Hole Oceanographic Institution – Gi
Le Voyer, Marion, (202) 633-1817 levoyerm@si.edu, Smithsonian Institution / National Museum of Natural History – Gz
Lea, David W., (805) 893-8665 lea@geol.ucsb.edu, Univ of California, Santa Barbara – PeCm
Lea, Peter D., (207) 725-3439 plea@bowdoin.edu, Bowdoin Coll – Gl
Leach, Harry, +44 0151 794 4097 Leach@liverpool.ac.uk, Univ of Liverpool – Og
Leadbetter, Jared R., 626.395.4182 jleadbetter@caltech.edu, California Inst of Tech – Pm
Leake, Bernard E., 00442920876421 leakeb@cf.ac.uk, Cardiff Univ – GipGz
Leake, Martha A., (229) 333-5756 mleake@valdosta.edu, Valdosta State Univ – Xg
Leap, Darrell I., (765) 494-3699 mountains2oceans@comcast.net, Purdue Univ – Hy
Lear, C, carrie@earth.cf.ac.uk, Cardiff Univ – Ou
Lear, Caroline, +44(0)29 208 79004 LearC@cardiff.ac.uk, Univ of Wales – Og
Leatham, W. Britt, (909) 537-5322 bleatham@csusb.edu, California State Univ, San Bernardino – PsmOg
Leather, Kimberly, Kimberley.Leather@manchester.ac.uk, Univ of Manchester – Oa
Leavell, Daniel N., (740) 366-9342 leavell.6@osu.edu, Ohio State Univ – Ge
Leavens, Peter B., 302-831-8106 pbl@udel.edu, Univ of Delaware – Gz
Leavitt, Steven W., (520) 621-6468 sleavitt@ltrr.arizona.edu, Univ of Arizona – Csc
Lebedev, Maxim, +61 8 9266 3519 M.Lebedev@curtin.edu.au, Curtin Univ – Yx
Lebofsky, Larry A., (520) 621-6947 lebofsky@lpl.arizona.edu, Univ of Arizona – OngXm

Lebold, Joe, (304) 293-0749 joe.lebold@mail.wvu.edu, West Virginia Univ – PeGrg
Lechler, Paul J., (775) 682-8773 plechler@unr.edu, Univ of Nevada – Cg
Leckie, R. Mark, (413) 545-1948 mleckie@geo.umass.edu, Univ of Massachusetts, Amherst – PmGru
Lee, Alexis, (910) 962-3736 leea@uncw.edu, Univ of North Carolina Wilmington – On
Lee, Alyce, 304-256-0270 alee@concord.edu, Concord Univ – Og
Lee, Arthur C., (865) 220-9145 leea@roanestate.edu, Roane State Community Coll - Oak Ridge – Ges
Lee, Chung M., (801) 581-8218 chunglee@geog.utah.edu, Univ of Utah – Ou
Lee, Cin-Ty A., 713.348.4652 ctlee@rice.edu, Rice Univ – Cg
Lee, Cindy, (632) 220-2101 cindy.lee@stonybrook.edu, SUNY, Stony Brook – OcCo
Lee, Cindy M., (864) 656-0672 lc@clemson.edu, Clemson Univ – GeClOe
Lee, Craig M., (206) 685-7656 craig@apl.washington.edu, Univ of Washington – Op
Lee, Daphne E., +64 3 479-7525 daphne.lee@otago.ac.nz, Univ of Otago – Pib
Lee, Eung Seok, (740) 593-1101 leee1@ohio.edu, Ohio Univ – Hw
Lee, Hung, (519) 824-4120 Ext.53828 hlee@uoguelph.ca, Univ of Guelph – On
Lee, In Young, (630) 252-8724 Argonne National Laboratory – Oa
Lee, Jaeheon, 520-626-4967 jaeheon@email.arizona.edu, Univ of Arizona – Nx
Lee, Jeffrey, (509) 963-2801 jeff@geology.cwu.edu, Central Washington Univ – Gct
Lee, Jeffrey A., 806-834-8228 jeff.lee@ttu.edu, Texas Tech Univ – Oy
Lee, Jejung, leej@umkc.edu, Univ of Missouri-Kansas City – GqHw
Lee, Kanani K., (203) 432-4354 kanani.lee@yale.edu, Yale Univ – Gy
Lee, Keenan, (303) 273-3808 klee@mines.edu, Colorado School of Mines – Or
Lee, Martin, +4401413302634 Martin.Lee@glasgow.ac.uk, Univ of Glasgow – Gz
Lee, Meehye, 82-2-3290-3178 meehye@korea.ac.kr, Korea Univ – OagCm
Lee, Michael D., 510-885-3155 michael.lee@csueastbay.edu, California State Univ, East Bay – Hg
Lee, Ming-Kuo, (334) 844-4898 leeming@auburn.edu, Auburn Univ – HwGe
Lee, Rachel J., (315) 312-5506 rachel.lee@oswego.edu, SUNY, Oswego – GvOr
Lee, Sukyoung, (814) 863-1587 slg9@psu.edu, Pennsylvania State Univ, Univ Park – Ow
Lee, Wook, (814) 732-2291 wlee@edinboro.edu, Edinboro Univ of Pennsylvania – Oi
Lee, Young Jae, 82-2-3290-3181 youngjlee@korea.ac.kr, Korea Univ – ScGze
Lee, Zhongping , zhongping.lee@umb.edu, Univ of Massachusetts, Boston – Org
Lee-Gorishti, Yolanda, 203-392-6647 yolanda.lee-gorishti@uconn.edu, Southern Connecticut State Univ – GaOe
Leech, Mary L., (415) 338-1144 leech@sfsu.edu, San Francisco State Univ – GptGz
Leeman, William P., (713) 348-4892 leeman@rice.edu, Rice Univ – Gi
Lees, Jonathan M., (919) 962-0695 jonathan.lees@unc.edu, Univ of North Carolina, Chapel Hill – YsGv
Leetaru, Hannes E., 217-333-5058 leetaru@isgs.uiuc.edu, Illinois State Geological Survey – Eo
Lefebvre, Rene, (418) 654-2651 rene.lefebvre@ete.inrs.ca, Universite du Quebec – Hw
LeFever, Julie A., jlefever@nd.gov, North Dakota Geological Survey – Gg
LeFever, Richard D., (701) 777-3014 richard.lefever@engr.und.edu, Univ of North Dakota – GsrGo
Lefticariu, Liliana, (618) 453-7373 lefticariu@geo.siu.edu, Southern Illinois Univ Carbondale – CslGe
Leger, Carol, 603.358.2570 cleger@keene.edu, Keene State Coll – Og
Leggitt, Leroy, lleggitt@llu.edu, Loma Linda Univ – Pi
Legore, Virginia L., (509) 376-5019 virginia.legore@pnl.gov, Pacific Northwest National Laboratory – Cg
Lehane, Mary, +353 21 4902764 m.lehane@ucc.ie, Univ Coll Cork – On
Lehman, Thomas M., 806-834-3148 tom.lehman@ttu.edu, Texas Tech Univ – Gs
Lehrberger, Gerhard, +49 89 289 25832 lehrberger@tum.de, Technische Universitaet Muenchen – EgGxg
Lehre, Andre K., (707) 826-3165 akl1@humboldt.edu, Humboldt State Univ – GmHs
Lehrmann, Daniel J., 210-999-7654 dlehrmann@trinity.edu, Trinity Univ – PiGs
Lehto, Heather L., (325) 486-6990 heather.lehto@angelo.edu, Angelo State Univ – GvYsOe
Leichmann, Jaromir, +420 549 49 5559 leichman@sci.muni.cz, Masaryk Univ – GxpGi
Leichter, James J., (858) 822-5330 jleichter@ucsd.edu, Univ of California, San Diego – Ob
Leier, Andrew L., 803-777-9941 aleier@geol.sc.edu, Univ of South Carolina – Gs
Leighton, Henry G., (514) 398-3766 henry.leighton@mcgill.ca, McGill Univ – Oa
Leighton, Lindsey, 780-492-3983 lleighto@ualberta.ca, Univ of Alberta – Pi
Leighty, Robert S., (480) 461-7021 rleighty@mesacc.edu, Mesa Community Coll – Og
Leimer, H. Wayne, (931) 372-3522 hwleimer@tntech.edu, Tennessee Tech Univ – Gz
Leinen, Margaret, (401) 874-6222 mleinen@nsf.gov, Univ of Rhode Island – Ou
Leinen, Margaret, 858-534-2827 mleinen@ucsd.edu, Univ of California, San Diego – OgPe
Leinfelder, Reinhold, 089/2180 6629 r.leinfelder@lrz.uni-muenchen.de, Ludwig-Maximilians-Universitaet Muenchen – Pg
Leitch, Alison, (709) 737-3306 aleitch@mun.ca, Memorial Univ of Newfoundland – Yg
Leite, Michael B., (308) 432-6377 mleite@csc.edu, Chadron State Coll – Gg
Leith, Kerry, +49 (89) 289 - 25867 kerry.leith@tum.de, Technische Universitaet Muenchen – GmNrOi
Leithold, Elana L., (919) 515-7282 lonnie_leithold@ncsu.edu, North Carolina State Univ – Gs
Leitz, Robert E., (303) 556-3072 Metropolitan State Coll of Denver – Gz
Lekan, Thomas, lekan@sc.edu, Univ of South Carolina – On
Lekic, Vedran, (301) 405-4086 ved@umd.edu, Univ of Maryland – Ysg
Leland, John, 775-784-6670 jleland@unr.edu, Univ of Nevada, Reno – On
Lemay, Phillip W., (225) 388-6922 philip@lgs.bri.lsu.edu, Louisiana State Univ – Gg
Lemiszki, Peter J., (865) 594-5596 peter.lemiszki@tn.gov, Pellissippi State Community Coll – GcgOi
Lemke, Karen A., (715) 346-2709 klemke@uwsp.edu, Univ of Wisconsin, Stevens Point – GmOy
Lemke, Lawrence D., (989) 774-1144 l.d.lemke@cmich.edu, Central Michigan Univ – HwGes
Lemmon, Mark, 979-458-8098 lemmon@tamu.edu, Texas A&M Univ – Oa
Lemmond, Peter C., (508) 289-2457 plemmond@whoi.edu, Woods Hole Oceanographic Institution – On
Lempe, Bernhard, +49 89289 25862 lempe@tum.de, Technische Universitaet Muenchen – GlgNg
Lenardic, Adrian, (713) 348-4883 ajns@rice.edu, Rice Univ – Gc
Lenczewski, Melissa E., (815) 753-7937 lenczewski@niu.edu, Northern Illinois Univ – Ge
Lene, Gene W., (210) 436-3011 glene@stmarytx.edu, Saint Mary's Univ – Ge
Lener, Edward, (540) 231-9249 lener@vt.edu, Virginia Polytechnic Inst & State Univ – On
Lenhart, Stephen W., (540) 831-5257 slenhart@radford.edu, Radford Univ
Lenkey, László, lenkey@pangea.elte.hu, Eotvos Lorand Univ – YhHwGa
Lennox, Paul G., + 61 2 9385 8096 p.lennox@unsw.edu.au, Univ of New South Wales – GcgOe
Lentz, David R., (506) 547-2070 Univ of New Brunswick – Cg
Lentz, Leonard J., 717–702–2040 lelentz@pa.gov, Pennsylvania Bureau of Topographic & Geologic Survey – Ec
Lenz, Alfred C., (519) 661-3195 aclenz@uwo.ca, Western Univ – PiePi
Lenz, Petra H., 808-956-8003 petra@pbrc.hawaii.edu, Univ of Hawai'i, Manoa – Og
Leonard, Eric M., (719) 389-6513 eleonard@coloradoColl.edu, Colorado Coll – Gl
Leonard, Lynn A., 910-962-2338 lynnl@uncw.edu, Univ of North Carolina Wilmington – Gu
Leonard, Meredith L., 818.778.5595 leonarml@lavc.edu, Los Angeles Valley Coll – OweOi
Leone, James, 518-473-9988 jleone@mail.nysed.gov, New York State Geological Survey – EoGoCg
Leonhart, Leo S., (520) 881-7300 Univ of Arizona – Hw
Leorri, Eduardo, 252 737 2529 leorrie@ecu.edu,

East Carolina Univ – GsPm
Lepain, David, 907-451-5085 david.lepain@alaska.gov, Alaska Division of Geological & Geophysical Surveys – EoGos
Lepper, Kenneth E., (701) 231-6746 ken.lepper@ndsu.edu, North Dakota State Univ – CcGml
Leppert, Ken, (318) 342-1918 Univ of Louisiana, Monroe – Oa
Lerach, David G., (970) 351-2853 David.Lerach@unco.edu, Univ of Northern Colorado – Owa
Lerch, Derek, dlerch@frc.edu, Feather River Coll – Og
Lerch, Robert N., 573-882-1402 Univ of Missouri, Columbia – Os
Lerczak, Jim, 541-737-6128 jlerczak@coas.oregonstate.edu, Oregon State Univ – Op
Lerdau, Manuel, (434) 924-3325 mtl5g@virginia.edu, Univ of Virginia – On
Lerman, Abraham, 847-491-7385 abe@earth.northwestern.edu, Northwestern Univ – CgGnCm
Lermurier, Nathalie, +33(0)3 44062540 nathalie.lermurier@lasalle-beauvais.fr, Institut Polytechnique LaSalle Beauvais (ex-IGAL) – On
Lerner-Lam, Arthur L., (845) 365-8356 lerner@ldeo.columbia.edu, Columbia Univ – Ys
Lesher, Charles E., (530) 752-9779 celesher@ucdavis.edu, Univ of California, Davis – Gi
Lesher, Michael, 7056751151 x2276 MLesher@laurentian.ca, Laurentian Univ, Sudbury – Em
Lesht, Barry, 312-413-3176 blesht@uic.edu, Univ of Illinois at Chicago – Op
Lesht, Barry M., (630) 252-4208 bmlesht@anl.gov, Argonne National Laboratory – Op
Lessard, Evelyn J., (206) 543-8795 elessard@u.washington.edu, Univ of Washington – Ob
Leszczynski, Raymond F., 3152948613 ext. 2313 Ray@cayuga-cc.edu, Cayuga Community Coll – GglGm
Letelier, Ricardo, (541) 737-3890 letelier@coas.oregonstate.edu, Oregon State Univ – Obr
Lethbridge, Mark, mark.lethbridge@flinders.edu.au, Flinders Univ – Oiy
Letsinger, Sally L., (812) 855-1356 sletsing@indiana.edu, Indiana Univ – HqOir
Leung, Irene S., (718) 960-8572 Graduate School of the City Univ of New York – Gz
Levander, Alan, (713) 348-6064 alan@rice.edu, Rice Univ – Ys
Levas, Stephen J., 262-472-6200 levass@uww.edu, Univ of Wisconsin, Whitewater – CmGe
Levasseur, Emile , levasseure@easternct.edu, Eastern Connecticut State Univ – Og
Leventer, Amy, (315) 228-7213 aleventer@colgate.edu, Colgate Univ – Ou
Leventon, Julia, +44(0) 113 34 31635  J.Leventon@leeds.ac.uk, Univ of Leeds – Ge
Lever, Helen, +44 (0) 131 451 4057 h.lever@hw.ac.uk, Heriot-Watt Univ – Go
Leverington, David W., 806-834-5310 david.leverington@ttu.edu, Texas Tech Univ – Oiy
Leveson, David J., (718) 951-5330 Graduate School of the City Univ of New York – On
Levesque, Andre, levesque.andre@ggl.ulaval.ca, Universite Laval – Gz
Levey, Raymond A., (801) 585-3826 rlevey@egi.utah.edu, Univ of Utah – Eo
Levin, Lisa A., (858) 534-3579 llevin@ucsd.edu, Univ of California, San Diego – Ob
Levin, Naomi, 410-516-4317 nlevin3@jhu.edu, Johns Hopkins Univ – Gs
Levin, Vadim, 848-445-5415 vlevin@eps.rutgers.edu, Rutgers, The State Univ of New Jersey – Ysg
Levine, Norman S., (843) 953-5308 levinen@cofc.edu, Coll of Charleston – OiGeNg
Levine, Rebekah, 406-683-7134 rebekah.levine@umwestern.edu, Univ of Montana Western – HgGmOa
Levino, Lynn J., 412-442-4299 Pennsylvania Bureau of Topographic & Geologic Survey – Go
Levinson, Alfred A., (403) 220-5846 Univ of Calgary – Eg
Levinton, Jeffrey, (631) 632-8602 jeffrey.levinton@stonybrook.edu, SUNY, Stony Brook – Ob
Levson, Victor M., (250) 952-0391 vic.levson@gems9.gov.bc.ca, Univ of Victoria – Gl
Levy, David, david.levy@umb.edu, Univ of Massachusetts, Boston – On
Levy, Jonathan, (513) 529-1947 levyj@MiamiOh.edu, Miami Univ – Hw
Levy, Melissa H., (916) 484-8684 levym@arc.losrios.edu, American River Coll – GgOgy
Levy, Schon S., (505) 667-9504 sslevy@lanl.gov, Los Alamos National Laboratory – Gi
Lew, Alan A., (928) 523-6567 alan.lew@nau.edu, Northern Arizona Univ – On
Lew, Jeffrey, (310) 825-3023 lew@atmos.ucla.edu, Univ of California, Los Angeles – Oa
Lewis, Alan G., (604) 822-3626 alewis@eos.ubc.ca, Univ of British Columbia – Ob
Lewis, Brian T. R., (206) 543-7419 blewis@u.washington.edu, Univ of Washington – Yr
Lewis, Chris, cjlewis@ccsf.edu, City Coll of San Francisco – GgEmOs
Lewis, Gerald L., (626) 585-7137 gllewis@pasadena.edu, Pasadena City Coll – Ps
Lewis, John S., (360) 873-8781 jsl@u.arizona.edu, Univ of Arizona – Xcm
Lewis, Jon C., (724) 357-5624 jclewis@iup.edu, Indiana Univ of Pennsylvania – Gc
Lewis, Katie, 806-746-6101 krothlisberger@tamu.edu, Texas A&M Univ – Sco
Lewis, Marlon R., (902) 494-3513 marlon.lewis@dal.ca, Dalhousie Univ – Ob
Lewis, Mary, 5102357800 x.4284 mlewis@contracosta.edu, Contra Costa Coll – Gg
Lewis, Reed S., (208) 885-7472 reedl@uidaho.edu, Univ of Idaho – GgiEg
Lewis, Ronald D., (334) 844-4886 lewisrd@auburn.edu, Auburn Univ – Pi
Lewis, Stephen D., (559) 278-6956 slewis@csufresno.edu, California State Univ, Fresno – Yg
Leybourne, Matthew I., 7056751151 x2263 mleybourne@laurentian.ca, Laurentian Univ, Sudbury – Cg
Leyden, Barbara W., (813) 974-0324 Univ of South Florida, Tampa – Pl
Leyrit, Hervé, +33(0)3 44068998 herve.leyrit@lasalle-beauvais.fr, Institut Polytechnique LaSalle Beauvais (ex-IGAL) – Gv
Li, Aibing, 713-743-2878 ali2@uh.edu, Univ of Houston – Ys
Li, Baosheng, 631 632-9642 baosheng.li@sunysb.edu, SUNY, Stony Brook – Gy
Li, Chusi, (812) 855-1558 cli@indiana.edu, Indiana Univ, Bloomington – GiEm
Li, Dan, lidan@bu.edu, Boston Univ – HqOa
Li, Gary, 510-885-3165 gary.li@csueastbay.edu, California State Univ, East Bay – Oi
Li, Junran, 918-631-2517 junran-li@utulsa.edu, The Univ of Tulsa – Gme
Li, Liangping, (605) 394-2461 liangping.li@sdsmt.edu, South Dakota School of Mines & Tech – Hqw
Li, Lin, ll3@iupui.edu, Indiana Univ / Purdue Univ, Indianapolis – XgOrg
Li, Long, 780-492-9288 long4@ualberta.ca, Univ of Alberta – CsGpv
Li, Peng, (501) 683-0114 peng.li@arkansas.gov, Arkansas Geological Survey – Gog
Li, Ping-Chi, (931) 372-3752 PLI@TnTech.edu, Tennessee Tech Univ – Oi
Li, Rong-Yu, 204-727-9684 lir@brandonu.ca, Western Univ – Pie
Li, Tim, timli@hawaii.edu, Univ of Hawai'i, Manoa – Oa
Li, Wenhong, 919 684-5015 wl66@duke.edu, Duke Univ – On
Li, William K., (902) 426-6349 lib@mar.dfo-mpo.gc.ca, Dalhousie Univ – Ob
Li, Xiangshan, 713-743-0742 xli10@uh.edu, Univ of Houston – Oaw
Li, Yaoguo, (303) 273-3510 ygli@mines.edu, Colorado School of Mines – Yev
Li, Yong-Gang, (213) 740-3556 hylirenko@marshall.usc.edu, Univ of Southern California – Ys
Li, Yuan-Hui, 808-956-6297 yhli@soest.hawaii.edu, Univ of Hawai'i, Manoa – Cm
Li, Yuncong, (305) 246-7000 yunli@ufl.edu, Univ of Florida – ScOs
Li, Zhenhong, +44 (0) 191 208 5704 zhenhong.li@ncl.ac.uk, Univ of Newcastle Upon Tyne – Yd
Liang, George, (785) 532-6101 gliang@ksu.edu, Kansas State Univ – So
Liang, Liyuan, 865-241-3933 2ll@ornl.gov, Oak Ridge National Laboratory – Cl
Liang, Yan, (401) 863-9477 Yan_Liang@Brown.edu, Brown Univ – Cp
Liauw, Henri L., hliauwap@broward.edu, Broward Coll – Gg
Libarkin, Julie C., (517) 355-8369 libarkin@msu.edu, Michigan State Univ – Oen
Liberty, Lee M., (208) 426-1166 lml@cgiss.boisestate.edu, Boise State Univ – Ys
Libra, Robert D., Robert.Libra@dnr.iowa.gov, Iowa Dept of Natural Resources – Gg

Licciardi, Joseph M., 603-862-1718 joe.licciardi@unh.edu, Univ of New Hampshire – Gl
Licht, Alexis, licht@uw.edu, Univ of Washington – Gsn
Licht, Kathy J., (317) 278-1343 Indiana Univ / Purdue Univ, Indianapolis – Gl
Lichtenberger, János, lityi@sas.elte.hu, Eotvos Lorand Univ – OarXy
Lichtner, Peter C., (505) 667-3420 lichtner@lanl.gov, Los Alamos National Laboratory – On
Liddell, W. David, (435) 797-1261 dave.liddell@usu.edu, Utah State Univ – GsPo
Lidgard, Scott H., (312) 665-7625 slidgard@fieldmuseum.org, Field Museum of Natural History – Pi
Lidiak, Edward G., (412) 624-8871 egl@pitt.edu, Univ of Pittsburgh – Gi
Lidicker, Jr., William Z., (510) 642-3059 Univ of California, Berkeley – Pv
Liebe, Richard M., (585) 395-5100 (Ext. 7524) rliebe@weather.brockport.edu, SUNY, The Coll at Brockport – Ps
Liebens, Johan, 850-474-2065 liebens@uwf.edu, Univ of West Florida – So
Lieberman, Bruce S., (785) 864-2741 blieber@ku.edu, Univ of Kansas – Poi
Lieberman, Robert C., (631) 632-8214 robert.liebermann@sunysb.edu, SUNY, Stony Brook – GyYsx
Liebling, Richard, 516 463-6545 georsl@hofstra.edu, Hofstra Univ – Gg
Liebling, Richard S., (212) 772-5412 Graduate School of the City Univ of New York – Gz
Lierman, Robert T., (859) 622-1278 tom.lierman@eku.edu, Eastern Kentucky Univ – Gsr
Lifton, Nathaniel A., (765) 494-0754 nlifton@purdue.edu, Purdue Univ – GmCc
Light, Bonnie, 206-543-9824 bonnie@apl.washington.edu, Univ of Washington – Oa
Lightbody, Anne F., 603-862--0711 anne.lightbody@unh.edu, Univ of New Hampshire – HgsOn
Likos, William J., 608-890-2662 likos@wisc.edu, Univ of Wisconsin, Madison – NgSp
Lilley, Marvin D., (206) 543-0859 lilley@ocean.washington.edu, Univ of Washington – Ob
Lillie, Robert J., lillier@geo.oregonstate.edu, Oregon State Univ – Ye
Lilly, Troy, 325-574-7922 tlilly@wtc.edu, Western Texas Coll – GgOg
Lima, Eduardo A., 617-324-2829 limaea@mit.edu, Massachusetts Inst of Tech – Ym
Lima, Ivan D., ilima@whoi.edu, Woods Hole Oceanographic Institution – Oc
Limp, W. F., 479-575-7909 flimp@uark.edu, Univ of Arkansas, Fayetteville – Oi
Lin, Douglas, (831) 459-2732 lin@lick.ucsc.edu, Univ of California, Santa Cruz – Xc
Lin, Fan-Chi, (801) 581-4373 fanchi.lin@utah.edu, Univ of Utah – Ygs
Lin, Gong-yuh, gongyuh.lin@csun.edu, California State Univ, Northridge – Oa
Lin, Hsing K., (907) 474-6347 hklin@alaska.edu, Univ of Alaska, Fairbanks – On
Lin, Jialin, (614) 292-6634 lin.789@osu.edu, Ohio State Univ – Oa
Lin, Jian, (508) 289-2576 jlin@whoi.edu, Woods Hole Oceanographic Institution – Gt
Lin, John C., 801-581-7530 john.lin@utah.edu, Univ of Utah – Oa
Lin, Jung-Fu, 512-471-8054 afu@jsg.utexas.edu, Univ of Texas at Austin – GyYm
Lin, Senjie, (860) 405-9168 senjie.lin@uconn.edu, Univ of Connecticut – Ob
Lin, Shoufa, (519) 888-4567 (Ext. 6557) shoufa@uwaterloo.ca, Univ of Waterloo – On
Lin, Wuyin, (631) 632-3141 Wuyin.Lin@stonybrook.edu, SUNY, Stony Brook – Oa
Lin, Xiaomao, (785) 532-6816 xlin@ksu.edu, Kansas State Univ – On
Lin, Yu-Feng F., 217-333-0235 yflin@illinois.edu, Univ of Illinois, Urbana-Champaign – Hg
Linares, Rogelio, ++935811259 rogelio.linares@uab.cat, Universitat Autonoma de Barcelona – NgYx
Lincoln, Beth Z., (517) 629-0331 blincoln@albion.edu, Albion Coll – Gc
Lincoln, Jonathan M., (973) 655-7273 lincolnj@mail.montclair.edu, Montclair State Univ – Gr
Lincoln, Timothy N., 517-629-0486 tlincoln@albion.edu, Albion Coll – ClHw
Lindberg, David R., (510) 642-3926 Univ of California, Berkeley – Pi
Lindberg, David R., drl@berkeley.edu, Univ of California, Berkeley – Po
Lindberg, Jonathan W., (509) 376-5005 jon.lindberg@pnl.gov, Pacific Northwest National Laboratory – Ec
Lindberg, Steven E., partnerships@ornl.gov, Oak Ridge National Laboratory – Cl
Lindbo, David, (919) 793-4428 North Carolina State Univ – Sd
Linde, Alan T., (202) 478-8835 alinde@carnegiescience.edu, Carnegie Institution of Washington – Ys
Lindemann, Richard H., rlindemann@skidmore.edu, Skidmore Coll – Pi
Lindemann, William C., (505) 646-3405 New Mexico State Univ, Las Cruces – Sb
Lindenmeier, Clark W., (509) 376-8419 clark.lindenmeier@pnl.gov, Pacific Northwest National Laboratory – Cg
Lindgren, Paula, 441413305442 Paula.Lindgren@glasgow.ac.uk, Univ of Glasgow – Oa
Lindley, Stacy, (916) 278-6337 stacy.lindley@csus.edu, California State Univ, Sacramento – On
Lindline, Jennifer, (505) 426-2046 lindlinej@nmhu.edu, New Mexico Highlands Univ – GizOe
Lindo Atichati, David, (718) 982-2919 david.Lindo@csi.cuny.edu, Coll of Staten Island – Gue
Lindquist, Anna, 906-635-2140 alindquist1@lssu.edu, Lake Superior State Univ – YmGz
Lindsay, Everett H., (520) 621-6024 ehlind@geo.arizona.edu, Univ of Arizona – Pv
Lindsay, Matthew B., (306) 966-5693 matt.lindsay@usask.ca, Univ of Saskatchewan – ClGg
Lindsey, Kassandra, 303-384-2660 kolindsey@mines.edu, Colorado School of Mines – NgGmOi
Lindsley, Donald H., (631) 632-8195 donald.lindsley@sunysb.edu, SUNY, Stony Brook – Cp
Lindsley-Griffin, Nancy, (402) 472-2629 nlg@unl.edu, Univ of Nebraska, Lincoln – Gt
Lindzen, Richard S., (617) 253-2432 rlindzen@mit.edu, Massachusetts Inst of Tech – Ow
Lineback, Neal G., (828) 262-3000 linebackng@appstate.edu, Appalachian State Univ – Oy
Liner, Christopher , 479-575-4835 liner@uark.edu, Univ of Arkansas, Fayetteville – YseGg
Lines, Larry R., (403) 220-5841 Univ of Calgary – Ye
Lini, Andrea, (802) 656-0245 andrea.lini@uvm.edu, Univ of Vermont – CsGn
Link, Curtis A., (406) 496-4165 clink@mtech.edu, Montana Tech – Ye
Link, Paul K., (208) 282-3365 linkpaul@isu.edu, Idaho State Univ – Gs
Linky, Edward, (212) 772-5265 linky.edward@epamail.epa.gov, Hunter Coll (CUNY) – On
Linn, Anne M., (202) 334-2744 alinn@nas.edu, National Academy of Sciences/National Research Council – Gs
Linn, Rodman, (505) 665-6254 rrl@lanl.gov, Los Alamos National Laboratory – Ng
Linneman, Scott R., (360) 650-7207 Scott.Linneman@wwu.edu, Western Washington Univ – GmOe
Linnen, Robert, 519-661-2111 x89207 rlinnen@uwo.ca, Western Univ – EmCe
Linol, Bastien, bastien.aeon@gmail.com, Nelson Mandela Metropolitan Univ – Gs
Liou, Juhn G., (650) 723-2716 liou@pangea.stanford.edu, Stanford Univ – Gp
Lipeles, Maxine I., (314) 935-5482 milipeles@seas.wustl.edu, Washington Univ in St. Louis – On
Lippelt, Irene D., (608) 262-7430 irene.lippelt@wgnhs.uwex.edu, Univ of Wisconsin - Extension – Gg
Lippert, Peter C., (801) 581-4599 pete.lippert@utah.edu, Univ of Utah – YmGtPe
Lippmann, Thomas C., 603-862-4450 lippmann@ccom.unh.edu, Univ of New Hampshire – On
Lips, Elliott W., (801) 581-8218 elliott.lips@geog.utah.edu, Univ of Utah – Gm
Lisenbee, Alvis L., (605) 394-2461 Alvis.Lisenbee@sdsmt.edu, South Dakota School of Mines & Tech – Gc
Lisichenko, Richard, 785-628-4159 rlisiche@fhsu.edu, Fort Hays State Univ – Oi
Lisiecki, Lorraine, (805) 893-4437 lisiecki@geol.ucsb.edu, Univ of California, Santa Barbara – Gn
Lisle, R J., lisle@cf.ac.uk, Cardiff Univ – Gc
Lisle, Thomas, (707) 825-2930 tel7001@humboldt.edu, Humboldt State Univ – Hs
Liss, Peter, +44 (0)1603 59 2563 p.liss@uea.ac.uk,

Univ of East Anglia – Co
Lithgow-Bertelloni, Carolina, +44 020 7679 37220 c.lithgow-bertelloni@ucl.ac.uk, Univ Coll London – Yg
Little, Crispin, +44(0) 113 34 36621 earctsl@leeds.ac.uk, Univ of Leeds – Pg
Little, Jonathan, jlittle@monroecc.edu, Monroe Community Coll – OylOi
Little, Tim, +64 4 463 6198 Tim.Little@vuw.ac.nz, Victoria Univ of Wellington – Gct
Little, William W., (208) 496-7679 littlew@byui.edu, Brigham Young Univ - Idaho – GsdGm
Littler, Kate L., 01326255725 k.littler@exeter.ac.uk, Exeter Univ – PeCl
Liu, Chuntao, 361-825-3845 Chuntao.Liu@tamucc.edu, Texas A&M Univ, Corpus Christi – Oa
Liu, Dantong, dantong.liu@manchester.ac.uk, Univ of Manchester – Oa
Liu, Gaisheng, 785-864-2115 gliu@kgs.ku.edu, Univ of Kansas – Hg
Liu, Jian G., +44 20 759 46418 j.g.liu@imperial.ac.uk, Imperial Coll – Or
Liu, Jian-yi, (406) 994-6905 uesjl@montana.edu, Montana State Univ – On
Liu, Jingpu P., 919-515-3711 jpliu@ncsu.edu, North Carolina State Univ – Gu
Liu, Jiping, jiping.liu@eas.gatech.edu, Georgia Inst of Tech – Oa
Liu, Lanbo, (860) 486-1388 lanbo.liu@uconn.edu, Univ of Connecticut – Yg
Liu, Lijun, (217) 300-0378 ljliu@illinois.edu, Univ of Illinois, Urbana-Champaign – YgGtm
Liu, Mian, (573) 882-3784 LiuM@missouri.edu, Univ of Missouri – Yg
Liu, Paul Hiaibao, Paul.Liu@dnr.iowa.gov, Iowa Dept of Natural Resources – Gg
Liu, Ping, (631) 632-3195 Ping.Liu@stonybrook.edu, SUNY, Stony Brook – Oa
Liu, Shimin, (814) 863-4491 szl3@psu.edu, Pennsylvania State Univ, Univ Park – Nm
Liu, Weibo, 561 297-4965 liuw@fau.edu, Florida Atlantic Univ – On
Liu, Xiaoming, (919) 962-0675 xiaomliu@unc.edu, Univ of North Carolina, Chapel Hill – Cgl
Liu, Yajing, (514) 398-4085 yajing.liu@mcgill.ca, McGill Univ – Ygs
Liu, Yongqiang, (706) 559-4240 yliu@fs.fed.us, Georgia Inst of Tech – Ow
Liu, Zhanfei, 361-749-6772 zhanfei.liu@utexas.edu, Univ of Texas at Austin – Og
Liu, Zheng-yu, (608) 262-0777 Univ of Wisconsin, Madison – Op
Liu, Zhenxian, (314) 882-3784 zxliu@bnl.gov, Carnegie Institution of Washington – Gy
Liutkus, Cynthia, 828-262-6933 liutkuscm@appstate.edu, Appalachian State Univ – Gd
Livaccari, Richard F., (970) 248-1081 rlivacca@coloradomesa.edu, Colorado Mesa Univ – Gc
Lively, Rich, (612) 626-3103 lively@umn.edu, Univ of Minnesota – Cc
Livens, Francis, +44 0161 275-4647 francis.livens@manchester.ac.uk, Univ of Manchester – Cg
Livermore, Phil, +44(0) 113 34 30379 p.w.livermore@leeds.ac.uk, Univ of Leeds – Ym
Livingstone, Daniel A., (919) 684-3264 livingst@duke.edu, Duke Univ – Pe
Lizarralde, Daniel, 508-289-2942 dlizarralde@whoi.edu, Woods Hole Oceanographic Institution – Ys
Llanos, Hilario, hilario.llanos@ehu.eus, Univ of the Basque Country UPV/EHU – Hgy
Llewellin, Ed, +44 (0) 191 33 42336 ed.llewellin@durham.ac.uk, Durham Univ – Gv
Lloyd, Glenn D., 573/368-2176 glenn.lloyd@dnr.mo.gov, Missouri Dept of Natural Resources – On
Lloyd, Graeme, +44 (1865) 272056 graeme.lloyd@earth.ox.ac.uk, Univ of Oxford – Gs
Lloyd, Jonathan, +44 0161 275-7155 Jon.Lloyd@manchester.ac.uk, Univ of Manchester – Ge
Loague, Keith, (650) 723-0847 keith@pangea.stanford.edu, Stanford Univ – Hq
Lobben, Amy, (541) 346-4566 lobben@uoregon.edu, Univ of Oregon – OinOn
Lobegeier, Melissa, 615-898-2403 Melissa.Lobegeier@mtsu.edu, Middle Tennessee State Univ – Pmg
Locat, Jacques E., (418) 656-2179 locat@ggl.ulaval.ca, Universite Laval – Ng
Lock, Brian E., (337) 482-6823 belock@louisiana.edu, Univ of Louisiana at Lafayette – Gs
Locke, Daniel B., (207) 287-7171 daniel.b.locke@maine.gov, Dept of Agriculture, Conservation, and Forestry – Hw
Locke, David C., (718) 997-3271 Graduate School of the City Univ of New York – Ca
Locke, Erika , Erika.Locke@tamucc.edu, Texas A&M Univ, Corpus Christi – Ggo
Locke, James L., (415) 485-9526 jim@marin.edu, Coll of Marin – Og
Locke, Randall, (217) 333-3866 rlocke@illinois.edu, Univ of Illinois, Urbana-Champaign – Cl
Locker, Stanley D., (727) 553-1502 stan@marine.usf.edu, Univ of South Florida – Ou
Lockley, Martin G., (303) 556-4884 Univ of Colorado, Denver – Pi
Lockwood, John P., (808) 967-8579 jplockwood@volcanologist.com, Univ of Hawai'i, Hilo – Gvg
Lockwood, Rowan, (757) 221-2878 rxlock@wm.edu, Coll of William & Mary – Po
Lodders-Fegley, Katharina, (314) 935-4851 lodders@levee.wustl.edu, Washington Univ in St. Louis – Xc
Lodge, Robert, 715-836-4361 lodgerw@uwec.edu, Univ of Wisconsin, Eau Claire – Gc
Loeb, Valerie, 831-771-4400 loeb@mlml.calstate.edu, Moss Landing Marine Laboratories – On
Lofthouse, Stephen T., (212) 346-1760 slofthouse@fsmail.pace.edu, Pace Univ, New York Campus – Og
Logan, Alan, (506) 648-5715 logan@unbsj.ca, Univ of New Brunswick Saint John – Ob
Logan, John M., (541) 346-4587 jmllogan@aol.com, Univ of Oregon – Nr
Logsdon, Miles G., (206) 543-5334 mlog@u.washington.edu, Univ of Washington – Oi
Lohan, Maeve, +44 (0)23 80596449 M.Lohan@soton.ac.uk, Univ of Southampton – Ob
Loheide, Steven P., (608) 265-5277 loheide@wisc.edu, Univ of Wisconsin, Madison – HwSp
Lohman, Rowena B., rbl62@cornell.edu, Cornell Univ – YsOrGt
Lohmann, George P., (508) 289-2840 glohmann@whoi.edu, Woods Hole Oceanographic Institution – Pe
Lohmann, Kyger C., (734) 763-2298 kacey@umich.edu, Univ of Michigan – CsGsCl
Lohrengel, II, C. Frederick, 435-586-7941 lohrengel@suu.edu, Southern Utah Univ – GsrPs
Lokau, Katja, +49 89 289 25857 katja.lokau@tum.de, Technische Universitaet Muenchen – Ng
Lomaga, Margaret, lomagam@sunysuffolk.edu, Suffolk County Community Coll, Ammerman Campus – Og
Lombardo, Kelly, 860-405-9256 kelly.lombardo@uconn.edu, Univ of Connecticut – Op
London, David, (405) 325-3253 dlondon@ou.edu, Univ of Oklahoma – CpGiz
Lonergan, Lidia, +44 20 759 46465 l.lonergan@imperial.ac.uk, Imperial Coll – GtcGs
Long, Ann D., (217) 244-6172 annlong@illinois.edu, Univ of Illinois, Urbana-Champaign – Gm
Long, Austin, (520) 621-8888 along@geo.arizona.edu, Univ of Arizona – Cl
Long, Colleen, longcm@illinois.edu, Univ of Illinois, Urbana-Champaign – Gg
Long, David T., (517) 353-9618 long@msu.edu, Michigan State Univ – ClGeb
Long, Leon E., (512) 459-7838 leonlong@jsg.utexas.edu, Univ of Texas at Austin – Cc
Long, Lisa, 614 265 6590 lisa.long@dnr.state.oh.usm, Ohio Dept of Natural Resources – On
Long, Maureen D., (203) 432-5031 maureen.long@yale.edu, Yale Univ – Ys
Long , Sean P., 509-335-8868 sean.p.long@wsu.edu, Washington State Univ – Gc
Longerich, Henry, (709) 737-8380 henry@sparky2.esd.mun.ca, Memorial Univ of Newfoundland – Cg
Longnecker, Krista, (508) 289-2824 klongnecker@whoil.edu, Woods Hole Oceanographic Institution – Oc
Longoria, Jose F., (305) 348-3614 longoria@fiu.edu, Florida International Univ – Ps
Longstaffe, Frederick J., (519) 661-3177 flongsta@uwo.ca, Western Univ – Cs
Longworth, Brett, 508-289-3559 blongworth@whoi.edu, Woods Hole Oceanographic Institution – Ou
Lonn, Jeff, (406) 496-4177 jlonn@mtech.edu, Montana Tech of The Univ of Montana – GgtGc

Lonsdale, Darcy J., (631) 632-8712 darcy.lonsdale@stonybrook.edu, SUNY, Stony Brook – Ob

Lonsdale, Peter F., (858) 534-2855 plonsdale@ucsd.edu, Univ of California, San Diego – Gu

Looney, Brian, 803-725-3692 brian02.looney@srnl.doe.gov, Clemson Univ – Hq

Loope, David B., (402) 472-2647 dloope1@unl.edu, Univ of Nebraska, Lincoln – Gsg

Loope, Henry, (812) 856-3117 hloope@indiana.edu, Indiana Univ – Gls

Looy, Cynthia, 510-642-1607 looy@berkeley.edu, Univ of California, Berkeley – Pb

Lopez, Alberto, (787) 265-3845 alberto.lopez3@upr.edu, Univ of Puerto Rico – YdGtYs

Lopez, Dina L., (740) 593-9435 lopezd@ohio.edu, Ohio Univ – CgGeq

Lopez, Glenn R., (631) 632-8660 glenn.lopez@stonybrook.edu, SUNY, Stony Brook – Ob

Lopez, Margarita, marlopez@cicese.mx, Centro de Investigación Científica y de Educación Superior de Ensenada – Cc

Lord, Mark L., (828) 227-2271 mlord@wcu.edu, Western Carolina Univ – HwGmOe

Lorenz, Ralph, (520) 621-5585 rlorenz@lpl.arizona.edu, Univ of Arizona – Xg

Lorenzo, Juan M., (225) 388-2497 juan@geol.lsu.edu, Louisiana State Univ – Ys

Lorenzoni, Irene, +44 (0)1603 59 3173 i.lorenzoni@uea.ac.uk, Univ of East Anglia – Eg

Lorinczi, Piroska, +44(0) 113 34 39245 earpl@leeds.ac.uk, Univ of Leeds – Np

Loring, Arthur P., (718) 262-2079 York Coll (CUNY) – Gm

Losh, Steven, (507) 389-6323 steven.losh@mnsu.edu, Minnesota State Univ – GzxGo

Losos, Zdenek, ++420 549 49 5623 losos@sci.muni.cz, Masaryk Univ – Gzy

Lott, Dempsey E., (508) 289-2929 dlott@whoi.edu, Woods Hole Oceanographic Institution – Oc

Lotterhos, Kathleen, (781) 581-7370 k.lotterhos@neu.edu, Northeastern Univ – On

Lottermoser, Bernd, +44 01326 255973 B.Lottermoser@exeter.ac.uk, Exeter Univ – Ca

Loucks, Bob, (512) 471-0366 bob.loucks@beg.utexas.edu, Univ of Texas at Austin, Jackson School of Geosciences – Go

Louden, Keith E., (902) 494-3452 keith.louden@dal.ca, Dalhousie Univ – Yr

Lough, Amanda, (215) 895 6456 amanda.c.lough@drexel.edu, Drexel Univ – Ysg

Louie, John, (775) 784-4219 louie@unr.edu, Univ of Nevada, Reno – Yg

Lounsbury, Diane E., (585) 245-5291 lounsbur@geneseo.edu, SUNY, Geneseo – On

Lourenço, José M., 00351259350280 martinho@utad.pt, Universidade de Trás-os-Montes e Alto Douro – YeOin

Lourenco, Sergio, lourencosd@cf.ac.uk, Cardiff Univ – Sp

Love, David W., (505) 835-5146 dave@gis.nmt.edu, New Mexico Inst of Mining & Tech – Ge

Love, Gordon, gordon.love@ucr.edu, Univ of California, Riverside – Cog

Lovekin, Jonathan, 303-384-2654 jlovekin@mines.edu, Colorado School of Mines – NgOuGs

Loveland, Andrea M., (307) 766-2286 andrea.loveland@wyo.gov, Wyoming State Geological Survey – Gct

Loveland, Karen, 573-368-2142 karen.loveland@dnr.mo.gov, Missouri Dept of Natural Resources – On

Loveless, Jack, (413) 585-2657 jloveles@smith.edu, Smith Coll – Gtc

Lovell, Mark D., (208) 496-1903 lovellm@byui.edu, Brigham Young Univ - Idaho – HyOiGo

Lovell, Mike, +440116 252 3798 mike.lovell@le.ac.uk, Leicester Univ – Yg

LoVetere, Crystal, clovetere@cerritos.edu, Cerritos Coll – Oy

Lovett, Andrew A., +44 (0)1603 59 3126 a.lovett@uea.ac.uk, Univ of East Anglia – Oiu

Lovett, Cole, (269) 927-8744 lovett@lakemichiganColl.edu, Lake Michigan Coll – Gg

Lowe, Donald R., (650) 725-3040 lowe@pangea.stanford.edu, Stanford Univ – Gd

Lowe, Douglas, Douglas.Lowe@manchester.ac.uk, Univ of Manchester – Oa

Lowe, John C., (202) 994-6188 George Washington Univ – Og

Lowe, Mike, (801) 537-3389 mikelowe@utah.gov, Utah Geological Survey – Hw

Lowell, Robert P., (540) 231-6004 rlowell@vt.edu, Virginia Polytechnic Inst & State Univ – Yrh

Lowell, Thomas V., (513) 556-4165 thomas.lowell@uc.edu, Univ of Cincinnati – Gl

Lowenstein, Tim K., (607) 777-4254 lowenst@binghamton.edu, Binghamton Univ – Cl

Lower, Steven K., 614 292-1571 lower.9@osu.edu, Ohio State Univ – Py

Lowery, Birl, (608) 262-2752 blowery@wisc.edu, Univ of Wisconsin, Madison – Sp

Lowery, Mary Sue, (619) 260-4600 slowery@sandiego.edu, Univ of San Diego – Ob

Lowry, Anthony R., (435) 797-7096 tony.lowry@usu.edu, Utah State Univ – Yds

Lowry, Christopher S., 716-645-4266 cslowry@buffalo.edu, SUNY, Buffalo – Hw

Lowry, David, +44 1784 443105 D.Lowry@rhul.ac.uk, Univ of London, Royal Holloway & Bedford New Coll – Cs

Lowry, Wallace D., (540) 231-6521 Virginia Polytechnic Inst & State Univ – Gc

Lowry, William R., (314) 935-5821 Washington Univ in St. Louis – On

Loxsom, Fred, loxsomf@easternct.edu, Eastern Connecticut State Univ – On

Loyd, Sean, 657-278-4537 sloyd@fullerton.edu, California State Univ, Fullerton – CgGe

Loydell, David, +44 023 92 842698 david.loydell@port.ac.uk, Univ of Portsmouth – Gr

Loynachan, Thomas E., (515) 290-7154 teloynac@iastate.edu, Iowa State Univ of Science & Tech – So

Lozier, M. Susan, (919) 681-8199 s.lozier@duke.edu, Duke Univ – Op

Lozinsky, Richard P., (714) 992-7445 rlozinsky@fullcoll.edu, Fullerton Coll – GreGm

Lozowski, Edward P., (780) 492-0385 edward.lozowski@ualberta.ca, Univ of Alberta – Oa

Lu, Youyu, (902) 426-7780 youyu.lu@ec.gc.ca, Dalhousie Univ – Op

Lu, Yuehan, 205-348-1882 yuehanlu@as.ua.edu, Univ of Alabama – GgOuGe

Lu, Zhong, 214-768-0101 Southern Methodist Univ – Or

Lu, Zunli, 315-443-0281 zunlilu@syr.edu, Syracuse Univ – Cg

Lucas, Cathy, +44 (0)23 80596617 cathy.lucas@noc.soton.ac.uk, Univ of Southampton – Ob

Lucas, Franklin A., frank.lucas@uniben.edu, Univ of Benin – PmlGr

Lucas-Clark, Joyce, 415-452-5046 jluclark@comcast.net, City Coll of San Francisco – PgGg

Lucey, Paul G., (808) 956-3137 Univ of Hawai'i, Manoa – Xy

Lucia, F. J., 512-471-1534 jerry.lucia@beg.utexas.edu, Univ of Texas at Austin – Eo

Lucia, F. Jerry, (512) 471-7367 jerry.lucia@beg.utexas.edu, Univ of Texas at Austin, Jackson School of Geosciences – Go

Luciano, Katherine E., (843) 953-6843 lucianok@dnr.sc.gov, Dept of Natural Resources – GusGm

Lucotte, Marc Michel, 514-987-3000 #3767 lucotte.marc_michel@uqam.ca, Universite du Quebec a Montreal – Oc

Lucus, Beth, (540) 231-4595 blucas06@vt.edu, Virginia Tech – On

Luczaj, John A., (920) 465-5139 luczajj@uwgb.edu, Univ of Wisconsin, Green Bay – HwGsCl

Ludman, Allan, (718) 997-3271 Graduate School of the City Univ of New York – Gg

Ludman, Allan, (718) 997-3300 allan.ludman@qc.cuny.edu, Queens Coll (CUNY) – GgtGr

Ludvigson, Greg A., (785) 864-2734 gludvigson@ku.edu, Univ of Kansas – GrCsPe

Ludwikoski, David J., (443) 840-4216 dludwikoski@ccbcmd.edu, Community Coll of Baltimore County, Catonsville – OegOn

Lueth, Virgil L., 575-835-5140 vwlueth@nmt.edu, New Mexico Inst of Mining and Tech – GzEg

Lukas, Roger, (808) 956-7896 rlukas@soest.hawaii.edu, Univ of Hawai'i, Manoa – Op

Lukinbeal, Christopher, clukinbe@email.arizona.edu, Univ of Arizona – Oi

Lumpkin, Thomas A., (509) 335-2726 christie.lumpkin@wsu.edu, Washington State Univ – On

Lumsden, David N., (901) 678-4359 dlumsden@memphis.edu, Univ of Memphis – Gd

Lund, David, 860-405-9331 david.lund@uconn.edu,

Univ of Connecticut – Oc
Lund, Jay R., (916) 752-5671 Univ of California, Davis – Hs
Lund, Steven P., (213) 740-5835 slund@usc.edu,
Univ of Southern California – Ym
Lund, William R., (435) 865-9034 billlund@utah.gov,
Utah Geological Survey – Ng
Lundblad, Steven P., 808-974-7641 slundbla@hawaii.edu,
Univ of Hawai'i, Hilo – GraGe
Lundelius, Ernest L., (512) 232-5513 erniel@utexas.edu,
Univ of Texas at Austin – PvePo
Lundin, Robert F., (480) 965-3514 robert.lundin@asu.edu,
Arizona State Univ – Pm
Lundquist, Jessica D., 206-685-7594 jdlund@uw.edu,
Univ of Washington – Oa
Lundstrom, Craig C., (217) 244-6293 lundstro@illinois.edu,
Univ of Illinois, Urbana-Champaign – Cp
Lundstrom, Elizabeth, elundstr@umn.edu,
Univ of Minnesota, Twin Cities – Ca
Lunine, Jonathan I., (520) 621-2789 jlunine@lpl.arizona.edu,
Univ of Arizona – Xc
Luo, Chao, chao.luo@eas.gatech.edu, Georgia Inst of Tech – Oa
Luo, Gang, 512-475-6613 gangluo66@gmail.com,
Univ of Texas at Austin – Gtc
Luo, Jun, (417) 836-4273 junluo@missouristate.edu,
Missouri State Univ – Oi
Luo, Zhexi, (412) 622-6578 calleryb@CarnegieMNH.org,
Carnegie Museum of Natural History – Pv
Lupankwa, Mlindelwa , 012 382 6213 lupankwam@tut.ac.za,
Tshwane Univ of Tech – HywGg
Lupia, Richard, (405) 325-7229 rlupia@ou.edu,
Univ of Oklahoma – Pm
Lupo, Anthony R., (573) 884-1638 lupoa@missouri.edu,
Univ of Missouri, Columbia – Oa
Lupulescu, Marian V., (518) 474-1432 marian.lupulescu@nysed.gov,
New York State Geological Survey – GzxEm
Lusardi, Barb, (612) 626-4791 lusar001@umn.edu, Univ of Minnesota – Gl
Lusk, Braden , 859-257-1105 braden.lusk@uky.edu,
Univ of Kentucky – Nm
Luth, Robert W., (780) 492-2740 robert.luth@ualberta.ca,
Univ of Alberta – Gi
Luther, Amy, (225) 578-2337 aluther@lsu.edu, Louisiana State Univ – Gc
Luther, Douglas S., (808) 956-5875 dluther@soest.hawaii.edu,
Univ of Hawai'i, Manoa – Op
Luther, Mark E., (727) 553-1528 mluther@marine.usf.edu,
Univ of South Florida – Op
Luther, III, George W., (302) 645-4208 luther@udel.edu,
Univ of Delaware – Cm
Luttge, Andreas, (713) 348-6304 aluttge@rice.edu, Rice Univ – Cg
Luttrell, Karen, (225) 578-5620 kluttrell@lsu.edu,
Louisiana State Univ – Yg
Lutz, Alexandra, 775.673.7418 alexandra.lutz@dri.edu,
Univ of Nevada, Reno – Hw
Lutz, Pascale, +33(0)3 44068990 pascale.lutz@lasalle-beauvais.fr,
Institut Polytechnique LaSalle Beauvais (ex-IGAL) – YgxYd
Lutz, Timothy M., (610) 436-3498 tlutz@wcupa.edu,
West Chester Univ – Gq
Lux, Daniel R., (207) 581-4494 dlux@maine.edu, Univ of Maine – Gi
Luyendyk, Bruce P., (805) 893-2827 Univ of California, Santa Barbara – Yr
Luzincourt, Marc R., (418) 654-3715 mluzinco@nrcan.gc.ca,
Universite du Quebec – Cs
Lužar-Oberiter, Borna, +38514606115 bluzar@geol.pmf.hr,
Univ of Zagreb – Gg
Lwiza, Kamazima M., (631) 632-7309 kamazima.lwiza@stonybrook.edu,
SUNY, Stony Brook – Oc
Lübbe, Maike, 089/2180 4337 Meike.Luebbe@lrz.uni-muenchen.de,
Ludwig-Maximilians-Universitaet Muenchen – Gz
Lyle, Mike, 757-822-7189 tclylem@tcc.edu, Tidewater Community Coll – Gg
Lynch, Michael J., (859) 323-0561 mike.lynch@uky.edu,
Univ of Kentucky – On
Lynch-Stieglitz, Jean, (404) 894-3944 jean@eas.gatech.edu,
Georgia Inst of Tech – PeCsOg
Lynds, Ranie, (307) 766-2286 x235 ranie.lynds@wyo.gov,
Wyoming State Geological Survey – Eo
Lynds, Ranie, (307) 766-2286 ranie.lynds@wyo.gov,
Univ of Wyoming – Gsr
Lynn, Resler M., (540) 231-5790 resler@vt.edu,
Virginia Polytechnic Inst & State Univ – Oy
Lyon, Elizabeth C., 717.702.2063 elyon@pa.gov,
Pennsylvania Bureau of Topographic & Geologic Survey – On
Lyon, Ian, Ian.Lyon@manchester.ac.uk, Univ of Manchester – Xc
Lyon, Linda M., (406) 683-7075 linda.lyon@umwestern.edu,
Univ of Montana Western – Ou
Lyon, Ronald J. P., (650) 725-8077 lyon@pangea.stanford.edu,
Stanford Univ – Or
Lyons, Lawrence, (310) 206-7876 larry@atmos.ucla.edu,
Univ of California, Los Angeles – Oa
Lyons, Timothy W., (951) 827-3106 timothy.lyons@ucr.edu,
Univ of California, Riverside – CgOcCl
Lyons, Walter A., 970-351-2647 walyons@frii.com,
Univ of Northern Colorado – Oaw
Lyons, William B., ( 614) 688-3241 lyons.142@osu.edu,
Ohio State Univ – Cl
Lyson, Tyler, 303-370-6328 Tyler.Lyson@dmns.org,
Denver Museum of Nature & Science – Pv
Lytwyn, Jennifer N., 713-743-3296 jlytwyn@uh.edu, Univ of Houston – Gi

## M

Ma, Chong, (334) 844-4203 chongma@auburn.edu, Auburn Univ – GctGg
Ma, Jingsheng, +44 (0)131 451 8296 jingsheng.ma@pet.hw.ac.uk,
Heriot-Watt Univ – Np
Ma, Lena Q., (352) 392-9063 lqma@ufl.edu, Univ of Florida – Sc
Ma, Lin, 915-747-5218 lma@utep.edu, Univ of Texas, El Paso – Hg
Ma, Shuo, 619-594-3091 sma@mail.sdsu.edu, San Diego State Univ – Ys
Ma, Yanxia, yma@lsu.edu, Louisiana State Univ – Og
Maa, Jerome P-Y., (804) 684-7270 maa@vims.edu,
Coll of William & Mary – On
Maantay, Juliana, 718-960-8574 maantay@aol.com,
Lehman Coll (CUNY) – Oi
Maas, Regan, 818-677-3515 regan.maas@csun.edu,
California State Univ, Northridge – Oi
Maasch, Kirk A., (207) 581-2197 kirk@iceage.umeqs.maine.edu,
Univ of Maine – Oa
Mabee, Stephen B., (413) 545-4814 sbmabee@geo.umass.edu,
Massachusetts Geological Survey – Hw
Maboko, Makenya A., +255222410013 mmaboko@uccmail.co.tz,
Univ of Dar es Salaam – GpCgGx
Mac Eachern, James A., (604) 291-5388 jmaceach@sfu.ca,
Simon Fraser Univ – Gs
Mac Niocaill, Conall, +44 (0)1865 282135 conallm@earth.ox.ac.uk,
Univ of Oxford – GtgYm
Macadam, John, J.D.Macadam@exeter.ac.uk, Exeter Univ – Ge
Macalady, Jennifer L., (814) 865-6330 jlm80@psu.edu,
Pennsylvania State Univ, Univ Park – PyClSb
MacAyeal, Douglas R., (773) 702-8027 drm7@midway.uchicago.edu,
Univ of Chicago – Yr
MacCarthy, Ivor, +353 21 4902152 i.maccarthy@ucc.ie,
Univ Coll Cork – Gs
MacCarthy, Patrick, pmaccart@mines.edu, Colorado School of Mines – Co
MacCready, Parker, (206) 685-9588 parker@ocean.washington.edu,
Univ of Washington – Op
Macdonald, Francis, fmacdon@fas.harvard.edu, Harvard Univ – Pe
MacDonald, Iain, +44(0)29 208 77302 McdonaldI1@cardiff.ac.uk,
Univ of Wales – Ca
Macdonald, R. Heather, (757) 221-2443 rhmacd@wm.edu,
Coll of William & Mary – GsOe
MacDonald, William D., (607) 777-2863 Binghamton Univ – Gc
Macdougall, J. Douglas, jdmacdougall@ucsd.edu,
Univ of California, San Diego –
Mace, Gerald, 801-585-9489 jay.mace@utah.edu, Univ of Utah – Oa
MacFadden, Bruce J., 3523921721 x496 bmacfadd@flmnh.ufl.edu,
Univ of Florida – Pv
Macfarlane, Andrew W., (305) 348-3980 macfarla@fiu.edu,
Florida International Univ – Em
MacGregor, Kelly, (651) 696-6441 macgregor@macalester.edu,
Macalester Coll – GmOl
Machel, Hans G., (780) 492-5659 hans.machel@ualberta.ca,
Univ of Alberta – Go
Machovina, Brett , 719-333-3080 brett.machovina@usafa.edu,
United States Air Force Academy – Oir
Macias, Steve E., (360) 475-7711 smacias@olympic.edu, Olympic Coll – Gg
Maciha, Mark J., (928) 523-8242 Mark.Maciha@nau.edu,
Northern Arizona Univ – OnnOn

MacInnes, Breanyn, 509-963-2827 macinnes@geology.cwu.edu, Central Washington Univ – GseGu
MacInnes, Michael, (505) 665-2154 Los Alamos National Laboratory – On
MacIntyre, Hugh, (902) 494-2932 hugh.macintyre@dal.ca, Dalhousie Univ – Ob
Macintyre, Ian G., (202) 633-1339 Smithsonian Institution / National Museum of Natural History – Gs
Mack, John E., (716) 878-5006 mackje@buffalostate.edu, Buffalo State Coll – Xg
Mackaness, William A., +44 (0) 131 650 8163 william.mackaness@ed.ac.uk, Edinburgh Univ – Oi
Mackas, David L., (250) 363-6442 mackas@ios.bc.ca, Univ of Victoria – Ob
Mackenzie, Fred T., (808) 956-6344 fredm@soest.hawaii.edu, Univ of Hawai'i, Manoa – Cm
Macko, Stephen A., (434) 924-6849 sam8f@virginia.edu, Univ of Virginia – CosOc
Mackowiak, Cheryl, 850-875-7126 Echo13@ufl.edu, Univ of Florida – So
MacLachlan, Ian R., (403) 329-2076 maclachlan@uleth.ca, Univ of Lethbridge – Oe
MacLachlan, John, (905) 525-9140 (Ext. 24195) maclacjc@mcmaster.ca, McMaster Univ – Gs
MacLaughlin, Mary M., (406) 496-4655 mmaclaughlin@mtech.edu, Montana Tech of the Univ of Montana – Nr
MacLean, John S., 435.586.1937 johnmaclean@suu.edu, Southern Utah Univ – GctOe
MacLean, Wallace H., whm@eps.mcgill.ca, McGill Univ – Em
Maclennan, John, +44 (0) 1223 761602 jmac05@esc.cam.ac.uk, Univ of Cambridge – Gi
MacLeod, C, macleod@cf.ac.uk, Cardiff Univ – Gu
MacLeod, Chris, +44(0)29 208 74332 MacLeod@cardiff.ac.uk, Univ of Wales – Og
MacLeod, Kenneth A., (573) 884-3118 macleodk@missouri.edu, Univ of Missouri – Pg
MacMahan, Jamie, jhMacMah@nps.edu, Naval Postgraduate School – Onp
MacNish, Robert, (520) 621-5082 macnish@hwr.arizona.edu, Univ of Arizona – Hw
Macomber, Richard, (718) 460-1476 Long Island Univ, Brooklyn Campus – Pg
Macpherson, Colin G., 0191 3342283 colin.macpherson@durham.ac.uk, Durham Univ – CgGiCs
MacPherson, Glenn J., (202) 633-1803 macphers@si.edu, Smithsonian Institution / National Museum of Natural History – XcGi
Macpherson, Gwen L., (785) 864-2742 glmac@ku.edu, Univ of Kansas – Hw
MacQuarrie, Pamela, (403) 440-6176 pmacquarrie@mtroyal.ca, Mount Royal Univ – Oy
Madadi, Mahyar, +61 8 9266-2324 Mahyar.Madadi@curtin.edu.au, Curtin Univ – Ye
Maddock, III, Thomas, (520) 621-7120 maddock@hwr.arizona.edu, Univ of Arizona – Hq
Maddocks, Rosalie F., 713-743-2772 rmaddocks@uh.edu, Univ of Houston – Pm
Madej, Mary Ann, mary_ann_madej@usgs.gov, Humboldt State Univ – Gm
Madin, Ian P., (971) 673-1555 ian.madin@dogami.state.or.us, Oregon Dept of Geology & Mineral Industries – Gg
Madison, Frederick W., (608) 263-4004 fredmad@wisc.edu, Univ of Wisconsin, Madison – Sd
Madsen, David B., (801) 537-3314 nrugs.dmadsen@state.ut.us, Utah Geological Survey – Ga
Madsen, John A., (302) 831-1608 jmadsen@udel.edu, Univ of Delaware – YrOue
Maerz, Christian, +44 (0) 1133431504 c.maerz@leeds.ac.uk, Univ of Leeds – CmOuEm
Magaard, Lorenz, 808-956-7509 lorenz@hawaii.edu, Univ of Hawai'i, Manoa – Op
Magloughlin, Jerry F., (970) 491-1812 jerrym@cnr.colostate.edu, Colorado State Univ – Gpz
Magnani, M. Beatrice, 214-768-1751 Southern Methodist Univ – YsGtYe
Magnani, Maria B., (901) 678-2007 mmagnani@memphis.edu, Univ of Memphis – YesYg
Magnusdottir, Gudrun, (949) 824-3520 gudrun@uci.edu, Univ of California, Irvine – Oa
Magondu, Moffat G., mmagondu@jkuat.ac.ke, Jomo Kenyatta Univ of Agriculture & Tech – OirOg
Magsino, Sammantha L., (202) 334-2744 smagsino@nas.edu, National Academy of Sciences/National Research Council – GvNg
Magson, Justine, +27 (0)51 401 2373 markramj1@ufs.ac.za, Univ of the Free State – CgGi
Mahaffee, Tina, 478-934-3400 tmahaffee@mgc.edu, Middle Georgia Coll – Gg
Mahaffey, Claire, +44 0151 794 4090 Claire.Mahaffey@liverpool.ac.uk, Univ of Liverpool – Oc
Mahan, Kevin H., kevin.mahan@colorado.edu, Univ of Colorado – Gpc
Mahat, Vinod, (630) 252-6407 vmahat@anl.gov, Argonne National Laboratory – Hs
Mahatsente, Rezene, 205-348-5772 rmahatsente@ua.edu, Univ of Alabama – YgmYv
Maher, John, 9134698500 x5953 jmaher1@jccc.edu, Johnson County Community Coll – Oy
Maher, Kierran, 575-835-6354 kmaher@nmt.edu, New Mexico Inst of Mining and Tech – Eg
Maher, Jr., Harmon D., 402-554-4807 harmon_maher@unomaha.edu, Univ of Nebraska at Omaha – GcsGt
Maher, Jr., Louis J., (608) 262-9595 maher@geology.wisc.edu, Univ of Wisconsin, Madison – Pl
Mahlen, Nancy J., (585) 245-5016 mahlen@geneseo.edu, SUNY, Geneseo – GgpCc
Mahler, Robert L., (208) 885-7025 Univ of Idaho – On
Mahmood, Rezaul, (270) 745-5979 rezaul.mahmood@wku.edu, Western Kentucky Univ – Oa
Mahmoud, Sara A., 002-03-3921595 geo_soso2006@yahoo.com, Alexandria Univ – Gg
Mahoney, J. Brian, (715) 836-4952 mahonej@uwec.edu, Univ of Wisconsin, Eau Claire – GstEg
Mahood, Gail A., (650) 723-1429 gail@pangea.stanford.edu, Stanford Univ – Gi
Mahowald, Natalie M., nmm63@cornell.edu, Cornell Univ – Oa
Maier, Raina M., (520) 621-7231 rmaier@ag.arizona.edu, Univ of Arizona – On
Maillol, Jean-Michel, (403) 220-8393 maillol@ucalgary.ca, Univ of Calgary – Ym
Main, Ian G., +44 (0) 131 650 4911 Ian.Main@ed.ac.uk, Edinburgh Univ – Ys
Maio, Chris, (907) 474-5651 cvmaio@alaska.edu, Univ of Alaska, Fairbanks – Oy
Maisey, John G., (212) 769-5811 American Museum of Natural History – Pv
Majodina, Thando, (012) 382-6283 majodinato@tut.ac.za, Tshwane Univ of Tech – GgzGe
Major, Penni, 281-765-7865 penny.westerfeld@nhmccd.edu, Lonestar Coll - North Harris – Gg
Major, Ruth H., 518-629-7131 Hudson Valley Community Coll – Ggh
Maju-Oyovwikowhe, Efetobore G., efetobore.maju-oyovwikowhe@uniben.edu, Univ of Benin – GszGo
Majzlan, Juraj, +49(0)3641 948700 juraj.majzlan@uni-jena.de, Friedrich-Schiller-Univ Jena – Gz
Mak, John E., (631) 632-8673 john.mak@stonybrook.edu, SUNY, Stony Brook – OaCas
Mak, Mankin, (217) 333-8071 mak@atmos.uiuc.edu, Univ of Illinois, Urbana-Champaign – Oa
Makkawi, Mohammad, +966138602621 makkawi@kfupm.edu.sa, King Fahd Univ of Petroleum and Minerals – HwGoq
Makovicky, Peter J., (312) 665-7633 pmakovicky@fieldmuseum.org, Field Museum of Natural History – Pv
Malahoff, Alexander, (808) 956-6802 malahoff@hawaii.edu, Univ of Hawai'i, Manoa – Cm
Malanotte-Rizzoli, Paola M., (617) 253-2451 rizzoli@mit.edu, Massachusetts Inst of Tech – Op
Malcolm, Elizabeth, 757-233-8751 emalcolm@vwc.edu, Virginia Wesleyan Coll – OaCgOg
Malcuit, Robert J., malcuit@denison.edu, Denison Univ – Gx
Malega, Ron, 417-836-4556 rmalega@missouristate.edu, Missouri State Univ – On
Malhotra, Renu, (520) 626-5899 renu@lpl.arizona.edu, Univ of Arizona –
Malin, Gill, +44 (0)1603 59 2531 g.malin@uea.ac.uk, Univ of East Anglia – Obc
Malin, Peter E., (919) 681-8889 p.malin@auckland.ac.nz, Duke Univ – Ys
Malinconico, Lawrence L., (610) 330-5195 malincol@lafayette.edu, Lafayette Coll – Yg
Malinowski, Jacalyn W., 217-244-8035 jwittm2@illinois.edu,

Univ of Illinois, Urbana-Champaign – PgGsPy
Malinowski, Jon C., (845) 938-4673 mal@usma.edu, United States Military Academy – On
Malinverno, Alberto , (845) 365-8577 alberto@ldeo.columbia.edu, Columbia Univ – Gug
Mallard, Laura, mallardl@appstate.edu, Appalachian State Univ – GtOe
Mallarino, Antonio W., (515) 294-6200 apmallar@iastate.edu, Iowa State Univ of Science & Tech – So
Mallick, Subhashis, 307-766-2884 smallick@uwyo.edu, Univ of Wyoming – YseYg
Mallinson, David, (252) 328-1344 mallinsond@ecu.edu, East Carolina Univ – Gu
Malo, Michel, michel.malo@ete.inrs.ca, Universite du Quebec – GctGs
Malone, Andrew, 920-832-7396 andrew.g.malone@lawrence.edu, Lawrence Univ – Ol
Malone, David H., (309) 438-7649 dhmalon@ilstu.edu, Illinois State Univ – GcsEg
Malone, Shawn J., 765-285-8263 sjmalone@bsu.edu, Ball State Univ – GtxGm
Malone, Stephen D., 206-685-3811 steve@ess.washington.edu, Univ of Washington – YsGt
Maloney, Eric, emaloney@atmos.colostate.edu, Colorado State Univ – Oa
Maloof, Adam C., (609) 258-4101 maloof@princeton.edu, Princeton Univ – Ym
Malservisi, Rocco, 089/2180 4202 malservisi@rsmas.miami.edu, Ludwig-Maximilians-Universitaet Muenchen – Yg
Malzer, Gary L., (612) 625-6728 gmalzer@soils.umn.edu, Univ of Minnesota, Twin Cities – Sc
Mamot, Philipp, +49 (89) 289 25895 philipp.mamot@tum.de, Technische Universitaet Muenchen – GmOg
Mana, Sara , smana@salemstate.edu, Salem State Univ – Gci
Manabe, Syukuro, (609) 258-2790 manabe@princeton.edu, Princeton Univ – Oa
MANCA, Prof. Pierpaolo P., 070-675-5529 ppmanca@unica.it, Universita di Cagliari – Nm
Manchester, Steven R., 3523921721 x495 steven@flmnh.ufl.edu, Univ of Florida – Pb
Mancini, Ernest A., (205) 348-4319 emancini@as.ua.edu, Univ of Alabama – Gro
Mand, Arlene, (215) 573-3164 amand@sas.upenn.edu, Univ of Pennsylvania – On
Manda, Alex K., (252) 328-9403 mandaa@ecu.edu, East Carolina Univ – HwgHq
Mandel, Rolfe, (785) 864-2171 mandel@ku.edu, Univ of Kansas – Gam
Mandelman, John, jmandelman@neaq.org, Univ of Massachusetts, Boston – Ob
Mandia, Scott, 631-451-4104 mandias@sunysuffolk.edu, Suffolk County Community Coll, Ammerman Campus – Ow
Mandra, York T., (415) 338-2061 ytmandra@sfsu.edu, San Francisco State Univ – Pg
Mandrone, Giuseppe, giuseppe.mandrone@unito.it, Università di Torino – Ng
Manduca, Cathryn A., (507) 222-7096 cmanduca@carleton.edu, Carleton Coll – Gi
Mandziuk, William S., (204) 474-7826 mandziu0@cc.umanitoba.ca, Univ of Manitoba – Gg
Manecan, Teodosia, (212) 772-5265 tmanecan@hunter.cuny.edu, Hunter Coll (CUNY) – Gp
Maneta, Marco, 406-243-2454 marco.maneta@mso.umt.edu, Univ of Montana – Hsg
Manfrino, Carrie M., (908) 737-3697 cmanfrin@kean.edu, Kean Univ – GuOb
Manga, Michael, (510) 643-8532 manga@seismo.berkeley.edu, Univ of California, Berkeley – GvXy
Manganini, Steven J., (508) 289-2778 smanganini@whoi.edu, Woods Hole Oceanographic Institution – Ou
Mangano, Gabriela, 306-966-5730 gabriela.mangano@usask.ca, Univ of Saskatchewan – PsGs
Mangel, Mark S., (831) 459-5785 msmangel@cats.ucsc.edu, Univ of California, Santa Cruz – On
Manger, Walter, 479-575-3355 wmanger@uark.edu, Univ of Arkansas, Fayetteville – Gr
Manghnani, Murli H., (808) 956-7825 murli@soest.hawaii.edu, Univ of Hawai'i, Manoa – Yx
Mango, Helen N., (802) 468-1478 helen.mango@castleton.edu, Castleton Univ – CgGe
Mangriotis, Maria-Daphne , +44 (0) 131 451 3565 m.mangriotis@hw.ac.uk, Heriot-Watt Univ – Ys
Mankiewicz, Carol, (608) 363-2371 mankiewi@beloit.edu, Beloit Coll – Gs
Manley, Patricia L., (802) 443-5430 patmanley@middlebury.edu, Middlebury Coll – YrGu
Manley, Thomas O., (802) 443-3114 tmanley@middlebury.edu, Middlebury Coll – Op
Mann, Dan, dhmann@alaska.edu, Univ of Alaska, Fairbanks – Oy
Mann, Daniel H., (907) 474-5872 Univ of Alaska, Fairbanks – Pe
Mann, Keith O., (740) 368-3620 komann@owu.edu, Ohio Wesleyan Univ – Pi
Mann, Michael E., (814) 863-4075 mann@meteo.psu.edu, Pennsylvania State Univ, Univ Park – Oa
Mann, Paul, (512) 471-0452 paulm@ig.utexas.edu, Univ of Texas at Austin – Gt
Manning, Andrew, a.manning@uea.ac.uk, Univ of East Anglia – Oa
Manning, Christina, +44 1784 443835 c.manning@es.rhul.ac.uk, Univ of London, Royal Holloway & Bedford New Coll – Cg
Manning, Craig E., (310) 206-3290 manning@epss.ucla.edu, Univ of California, Los Angeles – GxCp
Manon, Matthew R., (518) 388-8015 manonm@union.edu, Union Coll – Gp
Manos, Leah D., (660) 562-1385 lmanos@nwmissouri.edu, Northwest Missouri State Univ – Oy
Manship, Lori L., 432 552 2245 manship_l@utpb.edu, Univ of Texas, Permian Basin – PioOi
Mansour, Ahmed S., 002-03-3921595 ah_sadek@hotmail.com, Alexandria Univ – Ggs
Manspeizer, Warren, (973) 353-5100 mansp@andromeda.rutgers.edu, Rutgers, The State Univ of New Jersey, Newark – Gr
Mantei, Erwin J., (417) 836-5446 emantei@missouristate.edu, Missouri State Univ – Ct
Mantilla Figueroa, Luis C., lcmantil@uis.edu.co, Universidad Industrial de Santander – EgCec
Manton, William I., (972) 883-2441 manton@utdallas.edu, Univ of Texas, Dallas – Cc
Mantua, Nathan J., nmantua@u.washington.edu, Univ of Washington – Oa
Manzi, Anthony, 718-409-7371 amanzi@sunymaritime.edu, SUNY, Maritime Coll – Ow
Manzocchi, Tom, (+353) 1 716 2605 tom.manzocchi@ucd.ie, Univ Coll Dublin – Goc
Mao, Ho-kwang, 202-478-8960 hmao@carnegiescience.edu, Carnegie Institution of Washington – Yx
Mao, Ho-kwang (David), (773) 702-8136 Univ of Chicago – Yg
Mapani, Benjamin S., +264-61-2063745 bmapani@unam.na, Univ of Namibia – GtcGp
Mapholi, Thendo, +27 (0)51 401 2724 mapholit@ufs.ac.za, Univ of the Free State – Ggp
Marcano, Eugenio J., (413) 538-3237 emarcano@mtholyoke.edu, Mount Holyoke Coll – OiSoOn
Marcantonio, Franco, 979-845-9240 marcantonio@geo.tamu.edu, Texas A&M Univ – Ct
Marchal, Olivier, (508) 289-3374 omarchal@whoi.edu, Woods Hole Oceanographic Institution – Pe
Marchand, Gerard, (413) 538-3173 gmarchan@mtholyoke.edu, Mount Holyoke Coll – Gga
Marchand, Roger T., (206) 685-3757 rojmarch@u.washington.edu, Univ of Washington – Oaa
Marchant, David R., (617) 353-3236 marchant@bu.edu, Boston Univ – Gm
Marchetti, David W., (970) 943-2367 dmarchetti@western.edu, Western State Colorado Univ – GmClc
Marchisin, John, (212) 346-1502 Pace Univ, New York Campus – Og
Marchisin, John, (201) 200-3161 jmarchisin@njcu.edu, New Jersey City Univ – Gg
Marchitto, Thomas M., 303-492-7739 tom.marchitto@colorado.edu, Univ of Colorado – Oc
Marco, Shmulik, shmulikm@tau.ac.il, Tel Aviv Univ – Ggc
Marenco, Katherine N., (610) 526-5627 kmarenco@brynmawr.edu, Bryn Mawr Coll – PiePo
Marenco, Pedro J., (610) 526-7580 pmarenco@brynmawr.edu, Bryn Mawr Coll – PgCsPo
Mareschal, Jean-Claude, 514-987-3000 #6864 mareschal.jean-claude@uqam.ca, Universite du Quebec a Montreal – Yg
Marfurt, Kurt J., (405) 325-5669 kmarfurt@ou.edu, Univ of Oklahoma – Ys

Margot, Jean-Luc, (310) 206-8345 jlm@epss.ucla.edu, Univ of California, Los Angeles – XyYdv
Margrave, Gary F., (403) 220-4606 Univ of Calgary – Ye
Maria, Tony, 812-461-5326 ahmaria@usi.edu, Univ of Southern Indiana – Gv
Mariani, Elisabeth, +440151 794 5180 Mariani@liverpool.ac.uk, Univ of Liverpool – Nr
Marinelli, Roberta, 541-737-5195 marinelr@oregonstate.edu, Oregon State Univ – GuOcGe
Marino, Miguel A., (530) 752-0684 mamarino@ucdavis.edu, Univ of California, Davis – Hw
Marion, Giles, (775) 673-7349 gmarion@dri.edu, Desert Research Inst – Sc
Maritan, Lara, +390498279143 lara.maritan@unipd.it, Università degli Studi di Padova – Gx
Marjanac, Tihomir, +38514606109 marjanac@geol.pmf.hr, Univ of Zagreb – GsgYg
Mark, Bryan G., (614) 247-6180 mark.9@osu.edu, Ohio State Univ – Oa
Mark, Tucker, 802-626-6328 mark.tucker@lyndonstate.edu, Lyndon State Coll – Oa
Market, Patrick S., 573-882-1496 marketp@missouri.edu, Univ of Missouri, Columbia – Oa
Markey, Richard, richard.markey@colostate.edu, Colorado State Univ – Cca
Markley, Michelle J., (413) 538-2814 mmarkley@mtholyoke.edu, Mount Holyoke Coll – GcgGt
Marko, Wayne, 847-491-3459 wayne@earth.northwestern.edu, Northwestern Univ – Gp
Markowitz, Steven, 718-6704180 Queens Coll (CUNY) – GbOn
Markowski, Antonette K., (717) 702-2038 amarkowski@pa.gov, Pennsylvania Bureau of Topographic & Geologic Survey – Eco
Marks, Dennis W., 229-244-5159 Valdosta State Univ – On
Markun, Francis J., (630) 252-7521 Argonne National Laboratory – Cc
Markwith, Scott H., 561 297-2102 smarkwit@fau.edu, Florida Atlantic Univ – Oy
Marley, Nancy A., (630) 252-5014 namarley@anl.gov, Argonne National Laboratory – Oa
Marlowe, Brian W., 813-253-7685 bmarlowe2@hccfl.edu, Hillsborough Community Coll – Og
Marobhe, Isaac M., 255222410013 marobhe@udsm.ac.tz, Univ of Dar es Salaam – YegYx
Marone, Chris, (814) 865-7964 cjm38@psu.edu, Pennsylvania State Univ, Univ Park – Yx
Marquez, L. Lynn, (717) 871-4339 lynn.marquez@millersville.edu, Millersville Univ – Cg
Marr, Paul G., (717) 477-1656 pgmarr@ship.edu, Shippensburg Univ – Oi
Marra, John, (718) 951-5000 x2594 jfm7780@brooklyn.cuny.edu, Brooklyn Coll (CUNY) – Ob
Marrett, Randall A., (512) 471-2113 marrett@mail.utexas.edu, Univ of Texas at Austin – Gc
Marrs, Rob, +44 0151 795 5172 Calluna@liverpool.ac.uk, Univ of Liverpool – Os
Marsaglia, Kathleen M., (818) 677-6309 kathie.marsaglia@csun.edu, California State Univ, Northridge – Gs
Marsan, Yvonne, 910-962-7288 marsany@uncw.edu, Univ of North Carolina Wilmington – Oi
Marschik, Robert, 089/2180 6527 robert.marschik@iaag.geo.uni-muenchen.de, Ludwig-Maximilians-Universitaet Muenchen – Gg
Marsellos, Antonios, (516) 463-5567 antonios.e.marsellos@hofstra.edu, Hofstra Univ – GtqOi
Marsh, Julian S., +27 (0)46-603-7394 goonie.marsh@ru.ac.za, Rhodes Univ – GiCgGv
Marsh, Robert, +44 (0)23 8059 6214 rma@noc.soton.ac.uk, Univ of Southampton – Op
Marshak, Stephen, (217) 333-7705 smarshak@illinois.edu, Univ of Illinois, Urbana-Champaign – Gc
Marshall, Charles R., 510- 642-1821 crmarshall@berkeley.edu, Univ of California, Berkeley – PqoPi
Marshall, Claude Monte, (619) 594-1395 mmarshall@mail.sdsu.edu, San Diego State Univ – Ym
Marshall, Craig, (785) 864-6029 cpmarshall@ku.edu, Univ of Kansas – GzOm
Marshall, Dan D., (604) 291-5387 Simon Fraser Univ – Ca
Marshall, Jeffrey S., 909-869-3461 marshall@csupomona.edu, California State Polytechnic Univ, Pomona – GmtHs
Marshall, Jim, +44-151-794-5177 Isotopes@liverpool.ac.uk, Univ of Liverpool – Cs
Marshall, John C., (617) 253-9615 marshall@gulf.mit.edu, Massachusetts Inst of Tech – Op
Marshall, John E., +44 (0)23 80592015 jeam@noc.soton.ac.uk, Univ of Southampton – PlGro
Marshall, Katherine, (612) 626-8811 kjmarsha@umn.edu, Univ of Minnesota – Gl
Marshall, Scott, 828-265-8680 marshallst@appstate.edu, Appalachian State Univ – YxGc
Marshall, Stephen, (519) 824-4120 Ext.52720 samarsha@uoguelph.ca, Univ of Guelph – On
Marshall, Thomas R., 605-677-6164 tom.marshall@usd.edu, South Dakota Dept of Environment and Natural Resources – GgEo
Marsham, John, +44(0) 113 34 36422 j.marsham@leeds.ac.uk, Univ of Leeds – Oa
Marston, Sallie A., marston@email.arizona.edu, Univ of Arizona – Ou
Martel, Richard, (418) 654-2683 rene.martel@ete.inrs.ca, Universite du Quebec – Hw
Martel, Stephen J., (808) 956-7797 martel@soest.hawaii.edu, Univ of Hawai'i, Manoa – Ng
Martens, Hilary R., (406) 243-6855 hilary.martens@umontana.edu, Univ of Montana – YgsYd
Martill, David, +44 023 92 842256 david.martill@port.ac.uk, Univ of Portsmouth – Po
Martin, Anthony J., (740) 368-3621 geoam@learnlink.emory.edu, Emory Univ – Pe
Martin, Arturo, amartin@cicese.mx, Centro de Investigación Científica y de Educación Superior de Ensenada – Gd
Martin, Barton S., (740) 368-3621 bsmartin@owu.edu, Ohio Wesleyan Univ – GxvCg
Martin, Beth, 314-935-4136 martin@wustl.edu, Washington Univ in St. Louis – Ng
Martin, Candace E., +64 3 479-7526 candace.martin@otago.ac.nz, Univ of Otago – ClcGz
Martin, Charles W., (765) 983-1672 martich@earlham.edu, Earlham Coll – Gg
Martin, Ellen E., (352) 392-2141 eemartin@ufl.edu, Univ of Florida – Cl
Martin, Gale D., 702-651- 3141 Gale.Martin@csn.edu, Coll of Southern Nevada - West Charleston Campus – Gg
Martin, James E., (337) 482-6468 geology@louisiana.edu, Univ of Louisiana at Lafayette – PvGg
Martin, James E., (605) 394-2461 James.Martin@sdsmt.edu, South Dakota School of Mines & Tech – Ps
Martin, John W., (408) 298-2181 San Jose City Coll – Og
Martin, Jonathan B., (352) 392-6219 jbmartin@ufl.edu , Univ of Florida – ClHg
Martin, Jonathan E., (608)262-9845 jon@aos.wisc.edu, Univ of Wisconsin, Madison – Ow
Martin, Mona, (618) 453-4362 monamartin@siu.edu, Southern Illinois Univ Carbondale – On
Martin, Pamela, (773) 834-5245 Univ of Chicago – Pe
Martin, Paul F., (509) 376-2519 paul.martin@pnl.gov, Pacific Northwest National Laboratory – Cg
Martin, Randal S., (505) 835-5469 New Mexico Inst of Mining and Tech – Oa
Martin, Robert F., (514) 324-2579 robert.martin@mcgill.ca, McGill Univ – GziEm
Martin, Ronald E., (302) 831-2569 daddy@udel.edu, Univ of Delaware – PmePg
Martin, Scot T., 617-495-7620 smartin@seas.harvard.edu, Harvard Univ – Cg
Martin, Scott C., 330-941-3026 scmartin@ysu.edu, Youngstown State Univ – Ge
Martin, Seelye, (206) 543-6438 seelye@ocean.washington.edu, Univ of Washington – Op
Martin, Silvana, +390498279104 silvana.martin@unipd.it, Università degli Studi di Padova – Gc
Martin, Timothy J., (630) 252-8708 tjmartin@anl.gov, Argonne National Laboratory – Oa
Martin, Walter, 704-687-5954 wemartin@uncc.edu, Univ of North Carolina, Charlotte – Oa
Martin, William R., (508) 289-2836 wmartin@whoi.edu, Woods Hole Oceanographic Institution – Oc
Martin-Hayden, James, (419) 530-2634 jhayden@geology.utoledo.edu, Univ of Toledo – Hq

Martinez, Fernando, (808) 956-6882 martinez@soest.hawaii.edu, Univ of Hawai'i, Manoa – Yr
Martinez, Judy, (801) 581-6553 judy.martinez@utah.edu, Univ of Utah – On
Martínez Fernández, Francisco, ++935811513 francisco.martinez@uab.cat, Universitat Autonoma de Barcelona – GpCp
Martínez Ribas, Ricard, ++935811464 ricard.martinez@uab.cat, Universitat Autonoma de Barcelona – Pi
Martinez Torres, Luis Miguel, luismiguel.martinez@ehu.eus, Univ of the Basque Country UPV/EHU – Ggt
Martinez-Hackert, Bettina, 716-878-6731 martinb@buffalostate.edu, Buffalo State Coll – Gvm
Martini, Anna M., (413) 542-2067 ammartini@amherst.edu, Amherst Coll – Cl
Martini, I. Peter, (519) 824-4120 x52488 pmartini@uoguelph.ca, Univ of Guelph – Gsl
Martino, Ronald L., (304) 696-2715 martinor@marshall.edu, Marshall Univ – Gs
Martins, Zita, +44 20 759 49982 z.martins@imperial.ac.uk, Imperial Coll – Xm
Martinson, Douglas G., (845) 365-8830 Columbia Univ – Op
Martinuš, Maja, +38514606089 majamarti@geol.pmf.hr, Univ of Zagreb – Gg
Martiny, Adam, 949-824-9713 amartiny@uci.edu, Univ of California, Irvine – Ob
Martire, Luca, luca.martire@unito.it, Università di Torino – Gd
Marty, Kevin, 760-355-5761 kevin.marty@imperial.edu, Imperial Valley Coll – Gg
Martz, Todd, (858) 534-7466 trmartz@ucsd.edu, Univ of California, San Diego – Gg
Marvinney, Robert G., (207) 287-2804 robert.g.marvinney@maine.gov, Dept of Agriculture, Conservation, and Forestry – Gg
Marzen, Luke J., (334) 844-3462 marzelj@auburn.edu, Auburn Univ – Oir
Marzolf, John E., (618) 453-7372 marzolf@geo.siu.edu, Southern Illinois Univ Carbondale – Gs
Marzoli, Andrea, +390498279154 andrea.marzoli@unipd.it, Università degli Studi di Padova – Cg
Marzulli, Walter, (609) 292-2576 walt.marzulli@dep.state.nj.us, New Jersey Geological and Water Survey – Oy
Masch, Ludwig, 089/2180 4273 masch@min.uni-muenchen.de, Ludwig-Maximilians-Universitaet Muenchen – Gz
Masciocco, Luciano, luciano.masciocco@unito.it, Università di Torino – Ge
Masek, Ondrej, +44 (0) 131 650 5095 ondrej.masek@ed.ac.uk, Edinburgh Univ – Ng
Masiello, Caroline A., 713.348.5234 masiello@rice.edu, Rice Univ – On
Maslowski, Wieslaw, (831) 656-3162 maslowsk@nps.edu, Naval Postgraduate School – Op
Mason, Allen S., (505) 667-4140 Los Alamos National Laboratory – Oa
Mason, Charles E., (606) 783-2166 c.mason@moreheadstate.edu, Morehead State Univ – PiGrs
Mason, Owen, (907) 474-6293 Univ of Alaska, Fairbanks – Ga
Mason, Philippa J., +44 20 759 46528 p.j.mason@imperial.ac.uk, Imperial Coll – Or
Mason, Robert, (860) 405-9129 robert.mason@uconn.edu, Univ of Connecticut – OcCtm
Mason, Roger A., (709) 737-4385 rmason@sparky2.esd.mun.ca, Memorial Univ of Newfoundland – Gz
Mason, Sherri A., 716.673.3292 mason@fredonia.edu, SUNY Fredonia – OgGe
Mass, Clifford F., cliff@atmos.washington.edu, Univ of Washington – Oa
Massare, Judy A., (585) 395-2419 jmassare@brockport.edu, SUNY, The Coll at Brockport – Pv
Massey, Michael, michael.massey@csueastbay.edu, California State Univ, East Bay – OnGeCg
Massironi, Matteo, +390498279116 matteo.massironi@unipd.it, Università degli Studi di Padova – XgOr
Mastalerz, Maria, (812) 855-5412 mmastale@indiana.edu, Indiana Univ – Ec
Mastalerz, Maria D., 812-855-7416 mmastale@indiana.edu, Indiana Univ, Bloomington – Ec
Masterlark, Timothy L., (605) 394-2461 Timothy.Masterlark@sdsmt.edu, South Dakota School of Mines & Tech – Yg
Masterman, Steven S., (907) 451-5001 steve.masterman@alaska.gov, Alaska Division of Geological & Geophysical Surveys – EgNg
Masters, T. Guy, (858) 534-4122 tmasters@ucsd.edu, Univ of California, San Diego – YsGy
Mata, Scott, (657) 278-7096 smata@fullerton.edu, California State Univ, Fullerton – Gg
Matano, Ricardo, (541) 737-2212 rmatano@coas.oregonstate.edu, Oregon State Univ – Oap
Matheney, Ronald K., (701) 777-4569 ronald.matheney@engr.und.edu, Univ of North Dakota – Cs
Mathenge, Christine, 931-221-6434 mathengec@apsu.ecu, Austin Peay State Univ – Oy
Mather, Tamsin A., +44 (0)1865 282125 tamsin.mather@earth.ox.ac.uk, Univ of Oxford – Gv
Mathers, Hannah, hannah.mathers@glasgow.ac.uk, Univ of Glasgow – GmlGg
Mathewson, Christopher C., (979) 845-2488 mathewson@geo.tamu.edu, Texas A&M Univ – NgHwEg
Mathez, Edmond A., (212) 769-5379 Graduate School of the City Univ of New York – Gi
Mathias, Simon, s.a.mathias@durham.ac.uk, Durham Univ – Oi
Mathison, Mark E., (515) 294-9686 mathison@iastate.edu, Iowa State Univ of Science & Tech – Oe
Mathur, Dr R., +919347303173 ramrajm@rediffmail.com, Univ Coll of Science, Osmania Univ – YemYx
Mathur, Ryan, 814-641-3725 mathur@juniata.edu, Juniata Coll – CeYgHw
Matisoff, Gerald, (216) 368-3677 gerald.matisoff@case.edu, Case Western Reserve Univ – Cl
Matney, Timothy, 330 972-6892 matney@uakron.edu, Univ of Akron – Ga
Matson, Pamela A., (650) 725-6812 matson@pangea.stanford.edu, Stanford Univ – Sb
Matson, Sam, 208-426-3645 sammatson@boisestate.edu, Boise State Univ – GrPv
Matsumoto, Katsumi, 612-624-0275 katsumi@umn.edu, Univ of Minnesota, Twin Cities – Oc
Matter, Juerg M., +44 (0)23 80593042 J.Matter@southampton.ac.uk, Univ of Southampton – Ng
Mattey, Dave, +44 1784 443587 D.Mattey@rhul.ac.uk, Univ of London, Royal Holloway & Bedford New Coll – Cs
Mattheus, Paul E., (907) 474-7758 Univ of Alaska, Fairbanks – Po
Mattheus, Robin, 906-635-2155 crmattheus@gmail.com, Lake Superior State Univ – GsOn
Matthews, Adrian, +44 (0)1603 59 3733 a.j.matthews@uea.ac.uk, Univ of East Anglia – Ow
Matthews, Robert A., (916) 752-0179 Univ of California, Davis – Ge
Mattigod, Shas V., (509) 376-4311 shas.mattigod@pnl.gov, Pacific Northwest National Laboratory – Cl
Mattinson, Chris, 509.963.1628 mattinson@geology.cwu.edu, Central Washington Univ – Gzp
Mattioli, Glen, (817) 272-2987 mattioli@uta.edu, Univ of Texas, Arlington – GitYd
Mattison, Katherine W., (573) 341-4616 kmattisn@umr.edu, Missouri Univ of Science and Tech – On
Mattox, Stephen R., (616) 331-3734 mattoxs@gvsu.edu, Grand Valley State Univ – OeGv
Mattox, Tari, 616-234-2119 tmattox@grcc.edu, Grand Rapids Community Coll – GgvCg
Mattson, Peter H., (718) 997-3300 Graduate School of the City Univ of New York – Gc
Mattson, Peter H., (718) 997-3335, Queens Coll (CUNY) – Gc
Matty, David J., 801-626-6201 dmatty@weber.edu, Weber State Univ – Gig
Matyjasik, Basia, (801) 537-3122 basiamatyjasik@utah.gov, Utah Geological Survey – Gg
Matyjasik, Marek, (801) 626-7726 mmatyjasik@weber.edu, Weber State Univ – HwGe
Matzke, David, (313) 593-5036 dmatzke@umich.edu, Univ of Michigan, Dearborn – Xy
Matzke, Gordon E., matzkeg@geo.oregonstate.edu, Oregon State Univ – Og
Maul, George A., (321) 674-8096 gmaul@fit.edu, Florida Inst of Tech – Op
Maulud, Mat Ruzlin, 03-79674144 matruzlin@um.edu.my, Univ of Malaya – On
Maurice, Patricia, Patricia.A.Maurice.3@nd.edu, Univ of Notre Dame – Cl
Maurrasse, Florentin J., (305) 348-2350 maurrass@fiu.edu, Florida International Univ – PsGsn
Mauzerall, Denise L., (609) 258-2498 mauzeral@princeton.edu,

Princeton Univ – Oa
Mavko, Gerald M., (650) 723-9438 Stanford Univ – Yg
Mawalagedara, Rachindra, 515-294-8633 rmawala@iastate.edu, Iowa State Univ of Science & Tech – Oa
Maxwell, Arthur E., 512-471-0411 art@ig.utexas.edu, Univ of Texas at Austin – Og
Maxwell, Michael, (604) 296-4218 Michael_Maxwell@golder.com, Univ of British Columbia – Yg
Maxwell, Reed M., 303-384-2456 rmaxwell@mines.edu, Colorado School of Mines – Hs
May, Cynthia, (715) 682-1499 cmay@northland.edu, Northland Coll – OirOy
May, Daniel J., (203) 932-7262 dmay@newhaven.edu, Univ of New Haven – GtxGe
May, Diane M., (417) 836-6900 dianemay@missouristate.edu, Missouri State Univ – On
May, Fred E., (801) 581-8218 pscem.fmay@state.ut.us, Univ of Utah – On
May, James, (662) 325-5806 Mississippi State Univ – Hw
May, Michael, (502) 745-4555 michael.may@wku.edu, Western Kentucky Univ – Ge
May, S J., (972) 377-1635 smay@collin.edu, Collin Coll - Preston Ridge Campus – GgoGc
Mayborn, Kyle R., (309) 298-1577 KR-Mayborn@wiu.edu, Western Illinois Univ – Gic
Mayer, Alex S., (906) 487-3372 asmayer@mtu.edu, Michigan Technological Univ – Hy
Mayer, Bernhard, (403) 220-5389 bmayer@ucalgary.ca, Univ of Calgary – Cs
Mayer, Christine M., (419) 530-8377 christine.mayer@utoledo.edu, Univ of Toledo – On
Mayer, James R., (678) 839-4055 jmayer@westga.edu, Univ of West Georgia – Hw
Mayer, Larry A., 603-862-2615 larry.mayer@unh.edu, Univ of New Hampshire – Gu
Mayer, Margaret, 928-724-6722 Dine' Coll – GeOn
Mayer, Ulrich, (604) 822-1539 umayer@eos.ubc.ca, Univ of British Columbia – Nm
Mayes, Melanie A., 865-574-7336 mayesma@ornl.gov, Univ of Tennessee, Knoxville – ClGeHw
Mayewski, Paul A., (207) 581-3019 paul.mayewski@maine.edu, Univ of Maine – Pe
Mayfield, Michael W., (828) 262-7058 mayfldmw@appstate.edu, Appalachian State Univ – Hs
Maynard, J. Barry, (513) 556-5034 maynarjb@uc.edu, Univ of Cincinnati – Cl
Mayne, Rhiannon G., 817-257-4172 r.mayne@tcu.edu, Texas Christian Univ – XmgXc
Mayo, Alan L., 801-422-2338 alan_mayo@byu.edu, Brigham Young Univ – Hw
Mayor, Shane D., 530-898-6337 sdmayor@csuchico.edu, California State Univ, Chico – Oa
Mazariegos, Ruben A., (210) 381-3523 Univ of Texas, Pan American – Ye
Mazer, James J., (630) 252-7362 Argonne National Laboratory – Cl
Mazzella, Prof. Antonio A., 070-675-5542 mazzella@unica.it, Universita di Cagliari – Gq
Mazzocco, Elizabeth, emazzocc@ashland.edu, Ashland Univ – Oa
Mazzoli, Claudio, +390498279144 claudio.mazzoli@unipd.it, Università degli Studi di Padova – Gx
Mazzullo, Salvatore J., (316) 978-3140 Wichita State Univ – Go
Mbila, Monday O., 256-372-4185 monday.mbila@aamu.edu, Alabama A&M Univ – Sd
Mbuh, Mbongowo J., 701-777-4587 mbongowo.mbuh@und.edu, Univ of North Dakota – OrHs
Mc Mahon, Michelle, 281-765-7865 mmckubota@aol.com, Lonestar Coll - North Harris – Go
McAfee, Gerald B., (432) 335-6558 bmcafee@odessa.edu, Odessa Coll – Pi
McAlester, A. Lee, (214) 768-2623 Southern Methodist Univ – Pe
McAllister, Arnold L., (506) 453-4804 Univ of New Brunswick – Em
McAndrews, John H., (416) 586-5609 Univ of Toronto – Pl
McArthur, John M., +44 020 7679 2376 j.mcarthur@ucl.ac.uk, Univ Coll London – ClHwPs
McArthur, Russell, (415) 338-1755 San Francisco State Univ – On
McArthur, Russell , rmcarthur@ccsf.edu, City Coll of San Francisco – Gg
McAtee, Mike, 909-869-3648 lmmcatee@csupomona.edu, California State Polytechnic Univ, Pomona – On
McBeth, Joyce M., joyce.mcbeth@usask.ca, Univ of Saskatchewan – PyGeSb
McBirney, Alexander R., (541) 344-2539 Univ of Oregon – Gv
McBride, Earle F., (512) 471-1905 efmcbride@mail.utexas.edu, Univ of Texas at Austin – Gd
McBride, John H., (801) 422-5219 john_mcbride@byu.edu, Brigham Young Univ – YeGoe
McBride, Randolph, (703) 993-1642 rmcbride@gmu.edu, George Mason Univ – On
McBride, Raymond G., (519) 824-4120 (Ext. 52492) rmcbride@uoguelph.ca, Univ of Guelph – Os
McCaffrey, Bill, +44(0) 113 34 36625 w.d.mccaffrey@leeds.ac.uk, Univ of Leeds – Gs
McCaffrey, Kenneth J., 0191 3742523 k.j.mccaffrey@durham.ac.uk, Durham Univ – Gc
McCair, Andrew, +44(0) 113 34 35219 a.m.mccaig@leeds.ac.uk, Univ of Leeds – Gc
McCall, Peter L., (216) 368-3676 plm4@case.edu, Case Western Reserve Univ – Po
McCallister, Robert, (608) 758-6556 robert.mccallister@uwc.edu, Univ of Wisconsin Colls – OyGgOn
McCallum, I. Stewart (Stu), mccallum@uw.edu, Univ of Washington – GivGz
McCarter, Michael K., (801) 581-8603 k.mccarter@utah.edu, Univ of Utah – Nm
McCarthy, Daniel, 904 256-7369 Jacksonville Univ – Og
McCarthy, Daniel P., (905) 688-5550 (Ext. 3864) dmccarthy2@brocku.ca, Brock Univ – Gl
McCarthy, Francine G., (905) 688-5550 (Ext. 4286) Francine@brocku.ca, Brock Univ – Pm
McCarthy, James T., 617-495-2330 jmccarthy@oeb.harvard.edu, Harvard Univ – Ob
McCarthy, Michael P., 206-685-2543 mccarthy@ess.washington.edu, Univ of Washington – XyYx
McCarthy, Paul, (907) 474-6894 ffpjm@uaf.edu, Univ of Alaska, Fairbanks – Gs
McCartney, Kevin, (207) 768-9482 kevin.mccartney@umpi.edu, Univ of Maine, Presque Isle – Pm
McCartney, M. C., (608) 263-7393 carol.mccartney@wgnhs.uwex.edu, Univ of Wisconsin - Extension – HwGle
McCaslin, Bobby D., (505) 646-3405 New Mexico State Univ, Las Cruces – Sc
McCauley, Marlene, (336) 316-2236 mmccauley@guilford.edu, Guilford Coll – Cp
McCauley, Steven, 214-860-2398 smccauley@dcccd.edu, El Centro Coll - Dallas Community Coll District – GgOw
McCausland, Phil J., 519-661-2111 x88008 pmccausl@uwo.ca, Western Univ – YmXm
McClain, James S., (916) 752-7093 Univ of California, Davis – Yr
McClaughry, Jason, (541) 523-3133 jason.mcclaughry@dogami.state.or.us, Oregon Dept of Geology & Mineral Industries – Gg
McClay, Ken R., +44 1784 443618 k.mcclay@rhul.ac.uk, Univ of London, Royal Holloway & Bedford New Coll – GctGo
McClellan, Elizabeth, emcclellan@radford.edu, Radford Univ – GipGd
McClellan, Guerry H., (352) 392-2231 guerry@ufl.edu, Univ of Florida – En
McClelland, James W., (361) 749-6756 jimm@utexas.edu, Univ of Texas at Austin – CgHsCs
McClelland, Lori, (775)784-4977 mcclella@unr.edu, Univ of Nevada, Reno – Ys
McClelland, William C., 319-335-1827 bill-mcclelland@uiowa.edu, Univ of Iowa – GtCcGc
McClenaghan, Seán h., + 353 1 8961585 mcclens@tcd.ie, Trinity Coll – Eg
McCluskey, James, (715) 261-6362 jim.mccluskey@uwc.edu, Univ of Wisconsin Colls – OinOn
McCollor, Doug, doug.mccollor@bchydro.com, Univ of British Columbia – Oaw
McCollough, Cherie, 361-825-3166 Cherie.McCollough@tamucc.edu, Texas A&M Univ, Corpus Christi – On
McCollum, Linda B., (509) 359-7473 lmccollum@ewu.edu, Eastern Washington Univ – Po
McConaugha, John R., (757) 683-4698 jmcconau@odu.edu, Old Dominion Univ – Ob
McConnaughhay, Mark, 845-431-8536 mcconn@sunydutchess.edu, Dutchess Community Coll – OgwEo
McConnell, David, (919) 515-0381 david_mcconnell@ncsu.edu, North Carolina State Univ – GgOeGc

McConnell, Joseph, (775) 673-7348 jmcconn@dri.edu, Univ of Nevada, Reno – On
McConnell, Robert L., rmconnel@umw.edu, Univ of Mary Washington – Ge
McConnell, Vicki S., (971) 673-1543 vicki.mcconnell@dogami.state.or.us, Oregon Dept of Geology & Mineral Industries – Gv
McCord, James T., (505) 844-5157 jtmccor@somnet.sandia.gov, New Mexico Inst of Mining and Tech – Hq
McCorkle, Daniel C., (508) 289-2949 dmccorkle@whoi.edu, Woods Hole Oceanographic Institution – Cm
McCormack, John K., (775) 784-4518 mccormac@mines.unr.edu, Univ of Nevada, Reno – Gz
McCormick, George R., erie-@msn.com, Univ of Iowa – Gz
McCormick, Michael L., 315-859-4832 mmccormi@hamilton.edu, Hamilton Coll – Py
McCoy, Edward L., (330) 263-3884 mccoy.13@osu.edu, Ohio State Univ – Sp
McCoy, Floyd W., 236-9115 fmccoy@hawaii.edu, Windward Community Coll – GgaGu
McCoy, Kevin, 303-384-2632 kemccoy@mines.edu, Colorado School of Mines – NgOiHw
Mccoy, Roger M., (801) 581-8218 Univ of Utah – Or
McCoy, Scott W., 775-682-7205 scottmccoy@unr.edu, Univ of Nevada, Reno – GmNg
McCoy, Timothy, 202-633-2206 mccoyt@si.edu, Univ of Tennessee, Knoxville – Xm
McCoy, William D., (413) 545-1535 wdmccoy@geo.umass.edu, Univ of Massachusetts, Amherst – Oy
McCoy, Jr., Floyd W., (808) 235-7318 Univ of Hawai'i, Manoa – Gu
McCraw, David J., (505) 835-5594 djmc@mailhost.nmt.edu, New Mexico Inst of Mining & Tech – Oy
McCreary, Julian P., (808) 956-2216 jay@soest.hawaii.edu, Univ of Hawai'i, Manoa – Op
McCrink, Timothy P., (916) 324-2549 California Geological Survey – Ng
McCulley, Dawn, (209) 667-3466 dmcculley@csustan.edu, California State Univ, Stanislaus – On
McCulloh, Richard P., (225) 578-5327 mccullo@lsu.edu, Louisiana State Univ – Gg
McCullough, Jr., Edgar J., (520) 621-6024 Univ of Arizona – Ge
McCurdy, Maureen, (318) 257-3165 mm@coes.latech.edu, Louisiana Tech Univ – Hw
McCurry, Michael O., (208) 282-3960 mccumich@isu.edu, Idaho State Univ – Gz
McCutcheon, Karen Sue, 307-472 5359 kmccutcheon@casperColl.edu, Casper Coll – Oeg
McDade, Ian C., (416)736-2100 #22859 mcdade@yorku.ca, York Univ – Or
McDaniel, Paul A., (208) 885-7012 Univ of Idaho – Sd
McDermott, Christopher I., +44 (0) 131 650 5931 cmcdermo@staffmail.ed.ac.uk, Edinburgh Univ – Hg
McDermott, Frank, (+353) 1 716 2327 frank.mcdermott@ucd.ie, Univ Coll Dublin – GgCgOg
McDermott, Thomas M., (585) 395-5718 SUNY, The Coll at Brockport – On
McDonald, Andrew M., 7056751151 x2266 amcdonald@laurentian.ca, Laurentian Univ, Sudbury – Gz
McDonald, Brenna, (573) 368-2163 brenna.mcdonald@dnr.mo.gov, Missouri Dept of Natural Resources – Ge
McDonald, Catherine, (406) 496-4883 kmcdonald@mtech.edu, Montana Tech of The Univ of Montana – Grs
McDonald, Eric, (775) 673-7302 emcdonald@dri.edu, Univ of Nevada, Reno – Sp
McDonald, Gregory N., (801) 537-3383 gregmcdonald@utah.gov, Utah Geological Survey – Ng
McDonald, Kyle C., (818) 354-3440 kyle.mcdonald@jpl.nasa.gov, Jet Propulsion Laboratory – On
McDonald, Lynn, (505) 667-1582 lmcdonald@lanl.gov, Los Alamos National Laboratory – On
McDonough, William F., (301) 405-5561 mcdonoug@geol.umd.edu, Univ of Maryland – Cg
McDougall, Jim, 253-566-5060 JMcDougall@tacomacc.edu, Tacoma Community Coll – Og
McDowell, Fred W., (512) 471-1672 mcdowell@mail.utexas.edu, Univ of Texas at Austin – Cc
McDowell, Patricia F., (541) 346-4567 pmcd@uoregon.edu, Univ of Oregon – Gm
McDowell, Rob J., 770-274-5466 rmcdowell@gsu.edu, Georgia State Univ, Perimeter Coll, Dunwoody Campus – Gts
McDowell, Ronald, (304) 594-2331 mcdowell@geosrv.wvnet.edu, West Virginia Univ – GsPiCe
McDuff, Russell E., (206) 545-1947 mcduff@ocean.washington.edu, Univ of Washington – Oc
McElroy, Anne, (631) 632-8488 anne.mcelroy@stonybrook.edu, SUNY, Stony Brook – Oc
McElroy, Brandon, (307) 766-3601 bmcelroy@uwyo.edu, Univ of Wyoming – Gsm
McElroy, Michael B., (617) 495-9261 mbm@io.harvard.edu, Harvard Univ – Oa
McElroy, Tom, (416) 736-2100 x22113 tmcelroy@yorku.ca, York Univ – Oa
McElwain, Jenny, (312) 665-7635 jennifer.mcelwain@gmail.com, Field Museum of Natural History – Pb
McElwaine, Jim, +44 (0) 191 33 42286 Durham Univ – Yg
McElwee, Carl D., (785) 864-3965 cmcelwee@ku.edu, Univ of Kansas – Hg
McEvoy, Jamie , (406) 994-4069 jamie.mcevoy@montana.edu, Montana State Univ – Oe
McEwen, Alfred, (520) 621-4573 mcewen@lpl.arizona.edu, Univ of Arizona – Xg
McFadden, Bruce J., (352) 846-2000 bmacfadd@flmnh.ufl.edu, Univ of Florida – Pv
McFadden, Jennifer, 717-361-1392 mcfaddenj@etown.edu, Elizabethtown Coll – OnnOn
McFadden, Leslie M., (505) 277-6121 lmcfadnm@unm.edu, Univ of New Mexico – Sd
McFarlane, Norman, (250) 363-8227 norm.mcfarlane@ec.gc.ca, Univ of Victoria – Oa
McFarquhar, Greg M., (217) 265-5458 mcfarq@illinois.edu, Univ of Illinois, Urbana-Champaign – Oaw
McFiggans, Gordon, +44 0161 306-3954 Gordon.B.Mcfiggans@manchester.ac.uk, Univ of Manchester – Oa
McGahan, Donald, (254) 968-9701 mcgahan@tarleton.edu, Texas A&M Univ – Sdc
McGarvie, Dave, +44 (0) 131 549 7140 x 71140 dave.mcgarvie@open.ac.uk, The Open Univ – GviCt
McGauley, Michael G., 305-237-2687 mmcgaule@mdc.edu, Miami Dade Coll (Kendall Campus) – Owp
McGeary, Susan, (302) 831-8174 smcgeary@udel.edu, Univ of Delaware – Oe
McGee, David, (617) 324-3545 davidmcg@mit.edu, Massachusetts Inst of Tech – Cl
McGee, Tara, 780-492-3042 tmcgee@ualberta.ca, Univ of Alberta – On
McGehee, Richard V., rmcgehee@austincc.edu, Austin Community Coll District – GgOn
McGehee, Thomas L., (512) 595-3590 kftlm00@tamuk.edu, Texas A&M Univ, Kingsville – Cl
McGhee, Jr., George R., (848) 445-8523 mcghee@rci.rutgers.edu, Rutgers, The State Univ of New Jersey – PoqPi
McGill, George E., (413) 545-0140 Univ of Massachusetts, Amherst – Gc
McGill, Sally F., (909) 537-5347 smcgill@csusb.edu, California State Univ, San Bernardino – GtYdGm
McGinn, Chris, (919) 530-6269 cmcginn@nccu.edu, North Carolina Central Univ – Ogi
McGinnis, Lyle D., (630) 252-8722 Argonne National Laboratory – Yg
McGivern, Tiffany, (315)792-3134 Utica Coll – Geg
McGlade, Jacqueline, +44 020 7679 32839 jacquie.mcglade@ucl.ac.uk, Univ Coll London – Ge
McGlathery, Karen, (804) 924-0558 kjm4k@virginia.edu, Univ of Virginia – Ob
McGlue, Michael M., 859-257-3758 michael.mcglue@uky.edu, Univ of Kentucky – Gso
McGoldrick, Peter J., 61 3 6226 7209 Univ of Tasmania – Ce
Mcgowan, Alistair, +4401413305449 Alistair.McGowan@glasgow.ac.uk, Univ of Glasgow – Pg
McGowan, Eileen, 413-586-8305 emcgowan@geo.umass.edu, Univ of Massachusetts, Amherst – Xg
McGowan, John A., (858) 534-2074 jmcgowan@ucsd.edu, Univ of California, San Diego – Ob
McGrail, Bernard P., (509) 376-9193 pete.mcgrail@pnl.gov, Pacific Northwest National Laboratory – Cg
McGrath, Andrew, andrew.mcgrath@flinders.edu.au, Flinders Univ – Ora
McGrath, Steve F., 406-496-4767 smcgrath@mtech.edu,

Montana Tech of The Univ of Montana – CaEgCe
McGrath, Jr., Dorn C., (202) 994-6185 dornmcg@gwu.edu, George Washington Univ – Oy
McGraw, Maureen A., (505) 665-8128 mcgraw@lanl.gov, Los Alamos National Laboratory – On
McGregor, Kent M., (940) 565-2380 kent.mcgregor@unt.edu, Univ of North Texas – Oag
McGregor, Stuart W., 205-247-3629 smcgregor@gsa.state.al.us, Geological Survey of Alabama – Hs
McGrew, Allen J., (937) 229-5638 amcgrew1@udayton.edu, Univ of Dayton – GtcGp
McGuire, Angela, 509-527-5696 cotaac@whitman.edu, Whitman Coll – Gi
McGuire, Bill, +44 020 7679 33449 w.mcguire@ucl.ac.uk, Univ Coll London – Gv
McGuire, Jeffrey J., (508) 289-3290 jmcguire@whoi.edu, Woods Hole Oceanographic Institution – Ys
McGuire, Jennifer, 651-962-5254 jtmcguire@stthomas.edu, Univ of Saint Thomas – CgHsGb
McGwire, Kenneth, 775.673.7324 ken.mcgwire@dri.edu, Univ of Nevada, Reno – Or
McHargue, Timothy, timmchar@stanford.edu, Univ of Missouri – Gs
McHenry, Lindsay J., (414) 229-3951 lmchenry@uwm.edu, Univ of Wisconsin, Milwaukee – GzaXg
McHugh, Cecilia, (718) 997-3330 cecilia.mchugh@qc.cuny.edu, Queens Coll (CUNY) – Ou
McHugh, Cecilia M., (718) 997-3322 Graduate School of the City Univ of New York – Ou
McHugh, Julia, 970-248-1993 jumchugh@coloradomesa.edu, Colorado Mesa Univ – Pgv
McIlvin, Matt, 508-289-2884 mmcilvin@whoi.edu, Woods Hole Oceanographic Institution – Oc
McInnes, Kevin J., (979) 845-5986 k-mcinnes@tamu.edu, Texas A&M Univ – Sp
McIntosh, Jennifer, 520-626-2282 mcintosh@hwr.arizona.edu, Univ of Arizona – Hy
McIntosh, Kirk D., (512) 471-0480 kirk@ig.utexas.edu, Univ of Texas at Austin – Yr
McIntosh, William C., (505) 835-5324 mcintosh@nmt.edu, New Mexico Inst of Mining & Tech – Cc
McIntyre, Andrew, (718) 997-3329 Queens Coll (CUNY) – Pe
McIsaac, Gregory F., (217) 333-9411 Univ of Illinois, Urbana-Champaign – Hs
McKay, Jennifer L., 541-737-4054 mckay@coas.oregonstate.edu, Oregon State Univ – Cm
McKay, Larry D., (865) 974-5498 lmckay@utk.edu, Univ of Tennessee, Knoxville – HyGe
McKay, Matthew P., (417) 836-5318 MatthewMcKay@missouristate.edu, Missouri State Univ – Gct
McKay, Robert M., Robert.McKay@dnr.iowa.gov, Iowa Dept of Natural Resources – Gg
McKay III, E. Donald, (217) 333-0044 emckay@illinois.edu, Univ of Illinois, Urbana-Champaign – GlsGr
McKean, Adam, (801) 537-3386 adammckean@utah.gov, Utah Geological Survey – Gmg
McKean , Rebecca, 920-403-3227 rebecca.mckean@snc.edu, Saint Norbert Coll – GdPgGh
McKee, James W., (414) 424-4460 mckee@athenet.net, Univ of Wisconsin, Oshkosh – Gr
McKeegan, Kevin D., (310) 825-3580 mckeegan@epss.ucla.edu, Univ of California, Los Angeles – XcCsc
McKenna, Thomas E., 302-831-8257 mckennat@udel.edu, Univ of Delaware – Hg
McKenney, Rosemary, 253-535-8726 mckennra@plu.edu, Pacific Lutheran Univ – Gm
McKenzie, Charlotte, 406-496-4180 cmckenzie@mtech.edu, Montana Tech of The Univ of Montana – On
McKenzie, Connie, mckenzie@coes.latech.edu, Louisiana Tech Univ – On
McKenzie, Jeffrey M., (514) 398-6767 jeffrey.mckenzie@mcgill.ca, McGill Univ – Hw
McKenzie, Phyllis, (202) 633-1860 mckenzie@si.edu, Smithsonian Institution / National Museum of Natural History – On
McKenzie, Ross, ross.mckenzie@agric.gov.ab.ca, Univ of Lethbridge – On
McKenzie, Scott, (814) 824-2382 smckenzie@mercyhurst.edu, Mercyhurst Univ – PgXmOe
McKibben, Michael A., (951) 827-3444 michael.mckibben@ucr.edu, Univ of California, Riverside – Cg
McKinley, Galen, (608) 262-4817 galen@aos.wisc.edu, Univ of Wisconsin, Madison – OcpOg
McKinney, D. Brooks, (315) 781-3304 dbmck@hws.edu, Hobart & William Smith Colls – Gx
McKinney, Frank K., (828) 262-2748 mckinneyfk@appstate.edu, Appalachian State Univ – Pi
McKinney, Mac, 205-247-3549 mmckinney@gsa.state.al.us, Geological Survey of Alabama – Gg
McKinney, Marg J., (828) 262-2747 mckinnymj@appstate.edu, Appalachian State Univ – Pg
McKinney, Michael L., (865) 974-6359 mmckinne@utk.edu, Univ of Tennessee, Knoxville – GePo
McKinney, Nathan, 850-474-3207 nmckinney@uwf.edu, Univ of West Florida – Oi
McKinnon, William B., (314) 935-5604 mckinnon@wustl.edu, Washington Univ in St. Louis – Xg
McKnight, Brian K., (920) 233-4595 briankmcknight@yahoo.com, Univ of Wisconsin, Oshkosh – Gdu
McLafferty, Sara L., (212) 772-5224 slm@everest.hunter.cuny.edu, Graduate School of the City Univ of New York – On
McLaskey, Greg C., gcm8@cornell.edu, Cornell Univ – YsGc
McLaughlin, Patrick I., 812-855-1350 pimclaug@iu.edu, Indiana Univ – GrClGs
McLaughlin, Patrick I., (608) 262-8658 pimclaughlin@wisc.edu, Univ of Wisconsin, Extension – Gr
McLaughlin, Peter P., 302-831-8263 ppmclau@udel.edu, Univ of Delaware – HgPm
McLaughlin, Richard, 361-825-2010 Richard.Mclaughlin@tamucc.edu, Texas A&M Univ, Corpus Christi – On
McLaughlin, Richard A., (919) 515-7306 North Carolina State Univ – Sc
McLaughlin, Jr., Peter P., 302-831-2833 ppmclau@udel.edu, Univ of Delaware – Gr
McLaurin, Brett T., 570-389-4142 bmclauri@bloomu.edu, Bloomsburg Univ – GrsGd
McLean, Dewey M., (540) 231-6521 dmclean@vt.edu, Virginia Polytechnic Inst & State Univ – Pe
McLean, Noah, noahmc@ku.edu, Univ of Kansas – CcGq
McLemore, Virginia, (505) 835-5521 ginger@gis.nmt.edu, New Mexico Inst of Mining & Tech – Em
McLennan, John D., (801) 587-7925 jmclennan@egi.utah.edu, Univ of Utah – Nrp
McLennan, Scott M., (631) 632-8194 scott.mclennan@sunysb.edu, SUNY, Stony Brook – Cg
Mcleod, Andrew T., +44 (0) 131 650 5434 Andy.McLeod@ed.ac.uk, Edinburgh Univ – Og
McLeod, Claire, 513-529-9662 mcleodcl@miamioh.edu, Miami Univ – Gx
McLeod, Clara, cpmcleod@wustl.edu, Washington Univ in St. Louis – On
McLeod, Samuel A., (213) 763-3325 smcleod@ref.usc.edu, Los Angeles County Museum of Natural History – Pv
McManus, Dean A., (206) 543-0587 mcmanus@u.washington.edu, Univ of Washington – Ou
McManus, George B., 860-405-9164 george.mcmanus@uconn.edu, Univ of Connecticut – Ob
McManus, Jerry F., (508) 289-3328 jmcmanus@whoi.edu, Woods Hole Oceanographic Institution – Pe
McManus, Jerry F., (845) 365-8722 jmcmanus@ldeo.columbia.edu, Columbia Univ – PeOu
McManus, Margaret A., (831) 459-4736 Univ of California, Santa Cruz – On
McManus, Margaret Anne, (808) 956-8623 mamc@hawaii.edu, Univ of Hawai'i, Manoa – Op
McMechan, George A., (972) 883-2419 mcmec@utdallas.edu, Univ of Texas, Dallas – Ys
McMenamin, Mark, 413-538-2280 mmcmenam@mtholyoke.edu, Mount Holyoke Coll – PgGst
McMillan, Margaret E., (501) 569-3024 memcmillan@ualr.edu, Univ of Arkansas at Little Rock – Gm
McMillan, Nancy J., 575-646-5000 nmcmilla@nmsu.edu, New Mexico State Univ, Las Cruces – Gi
McMillan, Robert S., (520) 621-6968 bob@lpl.arizona.edu, Univ of Arizona – OnnOn
McMonagle, Julie, (570) 408-4604 julie.mcmonagle@wilkes.edu, Wilkes Univ – Gge
McMullin, David W., (902) 585-1276 david.mcmullin@acadiau.ca, Acadia Univ – Oe
McMurdie, Lynn A., 206-685-9405 mcmurdie@atmos.washington.edu, Univ of Washington – Oa

McMurtry, Gary M., (808) 956-6858 garym@soest.hawaii.edu, Univ of Hawai'i, Manoa – CcGvu
McNair, Laurie A., (505) 665-3328 mcnair@lanl.gov, Los Alamos National Laboratory – On
McNally, Karen C., kmcnally@pmc.ucsc.edu, Univ of California, Santa Cruz – Ys
McNamara, Allen, 480-965-1733 allen.mcnamara@asu.edu, Arizona State Univ – Yx
McNamara, James P., (208) 426-1354 jmcnamar@boisestate.edu, Boise State Univ – Hg
McNamara, Jodi, (315) 228-7201 jmcnamara@colgate.edu, Colgate Univ – On
McNamara, Kenneth, +44 (0) 1223 333410 kmcn07@esc.cam.ac.uk, Univ of Cambridge – Po
McNaught, Mark A., (330) 829-8226 mcnaugma@mountunion.edu, Mount Union Coll – Gcg
McNeal, Karen S., (334) 844-4282 ksm0041@auburn.edu, Auburn Univ – OeGeCm
McNeill, Donald F., (305) 284-3360 d.mcneill@miami.edu, Univ of Miami – GsrGu
McNeill, Lisa, +44 (0)23 80593640 lcmn@noc.soton.ac.uk, Univ of Southampton – GtYr
McNichol, Ann P., (508) 289-3394 amcnichol@whoi.edu, Woods Hole Oceanographic Institution – Oc
McNicol, Barbara, (403) 440-6175 bmcnicol@mtroyal.ca, Mount Royal Univ – Og
McNulty, Brendan A., (310) 243-3412 bmcnulty@csudh.edu, California State Univ, Dominguez Hills – Gc
McPhaden, Michael J., (206) 526-6783 mcphaden@pmel.noaa.gov, Univ of Washington – Og
McPhail, D C "Bear", +61 2 6125 2776 bear@ems.anu.edu.au, Australian National Univ – Cl
McPherron, Robert L., (310) 825-1882 rmcpherr@igpp.ucla.edu, Univ of California, Los Angeles – Xy
McPherron, Robert L., (818) 887-6220 rmcpherron@igpp.ucla.edu, Univ of California, Los Angeles – Xy
McPherson, Brian, 801-581-5634 bmcpherson@egi.utah.edu, Univ of Utah – Ye
McPherson, Robert C., (707) 826-5828 rm4@humboldt.edu, Humboldt State Univ – Gt
McPhie, Jocelyn, 61 3 6226 2892 Jocelyn.McPhie@utas.edu.au, Univ of Tasmania – Gv
McQuaid, Jim, +44(0) 113 34 36724 j.b.mcquaid@leeds.ac.uk, Univ of Leeds – Oa
McQuarrie, Nadine, (412) 624-8870 nmcq@pitt.edu, Univ of Pittsburgh – Gct
McRivette, Michael, 517-629-0276 mmcrivette@albion.edu, Albion Coll – GtOir
McRoberts, Christopher A., (607) 753-2925 mcroberts@cortland.edu, SUNY, Cortland – Pg
McSween, Jr., Harry Y., (865) 974-6359 mcsween@utk.edu, Univ of Tennessee, Knoxville – XcGi
McSweeney, Kevin, (608) 262-0331 kmcsween@wisc.edu, Univ of Wisconsin, Madison – SdaSf
McWethy, David B., (406) 994-6915 dmcwethy@montana.edu, Montana State Univ – PeSb
McWilliams, James, (310) 206-2829 jcm@atmos.ucla.edu, Univ of California, Los Angeles – Oga
McWilliams, Michael O., (650) 723-3718 mcwilliams@stanford.edu, Stanford Univ – CcYme
McWilliams, Robert G., mcwillrg@miamioh.edu, Miami Univ – Ps
MdNamee, Brittani D., (828) 350-4554 bmcnamee@unca.edu, Univ of North Carolina, Asheville – GzxEg
Mead, James I., mead@etsu.edu, Northern Arizona Univ – Po
Meade, Brendan, 617-495-8921 meade@fas.harvard.edu, Harvard Univ – Yg
Meador, Cindy D., 806-651-2582 cmeador@wtamu.edu, West Texas A&M Univ – Og
Meadows, Guy A., 906/487-1106 gmeadows@mtu.edu, Michigan Technological Univ – Gu
Meadows, Wayne R., (505) 665-0291 wmeadows@lanl.gov, Los Alamos National Laboratory – Yg
Means, Guy H., (850) 617-0312 Florida Geological Survey – PgGe
Measures, Christopher, (808) 956-8693 chrism@soest.hawaii.edu, Univ of Hawai'i, Manoa – Oc
Measures, Elizabeth A., (432) 837-8117 measures@sulross.edu, Sul Ross State Univ – Gdq
Mechoso, Carlos R., (310) 825-3057 mechoso@atmos.ucla.edu, Univ of California, Los Angeles – Oa
Meckel, Timothy A., 512-471-4306 tip.meckel@beg.utexas.edu, Univ of Texas at Austin – Gr
Mecklenburgh, Julian, +440161 275-3821 Julian.Mecklenburgh@manchester.ac.uk, Univ of Manchester – Gc
Medaris, Jr., Levi G., (608) 833-4258 medaris@geology.wisc.edu, Univ of Wisconsin, Madison – GipSa
Meddaugh, W. S., (940) 397-4469 scott.meddaugh@mwsu.edu, Midwestern State Univ – GoEgCg
Medina, Cristian R., 855-9992 crmedina@indiana.edu, Indiana Univ – GoeHw
Medina Elizalde, Martin A., (334) 844-4966 mam0199@auburn.edu, Auburn Univ – PeHyCl
Medlin, Peggy, (931) 372-3121 pmedlin@tntech.edu, Tennessee Tech Univ – On
Meduniæ, Gordana, +38514605909 gpavlovi@inet.hr, Univ of Zagreb – CgScGe
Meehan, Katharine, 541-346-4521 meehan@uoregon.edu, Univ of Oregon – On
Meentemeyer, Ross, 704-687-5944 rkmeente@email.uncc.edu, Univ of North Carolina, Charlotte – Oi
Meere, Pat, +353 21 4903056 p.meere@ucc.ie, Univ Coll Cork – Gc
Meert, Joseph G., (352) 392-2231 jmeert@ufl.edu, Univ of Florida – YgGt
Meesters, Antoon, +31 20 59 87362 a.g.c.a.meesters@vu.nl, VU Univ Amsterdam – OwHg
Megarry, Will, w.megarry@qub.ac.uk, Queens Univ Belfast – Gae
Megdal, Sharon B., (520) 792-9591 smegdal@ag.arizona.edu, Univ of Arizona – Hg
Mehes, Roxane J., (705) 673-6575 rmehes@laurentian.ca, Laurentian Univ, Sudbury – On
Mehnert, Edward, (217) 244-2765 emehnert@illinois.edu, Univ of Illinois, Urbana-Champaign – Hyw
Mehra, Aradhana, +44 01332 591133 a.mehra@derby.ac.uk, Univ of Derby – Co
Mehrtens, Charlotte J., (802) 656-0243 cmehrten@uvm.edu, Univ of Vermont – Gs
Mei, Shenghua, 612-626-0572 meixx002@umn.edu, Univ of Minnesota, Twin Cities – Yx
Meigs, Andrew J., (541) 737-1214 meigsa@geo.oregonstate.edu, Oregon State Univ – Gc
Meisner, Caroline B., (775) 340-7341 caroline.meisner@gbcnv.edu, Great Basin Coll – OgSo
Meister, Paul A., (309) 438-7479 pameist@ilstu.edu, Illinois State Univ – GgHgOn
Meisterernst, Götz, 089/2180 4356 goetz.meisterernst@lrz.uni-muenchen.de, Ludwig-Maximilians-Universitaet Muenchen – Gz
Meixner, Thomas, 520-626-1532 tmeixner@hwr.arizona.edu, Univ of Arizona – Cg
Mekik, Figen A., (616) 331-2811 mekikf@gvsu.edu, Grand Valley State Univ – Gxc
Meko, David M., (520) 621-3457 dmeko@ltrr.arizona.edu, Univ of Arizona – Hq
Melas, Faye, (212) 772-5265 ffmelas@yahoo.com, Hunter Coll (CUNY) – Gs
Melbourne, Timothy I., (509) 963-2799 tim@geology.cwu.edu, Central Washington Univ – Yd
Melcher, Frank, frank.melcher@unileoben.ac.at, Univ Leoben – GgEg
Melchin, Michael, (902) 867-5177 Dalhousie Univ – Pg
Melchiorre, Erik, (909) 537-7754 emelch@csusb.edu, California State Univ, San Bernardino – EmHwCl
Meldahl, Keith H., (760) 757-2121, x6412 kmeldahl@miracosta.edu, MiraCosta Coll – Gg
Melendez, Christyanne, 828-262-7859 melendezc@appstate.edu, Appalachian State Univ – Ggv
Melichar, Rostislav, +420 549 49 5812 melda@sci.muni.cz, Masaryk Univ – GtcGr
Melim, Leslie A., (309) 298-1377 LA-Melim@wiu.edu, Western Illinois Univ – Gd
Mellor, George, (609) 258-6570 glmellor@princeton.edu, Princeton Univ – Op
Melosh, H. J., (765) 494-3290 jmelosh@purdue.edu, Purdue Univ – XmgXy
Melosh, Henry J., (520) 621-2806 jmelosh@lpl.arizona.edu, Univ of Arizona – Gt

Melson, William G., (202) 633-1809 melsonw@si.edu, Smithsonian Institution / National Museum of Natural History – GvtGa
Melton, Greg, meltong@byui.edu, Brigham Young Univ - Idaho – OgEgGp
Meltzer, Anne S., (610) 758-3673 asm3@lehigh.edu, Lehigh Univ – Ys
Melville, W. Kendall, (858) 534-0478 kmelville@ucsd.edu, Univ of California, San Diego – Op
Memeti, Valbone, (657) 278-2036 vmemeti@fullerton.edu, California State Univ, Fullerton – GivCa
Mendelson, Carl V., (608) 363-2223 mendelsn@beloit.edu, Beloit Coll – Pm
Mendelssohn, Irving A., (225) 388-6425 Louisiana State Univ – Ob
Mendoza, Carl, (780) 492-2664 carl.mendoza@ualberta.ca, Univ of Alberta – Hw
Mendoza, Luis H., lmendoza@cicese.mx, Centro de Investigación Científica y de Educación Superior de Ensenada – Ne
Menegon, Luca, +44 1752 584931 luca.menegon@plymouth.ac.uk, Univ of Plymouth – Gct
Meng, Jin, (212) 496-3337 American Museum of Natural History – Pv
Meng, Lingsen, meng@epss.ucla.edu, Univ of California, Los Angeles – Ys
Meng, Qingmin, 662-268-1032 Ext 240 Mississippi State Univ – Oig
Mengel, David, (785) 532-2166 dmengel@ksu.edu, Kansas State Univ – So
Menke, William H., (845) 365-8438 Columbia Univ – Ys
Menking, Kirsten M., 845-437-5545 kimenking@vassar.edu, Vassar Coll – GmPeGc
Menninga, Clarence, menn@calvin.edu, Calvin Coll – CcGg
Menold, Carrie A., 517-629-0312 cmenold@albion.edu, Albion Coll – Gpz
Mensah, Emmanuel, +233 24 4186 193 emmamensah6@yahoo.com, Kwame Nkrumah Univ of Science and Tech – EgGg
Menschik, Florian M., +49 (89) 289 25883 menschik@tum.de, Technische Universitaet Muenchen – NgOiNr
Mensing, Scott A., (775) 784-6346 smensing@unr.edu, Univ of Nevada, Reno – Oy
Menuge, Julian F., (+353) 1 716 2141 j.f.menuge@ucd.ie, Univ Coll Dublin – EmCcs
Menzies, John, (905) 688-5550 x3865 jmenzies@brocku.ca, Brock Univ – GlmGs
Mercer, Andrew, 662-268-1032 Ext 231 mercer@GRI.MsState.Edu, Mississippi State Univ – Owa
Merchant, James W., (402) 472-7531 jmerchant@unl.edu, University of Nebraska - Lincoln – Oru
Mercier, Michael , (905) 525-9140 (Ext. 27597) mercieme@mcmaster.ca, McMaster Univ – Ou
Merck, Jr., John W., (301) 405-2808 jmerck@umd.edu, Univ of Maryland – Pv
Meredith, Philip, +44 020 7679 37824 p.meredith@ucl.ac.uk, Univ Coll London – Nr
Meretsky, Vicky J., (812) 855-5971 meretsky@indiana.edu, Indiana Univ, Bloomington – Sf
Mereu, Robert F., (519) 661-3605 a424@uwo.ca, Western Univ – Ys
Merguerian, Charles M., 516 463-5567 geocmm@hofstra.edu, Hofstra Univ – Gc
Merifield, Paul M., (310) 794-5019 pmerifie@ucla.edu, Univ of California, Los Angeles – Ng
Merino, Enrique, (812) 855-5088 merino@indiana.edu, Indiana Univ, Bloomington – Cl
Merkel, Timo Casjen, 089/2180 4337 casjen.merkel@lrz.uni-muenchen.de, Ludwig-Maximilians-Universitaet Muenchen – Gz
Merlis, Timothy, 514-398-3140 timothy.merlis@mcgill.ca, McGill Univ – Oaw
Merriam, Daniel F., (785) 864-2127 dmerriam@kgs.ku.edu, Univ of Kansas – Gg
Merriam, James B., (306) 966-5716 jim.merriam@usask.ca, Univ of Saskatchewan – Yg
Merrifield, Mark A., (808) 956-6161 markm@soest.hawaii.edu, Univ of Hawai'i, Manoa – Op
Merrill, Glen K., (713) 221-8168 merrillg@uhd.edu, Univ of Houston Downtown – Pi
Merrill, John T., (401) 874-6715 jmerrill@boreas.gso.uri.edu, Univ of Rhode Island – Oc
Merrill, Ronald T., (206) 543-6686 Univ of Washington – Ym
Merritt, Andrew, +44 1752 584702 andrew.merritt@plymouth.ac.uk, Univ of Plymouth – Ng
Merritt, Dare, 252 328 6360 merrittd@ecu.edu, East Carolina Univ – On
Merritts, Dorothy J., (717) 291-4398 dorothy.merritts@fandm.edu, Franklin and Marshall Coll – Gm
Mertzman, Stanley A., (717) 291-3818 stan.mertzman@fandm.edu, Franklin and Marshall Coll – GizOm
Meskhidze, Nicholas, 919-515-7243 nicholas_meskhidze@ncsu.edu, North Carolina State Univ – Oa
Messer, Sharon, 740-351-3456 smesser@shawnee.edu, Shawnee State Univ – On
Messina, Paula, (408) 924-5027 paula.messina@sjsu.edu, San Jose State Univ – OeGm
Metaxas, Anna, (902) 494-3021 anna.metaxas@dal.ca, Dalhousie Univ – Ob
Metcalf, Meredith, (860) 465-4370 metcalfm@easternct.edu, Eastern Connecticut State Univ – OirHw
Metcalf, Rodney V., (702) 895-4442 kim.metcalf@unlv.edu, Univ of Nevada, Las Vegas – Gp
Metcalfe, Ian, +61-2-67733499 imetcal2@une.edu.au, Univ of New England – PmGtg
Metesh, John J., (406) 496-4159 jmetesh@mtech.edu, Montana Tech of The Univ of Montana – HwCa
Metz, Nicholas, (315) 781-3615 metz@hws.edu, Hobart & William Smith Colls – Oa
Metz, Robert, (908) 737-3687 rmetz@kean.edu, Kean Univ – Gr
Metzger, Ellen P., (408) 924-5048 ellen.metzger@sjsu.edu, San Jose State Univ – GpOe
Metzger, Marc J., +44 (0) 131 651 4446 mmetzger@staffmail.ed.ac.uk, Edinburgh Univ – Ge
Metzger, Ronald A., (541) 888-7216 rmetzger@socc.edu, Southwestern Oregon Community Coll – PmOePs
Metzler, Christopher V., (760) 944-4449, x7738 cmetzler@miracosta.edu, MiraCosta Coll – Gg
Meyer, Brian, bmeyer2@gsu.edu, Georgia State Univ – GesHw
Meyer, David L., (513) 556-4530 david.meyer@uc.edu, Univ of Cincinnati – Pi
Meyer, Franz J., 907-474-7767 fmeyer@gi.alaska.edu, Univ of Alaska, Fairbanks – Or
Meyer, Gary, (612) 626-1741 meyer015@umn.edu, Univ of Minnesota – Gl
Meyer, Grant A., (505) 277-4204 gmeyer@unm.edu, Univ of New Mexico – Gm
Meyer, Jeffrey W., (805) 965-0531 (Ext. 4270) meyerj@sbcc.edu, Santa Barbara City Coll – GgzGx
Meyer, Judith, (417) 836-5604 judithmeyer@missouristate.edu, Missouri State Univ – On
Meyer, Lewis, +44 01326 253766 l.h.i.meyer@exeter.ac.uk, Exeter Univ – Nmr
Meyer, Philip D., (503) 417-7552 philip.meyer@pnl.gov, Pacific Northwest National Laboratory – Hw
Meyer, Rebecca A., (812) 855-2687 reameyer@indiana.edu, Indiana Univ – Ec
Meyer, Scott C., (217) 333-5382 Illinois State Water Survey – Hw
Meyer, Steven J., (920) 465-5022 meyers@uwgb.edu, Univ of Wisconsin, Green Bay – OawOg
Meyer, W. Craig, (818) 710-4241 meyerwc@pierceColl.edu, Los Angeles Pierce Coll – GeuPm
Meyer, William T., (630) 969-6586 meyer@iastate.edu, Argonne National Laboratory – Cg
Meyer Dombard, DArcy, 312-996-2423 drmd@uic.edu, Univ of Illinois at Chicago – Py
Meyers, Jamie A., 507-457-5266 jmeyers@winona.edu, Winona State Univ – Gsd
Meyers, Philip A., (734) 764-0597 pameyers@umich.edu, Univ of Michigan – Co
Meylan, Anne, (727) 896-8626 Univ of South Florida – Ob
Meylan, Maurice A., (601) 266-4527 mmeylan@otr.usm.edu, Univ of Southern Mississippi – Gu
Meyzen, Christine M., +390498279153 christine.meyzen@unipd.it, Università degli Studi di Padova – Cg
Mezga, Aleksandar, +38514606116 amezga@geol.pmf.hr, Univ of Zagreb – Pg
Mezger, Jochen E., (907) 474-7809 jemezger@alaska.edu, Univ of Alaska, Fairbanks – GcpCc
Miah, Khalid, (406) 496-4888 kmiah@mtech.edu, Montana Tech – YexYs
Miall, Andrew D., (416) 978-8841 miall@es.utoronto.ca, Univ of Toronto – Gso
Miao, Xiaodong, (217) 244-2516 miao@illinois.edu, Univ of Illinois, Urbana-Champaign – GmsGl

Miao, Xin, (417) 836-5173 xinmiao@missouristate.edu, Missouri State Univ – Or
Micallef, Aaron, aaron.micallef@um.edu.mt, Univ of Malta – GumOu
Michael, Holly A., (302) 831-4197 hmichael@udel.edu, Univ of Delaware – Hwq
Michael, Peter J., (918) 631-3017 pjm@utulsa.edu, The Univ of Tulsa – GivGz
Michaels, Patrick J., (804) 924-0549 pmichaels@cato.org, Univ of Virginia – Oa
Michaels, Paul, (208) 426-1929 pm@cgiss.boisestate.edu, Boise State Univ – Ne
Michalek, Thomas E., 406-496-4405 tmichalek@mtech.edu, Montana Tech of The Univ of Montana – HgwHs
Michalski, Greg, (765) 494-3704 gmichals@purdue.edu, Purdue Univ – CsGeOa
Michaud, Jene D., 808-974-7411 jene@hawaii.edu, Univ of Hawai'i, Hilo – HqwGm
Michaud, Yves, (418) 654-2647 Universite du Quebec – Gm
Michel, Fred A., fmichel@ccs.carleton.ca, Carleton Univ – Hw
Michel, Jacqueline, 504-280-6325 jmichel@uno.edu, Univ of New Orleans – Cg
Michel, Lauren, (931) 372-3188 lmichel@tntech.edu, Tennessee Tech Univ – PeCs
Michel, Suzanne, 6196447454 x3028 Suzanne.Michel@gcccd.edu, Cuyamaca Coll – Gg
Michelfelder, Gary, (417) 836-3171 garymichelfelder@missouristate.edu, Missouri State Univ – GiCgGv
Mickelson, Andrew M., 901-678-4505 amicklsn@memphis.edu, Univ of Memphis – Ga
Mickelson, David M., (608) 262-7863 mickelson@geology.wisc.edu, Univ of Wisconsin, Madison – Gl
Mickus, Kevin L., (417) 836-6375 kevinmickus@missouristate.edu, Missouri State Univ – Yg
Miclaus, Crina, 00402324095 miclaus@uaic.ro, Al. I. Cuza Univ of Iasi – Gsr
Middlemiss, Lucie, +44(0) 113 34 35246 l.k.middlemiss@leeds.ac.uk, Univ of Leeds – On
Middleton, Carrie A., (303) 293-5019 cmiddleton@osmre.gov, Department of Interior Office of Surface Mining Reclamation and Enforcement – OrGfOi
Middleton, Larry T., (928) 523-2429 larry.middleton@nau.edu, Northern Arizona Univ – Gs
Mies, Jonathan W., (423) 425-4606 Jonathan-Mies@utc.edu, Univ of Tennessee at Chattanooga – GctHg
Miguel, Carlos, (598) 25251552 cmiguel@fcien.edu.uy, Universidad de la Republica Oriental del Uruguay (UDELAR) – OirSf
Mikesell, Dylan, dylanmikesell@boisestate.edu, Boise State Univ – Yg
Mikhaltsevitch, Vassily, +61 8 9266-4976 V.Mikhaltsevitch@curtin.edu.au, Curtin Univ. – Ye
Mikkelsen, Paula, mikkelsen@museumoftheearth.org, Cornell Univ – Pi
Mikkelsen, Paula, 607-273-6623 x 20 pmm37@cornell.edu, Paleontological Research Institution – Pi
Mikulic, Donald G., 217-244-2518 mikulic@isgs.uiuc.edu, Illinois State Geological Survey – Ps
Mikulich, Matthew J., 719-395-6794 mjmikulich@msn.com, Virginia Polytechnic Inst & State Univ – Ye
Milam, Keith A., (740) 593-1106 milamk@ohio.edu, Ohio Univ – Xg
Milan, Luke, +61267732019 lmilan@une.edu.au, Univ of New England – GtcGp
Miles, Randall J., 573-882-6607 Univ of Missouri, Columbia – Sd
Milewski, Adam, (706) 542-2652 Univ of Georgia – Hw
Militzer, Burkhard, militzer@seismo.berkeley.edu, Univ of California, Berkeley – Gy
Milkereit, Bernd, (416) 978-2466 bm@physics.utoronto.ca, Univ of Toronto – Ye
Millan, Christina, 614-292-0863 millan.2@osu.edu, Ohio State Univ – Ggc
Millen, Timothy M., 847-697-1000 tmillen@elgin.edu, Elgin Community Coll – Gg
Miller, Arnold I., (513) 556-4022 arnold.miller@uc.edu, Cincinnati Museum Center – Pq
Miller, Barry W., (865) 594-5599 Barry.Miller@tn.gov, Tennessee Geological Survey – EcOiGg
Miller, Brent, (979) 458-3671 bvmiller@geo.tamu.edu, Texas A&M Univ – Cc
Miller, Calvin F., 615-322-2232 calvin.miller@vanderbilt.edu, Vanderbilt Univ – Gi
Miller, Charles M., (505) 667-8415 Los Alamos National Laboratory – Cc
Miller, Christian A., 808-956-9607 cmiller3@hawaii.edu, Univ of Hawai'i, Manoa – CslCm
Miller, David, dmiller37@csub.edu, California State Univ, Bakersfield – GgtGc
Miller, David S., (630) 252-7191 Argonne National Laboratory – Ge
Miller, Donald S., (518) 478-0758 milled2@rpi.edu, Rensselaer Polytechnic Inst – Cc
Miller, Doug, 828-232-5158 dmiller@unca.edu, Univ of North Carolina at Asheville – OanOn
Miller, Douglas C., (302) 645-4277 dmiller@udel.edu, Univ of Delaware – Ob
Miller, Elizabeth L., (650) 723-1149 miller@pangea.stanford.edu, Stanford Univ – Gc
Miller, Geoffrey G., (506) 643-2361 Los Alamos National Laboratory – Cc
Miller, Gerald A., (515) 291-3442 soil@iastate.edu, Iowa State Univ of Science & Tech – SdaSo
Miller, Gifford H., 303-492-6962 gmiller@colorado.edu, Univ of Colorado – Cc
Miller, Harvey J., (801) 585-3972 miller.81@osu.edu, Ohio State Univ – Oi
Miller, Hugh, (303) 273-3558 hbmiller@mines.edu, Colorado School of Mines – Nmg
Miller, Ian, 303-370-8351 Ian.Miller@dmns.org, Denver Museum of Nature & Science – PbGg
Miller, James D., 218-726-6582 mille066@tc.umn.edu, Univ of Minnesota, Duluth – Gi
Miller, James D., 218-726-8385 mille066@d.umn.edu, Univ of Minnesota, Twin Cities – Gx
Miller, James F., (417) 836-5800 Missouri State Univ – Ps
Miller, Jerry R., (828) 227-2269 jmiller@wcu.edu, Western Carolina Univ – GmeGf
Miller, John, (416)736-5245 jrmiller@yorku.ca, York Univ – Or
Miller, Jonathan S., (408) 924-5015 jonathan.miller@sjsu.edu, San Jose State Univ – GiCc
Miller, Joshua H., josh.miller@uc.edu, Cincinnati Museum Center – Pe
Miller, Kate, (979) 845-3651 kcmiller@tamu.edu, Texas A&M Univ – Ys
Miller, Keith B., (785) 532-2250 kbmill@ksu.edu, Kansas State Univ – Pe
Miller, Kenneth G., (848) 445-3622 kgm@rci.rutgers.edu, Rutgers, The State Univ of New Jersey – GurPm
Miller, M. Meghan, (509) 963-2825 meghan@cwu.edu, Central Washington Univ – Yd
Miller, Mark, 601-266-4729 m.m.miller@usm.edu, Univ of Southern Mississippi – On
Miller, Marli G., (541) 346-4410 millerm@uoregon.edu, Univ of Oregon – Gc
Miller, Marvin R., (406) 496-4155 mmiller@mtech.edu, Montana Tech of The Univ of Montana – Hw
Miller, Max, 303-404-5415 max.miller@frontrange.edu, Front Range Community Coll - Westminster – OiyOe
Miller, Michael B., (225) 388-3412 byron@lgs.bri.lsu.edu, Louisiana State Univ – Go
Miller, Molly F., 615-322-3528 molly.miller@vanderbilt.edu, Vanderbilt Univ – Pe
Miller, Murray H., (519) 824-4120 (Ext. 53758) jmmiller7@sympatico.ca, Univ of Guelph – So
Miller, Randall F., (506) 643-2361 Univ of New Brunswick – Pg
Miller, Raymond M., (630) 252-3395 rmmiller@anl.gov, Argonne National Laboratory – Sb
Miller, Richard, 252 328 9372 millerri@ecu.edu, East Carolina Univ – Or
Miller, Richard D., (785) 864-2091 rmiller@kgs.ku.edu, Univ of Kansas – Ye
Miller, Richard H., (619) 594-5118 rmiller@geology.sdsu.edu, San Diego State Univ – Ps
Miller, Robert B., (408) 924-5025 robert.b.miller@sjsu.edu, San Jose State Univ – Gct
Miller, Robert N., (541) 737-4555 miller@coas.oregonstate.edu, Oregon State Univ – OpnOa
Miller, Ronald L., (212) 678-5577 rmiller@giss.nasa.gov, Columbia Univ – Oa
Miller, Steven F., (506) 643-2361 Argonne National Laboratory – Gg
Miller, Ted R., 605-677-6867 ted.miller@usd.edu, South Dakota Dept of Environment and Natural Resources – On
Miller, Wade E., 801 422-2321 wem@geology.byu.edu, Brigham Young Univ – Pv

Miller, William C., (707) 826-3110 wm1@humboldt.edu, Humboldt State Univ – Pi

Miller, William P., (601) 325-2912 Univ of Georgia – Sc

Miller-Hicks, Bryan, Bryan.Miller-Hicks@gcccd.edu, Cuyamaca Coll – Gg

Millet, Dylan B., 612-626-3259 dbm@umn.edu, Univ of Minnesota, Twin Cities – Oa

Millette, Thomas L., (413) 538-2813 tmillett@mtholyoke.edu, Mount Holyoke Coll – OriGm

Milligan, Mark R., (801) 537-3326 nrugs.milliga@state.ut.us, Utah Geological Survey – Gg

Milligan, Timothy, (902) 426-3273 milligant@dfo-mpo.gc.ca, Dalhousie Univ – Gs

Milliken, Kitty L., (512) 471-6082 kittym@mail.utexas.edu, Univ of Texas at Austin – Gd

Millington, Andrew, andrew.millington@flinders.edu.au, Flinders Univ – OuyOr

Mills, Aaron L., (804) 924-0564 alm7d@virginia.edu, Univ of Virginia – So

Mills, Eric L., (902) 471-2016 e.mills@dal.ca, Dalhousie Univ – ObnOn

Mills, James G., (765) 658-4669 jmills@depauw.edu, DePauw Univ – Gi

Mills, Jon, ++44 (0)191 208 5393 jon.mills@ncl.ac.uk, Univ of Newcastle Upon Tyne – OrYdOg

Mills, Rachel A., +44 (0)23 80592678 Rachel.Mills@soton.ac.uk, Univ of Southampton – Cm

Mills, Stephanie, +44 020 8417 2950 S.Mills@kingston.ac.uk, Kingston Univ – Ol

Mills, Suzanne, (905) 525-9140 smills@mcmaster.ca, McMaster Univ – On

Milne, Glenn A., 613-562-5800 6424 gamilne@uottawa.ca, Univ of Ottawa – YgOag

Milner, Lloyd R., (225) 578-3410 lmilne1@lsu.edu, Louisiana State Univ – Gg

Milroy, Scott, (228) 688-2325 scott.milroy@usm.edu, Univ of Southern Mississippi – Ob

Milstead, Terence, (828) 262-7057 milsteadtm@appstate.edu, Appalachian State Univ – On

Min, Doo-Hong, dmin@ksu.edu, Kansas State Univ – So

Min, Kyoungwon, (352) 392-2720 kmin@ufl.edu, Univ of Florida – Cc

Minarik, William G., 514-398-2596 william.minarik@mcgill.ca, McGill Univ – Cp

Mincer, Tracy, (508) 289-3640 tmincer@whoi.edu, Woods Hole Oceanographic Institution – Cm

Mindell, Randal, 604-527-5226 mindellr@douglasColl.ca, Douglas Coll – PbGsg

Miner, Andy, 509.963.2822 minera@geology.cwu.edu, Central Washington Univ – Yd

Miner, James J., 217-244-5786 miner@illinois.edu, Univ of Illinois – Sf

Minium, Deborah, (425) 564-5120 deborah.minium@bellevueColl.edu, Bellevue Coll – Gg

Minnich, Richard A., (951) 827-5515 richard.minnich@ucr.edu, Univ of California, Riverside – Oy

Minor, Timothy B., (775) 673-7477 tminor@dri.edu, Desert Research Inst – Or

Minshull, Timothy A., +44 (0)23 80596569 tmin@noc.soton.ac.uk, Univ of Southampton – Yr

Minster, J. Bernard H., (858) 945-0693 jbminster@ucsd.edu, Univ of California, San Diego – Ys

Minter, Nicholas, +44 023 92 842288 nic.minter@port.ac.uk, Univ of Portsmouth – Gs

Minton, David, (765) 494-3292 daminton@purdue.edu, Purdue Univ – Xg

Minzoni, Marcello, (205)348-0768 marcello.minzoni@ua.edu, Univ of Alabama – GosEo

Minzoni, Rebecca T., (205)348-6050 rebecca.minzoni@ua.edu, Univ of Alabama – GsPme

Miot da Silva, Graziela, graziela.miotdasilva@flinders.edu.au, Flinders Univ – Gug

Miranda, Elena A., 818-677-4671 elena.miranda@csun.edu, California State Univ, Northridge – Gc

Mirnejad, Hassan , 513-529-3216 mirnejh@miamioh.edu, Miami Univ – GxCg

Mishler, Brent, (510) 642-6810 bmishler@berkeley.edu, Univ of California, Berkeley – Pe

Mishra, Umakant, 630-252-1108 umishra@anl.gov, Argonne National Laboratory – Os

Misner, Tamara, 814-732-1352 tmisner@edinboro.edu, Edinboro Univ of Pennsylvania – GmHs

Misra, Debsmita, (907) 474-5339 dmisra@alaska.edu, Univ of Alaska, Fairbanks – HwOrNg

Misra, Kula C., (865) 974-6020 kmisra@utk.edu, Univ of Tennessee, Knoxville – Eg

Misra, Siddharth, 405-325-6787 misra@ou.edu, Univ of Oklahoma – Np

Mitasova, Helena, 919-513-1327 helena_mitasova@ncsu.edu, North Carolina State Univ – Oi

Mitchell, Charles C., (334) 844-5489 mitchc1@auburn.edu, Auburn Univ – Sc

Mitchell, Charles E., (716) 645-4290 cem@geology.buffalo.edu, SUNY, Buffalo – Po

Mitchell, Jonathan, (310) 825-2970 mitch@epss.ucla.edu, Univ of California, Los Angeles – OaXg

Mitchell, Martin D., (507) 389-1610 martin.mitchell@mnsu.edu, Minnesota State Univ – Ou

Mitchell, Neil C., neil.mitchell@manchester.ac.uk, Univ of Manchester – GumYg

Mitchell, Robert J., (360) 650-3591 robert.mitchell@geol.wwu.edu, Western Washington Univ – Hw

Mitchell, Roger H., (807) 343-8287 roger.mitchell@lakeheadu.ca, Lakehead Univ – Gi

Mitchell, Simon F., 876-927-2728 geoggeol@uwimona.edu.jm, Univ of the West Indies Mona Campus – GsgGs

Mitchell, Tom, +44 (0)20 7679 7361 tom.mitchell@ucl.ac.uk, Univ Coll London – Gc

Mitchneck, Beth A., bethm@email.arizona.edu, Univ of Arizona – On

Mitchum, Gary T., (727) 553-3941 gmitchum@marine.usf.edu, Univ of South Florida – Op

Mitra, Chandana, (334) 844-4229 czm0033@auburn.edu, Auburn Univ – OgaOi

Mitra, Gautam, (585) 275-5816 gautam.mitra@rochester.edu, Univ of Rochester – GctNr

Mitra, Shankar, (405) 325-4462 smitra@ou.edu, Univ of Oklahoma – Gc

Mitra, Siddhartha, 252 328 6611 mitras@ecu.edu, East Carolina Univ – Co

Mitri, Hani, (514) 398-4755 McGill Univ – Nm

Mitrovica, Jerry X., 617-496-2732 jxm@eps.harvard.edu, Harvard Univ – Yg

Mittelstaedt, Eric L., (208) 885-2045 emittelstaedt@uidaho.edu, Univ of Idaho – Yr

Mitterer, Richard M., (972) 883-2462 mitterer@utdallas.edu, Univ of Texas, Dallas – Col

Mittlefehldt, David, (281) 483-5043 david.w.mittlefehldt@nasa.gov, Univ of Tennessee, Knoxville – Xm

Mix, Alan C., 541-737-5212 amix@coas.oregonstate.edu, Oregon State Univ – Cs

Miyagi, Lowell, (801) 581-6619 lowell.miyagi@utah.edu, Univ of Utah – Gz

Miyares, Ines, (212) 772-5443 imiyares@hunter.cuny.edu, Hunter Coll (CUNY) – On

Moayyed, Mohsen -., +98 (411) 339 2717 moayyed@tabrizu.ac.ir, Univ of Tabriz – Gig

Moazzen, Mohssen -., +98 (411) 339 2679 moazzen@tabrizu.ac.ir, Univ of Tabriz – GpCgGz

Mobbs, Stephen, +44(0) 113 34 35158 s.d.mobbs@leeds.ac.uk, Univ of Leeds – Oa

Moberly, Ralph, (808) 956-8765 ralph@soest.hawaii.edu, Univ of Hawai'i, Manoa – Gu

Mobley, Curtis D., (206) 230-8166 Univ of Washington – Ob

Mock, R. Stephen, (406) 683-7261 steve.mock@umwestern.edu, Univ of Montana Western – Ca

Mock, Thomas, +44 (0)1603 59 2566 t.mock@uea.ac.uk, Univ of East Anglia – Py

Mock, Timothy D., 202-478-8466 tmock@carnegiescience.edu, Carnegie Institution of Washington – Cc

Mockler, Theodore, (505) 667-4318 mockler@lanl.gov, Los Alamos National Laboratory – On

Mode, William N., (920) 424-7004 mode@uwosh.edu, Univ of Wisconsin, Oshkosh – Gl

Modzelewski, Henryk, (604) 822-3591 hmodzelewski@eos.ubc.ca, Univ of British Columbia – Oa

Moe-Hoffman, Amy P., 303-492-0018 APM105@MSSTATE.EDU, Univ of Colorado – Pg

Moe-Hoffman, Amy P., 662-268-1032 Ext 234 paleomoe@gmail.com, Mississippi State Univ – On

Moecher, David P., (859) 257-6939 moker@uky.edu, Univ of Kentucky – GxtCg

Moeck, Inga, inga.moeck@tum.de, Technische Universitaet Muenchen – GcgOn
Moeglin, Thomas D., (417) 836-5800 Missouri State Univ – Ng
Moersch, Jeffery E., (865) 974-2366 jmoersch@utk.edu, Univ of Tennessee, Knoxville – XgOr
Moffett, James, 213-740-5626 jmoffett@usc.edu, Univ of Southern California – Oc
Mofjeld, Harold O., (206) 526-6819 mofjeld@pmel.noaa.gov, Univ of Washington – Op
Moghanloo, Rouzbeh, rouzbeh.gm@ou.edu, Univ of Oklahoma – Np
Mogilevskaya, Sonia, 612-625-4810 mogil003@umn.edu, Univ of Minnesota, Twin Cities – Ng
Mogk, David W., (406) 994-6916 mogk@montana.edu, Montana State Univ – Gp
Mohammad Reza, Hosseinzadeh, +98 (411) 339 2697 Univ of Tabriz – Egm
Mohapatra, Rama P., (507) 389-2617 rama.mohapatra@mnsu.edu, Minnesota State Univ – OirOg
Mohr, Marcus, 089/2180 4230 marcus.mohr@geophysik.uni-muenchen.de, Ludwig-Maximilians-Universitaet Muenchen – Yg
Mohrig, David, (512) 471-2282 mohrig@jsg.utexas.edu, Univ of Texas at Austin – GsmGr
MOINE, Bertrand N., 33-477481513 moineb@univ-st-etienne.fr, Université Jean Monnet, Saint-Etienne – CgtGx
Mojzsis, Stephen J., (303) 492-5014 stephen.mojzsis@colorado.edu, Univ of Colorado – XcCcGp
Mokos, Jennifer T., 740-368-3624 jtmokos@owu.edu, Ohio Wesleyan Univ – On
Moldowan, J. Michael, (650) 725-0913 moldowan@pangea.stanford.edu, Stanford Univ – Co
Moldwin, Mark B., (734) 647-3370 mmoldwin@umich.edu, Univ of Michigan – On
Molina, Jean-Alex E., (651) 647-9865 jamolina@umn.edu, Univ of Minnesota, Twin Cities – SbCg
Molina, Mario J., (858) 534-1696 mjmolina@ucsd.edu, Univ of California, San Diego – Oa
Molinari, John E., (518) 442-4562 molinari@atmos.albany.edu, SUNY, Albany – Oa
Moll, Nancy E., 760-776-7272 nmoll@Collofthedesert.edu, Coll of the Desert – Gge
Mollner, Daniel, (507) 933-7569 Gustavus Adolphus Coll – On
Molnár, Gábor, molnar@sas.elte.hu, Eotvos Lorand Univ – OrYgOi
Molnar, Peter, 303-492-4936 peter.molnar@colorado.edu, Univ of Colorado – Gt
Molnar, Sheri, 519-661-2111, ext.87031 smolnar8@uwo.ca, Western Univ – Ys
Molnia, Bruce F., (703) 648-4120 bmolnia@usgs.gov, Duke Univ – Or
Moloney, Marguerite M., (985) 448-4878 marguerite.moloney@nicholls.edu, Nicholls State Univ – GeOnn
Molotch, Noah, Noah.Molotch@colorado.edu, Univ of Colorado – Hg
Momayez, Moe, (520) 621-6580 moe.momayez@arizona.edu, Univ of Arizona – NmYxNr
Momen, Nasim, (617) 353-5679 Boston Univ – On
Momm, Henrique G., (615) 904-8372 henrique.momm@mtsu.edu, Middle Tennessee State Univ – OiHsOr
Momohara, Kristin, (808) 956-2910 kristinu@soest.hawaii.edu, Univ of Hawai'i, Manoa – On
Monahan, Adam, 250-721-6120 monahana@uvic.ca, Univ of Victoria – On
Monahan, Edward C., (860) 405-9110 edward.monahan@uconn.edu, Univ of Connecticut – Oa
Monari, Stefano, +390498279171 stefano.monari@unipd.it, Università degli Studi di Padova – Pg
Moncrief, John F., (612) 625-2771 moncr001@umn.ed, Univ of Minnesota, Twin Cities – Sp
Moncrieff, John B., +44 (0) 131 650 5402 J.Moncrieff@ed.ac.uk, Edinburgh Univ – Oa
Monecke, Thomas, 303-273-3841 tmonecke@mines.edu, Colorado School of Mines – Em
Monet, Julie, 530-898-3460 jmonet@csuchico.edu, California State Univ, Chico – GgNg
Monger, Bruce, bcm3@cornell.edu, Cornell Univ – Ob
Monger, H. C., (505) 646-3405 New Mexico State Univ, Las Cruces – Sa
Monson, Jessica, 217-265-6895 jlbm@illinois.edu, Univ of Illinois, Urbana-Champaign – Gg
Montagna, Paul, 361-825-2040 Paul.Montagna@tamucc.edu, Texas A&M Univ, Corpus Christi – On
Montagne, Cliff, (406) 599-7755 montagne@montana.edu, Montana State Univ – Sd
Montana, Carlos J., 915-747-5498 montana@utep.edu, Univ of Texas, El Paso – Yg
Montañez, Isabel P., (530) 754-7823 ipmontanez@ucdavis.edu, Univ of California, Davis – Gd
Montayne, Simone, (907) 451-5036 simone.montayne@alaska.gov, Alaska Division of Geological & Geophysical Surveys – Gg
Monteleone, Brian D., (508) 289-2405 bmonteleone@whoi.edu, Woods Hole Oceanographic Institution – Cc
Montenari, Michael, (+44) 01782 733162 m.montenari@keele.ac.uk, Keele Univ – PmgPs
Montenegro, Alvaro, (614) 688-5451 montenegro.8@osu.edu, Ohio State Univ – OgaOy
Montesi, Laurent G., (301) 405-7534 montesi@umd.edu, Univ of Maryland – YgXyGt
Monteverde, Donald H., (609) 292-2576 don.monteverde@dep.state.nj.us, New Jersey Geological and Water Survey – Grg
Monteverdi, John P., (415) 338-7728 montever@sfsu.edu, San Francisco State Univ – Oa
Montgomery, Carla W., (815) 753-1943 datarock@niu.edu, Northern Illinois Univ – Cc
Montgomery, David R., 206-685-2560 bigdirt@uw.edu, Univ of Washington – Gm
Montgomery, Homer A., 972-883-2496 mont@utdallas.edu, Univ of Texas, Dallas – Pg
Montgomery, Keith, (715) 845-9602 keith.montgomery@uwc.edu, Univ of Wisconsin Colls – OyGg
Montgomery, Michael T., (831) 656-2296 mtmontgo@nps.edu, Naval Postgraduate School – OwaOg
Montgomery, Tamra S., 217-333-5105 tmntgmry@illinois.edu, Univ of Illinois, Urbana-Champaign – On
Montgomery, William W., (201) 200-3161 wmontgomery@njcu.edu, New Jersey City Univ – Hw
Montoya, Joseph, (404) 385-0479 Georgia Inst of Tech – Cm
Montoya, Judith, (928) 523-8523 judith.montoya@nau.edu, Northern Arizona Univ – Oy
Montwill, Gail F., (909) 946-1796 Santiago Canyon Coll – Gg
Moodie, T. Bryant, (780) 492-5742 bryant.moodie@ualberta.ca, Univ of Alberta – On
Moody, Jennie L., (804) 924-0592 jlm8h@virginia.edu, Univ of Virginia – Oa
Mooers, Howard, (218) 726-7239 hmooers@d.umn.edu, Univ of Minnesota, Duluth – GleHw
Mookerjee, Matty, (707) 664-2002 matty.mookerjee@sonoma.edu, Sonoma State Univ – GcYgOi
Moon, Charlie, +44 01326 371822 C.J.Moon@exeter.ac.uk, Exeter Univ – Nx
Moon, Wooil, (204) 474-9833 wmoon@cc.umanitoba.ca, Univ of Manitoba – Ys
Mooney, Phillip, (707) 664-2328 mooneyp@sonoma.edu, Sonoma State Univ – OnGc
Moorberg, Colby, (785) 532-7207 Kansas State Univ – So
Moore, Andrew, (765) 983-1672 moorean@earlham.edu, Earlham Coll – Gm
Moore, Bradley S., (858) 822-6650 bsmoore@ucsd.edu, Univ of California, San Diego – Cm
Moore, Christopher M., +44 (0)23 80594801 cmm297@noc.soton.ac.uk, Univ of Southampton – Ob
Moore, Daniel K., (208) 496-1902 moored@byui.edu, Brigham Young Univ - Idaho – GxiCp
Moore, Dennis, dmoore@pmel.noaa.gov, Univ of Hawai'i, Manoa – Og
Moore, Dennis W., (734) 763-0202 dennis.w.moore@noaa.gov, Univ of Washington – Op
Moore, Duane M., 505-277-4204 dewey33@unm.edu, Univ of New Mexico – Sc
Moore, Gregory F., (808) 956-6854 gmoore@hawaii.edu, Univ of Hawai'i, Manoa – Gt
Moore, J. Casey, (831) 459-2574 cmoore@es.ucsc.edu, Univ of California, Santa Cruz – Gc
Moore, Jefferson K., 949-824-5391 jkmoore@uci.edu, Univ of California, Irvine – Op
Moore, Jeffrey, (801) 585-0491 jeff.moore@utah.edu, Univ of Utah – NgrGm
Moore, Joel, (410) 704-4245 moore@towson.edu, Towson Univ – ClsSc
Moore, Joseph N., (801) 585-6931 jmoore@egi.utah.edu, Univ of Utah – Gv

Moore, Kathryn, +44 (0)1326 255693 K.Moore@exeter.ac.uk, Exeter Univ – Ge
Moore, Laura J., (919) 962-5960 moorelj@email.unc.edu, Univ of North Carolina, Chapel Hill – Gme
Moore, Michael E., 717.702.2024 michmoore@pa.gov, Pennsylvania Bureau of Topographic & Geologic Survey – HwOi
Moore, Phillip, +44 (0) 191 208 5040 philip.moore@ncl.ac.uk, Univ of Newcastle Upon Tyne – Yd
Moore, Richard W., 831-656-1041 rwmoor1@nps.edu, Naval Postgraduate School – Ow
Moore, Robert M., (902) 494-3871 robert.moore@dal.ca, Dalhousie Univ – Oc
Moore, Theodore C., (734) 763-0202 tedmoore@umich.edu, Univ of Michigan – Ou
Mooreman, Thomas B., (515) 294-2308 tom.moorman@ars.usda.gov, Iowa State Univ of Science & Tech – So
Moores, Eldridge M., (916) 752-0352 Univ of California, Davis – Gt
Moorhead, Daryl L., (419) 530-2017 daryl.moorhead@utoledo.edu, Univ of Toledo – On
Moorhead, Kevin K., (828) 232-5183 moorhead@unca.edu, Univ of North Carolina, Asheville – Sb
Moorkamp, Max, +440116 252 3632 mm489@le.ac.uk, Leicester Univ – Yg
Moorman, Brian J., (403) 220-4835 moorman@ucalgary.ca, Univ of Calgary – Gm
Mora-Klepeis, Gabriela, 802-656-0246 gmora@uvm.edu, Univ of Vermont – CacGi
Morabia, Alfredo, 718-670-4180 Queens Coll (CUNY) – GbOn
Morales, Michael A., (620) 794-0191 mmorales@emporia.edu, Emporia State Univ – PvGrg
Morales, Tomas, tomas.morales@ehu.eus, Univ of the Basque Country UPV/EHU – NgHg
Moran, Dawn, (508) 289-4918 dmoran@whoi.edu, Woods Hole Oceanographic Institution – Ob
Moran, Jean E., 510-885-2491 jean.moran@csueastbay.edu, California State Univ, East Bay – HwCl
Moran, S. Bradley, (401) 874-6530 moran@gso.uri.edu, Univ of Rhode Island – Oc
Moran, Susan, (520) 670-6380 (Ext. 171) susan.moran@ars.usda.gov, Univ of Arizona – So
Moran-Zenteno, Dante J., dante@tonatiuh.igeofcu.unam.mx, Universidad Nacional Autonoma de Mexico – Cc
Morand, Vincent J., +61 3 9479 5641 v.morand@latrobe.edu.au, La Trobe Univ – GgcGt
Morealli, Sarah A., 540 654-1402 smoreall@umw.edu, Univ of Mary Washington – Gg
Morehouse, Barbara, morehoub@email.arizona.edu, Univ of Arizona – On
Morel, Francois M M., (609) 258-2416 morel@princeton.edu, Princeton Univ – Cl
Morel-Kraepiel, Anne, 609-258-7415 kraepiel@Princeton.edu, Princeton Univ – Em
Moreno, Rafael, (303) 556-8477 Metropolitan State Coll of Denver – Og
Moreton, Kim, (+44) (0) 7854825368 k.moreton@exeter.ac.uk, Exeter Univ – GeEmOu
Morgan, Craig D., (801) 537-3370 craigmorgan@utah.gov, Utah Geological Survey – Go
Morgan, Cristine L., (979) 845-3603 cmorgan@tamu.edu, Texas A&M Univ – Spd
Morgan, Daniel, +44(0) 113 34 35202 d.j.morgan@leeds.ac.uk, Univ of Leeds – Giv
Morgan, Daniel J., (615) 343-3141 dan.morgan@vanderbilt.edu, Vanderbilt Univ – GmCc
Morgan, Emory, (757) 727-5783 emory.morgan@hampton.edu, Hampton Univ – Oe
Morgan, F D., (617) 253-7857 morgan@erl.mit.edu, Massachusetts Inst of Tech – Yg
Morgan, Gary, 757.727.5783 gary.morgan@hamptonu.edu, Hampton Univ – OnnOn
Morgan, George B., (405) 325-2642 gmorgan@ou.edu, Univ of Oklahoma – Gi
Morgan, Jason, +44 1784 443606 Jason.Morgan@rhul.ac.uk, Univ of London, Royal Holloway & Bedford New Coll – Yg
Morgan, Joanna, +44 20 759 46423 j.v.morgan@imperial.ac.uk, Imperial Coll – Yg
Morgan, John D., 850-474-2224 jmorgan3@uwf.edu, Univ of West Florida – Oi
Morgan, Julia K., (713) 348-6330 morganj@rice.edu, Rice Univ – Gt
Morgan, Kelly, 239-658-3400 conserv@ufl.edu, Univ of Florida – So
Morgan, Ken M., (817) 257-7721 k.morgan@tcu.edu, Texas Christian Univ – OrGeOi
Morgan, Matt, 303-384-2632 mmorgan@mines.edu, Colorado School of Mines – GgmXm
Morgan, Michael C., (608)265-8159 morgan@aurora.aos.wisc.edu, Univ of Wisconsin, Madison – Ow
Morgan, Paul, 303-384-2648 morgan@mines.edu, Colorado School of Mines – YhGtYg
Morgan, Paul, paul.morgan@nau.edu, Northern Arizona Univ – Yh
Morgan, Siobahn M., (319) 273-2389 siobahn.morgan@uni.edu, Univ of Northern Iowa – On
Morgan, Sven S., (989) 774-1082 morga1ss@cmich.edu, Central Michigan Univ – Gc
Morgan, Tamie, (817) 257-7743 tamie.morgan@tcu.edu, Texas Christian Univ – Oi
Morgan, Terrance L., (505) 667-0837 Los Alamos National Laboratory – Ng
Morgan, William, +44 0161 306-6586 Will.Morgan@manchester.ac.uk, Univ of Manchester – Oa
Morin, Paul, 612-626-0505 lpaul@umn.edu, Univ of Minnesota, Twin Cities – On
Morison, James H., (206) 543-1394 morison@apl.washington.edu, Univ of Washington – Op
Moritz, Wolfgang, 089/2180 4336 w.moritz@lrz.uni-muenchen.de, Ludwig-Maximilians-Universitaet Muenchen – Gz
Moro, Alan, +38514606093 amoro@geol.pmf.hr, Univ of Zagreb – PgsGg
Morozov, Igor B., (306) 966-2761 igor.morozov@usask.ca, Univ of Saskatchewan – YsGy
Morra, Matthew J., (208) 885-6315 Univ of Idaho – Sb
Morrin, M. Elizabeth, 973-353-5100 morrin@andromeda.rutgers.edu, Rutgers, The State Univ of New Jersey, Newark – On
Morris, Antony, +44 1752 584766 a.morris@plymouth.ac.uk, Univ of Plymouth – YmGtu
Morris, Billy, 706-368-7528 bmorris@highlands.edu, Georgia Highlands Coll – Gg
Morris, Brenda, (760) 744-1150 ext. 2512 bmorris@palomar.edu, Palomar Coll – On
Morris, David, (314) 935-6926 Los Alamos National Laboratory – Cc
Morris, Donald P., (610) 758-5175 dpm2@lehigh.edu, Lehigh Univ – On
Morris, Geoffrey, (785) 532-3397 gpmorris@ksu.edu, Kansas State Univ – So
Morris, John A., 662-268-1032 Ext 235 jam16@msstate.edu, Mississippi State Univ – OirOw
Morris, Kalon, (949) 582-4649 kmorris@saddleback.edu, Saddleback Community Coll – OgpOw
Morris, Robert W., rmorris@wittenberg.edu, Wittenberg Univ – PiGg
Morris, Simon C., +44 (0) 1223 333414 sc113@esc.cam.ac.uk, Univ of Cambridge – Po
Morris, Thomas H., (801) 422-3761 tom_morris@byu.edu, Brigham Young Univ – Gr
Morris, William A., (905) 525-9140 (Ext. 24195) morriswa@mcmaster.ca, McMaster Univ – Yg
Morrow, Robert H., 803.896.1214 morrowr@dnr.sc.gov, Dept of Natural Resources – Gtp
Morschauser, Lindsey, 662-268-1032 Ext 243 lcm193@msstate.edu, Mississippi State Univ – Ow
Morse, David L., (512) 232-3241 Univ of Texas at Austin – Yg
Morse, Linda D., (757) 221-2444 ldmors@wm.edu, Coll of William & Mary – GeOg
Morse, Stearns A., (413) 545-0175 Univ of Massachusetts, Amherst – Gi
Mortensen, James K., (604) 822-6208 jmortensen@eos.ubc.ca, Univ of British Columbia – Cc
Mortenson, Kristine B., 801-422-3919 kris_mortenson@byu.edu, Brigham Young Univ – On
Mortimer, Robert, +44(0) 113 34 35251 r.j.g.mortimer@leeds.ac.uk, Univ of Leeds – Ge
Mortlock, Richard , (848) 445-3423 rmortloc@rci.rutgers.edu, Rutgers, The State Univ of New Jersey – CgmCs
Morton, Allan E., (602) 426-4351 Central Arizona Coll – Oe
Morton, Bruce, mortonb@easternct.edu, Eastern Connecticut State Univ – Og
Morton, Douglas M., (951) 276-6397 scamp@ucrac1.ucr.edu, Univ of California, Riverside – Gp
Morton, Robert, (813) 974-2773 Univ of South Florida, Tampa – Gs
Morton, Roger D., (780)-964-1935 rogermorton1@me.com, Univ of Alberta – Em

Morton, Ronald, rmorton@d.umn.edu, Univ of Minnesota, Duluth – EgGv
Moscardelli, Lorena G., 512-471-4971 lorena.moscardelli@beg.utexas.edu, Univ of Texas at Austin – GmYs
Mosenfelder, Jed, jmosenfe@umn.edu, Univ of Minnesota, Twin Cities – Cp
Moser, Desmond, 519-661-4214 dmoser22@uwo.ca, Western Univ – GtCcXg
Moser, Desmond E., (801) 585-3782 demoser@mines.utah.edu, Univ of Utah – Cc
Mosher, David C., 603-862-5493 dmosher@ccom.unh.edu, Univ of New Hampshire – Og
Mosher, David C., (902) 426-3149 mosher@agc.bio.ns.ca, Dalhousie Univ – Ou
Mosher, Sharon, (512) 471-4135 mosher@mail.utexas.edu, Univ of Texas at Austin – Gc
Moshier, Stephen O., (630) 752-5856 stephen.moshier@wheaton.edu, Wheaton Coll – Gd
Moskalski, Susanne, susanne.moskalski@stockton.edu, Stockton Univ – GsOn
Moskowitz, Bruce M., 612-624-1547 bmosk@umn.edu, Univ of Minnesota, Twin Cities – Ym
Mosley-Thompson, Ellen E., (614) 292-2580 thompson.4@osu.edu, Ohio State Univ – Oa
Moslow, Thomas F., (403) 269-6911 Univ of Calgary – Co
Moss, Neil E., 205-247-3557 nmoss@gsa.state.al.us, Geological Survey of Alabama – Hw
Moss, Patti, 509-527-5225 mosspm@whitman.edu, Whitman Coll – On
Mossa, Joann, (352) 294-7510 mossa@ufl.edu, Univ of Florida – GmHsOy
Mossman, David J., (506) 364-2326 dmossman@mta.ca, Mount Allison Univ – EmOn
Mossop, John W., (514) 398-4755 McGill Univ – Nm
Motani, Ryosuke, 530-754-6284 rmotani@ucdavis.edu, Univ of California, Davis – Pv
Motavalli, Peter P., 573-884-3212 motavallip@missouri.edu, Univ of Missouri, Columbia – Os
Mote, Philip, 541-737-5694 pmote@coas.oregonstate.edu, Oregon State Univ – Oag
Mote, Philip W., pmote@coas.oregonstate.edu, Univ of Washington – Oa
Mottl, Michael J., (808) 956-7006 mmottl@soest.hawaii.edu, Univ of Hawai'i, Manoa – Ou
Motyka, James, motykaj@easternct.edu, Eastern Connecticut State Univ – Og
Mouat, David A., (775) 673-7402 dmouat@dri.edu, Desert Research Inst – Ge
Moucha, Robert, 315-443-6239 rmoucha@syr.edu, Syracuse Univ – GtYeg
Mouginis-Mark, Peter J., (808) 956-3147 Univ of Hawai'i, Manoa – Xg
Moulis, Anastasia, 617-552-8300 anastasia.macherides@bc.edu, Boston Coll – Ys
Moum, James N., 541-737-2553 jmoum@coas.oregonstate.edu, Oregon State Univ – Op
Mound, Jon, +44(0) 113 34 35216 earjem@leeds.ac.uk, Univ of Leeds – Gt
Mount, Gregory, 724-357-7662 gregory.mount@iup.edu, Indiana Univ of Pennsylvania – Hy
Mount, Jeffrey F., jfmount@ucdavis.edu, Univ of California, Davis – Gm
Mountain, Carol S., (845) 356-8551 carolm@ldeo.columbia.edu, Columbia Univ – On
Mountain, Gregory S., 848-445-0817 gmtn@rci.rutgers.edu, Rutgers, The State Univ of New Jersey – GuYrGr
Mountney, Nigel, +44(0) 113 34 35249 n.p.mountney@leeds.ac.uk, Univ of Leeds – Gs
Mousset-Jones, Pierre, (775) 784-6959 mousset@mines.unr.edu, Univ of Nevada, Reno – Nm
Mower, Richard N., 989-774-3821 mower1rn@cmich.edu, Central Michigan Univ – Ow
Mowrer, Jake, 979-845-5366 jake.mowrer@tamu.edu, Texas A&M Univ – Sc
Moy, Christopher M., 64-3-479-5279 chris.moy@otago.ac.nz, Univ of Otago – Gus
MOYEN, Jean-François, (+33)477481510 jean.francois.moyen@univ-st-etienne.fr, Université Jean Monnet, Saint-Etienne – GiCtGp
Moyer, Elisabeth, (773) 834-2992 Univ of Chicago – Oa
Moyer, Kerry A., (814) 732-2454 kmoyer@edinboro.edu, Edinboro Univ of Pennsylvania – Oa
Moysey, Stephen M., (864) 656-5019 smoysey@clemson.edu, Clemson Univ – Hw
Mozley, Peter S., (575) 835-5311 mozley@nmt.edu, New Mexico Inst of Mining and Tech – Gs
Mozzachiodi, Riccardo, 361-825-3634 Riccardo.Mozzachiodi@tamucc.edu, Texas A&M Univ, Corpus Christi – On
Mozzi, Paolo, +390498279190 paolo.mozzi@unipd.it, Università degli Studi di Padova – Gm
Mrinjek, Ervin, +38514606057 ervin.mrinjek@zg.t-com.hr, Univ of Zagreb – GscGg
Mroz, Eugene J., (505) 667-7758 Los Alamos National Laboratory – Oa
Mshiu, Elisante E., mshiutz@udsm.ac.tz, Univ of Dar es Salaam – GgCeOr
Mucci, Alfonso, (514) 398-4892 alfonso.mucci@mcgill.ca, McGill Univ – CmlOc
Mucciardi, Frank, (514) 398-4755 McGill Univ – Nx
Muchez, Philippe, philippe.muchez@ees.kuleuven.be, Katholieke Universiteit Leuven – EmCe
Mucsi, lászló, +3662544156 mucsi@geo.u-szeged.hu, Univesity of Szeged – OirOy
Mudd, Simon N., +44 (0) 131 650 2535 simon.m.mudd@ed.ac.uk, Edinburgh Univ – Hg
Mudrick, Stephen E., (573) 882-6721 Univ of Missouri, Columbia – Oa
Muehlenbachs, Karlis, (780) 492-2827 karlis.muehlenbachs@ualberta.ca, Univ of Alberta – Cs
Mueller, Amy, 617.373.8131 a.mueller@northeastern.edu, Northeastern Univ – On
Mueller, Erich M., (334) 460-7136 em256@cornell.edu, Univ of South Alabama – Ob
Mueller, Karl J., (303) 492-7336 karl.mueller@colorado.edu, Univ of Colorado – GtcGm
Mueller, Paul A., (352) 392-2231 pamueller@ufl.edu, Univ of Florida – Cc
Mueller, Raymond G., (609) 652-4209 ray.mueller@stockton.edu, Richard Stockton Coll of New Jersey – SoGam
Mueller, Thomas, (724) 938-4255 mueller@calu.edu, California Univ of Pennsylvania – Oi
Muggeridge, Ann H., +44 20 759 47379 a.muggeridge@imperial.ac.uk, Imperial Coll – Np
Muhleman, Duane O., (626) 395-6186 dom@gps.caltech.edu, California Inst of Tech – Xy
Muir, William, 909-387-1603 wmuir@sbccd.cc.ca.us, San Bernardino Valley Coll – Og
Mukasa, Samuel B., (734) 936-3227 mukasa@umich.edu, Univ of Michigan – Cc
Mukhopadhyay, Sujoy, (530) 752-4711 sujoy@ucdavis.edu, Univ of California, Davis – Cg
Mulcahy, Sean R., 360-650-3645 sean.mulcahy@wwu.edu, Western Washington Univ – GpzGp
Muldoon, Maureen A., (920) 424-4461 muldoon@uwosh.edu, Univ of Wisconsin, Oshkosh – Hw
Mulholland, Margaret, (757) 683-3972 mmulholl@odu.edu, Old Dominion Univ – Ob
Mulibo, Gabriel D., gmbelwa@yahoo.com, Univ of Dar es Salaam – Ysg
Mulla, David J., (612) 625-6721 mulla003@umn.edu, Univ of Minnesota, Twin Cities – Sp
Mullen, Steven L., (520) 621-6842 mullen@atmo.arizona.edu, Univ of Arizona – Oa
Muller, Andrew C., (410) 293-6569 amuller@usna.edu, United States Naval Academy – OpnOu
Muller, Dietmar, (029) 351-3244 d.muller@usyd.edu.au, Univ of Sydney – Yrg
Muller, Otto H., (607) 871-2208 Alfred Univ – Gc
Muller, Peter, (808) 956-8081 pmuller@soest.hawaii.edu, Univ of Hawai'i, Manoa – Op
Muller, Wolfgang, +44 1784 443584 wolfgang.muller@rhul.ac.uk, Univ of London, Royal Holloway & Bedford New Coll – CcPeGa
Muller-Karger, Frank E., (727) 553-3335 Univ of South Florida – Or
Mulligan, Kevin R., 806-834-0391 kevin.mulligan@ttu.edu, Texas Tech Univ – OiGmSo
Mullins, Gregory L., (540) 231-4383 gmullins@vt.edu, Virginia Polytechnic Inst & State Univ – Sc
Mulrooney, Timothy, (919) 530-6269 tmulroon@nccu.edu, North Carolina Central Univ – Oir
Mulvaney, Richard L., (217) 333-9467 Univ of Illinois, Urbana-Champaign – Sc
Mulvany, Patrick S., (573) 341-4616 mulvany@umr.edu, Missouri Univ of Science and Tech – Gg
Mumin, A. Hamid, 204-727-9685 mumin@brandonu.ca, Brandon Univ – EgGzt
Munasinghe, Tissa, munasit@lahc.edu, Los Angeles Harbor Coll – Og
Mundie, Ben, (541) 967-2039 Ben.A.Mundie@mlrr.oregongeology.com,

Oregon Dept of Geology & Mineral Industries – Ge
Mungall, James E., (416) 978-2975 mungall@es.utoronto.ca, Univ of Toronto – GiEgCg
Munguia, Luis, lmunguia@cicese.mx, Centro de Investigación Científica y de Educación Superior de Ensenada – Ys
Munk, LeeAnn, 907 786-6895 lamunk@uaa.alaska.edu, Univ of Alaska, Anchorage – Cle
Munk, Walter H., (858) 534-2877 wmunk@ucsd.edu, Univ of California, San Diego – Op
Munn, Barbara J., (916) 278-6811 California State Univ, Sacramento – Ggp
Munroe, Jeffrey S., (802) 443-3446 jmunroe@middlebury.edu, Middlebury Coll – Gln
Munski, Douglas C., (701) 777-4591 douglas.munski@und.edu, Univ of North Dakota – Oe
Muntean, John, (775) 682-8748 munteanj@unr.edu, Univ of Nevada – Eg
Muntean, Thomas, 517-264-3943 tmuntean@adrian.edu, Adrian Coll – Gde
Murchie, Scott, 240-228-6235 scott.murchie@jhuapl.edu, Univ of Tennessee, Knoxville – Xg
Murdoch, Lawrence C., (864) 656-2597 lmurdoc@clemson.edu, Clemson Univ – Hw
Murgulet, Dorina, 361-825-2309 Dorina.Murgulet@tamucc.edu, Texas A&M Univ, Corpus Christi – Hw
Murgulet, Valeriu, 361-825-6023 Valeriu.Murgulet@tamucc.edu, Texas A&M Univ, Corpus Christi – Cg
Murowchick, James B., (816) 235-2979 murowchickj@umkc.edu, Univ of Missouri-Kansas City – CgGz
Murphree, James Thomas, 831-656-2723 murphree@nps.edu, Naval Postgraduate School – Ow
Murphy, Alexander B., (541) 346-4571 abmurphy@uoregon.edu, Univ of Oregon – On
Murphy, Cindy, (902) 867-2299 cmurphy@stfx.ca, Saint Francis Xavier Univ – Gg
Murphy, David T., 61 07 31382329 david.murphy@qut.edu.au, Queensland Univ of Tech – CgEgCc
Murphy, Edward C., (701) 328-8002 emurphy@nd.gov, North Dakota Geological Survey – On
Murphy, J. Brendan, (902) 867-2481 Dalhousie Univ – Gx
Murphy, Michael, 713-743-3564 mmurphy@central.uh.edu, Univ of Houston – Gc
Murphy, Michael A., michael.murphy@ucr.edu, Univ of California, Riverside – Ps
Murphy, Todd, (318) 342-3428 murphy@ulm.edu, Univ of Louisiana, Monroe – Oa
Murphy, Vincent, 617-552-8300 Boston Coll – Yg
Murphy, William M., 530-898-5163 wmurphy@csuchico.edu, California State Univ, Chico – Cl
Murray, A. Bradshaw, (919) 684-5847 abmurray@duke.edu, Duke Univ – Ou
Murray, Alison E., (775) 673-7361 alison@dri.edu, Desert Research Inst – Py
Murray, Andrew, 251-460-7325 amurray@southalabama.edu, Univ of South Alabama – Ow
Murray, Christopher J., (509) 376-5848 chris.murray@pnl.gov, Pacific Northwest National Laboratory – Gq
Murray, Daniel P., (401) 874-2197 dpmurray@uri.edu, Univ of Rhode Island – GpOe
Murray, David, (401) 863-3531 David_Murray@Brown.edu, Brown Univ – Ou
Murray, James W., (206) 543-4730 jmurray@u.washington.edu, Univ of Washington – Oc
Murray, John, + 353 (0)91 495 095 john.murray@nuigalway.ie, National Univ of Ireland Galway – PgGs
Murray, Kent S., (313) 436-9129 kmurray@umich.edu, Univ of Michigan, Dearborn – Hw
Murray, Kyle E., (405) 325-7502 kyle.murray@ou.edu, Univ of Oklahoma – HwOiGo
Murray, Mark, 575.835.6930 murray@ees.nmt.edu, New Mexico Inst of Mining and Tech – Yds
Murray, Richard, (617) 353-6532 rickm@bu.edu, Boston Univ – Cm
Murrell, Coling, +44 (0)1603 59 2959 j.c.murrell@uea.ac.uk, Univ of East Anglia – Oa
Murrell, Michael T., (505) 667-0967 Los Alamos National Laboratory – Cc
Murthy, Prahlad N., 570-408-4617 prahlad.murthy@wilkes.edu, Wilkes Univ – OanOn
Murtugudde, Raghuram G., (301) 314-2622 ragu@essic.umd.edu, Univ of Maryland – Obr
Mushkin, Amit, 972(02)5314254 mushkin@uw.edu, Univ of Washington – OrCcGm
Musil, Rudolf, +420 549 49 5997 rudolf@sci.muni.cz, Masaryk Univ – PgvGe
Muskatt, Herman, 792-3028 ewelch@utica.edu, Utica Coll – GgrPg
Musolf, Gene E., (715) 845-9602 Univ of Wisconsin Colls – Oy
Musselman, Zachary A., (601) 974-1344 musseza@millsaps.edu, Millsaps Coll – Gm
Mustard, John F., (401) 863-1264 john_mustard@brown.edu, Brown Univ – Or
Mustard, Peter S., (604) 291-5389 pmustard@sfu.ca, Simon Fraser Univ – Gs
Mustart, David A., (415) 338-7729 mustart@sfsu.edu, San Francisco State Univ – GizGv
Muszer, Antoni, antoni.muszer@uwr.edu.pl, Univ of Wroclaw – Eg
Muthukrishnan, Suresh, (864) 294-3361 suresh.muthukrishnan@furman.edu, Furman Univ – OiGmOr
Mutis-Duplat, Emilio, (432) 552-2243 mutis_e@utpb.edu, Univ of Texas, Permian Basin – Gp
Mutter, John C., (845) 365-8730 Columbia Univ – Yr
Mutter, John C., (646) 269-6942 jcm@ldeo.columbia.edu, Columbia Univ – YrGtYs
Muxworthy, Adrian, +44 20 759 46442 adrian.muxworthy@imperial.ac.uk, Imperial Coll – Ym
Muza, Jay P., (954) 201-6771 jmuza@broward.edu, Broward Coll, Central Campus – OuPme
Mwangi, Nancy , nwmwangi@jkuat.ac.ke, Jomo Kenyatta Univ of Agriculture & Tech – OiiOr
Myer, George H., (215) 204-7173 gmyer@temple.edu, Temple Univ – Gz
Myers, Alan R., 217-300-2570 amyers76@illinois.edu, Univ of Illinois, Urbana-Champaign – EcGo
Myers, Clifford D., (413) 236-4601 cmyers@berkshirecc.edu, Berkshire Community Coll – On
Myers, Corinne E., 505.277.4204 cemyers@unm.edu, Univ of New Mexico – Po
Myers, James D., (307) 766-2203 magma@uwyo.edu, Univ of Wyoming – Gi
Myers, Jeffrey A., 503-838-8365 myersj@wou.edu, Western Oregon Univ – Gs
Myers, Orrin B., (505) 665-3742 obm@lanl.gov, Los Alamos National Laboratory – Py
Myers, Paul, 780-492-6706 pmyers@ualberta.ca, Univ of Alberta – Og
Myers, Paul B., (610) 758-3665 pbm1@lehigh.edu, Lehigh Univ – Hw
Myers, Tammy, (717) 477-1685 tlmyers@ship.edu, Shippensburg Univ – On
Mylavarapu, S R., 352-392-1951 ext 202 raom@ufl.edu, Univ of Florida – Sc
Müller, Lena, 089/2180 4340 lena-mueller@lrz.uni-muenchen.de, Ludwig-Maximilians-Universitaet Muenchen – Gz
Müller-Sohnius, Dieter, 089/2180 4259 mueso@min.uni-muenchen.de, Ludwig-Maximilians-Universitaet Muenchen – Gz
Mylroie, John E., 662-268-1032 Ext 237 mylroie@geosci.msstate.edu, Mississippi State Univ – Gm
Myneni, Satish C B., (609) 258-5848 smyneni@princeton.edu, Princeton Univ – Sc
Myrbo, Amy, 612-626-7889 amyrbo@umn.edu, Univ of Minnesota, Twin Cities – PeOe
Myrow, Paul M., (719) 389-6790 pmyrow@coloradoColl.edu, Colorado Coll – Gs
Mysak, Lawrence A., (514) 398-3768 lawrence.mysak@mcgill.ca, McGill Univ – Op
Mysen, Bjorn O., (202) 478-8975 bmysen@carnegiescience.edu, Carnegie Institution of Washington – Cp
Möller, Andreas, 785864358 amoller@ku.edu, Univ of Kansas – Gt

## N

Naar, David F., (727) 553-1637 dnaar@usf.edu, Univ of South Florida – Ou
Nabelek, John L., 541-737-2757 nabelek@coas.oregonstate.edu, Oregon State Univ – GtYs
Nabelek, Peter I., (573) 884-6463 NabelekP@missouri.edu, Univ of Missouri – Ct
Nabighian, Misac, (303) 273-3933 mnabighi@mines.edu, Colorado School of Mines – Ye
Nadeau, Olivier, 613-558-8395 onadeau@uottawa.ca, Univ of Ottawa – Gx

Nadeau, Patricia, pnadeau@salemstate.edu, Salem State Univ – Gv
Nadiga, Balu, (505) 667-9466 balu@lanl.gov, Los Alamos National Laboratory – On
Nadin, Elisabeth S., (907) 474-5181 enadin@alaska.edu, Univ of Alaska, Fairbanks – Gtc
Nadiri, Ata Allah, +989143003725 nadiri@tabrizu.ac.ir, Univ of Tabriz – HwGqe
Nadon, Gregory C., (740) 593-4212 nadon@ohio.edu, Ohio Univ – Gs
Naehr, Thomas, (361) 825-2470 Thomas.Naehr@tamucc.edu, Texas A&M Univ, Corpus Christi – Gu
Naftz, David L., (801) 975-3389 dlnaftz@usgs.gov, Univ of Utah – Cg
Nagaoka, Lisa A., (940) 565-2510 lisa.nagaoka@unt.edu, Univ of North Texas – Ga
Nagel, Athena, (662) 268-1032 ext 238 amo58@msstate.edu, Mississippi State Univ – OgiOr
Nagel-Myers, Judith, 315-229-5239 jnagel@stlawu.edu, St. Lawrence Univ – PisPe
Nagihara, Seiichi, 806-834-4481 seiichi.nagihara@ttu.edu, Texas Tech Univ – Oi
Nagy, Kathryn L., 312-355-3276 klnagy@uic.edu, Univ of Illinois at Chicago – Cg
Nagy-Shadman, Elizabeth, 626 585-3369 eanagy-shadman@pasadena.edu, Pasadena City Coll – Gg
Nair, Vimala D., (352) 392-1803 Ext. 324 vdn@ufl.edu, Univ of Florida – Sc
Najjar, Raymond G., (814) 863-1586 najjar@meteo.psu.edu, Pennsylvania State Univ, Univ Park – Ow
Nakagawa, Masami, 303-384-2132 mnakagaw@mines.edu, Colorado School of Mines – Oe
Nakamura, Noboru, (773) 702-3802 nnn@bethel.uchicago.edu, Univ of Chicago – Oa
Nakamura, Yosio, (512) 471-0428 yosio@ig.utexas.edu, Univ of Texas at Austin – Ys
Naldrett, Anthony J., 44 (0)1243 5318 ajnaldrett@yahoo.com, Univ of Toronto – EmGiCp
Nalin, Ronald, (909) 558-7151 rnalin@llu.edu, Loma Linda Univ – Gsr
Namikas, Steven, (225) 578-6142 snamik1@lsu.edu, Louisiana State Univ – Gm
Nance, Hardie S., 512-471-6285 seay.nance@beg.utexas.edu, Univ of Texas at Austin – Grc
Nance, R. Damian, (740) 593-1107 nance@ohio.edu, Ohio Univ – Gtc
Nance, Seay, 512-471-6285 seay.nance@beg.utexas.edu, Univ of Texas at Austin, Jackson School of Geosciences – Gg
Napieralski, Jacob, 313.593.5157 jnapiera@umd.umich.edu, Univ of Michigan, Dearborn – GmlOi
Naples, Virginia, 815-753-7820 vlnaples@niu.edu, Northern Illinois Univ – Pv
Napora, Theodore A., (401) 874-6222 Univ of Rhode Island – Ob
Napp, Jeffrey, (206) 526-4148 jeff.napp@noaa.gov, Univ of Washington – Ob
Naranjo, Ramon, 775-887=7659 rnaranjo@usgs.gov, Univ of Nevada, Reno – CgHw
Narbonne, Guy M., (613) 533-6168 narbonne@queensu.ca, Queen's Univ – PiyGs
Narod, Barry, (604) 822-2267 narod@eos.ubc.ca, Univ of British Columbia – Yx
Nash, Barbara P., (801) 581-8587 barb.nash@utah.edu, Univ of Utah – Gi
Nash, David, david.nash@uc.edu, Univ of Cincinnati – Gm
Nash, Greg, 801-585-9986 gnash@egi.utah.edu, Univ of Utah – Gt
Nash, Jonathan, 541-737-4573 nash@coas.oregonstate.edu, Oregon State Univ – Op
Nash, T A., 614 265 6582 thomas.nash@dnr.state.oh.us, Ohio Dept of Natural Resources – Gl
Nashold, Barney W., (630) 252-7698 bwnashold@anl.gov, Argonne National Laboratory – On
Nasir, Sobhi J., 00974-4852207 nasir54@qu.edu.qa, Univ of Qatar – Gx
Naslund, H. Richard, (607) 777-4313 Binghamton Univ – Gi
Nasraoui, Mohamed, +33 (0)344063813 mohmed.nasraoui@lasalle-beauvais.fr, Institut Polytechnique LaSalle Beauvais (ex-IGAL) – EgGzEm
Nater, Edward A., (612) 625-1725 enater@umn.edu, Univ of Minnesota, Twin Cities – Sd
Nathan , Stephen, 860-465-5579 nathans@easternct.edu, Eastern Connecticut State Univ – PmEo
Nathan, Stephen, 413-967-5448 snathan@geo.umass.edu, Univ of Massachusetts, Amherst – PmOg
Nathan, Terrence R., (530) 752-1609 trnathan@ucdavis.edu, Univ of California, Davis – Oaw
Nations, Jack D., dale.nations@nau.edu, Northern Arizona Univ – Pm
Natland, James, 305 421-4123 jnatland@rsmas.miami.edu, Univ of Miami – Gi
Naumann, Malik, 089/2180 6714 Ludwig-Maximilians-Universitaet Muenchen – Pg
Naumann, Terry R., trnaumann@uaa.alaska.edu, Univ of Alaska, Anchorage – Gi
Navarre-Sitchler, Alexis, 303-384-2219 asitchle@mines.edu, Colorado School of Mines – Ca
Naveira Garabato, Alberto, +44 (0)23 80592680 acng@noc.soton.ac.uk, Univ of Southampton – Op
Navrotsky, Alexandra, (530) 752-3292 anavrotsky@ucdavis.edu, Univ of California, Davis – Om
Naylor, Mark, +44 (0) 131 650 4918 Mark.Naylor@ed.ac.uk, Edinburgh Univ – Gt
Naylor, Shawn, 812-855-2504 snaylor@indiana.edu, Indiana Univ – Hw
Naymik, Thomas G., 614-433-9280 tnaymik@geosyntec.com, Ohio State Univ – Hw
Nduati, Eunice W., enduati@jkuat.ac.ke, Jomo Kenyatta Univ of Agriculture & Tech – OrrOi
Neal, Clive R., (574) 631-8328 neal.1@nd.edu, Univ of Notre Dame – Gi
Neal, Donald W., (252) 328-4392 neald@ecu.edu, East Carolina Univ – Gr
Neal, William J., (616) 331-3381 nealw@gvsu.edu, Grand Valley State Univ – GsdGr
Nealson, Kenneth, (213) 821-2271 knealson@usc.edu, Univ of Southern California – Py
Neboga, Victoria V., 717–702–2026 vneboga@pa.gov, Pennsylvania Bureau of Topographic & Geologic Survey – Hw
Nechaev, Dmitri, (228) 688-2573 dmitri.nechaev@usm.edu, Univ of Southern Mississippi – Op
Nedunuri, Krishna K., 937 376 6455 knedunuri@centralstate.edu, Central State Univ – HwCaSo
Needham, Tim, +44(0) 113 34 39104 D.T.Needham@leeds.ac.uk, Univ of Leeds – Gc
Neelin, J. David, (310) 206-3734 neelin@atmos.ucla.edu, Univ of California, Los Angeles – Oa
Neely, Haly L., 979-845-3041 hneely@tamu.edu, Texas A&M Univ – Spd
Neethling, Stephen, +44 20 759 49341 s.neethling@imperial.ac.uk, Imperial Coll – Gz
Negrini, Robert M., (661) 654-2185 rnegrini@csub.edu, California State Univ, Bakersfield – YmgGn
Nehru, Cherukupalli E., (718) 951-5417 Graduate School of the City Univ of New York – Gi
Nehyba, Slavomír, +420 549 49 6067 slavek@sci.muni.cz, Masaryk Univ – GsdGg
Neighbors, Corrie, (505) 538-6352 neighborsc@wnmu.edu, Western New Mexico Univ – Ysr
Neill, Owen K., (509) 335-6770 owen.neill@wsu.edu, Washington State Univ – Ca
Neill, Simon P., 01248 383938 s.p.neill@bangor.ac.uk, Univ of Wales – OnpOg
Neinow, Peter W., +44 (0) 131 650 9139 Peter.Nienow@ed.ac.uk, Edinburgh Univ – Ol
Neish, Catherine, 519-661-2111, ext.83188 cneish@uwo.ca, Western Univ – Xg
Nekvasil, Hanna, 631-632-8201 hanna.nekvasil@sunysb.edu, SUNY, Stony Brook – Cp
Nelsen, Lori, (864)294-3481 lori.nelsen@furman.edu, Furman Univ – Ca
Nelson, Bruce K., (206) 543-4434 bnelson@u.washington.edu, Univ of Washington – Cc
Nelson, Bruce W., bwn@virginia.edu, Univ of Virginia – Gs
Nelson, Daren T., (910) 775-4589 daren.nelson@uncp.edu, Univ of North Carolina, Pembroke – GmHwOe
Nelson, Kenneth A., (785) 864-2164 nelson@kgs.ku.edu, Univ of Kansas – Oy
Nelson, Marc A., (501) 575-3964 manelson@comp.uark.edu, Univ of Arkansas, Fayetteville – Cl
Nelson, Michael G., (801) 585-3064 nelsonelson@aol.com, Univ of Utah – Nm
Nelson, Nathan, (785) 532-5115 nonelson@ksu.edu, Kansas State Univ – So
Nelson, Priscilla P., 303-273-3700 pnelson@mines.edu, Colorado School of Mines – NrgNm
Nelson, Robert E., (207) 859-5800 Colby Coll – Pe
Nelson, Robert K., (508) 289-2656 rnelson@whoi.edu, Woods Hole Oceanographic Institution – Co

Nelson, Robert M., (818) 354-1797 Jet Propulsion Laboratory – Xg
Nelson, Robert S., (309)438-7808 rsnelso@ilstu.edu, Illinois State Univ – Gm
Nelson, Stephen A., (504) 862-3194 snelson@tulane.edu, Tulane Univ – Gi
Nelson, Stephen T., 801-422-8688 steve_nelson@byu.edu, Brigham Young Univ – Cs
Nelson, Wendy, (410) 704-3133 wrnelson@towson.edu, Towson Univ – GiCc
Nelwamondo, T. M., +27 15 962 8582 tshililo.nelwamondo@univen.ac.za, Univ of Venda – GeOuy
Nemcok, Michal, +421-2-5463-0337 mnemcok@egi.utah.edu, Univ of Utah – GctGo
Nemon, Amy, (270) 745-6952 amy.nemon@wku.edu, Western Kentucky Univ – Oig
Nenes, Athanasios, (404) 894-9225 nenes@eas.gatech.edu, Georgia Inst of Tech – Oa
Nesbitt, Alex, (775) 682-8762 anesbitt@unr.edu, Univ of Nevada – On
Nesbitt, Carl, 775-784-8287 carln@unr.edu, Univ of Nevada, Reno – Nx
Nesbitt, Elizabeth (Liz), (206) 543-5949 lnesbitt@uw.edu, Univ of Washington – Pim
Nesbitt, H W., (519) 661-3194 hwn@uwo.ca, Western Univ – Cl
Nesbitt, Stephen W., (217) 244-3740 snesbitt@illinois.edu, Univ of Illinois, Urbana-Champaign – Oa
Nesheim, Tim, tonesheim@nd.gov, North Dakota Geological Survey - Gg
Nesse, William D., (970) 351-2830 w.d.nesse@gmail.com, Univ of Northern Colorado – Gxz
Nestell, Galina P., (817) 272-2987 gnestell@exchange.uta.edu, Univ of Texas, Arlington – Pms
Nestola, Fabrizio, +390498279160 fabrizio.nestola@unipd.it, Università degli Studi di Padova – Gz
Netoff, Dennis I., 936 294-1454 geo_din@shsu.edu, Sam Houston State Univ – Gm
Nettles, Meredith, (845) 365-8613 nettles@ldeo.columbia.edu, Columbia Univ – YsGl
Neubaum, John C., (717) 702-2039 jneubaum@pa.gov, Pennsylvania Bureau of Topographic & Geologic Survey – EcnGg
Neubeck, William S., (518) 388-6519 neubeckw@union.edu, Union Coll – Gm
Neuberg, Jurgen, +44(0) 113 34 36769 j.neuberg@leeds.ac.uk, Univ of Leeds – Gv
Neufeld, Jerome A., +44 (0) 1223 765709 jn271@cam.ac.uk, Univ of Cambridge – GoYgGt
Neuhauser, Kenneth R., (785) 628-5349 kneuhaus@fhsu.edu, Fort Hays State Univ – Gc
Neuman, Andrew G., (403) 823-7707 Royal Tyrrell Museum of Palaeontology – Pv
Neuman, Dennis R., (406) 994-4822 dneuman@montana.edu, Montana State Univ – On
Neuman, Shlomo P., (520) 621-7114 neuman@hwr.arizona.edu, Univ of Arizona – HwqGq
Neumann, A. Conrad, (919) 962-0190 neumann@marine.unc.edu, Univ of North Carolina, Chapel Hill – Gu
Neumann, Klaus, 765-285-8262 kneumann@bsu.edu, Ball State Univ – Cl
Neumann, Patricia, (310) 660-3593 pneumann@elcamino.edu, El Camino Coll – Og
Neuweiler, Fritz, (418) 656-7479 fritz.neuweiler@ggl.ulaval.ca, Universite Laval – GsdPo
Neves, Douglas, (310) 660-3593 dneves@elcamino.edu, El Camino Coll – Og
Nevins, Joseph, 845-437-7823 jonevins@vassar.edu, Vassar Coll – Ou
Nevins, Susan K., (607) 753-2815 GeologyDept@cortland.edu, SUNY, Cortland – On
Newberry, Rainer J., ffrn@uaf.edu, Univ of Alaska, Fairbanks – Em
Newbold, K. B., (905) 525-9140 x27948 newbold@mcmaster.ca, McMaster Univ – Ge
Newcomer, Darrell R., (509) 376-1054 darrell.newcomer@pnl.gov, Pacific Northwest National Laboratory – Hw
Newell, Charles J., (713) 663-6600 Rice Univ – Hw
Newell, Dennis L., dennis.newell@usu.edu, Utah State Univ – Cls
Newell, Kerry D., (785) 864-2183 dnewell@kgs.ku.edu, Univ of Kansas – Go
Newland, Leo, l.newland@tcu.edu, Texas Christian Univ – CaOs
Newman, Andrew V., (404) 894-3976 anewman@gatech.edu, Georgia Inst of Tech – YdsYg
Newman, Brent D., (505) 667-3021 bnewman@lanl.gov, Los Alamos National Laboratory – Gg
Newman, Dianne K., (636) 395-3543 dkn@caltech.edu, California Inst of Tech – Py
Newman, Jamie, 212-769-5386 jamien@amnh.org, American Museum of Natural History – Gg
Newman, Jonathan, (519) 824-4120 Ext.52147 jnewma01@uoguelph.ca, Univ of Guelph – On
Newman, Julie , (979) 845-9283 julie-newman@geos.tamu.edu, Texas A&M Univ – Gct
Newman, Steven B., (860) 832-2940 ccsuwxman@gmail.com, Central Connecticut State Univ – Oa
Newman, William A., (858) 534-2150 wnewman@ucsd.edu, Univ of California, San Diego – Ob
Newman, William I., (310) 825-3912 win@ucla.edu, Univ of California, Los Angeles – Xy
Newsom, Horton E., (505) 277-0375 newsom@unm.edu, Univ of New Mexico – XgcXm
Newton, Anthony J., +44 (0) 131 650 2546 Anthony.Newton@ed.ac.uk, Edinburgh Univ – On
Newton, Cathryn R., 315-443-2672 crnewton@syr.edu, Syracuse Univ – Po
Newton, Jan, (206) 616-3641 newton@ocean.washington.edu, Univ of Washington – Ob
Newton, Rob, +44(0) 113 34 31631 d.oneill@leeds.ac.uk, Univ of Leeds – Ge
Newton, Robert C., (310) 206-2917 rcnewton@ucla.edu, Univ of California, Los Angeles – CpGp
Newton, Robert M., (413) 585-3946 rnewton@smith.edu, Smith Coll – Gm
Newton, Seth A., 205-247-3708 snewton@gsa.state.al.us, Geological Survey of Alabama – OnGg
Nex, Paul A., 011 717 6563 Paul.Nex@wits.ac.za, Univ of the Witwatersrand – Eg
Nezat, Carmen A., (509) 359-7959 cnezat@ewu.edu, Eastern Washington Univ – ClGeSc
Ng, Crystal, 612-624-9243 gcng@umn.edu, Univ of Minnesota, Twin Cities – Hy
Ngigi, Thomas G., tgngigi@jkuat.ac.ke, Jomo Kenyatta Univ of Agriculture & Tech – OriOr
Ngo, Thanh X., +84-9782 24246 ngoxuanthanh@humg.edu.vn, Hanoi Univ of Mining & Geology – GtcGi
Ngoy, Kikombo, (908) 737-3718 kngoy@kean.edu, Kean Univ – Oy
Nguyen, Maurice, 612-626-8739 nguyenm@umn.edu, Univ of Minnesota – Gl
Ngwenya, Bryne T., +44 (0) 131 650 8507 Bryne.Ngwenya@ed.ac.uk, Edinburgh Univ – Co
Ni, James, (512) 550-4620 jni@nmsu.edu, Univ of Missouri – Ygs
Ni, James F., (505) 646-1920 jni@nmsu.edu, New Mexico State Univ, Las Cruces – YsGtYg
Ni-Meister, Wenge, (212) 772-5321 wenge.ni-meister@hunter.cuny.edu, Hunter Coll (CUNY) – Or
Nicholas, Chris, +353 1 8962176 nicholyj@tcd.ie, Trinity Coll – Eo
Nicholas, Joseph W., (540) 654-1470 jnicola@mwc.edu, Mary Washington Coll – Oy
Nicholl, Amy, 970-351-1239 Amy.Nicholl@unco.edu, Univ of Northern Colorado – GgOge
Nicholls, James W., (403) 220-7127 Univ of Calgary – Gi
Nichols, Caitlyn, caitlyn.nichols@csi.cuny.edu, Coll of Staten Island – Ge
Nichols, Kyle K., 518-580-5194 knichols@skidmore.edu, Skidmore Coll – Gm
Nichols, Liz, 949-451-5561 lnichols@ivc.edu, Irvine Valley Coll – On
Nichols, Terry E., (501) 575-7317 tenichol@comp.uark.edu, Univ of Arkansas, Fayetteville – On
Nicholson, David, (508) 289-3547 dnicholson@whoi.edu, Woods Hole Oceanographic Institution – Gu
Nicholson, Kirsten N., 765-285-8268 knichols@bsu.edu, Ball State Univ – Gi
Nicholson, Nanette, 212-769-5390 nnicholson@amnh.org, American Museum of Natural History – On
Nick, Kevin, (909) 558-4530 knick@llu.edu, Loma Linda Univ – GsbGr
Nickmann, Marion, +49 89 289 25853 nickmann@tum.de, Technische Universitaet Muenchen – Ng
Nicolas, Michelle P., (204) 945-6571 michelle.nicolas@gov.mb.ca, Manitoba Geological Survey – GrsGo
Nicolaysen, Kirsten P., 509-527-4934 nicolakp@whitman.edu, Whitman Coll – GiCgGa
Nicolescu, Stefan, 203-432-3141 stefan.nicolescu@yale.edu, Yale Univ – CtGpz
Nicoletti, Jeremy D., (603) 271-5762 jeremy.nicoletti@des.nh.gov, New Hampshire Geological Survey – GgHg

Nicoll, Kathleen, (801) 581-8218 kathleen.nicoll@geog.utah.edu, Univ of Utah – Gs
Nicot, Jean-Philippe, 512-471-6246 jp.nicot@beg.utexas.edu, Univ of Texas at Austin – Hw
Niculescu, Bogdan M., +40-21-3125024 bogdan.niculescu@gg.unibuc.ro, Univ of Bucharest – Yge
Niedzielski, Michael A., michael.niedzielski@und.edu, Univ of North Dakota – Oi
Nielsen, Donald R., (530) 753-5760 drnielsen@ucdavis.edu, Univ of California, Davis – Sp
Nielsen, Gerald A., (406) 994-5075 nielsenmontana@aol.com, Montana State Univ – Sd
Nielsen, Gregory B., (801) 626-6394 gnielsen@weber.edu, Weber State Univ – Gg
Nielsen, Kurt E., 831-656-2295 nielsen@nps.edu, Naval Postgraduate School – Ow
Nielsen, Mary, 605-677-5649 Mary.Nielsen@usd.edu, Univ of South Dakota – On
Nielsen, Peter A., (603) 358-2553 pnielsen@keene.edu, Keene State Coll – GxzGc
Nielsen, Peter J., (801) 537-3359 peternielsen@utah.gov, Utah Geological Survey – Eo
Nielsen, Roger L., (541) 737-1235 nielsenr@geo.oregonstate.edu, Oregon State Univ – GizGv
Nielsen, Sune G., (508) 289-2837 snielsen@whoi.edu, Woods Hole Oceanographic Institution – Cs
Nielsen-Gammon, John, (979) 862-2248 n-g@tamu.edu, Texas A&M Univ – Oa
Niem, Alan R., niema@geo.oregonstate.edu, Oregon State Univ – GdsGr
Niemann, William L., (304) 696-6721 niemann@marshall.edu, Marshall Univ – Ng
Niemi, Nathan, 734-764-6377 naniemi@umich.edu, Univ of Michigan – Gc
Niemi, Tina M., (816) 235-5342 niemit@umkc.edu, Univ of Missouri-Kansas City – Gt
Niemitz, Jeffery W., niemitz@dickinson.edu, Dickinson Coll – ClOuGn
Nieto, Antonio, (814) 769-0657 antonionieto@gmail.com, Pennsylvania State Univ, Univ Park – NmOeGq
Niewendorp, Clark, (971) 673-1555 clark.niewendorp@dogami.state.or.us, Oregon Dept of Geology & Mineral Industries – Eg
Nikitina, Daria L., (610) 436-3103 dnikitina@wcupa.edu, West Chester Univ – Gm
Nikolinakou, Maria-Aikaterini, 512-475-6613 mariakat@austin.utexas.edu, Univ of Texas at Austin – Np
Nikulin, Alex, (607) 777-2518 anikulin@binghamton.edu, Binghamton Univ – Ys
Nimis, Paolo, +390498279161 paolo.nimis@unipd.it, Università degli Studi di Padova – Eg
Ninesteel, Judy J., (304) 788-6956 JJNinesteel@mail.wvu.edu, Potomac State Coll – Hg
Ning, Liang, 814-8760987 lun115@psu.edu, Univ of Massachusetts, Amherst – Ow
Nisbet, Euan, +44 1784 443809 E.Nisbet@rhul.ac.uk, Univ of London, Royal Holloway & Bedford New Coll – Og
Nissen-Meyer, Targe, +44 (1865) 282149 tarje.nissen-meyer@earth.ox.ac.uk, Univ of Oxford – Ys
Nitecki, Matthew H., (312) 665-7093 mnitecki@fmnh.org, Field Museum of Natural History – Pi
Nittler, Larry R., (202) 478-8460 lnittler@carnegiescience.edu, Carnegie Institution of Washington – Xc
Nittrouer, Charles A., 206-543-5099 nittroue@ocean.washington.edu, Univ of Washington – GuYr
Nittrouer, Jeffrey A., 713.348.4886 nittrouer@rice.edu, Rice Univ – Gms
Niu, Fenglin, (713) 348-4122 niu@rice.edu, Rice Univ – Ysg
Niu, Guo-Yue, niu@geo.utexas.edu, Univ of Texas at Austin – SpOa
Niu, Yaoling, +44 (0) 191 33 42311 yaoling.niu@durham.ac.uk, Durham Univ – On
Nixon, Cheryl, cheryl.nixon@umb.edu, Univ of Massachusetts, Boston – On
Nixon, R. Paul, 801-422-4657 paul_nixon@byu.edu, Brigham Young Univ – Go
Nixon, Scott W., (401) 874-6803 Univ of Rhode Island – Ob
Niyogi, Dev, (765) 494-9531 climate@purdue.edu, Purdue Univ – Oa
Njau, Jackson K., (812) 856-3170 jknjau@indiana.edu, Indiana Univ, Bloomington – Pv
Njoku, Eni G., (818) 354-3693 eni.g.njoku@jpl.nasa.gov, Jet Propulsion Laboratory – Or
Nkedi-Kizza, Peter, (352) 392-1951 kizza@ufl.edu , Univ of Florida – Sp
Nkotagu, Hudson H., nkotaguh@yahoo.com, Univ of Dar es Salaam – HwgGe
Nkuna, Tinyiko R., +27 15 962 8017 tinyiko.nkuna@univen.ac.za, Univ of Venda for Science & Tech – HgwOa
Noakes, John E., (706) 542-1395 Univ of Georgia – Cc
Noble, Paula J., 775-784-6211 noblepj@unr.edu, Univ of Nevada, Reno – PmGnPs
Noblett, Jeffrey B., (719) 389-6516 jnoblett@coloradoColl.edu, Colorado Coll – Gxv
Noe, Garry, (757) 455-3284 gnoe@vwc.edu, Virginia Wesleyan Coll – HsOin
Noffke, Nora K., (757) 683-3313 nnoffke@odu.edu, Old Dominion Univ – GsPyGu
Nogueira, Ricardo, 404-413-5791 rnogueira@gsu.edu, Georgia State Univ – Oaw
Nolan, Robert P., (718) 951-4242 Graduate School of the City Univ of New York – Co
Nolin, Anne, (541) 737-8051 nolina@geo.oregonstate.edu, Oregon State Univ – Or
Noll, Mark R., (585) 395-5717 mnoll@esc.brockport.edu, SUNY, The Coll at Brockport – Cl
Noll, Michael G., (229) 333-7143 mgnoll@valdosta.edu, Valdosta State Univ – On
Noltimier, Hallan C., (614) 292-9796 noltimier.2@osu.edu, Ohio State Univ – Ym
Nondorf, Lea, (501) 683-0110 lea.nondorf@arkansas.gov, Arkansas Geological Survey – Cg
Noonan, Mathew T., 605-677-6152 matthew.noonan@usd.edu, South Dakota Dept of Environment and Natural Resources – Hw
Noone, David, (541) 737-3629 dcn@coas.oregonstate.edu, Oregon State Univ – Owg
Nord, Julia, (703) 993-3395 jnord@gmu.edu, George Mason Univ – Gzg
Nordeng, Stephan, (701) 777-3455 stephan.nordeng@engr.und.edu, Univ of North Dakota – Go
Nordstrom, Greg, 662-268-1032 Ext 248 gjn2@msstate.edu, Mississippi State Univ – Ow
Nordt, Lee C., (254) 710-4288 lee_nordt@baylor.edu, Baylor Univ – SdGa
Nordwald, Dan, 573-368-2451 dan.nordwald@dnr.mo.gov, Missouri Dept of Natural Resources – On
Norell, Mark A., (212) 769-5804 American Museum of Natural History – Pv
Noren, Anders, 612-626-3298 noren021@umn.edu, Univ of Minnesota, Twin Cities – Gn
Norford, Brian, (403) 292-7000 Univ of Calgary – Gr
Norman, David, +44 (0) 1223 333426 dn102@esc.cam.ac.uk, Univ of Cambridge – Po
Norman, David K., (360) 902-1439 dave.norman@dnr.wa.gov, Washington Division of Geology & Earth Resources – Gg
Norman, John M., (608)262-4576 jmnorman@wisc.edu, Univ of Wisconsin, Madison – So
Norman, Ralph R., (205) 247-3587 rnorman@gsa.state.al.us, Geological Survey of Alabama – Hwg
Norris, Christopher A., 203-432-3748 christopher.norris@yale.edu, Yale Univ – Pv
Norris, Dean R., (321) 674-8096 norris@fit.edu, Florida Inst of Tech – Ob
Norris, Geoffrey, (416) 978-4851 norris@quartz.geology.utoronto.ca, Univ of Toronto – Pl
Norris, Joel R., (858) 822-4420 jnorris@ucsd.edu, Univ of California, San Diego – Oa
Norris, Richard D., (858) 822-1868 rnorris@ucsd.edu, Univ of California, San Diego – Gu
Norris-Tull, Delena, 406-683-7043 delena.norris@umwestern.edu, Univ of Montana Western – Oe
Norrish, Winston, 509.963.2192 norrishw@geology.cwu.edu, Central Washington Univ – GgeGo
Norry, Mike, +440116 252 3803 nah@le.ac.uk, Leicester Univ – Pe
North, Gerald R., (979) 845-8083 g-north@tamu.edu, Texas A&M Univ – Ow
North, Leslie, leslie.north@wku.edu, Western Kentucky Univ – Og
Northrup, Clyde J., (208) 426-1581 cjnorth@boisestate.edu, Boise State Univ – Gc
Norton, Stephen A., (207) 581-2156 norton@maine.edu, Univ of Maine – Cl
Norton-Krane, Abby N., (216) 987-5227 Abby.Norton-Krane@tri-c.edu,

Cuyahoga Community Coll - Western Campus – Og
Norwood, James, 334-844-3414 jan0003@auburn.edu, Auburn Univ – Oy
Nosal, Thomas E., 86004243590 thomas.nosal@ct.gov, Dept of Energy and Environmental Protection – On
Nothdurft, Luke, +61 7 3138 1531 l.nothdurft@qut.edu.au, Queensland Univ of Tech – GdCmGg
Nourse, Jonathan A., 909-869-3460 janourse@csupomona.edu, California State Polytechnic Univ, Pomona – GctEm
Novacek, Michael J., (212) 769-5805 American Museum of Natural History – Pv
Novak, Milan , +420 549 49 6188 mnovak@sci.muni.cz, Masaryk Univ – GzpGi
Novak, Thomas, 859-257-3818 thomas.novak@uky.edu, Univ of Kentucky – Nm
Nowack, Robert L., (765) 494-5978 nowack@purdue.edu, Purdue Univ – Ys
Nowell, Arthur R. M., (206) 543-6605 nowell@cofs.washington.edu, Univ of Washington – Ou
Nowlin, Jr., Worth D., (979) 846-6747 wnowlin@tamu.edu, Texas A&M Univ – Opn
Nowotarski, Christopher J., (979) 845-3305 cjnowotarski@tamu.edu, Texas A&M Univ – Owa
Noyes, Jim, (310) 660-3593 tnoyes@elcamino.edu, El Camino Coll – Og
Noyes, Joanne M., (605) 394-6972 joanne.noyes@state.sd.us, South Dakota Dept of Environment and Natural Resources – Hw
Ntarlagiannis, Dimitrios, 973-353-5189 dimntar@andromeda.rutgers.edu, Rutgers, The State Univ of New Jersey, Newark – Yg
Nton, M. E., aa.elueze@mail.ui.edu.ng, Univ of Ibadan – Go
Ntsaluba, Bantubonke I., +27 (0)46-603-8309 b.ntsaluba@ru.ac.za, Rhodes Univ – Gi
Nudds, John, +44 0161 275-7861 john.nudds@manchester.ac.uk, Univ of Manchester – Pg
NUDE, Prosper M., +233-244-116879 pmnude@ug.edu.gh, Univ of Ghana – GipGx
Nuester, Jochen, jnuester@csuchico.edu, California State Univ, Chico – Cm
Null, E. Jan, (415) 338-2061 San Francisco State Univ – Ge
Nunn, Jeffrey A., (225) 388-6657 jeff@geol.lsu.edu, Louisiana State Univ – Yg
Nur, Amos M., (650) 723-9526 Stanford Univ – Yg
Nusbaum, Robert L., (843) 953-5596 nusbaumr@cofc.edu, Coll of Charleston – Gz
Nuss, Wendell A., 831-656-2308 nuss@nps.edu, Naval Postgraduate School – Ow
Nussear, Ken, 775-784-6995 Univ of Nevada, Reno – On
Nuttall, Brandon C., 859-323-0544 bnuttall@uky.edu, Univ of Kentucky – Eo
Nuwategeka, Expedito, +256 782 889 985 nuwategeka@gmail.com, Gulu Univ – OyyGv
Nwachukwu, J. I., 234-803-725-8122 jnwachuk@oauife.edu.ng, Obafemi Awolowo Univ – GoCg
Nyblade, Andrew A., (814) 863-8341 andy@geosc.psu.edu, Pennsylvania State Univ, Univ Park – Yg
Nye, Janet, (631) 632-3187 janet.nye@stonybrook.edu, SUNY, Stony Brook – Ob
Nyhan, John W., (505) 667-3163 Los Alamos National Laboratory – So
Nyland, Edo, (780) 492-5502 edo@phys.ualberta.ca, Univ of Alberta – Ys
Nyquist, Jonathan, (215) 204-7484 nyq@temple.edu, Temple Univ – Yg
Nystuen, Jeffrey A., (206) 543-1343 nystuen@apl.washington.edu, Univ of Washington – Op
Nzengung, Valentine A., (706) 542-2699 vnzengun@uga.edu, Univ of Georgia – ClGe

## O

O'Brien, Arnold L., (978) 934-3902 arnold_obrien@uml.edu, Univ of Massachusetts Lowell – Hw
O'Brien, John M., (201) 200-3161 jobrien@njcu.edu, New Jersey City Univ – Gs
O'Brien, Lawrence E., (914) 341-4570 lobrien@sunyorange.edu, Orange County Community Coll – Gg
O'Brien, Neal R., (315) 267-2287 obriennr@potsdam.edu, SUNY Potsdam – Gs
O'Brien, Rachel, (814) 332-2875 robrien@allegheny.edu, Allegheny Coll – HwGe
O'Brien, Suzanne R., (508) 531-1390 s6obrien@bridgew.edu, Bridgewater State Univ – Og
O'Callaghan, Mick, +353 21 4902662 mick.ocallaghan@ucc.ie, Univ Coll Cork – On
O'Connor, George A., (352) 392-1803 (Ext. 329) gao@ufl.edu, Univ of Florida – Sc
O'Connor, Yuet-Ling, yloconnor@pasadena.edu, Pasadena City Coll – Ge
O'Donnell, James, 860-405-9171 james.odonnell@uconn.edu, Univ of Connecticut – Op
O'Driscoll, Michael A., 252 328 5578 odriscollm@ecu.edu, East Carolina Univ – Hg
O'Driscoll, Nelson, 902-585-1679 nelson.odriscoll@acadiau.ca, Acadia Univ – CgGgOn
O'Farrell, Keely A., 859-323-4876 k.ofarrell@uky.edu, Univ of Kentucky – Ygd
O'Geen, Toby, 530-752-2155 atogeen@ucdavis.edu, Univ of California, Davis – Os
O'Gorman, Paul, (617) 452-3382 pog@mit.edu, Massachusetts Inst of Tech – Oa
O'Halloran, Ivan, (519) 674-1635 iohallo@ridgetownc.uoguelph.ca, Univ of Guelph – So
O'Hara, Kieran D., (859) 257-6931 geokoh@uky.edu, Univ of Kentucky – Gc
O'Hara, Matthew J., (509) 373-1671 matt.ohara@pnl.gov, Pacific Northwest National Laboratory – Cg
O'Keefe, Jennifer, (606) 783-2349 j.okeefe@moreheadstate.edu, Morehead State Univ – PlEcOe
O'Keeffe, Mike, 303-384-2637 okeeffe@mines.edu, Colorado School of Mines – GgeEg
O'Leary, Maureen, 631-444-3730 maureen.oleary@sunysb.edu, SUNY, Stony Brook – Pv
O'Meara, Stephanie, (970) 225-3584 Stephanie_O'Meara@partner.nps.gov, Colorado State Univ – Gc
O'Melia, Charles R., (410) 516-7102 omelia@jhu.edu, Johns Hopkins Univ – On
O'Mullan, Gregory, 718-997-3452 gomullan@qc.cuny.edu, Queens Coll (CUNY) – PyOb
O'Neil, Caron E., (717) 702-2042 coneil@pa.gov, Pennsylvania Bureau of Topographic & Geologic Survey – GgOn
O'Neil, James, (734) 764-1435 jro@umich.edu, Univ of Michigan – Cs
O'Neil, Jennifer, 775-682-8747 joneil@unr.edu, Univ of Nevada, Reno – On
O'Neil, Jonathan, 613-562-5800 6273 Jonathan.oneil@uottawa.ca, Univ of Ottawa – Cgc
O'Neil, Patrick E., 205-247-3586 poneil@gsa.state.al.us, Geological Survey of Alabama – Hs
O'Neill, Larry, (541) 737-2064 loneill@coas.oregonstate.edu, Oregon State Univ – Owg
O'Neill, Patrick M., (504) 388-2681 pat@lgs.bri.lsu.edu, Louisiana State Univ – Sf
O'Neill, Patrick M., (225) 578-8590 poneil2@lsu.edu, Louisiana State Univ – Sf
O'Reilly, Andrew M., 662-915-2483 aoreilly@olemiss.edu, Univ of Mississippi – Hwq
O'Reilly, Catherine M., 309-438-3493 cmoreil@ilstu.edu, Illinois State Univ – HgCgGg
O'Rourke, Thomas D., (607) 255-6470 tdo1@cornell.edu, Cornell Univ – Nge
Oakes-Miller, Hollie, hollie.oakesmiller@pcc.edu, Portland Community Coll - Sylvania Campus – Gg
Oakley, Adrienne, 484-646-4334 oakley@kutztown.edu, Kutztown Univ of Pennsylvania – YrOun
Oakley, Bryan, (860) 465-0418 oakleyb@easternct.edu, Eastern Connecticut State Univ – OnGml
Oakley, Bryan A., boakley@my.uri.edu, Univ of Rhode Island – Gsl
Oaks, Jr., Robert Q., (435) 752-0867 boboaks@comcast.net, Utah State Univ – GsHwYv
Oberdorfer, June A., (408) 924-5026 june.oberdorfer@sjsu.edu, San Jose State Univ – HwGeHy
Obermeyer, Nancy J., 812-237-4351 njobie28@kiva.net, Indiana State Univ – Oi
Obia, Godson C., 217-581-3328 gcobia@eiu.edu, Eastern Illinois Univ – On
Obolewicz, Dave, dobolewicz@keene.edu, Keene State Coll – GgEmOw
Obour, Augustine, (785) 625-3425 aobour@ksu.edu, Kansas State Univ – So
Obrad, Jennifer M., 217-333-8741 jobrad@illinois.edu, Univ of Illinois, Urbana-Champaign – EcGo

Obreza, Thomas A., 352-392-1951 ext 243 obreza@ufl.edu, Univ of Florida – So

Obrist, Daniel, (978) 934-3900 Daniel_Obrist@uml.edu, Univ of Massachusetts Lowell – Oa

Ocan, O. O., 234-803-705-8796 Obafemi Awolowo Univ – GtzGp

Occhiuzzi, Tony, 4807528888 x 81369 tocchiuzzi.cds@tuhsd.k12.az.us, Mesa Community Coll – On

Oches, Rick, (781) 891-2937 roches@bentley.edu, Bentley Univ – GeOePe

OConnell, Suzanne B., (860) 685-2262 soconnell@wesleyan.edu, Wesleyan Univ – GsuGe

Odendaal, Adriaan, +27 (0)51 401 9928 odendaalai@ufs.ac.za, Univ of the Free State – GgsGr

Odera, Patroba A., podera@jkuat.ac.ke, Jomo Kenyatta Univ of Agriculture & Tech – YdOge

Odeyemi, Idowu O., 08033436110 Federal Univ of Tech, Akure – GgxOr

Odland, Sarah K., 845-365-8633 odland@ldeo.columbia.edu, Columbia Univ – On

Odling, Noelle, +44(0) 113 34 32806 n.e.odling@leeds.ac.uk, Univ of Leeds – Hg

Odogba, Ismaila, 715-346-4451 iodogba@uwsp.edu, Univ of Wisconsin, Stevens Point – Oui

Odokuma-Alonge, Ovie, ovie.odokuma-alonge@uniben.edu, Univ of Benin – CgGip

Odom, John W., (334) 844-3966 jodom@acesag.auburn.edu, Auburn Univ – Sc

Odom, LeRoy A., (850) 644-6706 odom@magnet.fsu.edu, Florida State Univ – Cg

Odom, Robert I., 206-685-3788 odom@apl.washington.edu, Univ of Washington – YrGu

Oduneye, Olusola C., +2348066744769 ocoduneye@lautech.edu.ng, Ladoke Akintola Univ of Tech – GooGo

Oduor, Peter, (701) 231-7145 Peter.Oduor@ndsu.edu, North Dakota State Univ – Oi

Oechel, Walter, walter.oechel@open.ac.uk, The Open Univ – Oa

Oelkers, Eric, +44 020 7679 37870 e.oelkers@ucl.ac.uk, Univ Coll London – ClGze

Oestreich, William, (508) 289-3685 woestreich@whoi.edu, Woods Hole Oceanographic Institution – Oc

Oganov, Artem, 631-632-1429 artem.oganov@sunysb.edu, SUNY, Stony Brook – Gz

Ogard, Allen E., (505) 667-6344 Los Alamos National Laboratory – Hw

Ogata, Kei, +31 20 59 87288 k.ogata@vu.nl, VU Univ Amsterdam – Gt

Ogbahon, Osazuwa A., 08064253825 oaogbahon@futa.edu.ng, Federal Univ of Tech, Akure – Ggo

Ogbamikhumi, Alexander, alexander.ogbamikhumi@uniben.edu, Univ of Benin – YsgGg

Ogg, James G., (765) 494-8681 jogg@purdue.edu, Purdue Univ – PsGsYm

Ogiesoba, Osareni C., 512-471-6250 osareni.ogiesoba@beg.utexas.edu, Univ of Texas at Austin – Ygs

Oglesby, David D., (951) 827-2036 david.oglesby@ucr.edu, Univ of California, Riverside – Ys

Oglesby, Elizabeth, eoglesby@email.arizona.edu, Univ of Arizona – On

Oglesby, Robert J., (402) 472-1507 roglesby2@unl.edu, Univ of Nebraska, Lincoln – Oa

Ogram, Andrew V., 3523921951 (Ext 211) aogram@ufl.edu, Univ of Florida – Sb

Ogston, Andrea S., (206) 543-0768 ogston@ocean.washington.edu, Univ of Washington – Ou

Ogungbesan, Gbenga O., +2348053432408 googungbesan@lautech.edu.ng, Ladoke Akintola Univ of Tech – GooGo

Ogunsanwo, O., femiogunsanwo2003@yahoo.com, Univ of Ilorin – Ng

Ohman, Mark D., (858) 534-2754 mohman@ucsd.edu, Univ of California, San Diego – Ob

Ohmoto, Hiroshi, (814) 865-4074 ohmoto@geosc.psu.edu, Pennsylvania State Univ, Univ Park – Cs

Oikawa, Patricia, patty.oikawa@csueastbay.edu, California State Univ, East Bay – SbCoSf

Ojakangas, Richard W., (218) 726-7923 rojakang@d.umn.edu, Univ of Minnesota, Duluth – Gd

Ojala, Carl F., (313) 487-0218 Eastern Michigan Univ – Oa

Ojo, S. B., 234-803-703-7226 sbojo@oauife.edu.ng, Obafemi Awolowo Univ – Yg

Okafor, Florence A., (256) 372-4926 florence.okafor@aamu.edu, Alabama A&M Univ – OnEn

Okal, Emile A., (847) 491-3194 emile@earth.northwestern.edu, Northwestern Univ – YsgYr

Okalebo, Jane, jane.akalebo@gmail.com, Univ of Nebraska, Lincoln – Oa

Okaya, David A., (213) 740-7452 okaya@earth.usc.edu, Univ of Southern California – Ys

Okulewicz, Steven C., (516) 463-6545 geosco@hofstra.edu, Hofstra Univ – CgGue

Okumura, Yuko, 512-471-0383 yukoo@ig.utexas.edu, Univ of Texas at Austin – OaPe

Okunade, Samuel, (937) 376-6455 Central State Univ – Gm

Okunlola, Ougbenga. A., +2348033814070 o.okunlola@mail.ui.edu.ng, Univ of Ibadan – EgCgGe

Okunuwadje, Sunday E., sunday.okunuwadje@uniben.edu, Univ of Benin – GsoGd

Ola, P S., 08035927594 psola@futa.edu.ng, Federal Univ of Tech, Akure – GosGg

Olabode, Solomon O., 08033783498 Federal Univ of Tech, Akure – GgoGs

Oladunjoye, M. A., ma.oladunjoye@mail.ui.edu.ng, Univ of Ibadan – Yg

Olafsen-Lackey, Susan, (402) 371-6512 slackey@unl.edu, Unversity of Nebraska - Lincoln – Hw

Olanrewaju, Johnson, 814-871-7453 olanrewa001@gannon.edu, Gannon Univ – Cg

Olarewaju, V. O., 234-803-403-2030 volarewa@oauife.edu.ng, Obafemi Awolowo Univ – GozCg

Olariu, Cornel, 512-471-1519 cornelo@jsg.utexas.edu, Univ of Texas at Austin – Gr

Olariu, Mariana, iulialet@yahoo.com, Univ of Texas at Austin – Gr

Olatunji, A. S., +2348055475865 as.olatunji@mail.ui.edu.ng, Univ of Ibadan – CgGeCt

Olawuyi, Kehinde A., +2348107331222 gideonola2001@yahoo.com, Univ of Ilorin – YeGge

Olayinka, A. I., aa.elueze@mail.ui.edu.ng, Univ of Ibadan – Yg

Olayiwola, M. A., 234-806-448-7416 mayiwola@oauife.edu.ng, Obafemi Awolowo Univ – GugPi

Olcott Marshall, Alison, (785) 864-1917 olcott@ku.edu, Univ of Kansas – Py

Oldenburg, Douglas W., (604) 822-5406 doldenburg@eos.ubc.ca, Univ of British Columbia – Yg

Oldershaw, Alan E., (403) 220-3258 oldersha@geo.ucalgary.ca, Univ of Calgary – Gs

Oldfield, Jonathan, +44 (0)121 414 2943 j.d.oldfield@bham.ac.uk, Univ of Birmingham – Oy

Oldham, Richard L., 568-3100 (12147) OldhamR@arc.losrios.edu, American River Coll – Gg

Oldow, John S., (972) 883-2403 oldow@utdallas.edu, Univ of Texas, Dallas – Gc

Olea, Ricardo A., (703) 648-6414 rolea@usgs.gov, Univ of Kansas – GqoEc

Oleinik, Anton, (561) 297-3297 aoleinik@fau.edu, Florida Atlantic Univ – Pe

Olesik, John W., (614) 292-6954 olesik.2@osu.edu, Ohio State Univ – Ca

Oleynik, Sergey, (609) 258-2390 soleynik@princeton.edu, Princeton Univ – Cs

Olhoeft, Gary R., (303) 273-3458 golhoeft@mines.edu, Colorado School of Mines – Yg

Olivares, Jose A., (505) 665-2643 Los Alamos National Laboratory – Cs

Oliveira, Alcino S., soliveir@utad.pt, Universidade de Trás-os-Montes e Alto Douro – Hww

Oliver, Adolph A., (415) 786-6865 Chabot Coll – Ys

Oliver, Kevin, +44 (0)23 80596490 K.Oliver@noc.soton.ac.uk, Univ of Southampton – Op

Oliver, Matthew J., 302-645-4079 moliver@udel.edu, Univ of Delaware – ObrOg

Oliver, Ronald D., (702) 794-7095 ron_oliver@lanl.gov, Los Alamos National Laboratory – Yg

Oliveras Castro, Valentí, ++935813092 valenti.oliveras@uab.cat, Universitat Autonoma de Barcelona – Gvi

Olivo, Gema, 613 533-6998 olivo@queensu.ca, Queen's Univ – Eg

Ollerhead, Jeffery W., (506) 364-2428 jollerhead@mta.ca, Mount Allison Univ – Yr

Olmsted, Wayne, (406) 496-4151 Montana Tech of the Univ of Montana – Ca

Olney, Jessica L., 813-253-7647 jolney2@hccfl.edu, Hillsborough Community Coll – OgGg

Olney, Matthew P., molney@hccfl.edu, Hillsborough Community Coll – OgGg

Olorunfemi, A. O., 234-803-392-1712 akinrewa@oauife.edu.ng, Obafemi Awolowo Univ – Go

Olorunfemi, M. O., 234-903-719-2169 mlorunfe@oauife.edu.ng, Obafemi Awolowo Univ – Yg
Olsen, Amanda A., 207 581-2194 amanda.olsen@umit.maine.edu, Univ of Maine – Cl
Olsen, Khris B., (509) 376-4114 kb.olsen@pnl.gov, Pacific Northwest National Laboratory – Cg
Olsen, Kim B., 619-594-2649 kbolsen@mail.sdsu.edu, San Diego State Univ – Ye
Olsen, Paul E., (845) 365-8491 Columbia Univ – Gm
Olson, Hillary C., (512) 471-0455 olson@utig.ig.utexas.edu, Univ of Texas at Austin – Ps
Olson, James R., 605-677-6866 jim.olson@usd.edu, South Dakota Dept of Environment and Natural Resources – On
Olson, Kenneth R., (217) 333-9639 Univ of Illinois, Urbana-Champaign – Sc
Olson, Kimberly, olsonk@arc.losrios.edu, American River Coll – Oye
Olson, Neil F., (603) 271-2875 neil.olson@des.nh.gov, New Hampshire Geological Survey – GeHy
Olson, Peter L., (410) 516-7707 olson@jhu.edu, Johns Hopkins Univ – Yg
Olson, Ted L., 435 283-7533 ted.olson@snow.edu, Snow Coll – Ys
Olsson, Richard K., (848) 445-3043 olsson@rci.rutgers.edu, Rutgers, The State Univ of New Jersey – PmGr
Olszewski, Kathy, (718) 409-7366 kolszewski@sunymaritime.edu, SUNY, Maritime Coll – Cg
Olszewski, Thomas, 979-845-2465 tomo@geo.tamu.edu, Texas A&M Univ – Pe
Oltman-Shay, Joan M., (425) 644-9660 Univ of Washington – Op
Olugboji, Tolulope M., olugboji@umd.edu, Univ of Maryland – Ysg
Olyphant, Greg A., (812) 855-5154 olyphant@indiana.edu, Indiana Univ, Bloomington – Hw
Omar, Gomaa I., (215) 898-6908 gomar@sas.upenn.edu, Univ of Pennsylvania – Cc
Omelon, Christopher, 512-471-5440 omelon@jsg.utexas.edu, Univ of Texas at Austin – Cg
Omotoso, O. A., deletoso2002@yahoo.com, Univ of Ilorin – Ge
Oms Llobet, Oriol, ++34935811218 joseporiol.oms@uab.cat, Universitat Autonoma de Barcelona – GrsYm
Onasch, Charles M., (419) 372-7197 conasch@bgnet.bgsu.edu, Bowling Green State Univ – Gc
Onderdonk, Nate, (562) 985-2654 nate.onderdonk@csulb.edu, California State Univ, Long Beach – Gtm
ONeal, Michael, 302-831-8273 michael@udel.edu, Univ of Delaware – Gm
Ono, Shuhei, (617) 253-0474 sono@mit.edu, Massachusetts Inst of Tech – PyCs
Onstott, Tullis C., (609) 258-6898 tullis@princeton.edu, Princeton Univ – Py
Oommen, Thomas, (906) 487-2045 toommen@mtu.edu, Michigan Technological Univ – NeOiNg
Oona, Hain, (505) 667-5685 Los Alamos National Laboratory – On
Oostdam, Bernard L., 717-871-9928 boostdam@hotmail.com, Millersville Univ – Ou
Oostrom, Martinus, (509) 372-6044 mart.oostrom@pnl.gov, Pacific Northwest National Laboratory – Hq
Opdyke, Neil D., (352) 392-6127 drno@ufl.edu, Univ of Florida – Ym
Opeloye, S A., 08060981972 Federal Univ of Tech, Akure – GgoGs
Ophori, Duke U., (973) 655-7558 ophorid@mail.montclair.edu, Montclair State Univ – Hw
Oppenheimer, Michael, (609) 258-2338 omichael@princeton.edu, Princeton Univ – OaGe
Opper, Carl, 727-791-2536 opper.carl@spColl.edu, Saint Petersburg Coll, Clearwater – Hw
Opperman, William, wopperma@broward.edu, Broward Coll – Gg
Oppo, Delia W., 508-289-2681 doppo@whoi.edu, Woods Hole Oceanographic Institution – Pm
Oppong, Joseph R., (940) 565-2181 oppong@unt.edu, Univ of North Texas – On
Orchard, Michael, 604-669-0909 morchard@nrcan.gc.ca, Univ of British Columbia – PgGg
Orcutt, John A., (858) 534-2887 jorcutt@ucsd.edu, Univ of California, San Diego – YrsYd
Orf, Leigh, 989-774-1923 leigh.orf@cmich.edu, Central Michigan Univ – Ow
Orians, Kristin J., (604) 822-6571 korians@eoas.ubc.ca, Univ of British Columbia – Cm
Origlia, Hector D., +5435846761998 doriglia@exa.unrc.edu.ar, Universidad Nacional de Rio Cuarto – NgOn
Orlandini, Kent A., (630) 252-4236 Argonne National Laboratory – Cc
Orlove, Benjamin S., bsorlove@ucdavis.edu, Columbia Univ – Oa
Orme, Amalie, (818) 677-3532 amalie.orme@csun.edu, California State Univ, Northridge – Gm
Orndorff, Richard L., (509) 359-2855 rorndorff@ewu.edu, Eastern Washington Univ – Ge
Orphan, Victoria, (626) 395-1786 vorphan@gps.caltech.edu, California Inst of Tech – Py
Orr, Robert D., (209) 376-2481 robert.orr@pnl.gov, Pacific Northwest National Laboratory – Ca
Orr, William , orrw@pdx.edu, Portland State Univ – GgOg
Orr, William N., (541) 346-4577 worr@darkwing.uoregon.edu, Univ of Oregon – Pm
Orrock, John, orrock@wustl.edu, Washington Univ in St. Louis – On
Orsi, Alejandro H., (979) 845-4014 aorsi@ocean.tamu.edu, Texas A&M Univ – Op
Ort, Michael H., (928) 523-9363 michael.ort@nau.edu, Northern Arizona Univ – Gv
Ortega-Ariza, Diana, 706-507-8090 ortegaariza_diana@columbusstate.edu, Columbus State Univ – GdgGr
Ortegren, Jason, 850-474-2491 jortegren@uwf.edu, Univ of West Florida – PeOu
Ortiz, Joseph D., 330-672-2225 jortiz@kent.edu, Kent State Univ – Gs
Ortmann, Anthony L., 270-809-6755 aortmann@murraystate.edu, Murray State Univ – OnYg
Ortt, Jr., Richard A., (410) 554-5503 richard.ortt@maryland.gov, Maryland Department of Natural Resources – Op
Orville, Richard E., (979) 845-7671 rorville@tamu.edu, Texas A&M Univ – Oa
Osagiede, Edoseghe E., edoseghe.osagiede@uniben.edu, Univ of Benin – GctGx
Osborn, Gerald D., (403) 220-6448 Univ of Calgary – Gm
Osborn, Joe, (651) 779-3434 joe.osborn@century.edu, Century Coll – Og
Osborn, Stephen G., (909)869-3009 sgosborn@csupomona.edu, California State Polytechnic Univ, Pomona – HwClSp
Osborn, Timothy J., +44 (0)1603 59 2089 t.osborn@uea.ac.uk, Univ of East Anglia – Oa
Osborne, Todd, 352-392-1803 osbornet@ufl.edu, Univ of Florida – Sf
Osburn, Chris, 919-515-0382 chris_osburn@ncsu.edu, North Carolina State Univ – Cms
Osburn, Magdalena , 847-491-4254 maggie@earth.northwestern.edu, Northwestern Univ – Py
Oschlies, Andreas, aoschlies@geomar.de, Dalhousie Univ – Ob
Oser, Jens, 089/2180 4208 oeser@geophysik.uni-muenchen.de, Ludwig-Maximilians-Universitaet Muenchen – Yg
Oshun, Jasper, 707-826-3112 jo1196@humboldt.edu, Humboldt State Univ – HyGmg
Osi, Attila, hungaros@freemail.hu, Eotvos Lorand Univ – PviPe
Osinski, Gordon R., 519-661-2111 ext.84208 gosinbski@uwo.ca, Western Univ – XgcGp
Oskin, Michael, (530) 752-3993 meoskin@ucdavis.edu, Univ of California, Davis – GmtGc
Osleger, David A., (530) 754-7824 daosleger@ucdavis.edu, Univ of California, Davis – Gr
Osman, Mutasim S., +966138603184 mutasimsami@kfupm.edu.sa, King Fahd Univ of Petroleum and Minerals – GosGc
Osmond, Deanna L., (919) 515-7303 deanna_osmond@ncsu.edu, North Carolina State Univ – Sb
Osmond, John K., (850) 644-5860 osmond@gly.fsu.edu, Florida State Univ – Cc
Ossolinski, Justin, (508) 289-4995 jossolinski@whoi.edu, Woods Hole Oceanographic Institution – Oc
Oster, Jessica L., 615-322-1461 jessica.l.oster@vanderbilt.edu, Vanderbilt Univ – Cls
Osterberg, Erich C., erich.c.osterberg@dartmouth.edu, Dartmouth Coll – OaCl
Oswalt, Ginny L., (501) 569-3546 vlmartin@ualr.edu, Univ of Arkansas at Little Rock – On
Otamendi, Juan E., +543584676198 jotamendi@exa.unrc.edu.ar, Universidad Nacional de Rio Cuarto – CgaCp
Othberg, Kurt L., othberg@uidaho.edu, Univ of Idaho – GmgGl
Otis, Gard, (519) 814-4120 Ext.52478 gotis@uoguelph.ca, Univ of Guelph – On
Ott, Kevin C., (505) 667-4600 Los Alamos National Laboratory – On
Ottavi-Pupier, Elsa, +33 (0)3 44069358 elsa.ottavi-pupier@lasalle-

beauvais.fr,
Institut Polytechnique LaSalle Beauvais (ex-IGAL) – GvzGi
Otter, Ryan, 615-898-2063 rrotter@mtsu.edu,
Univ of Tennessee, Knoxville – On
Otterstrom, Samuel M., (801) 422-7751 samuel_otterstrom@byu.edu,
Brigham Young Univ – Ou
Otto, LeeAnn, (619) 260-4600 lotto@sandiego.edu, Univ of San Diego – On
Otvos, Ervin G., 228-872-4235 ervin.otvos@usm.edu,
Univ of Southern Mississippi – On
Ouellette, Vicki L., (970) 351-2647 vicki.ouellette@unco.edu,
Univ of Northern Colorado – On
Oughton, John, (651) 748-2637 john.oughton@century.edu,
Century Coll – Og
Ouimet, William, 860-486-3772 william.ouimet@uconn.edu,
Univ of Connecticut – GmOy
Ouimette, Mark A., (325) 670-1383 ouimette@hsutx.edu,
Hardin-Simmons Univ – GieGt
Over, D. Jeffrey, (585) 245-5294 over@geneseo.edu, SUNY, Geneseo – Ps
Overland, James E., james.e.overland@noaa.gov,
Univ of Washington – Oa
Overpeck, Jonathan T., 520-626-4345 jto@email.arizona.edu,
Univ of Arizona – Oa
Oviatt, Candace, (401) 874-6132 coviatt@gso.uri.edu,
Univ of Rhode Island – Ob
Oviatt, Charles G., (314) 288-7264 joviatt@ksu.edu,
Kansas State Univ – Gm
Oware, Erasmus K., erasmuso@buffalo.edu, SUNY, Buffalo – YgHq
Owen, Alan W., +4401413305461 alan.owen@glasgow.ac.uk,
Univ of Glasgow – PgsPo
Owen, Donald E., (409) 880-8234 donald.owen@lamar.edu,
Lamar Univ – Gr
Owen, Lewis, 513-556-42-3 Lewis.Owen@uc.edu,
Univ of Cincinnati – GmlGt
Owen, Robert M., (734) 763-4593 rowen@umich.edu,
Univ of Michigan – OcGu
Owens, Brent E., (757) 221-1813 beowen@wm.edu,
Coll of William & Mary – Gx
Owens, Tamera, (231) 777-0289 tamera.owens@muskegoncc.edu,
Muskegon Community Coll – On
Owens, Thomas J., 803-777-4530 owens@seis.sc.edu,
Univ of South Carolina – Ys
Owensby, Clenton E., (785) 532-7232 owensby@ksu.edu,
Kansas State Univ – Sf
Owoseni, Joshua O., 08060719411 Federal Univ of Tech, Akure – GgHw
Oxford, Jeremiah, 503-838-8680 oxfordj@wou.edu,
Western Oregon Univ – Og
Oyawale, A. A., 234-803-331-2150 aoyawale@oauife.edu.ng,
Obafemi Awolowo Univ – GeoCg
Oyediran, I. A., ia.oyediran@mail.ui.edu.ng, Univ of Ibadan – Ng
Oza, Rupal, 212-650-3035 rupal.oza@hunter.cuny.edu,
Hunter Coll (CUNY) – On
Ozbay, M. Ugur, (303) 273-3710 mozbay@mines.edu,
Colorado School of Mines – Nr
Ozdenerol, Esra, 901-678-2787 eozdenrl@memphis.edu,
Univ of Memphis – Oi
Ozel Ym, Esra , 00904242370000-5953 eozel@firat.edu.tr, Firat Univ – Gxi
Ozkan-Haller, Tuba, 541-737-9170 ozkan@coas.oregonstate.edu,
Oregon State Univ – On
Ozsvath, David L., (715) 346-2287 dozsvath@uwsp.edu,
Univ of Wisconsin, Stevens Point – Hw
Oztekin Okan, Ozlem, 00904242370000-5983 ooztekin@firat.edu.tr,
Firat Univ – HgwHs
Ozturk, Nevin, 00904242370000-5956 nevinozturk@firat.edu.tr,
Firat Univ – Cg

## P

Paavola, Jouni, +44(0) 113 34 36787 j.paavola@leeds.ac.uk,
Univ of Leeds – Ge
Pace, Michael L., (434) 924-6541 mlp5fy@virginia.edu,
Univ of Virginia – On
Pacheco, Fernando A., fpacheco@utad.pt,
Universidade de Trás-os-Montes e Alto Douro – Hq
Pachut, Jr., Joseph F., (317) 274-7785 jpachut@iupui.edu,
Indiana Univ / Purdue Univ, Indianapolis – Po
Pacia, Christina, (908) 737-3738 cpacia@kean.edu, Kean Univ – On
Padden, Maureen, (905) 525-9140 (Ext. 20118) paddenm@mcmaster.ca,
McMaster Univ – Cg
Padian, Kevin, (510) 642-7434 kpadian@berkeley.edu,
Univ of California, Berkeley – Pv
Paduan, Jeffrey D., (831) 656-3350 paduan@nps.edu,
Naval Postgraduate School – Op
Paez, H. A., 905-525-9140 (Ext. 26099) paezha@mcmaster.ca,
McMaster Univ – Eg
Page, David, 775-673-7110 dave.page@dri.edu, Desert Research Inst – Ga
Page, F Zeb, (440) 775-6701 zeb.page@oberlin.edu, Oberlin Coll – GpCsg
Page, Kevin, +44 1752 584750 kevin.page@plymouth.ac.uk,
Univ of Plymouth – Gr
Page, Philippe, philippe_page@uqac.ca,
Universite du Quebec a Chicoutimi – GiCgEm
Pagiatakis, Spiros, (416)736-2100 #77757 spiros@yorku.ca,
York Univ – Yd
Pagliari, Paulo H., 507-752-5065 pagli005@umn.edu,
Univ of Minnesota, Twin Cities – Sc
Paglione, Timothy, 718-262-2654 tpaglione@york.cuny.edu,
York Coll (CUNY) – Xm
Pagnac, Darrin C., 605-394-2469 Darrin.Pagnac@sdsmt.edu,
South Dakota School of Mines & Tech – Pv
Paige, David A., (310) 825-4268 dap@epss.ucla.edu,
Univ of California, Los Angeles – Xy
Pain, Christopher, +44 20 759 49322 c.pain@imperial.ac.uk,
Imperial Coll – Yg
Paine, Jeffrey G., 512-471-1260 jeff.paine@beg.utexas.edu,
Univ of Texas at Austin – On
Paine, Jeffrey G., (512) 475-9524 jeff.paine@beg.utexas.edu,
Univ of Texas at Austin, Jackson School of Geosciences – Gr
Pair, Donald, (937) 229-2936 Don.Pair@notes.udayton.edu,
Univ of Dayton – Gl
Pakhomov, Evgeny, (604) 827-5564 epakhomov@eos.ubc.ca,
Univ of British Columbia – Ob
Paktunc, Dogan, 613-947-7061 Dogan.Packtunc@nrcan-rncan.gc.ca,
Univ of Ottawa – Gz
Palace, Michael W., 603-862-4193 mike.palace@unh.edu,
Univ of New Hampshire – On
Palamartchouk, Kirill, +44 191 208 6421 kirill.palamartckirill.palamartchouk@ncl.ac.uk, Univ of Newcastle Upon Tyne – Yd
Palenik, Brian, (858) 534-7505 bpalenik@ucsd.edu,
Univ of California, San Diego – Ob
Palin, J. Michael, +64 3 479-9083 michael.palin@otago.ac.nz,
Univ of Otago – CcGip
Palinkaš, Ladislav, +38514605971 lpalinka@geol.pmf.hr,
Univ of Zagreb – CgeCp
Palladino, Steve D., (805) 654-6400 (Ext. 1365) Ventura Coll – Oi
Pallis, Ted J., (609) 984-6587 ted.pallis@dep.state.nj.us,
New Jersey Geological and Water Survey – Oi
Pallister, John S., (360) 993-8964 jpallist@usgs.gov,
Univ of Pittsburgh – GviGg
Palluconi, Frank D., (818) 354-8362 frank.d.palluconi@jpl.nasa.gov,
Jet Propulsion Laboratory – Or
Palm, Risa I., (404) 413-2574 risapalm@gsu.edu,
Georgia State Univ – On
Palma, Miclelle, (262) 521-5542 michelle.palma@uwc.edu,
Univ of Wisconsin Colls – OnnOn
Palmer, Arthur N., (607) 436-3064 palmeran@oneonta.edu,
SUNY, Oneonta – Hg
Palmer, Christina, (909) 537-5336 cpalmer@csusb.edu,
California State Univ, San Bernardino – On
Palmer, Clare, cpalmer@wustl.edu, Washington Univ in St. Louis – On
Palmer, Derecke, 612 9385 8719 d.palmer@unsw.edu.au,
Univ of New South Wales – Ye
Palmer, Donald F., (330) 672-2680 dpalmer@kent.edu,
Kent State Univ – Yg
Palmer, Evan, 719-333-3080 ronald.palmer@usafa.edu,
United States Air Force Academy – Oig
Palmer, H. C., (519) 661-2111 x86749 cpalmer@uwo.ca,
Western Univ – YmGv
Palmer, Martin R., +44 (0)23 80596607 pmrp@noc.soton.ac.uk,
Univ of Southampton – Cg
Palmer, Paul, +44 (0) 131 650 7724 paul.palmer@ed.ac.uk,
Edinburgh Univ – Oa
Palmquist, John C., (414) 832-6732 palmquij@lawrence.edu,
Lawrence Univ – GcpGg
Palter, Jaime, jpalter@uri.edu, McGill Univ – OagOe

Palumbo, Tony V., (423) 576-8002 avp@ornl.gov, Oak Ridge National Laboratory – Sb
Palutoglu, Mahmut, 00904242370000-5954 mpalutoglu@firat.edu.tr, Firat Univ – YgsGc
Pamukcu, Alya, (609) 258-0836 alya@princeton.edu, Princeton Univ – NgGe
Pan, Feifei, 940-369-5109 feifei/pan@unt.edu, Univ of North Texas – Hw
Pan, William L., (509) 335-3611 wlpan@wsu.edu, Washington State Univ – Sb
Pan, Yuanming, (706) 966-5699 yuanming.pan@usask.ca, Univ of Saskatchewan – Gx
Pan, Zaitao, (314) 977-3114 panz@eas.slu.edu, Saint Louis Univ – Oa
Panagiotakopulu, Eva, +44 (0) 131 650 2531 Eva.P@ed.ac.uk, Edinburgh Univ – Pe
Panah, Assad I., (814) 362-7569 aap@pitt.edu, Univ of Pittsburgh, Bradford – GorGc
Pancha, Aasha, (775) 784-4254 pancha@seismo.unr.edu, Univ of Nevada, Reno – Ys
Panero, Wendy R., 614 292 6290 panero.1@osu.edu, Ohio State Univ – Yg
Panetta, Richard L., (979) 845-1386 r-panetta@tamu.edu, Texas A&M Univ – Ow
Pangle, Luke, lpangle@gsu.edu, Georgia State Univ – HqCs
Pani, Eric A., (318) 342-1878 Univ of Louisiana, Monroe – Oa
Paniconi, Claudio, claudio.paniconi@ete.inrs.ca, Universite du Quebec – Hw
Panish, Peter T., (413) 545-2593 panish@geo.umass.edu, Univ of Massachusetts, Amherst – Gp
Pankow, Kristine L., 801-585-6484 pankow@seis.utah.edu, Univ of Utah – Ys
Panning, Mark, (352) 392-2634 mpanning@ufl.edu, Univ of Florida – Ys
Panno, Samuel V., (217) 244-2456 s-panno@illinois.edu, Univ of Illinois, Urbana-Champaign – ClGgHw
Pant, Hari, 718-960-5859 hari.pant@lehman.cuny.edu, Lehman Coll (CUNY) – Cg
Panter, Kurt S., (419) 372-7337 kpanter@bgnet.bgsu.edu, Bowling Green State Univ – Gv
Paola, Christopher, 612-624-8025 cpaola@umn.edu, Univ of Minnesota, Twin Cities – Gs
Papaleo, Silvanna, spapaleo@zircon.geology.utoronto.ca, Univ of Toronto – On
Papanastassiou, D.A. (Dimitri), (626) 395-6179 dap@gps.caltech.edu, California Inst of Tech – Cc
Papcun, George, 12012993161 gpapcun@njcu.edu, New Jersey City Univ – Oe
Papendick, Robert I., (509) 335-1552 papendick@wsu.edu, Washington State Univ – Sp
Papike, James J., (505) 277-1646 jpapike@unm.edu, Univ of New Mexico – Gz
Papineau, Dominic, d.papineau@ucl.ac.uk, Univ Coll London – Cg
Papp, Kenneth R., (907) 696-0079 kenneth.papp@alaska.gov, Alaska Division of Geological & Geophysical Surveys – Gg
Pappas, Matthew, 631-451-4301 pappasm@sunysuffolk.edu, Suffolk County Community Coll, Ammerman Campus – Og
Paquette, Jeanne, (514) 398-4402 jeanne.paquette@mcgill.ca, McGill Univ – Gz
Paradise, Thomas R., (479) 575-3159 paradise@uark.edu, Univ of Arkansas, Fayetteville – OgyGm
Parai, Rita, 314-935-3974 rparai@levee.wustl.edu, Washington Univ in St. Louis – Cs
Parashar, Rishi, 775.673-7496 rishi.parashar@dri.edu, Univ of Nevada, Reno – Hq
Parcell, William C., (316)978- 3140 Wichita State Univ – Gr
Pardi, Richard R., (201) 595-2695 William Paterson Univ – Cc
Parendes, Laurie A., (814) 732-2454 lparendes@edinboro.edu, Edinboro Univ of Pennsylvania – On
Parent, Michel, (418) 654-2657 Universite du Quebec – Gl
Pares, Josep M., (734) 615-0472 jmpares@umich.edu, Univ of Michigan – Ym
Parham, James, (657) 278-2043 jparham@fullerton.edu, California State Univ, Fullerton – PvoGg
Parise, John B., (631) 632-8196 john.parise@sunysb.edu, SUNY, Stony Brook – Gz
Pariseau, William G., (801) 581-5164 w.pariseau@utah.edu, Univ of Utah – Nr
Parish, Cynthia L., cparish@gt.rr.com, Lamar Univ – Gg
Parizek, Richard R., (814) 865-3012 parizek@ems.psu.edu, Pennsylvania State Univ, Univ Park – Hw
Park, Jeffrey J., (203) 432-3172 jeffrey.park@yale.edu, Yale Univ – YsOg
Park, Myung-Sook, 831-656-2858 mpark@nps.edu, Naval Postgraduate School – Ow
Park, Sohyun, 089/2180 4333 soohyun.park@physik.uni-muenchen.de, Ludwig-Maximilians-Universitaet Muenchen – Gz
Park, Stephen K., (951) 827-4501 magneto@ucrmt.ucr.edu, Univ of California, Riverside – Ym
Parker, David B., 806-651-4099 dparker@wtamu.edu, West Texas A&M Univ – Sbf
Parker, Don, 806-291-1121 Don.parker@wbu.edu, Wayland Baptist Univ – GivCg
Parker, Don F., (254) 710-2192 don_parker@baylor.edu, Baylor Univ – Giv
Parker, Doug, +44(0) 113 34 36739 d.j.parker@leeds.ac.uk, Univ of Leeds – Ow
Parker, Gary, 217-244-6161 parkerg@illinois.edu, Univ of Illinois, Urbana-Champaign – Gm
Parker, Jack R., (705) 670-5924 jack.parker@ontario.ca, Ontario Geological Survey – GgEgOe
Parker, Jim, 713-743-6750 jlparker9@uh.edu, Univ of Houston – On
Parker, Joan, 831-771-4400 parker@mlml.calstate.edu, Moss Landing Marine Laboratories – On
Parker, Kent E., (509) 373-6337 kent.parker@pnl.gov, Pacific Northwest National Laboratory – Cg
Parker, Marjorie, (207) 725-3628 mparker@bowdoin.edu, Bowdoin Coll – On
Parker, Matthew, 919-513-4367 mdparker@ncsu.edu, North Carolina State Univ – Ow
Parker, Richard M., 361-593-3590 Richard.Parker@tamuk.edu, Texas A&M Univ, Kingsville – GgPg
Parker, Robert L., (858) 534-2475 rlparker@ucsd.edu, Univ of California, San Diego – Ym
Parker, Stephen R., (406) 496-4185 sparker@mtech.edu, Montana Tech of the Univ of Montana – On
Parker, William C., (850) 644-1568 parker@gly.fsu.edu, Florida State Univ – Pi
PArkes, R. J., +44(0)29 208 70058 ParkesRJ@cardiff.ac.uk, Univ of Wales – Cg
Parkin, Gary, (519) 824-4120 (Ext. 52452) gparkin@uoguelph.ca, Univ of Guelph – Hw
Parkin, Geoffrey, +44 (0) 191 208 6146 geoff.parkin@ncl.ac.uk, Univ of Newcastle Upon Tyne – Hg
Parkinson, Christopher D., 504-280-6792 cparkins@uno.edu, Univ of New Orleans – Gp
Parks, Carlton R., (407) 868-0508 Florida Inst of Tech – Ow
Parks, George K., (510) 643-5512 parks@ssl.berkeley.edu, Univ of Washington – YxXy
Parlak, Osman, 03223387081 parlak@cu.edu.tr, Cukurova Universitesi – GipCc
Parman, Stephen, (401) 863-3352 Stephen_Parman@brown.edu, Brown Univ – CpgCt
Parmentier, E. Marc, (401) 863-1700 EM_Parmentier@Brown.edu, Brown Univ – Yg
Parnell, Roderic A., (928) 523-3329 roderic.parnell@nau.edu, Northern Arizona Univ – Cl
Parnella, Bill, 302-831-3156 parnella@udel.edu, Univ of Delaware – Gg
Parr, Kate, +44 0151 795 4640 Kate.Parr@liverpool.ac.uk, Univ of Liverpool – Pe
Parrick, Brittany, 614 265 6581 Brittany.Parrick@dnr.state.oh.us, Ohio Dept of Natural Resources – Gg
Parris, Thomas M., 859-323-0527 mparris@uky.edu, Univ of Kentucky – Eo
Parrish, Christopher E., 603-862-3438 cparrish@ccom.unh.edu, Univ of New Hampshire – Or
Parrish, Pia, (619) 594-5587 pparrish@mail.sdsu.edu, San Diego State Univ – On
Parrish, Randy, +440116 252 3315 rrp@nigl.nerc.ac.uk, Leicester Univ – Cs
Parry, William T., (801) 581-6217 wparry@comcast.net, Univ of Utah – CgGz
Parsekian, Andrew D., (307) 766-3603 aparseki@uwyo.edu, Univ of Wyoming – Yg
Parsen, Mike J., 608629419 mike.parsen@wgnhs.uwex.edu, Univ of Wisconsin - Extension – Hw
Parsons, Barry, +44 (1865) 272017 barry.parsons@earth.ox.ac.uk, Univ of Oxford – Yd

Faculty Index -P

Parsons, Gail, (309) 794-7318 GailParsons@augustana.edu, Augustana Coll – On
Parsons, Jeffrey D., (206) 221-6627 parsons@ocean.washington.edu, Univ of Washington – Ou
Parsons-Hubbard, Karla M., (440) 775-8353 karla.hubbard@oberlin.edu, Oberlin Coll – PoGu
Partin, Camille, camille.partin@usask.ca, Univ of Saskatchewan – Gtc
Paschert, Karin, 089/2180 6573 karin.paschert@lmu.de, Ludwig-Maximilians-Universitaet Muenchen – Gg
Pascoe, Richard, +4401326 371838 R.D.Pascoe@exeter.ac.uk, Exeter Univ – Gz
Pascussi, Michael, (518) 486-4820 mpascucc@mail.nysed.gov, New York State Geological Survey – Eo
Pashin, Jack C., (205) 349-2852 jpashin@gsa.state.al.us, Mississippi State Univ – GoEcGs
Pasicznyk, David L., (609) 984-6587 dave.pasicznyk@dep.state.nj.us, New Jersey Geological and Water Survey – Yg
Pasken, Robert W., (314) 977-3125 paskenrw@slu.edu, Saint Louis Univ – Ow
Passey, Benjamin, (410) 340-9473 passey@umich.edu, Univ of Michigan – CsaCl
Pasteris, Jill D., (314) 935-5434 pasteris@levee.wustl.edu, Washington Univ in St. Louis – GbzGe
Pasteris, Jill D., pasteris@wustl.edu, Washington Univ in St. Louis – On
Pasternack, Gregory B., (530) 754-9243 gpast@ucdavis.edu, Univ of California, Davis – GmHg
Pasternack, Gregory B., (530) 302-5658 gpast@ucdavis.edu, Univ of California, Davis – GmHg
Pate, John, (573) 368-2148 john.pate@dnr.mo.gov, Missouri Dept of Natural Resources – Ge
Patel, Vinodkumar A., 217-244-0639 patel@isgs.uiuc.edu vapatel@illinois.edu, Illinois State Geological Survey – On
Patenaude, Genevieve, +44 (0) 131 651 4472 Genevieve.Patenaude@ed.ac.uk, Edinburgh Univ – Ge
Patera, Edward S., (505) 667-4457 Los Alamos National Laboratory – Cg
Paterson, Colin J., (605) 394-2461 Colin.Paterson@sdsmt.edu, South Dakota School of Mines & Tech – Eg
Paterson, John R., +61-2-67732101 jpater20@une.edu.au, Univ of New England – PioPs
Paterson, Scott R., (213) 740-6103 paterson@usc.edu, Univ of Southern California – Gc
Patino-Douce, Alberto E., (706) 542-2394 alpatino@uga.edu, Univ of Georgia – OnnOn
Patino-Douce, Marta, (706) 542-2399 mapatino@uga.edu, Univ of Georgia – Gi
Paton, Douglas, +44(0) 113 34 35238 D.A.Paton@leeds.ac.uk, Univ of Leeds – Gc
Patoux, Jerome, 206-685-1736 jerome@atmos.washington.edu, Univ of Washington – Oa
Patrick, David M., (601) 266-4530 Univ of Southern Mississippi – Ng
Patronas, Dennis, dpatronas@disl.org, Dauphin Island Sea Lab – On
Patterson, Jodi T., (801) 537-3310 jpatters@utah.gov, Utah Geological Survey – On
Patterson, Mark, (781) 581-7370 m.patterson@neu.edu, Northeastern Univ – On
Patterson, Molly , (607) 777-2831 patterso@binghamton.edu, Binghamton Univ – CmOl
Patterson, R. Timothy, tpatters@ccs.carleton.ca, Carleton Univ – Pm
Patterson, William P., (306) 966-5691 bill.patterson@usask.ca, Univ of Saskatchewan – CsPeGn
Pattison, David R., (403) 220-3263 pattison@geo.ucalgary.ca, Univ of Calgary – Gp
Pattison, Simon A., 204-727-7468 pattison@brandonu.ca, Brandon Univ – GsoGd
Patton, Howard J., (505) 667-1003 patton@lanl.gov, Los Alamos National Laboratory – Yg
Patton, James L., (510) 642-3059 Univ of California, Berkeley – Pv
Patton, Jason A., 479-968-0676 jpatton@atu.edu, Arkansas Tech Univ – GgeGc
Patton, Jason R., jrp2@humboldt.edu, Humboldt State Univ – GuOg
Patton, Peter C., (860) 685-2268 ppatton@wesleyan.edu, Wesleyan Univ – Gm
Patton, Regan L., (509) 534-3670 rpatton@wsu.edu, Washington State Univ – Gc
Patton, Terri, 630-252-3294 tlpatton@anl.gov, Argonne National Laboratory – Gg
Patton, Thomas W., 406-496-4153 tpatton@mtech.edu, Montana Tech of The Univ of Montana – Hw
Pattrick, Richard, richard.pattrick@manchester.ac.uk, Univ of Manchester – Gz
Patwardhan, Kaustubh, 845-257-3738 patwardk@newpaltz.edu, SUNY, New Paltz – Ggi
Paty, Carol M., (404) 894-2860 carol.paty@eas.gatech.edu, Georgia Inst of Tech – YmgOn
Patzkowsky, Mark E., (814) 863-1959 brachio@geosc.psu.edu, Pennsylvania State Univ, Univ Park – Po
Paul, John H., (727) 553-1168 jpaul@marine.usf.edu, Univ of South Florida – Ob
Paulsell, Robert L., (225) 578-8655 rpaulsell@lsu.edu, Louisiana State Univ – Oy
Paulsen, Timothy S., (920) 424-7002 paulsen@uwosh.edu, Univ of Wisconsin, Oshkosh – Gc
Paulson, Suzanne, (310) 206-4442 paulson@atmos.ucla.edu, Univ of California, Los Angeles – Oa
Pavelsky, Tamlin M., (919) 962-4239 pavelsky@unc.edu, Univ of North Carolina, Chapel Hill – Hg
Pavia, Giulio, giulio.pavia@unito.it, Università di Torino – Pg
Pavlis, Gary L., (812) 855-5141 pavlis@indiana.edu, Indiana Univ, Bloomington – YsGtYe
Pavlis, Terry L., (504) 280-6797 tpavlis@uno.edu, Univ of New Orleans – Gt
Pavlis, Terry L., 915-747-5570 tlpavlis@utep.edu, Univ of Texas, El Paso – Gc
Pavloski, Charles, (814) 865-0478 pavloski@essc.psu.edu, Pennsylvania State Univ, Univ Park – Ow
Pavlovskaya, Marianna, (212) 772-5320 mpavlov@hunter.cuny.edu, Hunter Coll (CUNY) – Oi
Pavlowsky, Robert T., (417) 836-8473 robertpavlowsky@missouristate.edu, Missouri State Univ – Gm
Pawley, Alison, +44 0161 275-3944 alison.pawley@manchester.ac.uk, Univ of Manchester – Gc
Pawloski, Gayle A., (925) 423-0437 pawloski1@llnl.gov, Lawrence Livermore National Laboratory – Gg
Pawlowicz, Richard A., (604) 822-1356 rpawlowicz@eos.ubc.ca, Univ of British Columbia – Op
Paysen-Petersen, Lukas, +49 (89) 289 25895 lukas.paysen-petersen@tum.de, Technische Universitaet Muenchen – NgrGq
Paytan, Adina, (650) 724-4073 apaytan@pangea.stanford.edu, Stanford Univ – Oc
Paz Gonzalez, Antonio, 34 981167000 tucho@udc.es, Coruna Univ – Sd
Pazzaglia, Frank J., (610) 758-3667 fjp3@lehigh.edu, Lehigh Univ – Gm
Pe-Piper, Georgia, (902) 420-5744 Dalhousie Univ – Gv
Peace, Walter G., (905) 525-9140 (Ext. 23517) peacew@mcmaster.ca, McMaster Univ – Ou
Peacock, Caroline, +44(0) 113 34 37877 C.L.Peacock@leeds.ac.uk, Univ of Leeds – Sb
Peacock, Simon M., simon.peacock@ubc.ca, Univ of British Columbia – GtpYs
Peacor, Donald R., (734) 764-1452 drpeacor@umich.edu, Univ of Michigan – Gz
Peakall, Jeff, +44(0) 113 34 35205 j.peakall@leeds.ac.uk, Univ of Leeds – Gs
Peake, Jeffrey S., (402) 554-2662 jpeake@unomaha.edu, Univ of Nebraska at Omaha – OyaOa
Pearce, J A., PearceJA@cf.ac.uk, Cardiff Univ – Cg
Pearce, Jamie R., +44 (0) 131 650 2294 jamie.pearce@ed.ac.uk, Edinburgh Univ – Ge
Pearson, Ann, (617) 384-8392 pearson@eps.harvard.edu, Harvard Univ – Cm
Pearson, David A., 7056751151 x2336 dpearson@laurentian.ca, Laurentian Univ, Sudbury – Oe
Pearson, David M., 208-282-3486 peardavi@isu.edu, Idaho State Univ – Gc
Pearson, Eugene F., (209) 946-2926 epearson@pacific.edu, Univ of the Pacific – GsPgOg
Pearson, Graham, (780) 492-4156 gdpearson@ualberta.ca, Univ of Alberta – Cec
Pearson, P N., PearsonP@cf.ac.uk, Cardiff Univ – Pe
Pearson, Paul, +44(0)29 208 74579 PearsonP@cardiff.ac.uk, Univ of Wales – PmCg
Pearthree, Philip A., 520-621-2470 Pearthree, Phpearthre@email.arizona.

edu,
Univ of Arizona – Ge
Peate, David W., (319) 335-0567 david-peate@uiowa.edu,
Univ of Iowa – CgGi
Peavy, Samuel T., 229-931-2330 speavy@canes.gsw.edu,
Georgia Southwestern State Univ – Yg
Pec, Matej, mpec@umn.edu, Univ of Minnesota, Twin Cities – Yx
Pechmann, James C., (801) 581-3858 pechmann@seis.utah.edu,
Univ of Utah – Ys
Peck, John A., (330) 972-7659 jpeck@uakron.edu,
Univ of Akron – GseGn
Peck, Robert, 304-384-5327 Concord Univ – Pg
Peck, Theodore R., (217) 333-9486 Univ of Illinois, Urbana-Champaign – Sc
Peck, William H., (315) 228-7698 wpeck@colgate.edu,
Colgate Univ – GpiCs
Peddle, Derek R., derek.peddle@uleth.ca, Univ of Lethbridge – Or
Pedemonte, Virginia, (598) 25251552 vpedemonte@fcien.edu.uy,
Universidad de la Republica Oriental del Uruguay (UDELAR) – OirOu
Pedentchouk, Nikolai, +44 (0)1603 59 3395 n.pedentchouk@uea.ac.uk,
Univ of East Anglia – Cs
Pedersen, Joel A., (608) 263-4971 joelpedersen@wisc.edu,
Univ of Wisconsin, Madison – So
Pedersen, Thomas F., tfp@uvic.ca, Univ of Victoria – CmOcCs
Pederson, Darryll T., (402) 472-7563 dpederson2@unl.edu,
Univ of Nebraska, Lincoln – Hw
Pederson, Joel L., (435) 797-7097 joel.pederson@usu.edu,
Utah State Univ – Gm
Pedley, Kate, +64 3 3667001 Ext 3892 kate.pedley@canterbury.ac.nz,
Univ of Canterbury – Og
Pedone, Vicki A., (818) 677-6694 vicki.pedone@csun.edu,
California State Univ, Northridge – GdCs
Peebles, Ernst B., (727) 553-3983 epeebles@mail.usf.edu,
Univ of South Florida – ObCs
Peele, R Hampton, (225) 328-6651 hampton@lsu.edu,
Louisiana State Univ – OirOg
Peeters, Frank J., +31 20 5987419 f.j.c.peeters@vu.nl,
VU Univ Amsterdam – PmOg
Pegion, Kathy, 703-993-5727 kpegion@gmu.edu,
George Mason Univ – Oap
Peirce, Christine, 0191 3742515 christine.peirce@durham.ac.uk,
Durham Univ – Yg
Pekar, Stephen, (718) 997-3305 stephen.pekar@qc.cuny.edu,
Queens Coll (CUNY) – PeGrOu
Pellenbarg, Robert, rpellenbarg@Collofthedesert.edu,
Coll of the Desert – OcGu
Pellerin, Jocelyne, (418) 724-1704 jocelyne_pellerin@uqar.uquebec.ca,
Universite du Quebec a Rimouski – Ob
Pelletier, Emilien, (418) 723-1986 (Ext. 1764) emilien_pelletier@uqar.ca,
Universite du Quebec a Rimouski – Oc
Pelletier, Jon D., (520) 626-2126 jdpellet@email.arizona.edu,
Univ of Arizona – Gm
Pellin, Michael J., (630) 972-3510 Argonne National Laboratory – Om
Pellowski, Christopher J., (605) 394-2465 Christopher.Pellowski@sdsmt.edu,
South Dakota School of Mines & Tech – Gg
Pelton, John R., (208) 426-3640 jpelton@boisestate.edu,
Boise State Univ – YgeYs
Peltzer, Gilles, (310) 206-2156 peltzer@epss.ucla.edu,
Univ of California, Los Angeles – OrGt
Pemberton, S. George, (780) 492-2044 george.pemberton@ualberta.ca,
Univ of Alberta – Gr
Pendleton, Daniel, dpendlton@neaq.org,
Univ of Massachusetts, Boston – Ob
Peng, Lee Chai, 03-79674233 leecp@um.edu.my, Univ of Malaya – Pg
Peng, Yongbo, 225-578-3413 ypeng@lsu.edu, Louisiana State Univ – Cg
Peng, Zhigang, 404-894-0231 zhigang.peng@eas.gatech.edu,
Georgia Inst of Tech – YsgOn
Penna, Nigel, +44 (0) 191 208 8747 nigel.penna@ncl.ac.uk,
Univ of Newcastle Upon Tyne – Gq
Pennacchioni, Giorgio, +390498279106 giorgio.pennacchioni@unipd.it,
Università degli Studi di Padova – Gc
Penney, Paulette, (254) 710-2361 paulette_penney@baylor.edu,
Baylor Univ – On
Pennington, Deana, 915-747-5867 ddpennington@utep.edu,
Univ of Texas, El Paso – Oi
Pennington, Wayne D., (906) 487-2005 wayne@mtu.edu,
Michigan Technological Univ – YsNe
Penniston-Dorland, Sarah, (301) 405-4087 sarahpd@geol.umd.edu,
Univ of Maryland – Gp
Pennock, Jonathan R., (334) 861-7531 jonathan.pennock@unh.edu,
Dauphin Island Sea Lab – Oc
Pennuto, Christopher, pennutcm@buffalostate.edu, SUNY, Buffalo – Ng
Penny, Andrew, 831-656-3101 abpenny@nps.edu,
Naval Postgraduate School – Ow
Pentcheva, Rossitza, 089/2180 4352 pentcheva@lrz.uni-muenchen.de,
Ludwig-Maximilians-Universitaet Muenchen – Gz
Penzo, Michael A., (508) 531-1390 mpenzo@bridgew.edu,
Bridgewater State Univ – Ge
Pepino, Richard V., 717 358-4555 richard.pepino@fandm.edu,
Franklin and Marshall Coll – Ou
Peppe, Daniel J., (254) 710-2629 daniel_peppe@baylor.edu,
Baylor Univ – PbYmPe
Pepper, Ian L., (520) 621-7234 ipepper@ag.arizona.edu,
Univ of Arizona – On
Peprah, Ebenezer, (310) 660-3593 epeprah@elcamino.edu,
El Camino Coll – Og
Perault, David R., (434) 544-8370 perault@lynchburg.edu,
Lynchburg Coll – Ge
Percell, Peter, 713-743-6724 ppercell@uh.edu, Univ of Houston – Oa
Percival, Carl, C.Percival@manchester.ac.uk, Univ of Manchester – Oa
Percy, David, (503) 725-3373 Portland State Univ – Oi
Perdrial, Julia, (802) 656-0665 julia.perdrial@uvm.edu,
Univ of Vermont – ClSc
Perdrial, Nicolas, Nicolas.Perdrial@uvm.edu, Univ of Vermont – GezCl
Perdue, Edward Michael, (404) 894-3942 michael.perdue@eas.gatech.edu,
Georgia Inst of Tech – Co
Pereira, Alcides C., +351-239 860 563 apereira@dct.uc.pt,
Universidade de Coimbra – GebOr
Peres, Carlos, +44 (0)1603 59 2549 c.peres@uea.ac.uk,
Univ of East Anglia – Sf
Perez, Adriana, 915-831-8875 aperez28@epcc.edu,
El Paso Community Coll – Gg
Perez, Florante, (213) 620-5026 California Geological Survey – Ng
Perez, Marco A., mperez@cicese.mx,
Centro de Investigación Científica y de Educación Superior de Ensenada – Ye
Perez, Richard R., (518) 437-8751 perez@asrc.cestm.albany.edu,
SUNY, Albany – Oa
Perez-Huerta, Alberto, (205) 348-8382 aphuerta@ua.edu,
Univ of Alabama – PgClGz
Perfect, Edmund, (865) 974-2366 eperfect@utk.edu,
Univ of Tennessee, Knoxville – SpHwGq
Perfit, Michael R., (352) 392-2128 mperfit@ufl.edu, Univ of Florida – Giu
Peri, Francesco, Francesco.Peri@umb.edu,
Univ of Massachusetts, Boston – On
Perkins, Ronald D., (919) 684-3376 rperkins@duke.edu,
Duke Univ – Gdo
Perkins, William A., (509) 372-6131 wa_perkins@pnl.gov,
Pacific Northwest National Laboratory – Hg
Perkins, III, Dexter, (701) 777-2991 dexter.perkins@engr.und.edu,
Univ of North Dakota – Gp
Perkis, Bill, bill.perkis@gogebic.edu,
Gogebic Community Coll – Gg
Perret, Didier H., (418) 654-2686 didier.perret@canada.ca,
Universite du Quebec – NgeOu
Perrie, William, (902) 426-3985 william.perrie@dfo-mpo.gc.ca,
Dalhousie Univ – Opw
Perrin, Richard E., (505) 667-4755 Los Alamos National Laboratory – Cc
Perron, Taylor, (617) 253-5735 perron@mit.edu,
Massachusetts Inst of Tech – Gm
Perrouty, Stephane, sperrout@uwo.ca, Western Univ – EgGcYe
Perry, Baker, (828) 262-7597 perrylb@appstate.edu,
Appalachian State Univ – Og
Perry, David J., (415) 786-6887 Chabot Coll – Op
Perry, Eugene C., (815) 753-1943 eperry@niu.edu,
Northern Illinois Univ – Cs
Perry, Frank V., (505) 667-1033 fperry@lanl.gov,
Los Alamos National Laboratory – Gv
Perry, Kevin D., (801) 581-6138 kevin.perry@utah.edu,
Univ of Utah – Oa
Perry , Kyle, 859-257-0133 kyle.perry@uky.edu,
Univ of Kentucky – Nr
Perry, Randall, +44 20 759 46425 r.perry@imperial.ac.uk, Imperial Coll

– Og
Persaud, Naraine, (540) 231-3817 npers@vt.edu, Virginia Polytechnic Inst & State Univ – Sp
Persaud, Patricia, 225-578-5676 ppersaud@lsu.edu, Louisiana State Univ – Ys
Perscio, Lyman, 814-824-2076 lpersico@mercyhurst.edu, Mercyhurst Univ – GsSpHg
Persico, Lyman P., 509-527-5157 persiclp@whitman.edu, Whitman Coll – Gm
Person, Arthur, (814) 863-8568 Pennsylvania State Univ, Univ Park – Ow
Person, Jeff, jjperson@nd.gov, North Dakota Geological Survey – Pg
Person, Mark A., (575) 835-6506 mperson@ees.nmt.edu, New Mexico Inst of Mining and Tech – HyYhg
Perumal, Ram, (785) 625-3425 perumal@ksu.edu, Kansas State Univ – So
Pervunina, Aelita, +78142766039 aelita@krc.karelia.ru, Inst of Geology (Karelia, Russia) – GpvEm
Peryea, Frank J., (509) 663-8181 Washington State Univ – Sc
Pesavento, Jim, 7607441150 x2516 jpesavento@palomar.edu, Palomar Coll – GgOg
Pessagno, Jr., Emile A., (972) 883-2444 pessagno@utdallas.edu, Univ of Texas, Dallas – Pm
Pesses, Michael, mpesses@avc.edu, Antelope Valley Coll – Oyi
Pestrong, Raymond, (415) 338-2080 rayp@sfsu.edu, San Francisco State Univ – Ng
Petcovic, Heather L., (269) 387-5488 heather.petcovic@wmich.edu, Western Michigan Univ – Oe
Peteet, Dorothy M., (845) 365-8420 peteet@ldeo.columbia.edu, Columbia Univ – PleOn
Peterman, Emily M., 207-725-3846 epeterma@bowdoin.edu, Bowdoin Coll – Gtp
Peters, Catherine A., (609) 258-5645 cap@phoenix.princeton.edu, Princeton Univ – Hw
Peters, Lisa, (505) 835-5217 lisa@nmt.edu, New Mexico Inst of Mining & Tech – Gg
Peters, Mark T., (202) 586-9279 Mark_Peters@notes.ymp.gov, Los Alamos National Laboratory – Yg
Peters, Roger M., roger.peter@uwex.edu, Univ of Wisconsin - Extension – Gg
Peters, Shanan, 608-262-5987 peters@geology.wisc.edu, Univ of Wisconsin, Madison – Gs
Peters, Stephen C., 610/758-3660 scp2@lehigh.edu, Lehigh Univ – Cl
Petersen, Bruce, (314) 935-5643 petersen@wustl.edu, Washington Univ in St. Louis – On
Petersen, Erich U., (801) 581-7238 erich.petersen@utah.edu, Univ of Utah – Em
Petersen, Gary W., (814) 865-1540 gwp2@psu.edu, Pennsylvania State Univ, Univ Park – Sd
Petersen, Nikolai, 089/2180 4233 petersen@geophysik.uni-muenchen.de, Ludwig-Maximilians-Universitaet Muenchen – Yg
Petersen, Ulrich, (617) 495-2353 speters@fas.harvard.edu, Harvard Univ – Eg
Peterson, Bradley, Bradley.Peterson@stonybrook.edu, SUNY, Stony Brook – Ob
Peterson, Dallas, (785) 532-0405 dpeterso, Kansas State Univ – So
Peterson, Eric W., (309) 438-7865 ewpeter@ilstu.edu, Illinois State Univ – Hws
Peterson, Eugene J., (505) 667-5182 Los Alamos National Laboratory – Ct
Peterson, Gary L., (619) 594-5594 gpeterson@geology.sdsu.edu, San Diego State Univ – Gr
Peterson, Holly, 336-316-2263 petersonhe@guilford.edu, Guilford Coll – HwGe
Peterson, Jon W., peterson@hope.edu, Hope Coll – Ge
Peterson, Joseph E., (847) 697-1000 jpeterson@elgin.edu, Elgin Community Coll – Gg
Peterson, Joseph E., 920-424-4463 petersoj@uwosh.edu, Univ of Wisconsin, Oshkosh – PgGr
Peterson, Larry C., 305-284-6821 l.peterson@miami.edu, Univ of Miami – Gr
Peterson, Larry C., 305 421-4692 lpeterson@rsmas.miami.edu, Univ of Miami – Gu
Peterson, Peter A., (515) 294-9652 pap@iastate.edu, Iowa State Univ of Science & Tech – On
Peterson, Richard E., 806-834-3418 richard.peterson@ttu.edu, Texas Tech Univ – Oa
Peterson, Ronald C., (613) 533-6180 peterson@queensu.ca, Queen's Univ – Gz
Peterson, Virginia L., (616) 331-2811 petersvi@gvsu.edu, Grand Valley State Univ – Gxc
Pethick, Andrew, +618 9266-2297 Andrew.Pethick@curtin.edu.au, Curtin Univ – YevYg
Petit, Martin A., (309) 694-5327 Illinois Central Coll – Oe
Petr, Vilem, 303-273-3222 vpetr@mines.edu, Colorado School of Mines – Nm
Petrie, Elizabeth S., (970) 943-2117 epetrie@western.edu, Western State Colorado Univ – GcoYe
Petrinec, Zorica, +38514605969 zoricap@geol.pmf.hr, Univ of Zagreb – GipGz
Petronis, Michael S., (505) 454-3513 mspetro@nmhu.edu, New Mexico Highlands Univ – YmGcv
Petrovay, Kristof, +36 20 3722500/6621 K.Petrovay@astro.elte.hu, Eotvos Lorand Univ – Xy
Petruncio, Emil T., 410-293-6564 petrunci@usna.edu, United States Naval Academy – Opr
Petsch, Steven, (413) 545-4413 spetsch@geo.umass.edu, Univ of Massachusetts, Amherst – Col
Pettengill, Gordon H., (617) 253-6384 hgp@mit.edu, Massachusetts Inst of Tech – Og
Petters, Markus, 919-515-7144 markus_petters@ncsu.edu, North Carolina State Univ – Oa
Pettijohn, J. C., 217-333-3540 jcpettij@illinois.edu, Univ of Illinois, Urbana-Champaign – Hg
Pettinga, Jarg R., (03) 3667-001 x7716 jarg.pettinga@canterbury.ac.nz, Univ of Canterbury – Gtc
Pettit, Erin C., 907-474-5389 ecpettit@alaska.edu, Univ of Alaska, Fairbanks – Ol
Petty, Grant W., (608 )263-3265 gpetty@aos.wisc.edu, Univ of Wisconsin, Madison – OarOw
Pettyjohn, Wayne, 405-372-1981 Oklahoma State Univ – Hw
Petuch, Edward J., (561) 297-2398 epetuch@fau.edu, Florida Atlantic Univ – Pg
Peucker-Ehrenbrink, Bernhard, (508) 289-2518 bpeucker@whoi.edu, Woods Hole Oceanographic Institution – CmlHs
Pevzner, Roman, +618 9266-9805 R.Pevzner@curtin.edu.au, Curtin Univ – Yxx
Pezelj, Đurðica, +38514606117 durpezelj@yahoo.com, Univ of Zagreb – PgbPm
Pfannkuch, Hans O., (612) 624-1620 h2olafpf@umn.edu, Univ of Minnesota, Twin Cities – HwGeh
Pfefferkorn, Hermann W., (215) 898-5156 hpfeffer@sas.upenn.edu, Univ of Pennsylvania – PbePs
Pfirman, Stephanie L., (212) 854-5120 pfirman@ldeo.columbia.edu, Columbia Univ – OpGe
Pfuhl, Helen, 089/2180 4230 helen.pfuhl@geophysik.uni-muenchen.de, Ludwig-Maximilians-Universitaet Muenchen – Yg
Pheifer, Raymond N., 217-581-2626 rnpheifer@eiu.edu, Eastern Illinois Univ – EcPg
Phelps, Tommy, 865-574-7290 phelpstj1@ornl.gov, Univ of Tennessee, Knoxville – On
Phelps, William, 951-222-8350 william.phelps@rcc.edu, Riverside City Coll – GgOpPg
Philander, S. George H., (609) 258-5683 gphlder@princeton.edu, Princeton Univ – Op
Phillippe, Jennifer, 304-384-5157 elyon@concord.edu, Concord Univ – GegOe
Phillippi, Nathan E., 910-521-6588 nathan.phillippi@uncp.edu, Univ of North Carolina, Pembroke – Oni
Phillips, Andrew C., (217) 333-2513 aphillps@illinois.edu, Univ of Illinois, Urbana-Champaign – GsmOi
Phillips, Brian L., (631) 632-6853 brian.phillips@stonybrook.edu, SUNY, Stony Brook – Gz
Phillips, Dennis, (505) 667-4253 Los Alamos National Laboratory – Co
Phillips, Fred M., (575) 835-5540 phillips@nmt.edu, New Mexico Inst of Mining and Tech – Hw
Phillips, Michael, (815) 224-0394 mike_phillips@ivcc.edu, Illinois Valley Community Coll – GgeGm
Phillips, Paul, (785) 628-5969 pphillip@fhsu.edu, Fort Hays State Univ – Oe
Phillips, Richard, +44(0) 113 34 31728 R.J.Phillips@leeds.ac.uk, Univ of Leeds – Gt
Phillips, William M., (208) 885-8928 phillips@uidaho.edu, Univ of Idaho – Gm
Phillips, William R., (801) 378-4545 Brigham Young Univ – Gz
Phillips, William S., (505) 667-8106 wsp@lanl.gov,

Los Alamos National Laboratory – Yg
Philp, R. Paul, (405) 325-4469 pphilp@ou.edu, Univ of Oklahoma – Co
Philpotts, Anthony R., (860) 486-1394 anthony.philpotts@uconn.edu, Univ of Connecticut – Gi
Phinney, Robert A., (609) 258-4118 rphinney@princeton.edu, Princeton Univ – Ys
Phipps, Stephen P., (215) 898-4602 sphipps@sas.upenn.edu, Univ of Pennsylvania – Gc
Phipps Morgan, Jason, jp369@cornell.edu, Cornell Univ – GvCmPe
Piazzoni, Antonio Sebastian, 089/2180 4230 antonio.piazzoni@geophysik.uni-muenchen.de, Ludwig-Maximilians-Universitaet Muenchen – Yg
Picard, M. Dane, (801) 581-7991 dane.picard@utah.edu, Univ of Utah – Gd
Picardal, Flynn W., (812) 855-0732 picardal@indiana.edu, Indiana Univ, Bloomington – So
Piccinini, Leonardo, +390498279124 leonardo.piccinini@unipd.it, Università degli Studi di Padova – Hg
Piccoli, Philip M., (301) 405-6966 piccoli@umd.edu, Univ of Maryland – Cg
Pichevin, Laetitia, +44 (0) 131 650 5980 Laetitia.Pichevin@ed.ac.uk, Edinburgh Univ – Cg
Pichler, Thomas, (813) 974-0321 pichler@cas.usf.edu, Univ of South Florida, Tampa – Cl
Pickard, Megan, (208) 486-7678 pickardm@byui.edu, Brigham Young Univ - Idaho – GiOeGe
Pickering, Ingrid J., (306) 966-5706 ingrid.pickering@usask.ca, Univ of Saskatchewan – CgtOn
Pickering, Kevin, +44 020 7679 31325 kt.pickering@ucl.ac.uk, Univ Coll London – Gs
Pickett, Nicki, 954-201-6677 npickett@broward.edu, Broward Coll, Central Campus – On
Pickup, Gillian E., +44 (0) 131 451 3168 g.pickup@hw.ac.uk, Heriot-Watt Univ – Hw
Pier, Stanley M., (713) 522-0633 Rice Univ – Co
Pierce, David, 440-525-7341 dpierce@lakelandcc.edu, Lakeland Community Coll – GgHsOw
Pierce, Heather, (585) 292-2426 hpierce@monroecc.edu, Monroe Community Coll – Oyi
Pierce, Jeff, jeffrey.pierce@colostate.edu, Colorado State Univ – Oa
Pierce, Jennifer L., (208) 426-5380 jenpierce@boisestate.edu, Boise State Univ – GmPe
Pierce, Larry, (573) 368-2191 larry.pierce@dnr.mo.gov, Missouri Dept of Natural Resources – On
Pieren, Agustin P., 34 913944819 vdposgrado@geo.ucm.es, Univ Complutense de Madrid – GrEgg
Pieri, David C., (818) 354-6299 Jet Propulsion Laboratory – Gt
Pierrehumbert, Raymond T., (773) 702-8811 Univ of Chicago – Oa
Pierzynski, Gary M., 785-532-6101 gmp@ksu.edu, Kansas State Univ – Sc
Pieters, Carle M., (401) 863-2417 Carle_Pieters@Brown.edu, Brown Univ – Or
Pieters, Roger, (604) 822-4297 rpieters@eos.ubc.ca, Univ of British Columbia – Op
Pietrafesa, Leonard J., (919) 515-3717 leonard_pietrafesa@ncsu.edu, North Carolina State Univ – Op
Pietranik, Anna, anna.pietranik@uwr.edu.pl, Univ of Wroclaw – CgGi
Pietras, Jeff, (607) 777-3348 jpietras@binghamton.edu, Binghamton Univ – GsEo
Pietro, Kathryn R., (508) 289-3862 kpietro@whoi.edu, Woods Hole Oceanographic Institution – Gu
Pietrzak-Renaud, Natalie, 519-473-3766 npietrz@uwo.ca, Western Univ – Eg
Pigott, John D., (405) 325-4498 jpigott@ou.edu, Univ of Oklahoma – GoYeCg
Pike, J, PikeJ@cf.ac.uk, Cardiff Univ – On
Pike, Jenny, +44(0)29 208 75181 PikeJ@cardiff.ac.uk, Univ of Wales – Og
Pike, Scott, (503) 370-6587 spike@willamette.edu, Willamette Univ – Gag
Pike, Steven M., (508) 289-2350 spike@whoi.edu, Woods Hole Oceanographic Institution – Ca
Pikelj, Kristina, +38514606113 kpikelj@geol.pmf.hr, Univ of Zagreb – Gs
Pikitch, Ellen K., (631) 632-9599 ellen.pikitch@stonybrook.edu, SUNY, Stony Brook – Ob
Pilarczyk, Jessica E., (228) 688-3177 jessica.pilarczyk, Univ of Southern Mississippi – Ou
Pilkey, Jr., Orrin H., (919) 684-4238 opilkey@geo.duke.edu, Duke Univ – Ou
Pilson, Michael E., (401) 874-6104 pilson@gso.uri.edu, Univ of Rhode Island – Oc
Pinan-Llamas, Aranzazu, 260-481-6253 pinana@ipfw.edu, Indiana Univ / Purdue Univ, Fort Wayne – GcdGp
Pinckney, James, pinckney@sc.edu, Univ of South Carolina – Ob
Pinet, Paul, (315) 228-7656 ppinet@colgate.edu, Colgate Univ – Ou
Pingitore, Jr., Nicholas E., 915-747-5754 npingitore@utep.edu, Univ of Texas, El Paso – Cl
Piniella Febrer, Juan Francesc, ++34935813088 juan.piniella@uab.cat, Universitat Autonoma de Barcelona – Gz
Pinkel, Robert, (858) 534-2056 rpinkel@ucsd.edu, Univ of California, San Diego – Op
Pinnt, Todd, todd.pinnt@azwestern.edu, Arizona Western Coll – Oyi
Pinter, Nicholas, (530) 754-1041 npinter@ucdavis.edu, Univ of California, Davis – Gm
Pinti, Daniele Luigi, 514-987-3000 #2572 pinti.daniele@uqam.ca, Universite du Quebec a Montreal – Cs
Pintilei, Mitica, 0040232401494 mpintilei@gmail.com, Al. I. Cuza Univ of Iasi – CgeCt
Piotrowski, Alexander, +44 (0) 1223 333473 apio04@esc.cam.ac.uk, Univ of Cambridge – Ge
Piper, David J., (902) 426-6580 david.piper@canada.ca, Dalhousie Univ – GusGo
Piper, John, +44 0151 794 3461 Sg04@liverpool.ac.uk, Univ of Liverpool – YmGt
Pipkin, Bernard W., (213) 740-6106 pipkin@usc.edu, Univ of Southern California – Ng
Pires, E M., (516) 299-2318 mark.pires@liu.edu, Long Island Univ, C.W. Post Campus – Og
Pirie, Diane H., (305) 348-2876 Florida International Univ – Gg
Pisias, Nicklas, npisias@coas.oregonstate.edu, Oregon State Univ – Gu
Pitlick, John, (303) 492-5906 pitlick@colorado.edu, Univ of Colorado – Gm
Pittman, Jason, (916) 608-6668 PittmaJ@flc.losrios.edu, Folsom Lake Coll – OgiOi
Pizzuto, James E., (302) 831-2710 pizzuto@udel.edu, Univ of Delaware – Gm
Placzek, Christa, 0747 814 756 christa.placzek@jcu.edu.au, James Cook Univ – CclGe
Plane, David A., plane@email.arizona.edu, Univ of Arizona – On
Plank, Gabriel, (775) 784-7039 gabe@seismo.unr.edu, Univ of Nevada, Reno – Ys
Plank, Owen C., (706) 542-9072 Univ of Georgia – Sc
Plank, Terry A., (845) 365-8410 tplank@ldeo.columbia.edu, Columbia Univ – GxCg
Plankell, Eric T., 217-265-8029 eplankel@illinois.edu, Univ of Illinois, Urbana-Champaign – Hw
Plant, Jane, +44 20 759 47416 jane.plant@imperial.ac.uk, Imperial Coll – Ge
Plant, Jeffrey J., (818) 393-3799 plant@jpl.nasa.gov, Jet Propulsion Laboratory – On
Plante, Alain F., (215) 898-9269 aplante@sas.upenn.edu, Univ of Pennsylvania – SbOsCo
Plante, Martin, (418) 656-8121 martin.plante@ggl.ulaval.ca, Universite Laval – Cg
Plasienka, Dusan, 0042160296529 plasienka@fns.uniba.sk, Comenius Univ in Bratislava – Gt
Platnick, Steven, steven.platnick@nasa.gov, Univ of Wisconsin, Madison – Or
Platt, Brian F., 662-915-5440 bfplatt@olemiss.edu, Univ of Mississippi – GsPe
Platt, John P., (213) 821-1194 jplatt@usc.edu, Univ of Southern California – GctGp
Platt, Rutherford H., (413) 545-2499 rplatt@geo.umass.edu, Univ of Massachusetts, Amherst – Ou
Plattner, Christina, 089/21804220 christina.plattner@geophysik.uni-muenchen.de, Ludwig-Maximilians-Universitaet Muenchen – Yg
Plescia, Jeffrey, (202) 358-0295 Jet Propulsion Laboratory – On
Plewe, Brandon, (801) 378-4161 brandon_plewe@byu.edu, Brigham Young Univ – Oi
Plink-Bjorklund, Piret, (303) 384-2042 pplink@mines.edu, Colorado School of Mines – Gs
Plint, A G., (519) 661-3179 gplint@uwo.ca, Western Univ – Gs
Plotkin, Pamela, 979-845-3902 plotkin@tamu.edu, Texas A&M Univ – Ob
Plotnick, Roy E., (312) 996-2111 plotnick@uic.edu, Univ of Illinois at Chicago – PoiGq

Plug, Lawrence, 902-494-1200 lplug@is.dal.ca, Dalhousie Univ – Gm
Pluhar, Chris, (559) 278-1128 cpluhar@csufresno.edu, California State Univ, Fresno – GtYmNg
Plumb, Raymond A., (617) 253-6281 plumb@mit.edu, Massachusetts Inst of Tech – Ow
Plummer, Charles C., plummercc@csus.edu, California State Univ, Sacramento – Gp
Plummer, Rebecca, (301) 405-6980 rplummer@umd.edu, Univ of Maryland – Csa
Plymate, Thomas G., (417) 836-4419 tomplymate@missouristate.edu, Missouri State Univ – Gx
Poch Serra, Joan, ++935811085 joan.poch@uab.cat, Universitat Autonoma de Barcelona – Gr
Pociask, Geoff, 217-265-8212 pociask@illinois.edu, Univ of Illinois, Urbana-Champaign – Hw
Podolak, Morris, (052) 838-0976 morris@post.tau.ac.il, Tel Aviv Univ – OnnOn
Poeter, Eileen P., epoeter@mines.edu, Colorado School of Mines – Ng
Pogge von Strandmann, Phillip, +44 020 7679 33637 p.strandmann@ucl.ac.uk, Univ Coll London – Csl
Pogue, Kevin R., (509) 527-5955 pogue@whitman.edu, Whitman Coll – Gc
Pohll, Greg, 775-682-6349 greg.pohll@dri.edu, Univ of Nevada, Reno – Hwq
Poirier, Andre, 514-987-3000 #1718 Universite du Quebec a Montreal – Cs
Pojeta, John, (202) 633-1347 Smithsonian Institution / National Museum of Natural History – Pi
Pokorny, Eugene W., (702) 295-7496 eugene_pokorny@lanl.gov, Los Alamos National Laboratory – Nm
Pokras, Edward M., epokras@keene.edu, Keene State Coll – GgPmOb
Polat, Ali, 519-253-3000 x2495 polat@uwindsor.ca, Univ of Windsor – GixCc
Polet, Jascha, 909-869-3459 jpolet@csupomona.edu, California State Polytechnic Univ, Pomona – Ysg
Polito, Thomas A., (515) 294-0513 tpolito@iastate.edu, Iowa State Univ of Science & Tech – So
Polk, Jason, jason.polk@wku.edu, nWestern Kentucky Univ – Gm
Pollack, Henry N., (734) 763-0084 hpollack@umich.edu, Univ of Michigan – Yg
Pollack, Jennifer, 361-825-2041 Jennifer.Pollack@tamucc.edu, Texas A&M Univ, Corpus Christi – Ob
Pollak, Robert, pollak@wustl.edu, Washington Univ in St. Louis – On
Pollard, David D., (650) 723-4679 dpollard@pangea.stanford.edu, Stanford Univ – Gc
Pollock, Meagen, (330) 263-2202 mpollock@wooster.edu, Coll of Wooster – Gi
Polly, P D., (812) 855-7994 pdpolly@indiana.edu, Indiana Univ, Bloomington – PvgPq
Polovina, Jeffrey J., (808) 983-5390 jeffrey.polovina@noaa.gov, Univ of Hawai'i, Manoa – On
Polvani, Lorenzo M., (212) 854-7331 lmp3@columbia.edu, Columbia Univ – OwaGq
Polyak, Victor J., (505) 277-4204 polyak@unm.edu, Univ of New Mexico – CcGzm
Polzin, Dierk T., dtpolzin@wisc.edu, Univ of Wisconsin, Madison – Oa
Pomeroy, George M., (717) 477-1776 gmpome@ship.edu, Shippensburg Univ – Ou
Pommier, Anne, (858) 822-5025 pommier@ucsd.edu, Univ of California, San Diego – Cg
Pond, Lisa G., (225) 578-0401 lgpond@lsu.edu, Louisiana State Univ – Og
Pond, Stephen G., (604) 822-2205 spond@eos.ubc.ca, Univ of British Columbia – Op
Ponette-Gonzalez, Alexandra, 940-565-4012 alexandra.ponette@unt.edu, Univ of North Texas – On
Pons Muñoz, Josep Maria, ++935811054 josepmaria.pons@uab.cat, Universitat Autonoma de Barcelona – Pv
Poole, T. Craig, (559) 442-4600 craig.poole@fresnocityColl.edu, Fresno City Coll – Gg
Pope, Gregory A., (973) 655-7385 popeg@mail.montclair.edu, Montclair State Univ – GmaOy
Pope, Jeanette K., 765-658-4105 jpope@depauw.edu, DePauw Univ – Ge
Pope, Lin F., 601-266-6054 lin.pope@usm.edu, Univ of Southern Mississippi – Gz
Pope, Michael , (979) 845-4376 mcpope@geo.tamu.edu, Texas A&M Univ – GrCc
Pope, Richard, +44 (0)1332 591751 r.j.pope@derby.ac.uk, Univ of Derby – Gam
Popp, Brian N., (808) 956-6206 popp@soest.hawaii.edu, Univ of Hawai'i, Manoa – Cs
Popp, Christoph, (202) 633-1837 poppc@si.edu, Smithsonian Institution / National Museum of Natural History – Gv
Porcelli, Don, +44 (1865) 282121 don.porcelli@earth.ox.ac.uk, Univ of Oxford – Cg
Porch, William M., (505) 667-0971 wporch@lanl.gov, Los Alamos National Laboratory – Yg
Poreda, Robert J., (585) 275-0051 robert.poreda@rochester.edu, Univ of Rochester – Cg
Porlas, Dustin, (801) 581-7162 Dustin.porlas@utah.edu, Univ of Utah – On
Portell, Roger W., 3523921721 x258 portell@flmnh.ufl.edu, Univ of Florida – Pi
Porter, Claire, 612-626-0505 porte254@umn.edu, Univ of Minnesota, Twin Cities – Oi
Porter, Dwayne, porter@sc.edu, Univ of South Carolina – OgHs
Porter, John H., (804) 924-8999 jhp7e@virginia.edu, Univ of Virginia – Or
Porter, Stephen C., scporter@uw.edu, Univ of Washington – Gl
Porter, Susannah, (805) 893-8954 porter@geol.ucsb.edu, Univ of California, Santa Barbara – Gn
Portillo, Danny, (719) 333-3080 United States Air Force Academy – Oi
Portner, Ryan, ryan.portner@sjsu.edu, San Jose State Univ – Gdv
Posiloviæ, Hrvoje, +38514606087 posilovic@geol.pmf.hr, Univ of Zagreb – Gg
Posler, Gerry L., (785) 532-6101 gposler@ksu.edu, Kansas State Univ – Sf
Pospelova, Vera, (250) 721-6314 vpospe@uvic.ca, Univ of Victoria – PelOb
Post, Donald F., (520) 621-1262 postdf@ag.arizona.edu, Univ of Arizona – Sd
Post, Jeffrey E., (202) 633-1814 postj@si.edu, Smithsonian Institution / National Museum of Natural History – Gz
Post, Vincent , vincent.post@flinders.edu.au, Flinders Univ – Hw
Poteet, Mary F., 512-471-5209 mpoteet@jsg.utexas.edu, Univ of Texas at Austin – Pg
Potel, Sébastien, +33(0)3 44068999 sébastien.potel@lasalle-beauvais.fr, Institut Polytechnique LaSalle Beauvais (ex-IGAL) – GpzGi
Potess, Marla, (325) 670-1395 marla.potess@hsutx.edu, Hardin-Simmons Univ – Ogu
Poths, Jane, (505) 667-1506 Los Alamos National Laboratory – Cc
Potra, Adriana, 479-575-6419 potra@uark.edu, Univ of Arkansas, Fayetteville – EgCcGg
Potter, David, (780) 2481518 dkpotter@ualberta.ca, Univ of Alberta – Ye
Potter, Eric C., (512) 471-7090 eric.potter@beg.utexas.edu, Univ of Texas at Austin, Jackson School of Geosciences – Eo
Potter, Kenneth W., (608) 262-0400 potter@engr.wisc.edu, Univ of Wisconsin, Madison – Hg
Potter, Lee S., 319-273-2759 lspotter@earthlink.net, Univ of Northern Iowa – Gz
Potter, Paul E., (513) 556-3732 Univ of Cincinnati – Gs
Potter, Jr., Donald B., bpotter@sewanee.edu, Sewanee: Univ of the South – Gc
Potts, Donald C., (831) 459-4417 potts@biology.ucsc.edu, Univ of California, Santa Cruz – Ob
Poty, Edouard, 32 4 366 52 83 e.poty@ulg.ac.be, Universite de Liege – PiGsPe
Poudel, Durga, 337-482-6163 ddpoudel@louisiana.edu, Univ of Louisiana at Lafayette – Ge
Poulsen, Christopher J., (734) 615-2236 poulsen@umich.edu, Univ of Michigan – Pe
Poulson, Simon, 775.784.1833 poulson@mines.unr.edu, Univ of Nevada, Reno – Cls
Poulson, Simon R., (775)784-1104 poulson@mines.unr.edu, Univ of Nevada, Reno – Cs
Poulton, Mary, 520-621-8391 mpoulton@email.arizona.edu, Univ of Arizona – Ng
Poulton, Simon, +44(0) 113 34 35237 s.poulton@leeds.ac.uk, Univ of Leeds – Cls
Pound, Kate S., 320-308-2014 kspound@stcloudstate.edu, Saint Cloud State Univ – GeOe
Pourmand, Ali, apourmand@rsmas.miami.edu, Univ of Miami – Cga
Pourret, Olivier, +33(0)3 44068979 olivier.pourret@lasalle-beauvais.fr, Institut Polytechnique LaSalle Beauvais (ex-IGAL) – CgaGe
Powell, Brian, (808) 956-6724 powellb@hawaii.edu, Univ of Hawai'i, Manoa – Op
Powell, Brian A., 864656-1004 bpowell@clemson.edu, Clemson Univ – Cg

Powell, Christine A., (901) 678-2007 capowell@memphis.edu, Univ of Memphis – Ys
Powell, J. Mark, (608) 890-0070 jmpowel2@wisc.edu, Univ of Wisconsin, Madison – So
Powell, Jane, +44 (0)1603 59 2822 j.c.powell@uea.ac.uk, Univ of East Anglia – Ge
Powell, Matthew G., (814) 641-3602 powell@juniata.edu, Juniata Coll – PieGs
Powell, Ross D., (815) 753-7952 rpowell@niu.edu, Northern Illinois Univ – Gs
Powell, Thomas M., (510) 642-7455 zackp@socrates.berkeley.edu, Univ of California, Berkeley – Op
Powell, Wayne G., (718) 951-5761 wpowell@brooklyn.cuny.edu, Graduate School of the City Univ of New York – EmGaCs
Powell, Wayne G., (718) 951-5416 wpowell@brooklyn.cuny.edu, Brooklyn Coll (CUNY) – GaCsOe
Power, Mary E., (510) 643-7776 Univ of California, Berkeley – Pe
Powers, Roger W., (907) 474-6188 Univ of Alaska, Fairbanks – On
Powers, Sean, spowers@southalabama.edu, Univ of South Alabama – Ob
Poyla, David, +44 0161 275-3818 david.polya@manchester.ac.uk, Univ of Manchester – Cm
Poynton, Helen, helen.poynton@umb.edu, Univ of Massachusetts, Boston – On
Pradhanang, Soni M., (401) 874-5980 spradhanang@uri.edu, Univ of Rhode Island – HsqHq
Prahl, Fredrick G., 541-737-3969 fprahl@coas.oregonstate.edu, Oregon State Univ – Oc
Prakash, Anupma, 907-474-1897 prakash@gi.alaska.edu, Univ of Alaska, Fairbanks – Or
Pranter, Matthew J., (405) 325-3253 matthew.pranter@ou.edu, Univ of Oklahoma – Go
Pranter, Matthew J., 405-325-4451 matthew.pranter@ou.edu, Univ of Oklahoma – Gos
Prasad, Vara , (785) 532-3746 vara@ksu.edu, Kansas State Univ – So
Prather, Kimberly A., (858) 822-5312 kprather@ucsd.edu, Univ of California, San Diego – OaCmHg
Prather, Michael J., (949) 824-5838 mprather@uci.edu, Univ of California, Irvine – Oa
Pratson, Lincoln F., (919) 681-8077 lincoln.pratson@duke.edu, Duke Univ – Gs
Pratt, Allyn R., (505) 667-4308 pratt_allyn_r@lanl.gov, Los Alamos National Laboratory – Ge
Pratt, Brian R., (306) 966-5725 brian.pratt@usask.ca, Univ of Saskatchewan – PgGsr
Pratt, Lisa M., (812) 855-9203 prattl@indiana.edu, Indiana Univ, Bloomington – Co
Pratt, R. Gerhard, 519-661-2111 ext. 86690 gpratt2@uwo.ca, Western Univ – Yeg
Pratte, Darrell D., 573 368 2300 darrell.pratte@mda.mo.gov, Missouri Dept of Agriculture – On
Prave, Tony, +44 (0)1334462381 ap13@st-andrews.ac.uk, Univ of St. Andrews – GgrGs
Precedo, Laura, 954-201-6674 lprecedo@broward.edu, Broward Coll, Central Campus – Oc
Precht, Francis L., (301) 687-4440 fprecht@frostburg.edu, Frostburg State Univ – Oy
Prell, Warren L., (401) 863-3221 Warren_Prell@Brown.edu, Brown Univ – Ou
Prencipe, Mauro, mauro.prencipe@unito.it, Università di Torino – Gz
Presley, Bobby J., (979) 845-5136 bpresley@ocean.tamu.edu, Texas A&M Univ – Oc
Presley, DeAnn, 785532128 deann@ksu.edu, Kansas State Univ – So
Presnall, Dean C., (972) 883-2444 presnall@utdallas.edu, Univ of Texas, Dallas – Cp
Prestegaard, Karen L., (301) 405-6982 kpresto@geol.umd.edu, Univ of Maryland – Hgw
Preston, Thomas C., 514-398-3766 thomas.preston@mcgill.ca, McGill Univ – OanOn
Preston, William L., (805) 756-2210 wpreston@calpoly.edu, California Polytechnic State Univ – Oe
Preto, Nereo, +390498279174 nereo.preto@unipd.it, Università degli Studi di Padova – Gs
Prevec, Steve, +27 (0)46-603-8309 s.prevec@ru.ac.za, Rhodes Univ – Cg
Prevost, Jean-Herve, (609) 258-5424 Princeton Univ – Ng
Prewett, Jerry L., (573) 368-2101 jerry.prewett@dnr.mo.gov, Missouri Dept of Natural Resources – GgHw
Prezant, Robert S., (718) 997-4105 RPrezant@QC1.QC.EDU, Graduate School of the City Univ of New York – Py
Price, Alan P., (262) 521-5498 paul.price@uwc.edu, Univ of Wisconsin Colls – OyGmOg
Price, Colin G., +97236406029 cprice@flash.tau.ac.il, Tel Aviv Univ – Oa
Price, David, +44 (0)2076798581 d.price@ucl.ac.uk, Univ Coll London – GzYgPm
Price, Douglas M., 330-941-3019 dmprice@ysu.edu, Youngstown State Univ – Cg
Price, Gregory, +44 1752 584771 G.Price@plymouth.ac.uk, Univ of Plymouth – Gsr
Price, Jason R., 717-872-3005 Jason.Price@millersville.edu, Millersville Univ – Gs
Price, Jonathan D., (940) 397-4288 jonathan.price@mwsu.edu, Midwestern State Univ – GiCgGz
Price, Jonathan G., 775/682-8766 jprice@unr.edu, Univ of Nevada – Gg
Price, Katie, 404-413-5780 kprice@gsu.edu, Georgia State Univ – HsGm
Price, Kevin, (785) 532-6101 kpprice@ksu.edu, Kansas State Univ – Or
Price, L G., (575) 835-5752 greer@nmbg.nmt.edu, New Mexico Inst of Mining & Tech – Gg
Price, Maribeth H., (605) 394-2468 Maribeth.Price@sdsmt.edu, South Dakota School of Mines & Tech – Oir
Price, Marie D., (202) 994-6187 George Washington Univ – Oy
Price, Nancy A., 503-725-3398 naprice@pdx.edu, Portland State Univ – GcOeGt
Price, Peter, (573) 368-2131 peter.price@dnr.mo.gov, Missouri Dept of Natural Resources – GeHy
Price, Peter E., 281-765-7764 Peter.E.Price@nhmccd.edu, Lonestar Coll - North Harris – Oi
Price, Raymond A., (613) 533-6542 pricera@queensu.ca, Queen's Univ – GtcGe
Price, Rene, (305) 348-3119 pricer@fiu.edu, Florida International Univ – Hw
Prichard, Hazel, +44(0)29 208 74323 prichard@cardiff.ac.uk, Univ of Wales – EgGze
Prichard, Terry L., (559) 468-2086 tlprichard@ucdavis.edu, Univ of California, Davis – On
Prichonnet, Gilbert P., 514-987-3000 #3383 prichonnet.gilbert@uqam.ca, Universite du Quebec a Montreal – Pi
Prichystal, Antonín, +420 549 49 6699 prichy@sci.muni.cz, Masaryk Univ – GgaGv
Pride, Douglas E., 614-329-0941 pride.1@osu.edu, Ohio State Univ – Em
Pride, Steven, (510) 495-2823 SRPride@lbl.gov, Univ of California, Berkeley – Ys
Pridmore, Cindy L., (916) 925-6902 California Geological Survey – Ng
Priesendorf, Carl, (816) 604-2549 carl.priesendorf@mcckc.edu, Metropolitan Community Coll-Kansas City – GgOy
Priest, Eric, (847) 543-2585 epriest@clcillinois.edu, Coll of Lake County – Owe
Priest, George R., (541) 574-6642 george.priest@dogami.state.or.us, Oregon Dept of Geology & Mineral Industries – Gg
Priestley, Keith, +44 (0) 1223 337195 Univ of Cambridge – YgGt
Primeau, Francois, 949-824-9435 fprimeau@uci.edu, Univ of California, Irvine – Op
Pringle, James M., (603) 862-5000 jpringle@cisunix.unh.edu, Univ of New Hampshire – Op
Pringle, Jamie, (+44) 01782 733163 j.k.pringle@keele.ac.uk, Keele Univ – Yg
Pringle, Patrick, 360-736-9361 ppringle@centralia.edu, Centralia Coll – OgGvOn
Prinn, Ronald G., 617-253-2452 rprinn@mit.edu, Massachusetts Inst of Tech – Oa
Prins, Maarten A., +31 20 59 83635 m.a.prins@vu.nl, VU Univ Amsterdam – GsOg
Prinz, Martin, (212) 769-5381 Graduate School of the City Univ of New York – Gi
Prior, William L., 501 683-0117 bill.prior@arkansas.gov, Arkansas Geological Survey – GggEc
Pristas, Ronald, (609) 984-6587 ron.pristas@dep.state.nj.us, New Jersey Geological and Water Survey – Oi
Pritchard, Chad, 509-359-7026 cpritchard@ewu.edu, Eastern Washington Univ – Gct
Pritchard, Matthew E., mp337@cornell.edu, Cornell Univ – YdGvOl
Pritchard, Matthew E., (607) 255-4870 mp337@cornell.edu, Cornell Univ – YdGvOl
Pritchett, Brittany, 405-325-7331 brittanyp@ou.edu,

Univ of Oklahoma – Go
Privette, David, (704)330-6750 David.Privette@cpcc.edu, Central Piedmont Community Coll – Oy
Prohiæ, Esad, +38514605963 eprohic@geol.pmf.hr, Univ of Zagreb – CglGe
Proudhon, Benoit, +33(0)3 44068996 benoit.proudhon@lasalle-beauvais.fr, Institut Polytechnique LaSalle Beauvais (ex-IGAL) – GgcGt
Provin, Tony L., (979) 862-4955 t-provin@tamu.edu, Texas A&M Univ – Sc
Pruell, Richard J., (401) 782-3000 Univ of Rhode Island – Co
Pruss, Sara B., (413) 585-3948 spruss@smith.edu, Smith Coll – Pg
Pryor, Sara C., 607-255-3376 sp2279@cornell.edu, Cornell Univ – Oan
Prytulak, Julie, +44 20 759 46474 j.prytulak@imperial.ac.uk, Imperial Coll – Gg
Ptacek, Anton D., ptacek2@juno.com, San Diego State Univ – Gx
Pu, Zhaoxia, 801-585-3864 zhaoxia.pu@utah.edu, Univ of Utah – Oa
Puchtel, Igor, (301) 405-4054 ipuchtel@umd.edu, Univ of Maryland – CcgGi
Puckette, James, 405-744-6358 jim.puckette@okstate.edu, Oklahoma State Univ – Go
Puelles, Pablo, pablo.puelles@ehu.eus, Univ of the Basque Country UPV/EHU – Gct
Puente, Carlos E., (916) 752-0689 Univ of California, Davis – Hg
Pufahl, Peir K., 902-585-1858 peir.pufahl@acadiau.ca, Acadia Univ – Gs
Puffer, John H., (973) 353-5100 jpuffer@andromeda.rutgers.edu, Rutgers, The State Univ of New Jersey, Newark – GivGe
Pugh, Teresa, 615 322 2975 teri.pugh@vanderbilt.edu, Vanderbilt Univ – On
Pujana, Ignacio, (972) 883-2461 pujana@utdallas.edu, Univ of Texas, Dallas – Pm
Pujol, Jose, (901) 678-4827 jpujol@memphis.edu, Univ of Memphis – Ye
Pullammanappallil, Satish, (775) 784-613 satish@seismo.unr.edu, Univ of Nevada, Reno – Ys
Pulliam, Robert J., (512) 471-6156 jay@ig.utexas.edu, Univ of Texas at Austin – Ys
Pulliam, Robert J., (254) 710-2183 jay_pulliam@baylor.edu, Baylor Univ – YsgYe
Pumphrey, Hugh C., +44 (0) 131 650 6026 Hugh.Pumphrey@ed.ac.uk, Edinburgh Univ – Oa
Pun, Aurora, (505) 277-5629 apun@unm.edu, Univ of New Mexico – Ca
Pundsack, Jonathan, 612-626-0505 pundsack@umn.edu, Univ of Minnesota, Twin Cities – On
Punyasena, Surangi W., (217) 244-8049 punyasena@life.illinois.edu, Univ of Illinois, Urbana-Champaign – PblPe
Purcell, Rita, 781-283-3151 rpurcell@wellesley.edu, Wellesley Coll – On
Purchase, Megan, +27 (0)51 401 7158 purchasemd@ufs.ac.za, Univ of the Free State – Gx
Purdie, Duncan A., +44 (0)23 80592263 duncan.purdie@noc.soton.ac.uk, Univ of Southampton – Ob
Purdom, William B., (541) 552-6494 purdom@sou.edu, Southern Oregon Univ – Gx
Purdy, Ann, (248) 628-1562 Wayne State Univ – Gg
Purdy, G. Michael, (845) 365-8348 mpurdy@ldeo.columbia.edu, Columbia Univ – Yr
Purkiss, Robert, 325-486-6987 robert.purkiss@angelo.edu, Angelo State Univ – Gg
Purnell, Mark, +440116 252 3645 map2@le.ac.uk, Leicester Univ – Po
Purtle, Jennifer M., (501) 575-7317 jms14@comp.uark.edu, Univ of Arkansas, Fayetteville – On
Pusede, Sally, (434) 924-4544 sep6a@virginia.edu, Univ of Virginia – Oa
Putirka, Keith D., (559) 278-4524 kputirka@csufresno.edu, California State Univ, Fresno – GviGp
Putkonen, Jaakko, (701) 777-3213 jaakko.putkonen@engr.und.edu, Univ of North Dakota – Gml
Putnam, Aaron E., 207-581-2186 aaron.putnam@maine.edu, Univ of Maine – Gll
Putnam, Peter E., (403) 215-7850 Univ of Calgary – Gs
Putnam, Roger, RPutnam1@csustan.edu, California State Univ, Stanislaus – Ggi
Pyle, Eric J., 540-568-7115 pyleej@jmu.edu, James Madison Univ – Oe
Pylypenko, Svitlana, (303) 384-2140 spylypen@mines.edu, Colorado School of Mines – Omn
Pyrtle, Ashanti J., apyrtle@seas.marine.usf.edu, Univ of South Florida – Oc
Pysklywec, Russell N., (416) 978-4852 russ@geology.utoronto.ca, Univ of Toronto – Yg
Pöllmann, Herbert, +345 5526110 herbert.poellman@geo.uni-halle.de, Martin-Luther-Universitaet Halle-Wittenberg – Gz

## Q

Qi, Feng, (908) 737-3702 fqi@kean.edu, Kean Univ – OiyOg
Qiu, Bo, (808) 956-4098 bo@soest.hawaii.edu, Univ of Hawai'i, Manoa – Op
Qiu, Xiaomin, (417) 836-3129 qiu@missouristate.edu, Missouri State Univ – Oi
Qu, Deyang, deyang.qu@umb.edu, Univ of Massachusetts, Boston – Om
Quade, Jay, 520-626-3223 quadej@email.arizona.edu, Univ of Arizona – Sc
Qualls, Robert G., (775) 327-5014 qualls@unr.edu, Univ of Nevada, Reno – OsHsSf
Quan, Tracy, 405-744-6358 tracy.quan@okstate.edu, Oklahoma State Univ – Og
Quattro, Joseph, Quattro@biol.sc.edu, Univ of South Carolina – Ob
Quay, Paul D., (206) 545-8061 pdquay@u.washington.edu, Univ of Washington – Oc
Quick, Thomas J., (330) 972-6935 tquick@uakron.edu, Univ of Akron – GgCaYe
Quinn, Claire, +44(0) 113 34 38700 c.h.quinn@leeds.ac.uk, Univ of Leeds – Ge
Quinn, Courtney, (864)294-3655 courtney.quinn@furman.edu, Furman Univ – On
Quinn, Heather A., (410) 554-5522 heather.quinn@maryland.gov, Maryland Department of Natural Resources – GgHg
Quinn, James G., (401) 874-6219 jgquinn@gso.uri.edu, Univ of Rhode Island – Oc
Quinn, John, (630) 252-5357 quinnj@anl.gov, Argonne National Laboratory – Hw
Quinn, Paul, +44 (0) 191 208 5773 p.f.quinn@ncl.ac.uk, Univ of Newcastle Upon Tyne – Hg
Quinn, Terry, 512-471-0377 quinn@ig.utexas.edu, Univ of Texas at Austin – Pe
Quintanilla Terminal, Alejandra, aquintan@umn.edu, Univ of Minnesota, Twin Cities – Yx
Quintero, Sylvia, (520) 621-6025 squinter@email.arizona.edu, Univ of Arizona – On
Quinton, Page C., (315) 267-2815 quintopc@potsdam.edu, SUNY Potsdam – PeCsPm

## R

Rabalais, Nancy N., (225) 851-2800 Louisiana State Univ – Ob
Raber, George, 601-266-5807 george.raber@usm.edu, Univ of Southern Mississippi – Oi
Raczkowski, Charles, (336) 285-4847 raczkowc@ncat.edu, North Carolina Agricultural & Tech State Univ – So
Radakovich, Amy, (612) 626-4434 rada0042@d.umn.edu, Univ of Minnesota – Gp
Radcliffe, David E., (706) 542-0897 Univ of Georgia – Sp
Radcliffe, Dennis, 516-463-6545 dennis.radcliffe@hofstra.edu, Hofstra Univ – Gzx
Radebaugh, Jani, 801 422-9127 jani.radebaugh@byu.edu, Brigham Young Univ – Xg
Rademacher, Laura K., (209) 946-7351 lrademacher@pacific.edu, Univ of the Pacific – ClHyOs
Radic, Valentina, 604-827-1446 vradic@eos.ubc.ca, Univ of British Columbia – OlaOg
Radke, Lawrence F., radke@ucar.edu, Univ of Washington – Oa
Radko, Timour, (831) 656-3318 tradko@nps.edu, Naval Postgraduate School – Op
Rae, James, +44 01334 464948 jwbr@st-andrews.ac.uk, Univ of St. Andrews – Cg
Raef, Abdelmoneam E., (785) 532-2240 abraef@ksu.edu, Kansas State Univ – YesOr
Raeside, Robert, (902) 585-1583 Dalhousie Univ – Gx
Raeside, Robert P., (902) 585-1323 rob.raeside@acadiau.ca, Acadia Univ – GptGc
Rafferty, Milton D., (417) 836-5800 Missouri State Univ – On
Ragland, Paul C., (850) 644-5018 ragland@magnet.fsu.edu, Florida State Univ – Ca
Ragotzkie, Robert A., ragotzkie@wisc.edu, Univ of Wisconsin, Madison – Ob
Rahaman, M. A., 234-800-337-7441 mrahaman@oauife.edu.ng, Obafemi Awolowo Univ – GpcGi
Rahl, Jeffrey, 540-458-8101 rahlj@wlu.edu, Washington & Lee Univ – Gt

Rahn, Kenneth A., (401) 874-6713 krahn@uri.edu, Univ of Rhode Island – Oc
Rahn, Perry H., (605) 394-2527 perry.rahn@sdsmt.edu, South Dakota School of Mines & Tech – Ng
Rai, Chandra S., (405) 325-6866 crai@ou.edu, Univ of Oklahoma – Np
Railsback, L. Bruce, (706) 542-3453 rlsbk@gly.uga.edu, Univ of Georgia – PeGdCl
Raine, Robin, +353 (0)91 492271 robin.raine@nuigalway.ie, National Univ of Ireland Galway – Yg
Rains, Daniel S., (501) 683-0153 daniel.rains@arkansas.gov, Arkansas Geological Survey – GgOi
Rains, Mark C., 974-0323 Univ of South Florida, Tampa – Hw
Raj, John K., 03-79674225 jkraj@um.edu.my, Univ of Malaya – Ng
Rakovan, John F., (513) 529-3245 rakovajf@miamioh.edu, Miami Univ – GzCl
Rakowski, Cynthia L., (509) 375-6927 cindy.rakowski@pnl.gov, Pacific Northwest National Laboratory – On
Rallis, Donald N., (540) 654-1492 drallis@mwcgw.mwc.edu, Mary Washington Coll – On
Ramage, Joan, (610) 758-6410 jmr204@lehigh.edu, Lehigh Univ – OrGlm
Ramagwede, Fhatuwani L., 0027128411911 info@geoscience.org.za, Geological Survey of South Africa – EgGzEg
Raman, Sethu S., (919) 515-7144 sethu_raman@ncsu.edu, North Carolina State Univ – Oa
Ramanathan, V., (858) 534-0219 vramanathan@ucsd.edu, Univ of California, San Diego – Oa
Rambaud, Fabienne M., frambaud@austincc.edu, Austin Community Coll District – Eg
Ramelli, Alan R., (775) 784-4151 ramelli@unr.edu, Univ of Nevada – Ng
Ramirez, Abelardo L., (925) 422-6919 ramirez3@llnl.gov, Lawrence Livermore National Laboratory – Ng
Ramirez, Pedro C., (323) 343-2417 pramire@calstatela.edu, California State Univ, Los Angeles – Gs
Ramirez, Wilson R., (787) 265-3845 wilson.ramirez1@upr.edu , Univ of Puerto Rico – GudOu
Ramos, Frank C., 575-646-2511 framos@nmsu.edu, New Mexico State Univ, Las Cruces – CgGg
Rampedi, Isaac T., +27115592429 isaacr@uj.ac.za, Univ of Johannesburg – PySdHg
Ramsey, Kelvin W., (302) 831-3586 kwramsey@udel.edu, Univ of Delaware – Oa
Ramsey, Kelvin W., 302-831-2833 kwramsey@udel.edu, Univ of Delaware – Gs
Ramsey, Michael S., (412) 624-8772 ramsey@ivis.eps.pitt.edu, Univ of Pittsburgh – Or
Ramspott, Matthew E., 301/687-4412 meramspott@frostburg.edu, Frostburg State Univ – Or
Ranalli, Giorgio, granalli@ccs.carleton.ca, Carleton Univ – Yg
Rancan, Helen L., 609-984-6587 helen.rancan@dep.state.nj.us, New Jersey Geological and Water Survey – Hs
Randall, David A., randall@atmos.colostate.edu, Colorado State Univ – Oa
Randall, George, (505) 667-9483 Los Alamos National Laboratory – Yg
Randall, Gyles W., (507) 835-3620 Univ of Minnesota, Twin Cities – So
Randazzo, Anthony F., (352) 392-2634 afrgeohazards@bellsouth.net, Univ of Florida – Gd
Randlett, Victoria, (775) 327-5078 randlett@unr.edu, Univ of Nevada, Reno – Ou
Ranhofer, Melissa, (864)294-3647 melissa.ranhofer@furman.edu, Furman Univ – Gg
Rankey, Gene, (785) 864-6028 grankey@ku.edu, Univ of Kansas – GsrGu
Rankin, Robert, (780) 492-5082 rankin@phys.ualberta.ca, Univ of Alberta – Xy
Ransom, Michel D., (785) 532-7203 mdransom@ksu.edu, Kansas State Univ – SdcGm
Ranson, William A., (864) 294-3364 bill.ranson@furman.edu, Furman Univ – GxzGp
Ranville, James, (303) 273-3004 jranvill@mines.edu, Colorado School of Mines – Ca
Raol, Marcie, Marcie.Raol@fairmontstate.edu, Fairmont State Univ – Oe
Rapley, Chris, +44 020 7679 33560 christopher.rapley@ucl.ac.uk, Univ Coll London – OgrOl
Rapp, George R., 218-726-7629 grapp@d.umn.edu, Univ of Minnesota, Duluth – GaeCg
Rappe, Michael S., (808) 236-7464 rappe@hawaii.edu, Univ of Hawai'i, Manoa – Og
Rappenglueck, Bernhard, 713-743-2469 brappenglueck@uh.edu, Univ of Houston – Oa
Rasbury, E. Troy, 631-632-1488 troy.rasbury@sunysb.edu, SUNY, Stony Brook – Cc
Rashed, Mohamed A., 002-03-3921595 Rashedmohamed@yahoo.com, Alexandria Univ – GgOs
Rasmussen, Craig, (520) 621-7223 crasmuss@cals.arizona.edu, Univ of Arizona – GmCl
Rasmussen, Kenneth, (703) 323-2139 krasmussen@nvcc.edu, Northern Virginia Community Coll - Annandale – Gus
Rasmussen, Kristen L., kristenr@rams.colostate.edu, Colorado State Univ – Oa
Rasmussen, Pat E., (613) 868-8609 pat.rasmussen@canada.gc.ca, Univ of Ottawa – GeOaGb
Rasmussen, Steen, (505) 665-0052 steen@lanl.gov, Los Alamos National Laboratory – Yg
Rasmussen, Tab, (314) 935-4844 dtrasmus@wustl.edu, Washington Univ in St. Louis – On
Ratajeski, Kent, (859) 257-4444 krata2@email.uky.edu, Univ of Kentucky – Gi
Ratchford, M. E., (501) 683-0118 ed.ratchford@arkansas.gov, Arkansas Geological Survey – GocEo
Ratchford, Michael , 208-885-7991 eratchford@uidaho.edu, Univ of Idaho – GcoEg
Rath, Carolyn, (657) 278-7096 crath@fullerton.edu, California State Univ, Fullerton – Gg
Rathburn, Anthony, 812-237-2269 trathburn@gmail.com, Indiana State Univ – PyOu
Rathburn, Sara L., (970) 491-6956 rathburn@warnercnr.colostate.edu, Colorado State Univ – Ggm
Raub, Tim, +44 01334 464012 timraub@st-andrews.ac.uk, Univ of St. Andrews – Py
Rauber, Robert M., (217) 333-2835 r-rauber@illinois.edu, Univ of Illinois, Urbana-Champaign – Oa
Rauch, Henry W., (304) 293-2187 hrauch@wvu.edu, West Virginia Univ – HwGe
Rauch, Marta, marta.rauch@uwr.edu.pl, Univ of Wroclaw – Gc
Raudsepp, Mati, (604) 822-6396 mraudsepp@eos.ubc.ca, Univ of British Columbia – GzeOm
Ravat, Dhananjay, (859) 257-4726 dhananjay.ravat@uky.edu, Univ of Kentucky – YgmYv
Ravelo, Ana C., 831-459-3722 acr@pmc.ucsc.edu, Univ of California, Santa Cruz – OcGe
Ravishankara, A.R., a.r.ravishankara@colostate.edu, Colorado State Univ – Oag
Ravizza, Gregory, 808-956-2916 gravizza@soest.hawaii.edu, Univ of Hawai'i, Manoa – Cm
Rawling, Geoffrey, (505) 366-2535 geoff@gis.nmt.edu, New Mexico Inst of Mining & Tech – Gg
Rawling, J. E., (608) 263-6839 elmo.rawling@wgnhs.uwex.edu, Univ of Wisconsin - Extension – Gl
Rawlins, Michael, 413 545-0659 rawlins@geo.umass.edi, Univ of Massachusetts, Amherst – Oa
Rawlinson, Nick, nr441@cam.ac.uk, Univ of Cambridge – Ygg
Ray, G. Carleton, (804) 924-0551 cr@virginia.edu, Univ of Virginia – Ob
Ray, Pallav, 321-674-7191 pray@fit.edu, Florida Inst of Tech – Oa
Ray, Waverly, 6196447454 x3018 Waverly.Ray@gcccd.edu, Cuyamaca Coll – Gg
Rayburn, John A., 845-257-3767 rayburnj@newpaltz.edu, SUNY, New Paltz – Gml
Raymond, Anne, (979) 845-0644 raymond@geo.tamu.edu, Texas A&M Univ – Pb
Raymond, Carol A., (818) 354-8690 Jet Propulsion Laboratory – Yr
Raymond, Charles F., 206-685-9697 charlie@ess.washington.edu, Univ of Washington – Ol
Raymond, Loren A., (828) 262-2749 raymondla@appstate.edu, Appalachian State Univ – Gx
Rayne, Todd W., (315) 859-4698 trayne@hamilton.edu, Hamilton Coll – Hy
Rayner, John N., (614) 292-2514 Ohio State Univ – Oa
Rea, David K., (734) 936-0521 davidrea@umich.edu, Univ of Michigan – Gu
Read, Adam S., (505) 366-2533 adamread@gis.nmt.edu, New Mexico Inst of Mining & Tech – Gg
Read, J. Fred, (540) 231-5124 jread@vt.edu, Virginia Polytechnic Inst & State Univ – Gs
Reading, Anya, 61 3 6226 2477 anya.reading@utas.edu.au, Univ of Tasmania – Ysg
Reagan, Mark K., (319) 335-1802 mark-reagan@uiowa.edu,

Univ of Iowa – Gi
Reams, Max W., (815) 939-5394 mreams@olivet.edu, Olivet Nazarene Univ – GsmPg
Reaven, Sheldon, (631) 632-8765 sheldon.reaven@stonybrook.edu, SUNY, Stony Brook – On
Reavy, John, +353 21 4902886 j.reavy@ucc.ie, Univ Coll Cork – Gi
Reay, David, +44 (0) 131 650 7723 david.reay@ed.ac.uk, Edinburgh Univ – Oa
Reay, William G., 804-684-7119 wreay@vims.edu, Coll of William & Mary – Hg
Reber, Jacqueline, (515) 294-7513 jreber@iastate.edu, Iowa State Univ of Science & Tech – Gct
Reboulet, Edward, 785-864-2173 reboulet@kgs.ku.edu, Univ of Kansas – Hy
Rech, Jason, 513-529-1935 rechja@MiamiOh.edu, Miami Univ – Gm
Rechcigl, John E., 813-633-4111 rechcigl@ufl.edu, Univ of Florida – So
Reche Estrada, Joan, ++935811781 joan.reche@uab.cat, Universitat Autonoma de Barcelona – GpCp
Rechtien, Richard D., (573) 341-4616 Missouri Univ of Science and Tech – Ye
Rector, James W., (510) 643-7820 Univ of California, Berkeley – Ye
Redalje, Donald G., (228) 688-1174 donald.redalje@usm.edu, Univ of Southern Mississippi – Ob
Redden, Jack A., (605) 394-2461 South Dakota School of Mines & Tech – Gp
Redden, Marcella, 205-247-3654 mmcintyre@gsa.state.al.us, Geological Survey of Alabama – Gg
Reddy, Christopher M., (508) 289-2316 creddy@whoi.edu, Woods Hole Oceanographic Institution – Co
Reddy, K R., (352) 392-1803 krr@ufl.edu, Univ of Florida – Sbc
Redfern, Jonathan, +440161 275-3773 jonathan.redfern@manchester.ac.uk, Univ of Manchester – GsoGl
Redfern, Simon, +44 (0) 1223 333475 satr@cam.ac.uk, Univ of Cambridge – Gz
Redmond, Brian T., (570) 408-4803 brian.redmond@wilkes.edu, Wilkes Univ – Gs
Ree, Jin-Han, 82-2-3290-3175 reejh@korea.ac.kr, Korea Univ – Gct
Reece, Julia S., 979-458-2728 jreece@geos.tamu.edu, Texas A&M Univ – SpGso
Reed, Denise J., (504) 280-7395 djreed@uno.edu, Univ of New Orleans – Gm
Reed, Donald L., (408) 924-5036 donald.reed@sjsu.edu, San Jose State Univ – GuYg
Reed, Mark H., (541) 346-5587 mhreed@uoregon.edu, Univ of Oregon – CgEm
Reed, Robert M., (512) 471-0356 rob.reed@beg.utexas.edu, Univ of Texas at Austin, Jackson School of Geosciences – GxoGc
Reed, Wallace E., wer@virginia.edu, Univ of Virginia – Or
Reeder, Richard J., (631) 632-8208 rjreeder@stonybrook.edu, SUNY, Stony Brook – ClGz
Reedy, Robert C., (512) 471-7244 bob.reedy@beg.utexas.edu, Univ of Texas at Austin, Jackson School of Geosciences – Hg
Reedy, Robert C., (505) 277-0300 rreedy@unm.edu, Univ of New Mexico – Xcy
Rees, Margaret N., (702) 895-3890 peg.rees@unlv.edu, Univ of Nevada, Las Vegas – Gs
Reese, Andy, 601-266-4729 andy.reese@usm.edu, Univ of Southern Mississippi – Oy
Reese, Brandi, 361-825-3022 Brandi.Reese@tamucc.edu, Texas A&M Univ, Corpus Christi – Ob
Reese, Joseph F., (814) 732-2529 jreese@edinboro.edu, Edinboro Univ of Pennsylvania – GctOe
Reese, Stuart O., 717.702.2028 sreese@pa.gov, Pennsylvania Bureau of Topographic & Geologic Survey – Hw
Reesman, Arthur L., 615-322-2976 Vanderbilt Univ – Gg
Reeve, Andrew S., (207) 581-2353 asreeve@maine.edu, Univ of Maine – Hw
Reeves, Claire, +44 (0)1603 59 3625 c.reeves@uea.ac.uk, Univ of East Anglia – Cg
Reeves, Donald Matt, (907) 786-1372 dmreeves2@uaa.alaska.edu, Univ of Alaska, Anchorage – GeHgOn
Reeves, Donald Matthew M., (269) 387-5493 matt.reeves@wmich.edu, Western Michigan Univ – HwqNr
Refenes, James L., (734) 995-7594 james.refenes@cuaa.edu, Concordia Univ – Oe
Refsnider, Kurt, (928) 350-2256 kurt.refsnider@prescott.edu, Prescott Coll – GmlPe
Regalla, Christine, cregalla@bu.edu, Boston Univ – GtcGm
Reganold, John P., (509) 335-8856 reganold@wsu.edu, Washington State Univ – SoOs
Regehr, David, (785) 532-6101 dregehr@ksu.edu, Kansas State Univ – So
Reháková, Daniela, +421260296700 rehakova@fns.uniba.sk, Comenius Univ in Bratislava – Pms
Rehkamper, Mark, +44 20 759 46391 markrehk@imperial.ac.uk, Imperial Coll – CsaCc
Rehm, George W., (612) 625-6210 grehm@soils.umn.edu, Univ of Minnesota, Twin Cities – Sc
Reichard, James S., (912) 478-1153 JReich@GeorgiaSouthern.edu, Georgia Southern Univ – Hw
Reichard, Ronnal, (321) 674-7522 Florida Inst of Tech – Oo
Reichenbacher, Bettina, 089/2180 6603 b.reichenbacher@lrz.uni-muenchen.de, Ludwig-Maximilians-Universitaet Muenchen – Pg
Reichler, Thomas, 801-585-0040 thomas.reichler@utah.edu, Univ of Utah – Oa
Reid, Arch M., (713) 893-1327 archreid@gmail.com, Univ of Houston – GizXm
Reid, Brian, +44 (0)1603 59 2357 b.reid@uea.ac.uk, Univ of East Anglia – GeCg
Reid, Catherine, 3667001 Ext 7764 catherine.reid@canterbury.ac.nz, Univ of Canterbury – Po
Reid, Jeffrey C., (919) 707-9205 Jeff.Reid@ncdenr.gov, North Carolina Geological Survey –
Reid, Joshua, josh.reid@umb.edu, Univ of Massachusetts, Boston – Ga
Reid, Mary R., (928) 523-7200 mary.reid@nau.edu, Northern Arizona Univ – Gi
Reid, Ruth P., 305 421-4606 preid@rsmas.miami.edu, Univ of Miami – Gs
Reid, Steven K., 606-783-5293 s.reid@morehead-st.edu, Morehead State Univ – Gs
Reidel, Stephen P., (509) 376-9932 sreidel@wsu.edu, Washington State Univ – GivGt
Reidenbach, Matthew A., (434) 243-4937 mar5jj@virginia.edu, Univ of Virginia – Hg
Reif, Samantha, 217-786-2764 Samantha.Reif@llcc.edu, Lincoln Land Community Coll – Gg
Reilinger, Robert E., (617) 253-7860 reilinge@erl.mit.edu, Massachusetts Inst of Tech – Yg
Reimer, Andreas, +49 (0)551 392164 areimer@gwdg.de, Georg-August Univ of Goettingen – OcCmm
Reimers, Clare, 541-737-0220 creimers@coas.oregonstate.edu, Oregon State Univ – Oc
Reinen, Linda A., (909) 621-8672 lreinen@pomona.edu, Pomona Coll – Gct
Reiners, Peter, 520-626-2236 reiners@email.arizona.edu, Univ of Arizona – Cg
Reinfelder, John R., (848) 932-8013 reinfelder@envsci.rutgers.edu, Rutgers, The State Univ of New Jersey – ClObCs
Reinfelder, Ying Fan, (848) 445-2044 yingfan@rci.rutgers.edu, Rutgers, The State Univ of New Jersey – HwyOw
Reinhardt, Edward G., (905) 525-9140 (Ext. 27594) ereinhar@mcmaster.ca, McMaster Univ – Pm
Reisner, Jon M., (505) 665-1889 reisner@lanl.gov, Los Alamos National Laboratory – Oa
Reiss, Nathan, (212) 346-1502 Pace Univ, New York Campus – Oa
Reiten, Jon C., 406-657-2630 jreiten@mtech.edu, Montana Tech of The Univ of Montana – Hw
Reiter, Marshall A., (575) 835-5306 mreiter@nmt.edu, New Mexico Inst of Mining and Tech – Yh
Reitner, Joachim, +49 (0)551 397950 jreitne@gwdg.de, Georg-August Univ of Goettingen – PyyPy
Reitz, Elizabeth J., (706) 542-1464 ereitz@arches.uga.edu, Univ of Georgia – Ga
Remacha Grau, Eduard, ++935811603 eduard.remacha@uab.cat, Universitat Autonoma de Barcelona – Gso
Remenda, Victoria H., (613) 533-6594 remendav@queensu.ca, Queen's Univ – Hw
Rempel, Alan W., (541) 346-6316 rempel@uoregon.edu, Univ of Oregon – YgNrOl
Remson, Irwin, (650) 723-9191 Stanford Univ – Hw
Renard, Robert J., 831-375-2354 bobandotty@aol.com, Naval Postgraduate School – Ow
Renaut, Robin W., (306) 966-5705 robin.renaut@usask.ca, Univ of Saskatchewan – Gs

Reneau, Raymond B., (540) 231-9779 reneau@vt.edu, Virginia Polytechnic Inst & State Univ – Sc
Reneau, Steven L., (505) 665-3151 sreneau@lanl.gov, Los Alamos National Laboratory – Gm
Renfrew, Ian, +44 (0)1603 59 2557 i.renfrew@uea.ac.uk, Univ of East Anglia – Ow
Renfrew, Melanie, renfremp@lahc.edu, Los Angeles Harbor Coll – Og
Renne, Paul, (510) 644-1350 prenne@bgc.org, Univ of California, Berkeley – Cc
Rennie, Colette, (902) 867-2299 crennie@stfx.ca, Saint Francis Xavier Univ – Gg
Rennie, Tim, (613) 258-8336 Ext. 61286 trennie@kemptvillec.uoguelph.ca, Univ of Guelph – On
Rennie, Tom, (775) 682-6088 tom@seismo.unr.edu, Univ of Nevada, Reno – Ys
Renock, Devon, Devon.J.Renock@Dartmouth.edu, Dartmouth Coll – GzCa
Renshaw, Carl E., (603) 646-3365 carl.renshaw@dartmouth.edu, Dartmouth Coll – HgGc
Rentel, Raimund, +27 (0)51 401 2414 rentelr@ufs.ac.za, Univ of the Free State – Gz
Renton, John J., (304) 293-5603 jrenton@wvu.edu, West Virginia Univ – Ec
Renwick, William, (513) 529-5010 renwicwh@miamioh.edu, Miami Univ – GmHg
Repeta, Daniel J., (508) 289-2635 drepeta@whoi.edu, Woods Hole Oceanographic Institution – Co
Repka, James, (949) 582-4694 jrepka@saddleback.edu, Saddleback Community Coll – GgmOe
Reshef, Moshe, 972-36406880 mosher@post.tau.ac.il, Tel Aviv Univ – Yeg
Resing, Joseph A., (206) 526-6184 resing@pmel.noaa.gov, Univ of Washington – Oc
Resio, D. R., (407) 768-8000 Florida Inst of Tech – Op
Resnic, Victor S., 281-618-5800 rvictor@mcleodusa.net, Lonestar Coll - North Harris – Go
Resnichenko, Yuri, (598) 2525 1552 yresni@fcien.edu.uy, Universidad de la Republica Oriental del Uruguay (UDELAR) – OirOn
Resor, Phillip G., 860 6853139 presor@wesleyan.edu, Wesleyan Univ – Gct
Ressel, Mike, (775) 682-7844 mressel@unr.edu, Univ of Nevada – Eg
Ressler, Daniel E., 570-372-4216 resslerd@susqu.edu, Susquehanna Univ – Sp
Restrepo, Jorge I., (561) 297-2795 restrepo@fau.edu, Florida Atlantic Univ – Hq
Retallack, Gregory J., (541) 346-4558 gregr@uoregon.edu, Univ of Oregon – PbOsGa
Retelle, Michael J., (207) 786-6155 mretelle@bates.edu, Bates Coll – Gl
Rettig, Andrew, 937-229-2261 arettig1@udayton.edu, Univ of Dayton – Oy
Retzler, Andrew J., (612) 626-3895 aretzler@umn.edu, Univ of Minnesota – GrsPi
Reusch, David B., 575-835-5404 dreusch@ees.nmt.edu, New Mexico Inst of Mining and Tech – On
Reusch, Douglas N., 207-778-7463 info@reuschlaw.de, Univ of Maine - Farmington – GtCgOu
Reuss, Robert L., 617-627-3494 bert.reuss@tufts.edu, Tufts Univ – Gzi
Reuter, Gerhard, (780) 492-0358 gerhard.reuter@ualberta.ca, Univ of Alberta – Ow
Revelle, Douglas O., (505) 667-1256 revelle@lanl.gov, Los Alamos National Laboratory – Oa
Revenaugh, Justin, 612-624-7553 justinr@umn.edu, Univ of Minnesota, Twin Cities – Ys
Revetta, Frank A., (315) 267-2289 revettfa@potsdam.edu, SUNY Potsdam – Yg
Rex, Arthur B., (828) 262-6911 rexab@appstate.edu, Appalachian State Univ – Oi
Rexius, James E., 734-462-4400 jrexius@schoolcraft.edu, Schoolcraft Coll – Gl
Reyes, Alfonso, reyeszca@cicese.mx, Centro de Investigación Científica y de Educación Superior de Ensenada – Ne
Reynolds, Barbara C., (828) 232-5048 kreynolds@unca.edu, Univ of North Carolina, Asheville – On
Reynolds, James H., (828) 884-8377 reynoljh@brevard.edu, Brevard Coll – GrvGe
Reynolds, Robert W., (541) 383-7557 Central Oregon Community Coll – Gv
Reynolds, Stephen J., (480) 965-9049 sreynolds@asu.edu, Arizona State Univ – GctOe
Rezaie-Boroon, Mohammad H., (323) 343-2406 mrezaie@calstatela.edu, California State Univ, Los Angeles – GeCg
Rheuban, Jennie, (508) 289-3782 jrheuban@whoi.edu, Woods Hole Oceanographic Institution – Oc
Rhines, Peter B., 206-543-0593 rhines@atmos.washington.edu, Univ of Washington – Op
Rhoads, Bruce, brhoads@illinois.edu, Univ of Illinois, Urbana-Champaign – Gm
Rhoads, James, (480) 727-7133 james.rhoads@asu.edu, Arizona State Univ – On
Rhode, David E., (775) 673-7310 dave@dri.edu, Desert Research Inst – Pe
Rhodes, Amy L., (413) 585-3947 arhodes@smith.edu, Smith Coll – ClGe
Rhodes, Carol J., (979) 845-3001 cj-rhodes@tamu.edu, Texas A&M Univ – On
Rhodes, Dallas D., (912) 478-5361 Georgia Southern Univ – Gm
Rhodes, Edward J., (310) 825-3880 erhodes@epss.ucla.edu, Univ of California, Los Angeles – CcGma
Rhodes, Frank H. T., (607) 255-6233 Cornell Univ – Pi
Rhodes, J. Michael, (413) 545-2841 jmrhodes@geo.umass.edu, Univ of Massachusetts, Amherst – GviCa
Rhyne, Andrew, arhyne@neaq.org, Univ of Massachusetts, Boston – Ob
Rial, Jose A., (919) 966-4553 jose_rial@unc.edu, Univ of North Carolina, Chapel Hill – Ys
Ribbe, Paul H., (540) 231-6880 ribbe@vt.edu, Virginia Polytechnic Inst & State Univ – Gz
Ricciardi, Karen, karen.ricciardi@umb.edu, Univ of Massachusetts, Boston – Gq
Rice, Benjamin, 847-491-8190 ben@earth.northwestern.edu, Northwestern Univ – On
Rice, Chuck W., (785) 532-7217 cwrice@ksu.edu, Kansas State Univ – Sb
Rice, James R., (617) 495-3445 rice@esag.harvard.edu, Harvard Univ – Yg
Rice, Karen C., (434) 243-3429 kcr4y@virginia.edu, Univ of Virginia – HgCg
Rice, Keith W., 715-346-4454 krice@uwsp.edu, Univ of Wisconsin, Stevens Point – Oir
Rice, Melissa S., (360) 650-3592 melissa.rice@wwu.edu, Western Washington Univ – XgGsm
Rice, Murray, 940-565-3861 murray.rice@unt.edu, Univ of North Texas – On
Rice, Pamela J., (612) 625-1909 price@soils.umn.edu, Univ of Minnesota, Twin Cities – So
Rice, Thomas J., (715) 889-9401 trice@calpoly.edu, California Polytechnic State Univ – Sd
Rice, Thomas L., (937) 766-6140 trice@cedarville.edu, Cedarville Univ – GemEo
Rice-Snow, R. Scott, 765-285-8269 ricesnow@bsu.edu, Ball State Univ – Gm
Rich, Fredrick J., (912) 478-0849 frich@georgiasouthern.edu, Georgia Southern Univ – Pl
Rich, Thomas B., (605) 677-6150 tom.rich@usd.edu, South Dakota Dept of Environment and Natural Resources – Hw
Richard, Benjamin H., benjamin.richard@wright.edu, Wright State Univ – Ye
Richard, Gigi, 970-248-1689 grichard@coloradomesa.edu, Colorado Mesa Univ – Hsg
Richard, Robert, (310) 825-6663 rrichard@igpp.ucla.edu, Univ of California, Los Angeles – Yg
Richards, James H., (530) 752-0170 jhrichards@ucdavis.edu, Univ of California, Davis – Os
Richards, Jeremy P., (780) 492-3430 jeremy.richards@ualberta.ca, Univ of Alberta – Em
Richards, Kelvin J., 808-956-5399 rkelvin@hawaii.edu, Univ of Hawai'i, Manoa – Op
Richards, Laura, +44 0161 306-0361 laura.richards@manchester.ac.uk, Univ of Manchester – Ge
Richards, Mark A., Mark_Richards@berkeley.edu, Univ of California, Berkeley – Yg
Richards, Paul G., (845) 365-8389 Columbia Univ – Ys
Richards, Paul L., prichard@brockport.edu, SUNY, The Coll at Brockport – Hg
Richards-McClung, Bryony, 801-585-0599 bmcclung@egi.utah.edu, Univ of Utah – Go

Richardson, Eliza, (814) 863-2507 eliza@psu.edu, Pennsylvania State Univ, Univ Park – YsOe
Richardson, Mary J., (979) 845-7966 mrichardson@ocean.tamu.edu, Texas A&M Univ – Op
Richardson, Randall M., (520) 621-4950 rmr@email.arizona.edu, Univ of Arizona – Yg
Richardson, Steve, 021-650-2921 steve.richardson@uct.ac.za, Univ of Cape Town – Cg
Richardson, Tammi, richardson@biol.sc.edu, Univ of South Carolina – Ob
Richaud, Mathieu, (559) 278-4557 mathieu@csufresno.edu, California State Univ, Fresno – GusCs
Richey, Jeffrey E., (206) 543-7339 jrichey@u.washington.edu, Univ of Washington – Oc
Richmond, Marshall C., (509) 372-6241 mc_richmond@pnl.gov, Pacific Northwest National Laboratory – Hs
Richter, Carl, (337) 482-5353 richter@louisiana.edu, Univ of Louisiana at Lafayette – Ym
Richter, Daniel D., 919-613-8031 drichter@duke.edu, Duke Univ – Os
Richter, David, David.Richter.26@nd.edu, Univ of Notre Dame – Oa
Richter, Frank M., (773) 702-8118 richter@geosci.uchicago.edu, Univ of Chicago – Gt
Richter, Suzanna L., (717) 358-5843 suzanna.richter@fandm.edu, Franklin and Marshall Coll – Ge
Ricka, Adam, +420 549 49 6605 ricka@sci.muni.cz, Masaryk Univ – Hgy
Rickaby, Ros, +44 (1865) 272034 rosalind.rickaby@earth.ox.ac.uk, Univ of Oxford – Gz
Rickard, D, rickard@cf.ac.uk, Cardiff Univ – Cg
Ricketts, Hugo, +44 0161 306-3911 h.ricketts@manchester.ac.uk, Univ of Manchester – Oa
Ricketts, Jason, (915) 747-5599 jricketts@utep.edu, Univ of Texas, El Paso – GctGi
Ridd, Merrill K., (801) 581-7939 merrill.ridd@geog.utah.edu, Univ of Utah – Or
Ridenour, Gregory D., (931) 221-7454 ridenourg@apsu.edu, Austin Peay State Univ – HsOgn
Ridenour, Wendy M., 406 683-7264 wendy.ridenour@umwestern.edu, Univ of Montana Western – Pe
Ridge, John C., (617) 627-3494 jack.ridge@tufts.edu, Tufts Univ – GlnGg
Ridgway, Kenneth D., (765) 494-3269 ridge@purdue.edu, Purdue Univ – Gs
Ridgwell, Andy, (951) 827-3186 andy@seao2.org, Univ of California, Riverside – On
Riding, Robert, 865-974-2366 rriding@utk.edu, Univ of Tennessee, Knoxville – Gs
Ridky, Alice M., (207) 859-5800 amridky@colby.edu, Colby Coll – On
Ridley, John R., (970) 491-5943 jridley@cnr.colostate.edu, Colorado State Univ – Eg
Ridley, Moira K., (806) 834-0627 moira.ridley@ttu.edu, Texas Tech Univ – Cl
Riebe, Clifford S., 307-766-3965 criebe@uwyo.edu, Univ of Wyoming – ClGme
Riebesell, John, (313) 593-5132 jriebese@umich.edu, Univ of Michigan, Dearborn – Ou
Riedel, Oliver, 089/2180 4335 oliver.riedl@lrz.uni-muenchen.de, Ludwig-Maximilians-Universitaet Muenchen – Gz
Rieder, Michael M., +49 89 28925864 rieder@tum.de, Technische Universitaet Muenchen – GgxGs
Riediger, Cynthia L., (403) 220-8783 riediger@geo.ucalgary.ca, Univ of Calgary – Co
Riedinger, Natascha, 405-744-6358 natascha.riedinger@okstate.edu, Oklahoma State Univ – CgGu
Rieger, Duayne, 814-341-1674 drieger@ccri.edu, Community Coll of Rhode Island – Ys
Rieke, George H., (520) 621-2832 grieke@as.arizona.edu, Univ of Arizona – On
Rieken, Eric R., (210) 381-3526 Univ of Texas, Pan American – Gt
Riemer, Nicole, 217-244-2844 nriemer@illinois.edu, Univ of Illinois, Urbana-Champaign – Oa
Riemersma, Peter E., 616-331-3553 riemersp@gvsu.edu, Grand Valley State Univ – Hw
Rieppel, Olivier C., (312) 665-7630 orieppel@fieldmuseum.org, Field Museum of Natural History – Pv
Ries, Justin, 7815817370 x342 j.ries@neu.edu, Northeastern Univ – Cm
Riess, Carolyn M., (512) 468-1832 criess@austincc.edu, Austin Community Coll District – GoeGg
Rietbrock, Andreas, +44-151-794-5181 A.Rietbrock@liverpool.ac.uk, Univ of Liverpool – YsGv
Rietmeijer, Frans J., (505) 277-5733 fransjmr@unm.edu, Univ of New Mexico – Gp
Rigby, John, 61 7 3138 1638 j.rigby@qut.edu.au, Queensland Univ of Tech – Pb
Riggs, Eric, 9798453651 emriggs@geos.tamu.edu, Texas A&M Univ – Gy
Riggs, Nancy, (928) 523-9362 nancy.riggs@nau.edu, Northern Arizona Univ – Gv
Riggs, Stanley R., (252) 328-6015 riggss@ecu.edu, East Carolina Univ – Gu
Rignot, Eric, (949) 824-3739 erignot@uci.edu, Univ of California, Irvine – OlpOr
Rigo, Manuel, +390498279175 manuel.rigo@unipd.it, Università degli Studi di Padova – Gs
Rigsby, Catherine A., (252) 328-4297 rigsbyc@ecu.edu, East Carolina Univ – Gs
Riha, Susan, (607) 255-1729 sjr4@cornell.edu, Cornell Univ – Sf
Riker-Coleman, Kristin E., 715-394-8410 krikerco@uwsuper.edu, Univ of Wisconsin, Superior – Gg
Riley, James, (520) 626-6681 jjriley@ag.arizona.edu, Univ of Arizona – On
Riley, Rhonda, riley@suu.edu, Southern Utah Univ – On
Rimmer, Susan M., (618) 453-7369 srimmer@siu.edu, Southern Illinois Univ Carbondale – EcCgGo
Rimstidt, J. Donald, (540) 231-6894 jdr02@vt.edu, Virginia Polytechnic Inst & State Univ – Clg
Rinae, Makhadi, +27 (0)51 401 9008 makhadi@ufs.ac.za, Univ of the Free State – Gge
Rind, David H., (212) 678-5593 drind@giss.nasa.gov, Columbia Univ – Oar
Rindsberg, Andrew K., 205 652 3416 arindsberg@uwa.edu, Univ of West Alabama – PeGePi
Rink, W. J., (905) 525-9140 (Ext. 24178) rinkwj@mcmaster.ca, McMaster Univ – Cc
Rinterknecht, Vincent, +33 (0)1 45 07 55 81 vincent.rinterknecht@lgp.cnrs.fr, Univ of St. Andrews – CcGma
Riordan, Allen J., (919) 515-7973 al_riordan@ncsu.edu, North Carolina State Univ – Oa
Riordan, Jean, (907) 696-0079 jean.riordan@alaska.gov, Alaska Division of Geological & Geophysical Surveys – Gg
Rios-Sanchez, Miriam, 218-755-2595 mriossanchez@bemidjistate.edu, Bemidji State Univ – HwOr
Ripley, Edward M., (812) 855-1196 ripley@indiana.edu, Indiana Univ, Bloomington – Em
Riser, Stephen C., (206) 543-1187 riser@ocean.washington.edu, Univ of Washington – Op
Risk, Dave A., drisk@stfx.ca, Saint Francis Xavier Univ – Osr
Risk, Michael J., (905) 525-9140 riskmj@mcmaster.ca, McMaster Univ – Po
Ritchie, Alexander W., (843) 953-5591 ritchiea@cofc.edu, Coll of Charleston – Gc
Ritchie, Harold C., (902) 494-5192 hritchie@phys.ocean.dal.ca, Dalhousie Univ – Ow
Ritsche, Michael, 630-252-1554 mtritsche@anl.gov, Argonne National Laboratory – Oa
Ritsema, Jeroen, (734) 615-6405 jritsema@umich.edu, Univ of Michigan – Ys
Rittenour, Tammy M., (435) 797-1273 tammy.rittenour@usu.edu, Utah State Univ – GmCcGa
Ritter, Charles J., (937) 229-2953 Univ of Dayton – Ct
Ritter, Joachim R., +49-721-60844539 joachim.ritter@kit.edu, Karlsruhe Inst of Tech – YsGtv
Ritter, John B., (937) 327-7332 jritter@wittenberg.edu, Wittenberg Univ – Gm
Ritter, Leonard, (519) 824-4120 Ext.52980 lritter@uoguelph.ca, Univ of Guelph – On
Ritter, Michael E., 715-346-4449 mritter@uwsp.edu, Univ of Wisconsin, Stevens Point – Ow
Ritter, Paul, (309) 268-8640 paul.ritter@heartland.edu, Heartland Community Coll – Og
Ritter, Scott M., 801-4224239 scott_ritter@byu.edu, Brigham Young Univ – Ps
Ritterbush, Linda A., (805) 493-3265 ritterbu@clunet.edu, California Lutheran Univ – Pi
Ritz, Richard, (402) 280-2461 richard.ritz@afwa.af.mil,

Creighton Univ – Ow
Ritzi, Jr., Robert W., 937 775-3455 robert.ritzi@wright.edu, Wright State Univ – Hw
Rius, Marc, +44 (0)23 8059 3275 m.rius@soton.ac.uk, Univ of Southampton – Onn
Rivard, Benoit, (780) 492-0345 benoit.rivard@ualberta.ca, Univ of Alberta – Or
Rivera, Edna L., (312) 996-6123 eriver15@uic.edu, Univ of Illinois at Chicago – On
Rivera, Mark, 907-786-1235 marivera@uaa.alaska.edu, Univ of Alaska, Anchorage – Gg
Rivers, Toby C. J. S., (709) 737-8392 trivers@sparky2.esd.mun.ca, Memorial Univ of Newfoundland – Gp
Rizeli, Mustafa Eren, 00904242370000-5961 merizeli@firat.edu.tr, Firat Univ – Gxi
Rizoulis, Athanasios, +44 0161 275-0311 A.Rizoulis@manchester.ac.uk, Univ of Manchester – Ge
Roach, Michael, 61 3 6226 2474 Univ of Tasmania – Yg
Roadcap, George S., (217) 333-7951 Illinois State Water Survey – Hw
Robarge, Wayne P., (919) 515-1454 North Carolina State Univ – Sc
Robas, Sheryl A., (609) 258-6144 srobas@princeton.edu, Princeton Univ – On
Robbins, Debra C., (828) 251-6441 drobbins@unca.edu, Univ of North Carolina, Asheville – On
Robbins, Gary A., (860) 486-2448 gary.robbins@uconn.edu, Univ of Connecticut – Hw
Robert, Genevieve, (207) 786-6105 grobert@bates.edu, Bates Coll – CpGv
Robert, Sanborn, (262) 335-5263 robert.sanborn@uwc.edu, Univ of Wisconsin Colls – GgOyn
Roberts, A. Lynn, (410) 516-4387 lroberts@jhu.edu, Johns Hopkins Univ – On
Roberts, Charles E., (561) 297-3254 croberts@fau.edu, Florida Atlantic Univ – Ori
Roberts, Eric M., +61747816947 eric.roberts@jcu.edu.au, James Cook Univ – GsPvCc
Roberts, Frank, (504) 388-2964 Montgomery County Community Coll – Gp
Roberts, Gerald, +44 020 3073 8033 gerald.roberts@ucl.ac.uk, Birkbeck Coll – Ne
Roberts, Harry H., (225) 388-2964 harry@antares.esl.lsu.edu, Louisiana State Univ – Gs
Roberts, Jennifer A., (785) 864-4997 jenrob@ku.edu, Univ of Kansas – PyCl
Roberts, Mark L., (508) 289-3654 mroberts@whoi.edu, Woods Hole Oceanographic Institution – Yg
Roberts, Paul H., (310) 206-2707 roberts@math.ucla.edu, Univ of California, Los Angeles – Ym
Roberts, Peter, (505) 667-1199 proberts@lanl.gov, Los Alamos National Laboratory – Ys
Roberts, Sarah K., (925) 423-4112 roberts28@llnl.gov, Lawrence Livermore National Laboratory – Cl
Roberts, Sheila J., (419) 372-0354 sjrober@bgnet.bgsu.edu, Bowling Green State Univ – CgHw
Roberts, Sheila M., (406) 683-7017 sheila.roberts@umwestern.edu, Univ of Montana Western – Ge
Roberts, Stephen, +44 ()023 80593246 steve.roberts@noc.soton.ac.uk, Univ of Southampton – Cg
Robertson, Alastair H., +44 (0) 131 650 8546 alastair.robertson@ed.ac.uk, Edinburgh Univ – GtgGs
Robertson, Andrew W., 845-680-4491 awr@iri.columbia.edu, Columbia Univ – Oa
Robertson, Charles E., (573) 341-4616 Missouri Univ of Science and Tech – Gc
Robertson, Daniel E., 585-292-2422 drobertson@monroecc.edu, Monroe Community Coll – Eg
Robertson, James M., (608) 263-7384 james.robertson@uwex.edu, Univ of Wisconsin - Extension – Eg
Robin, Michel R., 613-562-5800 Ext 6852 mrobin@uottawa.ca, Univ of Ottawa – Hw
Robin, Pierre-Yves F., 905 828-5419 Univ of Toronto – Gc
Robinson, Alexander, 713-743-2547 acrobinson@uh.edu, Univ of Houston – Gc
Robinson, Bruce A., (505) 667-1910 robinson@lanl.gov, Los Alamos National Laboratory – On
Robinson, Carol, +44 (0)1603 59 3174 carol.robinson@uea.ac.uk, Univ of East Anglia – Ob
Robinson, Clare, +44 0161 275-3296 A.Rizoulis@manchester.ac.uk, Univ of Manchester – Co
Robinson, Cordula, cordula@crsa.bu.edu, Boston Univ – Gm
Robinson, David, +44 (0)1908 653493 x 53493 david.robinson@open.ac.uk, The Open Univ – On
Robinson, Delores, dmr@ua.edu, Univ of Alabama – Gc
Robinson, Edward, (305) 348-3572 draper@fiu.edu, Florida International Univ – Pm
Robinson, Edwin S., (540) 231-6521 esrobinson@vt.edu, Virginia Polytechnic Inst & State Univ – Yg
Robinson, Francis J., (203) 432-2033 francis.robinson@yale.edu, Yale Univ – Gg
Robinson, George W., grobinson@stlawu.edu, St. Lawrence Univ – Gz
Robinson, Ian S., +44 (0)23 80593438 isr@noc.soton.ac.uk, Univ of Southampton – Or
Robinson, Judith, 973-353-1976 judy.robinson@rutgers.edu, Rutgers, The State Univ of New Jersey, Newark – Yg
Robinson, Kevin, (619) 594-1386 rockrobinson@gmail.com, San Diego State Univ – Gc
Robinson, Leonie, +44 0151 795 4387 Leonie.Robinson@liverpool.ac.uk, Univ of Liverpool – Ob
Robinson, Lori, (612) 626-7429 robin126@umn.edu, Univ of Minnesota – Oi
Robinson, Mark, (480) 727-9691 mark.s.robinson@asu.edu, Arizona State Univ – On
Robinson, Michael, (912) 598-3310 mike.robinson@skio.usg.edu, Georgia Southern Univ – Oin
Robinson, Michael A., marobinson3@sbcc.edu, Santa Barbara City Coll – OyiOw
Robinson, Paul D., (618) 453-7373 robinson@geo.siu.edu, Southern Illinois Univ Carbondale – Gz
Robinson, Paul T., (902) 494-2361 Dalhousie Univ – Gv
Robinson, Peter, (413) 545-2286 Univ of Massachusetts, Amherst – Gc
Robinson, Peter, (303) 492-5108 peter.robinson@colorado.edu, Univ of Colorado – Pv
Robinson, R. Craig, (860) 832-2950 Central Connecticut State Univ – Xy
Robinson, Richard, robinson_richard@smc.edu, Santa Monica Coll – Gg
Robinson, Ruth, +44 01334 463996 rajr@st-andrews.ac.uk, Univ of St. Andrews – Gs
Robinson, Sarah, (719) 333-9287 sarah.robinson@usafa.edu, United States Air Force Academy – GgOi
Robinson, Steve, j.s.robinson@reading.ac.uk, Univ of Reading – Sf
Robinson, Stuart, +44 (1865) 272058 stuartr@earth.ox.ac.uk, Univ of Oxford – PeGr
Robinson, Walter, 919-515-7002 walter_robinson@ncsu.edu, North Carolina State Univ – Oa
Robinson, Walter A., (217) 333-2292 robinson@atmos.uiuc.edu, Univ of Illinois, Urbana-Champaign – Oa
Robinson, William, william.robinson@umb.edu, Univ of Massachusetts, Boston – On
Robison, Richard A., (785) 864-2739 rrobisn@ku.edu, Univ of Kansas – PiGr
Robl, Thomas , tom.robl@uky.edu, Univ of Kentucky – EcCo
Rocha, Guillermo, 7189515000 x2887 grocha@brooklyn.cuny.edu, Brooklyn Coll (CUNY) – GgeCg
Roche, Didier, +31 20 59 83077 didier.roche@vu.nl, VU Univ Amsterdam – Pe
Roche, James E., (225) 388-2707 jroche@geol.lsu.edu, Louisiana State Univ – Gg
Rochester, Michael G., (709) 737-7565 mrochest@morgan.ucs.mun.ca, Memorial Univ of Newfoundland – Yg
Rochette, Elizabeth A., (603) 862-0713 Univ of New Hampshire – Sc
Rochette, Scott M., (585) 395-2603 srochett@brockport.edu, SUNY, The Coll at Brockport – Ow
Rocholl, Alexander, 089/2180 4293 rocholl@min.uni-muenchen.de, Ludwig-Maximilians-Universitaet Muenchen – Gz
Rock, Jessie, 701-231-7951 jessie.rock@ndsu.edu, North Dakota State Univ – Og
Rockaway, John D., (859) 572-5412 rockawayj@nku.edu, Northern Kentucky Univ – Ng
Rockwell, Thomas K., (619) 594-4441 trockwell@mail.sdsu.edu, San Diego State Univ – Gm
Rockwood, Anthony A., (303) 556-8399 Metropolitan State Coll of Denver – Ow
Rocque, David, 207-287-2666 david.rocque@maine.gov, Dept of Agriculture, Conservation, and Forestry – Sd
Rodbell, Donald T., (518) 388-6034 rodbelld@union.edu, Union Coll – GlmGe

Roden, Eric E., (608) 260-0724 eroden@geology.wisc.edu, Univ of Wisconsin, Madison – PySbCl
Roden, Gunnar I., (206) 543-5627 giroden@u.washington.edu, Univ of Washington – Op
Roden, Michael F., (706) 542-2416 mroden@uga.edu, Univ of Georgia – Gi
Rodgers, David W., (208) 282-3460 rodgdavi@isu.edu, Idaho State Univ – Gct
Rodgers, Jim, (307) 766-2286 Ext. 255 james.rodgers@wyo.gov, Wyoming State Geological Survey – Gg
Rodgers, John C., (662) 325-0732 jcr100@msstate.edu, Mississippi State Univ – Oy
Rodgers, N, RodgersN@cf.ac.uk, Cardiff Univ – Ow
Rodgers, Nick, +44(0)29 208 79064 RodgersN@cardiff.ac.uk, Univ of Wales – Xm
Rodi, William, (617) 253-7855 rodi@mit.edu, Massachusetts Inst of Tech – Yg
Rodland, David L., (740) 826-8425 drodland@muskingum.edu, Muskingum Univ – PgiPe
Rodolfo, Kelvin S., krodolfo@uic.edu, Univ of Illinois at Chicago – GueGs
Rodrigues, Cyril G., 519-253-3000 ext. 2499 cgr@uwindsor.ca, Univ of Windsor – Pm
Rodriguez, Joaquin, (212) 772-5321 Graduate School of the City Univ of New York – Pi
Rodriguez, Lizzette A., (787) 265-3845 lizzette.rodriguez1@upr.edu, Univ of Puerto Rico – GviOr
Rodriguez, Vanessa del S., (787) 722-2526 Puerto Rico Bureau of Geology – On
Rodriguez-Blanco, Juan Diego, +353 1 8961691 j.d.rodriguez-blanco@tcd.ie, Trinity Coll – GzClOm
Roe, Carol, 757-221-2440 crroex@wm.edu, Coll of William & Mary – On
Roe, Gerard H., 206-697-3298 gerard@ess.washington.edu, Univ of Washington – OaGmOl
Roecker, Steven W., (518) 276-6773 roecks@rpi.edu, Rensselaer Polytechnic Inst – Yg
Roegiers, Jean-Claude, (405) 255-5459 jroegiers@ou.edu, Univ of Oklahoma – Nr
Roelofse, Frederick, +27 (0)51 401 9001 roelofsef@ufs.ac.za, Univ of the Free State – Gip
Roemer, Elizabeth, (520) 621-2897 eroemer@pirlmail.lpl.arizona.edu, Univ of Arizona – On
Roemmich, Dean H., (858) 534-2307 droemmich@ucsd.edu, Univ of California, San Diego – Op
Roering, Joshua J., (541) 346-5574 jroering@uoregon.edu, Univ of Oregon – Gm
Roeske, Sarah M., (530) 752-4933 smroeske@ucdavis.edu, Univ of California, Davis – GtcGp
Roesler, Collin, 207-725-3842 croesler@bowdoin.edu, Bowdoin Coll – OgpOr
Roethel, Frank J., (631) 632-8732 frank.roethel@stonybrook.edu, SUNY, Stony Brook – Oc
Rogers, Garry C., (250) 363-6450 Univ of Victoria – Ys
Rogers, Jefferson S., (731) 881-7442 jrogers@utm.edu, Univ of Tennessee, Martin – On
Rogers, Jeffery C., (614) 292-0148 rogers.21@osu.edu, Ohio State Univ – Oa
Rogers, Joe D., 806-651-2570 West Texas A&M Univ – Ga
Rogers, Karyn L., (518) 276-2372 rogerk5@rpi.edu, Rensselaer Polytechnic Inst – PyCl
Rogers, Pamela Z., (505) 667-1765 Los Alamos National Laboratory – Cl
Rogers, Raymond R., (651) 696-6434 rogersk@macalester.edu, Macalester Coll – Gs
Rogers, Robert D., (209) 667-3466 rrogers1@csustan.edu, California State Univ, Stanislaus – GcmGt
Rogers, Steven, (+44) 01782 733752 s.l.rogers@keele.ac.uk, Keele Univ – Gg
Rogers, William J., 806-651-2581 West Texas A&M Univ – On
Rogerson, Robert J., (403) 329-5117 rogerson@uleth.ca, Univ of Lethbridge – Gm
Roggenthen, William M., (605) 394-2461 William.Roggenthen@sdsmt.edu, South Dakota School of Mines & Tech – NgYg
Rogowski, Andrew S., (814) 863-8758 asr@psu.edu, Pennsylvania State Univ, Univ Park – Sp
Rohay, Alan C., (509) 376-6925 alan.rohay@pnl.gov, Pacific Northwest National Laboratory – Ys
Rohe, Randall, randall.rohe@uwc.edu, Univ of Wisconsin Colls – OyGgOn
Rohr, David M., (432) 837-8167 drohr@sulross.edu, Sul Ross State Univ – PiGd
Rohs, C. Renee, (660) 562-1201 rrohs@nwmissouri.edu, Northwest Missouri State Univ – Cg
Roinstad, Lori L., 605-677-6154 lori.roinstad@usd.edu, South Dakota Dept of Environment and Natural Resources – On
Rokop, Donald J., (505) 667-4299 Los Alamos National Laboratory – Cc
Rollins, Harold B., (412) 624-8780 snail@pitt.edu, Univ of Pittsburgh – Pe
Rollins, Kyle, (573) 368-2171 kyle.rollins@dnr.mo.gov, Missouri Dept of Natural Resources – OyGg
Rollinson, Hugh, +44 01332 591786 h.rollinson@derby.ac.uk, Univ of Derby – Cg
Rollinson, Paul A., (417) 836-5688 paulrollinson@missouristate.edu, Missouri State Univ – On
Rolston, Dennis E., (916) 752-2113 Univ of California, Davis – So
Roman, Aubrecht, aubrecht@fns.uniba.sk, Comenius Univ in Bratislava – GrXg
Roman, Charles T., (401) 874-6885 croman@gso.uri.edu, Univ of Rhode Island – Ob
Roman, Diana C., 202-478-8834 droman@carnegiescience.edu, Carnegie Institution of Washington – Gv
Roman, Eric W., (609) 984-6587 eric.roman@dep.state.nj.us, New Jersey Geological and Water Survey – Hw
Romanak, Katherine D., (512) 471-6136 katherine.romanak@beg.utexas.edu, Univ of Texas at Austin, Jackson School of Geosciences – CgScGe
Romanovsky, Vladimir, (907) 474-7459 ffver@uaf.edu, Univ of Alaska, Fairbanks – Yg
Romanowicz, Barbara A., 510-643-5690 barbara@seismo.berkeley.edu, Univ of California, Berkeley – Ys
Romanowicz, Edwin A., (518) 564-2152 romanoea@plattsburgh.edu, Plattsburgh State Univ (SUNY) – Hw
Ronayne, Michael J., (970) 491-0666 Michael.Ronayne@colostate.edu, Colorado State Univ – Hwq
Rondot, Beth, (516) 299-2318 beth.rondot@liu.edu, Long Island Univ, C.W. Post Campus – On
Rood, Richard B., (734) 647-3530 rbrood@umich.edu, Univ of Michigan – Owa
Rooney, Neil, (519) 824-4120 Ext.52573 nrooney@uoguelph.ca, Univ of Guelph – On
Rooney, Tyrone, 517-432-5522 rooneyt@msu.edu, Michigan State Univ – Gi
Root, Tara L., 561 297-3253 troot@fau.edu, Florida Atlantic Univ – Hg
Roozeboom, Kraig, (785) 532-3781 kraig@ksu.edu, Kansas State Univ – So
Rose, Arthur W., (814) 238-2838 awr1@psu.edu, Pennsylvania State Univ, Univ Park – Cge
Rose, Candace M., (630) 252-3499 cmrose@anl.gov, Argonne National Laboratory – On
Rose, Catherine V., +353 1 8961165 crose@tcd.ie, Trinity Coll – Cg
Rose, Dan, +353 21 4902189 Univ Coll Cork – On
Rose, I. S., (678) 839-4067 srose@westga.edu, Univ of West Georgia – OyaOw
Rose, Peter E., (801) 585-7785 prose@egi.utah.edu, Univ of Utah – Np
Rose, Seth E., 404413-5750 geoser@langate.gsu.edu, Georgia State Univ – Hw
Rose, Timothy, (202) 633-1398 roset@si.edu, Smithsonian Institution / National Museum of Natural History – CaGv
Rose, William I., (906) 487-2367 raman@mtu.edu, Michigan Technological Univ – Gv
Roselle, Gregory T., (208) 496-7683 roselleg@byui.edu, Brigham Young Univ - Idaho – GpCaHg
Rosen, Carl J., (612) 625-8114 crosen@umn.edu, Univ of Minnesota, Twin Cities – Sc
Rosen, Peter S., p.rosen@neu.edu, Northeastern Univ – Onu
Rosenberg, Gary, 215.299.1033 gr347@drexel.edu, Drexel Univ – Pi
Rosenberg, Gary D., (317) 274-7468 grosenbe@iupui.edu, Milwaukee Public Museum – GhPg
Rosenberg, Philip E., (509) 335-4368 rosenberg@wsu.edu, Washington State Univ – GzCg
Rosenfeld, Carla, rose0859@umn.edu, Univ of Minnesota, Twin Cities – Py
Rosenfeld, John L., (310) 825-1505 rosenfel@ucla.edu, Univ of California, Los Angeles – Gp

Rosengaus, Rebeca, (617) 373-7032 r.rosengaus@neu.edu, Northeastern Univ – On
Rosenthal, Yair, 848-932-6555X227 rosentha@marine.rutgers.edu, Rutgers, The State Univ of New Jersey – CmlOc
Ross, Andrew, +44(0) 113 34 37590 a.n.ross@leeds.ac.uk, Univ of Leeds – Ow
Ross, David A., (508) 289-2578 dross@whoi.edu, Woods Hole Oceanographic Institution – GuOg
Ross, Gerald M., (403) 292-7000 Univ of Calgary – Gt
Ross, Kirstin, kirstin.ross@flinders.edu.au, Flinders Univ – On
Ross, Martin E., (617) 373-3263 m.ross@neu.edu, Northeastern Univ – Gie
Ross, Nancy L., (540) 231-6356 nross@vt.edu, Virginia Polytechnic Inst & State Univ – Gz
Ross, Robert M., (607) 273-6623 x18 rmr16@cornell.edu, Paleontological Research Institution – PoOe
Ross, Tetjana, (902) 494-1327 tetjana.ross@dal.ca, Dalhousie Univ – Op
Ross, Thomas E., (910) 521-6218 tom.ross@uncp.edu, Univ of North Carolina, Pembroke – Oy
Rossby, Hans T., (401) 874-6521 trossby@gso.uri.edu, Univ of Rhode Island – Op
Rosscoe, Steven, (325) 670-1387 srosscoe@hsutx.edu, Hardin-Simmons Univ – PmGsPs
Rossell, Irene M., (828) 232-5185 irossell@unca.edu, Univ of North Carolina, Asheville – On
Rossetti, Piergiorgio, piergiorgio.rossetti@unito.it, Università di Torino – Eg
Rossman, George R., (626) 395-6471 grr@gps.caltech.edu, California Inst of Tech – GzCa
Rost, Sebastian, +44(0) 113 34 35212 s.rost@leeds.ac.uk, Univ of Leeds – Ys
Rostam-Abadi, Massoud, 217-244-4977 massoud@isgs.uiuc.edu, Illinois State Geological Survey – On
Rostami, Jamal, 814-863-7606 jur17@psu.edu, Pennsylvania State Univ, Univ Park – Nm
Rostoker, Gordon, (780) 492-5286 rostoker@space.ualberta.ca, Univ of Alberta – Xy
Rostron, Ben, (780) 492-2178 ben.rostron@ualberta.ca, Univ of Alberta – On
Roth, Gregory W., (814) 863-1018 gwr@psu.edu, Pennsylvania State Univ, Univ Park – So
Roth, Leonard T., troth@hccfl.edu, Hillsborough Community Coll – Og
Roth, Peter H., (801) 581-6704 peter.roth@utah.edu, Univ of Utah – Pm
Rothberg, Maryann, (609) 258-4655 mar@cedr.princeton.edu, Princeton Univ – On
Rothman, Daniel H., (617) 253-7861 dhr@mit.edu, Massachusetts Inst of Tech – Yg
Rothrock, David A., (206) 545-2262 rothrock@apl.washington.edu, Univ of Washington – Op
Rothstein, Lewis, (401) 874-6517 lrothstein@gso.uri.edu, Univ of Rhode Island – Op
Rotjan, Randi, rrotjan@neaq.org, Univ of Massachusetts, Boston – Ob
Rouff, Ashaki, 973-353-2511 Rutgers, The State Univ of New Jersey, Newark – Cl
Rouff, Ashaki, 718-997-3073 ashaki.rouff@qc.cuny.edu, Queens Coll (CUNY) – Cac
Roughgarden, Joan, (650) 723-3648 Stanford Univ – Ob
Rougvie, James R., 608-363-2268 rougviej@beloit.edu, Beloit Coll – GxpGz
Rouhani, Farhang, (540) 654-1895 frouhani@mwc.edu, Mary Washington Coll – On
Rouleau, Alain, 4185455011x5213 arouleau@uqac.ca, Universite du Quebec a Chicoutimi – Hw
Rouleau, Pierre M., (604) 637-6294 Sir Wilfred Grenfell Coll – Yg
Rounds, Steven W., (916) 278-7828 rounds@csus.edu, California State Univ, Sacramento – Gg
Rounsevell, Mark D., +44 (0) 131 651 4468 mark.rounsevell@ed.ac.uk, Edinburgh Univ – Ou
Rouse, Gregory W., (858) 534-7943 grouse@ucsd.edu, Univ of California, San Diego – Ob
Rouse, Jesse, 910-521-6387 jesse.rouse@uncp.edu, Univ of North Carolina, Pembroke – Oni
Rouse, Joseph D., 16717352685 jdrouse@uguam.uog.edu, Univ of Guam – Hg
Rouse, Roland C., (734) 763-0952 rouserc@umich.edu, Univ of Michigan – Gz
Rouse, Jr., Lawrence J., (225) 388-2953 Louisiana State Univ – Op
Rousell, Don H., 7056751151 x2265 drousell@laurentian.ca, Laurentian Univ, Sudbury – Gc
Roussel-Dupre, R., (505) 667-9228 rroussel-dupre@lanl.gov, Los Alamos National Laboratory – On
Roussenov, Vassil, +44 0151 794 4099 V.Roussenov@liverpool.ac.uk, Univ of Liverpool – Og
Rovey, Charles W., (417) 836-6890 charlesrovey@missouristate.edu, Missouri State Univ – HyGl
Rowan, Christopher J., 330-672-7428 crowan5@kent.edu, Kent State Univ – Gt
Rowden, Robert, Robert.Rowden@dnr.iowa.gov, Iowa Dept of Natural Resources – Gg
Rowe, Charlotte A., 505-665-6404 char@lanl.gov, New Mexico Inst of Mining and Tech – Ys
Rowe, Christie, (514) 398-2769 christie.rowe@mcgill.ca, McGill Univ – GctEm
Rowe, Clinton M., (402) 472-1946 crowe1@unl.edu, Univ of Nebraska, Lincoln – Oa
Rowe, Gilbert T., (409) 740-4458 roweg@tamug.edu, Texas A&M Univ – Ob
Rowe, Timothy B., (512) 471-1725 rowe@mail.utexas.edu, Univ of Texas at Austin – Pv
Rowell, Albert J., (785) 864-2747 arowell@ku.edu, Univ of Kansas – Pi
Rowland, Scott K., (808) 956-3150 Univ of Hawai'i, Manoa – On
Rowland, Stephen, (702) 895-3625 steve.rowland@unlv.edu, Univ of Nevada, Las Vegas – Pi
Rowley, David B., (773) 702-8146 rowley@geosci.uchicago.edu, Univ of Chicago – GtgGc
Rowley, Rex J., 309-438-7832 rjrowle@ilstu.edu, Illinois State Univ – Oi
Roy, Denis W., (418) 545-5011 dwroy@uqac.ca, Universite du Quebec a Chicoutimi – Gc
Roy, Martin, 514-987-3000 #7619 roy.matin@uqam.ca, Universite du Quebec a Montreal – Gl
Roy, Suzanne, (418) 723-1986 x1748 suzanne_roy@uqar.ca, Universite du Quebec a Rimouski – Ob
Roy, William R., 217-333-1197 roy@isgs.uiuc.edu, Illinois State Geological Survey – Sc
Roychoudhury, Alakendra N., +27 21 808 3124 roy@sun.ac.za, Stellenbosch Univ – ClmOc
Royden, Leigh H., (617) 253-1292 lhroyden@mit.edu, Massachusetts Inst of Tech – Gt
Royer, Dana, (860) 685-2836 droyer@wesleyan.edu, Wesleyan Univ – PeoPb
Royer, Thomas, (757) 683-5547 royer@ccpo.odu.edu, Old Dominion Univ – Op
Royer, Todd, 812-855-0563 Indiana Univ, Bloomington – HsOg
Rozmus, Wojciech, (780) 492-8486 rozmus@phys.ualberta.ca, Univ of Alberta – On
Ruan, Youyi, 609-258-5031 youyir@princeton.edu, Princeton Univ – Ysg
Ruark, Matthew D., (608) 263-2889 mdruark@wisc.edu, Univ of Wisconsin, Madison – So
Rubbo, Marco, marco.rubbo@unito.it, Università di Torino – Gz
Rubin, Alan E., (310) 825-3202 rubin@igpp.ucla.edu, Univ of California, Los Angeles – Xm
Rubin, Allan M., (609) 258-1506 arubin@princeton.edu, Princeton Univ – Yg
Rubin, Charles M., (509) 963-2827 rbeling@wvu.edu, Central Washington Univ – Gt
Rubin, Kenneth H., (808) 956-8973 krubin@hawaii.edu, Univ of Hawai'i, Manoa – CcGvCg
Rubio-Sierra, Javier, 089/2180 4317 rubio@lrz.uni-muenchen.de, Ludwig-Maximilians-Universitaet Muenchen – Gz
Rucklidge, John C., (416) 978-2061 jcr@quartz.geology.utoronto.ca, Univ of Toronto – Ge
Ruddick, Barry R., (902) 494-2505 barry.ruddick@dal.ca, Dalhousie Univ – Op
Ruddiman, William F., 540-348-1963 wfr5c@virginia.edu, Univ of Virginia – Gu
Rudge, John, +44 (0) 1223 765545 jfr23@cam.ac.uk, Univ of Cambridge – YgGt
Rudnick, Daniel L., (858) 534-7669 drudnick@ucsd.edu, Univ of California, San Diego – Op
Rudnick, Roberta L., rudnick@geol.umd.edu, Univ of Maryland – CgsCt
Rudnicki, Ryan E., 210-486-0061 rrudnicki@alamo.edu, San Antonio Community Coll – Oyr
Rueger, Bruce F., (207) 859-5806 bfrueger@colby.edu, Colby Coll – Pl

Ruehr, Thomas A., (805) 756-2552 truehr@calpoly.edu, California Polytechnic State Univ – Sb
Ruff, Larry J., (734) 763-9301 ruff@umich.edu, Univ of Michigan – Ys
Ruffel, Alice, (214) 507-9014 aruffel@dcccd.edu, El Centro Coll - Dallas Community Coll District – Gg
Ruffman, Alan, (902) 422-6482 Dalhousie Univ – Ys
Ruggiero, Peter, (541) 737-1239 ruggierp@science.oregonstate.edu, Oregon State Univ – On
Ruhl, Laura S., 501-683-4197 lsruhl@ualr.edu, Univ of Arkansas at Little Rock – GeClGb
Ruina, Andy L., (607) 255-7108 alr3@cornell.edu, Cornell Univ – Yx
Ruiz, Joaquin, (520) 621-4090 jruiz@email.arizona.edu, Univ of Arizona – Cgc
Ruiz-Diaz, Dorivar, (785) 532-6183 ruizdiaz@ksu.edu, Kansas State Univ – So
Rule, Joseph H., (757) 683-3274 jrule@odu.edu, Old Dominion Univ – ClOsSc
Rumble, III, Douglas, 202-478-8990 drumble@carnegiescience.edu, Carnegie Institution of Washington – Gp
Rumrill, Julie, 203-392-5842 rumrillj1@southernct.edu, Southern Connecticut State Univ – GgeGl
Rumstay, Kenneth S., (229) 333-5754 krumstay@valdosta.edu, Valdosta State Univ – Onn
Rundberg, Robert S., (505) 667-4559 Los Alamos National Laboratory – Gq
Rundle, John, jbrundle@ucdavis.edu, Univ of California, Davis – Yd
Rundquist, Bradley C., (701) 777-4246 bradley.rundquist@und.edu, Univ of North Dakota – OriOy
Rundquist, Donald C., (402) 472-3471 drundquist@unl.edu, Unversity of Nebraska - Lincoln – Ori
Runge, Edward C. A., (979) 845-3041 e-runge@tamu.edu, Texas A&M Univ – Sd
Runkel, Anthony, 612-627-4780 (Ext. 222) runke001@umn.edu, Univ of Minnesota, Twin Cities – Gs
Runnegar, Bruce, (310) 206-1738 Univ of California, Los Angeles – Po
Runyon, Cassandra R., (843) 953-8279 runyonc@cofc.edu, Coll of Charleston – XgOse
Rupert, Gerald B., (573) 341-4616 Missouri Univ of Science and Tech – Ye
Rupp, David, 541-737-5222 drupp@coas.oregonstate.edu, Oregon State Univ – Oap
Rupp, John A., (812) 855-1323 rupp@indiana.edu, Indiana Univ – Go
Ruppel, Stephen C., (512) 471-2965 stephen.ruppel@beg.utexas.edu, Univ of Texas at Austin, Jackson School of Geosciences – Gd
Rupper, Summer, 801 422-6946 summer_rupper@byu.edu, Brigham Young Univ – OlPe
Ruppert, Kelly R., (657) 278-3561 kruppert@fullerton.edu, California State Univ, Fullerton – Gg
Ruprecht, Philipp P., 775-682-6084 pruprecht@unr.edu, Univ of Nevada, Reno – Giv
Rush, Jason, (785) 864-2178 rush@kgs.ku.edu, Univ of Kansas – Gg
Rusmore, Margaret E., 323 259 2565 rusmore@oxy.edu, Occidental Coll – Gc
Russ, Tom, (301) 934-7814 Truss@csmd.edu, Coll of Southern Maryland – GgSo
Russell, Ann D., (530) 752-3311 adrussell@ucdavis.edu, Univ of California, Davis – Cm
Russell, Armistead G., (404) 894-3079 ted.russell@ce.gatech.edu, Georgia Inst of Tech – Oa
Russell, Christopher T., (310) 825-3188 ctrussell@igpp.ucla.edu, Univ of California, Los Angeles – Xy
Russell, Dale A., (919) 515-1339 dale_russell@ncsu.edu, North Carolina State Univ – Pv
Russell, James K., (604) 822-2703 krussell@eos.ubc.ca, Univ of British Columbia – GviCg
Russell, James M., (401) 863-6330 James_Russell@Brown.edu, Brown Univ – Gn
Russell, Joellen, 520-626-2194 jrussell@email.arizona.edu, Univ of Arizona – Oc
Russell, Lynn M., (858) 534-4852 lmrussell@ucsd.edu, Univ of California, San Diego – Oa
Russell, R. Doncaster, (604) 822-2551 drussell@eos.ubc.ca, Univ of British Columbia – Yg
Russell, Ron, (512) 471-8831 ron.russell@beg.utexas.edu, Univ of Texas at Austin, Jackson School of Geosciences – On
Russell, Sally, +44(0) 113 34 35279 S.Russell@leeds.ac.uk, Univ of Leeds – Ge
Russell, Terry P., (250) 721-6184 trussell@uvic.ca, Univ of Victoria – On
Russell, Theresa J., (501) 575-4403 trussell@comp.uark.edu, Univ of Arkansas, Fayetteville – On
Russelle, Michael P., (612) 625-8145 russelle@soils.umn.edu, Univ of Minnesota, Twin Cities – Sc
Russo, Mary Rose, 609 258-4101 mrusso@princeton.edu, Princeton Univ – On
Russo, Raymond, 352-392-6766 rrusso@ufl.edu, Univ of Florida – Yg
Russo, Tess A., 814-865-7389 russo@psu.edu, Pennsylvania State Univ, Univ Park – Hw
Rust, Derek, +44 023 92 842298 derek.rust@port.ac.uk, Univ of Portsmouth – Gt
Rustad, James R., jrrustad@ucdavis.edu, Univ of California, Davis – Cl
Rutberg, Randye L., (212) 772-5326 randye.rutberg@hunter.cuny.edu, Hunter Coll (CUNY) – Cc
Rutford, Robert H., (972) 883-6470 rutford@utdallas.edu, Univ of Texas, Dallas – Gl
Rutherford, Malcolm J., (401) 863-1927 Malcolm_Rutherford@Brown.edu, Brown Univ – Cp
Rutherford, Michael P., (604) 637-6328 Sir Wilfred Grenfell Coll – Sb
Rutledge, Steven A., rutledge@atmos.colostate.edu, Colorado State Univ – Oa
Rutstein, Martin S., (914) 257-3763 SUNY, New Paltz – Gz
Ruttan, Lore, 404 7274217 lruttan@emory.edu, Emory Univ – Ob
Ruttenberg, Kathleen, 808-956-9371 kcr@soest.hawaii.edu, Univ of Hawai'i, Manoa – Cg
Rutter, Ernest, +44 0161 275-3945 e.rutter@manchester.ac.uk, Univ of Manchester – NrGct
Rutter, Nathaniel W., (780) 492-3085 nat.rutter@ualberta.ca, Univ of Alberta – Ge
Ruzicka, Alexander (Alex) M., (503) 725-3372 ruzickaa@pdx.edu, Portland State Univ – Xm
Ruzicka, Jaromir, jaromirr@hawaii.edu, Univ of Hawai'i, Manoa – On
Ryall, Patrick J., (902) 494-3465 pryall@is.dal.ca, Dalhousie Univ – Yg
Ryan, Cathy, (403) 220-2793 ryan@geo.ucalgary.ca, Univ of Calgary – Hw
Ryan, Jeffrey G., (813) 974-1598 ryan@shell.cas.usf.edu, Univ of South Florida, Tampa – Ct
Ryan, Peter C., (802) 443-2557 pryan@middlebury.edu, Middlebury Coll – Cl
Ryan, Susan, (724) 938-4531 ryan@calu.edu, California Univ of Pennsylvania – Og
Ryan, William F., (814) 865-0478 Pennsylvania State Univ, Univ Park – Ow
Ryan, William B. F., (845) 365-8312 Columbia Univ – Yr
Rychert, Catherine A., +44 (0)23 80598663 C.Rychert@soton.ac.uk, Univ of Southampton – Yg
Ryder, Isabelle, +44-151-794-5143 I.Ryder@liverpool.ac.uk, Univ of Liverpool – Gt
Ryder, Roy, rryder@southalabama.edu, Univ of South Alabama – SdOry
Rye, Danny M., (203) 432-3174 danny.rye@yale.edu, Yale Univ – Cs
Rygel, Michael C., (315) 267-3401 rygelmc@potsdam.edu, SUNY Potsdam – GsrGo
Rykaczewski, Ryan, ryk@sc.edu, Univ of South Carolina – Ob
Rysgaard, Soren, 204-272-1611 rysgaard@umanitoba.ca, Univ of Manitoba – Gl
Rößner, Gertrud, 089/2180 6612 g.roessner@lrz.uni-muenchen.de, Ludwig-Maximilians-Universitaet Muenchen – Pg

## S

Sá, Artur A., asa@utad.pt, Universidade de Trás-os-Montes e Alto Douro – PgiPs
Saal, Alberto E., (401) 863-7238 Alberto_Saal@Brown.edu, Brown Univ – Cg
Saalfeld, Alan J., 614 292 6665 saalfeld.1@osu.edu, Ohio State Univ – Yd
Saar, Martin O., 612-625-7332 saar@umn.edu, Univ of Minnesota, Twin Cities – Hg
Sabala-Foreman, Susan M., (928) 523-4561 Northern Arizona Univ – On
Sabine, Christopher L., (206) 526-4809 chris.sabine@noaa.gov, Univ of Washington – Oc
Sabol, Martin, sabol@fns.uniba.sk, Comenius Univ in Bratislava – Pvo
Sabra, Karim, (404) 385-6193 Georgia Inst of Tech – Yg
Sacchi, Mauricio D., (780) 492-1060 sacchi@phys.ualberta.ca, Univ of Alberta – Ys
Saccocia, Peter J., psaccocia@bridgew.edu, Bridgewater State Univ – Cm
Sack, Richard, 425-880-4418 rosack@uw.edu, Univ of Washington – CgOn
Sacks, I. Selwyn, (202) 478-8839 ssacks@carnegiescience.edu,

Carnegie Institution of Washington – Ys
Sacramentogrilo, Isabelle, 619-594-5607 isacramentogrilo@mail.sdsu.edu, San Diego State Univ – Gg
Sadd, James L., 323 259 2518 jsadd@oxy.edu, Occidental Coll – Oi
Sadiq, Abdulali A., sadiqa@qu.edu.qa, Univ of Qatar – Or
Sadler, Peter M., (951) 827-5616 peter.sadler@ucr.edu, Univ of California, Riverside – GrPs
Sadowsky, Michael J., (612) 624-2706 sadowsky@soils.umn.edu, Univ of Minnesota, Twin Cities – Sb
Saeidi, Ali, (418) 545-5011 x2561 asaeidi@uqac.ca, Universite du Quebec a Chicoutimi – Nrm
Saffer, Demian M., 814-865-7965 dsaffer@geosc.psu.edu, Pennsylvania State Univ, Univ Park – Hw
Sagebiel, J. C., 909 307-2669 237 jsagebiel@sbcm.sbcounty.gov, San Bernardino County Museum – Pv
Sageman, Bradley B., (847) 467-2257 brad@earth.northwestern.edu, Northwestern Univ – PsGsPe
Sager, William W., (979) 845-9828 wsager@ocean.tamu.edu, Texas A&M Univ – Ou
Sager, William W., (713) 743-3108 wwsager@uh.edu, Univ of Houston – GtYrm
Sagiroglu, Ahmet, 00904242370000-5990 sagiroglu@firat.edu.tr, Firat Univ – Egm
Sahagian, Dork, 610-758-6379 dork.sahagian@lehigh.edu, Lehigh Univ – PeGvr
Sahay, Pratap, pratap@cicese.mx, Centro de Investigación Científica y de Educación Superior de Ensenada – Ys
Sahoo, Prasanta, (321) 674-8147 Florida Inst of Tech – Oo
Sahr, John D., (206) 616-7175 jdsahr@ee.washington.edu, Univ of Washington – OrXym
Saikia, Udoy, udoy.saikia@flinders.edu.au , Flinders Univ – On
Saila, Saul B., (401) 874-6485 saila@gso.uri.edu, Univ of Rhode Island – Ob
Saillet, Elodie, +33(0)3 44067563 elodie.saillet@lasalle-beauvais.fr, Institut Polytechnique LaSalle Beauvais (ex-IGAL) – GctGo
Saini-Eidukat, Bernhardt, (701) 231-8785 bernhardt.saini-eidukat@ndsu.edu, North Dakota State Univ – GiCg
Saint, Prem K., psaint@fullerton.edu, California State Univ, Fullerton – Hw
Saito, Mak A., 508-289-3696 msaito@whoi.edu, Woods Hole Oceanographic Institution – Oc
Saja, David B., 2162314600 x3229 dsaja@cmnh.org, Cleveland Museum of Natural History – Gdc
Sak, Peter B., (717) 245-1423 sakp@dickinson.edu, Dickinson Coll – Gcm
Sakhaee-Pour, Ahmad, 405-325-3306 sakhaee@ou.edu, Univ of Oklahoma – Np
Saku, James C., (301) 687-4724 jsaku@frostburg.edu, Frostburg State Univ – On
Salami, B. M., 234-803-321-9685 salamibm@oauife.edu.ng, Obafemi Awolowo Univ – Hg
Salami, Sikiru A., sikiru.salami@uniben.edu, Univ of Benin – YgeGg
Salaun, Pascal, +44-151-794-4101 Pascal.Salaun@liverpool.ac.uk, Univ of Liverpool – Em
Salaun, Rachel, +440151 795 4649 Rachel.Jeffreys@liverpool.ac.uk, Univ of Liverpool – Oc
Saleeby, Jason B., (626) 395-6141 jason@gps.caltech.edu, California Inst of Tech – Gc
Salisbury, Joseph, 603-862-0849 joe.salisbury@unh.edu, Univ of New Hampshire – On
Salje, Ekhard, +44 (0) 1223 768321 ekhard@esc.cam.ac.uk, Univ of Cambridge – Gy
Salle, Bethan, (214) 578-1914 bsalle@dcccd.edu, El Centro Coll - Dallas Community Coll District – Gg
Sallu, Susannah, +44(0) 113 34 31641 s.sallu@leeds.ac.uk, Univ of Leeds – Ga
Salmon, Richard L., (858) 534-2090 rsalmon@ucsd.edu, Univ of California, San Diego – Op
Salmun, Haydee, (212) 772-4159 hsalmun@hunter.cuny.edu, Hunter Coll (CUNY) – OpgOa
Salters, Vincent J., (850) 644-1934 salters@magnet.fsu.edu, Florida State Univ – Cg
Saltzman, Eric S., 949-824-3936 esaltzma@uci.edu, Univ of California, Irvine – Oa
Saltzman, Matthew R., 614 292-0481 saltzman.11@osu.edu, Ohio State Univ – Gs
Salviulo, Gabriella, +390498279157 gabriella.salviulo@unipd.it, Università degli Studi di Padova – Gz
Salvucci, Guido D., (617) 353-8344 gdsalvuc@bu.edu, Boston Univ – Hq
Salyards, Stephen L., (310) 825-3043 salyards@epss.ucla.edu, Univ of California, Los Angeles – Ys
Salzer, Matthew W., (520) 621-2946 msalzer@ltrr.arizona.edu, Univ of Arizona – PeGav
Sambrotto, Raymond N., (845) 365-8402 Columbia Univ – Ob
Samelson, Roger, 541-737-4752 rsamelson@coas.oregonstate.edu, Oregon State Univ – Op
Sammarco, Paul W., (225) 851-2800 Louisiana State Univ – On
Sammis, Charles G., (213) 740-5836 sammis@usc.edu, Univ of Southern California – Yg
Sammis, Theodore W., (505) 646-3405 New Mexico State Univ, Las Cruces – Sp
Samonds, Karen, 815-753-3201 ksamonds@niu.edu, Northern Illinois Univ – Pv
Sample, James C., (928) 523-0881 james.sample@nau.edu, Northern Arizona Univ – Cl
Samson, Iain M., 519-253-3000 ext. 2489 ims@uwindsor.ca, Univ of Windsor – EgCg
Samson, John C., (780) 492-3616 samson@space.ualberta.ca, Univ of Alberta – Xy
Samson, Scott D., 315-443-2672 sdsamson@syr.edu, Syracuse Univ – Cc
Samsonov, Sergey V., (613) 656-1534 sergey.samsonov@canada.ca, Natural Resources Canada – YdOrYg
Samuelson, Alan C., (765) 285-8270 Ball State Univ – Hw
Sanabia, Elizabeth R., 410-293-6556 sanabia@usna.edu, United States Naval Academy – Ow
Sanchez, Charles, (520) 782-3836 sanchez@ag.arizona.edu, Univ of Arizona – On
Sanchez, Marcelo, 979.862.6604 msanchez@civil.tamu.edu, Texas A&M Univ – SpNr
Sanchez, Veronica I., (361) 593-3590 veronica.sanchez@tamuk.edu, Texas A&M Univ, Kingsville – GctGm
Sanchez Roman, Monica, m.sanchezroman@vu.nl, VU Univ Amsterdam – GdsGg
Sanchez-Azofeifa, G. Arturo, (780) 492-1822 arturo.sanchez@ualberta.ca, Univ of Alberta – OirOg
Sandau, Ken L., 406-496-4151 ksandau@mtech.edu, Montana Tech of The Univ of Montana – Oi
Sandel, Bill R., (520) 621-4305 sandel@vega.lpl.arizona.edu, Univ of Arizona – Oa
Sanden, Eric M., (715) 425-3729 eric.m.sanden@uwrf.edu, Univ of Wisconsin, River Falls – Or
Sanders, Laura L., (773) 442-6051 L-Sanders@neiu.edu, Northeastern Illinois Univ – HwGe
Sanders, Ronald S., (818) 354-2867 sanders@jpl.nasa.gov, Jet Propulsion Laboratory – Gr
Sandhu, Harpinder, harpinder.sandhu@flinders.edu.au, Flinders Univ – On
Sandlin, Stephen H., 909-389-8644 ssandlin@sbccd.cc.ca.us, San Bernardino Valley Coll – Oy
Sandor, Jonathan A., (515) 294-2209 jasandor@iastate.edu, Iowa State Univ of Science & Tech – Sd
Sandoval, Samuel, (530) 754-9646 samsandoval@ucdavis.edu, Univ of California, Davis – Hg
Sandoval Solis, Samuel, 530-750-9722 samsandoval@ucdavis.edu, Univ of California, Davis – Hg
Sandvol, Eric A., (573) 884-9616 sandvole@missouri.edu, Univ of Missouri – Ys
Sandwell, David T., (858) 534-7109 dsandwell@ucsd.edu, Univ of California, San Diego – Yr
Sandwith, Zoe, (508) 289-3942 zsandwith@whoi.edu, Woods Hole Oceanographic Institution – Oc
Sandy, Michael R., (937) 229-3436 Univ of Dayton – Pi
Sanford, Allan R., sanford@nmt.edu, New Mexico Inst of Mining and Tech – Ys
Sanford, Lawrence P., (410) 221-8429 lsanford@hpl.umces.edu, Univ of Maryland – Op
Sanford, Robert A., 217-244-7250 rsanford@illinois.edu, Univ of Illinois, Urbana-Champaign – Py
Sanford, Thomas B., (206) 543-1365 sanford@apl.washington.edu, Univ of Washington – Op
Sanford, William E., (970) 491-5929 bills@cnr.colostate.edu, Colorado State Univ – Hw
Sanislav, Ioan, ioan.sanislav@jcu.edu.au, James Cook Univ – GctGp
Sanjurjo, Jorge, 0034981167000 jsanjurjo@udc.es, Coruna Univ – GaCcPy

Sankey, Julia, (209) 667-3466 jsankey@csustan.edu, California State Univ, Stanislaus – Pve

Sansone, Francis J., 808-956-2912 sansone@hawaii.edu, Univ of Hawai'i, Manoa – Cm

Santana, Vicente, vicentejose.santana@ehu.eus, Univ of the Basque Country UPV/EHU – GgYm

Santander, Erma, (520) 621-7120 erma@hwr.arizona.edu, Univ of Arizona – On

Santelli, Cara, santellic@si.edu, Smithsonian Institution / National Museum of Natural History – Py

Santelmann, Mary V., (541) 737-1215 santelmm@geo.oregonstate.edu, Oregon State Univ – Oy

Santi, Paul M., 303-273-3108 psanti@mines.edu, Colorado School of Mines – Ng

Santillan, Marcelo, (509) 963-1107 marcelo@geology.cwu.edu, Central Washington Univ – Yd

Santos, Eduardo, (785) 532-6932 esantos@ksu.edu, Kansas State Univ – So

Santos, Hernan, (787) 265-3845 hernan.santos@upr.edu, Univ of Puerto Rico – PiGdr

Santschi, Peter H., (409) 740-4476 santschip@tamug.edu, Texas A&M Univ – Oc

Sanudo, Sergio, 213-821-1302 sanudo@usc.edu, Univ of Southern California – Py

Sapigao, Gladys, 718-997-3300 gladys.sapigao@qc.cuny.edu, Queens Coll (CUNY) – OnnOn

Sar, Abdullah, 00904242370000-5961 asasr@firat.edu.tr, Firat Univ – GviGp

Sarachik, Edward, 206-543-6720 sarachik@atmos.washington.edu, Univ of Washington – Oa

Sarah, Willig B., (215) 898-5724 sawillig@verizon.net, Univ of Pennsylvania – Ge

Saravanan, R., 979-845-0175 r.saravanan@tamu.edu, Texas A&M Univ – Oa

Sargent, Kenneth A., (864) 294-3362 ken.sargent@furman.edu, Furman Univ – Hy

Saripalli, Srikanth, (480) 727-0023 srikanth.saripalli@asu.edu, Arizona State Univ – On

Sarmiento, Jorge L., (609) 258-6585 jls@princeton.edu, Princeton Univ – Oc

Sarrionandia, Fernando, fernando.sarrionandia@ehu.eus, Univ of the Basque Country UPV/EHU – GigGv

Sartain, Jerry B., (352) 392-1803 (Ext. 330) sartain@ufl.edu, Univ of Florida – So

Sarwar, A. K. Mostofa, (504) 280-6717 asarwar@uno.edu, Univ of New Orleans – Ye

Sasmaz, Ahmet, +905323406642 asasmaz@firat.edu.tr, Firat Univ – EgGeCe

Sasowsky, Ira D., (330) 972-5389 ids@uakron.edu, Univ of Akron – HwGmo

Sasowsky, Kathryn, (216) 987-5227 Kathryn.Sasowsky@tri-c.edu, Cuyahoga Community Coll - Western Campus – Og

Sass, Henrik, +44(0)29 208 76001 SassH@cardiff.ac.uk, Univ of Wales – Co

Sass, John H., (928) 556-7226 Northern Arizona Univ – Yg

Sassenrath, Gretchen, (620) 421-4826 gsassenrath@ksu.edu, Kansas State Univ – So

Sassi, Raffaele, +390498279149 raffaele.sassi@unipd.it, Università degli Studi di Padova – Ggx

Sato, Yoko, (973) 655-4448 satoyo@mail.montclair.edu, Montclair State Univ – Gg

Satow, Christopher, +44 208 417 2950 C.Satow@kingston.ac.uk, Kingston Univ – Ge

Satterfield, Joseph I., 325-486-6766 joseph.satterfield@angelo.edu, Angelo State Univ – Gc

Sauck, William A., (269) 387-4991 bill.sauck@wmich.edu, Western Michigan Univ – YgeYv

Sauer, Nancy N. S., (505) 665-3759 Los Alamos National Laboratory – On

Sauer, Peter, 812-855-6591 pesauer@indiana.edu, Indiana Univ, Bloomington – Pe

Saunders, Andy D., +440116 252 3923 ads@le.ac.uk, Leicester Univ – GiCgGe

Saunders, Charles, CharlesSaunders@rmc.edu, Randolph-Macon Coll – Ghx

Saunders, James A., (334) 844-4884 saundja@auburn.edu, Auburn Univ – Cl

Saunders, Kate, +44 (0) 131 650 2544 kate.saunders@ed.ac.uk, Edinburgh Univ – GviGz

Saunders, Ralph, (310) 243-3284 rsaunders@csudh.edu, California State Univ, Dominguez Hills – Og

Saunders, W. Bruce, (610) 526-5114 wsaunder@brynmawr.edu, Bryn Mawr Coll – Pi

Sautter, Leslie R., (843) 953-5586 sautterl@cofc.edu, Coll of Charleston – Ob

Sava, Diana, 720-981-0522 diana.sava@beg.utexas.edu, Univ of Texas at Austin, Jackson School of Geosciences – Yg

Sava, Diana C., 512-475-9507 diana.sava@beg.utexas.edu, Univ of Texas at Austin – Yg

Sava, Paul, 303-384-2362 psava@mines.edu, Colorado School of Mines – Ye

Savage, Brian, (401) 874-5392 savage@uri.edu, Univ of Rhode Island – Ys

Savage, Norman M., (541) 346-4585 nmsavage@uoregon.edu, Univ of Oregon – Pi

Savard, Dany, 418-545-5011 ddsavard@uqac.ca, Universite du Quebec a Chicoutimi – Ca

Savigny, K. Wayne, (604) 684-5900 Univ of British Columbia – Ng

Savin, Samuel M., (216) 368-6592 sms7@po.cwru.edu, Case Western Reserve Univ – Cs

Savina, Mary E., 507-222-4404 msavina@carleton.edu, Carleton Coll – Gm

Savov, Ivan, +44(0) 113 34 35199 earis@leeds.ac.uk, Univ of Leeds – Cg

Savrda, Charles E., (334) 844-4887 savrdce@auburn.edu, Auburn Univ – GdPeGr

Sawin, Robert S., (785) 864-2099 bsawin@kgs.ku.edu, Univ of Kansas – Gg

Sawyer, Carol F., 251-460-6169 sawyer@southalabama.edu, Univ of South Alabama – GmOye

Sawyer, Dale S., (713) 348-5106 dale@rice.edu, Rice Univ – Yr

Sawyer, Edward W., (418) 545-5011 x5636 edward-w_sawyer@uqac.ca, Universite du Quebec a Chicoutimi – GptGi

Sawyer, J. F., (605) 394-2462 Foster.Sawyer@sdsmt.edu, South Dakota School of Mines & Tech – NgGo

Saxena, Surenda K., (718) 951-5416 Graduate School of the City Univ of New York – Cg

Saxena, Surendra K., (305) 348-3030 saxenas@fiu.edu, Florida International Univ – Gy

Sayles, Frederick L., (508) 289-2561 fsayles@whoi.edu, Woods Hole Oceanographic Institution – Oc

Saylor, Beverly Z., 216 368 3763 bzs@case.edu, Case Western Reserve Univ – Gs

Sbar, Marc, msbar@email.arizona.edu, Univ of Arizona – Yes

Scanlin, Michael A., 717-361-1323 jte2@psu.edu, Elizabethtown Coll – Ye

Scanlon, Bridget R., 512-471-8241 bridget.scanlon@beg.utexas.edu, Univ of Texas at Austin – Hg

Scanlon, Todd M., (434) 924-3382 tms2v@virginia.edu, Univ of Virginia – Hg

Scannapieco, Evan, (480) 727-6788 evan.scannapieco@asu.edu, Arizona State Univ – On

Scapelli, Krista, (817) 257-7270 k.l.scarpelli@tcu.edu, Texas Christian Univ – On

Scarnato, Barbara V., bvscarna@nps.edu, Naval Postgraduate School – Ow

Scarselli, Nicola, +44 1784 443597 nicola.scarselli@rhul.ac.uk, Univ of London, Royal Holloway & Bedford New Coll – Go

Schaal, Barbara, (314) 935-6822 schaal@wustl.edu, Washington Univ in St. Louis – On

Schaap, Marcel, (520) 626-4532 mschapp@cals.arizona.edu, Univ of Arizona – So

Schade, Gunnar, (979) 845-7671 schade@ariel.met.tamu.edu, Texas A&M Univ – Oa

Schaef, Herbert T., (509) 373-9949 todd.schaef@pnl.gov, Pacific Northwest National Laboratory – Cg

Schaefer, Janet R. G., (907) 451-5005 janet.schaefer@alaska.gov, Alaska Division of Geological & Geophysical Surveys – Gv

Schaefer, Joerg, 845-365-8703 Columbia Univ – Pe

Schafer, Carl M., (586) 286-2154 schaferc@macomb.edu, Macomb Community Coll, Center Campus – Gg

Schafer, Tom, 785-628-5969 tschafer@fhsu.edu, Fort Hays State Univ – Oyi

Schafer, Jr., John W., (515) 294-3063 jschafer@iastate.edu, Iowa State Univ of Science & Tech – So

Schaffrin, Burkhard A., 614 292-0502 schaffrin.1@osu.edu, Ohio State Univ – Yd

Schaller, Mirjam, 734-615-4286 mirjam@umich.edu, Univ of Michigan – Gm

Schaller, Morgan F., (518) 276-3358 schall@rpi.edu, Rensselaer Polytechnic Inst – CsPe
Schapaugh, Bill T., (785) 532-7242 wts@ksu.edu, Kansas State Univ – So
Schardt, Christian, (218) 726-7899 cschardt@d.umn.edu, Univ of Minnesota, Duluth – EgCg
Schardt, Lawrence A., (814) 863-7655 las233@psu.edu, Pennsylvania State Univ, Univ Park – Sd
Scharer, Katherine, 828-262-6739 scharerkm@appstate.edu, Appalachian State Univ – Gc
Scharman, Mitchell, (304) 696-5435 scharman@marshall.edu, Marshall Univ – GctOg
Scharnberger, Charles K., Charles.scharnberger@millersville.edu, Millersville Univ – Gc
Schauble, Edwin A., (310) 825-3880 schauble@ucla.edu, Univ of California, Los Angeles – Cgs
Schauss, Kim E., (812) 464-1701 keschauss@usi.edu, Univ of Southern Indiana – On
Scheel, Patrick, 573-368-2243 patrick.scheel@dnr.mo.gov, Missouri Dept of Natural Resources – On
Scheer, Clemens, clemens.scheer@qut.edu.au, Queensland Univ of Tech – SbGe
Scheibe, Timothy D., (509) 372-6065 tim.scheibe@pnl.gov, Pacific Northwest National Laboratory – Hw
Scheidemen, Kathy J., (805) 893-7615 kathys@icess.ucsb.edu, Univ of California, Santa Barbara – On
Scheidt, Brian, 573-518-2314 bscheidt@MineralArea.edu, Mineral Area Coll – Hw
Schell, Marie, (519) 661-3191 mschell@uwo.ca, Western Univ – On
Schellart, Wouter , +31 20 59 8610 w.p.schellart@vu.nl, VU Univ Amsterdam – Gt
Schellenberg, Stephen A., 61959421039 saschellenberg@mail.sdsu.edu, San Diego State Univ – Pe
Schelske, Claire L., (352) 392-9617 Univ of Florida – Pe
Schenck, William S., (302) 831-2833 rockman@udel.edu, Univ of Delaware – GgpGc
Scher, Howie, 803-777-2410 hscher@geol.sc.edu, Univ of South Carolina – GuCl
Scherer, Reed P., (815) 753-7951 reed@niu.edu, Northern Illinois Univ – Pm
Schermer, Elizabeth R., (360) 650-3658 schermer@geol.wwu.edu, Western Washington Univ – Gt
Schiappa, Tamra A., (724) 738-2829 tamra.schiappa@sru.edu, Slippery Rock Univ – PiGrOe
Schieber, Juergen, (812) 856-4740 jschiebe@indiana.edu, Indiana Univ, Bloomington – Gs
Schiebout, Judith A., (225) 578-2717 schiebout@geol.lsu.edu, Louisiana State Univ – Pv
Schiefer, Erik, (928) 523-6535 Erik.Schiefer@nau.edu, Northern Arizona Univ – GmOyi
Schiffbauer, James, (573) 882-9501 schiffbauerj@missouri.edu, Univ of Missouri – Poi
Schiffman, Peter, (530) 752-3669 pschiffman@ucdavis.edu, Univ of California, Davis – Gp
Schilling, Jean-Guy, (401) 874-6628 jgs@gso.uri.edu, Univ of Rhode Island – Ou
Schilling, Keith, Keith.Schilling@dnr.iowa.gov, Iowa Dept of Natural Resources – Gg
Schilling, Keith E., (319) 335-1575 keith.schilling@dnr.iowa.gov, Univ of Iowa – Hgs
Schimmelmann, Arndt, (812) 855-7645 aschimme@indiana.edu, Indiana Univ, Bloomington – CsGsPe
Schimmrich, Steven, (845) 687-7683 schimmrs@sunyulster.edu, SUNY, Ulster County Community Coll – Og
Schincariol, Robert A., (519) 661-3732 schincar@uwo.ca, Western Univ – HwsNg
Schindler, Michael, 7056751151 x2368 mschindler@laurentian.ca, Laurentian Univ, Sudbury – Gz
Schirmer, Ron, (507) 389-6929 ronald.schirmer@mnsu.edu, Minnesota State Univ – GaPb
Schlautman, Mark, (864) 656-4059 mschlau@clemson.edu, Clemson Univ – ClHgSc
Schlegel, Alan, (620) 376-4761 schlegel@ksu.edu, Kansas State Univ – So
Schleifer, Stanley, 718-262-2726 sschleifer@york.cuny.edu, York Coll (CUNY) – Ge
Schlesinger, Michael E., (217) 333-2192 schlesin@illinois.edu, Univ of Illinois, Urbana-Champaign – Oa
Schlesinger, William H., (919) 613-8004 schlesin@duke.edu, Duke Univ – On
Schlichting, Hilke, (626) 316-3629 hilke@ucla.edu, Univ of California, Los Angeles – Xy
Schlische, Roy W., (848) 445-3142 schlisch@rci.rutgers.edu, Rutgers, The State Univ of New Jersey – Gct
Schlosser, C. Adam, 617-253-3983 casch@mit.edu, Massachusetts Inst of Tech – Oa
Schlosser, Peter, (845) 365-8707 Columbia Univ – Hw
Schlue, John W., jwschlue@yahoo.com, New Mexico Inst of Mining and Tech – YsgOg
Schlumpberger, Debbie, 320-308-3260 dmschlumpberger@stcloudstate.edu, Saint Cloud State Univ – On
Schlögl, Ján , schlogl@fns.uniba.sk, Comenius Univ in Bratislava – Piy
Schmahl, Wolfgang, 089/2180 4311 Wolfgang.Schmahl@lrz.uni-muenchen.de, Ludwig-Maximilians-Universitaet Muenchen – Gz
Schmandt, Brandon, 505.277.4204 bschmandt@unm.edu, Univ of New Mexico – Yg
Schmerr, Nicholas, 301-405-4385 nschmerr@umd.edu, Univ of Maryland – XyYs
Schmid, Dieter, 089/2180 6635 d.schmid@lrz.uni-muenchen.de, Ludwig-Maximilians-Universitaet Muenchen – Pg
Schmid, Ginger L., (507) 389-2824 ginger.schmid@mnsu.edu, Minnesota State Univ – SdGmSf
Schmid, Katie, 412-442-4232 kschmid@pa.gov, Pennsylvania Bureau of Topographic & Geologic Survey – Go
Schmidt, Amanda H., 440-775-8351 amanda.schmidt@oberlin.edu, Oberlin Coll – Gm
Schmidt, Bennetta, bennetta.schmidt@lamar.edu, Lamar Univ – Gg
Schmidt, Dale R., 217-300-1169 schmidt2@illinois.edu, Univ of Illinois, Urbana-Champaign – Ge
Schmidt, David , dasc@uw.edu, Univ of Washington – Gt
Schmidt, David, 937 775-3539 david.schmidt@wright.edu, Wright State Univ – PiGd
Schmidt, Jonathan, (519) 824-4120 Ext.53966 jonschm@uoguelph.ca, Univ of Guelph – On
Schmidt, Keegan L., (208) 790-2283 klschmidt@lcsc.edu, Lewis-Clark State Coll – Gc
Schmidt, Lisa, lschmidt@sbccd.cc.ca.us, San Bernardino Valley Coll – Oy
Schmidt, Matthew, 757-683-4285 mwschmid@odu.edu, Old Dominion Univ – Ou
Schmitt, Danielle M., (609) 258-7015 dschmitt@princeton.edu, Princeton Univ – Oe
Schmitt, Douglas R., (780) 492-3985 dschmitt@ualberta.ca, Univ of Alberta – YxNrYe
Schmitt, James G., (406) 994-6903 jschmitt@montana.edu, Montana State Univ – Gr
Schmitt, Michael A., (612) 625-7017 mschmitt@soils.umn.edu, Univ of Minnesota, Twin Cities – Sc
Schmittner, Andreas, 541-737-9952 aschmittner@coas.oregonstate.edu, Oregon State Univ – OaYr
Schmitz, Darrel W., 662-268-1032 Ext 241 schmitz@geosci.msstate.edu, Mississippi State Univ – Hw
Schmitz, Mark D., 208-426-5907 markschmitz@boisestate.edu, Boise State Univ – CcaCg
Schmutz, Phillip P., 850-474-3418 pschmutz@uwf.edu, Univ of West Florida – Gm
Schneider, David, (613) 562-5800 x6155 david.schneider@uottawa.ca, Univ of Ottawa – GtCcGg
Schneider, Edwin K., (703) 993-5364 eschnei1@gmu.edu, George Mason Univ – Oa
Schneider, Jim, 7144320202 x21317 jschneider@occ.cccd.edu, Orange Coast Coll – Geu
Schneider, John F., (630) 252-8923 Argonne National Laboratory – Ca
Schneider, Julius, 089/2180 4354 julius.schneider@lrz.uni-muenchen.de, Ludwig-Maximilians-Universitaet Muenchen – Gz
Schneider, Niklas, (808) 956-8383 nschneid@hawaii.edu, Univ of Hawai'i, Manoa – Og
Schneider, Robert J., 508-289-2756 rschneider@whoi.edu, Woods Hole Oceanographic Institution – Yr
Schneider, Robert V., 361-593-3589 Robert.Schneider@tamuk.edu, Texas A&M Univ, Kingsville – GoYes
Schneider, Tapio, (626) 395-6143 tapio@caltech.edu, California Inst of Tech – OawOg
Schneiderman, Jill S., 845-437-5542 schneiderman@vassar.edu, Vassar Coll – Gs
Schnetzer, Astrid, (919) 515-7837 aschnet@ncsu.edu,

North Carolina State Univ – Ob
Schnoebelen, Douglas J., (319) 358-3617 douglas-schnoebelen@uiowa.edu,
Univ of Iowa – Hg
Schoenberger, Erica, (410) 516-6158 ericas@jhu.edu,
Johns Hopkins Univ – On
Schoene, R B., (609) 258-5747 bschoene@princeton.edu,
Princeton Univ – Cc
Schoenemann, Spruce W., (406) 683-7624 spruce.schoenemann@umwestern.edu,
Univ of Montana Western – GeCgs
Scholtz, Theresa C., (630) 252-6499 Argonne National Laboratory – Ge
Scholz, Christopher A., 315-443-2672 cascholz@syr.edu,
Syracuse Univ – Gs
Scholz, Christopher H., (845) 365-8360 Columbia Univ – Ys
Schoof, Christian, (604) 822-3063 cschoof@eos.ubc.ca,
Univ of British Columbia – Ol
Schoonen, Martin A., (631) 632-8007 martin.schoonen@sunysb.edu,
SUNY, Stony Brook – Cl
Schoonmaker, Adam, 792-2577 adschoonmaker@utica.edu,
Utica Coll – GczGx
Schoonmaker, Jane E., (808) 956-9935 jane@soest.hawaii.edu,
Univ of Hawai'i, Manoa – Cl
Schopf, J. William, (310) 825-1170 schopf@epss.ucla.edu,
Univ of California, Los Angeles – Po
Schopf, Paul S., (703) 993-5394 pschopf@gmu.edu,
George Mason Univ – Opa
Schouten, Hans, (508) 289-2574 hschouten@whoi.edu,
Woods Hole Oceanographic Institution – Yr
Schrag, Daniel P., (617) 495-7676 schrag@eps.harvard.edu,
Harvard Univ – Cg
Schrage, Jon M., (402) 280-5759 schragej@gmail.com,
Creighton Univ – Ow
Schrank, Christoph, +61 7 3138 1583 christoph.schrank@qut.edu.au,
Queensland Univ of Tech – GcqGt
Schreiber, B. Charlotte, (718) 997-3300 Queens Coll (CUNY) – Gd
Schreiber, Charlotte, 206-297-1454 geologo1@uw.edu,
Univ of Washington – Gsd
Schreiber, Madeline E., (540) 231-3377 Virginia Polytechnic Inst & State Univ – Hw
Schreier, Hans D., (604) 822-4401 Univ of British Columbia – Og
Schrenk, Matt, (517) 884-7966 schrenkm@msu.edu,
Michigan State Univ – Py
Schrieber, B. Charlotte, (718) 997-3300 Graduate School of the City Univ of New York – Gd
Schriver, David, (310) 825-6663 dave@igpp.ucla.edu,
Univ of California, Los Angeles – Yg
Schroder-Adams, Claudia, csadams@ccs.carleton.ca, Carleton Univ – Pm
Schroeder, Dustin M., (650) 725-7861 dustin.m.schroeder@stanford.edu,
Stanford Univ – Gl
Schroeder, John L., 806-834-5678 john.schroeder@ttu.edu,
Texas Tech Univ – Oa
Schroeder, Kathleen, (828) 262-7055 schroederk@appstate.edu,
Appalachian State Univ – On
Schroeder, Norman C., (505) 667-0967 Los Alamos National Laboratory – Cc
Schroeder, Paul A., (706) 542-2384 schroe@uga.edu,
Univ of Georgia – GzEnCl
Schroeder, Stefan, +44 0161 306-6870 stefan.schroeder@manchester.ac.uk,
Univ of Manchester – Go
Schroeder, William W., (334) 861-7528 wschroeder@disl.org,
Dauphin Island Sea Lab – Og
Schroth, Andrew W., (802)656 3481 aschroth@uvm.edu,
Univ of Vermont – ClGz
Schubert, Brian, 337-482-6967 bas9777@louisiana.edu,
Univ of Louisiana at Lafayette – Cg
Schubert, Gerald, (310) 825-4577 Univ of California, Los Angeles – Yg
Schuberth, Bernhard, 089/2180 4220 bernhard@geophysik.uni-muenchen.de, Ludwig-Maximilians-Universitaet Muenchen – Yg
Schulingkamp, Arren, 225-578-3412 warrenii@lsu.edu,
Louisiana State Univ – Gg
Schulmeister, Marcia K., (620) 341-5983 mschulme@emporia.edu,
Emporia State Univ – HwClGe
Schulp, Nynke , +31 20 59 83082 nynke.schulp@vu.nl,
VU Univ Amsterdam – Oy
Schulte, Kimberly, 770-274-5083 kschulte1@gsu.edu,
Georgia State Univ, Perimeter Coll, Dunwoody Campus – Ge
Schultz, Adam, 541-737-9832 adam@coas.oregonstate.edu,
Oregon State Univ – Yg
Schultz, David, +44 0161 306-3909 david.schultz@manchester.ac.uk,
Univ of Manchester – Ow
Schultz, Gerald E., 806-651-2580 gschultz@wtamu.edu,
West Texas A&M Univ – PvGzOg
Schultz, Jan, (805) 965-0581 (Ext. 2313) schultz@sbcc.edu,
Santa Barbara City Coll – GgePg
Schultz, Peter H., (401) 863-2417 Peter_Schultz@Brown.edu,
Brown Univ – Xg
Schulz, Layne D., 605-67-76161 layne.schulz@usd.edu,
South Dakota Dept of Environment and Natural Resources – Glg
Schulz, William J., 410-293-6563 schulz@usna.edu,
United States Naval Academy – Opw
Schulze, Anja, 409 740 4540 schulzea@tamug.edu,
Texas A&M Univ – Ob
Schulze, Daniel J., 905-8283970 dschulze@utm.utoronto.ca,
Univ of Toronto – Gi
Schulze, Karl, (630) 466-2652 kschulze@waubonsee.edu,
Waubonsee Community Coll – OgwOa
Schulze-Makuch, Dirk, 509-335-1180 dirksm@wsu.edu,
Washington State Univ – Hw
Schumacher, Courtney, (979) 845-5522 cschu@tamu.edu,
Texas A&M Univ – Oa
Schumacher, Matthew, mschumac@stfx.ca, Saint Francis Xavier Univ – Og
Schumann, Arnold W., (863) 956-1151 schumaw@ufl.edu,
Univ of Florida – So
Schumer, Rina, (775) 673-7414 rina.schumer@dri.edu,
Univ of Nevada, Reno – HqwGm
Schuster, Gerard T., (801) 581-4373 j.schuster@utah.edu,
Univ of Utah – Ye
Schutt, Derek L., (970) 491-5786 schutt@warnercnr.colostate.edu,
Colorado State Univ – Ysg
Schwab, Brandon E., bes21@humboldt.edu,
Humboldt State Univ – GizGv
Schwab, Brandon E., (828) 227-7495 beschwab@wcu.edu,
Western Carolina Univ – GizGv
Schwab, Fred, (310) 825-3123 schwab@igpp.ucla.edu,
Univ of California, Los Angeles – Ys
Schwab, Frederick, (540) 458-5830 schwabf@wlu.edu,
Washington & Lee Univ – Gdg
Schwab, James J., (518) 437-8754 schwab@asrc.cestm.albany.edu,
SUNY, Albany – Oa
Schwab, Paul, 979-845-3663 pschwab@tamu.edu,
Texas A&M Univ – Sc
Schwankl, Larry J., (530) 752-1130 ljschwankl@ucdavis.edu,
Univ of California, Davis – On
Schwarcz, Henry P., (905) 525-9140 (Ext. 24186) schwarcz@mcmaster.ca,
McMaster Univ – Cs
Schwartz, David, (408) 479-6495 daschwar@cabrillo.edu,
Cabrillo Coll – Gu
Schwartz, Frank W., (614) 292-6196 schwartz.11@osu.edu,
Ohio State Univ – Hg
Schwartz, Hilde, (831) 459-5429 hschwartz@pmc.ucsc.edu,
Univ of California, Santa Cruz – Pv
Schwartz, Joshua J., (818) 677-5813 joshua.schwartz@csun.edu,
California State Univ, Northridge – GiCc
Schwartz, Matthew C., 850-474-3469 mschwartz@uwf.edu,
Univ of West Florida – OcCl
Schwartz, Susan Y., (831) 459-3133 sschwartz@pmc.ucsc.edu,
Univ of California, Santa Cruz – Ys
Schwartz, Theresa M., (814) 332-2873 tschwartz@allegheny.edu,
Allegheny Coll – Gs
Schwarz, Karen M., (610) 436-2788 kvanlandingham@wcupa.edu,
West Chester Univ – On
Schwarzschild, Arthur C., (757) 331-1246 acs7q@virginia.edu,
Univ of Virginia – On
Schweitzer, Carrie E., 330-244-3303 cschweit@kent.edu,
Kent State Univ at Stark – PiGg
Schwerdtner, Walfried M., (416) 978-5080 fried@quartz.geology.utoronto.ca, Univ of Toronto – GctGq
Schwert, Donald P., (701) 231-7496 donald.schwert@ndsu.edu,
North Dakota State Univ – PeGe
Schwimmer, David R., 706 569-3028 schwimmer_david@columbusstate.edu, Columbus State Univ – Pv
Schwimmer, Reed A., 609-896-5346 rschwimmer@rider.edu,

Rider Univ – GsOgn
Schwob, Stephanie L., 214-768-2770 Southern Methodist Univ – On
Sclater, John G., (858) 534-3051 jsclater@ucsd.edu, Univ of California, San Diego – Yh
Scoates, James S., (604) 822-3667 jscoates@eos.ubc.ca, Univ of British Columbia – Gi
Scotese, Christopher R., (817) 272-2987 cscotese@exchange.uta.edu, Univ of Texas, Arlington – Gt
Scott, Andrew C., +44 1784 443608 A.Scott@rhul.ac.uk, Univ of London, Royal Holloway & Bedford New Coll – Pb
Scott, Christopher A., cascott@email.arizona.edu, Univ of Arizona – Hw
Scott, Craig, (403) 823-7707 Royal Tyrrell Museum of Palaeontology – Pv
Scott, Darren M., (905) 525-9140 (Ext. 24953) scottdm@mcmaster.ca, McMaster Univ – Oi
Scott, David B., (902) 494-3604 Dalhousie Univ – Pm
Scott, Edward R., (808) 956-3955 Univ of Hawai'i, Manoa – Xm
Scott, Kathy, (604) 822-5606 kscott@eos.ubc.ca, Univ of British Columbia – On
Scott, Larry, 303-384-2631 lmscott@mines.edu, Colorado School of Mines – Oy
Scott, Robert B., 512-471-0375 rscott@ig.utexas.edu, Univ of Texas at Austin – Og
Scott, Robert W., (918) 230-3436 rwscott@cimtel.net, The Univ of Tulsa – GsrPi
Scott, Steven D., (416) 978-5424 scottsd@es.utoronto.ca, Univ of Toronto – GuEmCm
Scott, Thomas A., (951) 827-5115 thomas.scott@ucr.edu, Univ of California, Riverside – Oy
Scott, Tim, (401) 254-3108 Roger Williams Univ – Ob
Scott, Verne H., (530) 752-0453 vhscott@ucdavis.edu, Univ of California, Davis – Hs
Scott, Vernon, 405-744-6358 vscott@okstate.edu, Oklahoma State Univ – Oe
Scott Smith, Barbara H., (604) 984-9609 bhssmith@allstream.net, Univ of British Columbia – Gz
Scott-Dupree, Cynthia, (519) 824-4120 Ext.52477 cscottdu@uoguelph.ca, Univ of Guelph – On
Scotton, Paolo, +390498279120 paolo.scotton@unipd.it, Università degli Studi di Padova – Hg
Scow, Kate, (530) 752-4632 kmscow@ucdavis.edu, Univ of California, Davis – Os
Scranton, Mary I., (631) 632-8735 mary.scranton@stonybrook.edu, SUNY, Stony Brook – Oc
Screaton, Elizabeth J., (352) 3924612 screaton@ufl.edu, Univ of Florida – Hg
Scudder, Sylvia J., 3523921721 x246 scudder@flmnh.ufl.edu, Univ of Florida – Pg
Scuderi, Louis A., (505) 277-4204 tree@unm.edu, Univ of New Mexico – PeGms
Scyphers, Steven, 781.581.7370 s.scyphers@northeastern.edu, Northeastern Univ – On
Seacrest, Tyler, 406 683-7263 tyler.seacrest@umwestern.edu, Univ of Montana Western – Gq
Seager, Richard, (845) 365-8743 Columbia Univ – Ow
Seager, Sara, 617-253-6775 seager@mit.edu, Massachusetts Inst of Tech – Xy
Seal, Thom, (775) 682-8813 tseal@unr.edu, Univ of Nevada, Reno – NxEmNm
Seaman, Nelson L., (814) 863-1583 Pennsylvania State Univ, Univ Park – Ow
Seaman, Sheila J., (413) 545-2822 sjs@geo.umass.edu, Univ of Massachusetts, Amherst – Gi
Searle, Mike, +44 (1865) 272022 mike.searle@earth.ox.ac.uk, Univ of Oxford – Gt
Searls, Mindi L., 402-472-6934 Univ of Nebraska, Lincoln – YgGg
Sears, Heather, 7815817370 x316 h.sears@neu.edu, Northeastern Univ – On
Sears, James W., (406) 243-5251 james.sears@umontana.edu, Univ of Montana – Gc
Sebol, Lesley, (303) 384-2633 lsebol@mines.edu, Colorado School of Mines – HwGeOg
Secco, Luciano, +390498279158 luciano.secco@unipd.it, Università degli Studi di Padova – Gz
Secco, Richard A., (519) 661-4079 secco@uwo.ca, Western Univ – Gy
Secord, Ross, (402) 472-2663 rsecord2@unl.edu, Univ of Nebraska, Lincoln – PvCs
Sediek, Kadry N., 002-03-3921595 kknsed@yahoo.com, Alexandria Univ – GgsGd
Sedivy, Robert, 402-465-9021 rasedivy@anl.gov, Argonne National Laboratory – Hw
Sedivy, Robert A., (630) 252-1897 rasedivy@anl.gov, Argonne National Laboratory – Hw
Sedlacek, Alexa, 319-273-3072 alexa.sedlacek@uni.edu, Univ of Northern Iowa – Cs
Seeber, Leonardo, (845) 365-8385 Columbia Univ – Ys
Seedorff, Eric, (520) 626-3921 seedorff@email.arizona.edu, Univ of Arizona – Eg
Seeger, Cheryl M., (573) 368-2184 cheryl.seeger@dnr.mo.gov, Missouri Dept of Natural Resources – Gig
Seeley, Mark W., (612) 625-4724 mseeley@umn.edu, Univ of Minnesota, Twin Cities – Oa
Seewald, Jeffrey S., (508) 289-2966 jseewald@whoi.edu, Woods Hole Oceanographic Institution – Cp
Segall, Marylin, 801-585-5730 mpsegall@egi.utah.edu, Univ of Utah – GeOu
Segall, Paul, (650) 725-7241 Stanford Univ – Yg
Segars, William P., (706) 542-9072 Univ of Georgia – Sc
Seibel, Erwin, (415) 338-2061 San Francisco State Univ – On
Seibt, Ulrike, (310) 206-4442 useibt@ucla.edu, Univ of California, Los Angeles – PyCg
Seid, Mary J., 217-244-8171 maryseid@illinois.edu, Univ of Illinois, Urbana-Champaign – Gz
Seidemann, David E., (718) 951-5761 Graduate School of the City Univ of New York – Cc
Seidemann, David E., 7189515000 x2882 dseidemann@earthlink.net, Brooklyn Coll (CUNY) – Cc
Seifert, Karl E., (515) 294-5265 kseifert@iastate.edu, Iowa State Univ of Science & Tech – Ct
Seifoullaev, Roustam K., (512) 232-3223 roustam@utig.ig.utexas.edu, Univ of Texas at Austin – Ye
Seigley, Lynette S., 319-335-1598 lynette.seigley@dnr.iowa.gov, Iowa Dept of Natural Resources – Gg
Seitz, Jeffery C., (510) 885-3438 jeff.seitz@csueastbay.edu, California State Univ, East Bay – CgOeGx
Selby, Dave, +44 (0) 191 33 42294 david.selby@durham.ac.uk, Durham Univ – Cc
Selden, Paul A., (785) 864-2751 selden@ku.edu, Univ of Kansas – Pio
Selim, Hussein M., (504) 388-2110 Louisiana State Univ – Sp
Selin, Noelle, (617) 324-2592 selin@mit.edu, Massachusetts Inst of Tech – Oa
Selleck, Bruce, (315) 228-7949 bselleck@colgate.edu, Colgate Univ – Gs
Sellmeier, Bettina, +49 (89) 289 25822 sellmeier@tum.de, Technische Universitaet Muenchen – Ngr
Selph, Karen E., (808) 956-7941 selph@hawaii.edu, Univ of Hawai'i, Manoa – Ob
Selverstone, Jane E., (505) 277-6528 selver@unm.edu, Univ of New Mexico – Gpt
Semazzi, Fred H. M., (919) 515-1434 fred_semazzi@ncsu.edu, North Carolina State Univ – Oa
Semken, Steven, (480) 965-7965 semken@asu.edu, Arizona State Univ – OenGg
Semken, Jr., Holmes A., (319) 335-1830 holmes-semken@uiowa.edu, Univ of Iowa – Pv
Semtner, Albert J., sbert@nps.edu, Naval Postgraduate School – Op
Sen, Gautam, (305) 348-2299 seng@fiu.edu, Florida International Univ – Gi
Sen, Mrinal K., (512) 471-0466 Univ of Texas at Austin – Ye
Sen, Pragnyadipta, Pragnyadipta.Sen@oneonta.edu, SUNY, Oneonta – Gct
Sen Gupta, Barun K., (225) 388-5984 barun@geol.lsu.edu, Louisiana State Univ – Pm
Send, Uwe, (858) 822-6710 usend@ucsd.edu, Univ of California, San Diego – Op
Senko, John M., 330 972-8047 senko@uakron.edu, Univ of Akron – Cg
Sennert, Sally K., (202) 633-1805 kuhns@si.edu, Smithsonian Institution / National Museum of Natural History – GvOr
Seong, Jeong C., (678) 839-4069 jseong@westga.edu, Univ of West Georgia – Oir
Sephton, Mark, +44 20 759 46542 m.a.sephton@imperial.ac.uk, Imperial Coll – Xm
Sepúlveda, Julio C., jsepulveda@Colorado.edu, Univ of Colorado – Co
Seramur, Keith C., (828) 262-3049 Appalachian State Univ – Hw
Serenko, Thomas J., (614) 265-6598 thomas.serenko@dnr.state.oh.us, Ohio Dept of Natural Resources – EgnEm
Sericano, Jose L., (979) 862-2323 jose@gerg.tamu.edu,

Faculty Index -S

Texas A&M Univ – Cm
Serne, R. Jeffrey, (360) 539-6162 jeff.serne@pnnl.gov, Pacific Northwest National Laboratory – CgSoCl
Serpa, Laura F., (504) 280-6801 lserpa@uno.edu, Univ of New Orleans – Ys
Serpa, Laura F., (915) 747-6058 lfserpa@utep.edu, Univ of Texas, El Paso – Ye
Serrano, Carmen, (207) 872-3244 serrano@fit.edu, Florida Inst of Tech – On
Serrano, Carmen, (321) 674-8096 serrano@fit.edu, Florida Inst of Tech – On
Serreze, Mark, mark.serreze@colorado.edu, Univ of Colorado – Oa
Sertich, Joseph, 303-370-6331 Joe.Sertich@dmns.org, Denver Museum of Nature & Science – Pv
Sessions, Alex L., 626.395.6445 als@gps.caltech.edu, California Inst of Tech – Co
Sethi, Parvinder S., (540) 831-5619 psethi@radford.edu, Radford Univ – Gz
Setterholm, Dale, (612) 626-5119 sette001@umn.edu, Univ of Minnesota – Gs
Sevellec, Florian, +44 (0)23 80594850 Florian.Sevellec@noc.soton.ac.uk, Univ of Southampton – Op
Severinghaus, Jeffrey P., (858) 822-2483 jseveringhaus@ucsd.edu, Univ of California, San Diego – Pe
Severs, Matthew R., 609-626-6857 matthew.severs@stockton.edu, Stockton Univ – GxCgEg
Sewall, Jacob, 484-646-5864 sewall@kutztown.edu, Kutztown Univ of Pennsylvania – GeOa
Sexton, John L., (618) 453-7374 sexton@geo.siu.edu, Southern Illinois Univ Carbondale – Ye
Sexton, Philip, philip.sexton@open.ac.uk, The Open Univ – Ge
Seyfang, Gill, +44 (0)1603 59 2956 g.seyfang@uea.ac.uk, Univ of East Anglia – Og
Seyfried, Jr., William E., 612-624-1333 wes@umn.edu, Univ of Minnesota, Twin Cities – Cm
Seyler, Beverly, 217-244-2389 seyler@isgs.uiuc.edu, Illinois State Geological Survey – Eo
Seymour, Kevin L., kevins@rom.on.ca, Royal Ontario Museum – PvoPg
Seyoum, Wondwosen M., (309) 438-2833 wmseyou@ilstu.edu, Illinois State Univ – HwqHs
Shaaban, Mohamad N., 002-03-3921595 Moshaaban@yahoo.com, Alexandria Univ – GgsGd
Shaak, Graig D., 3523921721 x257 gdshaak@flmnh.ufl.edu, Univ of Florida – Pe
Shackley, Simon J., +44 (0) 131 650 7862 Simon.Shackley@ed.ac.uk, Edinburgh Univ – Eg
Shade, Harry, 4087412045 x 3678 geology1@earthlink.net, West Valley Coll – Gg
Shade, Janet, 814-362-7560 jas144@pitt.edu, Univ of Pittsburgh, Bradford – On
Shadwick, Elizabeth H., 804-684-7247 shadwick@vims.edu, Coll of William & Mary – Oc
Shah, Subhash N., (405) 325-6871 subhash@ou.edu, Univ of Oklahoma – Np
Shahar, Anat, (202) 478-8929 ashahar@carnegiescience.edu, Carnegie Institution of Washington – CgYxCs
Shail, Robin, +44 01326 371826 R.K.Shail@exeter.ac.uk, Exeter Univ – Gtc
Shakal, Anthony F., (916) 322-7481 California Geological Survey – Ys
Shakoor, Abdul, (330) 672-2968 ashakoor@kent.edu, Kent State Univ – Ng
Shakun, Jeremy D., 617-552-1625 jeremy.shakun@bc.edu, Boston Coll – Gg
Shalimba, Ester, +264-61-2063839 eshalimba@unam.na, Univ of Namibia – Ggp
Shamberger, Kathryn E., (979) 845-5752 katie.shamberger@tamu.edu, Texas A&M Univ – Oc
Shams, Asghar, +44 (0)131 451 3904 a.shams@hw.ac.uk, Heriot-Watt Univ – Yg
Shanafield, Margaret, margaret.shanafield@flinders.edu.au, Flinders Univ – Hw
Shanahan, Timothy M., 512-232-7051 tshanahan@jsg.utexas.edu, Univ of Texas at Austin – PeGsCg
Shane, Tyrrell, +353 (0)91 494387 shane.tyrrell@nuigalway.ie, National Univ of Ireland Galway – Gs
Shang, Congxiao, +44 (0)1603 59 3123 c.shang@uea.ac.uk, Univ of East Anglia – Hw
Shank, Gerard C., 512-550-7139 cshank@utexas.edu, Univ of Texas at Austin – Ob
Shank, Stephen G., 717-702-2021 stshank@pa.gov, Pennsylvania Bureau of Topographic & Geologic Survey – GziGp
Shankland, Thomas J., (505) 667-4907 shanklan@lanl.gov, Los Alamos National Laboratory – Yx
Shannon, Jack D., (630) 252-5807 jack_shannon@anl.gov, Argonne National Laboratory – Ow
Shannon, Jeremy, (906) 487-3573 jmshanno@mtu.edu, Michigan Technological Univ – Gg
Shapiro, Russell, (530) 898-4300 rsshapiro@csuchico.edu, California State Univ, Chico – PyGdPg
Shapley, Mark, shap0029@umn.edu, Univ of Minnesota, Twin Cities – Gn
Sharkey, Debra, (916) 691-7210 Cosumnes River Coll – Oy
Sharma, Govind, (205) 851-5462 aamaxs01@aamu.edu, Alabama A&M Univ – On
Sharma, Mukul, 603-646-0024 Mukul.Sharma@dartmouth.edu, Dartmouth Coll – Ge
Sharma, Shikha, 304-293-5603 Shikha.Sharma@mail.wvu.edu, West Virginia Univ – Csa
Sharma, Shiv K., (808) 956-8476 Univ of Hawai'i, Manoa – Gx
Sharma, Suresh, ssharma@ou.edu, Univ of Oklahoma – Np
Sharp, Jonathan H., (302) 645-4259 jsharp@udel.edu, Univ of Delaware – Oc
Sharp, Martin J., (780) 492-4156 martin.sharp@ualberta.ca, Univ of Alberta – Gl
Sharp, Patricia S., (936) 468-2095 psharp@sfasu.edu, Stephen F. Austin State Univ – Gg
Sharp, S. L., 540-231-4080 llyn@vt.edu, Virginia Polytechnic Inst & State Univ – Oe
Sharp, Thomas G., 480-965-3071 tom.sharp@asu.edu, Arizona State Univ – Gz
Sharp, Zachary D., (505) 277-4204 zsharp@unm.edu, Univ of New Mexico – Cs
Sharp, Jr., John M., (512) 471-3317 jmsharp@jsg.utexas.edu, Univ of Texas at Austin – HqwGe
Sharples, Jonathan, +44 0151 794 4093 Jonathan.Sharples@liverpool.ac.uk, Univ of Liverpool – Ocp
Shaulis, James R., 717.702.2037 jshaulis@pa.gov, Pennsylvania Bureau of Topographic & Geologic Survey – EcOe
Shaver, Stephen A., (931) 598-1116 sshaver@sewanee.edu, Sewanee: Univ of the South – Eg
Shaw, Bruce, (845) 365-8380 Columbia Univ – Ys
Shaw, Christopher A., (323) 857-6317 chrissha@bcf.usc.edu, Los Angeles County Museum of Natural History – Pv
Shaw, Cliff S., 506-447-3195 cshaw@unb.ca, Univ of New Brunswick – Gi
Shaw, Colin, 406 994 6760 colin.shaw1@montana.edu, Montana State Univ – GcEmGp
Shaw, Fred, 573-368-2479 fred.shaw@dnr.mo.gov, Missouri Dept of Natural Resources – On
Shaw, Frederick C., (212) 642-2202 Lehman Coll (CUNY) – Ps
Shaw, Frederick C., (718) 960-8565 fcshaw@verizon.net, Graduate School of the City Univ of New York – Pi
Shaw, George H., (518) 388-6310 shawg@union.edu, Union Coll – Yx
Shaw, Glenn D., 406-496-4809 gshaw@mtech.edu, Montana Tech of the Univ of Montana – HwCsHs
Shaw, Joey N., (334) 844-3957 shawjo1@auburn.edu, Auburn Univ – Sd
Shaw, John, (780) 492-3573 john.shaw@ualberta.ca, Univ of Alberta – Gl
Shaw, John, (617) 495-8008 shaw@eps.harvard.edu, Harvard Univ – Gt
Shaw, John B., 479-575-7489 shaw84@uark.edu, Univ of Arkansas, Fayetteville – GsrGg
Shaw, Katy, (253) 833-9111 kshaw@greenriver.edu, Green River Community Coll – OggGg
Shaw, Kenneth L., (212) 642-2202 kshaw@egi.utah.edu, Univ of Utah – Ye
Shaw, Liz, e.j.shaw@reading.ac.uk, Univ of Reading – Sb
Shaw, Ping-Tung, (919) 515-7276 ping-tung_shaw@ncsu.edu, North Carolina State Univ – Op
Shaw, Richard F., (225) 388-6734 Louisiana State Univ – Ob
Shaw, Tiffany A., (845) 365-8571 tas@ldeo.columbia.edu, Columbia Univ – Oa
Shaw Faulkner, Melinda, (936) 468-2236 mgshaw@sfasu.edu, Stephen F. Austin State Univ – GeClEg
Shcherbakov, Robert, 519-661-2111 ext.84212 rshcherb@uwo.ca, Western Univ – YgsYd
Shchukin, Eugene D., (410) 516-5079 shchukin@jhuvms.hcf.jhu.edu,

Johns Hopkins Univ – On
Shea, Erin, 907-786-6846 eshea2@uaa.alaska.edu, Univ of Alaska, Anchorage – CcGcCg
Shearer, Peter M., (858) 534-2260 pshearer@ucsd.edu, Univ of California, San Diego – Ysg
Shearer, Jr., Charles K., (505) 277-9159 cshearer@unm.edu, Univ of New Mexico – Gi
Shearman, Kipp, 541-737-1866 shearman@coas.oregonstate.edu, Oregon State Univ – Opn
Shears, Andrew, ashears@mansfield.edu, Univ of Wisconsin Colls – Oin
shedied, ahmad g., 01005856505 ags00@fayoum.edu.eg, Fayoum Univ – GgHww
Sheehan, Anne F., (303) 492-4597 anne.sheehan@colorado.edu, Univ of Colorado – Yse
Sheehan, Peter M., (414) 278-2741 sheehan@mpm.edu, Milwaukee Public Museum – Pi
Sheehan, Peter M., (414) 988-9128 sheehan@uwm.edu, Univ of Wisconsin, Milwaukee – Pie
Sheets, H. David, hsheets@buffalo.edu, SUNY, Buffalo – Gg
Sheldon, Amy L., (585) 245-5988 sheldon@geneseo.edu, SUNY, Geneseo – Hw
Sheldon, Amy L., sheldon@geneseo.edu, SUNY, Buffalo – CgGg
Shell, Karen M., (541) 737-0980 kshell@coas.oregonstate.edu, Oregon State Univ – Oa
Shellito, Lucinda, (970) 351-2491 lucinda.shellito@unco.edu, Univ of Northern Colorado – OaPeOw
Shelton, Dale W., (410) 554-5505 dale.shelton@maryland.gov, Maryland Department of Natural Resources – OeGg
Shelton, Kevin L., (573) 882-1004 SheltonKL@missouri.edu, Univ of Missouri – EmCs
Shelton, Sally Y., (605) 394-2487 Sally.Shelton@sdsmt.edu, South Dakota School of Mines & Tech – Pg
Shem, Linda M., (630) 252-3857 Argonne National Laboratory – Ge
Shen, Jian, (804) 684-7359 shen@vims.edu, Coll of William & Mary – Op
Shen, Po-Yu (Paul), (519) 661-3142 pys@uwo.ca, Western Univ – Yh
Shen, Samuel S., (780) 492-0216 shen@ualberta.ca, Univ of Alberta – Oa
Shen, Xinhua, (319) 273-2536 xinhua.shen@uni.edu, Univ of Northern Iowa – Oaw
Shen, Yang, (401) 874-6848 yshen@gso.uri.edu, Univ of Rhode Island – Ou
Sheng, Jinyu, (902) 494-2718 jinyu.sheng@dal.ca, Dalhousie Univ – Op
Shepard, Michael K., (570) 389-4568 mshepard@bloomu.edu, Bloomsburg Univ – XgYgOr
Shepardson, Daniel P., (765) 494-5284 dshep@purdue.edu, Purdue Univ – Oe
Shepherd, Gordon G., 4167362100 x33221 gordon@yorku.ca, York Univ – OarOn
Shepherd, John G., jgs@noc.soton.ac.uk, Univ of Southampton – Oa
Shepherd, Mark A., (512) 574-4227 mark.shepherd@austincc.edu, Austin Community Coll District – Oa
Shepherd, Stephanie L., 334-844-4926 sls0070@auburn.edu, Auburn Univ – GmOi
Sheppard, Scott S., (202) 478-8854 ssheppard@carnegiescience.edu, Carnegie Institution of Washington – Xg
Shepson, Paul B., (765) 494-7441 pshepson@purdue.edu, Purdue Univ – Oa
Sheridan, Michael F., (716) 645-5345 mfs@geology.buffalo.edu, SUNY, Buffalo – Gv
Sheridan, Robert E., (848) 445-2015 rsheridan@rci.rutgers.edu, Rutgers, The State Univ of New Jersey – YrGu
Sherman-Morris, Kathleen M., (662) 268-1032 x242 kms5@geosci.msstate.edu, Mississippi State Univ – Oa
Sherr, Barry, (541) 231-0301 sherrb@coas.oregonstate.edu, Oregon State Univ – Ob
Sherr, Evelyn, sherre@coas.oregonstate.edu, Oregon State Univ – Ob
Sherrell, Robert M., (848) 932-6555 (Ext. 252) sherrell@ahab.rutgers.edu, Rutgers, The State Univ of New Jersey – CmOu
Sherriff, Barbara L., (204) 474-9786 bl_sherriff@umanitoba.ca, Univ of Manitoba – Gz
Sherrod, Brian L., 253-653-8358 bsherrod@ess.washington.edu, Univ of Washington – YsGt
Sherrod, Laura, (484) 646-4113 sherrod@kutztown.edu, Kutztown Univ of Pennsylvania – YgHw
Shervais, John W., (435) 797-1274 john.shervais@usu.edu, Utah State Univ – Gi
Sherwood, William C., (540) 568-6473 sherwowc@jmu.edu, James Madison Univ – So
Sherwood Lollar, Barbara, 416-978-0770 bslollar@chem.utoronto.ca, Univ of Toronto – Cs
Shevenell, Lisa, (775) 784-1779 lisaas@unr.edu, Univ of Nevada, Reno – Hw
Shevenell, Lisa, (775) 784-6691 lisas@atlasgeoinc.com, Univ of Nevada – Cg
Shew, Roger D., 910-962-7676 shewr@uncw.edu, Univ of North Carolina Wilmington – Eo
Shi, Wei, 919-513-4641 wei_shi@ncsu.edu, North Carolina State Univ – Sb
Shi, Xuan, 479-575-7906 xuanshi@uark.edu, Univ of Arkansas, Fayetteville – Ori
Shiaris, Michael, michael.shiaris@umb.edu, Univ of Massachusetts, Boston – On
Shieh, Sean R., 5196612111 x82467 sshieh@uwo.ca, Western Univ – YxGzOm
Shieh, Yuch-Ning, (765) 494-3272 ynshieh@purdue.edu, Purdue Univ – CsGpv
Shiel, Alyssa, (541) 737-5209 ashiel@coas.oregonstate.edu, Oregon State Univ – Cgs
Shields, Robin, +44 (0) 131 451 8215 robin.shields@hw.ac.uk, Heriot-Watt Univ – Cg
Shields-Zhou, Graham, +44 020 7679 7821 g.shields@ucl.ac.uk, Univ Coll London – Cg
Shiller, Alan M., (228) 688-1178 alan.shiller@usm.edu, Univ of Southern Mississippi – OcCtHs
Shilpakar, Prabin ., 972-883-2408 shilpakar@utdallas.edu, Univ of Texas, Dallas – GcYdGc
Shilts, William W., (217) 333-5111 shilts@illinois.edu, Univ of Illinois, Urbana-Champaign – Gs
Shim-Chim, Richard, (609) 984-6587 rich.shim-chim@dep.state.nj.us, New Jersey Geological and Water Survey – Hw
Shimizu, Melinda, 503-838-9320 shimizum@wou.edu, Western Oregon Univ – OirOg
Shimizu, Nobumichi, (508) 289-2963 Woods Hole Oceanographic Institution – Ca
Shinn, Eugene, (727) 893-3100 Univ of South Florida – Gg
Shinoda, Toshiaki, 361-825-3636 Toshiaki.Shinoda@tamucc.edu, Texas A&M Univ, Corpus Christi – Oa
Shipley, Thomas H., (512) 471-0430 tom@ig.utexas.edu, Univ of Texas at Austin – Yr
Shirer, Hampton N., (814) 863-1992 hns@psu.edu, Pennsylvania State Univ, Univ Park – Ow
Shirey, Steven B., (202) 478-8473 sshirey@carnegiescience.edu, Carnegie Institution of Washington – CcGiz
Shirgaokar, Manish, 780.492.2802 shirgaokar@ualberta.ca, Univ of Alberta – Ou
Shirley, Terry, 704-687-5925 trshirle@uncc.edu, Univ of North Carolina, Charlotte – Oaw
Shock, Everett, 480-965-0631 everett.shock@asu.edu, Arizona State Univ – On
Shoemaker, Kurt A., (740) 351-3395 kshoemaker@shawnee.edu, Shawnee State Univ – GxzGm
Shofner, Gregory A., 410-704-2220 gshofner@towson.edu, Towson Univ – Cg
Sholkovitz, Edward R., (508) 289-2346 esholkovitz@whoi.edu, Woods Hole Oceanographic Institution – Oc
Shorey, Christian V., (303) 273-3556 cshorey@mines.edu, Colorado School of Mines – Gg
Short, Heather, hashort@eps.mcgill.ca, McGill Univ – Gc
Shortt, Niamh K., +44 (0) 131 651 7130 Niamh.Shortt@ed.ac.uk, Edinburgh Univ – Eg
Shoup, Doug, (620) 421-1530 dshoup@ksu.edu, Kansas State Univ – So
Showers, William J., (919) 515-7143 w_showers@ncsu.edu, North Carolina State Univ – Cs
Showman, Adam, (520) 621-4021 showman@lpl.arizona.edu, Univ of Arizona – On
Shreve, Ronald L., (310) 825-5273 shreve@ess.ucla.edu, Univ of California, Los Angeles – Gm
Shreve, Ronald L., shreve@ess.ucla.edu, Univ of Washington – Yr
Shroat-Lewis, Rene A., 501-683-7743 rashroatlew@ualr.edu, Univ of Arkansas at Little Rock – PeOe
Shroba, Cynthia S., 702-651- 7427 cindy.shroba@csn.edu, Coll of Southern Nevada - West Charleston Campus – Gg

Shroder, Jr., John F., (402) 554-2770 jshroder@unomaha.edu, Univ of Nebraska at Omaha – GmlOy
Shropshire, K. Lee, (970) 351-2285 leeshrop@att.net, Univ of Northern Colorado – Pg
Shroyer, Emily L., 541-737-1298 eshroyer@coas.oregonstate.edu, Oregon State Univ – On
Shroyer, Jim, (785) 532-6101 jshroyer@ksu.edu, Kansas State Univ – So
Shu, Jinfu, (202) 478-8963 j.shu@gl.ciw.edu, Carnegie Institution of Washington – Yx
Shuford, James W., (205) 851-5462 jshufard@aamu.edu, Alabama A&M Univ – So
Shugart, Jr., Herman H., (804) 924-7642 hhs@virginia.edu, Univ of Virginia – Og
Shuib, Mustaffa Kamal, 03-79674227 mustaffk@um.edu.my, Univ of Malaya – Gc
Shukla, Jagadish, (703) 993-1983 jshukla@gmu.edu, George Mason Univ – Oa
Shuller-Nickles, Lindsay C., 864656-1448 lshulle@clemson.edu, Clemson Univ – Gz
Shulski, Martha D., mshulski3@unl.edu, Univ of Nebraska, Lincoln – Oa
Shum, CK, 614 292 7118 ckshum@osu.edu, Ohio State Univ – Yd
Shumaker, Robert C., (304) 293-5603 rshumaker@wvu.edu, West Virginia Univ – Go
Shuman, Bryan N., 307-766-6442 bshuman@uwyo.edu, Univ of Wyoming – PeGne
Shuman, Larry M., (404) 228-7276 Univ of Georgia – Sc
Shuman, Randy, (206) 296-8243 rshuman531@gmail.com, Univ of Washington – Ob
Shumway, Matthew J., (801) 422-2707 jms7@byu.edu, Brigham Young Univ – On
Shumway, Sandra, (860) 405-9282 sandra.shumway@uconn.edu, Univ of Connecticut – Ob
Shuster, Robert D., (402) 554-2457 rshuster@unomaha.edu, Univ of Nebraska at Omaha – GiaOe
Shuttleworth, W. James, (520) 621-8787 shuttle@hwr.arizona.edu, Univ of Arizona – Hs
Siahcheshm, Kamal, +98 (411) 339 2699 Univ of Tabriz – Egm
Sibeko, Skhumbuzo, 012 382 6271 sibekosg@tut.ac.za, Tshwane Univ of Tech – GggGg
Sibert, John R., sibert@hawaii.edu, Univ of Hawai'i, Manoa – Ob
Sibley, Paul, (519) 824-4120 Ext.52707 psibley@uoguelph.ca, Univ of Guelph – On
Sibray, Steven S., (308) 632-1382 ssibray@unl.edu, Unversity of Nebraska - Lincoln – HwGg
Sibson, Rick H., +64 3 479-7506 rick.sibson@otago.ac.nz, Univ of Otago – Gc
Sicard, Karri, 907-451-5040 karri.sicard@alaska.gov, Alaska Division of Geological & Geophysical Surveys – Gg
Siddoway, Christine S., (719) 389-6717 csiddoway@coloradoColl.edu, Colorado Coll – Gc
Sidle, Roy, 828-262-2747 sidle.roy@epa.gov, Appalachian State Univ – Hy
Sidor, Christian A., 206-221-3285 casidor@uw.edu, Univ of Washington – Pv
Siegel, Coralie, 3138 7726 c.siegel@qut.edu.au, Queensland Univ of Tech – GiCc
Siegel, Donald I., 315-443-2672 disiegel@syr.edu, Syracuse Univ – Hw
Siegrist, jr, Henry G., 410 827 8095 g.a.siegrist@gmail.com, Univ of Guam – GdzHy
Siemens, Michael A., (573) 368-2134 mike.siemens@dnr.mo.gov, Missouri Dept of Natural Resources – GgHgOi
Siesser, William G., 615-322-2984 william.g.siesser@vanderbilt.edu, Vanderbilt Univ – Pm
Siewers, Fredrick D., (270) 745-5988 fred.siewers@wku.edu, Western Kentucky Univ – Pg
Sigler, Jeffrey M., 504-862-3257 jsigler@tulane.edu, Tulane Univ – GgOwa
Sigler, William V., (419) 530-2897 Univ of Toledo – On
Sigloch, Karin, +44 (1865) 272000 karin.sigloch@earth.ox.ac.uk, Univ of Oxford – YgsGt
Sigman, Daniel M., (609) 258-2194 sigman@princeton.edu, Princeton Univ – Cg
Sigurdsson, Haraldur, (401) 874-6596 haraldur@gso.uri.edu, Univ of Rhode Island – Ou
Sikora, Todd D., 717-872-3292 Todd.Sikora@millersville.edu, Millersville Univ – Oa
Silliman, James, (361) 825-3718 James.Silliman@tamucc.edu, Texas A&M Univ, Corpus Christi – Co
Silva, Lucas C., lcsilva@ucdavis.edu, Univ of California, Davis – Sb
Silva, Michael A., (916) 324-0768 California Geological Survey – Ng
Silva-Castro, Jhon, 859-257-1173 john.silva@uky.edu, Univ of Kentucky – Nm
Silvano, Janet, (617) 627-3494 Tufts Univ – On
Silveira, Maria, 8637351314 x209 mlas@ufl.edu, Univ of Florida – So
Silver, Eli A., (831) 459-2266 esilver@pmc.ucsc.edu, Univ of California, Santa Cruz – Yr
Silver, Leon T., (626) 395-6490 lsilver@gps.caltech.edu, California Inst of Tech – Cc
Silver, Mary W., (831) 459-2908 msilver@cats.ucsc.edu, Univ of California, Santa Cruz – Ob
Silvertooth, Jeffrey C., (520) 621-7145 silver@ag.arizona.edu, Univ of Arizona – So
Silvestri, Alberta, +390498279142 alberta.silvestri@unipd.it, Università degli Studi di Padova – Gz
Simandl, George J., (250) 952-0413 gsimandle@galaxy.gov.bc.ca, Univ of Victoria – En
Simard, Suzanne, suzanne.simard@ubc.ca, Univ of British Columbia – Sb
Simila, Gerald W., (818) 677-3543 gerry.simila@csun.edu, California State Univ, Northridge – Yg
Simjouw, Jean-Paul, 203-932-1253 JSimjouw@newhaven.edu, Univ of New Haven – OcCma
Simmons, Craig, craig.simmons@flinders.edu.au, Flinders Univ – Hw
Simmons, F. William, (217) 333-4424 Univ of Illinois, Urbana-Champaign – Sp
Simmons, Lizanne V., (949) 589-0562 Santiago Canyon Coll – Gr
Simmons, M. G., (617) 253-6393 gene@hager-richter.com, Massachusetts Inst of Tech – Yg
Simmons, Peter, +44 (0)1603 59 3122 p.simmons@uea.ac.uk, Univ of East Anglia – Eg
Simmons, Stuart, 801-581-4122 ssimmons@egi.utah.edu, Univ of Utah – Gz
Simmons, William B., (504) 280-6791 wsimmons@uno.edu, Univ of New Orleans – Gz
Simmons, William B., 7342866325 , 5042806325 wsimmons@uno.edu, Univ of Michigan – Gz
Simms, Alexander, (805) 893-7292 asimms@geol.ucsb.edu, Univ of California, Santa Barbara – GsuGm
Simms, Janet E., (601) 634-3493 Mississippi State Univ – Yg
Simon, Kathy, kasimon@calpoly.edu, California Polytechnic State Univ – On
Simons, Frederik J., (609) 258-2598 fjsimons@princeton.edu, Princeton Univ – YsdYm
Simons, Mark, (626) 395-6984 simons@caltech.edu, California Inst of Tech – Ys
Simonson, Bruce M., (440) 775-8347 bruce.simonson@oberlin.edu, Oberlin Coll – GsXm
Simony, Philip S., (403) 220-6679 Univ of Calgary – Gc
Simovich, Marie, (619) 260-4600 scroom@sandiego.edu, Univ of San Diego – On
Simpkins, William W., (515) 294-7814 bsimp@iastate.edu, Iowa State Univ of Science & Tech – HwGlCs
Simpson, Edward L., (610) 683-4447 simpson@kutztown.edu, Kutztown Univ of Pennsylvania – Gs
Simpson, Frank, 519-253-3000 ext. 2487 franks@uwindsor.ca, Univ of Windsor – Gs
Simpson, Myrna Joyce S., 416 2877234 msimpson@utsc.utoronto.ca, Univ of Toronto – Ge
Simpson, Robert M., (731) 881-7439 msimpson@utm.edu, Univ of Tennessee, Martin – OawOi
Simpson, Jr., H. James, (703) 779-2043 simpsonj@ldeo.columbia.edu, Columbia Univ – Cm
Sims, Albert L., (218) 281-8619 simsx008@umn.edu, Univ of Minnesota, Twin Cities – Sc
Sims, Douglas , 702-651-4840 douglas.sims@csn.edu, Coll of Southern Nevada - West Charleston Campus – GeSp
Sims, Kenneth W., 307-766-3386 ksims7@uwyo.edu, Univ of Wyoming – CgGvCa
Sinclair, Alastair J., (604) 822-3086 asinclair@eos.ubc.ca, Univ of British Columbia – EmGg
Sinclair, Hugh D., +44 (0) 131 650 2518 Hugh.Sinclair@ed.ac.uk, Edinburgh Univ – Os
Singer, Bradley S., (608) 262-6366 bsinger@geology.wisc.edu, Univ of Wisconsin, Madison – Cc
Singer, David M., 330-672-3006 dsinger4@kent.edu, Kent State Univ –

Gze
Singer, Jared W., (518) 276-4095 singej2@rpi.edu,
Rensselaer Polytechnic Inst – CaOm
Singer, Jill K., (716) 878-4724 singerjk@buffalostate.edu,
Buffalo State Coll – GsOp
Singer, Michael, +44 01334 462874 michael.singer@st-andrews.ac.uk,
Univ of St. Andrews – Hg
Singerling, Sheryl, singerlings@si.edu,
Smithsonian Institution / National Museum of Natural History – Xm
Singh, Hanumant, 617.373.7286 ha.singh@neu.edu,
Northeastern Univ – On
Singh, Harbans, (973) 655-7383 singhh@mail.montclair.edu,
Montclair State Univ – On
Singha, Kamini, 303-273-3822 ksingha@mines.edu,
Colorado School of Mines – Hw
Singler, Charles R., 330-941-3611 crsingler@ysu.edu,
Youngstown State Univ – Gs
Singleton, John, (970) 491-0740 john.singleton@colostate.edu,
Colorado State Univ – Gct
Sinha, A. Krishna, (540) 231-5580 pitlab@vt.edu,
Virginia Polytechnic Inst & State Univ – Gt
Sinha, Ashish, (310) 243-3166 asinha@csudh.edu,
California State Univ, Dominguez Hills – PeCs
Sinton, John M., (808) 956-7751 sinton@hawaii.edu,
Univ of Hawai'i, Manoa – Gi
Sintubin, Manuel, manuel.sintubin@ees.kuleuven.be,
Katholieke Universiteit Leuven – GtcGa
Sipos, György, +3662544156 gysipos@geo.u-szeged.hu,
Univesity of Szeged – HqYm
Sirbescu, Mona, 989-774-4497 sirbe1mc@cmich.edu,
Central Michigan Univ – Gi
Sirk, Robert A., (931) 221-7473 sirkr@apsu.edu,
Austin Peay State Univ – GeOyu
Sisson, Virginia, 713-743-7634 vbsisson@uh.edu,
Univ of Houston – Gi
Sisterson, Doug, 630-252-5836 dlsisterson@anl.gov,
Argonne National Laboratory – Oa
Sit, Stefany , 312-413-1868 ssit@uic.edu, Univ of Illinois at Chicago – Ygs
Sitwell, O.F. George, sitwell@telus.net, Univ of Alberta – On
Size, William B., (404) 727-0203 wsize@emory.edu, Emory Univ – Gi
Sjostrom, Derek, (406) 238-7387 derek.sjostrom@rocky.edu,
Rocky Mountain Coll – ClGsHs
Skalbeck, John D., 262-595-2490 skalbeck@uwp.edu,
Univ of Wisconsin, Parkside – HwYgGg
Skarke, Adam, (662) 268-1032 ext 258 adam.skarke@msstate.edu,
Mississippi State Univ – GuYrOn
Skehan, James W., 617-552-8312 james.skehan@bc.edu, Boston Coll – Gc
Skelton, Lawrence H., mskelton5@cox.net, Univ of Kansas – Gg
Skemer, Philip, 314-935-3584 pskemer@levee.wustl.edu,
Washington Univ in St. Louis – GcCp
Skidmore, Mark L., (406) 994-7251 skidmore@montana.edu,
Montana State Univ – ClPyOl
Skiles, S. McKenzie, (801) 863-9096 mckenzie.skiles@uvu.edu,
Utah Valley Univ – HsOri
Skillen, James, 616-526-7546 jrs39@calvin.edu, Calvin Coll – Ou
Skinner, Brian J., (203) 432-3175 brian.skinner@yale.edu,
Yale Univ – CgEmGz
Skinner, H. Catherine W., (203) 432-3787 Yale Univ – Gz
Skinner, Luke C., +44 (0) 1223 764912 luke00@esc.cam.ac.uk,
Univ of Cambridge – GeCmOg
Skinner, Randall, 801-422-6083 randy_skinner@byu.edu,
Brigham Young Univ – Gg
Skippen, George B., gskippen@ccs.carleton.ca, Carleton Univ – Cg
Sklar, Leonard, 415-338-1204 leonard@sfsu.edu,
San Francisco State Univ – Gm
Skoda, Radek, +420 549 49 7392 rskoda@sci.muni.cz,
Masaryk Univ – GzOmGi
Skoog, Annelie, 860-405-9220 annelie.skoog@uconn.edu,
Univ of Connecticut – Oc
Skubis, Steven T., (315) 312-2799 SUNY, Oswego – Ow
Skyllingstad, Eric, 541-737-5697 skylling@coas.oregonstate.edu,
Oregon State Univ – Ow
Slade, Jr., Raymond M., rslade@austincc.edu,
Austin Community Coll District – HgsHq
Sladek, Chris, (775) 784-6970 csladek@unr.edu,
Univ of Nevada, Reno – Gg
Slater, Brian, 518-473-9988 bslater@mail.nysed.gov,
New York State Geological Survey – GoEoGr
Slater, Brian K., (614) 292-5891 slater.39@osu.edu,
Ohio State Univ – Sd
Slater, David, (775) 784-4893 dslater@seismo.unr.edu,
Univ of Nevada, Reno – Ys
Slater, Gregory F., (905) 525-9140 (Ext. 26388) gslater@mcmaster.ca,
McMaster Univ – Cg
Slater, Lee S., 973-353-5109 lslater@andromeda.rutgers.edu,
Rutgers, The State Univ of New Jersey, Newark – Yg
Slater, Tom, +44 (0) 131 650 9506 tom.slater@ed.ac.uk,
Edinburgh Univ – On
Slatkin, Montgomery, (510) 642-6300 Univ of California, Berkeley – Po
Slatt, Roger M., (405) 325-3253 sgg@hoth.gcn.ou.edu,
Univ of Oklahoma – Go
Slattery, Michael C., (817) 257-7506 m.slattery@tcu.edu,
Texas Christian Univ – HgSpOy
Slattery, William, 937 775-3455 william.slattery@wright.edu,
Wright State Univ – OeGr
Slaughter, Richard, 608/262-2399 rich@geology.wisc.edu,
Univ of Wisconsin, Madison – Pg
Slavetskas, Carol, cslavets@binghamton.edu, Binghamton Univ – On
Slawinski, Michael A., (709) 737-7541 mslawins@mun.ca,
Memorial Univ of Newfoundland – Ys
Sledzinski, Grazyna, 585 3232313 msledzin@aol.com,
Wayne State Univ – Gg
Sleep, Norman H., (650) 723-0882 Stanford Univ – Yg
Sleezer, Richard O., rsleezer@emporia.edu, Emporia State Univ – Os
Sletten, Ronald S., 206-543-0571 sletten@uw.edu,
Univ of Washington – OsSb
Slingerland, Rudy L., (814) 865-6892 sling@geosc.psu.edu,
Pennsylvania State Univ, Univ Park – Gs
Sloan, Doris, (510) 527-5710 dsloan@berkeley.edu,
Univ of California, Berkeley – Pme
Sloan, Heather, 718-960-8008 Heather.Sloan@lehman.cuny.edu,
Lehman Coll (CUNY) – Yr
Sloan, Jon R., (818) 677-4880 jon.sloan@csun.edu,
California State Univ, Northridge – Pm
Sloan, Lisa C., (831) 459-3693 lsloan@pmc.ucsc.edu,
Univ of California, Santa Cruz – Pe
Slobodnik, Marek, +420 549 49 7055 marek@sci.muni.cz,
Masaryk Univ – GeEg
Sloss, Craig, 610731382610 c.sloss@qut.edu.au,
Queensland Univ of Tech – GsrGm
Slovinsky, Peter, (207) 287-7173 peter.a.slovinsky@maine.gov,
Dept of Agriculture, Conservation, and Forestry – Gu
Slowey, Niall C., (979) 845-8478 nslowey@ocean.tamu.edu,
Texas A&M Univ – Ou
Smaglik, Suzanne M., 307-855-2146 ssmaglik@cwc.edu,
Central Wyoming Coll – GgOeCg
Small, Christopher, (845) 365-8354 Columbia Univ – Or
Small, Lawrence, smalll@coas.oregonstate.edu, Oregon State Univ – Ob
Smalley, Gabriela W., 609-896-5097 gsmalley@rider.edu,
Rider Univ – ObcOp
Smalley, Glendon W., (615) 598-5714 Sewanee: Univ of the South – Sf
Smalley, Jr., Robert, (901) 678-2007 rsmalley@memphis.edu,
Univ of Memphis – Yd
Smallwood, Shane, 614 265 6595 shane.smallwood@dnr.state.oh.us,
Ohio Dept of Natural Resources – Ec
Smart, Christopher, +44 1752 584764 C.Smart@plymouth.ac.uk,
Univ of Plymouth – Pg
Smart, Katie A., 011 717 6549 katie.smart2@wits.ac.za,
Univ of the Witwatersrand – CsGx
Smay, Jessica J., 4082982181 x3933 jessica.smay@sjcc.edu,
San Jose City Coll – OeGm
Smayda, Theodore J., (401) 874-6171 tsmayda@gso.uri.edu,
Univ of Rhode Island – Ob
Smedes, Harry W., (541) 552-6479 Southern Oregon Univ – Gi
Smedley, Andrew, +44 0161 306-8770 Andrew.Smedley@manchester.ac.uk,
Univ of Manchester – Oa
Smee, Delbert L., 361-825-3637 Lee.Smee@tamucc.edu,
Texas A&M Univ, Corpus Christi – On
Smethie, William, (845) 365-8566 Columbia Univ – Cg
Smilnak, Roberta A., (303) 556-3144 Metropolitan State Coll of Denver – Og
Smirnov, Aleksey K., (906) 487-2365 asmirnov@mtu.edu,
Michigan Technological Univ – Yv

Smirnov, Anna, (418) 654-3711 asmirnov@nrcan.gc.ca, Universite du Quebec – Cs
Smith, Alan L., (909) 537-5409 alsmith@csusb.edu, California State Univ, San Bernardino – GviGz
Smith, Alison J., (330) 672-3709 alisonjs@kent.edu, Kent State Univ – Gn
Smith, Arthur, (610) 436-3335 West Chester Univ – Oe
Smith, Betty E., 217-581-6340 besmith@eiu.edu, Eastern Illinois Univ – Oi
Smith, Brianne, 718-951-5000 x2689 Brianne.Smith43@brooklyn.cuny.edu, Brooklyn Coll (CUNY) – Hs
Smith, C. K., 210-486-0062 csmith55@alamo.edu, San Antonio Community Coll – Oy
Smith, C. Ken, (931) 598-3219 ksmith@sewanee.edu, Sewanee: Univ of the South – Sf
Smith, Catherine H., (505) 667-0113 chsmith@lanl.gov, Los Alamos National Laboratory – Ca
Smith, Charles K., csmith55@alamo.edu, Alamo Colls - San Antonio Coll – Oy
Smith, Craig R., (808) 956-7776 csmith@soest.hawaii.edu, Univ of Hawai'i, Manoa – Ob
Smith, Dan, +440116 252 5355 djs40@le.ac.uk, Leicester Univ – Ge
Smith, David C., (401) 874-6172 dcsmith@gsosun1.gso.uri.edu, Univ of Rhode Island – Ob
Smith, David E., (804) 982-3058 des3e@virginia.edu, Univ of Virginia – Ob
Smith, David R., 410-293-6553 drsmith@usna.edu, United States Naval Academy – Ow
Smith, Deborah K., (508) 289-2472 dsmith@whoi.edu, Woods Hole Oceanographic Institution – Yr
Smith, Dena M., 303-735-2011 dena.smith@colorado.edu, Univ of Colorado – Pg
Smith, Diane R., (210) 999-7656 dsmith@trinity.edu, Trinity Univ – Gi
Smith, Donald R., (831) 459-5041 dsmith@scipp.ucsc.edu, Univ of California, Santa Cruz – On
Smith, Douglas, (512) 471-4261 doug@maestro.geo.utexas.edu, Univ of Texas at Austin – Gi
Smith, Douglas L., (210) 458-5751 Univ of Florida – Yh
Smith, Dwight E., 410-293-6566 United States Naval Academy – Owg
Smith, Elizabeth Y., 702-895-4065 elizabeth.smith@unlv.edu, Univ of Nevada, Las Vegas – OnnOn
Smith, Eugene I., (702) 895-3971 gsmith@ccmail.nevada.edu, Univ of Nevada, Las Vegas – Gi
Smith, Florence P., (630) 252-7980 Argonne National Laboratory – Ca
Smith, Gerald J., stratigrapher@msn.com, SUNY, Buffalo – Gs
Smith, Gerald R., (734) 764-0491 grsmith@umich.edu, Univ of Michigan – Ps
Smith, Gordon W., (765) 494-7681 McGill Univ – Nx
Smith, Grant, 503-838-8862 smithg@wou.edu, Western Oregon Univ – Og
Smith, H D., SmithHD@cf.ac.uk, Cardiff Univ – On
Smith, J. Leslie, (604) 822-4108 lsmith@eos.ubc.ca, Univ of British Columbia – Hw
Smith, James A., (609) 258-4615 jsmith@princeton.edu, Princeton Univ – Hg
Smith, James E., (905) 525-9140 (Ext. 24534) smithja@mcmaster.ca, McMaster Univ – Hw
Smith, Janet S., (717) 477-1757 jssmit@ship.edu, Shippensburg Univ – Oi
Smith, Jason J., 607-778-5116 smithjj@sunybroome.edu, Broome Community Coll – Ggs
Smith, Jeffery L., (509) 335-7648 Washington State Univ – Sb
Smith, Jen R., jensmith@wustl.edu, Washington Univ in St. Louis – Ga
Smith, Jennifer E., (858) 246-0803 smithj@ucsd.edu, Univ of California, San Diego – Obn
Smith, Jennifer R., 314-935-9451 jensmith@levee.wustl.edu, Washington Univ in St. Louis – Ga
Smith, Jim, j.smith@flinders.edu.au, Flinders Univ – On
Smith, Jim, +353 21 4902151 j.smith@ucc.ie, Univ Coll Cork – Pl
Smith, Jim, +44 023 92 842416 jim.smith@port.ac.uk, Univ of Portsmouth – Ge
Smith, Jon, 785-864-2179 jjsmith@kgs.ku.edu, Univ of Kansas – GsPe
Smith, Joseph P., 410-293-6568 jpsmith@usna.edu, United States Naval Academy – CmOg
Smith, K. L., (702) 895-3971 Wichita State Univ – On
Smith, Karen, 630-252-0136 smithk@anl.gov, Argonne National Laboratory – Hw
Smith, Karl A., (805) 756-2262 ksmith@tc.umn.edu, Univ of Minnesota, Twin Cities – Nx
Smith, Kathlyn M., (912) 478-5398 KSmith@GeorgiaSouthern.edu, Georgia Southern Univ – Pv
Smith, Ken D., (775) 784-4218 ken@seismo.unr.edu, Univ of Nevada, Reno – YsGt
Smith, Langhorne, 518-473-6262 lsmith@remove@this.mail.nysed.gov, New York State Geological Survey – Go
Smith, Larry N., (406) 496-4859 lsmith@mtech.edu, Montana Tech of the Univ of Montana – GsoGm
Smith, Laurence C., (310) 825-3154 lsmith@geog.ucla.edu, Univ of California, Los Angeles – Hs
Smith, Matthew C., (352) 392-2106 mcsmith@ufl.edu, Univ of Florida – Gi
Smith, Michael, 205-247-3724 msmith@gsa.state.al.us, Geological Survey of Alabama – Hg
Smith, Michael S., (910) 962-3496 smithms@uncw.edu, Univ of North Carolina Wilmington – GzpGa
Smith, Mike, +44020 8417 2500 Michael.Smith@kingston.ac.uk, Kingston Univ – Or
Smith, Mike, 970-204-8134 Mike.Smith@frontrange.edu, Front Range Community Coll - Larimer – GgOe
Smith, Ned P., (407) 674-8096 Florida Inst of Tech – Op
Smith, Norman D., (402) 472-5362 nsmith3@unl.edu, Univ of Nebraska, Lincoln – Gsm
Smith, Paul H., (505) 667-7494 Los Alamos National Laboratory – On
Smith, Paul L., (604) 822-6456 psmith@eos.ubc.ca, Univ of British Columbia – Pg
Smith, Paul L., 605-394-1990 Paul.Smith@sdsmt.edu, South Dakota School of Mines & Tech – Oa
Smith, Peter, (520) 621-2725 psmith@lpl.arizona.edu, Univ of Arizona – Or
Smith, Peter C., (902) 426-3474 smithp@mar.dfo-mpo.gc.ca, Dalhousie Univ – Op
Smith, Peter J., peterjsmith@shaw.ca, Univ of Alberta – On
Smith, Phillip J., (765) 494-3286 pjsmith@purdue.edu, Purdue Univ – Oa
Smith, Richard S., 7056751151 x2364 rsmith@laurentian.ca, Laurentian Univ, Sudbury – Ye
Smith, Robert B., (801) 581-7129 robert.b.smith@utah.edu, Univ of Utah – Ys
Smith, Robert L., (541) 752-6597 rsmith@coas.oregonstate.edu, Oregon State Univ – OpgOn
Smith, Ronald B., (203) 432-3129 ronald.smith@yale.edu, Yale Univ – Oa
Smith, Ronald M., (509) 376-5831 rmsmith@pnl.gov, Pacific Northwest National Laboratory – Hg
Smith, Russell, 915-831-8875 RussellS@epcc.edu, El Paso Community Coll – Gg
Smith, Sean, 207-581-2198 sean.m.smith@maine.edu, Univ of Maine – Hs
Smith, Shane V., 330-941-1752 svsmith@ysu.edu, Youngstown State Univ – Gs
Smith, Stephen V., svsmith@hawaii.edu, Univ of Hawai'i, Manoa – Ou
Smith, Steve, +44 1784 443635 Steve.Smith-1@rhul.ac.uk, Univ of London, Royal Holloway & Bedford New Coll – Ge
Smith, Steven C., 916-435-9180 smithsc@arc.losrios.edu, American River Coll – OynOn
Smith, Steven J., 319-273-6495 smithsj@uni.edu, Univ of Northern Iowa – Og
Smith, Susan M., 406-496-4173 ssmith@mtech.edu, Montana Tech of The Univ of Montana – On
Smith, Terence E., tsmith@uwindsor.ca, Univ of Windsor – Gi
Smith, Terry L., (805) 756-2262 tsmith@calpoly.edu, California Polytechnic State Univ – So
Smith, Thomas M., (804) 924-3107 tms9a@virginia.edu, Univ of Virginia – Og
Smith, William H., (314) 935-5638 whsmith@dasi.wustl.edu, Washington Univ in St. Louis – Xc
Smith-Engle, Jennifer M., (361) 825-2436 Jennifer.Smith-Engle@tamucc.edu, Texas A&M Univ, Corpus Christi – Gs
Smith-Konter, Bridget R., 808-956-3618 brkonter@hawaii.edu, Univ of Hawai'i, Manoa – YdXy
Smithson, Jayne, 5102357800 x.4284 jsmithson@contracosta.edu, Contra Costa Coll – Gg
Smithyman, Brenden, bsmithym@uwo.ca, Western Univ – Ye
Smosna, Richard A., (304) 293-5603 rsmosna@wvu.edu, West Virginia Univ – GrEo
Smrekar, Suzanne E., (818) 354-4192 Jet Propulsion Laboratory – Xy
Smukler, Sean, sean.smukler@ubc.ca, Univ of British Columbia – Os
Smylie, Douglas E., (416) 736-2100 #66438 doug@core.yorku.ca,

York Univ – Yg
Smyth, Joseph R., 303-492-5521 joseph.smyth@colorado.edu, Univ of Colorado – Gz
Smyth, Rebecca C., (512) 471-0232 rebecca.smyth@beg.utexas.edu, Univ of Texas at Austin, Jackson School of Geosciences – Hg
Smyth, Thomas J., (919) 515-2838 North Carolina State Univ – Sc
Smyth, William D., (541) 737-3029 smyth@coas.oregonstate.edu, Oregon State Univ – Op
Smythe, William, (310) 825-2434 Univ of California, Los Angeles – Ys
Snead, John I., (225) 578-3454 snead@lsu.edu, Louisiana State Univ – Gm
Snider, Henry I., (860) 228-9815 snider@easternct.edu, Eastern Connecticut State Univ – Ge
Snieder, Roel, 303-273-3456 rsnieder@mines.edu, Colorado School of Mines – Ye
Snodgrass, Eric R., (217) 333-3537 snodgrss@illinois.edu, Univ of Illinois, Urbana-Champaign – Oa
Snoeckx, Hilde, (850) 474-3377 snoeckx@uwf.edu, Univ of West Florida – Oug
Snoke, J. Arthur, (540) 231-6028 snoke@vt.edu, Virginia Polytechnic Inst & State Univ – Ys
Snow, Eleanour, (813) 974-0319 snow@usf.edu, Univ of South Florida, Tampa – Gz
Snow, Jonathan, 713-743-3298 jesnow@uh.edu, Univ of Houston – Ca
Snow, Julie A., (724) 738-2503 julie.snow@sru.edu, Slippery Rock Univ – Oa
Snyder, Daniel, 478-274-7806 dsnyder@mgc.edu, Middle Georgia Coll – Gg
Snyder, Jeffrey A., (419) 372-0533 jasnyd@bgnet.bgsu.edu, Bowling Green State Univ – Gm
Snyder, Jennifer L., (610) 359-5291 jsnyder2@dccc.edu, Delaware County Community Coll – Og
Snyder, Lori D., (715) 836-5086 snyderld@uwec.edu, Univ of Wisconsin, Eau Claire – Gx
Snyder, Noah, 617-552-0839 snyderno@bc.edu, Boston Coll – GgHg
Snyder, Peter K., 612-625-8207 pksnyder@umn.edu, Univ of Minnesota, Twin Cities – Oa
Snyder, Richard L., rlsnyder@ucdavis.edu, Univ of California, Davis – Oa
Snyder, Walter S., (208) 426-3645 wsnyder@boisestate.edu, Boise State Univ – GrcGt
Sobel, Adam H., (212) 854-6587 ahs129@columbia.edu, Columbia Univ – Oa
Sobotka, Jerzy, jerzy.sobotka@uwr.edu.pl, Univ of Wroclaw – Yg
Sohi, Sran P., +44 (0) 131 651 4471 Saran.Sohi@ed.ac.uk, Edinburgh Univ – Os
Sohn, Robert A., (508) 289-3616 rsohn@whoi.edu, Woods Hole Oceanographic Institution – Yr
Soja, Constance M., (315) 228-7200 csoja@colgate.edu, Colgate Univ – Pi
Sokolik, Irina, (404) 894-6180 isokolik@eas.gatech.edu, Georgia Inst of Tech – Oar
Sokolova, Elena, (204) 747-8252 elena_sokolova@umanitoba.ca, Univ of Manitoba – Gz
Solan, Martin, +44 (0)23 80593755 M.Solan@soton.ac.uk, Univ of Southampton – Ob
Solana, Camen, +44 023 92 842394 carmen.solana@port.ac.uk, Univ of Portsmouth – Gv
Solar, Gary S., (716) 878-4900 solargs@buffalostate.edu, Buffalo State Coll – GcpGt
Soldat, Douglas J., (608) 263-3631 djsoldat@wisc.edu, Univ of Wisconsin, Madison – So
Solecki, William, (973) 655-5129 soleckiw@mail.montclair.edu, Montclair State Univ – Ou
Solecki, William, 212-7724536 wsolecki@hunter.cuny.edu, Hunter Coll (CUNY) – Ou
Solferino, Giulio, +44 1784 443585 giulio.solferino@rhul.ac.uk, Univ of London, Royal Holloway & Bedford New Coll – Gg
Solis, Michael P., 614 265 6597 michael.solis@dnr.state.oh.us, Ohio Dept of Natural Resources – GgcGt
Soll, Wendy E., (505) 665-6930 weasel@lanl.gov, Los Alamos National Laboratory – Hw
Solomatov, Viatcheslav S., 314-935-7882 slava@wustl.edu, Washington Univ in St. Louis – Yg
Solomon, Douglas K., (801) 581-7231 kip.solomon@utah.edu, Univ of Utah – Hw
Solomon, Keith, (519) 824-4120 Ext 58792 ksolomon@uoguelph.ca, Univ of Guelph – On
Solomon, Sean , 845-365-8714 solomon@ldeo.columbia.edu, Columbia Univ – Xy
Solomon, Sean C., ssolomon@carnegiescience.edu, Carnegie Institution of Washington – YgsGt
Somarin, Alireza, 204-727-9680 somarina@brandonu.ca, Brandon Univ – Gig
Somasundaran, Ponisseril, (212) 854-2926 ps24@columbia.edu, Columbia Univ – Nx
Somayazulu, Maddury, (202) 478-8911 zulu@gl.ciw.edu, Carnegie Institution of Washington – Yx
Somenhally, Anil, ASomenahally@tamu.edu, Texas A&M Univ – Sb
Somers, Jr., Arnold E., (229) 333-5664 gsomers@valdosta.edu, Valdosta State Univ – On
Somerville, James M., +44 (0) 131 451 3162 j.m.somerville@hw.ac.uk, Heriot-Watt Univ – Np
Somerville, Richard C., (858) 534-4644 rsomerville@ucsd.edu, Univ of California, San Diego – Oa
Sommer, Ulrich , usommer@geomar.de, Dalhousie Univ – Ob
Sommerfield, Christopher K., 302-645-4255 cs@udel.edu, Univ of Delaware – On
Sonder, Leslie J., 603-646-2372 Leslie.Sonder@dartmouth.edu, Dartmouth Coll – Gq
Sondergeld, Carl H., (405) 325-6870 csondergeld@ou.edu, Univ of Oklahoma – NpYex
Sone, Hiroki, (608) 890-0531 hsone@wisc.edu, Univ of Wisconsin, Madison – NrYx
Sonett, Charles P., (520) 621-6935 sonett@dakotacom.net, Univ of Arizona – Yg
Song, Alex, +44 02076973158 alex.song@ucl.ac.uk, Univ Coll London – Ys
Song, Xiaodong, (217) 333-1841 xsong@illinois.edu, Univ of Illinois, Urbana-Champaign – Ys
Sonnenberg, Stephen A., 303-384-2182 ssonnenb@mines.edu, Colorado School of Mines – Go
Sonntag, Mark S., 325-942-2136 mark.sonntag@angelo.edu, Angelo State Univ – Xg
Sonzogni, William C., (608) 262-8062 Univ of Wisconsin, Madison – Oc
Soreghan, Gerilyn S., (405) 325-4482 lsoreg@ou.edu, Univ of Oklahoma – Gs
Soreghan, Michael J., (405) 325-3393 msoreg@ou.edu, Univ of Oklahoma – Gs
Sorensen, Sorena, sorena@volcano.si.edu, Univ of Maryland – Gp
Sorensen, Sorena S., (202) 633-1820 sorensens@si.edu, Smithsonian Institution / National Museum of Natural History – Gp
Sorenson, Mary Clare, 715-346-2629 msorenso@uwsp.edu, Univ of Wisconsin, Stevens Point – On
Sorkhabi, Rasoul, 801-581-9070 rsorkhabi@egi.utah.edu, Univ of Utah – GctGc
Sorlien, Christopher C., 845-359-7631 chris@crustal.ucsb.edu, Univ of California, Santa Barbara – Yr
Sorooshian, Soroosh, (949) 824-8825 soroosh@uci.edu, Univ of California, Irvine – Hg
Sorooshian, Soroosh, soroosh@uci.edu, Univ of Arizona – Hg
Soster, Frederick M., 765-658-4670 fsoster@depauw.edu, DePauw Univ – Gs
Sottile, Jr., Joseph, 859-257-4616 joe.sottile@uky.edu, Univ of Kentucky – Nm
Souch, Catherine J., (317) 274-1103 Indiana Univ, Indianapolis – Gm
Soule, Melissa, (508) 289-2879 msoule@whoi.edu, Woods Hole Oceanographic Institution – Oc
Soule, Peter T., (828) 262-7056 soulept@appstate.edu, Appalachian State Univ – Oa
Soule, S. Adam, (508) 289-3213 ssoule@whoi.edu, Woods Hole Oceanographic Institution – Gvu
Sousa, Luís M., lsousa@utad.pt, Universidade de Trás-os-Montes e Alto Douro – EnGgNg
Sousa, Wayne P., (510) 642-2435 Univ of California, Berkeley – Ob
Southam, Gordon, (519) 661-3197 gsoutham@uwo.ca, Western Univ – Py
Southam, John R., (305) 284-1898 jsoutham@miami.edu, Univ of Miami – Op
Southard, John B., (617) 253-3397 southard@mit.edu, Massachusetts Inst of Tech – Gs
Southard, Randal J., (530) 752-2199 rjsouthard@ucdavis.edu, Univ of California, Davis – Os
Southon, John, (949) 824-3674 jsouthon@uci.edu, Univ of California, Irvine – CcOcPe
Southwell, Benjamin, (906) 635-2076 bsouthwell@lssu.edu,

Lake Superior State Univ – On
Sowers, Todd, (814) 863-8093 sowers@geosc.psu.edu, Pennsylvania State Univ, Univ Park – Cs
Spaeth, Matthew P., 217-265-6578 spaeth@illinois.edu, Univ of Illinois, Urbana-Champaign – Ge
Spahr, Paul, 614 265 6577 paul.spahr@dnr.state.oh.us, Ohio Dept of Natural Resources – GeHwOi
Spalding, Brian P., bps@ornl.gov, Oak Ridge National Laboratory – Sc
Spane, Frank A., (509) 376-8329 frank.spane@pnl.gov, Pacific Northwest National Laboratory – Hg
Spangler, Daniel P., (352) 392-2106 Univ of Florida – Hw
Spanos, T.J.T (Tim), (780) 435-5245 tim@phys.ualberta.ca, Univ of Alberta – YgsEo
Sparks, David, (979) 458-1051 david-w-sparks@geos.tamu.edu, Texas A&M Univ – YdGq
Sparks, Thomas N., 859-323-0552 sparks@uky.edu, Univ of Kentucky – Gg
Spayd, Steven E., (609) 984-6587 steve.spayd@dep.state.nj.us, New Jersey Geological and Water Survey – HwGb
Spaziani, Amy, 740 5487348 amy.spaziani@dnr.state.oh.us, Ohio Dept of Natural Resources – Ou
Spear, Frank S., (518) 276-6103 spearf@rpi.edu, Rensselaer Polytechnic Inst – GpCpGc
Spears, Ellen, espears@emory.edu, Univ of Alabama – On
Speck, Angela, (573) 882-8371 speckan@missouri.edu, Univ of Missouri – Xc
Specker, Jennifer, (401) 874-6858 jspecker@gsosun1.gso.uri.edu, Univ of Rhode Island – Ob
Speece, Marvin A., (406) 496-4188 mspeece@mtech.edu, Montana Tech – Ye
Speed, Don, (602) 285-7244 d.speed@phoenixColl.edu, Phoenix Coll – Gg
Speer, James, (812) 237-2257 jim.speer@indstate.edu, Indiana State Univ – PeOyGe
Speidel, David H., (718) 997-3323 Graduate School of the City Univ of New York – Cg
Speijer, Robert, robert.speijer@ees.kuleuven.be, Katholieke Universiteit Leuven – Pym
Spell, Terry L., (702) 895-1171 terry.spell@unlv.edu, Univ of Nevada, Las Vegas – CcGvi
Spence, George D., (250) 721-6187 gspence@uvic.ca, Univ of Victoria – Ys
Spencer, Edgar W., (540) 458-8866 spencere@wlu.edu, Washington & Lee Univ – GcOge
Spencer, Jeremy M., 330 972-2394 jspencer@uakron.edu, Univ of Akron – Oya
Spencer, Joel Q., (785) 532-2249 joelspen@ksu.edu, Kansas State Univ – CcGs
Spencer, Larry T., (804) 683-5189 lts@oz.plymouth.edu, Plymouth State Univ – Gg
Spencer, Mary R., 859-257-8359 mary.spencer@uky.edu, Univ of Kentucky – On
Spencer, Matt, (906) 635-2085 mspencer@lssu.edu, Lake Superior State Univ – Gl
Spencer, Matthew, +44 0151 795 4399 M.Spencer@liverpool.ac.uk, Univ of Liverpool – Og
Spencer, Patrick K., (509) 527-5222 spencerp@whitman.edu, Whitman Coll – Pg
Spencer, Ronald J., (403) 220-6447 spencer@geo.ucalgary.ca, Univ of Calgary – Cg
Spera, Frank J., (805) 893-4880 Univ of California, Santa Barbara – Cp
Sperazza, Michael, 631-632-1687 michael.sperazza@sunysb.edu, SUNY, Stony Brook – On
Spero, Howard J., (530) 752-3307 hjspero@ucdavis.edu, Univ of California, Davis – PeCs
Spetzler, Hartmut A., 303-492-6715 spetzler@colorado.edu, Univ of Colorado – Ys
Spiegelman, Marc, (845) 365-8425 Columbia Univ – Ys
Spieler, Oliver, 089/2180 4221 spieler@min.uni-muenchen.de, Ludwig-Maximilians-Universitaet Muenchen – Gz
Spiess, Richard, +390498279150 richard.spiess@unipd.it, Università degli Studi di Padova – Gp
Spigel, Lindsay, (207) 287-7177 lindsay.spigel@maine.gov, Dept of Agriculture, Conservation, and Forestry – Gm
Spikes, Kyle T., 512-471-7674 kyle.spikes@jsg.utexas.edu, Univ of Texas at Austin – Ye
Spilde, Michael N., (505) 277-5430 mspilde@unm.edu, Univ of New Mexico – Gz
Spilker, Linda J., (818) 354-1647 linda.j.spilker@jpl.nasa.gov, Jet Propulsion Laboratory – Xy
Spindel, Robert C., (206) 543-1310 spindel@apl.washington.edu, Univ of Washington – Op
Spinelli, Glenn, 575.835.6512 spinelli@nmt.edu, New Mexico Inst of Mining and Tech – Hw
Spinler, Joshua C., (501) 569-3544 jxspinler@ualr.edu, Univ of Arkansas at Little Rock – YdGt
Spinosa, Claude, (208) 426-5905 cspinosa@boisestate.edu, Boise State Univ – Pi
Spitz, Yvette H., 541-737-3227 yspitz@coas.oregonstate.edu, Oregon State Univ – Yr
Spivak, Amanda C., (508) 289-4847 aspivak@whoi.edu, Woods Hole Oceanographic Institution – CmoOn
Spokas, Kurt, 612-626-2834 kurt.spokas@ars.usda.gov, Univ of Minnesota, Twin Cities – Spo
Spongberg, Alison L., (419) 530-4091 alison.spongberg@utoledo.edu, Univ of Toledo – Co
Spooner, Alecia, 425-388-9003 aspooner@everettcc.edu, Everett Community Coll – Og
Spooner, Edward T. C., (416) 978-3280 etcs@geology.utoronto.ca, Univ of Toronto – En
Spooner, Ian S., (902) 585-1312 ian.spooner@acadiau.ca, Acadia Univ – Ge
Spotila, James A., (540) 231-2109 spotila@vt.edu, Virginia Polytechnic Inst & State Univ – Gt
Spracklen, Dominick, +44(0) 113 34 37488 d.v.spracklen@leeds.ac.uk, Univ of Leeds – On
Spratt, Deborah A., (403) 220-6446 spratt@geo.ucalgary.ca, Univ of Calgary – Gc
Spray, John G., (506) 453-3550 jgs@unb.ca, Univ of New Brunswick – Gp
Spreng, Alfred C., (573) 341-4669 aspreng@umr.edu, Missouri Univ of Science and Tech – Gr
Sprenke, Kenneth F., (208) 885-5791 ksprenke@uidaho.edu, Univ of Idaho – Ye
Springer, Abraham E., (928) 523-7198 abe.springer@nau.edu, Northern Arizona Univ – Hw
Springer, Dale A., (570) 389-4747 dspringe@bloomu.edu, Bloomsburg Univ – PivOe
Springer, Everett P., (505) 667-0569 everetts@lanl.gov, Los Alamos National Laboratory – Hq
Springer, Gregory S., (740) 593-9431 springeg@ohio.edu, Ohio Univ – Gm
Springer, Kathleen B., (909) 307-2669 242 kspringer@sbcm.sbcounty.gov, San Bernardino County Museum – Og
Springer, Robert K., springer@brandonu.ca, Brandon Univ – Gi
Springston, George E., (802) 485-2734 gsprings@norwich.edu, Norwich Univ – GmOiYg
Sprinkel, Douglas A., (801) 391-1977 douglassprinkel@utah.gov, Utah Geological Survey – GosGg
Sprinkle, James T., (512) 471-4264 echino@jsg.utexas.edu, Univ of Texas at Austin – PioGr
Spruill, Richard K., (252) 328-4399 spruillr@ecu.edu, East Carolina Univ – Hw
Spry, Paul G., (515) 294-9637 pgspry@iastate.edu, Iowa State Univ of Science & Tech – Em
Spurr, Aaron, (319) 273-3789 aaron.spurr@uni.edu, Univ of Northern Iowa – Oeg
Squelch, Andrew P., +61 8 6436 8725 A.Squelch@curtin.edu.au, Curtin Univ – NmYe
Squires, Richard L., (818) 677-2514 richard.squires@csun.edu, California State Univ, Northridge – Pi
Squyres, Steven W., (607) 255-3508 sws6@cornell.edu, Cornell Univ – Xg
Sremac, Jasenka, +38514606108 jsremac@yahoo.com, Univ of Zagreb – PesPb
Srimal, Neptune, (305) 919-5969 srimal@fiu.edu, Florida International Univ – Gt
Srinivasan, Balakrishnan, +914132655008 sbala.esc@pondiuni.edu.in, Pondicherry Univ – CcGiz
Srinivasan, Gopalan, (416) 946-0278 srini@geology.utoronto.ca, Univ of Toronto – Ow
Sritharan, Subramania I., (937) 376-6275 sri@centralstate.edu, Central State Univ – Hg
SRIVASTAVA, HARI B., +919415353606 hbsrivastava@gmail.com, Banaras Hindu Univ – GctGp
Srivastava, Ramesh C., (312) 502-7139 srivast@geosci.uchicago.edu, Univ of Chicago – OarOw
Sriver, Ryan, 217-300-0364 rsriver@illinois.edu,

Univ of Illinois, Urbana-Champaign – Oa
Srnka, Len, (858) 822-1510 lsrnka@ucsd.edu, Univ of California, San Diego – Ym
Srogi, LeeAnn, (610) 436-2721 esrogi@wcupa.edu, West Chester Univ – Gp
St, Achim, (979) 862-4170 astoessel@ocean.tamu.edu, Texas A&M Univ – OpwOl
St. Amour, Natalie , nstamour@uwo.ca, Western Univ – Cs
St. Jean, Joseph, (919) 966-4516 jstjean@email.unc.edu, Univ of North Carolina, Chapel Hill – Pm
St. John, James C., (404) 894-1754 jim.stjohn@eas.gatech.edu, Georgia Inst of Tech – Oa
St. John, Kristen E., 540-568-6675 stjohnke@jmu.edu, James Madison Univ – Ou
Stachel, Thomas, (780) 492-0865 thomas.stachel@ualberta.ca, Univ of Alberta – GiCs
Stachnik, Joshua, 610-758-2581 jcs612@lehigh.edu, Lehigh Univ – Ys
Stack, Andrew, (404) 894-3895 stackag@ornl.gov, Georgia Inst of Tech – Cg
Stadnyk, Leona, (403) 440-6165 lstadnyk@mtroyal.ca, Mount Royal Univ – On
Stafford, C. Russell, (812) 237-3989 Russell.Stafford@indstate.edu, Indiana State Univ – Gam
Stafford, Emily, (828) 227-7367 esstafford@wcu.edu, Western Carolina Univ – PgGg
Stafford, James, (307) 766-2286 x252 james.stafford@wyo.gov, Wyoming State Geological Survey – HgyEg
Stafford, Kevin W., (936) 468-2429 staffordk@sfasu.edu, Stephen F. Austin State Univ – Hw
Stahl, Terry L., (202) 478-8870 stahl@dtm.ciw.edu, Carnegie Institution of Washington – On
Stahle, David W., (479) 575-3703 dstahle@uark.edu, Univ of Arkansas, Fayetteville – GegGg
Stahlman, Phillip, (785) 625-3425 stahlman@ksu.edu, Kansas State Univ – So
Stahmann, Paul, (815) 479-7593 pstahmann@mchenry.edu, McHenry County Coll – Oyg
Stakes, Debra, (805) 546-3100 dstakes@cuesta.edu, Cuesta Coll – Gg
Staley, Amie, (612) 626-4819 astaley@umn.edu, Univ of Minnesota – Gl
Staley, Andrew, (410) 260-8818 andrewstaley@maryland.gov, Maryland Department of Natural Resources – Hw
Stamm, Alfred J., (315) 312-2806 stamm@oswego.edu, SUNY, Oswego – Ow
Stampone, Mary D., 603-862-3136 mary.stampone@unh.edu, Univ of New Hampshire – On
Stan, Cristiana, (703) 993-5391 cstan@gmu.edu, George Mason Univ – Oa
Stan, Oana, 0040232201467 cristina.stan@uaic.ro, Al. I. Cuza Univ of Iasi – Ge
Standridge, Debbie, Deborah.Standridge@gsw.edu, Georgia Southwestern State Univ – On
Stanford, Loudon R., 208-885-7479 stanford@uidaho.edu, Univ of Idaho – GlmGg
Stanford, Scott D., (609) 292-2576 scott.stanford@dep.nj.gov, New Jersey Geological and Water Survey – Gml
Stanley, Clifford R., (902) 585-1344 cliff.stanley@acadiau.ca, Acadia Univ – Cg
Stanley, Daniel J., (202) 633-1354 Smithsonian Institution / National Museum of Natural History – Ou
Stanley, George R., (210) 486-0045 gstanley@alamo.edu, San Antonio Community Coll – Og
Stanley, Thomas M., (405) 325-7281 tmstanley@ou.edu, Univ of Oklahoma – PiGsc
Stanley, Val L., (608) 263-4004 val.stanley@wgnhs.uwex.edu, Univ of Wisconsin - Extension – GmpOi
Stanley, Jr., George D., (406) 243-5693 george.stanley@umontana.edu, Univ of Montana – Pi
Stansell, Nathan D., 815-753-1943 nstansell@niu.edu, Northern Illinois Univ – Gl
Stanton, Kathryn, 916-558-2343 stantok@scc.losrios.edu, Sacramento City Coll – GgPg
Stanton, Kelsay, kstanton@wvc.edu, Wenatchee Valley Coll – GgOe
Stanton, Stephen, (573) 882-4860 Univ of Missouri – On
Stanton, Timothy P., (831) 656-3144 stanton@nps.edu, Naval Postgraduate School – Op
Stapleton, Michael G., (724) 738-2495 micheal.stapleton@sru.edu, Slippery Rock Univ – So
Starek, Michael , 361-825-3978 Michael.Starek@tamucc.edu, Texas A&M Univ, Corpus Christi – Oi
Stark, Colin, (845) 365-8742 Columbia Univ – Gg
Stark, Robert, 089/2180 4329 stark@lrz.uni-muenchen.de, Ludwig-Maximilians-Universitaet Muenchen – Gz
Starr, Richard, 831-771-4400 starr@mlml.calstate.edu, Moss Landing Marine Laboratories – On
Starrfield, Sumner, (480) 965-7569 sumner.starrfield@asu.edu, Arizona State Univ – YhsYv
Starrs, Paul F., (775) 784-6930 starrs@unr.edu, Univ of Nevada, Reno – Oy
Stasko, Stanislaw, stanislaw.stasko@uwr.edu.pl, Univ of Wroclaw – Hg
Staten, Paul W., (812) 856-5135 pwstaten@indiana.edu, Indiana Univ, Bloomington – Oa
Statham, Peter, +44 (0)23 80592679 Univ of Southampton – Ob
Staub, James R., (406) 243-4953 james.staub@umontana.edu, Univ of Montana – GsrEo
Stauffer, David R., (814) 863-3932 stauffer@essc.psu.edu, Pennsylvania State Univ, Univ Park – Ow
Stauffer, Mel R., (306) 966-5708 mel.stauffer@usask.ca, Univ of Saskatchewan – Gc
Stead, Douglas, (604) 291-5387 Simon Fraser Univ – Ng
Steadman, David W., (352) 273-1969 dws@flmnh.ufl.edu, Univ of Florida – PvGaPs
Steadman, Todd A., 864-656-2536 tsteadm@clemson.edu, Clemson Univ – OnnOn
Stearley, Ralph F., (616) 526-6370 rstearle@calvin.edu, Calvin Coll – PvOgGh
Stearman, Will, +61 7 3138 4165 w.stearman@qut.edu.au, Queensland Univ of Tech – Geg
Stearns, David W., (405) 325-3253 mtdstearns@aol.com, Univ of Oklahoma – Gc
Stearns, Leigh, (785) 864-4202 stearns@ku.edu, Univ of Kansas – Ol
Stearns, Richard G., 615-322-2976 stearnrg@ctrvax.vanderbilt.edu, Vanderbilt Univ – Yg
Steart, David, +61 3 9479 5641 D.Steart@latrobe.edu.au, La Trobe Univ – GgPb
Stebbins, Jonathan F., (650) 723-1140 stebbins@pangea.stanford.edu, Stanford Univ – Cg
Stechmann, Samuel, 608-263-4351 stechmann@wisc.edu, Univ of Wisconsin, Madison – Ona
Steck, Lee, (505) 665-3528 Los Alamos National Laboratory – Gg
Steckler, Michael, (845) 365-8479 Columbia Univ – Yv
Steel, Ronald J., (512) 471-0954 rsteel@mail.utexas.edu, Univ of Texas at Austin – Gs
Steele, Andrew, 202-478-8974 asteele@carnegiescience.edu, Carnegie Institution of Washington – Xy
Steele, Ken, 479-575-6790 ksteele@uark.edu, Univ of Arkansas, Fayetteville – HgCgGg
Steele, Kenneth F., (501) 575-4403 ksteele@comp.uark.edu, Univ of Arkansas, Fayetteville – Cg
Steele, Michael A., (570) 408-4763 michael.steele@wilkes.edu, Wilkes Univ – Py
Steele-MacInnes, Matthew, 520-621-1385 steelemacinnis@email.arizona.edu, Univ of Arizona – Eg
Steen, Andrew, 865-974-0821 asteen1@utk.edu, Univ of Tennessee, Knoxville – CoOb
Steenberg, Julia, (612) 626-1830 and01006@umn.edu, Univ of Minnesota – Gs
Steenburgh, Jim, 801-581-8727 jim.steenburgh@utah.edu, Univ of Utah – Oa
Steenhuis, Tammo S., (607) 255-2489 tss1@cornell.edu, Cornell Univ – HgOs
Steeples, Don W., (785) 737-3399 don@ku.edu, Univ of Kansas – YseGe
Steer, David N., 330 972-2099 steer@uakron.edu, Univ of Akron – Yse
Stefani, Cristina, +390498279195 cristina.stefani@unipd.it, Università degli Studi di Padova – Gd
Stefano, Christopher J., 906-487-3028 cjstefano@mtu.edu, Michigan Technological Univ – GzxCg
Stefanova, Ivanka, 651-646-0665 stefa014@umn.edu, Univ of Minnesota, Twin Cities – Pl
Steffens, Katja, 089/2180 6533 katja.steffens@lmu.de, Ludwig-Maximilians-Universitaet Muenchen – Gg
Steger, John M., 305-237-2609 jsteger@mdc.edu, Miami Dade Coll (Kendall Campus) – OgwOg
Stegman, Dave, (858) 822-0767 dstegman@ucsd.edu, Univ of California, San Diego – Yd
Steidl, Gregg M., (609) 984-6587 gregg.steidl@dep.state.nj.us,

New Jersey Geological and Water Survey – Ge
Steidl, Jamison H., (805) 893-4905 steidl@eri.ucsb.edu, Univ of California, Santa Barbara – YsNeYg
Steig, Eric J., (206) 685-3715 steig@uw.edu, Univ of Washington – CsOla
Steiger, Scott, (315) 312-2802 steiger@oswego.edu, SUNY, Oswego – Ow
Stein, Carol A., (312) 996-9349 cstein@uic.edu, Univ of Illinois at Chicago – Yg
Stein, Seth, (847) 491-5265 seth@earth.northwestern.edu, Northwestern Univ – YsdYg
Steinacker, Reinhold, 0043 1 4277 53730 reinhold.steinacker@univie.ac.at, Univ of Vienna – Owa
Steinberg, Roger T., (361) 698-1665 rsteinb@delmar.edu, Del Mar Coll – GgoPg
Steinberger, Julia, +44 (0)785 607 9625 J.K.Steinberger@leeds.ac.uk, Univ of Leeds – Ou
Steinen, Randolph P., (860) 933-2590 randolph.steinen@ct.gov, Dept of Energy and Environmental Protection – Gs
Steiner, Jeffrey, (212) 650-6465 City Coll (CUNY) – Cp
Steinker, Don C., (419) 372-7200 Bowling Green State Univ – Po
Steinman, Byron A., 218-726-7435 bsteinma@d.umn.edu, Univ of Minnesota, Duluth – Gn
Steinmetz, John C., (812) 855-5067 jsteinm@indiana.edu, Indiana Univ – Pm
Stelck, Charles R., (780) 492-3085 Univ of Alberta – Ps
Steller, Diana, 831-771-4400 dsteller@mlml.calstate.edu, Moss Landing Marine Laboratories – On
Stelling, Pete, 360-650-4095 pete.stelling@wwu.edu, Western Washington Univ – GxEgGv
Steltenpohl, Mark G., (334) 844-4893 steltmg@auburn.edu, Auburn Univ – Gtc
Stensrud, David J., (814) 863-7714 djs78@psu.edu, Pennsylvania State Univ, Univ Park – Ow
Stephen, Daniel, (801) 863-8584 Daniel.Stephen@uvu.edu, Utah Valley Univ – PiGsPe
Stephen, Ralph A., (508) 289-2583 rstephen@whoi.edu, Woods Hole Oceanographic Institution – Ys
Stephens, Jason H., (512) 484-0874 jason.stephens@austincc.edu, Austin Community Coll District – YrGuYe
Stephenson, Gerry, gerry.stephenson@rogers.com, Univ of Guelph – On
Stephenson, Randell A., +44 (0)1224 274817 r.stephenson@abdn.ac.uk, Univ of Aberdeen – YgGtOg
Stepien, Carol A., 419-530-8362 carol.stepien@utoledo.edu, Univ of Toledo – On
Steppe, Cecily N., 410-293-6558 natunewi@usna.edu, United States Naval Academy – Ob
Stermer, Ed, estermer@icc.cc.il.us, Illinois Central Coll – Og
Stern, Charles R., 303-492-7170 charles.stern@colorado.edu, Univ of Colorado – Gi
Stern, Herschel I., hstern@miracosta.edu, MiraCosta Coll – Oy
Stern, Robert J., (972) 883-2442 rjstern@utdallas.edu, Univ of Texas, Dallas – Gt
Sternberg, Ben K., (520) 621-2439 bkslasi@u.arizona.edu, Univ of Arizona – Ye
Sternberg, Richard W., (206) 543-6487 rws@ocean.washington.edu, Univ of Washington – Ou
Sternberg, Robert S., (717) 291-4133 rob.sternberg@fandm.edu, Franklin and Marshall Coll – YmGa
Sternberg, Rolf, (973) 655-7386 Montclair State Univ – On
Stetler, Larry D., (605) 394-2464 Larry.Stetler@sdsmt.edu, South Dakota School of Mines & Tech – Ng
Stevens, Calvin H., (408) 924-5029 calvin.stevens@sjsu.edu, San Jose State Univ – PsGrs
Stevens, De Anne S., (907) 451-5014 deanne.stevens@alaska.gov, Alaska Division of Geological & Geophysical Surveys – Ng
Stevens, Duane E., dstevens@hawaii.edu, Univ of Hawai'i, Manoa – Ow
Stevens, Joan M., (805) 756-2261 jstevens@calpoly.edu, California Polytechnic State Univ – On
Stevens, Liane M., (936) 468-2024 stevenslm@sfasu.edu, Stephen F. Austin State Univ – Gpc
Stevens, Michael, Michael.Stevens@uni.edu, Univ of Northern Iowa – Oe
Stevens, Philip S., (812) 855-0732 pstevens@indiana.edu, Indiana Univ, Bloomington – Oa
Stevens, Robert G., (509) 786-9231 stevensr@wsu.edu, Washington State Univ – Sb
Stevens, Stan, 413-545-0773 sstevens@geo.umass.edu, Univ of Massachusetts, Amherst – Oy
Stevens (Landon), Lora R., 562-985-4817 lsteven2@csulb.edu, California State Univ, Long Beach – Gn
Stevenson, David, +44 (0) 131 650 6750 David.S.Stevenson@ed.ac.uk, Edinburgh Univ – Oa
Stevenson, David J., (626) 395-6108 djs@gps.caltech.edu, California Inst of Tech – Xy
Stevenson, John A., +44 (0) 131 650 7526 john.stevenson@ed.ac.uk, Edinburgh Univ – Gv
Stevenson, Ross, 514-987-3000 #7835 stevenson.ross@uqam.ca, Universite du Quebec a Montreal – Cg
Steward, Grieg F., 808-956-6775 grieg@hawaii.edu, Univ of Hawai'i, Manoa – Ob
Stewart, Alexander K., (315) 229-5087 astewart@stlawu.edu, St. Lawrence Univ – GlmHw
Stewart, Brian W., (412) 624-8883 bstewart@pitt.edu, Univ of Pittsburgh – CclGe
Stewart, Dion C., 678-240-6227 dstewart29@gsu.edu, Georgia State Univ, Perimeter Coll, Alpharetta Campus – Gzi
Stewart, Esther K., (608) 263-3201 esther.stewart@wgnhs.uwex.edu, Univ of Wisconsin - Extension – Gg
Stewart, Gary, 405-372-6063 Oklahoma State Univ – Go
Stewart, Gillian, (718) 997-3104 gillian.stewart@qc.cuny.edu, Queens Coll (CUNY) – OcCm
Stewart, Kevin G., (919) 962-0683 kgstewar@email.unc.edu, Univ of North Carolina, Chapel Hill – Gc
Stewart, Mark T., (813) 974-8749 mark@usf.edu, Univ of South Florida, Tampa – Hw
Stewart, Michael A., (217) 244-5025 stewart1@illinois.edu, Univ of Illinois, Urbana-Champaign – Gi
Stewart, Robert, 713-743-3081 rrstewart@uh.edu, Univ of Houston – Ye
Stewart, Robert H., rstewart@ocean.tamu.edu, Texas A&M Univ – Op
Stewart, Robert R., 713-743-3399 rrstewart@uh.edu, Univ of Houston – Ye
Stewart, Robert R., (403) 220-3265 stewart@ucalgary.ca, Univ of Calgary – Ye
Steyn, Douw G., (604) 364-1266 dsteyn@eos.ubc.ca, Univ of British Columbia – Oaw
Stickel, Robert, (404) 385-4413 robert.stickel@eas.gatech.edu, Georgia Inst of Tech – Oa
Stickney, Michael C., (406) 496-4332 mstickney@mtech.edu, Montana Tech of The Univ of Montana – YsGtm
Stickney, Robert R., (979) 845-3854 stickney@tamu.edu, Texas A&M Univ – Ob
Stidham, Christiane W., (631) 632-8059 christiane.stidham@sunysb.edu, SUNY, Stony Brook – YgGe
Stieglitz, Marc, (404) 385-6530 marc.stieglitz@ce.gatech.edu, Georgia Inst of Tech – Hsg
Stieglitz, Ronald D., (920) 465-2371 stieglir@uwgb.edu, Univ of Wisconsin, Green Bay – Gr
Stierle, Andrea, (406) 496-4117 andrea.stierle@umontana.edu, Montana Tech of the Univ of Montana – On
Stierman, Donald J., (419) 530-2860 donald.stierman@utoledo.edu, Univ of Toledo – YgGeYs
Stigall, Alycia L., (740) 593-0393 stigall@ohio.edu, Ohio Univ – Po
Stiles, Lynn F., (609) 652-4677 stilesl@pollux.stockton.edu, Richard Stockton Coll of New Jersey – On
Stillings, Lisa, (775) 784-5803 Univ of Nevada, Reno – Cm
Stimac, John P., (217) 581-6245 jpstimac@eiu.edu, Eastern Illinois Univ – Gct
Stimer, Debra, 330-244-3511 Kent State Univ at Stark – On
Stimpson, Ian G., (+44) 01782 733182 i.g.stimpson@keele.ac.uk, Keele Univ – YesYg
Stine, Alexander, 415-338-1209 stine@sfsu.edu, San Francisco State Univ – Gg
Stine, Scott W., 510-885-3159 scott.stine@csueastbay.edu, California State Univ, East Bay – Gm
Stinger, Lindsay C., +44(0) 113 34 37530 l.stringer@leeds.ac.uk, Univ of Leeds – On
Stinson, Amy L., 949-451-5622 astinson@ivc.edu, Irvine Valley Coll – Gc
Stinson, Amy L., (949) 361-1260 mesiem@aol.com, Santiago Canyon Coll – Gc
Stix, John, (514) 398-5391 john.stix@mcgill.ca, McGill Univ – Gv
Stixrude, Lars, +44 020 7679 37929 l.stixrude@ucl.ac.uk, Univ Coll London – Yg
Stock, Carl W., 205-348-1883 cstock@geo.ua.edu, Univ of Alabama – Pi
Stock, Joann M., (626) 395-6938 jstock@gps.caltech.edu, California Inst of Tech – Gt
Stockli, Daniel, 512-475-6037 stockli@jsg.utexas.edu,

Univ of Texas at Austin – Gc
Stocks, Lee, (570) 662-4612 lstocks@mansfield.edu, Mansfield Univ – OyGgm
Stockton, Charles W., (520) 621-7680 stockton@ltrr.arizona.edu, Univ of Arizona – Hq
Stoddard, Edward F., (919) 515-7939 skip_stoddard@ncsu.edu, North Carolina State Univ – Gp
Stoddard, Paul R., (815) 753-7929 pstoddard@niu.edu, Northern Illinois Univ – Gt
Stoecker, Diane, (410) 221-8407 stoecker@hpl.umces.edu, Univ of Maryland – Ob
Stoelinga, Mark T., 206-708-8588 mstoelinga@3tier.com, Univ of Washington – Oa
Stoessel, Marion, 979 845 7662 mstoessel@ocean.tamu.edu, Texas A&M Univ – Op
Stoessell, Ronald K., (504) 280-6795 rstoesse@uno.edu, Univ of New Orleans – Cl
Stofan, Ellen R., (818) 354-2076 ellen.r.stofan@jpl.nasa.gov, Jet Propulsion Laboratory – Xg
Stoffa, Paul L., (512) 471-6405 pauls@ig.utexas.edu, Univ of Texas at Austin – Ye
Stohr, Christopher J., (217) 244-2186 cstohr@illinois.edu, Univ of Illinois, Urbana-Champaign – OrGeNg
Stokes, Martin, +44 1752 584772 M.Stokes@plymouth.ac.uk, Univ of Plymouth – Gm
Stokes, Patricia, (801) 537-3320 nrugs.pstokes@state.ut.us, Utah Geological Survey – On
Stolper, Daniel, (609) 258-1052 dstolper@princeton.edu, Princeton Univ – Pe
Stolper, Edward M., (626) 395-6504 ems@gps.caltech.edu, California Inst of Tech – Cp
Stone, Alan T., (410) 516-8476 astone@jhu.edu, Johns Hopkins Univ – ClSc
Stone, Glenn D., (314) 935-5239 stone@artsci.wustl.edu, Washington Univ in St. Louis – On
Stone, Jim, 970-339-6664 jim.stone@aims.edu, Aims Community Coll – Og
Stone, John O., 206-221-6332 stn@uw.edu, Univ of Washington – Cca
Stone, Loyd, (785) 532-6101 stoner@ksu.edu, Kansas State Univ – Sp
Stone, Peter H., (617) 253-2443 phstone@mit.edu, Massachusetts Inst of Tech – Oa
Stoner, Joseph, 541-737-9002 jstoner@coas.oregonstate.edu, Oregon State Univ – Gsr
Storey, Craig D., +44 023 92 842245 craig.storey@port.ac.uk, Univ of Portsmouth – GptCc
Stork, Allen L., (970) 943-3044 astork@western.edu, Western State Colorado Univ – Giv
Stormer, Jr., John C., (904) 456-7884 Rice Univ – Gi
Storrs, Glenn W., (513) 345-8500 Univ of Cincinnati – Pv
Storrs, Glenn W., (513) 455-7141 gstorrs@cincymuseum.org, Cincinnati Museum Center – Pv
Stotler, Randy, (785) 864-6048 rstotler@ku.edu, Univ of Kansas – Hw
Stott, Lowell D., (213) 740-5120 stott@usc.edu, Univ of Southern California – Pm
Stoudt, Emily L., (432) 552-2244 stoudt_e@utpb.edu, Univ of Texas, Permian Basin – GdPsi
Stout, James H., (612) 624-4344 jstout@umn.edu, Univ of Minnesota, Twin Cities – GpzGt
Stover, Susan G., 785-864-2063 sstover@kgs.ku.edu, Univ of Kansas – Gg
Stowell, Harold H., (205) 348-5098 hstowell@ua.edu, Univ of Alabama – GpCcGt
Stowell Gale, Julia, 512-232-7957 julia.gale@beg.utexas.edu, Univ of Texas at Austin, Jackson School of Geosciences – Gc
Stoykova, Kristalina C., +35929792213 stoykova@geology.bas.bg, Bulgarian Academy of Sciences – PemPi
Strachan, Rob, +44 023 92 842279 rob.strachan@port.ac.uk, Univ of Portsmouth – Cc
Strachan, Scotty, strachan@unr.edu, Univ of Nevada, Reno – Pe
Stracher, Glenn B., (478) 289-2073 stracher@ega.edu, East Georgia State Coll – GycEc
Strack, Otto D., (612) 625-3009 strac001@tc.umn.edu, Univ of Minnesota, Twin Cities – Hw
Straffin, Eric, (814) 732-1574 estraffin@edinboro.edu, Edinboro Univ of Pennsylvania – GmsOs
Strahler, Alan, (617) 353-5984 Boston Univ – Or
Straight, William, (703) 948-7750 wstraight@nvcc.edu, Northern Virginia Community Coll - Loudoun Campus – GgPv
Strain, Priscilla L., 202-633-2481 Smithsonian Institution / National Air & Space Museum – Or
Stramler, Kirstie L., (415) 452-5046 City Coll of San Francisco – OagGg
Stramski, Dariusz, (858) 534-3353 dstramski@ucsd.edu, Univ of California, San Diego – Op
Strangeway, Robert J., (310) 206-6247 strange@igpp.ucla.edu, Univ of California, Los Angeles – On
Strasser, Jeffrey C., (309) 794-7218 JeffreyStrasser@augustana.edu, Augustana Coll – Gm
Strasser, Stefan, 089/2180 4340 stefan.strasser@vr-web.de, Ludwig-Maximilians-Universitaet Muenchen – Gz
Stratton, James F., (217) 581-2626 jfstratton@eiu.edu, Eastern Illinois Univ – Pg
Straub, David, (514) 398-8995 david.straub@mcgill.ca, McGill Univ – Op
Straub, Derek J., 570-372-4767 straubd@susqu.edu, Susquehanna Univ – Oa
Straub, Katherine H., (570) 372-4318 straubk@susqu.edu, Susquehanna Univ – Ow
Straub, Kyle M., 504-862-3273 kmstraub@tulane.edu, Tulane Univ – GgCg
Straus, David M., (703) 993-5719 dstraus@gmu.edu, George Mason Univ – Oa
Strauss, Harald, +49-251-83-33932 hstrauss@uni-muenster.de, Universitaet Muenster – CsGg
Strauss, Justin V., (603) 646-6954 justin.v.strauss@dartmouth.edu, Dartmouth Coll – GsCgGt
Straw, Byron, (970) 351-2470 byron.straw@unco.edu, Univ of Northern Colorado – GlmYg
Strawn, Daniel G., (208) 885-2713 dgstrawn@uidaho.edu, Univ of Idaho – Sc
Strayer, Luther M., (510) 885-3083 luther.strayer@csueastbay.edu, California State Univ, East Bay – Gc
Streck, Martin J., (503) 725-3379 streckm@pdx.edu, Portland State Univ – GivCt
Strecker, Manfred, strecker@geo.uni-potsdam.de, Cornell Univ – Gt
Streepey Smith, Meg, (765) 973-2168 streeme@earlham.edu, Earlham Coll – Gt
Streig, Ashley, 503-725-3371 streig@pdx.edu, Portland State Univ – YsGt
Strick, James E., (717) 291-3856 james.strick@fandm.edu, Franklin and Marshall Coll – Ghe
Strickland, Richard M., (206) 543-3131 strix@ocean.washington.edu, Univ of Washington – ObgOg
Stright, Lisa, (970) 491-4296 lisa.stright@colostate.edu, Colorado State Univ – GoNpYe
Stright, Lisa, (801) 585-5461 lisa.stright@utah.edu, Univ of Utah – Ggo
Strmiæ Palinkaš, Sabina, +38514605961 sabina.strmic@inet.hr, Univ of Zagreb – CgeCc
Strobel, Darrell F., (410) 516-7829 strobel@jhu.edu, Johns Hopkins Univ – OanOn
Strobel, Timothy A., 202-478-8943 tstrobel@carnegiescience.edu, Carnegie Institution of Washington – Om
Strock, Jeffrey S., (507) 752-7372 stroc001@umn.edu, Univ of Minnesota, Twin Cities – So
Stroeve, Julienne, j.stroeve@ucl.ac.uk, Univ Coll London – Olr
Strole, Torie L., 217-244-2392 strole@illinois.edu, Univ of Illinois, Urbana-Champaign – On
Strom, Kyle, (540) 231-0979 strom@vt.edu, Virginia Tech – HsGs
Strom, Robert G., 520- 621-2720 rstrom@LPL.arizona.edu, Univ of Arizona – Xg
Strong, Courtenay, 801-585-0049 court.strong@utah.edu, Univ of Utah – Oa
Strong, Ellen, stronge@si.edu, Smithsonian Institution / National Museum of Natural History – OnPi
Strother, Paul, 617-552-8395 paul.strother@bc.edu, Boston Coll – Pb
Strub, Ted P., 541-737-3015 tstrub@coas.oregonstate.edu, Oregon State Univ – Oa
Struzhkin, Viktor V., 202-478-8952 vstruzhkin@carnegiescience.edu, Carnegie Institution of Washington – Gy
Stuart, Graham, +44(0) 113 34 35217 g.w.stuart@leeds.ac.uk, Univ of Leeds – Ys
Stuart, Neil, +44 (0) 131 650 2549 N.Stuart@ed.ac.uk, Edinburgh Univ – Oi
Stubler, Craig, (805) 756-2188 cstubler@calpoly.edu, California Polytechnic State Univ – So
Stucker, James D., (614) 265-6601 james.stucker@dnr.state.oh.us, Ohio Dept of Natural Resources – EgCg
Stucki, Joseph W., (217) 333-9636 Univ of Illinois, Urbana-Champaign

– Sc
Student, James J., (989) 774-2295 stude1jj@cmich.edu,
Central Michigan Univ – CaGiz
Stuiver, Minze, minze@uw.edu, Univ of Washington – Cc
Stull, Robert J., (323) 343-2408 rstull@calstatela.edu,
California State Univ, Los Angeles – Gi
Stull, Roland B., 604-822-5901 rstull@eos.ubc.ca,
Univ of British Columbia – Oa
Stumbea, Dan, +40 232 201 464 dan.stumbea@uaic.ro,
Al. I. Cuza Univ of Iasi – Gez
Stump, Brian W., (214) 768-1223 Southern Methodist Univ – Ys
Stumpf, Andrew J., 217-244-6462 astumpf@illinois.edu,
Univ of Illinois, Urbana-Champaign – Gl
Stunz, Greg, 361-825-3254 Greg.Stunz@tamucc.edu,
Texas A&M Univ, Corpus Christi – On
Stupazzini, Marco, 089/2180 4143 stupa@geophysik.uni-muenchen.de,
Ludwig-Maximilians-Universitaet Muenchen – Yg
Sturchio, Neil C., (630) 252-3986 Argonne National Laboratory – Cg
Sturchio, Neil C., (302) 831-8706 sturchio@udel.edu,
Univ of Delaware – ClcCa
Sturges, Bill, +44 (0)1603 59 2018 w.sturges@uea.ac.uk,
Univ of East Anglia – Oa
Sturm, Diana, 251-460-6381 dsturm@southalabama.edu,
Univ of South Alabama – Geg
Sturz, Anne A., (619) 260-4600 asturz@sandiego.edu,
Univ of San Diego – Cm
Stute, Martin, (845) 365-8704 martins@ldeo.columbia.edu,
Columbia Univ – CsHgGe
Stutsman, Sam, 251-461-1508 sstutsman@southalabama.edu,
Univ of South Alabama – Oiy
Stutz, Jochen P., (310) 825-1217 jochen@atmos.ucla.edu,
Univ of California, Los Angeles – Oa
Su, Haibin, 361-593-3590 Haibin.Su@tamuk.edu,
Texas A&M Univ, Kingsville – Oir
Su, Xiaobo, (541) 346-4568 xiaobo@uoregon.edu,
Univ of Oregon – On
Suarez, Celina, (479) 575-4866 casuarez@uark.edu,
Univ of Arkansas, Fayetteville – ClPeGg
Sublette, Kerry, 918-631-3085 kerry-sublette@utulsa.edu,
The Univ of Tulsa – Ge
Suchy, Daniel R., (785) 864-2160 datares@kgs.ku.edu,
Univ of Kansas – Gg
Suen, C. J., (559) 278-8656 john_suen@csufresno.edu,
California State Univ, Fresno – HwCsGe
Sugarman, Peter J., (609) 292-2576 pete.sugarman@dep.state.nj.us,
New Jersey Geological and Water Survey – Gr
Sugarman, Peter P., (609) 292-6842 Rutgers, The State Univ of New Jersey
– Gr
Sugihara, George, (858) 534-5582 gsugihara@ucsd.edu,
Univ of California, San Diego – Ob
Sui, Daniel Z., (409) 845-7154 sui@geog.tamu.edu,
Texas A&M Univ – Or
Sukop, Michael C., (305) 348-3117 sukopm@fiu.edu,
Florida International Univ – HwqSp
Sulanowska, Margaret, (508) 289-2306 msulanowska@whoi.edu,
Woods Hole Oceanographic Institution – Cm
Sullivan, Bill, (207) 859-5800 wasulliv@colby.edu, Colby Coll – Gct
Sullivan, Donald G., donald.sullivan@du.edu, Univ of Denver – Og
Sullivan, Jack B., (301) 405-0106 jsull@umd.edu, Univ of Maryland – Ou
Sullivan, Raymond, (415) 338-2061 sullivan@sfsu.edu,
San Francisco State Univ – Go
Sullivan-Watts, Barbara K., (401) 874-6659 bsull@gso.uri.edu,
Univ of Rhode Island – Og
Sultan, Mohamed, (630) 252-1929 Argonne National Laboratory – Cg
Sultan, Mohamed, (269) 387-5487 mohamed.sultan@wmich.edu,
Western Michigan Univ – OriGe
Sumida, Stuart S., (909) 537-5346 California State Univ, San Bernardino
– Pv
Summers, Robert, 780-492-0342 robert.summers@ualberta.ca,
Univ of Alberta – On
Summers, Sara, (801) 626-6208 sarasummers@weber.edu,
Weber State Univ – Gzg
Summons, Roger, (617) 452-2791 rsummons@mit.edu,
Massachusetts Inst of Tech – PyCo
Sumner, Dawn Y., (530) 752-5353 dysumner@ucdavis.edu,
Univ of California, Davis – Gs
Sumner, Esther, +44 (0)23 80592067 E.J.Sumner@soton.ac.uk,
Univ of Southampton – Gs
Sumrall, Colin, (865) 974-2366 csumrall@utk.edu,
Univ of Tennessee, Knoxville – Pi
Sun, Alexander, 512-475-6190 alex.sun@beg.utexas.edu,
Univ of Texas at Austin – Gq
Sun, Hongbing, 609-895-5185 hsun@rider.edu, Rider Univ – HwCgSc
Sun, Wen-Yih, (765) 494-7681 wysun@purdue.edu, Purdue Univ – Oa
Sun, Yifei, (818) 677-3532 yifei.sun@csun.edu,
California State Univ, Northridge – Oi
Sun, Yuefeng, (979) 845-0635 sun@geos.tamu.edu,
Texas A&M Univ – GoYe
Sundareshwar, P. V., 605-394-2492 psundareshwar@usaid.gov,
South Dakota School of Mines & Tech – OeCgSb
Sundby, Bjorn, (418) 723-1986 (Ext. 1767) b.sundby@uquebec.ca,
Universite du Quebec a Rimouski – Oc
Sundby, Bjorn, 514-398-4883 bjorn.sundby@mcgill.ca, McGill Univ – Cm
Sundell, Ander, 208.562.3354 andersundell@cwidaho.cc,
Coll of Western Idaho – Ggc
Sundell, Kent A., (307) 268-2498 ksundell@casperColl.edu,
Casper Coll – Gg
Sunderlin, David, 610-330-5198 sunderld@lafayette.edu ,
Lafayette Coll – Gg
Suneson, Neil , (405) 325-1472 nsuneson@ou.edu, Univ of Oklahoma –
Grc
Suntharalingam, Parvadha, +44 (0)1603 59 1423 p.suntharalingam@uea.
ac.uk, Univ of East Anglia – Oc
Superchi-Culver, Tonia, 303-492-5211 toni.culver@colorado.edu,
Univ of Colorado – Pg
Suppe, John, jsuppe@uh.edu, Univ of Houston – Gtc
Surge, Donna M., 919-843-1994 donna64@unc.edu,
Univ of North Carolina, Chapel Hill – Pe
Surian, Nicola, +390498279125 nicola.surian@unipd.it,
Università degli Studi di Padova – Gm
Surpless, Benjamin E., (210) 999-7110 bsurples@trinity.edu,
Trinity Univ – GctGi
Surpless, Kathleen D., (210) 999-7365 ksurples@trinity.edu,
Trinity Univ – Gs
Susak, Nicholas J., (506) 453-4803 nsusak@unb.ca,
Univ of New Brunswick – Cg
Sushama, Laxmi, 514-987-3000 #2414 sushama.laxmi@uqam.ca,
Universite du Quebec a Montreal – Hy
Sussman, Robert W., (314) 935-5264 rwsussma@artsci.wustl.edu,
Washington Univ in St. Louis – On
Suszek, Thomas J., (920) 424-2268 suszek@uwosh.edu,
Univ of Wisconsin, Oshkosh – Gg
Sutherland, Bruce, 780-492-0573 bruce.sutherland@ualberta.ca,
Univ of Alberta – Cm
Sutherland, Dave, 541-346-8753 dsuth@uoregon.edu, Univ of Oregon –
Op
Sutherland, Mary K., (406) 496-4410 msutherland@mtech.edu,
Montana Tech of The Univ of Montana – Hws
Sutherland, Stuart, (604) 822-0176 ssutherland@eos.ubc.ca,
Univ of British Columbia – Pg
Sutherland, Wayne, (307) 766-2286 Ext. 247 wayne.sutherland@wyo.gov,
Wyoming State Geological Survey – En
Suttle, Curtis, (604) 822-8610 csuttle@eos.ubc.ca,
Univ of British Columbia – Ob
Suttner, Lee J., (812) 855-4957 suttner@indiana.edu,
Indiana Univ, Bloomington – Gd
Sutton, Mark, +44 20 759 47487 m.sutton@imperial.ac.uk,
Imperial Coll – Pg
Sutton, Paul C., (303) 871-2399 psutton@du.edu, Univ of Denver – Oi
Sutton, Sally J., (970) 491-5995 sallys@warnercnr.colostate.edu,
Colorado State Univ – GdCl
Suyker, Andrew E., asuyker@unl.edu, Univ of Nebraska, Lincoln – Oa
Sverdrup, Keith A., (414) 229-4017 sverdrup@uwm.edu,
Univ of Wisconsin, Milwaukee – YsGt
Sverjensky, Dimitri A., (410) 516-8568 sver@jhu.edu,
Johns Hopkins Univ – Cl
Svitra, Zita V., (505) 667-7616 Los Alamos National Laboratory – On
Svoma, Bohumil, svomab@missouri.edu, Univ of Missouri, Columbia –
Oa
Swain, Geoffrey W., (321) 674-8096 Florida Inst of Tech – Ob
Swain, Geoffry, (321) 674-7129 Florida Inst of Tech – Oo
Swanger, Kate, (978) 934-2664 kate_swanger@uml.edu,
Univ of Massachusetts Lowell – Gl
Swann, Abigail L., 206-616-0486 aswann@atmos.washington.edu,

Univ of Washington – Oan
Swanson, Basil I., (505) 667-5814 Los Alamos National Laboratory – On
Swanson, Donald A., (808) 967-8863 donswan@usgs.gov, Univ of Hawai'i, Manoa – Gv
Swanson, Karen, (201) 595-2589 William Paterson Univ – Cc
Swanson, R. L., (631) 632-8704 larry.swanson@stonybrook.edu, SUNY, Stony Brook – Og
Swanson, Sherman, (775) 784-4057 sswanson@agnt1.ag.unr.edu, Univ of Nevada, Reno – Hs
Swanson, Susan K., (608) 363-2132 swansons@beloit.edu, Beloit Coll – Hw
Swanson, Terry W., tswanson@uw.edu, Univ of Washington – CcGe
Swap, Robert J., (434) 924-7714 rjs8g@virginia.edu, Univ of Virginia – On
Swapp, Susan M., (307) 766-2513 swapp@uwyo.edu, Univ of Wyoming – Gp
Swarr, Gretchen, (508) 289-2558 gswarr@whoi.edu, Woods Hole Oceanographic Institution – Oc
Swart, Peter K., (305) 421-4103 pswart@rsmas.miami.edu, Univ of Miami – ClPsCs
Swartwood, Jade L., (570) 389-4108 jswartwo@bloomu.edu, Bloomsburg Univ – On
Swaters, Gordon E., (780) 492-7159 gordon.swaters@ualberta.ca, Univ of Alberta – Op
Sweeney, Mark D., (509) 373-0703 mark.sweeney@pnl.gov, Pacific Northwest National Laboratory – Yg
Sweeney, Mark R., (605) 677-6142 mark.sweeney@usd.edu, Univ of South Dakota – Gms
Sweeny, Daniel, (620) 421-4826 dsweeney@ksu.edu, Kansas State Univ – So
Sweet, Alisan C., 806-834-2398 alisan.sweet@ttu.edu, Texas Tech Univ – Gs
Sweet, Dustin E., (806) 834-8390 dustin.sweet@ttu.edu, Texas Tech Univ – Gsd
Sweetman, Steve, +44 023 9284 2257 steve.sweetman@port.ac.uk, Univ of Portsmouth – Pg
Swennen, Rudy, rudy.swennen@ees.kuleuven.be, Katholieke Universiteit Leuven – Gso
Swenson, John B., 218-726-6844 jswenso2@d.umn.edu, Univ of Minnesota, Duluth – Gr
Swetnam, Thomas W., tswetnam@email.arizona.edu, Univ of Arizona – On
Swetnam, Thomas W., (520) 621-2112 tswetnam@ltrr.arizona.edu, Univ of Arizona – Pe
Swett, Keene, (319) 351-4644 keene-swett@uiowa.edu, Univ of Iowa – Gs
Swift, Donald J. P., (757) 683-4937 dswift@odu.edu, Old Dominion Univ – Ou
Swift, Elijah V., (401) 874-6146 lige@gso.uri.edu, Univ of Rhode Island – Ob
Swift, Robert P., (505) 665-7871 bswift@lanl.gov, Los Alamos National Laboratory – Nr
Swift, Stephen A., (508) 289-2626 sswift@whoi.edu, Woods Hole Oceanographic Institution – Yr
Swindle, Timothy, 520-621-4128 tswindle@lpl.arizona.edu, Univ of Arizona – Xc
Swisher III, Carl C., 848-445-5363 cswish@rci.rutgers.edu, Rutgers, The State Univ of New Jersey – Cc
Swope, R. J., (317) 278-0132 rjswope@iupui.edu, Indiana Univ / Purdue Univ, Indianapolis – Gz
Swyrtek, Sheila, (810)232-9312 sheila.swyrtek@mcc.edu, Charles Stewart Mott Community Coll – Ga
Sydora, Richard D., (780) 492-3624 rsydora@phys.ualberta.ca, Univ of Alberta – Xy
Sykes, Lynn R., (845) 359-7428 sykes@ldeo.columbia.edu, Columbia Univ – YsGtOn
Sylva, Sean, 508-289-3546 ssylva@whoi.edu, Woods Hole Oceanographic Institution – Oc
Sylvan, Jason B., (979) 845-5105 jasonsylvan@tamu.edu, Texas A&M Univ – ObPy
Sylvester, Paul J., (709) 737-4736 sylvester@sparky2.esd.mun.ca, Memorial Univ of Newfoundland – Ca
Sylvester, Steven, (717) 291-3821 steve.sylvester@fandm.edu, Franklin and Marshall Coll – Ca
Sylvia, Elizabeth R., 410 554-5542 elizabeth.sylvia@maryland.gov, Maryland Department of Natural Resources – Ge
Symbalisty, E.M.D., (505) 667-9670 esymbalisty@lanl.gov, Los Alamos National Laboratory – On
Symes, William S., (713) 348-5997 symes@caam.rice.edu, Rice Univ – On
Symons, David T., (519) 253-3000 x2493 dsymons@uwindsor.ca, Univ of Windsor – YmGtEm
Syrett, William, (814) 865-6172 wjs1@psu.edu, Pennsylvania State Univ, Univ Park – Ow
Syrup, Krista A., (708) 974-5615 syrup@morainevalley.edu, Moraine Valley Community Coll – Cs
Syverson, Kent M., (715) 836-3676 syverskm@uwec.edu, Univ of Wisconsin, Eau Claire – Gl
Syvitski, Jai P., (303) 492-7909 james.syvitski@colorado.edu, Univ of Colorado – GsuGq
Szabo, Csaba, cszabo@elte.hu, Virginia Polytechnic Inst & State Univ – Gi
Szatmári, József, +3662544156 szatmari@geo.u-szeged.hu, Univesity of Szeged – OriOy
Szczepanski, Jacek, jacek.szczepanski@uwr.edu.pl, Univ of Wroclaw – Ggc
Szecsody, James E., (509) 372-6080 jim.szecsody@pnl.gov, Pacific Northwest National Laboratory – So
Székely, Balázs, balazs.szekely@ttk.elte.hu, Eotvos Lorand Univ – GmOri
Szekielda, Karl H., (212) 772-4019 szekielda@aol.com, Hunter Coll (CUNY) – Org
Szeliga, Walter, 509-963-2705 walter@geology.cwu.edu, Central Washington Univ – YgsYd
Szente, Istvan, szente@ludens.elte.hu, Eotvos Lorand Univ – Pig
Szeto, Anthony M. K., (416)736-2100 #77703 szeto@yorku.ca, York Univ – Yg
Szilagyi, Jozsef, jszilagyi@unl.edu, Unversity of Nebraska - Lincoln – HqOa
Szlavecz, Katalin, (410) 516-8947 szlavecz@jhu.edu, Johns Hopkins Univ – Pi
Szulczewski, Melanie, (540) 654-2345 mszulcze@umw.edu, Univ of Mary Washington – ScClOg
Szunyogh, Istvan, (979) 458-0553 szunyogh@tamu.edu, Texas A&M Univ – Oa
Szymanski, David, (781) 891-2901 dszymanski@bentley.edu, Bentley Univ – GvfCg
Szymanski, Jason, 585-292-2423 jszymanski@monroecc.edu, Monroe Community Coll – GlPe
Szynkiewicz, Anna, 865-974-6006 aszynkie@utk.edu, Univ of Tennessee, Knoxville – Cs
Söllner, Frank, 089/2180 6519 fank.soellner@iaag.geo.uni-muenchen.de, Ludwig-Maximilians-Universitaet Muenchen – Gg

## T

Tabidian, M. Ali, (818) 677-2536 ali.tabidian@csun.edu, California State Univ, Northridge – Hw
Taboada Castro, María T., 00 34 981 167000 teresat@udc.es, Coruna Univ – Sd
Taboga, Karl, (307) 766-2286 x226 karl.taboga@wyo.gov, Wyoming State Geological Survey – HwGe
Tabrizi, Azam, 757-822-5020 ATabrizi@tcc.edu, Tidewater Community Coll – Pm
Tabrum, Alan R., (412) 622-3265 tabruma@carnegiemnh.org, Carnegie Museum of Natural History – Pv
Tacinelli, John C., (507) 285-7501 john.tacinelli@rctc.edu, Rochester Community & Technical Coll – Gig
Tackett, Lydia S., 701-231-6164 lydia.tackett@ndsu.edu, North Dakota State Univ – PiGsPo
Tadesse, Tsegaye, ttadesse2@unl.edu, Univ of Nebraska, Lincoln – Oar
Taggart, Christopher T., (902) 494-7144 chris.taggart@dal.ca, Dalhousie Univ – Ob
Taggart, Ralph E., 517-353-5175 taggart@msu.edu, Michigan State Univ – Pb
Tagliabue, Alessandro, +44 0151 794 4651 A.Tagliabue@liverpool.ac.uk, Univ of Liverpool – Oc
Taib, Samsudin Hj., 03-79674235 samsudin@um.edu.my, Univ of Malaya – Yg
Tailby, Nichlos D., (518) 276-3247 tailbn@rpi.edu, Rensselaer Polytechnic Inst – Cpt
Taillefert, Martial, (404) 894-6043 mtaillef@eas.gatech.edu, Georgia Inst of Tech – ClmCa
Tait, C. Drew, (505) 667-7603 Los Alamos National Laboratory – Cp
Tait, Kim, ktait@rom.on.ca, Western Univ – Xm
Tait, Kimberly T., (416) 586-5820 ktait@rom.on.ca, Royal Ontario Museum – GzyXm

Tajik, Atieh, 404 413 5790 geoatt@langate.gsu.edu, Georgia State Univ – Gg
Takahashi, Taro, (845) 365-8537 Columbia Univ – Cm
Takeuchi, Akira, 81-76-445-6654 takeuchi@sci.u-toyama.ac.jp, Univ of Toyama – GtcYe
Takle, Eugene S., (515) 294-9871 gstakle@iastate.edu, Iowa State Univ of Science & Tech – Oaw
Talbot, Helen, +44 (0) 191 208 6426 helen.talbot@ncl.ac.uk, Univ of Newcastle Upon Tyne – Co
Talbot, James L., (360) 733-4282 talbot@wwu.edu, Western Washington Univ – Gc
Talley, John H., 302-831-2833 waterman@udel.edu, Univ of Delaware – Oa
Talley, Lynne D., (858) 534-6610 ltalley@ucsd.edu, Univ of California, San Diego – Op
Talwani, Manik, (713) 348-6067 manik@rice.edu, Rice Univ – Yr
Talwani, Pradeep, talwani@geol.sc.edu, Univ of South Carolina – Ys
Tamish, Mohamed M., 002-03-3921595 mtamish@hotmail.com, Alexandria Univ – GgsCg
Tan, Chunyang, tanc@umn.edu, Univ of Minnesota, Twin Cities – Cm
Taney, R. Marieke, (928) 523-2384 Marieke.Taney@nau.edu, Northern Arizona Univ – OnnOn
Tanhua, Toste, ttanhua@geomar.de, Dalhousie Univ – Oc
Tanimoto, Toshiro, (805) 893-8375 toshiro@geol.ucsb.edu, Univ of California, Santa Barbara – Ys
Tanner, Benjamin R., (828) 227-3915 btanner@wcu.edu, Western Carolina Univ – Cos
Tapanila, Leif, 208-282-3871 tapaleif@isu.edu, Idaho State Univ – Pi
Tapanila, Lori, 208-282-5024 tapalori@isu.edu, Idaho State Univ – Gg
Tape, Carl, 907-474-5456 carltape@gi.alaska.edu, Univ of Alaska, Fairbanks – Ys
Tapp, J. B., 918-631-3018 jbt@utulsa.edu, The Univ of Tulsa – Gc
Tarduno, John A., (585) 275-5713 john@earth.rochester.edu, Univ of Rochester – YmGtXm
Tarka, Robert, robert.tarka@uwr.edu.pl, Univ of Wroclaw – Hg
Tarr, Alexander, 508 929-8474 alexander.tarr@worcester.edu, Worcester State Univ – Og
Tarshizi, Ebrahim, 906/487-2582 tarshizi@mtu.edu, Michigan Technological Univ – Nm
Tary, Anna K., (781) 891-2236 atary@bentley.edu, Bentley Univ – Gl
Taskey, Ronald D., (805) 756-1160 rtaskey@calpoly.edu, California Polytechnic State Univ – Sf
Tassier-Surine, Stephanie, Stephanie.Surine@dnr.iowa.gov, Iowa Dept of Natural Resources – Gg
Tatabatai, M. A., (515) 294-7848 malit@iastate.edu, Iowa State Univ of Science & Tech – Sc
Tatarskii, Viatcheslav, (404) 894-9224 vvt@eas.gatech.edu, Georgia Inst of Tech – Oa
Tate, Garrett W., garrett.tate@vanderbilt.edu, Vanderbilt Univ – Gc
Tatham, Robert H., (512) 471-9129 tatham@mail.utexas.edu, Univ of Texas at Austin – Yg
Tausig, Heather, htausig@neaq.gov, Univ of Massachusetts, Boston – Og
Tauxe, John D., (423) 574-5348 Oak Ridge National Laboratory – Hw
Tauxe, Lisa, (858) 534-6084 ltauxe@ucsd.edu, Univ of California, San Diego – Ym
Tavener, Kristi, (807) 343-8677 ktavener@lakeheadu.ca, Lakehead Univ – On
Tawabini, Bassam S., +96638607643 bassamst@kfupm.edu.sa, King Fahd Univ of Petroleum and Minerals – GeHsCa
Taylor, Brian, (808) 956-6649 taylorb@hawaii.edu, Univ of Hawai'i, Manoa – Gt
Taylor, Carolyn, carolyn.taylor@mcmail.maricopa.edu, Mesa Community Coll – Gg
Taylor, Chuck J., 859-323-0523 charles.taylor@uky.edu, Univ of Kentucky – Gg
Taylor, Danny L., (775) 784-6922 dtaylor@mines.unr.edu, Univ of Nevada, Reno – Nm
Taylor, Donavon, (715) 425-3395 donavon.h.taylor@uwrf.edu, Univ of Wisconsin, River Falls – Sp
Taylor, Edith, 785-864-3621 etaylor@ku.edu, Univ of Kansas – Pb
Taylor, Elwynn, (808) 956-3899 setaylor@iastate.edu, Iowa State Univ of Science & Tech – Oa
Taylor, Frank J. R., (604) 822-4587 mtaylor@eos.ubc.ca, Univ of British Columbia – Ob
Taylor, Frederick W., (512) 471-0453 fred@ig.utexas.edu, Univ of Texas at Austin – GtmGe
Taylor, G. Jeffrey, (808) 956-3899 Univ of Hawai'i, Manoa – Xm
Taylor, Gordon T., (631) 632-8688 gordon.taylor@stonybrook.edu, SUNY, Stony Brook – Obc
Taylor, Graeme, +44 1752 584770 G.Taylor@plymouth.ac.uk, Univ of Plymouth – Yg
Taylor, Gregory R., (530) 898-6369 taylgr@shasta.csuchico.edu, California State Univ, Chico – Oa
Taylor, Hugh P., (626) 395-6116 hptaylor@gps.caltech.edu, California Inst of Tech – Cs
Taylor, John F., (724) 357-4469 jftaylor@iup.edu, Indiana Univ of Pennsylvania – PsiGs
Taylor, Kenneth B., (919) 707-9211 kenneth.b.taylor@ncdenr.gov, North Carolina Geological Survey – YsGgEo
Taylor, Kevin, +44 0161 275-8557 kevin.taylor@manchester.ac.uk, Univ of Manchester – Go
Taylor, Lansing, 801-581-8430 ltaylor@egi.utah.edu, Univ of Utah – Gc
Taylor, Lawrence A., (865) 974-6013 lataylor@utk.edu, Univ of Tennessee, Knoxville – GiCpXg
Taylor, Lawrence D., (517) 629-0308 ltaylor@albion.edu, Albion Coll – Gl
Taylor, Matthew, 303-871-2656 mtaylor7@du.edu, Univ of Denver – Og
Taylor, Michael, michael.taylor@flinders.edu.au, Flinders Univ – On
Taylor, Michael H., 785-864-5828 mht@ku.edu, Univ of Kansas – Gt
Taylor, Nathan H., 501 683-1085 nathan.taylor@arkansas.gov, Arkansas Geological Survey – OiyGe
Taylor, Penny M., 413-538-3236 pmtaylor@mtholyoke.edu, Mount Holyoke Coll – Gg
Taylor, Peter, +44(0) 113 34 37169 P.G.Taylor@leeds.ac.uk, Univ of Leeds – Ge
Taylor, Peter A., (416)736-2100 #77707 pat@yorku.ca, York Univ – Oa
Taylor, Rex N., +44 (0)23 80592007 rex@noc.soton.ac.uk, Univ of Southampton – GvCa
Taylor, Richard P., richard_taylor@carleton.ca, Carleton Univ – Cg
Taylor, Robert W., (973) 655-4129 Montclair State Univ – On
Taylor, Ryan W., 914 251 6652 ryan.taylor@purchase.edu, SUNY, Purchase – OiyOg
Taylor, Sid, (902) 867-2299 staylor@stfx.ca, Saint Francis Xavier Univ – Gg
Taylor, Stephen B., 503-838-8398 taylors@wou.edu, Western Oregon Univ – Gm
Taylor, Steven R., (505) 667-1007 taylor@lanl.gov, Los Alamos National Laboratory – Ys
Taylor, Ta-Shana A., (305) 284-4254 t.taylor2@miami.edu, Univ of Miami – OeGePv
Taylor, Thomas N., taylor.13@osu.edu, Ohio State Univ – Pb
Taylor, Thomas N., 785-864-3625 tntaylor@ku.edu, Univ of Kansas – Pb
Taylor, Wanda J., (702) 895-4615 wanda.taylor@unlv.edu, Univ of Nevada, Las Vegas – Gc
Taylor, Wayne A., (505) 667-4253 Los Alamos National Laboratory – Cl
Tchakerian, Vatche P., (409) 845-7997 Texas A&M Univ – Gm
Teagle, Damon, +44 (0)23 80592727 dat@noc.soton.ac.uk, Univ of Southampton – Cg
Teasdale, Rachel, 530-898-5547 rteasdale@csuchico.edu, California State Univ, Chico – Gv
Tedesco, Lenore P., (317) 274-7154 ltedesco@iupui.edu, Indiana Univ / Purdue Univ, Indianapolis – Gs
Tedford, Richard H., (212) 769-5809 American Museum of Natural History – Pv
Teed, Rebecca, 937 775-3446 rebecca.teed@wright.edu, Wright State Univ – OePe
Teeuw, Richard, +44 023 92 842267 richard.teeuw@port.ac.uk, Univ of Portsmouth – OaGmOi
Tefend, Karen S., ktefend@westga.edu, Univ of West Georgia – CgGe
Teixell Cácharo, Antoni, ++935811163 antonio.teixell@uab.cat, Universitat Autonoma de Barcelona – GtcYx
Tellam, John, +44 (0)121 41 46138 j.h.tellam@bham.ac.uk, Univ of Birmingham – Hw
Teller, James T., (204) 474-9270 tellerjt@ms.umanitoba.ca, Univ of Manitoba – Gs
Telmer, Kevin, ktelmer@uvic.ca, Univ of Victoria – Cl
Telyakovskiy, Aleksey, 775.784.1364 alekseyt@unr.edu, Univ of Nevada, Reno – Hq
Tempel, Gina, (775) 784-4706 gina@mines.unr.edu, Univ of Nevada, Reno – Cl
Temples, Tommy, (803) 348-0472 ttemples@sc.rr.com, Clemson Univ – GoYg
Templeton, Alan R., (314) 935-6868 temple_a@wustl.edu, Washington Univ in St. Louis – On

Templeton, Alexis, 303-492-6069 alexis.templeton@colorado.edu, Univ of Colorado – ClGe
Templeton, Jeffrey H., (503) 838-8858 templej@wou.edu, Western Oregon Univ – GviOe
Ten Brink, Norman W., tenbrinn@gvsu.edu, Grand Valley State Univ – Gm
Tencate, James, (505) 665-6667 tencate@lanl.gov, Los Alamos National Laboratory – On
Teng, Fangzhen, 206-543-7615 fteng@uw.edu, Univ of Washington – Cg
Teng, Ta-liang, (213) 740-5838 lteng@usc.edu, Univ of Southern California – YsGtYe
Tenorio, Victor O., (520) 621-3858 vtenorio@email.arizona.edu, Univ of Arizona – Nmx
Teodoriu, Catalin, 405-325-6872 cteodoriu@ou.edu, Univ of Oklahoma – Np
Tepley, III, Frank J., 541-737-2064 ftepley@coas.oregonstate.edu, Oregon State Univ – GiCs
Teplitski, Max, 352-392-1951 maxtep@ufl.edu, Univ of Florida – On
Tepper, Jeffrey H., (253) 879-3820 jtepper@pugetsound.edu, Univ of Puget Sound – GiCgGv
ter Voorde, Marlies, +31 20 59 87343 m.ter.voorde@vu.nl, VU Univ Amsterdam – Gqt
Ter-Simonian, Vardui, (213) 740-6106 tersimon@usc.edu, Univ of Southern California – On
Tera, Fouad, (202) 478-8472 ftera@carnegiescience.edu, Carnegie Institution of Washington – CcgXm
Terkla, David, david.terkla@umb.edu, Univ of Massachusetts, Boston – Onn
Terry, Dennis O., (215) 204-8226 doterry@temple.edu, Temple Univ – Gr
Tertyshnikov, Konstantin, +61 8 9266 2297 Konstantin.Tertyshnikov@curtin.edu.au, Curtin Univ – Ye
Terwey, Wes, terwey@southalabama.edu, Univ of South Alabama – Ow
Tesfaye, Samson, (573) 681-5586 Tesfayes@lincolnU.edu, Univ of Missouri – GtcOr
Tesso, Tesfaye, (785) 532-7238 ttesso@ksu.edu, Kansas State Univ – So
Tetrault, Denis, 519-253-3000 ext. 2495 deniskt@uwindsor.ca, Univ of Windsor – Gg
Tett, Simon F., +44 (0) 131 650 5341 Simon.Tett@ed.ac.uk, Edinburgh Univ – Ow
Tettenhorst, Rodney T., 614 247-4246 tettenhorst.2@osu.edu, Ohio State Univ – Gz
Tew , Berry H., 205-247-3679 ntew@gsa.state.al.us, Geological Survey of Alabama – GroGs
Tew, Nick, 205.348.4558 bhtew@ua.edu, Univ of Alabama – EoGro
Tewksbury, Barbara J., (315) 859-4713 btewksbu@hamilton.edu, Hamilton Coll – Gc
Textoris, Daniel A., (919) 962-0690 dtextori@email.unc.edu, Univ of North Carolina, Chapel Hill – Gd
Teyssier, Christian P., (612) 624-6801 teyssier@umn.edu, Univ of Minnesota, Twin Cities – GctGg
Tezcan, Levent, +90 (312) 2977750 tezcan@hacettepe.edu.tr, Hacettepe Univ – Hw
Thackray, Glenn D., (208) 282-3565 thacglen@isu.edu, Idaho State Univ – Gl
Thaisen, Kevin G., 616-331-9219 thaisenk@gvsu.edu, Grand Valley State Univ – XgGxOi
Thakurta, Joyashish, (269) 387-3667 joyashish.thakurta@wmich.edu, Western Michigan Univ – GiEg
Thale, Paul R., 406-496-4653 pthale@mtech.edu, Montana Tech of The Univ of Montana – Oi
Tharp, Thomas M., (765) 494-8678 ttharp1@purdue.edu, Purdue Univ – Nm
Thatje, Sven, +44 (0)23 80592009 svth@noc.soton.ac.uk, Univ of Southampton – Ob
Thayer, Paul A., (910) 520-8719 thayer@uncw.edu, Univ of North Carolina Wilmington – GdoHg
Thayn, Jonathan B., 309-438-8112 jthayn@ilstu.edu, Illinois State Univ – Oir
Theiling, Bethany P., 918-631-2754 bethany-theiling@utulsa.edu, The Univ of Tulsa – Cs
Theis, Karen, +44 0161 275-0407 Karen.Theis@manchester.ac.uk, Univ of Manchester – On
Theissen, Kevin, 651-962-5243 kmtheissen@stthomas.edu, Univ of Saint Thomas – GnOg
Themelis, Nickolas J., (212) 854-2138 njt1@columbia.edu, Columbia Univ – Nx
Thériault, Julie Mireille, 514 987-3000 #4276 theriault.julie@uqam.ca, Universite du Quebec a Montreal – On
Therrien, Francois, (403) 823-7707 Royal Tyrrell Museum of Palaeontology – Pv
Therrien, Pierre, therrien@ggl.ulaval.ca, Universite Laval – Gq
Therrien, Rene, (418) 656-5400 rene.therrien@ggl.ulaval.ca, Universite Laval – HwGe
Thibault, Yves, 613-992-1376 yves.thibault@nrcan.gc.ca, Western Univ – Gz
Thibodeau, Alyson M., (717) 245-8337 thibodea@dickinson.edu, Dickinson Coll – CcGaCs
Thiel, Dr., Volker, +49 (0)551 3914395 vthiel@gwdg.de, Georg-August Univ of Goettingen – CooCo
Thieme, Donald, (229) 333-5752 dmthieme@valdosta.edu, Valdosta State Univ – GaSdGm
Thien, Steve J., (785) 532-7207 sjthien@ksu.edu, Kansas State Univ – Sc
Thigpen, Ryan, 859-2181532 ryan.thigpen@uky.edu, Univ of Kentucky – Gtc
Thirlwall, Matthew, +44 1784 443609 M.Thirlwall@rhul.ac.uk, Univ of London, Royal Holloway & Bedford New Coll – Cs
Thiruvathukal, John V., (973) 655-4417 Montclair State Univ – Yg
Thole, Jeffrey T., (651) 696-6426 thole@macalester.edu, Macalester Coll – GgCaHw
Thomas, Amanda, amthomas@uoregon.edu, Univ of Oregon – Ys
Thomas, Andrew D., +44 (0)1970 622 781 ant23@aber.ac.uk, Univ of Wales – SbGmSf
Thomas, Debbie, 979 862 7742 dthomas@ocean.tamu.edu, Texas A&M Univ – Ou
Thomas, Donald M., (808) 956-6482 dthomas@soest.hawaii.edu, Univ of Hawai'i, Manoa – Ca
Thomas, Elizabeth K., ekthomas@buffalo.edu, SUNY, Buffalo – Cos
Thomas, Ellen, (860) 685-2238 Yale Univ – Pm
Thomas, Florence, (808) 236-7418 fithomas@hawaii.edu, Univ of Hawai'i, Manoa – Ob
Thomas, Helmuth, (902) 494-7177 helmuth.thomas@dal.ca, Dalhousie Univ – Oc
Thomas, Jay, (315)443-7631 jthom102@syr.edu, Syracuse Univ – Gi
Thomas, Jim, (775) 887-7648 tom_j_smith@usgs.gov , Univ of Nevada, Reno – Cg
Thomas, John, 352-392-1951 ext 216 thomas@ufl.edu, Univ of Florida – Sp
Thomas, Kimberly W., (505) 667-4379 Los Alamos National Laboratory – Ct
Thomas, Margaret A., (860) 424-3583 margaret.thomas@ct.gov, Dept of Energy and Environmental Protection – Gg
Thomas, Mark, +44(0) 113 34 35233 m.e.thomas@leeds.ac.uk, Univ of Leeds – Ng
Thomas, Megan D., 410-293-6574 mdthomas@usna.edu, United States Naval Academy – Og
Thomas, Peter, 361749-6768 peter.thomas@utexas.edu, Univ of Texas at Austin – Ob
Thomas, Ray G., (352) 392-7984 rgthomas@ufl.edu, Univ of Florida – Og
Thomas, Robert C., (406) 683-7615 rob.thomas@umwestern.edu, Univ of Montana Western – GseOe
Thomas, Roger D. K., (717) 358-4135 roger.thomas@fandm.edu, Franklin and Marshall Coll – PoGh
Thomas, Valerie, (404) 385-7254 valerie.thomas@isye.gatech.edu, Georgia Inst of Tech – Ou
Thomas, William A., (205) 247-3547 geowat@uky.edu, Geological Survey of Alabama – GtcGr
Thomason, Jason F., 217-244-2508 jthomaso@illinois.edu, Univ of Illinois, Urbana-Champaign – Hw
Thomasson, Joseph R., (785) 628-5665 Fort Hays State Univ – Pb
Thomka, James R., 330 972-5749 jthomka@uakron.edu, Univ of Akron – PogGg
Thompson, Allan M., (302) 831-2585 thompson@udel.edu, Univ of Delaware – Gx
Thompson, Andrew F., 626.395.8345 andrewt@caltech.edu, California Inst of Tech – Op
Thompson, Anu, +440151 794 4095 Anu@liverpool.ac.uk, Univ of Liverpool – Oc
Thompson, Carol A., (254) 968-9739 cthompson@tarleton.edu, Tarleton State Univ – HwGem
Thompson, Christopher J., (509) 376-6602 chris.thompson@pnl.gov, Pacific Northwest National Laboratory – Ca
Thompson, Curtis, (785) 532-5776 cthompso@ksu.edu, Kansas State Univ – So

Thompson, David W. J., davet@atmos.colostate.edu, Colorado State Univ – Yr
Thompson, Geoffrey, (508) 289-2397 gthompson@whoi.edu, Woods Hole Oceanographic Institution – Cm
Thompson, George A., (650) 723-3714 Stanford Univ – Yg
Thompson, Glennis, (510) 642-7025 Univ of California, Berkeley – On
Thompson, Jann W. M., (202) 633-1357 Smithsonian Institution / National Museum of Natural History – Pg
Thompson, Joel B., (727) 864-8991 thompsjb@eckerd.edu, Eckerd Coll – Py
Thompson, John F., jft66@cornell.edu, Cornell Univ – EgNx
Thompson, John F. H., (604) 687-1117 Univ of British Columbia – Em
Thompson, Joseph L., (505) 667-4559 Los Alamos National Laboratory – Ct
Thompson, Keith R., (902) 494-3491 keith.thompson@dal.ca, Dalhousie Univ – Op
Thompson, Kenneth W., 620-341-5985 kthompso@emporia.edu, Emporia State Univ – Oe
Thompson, Lonnie G., (614) 292-6652 thompson.3@osu.edu, Ohio State Univ – Gl
Thompson, LuAnne, (206) 543-9965 luanne@ocean.washington.edu, Univ of Washington – Op
Thompson, Margaret D., (781) 283-3029 mthompson@wellesley.edu, Wellesley Coll – Gc
Thompson, Michael D., (630) 252-9269 Argonne National Laboratory – Yg
Thompson, Michael L., (515) 294-2415 mlthomps@iastate.edu, Iowa State Univ of Science & Tech – Sc
Thompson, Todd A., (812) 855-7428 tthomps@indiana.edu, Indiana Univ – GsmGu
Thompson, Tommy B., (775) 327-5146 tommyt@mines.unr.edu, Univ of Nevada, Reno – Eg
Thompson, Wiley C., 845-938-2305 wiley.thompson@usma.edu, United States Military Academy – Oy
Thompson, William G., (508) 289-2630 wthompson@whoi.edu, Woods Hole Oceanographic Institution – PeCc
Thoms, Richard E., (503) 725-3379 dick@pdx.edu, Portland State Univ – Ps
Thomsen, Charles E., (916) 484-8184 thomsec@arc.losrios.edu, American River Coll – Oy
Thomsen, Laurenz A., l.thomsen@iu-bremen.de, Univ of Washington – Ob
Thomsen, Leon, 713-743-3386 lathomsen@uh.edu, Univ of Houston – Ye
Thomson, Cary, 604-822-0653 cthomson@eos.ubc.ca, Univ of British Columbia – On
Thomson, Cynthia, 212-854-9896 cthomson@iri.columbia.edu, Columbia Univ – On
Thomson, Dennis W., dwt2@psu.edu, Pennsylvania State Univ, Univ Park – Ow
Thomson, Jack, +44 0151 794 3594 Jack.Thomson@liverpool.ac.uk, Univ of Liverpool – Ob
Thomson, Jennifer A., (509) 359-7478 jthomson@ewu.edu, Eastern Washington Univ – Gp
Thomson, Richard E., (250) 363-6555 ThomsonR@pac.dfo-mpo.gc.ca, Univ of British Columbia – Op
Thomson, Vivian E., (434) 924-3964 vet4y@virginia.edu, Univ of Virginia – On
Thorbjarnarson, Kathryn W., (619) 594-5586 kthorbjarnarson@mail.sdsu.edu, San Diego State Univ – Hq
Thorkelson, Derek J., (604) 291-5390 dthorkel@sfu.ca, Simon Fraser Univ – Gt
Thorleifson, Harvey, (612) 626-2150 thorleif@umn.edu, Univ of Minnesota – Gl
Thorleifson, Harvey, 612-627-4780 ext 224 thorleif@umn.edu, Univ of Minnesota, Twin Cities – Gg
Thornberry-Ehrlich, Trista L., (970) 225-3584 tthorn@warnercnr.colostate.edu, Colorado State Univ – GgcGx
Thorncroft, Christopher D., (518) 442-4555 SUNY, Albany – Oa
Thorne, Michael, (801) 585-9792 michael.thorne@utah.edu, Univ of Utah – YsxYv
Thorne, Paul D., (509) 372-4482 paul.thorne@pnl.gov, Pacific Northwest National Laboratory – HgyYg
Thorne, Richard M., (310) 614-6630 rmt@atmos.ucla.edu, Univ of California, Los Angeles – Oa
Thornton, Daniel C., 979 845 4092 dthornton@ocean.tamu.edu, Texas A&M Univ – Ob
Thornton, Edward B., (831) 656-2847 thornton@nps.edu, Naval Postgraduate School – On
Thornton, Joel A., 206-543-4010 thornton@atmos.washington.edu, Univ of Washington – Oa
Thorson, Robert M., (860) 486-1396 robert.thorson@uconn.edu, Univ of Connecticut – Gm
Thosteson, Eric D., (321) 674-8096 thosteson@fit.edu, Florida Inst of Tech – On
Thul, David, 801-585-7013 dthul@egi.utah.edu, Univ of Utah – Gg
Thunell, Robert C., (803) 777-7593 thunell@geol.sc.edu, Univ of South Carolina – Pe
Thurber, Clifford, (608) 262-6027 thurber@geology.wisc.edu, Univ of Wisconsin, Madison – Ysx
Thurber, David L., (718) 997-3300 Graduate School of the City Univ of New York – Cl
Thurber, David L., (718) 997-3325 dlthumper@aol.com, Queens Coll (CUNY) – Cl
Thurnherr, Andreas M., (845) 365-8816 ant@ldeo.columbia.edu, Columbia Univ – Op
Thuro, Kurosch, +498928925850 thuro@tum.de, Technische Universitaet Muenchen – NgrGg
Thurow, Juergen, +44 020 7679 32416 j.thurow@ucl.ac.uk, Univ Coll London – Gs
Thurston, Phillips C., 7056751151 x2372 pthurston@laurentian.ca, Laurentian Univ, Sudbury – OgCgGr
Thurtell, George W., (519) 824-4120 (Ext. 52453) gthurtel@uoguelph.ca, Univ of Guelph – Oa
Thy, Peter, pthy@ucdavis.edu, Univ of California, Davis – Gx
Tiampo, Kristy F., 519-661-3188 ktiampo@uwo.ca, Western Univ – Yd
Tibljaš, Darko, +38514605970 dtibljas@geol.pmf.hr, Univ of Zagreb – GzScGx
Tibuleac, Ileana, (775) 784-6256 ileana@seismo.unr.edu, Univ of Nevada, Reno – Ys
Tice, Mike, 979-845-3138 tice@geo.tamu.edu, Texas A&M Univ – Py
Tick, Geoffrey, gtick@ua.edu, Univ of Alabama – Hw
Tidwell, Allan, (850) 526-2761 tidwella@chipola.edu, Chipola Coll – Og
Tidwell, David, 205-247-3698 dtidwell@gsa.state.al.us, Geological Survey of Alabama – Gm
Tidwell, Vincent C., (505) 844-6025 vctidwe@sandia.gov, New Mexico Inst of Mining and Tech – Hwq
Tierney, Jessica, 520-621-5377 jesst@email.arizona.edu, Univ of Arizona – Co
Tierney, Kate, 319.335.0670 kate-tierney@uiowa.edu, Univ of Iowa – GrCsGg
Tierney, Kate E., 740-587-6487 tierneyk@denison.edu, Denison Univ – CgGrOg
Tiffney, Bruce H., (805) 893-2959 bruce.tiffney@ccs.ucsb.edu, Univ of California, Santa Barbara – Pb
Tijani, Moshood N., +2348023252339 mn.tijani@mail.ui.edu.ng, Univ of Ibadan – HwCgGe
Tikoff, Basil, 608-262-4678 basil@geology.wisc.edu, Univ of Wisconsin, Madison – Gc
Tikoo-Schanz, Sonia, (848) 445-3444 sonia.tikoo@rutgers.edu, Rutgers, The State Univ of New Jersey – YmXy
Tillman, James E., mars@atmos.washington.edu, Univ of Washington – Oa
Tilt, Jenna, 541-737-1232 tiltj@oregonstate.edu, Oregon State Univ – Ou
Tilton, Eric E., 303-735-5033 eric.small@colorado.edu, Univ of Colorado – Hw
Timár, Gábor, +36 1 3722700/1762 timar@caesar.elte.hu, Eotvos Lorand Univ – YgdOi
Timmer, Jaqueline R., 406-496-4842 jtimmer@mtech.edu, Montana Tech of The Univ of Montana – Ca
Timmermann, Axel, (808) 956-2720 axel@hawaii.edu, Univ of Hawai'i, Manoa – Op
Timmermans, Mary-Louise, 203 432-3167 mary-louise.timmermans@yale.edu, Yale Univ – Oc
Timmons, David, david.timmons@umb.edu, Univ of Massachusetts, Boston – On
Timmons, J M., 575-835-5237 mtimmons@nmbg.nmt.edu, New Mexico Inst of Mining and Tech – GctGs
Timmons, Stacy, (575) 835-6951 stacyt@nmbg.nmt.edu, New Mexico Inst of Mining & Tech – Gg
Tindall, Sarah E., (610) 683-4446 tindall@kutztown.edu, Kutztown Univ of Pennsylvania – Gc
Ting, Mingfang, (845) 365-8374 ting@ldeo.columbia.edu, Columbia Univ – Oa
Tingey, David G., 801-422-7752 david_tingey@byu.edu, Brigham Young Univ – On

Tingley, Joseph V., jtingley@unr.edu, Univ of Nevada – Em
Tinjum, James M., 608/262-0785 tinjum@epd.engr.wisc.edu, Univ of Wisconsin, Madison – Ng
Tinker, Scott W., (512) 471-0209 scott.tinker@beg.utexas.edu, Univ of Texas at Austin, Jackson School of Geosciences – Gor
Tinkham, Doug, 7056751151 x2270 dtinkham@laurentian.ca, Laurentian Univ, Sudbury – Gp
Tinkler, Dorothy, 541-881-5967 dtinkler@tvcc.cc, Treasure Valley Community Coll – Oi
Tinnon, Vicki, 228-214-3335 vicki.tinnonbrock@usm.edu, Univ of Southern Mississippi – On
Tinsley, Mark, (434) 832-7708 TinsleyM@cvcc.vccs.edu, Central Virginia Community Coll – Og
Tipper, Ed, ett20@cam.ac.uk, Univ of Cambridge – Cl
Tipping, Robert, (612) 626-5437 tippi001@umn.edu, Univ of Minnesota – Hw
Tissot, Philippe , 361-825-3776 Philippe.Tissot@tamucc.edu, Texas A&M Univ, Corpus Christi – On
Titley, Spencer R., (520) 621-6018 stitley@email.arizona.edu, Univ of Arizona – Eg
Titus, Robert C., (607) 431-4733 titusr@hartwick.edu, Hartwick Coll – Ps
Titus, Sarah J., (507) 222-4419 stitus@carleton.edu, Carleton Coll – Gc
Tivey, Margaret K., (508) 289-3362 mtivey@whoi.edu, Woods Hole Oceanographic Institution – Cm
Tivey, Maurice A., (508) 289-2265 mtivey@whoi.edu, Woods Hole Oceanographic Institution – Gt
Tiwari, Chetan, 940-369-8103 chetan.tiwari@unt.edu, Univ of North Texas – Oi
Tlusty, Michael, mtlusty@neaq.org, Univ of Massachusetts, Boston – Ob
Tobias, Craig, 860-405-9140 craig.tobias@uconn.edu, Univ of Connecticut – Oc
Tobin, Harold, 608-265-5796 htobin@geology.wisc.edu, Univ of Wisconsin, Madison – Yg
Tobin, Harold J., (608)265-5796 htobin@wisc.edu, Univ of Wisconsin, Madison – Yr
Tobin, Tom S., (205) 348-1878 ttobin@ua.edu, Univ of Alabama – PsCsPe
Tobisch, Othmar T., otobisch@yahoo.com, Univ of California, Santa Cruz – Gct
Todd, Brenda, (402) 554-2662 btodd@unomaha.edu, Univ of Nebraska at Omaha – On
Todd, Claire E., (253) 535-5163 toddce@plu.edu, Pacific Lutheran Univ – Glm
Todd, Steve, stephen.todd15@pcc.edu, Portland Community Coll - Sylvania Campus – Ow
Todhunter, Paul, (701) 777-4593 paul.todhunter@und.edu, Univ of North Dakota – Oy
Toke', Nathan, (801) 863-8117 nathan.toke@uvu.edu, Utah Valley Univ – GtmOi
Tokmakian, Robin T., (831) 656-3255 rtt@nps.edu, Naval Postgraduate School – Op
Toksoz, M N., (617) 253-7852 toksoz@mit.edu, Massachusetts Inst of Tech – Ys
Tolhurst, Trevor, +44 (0)1603 59 3124 t.tolhurst@uea.ac.uk, Univ of East Anglia – Sb
Tolstoy, Maria, (845) 365-8791 Columbia Univ – Yr
Tomascak, Paul B., (315) 312-2285 tomascak@oswego.edu, SUNY, Oswego – CgGzi
Tomasko, Martin G., (520) 621-6969 mtomasko@lpl.arizona.edu, Univ of Arizona – Oa
Tomaso, Matthew S., (973) 655-7990 thomasw@mail.montclair.edu, Montclair State Univ – Ga
Tomašiæ, Nenad, +38514605968 ntomasic@geol.pmf.hr, Univ of Zagreb – GziCg
Tomiæ, Vladimir, +38514606094 vtomic@geol.pmf.hr, Univ of Zagreb – GgPgOi
Tomkiewicz, Warren, 603.535.2573 warrent@plymouth.edu, Plymouth State Univ – OggOe
Tomkin, Jonathan H., 217-244-2928 tomkin@illinois.edu, Univ of Illinois, Urbana-Champaign – Gm
Tomlinson, Emma L., + 353 1 8963856 tomlinse@tcd.ie, Trinity Coll – Cc
Tomlinson, Jaime L., 302-831-2649 jaimet@udel.edu, Univ of Delaware – Hg
Tomlinson, Peter, (785) 532-3198 ptomlin@ksu.edu, Kansas State Univ – So
Tompson, Andrew F., (925) 422-6348 tompson1@llnl.gov, Lawrence Livermore National Laboratory – Ng
Tomson, Mason B., (713) 527-6048 mtomson@rice.edu, Rice Univ – Cg
Toner, Brandy M., (612) 624-1362 toner@umn.edu, Univ of Minnesota, Twin Cities – ClmSc
Toner, Rachel, (307) 766-2286 Ext. 248 rachel.toner@wyo.gov, Wyoming State Geological Survey – Eo
Tong, Daoqin, daoqin@email.arizona.edu, Univ of Arizona – Oi
Tong, Vincent C., +44020 3073 8035 vincent.tong@ucl.ac.uk, Birkbeck Coll – Yg
Tongdee, Poetchanaporn, ptongdee@hccfl.edu, Hillsborough Community Coll – Og
Toni, Rousine T., 002-03-3921595 dr_rosine@hotmail.com, Alexandria Univ – GgPg
Tonon, Marco Davide, marco.tonon@unito.it, Università di Torino – Oe
Toomey, Douglas R., (541) 346-5576 drt@uoregon.edu, Univ of Oregon – YsGt
Toon, Sam, (+44) 01782 733698 s.m.toon@keele.ac.uk, Keele Univ – Ys
Toonen, Robert, (808) 236-7425 toonen@hawaii.edu, Univ of Hawai'i, Manoa – Ob
Toor, Gurpal, 813-6334152 gstoor@ufl.edu, Univ of Florida – Sc
Toran, Laura, (215) 204-2352 ltoran@temple.edu, Temple Univ – Hw
Torcellini, Paul, 860-465-0368 torcellinip@easternct.edu, Eastern Connecticut State Univ – On
Torlaschi, Enrico, 514-987-3000 #6848 torlaschi.enrico@uqam.ca, Universite du Quebec a Montreal – Ow
Tornabene, Livio, ltornabe@uwo.ca, Western Univ – XgOr
Toro, Jaime, 3042935603 ext. 4327 jtoro@wvu.edu, West Virginia Univ – Gc
Torreano, Scott, (615) 598-1271 storrean@sewanee.edu, Sewanee: Univ of the South – Sf
Torrents, Alba, (301) 405-1979 alba@eng.umd.edu, Univ of Maryland – Sb
Torres, Joseph J., (727) 553-1169 jtorres@marine.usf.edu, Univ of South Florida – Ob
Torres, Marta E., 541-737-2902 mtorres@coas.oregonstate.edu, Oregon State Univ – Cm
Torres, Raymond, (803) 777-4506 torres@geol.sc.edu, Univ of South Carolina – Hy
Torres Parisian, Cathleen, ctorresp@umn.edu, Univ of Minnesota, Twin Cities – Or
Tortell, Philippe, (604) 822-4728 ptortell@eos.ubc.ca, Univ of British Columbia – Ob
Torvela, Taija, +44(0) 113 34 36620 t.m.torvela@leeds.ac.uk, Univ of Leeds – GctEg
Tosdal, Richard, rtosdal@eos.ubc.ca, Univ of British Columbia – Em
Toth, Charles K., 614-292-7681 toth.2@osu.edu, Ohio State Univ – OrYd
Toth, Emoke, tothemoke.pal@gmail.com, Eotvos Lorand Univ – PmsPb
Tóth, József, (780) 492-1115 joe.toth@ualberta.ca, Univ of Alberta – Hw
Totsche, Kai U., +49(0)3641 948650 kai.totsche@uni-jena.de, Friedrich-Schiller-Univ Jena – HwClOs
Totten, Matthew W., (785) 532-2227 mtotten@ksu.edu, Kansas State Univ – Gso
Totten, Stanley M., (812) 866-7245 totten@hanover.edu, Hanover Coll – Gl
Touchton, Joseph T., (334) 844-3952 jtouchto@acesag.auburn.edu, Auburn Univ – Sc
Toullec, Renaud, +33(0)3 44069333 renaud.toullec@lasalle-beauvais.fr, Institut Polytechnique LaSalle Beauvais (ex-IGAL) – GsoEo
Towery, Brooke L., (850) 484-2056 btowery@pensacolastate.edu, Pensacola Junior Coll – Gg
Towner, Ronald, (520) 621-6465 rht@email.arizona.edu, Univ of Arizona – Gae
Townsend-Small, Amy, amy.townsend-small@uc.edu, Univ of Cincinnati – Co
Toy, Virginia G., 64 3 479 7506 virginia@geology.co.nz, Univ of Otago – Gct
Tracy, Matthew, matthew.tracy@usafa.edu, United States Air Force Academy – On
Tracy, Robert J., (540) 231-5980 rtracy@vt.edu, Virginia Polytechnic Inst & State Univ – Gp
Tracy, Sally J., (609) 258-3401 sjtracy@princeton.edu, Princeton Univ – Gy
Tran, Linda C., 281-765-7865 Linda.C.Tran@nhmccd.edu, Lonestar Coll - North Harris – Oi
Trandafir, Aurel, 801-581-7062 a.trandafir@utah.edu, Univ of Utah – Nr
Tranel, Lisa M., (309) 438-7966 ltranel@ilstu.edu, Illinois State Univ –

Gmt
Trapp, Robert J., 217-300-0967 jtrapp@illinois.edu, Univ of Illinois, Urbana-Champaign – Oa
Travers, Steven, (309) 268-8640 steve.travers@heartland.edu, Heartland Community Coll – Og
Travis, Bryan J., (505) 667-1254 bjtravis@lanl.gov, Los Alamos National Laboratory – Hq
Treadwell-Steitz, Carol, (518) 564-2028 Plattsburgh State Univ (SUNY) – Gg
Tredoux, Marian, +27 (0)51 401 9016 mtredoux@ufs.ac.za, Univ of the Free State – Cga
Trefry, John H., (321) 674-7305 jtrefry@fit.edu, Florida Inst of Tech – Cm
Trehu, Anne M., 541-737-2655 trehu@coas.oregonstate.edu, Oregon State Univ – Yr
Treloar, Peter, +44020 8417 2525 P.Treloar@kingston.ac.uk, Kingston Univ – Gc
Trembanis, Arthur C., 302-831-2498 art@udel.edu, Univ of Delaware – On
Tremblay, Alain, 514-987-3000 #1397 tremblay.alain@uqam.ca, Universite du Quebec a Montreal – Gct
Tremblay, Bruno, (514) 398-4369 bruno.tremblay@mcgill.ca, McGill Univ – Op
Tremblay, Thomas A., (512) 475-9537 tom.tremblay@beg.utexas.edu, Univ of Texas at Austin, Jackson School of Geosciences – Oy
Trenhaile, Alan S., 519-253-3000 ext. 2184 tren@uwindsor.ca, Univ of Windsor – Gm
Trentham, Robert C., 432 552-2432 trentham_r@utpb.edu, Univ of Texas, Permian Basin – Go
Treude, Tina, (310) 267-5213 ttreude@g.ucla.edu, Univ of California, Los Angeles – ObPyCm
Trevino, Ramon H., (512) 471-3362 ramon.trevino@beg.utexas.edu, Univ of Texas at Austin, Jackson School of Geosciences – GsoGe
Trevors, Jack, (519) 824-4120 Ext.53367 jtrevors@uoguelph.ca, Univ of Guelph – On
Treworgy, Janis D., (618) 374-5294 janis.treworgy@principia.edu, Principia Coll – GgPvGe
Trexler, James H., (775) 784-1504 trexler@mines.unr.edu, Univ of Nevada, Reno – Gs
Triay, Ines, (505) 665-1755 Los Alamos National Laboratory – Hg
Trifonoff, Karen M., (570) 389-4569 ktrifono@bloomu.edu, Bloomsburg Univ – Oi
Tripati, Aradhna, ripple@epss.ucla.edu, Univ of California, Los Angeles – CmPeCs
Triplehorn, Donald M., (907) 474-6891 Univ of Alaska, Fairbanks – Gs
Triplett, Laura, (507) 933-7442 ltriplet@gustavus.edu, Gustavus Adolphus Coll – GemCl
Tripoli, Gregory J., (608)262-3700 tripoli@aos.wisc.edu, Univ of Wisconsin, Madison – Ow
Trixler, Frank, 089/2179 509 trixler@lrz.uni-muenchen.de, Ludwig-Maximilians-Universitaet Muenchen – GzOmXc
Troch, Peter A., (520) 626-1277 patroch@hwr.arizona.edu, Univ of Arizona – Hs
Trocine, Robert P., (407) 674-8096 Florida Inst of Tech – Cm
Trofimovs, Jessica, +61 7 3138 2766 jessica.trofimovs@qut.edu.au, Queensland Univ of Tech – GsvGu
Tromp, Jeroen, (609) 258-4128 jtromp@princeton.edu, Princeton Univ – Yss
Troost, Kathy G., 206-221-1770 ktroost@uw.edu, Univ of Washington – Oun
Trop, Jeffrey M., (570) 577-3027 jtrop@bucknell.edu, Bucknell Univ – Gs
Trost, G K., (501) 575-7317 garrett@dicksonstreet.com, Univ of Arkansas, Fayetteville – On
Trouet, Valerie, (520) 626-8004 trouet@email.arizona.edu, Univ of Arizona – PeOya
Trout, Jennifer, (269) 387-8633 jennifer.l.trout@wmich.edu, Western Michigan Univ – OnnOn
Troy, Marleen, (570) 408-4615 marleen.troy@wilkes.edu, Wilkes Univ – Hw
Trudgill, Bruce D., (303) 273-3883 btrudgil@mines.edu, Colorado School of Mines – Gc
Trueman, Clive, +44 (0)23 80596571 trueman@noc.soton.ac.uk, Univ of Southampton – Cg
Trujillo, Alan P., 7607441150x2734 atrujillo@palomar.edu, Palomar Coll – OguOe
Trullenque, Ghislain, +33(0)3 44069327 ghislain.trullenque@laslle-beauvais.fr, Institut Polytechnique LaSalle Beauvais (ex-IGAL) – GcNrGg
Trumbore, Susan E., (949) 824-6142 setrumbo@uci.edu, Univ of California, Irvine – Sc
Trussell, Geoffrey, 7815817370 x300 g.trussell@neu.edu, Northeastern Univ – On
Trust, Michael, (617) 287-5287 Univ of Massachusetts, Boston – Or
Tsai, Victor, 626.395.6993 tsai@caltech.edu, California Inst of Tech – Yg
Tshudy, Dale, (814) 732-2453 dtshudy@edinboro.edu, Edinboro Univ of Pennsylvania – Pi
Tsikos, Hari, +27 (0)46-603-8511 h.tsikos@ru.ac.za, Rhodes Univ – CgEg
Tso, Jonathan L., (540) 831-5638 jtso@radford.edu, Radford Univ – Gcp
Tsoflias, George P., (785) 864-4584 tsoflias@ku.edu, Univ of Kansas – YgeGe
Tsujita, Cameron J., (519) 661-2111 (Ext. 86740) ctsujita@uwo.ca, Western Univ – Pi
Tsvankin, Ilya D., (303) 273-3060 ilya@dix.mines.edu, Colorado School of Mines – Ye
Tu, Wei, (912) 478-5233 wtu@georgiasouthern.edu, Georgia Southern Univ – Oi
Tubía, Jose M., jm.tubia@ehu.eus, Univ of the Basque Country UPV/EHU – GctYg
Tucholke, Brian E., (508) 289-2494 btucholke@whoi.edu, Woods Hole Oceanographic Institution – Gut
Tucker, Carla M., (409) 880-8236 ctucker@lamar.edu, Lamar Univ – Hw
Tucker, Gregory E., (303) 492-6985 gregory.tucker@colorado.edu, Univ of Colorado – Gmt
Tudhope, Alexander W., +44 (0) 131 650 8508 Sandy.Tudhope@ed.ac.uk, Edinburgh Univ – Cs
Tufford, Dan, tufford@sc.edu, Univ of South Carolina – On
Tulaczyk, Slawek, (831) 459-5207 stulaczy@ucsc.edu, Univ of California, Santa Cruz – GlmNg
Tull, James F., (850) 644-1448 jtull@fsu.edu, Florida State Univ – Gct
Tuller, Markus, (520) 621-7225 mtuller@cals.arizona.edu, Univ of Arizona – On
Tullis, Jan A., (401) 863-1921 jan_tullis@brown.edu, Brown Univ – Gcy
Tullis, Jason A., (479) 575-8784 jatullis@uark.edu, Univ of Arkansas, Fayetteville – Ori
Tullis, Terry E., (401) 863-3829 Terry_Tullis@Brown.edu, Brown Univ – Yx
Tung, Ka-Kit, tung@amath.washington.edu, Univ of Washington – Oa
Tung, Wen-wen, (765) 494-0272 wwtung@purdue.edu, Purdue Univ – Oa
Tunnicliffe, Verena, (250) 721-7135 verenat@uvic.ca, Univ of Victoria – Ob
Turbeville, John, (760) 944-4449, x6413 JTurbeville@miracosta.edu, MiraCosta Coll – GgOg
Turchyn, Alexandra, +44 (0) 1223 333479 atur07@esc.cam.ac.uk, Univ of Cambridge – Ge
Turco, Richard, (310) 825-6936 turco@atmos.ucla.edu, Univ of California, Los Angeles – Oa
Turcotte, Donald L., 530-752-6808 dlturcotte@ucdavis.edu, Univ of California, Davis – Yg
Turek, Andrew, (519) 969-6710 turek@uwindsor.ca, Univ of Windsor – Cc
Turk, Judith , 609 652 4209 judith.turk@stockton.edu, Richard Stockton Coll of New Jersey – Sfo
Turk, Judy, Judith.Turk@stockton.edu, Stockton Univ – SpGmSo
Turnbull, William D., (312) 665-7092 collections@fieldmuseum.org, Field Museum of Natural History – Pv
Turner, A. Keith, kturner@mines.edu, Colorado School of Mines – Ng
Turner, Derek, 604-777-6568 turnerd1@douglasColl.ca, Douglas Coll – Gem
Turner, Elizabeth C., 7056751151 x2267 eturner@laurentian.ca, Laurentian Univ, Sudbury – GrdPo
Turner, Eugene, eugene.turner@csun.edu, California State Univ, Northridge – Or
Turner, Grenville, +44 0161 275-0401 grenville.turner@manchester.ac.uk, Univ of Manchester – Xc
Turner, Jay, (314) 935-5480 jrturner@wustl.edu, Washington Univ in St. Louis – On
Turner, Jenni, +44 (0)1603 59 3109 jenni.turner@uea.ac.uk, Univ of East Anglia – Gt
Turner, Kerry, +44 (0)1603 59 2551 r.k.turner@uea.ac.uk, Univ of East Anglia – GeEg
Turner, Mark, (972) 881-5936 mturner@collin.edu, Collin Coll - Spring Creek Campus – Gg
Turner, Robert E., (225) 388-6454 Louisiana State Univ – Ob
Turner, Robert S., rtt@ornl.gov, Oak Ridge National Laboratory – Cl
Turner, Wesley L., (936) 468-1049 turnerwl@sfasu.edu,

Stephen F. Austin State Univ – Gg
Turner, III, Henry, (479) 575-7295 hturner@uark.edu, Univ of Arkansas, Fayetteville – GgtGc
Turnock, Allan C., (204) 474-6911 ac_turnock@umanitoba.ca, Univ of Manitoba – Gp
Turpening, Roger M., (906) 487-1784 roger@mtu.edu, Michigan Technological Univ – Ye
Turrin, Brent D., 848-445-3177 bturrin@rci.rutgers.edu, Rutgers, The State Univ of New Jersey – CcGvYm
Turski, Mark P., (603) 535-2749 markt@plymouth.edu, Plymouth State Univ – Og
Turtle, Elizabeth, (520) 621-8284 turtle@lpl.arizona.edu, Univ of Arizona – Xg
Tuttle, Samuel , stuttle@mtholyoke.edu, Mount Holyoke Coll – Gq
Tvelia, Sean, (631) 451-4303 tvelias@sunysuffolk.edu, Suffolk County Community Coll, Ammerman Campus – Ggl
Twelker, Evan, 907-451-5086 evan.twelker@alaska.gov, Alaska Division of Geological & Geophysical Surveys – Eg
Twine, Tracy E., 612-625-7278 twine@umn.edu, Univ of Minnesota, Twin Cities – Oa
Twiss, Robert J., (530) 752-1860 rjtwiss@ucdavis.edu, Univ of California, Davis – Gc
Tyburczy, James A., (480) 965-2637 jim.tyburczy@asu.edu, Arizona State Univ – GyzYx
Tyce, Robert, (401) 874-6879 tyce@oce.uri.edu, Univ of Rhode Island – Ou
Tyler, Carrie, 513-529-8311 tylercl@miamioh.edu, Miami Univ – Pb
Tyler, E. Jerry, (608) 262-0853 ejtyler@wisc.edu, Univ of Wisconsin, Madison – Sd
Tyler, Scott, (775) 784-6250 styler@unr.edu, Univ of Nevada, Reno – Hg
Tynan, Cynthia T., (206) 860-6793 ctynan@coastalstudies.org, Univ of Washington – Ob
Tyning, Thomas F., (413) 236-4502 ttyning@berkshirecc.edu, Berkshire Community Coll – On
Tyrrell, Toby, +44 (0)23 80596110 toby.tyrrell@soton.ac.uk, Univ of Southampton – OgPe
Tziperman, Eli, (617) 384-8381 eli@eps.harvard.edu, Harvard Univ – Op
Törnqvist, Torbjörn E., 504-314-2221 tor@tulane.edu, Tulane Univ – Gs

## U

Uahengo, Collen, +8613021208022 cuahengo@unam.na, Univ of Namibia – GgoEg
Uchupi, Elazar, (508) 289-2830 Woods Hole Oceanographic Institution – Ou
Uddin, Ashraf, (334) 844-4885 uddinas@auburn.edu, Auburn Univ – GdOn
Ufkes, Els , +31 20 5989953 els.ufkes@falw.vu.nl, VU Univ Amsterdam – PmOgGu
Ugland, Richard, 541-552.6479 lane@sou.edu, Southern Oregon Univ – Gg
Uhen, Mark, (703) 993-5264 muhen@gmu.edu, George Mason Univ – Pv
Ukstins Peate, Ingrid, 319-335-1824 ingrid-peate@uiowa.edu, Univ of Iowa – GviXg
Ulanski, Stanley L., (540) 568-6130 ulanskl@jmu.edu, James Madison Univ – Og
Ullman, David J., 715.682.1312 dullman@northland.edu, Northland Coll – GlPeGm
Ullman, William J., (302) 645-4302 ullman@udel.edu, Univ of Delaware – Cm
Ulmer, Gene C., (215) 204-7171 gulmer@temple.edu, Temple Univ – Cp
Ulmer-Scholle, Dana S., (575) 835-5673 dana.ulmer-scholle@nmt.edu, New Mexico Inst of Mining and Tech – GsdCl
Umhoefer, Paul J., (928) 523-6464 paul.umhoefer@nau.edu, Northern Arizona Univ – Gt
Underwood, Ben, bsunderw@umail.iu.edu, Indiana Univ, Bloomington – Ca
Underwood, Charlie, +44 020 3073 8036 Birkbeck Coll – Pg
Underwood, Michael B., (573) 882-4685 underwoodm@missouri.edu, Univ of Missouri – GsuGt
Underwood, Stephen J., (912) 478-5361 SJUnderwood@GeorgiaSouthern.edu, Georgia Southern Univ – Ow
Unger, Corey, (801) 538-4810 coreyunger@utah.gov, Utah Geological Survey – Oi
Unkefer, Clifford J., (505) 665-2560 Los Alamos National Laboratory – On
Unrug, Kot F., (606) 257-1883 Univ of Kentucky – Nr
Unsworth, Martyn, (780) 492-3041 martyn.unsworth@ualberta.ca, Univ of Alberta – GtvYe
Unsworth, Martyn, unsworth@ualberta.ca, Cornell Univ – Ye
Upchurch, Paul, +44 020 7679 37947 p.upchurch@ucl.ac.uk, Univ Coll London – Po
Ural, Melek, 00904242370000-5960 melekural@firat.edu.tr, Firat Univ – GivCc
Urban-Rich, Juanita, juanita.urban-rich@umb.edu, Univ of Massachusetts, Boston – Ob
Urbanczyk, Kevin, (432) 837-8110 kevinu@sulross.edu, Sul Ross State Univ – GiOi
Uriarte, Jesus Angel, jesusangel.uriarte@ehu.eus, Univ of the Basque Country UPV/EHU – NgHg
Urosevic, Milovan, +61 8 9266-2296 M.Urosevic@curtin.edu.au, Curtin Univ – YerYx
Urquhart, Mary, 972-883-2496 urquhart@utdallas.ede, Univ of Texas, Dallas – Xy
Urzua, Alfredo, 617-552-8339 alfredo.urzua@bc.edu, Boston Coll – Ng
Ustaszewski, Kamil, 0049(0)3641/948623 kamil.u@uni-jena.de, Friedrich-Schiller-Univ Jena – Gc
Ustin, Susan L., (530) 752-0621 slustin@ucdavis.edu, Univ of California, Davis – Or
Ustunisik, Gokce K., (605) 394-2461 Gokce.Ustunisik@sdsmt.edu, South Dakota School of Mines & Tech – Cp
Utgard, Russell O., 614 247-4246 utgard.1@osu.edu, Ohio State Univ – Ge
Utley, Tom, 321-674-8120 tutley@fit.edu, Florida Inst of Tech – Ow
Utter, James M., (914) 251-6642 SUNY, Purchase – On
Uzochukwu, Godfrey A., (336) 285-4866 uzo@ncat.edu, North Carolina Agricultural & Tech State Univ – Gz
Uzunlar, Nuri, (605) 394-2494 nuri.uzunlar@sdsmt.edu, South Dakota School of Mines & Tech – EmGcEo

## V

Vacher, H. Leonard, (813) 974-5267 vacher@chuma.cas.usf.edu, Univ of South Florida, Tampa – Hw
Vacquier, Victor D., (858) 534-4803 vvacquier@ucsd.edu, Univ of California, San Diego – Ob
Vaezi, Reza, +98 (411) 339 2721 vaezi@tabrizu.ac.ir, Univ of Tabriz – GeHgs
Vail, Lance W., (509) 372-6237 lance.vail@pnl.gov, Pacific Northwest National Laboratory – Hg
Vail, Peter R., (713) 348-4888 vail@rice.edu, Rice Univ – Gr
Vaillancourt, Robert, 717-872-3294 Robert.Vaillancourt@millersville.edu, Millersville Univ – Obc
Valdes, Juan B., (520) 621-2266 jvaldes@email.arizona.edu, Univ of Arizona – Hs
Valencia, Victor A., (520) 300-1605 victor.valencia@wsu.edu, Washington State Univ – OgGit
Valenti, Christine, (973) 655-4448 Montclair State Univ – Gg
Valentine, David, valentine@geol.ucsb.edu, Univ of California, Santa Barbara – Cm
Valentine, Gregory, (716) 645-4295 gav4@buffalo.edu, SUNY, Buffalo – Gv
Valentine, James W., (510) 643-5791 Univ of California, Berkeley – Po
Valentine, Michael , mvalentine@highline.edu, Highline Coll – Gt
Valentine, Michael, 253-566-5060 mvalentine@tacomacc.edu, Tacoma Community Coll – Og
Valentine, Michael J., (253) 879-3129 mvalentine@pugetsound.edu, Univ of Puget Sound – YmGc
Valentino, David W., (315) 312-2798 dvalenti@oswego.edu, SUNY, Oswego – GtcGp
Valley, John W., (608) 263-5659 valley@geology.wisc.edu, Univ of Wisconsin, Madison – CsGxz
Valsami-Jones, Eugenia, +44 (0)121 414 5537 e.valsamijones@bham.ac.uk, Univ of Birmingham – ClGeCa
van Alstine, James, +44(0) 113 34 37531 J.VanAlstine@leeds.ac.uk, Univ of Leeds – Ge
Van Alstine, James B., (320) 589-6313 vanalstj@mrs.umn.edu, Univ of Minnesota, Morris – Pg
Van Arsdale, Roy B., (901) 678-4356 rvanrsdl@memphis.edu, Univ of Memphis – Gcm
Van Avendonk, Harm, 512-471-0429 harm@ig.utexas.edu, Univ of Texas at Austin – Yg
van Balen, Ronald T., +31 20 59 87324 r.t.van.balen@vu.nl, VU Univ Amsterdam – GgmGt

van Bever Donker, Jan M., +27 021 959 3263 jvanbeverdonker@uwc.ac.za, Univ of the Western Cape – GcpGt
Van Brocklin, Matthew F., (315) 229-5197 mvanbrocklin@stlawu.edu, St. Lawrence Univ – GgOgn
Van Buer, Nicholas J., 909-869-3457 njvanbuer@csupomona.edu, California State Polytechnic Univ, Pomona – GxtGz
van de Flierdt, Tina, +44 20 759 41290 tina.vandeflierdt@imperial.ac.uk, Imperial Coll – Cs
van de Gevel, Saskia, (828) 262-7028 gevelsv@appstate.edu, Appalachian State Univ – Og
Van De Poll, Henk W., (506) 453-4804 Univ of New Brunswick – Gs
Van de Water, Peter, 559-278-2912 pvandewater@csufresno.edu, California State Univ, Fresno – GrPbGf
van den Akker, Ben, ben.vandenakker@flinders.edu.au, Flinders Univ – On
Van Den Broeke, Matthew S., (402) 472-2418 mvandenbroeke2@unl.edu, Univ of Nebraska, Lincoln – Oar
van den Heever, Sue, sue@atmos.colostate.edu, Colorado State Univ – Oa
Van Den Hoek, Jamon, 541-737-1229 vandenhj@oregonstate.edu, Oregon State Univ – Our
van der Berg, Stan, +440151 794 4096 Vandenberg@liverpool.ac.uk, Univ of Liverpool – Em
Van Der Flier-Keller, Eileen, (250) 472-4019 fkeller@uvic.ca, Univ of Victoria – Cl
van der Hilst, Robert, (617) 253-6977 hilst@mit.edu, Massachusetts Inst of Tech – Ys
van der Horst, Dan, +44 (0) 131 651 4467 Dan.vanderHorst@ed.ac.uk, Edinburgh Univ – Eg
van der Land, Crees, +44 (0) 191 208 6513 cees.van.der.land@ncl.ac.uk, Univ of Newcastle Upon Tyne – Go
van der Lee, Suzan, (847) 491-8183 suzan@earth.northwestern.edu, Northwestern Univ – YsxYg
van der Lubbe, Jeroen, +31 20 59 87366 h.j.l.vander.lubbe@vu.nl, VU Univ Amsterdam – Gs
van der Pluijm, Ben, (734) 763-0373 vdpluijm@umich.edu, Univ of Michigan – GctOe
van der Velde, Ype, +31 20 59 87402 y.vander.velde@vu.nl, VU Univ Amsterdam – Hg
van der Voo, Rob, (734) 764-8322 voo@umich.edu, Univ of Michigan – Ym
van der Werf, Guido R., +31 20 59 85687 guido.vander.werf@vu.nl, VU Univ Amsterdam – Sb
van der Zande, Emma, +31 20 59 86517 emma.vander.zanden@vu.nl, VU Univ Amsterdam – Oy
van Dijk, Deanna, (616) 526-6510 dvandijk@calvin.edu, Calvin Coll – GmOny
van Dongen, Bart, +44 0161 306-7460 Bart.VanDongen@manchester.ac.uk, Univ of Manchester – Co
Van Eerd, Laura, (519) 674-1644 lvaneerd@ridgetownc.uoguelph.ca, Univ of Guelph – On
van Gardingen, Paul R., +44 (0) 131 650 7253 P.Vangardingen@ed.ac.uk, Edinburgh Univ – On
Van Geen, Alexander, (845) 365-8644 Columbia Univ – Cg
van Hees, Edmond H., (313) 577-9436 midas@wayne.edu, Wayne State Univ – GzxEg
van Hinsberg, Vincent, vincent.vanhinsberg@mcgill.ca, McGill Univ – EgCg
Van Horn, Stephen R., (740) 826-8306 svanhorn@muskingum.edu, Muskingum Univ – GeOiEo
van Huissteden, Ko, +31 20 59 87354 j.van.huissteden@vu.nl, VU Univ Amsterdam – CgSb
van Hunen, Jeroen, +44 (0) 191 33 42293 jeroen.van-hunen@durham.ac.uk, Durham Univ – Gt
Van Iten, Heyo, (812) 866-7303 vaniten@hanover.edu, Hanover Coll – Pg
van Keken, Peter J., (734) 764-1497 keken@umich.edu, Univ of Michigan – Yg
Van Kooten, Gerald K., (616) 526-6374 gvankooten1@gmail.com, Calvin Coll – GoCePi
van Kranendonk, Martin, m.vankranendonk, Univ of New South Wales – GxPeGd
Van Loon, Lisa, 306-657-3585 lisa.vanloon@lightsource.ca, Western Univ – Ca
Van Mooy, Benjamin, (508) 289-2740 bvanmooy@whoi.edu, Woods Hole Oceanographic Institution – Oc
Van Nest, Julieann, (518) 474-5814 julieann.vannest@nysed.gov, New York State Geological Survey – GamOi
Van Niewenhuise, Donald, 713-743-3423 donvann@uh.edu, Univ of Houston – Ps
van Norden, Maxim F., (228) 688-7123 maxim.vannorden@usm.edu, Univ of Southern Mississippi – Op
Van Oostende, Nicolas, 609-258-1052 oostende@princeton.edu, Princeton Univ – Ob
Van Orman, James A., jav12@case.edu, Case Western Reserve Univ – CgGyCp
van Oss, Carel J., 716-829-2900 SUNY, Buffalo – Gy
Van Roosendaal, Susan, (801) 581-8218 Univ of Utah – On
Van Ry, Michael, mvanry@occ.cccd.edu, Orange Coast Coll – Gv
Van Ryswick, Stephen, 410-554-5544 stephen.vanryswick@maryland.gov, Maryland Department of Natural Resources – Ge
Van Schmus, W. Randall, (785) 864-2727 rvschmus@ku.edu, Univ of Kansas – Cc
Van Stan, John, (912) 478-8040 jvanstan@georgiasouthern.edu, Georgia Southern Univ – HgClSb
Van Straaten, H. Peter, (519) 824-4120 x52454 pvanstra@uoguelph.ca, Univ of Guelph – En
van Teeffelen, Astrid, +31 20 59 82526 astrid.van.teeffelen@vu.nl, VU Univ Amsterdam – Oy
Van Tongeren, Jill A., (848) 445-5363 jvantongeren@eps.rutgers.edu, Rutgers, The State Univ of New Jersey – GpcGt
Van Vleet, Edward S., (727) 553-1165 vanvleet@usf.edu, Univ of South Florida – Co
van Vliet, Jasper, +31 20 59 83052 jasper.van.vliet@vu.nl, VU Univ Amsterdam – Oy
van Westrenen, Wim, w.van.westrenen@vu.nl, VU Univ Amsterdam – CpGz
Van Wijk, Jolante, (575) 835-6661 jolante.vanwijk@nmt.edu, New Mexico Inst of Mining and Tech – GtoYe
Vanacore, Elizabeth, (787) 265-3845 elizabeth.vanacore@upr.edu, Univ of Puerto Rico – Yxs
VanAller-Hernick, Linda A., (518) 486-3699 lhernick@mail.nysed.gov, New York State Geological Survey – Pgb
Vancas, Tina, (814) 865-2622 tqs5@psu.edu, Pennsylvania State Univ, Univ Park – On
Vance, Robert K., (912) 478-5353 rkvance@georgiasouthern.edu, Georgia Southern Univ – GgEgGi
Vandeberg, Gregory S., 701-777-4588 gregory.vandeberg@und.edu, Univ of North Dakota – GmOiGl
Vandemark, Douglas C., 603-862-0195 doug.vandemark@unh.edu, Univ of New Hampshire – Or
Vandenbroucke, Thijs, thijs.vandenbroucke@ugent.be, Ghent Univ – Ps
Vander Auwera, Jacqueline, 32 4 366 22 53 jvdauwera@ulg.ac.be, Universite de Liege – Giv
Vanderkluysen, Loyc, (215) 571-4673 loyc@drexel.edu, Drexel Univ – GviGz
Vanderlip, Richard, (785) 532-6101 vanderrl@ksu.edu, Kansas State Univ – So
VanDervoort, Dane S., (205) 247-3626 dvandervoort@gsa.state.al.us, Geological Survey of Alabama – GcpOi
VanDerwerker, Tiffany J., 410-554-5558 tiffany.vanderwerker@maryland.gov, Maryland Department of Natural Resources – Hw
Vandike, James E., (573) 341-4616 Missouri Univ of Science and Tech – Hg
VanDorpe, Paul E., Paul.VanDorpe@dnr.iowa.gov, Iowa Dept of Natural Resources – Gg
VanGundy, Robert D., (276) 376-4656 rdv4v@uvawise.edu, Univ of Virginia Coll, Wise – GeHgOg
VanHorn, Jason, (616) 526-7623 jev35@calvin.edu, Calvin Coll – Oi
Vaniman, David T., (505) 667-1863 vaniman@lanl.gov, Los Alamos National Laboratory – Gx
vanKeken, Peter E., pvankeken@carnegiescience.edu, Carnegie Institution of Washington – GtYsCg
Vanko, David A., (410) 704-2121 dvanko@towson.edu, Towson Univ – Gxz
Vann, David R., (215) 898-4906 drvann@sas.upenn.edu, Univ of Pennsylvania – GeSfXm
Vannier, Ryan G., 616-331-3164 vannierr@gvsu.edu, Grand Valley State Univ – CgGn
Vannucchi, Paola, +44 01784 443616 Paola.Vannucchi@rhul.ac.uk, Univ of London, Royal Holloway & Bedford New Coll – Gu
Vanos, Jennifer K., (806) 786-2577 jkvanos@ucsd.edu, Univ of California, San Diego – Oa
Vaqueiro Rodriguez, Marcos, 0034637756150 mvaqueiro@frioya.es,

Coruna Univ – GcmYd
Varadi, Ferenc D., (310) 206-6484 Univ of California, Los Angeles – On
Varady, Robert G., (520) 626-4393 rvarady@email.arizona.edu, Univ of Arizona – On
Varekamp, Johan C., (860) 685-2248 jvarekamp@wesleyan.edu, Wesleyan Univ – CgGzv
Varela, Diana, 250-721-6120 dvarela@uvic.ca, Univ of Victoria – Ob
Vargo, Gabriel A., (727) 553-1167 gvargo@marine.usf.edu, Univ of South Florida – Ob
Varner, Ruth K., (603) 862-0853 ruth.varner@unh.edu, Univ of New Hampshire – Cl
Varricchio, David J., (406) 994-6907 djv@montana.edu, Montana State Univ – Pv
Vassiliou, Andreas H., 973-353-5100 ahvass@andromeda.rutgers.edu, Rutgers, The State Univ of New Jersey, Newark – Gz
Vaudrey, Jamie, 860-405-9149 jamie.vaudrey@uconn.edu, Univ of Connecticut – Ob
Vaughan, David, +44 0161 275-3935 david.vaughan@manchester.ac.uk, Univ of Manchester – GzClGe
Vaughan, Geraint, +44 0161 306-3931 Geraint.Vaughan@manchester.ac.uk, Univ of Manchester – Oa
Vaughan, Michael T., 631 632-8030 michael.vaughan@sunysb.edu, SUNY, Stony Brook – Yx
Vaughan, Naomi, +44 (0)1603 59 3904 n.vaughan@uea.ac.uk, Univ of East Anglia – Ow
Vautier, Yannick, +33(0)3 44068991 yannick.vautier@lasalle-beauvais.fr, Institut Polytechnique LaSalle Beauvais (ex-IGAL) – GodCo
Vaz, Nuno M., nunovaz@utad.pt, Universidade de Trás-os-Montes e Alto Douro – PmlYg
Veblen, Thomas T., (303) 492-8528 veblen@colorado.edu, Univ of Colorado – Og
Veeder, Glenn J., (818) 354-7388 Jet Propulsion Laboratory – On
Veeger, Anne I., (401) 874-2187 veeger@uri.edu, Univ of Rhode Island – HwCl
Vega, Anthony J., (814) 226-2023 avega@clarion.edu, Clarion Univ – OawOp
Vegas, Nestor, 00346015374 nestor.vegas@ehu.eus, Univ of the Basque Country UPV/EHU – GcYmGt
Veizer, Jan, 613-562-5800 6461 Veizer@uottawa.ca, Univ of Ottawa – Cl
Velasco, Aaron A., (915) 747-5101 aavelasco@utep.edu, Univ of Texas, El Paso – Yx
Velbel, Michael A., (517) 355-4626 velbel@msu.edu, Michigan State Univ – ClGdSc
Velicogna, Isabella, 949-824-2685 isabella.velicogna@gmail.com, Univ of California, Irvine – Or
Velinsky, David, 215 299 1147 velinsky@drexel.edu, Drexel Univ – OcCms
Velli, Marco, 310-206-4892 mvelli@ucla.edu, Univ of California, Los Angeles – Yg
Vengosh, Avner, 919 684 5847 vengosh@duke.edu, Duke Univ – Hw
Venkatesan, M I., (310) 206-2561 Univ of California, Los Angeles – Co
Venn, Cynthia, (570) 389-4141 cvenn@bloomu.edu, Bloomsburg Univ – OgClOc
Venterea, Rodney T., (612) 624-7842 venterea@soils.umn.edu, Univ of Minnesota, Twin Cities – Sp
Venti, Nicholas, 413-687-3515 nventi@geo.umass.edu, Univ of Massachusetts, Amherst – Og
Vento, Frank, 814-824-2581 fvento@mercyhurst.edu, Mercyhurst Univ – GaOsGm
Ventura, Stephen J., (608) 262-6416 sventura@wisc.edu, Univ of Wisconsin, Madison – OrSo
Venzke, Edward, (202) 633-1822 venzkee@si.edu, Smithsonian Institution / National Museum of Natural History – Gv
Vepraskas, Michael J., (919) 515-1458 North Carolina State Univ – Sd
Ver Straeten, Charles, (518) 486-2004 cverstrae@mail.nysed.gov, New York State Geological Survey – Grs
Verardo, Stacey, (703) 993-1045 sverardo@gmu.edu, George Mason Univ – Pei
Veraverbeke, Sander, +31 20 59 82975 s.n.n.veraverbeke@vu.nl, VU Univ Amsterdam – Hg
Verburg, Paul, 775-784-4511 pverburg@cabnr.unr.edu, Univ of Nevada, Reno – Sp
Verburg, Peter, +31 20 59 83594 peter.verburg@vu.nl, VU Univ Amsterdam – OyGe
Verlinde, Johannes, (814) 863-9711 verlinde@essc.psu.edu, Pennsylvania State Univ, Univ Park – Ow
Vermeesch, Pieter, +44 02076792418 p.vermeesch@ucl.ac.uk, Univ Coll London – Cc
Vermeij, Geerat J., (530) 752-2234 gjvermeij@ucdavis.edu, Univ of California, Davis – Pe
Vermeul, Vincent R., (509) 376-8316 vince.vermeul@pnl.gov, Pacific Northwest National Laboratory – Hg
Vernhes, Jean-David, +33(0)3 44062547 jean-david.vernhes@lasalle-beauvais.fr, Institut Polytechnique LaSalle Beauvais (ex-IGAL) – YgNgr
Veronis, George, (203) 432-3148 george.veronis@yale.edu, Yale Univ – Op
Verosub, Kenneth L., (530) 752-6911 klverosub@ucdavis.edu, Univ of California, Davis – Ym
Versteeg, Roelof, versteeg@ldeo.columbia.edu, Columbia Univ – Yg
Vervoort, Jeffrey D., (509) 335-5597 vervoort@wsu.edu, Washington State Univ – CcaGt
Vesovic, Velisa, +44 20 759 47352 v.vesovic@imperial.ac.uk, Imperial Coll – Np
Vesper, Dorothy, 304 293 5603 dorothy.vesper@mail.wvu.edu, West Virginia Univ – Hw
Vetter, Scott K., (318) 869-5055 svetter@centenary.edu, Centenary Coll of Louisiana – Gi
Vice, Mari A., (608) 342-1055 vice@uwplatt.edu, Univ of Wisconsin, Platteville – Gd
Vicens Batet, Enric, ++935811783 enric.vicens@uab.cat, Universitat Autonoma de Barcelona – Pg
Vickery, Nancy, nvicker2@une.edu.au, Univ of New England – GgeYg
Vidal Romaní, Juan Ramon, 00 34 981 167000 xemoncho@udc.es, Coruna Univ – GmlGc
Vidale, John E., (206) 543-6790 vidale@uw.edu, Univ of Washington – YsGtYg
Videtich, Patricia E., 616-331-3887 videticp@gvsu.edu, Grand Valley State Univ – Gd
Vidoviæ, Jelena, +38514606090 jelena.vidovic@geol.pmf.hr, Univ of Zagreb – Gg
Viegas, Anthony V., +91-832-6519332 anthonyviegas@yahoo.com, Goa Univ – Gig
Vieira, David J., (505) 667-7231 Los Alamos National Laboratory – On
Viens, Rob, (425) 564-3158 rob.viens@bellevueColl.edu, Bellevue Coll – GgOg
Vierrether, Chris, 573-368-2370 chris.vierrether@dnr.mo.gov, Missouri Dept of Natural Resources – OnEc
Viertel, Dave, 217-581-6244 dviertel@eiu.edu, Eastern Illinois Univ – Or
Vietti, Laura, (307) 314-2024 lvietti@uwyo.edu, Univ of Wyoming – Pv
Vig, Pradeep K., 618.545.3373 pvig@kaskaskia.edu, Kaskaskia Coll – Gg
Vigouroux-Caillibot, Nathalie, 604-527-5860 vigourouxcaillibotn@douglasColl.ca, Douglas Coll – GvzGg
Vilcaez, Javier, 405744-6358 vilcaez@okstate.edu, Oklahoma State Univ – HwNpCl
Villalard-Bohnsack, Martine, (401) 254-3243 Roger Williams Univ – Ob
Villalobos, Joshua, (915) 831-7001 jvillal6@epcc.edu, El Paso Community Coll – Gg
Villareal, Tracy A., (361) 749-6732 tracyv@austin.utexas.edu, Univ of Texas at Austin – Ob
Villegas, Monica B., 543584676198 mvillegas@exa.unrc.edu.ar, Universidad Nacional de Rio Cuarto – Gsc
Villeneuve, Marlene C., +64 3 3667001 Ext 45682 marlene.villeneuve@canterbury.ac.nz, Univ of Canterbury – Ng
Vimont, Daniel J., (608)262-2828 dvimont@wisc.edu, Univ of Wisconsin, Madison – Oa
Vincent, Dayton G., dvincent@purdue.edu, Purdue Univ – Oa
Vincent, Paul, (229) 333-5752 pvincent@valdosta.edu, Valdosta State Univ – Oig
Vincent, Robert K., (419) 372-0160 rvincen@bgnet.bgsu.edu, Bowling Green State Univ – Yg
Vinciguerra, Sergio, +44 0116 252 3634 sv127@le.ac.uk, Leicester Univ – Nr
Vinciguerra, Sergio Carmelo, sergiocarmelo.vinciguerra@unito.it, Università di Torino – Nr
Vinton, Bonita L., (724) 738-2048 bonita.vinton@sru.edu, Slippery Rock Univ – Gg
Visaggi, Christy, cvissagi@gsu.edu, Georgia State Univ – PiOe
Viskupic, Karen, (208)426-3658 karenviskupic@boisestate.edu, Boise State Univ – OeGc

Visona', Dario, +390498279147 dario.visona@unipd.it, Università degli Studi di Padova – Gi
Visscher, Pieter, (860) 486-4434 pieter.visscher@uconn.edu, Univ of Connecticut – Co
Visscher, Pieter T., 860-405-9159 pieter.visscher@uconn.edu, Univ of Connecticut – Co
Visser, Jenneke M., (337) 482-6966 jvisser@louisiana.edu, Univ of Louisiana at Lafayette – Ge
Vitek, John D., (405) 780-0623 jvitek@neo.tamu.edu, Texas A&M Univ – GmOe
Vitek, John D., 405-744-6358 jvitek@okstate.edu, Oklahoma State Univ – Gm
Viti, Cecilia, cecilia.viti@unisi.it, Univ of Siena – Gz
Vlahos, Penny, (860) 405-9269 penny.vlahos@uconn.edu, Univ of Connecticut – OcGeu
Vlahovic, Gordana, (919) 530-5172 gvlahovic@nccu.edu, North Carolina Central Univ – YsGtOe
Vocadlo, Lidunka, +44 020 7679 37919 l.vocadlo@ucl.ac.uk, Univ Coll London – Gy
Voelker, Bettina, (303) 273-3152 voelker@mines.edu, Colorado School of Mines – Ca
Vogel, Eve, 413-545-0778 evevogel@geo.umass.edu, Univ of Massachusetts, Amherst – Oy
Vogl, Jim, (352) 392-6987 jvogl@ufl.edu, Univ of Florida – Gct
Voglesonger, Kenneth M., 773-442-6053 K-Voglesonger@neiu.edu, Northeastern Illinois Univ – ClGe
Vogt, Steven, (831) 459-2151 vogt@ucolick.org, Univ of California, Santa Cruz – Xc
Voice, Peter J., (269) 387-5446 peter.voice@wmich.edu, Western Michigan Univ – GsOn
Voicu, Gabriel-Constantin, 514-987-3000 #1648 gvoicu@videotron.ca, Universite du Quebec a Montreal – GcCg
Voight, Barry, (814) 238-4431 voight@ems.psu.edu, Pennsylvania State Univ, Univ Park – GveGg
Voigt, Vicki, 573-368-2128 vicki.voigt@dnr.mo.gov, Missouri Dept of Natural Resources – Gg
Vojtko, Rastislav , vojtko@fns.uniba.sk, Comenius Univ in Bratislava – GctGm
Volborth, Alexis, (406) 496-4134 Montana Tech of the Univ of Montana – Ca
Voller, Vaughan R., (612) 625-0764 volle001@tc.umn.edu, Univ of Minnesota, Twin Cities – Om
Vollmer, Frederick W., 845-257-3760 vollmerf@newpaltz.edu, SUNY, New Paltz – GctGx
Vollmer, Steve, 7815817370 x312 s.vollmer@neu.edu, Northeastern Univ – On
von Bitter, Peter H., 416-586-5591 peterv@rom.on.ca, Univ of Toronto – Pm
von der Handt, Anette, 612-624-7370 avdhandt@umn.edu, Univ of Minnesota, Twin Cities – GiCt
von Frese, Ralph R., (614) 292-5635 von-frese.3@osu.edu, Ohio State Univ – Ye
von Glasow, Roland, +44 (0)1603 59 3204 r.von-glasow@uea.ac.uk, Univ of East Anglia – Yg
Von Reden, Karl F., (508) 289-3384 kvonreden@whoi.edu, Woods Hole Oceanographic Institution – Ct
von Salzen, Knut, knut.vonsalzen@canada.ca, Univ of Victoria – Oa
von Salzen, Knut, (250) 363-8287 knut.vonsalzen@canada.ca, Univ of British Columbia – Oa
von Seggern, David, (775) 784-4242 vonseg@seismo.unr.edu, Univ of Nevada, Reno – Ys
Vondra, Carl F., (515) 294-5867 cfvondra@iastate.edu, Iowa State Univ of Science & Tech – Gr
Vong, Richard J., 541-737-5693 vong@coas.oregonstate.edu, Oregon State Univ – Oa
Vonk, Jorien, +31 20 59 87336 j.e.vonk@vu.nl, VU Univ Amsterdam – Hg
Voorhees, David H., (630) 466-2783 dvoorhees@waubonsee.edu, Waubonsee Community Coll – OgGg
Voorhees, Kent J., kvoorhee@mines.edu, Colorado School of Mines – Co
Vopson, Melvin M., +44 023 92 842246 melvin.vopson@port.ac.uk, Univ of Portsmouth – Yg
Voroney, R. Paul, (519) 824-4120 (Ext. 53057) pvoroney@uoguelph.ca, Univ of Guelph – Sb
Vorwald, Brian, vorwalb@sunysuffolk.edu, Suffolk County Community Coll, Ammerman Campus – Og
Voss, Regis D., (515) 294-1923 Iowa State Univ of Science & Tech – So
Votaw, Robert, rvotaw@iun.edu, Indiana Univ Northwest – PiGrs
Voulgaris, George, (803) 777-2549 gvoulgaris@geol.sc.edu, Univ of South Carolina – On
Vrba, Elisabeth S., 203-432-5008 elisabeth.vrba@yale.edu, Yale Univ – Pvo
Vroon, Pieter, p.z.vroon@vu.nl, VU Univ Amsterdam – GvCcl
Vuke, Susan M., (406) 496-4326 svuke@mtech.edu, Montana Tech of The Univ of Montana – GgrGs
Vukovich, George, (416)736-2100 #30090 vukovich@yorku.ca, York Univ – Oa
Vulava, Vijay M., (843) 953-1922 vulavav@cofc.edu, Coll of Charleston – CgHwSc

## W

Waag, Charles J., (208) 426-3658 cwaag@boisestate.edu, Boise State Univ – Gc
Wach, Grant D., (902) 494-8019 grant.wach@dal.ca, Dalhousie Univ – GorGs
Wada, Ikuko, (612) 301-9535 iwada@umn.edu, Univ of Minnesota, Twin Cities – OnYgGt
Waddington, David C., (604) 527-5230 waddingtond@douglasColl.ca, Douglas Coll – OgGuc
Waddington, Edwin D., 206-543-4585 edw@uw.edu, Univ of Washington – Ol
Waddington, J. M., (905) 525-9140 (Ext. 23217) wadding@mcmaster.ca, McMaster Univ – Pe
Wade, Bridget S., +44 020 7679 32423 b.wade@ucl.ac.uk, Univ Coll London – PmeCl
Wade, Jeff, (773) 442-6050 j-wade@neiu.edu, Northeastern Illinois Univ – OnnOn
Wade, Phillip, 503-838-8225 wadep@wou.edu, Western Oregon Univ – Oge
Wade, Terry L., (979) 862-2325 t-wade@geos.tamu.edu, Texas A&M Univ – OcCao
Wadhwa, Meenakshi, (480) 965-0796 meenakshi.wadhwa@asu.edu, Arizona State Univ – On
Wadhwa, Meenakshi, (312) 665-7639 mwadhwa@fieldmuseum.org, Field Museum of Natural History – Xm
Wadsworth, William B., (562) 907-4200 Whittier Coll – Gi
Waff, Harve S., (541) 346-5577 waff@uoregon.edu, Univ of Oregon – Yg
Wagger, Michael G., (919) 515-4269 North Carolina State Univ – Sb
Waggoner, Karen, 432-685-5540 kwaggoner@midland.edu, Midland Coll – GghGc
Wagner, John R., (864) 656-5024 jrwgnr@clemson.edu, Clemson Univ – Oe
Wagner, Kaleb, (612) 626-6901 kewagner@umn.edu, Univ of Minnesota – Gl
Wagner, Lara S., 202-478-8838 lwagner@carnegiescience.edu, Carnegie Institution of Washington – YsgGt
Wagner, Peter J., 312-665-7634 pwagner@fieldmuseum.org, Field Museum of Natural History – Pi
Wagner, Rick, 205-247-3622 rwagner@gsa.state.al.us, Geological Survey of Alabama – Cg
Wagner, Timothy J., (402) 280-2239 timothywagner@creighton.edu, Creighton Univ – Oa
Wagner Riddle, Claudia, (519) 824-4120 (Ext. 2787) cwagnerr@uoguelph.ca, Univ of Guelph – Ow
Wagoner, Jeffrey L., (925) 422-1374 wagoner1@llnl.gov, Lawrence Livermore National Laboratory – Gs
Waheed, Umair, (609) 258-2921 uwaheed@princeton.edu, Princeton Univ – Ys
Wahl, Tim, (612) 626-0992 tewahl@umn.edu, Univ of Minnesota – Oi
Waid, Christopher, 614 265 6627 christopher.waid@dnr.state.oh.us, Ohio Dept of Natural Resources – Eo
Waite, Cynthia, (213) 740-6109 waite@usc.edu, Univ of Southern California – On
Waite, Gregory P., (906) 487-3554 gpwaite@mtu.edu, Michigan Technological Univ – YsGv
Waithaka, Hunja, hunja@eng.jkuat.ac.ke, Jomo Kenyatta Univ of Agriculture & Tech – YdOgi
Wakabayashi, John, 559-278-6459 jwakabayashi@csufresno.edu, California State Univ, Fresno – GmcGt
Wake, Cameron P., (603) 862-2329 cameron.wake@unh.edu, Univ of New Hampshire – Gl
Wake, David B., (510) 642-3059 wakelab@uclink.berkeley.edu, Univ of California, Berkeley – Po
Wake, Marvalee H., (510) 642-4743 Univ of California, Berkeley – Po

Wakefield, Kelli, kelliwakefield@hotmail.com, Mesa Community Coll – Gg
Wala, Andrew M., 859-257-2959 Univ of Kentucky – Nm
Walcek, Christopher J., (518) 437-8720 SUNY, Albany – Oa
Waldbusser, George, 541-737-8964 waldbuss@coas.oregonstate.edu, Oregon State Univ – Ob
Walden, John, +44 01334 463688 jw9@st-andrews.ac.uk, Univ of St. Andrews – Ym
Waldhauser, Felix, (845) 365-8538 felixw@ldeo.columbia.edu, Columbia Univ – YsgGt
Waldron, John W., john.waldron@ualberta.ca, Univ of Alberta – Gc
Walford, Nigel, +44020 8417 2512 nwalford@kingston.ac.uk, Kingston Univ – Oi
Walker, Charles T., (562) 985-4818 California State Univ, Long Beach – Cl
Walker, David, (845) 365-8658 Columbia Univ – Gx
Walker, J. Douglas, (785) 864-7711 jdwalker@ku.edu, Univ of Kansas – Gc
Walker, James A., (815) 753-7936 jwalker@niu.edu, Northern Illinois Univ – Giv
Walker, Jeffrey R., 845-437-5546 jewalker@vassar.edu, Vassar Coll – GzvGp
Walker, John L., (630) 252-6803 jlwalker@anl.gov, Argonne National Laboratory – Cg
Walker, Kenneth R., (865) 974-6017 kwalker5@utk.edu, Univ of Tennessee, Knoxville – Pe
Walker, Mark, (775) 784-1938 mwalker@cabnr.unr.edu, Univ of Nevada, Reno – On
Walker, Nan D., (225) 388-5331 Louisiana State Univ – Og
Walker, Peter A., (541) 346-4541 pwalker@oregon.uoregon.edu, Univ of Oregon – On
Walker, Raymond J., (310) 825-7685 rwalker@igpp.ucla.edu, Univ of California, Los Angeles – On
Walker, Richard, +440116 252 3628 Leicester Univ – Gs
Walker, Richard, +44 (1865) 282115 richard.walker@earth.ox.ac.uk, Univ of Oxford – Or
Walker, Richard J., (301) 405-4089 rjwalker@geol.umd.edu, Univ of Maryland – CcXcCg
Walker, Sally E., (706) 542-2396 swalker@gly.uga.edu, Univ of Georgia – Pm
Wallace, Adam F., (302) 831-1950 afw@udel.edu, Univ of Delaware – ClGzOm
Wallace, Davin, (228) 688-3060 davin.wallace@usm.edu, Univ of Southern Mississippi – Ou
Wallace, Douglas, (902) 494-4132 douglas.wallace@dal.ca, Dalhousie Univ – Oc
Wallace, Douglas W. R., (631) 344-2945 wallace@notes.cc.sunysb.edu, SUNY, Stony Brook – Oc
Wallace, Janae, (801) 537-3387 nrugs.jwallace@state.ut.us, Utah Geological Survey – Hw
Wallace, John M., 206-543-7390 wallace@atmos.washington.edu, Univ of Washington – Oa
Wallace, Laura, lwallace@ig.utexas.edu, Univ of Texas at Austin – GtYdOr
Wallace, Paul, (541) 346-5985 pwallace@uoregon.edu, Univ of Oregon – GviCg
Wallace, Tim, 662-268-1032 Ext 244 tjw5@msstate.edu, Mississippi State Univ – Ow
Wallace, William G., (718) 982-3876 william.wallace@csi.cuny.edu, Graduate School of the City Univ of New York – Og
Wallace, Jr., Terry C., (505) 667-3644 wallacet@lanl.gov, Los Alamos National Laboratory – Ys
Wallender, Wes W., 530.752.0688 wwwallender@ucdavis.edu, Univ of California, Davis – Hg
Waller, Thomas R., (202) 357-2127 Smithsonian Institution / National Museum of Natural History – Pi
Wallis, Ilka, ilka.wallis@flinders.edu.au, Flinders Univ – HqCg
Walrod, Amanda G., (501) 575-7317 awalrod@comp.uark.edu, Univ of Arkansas, Fayetteville – On
Walsh, Christopher, (301) 405-4351 cswalsh@umd.edu, Univ of Maryland – On
Walsh, Daniel E., (907) 474-6746 dewalsh@alaska.edu, Univ of Alaska, Fairbanks – On
Walsh, Ellen c., (920) 832-6739 ellen.c.walsh@lawrence.edu, Lawrence Univ – On
Walsh, Emily O., (319) 895-4302 ewalsh@cornellColl.edu, Cornell Coll – GptCc
Walsh, J.P., 252 3285431 walshj@ecu.edu, East Carolina Univ – Gs
Walsh, John E., (217) 333-7521 walsh@atmos.uiuc.edu, Univ of Illinois, Urbana-Champaign – Oa
Walsh, John J., (727) 553-1164 jwalsh@marine.usf.edu, Univ of South Florida – Ob
Walsh, Maud, (225) 578-1211 evwals@lsu.edu, Louisiana State Univ – GeOg
Walsh, Tim R., 806-291-1123 Wayland Baptist Univ – GsPmGo
Walter, Lynn M., (734) 763-4590 lmwalter@umich.edu, Univ of Michigan – Cl
Walter , Michael , 0117 9515007 m.j.walter@bristol.ac.uk, Univ of Bristol – GiCpg
Walter, Nathan A., (678) 839-4070 awalter@westga.edu, Univ of West Georgia – Og
Walter, Robert C., 717 358-7198 robert.walter@fandm.edu, Franklin and Marshall Coll – Cc
Walter, Thomas, (212) 772-5457 twalter@hunter.cuny.edu, Hunter Coll (CUNY) – OyeOw
Walter-Shea, Elizabeth A., (402) 472-1553 Univ of Nebraska, Lincoln – Oa
Walters, James C., james.walters@uni.edu, Univ of Northern Iowa – Gm
Waltham, Dave, +44 1784 443617 D.Waltham@rhul.ac.uk, Univ of London, Royal Holloway & Bedford New Coll – Yg
Walther, Ferdinand, 089/2180 4346 macferdi@gmx.org, Ludwig-Maximilians-Universitaet Muenchen – Gz
Walther, John V., (214) 768-3174 Southern Methodist Univ – Cg
Walton, Anthony W., (785) 864-2726 twalton@ku.edu, Univ of Kansas – GsEoGv
Walton, Gabriel, (303) 273-2235 gwalton@mines.edu, Colorado School of Mines – NgrYg
Walton, Ian, 801-581-8497 iwalton@egi.utah.edu, Univ of Utah – Gq
Walton, Nick, +44 023 92 842263 nick.walton@port.ac.uk, Univ of Portsmouth – Hw
Walworth, James, (520) 626-3364 walworth@ag.arizona.edu, Univ of Arizona – Soc
Wampler, J. Marion, kayargon@earthlink.net, Georgia State Univ – Cc
Wampler, Peter J., 616-331-2834 wamplerp@gvsu.edu, Grand Valley State Univ – HsOiGm
Wanamaker, Alan D., (515) 294-5142 adw@iastate.edu, Iowa State Univ of Science & Tech – CsPeOg
Wang, Alian, (314) 935-5671 alianw@levee.wustl.edu, Washington Univ in St. Louis – Ca
Wang, Bin, wangbin@hawaii.edu, Univ of Hawai'i, Manoa – Ow
Wang, Chi-Yuen, (510) 642-2288 chiyuen@seismo.berkeley.edu, Univ of California, Berkeley – Yg
Wang, Chien, (617) 253-5432 wangc@mit.edu, Massachusetts Inst of Tech – Oa
Wang, Dongmei, (701) 777-6143 dongmei.wang@engr.und.edu, Univ of North Dakota – Ng
Wang, Enru, enru.wang@und.edu, Univ of North Dakota – Ogi
Wang, Harry, (804) 684-7215 wang@vims.edu, Coll of William & Mary – Op
Wang, Herbert F., (608) 262-5932 hfwang@wisc.edu, Univ of Wisconsin, Madison – YgHwYx
Wang, Hong, 217-244-7692 hongwang@illinois.edu, Univ of Illinois, Urbana-Champaign – Cg
Wang, Hsiang-Jui, (404) 894-3748 raywang@eas.gatech.edu, Georgia Inst of Tech – Oa
Wang, Jeen-Hwa, 886-2-27839910-326 jhwang@earth.sinica.edu.tw, Academia Sinica – Ys
Wang, Jianhua, (202) 478-8457 jwang@carnegiescience.edu, Carnegie Institution of Washington – Cs
Wang, Jianwei, (225) 578-5532 jianwei@lsu.edu, Louisiana State Univ – ClOmGy
Wang, Jim, (225) 578-1360 jjwang@agcenter.lsu.edu, Louisiana State Univ – Sc
Wang, Kun, 314-935-3855 kunwang@levee.wustl.edu, Washington Univ in St. Louis – Xc
Wang, Lillian T., 302-831-1096 lillian@udel.edu, Univ of Delaware – Oi
Wang, Pao-Kuan, (608)263-6479 pao@windy.aos.wisc.edu, Univ of Wisconsin, Madison – Oa
Wang, Qing, 831-656-7716 qwang@nps.edu, Naval Postgraduate School – Ow
Wang, Wei-Chyung, (518) 437-8708 wcwang@albany.edu, SUNY, Albany – Oa
Wang, Weihong, (801) 863-7607 Weihong.Wang@uvu.edu, Utah Valley Univ – CmGeOi
Wang, Yang, (850) 644-1121 ywang@magnet.fsu.edu, Florida State Univ – CgsPe
Wang, Yanghua, +44 20 759 41171 yanghua.wang@imperial.ac.uk,

Imperial Coll – Yg
Wang, Yuhang, (404) 894-3995 yuhang.wang@eas.gatech.edu, Georgia Inst of Tech – Oa
Wang, Yumei, (971) 673-1555 meimei.wang@state.or.us, Oregon Dept of Geology & Mineral Industries – Ne
Wang, Yuqing, yuqing@hawaii.edu, Univ of Hawai'i, Manoa – Oa
Wang, Yuxuan, (409) 740-4829 wangyx@tamug.edu, Texas A&M Univ – Oa
Wang, Yuxuan, 713-743-9049 ywang140@uh.edu, Univ of Houston – Oa
Wang, Z. Aleck, (508) 289-3676 zawang@whoi.edu, Woods Hole Oceanographic Institution – Cm
Wang, Zhankun, 979 458 3464 zhankunwang@tamu.edu, Texas A&M Univ – Op
Wang, Zhenming, 859-323-0564 zmwang@uky.edu, Univ of Kentucky – Ys
Wang, Zhenming, (859) 257-5500 x142 zmwang@uky.edu, Univ of Kentucky – Yse
Wang, Zhi (Luke), (559) 278-4427 zwang@csufresno.edu, California State Univ, Fresno – SpHwOi
Wang, Zhuo, 217-244-4270 zhuowang@illinois.edu, Univ of Illinois, Urbana-Champaign – Oa
Wanke, Ansgar, awanke@unam.na, Univ of Namibia – GsCl
Wankel, Scott, (508) 289-3944 sdwankel@whoi.edu, Woods Hole Oceanographic Institution – Cm
Wanless, Dorsey, dwanless@boisestate.edu, Boise State Univ – Gi
Wanless, Harold R., (305) 284-2697 hwanless@miami.edu, Univ of Miami – Gse
Wannamaker, Phillip E., (801) 581-3547 pewanna@egi.utah.edu, Univ of Utah – Ye
Warburton, David L., (561) 297-3312 warburto@fau.edu, Florida Atlantic Univ – CgGe
Ward, Bess B., (609) 258-5150 bbw@princeton.edu, Princeton Univ – Ob
Ward, Brent C., (604) 291-4229 bcward@sfu.ca, Simon Fraser Univ – Ge
Ward, Calvin H., (713) 348-4086 wardch@rice.edu, Rice Univ – Og
Ward, Colin R., + 61 2 9385 4718 c.ward@unsw.edu.au, Univ of New South Wales – EcGd
Ward, David M., (406) 994-3401 umbdw@montana.edu, Montana State Univ – On
Ward, Dylan, dylan.ward@uc.edu, Univ of Cincinnati – GmqOi
Ward, J. Evan, 860-405-9073 evan.ward@uconn.edu, Univ of Connecticut – Ob
Ward, James W., (325 )486-6767 James.Ward@angelo.edu, Angelo State Univ – HwClOn
Ward, Larry G., (603) 862-5132 larry.ward@unh.edu, Univ of New Hampshire – Gu
Ward, Marie D., (604) 291-4229 Univ of Alaska, Fairbanks – On
Ward, Peter D., 206-543-2962 swift@ocean.washington.edu, Univ of Washington – Po
Ward, Steven N., (831) 459-2480 sward@es.ucsc.edu, Univ of California, Santa Cruz – Ys
Wardlaw, Norman C., (403) 220-6429 nwardlaw@geo.ucalgary.ca, Univ of Calgary – Go
Waren, Kirk B., 406-496-4866 kwaren@mtech.edu, Montana Tech of The Univ of Montana – HwsOe
Warger, Jane, (909) 652-6485 jane.warger@chaffey.edu, Chaffey Coll – OgGg
Warhaft, Zellman, 607-255-3898 zw16@cornell.edu, Cornell Univ – On
Warke, Patricia, p.warke@qub.ac.uk, Queens Univ Belfast – GmOym
Warkentin, Alicia, (604) 827-5284 awarkentin@eoas.ubc.ca, Univ of British Columbia – On
Warland, Jon, (519) 824-4120 (Ext. 6374) jwarland@uoguelph.ca, Univ of Guelph – Oa
Warme, John E., jwarme@mines.edu, Colorado School of Mines – Pi
Warner, Mark J., (206) 543-0765 warner@u.washington.edu, Univ of Washington – Op
Warner, Michael, +44 20 759 46535 m.warner@imperial.ac.uk, Imperial Coll – Ys
Warner, Nicholas H., 585-245-5291 warner@geneseo.edu, SUNY, Geneseo – XgGrHg
Warner, Richard D., (864) 656-5023 wrichar@ces.clemson.edu, Clemson Univ – Gz
Warner, Timothy A., 3042935603x4328 tim.warner@mail.wvu.edu, West Virginia Univ – Or
Warnock, Jonathan P., (724) 357-5627 jwarnock@iup.edu, Indiana Univ of Pennsylvania – PemPv
Warny, Sophie, (225) 578-5089 swarny@lsu.edu, Louisiana State Univ – Pl
Warren, Jessica, 302-831-2569 warrenj@udel.edu, Univ of Delaware – GxCpGt
Warren, Joseph, (631) 632-5045 Joe.Warren@stonybrook.edu, SUNY, Stony Brook – Ob
Warren, Lesley A., (905) 525-9140 (Ext. 27347) warrenl@mcmaster.ca, McMaster Univ – CgHg
Warren, Linda, (314) 977-3197 lwarren8@slu.edu, Saint Louis Univ – Yg
Warren, Paul, (310) 825-2015 Univ of California, Los Angeles – Xm
Warren, Rachel, +44 (0)1603 59 3912 r.warren@uea.ac.uk, Univ of East Anglia – Eg
Warren, Richard G., (505) 667-7063 rgw@lanl.gov, Los Alamos National Laboratory – Gi
Warren, Richard L., (520) 621-2320 Univ of Arizona – On
Warren, Stephen G., (206) 543-7230 sgw@atmos.washington.edu, Univ of Washington – Oal
Warrick, Arthur W., (520) 621-1516 aww@ag.arizona.edu, Univ of Arizona – Sp
Wartes, Marwan A., (907) 451-5056 marwan.wartes@alaska.gov, Alaska Division of Geological & Geophysical Surveys – GsoGt
Warwick, Phillip, +44 (0)23 80592780 phil.warwick@noc.soton.ac.uk, Univ of Southampton – CaGe
Wash, Carlyle H., 831-656-7776 wash@nps.edu, Naval Postgraduate School – Ow
Washburn, Robert H., (814) 641-3600 Juniata Coll – Gs
Washburne, James C., (520) 626-4107 jwash@hwr.arizona.edu, Univ of Arizona – HsOge
Wasomi, Charles B., cwasomi@jkuat.ac.ke, Jomo Kenyatta Univ of Agriculture & Tech – OrrOi
Wassenaar, Len, (306) 239-2270 Univ of Saskatchewan – Hw
Wasserburg, Gerald J., (626) 395-6139 gjw@gps.caltech.edu, California Inst of Tech – Cc
Wassermann, Joachim, 08141/5346762 jowa@geophysik.uni-muenchen.de, Ludwig-Maximilians-Universitaet Muenchen – Yg
Wasson, John T., (310) 825-1986 jtwasson@ucla.edu, Univ of California, Los Angeles – Xm
Wasylenki, Laura E., (812) 855-7508 lauraw@indiana.edu, Indiana Univ, Bloomington – Cl
Waszek, Lauren, lwaszek@nmsu.edu, New Mexico State Univ, Las Cruces – Ys
Watanabe, Tohru, 81-76-445-6650 twatnabe@sci.u-toyama.ac.jp, Univ of Toyama – YxsGv
Waters, Dave, +44 (1865) 282457 dave.waters@earth.ox.ac.uk, Univ of Oxford – Gp
Waters, Johnny, 828-262-7820 watersja@appstate.edu, Univ of Tennessee, Knoxville – Pg
Waters, Johnny A., watersja@appstate.edu, Appalachian State Univ – Pi
Waters, Michael R., (979) 845-5246 mwaters@tamu.edu, Texas A&M Univ – Ga
Waters-Tormey, Cheryl, (828) 227-3696 cherylwt@wcu.edu, Western Carolina Univ – Gct
Waterstone, Marvin, (520) 621-1478 marvinw@u.arizona.edu, Univ of Arizona – On
Watkins, David, D.C.Watkins@exeter.ac.uk, Exeter Univ – Hg
Watkins, David K., (402) 472-2177 dwatkins1@unl.edu, Univ of Nebraska, Lincoln – PmGu
Watkins, James, watkins4@uoregon.edu, Univ of Oregon – Cg
Watkinson, A. John, (509) 335-2470 watkinso@mail.wsu.edu, Washington State Univ – Gc
Watkinson, Andrew, +44 (0)1603 59 2267 a.watkinson@uea.ac.uk, Univ of East Anglia – Ge
Watkinson, David H., dwatkson@ccs.carleton.ca, Carleton Univ – Em
Watkinson, Ian, +44 1784 414046 i.watkinson@es.rhul.ac.uk, Univ of London, Royal Holloway & Bedford New Coll – Gt
Watkinson, Matthew, +44 1752 584765 M.P.Watkinson@plymouth.ac.uk, Univ of Plymouth – Gs
Watney, Lynn W., (785) 864-2184 lwatney@kgs.ku.edu, Univ of Kansas – Gr
Watson, Alan, (519) 824-4120 Ext.52356 awatson@uoguelph.ca, Univ of Guelph – On
Watson, David B., (865) 241-4749 v6i@ornl.gov, Oak Ridge National Laboratory – Hw
Watson, E. Bruce, (518) 276-8838 watsoe@rpi.edu, Rensselaer Polytechnic Inst – CpcCt
Watson, Kelly, 859-622-1419 kelly.watson@eku.edu,

Faculty Index -W

Eastern Kentucky Univ – Or
Watters, Robert J., (775) 784-6069 watters@mines.unr.edu, Univ of Nevada, Reno – Nrg
Watters, Thomas R., 202-633-2483 Smithsonian Institution / National Air & Space Museum – Xg
Wattrus, Nigel J., (218) 726-7154 nwattrus@d.umn.edu, Univ of Minnesota, Duluth – Yr
Watts, Anthony B., +44 (1865) 272032 tony@earth.ox.ac.uk, Univ of Oxford – GutYv
Watts, Chester F., (540) 831-5637 cwatts@radford.edu, Radford Univ – NgrHw
Watts, D. Randolph, (401) 874-6507 rwatts@gso.uri.edu, Univ of Rhode Island – Op
Watts, Doyle, 937 775-3455 doyle.watts@wright.edu, Wright State Univ – YeOr
Waugh, Darryn W., (410) 516-8344 waugh@jhu.edu, Johns Hopkins Univ – Oa
Waugh, John, 757-822-7436 tcwaugj@tcc.edu, Tidewater Community Coll – Gg
Waugh, Richard A., (608) 342-1386 waugh@uwplatt.edu, Univ of Wisconsin, Platteville – Gc
Waugh, Truman, 785-864-2119 twaugh@ku.edu, Univ of Kansas – Ca
Wauthier, Christelle, 814-865-6711 cuw25@psu.edu, Pennsylvania State Univ, Univ Park – On
Wax, Charles L., (662) 325-3915 wax@geosci.msstate.edu, Mississippi State Univ – Oa
Wayne, William J., wwayne3@unl.edu, Univ of Nebraska, Lincoln – Gm
Wdowinski, Shimon, (305) 348-6826 shimon.wdowinski@fiu.edu, Univ of Miami – YdGt
Weaver, Andrew J., (250) 472-4001 weaver@ocean.seos.uvic.ca, Univ of Victoria – Op
Weaver, Barry L., (405) 325-4492 bweaver@ou.edu, Univ of Oklahoma – Ct
Weaver, Douglas J., (702) 295-5916 douglas_weaver@lanl.gov, Los Alamos National Laboratory – On
Weaver, John T., (250) 721-6155 weaver@phys.uvic.ca, Univ of Victoria – Ym
Weaver, Justin E., 806-834-4610 justin.e.weaver@ttu.edu, Texas Tech Univ – Oa
Weaver, Robert, (321) 674-7273 Florida Inst of Tech – Oo
Weaver, Stephen G., (719) 389-6954 sweaver@coloradoColl.edu, Colorado Coll – Gx
Weaver, Thomas A., (505) 667-8464 tweaver@lanl.gov, Los Alamos National Laboratory – Yg
Weaver Bowman, Kristin, (657) 278-3331 kweaver-bowman@fullerton.edu, California State Univ, Fullerton – Oe
Webb, Christine, (410) 507-3070 webbc@si.edu, Smithsonian Institution / National Museum of Natural History – GzgOn
Webb, Craig A., cwebb@mtsac.edu, Mt. San Antonio Coll – OgwOg
Webb, Elizabeth A., 5196612111 x80208 ewebb5@uwo.ca, Western Univ – CslSa
Webb, Fred, (828) 262-2166 webbfj@appstate.edu, Appalachian State Univ – Gr
Webb, John A., +61 3 9479 1273 john.webb@latrobe.edu.au, La Trobe Univ – GeHwGa
Webb, Laura E., (802) 656-8136 lewebb@uvm.edu, Univ of Vermont – GtCcGc
Webb, Nathan D., (217) 244-2426 ndwebb2@illinois.edu, Univ of Illinois, Urbana-Champaign – EoGo
Webb, Peter N., (614) 292-7285 webb.3@osu.edu, Ohio State Univ – Pm
Webb, Robert H., (520) 626-3293 rhwebb@usgs.gov, Univ of Arizona – Gm
Webb, Spahr, (845) 365-8439 Columbia Univ – Yr
Webb, III, Thompson, (401) 863-3128 thompson_webb_iii@brown.edu, Brown Univ – PeIOa
Webber, Andrew, (513) 455-7160 Cincinnati Museum Center – Pi
Webber, Jeffrey R., (609) 652-4213 jeffrey.webber@stockton.edu, Stockton Univ – GcpGt
Webber, Karen L., (504) 280-7395 kwebber@uno.edu, Univ of Michigan – Gv
Webber, Karen L., (504) 280-6791 kwebber@uno.edu, Univ of New Orleans – Gv
Weber, Bodo, bweber@cicese.mx, Centro de Investigación Científica y de Educación Superior de Ensenada – Gp
Weber, Gerald E., (831) 459-5429 gweber@pmc.ucsc.edu, Univ of California, Santa Cruz – NgGmo
Weber, John C., 616-331-3191 weberj@gvsu.edu, Grand Valley State Univ – Gc
Weber, Karrie A., (402) 472-2739 kweber@unl.edu, Univ of Nebraska, Lincoln – PyCl
Weber, Keith, (208) 282-2757 webekeit@isu.edu, Idaho State Univ – Oi
Weber, Rodney J., (404) 894-1750 rweber@eas.gatech.edu, Georgia Inst of Tech – Oa
Weber-Diefenbach, Klaus, 089/2180 6549 klaus.diefenbach@iaag.geo.uni-muenchen.de, Ludwig-Maximilians-Universitaet Muenchen – Gg
Weborg-Benson, Kimberly, (716) 673-3293 kim.weborg-benson@fredonia.edu, SUNY Fredonia – GgOaPy
Webre, Cherri B., (225) 388-8328 cherri@lgs.bri.lsu.edu, Louisiana State Univ – On
Webster, Ferris, (302) 645-4266 Univ of Delaware – Op
Webster, Gary D., (509) 335-4369 webster@wsu.edu, Washington State Univ – Pis
Webster, James D., 212-769-5401 jdw@amnh.org, American Museum of Natural History – Eg
Webster, John R., (701) 858-3873 john.webster@minotstateu.edu, Minot State Univ – Giz
Webster, Peter J., (404) 894-1748 pjw@eas.gatech.edu, Georgia Inst of Tech – Oap
Wedding, William C., 859-257-1883 chad.wedding@uky.edu, Univ of Kentucky – Nm
Weddle, Thomas K., (207) 287-7170 thomas.k.weddle@maine.gov, Dept of Agriculture, Conservation, and Forestry – GlHw
Weeden, Lori, (978) 934-3344 lori_weeden@uml.edu, Univ of Massachusetts Lowell – Gge
Weeraratne, Dayanthie, (818) 677-2046 dsw@csun.edu, California State Univ, Northridge – Yg
Weertman, Johannes, (847) 491-3197 j_weertman2@northwestern.edu, Northwestern Univ – Om
Weglein, Arthur B., (713) 743-3848 aweglein@uh.edu, Univ of Houston – Ye
Wegmann, Karl, 919-515-0380 karl_wegmann@ncsu.edu, North Carolina State Univ – Gm
Wegner, John, 404 727 4206 jwegner@emory.edu, Emory Univ – Sf
Wehmiller, John F., (302) 831-2926 jwehm@udel.edu, Univ of Delaware – Cl
Wehner, Peter J., (512) 223-4276 pwehner@austincc.edu, Austin Community Coll District – Gv
Wehrmann, Laura, 089/2180 Ludwig-Maximilians-Universitaet Muenchen – Pg
Wei, Xiaofang, 937 376 6193 xwei@centralstate.edu, Central State Univ – OrgOe
Weibel, C. P., 217-333-5108 c-weibel@illinois.edu, Univ of Illinois, Urbana-Champaign – Gl
Weiblen, Paul W., 612-625-3477 pweib@umn.edu, Univ of Minnesota, Twin Cities – Gi
Weidner, Donald J., (631) 632-8211 donald.weidner@sunysb.edu, SUNY, Stony Brook – Gy
Weil, Arlo B., (610) 526-5113 aweil@brynmawr.edu, Bryn Mawr Coll – GctYm
Weiland, Thomas J., tjw@canes.gsw.edu, Georgia Southwestern State Univ – Gi
Weimer, Paul, (303) 492-3809 paul.weimer@colorado.edu, Univ of Colorado – Gor
Weinberger, Alycia J., (202) 478-8820 aweinberger@carnegiescience.edu, Carnegie Institution of Washington – Xg
Weinstein, Charles E., (413) 236-4556 cweinste@berkshirecc.edu, Berkshire Community Coll – On
Weintraub, Michael N., 419-530-2585 michael.weintraub@utoledo.edu, Univ of Toledo – On
Weirich, Frank H., (319) 335-0156 frank-weirich@uiowa.edu, Univ of Iowa – Gm
Weis, Dominique, (604) 822-1697 dweis@eos.ubc.ca, Univ of British Columbia – Cc
Weisberg, Robert H., (727) 553-1568 rweisberg@marine.usf.edu, Univ of South Florida – Op
Weisener, Christopher, 5192533000 x3753 weisener@uwindsor.ca, Univ of Windsor – CoGePy
Weisenfluh, Gerald A., (859) 323-0505 jerryw@uky.edu, Univ of Kentucky – EcOi
Weislogel, Amy L., 304-293-5603 Amy.Weislogel@mail.wvu.edu, West Virginia Univ – Gs

Weisman, Robert A., 320-308-3247 raweisman@stcloudstate.edu, Saint Cloud State Univ – Ow
Weiss, Alfred W., 630.466.2720 aweiss@waubonsee.edu, Waubonsee Community Coll – Oiy
Weiss, Benjamin, (626) 395-6863 bpweiss@mit.edu, Massachusetts Inst of Tech – Ym
Weiss, Chester J., 540-231-6521 cjweiss@unm.edu, Univ of New Mexico – Ye
Weiss, Christopher C., 806-834-4712 chris.weiss@ttu.edu, Texas Tech Univ – OaaOa
Weiss, Dennis, (212) 650-6849 Graduate School of the City Univ of New York – Pg
Weiss, Dominik, +44 20 759 46383 d.weiss@imperial.ac.uk, Imperial Coll – CgsCl
Weiss, Ray F., (858) 534-2598 rfweiss@ucsd.edu, Univ of California, San Diego – Cm
Weissmann, Gary, 505-277-4204 weissman@unm.edu, Univ of New Mexico – Hw
Welby, Charles W., (919) 515-7158 cwwelby@unity.ncsu.edu, North Carolina State Univ – Hw
Welch, Steve M., (785) 532-7236 welchsm@ksu.edu, Kansas State Univ – Sp
Welch, Susan A., 614-292-9059 welch.318@osu.edu, Ohio State Univ – ClGeCs
Weldeab, Syee, 805-893-4903 weldeab@geol.ucsb.edu, Univ of California, Santa Barbara – PeCm
Weldon, Elise M., (541) 346-4647 lili@uoregon.edu, Univ of Oregon – Gmt
Weldon, Ray J., (541) 346-4584 ray@uoregon.edu, Univ of Oregon – Gc
Welford, Mark R., (912) 478-5943 mwelfgeog@gsvms2.cc.gasou.edu, Georgia Southern Univ – Oy
Welhan, John A., (208) 282-4254 welhjohn@isu.edu, Idaho State Univ – Hw
Welling, Tim, 845-431-8536 welling@sunydutchess.edu, Dutchess Community Coll – GeOs
Wellner, Julia, 713-743-0214 jwellner@uh.edu, Univ of Houston – Gs
Wells, David E., (228) 688-3389 dew@unb.ca, Univ of Southern Mississippi – YdOi
Wells, Michael L., (702) 895-0828 michael.wells@unlv.edu, Univ of Nevada, Las Vegas – Gc
Wells, Neil, +44 (0)23 80592428 n.c.wells@soton.ac.uk, Univ of Southampton – OpwOn
Wells, Neil A., (330) 672-2951 nwells@kent.edu, Kent State Univ – Gs
Wells, Randy, (941) 388-4441 Univ of South Florida – Ob
Wells, Steve G., (775) 673-7470 sgwells@dri.edu, Univ of Nevada, Reno – On
Welp, Lisa, (765) 496-6896 lwelp@purdue.edu, Purdue Univ – Cs
Welsh, James L., 507-933-7335 welsh@gac.edu, Gustavus Adolphus Coll – Gxc
Welzenbach, Linda, (202) 633-1825 welzenbl@si.edu, Smithsonian Institution / National Museum of Natural History – Xm
Wen, Lianxing, (631) 632-1726 lianxing.wen@sunysb.edu, SUNY, Stony Brook – Ys
Wen, Xianyun, +44(0) 113 34 31425 x.wen@leeds.ac.uk, Univ of Leeds – Og
Wen, Yuming, 16717352685 ywen@uguam.uog.edu, Univ of Guam – Oir
Wendlandt, Richard F., (303) 273-3809 rwendlan@mines.edu, Colorado School of Mines – GizCp
Weng, Kevin, 808-956-4109 kevin.weng@hawaii.edu, Univ of Hawai'i, Manoa – Ob
Weng, Qihao, 812-237-2255 qweng@indstate.edu, Indiana State Univ – Oru
Wenk, Hans-Rudolf, (510) 642-7431 wenk@berkeley.edu, Univ of California, Berkeley – Gz
Wennberg, Paul O., (626) 395-2447 wennberg@gps.caltech.edu, California Inst of Tech – Oa
Wenner, David B., (706) 542-2393 dwenner@uga.edu, Univ of Georgia – Cs
Wenner, Jennifer M., (920) 424-7003 wenner@uwosh.edu, Univ of Wisconsin, Oshkosh – Gi
Wenzel, Christopher, (307) 532-8293 chris.wenzel@ewc.wy.edu, Eastern Wyoming Coll – SbfHg
Werdon, Melanie B., (907) 451-5082 melanie.werdon@alaska.gov, Alaska Division of Geological & Geophysical Surveys – Eg
Werhner, Matthew J., (813) 253-7263 mwerhner@hccfl.edu, Hillsborough Community Coll – Gg
Werne, Josef, (412) 624-8775 jwerne@pitt.edu, Univ of Pittsburgh – Co
Werner, Adrian , adrian.werner@flinders.edu.au, Flinders Univ – Hw
Werner, Alan, (413) 538-2134 awerner@mtholyoke.edu, Mount Holyoke Coll – GlOuGm
Werner, Bradley T., (858) 534-0583 bwerner@ucsd.edu, Univ of California, San Diego – Gm
Werner, Christopher L., christopher.werner@shell.com, McMaster Univ – GsoHq
Wernicke, Brian P., (626) 395-6192 brian@gps.caltech.edu, California Inst of Tech – Gtc
Werts, Scott P., 803-323-4930 wertss@winthrop.edu, Winthrop Univ – ScbPy
Wesely, Marvin L., (630) 252-5827 mlwesely@anl.gov, Argonne National Laboratory – Oa
Wesler, Kit W., (270) 809-3457 kwesler@murraystate.edu, Murray State Univ – Oni
Wesley Wood, Charley, 850-9837126 woodwes@ufl.edu, Univ of Florida – Os
Wesnousky, Steve, (775) 784-6067 stevew@seismo.unr.edu, Univ of Nevada, Reno – Ys
Wessel, Paul, (808) 956-4778 pwessel@hawaii.edu, Univ of Hawai'i, Manoa – Yr
West, A. Joshua (Josh), (213) 740-6736 joshwest@usc.edu, Univ of Southern California – ClGmHs
West, David P., (802) 443-3476 dwest@middlebury.edu, Middlebury Coll – Gc
West, Jared, +44(0) 113 34 35253 L.J.West@leeds.ac.uk, Univ of Leeds – Hg
West, Keith, (715) 735-4352 keith.west@uwc.edu, Univ of Wisconsin Colls – OynOn
West, Larry T., (706) 542-0906 Univ of Georgia – Sd
West, Olivia M., (615) 576-0505 qm5@ornl.gov, Oak Ridge National Laboratory – Ng
West, Robert, 323-260-8115 westrb@elac.edu, East Los Angeles Coll – GgmSd
West, Ronald R., (785) 532-2248 rrwest@ksu.edu, Kansas State Univ – PoeGs
West, Terry R., (765) 494-3296 trwest@purdue.edu, Purdue Univ – Ng
Westerink, Joannes J., (574) 631-6475 Univ of Notre Dame – On
Westerman, David S., (802) 485-2337 westy@norwich.edu, Norwich Univ – Gxi
Westgate, James W., (409) 880-8237 james.westgate@lamar.edu, Lamar Univ – Pev
Westgate, John A., (416) 287-7235 Univ of Toronto – Gl
Westhoff, Martijn C., +31 20 59 85741 m.c.westhoff@vu.nl, VU Univ Amsterdam – Hg
Westrop, Stephen R., (405) 325-0542 swestrop@ou.edu, Univ of Oklahoma – Pi
Westwood, Rachel, (+44) 01782 734309 r.f.westwood@keele.ac.uk, Keele Univ – Yx
Wetherelt, Andrew, +44 01326 371827 A.Wetherelt@exeter.ac.uk, Exeter Univ – Nm
Wettlaufer, John S., (203) 432-0892 john.wettlaufer@yale.edu, Yale Univ – Yg
Wettstein, Justin, (541) 737-5177 justinw@coas.oregonstate.edu, Oregon State Univ – Oa
Wetz, Michael, 361-825-2132 Michael.Wetz@tamucc.edu, Texas A&M Univ, Corpus Christi – Ob
Wetzel, Dan L., (907) 488-3746 Univ of Alaska, Fairbanks – On
Wetzel, Laura R., (727) 864-8484 wetzellr@eckerd.edu, Eckerd Coll – YrOuGg
Whalen, Kristen, (508) 289-3627 kwhalen@whoi.edu, Woods Hole Oceanographic Institution – Ob
Whalen, Michael T., 907-474-5302 ffmtw@uaf.edu, Univ of Alaska, Fairbanks – Gs
Whaler, Kathryn A., +44 (0) 131 650 4904 kathy.whaler@ed.ac.uk, Edinburgh Univ – Ym
Whalley, John, +44 023 9284 2257 john.whalley@port.ac.uk, Univ of Portsmouth – Gc
Whatley, Robin L., 312-344-8604 rwhatley@colum.edu, Columbia Coll Chicago – Gg
Whattam, Scott A., 82-2-3290-3182 whattam@korea.ac.kr, Korea Univ – GiCtg
Wheatcraft, Steve, (775) 784-1973 steve@wheatcraftandassociates.com, Univ of Nevada, Reno – Hw
Wheatcroft, Rob, 541-737-3891 raw@coas.oregonstate.edu, Oregon State Univ – OgGg

Wheaton, John R., (406) 272-1603 jwheaton@mtech.edu, Montana Tech of The Univ of Montana – HwEc
Wheeler, Andy J., +353 21 4903951 a.wheeler@ucc.ie, Univ Coll Cork – Gu
Wheeler, Betty, whee0023@umn.edu, Univ of Minnesota, Twin Cities – Hw
Wheeler, Greg, 916-278-6337 wheelergr@csus.edu, California State Univ, Sacramento – Em
Wheeler, John, +440151 794 5172 Johnwh@liverpool.ac.uk, Univ of Liverpool – Em
Wheeler, Patricia, pwheeler@coas.oregonstate.edu, Oregon State Univ – Cm
Wheeler, Paul, +44 01326 255778 P.Wheeler@exeter.ac.uk, Exeter Univ – Nx
Wheeler, Richard F., (931) 221-1004 wheelerr@apsu.edu, Austin Peay State Univ – Ggo
Whelan, Jean K., (508) 289-2819 jwhelan@whoi.edu, Woods Hole Oceanographic Institution – Co
Whelan, Peter M., (320) 589-6315 whelan@mrs.umn.edu, Univ of Minnesota, Morris – Gx
Whiley, Harriet, harriet.whiley@flinders.edu.au, Flinders Univ – On
Whipkey, Charles, (540) 654-1428 cwhipkey@umw.edu, Univ of Mary Washington – Cg
Whipple, Kelin, 480-965-9508 kxw@asu.edu, Arizona State Univ – Gt
Whisonant, Robert C., (540) 831-5224 rwhisona@radford.edu, Radford Univ – Gsa
Whitaker, Brenda, blw@uvawise.edu, Univ of Virginia Coll, Wise – On
Whitaker, Rodney W., (505) 667-7672 rww@lanl.gov, Los Alamos National Laboratory – Oa
White, Angelicque, 541-737-6397 awhite@coas.oregonstate.edu, Oregon State Univ – Ob
White, Bekki C., 501 296-1877 bekki.white@arkansas.gov, Arkansas Geological Survey – GoEoGg
White, Craig M., (208) 426-3633 cwhite@boisestate.edu, Boise State Univ – Giv
White, George W., (301) 687-4264 gwhite@frostburg.edu, Frostburg State Univ – On
White, James D., +64 3 479-9009 james.white@otago.ac.nz, Univ of Otago – Gvs
White, James W. C., 303-492-5494 jwhite@colorado.edu, Univ of Colorado – Cs
White, Jeffrey, (812) 855-0731 whitej@indiana.edu, Indiana Univ, Bloomington – Hs
White, Jeffrey G., (919) 515-2389 Jeffrey_White@ncsu.edu, North Carolina State Univ – Or
White, Jeffrey R., (812) 855-0731 whitej@indiana.edu, Indiana Univ, Bloomington – Hg
White, John C., (859) 622-1276 john.white@eku.edu, Eastern Kentucky Univ – GiCgt
White, Jonathan, (303) 384-2650 jwhite@mines.edu, Colorado School of Mines – NgOuGg
White, Joseph C., (506) 453-4804 clancy@unb.ca, Univ of New Brunswick – Gc
White, Joseph D., (254) 710-2141 Joseph_D_White@baylor.edu, Baylor Univ – Or
White, Lisa D., 510-664-4966 ldwhite@berkeley.edu, Univ of California, Berkeley – OePim
White, Lisa D., (415) 338-1963 lwhite@sfsu.edu, San Francisco State Univ – Pg
White, Mark D., (509) 372-6070 mark.white@pnl.gov, Pacific Northwest National Laboratory – Hy
White, Martin, +353 (0)91 493 214 martin.white@nuigalway.ie, National Univ of Ireland Galway – Op
White, Nicky, +44 (0) 1223 337063 njw10@cam.ac.uk, Univ of Cambridge – YgGt
White, Robert, +44 (0) 1223 337187 rswl@cam.ac.uk, Univ of Cambridge – YgGt
White, Sarah Jane, 609-258-4101 mrusso@princeton.edu, Princeton Univ – Ge
White, Scott, (970) 247-7475 white_s@fortlewis.edu, Fort Lewis Coll – Oy
White, Scott M., (803) 777-6304 swhite@geol.sc.edu, Univ of South Carolina – Yr
White, Susan, just19@earthlink.net, Los Angeles Harbor Coll – Og
White, Susan Q., 61 3 9479 5641 susanqwhite@netspace.net.au, La Trobe Univ – GmgGe
White, William B., (814) 667-2709 wbw2@psu.edu, Pennsylvania State Univ, Univ Park – CgGzy
White, William M., (607) 255-7466 wmw4@cornell.edu, Cornell Univ – CctCa
Whiteaker, Timothy L., 512-471-0570 twhit@mail.utexas.edu, Univ of Texas at Austin – Oi
Whitehead, James, (506) 453-4593 Univ of New Brunswick – Gg
Whitehead, James, +44 0161 306-3953 James.Whitehead@manchester.ac.uk, Univer sity of Manchester – Oa
Whitehead, Peter W., +61 742321200 peter.whitehead@jcu.edu.au, James Cook Univ – Ggv
Whitehead, Robert E., rwhitehead@laurentian.ca, Laurentian Univ, Sudbury – Ce
Whitehill, Matthew, 218-733-5981 m.whitehill@lsc.edu, Lake Superior Coll – Gg
Whiteley, Martin, +44 01332 593752 m.whiteley@derby.ac.uk, Univ of Derby – Go
Whiteside, Jessica H., +44 (0)23 80593199 J.Whiteside@soton.ac.uk, Univ of Southampton – Pe
Whiteway, James, (416)736-2100 #22310 whiteway@yorku.ca, York Univ – Oa
Whitfield, Thomas G., (717) 702-2023 twhitfield@pa.gov, Pennsylvania Bureau of Topographic & Geologic Survey – Oi
Whiticar, Michael J., (250) 721-6514 whiticar@uvic.ca, Univ of Victoria – CosGo
Whiting, Peter J., (216) 368-3989 peter.whiting@case.edu, Case Western Reserve Univ – Gm
Whitlock, Cathy, (406) 994-6910 whitlock@montana.edu, Montana State Univ – Pey
Whitman, Brian E., (570) 408-4882 brian.whitman@wilkes.edu, Wilkes Univ – HqwSb
Whitman, Dean, (305) 348-3089 whitmand@fiu.edu, Florida International Univ – YgGet
Whitman, Jill M., (253) 535-8720 whitmaj@plu.edu, Pacific Lutheran Univ – OuGuOe
Whitmarsh, Robert B., +44 (0)23 80596564 rbw@noc.soton.ac.uk, Univ of Southampton – Yr
Whitmeyer, Steven J., whitmesj@jmu.edu, James Madison Univ – GctOi
Whitmore, John H., (937) 766-7947 johnwhitmore@cedarville.edu, Cedarville Univ – PgGsg
Whitney, D.A., (785) 532-6101 whitney@ksu.edu, Kansas State Univ – So
Whitney, Donna L., (612) 626-7582 dwhitney@umn.edu, Univ of Minnesota, Twin Cities – Gpt
Whitney, Earl M., (505) 667-3595 whitney@lanl.gov, Los Alamos National Laboratory – On
Whitney, James A., (706) 548-6894 jamesawhitney@hotmail.com, Univ of Georgia – Giv
Whitney, Michael, 860-405-9157 michael.whitney@uconn.edu, Univ of Connecticut – Op
Whittaker, Alex, +44 20 759 47491 a.whittaker@imperial.ac.uk, Imperial Coll – Gt
Whittaker, Amber, 207-287-2803 amber.whittaker@maine.gov, Dept of Agriculture, Conservation, and Forestry – OiGcCg
Whittecar, Jr., G. Richard, (757) 683-5197 rwhittec@odu.edu, Old Dominion Univ – GmHwOe
Whittemore, Donald O., (785) 864-2182 donwhitt@kgs.ku.edu, Univ of Kansas – Hw
Whittemore, Donald O., 785-864-3965 donwhitt@kgs.ku.edu, Univ of Kansas – Cl
Whittier, Michael, MWhittier@csustan.edu, California State Univ, Stanislaus – Ggz
Whittington, Alan, (573) 884-7625 whittingtona@missouri.edu, Univ of Missouri – GivCp
Whittington, Carla, (206) 878-3710 cwhittin@highline.edu, Highline Coll – GgvOg
Whitton, Mark, +44 023 9284 2257 mark.witton@port.ac.uk, Univ of Portsmouth – Pg
Wiberg, Patricia L., (434) 924-7546 pw3c@virginia.edu, Univ of Virginia – Og
Wicander, Reed, (989) 774-3179 wican1r@cmich.edu, Central Michigan Univ – Pl
Wickert, Andrew D., (612) 625-6878 awickert@umn.edu, Univ of Minnesota, Twin Cities – Gml
Wickham, John S., (817) 272-2987 wickham@uta.edu, Univ of Texas, Arlington – Gc
Wickham, Thomas, (724) 938-4180 wickham@calu.edu, California Univ of Pennsylvania – Ou
Wicks, Carol M., (225) 578-2692 cwicks@lsu.edu,

Louisiana State Univ – HwClHy
Wicks, Frederick J., (416) 978-5395 fredw@rom.on.ca, Univ of Toronto – Gz
Wicks, Frederick J., fredw@rom.on.ca, Royal Ontario Museum – Gz
Wicks, June, 609-258-4101 jwicks@princeton.edu, Princeton Univ – Cg
Widanagamage, Inoka, 662-915-2154 ihwidana@olemiss.edu, Univ of Mississippi – ClGep
Widdowson, Mark A., (540) 231-7153 Virginia Tech – HwSpHq
Widom, Elisabeth, (513) 529-5048 widome@miamioh.edu, Miami Univ – CcGve
Widory, David, 514-987-3000 #1968 widory.david@uqam.ca, Universite du Quebec a Montreal – Cs
Wiedenmann, Jorg, +44 (0)23 80596497 joerg.wiedenmann@noc.soton.ac.uk, Univ of Southampton – Ob
Wiederspahn, Mark, (512) 471-0406 markw@ig.utexas.edu, Univ of Texas at Austin – On
Wiederwohl, Chrissy, 979-845-7191 chrissyw@tamu.edu, Texas A&M Univ – Op
Wielicki, Matthew, (205) 348-0548 mmwielicki@ua.edu, Univ of Alabama – CcgGt
Wiens, Douglas A., (314) 935-6517 doug@wustl.edu, Washington Univ in St. Louis – YsrYg
Wierenga, Peter J., (520) 792-9591 wierenga@ag.arizona.edu, Univ of Arizona – Sp
Wiese, Katryn, (415) 452-5061 katryn.wiese@mail.ccsf.edu, City Coll of San Francisco – GiOg
Wiesenburg, Denis A., (601) 266-4937 denis.wiesenburg@usm.edu, Univ of Southern Mississippi – Oc
Wieser, Carola, +49 (89) 289 25885 carola.wieser@tum.de, Technische Universitaet Muenchen – Ngr
Wiesner, Mark R., (713) 348-5129 wiesner@rice.edu, Rice Univ – Og
Wiggert, Jerry, (228) 688-3491 jerry.wiggert@usm.edu, Univ of Southern Mississippi – Op
Wiggin, Jack, jack.wiggin@umb.edu, Univ of Massachusetts, Boston – Ou
Wigley, Rochelle, 603-862-1135 rochelle@ccom.unh.edu, Univ of New Hampshire – CgGu
Wigmosta, Mark S., (509) 372-6238 mark.wigmosta@pnl.gov, Pacific Northwest National Laboratory – Hs
Wignall, Paul, +44(0) 113 34 35247 P.B.Wignall@leeds.ac.uk, Univ of Leeds – PgGs
Wijbrans, Jan, j.r.wijbrans@vu.nl, VU Univ Amsterdam – CcGtv
Wijesinghe, Ananda M., (925) 423-0605 Lawrence Livermore National Laboratory – Nr
Wikgren, Brooke, bwikgren@neaq.org, Univ of Massachusetts, Boston – Oi
Wiklund, Marie-Louise M., +358 (0)29 503 2544 marie-louise.wiklund@gtk.fi, Geological Survey of Finland – OnnOn
Wilbur, Bryan, 626 585-3118 bcwilbur@pasadena.edu, Pasadena City Coll – Gg
Wilch, Thomas I., (517) 629-0759 twilch@albion.edu, Albion Coll – GlvGm
Wilcock, Peter W., (410) 516-5421 Johns Hopkins Univ – Gm
Wilcock, William S., (206) 543-6043 wilcock@u.washington.edu, Univ of Washington – Ou
Wilcox, Andrew, 406-243-4761 andrew.wilcox@mso.umt.edu, Univ of Montana – Gm
Wilcox, Jeffrey D., (828) 232-5184 jwilcox@unca.edu, Univ of North Carolina, Asheville – HwClGg
Wildeman, Thomas R., twildema@mines.edu, Colorado School of Mines – Ct
Wilder, Lee, (603) 271-1976 leland.wilder@des.nh.gov, New Hampshire Geological Survey – GgOee
Wilder, Margaret, mwilder@email.arizona.edu, Univ of Arizona – On
Wilderman, Candie, (717) 245-1573 wilderma@dickinson.edu, Dickinson Coll – Hs
Wiles, Gregory C., (330) 263-2298 gwiles@wooster.edu, Coll of Wooster – Gl
Wiley, Thomas J., (541) 476-2496 tom.wiley@dogami.state.or.us, Oregon Dept of Geology & Mineral Industries – Gg
Wilf, Peter D., (814) 865-6721 pwilf@geosc.psu.edu, Pennsylvania State Univ, Univ Park – Py
Wilfing, Lisa, +49 (89) 289 25857 lisa.wilfing@tum.de, Technische Universitaet Muenchen – Ngr
Wilhelmson, Robert B., (217) 333-8651 bw@ncsa.uiuc.edu, Univ of Illinois, Urbana-Champaign – Oa
Wilhelmy, Jerry B., (505) 665-3188 Los Alamos National Laboratory – On
Wilhite, Donald A., (402) 472-4270 dwilhite2@unl.edu, Univ of Nebraska, Lincoln – OayOn
Wilkerson, Christine M., (801) 537-3332 nrugs.cwilkers@state.ut.us, Utah Geological Survey – Gg
Wilkerson, Forrest, (507) 389-2617 forrest.wilkerson@mnsu.edu, Minnesota State Univ – OawGm
Wilkerson, M. S., (765) 658-4666 mswilke@depauw.edu, Univ of Illinois, Urbana-Champaign – Gc
Wilkey, Patrick L., (630) 252-6258 Argonne National Laboratory – Ng
Wilkie, Ann C., (352) 392-8699 acwilkie@ifas.ufl.edu, Univ of Florida – Sb
Wilkie, Richard W., (413) 253-5752 rwilkie@geo.umass.edu, Univ of Massachusetts, Amherst – OynOn
Wilkins, Colin, +44 1752 584773 C.Wilkins@plymouth.ac.uk, Univ of Plymouth – Eg
Wilkins, David E., (208) 426-2390 dwilkins@boisestate.edu, Boise State Univ – OyGm
Wilkins, Michael, (614) 688-2134 wilkins.231@osu.edu, Ohio State Univ – PyCl
Wilkinson, Bruce, 315-443-3869 eustasy@syr.edu, Syracuse Univ – Gs
Wilkinson, Bruce H., (734) 764-1435 eustasy@umich.edu, Univ of Michigan – GseGt
Wilkinson, Jamie, +44 20 759 46415 j.wilkinson@imperial.ac.uk, Imperial Coll – Em
Wilks, Daniel, (607) 255-1750 dsw5@cornell.edu, Cornell Univ – Oa
Wilks, Maureen, (505) 835-5322 mwilks@gis.nmt.edu, New Mexico Inst of Mining & Tech – Gp
Willahan, Duane, (408) 848-4702 Gavilan Coll – Gg
Willenbring, Jane, 215-746-8197 erosion@sas.upenn.edu, Univ of Pennsylvania – GmCcOl
William, Nancy, (785) 532-7257 nkw@ksu.edu, Kansas State Univ – On
Williams, Aaron, (251) 460-6915 bwilliams@southalabama.edu, Univ of South Alabama – Ow
Williams, Allison M., (905) 525-9140 (Ext. 24334) awill@mcmaster.ca, McMaster Univ – On
Williams, Amy, 410-704-2744 Towson Univ – Ca
Williams, Bruce A., (509) 372-3799 bruce.williams@pnl.gov, Pacific Northwest National Laboratory – Ng
Williams, Carolyn S., (540) 231-6894 wilcar@vt.edu, Virginia Polytechnic Inst & State Univ – On
Williams, Christopher J., (717) 291-3814 chris.williams@fandm.edu, Franklin and Marshall Coll – PeSb
Williams, Curtis J., (714) 484-7000 (Ext. 48181) Cypress Coll – Gg
Williams, Dave E., williams.david@interchange.ubc.ca, Univ of British Columbia – Oc
Williams, David A., (270) 827-3414 williams@uky.edu, Univ of Kentucky – Eg
Williams, Erik, (775) 784-1396 eswilliams@unr.edu, Univ of Nevada, Reno – Ys
Williams, Forrest, 831-656-3274 fwillia@comcast.net, Naval Postgraduate School – Ow
Williams, Graham, (902) 426-5657 Dalhousie Univ – Pl
Williams, Harris, (919) 530-7127 hewhew@nccu.edu, North Carolina Central Univ – Ogg
Williams, Harry F. L., (940) 565-3317 harryf.williams@unt.edu, Univ of North Texas – Gm
Williams, Helen, +44 (0) 191 33 42546 h.m.williams2@durham.ac.uk, Durham Univ – Cs
Williams, Ian S., (709) 737-8395 ian.williams@uwrf.edu, Univ of Wisconsin, River Falls – Yg
Williams, James H., (573) 341-4616 Missouri Univ of Science and Tech – Gg
Williams, Jeremy C., (330) 672-1459 jwill243@kent.edu, Kent State Univ – Cg
Williams, John W., (408) 924-5050 john.williams@sjsu.edu, San Jose State Univ – Ng
Williams, John W., (608) 265-5537 jww@geography.wisc.edu, Univ of Wisconsin, Madison – PeOyPl
Williams, Kaj, (406) 994-3338 kaj.williams@montana.edu, Montana State Univ – Oa
Williams, Kay R., (717) 477-1602 krwill@ship.edu, Shippensburg Univ – Oa
Williams, Kevin K., 716-878-5116 williakk@buffalostate.edu, Buffalo State Coll – GmXg
Williams, Kevin K., 716-878-4911 williakk@buffalostate.edu, Buffalo State Coll – Xg

Williams, Kevin L., (206) 543-3949 williams@apl.washington.edu, Univ of Washington – Op
Williams, Kim R., (303) 273-3245 krwillia@mines.edu, Colorado School of Mines – Ca
Williams, Mark, +44 0116 252 3642 mri@le.ac.uk, Leicester Univ – Po
Williams, Mark W., (303) 492-8830 Univ of Colorado – Og
Williams, Mathew, +44 (0) 131 650 7776 Mat.Williams@ed.ac.uk, Edinburgh Univ – Oa
Williams, Michael L., (413) 545-0745 mlw@geo.umass.edu, Univ of Massachusetts, Amherst – Gc
Williams, Paul, +44 0161 306-3905 paul.i.williams@manchester.ac.uk, Univ of Manchester – Oa
Williams, Paul F., (506) 453-5185 pfw@unb.ca, Univ of New Brunswick – Gc
Williams, Quentin, (831) 459-3132 qwilliams@pmc.ucsc.edu, Univ of California, Santa Cruz – Gy
Williams, Ric, +44-151-794-5136 Ric@liverpool.ac.uk, Univ of Liverpool – Og
Williams, Roger T., rtwillia@nps.edu, Naval Postgraduate School – Ow
Williams, Stanley N., (480) 965-1438 stanley.williams@asu.edu, Arizona State Univ – Gv
Williams, Wayne K., (423) 425-4427 wayne-williams@utc.edu, Univ of Tennessee at Chattanooga – Gdg
Williams, Wyn, +44 (0) 131 650 4909 Wyn.Williams@ed.ac.uk, Edinburgh Univ – Ym
Williams, II, Richard T., (865) 974-6169 rwilliams@utk.edu, Univ of Tennessee, Knoxville – Yg
Williams-Bruinders, Leizel, +27 (0)41 504 4361 leizel.williams@nmmu.ac.za, Nelson Mandela Metropolitan Univ – On
Williams-Jones, Anthony E., (514) 398-1676 willyj@eps.mcgill.ca, McGill Univ – Ce
Williams-Jones, Glyn, (778) 782-3306 glynwj@sfu.ca, Simon Fraser Univ – GvCeYe
Williamson, Ben, +44 01326 371856 B.J.Williamson@exeter.ac.uk, Exeter Univ – EmGv
Williamson, Douglas, 212-772-5265 douglas.williamson@hunter.cuny.edu, Hunter Coll (CUNY) – Oi
Willis, Grant C., (801) 537-3355 grantwillis@utah.gov, Utah Geological Survey – Gg
Willis, Julie B., 208-496-1905 willisj@byui.edu, Brigham Young Univ - Idaho – GtcOi
Willis, Marc, 714-992-7446 mwillis@fullcoll.edu, Fullerton Coll – Gg
Willoughby, Hugh E., 305-348-0243 hugh.willoughby@fiu.edu, Florida International Univ – Oa
Wills, Christopher J., (415) 557-1668 California Geological Survey – Ng
Wills, William V., 813-253-7809 wwills@hccfl.edu, Hillsborough Community Coll – Og
Willsey, Shawn P., (208) 732-6421 swillsey@csi.edu, Coll of Southern Idaho – GgcGv
Willson, Lee, 713.348.6219 Rice Univ – On
Wilmoth, Valorie, vthorson@slu.edu, Saint Louis Univ – Ow
Wilmut, Michael, (250) 472-4343 Univ of Victoria – On
Wilson, Alicia M., (803) 777-1240 awilson@geol.sc.edu, Univ of South Carolina – Hw
Wilson, Blake, (785) 864-2118 bwilson@kgs.ku.edu, Univ of Kansas – Oi
Wilson, Carol A., carolw@lsu.edu, Louisiana State Univ – GsOnCc
Wilson, Charlie, +44 (0)1603 59 1386 charlie.wilson@uea.ac.uk, Univ of East Anglia – Ge
Wilson, Clark R., (512) 471-5008 crwilson@jsg.utexas.edu, Univ of Texas at Austin – Yg
Wilson, Fred L., (325) 486-6984 fwilson@angelo.edu, Angelo State Univ – Gm
Wilson, Gary S., +64 3 479-7519 gary.wilson@otago.ac.nz, Univ of Otago – GuYmGs
Wilson, Greg, 541-737-4015 wilsongr@coas.oregonstate.edu, Oregon State Univ – Onp
Wilson, Greg C., 616-331-2392 wilsong@gvsu.edu, Grand Valley State Univ – Gm
Wilson, James R., jwilson@weber.edu, Weber State Univ – Gz
Wilson, Jeffey S., (317) 274-1128 jeswilso@iupui.edu, Indiana Univ, Indianapolis – Oy
Wilson, Jeffrey A., (734) 647-7461 wilsonja@umich.edu, Univ of Michigan – Gg
Wilson, John D., (780) 492-0353 Univ of Alberta – Ow
Wilson, John L., (575) 835-5308 jwilson@nmt.edu, New Mexico Inst of Mining and Tech – Hw
Wilson, John R., (610) 330-5197 wilsonj@lafayette.edu, Lafayette Coll – Oi
Wilson, Laura E., (785) 639-6192 lewilson6@fhsu.edu, Fort Hays State Univ – PveGs
Wilson, Lorne G., (520) 621-9108 lorne@email.arizona.edu, Univ of Arizona – Hw
Wilson, Lucy A., (506) 648-5607 lwilson@unbsj.ca, Univ of New Brunswick Saint John – Ga
Wilson, Mark A., (330) 263-2247 mwilson@wooster.edu, Coll of Wooster – Pi
Wilson, Merwether, +44 (0) 131 650 8636 Meriwether.Wilson@ed.ac.uk, Edinburgh Univ – Ob
Wilson, Michael C., wilsonmi@douglasColl.ca, Douglas Coll – GaPgGs
Wilson, P. Christopher, 772-468-3922 ext 119 pcwilson@ufl.edu, Univ of Florida – So
Wilson, Paul A., +44 (0)23 80596164 paul.wilson@noc.soton.ac.uk, Univ of Southampton – GsCg
Wilson, Rick I., (709) 737-8386 California Geological Survey – Ge
Wilson, Robert, +44 01334 463914 rjsw@st-andrews.ac.uk, Univ of St. Andrews – On
Wilson, Robert E., (631) 632-8689 robert.wilson@stonybrook.edu, SUNY, Stony Brook – Op
Wilson, Roy R., (860) 465-4370 wilsonr@easternct.edu, Eastern Connecticut State Univ – Oi
Wilson, Sarah, (540) 458-8800 wilsons@wlu.edu, Washington & Lee Univ – On
Wilson, Steven D., (217) 333-0956 Illinois State Water Survey – Hw
Wilson, Terry J., 614-292-0723 wilson.43@osu.edu, Ohio State Univ – Gc
Wilson, Thomas, +64 3 3667001 Ext 45511 thomas.wilson@canterbury.ac.nz, Univ of Canterbury – Gv
Wilson, Thomas B., (520) 621-9308 twilson@ag.arizona.edu, Univ of Arizona – On
Wilson, Thomas H., (304) 293-6431 tom.wilson@mail.wvu.edu, West Virginia Univ – YeGcYg
Wilton, Derek H., (709) 737-8389 dwilton@sparky2.esd.mun.ca, Memorial Univ of Newfoundland – EgCae
Wilton, Robert D., (905) 525-9140 (Ext. 24536) wiltonr@mcmaster.ca, McMaster Univ – On
Wiltshire, John C., (808) 956-6042 johnw@soest.hawaii.edu, Univ of Hawai'i, Manoa – Og
Wimbush, Mark, (401) 874-6515 m.wimbush@gso.uri.edu, Univ of Rhode Island – Op
Winant, Clinton D., (858) 534-2067 cwinant@ucsd.edu, Univ of California, San Diego – On
Winberry, Paul, winberry@geology.cwu.edu, Central Washington Univ – Yg
Winchell, Robert E., (562) 985-4920 California State Univ, Long Beach – Gz
Winckler, Gisela, 845-365-8756 Columbia Univ – Cg
Windsor, John G., (321) 674-8096 jwindsor@fit.edu, Florida Inst of Tech – Oc
Wine, Paul H., (404) 894-3425 pw7@prism.gatech.edu, Georgia Inst of Tech – Oa
Winebrenner, Dale P., 206-543-1393 dpw@apl.washington.edu, Univ of Washington – Olr
Wing, Scott L., (202) 357-2649 Smithsonian Institution / National Museum of Natural History – Pb
Winglee, Robert M., 206-685-8160 winglee@ess.washington.edu, Univ of Washington – XyYx
Winguth, Arne M., 817 272 2987 awinguth@uta.edu, Univ of Texas, Arlington – Opa
Winkler, Dale A., (214) 768-2750 Southern Methodist Univ – Pv
Winklhofer, Michael, 089/2180 4143 michael@geophysik.uni-muenchen.de, Ludwig-Maximilians-Universitaet Muenchen – Yg
Winslow, Margaret S., (212) 650-6984 Graduate School of the City Univ of New York – Gc
Winsor, Roger A., (828) 262-7053 winsorra@appstate.edu, Appalachian State Univ – On
Winston, Barbara, (314) 935-7047 Washington Univ in St. Louis – On
Winterbottom, Wesley, winterbottomw@easternct.edu, Eastern Connecticut State Univ – Og
Winterkamp, Judith L., (505) 667-1264 judyw@lanl.gov, Los Alamos National Laboratory – On
Winton, Alison, (806) 834-0497 alison.winton@ttu.edu,

Texas Tech Univ – OnnOn
Wintsch, Robert P., (812) 855-4018 wintsch@indiana.edu, Indiana Univ, Bloomington – Gp
Wirth, Karl R., (651) 696-6449 wirth@macalester.edu, Macalester Coll – GiOn
Wisdom, Jack, (617) 253-7730 wisdom@mit.edu, Massachusetts Inst of Tech – Og
Wise, Michael A., (202) 633-1826 wisem@si.edu, Smithsonian Institution / National Museum of Natural History – Gz
Wise, Jr, Sherwood W., (850) 644-6265 swise@fsu.edu, Florida State Univ – PmGu
Wisely, Beth , 307-268 2233 bwisely@casperColl.edu, Casper Coll – YgHwGc
Wishart, De Bonne N., dwishart@centralstate.edu, Central State Univ – YgCa
Wishner, Karen, (401) 874-6402 kwishner@gso.uri.edu, Univ of Rhode Island – Ob
Withers, Kim, 361-825-5907 Kim.Withers@tamucc.edu, Texas A&M Univ, Corpus Christi – On
Withers, Mitchell M., (901) 678-2007 mwithers@memphis.edu, Univ of Memphis – Ys
Withers, Tony, 519-661-2111 x.88627 tony.withers@uwo.ca, Western Univ – Cp
Withjack, Martha O., (848) 445-3445 drmeow3@yahoo.com, Rutgers, The State Univ of New Jersey – Gct
Witkowski, Christine, (860) 343-5781 cwitkowski@mxcc.commnet.edu, Middlesex Community Coll – GeOg
Witt, Emma, emma.witt@stockton.edu, Stockton Univ – Hgs
Witte, Ronald W., (609) 292-2576 ron.witte@dep.state.nj.us, New Jersey Geological and Water Survey – Gl
Witter, Rob, 541-574-7969 rwitter@usgs.gov, Oregon Dept of Geology & Mineral Industries – Og
Wittke, James, 9285239565/9044 james.wittke@nau.edu, Northern Arizona Univ – Gi
Wittke, Seth, (307) 766-2286 Ext. 244 seth.wittke@wyo.gov, Wyoming State Geological Survey – Ng
Wittkop, Chad, (507) 389-6929 chad.wittkop@mnsu.edu, Minnesota State Univ – GsCgGl
Witzke, Brian J., (319) 335-1590 brian-witzke@uiowa.edu, Univ of Iowa – Ps
Wixman, Ronald, (541) 346-4568 rwixman@uoregon.edu, Univ of Oregon – On
Wobus, Reinhard A., (413) 597-2470 reinhard.a.wobus@williams.edu, Williams Coll – Giz
Woerheide, Gert, geobiologie@geo.lmu.de, Ludwig-Maximilians-Universitaet Muenchen – Pyo
Wofsy, Steven C., (617) 495-4566 scw@io.harvard.edu, Harvard Univ – Oa
Wogelius, Roy, (+44)-(0)161-275 3841 Roy.Wogelius@manchester.ac.uk, Univ of Manchester – Cg
Wohl, Ellen E., (970) 491-5298 ellen.wohl@colostate.edu, Colorado State Univ – Gm
Wohletz, Kenneth H., (505) 667-9202 wohletz@lanl.gov, Los Alamos National Laboratory – Gv
Wojewoda, Jurand, jurand.wojewoda@uwr.edu.pl, Univ of Wroclaw – Gs
Wojtal, Steven F., (440) 775-8352 steven.wojtal@oberlin.edu, Oberlin Coll – Gc
Wolak, Jeannette, (931) 372-3695 jwolak@tntech.edu, Tennessee Tech Univ – Gsr
Wolaver, Brad, (512) 471-1368 brad.wolaver@beg.utexas.edu, Univ of Texas at Austin – HwGe
Wolcott, Donna L., (919) 515-7866 donna_wolcott@ncsu.edu, North Carolina State Univ – Ob
Wolcott, Ray, 6196447454 x3099 rwolcott@palomar.edu, Cuyamaca Coll – Og
Wolcott, Thomas G., (919) 515-7866 tom_wolcott@ncsu.edu, North Carolina State Univ – Ob
WoldeGabriel, Giday, 505-667-8749 wgiday@lanl.gov, Los Alamos National Laboratory – Gx
Wolf, Aaron T., (541) 737-2722 wolfa@geo.oregonstate.edu, Oregon State Univ – Hg
Wolf, Lorraine W., (334) 844-4878 wolflor@auburn.edu, Auburn Univ – Ysg
Wolf, Michael B., (309) 794-7304 MichaelWolf@augustana.edu, Augustana Coll – Gi
Wolfe, Amy, 513-529-5009 wolfeal4@miamioh.edu, Miami Univ – Cs
Wolfe, Ben, (816) 604-6622 ben.wolfe@mcckc.edu, Metropolitan Community Coll-Kansas City – GgOyg
Wolfe, Christopher, christopher.wolfe@stonybrook.edu, SUNY, Stony Brook – Op
Wolfe, Karen M., (615) 898-2726 karen.wolfe@mtsu.edu, Middle Tennessee State Univ – On
Wolfe, Paul J., 937 775-2201 paul.wolfe@wright.edu, Wright State Univ – Ye
Wolff, Eric W., +44 (0) 1223 333486 ew428@cam.ac.uk, Univ of Cambridge – OlPe
Wolff, George, +44-151-794-4094 Wolff@liverpool.ac.uk, Univ of Liverpool – Og
Wolff, John A., (509) 335-2825 jawolff@wsu.edu, Washington State Univ – Giv
Wolfgram, Diane, (406) 496-4353 dwolfgram@mtech.edu, Montana Tech of the Univ of Montana – EgoNm
Wolfsberg, Andrew V., (505) 667-3599 awolf@lanl.gov, Los Alamos National Laboratory – Hw
Wolfsberg, Kurt, (505) 667-4464 Los Alamos National Laboratory – Cc
Wolken, Gabriel J., (907) 451-5018 gabriel.wolken@alaska.gov, Alaska Division of Geological & Geophysical Surveys – Gm
Wollan, Jacinda, mandy.looser@ndsu.edu, North Dakota State Univ – On
Wolny, Dave, (970) 248-1154 dwolmy@mesastate.edu, Colorado Mesa Univ – Ys
Wolosz, Thomas H., (518) 564-4031 woloszth@plattsburgh.edu, Plattsburgh State Univ (SUNY) – Pe
Woltemade, Christopher J., (717) 477-1143 cjwolt@ship.edu, Shippensburg Univ – HgGmHs
Wolter, Calvin, Calvin.Wolter@dnr.iowa.gov, Iowa Dept of Natural Resources – GgOi
Wolverton, Steve, 940-565-4987 steven.wolverton@unt.edu, Univ of North Texas – Ga
Wong, Chi S., (250) 363-6407 WongCS@pac.dfo-mpo.gc.ca, Univ of British Columbia – Cm
Wong, Cindy, (716) 878-6731 solargs@buffalostate.edu, Buffalo State Coll – On
Wong, Corinne I., 617-553-1817 wongcw@bc.edu, Boston Coll – PeCsHw
Wong, George T. F., (757) 683-4932 gwong@odu.edu, Old Dominion Univ – Oc
Wong, Martin, (315) 228-7203 mswong@colgate.edu, Colgate Univ – Gtc
Wong, Teng-fong, (631) 632-8212 teng-fong.wong@stonybrook.edu, SUNY, Stony Brook – Yx
Woo, David, 510-885-3160 david.woo@csueastbay.edu, California State Univ, East Bay – Or
Woo, Ming-Ko, (905) 525-9140 woo@mcmaster.ca, McMaster Univ – Hg
Wood, Aaron R., (515) 294-8862 awood@iastate.edu, Iowa State Univ of Science & Tech – PveGr
Wood, Bernard, +44 (1865) 272014 Bernie.Wood@earth.ox.ac.uk, Univ of Oxford – Gz
Wood, Charles W., (334) 844-3997 woodcha@auburn.edu, Auburn Univ – Sc
Wood, Craig B., 401-865-2585 cbwood@providence.edu, Providence Coll – Pv
Wood, David A., 210-486-0063 dwood30@alamo.edu, San Antonio Community Coll – Onn
Wood, Eric F., (609) 258-4675 efwood@princeton.edu, Princeton Univ – HgOrHq
Wood, Howard, (361) 825-3335 tony.wood@tamucc.edu, Texas A&M Univ, Corpus Christi – OeHsOn
Wood, Ian, +44 020 7679 32405 ian.wood@ucl.ac.uk, Univ Coll London – Gz
Wood, Jacqueline, (504) 671-6485 jwood@lakelandcc.edu, Delgado Community Coll – Gg
Wood, James R., (906) 487-2894 jrw@mtu.edu, Michigan Technological Univ – Cl
Wood, Kim, 662-268-1032 kimberly.wood@msstate.edu, Mississippi State Univ – OawOr
Wood, Lesli, lwood@mines.edu, Colorado School of Mines – GsoGm
Wood, Lesli J., 512-471-0328 lesli.wood@beg.utexas.edu, Univ of Texas at Austin – Eo
Wood, Neill, +44 01326 255163 n.a.wood@exeter.ac.uk, Exeter Univ – YeOe
Wood, Robert, 206-543-1203 robwood@atmos.washington.edu, Univ of Washington – Oa
Wood, Spencer H., (208) 426-3629 swood@boisestate.edu, Boise State Univ – Gmm
Wood, Stephen E., 206-543-0900 sewood@ess.washington.edu, Univ of Washington – On
Wood, Thomas R., (208) 533-8164 twood@uidaho.edu,

Univ of Idaho – Hw
Woodall, Debra W., (386) 506-3765 WoodalD@daytonastate.edu, Daytona State Coll – GgOg
Woodard, Gary C., (520) 621-5399 gwoodard@sahra.arizona.edu, Univ of Arizona – On
Woodburne, Michael O., (909) 787-5028 michael.woodburne@ucr.edu, Univ of California, Riverside – Pv
Woodcock, Curtis, curtis@bu.edu, Boston Univ – Or
Woodcock, Nigel, +44 (0) 1223 333430 nhw1@esc.cam.ac.uk, Univ of Cambridge – GcsGt
Woodgate, Rebecca A., (206) 221-3268 woodgate@apl.washington.edu, Univ of Washington – Op
Woodhouse, Connie, 520-626-0235 conniew1@email.arizona.edu, Univ of Arizona – Pe
Woodhouse, Iain H., +44 (0) 131 650 2527 i.h.woodhouse@ed.ac.uk, Edinburgh Univ – Oi
Woodhouse, John, +44 (1865) 272021 john.woodhouse@earth.ox.ac.uk, Univ of Oxford – Yg
Woodin, Sarah, woodin@biol.sc.edu, Univ of South Carolina – Ob
Wooding, Frank B., (508) 289-3334 Woods Hole Oceanographic Institution – Yr
Woodland, Bertram G., (312) 665-7648 Field Museum of Natural History – Gp
Woodley, Teresa, (604) 822-3146 twoodley@eos.ubc.ca, Univ of British Columbia – On
Woodruff, Jonathan D., 413-577-3831 woodruff@geo.umass.edu, Univ of Massachusetts, Amherst – Gs
Woodruff, William H., (505) 665-2557 Los Alamos National Laboratory – On
Woods, Adam D., (657) 278-2921 awoods@fullerton.edu, California State Univ, Fullerton – GsPe
Woods, Andy, +44 (0) 1223 765702 andy@bpi.cam.ac.uk, Univ of Cambridge – YgGt
Woods, Karen M., (409) 880-2251 karen.woods@lamar.edu, Lamar Univ – On
Woods, Neal, woodsn@mailbox.sc.edu, Univ of South Carolina – On
Woods, Rachel A., +44 (0) 131 650 6014 Rachel.Wood@ed.ac.uk, Edinburgh Univ – Gs
Woodward, Lee A., (505) 277-5309 Univ of New Mexico – Gc
Woodward, Mac B., 501 683-0113 mac.woodward@arkansas.gov, Arkansas Geological Survey – Gog
Woodwell, Grant R., (540) 654-1427 gwoodwel@mwc.edu, Univ of Mary Washington – Gc
Wooldridge, C F., wooldridge@cf.ac.uk, Cardiff Univ – On
Woolery, Edward W., (859) 257-3016 woolery@uky.edu, Univ of Kentucky – Ys
Worcester, Peter A., (812) 866-7306 worcestr@hanover.edu, Hanover Coll – Gi
Worden, Richard, 0151 794 5184 R.Worden@liverpool.ac.uk, Univ of Liverpool – Gs
Worrall, Fred, 0191 3742525 fred.worrall@durham.ac.uk, Durham Univ – Cg
Wortel, Matthew J., (319) 335-3992 matthew-wortel@uiowa.edu, Univ of Iowa – Gx
Worthington, Lindsay L., 505.277.4204 lworthington@unm.edu, Univ of New Mexico – Yg
Wortmann, Ulrich B., 416-978-2084 Univ of Toronto – Gg
Wraith, Jon M., (406) 994-1997 jean.dixon@montana.edu, Montana State Univ – Sp
Wright, Alan, 561-992-1555 alwr@ufl.edu, Univ of Florida – Sc
Wright, Carrie L., (812) 465-1145 clwright@usieagles.org, Univ of Southern Indiana – Oe
Wright, Eric S., (406) 683-7274 eric.wright@umwestern.edu, Univ of Montana Western – Gq
Wright, James D., (848) 445-5722 jdwright@rci.rutgers.edu, Rutgers, The State Univ of New Jersey – CsOuPm
Wright, James (Jim) E., (706) 542-4394 jwright@gly.uga.edu, Univ of Georgia – Gt
Wright, Kathyrn, (269) 387-5486 kathyrn.wright@wmich.edu, Western Michigan Univ – On
Wright, Stephen F., (802) 656-4479 swright@uvm.edu, Univ of Vermont – Gc
Wright, Tim, +44(0) 113 34 35258 t.j.wright@leeds.ac.uk, Univ of Leeds – Yd
Wright, V P., wrightVP@cf.ac.uk, Cardiff Univ – Gs
Wronkiewicz, David J., (630) 252-4385 Argonne National Laboratory – Cg
Wu, Charles T., (519) 661-3791 ctwu@uwo.ca, Western Univ – Ca
Wu, Chin, (608) 263-3078 chinwu@engr.wisc.edu, Univ of Wisconsin, Madison – Onp
Wu, David T., (303) 273-2066 dwu@mines.edu, Colorado School of Mines – On
Wu, Francis T., (607) 777-2512 Binghamton Univ – Ys
Wu, Jonny, (713) 743-9624 jwu40@central.uh.edu, Univ of Houston – GtcGo
Wu, Patrick, (403) 220-7855 ppwu@ucalgary.ca, Univ of Calgary – Ys
Wu, Ru-shan, (831) 459-5135 wrs@es.ucsc.edu, Univ of California, Santa Cruz – Ys
Wu, Shiliang, (906) 487-2590 slwu@mtu.edu, Michigan Technological Univ – OaCg
Wu, Shuang-Ye, (937) 229-1720 Shuang-Ye.Wu@notes.udayton.edu, Univ of Dayton – Oy
Wu, Xiaoqing, (515) 294-9872 wuxq@iastate.edu, Iowa State Univ of Science & Tech – Oa
Wu, Xingru, xingru.wu@ou.edu, Univ of Oklahoma – Np
Wu, Yutian, (765) 494-8677 wu640@purdue.edu, Purdue Univ – Oa
Wuebbles, Donald J., (217) 244-1568 wuebbles@illinois.edu, Univ of Illinois, Urbana-Champaign – Oa
Wuerthele, Norman, (412) 622-3265 wuerthelen@carnegiemnh.org, Carnegie Museum of Natural History – Pv
Wulamu, Wasit, (314) 977-5156 awulamu@slu.edu, Saint Louis Univ – Oi
Wulff, Andrew, (270) 745-5976 andrew.wulff@wku.edu, Western Kentucky Univ – Gv
Wulff, Andrew H., (562) 907-4220 Whittier Coll – Gi
Wunderman, Richard, (202) 633-1827 wunderma@si.edu, Smithsonian Institution / National Museum of Natural History – Gv
Wunsch, Carl, (617) 496-2732 cwunsch@fas.harvard.edu, Harvard Univ – Op
Wunsch, Carl I., (617) 253-5937 cwunsch@mit.edu, Massachusetts Inst of Tech – Op
Wunsch, David R., (302) 831-8258 dwunsch@udel.edu, Univ of Delaware – ClHwNg
Wust-Bloch, Gilles H., (03-) 640-5475 l, Tel Aviv Univ – Ys
Wyckoff, John W., (303) 556-2590 john.wyckoff@ucdenver.edu, Univ of Colorado, Denver – Oy
Wyckoff, William K., (406) 994-6914 uesww@montana.edu, Montana State Univ – On
Wylie, Ann G., (301) 405-4079 awylie@umd.edu, Univ of Maryland – Gz
Wyllie, Peter J., (626) 395-6461 wyllie@gps.caltech.edu, California Inst of Tech – Cp
Wyman, Derek A., (618) 938-0117 Univ of Saskatchewan – Cg
Wyngaard, John C., wyngaard@ems.psu.edu, Pennsylvania State Univ, Univ Park – Ow
Wynn, Elizabeth A., (205) 247-3671 awynn@gsa.state.al.us, Geological Survey of Alabama – HsOyi
Wynn, Thomas C., (570) 484-2081 twynn@lhup.edu, Lock Haven Univ – GsPiEo
Wypych, Alicja, 907-451-5010 alicja.wypych@alaska.gov, Alaska Division of Geological & Geophysical Surveys – Gg
Wyse Jackson, Patrick N., + 353-1-8961477 wysjcknp@tcd.ie, Trinity Coll – PiGh
Wysession, Michael E., (314) 935-5625 michael@wucore.wustl.edu, Washington Univ in St. Louis – Ys
Wysocki, Mark, (607) 255-2568 mww3@cornell.edu, Cornell Univ – Oa
Wysong, Jr., James F., (813) 253-7805 jwysong@hccfl.edu, Hillsborough Community Coll – Ow
Wyss, Andre R., (805) 893-8628 wyss@geol.ucsb.edu, Univ of California, Santa Barbara – Pv

## X

Xia, Renjie, (217) 244-6166 Illinois State Water Survey – Hs
Xiao, Shuhai, 540-231-1366 xiao@vt.edu, Virginia Polytechnic Inst & State Univ – Pi
Xie, Feiqin, (361) 825-3229 feiqin.xie@tamucc.edu, Texas A&M Univ, Corpus Christi – OarOw
Xie, Lian, (919) 515-1435 xie@ncsu.edu, North Carolina State Univ – Oa
Xie, Xiangyang, (817) 257-4395 x.xie@tcu.edu, Texas Christian Univ – GoEog
Xie, Xiao-bi, (831) 459-5094 xie@es.ucsc.edu, Univ of California, Santa Cruz – Ys
Xie, Zhixiao, (561) 297-2852 xie@fau.edu, Florida Atlantic Univ – OirOy
Xu, Huifang, 608/265-5587 hfxu@geology.wisc.edu, Univ of Wisconsin, Madison – Gz
Xu, Li, (508)289-3673 lxu@whoi.edu,

Woods Hole Oceanographic Institution – Yr
Xu, Shangping, 414-229-6148 xus@uwm.edu, Univ of Wisconsin, Milwaukee – Hg
Xu, Wei, wei.xu@uleth.ca, Univ of Lethbridge – Oi
Xuan, Chuang, +44 (0)23 80596401 C.Xuan@soton.ac.uk, Univ of Southampton – Pe

## Y

Ysmail, 00904242370000-5970 iyildirim@firat.edu.tr, Firat Univ – ScGix
Yabusaki, Steven B., (509) 372-6095 steve.yabusaki@pnl.gov, Pacific Northwest National Laboratory – Hg
Yacobucci, Peg M., (419) 372-7982 mmyacob@bgsu.edu, Bowling Green State Univ – Po
Yacucci, Mark, (309) 268-8640 mark.yacucci@heartland.edu, Heartland Community Coll – Og
Yacucci, Mark A., 217-265-0747 yacucci@illinois.edu, Univ of Illinois – Oi
Yalcin, Kaplan, (541) 737-1230 yalcink@geo.oregonstate.edu, Oregon State Univ – GgeOe
Yalcin, Rebecca, yalcinr@onid.orst.edu, Oregon State Univ – Gg
Yalda, Sepideh, 717-872-3293 Sepi.Yalda@millersville.edu, Millersville Univ – Oa
Yamanaka, Tsuyuko, +44 0151 795 5291 T.Yamanaka@liverpool.ac.uk, Univ of Liverpool – Ob
Yan, Eugene, (630) 252-6322 eyan@anl.gov, Argonne National Laboratory – Hw
Yan, Jun, (270) 745-8952 jun.yan@wku.edu, Western Kentucky Univ – Oi
Yan, Xiao-Hai, (302) 831-3694 Univ of Delaware – Or
Yan, Y E., (630) 252-6322 eyan@anl.gov, Argonne National Laboratory – Gg
Yancey, Thomas E., (979) 845-0643 yancey@geo.tamu.edu, Texas A&M Univ – Pg
Yandle, Tracy, 404 727 5652 tyandle@emory.edu, Emory Univ – Ob
Yang, Changbing, 512-471-4364 changbing.yang@beg.utexas.edu, Univ of Texas at Austin – Hw
Yang, Gang, (970)-491-3789 gang.yang@colostate.edu, Colorado State Univ – Cc
Yang, Jianwen, 519-253-3000 x2181 jianweny@uwindsor.ca, Univ of Windsor – HwYg
Yang, Panseok, (204) 474-6910 panseok.yang@umanitoba.ca, Univ of Manitoba – CaGp
Yang, Ping, (979) 845-7679 pyang@tamu.edu, Texas A&M Univ – Oa
Yang, Wan, 316 -978-3140 Wichita State Univ – Gs
Yang, Y, YangY6@cf.ac.uk, Cardiff Univ – Hg
Yang, Zhiming, (919) 530-5296 zyang@nccu.edu, North Carolina Central Univ – Ory
Yang, Zong-Liang, (512) 471-3824 liang@mail.utexas.edu, Univ of Texas at Austin – OwaHq
Yankovsky, Sasha, (803) 777-3550 ayankovsky@geol.sc.edu, Univ of South Carolina – OpnOw
Yao, Wensheng, (727) 553-3922 Univ of South Florida – Oc
Yao, Yon, +27 (0)46-603-7393 y.yao@ru.ac.za, Rhodes Univ – Eg
Yapp, Crayton J., (214) 768-3897 Southern Methodist Univ – Csl
Yarbrough, Lance D., 662-915-7499 ldyarbro@olemiss.edu, Univ of Mississippi – Ng
Yarbrough, Robert A., 912-478-0846 RYarbrough@GeorgiaSouthern.edu, Georgia Southern Univ – On
Yardley, Bruce, +44(0) 113 34 35227 B.W.D.Yardley@leeds.ac.uk, Univ of Leeds – Cg
Yarger, Douglas N., (515) 294-9872 doug@iastate.edu, Iowa State Univ of Science & Tech – Oa
Yarger, Douglas N., doug@iastate.edu, Iowa State Univ of Science & Tech – Oa
Yassin, Barbara E., (601) 961-5571 barbara_yassin@mdeq.ms.gov, Mississippi Office of Geology – OyiGm
Yates, Martin G., (207) 581-2154 yates@maine.edu, Univ of Maine – EgGzx
Yates, Mary Anne, (505) 667-7090 Los Alamos National Laboratory – On
Yau, Man Kong, (514) 398-3719 peter.yau@mcgill.ca, McGill Univ – Oa
Ye, Hengchun, (323) 343-2229 hye2@calstatela.edu, California State Univ, Los Angeles – OyHs
Ye, Ming, mingye@scs.fsu.edu, Florida State Univ – Hg
Yeager, Kevin M., (859) 257-5431 kevin.yeager@uky.edu, Univ of Kentucky – GsCmc
Yeats, Robert S., yeatsr@geo.oregonstate.edu, Oregon State Univ – GtYrNg
Yee, Nathan, 848-932-5714 nyee@envsci.rutgers.edu, Rutgers, The State Univ of New Jersey – PyCo
Yegulalp, Tuncel M., (212) 854-2984 yegulalp@columbia.edu, Columbia Univ – Nm
Yeh, Jim, (520) 621-5943 yeh@hwr.arizona.edu, Univ of Arizona – Sp
Yeh, Joseph S., (512) 471-3323 joseph.yeh@beg.utexas.edu, Univ of Texas at Austin, Jackson School of Geosciences – Gm
Yeh, Tian-Chyi J., (520) 621-5943 ybiem@hwr.arizona.edu, Univ of Arizona – Hq
Yelderman, Jr., Joe C., (254) 710-2185 joe_yelderman@baylor.edu, Baylor Univ – HwgOu
Yelisetti, Subbarao, 361-593-4894 Subbarao.yelisetti@tamuk.edu, Texas A&M Univ, Kingsville – Ygs
Yelle, Roger, (520) 621-6243 yelle@lpl.arizona.edu, Univ of Arizona – On
Yellich, John, (269) 387-8649 john.a.yellich@wmich.edu, Western Michigan Univ – EgGeHw
Yeung, Laurence Y., (713) 348-6304 ly19@rice.edu, Rice Univ – CsOaCl
Yi, Chuixiang, 718-997-3366 chuixiang.yi@qc.cuny.edu, Queens Coll (CUNY) – Oaw
Yi, Yuchan, 614 292 6005 yi.3@osu.edu, Ohio State Univ – Ydv
Yiannakoulias, Niko, (905) 525-9140 (Ext. 20117) yiannan@mcmaster.ca, McMaster Univ – Ge
Yilmaz, Alper, 614-247-4323 yilmaz.15@osu.edu, Ohio State Univ – Or
Yin, An, (310) 825-8752 yin@epss.ucla.edu, Univ of California, Los Angeles – Gt
Yin, Jianjun, 520-626-7453 yin@email.arizona.edu, Univ of Arizona – Og
Yin, Qing-zhu, 530-752-0934 qyin@ucdavis.edu, Univ of California, Davis – Cc
Yoakum, Barbara, 410-293-6928 yoakum@usna.edu, United States Naval Academy – On
Yogodzinkski, Gene M., 803-777-9524 gyogodzin@geol.sc.edu, Univ of South Carolina – Gi
Yoh, Shing, (908) 737-3692 syoh@kean.edu, Kean Univ – Ow
Yon, Lisa, 7607441150 x2369 lyon@palomar.edu, Palomar Coll – Ogg
Yong, Wenjun, 519-661-2111 ext 86628 wyong4@uwo.ca, Western Univ – Yx
Yonkee, W. A., (801) 626-7419 ayonkee@weber.edu, Weber State Univ – Gct
Yoo, Kyungsoo, 612-624-7784 kyoo@umn.edu, Univ of Minnesota, Twin Cities – Sd
Yool, Stephen R., yools@email.arizona.edu, Univ of Arizona – Or
Yoshida, Glenn, (323) 241-5296 yoshidgy@lasc.edu, Los Angeles Southwest Coll – Gg
Yoshinobu, Aaron S., (806) 834-7715 aaron.yoshinobu@ttu.edu, Texas Tech Univ – GctGu
Yoskowitz, David , 361-825-2966 david.yoskowitz@tamucc.edu, Texas A&M Univ, Corpus Christi – On
You, Jinsheng, jyou2@unl.ed, Univ of Nebraska, Lincoln – Hg
Young, Donald W., (602) 542-5025 Univ of Arizona – On
Young, Edward D., (310) 267-4930 eyoung@epss.ucla.edu, Univ of California, Los Angeles – CsXc
Young, George S., (814) 863-4228 young@ems.psu.edu, Pennsylvania State Univ, Univ Park – Ow
Young, Glen, (573) 368-2333 glen.young@dnr.mo.gov, Missouri Dept of Natural Resources – Gg
Young, Graham A., (204) 988-0648 gyoung@cc.umanitoba.ca, Univ of Manitoba – Pi
Young, Grant M., (519) 473-5692 gyoung@uwo.ca, Western Univ – GmGt
Young, Harvey R., 204-727-9798 young@brandonu.ca, Brandon Univ – Gd
Young, James E., (828) 262-8482 youngje@appstate.edu, Appalachian State Univ – Oe
Young, Jeffrey, (204) 474-8863 jeff.young@umanitoba.ca, Univ of Manitoba – GcOe
Young, Jeri J., 602.708.8558 jeri.young@azgs.az.gov, Univ of Arizona – Ys
Young, John A., (608)262-5963 jayoung@wisc.edu, Univ of Wisconsin, Madison – Oa
Young, Michael H., (512) 475-8830 michael.young@beg.utexas.edu, Univ of Texas at Austin – GeHw
Young, Patrick, (480) 727-6581 patrick.young.1@asu.edu, Arizona State Univ – On
Young, Priscilla E., 605-677-6144 priscilla.young@usd.edu, South Dakota Dept of Environment and Natural Resources – On
Young, Richard A., (585) 245-5296 young@geneseo.edu, SUNY, Geneseo – GmXgOr
Young, Richard E., (808) 956-7024 ryoung@hawaii.edu,

Univ of Hawai'i, Manoa – Ob
Young, Robert S., (828) 227-3822 ryoung@wcu.edu, Western Carolina Univ – GsOn
Young, Tim, 217-244-2772 young@illinois.edu, Univ of Illinois, Urbana-Champaign – Hw
Young, William, +44(0) 113 34 31640 c.w.young@leeds.ac.uk, Univ of Leeds – Ge
Young, William R., (858) 534-1380 wryoung@ucsd.edu, Univ of California, San Diego – Op
Yow, Donald M., (859) 622-1420 don.yow@eku.edu, Eastern Kentucky Univ – OwyOg
Yow, Sonja H., (859) 622-1424 sonja.yow@eku.edu, Eastern Kentucky Univ – On
Yu, Jin-Yi, (949) 824-3878 jyyu@uci.edu, Univ of California, Irvine – OapOw
Yu, Qian, (413) 545-2095 qyu@geo.umass.edu, Univ of Massachusetts, Amherst – Ori
Yu, Wen-che, fgsyw@earth.sinica.edu.tw, Academia Sinica – Ys
Yu, Zicheng, 610-758-6751 ziy2@lehigh.edu, Lehigh Univ – Pe
Yuan, Fei, (507) 389-2617 fei.yuan@mnsu.edu, Minnesota State Univ – Ori
Yuan, Xiaojun, (845) 365-8820 Columbia Univ – Op
Yue, Stephen, (514) 398-4755 McGill Univ – Nx
Yuen, Cheong-yip R., (630) 252-4869 yuenr@anl.gov, Argonne National Laboratory – Ge
Yuen, David A., 612-624-1868 davey@umn.edu, Univ of Minnesota, Twin Cities – Yg
Yule, J. Douglas, (818) 677-6238 j.d.yule@csun.edu, California State Univ, Northridge – Gt
Yun, Misuk, (204) 474-8870 yun@cc.umanitoba.ca, Univ of Manitoba – Cs
Yun, Seong-Taek, 82-2-3290-3176 styun@korea.ac.kr, Korea Univ – ClHyCs
Yung, Yuk L., (626) 395-6940 yly@gps.caltech.edu, California Inst of Tech – Oa
Yunwei, Sun, (925)422-1587 sun4@llnl.gov, Lawrence Livermore National Laboratory – Ng
Yuretich, Richard F., (413) 545-0538 yuretich@geo.umass.edu, Univ of Massachusetts, Amherst – Cl
Yurkovich, Steven P., (828) 227-7367 yurkovich@wcu.edu, Western Carolina Univ – Gx
Yurtsever, Ayhan, 089/2180 4346 ayhan.yurtsever@ltz.uni-uemchen.de, Ludwig-Maximilians-Universitaet Muenchen – Gz
Yusoff, Ismail, 03-79674153 ismaily70@um.edu.my, Univ of Malaya – Hg
Yuter, Sandra, 919-513-7963 sandra_yuter@ncsu.edu, North Carolina State Univ – Oa
Yuter, Sandra E., seyuter@ncsu.edu, Univ of Washington – Oa
Yutzy, Gale, (301) 687-4369 gyutzy@frostburg.edu, Frostburg State Univ – On
Yvon-Lewis, Shari A., 979 458 1816 syvonlewis@ocean.tamu.edu, Texas A&M Univ – Oc

## Z

Zabielski, Victor, (703) 845-6507 vzabielski@nvcc.edu, Northern Virginia Community Coll - Alexandria – GgOg
Zaccaria, Daniele, 530-219-7502 dzaccaria@ucdavis.edu, Univ of California, Davis – Hg
Zachos, James C., (831) 459-4644 jzachos@pmc.ucsc.edu, Univ of California, Santa Cruz – Cm
Zachos, Louis, (662) 915-8827 lgzachos@olemiss.edu, Univ of Mississippi – Ger
Zachry, Doy L., (479) 575-2785 dzachry@uark.edu, Univ of Arkansas, Fayetteville – Gr
Zafiriou, Oliver C., (508) 289-2342 ozafiriou@whoi.edu, Woods Hole Oceanographic Institution – Oc
Zahm, Christopher K., 512-471-3159 chris.zahm@beg.utexas.edu, Univ of Texas at Austin – Yg
Zaitchik, Benjamin, 410-516-7135 bzaitch1@jhu.edu, Johns Hopkins Univ – Oa
Zaja, Annalisa, +390498279193 annalisa.zaja@unipd.it, Università degli Studi di Padova – Yg
Zajac, Roman N., (203) 932-7114 rzajac@newhaven.edu, Univ of New Haven – Oni
Zakardjian, Bruno, (418) 723-1986 (Ext. 1570) bruno_zakardjian@uqar.qc.ca, Universite du Quebec a Rimouski – Ob
Zakrzewski, Richard J., (785) 628-5389 rzakrzew@fhsu.edu, Fort Hays State Univ – PvsPe
Zalasiewicz, Jan, +440116 252 3928 jaz1@le.ac.uk, Leicester Univ – Po
Zaleha, Michael J., (937) 327-7331 mzaleha@wittenberg.edu, Wittenberg Univ – Gs
Zaleha, Robert, (216) 987-5278 Robert.Zaleha@tri-c.edu, Cuyahoga Community Coll - Western Campus – Og
Zaman, Musharraf, 405-325-2626 zaman@ou.edu, Univ of Oklahoma – Np
Zamani, Behzad, +98 (411) 339 2700 zamani@tabrizu.ac.ir, Univ of Tabriz – GtYs
Zambito, James J., 608.262.3385 jay.zambito@wgnhs.uwex.edu, Univ of Wisconsin - Extension – Gsg
Zampieri, Dario, +390498279179 dario.zampieri@unipd.it, Università degli Studi di Padova – Gc
Zamzow, Craig E., czamzow@clarion.edu, Clarion Univ – Gi
Zanazzi, Alessandro, (801) 863-5395 alessandro.zanazzi@uvu.edu, Utah Valley Univ – CgsPe
Zandt, George, (520) 621-2273 gzandt@email.arizona.edu, Univ of Arizona – Ys
Zanella, Elena, elena.zanella@unito.it, Università di Torino – Ym
Zanetti, Kathleen, (702) 895-4789 Univ of Nevada, Las Vegas – Gg
Zaneveld, J. R., ron@wetlabs.com, Oregon State Univ – Yr
Zappa, Christopher J., 845-365-8547 zappa@ldeo.columbia.edu, Columbia Univ – Op
Zarillo, Gary A., zarillo@fit.edu, Florida Inst of Tech – Ou
Zarnetske, Jay, (517) 353-3249 jpz@msu.edu, Michigan State Univ – Hwg
Zarroca Hernández, Mario, ++935812033 mario.zarroca.hernandez@uab.cat, Universitat Autonoma de Barcelona – YgNg
Zaspel, Craig E., (406) 683-7366 craig.zaspel@umwestern.edu, Univ of Montana Western – Yg
Zattin, Massimiliano, +390498279186 massimiliano.zattin@unipd.it, Università degli Studi di Padova – Gs
Zavada, Michael S., (401) 865-2163 mzavada@providence.edu, Providence Coll – On
Zawadzki, Isztar I., (514) 398-1034 isztar.zawadzki@mcgill.ca, McGill Univ – Oar
Zawiskie, John M., (248) 645-3252 Wayne State Univ – Gg
Zawislak, Ronald L., (615)898-5609 Ron.Zawislak@mtsu.edu, Middle Tennessee State Univ – Ye
Zayac, John M., (818) 710-2218 zayacjm@pierceColl.edu, Los Angeles Pierce Coll – Ggv
Zebker, Howard A., (650) 723-8067 zebker@stanford.edu, Stanford Univ – Yx
Zeebe, Richard E., (808) 956-6473 zeebe@soest.hawaii.edu, Univ of Hawai'i, Manoa – Ou
Zehnder, Joseph A., 402-280-2448 zehnder@creighton.edu, Creighton Univ – Oa
Zehr, Jonathan P., (831) 459-4009 zehrj@cats.ucsc.edu, Univ of California, Santa Cruz – Ob
Zeidouni, Mehdi, mehdi.zeidouni@beg.utexas.edu, Univ of Texas at Austin – Eo
Zeiger, Elaine, (312) 665-7627 Field Museum of Natural History – On
Zeigler, E. Lynn, 678-891-3767 ezeigler1@gsu.edu, Georgia State Univ, Perimeter Coll, Clarkston Campus – Gd
Zeitler, Peter K., (610) 758-3671 pkz0@lehigh.edu, Lehigh Univ – Cc
Zeitlhöfler, Matthias, 089/2180 6570 matthias@zeitlhoefler.de, Ludwig-Maximilians-Universitaet Muenchen – Gg
Zelazny, Lucian W., (540) 231-9781 Virginia Polytechnic Inst & State Univ – Sc
Zelizer, Nora, 609 258-5809 nzelizer@princeton.edu, Princeton Univ – On
Zelt, Colin A., (713) 348-4757 czelt@rice.edu, Rice Univ – Yg
Zeman, Josef, +420 549 49 8295 jzeman@sci.muni.cz, Masaryk Univ – CgpCe
Zender, Charles, 949-824-2987 zender@uci.edu, Univ of California, Irvine – Oa
Zeng, Hongliu, 512-475-6382 hongliu.zeng@beg.utexas.edu, Univ of Texas at Austin – YsGs
Zeng, Ning, (301) 405-5377 zeng@umd.edu, Univ of Maryland – OaCgGl
Zentilli, Marcos, (902) 494-3873 Dalhousie Univ – Eg
Zentner, Nick, (509) 963-2828 nick@geology.cwu.edu, Central Washington Univ – Gg
Zerkle, Aubrey, +44 01334 464949 az29@st-andrews.ac.uk, Univ of St. Andrews – Co
Zhai, Xiaoming, 847 543-2504 Coll of Lake County – Gp
Zhai, Xiaoming, +44 (0)1603 59 3762 xiaoming.zhai@uea.ac.uk, Univ of East Anglia – Og

Zhan, Hongbin, (979) 862-7961 zhan@geo.tamu.edu, Texas A&M Univ – Hw
Zhan, Zhongwen, 626-395-6906 zwzhan@caltech.edu, California Inst of Tech – Ys
Zhang, Bo, 205-348-4544 bzhang33@ua.edu, Univ of Alabama – YeGoEo
Zhang, Caiyun, 561-297-2648 czhang3@fau.edu, Florida Atlantic Univ – Or
Zhang, Chi, 785-864-4974 chizhang@ku.edu, Univ of Kansas – Ygx
Zhang, Chunfu, (785) 628-5348 c_zhang35@fhsu.edu, Fort Hays State Univ – Ggz
Zhang, Guorong, (785) 625-3425 gzhang@ksu.edu, Kansas State Univ – So
Zhang, Henian, (404) 894-1738 henian.zhang@eas.gatech.edu, Georgia Inst of Tech – Oa
Zhang, Huan, 860-405-9237 huan.zhang@uconn.edu, Univ of Connecticut – Ob
Zhang, Jin, jinzhang@unm.edu, Univ of New Mexico – Gy
Zhang, Jinhong, (520) 626-9656 jhzhang@email.arizona.edu, Univ of Arizona – Nm
Zhang, Lin, 361-825-2436 Lin.Zhang@tamucc.edu, Texas A&M Univ, Corpus Christi – Cs
Zhang, Max, 607-254-5402 kz33@cornell.edu, Cornell Univ – On
Zhang, Minghua , 530-752-4953 mhzhang@ucdavis.edu, Univ of California, Davis – Hg
Zhang, Minghua, (631) 632-8318 minghua.zhang@stonybrook.edu, SUNY, Stony Brook – Oa
Zhang, Ning, 937 376 6043 nzhang@centralstate.edu, Central State Univ – Emo
Zhang, Qiaofeng (Robin), (270) 809-6760 qzhang@murraystate.edu, Murray State Univ – OirOu
Zhang, Ren, (254)7102496 Ren_Zhang@baylor.edu, Baylor Univ – Cs
Zhang, Renyi, (979) 845-7671 renyi-zhang@geos.tamu.edu, Texas A&M Univ – Oa
Zhang, Rong, (609) 987-5061 rong.zhang@noaa.gov, Princeton Univ – Op
Zhang, Tongwei, 512-232-1496 tongwei.zhang@beg.utexas.edu, Univ of Texas at Austin – Cgs
Zhang, Wei, wei2.zhang@umb.edu, Univ of Massachusetts, Boston – On
Zhang, Xi, (831) 502-8126 xiz@ucsc.edu, Univ of California, Santa Cruz – OaXy
Zhang, Xinning, 609-258-7438 Xinningz@princeton.edu, Princeton Univ – Ou
Zhang, Y. J., 804-684-7466 yjzhang@vims.edu, Coll of William & Mary – Op
Zhang, Yang, 919-515-9688 yang_zhang@ncsu.edu, North Carolina State Univ – Oa
Zhang, Ye, (307) 766-3386 yzhang9@uwyo.edu, Univ of Wyoming – Hy
Zhang, Yige, yigezhang@fas.harvard.edu, Texas A&M Univ – Co
Zhang, Yong, (205) 348-3317 yzhang264@ua.edu, Univ of Alabama – Hqw
Zhang, You-Kuan, (319) 335-1806 you-kuan-zhang@uiowa.edu, Univ of Iowa – Hw
Zhang, Youxue, (734) 763-0947 youxue@umich.edu, Univ of Michigan – Cp
Zhao, Bo, 541-737-3497 bzhao@coas.oregonstate.edu, Oregon State Univ – Oi
Zhao, Li, 886-02-27839910-320 zhaol@earth.sinica.edu.tw, Academia Sinica – YsgGt
Zhao, Xixi, (831) 459-4847 xzhao@es.ucsc.edu, Univ of California, Santa Cruz – Ym
Zhaohui, Li, (262) 595-2487 li@uwp.edu, Univ of Wisconsin, Parkside – ClGzHw
Zhdanov, Michael S., (801) 581-7750 michael.zhdanov@utah.edu, Univ of Utah – Ye
Zheng, Chunmiao, 205-348-0579 czheng@ua.edu, Univ of Alabama – Hw
Zheng, Yan, (718) 997-3329 Queens Coll (CUNY) – CgHwCm
Zhou, Hua-Wei, (713) 743-3440 hzhou@uh.edu, Univ of Houston – YseYg
Zhou, Meng, meng.zhou@umb.edu, Univ of Massachusetts, Boston – Op
Zhou, Wendy W., (303) 384-2181 wzhou@mines.edu, Colorado School of Mines – NgOir
Zhou, Xiaobing, (406) 496-4350 xzhou@mtech.edu, Montana Tech – Or
Zhou, Xiaolu, (912) 478-5943 xzhou@georgiasouthern.edu, Georgia Southern Univ – Or
Zhou, Ying, 540-231-6521 yingz@vt.edu, Virginia Polytechnic Inst & State Univ – Ys
Zhou, Yu, 845-437-5543 yuzhou@vassar.edu, Vassar Coll – Oun
Zhou, Yuyu, (515) 294-2842 yuyuzhou@iastate.edu, Iowa State Univ of Science & Tech – Oyi
Zhu, Chen, (812) 856-1884 chenzhu@indiana.edu, Indiana Univ, Bloomington –
Zhu, Junfeng, 859-323-0530 junfeng.zhu@uky.edu, Univ of Kentucky – Hw
Zhu, Lupei, (314) 977-3118 zhul@slu.edu, Saint Louis Univ – YsGt
Zhu, Ping, 305-348-7096 zhup@fiu.edu, Florida International Univ – Oa
Zhu, Qingzhi, (631) 632-8747 Qing.Zhu@stonybrook.edu, SUNY, Stony Brook – Oc
Zhu, Wenlu, (301) 405-1831 wzhu@umd.edu, Univ of Maryland – YrNrHg
Zhuang, Guangsheng, gzhuang@lsu.edu, Louisiana State Univ – CocGs
Zhuang, Qianlai, (765) 494-9610 qzhuang@purdue.edu, Purdue Univ – Oa
Zieg, Michael J., (724) 738-2501 michael.zieg@sru.edu, Slippery Rock Univ – Gi
Ziegler, Karen, (505) 277-1437 kziegler@unm.edu, Univ of New Mexico – CsXc
Ziemer, Robert R., rrz7001@humboldt.edu, Humboldt State Univ – HgGmOn
Zierenberg, Robert A., (530) 752-1863 razierenberg@ucdavis.edu, Univ of California, Davis – Cs
Zimbelman, James R., 202-633-2471 Smithsonian Institution / National Air & Space Museum – Xg
Zimmer, Brian, 828-262-7517 zimmerbw@appstate.edu, Appalachian State Univ – Gg
Zimmerman, Aaron, (970) 491-3789 aaron.zimmerman@colostate.edu, Colorado State Univ – Cc
Zimmerman, Andrew, 352-392-0070 azimmer@ufl.edu, Univ of Florida – Co
Zimmerman, Brian S., (814) 732-2529 bzimmerman@edinboro.edu, Edinboro Univ of Pennsylvania – EgGzi
Zimmerman, Mark, 612-626-0572 zimme030@umn.edu, Univ of Minnesota, Twin Cities – Yx
Zimmerman, R. Eric, (630) 252-6816 Argonne National Laboratory – Ng
Zimmerman, Robert W., +44 20 7594 7412 r.w.zimmerman@imperial.ac.uk, Imperial Coll – NrHwNp
Zimmerman, Ronald K., (225) 388-8302 ron@lgs.bri.lsu.edu, Louisiana State Univ – Go
Zimmerman, Jr., Jay, (618) 453-3351 zimmerman@geo.siu.edu, Southern Illinois Univ Carbondale – Gc
Zimmermann, George, (609) 652-4308 george.zimmermann@stockton.edu, Richard Stockton Coll of New Jersey – On
Zink, Albert, 089/2180 4340 Ludwig-Maximilians-Universitaet Muenchen – Gz
Zinsmeister, William J., (765) 494-0279 wjzins@purdue.edu, Purdue Univ – Ps
Ziolkowski, Anton M., +44 (0) 131 650 8511 anton.ziolkowski@ed.ac.uk, Edinburgh Univ – YesYg
Ziolkowski, Lori A., 803-777-0035 lziolkowski@geol.dc.edu, Univ of South Carolina – CoOcCs
Zipf, Karl, 303-384-2371 rzipf@mines.edu, Colorado School of Mines – Nm
Zipper, Carl E., (540) 231-9782 czip@vt.edu, Virginia Polytechnic Inst & State Univ – Oe
Zipser, Edward, 801-585-0467 ed.zipser@utah.edu, Univ of Utah – Oa
Ziramov, Sasha, +61 8 9266-4973 Sasha.Ziramov@curtin.edu.au, Curtin Univ – Ye
Zlotkin, Howard, 12012003161 hzlotkin@njcu.edu, New Jersey City Univ – Oe
Zlotnik, Vitaly A., (402) 472-2495 vzlotnik1@unl.edu, Univ of Nebraska, Lincoln – HwyHq
Zoback, Mark D., (650) 725-9295 zoback@stanford.edu, Stanford Univ – Yg
Zodrow, Erwin L., erwin_zodrow@uccb.ca, Cape Breton Univ – Pi
Zois, Constantine S., (908) 737-3693 czois@kean.edu, Kean Univ – Ow
Zollner, Stefan, (575) 646-7627 zollner@nmsu.edu, New Mexico State Univ, Las Cruces – Om
Zollweg, James A., (585) 395-2352 jzollweg@brockport.edu, SUNY, The Coll at Brockport – Hs
Zolotov, Mikhail Y., (480) 965-4739 zolotov@asu.edu, Arizona State Univ – Cg
Zonneveld, John-Paul, 780-492-3287 zonneveld@ualberta.ca, Univ of Alberta – Go
Zou, Haibo, (334) 844-4315 haibo.zou@auburn.edu, Auburn Univ – CcGiv
Zouhri, Lahcen, +33(0)3 44068976 lahcen.zouhri@lasalle-beauvais.fr,

Institut Polytechnique LaSalle Beauvais (ex-IGAL) – HwgGe
Zreda, Marek, 520-621-4072 marek@hwr.arizona.edu, Univ of Arizona – Cc
Zuber, Maria T., 617-253-6397 zuber@mit.edu, Massachusetts Inst of Tech – Xy
Zuend, Andreas, (514) 398-7420 andreas.zuend@mcgill.ca, McGill Univ – Oaw
Zume, Joseph T., (717) 477-1548 jtzume@ship.edu, Shippensburg Univ – HgYg
Zumwalt, Gary S., gzumwalt@latech.edu, Louisiana Tech Univ – Pi
Zuppann, Charles W., 812 855 9199 czuppann@indiana.edu, Indiana Univ – Go
Zurawski, Ronald P., (615) 532-1502 ronald.zurawski@tn.gov, Tennessee Geological Survey – GgEg
Zurevinski, Shannon, (807) 343-8015 shannon.zurevinski@lakeheadu.ca, Lakehead Univ – Gz
Zurick, David, (859) 622-1427 david.zurick@eku.edu, Eastern Kentucky Univ – On
Zuza, Andrew V., (775) 682-8752 azuza@unr.edu, Univ of Nevada – Gc
Zwiefelhofer, Luke, 507-457-2778 lzwiefelhofer@winona.edu, Winona State Univ – On
Zyvoloski, George A., (505) 667-1581 Los Alamos National Laboratory – Ng

www.ingramcontent.com/pod-product-compliance
Lightning Source LLC
Chambersburg PA
CBHW082117210326
41599CB00031B/5789